Patrick J. Pinnell
5/21/2016

Time-Saver Standards for Housing and Residential Development

Other McGraw-Hill Books of Interest

Time-Saver Standards for Housing and Residential Development

SECOND EDITION

Edited by

JOSEPH DE CHIARA
JULIUS PANERO
MARTIN ZELNIK

McGraw-Hill, Inc.

New York   San Francisco   Washington, D.C.   Auckland   Bogotá
Caracas   Lisbon   London   Madrid   Mexico City   Milan
Montreal   New Delhi   San Juan   Singapore
Sydney   Tokyo   Toronto

Library of Congress Cataloging-in-Publication Data

Time-saver standards for housing and residential development / [edited
    by] Joseph De Chiara, Julius Panero, Martin Zelnik.—2nd ed.
            p.   cm.
        Rev. ed. of: Time-saver standards for residential development.   1st
    ed. © 1984.
        Includes index.
        ISBN 0-07-016301-4 (acid-free paper)
        1. Homesites—Planning—Handbooks, manuals, etc.   2. Dwellings—
    Designs and plans—Handbooks, manuals, etc.   I. De Chiara, Joseph,
    date.   II. Panero, Julius.   III. Zelnik, Martin, date.
    IV. Time-saver standards for residential development.
    NA9051.T55   1995
    711'.58—dc20   94-5366
                                                                    CIP

1 2 3 4 5 6 7 8 9 0   KGP/KGP   9 0 9 8 7 6 5 4

ISBN 0-07-016301-4

*The sponsoring editor for this book was Joel Stein, the
editing supervisor was Stephen M. Smith, and the
production supervisor was Suzanne W. Babeuf. It was
set in Univers Regular by North Market Street Graphics.*

Printed and bound by Arcata Graphics/Kingsport.

This book is printed on acid-free paper.

Contents

Contents

Contents

Preface to Second Edition

Housing standards and criteria are by no means static. They are, in fact, constantly evolving to reflect changing economic conditions, social values, and technological innovations. It is in the attempt to keep current with these changes that this Second Edition of *Time-Saver Standards for Housing and Residential Development* has been put forth.

The dwelling unit, in the form of either the single-family house or the apartment, is considered the basic building block of the community. Over the years significant changes to the dwelling unit have occurred, from within by technological developments and from without by social and economic changes.

Inside the dwelling unit, we have seen the expansion of functions in both the kitchen and bathroom areas. The advent of the computer/cable/information highway tidal wave is rapidly changing the entertainment and recreation activities of the home. As a result, the traditional living room and recreation room are in transition. New multimedia rooms are being created. Another significant development has been the explosion in the use of the home as the workplace. This has resulted in the home office and its related communications center. Accessibility for the handicapped has been a third strong influence in the design of housing.

Outside the dwelling unit, several evolving trends are affecting the design of apartment buildings and the organization of the neighborhood. The need for greater security owing to the escalating crime rate has become a major design concern. The harming of the environment by destruction of the natural habitat, air and water pollution, and overdevelopment is another major design concern. Conservation measures are now, and will continue to be, strong influences in site selection and housing planning.

In this Second Edition, standards and criteria were reviewed and updated. Several chapters underwent major revision and new graphic material, illustrations, and architectural plans were added, which increased this book's page count by over 200 pages. As a result of all these changes, we strongly feel that the quality and usefulness of this unique reference source have been enhanced, and that it will continue to serve the needs of the many design professions.

Joseph De Chiara
Julius Panero
Martin Zelnik

Preface to First Edition

Residential development, which includes land subdivision, tract development, large-scale housing, and rehabilitation, is a vital and intricate part of our society and economy. It encompasses a broad, loosely knit network of significant components including planners, architects, bankers, lawyers, social scientists, politicians, as well as builders and developers. The conception and execution of any one kind of residential project is a long and complex process. This book deals with one aspect of this process, that of the architectural planning and design of the physical space.

Specifically, the aim of this book is to provide a comprehensive source of architectural design data for housing and residential development. Although there is presently a great deal of this type of information available, it is fragmented and scattered among many government agencies, research foundations, housing groups, and publications. From this wealth of material, we have assembled and selected what we feel to be the most appropriate and useful by the design professions. It is intended primarily for those involved with the architectural development of proposals, programs, preliminary studies, and analyses. This would mainly include architects, planners, landscape architects, designers, and students. The material, however, will also be of significant value to the many other related professionals. The material selected is strongly weighted towards the more technical and practical, rather than the theoretical, aspects of housing. However, there is no inclusion of engineering design data, that is, structural and mechanical, because these aspects are thoroughly presented in other available publications. Also, the selection of material does not represent any specific philosophy or viewpoint of housing, such as high-rise versus low-rise development, high-density versus low-density projects, or public or private development. The material is that which we feel will enable the designer to evolve better and more meaningful living environments.

Housing standards and criteria are by no means static or etched in stone. They are in a constant state of flux, and, over the years, gradually evolved to reflect the changing economic and social values of our society. Therefore, the user of this book is cautioned to use the material presented only as a guide in arriving at specific decisions. The book is not an attempt to provide definitive formulae for a highly complex process. Each situation, even though it may appear similar, must by reviewed and analyzed on its own merits.

The format of this book follows the tradition of the well-known *Time-Saver Standards* series, which includes *Time-Saver Standards for Architectural Design Data* and *Time-Saver Standards for Building Types*. It is intended to be comprehensive in scope and essentially graphic in presentation.

The editor wishes to acknowledge the generous support and assistance of the many agencies, organizations, publishers, and practitioners in allowing the use of their material. In particular, we are indebted to *Architectural Record, Progressive Architecture, House and Home,* and *Architectural Forum.* Also, I wish to thank my good friend and co-author, Dr. Lee Koppelman, for use of some of the material from one of our early publications entitled *Housing/Planning and Design.*

Joseph De Chiara

**Time-Saver
Standards for
Housing and
Residential
Development**

General Planning and Neighborhood Organization

PLANNING UNITS—TYPES AND FUNCTIONS

Types of Planning Units

In planning a system of interrelated areas and facilities, each type of geographic area—neighborhood, community, school district, city, county, region, state, and nation—must be considered. In some instances, the entire state must be included and, when accessible, areas and facilities provided by the federal government must be taken into account.

The Neighborhood

The neighborhood is a residential area with homogeneous characteristics, of a size comparable to that usually served by an elementary school. A typically ideal neighborhood for planning purposes would be an area ¾ to 1 mile square and containing about 6000 to 8000 people.

Neighborhoods occur in various shapes and sizes. Population densities vary from a few thousand to many thousands per square mile, and there is also a wide variation in the numbers of children. Therefore, each neighborhood must be studied carefully. Because most residents live within a short distance of the school or playground, they walk to it and tend to use it frequently, often for shorter periods than in the centers planned for a larger geographic unit.

The Community

The community is a section of a city, primarily a residential area. It usually represents the service area of a high school, contains a large business center, and commonly constitutes a section of the city measuring 2 or 3 miles across. It can be thought of as a "community of neighborhoods" because it is usually composed of three to five neighborhoods. Consequently, the population varies, but on the average is 3 to 5 times that of a neighborhood, or from 20,000 to 40,000 people. It may have a less pronounced homogeneity than the neighborhood but should not be so dissimilar as to make unified planning impossible. If the dissimilarities are pronounced, the community may need to be subdivided for planning.

The City or School District

The area designated as the city, town, borough, or village lends itself to the provision of areas and facilities for use by the entire population of the political subdivision. Major parks, golf courses, camps, museums, and botanical gardens, which cannot be provided in each neighborhood and community, are typical citywide areas. In small localities comprising one community and with a single high school, citywide planning is largely comparable with planning for a single community as described above, although some facilities commonly provided in larger citywide areas are included.

School districts vary widely in size and population, but districtwide school planning involves primarily neighborhoods and communities. Some of the large school districts provide districtwide facilities for an outdoor education-recreation complex, interscholastic activities, consolidated educational programs, and some type of post-high school center or community college for day pupils.

Larger Units

The county or the region, which is a geographic area that sometimes includes parts of more than one county, is increasingly used as a unit for planning. Many such planning units are located in close proximity to a metropolitan city and include both the city and the surrounding region. Others are primarily rural in nature and are composed primarily of unincorporated areas. Planning on a regional or district basis lends itself to the provision of extensive properties usable for family outings, winter and water sports, and other activities requiring large land and water areas. Since these properties are distant from dense population centers, people require transportation in order to reach them and consequently tend to use them less frequently but for longer periods than is true of the areas in the smaller planning units. The increase in statewide planning makes it important that plans developed for smaller geographic units take into account existing and proposed facilities for statewide use.

Unit Relationships

Even though area and facility planning is done for various geographic units and political subdivisions and involves a great variety of school and recreational properties, there is a relationship among the resultant plans. The areas and facilities provided in one unit or subdivision often influence the need for them elsewhere. Areas and facilities in the larger units supplement those in the smaller ones and are used by the people living in all these subdivisions. Therefore, cooperation among the agencies involved in planning at different levels is essential in order to achieve coordinated programs.

LOCATION

In planning a housing development, a careful study and analysis of all the factors that eventually determine rentals or sales must be made. Land costs, building costs, operating and maintenance expenses, taxes, insurance rates, interest rates, and all other items that enter into the cost of producing the rental property and maintaining the services of shelter for which it is planned must sum up to a figure consistent with the rental income expectancy of the completed property.

In order to be acceptable, a housing development must meet certain fundamental requisites with respect to location and planning. First, the project should be located in or near a city or town where there is a definite prospect of continuing demand for housing at the proposed rental rates; it should be located in a neighborhood where the possibilities of future deterioration are at the minimum; and the site itself should be suitable for the development of a project of the type and magnitude contemplated. Second, the plans of the entire development should embody qualities of design and construction in terms of open space, lawns and planting, light and air, convenience and privacy of the dwelling units, and other amenities of family living, to the end that appeal may be enhanced and the factors of obsolescence may be minimized.

Rental income or selling price, to a great extent limited by the conditions prevailing in the market, is the controlling factor in the planning of a project. A fair return on the investment is assured only when costs, both construction and operating, are geared to the rental income expectancy. Planning affects not only the original cost of construction, but also, and equally important, the cost of operating and maintaining the property. But planning alone cannot be relied upon as the sole means of attaining these goals, nor is it to be inferred that, under varying conditions and circumstances, a given plan will result in the same rental rate. Experience has indicated, however, that under similar conditions of land costs, financing costs, and taxes, certain types of planning permit lower rentals than do other types. When the fullest possible advantage is taken of economies of layout and construction, less rental income is required to sustain a given project.

In the search for methods of reducing building and operating costs, there are certain essentials that must not be sacrificed. Substantial, well-built structures, located in good residential neighborhoods, with adequate space and equipment must be provided. A garden environment with play space for children is desirable, and ordinarily will be required. Plans must provide for light and ventilation, and each unit should incorporate the highest attainable degree of privacy and other amenities of family living. There must be adequate provision for automobiles, in either garages or parking spaces. Due attention must be given to the problem of servicing the dwelling units and the removal of waste.

COMPATIBILITY OF OTHER LAND USES WITH HOUSING

LISTING	GENERAL DESCRIPTION	COMPATIBILITY WITH HOUSING
Parks and Playgrounds		Desirable
Elementary School	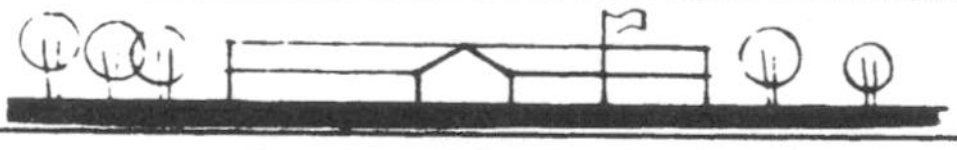	Desirable
Churches and Synagogues		Desirable
Housing in Good Condition		Desirable
Local Shopping	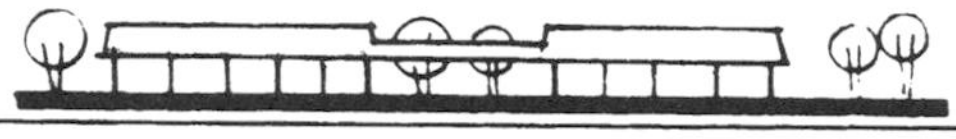	Desirable
Medical Facilities	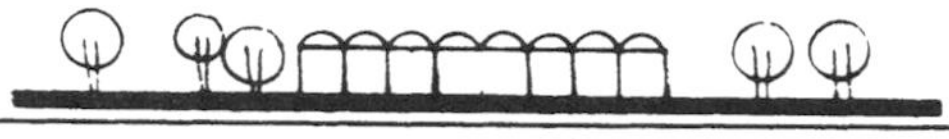	Desirable
Stores or Shops		Acceptable
Highway with Buffer Strips.		Acceptable
Housing in Fair Condition		Acceptable
Industrial Park		Acceptable
High School		Acceptable
Industrial Uses—Not Properly Screened		Not Acceptable
Airports		Not Acceptable
Highway without Buffer Strips		Not Acceptable
Warehouses, Railroad Tracks and Yards		Not Acceptable
Deteriorated or Dilapidated Housing		Not Acceptable

Fig. 1

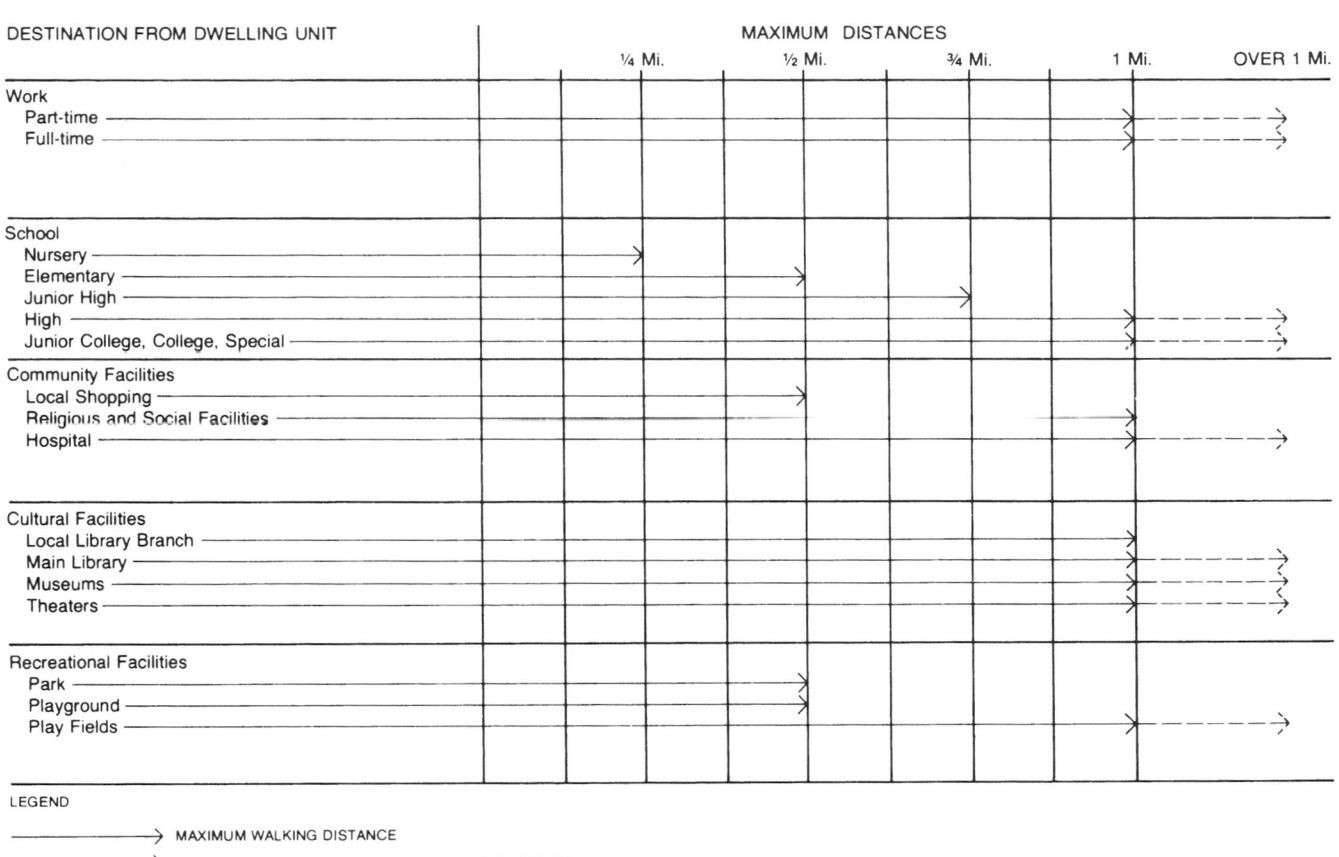

DESTINATION FROM DWELLING UNIT	MAXIMUM DISTANCES				
	¼ Mi.	½ Mi.	¾ Mi.	1 Mi.	OVER 1 Mi.
Work					
Part-time				→	- - →
Full-time				→	- - →
School					
Nursery	→				
Elementary		→			
Junior High			→		
High				→	- - →
Junior College, College, Special				→	- - →
Community Facilities					
Local Shopping		→			
Religious and Social Facilities				→	
Hospital				→	→
Cultural Facilities					
Local Library Branch				→	
Main Library				→	- - →
Museums				→	- - →
Theaters				→	- - →
Recreational Facilities					
Park		→			
Playground		→			
Play Fields				→	- - →

LEGEND

———————→ MAXIMUM WALKING DISTANCE

– – – – – –→ RIDING DISTANCE (CAR OR PUBLIC TRANSPORTATION)

Fig. 1

RELATIONSHIP OF EMPLOYMENT AND CENTRAL FACILITIES TO HOUSING SITE

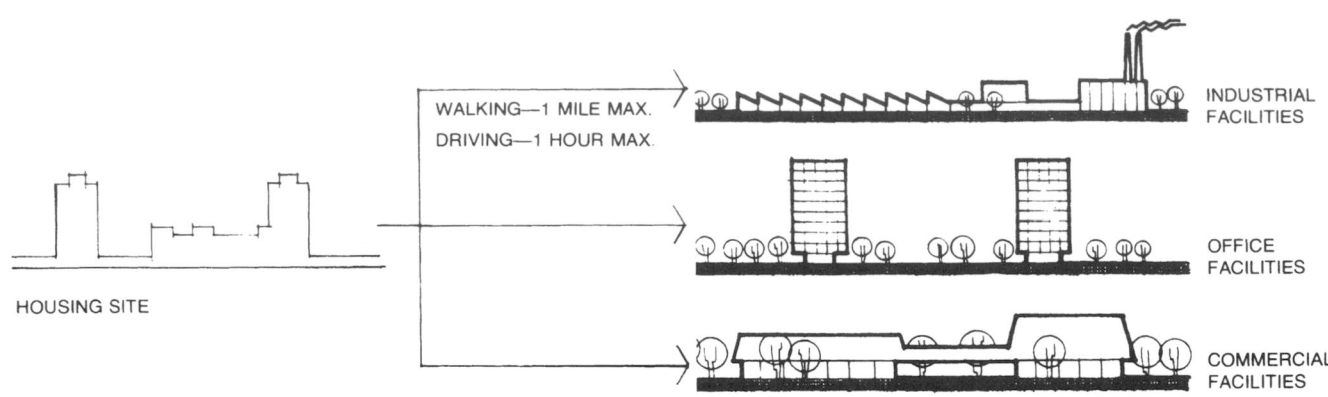

Fig. 1   Accessibility of employment facilities.

The selection of the community in which to build is the first problem confronting the sponsor of a project. To be acceptable, a project must be located in or near a city or town where there are adequate sources of employment, preferably in diversified occupations rather than in one or a few principal industries. Above all, it must be located in an economically stable community where there is evidence of a defi-

nite and continuing demand for housing at rentals sufficient to cover the requirements of the project as a sound business enterprise.

Within a given community, a location meriting approval should be readily accessible to places of employment and satisfactory transportation facilities should be available. It should be conveniently situated with respect to schools, churches, shopping centers, and the recreational

facilities of the community. The site must be suitable for residential development, free from the hazards of floods, fog, smoke, noxious odors, nuisance industries, and the like. It must be located in a neighborhood where zoning or other types of protective regulation will permit the sort of development contemplated, and it must conform with city, county, or regional planning where such planning is in force.

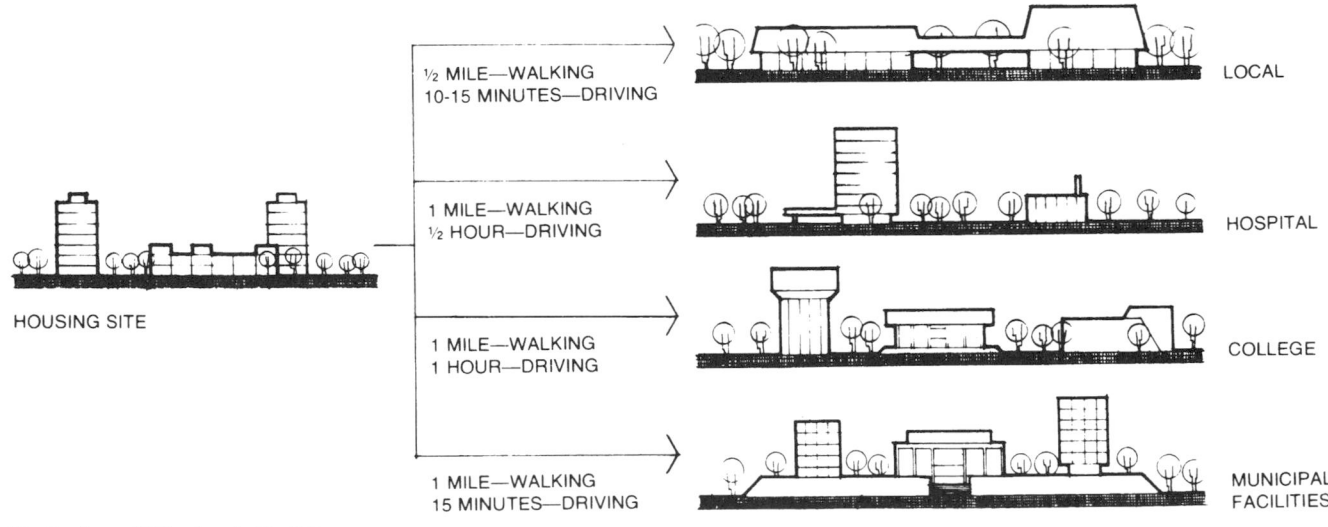

Fig. 2   Accessibility of central facilities.

RELATIONSHIP OF RECREATIONAL AND CULTURAL FACILITIES TO HOUSING SITE

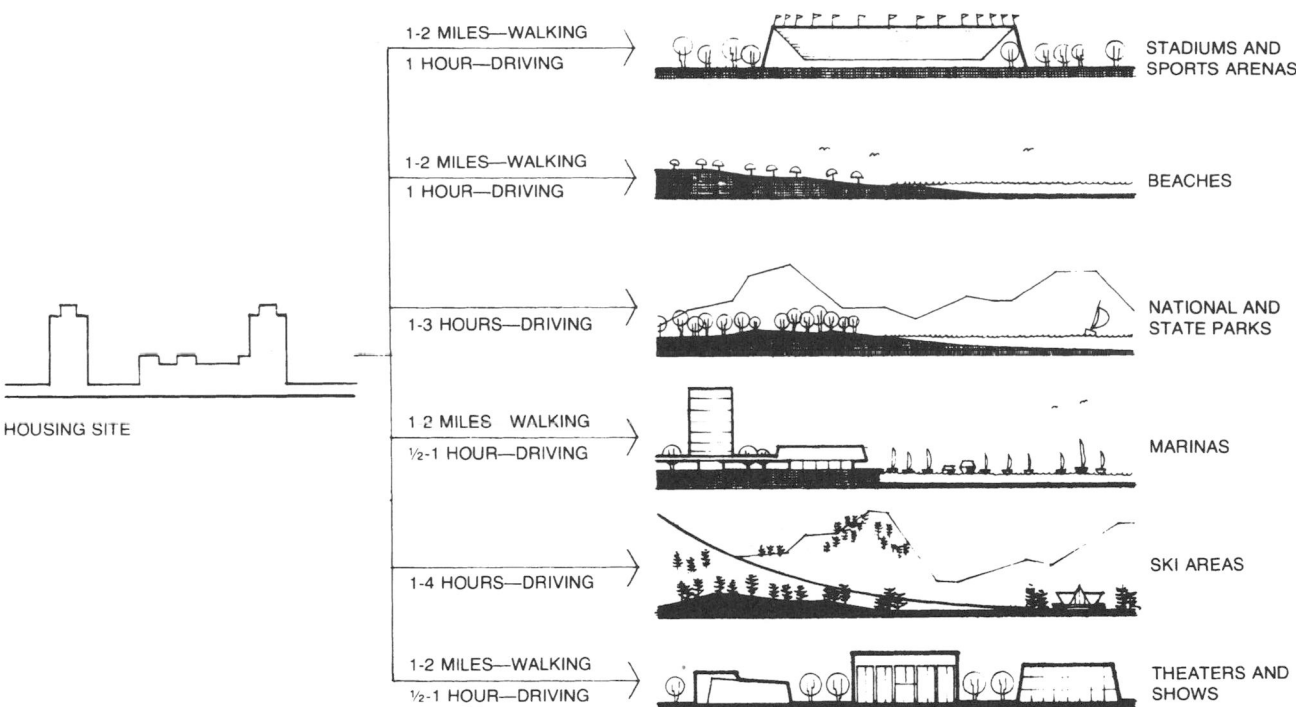

Fig. 1   Accessibility of recreational and cultural facilities.

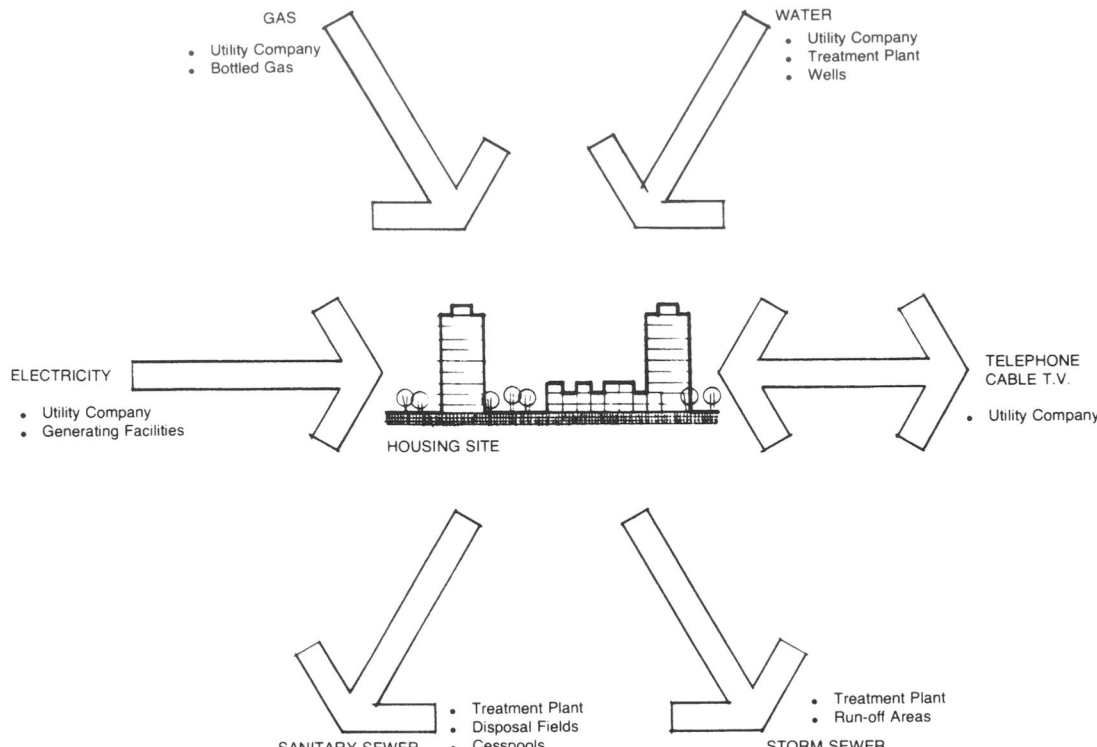

Fig. 1

Designation	Location within the street right of way	Function	Planning considerations
Sanitary sewers	Underground below the frost line and all other utilities. Between the water main and the storm sewer.	Intended to carry off water and other liquids containing organic materials, materials subject to decomposition and other waste, but not storm or surface water.	Essential to handle large volume of sewerage; if soil has poor absorption rate, sewers are necessary; local treatment plant can be utilized.
Storm sewers	Underground below the frost line and all other utilities. 21 to 22 ft either side of the street right-of-way center line. Generally adjacent to the curb on narrow streets and under the roadway on wide streets.	Intended to carry off storm, surface, and any other clear water or liquid not containing organic materials or other materials subject to decomposition.	In built-up areas, storm water must be removed from site; in rural or suburban areas, storm water may drain to adjacent streams or lands.
Water	Underground below the frost line but above the sewers. 20 to 30 ft either side of the street right-of-way center line. Generally under the sidewalk on narrow streets and adjacent to the curb on wide streets.	Intended to deliver a supply of potable water to a community for public or private use.	An adequate and sustained supply of water is essential to any housing development.
Gas	Underground below the frost line but above water and sewer. 30 to 34 ft either side of the street right-of-way center line. Generally under the sidewalk on narrow streets and adjacent to the curb on wide streets	Intended to deliver a supply of combustible gas to a community for public or private use.	If gas is not available, electricity may be substituted; bottled gas is another alternative.
Electricity (conduit)	Underground below the frost line but above water and sewer. 15 ft either side of the street right-of-way center line. Generally adjacent to or under the roadway pavement.	Intended to deliver a supply of electrical energy to a community for public or private use.	An adequate source of electricity is critical; source should be able to supply expanded future needs also.
Telephone, TV Cable	Underground below the frost line but above water and sewer. 15 ft either side of the street right-of-way center line. Generally adjacent to or under the roadway pavement.	Intended to facilitate and maintain a communication network for public or private use on an intra- or intercommunity basis.	These services are becoming increasingly more important; expansion of such facilities can be expected.

Classification	Cause	Effects	Solutions

Air, smoke, and dust	Incinerators
	Industry
	Vehicular traffic
	Refuse dumps
		Soiling of property
		Nuisance
		Visibility reduction
		Corrosion and other material
		  decomposition
		Plant damage
		Toxic effects
		Health hazards
		May depreciate property value
			Use of:
			Cyclones
			Bag filters
			Electrostatic precipitators
			Washers, etc.
			More efficient fuels
			Automotive pollution-control
			  devices

Noise	Airports Industry Railroads Recreational activities Trucks Vehicular traffic	Nuisance Health hazards May depreciate property value	Use of: Landscaping as buffer More effective vehicle noise-control devices Open-space separation
Odor	Disposal areas Dumps Industry Sewage-treatment plants Swamps	Nuisance May indicate other air pollution May depreciate property value	Use of: Good housekeeping Chemical control Masking Counteractant Etc.
Visual	Buildings in deteriorated condition Overhead wires Signs Poor landscaping Poor maintenance	Nuisance May depreciate property value	Use of: Landscaping as barrier Good building and property maintenance Sensitive selection of advertising graphics Concealment of utility lines High standards of landscaping

Water	Industry
	Recreational uses
	Polluted streams and lakes
		Soiling of property
		Nuisance
		Corrosion and other material
		  decomposition
		Wildlife damage
		Health hazards
		May depreciate property value
			Use of:
			Controlled discharge of
			  waste and chemicals

Fig. 1

POPULATION LEVELS REQUIRED TO SUPPORT URBAN ACTIVITIES

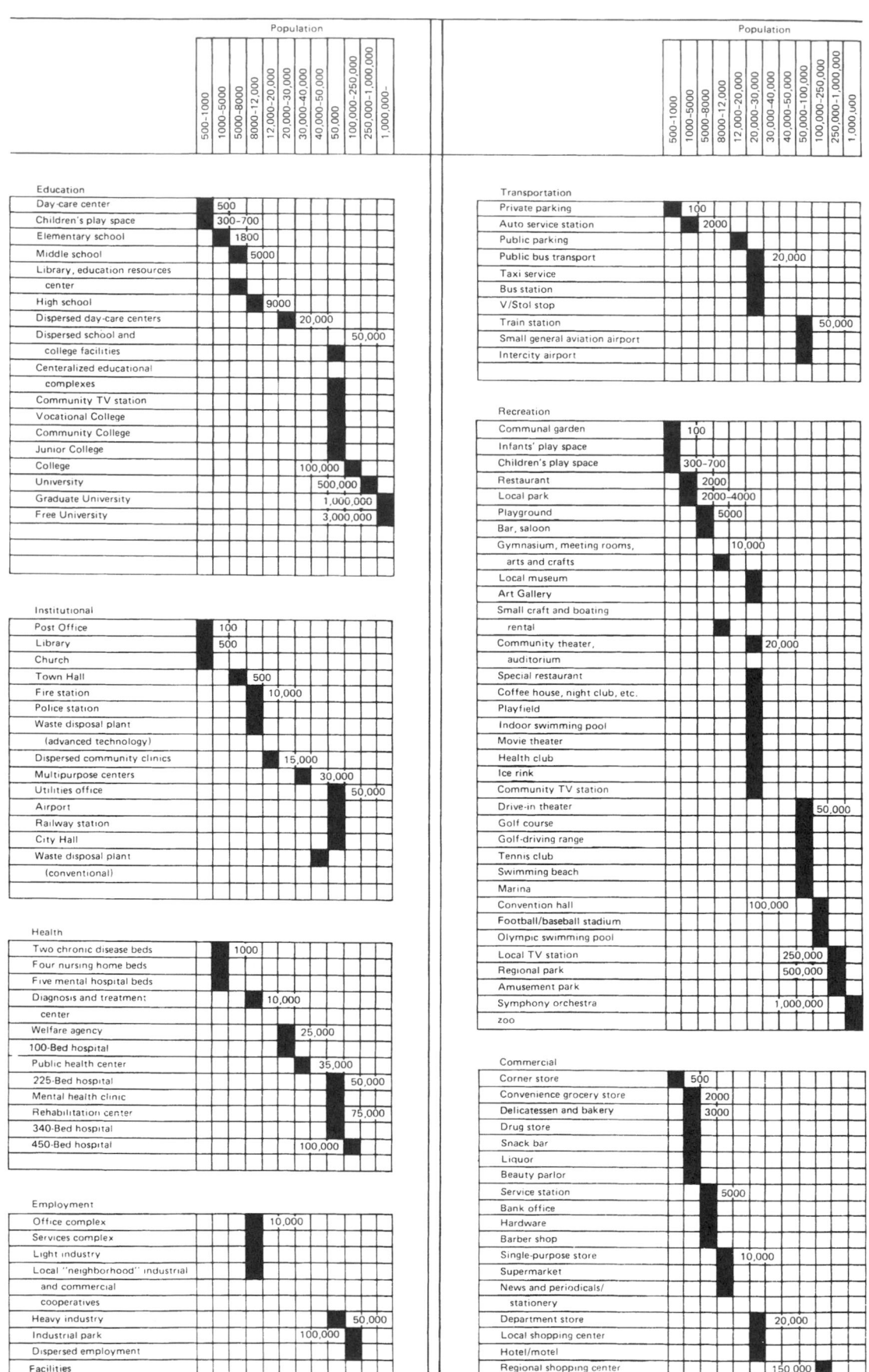

Fig. 1

TYPES OF NEIGHBORHOODS

A man's home is his castle. This is true whether the castle is the traditional single-family detached dwelling or a modern apartment high in the sky. It is true even if the castle is surrounded by a moat of uncut grass, a fleet of scooters, tricycles, and skates. But most families and individuals do not live entirely within their castles. They live on a street in a neighborhood.

What makes a neighborhood? In addition to individual homes, a neighborhood contains schools, churches, parks, and business centers. Some things are the result of joint effort—the streets, storm drainage system, water supply, electricity, telephone, power, gas, and the sewage disposal system. Even the street names and house addresses are a part of the neighborhood as well as of the individual residence. Letter carriers, milk delivery people, police officers, and delivery people are a few of the inhabitants who work in the neighborhood but who do not live there. A man may live in his castle, but he does not live alone.

The first step in planning for housing is the determination of the location, size, and characteristics of the existing residential neighborhood. It is also important to know the location of potential residential areas and the potential market for residential housing.

It is important to know something about each of the residential areas in a neighborhood sense—in what ways the neighborhood is growing, declining, or changing character. This is not revealed entirely by exterior inspection. It is necessary to go beneath the surface and understand the underlying economic facts about the area. Is the type of home in this area suitable for modern accommodations? Some communities that were built as recently as the post-Korean period consist predominantly of two-bedroom units. Two-bedroom units have proved to be inadequate for the larger postwar family in some localities. Conversely, there are many older towns with houses of seven and eight bedrooms that are far too large for most families today. Does the neighborhood need rejuvenation through new building, or protection from commercialization and overcrowding?

What are the different types of neighborhoods, and what are their special needs? Residential neighborhoods may be broadly classified into five types.

At one end of the scale, there is the older neighborhood of fine old mansions that have survived several generations, perhaps dating from before the turn of the century. Here the homes and lots are large. Land and building values have remained consistently high for several generations. These neighborhoods were well laid out, the buildings well constructed. They have sturdily survived to this day, but what of their future? Among the dangers this kind of neighborhood may face are (1) overcrowding of structures (doubling up of families, conversion of homes to apartments), (2) undermaintenance of structures, (3) aging of public schools and utilities, and (4) commercialization of properties (partial conversion to business use).

What can help this type of older residential neighborhood? Zoning may help by preventing the incursion of new, adverse land uses such as manufacturing, commercial uses, and warehousing. Such uses are likely to appear in older residential neighborhoods located next to commercial areas. First, it is necessary to prevent this kind of land-use invasion by establishing a clearly defined boundary around the neighborhood and by permitting only residentially com-patible uses. Zoning may help by providing a favorable climate for continued residential occupancy of the area. This may be done by recognizing that the older type of home found in the area may no longer be practical for modern accommodations, even if it is still structurally sound. What can be done if it is found, from real estate market analysis, that an older area is declining because of its inability to compete with the modern suburban home? It may be well to recognize the fact that such older areas are open to a greater threat from the introduction of commercial uses than they are from the introduction of different residential building types. It is sometimes better in such communities to permit scattered apartment clusters when individual older residences have deteriorated and have outlived their economic life.

A second general type of residential neighborhood is the moderately old single-family residential area. This is the residential area that was developed perhaps in the 1920s and early 1930s. Such areas are often characterized by spacious lots, two-level single-family detached dwellings, and detached garages. Often this type of neighborhood was originally a tract development created by a single realtor. It may have many advantages such as complete public facilities, a relatively modern school, and adequate park and playground property. It may also be farther from the core of the community, so that it stands less chance of becoming mixed with industrial and commercial uses.

A third type of residential neighborhood is one that is in transition from residential to commercial uses. It is usually an older area, but new areas are also affected. It is a neighborhood that already has mixed land uses (some business or industrial as well as residential uses) and mixed residential building types (duplexes and apartments as well as single-family homes). Such neighborhoods are found both close to the downtown core and on the fringe of the community. In the very old area close to downtown where there has been no zoning protection, older residential homes may have been demolished for business and industrial uses in a haphazard pattern so that today homes are next door to businesses. In the newer, outlying areas, vacant land may have been developed at different times for different land uses, with modern residential tract developments going up next to older commercial and industrial establishments. The buildings of the transitional neighborhood are usually old and undermaintained. In some places, modern developments are pushing out older ones. In any case, the transitional neighborhood is changing, and the problem is to direct the change.

The fourth kind of residential neighborhood is the new residential neighborhood located in the outlying fringe areas—the emerging neighborhood. This neighborhood may be only partially developed at this point. Its problems often are merely ones of providing full public facilities such as streets, curbs, gutters, storm drainage, schools, and recreation areas. The emerging neighborhood with vacant land for development is often the place for a new type of urban development that permits the intermingling of different building types.

The fifth general type of neighborhood is the residential "remnant" neighborhood. Remnant neighborhoods are the residential "pockets" of two or three blocks that are found scattered throughout a community. Often these are the remaining segments of once-flourishing communities. Within almost all communities, there is a section "on the other side of the tracks" that is a small slum remnant of a once-thriving com-munity. The district may be completely surrounded by industrial uses. Some blocks may have two or three kinds of land uses. This kind of residential pocket may be slated for ultimate redevelopment by either private or public uses over a long period of time.

What might be called a sixth "neighborhood," but is not really a neighborhood, is the isolated housing unit found scattered throughout the community. These residential fragments are generally not zoned for residential use. They are a part of the industrial or commercial area.

LOCATION

Location criteria include factors of health, safety, convenience, amenity, and economy, all related to a residential lifestyle. In general, three basic examinations for any housing site should be undertaken. These are:

1. The environmental setting
2. The neighborhood setting
3. The ancillary setting

The Environmental Setting

Residential communities should be compatible with the carrying capacity of the natural environment. This means that topography and soil conditions must be considered to ensure the stability of the physical structures as well as to identify and preserve unique natural features for protection and preservation. In addition, the density of development should relate well to the availability of a continuous safe yield of potable water and to the capacity of the environment to absorb the soil and sewage waste of the community. Wind directions should be considered vis-à-vis potential smoke and fume pollution.

The Neighborhood Setting

Whether the new housing is to be built in an existing community or is part of a newly created one, the primary awareness must be of the general neighborhood condition. In existing communities the significant indexes include the age and condition of buildings, the relative stability and/or direction of growth, accessibility, the mix of housing types, and the existing nonresidential uses. In new communities the two major factors are accessibility and compatibility with the surrounding land uses.

The Ancillary Setting

The prime concern is the adequacy of community facilities for work, shopping, and community activity including education, recreation, religion, health services, and protection. The nonresidential uses should be in compatible settings, with sufficient buffers to ensure protection from noise, air, and visual pollution. The following list typifies compatible and noncompatible uses.

It is not usually necessary or practical for all services to be within close physical proximity to the residential neighborhood. Therefore, accessibility in terms of convenience and safety is an important community feature. The following pages illustrate general standards and criteria for location, compatibility, and accessibility.

Compatibility

The neighborhood or immediate surrounding area must also be carefully considered with the new housing. The condition of structures and character of the neighborhood are significant

TYPES OF NEIGHBORHOODS

indications. Most important to determine would be the prevailing trend of the neighborhood, for example: An existing so-called stable neighborhood may be deteriorating and on its way to becoming a slum area, or vice versa. Since housing is expected to be economically viable for a 40- to 50-year period, the eternal factors acting upon a neighborhood are essential to identify.

Listed below are some general land uses and their degree of desirability as neighbors for a housing development:

Desirable	Acceptable	Not acceptable
Parks and Playgrounds	Stores or shops	Industrial uses, not properly screened
Elementary Schools	Highways with buffer strips	Airports
Churches	Housing in fair condition	Highways without buffer strips
Housing, in Good Condition	Industrial park	Warehouses
Local Shopping	High school	Railroad tracks and yards
Medical Facilities		Deteriorated or dilapidated housing

Private Automobile

Currently the automobile is the most convenient way to travel. However, it is expensive and highly inefficient as a means of moving large numbers of people from their homes to work or other facilities. Automobiles require highways to get to a destination and once there need parking space. Both of these requirements commit large amounts of land that can be utilized more productively. Also, in recent years the automobile has become the major culprit in atmospheric pollution.

The figure on p. 5 indicates the acceptable accessibility criteria for various destinations.

Accessibility

The housing development must have convenient routes of access to employment, shopping, institutional needs, and recreation. Location near major highways or public transportation is desirable.

The three methods of access are:
1. Walking
2. Public transportation
3. Private automobile

Walking

The simplest method of access to any facility would be by walking. Even though many facilities cannot be located within walking distance, effort should be made to have as many as possible of the facilities used most frequently nearby.

A variation of walking is the use of bicycles. This greatly increases the range of accessibility but is still highly economical in time and money. Provisions should be made for the maximum utilization of bicycles within the housing development and the surrounding circulation network. This can be achieved by the use of special bike lanes parallel to the street or completely independent bikeways away from the vehicular traffic.

Public Transportation

Aside from walking, planners generally agree that mass transportation modes provide the most rational means of circulation within urban and suburban settings. This includes trains, buses, monorails, subways, minibuses, and many other variations. Public transportation is more efficient and can move greater numbers of people than any other means. Also, mass transportation uses less land for right-of-ways and reduces air pollution in sharp contrast to the automobile.

Necessary utilities must be readily available to the site if the development is to be built. These include water, sanitary and storm sewers, gas, and electricity.

Water

The best source for water is to have a public water supply system. This usually means water of sufficient quantity and quality as to present no difficulties. However, if no public system exists, it will be necessary to provide a private system utilizing well water, if it is readily available. In dealing with detached houses, use of individual wells may be explored. Not only does the use of wells add cost but additional precautions are necessary to ensure protection of the water system from contamination from septic tanks or sewers. In addition, storage tanks most likely will be required to meet the demand of peak-hour consumption, constant pressure, and a 2-h fire-protection reserve.

Sanitary Sewers

The best method is to have a public system available to connect with the new sewer lines. This will generally provide proper sewage treatment and disposal at a distant plant. In lieu of an available public system, it will be necessary to provide a private self-contained package type of sewage-treatment plant. Despite the initial cost, this method is acceptable. Provisions should be made to hook up to a public system, if and when one does become available. The least recommended method is to use septic tanks. The disadvantage of septic tanks is the possible contamination of nearby wells or of the water supply and the need for drainage fields. If the soil does not have proper percolation, this may seriously limit the use of the land.

Storm Sewers

Storm sewers are generally recommended to be separated from sanitary sewers. As with the other utilities, it is best that a public system of storm sewers exist. If none exists, discharge into adjacent lakes or streams may be satisfactory. Precautions are necessary to protect these natural water courses from becoming polluted. Permission to discharge into them may also be required from local and state authorities.

Electricity

An adequate supply of power both for present and for future use is essential. The amount of electricity used has steadily increased over the years and, from all indications, will continue to do so in the near future. Consultation and planning with local utility companies is recommended. Service lines should be placed underground to minimize disruptions from bad weather and to improve the visual aspects of the landscape.

Gas

Gas mains, if required, should be carefully located and protected from possible damage.

Cost analysis should be made for installing new gas mains as opposed to utilizing an all-electric system.

Density

Density of residential development is the ratio of occupancy (dwellings, persons, families or habitable rooms, etc.) to land area (acre, hectare, or square mile, etc.). It can be expressed in different ways, according to the choice of terms for occupancy or land area. Of all measures of intensity, density is the one most commonly accepted as reflecting the livability of multifamily housing.

In the United States the most consistently used ratio is number of dwellings (or families) per acre. Usually no distinctions are made regarding the size of individual dwellings and/or number of persons occupying them. Project size acreage is generally expressed in net acres. Areas devoted to commercial activities, community facilities, public recreation, and major roads are excluded from net acreage computations, but these are included in gross acre computations. The most frequent variation in the expression of density consists of substituting persons for dwellings. This is often used when the housing is not occupied by families of average size. For instance, in stating the density of a dormitory group, persons per acre is more meaningful than either dwelling units (since the latter are frequently single sleeping rooms) or families (which might consist of one person).

The two components of the density ratio, occupancy and size of site, are also meaningful when considered independently. Occupancy is a measure of activity on a site and is also significant in an appraisal of the adequacy of community facilities. The provision of schools, recreation space, shops, and similar facilities is directly related to number of persons. Site size, as a separate component, influences the quality of a housing project. While density can be described in a straightforward fashion, degrees of density are difficult to state in absolute terms. What constitutes high, medium, or low density is relative to a number of factors, namely, a county's or city's tradition of residential development, location of housing, and building type. For example, high-density housing on the fringe of a city may be considered medium or even low density at its core. Fifteen dwelling units to the acre is regarded as high for single-family detached structures, medium for row houses, and low for multistory structures.

Coverage

Coverage is the percentage of land occupied by structures. The higher the coverage, the less open land for outdoor recreation, gardens, parking spaces, and other needs.

Coverage is a somewhat misleading measure for large projects with a variety of building types because it is a percentage for the entire project and may not represent the coverage of any specific building group. It would be more accurate to measure coverage separately for each subarea of similar building types.

Floor Area Ratio

Floor area ratio is defined by the American Public Health Association as "the total floor area of all stories used for residential purposes, divided by the area of residential land." It is a measure of building bulk, and is often preferred over coverage because the latter fails to reflect above-

TABLE 1   Comparison of Housing Formation and Rate of Population Growth, U.S., 1920–1970

	No. of households	Percent increase	Population	Percent increase	Average household size
1920	17.6	24.5	74.1		4.2
1930	23.3	32.4	92.6	25.0	4.0
1940	27.7	19.1	110.5	9.5	3.67
1950	37.1	33.7	127.6	25.8	3.37
1960	47.5	11.6		10.1	3.33
1970	56.2	8.8			3.17

ground development. On the other hand, floor area ratio does not reveal the amount of open space available on a site. A one-story building that covers 100 percent of a site and a two-story building that covers 50 percent of a site both have a floor area ratio of 1.0. Because neither floor area ratio nor coverage alone describes the characteristics of residential development, they are often used in combination.

Since architects, to date, have not yet designed a truly flexible house that can expand and contract as the needs of the family change, the solution must be achieved in a different manner.

The only other solution to retain neighborhood stability is to provide, within the neighborhood, sufficient variety of living units that will enable most of the families to provide for their needs at different stages of their development. The chief factors in planning for housing center around the "family" and the "household" rather than the individual. This results from the fact that the family represents the basic social unit in our society today and that a house or apartment is generally occupied by a "family."

The difference between a "household" and a "family," as used by the Census Bureau, is highly significant. A household is defined as all the persons who occupy a housing unit. A household includes the related family members and all the unrelated persons who share the housing unit. A "family" refers to a group of two persons or more related by blood, marriage, or adoption and residing together.

The rate of household formation in the United States has consistently been greater than the rate of population growth. This can be seen in Table 1.

The rate of household formation is a highly important factor in housing demand. Also, most important for housing is that the average size of households in the United States has continually decreased. In 1920 the average household size was 4.2 persons, but in 1970 it was reduced to 3.17 persons. The decrease in number of persons per household during the last decade resulted from an increase in the proportion of households maintained by both young and older adults living alone and from a decrease in the average number of children under 18 years of age per household.

Rental

The rental type of housing provides for less permanent type of occupancy. It also provides smaller housing units for newly married couples and for elderly people. The major criticism is that it provides transient-type occupancy with people who have no permanent roots to the community. However, it does give flexibility in providing a range of apartment types and sizes. The number of school-age children depends on the size of apartments.

Cooperative

This type of housing provides a type of occupancy that encourages participation in the ownership of the project. It creates a sense of communal ownership and belonging to an identifiable group. It makes for permanent residency.

Condominium

This provides ownership of housing units with all its related considerations. It avoids all the maintenance chores, yet enables occupants to remain aloof of community participation. It is a relatively permanent form of occupancy. The number of school-age children depends on the size of apartments.

HOUSING TYPES AND SERVICES

Type	Individual capability	Housing and advantages for the individual	Housing problems for the individual	Population most likely housed	Approximate percentage of persons housed
Fully independent household:					
Single-family house	Self-contained, self-sufficient household; residents capable of own personal care, housekeeping and cooking	Familiar surroundings, low monthly rent	Property taxes, maintenance costs, freezes property equity for alternative uses of these funds	All socioeconomic and racial groups have an overwhelming preference for this type of housing; primary groups are married couples	69.5% total; 59.8% (estimated) are fully independent
Apartment house		Freezes assets from previously owned home, few building maintenance responsibilities	Relatively higher monthly rent; little management provision for supportive services; tenants of low capability not acceptable and will be moved out	All socioeconomic and racial groups, usually in age ranges from late sixties to late seventies; about two-thirds are single persons	21.3% total (includes an estimated 3% federally subsidized housing)
Retirement community		Exposure/emergence in active age peer social system, many recreational amenities, few building maintenance responsibilities	High rent and activity costs; little management provision for supportive services; tenants of low capability not acceptable and will be moved out	Mainly upper-middle to upper income, Anglo populations, usually these people are relatively younger (55–70), more active, and generally couples	About 1%
Mobile home		Relatively low cost new housing, some recreational amenities, few building maintenance responsibilities	Little management provision for supportive services; tenants of low capability are not acceptable and will be moved out; location usually in a noncentral area	Mainly middle-lower to middle income, Anglo, in middle sixties to late seventies; both couples and singles	About 1%
Semi-independent household:					
Single-family house	Not self-contained, not self-sufficient; residents are capable of own personal care but may require assistance in cooking and housekeeping	Same as single-family house above	Same as single-family house above; supportive services required	Same as single-family homes above	Estimated 8.7% are semi-independent
Apartment house		Same as apartment above	Same as apartment above; supportive services required may not be available	Same as apartments above	Estimated 0.5%
Boarding house		Meals, housekeeping provided; extensive social interaction available; frees assets from previous home	Moderate cost; regimentation to routine and regulations of housing unit; zoning for such facilities may place them in unfamiliar and undesirable setting	Primarily single men, low to low-middle income	2%
Residence with children, other relatives		Low cost for older person; availability of intimate support groups; few building maintenance responsibilities; may permit freeing assets in former residence	Loss of much individual freedom and privacy; stressful on interpersonal relationships	Principally single elderly women, middle income and below; slight variation among racial and ethnic groups	11.2% total estimated 2.2% semi-independent
Retirement hotels or apartments	Can still be self-contained, but is less self-sufficient; residents capable of personal care; cooking and housekeeping tasks are incorporated into the housing program; health care is available on emergency basis	Meals, housekeeping services provided; extensive social interaction possible; limited health services coverage; project usually does not provide life care	Low to high cost depending on quality of accommodations; regimentation of meal and housing rules; frequently very incapacitated will have to relocate to more supportive environment; building may not be well designed (remodeled) to accommodate the needs of the tenant population; also may be poorly located	Population ranges over all income groups because of the variance in the cost and quality of supportive services; tenants generally are in the early to middle eighties; sex balance; mainly single	Estimated 0.5%
Homes for the aged		Same as above, except provides life care	High rent and monthly service cost, many also have high enrollment cost; regimentation to routine of meals and other housing rules	Principally middle- to upper-income Anglos, both couples and singles (Section 202 housing does permit limited access to low-middle income); average tenant ages range from middle seventies to late eighties	1% total; estimated 0.8% at least semi-independent
Dependent household:					
Intermediate-care housing	Neither self-contained nor self-sufficient; help given if needed in getting about; personal care, grooming, etc., in addition to cooking and household tasks; short-term health care available if needed	Same as above, large proportion of costs are payable under Medicare and other health coverage	Moderate to high cost; many of these facilities are converted nursing homes and may have the appearance and orientation of these institutions	Same as above, except regarding sex balance—this will tend to be dominated by females; income range is explained because many higher-income people will be receiving this care under the life care plans of their home for the aged	Estimated 0.2%
Nursing homes	Neither self-contained nor self-sufficient; total care available for health, personal, and household functions	Full range of medical and personal services; large proportion of costs are payable under Medicare and other health coverage	High cost, institutional environmental, negatively perceived as terminal-care facility	Same as above	5%

Fig. 1

LARGE-SCALE DEVELOPMENTS

A project involves more than one building on a large site, usually a superblock. The type of housing can be either low- or high-rise. The site is characterized by low land coverage (20 to 40 percent) and provisions for basic community facilities, such as play areas and sitting areas. Construction is dependent upon height of building. Because of low lot coverage, the project often has extensive landscaping and open areas.

The project usually is under one ownership, and the dwelling units are rented. In recent years there has been an increase in cooperative and condominium ownership.

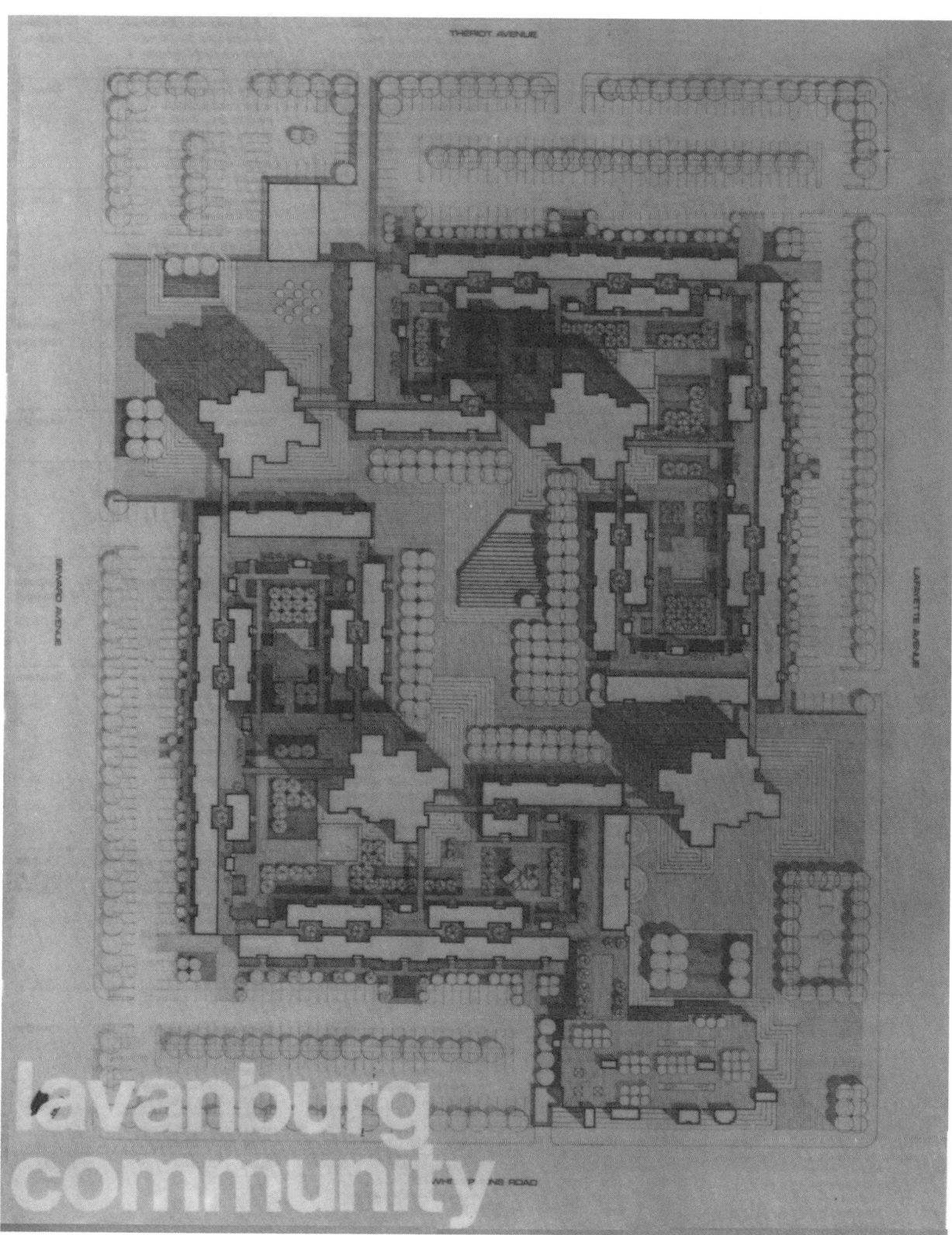

Fig. 1   Site plan. Conklin and Rossant, Architects.

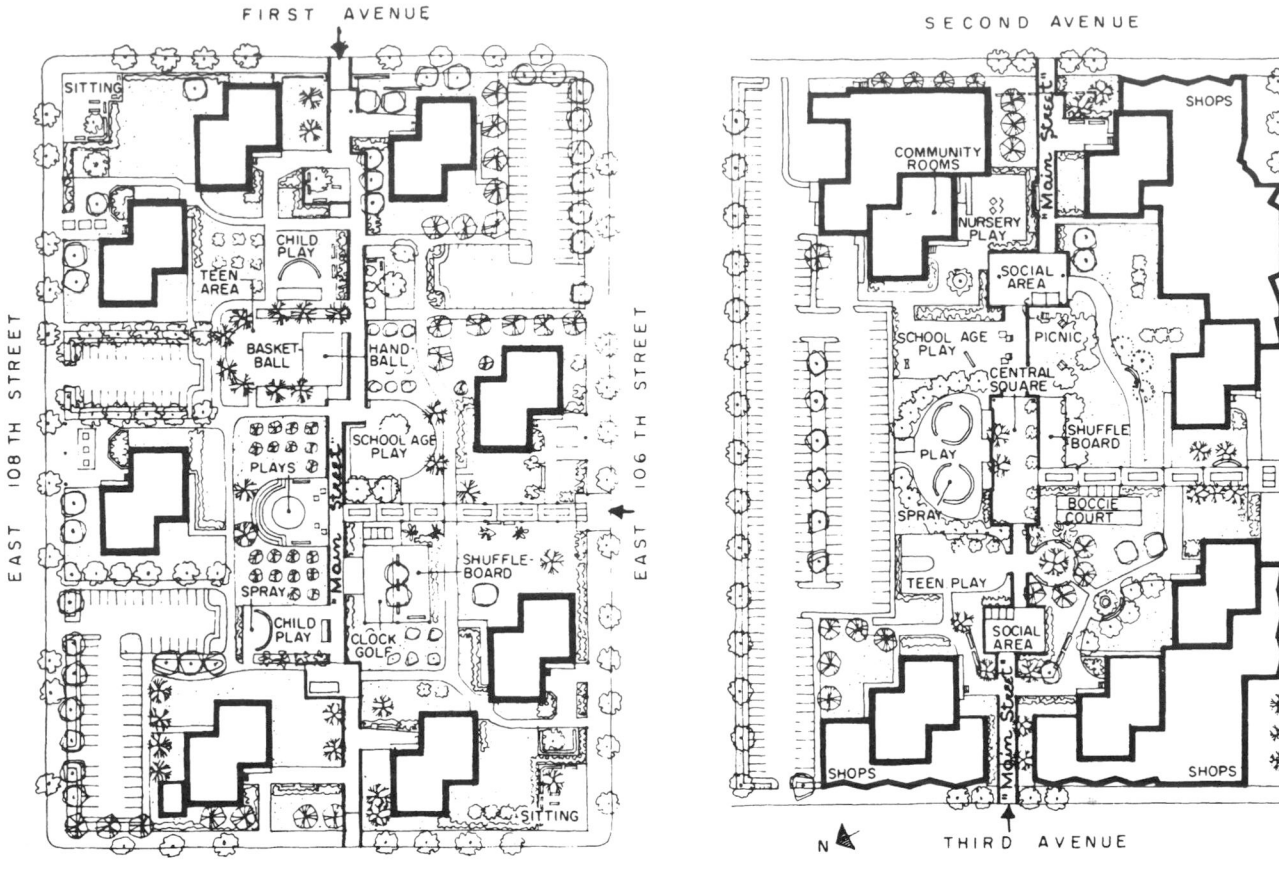

Fig. 2   Franklin Plaza, New York City.

The Superblock

The superblock is a relatively large area, usually containing one or more common open spaces, and bounded on all sides by major arterial roads. Secondary and cul-de-sac streets branch out from the major arterial road to provide internal access and service to the residential and nonresidential elements. No through traffic is permitted. Many variations on the superblock are possible. The area of the superblock frequently is used as the basis for the physical organization of the neighborhood or subneighborhood. The superblock is characterized by an articulated street system that will provide proper circulation and access to all buildings, service areas, and parking facilities. Another characteristic of the superblock is the provision of a variety of open spaces and their integration with the residential uses. This would include play lots, playgrounds, and sitting areas. Another important element is a functional layout of pedestrian walks and paths, one that will provide for the complete separation of vehicular and pedestrian circulation.

Air rights are simply rights to use the space above a particular area, without disturbing the primary use of the land itself. The most logical areas that lend themselves to such use are open or one-story uses, such as highway, railroads, streets, parking lots, and low structures.

Such use enables the land to be fully utilized. It brings a greater return to the owner and provides greater concentration of development.

In a time when land is becoming increasingly scarce, especially in urban areas, the use of air rights provides an important source of sites. In most cases, the uses will be different and the question of compatibility arises. Care must be taken to protect the individual functions so that they do not interfere with each other.

In recent years, by utilizing air rights, housing has been constructed or proposed over freeways, railroads, schools, and commercial development.

From a planning standpoint, the utilization of such air rights would provide additional opportunity for necessary development in the community. It could provide greater concentration and efficient functional relationship of the different uses. Such air rights would also make property more valuable and thereby increase taxes on these properties.

Air rights would enable some community development to occur vertically rather than constantly to seek new horizontal expansion.

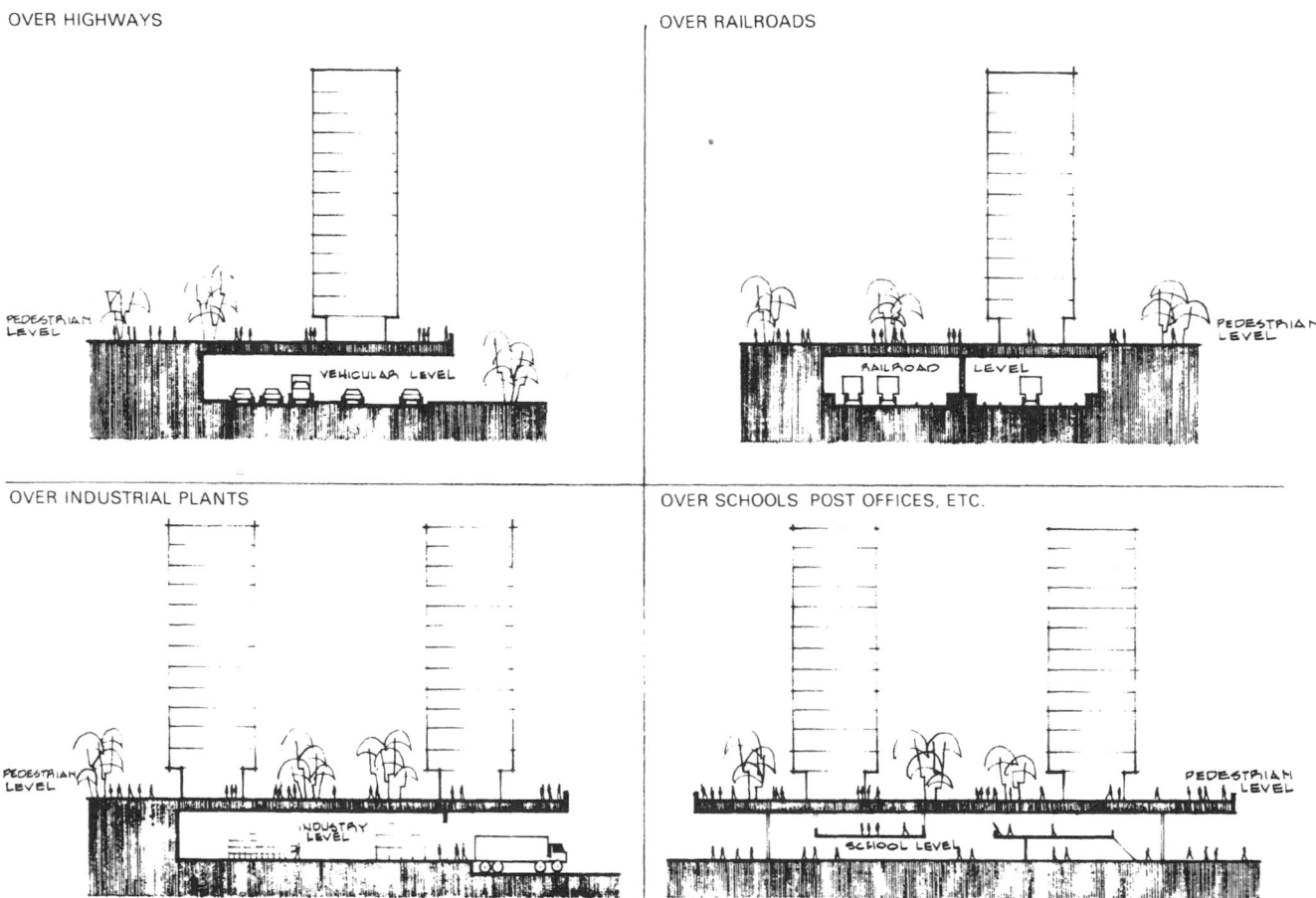

Fig. 1 Air rights/residential development.

DEVELOPMENT OVER WATER

The scarcity of land for new housing sites has led to building over bodies of water. This takes the form of filling in low swamp and marshlands or simply extending the shoreline out into a body of water with proper fill. Recently, various proposals have been made that would create sufficient platform areas over water to build an entire city. Figures 1 to 4 are the four practical methods as to how this can be achieved.

Land filling, as shown in Fig. 1, is the basic method for building in the water. Involving simply the displacement of water and other materials, this method is practical at shallow depths when fill is readily available. But because fill is paid for by the cubic yard, the cost rises directly as the depth of the water increases. The island is used for a hypothetical generating plant; in this instance the great weights of the buildings are carried by caisson to the firm underbottom. The sides of the island are armored with stone to resist the abrasion of the waves and currents. An actual 16-acre island was built by Detroit Edison Co. for a power plant at Harbor Beach, Michigan, in Lake Huron. The water ranges from 6 to 20 ft deep, a very practical depth for fill. A protective breakwater runs behind the island.

Fig. 1

Polders of dry land are made by putting up dikes to restrain the waters, and exposing the former sea bottom or lake bottom, as shown in Fig. 2. The dikes themselves are built much like long strips of landfill, except that they must be put together somewhat more carefully, with an eye to restraining the inevitable seepage. This seepage, sometimes treated for pollutants, is pumped out along with rain water. Polders are generally more economical than landfill for reclaiming large areas because less material is used. In the above example, an office building stands within the polder. The Dutch have seized more than 1,600,000 acres of land from the sea since the thirteenth century and today are reclaiming land for about $2000 per acre. So far, the deepest polder in the Netherlands, near Rotterdam, is 21 ft below sea level.

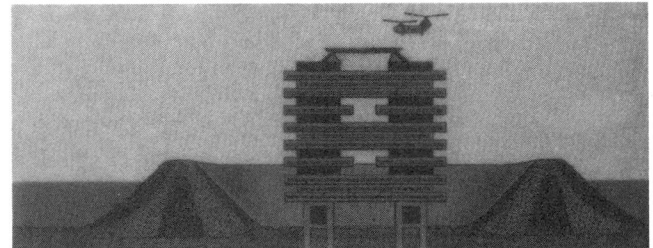

Fig. 2

Piles are used to support many water structures, from the wooden poles on which most of Venice rests to the gigantic steel legs that support offshore oil-drilling platforms. Piles are made also of concrete, but whatever their material, they must be driven into firm bottom soil if, as in Fig. 3, they are intended to hold up any kind of permanent structure. Piling has two important advantages over the methods shown above. Structures can be erected at greater practical depths, and instead of blocking normal water currents, they permit them to pass underneath. The deepest known fixed piling installation is an oil platform in 340 ft of water in the Gulf of Mexico. Freeport Sulphur Co.'s Caminada mine is located 6 miles off the Louisiana shore in 50 ft of water. A heated, insulated underwater pipeline was trenched into the Gulf floor to carry the molten sulfur to the mainland.

Fig. 3

Floating structures have been designed to support airports as shown in Fig. 4, oil rigs and housing projects. The most flexible form of building on water, floats can be moored at varying depths without increasing their cost and can be moved from one location to another. As a rule, floating structures cost more than other methods of building on water. They must be even stronger than most ships because they are not intended to ride with the force of the waves, but to resist it. Figure 4 illustrates two methods of improving seaworthiness. First, the flotation chambers that make the airport buoyant are submerged so that they will be below the greatest turbulence in storms, permitting waves to pass through a fairly open framework. Second, a protective breakwater of large bags, partially filled with water, is moored around the airport to absorb some of the force of the waves.

Fig. 4

RESIDENTIAL DENSITY

DENSITY OF RESIDENTIAL DEVELOPMENT

This section deals with selected methods of measuring the density of development within residential areas of a neighborhood, and of determining proper limits for such density.

Governing Criteria for Density

The intensity of land use should not be so great as to cause congestion of buildings or to preclude the amenities of good housing. Specifically, densities should be limited to provide:

1. Adequate daylight, sunlight, air, and usable open space for all dwellings
2. Adequate space for all community facilities
3. A general feeling of openness and privacy

Densities should have a reasonable relationship to land and improvement costs. Two types of density measurement are needed:

1. Density measures for residential areas of the neighborhood (called residential or dwelling densities) to ensure adequate open space, light, and air for residential facilities
2. Density measures for the entire neighborhood (termed neighborhood densities), taking all land uses into account, to ensure provision of adequate community facilities in relation to population load

Site Planning Characteristics Reflected by Residential Density

The importance of density measurement as a planning tool arises from the fact that densities reflect with a certain degree of accuracy important characteristics of site planning. Densities show the crowding of people and structures on the land and the amount of open space available to the families. For example, the percent of land covered by buildings reflects in general the amount of open space available for gardens, children's play, outdoor living, the drying of laundry, and the like.

Since densities bear an obvious relation to the spacing of buildings and their height, another important factor is measured by densities, namely, the approximate amount of light and air admitted to dwellings.

Density standards are useful as a guide for preliminary design schemes, and for estimating population loads and required areas of land. Density measurements provide a uniform and objective method of comparison of site plans for general openness, amenity, and livability.

Density standards have major value as controls in zoning ordinances, subdivision regulations, and the like. Proper standards, carried out through competent design, give assurance that land crowding, encroachments on daylight, and similar blight-inducing factors will be controlled.

It must be recognized that density figures, no matter how accurately computed, are but a crude index of the design quality of a site plan. Being rigid mathematical ratios for relatively large areas, they cannot properly reflect all factors of design. For example, suitable average densities for large tracts of land will not necessarily ensure that buildings are not crowded together in some parts of the development area. The amount of open space established by density standards has limited meaning unless that space is properly distributed and designed for usability.

Good design practice can provide adequate open space for all outdoor functions of family life at relatively high densities. On the other hand, poor site planning may create land crowding and lack of usable open space even at low densities. In addition to meeting density standards, therefore, residential areas must also comply with all standards for spacing of structures, orientation, and other features of site layout.

Measures of Density

The intensity of residential use can be expressed by different types of density calculations, showing mathematical relationships between the area of a given piece of land and the population load or building bulk. Area measurements are usually given in acres, population load as number of persons or families, and building bulk in terms of ground area covered or total floor area. Thus, for example, population density is expressed as the number of persons (or families) per acre of land, or as acres of land per 1000 persons, and dwelling density as the number of dwelling units per acre of land or as the number of acres (or square feet) of land per dwelling unit.

A complete discussion of the many methods of density measurement used for planning or regulatory purposes is beyond the scope of this book. A limited number of density factors relating to residential land use have been selected for further discussion on the basis that they seem to best reflect the characteristics of the site plan.

Net dwelling density The number of dwelling units per acre of net residential land (land devoted to residential buildings and accessory uses on the same lots, such as informal open space, drives, and service areas, but excluding land for streets, public parking, playgrounds, and nonresidential buildings).[1]

Building coverage The proportion of net or gross residential land taken up by buildings.

Building bulk (floor area ratio) The total floor area of all stories used for residential purposes, divided by the area of residential land.

Application of each of these measures of residential density is discussed in later paragraphs of this section.

Useful as they are in planning residential land, residential densities are not an adequate measure of land use in the neighborhood as a whole. Requirements as to light and air, for instance, can be met in terms of residential densities that may still overtax the available schools, playgrounds, streets, or other community facilities. The building up of one tract after another on the basis of maximum residential densities alone, without regard for these neighborhood elements, will lead to most serious land crowding.

A further type of density measurement is therefore needed:

Neighborhood density The number of dwelling units per acre of total neighborhood land (new residential land plus streets and land used for schools, recreation, shopping, and other neighborhood community purposes).[2]

[1]Gross dwelling density, a measurement much used in the past, is not employed in this section. Gross density is the number of dwelling units per acre of gross residential land (land as described above, plus bordering streets up to limited distances—ordinarily to the center of the street).

It is one purpose of this book to encourage replacement of the gross dwelling density concept by that of overall neighborhood density.

[2]Neighborhood land excludes nonneighborhood uses and unusable land within the neighborhood boundaries.

In addition to residential densities discussed in this chapter, neighborhood density standards must be met in order not to overload playgrounds, schools, and other community-, district- or citywide facilities.

Net Dwelling Densities: Basis of Calculation

Table 1 gives recommended area (net residential land) allowances per family with the various dwelling types.

For one- and two-family dwellings, only the recommended total lot area is shown. This is based on the sizes assumed below:

Dwelling type	Lot size or equivalent, ft	Net residential area per family, ft^2
One-family detached	60 × 100	6000
One-family semidetached	80 × 100 for two families	4000
Two-family detached	80 × 100 for two families	4000
One-family attached (row)	20 × 100 plus 40-ft side yard between each 10 units*	2400
Two-family semidetached	48 × 100 for two families	2400

*Figures are for two-story, 25-ft-minimum lot width is recommended.

For multifamily dwellings, total land area is derived from its component parts: (1) area covered by buildings, (2) outdoor living space, (3) area for service, laundry drying, walks, and setbacks, (4) off-street residential parking areas. These together constitute the net residential land area.

The area covered by multifamily buildings has been assumed on the basis of floor area allowances per family under normal contempory design and construction practice, as shown below. Total floor area of buildings (including shared circulation space) is divided by the number of stories per building. Gross floor area per family is assumed to increase as height increases, because of the need for added interior service and circulation space.

Height of building (stories)	Assumed gross floor area, ft^2	Area covered by building, ft^2
2	870 per family	435 per family
3	870 per family	290 per family
6	870 per family	145 per family
9	945 per family	105 per family
13	945 per family	75 per family (approx)

Since floor areas may vary with local design practice for various types of multifamily dwellings, the figures above should be adjusted where necessary. The effect of such adjustments on Table 1 and later tables should then be checked before these are applied to local plan solutions.

Allowances for outdoor living space in Table 1 are based on established standards, and should be complied with. Areas for service, walks, setbacks, and off-street parking are the most diffi-

TABLE 1   Allocation of Net Residential Land to Major Dwelling Uses

(Recommended allowance per family, by dwelling type and by component uses*)

Dwelling type	Land area, ft² per family†				
	Total	Covered by buildings	Outdoor living‡	Service, walks, and setback	Off-street parking
One- and two-family (individual access and services):					
1-family detached	6000	Varies	within	lot	area
1-family semidetached or 2-family detached	4000	Varies	within	lot	area
1-family attached (row) or 2-family semidetached	2400	Varies	within	lot	area
Multifamily (common access and services):					
2-story	1465	435	415	455	160
3-story	985	290	315	220	160
6-story	570	145	215	50	160
9-story	515	105	215	35	160
13-story	450	75	215	36	125

*The standards of this table apply only to net residential land. Plans for a development must comply, in addition, with neighborhood density standards for streets and community facilities.

†For basis of allowance, see text: Net Dwelling Densities: Basis of Calculation.

‡Including playlot for small children.

TABLE 2   Net Dwelling Densities and Building Coverage

(Recommended standard values, by dwelling type*)

Dwelling type	Net dwelling density, units per acre of net residential land		Net building coverage, % of net residential land built over
	Standard: desirable	Standard: maximum	Standard: maximum
One- and two-family:			
1-family detached	5	7	0
1-family semidetached or 2-family detached	10	12	0
1-family attached (row) or 2-family semidetached	16	19	30
Multifamily:			
2-story	25	30	30
3-story	40	45	30
6-story	65	75	25
9-story	75	85	20
13-story	85	95	17

*In addition to meeting the standards of this table, plans for a development must comply with neighborhood density standards for streets and community facilities.

cult to assess. The figures shown are based on generally accepted servicing and layout practice for different dwelling types. Off-street parking is calculated at 240 ft² per car, with ½ to ⅔ car per family (in multiple dwellings).

It is recognized that the figures given can serve only as a guide and that satisfactory design solutions may be achieved with different area allowances.

Net Dwelling Densities: One- and Two-Family Houses

Table 2 translates the above lot sizes and other net residential area requirements into recommended net dwelling densities for one-, two- and multifamily dwellings.

Recommended lot sizes for one- and two-family houses will result in maximum densities of 7 units per net acre of residential land for detached one-family houses, and 12 units per acre for semidetached houses of this type. One-family row houses should not normally exceed 19 dwellings per net acre. Although higher densities for these dwelling types may be compatible with standards for light and air, it is doubtful whether densities beyond these maxima will permit sufficient flexibility in design to ensure privacy and other amenities that should be obtained with one- and two-family dwellings.[3]

Although the above dwelling densities are approved as standard, lower densities (shown in Table 2) should be the goal, especially in an unfavorable location. They will permit flexibility in site layout where poor topography reduces the amount of usable space attached to the house, or where larger than normal setbacks are needed for noise reduction. Lower densities are also desirable to permit increased lot widths for privacy.

Net Dwelling Densities: Multifamily Buildings

Apartment layout makes possible the shared use of service areas, approaches, playlots, and other residential land by a number of families

[3]For instance, spacing standards for sunlight may require a minimum distance between facing rows of buildings of two times the height of the building. For one-story buildings this might permit a minimum of 20 ft between buildings. Yet 20 ft is too little to give an adequate sense of space or privacy in backyards, and cannot be considered acceptable.

and thereby permits some reduction of area allowances per family as compared with layouts in individual lots. Greater sharing of outdoor areas is possible as the number of families increases. Therefore, space allowances per family can be decreased somewhat for taller apartments housing a more concentrated population, without impairing livability.

It should also be remembered that the more stories a building has, the less ground area per family is covered by the building. Assuming, for instance, the same floor area for each family, a six-story apartment housing x families will cover only one-half of the ground covered by two three-story buildings housing (together) the same number of families. These considerations permitting higher densities as the number of stories increases, without detriment to health or amenity, are reflected in the figures of Table 2.

Densities of multifamily buildings should be kept within the desirable range of the table: from 25 units per net residential acre for two-story apartments to 85 dwellings per net residential acre for 13-story elevator apartments. Although somewhat higher densities may be attainable, it is doubtful whether satisfactory site layouts meeting all standards can be devised except under especially favorable conditions. In no case should net dwelling densities exceed the maximum figure shown in Table 2.

Net Dwelling Densities in Relation to Population Densities

Dwelling densities have the limitation that they do not measure the exact population load on residential land. The number of persons per room is likely to decrease, and floor area per person is likely to increase, from low- to high-income families. If the dwelling count is to represent the actual population load, both the dwelling sizes (number of rooms per dwelling) and occupancy condition (number of persons per room) must be taken into account.

As far as housing environment is concerned, the number of persons per acre is particularly useful as an index of the population load on the various community facilities. For this reason, standards for population density are most usefully applied on a neighborhoodwide basis. However, population load has a direct effect on the amount of residential land required for multiple dwellings. Net population densities, therefore, are useful as a guide to residential land-area

21

RESIDENTIAL DENSITY

requirements in multiple-dwelling developments.[4] There has not been sufficient research to determine the exact population densities that conform to the required amount of usable open space. Population densities should under no circumstances be so high that the outdoor residential space requirements cannot be met.

Building Coverage

Building coverage is the proportion of net or gross residential land area taken up by buildings. Thus, for instance, 40 percent net coverage means that 40 percent of the residential land area is covered by buildings, leaving 60 percent in open land for residential outdoor uses.

While building coverage bears an obvious relationship to population density, it is nonetheless a separate matter that must be considered on its own merits. Even if, by using low buildings, a low density is maintained, it is obvious that if these buildings cover too large a percentage of the land, insufficient outdoor space will remain for various uses conducive to health, and this lack of space may also result in inadequate arrangements for circulation.

Figures for building coverage are more tangible standards than those described for light and air and for other criteria that would affect building spacing, and are therefore useful in municipal regulation. However, such figures are a means of achieving an end, rather than the end itself. Poorly located buildings covering only 25 percent of the net residential land may easily admit less light to living and sleeping rooms than well-designed ones with 35 percent coverage. Coverage and height are closely interrelated, and can be established only in the process of design. At the present time, 20 to 30 percent coverage of land within property lines appears to be practical and to permit conformity with standards for light, air, and open spaces. Controls that set maximum net coverages exceeding 35 percent may fail to provide sufficient open space and may lead to overcrowding of people on the land.

In the author's opinion, no designs for arrangement of multiple dwellings have yet been published that provide for adequate sunlight (at least in latitudes of the temperate zone) and at the same time show net building cover-

age in excess of 40 percent. The lower values in Table 2 are believed in line with progressive current practice.

Net building coverage by itself is the crudest measure of residential density, and unless it is related to building height and population density, it measures no more than the approximate overall amount of outdoor space available for gardens, children's play, adults' recreation, laundry drying, driveways, private garages, etc. The usability of outdoor space will depend on its good design, and the amount of space available to each family will depend on the population load put on the land.

Building Bulk (Floor Area Ratio)

The measurement of building bulk in terms of "floor area ratios" has been found so useful as a density control that it is being applied increasingly by planners both in the United States and Britain. The floor area ratio is a comparatively recent concept that requires a clear understanding.

Floor area ratio is the total floor area of all stories[5] used for residential purposes, divided by the area of residential land.

For example, a floor area ratio of 1.00 means that the combined floor area of buildings equals the residential land area. This corresponds to a building coverage of 25 percent by four-story buildings or a coverage of 50 percent by two-story buildings. A floor area ratio of 1.20 may mean that 30 percent of the area of the land is covered by four-story structures or that 15 percent of the land is covered by eight-story structures.

Although in current zoning ordinances floor area ratios refer to net residential area, figures for floor area ratios are given here in relation to gross residential site areas (including land for streets), because, from the point of view of spacing buildings for sunlight and daylight penetration, it does not make any difference whether streets occupy some of the intervening open spaces.

Because floor area ratio establishes a mathematical relation between the land area, the floor area of the building, and its height, it is considered among the most accurate indexes for adequacy of light and air.[6] This becomes clear when floor area ratio is related to the spacing of buildings and their height. If, for instance, parallel rows of six-story buildings are spaced two and one-half times their height to permit proper sunlight admission, the floor area ratio must be approximately 1.14 with normal story height and depth of building.[7]

Based on similar computations, floor area ratios required to enable rows of buildings of different height to be spaced two and one-half times their height will range from 0.86 for three-story apartments[8] to 1.27 for nine-story elevator apartments. Apartments of thirteen stories will require a floor area ratio of 1.34.

If the above floor area ratios are used as density controls, they will generally assure adequate admission of sunshine, daylight, and air to dwellings. However, residential areas should also meet standards for dwelling densities based on usability of residential land, and they must also comply with neighborhood densities.

It should also be noted that floor area ratios do not reflect population densities, because floor area per person varies (usually increasing as income increases). In order to measure population loads, an additional index of floor area per person should be used. This makes it impossible to relate density in terms of floor area ratios to population density.

---

[4]Population density for new developments may be approximated by multiplying the net dwelling density by the average size of family (based on proposed dwelling size).

[5]The ground area of the building multiplied by the number of stories gives the total floor area (except where there are setbacks in upper stories). For instance, the total floor area of a two-story building covering 800 ft^2 of ground is 1600 ft^2 (2 × 800). For a four-story building having the same ground area of 800 ft^2, the total floor area of all stories is 3200 ft^4 (4 × 800).

[6]The mathematical relationship of floor area ratio to building coverage and height is expressed by the following formula:

$$F = \frac{G \times S}{L} = B \times S$$

where F = floor area ratio
G = ground area of building
S = number of stories
L = area of land
B = building coverage (ground area of building divided by area of land)

[7]Assuming a 10-ft story height, the distance between buildings will be 2½ × 60 ft equals 150 ft. If, furthermore, the buildings are assumed to be 35 ft deep, their coverage will be 35/185, or 19 percent. The floor area ratio will be the coverage times number of stories, or 0.19 × 6 equals 1.14.

[8]Assumed height of stories: 10 ft per story for first six stories; 85 ft for nine stories; 122 ft for thirteen stories. Assumed depth of buildings: 35 ft.

Type	Occupancy relationship	Legal control	Characteristics	Planning considerations
Rental	Tenant	Lease	An owner builds and finances the building or complex; the occupants rent their dwelling units (apartments); utilities, appliances and furnishings may be included in the rental charge; enclosed or open off-street parking may be provided at an additional charge to the tenant; maintenance and operating costs are almost always the responsibility of the owner; time of leases varies, but three years is most common; tenant families tend to be somewhat transient; rental projects are erected primarily for investment.	Tenants tend to be more transient than other types of occupants. Large apartments generate more children than small apartments. Smaller apartments will generally be occupied by single people, young married couples, or the elderly.
Cooperative	Tenant	Stock	Tenant-owner corporation own the building or complex; tenants own stock in the building or complex in proportion to the value of their dwelling units; depending upon the lease conditions, a tenant-owner may sell his stock either back to the corporation or to a new tenant-owner when he moves; mortgage, operating, maintenance and any other costs for the building or complex are paid by the tenant-owner corporation.	Families have a vested interest, will tend toward a stable occupancy; greater interest and participation will occur both in project and community affairs.
Condominium	Ownership	Ownership	A form of cooperative; occupant owns outright his dwelling unit upon which there are no restrictions as to sale, rental, or transfer; the owner-occupant is responsible for the mortgage (if he has one), operating, maintenance, and any other costs only insofar as they pertain to his dwelling unit; all spaces beyond the individually owned dwelling units are held in common ownership.	Owner will generally react as any homeowner in the community; since the unit is owned outright, the owner will invest additional funds for maintenance and upkeep; no control over buying and selling of units.

Fig. 1

CONDOMINIUM/COOPERATIVE OWNERSHIP

The purpose of the schematic diagrams and explanations set forth below is to establish a basis for the reader to understand the relationship between various housing structures and forms of ownership.

Types of Residential Structures

1. Single-unit/single-family detached

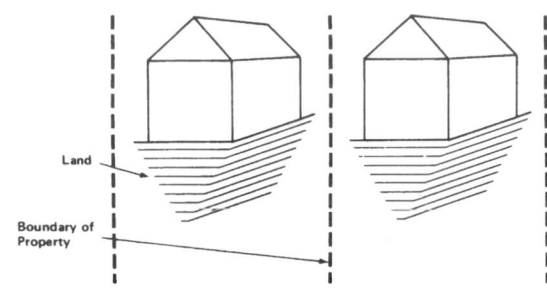

Fig. 1a

2. Single-unit/single-family attached (townhouse) (row house)

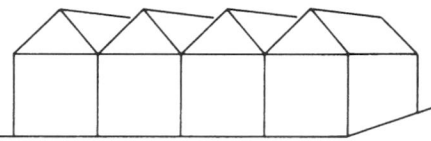

Fig. 1b

3. Fourplex

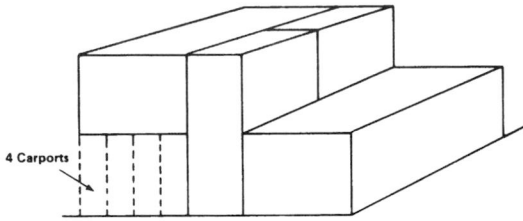

Fig. 1c

4. Garden

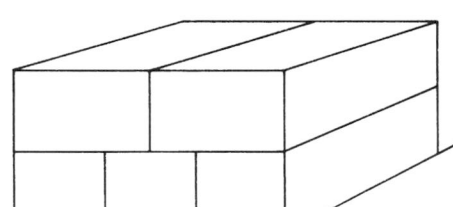

Fig. 1d

5. High-rise or low-rise, depending upon number of stories

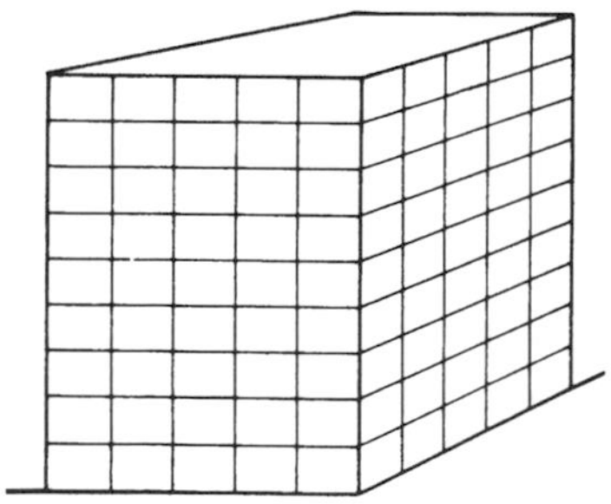

Fig. 1e

Ownership

1. *Traditional form of ownership of single-unit/single-family structures*

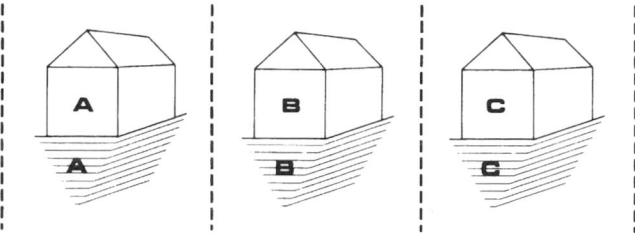

   a. Detached single-unit/single-family structures. Note: Letters on structures in Figs.
   2 through 13 indicate owners of structure. Letters on land indicate owners of
   land.

Fig. 2

   b. Attached single-unit/single-family structures (townhouse) (row house).

Fig. 3

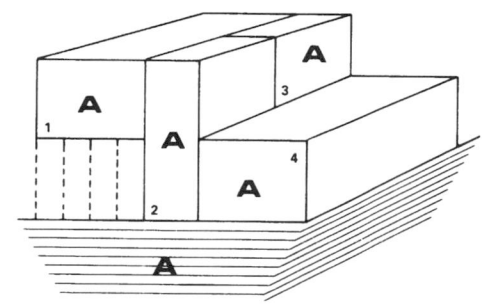

2. *Traditional form of ownership of multiu-
nit/multifamily structures:* Single-deed ownership in fee simple of multiunit structure and real
property extending to boundary of property.
Structure held for rental purposes.
   a. Multiunit/multifamily fourplex structure. Note: Numbers indicate rental
   occupants of structure owned by A.

Fig. 4

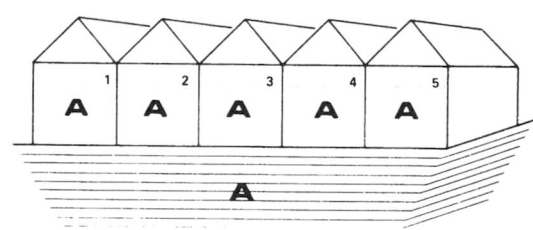

   b. Attached multiunit/multifamily structure, rental (townhouse) (row house).
   Note: Numbers in Fig. 5 indicate rental
   occupants of structure owned by A.

Fig. 5

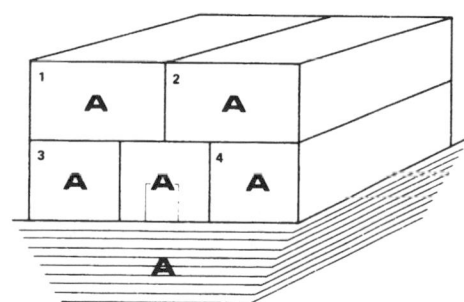

   c. Multiunit/multifamily structure, rental
   (garden). Numbers in Fig. 6 indicate
   rental occupants of structure owned
   by A.

Fig. 6

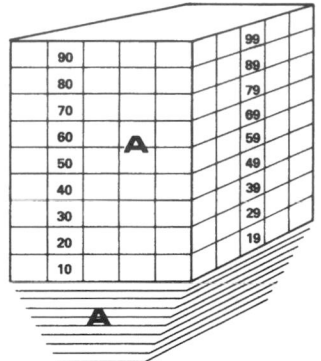

d. Multiunit/multifamily structure, rental (high-rise).

Fig. 7

3. *Development of homes with traditional form of ownership as a community (also called PUD):* Single-deed ownership in fee simple of individual home and real property extending to boundary of property. Fee-simple owners must be members of incorporated homeowners association (HOA).

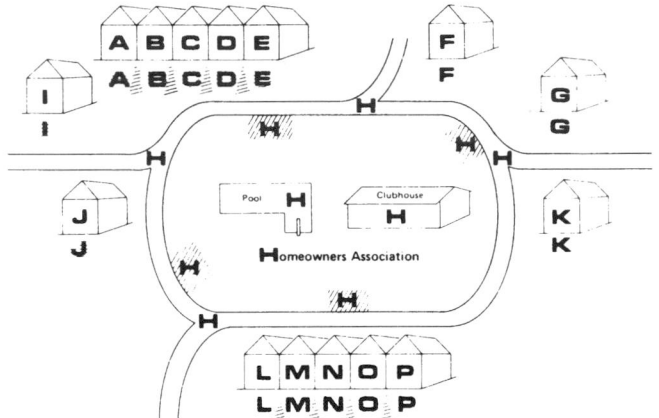

a. Usual configuration:[1] H owns title to community facilities and all land, except land directly under lots on which homes are located. This land and the homes are owned by individual homeowners (A, B, C, D, E, etc.). H also owns title to streets which are not publicly owned. A, B, C, D, E, etc., must be members of corporation H.

Fig. 8

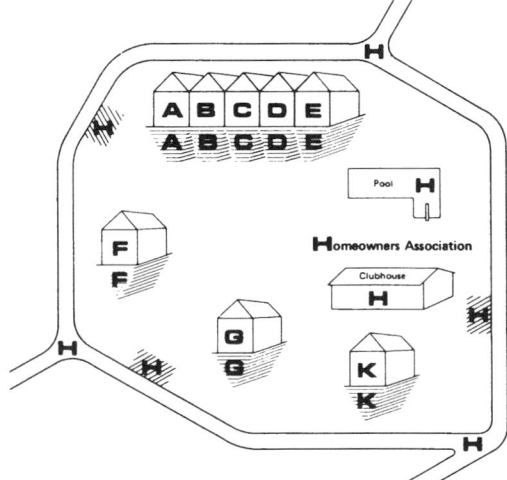

b. Zero lot line (alternate configuration): H owns title to all community facilities and land, except land directly under each structure, which is owned by owner (A, B, C, D, E, F, G, or K) of structure. H also owns title to streets which are not publicly owned. A, B, C, D, E, F, G, and K must be members of corporation H.

Fig. 9

[1]Figures 8 and 9 show single-unit/single-family detached and single-unit/single-family attached structures which are the types of structures most commonly used. However, any type of structure may be used and the legal owner of the structure is a member of the HOA. These configurations are most commonly called planned unit developments (PUDs).

4. *Cooperative corporation ownership:* Single-deed fee-simple ownership by a cooperative corporation of structures and real property (including community facilities) extending to boundary of property (streets). All stockholders (or members) in the cooperative corporation are given exclusive right to use the unit and share use of the community facilities.

Figure 10 shows different types of residential structures to display a cooperative community. It should be noted that it is most common for a cooperative to be composed of a single type of residential structure. In the HUD mortgage insurance program there must be at least five dwelling units; therefore, single-unit/single-family detached structures or single-unit/single-family attached structures must consist of at least five dwelling units. A cooperative cannot consist of a single fourplex structure (i.e., less than five units). Numbers in Fig. 10 indicate stockholders in cooperative corporation A, which owns title to all structures, streets, and community facilities.

5. *Condominium ownership:* Single-deed fee-simple ownership of individual units and an undivided interest in a fee representing the common elements (i.e., purchasers are owners of individual condominium units and partial owners of the common elements).

Condominium (or condominium development or condominium project): means the real property including structures and community facilities as recorded under condominium law.

Condominium unit: means a unit owned in fee together with an undivided interest in the common elements and areas including community facilities.

Unit: means that portion of the condominium designated for exclusive residential or commercial use.

Common elements: means the land and all portions of the condominium structures other than the units.

Developer: means the entity which causes a development to be constructed and recorded as a condominium.

Figure 11 shows different types of residential structures to display a condominium community. It should be noted that it is most common for a condominium to be composed of a single type of residential structure. In the HUD mortgage insurance program there must be at least four dwelling units; therefore, single-unit/single-family detached structures or single-unit/single-family attached structures must consist of at least four dwelling units. In this case a condominium may consist of a single fourplex structure. A–Z each own an individual unit and have an undivided interest in the common elements (including roofs and other structural elements).

    a. Expandable or add-on type of condominium: The expandable type of condominium is, as the name suggests, a single condominium development which is built in phases. The phases are separated by temporary property lines which are not lifted until after each of the subsequent phases is completed and sold. The total number of units in the ultimate condominium project must be known. For this example it is assumed that there is a total of 90 units constructed in three phases of 30 units each. Using a ratio of 1/total number of units, a purchaser of a unit in the first phase will progressively own 1/30, 1/60, then 1/90 undivided interest in the common area. Although the fraction of each first-phase purchaser decreases as each phase is added on, the smaller fraction

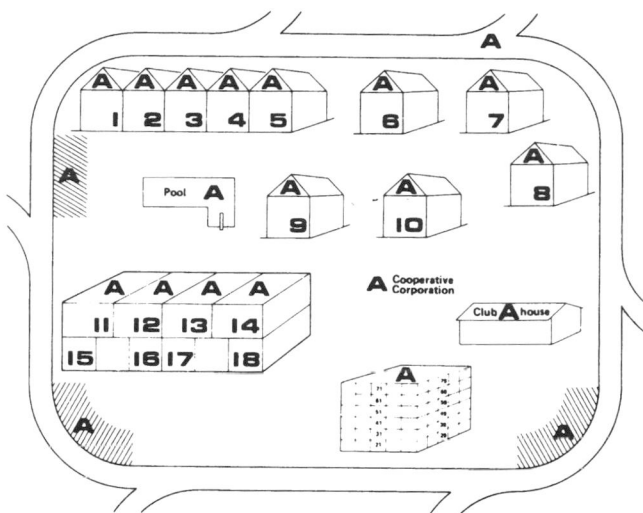

Fig. 10

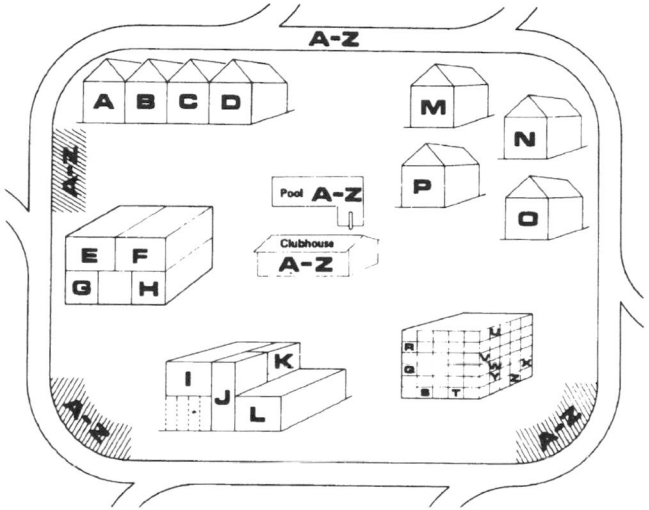

Fig. 11

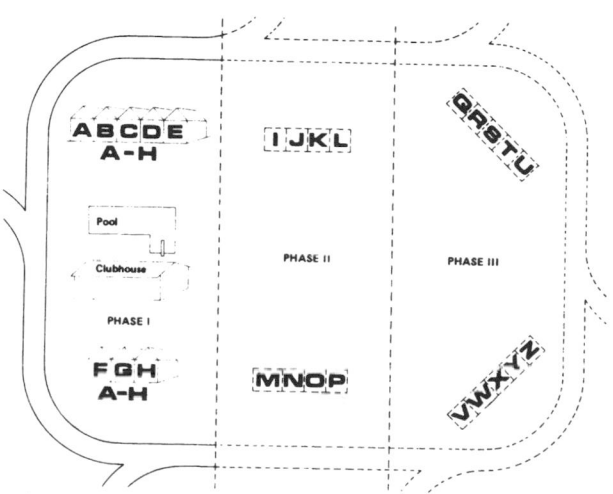

Fig. 12

CONDOMINIUM/COOPERATIVE OWNERSHIP, FAMILY CYCLE

will reflect an interest in a larger common area. The units in Fig. 12 could be structures other than townhouses, but if the units in the structures differ in size a ratio other than 1/total will have to be used (such as unit area/total area).

b. Multiple condominium (series) developments: In this type of development a series of condominiums, usually with 20 to 50 units, is constructed so they can be more readily marketed and titles can be more quickly conveyed. It should not be confused with the add-on or expandable condominium, which is only one condominium. Each of these condominiums is on a separate parcel of property in which each unit owner in that condominium has an undivided interest in its common area. No unit owner in one condominium has an undivided interest in any other condominium. The unit owners of all condominiums, however, are automatically members of the community facilities corporation. In most cases the individual condominium boards of directors delegate their rights to select contractors for maintenance of lawns, trash and snow removal, etc., to the community facilities corporation. This gives the effect that there is a master association, which is not true. An association or corporation on a separate parcel of property can only be the master of the property to which it holds title. The condominiums sketched in Fig. 13 could be made up of other types of structures than the townhouses depicted.

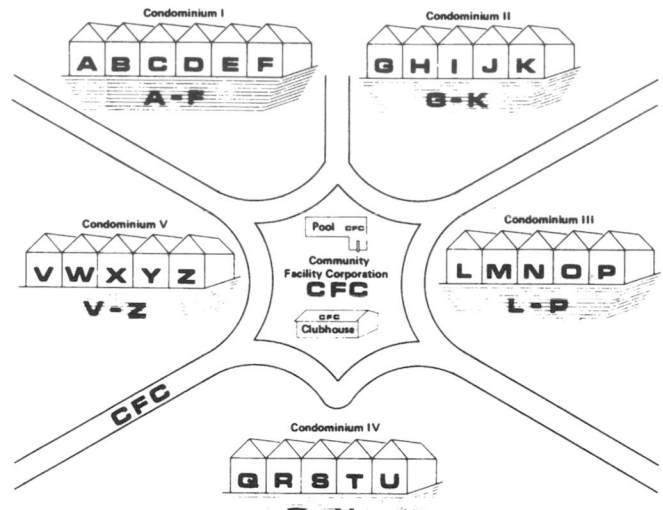

Fig. 13

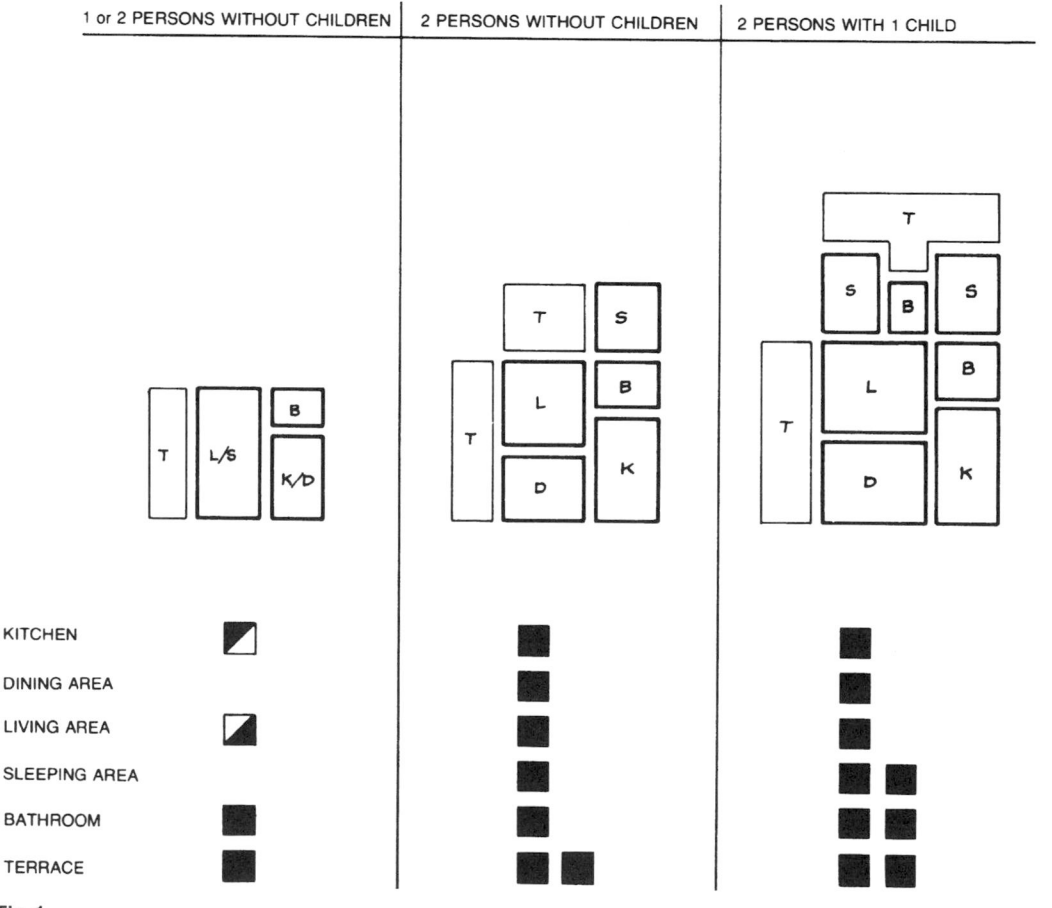

FAMILY LIFE CYCLE

One of the basic factors in housing design is the continually changing family size, e.g., organization, composition, age, and size. As a result of this continuous change, the physical space requirements also change. Most often it is a gradual process over the years.

The conflict occurs with the ever-fluctuating family organization and the inflexible physical space they occupy at a particular period in time. For example, when a family needs an additional bedroom or more recreational space, it cannot easily increase or expand its space. This is possible with detached single-family houses but becomes quite difficult to accomplish with any other type of living unit. In like manner, when the family is getting smaller, the physical space requirements will contract substantially. Again, a situation exists where the physical space does not match the family needs.

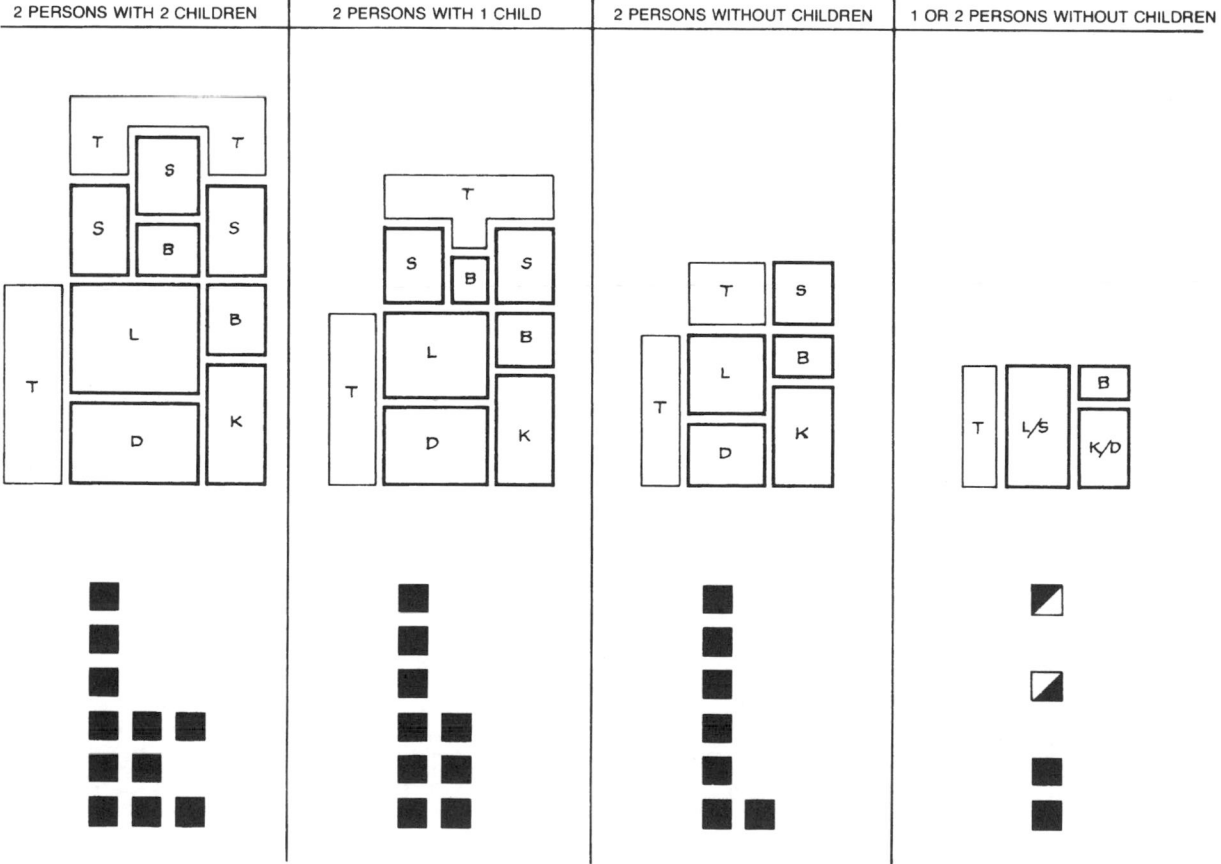

Fig. 1 *(Continued)*

INTENSITY OF DEVELOPMENT (LAND-USE INTENSITY)

The intensity of development of land for residential use has been measured throughout the country in a variety of ways. The land-use intensity (LUI) method has won increased acceptance as a comprehensive technical approach to this subject. The LUI method relates the number of living units by floor area to a recommended amount of site area according to their location and the type and size of housing. It includes useful guidance material as to needed parking space, recreation space, livability space, etc., as related to floor area.

The correlation of these elements shown in Table 4 indicates, for each LUI number, the maximum square foot amount of floor area (FA) and the minimum ratios of open space (OS), livability space (LS), and recreation space (RS), for each square foot of site or land area (LA). Also indicated are the minimum number of occupant car (OC), and total car (TC) parking spaces per living unit.

DEFINITIONS

Land Area (LA)

Land area for LUI computations is the sum of site land area for residential use plus one-half of the area of any abutting walk, alley, or street right-of-way plus one-half of the area of any abutting beneficial open space with reasonable expectancy of permanence such as streams or park land. (Countable area for abutting beneficial open space should have a maximum width of 35 ft for single-family building types, 50 ft for walkup apartments, and 70 ft for high-rise buildings, measured at right angles to the property line.)

Reduced Site Area for Steep Slopes

For nonelevator buildings on sites with 10 percent or more of their original area containing existing slopes of 20 percent or more, the land area for LUI assignment considerations should be reduced 1 percent for each total percentage point of average slope within the steep-sloped portion of the site area.

Total site area, 45,000 ft^2
Area of average 30% slope, 26,250 ft^2
30% × 26,250 ft^2, 7,875 ft^2

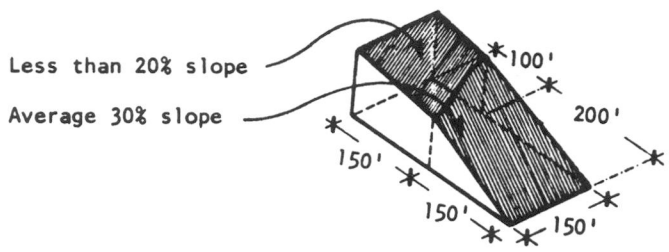

Total site area for LUI assignment, 45,000 ft^2 − 7,875 ft^2 = 37,125 ft^2

Building Area (BA)

Building area is the total land area covered by residential buildings, measured horizontally from the faces of the exterior walls (or the exterior lines of omitted walls) at main grade level. Entrance platforms, steps, and terraces are not countable as building area.

A— Single Family Detached
B— Town Houses
C— Garden Apartments
D— High Rise Apartments

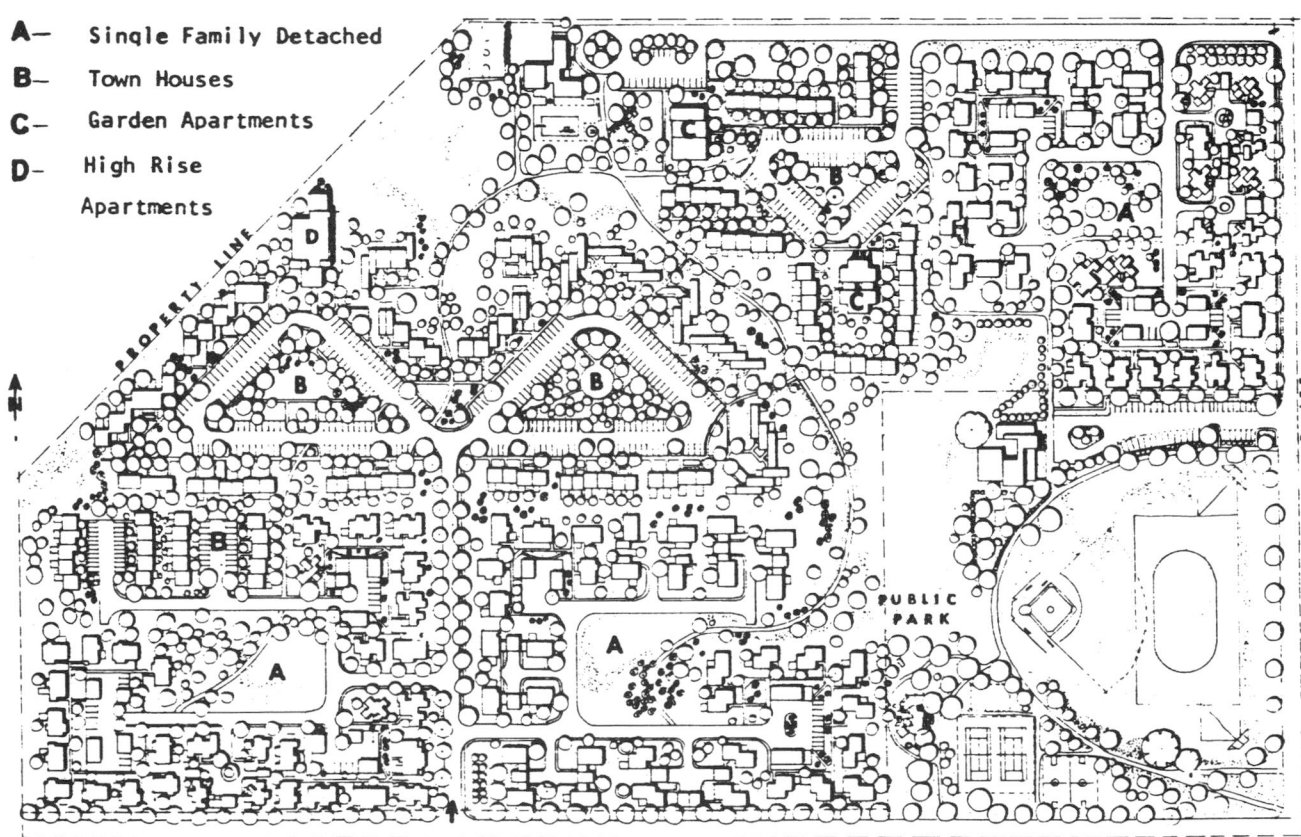

Fig. 1 Variety of building types in combination.

Floor Area (FA) and Floor Area Ratio (FAR)

Floor area is the total floor area for residential use on all floors of a building or buildings, measured from the outside faces of the exterior walls, including halls, lobbies, stairways, elevator shafts, enclosed porches, balconies, and below-grade floor areas used for habitation and residential access.

Not countable: (1) open terrace, patio, atrium, or balcony; (2) carport, garage, breezeway, or tool shed; (3) special-purpose areas for the common use of all the occupants, such as a recreation room or social hall; (4) staff space for therapy or examination in care housing; (5) basement spaces not used for living accommodations, or (6) any commercial or other nonresidential space.

The floor area ratio (FAR) times the land area (LA) equals the *maximum* amount of floor area (FA) acceptable for the development of a property. $FAR \times LA = FA$, or $FA/LA = FAR$

Open Space (OS) and Open Space Ratio (OSR)

Open space is the sum of the uncovered open space and one-half of the covered open space.

Uncovered open space: The horizontal area of the site not covered by building area (BA), plus open exterior balconies and roof area improved as recreation space (RS).

Covered open space (COS): The usable open space that is closed to the sky, having two clear unobstructed open or partially opened sides (minimum 50 percent open). The square foot amount countable as covered open space may not exceed the square foot amount of the open sides. Examples: covered balconies, covered portions of improved roof area or spaces under buildings supported by posts, columns, or cantilevers.

The open space ratio (OSR) times the land area (LA) equals the minimum amount of open space acceptable for the development of a property. $OSR \times LA = OS$, or $OS/LA = OSR$

Livability Space (LS) and Livability Space Ratio (LSR)

Livability space is nonvehicular open space, including lawns, planting space, walks, paved terraces, and sitting areas and the unpaved portions of street right-of-ways. No paved areas for car traffic or parking can be included as livability space.

The livability space ratio (LSR) times the land area (LA) equals the minimum amount of livability space acceptable for the development of a property. $LSR \times LA = LS$, or $LS/LA = LSR$

Recreation Space (RS) and Recreation Space Ratio (RSR)

Recreation space is a public or private exterior area improved for recreation of all residents, having at least dimension of 50 ft, an average dimension of 100 ft, and a minimum area of 10,000 ft².

A smaller least dimension is acceptable if the recreation space is usable improved roof area.

A smaller dimension and area are acceptable if 10,000 ft² is more than the total needed.

Countable recreation space should be a minimum of 20 ft from any residential wall containing a window on the ground floor.

The recreation space ratio (RSR) times the land area (LA) equals the minimum amount of recreation space acceptable for the development of a property. $RSR \times LA = RS$, or $RS/LA = RSR$

Occupant Car Space (OCS) and Occupant Car Ratio (OCR)

Occupant car space is garage, carport, or other parking space available to the residents without time limits.

The occupant car ratio (OCR) times the number of living units (LU) equals the minimum number of car parking spaces (CPS) for residents in the development of a property. $OCR \times LU = CPS$, or $CPS/LU = OCR$

Total Car Space (TCS) and Total Car Ratio (TCR)

Total car space is occupant car space plus other parking space that is available for unlimited or seldom limited time periods (primarily for guests).

The total car ratio (TCR) times the number of living units (LU) equals the minimum number of car parking spaces acceptable for a development including space for guest cars. $TCR \times LU = TCS$, or $TCS/LU = TCR$

LAND-USE INTENSITY SCALE

To rate or measure, it is necessary to have a measurement scale. For LUI, the rating scale is based first and most directly on the relationship of total floor area (FA) to total land area (LA).

As shown on the basic scale in Table 1, the LUI scale starts with an FAR of 0.025 for a LUI number of 1.0. The FAR doubles at each succeeding full LUI number on the scale. The floor area ratio (FAR) of 0.025 indicates that the maximum floor area (FA) desired at a LUI rating of 1.0 is 1089 ft² per acre. (43,560 × 0.025) For a LUI rating of 2 the FAR would be 0.05 and the FA would be 2178 ft².

TABLE 1 Basic Scale

Land-use intensity number		Floor area ratio
LUI	1.0	0.025
LUI	2.0	0.050
LUI	3.0	0.100
LUI	4.0	0.200
LUI	5.0	0.400
LUI	6.0	0.800
LUI	7.0	1.60
LUI	8.0	3.20

Fig. 2 Combined cluster housing and parking court site plan.

Parking court and housing cluster

INTENSITY OF DEVELOPMENT (LAND-USE INTENSITY)

TABLE 2   Expanded Scale

Land-use intensity number		Floor area ratio
LUI	3.0	0.100
LUI	3.1	0.107
LUI	3.2	0.115
LUI	3.3	0.123
LUI	3.4	0.132
LUI	3.5	0.141
LUI	3.6	0.152
LUI	3.7	0.162
LUI	3.8	0.174
LUI	3.9	0.187
LUI	4.0	0.200

TABLE 3   Land-Use Intensity Scale (Expanded)

Land-use intensity (LUI)	Floor area ratio (FAR)	Land-use intensity (LUI)	Floor area ratio (FAR)	Land-use intensity (LUI)	Floor area ratio (FAR)
3.0	0.100	4.7	0.325	6.4	1.06
3.1	0.107	4.8	0.348	6.5	1.13
3.2	0.115	4.9	0.373	6.6	1.21
3.3	0.123	5.0	0.400	6.7	1.30
3.4	0.132	5.1	0.429	6.8	1.39
3.5	0.141	5.2	0.459	6.9	1.49
3.6	0.152	5.3	0.492	7.0	1.60
3.7	0.162	5.4	0.528	7.1	1.72
3.8	0.174	5.5	0.566	7.2	1.84
3.9	0.187	5.6	0.606	7.3	1.97
4.0	0.200	5.7	0.650	7.4	2.11
<u>4.1</u>	<u>0.214</u>	5.8	0.696	7.5	2.26
4.2	0.230	5.9	0.746	7.6	2.42
4.3	0.246	6.0	0.800	7.7	2.60
4.4	0.264	6.1	0.857	7.8	2.79
4.5	0.283	6.2	0.919	7.9	2.99
4.6	0.303	6.3	0.985	8.0	3.20

Each full number on the LUI scale has 10 subdivisions, such as 3.1, 3.2, and 3.3, for intensities between the full numbers as shown on the expanded scale in Table 2.

Computed FARs between FARs shown on the LUI scale are raised to the FAR of the next higher LUI number on the scale, and standards for that LUI number apply. FARs 0.188 through 0.200 are classified as LUI 4.0.

As intensities less than LUI 3.0 are seldom used in HUD projects, LUIs 1.0 through 2.9 are not included in the tables.

LAND-USE INTENSITY NUMBER

The LUI number assigned for the development of a site is determined by the average living unit size and the number of living units for the site that are agreed to by the developer, the local authority, and HUD. It should provide an intensity of development that is appropriate to the characteristic of the site and its location in the community.

To find the LUI number, the average living-unit size (1250 ft^2 in the example below) is multiplied by the number of living units (8 in the example below) to determine the floor area (FA). The floor area is then divided by the gross site or land area (LA) (48,500 ft^2 in the example below) to determine the floor area ratio (FAR).

$$FA/LA = FAR$$

8 × 1250 = FA 10,000/LA 48,500 = FAR 206. The LUI number whose FAR is just above the FAR determined (0.206) is the LUI number 4.1 for the site as shown underlined in Table 3.

USE OF LUI CRITERIA

Reading horizontally along the line for LUI 3.8 in Table 4 are the ratio amounts of floor area, open space, livability space, recreation space, and the number of car parking spaces per living unit required for the project.

LAND-USE INTENSITY RANGE

The land-use intensity range indicated in Table 5 for each building type is the range that has proved to be the most favorable for that building type. Intensities higher or lower than the range indicated tend to over- or underdevelop a property. The intensity for a building type alone on a property or in a group mixed with groups or individual specimens of other building types should usually be within the range indicated.

TABLE 4

Land-use intensity (LUI)	Floor area ratio (FAR)	Open space ratio (OSR)	Livability space ratio (LSR)	Recreation space ratio (RSR)	Occupant car ratio (OCR)	Total car ratio (TCR)
3.0	0.100	0.80	0.65	0.025	2.0	2.2
3.1	0.107	0.80	0.62	0.026	1.9	2.1
3.2	0.115	0.79	0.60	0.026	1.9	2.1
⋮	⋮	⋮	⋮	⋮	⋮	⋮
3.7	0.162	0.77	0.53	0.032	1.6	1.8
3.8	0.174	0.77	0.52	0.033	1.5	1.7
3.9	0.187	0.77	0.52	0.036	1.5	1.7
4.0	0.200	0.76	0.52	0.036	1.4	1.6

TABLE 5   Favorable Land-Use Intensity Ranges for Various Building Types

Building type	Range of land-use intensity
Single-Family Types	
1-story detached	From LUI 1.0 to LUI 3.8
1-story townhouse	From LUI 2.0 to LUI 3.9
2-story detached	From LUI 2.0 to LUI 4.0
2-story townhouse	From LUI 3.7 to LUI 4.8
Walk-up Apartments	
2-story garden apartment	From LUI 3.9 to LUI 5.0
3-story apartment	From LUI 4.9 to LUI 6.0
4-story apartment	From LUI 5.5 to LUI 6.5
Elevator Buildings	
6-story apartment	From LUI 5.9 to LUI 6.9
8-story apartment	From LUI 6.2 to LUI 7.2
10-story apartment	From LUI 6.5 to LUI 7.5
12-story apartment	From LUI 6.8 to LUI 7.9
18-story apartment	From LUI 7.2 to LUI 8.4
24-story or more	From LUI 7.7 to LUI 9.4

When the acceptable land-use intensity number has been determined, the developer may use the amount of floor area provided by the floor area ratio for that number in many different ways. Figure 4 shows how the same amount of floor area can be used in 2-, 5-, or 10-story buildings.

example To find the LUI number for a development contemplating living units of approximately 1200 ft^2 and a desired density of 6 living units per gross acre, follow the horizontal line from 1200 to its intersection with the vertical line below 6, to determine a LUI of 3.8. For combinations of living-unit size and number per acre not

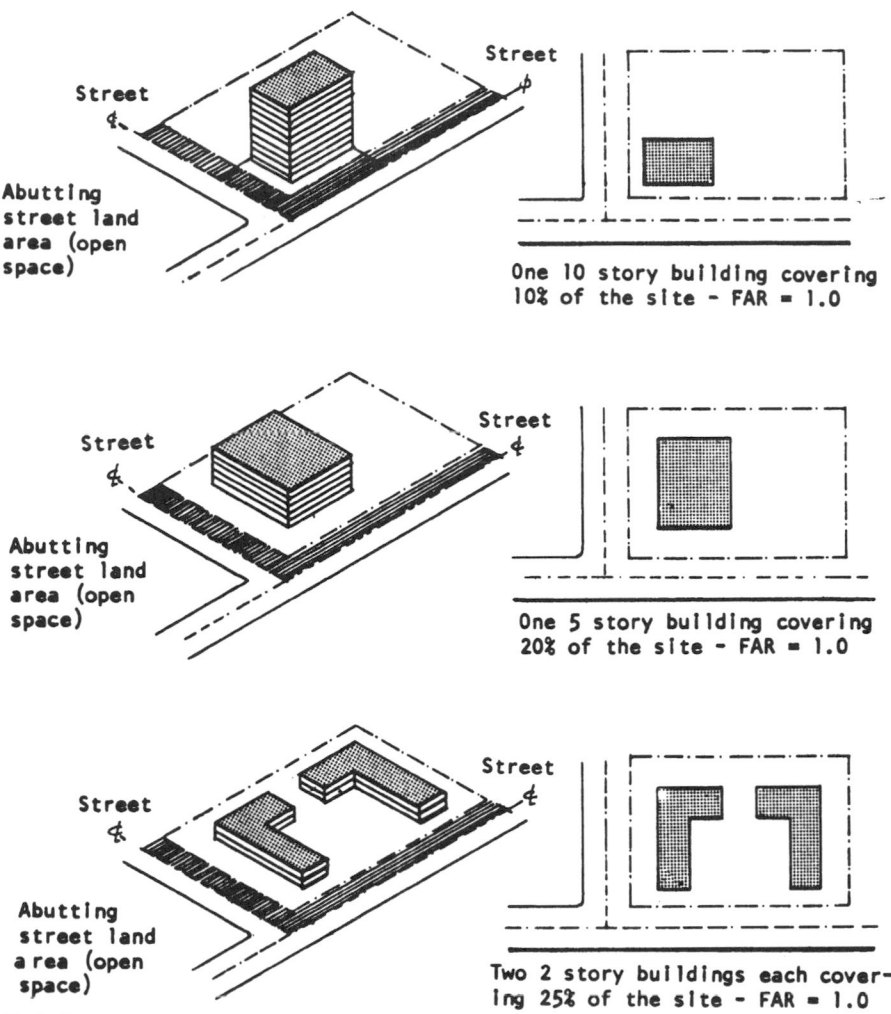

One 10 story building covering 10% of the site - FAR = 1.0

One 5 story building covering 20% of the site - FAR = 1.0

Two 2 story buildings each covering 25% of the site - FAR = 1.0

Fig. 3   Floor area ratios.

TABLE 6   Single-Family-Dwelling Land-Use Intensity Numbers

Net living unit size	Number of living units per gross acre										
	4	5	6	8	10	12	14	16	18	20	25
600 ft²				3.2	3.5	3.8	4.0	4.2	4.4	4.5	4.8
700 ft²			3.0	3.4	3.7	4.0	4.2	4.4	4.6	4.7	5.1
800 ft²		3.0	3.2	3.6	3.9	4.2	4.4	4.6	4.8	4.9	5.3
900 ft²		3.1	3.4	3.8	4.1	4.4	4.6	4.8	4.9	5.1	5.4
1000 ft²	3.0	3.2	3.5	3.9	4.2	4.5	4.7	4.9	5.1	5.3	
1100 ft²	3.1	3.4	3.6	4.1	4.4	4.7	4.9	5.1	5.2	5.4	
1200 ft²	3.2	3.5	3.8	4.2	4.5	4.8	5.0	5.2	5.4		
1300 ft²	3.3	3.6	3.9	4.3	4.6	4.9	5.1	5.3			
1400 ft²	3.4	3.7	4.0	4.4	4.7	5.0	5.2	5.4			
1500 ft²	3.5	3.8	4.1	4.5	4.8	5.1	5.3				
1600 ft²	3.6	3.9	4.2	4.6	4.9	5.2	5.4				
1700 ft²	3.7	4.0	4.3	4.7	5.0	5.3					
1800 ft²	3.8	4.1	4.4	4.8	5.1	5.4					
1900 ft²	3.9	4.2	4.4	4.9	5.2	5.4					
2000 ft²	3.9	4.2	4.5	4.9	5.3						

INTENSITY OF DEVELOPMENT (LAND-USE INTENSITY)

shown in Tables 7 through 10, LUI numbers can be calculated as described under Land-Use Intensity Number. Allowance for common space can be calculated as described under Adjusted Floor Area for Walk-up Apartments or Adjusted Floor Area for High-Rise Apartments. If exact figures for percentage of common space are available, they can be substituted for the 11 or 20 percent averages shown here.

ADJUSTED FLOOR AREA FOR WALK-UP APARTMENTS

The floor area (FA) for individual garden apartment units should be increased by 11 percent before computing the FAR, to allow approximately 10 percent of the total floor area for common-use halls, stairways, etc.

ADJUSTED FLOOR AREA FOR HIGH-RISE APARTMENTS

The floor area (FA) for individual high-rise apartment units should be increased by 20 percent before computing the FAR, to allow approximately 17 percent of the total floor area for lobbies, elevator shafts, stairways, halls, etc.

MAXIMUM NUMBER OF LIVING UNITS FOR A PROJECT

To find the maximum number of living units for a project in compliance with the LUI standards, multiply the acceptable number of living units per acre used in identifying the LUI number by the size of the project or land area.

example 6 LU per acre × 60 acres = 360 project LU.

MINIMUM LAND AREA FOR A PROJECT

To find the minimum amount of land or gross site area for a project in compliance with the LUI standards, divide the total number of living units needed for a project by the acceptable number of living units per acre.

example A 160-unit townhouse development is needed for low-rent housing, to average 1200 ft² per living unit at a density of 8 living units per acre. The land-use intensity (LUI) number for this living unit density and size is 4.2 (see Table 6).
160/8 = 20 acres needed, in a location where LUI 4.2 would be appropriate.

LAND AREA RANGE DETERMINATION WHERE DENSITY IS NOT KNOWN

To find the most favorable range of land area which will comply with LUI guidelines when only the number of units, the floor area and the building type are known, find the LUI range for the building type in Table 5. Find the related floor area ratios in Tables 7, 9, and 11. To find the most favorable land area range, divide the total proposed residential floor area by the floor area ratio for the highest and lowest LUI number shown for the proposed building type in Table 5.

TABLE 7 Single-Family-Dwelling Land-Use Intensity Criteria

Land-use intensity (LUI)	Floor area ratio (FAR)	Open space ratio (OSR)	Livability space ratio (LSR)	Recreation space ratio (RSR)	Occupant car ratio (OCR)	Total car ratio (TCR)
3.0	0.100	0.80	0.65	0.025	2.0	2.2
3.1	0.107	0.80	0.62	0.026	1.9	2.1
3.2	0.115	0.79	0.60	0.026	1.9	2.1
3.3	0.123	0.79	0.58	0.028	1.8	2.0
3.4	0.132	0.78	0.55	0.029	1.7	1.9
3.5	0.141	0.78	0.54	0.030	1.7	1.9
3.6	0.152	0.78	0.53	0.030	1.6	1.8
3.7	0.162	0.77	0.53	0.032	1.6	1.8
3.8	0.174	0.77	0.52	0.033	1.5	1.7
3.9	0.187	0.77	0.52	0.036	1.5	1.7
4.0	0.200	0.76	0.52	0.036	1.4	1.6
4.1	0.214	0.76	0.51	0.039	1.4	1.6
4.2	0.230	0.75	0.51	0.039	1.4	1.5
4.3	0.246	0.75	0.49	0.039	1.3	1.5
4.4	0.264	0.74	0.48	0.042	1.3	1.5
4.5	0.283	0.74	0.48	0.042	1.2	1.4
4.6	0.303	0.73	0.46	0.046	1.2	1.4
4.7	0.325	0.73	0.46	0.046	1.2	1.3
4.8	0.348	0.73	0.45	0.049	1.1	1.3
4.9	0.373	0.72	0.45	0.052	1.1	1.3
5.0	0.400	0.72	0.44	0.052	1.1	1.2
5.1	0.429	0.72	0.43	0.055	1.0	1.2
5.2	0.459	0.72	0.42	0.056	1.0	1.2
5.3	0.492	0.71	0.41	0.059	0.99	1.1
5.4	0.528	0.71	0.41	0.062	0.96	1.1

TABLE 8 Walk-up Apartment—Land-Use Intensity Numbers

Net living unit size*	Number of living units per gross acre										
	10	12	14	16	20	25	30	40	50	60	80
400 ft²	3.6	3.8	4.1	4.4	4.7	5.1	5.4	5.7	6.1		
500 ft²	3.7	3.9	4.1	4.4	4.7	5.0	5.4	5.7	6.0	6.4	
600 ft²	3.7	3.9	4.2	4.3	4.7	5.0	5.2	5.7	6.0	6.2	
700 ft²	3.9	4.2	4.4	4.6	4.8	5.2	5.5	5.9	6.2	6.5	
800 ft²	4.1	4.3	4.6	4.8	5.1	5.4	5.7	6.1	6.4		
900 ft²	4.2	4.5	4.7	4.9	5.1	5.6	5.8	6.2			
1000 ft²	4.4	4.7	4.9	5.1	5.4	5.7	6.0	6.4			
1100 ft²	4.5	4.8	5.0	5.2	5.4	5.9	6.1	6.5			
1200 ft²	4.7	4.9	5.0	5.3	5.7	6.0	6.1				
1300 ft²	4.8	5.0	5.3	5.5	5.8	6.1	6.4				
1400 ft²	4.9	5.0	5.4	5.6	5.9	6.2	6.5				
1500 ft²	5.0	5.1	5.5	5.7	6.0	6.3					
1600 ft²	5.1	5.3	5.7	5.8	6.1	6.3					
1700 ft²	5.2	5.4	5.6	5.8	6.2	6.5					
1800 ft²	5.2	5.5	5.6	5.9	6.1						

*Total floor areas shown have been increased by 11 percent as directed above to include common space in deriving correct LUI numbers.

example Proposed 100 each 1200-ft² 2-story townhouses. Problem, how much land area is needed to comply with LUI standards.

Table 5 shows for 2-story townhouses a LUI range of 3.7 to 4.8. Table 7 shows LUI 3.7 = FAR 0.162 and LUI 4.8 = FAR 0.348.

Total proposed floor area = 100 LU × 1200 ft² = 120,000 ft²

$$\text{Minimum site size} = \frac{120,000 \text{ ft}^2}{0.348} = 344,828 \text{ ft}^2$$
or 7.9 acres

$$\text{Maximum site size} = \frac{120,000 \text{ ft}^2}{0.162} = 740,740 \text{ ft}^2$$
or 17.0 acres

Site area range needed to comply with LUI standards = 7.9 to 17 acres.

TABLE 9 Walk-up Apartment—Land-Use Intensity Criteria

Land-use intensity (LUI)	Floor area ratio (FAR)	Open space ratio (OSR)	Livability space ratio (LSR)	Recreation space ratio (RSR)	Occupant car ratio (OCR)	Total car ratio (TCR)
3.6	0.152	0.78	0.53	0.030	1.6	1.8
3.7	0.162	0.77	0.53	0.032	1.6	1.8
3.8	0.174	0.77	0.52	0.033	1.5	1.7
3.9	0.187	0.77	0.52	0.036	1.5	1.7
4.0	0.200	0.76	0.52	0.036	1.4	1.6
4.1	0.214	0.76	0.51	0.039	1.4	1.6
4.2	0.230	0.75	0.51	0.039	1.6	1.4
4.3	0.246	0.75	0.49	0.039	1.3	1.5
4.4	0.264	0.74	0.48	0.042	1.3	1.5
4.5	0.283	0.74	0.48	0.042	1.2	1.4
4.6	0.303	0.73	0.46	0.046	1.2	1.4
4.7	0.325	0.73	0.46	0.046	1.2	1.3
4.8	0.348	0.73	0.45	0.049	1.1	1.3
4.9	0.373	0.72	0.45	0.052	1.1	1.3
5.0	0.400	0.72	0.44	0.052	1.1	1.2
5.1	0.429	0.72	0.43	0.055	1.0	1.2
5.2	0.459	0.72	0.42	0.056	1.0	1.2
5.3	0.492	0.71	0.41	0.059	0.99	1.1
5.4	0.528	0.71	0.41	0.062	0.96	1.1
5.5	0.566	0.71	0.40	0.062	0.93	1.1
5.6	0.606	0.70	0.40	0.065	0.90	1.0
5.7	0.650	0.70	0.40	0.065	0.87	1.0
5.8	0.696	0.69	0.40	0.070	0.84	0.99
5.9	0.746	0.69	0.40	0.075	0.82	0.96
6.0	0.800	0.68	0.40	0.080	0.79	0.93
6.1	0.857	0.68	0.40	0.080	0.77	0.90
6.2	0.919	0.68	0.40	0.083	0.74	0.85
6.3	0.985	0.68	0.40	0.085	0.72	0.85
6.4	1.06	0.68	0.40	0.085	0.70	0.83
6.5	1.13	0.67	0.41	0.090	0.68	0.81

TABLE 10 High-Rise-Apartment Land-Use Intensity Numbers

Net living unit size*	Number of living units per gross acre											
	30	40	50	60	80	100	120	140	160	180	200	240
400 ft²					6.2	6.5	6.8	7.0	7.2	7.4	7.5	7.8
500 ft²				6.1	6.5	6.8	7.1	7.3	7.5	7.7	7.8	
600 ft²			6.1	6.3	6.8	7.1	7.4	7.6	7.8	7.9		
700 ft²		6.0	6.3	6.6	7.0	7.3	7.5	7.8	8.0			
800 ft²		6.2	6.5	6.8	7.2	7.5	7.8	8.0				
900 ft²	5.9	6.4	6.7	6.9	7.4	7.7	7.9					
1000 ft²	6.1	6.5	6.8	7.1	7.5	7.8						
1100 ft²	6.2	6.6	7.0	7.2	7.7	8.0						
1200 ft²	6.4	6.8	7.1	7.4	7.8							
1300 ft²	6.5	6.9	7.2	7.4	7.9							
1400 ft²	6.6	7.0	7.3	7.6	8.0							
1500 ft²	6.7	7.1	7.4	7.7								
1600 ft²	6.8	7.2	7.5	7.8								
1700 ft²	6.9	7.3	7.6	7.9								
1800 ft²	6.9	7.4	7.7	7.9								

*Total floor areas shown have been increased by 20 percent as directed above to include common space in deriving correct LUI numbers.

TABLE 11 High-Rise-Apartment Land-Use Intensity Criteria

Land-use intensity (LUI)	Floor area ratio (FAR)	Open space ratio (OSR)	Livability space ratio (LSR)	Recreation space ratio (RSR)	Occupant car ratio (OCR)	Total car ratio (TCR)
5.9	0.746	0.69	0.40	0.075	0.82	0.96
6.0	0.800	0.68	0.40	0.080	0.79	0.93
6.1	0.857	0.68	0.40	0.080	0.77	0.90
6.2	0.919	0.68	0.40	0.083	0.74	0.87
6.3	0.985	0.68	0.40	0.085	0.72	0.85
6.4	1.06	0.68	0.40	0.085	0.70	0.83
6.5	1.13	0.67	0.41	0.090	0.68	0.81
6.6	1.21	0.67	0.41	0.097	0.66	0.79
6.7	1.30	0.67	0.42	0.104	0.64	0.77
6.8	1.39	0.68	0.42	0.104	0.62	0.75
6.9	1.49	0.68	0.43	0.104	0.60	0.73
7.0	1.60	0.68	0.43	0.112	0.58	0.71
7.1	1.72	0.68	0.45	0.115	0.57	0.69
7.2	1.84	0.69	0.46	0.115	0.56	0.67
7.3	1.97	0.70	0.47	0.118	0.54	0.65
7.4	2.11	0.71	0.49	0.127	0.52	0.63
7.5	2.26	0.72	0.50	0.136	0.50	0.61
7.6	2.42	0.75	0.51	0.145	0.49	0.60
7.7	2.60	0.76	0.52	0.145	0.47	0.58
7.8	2.79	0.81	0.56	0.145	0.46	0.56
7.9	2.99	0.83	0.57	0.150	0.45	0.55
8.0	3.20	0.86	0.61	0.160	0.44	0.54

EVALUATION OF SITE EXPOSURE TO AIRCRAFT NOISE

EVALUATION OF SITE EXPOSURE TO AIRCRAFT NOISE

If noise exposure forecast (NEF) or composite noise rating (CNR) contours are available, locate the site by referring to the marked scale. Also locate a point roughly in the center of the area covered by the principal runways. If the site lies outside the NEF-30 (CNR-100) contour, draw a straight line to connect these two points. Measure along this line the distances between (1) the NEF-40 (CNR-115) and NEF-30 (CNR-100) contours and (2) the NEF-30 (CNR-100) contour and the site. Now use Table 1 to evaluate the site's exposure to aircraft noise.

If NEF or CNR contours are not available, determine the effective number of operations for the airport as follows. Multiply the number of nighttime jet operations by 17. Then add the number of daytime jet operations to obtain an effective total. *Any* supersonic jet operation automatically places an airport in the largest category of Table 2, which governs noise acceptability.

On a map of the area that shows the principal runways, mark the locations of the site and of the center of the area covered by the principal runways. Then, using the distances below, you can construct approximate NEF-40 and NEF-30 contours for the major runways and flight paths most likely to affect the site. Again use Table 1 to evaluate the site's exposure to aircraft noise.

examples

Example 1: The illustration at the top shows two sites located on a map that has NEF contours. We draw a line from each of these sites to a point roughly in the center of the area covered by the principal runways.

Measuring along these lines, we find that site 1 lies outside the NEF-30 contour at a distance greater than that between the NEF-30 and NEF-40 contours and that site 2 lies outside the NEF-30 contour at a distance less than that between the NEF-30 and NEF-40 contours.

Therefore, the exposure of site 1 to aircraft noise is clearly acceptable and the exposure of site 2 is normally acceptable.

Example 2: The illustration at the bottom of the page shows an airport for which NEF or CNR

TABLE 1 Site Exposure to Aircraft Noise

Distance from site to the center of the area covered by the principal runways	Acceptability category
Outside the NEF-30 (CNR-100) contour, at a distance greater than or equal to the distance between the NEF-30 and NEF-40 (CNR-100, CNR-115) contours	Clearly acceptable
Outside the NEF-30 (CNR-100) contour, at a distance less than the distance between the NEF-30 and NEF-40 (CNR-100, CNR-115) contours	Normally acceptable
Between the NEF-30 and NEF-40 (CNR-100, CNR-115) contours	Normally unacceptable
Within the NEF-40 (CNR-115) contour	Clearly unacceptable

TABLE 2 Distances for Approximate NEF Contours

Effective number of operations	Distances to NEF 30 contour		Distances to NEF 40 contour	
	①	②	①	②
0–50	1000 ft	1 mile	0	0
51–500	½ mile	3 miles	1000 ft	1 mile
501–1300	1½ miles	6 miles	2000 ft	2½ miles
More than 1300 or any supersonic jet operations	2 miles	10 miles	3000 ft	4 miles

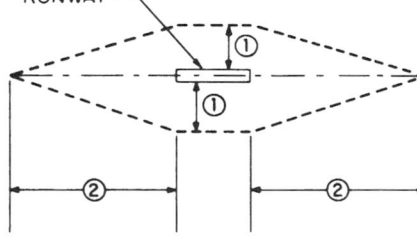

Fig. 1 Construction of approximate NEF contours using the distances in Table 2.

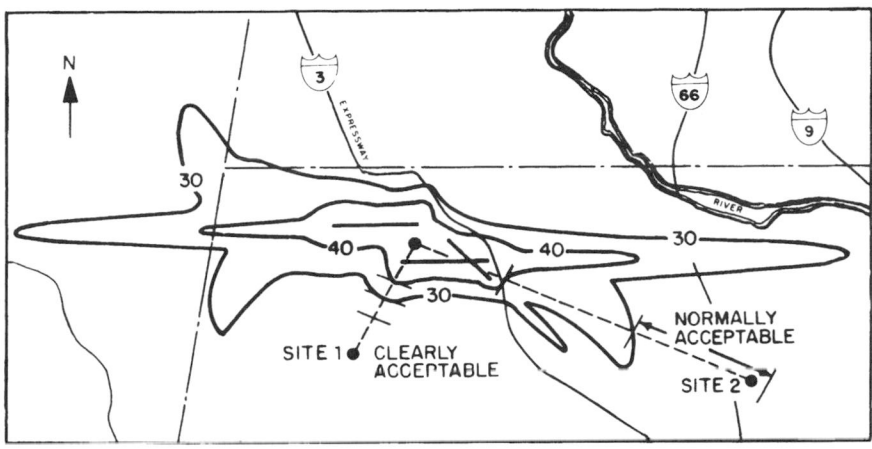

Fig. 2   Example of NEF contours.

contours are not available. The airport has 20 nighttime and 125 daytime jet operations. There are no supersonic flights, and so we determine the effective number of operations as follows:

$$20 \text{ (nighttime)} \times 17 = 340$$

Add to this the actual number of daytime operations:

$$340 + 125 \text{ (daytime)} = 465$$

Using the distances in Table 2, we construct approximate NEF contours and then draw a line from the site to a point roughly in the center of the area covered by the principal runways. Measuring along this line, we find that the site lies outside the NEF-30 contour at a distance greater than that between the NEF-30 and NEF-40 contours. Therefore, the site's exposure to aircraft noise is clearly acceptable.

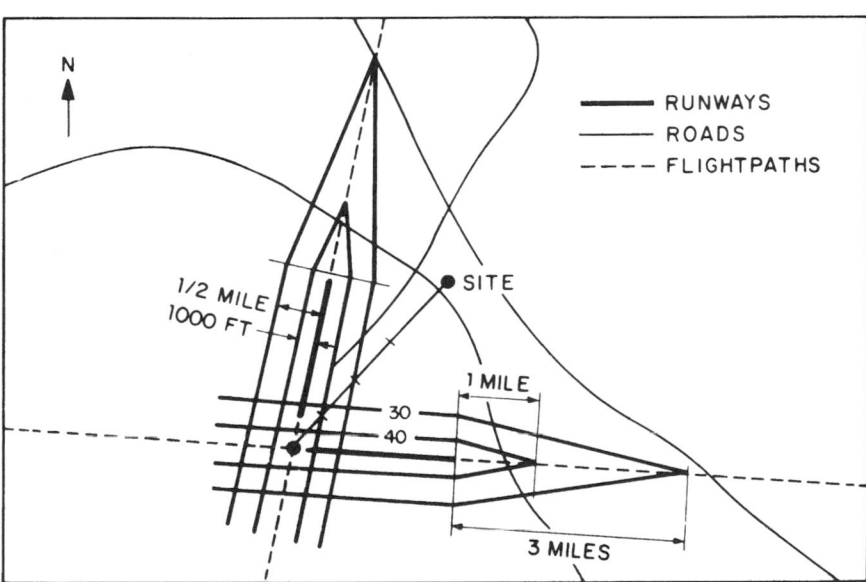

Fig. 3   Example of approximate NEF contours drawn for an airport with an effective number of operations between 51 and 500.

EVALUATION OF SITE EXPOSURE TO ROADWAY NOISE

EVALUATION OF SITE EXPOSURE TO ROADWAY NOISE

Traffic surveys show that the level of roadway noise depends on the percentage of trucks in the total traffic volume. To account for this effect, these guidelines provide for separate evaluation of automobile and truck traffic.

Before proceeding with these separate evaluations, however, determine the effective distance from the site to each road by locating the distances from the site to the centerlines of the nearest and farthest lanes of traffic. Now lay a straightedge to connect these two distances and read off the value at the point where the straightedge crosses the middle scale. This value is the effective distance to the road.

Automobile Traffic

The numbers in Fig. 1, which is used to evaluate the site's exposure to automobile noise, were arrived at with the following assumptions:
- There is no traffic signal or stop sign within 800 ft of the site.
- The mean automobile traffic speed is 60 mi/h.
- There is line-of-sight exposure from the site to the road; i.e., there is no barrier that effectively shields the site from the road.

If a road meets these three conditions, proceed to the figure for an immediate evaluation of the site's exposure to the automobile noise from that road. But if any of these conditions are different, make the necessary adjustment(s) and *then* use the figure for the evaluation.

Adjustments for Automobile Traffic

Stop-and-go traffic If there is a traffic signal or stop sign within 800 ft of the site, multiply the total number of automobiles per hour by 0.1.

Mean traffic speed If there is no traffic signal or stop sign within 800 ft of the site *and* the mean automobile speed is other than 60 mi/h, multiply the total number of automobiles by the appropriate adjustment factor.

TABLE 1

Mean traffic speed, mi/h	Adjustment factor
20	0.12
25	0.18
30	0.25
35	0.32
40	0.40
45	0.55
50	0.70
55	0.85
60	1.00
65	1.20
70	1.40

Barrier adjustment This adjustment affects distance and applies equally to automobiles and trucks on the same road. Therefore, instructions for this adjustment appear after those for truck traffic.

Truck Traffic

The numbers in Fig. 3, which is used to evaluate the site's exposure to truck noise, were arrived at with the following assumptions:
- There is a road gradient of less than 3 percent.

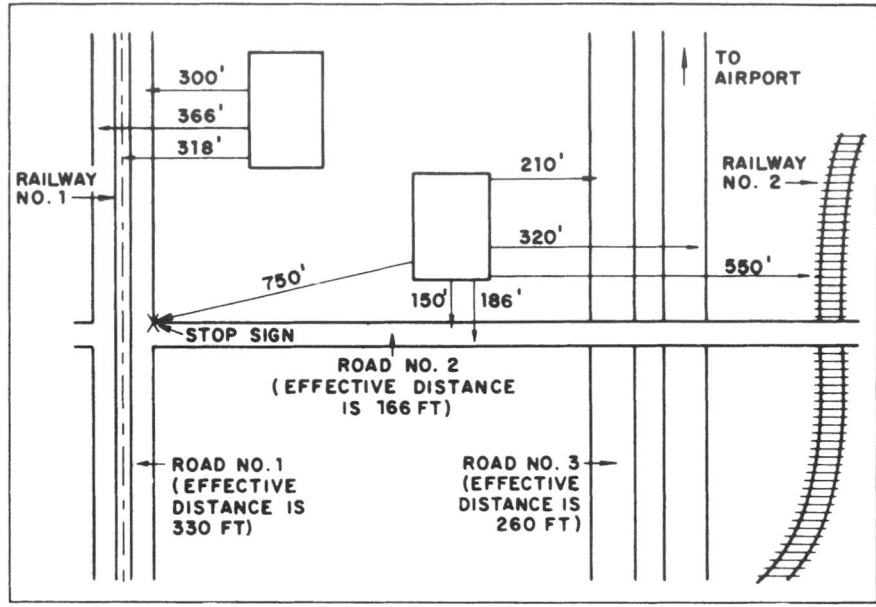

Fig. 1 Plan view of site showing how distances should be measured from the location of the dwelling nearest to the source.

- There is no traffic signal or stop sign within 800 ft of the site.
- The mean truck traffic speed is 30 mi/h.
- There is line-of-sight exposure from the site to the road; i.e., there is no barrier that effectively shields the site from the road.

If a road meets these four conditions, proceed to Fig. 4 for an immediate evaluation of the site's exposure to truck noise from that road. But if any of the conditions are different, make the necessary adjustment(s) listed below and *then* use Fig. 3 for the evaluation.

Adjustments for Truck Traffic

Road gradient If there is a gradient of 3 percent or more, multiply the number of trucks per hour in the uphill direction by the appropriate adjustment factor.

TABLE 2

Percent of gradient	Adjustment factor
3–4	1.4
5–6	1.7
More than 6	2.5

Add to this adjusted figure the number of trucks per hour in the downhill direction.

Stop-and-go traffic If there is a traffic signal or stop sign within 800 ft of the site, multiply by 5 the total number of trucks.

Mean traffic speed Make this adjustment only if there is no traffic signal or stop sign within 800 ft of the site *and* the mean speed is not 30 mi/h.

If the mean truck speed differs with direction, treat the uphill and downhill traffic separately.

Multiply each by the appropriate adjustment factor.

TABLE 3

Mean traffic speed, mi/h	Adjustment factor
20	1.60
25	1.20
30	1.00
35	0.88
40	0.75
45	0.69
50	0.63
55	0.57
60	0.50
65	0.46
70	0.43

examples The site shown is exposed to noise from three major roads: Road 1 has four lanes, each 12 ft wide, and a 30-ft-wide median strip that accommodates a rapid-transit line. Road 2 has four lanes, each 12 ft wide. Road 3 has six lanes, each 15 ft wide, and a median strip 35 ft wide.

Example 1: Road 1. The distance from the site to the center of the nearest lane of traffic is 300 ft. The distance to the centerline of the farthest lane of traffic is 366 ft. Figure 1 shows that the effective distance from the site to this road is 330 ft.

Road 2. The distance to the center of the nearest lane of traffic is 150 ft. The distance to the centerline of the farthest lane of traffic is 186 ft. Figure 1 shows that the effective distance from the site to this road is 166 ft.

Road 3. The distance to the centerline of the nearest lane of traffic is 210 ft. The distance to the centerline of the farthest lane of traffic is 320 ft. Figure 1 shows that the effective distance from the site to this road is 260 ft.

Example 2: Road 1 meets the three conditions that allow for an immediate evaluation. In obtaining the information necessary for this

EVALUATION OF SITE EXPOSURE TO ROADWAY NOISE

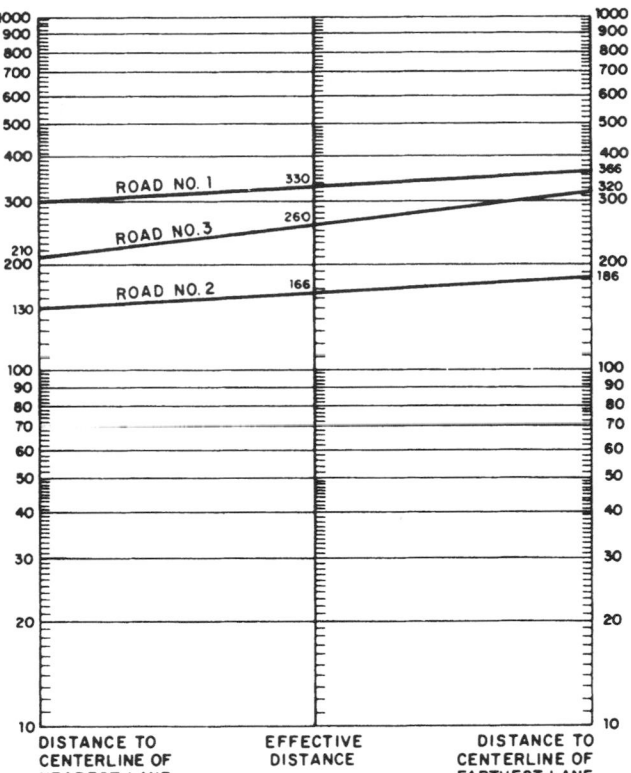

Fig. 2 Example of how Fig. 1 is used to determine effective distances.

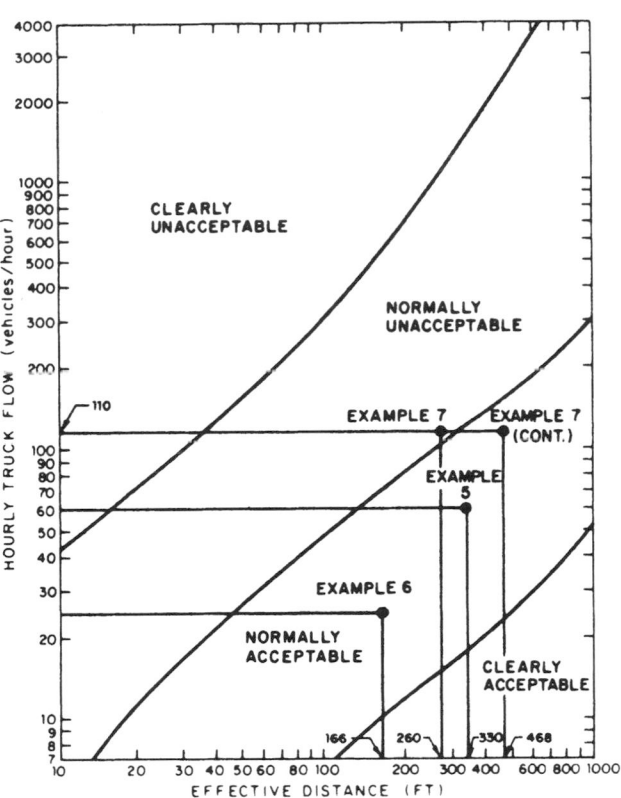

Fig. 3 Example of how Fig. 2 is used to evaluate site exposure to automobile noise.

evaluation, we found that the hourly automobile flow is 800 vehicles. In Fig. 2, we locate on the vertical scale the point representing 800 vehicles per hour and on the horizontal scale the point representing 330 ft. (Note that we must estimate the location of this point.) Using a straightedge, we draw lines to connect these two values and find that the site's exposure to automobile noise from this road is normally acceptable.

Example 3: Road 2 has a stop sign at 750 ft from the site. The hourly automobile flow is reported as being 900 vehicles. We adjust for stop-and-go traffic:

$$900 \times 0.1 = 90 \text{ vehicles}$$

and find from Fig. 2 that the exposure to automobile noise is clearly acceptable.

Example 4: Road 3 is a depressed highway. There is no traffic signal or stop sign and the mean speed is 60 mi/h. The hourly automobile flow is 1200 vehicles. The road profile shields all residential levels of the housing from line of sight to the traffic. The only adjustment that can be made is the barrier adjustment. This adjustment is necessary, however, only when the site's exposure to noise has been found clearly or normally unacceptable. Figure 2 shows that the exposure to automobile noise is normally acceptable. Therefore, no adjustment for barrier is necessary.

Example 5: Road 1 meets the four conditions that allow for an immediate evaluation. The hourly truck flow is 60 vehicles. Figure 3 shows that the site's exposure to truck noise from this road is normally acceptable.

Example 6: Road 2 has a stop sign at 750 ft from the site. There is also a road gradient of 4 percent. No trucks are allowed on this road, but 4 buses per hour are scheduled, 2 in each direction.

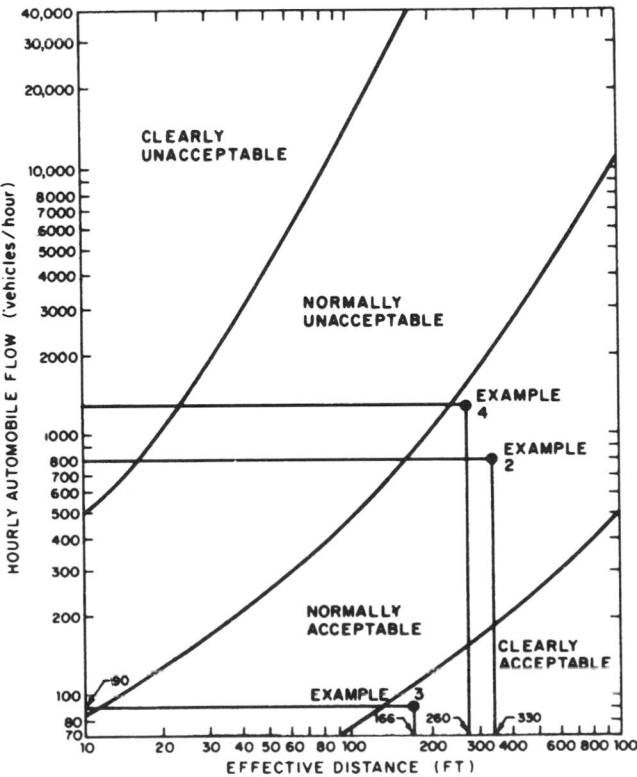

Fig. 4 Example of how Fig. 3 is used to evauate the site's exposure to truck noise.

EVALUATION OF SITE EXPOSURE TO ROADWAY NOISE

We adjust first for gradient:

$$2 \times 1.4 =$$

Uphill:	2.8 vehicles
Downhill:	2.0 vehicles
Total flow:	4.8 vehicles

And then adjust for stop-and-go traffic:

$$4.8 \times 5 = 24 \text{ vehicles (per hour)}$$

Figure 3 shows that the exposure to truck (bus) noise from this road is normally acceptable.

Example 7: The profile of road 3 shields all residential levels of the housing from line of sight to the traffic. The mean truck speed is 50 mi/h. The hourly truck flow is 175 vehicles. We adjust for mean speed:

$$175 \times 0.63 = 110.25$$
$$= 110 \text{ vehicles}$$

and find from Fig. 4 that exposure to truck noise is normally unacceptable. Therefore, we proceed with the barrier adjustment.

Road 3 has been depressed 25 ft from the 150-ft elevation of the natural terrain. The actual road elevation, therefore, is 125 ft. We find the effective road elevation to be

$$125 + 5 = 130 \text{ ft}$$

Six stories are planned for the housing, which is located at an elevation of 130 ft. The effective site elevation for the highest story is

$$6 \times 10 = 60 + 130 - 5 = 185 \text{ ft}$$

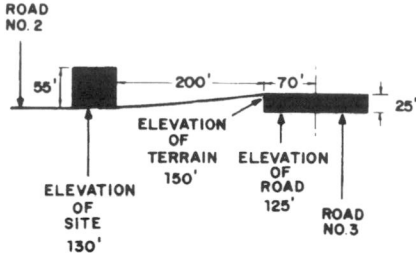

Fig. 5 Detail of site showing the measurements necessary for a barrier adjustment.

TABLE 1

Type	Size	Area per 1000 persons	Service area	Location	Usual facilities and remarks
Neighborhood park	3-acre minimum	1 to 2 acres per 1000 persons depending upon shape and intensity of development	Approximately ½ mile radius similar to elementary school service area	Preferably adjoining elementary school near center of neighborhood unit	Unsupervised sports, play equipment, multiple-use paved areas, turf area, and planting, some passive areas desirable, minimum of auto parking. Summer playground program with small shelter desirable
Playfield	10–25 acres. May be part of larger scenic area if location provides convenient service	½ acre per 1000 persons with at least 1 active play area per 1000 people	Approximately 1 mile radius, similar to high school service area	At or near the intersection of major or secondary thoroughfares near center of service area	Baseball, football, soft-ball, tennis, and other active athletic areas. Some facilities may be lighted for night use, and substantial auto parking required. May include playground-type area
Community park	25–100 acres	2 acres per 1000 persons	Approximately 1 mile radius. Similar to high school service area	At or near intersection of major or secondary thoroughfares near center of service area	Similar to playfield but at least ½ area for picnicking, and family activity. May include community center, swimming pool, and water activities such as fishing. Off-street parking required and passive area desirable
Large park	Minimum of 100 acres, preferably several hundred acres	Approximately 5 acres per 1000 people	3 miles or more radius with good accessibility by auto	Where appropriate sites can be obtained incorporating natural features, one area for each 50,000 to 100,000 persons desirable within urbanized area or on the periphery	Active athletic areas similar to playfield but at least one-half the area should be rustic and provide picnicking, hiking, camping, archery, etc.; golf courses, fishing, boating, and water sports may be included. Much off-street parking required, with interior roadways, shelter, swimming pools, and quiet; passive areas desirable
Parkways, ornamental areas	Size varies depending on conditions and nature of area	Where sites dictate development	No specific service area as most serve entire urban area	Along waterways or as aesthetic treatment	Largely scenic areas but may include picnicking
Special-use areas	Size varies depending on conditions and nature of area	Specific facilities will dictate area per 1000	No specific service area, as most serve entire urban area	Near center of urban area	Combination of two or more classifications such as zoo, botanical garden, or exhibition area within a community park, or playfield
Reservations and preserves	Several hundred to a thousand acres or more	10 acres per 1000 persons. May include some close-in recreation areas	Entire urban area	Usually on fringe of urban development at appropriate sites	Rustic and wild areas, camping, nature, and hiking trails, bridle paths, bird sanctuary, boating, fishing, and similar uses not requiring intensive development
Regional recreation areas	Several thousand acres	No specific standard. May be partially included in area of preserves and reservations	Entire region	Within 1–3 hours driving time of urban center	Lake, river, or reservoir providing fishing, boating, water sports, picnicking, hunting, camping, and similar facilities
Tennis court, outdoor basketball, and the other court sports	2 acres is ideal	1 acre for every 5000 people	Approximately 1 mile radius	Located in playfields or community park	May be in community park or large park
Baseball diamond	Regulation Junior diamonds	1 per 30,000 people. 1 per 3000 of ages 5 to 14 years. 1 per 10,000 people	Approximately 1 mile radius	Located in playfields preferably	May be in community park or large park
Fishing (no boats)	Minimum of 3 surface acres	5% instant capacity of population		Located within an hour's drive or 50 miles and within 5 to 10 miles of an all-weather highway	This standard is for only small city lakes such as those developed from playa lakes

(continued)

RECOMMENDED LAND STANDARDS

TABLE 1 (*Continued*)

Type	Size	Area per 1000 persons	Service area	Location	Usual facilities and remarks
Community swimming pool	4500 ft^2 of water surface	1 pool for every 30,000 people	Serves 150 people at a time	Located in community park or large park	Deck area should always equal or exceed square footage of water surface area, since not more than one-half of the swimmers will be in the water at any time
Golf course	18 holes—110 acre minimum. 150 acres for a good course	18 holes for first 20,000 people. 18 additional holes for every 30,000 people	Can accommodate 500 to 550 persons per day	Gently rolling area with some trees is preferable	May be located within a community park, play-field, or large park
Picnic area	2 acres per 1000 people	8 units per acre	5 mile radius average driving distance. Each unit serves 8 people	In neighborhood or community parks	Planned on a walk-in basis with multicar parking areas. Trees and shade should be provided as well as sanitary facilities within the area
Community center building	7500–10,000 ft^2	1 for every 25,000 people	½ to ¾ mile radius	Within large park or possibly community park	Preferably in conjunction with a swimming pool and/or party house
Playgrounds	½ acre	½ acre for every 5000 people	½ mile radius	Preferably within neighborhood park	
Summer playground	200 ft^2 minimum	1 per neighborhood park	½ mile radius	Within neighborhood park	Shelter for storage of supplies and playing of games for summer recreation program

CHECKLIST FOR EXAMINATION AND COMPARISON OF SITES

1. *Conformance with urban pattern*
   a. Conformance with accepted urban development plans, or tentative plans, or probable trends in land use
   b. Present zoning; possible changes
   c. Approval of city planning bodies
   d. Possibility of closing existing streets, dedicating new streets
   e. Effect of building codes and possibility of modification

2. *Slum clearance considerations*
   a. Number, character, and condition of existing buildings on site
   b. Number of families housed at present
   c. Relocation of present residents
   d. Equivalent elimination

3. *Characteristics of site and environment*
   a. Area of site compared with area needed for buildings and project facilities
   b. Shape of site; parcels necessarily excluded; deed restrictions; easements
   c. Topography as it affects livability of the site plan; favorable features such as existing shade trees, pleasing outlook, desirable slopes
   d. Quality of neighborhood; extent of nonresidential land use; suitability of neighborhood for dwelling type desired
   e. Effect of project on neighborhood
   f. Hazards; possibility of flooding, slides, or subsidence. Proximity to railroads, high-speed trafficways, high embankments, unprotected bodies of water; presence of insect or rodent breeding places; or high groundwater level that might cause dampness in building
   g. Nuisances; nearness to industrial plants, railroads, switchyards, heavy-traffic streets, airports, etc., causing noise, smoke, dust, odor, vibrations

4. *Availability of special municipal services*
   a. Garbage and rubbish collection
   b. Fire protection as affected by site location and street access
   c. Streets: lighting, cleaning, maintenance, snow removal, tree planting and maintenance, etc.
   d. Police protection and other municipal services

5. *Civic and community facilities*
   a. Public transportation facilities: means, routes, adequacy and expense of transportation to employment, schools, central business district, etc.
   b. Accessibility to paved thoroughfares
   c. Amount and character of employment within walking distance and within reasonable travel radius
   d. Stores and markets; kinds and locations; need for additional facilities as part of project development
   e. Schools—grade, junior high, and high: locations, capacities, adequacy; probability of enlargement, if needed
   f. Parks and playgrounds: locations, facilities provided, adequacy, maintenance and supervision supplied; possible additions
   g. Churches, theaters, clinics

6. *Appropriateness of project design to site, with reference to livability*
   a. Type or types of dwellings
   b. Project density
   c. Utility selection

7. *Elements of project development cost*
   a. Land costs, including site acquisition, expense, and unpaid special assessments
   b. Effect of soil conditions, topographic features, project density appropriate to the neighborhood, availability of utilities, extent of existing street improvements, recreational facilities and additions to be provided by municipality or utility companies, etc.
   c. Building types, utility selection, site conditions, and requirements for nondwelling structures

8. *Project maintenance and operating costs*
   a. Differences in costs of utilities appropriate to the respective sites
   b. Differentials in grounds maintenance costs due to topography
   c. Differences in estimated payments in lieu of taxes

PHYSICAL ELEMENTS IN THE ORGANIZATION OF THE NEIGHBORHOOD

Residential Element	Area Served (radius in ft. or mi.)	Number of Families Served	Types of Open Space and Community Facilities Required	Plan Relationships
Single-family detached home	0-40 ft.	1	Patio, outdoor recreation area, family room	
Single dwelling unit in a multi-unit building	0-40 ft.	1	Terrace or balcony, open corridor, outdoor living room	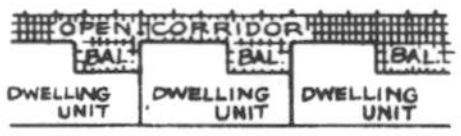
Typical floor in a multi-unit building	40-200 ft.	4-10	Enclosed play space, enclosed sitting area	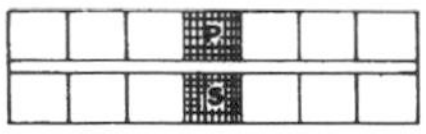
Apartment building	200-400 ft.	10-150	Outdoor areas for play and sitting, roof deck, pool, community room, tot lot	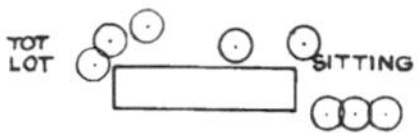
Complex of apartment buildings or a residential block or street	400-800 ft.	30-500	Outdoor areas for play and sitting, pool or pools, small community building	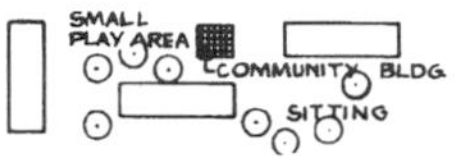
Hamlet or cluster of blocks	800-4000 ft.	90-1,500	Outdoor areas for sports, play (playground) and sitting, pool or pools, community building	
Single neighborhood	¼-½ mi.	1,000-5,000	Play field (sports), playground, sitting and picnic areas, pools, large community building	
Cluster of neighborhoods	½-2 mi.	3,000-15,000	Play fields, playgrounds, sitting and picnic areas possibly with a lake, pools, recreation and community center	

Fig. 1   Neighborhood organization.

BASIC PLANNING UNIT

Single-Family Detached Home

The single-family house and the single apartment unit are the basic planning units containing the family as a social entity. The type and character of this basic planning unit play a strong part in establishing the nature and quality of the community as a whole.

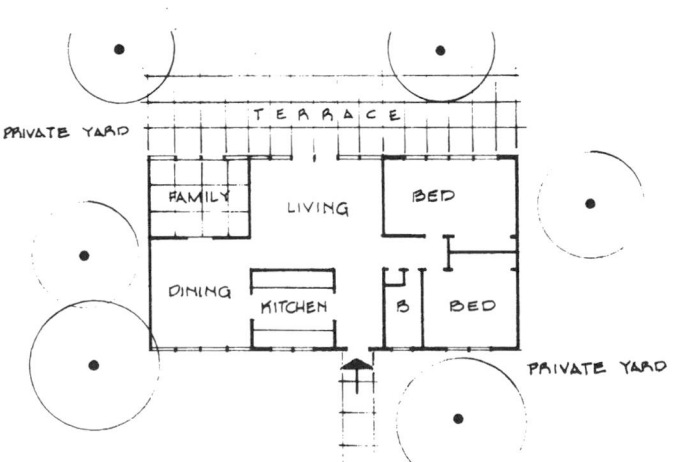

Fig. 2

Single Dwelling Unit (in a Multiunit Building)

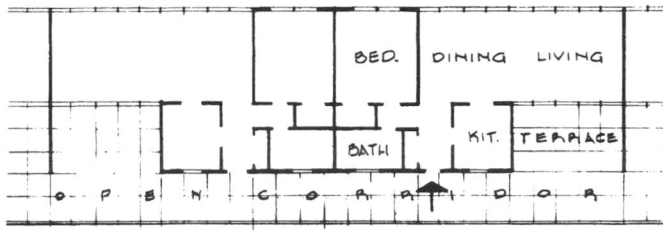

Fig. 3

TYPICAL FLOOR/APARTMENT BUILDING

Cluster—Typical Floor (in a Multiunit Building)

Outside of the family, this grouping of several dwelling units forms the most intimate of associations. There is a strong personal identification among all individuals. Physical proximity is an important element.

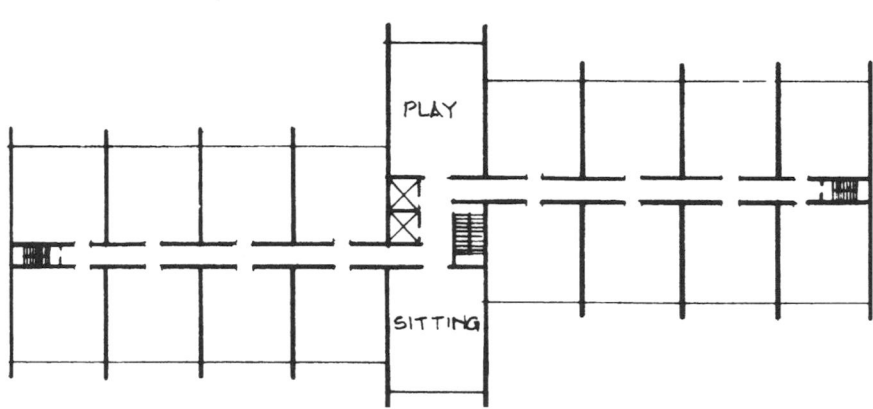

Fig. 4

Apartment Building (Roof and Ground Plan Shown)

The apartment building contains several clusters. It can support a wider range of facilities. Personal identity and close proximity are significant in this relationship.

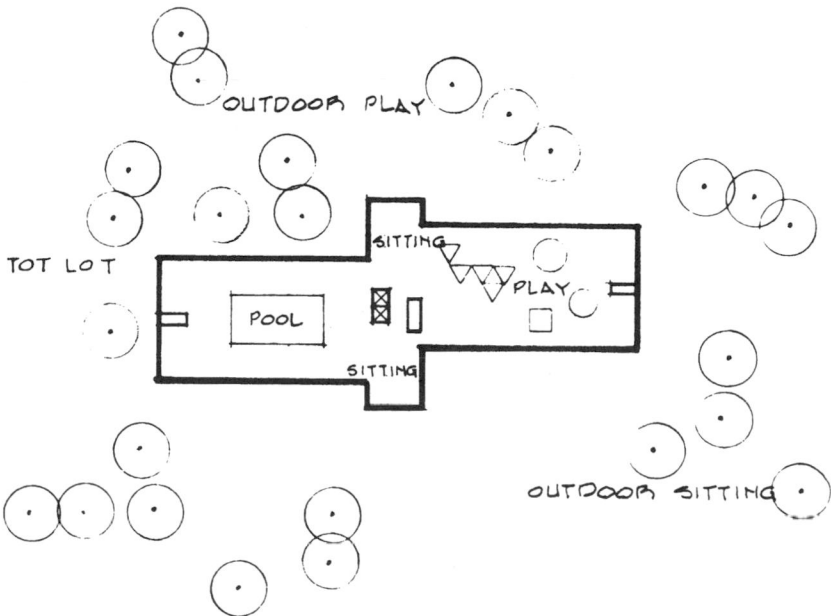

Fig. 5

PHYSICAL ELEMENTS IN THE ORGANIZATION OF THE NEIGHBORHOOD

APARTMENT BUILDING COMPLEX

Complex of Apartment Buildings

As the grouping becomes larger, relationships are more selective and based on special interests. Less personal contact and physical proximity.

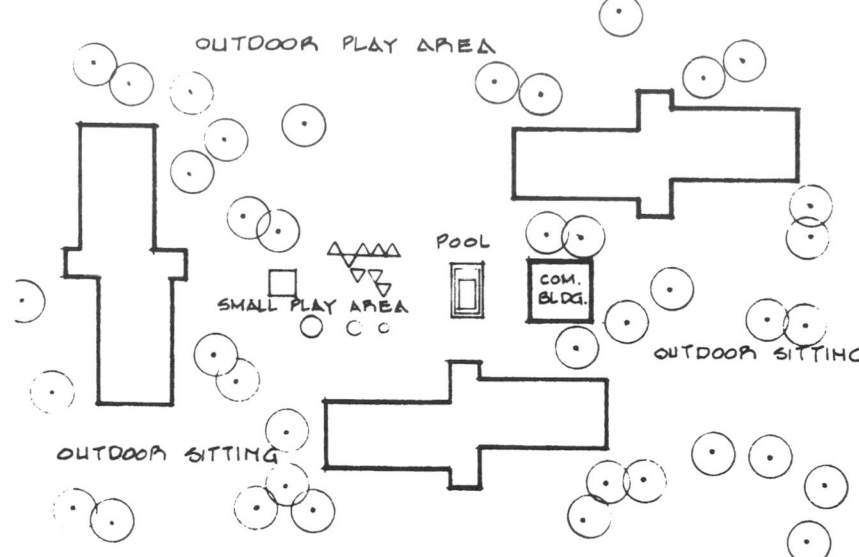

Fig. 6

Hamlet or Cluster of Blocks

This grouping is similar to a neighborhood, except in number. It contains a wide range of families. There is limited personal contact and a wider assortment of facilities.

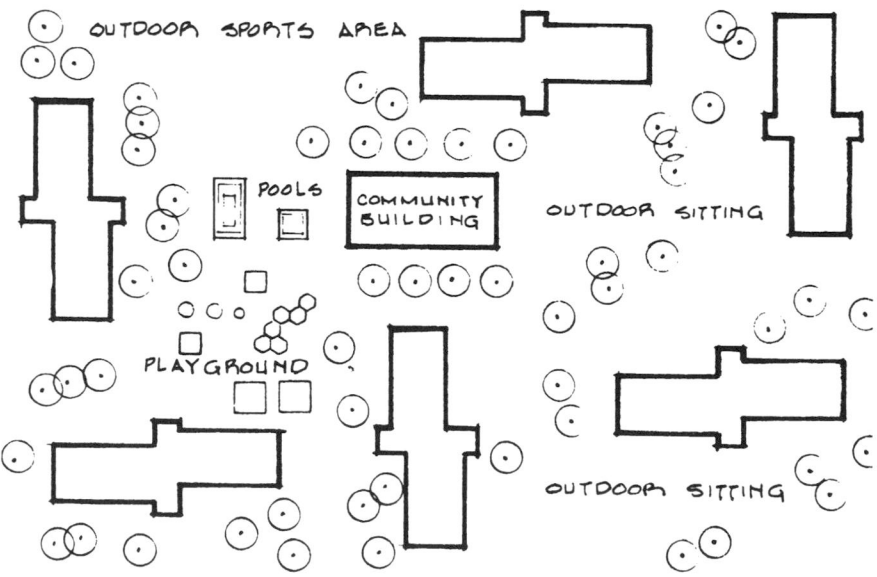

Fig. 7

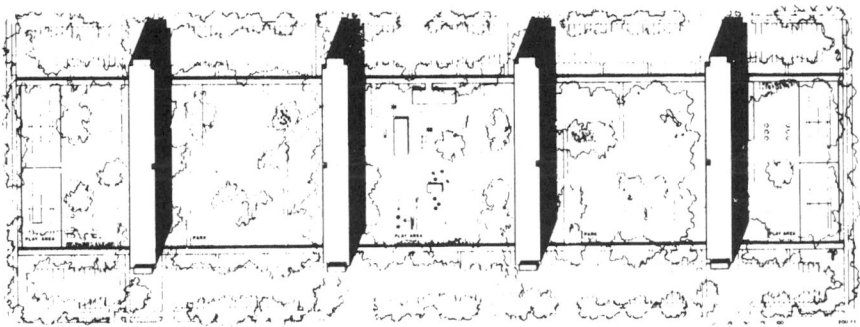

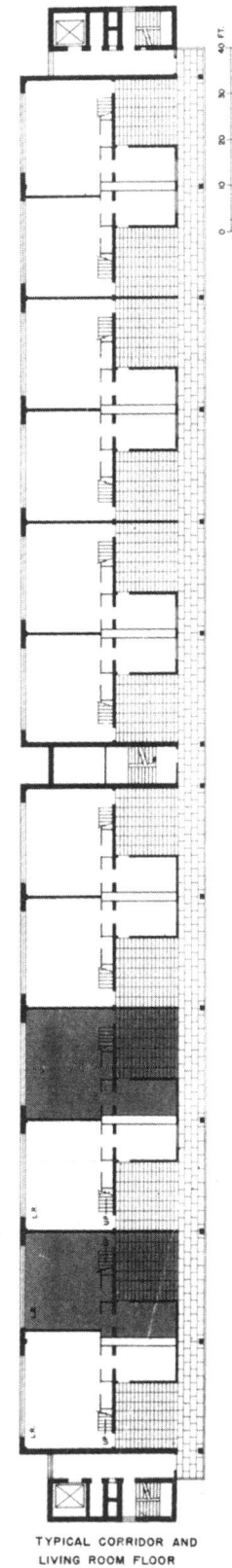

Cluster of Blocks

A full flood of *community living* is in this imaginative design for a luxurious vertical neighborhood. All ground floor area is for use of all tenants, including the enclosed first floor of the apartment units. In the plot for four of these apartment units, communal land is subtended into parks and play areas, and fenced by parking areas down the long sides of the rectangular plot. Short ends are fenced with tennis courts.

The architects of this apartment type also gave deep and obvious attention to the more intimate undertow of *family life,* in addition to the community life of the ground level, and the intermediate porch life on the terraces adjoining the sidewalks in the sky. Within the apartments in this ingenious design the family can be alone without being closed in; and even in the one-bedroom apartments there is the duplex arrangement, guaranteeing further privacy. Three-bedroom and one-bedroom apartments are created simply by transferring proprietorship of one of the two bedrooms in the basic arrangement.

Note generous size of terraces off gallery.

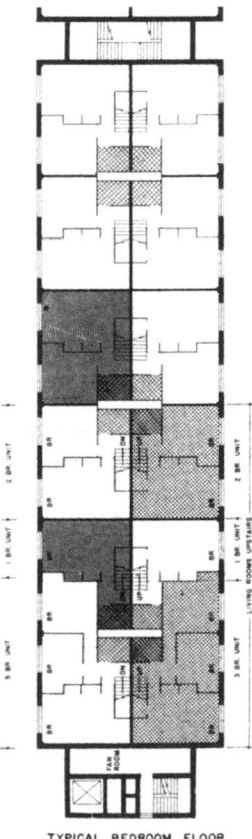

TYPICAL BEDROOM FLOOR

Fig. 8

TYPICAL CORRIDOR AND
LIVING ROOM FLOOR

PHYSICAL ELEMENTS IN THE ORGANIZATION OF THE NEIGHBORHOOD

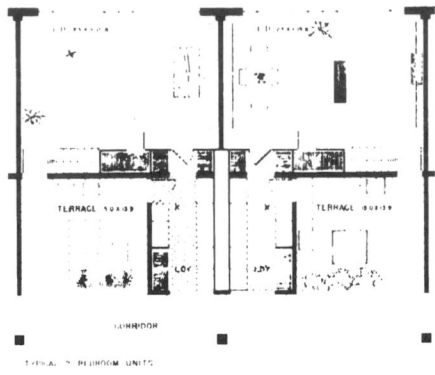

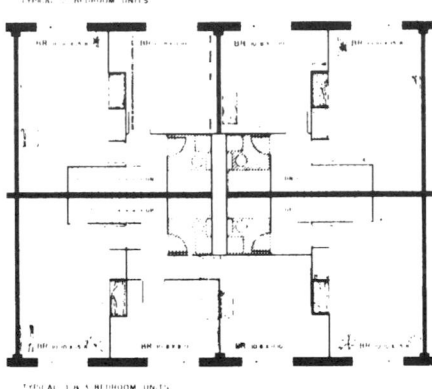

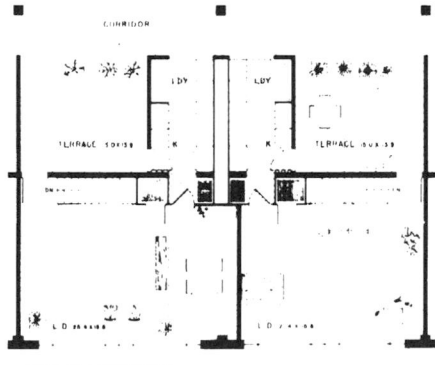

Fig. 9 Site plan. Leinweber, Yamasaki & Hellmuth, Architects.

Entire floors of bedrooms occupy every third level above ground floor (which is devoted to community space, and contains no apartments). At the other end of connecting stairways within the apartments are living room and traffic floors.

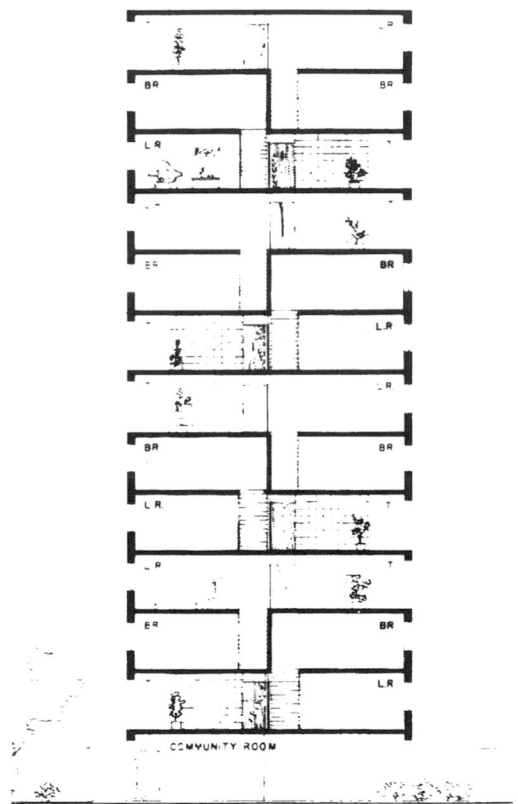

PHYSICAL ELEMENTS IN THE ORGANIZATION OF THE NEIGHBORHOOD

NEIGHBORHOOD COMPLEX

Single Neighborhood

The complex in Fig. 10 is sufficiently large to support an elementary school, local shopping, and a range of recreational facilities. It is small enough to identify with personally yet large enough to sustain a variety of interests and friendships. It is a major planning tool in organizing larger physical areas.

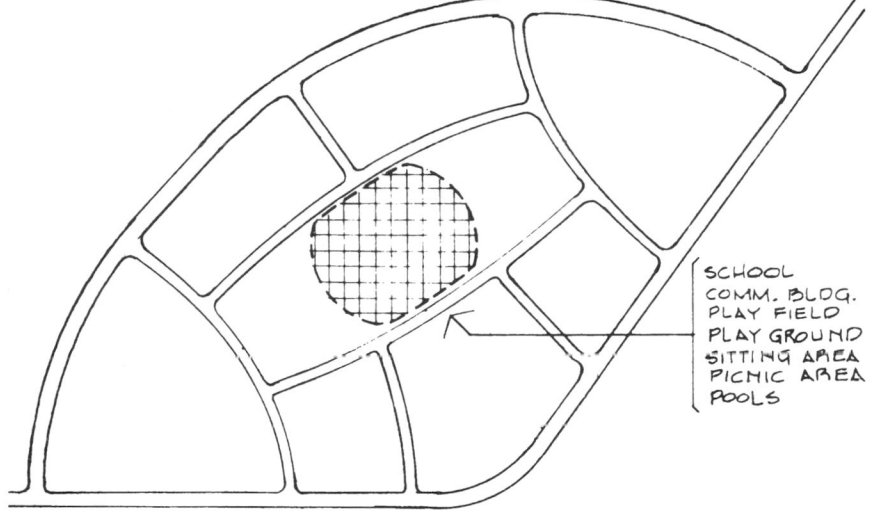

SCHOOL
COMM. BLDG.
PLAY FIELD
PLAY GROUND
SITTING AREA
PICNIC AREA
POOLS

Fig. 10

Cluster of Neighborhoods

The cluster in Fig. 11 contains an adequate number of people to support a full range of educational, social, and economic facilities. It may be considered a small town or village.

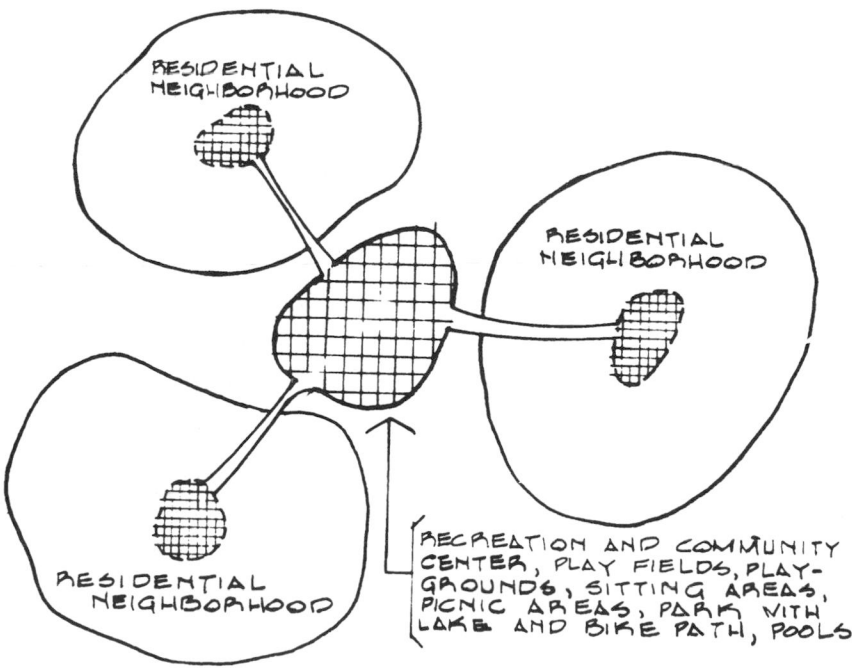

RESIDENTIAL NEIGHBORHOOD

RESIDENTIAL NEIGHBORHOOD

RESIDENTIAL NEIGHBORHOOD

RECREATION AND COMMUNITY CENTER, PLAY FIELDS, PLAY-GROUNDS, SITTING AREAS, PICNIC AREAS, PARK WITH LAKE AND BIKE PATH, POOLS

Fig. 11

PHYSICAL ELEMENTS IN THE ORGANIZATION OF THE NEIGHBORHOOD

The Garden City, Ebenezer Howard

Ebenezer Howard put forth his concept of a garden city in a book entitled *Tomorrow: A Peaceful Path to Real Reform* in 1898. The basic goal was to combine the advantages of town life with that of the country. He advocated the building of "towns designed for healthy living and industry of a size that makes possible a full measure of social life but not larger; surrounded by a rural belt; the whole of the land being in public ownership, or held in trust for the community."

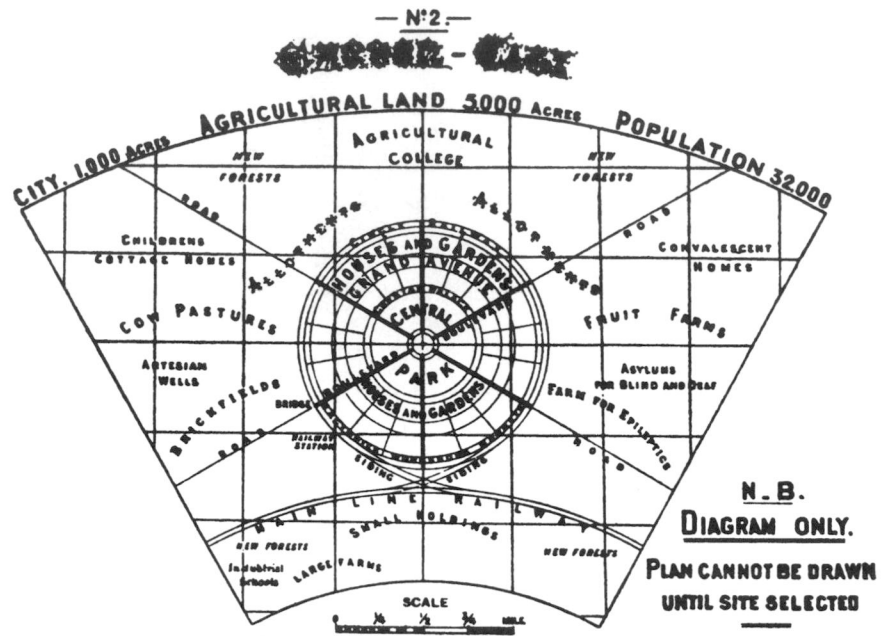

Fig. 12

Ward and Center of Garden City

Total area of city, 6000 acres
Built-up area, 1000 acres
Permanent greenbelt, 5000 acres
Total population, 32,000 people
City organization:
Center: Civic buildings
1st ring: Central park
2d ring: Housing of various types bisected by Grand Ave.
3d ring: Crystal palace or covered promenades
4th ring: Factories and warehouses
Green belt: Permanent open space

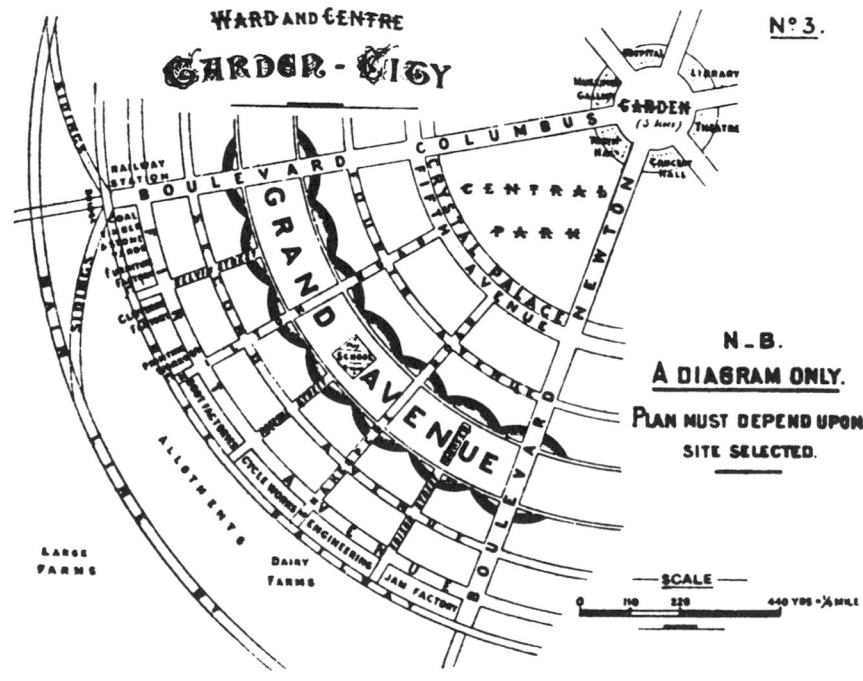

Fig. 13

PHYSICAL ELEMENTS IN THE ORGANIZATION OF THE NEIGHBORHOOD

NEIGHBORHOOD UNIT

Clarence Perry

In a preliminary study in 1926 and in a report published by the Committee on the Regional Plan of New York and Its Environs in 1929, Perry enunciated his neighborhood theory. Its basic principles were:

1. Major arterials and through traffic routes should not pass through residential neighborhoods. Instead, these streets should provide the boundaries of the neighborhood.

2. Interior street patterns should be designed and constructed through use of cul-de-sacs, curved layout and light-duty surfacing so as to encourage a quiet, safe, low-volume traffic movement and preservation of the residential atmosphere.

3. The population of the neighborhood should be that which is necessary to support its elementary school. (When Perry formulated his theory, this population was estimated at about 5000 persons; current elementary school size standards probably would lower the figure to 3000 to 4000 persons.)

4. The neighborhood focal point should be the elementary school centrally located on a common or green, along with other institutions that have service areas coincident with the neighborhood boundaries.

5. The neighborhood would occupy approximately 160 acres with a density of 10 families per acre. The shape would be such that no child would walk more than ½ mile to school.

6. The unit would be served by shopping facilities, churches, a library, and a community center located near the elementary school.

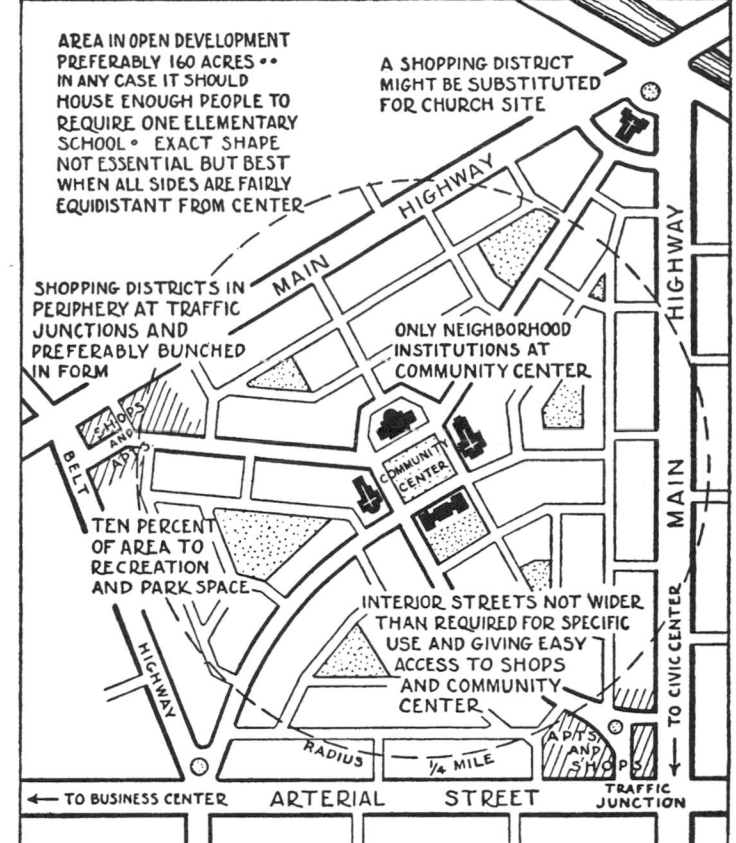

Fig. 14

The Neighborhood Unit—Clarence Stein

The elementary school is the center of the unit and within a ½ mile radius of all residents in the neighborhood (Fig. 15). A small shopping center for daily needs is located near the school. Most residential streets are suggested as cul-de-sac or "dead-end" roads to eliminate through traffic, and park space flows through the neighborhood in a manner reminiscent of the Radburn plan (see Figs. 17 through 20).

The grouping of three neighborhood units is served by a high school and one or two major commercial centers, the radius for walking to these facilities being 1 mile.

N. L. Englehardt, Jr.

N. L. Englehardt, Jr., has presented a comprehensive pattern of the neighborhood as a component of the successively larger segments in a city structure (Fig. 16). The neighborhood unit includes the elementary school, a small shopping district, and a playground. These facilities are grouped near the center of the unit so that the walking distance between them and the home does not exceed ½ mile. An elementary school with a standard enrollment of between 600 and 800 pupils will represent a population of about 1700 families in the neighborhood unit.[1]

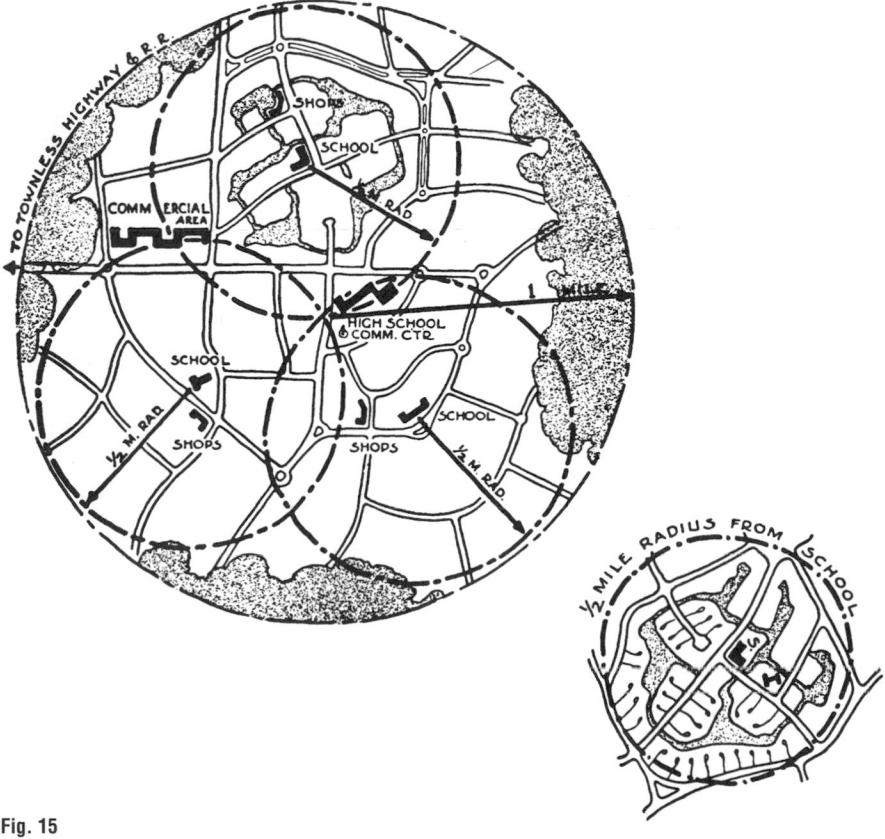

[1] N. L. Englehardt, Jr., The School-Neighborhood Nucleus, *Architectural Forum*, October 1943.

Fig. 15

PHYSICAL ELEMENTS IN THE ORGANIZATION OF THE NEIGHBORHOOD

Two such units (3400 families) will support a junior high school with a recreation center in conjunction; the walking distance does not exceed 1 mile from the center to the most remote home. Four units (6800 families) will require a senior high school and a commercial center. It will also be an appropriate size for a major park and recreation area. This grouping of four neighborhood units forms a "community" with a population of about 24,000 people. The component parts of this community pattern are integrated, and such communities may be arranged in whatever combinations the sources of employment and communication to and from them may require.

A Neighborhood Unit by José Sert

Figure 17 illustrates an organization of neighborhood units suggested by José Sert. While some authorities have stated that the maximum walking distance from home to the elementary school should be ½ mile, this diagram indicates a maximum distance of about ¼ mile, which is the standard accepted by a number of communities. In contrast to a population density of 20 to 25 persons assumed as a desirable average in many communities, Sert assumes a density of 2 or 3 times this number, which may account for the shorter walking distances he proposes from homes to the several schools in his scheme.

The elementary school occupies a central position in the neighborhood unit, and a group of these units—six to eight in number—constitutes a "township" with a population of between 56,000 and 80,000 people. A junior high school serves four neighborhoods; a senior high school serves the eight units; these facilities are situated within a township center surrounded by a greenbelt. The neighborhood unit includes the elementary school, preschool play lots, playground, church, shopping center, library, and emergency clinic. The township center includes the junior and senior high schools, community auditorium and meeting rooms, concert hall, theaters, main shopping center, recreation and administrative center.

Traffic ways bypass the neighborhood units and connect them with the civic center, which includes the regional facilities for administration, education, hotels, trade and recreation, and transportation stations on one side, and on the other side are the locations for light industrial plants. All these elements are separated from each other by greenbelts, and the open countryside is accessible to all the people.

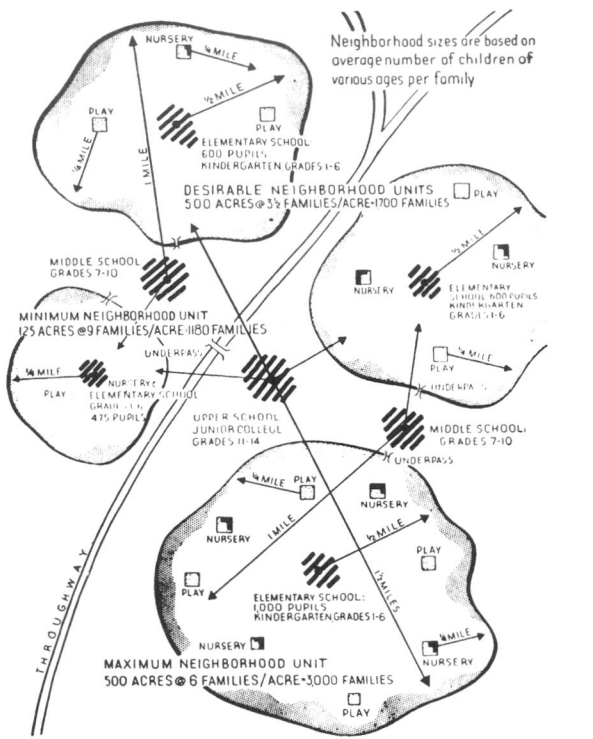

Fig. 16

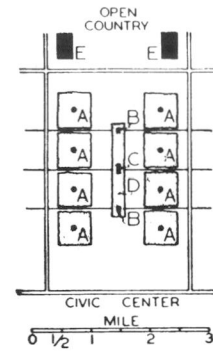

A Neighborhood Unit
B Junior High School
C Senior High School
D Township Center
E Light Industry

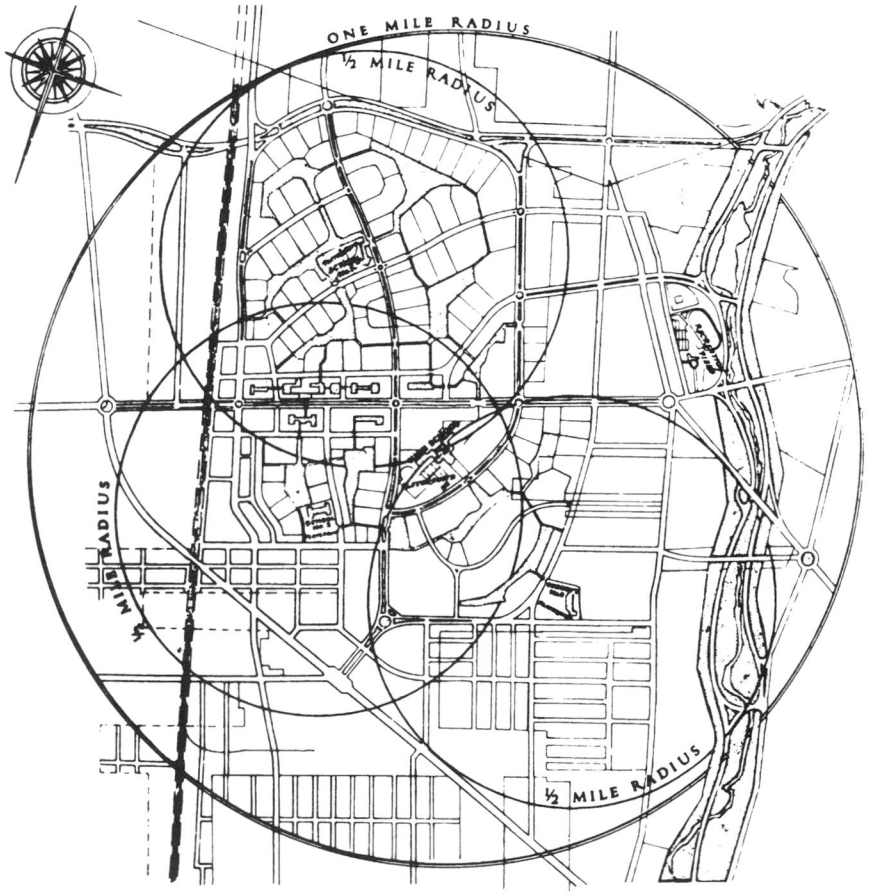

Fig. 17

General Plan Showing Neighborhoods

In their design of the suburb of Radburn in New Jersey, C. S. Stein and Henry Wright introduced a new approach to residential planning. They originated the superblock idea, the main feature of which is the separation of pedestrian and automobile traffic (Fig. 18). At Radburn, houses are grouped around a series of cul-de-sacs that are linked by walkways with the park, the school, and the shops, all of which are located in the interior of the superblock. The superblock is considered an ideal solution to the circulation problem, since it provides a means of locating the houses off the main road.

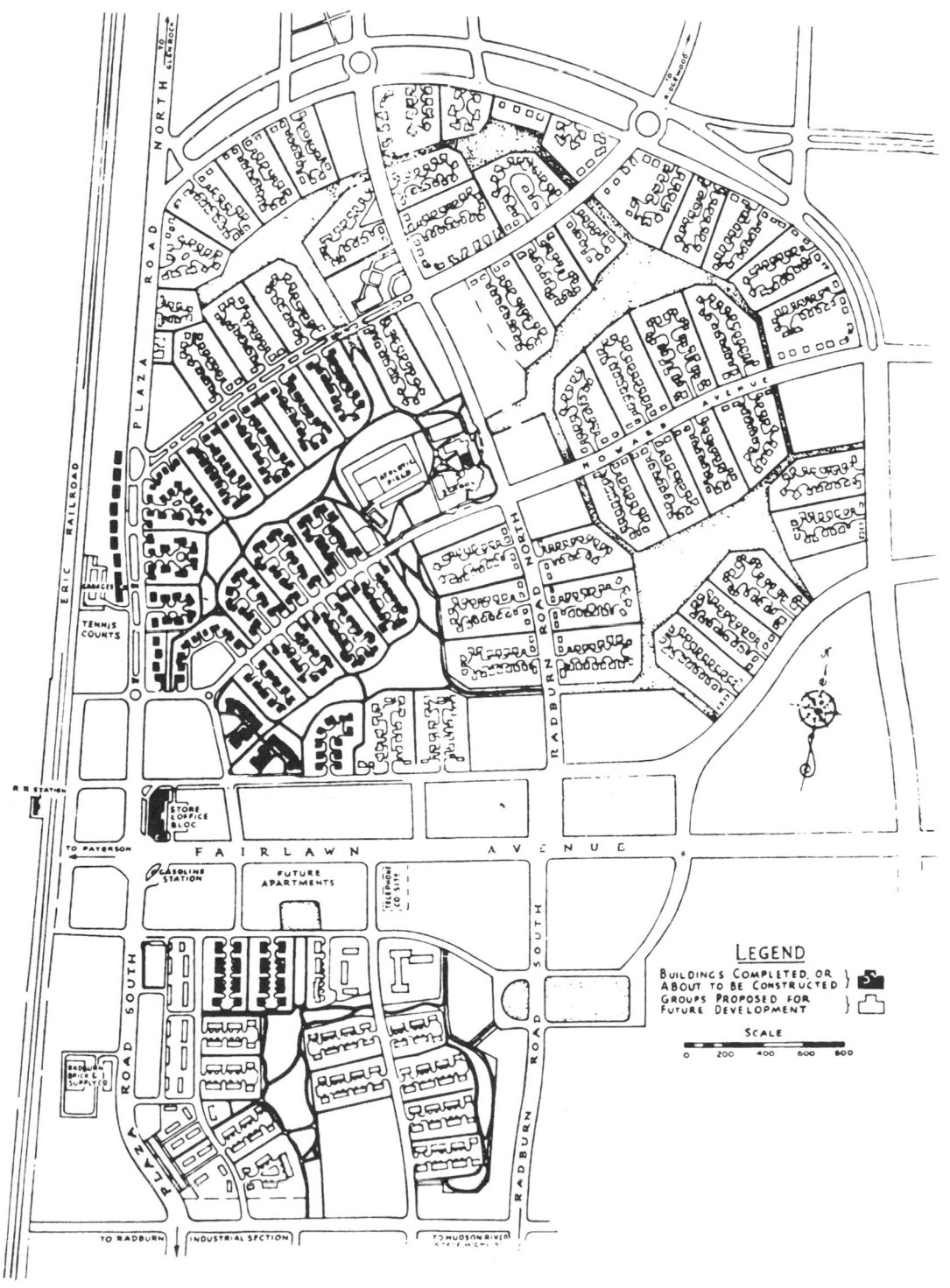

Fig. 18   Plan of Radburn.

PHYSICAL ELEMENTS IN THE ORGANIZATION OF THE NEIGHBORHOOD

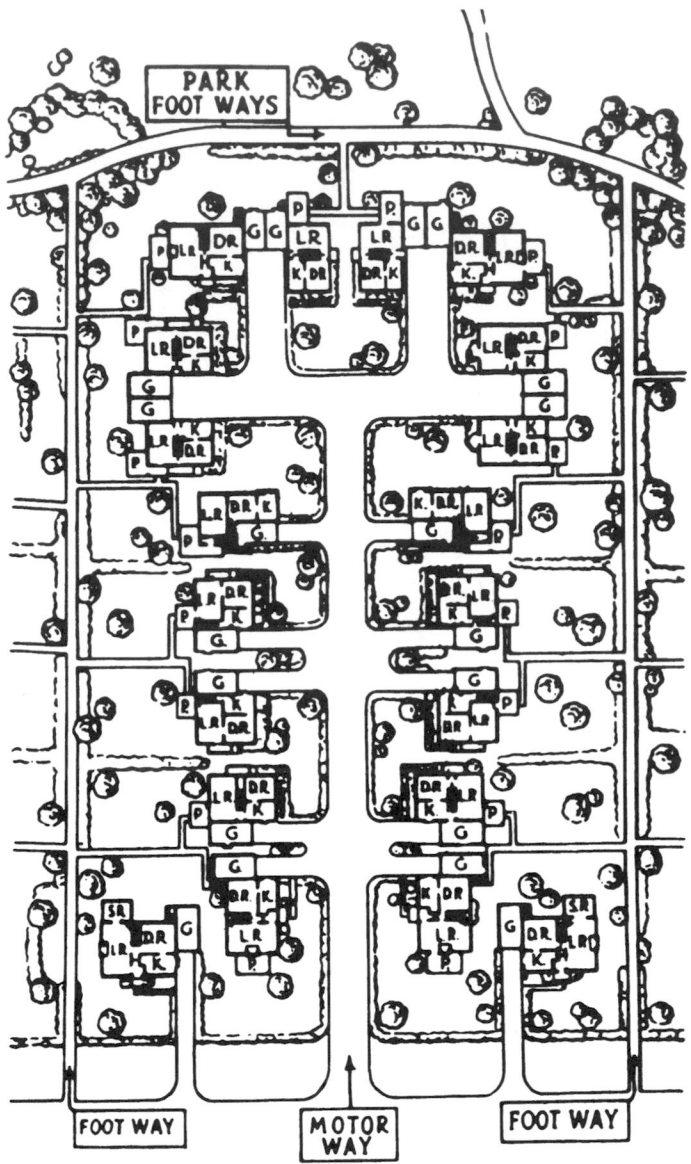

Northwest Neighborhood

Figure 19 shows a typical cul-de-sac street employed at Radburn. Its characteristics may be summarized as follows: The short cul-de-sac acts as a service lane only; it provides vehicular access to houses and garages, permitting delivery and other services, and it also serves for most of the parking; footways located on the perimeter of each cul-de-sac house group serve as sidewalks. As opposed to established planning practices, houses have been "turned around," the living rooms, porches, and as many bedrooms as possible facing the gardens at the rear of dwellings, and kitchens and cellar storage facing the service lane.

The dwellings are loosely disposed around the dead-end streets and, as a group, they show little of formal architectural discipline. The landscaping, judiciously planned, undoubtedly is the most important uniting element in the composition. Other uniting elements are the consistency in the use of building materials and the continuity in roof lines. Also, by joining houses by means of coupling their garages, the usual disorderly appearance of the free-standing houses in relation to each other has been eliminated and sufficient space left on either side of the buildings. The architectural informality of the Radburn cul-de-sac distinguishes it from the British dead-end street in which a formal correlation of the houses predominates.

Fig. 19   Plan of a typical "lane" at Radburn. The park in the center of the superblock is shown at the top; the motorways to the houses are at right angles to the park.

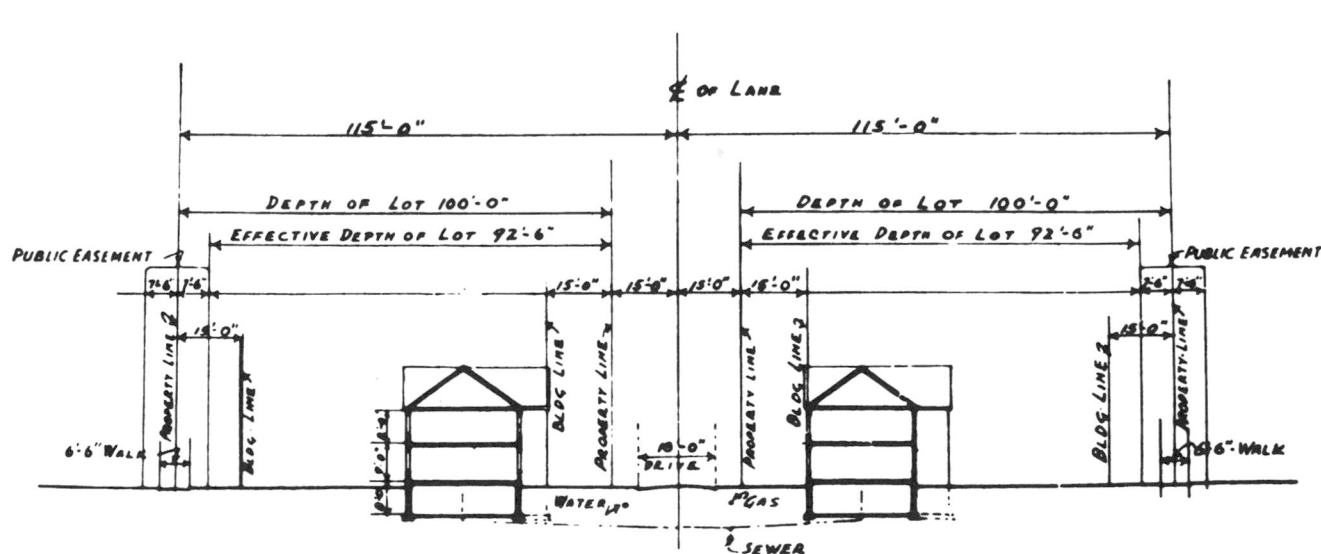

Fig. 20   Typical transverse section of a "lane" in the first unit of Radburn.

Site Considerations and Site Planning

SITE PLANNING DEFINED

Site planning is a broad term that embraces selection of sites; location of buildings in functional relation to each other, to the shape and topography of the site, and to the environment; provision within the site of suitable circulation routes well related to existing or proposed streets and walks; determination of land use to complement the buildings, such as private yards, parking space and recreation areas. These and many other things are included within the scope of site planning.

There has been at least some evidence that, in the minds of inexperienced or thoughtless designers, the site plan is looked upon as the arrangement of a group of buildings into a pattern, pleasing in its two-dimensional qualities, or as a simple scattering of buildings. In either case a few details only, such as the relationship of existing topography, street grades, and sewer depths, seem to have been considered as complicating features.

The site plan is a complex thing, and any underestimation of its importance risks the success of a project. The site plan is shaped by climate, by local housing customs, economic conditions, and laws; by the location of the site with respect to employment, transportation, utilities, and social institutions; by the cost of the land, the relative cost of various forms of construction, and the cost of utilities and maintenance; by the habits, incomes, and composition of the families to be housed. It is influenced by the area, shape, and topography of the site; the number of dwelling units proposed and whether these are to be apartments, flats, row or town houses; the orientation and spacing of the buildings; the method of waste collection and disposal and the landscape development and the preservation of existing trees. All these factors must be correlated to produce a simple, livable, economical pattern of land use in which the land and buildings are integrated and so organized as to serve the needs of the families to be housed. The organization of the plan, if satisfactory, will also harmonize, not conflict, with the character of the land.

The physical site characteristics contained here fall into five general groups. They include natural characteristics, orientation, circulation, parking, and utilities. The first group contains the major constraints vis-à-vis grades and soil conditions. Availability of potable water, energy supplies, and adequate municipal services for waste disposal are not included in this discussion, although they certainly act as constraints to site development.

Orientation is an important consideration for aesthetic and practical reasons. Despite controlled atmospheres within the dwelling unit, proper attention to placement and orientation can add greatly to the efficiency and comfort of the inhabitants.

Circulation and parking are increasing in importance as integral parts of site design for residential communities, including the higher-density urban communities, because of the trend toward heavier reliance on the private automobile.

The location of utilities is also receiving increased attention commensurate with the rise In energy consumption. Placement of equipment, lines, and generating facilities is of aesthetic concern and relates directly to reliability of service as affected by storm conditions.

Every housing site represents a special situation and should be approached and designed with this in mind. It is meaningless and even harmful, therefore, to try to establish rigid rules for universal application.

A SITE AND ITS IMMEDIATE NEIGHBORHOOD

Initial site selection involves many factors, an important one being location with respect to the general neighborhood. Frequently the recommendations of professional planners who have assisted in the preparation of long-range development plans and zoning maps aid in site selection. Such recommendations consider land-use compatibility, availability of community facilities, etc.

There are no hard and fast rules regarding mixtures of land uses. Where rules have been formulated (as in most zoning ordinances), they exist to protect residential areas from the nuisances commonly associated with nonresidential uses. A more realistic approach to mixing is to evaluate each proposed use on its merits. For example, a nonresidential area is not a priori an acceptable or nonacceptable neighbor to a housing site. The compatibility or noncompatibility of the two uses should be carefully examined. If they conflict, a generous buffer between the two is called for. If they do not conflict, there is no cause for alarm. In zoning this reasoning has resulted in the adoption of performance standards to replace conventional use districts. These performance standards set measurable maximum limits for certain nuisance factors associated with many uses. Among these are noise, smoke, and traffic. Some of these problems can be alleviated, if not entirely eliminated, through careful site planning if planners have a thorough knowledge of existing off-site conditions in advance of their work. They cannot, of course, anticipate all problems and future changes.

Generally sources of noise are easy to predict. Large playgrounds or other public recreation places, commercial and industrial complexes, and transportation facilities are the most common ones. Buffers of open space and sound-absorbent plant materials can help to control such noises. In the case of smoke and other annoying atmospheric pollutants, two choices are open. One is simply not to build housing on a site which is downwind of a known air-pollution source. The second is to force the offender to install air-pollution control devices. The latter is not a job for housing site planners alone, although they can contribute to public awareness of the menace.

Traffic hazards are associated with high speed, heavy circulation, and dangerous roads and intersections. Again, these factors are usually predictable for any site in a built-up portion of a city or in sections that are in transition. If traffic hazards cannot be eliminated through external controls, they can at least be substantially reduced by careful internal planning of a site. Techniques such as locating dwelling units to focus on internal site spaces, using structures or plants to protect the vulnerable edges, and separating the pedestrian paths and vehicular roadway systems are all effective.

Performance standards in themselves can do little if anything to combat invasion of privacy. Here again, a thorough examination of neighborhood characteristics is essential in order to pinpoint any real or potential threats to privacy. From such a survey, it is possible to predict natural traffic routes by examining their generation from points outside the site. A skillful site design can do a great deal to prevent the unwarranted use of a site by nonresidents. If a site is used as a shortcut by outsiders, pedestrians, and automobile drivers alike, this would indicate that the plan itself is at fault, since it fails to discourage and may, in fact, encourage such traffic. The all-too-frequently used sign "Trespassers Keep Out" is eloquent testimony to this failure.

The planner is faced with a more complex problem if the site is conceived as a neighborhood focal point or if it is to incorporate shops and other community services. In these cases, loss of privacy can be prevented by designing separate public and private areas and laying out communal paths and gathering places so that they are at adequate distances from dwelling units.

In extreme cases, when it is almost impossible to regulate these neighborhood nuisances or to abandon a site in favor of a better location, the only alternative may be to erect a physical barrier that insulates the housing site from its surroundings. Even this alternative is ineffective against air pollution.

SITE SIZE AND SHAPE

There are no meaningful maximums or minimums in the abstract that can be applied to the size of a housing site. It is the choice of what is built upon the land and its relationship to the total community that determines site size suitability. Unfortunately, site and program of development are often chosen independently of one another, causing a misfit that could have been prevented. A developer decides upon a specific type of housing in advance of land acquisition, resulting in either overcrowding or wasting of land and an inferior plan for designated uses.

As is true for size, there are no abstract optimums for the shape of a parcel of land. Certain shapes, particularly very narrow or irregular properties, will severely restrict the choice of possible building types and/or placement on the site. Parcels that are square or possess sufficient width for an arrangement of buildings in adequately spaced parallel rows permit a greater choice. In practice the minimum width of a site is set either directly by zoning (40 ft, 50 ft, etc.) or indirectly by setback requirements from lot lines. The effects of adjoining property and activity often vary according to a site's physical configuration. For example, the narrower the site, the greater the influence that adjacent buildings have on its supply of light and air. Conversely, the wider the site, the weaker the influence.

SPECIAL CONSIDERATIONS FOR LARGE SITES

As a result of economies of large-scale building and the current interest in planned communities and new towns, bigger and bigger sites are being developed. Therefore, this facet of planning warrants some special consideration. At the outset, it is necessary to distinguish between composite sites and single sites. Composite sites consist of numbers of parcels that are usually sold and developed as separate projects that may or may not be part of a coordinated design. Single sites are planned and built according to a coordinated site plan. There is not always a clear-cut line between the two types of developments—particularly when composite site parcels are sold individually but are nevertheless subjected to review for conformity to an overall development plan. In either situation, one of the most desirable attributes of a site is variety. It follows that one of the least desirable characteristics is monotony. Monotony can have a deadening effect. Even though great care may have been taken to develop a diversity of housing types and site details, a large site often seems monotonous just by virtue of size alone. To counteract this, several design techniques are possible. The shape of a site can be adjusted. The massiveness of a large site is camouflaged somewhat if its shape is irregular or narrow. This serves to prohibit site development which looks the same from all directions.

The visual field can also be broken up by exploiting a site's natural amenities, such as rock outcroppings, water courses, varied topography, or attractive vegetation. Where such amenities are lacking, it is imperative to avoid layouts whereby the whole site may be seen at a single glance.

Allowance for visual breaks and contrast is a vital factor in site design. Even when building designs must be repeated to hold down costs, as is necessary in the case of low-rent housing, scattering the units throughout a community rather than grouping them together can relieve the tedious sameness produced by masses of identical buildings.

TOPOGRAPHY AND CLIMATE

Topography and climate play a most important part in determining the appropriateness of the location and design of a site. Gently rolling land offers greater opportunities for variety in site planning and architectural design than does flat land. Grade changes permit more imaginative determinations of building-to-building relationships, automobile storage, and outdoor passive and active recreation areas than do sites lacking in irregularities. The latter are dependent on excavations, manmade hills and water bodies, and architectural forms to create interest.

Today, "view" lots are prized. A high site generally has more inherent possibilities for a broad vista than a low site does. Also, a high site that can be seen from a distance—the easily identifiable focal point of its surroundings—

has a definite psychological attraction. From a practical standpoint, high ground is also appealing because liquid wastes and unpleasant odors do not collect there and because it is relatively secure from floods—all problems of low land. However, low-land sites are not always a second choice. Those bordering water are often very desirable. Usually a substantial investment, in the form of retaining walls, expensive drainage systems, and landfill, for example, is required to make such land suitable for building. The costs of these improvements are usually offset by the advantages a water site provides—spectacular views, sports and recreation, permanent privacy on at least one side.

A low site can also be desirable in areas of climate extremes, for hollows tend to be protected from both excessive heat and cold. This factor is becoming less and less important since local climatic conditions can be influenced artificially through the use of vegetation, structural barriers, manmade topography, and central heating and cooling systems. With technological advances, the livability of low land versus high land is becoming more and more a simple question of taste.

BUILDABLE QUALITIES

Three factors are of major concern under this general heading—soils, slopes, and vegetation. For many regions of the United States, natural surveys identify and record soil types. At specific sites, soils are examined by engineering surveys. They are tested for drainage, water-table level, and load-bearing capacities—all of which should affect both the type and location of buildings on any given site. A knowledge of soil conditions or test borings is necessary for all but the most modest structures. Rock outcroppings can complicate initial site preparation, but if they exist in limited amounts they may be exposed and become assets in site appearance. Soils not well suited to the growth of plant materials can be treated or covered or replaced with others that are.

Slopes are conducive to imaginative site planning. Gentle slopes are preferable to very rugged terrain for most residential building

purposes. Besides aiding in drainage they are less costly to build on than steep slopes. The latter require elaborate footings and a great deal of earth moving to create usable ground areas. The Federal Housing Administration requires a minimum slope of 1 percent and a maximum of 8 percent around buildings. Where sites are flat, considerable success has been achieved by the building of artificial hills. To be sure, many examples can be found of spectacular housing built on the edge or down the slopes of very steep hillsides. This raw land is cheap to buy but expensive to develop and may not allow for outdoor recreation. When on-site parking is required, it must be built directly into the dwelling structure, an additional complication and expense. Parking on such slopes is often dangerous and unsightly. Yet the magnificent vistas and the possible proximity to the central city are compensating factors.

Serious erosion and drainage problems have resulted from poor handling of steep land—improper terracing for building sites, careless placing of buildings, inadequate underground tiles to augment natural seepage, and neglect of the banks of drainage channels that carry away water during periods of heavy, seasonal rainfall. If the site is large enough and has varied topography, it is best to leave the very steep portions in their natural state and build on the more gentle slopes. These gentle slopes can be protected through the use of plant materials. Any remaining flat land can be used for recreational purposes.

The value of vegetation as an aspect of site quality cannot be overestimated. There is no question that trees and shrubs enhance the livability of housing areas. If an undeveloped site contains healthy and attractive vegetation, it should be preserved. If a site lacks vegetation, planting should be undertaken at the earliest possible date. In cases where soils or subsurface conditions are not conducive to growth and the introduction of new soil is not feasible, an alternative is to plant trees and shrubs in boxes. The task of creating and maintaining a green site can be accomplished only if occupant needs are taken into account before a site is acquired, or once it is acquired, before new structures are placed on it.

TOPOGRAPHY

Topography is an important element in determining the acceptability or value of a site. It greatly affects the layout of buildings and how they can be placed upon it, and it affects the cost of foundations and utility lines.

In order to make a proper judgment, detailed information in the form of accurate surveys showing contours is necessary. Such information must be interpreted by architects and engineers.

The best type of topography for housing is generally considered to be level or gently rolling terrain with slopes less than 10 to 20 percent. For single-family detached housing, the lot size should be increased in relation to the slope. It should also be high ground with good drainage.

However, a site should not be discarded because of rugged contours. Such features may, by careful study and imaginative design, be turned into an advantage and add features that would not be available on a level site.

Some Common Topographic Positions (Fig. 1)

Area 1 is a floodplain. It is subject to flooding during heavy storms.

Area 2 is an alluvial fan. The soil has been forming over the years as a result of water eroding material from the watershed above and depositing it near the mouth of the waterway. An alluvial fan can be hard hit by flash floods after heavy rains unless an adequate water-disposal system has been provided to control the runoff from the watershed above.

Area 3 is an upland waterway where water flowing from the higher surrounding land will concentrate. Natural waterways should not be used unless an adequate ditch or diversion terrace has been constructed to divert water from the site.

Area 4 is a low depressed area where water accumulates from higher surrounding areas. These soils remain wet and spongy for long periods.

Area 5 is a steep hillside. Many soils on steep slopes are shallow to rock. Some are subject to severe slippage. On all slopes, one must be careful of soil movement through gravity or by water erosion. Yet some steep hillsides can be used safely as building sites. The problem can be solved by studying the soils and avoiding the bad ones.

Area 6 is a deep, well-drained soil found on ridgetops and gently sloping hillsides. Generally these areas have the smallest water-management problems. They are the best building sites, other things being equal.

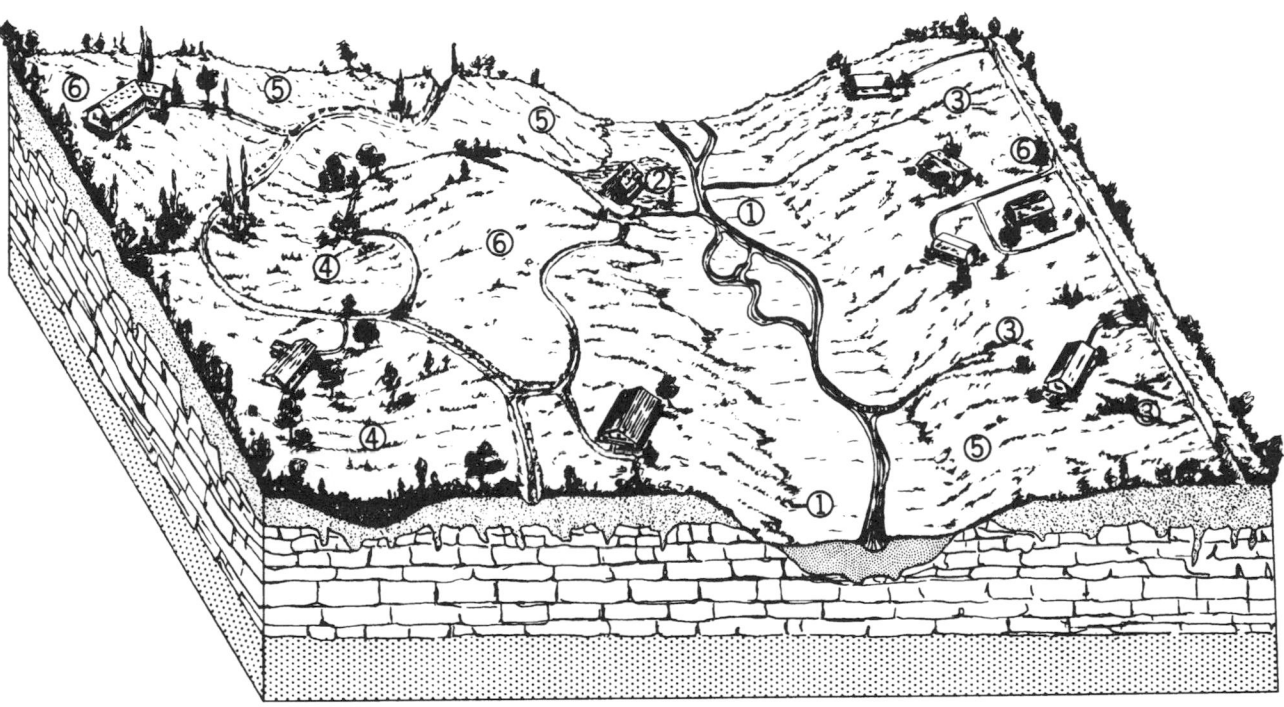

Fig. 1

HOUSING SITE

1. *A gently sloping site is preferable to one presenting serious topographical difficulties.* This will be apparent by an examination of Figs. 2 and 3 in which the same area has been assumed in both cases, the same access streets, and the same number of dwelling units, but with different topography. A study of these plans will make it evident that in Fig. 2 there is a greater likelihood of expensive cut and fill. The steep slope also requires added provision for surface drainage to prevent heavy accumulation of rain water. Culverts and large-sized storm sewers may be required to remove it.

If the buildings are on a steep slope, they may be more costly because of added exterior walls

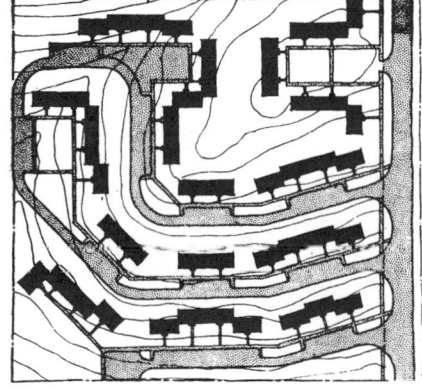

Fig. 2

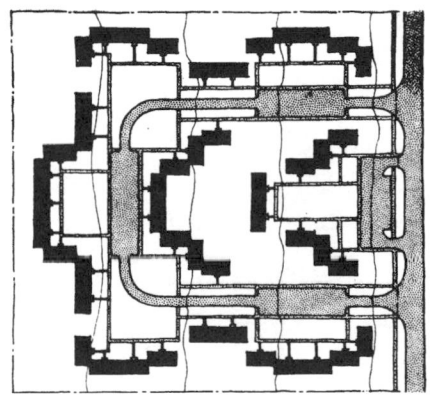

Fig. 3

Fig. 4   Site plan. Westchester Development Corp., Pokorny & Pertz, Architects.

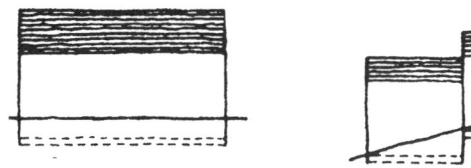

Fig. 5

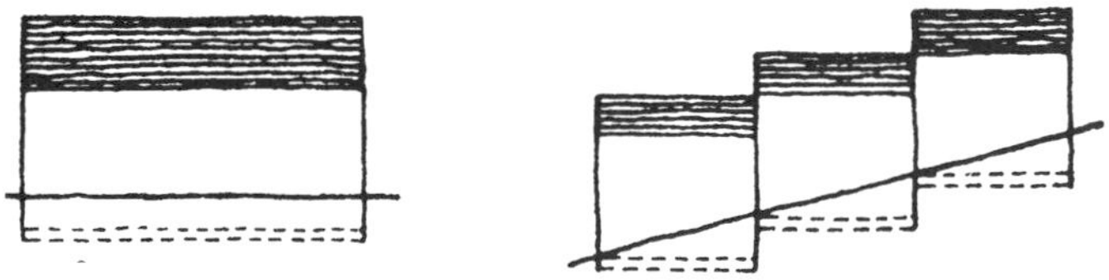

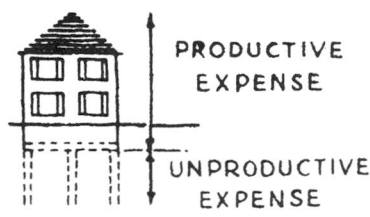

Fig. 6

Fig. 7

PRODUCTIVE EXPENSE

UNPRODUCTIVE EXPENSE

(Fig. 5). This is especially true where basements are omitted.

Where a long building runs perpendicular to the contours, it may be necessary to vary the floor and roof levels (Fig. 6). This means added costs of roof and finishing the stepped gable ends.

A comparison of the two plans in Figs. 2 and 3 will also demonstrate the added flexibility of planning in the level site due to the unrestricted possibilities of placing the buildings. The difference in length and cost of road improvements to produce similar ease of access will also be apparent.

2. *Sites containing soft ground, heavy uncompacted fill, or outcroppings of rock should be avoided* (Fig. 7). Preparation of the site to obviate these objectionable features is expensive and adds nothing to rental value.

3. *Choose a site where heavy-duty road construction will not be required.* The traffic tributary to the average housing development is not heavy, and comparatively light hard-surfaced roads of moderate widths will suffice. If the needs of urban or through traffic require the construction of heavy-duty roads, either boundary or internal, at the expense of the project, an unproductive burden of cost is saddled on the enterprise.

4. *Sites remote from public roads and utilities are less desirable than those where these facilities are immediately available.* If roads and utilities must be brought from a great distance, low-priced land may prove to be prohibitive in final cost.

INLAND AND SHORE CONSIDERATIONS OF HOUSING SITES

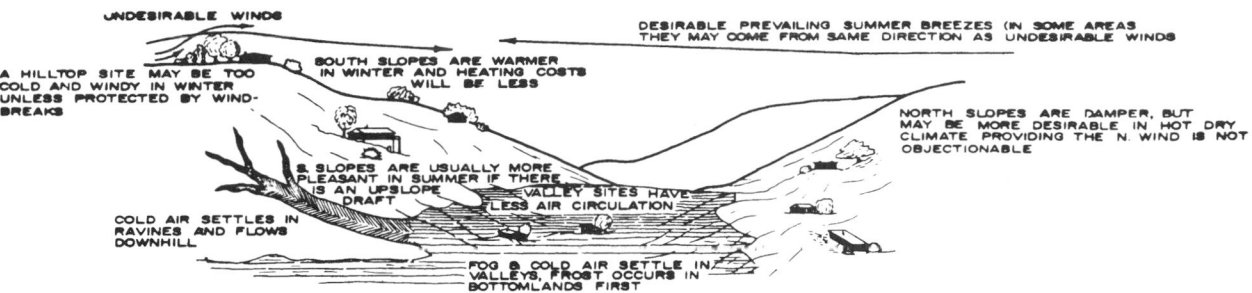

Fig. 8 Inland home sites.

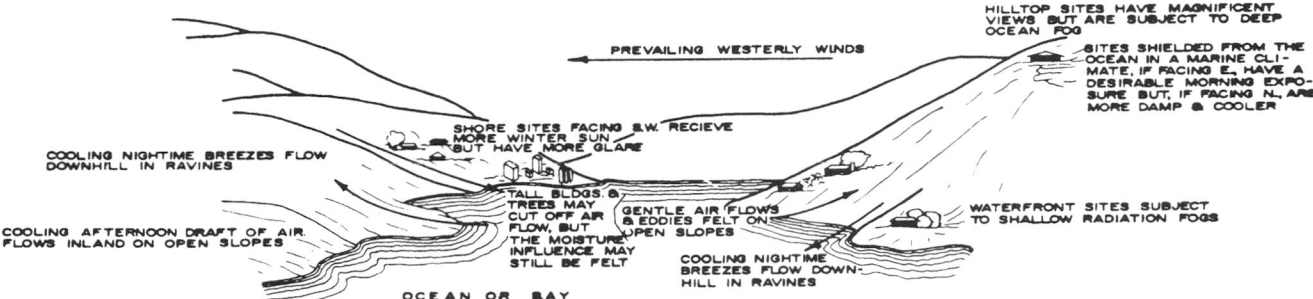

Fig. 9 Shore home sites.

BUILDABLE AREA

No site should be given even tentative consideration unless the amount of buildable area it contains is known. If the site includes steeply sloping land, at least a sketch topography should be available. Data on soil conditions, particularly where there are areas of poor bearing due to natural conditions or to artificial fill, should also be obtained. A site engineer should cooperate with the planner in laying out topographically difficult sites. Runs and depths of sewer cuts constitute an important element of cost. Unbuildable areas of poor-bearing soil may often be used for parking or recreation areas, and thus need not cause a serious loss of useful area. Land that is unbuildable because it is so steep that construction cost becomes excessive is ordinarily of little use for other purposes, but all land may be of value to the project in giving more light and air to the houses. At the periphery of a project open area may provide useful protection against undesirable factors in the environment, acting as a miniature "greenbelt."

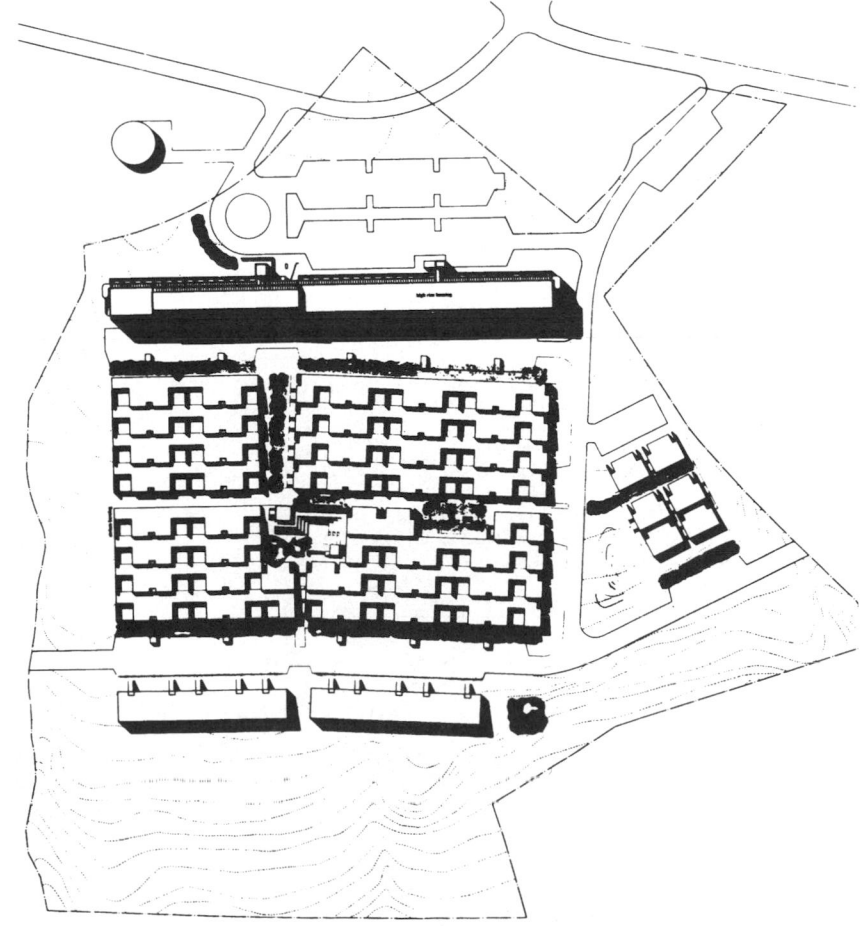

Fig. 10 Site plan. Ithaca, N.Y., Werner, Seligmann & Associates, Architects.

SITE CONFIGURATION

SITE CONFIGURATION

It will be evident from Fig. 1 that case A will afford frontages toward which the building units may be faced, and little road development inside the site will be necessary. Case B, with a narrow frontage and a depth of almost ⅓ mile, will of necessity require interior roads.

In this connection, it may be stated as a general principle that the narrow and deep site presents problems of site planning similar to the difficulties encountered in planning a single dwelling unit on a narrow and deep lot.

1. *Where surrounding roads must be constructed and paid for directly or by assessment, the site most nearly square is preferable.* This is a simple matter of geometry and is illustrated by Fig. 2.

2. *Utilities.* The knowledge that utilities are available to a given site is insufficient evidence on which to proceed. The adequacy of such utilities to bear the added loads that will be created by the proposed project must be satisfactorily determined. If water mains and sewers must be replaced, they might as well not be there.

Where access to a sewerage system is not possible and septic tanks are resorted to, the site should be carefully studied to determine that:

    a. There will be an available disposal field of adequate area.

    b. The soil will absorb the outflow water from the tanks.

    c. Public authorities will approve such installation.

3. *Fire protection.* Careful investigation should be made of the rate of fire insurance. If nonfireproof structures have been contemplated, the sponsors should investigate whether a differential in insurance cost would warrant the adoption of fireproof construction, or whether a different site should be chosen.

SHAPE

The shape of a site is a critical factor and influences usability of the site. Therefore, in assessing a site's usability, the following factors must be evaluated as a consequence of site shape:

1. Size
2. Accessibility
3. Visibility

To evaluate the effect of the shape of the site on the size of potential building locations, all applicable setback requirements in the local zoning ordinance must be defined and located. The space which remains is the area on which the building may be designed. A site with irregular proportions can sometimes be rendered useless after the setback requirements are subtracted. For example, Fig. 3 shows a site with 70 ft of frontage and 365 ft of depth. Once the setback requirements are taken into account, this site has greatly limited usability.

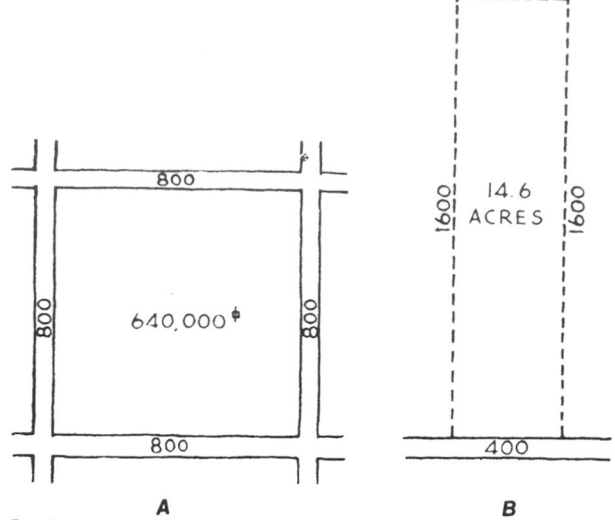

Fig. 1

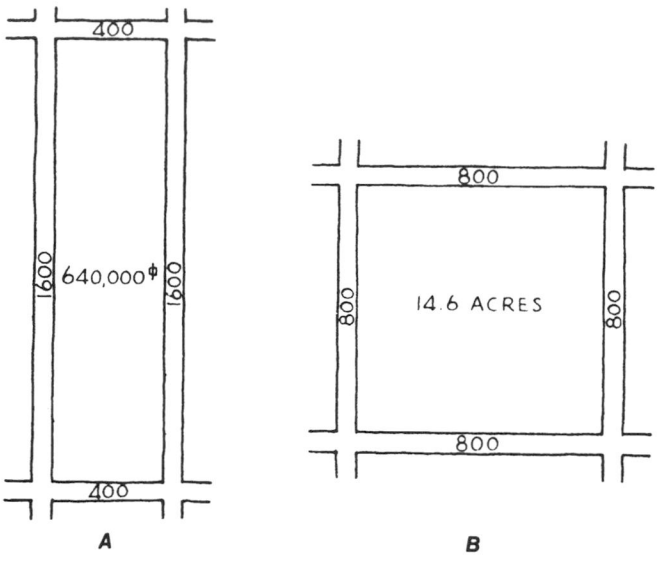

Fig. 2

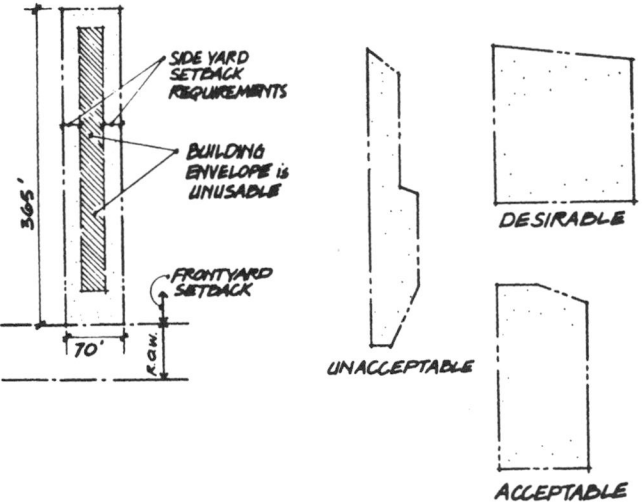

Fig. 3

SLOPES AND GRADES

Figure 1 gives desirable limits for slopes on different types of areas. Deviations may be warranted by especially favorable conditions, such as porous soils, mild climates, or light rainfall; also if local experience indicates that other gradients are satisfactory.

Failure to provide positive pitch away from buildings and to give open areas adequate slopes has necessitated costly regrading and reconstruction work on numerous projects. The trouble has been due in part to inaccurate construction, but incomplete or poorly conceived plans have been a contributing cause.

Of two basic design methods, one provides for drainage mainly across grassed areas, generally through "swales," until the water reaches streets, drives, or storm sewer inlets. This scheme, requiring the flow of water from walks onto lawns, is not altogether effective when slopes are inadequate and finished grading is not accurately executed, or if the turf is above the walk level. Swale drainage occasionally is carried under walks by small culverts (6- to 8-in pipes or boxes). These are slight hazards and frequently become stopped. The other method employs walks to a considerable extent as drainage channels. This scheme has met some objection; nevertheless, it generally is more economical and practical than the use of swales, and it has been used far more widely. Moreover, when walks have been given proper cross and longitudinal slopes, with sewer inlets provided at points of concentrated storm-water flow, there has been no serious inconvenience or complications.

AREA	FUNCTION	SLOPE IN PERCENT MAX.	SLOPE IN PERCENT MIN.
STREETS, SERVICE DRIVES AND PARKING AREAS		8.0	0.5
COLLECTOR AND APPROACH WALKS		10.0	0.5
ENTRANCE WALKS		4.0	1.0
RAMPS		15.0	—
PAVED PLAY AND SITTING AREAS		2.0	0.5
LAWN AREAS		25.0	1.0
GRASSED PLAYGROUNDS		4.0	0.5
SWALES		10.0	1.0
GRASSED BANKS		4 TO 1 SLOPE (3 TO 1 PREF.)	
PLANTED BANKS		2 TO 1 SLOPE (3 TO 1 PREF.)	

Fig. 1 Desirable slopes.

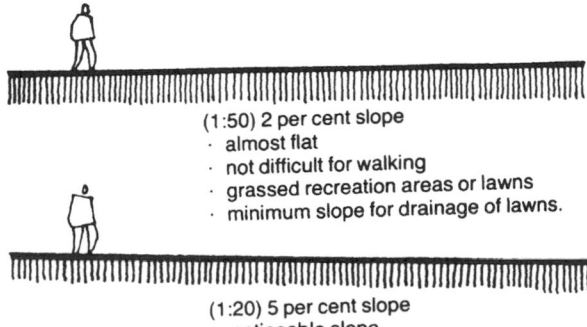

(1:50) 2 per cent slope
· almost flat
· not difficult for walking
· grassed recreation areas or lawns
· minimum slope for drainage of lawns.

(1:20) 5 per cent slope
· noticeable slope
· a little difficult for walking
· steepest slope for parking areas.

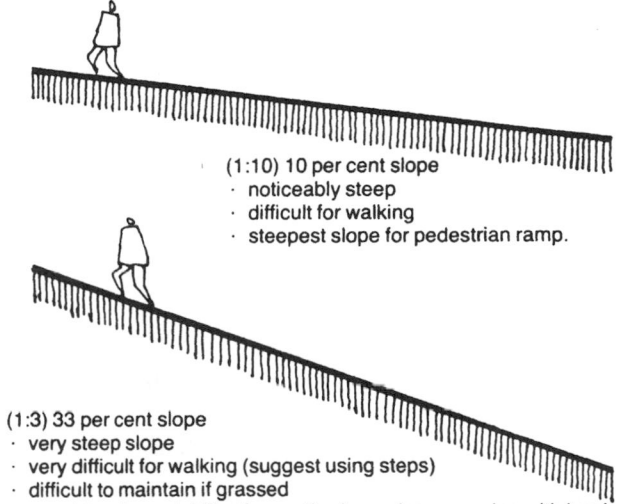

(1:10) 10 per cent slope
· noticeably steep
· difficult for walking
· steepest slope for pedestrian ramp.

(1:3) 33 per cent slope
· very steep slope
· very difficult for walking (suggest using steps)
· difficult to maintain if grassed
· consider slope stabilization methods, such as covering with hard materials, or use retaining walls.

Fig. 2 Examples of commonly used gradients. *Note*: 10 percent (1:10) slopes and steeper are especially dangerous in winter conditions.

SLOPES AND GRADES

Fig. 3 Lawn areas slope to street or drainage channel.

Fig. 4 Crowned sections are used for athletic fields, play areas, and traffic islands.

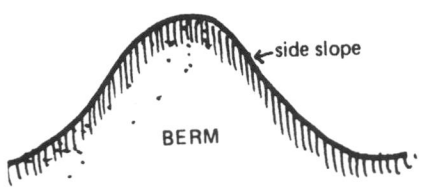

Fig. 5 Berm side slopes and planted slopes must be carefully selected, depending on plant cover and soil.

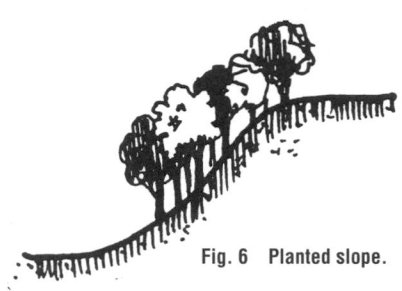

Fig. 6 Planted slope.

TABLE 1   Slopes of General Landscape Areas

Condition	Maximum		Minimum		Preferred
Lawns and grass areas	25%	4:1*	1%	100:1	1.5–10%
Grassed athletic fields	2%	50:1	0.5%	200:1	1%
Berms and mounds	20%	5:1	5%	20:1	10%
Mowed slopes	25%	4:1*	...		20%
Unmowed grass banks	Material of repose		...		25%
Planted slopes and beds	10%	10:1	0.5%	200:1	3–5%

*The maximum slope that mowing machinery can work on is approximately 25%.

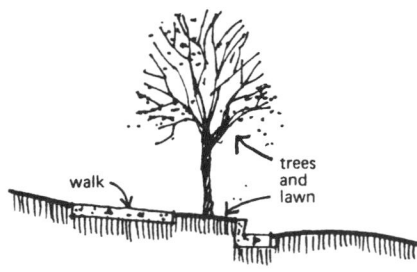

Fig. 7 Trails, rural roads, and highways.

TABLE 2   Slopes of Streets and Ways

Condition	Maximum		Minimum		Preferred	
Crown of improved streets	3%	33⅓:1	1%	100:1	2%	50:1
Crown of unimproved streets	3%	33⅓:1	2%	50:1	2.5%	40:1
Slide slope on walks	4%	25:1	1%	100:1	1–2%	
Tree lawns	20%	5:1	1%	100:1	2–3%	
Slope of shoulders	15%	66⅔:1	1%	100:1	2–3%	
Longitudinal slope of streets	20%	5:1	0.5%	200:1	1–10%	
Longitudinal slope of driveways	20%	5:1	0.25%	400:1	1–10%	
Longitudinal slope of parking areas	5%	20:1	0.25%	400:1	2–3%	
Longitudinal slope of sidewalks	10%	10:1	0.5%	200:1	1–5%	
Longitudinal slope of valleyed section	5%	25:1	0.5%	200:1	2–3%	

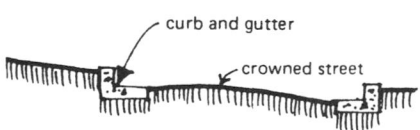

Fig. 8 Local and feeder streets.

Fig. 9 Urban streets and public parking.

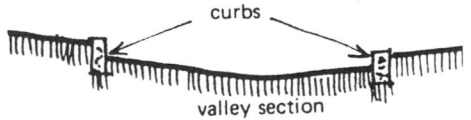

Fig. 10 Parking areas and utility and service roads.

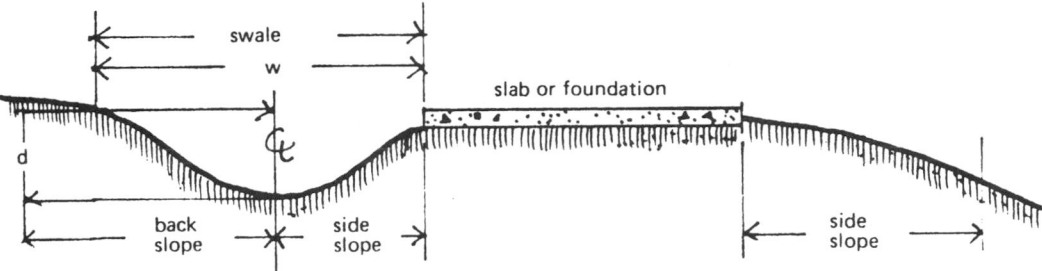

Fig. 11

TABLE 3   Slopes adjacent to Buildings

Condition	Maximum		Minimum		Range preferred
Side slopes with vehicular access	10%	10:1	0.5%	200:1	1–3%
Back slopes with vehicular access	15%	6.66:1	0.5%	200:1	1–5%
Side slopes without vehicular access	15%	6.66:1	0.5%	200:1	1–10%
Back slopes without vehicular access	20%	5:1	0.5%	200:1	1–10%

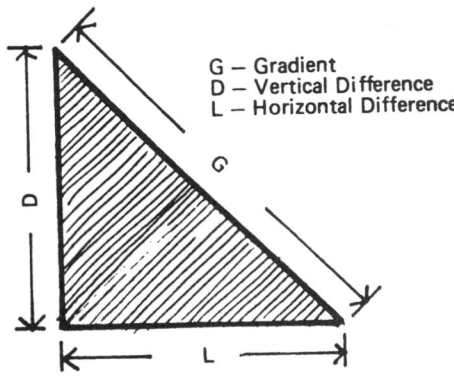

G — Gradient
D — Vertical Difference
L — Horizontal Difference

Fig. 12

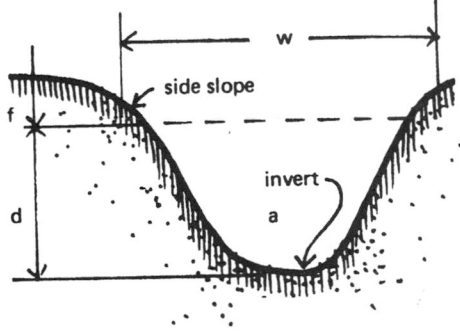

Fig. 13   Ditch. w = width of the channel; d = depth of the channel; a = area of the cross-section; p = wetted perimeter; f = freeboard; invert = bottom elevation of the channel.

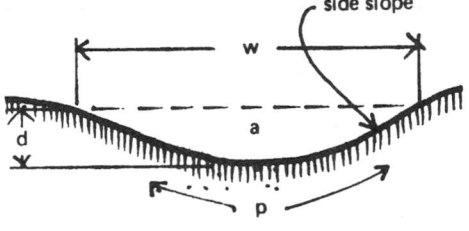

Fig. 14   Swale. w = width of the channel; d = depth of the channel; a = area of the cross-section; p = wetted perimeter; f = freeboard; invert = bottom elevation of the channel.

TABLE 4   Slopes of Drainage Channels

Condition	Maximum		Minimum		Preferred
Swale side slopes	10%	10:1	1%	100:1	2%
Longitudinal slope of swales					
Grass invert	8%	12.5:1	1%	100:1	1.5–2%
Paved invert	12%	8.33:1	0.5%	100:1	4–6%
Ditchside slopes	of repose		...		20–25%
Grass invert	8%	12.5:1	1%	100:1	2–3%
Paved invert	10%		...		5–6%

SOIL CONDITIONS

A thorough investigation of soil conditions is essential. The soil must be such that it can reasonably sustain the weight of the proposed buildings and not cause any other problems. A rocky base will result in expensive foundation work, difficult site development, and drainage problems. Installation of underground utility lines, such as water, gas, and sewers, would be made extremely difficult and costly.

A swampy condition will result in the use of piles to support the buildings and possible flooding conditions. Special waterproofing would be required for foundations, basements, and underground garages. The nature of the soil will also determine the effectiveness of its ability to grow grass, trees, and other vegetation.

TABLE 1 Soil Conditions

Designation	Value as foundation below frost line	Potential frost action	Compressibility and expansion	Drainage characteristics	Planning considerations
Gravel and gravelly soils	Good to excellent	None to medium	Almost none to slight	Excellent to practically impervious	Best location for buildings; excellent for accessory buildings, play areas, parking areas
Sand and sandy soils	Fair to good	None to high	Almost none to medium	Excellent to practically impervious	Good location for main buildings, accessory buildings, and active recreational areas
Low compressibility, fine-grained soils	Fair to poor	Medium to very high	Slight to high	Fair to practically impervious	Fair location for low or accessory buildings; good for play areas and parking
High-compressibility, fine-grained soils	Poor to very poor	Medium to very high	High	Fair to practically impervious	Not good for building; retain as permanent greenbelt or open space
Peat and other fibrous organic soils	Not suitable	Slight	Very high	Fair to poor	Excavation in this material is difficult and expensive; poor location for structures; retain as open space or park area
Rock	Good to excellent	Very high	Almost none	Impervious	Excavation in this material is difficult and expensive; poor location for structures; retain as open space or park area

TABLE 2 Nominal Values of Allowable Bearing Pressures for Spread Foundations

Type of bearing material	Consistency in place	Allowable bearing pressure (tons per square foot)	
		Ordinary range	Recommended value for use
Massive crystalline igneous and metamorphic rock: granite, diorite, basalt, gneiss, thoroughly cemented conglomerate (sound condition allows minor cracks)	Hard, sound rock	60–100	80
Foliated metamorphic rock: slate, schist (sound condition allows minor cracks)	Medium-hard sound rock	30–40	35
Sedimentary rock: hard cemented shales, siltstone, sandstone, limestone without cavities	Medium-hard sound rock	15–25	20
Weathered or broken bedrock of any kind except highly argillaceous rock (shale)	Soft rock	8–12	10
Compaction shale or other highly argillaceous rock in sound condition	Soft rock	8–12	10
Well-graded mixture of fine- and coarse-grained soil: glacial till, hardpan, boulder clay (GW-GC, GC, SC)	Very compact	8–12	10
Gravel, gravel-sand mixtures, boulder-gravel mixtures (GW, GP, SW, SP)	Very compact	7–10	8
	Medium to compact	5–7	6
	Loose	3–6	4
Coarse to medium sand, sand with little gravel (SW, SP)	Very compact	4–6	4
	Medium to compact	3–4	3
	Loose	2–3	2
Fine to medium sand, silty or clayey medium to coarse sand (SW, SM, SC)	Very compact	3–5	3
	Medium to compact	2–4	2.5
	Loose	1–2	1.5
Fine sand, silty or clayey medium to fine sand (SP, SM, SC)	Very compact	3–4	3
	Medium to compact	2–3	2
	Loose	1–2	1.5
Homogeneous inorganic clay, sandy or silty clay (CL, CH)	Very stiff to hard	3–6	4
	Medium to stiff	1–3	2
	Soft	0.5–1	0.5
Inorganic silt, sandy or clayey silt, varved silt-clay–fine sand (ML, MH)	Very stiff to hard	2–4	3
	Medium to stiff	1–3	1.5
	Soft	0.5–1	0.5

Under the Casagrande system of classification, C = clay, G = gravel, H = high compressibility, L = low to medium compressibility, M = silt, P = poorly graded, S = sand, and W = well graded.

In considering the visual assets of a site, extensive field observation is necessary. Features observed in the field can be mapped (see Fig. 1) and considered along with the site's physical features in compiling the development plan. Visual characteristics to be considered include ridge tops and valley bottoms, brooks and streams, ledges, stone walls, views and vistas, significant vegetation (such as hemlocks and other evergreens, wetlands plants, and wildflowers), and other aesthetic assets such as waterfalls and historic buildings.

The remainder of the analysis is conducted by considering each of the major natural resource characteristics of the site in relation to land uses proposed for the site. For a typical subdivision, the major land uses would be:

- Water supply
- Septic systems or sewers
- Buildings and dwellings
- Roads and parking areas

The primary natural resource factors affecting (and affected by) the land uses are:

- Depth to water table
- Earth material characteristics (i.e., soil percolation rates, susceptibility to erosion, etc.)
- Slope
- Depth to bedrock
- Flood-prone and storm-prone areas

Depending on the site, its proposed use, and the level of detail of the analysis, additional natural resource factors can be considered. These include vegetation, wildlife value, wetlands, drainage areas, availability of groundwater, bedrock type, agricultural capability, and other factors.

Figures 3 through 8 evaluate two of the proposed land uses—septic systems and buildings—in terms of four resource factors—depth to bedrock, depth to water table, earth materials, and slope. The degree to which a natural resource factor limits the proposed land use will vary from location to location. For example, in areas where bedrock is more than 10 ft below the surface, bedrock usually will not limit or make special design necessary for dwellings and septic systems. In areas where bedrock is somewhat closer to the surface, some special design measures may be required. The most severe limitations will be imposed where bedrock is shallow and outcrops are frequent. These varying conditions are designated on the charts as optimum, marginal, and critical, respectively.

The site has been mapped for each of the four major resource characteristics, with shaded areas designating portions of the site where design or development restrictions are imposed. When these maps are combined (Fig. 9), overall development opportunities and limitations are revealed.

One of the greatest limitations to conducting an analysis of this type is lack of data. Even in mapped areas, information may not be detailed enough to be useful in site analyses. However, these inventory maps can be used to determine the specific resource concerns that should lead to further site investigation. Field observation at the site can provide information that is not otherwise available; field work is also important in confirming existing data and compiling information on unmapped features such as vegetation.

The site is a tract of approximately 600 acres in rural Connecticut that contains many features typical of suburban and rural areas. A stream runs through the southeastern portion of the site and is fed by a spring and a small red maple swamp. The areas immediately adjacent to the streambanks are subject to occasional flooding. The land was formerly farmed, and consists primarily of second-growth forest. Elevation varies from 325 to 600 ft above sea level.

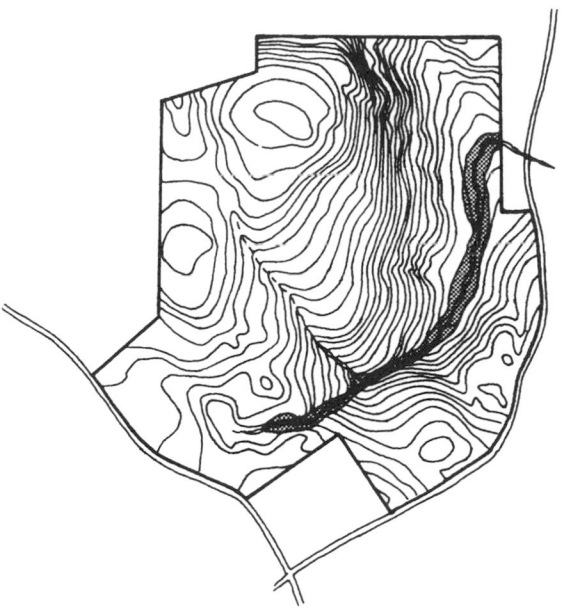

 Contours (vertical distance between contour lines is 10 feet)

 Flood-prone areas (areas where there is a 1% chance of flooding in any given year)

Fig. 1

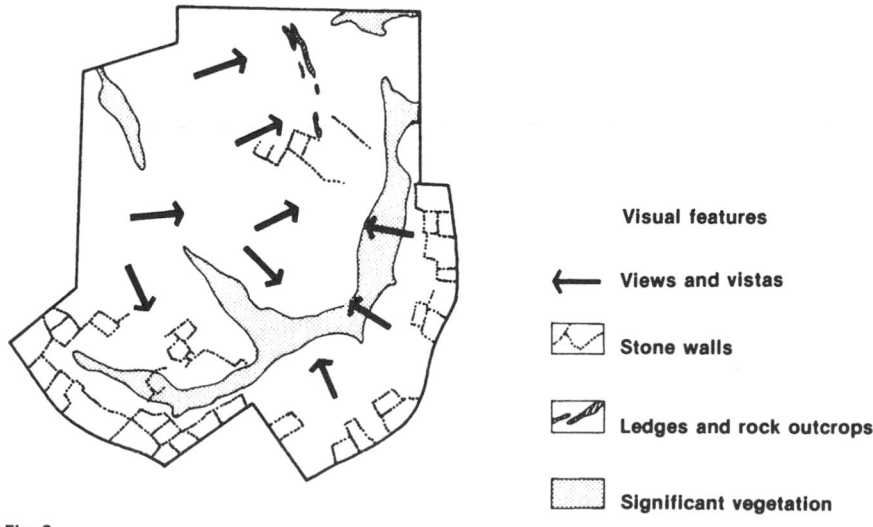

Fig. 2

Visual features

← **Views and vistas**

Stone walls

Ledges and rock outcrops

Significant vegetation

SITE EVALUATION

DEPTH TO WATER TABLE AND
EARTH MATERIALS (PERCOLATION)

Illustrated below are physical conditions commonly encountered at development sites, accompanied by brief descriptions of engineering measures (for foundations and septic systems) required for development in areas where these conditions are present. Development under "critical conditions" is usually prohibitively expensive as well as environmentally damaging.

	OPTIMUM CONDITIONS	MARGINAL CONDITIONS	CRITICAL CONDITIONS
DEPTH TO WATER TABLE	**Greater than 10 feet**	**fluctuates from 3 to 10 feet**	**Permanently high: wetlands***
	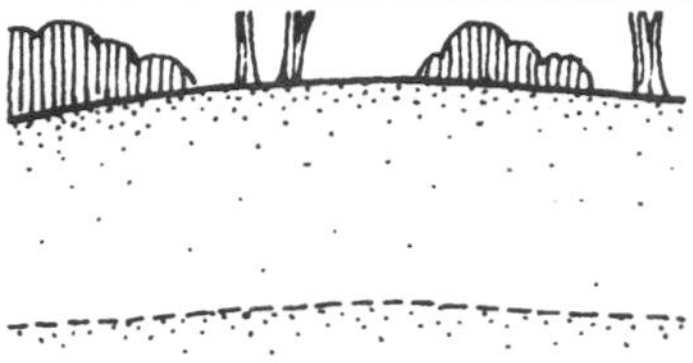 Conventionally designed basements and septic systems will not be flooded.	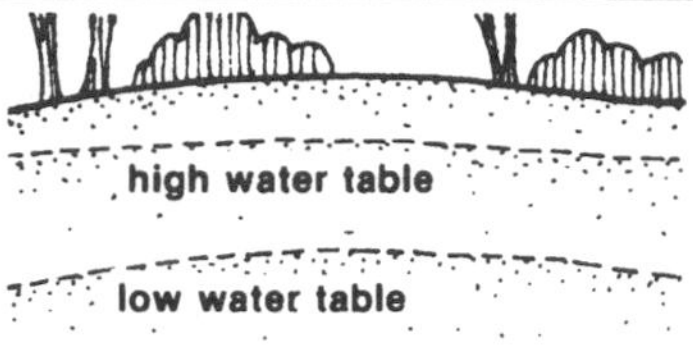 Building footings should be properly drained; shallow foundations may be necessary. Septic systems require curtain drain and/or use of fill material similar to existing soil.	 Severe limitations on development. Shallow foundations required: site preparation includes removal of organic material and replacement with clean fill. Elaborate drainage and fill necessary for septic systems.
EARTH MATERIALS** (PERCOLATION)	**Percolation up to 20 minutes per inch**	**Percolation 20-60 minutes per inch**	**Percolation more than 60 minutes per inch**
	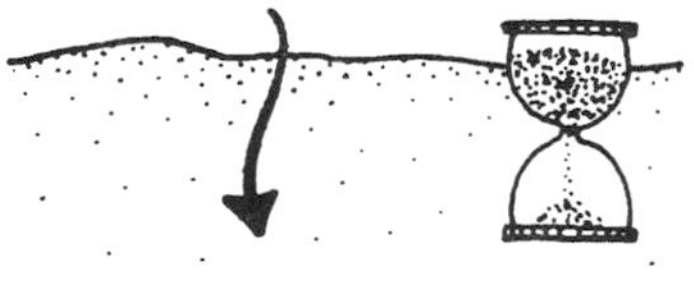 Conventional building design adequate. Standard septic systems adequate in most cases; special design needed where percolation exceeds .5 minutes/inch, which is too fast for adequate renovation.	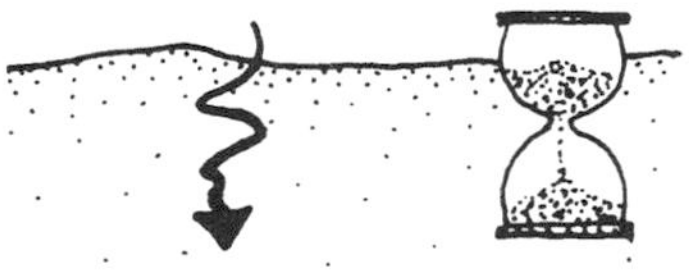 Standard building design; larger leaching area required for septic systems, with standard or special trench design. Water mounding may occur because of slow percolation.	 Standard building design; severe limitations on septic systems, with extensive leaching fields, fill, and/or aboveground systems required.

Fig. 3

* Wetlands permit required

** The percolation rate, or the speed at which water can flow through the soil, is one of several soil characteristics that affect development. Another is the ability of various soil types to support foundations. Most upland soils (glacial till) and sand and gravel deposits will adequately support foundations. Problems are frequent in clays, peat deposits and other wetland soils, where special measures are necessary to prevent buildings from settling.

Depth to Water Table

This information is not commonly available on maps. An indication of water table depth can be obtained from soils maps compiled by the Soil Conservation Service (SCS). Soils maps are available from SCS Field Offices and DEP's Natural Resources Center. Soils maps showing regulated inland wetlands may be obtained from the Water Resources Unit, DEP. SCS publications include tables which indicate saturated soils and soils with groundwater within 3 ft of the surface. These tables can be used along with soils maps (and field testing where necessary) to complete a general map of groundwater characteristics for development sites.

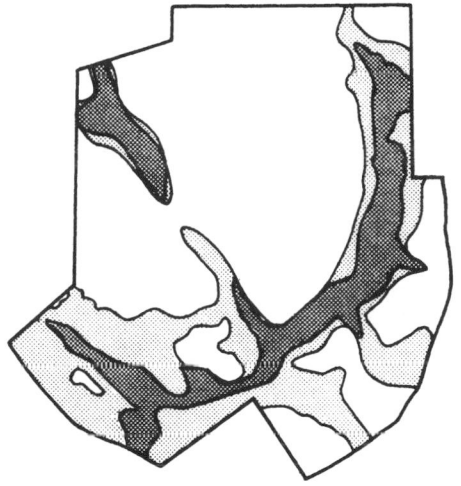

Saturated soils (regulated inland wetlands)

Water table within three feet of surface

Optimum ground water conditions for development

Fig. 4

Earth Materials: Percolation Probability

Percolation rates can be estimated from information compiled by the SCS. Detailed soils maps can be used in conjunction with SCS keys which indicate percolation rate probabilities for the various soil types. Each soil type has been placed into one of four categories: fast, probably fast, probably slow, and slow. Field testing will be necessary at possible septic system sites to provide more accurate data. The cross-hatched portions of the map represent areas where other resource characteristics (wetlands and shallow bedrock) preclude the use of percolation probability.

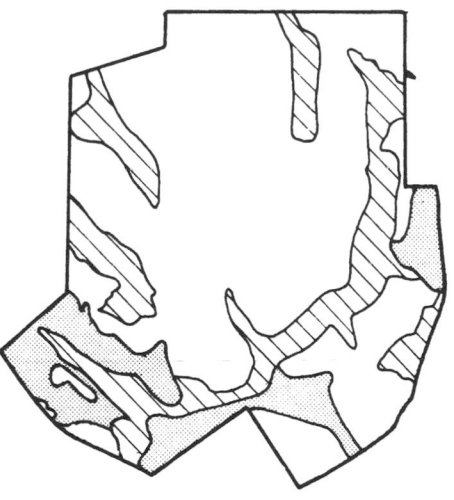

Slow percolation (80% chance that water in a test hole will take longer than 30 minutes to fall one inch)

Probably fast percolation (50% chance that water in a test hole will take 10 to 20 minutes to fall one inch)

Areas where other resource characteristics (shallow water table and bedrock) **preclude the use of percolation probability**

Fig. 5

SITE EVALUATION

SLOPE AND DEPTH TO BEDROCK

Frequently encountered physical conditions of development sites are outlined here along with the engineering measures commonly used in response to those conditions. Expense of site preparation and septic system installation is considerably greater on steep slopes and in areas where bedrock is near the surface than in areas where conditions are less restrictive for development.

	OPTIMUM CONDITIONS	MARGINAL CONDITIONS	CRITICAL CONDITIONS
SLOPE	**Less than 10 percent**	**10 to 15 percent**	**Greater than 15 percent**
	Suitable for construction of buildings and septic systems using conventional design.	Some grading may be necessary to prepare building sites; septic system trench design should be adjusted to accommodate slope.	Hazardous for heavy equipment. Considerable grading necessary at building sites, requiring precautions against erosion and soil slumping. Extreme difficulty in septic system installation with use of grading and fill.*
DEPTH TO BEDROCK	**Greater than 10 feet**	**LESS THAN 10 FEET**	**Numerous outcrops**
	Conventional building and septic system design is adequate. Trenches should be 3-6 feet below surface and at least 4 feet above bedrock.	Septic system trenches should be 4 feet above bedrock and covered with fill to proper depth. Some removal of bedrock may be necessary for building sites; foundations should rest on same material throughout.	Fill required for septic system installation.* Blasting required for building site preparation; foundations should rest on gravel cushions to prevent uneven settling.

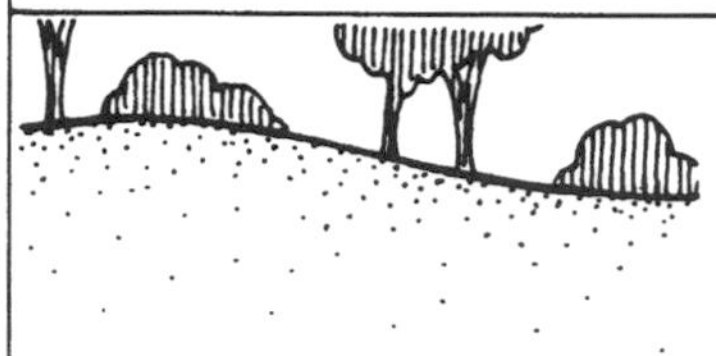

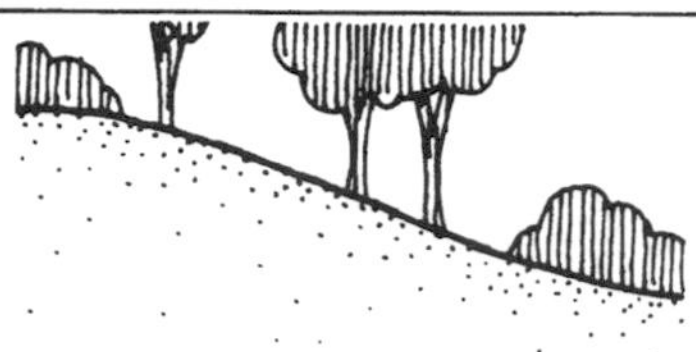

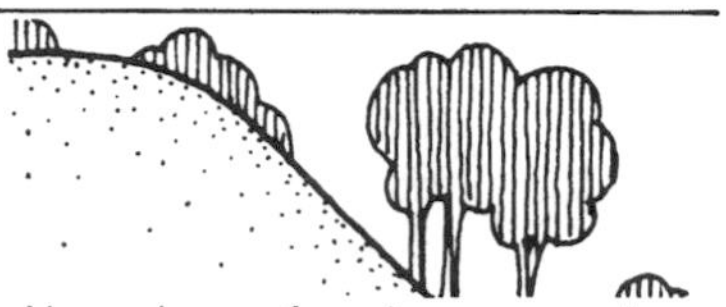

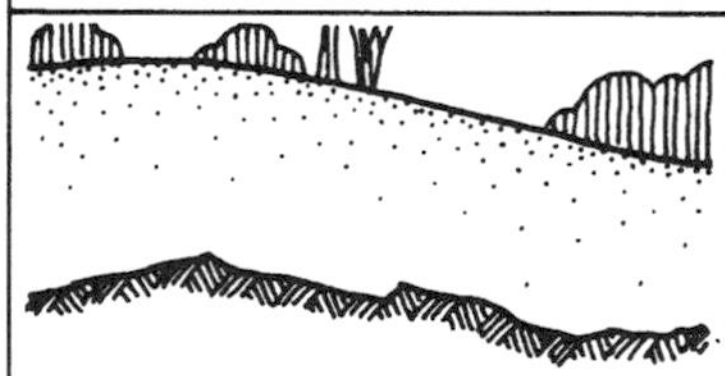

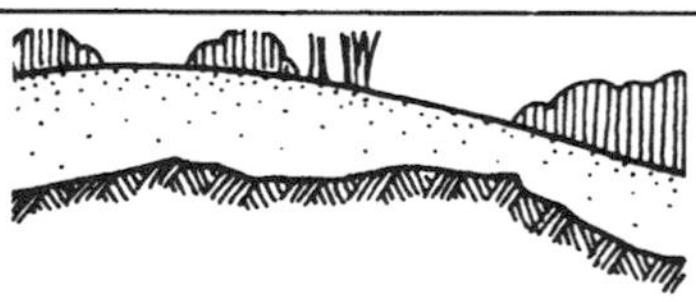

Fig. 6

* Relatively complex and expensive measures are required for effective septic system operation where natural resource conditions are critical. These include techniques (such as fill and subsurface drainage systems) to make conventional systems function under unusual conditions, as well as the use of special equipment and methods such as leaching galleries and above-ground systems. In either case, a professional engineer and the Water Compliance Unit of DEP should be consulted.

Slope

Maps showing steep slopes can be compiled from topographic maps, SCS maps, and observations in the field. The Natural Resources Center offers technical assistance in the compilation of these maps. For smaller sites, field observations may be adequate for delineating areas with steep slopes.

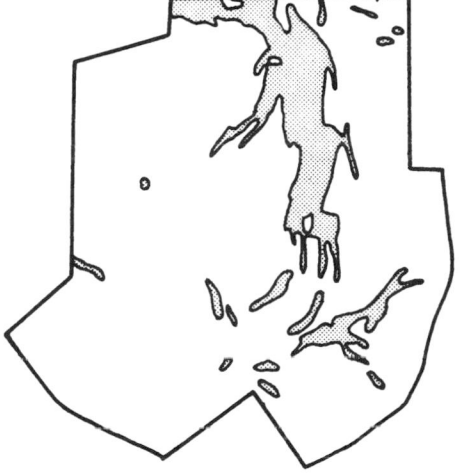

░░░	**Slope 15% or greater** (elevation change of at least 15 feet per 100 feet horizontal distance)
☐	**Moderate: slope less than 15%**

Fig. 7

Depth to Bedrock

Bedrock outcrops can be mapped through field observations and may be visible in some aerial photographs. SCS maps and surficial geology maps can be used to generally determine shallow bedrock areas. Surficial geology maps show areas where bedrock is within 10 ft of the surface; SCS soils maps can be interpreted to show areas where soil is rocky or where bedrock is 2 ft or less below the surface.

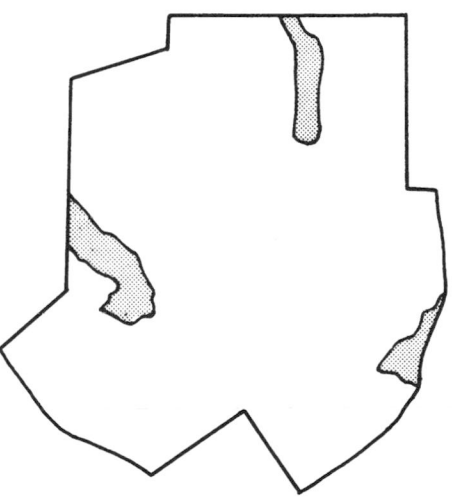

░░░	**Bedrock close to surface; frequent outcrops**
☐	**Optimum depth to bedrock conditions for development**

Fig. 8

SITE EVALUATION

THE COMPOSITE MAP

The most important use of the composite map is the identification of the specific types of problems that exist on various parts of the site. Knowing these problems early in the planning stage allows the developer to use the characteristics of the site to best advantage and to avoid the necessity of making expensive changes later in the planning or construction processes.

Figures 3 through 8 outline the limitations imposed by individual resource factors. In the final analysis, however, all the factors will have to be considered together. Listed below are the combinations of resource characteristics that occur most frequently at the site, and the engineering measures required for development of areas with those characteristics.

1. *Conditions:* Bedrock 10 ft or more below surface, soil percolation probably fast, slope less than 15 percent, water table 10 ft or more below surface.

These areas present the best conditions for development. Conventional construction methods and design can be used for buildings and septic systems, and little or no grading is required for roads and driveways.

2. *Conditions:* Bedrock 10 ft or more below surface, soil percolation slow, slope less than 15 percent, high groundwater 1 to 2 months per year.

Seasonally high groundwater will require special septic system design, possibly including the use of curtain drains or fill. Shallow foundations may be necessary for buildings; basements should be adequately drained.

3. *Conditions:* Bedrock 10 ft or more below surface, soil percolation fast or probably fast, slope 15 percent or greater, water table 10 ft or more below surface.

Grading for buildings and roads could cause erosion problems; proper fill should be used to prevent slumping or settling of foundations. Restrictions on septic systems are severe: elaborate leaching field design and fill may be necessary, and machine operation may be unsafe.

4. *Conditions:* Bedrock 10 ft or more below surface, soil percolation slow, slope less than 15 percent, high groundwater 1 to 2 months per year.

Buildings should be properly drained; shallow foundations may be necessary. Septic system design requirements are complex and expensive, requiring increased leaching field size or leaching galleries to compensate for slow percolation, and the use of curtain drains or fill to prevent failure of systems due to seasonal high groundwater.

5. *Conditions:* Bedrock near surface with numerous outcrops, slope 15 percent or greater, water table 10 ft or more below surface.

These areas are unsuitable for septic systems and impose severe restrictions on building foundation with blasting, grading, and/or fill required.

6. *Conditions:* Bedrock near surface with numerous outcrops, slopes less than 15 percent, seasonally high groundwater.

Very severe restrictions are imposed upon septic systems with fill and subsurface drainage system probably necessary to prevent failure. Shallow foundations will be necessary for buildings with care taken to ensure that buildings rest on the same material (such as gravel backfill) throughout to prevent uneven settling.

7. *Conditions:* Bedrock 10 ft or more below surface slope less than 15 percent, permanently high water table (wetlands).

Wetlands are unsuitable for septic systems without extensive excavation, filling, and drainage. Restrictions on buildings are severe, with removal of organic material and replacement with clean and compacted soil is necessary. Wetlands permit required.

Fig. 9

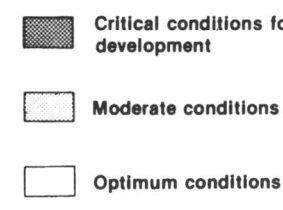

Critical conditions for development

Moderate conditions

Optimum conditions

Scale: 1 inch = 1,000 feet

A SUGGESTED DEVELOPMENT PLAN

The composite map identified portions of the site where conditions for development are optimum. Buildings and roads were concentrated in these areas, leaving sensitive areas such as wetlands and steep slopes largely undisturbed. This serves the dual purpose of minimizing environmental disturbance and preserving the visual quality of the site, since these areas are often of the greatest scenic value.

Clustered single-family dwellings, townhouses, and walk-up apartments are included in the plan to illustrate the various options open to the developer. (This plan is intended to give a general indication of how a site of this type could be developed, and the exact types and positions of buildings and roads are not important.) Although

lot sizes are somewhat smaller than normal, adjacent open space increases land area available for recreational use by residents.

A continual ribbon of open space—in areas where soil types pose limitations on development—provides pedestrian circulation that is separated from vehicular traffic. A buffer of open space on the southwestern portion of the

site increases privacy and shields residential areas from traffic noise.

Roads, instead of being in conflict with the topography, are parallel or at an oblique angle to the site's contours. Residential development is located on loops and cul-de-sacs where traffic is light; and road layout discourages outside through traffic.

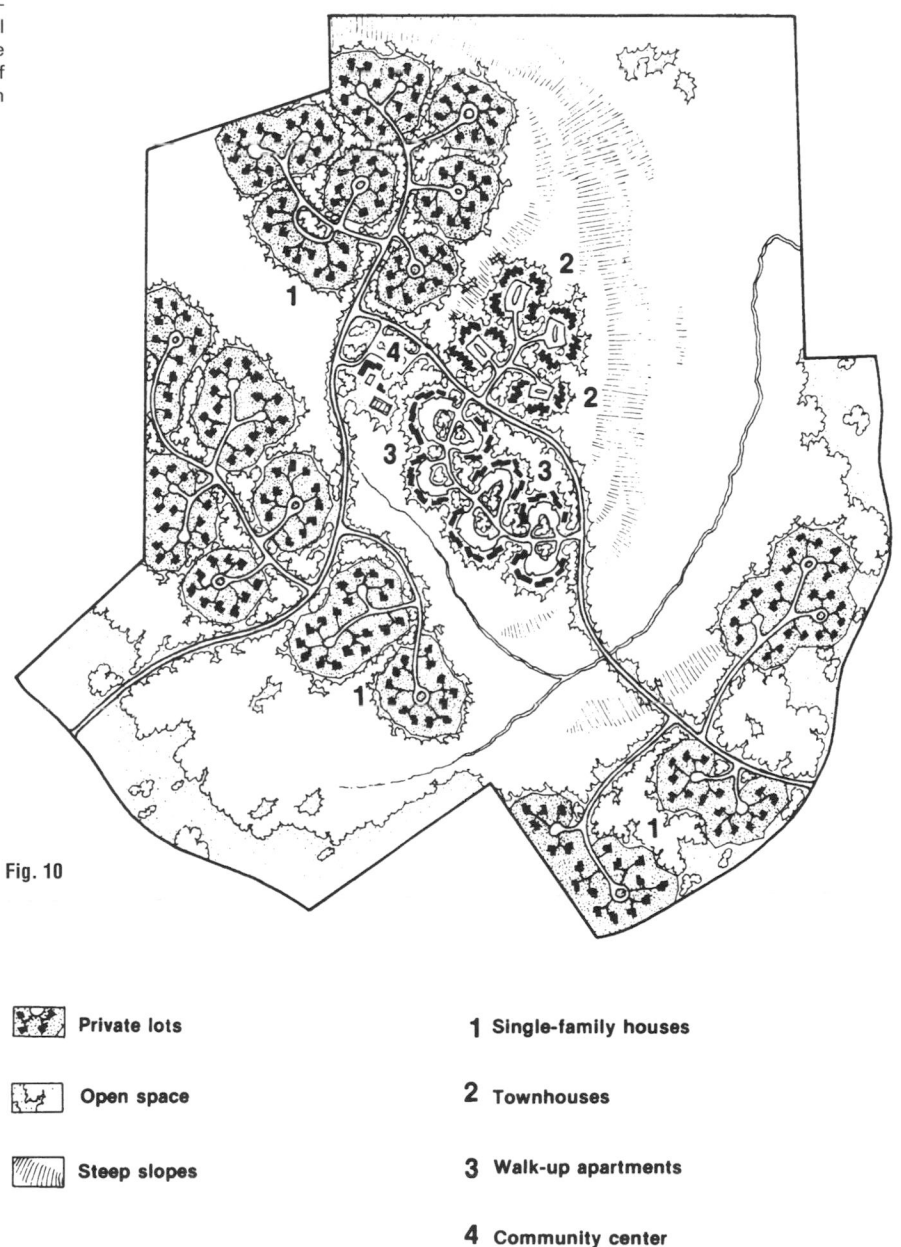

Fig. 10

▨ Private lots	**1** Single-family houses
▨ Open space	**2** Townhouses
▨ Steep slopes	**3** Walk-up apartments
	4 Community center

BUILDING ORIENTATION/SUN

ORIENTATION—SUN

Orientation is the placement of a building or apartment so that it may obtain the best advantages in relation to its physical location. The major considerations in orientation of a building or dwelling unit are:
1. Sunlight
2. Prevailing breezes
3. Views

SUNLIGHT

The objective of orientation for the sun is to obtain sunlight when it is desired and to block out sunlight when it is not desired.

Since the United States has a great variety of climatic conditions, it is difficult to make universal assumptions in this respect. However, it can be generally stated that the objective is to have sun when it is desired and to avoid it when it is not wanted. In the wintertime, it is desirable to have a maximum of sunlight. The hot summer sun, particularly at noon and late afternoon, is not desirable. Sunlight in the morning is delightful throughout the entire year. People will differ in their individual choices of how a dwelling unit should be oriented, but there are some considerations that are generally accepted:

1. Each apartment should get some sun at some time of the day.

2. Since people express a variety of desires as to amount and exposure to sunlight, it is advantageous for dwelling units to have different sun orientations.

3. No apartment should be oriented completely toward the north, because any dwelling unit facing north will get no sun.

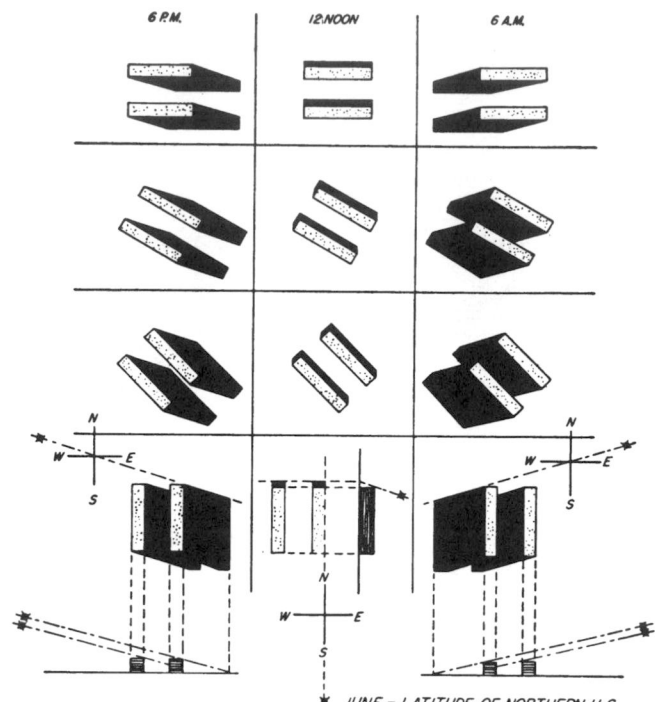

Fig. 1   Summer orientation.

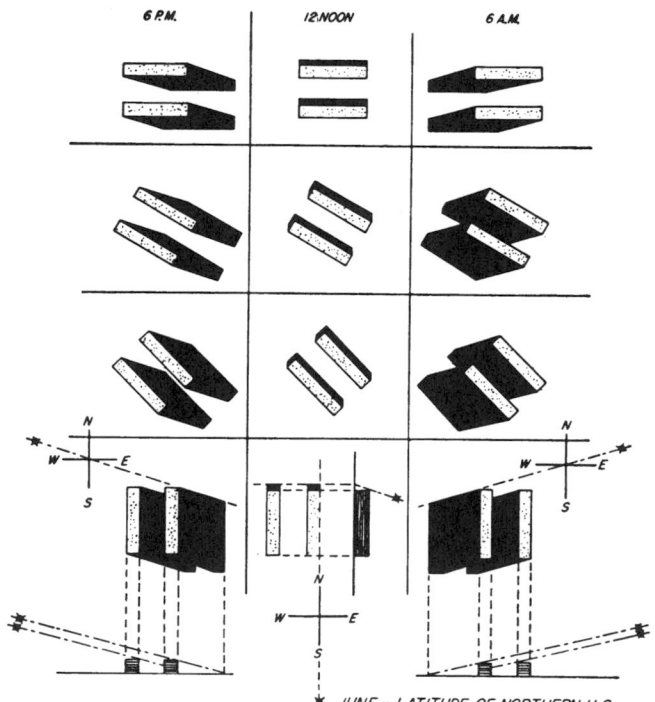

Fig. 2   Winter orientation.

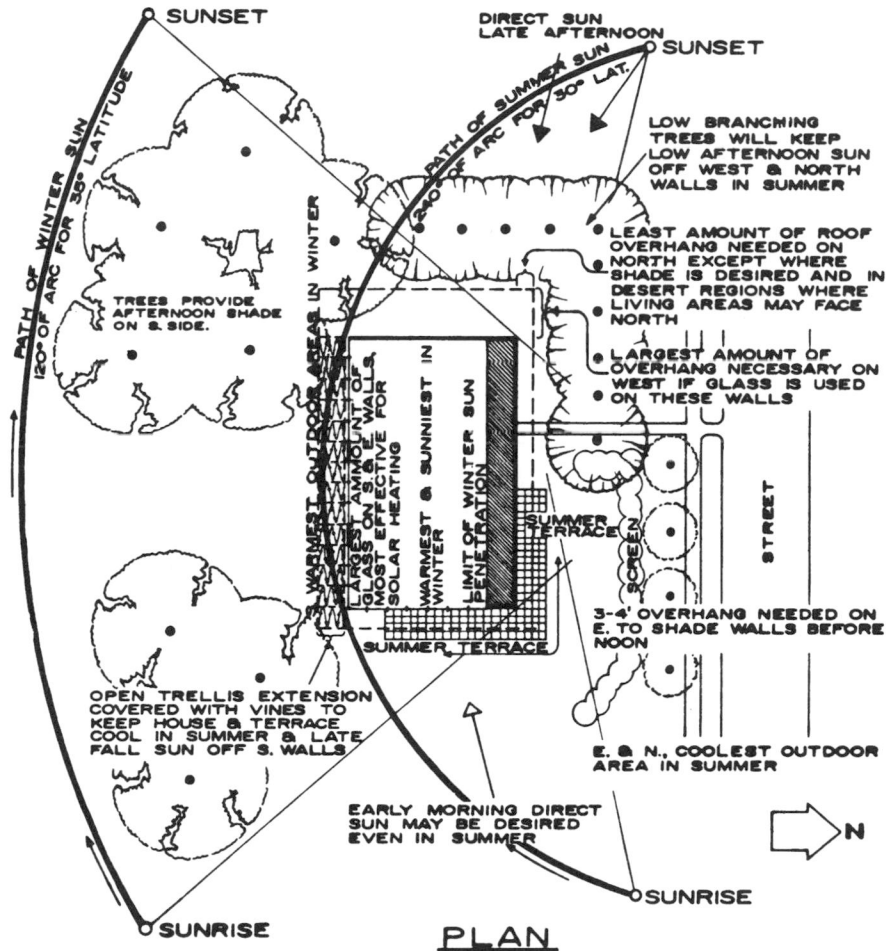

PLAN

Fig. 3

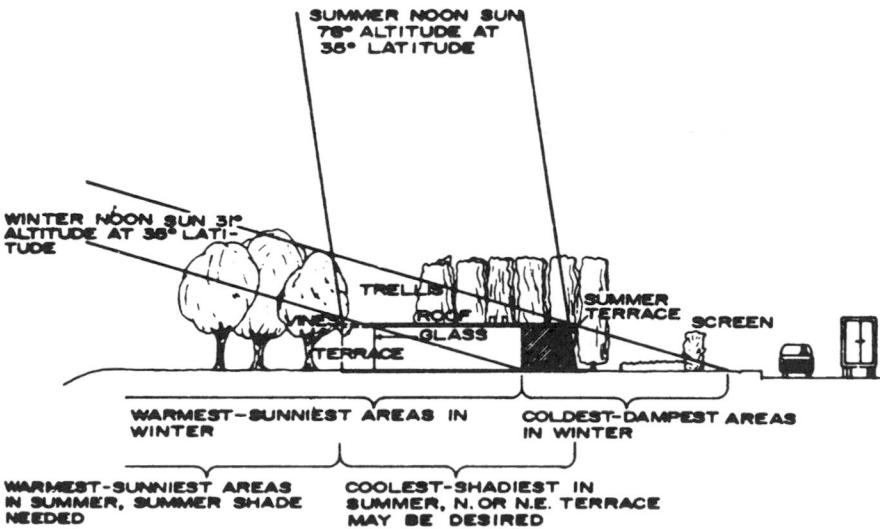

ELEVATION

Fig. 4

BUILDING ORIENTATION

The following is a list of orientations and their resulting effects.

South Orientation

Sunlight will occur from late morning to early afternoon. During the summer months the sun will appear high in the sky and during the winter months the sun will appear low in the sky.

South orientation is the best to obtain the maximum sunlight during the day.

The hot summer sun can easily be controlled by properly designed overhangs. The low angle of the sun during the winter months will enable the rays to penetrate deep into the room.

East Orientation

Sunlight will occur only in the morning hours. During the summer months, when the sun rises in the east, the early morning hours will have sunlight. The sun will be very low in the sky and the sun will generally not be too intense. During the winter months, the sun will rise more toward the southeast, thus providing a shorter period of sunlight.

Southeast Orientation

Sunlight will occur from early morning to late morning, or possibly to noon. At midmorning the sun will be reasonably high in the sky and will provide a moderately intense sunlight.

Southwest Orientation

Sunlight will occur from early afternoon to late afternoon. The sun will be reasonably high in the sky. The rays of the sun will be much more intense than the morning sun. In some areas during the winter months the sun will set in the southwest.

West Orientation

Sunlight will be present from midafternoon to late afternoon. During the summer months, the west sun will be very intense. It will set generally in the west or northwest. During the winter months, the sun will generally set in the southwest.

North Orientation

No sunlight will be obtained from a direct north orientation.

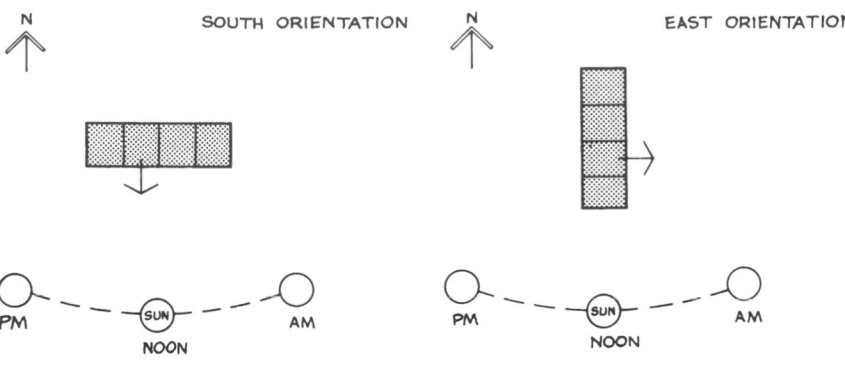

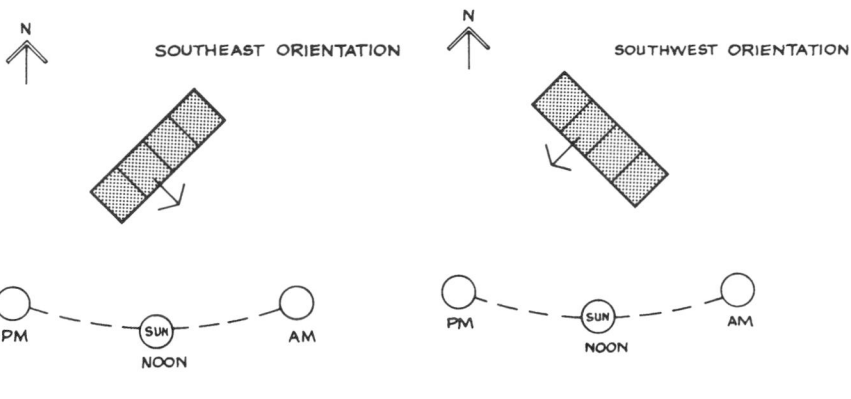

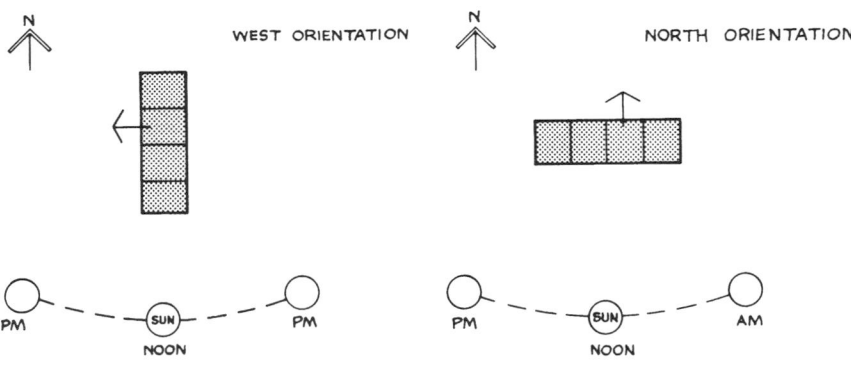

Fig. 5

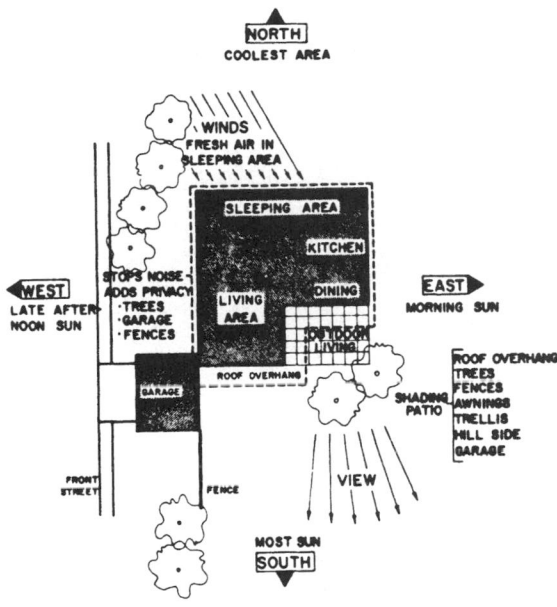

Fig. 6 Factors affecting orientation.

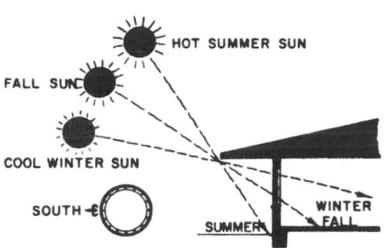

Fig. 7 Seasonal sun angles.

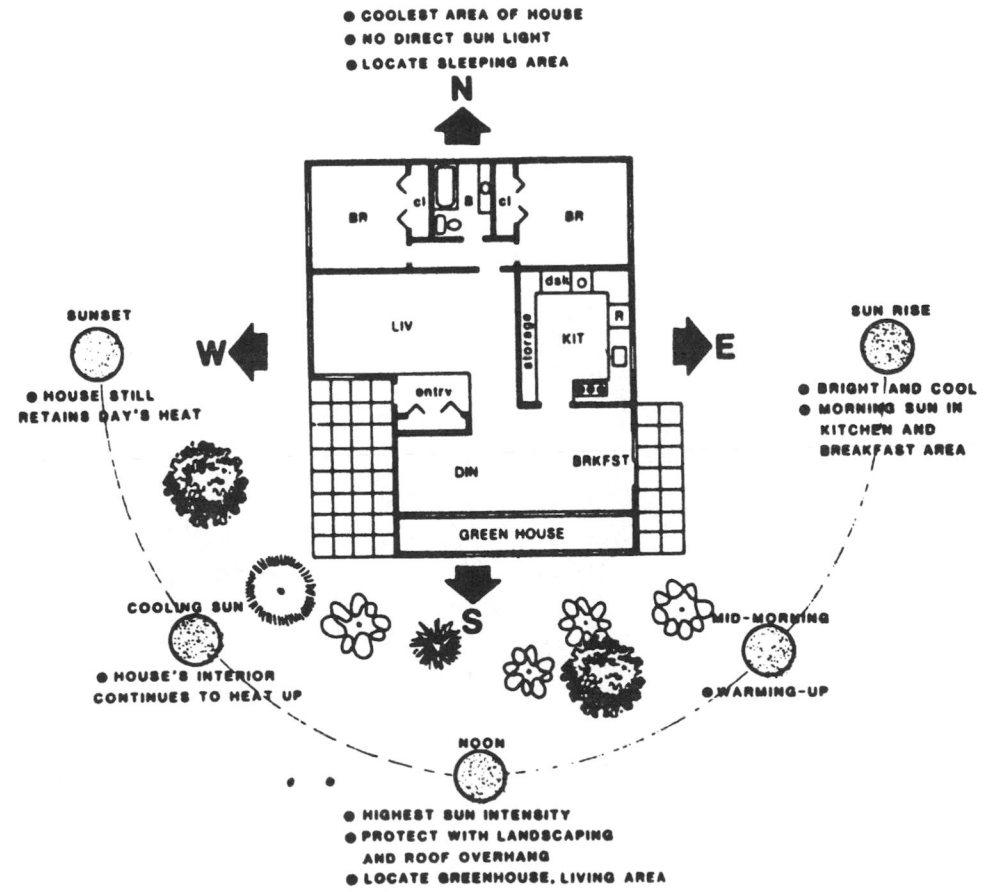

Fig. 8 Passive solar consideration.

BUILDING ORIENTATION/BREEZES AND VENTILATION

BREEZES

Landscaping plans can be custom designed so that trees and other plantings buffer a home from hot sun and cold winds, and also channel cooling breezes toward it.

Figure 1 shows how to work out an energy-saving planting scheme by using local solar data and relating it to a specific site, structural characteristics of the home, and wind patterns in the area.

This landscaping plan is for a hypothetical home in Miami, Fla. The house is rectangular, has an overhang, is built on a quarter-acre lot, and is oriented 5° east of south. It was necessary that prevailing winds from the east not be blocked at ground level and the roof surface be unobstructed for solar collection.

In the study, solar data was gathered during 12 hours (from 6 a.m. through 5 p.m.) on June 22, the longest and highest radiation day of the year. The data was translated into schematics that show how shifts in the sun's position during the day direct radiation to different areas of the home, thus indicating where trees are needed for shading.

The small drawings in Fig. 1 represent key hours in the day-long study. From left (*running counterclockwise*) you can see how shading requirements change between 6 a.m., when the solar load is on east and north walls, and 5 p.m., when the low sun angle sends radiation to west and north walls.

The large drawing at the bottom of Fig. 1 represents the final landscape plan, which incorporates some additional planting elements to improve wind control.

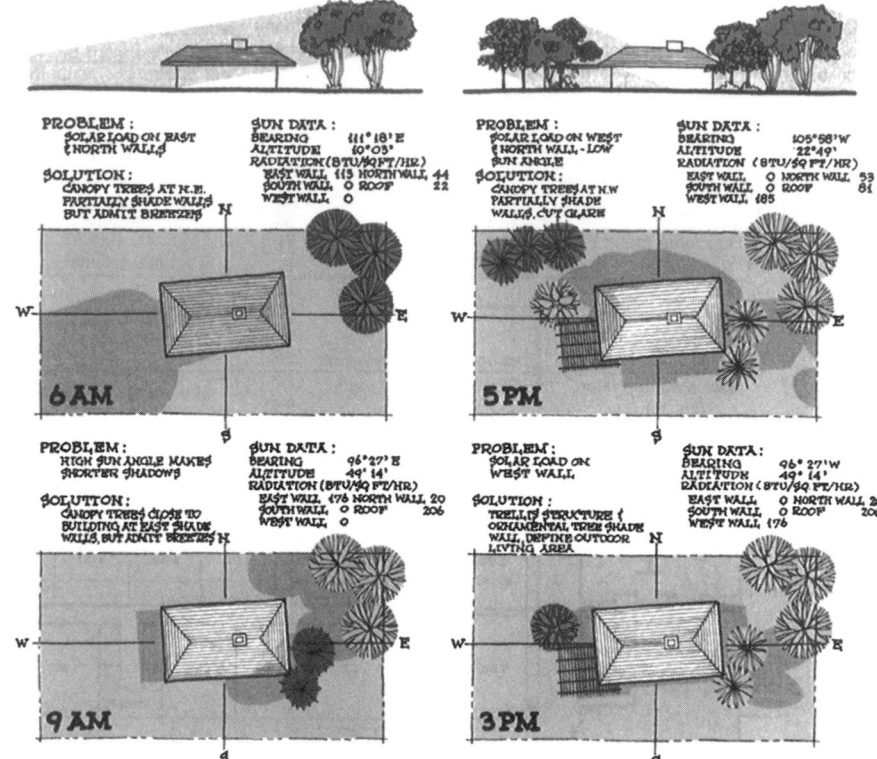

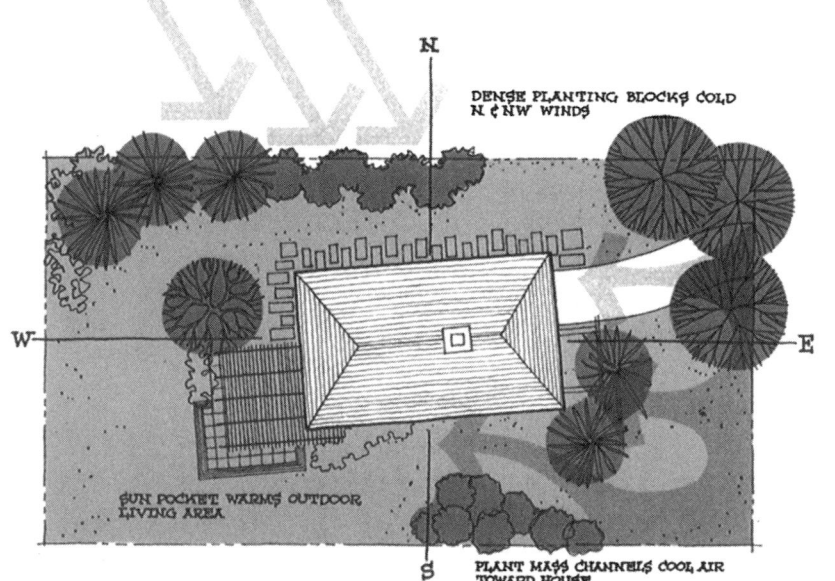

Fig. 1

BUILDING ORIENTATION/BREEZES AND VENTILATION

The study in Fig. 2 suggests energy-saving siting and orientation treatments for both single- and multifamily projects. Here, the primary goal is to reduce cooling loads since the projects are in a hot, humid region.

Each home in the single-family project (*top plan*) is oriented on its lot so it will receive maximum cooling from the wind and minimum heat from the sun. Local streets are laid out on an east-west axis to channel cooling westerly breezes through the project.

The center plan—a small segment of the same single-family project—shows how trees can be placed on common property lines so the same trees shade adjoining homes at different times of the day.

In the multifamily project (*bottom plan*), linear clustering of townhouses or apartment buildings is suggested. This allows the long side of most building groups to be oriented on an east-west axis, minimizing heating effects from the sun on wall surfaces. Cluster ends are kept open to channel prevailing breezes through the wide, open areas between building groups.

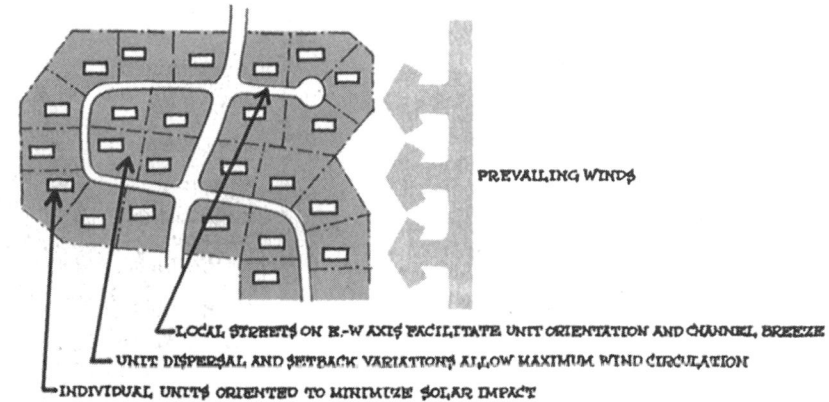

PREVAILING WINDS

LOCAL STREETS ON E-W AXIS FACILITATE UNIT ORIENTATION AND CHANNEL BREEZE

UNIT DISPERSAL AND SETBACK VARIATIONS ALLOW MAXIMUM WIND CIRCULATION

INDIVIDUAL UNITS ORIENTED TO MINIMIZE SOLAR IMPACT

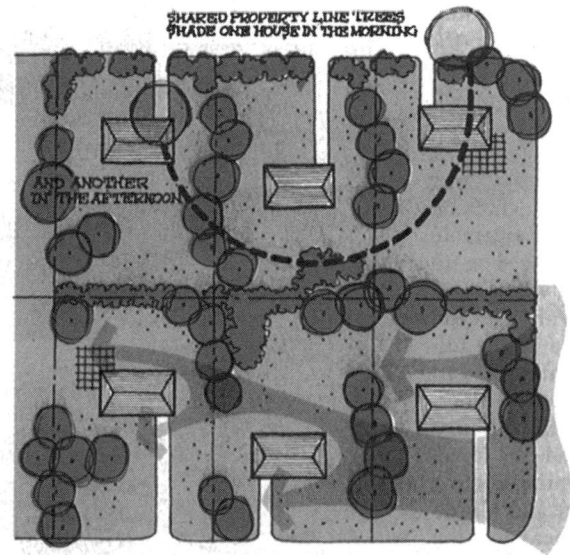

SHARED PROPERTY LINE TREES SHADE ONE HOUSE IN THE MORNING

AND ANOTHER IN THE AFTERNOON

BREEZES FLOW BENEATH CANOPY TREES & BETWEEN STAGGERED STRUCTURES

WIDE LINEAR OPEN SPACES CHANNEL BREEZE THROUGH DEVELOPMENT AND CREATE A SCENIC AMENITY FOR EACH DWELLING

UNITS FORM LINEAR CLUSTERS, ORIENTED ON AN E-W AXIS FOR MINIMUM SOLAR EXPOSURE OF WALL SURFACES

CLUSTER ENDS ARE KEPT LOOSE IN ORDER TO GUIDE WINDS INTO & THROUGH THE CENTER

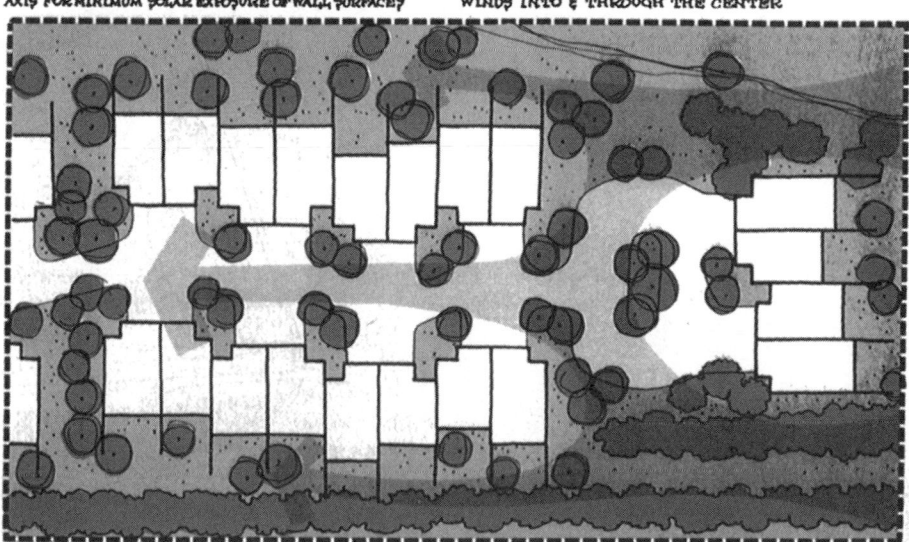

Fig. 2

BUILDING ORIENTATION/BREEZES AND VENTILATION

Design, siting, and landscaping features can help reduce energy loads for housing in colder climates. They include:

■ A cluster and landscaping plan that blocks cold winter winds, but channels cooling summer breezes through the clusters (Fig. 3).

■ A suggestion for a steep-slope design that channels cold winds up and away from the houses.

■ A house design that, among other things, uses its attached garage as a wind buffer.

■ A window design for a southeastern exposure that speeds morning heating during cold weather.

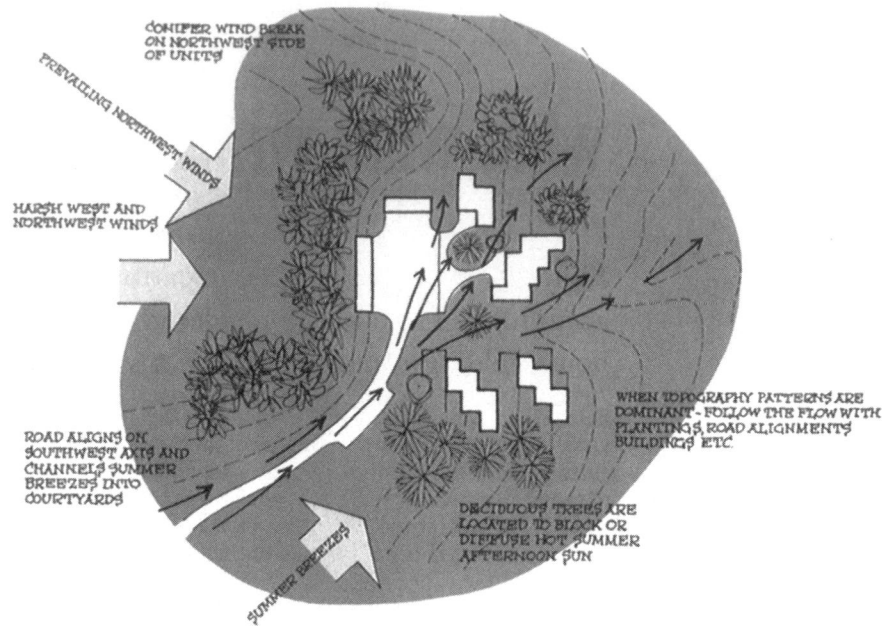

Fig. 3 A cluster design for sun and wind protection.

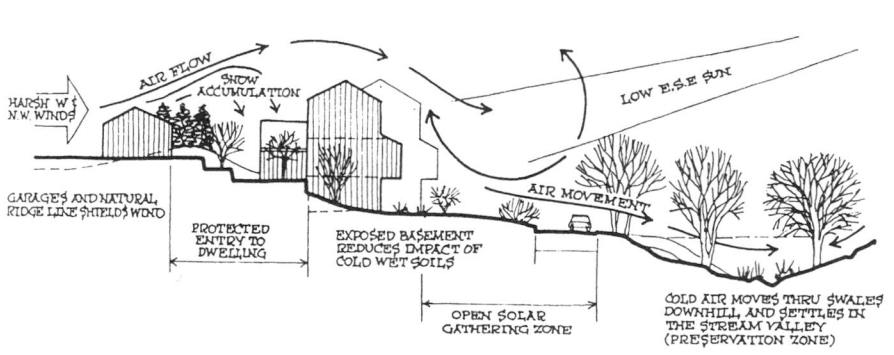

Fig. 4 A steep-slope design for wind protection.

Fig. 5 A window design for heating help.

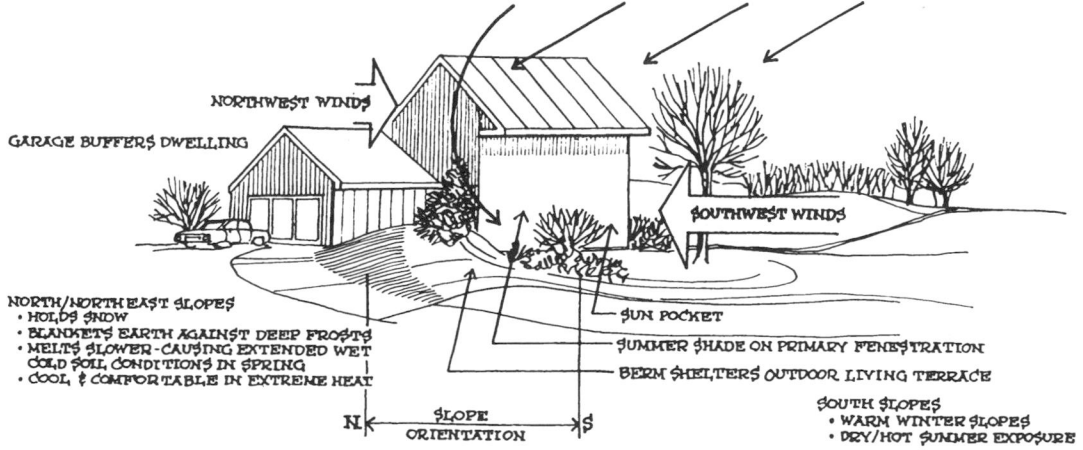

Fig. 6 A unit design for sun and wind protection.

VENTILATION

Buildings, especially in warmer climates, should be oriented toward the prevailing breezes, in addition to the sun orientation. However, this may not always be possible. Generally, orientation of the building in relation to sunlight is considered more important than orientation relating to the prevailing breezes. When it is possible to take advantage of prevailing breezes, the long side of the building should be faced toward the breeze. Recent developments in the widespread use of air conditioning and the awareness of air pollution have deemphasized the use of natural breezes for ventilation.

Building Ventilation

Building location with respect to prevailing breezes is an important planning factor, especially in warmer climates and during the hot seasons of the year. In Fig. 7 A, B, and C are plan views indicating the percentage of ventilation for different building arrangements, assuming they are placed perpendicular to the prevailing wind. In Fig. 7 D, E, and F show the cross section of buildings and the effect the placement to each other has on the prevailing wind.

Dwelling-Unit Ventilation

Figure 8 shows how the dwelling unit layout affects ventilation. A and B are plan views of typical corner rooms. C and D are sections and indicate the result of window heights on ventilation. Plan E illustrates through-ventilation, which is the best; corner ventilation in plan F is also good. Plans G and H indicate poor or no ventilation.

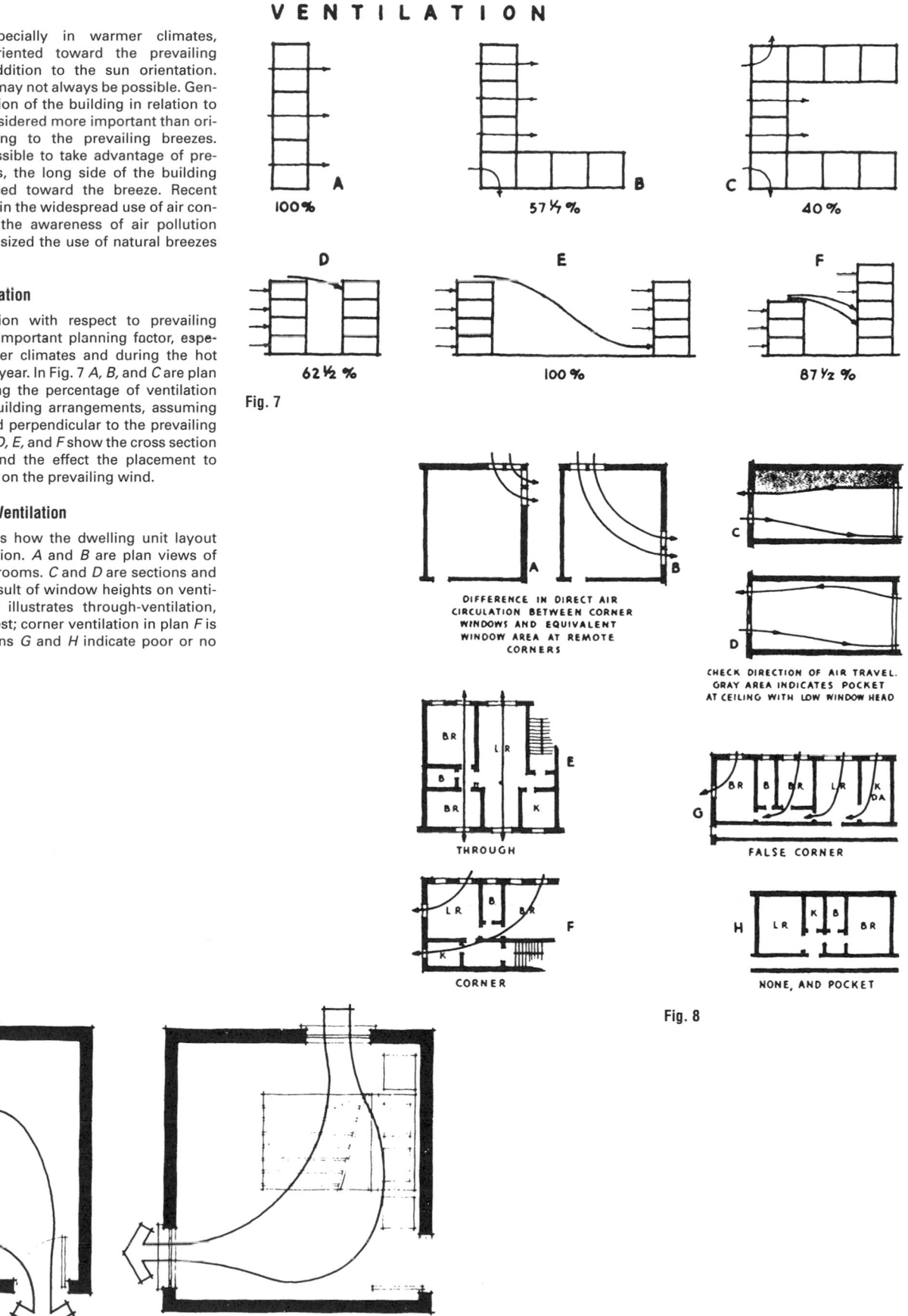

Fig. 7

Fig. 8

Fig. 9 Cross-ventilation may be obtained by placing windows in opposite corners of the same room or by an open door and window.

BUILDING ORIENTATION/VIEWS AND VISTAS

VIEWS AND VISTAS

When the situation presents itself, maximum advantage should be taken of any possible views. Natural vistas, such as rivers, mountains, parks, and lakes, are the most desirable. However, many manmade views are equally interesting, such as bridges, golf courses, and recreational areas. Care should be exercised in the placement of buildings to each other so that the view is not blocked by another building. Buildings should be staggered to provide maximum exposure for the most buildings.

VIEW PROTECTION

0-2% GRADE: NO VIEW EXCEPT NEIGHBORS, STREET; VARIED SETBACKS, ANGLED HOUSES AND VARIED LOT WIDTHS MIGHT DECREASE MONOTONY OF AREA

2-10% GRADE: SAME VIEW IF ALL HOMES ARE KEPT SINGLE STORY, LOW ROOF LINES; FENCING IS MODERATE; CONSIDERATION IS GIVEN TO WAIVING SETBACKS IF SUPERIOR VIEW IS GAINED.

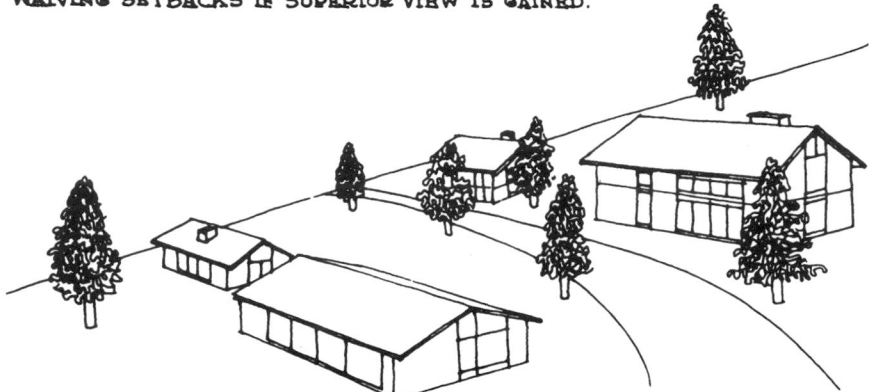

10-20% GRADE: EXCELLENT VIEW SITUATION, BUT ROADS SHOULD FOLLOW CONTOURS; VARIED SET-BACKS, 2-STORY HOMES ON HIGH SIDE PERMITTED IF LOTS ARE ADEQUATE TO INSURE MAJORITY GOOD VIEWS

Fig. 1

ROOM ORIENTATION

Individual room orientation to the sun is, to a degree, a personal choice. However, there are generally accepted orientations for different rooms within the dwelling unit. Most frequently, it is very difficult to achieve the ideal orientation for each room and some compromises must be made. It is beneficial to obtain some sunlight in the dwelling during the day. A dwelling that is facing directly north and receiving no sunlight tends to be a cool or dreary dwelling. Figure 1 and the following discussion for the orientation of each individual room should be used as a guide rather than a rule.

Bedrooms

Sunlight streaming into the bedroom upon waking up in the morning is a pleasant feeling. This is achieved by an easterly or southeasterly exposure. West exposure should be avoided because the west sun during the summer months is strong and heats up the room during the late afternoon and early evening.

Living Room

The living room should have a southerly exposure. In hot climates, a properly designed overhang will prevent direct sunlight from entering the room. In wintertime the low angle of the sun will allow sunlight to penetrate the depth of the room.

Dining Area

An easterly exposure will allow sunlight to enter during the morning for breakfast meals. If, however, it is desirable to view a sunset or have sunlight during the evening meal, a westerly or southwesterly exposure should be used. Since lunch is a minimal meal, this usually is not a major consideration.

Kitchen

North is generally considered to be the best orientation for kitchens and laundry areas because it provides an even, nonglare light. However, many people prefer some sunlight in the kitchen. A great deal would depend on the personal preference of the individual.

Multipurpose Area

This most often is a major activity area involving many members of the family. It should have a similar orientation to the living room.

Library

The library should be oriented toward the north. This will provide an even light source and will prevent any possible damage to books by the direct rays of the sun.

Fig. 1

BUILDING ORIENTATION TO STREET AND BUILDING GROUPING

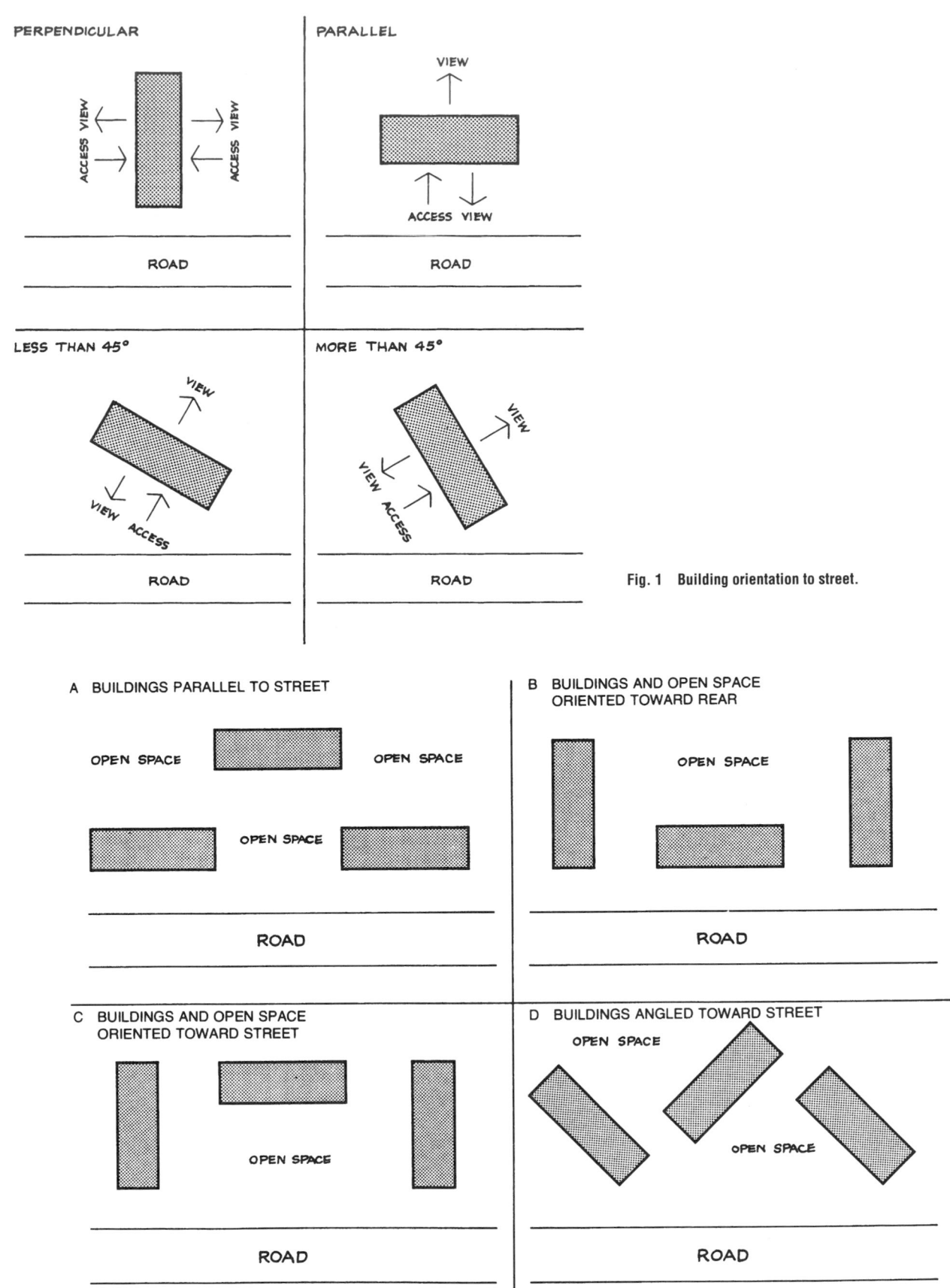

Fig. 1 Building orientation to street.

Fig. 2 Building grouping.

GENERAL

On housing sites it is important to plan for easy and direct movement of pedestrians and vehicles. Convenience of circulation and safety must be considered and planned together. Pedestrians generally prefer to walk in direct, straight lines. When they must use indirect or awkwardly placed walkways, they may take unauthorized routes, often trampling grass, shrubs, and other plants. Paths should follow topography and natural lines of movement, widening as traffic increases and narrowing in less-used areas. The pedestrian circulation system should also be designed to distinguish between the front and rear entrances of buildings. The quality of a multifamily dwelling appears to decrease when one entrance has to accommodate the removal of garbage and the entrance of guests, for instance. In part, this is a matter of interior space planning, but it is also a concern of site design.

Some pedestrian traffic considerations also apply to vehicular traffic: automobiles, scooters, service trucks, and in some cases, bicycles. It is necessary that minor roads come close to buildings to facilitate delivery of goods, to give protection in inclement weather, and to provide access for emergency vehicles.

Vehicles should be able to approach residential buildings, but need not remain close and conflict with pedestrian movement. The ideal solution seems to be the vertical separation of pedestrians and vehicles. However, this type of separation is generally limited to central city locations where heavy traffic volume justifies the great expense. At outlying locations horizontal separation is much more common. Many of the site plans included in this book show the separation of pedestrians and vehicles and most frequently accomplish it by restricting automobiles to the periphery of a site and allowing free pedestrian movement in the center.

VEHICULAR

A housing site should be provided with a periphery road system sufficiently wide to allow the maximum number of vehicles to move freely at all times. This road system should accommodate all through traffic that is bypassing the site and all traffic generated by the housing development.

In addition, adequate vehicular access into and out of the site is required. Such access will include delivery, service, and emergency vehicles. Access roads should be of sufficient width for all types of vehicles and also be provided with proper turning space. The following are based on FHA standards for vehicular access:

1. Access and circulation for fire-fighting equipment, furniture moving vans, fuel trucks, garbage collection, deliveries, and snow removal must be planned for efficient operation and in accordance with local custom.

2. Each property must be provided with vehicular access by an abutting public or private street.

3. Streets must be provided on the site where necessary to furnish principal trafficways for convenient access to the living units and other important facilities on the property.

4. The width and construction of the required street and provisions for its continued maintenance must provide safe and suitable vehicular access to and from the property at all times. Dead-end streets include adequate vehicular turning space.

5. The street pattern should discourage unnecessary through traffic.

6. The street system must connect with adjoining plotted streets except where topography does not permit or where such street connections would adversely affect the property.

7. The street system must provide convenient circulation by means of minor streets and properly located collector and arterial streets.

8. Cul-de-sacs must be provided with adequate paved turning space, usually a turning circle of at least 80 ft in diameter (100 ft is preferable).

9. Street rights-of-way must be of adequate width to accommodate the contemplated parking and traffic load in accordance with the type of street.

10. Proper street alignment and gradients are necessary. Streets must be adapted to the topography, preserving to the extent possible the natural contour and site features. They must have a suitable alignment and gradients for safety of traffic, satisfactory surface and groundwater drainage, and proper functioning of sanitary and storm sewer systems.

11. Recognition of existing facilities is required. The street system must be designed to recognize existing easements, utility lines, etc., which are to be preserved, and must be designed to permit connection to existing facilities where necessary for the proper functioning of the drainage and utility systems.

12. Well-designed street intersections are essential. Street intersections must generally be at right angles. Intersections of more than two streets and offsets at junctions of less than 125 ft should be avoided.

13. Driveways must be provided on the site where necessary for convenient access to the living units, garage compounds, parking areas, service entrances of buildings, collection of refuse, and all other necessary services.

14. Driveways must be planned for convenient circulation suitable for traffic needs and safety. Cul-de-sacs must be provided with adequate paved vehicular turning space, usually a turning circle at least 80 ft in diameter, except for short, straight-service driveways with light traffic.

PEDESTRIAN

A safe and convenient system of pedestrian walks is essential. It should be functionally organized and follow the natural traffic patterns of pedestrians. Walks should be wide enough to accommodate two-way traffic. Paved areas, especially at main entrances, must be of sufficient area to accommodate anticipated activities.

The following standards are based on FHA standards for pedestrian circulation:

1. Access to the dwellings and circulation between buildings and other important project facilities for vehicular and pedestrian traffic must be comfortable and convenient for the occupants.

2. Walking distance from the main entrances of buildings to a street, driveway, or parking court must usually be less than 100 ft; exception to this standard should be reasonably justified by compensating advantages, such as desirable views and site preservation through adaptation to topography. In no case must the distance exceed 250 ft.

3. Street sidewalks and on-site walks must be provided for convenient and safe access to all living units from streets, driveways, parking courts, or garages and for convenient circulation and access to all project facilities.

4. Width, alignment, and gradient of walks must provide safety, convenience, and appearance suitable for pedestrian traffic, shopping carts, and for moving of furniture. Small jogs in the alignments shall be avoided.

5. Steps and stepped ramps must be avoided if possible in order to facilitate servicing with wheeled vehicles.

6. An open and unobstructed passageway must be provided at grade level of each inner court. Such passageways must have a cross-sectional area of not less than 40 ft^2 and sufficient headroom to permit the passage of nonvehicular fire-fighting equipment, and must be continuous from the inner court of a yard, or an unobstructed open area between buildings.

Planning the Walks

The total walk area per dwelling unit varies considerably in different projects. In some, walks surround every building; in others, service drives provide the only pedestrian access. In projects of high density there are usually numerous walks. Although the need for economy must be kept in mind, a good walk system that promotes convenience is a sound investment in all projects.

A classification of walks is useful to provide uniform terminology and to serve as a check on the scope of this feature of planning:

1. Sidewalks: parallel to city and project streets

2. Collector walks: not parallel to streets; designed for general circulation

3. Approach walks: leading to buildings or groups of buildings from other walks, streets, or drives

4. Entrance walks: leading directly to dwelling or building entrances

The walk plan should be functional, built up of primary, secondary, and tertiary elements, each adjusted in location, width, and material to serve its purpose. Directness of access is essential; otherwise most people seem inclined to shortcut, unless they are funneled into the intended paths by planting or barriers. Adults are the most difficult to cope with; they make and then follow beaten tracks, whereas children either scatter in every direction or play on the paved areas.

Walks must be laid out so that they follow the natural path of circulation. They should be functional rather than formal in design and layout.

Walks and Service Circulation

A well-planned system of walks is needed for convenient access to recreation areas, service areas, parking areas, and other common facilities. Walks serve many purposes. They provide access and circulation to and between homes and facilities. They are a place to enjoy a pleasant walk, a place to push the baby carriage, a place to play hopscotch or marbles or ride a small vehicle. Thoughtfully designed, they are not only a utility; they add beauty, especially if combined with pleasant sitting areas which people can enjoy.

Walks are utilitarian as well, and when such service as garbage and trash collection is anticipated, using small motor carts or wagons on the walk system, their width should be adequate for this purpose.

Avoid steps when possible. Ramps for easy change in grade should be used, as they make the transition easier for service vehicles and electric carts, as well as elderly residents.

STREET PARKING

STREET PARKING

Shown in Figs. 1 and 2 are the minimum dimensions for three types of on-street parking and two types of bay parking. Parallel, 45°, and 60° parking at curb are the most usual methods. Since it is a natural tendency for tenants to park in front of their homes, it is wise to provide for it. The dimensions indicated are for streets in average community developments. Only parallel parking is permitted on main (collector) streets. Diagonal parking is permitted on minor streets. Where traffic is heavy, streets should be increased in width. Sidewalks that are adjacent to the curb must be widened to allow for overhang of bumpers.

Parking bays, both directly off the street and within property lines, offer a solution where parking is not permitted on a public way. Parking bays directly off the street are not only more convenient for the tenants but are less expensive to construct and more economical to maintain. These should be used, however, only on minor streets. The illustrations are for streets with two-way traffic.

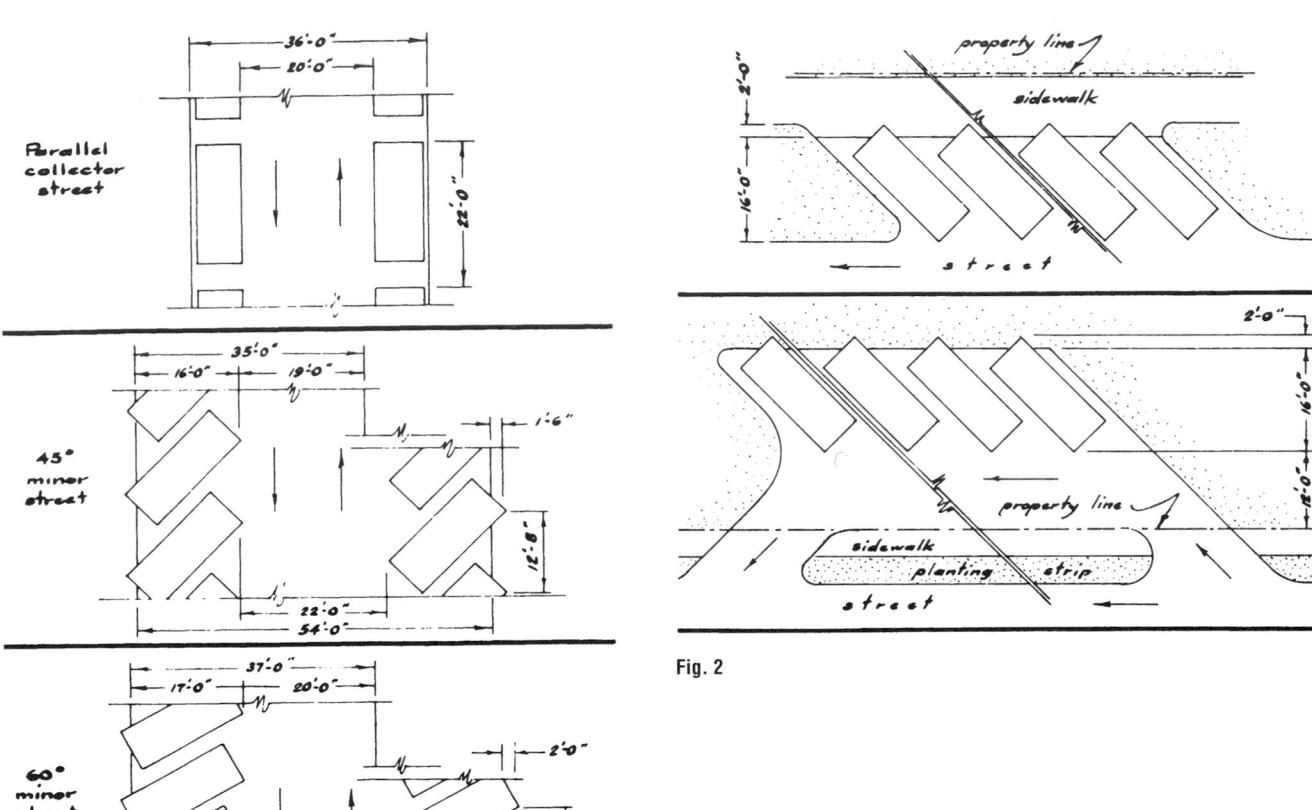

Fig. 1

Fig. 2

CAR CIRCULATION AND ACCESS

Traffic circulation should be around a residential development, not through it. Major traffic arteries, existing or assured, should provide for fast and convenient access to the development. When the project has been reached, however, safety, convenience, and pleasant living for the residents take precedence over traffic speed and shortcuts through the project.

Arterial streets should not be used for car access to the homesites. The backing of cars off the on-lot parking spaces into fast-moving traffic is too hazardous.

Collector streets of ample width and flowing alignment should feed traffic between the arterial streets and to a network of minor access streets on which most of the homesites are located. Location of collector streets near the perimeter of a development is often advantageous.

Short loop streets and cul-de-sacs are best for the minor streets because they provide the safest access to and from homesites in small housing groups. Superblocks made up of several or many such minor elements are very appropriate to the cluster subdivision and townhouse-on-the-green concept. With careful design, the common open space and facilities can be accessible on foot from all homes without crossing any street.

The width of minor streets may be limited to two traffic lanes if car parking for owners and guests is adequately and conveniently provided elsewhere. Through careful design, this can produce savings in grading, drainage, and street construction.

Type of Facility	Function and Design Features	Spacing	Widths		Desirable Maximum Grades	Speed	Other Features
			R.O.W.	Pavement			
Freeways	Provide regional and metropolitan continuity and unity. Limited access: no grade crossing; no traffic stops.	Variable; related to regional pattern of population and industrial centers	200-300'	Varies; 12' per lane; 8-10' shoulders both sides of each roadway; 8'-60' median strip.	3%	60 mph	Depressed, at grade, or elevated. Preferably depressed though urban areas. Require intensive landscaping, service roads, or adequate rear lot building setback lines (75') where service roads are not provided.
Expressways	Provide metropolitan and city continuity and unity. Limited access; some channelized grade crossings and signals at major intersections. Parking prohibited.	Variable; generally radial or circumferential	200-250'	Varies; 12' per lane; 8-10' shoulders; 8-30' median strip.	4%	50 mph	Generally at grade. Requires landscaping and service roads or adequate rear lot building setback lines (75') where service roads are not provided.
Major Roads (Major Arterials)	Provide unity throughout contiguous urban area. Usually form boundaries for neighborhoods. Minor access control; channelized intersection; parking generally prohibited.	1½ to 2 miles	120-150'	84' maximum for 4 lanes, parking and median strip.	4%	35-45 mph	Require 5' wide detached sidewalks in urban areas, planting strips (5'-10' wide or more) and adequate building setback lines (30') for buildings fronting on street; 60' for buildings backing on street.
Secondary Roads (Minor Arterials)	Main feeder streets. Signals where needed; stop signs on side streets. Occasionally form boundaries for neighborhoods.	¾ to 1 mile	80'	60'	5%	35-40 mph	Require 5' wide detached sidewalks, planting strips between sidewalks and curb 5' to 10' or more, and adequate building setback lines (30').
Collector Streets	Main interior streets. Stop signs on side streets.	¼ to ½ mile	64'	44' (2-12' traffic lanes; 2-10' parking lanes)	5%	30 mph	Require at least 4' wide detached sidewalks; vertical curbs; planting strips are desirable; building setback lines 30' from right of way.
Local Streets	Local service streets. Nonconducive to through traffic.	at blocks	50'	36' where street parking is permitted	6%	25 mph	Sidewalks at least 4' in width for densities greater than 1 d.u./acre, and curbs and gutters.
Cul-de-sac	Street open at only one end, with provision for a practical turnaround at the other.	only wherever practical	50' (90' dia. turnaround)	30'-36' (75' turnaround)	5%		Should not have a length greater than 500 feet.

Fig. 1

STREET STANDARDS

One set of design standards that can be used as a basis for a comprehensive street plan is that developed by the National Committee for Traffic Safety in 1961. These standards have been extensively used and accepted. Over the years, these standards have resulted in acceptable and functional street systems for both vehicle movement and pedestrian safety. It must be understood that these standards are minimum and should be utilized only as a guide to meet existing conditions.

Similar standards are available from the Institute of Traffic Engineers.

TABLE 1 Recommendations of the National Committee for Traffic Safety

Design of residential streets			Design of feeder or collector streets			
	Single-family	Multifamily				
Street width	50 ft	60 ft	Street width	60 ft	Horizontal alignment	90° intersections preferred; less than 60° unduly hazardous
Pavement width curbs	26 ft	32 ft	Pavement width	36 ft		
	Straight curb recommended	Same	Curbs	Straight curb recommended		12 ft curb radius for local and feeder streets
Sidewalks:			Sidewalks:			
Width	4 ft minimum		Width	4 ft minimum		50 ft curb radius for feeder street intersecting main highway
Setback	3 ft minimum if no trees; 7 ft minimum with trees	Same	Setback	3 ft mi.nimum if no trees, 7 ft minimum with trees		
Horizontal alignment	200 ft minimum sight distance	Same	Horizontal alignment	Same as for local residential streets	Design of sidewalks:	
Vertical alignment	6–8% maximum grade desirable; 3–4% per 100 ft maximum rate of change	Same	Vertical alignment	Same as for local residential streets	Placement	Setback should be minimum of 7 ft where trees are planted between curb and sidewalk; minimum of 3 ft if no trees
			Pavement surface	Same as for local residential streets		
Cul-de-sac	400–500 ft maximum length	Same	Design of intersections:			
			Sight distance	Such that each vehicle is visible to the other driver when each is 75 ft from the intersection for 25 mi/h maximum speed. No building or other sight obstruction within sight triangle	Width	4 ft minimum (4½–5 ft minimum near shopping centers)
Turnarounds	40 ft minimum curb radius without parking; 50 ft minimum curb radius with parking	Same				
Pavement surface	Nonskid with strength to carry traffic load	Same	Vertical alignment	Flat grade within intersection		
				Flat section preferred from 50 to 100 ft each way from intersection, but in no case over 3–5% grade; 6% maximum between 100 and 150 ft of intersection		

NOTE: The table is based on a maximum speed of 25 mi/h, determined by the Uniform Vehicle Code. Recommendations will be reasonably satisfactory even if some speeds moderately exceed 25 mi/h.

STREET CLASSIFICATION

The overall street system for a housing development must conform to the circulation requirements of the master plan for the community. This will provide maximum accessibility to all parts of the community and ensure proper coordination with proposed circulation changes.

Direct access to a major arterial highway is essential. Such intersections must be adequately controlled with lights or other means. The practical minimum distance between intersections on the major arterial highway should be 800 to 1000 ft. No through streets should be provided. All circulation should be around the periphery of the development to the major arterial highway.

Each lane of traffic will carry from 600 to 800 cars per hour. Horizontal alignment of all collector, minor, loop, and access streets should provide for a minimum of 200 ft clear sight distance. The vertical alignment should not exceed 6 to 8 percent grade differential. Sidewalks, when used, should be a minimum of 4 ft wide. When trees are planted between the curb and sidewalk, the sidewalk should be set back approximately 8 ft. If no trees are used, the setback should be 4 ft.

Fig. 1 Types of streets.

STREET CLASSIFICATION

DESIGNATION	RIGHT-OF-WAY WIDTH	FUNCTION
Arterial Streets and Highways	80-120 feet	Primarily devoted to the movement of high volumes of traffic at relatively high speed; only rarely interrelated with adjacent land areas; vehicular access is almost always limited.
Marginal Access Streets	40 feet	These are minor streets that are parallel to and adjacent to arterial streets and highways; and which provide access to abutting properties and protection from through traffic.
Collector Streets	60-80 feet pavement width— 32 feet min.	These carry traffic from minor streets to the major system of arterial streets and highways, including the principal entrance streets of a residential development and streets for circulation within such a development. They permit access to adjacent land areas, but generally do not permit long-distance through traffic.
Minor Streets	50-60 feet pavement width— 24-32 feet	These carry traffic from collector streets to the individual land parcels within any given area. The primary function of these streets is to provide access to abutting properties.
Loop	50-60 feet pavement width— 24-32 feet	Same as minor street.
Cul-de-sac	800 feet max. 40 feet min. curb radius without parking	Dead-end street with proper turning radius at end. Provides quiet residential street with no through traffic. Also helps solve difficult site problems with restricted access.
Alleys	20 feet	These are minor ways that are used primarily for vehicular service access to the back or the side of properties otherwise abutting a street. May be necessary in group, row houses, or apartment developments. Not recommended in single-family developments.

Fig. 2

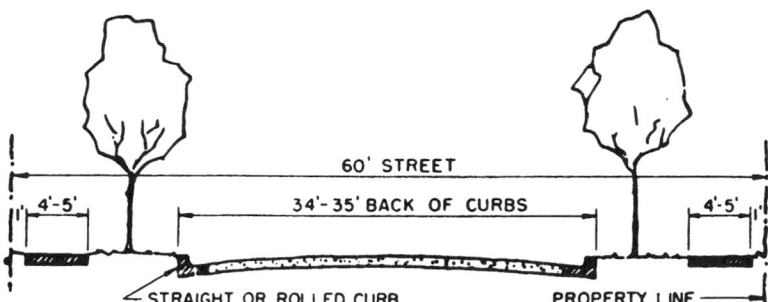

Fig. 3 Typical 60-ft street cross section.

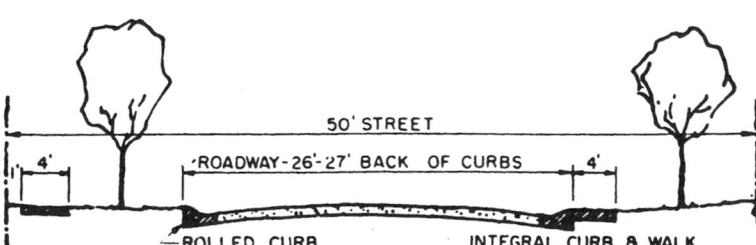

Fig. 4 Typical 50-ft street cross section.

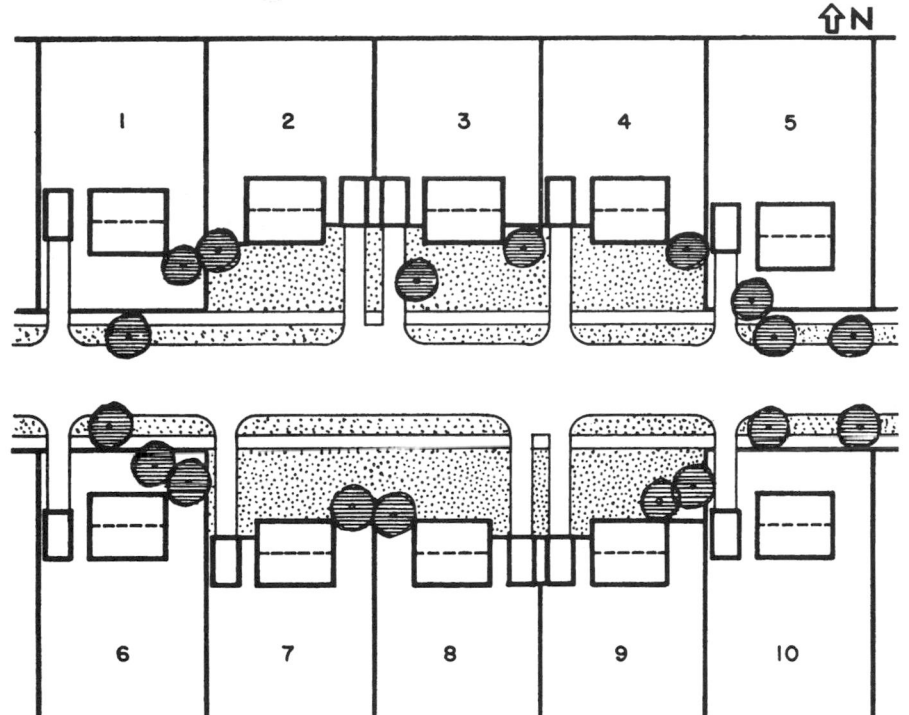

Fig. 1   The straight street—improved design.

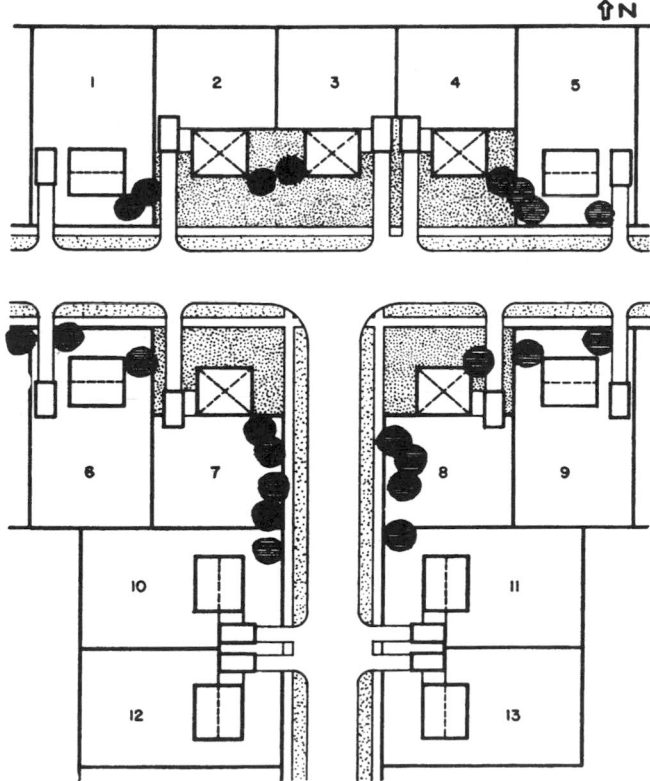

Fig. 2   The T junction—improved design.

TYPICAL STREET ARRANGEMENTS

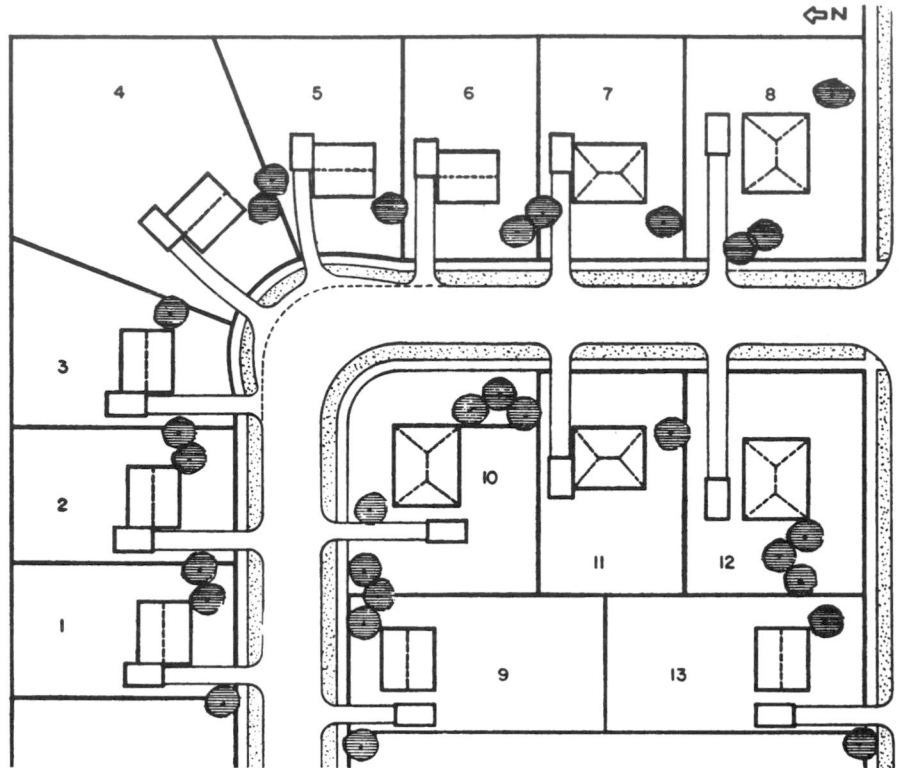

Fig. 3   The loop.

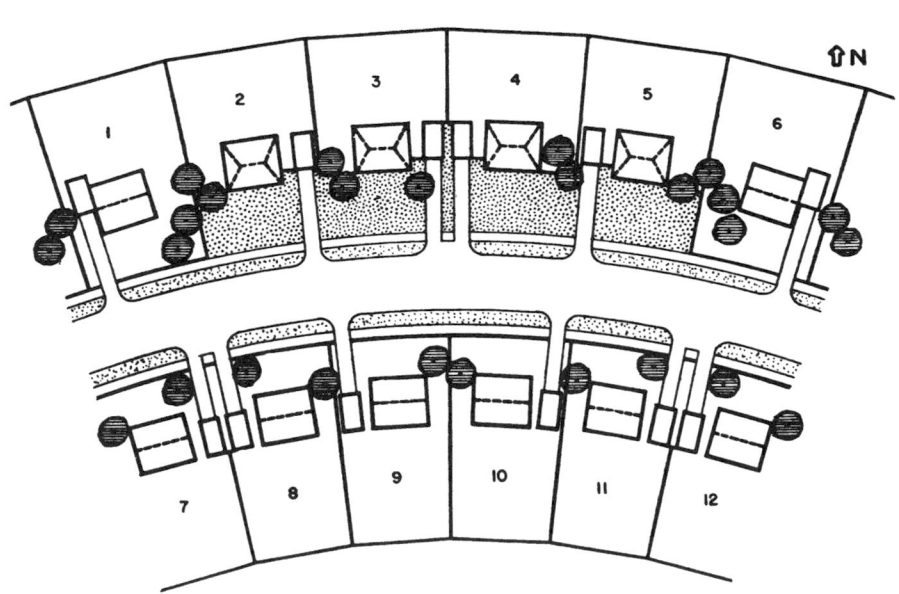

Fig. 4   The curved street—improved design.

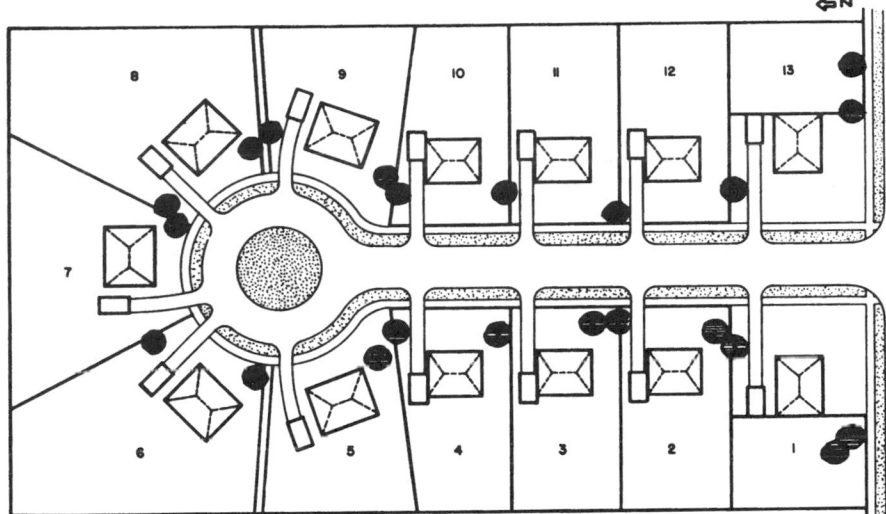

Fig. 5   Cul-de-sac.

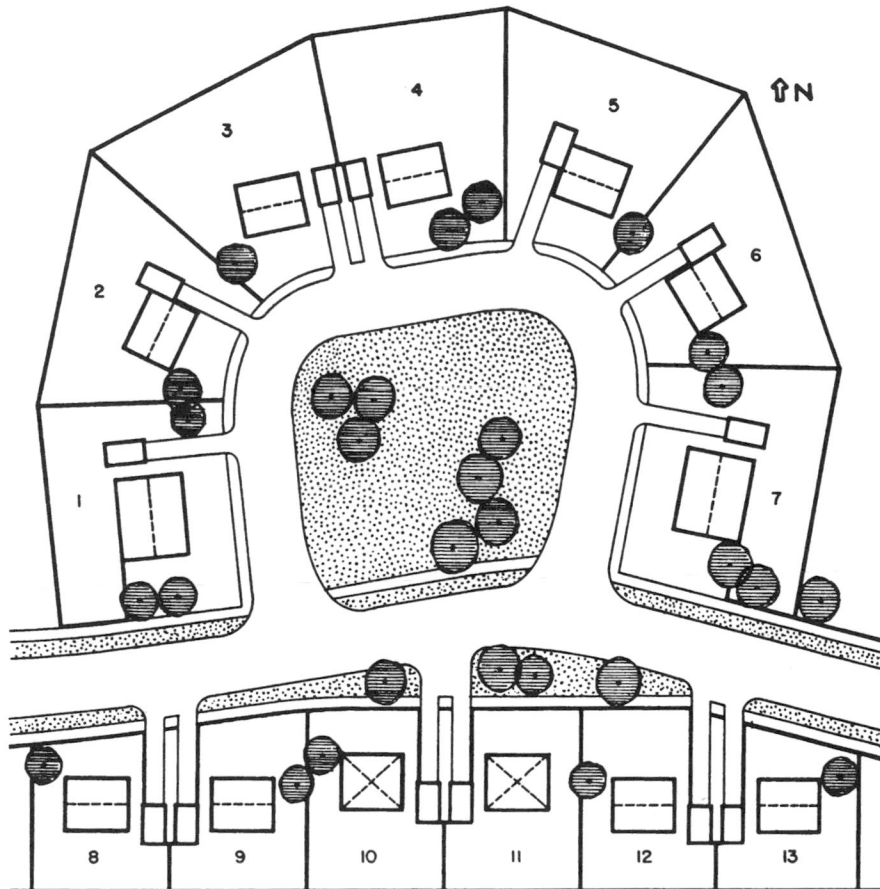

Fig. 6   A group around a green.

ACCESSORY PARKING

PARKING—GENERAL

Sufficient off-street parking should be provided for the housing site. In developments convenient to public transportation, efforts should be made to minimize parking areas in excess of actual needs. In general, several moderate-sized parking areas are preferable to one or two large areas. Parking should never be more than 200 ft from the dwellings it serves. A shorter distance is much more desirable.

The following are some of the FHA standards for parking:

1. Paved parking areas and courts must be provided to meet the needs of the residents and their guests without interference with normal traffic.

2. Parking areas and courts must be located for convenient access to the living units without impairing the views from living rooms, entrances, or front yards.

3. Dimensions of parking areas and courts must be adequate for convenient use for occupant parking.

4. Where necessary to provide for bumper clearance and suitable screen planting, parking facilities must not be nearer than 5 ft to any street, property line, or project facility.

5. Driveways must have two traffic lanes for their entire length, usually 18 ft in addition to any parking space, except that a single lane may be used for short, straight-service driveways where two-way traffic is not anticipated.

6. Garages, carports, and parking bays must be set back at least 8 ft from the nearest edge of any moving traffic lane to the extent necessary to provide sight lines for safe entry into the trafficway.

AUTOMOBILE STORAGE

The tremendous increase in automobile ownership has had a profound influence on the design and location of residential districts. The problem confronting site planners is to create a

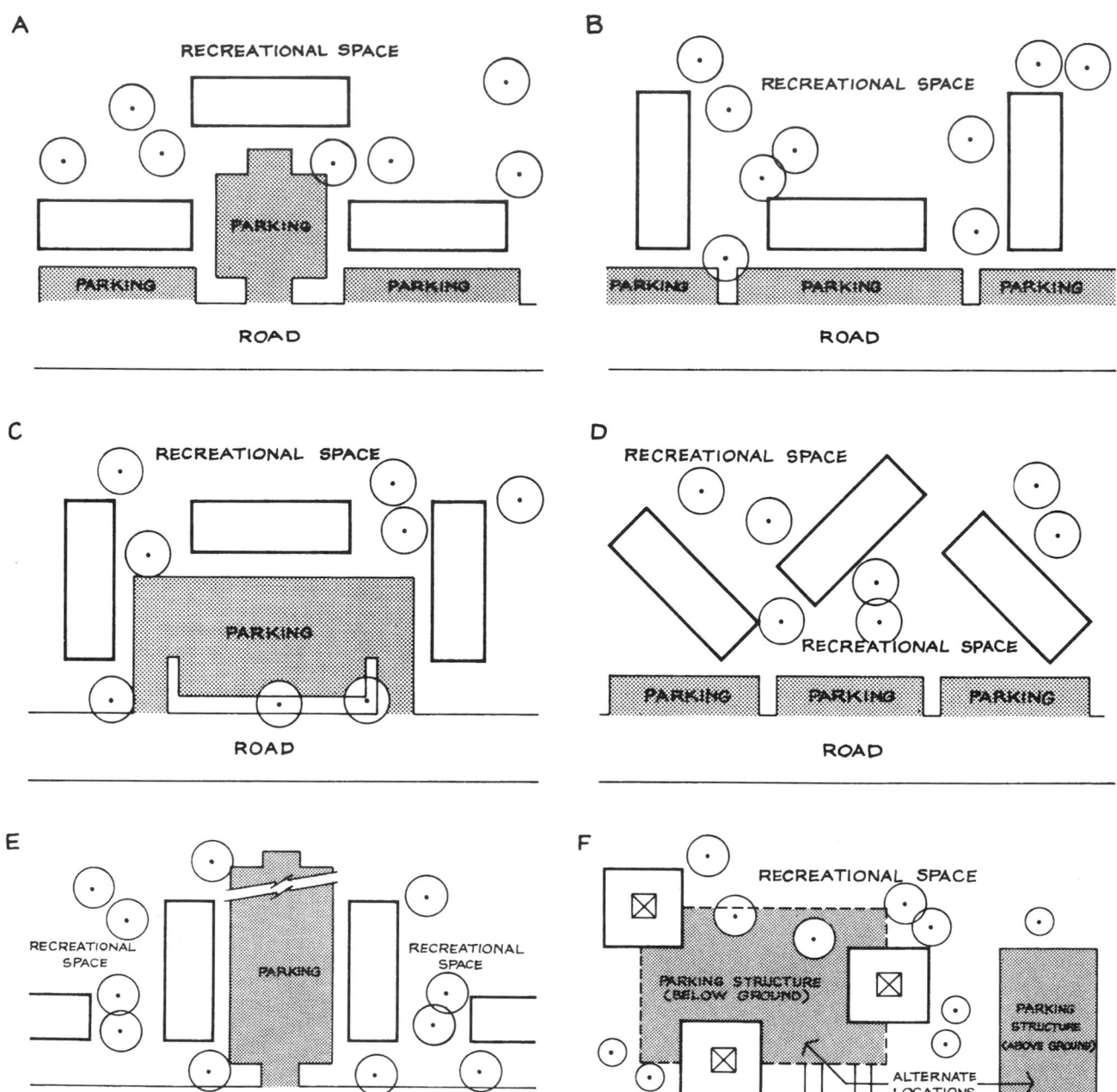

Fig. 1   Building arrangements with parking.

balance between the allocation of space for automobile storage and for other outdoor uses. This problem is made more acute by the pressure to satisfy occupants' demands for parking as close as possible to their dwellings.

The ownership of automobiles is reaching or has already reached a ratio of one car per family. This ratio seems to prevail for all projects regardless of occupants' incomes, but many sites are not being built to this standard. If ownership of cars increases, the inadequacy of accommodations will be felt even more. Unlike other site facilities, parking space cannot be added in stages without destroying some aspect of quality. In high-income housing projects, the ratio can be as high as two cars per family. It follows that for all new housing developments, at least one parking space is needed for each dwelling unit. The exceptions might be for housing for the elderly and central-city housing where the cost of providing parking

facilities is prohibitive and the ownership and operation of a motor vehicle is limited because of other factors. At many sites there is a need to allocate space for visitors' cars in addition to those belonging to residents.

If parking needs are not met on-site, crowding of adjacent streets is likely to result. However, if parking needs are met on-site, the residential buildings sometimes look as though they are built on huge parking lots. The automobile then dominates the site and infringes on privacy. The net effect is a reduction, if not the disappearance, of site quality.

The most satisfactory solution is to store cars in areas hidden from view, preferably underground. In this way, automobiles would be convenient to dwellings, greater pedestrian safety would be ensured, and open space would be preserved for other uses. However, underground parking is prohibitive in cost for most developments. The next best solution is the

construction of parking structures, preferably low ones whose roofs can be used for playgrounds, laundry areas, sundecks, etc. Some recent multiple housing projects have been designed with parking on lower floors and apartments on upper floors of the same structures. Solutions involving complete separation of pedestrians and automobiles, though costly, may become mandatory for central-city, high-density housing. In outlying areas and sites of lower intensity development, it is possible to satisfy parking demands without vertical separation if careful site planning is followed. Usually, parking space is provided by a number of small lots, screened from active pedestrian parts of the site. A number of small lots are aesthetically preferable to one huge paved area and, in addition, generally permit the majority of cars to be stored reasonably close to the individual living units.

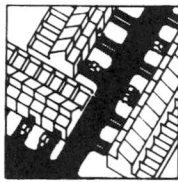

a. individual parking on grade adjacent to dwelling unit.

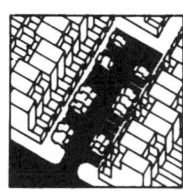

b. common parking on grade, adjacent to and shared by groups of dwelling units.

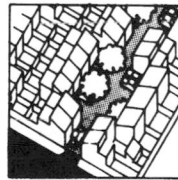

c. common parking on grade, integrated with common open space, adjacent/near to and shared by groups of dwelling units.

d. common parking on grade, near to and shared by groups of dwelling units.

e. common parking in building structure below housing, shared by groups/all dwelling units.

f. common parking on grade, separate from and shared by all dwelling units.

g. common parking in separate building structure, shared by all dwelling units.

Fig. 2

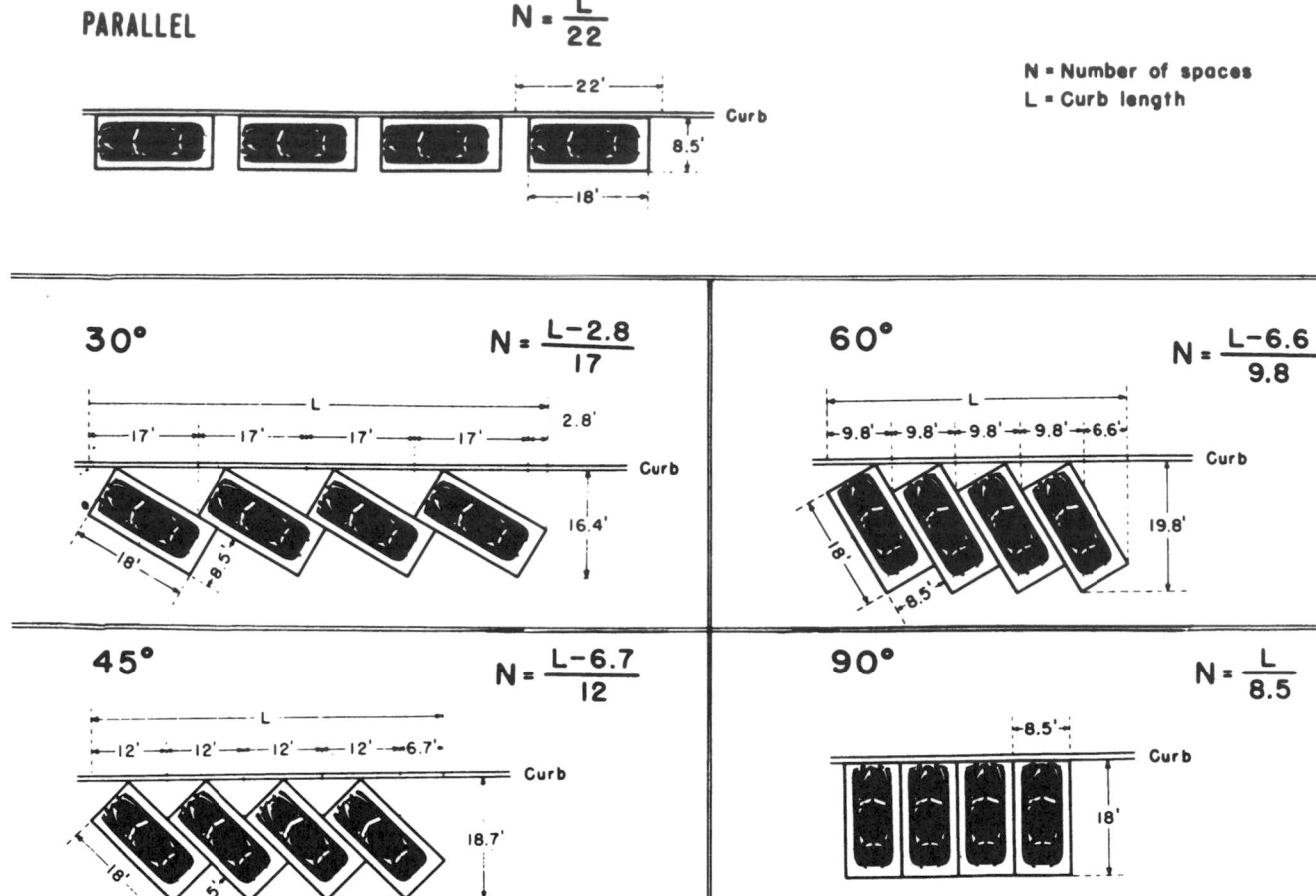

Fig. 1   Curb parking—space requirements for curb parking at various angles.

Parking

BAY

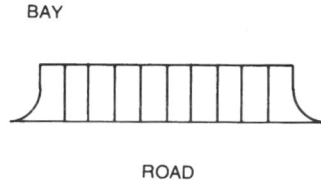

ROAD

The parking bay system is the most efficient and least expensive arrangement. It occupies a minimal area and may serve individual buildings or groups of buildings along minor residential streets. However, vehicles backing out into traffic may create hazardous conditions.

BUFFER

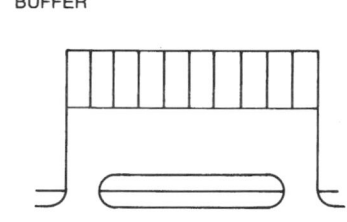

ROAD

The buffer system is similar to the parking bay, except that this arrangement separates parking maneuvering from roadway traffic. It occupies more area than the parking bay and may also serve individual buildings or groups of buildings. It may be located on a fairly active street.

PERPENDICULAR COURT

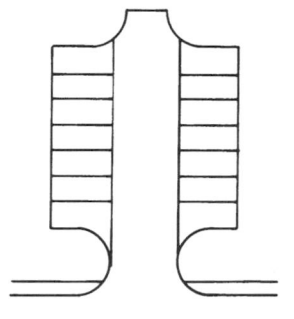

ROAD

While this is one of the safest and most attractive systems, its relatively great depth penetrates the site and may absorb area that might best be used in other ways. In some cases, buildings may be located around the parking court, thus integrating the parking space with surrounding spaces. It may be located on a fairly active street.

PARALLEL

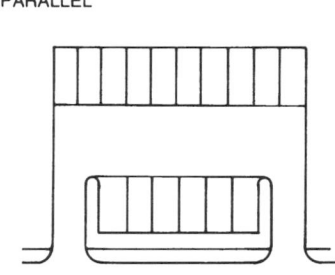

ROAD

Parallel parking is similar to the buffer system in all respects except that this design is more efficient in so far as it uses a double-loaded parking aisle. Like the parking court, it also penetrates the site fairly deeply, but may also be integrated with the open spaces around it. It may be located on a fairly active street.

Fig. 1 Parking.

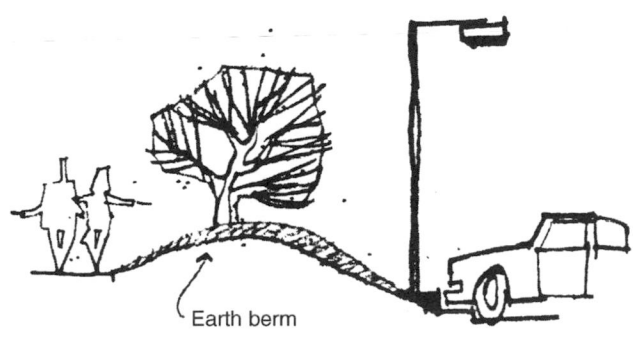

Fig. 2 Earth berm—helps screen parking.

PARKING GARAGES

Using the basement of apartment buildings for parking purposes presents problems that must be weighed against each other to determine the most economical manner of providing space for the required number of cars. As is shown by Fig. 1, the type of parking and the number of cars that can be accommodated depend upon the dimensions of the space and location of columns in the area assigned to parking. To achieve the ideal arrangement may prove too costly, and it may be more economical to use more floor space for a maximum number of cars. Each individual case must be studied to obtain the best net results.

Where local regulations permit and space is available, parking should be provided in or just off the street.

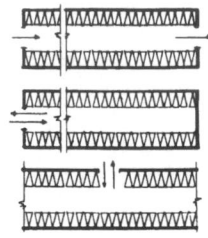

Open basement space with clear span and no columns is the most desirable but may be too costly to construct. The three diagrams illustrate three methods of circulation in open basements.

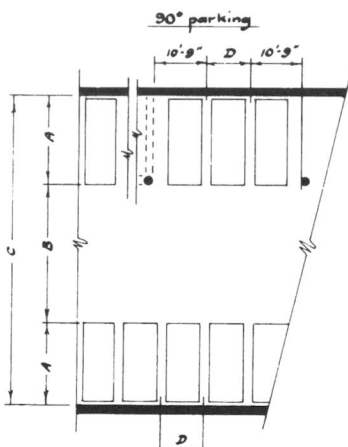

Fig. 1

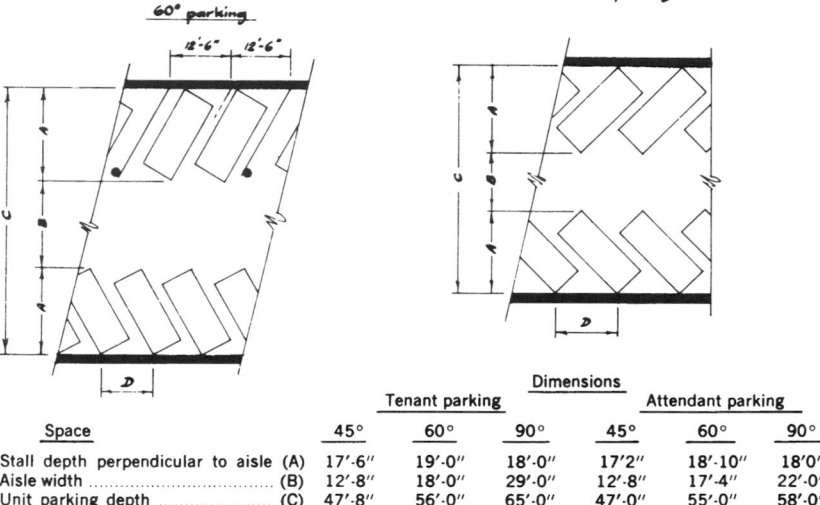

	Dimensions					
	Tenant parking			Attendant parking		
Space	45°	60°	90°	45°	60°	90°
Stall depth perpendicular to aisle (A)	17'-6"	19'-0"	18'-0"	17'2"	18'-10"	18'0"
Aisle width .... (B)	12'-8"	18'-0"	29'-0"	12'-8"	17'-4"	22'-0"
Unit parking depth .... (C)	47'-8"	56'-0"	65'-0"	47'-0"	55'-0"	58'-0"
Stall width parallel to aisle .... (D)	12'-8"	10'-6"	9'-0"	11'-4"	9'-3"	8'-0"

Note: Where 45° and 60° parking is necessary, one way traffic should be planned.

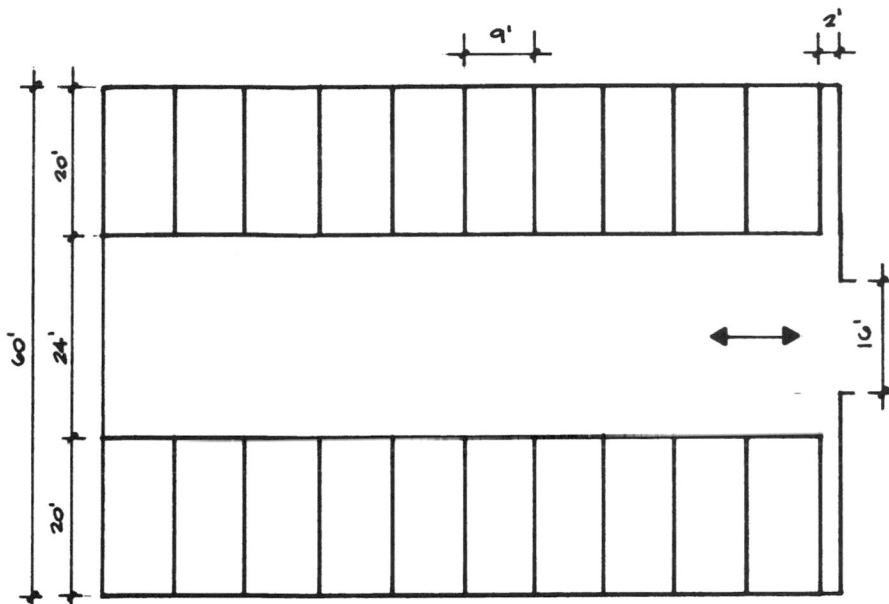

Fig. 1   90° perimeter parking.

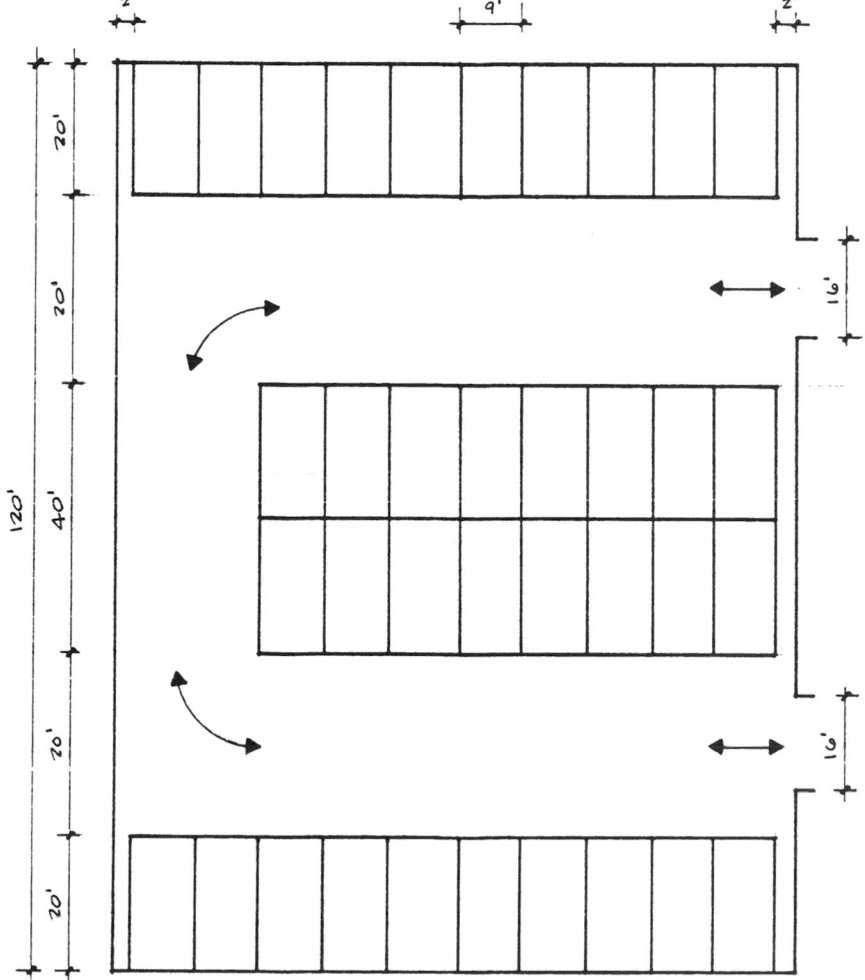

Fig. 2   90° perimeter and island parking.

45° PERIMETER AND ISLAND PARKING

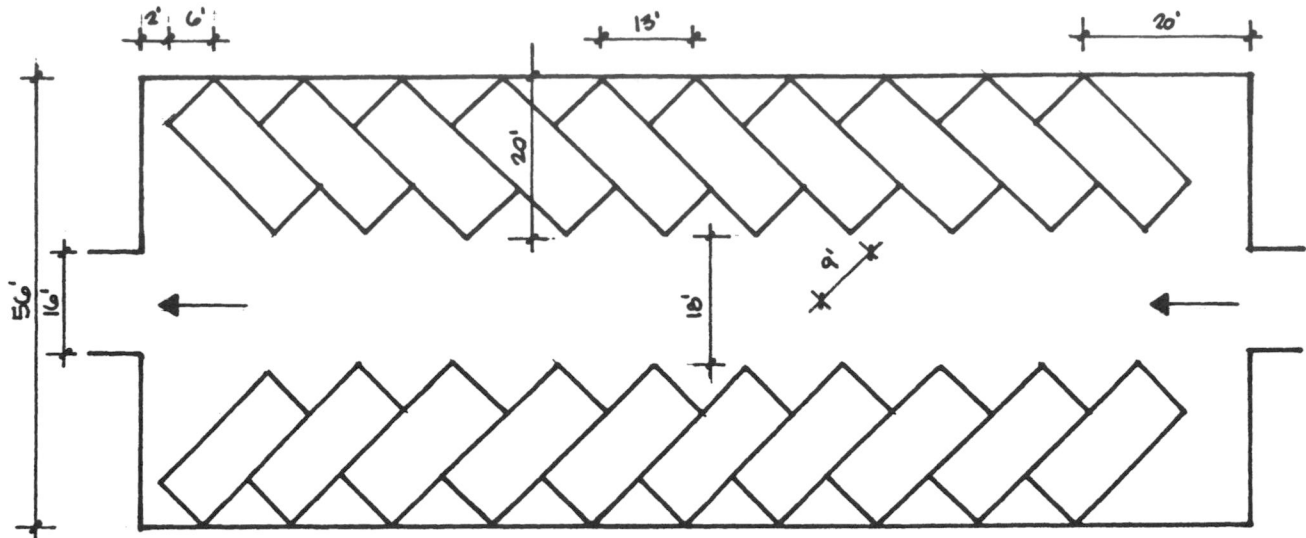

Fig. 1   45° perimeter single-unit parking.

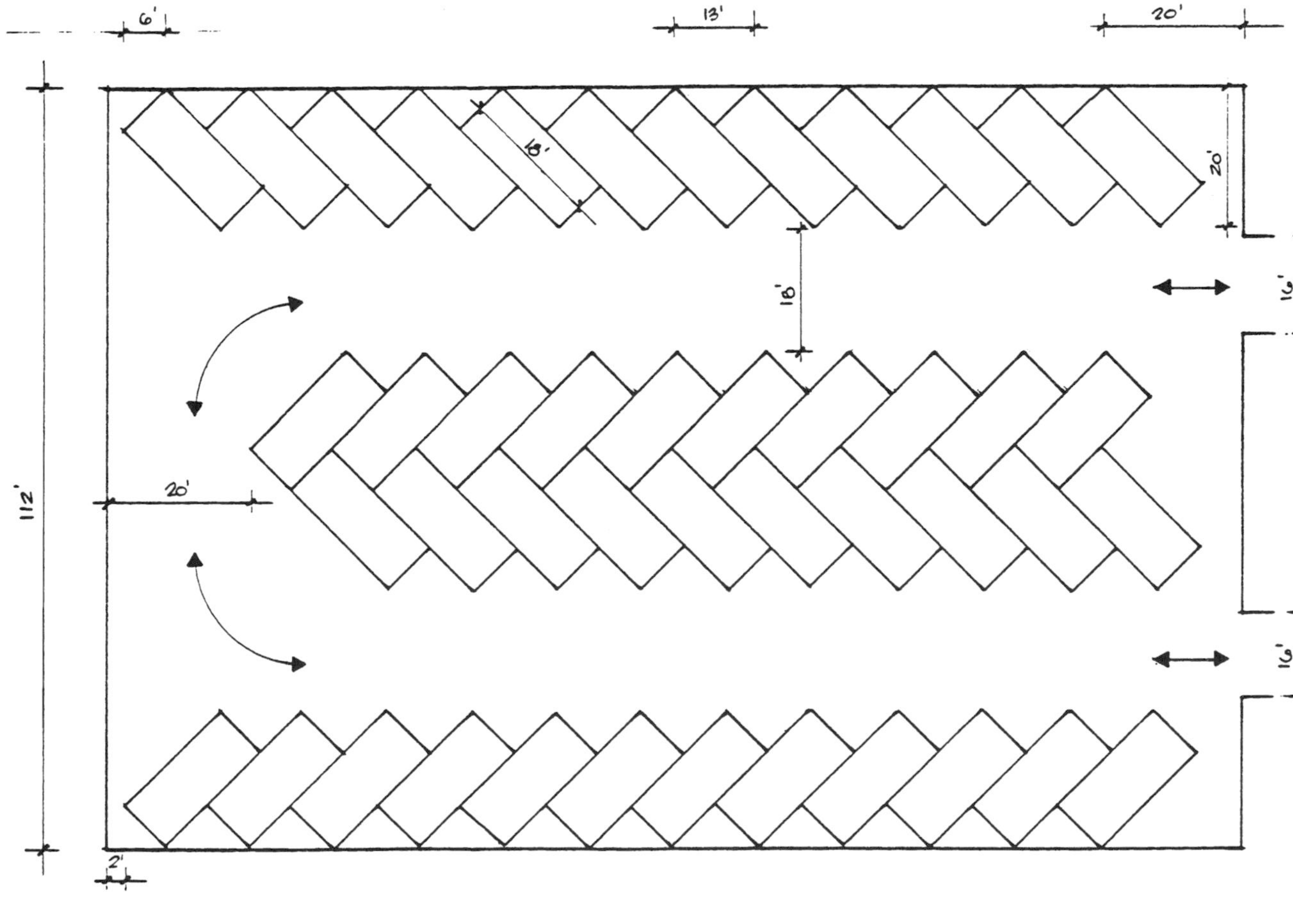

Fig. 2   45° perimeter and island single-unit parking.

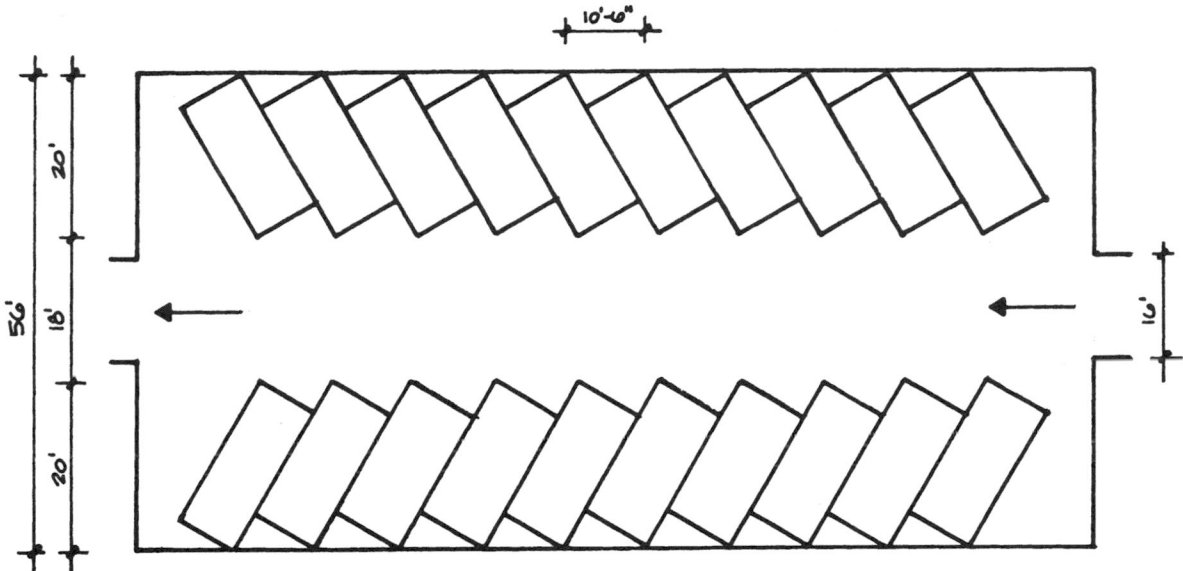

Fig. 1   60° perimeter parking.

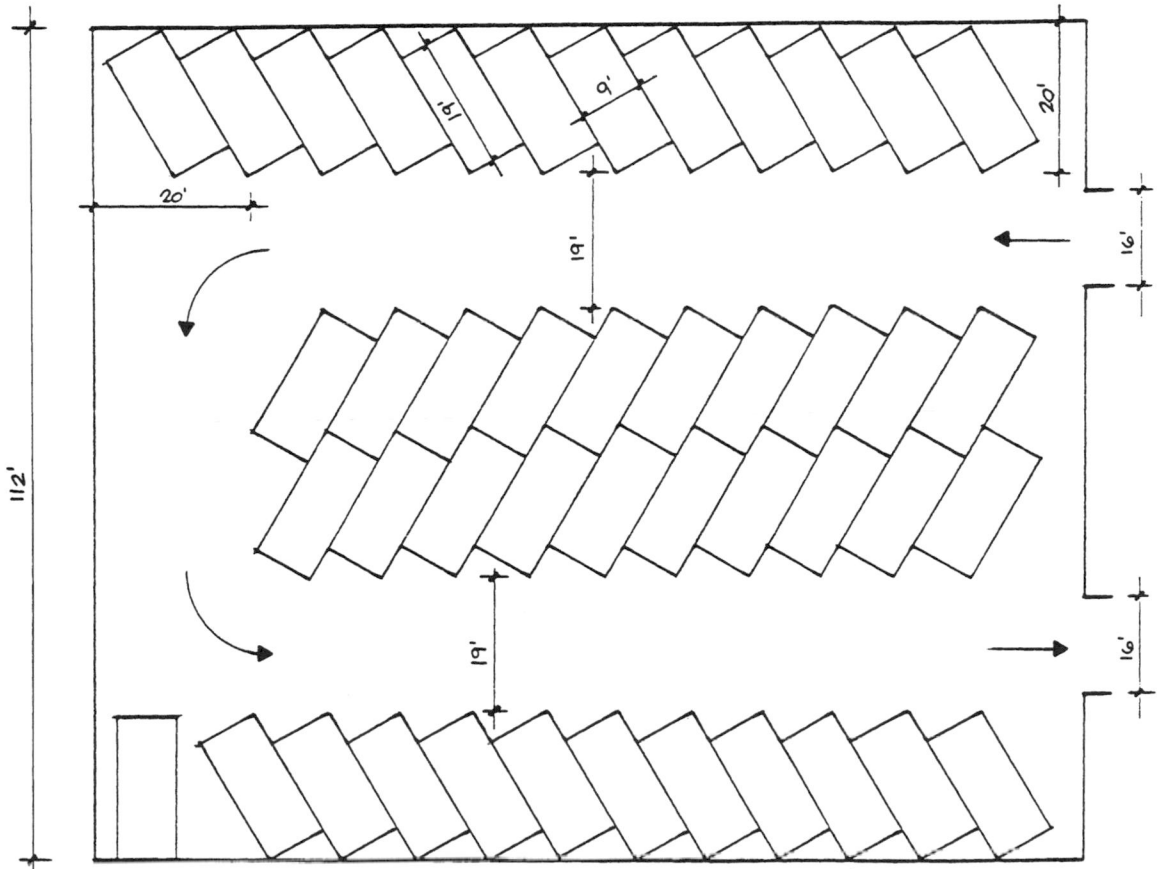

Fig. 2   60° head to head island and perimeter parking.

REQUIREMENTS FOR VEHICULAR TURNS

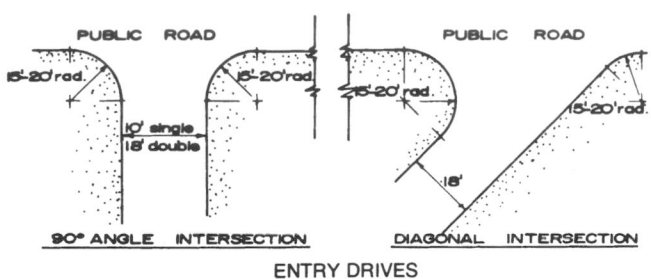

Fig. 1

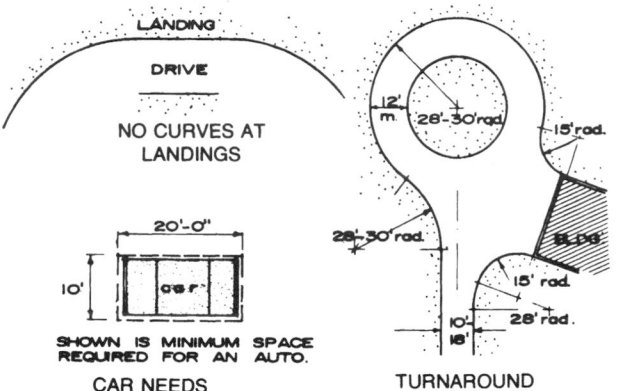

Fig. 2

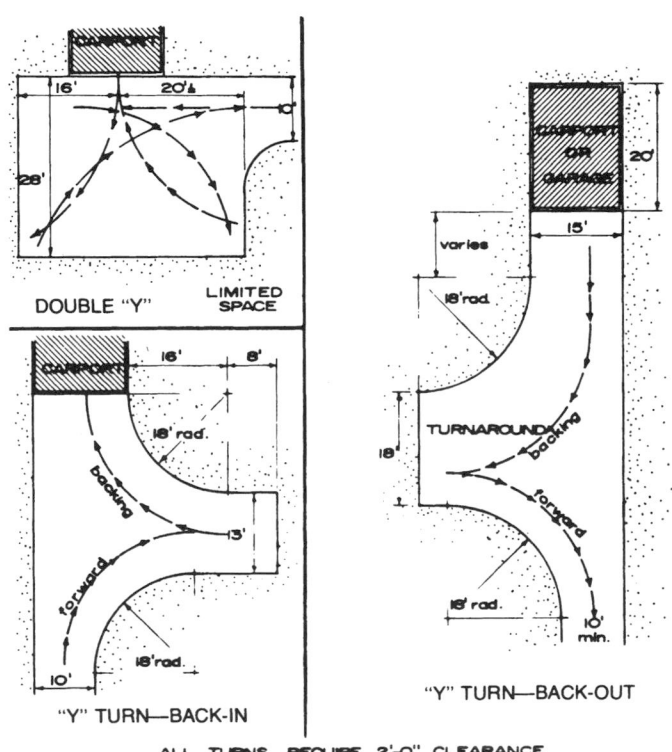

Fig. 3

a. private open space on grade adjacent to dwelling unit; common open space reduced to access.

b. private open space on/in building structure, adjacent to dwelling unit; common open space reduced to access.

c. private open space on grade or on/in building structure, adjacent to dwelling unit; common open space shared by groups of dwelling units.

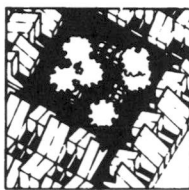

d. private open space on grade or on/in building structure, adjacent to dwelling unit; common open space, integrated with parking, shared by groups of dwelling units.

e. common open space shared by groups of dwelling units.

f. common open space, integrated with parking, shared by groups of dwelling units.

g. common open space shared by all dwelling units.

Fig. 1

LOCATION OF UTILITIES

From a planning standpoint, the location of utility lines always presents problems. First, each utility company or agency has its own standards and requirements, which very rarely relate to the other utilities. Servicing or installation of new lines frequently requires breaking up of the street and disruption of traffic. Overhead electrical and telephone lines are unsightly and are subject to disruption by severe weather conditions.

In any new housing development, the location and interrelationship of all utility lines must be carefully studied for efficiency and appearance.

Water Supply

Water supply mains may be located under the sidewalk, in the planting strip, or under the street. Minimum design requirements will locate them at least 10 ft from the nearest sewer or gas main and above the highest sewer or gas main. Some engineers place water mains on the north side of the east-west street, and on the east side of a north-south street, so that the rays of the sun will be more effective in preventing freezing.

If wells are used, they should be located sufficiently distant from septic tanks, sewers, cesspools, and drainage fields. The usual recommended minimum distances are 50 ft from septic tanks and sewers, 100 ft from drainage fields, and 150 ft from cesspools.

Sanitary Sewer

The sanitary sewer mains are generally located on the centerline of the road. The line is a clay tile pipe. If it were located in the planting strip, the roots of the trees might cause breaks in the pipes. The centerline location also locates the pipe equidistant from building lines on both sides of the street. The sewer line should be located below the water supply mains.

Storm Sewer

Storm sewers are generally located one-third the distance from the curb line to the centerline of street. It is always located on the opposite side of the street from the waterline. This is to prevent any possible contamination.

Electricity

Though best located in an underground conduit, overhead power lines are often situated above planting strips, causing interference with trees, the danger of falling wires, and unsightly appearance.

An alternate location for electric power lines is at the rear of the lots, either above or below ground, and then service lines are brought into the house. When this is done, proper easements are necessary for servicing of the lines when required.

The trend is, despite additional cost, to place electric power lines underground for two reasons. First, it reduces the chances of power failures and second, it eliminates unsightly clutter in the landscape, adding to the aesthetic appeal.

Telephone, TV Cable

Similar to electric power, telephone lines can be located either above or below ground. In the past almost all lines were above ground and either utilized the electric line poles or set up an additional line. In either case, they are unsightly and subject to disruption by the weather.

Telephone lines, TV cables, and other special lines should all be located underground. Some attempts have been made to combine all electrical, telephone, and TV cables into a common underground trench that would simplify additional installations and maintenance. These lines may also be located at the rear-lot easement, if necessary.

Gas

Gas mains are generally located under the sidewalk or in the planting strip. They normally do not have any special requirements.

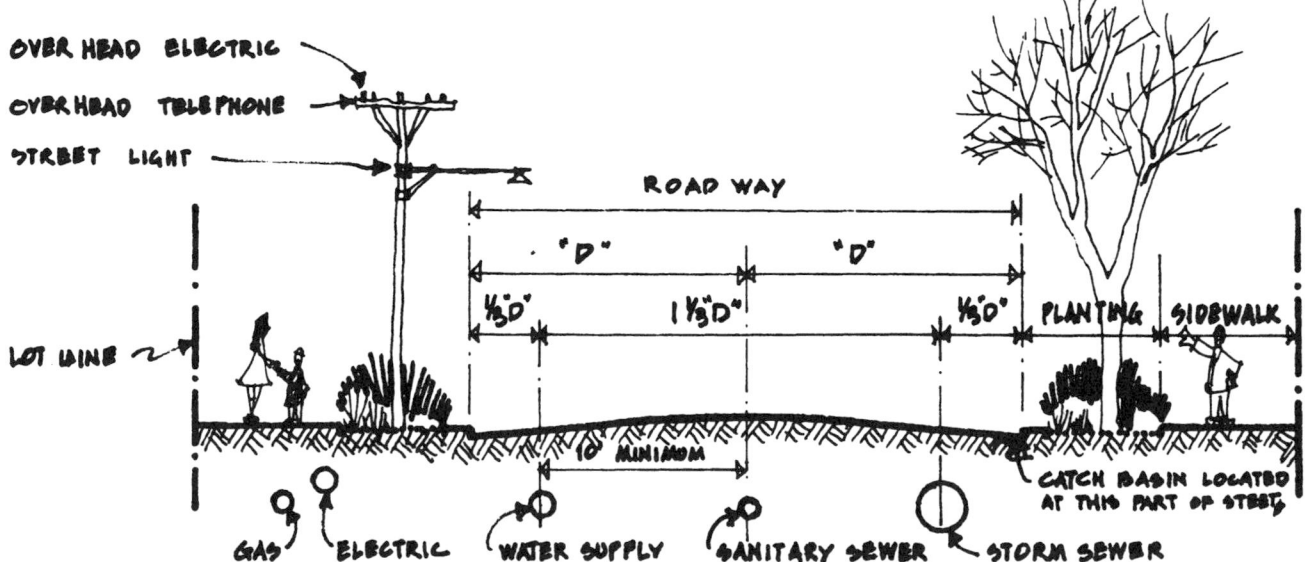

Fig. 1   The street cross section is generalized. Actual conditions can have all or some of the elements indicated.

SITE SELECTION

The best way to control noise is to avoid it. This can most effectively be done if the first design task is that of selecting the building site, because careful selection can steer the proposed building away from noise problems. Likewise, for some projects it may be cheaper to abandon a site already selected and relocate to a quieter area than to make extensive revisions to an acoustically unacceptable scheme. For many projects, however, other considerations than sound may preclude the selection of a quiet site or relocation to a quiet site.

Keep in mind that acoustic conditions are rarely stagnant and you should consult zoning and planning authorities to determine future plans for the surrounding area. A seemingly suitable site can later be surrounded by industrial areas or traffic arteries, or subjected to aircraft overflights greatly increasing on-site levels. Therefore, it would be wise to attempt to predict future noise levels at your site to determine the impact of plans for the surrounding area.

In selecting a quiet site, refer to Table 1, which gives desirable minimums for the distances from transportation system sound sources to a building or site. Also, look for exist-

ing natural and manmade sound barriers, examples of which are shown in Figs. 1 to 5 and in Figs. 6 and 7. Sites on rolling terrain separated from railways and highways by heavy, wide stands of trees are generally quieter than sites located in hollows or on flat, open ground.

Give preference to sites which are predominantly upwind of noise sources. At large distances the upwind side is generally quieter than the downwind side of a noise source. The wind tends to bend the sound path upward, as shown in Fig. 8, thereby reducing the sound energy that impinges on an upwind site.

Sites near hills or traffic intersections are generally unfavorable because of the acceleration, deceleration, and braking of vehicles, especially if the traffic includes heavy truck traffic like that in Fig. 9. Most of all, congested areas of heavy traffic should be avoided, as shown in Fig. 10.

TABLE 1 Distance Criteria for Elimination of Nonintrusive Transportation System Noise Sources

Noise source	There is a possibility of excessive noise due to this source if the building is:				
Highway	Within 1000 ft of any major roadway*				
Railroad	Within 3000 ft of any railway line				
Aircraft	Within the distances given below:				
International airport		Commercial or military airport		General aviation airport	
Distance to side of runway	Distance to end of runway	Distance to side of runway	Distance to end of runway	Distance to side of runway	Distance to end of runway
3½ miles	16 miles	2½ miles	9½ miles	1 mile	5 miles

*A major roadway is one with traffic of more than 50 autos per hour or more than 5 trucks per hour.

Fig. 1 Shielding by a wall.

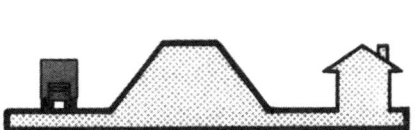

Fig. 2 Shielding by an earth berm.

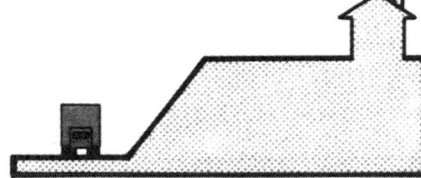

Fig. 3 Shielding by a depressed highway or railway.

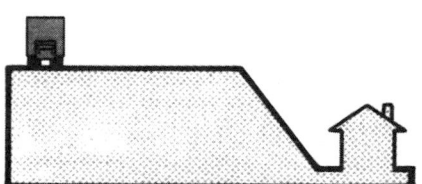

Fig. 4 Shielding by an elevated highway or railway.

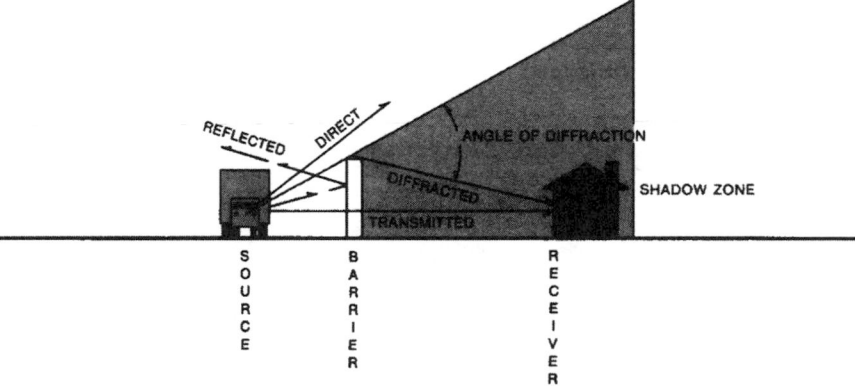

Fig. 5 Paths along which sound energy can travel from the source to the receiver. Also shown is the shadow zone and the angle of diffraction associated with the source-barrier-receiver geometry.

NOISE CONTROL

Fig. 6   Use of various noise barriers.

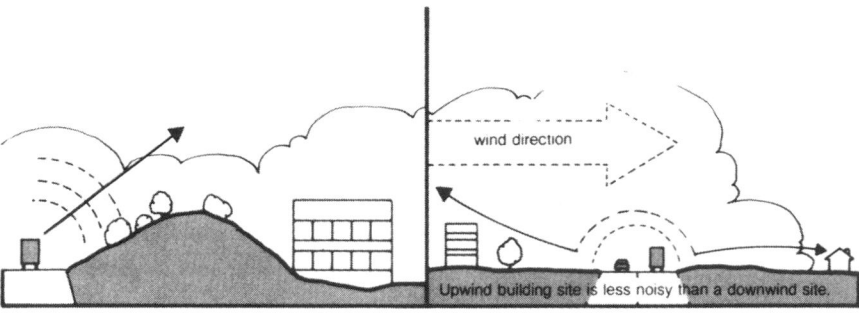

Fig. 7   Use of natural noise barriers.

Fig. 8   Selection of building sites relative to wind direction.

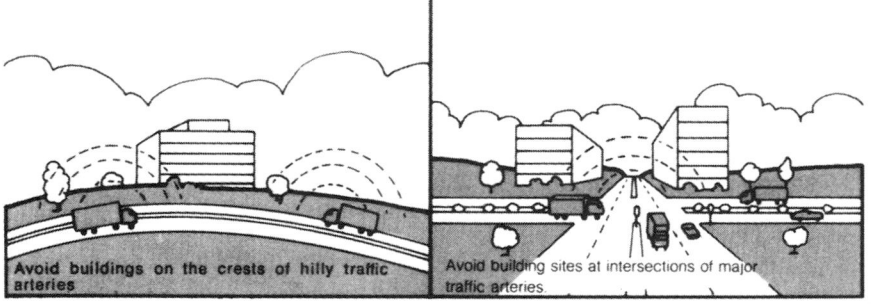

Fig. 9   Building sites near hilly traffic areas.

Fig. 10   Building sites near traffic junctions.

BUILDING LOCATION AND ORIENTATION

The thoughtful location and orientation of buildings on a site can aid in controlling noise. In order to determine whether or not there is apt to be a preferred "quiet" location on a site, it is generally necessary to consider several locations.

Buildings can be located in the quiet areas of a site with windowed facades facing the quiet areas, and with heavy, windowless walls facing the sources of sound. In general, the noise level near the facade of a building facing away from the predominant source of sound will be 3 dB less than near a facade facing the source. Acoustical shielding can be provided by the existing terrain, natural landscaping, or wooded areas.

If the building site is relatively close to a major highway or railway, and if the building is to be fairly long, two design concepts can be employed. One is to orient the building's major axis perpendicular to the direction of the highway or railway, and then to locate the noise-sensitive exterior rooms at the end of the building farthest from the roadway or track. A second design concept is to orient the building's axis parallel to the highway or railway and to provide materials having an extremely high sound intensity ratio (SIR) on the facade facing the noise source, while placing noise-sensitive rooms on the facade shielded by the building itself.

It is especially important that buildings not be parallel when located on both sides of an expressway in order to avoid multiple reflections of sound waves which increase sound levels. A random or staggered building layout or a cluster of buildings with no parallel building faces will avoid this problem of multiple reflections of sound waves between opposite buildings. Slightly curved buildings can be beneficial when the curvature is convex relative to the direction of the greatest noise source. U-shaped buildings or semienclosed courtyards provide areas for multiple reflections and should not be used as outdoor activity areas, because they tend to be quite reverberant and noisy. These layouts are shown in Fig. 11.

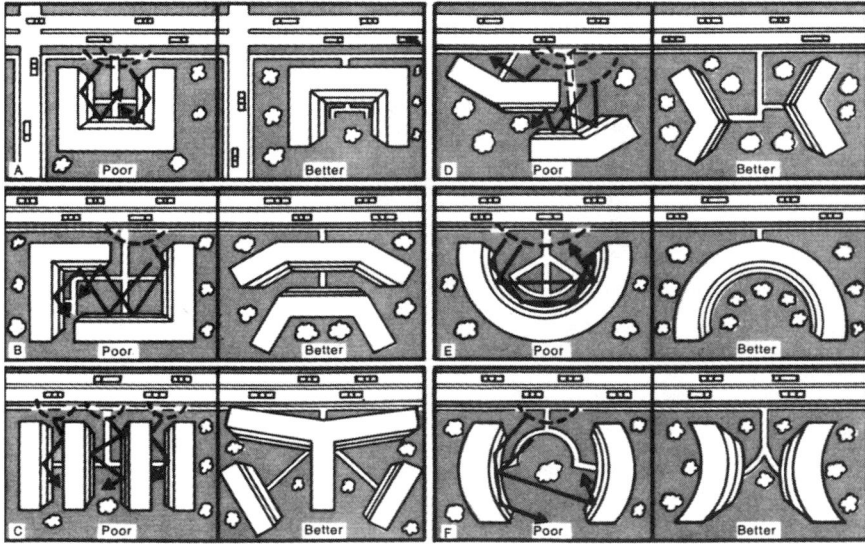

Fig. 11   Orientation of buildings on sites.

A variety of techniques can be used to control traffic noise in residential developments.

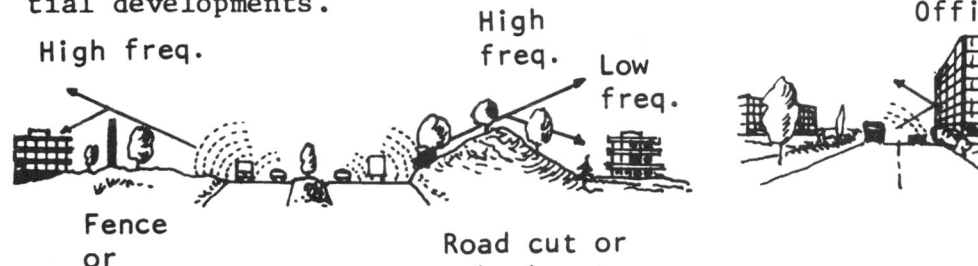

High freq.

High freq.

Low freq.

Office

Apartment

Fence or wall

Road cut or embankment

Buildings located in open areas are less noisy than in congested areas.

Traffic arteries between tall buildings are quite noisy.

Hollows or depressions are generally noisier than flat open land.

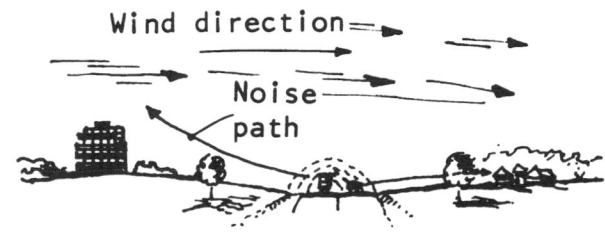

Wind direction

Noise path

Upwind building locations are less noisy than downwind locations.

Buildings located at intersections of major traffic arteries are extremely noisy due to accelerating, decelerating, and braking

Buildings located on the crests of hilly traffic arteries are very noisy due to low gear acceleration noise.

Fig. 12

NOISE CONTROL

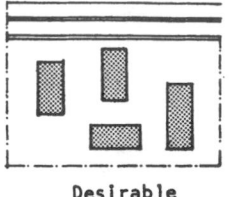 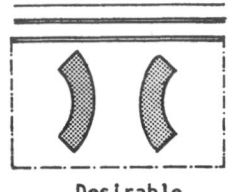 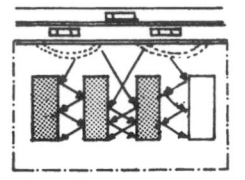

Desirable   Desirable   Undesirable

A layout which minimizes facing parallel building walls will significantly reduce unwanted noise.

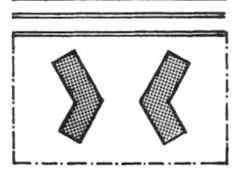 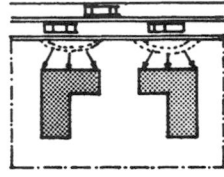 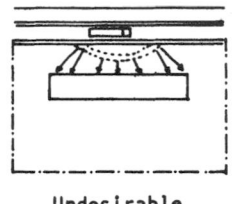

Desirable   Undesirable   Undesirable

Minimum building wall exposure to busy roads will reduce noise.

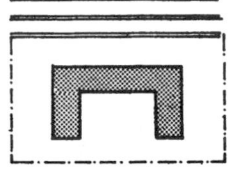 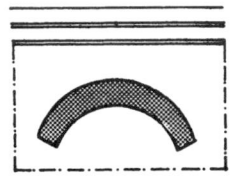 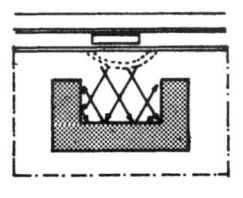

Desirable   Desirable   Undesirable

Orientation of building courts away from busy roads will reduce noise.

Fig. 13

BUILDING CONFIGURATION

Buildings can be arranged with noisy and quiet sides as previously mentioned, if the principal noise source is relatively near. Try to have rooms with low noise level criteria located on the quiet side, and rooms with high criteria located on the noisy side.

Often the most efficient methods of noise control in a development are available through appropriate site design techniques. Observance of the following principles will help to moderate or eliminate unwanted noise in residential developments.

Desirable   Undesirable

Fig. 14

Natural and manmade sound barriers should be utilized when they are available.

Fig. 15

Where there is ample space, trees can be used to moderate noise. A thick growth of leafy trees and underbrush reduces noise about 6 to 7 dB/100 ft (average over the audible frequency range). Low-frequency loss is 3 to 4 dB, high-frequency loss is 10 to 12 dB. A single row of trees is worthless as a noise barrier. Because of interreflection, multiple rows of trees are more effective. High-frequency loss is 3 to 4 dB.

WATER FOR FIRE PROTECTION

In any community, a strong reliable water supply is essential for fire fighting. The components which make up a water system include the source (wells, rivers, lakes), water works (filtration, purification, pumping), storage (ground level, elevated), and distribution (water mains, hydrants). It is assumed that an adequate and reliable local water system exists and the new development's distribution system will be connected to the local system.

The water system should as a minimum be able both to supply the domestic needs of the development as well as provide adequate flows for fire fighting.

The American Insurance Association (AIA) in its grading schedule has established fire flow requirements for the most congested high-value area of a community based on total population. Using this method, a community of 1000 population has a required fire flow of 1000 gal/min. Flows range upward to a maximum of 12,000 gal/min for a city of 200,000 population. Cities over 200,000 population are required to have an additional 2000 to 8000 gal/min available for a second fire. The fire flow requirements established by the American Insurance Association for other portions of a city are usually based upon the degree of congestion, the size of the buildings, occupancy, and the size of the fire department.

The American Insurance Association has established 500 gal/min as the minimum acceptable flow for the residential area provided a 30-ft separation between buildings is maintained. The flow requirements for residential areas are increased as the size and height of the buildings are increased and the separation is reduced.

Water for fire-fighting purposes should be available near all buildings constantly at pressures and quantities adequate for the anticipated hazard.

The following minimum fire flows should be available for the periods indicated for areas comprised of the following types of housing systems:

1. Single-family detached—separated by more than 30 ft—500 gal/min for 1 h.
2. Single-family detached, separated by less than 30 ft; single-family attached; multifamily low rise with not over six living units between fire walls if of combustible construction—750 gal/min for 1½ h.
3. All other residential buildings—1000 gal/min for 2 h.
4. Public buildings, depending on size, height, occupancy, and combustibility of structure and contents—750 to 2000 gal/min for 1½ to 4 h.

Fire flows should be available at a residual pressure of 20 lb/in^2 and should be in addition to peak domestic consumption demands.

Water mains, except to hydrants, should not be less than 8 in nominal inside diameter. Dead-end mains should not exceed 600 ft in length. Sectionalizing valves should be provided so that not more than 800 ft of piping will be affected by a break or shutdown.

Hydrants should not have less than a 6-in nominal inside diameter connection to the water main. A gate valve should be installed in the supply connection to each hydrant. Hydrants should be of a style and type in accordance with local practices and regulations.

Hydrants should be located at each street intersection with additional hydrants provided at midpoints along all streets, drives and cul-de-sacs, where the distance between intersections exceeds 500 ft. Spacing between hydrants serving multifamily dwellings and public buildings should be reduced to 300 ft.

A hydrant should be located within 200 ft of the standpipe siamese connection on all high-rise dwellings.

Hydrants should be placed within 5 to 10 ft of street or driveway pavement.

Hydrants should not be placed closer than 50 ft to the building being protected.

Water piping should be composed of, or lined with, a nontuberculating type of material in those geographic areas where tuberculation of water-main piping would reduce the effective water flow.

ACCESS TO SITE

One of the advantages of large-scale site development is the ability to predict the traffic volume more accurately based on land use and thus design the street for a specific level of traffic. For example, a street serving only single-family dwellings, with off-street parking and no through traffic, could be designed to minimal standards, as compared with a through street serving several high-rise clusters. Minimum standards for a street should take into consideration the possible need for fire apparatus to use the street for access to any of the buildings.

Streets not only serve the fire department as a means of access to buildings but also contribute to the separation between buildings of groups of buildings, thus limiting the potential for fire spread. The wider the street, the better the fire break it becomes. The system of streets serving a development must provide the public fire department with safe, reliable, and rapid access to all areas of the development as well as serve the day-to-day needs of the residents.

The Institute of Traffic Engineers has developed recommended practices for subdivision streets which express basic principles for street system layout as well as engineering standards and specifications for the individual design elements. Of the recommended principles, only a few specifically relate to fire department response and operations. Application of these principles to the design of a development site will result in a functional system meeting the needs of not only the residents but also the fire department.

The criteria established for site access recommend that a minimum of two separate roadways connect the development to public streets. This requirement not only will reduce the chance of all access to the development being accidentally blocked but also will permit emergency vehicles to enter the site from different directions. As an alternate, a single connecting roadway would be acceptable if the traffic lanes were separated by a median strip of sufficient width to reduce the chance of both lanes being blocked.

Streets should be planned, constructed, and maintained to permit unrestricted access into all parts of the developed area by emergency vehicles.

The developed area should be accessible from at least two separated connection roadways connected to one or more public streets or highways.

A single connection roadway may be used, provided it is of a divided design, i.e., opposing traffic lanes divided by a median strip. The median strip should be of sufficient width to allow passage of emergency vehicles when one of the roadways is blocked.

Every street within the developed area should be accessible from both of the connection roadways.

Streets leading to all parts of the developed area should be paved and be capable of supporting the heaviest axle loads permitted under applicable laws.

Width of pavement of streets should be not less than 22 ft where parking is prohibited on both sides, 29 ft where parallel curb parking is permitted on one side, and 36 ft where parallel curb parking is permitted on both sides. Off-street, diagonal, or 90° parking spaces should not encroach on the aforementioned street widths.

Street intersections should be lighted in accordance with the recommendations of the Illuminating Engineering Society. Minimum curb radius at intersections should be 20 ft. Intersections should meet at approximately a 90° angle.

Grades of streets should not exceed 10 percent, except that where winter icing is common, it should not exceed 8 percent. Grades up to 15 percent may be used for short distances, 600 ft or less, but should return to the lesser figures prior to street intersections.

All streets should be named or numbered, following the pattern established in the community. Duplication or close similarity to established street names should be avoided, unless the street is part of or an extension of an existing street. Street signs should be placed at each intersection and should be clearly readable both day and night.

Streets preferably should not be closed (terminate) at one end. Single egress-ingress (access) streets should be clearly marked at the entrance and should have a turnaround area with a radius of not less than 40 ft, plus 7 ft for each curb parking lane.

Large fire apparatus, particularly large water tankers, are likely to have individual axle loads in excess of 20,000 lb.

Dead-end streets are not recommended in other than low-density, single-family areas. Normal traffic densities may dictate circulating patterns, particularly in high-rise residential sections. The possibility of the only access being blocked by road repairs, fallen trees, or accidents should be considered wherever dead-end streets are used.

ACCESS TO BUILDINGS

Each residential structure and each structure used by the public must be located so as to assure unrestricted accessibility in case of emergency.

Driveways to be considered for fire department access should be at least 20 ft wide, paved, and designed for appropriate axle loads. On one-way drives, the 20 ft width may include one parallel parking lane.

Access to single-family dwellings, attached or detached, should be from a street or driveway which is not more than 100 ft from the dwelling.

Access to low-rise multifamily dwellings should be from a street or driveway which is not more than 75 ft from each street-level entrance. No low-rise multifamily dwelling should be located on a dead-end street or driveway more than 200 ft in length.

Access to high-rise multifamily dwellings should be from streets or driveways which are not more than 50 ft from each entrance. Access between such street or drive and the main entrances should not be restricted by parking spaces. No high-rise dwelling should be located on a dead-end street more than 200 ft in length.

Access to public buildings (schools, churches, community centers, stores, etc.) should be from two streets or driveways, one of

FIRE PROTECTION

which may be a dead end, or from a single, non-dead-end street or driveway. Streets and driveways needed for access should be within 50 ft of each main entrance.

Each building should be given a house num-ber, in conformity with local practice. The same number should not be used for more than one building on a street, and the numbering system should continue the pattern of the community. Where necessary to avoid confusion, each dwelling in single-family, attached housing should be given a separate house number. House numbers should be conspicuously posted on the street or access driveway side of the building.

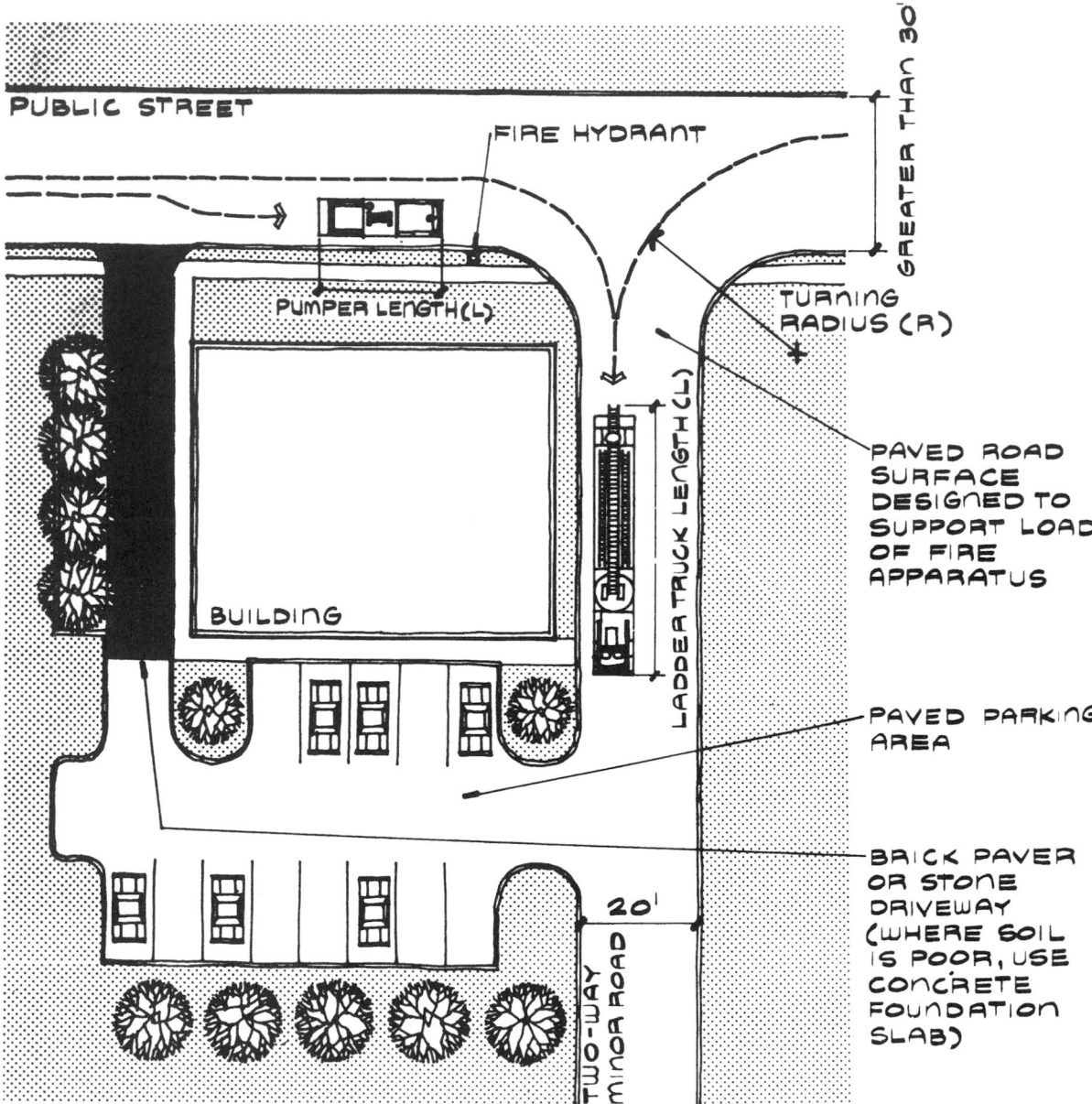

Note: Vertical clearance into below-grade parking areas, under pedestrian bridges, and the like should be at least 9 to 12 ft to provide access for fire apparatus.

Fig. 1 Site access for fire apparatus. The building site plan should provide adequate driveway widths, turning radii, and parking space on firm, level surfaces for fire apparatus. Avoid manmade and natural barriers that could interfere with movement of fire vehicles. Fire apparatus turning radii (R) typically vary from 28 to 40 ft and vehicle length (L) from 40 to 65 ft for ladder trucks and from 20 to 40 ft for pumpers.

SITE PLANNING FOR SOLAR ENERGY UTILIZATION

Site planning for the utilization of solar energy is concerned with two major issues: (1) access to the sun and (2) location of the building on the site to reduce its energy requirement. The placement and integration of the solar dwelling on the site in response to these concerns entails numerous decisions made at a variety of scales. The process may commence at a regional climatic and geographic scale and terminate at a specific location on the building site. At every scale, decisions regarding site selection, building orientation and placement, and site planning and design are made.

SITE SELECTION

At times a builder, developer, or designer may have the option of selecting a site or of determining the precise location on a larger site for the placement of the solar dwelling or dwellings. In such instances, the best site for effective solar energy utilization should be chosen by analyzing and evaluating carefully all of the following factors.
Geography of the area surrounding the site:
■ The daily and seasonal path of the sun across the site
■ The daily and seasonal windflow patterns around or through the site
■ The presence of earthforms which may block the sun or wind
■ The presence of low areas where cold air could settle
Topography of site:
■ Steepness of the slope—can it be built upon economically?

■ The presence of slopes beneficial or detrimental to energy conservation and solar energy utilization
Orientation of slopes on the site:
■ South-facing slopes for maximum solar exposure
■ West-facing slopes for maximum afternoon solar exposure
■ East-facing slopes for maximum morning solar exposure
■ North-facing slopes for minimum solar exposure
Geology underlying the site:
■ Depth and type of rock on the site
■ Unbuildable areas on the site
Existing soil potential and constraints:
■ Soils with engineering limitations unable to support structures
■ Soils with agricultural limitations, unable to support vegetation
Existing vegetation:
■ Size, variety, and location of vegetation which would impair solar collection
■ Building sites which would disturb existing vegetation to a minimum
■ Size, variety, and location of vegetation which would assist in energy conservation
Climatically protected areas on the site:
■ Areas protected at certain times of the day or year
■ Areas protected by topography
■ Areas protected by vegetation
Climatically exposed locations on the site:
■ Areas exposed to sun or wind
■ Areas exposed primarily in winter
■ Areas exposed primarily in summer

■ Areas exposed all seasons of the year
Natural access routes to and through the sites:
■ Adjacent streets for vehicular access to the site
■ Adjacent walkways for pedestrian access to the site
Solar radiation patterns on the site:
■ Daily and monthly
■ Seasonal
■ Impediments (e.g., vegetation that may cover the site or shadow buildable areas on the site)
Wind patterns on the site:
■ Daily and monthly
■ Seasonal
■ Impediments (e.g., thick vegetation or underbrush that may block air movement on or through the site)
Precipitation patterns on the site:
■ Fog movement, collection, or propensity patterns
■ Snow drift and collection patterns
■ Frost "pockets"
Temperature patterns on the site:
■ Daily and monthly
■ Seasonal
■ Warm areas
■ Cold areas
Water or air drainage patterns on or across the site:
■ Seasonal air or water flow patterns
■ Daily air or water flow patterns
■ Existing or natural impediments to air or water flow patterns
Tools for site analysis include air photos, topographic maps, climatic charts, or direct

ALTITUDE AND SCOPE

THE TOPOGRAPHY IS ANALYZED IN BOTH PLAN AND CROSS SECTION TO LOCATE BUILDABLE AREAS ON UPPER AND MIDDLE SLOPES.

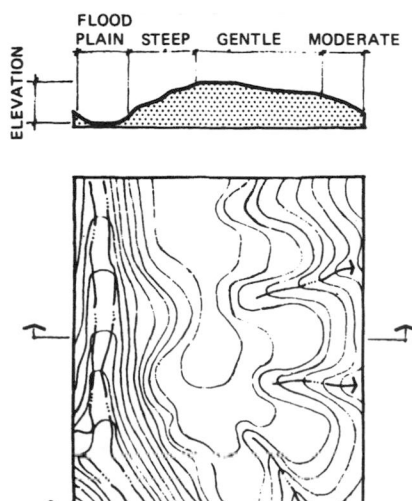

Fig. 1

ORIENTATION AND WINDS

THE SITE IS NEXT ASSESSED FOR AREAS ORIENTED IN A SOUTHERLY DIRECTION FOR MAXIMUM SOLAR EXPOSURE. ALSO, THE PREVAILING AND STORM WINDS WHICH MOVE REGULARLY OR OCCASIONALLY ACROSS THE SITE ARE PLOTTED.

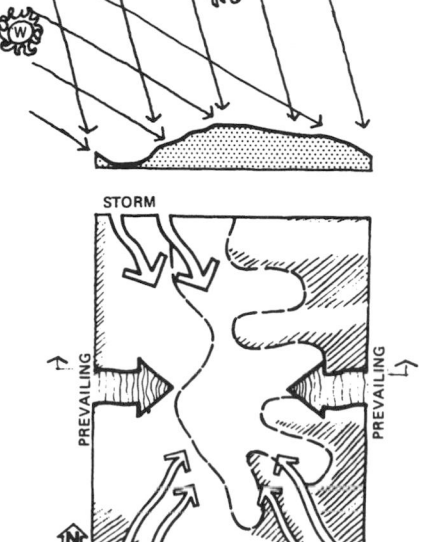

Fig. 2

VEGETATION AND MOISTURE

EXISTING VEGETATION AND MOISTURE PATTERNS ON THE SITE ARE RELATED TO THEIR POTENTIAL FOR ASSISTANCE IN THE CREATION OF SUN POCKETS AND FOR PROVIDING WIND PROTECTION. THE DENSITY AND TYPE OF VEGETATION ARE ANALYZED AND GRAPHICALLY DEPICTED IN ORDER TO GAIN AN UNDERSTANDING OF THE PATTERNS OF SHADE OR PROTECTION AND AIR OR MOISTURE FLOW.

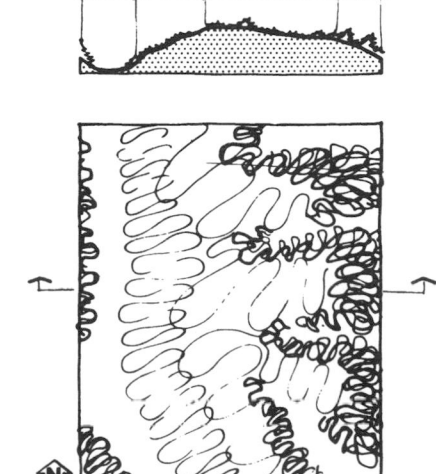

Fig. 3

SITE PLANNING FOR SOLAR ENERGY UTILIZATION

COMPOSITE SHOWING PREFERRED SITES

A COMPOSITE IS PREPARED FROM THE PRECEDING FACTORS SHOWING A RANKING OR A RATING OF THE PREFERRED SITES FOR PLACEMENT OF A SOLAR DWELLING ("a" BEING BEST, "b" NEXT BEST AND SO ON).

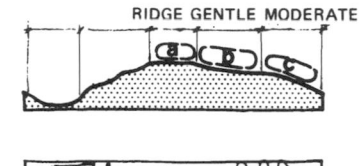

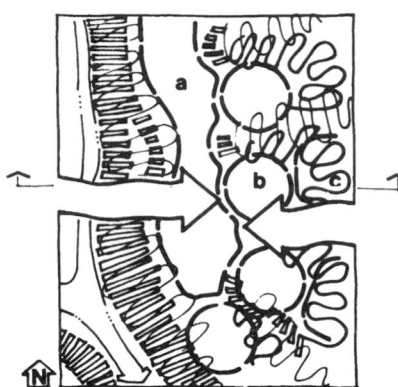

Fig. 4

observations on the site. Site selection at whatever scale must take into account the distinctive characteristics of the major climatic regions of the United States mentioned earlier. Once the data are collected and organized, they can be used to evaluate, rate, and eventually select a specific location or site for the placement of the dwelling, solar system, and other site-related activities.

A simplified example of the site analysis process for determining preferred locations for solar dwellings in western temperature climates is shown in Figures 1 through 4.

Siting and Orientation

Optimum solar energy utilization is achieved by the proper placement and integration of the dwelling, solar collectors, and other site-related activities and elements on the building site.

In addition to the dwelling, the most common activity areas found on residential sites include:
- Means of access (entrances to the site and to the dwelling)
- Means of service (service and storage areas)
- Areas for outdoor living (patios, terraces, etc.)
- Areas for outdoor recreation (play areas, pools, courts, etc.)

On sites where the dwelling(s) will be heated or cooled by solar energy, additional site planning factors must be considered for accommodating solar collection—by either dwelling or on-site collectors.

Each of the four major climatic regions in the United States has different siting and orientation considerations. The following is an overview of the major determinants for each region.

Cool Region

Maximum exposure of the dwelling and solar collector to the sun is the primary objective of site planning in cool regions. Sites with south-facing slopes are advantageous because they provide maximum exposure to solar radiation. Outdoor living areas should be located on the south sides of buildings to take advantage of the sun's heat. Exterior walls and fences can be used to create sun pockets and to provide protection from chilling winter winds.

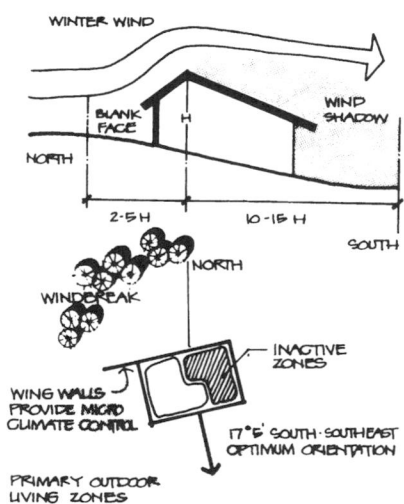

Fig. 5

Locating the dwelling on the leeward side of a hill or in an area protected from prevailing cold northwest winter winds—known as a window shadow—will conserve energy. Evergreen vegetation, earth mounds (berms), and windowless insulated walls can also be used to protect the north and northwest exterior walls of buildings from cold winter winds.

Structures can be built into hillsides or partially covered with earth and planting for natural insulation.

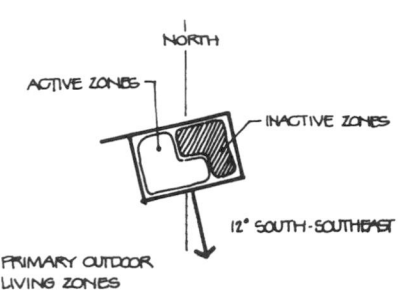

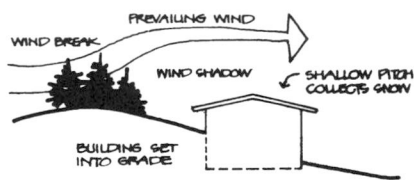

Fig. 6

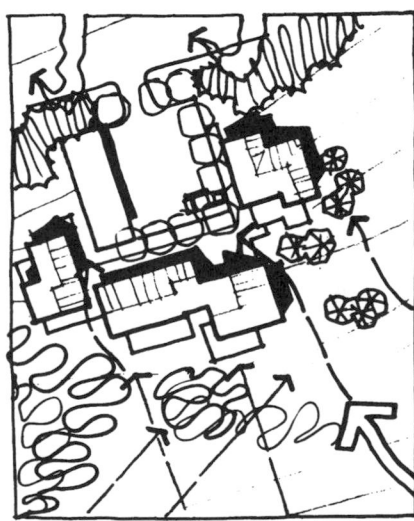

FOR CLUSTERED MULTI-FAMILY DWELLINGS, TERRACES AND OUTDOOR LIVING AREAS SHOULD BE INTEGRATED WITHIN THE BUILDING CLUSTERS. THIS WILL REDUCE COLD AIR MOVEMENT IN WINTER AND WILL CHANNEL AND DIRECT BREEZES IN SUMMER.

Fig. 7

Temperate Region

In the temperate region it is vital to assure maximum exposure of the solar collectors during the spring, fall, and winter months. To do so, the collector should be located on the middle to upper portion of any slope and should be oriented within an arc 10° either side of south. The primary outdoor living areas should be on the southwest side of the dwelling for protection from north or northwest winds. Only deciduous vegetation should be used on the south side of the dwelling, since this provides summer shade and allows for the penetration of winter sun.

The cooling impact of winter winds can be reduced by using existing or added landforms or vegetation on the north or northwest sides of the dwelling. The structure itself can be designed with steeply pitched roofs on the windward side, thus deflecting the wind and reducing the roof area affected by the winds. Blank walls, garages, or storage areas can be placed on the north sides of the dwelling. To keep cold winter winds out of the dwelling, north entrances should be protected with earth mounds, evergreen vegetation, walls, or fences. Outdoor areas used during warm weather should be designed and oriented to take advantage of the prevailing southwest summer breezes.

Hot, Humid Region

In hot, humid regions where the heating requirement is small, solar collectors for heating-only systems require maximum exposure to solar radiation, primarily during the winter months. During the remainder of the year air movement in and through the site and shading are the most important site design considerations. However, for solar cooling or domestic water heating, year-round solar collector exposure will be required. Collector orientation within an arc 10° either side of south is sufficient for efficient solar collection. Figures 11 through 13 illustrate a number of site planning and design considerations for solar energy utilization and energy conservation.

Hot, Arid Region

The objectives of siting, orientation and site planning in hot, arid regions are to maximize duration of solar radiation exposure on the collector and to provide shade for outdoor areas used in late morning or afternoon. To accomplish these objectives, the collector should be oriented south-southwest and the outdoor living areas should be located to the southeast of the dwelling in order to utilize early morning

STREETS AND PARKING AREAS SHADED WITH DECIDUOUS VEGETATION WILL ALSO CHANNEL SUMMER BREEZES AND REDUCE RADIATION REFLECTION WHILE ALLOWING THE SUN TO PENETRATE DURING THE WINTER.

Fig. 8

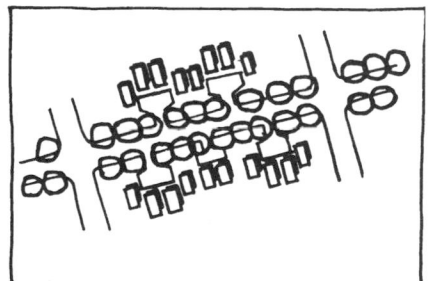

ROADWAYS CAN SERVE TO CHANNEL AND DIRECT DESIRABLE BREEZES OR BLOCK UNWANTED COLD WINDS. FOR TEMPERATE REGIONS, AN EAST-WEST STREET ORIENTATION CAN BEST SERVE THESE PURPOSES.

Fig. 9

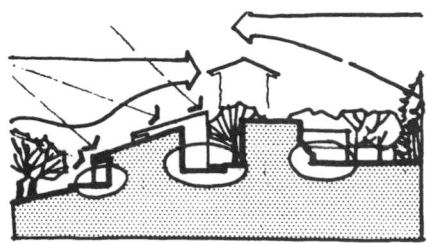

"SUN POCKETS" OR "SOLAR NOOKS" LOCATED ON THE SOUTH SIDES OF BUILDINGS MAY HELP EXTEND PERIODS OF SEDENTARY OUTDOOR LIVING DURING COOLER MONTHS.

Fig. 10

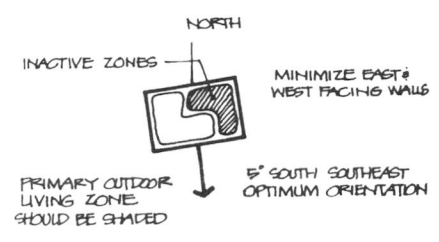

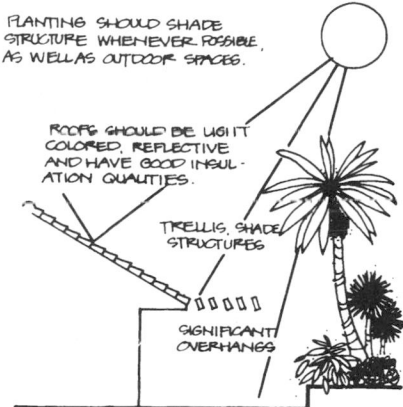

Fig. 11

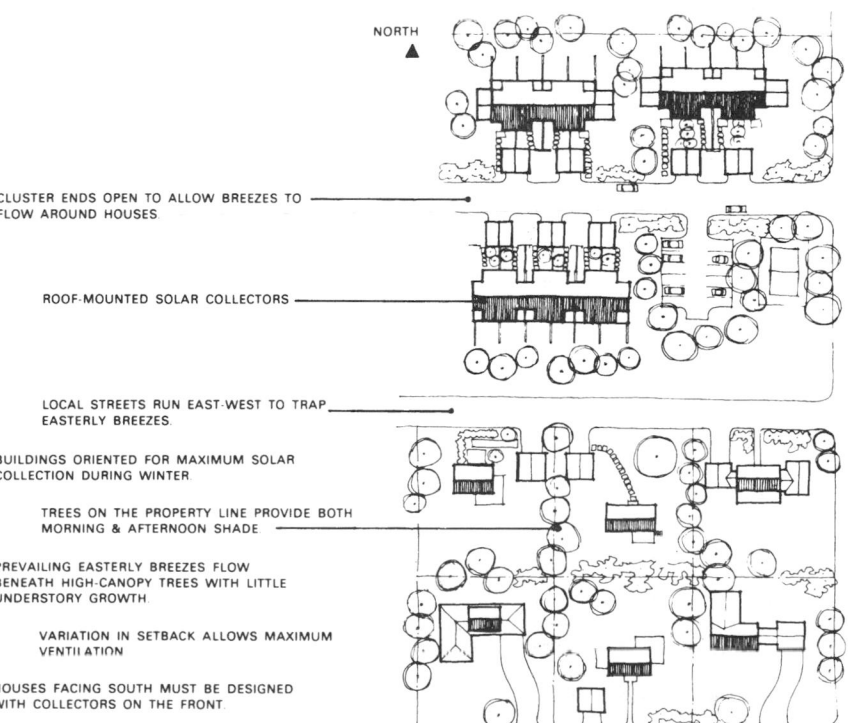

CLUSTER ENDS OPEN TO ALLOW BREEZES TO FLOW AROUND HOUSES.

ROOF-MOUNTED SOLAR COLLECTORS

LOCAL STREETS RUN EAST-WEST TO TRAP EASTERLY BREEZES.

BUILDINGS ORIENTED FOR MAXIMUM SOLAR COLLECTION DURING WINTER.

TREES ON THE PROPERTY LINE PROVIDE BOTH MORNING & AFTERNOON SHADE.

PREVAILING EASTERLY BREEZES FLOW BENEATH HIGH-CANOPY TREES WITH LITTLE UNDERSTORY GROWTH.

VARIATION IN SETBACK ALLOWS MAXIMUM VENTILATION.

HOUSES FACING SOUTH MUST BE DESIGNED WITH COLLECTORS ON THE FRONT.

Fig. 12   Site plan for hot, humid region illustrating principles of orientation and placement of buildings, streets, and planting.

SITE PLANNING FOR SOLAR ENERGY UTILIZATION

sun and take advantage of shade provided by the structure in the afternoon.

Indoor and outdoor activity areas should take maximum advantage of cooling breezes by increasing the local humidity level and lowering the temperature. This may be done by locating the dwelling on the leeward side of a lake, stream, or other bodies of water. Also, lower hillside sites will benefit from cooler natural air movement during early evening and warm air movement during early morning.

Excessive glare and radiation in the outdoor environment can be reduced by providing:

- Small shaded parking areas or carports
- Turf adjacent to the dwelling unit
- Tree-shaded roadways and parking areas
- Parking areas removed from the dwelling units
- East-west orientation of narrow roadways

Exterior wall openings should face south but should be shaded either by roof overhangs or by deciduous trees in order to limit excessive solar radiation into the dwelling. The size of the windows on the east and west sides of the dwelling should be minimized in order to reduce radiation heat gain into the house in early mornings and late afternoons. Multiple buildings are best arranged in clusters for heat

absorption, shading opportunities, and protection from east and west exposures.

Each climatic region has its own distinctive characteristics and conditions that influence site planning and dwelling design for solar energy utilization and for energy conservation. Table 1 suggests the general objectives of site planning and dwelling design for each climatic region as well as some methods for achieving these objectives. The chart reflects the seasonal trade-offs made between climatic optimums. In all cases, a detailed analysis should be undertaken to identify the site trade-offs between optimums for solar energy collection and optimums for energy conservation.

INTEGRATION OF THE BUILDING AND SITE

Ideally, a building is designed for the specific site on which it is to be placed. Commonly, however, a building design may be replicated with only minor changes on different sites and in different climates.

Site planning solutions are not as easy to replicate, because each site has a unique geography, geology, and ecology. The most appropriate way to integrate any building and its site

is first to analyze the site very carefully, and then to place the building on the site with a minimum of disruption and the greatest recognition and acceptance of the site's distinctive features.

It is possible, however, to provide general techniques for integrating buildings with their sites. Historically, a number of such techniques have evolved, among which are indigenous architectural characteristics adapted to local site conditions, architectural extensions to the building such as walls and covered walks, the use of native materials found on the site, and techniques for preserving or enhancing the native ecology.

In each climatic region, guidelines can be determined to help apply the many techniques available for integrating a building and its site in ways appropriate to the particular region. These guidelines can be particularly helpful in maximizing energy conservation and increasing the opportunity for successful use of solar heating and cooling.

Detailed Site Design

The detailed design of a site for optimum solar energy utilization and energy conservation entails the use of a variety of types of vegeta-

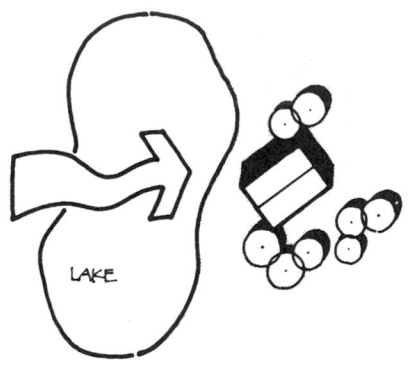

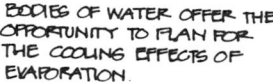

BODIES OF WATER OFFER THE OPPORTUNITY TO PLAN FOR THE COOLING EFFECTS OF EVAPORATION.

Fig. 13

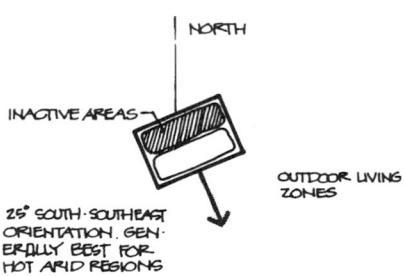

TABLE 1 Site Orientation Chart

Objectives	Cool	Temperate	Hot, humid	Hot arid
Adaptations	Maximize warming effects of solar radiation. Reduce impact of winter wind. Avoid local climatic cold pockets	Maximize warming effects of sun in winter. Maximize shade in summer. Reduce impact of winter wind but allow air circulation in summer	Maximize shade. Maximize wind	Maximize shade late morning and all afternoon. Maximize humidity Maximize air movement in summer
Position on slope	Low for wind shelter	Middle-upper for solar radiation exposure	High for wind	Low for cool air flow
Orientation on slope	South to southeast	South to southeast	South	East-southeast for afternoon shade
Relation to water	Near large body of water	Close to water, but avoid coastal fog	Near any water	On lee side of water
Preferred winds	Sheltered from north and west	Avoid continental cold winds	Sheltered from north	Exposed to prevailing winds
Clustering	Around sun pockets	Around a common, sunny terrace	Open to wind	Along east-west axis, for shade and wind
Building orientation*	Southeast	South to southeast	South, toward prevailing wind	South
Tree forms	Deciduous trees near building, evergreens for windbreaks	Deciduous trees nearby on west. No evergreens near on south	High canopy trees. Use deciduous trees near building	Trees overhanging roof if possible
Road orientation	Crosswise to winter wind	Crosswise to winter wind	Broad channel, east-west axis	Narrow, east-west
Materials coloration	Medium to dark	Medium	Light, especially for roof	Light on exposed surfaces, dark to avoid reflection

*Must be evaluated in terms of impact on solar collector, size, efficiency, and tilt.

SITE PLANNING FOR SOLAR ENERGY UTILIZATION

tion, paving, fences, walls, overhead canopies, and other natural and manmade elements. These elements are used to control the solar exposure, comfort, and energy efficiency of the site and the dwelling.

The materials used in site design have the ability to absorb, store, radiate, and deflect solar radiation as well as to channel warm or cool air flow. For instance, trees of all sizes and types block incoming and outgoing solar radiation, deflect and direct the wind, and moderate precipitation, humidity, and temperature in and around the site and dwelling. Shrubs deflect wind and influence site temperature and glare. Ground covers regulate absorption and radiation. Turf influences diurnal temperatures and is less reflective than most paving materials. Certain paving surfaces, fences, walls, canopies, trellises, and other site elements may be located on the site to absorb or reflect solar radiation, channel or block winds, and expose or cover the dwelling or solar collector.

DECIDUOUS TREES CAN BE USED FOR SUMMER SUN SHADING OF THE DWELLING AND YET ALLOW WINTER SUN PENETRATION THROUGH THEIR BARE BRANCHES FOR SOLAR COLLECTION. BARE BRANCHED DECIDUOUS TREES DO, HOWEVER, CAST A SUBSTANTIAL SHADOW AND WILL REDUCE COLLECTION EFFICIENCY. EVERGREENS SHADE COLLECTORS HEAVILY ALL YEAR.

Fig. 14

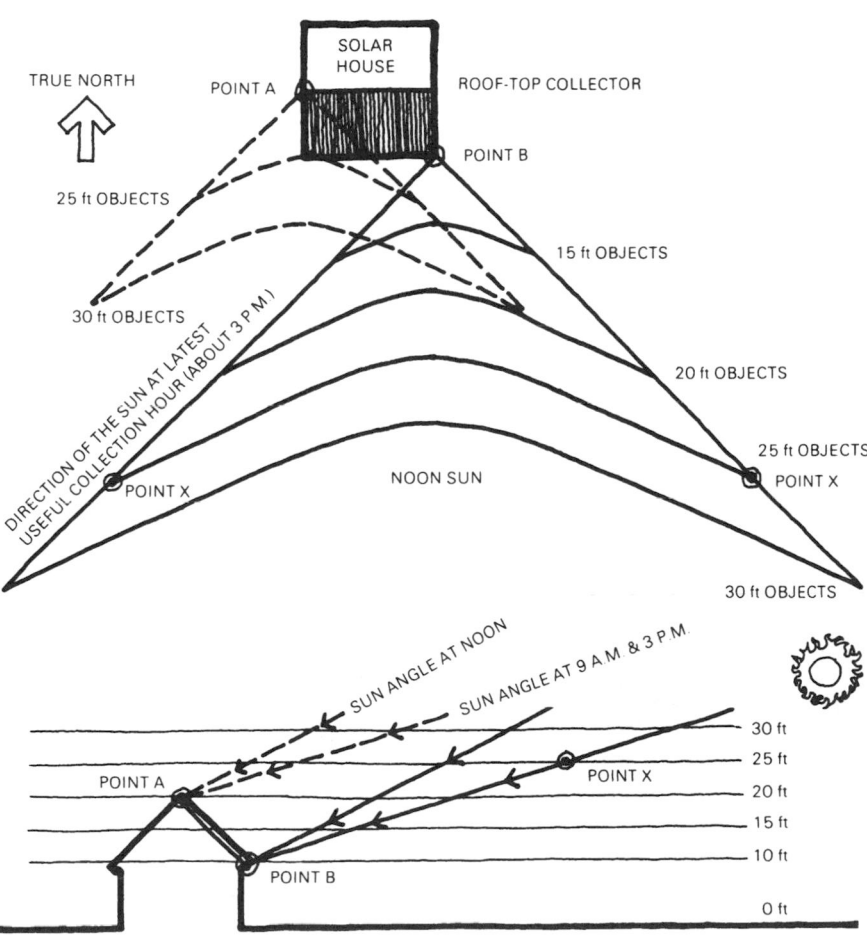

SOLAR INTERFERENCE BOUNDARIES OF INDIVIDUAL POINTS
LATITUDE 40
DECEMBER 21 WINTER SOLSTICE

EVERY POINT ON THE COLLECTOR FOR A GIVEN LATITUDE AND DAY OF THE YEAR, HAS A SET OF SOLAR INTERFERENCE BOUNDARIES. THESE DEFINE THE AREAS WITHIN WHICH OBJECTS OF A GIVEN HEIGHT ABOVE A FLAT SITE WILL CAST A SHADOW ON THE COLLECTOR. AREAS BEFORE AND AFTER USEFUL COLLECTION HOURS ARE NOT INCLUDED.

SOLAR INTERFERENCE BOUNDARIES ARE DRAWN BY PLOTTING IN PLAN THE POINTS OF INTERSECTION BETWEEN THE SUN ANGLES AND THE VARIOUS ELEVATIONS ABOVE THE ZERO GRADE. (SUCH AS POINT X)

Fig. 15

SITE PLANNING FOR SOLAR ENERGY UTILIZATION

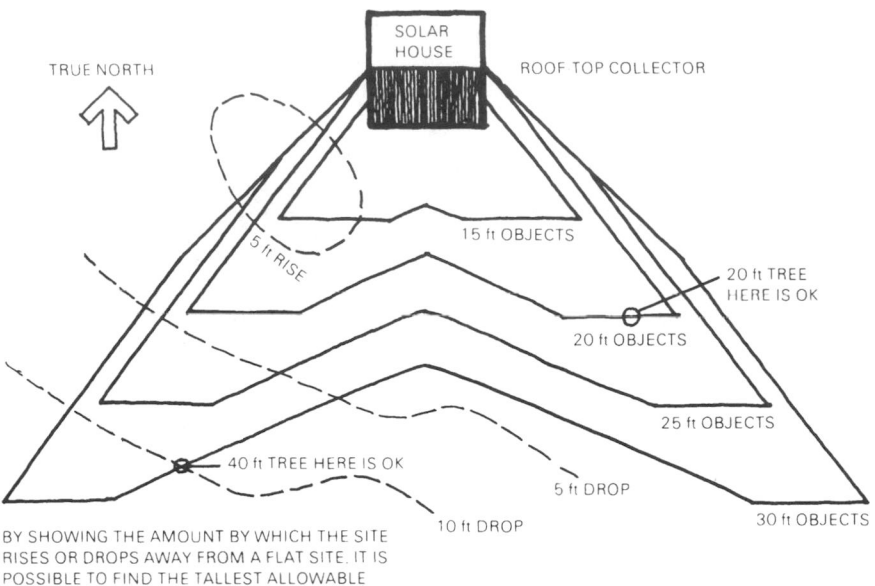

TRUE NORTH

SOLAR HOUSE

ROOF TOP COLLECTOR

5 ft RISE

15 ft OBJECTS

20 ft TREE HERE IS OK

20 ft OBJECTS

25 ft OBJECTS

40 ft TREE HERE IS OK

5 ft DROP

10 ft DROP

30 ft OBJECTS

BY SHOWING THE AMOUNT BY WHICH THE SITE RISES OR DROPS AWAY FROM A FLAT SITE. IT IS POSSIBLE TO FIND THE TALLEST ALLOWABLE OBJECT AT ANY POINT IN THE SITE

COMPOSITE SOLAR INTERFERENCE BOUNDARIES FOR ENTIRE COLLECTOR
LATITUDE 40
DECEMBER 21   WINTER SOLSTICE

A COMPOSITE PLAN OF THE SOLAR INTERFERENCE BOUNDARIES FOR EVERY POINT ON THE COLLECTOR CAN BE MADE RELATIVELY SIMPLY

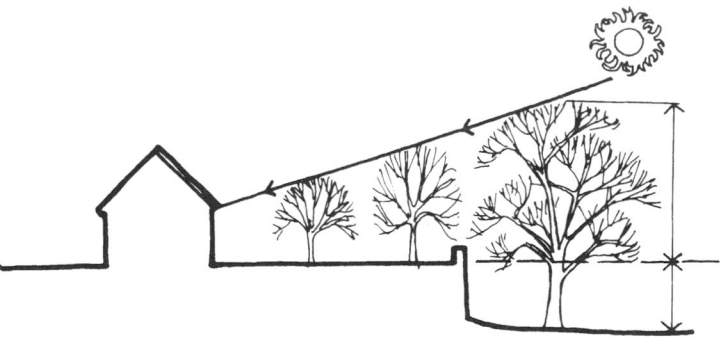

IF THE SITE FALLS AWAY TO THE SOUTH, LARGER TREES CAN BE PLANTED WITHOUT SHADING THE COLLECTOR

THE EXTRA HEIGHT ALLOWABLE CAN BE SHOWN IN PLAN

Fig. 16

MULTIPLE BRAKING EFFECT

Fig. 17   Multiple braking effect.

MULTI-LAYERED VEGETATION INCLUDING CANOPY TREES AND UNDERSTORY TREES OR SHRUBS PROVIDES A MULTIPLE BRAKING EFFECT, SUBSTANTIALLY DECREASING THE WIND VELOCITY MOVING OVER A SITE.

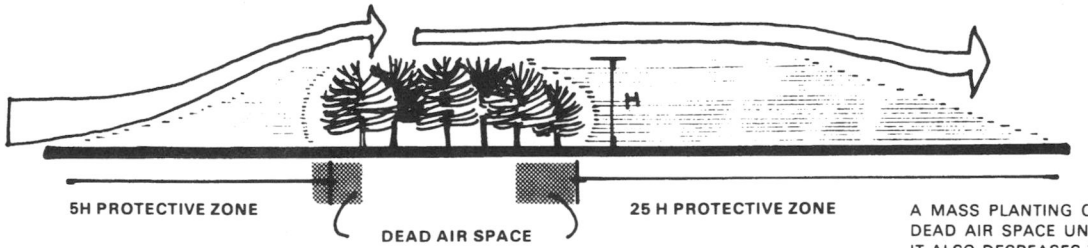

5H PROTECTIVE ZONE

DEAD AIR SPACE

25 H PROTECTIVE ZONE

A MASS PLANTING OF TREES PROVIDES A DEAD AIR SPACE UNDER AND AROUND ITSELF. IT ALSO DECREASES THE AIR VELOCITY 5 TIMES ITS HEIGHT TO WINDWARD AND 25 TIMES ITS HEIGHT TO LEEWARD OF THE PLANTING.

Fig. 18

SITE PLANNING FOR SOLAR ENERGY UTILIZATION

SHORT ZONE OF PROTECTION   LONG ZONE OF PROTECTION

Fig. 19

PLANTING ON THE LEEWARD SIDE OF A HILL
SUBSTANTIALLY INCREASES THE DOWNWIND
ZONE OF REDUCED AIR VELOCITY, WHILE
PLANTING ON THE WINDWARD SIDE
CORRESPONDING DECREASES THE ZONE.

FENCES, WALLS OR VEGETATION CAN BLOCK
NATURAL AIR FLOW PATTERNS. CARE MUST BE
TAKEN DURING SITE DESIGN TO PROVIDE THE
NECESSARY VISUAL CONTROL WHILE AVOIDING
ADVERSE CLIMATIC CONDITIONS. AS COOLER
AIR FLOWS DOWNHILL IN THE EVENING,
FENCES, WALLS OR PLANTINGS SHOULD NOT
UNINTENTIONALLY DAM THIS FLOW AND THUS
CREATE A COLD AIR POCKET WHERE IT IS NOT
WANTED.

VEGETATION MAY BE PRESERVED OR PLACED
IN SUCH A WAY AS TO CHANNEL OR BLOCK
DAILY OR SEASONAL AIR FLOW PATTERNS.

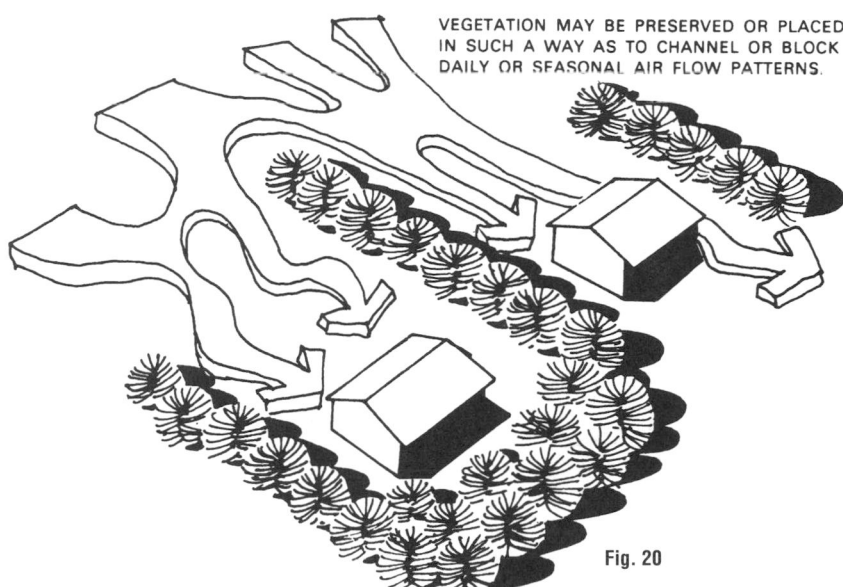

Fig. 20

VEGETATION PROPERLY PLACED CAN DEFLECT
RATHER THAN DAM COLD AIR FLOW

DWELLING UNPROTECTED FROM COLD AIR
FLOW

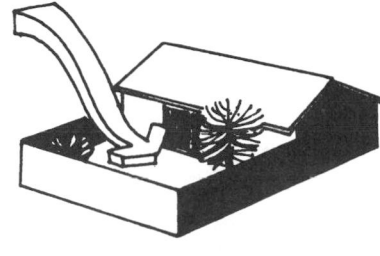

COLD AIR TRAPPED BY FENCE

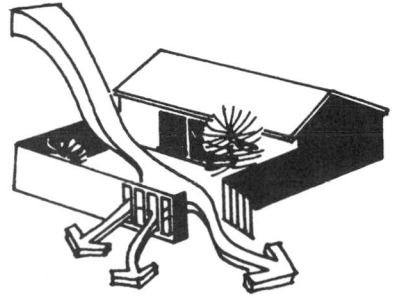

FENCE DESIGN CAN PROVIDE FOR COLD AIR
DRAINAGE

Fig. 21

SITE PLANNING FOR SOLAR ENERGY
UTILIZATION AND ENERGY CONSERVATION
BASED ON THE VEGETATION, TOPOGRAPHIC
AND CLIMATIC ANALYSIS OF THE SITE
SUGGESTS SOUTHERN EXPOSURE, NORTHERN
PROTECTION AND UNIMPAIRED AIR MOVEMENT
FOR MULTI-FAMILY HOUSING PROJECT
LOCATED IN A COOL CLIMATE.

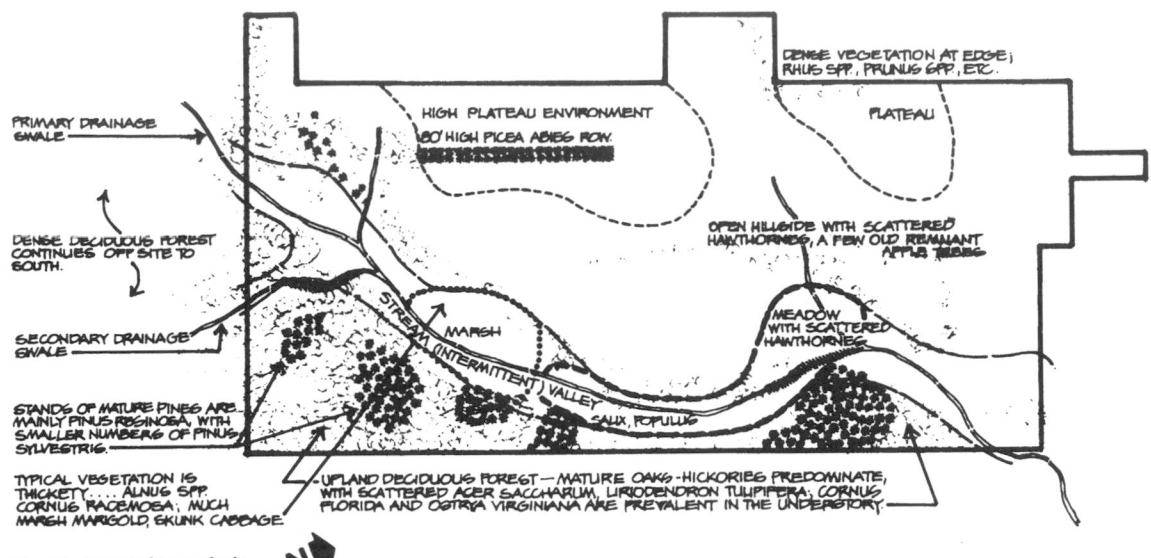

Fig. 22 Vegetation analysis.

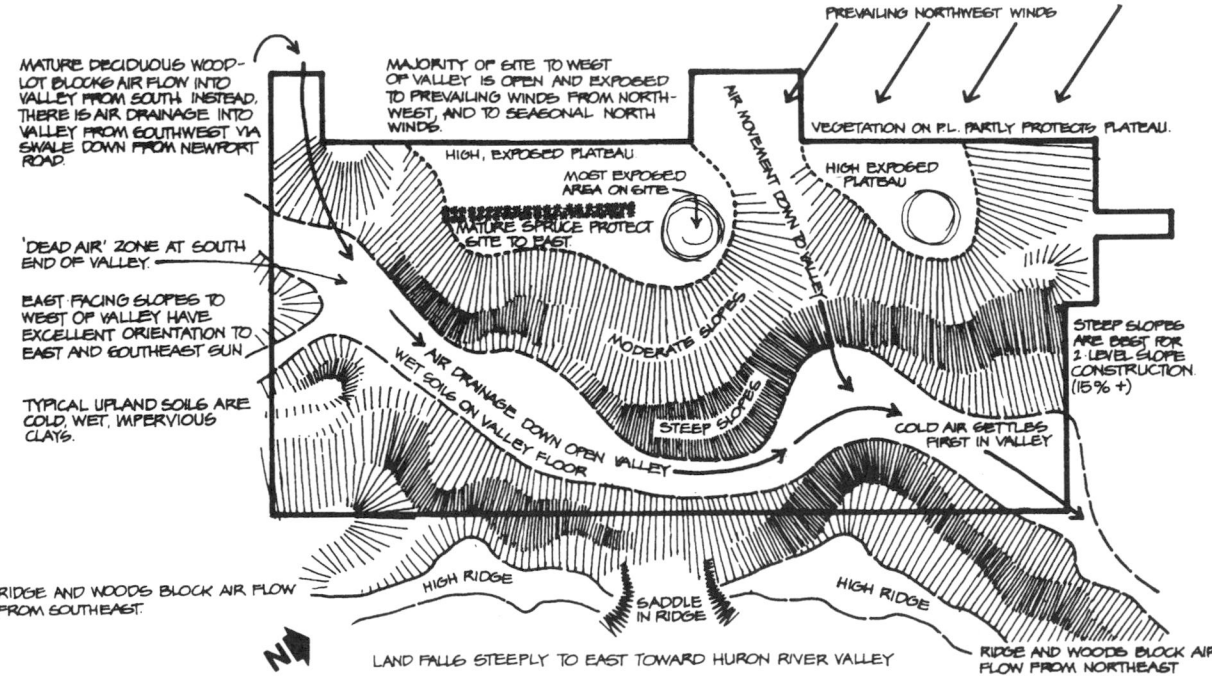

Fig. 23 Topographical analysis.

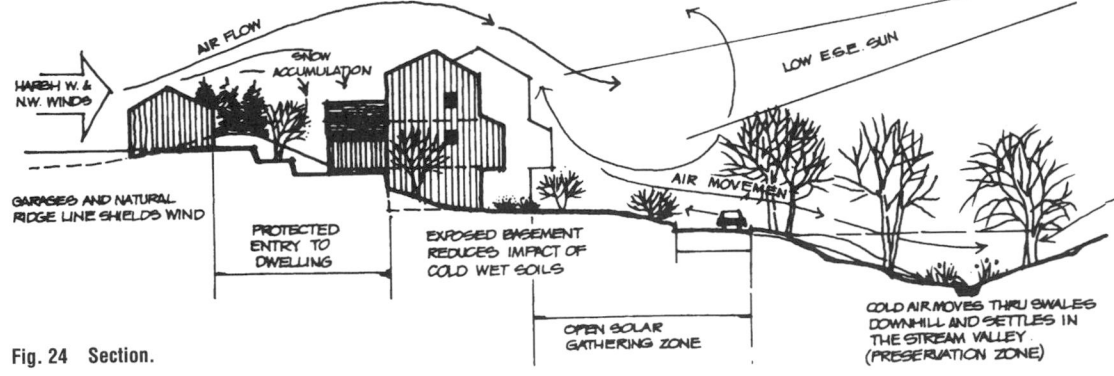

118   **Fig. 24** Section.

SITE PLANNING FOR SOLAR ENERGY UTILIZATION

The following site plan and dwelling design concept are indicative of the solar energy utilization and energy-conservation considerations for the cool and temperate regions. The techniques employed include:

■ The use of windbreak planting
■ The orientation of road alignment with planting on either side to channel summer breezes

■ The location of units in a configuration suggested by the topography
■ The use of the garage to buffer the dwelling from northwest winter winds
■ The use of berms to shelter outdoor living terraces
■ The use and location of deciduous trees to block or filter afternoon summer sun

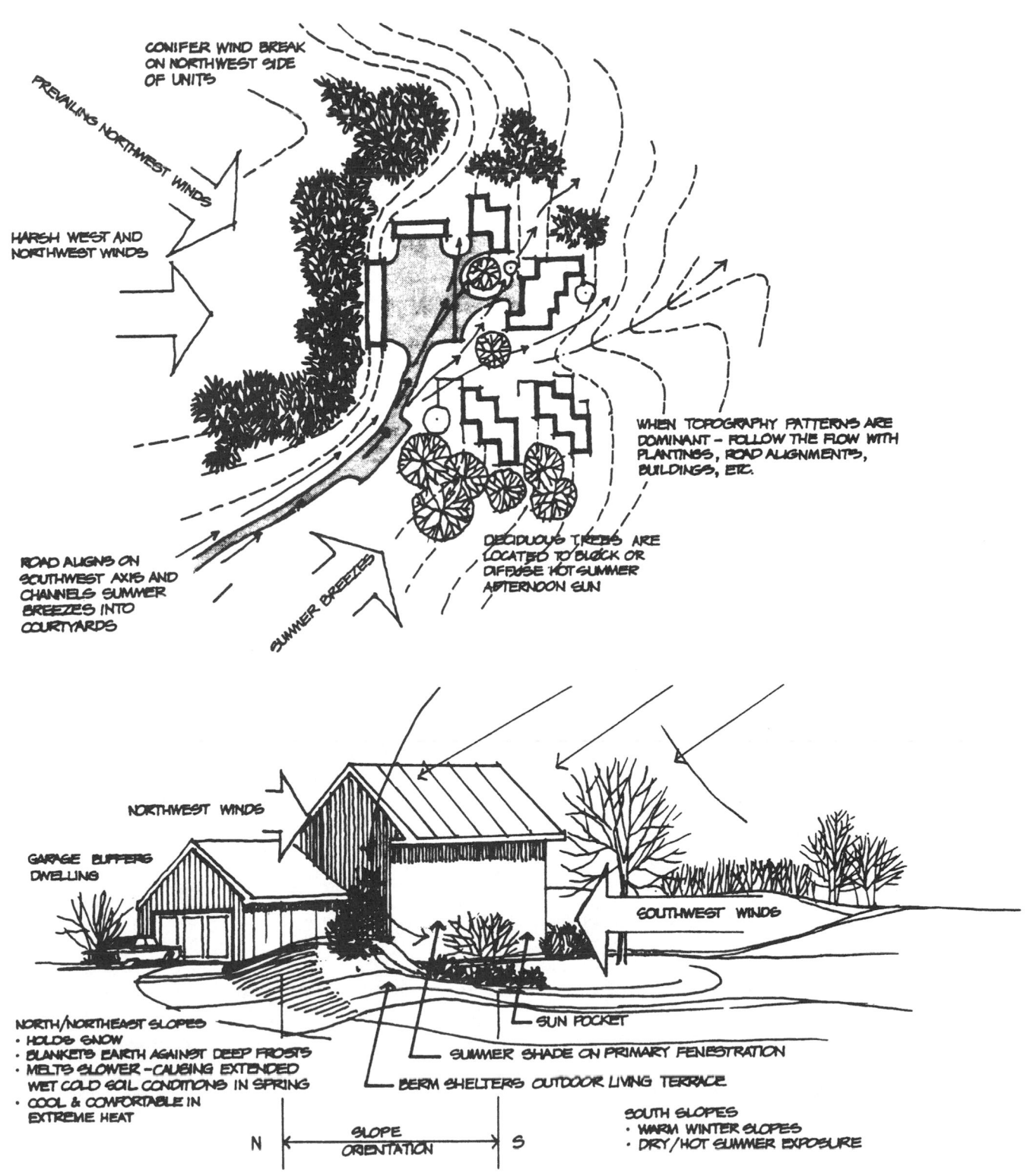

Fig. 25

SITE SECURITY ANALYSIS

This section presents the criteria that comprise the site analysis and the logic behind the analysis. These criteria can be used to determine the extent to which a site's design and development contribute to the vulnerability of residents by either increasing chances of victimization, causing residents to be fearful, or causing residents to alter their behavior by pulling back from their environment, thereby preventing the formation of the neighborhood ties that would make the community more resistant to crime.

In reviewing these criteria, it is important to realize that site design is only one of the factors that can contribute to a criminal act. The event itself is usually the result of a complex series of factors that may include not only deficiencies in the site but also the motivation of the criminal, features of the lifestyle of the victim that might place the person in special risk (such as heavy drinking, moving through the site at odd hours), and the absence of police protection or a supportive neighborhood.

Fear of crime is equally complex and is not the result of any one factor. Prior exposure to crime, a person's age, health, and general outlook on life can all play a role in how fearful one is about becoming a victim. This is also the case with altered behavior, the third measure of vulnerability. How people feel toward their neighbors and their general environment, and what they think the environment tells them about themselves all influence the degree to which they will choose to interact and to participate actively in their environment.

It is this complexity that led to the development of the comprehensive approach to security planning, an approach which recognizes that no one improvement or intervention can be expected to impact decisively on such a complicated event. The comprehensive approach is built on the awareness that the solution must be as complex as the problem, and it seeks to counteract each of the dimensions of the problem.

The site, therefore, is only one factor that affects a resident's vulnerability to crime. But it is an important one. The design, organization, and development of a site can determine the extent to which opportunities for victimizations occur and can evoke fear or offer reassurance to residents. A site can be organized so it encourages people to get to know one another and work together, or it can be laid out in such a way as to make the development of close relationships difficult, if not impossible.

The importance of site layout and design as a factor that can affect victimization rates, residents' fear of crime, and altered behavior has been confirmed many times by surveys.

THE CRITERIA

Six basic criteria comprise the site analysis. These criteria, discussed below, measure the extent to which a site contributes to a crime problem.

Penetrability

This criterion examines how access to the site is structured and controlled. In many public or low-income developments a security problem is created because access is uncontrolled. That is, there are no environmental suggestions as to how the site should be entered or how traffic should move through it: people enter and move through the site without crossing any barriers suggesting that they are entering someone's environment.

Ideally, the entrances to a site and its buildings should be structured and clearly marked. Where access is not desired, formal or symbolic barriers should exist. Generally, entrances to a site should be arranged so those coming and going can be easily seen. Such surveillance can be performed by a security guard or police patrol, but there should also be opportunity for casual surveillance—surveillance conducted by residents from their homes or from sitting areas around the entrances. The entryways should also be emphatic enough so that they clearly tell people coming in that they are entering someone's environment; that there is a difference between the street and the site.

Entrances to spaces inside the site are also important. A thoughtfully developed site usually has different spaces that are intended to be used for different purposes. Good design often strives to provide a hierarchy of space which moves from public spaces open to everyone, to semi-private spaces intended for specific groups of people, and finally to private spaces intended for individual households, all with well-defined entrances and boundaries. The significance here is that it is important not to overlook opportunities to establish them where they do not exist.

Many sites do not meet these standards. Many have no formal entrance points or boundaries that announce a residential environment. Many projects are penetrable from 360° on the compass. They can be entered from all sides and angles, and once on the site, there is no indication of how people are to move through the area. In these sites there is usually an absence of formal entryways, as well as barriers to restrain entry at undesired points.

The concept of penetrability aims at identifying these deficiencies. To use it properly, it is necessary to study a site extensively and to determine how people move onto the site and through it. By studying the streets and determining which ones are used the most and by whom, an opportunity may be seen to close some of them and channel traffic through a main entrance point. A look at the boundaries of the housing development may reveal whether there are any real or symbolic barriers that prevent people from entering. A study of worn areas and frequently used shortcuts that cut through people's private space may indicate that these should be closed and traffic forcefully redirected. In other instances, it may be best to yield to people's insistence on a route and formalize the path—particularly if it leads to important destinations such as bus stops or shopping areas that cannot realistically be changed.

From the design standpoint, a controlled entranceway usually has, as shown in Fig. 2, an outer lobby to which visitors may freely enter and an inner lobby to which visitors and residents are admitted after being checked for identification. A well-designed entranceway permits the guard to survey both the lobby and elevator waiting areas. Figures 3 and 4 provide additional examples of well-designed entrance and lobby areas.

Penetrability is only one factor that should be used to analyze a site, but it can be a critical one. For if anyone can enter a site without being informed by its layout and design that the site is a special environment that belongs to residents, then control over the whole site can be lost, making it difficult, if not impossible, to protect interior and private spaces.

Territoriality

Territoriality refers to the extent to which a housing development's design and layout encourage residents to take control of the site—in other words, to act on the common need to control the space upon which they live.

Good site design encourages territoriality. It invites residents to "claim" space adjacent to their units and, as a group, to assume control of semiprivate areas, such as courtyards. Inadequate site design and development does not do this. In these instances only two kinds of space may exist, public space, which anyone can occupy without challenge, or very private interior space located inside the unit. This means that the only line of defense, a fragile one at best, is the door or window. Ideally, there should be several lines of defense and definition beginning with public space, then semiprivate, and finally private space.

There are several ways to encourage territoriality. Semiprivate spaces, as discussed earlier, can be created by defining courtyards and structuring access to various parts of the site. It is also important to encourage residents to take control of space immediately adjacent to their unit.

Opportunities for Surveillance

This aspect of the site analysis involves assessing the site in terms of the extent to which activities occurring in public and semiprivate space can be observed. In assessing a site, it is important to recognize that there are two types of surveillance. The first is casual or informal surveillance—situations where the design of the site allows residents to casually or informally observe the activities of their neighbors or their families. Sites with good opportunities for casual surveillance usually avoid dark, labyrinthine pathways, instead favoring pathways that lead in front of houses where people are likely to be. Bus stops, lobbies, and entrances to elevators are all arranged so that people in these places can be observed by others. Good site design also provides opportunities for parents to observe their children at play. Kitchen windows overlooking play areas help accomplish this objective.

The significance of these features cannot be overestimated. They provide "eyes and ears" that can see or hear if help is needed; they reassure people that they are not alone and isolated, and this reduces fear. As a result, more people use the site, which in turn improves security because criminals will rarely act if they think they will be seen.

The other kind of surveillance is the formal surveillance undertaken by security guards and police. It is important that attention be paid to this element. A site should be carefully examined, in concert with the security guards and police who patrol it, to determine which design

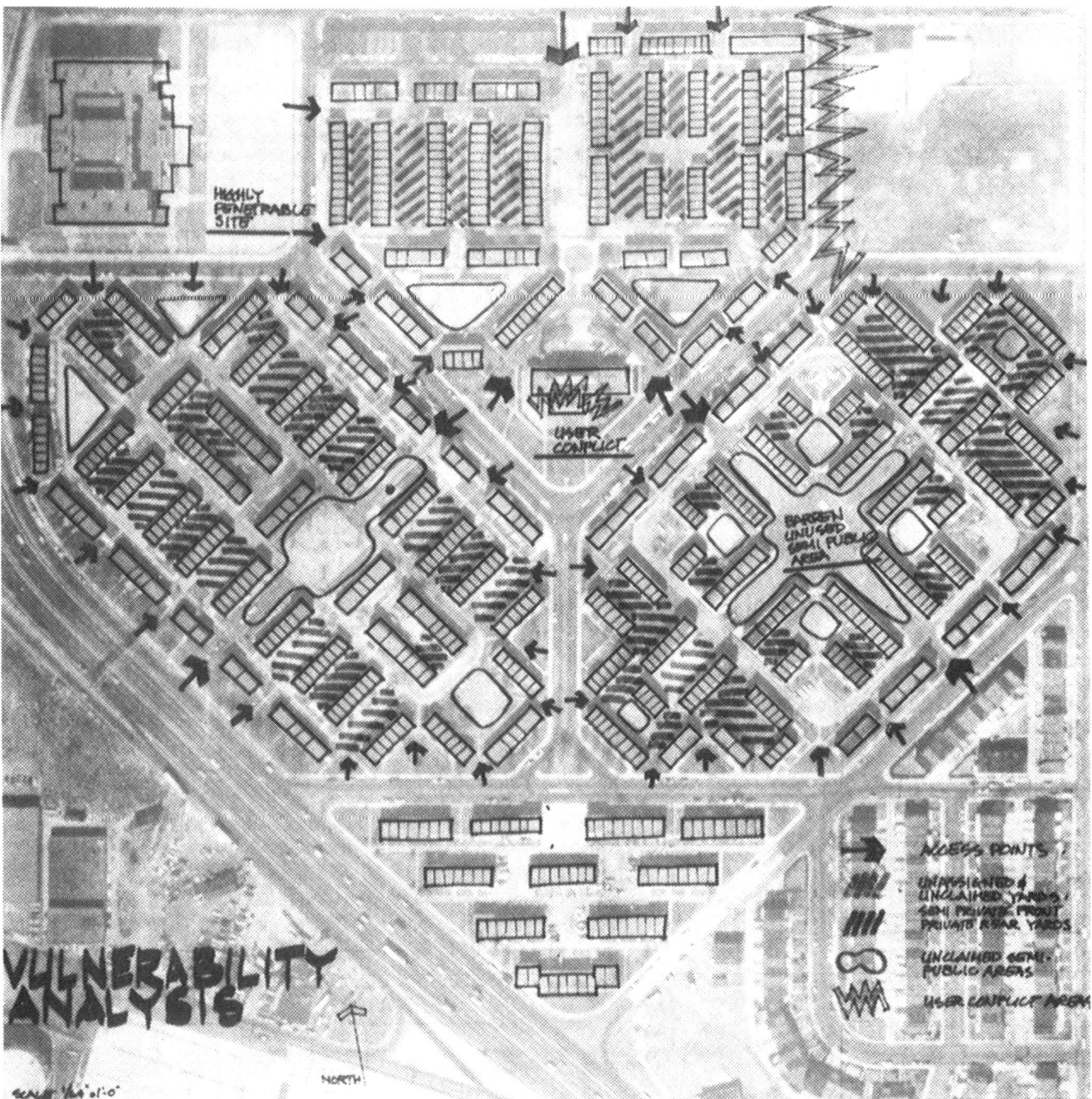

Fig. 1 Site vulnerability analysis.

or development features inhibit formal surveillance. Sometimes it can be high walls that block the view of patrolling guards or police, or provide a hiding place for burglars or troublemakers.

Another feature of some sites is that many have major areas that security guards and police officers cannot get to by car or reach by walking a reasonable distance from their cars. (It is a fact of life that police generally like to stay in, or at least near, their vehicle. They view it as their communication link to both

information and help.) In some large housing developments, an area that is frequently inaccessible is the ballfield. This was the case in Nickerson Gardens, Los Angeles, and similarly in East Terrace Homes, San Antonio, Texas, where a large open field in the center of the project could not be patrolled by car. In both of these cases, drug dealers would sell in the center of the large space, secure in the knowledge that they could see anyone coming in plenty of time to throw away their goods.

In Los Angeles, WBA proposed a service roadway to cut through the ballfield. As shown in Fig. 5, it was placed so it would not interfere with the layout of the ballfield yet would allow the police to patrol the field and to get their vehicles close to any part of the field.

If opportunities for informal and formal surveillance can be developed, fear of crime as well as people's sense of isolation can be reduced, and many criminal acts can be deterred. Because criminals rarely act where

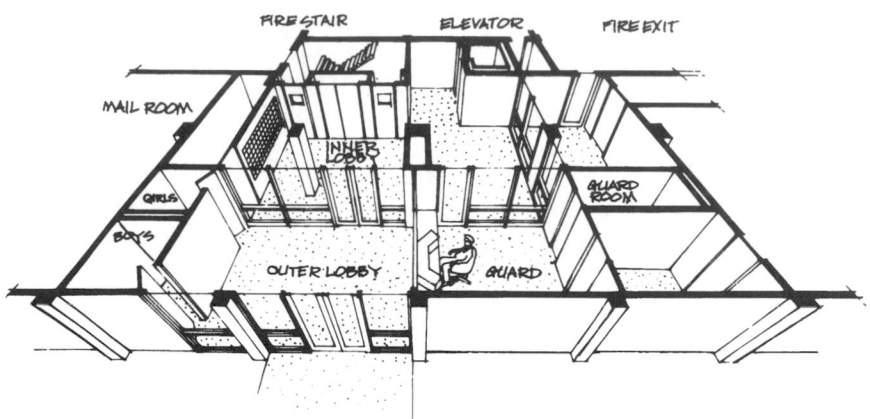

Fig. 2   Lobby design A.

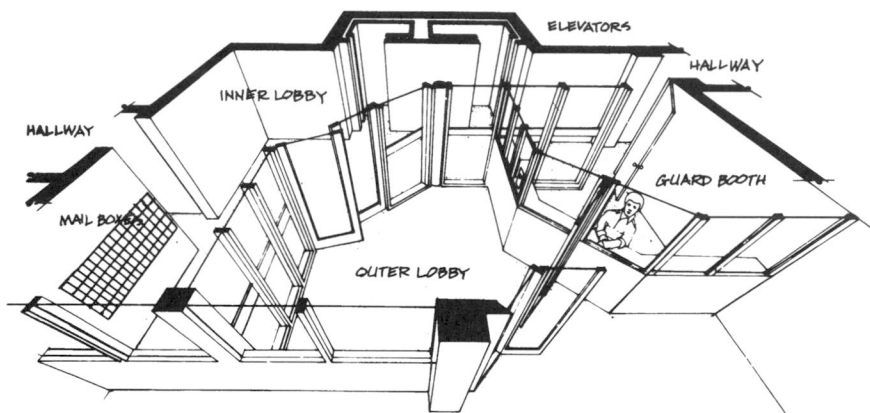

Fig. 3   Lobby design B.

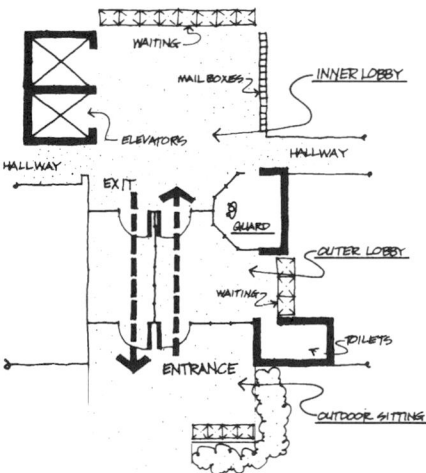

Fig. 4   Lobby design C.

they feel there is a good chance of being observed, surveillance is an important aspect of a site's layout and design.

Unassigned Space

Unassigned spaces are those which individuals or groups of residents have not been able to claim for their own use. Generally this is because these spaces lack environmental cues suggesting how the space is to be used and who should control it. Frequently there is no formal or informal supervision or control over these spaces, and their dimensions are poorly defined. Unassigned spaces may vary in size, location, and character; they may be front or rear yards that are unclaimed by tenants for their own use, or larger open spaces.

Large amounts of unassigned space can be a major vulnerability. Because these spaces are unprotected and uncared for, they provide opportunities for residents and outsiders to engage in mischievous and antisocial activities that would not be tolerated in situations where residents control and maintain their own territory.

Good design has little unassigned space. An effort is made to have several levels of space based on need. Spaces are clearly designated as public space, semiprivate space, and private space. Each of these kinds of space is organized so that it is clear who is to use it and for what purposes. The goal is thus to encourage residents to lay claim to space. Good design avoids creating large ambiguous, anonymous spaces that residents cannot control because such spaces frequently end up being surrendered to outsiders or to disruptive elements within the community or they simply become vacant eyesores. In such cases, these spaces, instead of enhancing the site, cause people to retreat from their environment and from each other.

In employing this criterion of the site analysis, it is important to look for spaces that are vacant or undeveloped, that no one seems to care for, or which are being used as gathering points for inappropriate behavior. As will be discussed in the next section, these areas must be identified and mapped, and alternatives developed to give them structure and a clearly assigned use.

One example of unassigned use was found in Arthur Capper Dwellings in Washington, D.C., a public housing development for which WBA prepared a comprehensive security plan. Part of the project consisted of low-rise buildings framed around large open spaces (see Fig. 6). Field observation revealed that many of these center spaces were completely unlit and unassigned. They were vacant wastelands in the center of more than one hundred family units.

The solution proposed to resolve the problem involved enriching these interior spaces and assigning them a clearly defined use. As shown in Fig. 7, enclosed yards were proposed for some of the units as well as parking, teen play, and sitting areas. Access was also structured to reduce penetrability.

The criterion of unassigned space involves identifying and cataloging areas on the site that lack definition and assignment and which as a consequence expose residents to fear and risk and inhibit their efforts to form neighborhood relationships.

Design Conflicts

Design conflicts occur when two incompatible activities are located next to one another without sufficient separation, or when two incompatible activities are forced to compete for the

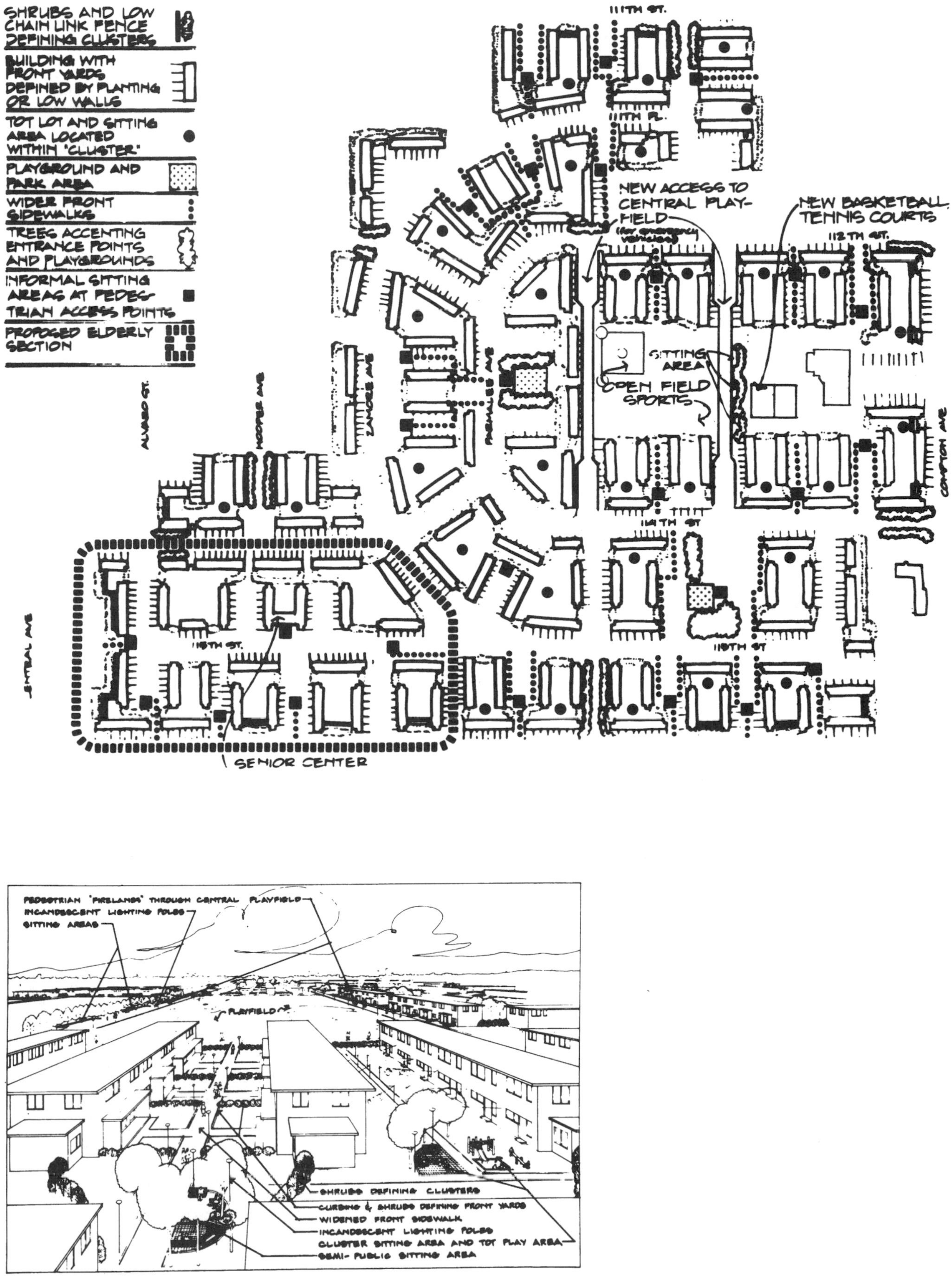

SHRUBS AND LOW CHAIN LINK FENCE DEFINING CLUSTERS

BUILDING WITH FRONT YARDS DEFINED BY PLANTING OR LOW WALLS

TOT LOT AND SITTING AREA LOCATED WITHIN "CLUSTER"

PLAYGROUND AND PARK AREA

WIDER FRONT SIDEWALKS

TREES ACCENTING ENTRANCE POINTS AND PLAYGROUNDS

INFORMAL SITTING AREAS AT PEDESTRIAN ACCESS POINTS

PROPOSED ELDERLY SECTION

111TH ST.
111TH PL.
112TH ST.

NEW ACCESS TO CENTRAL PLAYFIELD

NEW BASKETBALL TENNIS COURTS

SITTING AREA

OPEN FIELD SPORTS

ALVARO ST.
HOOPER AVE.
SANGER AVE.
PRESS AVE.
COMPTON AVE.
CENTRAL AVE.

114TH ST.
113TH ST.
115TH ST.
116TH ST.

SENIOR CENTER

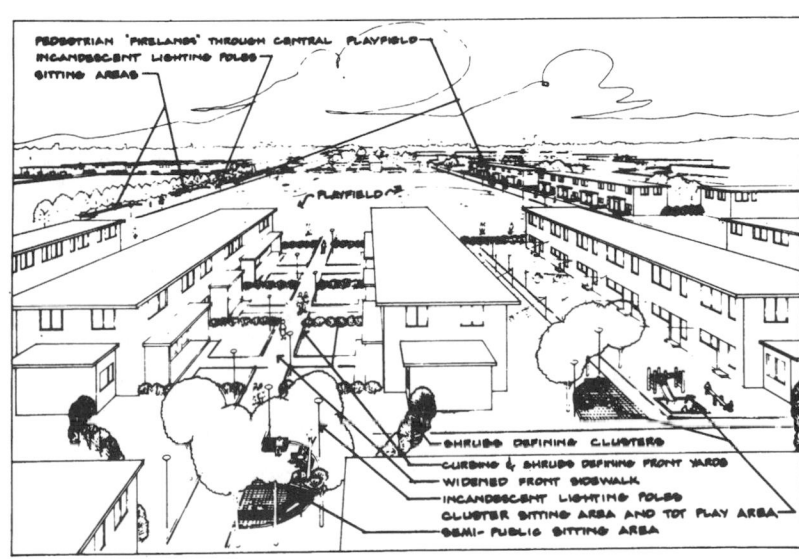

PEDESTRIAN "FIRELANES" THROUGH CENTRAL PLAYFIELD
INCANDESCENT LIGHTING POLES
SITTING AREAS

PLAYFIELD

SHRUBS DEFINING CLUSTERS
CURBING & SHRUBS DEFINING FRONT YARDS
WIDENED FRONT SIDEWALK
INCANDESCENT LIGHTING POLES
CLUSTER SITTING AREA AND TOT PLAY AREA
SEMI-PUBLIC SITTING AREA

Fig. 5   Proposed site plan.

SITE SECURITY ANALYSIS

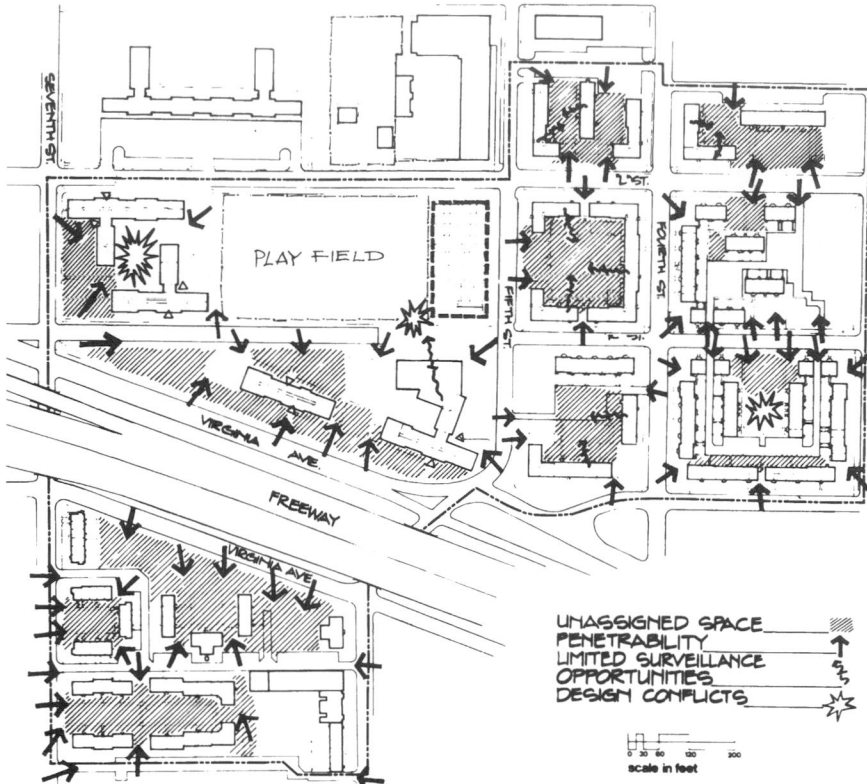

Fig. 6 Site security analysis.

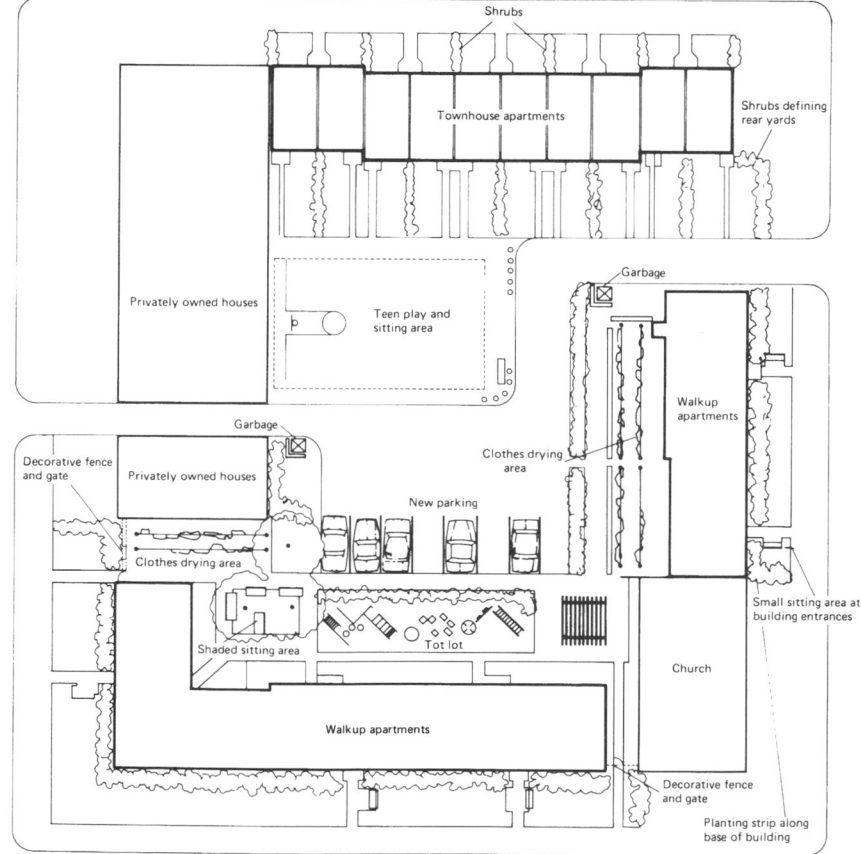

Fig. 7 Typical yard development.

same space, resulting in continued tension among residents.

The following examples illustrate the various kinds of design conflicts that can arise:

■ Pathways to a building for the elderly that lead them right next to an active teenage recreational area: These instances are particularly unfortunate because elderly persons frequently feel intimidated by rough play even though no threat is intended. The best solution in this case is to organize pathways so that older persons do not have to routinely pass close to teenage play areas to gain access to their homes unless they choose to. If that is impossible, buffers should be established between the walkways and the play areas.

■ A tot lot and basketball court located side by side: This situation can be extremely harmful to the best use of the site because teenagers have a tendency to expand their control over adjacent areas, with the result that younger children get pushed away. Tot lots next to teenage activity areas are also ill advised because the mothers of young children are frequently made uncomfortable by the rough talk and aggressive behavior of teenagers. In order to be successful, a tot lot must be not only fun for the child but comfortable for the mother as well; otherwise she will not bring her child there.

■ Entrances to high-rise buildings used as lounging areas: This frequently occurs where there are inadequate recreational facilities for both adults and young people or where entrances are not guarded or periodically surveyed by police. In such cases, residents are made uneasy while crossing these lounging or "hangout" areas, and it can be a source of tension in the community. The solution lies in establishing a controlled entranceway and in providing more attractive lounging and recreational spaces for residents.

■ A large ballfield borders directly on front yards of adjacent residences.

■ Elderly persons attending a lunch program at a community center use the same entrances as young people and must share lounge areas.

The effects of the above-noted design conflicts on crime, fear of crime, and altered behavior can be substantial. Competition over use can lead to quarrels which can end in assaults or worse. The same design conflict can also cause one user group to withdraw and do without needed facilities. In these cases, people are pulled apart rather than encouraged by the design of their environment to work together and coexist peacefully.

Design conflicts can also contribute to fear of crime whether entirely rational or not. An elderly person who must pass close to a rough play area cannot help but feel some sense of anxiety; the same is true of a mother who must pass through an entrance to her building that is crowded with lounging teenagers.

There are a number of solutions available for design conflicts. These will vary depending upon the site, the exact nature of the problem, and the resources available. In general, solutions should be directed at separating and containing competing users. This can be done by establishing buffers, such as walls, shrubs, or trees, or by providing an alternative site for one of the activities—relocating a conflicting activity, for example. Solutions can also include rerouting people around a design conflict. If the problem is outside the site, a change in the location of the entrance to the site might be considered. Petitions and inquiries to local zoning boards can also be initiated to prevent the location of businesses that would be in conflict with the needs of a residential environment.

Influence of Surrounding Neighborhoods

Neighborhood influences on a site are important to identify. Although neighborhoods can offer a setting which provides a variety of support services such as commercial stores, parks, and transportation lines, some neighborhoods generate pedestrian patterns or attract undesirable groups which impact negatively on the security of a site. Many times the location of a site in a neighborhood can make it especially vulnerable. Security in a housing development can be influenced if two different kinds of housing developments abut one another, such as an elderly next to a family project, a juxtaposition that could impact negatively since the vigorous activities and noise of the children in the family development could intrude upon the adjacent outdoor spaces of the elderly development and discourage their intended use by the elderly residents. A housing development located next to a public ballfield or basketball court may also be negatively impacted by the large groups of teenagers and young adults who are drawn to the facilities and spill over onto the site as they wait their turn or watch the games. The location of bus or subway stops can affect security on a site, sometimes by encouraging nonresidents to take shortcuts through the site; or, if located in poorly protected areas, these stops can be fear-evoking places of victimization.

Thus it is important to understand how the environment on a site is influenced by broader neighborhood factors. Neighborhood influences can determine how people do move through a site and what changes should be made to increase their security and that of the residents. If a site is located next to a crime-generating area, this factor must be understood and dealt with. If the site is isolated and people must move through pathways to other activities that expose them to crime or make them fearful, then these routes must be made safer.

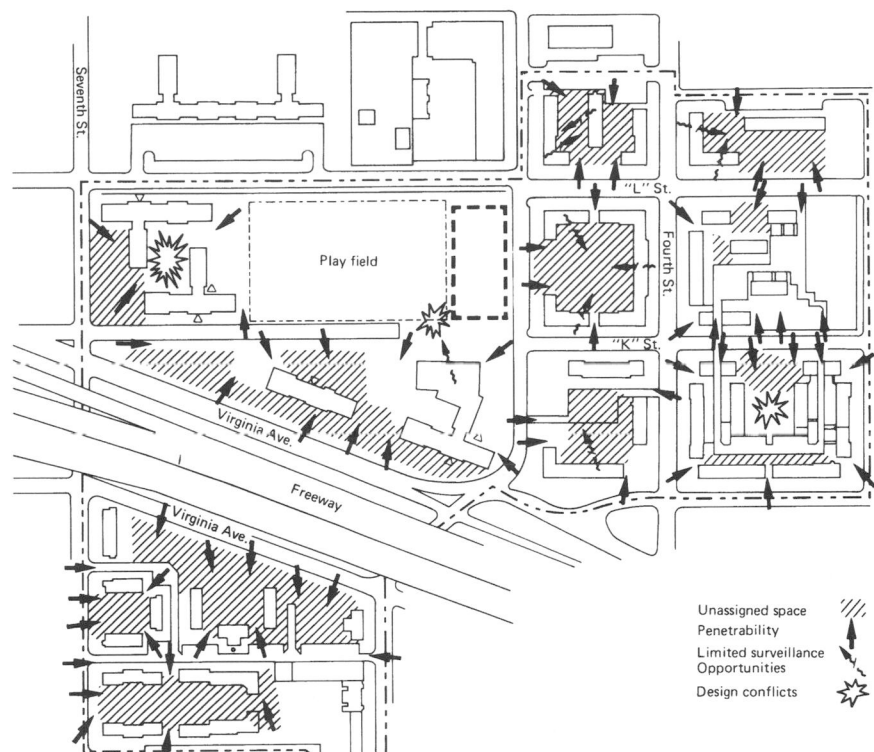

Unassigned space
Penetrability
Limited surveillance Opportunities
Design conflicts

Fig. 8a   Site security analysis/symbol graphics.

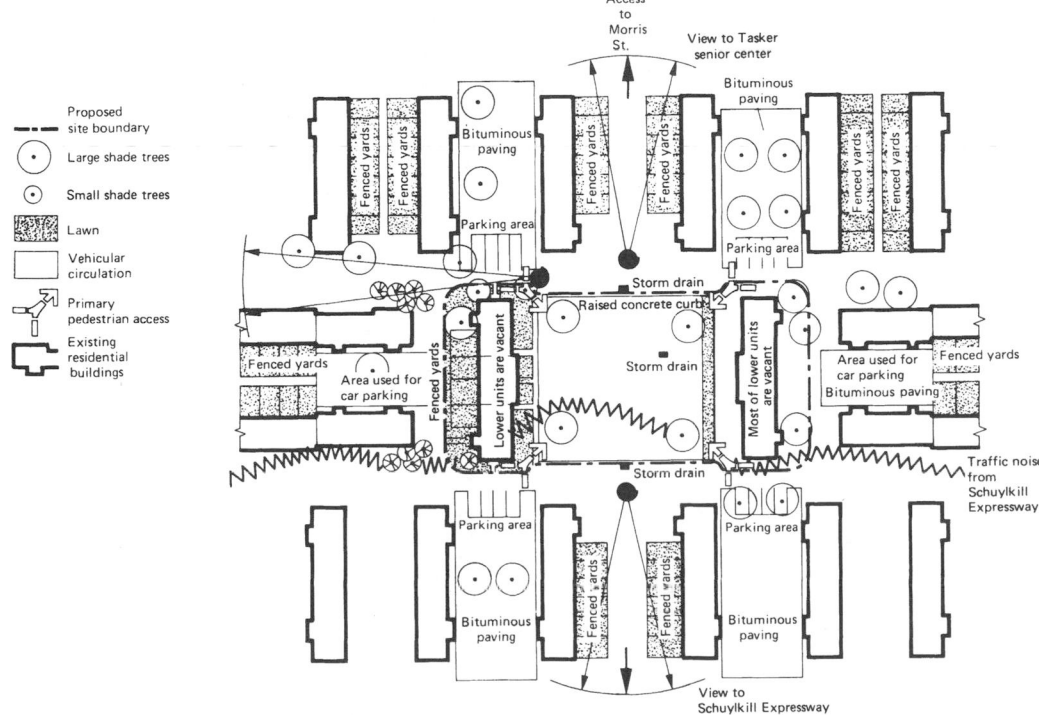

Proposed site boundary
Large shade trees
Small shade trees
Lawn
Vehicular circulation
Primary pedestrian access
Existing residential buildings

Fig. 8b   Family center site analysis/symbol graphics.

How will the site elements listed below contribute toward ➔	Eliminating Unassigned Space	
PLANT MATERIALS — Ground Covers	The qualities of a ground surface can indicate its intended use. Ground covers define yard, garden and lawn areas.	◖
Low Shrubs	Provide a residential character helpful in defining yard, garden and lawn areas and in developing shaded sitting areas.	◖
Mid-sized Shrubs	Effective at making areas semi-private to private, depending upon planting layout.	◖
Trees	Provide a residential character helpful in defining yard, garden and lawn areas and in developing shaded sitting areas.	◖
CONSTRUCTED ELEMENTS — Low Walls	Delineate and separate sub-areas of a project site if 18″–24″ in height. Can serve as sitting element contributing to use of site sub-areas.	◖
High Walls	Limit access and help delineate and separate sub-areas of the project site. Form private spaces, i.e. yards and patio areas.	○
Low Fences	Define and separate sub-areas of the project site. Can create semi-private areas such as individual yards.	⦶
High Fences	Minimal effect. Limit access and help delineate and separate sub-areas but do not actually contribute to use of space.	○
Gates	Assign space as entry and passage. (circulation)	⦶
Bollards	Prevent vehicle access, freeing space for pedestrian use. They define space and may serve as sitting elements.	◖
Paving Materials and Textures	Particular materials and patterns can indicate use and extent of sub-areas, and contribute to residents sense of territoriality.	⦶
Slopes and Berms	Can be used to define use areas and contribute to the development of play and sitting areas.	⦶
Stairs and Ramps	Define passage and pedestrian routes. Supplies access from one level or use area to another.	⦶
MANUFACTURED ELEMENTS — Site Furniture	These elements are useful in developing a space for assigned uses, such as sitting areas, game table areas, etc.	
Play Equipment Structures	These elements are useful in developing a play area to serve an assigned user group.	
Site Lighting	Extends the time period during which use areas can be actively used.	⦶

Fig. 1 Site elements security capabilities matrix.

Valuation Key

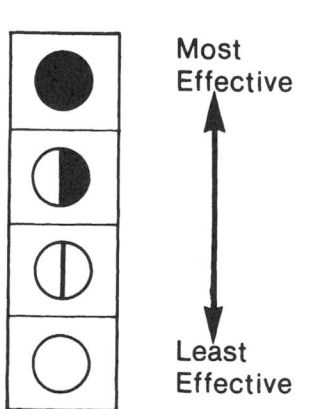

- ● Most Effective
- ↕
- ○ Least Effective

Minimizing Penetrability		Maximizing Surveillance		Minimizing Design Conflicts	
Form subtle symbolic barriers when mass planted in a planting bed or panel.	(circle, vertical line)	N.A.	(empty circle)	When mass planted, they subtly define areas, but do not effectively separate conflicting uses or groups.	(empty circle)
Capable of forming a symbolic barrier (dependent upon planting layout).	(circle, vertical line)	Excellent as a means to define areas that require visual surveillance.	(empty circle)	May provide adequate separation of use areas, however, mid-sized shrubs are more effective.	(circle, vertical line)
Form symbolic barriers that may develop into real barriers depending upon plants and layout.	(half-filled circle)	May substantially block surveillance of adjacent areas, depending upon planting layout.	(empty circle)	Excellent buffer/barrier to separate conflicting areas.	(half-filled circle)
Depending upon plant layout, can form symbolic barriers.	(circle, vertical line)	Most large trees will not hinder surveillance, though smaller flowering trees may.	(empty circle)	Mass plantings of small flowering trees can effectively separate use areas by forming screen barriers.	(half-filled circle)
Can define the project perimeter, and as symbolic barriers limit access to controllable points.	(half-filled circle)	If of sitting element height, they encourage use and activity in adjacent areas, thereby, contributing to surveillance.	(circle, vertical line)	Provide good symbolic separation of use areas.	(circle, vertical line)
Effective impenetrable barrier.	(filled circle)	Effectively block visual surveillance.	(empty circle)	Physically separate conflicting use areas. Can buffer noisy unattractive areas.	(filled circle)
Symbolic barriers that can be breached, but do minimize penetrability of the project site.	(circle, vertical line)	Do not hinder surveillance — if of metal picket or woven wire mesh fence types.	(empty circle)	Symbolically separate uses, such as an active play area and an adjacent walkway.	(circle, vertical line)
Excellent physical barrier, stopping penetrability of the project site.	(filled circle)	If of chain link or picket, construction, will only minimally interfere with surveillance.	(empty circle)	Separate conflicting uses and activity areas.	(filled circle)
Effective means to control access while stopping penetrability.	(circle, vertical line)	As an access point, gates concentrate traffic, thereby increasing surveillance, possibilities.	(circle, vertical line)	Do not contribute toward separating conflicting uses of areas.	(empty circle)
Prevent vehicular access, but permit free access for pedestrians and cyclists.	(circle, vertical line)	Do not hinder surveillance.	(empty circle)	Effectively separate auto and pedestrian traffic.	(circle, vertical line)
Form subtle symbolic barriers. Contrasting patterns and materials can define transition zones.	(half-filled circle)	N.A.	(empty circle)	Subtly define areas, but may not separate conflicts.	(empty circle)
Form symbolic barriers that discourage penetration of the project site.	(half-filled circle)	Should be sized, shaped and located so as not to block surveillance.	(empty circle)	Can separate conflicting uses as well as buffer noisy or unattractive areas.	(half-filled circle)
Can form symbolic barriers at the project perimeter or at an on-site use areas and at building entrances.	(circle, vertical line)	Due to change in grade, may limit surveillance from lower area.	(empty circle)	May separate conflicts, while not disrupting pedestrian traffic. At certain locations could be extra wide for sitting.	(circle, vertical line)
Capable of limiting penetrability of site sub-areas and encouraging outdoor activity of residents.	(circle, vertical line)	These elements encourage outdoor activity and residential use, thereby increasing surveillance.	(half-filled circle)	Should be grouped and located to identify with a particular group and function. Adequate facilities must be provided for other groups to avoid conflicts over use. (Applies to both site furniture and play equipment structures.)	(empty circle)
May function as symbolic barriers minimizing penetrability, particularly when actively used by residents.	(circle, vertical line)	Encourages increased surveillance of area if play elements are actively used.	(half-filled circle)		(empty circle)
May limit, at least initially, penetrability of site areas.	(circle, vertical line)	Effective and safe levels of lighting greatly aid night surveillance and residents' sense of security.	(filled circle)	N.A.	(empty circle)

Fig. 1 (*Continued*)

SITE ELEMENTS FOR SECURITY

TABLE 1   Ground Covers

Description	Ground covers are surface-growing plants that seldom achieve a height of more than 12 in. They include such plants as ivy, pachysandra, vinca, and myrtle.
Security aspects	These low-growing plants can be used to define a separate specific area of a site as well as to help establish the boundaries of a site. When planted in a bed or panel, they present an attractive symbolic barrier, or buffer area. When planted in mass, these plants can also be used to fill in large, vacant, and anonymous areas for which no specific use is practical. Used in this manner, they neutralize the space, leaving it available as a visual experience but clearly indicating it is not to be walked on.

TABLE 2   Recommended Ground Covers

Proper name	Common name	Type	Remarks	Hardiness zone
Hedera helix "Baltica"	Baltic ivy	Evergreen	Shade or sun, mows well	4–9
Euonymus fortunei	Wintercreeper	Semievergreen	Spray to prevent scale attack	5–10
Pachysandra terminalis	Japanese spurge	Evergreen	Shaded areas only	5–8
Vinca minor	Periwinkle, myrtle	Evergreen	Sun or partial shade, mows well	5–10

NOTE: There are also a number of low junipers (*Juniperus horizontalis, j. subina, j. chinensis,* and *j. conferta*) which make excellent ground covers in the height range of 1 to 2 ft (12- to 15-in plants should be planted 18-in on centers).

TABLE 3   Low Shrubs

Description	Low shrubs are bushy plants that do not grow beyond 3 ft in height. They consist of several woody stems rather than a single trunk, may be spreading, and are covered with either evergreen or deciduous foliage.
Security aspects	Low shrubs, when used properly, can be an important element in security planning as well as add warmth and texture to a site. They can assist in reducing a site's penetrability, in removing design conflicts, and in assigning space for particular uses. When planted closely together, they form a tightly knit symbolic barrier that does not limit surveillance. This barrier can be used to define garden and lawn areas and to buffer sitting areas from more active uses, thus minimizing conflicts over use. Low shrubs planted along a site's perimeter at points where access is not desired can structure access to the site. They are subtle in their intent as opposed to constructed architectural elements.

TABLE 4   Recommended Low Shrubs

Proper name	Common name	Type	Remarks	Hardiness zone
Berberis triacanthophora	Three-spine barberry	Evergreen	May be sheared	6
Euonymous alatus "Compactus"	Dwarf-winged euonymus	Deciduous	May be sheared	5–8
Ilex crenata	Japanese holly	Evergreen	Sheared periodically to keep within this height range	6–9
Juniperus horizontalis	Plumosa Andorra juniper	Evergreen		4–11
Taxus repandens baccata	Spreading English yew	Evergreen		4–7

TABLE 5   Mid-Sized Shrubs

Description	Mid-sized shrubs can reach a height of 6 to 10 ft within 5 to 10 years. Their foliage can be either deciduous or evergreen, and their woody stems may be thorny. Some of these shrubs may be used in their natural form in small groupings, or closely spaced and sheared into a hedge.
Security aspects	Shrubs form a substantial symbolic barrier with the potential of developing into a formal, impenetrable barrier. A few of these plants have thorns or spines which aid the plants in their own defense and establish a convincing barrier. Since shrubs with thorns can be a hazard, they should be planted at least 3 to 4 ft from walkways and other locations which are heavily used by residents. These shrubs may grow tall enough to substantially block visual surveillance of site areas but can be kept lower with pruning if they present a security hazard. Where space is at a premium, these plants may be grown as a hedge, effective at separating use areas over which there may be conflicts. Where more space is available, these shrubs may be used in their natural unsheared state to form an effective screen between use areas. Shrubs of this height may also be used to form screens defining semi-private to private front and rear yard areas. They are also effectively used in small groupings to landscape lawn, garden, and yard areas.

TABLE 6   Recommended Mid-Sized Shrubs

Proper name	Common name	Type	Remarks*	Hardiness zone
Berberis julianae	Wintergreen barberry	Evergreen	(H) spines	5–8
Euonymus alatus	Winged spindle tree	Deciduous	(H)	3–8
Ilex crenata microphylla	Little-leaf Japanese holly	Evergreen	(H)	
Ilex cornuta Burfordii	Burford Chinese holly	Evergreen	Berries, spines	6–7
Ligustrum ibolium	Ibolium privet	Deciduous	(H)	4–9
Prunus laruocerasus "Schipkaensis"	Cherry-laurel	Evergreen	(H)	6–9
Pyracantha coccinea lalandei	Scarlet firethorn	Deciduous	(H) thorns, flowering berries	5
Vibrunum dentatum	Arrowwood	Deciduous	(H)	2

*(H) indicates that the shrub is suitable for use in forming hedges.

TABLE 7   Trees

Description	Characterized by a single stem or trunk, trees are woody plants which grow to a height of at least 10 ft or taller. They may be divided into two categories—small trees, those under 25 ft in height (many of which have conspicuous flowers), and large trees, most of which mature to a height of 40 to 60 ft and are noted for their overhead canopy and shade-projecting capabilities.
Security aspects	Trees, with their heavy trunks and large canopies, can form an effective symbolic barrier when spaced 20 to 25 ft apart along a project property line. When the trees' vegetative canopies are above eye level, they do not hinder surveillance. Trees also enhance sitting areas, since their shade during the warm months of the year encourages people to use outdoor seating. This use provides neighbors with an opportunity to get to know and recognize one another and promotes informal surveillance on the site. Large trees are particularly well suited for defining the limits of areas such as on-site playfields and adjacent walkways. When so used, they should be kept 10 ft or so away from play equipment or play courts so that lower branches cannot impede the play. The shrubbier, smaller trees can also be used to define areas. They are particularly useful when a barrier or visual screen is desired, such as around a services area. Some species can even be sheared into a hedge form.

TABLE 8   Recommended Large Trees

Proper name	Common name	Remarks	Hardiness zone
Acer platanoides "Summershade"	Norway maple	Difficult to grow grass underneath	3–8
Liquidambar styraciflua	Sweet gum	Rather difficult to transplant in large sizes	4–9
Pyrus calleryana Bradford	Bradford pear		5–9
Quercus borealis	Northern red oak	Transplants easily	4–8
Sophora japonica (Regant)	Japanese pagoda tree	Flowers late summer, pods remain on tree*	4–9
Tilia cordata	Little-leaf linden (XP 110)	Very hardy	3–9
Zelkova serrata	Japanese zelkova	Grows fast, close in shape to elm	5–9

*Pagoda trees are used extensively in urban areas although the pods are considered poisonous. However, before falling, they deteriorate to the extent that they are not attractive to children as playthings.

TABLE 9   Recommended Small Trees

Proper name	Common name	Remarks	Hardiness zone
Crataegus lavallei	Lavalle hawthorn	Flowers late May, red fruits remain on tree through winter	4–7
Crataegus phaenopyrum	Washington hawthorn	Fruits remain all winter	4–7
Malus baccata	Siberian crab apple	White, early May, fruit red or yellow	2–9
Malus dorathea	Dorathea crab apple	Round shaped	4–9
Malus "Snowdrift"	Snowdrift crab apple	Vase shaped	4–9
Magnolia soulangiana	Saucer magnolia	Large flowers appear before leaves	5–10
Magnolia stellata	Star magnolia	Large flowers appear before leaves	5–9
Prunus cerasifera "Atropurpurea"	Purpleleaf plum		4–9

NOTE: The above trees are deciduous. Needle-leaf evergreen trees are difficult to grow under city conditions. (*Pinus nigra* and *Tsuga caroliniana* are the best choices.)

SITE ELEMENTS FOR SECURITY

TABLE 10 Low Walls

Description	Generally between 18 and 36 in in height; low walls are upright structures of masonry or wood construction. While they are not impenetrable, they do form excellent symbolic barriers.
Security aspects	Low walls can be used to define the project perimeter, channeling pedestrian movement and limiting access to predetermined, controllable points. As a barrier they may define and separate on-site use areas such as private yards, entry and sitting courts, and playgrounds, and they are especially appropriate for this function when space is limited. If constructed between 18 and 36 in in height, low walls can be casual sitting elements contributing to the use, activity, and surveillance of adjacent areas.
Examples	
Concrete wall (poured in place)	■ Requires minimal maintenance ■ Has the longest potential useful life—and the ability to withstand abuse ■ Initial construction costs are relatively high ■ Can be quite handsome in appearance
Brick wall	■ Attractive in appearance ■ With moderate care will serve a long useful life ■ A good value in terms of initial cost, appearance, and maintenance requirements
Concrete block wall (stucco surfaced or painted)	■ Low initial cost compared with other masonry walls ■ Appearance can be quite utilitarian (especially painted walls) ■ Requires more frequent maintenance

TABLE 11 High Walls

Description	Similar in construction and form to low walls, these upright structures are a minimum of 4 ft in height and are most frequently built at a height of 6 ft.
Security aspects	At a height of 4 ft, walls can form a substantial physical separation, while at 6 ft they form a relatively impenetrable barrier as well as offering visual privacy and a buffering of adjacent areas. With these characteristics, high walls can define the project perimeter and control penetrability into the project. They most effectively buffer off-site areas, such as noisy expressways or industry. Used less often within the project grounds, high walls can buffer and separate areas of conflicting use, such as a maintenance yard and a quiet sitting area for elderly residents. Walls can also provide needed separation between two intensively used play areas; such as one for elementary-school-age children and one for teenagers. In examples such as these, the buffering or separating walls must be located with care so that they do not create a security hazard by eliminating surveillance opportunities from adjacent residential buildings and pedestrian walkways.
Examples	
Concrete wall (poured in place)	■ Minimal maintenance required. ■ Will withstand abuse and has a long useful life potential ■ Construction costs are only slightly higher than for brick construction ■ Attractive appearance depending upon wall design
Brick wall	■ Attractive in appearance ■ Moderate maintenance requirements, usually performed on a 10- to 15-year cycle ■ With this moderate care, brick walls will serve a long useful life
Concrete block wall (painted finish)	■ Lowest initial cost of the masonry wall types ■ Requires more frequent maintenance ■ Utilitarian in appearance (can be improved by stucco surfacing) ■ Where walls are background elements, block walls may be the most economical and appropriate choice

TABLE 12 Low Fences

Description	Usually between 36 and 42 in in height, low fences are upright structures of wood or metal posts and rails with pickets or woven wire mesh. Traditionally, they have been used to indicate and maintain a property boundary, as well as serve as a barrier offering protection and/or confinement. Many types are now being standardized at 36 in in height.
Security aspects	Low fences can contribute to a site's security in several ways. They can be used to define a project's perimeter and to guide entry to fixed, observable points. They can also define and separate use areas on a site. Because of their low height, observation between separated areas is still possible.
Examples	
Wrought-iron fence	■ Attractive in appearance ■ Difficult to vandalize ■ Moderate maintenance requirements ■ High initial cost is offset by potentially long useful life
Tubular-steel fence	■ Reasonably attractive in appearance ■ A good value in terms of initial cost, low maintenance, and a reasonable resistance to vandalism ■ With periodic maintenance, useful life approximates that of wrought iron
Chain-link fence	■ Lowest initial cost ■ Least resistance to vandalism ■ Least attractive in appearance

TABLE 13 High Fences

Description	Identical in construction and form to low fences, these structures are a minimum of 4 ft in height, though they are usually constructed to a height of 6 to 8 ft, increasing their effectiveness as a relatively impenetrable barrier.
Security aspects	High fences are most frequently used to control access along a project perimeter by forcing those who wish to enter to pass to a controllable entranceway. High fences are used less frequently within the project grounds but may form partial screens or complete enclosures, to contain activity on basketball and handball courts, minimizing conflict with adjacent use areas. High fences also frequently enclose maintenance yards, keeping small children away from the hazards of the area and minimizing vandalism. For the above uses, metal picket or chain-link fences are most often used because they do not obstruct visibility and permit surveillance of adjacent areas. Where used to create private patios or terraces adjacent to low-rise residential buildings, they should be augmented to create a visual barrier as well.
Examples	
Wrought-iron fence	■ Attractive appearance ■ Difficult to vandalize ■ Moderate maintenance requirements ■ Potentially long useful life offsets high initial cost
Tubular-steel fence	■ Attractive in appearance ■ Characterized by reasonable initial cost, low maintenance requirements, and a satisfactory resistance to vandalism ■ Durability approximates that of wrought-iron fencing
Chain-link fence	■ Very low initial cost ■ Low resistance to vandalism ■ Least attractive in appearance

SITE ELEMENTS FOR SECURITY

TABLE 14   Gates

Description	The gates discussed here are designed to provide a substantial obstacle when closed.
Security aspects	Gates have several functions from a security standpoint. They can provide a difficult obstacle for someone who wishes to enter a protected area. Also, simply by the fact of being closed, they announce that there is a special area behind them to which entry is not permitted. Yet, when open, gates can provide a symbolic entry point indicating that those entering are moving into a particular environment.
	Gates intended to physically impede entry are not feasible unless such a policy can be conveniently formulated and implemented. Whatever the expected function of a gate, it is important to have a clear policy governing the opening and closing of the gate and who is to have authority so residents will not be greatly inconvenienced.
Examples of Pedestrian Gates and Vehicular Gates	
Wrought-iron gates	■ Attractive appearance ■ Difficult to vandalize ■ Moderate maintenance and a satisfactory resistance to vandalism
Tubular-steel gate	■ Reasonably attractive in appearance ■ Also characterized by low maintenance and a satisfactory resistance to vandalism ■ Initial cost is approximately the same as for gates of wrought-iron construction
Chain-link gate	■ Lowest initial cost ■ Least resistance to vandalism ■ Least attractive in appearance

TABLE 15   Bollards

Description	Bollards are small posts constructed of wood, metal, or concrete. They are usually 12 in or less in diameter and range from 24 to 30 in in height.
Security aspects	Bollards are used to separate and control vehicular and pedestrian traffic. Spaced as much as 5 ft apart, but never closer than 3 ft apart, they can bar vehicular access while permitting unhindered access for pedestrians and bicyclists.
	In addition to controlling vehicular access, they can be useful for defining on-site areas such as playgrounds and as casual sitting elements. Both uses should contribute to resident use and activity within an area.
Examples	
Wood bollard	■ Economical when 8 by 8 in square or under ■ Reasonably long useful life if pressure-treated ■ Can be quite attractive
Concrete bollard	■ Attractive in appearance ■ Minimal maintenance ■ Long useful life
Pipe bollard	■ Utilitarian in appearance ■ Durable and able to withstand abuse ■ Well suited for service and delivery area

TABLE 16 Paving Materials and Textures

Description	A variety of paving materials are suitable for use in housing projects. They fall, in general, into two categories: those that are poured in a liquid state and harden into place, such as concrete or bituminous asphalt, and pavings of small blocks or units such as brick, granite sets, and precast asphalt pavers.
Security aspects	Variations in paving material, texture, and color can be utilized to establish zones and use areas on a site. At the project perimeter and along entry and walkways, distinctive paving can indicate transitional areas between public streets and sidewalks and semiprivate and private residential areas. Within the project site, selection of a paving material and pattern for a particular use area can clearly indicate the boundary of that area and, by contrasting it with an adjacent area, minimize conflicts over use.
	Distinctive paving is one effective way to "assign" or associate site areas with a particular group of residential units or buildings. For example, a paved entry court of a particular material and pattern will give identity and unity to the buildings which front on the court, and contribute to the residents' sense of territoriality. Introducing a selected variety of paving materials and patterns, corresponding to and identifying the use areas and entry zones on the project site, will contribute to the residents' sense of security, as well as add visual interest and variety.

Examples		
Asphalt paving	■ Lowest in construction cost ■ Requires the most frequent maintenance in the form of periodic resurfacing ■ In most applications it is aesthetically unappealing and monotonous	Hexagonal pavers (asphalt)	■ Attractive pavers with a slightly resilient surface—pleasant for walking ■ Initial cost is somewhat more expensive than concrete but less than brick paving ■ Will withstand heavy usage and traffic
Concrete paving (broom finished scored on 3-ft grid)	■ Very moderate in cost with a long useful life expectancy ■ Minimal maintenance required ■ Offers a reasonably attractive surface ■ Offers a nonslip textured surface	Bomanite paving (concrete)	■ Similar in appearance to brick or stone paving ■ Less costly than brick or stone paving ■ May deteriorate over time and is somewhat difficult to clean "joints" and to patch surface
Concrete paving (exposed aggregate finish)	■ Presents a very attractive "pebbled" surface ■ Usually costs approximately twice as much as ordinary concrete work ■ Minimal maintenance required ■ Has a reasonably long useful life expectancy	Brick paving	■ A very attractive paving material ■ It ages gracefully and requires minimal maintenance ■ Its relatively high initial cost is offset by its potentially long useful life ■ An especially appropriate material in older urban areas
Precast pavers (concrete)	■ Cost is only slightly higher than ordinary concrete paving ■ Presents a very attractive textured surface similar to brick paving ■ Has a long useful life expectancy and requires only minimal maintenance	Brick grid (in concrete paving)	■ An economical and attractive way to incorporate brick into paving surfaces ■ Effective way to pave entrance plazas or courts at community or residential buildings ■ If properly constructed will require minimal maintenance and serve a long useful life

SITE ELEMENTS FOR SECURITY

TABLE 17 Slopes and Berms

Description	Slopes and berms are earthen barriers. A slope is an inclined surface which may be gentle, moderate, or steep, depending on its purpose and the site conditions. A berm is a mound of earth with sloping sides, located between areas of approximately the same elevation.
Security aspects	Berms and slopes are most effective on a housing project site in separating and buffering potentially conflicting use areas such as playground and an intensively used roadway.
	At the perimeter of a housing project, these earthen barriers can form an effective symbolic barrier that discourages penetration and directs pedestrians to proper entrances to the site.
	Within the project site, earthen slopes and berms can define areas set aside for quiet sitting, as well as active playfields. Berms can be incorporated into the design of a tot or elementary-school-age playground (for example, slides can be incorporated into the slope of a berm). When located adjacent to a teen play or athletic court or field, berms and slopes from excellent casual spectator seating, while buffering adjacent residential buildings from noise.
	Slopes and berms should be sized, shaped, and located with care so as not to block on-site surveillance.

Examples	
3-ft-height berm—2:1 slope (with ground-cover vegetation)	■ Serves as an excellent means of buffering sitting area from adjacent street and walkway
	■ With the ivy ground cover the berm forms an attractive setting for the seating area
	■ Ground cover vegetation reduces the frequency of maintenance
3-ft-height berm—3:1 slope (with grass vegetation)	■ Low, gentle berms give visual relief to uniformly "flat" urban areas
	■ Provides an attractive parklike setting
	■ Reduces the visual and noise impact of adjacent roadways
	■ Gentle slope permits normal mowing operations
6-ft-height slope—2:1 slope (with ground-cover vegetation)	■ Slope separates semiprivate residential areas from the public street and sidewalk
	■ Ground-cover vegetation reduces periodic maintenance

3

Subdivisions and Land Planning

Good subdivision requires the recognition and evaluation of the elements that will be of significance in creating functional, well-balanced, and aesthetically pleasing communities. In addition to requiring technical skill in laying out the subdivision, the creation of satisfactory development is also predicated on achieving coordinated action on the parts of the subdivision developer, planning board, and other municipal officials.

The general determinants of subdivision design include the following: the guidelines for community development as set forth in its comprehensive plan; the influence of existing peripheral development, and the effect of the physical characteristics of the site.

REQUIREMENTS OF THE MASTER PLAN

The desirability of a planning board's having an adopted master plan to use as a guide in reviewing proposals for land subdivision is obvious. This plan should include existing and proposed streets, parks, public reservations, sites for public buildings and structures, zoning districts, and routes for public utilities. The plan can be adopted by the planning board as its official guide, and after adoption can be changed by the board when conditions call for its amendment. The municipal governing body (town board, village board, or city council) in seeking advice from the planning board on development matters will obtain the benefits of the plan's guidance. Good procedure suggests that the governing body informally agree on the major elements of the plan before it is adopted by the planning board so that it will reflect the views of the elected officials in the community.

The board can refuse approval to a layout that is not properly related to the street layout shown on the master plan; that power gives the board its most important tool for implementing the street system planned for the community. More extensive use of this power in the past could have corrected many errors in street layout now obvious to planners and lay people.

SOME ALTERNATIVE WAYS OF ACHIEVING OPEN SPACE

The master plan may also show additions to existing park and school sites and locations for future parks and schools. Since the planning board is required to decide whether or not parks and playgrounds are needed, and if so, to require their reservation by the developer, the park and playgrounds shown on the community's master plan can be obtained (in part) as land is subdivided. In the case of school sites, the developer and planning board have in many instances reached agreement on substantial contributions of land at reasonable cost (or no cost) as a means of helping to provide the schools necessary to serve large new housing developments. More than one developer has found the provision of land for a new school a valuable sales aid when promoting a project.

Figure 1 shows a portion of a town master plan with the location of a proposed subdivision superimposed thereon. It should be noted that this master plan shows the location for a future park along the upper boundary of the proposed subdivision site and two collector streets (Orchard Road and Lincoln Road) for which additional rights-of-way are needed to allow for street widenings.

Figure 2 shows alternate ways of achieving open space.

EFFECT OF NEARBY DEVELOPMENT ON THE SITE

One obvious effect of existing development that may adjoin the site of a proposed subdivision comes from the need to provide for the extension of roads from the adjoining area into the new one. In some cases, the new development will need to employ the streets in the older one as the means of access to it, and in others the older subdivision streets will provide a second means of access to the new subdivision. Experience has shown that there are some basic principles that should not be violated when new streets are laid out adjacent to existing ones. One of these principles is that no "reserve strips" should be permitted at the end of a street so as to prohibit future access into land beyond it. The need for convenient traffic circulation throughout a community makes this protective device an obsolete method of providing "privacy" for a particular subdivision. Another principle is that the main means of access to a large new subdivision (say more than 10 lots) should be provided from a street designed to carry a fairly high traffic load and should not be provided through a local street designed only for light traffic. If the community does not have a master plan that shows how these traffic routes are to be laid out and coordinated as the area is developed, common sense will often indicate where through traffic or collector-street traffic is best routed. The planning board that has a master plan for traffic circulation will be in a better position to make sure that both new and existing development is not devalued by heavy or high-speed traffic.

When the subdivision design requires that a proposed street be continued to the edge of a presently undeveloped area to make provision for its future extension, it is desirable to require

a *temporary* turnaround at the end of the street to allow for convenient vehicular movement. Such excess right-of-way that may be required for the temporary turnaround can revert to the abutting lots when the street is extended.

Unless there is an existing or proposed street to be extended, it is generally undesirable to terminate a street at a property line. The problem of providing street access to the corner of a property can be solved by the provision of a short stub or "eyebrow" around which usable lots can be created (see Fig. 3 on p. 141).

When the new subdivision lies next to an area already provided with public services and utilities, the extension of these becomes an important factor in the layout. Water mains and hydrants can usually follow streets without serious problems, unless a significantly higher elevation is involved, which may call for some adjustment in water pressure. Gas mains are a similar utility, with pressure rarely a problem. Sanitary sewers, however, normally rely on gravity flow, and the grades of streets will very definitely affect the adequacy and cost of this service. In many cases, it is necessary to provide a sanitary sewer easement across lots to make the system workable. (It is good practice to have such easements follow lot lines where possible.) Pumping sewage should be avoided and in some areas will not be approved by health authorities. Storm water drainage is a comparable service; it requires careful analysis to relate its requirements to the street system, the slope of the individual lot, and the location of buildings.

Storm water drainage will need to be routed to some point or points at the boundary of the subdivision where it can be safely carried away (in some few cases the subdivision may include its own drainage "sump"). Where this water leaves the developer's property is a crucial

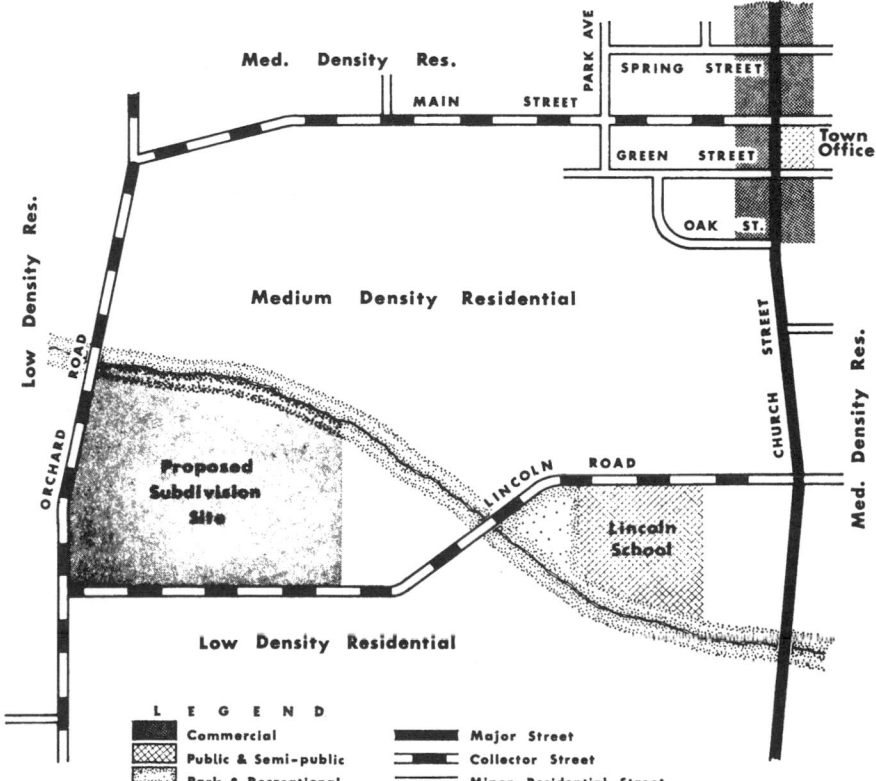

REQUIREMENTS OF THE MASTER PLAN

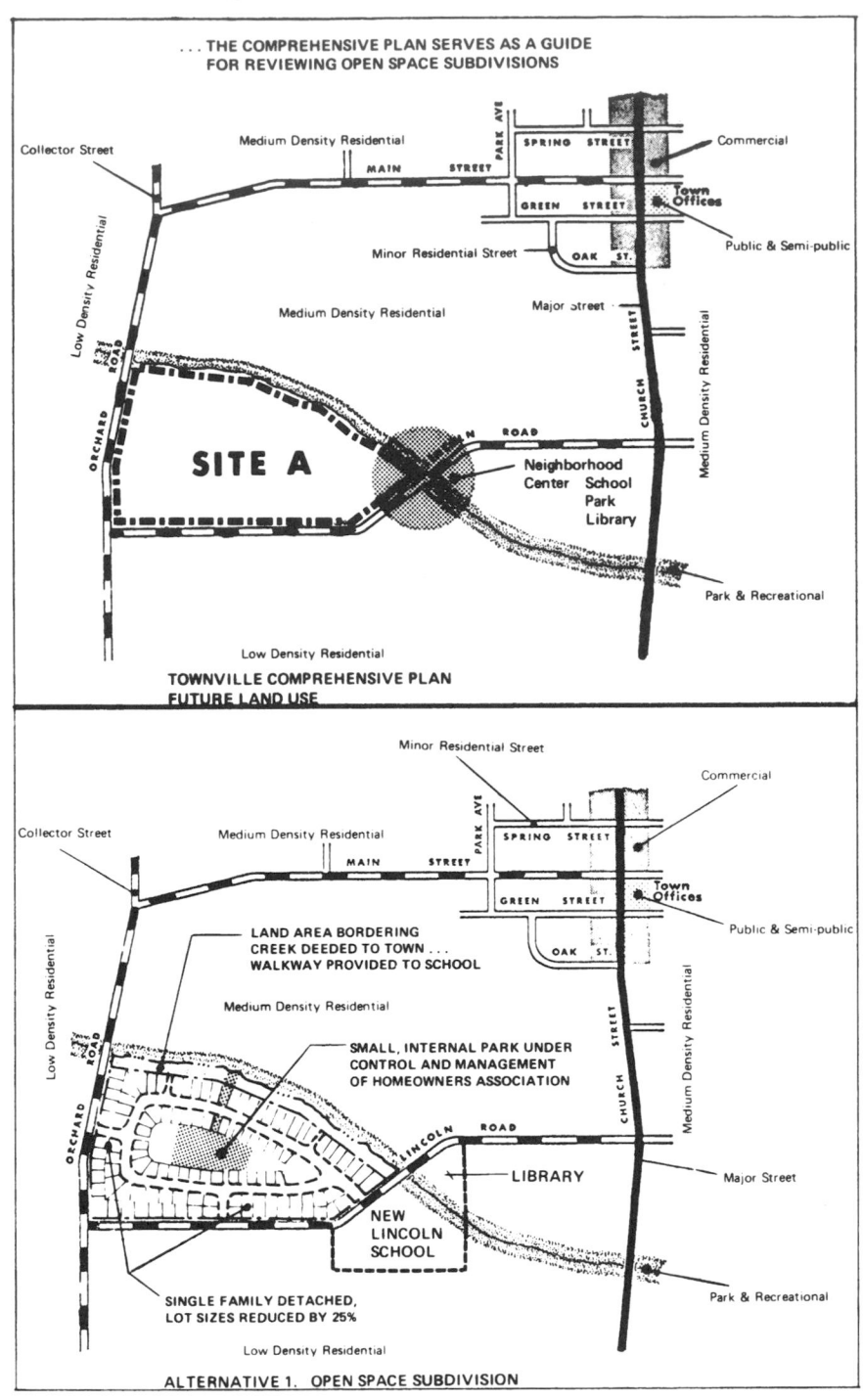

... THE COMPREHENSIVE PLAN SERVES AS A GUIDE
FOR REVIEWING OPEN SPACE SUBDIVISIONS

SITE A

TOWNVILLE COMPREHENSIVE PLAN
FUTURE LAND USE

ALTERNATIVE 1. OPEN SPACE SUBDIVISION

Fig. 2a

design matter. In the past it has caused a great deal of argument among landowners and often has been responsible for costly improvements by the municipality itself. Developments increase water runoff because the new lawns, roofs, driveways, and paved streets are less absorbent than vacant or farm land. This will increase loads on storm drains downstream, and means that the developer, adjoining owner, and municipality will have to cooperate in providing solutions. A master plan can include proposals for handling storm drainage on a long-term and coordinated basis and thus provide the planning board and developer with a guide to the solution of drainage problems.

Other connecting utilities and services needing study at the time of subdivision approval are electric power and street lighting, fire alarm boxes, street signs, and sidewalks.

The relation of the subdivision to a nearby school or park should be studied. Persons going to the park or children walking to school should be given a convenient and safe route, and the needs of persons who will be living in the future on land beyond the present subdivision itself will also require study. Again, a master plan will identify these needs and show how they can be met in the design of the new subdivision.

EFFECT OF THE PHYSICAL CHARACTERISTICS OF THE SITE

The effect of an area's physical characteristics is one of the most important factors to be considered in the design of any subdivision. When these characteristics are ignored, costs can go up and long-term values will be endangered. When selecting land for development, careful consideration should be given to its slope, drainage, and soils. In many cases these factors become so important that they will, in effect, dictate the type of development that is practical. A common example of this is the case where the site under consideration has a very steep slope, which makes intensive one-family housing impractical because of the cost involved in making small lots actually usable.

Developers should take advantage of trained engineering, surveying, and site design services when they begin to plan the layout of the subdivision. Engineers will normally need to make some sort of *topographic map* their basic tool in laying out streets at acceptable grades and in providing a storm water drainage system that is adequate. A topographic map shows the elevations of the site by use of contour lines and usually includes information about watercourses, rock outcrops, and the other physical features of the site. Since developers will normally need this type of map to make their plans, the planning board may reasonably require subdividers to supply it as part of their submission for planning board review. Many communities specify in their subdivision regulations that a topographic map be used as a basis for preparation of the subdivider's preapplication sketch (see Fig. 17 on p. 145) and preliminary plat (see Fig. 18 on p. 146). Where the land is steep, the topographic map will tell how steep it is and will show where roads should not be built. Where land is very flat, the topographic map will show where there is a need for careful design of the drainage system to avoid future flooding or stagnant water.

Specific data or information based on local experience or special study in the area of the site will be needed to show how the water table, type of soils, and underground rock structure will affect the proposed development. If the site is not to be served by water mains or a collective sewerage system, this type of data becomes absolutely essential, and it will be required by health authorities before they will approve any plans. It will also help the developer and planning board to decide on the most practical type of road system, since surface or subsurface rock can add greatly to the cost of road building, pipe laying, and building foundation works. A high water table may be equally difficult to handle.

Recent developments in aerial photography permit much of this kind of data to be obtained from air photos with a minimum of ground survey work. For large tracts, aerial photography will usually be a cheaper method of obtaining topographic and other information than detailed field work on the ground.

One of the most valuable characteristics of a site is the view it may have across neighboring lands or to the horizon. The property with a good view is desirable if the view is pleasing and the lot and houses are laid out to make use of the view. There are many examples of excellent views being wasted when the developer fails to locate the streets and houses in the way that will allow residents actually to see the view. While a planning board is not in the business of designing individual house layouts, it often can persuade the builder to use this valuable resource more fully. When the view is toward visually unattractive commercial or industrial areas, the lot layout should be modified to minimize this effect.

A common complaint about new subdivisions, particularly those that have a large number of new lots and houses, is that they are barren of trees. The preservation of existing healthy and well-suited trees that are already on the site is important in order to keep them as a future asset. Trees increase the value of the lots, as they make the new subdivision more attractive from the beginning. Many builders have found that good trees increase the market price of the lot or house by more than the saving obtained from "clean-sweep" bulldozing. In preserving trees it is important to realize that all trees have a limited useful life: many a handsome forest tree is actually nearing the end of its life and should be removed. The advice of a trained forester or landscape architect will be helpful in these matters. The planning board may require street trees as part of the improvements to be provided by the developer.

If the site under consideration has watercourses, or ponds, or other terrain features that can contribute to the beauty of its layout, it is well to take care to see that as much as possible of these gifts of nature is preserved. A planning board and developer should be able to maximize the use of these resources without unduly restricting the use of the property, as the solutions are usually matters of design detail rather than major land use.

Many sites for new subdivisions have formerly been active farms or had other rural uses that leave on the land, when they are abandoned, certain manmade features that can be turned to advantage at the time of subdivision. Examples of these are stone walls, fences, orchards, ponds, lanes, avenues of trees, and ornamental landscaping. Contrariwise, some of the existing manmade features may be a hindrance to an attractive subdivision, such as the existing buildings on the site, which may be obsolescent or out of character with the new buildings proposed. Old dwellings in this situation are likely to cause zoning variance problems, owing to their bulk, which will tend to make them unusable except for a multifamily purpose (or perhaps a nursing home, fraternal club, or even a commercial use). These uses in a new one-family area are a devaluating factor in many cases, and a good rule to follow under such conditions is to remove them and employ the site for a new use.

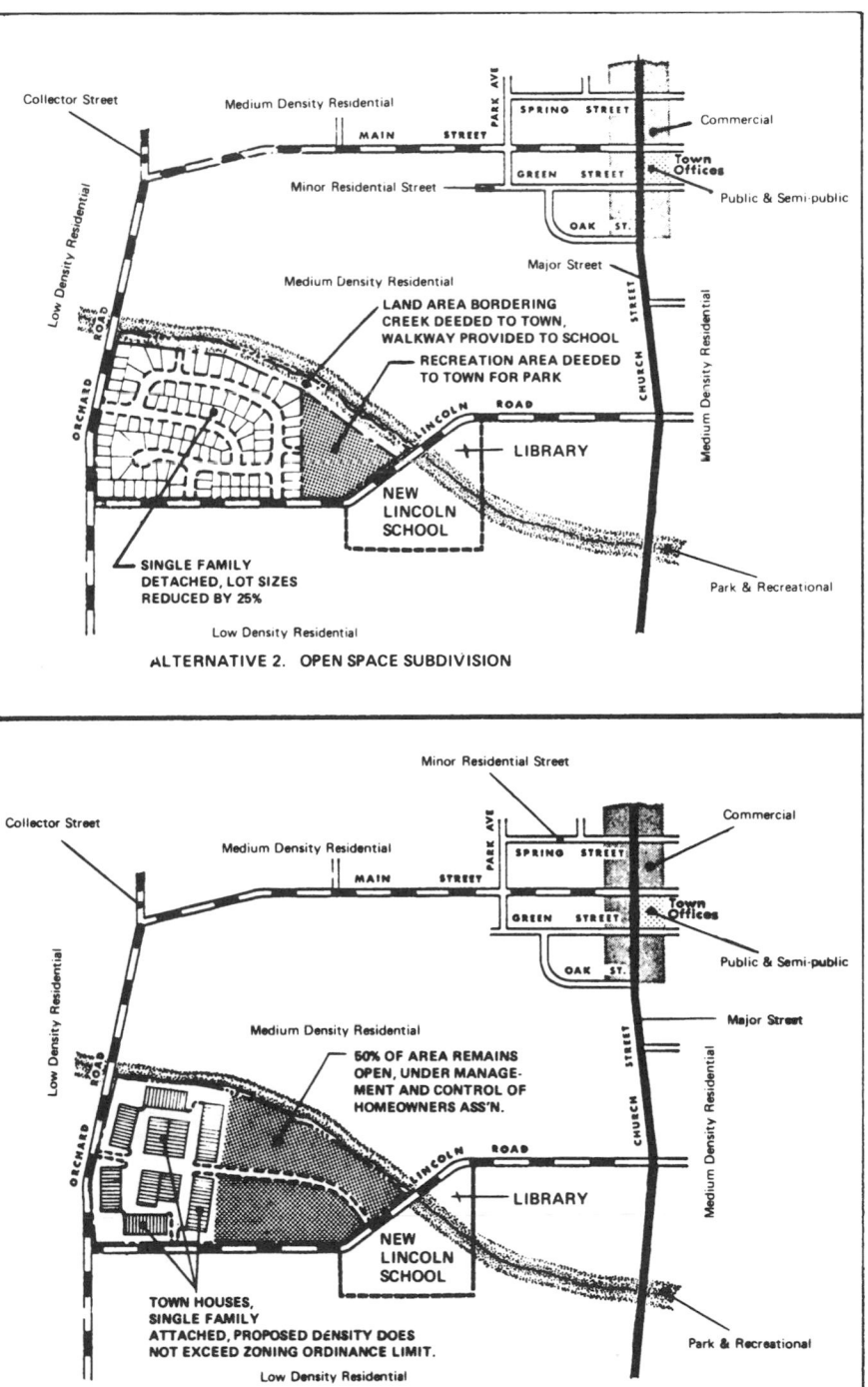

ALTERNATIVE 2. OPEN SPACE SUBDIVISION

ALTERNATIVE 3. OPEN SPACE SUBDIVISION

Fig. 2*b*

STREET AND LOT LAYOUT

The prime function of residential streets is to provide access to individual properties to accommodate their prospective traffic and to allow the convenient entry of fire-fighting, snow-removal and other road maintenance equipment. Streets should also be logically related to the topography and be coordinated into a system whereby each street performs the function for which it is intended.

The function that a street is intended to serve will determine both its right-of-way width and its pavement width. A minor residential street that serves a relatively low-density residential area may need less pavement width than if the same street served higher concentrations of residential development. This results both from the higher volumes of traffic on the street and from the resultant higher incidence of on-street parking. Collector streets and major streets carry progressively higher amounts of traffic than minor residential streets. This fact must be reflected in the criteria used for determining the street cross section. Other considerations affecting street right-of-way width are sidewalks, planting strips, and utilities, including street lights and fire hydrants.

The volume and speed of vehicular traffic on a street can be influenced by its particular design. An undifferentiated rectangular or grid street pattern usually does not include a collector or secondary street system and tends to make each local street as important as the next (see Fig. 4). This encourages through traffic at higher speeds on each street and also creates many potential traffic conflict points at the four-way intersections. One of the most trouble-free designs for a residential street is that of a "loop," which provides convenient access to each lot without encouraging through traffic (Fig. 5).

The dead-end or cul-de-sac street can also be used to advantage in residential subdivisions (Fig. 6). Through traffic is completely eliminated because there is only one entrance into the street. This creates an added sense of privacy, safety, and value to the lots fronting on this street. Two major drawbacks of cul-de-sac streets are that access to the interior lots can be impeded by a blockage at the open end and that traffic at the open end can become undesirably high if the street is too long and access to a large number of homes is provided. These streets should have paved turnarounds at their closed ends that are wide enough to permit

vehicles to negotiate the turn without the need for backing.

When residential development occurs along major streets and other highly traveled traffic arteries, special consideration must be given to its design. Lots should not front directly on or have direct access to such streets (Fig. 7a). When this occurs, the efficiency of these streets is reduced and they are no longer able to adequately perform the function for which they were designed. This problem can usually be solved by either building a marginal access street (Fig. 7b) or backing the lots up to the major street (Fig. 7c). The marginal access street provides frontage for the individual lots and greatly reduces the number of points of access to the major street. When the landscaped buffer strip is provided between the marginal access street and the major street, the traffic noise will be reduced and a more private environment created. Unless care is taken in designing the marginal access street, it may cause more traffic conflict at its entrances and exits than it is intended to solve. By maintaining a minimum safe distance between these entrances and exits and other intersections most of this traffic conflict can be avoided.

In cases where lots can be backed onto a major street, the land-use conflict can be reduced by requiring a landscaped buffer zone between the major street and the rear property line. In addition, a fence along the rear property line can provide for more privacy and a safer backyard.

Intersections are another important element of street design. When improperly designed, street intersections become potential traffic hazards. Streets should intersect at right angles (Fig. 9) and not at acute angles (Fig. 8). The centerlines of offset street intersections should be far enough apart so that traffic is deterred from cutting diagonally across them. Intersections should occur on straight sections of street instead of on curves, and should have gentle grades rather than steep slopes. Four-way intersections should be avoided except at the crossing of collector or major streets where traffic-control devices are utilized.

The blocks that make up a subdivision are inherently related to the street patterns. Although the number of intersections should be kept to a minimum, it is necessary to limit block length in order to permit adequate vehicular and pedestrian circulation within the subdivision. In situations where excessive block

lengths are unavoidable, such as under unusual topographic or drainage conditions, a right-of-way or easement for pedestrians should be provided across the block to break up its excessive length.

The lot layout and street arrangement in a subdivision are so closely interrelated that one cannot be planned without considering its effect on the other. Once the general lot size and dimension requirements have been determined, a street system can be designed to allow for the development of a desirable lot layout. In order to create a desirable home site that can be developed economically, several factors must be considered and certain general principles adhered to when lots are being laid out.

Good trees and other desirable natural growth should be preserved and the amount of grading kept to a minimum. Generally, it is preferable for the lot elevation to be somewhat higher than that of the abutting street. The grade between the street and the house location on the lot should not be excessive but should be enough to provide good surface drainage to the street and subsequently to a storm drainage system. Each lot should provide a desirable building site that allows adequate space for side yards and a driveway. It should be deep enough to allow for proper building setback and provide some space for outdoor activities.

The size and shape of the individual lot is often influenced by the type and size of dwelling contemplated for the development. This is especially true when the subdivider is also the home builder. Rectangular lots are generally the most usable. However, topography, street layout, and the shape of the original parcel often necessitate creation of lots that are not rectangular. When this occurs, odd-shaped lots with excessive jogs and corners should be avoided (Fig. 10). Whenever possible, side lot lines should be perpendicular to straight streets or radial to curved streets (Fig. 11). Corner lots that are too small do not provide an adequate building site (Fig. 12). Generally, corner lots should be larger than interior lots to allow for required setback from each street and provide a more usable backyard (Fig. 13).

When developing an odd-shaped parcel of land fronting on an existing road, creation of excessively deep lots should be avoided (Fig. 14). Use of a short cul-de-sac street can often facilitate development of the parcel into more desirable lots (Fig. 15).

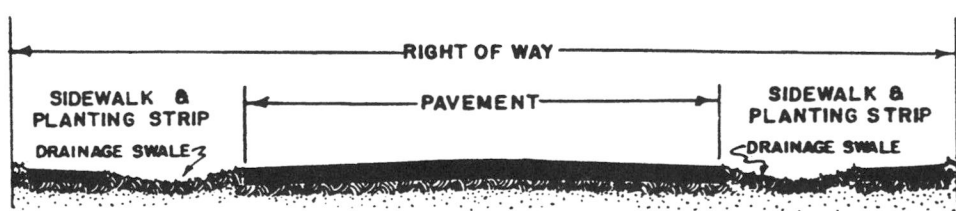

Fig. 1 Typical street.

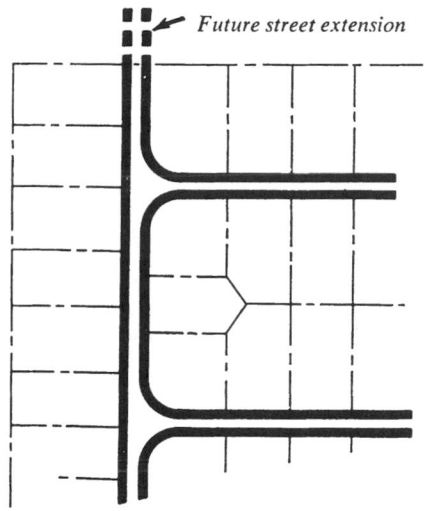

Fig. 2   Provision for future street extension.

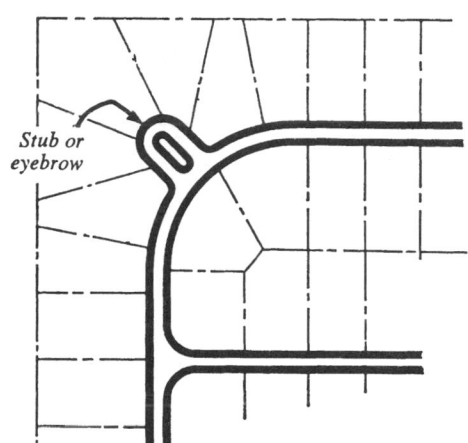

Fig. 3   Use of stub street or "eyebrow."

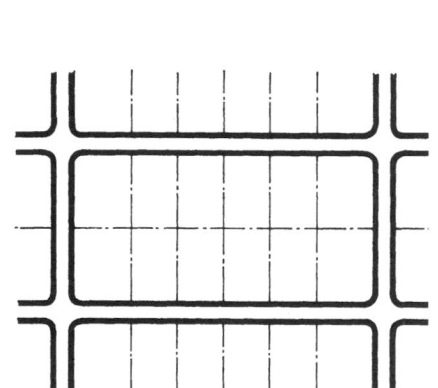

Fig. 4   A rectangular or grid street pattern.

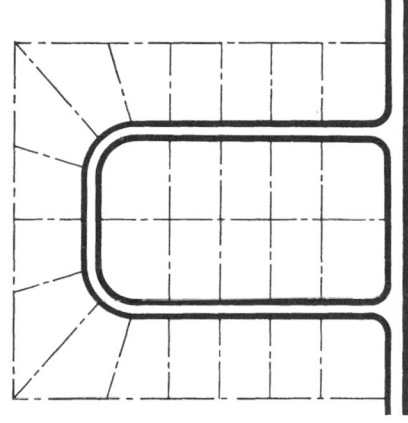

Fig. 5   A loop street.

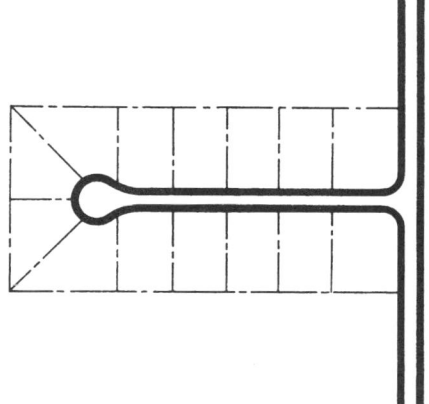

Fig. 6   A cul-de-sac street.

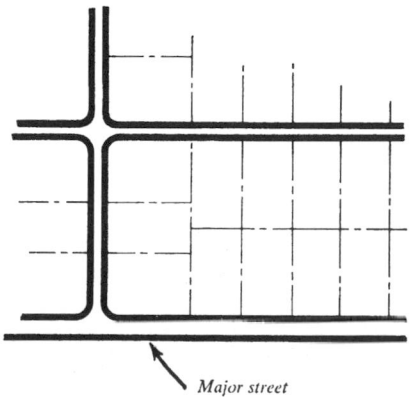

Fig. 7a   The practice of fronting lots directly on a major street is undesirable.

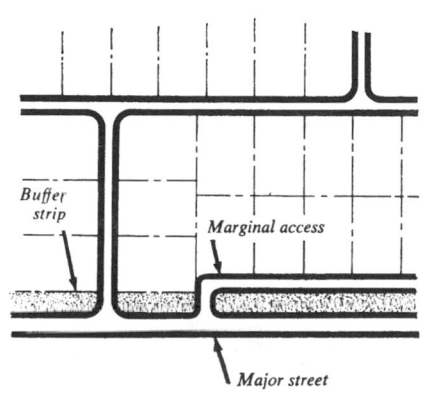

Fig. 7b   Use of a buffer strip and marginal access street is more desirable.

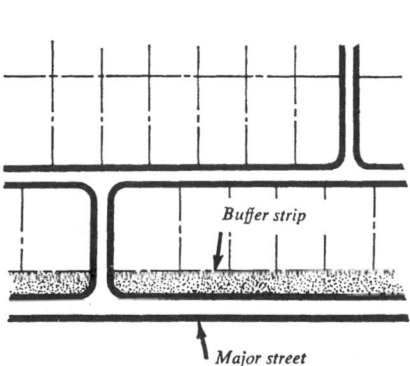

Fig. 7c   Use of a buffer strip when backing lots on a major street is desirable.

STREET AND LOT LAYOUT

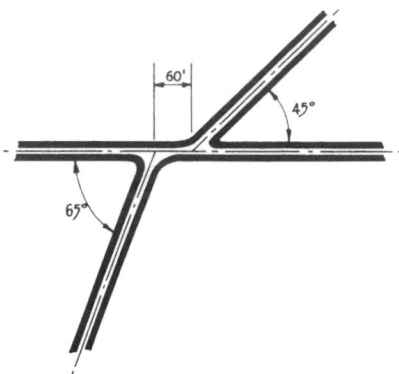

Fig. 8   Undesirable offset street intersection.

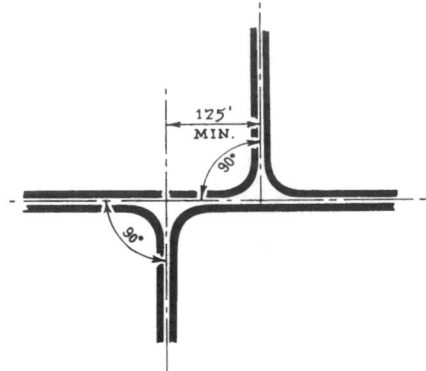

Fig. 9   More desirable street intersection.

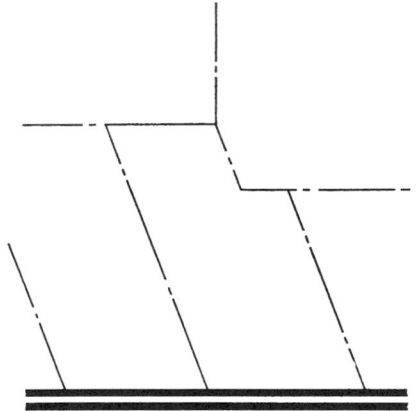

Fig. 10   Undesirable lot layout.

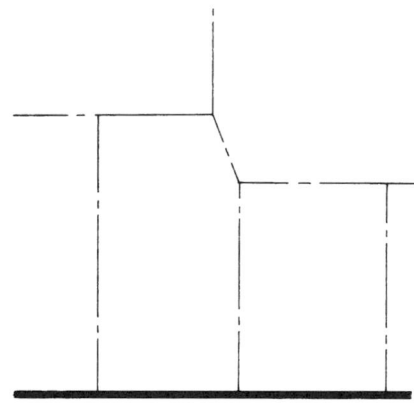

Fig. 11   More desirable lot layout.

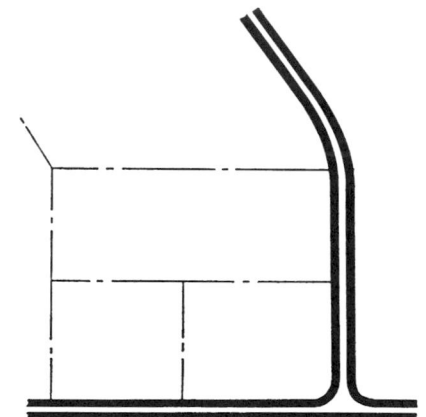

Fig. 12   Undesirable corner lot arrangement.

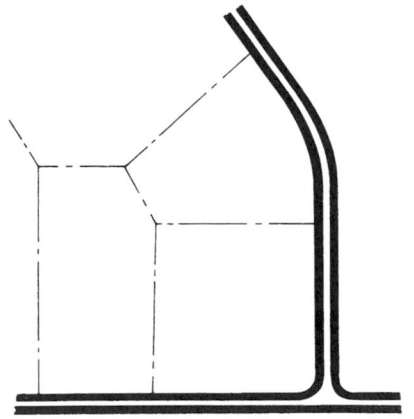

Fig. 13   More desirable corner lot arrangement.

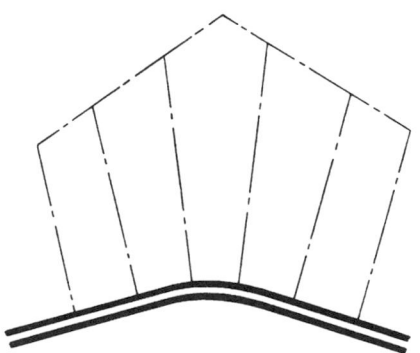

Fig. 14   Excessively deep lots.

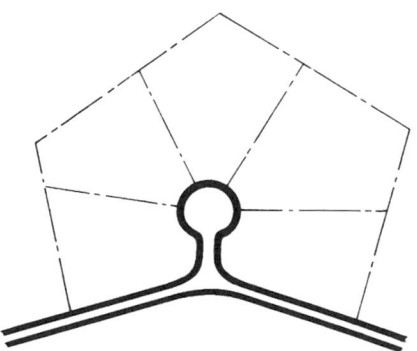

Fig. 15   More desirable lots.

DRAINAGE AND STREAMS

A subdivision site that is traversed by a small drainage way or a small stream often requires special consideration. A small stream may necessitate a different treatment from that used for a small drainage way. The lots should be laid out so that the drainage way will not be near the center of the lot (Fig. 16a). More desirable and usable lots (Fig. 16b) can be created by letting the side lot line follow the center of the drainage way and by providing an adequate easement on each side of this line for drainage purposes. The lot width should be increased to allow for the easement and still provide a suitable building site. When a small stream traverses a subdivision site, desirable lots can be created by providing a drainage right-of-way or easement on each side of the stream and backing the lots up to it (Fig. 16d). This treatment tends to preserve the stream bed in its natural state, provide continuous public or private open space, and eliminate the need for costly and undesirable driveway culverts that would be required if lots were fronted on the stream (Fig. 16c).

The development of a desirable street arrangement and lot layout is essential if the subdivision is to become an asset to the community. However, this alone is not enough. Adequate street improvements, utilities, and drainage facilities must be installed and certain community facilities provided.

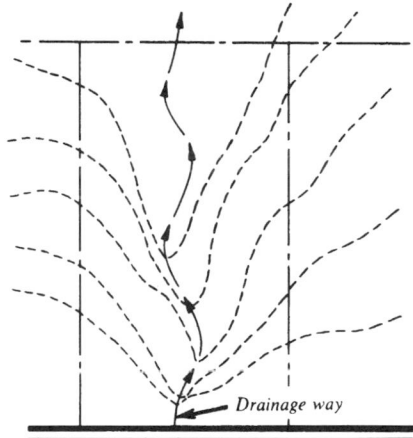

Fig. 16a Undesirable building site.

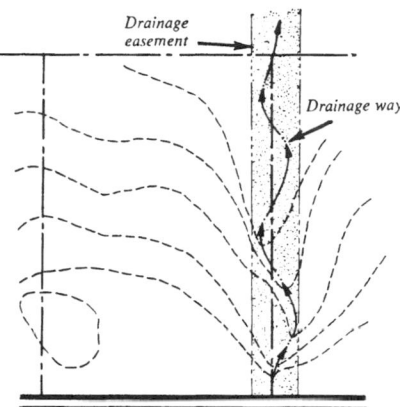

Fig. 16b More desirable building site.

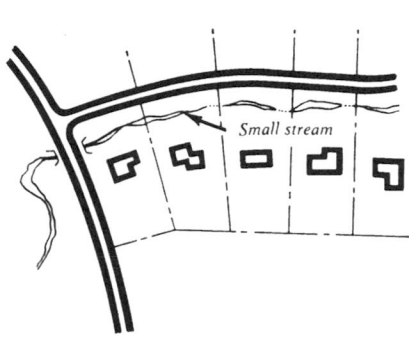

Fig. 16c Undesirable design.

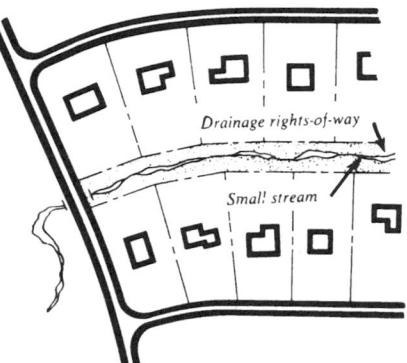

Fig. 16d More desirable design.

STREET AND LOT LAYOUT

CONTROLLING THE COST OF IMPROVEMENTS

Subdividers can be required to provide for acceptable street improvements in their subdivisions. If the proper standards of design have been used and the highest quality of construction maintained, the original cost and annual upkeep will, over the long run, prove cheaper than cutting original costs by inadequate design and construction. A low-cost street base and pavement, while reducing the developer's cost for an improved lot, will last only a few years, and the new taxpayers will join in paying the costs of future reconstruction or expensive annual maintenance.

The subdivider's provision of street lights, fire hydrants, fire alarm boxes, trees, and other items can be referred to the local officials involved for their approval. Private utility mains should be approved by the companies that will provide the service.

Many subdivision layouts are not carefully designed to ensure the most economical provision of street improvements. In fact, many planning boards have redesigned layouts to cut the length of street without sacrificing lots, a fact that points up the usefulness to the developer of obtaining experienced design assistance. Specific examples of uneconomic layout include excessive street pavement due to short blocks, excessive road construction costs due to steep grades requiring cuts, poor lot layouts resulting in unsalable lots such as corner lots too small, odd-shaped lots without a good building site, and the improper use of wet area, rock area, or land otherwise poorly suited to development.

Experience indicates that time and money spent, by both developer and planning board at the beginning of a development, on experienced technical assistance and complete site analysis will save money during construction and after completion. Rigorous application of the proper standards for improvements will return dividends in lower maintenance costs, greater contentment among the new residents, fewer burdens on the local municipal budget for the improvements finally needed, and a quality of development that will show higher and more stable tax values.

CONTROLLING MAINTENANCE COSTS

Proper layout of the new subdivision will obtain the most value for the least amount of street. This will reduce overall street maintenance costs—such as resurfacing, snow plowing, street cleaning, catch basin cleaning, hydrant and street light service, to name a few. Other direct savings will accrue to delivery vehicles through a reduction of the length of trip. Garbage and rubbish collections are similarly affected. In addition to street length, street grade will affect maintenance and servicing costs. Steep grades can force heavy vehicles to take circuitous routes, and can cause hazards amounting to virtual road blocks during adverse weather conditions. The cumulative savings by individuals from proper layout can amount to an impressive sum over a period of years.

PRELIMINARY PLAT

Experience indicates that it is generally not advisable for the subdivider to attempt to present a fully completed subdivision plat to the planning board before submitting a less detailed (and less costly) preliminary map for the board's review. There are usually changes needed after the subdivider and board go over the proposal, and making these changes at the preliminary stage can save costs. Because of this, a "preliminary plat" procedure is specified in many regulations. Town law makes provision for a town planning board to continually approve, with or without modifications, or disapprove a "preliminary plat."

At this preliminary stage, the subdivider is usually expected to present a carefully worked out plan for the development of the site but is not required to finalize this in expensive drawings. The preliminary plat (Fig. 18 on p. 146) should be at a suitable scale with accurate drafting so that all the characteristics of the final plat can be anticipated. Many boards require considerable supplementary data at this stage in addition to the developer's plan. Some of these supplementary requirements may be:

1. Affidavit by owner consenting to the application and submitting proof of ownership
2. Locational sketch showing how the proposed subdivision fits into the area around it
3. Preliminary plans and specifications for road construction, drainage, utilities, and other improvements
4. Temporary stakes along centerlines of roads to facilitate board's field inspection
5. Comments by health department officials on feasibility of water supply and sanitary wastes disposal
6. Comment by county, state, and federal agencies relating to public rights-of-way and sites for public development where applicable

Because of the detailed nature of the information required for the consideration of a preliminary plat, many regulations offer a "preapplication conference" procedure to subdividers where they are given opportunity to discuss the project with the board in order to determine its requirements before engaging technical help. This is a practical necessity where the project is a large one or where the developer is new in the community. Figure 17 on p. 145 shows a preapplication sketch used by a subdivider when discussing a project with the planning board, and suggested modifications and additional requirements were noted thereon by the board.

The action of the planning board, after its review of the subdivider's preliminary plat, should be such as to avoid any inference that the subdivider has, in fact, received approval of a *plat*, since this can be given only after a public hearing as required by law.

It is recommended that the board's action, including any changes it deems necessary before its final approval, be given to the subdivider in writing and be entered in its record. A written communication from the board to the subdivider is the most practical method of assuring that all parties have a clear understanding of the board's position at this stage.

For example, in taking action on the preliminary plat shown in Fig. 18, the planning board granted "conditional approval" subject to elimination of the four-way intersection by use of cul-de-sac and satisfactory adjustment of the lot layout.

HEALTH DEPARTMENT APPROVAL

Prior to the preparation of a final subdivision plat drawing, the applicant should check its design with the appropriate health department, so that the plat presented to the planning board for its approval will also be acceptable to the health officer and thus be suitable for filing with the county clerk.

Health officers recommend that subdividers engage engineering assistance for advice on the water supply and sewerage aspects of land proposed for development. In turn, engineers are encouraged to discuss their projects with the district health engineer prior to the preparation of detailed plans, since health requirements often necessitate changes in proposals, particularly with respect to the size of lots when individual sewerage systems are involved.

If water supply, other than a well for each lot, is proposed, the developer may need the approval of a local water district, or county or state authorities.

THE SUBDIVISION PLAT

After the preliminary plat has been brought into acceptable shape and health department requirements have been met, subdividers are ready to apply for final approval of the project. At this stage, many localities require a formal application (which may be supplementary to that required for the preliminary plat), an affidavit of ownership, and the payment of application fees. Figure 19 on p. 146 shows a subdivision plat.

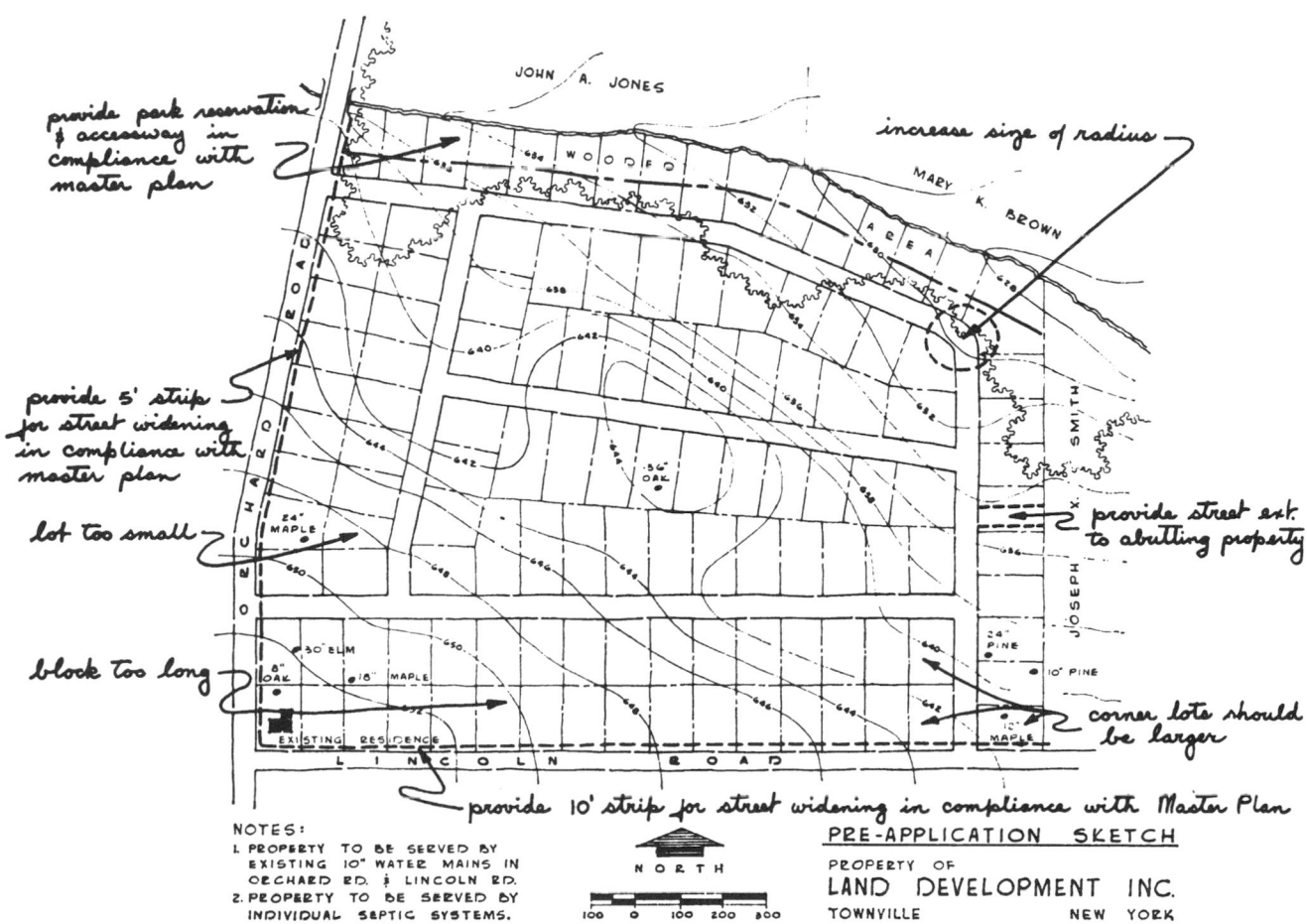

Fig. 17   A preapplication sketch of a subdivision with the planning board's review comments added.

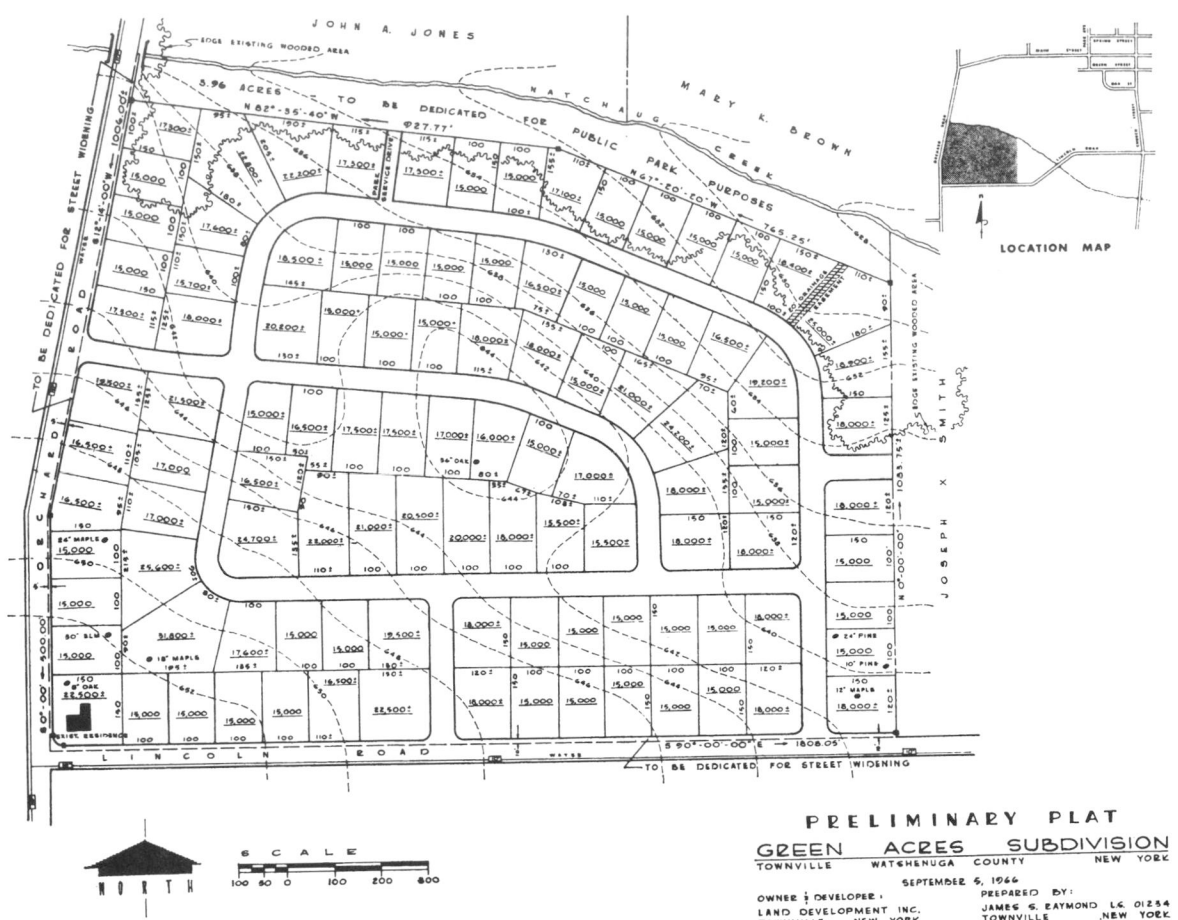

Fig. 18 A preliminary plat.

Fig. 19 A subdivision plat.

DISCOURAGE HEAVY THROUGH TRAFFIC

Minor streets should be so arranged as to make fast through travel impossible. Rapidly moving traffic on local residential streets results in an undue number of accidents and also unnecessarily increases the cost of pavement construction and maintenance.

The mixture of local and through traffic on a residential street creates a condition that tends toward a doubtful policy as to land use and neighborhood growth. Where lots have unlimited and direct access to a heavy traffic street there is a constant threat that the restrictive covenants and zoning ordinance may be broken down by pressure to convert detached dwelling lots into income properties.

The illustration at the left in Fig. 1 shows poor street planning that results in unfavorable conditions.

PLAN FOR EXTENSION OF MAJOR STREETS

In the development of a large subdivision, the relationship of the tract to the master city plan of a community, if such exists, should be ascertained. It is obvious that proposed major streets, transportation, recreation, and public utilities should be considered in planning a new segment of the city.

Where no official plans exist, however, provisions should be made for projecting major streets through the subdivision that now end at the boundary of the tract.

When major traffic streets are not planned as part of the subdivision, lots may be sold and houses built in the path of a future trafficway. Either the development of a through street is blocked or opening of the street later is an unnecessary expense.

TRAFFIC SHOULD FLOW TOWARD THOROUGHFARES

When traffic does not flow toward main thoroughfares, it causes an unnecessary use of local streets in order to reach the main trafficways. This excessive use of residential streets causes an added expense of pavement construction and maintenance. Local streets that carry unnecessary traffic form definite hazards to pedestrians and children.

The street design of a subdivision should be carefully planned to provide for all traffic demands and at the same time create a street arrangement that will make an attractive neighborhood. This will generally produce fewer streets than one that cuts up the land into numerous rectangles without consideration of proper traffic routing.

A monotonous street system of this type is generally extravagant, producing more streets than are needed.

MINOR STREETS SHOULD ENTER MAJOR STREETS AT RIGHT ANGLES

Streets should intersect each other as nearly at right angles as is practicable, and the number of streets converging upon a single point should be kept at a minimum. All minor streets approaching a major thoroughfare at acute angles should be turned so that for a distance of about 100 ft they will be at right angles to the major street.

When minor streets join a thoroughfare at raking angles, visibility is greatly impaired for both motorists and pedestrians. Drivers are also tempted to turn in and out of such streets without greatly reducing their speed.

The sketch plans at the left illustrate how hazardous traffic intersections can be improved by correct plotting to obtain streets crossing at right angles.

AVOID PLANNING OF DEAD-END STREETS

The practice has been, in the past, to place dead-end streets against railroad rights-of-way, open country, or some other permanent or temporary barrier. This should be avoided. The remote possibility, in many cases, of dead-end streets connecting with future streets in an adjoining tract has resulted in blighted property in that particular locality.

When there is a possibility of the street going on through, at some future time, the lot at the end of the street may be reserved for a given time and not sold or built on. If it develops that the street connection will not be necessary, the lot can become a building site and complete the design of the neighborhood. When conditions

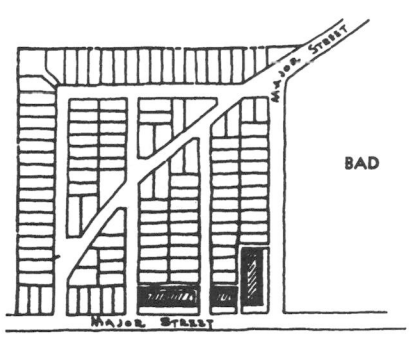
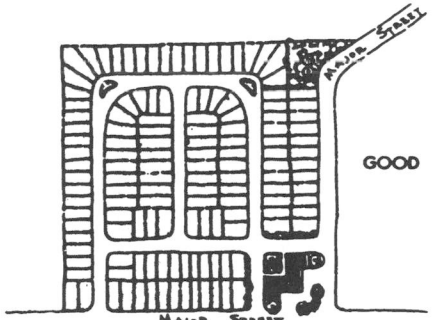

Fig. 1

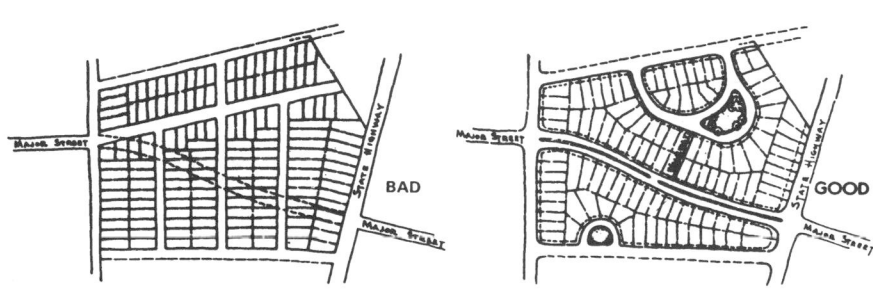

Fig. 2

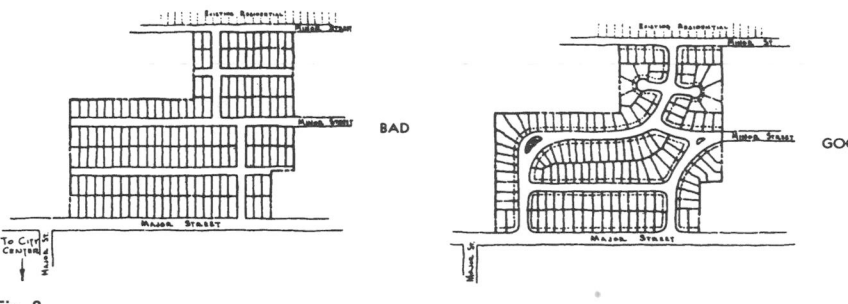

Fig. 3

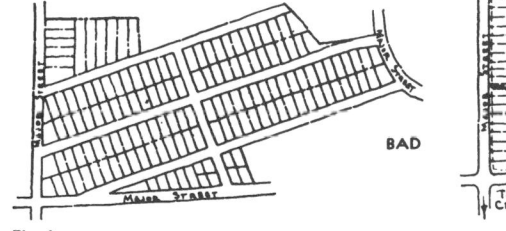

Fig. 4

DESIGN CONSIDERATIONS OF SUBDIVISIONS

make it impractical to avoid a dead-end street, it should be terminated by a turnaround. This circle should have a diameter of at least 100 ft and be at least the depth of one lot from the boundary line of the tract.

STREETS SHOULD FIT CONTOURS OF IRREGULAR LAND

When ground levels of a tract vary considerably, streets should be laid out to conform to natural conditions. Observations as to high and low ground are often adequate to determine the location of streets. If the land is rough, a topographical map should be made to obtain a complete representation of ground conditions.

Streets laid out to fit the contours of the land will avoid excessive grades and reduce construction cost. A subdivision plan based upon the topography of the site not only makes possible a better-designed development but also makes the installation of utilities more economical.

In locating streets, consideration should be given to the size and shape of lots and blocks in order to obtain the best use of the land.

SHORT BLOCKS ARE NOT ECONOMICAL

Figures 7 and 8 contrast two types of local street design—one an example of the rigid grid-iron pattern, the other planned to meet the requirements of local access and circulation.

Short blocks increase initial construction costs because of the large number of cross streets, and also increase traffic hazards and travel time through such districts. In the plan at the right better-shaped lots are secured and those facing the state highway are protected by a park strip. This plan also provides a local shopping center and a school site.

The platting of suburban residential blocks up to 1300 ft in length by two lot-depths wide, bounded by streets that are adjusted to topographic and traffic requirements, is recommended as being most economical.

LONG BLOCKS REQUIRE CROSSWALKS NEAR CENTER

The use of crosswalks through long blocks to afford more direct access to nearby community facilities is desirable because of the appeal and convenience that is lent to otherwise remotely situated residential lots. Such pedestrian ways near the middle of all blocks exceeding 1000 ft in length are recommended.

When a nearby shopping center, school, or park is so located that a large number of residents of a neighborhood are forced into circuitous routes in order that they may reach their destination, it is often desirable to provide crosswalks in shorter blocks—those over 750 ft in length. This often brings the playgrounds or grocery store as much as a ¼ mile nearer in walking distance to the doorsteps of many homes.

PLAN COMMERCIAL SITES WHERE NEEDED

Local shopping centers are definite assets to a community. They should be located within convenient and safe walking distance for the residents and designed to afford adequate off-street delivery and parking facilities.

Commercial structures should be concentrated at suitable centers adjoining a major thoroughfare and be accessible by way of local connecting residential streets. They should be designed together as a group and not as a series of unrelated separate stores.

To ascertain the amount of land needed for commercial use in a community, such factors as estimated per capita sales and the volume of business per store unit must be analyzed in determining the kind and number of stores that the neighborhood can profitably sustain.

PROVIDE SCHOOL AND CHURCH SITES

If a subdivision is large enough to warrant the consideration of all community requirements, locations should be provided for schools and churches. These sites should be centrally located for the convenience of all property owners and citizens in the vicinity. Adequate space should be provided for the parking of automobiles without interfering with private parking needs of those living near the school and church.

These buildings produce a favorable impression as to the stability of a community and therefore should form one of the early demonstrations of neighborhood growth. The

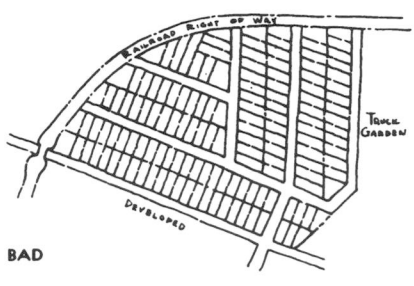

Fig. 5

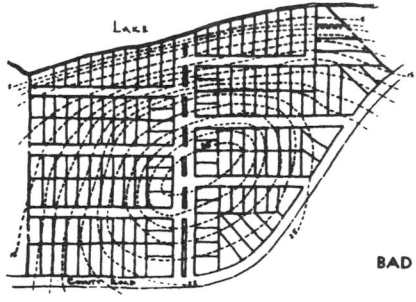
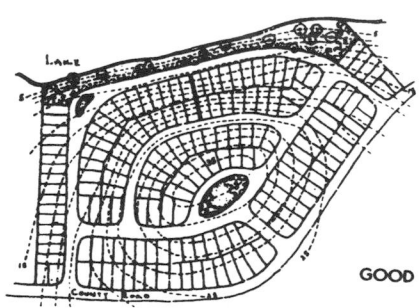

Fig. 6

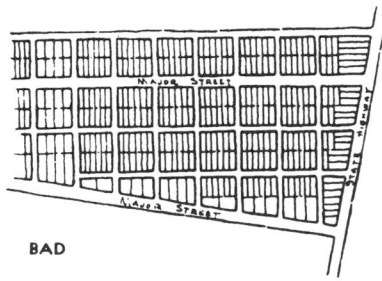
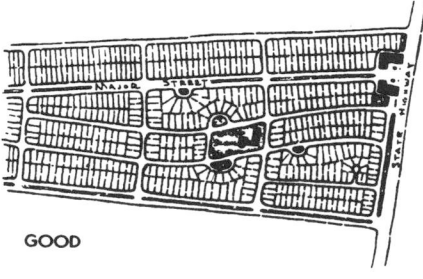

Fig. 7

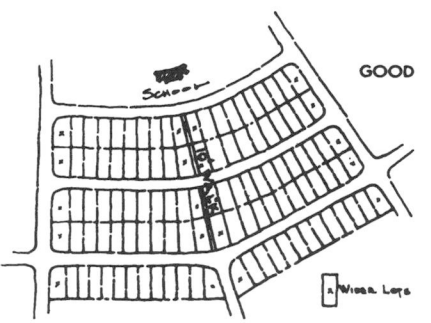

selection of convenient sites for such facilities as schools, churches, and local shopping centers will go far in increasing a subdivision's salability.

PARKS ARE A DEFINITE COMMUNITY ASSET

Rough wooded areas that are difficult to develop into economical dwelling sites are often well adapted for recreational use. Enhanced adjoining property values may exceed the cost of developing and dedicating such public open spaces. A well-located park also may offset the sales resistance of remotely situated lots and render the entire tract more marketable.

Parks are a definite asset to a community. They are a proper place for children and adults to enjoy the outdoors with safety. They reserve, for all time, natural features that all property owners in the vicinity can enjoy. They are as important to neighborhood development as any other general feature.

The improvement and maintenance of park areas should be handled in the same manner as street improvements and maintenance.

PRESERVE NATURAL FEATURES OF SITE FOR IMPROVED APPEARANCE

It is recommended that whenever possible all natural features of neighborhood should be preserved to add to the beauty of the tract.

In many cases valuable tree growth has been cut down and knolls have been removed in order to fill in lower ground. Frequently this is an unnecessary expense and only results in the ruination of what might be a more valuable residential property.

A more desirable neighborhood can be created when roads are located to fit the existing lay of the ground and placed in such a manner as to preserve, as far as possible, the native tree growth. The curving of streets to fit contours of the land and the saving of valuable trees add to the beauty of a development and reduce construction cost.

DEEP LOTS ARE WASTEFUL

Great depth in a residential lot generally does not increase its salability by virtue of its large area. This type of platting materially decreases the number of lots in a subdivision. Residential lots over 150 ft in depth are usually undesirable unless they are ¼ acre or more in size.

If consistent with economical land subdivision, residential lots of 50 or 60 ft in width should not greatly exceed 130 ft in depth. Lots of from 100 to 120 ft in depth will usually be found satisfactory for single-family dwellings. Lot sizes should be arrived at only after a careful study of local conditions and by an analysis of the relationship between front foot utility and street construction costs and the value of undeveloped land.

PLAN LOTS OF ADEQUATE WIDTH

The well-being of a neighborhood and the economic soundness of a project rest largely on the manner in which the land is divided into lots.

The width of these units should not be reduced beyond a minimum consistent with building coverage, and light and air requirements. Each developer should consider the question of lot width from the point of view of local regulations, the character and topography

of the site, the type of dwellings contemplated, and the ratio of raw (unimproved) land costs to the linear front-foot costs of local public utility improvements.

Practical building sites require lots at least 50 ft wide to provide adequate side yards for light, air, driveways, and to avoid crowding.

AVOID SHARP-ANGLED LOTS

Lots that have sharp-pointed corners are wasteful of land because the resulting wedge-shaped areas have little or no utility. Such lots also constitute poor building sites.

Sharp-angled lots can be avoided by planning streets to intersect at right angles and by making side lot lines perpendicular or radial to street lines.

In the adjacent sketches are contrasted an extravagant—though not unusual—type of subdivision plan and a suggested revised design that has 40 percent less street area, better-sized and shaped lots, and eliminates hazardous traffic intersections.

Attention also is directed to the manner in which deep lots are backed against the highway bounding one side of the tract, thus permitting all houses to face into the subdivision.

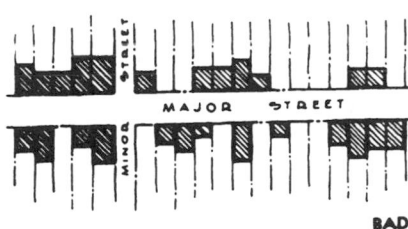
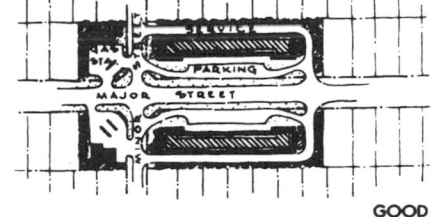

Fig. 9

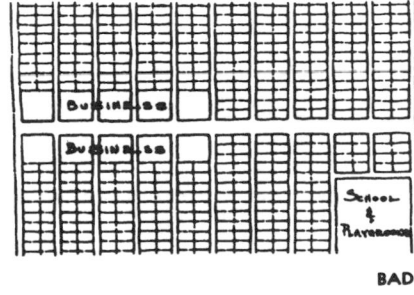
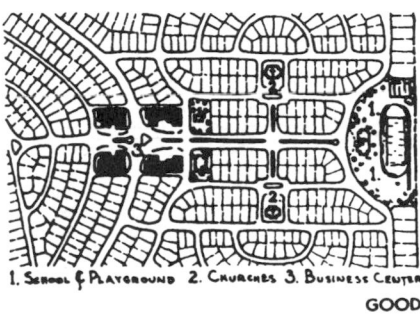

Fig. 10

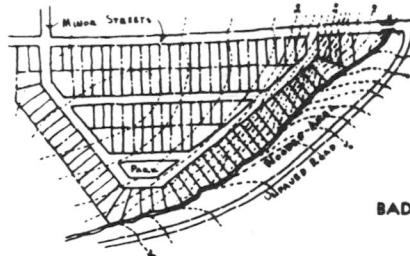
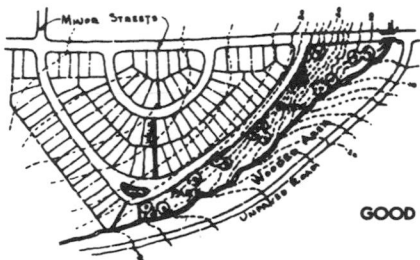

Fig. 11

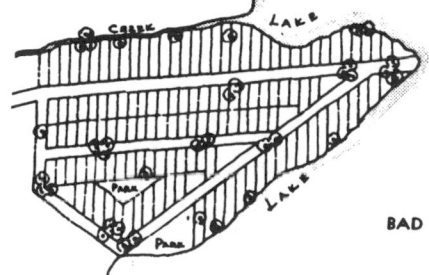
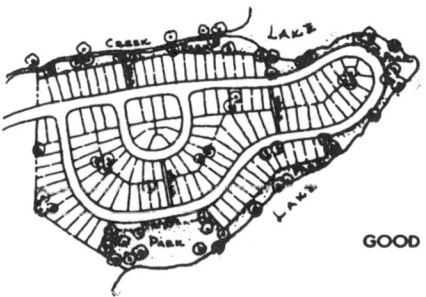

Fig. 12

DESIGN CONSIDERATIONS OF SUBDIVISIONS

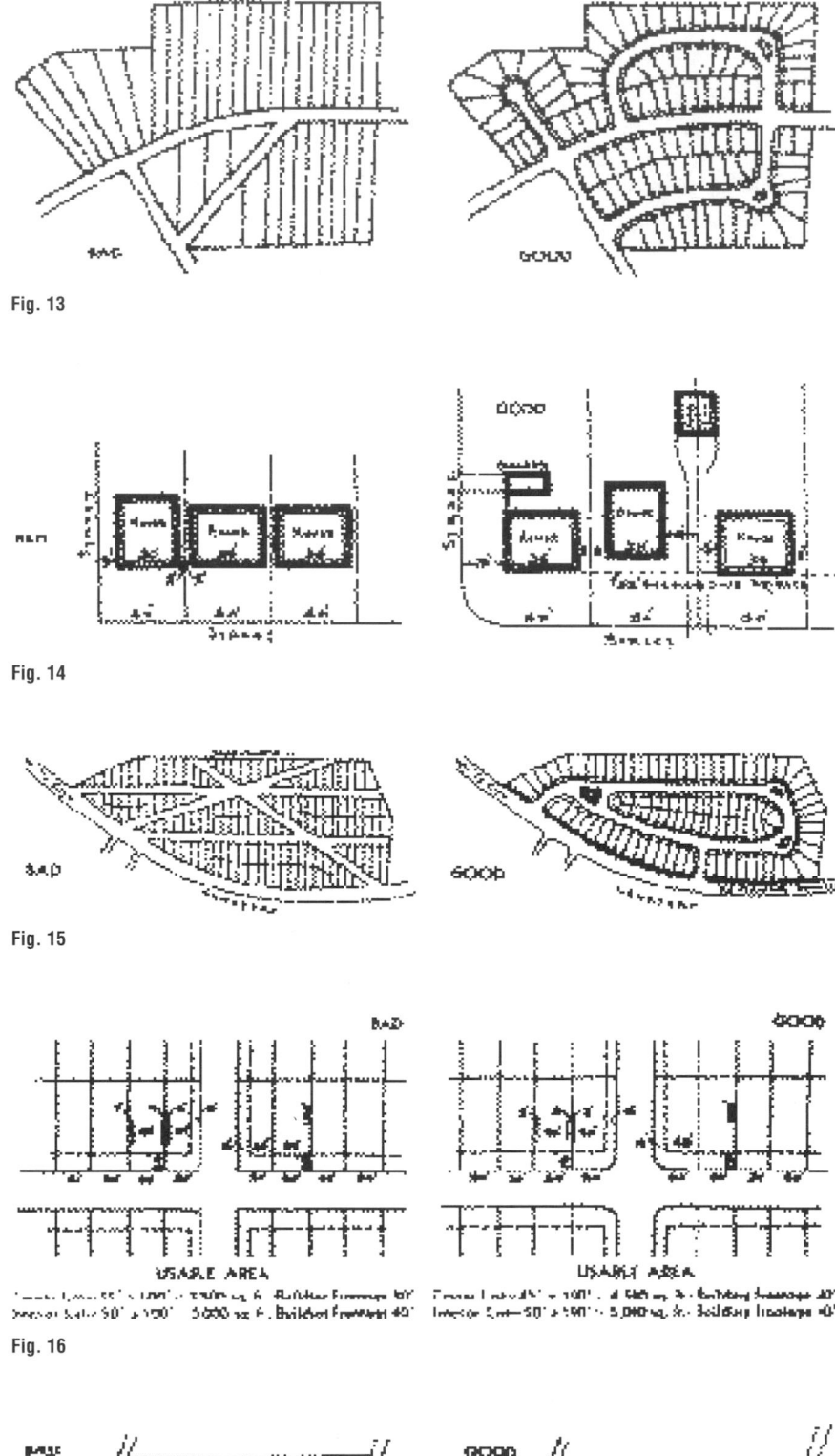

Fig. 13

Fig. 14

Fig. 15

Fig. 16

Fig. 17

PLAN WIDER CORNER LOTS

Every residential lot within a neighborhood should be sufficiently spacious to provide free area on all sides of the space to be covered by a dwelling. Because of the special requirements imposed upon corner lots by reason of necessary setbacks from two streets, it is recommended that corner lots be given extra width at least to the extent of the additional side yard demanded by the side street setback requirement. In the case of a normal corner lot with a side yard requirement of 15 ft, the width should be 10 ft wider than interior lots.

Regulations establishing minimum building line setbacks on the front, sides, and rear of dwellings must be considered by the subdivider at the time lot lines are established.

MAKE LOT LINES PERPENDICULAR TO STREET

In order that maximum use be obtained from all lots, it is suggested that the lot lines be kept perpendicular or radial to street lines. When this is not done, there is a tendency to build houses on lots so that the sides of the houses are parallel with the side lot lines. This creates an unattractive sawtooth arrangement and many times causes the front of one house to face into the side or rear of a neighboring house.

If a maximum use is to be made of every square foot of the lot area, it is important that the lot be well shaped. If lines are not kept perpendicular to the street, sharp-angled corners will result. These are difficult to utilize and give the area an undesirable appearance.

PLAN LOTS TO FACE DESIRABLE VIEWS

In laying out a subdivision the planner should take advantage of any natural or created beauty spot. Whenever possible, lots should be so faced that houses will look out over the park rather than face on side streets.

Developers should give consideration to the arrangement of lots so that the proposed dwellings will not overlook neighboring rear yards, face undeveloped and restricted property, or be exposed to the adverse effects of heavily traveled streets and adjacent nonconforming land uses.

Each lot within a new subdivision should not only constitute a good house site but also be so planned as to size, shape, and orientation that it takes full advantage of such desirable natural features as views, the slope of the land, sunlight, prevailing winds, shade trees, and adjoining public spaces.

PROTECT LOTS AGAINST ADJACENT NONCONFORMING USES

Residential lots should be arranged so that they will not be seriously affected by a nonconforming use of adjoining property. Objectionable properties can be blocked off by screen planting, or the lots backed against the nonconforming land so that houses built on them front away from the objectionable use. It is suggested that where possible the subdivision boundary line be along the rear of a lot rather than the center of a street.

The appearance and value of a building site are improved when it faces a similar site across the street. Correct location of lots, well-drawn restrictive covenants, and zoning ordinances are a protection against the blighting influence of adjoining nonconforming property uses.

PROTECT RESIDENTIAL LOTS
AGAINST MAJOR STREET TRAFFIC

When residential lots are located on a major thoroughfare, it is suggested that the through traffic be separated from local service by a planting strip about 20 ft wide.

An 18-ft local service roadway should be located inside of this planting protecting the residences against the noise and dust of traffic, and lessening the street dangers to children. Increase in the desirability of the lots will offset the cost of added street width, and the planting of trees and shrubs will add to its attractiveness.

In the past it has been the custom of developers of subdivisions to set aside all property on main thoroughfares for business or apartments because of the belief that a major highway was not a suitable place for a private dwelling. The result has been spotted developments with many vacant lots.

An excessive amount of street construction, the rigid and monotonous layouts of streets, the use of "butt" lots, and the subdividing of the wooded lakeshore, as shown in Fig. 21, would make the project costly to develop and difficult to market.

The revised plan (Fig. 22) has overcome these objections, and every lot has been made a desirable building site. Although this plan provides fewer lots, the changes permit a greater financial return and quicker sales for the developer and a better investment for the buyer.

The subdivision in Fig. 23 provides 101 desirable building sites. A majority of the houses face east or west and will therefore receive sunlight into their front rooms at some time during the day. In the preparation of the plat for recording, lots should be numbered consecutively throughout the entire tract.

The street plan is adapted to the topography and provides for surface water drainage. Although the number of entrances from the major thoroughfare is limited, the street pattern facilitates the flow of traffic from the principal approach. Curved streets create greater appeal than is possible in a gridiron plan. Blocks up to 1200 ft long are desirable and reduce expense for cross streets. This subdivision does not require its own system of major thoroughfares. However, recognition is made of the present and planned roadway pattern of the city in which it is located.

A subdivision of this size does not require provision for complete community facilities, such as stores, schools, and churches necessary in a larger neighborhood.

Complete information regarding the site and its relation to the town or city of which it is part is essential to the planning of a desirable residential neighborhood. Not only is it necessary to have a closed, true-boundary survey, but also complete topographical data, including locations of existing trees that might be preserved. The capacity of storm and sanitary sewers should be known. The adequacy of a safe water supply system and the existence of other essential utilities, and of transportation facilities, are important factors.

Residential subdivisions should be located where they will not be adversely affected by industrial expansion and other nonconforming uses. They should be in the trend of residential development of similar types of homes. To further assure stability, residential areas should be safeguarded by recorded protective covenants, and the establishment and enforcement of a zoning ordinance governing the use of the property and surrounding areas.

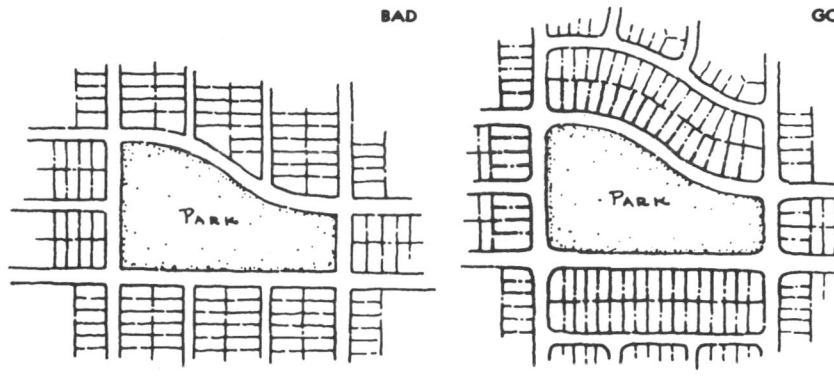

Fig. 18

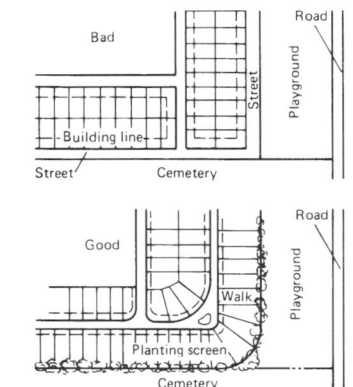

Fig. 19

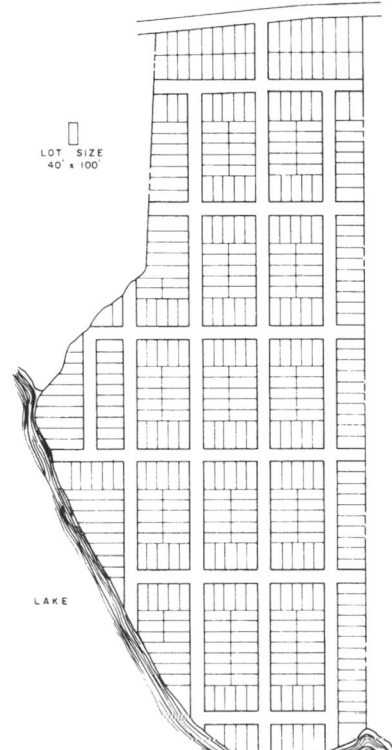

Fig. 21 Typical subdivision pattern. This type of subdivision plan is marked by excessive amounts of street construction, lots blocking the shoreline, and no open space.

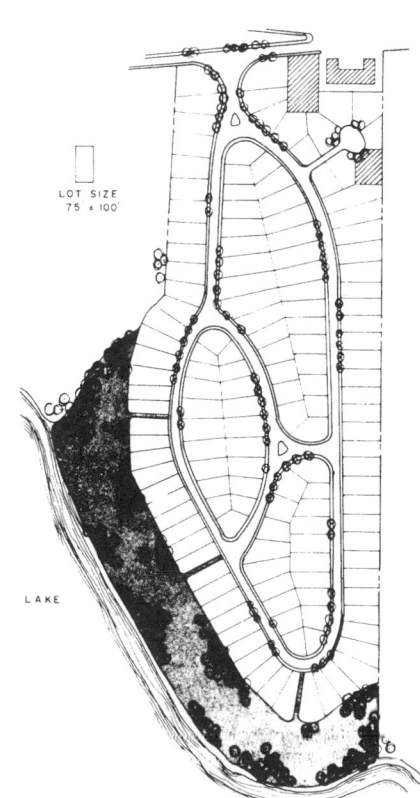

Fig. 20

Fig. 22 Revised pattern. This revised plan, though of fewer building lots, provides amenities of better building lots and preservation of the shoreline in community open space.

DESIGN CONSIDERATIONS OF SUBDIVISIONS

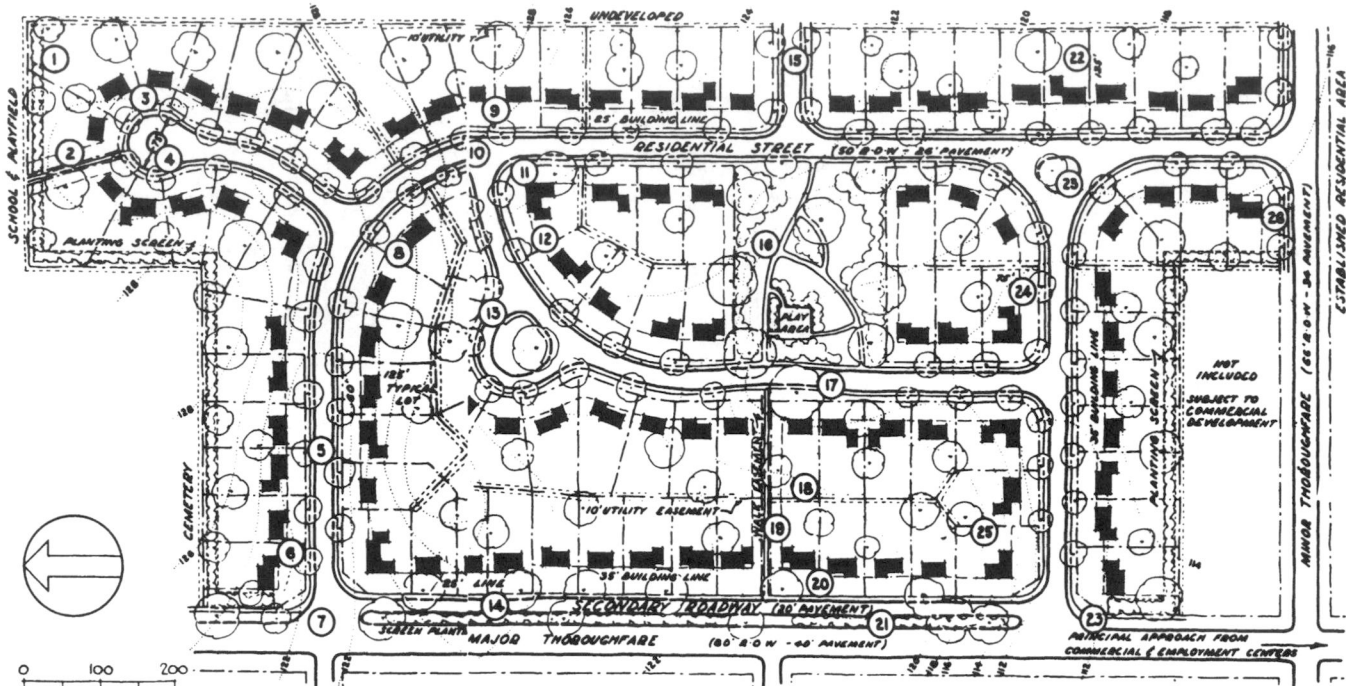

Fig. 23

Legend

1. 15-ft easement for planting screen to provide protection from nonresidential use.

2. 10-ft walk easement gives access to school.

3. Cul-de-sac utilizes odd parcel of land to advantage.

4. Turnaround right-of-way 100 ft in diameter.

5. Street trees planted approximately 50 ft apart where no trees exist.

6. Additional building setback improves subdivision entrance.

7. Street intersections at right angles reduce hazards.

8. Lot side line centered on street end to avoid car lights shining into residences.

9. Residences opposite street end set back farther to reduce glare from car lights.

10. Three-way intersections reduce hazards.

11. Property lines on 30-ft radii at corners.

12. Lot lines perpendicular to street right-of-way lines.

13. "Eyebrow" provides frontage for additional lots in deeper portion of block.

14. Secondary roadway eliminates hazard of entering major thoroughfare from individual driveways.

15. Provision for access to land now undeveloped.

16. Neighborhood park located near center of tract. Adjacent lots wider to allow for 15-ft protective side line setback.

17. Pavement shifted within right-of-way to preserve existing trees.

18. Above-ground utilities in rear line easements.

19. 10-ft walk easement provides access to

park. Adjacent lots wider to allow for 15-ft protective side line setback.

20. Variation of building line along straight street creates interest.

21. Screen planting gives protection from noise and lights on thoroughfare.

22. Lots backing to uncontrolled land given greater depth for additional protection.

23. Low planting at street intersections permits clear vision.

24. Wider corner lot permits equal building setback on each street.

25. Platting of block end to avoid siding properties to residences across street.

26. Lots sided to boundary street where land use across street is nonconforming.

SUMMARY COMPARISONS

Public Cul-de-Sacs

TABLE 1 Cost Comparisons—Single-Family Detached

	2.75	4	Typical standards 4	Zero lot line 5
Clearing and grubbing	$ 452	$ 350	$ 381	$ 330
Grading streets	332	276	392	257
Street pavement	593	498	731	459
Storm drainage	854	611	619	583
Sanitary sewer	943	827	923	801
Water distribution	552	468	531	447
Curbs and gutter	679	701	679	657
Driveways	760	700	700	500
Sidewalks	272	208	212	188
Street trees	366	306	306	294
Grading/seeding	1,157	741	768	523
Totals	$6960	$5686	$6242	$5039
Percent of 4 dwelling units per acre, typical standards plan	112	91	100	81

TABLE 2 Quantity Size Comparisons—Single-Family Detached

	2.75	4	Typical standards 4	Zero lot line 5
Square feet per unit	2,500	2,000	2,000	2,000
Average lot size, ft^2	14,363	9,611	9,675	7,390
Off-street parking	4	4	4	2+
Minor street width, ft	20	20	30	20
Minor street right-of-way, ft	28	28	50	28
Street pavement per dwelling unit	1,268	1,066	1,566	986
Linear feet of street per dwelling unit	61	51	51	49
Curbs and gutters per dwelling unit	93	96	93	90

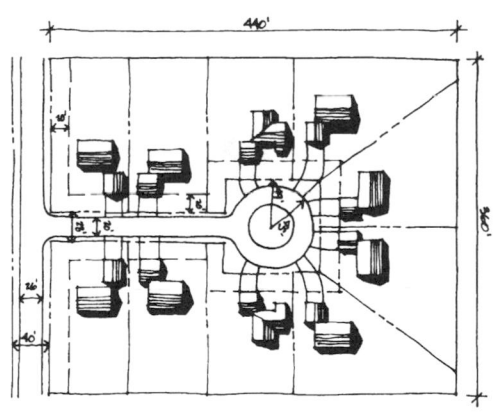

2.75   Net Density

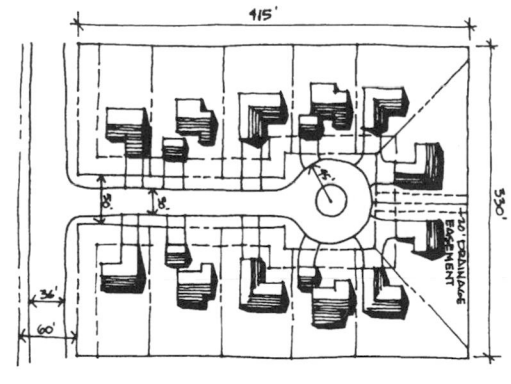

**4.0   Typical Standards
Net Density**

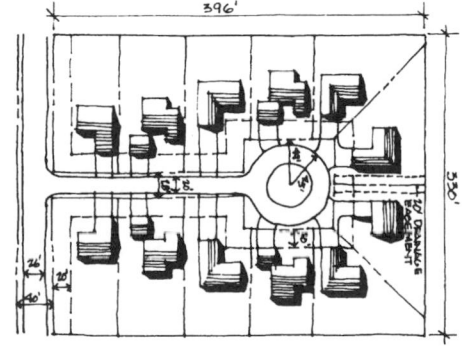

4.0   Net Density

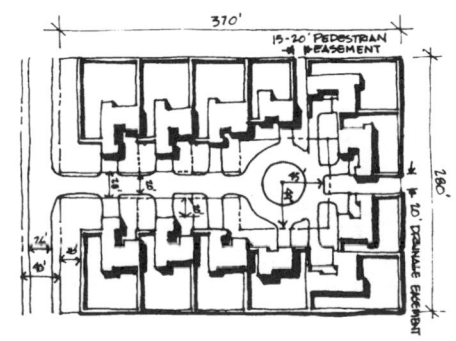

**5.0   Zero Lot Line
Net Density**

Fig. 1

153

COMPARISON OF TYPICAL CUL-DE-SAC DEVELOPMENT

TABLE 3 Cost Comparisons—Single-Family/Attached

	Duplex 5	Duplex 7.25	Triplex 6	Quadplex 8.25
Clearing and grubbing	$ 332	$ 253	$ 296	$ 215
Grading streets	198	160	183	137
Street pavement	356	287	333	250
Storm drainage	471	426	421	322
Sanitary sewer	711	633	669	607
Water distribution	375	321	348	300
Curbs and gutter	496	402	460	343
Driveways	370	320	320	398
Sidewalks	156	120	144	104
Street trees	216	180	204	150
Grading/seeding	594	331	486	312
Totals	$4275	$3433	$3664	$3138
Percent of dwelling units per acre, typical standards plan	68	55	62	50

TABLE 4 Quantity/Size Comparisons—Single-Family/Attached

	Duplex 5	Duplex 7.25	Triplex 6	Quadplex 8.25
Square feet per unit, ft^2	2450	2200	1800	900
Average lot size, ft^2	7854	4888	6375	4628
Off-street parking	4	2+	2	2
Minor street width, ft	20	20	20	20
Minor street right-of-way, ft	28	28	28	28
Street pavement per dwelling unit	761	615	715	535
Linear feet of street per dwelling unit	36	30	34	25
Curbs and gutters per dwelling unit	68	55	63	47

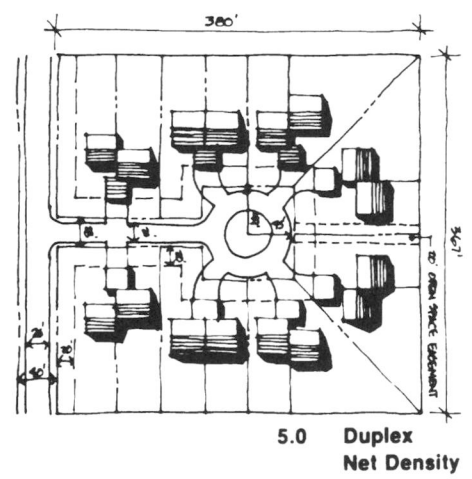

**5.0 Duplex
Net Density**

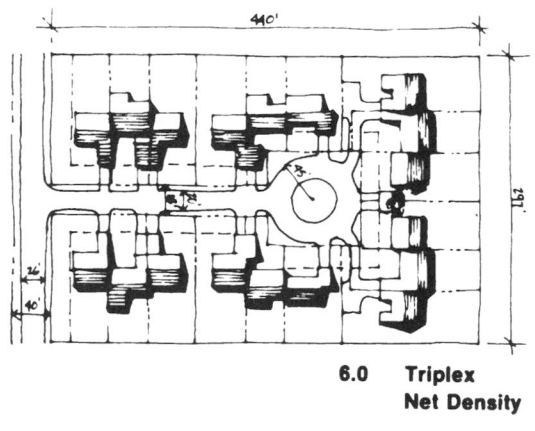

**6.0 Triplex
Net Density**

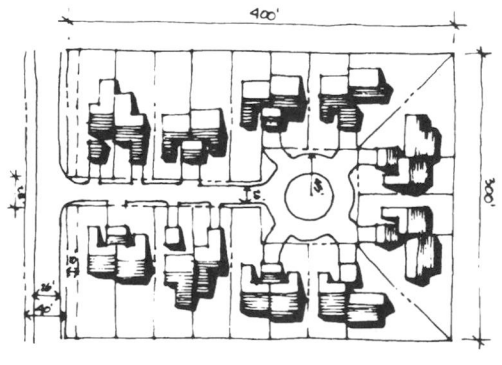

**7.25 Duplex
Net Density**

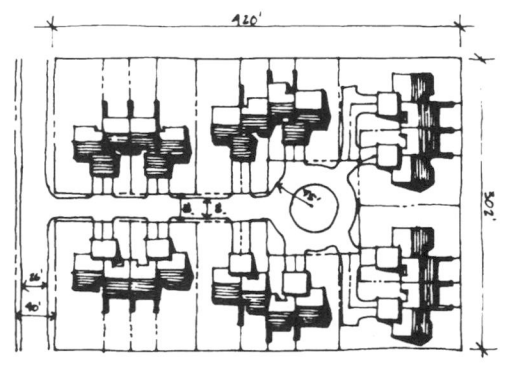

**8.25 Quadplex
Net Density**

Plan Efficiency

TABLE 5 Cost Comparisons of Four Dwelling Units per Acre Plans

	Public cul-de-sac	Public eyebrow	Public street	Typical standards cul-de-sac
Cleaning and grubbing	$ 350	$ 373	$ 339	$ 381
Grading streets	276	416	226	392
Street pavement	498	787	453	731
Storm drainage	611	997	763	619
Sanitary sewer	827	959	749	923
Water distribution	468	558	414	531
Curbs and gutter	701	861	533	679
Driveways	700	660	650	700
Sidewalks	208	268	272	212
Street trees	306	390	288	306
Grading/seeding	741	733	706	768
Totals	$5686	$7002	$5413	$6242
Percent of typical standards cul-de-sac	91	112	87	100

TABLE 6 Quantity/Size Comparisons of Four Dwelling Units per Acre Plans

	Public cul-de-sac	Public eyebrow	Public street	Typical standards cul-de-sac
Square feet per unit	2000	2000	2000	2000
Average lot size, ft^2	9611	9200	9425	9675
Off-street parking	4	4	4	4
Minor street width, ft	20	20	26*	30
Minor street right-of-way, ft	28	28	40*	50
Street pavement per dwelling unit	1066	1672	943	1566
Linear feet of street per dwelling unit	51	65	36	51
Curbs and gutter per dwelling unit	96	118	73	93

*Subcollector street dimensions.

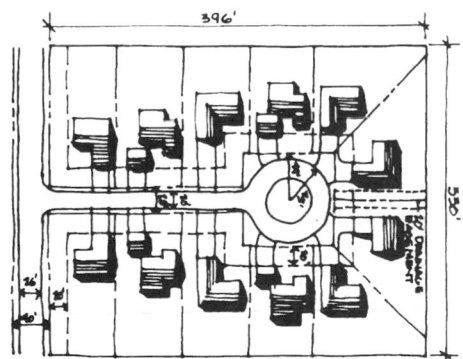

Public Cul-de-sac

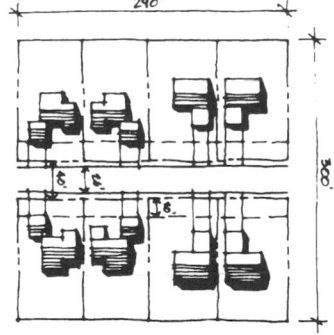

Public Street

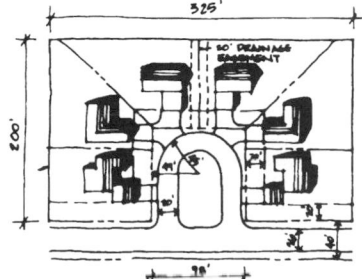

Public Eye Brow

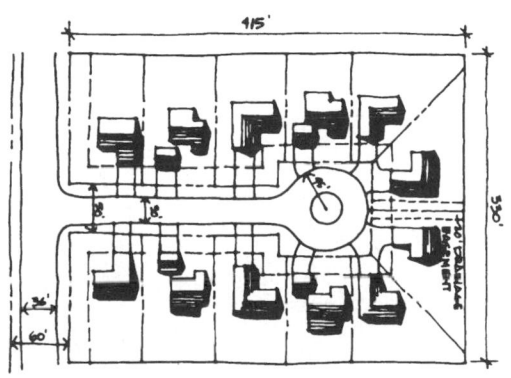

Typical Standards Cul-de-sac

Fig. 3

155

COMPARISON OF TYPICAL CUL-DE-SAC DEVELOPMENT

TABLE 7   Cost Comparisons of Four Dwelling Units per Acre Plans

	Private court	Commons court	Auto court
Clearing and grubbing	$ 386	$ 368	$ 361
Grading streets	326	427	468
Street pavement	696	756	875
Storm drainage	766	808	808
Sanitary sewer	1133	963	991
Water distribution	702	564	579
Curbs and gutter	511	1095	861
Driveways	440	500	500
Sidewalks	300	144	144
Street trees	152	144	144
Grading/seeding	860	751	687
Totals	$6272	$6520	$6418
Percent of typical standards cul-de-sac	100	104	103

TABLE 8   Quantity/Size Comparisons of Four Dwelling Units per Acre Plans

	Private court	Commons court	Auto court
Square feet per unit	2,000	2,100	2,100
Average lot size, ft²	10,875	9,114	844
Off-street parking	4	4	4
Minor street width, ft	18	20	20
Minor street right-of-way, ft	20	98	130
Street pavement per dwelling unit	1,470	1,622	1,881
Linear feet of street per dwelling unit	38	76	81
Curb and gutter per dwelling unit	70	150	118

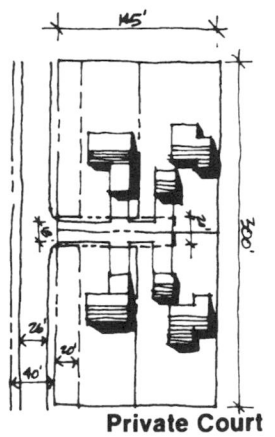

Private Court

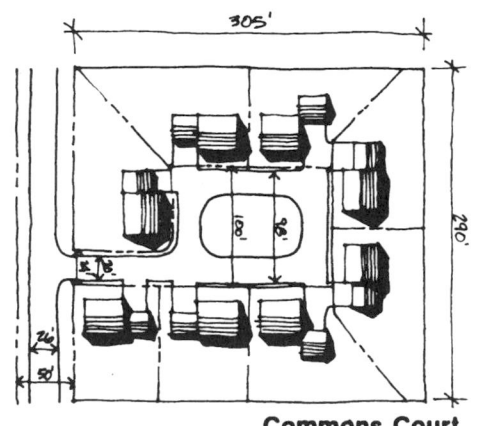

Commons Court

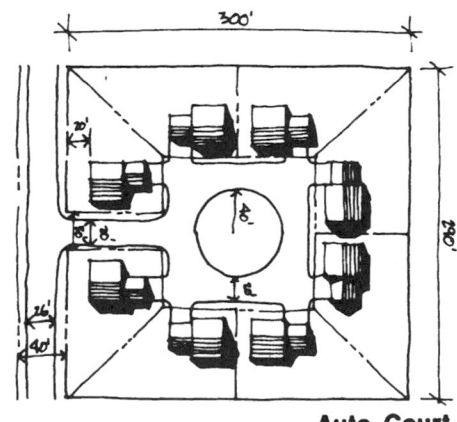

Auto Court

Fig. 4

Cul-de-Sac vs. the Court

TABLE 9   Cost Comparisons of Plans

	Zero lot line		Triplex	
	8/court	5/cul-de-sac	7.25/court	6/cul-de-sac
Clearing and grubbing	$ 216	$ 330	$ 237	$ 296
Grading streets	197	257	204	183
Street pavement	422	459	364	333
Storm drainage	490	583	396	421
Sanitary sewer	817	801	697	669
Water distribution	417	447	381	348
Curbs and gutter	540	657	445	460
Driveways	400	500	500	320
Sidewalks	168	188	156	144
Street trees	168	294	156	204
Grading/seeding	277	523	298	486
Totals	$4166	$5039	$3834	$3864

TABLE 10   Quantity/Size Comparisons of Plans

	Zero lot line		Triplex	
	8/court	5/cul-de-sac	7.25/court	6/cul-de-sac
Square feet per unit	1200	2000	1800	1800
Average lot size, ft^2	4299	7390	4583	6375
Off-street parking	3	2+	2	2
Minor street width, ft	20	20	18	20
Minor street right-of-way, ft	24	28	20	28
Street pavement per dwelling unit	986	915	785	715
Linear feet of street per dwelling unit	42	49	39	34
Curb and gutter per dwelling unit	74	90	61	63

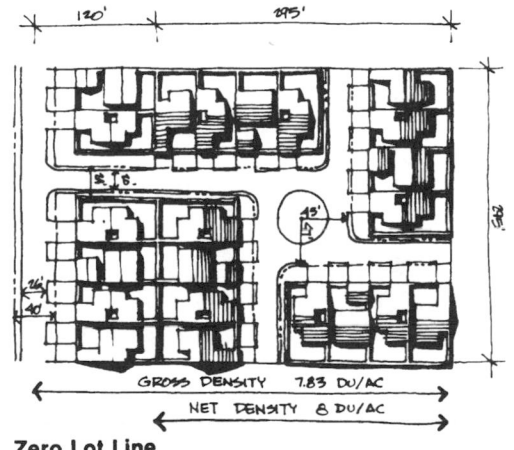

**Zero Lot Line
8/Court**

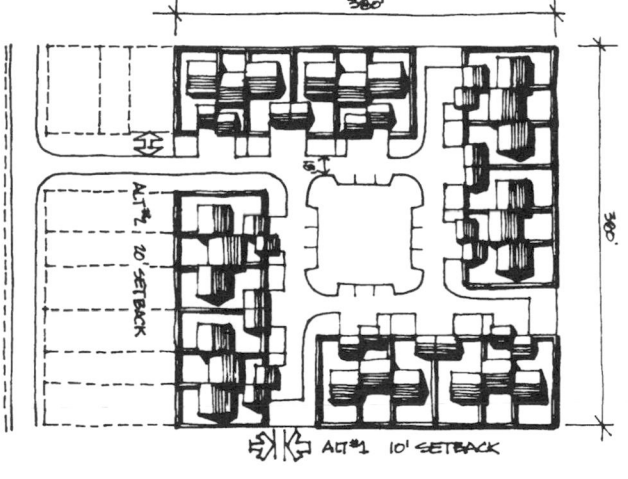

**Triplex
7.25/Court**

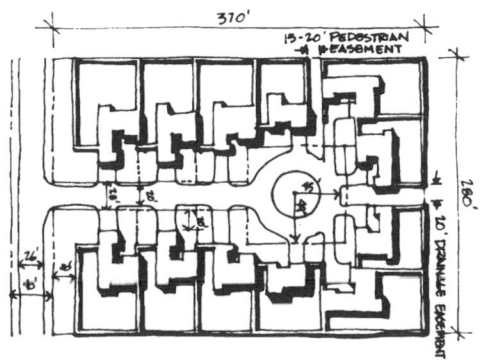

**Zero Lot Line
5/Cul-de-sac**

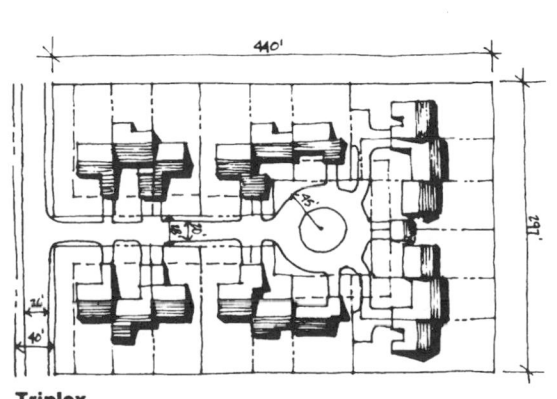

**Triplex
6/Cul-de-sac**

Fig. 5

COMPARISON OF TYPICAL CUL-DE-SAC DEVELOPMENT

TABLE 11   Cost Comparisons of Plans

	Duplex plans		
	5/cul-de-sac	7.25/cul-de-sac	6.8 private court
Cleaning and grubbing	$ 332	$ 253	$ 249
Grading streets	198	160	216
Street pavement	356	287	413
Storm drainage	471	426	449
Sanitary sewer	711	633	781
Water distribution	375	321	444
Curbs and gutter	496	402	627
Driveways	370	320	250
Sidewalks	156	120	188
Street trees	216	180	188
Grading/seeding	594	331	371
Totals	$4275	$3433	$4176

TABLE 12   Quantity/Size Comparisons of Plans

	Duplex plans		
	5/cul-de-sac	7.25/cul-de-sac	6.8 Private court
Square feet per unit	2450	2200	1800
Average lot size, ft²	7854	4888	5000
Off-street parking	4	2+	2
Minor street width, ft	20	20	18
Minor street right-of-way, ft	28	28	20
Street pavement per dwelling unit	761	615	889
Linear feet of street per dwelling unit	36	30	47
Curb and gutter per dwelling unit	68	55	86

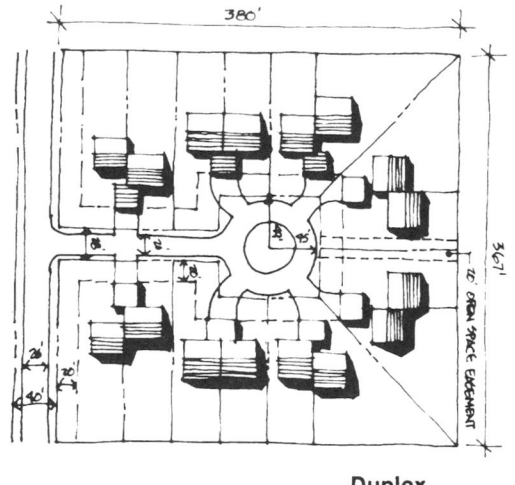

**Duplex
5/Cul-de-sac**

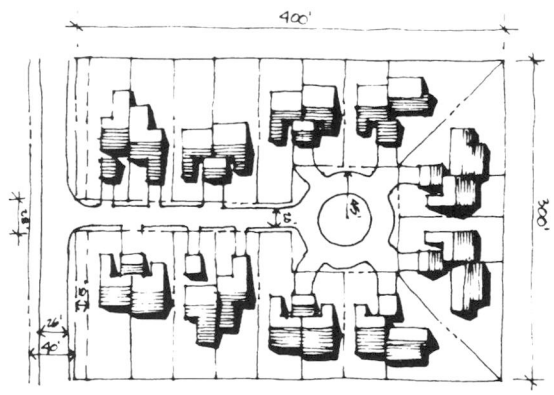

**Duplex
7.25/Cul-de-sac**

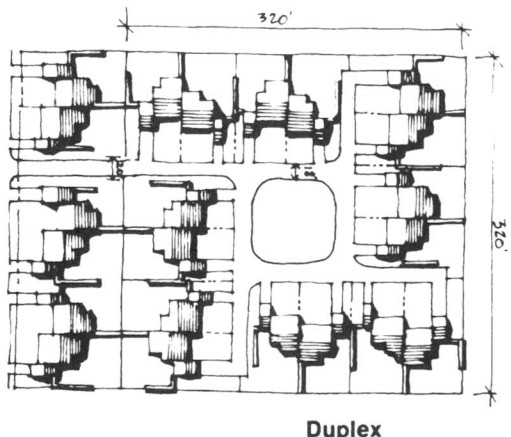

**Duplex
6.8/Private Court**

BUILDING ARRANGEMENTS

As buildings are first designed through a consideration of how they will ultimately fit into the site plan, site placement must be studied in detail. Some of the factors influencing the grouping of buildings are:

- The shape and slope of the buildable land areas
- The location of views and amenities
- The density to be achieved
- The character of the site's surroundings

Buildings may be placed on one side (single-loaded) or both sides (double-loaded) of vehicular access courts. The coordination of buildings on a site may involve one or more of four basic configurations which will be discussed in the following sections. Depending on the site's locale, streets and alleys may determine the site plan. Or existing streets or city blocks may serve as a formal site perimeter for newly designed interior court roads. On less constrained (often suburban) sites, buildings located on linear courts or in clusters become the primary planning approach.

Single-Loaded Courts

The single-loaded court works well with garage-court entry buildings on a site with a narrow buildable area. However, a lower density of houses is the typical result. This is acceptable if housing prices are high enough to accommodate the increased raw land cost per unit and a large enough site development budget exists to properly improve or preserve the across-court views. In more dense construction with multistory buildings, the single-loaded court road may be necessary to allow enough road frontage for adequate garage and/or parking. Where single-loaded courts are used with drive-under homes, homes may be designed with decks facing the green space beyond the pavement.

Double-Loaded Courts

A site planning technique to preserve open space and/or attain increased density will front buildings on both sides of a court road. Double-loaded courts concentrate buildings to disturb less of the site. The cross-court privacy problems make green-space entry drive-unders with decks built over the garages a less desirable solution. Moreover, double-loading drive-under homes on a site may result in canyonlike garage courts. Where this is necessary, focusing living spaces (and patios or decks) away from the paving area and other units is preferable.

Double-loaded courts permit densities in the upper ranges of 6 to 11 units per acre with green-space entry buildings. Garage-court entry buildings with entries facing the paving would yield a somewhat lower density because of the less regimented building configurations required for better livability. Buildable site area and site development economies are common reasons for double-loading roads.

However, improper use or overuse of this planning method is a main source of the garage court syndrome. A land plan that places too many units off a given amount of court road magnifies the feeling of density. Even with an abundance of surrounding open space, the first visual impact of the homes may send away prospective home buyers.

In urban infill areas, where the existing grid street pattern is fixed or inflexible ordinances force adherence to right-of-way and setback requirements, grid streets and alleys may be the only usable circulation system. Even with less restrictive controls on development, the orientation of adjacent structures will determine a building's appropriate positioning, setbacks, and street system, fitting it to the existing streetscape.

A somewhat larger infill site may provide enough interior open space for rear garages and parking courts. While street-sidewalk relationships would be maintained typical to other homes, the open space in the center of the block would give an interesting visual contrast. Parcels surrounding the open space could then be developed in individual phases. This could both break the linear alley feeling and provide some room for an intensive amenity, such as a tennis court or developed play area.

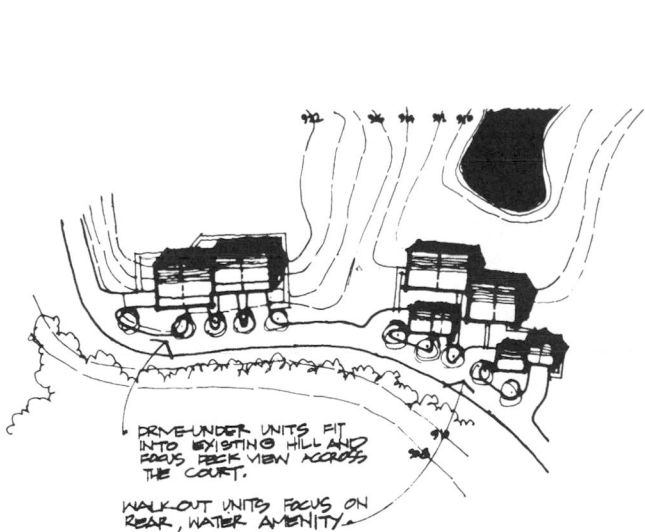

Fig. 1   The single-loaded court road.

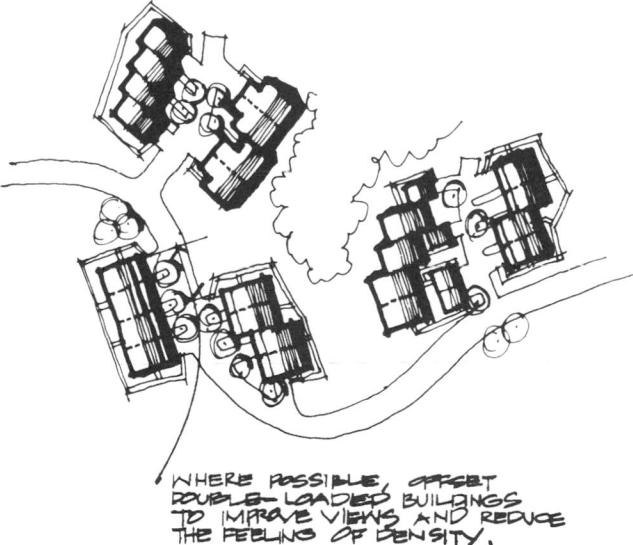

Fig. 2   Double loading of the court roads to preserve open space elsewhere and increase unit density.

BUILDING ARRANGEMENTS

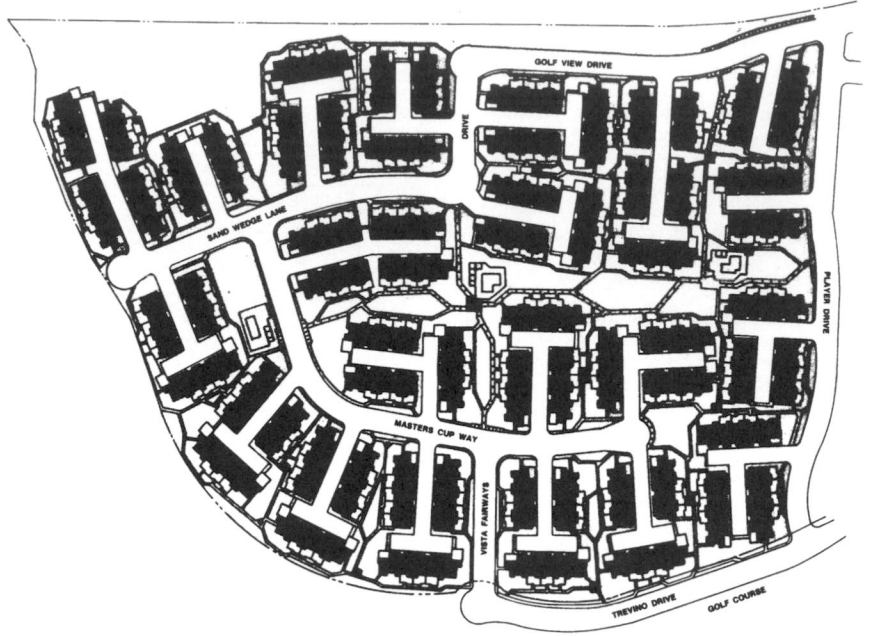

Fig. 3   Increased density of 10.5 units per acre achieved by double-loaded courts and T turnarounds.

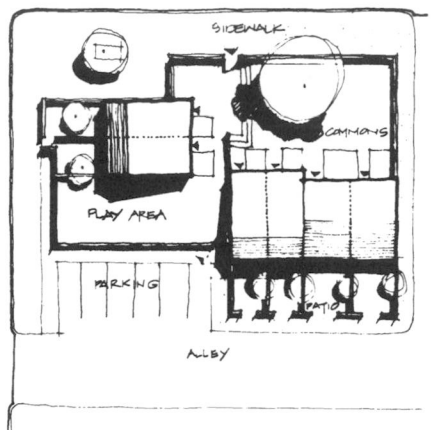

Fig. 4   A small urban infill development maintaining setbacks and regular orientation to adjacent buildings.

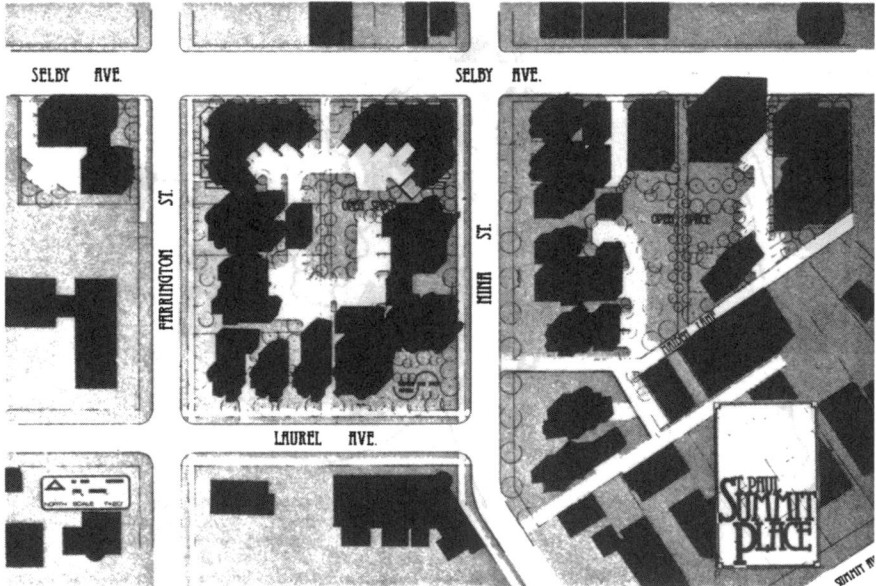

Fig. 5   A large urban infill site: maintaining the existing sidewalk/building relationship but turning the alley into an open space and parking court area.

Linear Groups

Linear court roads are typically used on sloping sites and best adapt to a narrow buildable land area or to increased densities. With either a single- or double-loaded form, minimum excavation occurs if the proper home types are used. Again, the garage-court feeling may easily result if too high a density is used; building type variations may help to break this appearance.

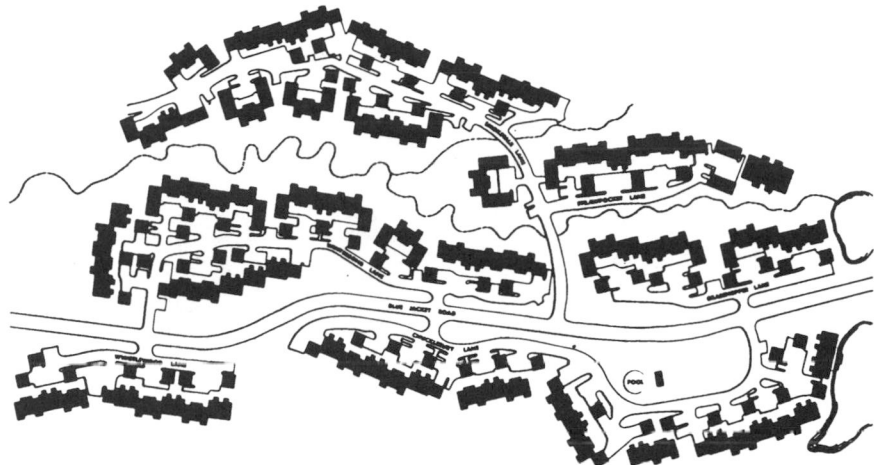

Fig. 6 This site plan shows 27 buildings of 3 to 12 units place off a major collector. Small parking bays or garages break the row house feel of this linear plan.

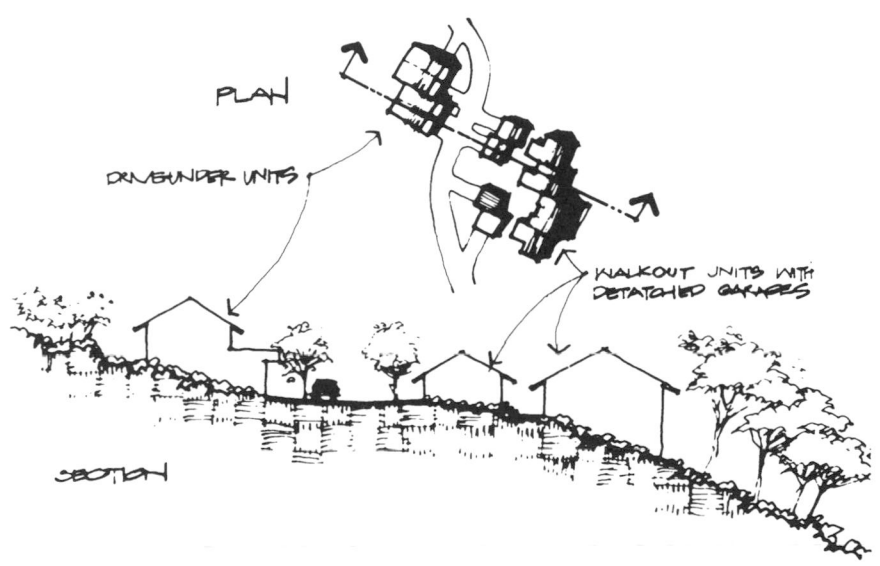

Fig. 7 A linear court will adapt to difficult terrain.

PARKING, GARAGES, AND STREET PLANNING

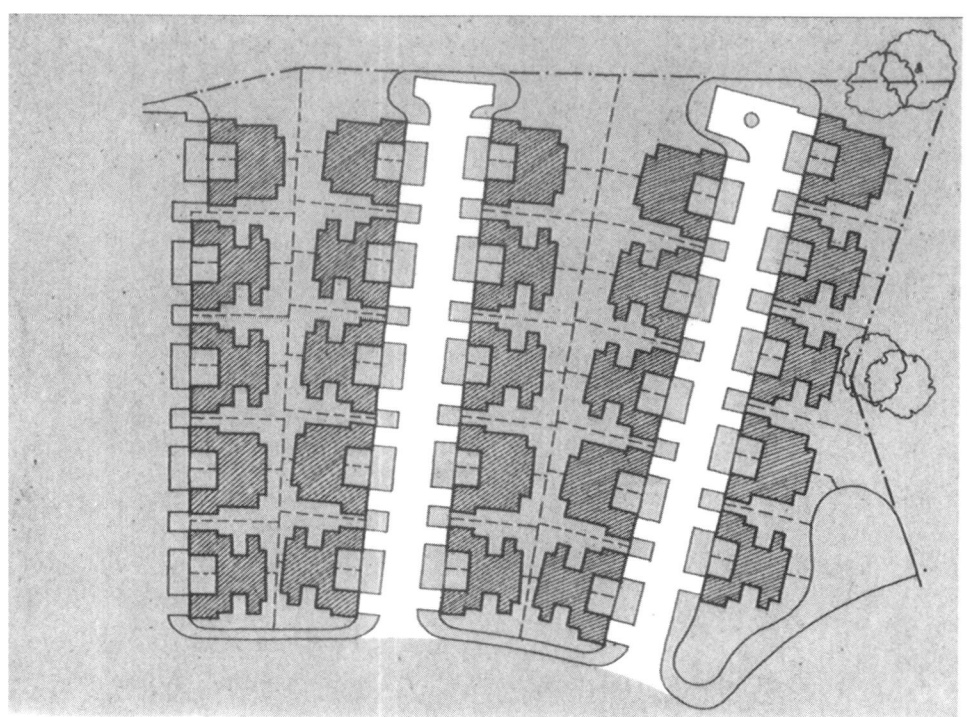

Fig. 1

Problem No. 1:
The garage-door streetscape

This is what happens when conventional single-family planning is used in medium-density situations. Instead of one house for every 100′ or so of frontage, this project has two houses, in the form of duplexes, and a net density of eight d.u. per acre. The result is a veritable canyon of garage doors, at best monotonous and at worst, when doors are left open, a showcase of all the junk that families keep in garages. The drives are just long enough to provide guest parking, and the oveall result is that the automobile dominates the entrances to the homes as well as the streetscape.

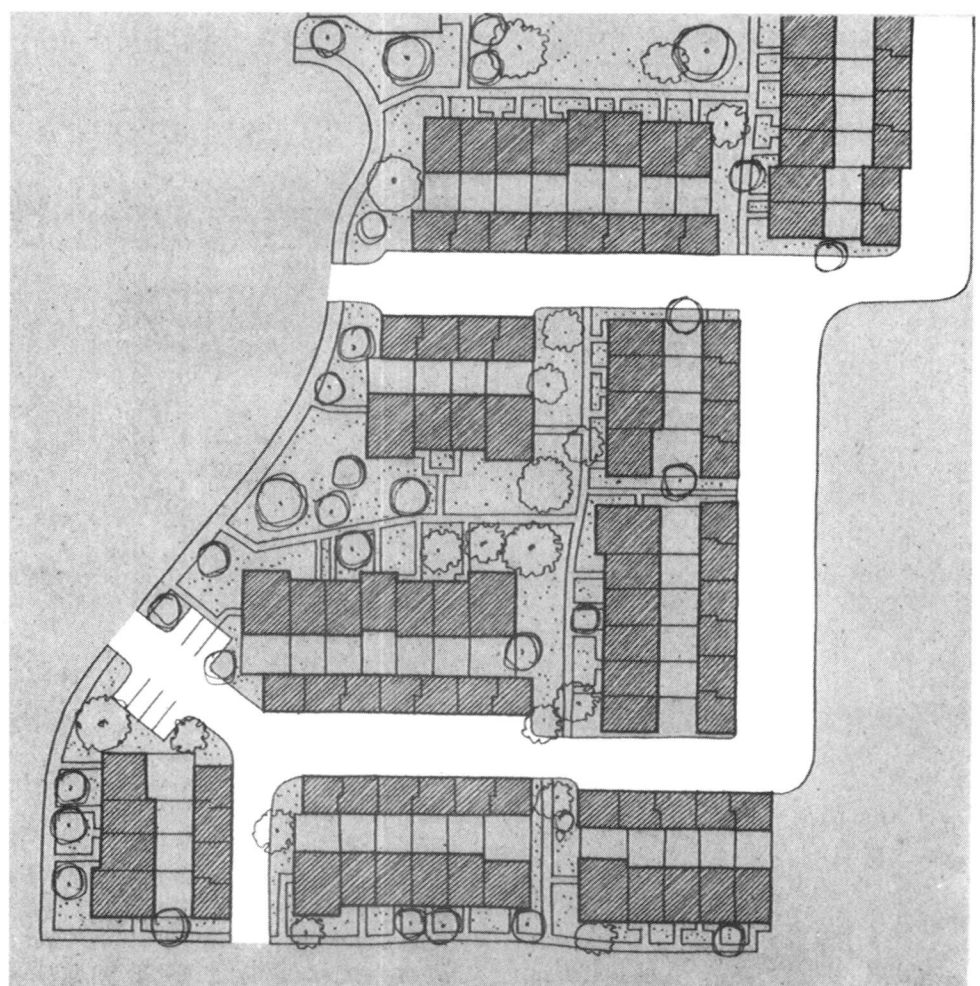

Problem No. 2:
Gasoline alley

One of the most widely used schemes to relieve the above problem is what might best be called the back-alley plan. An example is shown at left. The front entrances to these townhouses have been completely separated from the automobile access; nicely landscaped greenbelts lead from the street to the front doors, while on the other side of the units, garages open onto what might best be termed car service roads.

But while the front entrances are to some extent improved, other problems are created. Streets are even bleaker than those in the plan above; wall-to-wall garage doors and unrelieved road surface. Guests must approach the houses through these streets, and then they have an overly long walk, however attractive, to the unit entrances. Density of this project: ten d.u. per acre.

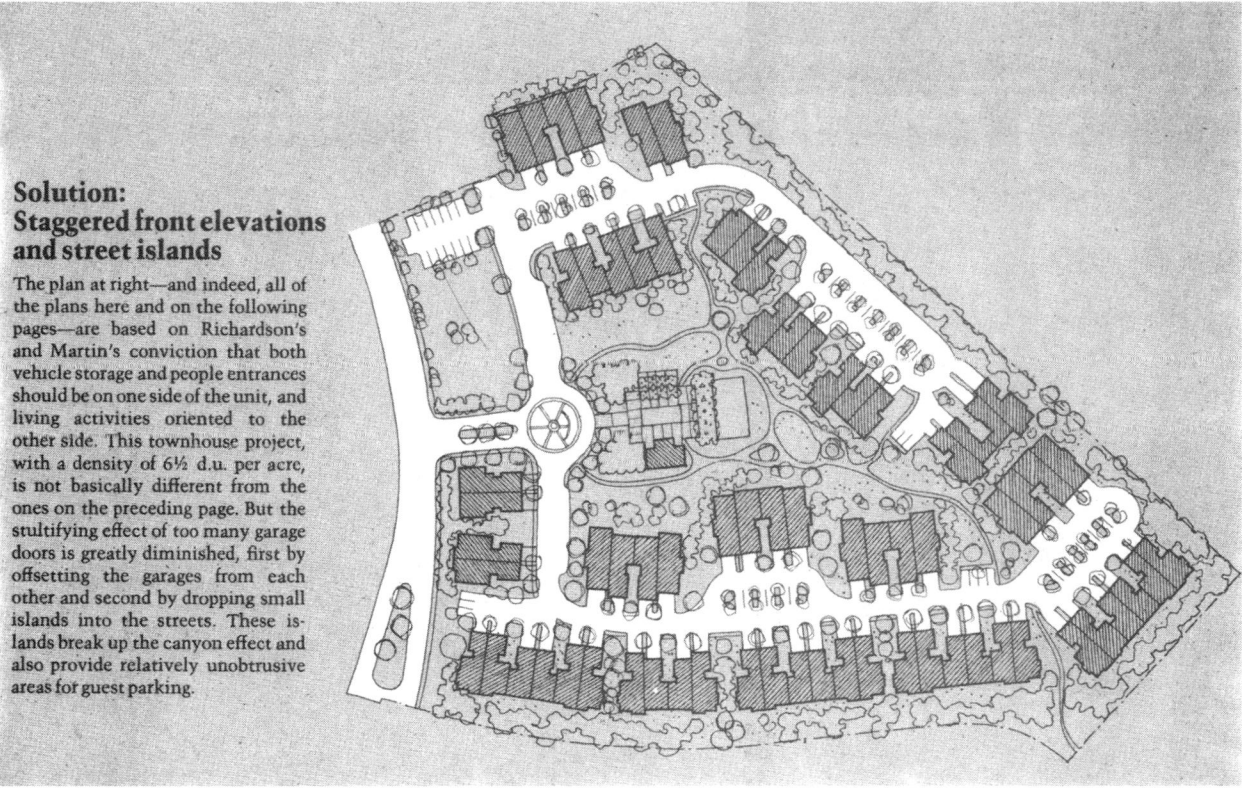

Solution: Staggered front elevations and street islands

The plan at right—and indeed, all of the plans here and on the following pages—are based on Richardson's and Martin's conviction that both vehicle storage and people entrances should be on one side of the unit, and living activities oriented to the other side. This townhouse project, with a density of 6½ d.u. per acre, is not basically different from the ones on the preceding page. But the stultifying effect of too many garage doors is greatly diminished, first by offsetting the garages from each other and second by dropping small islands into the streets. These islands break up the canyon effect and also provide relatively unobtrusive areas for guest parking.

Fig. 3

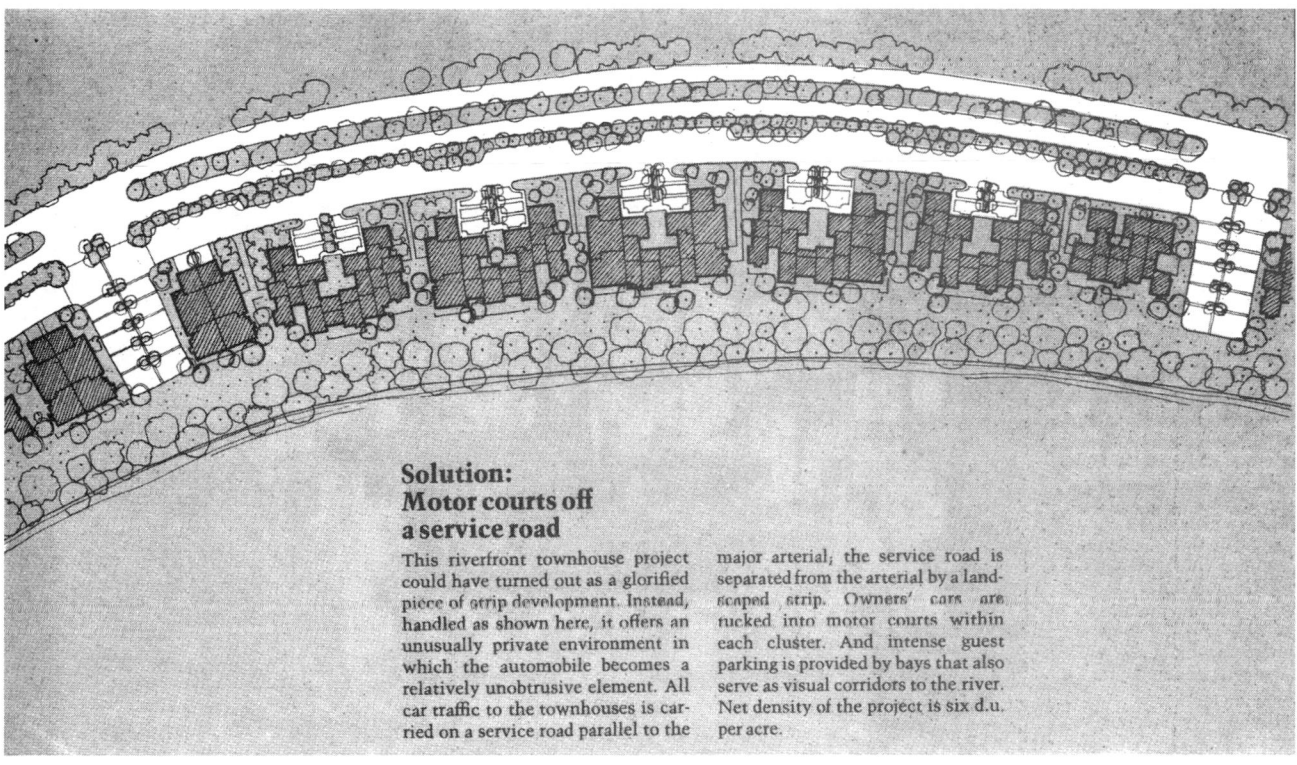

Solution: Motor courts off a service road

This riverfront townhouse project could have turned out as a glorified piece of strip development. Instead, handled as shown here, it offers an unusually private environment in which the automobile becomes a relatively unobtrusive element. All car traffic to the townhouses is carried on a service road parallel to the major arterial; the service road is separated from the arterial by a landscaped strip. Owners' cars are tucked into motor courts within each cluster. And intense guest parking is provided by bays that also serve as visual corridors to the river. Net density of the project is six d.u. per acre.

Fig. 4

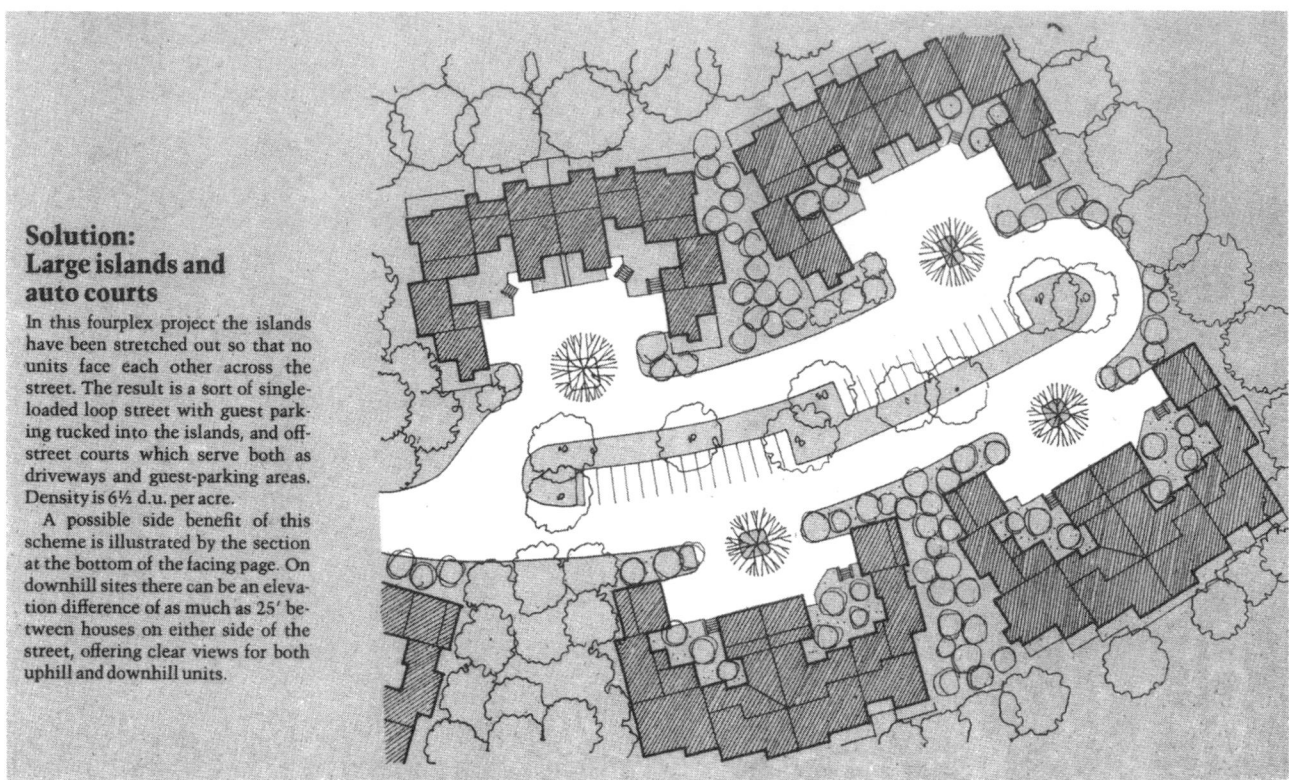

Solution: Large islands and auto courts

In this fourplex project the islands have been stretched out so that no units face each other across the street. The result is a sort of single-loaded loop street with guest parking tucked into the islands, and off-street courts which serve both as driveways and guest-parking areas. Density is 6½ d.u. per acre.

A possible side benefit of this scheme is illustrated by the section at the bottom of the facing page. On downhill sites there can be an elevation difference of as much as 25' between houses on either side of the street, offering clear views for both uphill and downhill units.

Fig. 5

Solution: pocket parks and dividers for patio houses...

Patio houses are great inside and a problem outside. They offer very private living both indoors and out. But the same feature that creates that privacy—the enclosure around the lot—can also turn the street into a canyon with walls broken only by a succession of garage doors or carports.

This patio-house site plan permits no such canyons. All the houses front on either a street island or a park, offering more open green space than many a conventional single-family project. And because the streets are single-loaded, the automobile intensity of the neighborhood is relatively low. Density is five d.u. per acre.

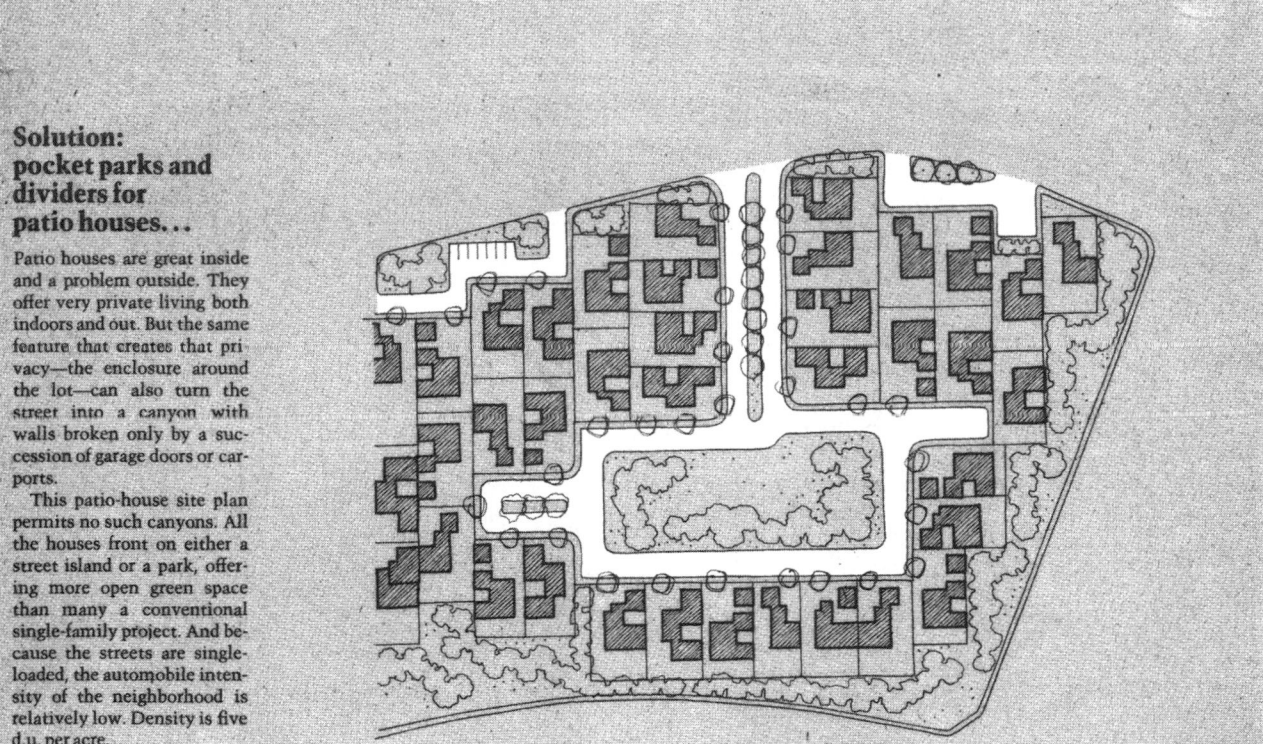

Fig. 7

... and the same, plus island parking, for townhouses

Essentially this is the same scheme as that for the patio-house layout above (it is, in fact, part of the same subdivision). The principle of a sort of super cul-de-sac wrapped around a small park is repeated. But, because the density is higher with townhouses (eight d.u. per acre), and because there is less frontage per unit, additional parking for guests is provided in bays on one side of the park.

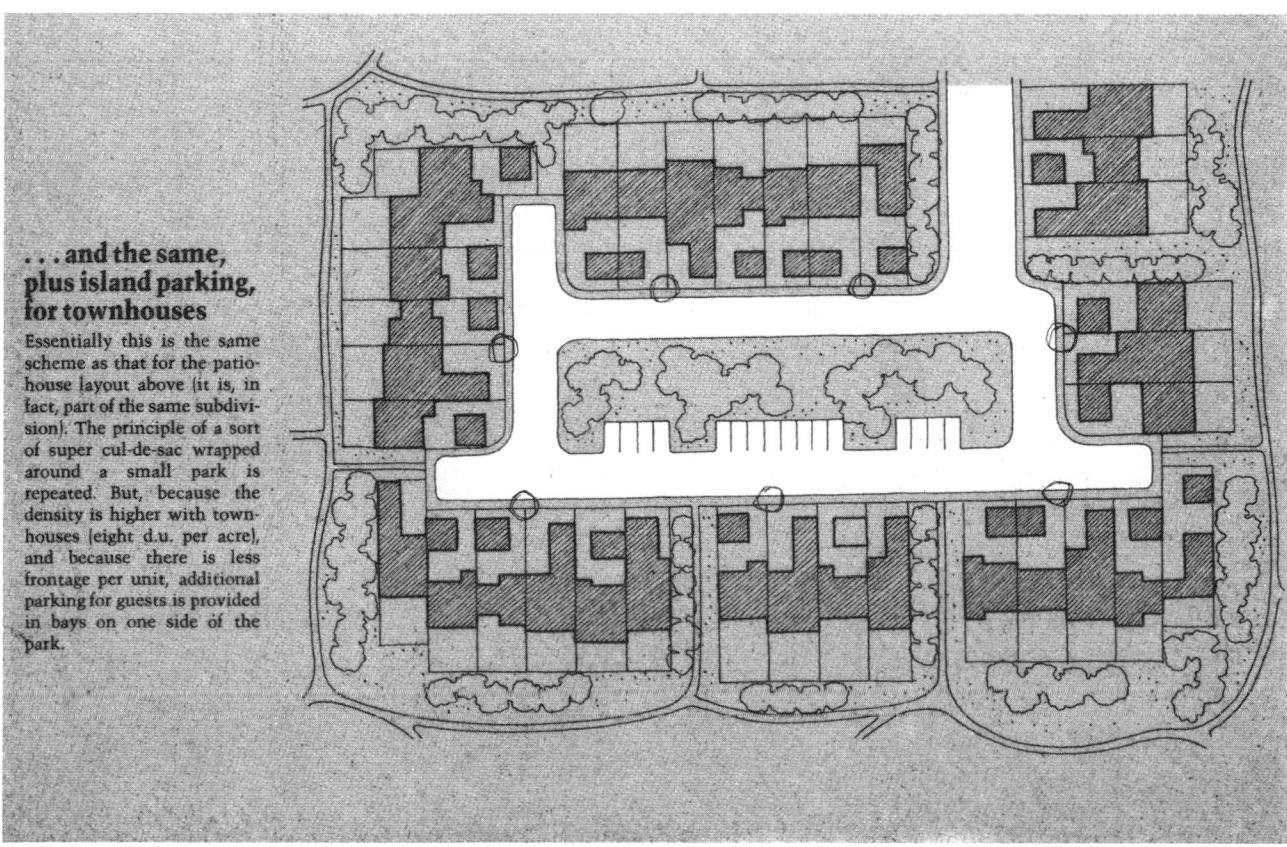

Fig. 8

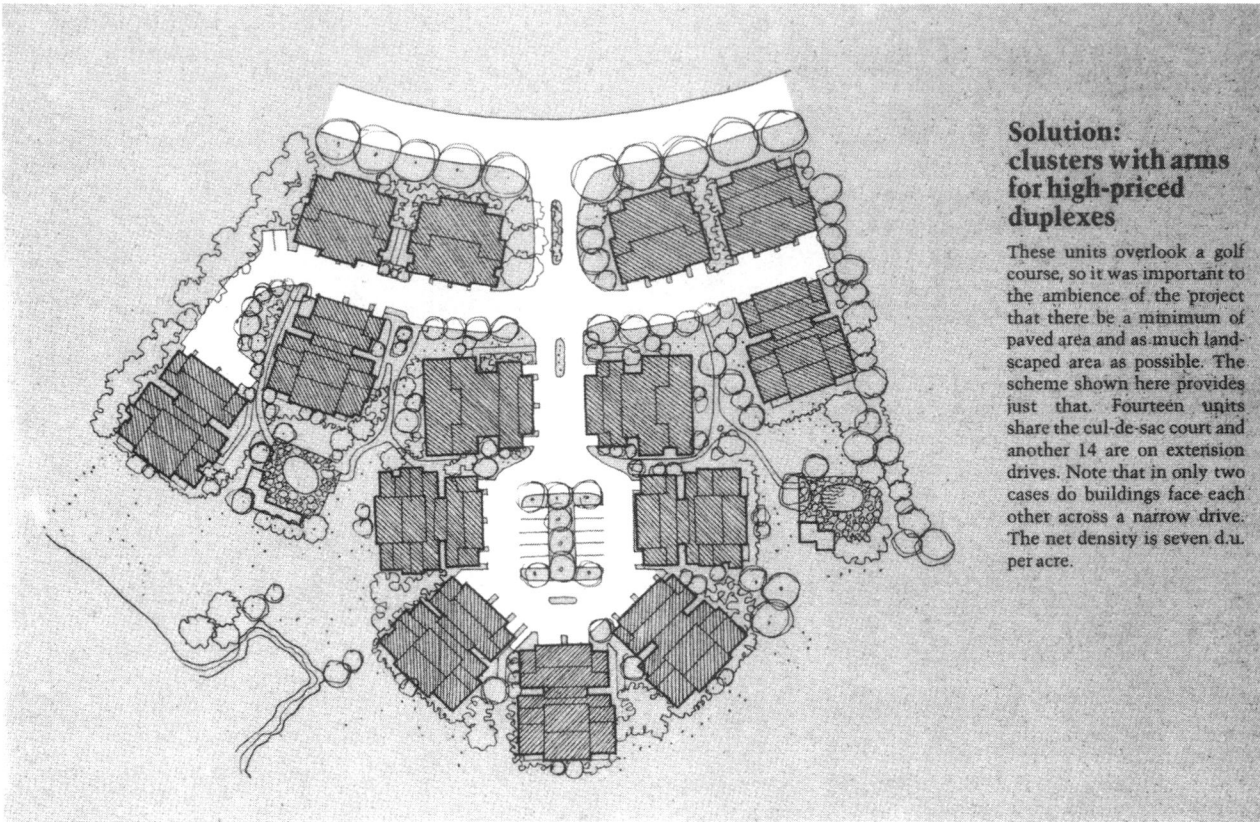

Fig. 9

Solution: clusters with arms for high-priced duplexes

These units overlook a golf course, so it was important to the ambience of the project that there be a minimum of paved area and as much land-scaped area as possible. The scheme shown here provides just that. Fourteen units share the cul-de-sac court and another 14 are on extension drives. Note that in only two cases do buildings face each other across a narrow drive. The net density is seven d.u. per acre.

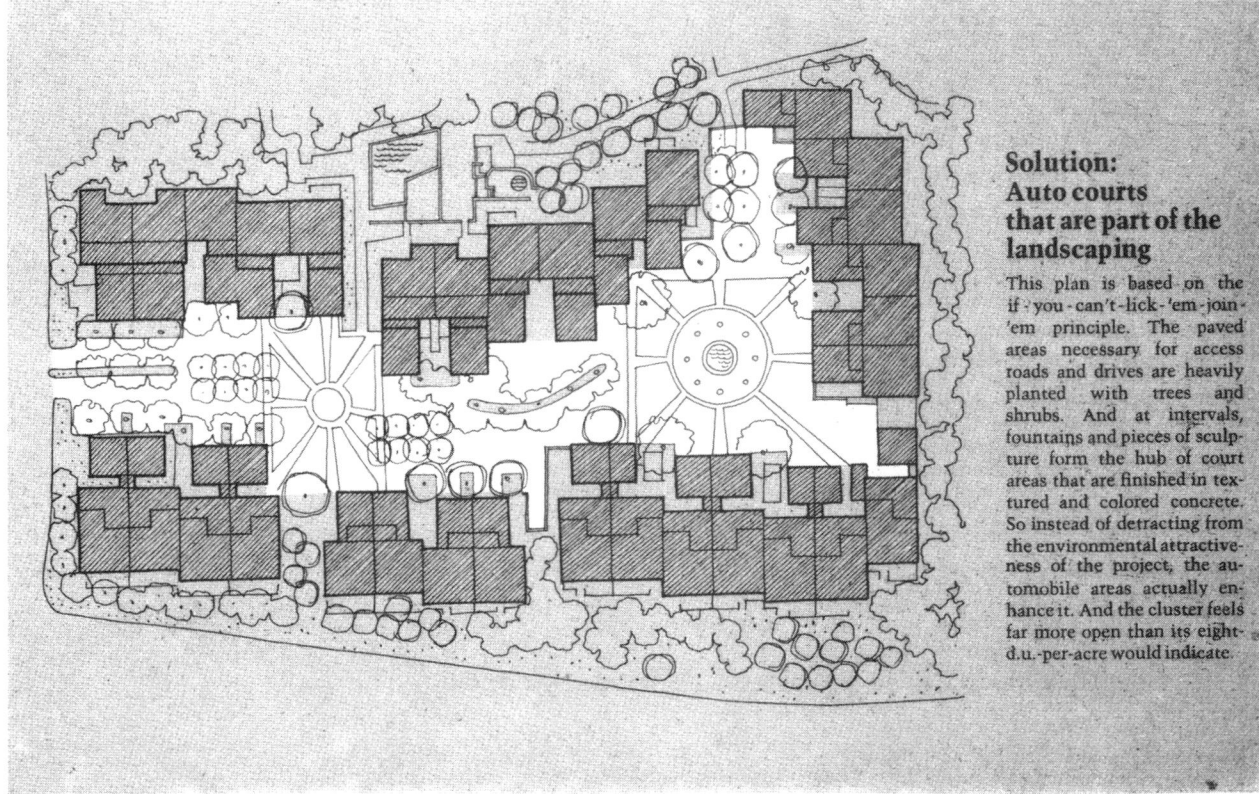

Solution: Auto courts that are part of the landscaping

This plan is based on the if-you-can't-lick-'em-join-'em principle. The paved areas necessary for access roads and drives are heavily planted with trees and shrubs. And at intervals, fountains and pieces of sculpture form the hub of court areas that are finished in textured and colored concrete. So instead of detracting from the environmental attractiveness of the project, the automobile areas actually enhance it. And the cluster feels far more open than its eight-d.u.-per-acre would indicate.

In studying a site for a possible subdivision, a critical determination is the number of lots per acre and the amount of open space for streets and park areas. Figure 1 provides a quick means of determining the relationship of number of lots per gross acre and percentages of open space.

It is assumed that the site is level and entirely buildable with no steep slopes, marshy land, or other obstructions.

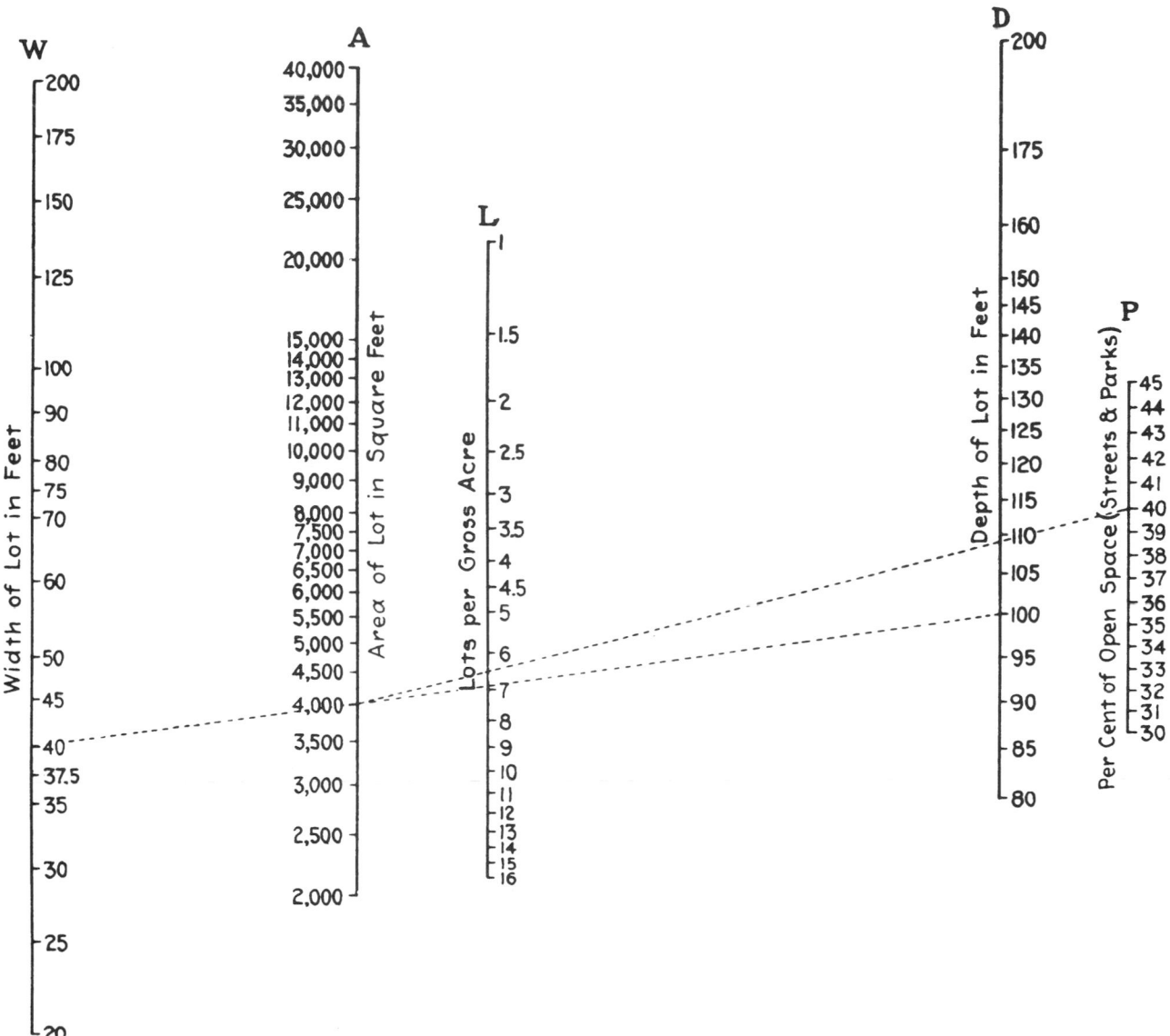

Fig. 1   Diagram for determining lots per gross acre for varying lot sizes and percentages of open space in streets and parks. Method of using diagram: Start with values on W and D scales; lay straightedge between them and read area of lot on A scale; choose value on P scale; lay straightedge between this value and determine on A scale; read required answer on L scale. In example shown, W = 40 ft and D = 100 ft; hence A = 4000 ft². With P = 40 percent, L = 6.5 lots per gross acre.

PLANNED UNIT DEVELOPMENT

PUD

Planned unit development (PUD) is a new way of designing residential neighborhoods that can provide a better environment for the people who live there and produce more profits for the developer and builder.

Under this approach, the planning board may waive technical requirements, such as yard regulations and height restrictions, to permit dwellings to be built together in clusters, leaving substantial land areas in a natural state. In addition, the governing body or planning board can grant bonuses of extra floor area to developers in return for good site plans or the provisions of common open space.

Some of the advantages for people living in a planned unit development are:

Larger houses for less money

More choice of house types

Preservation of natural features like ponds and trees

Community recreation space

Safe pedestrian ways and safer streets

More conveniently located schools and shops

Some of the advantages for the developer and builder are:

Less land used for streets

More efficient utility runs

Better drainage, less grading and site preparation

More varied house types that can reach a wider market

More dwelling units and bigger houses

The ability to include shops and stores

The following outline explains the most significant provisions of the new approach and shows how these advantages can be obtained.

Regulations that can be modified The planning board may authorize the following modifications to the zoning regulations, provided that the overall plan is satisfactory to the board and is not contrary to the master plan.

Bulk regulations

1. Floor area and dwelling units, rooms, or rooming units may be distributed without regard for zoning lot lines.
2. Open sapce may be distributed without regard to zoning lot lines.
3. Lot sizes may be reduced.
4. Yard regulations may be waived within a development.
5. Heigh regulations may be waived within a development, provided that regulations governing the spacing of buildings are satisfied.

Use regulations

1. Convenience shopping, restaurants, and certain other types of consumer services may be permitted within residential areas, provided that the board is satisfied that they represent an amenity, and provided that the total area devoted to such uses is no more than 2 to 5 percent of the overall floor area permitted in the development.
2. Outdoor swimming pools may be provided in the common open space, provided that their use is restricted to residents of the development and that the pool is located at a reasonable distance from the development's boundary and is adequately screened from the street.

Regulations that can be modified by special permit The planning board may grant special permits providing for minor variations in the requirements for front or rear yards, and in the regulations governing the height of buildings, along the boundaries of a planned unit development.

Bonuses given by special permit The planning board may grant bonuses of additional floor space to the developer for a good site plan with or without the provision of common open space. The bonus for a good site plan can be granted within the following limits:

1. The required open space may be reduced by 10 to 20 percent.
2. The required lot area per room, or the lot area required per dwelling unit, may be reduced by 5 to 10 percent.
3. The allowable floor area may be increased by 5 to 15 percent.

Larger Houses for Less Money

A large portion of the price paid for a house really goes to pay for land. Land prices in urban areas can be as much as 10 times as high as they would be in a suburb. Since planned unit development permits the builder to offer houses on smaller building lots, a house in a development where this approach has been used could cost substantially less than a comparable house on a larger lot. At the same time, the easing of yard restrictions in planned unit development permits the builder to construct a larger house than is now permitted on a conventional lot; so that a larger house, for less money, is a very real possibility under planned unit development.

More Choice of House Types

Where conventional development tends to produce street after street of the same type of dwelling, planned unit development encourages townhouses, garden apartments, detached houses, and atrium houses, all of varying sizes, to be built in the same development. This means more variety of family size and income level in any given area, and allows families to move from one type and size of house to another without leaving their old neighborhood.

Preservation of Natural Features

Instead of developing a whole section with paved streets and narrow, fenced-in yards, planned unit development permits as much as 30 percent of the land area to remain in its natural state, while housing the same number of families as conventional development, sometimes even more. This means that natural features, like ponds and rock outcroppings, as well as trees and streams, can be preserved near the places where people live.

At the same time, all houses continue to have their own private open space, which may well be larger than conventional backyards. The land saved for open space is land that would ordinarily have been devoted to unusable side yards and unnecessary streets.

Community Recreation Space

Open space created by planned unit development can be used for recreation areas like playing fields and swimming pools, and there can easily be extra open space for schools and community facilities. The new approach allows such facilities to be designed as an integral part of the residential neighborhood, instead of being in their own separate locations.

Safe Pedestrian Ways and Safer Streets

The community open space of planned unit development can also be used to create pedestrian greenways connecting houses with schools and larger open areas. Such greenways can be designed so that they cross few or no streets, providing safe routes for children to walk to school, or play space.

The intersection of two conventional "gridiron" streets creates as many as 16 potential places where a collison can take place. The neighborhood loop streets possible in planned unit development can have as few as three potential collision points. In addition, the clear distinction between through-traffic streets and neighborhood streets made possible by planned unit development provides a generally safer traffic pattern, with fewer cars moving more slowly, in the areas where people live.

More Convenience to Schools and Shops

In conventionally zoned areas, shops can only be placed in sections with commercial zoning. A planned unit development permits small groups of shops and restaurants in the middle of residential areas, giving the kind of convenience often found in the center of cities, but seldom in outlying residential districts.

In addition, by placing a school adjacent to community open space, it is likely to be far more centrally located than would be possible under conventional conditions.

Fewer and shorter Streets

Developers in large low-density tracts generally are responsible for building the streets themselves; therefore, the fewer and shorter streets needed for planned unit development mean a substantial saving for the developer. There may be as much as 30 percent less street area under planned unit development, which means not only less development cost but more valuable land available for housing.

More Efficient Utility Runs

Developers in a large tract normally must also build storm drains and the sewers for their development. Because of the more compact street system possible in planned unit development, the developer is likely to realize substantial savings in providing utilities.

Better Drainage, Less Site Preparation

Conventional gridiron street systems often work against the natural contours of the land, creating steep streets to which it is hard to relate houses, and low-lying areas susceptible to flooding. Planned unit development, by providing streets that are only for local vehicles, allows the builder to develop those parts of the site most suitable for housing, leaving hills and floodplain areas open as part of community open space. By not having to meet customary requirements for through and connecting streets, the builder frequently can realize a significant saving, as well as ending up with a far more satisfactory development.

More Sales Flexibility in House Types

Market conditions often change while an area is being developed. Mortage money may become easier or harder to come by; local conditions may cause a sudden influx of a new kind of house buyer. Conventional zoning and street maps tend to lock the builder into a single type of house with a very narrow variation in price range. The variety of house types possible under planned unit development allows the builder to appeal to a wider segment of the potential house market, and to switchs from detached houses to garden apartments, for example, as market conditions change.

More Dwelling Units and Bigger Houses

The bonus provisions of the planned unit development regulations can give the builder significantly more houses or apartments on a given piece of land, in addition to the extra buildings made possible by saving on the amount of land devoted to streets. The relaxation of yard requirements also permits bigger houses than are possible under conventional zoning, further enlarging the builder's flexibility in responding to the housing market.

Ability to Include Shops and Stores

Finally, developers and builders benefit from the opportunity provided by planned unit development to devote a portion of the floor area they build to space for shops and restaurants. Such commercial space has a high rental value and is not usually allowed in a residential development.

Streets

Street patterns are the most important element in establishing the character of a residential community. Most existing and mapped streets are based upon only two guiding principles: provision of maximum frontage of traditional lot size and maximum flow of all types of traffic on every street. The first step in dispelling the monotony caused by this system—particularly in low-density residential areas—is the establishment of a hierarchy of street types based on usage. Aside from expressways and highways, this hierarchy consists of three basic street types: Major collector streets, local collector streets, and local residential streets. Major collector streets are major arteries and interneighborhood streets; local collector streets pick up traffic from mocal residential streets in one neighborhood; and local residential streets are solely for the residential area served. Recommendations for specific characteristics of these three street types are:

Major collector streets

Traffic characteristics: All types of vehicles, through traffic
Pedestrian safety: Limitation of pedestrian crossover to a minimum of controlled point
Length: Unlimited
Width: 60–100 ft
Grades: 8 percent maximum, with other technical requirements conforming to the policies of the department of highways

Local collector streets

Traffic characteristics: Primarily private cars and service vehicles, through traffic discouraged
Pedestrian safety: Increased through limitation of traffic, but crossover points should be designated.
Length: Should be interrupted by intersections with major collector streets; intersections should be T-shaped in order to prevent local

traffic from crossing major collector
Width: 50 to 60 ft
Grades: 10 percent maximum, with other technical requirements conforming to the policies of the department of highways.

Local residential streets

Traffic characteristics: Private cars, except that service and emergency vehicles are permitted
Pedestrian safety: Problem minimized through restriction of traffic to residents of specific residential grouping; where possible, pedestrians should be able to pass beyond their own residential grouping without crossing any street. Furthermore, the pedestrian's path to local shops and elementary schools should cross as few streets as possible.
Length: Grid and modified grid system blocks should have a continuous frontage no longer than 800 ft, except that a 1200-ft-long block is permissible if a pedestrian access no narrower than 10 ft is provided near the midpoint of the block.
Cul-de-sacs should be no longer than 250 ft to the neck of the turnaround unless a connection to an adjacent street can be achieved by a 10-ft-wide paved pedestrian walk for emergency vehicle access.
P-loop streets should have a neck no longer than 700 ft and a loop circumference no longer than 2800 ft, measured at the centerline of the street. P-loop streets must have emergency vehicle access to an adjacent street.
Horseshoe-loop streets may be of varying lengths, depending on the number of dwelling units served.
Width: 40 to 50 ft
Specifications of street and sidewalk design and construction must conform to the standards of the department of highways.

Sidewalks and Pedestrian Ways

Sidewalks and pedestrian ways supplement and complete street systems in establishing the character of a residential environment. The pedestrian circulation system need not parallel the street system, but the following criteria must be observed:

1. A sidewalk must be provided on at least one side of a public street except where it can be demonstrated that such a sidewalk is not desirable.
2. Pedestrian circulation systems must be provided as convenient, safe, and attractive links between residential groupings, open space areas, recreational areas, schools, and local shopping areas.
3. The width of any sidewalk must be at least 4 ft.
4. Alternatives to the norms of asphalt or concrete pavement construction should be considered; surface treatment and forming methods can afford an opportunity to enhance the character of a residential environment.

Utility Placement

The requirements of utility locations generally follow the street pattern; however, easements through common open space augment the flexibility of utility placement. Easement requirements are as follows: 20 ft unobstructed width for one storm or sanitary line; 30 ft for two storm or sanitary lines. The feasibility of burying electrical and telephone lines should be studied, as well as the use of existing watercourses for storm drainage.

Site Characteristics

Preexisting site conditions have considerable importance in establishing the character of a residential development. Previous policy, again through street mapping and traditional lot sizes, has generally ignored the preservation of natural site characteristics. The planned unit development amendment not only permits but encourages flexible and positive responses to the natural assets of a site. Specific site assets that should be considered in a planned unit development are:

Trees Trees of 6-in diameter and larger are to be protected and saved wherever possible, particularly where a grouping of such trees exists; the feasibility of temporary removal and replacement of smaller trees should be considered.

Contours Responses to site profiles must be considered in planned unit developments; ridges, rock outcroppings, slopes, and hillocks all require that special consideration be given the siting of buildings.

Water Existing site water, in the form of watercourses, streams, marshes, and ponds should be considered as possible resources for the establishment of viable ponds, streams, or storm drainage courses.

Orientation The siting of a residential development should be assessed in terms of site profiles, views, sun, prevailing wind, and water resources.

Open Space Development

All of the above considerations should be considered with a view toward developing pleasant and usable open-space patterns throughout the residential community. This open space should be related to any existing parks or park plans.

Houses and Placement of Houses on Lots

The house is the most important item for each individual homeowner in the residential community. Past practice, dictated by the inflexibilities of street mapping and subdivision, was much too limited a range of choice for home buyers and developers alike. Typically, such practice resulted in deep, narrow houses on deep, narrow lots. The front yard in this situation is entirely given over to a paving network; sidewalks, driveways, and front walks. The side yard along which the largest dimension of the house must run is seldom much more than 8 ft wide and is virtually worthless. The rear yard becomes the only usable open space, and even here, since all houses are placed in a row, it is very difficult to establish any real privacy.

Without setting down standards for houses and their placement, it is the intention of the planned unit development regulations to introduce the kind of flexibility that will greatly improve the residential environment. Specifically, the individual house must be designed to relate the open area around each house to what occurs inside the house. The house has entrances for people and an entrance for automobiles; this fact has meaning in terms of both open area and the portion of the house that serves the entrance function; and a design response to this fact is expected. The house has indoor and outdoor living functions; this requires a design that relates the two, preferably resulting in increased privacy for outdoor living because of the way adjacent buildings are

PLANNED UNIT DEVELOPMENT

placed. Similar thought and design response should be shown in the arrangement of bedrooms, internal circulation, and service spaces.

Definitions

Planned unit: A land area that (1) has both individual building sites and common property such as a park, and (2) is designed and organized to be capable of satisfactory use and operation as a separate entity without necessarily having the participation of otherbuilding sites or other common property. The ownership of the common property may be either public or private.

Planned unit development A single planned unit as initially designed; or such a unit as expanded by annexation of additional land area; or a group of contiguous planned units, either operating as separate entities or merged into a single consolidated entity.

Homes association: An incorporated, non-profit organization operating under recorded land agreements through which (1) each lot owner in a planned unit or other described land area is automatically a member, and (2) each lot is automatically subject to a charge for a proportionate share of the expenses for the organization's activities, such as maintaining a common property.

Common property A parcel or parcels of land together with the improvements thereon, the use and enjoyment of which are shared by the owners and occupants of the individual building sites in the planned unit.

Regulations that can be modified The planning board may authorize modifications to the zoning regulations, provided that the overall plan is satisfactory to the board and is not contrary to the master plan.

ZONING OF PLANNED UNIT DEVELOPMENT

Traditionally, zones have been classified according to type of building. A single-family zone, for example, is based on the premise that all properties within the area should be developed for a single building of a similar type, specifically, the single-family home. In the traditional zoning ordinance, setbacks are required all the way around the house. The front yard setback is generally 20 to 25 ft. The side yard setback varies between 5 and 15 ft. The backyard traditionally is 25 to 30 ft. By maintaining these setbacks, a typical pattern of single-family detached dwellings is achieved. Also, a height limit of 30 to 40 ft (excluding chimneys and TV-radio antenna) and a minimum lot size are required. This type of residential zone has many advantages. It ensures that the property will be protected from undue encroachment and exclusion of light and air by the neighboring single-family residential building.

There are some disadvantages to traditional zoning that should be noted. This type of zoning often results in relatively monotonous development. Each street tends to become very much like the neighboring street. Moreover, this type of single-family detached residential zoning may be wasteful of land. Often the side yards are not large enough to be used for anything other than a path permitting someone to take the lawn mower from the front yard to the backyard. Modern planned unit development zoning encourages variety, full use of all open space, and the addition of residential amenities to ensure a long, useful life of the development. It does this by waiving the traditional restrictions, which allow only single-family houses on spacious lots. Under planned unit development zoning, the density of the area remains constant, but the developer may build an assortment of housing types—single-family units, duplexes, row houses, and apartments. This type of development, especially recommended for hilly terrain, has the added advantage of providing common parking facilities and play areas.

For example, in a 5-acre area zoned at a density of four families per acre (approximately 10,000 ft^2 minimum lot size), developers can build 20 units. They may choose to construct a row of 8 units plus 2 apartments of 6 units each, for a total of 20 units. This is the same number of units the developer could build under a traditional development of single-family homes. Under planned unit development, however, the developer may be able to achieve substantial savings in street and utility development. More important, the housing units can be laid out so that a substantial saving in land will be achieved. Many developers add special amenities to their developments, including swimming pools, golf courses, and community centers.

The planned unit development is usually incorporated into the zoning ordinance, not as a special zone but as a conditional use in any of the residential zones. Usually it must conform to the density of the zone in which it is located. In some communities, a 10 or 20 percent density bonus is offered as an incentive to encourage developers to use this method of building residential areas. Usually a 5-acre minimum is required, although in some communities there is an advantage in reducing this minimum to 1 or 2 acres to encourage planned unit development where an entire block in a built-up area is still vacant and under single ownership. The development plan is reviewed by the board of adjustment, which may be assisted by outside consultants. Attention should be paid to designing the zoning ordinance so that the review procedure does not discourage developers from taking advantage of the planned unit development.

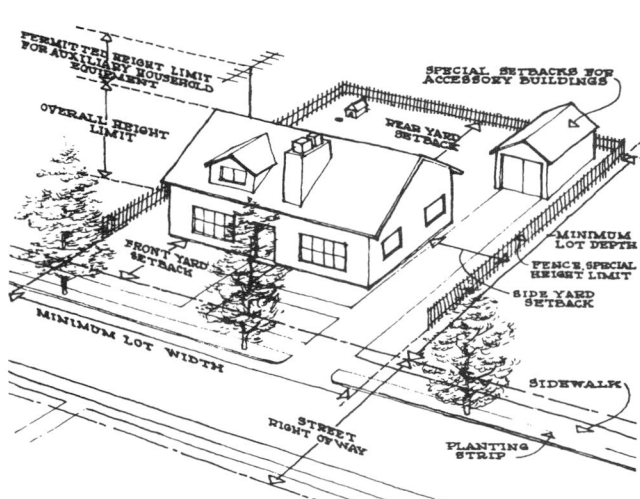

Fig. 1

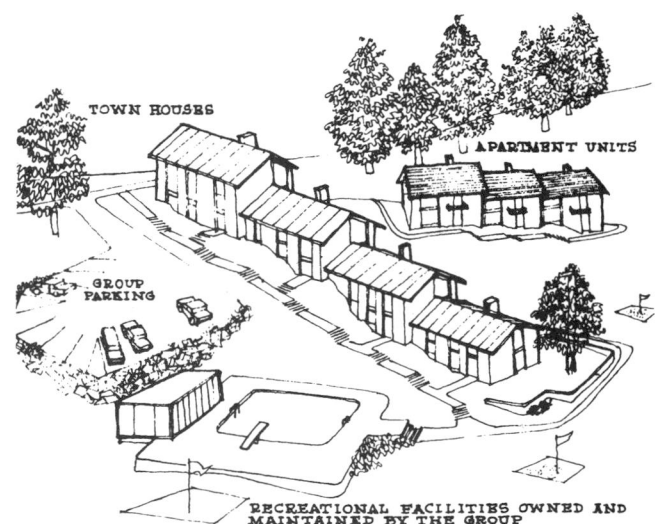

Fig. 2

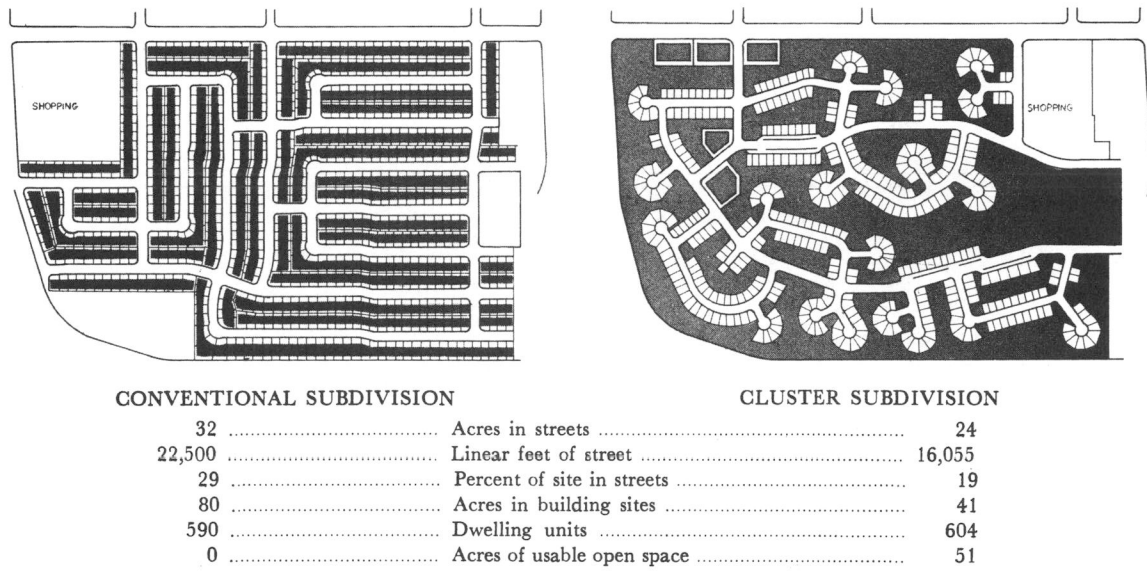

CONVENTIONAL SUBDIVISION		CLUSTER SUBDIVISION
32	Acres in streets	24
22,500	Linear feet of street	16,055
29	Percent of site in streets	19
80	Acres in building sites	41
590	Dwelling units	604
0	Acres of usable open space	51

Fig. 3 Comparison of a conventional subdivision and a cluster subdivision. Santa Clara County Planning Department.

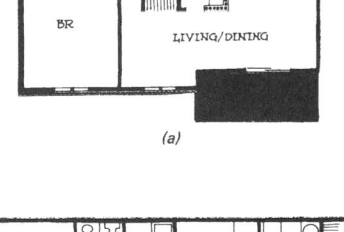

(a)

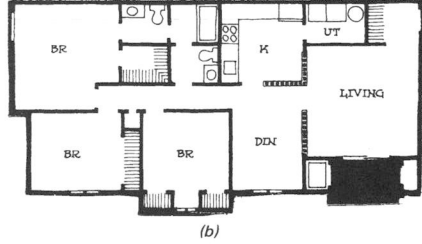

(b)

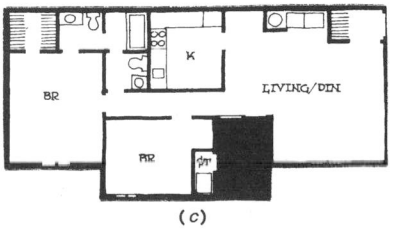

(c)

Fig. 4 Four housing types are included in the land plan for the 65-acre, 350-unit PUD. The 20-acre first phase has 18 zero-lot-line houses at a density of five per acre, 40 apartment units and 15 single-family homes on ½-acre lots, plus recreational center. The 21-acre second phase adds 58 fee-simple townhouses, 20 zero-lot-line homes, and 56 apartment units.

Fig. 5 Apartments are segregated by size in eight-unit buildings: (a) one-bedroom apartment; (b) three-bedroom apartment; (c) two-bedroom apartment.

SINGLE FAMILY
ZERO-LOT-LINE
APARTMENTS
TOWNHOUSES

PHASE I
LAKE AREA
RECREATION
PHASE II
0   100   200 F'T

CLUSTER DEVELOPMENT

Conventional Subdivision Design

In most subdivisions, the entire site is split up into single house lots of ½ acre or more. A large amount of roadway is required for access to the lots, and, since houses are dispersed, utility installation and maintenance costs are high. Lack of open space requires mixing of pedestrian and vehicular traffic, creating safety problems. Privacy is limited, and the landscape is often visually monotonous.

Cluster Design

In cluster developments, individual lot size is reduced in favor of common open space areas. Clustering allows for utilization of the best building sites while preserving environmentally sensitive areas. Concentration of buildings lowers installation costs for utilities, and reduces road-building requirements. Pedestrian and vehicular traffic can be separated; safety is increased by locating public recreational areas away from roads. Careful layout of open space can provide increased privacy and will help maintain the natural character of the site.

Clustering of single-family homes on private lots enables the benefits of private land ownership to be maintained. An alternative which provides larger open space areas and higher housing density is the construction of townhouses or apartments instead of individual homes. This type of development also allows for the most efficient layout of roads and utilities.

The zoning regulations of most towns contain no provisions for cluster development; therefore, variances would be necessary in most localities. Towns considering the adoption of cluster development ordinances should evaluate road width and surface water drainage standards to allow for narrower cul-de-sacs and drainage systems that, where possible, follow natural drainage patterns.

The approach is quite simple. Open space is preserved by permitting a subdivider to develop smaller lots than specified in the zoning ordinance, coupled with the requirement that the land saved by reserved for permanent open space. No increase in the number of units is allowed, thus retaining the original density prescribed by the zoning ordinance. For this reason the technique is sometimes referred to as "density zoning" or "density averaging."

A more common term for it is "cluster development," or "cluster zoning." While this is simple and easy to remember, it can be misinterpreted to mean houses tightly packed or clustered together. While this might be appropriate in some places, it is not a necessary feature of successful open space design, even though houses will be somewhat closer together than in a conventional development of the same acreage.

Still another term associated with the open space approach is "planned unti development" (PUD). The U.S. Department of Housing and Urban Development uses PUD specifically to mean a similar kind of residential development. However, planned unit development more commonly means a relatively large-scale development that includes commerical and public facilities, and sometimes industrial development, as well as housing, in the overall design. PUD also generally involves densities higher than permitted by existing zoning. PUD may incorporate the open-space concept in its design, but its uses are larger in scope.

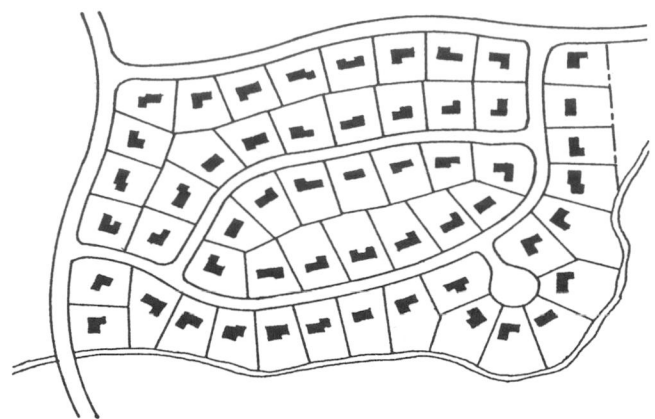

Fig. 1   Site: 30 acres; 54-lot subdivision.

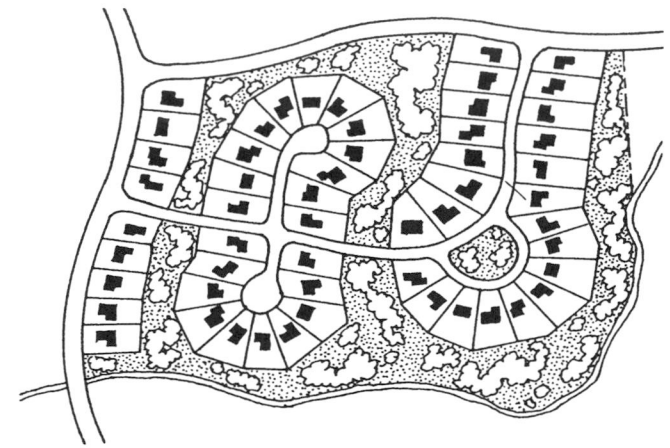

Fig. 2   Cluster development: 54 lots.

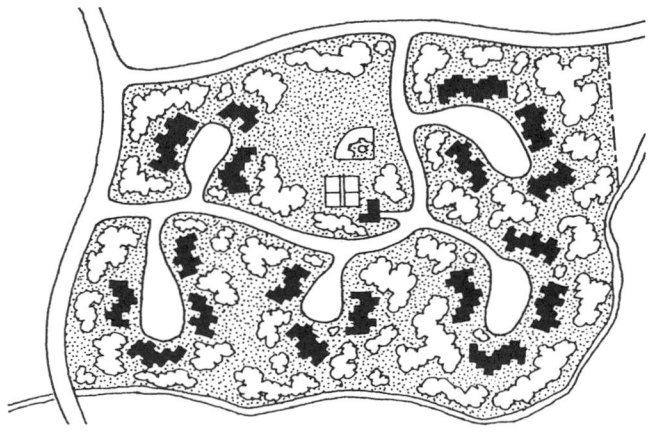

Fig. 3   Cluster development: 112 townhouses.

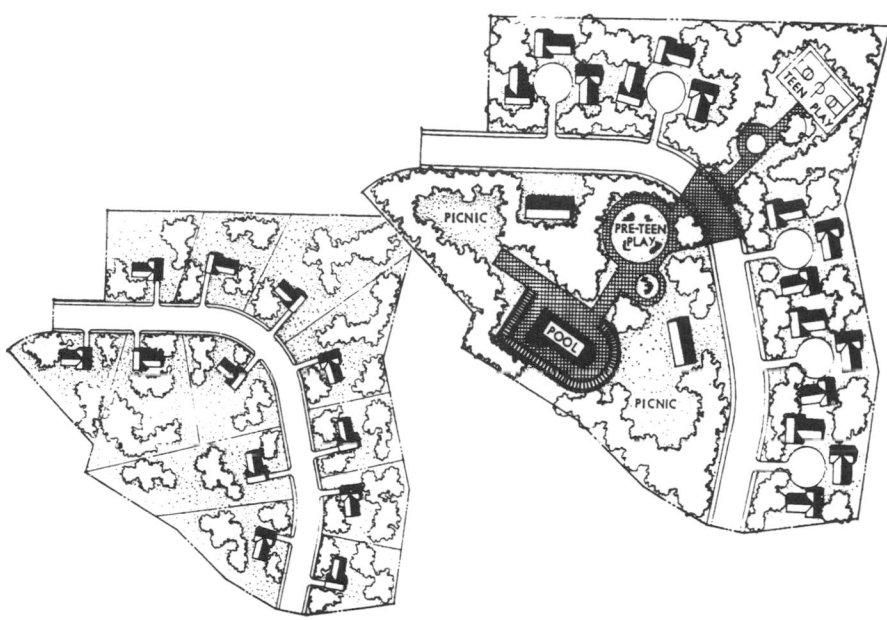

Fig. 4 A design for a cluster zone in Poughkeepsie, New York, by Burt Gold, shows how the land was originally divided into 12 parcels. The revised cluster plan has three additional houses. Strategic planning releases
sufficient land for a community park owned and maintained by the homeowners of the development. The recreational facilities include tennis courts, a pool, cabanas, and outdoor showers.

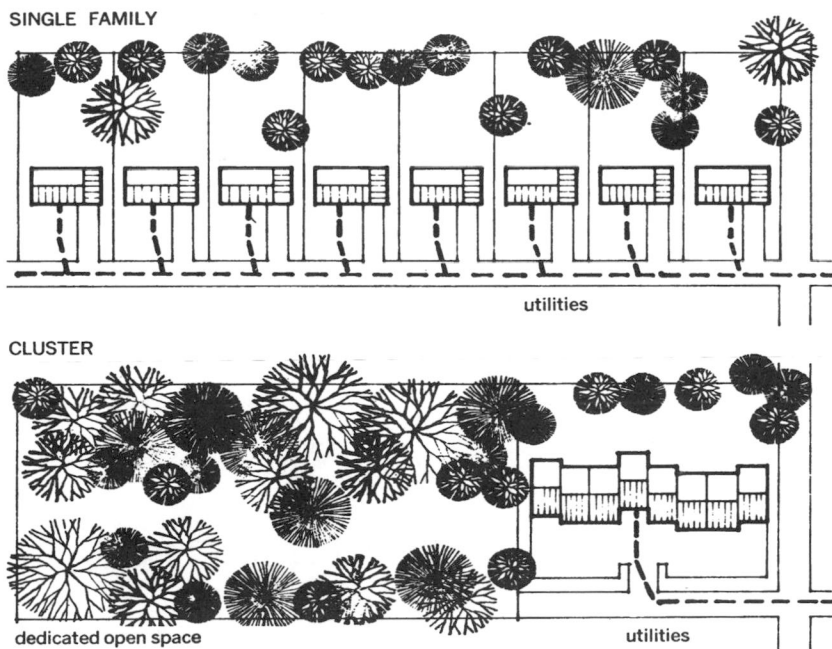

Fig. 5 Single family vs. cluster comparison shows open space gained by clustering homes.

CLUSTER DEVELOPMENT

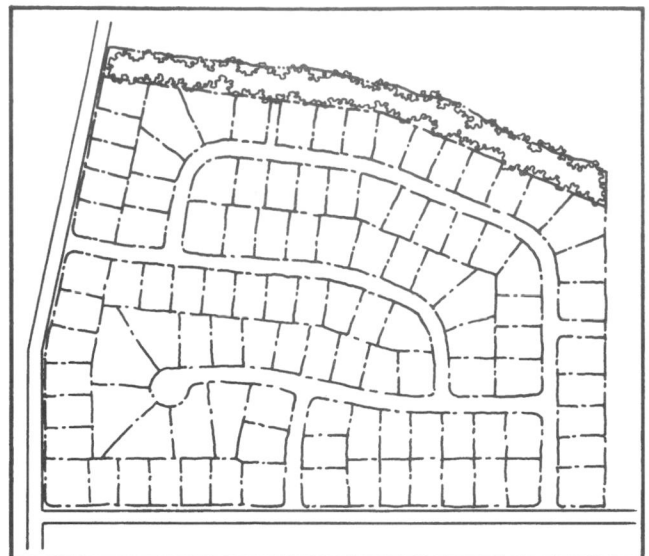

Fig. 6 Conventional development.

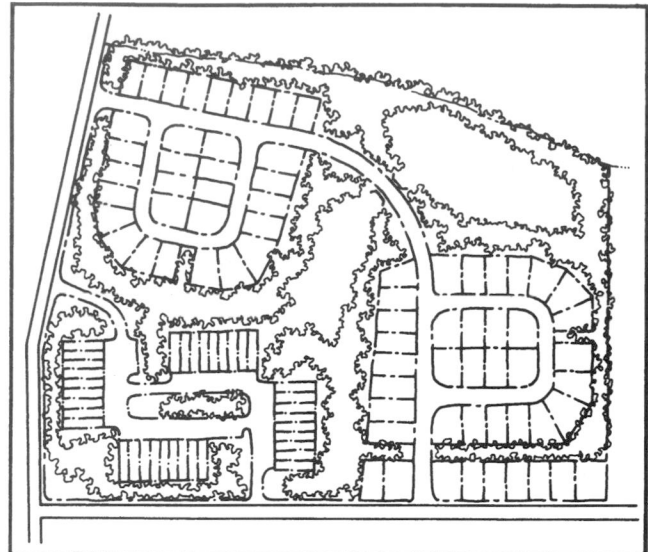

Fig. 7 Open-space development.

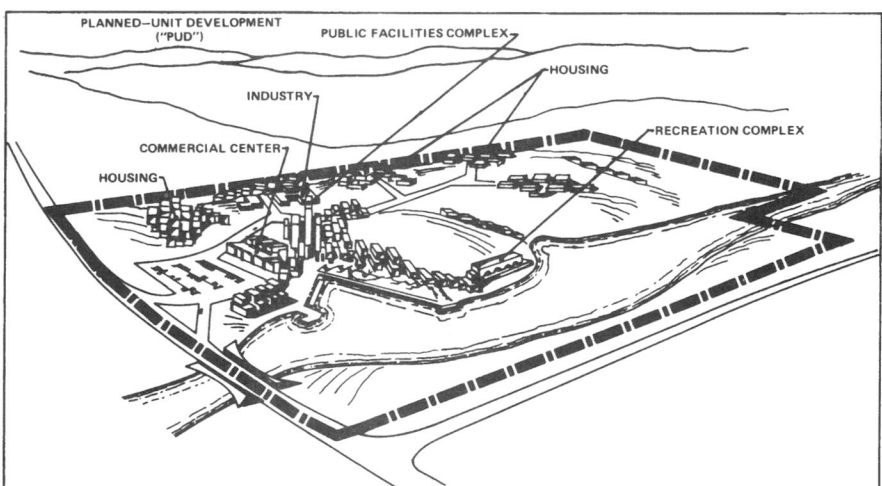

Fig. 8 Planned unit development.

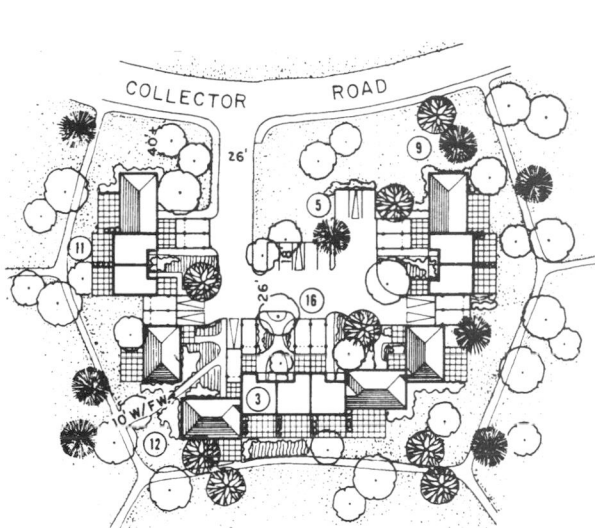

Fig. 9 A 13-unit cluster.

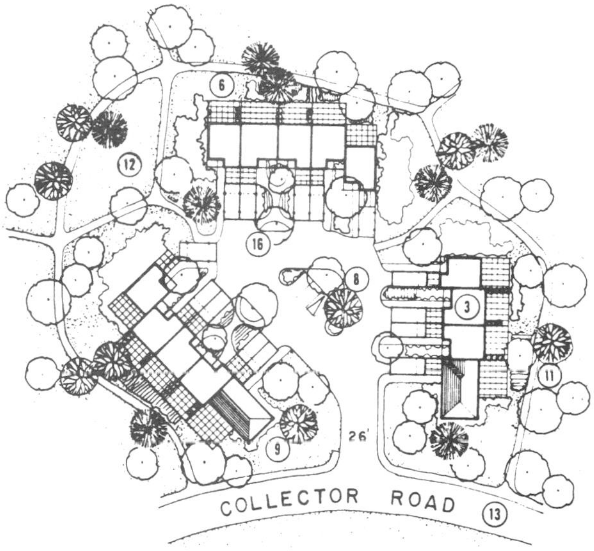

Fig. 10 An 18-unit cluster.

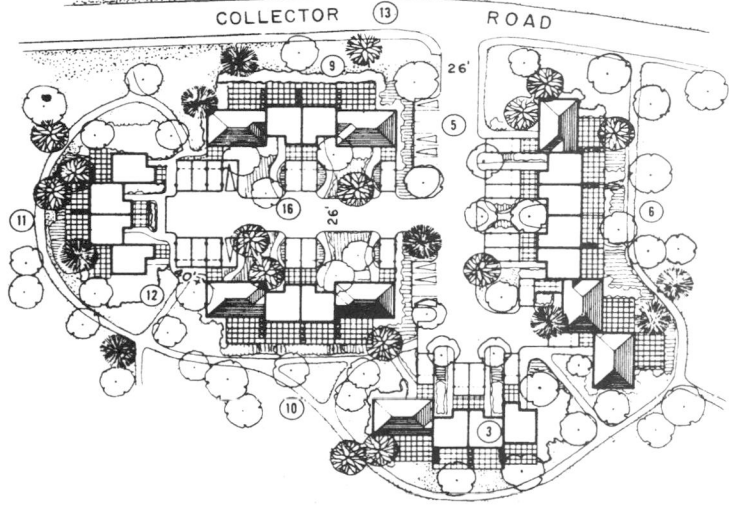

Fig. 11 A 23-unit cluster.

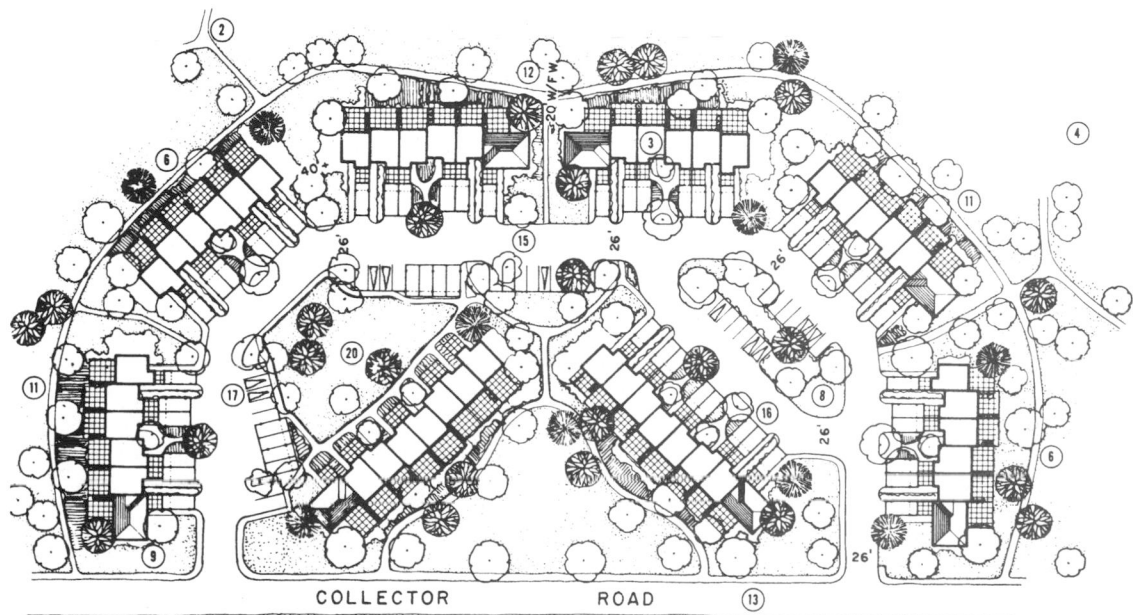

Fig. 12 A 52-unit cluster.

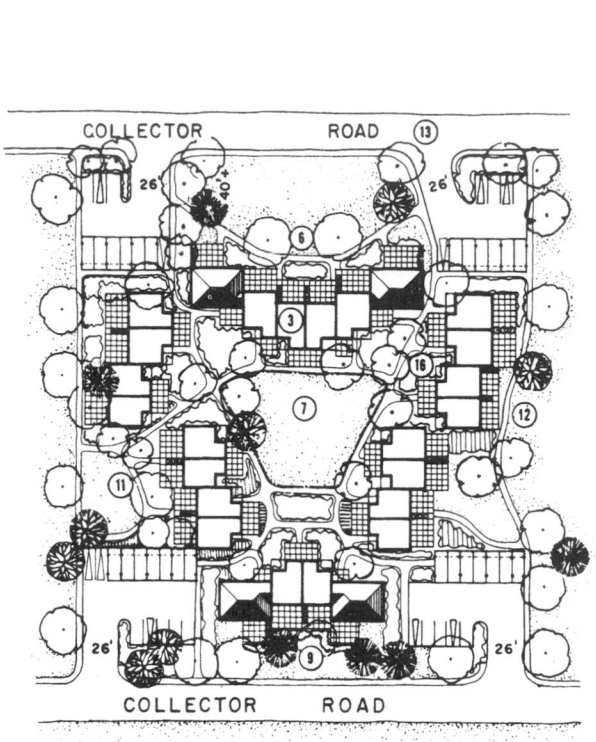

Fig. 13　A 26-unit cluster.

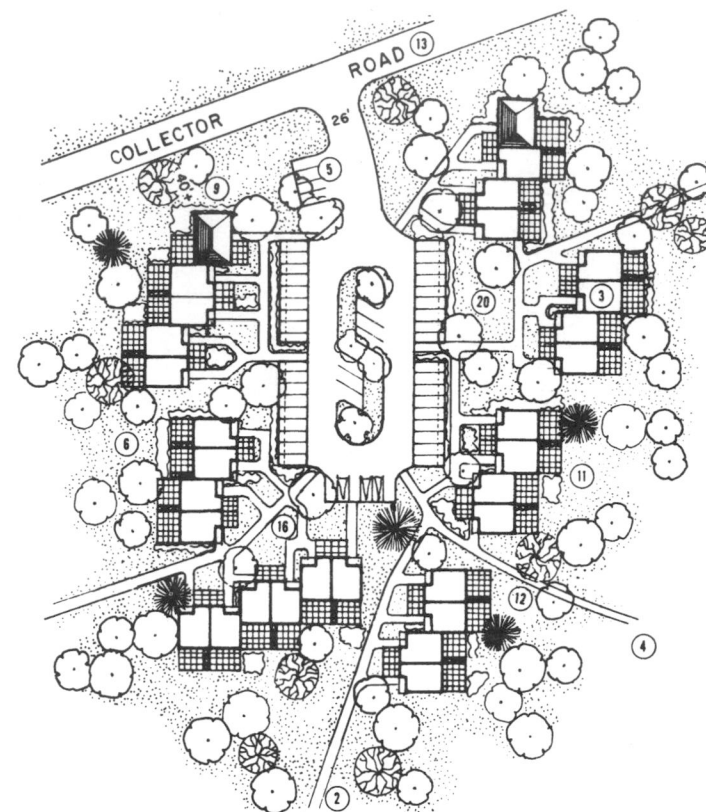

Fig. 14　A 31-unit cluster.

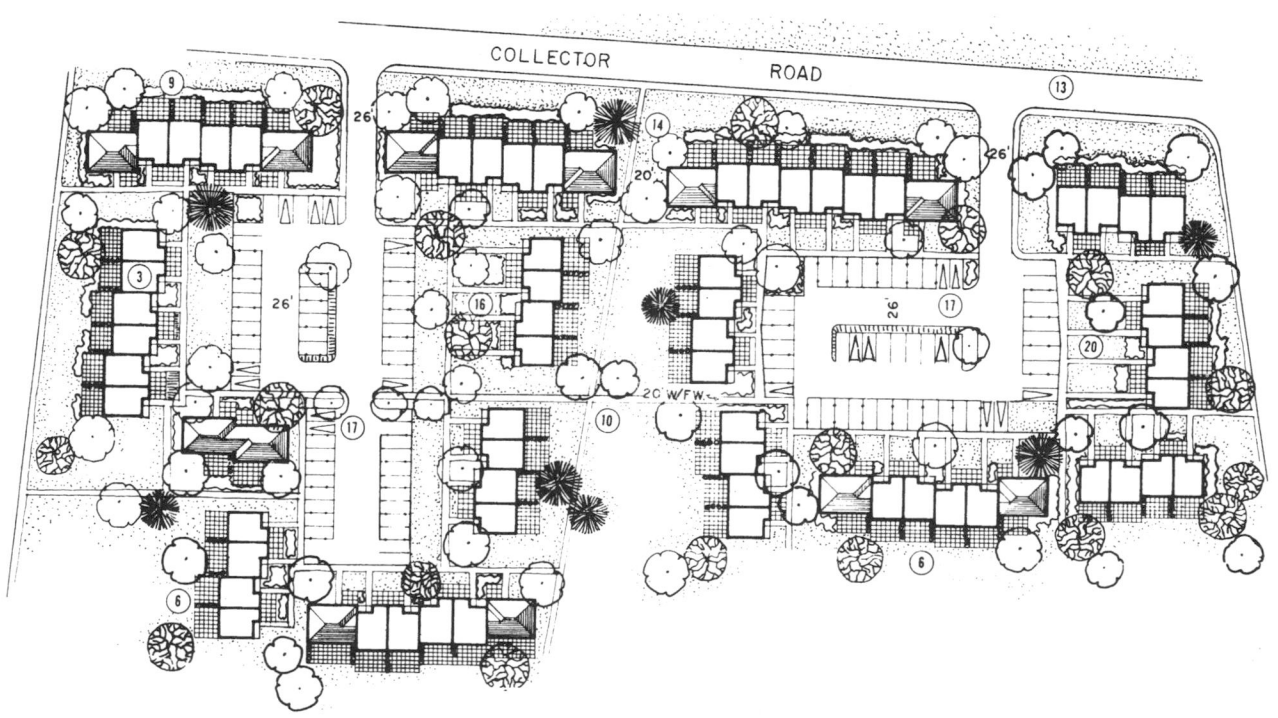

Fig. 15　A 72-unit cluster.

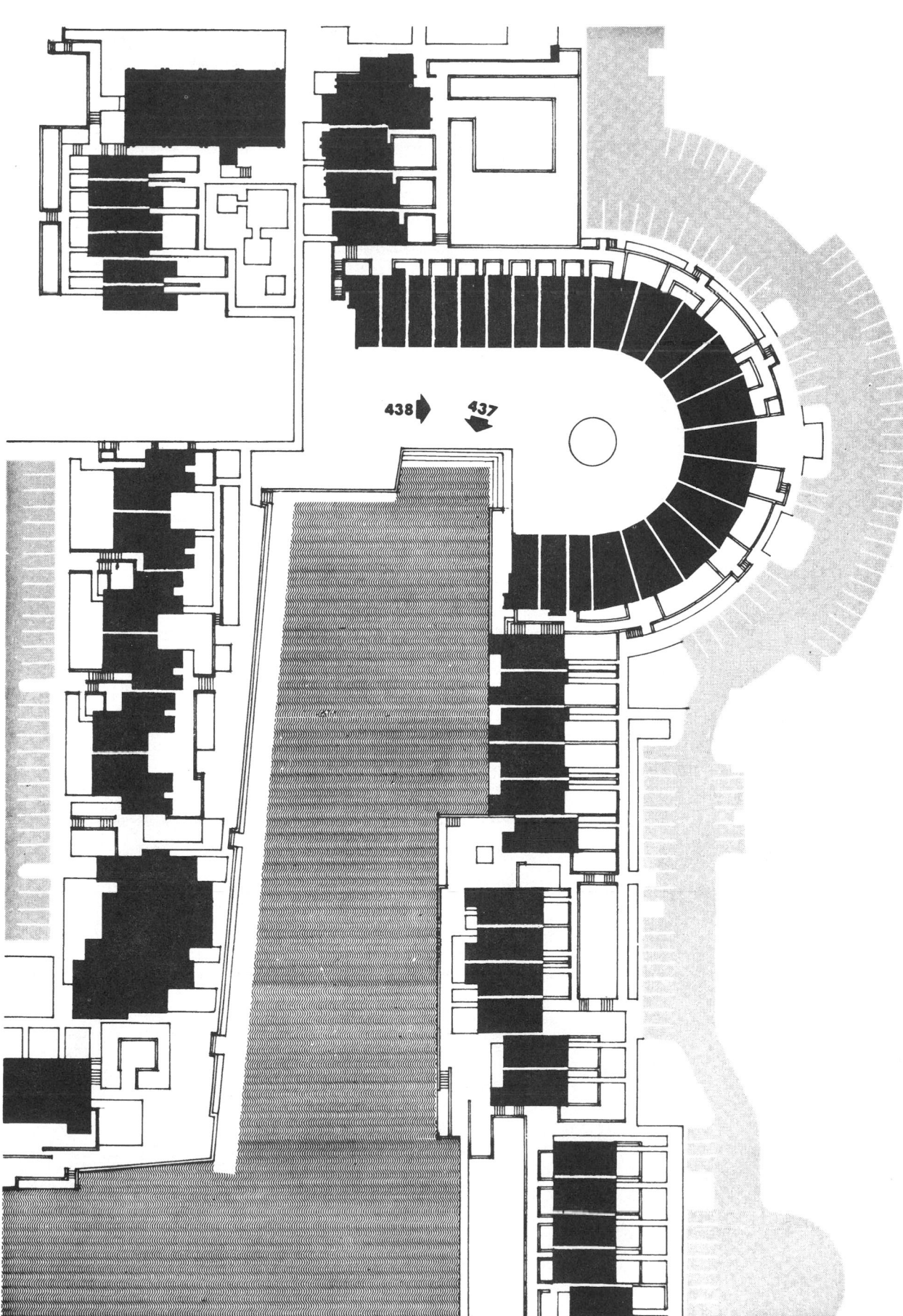

Fig. 16   Reston, Reston, Va. Whittlesey and Conklin, Architects.

DESIGN PRINCIPLES

The rear yard becomes the only usable open space, and even here, since all houses are placed in a row, it is very difficult to establish any real privacy.

Without setting down standards for houses and their placement, it is the intention of the planned unit development regulations to introduce the kind of flexibislity whichs will greatly improve the residential environment. Specifically, the individual house must be designed to relate the open area around each house to what occurs inside the house. The house has entrances for people and an entrance for automobiles; this fact has meaning in terms of both open area and the portion of the house which serves the entrance function; and a design response to this fact is expected. The house has indoor and outdoor living functions; this requires a design which relates the two, preferably resulting in increased privacy for outdoor living because of the way adjacent buildings are placed. Similar thought and design response should be shown in the arrangement of bedrooms, internal circulation, and service spaces.

Table shows land use and site utilization advantages of Planned Unit Development schemes (numbered 2, 3, and 4) over conventional subdivision scheme, 1.	GROSS SITE AREA	STREET AREA	STREET AREA % OF GROSS SITE AREA	NET SITE AREA
1 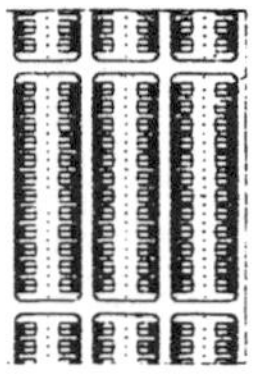	20 ACRES	6.3 ACRES	31.4%	13.7 ACRES
2 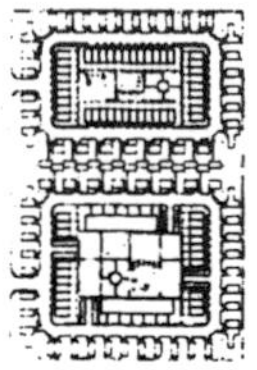	20 ACRES	5.6 ACRES	28%	14.4 ACRES
3 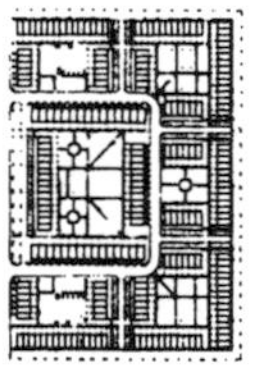	20 ACRES	4.1 ACRES	20.5%	15.9 ACRES
4 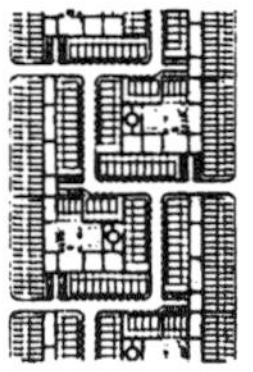	20 ACRES	5 ACRES	25%	15.0 ACRES

Fig. 1   Table of comparative advantages.

COMMON OPEN SPACE	NUMBER OF DWELLING UNITS	ALLOWABLE FLOOR AREA PER DWELLING UNIT	ALLOWABLE COVERAGE PER DWELLING UNIT	ALLOWABLE NUMBER OF ROOMS PER DWELLING UNIT	
NONE	semi-det: **198**	**1400 sq. ft.**	**700 sq. ft.**	**7.5**	Figures based on: typical zoning lot of 2800 sq. ft. F.A.R. (Floor Area Ratio) of .5 O.S.R. (Open Space Ratio) of 150 lot area per room of 375 sq. ft.
2.3 ACRES	detached: **59** semi-det: **23** townhouses: **62** garden apts: **56** total: **200**	**1840 sq. ft.**	**940 sq. ft.**	**9.5**	Figures based on: net site area divided by number of dwelling units, and application of full bonuses resulting in: F.A.R. (Floor Area Ratio) of .575 O.S.R. (Open Space Ratio) of 120 lot area per room of 337 sq. ft.
8.6 ACRES	townhouses: **213**	**1900 sq. ft.**	**980 sq. ft.**	**9.8**	
4.0 ACRES	townhouses: **210**	**1820 sq. ft.**	**975 sq. ft.**	**9.35**	

Fig. 1 (*Continued*)

DESIGN PRINCIPLES

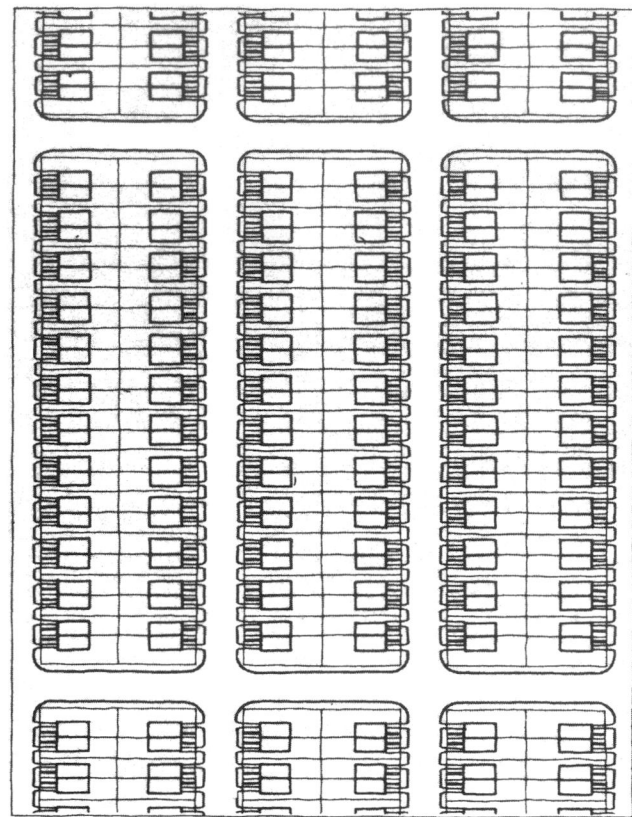

Fig. 2

1

Site plan shows maximum site utilization with semi-detached house types, typical of past standards. Note that all streets go straight through; no meaningful open space is provided, and the distance between adjacent structures is only 13 feet.

2

Site plan shows a variety of house types arranged around a loop street system. Note that streets serve only houses on the loops; that the houses are oriented toward their private yards; that large areas of common open space are provided in the interior of each loop; and that the distance between structures is never less than 20 feet.

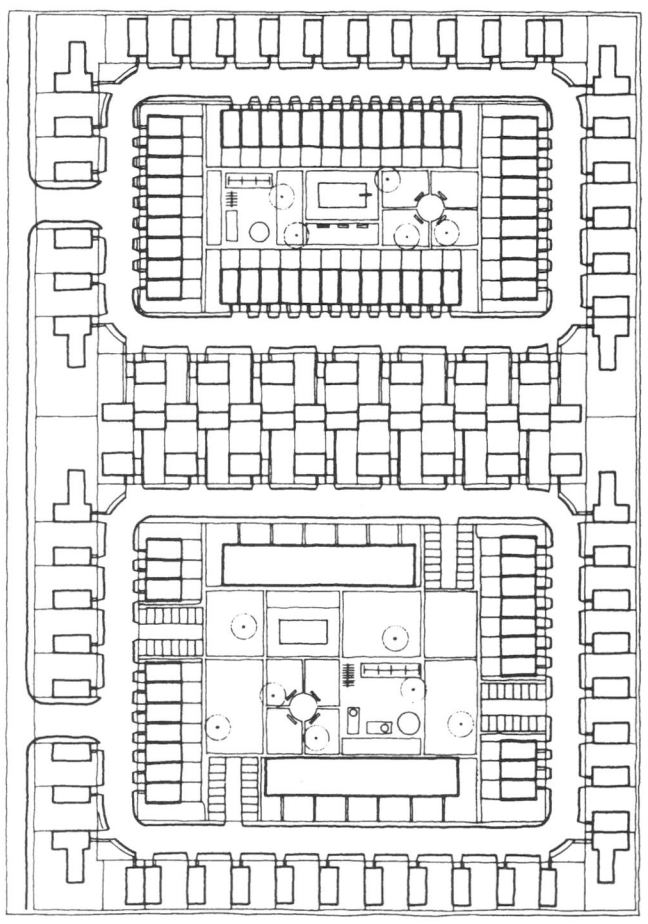

Fig. 3

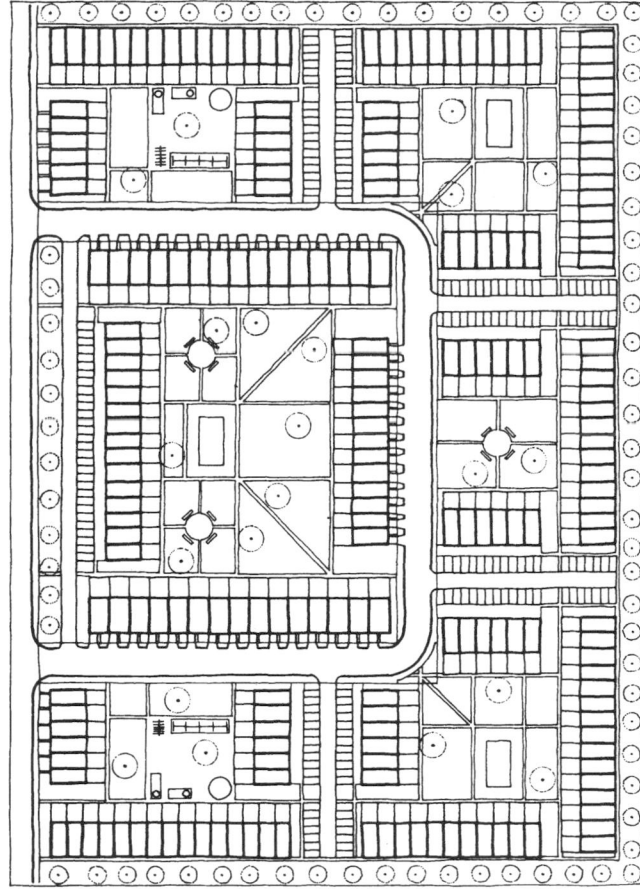

Fig. 4

3

Site plan shows townhouses arranged around a horseshoe loop and cul-de-sac street system. Note that the streets are solely for the use of the individual townhouse clusters; that each cluster encloses meaningful open space accessible through the private yards of individual houses; and spacing between building groups is very generous.

4

Site plan shows townhouses arranged on a modified grid street system. Note that, by making minor modifications in a typical grid system, through traffic can be eliminated; that the resulting building placement defines a series of open spaces, private and common; and that spacing between the buildings is generous.

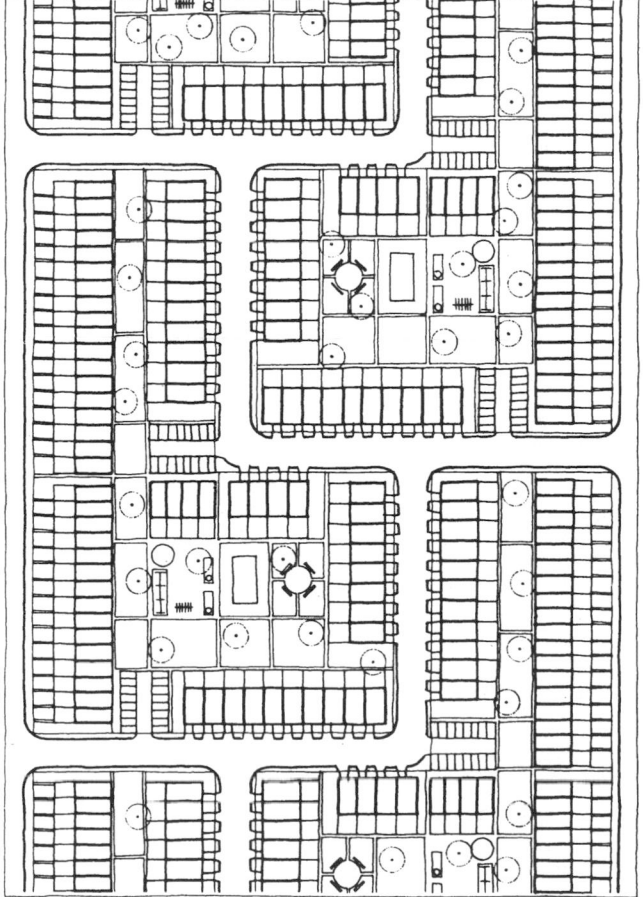

Fig. 5

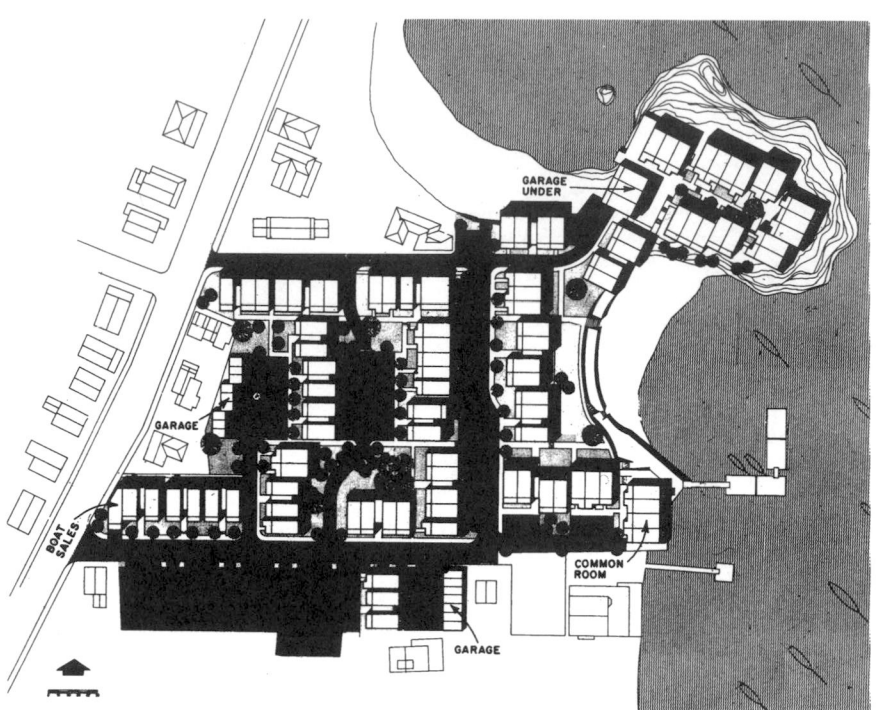

Fig. 1

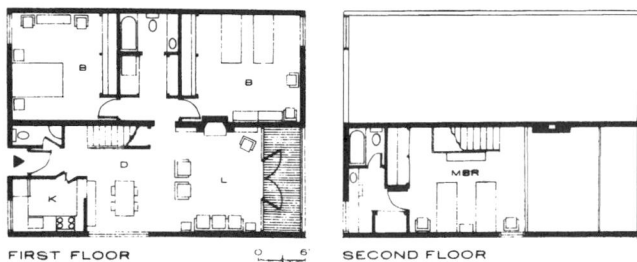

FIRST FLOOR   SECOND FLOOR

Fig. 2   Clover Landing, Marblehead, Mass. Chapman & Goyette, Architects.

OWNERSHIP AND MAINTENANCE OF OPEN SPACE

Two matters relating to open-space subdivision that often arouse apprehension on the part of both planning boards and the public are first, concern with the maintenance and control of open space, and second, the fear that the open space may someday be used for development, thus greatly increasing the total density. These are legitimate and sensitive issues. However, planning boards can ensure that developers not only prepare an appropriate physical design but can also provide proper legal safeguards for control and maintenance of the open space.

Two basic approaches are most commonly utilized. The open space can be dedicated to the community for use as a public facility, which would then mean that the municipality would maintain it; or it could be owned by a homeowners' association comprised of the residents of the subdivision and reserved for their use.

Most county districts are geared to assist developers, local groups, public agencies, and community associations in developing appropriate management plans for open space and natural areas.

Municipal Ownership

A number of communities require public dedication of open space.

Each method has its advantages and problems. With municipal ownership, of course, there is a firmer guarantee that the land will be used and cared for in compliance with the wishes of the larger community. It also can be a relatively inexpensive and painless way to add parks and open space resources for the community. However, should the open spaces be in an area not easily accessible, there is the danger of the municipality maintaining at public cost a facility that is, for all practical purposes, a private park.

The other side of the coin of municipal ownership is that prospective homeowners may not wish to live in or adjacent to a public park.

In the belief that any open space is better than none, many communities may be tempted to accept whatever land a builder is willing to dedicate to the municipality as open space. Care should be taken, however, that such land is appropriate for open space and compatible with community ownership and responsibility. The possibility of use for formal recreation such as ball fields is not necessarily a criterion. Much open space can serve a valuable function in its undeveloped state as a "wander space" for youngsters, as a visual amenity, or as a nature study area. However, in its eagerness to increase it supply of open space, a community

may find itself owning land which, through location, topography, or general condition, is not only unsuitable for formal recreational use but difficult to maintain and care for even in its undeveloped state. Or it may be too inconveniently located for use by a ny significant number of residents. Such land may turn into a dumping ground for autos and other wastes, and instead of benefiting the community, end up as a hazardous or unsightly area that can only be properly eliminated or supervised at great public expense.

Such land may be offered by developers because it is economically infeasible to build on, or because its slope, soil conditions or other characteristics do not satisfy the "buildability" criteria. They may often attempt to gain "credit" for such land, thus enabling them to build on the remaining piece to a higher density than would have been realistically possible on the total site.

To cope with this problem, a number of local governments require a builder to submit a conventional subdivision plan for the entire area, showing the lots that could be realistically created in terms of topography and costs under the existing zoning and in compliance with the subdivision regulations. The total number of lots arrived at in compliance with these qualifications establishes the maximum density for the open-space development.

A community can protect itself against the possibility of possessing land that is a liability rather than an asset by asking that developers "finish" the land before dedicating it to the municipality. As a result, the open space received is fully equipped and laid out for baseball and for other specific recreational uses.

The Homeowners' Association

Many of the problems associated with municipal ownership may be eliminated through the use of an alternative approach to the preservation of open space—the homeowners' association.

The homeowners' association is a nonprofit corporation made up of the residents to maintain the common open spaces and facilities in an open-space development. It is, in a sense, a small neighborhood government. Such associations may be voluntary or automatic. In voluntary associations, membership is optional, and while this idea may appeal to our democratic instincts, such an approach has many shortcomings. It can lead to administrative difficulties and to inequities among members and nonmembers in the use of land and facilities.

The automatic or mandatory homeowners' association is by far the more effective approach. Such an association should be legally established before sales in a development begin. As each lot is sold, the purchaser must become a

member of the association. This requirement "runs with the land"—that is, it is written into the deed of each individual lot in perpetuity.

The association is responsible for the care and maintenance of the open space and any developed facilities, such as ball fields, swimming pools, or meeting rooms that may be commonly owned. A monthly or yearly service charge is assessed against each member to cover the costs involved.

Developers retain membership in the association by virtue of their ownership of the unsold lots and in the early stages will have the majority membership and thus control the community facilities and open space.

A number of such associations have been operating for many years. In Radburn, New Jersey, where an association has existed since 1930, the annual fee is based on a prearranged percentage of the real estate tax paid to the township. In return, the association not only cares for the recreational facilities and the open space but also provides a library and a program of recreational and educational activities for all residents. The fee also covers the salary of a full-time manager and a small clerical staff, a necessity in a community as large as Radburn.

The municipality can require that such an association, if established, be set up by the developer according to prescribed standards. It may list a number of conditions for approval of such an association. These include the requirements that membership in the association be automatic for each lot owner and that the homes association gain title to all the common property and, once established, retain all responsibility for operation of the open space and common facilities.

The question of whether the open space should be municipally owned and maintained or whether it should become the property of a private homeowners' association is a decision that must be made by each municipality. Local circumstances, such as the need for public open space and the nature of existing development, will affect the decision.

Conclusion

Open-space subdivision should become an integral element of a municipality's strategy to achieve recreation and open space objectives. This subdivision technique will not, however, relieve a locality from having to acquire or in other ways obtain and preserve parklands. Though the community can facilitate, guide, and encourage its use, in the final analysis open space subdivision lies in the hands of private developers and is subject to the vagaries of the housing market. Yet as we have emphasized, maintaining open areas in new subdivisions is the most farsighted way of ensuring a sproper balance between people and nature.

LAND SUBDIVISION ANALYSIS

DESIGN PROCESS

In the design process, the data and interpretations resulting from the site analyses are used in evolving a final site plan. There are two aspects of the analyses that must be articulated in the design process: the identification of logical areas of the site and the identification of significant conditions. Actually each is inherent in the other, but this intrinsic relationship is easily lost in the dissecting process of analysis. The purpose of the analyses is to better understand the conditions of the site; thus a great deal of data about the site are produced. The purpose of site design is to work with the conditions of the site to achieve an environment that is consistent with these existing conditions and a physical form that is sympathetic to that which exists. The key to design is the organization of the data into manageable areas that pertain to how the site can be used. The design process involves the application of various criteria for planning and design to this basic environmental concept of the site.

The stages in development of a final plan and the inputs that are required for each stage are diagrammed in Fig. 1. The detailed analyses of the environment of the site result in maps of natural and manmade units. The first step in the design process is to combine these into one set of units which comprise all significant environmental factors. These are evaluated according to land management and development criteria to indicate the type of land use pattern the site can tolerate. When the developer's program requirements are added, it is possible to outline the realistic options for development of the site. After the developer and planner have settled on a course of action, basic design criteria are followed in alying out a conceptual or pre-preliminary site plan. This plan should be reviewed by the planning board, or preferably its staff, to determine if the development objectives for the strategic area will be accomplished. After this endorsement, the plan is refined to comply with all local regulations and conditions. This input results in the preliminary site plan, which is reviewed by the planning board and by the public at an open meeting or hearing. After preliminary approval by the planning board, the plan is fully detailed into working drawings which show how every aspect of the project will be built in compliance with all applicable design standards and construction codes. These drawings receive the planning board's final approval, and a building permit can then be issued.

The planning board's contribution to development within strategic areas is completed at the end of the preliminary site plan review. The production of final plans and supervision of construction should not require changes in the planning and design concepts for a site. For the most part, the final design and construction stages are concerned with details which have an effect on the architectural and internal quality of the site but have little impact on the community as a whole. These stages can be adequately handled by a building inspector cognizant of the planning board's objectives according to usual practice. Therefore, these stages are not treated here.

The results of the natural and manmade analyses are two sets of units which delineate all existing conditions and the extent or area of each. To begin the design process it is necessary to derive a common denominator for both natural and manmade considerations. This is achieved by superimposing the natural and manmade units, which produces a map combining all units. Each resulting area is a distinct set of conditions—a discrete unit of the environ-

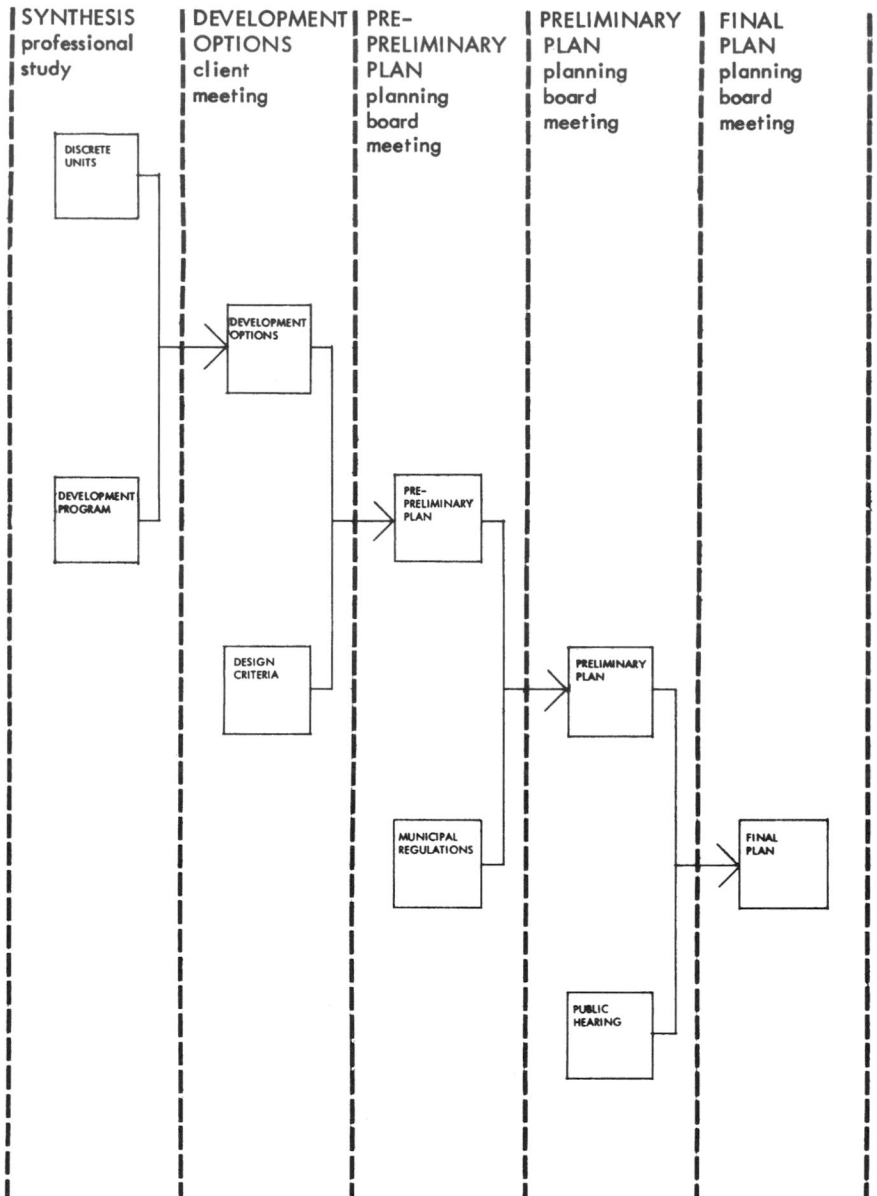

Fig. 1

ment. Discrete units are a synthesis of all environmental conditions expressed in areal terms. They are the environmental divisions of the site that reflect its existing functions and character. Therefore, they are the logical building blocks of a site plan that respect the environment.

The discrete units can be evaluated to determine how each should be used within the context of the total site environment. The most basic evaluation is the suitability of a unit for building. The intention of this evaluation is to protect the efficient functioning and character of the site which benefit the entire local area. In the public interest, the shaded units on the above maps are excluded from the potential building area of the site. The remaining units are generally suitable for building. They have either no serious constraints or problems, such as an occasional high water table or somewhat steep slopes, that are within the scope of normal site improvements. The factors which determine how they can be used are also derived from the natural and manmade analyses. They are given in Table 1.

Discrete Units

In applying these factors to the discrete units, a pattern of combinations was apparent. Five combinations were derived which enabled the units to be aggregated into larger coherent areas, each with a narrow range of conditions. The characteristics of each are listed below:

1. Flat and gently sloping, northwest-east aspect, steep drop-off to southeast, high visual exposure, access to town road, access to unbuildable area, access to building complex, no trees

2. Moderate slope, north aspect, high visual exposure, access to unbuildable area, scattered trees and boulders

3. Moderate to steep slope, southeast aspect, access to unbuildable area, no trees

4. Flat to moderate slope, northwest-north aspect, access to unbuildable area, no trees

5. Moderate to steep slope, north and southeast aspect, high visual exposure, small forested area

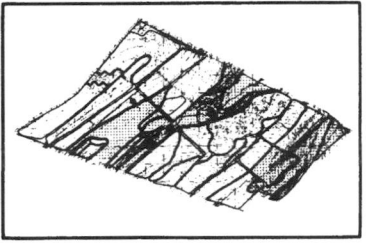

NATURAL UNITS

 drainage courses and areas subject to flooding
lowland areas subject to ponding
steeply sloping land (over 20%)

Fig. 2

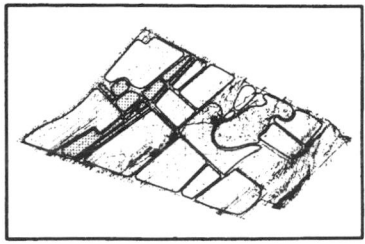

MAN-MADE UNITS

 power line right-of-way
farm house and barn

The discrete units distinguish all environmental resources at the site scale. They reveal the range and distribution of conditions but may be too detailed and fine in scale to be the basis for site design. Grouping the discrete units according to relative similarity of conditions provides more manageable units that are applicable to housing development. Although some generalization of detail is involved, the larger units remain consistent with the structure and the functions of the environment.

The functional limitations of the site are protected by the designation of unbuildable area, or open space. This comprises all environmental functions that are essential to the area and all serious limitations on building with the site. The open space pattern gives the design a basic envi-

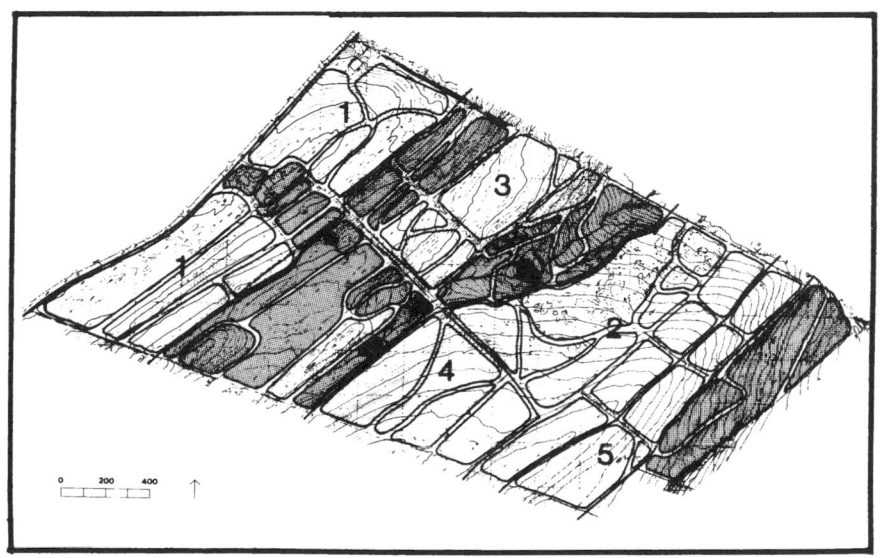

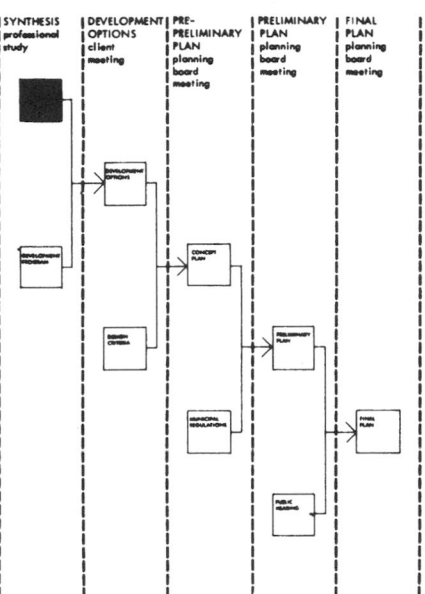

Fig. 3*b*

DISCRETE UNITS

 SUITABLE FOR DEVELOPMENT

 UNSUITABLE FOR DEVELOPMENT

Fig. 3*a*

TABLE 1

Natural	Manmade
Slope:	Access:
Aspect	Roads
Earthmoving and slope stability	Open space
	Views
Soil:	Service
Drainage conditions	Serviceability
Foundation conditions	Community facilities
Vegetation:	Edges:
Types	Stone walls and fence lines
Condition	Hedgerows
Visual:	Activity:
Exposure	Barn rehabilitation

ronmental appropriateness; however, it is not sufficient to express the character and identity of the site. This requires, in addition, a sympathy with the physical forms: the landforms, the indigenous plants, the organization of the land and the activities upon it, and the visual or spatial coherence that results from all these factors. This element of sympathy is provided by the structure and characterization of buildable areas. The characterization indicates the basic quality of the area with which the proposed use must be compatible; the edges of the areas are the main linear elements of the structure of the site. The development program discussed below can thus be related to the most appropriate areas of the site and organized according to the same elements that give the site its present character. By this means, the integrity of the environment will be continued as the site is developed for a new use.

The main objective of land developers is to achieve a return on their investment. In most cases, they have purchased land at a price based on the return to be expected under the local zoning designation. The zoning thus, in effect, requires the developer to build a minimum number of dwelling units in order to make a profit. Under the flexible provisions of a planned unit development, a mixture of housing types may be provided. Each type has different land requirements, and a different value in terms of its selling price for the developer. A mixture of housing types therefore requires a different number of dwelling units in order for the developer to realize the same value. Developers decide on the percentage of each housing type according to their evaluation of the housing market. In the case in Fig. 3 the most feasible mix was 30 percent apartments, 60 percent townhouses, and 10 percent single-family detached houses. The next step is to determine the number of each type according to the zoning and the conditions of the site.

The site has an area of 90 acres; it was zoned for 10,000-ft^2 single-family lots which yield about 2.5 dwelling units per acre. The zoning therefore permits 225 detached, single-family units on the site. As a PUD, the number of units of each type is determined by dividing the permitted number of units by the desired mix. In order to compensate the developer for the lower unit value of multifamily units, the number of garden apartments is increased by a factor of 3 and the number of townhouses is increased by a factor of 2. The total number of dwelling units is then 494.

DEVELOPMENT OPTIONS

Developers then consider the options for distributing this number on the site. By using the most suitable areas for each housing type and standard densities for each type, a total of 479 units was the maximum feasible number. This was acceptable to the developer; in other cases options such as greater than standard densities, major site improvements to enlarge the areas of suitable land or changes in the desired housing mix may be considered in reaching an acceptable program. By following the PUD option, the developer achieves a greater number of units, a desirable mix of housing types, and the neces-

TABLE 2

	Percent of permitted number of units	Number of units	PUD unit increase factor	Maximum number of units permitted
Apartments	30	67	3	201
Townhouses	60	135	2	270
Single-family houses	10	23	1	23
Total	100	225		494

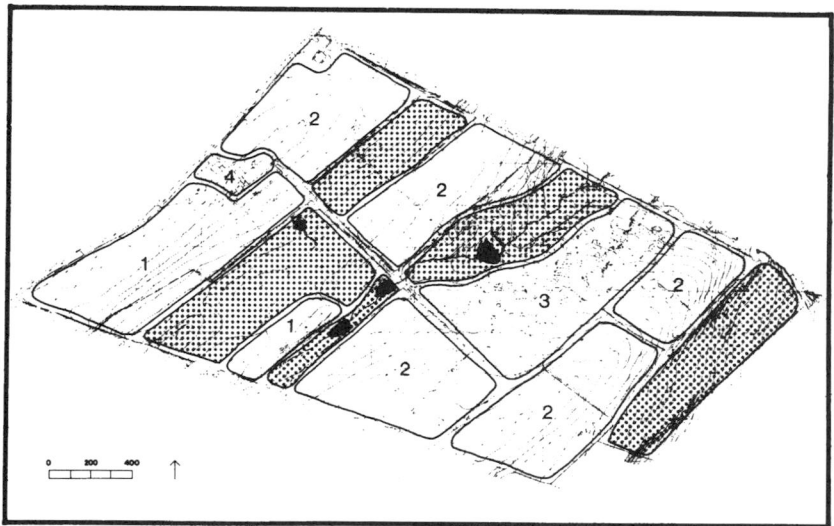

Fig. 4a

Fig. 4b

		ACRES	DENSITY FACTOR	# UNITS
1	APARTMENTS	13	12	156
2	TOWN HOUSES	36	8	288
3	SINGLE FAMILY HOUSES	14	2.5	35
4	RECREATION	1		
	OPEN SPACE	31		
	TOTAL	90	5.3	479

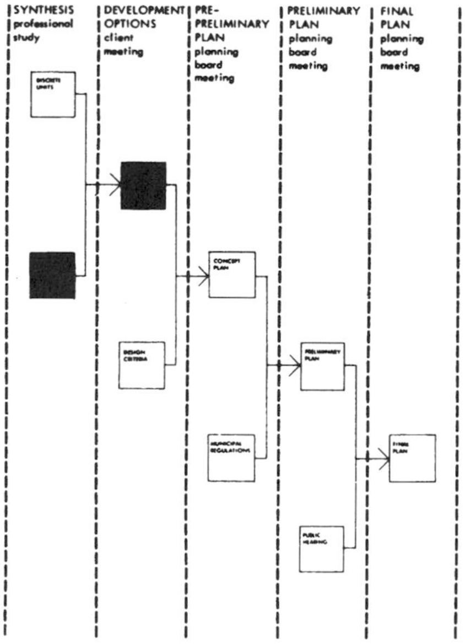

sary flexibility for a high quality of desing. The town achieves a relatively intensive development, appropriate for a strategic area. In addition an important segment of a townwide open space system is secured at no cost to the town.

The map resulting from the development options step provides a concept for the location and intensity of uses on the site. This concept for design of the site is developed into a desirable physical layout by the application of design criteria. Below is a list of criteria for determining building types and their location and orientation on a site. These are derived from basic principles of the effects of the environment on buildings and activities.

The design criteria in Fig. 7 cover basic principles of organization of buildings and access into a coherent system or pattern for a site. These principles should be combined with the site-derived criteria from Figs. 5 and 6 in the design of a conceptual site plan.

PRE-PRELIMINARY PLAN

The pre-preliminary plan is a schematic drawing showing the organization of all the major components of the development according to design criteria and the dictatesof the site. The size and relationship of all building and open space areas and the vehicular and pedestrian access routes are set. These provide the structure for the proposed development and ultimately determine its character.

The pre-preliminary plan is submitted to the planning board for review and comment. It does not require formal approval but should be the basis for discussion of the development concept. Following agreement on the concept, or a modification of it, the concept can be trans-

SUN

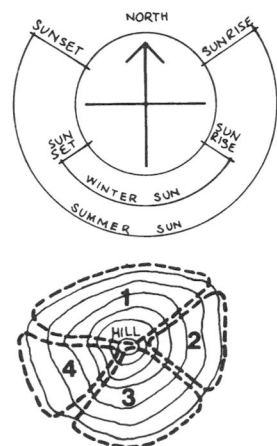

desirable slope orientation
■ winter   morning and afternoon south-east to south-west
■ summer   early morning and late afternoon east, west
undesirable slope orientation
■ winter   west, north, east
■ summer   south-east to south-west

aspects of the sides of a hill
1   cold side
2   cool side morning
3   hot side (summer)
4   warm side evening

WIND

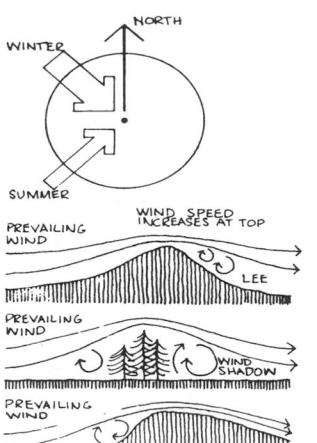

desirable   summer breezes
undesirable winter prevailing winds

buildings
insulate to protect from cold winds and open to allow winds to cool in summer

locate buildings on the lee side of the hill to protect them from winter prevailing winds

locate buildings or plant hedgerows or tree buffers to protect them from prevailing winter winds

the most desirable location for building with respect to wind and hills also depends on the shape of the hill

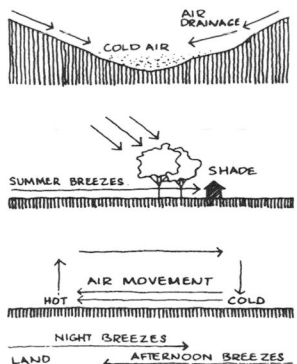

cold air accumulates in valley floors and then tends to move down the valley   this is particularly a night-time effect

plant trees to protect buildings from summer sun but allow for breezes

deciduous trees are desirable as they provide shade in summer while letting the winter sun through

hot air rises causing the air from colder areas to move in under the rising hot air   this process causes a circular movement with air moving from cold areas to hot areas

design buildings and layouts to take into account the daytime and night-time direction of air movement

Fig. 5a

LAND SUBDIVISION ANALYSIS

lated into a plan in which the actual components of the project are located on the site and designed.

The exercise of control over the development of land by a community is an application of the police power of the state. The fundamental document which sets forth the powers and provisions of land use control is the state enabling legislation which empowers local government to adopt land use controls. The most commonly used controls are the zoning ordinance and subdivision regulations.

Zoning Ordinance

The zoning ordinance is adopted by the community for the purpose of promoting the health, safety, morals, and general welfare of the community. The ordinance is a local law which regulates the use and development of all land, as follows:

- Dividing the community into districts
- Density of population
- Location and use of buildings, structures, and land
- Amount of open space
- Percentage of lot that may be occupied
- Size of yards and courts
- Size of buildings
- Building height and number of stories

VEGETATION

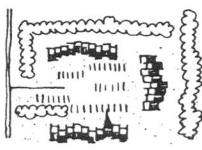

utilize existing tree patterns
- lines
  maintain scale, rhythm and form of hedgerows and other linear elements

- areas
  large scale structures relate well to large masses of trees, woods or forests

  areas can be cleared creating tree masses to relate to building complexes

- points
  scattered clumps and individual trees relate well to single family houses

  build in woods rather than treeless fields

DRAINAGE

maintain natural, surface drainage system

provide storage for excess runoff in lakes and temporary ponds

avoid building in flood plains (conform to the national flood insurance program)

locate structures in areas that are not prone to natural disasters or build structures that can withstand the natural disaster (e.g. hurricanes, tornados, earthquakes, tidal inundation, mudslides, erosion, floods, subsidance, etc.)

SOILS

structural quality of the soils should determine type of construction

maintain soil as a fertile medium for plant growth

minimize earthmoving requirements

restrict construction areas and movement of heavy machinery to avoid unnecessary soil compaction

stockpile topsoil from excavated areas for reuse after construction is completed

provide subgrade drainage of all wet soils, discharging into surface drainage system

maintain water balance and water table by avoiding compaction of soil, massive regrading and high coverage (causing excessive runoff)

Fig. 5b

TOPOGRAPHY

design and place buildings so that they follow contour lines and relate to the form of the terrain

build on the sides of hills not on the top so that the form of the natural feature will be protected

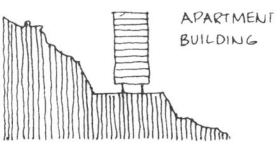

APARTMENT BUILDING

rest on a platform

TERRACED HOUSES

SINGLE FAMILY HOUSE

use slopes to create parking spaces under buildings

on steep slopes design buildings for that unique condition

imposed structures may hug the slope

stand completely free

avoid high land coverage on rough, unique and/or steep terrain   it is possible to keep the same floor area ratio while minimizing the coverage by building multi-story structures

unique soil conditions or fragile eco-logical system such as sand dunes require designs that do not disturb the natural system

preserve natural features such as rock outcrops, trees, etc. rather than "improving" them for ease of construction

	SLOPE	CONDITION AND SITE IMPROVEMENTS	SUGGESTED TYPE OF RESIDENTIAL USE
	flat 0%–5%	requires regrading and underground drainage system best left with no major development or construction	recreation open space
	gently sloping 3%–10%	most development can be sited with minor reshaping of land generally good for building	all housing types single family houses
	rolling 5%–20%	low retaining walls may be required around roads and parking areas generally good but less coverage	town houses
	steep 15%–30%	very high or tiered retaining walls may be necessary to accommodate grade changes – minimize land coverage	no parking lots apartments
	very steep over 30%	foundations and retaining structures are usually prohibitive in cost avoid building and construction of all types	open space

note   the suggested residential use is only meant as a guide and is based on land coverage and access only – combinations and/or changes can be made

Fig. 6*a*

VISUAL

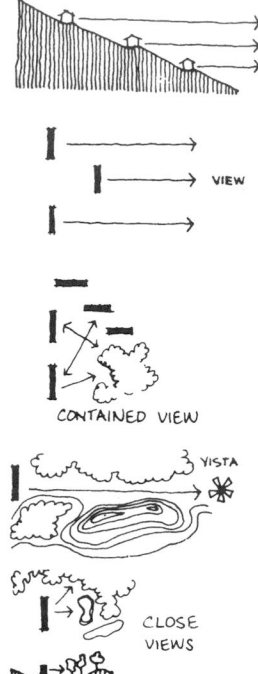

buildings located on a slope can take advantage of the view as long as the lower buildings are kept under the line of vision of the one above

buildings on flat terrain have to be placed so they do not block the view

when the terrain offers no view the buildings should be placed so that an internal or contained view becomes significant

place buildings to take advantage of vistas or to create vistas

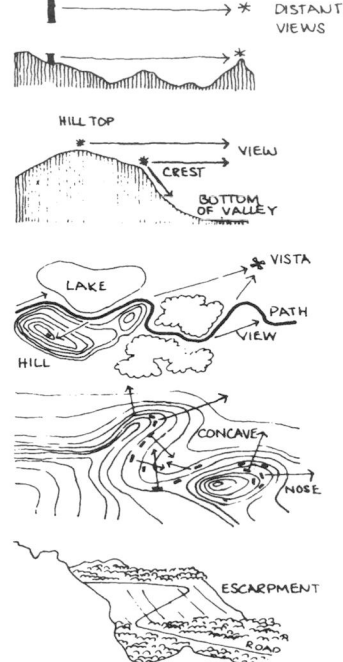

design and place buildings to take advantage of both close views and distant views

the brow or crest of a hill is a more critical location than the top of the hill as it offers views down the hill into the valley as well as distant views out

make use of visual elements when designing paths, roads, etc.

hollows or concave slopes are enclosed, sheltered, oriented internally or to a focused view

noses or convex slopes are exposed, expansive, oriented outward to a general view

avoid crossing prominent landforms with roads or other man-made structures therefore maintaining the identity or integrity of the features

SOUND

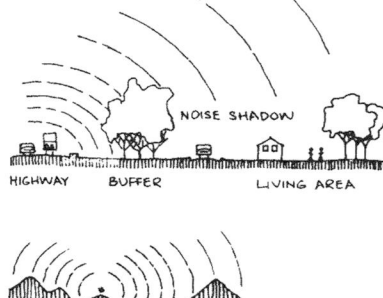

create a buffer between living areas and noise source

avoid prominent sites for projects that generate a lot of noise e.g. industry, highways, airports, etc.

noise reverberates through and along valleys

noise travels very well across water bodies

Fig. 6*b*

MOVEMENT PATTERN

create a hierarchy of roads each
with a clearly defined function
- limited access highway (interstate)
  inter-city, high speed, no development, grade separated interchanges

- arterial (state)
  intra-city or county
  no development

- local distributor (county or town)
  slow moving
  sidewalks
  some frontage access

- service road (minor street)
  building access
  sidewalks
  no through traffic

avoid through traffic in residential
neighborhoods by the correct selection
of street pattern

no hierarchy of roads created
through traffic is possible (except
within the pattern)

patterns o.k. for collector systems but
not so good for residential neighborhoods

hierarchy created but allow for too much
access i.e. causes strip development

through traffic not possible therefore
good for service roads in residential
neighborhoods

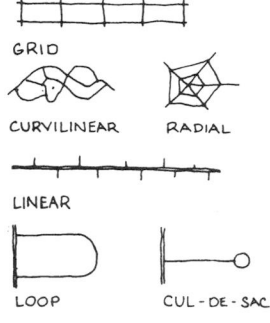

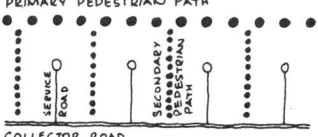

separate vehicular and pedestrian movements

separate the functional uses of roads and
paths (trucks, busses, bicycles, children,
idle strollers, etc.)

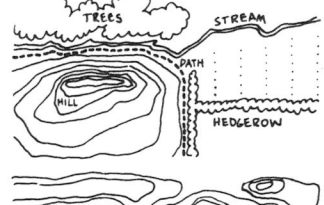

walkways and paths should follow
natural and/or man-made edges
and linear elements

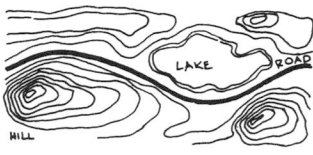

roads and paths should follow contour
lines and other natural features

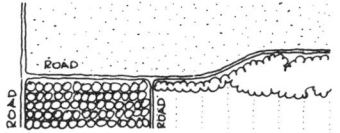

roads should follow established man-made
or natural patterns, forms, edges and lines

Fig. 7a

DEVELOPMENT PATTERN

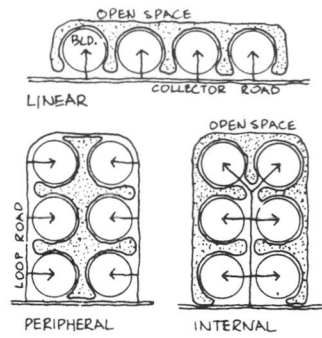

LINEAR

PERIPHERAL   INTERNAL

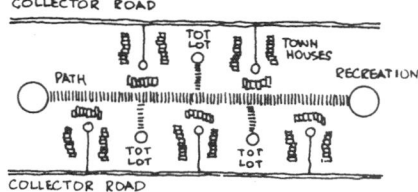

create an organized pattern of roads, open space, paths, land use and activities

distribute activities according to proximity, location and linkage

create a compatible environment

cluster like uses

create an overall mixture of uses and type of unit for variety

avoid a mixture of uses that are not compatible – e.g. heavy industry and residential

create an environment that is in sympathy and harmony with the natural and man-made resources of the area

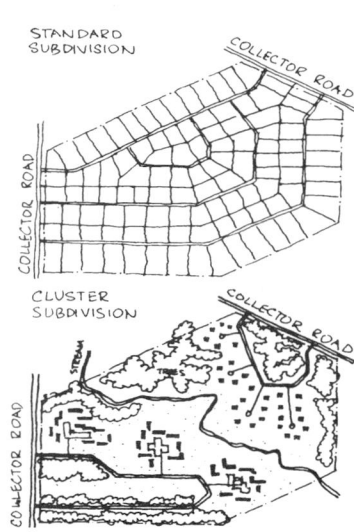

residential and other uses should be clustered to preserve natural features and to create a harmonious living and working environment

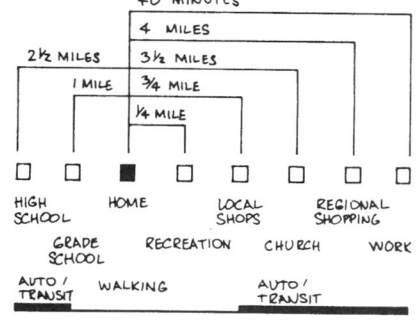

create residential developments with identifiable neighborhoods with a good relationship to all other activities

Fig. 7*b*

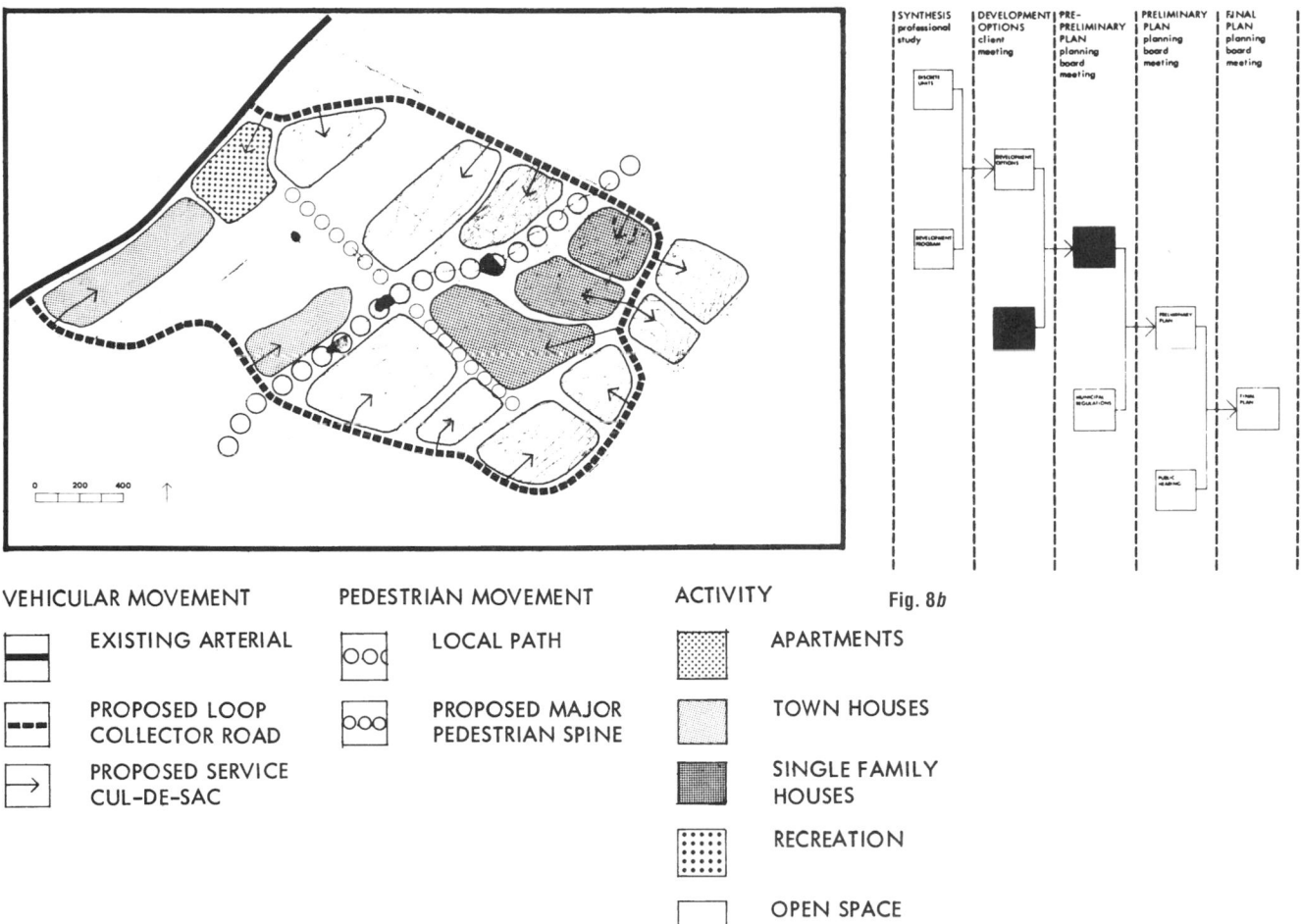

VEHICULAR MOVEMENT

▭ EXISTING ARTERIAL

▭ PROPOSED LOOP
COLLECTOR ROAD

→ PROPOSED SERVICE
CUL-DE-SAC

PEDESTRIAN MOVEMENT

ⓞⓞⓞ LOCAL PATH

ⓞⓞⓞ PROPOSED MAJOR
PEDESTRIAN SPINE

ACTIVITY

▦ APARTMENTS

▦ TOWN HOUSES

▦ SINGLE FAMILY
HOUSES

▦ RECREATION

☐ OPEN SPACE

Fig. 8b

Fig. 8a

Subdivision Regulations

The purpose of the regulations is to provide rules, standards, and procedures for the review of subdivisions by the planning board in order to promote the public health, safety, convenience, and general welfare of the community. It is administered to ensure orderly growth and development, the conservation, protection, and proper use of land and adequate provisions for circulation, utilities, and services. The regulations control:
- Procedure for submitting plans
- Design standards and specifications of streets, drainage, sewerage, water, street lights, fire hydrants, fire alarm boxes, trees
- Maximum block size
- Right-of-way width
- Pavement width
- Grades of streets
- Configuration of street intersections—angles and spacing
- Length of dead-end streets
- Provisions for fees and performance bonds

The zoning ordinance and subdivision regulations include standards for each of the above factors, and a proposed subdivision must conform in every respect. The planning unit development district is a special designation which exists in many communities. This permits mixtures of various housing and land use types. In order to achieve a more sympathetic and efficient layout, it also provides for adjustment of some of the usual standards to adapt to the conditions of the site. The preliminary plan in Fig. 9 shows the design of the site according to the town's policies.

PRELIMINARY PLAN

The preliminary plan for development of this site should be approved by the planning board. The board's decision should follow a public hearing, which may suggest modifications to the plan. The public hearing is held after the planning board has accepted the preliminary plan as complete, but before it has acted on the project. The hearing offers the only opportunity for the general public to voice its concerns or endorsement of the development proposal. Under the usual present arrangement, this is the public's only input into the planning process, and there is often, understandably, much confusion, resistance, and resentment. If a public input can be organized into the earlier stages of planning, much of this reaction could be avoided.

The hearing should satisfy the public that the planning board and other agencies have worked in the public interest. The developer and the planning board should demonstrate that the construction of utilities and provision of services have been well organized and coordinated with the rest of the community. The development of strategic areas should not cause economic hardship or degrade the quality of the environment in the community. The presentation of the proposal at the hearing should be organized to show the total impact on the environment. Table 3 includes a basic list of factors to be considered.

FINAL PLAN

The result of the public hearing is a final version of the preliminary plan that reflects all requirements of the planning board, the public, and the developer. If the planning process has been carried out properly and the community has made an adequate input, all of the necessary considerations in Table 3 will have been included. Otherwise, substantial revisions may be called for before the planning board can consider preliminary approval of the proposal. At this point in the process, the planning board should have sufficient information on the technical aspects of the proposal, its relationship to the environment, and its political implications.

Preliminary consideration by the planning board should not be a determination of whether the benefits of the project will outweigh the negative impacts; each should be judged independently. The negative impacts should be acceptable to the community and the benefits should be real. If the negative impact of "pollution of the environment" is unacceptable, the community will not be able to live with the development. If the benefits are not apparent, the potential of the strategic area has not been realized.

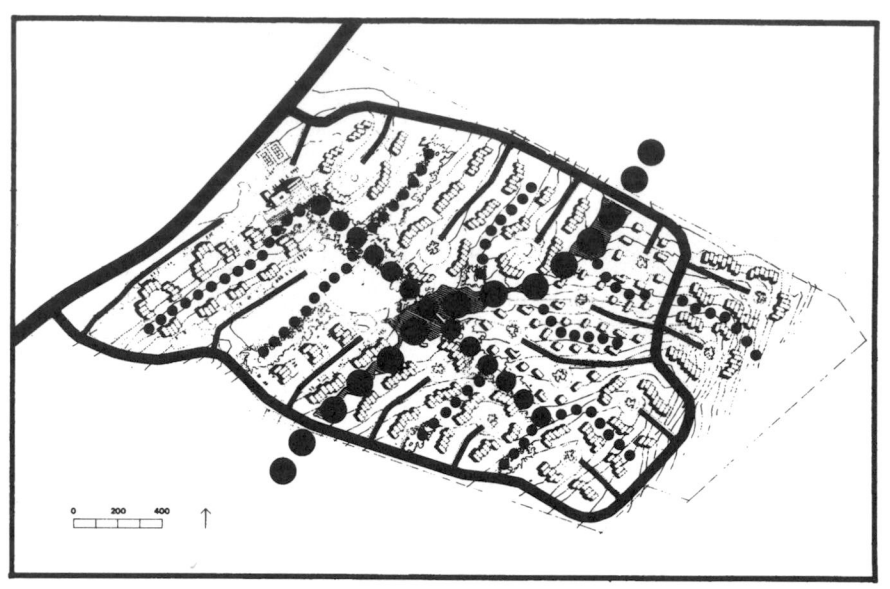

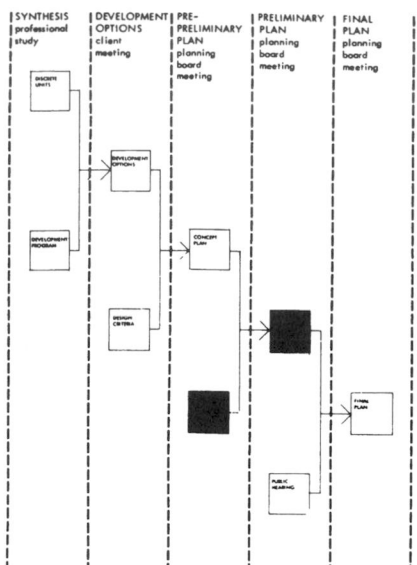

Fig. 9*b*

VEHICULAR MOVEMENT

	ARTERIAL
	COLLECTOR
	SERVICE

PEDESTRIAN MOVEMENT

●●	LOCAL PATH
●●●	MAJOR PATH
●●●	MINOR PATH

BUILDINGS

	APARTMENTS
	TOWN HOUSES
	SINGLE FAMILY
	RECREATION

	NUMBER OF UNITS	NUMBER OF ACRES
GARDEN APARTMENTS	152	17
TOWN HOUSES	289	39
SINGLE FAMILY HOUSES	37	11.4
OPEN SPACE AND RECREATION		22.6
TOTAL	478	90

SITE DENSITY   5.3 DWELLING UNITS PER ACRE

Fig. 9*a*

TABLE 3 Environmental Considerations

Natural resources	Manmade resources
Air:	Services:
Orientation to sun and wind	Demand for sewerage, water, schools, police, solid waste
Plantings to shade and screen	disposal, fire protection, recreation, health services
Air quality and sources of contamination: smoke, exhaust	Overburdening of utilities
fumes, dust, ambient air	Location of utility rights-of-way: underground lines, safety,
Water:	design, right-of-way management
Location in watershed	Movement:
Sources of water contamination: silt, sewage, fertilizer, road	Congestion
salt, refuse	Hazards to pedestrians
Water quality: dissolved oxygen, fish life, growths of algae	Parking and circulation facilities
and water weeds	Coordination of road pattern
Flood and drought cycle	Fumes; contaminated runoff
Coverage by buildings and paving: maintenance of water	Noise
table and aquifer recharge areas	Activity:
Multiple use of water resources	Incompatible land uses
Land:	Visual suitability: massing, materials, color
Organization of land use and buildings on terrain	Functional organization of land uses
Soil compaction	Adequacy of land areas: density and intensity of use
Soil erosion	Visual pollution: solid waste and debris
Disfiguring landscape by regrading	Housing maintenance
Dumping, litter, and sterilization of soil	

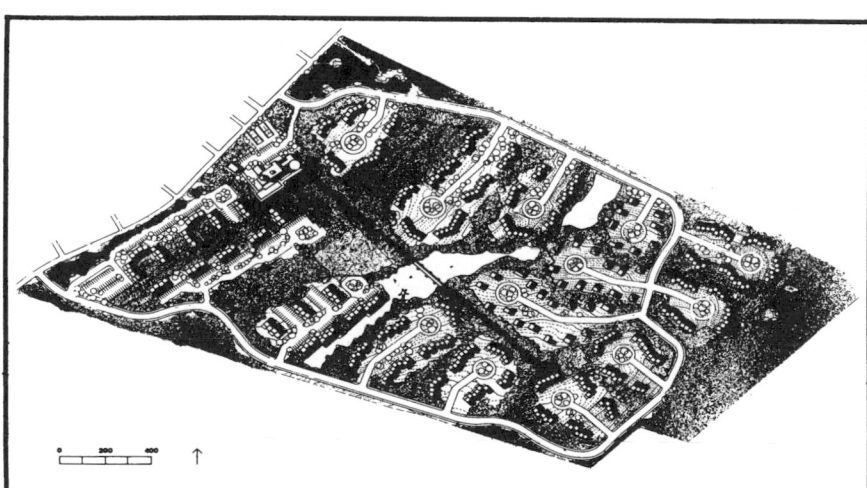

Fig. 10a

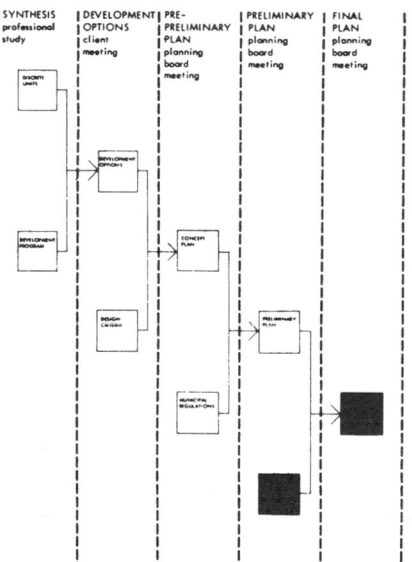

Fig. 10b

Fig. 11

TOWNHOUSE

Assumed density: **20 units/acre**

Lot area: 535 × 270 ft = 144,450 sf = 3.3 acres.
Number of units: 3.3 acres @ 20 = 66 units.
Parking: 66 @ 1.25 = 83 cars.

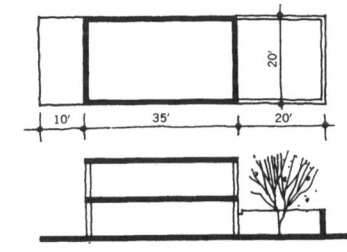

Ground utilization:

66 townhouses @ 700 sf =	46,200 sf
66 yards @ 600 sf	=	39,600 sf
83 cars @ 300 sf	=	24,900 sf
		110,700 sf

144,450 sf — 110,700 sf = 33,750 sf (25%).
Open area left for circulation and outdoor recreation (approximately half for recreation).

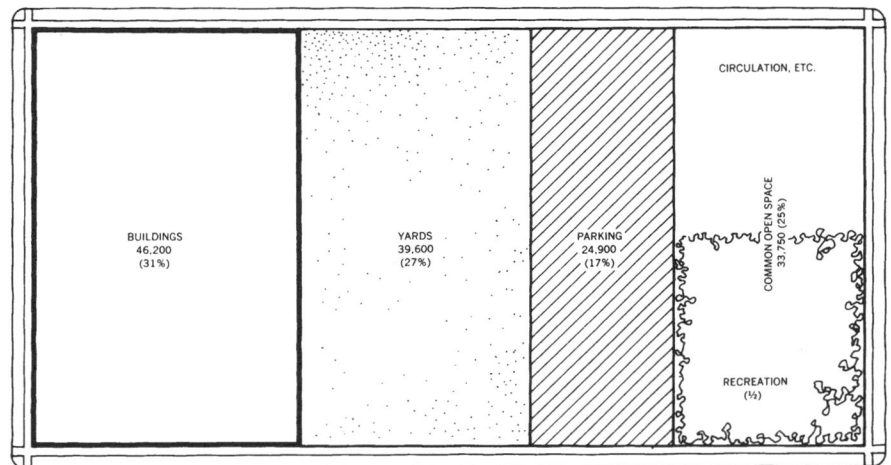

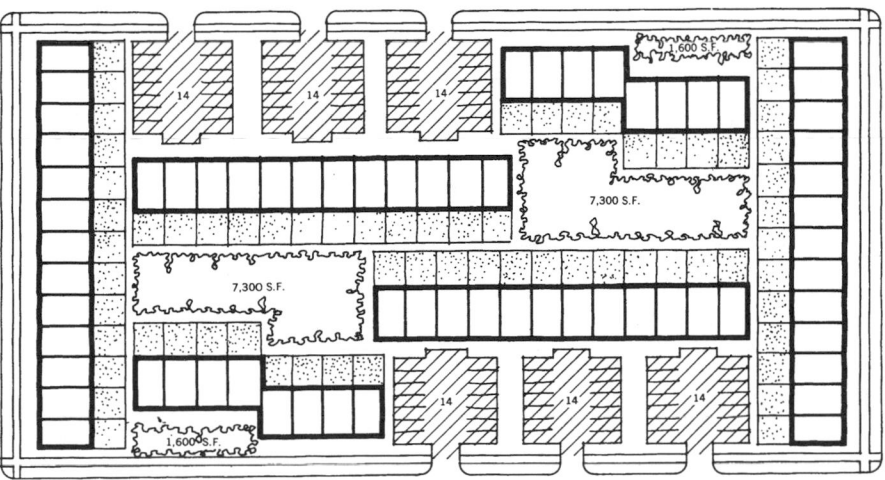

Fig. 1

TOWNHOUSE

To improve outdoor recreation space start with 35% of the land to be left open (or approximately 50,500 sf).

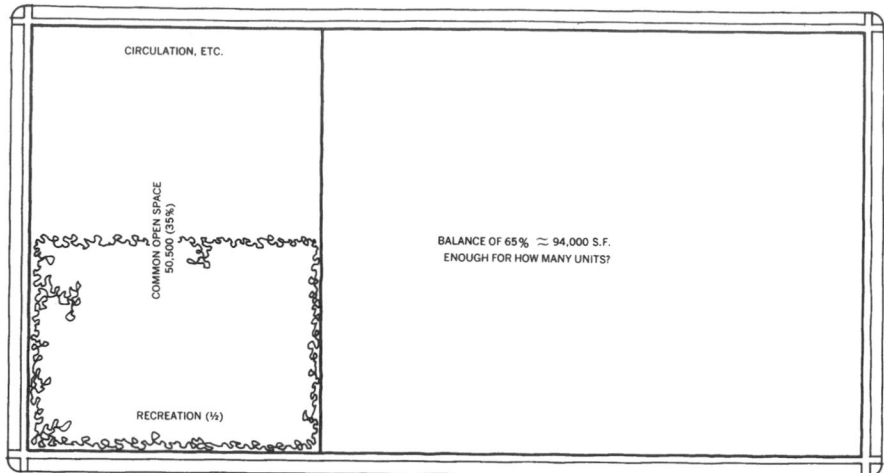

A single townhouse unit requires the following land:

Townhouse proper   =   700 sf
Yards   =   600 sf
Parking 1.25 @ 300 sf =   375 sf
                         1675 sf

94,000 sf ÷ 1675 sf = 56 units
56 ÷ 3.3 acres = **17 units/acre**

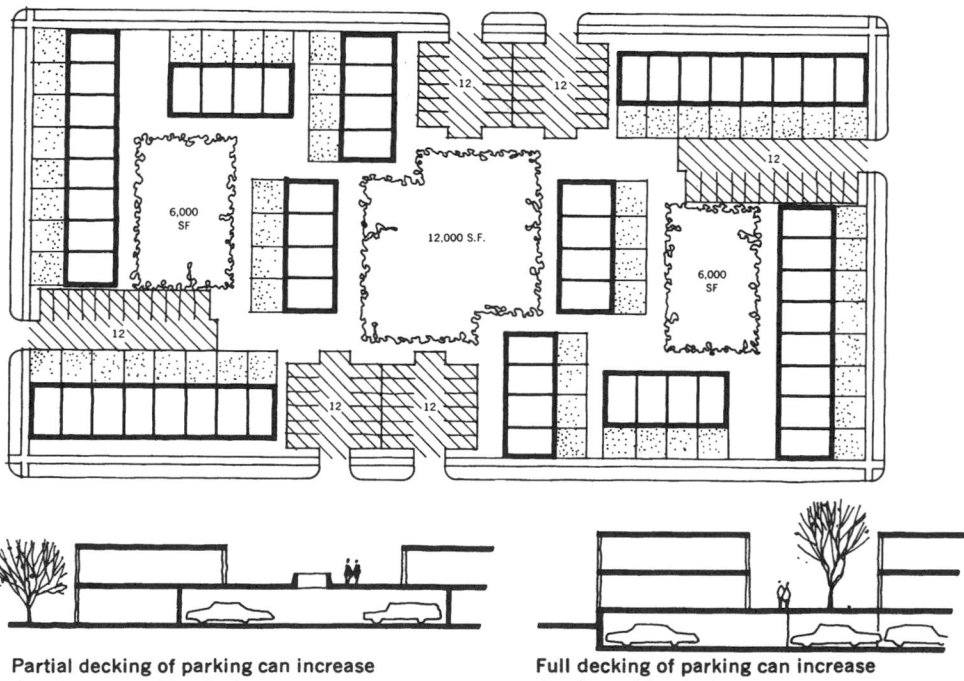

Partial decking of parking can increase density to over **20 units/acre.**

Full decking of parking can increase density to **25 units/acre.**

Fig. 2

THREE-STORY WALK-UP

Lot area: 535 sf × 270 sf	=	144,450 sf
Less 25% of lot for circulation and recreation	=	33,750 sf
Left for units and parking		110,700 sf

Each apartment needs for ground coverage
(114 × 48 ft) ÷ 12 (units per building) =   460 sf

Each apartment "carries" part of yards belonging to first floor apartments
(114 × 30 ft) ÷ 12 (units per building) =   280 sf
Parking 1.25 @ 300 sf   =   375 sf
                                              1,115 sf

110,700 sf ÷ 1115 sf = **99 units**
Because there are 12 units/building
  99 ÷ 12 = **8 buildings**
  8 buildings @ 12 = **96 units**
  96 ÷ 3.3 acres = **30 units/acre**

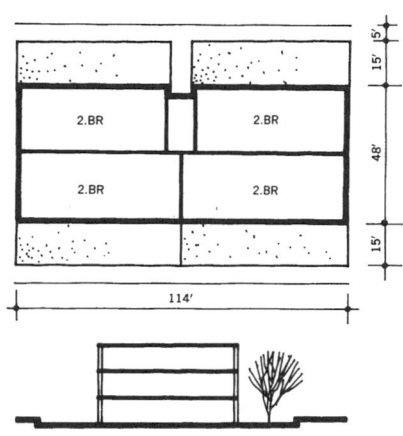

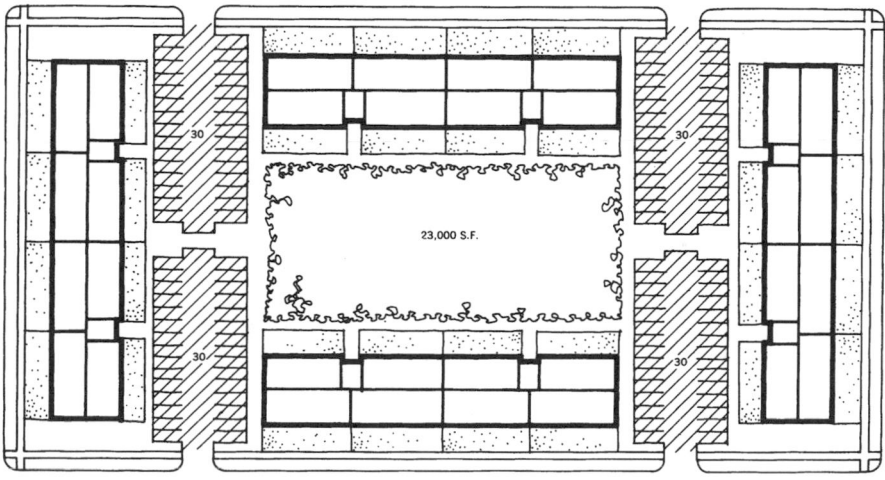

Fig. 3

DEVELOPMENT ANALYSIS

MIDRISE (Six Typical Floors)

To increase the outdoor recreation space lacking in the preceding study and to reduce ''sea of asphalt'' appearance, assume parking in two-level structure, the top of which is developed for recreation. Eliminating parking under the buildings, there are six typical apartment floors.

A. Unit:

6 apartment floors @ 40 units	=	240 units
Less 2 units per building for lobbies	=	—8 units
		232 units

232 ÷ 3.3 acres = **70 units/acre**

B. Parking:

2 parking levels in each garage @ 2 = 4 levels
4 @ 72 cars = ±290 cars = 232 apartments
@ 1.25

C. Recreation:

2 decks @ 21,000 sf	=	42,000 sf
On grade	=	11,000 sf
		53,000 sf

Required as per preceding study criteria (232 units @ 160 sf)	=	—37,120 sf
Overage		15,880 sf

If the 15,880 sf overage is used for 16 townhouses with private yards on top of the parking deck,

232 units + 16 units = 248 units
248 ÷ 3.3 acres = **75 units/acre**

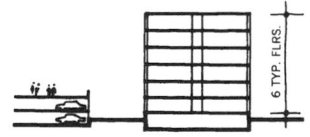

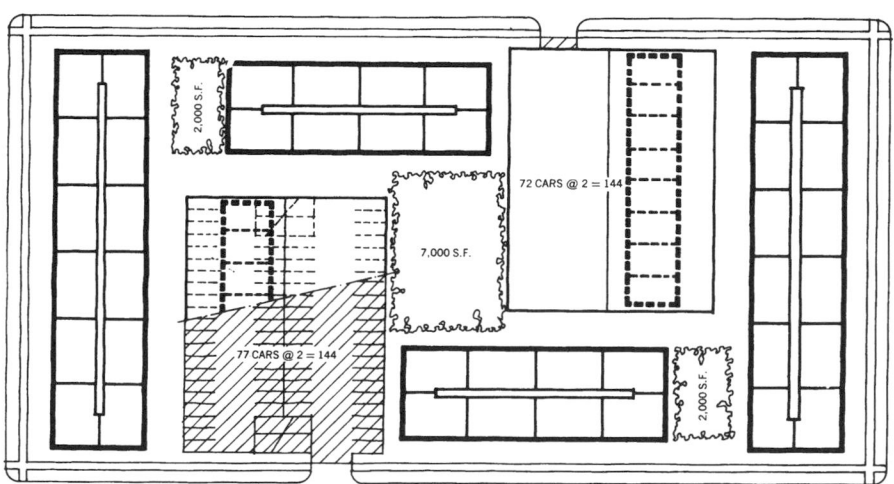

Fig. 4

HIGHRISE

Using parking as a determiner of ultimate density on the site

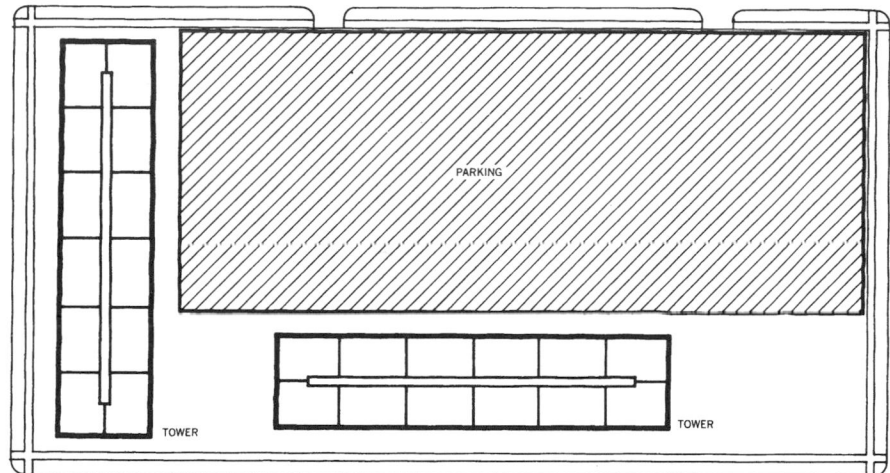

A. One-level parking in area shaded: 300 cars

Using 1:1.25 ratio, 300 cars serve 240 units
240 units ÷ 2 buildings = 120 units/building
120 units ÷ 12 units/floor = **10 floors**

240 ÷ 3.3 acres = **72 units/acre**

B. Two level parking in area shaded: 2 @ 260 = 520 cars

Using 1:1.25 parking ratio, 520 cars serve 416 units
416 units ÷ 2 buildings = 208 units/building
208 units ÷ 12 units/floor = 17.33, say **17 floors**
(17 @ 12 = 204 units/building
204 @ 2 = 408 units)

408 ÷ 3.3 acres = **123 units/acre**

C. Two-level parking over the entire site: 2 @ 400 = 800 cars

Using 1:1.25 parking ratio, 800 cars serve 640 units
640 units ÷ 2 buildings = 320 units/building
320 units ÷ 12 units/floor = 26.66, say **27 floors**
(27 @ 12 = 324 units/building
324 @ 2 = 648 units)

648 ÷ 3.3 acres = **200 units/acre**

. . . and so on upward depending on parking volume capacity.

Fig. 5

Community Facilities

MAXIMUM DISTANCES FOR COMMUNITY FACILITIES

The need for a total living environment surpassing the simple need for basic shelter has emerged as a significant development in recent years. Many housing developments and neighborhoods that appear stable and desirable are considered to owe a portion of credit to the ancillary community facilities for recreation and leisure, e.g., swimming pools, golf courses, and open green areas. In addition, the quality of community life is further enhanced by the proper amount and location of educational and sociocultural facilities.

This chapter is primarily concerned with facilities associated with neighborhood needs and design. However, a brief description is made of educational and recreational services that may serve a grouping of neighborhoods, e.g., junior and senior schools and large-scale parks. Each general grouping of facilities, educational, social and cultural, recreational and open-space, and neighborhood shopping, is presented as a self-contained unit. A summary chart of land-area requirements for the general run of facilities concludes the chapter.

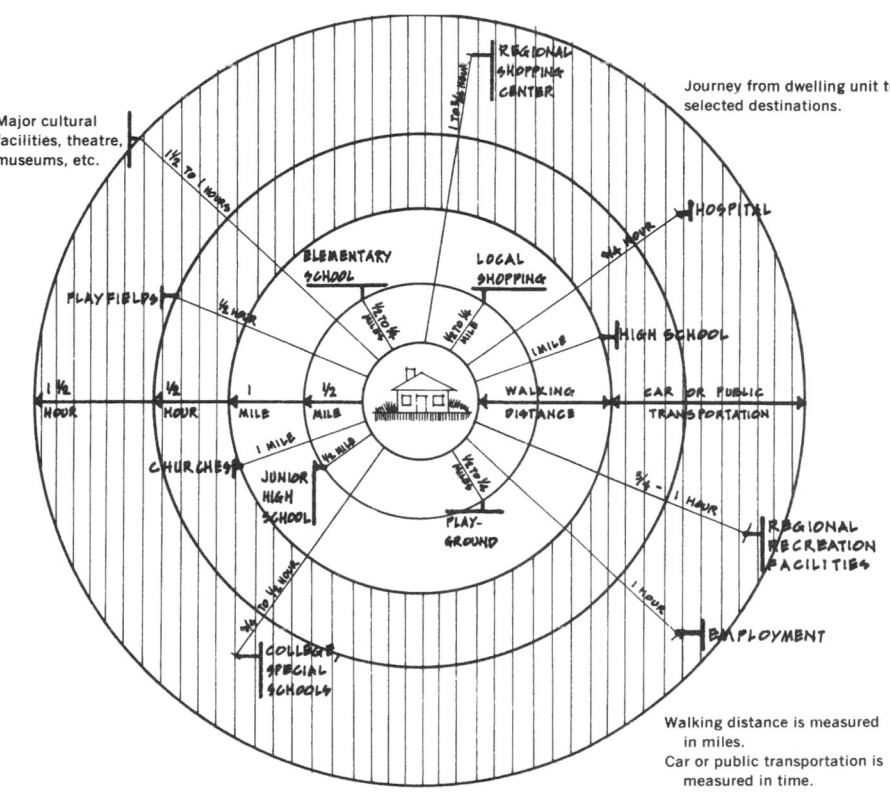

Fig. 1

SITE LOCATION

CRITICAL CRITERIA

Within ¼ mile (400 m) walking distance of elementary schools

Within ½ mile (800 m) walking distance of secondary schools

Within a ¼ mile (400 m) walking distance of public transportation

Within ¼ mile (400 m) walking distance of school-aged children's playground

Bounded by streets which are lightly trafficked

Not separated from child-oriented facilities by heavy traffic

Fig. 1

RECOMMENDED CRITERIA

Within ¼ mile walking distance from day-care facilities

Within ½ mile walking distance of low-cost child-oriented facilities

Buffered from natural hazards (railway lines)

Close to amenities, such as beaches, open spaces

Fig. 2

Site location is an important criterion of livability for families at high densities. For many families, their mobility and range of choice of activities may be limited by the presence of children or by limited incomes. Thus, it is important that the neighborhood provide for the family's basic needs and that the development not be isolated from its neighbors either physically or psychologically.

TABLE 1 Accessibility Standards

Infant Recreation		
Distance of preschool children's play area from dwelling	Adjacent	HIIS
Distance to play areas (2–6 years) from face of buildings	300–400 ft	APHA

Children's Recreation		
Maximum distance of children's (5–13 years) play area from dwelling	1300 ft	HIIS
Maximum walking distance to playground	¼–½ mile	APHA

Adult Recreation		
Maximum walking distance to athletic play field	1–1½ miles	APHA
Maximum walking distance to park	¼–½ mile	APHA

Nonresidential Built Space		
Maximum walking distance to community center	½ mile	APHA
Maximum distance to community facilities for old people (housing for aged)	½ mile	USA
Walking distance to schools	½ mile	APHA
Maximum walking distance to nursery school	¼ mile	APHA
Maximum walking distance to kindergarten	¼–½ mile	APHA
Maximum walking distance to elementary school	¼–½ mile	APHA
Walking distance to schools	½ mile	APHA
Maximum walking distance to elementary school	½ mile	CBH
Maximum walking distance to junior high school	¾–1 mile	APHA
Maximum travel time by public or special transit to junior high school	15–25 min	APHA
Maximum walking distance to grade school	1 mile	CBH
Maximum walking distance to senior high school	1–1½ mile	APHA
Maximum travel time by transit to senior high school	20–30 min	APHA
Maximum distance to high school	2½ miles	CBH
Maximum walking distance to shopping center	¼–½ mile	APHA
Maximum walking distance to local shopping center	¾ mile	CBH
Maximum distance to major shopping centers	4 miles	APHA
Maximum travel time by transport to district center (shopping, recreation, culture)	20–30 min	APHA
Maximum travel time to urban center	30–45 min	APHA
Maximum distance to health center	½ mile	APHA
Maximum travel time to health center by public transit	20 min	APHA

Roads (Residential)		
Maximum distance to setting down points from one- and two-person dwelling	100 ft	HIIS
Maximum distance to setting down points from three-person or more dwellings	150 ft	HIIS
Maximum carrying distance from car to house	300 ft	MOHLG
Maximum distance to vehicular way	200 ft	APHA
Desirable distance to dwelling from street or parking area	100 ft	PHA
Maximum distance to dwelling from street or parking area	200 ft	PHA
Maximum distance to dwelling from street or parking area	250 ft	FHA

Open Car Parking		
Minimum distance of parking areas from buildings	25 ft	PHA
Maximum distance of parking area from one- and two-person dwellings	150 ft	HIIS
Maximum distance of parking area from one- and two-person dwellings	100 ft	FHA
Maximum distance of parking area from three-person or more dwellings	250 ft	HIIS
Desirable distance to dwelling from street or parking area	100 ft	PHA
Maximum distance to dwelling from street or parking area	200 ft	PHA
Maximum distance to dwelling from street or parking area	250 ft	FHA
Minimum distance of parking areas from buildings	25 ft	PHA
Maximum distance of parking area from one- and two-person dwellings	150 ft	HIIS
Maximum distance of parking area from one- and two-person dwellings	100 ft	FHA
Maximum distance of parking area from three-person or more dwellings	250 ft	HIIS
Maximum distance of parking area from one- and two-person dwellings	150 ft	HIIS
Maximum distance of parking area from one- and two-person dwellings	100 ft	FHA
Maximum distance of parking area from three-person or more dwellings	250 ft	HIIS

NOTE:
HIIS: "Home in Its Setting," MOHLG (British).
APHA: American Public Health Association.
CBH: Community Builders Handbook.
MOHLG: Ministry of Housing and Local Government (British).

EDUCATIONAL FACILITIES

This category includes preschool and formal school services. In general, the neighborhood components will include a child-care center, nursery schools, and kindergartens in the preschool group, and elementary schools in the latter group.

These facilities must be within safe walking distance. Ideally, the children should have walking access without having to cross any vehicular streets. The maximum distance should not exceed ½ mile. Low-density areas require modification of these standards—usually met by the use of bus transportation.

DAY-CARE CENTERS

Day-care centers should be planned into developments housing 50 or more children between the ages of 3 and 5 years. These centers should contain a minimum of 5000 gross square feet. Under varying arrangements, full day care or any portion thereof (including extended custodial care) could be provided on a 12-month basis, with the teacher-pupil ratios not exceeding 8 to 1. The day-care centers shown could function adequately on ¾ acre of ground including ample fenced-in outdoor play spaces. In accordance with current educational practice for full day-care facilities, the learning spaces would be "open" and carpeted, and the program would follow a nongraded format. A kitchen would be included to enable every child to have a minimum of two snacks and one warm lunch each day.

NURSERY SCHOOL

The nursery school is for the care of children in the 4- to 5-year-old age group. The nursery school is planned for prekindergarten activities. The length of programs ranges from half-day to full-day and from several days a week to every day. Unlike the child-care center, the program is more limited in scope. The nursery school experience is also becoming increasingly common and considered essential by both parents and educators. Many believe that the nursery school should be part of the public school system.

The size of the center will be dependent upon the available number of children. The maximum size for a single group is considered to be 20 to 30 children. In no case should there be more than four groups per school.

The space needed will be approximately 600 to 1000 ft². State and federal regulations and recommendations of other recognized authorities for space standards must be complied with. The location should be on the first floor and relatively isolated from excessive noise, cars, and number of people. A nearby play area is essential and must be fenced off and separated from other outdoor uses.

The nursery school may be housed in a separate building or be part of a community center or other compatible type of building.

KINDERGARTEN

Kindergarten is for children 5 to 6 years old and is generally considered as preparatory to starting formal education. Over the years, kindergarten has become more and more essential as a required transitional period for the child. As such, kindergarten is more often part of the public school system and integrated with the elementary school. The size of the average kindergarten class should be about 20 to 25 children, assuming there will be a teacher and

an assistant. An outdoor play area is essential to the kindergarten.

ELEMENTARY SCHOOL

The elementary school is considered one of the basic organizational elements of the neighborhood. It is required by law that such a facility be provided to meet the basic need of education. As such, it has become the logical facility around which to develop the neighborhood. The type, size, and capacity of the elementary school facilities are determined by the school boards in accordance with state standards. Since there are no national standards on the organization of the elementary school, each community will differ in some aspects. One aspect is the number of grades in the school. Most school systems operate under one of the plans given in Table 1. The most common plan is the first one, the so-called 6–3–3 plan.

The kindergarten, when it is part of the public school system, is included within the elementary school complex. The minimum size for an elementary school is that it have at least one classroom per grade. For a K–6 school, this consists of a kindergarten and six classrooms, one for each grade, plus all the other necessary facilities.

The number of pupils per classroom varies in different school districts. The generally accepted maximum number per classroom is about 30

TABLE 1

Elementary school	Junior high school	Senior high school
Grades 1–6	Grades 7–9	Grades 10–12
Grades 1–6	Grades 7–8	Grades 9–12
Grades 1–8		Grades 9–12

pupils. Therefore, a minimum-sized school could contain about 180 pupils without a kindergarten, and about 210 pupils with a kindergarten. However, from the standpoint of administration, such a school may be considered too small, and the minimum size may be increased several times.

The land-area requirement for the elementary school will vary with land costs, available school board policies, and state requirements. For planning purposes, an area of 2 to 4 acres for larger schools would be appropriate, exclusive of outdoor recreation. Allow 5 to 6 acres more if the neighborhood playground is to be included.

SECONDARY EDUCATION

The junior and senior high school facilities may be part of the neighborhood pattern in high-density communities. In most suburban and moderate- to low-density areas, these facilities are normally associated with the broader community. In addition to serving the needs of the teenage student, these buildings are generally well suited for evening adult education and community recreational uses. Tables 2 to 6 contain general site and population served for the various school types.

MAXIMUM WALKING DISTANCES FOR STUDENTS

■ All distances given in Figs. 2 and 3 are considered to be maximum.
■ In high-density urban areas most schools are located within the maximum recommended walking distances.
■ In-low density rural areas many schools are located beyond maximum recommended walking distances. They must have bus service.

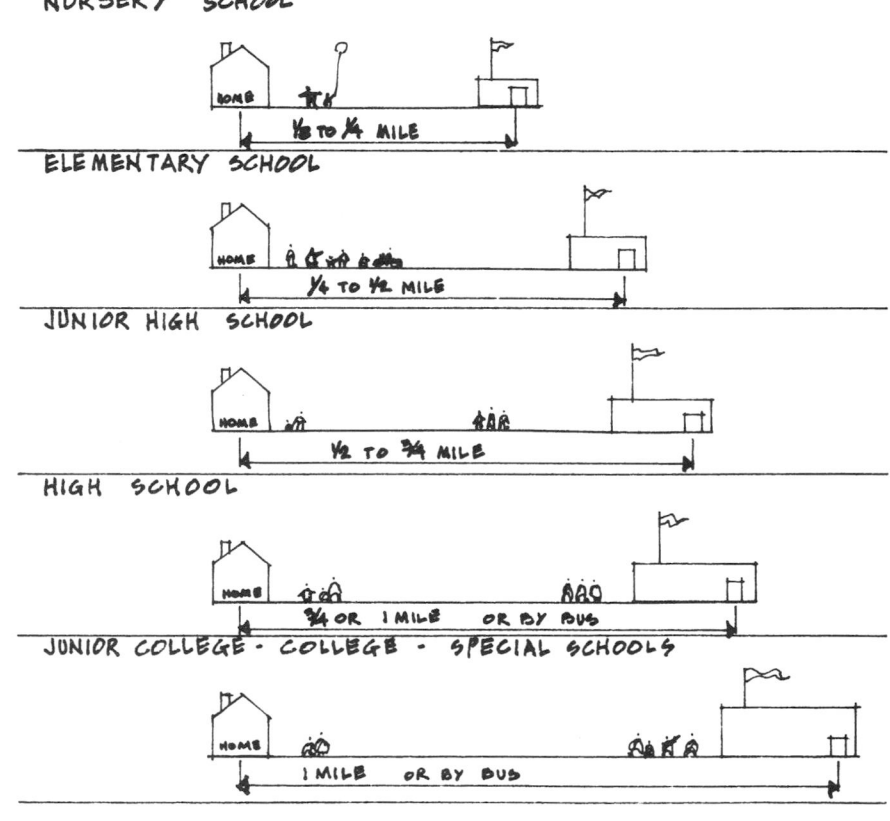

Fig. 1

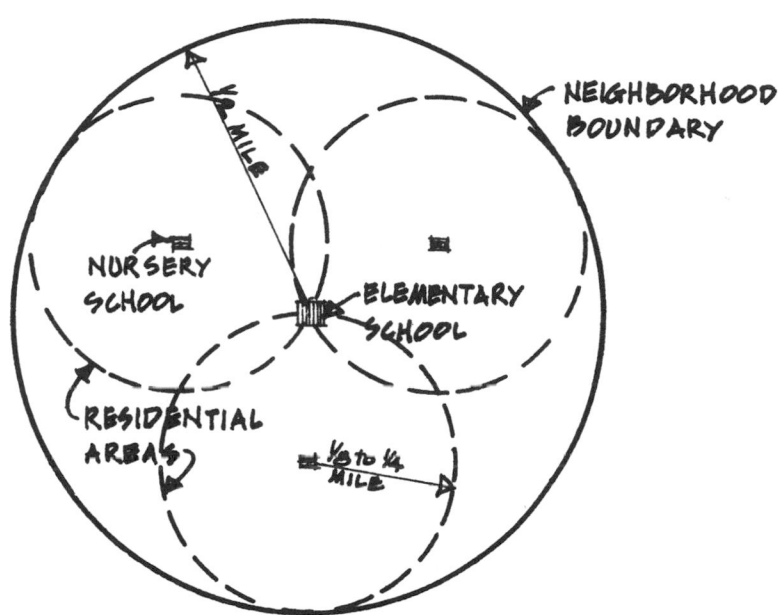

TYPICAL NEIGHBORHOOD ORGANIZATION

Fig. 2

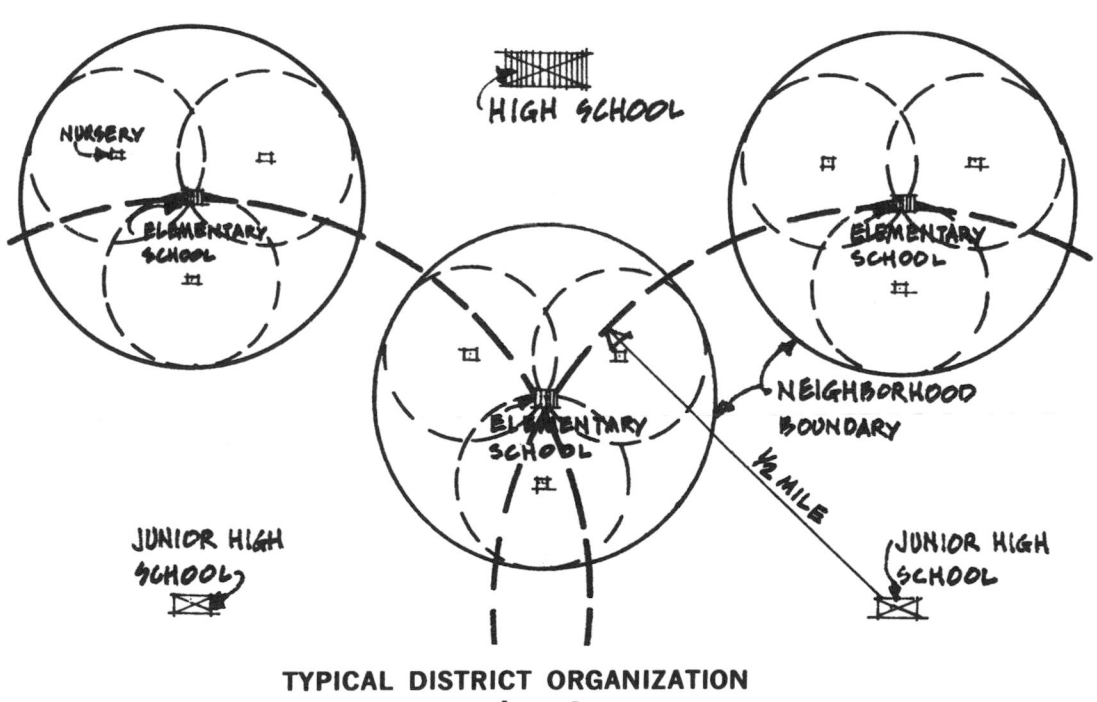

TYPICAL DISTRICT ORGANIZATION

Fig. 3

DAY-CARE CENTER
BUS STOP
COMMONS
BUS STOP

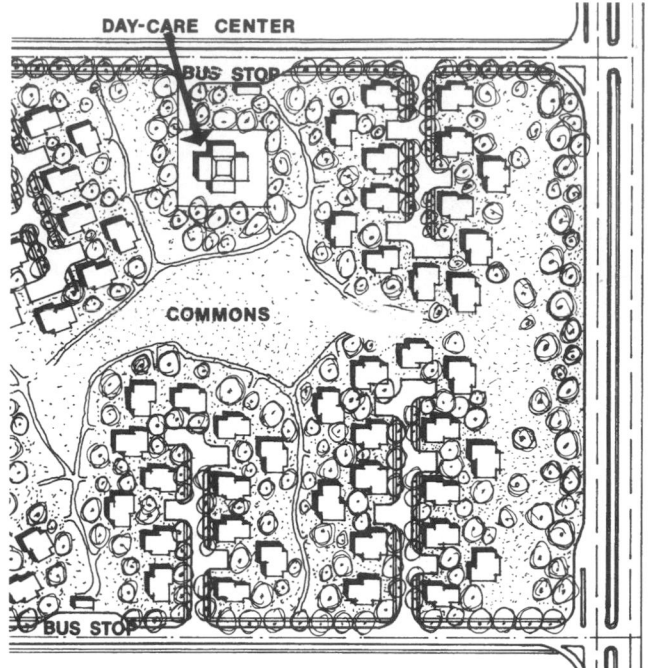

Fig. 4 Day-care center location plan.

ENVIRONMENTAL STUDY AREA
6' ENCLOSURE
SINK
CLOSET
PLAY-AREA
HEALTH UNIT
SHELF
A
A
PLAY-AREA
ADM.
ST.
SERVICE
TSR.
PLAY-AREA
KIT.
ST.
OUTDOOR PLAY-AREA

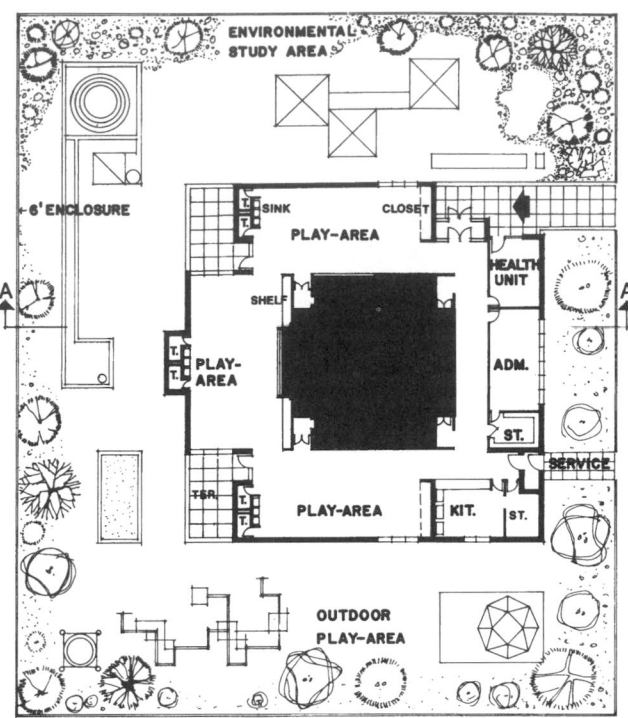

Fig. 5 Day-care center site plan.

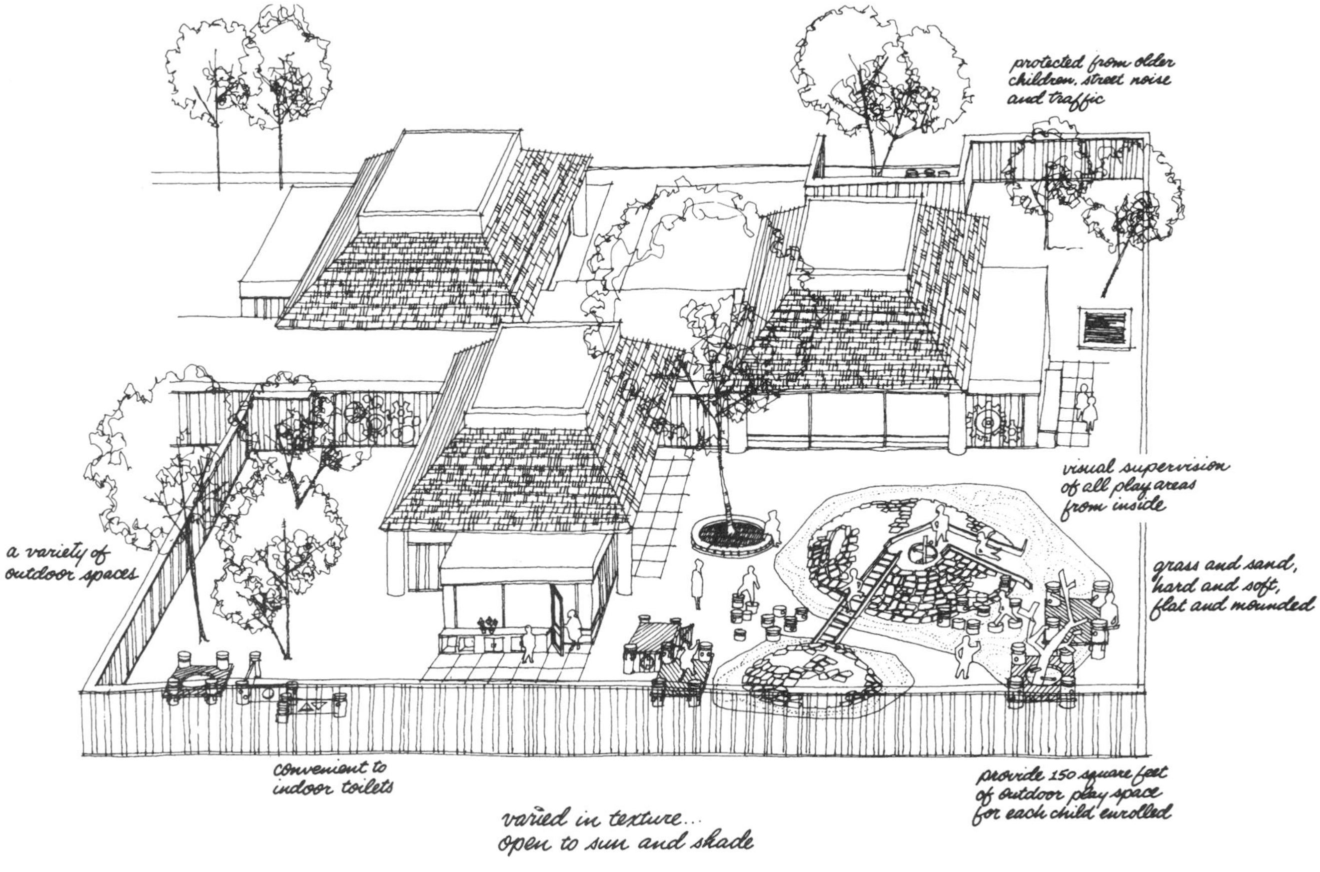

protected from older children, street noise and traffic
visual supervision of all play areas from inside
a variety of outdoor spaces
grass and sand, hard and soft, flat and mounded
convenient to indoor toilets
provide 150 square feet of outdoor play space for each child enrolled
varied in texture... open to sun and shade

Fig. 6 Child-care center.

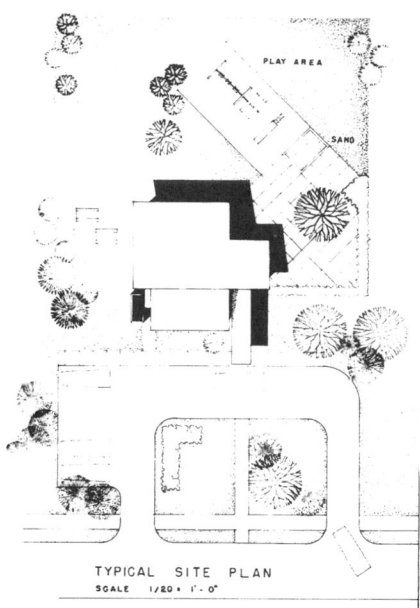

Fig. 7   Typical kindergarten site plan.

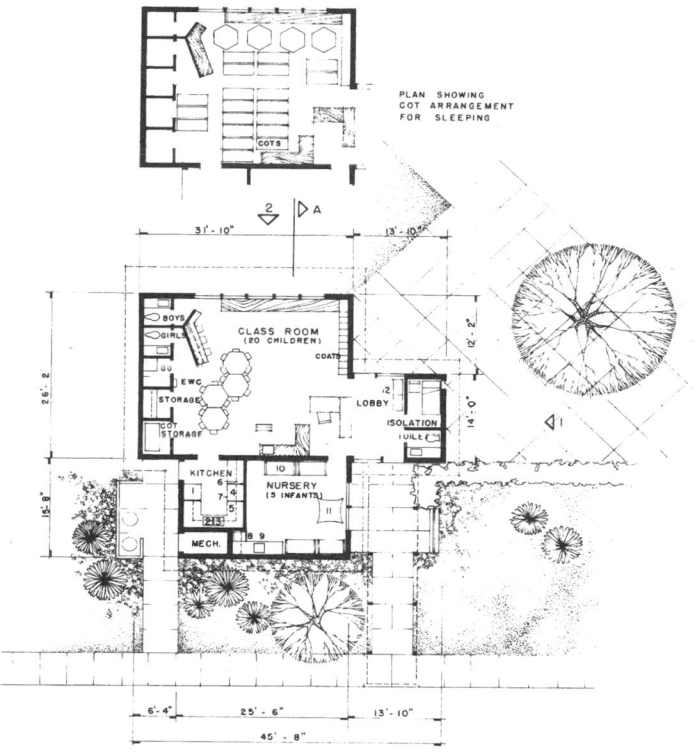

Fig. 8   Typical kindergarten floor plan.

TABLE 2   Nursery School—General Requirements

Assumed family size	3.25 persons per household	Area required	Two classes: 2000 ft²
			Four classes: 4000 ft²
			Six classes: 6000 ft²
Assumed population characteristics	50–60 children of nursery-school age per 1000 persons or 275–300 families	Accessory facilities	Playlot or children's play area with equipment. Play area should be completely fenced in from other activities.
Number of children of nursery-school age per family	0.15–0.25 children	Radius of area served	One to two blocks: desirable
			⅛ mile: maximum
Age of children served	2½ to 5 years old	Design features	Nursery school should be accessible by footpath from dwelling units without crossing any streets. If street must be crossed, it should be minor street.
Size of nursery school	Minimum: two classes (30 children)	General location	Near an elementary school, community center, or religious institution.
	Average: four classes (60 children)		
	Maximum: six classes (90 children)		
Population served	Four classes: 1000 persons, 275–300 families	Accessory parking	One space for each two classes.
	Six classes: 1500 persons, 425–450 families		
	Eight classes: 2000 persons, 550–600 families		

NOTE: These figures will vary for most areas. They are based on a full cross section of the population. Population figures should be checked for local age distribution and birth trends for any specific location.

EDUCATIONAL FACILITIES

TABLE 3 Elementary School—General Requirements

Assumed family size	3.25 persons per household	Area required	Minimum school: 7–8 acres
			Average school: 12–14 acres
			Maximum school: 16–18 acres
Assumed population characteristics	125–175 children of elementary-school age per 1000 persons or 275–300 families	Accessory facilities	Playground completely equipped for a wide range of activities. Playground area should be completely screened from street.
Number of children of elementary-school age per family	0.25–0.50 children	Radius of area served	¼ mile: desirable ½ mile: maximum
Age of children served	5–11 years	Design features	Elementary school should be accessible by footpath from dwelling units without crossing any streets. If street must be crossed, it should be a minor street.
Size of elementary school	Minimum: 250 pupils Average: 800 pupils Maximum: 1200 pupils	General location	Near center of residential area, near or adjacent to other community facilities, adjacent to playground.
Size of typical class	30–32 pupils	Accessory parking	One space per class plus three spaces.
Population served	Minimum school: 1500 persons Average school: 5000 persons Maximum school: 7000 persons		

NOTE: These figures will vary for most areas. They are based on a full cross section of the population. Population figures should be checked for local age distribution and birth trends for any specific location.

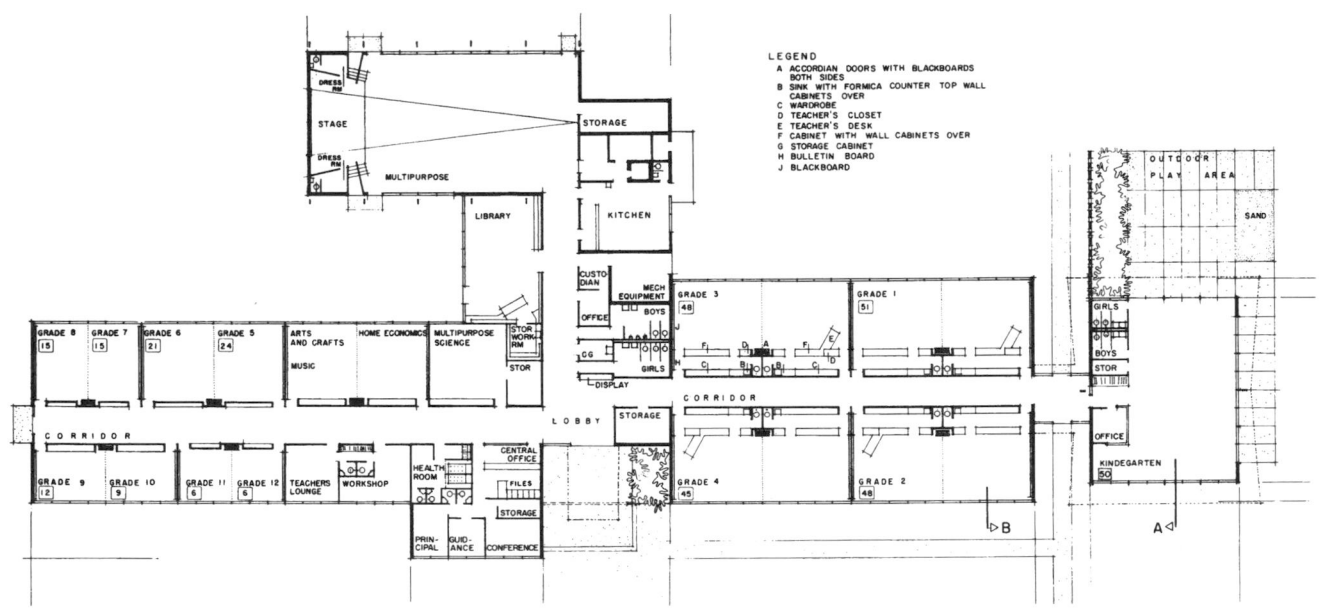

Fig. 9 Elementary school floor plan. Architects—Justement, Elam, Callmer and Kidd, W. C. Suite, Washington, D.C.

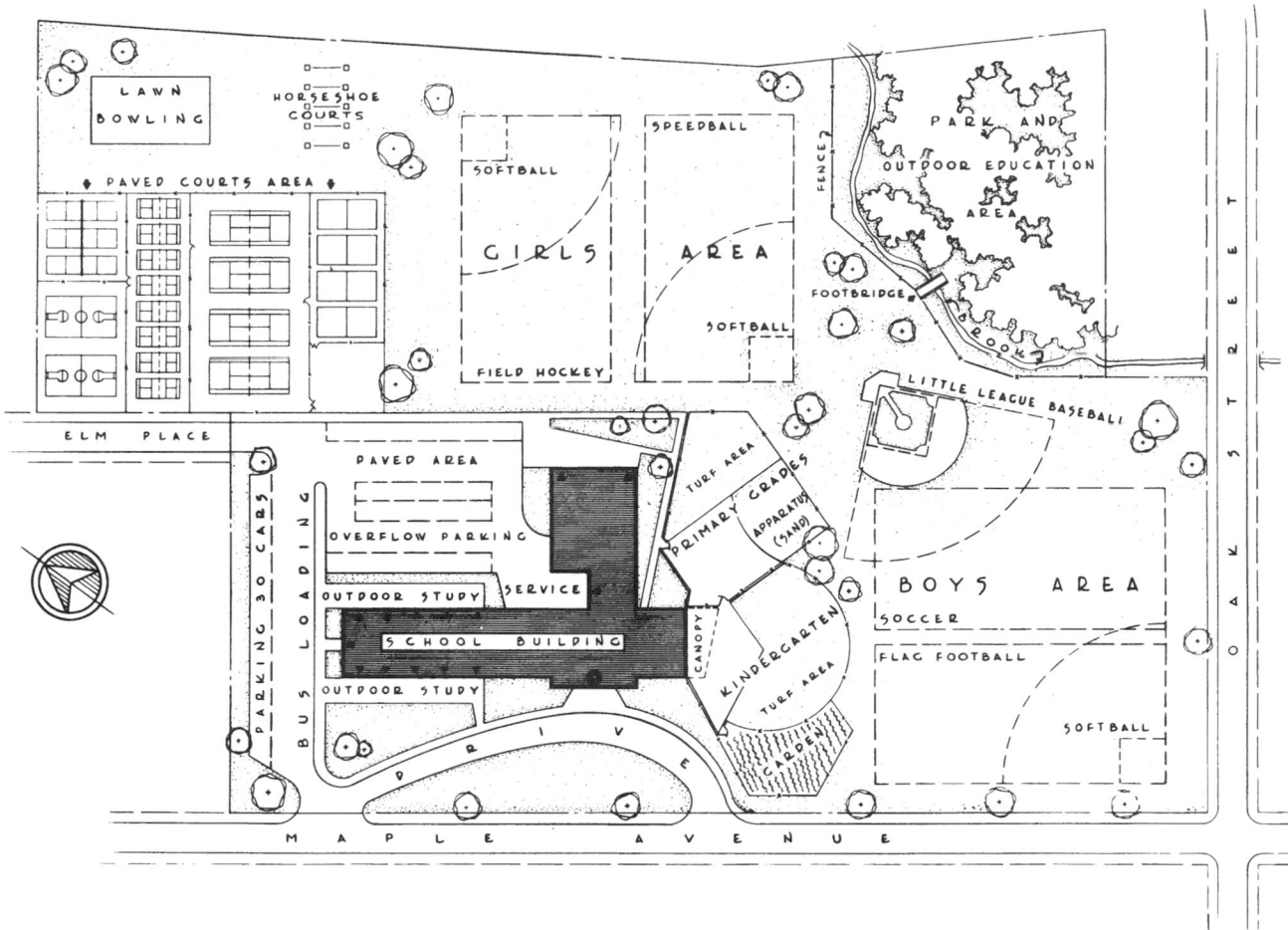

Fig. 10 Elementary school site showing distribution of outdoor play area.

TABLE 4 Junior High School—General Requirements

Assumed family size	3.25 persons per household	Area required	Minimum school: 18–20 acres
			Average school: 24–26 acres
			Maximum school: 30–32 acres
Assumed population characteristics	50–75 children of junior-high-school age per 1000 persons, 275–300 families	Accessory facilities	Playground and playfield completely equipped for a wide range of game activities.
Number of children of junior-high-school age per family	0.15–0.20 children	Radius of area served	½ mile: desirable
			¾ mile: maximum
Age of children served	12–14 years	Design features	School should be away from major arterial streets; pedestrian walkways from other areas should be provided; adjacent to park.
Size of junior high school	Minimum school: 800 pupils	General location	Located near concentration of dwelling units or near center of residential area.
	Average school: 1200 pupils		
	Maximum school: 1600 pupils		
Size of typical class	25–32 pupils	Accessory parking	One space per classroom plus six spaces.
Population served	Minimum school: 10,000 persons, 2750–3000 families		
	Average school: 16,000 persons, 4500–5000 families		
	Maximum school: 20,000 persons, 5800–6000 families		

NOTE: These figures will vary for most areas. They are based on a full cross section of the population. Population figures should be checked for local age distribution and birth trends for any specific location.

EDUCATIONAL FACILITIES

TABLE 5 Size of School Sites

School type	Minimum size, pupils	Ideal size, pupils	Maximum size, pupils	Site size, acres	Radius of area served, miles
Elementary	230	700	900	5 + 1 per 100 pupils	0.5
Junior high	750	1000	1500	15 + 1 per 100 pupils	1.0
Senior high	900	1500	2500	25 + 1 per 100 pupils	2.0
Elementary–junior high combination				15 + 1 per 100 pupils	1.0
Junior high–senior high combination				25 + 1 per 100 pupils	2.0
Elementary–park combination				8 + 1 per 100 pupils	0.5
Elementary–junior high–park combination				20 + 1 per 100 pupils	1.0
Junior high–senior high–park combination				18 + 1 per 100 pupils	1.0
Junior high–park combination				40 + 1 per 100 pupils	2.0
Senior high–park combination				35 + 1 per 100 pupils	2.0

Elementary School Site Size

School site area	Neighborhood population				
	1000 persons, 275 families, 90 pupils	2000 persons, 550 families, 180 pupils	3000 persons, 835 families, 270 pupils	4000 persons, 1100 families, 360 pupils	5000 persons, 1375 families, 450 pupils
Component uses					
Covered by building, ft^2		15,300	23,000	30,600	38,200
Service lawn and parking, ft^2		28,000	31,000	34,000	40,000
Margin for expansion (20% of 1 plus 2), ft^2		8,600	10,800	12,900	15,600
Total area					
Acres		1.2	1.5	1.8	2.2
Acres per 1000 persons		0.6	0.5	0.45	0.44
Square feet per family		94	78	70	68

SUMMARY OF EDUCATIONAL FACILITIES

TABLE 6 Neighborhood Facilities—Educational

Facility	Location and distance	Population served	Description	Function	Planning considerations
Child-care center	Central, 4–2 miles	500–1500	A facility integrally located within the housing complex; easily and speedily reached by walking; small-scale and residential in character.	To care for preschool-age children of 3 to 4 years of age.	Enables mothers to seek employment during the early years of motherhood. Provides a structured learning environment and social activity. Creates a minicore of community organization and identification.
Nursery	Central, 8–4 miles	2000–3000	Similar to the child-care center; greater emphasis on social adjustment and introduction to learning process.	To serve the social and educational needs of children from 2½ to 5 years of age.	Provides approx. 1000 ft^2 per class; a fenced-in play area with equipment; one parking space per each two classes. The nursery school should be accessible by footpath from dwelling unit without crossing street and may be near an elementary school or community center.
Kindergarten	Central, 4–2 miles	2000–3000	Formal instruction to educational process.	To serve the educational needs of children from 5 to 6 years of age. Preparation for elementary school.	Usually located within the elementary school complex; requires separate entrance and play areas.
Elementary school	Central, 4–2 miles	5000–8000	Usually the major community facility providing the center of activity for children and mothers.	To serve the educational needs of children from 6 to 11 years of age.	Provides 7–14 acres; a screened playground completely equipped for a wide range of activities; one parking space per class plus three spaces. The school should be accessible by footpath from dwelling units without crossing streets and near center of residential area, near or adjacent to other neighborhood facilities.

RELIGIOUS FACILITIES

Neighborhood churches frequently play an important role not only in the religious but also in the cultural and social activities of the community. They often serve as recreational and community centers as well.

The major problem in planning for churches and synagogues is that the religious makeup of the new community will not be known until the community is settled. If the population is of the same religious background, the problem is simplified. However, the probability is that there would be a wide variation in religious backgrounds, thereby making it almost impossible for neighborhood families to support a particular religious building. In that case, support would be required from several neighborhoods.

Membership Standards

C. A. Perry estimated that a population of 5000 persons could probably support three churches of about 1500 persons each. The Conference on Church Extension suggested one Protestant church for each 1500 to 2500 persons. It should be noted that on the national basis about 60 percent of the population may be expected to affiliate with a local church. In the western region this percentage is considerably lower. Thus, on the average, a population of 1500 might be expected to produce a church of 900 constituents, disregarding the churching of some people outside of the neighborhood, and internal heterogeneity of the population, which may cut the degree of affiliation to one specific church.

Many urban churchpeople believe that a church of about 500 members is the optimum size of a neighborhood institution, while a downtown church, to support a diversified staff and program, may need 1500 or 2000 or even more.

Area Requirements

There is no common uniform agreement as to the adequate size of a church site. Certainly the size of the site will depend upon the size of the church that is being projected, and the scope of the program. If an outdoor recreational program is desired, that will increase the size appreciably. If a parochial school is to be included, this will mean another appreciable increase in site needs. In most cases, landscaping and off-street parking should be provided for. Table 1 lists some standards that have been recommended.

Church Centers

The church and synagogue are still vital and important community facilities for many people. The role of these religious institutions has expanded into religious education and social activities. Parochial schools are common adjuncts to many churches and synagogues. In recent years they have expanded their activities to include all their constituents, especially the youth and the elderly. The result has been community centers built and run by the churches to meet the social and cultural needs. Such centers usually include meeting rooms, game rooms, shops, dining rooms, and some recreational facilities.

The location of the church center is usually adjacent to the religious institution, if sufficient land is available. Church parking areas can be used when there are no religious services.

PUBLIC LIBRARY

Branch Library

A branch library can play an important role as a cultural center. In addition to providing books, it can provide record and tape lending, music-listening facilities, visual-aid facilities, lecture series, and act as a general information center. With such an expanded role, the library or cultural center will be an important element in the neighborhood. The general standards for small public libraries are shown in Table 2.

Regardless of the size of the community, its library should provide access to enough books to cover the interests of the whole population.

TABLE 1

Source	Acres recommended for church site
Conference on Church Extension Standards based on:*	
0–400 membership	1 acre
400–800 membership	2 acres
800–1200 membership	3 acres
1200 or more members	4 acres
Presbyterian Board of Missions†	3 acres (on the average)
Urban Land Institute‡	3–5 acres (preferably near a shopping center)
Van Osdal§	5–6 acres (for a 600-seat church with 150 parking spaces)
	8 acres (for a Catholic church with a parochial school)

*Ross W. Sanderson, condensed report, conference on church extension, 1953, N.Y. National Council of Churches.
†Everett L. Perry, "Selections of a Church Site," *The City Church,* September 1953.
‡Urban Land Institute, *Community Builders Handbook,* 1954, p. 89.
§N. K. Van Osdal, Jr., "The Church and the Planned Community," *The City Church,* May 1952.

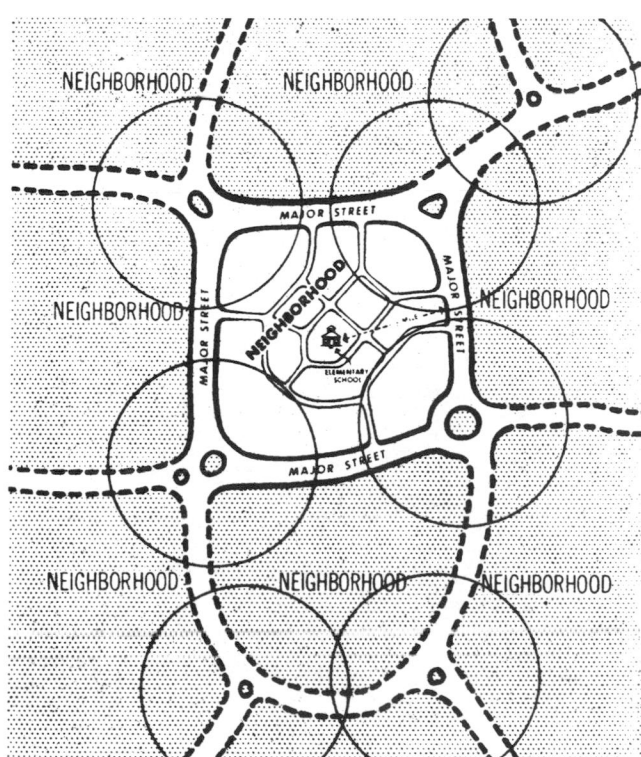

Fig. 1   Church serving one neighborhood only.

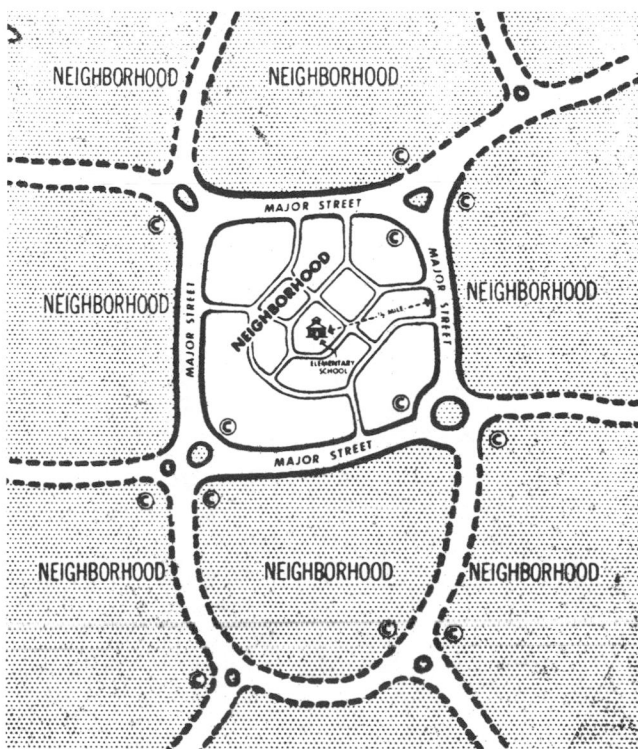

Fig. 2   Location of churches serving more than one neighborhood.

SOCIAL AND CULTURAL FACILITIES

1. Libraries serving populations from 5000 to 50,000 require a minimum of two books per capita.

2. Communities up to 5000 persons need access to a minimum of 10,000 volumes, or three books per capita, whichever is greater.

The library building should provide space for the full range of library services. All libraries should have designated areas for children's, young adult, and adult materials.

Multipurpose rooms should be provided for meeting, viewing, and listening by cultural, educational, and civic groups unless such facilities are readily available elsewhere in the community. They should be located for easy supervision so that they may be used for quiet reading and study when not needed by groups.

No single type of building is satisfactory for all public libraries. Each building is likely to be different, and its differences should be directly related to its service program.

The library building should be located in or near the community shopping center and at street level if possible. Adequate parking should be available nearby.

MULTISERVICE CENTER

One of the new social institutions to emerge in the community is the multiservice center. This facility is being encouraged by the federal government in Model Cities areas. It is to be similar to a neighborhood community center housing a multitude of federal, state, and local services. Also, it is to be the headquarters of local community action groups. At this neighborhood facility, assistance can be provided in solving problems such as employment service, job training, welfare, day care, voting, social security, and housing. In theory, each multiservice center should reflect the social, economic, and educational needs of each community. Its organization and operation must be flexible and multifunctional to adjust to constantly changing community needs. Each center, in essence, would be different from other centers.

Combined with these service facilities may be the more traditional social and recreational facilities, thereby making this center a truly community endeavor. The location, obviously, should be central and easily accessible to the entire community. Since many of the services would be for the elderly, disabled, or poor, consideration must be made for these groups. Some suggested locations would be adjacent to shopping areas, churches, or street level of multistory apartment buildings. Close proximity to bus stops and train stations would be essential.

Health Centers

Adequate medical services are essential on the local level. This should include medical, dental, and psychiatric services. This can be in the form of private and/or public facilities. A group practice providing a wide range of medical services would be one appropriate method. Medical and dental offices can easily be incorporated with the neighborhood shopping center. Provisions for accessibility to a hospital outside the neighborhood in emergencies are essential. Small preventive care and medical diagnostic facilities would also be appropriate.

COMMUNITY CENTERS

One way to provide activity space for both young and old is with a community center. It could provide meeting and recreational spaces, complete with a serving kitchen for catering. The lower (presumably more noisy) level could contain a teenage center with separate access and include game rooms, dance halls, etc. It should not be necessary for the teenagers to feel they must abandon the community to find activities especially suited to them; nor should their activities be in conflict with their parents' entertainment. The floor plan in Fig. 5 depicts an arrangement that is suitable for expansion to accommodate various neighborhood sizes.

Youth Center

The youth center may consist of a variety of building types. It may be a boys club, a YMCA or YWCA, or a settlement house. Sometimes, it

Fig. 3 M. Paul Friedberg & Associates, Landscape Architects.

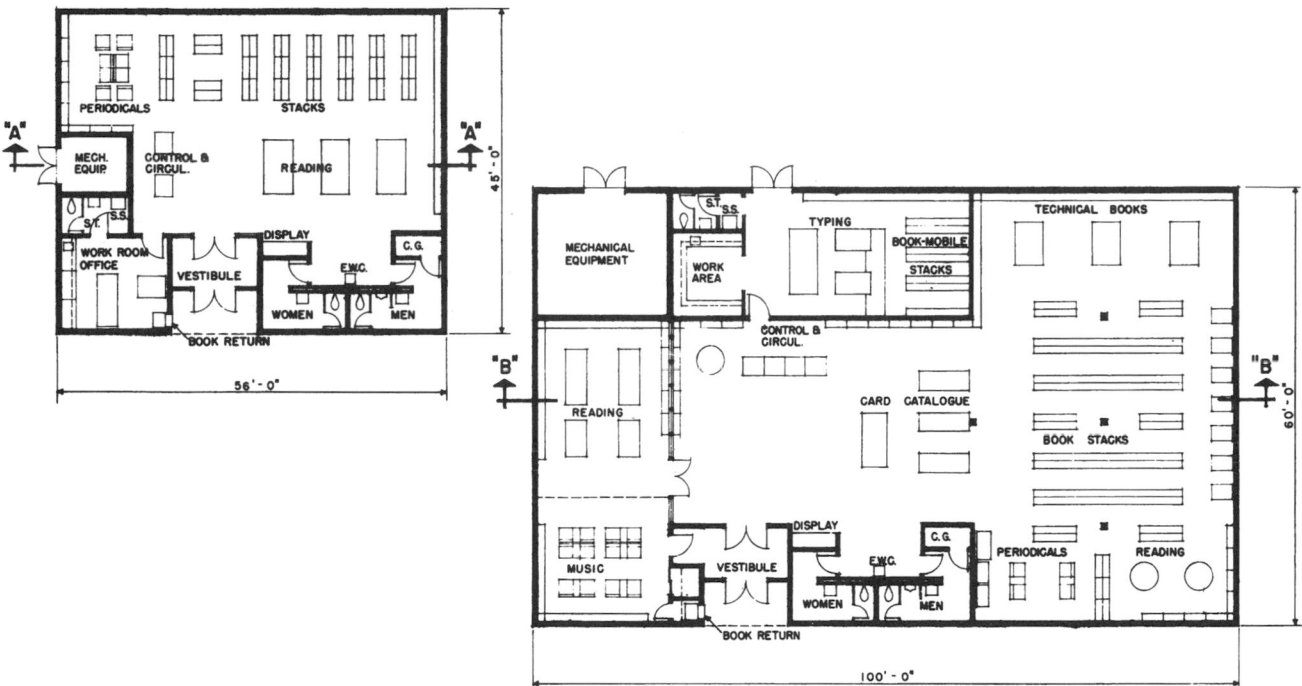

Fig. 4   Library buildings.

TABLE 2   Interim Standards for Small Public Libraries—Guidelines for Determining Minimum Space Requirements

Population served	Size of book collection	Shelving space,* lin ft of shelving†	Amount of floor space	Reader space	Staff work space	Estimated additional space needed‡	Total floor space
Under 2499	10,000 volumes	1300 lin ft	1000 ft²	Minimum 400 ft² for 13 seats, at 30 ft² per reader space	300 ft²	300 ft²	2000 ft²
2500–4999	10,000 volumes plus three books per capita for population over 3500	1300 lin ft; add 1 ft of shelving for every eight books over 10,000	1000 ft²; add 1 ft² for every 10 books over 10,000	Minimum 500 ft² for 16 seats; add five seats per thousand over 3500 population served, at 30 ft² per reader space	300 ft²	700 ft²	2500 ft² or 0.7 ft² per capita, whichever is greater
5000–9999	15,000 volumes plus two books per capita for population over 5000	875 lin ft; add 1 ft of shelving for every eight books over 15,000	1500 ft²; add 1 ft² for every 10 books over 15,000	Minimum 700 ft² for 23 seats; add four seats per thousand over 5000 population served, at 30 ft² per reader space	500 ft²; add 150 ft² for each full-time staff member over 3	1000 ft²	3500 ft² or 0.7 ft² per capita, whichever is greater
10,000–24,999	20,000 volumes plus two books per capita for populations over 10,000	2500 lin ft; add 1 ft of shelving for every eight books over 20,000	2000 ft²; add 1 ft² for every 10 books over 20,000	Min. 1200 ft² for 40 seats; add four seats per thousand over 10,000 population served, at 30 ft² per reader space	1000 ft²; add 150 ft² for each full-time staff member over 7	1800 ft²	7000 ft² or 0.7 ft² per capita, whichever is greater
25,000–49,000	50,000 volumes plus two books per capita for population over 25,000	6300 lin ft; add 1 ft of shelving for every eight books over 50,000	5000 ft²; add 1 ft² for every 10 books over 50,000	Minimum 2250 ft² for 75 seats; add three seats per thousand over 25,000 population served, at 30 ft² per reader space	1500 ft²; add 150 ft² for each full-time staff member over 13	5250 ft²	15,000 ft² or 0.6 ft² per capita, whichever is greater

*Libraries in systems need only to provide shelving for basic collection plus number of books on loan from resource center at any one time.
†A standard library shelf equals 3 lin ft.
‡Space for circulation desk, heating and cooling equipment, multipurpose room, stairways, janitors' supplies, toilets, etc., as required by community needs and the program of library services.

SOCIAL AND CULTURAL FACILITIES

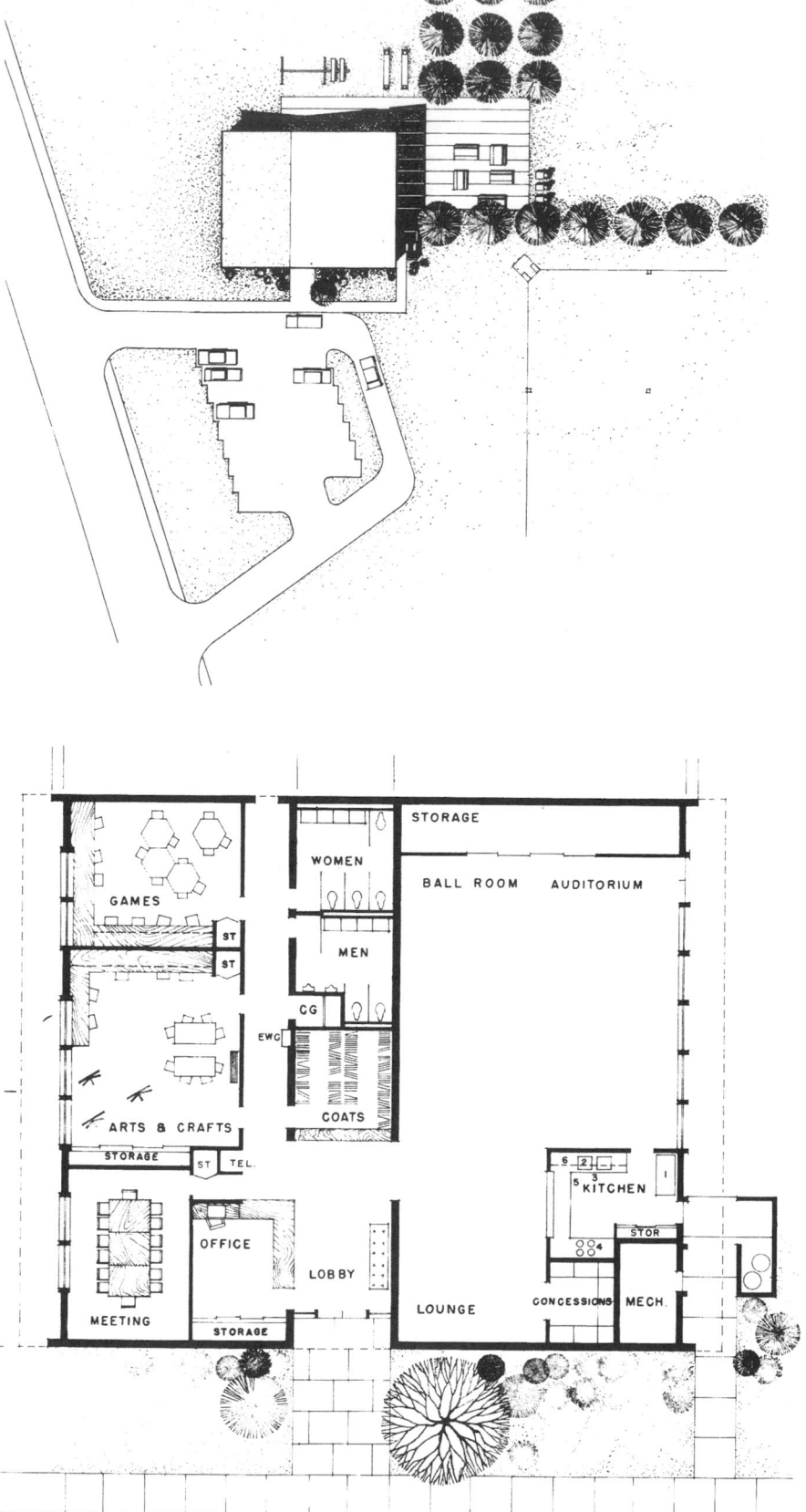

Fig. 5

may be a church or municipally sponsored facility. Essentially it caters to the young boys and girls with a range of social and recreational facilities. This would include meeting rooms, gym, swimming pools, shops, game rooms, and lounges.

Youth centers are similar to other community centers or recreation buildings except that they direct their activities to a restricted age group, usually under 21 years of age.

The location of such youth centers is very important. They should be easily accessible and convenient, preferably along major lines of circulation. The site should be large enough to provide for outdoor activities and future expansion. Adequate parking for staff and visitors is necessary. Hopefully many young people will either walk or use bicycles to get to the center.

RECREATION BUILDING

The recreation building is a community facility of a specialized type. It emerged as part of and usually is incorporated in the local park and recreation system. The building may contain a range of passive and active facilities from meeting rooms to gyms and indoor swimming pools. They are usually designed to function as community centers with strong citizen participation. Most recreation buildings are designed for both young and old on a year-round basis.

The programming of these facilities should be closely coordinated with existing or proposed facilities in other community buildings such as schools or private clubs. The location of the recreation building should be central to the population served and near to outdoor recreation areas. The integration of indoor-outdoor activities will provide a maximum year-round activities program. Adequate parking is essential for staff and visitors. A location near mass transit lines will remove somewhat the need for parking needs.

The building shown contains a full-sized gymnasium for regulation basketball, shower and locker rooms, storage areas, and health and administrative suites. Since the excessive height of a gymnasium might be out of scale with surrounding dwellings, the structure is shown partially recessed into the ground. The reinforced roof structure of the lower unit is flush with the adjoining grade and makes an excellent hard-surfaced play area with a backdrop wall for games such as handball. Access to the lower level is through a sunken courtyard.

SOCIAL AND CULTURAL FACILITIES

CENTERS FOR THE ELDERLY

These centers serve the senior citizens by providing them with a place to meet other people. The planned activities are primarily social. They usually include special-interest clubs, cultural groups, adult education programs, involvement in the arts, and centers for action groups. Space should be provided for passive recreation activities such as game rooms and shops.

The senior citizen center can be the location of an increasing number of social, health, welfare, and employment services for the elderly in the community. By placing these services at the center, they become readily available to a group that has become far less mobile than the general population. The activities of the center are highly flexible and are limited only by the people involved.

The physical space required should be flexible to meet the changing needs of the elderly. Most appropriate would be moderate-sized meeting rooms, shops, classrooms, and offices. No large specialized spaces such as gyms or swimming pools should be anticipated. The location must be considered for easy accessibility and a minimum of steps. Preferably it should be located in a separate one-story structure or the first floor of a high-rise apartment building. Parking requirements would be minimal.

SUMMARY OF FACILITIES

This category of institutional uses includes religious, health and public service centers, libraries, and community centers. Table 3 is a summary of the social and cultural facilities relative to people served, their function, and their planning considerations.

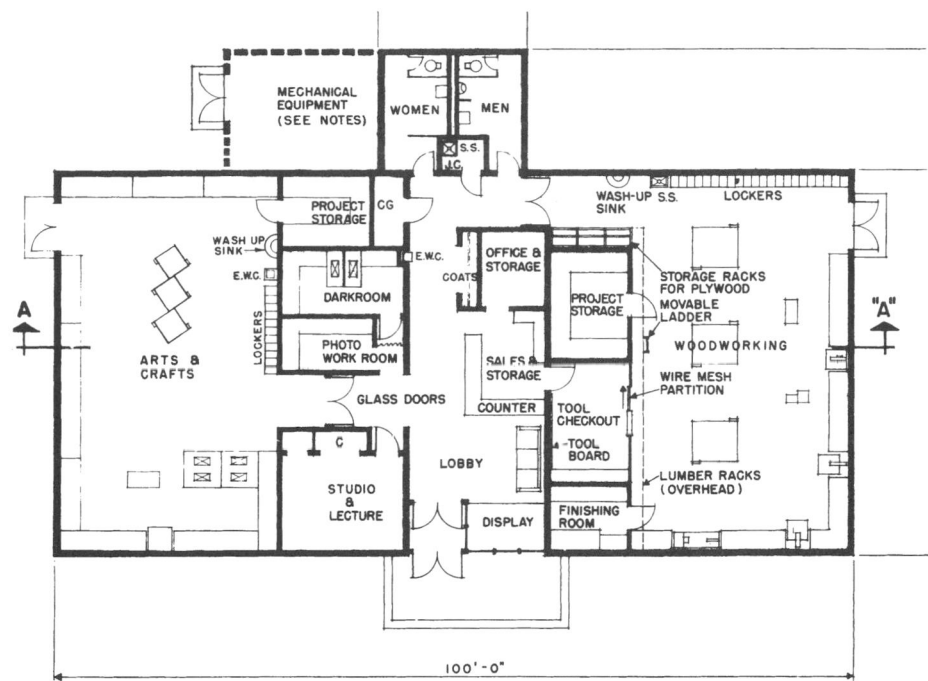

Fig. 6   Arts and crafts shop—youth center/center for the elderly.

Fig. 7   Lindsay-Bushwick housing, M. Paul Friedberg & Associates, Landscape Architects.

SOCIAL AND CULTURAL FACILITIES

TABLE 3 Social and Cultural Facilities

Facility	Location and distance	Number served	Description	Function	Planning considerations
Church or synagogue	Central to congregation, 1 mile max.	500–2500	A place of worship and often a social center for the congregation. If a school is included, it becomes the center for religious education.	A facility in which religious services and gatherings as well as social meetings and other neighborhood activities are held.	Should be easily accessible, on foot if possible, and located in a quiet area with sufficient space for landscaping and off-street parking.
Branch library	Central, 1 mile max.	Up to 5000	A small branch of a larger system. Should contain both children's and adult's reading rooms, and range of books, records, and magazines.	A facility to serve the limited reading and research needs of a neighborhood.	Should be easily accessible, preferably on a main thoroughfare, in a subshopping area, or near a neighborhood center. Ease of parking is desirable. A minimum of 80 years expansion should be possible.
Recreation center	¼–½ mile	Up to 5000	A multipurpose building. Usually includes a gym, swimming pool, and game rooms. Meeting rooms and other multipurpose areas.	To serve the recreational needs of a neighborhood including small-scale outdoor activities and a full range of indoor activities.	This facility may be part of a school or located in a park. It may generate noise and should thus be somewhat separated from other quieter facilities. Ease of parking is desirable.
Social center	¼–½ mile	Up to 5000	A multipurpose building with emphasis upon passive activities such as game rooms, reading and meeting rooms.	A facility to be used for social activities such as dances, banquets, meetings, special-interest shows, and exhibits.	This facility may be part of a school, should have some food-preparation facilities and should provide adequate off-street parking.
Health center	Central, ¼–½ mile	Up to 5000	Small medical offices that can provide basic diagnostic and treatment services. Usually a public facility but may be private group practice.	A facility to provide minor health services, to serve as a health information center and possibly act as a referral bureau.	This facility may be part of a school, should be easily accessible on foot with the provision for off-street parking.
Multiservice center	Central, ¼–½ mile	Up to 5000	Group of offices or one large flexible space that can be modified as needs change.	A facility to serve primarily s an information and acommunity-guidance center providing legal and other professional advice.	Provides a sense of contact with governmental organizations; brings the government closer to the people; provides an outlet for local participation.

TABLE 1   Types of Open Space in a Recreation System

Orientation	Function	Space, design, and service area	Example
Home-oriented space	Should meet aesthetic qualities and accommodate informal activities of an active and passive nature, i.e., sitting, reading, gardening, sunning, children's play, and family activity	Varies according to housing type; immediately adjacent or within 500 ft of each dwelling unit	Front- and backyards, driveways, sidewalks, porch, balconies, workshops, play rooms, recreation rooms
Home cluster or subneighborhood common space	Especially important in high-density areas, providing visual relief and aesthetic qualities for activities similar to those mentioned above, as well as meeting areas for small informal groups, walking, jogging, and dog walking	Must be visually accessible; varies from 500 ft^2 to 2 acres; designed to be as flexible and adaptable as possible; will serve an area of 100 yd to ¼ mile radius	Vacant lots, cul-de-sacs, boulevards, greenbelts, walkways, trails, playlots, rest areas, vest-pocket parks, parkettes
Neighborhood space	Should accommodate neighborhood interest preferences; may include sports areas for minor leagues, outdoor skating rinks, water play, as well as special events and informal passive activities	Space should be associated with an elementary school; varies from 4 to 20 acres; will serve 5000 people within an area of ¼ to ½ mile radius	Neighborhood parks or park-school combinations; playfields for baseball, soccer, and football; adventure playgrounds, wading pools, neighborhood centers
Community space	Should accommodate social, cultural, educational, and physical activities of particular interest to the community; multipurpose, year-round, day/night activities; low-level competitive sports with limited spectator space	Space should be associated with a secondary school; varies from 15 to 20 acres; will serve several neighborhoods or 15,000 to 25,000 people within a radius of ½ to 1½ miles; accessible by walking, cycling, and public transit	Community park or park-school combinations; facilities for playgrounds, recreation center, meeting rooms and library; track and field areas, sports fields, arena, and swimming pool

TABLE 2   General Standard for Open Space

Area	Acres per 1000 population	Service radius	Minimum size
Subneighborhood areas	Included in neighborhood and community parks	100 yd minimum	500 ft^2
Neighborhood park and elementary school combination*	4	¼–½ mile	10 acres
Community park and secondary school combination*	3	1–1½ mile	30 acres

*It is assumed that the park and the school are adjacent and completely accessible to each other. If they are not, then the acreage for the park and for the school should each be increased by 25 percent. These figures include the space occupied by the buildings on each site and the parking areas.

TABLE 3   Average Open Space Standards

Area	Acres per 1000 population	Service radius	Size
Tot lot (playlot, park lot)	0.25–0.5	⅛–¼ mile, usually ¼ mile	0.6–2.0 acres, usually 0.5 acre
Parkette (vest-pocket park)	0.5	⅛–¼ mile	0.6 to 1.0 acre, usually 0.5 acre
Neighborhood park (playground, local park, community park)	1.0–2.0	¼–½mile, usually ½ mile	¼–20 acres, usually 6 acres
Community park (playfield)	1.0–2.0	½–3 miles, usually 1 mile	4–100 acres, usually 8–25 acres

Every residential neighborhood requires a range of facilities for both children and adults that are easily accessible to the living units. Such facilities should include at least the following:

1. A preschool play area, usually called a playlot or a tot lot, for very young children who require close supervision. Such tot lots must be within easy reach of and visual observation from the living unit. With judicious arrangement of buildings, these spaces can usually be located between buildings or at the end of clusters, to form natural informal groups.

2. A centralized play space for grade-school children. This playground should be well-equipped and of adequate size to accommodate all the children that will use the facility.

3. Outdoor sitting areas should be provided for adults for passive activities. Such areas must be attractive and also provide a variety of visual experiences. They may include shaded areas, roof gardens, paved areas, and areas to view children at play.

4. More activity recreation areas should also be provided. These facilities can be used by a variety of age groups and require greater supervision and control. Such facilities may include tennis courts, boccie courts, bicycle riding paths, or archery ranges. The most popular facility in this category is probably the swimming pool. This can include a wading pool for young children, cabanas, and even a recreation building.

In addition, golf courses and open lands for passive recreation are desirable elements for the total community. Community recreation, where feasible, should stress year-round activities rather than limited seasonal ones.

The following pages present the salient aspects of each of the above-mentioned facilities. This section includes a summary description of the various facilities according to function, equipment, and the area served.

PLAYLOTS

The playlot should be provided in the central open area within each block or adjoining each cluster of dwellings. In developments serving families with children, playlots should be provided for preschool children up to 6 years of age.

They are a necessary element of housing developments to complement common open-space areas. Playlots may include (1) an enclosed area for play equipment and such special facilities as a sand area and a spray pool; (2) an open, turfed area for active play; and (3) a shaded area for quiet activities.

Location of Playlots

Playlots should be included as an integral part of the housing area design, and are desirably located within 300 to 400 ft of each living unit served. A playlot should be accessible without crossing any street, and the walkways thereto should have an easy gradient for pushing strollers and carriages. Playlots may be included in playgrounds close to housing areas, to serve the preschool age group in the adjoining neighborhood.

Size of Playlots

The enclosed area for play equipment and special facilities should be based on a minimum of 70 ft^2 per child, which is equivalent to 21 ft^2 per family. A minimum enclosed area of approximately 2000 ft^2 will serve some 30 preschool children (about 100 families). Such a size will accommodate only a limited selection of play equipment. To accommodate a full range of equipment and special facilities, including a spray pool, the minimum enclosed area should be about 1000 ft^2, which would serve up to 50 preschool children (about 165 families). Additional space is required to accommodate the elements of the playlot outside of the enclosed area. A turfed area at least 10 ft square should be provided for activity games.

Playlot Activity Spaces and Elements

A playlot should comprise the following basic activity spaces and elements:

1. An enclosed area with play equipment and special facilities including:
   a. Play equipment such as climbers, slides, swing sets, play walls and playhouses, and play sculpture
   b. A sand area
   c. A spray pool
2. An open, turfed area for running and active play
3. A shaded area for quiet activities
4. Miscellaneous elements—including benches for supervising parents; walks and other paved areas wide enough for strollers, carriages, tricycles, wagons, etc.; play space dividers (fences, walks, trees, shrubs), a step-up drinking fountain, trash containers, and landscape planting

Layout of Playlots

The specific layout and shape of each playlot will be governed by the existing site conditions and the facilities to be provided. General principles of layout can be described as follows:

1. The intensively used part of the playlot with play equipment and special facilities should be surrounded by a low enclosure with supplemental planting, and provided with one entrance-exit. This design will discourage intrusion by animals or older children, provide adequate and safe control over the children, and prevent the area from becoming a thoroughfare. Adequate drainage should be provided.

2. Equipment should be selected and arranged with adequate surrounding space in small, natural play groups. Traffic flow should be planned to encourage movement throughout the playlot in a safe, orderly manner. This traffic flow may be facilitated with walks, plantings, low walls, and benches.

3. Equipment that enables large numbers of children to play without taking turns (climbers, play sculpture) should be located near the entrance, yet positioned so that it will not cause congestion. With such an arrangement, children will tend to move more slowly to equipment that limits participation and requires turns (swings, slides), thereby modifying the load factor and reducing conflicts.

4. Sand areas, play walls, playhouses, and play sculpture should be located away from such pieces of equipment as swings and slides for safety and to promote a creative atmosphere for the child's world of make-believe. Artificial or natural shade is desirable over the sedentary play pieces, where children will play on hot days without immediate supervision. Play sculpture may be placed in the sand area to enhance its value by providing a greater variety of play opportunities. A portion of the area should be maintained free of equipment for general sand play that is not in conflict with traffic flow.

5. Swings or other moving equipment should be located near the outside of the equipment area, and should be sufficiently separated by walls or fences to discourage children from walking into them while they are moving. Swings should be oriented toward the best view and away from the sun. Sliding equipment should preferably face north, away from the summer sun. Equipment and metal surfaces should be located in available shade.

6. Spray pools should be centrally located, and step-up drinking fountains strategically placed for convenience and economy in relation to water supply and waste-disposal lines.

7. The open, turfed area for running and active play, and the shaded area for such quiet activities as reading and storytelling should be closely related to the enclosed equipment area and serve as buffer space around it.

8. Nonmovable benches should be conveniently located to assure good visibility and protection of the children at play. Durable trash containers should be provided and conveniently located to maintain a neat, orderly appearance.

Playlot Equipment for Preschool Children

Table 1 indicates types, quantities, and minimum play space requirements for various types of equipment totaling about 2800 ft^2; this area, plus additional space for circulation and play space dividers, will accommodate a full range of playlot equipment serving a neighborhood containing approximately 50 preschool children (about 165 families).

Smaller playlots may be developed to serve a neighborhood containing some 30 children (about 100 families) using a limited selection of equipment with play space requirements totaling about 1200 ft^2; this area, plus additional space for circulation and play space dividers, should consider the following desirable priorities: (1) a sand area, (2) a climbing device such as a climber, a play wall, or a piece of play sculpture, (3) a slide, and (4) a swing set. Where several playlots are provided, the equipment selections should be complementary, rather than all being the same type. For example, one playlot may include play walls or a playhouse, while another playlot may provide a piece of play sculpture. Also, such a costly but popular item as a spray pool may be justified in only one out of every two or three playlots provided.

TABLE 1

Equipment	Number of pieces	Play space requirements
Climber	1	10 × 25 ft
Junior swing set (four swings)	1	16 × 32 ft
Play sculpture	1	10 × 10 ft
Play wall or playhouse	1	15 × 15 ft
Sand area	1	15 × 15 ft
Slide	1	10 × 25 ft
Spray pool (including deck)	1	36 × 36 ft

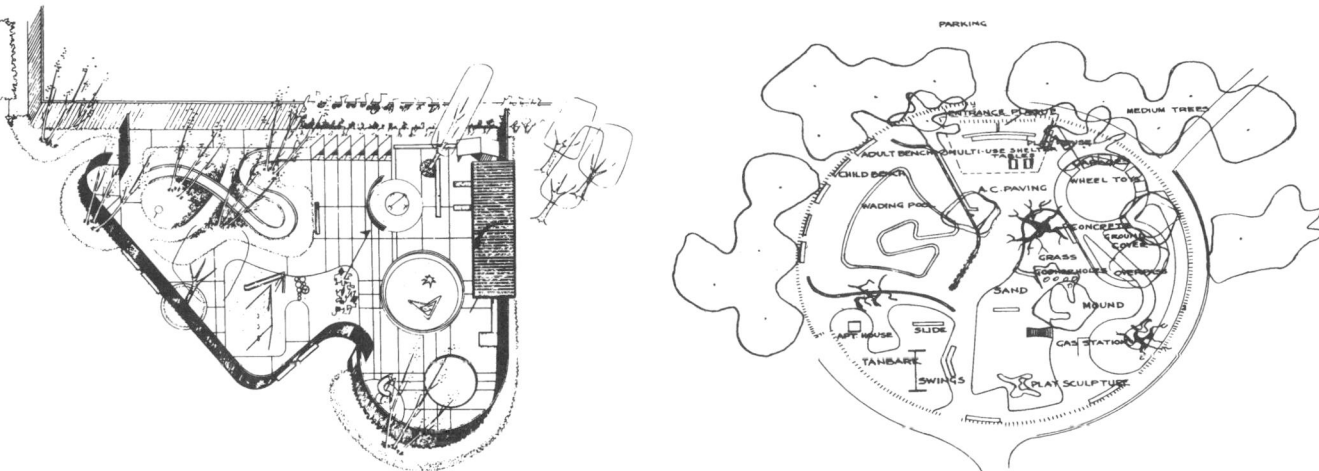

Fig. 1   Playlot site plan.

Fig. 2   Playlot layout.

Fig. 3   M. Paul Friedberg & Associates, Landscape Architects.

PLAYGROUNDS

Each housing development that generates several hundred children between the ages of 6 and 16 should provide play areas for these children.

The location of the playground should be at or near the center of the development or neighborhood so that it can be easily and safely reached from all the living units. Traditionally, the playground will be located at or adjacent to the elementary school and will serve as part of the neighborhood core of activities.

The playground at every elementary school should be of sufficient size and design, and properly maintained, to serve both the elementary educational program and the recreational needs of all age groups in the neighborhood. Since education and recreation programs complement each other in many ways, unnecessary duplication of essential outdoor recreation facilities should be avoided. Only where this joint function is not feasible should a separate playground be developed.

The planning and design of the playground are critical. It must provide for a variety of activities and serve the needs of both boys and girls of different ages. The main areas of the playground include:

1. An *apparatus area* for swings, slides, jungle gyms, and other equipment.
2. An *open area* for running, jumping, and informal play
3. *Courts and fields* for games of softball, soccer, tennis, handball, and volleyball
4. An *area for quiet games* such as crafts, checkers, and hobbies
5. A *wading or spray pool* for the younger children
6. A *sitting area* both for parents to observe the children and for the children to rest from play

The entire area should be enclosed and screened off from adjacent areas to prevent children from wandering off and to provide noise insulation for nearby living units.

The playground should be located to receive as much sunlight as possible during the day.

Although the maximum distance generally given for a child to walk to a playground is ¼ to ½ mile, it should be located as close as possible because it is most frequently and intensely used by children who live in the immediate vicinity. Ideally, no child should be required to cross any streets to get to the playground.

Size and Number of Playgrounds

Recommended size of a playground is a minimum of 6 to 8 acres, which would serve approximately 1000 to 1500 families. The smallest playground that will accommodate essential activity spaces is about 3 acres, serving approximately 250 families (about 110 elementary school children). This minimum area should be increased at the rate of 0.2 to 0.4 acre for each additional 50 families. More than one playground should be provided where: (1) a complete school playground is not feasible, (2) the population to be served exceeds 1500 families, or (3) the distance from the housing units is too great.

Playground Activity Spaces and Elements

A playground should contain the following basic activity spaces and elements:

1. A playlot, as described in the preceding section, with its equipment and surfacing as recommended.

2. An enclosed playground equipment area with supplemental planting for elementary school children, and with equipment as recommended.
3. An open, turfed area for informal active games for elementary school children.
4. Shaded areas for quiet activities such as reading, storytelling, quiet games, handicrafts, picnicking, and horseshoe pitching for both children and adults.
5. A paved and well-lighted multipurpose area large enough for:
   a. Activities such as roller skating, dancing, hopscotch, four-square, and captain ball.
   b. Games requiring specific courts, such as basketball, volleyball, tennis, handball, badminton, paddle tennis, and shuffleboard.
6. An area for field games, preferably well-lighted (including softball, junior baseball, touch or flag football, soccer, track and field activities, and other games), which will aslo serve for informal play of field sports and kite flying, and be used occasionally for pageants, field days, and other community activities.
7. Miscellaneous elements such as public shelter, storage space, toilet facilities, drinking fountains, walks, benches, trash containers, and buffer zones with planting.

Layout of Playgrounds

The layout of a playground will vary according to size of available area, its topography, and the specific activities desired. It should fit the site with maximum preservation of the existing terrain and such natural site features as large shade trees, interesting ground forms, rock outcrops, and streams. These features should be integrated into the layout to the maximum extent feasible for appropriate activity spaces as natural divisions of various use areas, and for landscape interest. Grading should be kept to a minimum consistent with activity needs, adequate drainage, and erosion control. General principles of layout are described as follows:

1. The playlot and the playground equipment area should be located adjacent to the school and to each other.
2. An open, turfed area for informal active play should be located close to the playlot and the playground equipment area for convenient use by all elementary school children.
3. Areas for quiet activities for children and adults should be somewhat removed from active play spaces and should be close to tree-shaded areas and other natural features of the site.
4. The paved multipurpose area should be set off from other areas by planting and so located near the school gymnaisum that it may be used for physical education without disturbing other school classes. All posts or net supports required on the courts should be con-

structed with sleeves and caps that will permit removal of the posts and supports.
5. The area for field games should be located on fairly level, well-drained land with finished grades not in excess of 2.5 percent: a minimum grade of 1 percent is acceptable on pervious soils having ground percolation for proper drainage.
6. In general the area of the playground may be divided as follows:
   a. Approximately half of the area should be parklike, including the open, turfed areas for active play, the shaded areas for quiet activities, and the miscellaneous elements as described in 7 below.
   b. The other half of the area should include ¾ to 1 acre for the playlot, playground equipment area, and the paved, multipurpose area, and 1¾ acres (for softball) to 1 acre (for baseball) for the field games area.
7. The playground site should be fully developed with landscape planting for activity control and traffic control, and for attractiveness. This site also should have accessible public shelter, storage for maintenance and recreation equipment, toilet facilities, drinking fountains, walks wide enough for strollers and carriages, bicycle paths, benches for adults and children, and trash containers.

Playground Equipment for Elementary School Children

Table 2 indicates types, quantities, and minimum play space requirements totaling about 6600 ft^2; this area, plus additional space for circulation, miscellaneous elements, and buffer zones, will accommodate a full range of playground equipment serving approximately 50 children at one time.

PLAYFIELDS

A large housing complex or residential development that will generate from 10,000 to 20,000 people, or the equivalent of several neighborhoods, will require a playfield. The playfield will provide a wide range of recreational activities for teenagers and adults.

The location of the playfield should be within ½ to 1 mile of every housing unit. The greater the population density, the closer the playfield. Ideally, it should be easily accessible by walking or bicycle. Usually, the playfield would be situated near or adjacent to a high school for obvious reasons.

The area requirements for each playfield will vary with the number and composition of the population to be served. The minimum area is considered to be about 10 acres, or 1 acre for each 800 to 1000 people served.

TABLE 2

Equipment	Number of pieces	Play space requirements
Balance beam	1	15 × 30 ft
Climbers	3	21 × 50 ft
Climbing poles	3	10 × 20 ft
Horizontal bars	3	15 × 30 ft
Horizontal ladder	1	15 × 30 ft
Merry-go-round	1	40 × 40 ft
Parallel bars	1	15 × 30 ft
Senior swing set (six swings)	1	30 × 45 ft
Slide	1	12 × 35 ft

Fig. 4

The most common recreational facilities provided in a playfield are:

1. *Major sports areas* for softball, baseball, football, and soccer
2. A *court game area* for games such as tennis, handball, and volleyball
3. A *swimming pool complex* for general swimming and diving activities
4. An *amphitheater or bandshell* for concerts, rallies, and other cultural activities
5. A *community recreation building* for indoor activities and inclement weather

NEIGHBORHOOD AREAS AND FACILITIES

Playlot

A playlot is a small recreation area designed for the safe play of preschool children.

Location

As an independent unit, the playlot is most frequently utilized in large housing projects or in other densely populated urban areas with high concentrations of preschool-age children. More often, it is incorporated as a vital feature of a larger recreation area. If a community is able to operate a neighborhood playground within ¼ mile of every home, playlots should be located at the playground sites. A location near a playground entrane, close to restrooms, and away from active game areas is desirable.

Size

The space devoted to a playlot depends upon the total open space available for development

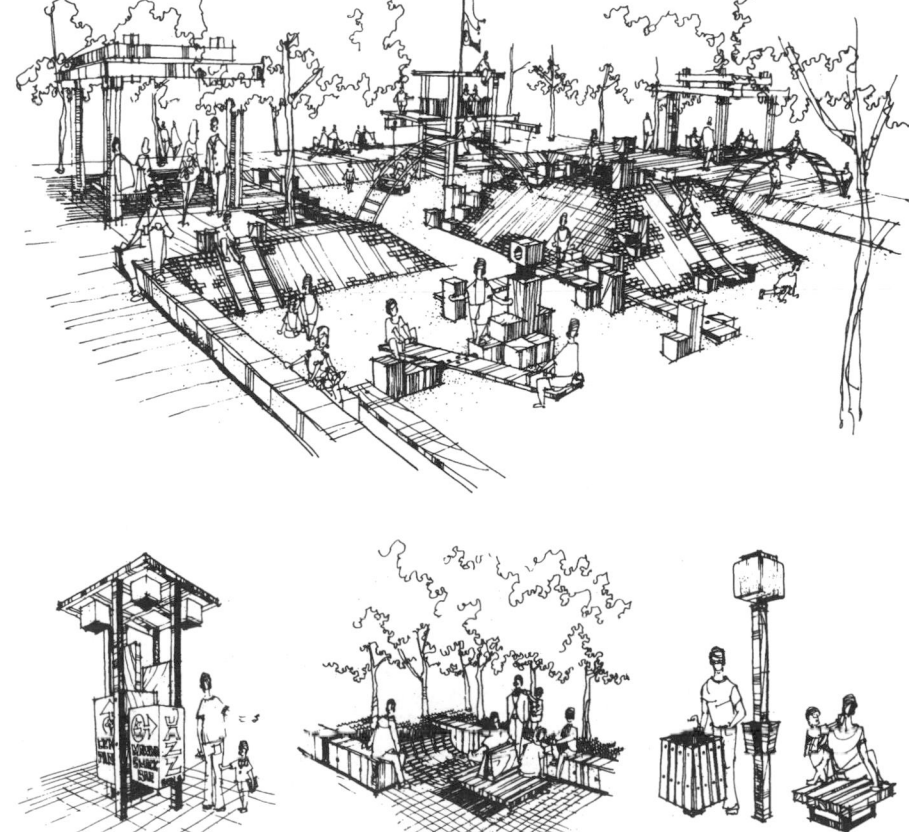

Fig. 5   M. Paul Friedberg & Associates, Landscape Architects.

Fig. 6   M. Paul Friedberg & Associates, Landscape Architects.

TABLE 3   Recommended Dimensions for Game Areas*

Games	Elementary school	Junior high school	High school (adults)	Area size (including buffer space), ft²
Basketball	40 × 60 ft	50 × 84 ft	50 × 84 ft	7,200
Basketball (college)			50 × 94 ft	8,000
Volleyball	25 × 50 ft	25 × 50 ft	30 × 60 ft	2,800
Badminton			20 × 44 ft	1,800
Paddle tennis			20 × 44 ft	1,800
Deck tennis			18 × 40 ft	1,250
Tennis		36 × 78 ft	26 × 78 ft	6,500
Ice hockey			85 × 200 ft	17,000
Field hockey			180 × 300 ft	64,000
Horseshoes		10 × 40 ft	10 × 50 ft	1,000
Shuffleboard			6 × 52 ft	640
Lawn bowling			14 × 110 ft	1,800
Boccie			15 × 75 ft	1,950
Tetherball	10-ft circle	12-ft circle	12-ft circle	400
Croquet	38 × 60 ft	38 × 60 ft	38 × 60 ft	2,200
Roque			30 × 60 ft	2,400
Handball (single-wall)	18 × 26 ft	18 × 26 ft	20 × 40 ft	1,200
Handball (four-wall)			23 × 46 ft	1,058
Baseball	210 × 210 ft	300 × 300 ft	400 × 400 ft	160,000
Archery		50 × 150 ft	50 × 300 ft	20,000
Softball (12-in ball)†	150 × 150 ft	200 × 200 ft	275 × 275 ft	75,000
Football			160 × 360 ft	80,000
Touch football		120 × 300 ft	160 × 360 ft	80,000
Six-man football			120 × 300 ft	54,000
Soccer (men) minimum			165 × 300 ft	65,000
maximum			240 × 360 ft	105,000
Soccer (women)			120 × 240 ft	40,000

*Table covers a single unit; many of above can be combined.
†Dimensions vary with size of ball used.

on a particular site. It may vary from 2500 to 10,000 ft².

General Features

The playlot should be enclosed with a low fence or solid plant materials in order to assist mothers or guardians in safeguarding their children. Careful thought should be given to placement of benches, with and without shade, for ease of supervision and comfort of parents and guardians. A drinking fountain with a step for tots will serve both children and adults.

Play equipment geared to the preschool child should combine attractive traditional play apparatus with creative, imaginative equipment. Such proven favorites as chair, bucket, and glider-type swings, 6-ft slides, and a small merry-go-round can be used safely. Hours of imaginative play will be enjoyed with such features as a simulated train, boat, airplane, and playhouse, and fiberglass or concrete animals. A small climbing structure should be included as well as facilities for sand play.

NEIGHBORHOOD PARK-SCHOOL (ELEMENTARY)

The neighborhood park-school is the primary unit in planning for physical education, recreation, and health education. This is a combination of an elementary school, neighborhood park, and playground. It is planned in such a manner that all areas and facilities are used to meet the educational and recreational needs of the people living in a neighborhood. It is essential that areas and facilities be cooperatively planned for the dual purpose of instruction and recreation, and that the school and community recreation programs be coordinated for maximum use of these areas and facilities by the entire neighborhood.

The park-school concept of combining education and recreation facilities on a single site has great merit. This combination makes possible a wider variety of opportunities on less acreage and at a lower cost than do separate installations. This approach is discussed here as it applies to areas at the neighborhood, community, and citywide levels.

Separately located recreation areas are also treated, since there may be certain circumstances under which the park-school may not be possible. It must be emphasized, however, that the combined approach is highly recommended.

Location

The neighborhood park-school should service an area with a maximum radius of ½ mile and a population of approximately 8000 people. any deviation in the population density (larger or smaller communities) may later change the service radius and/or acreage required for this installation.

Size

The minimum area recommended for a neighborhood park-school is 20 acres.

General Features

It is suggested that this area be developed as follows:

	Acres
School building	2.0
Parking	1.0
Playlot and apparatus	1.0
Hard-surface game courts and multiple-use area	2.5
Turf field-games area	5.5
Park area, including space for drama and quiet activities	5.5
Buffer zones and circulation	2.0
Recreation service building	0.2
Corner for senior citizens	0.3
Total	20.0

The school building should be at the edge of the area to provide for maximum development and utilization of the site, and playground equipment should be located far enough from the building to keep noise from interfering with class instruction.

A separate building containing the recreation leader's headquarters and public restroom facilities should be provided in close proximity to hard-surface and game areas.

Hard-surface areas should be contiguous to provide a larger area for recreational, recess, physical education, and intramural activities. The field area should be large enough for baseball and softball diamonds to accommodate all age levels, for various field games, and for special events. Paths and walks between areas should be placed so as to avoid traffic over lawns, and the arrangement of facilities and landscaping should make for ease of supervision.

NEIGHBORHOOD PLAYGROUND

Designed primarily to serve children under 14 years of age, the neighborhood playground should have additional features to interest teenagers and adults. The trend in recent years is for the neighborhood playground to become increasingly the center of activity for the wide variety of needs expressed by all residents. The more diversified interests of today's recreation consumer challenge the facility planner to pro-

vide for a broader program, with more attention devoted to multiple use by different age groups.

Modern planning for outdoor recreation at the neighborhood level places heavy emphasis on combining elementary-school needs with those of the community. This type of joint development is treated in the immediately preceding section on the neighborhood park-school.

Where elementary-school facilities are unavailable or inadequate, or joint development is impossible, a separate playground will be needed in each neighborhood.

Location

The neighborhood playground serves the recreation needs of the same population served by the neighborhood elementary school. Its maximum use radius will seldom exceed ½ mile, with most of the attendance originating within a ¼-mile distance. It should be located close to the center of the area to be served and away from heavily traveled streets and other barriers to easy and safe access.

Size

In order to have the desired features, the neighborhood playground would normally require a minimum of 8 acres. The particular facilities required will depend on the nature of the neighborhood, with space being allocated according to needs.

General Features

It is recommended that his area be developed as follows:

	Acres
Turf area for softball, touch football, soccer, speedball, and other field games	3.00
Hard-surface area for court games, such as netball, basketball, volleyball, and handball	0.50
Open space for informal play	0.50
Corner for senior citizens	0.30
Space for quiet games, storytelling, and crafts	0.20
Playlot	0.20
Children's outdoor theater	0.15
Apparatus area for elementary-age children	0.25
Service building for restrooms, storage, and equipment issue, or a small clubhouse with some indoor activity space	0.15
Circulation, landscaping, and buffer zones	2.00
Undesignated space	0.75
Total	8.00

Depending upon the relationship of the site to school and other recreation facilities in the neighborhood, optional features such as a recreation building, a park, tennis courts, or a swimming pool might be located at the neighborhood playground. If climatic conditions warrant, a spray or wading pool should be provided. The following space for optional features

	Acres
Recreation building	0.2
Park area (if there is no neighborhood park)	2.0
Swimming pool	0.5
Tennis courts	0.4
Total	3.1

should be added to the standards listed above:

The addition of optional features may require provision for off-street parking.

Equipment

The following types of equipment are recommended:

Several pieces of equipment designed as simulated stagecoaches, fire engines, boats, locomotives, etc.

Physical-fitness or obstacle-course features, such as a scaling wall, cargo net climber, etc.

Balance beam

Climbing structure, not to exceed 9 ft high

Horizontal ladder, not to exceed 7 ft high

Three horizontal bars with fixed heights, of rust-resistant metal

Straight slide 8 ft high or spiral slide 10 ft high

Six or more conventional swings, with low protective barriers

Pipe equipment formed into shapes

Sculptured forms

Merry-go-round, safety-type

The various apparatus groupings should be separated by plantings or attractive medium-height fencing.

NEIGHBORHOOD PARK

The neighborhood park is land set aside primarily for passive recreation. Ideally, it gives the impression of being rural, sylvan, or natural in its character. It emphasizes horticultural features, with spacious turf areas bordered by trees, shrubs, and sometimes floral arrangements. It is essential in densely populated areas, but not required where there is ample yard space attached to individual home sites.

Location

A park should be provided for each neighborhood. In many neighborhoods, it will be incorporated in the park-school site or neighborhood playground. A separate location is required if this combination is not feasible.

Size

A separately located neighborhood park normally requires 3 to 5 acres. As a measure of expediency, however, an isolated area as small as 1 or 2 acres may be used. Sometimes the neighborhood park function can be satisfactorily included as a portion of a community or citywide park.

General Features

The neighborhood park plays an important role in setting standards for community aesthetics. Therefore, it should include open lawn areas, plantings, and walks. Sculptured forms, pools, and fountains should also be considered for ornamentation. Creative planning will utilize contouring, contrasting surfaces, masonry, and other modern techniques to provide both eye appeal and utility.

COMMUNITY AREAS AND FACILITIES

Community Park-School (Junior High)

The community park-school (junior high), a joint development of school and community, provides an economical and practical approach to a communitywide facility for educational, cultural social, and recreational programs. This educational and recreational center generally refers to the combination of a junior high school and a community park.

Location

It is suggested that this facility provide service for an area with a radius of ½ to 1½ miles. Such an area will normally contain 20,000 to 30,000 people, but population density may modify the size of the area served.

Size:

Based upon current formulas for establishing junior-high-school and community-park sites, a minimum area of 35 acres is desirable.

General Features

It is suggested that the area be developed as follows:

	Acres
Buildings (school and community recreation)	5.00
Turf field-games area	8.00
Hard-surface games court and multiple-use area	2.75
Tennis courts	1.00
Football field with 440-yd track (220-yd straightaway)	4.00
Baseball field with hooded backstop	3.00
Playlot and apparatus	1.00
Park and natural areas	5.00
Parking	1.25
Buffer zones and circulation	4.00
Total	35.00

The following may be included as standard or optional features:

Swimming pool (usually related to the building)

Nature study trails and/or center

Day-camping center

There are many optional features that may be included in the community park-school. The inclusion of these is dependent upon the section of the country, available space, topography, community needs, climate, socioeconomic composition of the community, and other variables. The following may be included as optional features:

Archery range
Band shell
Boccie courts
Botanical garden
Croquet courts
Golf driving range
Golf putting course
Hard-surface area for dancing
Horseshoe pits
Ice-skating or roller-skating rink
Lake for boating
Lawn-bowling greens
Lighted courts and fields
Shuffleboard courts

In designing the community park-school, planners should consider the proper placement of apparatus and areas that serve multiple use, and also bear in mind appropriate safety features in the development of each area or facility.

COMMUNITY PARK-SCHOOL (SENIOR HIGH)

A community park-school (senior high) is planned to provide facilities for youth and adults to meet a wide range of educational and recreational needs and interests on a single site. It generally refers to a combination of a high school and a community park.

It is essential that coordination and cooperation be exercised by school and municipal authorities to ensure the maximum development and use of all facilities for instruction and recreation, both during and after school hours.

Location

It is suggested that the population density of the area as well as the total population of the community determine the scope and size of the area to be served by this facility. For example, the higher the population density, the smaller the service radius.

Size

Based on current formulas, a minimum area of 50 acres is suggested.

The site size should be based upon program needs, which will include: the physical education instructional program; school-supervised games, sports, and athletics; and school and community recreation activities during out-of-school hours.

General Features

It is suggested that the area be developed as follows:

	Acres
Buildings (including a gymnaisum and an aquatics center)	6.00
Turf field-games area for instruction, intramurals, interscholastic athletics practice, and recreation use	8.00
Hard-surface games court and multiple-use area	3.00
Tennis courts	1.50
Apparatus area for instructional use (optional)	0.12
Recreation area	5.00
Hard-surface area (for shuffleboard and outdoor bowling)	
Turf area (for horseshoes and croquet)	
Turf area (for golf and archery)	
Football field with bleachers and 440-yd track (220-yed straightaway)	6.00
Baseball field	3.50
Playlot and apparatus	0.50
Park and natural areas	5.00
Recreation building with senior-citizen center	0.50
Parking and driver-education range	6.00
Buffer zones and circulation	5.00
Total	50.12

For other features that may be incorporated into this facility, see the sections devoted to the community park-school (junior high) and the citywide or district park.

An adequate number of each kind of facility should be provided to permit full participation by the largest group that will be using the facility at any given time.

The total community park-school area should be landscaped to create a parklike setting that enhances and does not interfere with the instructional and recreational areas.

COMMUNITY PARK AND PLAYFIELD

The community park and playfield is designed to provide a variety of active and passive recreational services for all age groups of a community served by a large junior high school (20,000 to 30,000 residents). Primary requisites are outdoor fields for organized sports, indoor space for various activities, special facilities, and horticultural development.

Location

It is highly desirable that this facility be incorporated into the complex of a community park-school (junior high). Where this is not feasible, the community park and playfield should be located within ½ to 1½ miles of residents in its service area depending upon population density and ease of access.

Size

A separate community park and playfield requires an area of 15 to 20 acres. At least two-thirds of the area should be developed for active recreation purposes.

General Features

The following should be provided:

Fields for baseball, football, field hockey, soccer, and softball

Courts for tennis, basketball, boccie, volleyball, handball, horseshoes, shuffleboard, paddle tennis, and other games

Recreation building containing an auditorium, a gymnasium, and special-use rooms for crafts, dramatics, and social activities

Quiet recreation area

Hard-surface area for dodgeball and kickball

May include a neighborhood playground (see features under Neighborhood Playground)

CITYWIDE OR DISTRICT PARK

The citywide or district park serves a district of a larger city, or a total community of a smaller city. This facility should serve a population of from 50,000 to 100,000. It is designed to provide a wide variety of activities.

Location

This facility should be incorporated with a high school as a park-school development. Where this is not feasible, consideration should be given to placing the park as close as possible to the center of the population to be served. The land available will be a determining factor in site selection. While the service area will vary according to population density, a normal use radius is 2 to 4 miles.

Size

The citywide or district park may have from 50 to 100 acres.

General Features

Depending upon available acreage, topography, and natural features, the citywide or district park will contain a large number of different components. These would include, but not be limited to, the following:

A number of fields for baseball, football, soccer, and softball

Tennis center
Winter sports facilities
Day-camp center
Picnic areas (group and family)
Bicycling paths or tracks
Swimming pool
Lake for water sports
Pitch-and-putt golf course
Recreation building
Nature-centered trails
Skating rinks (ice and roller)
Playlot and apparatus
Parking areas
Outdoor theater

The above facilities should be separated by large turf and landscaped areas. Natural areas and perimeter buffers should be provided.

OFFICIAL BASEBALL

Babe Ruth Baseball (13 to 15 and 16 to 18 Years); Senior League Baseball (13 to 15 Years)

Recommended area Ground space is 3.0 to 3.85 acres minimum.

Size and dimension Baselines are 90'-0". Pitching distance is 60'-6". Pitcher's plate is 10 in above the level of home plate. Distance down foul lines is 320 ft minimum, 350 ft preferred. Outfield distance to center field is 400 ft+. For senior league baseball, recommended distance from home plate to outfield fence at all points is 300 ft².

Orientation Optimum orientation is to locate home palte so that the pitcher is throwing across the sun and the batter is not facing it. The line from home plate through the pitcher's mound and second base should run east-northeast.

Surface and drainage Surface is to be turf. Infield may be skinned, and shall be graded so that the baselines and home plate are level.

Special considerations Backstop is to be provided at a minimum distance of 40 ft or preferably 60 ft behind home plate.

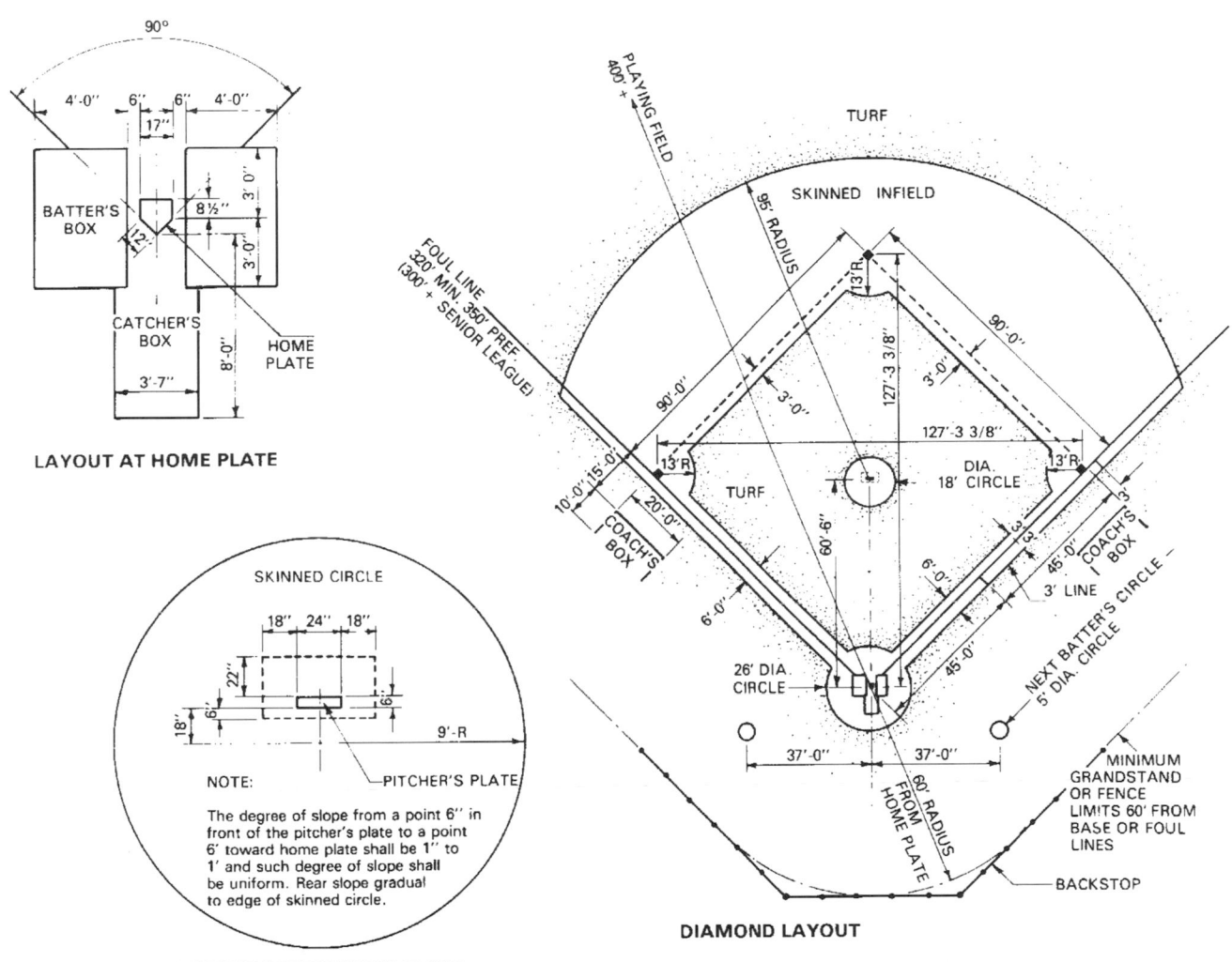

Fig. 7 Official baseball diamond. Foul lines, catcher's, batter's, and coach's boxes, next batter's circles, and 3-ft line shall be 2 to 3 in wide and marked with chalk or other white material. Caustic lime must not be used. Infield may be skinned.

BASEBALL

Little League (9 to 12 Years)

Recommended area Ground space is 1.2 acres minimum.

Size and dimension Baselines are 60'-0". Pitching distance is 46'-0". Pitcher's plate is 6 in above the level of home plate. Distance down foul line is 200 ft. Outfield distance to pocket in center field is 200 to 250 ft optional.

Orientation Optimum orientation is to locate home plate so that the pitcher is throwing across the sun and the batter is not facing it. The line from home plate through the pitcher's mound and second base should run east-northeast.

Surface and drainage Surface is to be turf. Infield may be skinned, and shall be graded so that the baselines and home plate are level.

Special considerations Backstop is to be provided at a recommended minimum distance of 25 ft behind home plate.

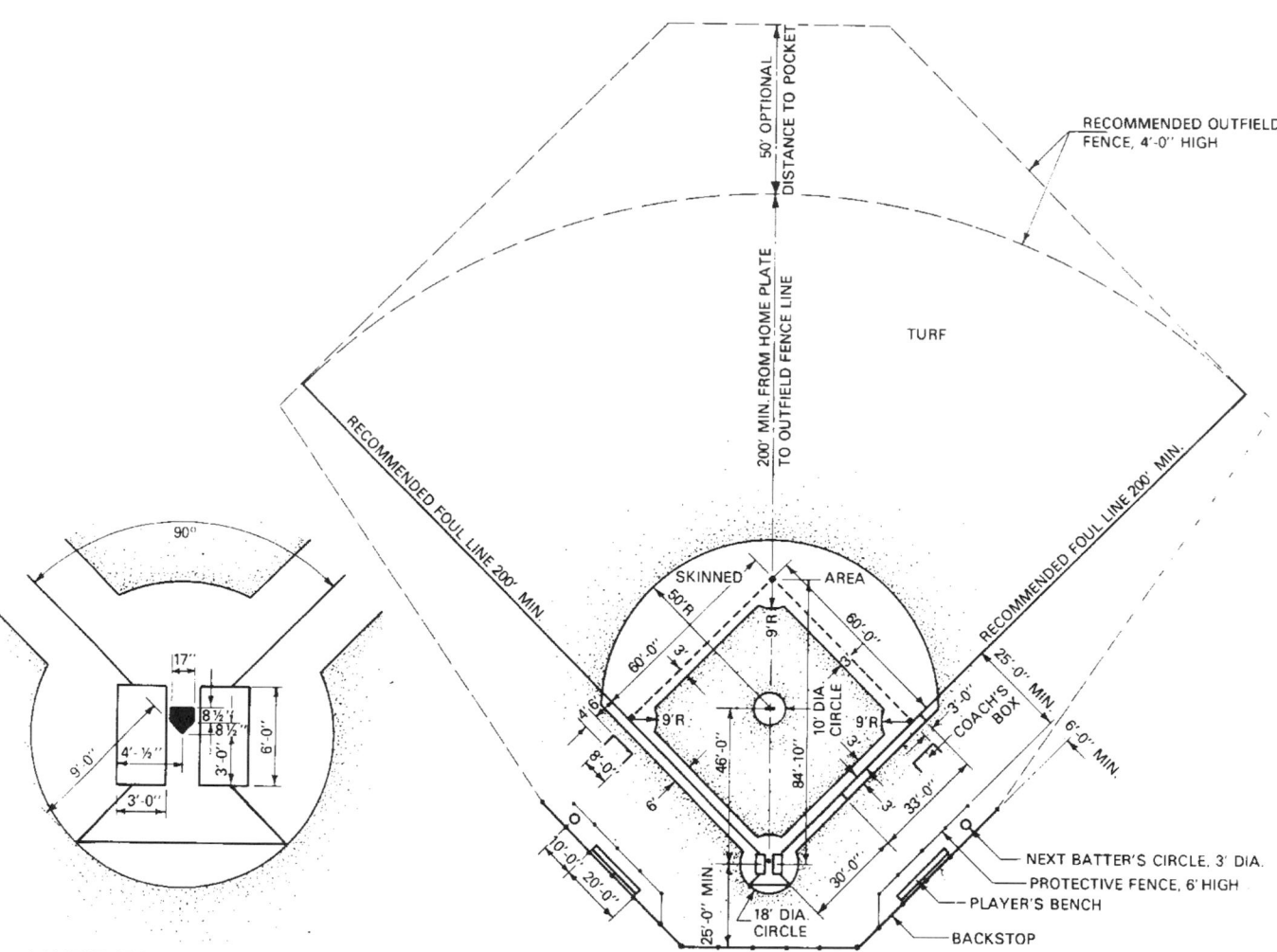

LAYOUT AT HOME PLATE

Fig. 8 Little league baseball diamond. Foul lines, catcher's, batter's, and coach's boxes, next batter's circles, and 3-ft restraining lines shall be 2 in wide and marked with white chalk or other white material. Caustic lime must not be used. Infield may be skinned.

SOFTBALL, 12-INCH

Fast and Slow Pitch

Recommended area Ground space is 62,500 ft² (1.5 acres) to 90,000 ft² (2.0 acres).

Size and dimension Baselines are 60'-0" for men and women, 45'-0" for uniors. Pitching dis-

tances are 46'-0" for men, 40'-0" for women, 35'-0" for juniors. Fast pitch playing field is 225-ft radius from home plate between foul lines for men and women. Slow pitch is 275-ft radius for men, 250-ft radius for women.

Orientation Optimum orientation is to locate home plate so that the pitcher is throwing across the sun and the batter is not facing it.

Surface and drainage Surface is to be turf. Infield may be skinned. The infield shall be graded so that the baselines and home plate are level.

Special considerations Backstop is to be located at a minimum distance of 25 ft behind home plate.

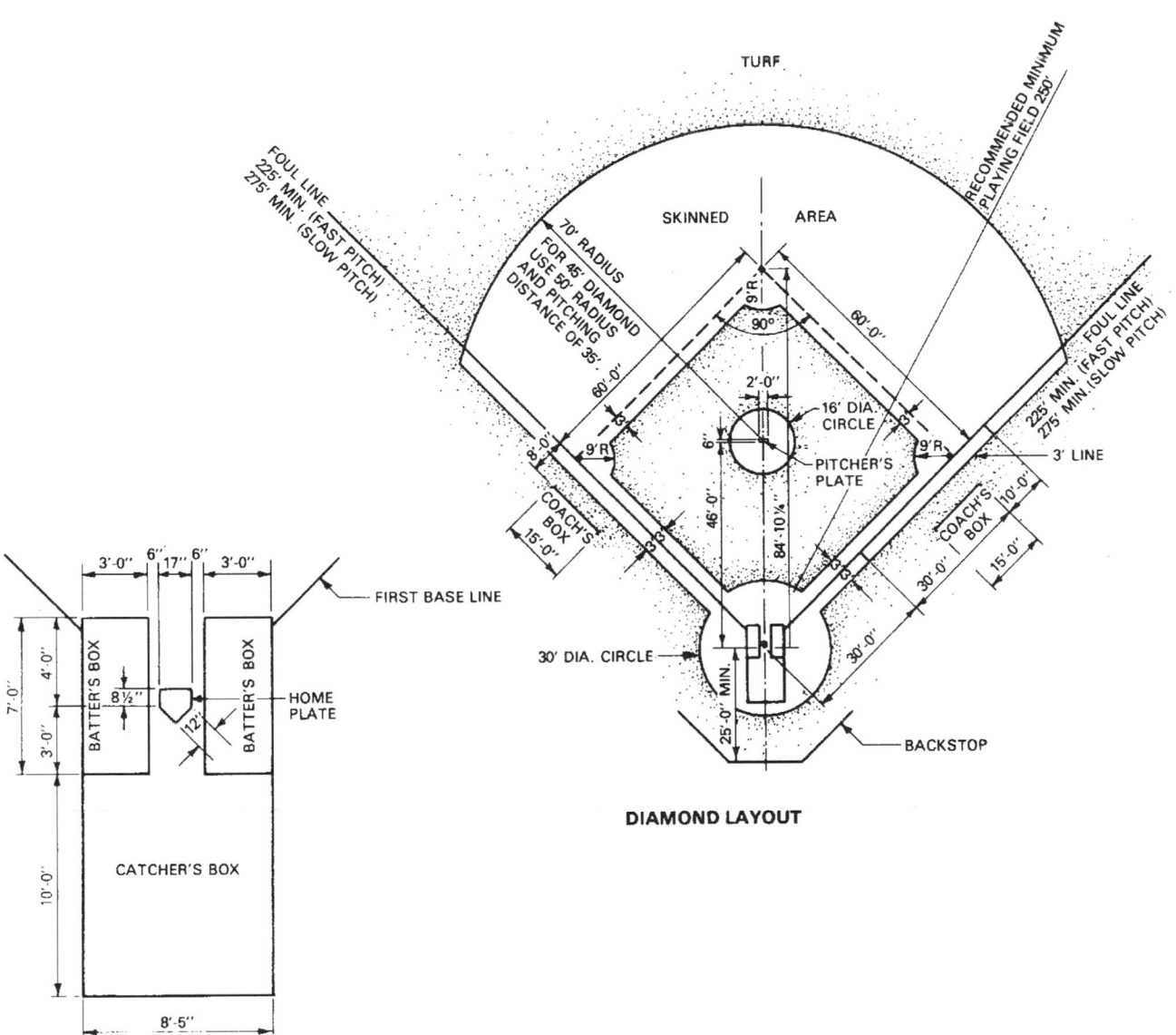

DIAMOND LAYOUT

LAYOUT AT HOME PLATE

Fig. 9 Diamond for 12-in softball. Foul lines, catcher's, batter's, and coach's boxes, and 3-ft lines are 2- to 3-in chalk lines. Pitching distance for women's softball to be 40'-0". For junior player (9 to 12 years) 45-ft distance between bases, 35-ft pitching distance.

RECREATION AREAS

SOCCER

Men's and Boys'

Recommended area Ground space is 75,250 ft² (1.7 acres) to 93,100 ft² (2.1 acres).

Size and dimension Playing field width is 195'-0" to 225'-0". Length is 330'-0" to 360'-0". Additional area recommended is 10'-0" minimum unobstructed psace on all sides.

Orientation Preferred orientation is for the long axis to be northwest-southeast to suit the angle of the sun in the fall playing season, or north-south for longer periods.

Surface and drainage Surface is to be turf. Preferred grading is a longitudinal crown with a 1 percent slope from center to each side and adequate underdrainage. Grading may be from side to side or corner to corner diagonally if conditions do not permit the preferred grading.

Special considerations Goal posts are to be provided at each end of the playing field.

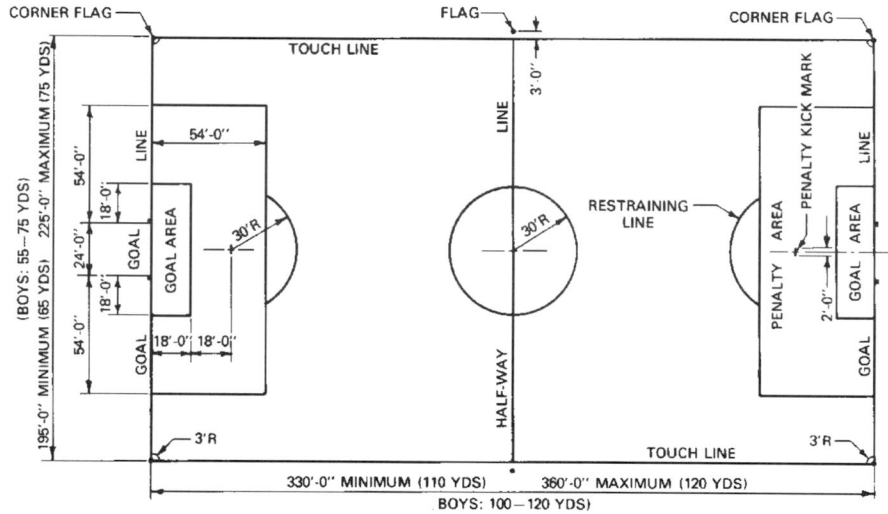

PLAYING FIELD LAYOUT

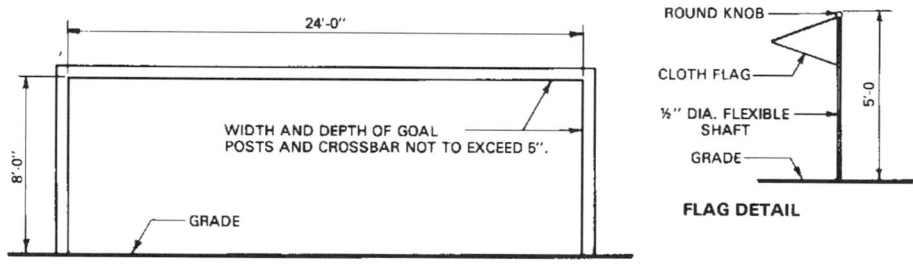

GOAL POSTS

FLAG DETAIL

Fig. 10 Playing field for men's and boys' soccer. Goal posts to be pressure treated with paintable, oil-borne preservative and painted above ground with three coats of white lead and oil. The goalposts and crossbar shall present a flat surface to the playing field, not less than 4 in nor more than 5 in. in width. Nets shall be attached to the posts, crossbar, and ground behind the goal. The top of the net must extend backward 2'-0" level with the crossbar. All dimensions are to the inside edge of lines. All lines shall be 2 in wide and marked with a white, nontoxic material which is not injurious to the eyes or skin.

BASKETBALL (NCAA)

Recommended area *High school:* ground space is 5040 ft² minimum to 7280 ft² maximum. *Collegiate:* ground space is 5600 ft² minimum to 7980 ft² maximum.

Size and dimension High school recommended court is 84 × 50 ft with a 10-ft unobstructed space

on all sides (3 ft minimum). Collegiate recommended court is 94 × 50 ft with a 10-ft unobstructed space on all sides (3 ft minimum).

Orientation Preferred orientation is for the long axis to be north-south.

Surface and drainage Surface is to be concrete or bituminous material with optional pro-

tective colorcoating. Drainage is to be end to end, side to side, or corner to corner diagonally at a minimum slope of 1 in. in 10 ft.

Special considerations Safety—backboard and goal support should have a minimum 32-in overhang and post may be padded if desired. Bottom edge and lower sides of rectangular backboard must be padded.

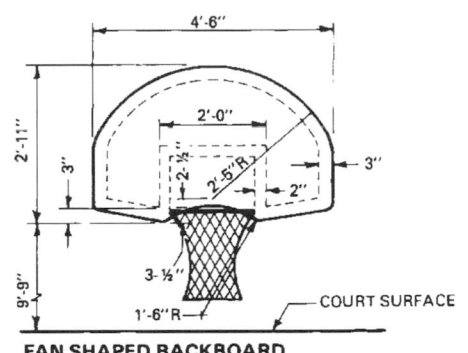

FAN SHAPED BACKBOARD

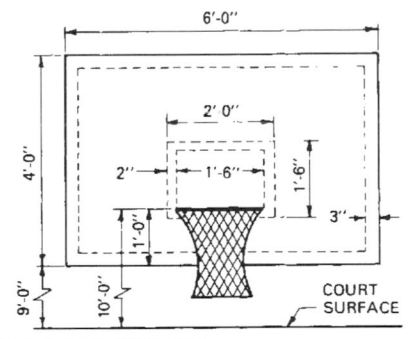

RECTANGULAR BACKBOARD

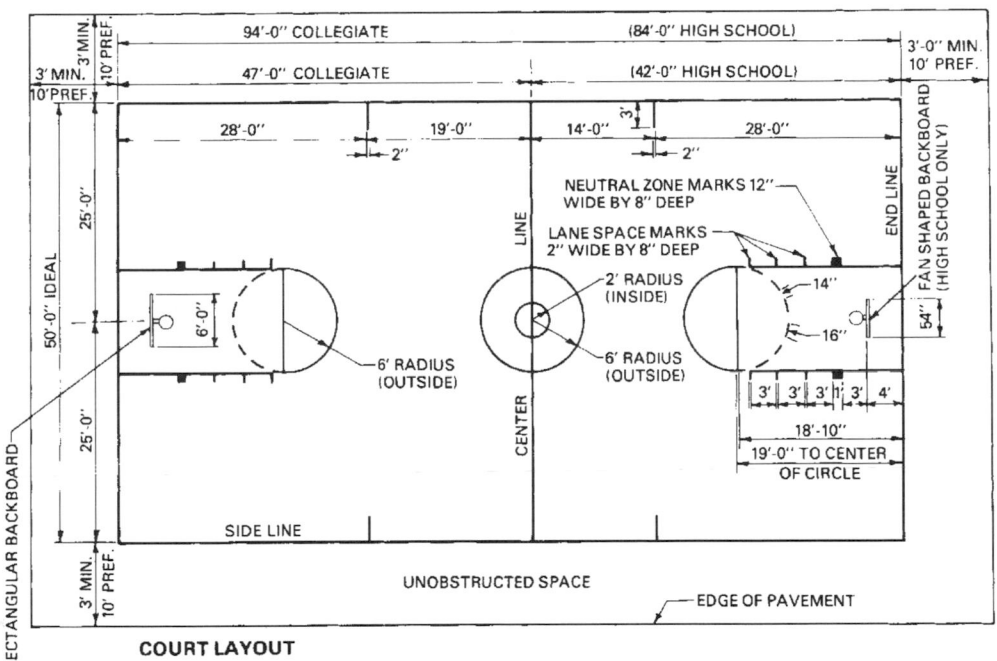

COURT LAYOUT

Fig. 11 NCAA basketball. The color of the lane space marks and neutral zone marks shall contrast with the color of the bounding lines. The mid-court marks shall be the same color as the bounding lines. All lines shall be 2 in wide (neutral zone excluded). All dimensions are to inside edge of lines except as noted. Backboard shall be of any rigid weather-resistant material. The front surface shall be flat and painted white unless it is transparent. If the backboard is transparent, it shall be marked with a 3-in wide white line around the border and an 18 × 24-in target area. If the backboard is transparent, it shall be marked with a 3-in wide white line around the border and an 18 × 24-in target area bounded with a 2-in wide white line.

RECREATION AREAS

ONE-WALL HANDBALL

Recommended area Ground space is 1665 ft^2 plus walls and footings.

Size and dimension Playing court is 20'-0" wide by 34'-0" long plus a required 11'-0" minimum width of surfaced area to the rear and a recommended 8'-6" minimum width on each side. Courts in battery are to be a minimum of 6'-0" between courts.

Orientation Preferred orientation is for the long axis to be north-south with the wall at the north end.

Surface and drainage Surface is to be smooth concrete with a minimum slope of 1 in. in 10 ft from the wall to the rear of the court.

Special considerations Fencing—court area preferably should be fenced with a 10-ft high chain link fence.

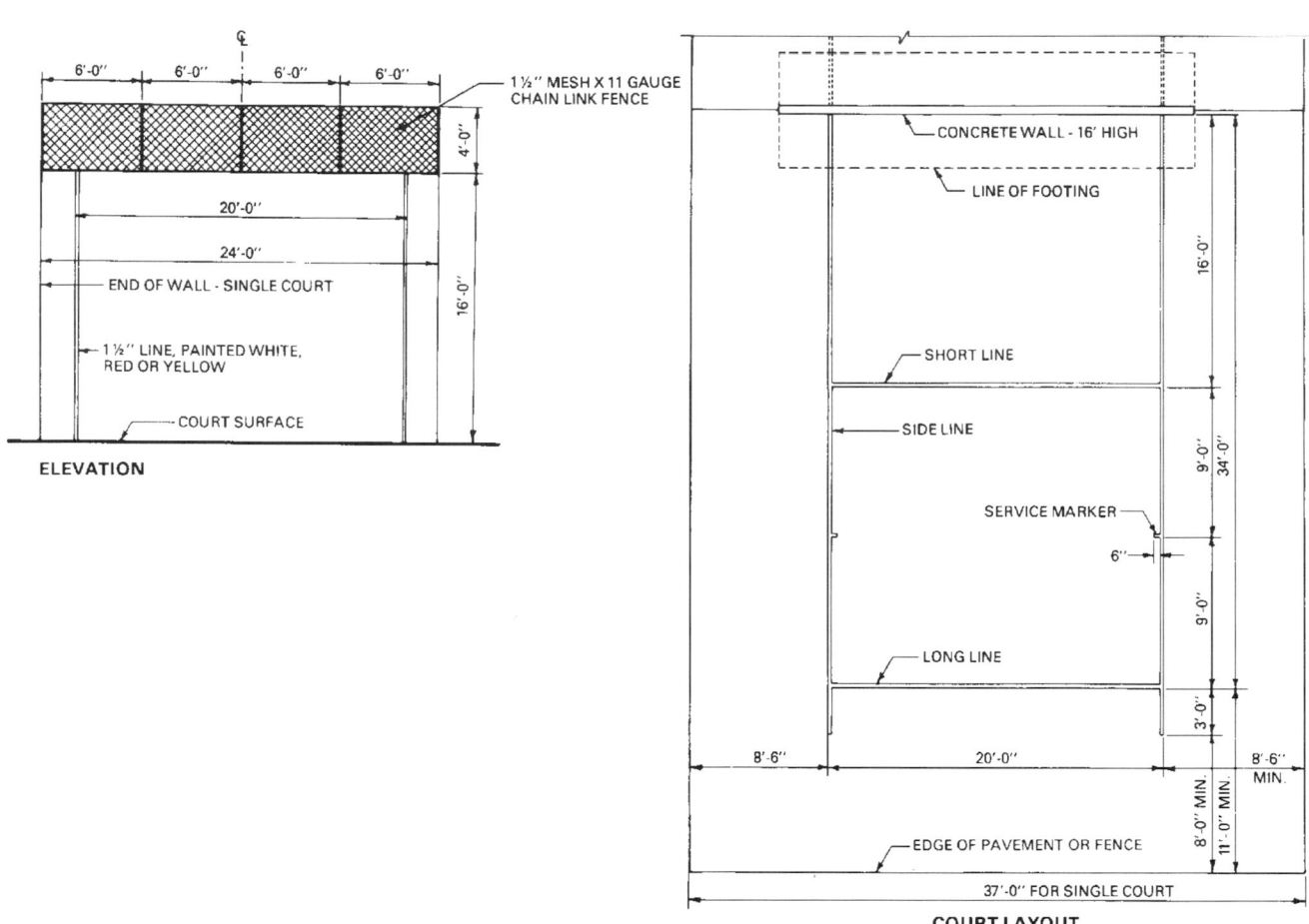

ELEVATION

COURT LAYOUT

Fig. 12   One-wall handball. Court markings 1½-in-wide lines painted white, red, or yellow.

TENNIS

Recommended area Ground space is 7200 ft² minimum.

Size and dimension Playing court is 36 × 78 ft plus at least 12 ft clearance on both sides or between courts in battery and 21 ft clearance on each end.

Orientation Orientation of long axis is to be north-south.

Surface and drainage Surface may be concrete or bituminous material with specialized protective colorcoating, or sand-clay. Drainage may be from end to end, side to side, or corner to corner diagonally at a minimum slope of 1 in. in 10 ft for pavement and level for sand-clay with underdrainage.

Special considerations Fencing—recommended 10-ft-high chain link fence on all sides.

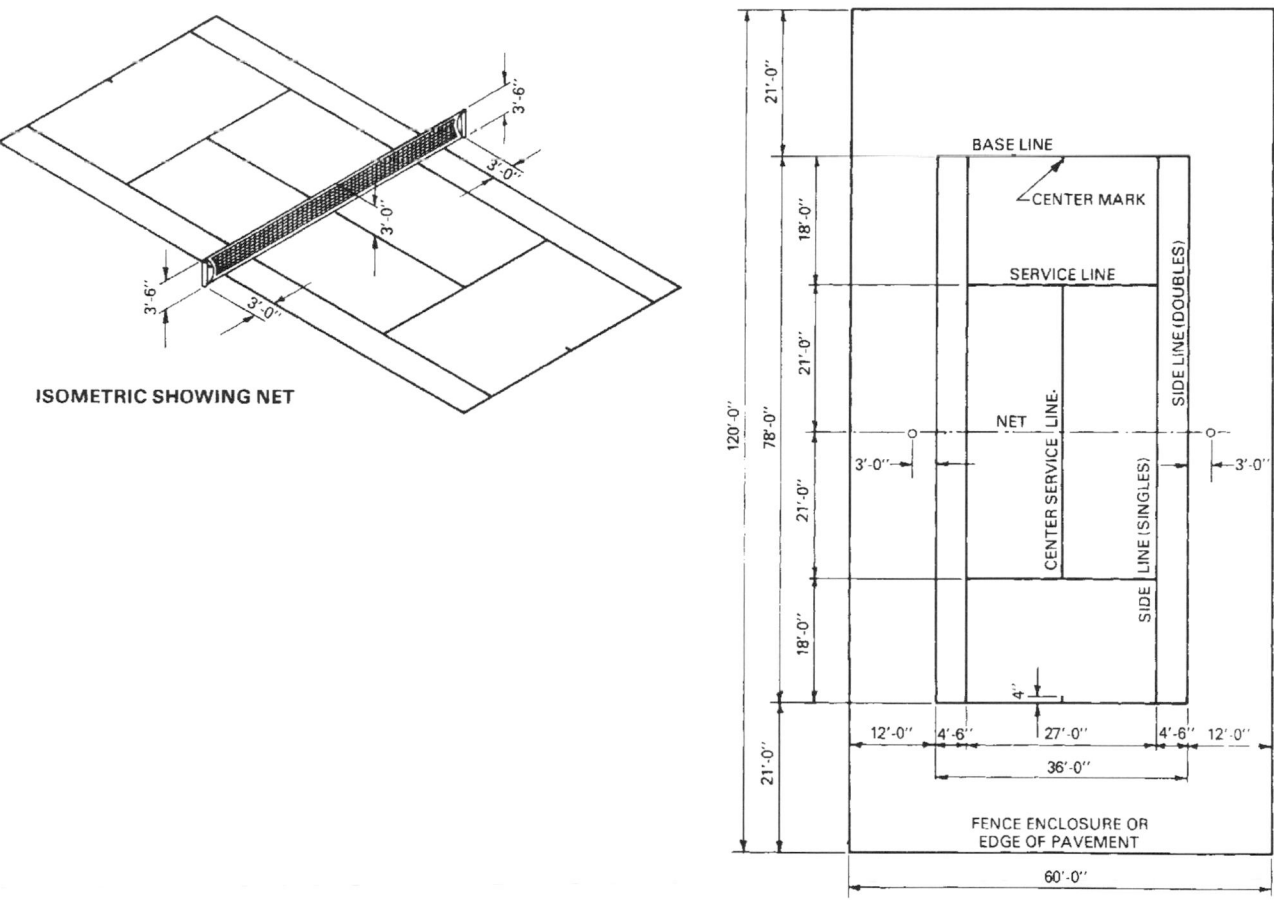

ISOMETRIC SHOWING NET

COURT LAYOUT

Fig. 13 Tennis court. All measurements for court markings are to the outside of lines except for those involving the center service line which is equally divided between the right and left service courts. All court markings to be 2 in wide. Fence enclosure, if provided, should be 10-ft-high, 11-gauge, 1¾-in mesh chain link. Minimum distance between sides of parallel courts to be 12′-0″.

VOLLEYBALL

Recommended area   Ground space is 4000 ft².

Size and dimension   Playing court is 30 × 60 ft plus 6 ft minimum, 10 ft preferred, unobstructed space on all sides.

Orientation   Preferred orientation is for the long axis to be north-south.

Surface and drainage   Recommended surface for intensive use is to be bituminous material or concrete, but sand-clay or turf may be used for informal play. Drainage is to be end to end, side to side, or corner to corner at a minimum slope of 1 in. in 10 ft.

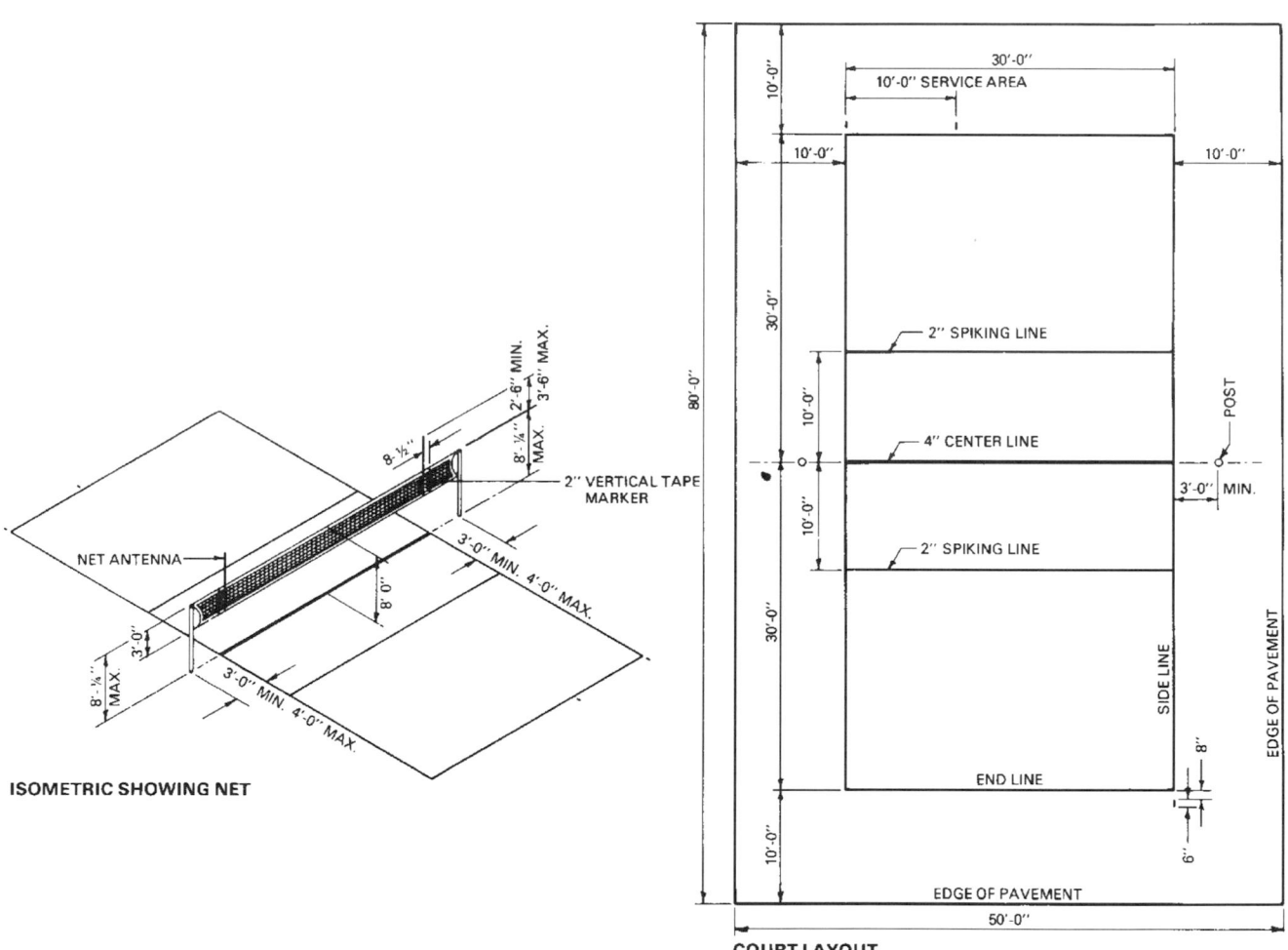

ISOMETRIC SHOWING NET

COURT LAYOUT

Fig. 14   Volleyball court. All measurements for court markings are to the outside of lines except for the centerline. All court markings to be 2 in wide except as noted. Net height at center to be: men 8'-0", women 7'-4¹⁄₄", high school 7'-0", elementary school 6'-6".

SHUFFLEBOARD

Recommended area   Ground space is 312 ft² minimum.

Size and dimension   Playing court is 6'-0" × 52'-0" plus a recommended minimum of 2'-0" on each side or 4'-0" between courts in battery.

Orientation   Recommended orientation is for the long axis to be north-south.

Surface and drainage   Surface is to be concrete with a burnished finish. Court surface is to be level with drainage away from the playing surface on all sides.

Special considerations   Secure covered storage for playing equipment should be provided near the court area.

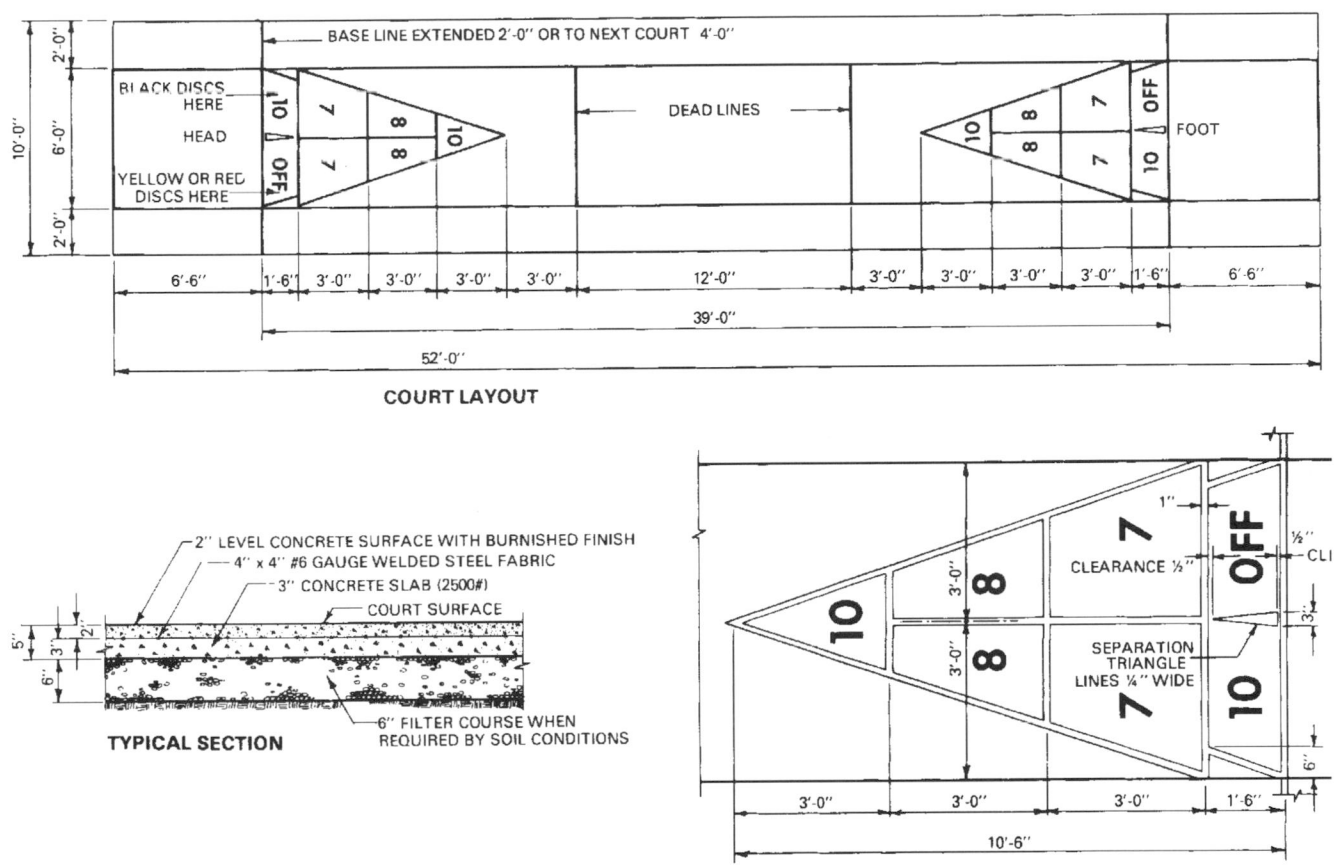

COURT LAYOUT

TYPICAL SECTION

COURT MARKING DETAIL

Fig. 15   Shuffleboard court. All dimensions are to centers of lines and to edge of court. Maximum line width 1½ in, minimum ¾ in. Lines and figures "10," "8," "7," and "10 OFF" should be marked with black shoe dye or black acrylic paint. Court to be constructed of concrete without expansion joints. A depressed alley at least 24 in wide, and not less than 4 in deep at midcourt, should be constructed between courts and on the outside of end courts. The alley should slope 1 in. in the first 6 ft of the length of the alley from each baseline, then slope to a minimum depth of 4 in at midcourt where a suitable water drain should be provided.

BOCCIE BALL

Recommended area Ground space is 1824 to 2816 ft².

Size and dimension Overall court dimensions are 13'-0" to 19'-6" wide by 78'-0" to 92'-0" long. Additional space of at least 3'-0" on each side

and 9'-0" on each end is recommended.

Orientation Preferred orientation is for the long axis to be north-south although it is of minor importance.

Surface and drainage Surface is to be preferably turf, although a mixture of sand and clay

may be used. Drainage may be in any direction at a recommended slope of 1 percent for turf and level for sand-clay with underdrainage.

Special considerations Optional low wooden barrier should be provided at each end and/or side of court.

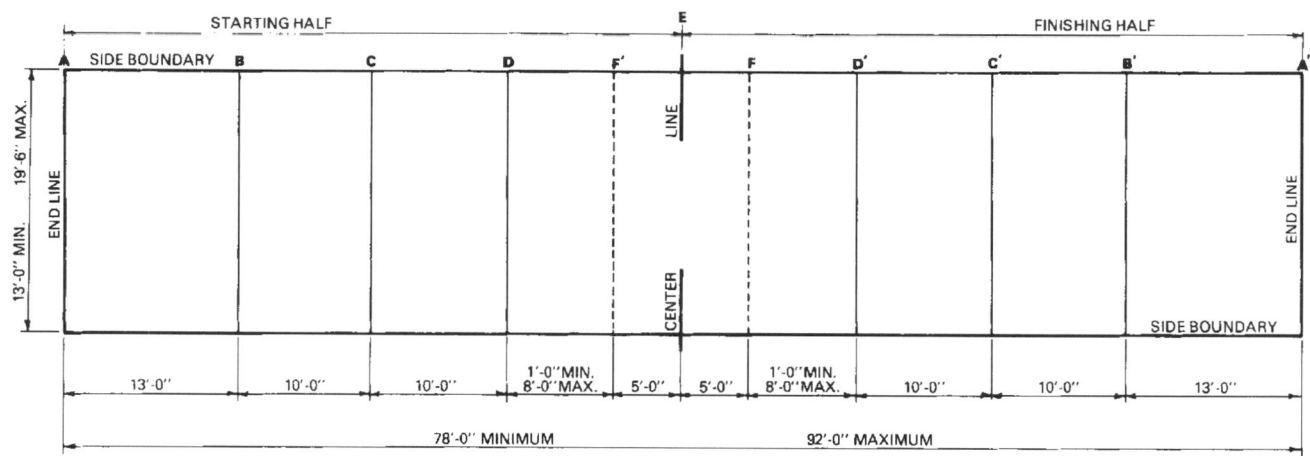

COURT LAYOUT

Fig. 16 Boccie. Court markings to be 2-in wide linen tape held in place with metal pins.

HORSEHOES

Recommended area Ground space is 1400 ft² including clear space.

Size and dimension Playing court is 10'-0" × 50'-0" plus a recommended 10-ft minimum unobstructed area on each end and a 5-ft (minimum) wide zone on each side.

Orientation Recommended orientation is for the long axis to be north-south.

Surface and drainage Surface of playing area, except for boxes and optional concrete walkways, should be turf. Area should be pitched to the side at a maximum slope of 2 percent. Elevation and slant of steel pegs should be between 2 and 3 in and equal.

Special considerations Boxes are to be filled with gummy potter's or blue clay. Safety—2'-0"-high backstop should be constructed at the end of the box to intercept overthrown or bounding shoes.

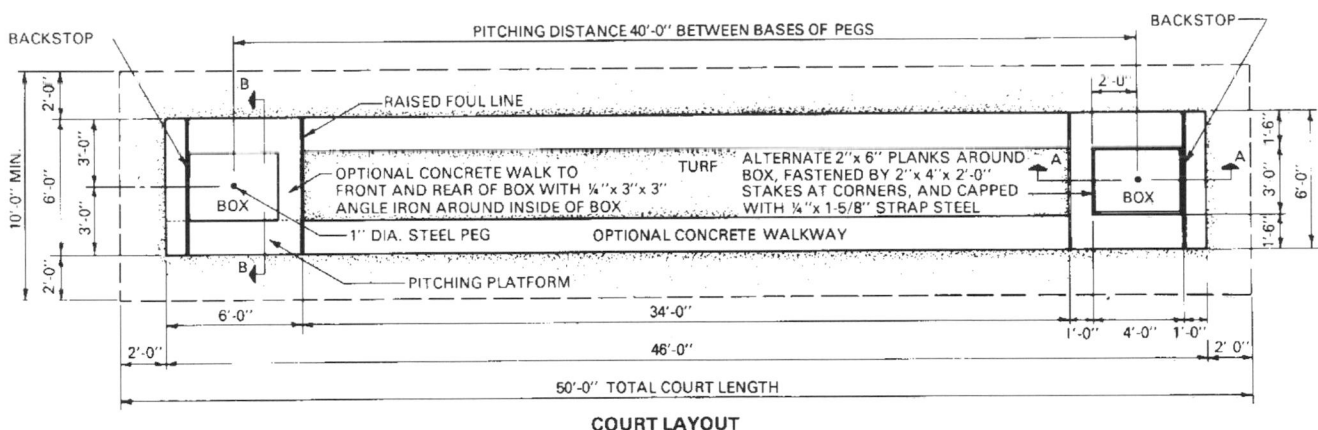

COURT LAYOUT

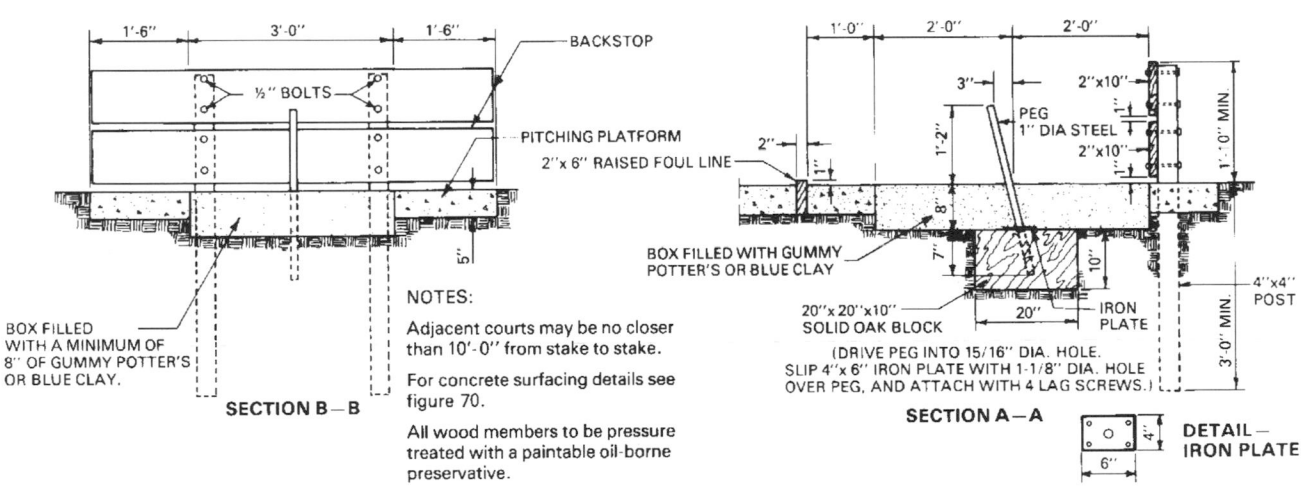

NOTES:

Adjacent courts may be no closer than 10'-0" from stake to stake.

For concrete surfacing details see figure 70.

All wood members to be pressure treated with a paintable oil-borne preservative.

Fig. 17 Horseshoe pitching court.

RECREATION AREAS

SWIMMING POOLS

Because of the promotion of swimming in high schools, colleges, and clubs, competitive swimming is receiving greater recognition throughout the United States. Junior championship meets require a pool at least 75 ft long and 42 ft wide. For senior championship meets, the ideal length is 55 yd long and 56 ft wide. The official short-course record requires a pool 75 ft in length and the long-course requires a pool 165 ft long. To fully meet these requirements it is wise to build a pool 1 in longer than the specified length, as the official distance cannot be even a fraction of an inch short.

The width of the pool is governed by the number and width of swimming lanes desired. The competitive pool should never have fewer than six lanes. The width of the lane represents the full spread of the arm in the breast stroke, and widths of 6, 7, and 8 ft have been designated as satisfactory. Lanes 5 ft wide are not recommended.

The slope of the pool floor should be as follows: No sudden changes of slope should be permitted in the area where the water is less than 5 ft deep. For depths less than 6 ft, the slope should be 1 ft in each 15 ft, and from the deep area in front of the diving board, the slope must not exceed 1 ft in each 15 ft.

Standards for the depth of swimming pools have been established by the Amateur Athletic Union (AAU), and these must be adhered to in connection with competitive swimming sanctioned by the organization. For the 75-ft pool, the shallow end should be at least 3 ft deep and slope not more than 1 ft each 15 ft and a minimum of 10 ft and preferably 12 ft under the diving boards. When high diving platforms (10 m) are used, the depth must be increased to 15 ft, maintaining other dimensions given. The National Collegiate Athletic Association (NCAA) recommends 3½ ft of water in the shallow end of the pool.

In any public pool the depth in the shallow end should never be more than 3 ft. For all commercial, municipal, and recreational pools, 80 to 85 percent of the total water area should be 5 ft or less in depth.

In planning for swimming pools, three factors must be considered. First consideration is the types of swimming pools. Second consideration is the location of the pools, and third, the size of the pools.

1. The three types of pools are:
    a. Wading pools for young children. The maximum depth of water in a wading pool is recommended not to exceed 24 in. Sufficient deck or circulation space for parents should be provided around the pool. this type of pool should be separated from a regular swimming pool because of safety and consideration for older swimmers.
    b. General swimming pool for older children and adults. The pool may be of any size or shape, but most frequently it is rectangular. The pool should be a minimum of 250 ft^2 with at least 25 ft^2 of water surface per bather at time of maximum load. The depth of the pool will vary from about 3 ft deep to about 10 ft deep. The deep end usually is used for diving. Not more than one-third of the pool area should have a water depth exceeding 5 ft.
    When only one pool is possible, the general pool serves both the swimming and diving functions. The shape of the pool should be such that there is clear distinction between the two activities.
    c. Diving pools are used for diving and deep-water activities. Whenever possible, this pool should be separated from the general swimming pool. The minimum depth of water in diving areas with a board 1 m or less above water should be 8 ft 6 in. If the board is over 1 m, but not over 3 m above the water, the depth should be 10 ft.

2. The location of pools should be such that they will receive a maximum of sunlight throughout the day. Avoid shadows from buildings and trees. The pools should have a southerly exposure. Avoid locating the pools too close to any dwelling units. An adequate buffer for noise and visual separation is essential. Avoid a location too close to the nearest lot lines.

3. The size of the pools should be sufficient to comfortably accommodate all swimmers at the peak time. After determining the maximum number of swimmers, allow at least 25 ft^2 per person.

Accessory facilities that are essential include a bathhouse, deck area for sunbathers, and an area for games, parking, and storage. If possible, the pool can be designed for winter activities such as ice skating and hockey.

PITCH-AND-PUTT GOLF

An undulating or steeply sloping area of land unsuitable for ordinary games can often be transformed into an excellent pitch-and-putt or approach course. When well laid out and maintained, this facility often proves extremely popular, particularly in densely populated urban areas. It is also one of the few games where the income can more than offset the cost of maintenance and management.

Space requirements vary according to the design of the course. Assuming, however, that only a mashie, niblick, and putter are to be used and the holes will not be less than 35 yd in length or more than 90 yd, 10 acres should be sufficient to provide 18 holes or half this area for 9 holes. The minimum width of any part of the course should not be less than 70 yd and greens should be from 350 to 500 yd^2 in area. There should be a clear putting space of not less than 10 yd in any direction around the hole.

There are two general types of golf courses in use today—9-hole and 18-hole.

The 18-hole course is the standard layout. The 9-hole course is a short course with the fairways and greens smaller than those of a regulation course but similar in every other way.

The average length of an 18-hole course is 6500 yd, while a 9-hole course is less than half of this lenght.

GOLF COURSES

The use of a golf course as the main feature of the open-space area in cluster or planned unit developments is becoming increasingly common. golf courses are attractive because they usually present a finely manicured, permanently green, well-landscaped vista. The course can meander over rolling terrain and fit into odd shapes, incorporating ponds and rock outcroppings.

A full 18-hole course requires from 160 to 200 acres of land. A 9-hole course needs about 100 acres. A par-3 course occupies about 30 acres. One drawback is that golf courses are expensive to build and to maintain. A golf course's main asset, from a developer's standpoint, is that land values adjacent to the fairways are substantially increased. More important, from a planning standpoint, is that permanent open space is created, which enhances the aesthetic atmosphere and physical attractiveness.

The total course length ranges from 5000 to 6000 yd. One fairway requires a 100-yd width. One method used to conserve land is to double the fairways, that is, use two parallel fairways. This requires a width of about 150 yd but reduces the total length. Another disadvantage in the use of a golf course is the relatively small numbers of people who can utilize it at any one time. Golf courses should be restricted to large developments where a range of other recreational facilities are available.

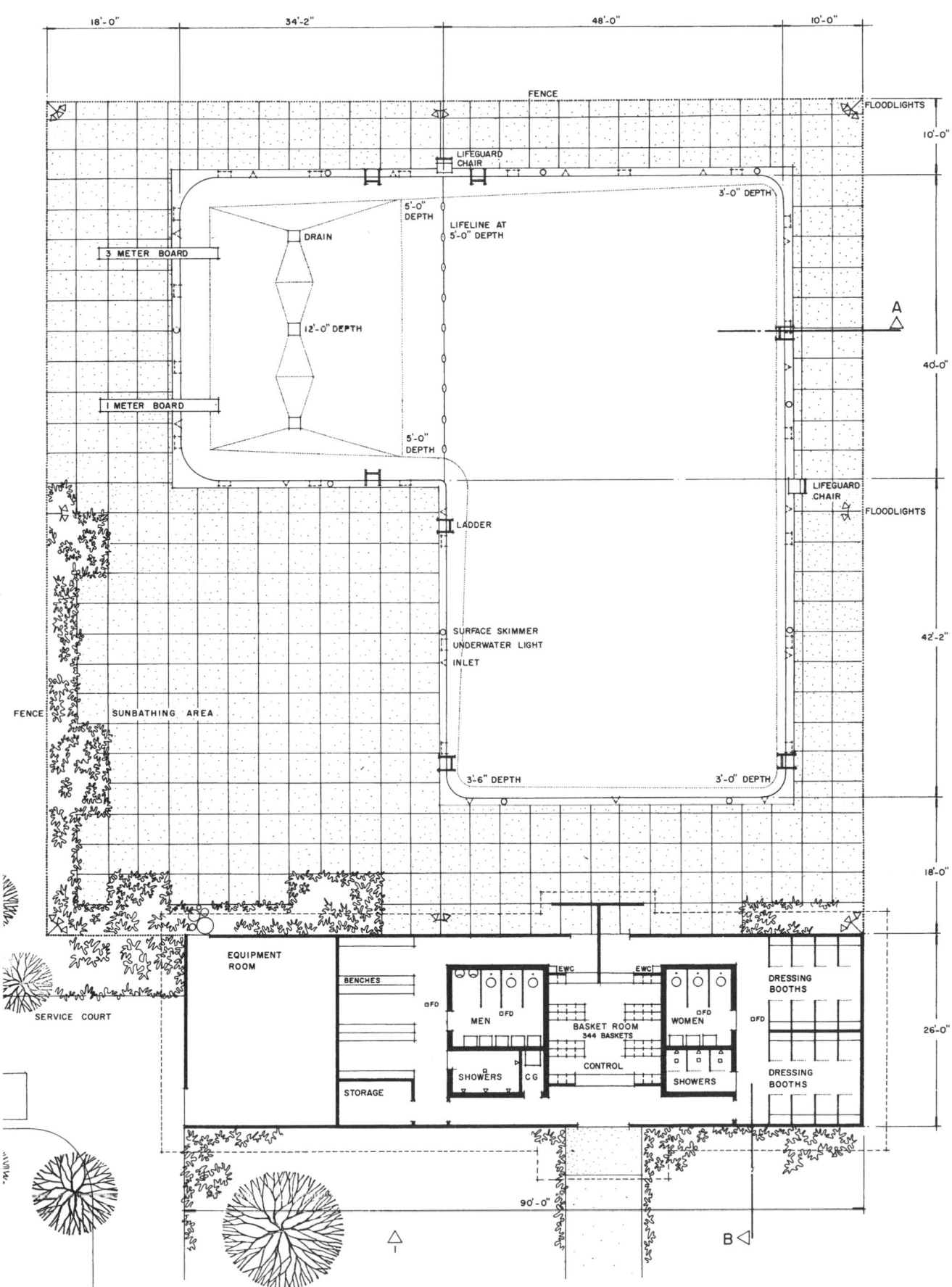

Fig. 18   A 25-m recreational swimming pool, 340 bathers.

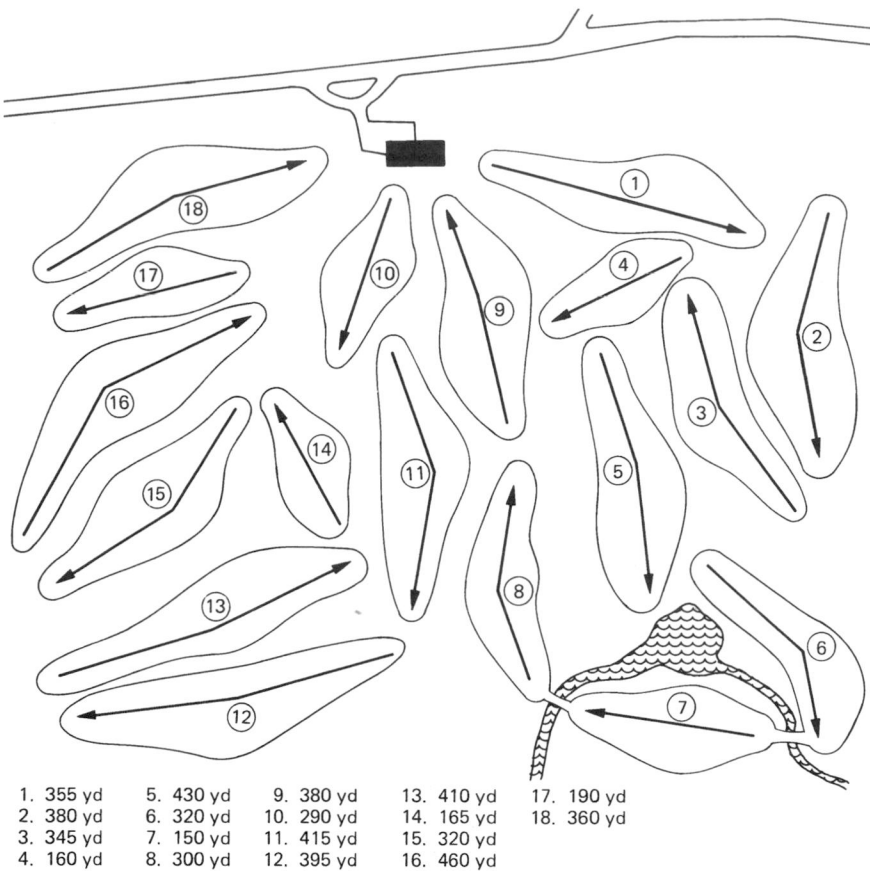

1. 355 yd   5. 430 yd   9. 380 yd   13. 410 yd   17. 190 yd
2. 380 yd   6. 320 yd   10. 290 yd   14. 165 yd   18. 360 yd
3. 345 yd   7. 150 yd   11. 415 yd   15. 320 yd
4. 160 yd   8. 300 yd   12. 395 yd   16. 460 yd

Fig. 19   Typical layout of an 18-hole golf course.

Four Basic Types of Golf Courses

The *par-3,* or pitch-and-putt course, has 18 short holes and can be built on about 45 to 50 acres. Most golfers do not consider this real golf.

The *nine-hole* course, with full-length fairways, can be built on 50 to 80 acres. It is real golf, but many golfers would prefer a regulation 18-hole course.

The *executive* course is the shortest course on which what is considered real 18-hole golf can be played. It is often used in retirement communities and takes up from 70 to 100 acres.

A *regulation* 18-hole golf course requires anywhere from 100 to 200 acres, depending on its length. If it plays more than 7000 yards from its back tees, it can legitimately be termed a *championship* course.

Most developers put in regulation courses; executive courses rank second. The selection of the latter usually results either from a special market, as noted above, or from a land squeeze; there are no great cost savings because there are still 18 tees and greens, the most expensive parts of a course to build.

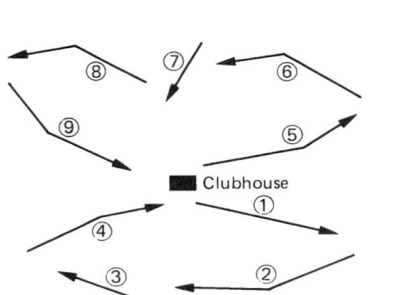

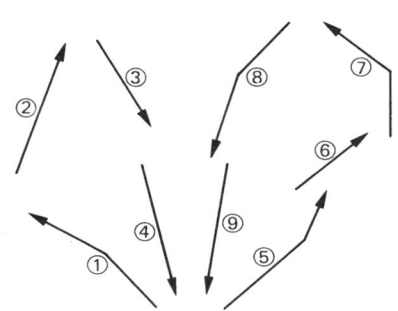

Fig. 20   Typical layout of a nine-hole golf course.

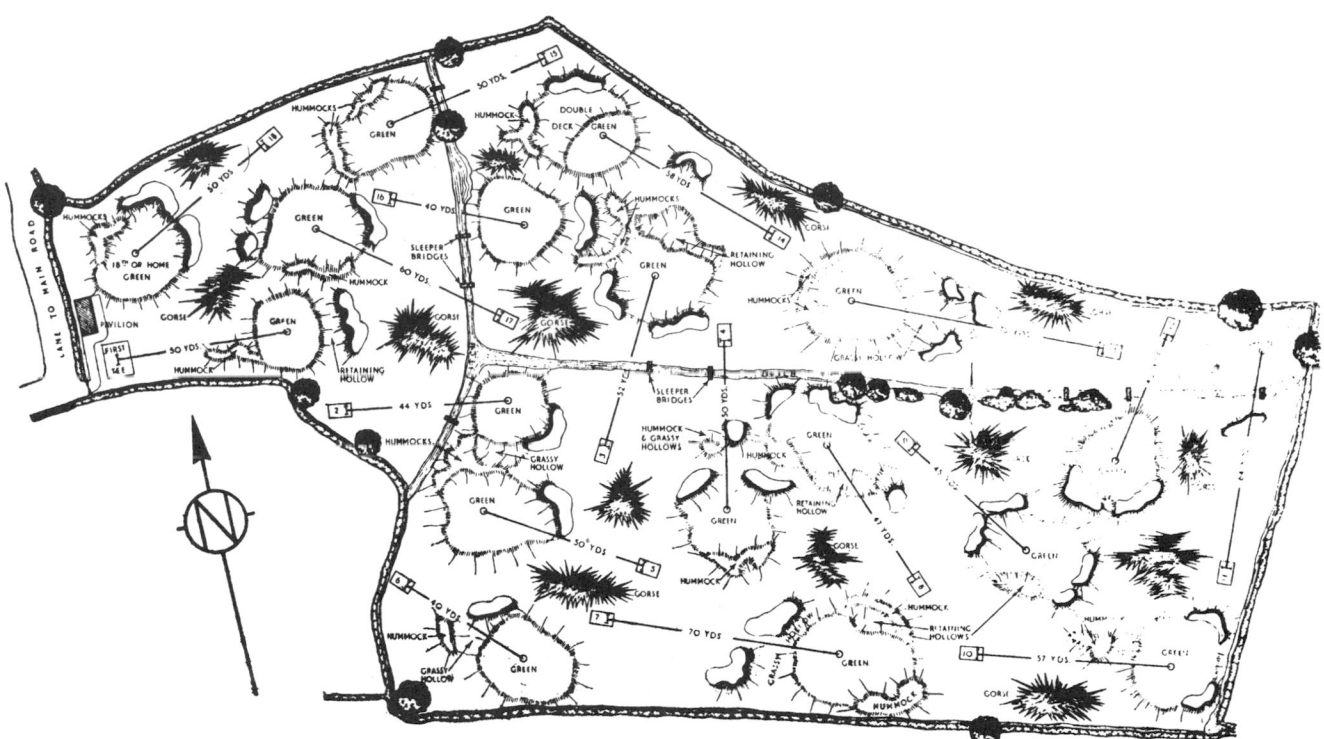

Fig. 21   A 13-acre 18-hole pitch-and-putt golf course.

Fig. 22   Par-3 golf course.

TABLE 4

	Min. area required	Max. area required	No. of parking spaces	Population served	Service radius	Average length
9-hole course	60 acres	80 acres	100 cars	One hole per 3000 or 27,000 persons	½–¾ h by car or public transportation	Approx. 2250 yd
18-hole course	120 acres	160 acres	200 cars	One hole per 1500 or 25,000 persons	1 h maximum by car or public transportation	6500 yd

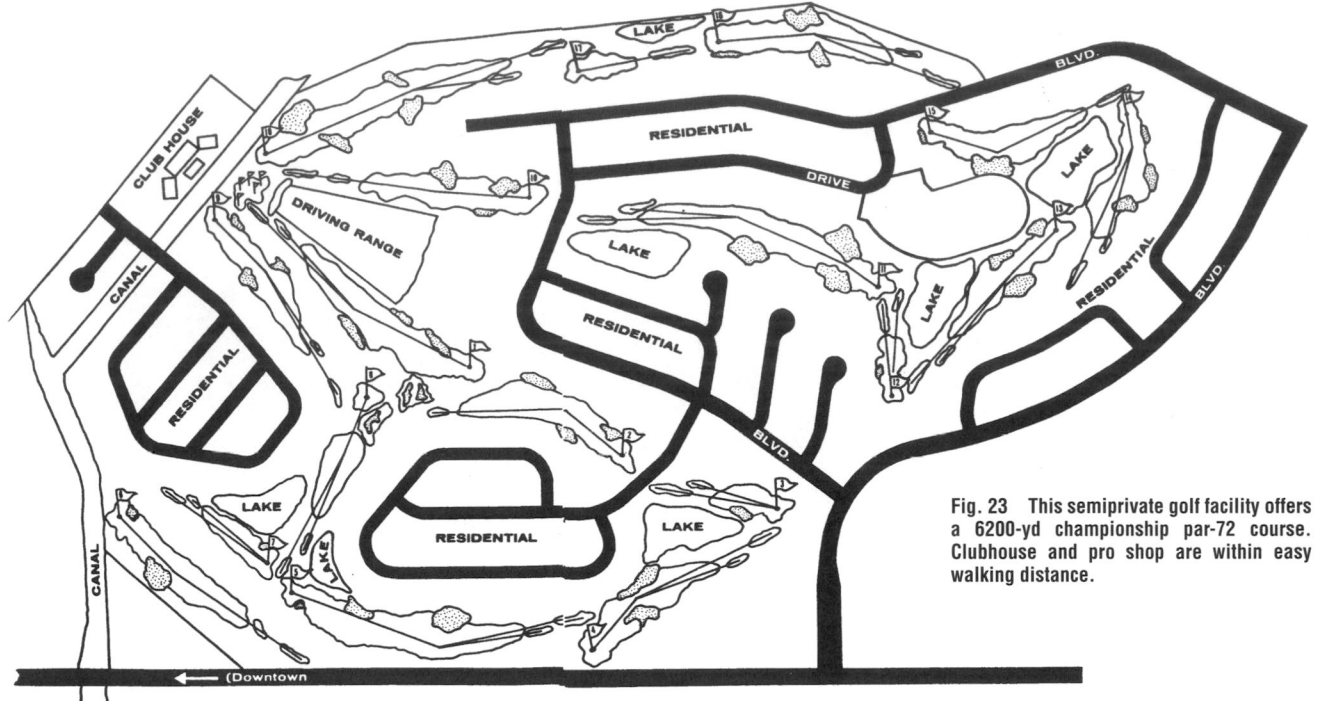

Fig. 23 This semiprivate golf facility offers a 6200-yd championship par-72 course. Clubhouse and pro shop are within easy walking distance.

Fig. 24 Condominiums overlook the fairways of the golf course. The buildings include both townhouse and apartment condominiums. The golf-course housing is worked in and around the course, yet fairways are grouped so as to provide long views in many directions.

SUMMARY OF RECREATIONAL FACILITIES

TABLE 5

Designation	Description and function	Facilities and equipment	Maximum distance served	No. of families served
Playlot	Generally a small area set aside for the play activity of young children under immediate adult supervision; usually located a short walking distance from the dwelling unit.	Generally includes children's slides, seesaw, sandboxes, climbing devices, swings. Drinking fountains and benches.	⅛ mile	10–20
Playground	An area set aside for the play activity of children and young adults with some adult supervision; may include some quiet areas usually located a reasonable walking distance from the dwelling unit.	Generally includes child's and adult' slides, seesaw, sandboxes, climbing devices, swings, tables for games. Drinking fountains and benches.	½ mile	40–100
Playfield	A large area set aside for active play and sports with the possibility of many persons participating at one time; usually containing spectator and preparation areas. May be more than 1 mile from the dwelling unit.	May include baseball and softball diamonds, football fields, tennis courts, running tracks. Parking lot. Drinking fountains; spectator seating	1 mile walking ½ h riding	2000–4000
Swimming pool	An area set aside for swimming and other low-scale water sports of all age groups with the possibility of spectators; may be more than 1 mile from the dwelling unit.	May include pools, shower rooms, diving boards. Parking lot. Drinking fountains; spectator seating	1 mile walking ½ h riding	2000–4000
Golf course	An area set aside specifically for the sport of golf.	May include up to 18-hole course, club house, driving range. Parking lot. Drinking fountains.	1 mile walking ½ h riding	2000–4000
Waterfront development	An area adjacent to a body of water set aside for pedestrians.	May include quiet sitting areas, bicycle paths, shops, restaurants.	1 mile	
Square, mall, plaza	An area, usually surrounded by buildings, set aside for pedestrians.	May include quiet sitting areas, bicycle paths, shops, restaurants.	¼ mile	
Park	An area, often well landscaped, set aside for various levels of recreation ranging from quiet sitting to fairly active play; several of these may exist within proximity of each other.	May include quiet sitting areas, bicycle paths, some play areas, drinking fountains, water areas.	½ mile	40–100

STANDARDS FOR PARK, RECREATION, AND OPEN SPACE

TABLE 6   Standards for Open Space and Recreation Areas

Area	Spatial standards, acres per 1000 resident population	Site size standards, acres	Service area radius, miles
Submetropolitan			
Minipark		1 or less	0.25
Neighborhood park	1.25	5*	0.5
Community park	1.50	20†	3.0
Other areas	1.75		
Total	4.50		
Metropolitan park		500‡	

*Minimum—2 additional acres for passive activity desirable.
†Minimum—10 additional acres for passive activity desirable.
‡Minimum.
SOURCE: Dade County, Florida.

Tables 6 and 7 are a summary of proposed standards with the suggested components for each type of recreation complex. At the neighborhood level, the individual components which comprise each major element, i.e., playground, playfield, recreation center building, and passive area, are shown.

RECREATION AREAS

TABLE 7   Park and Recreation Standards

Facility	Service distance	Population served (in thousands)	Acres per 1000 population

Neighborhood recreation complex
  Playground—play area and paved
  game courts
  Playfield—sportsfield and parking area
  Recreation center building—building,
  playlot, senior citizen and crafts area
  Passive—picnicking and park area
	¼ to ½ mile*	8–12	1.0

Community recreation complex
  Community center building, playlot,
  play area, paved game courts,
  sportsfield, swimming pool, skating-
  dancing circle, special events area,
  senior citizens, passive park and
  picnicking, and parking
	1 mile	Varies	0.5–1.0

Citywide recreation and parks
  Recreation park, amusement center
  (zoo, aquarium, etc.), sports center,
  parks and recreation administration
  center
	5 miles	Varies	5.0

*Service radius varies according to population density as follows: 49 persons per acre or less—½ mile (playfields 1 mile); 50 persons per acre or more—¼ mile (playfields ½ mile).
SOURCE: Baltimore, Maryland.

TABLE 8   Standards for Recreation Facilities

Facility	User group age	Spatial facility	Standards Per no. of persons in user group	Minimum facility	Facility and acres Within park total acres	Outside park total acres	Service area*	Miscellaneous
Lighted league play ball diamonds[†]	10–40	1	6,000	1	2.2	4.5	2–3 miles	Adjacent to high school, buffered from residential property
Lighted regulation tennis courts	12–64	1	4,000	6	3.5	4.5	2–3 miles	
Lighted shuffleboard courts	60+	1	1,000	12	0.5	1.0		Provide sheltered sitting area adjacent to courts; near public transit and major arterials
Regulation basketball courts	12–19	1	500	2	0.4	0.6	0.5 mile	
Swimming pools	6–15	1800 ft^2	1,000	1	0.6	0.8	1–1.5 miles	Optimum pool size 5000 ft^2; shallow wading areas separate from diving pool
Play apparatus areas	3–12	1	500	1	0.2	0.3	0.5 mile	Includes sitting areas, playground equipment for children, small open areas for free play
Golf courses (private included)[‡]	All ages	18 holes	50,000	1	110–150	110–150	30 min driving time	Rectangle most desirable shape extending north to south, gently rolling terrain preferred
Picnic grounds	All ages	1 acre	6,000				30 min maximum driving time	Located in natural beauty areas preferably adjacent to bodies of water, well-shaded, well-buffered from surrounding conflicting uses
Beaches[‡]	All ages	0.2 acre sand and 50 ft shoreline	1,000	1	2.5	2.5	30 min maximum driving time	
Boat ramps	All ages	2 lin ft	1,000	200 ft	6.5	8		Minimum 0.6 parking space per lin ft of ramp

*Includes acreage for auxiliary facilities.
[†]Primarily 60- to 75-ft combination diamonds.
[‡]Peak season tourist population as well as resident population.
SOURCE: Dade County, Florida.

TABLE 9   Park Area Standards

Neighborhood playlot	
Size	At least 1 acre preferred
Service area	¼ mile or less
Location	High-density neighborhoods where typical private yards do not exist
Usual facilities	Paved area, playground apparatus area for small children; usually private responsibility.
Neighborhood park	
Area per 1000 persons	3 acres
Size	5–10 acres, not including parking
Service area	½ mile or as limited by geographical barriers
Location	Preferably adjacent to elementary schools or near the center of the neighborhood
Usual facilities	Softball/baseball fields, multiple-use paved areas, playground apparatus areas, landscaped areas, and picnic areas. Small fieldhouse. Minimum of automobile parking.
District park	
Area per 1000 persons	3 acres
Size	10–30 acres
Service area	1 mile
Location	Preferably near the center of 4 or 5 neighborhoods
Usual facilities	Facilities of neighborhood park, tennis courts, football and soccer fields, lighting for evening use, community center/recreation buildings and swimming pool. Substantial parking areas. Skating rinks, sledding hill, and natural areas.

SOURCE: Elmhurst, Illinois.

TABLE 10   Area Standards

Type of facility	Minimum size	Maximum size	Acres per 1000*
Neighborhood			
Neighborhood park	5 acres	10 acres	2 per 1000
Neighborhood center			
Subcommunity			
Playfield	20 acres	30 acres	1.5 per 1000
Subcommunity center			
Community			
Area parks	150 acres	Variable	6.5 per 1000
Golf course—18 holes	140 acres	200 acres	18 holes per 50,000 population
Passive recreation area			Varies
Metropolitan			
Metropolitan park	1000 acres	Variable	10 per 1000
Greenbelt and open space			Varies
Special facilities			Varies

*The number of acres of parks required per 1000 population will vary in each neighborhood and community; the figures presented here are the composite base standards for Oklahoma City.
SOURCE: Oklahoma City, Oklahoma.

TABLE 11   Area and Quantity Standards

Public Open Space		
Percent of site for public open space including schools and churches	10%	CBH
Area of major park for each 100 population	3 acres	NRPA
Area of community park for each 100 population	1 acre	NRPA
Area of playfield for each 100 population	1–2 acres	NRPA
Area of neighborhood park for each 100 population	1 acre	NRPA
Area of neighborhood park per 1000 persons for one- and two-family neighborhood of		
1000 population	1.5 acres	NRPA
2000 population	1.0 acre	NRPA
3000 population	0.83 acre	NRPA
4000 population	0.75 acre	NRPA
5000 population	0.70 acre	NRPA
Area of neighborhood park per 1000 persons for multifamily neighborhood of		
1000 population	2.0 acres	APHA
2000 population	1.5 acres	APHA
3000 population	1.33 acres	NPHA
4000 population	1.25 acres	NPHA
5000 population	1.20 acres	NPHA
Area of neighborhood playground per 1000 persons for neighborhood of		
1000 population	2.75 acres	NPHA
2000 population	1.63 acres	NPHA
3000 population	1.33 acres	NPHA
4000 population	1.25 acres	NPHA
5000 population	1.20 acres	NPHA
Area of neighborhood park per family for one- or two-family neighborhood of		
1000 population	238 ft^2	NPHA
2000 population	158 ft^2	NPHA
3000 population	132 ft^2	NPHA
4000 population	119 ft^2	NPHA
5000 population	111 ft^2	
Area of neighborhood park per family for multifamily neighborhood of		
1000 population	318 ft^2	NPHA
2000 population	238 ft^2	NPHA
3000 population	208 ft^2	APHA
4000 population	198 ft^2	NPHA
5000 population	190 ft^2	NPHA
Area of neighborhood playground per family for neighborhood of		
1000 population	435 ft^2	NPHA
2000 population	258 ft^2	NPHA
3000 population	208 ft^2	NPHA
4000 population	198 ft^2	NPHA
5000 population	190 ft^2	NPHA
Minimum size for communal open space for nursery, school age children, and adults	3000 ft^2	FHA
For each additional family over 100 families	30 ft^2	FHA

TABLE 11   Area and Quantity Standards (*Continued*)

Infant Recreation		
Preschool children's play area per 100 dwellings	2500 ft^2	FHA
Preschool children's play area per bedroom	10 ft^2	PHA
Play areas (2–6 years) per child	50 ft^2	APHA
Play areas (2–6 years) per family	12–15 ft^2	APHA
Area of tot lot for each 3–800 population	1 acre	NRPA
Play space for 80 persons per acre schemes for each bed space	15–20 ft^2	MOHLG
Play space for 80 persons per acre schemes for each bed space	20–25 ft^2	PM

Children's Recreation		
Minimum size for playground areas for 2000 population	3.25 acres	CBH
Minimum size for playground areas for 3000 population	4.00 acres	CBH
Minimum size for playground areas for 4000 population	5.00 acres	CBH
Minimum size for playground areas for 5000 population	6.00 acres	CBH
Maximum for each bedroom for school age and adult recreation	2000 ft^2	PHA
For projects over 100 bedrooms add	50 ft^2	PHA
Minimum play area for 30 children	1500 ft^2	APHA
Maximum size of play area for 100 children	5000 ft^2	APHA
Play space for 80 persons per acre schemes for each bed space	15–20 ft^2	MOHLG
Play space for 80 persons per acre schemes for each bed space	20–25 ft^2	PM
Minimum size for neighborhood play lot	2500 ft^2	CBH

Adult Recreation		
Maximum for each bedroom for school age and adult recreation	2000 ft^2	PHA
For projects over 100 bedrooms add	50 ft^2	PHA

Private Open Space		
Area of land per family for outdoor living for two-story multifamily dwellings	415 ft^2	APHA
Area of land per family for outdoor living for three-story multifamily dwellings	315 ft^2	APHA
Area of land per family for outdoor living for 6-, 9-, and 13-story multifamily dwellings	215 ft^2	APHA

Roads		
Percent of site for streets	20%	CBH

Car Parking		
Area of off-street parking per family in two-, three-, six-, and nine-story multifamily schemes	160 ft^2	APHA
Area of off-street parking per family in 13-story multifamily schemes	125 ft^2	APHA
Number of car spaces per dwelling in single-family schemes	2	CBH
Number of car spaces per dwelling in multifamily schemes	1½	

Residential Space		
Maximum percent of site for residential use	35%	CBH

Detached Dwellings		
Area of land for single-family detached dwelling including outdoor space, service, and parking	6000 ft^2	APHA

TABLE 11 Area and Quantity Standards (*Continued*)

Attached Dwellings		
Area of land per family for single-family semidetached or two-family detached dwellings including outdoor space, service, and parking	4000 ft²	APHA
Area of land per family for single-family attached (row) or two-family semidetached dwellings including outdoor space, service, and parking	2400 ft²	APHA
Area of land per family covered by buildings for two-story multifamily dwellings	435 ft²	APHA
Area of land per family for multifamily two-story dwellings including outdoor space, service, and parking	1465 ft²	APHA
Walk-Up Apartments		
Area of land per family for multifamily three-story dwellings including outdoor space, service, and parking	985 ft²	APHA
Area of land per family covered by buildings for three-story multifamily dwellings	290 ft²	APHA
Elevator Apartments		
Area of land per family covered by buildings for six-story multifamily dwellings	145 ft²	APHA
Area of land per family covered by buildings for nine-story multifamily dwellings	105 ft²	APHA
Area of land per family covered by buildings for 13-story multifamily dwellings	75 ft²	APHA
Area of land per family for multifamily six-story dwellings including outdoor space, service, and parking	570 ft²	APHA
Area of land per family for multifamily nine-story dwellings including outdoor space, service, and parking	515 ft²	APHA
Area of land per family for multifamily 13-story dwellings including outdoor space, service, and parking	450 ft²	APHA
Nonresidential Built Space		
Percent of site for commercial use	5%	CBH
Minimum area for community rooms	1000 ft²	PSA
For projects with 100–300 units add	450 ft²	PSA
For projects with 300–500 units add	750 ft²	PSA
For projects with 500–1000 units add	900 ft²	PSA
Area of community room per family	4.5 ft²	PHA
Laundry facility for each 100 units	2500 ft²	FHA
Environmental Quality: Noise—Residential Space		
Maximum noise level in living areas of dwellings	50 db	APHA
Maximum noise level in study and sleeping areas of dwellings	30 db	APHA
Maximum noise level in proximity of bedrooms	40–50 db	APHA
Maximum noise level in proximity of any residential structure	50–60 db	APHA
Noise between dwellings should be no more than	45 db	PHA
Noise between dwellings should be no more than	50 db	FHA
Bedrooms should not be closer to group parking than	150 ft	HIIS

NOTE:
CBH: Community Builders Handbook.
NRPA: National Recreation and Park Association.
APHA: American Public Health Association.
FHA: Federal Housing Administration.
PHA: Public Housing Authority.
MOHLG: Ministry of Housing and Local Government (British).
PM: Parker Mains Report (MOHLG).
PSA: Play School Association.
HIIS: "Home in Its Setting," MOHLG.

TYPES OF SHOPS

Shopping and other commercial space may be necessary if the development is sufficiently large to support such facilities. This usually includes retail shops for convenient goods and the supply of basic services.

The location of the neighborhood shopping center is generally an arterial street at the intersection of a collector street. However, it may also be located more centrally within the neighborhood and closer to the other community facilities. Vehicular access for trucks is essential for deliveries of goods and other services. However, such vehicular access must not cross or interfere with pedestrian access to the shopping area.

The neighborhood shopping center generally includes 8 to 15 stores with an average gross floor area of about 40,000 ft². The site will vary from 1.5 to 4.0 acres, including parking area. A minimum of 800 to 1000 families in its trade area is needed to support this center.

Stores included in this category are stationery, laundry, bakery, hardware, service station, barber and beauty shops, small restaurants, drugstore, and a food market. Frequently, professional offices are included to provide medical and dental services.

In planning such facilities, it is most important that they are not larger than required by the development. This will either result in marginal business that cannot provide proper services or attract people from outside the development, causing undesired influx of people and cars.

Limited, but adequate, parking areas must be provided. The recommended standard for the amount of parking space to floor area of stores is a 2 to 1 ratio. That is, 2 ft² of parking space for every 1 ft² of sales area. Some standards suggest even more parking space, but this would negate the concept of pedestrian access to the center.

The location of the parking area should be between the stores and the street so that it will be away from the pedestrian access.

It is strongly recommended that the shopping center be in close proximity, not adjacent, to the school and play areas. This will encourage the multiple use of facilities and discourage the use of the automobile.

Two-story buildings may be utilized with medical, dental, or other services located there. The layout of store units should be as flexible as possible to make adjustments to meet changing community needs.

The entire complex needs sufficient buffer strips between shopping areas and adjacent residential uses. Such buffer strips for a small center should be at least 20 ft and have proper landscaping and fences. Also, it must be remembered that such shopping facilities must be clearly incidental to and compatible with the residential character of the property.

Food market: Should include specialty foods and delicatessen goods

Bakery shop: May be included in food market

Drugstore: Including reading matter, tobacco, and vanity goods

Restaurant: Including table service and take-out orders

Barber shop: Including shoeshine service

Beauty parlor: May be combined with barber shop

Laundry and dry-cleaning store: Combined service, including a laundromat

Hardware: Should include household goods

Service station: Including filling station, minor repairs, and auto accessories

NEIGHBORHOOD SHOPPING CENTERS

The typical neighborhood shopping center is a highly special merchandising entity. Far from being just a group of stores, it is a careful section of complementary stores, each making its own contribution to the drawing power of the group. Usually all are of the "service" type catering to people in their workaday shopping.

Its first appeal is the supermarket; its competitive advantage lies in its parking facilities.

Except for the very small center (Fig. 2a), which might serve a community of 250 to 300 families, a minimum of 500 families is considered necessary to support a center of from 10 to 12 shops, and 1000 to 3000 families to support a large one of from 25 to 40 shops.

With the supermarket are grouped other stores of the basic service type: a drugstore, dry cleaner, laundry agency, shoe repair shop, etc. As shopping centers grow in size, serving larger communities, other store types become practical (see plans of the larger developments).

The importance of highway approaches from other areas and transportation servicing the center should be stressed. Shopping centers often draw a sizable portion of their volume from areas outside what might be considered the normal or tributary area. This is particularly true if convenient parking is provided.

Adequate off-street parking should be an integral part of the shopping center, with good balance between front and rear parking. Best plan is to provide a moderate-sized parking area in front of the shops to take care of normal parking and a larger area to the sides and rear to accommodate automobiles during the peak period.

It has been found that if all parking is in front of the shops, store fronts must be an excessive distance from the street, presenting an unattractive paved area during off-peak periods. Conversely, if all parking is to the rear, the motorist is often discouraged in seeking parking space.

Under average conditions, 2 ft² of off-street parking space should be permanently reserved for each square foot of store area. Where the amount of pedestrian trade is expected to be relatively high because of adjacent multifamily developments or relatively low income groups, the ratio may be lowered somewhat. Where the drive-in trade will form the bulk of patronage, ratios up to 3 to 1 may be required for adequate parking.

TABLE 1   Neighborhood Shopping Center Size

Population served	Floor area required (sales area)	Customer parking area 2 to 1 ratio	Circulation, service, and planting areas, 25%	Total square feet	Total acres required	Square feet per family (gross)	Maximum walking distance
800 families, 2500 persons	20 ft² per family = 16,000 ft²	32,000 ft²	12,000 ft²	60,000	1.4	75	¼ mile
1600 families, 5000 persons	18 ft² per family = 28,800 ft²	57,600 ft²	36,400 ft²	172,800	4.0	100	½ mile

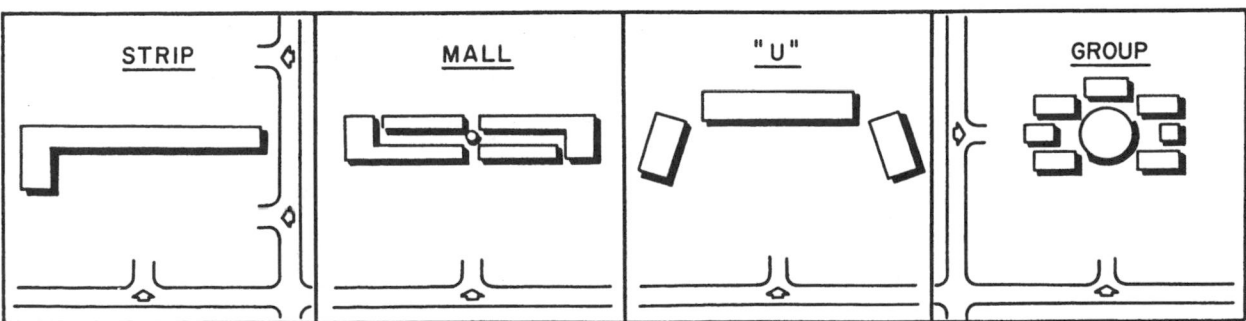

Fig. 1   There are four basic shopping center patterns. The strip and court forms are best for smaller centers.

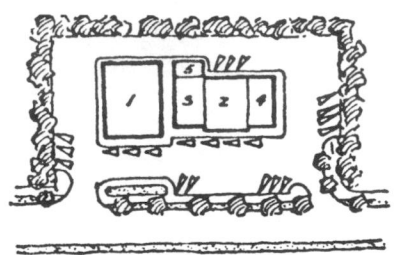

Store Group for 250 to 300 Families

(1) Grocery store, (2) drugs, (3) beauty and barber shop, (4) cleaning, laundry, and shoe repair, (5) service, heating, etc.

Fig. 2*a*

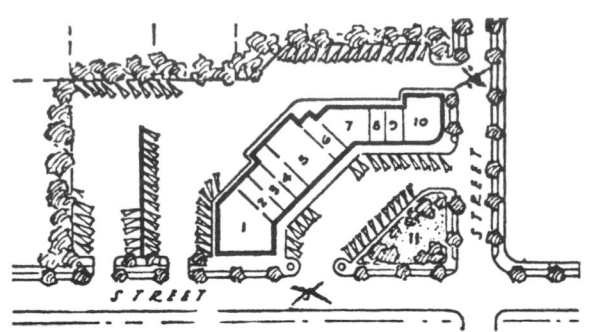

Store Group for 300 to 500 Families

(1) market, (2) cleaner and dyer, (3) ladies wear, (4) beauty shop, (5) variety store, (6) bakery, (7) delicatessen, (8) shoe repair & laundry, (9) barber shop, (10) drugs, (11) possible gas station

Fig. 2*b*

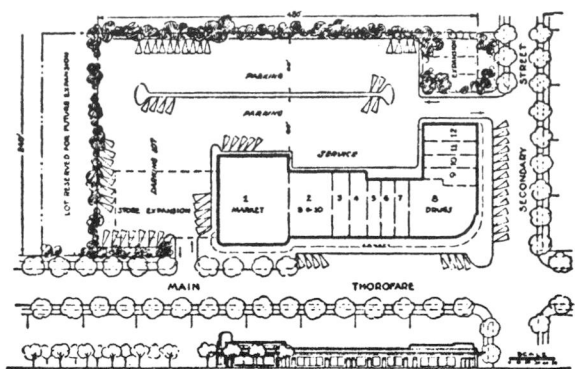

Shopping Center for 500 to 700 Families

(1) market, (2) variety store, (3) delicatessen, (4) ladies wear, (5) beauty shop, (6) florist, (7) gift shop, (8) drug store, (9) cleaner, (10) radio shop, (11) shoe repair, and (12) hardware

Fig. 2*c*

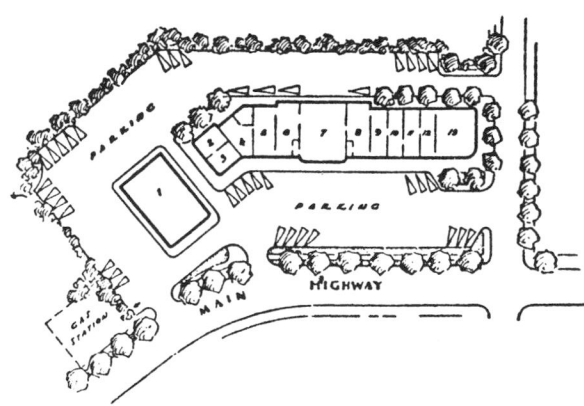

Shopping Center for 500 to 700 Families

(1) market, (2) shoe repair, (3) barber shop, (4) beauty shop, (5) ladies wear, (6) grocery store, (7) variety store, (8) delicatessen, (9) bakery, (10) cleaner and laundry, (11) radio and electrical shop, (12) florist, and (13) drug store

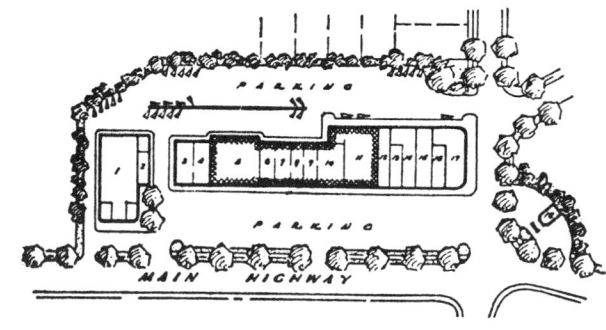

(Plan shows two stages of construction. The shaded portion is the initial installation.)

Shopping Center for 750 to 1000 Families

(1) theater, (2) barber shop, (3) radio and electrical, (4) liquor store, (5) supermarket, (6) gift shop, (7) beauty shop, (8) ladies wear, (9) haberdashery, (10) 5 & 10, (11) drug store, (12) shoe store, (13) cleaner and laundry, (14) bakery, (15) independent grocer, (16) books, (17) restaurant, and (18) gas station

Fig. 3*b*

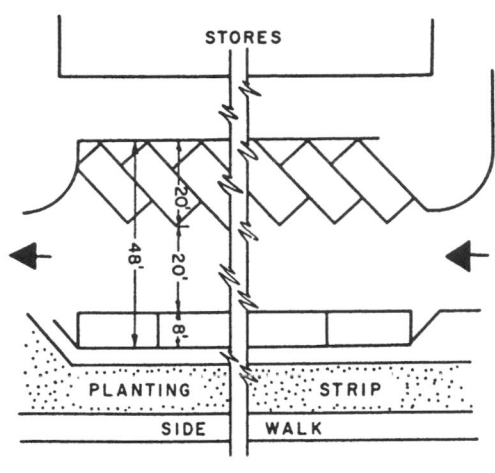

Parking in Restricted Widths

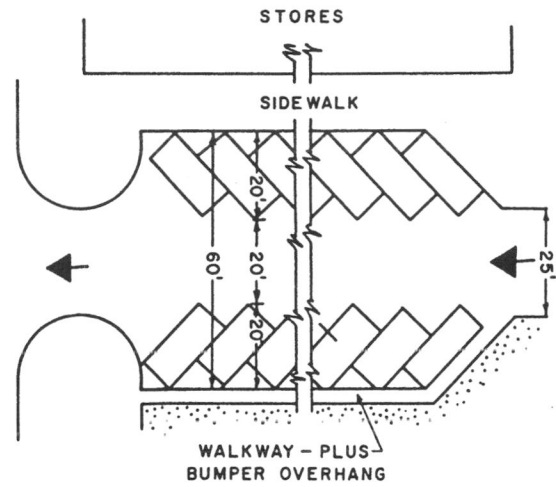

45 Degree Parking

45 degree stalls—width	8'-0"	9'-0"
No. of cars per 100 ft. of curb	9	8
Parallel stalls	20'-0"	20'-0"
No. of cars per 100 ft. of curb	5	5
No. of cars per acre excluding approaches	126	117

Width of car stalls	8'-0"	9'-0"
No. of cars per 100 ft. of curb	9	8
Curb occupied per car	11'-4"	12'-9"
No. of cars per acre excluding approaches	130	116

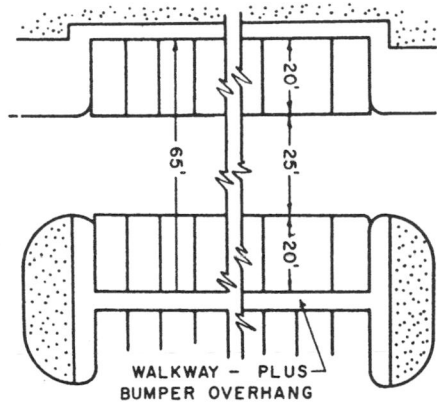

90 Degree - Right Angle Parking

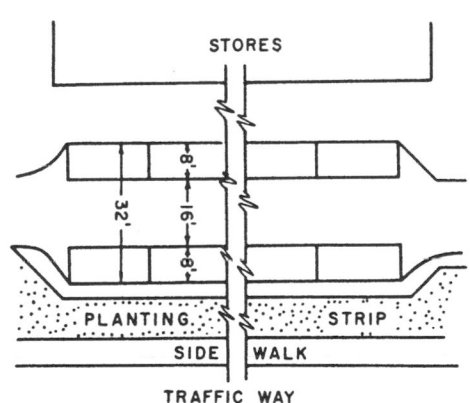

Parking in Narrow Widths

Width of car stalls	8'-0"	9'-0"
No. of cars per 100 ft. of curb	12	11
Curb occupied per car	8'-0"	9'-0"
No. of cars per acre excluding approaches	168	148

Note: Many developers feel the walkway between cars may be omitted.

Parallel stalls—width	20'-0"
No. of cars per 100 ft. of curb	5
No. of cars per acre excluding approaches	136

Note: Parallel parking space is seldom used as efficiently as is diagonal or right-angle parking with proper marking.

Fig. 4 Parking area layouts for the small shopping center.

NEIGHBORHOOD SHOPPING

The location of the neighborhood shopping center is generally located on the arterial street at the intersection of a collector street (see Fig. 5). Adequate parking in relationship to number of stores must be provided. The houses adjacent to the shopping center must be properly protected with planting or fences.

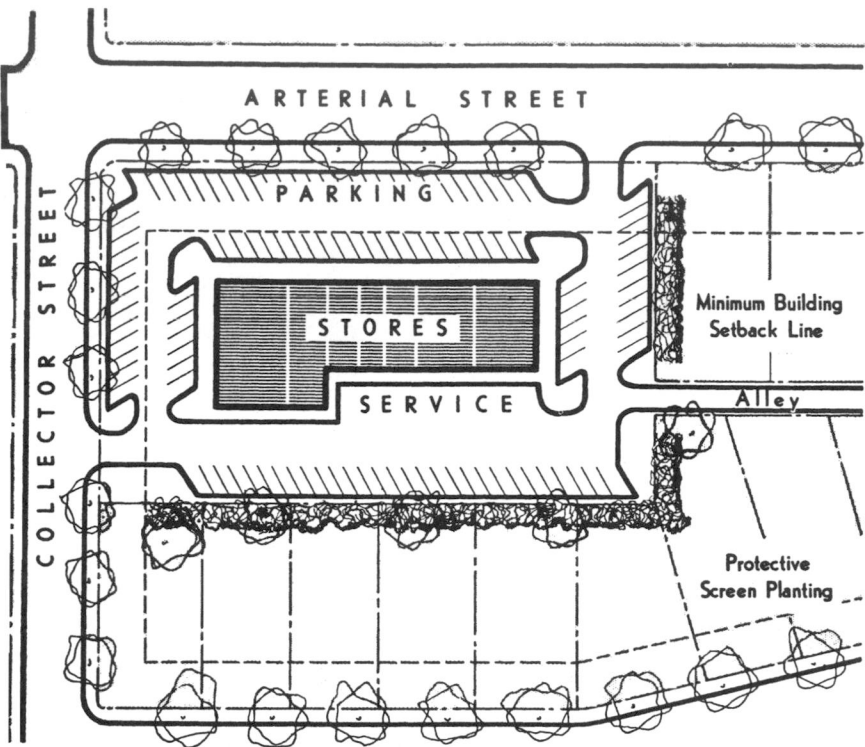

Fig. 5   Design principles applied to neighborhood shopping center.

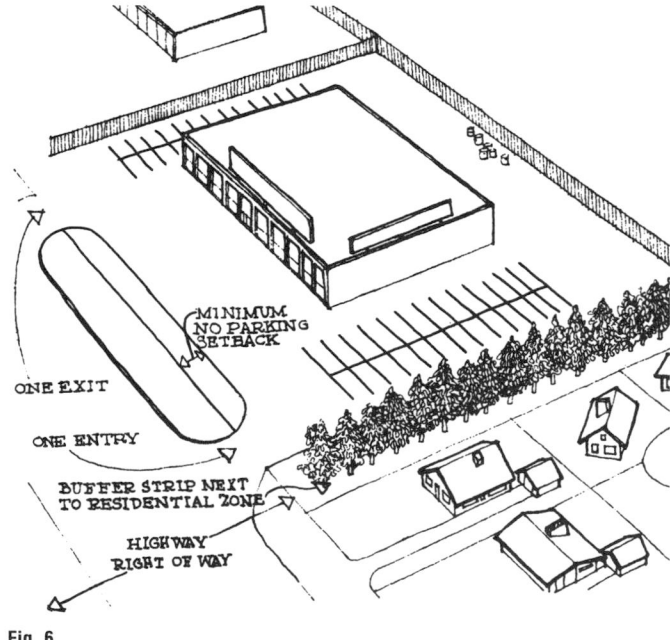

Fig. 6

The neighborhood center in Fig. 6 includes eight to fifteen stores with an average gross floor area of 40,000 ft^2 on a site of about four acres, including parking area. A minimum of 1000 families in its trade area is needed to support this center. Its principal tenants are the drugstore and supermarket, the rest including dry cleaner, beauty parlor, filling station, bakery, shoe repair, laundry agency, variety store, and barber shop.

In Fig. 7 the center's two major tenants, a bank and a supermarket, are located at each end of an L-shaped one-story building. In between are 13 retail shops.

In addition, there is a 500-seat movie theater, a 10,000-ft^2 medical building, 20,000 ft^2 of office space, and an 8000-ft^2 restaurant—all in free-standing buildings as shown on the site plan.

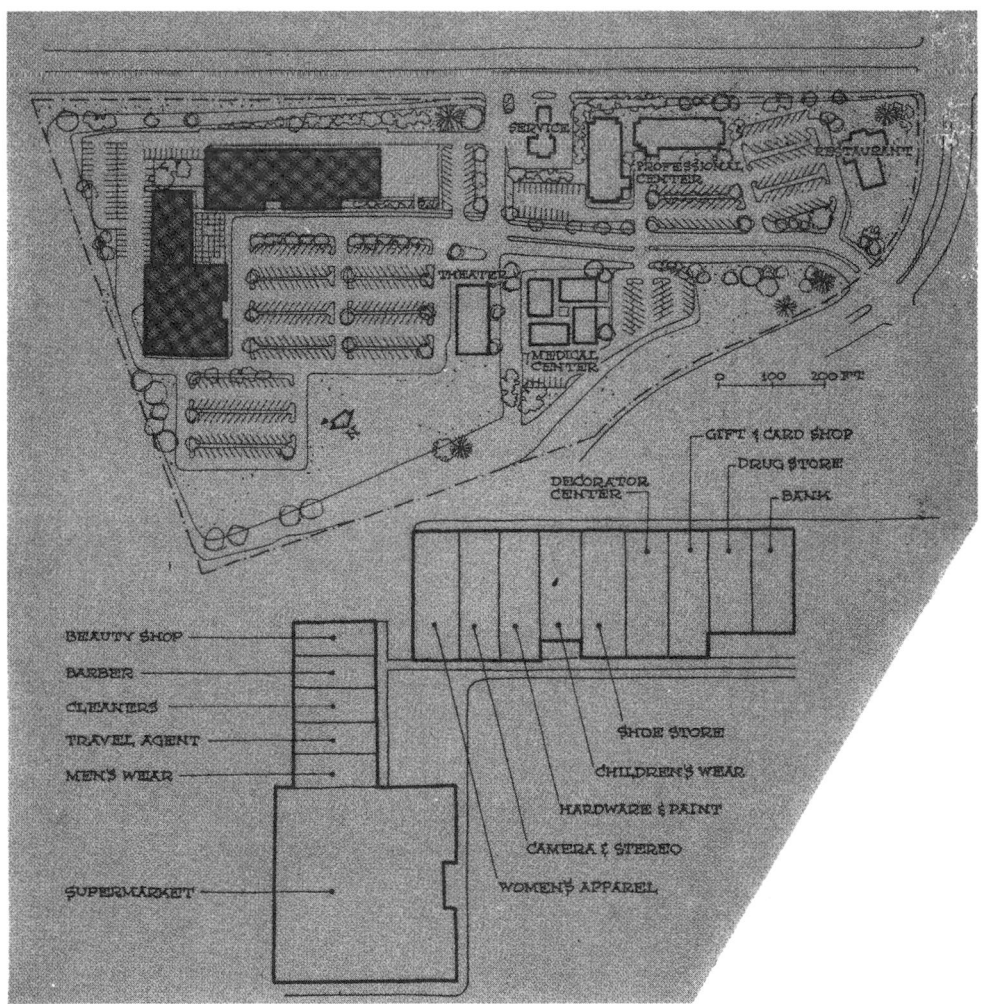

Fig. 7 Monach Bay Plaza, Laguna Miguel, California. Architects—Fernald, Nicol, Schiller.

LAND-AREA REQUIREMENTS FOR COMMUNITY FACILITIES

TABLE 1   Component Uses and Aggregate Area, by Type of Development and Population of Neighborhood

	Neighborhood population				
Type of development	1000 persons, 275 families	2000 persons, 550 families	3000 persons, 825 families	4000 persons, 1100 families	5000 persons, 1375 families
One- or two-family development*					
Area in component uses					
Acres in school site	1.20	1.20	1.50	1.80	2.20
Acres in playground	2.75	3.25	4.00	5.00	6.00
Acres in park	1.50	2.00	2.50	3.00	3.50
Acres in shopping center	0.80	1.20	2.20	2.60	3.00
Acres in general community facilities†	0.38	0.76	1.20	1.50	1.90
Aggregate area					
Acres: total	6.63	8.41	11.40	13.90	16.60
Acres per 1000 persons	6.63	4.20	3.80	3.47	3.32
Square feet per family	1,050	670	600	550	530
Multifamily development‡					
Area in component uses					
Acres in school site	1.20	1.20	1.50	1.80	2.20
Acres in playground	2.75	3.25	4.00	5.00	6.00
Acres in park	2.00	3.00	4.00	5.00	6.00
Acres in shopping center	0.80	1.20	2.20	2.60	3.00
Acres in general community facilities†	0.38	0.76	1.20	1.50	1.90
Aggregate area					
Acres: total	7.13	9.41	12.90	15.90	19.10
Acres per 1000 persons	7.13	4.70	4.30	3.97	3.82
Square feet per family	1130	745	680	630	610

*With private lot area of less than ¼ acre per family (for private lots of ¼ acre or more, park area may be omitted).

†Allowance for indoor social and cultural facilities (church, assembly hall, etc.) or separate health center, nursery school, etc., unallocated above. Need will vary locally.

‡Or other development predominantly without private yards.

COMMUNITY RECREATIONAL FACILITIES

CHILD	PROGRAM	TOT LOT	NURSERY/DAYCARE—PRIVATE	NURSERY/DAYCARE—PUBLIC	INTER. AGE PLAYGROUND—2nd CHOICE
MIN.	Minimum Size	1,500 SF. 25' min. dim.	600 SF.	600 SF.	2,500 SF. 35' min. dim.
	Minimum Area in Sunlight	25% for 2 hours	20% for 2 hours	20% for 2 hours	25% for 2 hours
LOCATION	within Private Outdoor Space	Yes	not applicable	not applicable	Yes
	within Private Open Space	Yes	not applicable	not applicable	Yes
	within Private Covered Space	15% max.	not applicable	not applicable	15% max.
	within Private Indoor Space	not applicable	Yes	Yes	not applicable
	within Semiprivate Outdoor Space	No	not applicable	not applicable	No
HEIGHT ABOVE & BELOW CURB	fronting on or within Private Outdoor Space	15' max. above At level of outdoor space	15' max. above at level of outdoor space	15' max. above at level of outdoor space	15' max. above at level of outdoor space
	fronting on or within Semiprivate Outdoor Space	not applicable	15' max. above 5' max. below	15' max. above 5' max. below	No
	within Private Indoor Space	not applicable	15' max. above at level of outdoor space	15' max. above at level of outdoor space	not applicable
	Height above and below nearest floor	± 5'	± 5'	± 5'	± 5'
ACCESS	Direct access from Private Outdoor Space	not applicable	Yes	No	not applicable
	Direct access from Lobby	Yes	Optional	No (fronting semiprivate space)	Yes
	Adjacent to Parking	Not within 30' min.	not applicable	not applicable	Not within 30' min.
WINDOWS	front on Private Outdoor Space	not applicable	Yes	Optional	not applicable
	front on Semiprivate or Public Outdoor Space	not applicable	Optional	Yes	not applicable
POLLUTION SOURCE	Air	35' min. horiz. at grade or +15' min. above grade	not applicable	not applicable	35' min. horiz. at grade or +15' min. above grade
	Noise	35' min. horiz. at grade or +15' min. above grade	No min.	No min.	35' min. horiz. at grade or +15' min. above grade
STANDARDS		1. for children 5-10 years, 2. equipment to be both kinetic—swings, slides, seesaws, etc., and static—sandboxes, wading pools, climbing apparatus, running and bike spaces, 3. benches for sitting at rate of 1/500 SF., 4. areas around equipment to not be surfaced with resilient material—rubber, elastaturf, sand, grass, 5. grade differences between lobby & facility to be accomplished by ramps, 6. when adjacent to intermediate playground, facilities should not be mixed	1. same elevation as tot lot ±2', 2. grade changes by ramp, 3. child's bathroom and adult bathroom, 4. long dimension of room on exterior wall, 5. for ages 1-5 years, 6. exterior wall on private outdoor space, min. 75% transparent	1-5—see NURSERY (private) 6. exterior wall on semiprivate and/or public open space to be 50% transparent	1. for children 5-10 years, 2. equipment to be both kinetic—swings, seesaws, merry-go-round, etc., and static—climbing apparatus, running and bike space, 3. benches for sitting at the rate of 1/800 SF., 4. areas around equipment to be surfaced with resilient material, 5. grade differences between lobby and facility to be accomplished by ramps, 6. when adjacent to Tot Lot, facilities should not be mixed

Fig. 1

COMMUNITY RECREATIONAL FACILITIES

	PROGRAM	MEETING/SOCIAL ROOM—PRIVATE	VOLLEYBALL	BASKETBALL	SWIMMING POOL	HANDBALL
MIN.	Minimum Size	600 SF.	Single court 2,830 SF.	1,550 SF. min. for half court	800 SF.	Single court 2,250 SF.
	Minimum Area in Sunlight	No min.	No min.	No min.	When outdoors 75% + 4 hours summer solstice	No min.
LOCATION	within Private *Outdoor* Space	not applicable	Yes	Yes	Yes	Yes
	within Private *Open* Space	not applicable	Yes	Yes	Yes	Yes
	within Private *Covered* Space	not applicable	Yes	Yes	No	Yes
	within Private *Indoor* Space	Yes	Yes	Yes	Yes	Yes
	within Semiprivate *Outdoor* Space	not applicable	No	No	No	No
HEIGHT ABOVE & BELOW CURB	fronting on or within Private Outdoor Space	No limit above / At level of outdoor space	Outdoors 100' max. above / At level of outdoor space	Outdoors 100' max. above / at level of outdoor space	Outdoors 150' max. above / 10' below	Outdoor 100' max. above / at level of outdoor space
	fronting on or within Semiprivate Outdoor Space	No limit above / 5' max. below	Outdoors 100' max. above / 5' max. below	Outdoors 100' max. above / 5' max. below	Outdoors 150' max. above / 10' below	Outdoors 100' max. above / 5'-0" below
	within Private Indoor Space	No limit above / At level of outdoor space	No limit above / No limit below	No limit above / No limit below	No limit above / No limit below	No limit above / No limit below
	Height above and below nearest floor	± 5'	± 5'	± 5'	± 5'	± 5'
ACCESS	Direct access from Private Outdoor Space	Yes	Yes when indoors and at grade ± 5'	Yes when indoors and at grade ±5'	Yes when indoors and at grade ± 5'-0"	Yes when indoors and at grade ± 5'-0"
	Direct access from Lobby	Yes	Yes when indoors	Yes when indoors	Yes when indoors	Yes when indoors
	Adjacent to Parking	not applicable	not applicable	not applicable	not applicable	not applicable
WINDOWS	front on Private Outdoor Space	Yes	Yes when indoors	Yes when indoors	Yes when indoors	Yes when indoors
	front on Semiprivate or Public Outdoor Space	Optional	Optional when indoors	Optional when indoors	Optional when indoors	Yes when indoors
POLLUTION SOURCE	Air	No min.	When outdoors 35' min. horiz. at grade or +15' min. above grade	When outdoors 35' min. horiz. at grade or +15' min. above grade	When outdoors 35' min. horiz at grade or +15' min. above grade	When outdoors 35' min. horiz at grade or +15' min. above grade
	Noise	No min.	When outdoors 35' min. horiz. at grade or +15' min. above grade	When outdoors 35' min. horiz. at grade or +15' min. above grade	When outdoors 35' min. horiz at grade or +15' min. above grade	When outdoors 35' min. horiz at grade or +15' min. above grade
STANDARDS		1. long dimension of room on exterior wall and to be 70% transparent, 2. transparency ratio of exterior walls to be 70%	1. minimum height throughout 20', 2. lighting level when in covered space or indoor space to meet current standards, 3. court dimensions to meet standards	1. minimum height 20' throughout, 2. lighting level in covered space to meet current standards, 3. court dimensions to meet standards	1. when indoors at least 50% of a wall that corresponds to the long dimension of the pool area should be transparent, 2. when indoors and open to the public, separate entry should be semiprivate or public provided from outdoor space	1. back wall and ceiling to be minimum 16'-0" high, 2. lighting level in covered or indoor space to meet current standards, 3. court dimensions to meet standard

MIXED

MIXED		PROGRAM	TENNIS COURT(S)	MEETING/SOCIAL ROOM—PUBLIC
MIN.		Minimum Size	7,200 SF. for single court and enclosure	600 SF.
		Minimum Area in Sunlight	No min.	No min.
LOCATION		within Private *Outdoor* Space	Yes	not applicable
		within Private *Open* Space	Yes	not applicable
		within Private *Covered* Space	Yes	not applicable
		within Private *Indoor* Space	Yes	Yes
		within Semiprivate *Outdoor* Space	No	not applicable
HEIGHT ABOVE & BELOW CURB		fronting on or within Private Outdoor Space	Outdoors 100' max. above at level of outdoor space	No limit above At level of outdoor space
		fronting on or within Semiprivate Outdoor Space	Outdoor 100' max. above 5' max. below	No limit above At level of outdoor space
		within Private Indoor Space	No limit above No limit below	No limit above At level of outdoor space
ACCESS		Height above and below nearest floor	± 5'	± 5'
		Direct access from Private Outdoor Space	Yes when indoors and at grade ± 5'	No
		Direct access from Lobby	Yes when indoors	No (fronting semiprivate space)
		Adjacent to Parking	not applicable	not applicable
WINDOWS		front on Private Outdoor Space	Yes when indoors	Optional
		front on Semiprivate or Public Outdoor Space	Yes when indoors	Yes
POLLUTION SOURCE		Air	When outdoors 35' min. horiz. at grade or +15' min. above grade.	No min.
		Noise	When outdoors 35' min. horiz. at grade or +5' min. above grade.	No min.
STANDARDS			1. when indoors under permanent or temporary structure, minimum height at edge of court 27' and 32' at center of net, 2. when indoors and open to the public, separate entry should be provided fronting semiprivate or public outdoor space, 3. lighting level in covered or indoor space to meet current stands, 4. court dimensions to meet standard	1. long dimension of room on exterior wall and to be 70% transparent, 2. transparency ratio of exterior walls to be 70%

Fig. 3

COMMUNITY RECREATIONAL FACILITIES

ADULT		PROGRAM	SHOP—CRAFTS	LAUNDRY ROOM
MIN.		Minimum Size	400 SF. 15' min. dim.	400 SF.
		Minimum Area in Sunlight	No min.	25% min.
LOCATION		within Private *Outdoor* Space	No	No
		within Private *Open* Space	No	No
		within Private *Covered* Space	No	No
		within Private *Indoor Space*	Yes	Yes
		within Semiprivate *Outdoor* Space	No	No
HEIGHT ABOVE & BELOW CURB		fronting on or within Private Outdoor Space	No limit above At level of outdoor space	No limit above At level of outdoor space
		fronting on or within Semiprivate Outdoor Space	No limit above 5' max. below	No limit above 5' max. below
		within Private Indoor Space	No limit above At level of outdoor space	No limit above At level of outdoor space
ACCESS		Height above and below nearest floor	± 5'	same level
		Direct access from Private Outdoor Space	Optional	Yes
		Direct access from Lobby	Yes	Yes
		Adjacent to Parking	not applicable	not applicable
WINDOWS		front on Private Outdoor Space	Yes	Yes
		front on Semiprivate or Public Outdoor Space	Optional	Optional
POLLUTION SOURCE		Air	No min.	No min.
		Noise	No min.	No min.
STANDARDS			1. must have bathroom and work-sink, 2. equipped with machine tools and/or sewing/weaving, and/or pottery, 3. exterior wall on private outdoor space at least 40% transparent, 4. be accessible to tenants 12 hrs. a day, 7 days a week, 5. for tenants only	1. long dimension of room on exterior wall, 2. at least 50% transparent, 3. seating, 4. for tenants only

Fig. 4

COMMUNITY RECREATIONAL FACILITIES

	PROGRAM	PASSIVE SPACE	TERRACE	ROOFTOP TERRACE	HEALTH CLUB TYPE FACILITIES	SHOP—AUTOMOTIVE
MIN.	Minimum Size	None	200 SF. min.; 400 max.	Up to 25% of adult SF. requirement max. 20' min. dim.	No min.	300 SF. 10' min. dim.
	Minimum Area in Sunlight	No min.	No min.	No min.	No min.	No min.
LOCATION	within Private Outdoor Space	Yes	Yes	Yes	Yes. As adjunct to indoor uses	No
	within Private Open Space	Yes	Yes	Yes	Yes. As adjunct to indoor uses	No
	within Private Covered Space	25% max.	25% max. by structure less than 18' above	25% max.	Yes. As adjunct to indoor uses	No
	within Private Indoor Space	No	No	No	Yes	Yes
	within Semiprivate Outdoor Space	Yes	25% max. by structure less than 18' above	No	Yes. As adjunct to indoor uses	No
HEIGHT BELOW CURB	fronting on or within Private Outdoor Space	15' max. above At level of outdoor space	No limit above Max. 5' below	40' min. to 200' max. above curb	No limit above At level of outdoor space	unlimited above
	fronting on or within Semiprivate Outdoor Space	5' max. above At level of outdoor space	No limit above Max. 5' below	40' min. to 200' max. above curb	No limit above 10' max. below	unlimited above
ABOVE CURB	within Private Indoor Space	not applicable	not applicable	not applicable	No limit	unlimited above
	Height above and below nearest floor	± 5'	No limit above	at same level	± 5'	± 5'
ACCESS	Direct access from Private Outdoor Space	not applicable	Optional	Optional	Optional	Optional
	Direct access from Lobby	Yes	not applicable	Yes	Yes	Optional
	Adjacent to Parking	Optional	not applicable	not applicable	not applicable	Yes
WINDOWS	front on Private Outdoor Space	not applicable	not applicable	not applicable	Yes	Yes
	front on Semiprivate or Public Outdoor Space	not applicable	not applicable		Optional	No
POLLUTION SOURCE	Air	No min.	When outdoors 35' min. horiz at grade or +15' min. above grade	100' from smokestack	No min.	No min.
	Noise	No min.	When outdoors 35' min. horiz at grade or +15' min. above grade	50' from cooling tower	No min.	No min.
STANDARDS		1. the passive space is to be apportioned as follows: a. amount of passive space in semiprivate outdoor space can be no more than 25% of total Adult requirement b. and cannot be greater in area than passive space in private outdoor space	1. only 1 apartment may front on a terrace. 2. when on grade (-5'—15') there can be a direct entry to the terrace when fronting on private outdoor space. 3. when fronting on semiprivate or public outdoor space, the entry from that space must have a locked entry	1. the terrace be surfaced with a durable paving material. 2. at least 15% of the terrace be covered either permanently or with awnings, etc., 3. that following facilities be provided: a. seating b. bathrooms c. drinking fountain	1. would include such facilities as: sauna, steam baths, various types of baths, swimming pools, showers, exercise equipment and lockers, 2. have a minimum height of 12'-0" throughout. 3. exterior wall to be a minimum of 50% transparent, 4. when open to public, entrance to be provided from semiprivate or public open space. 5. exterior spaces are not counted toward interior recreation space	1. equiped with bathroom 2. appropriately equipped with benches and pit 3. must be accessible to the tenants for 12 hrs a day, 7 days a week 4. for tenants only

ADULT

Fig. 5

5

Elements of the Dwelling Unit

TABLE 1   Mean Body Measurements

	Women	Men
Heights, standing, inches		
Top of head	64.0	69.8
Eye	59.5	64.9
Shoulder	52.3	57.3
Elbow	39.6	43.4
Palm	29.7	31.9
Thumb tip	26.6	28.8
Heights, seated, inches		
Seat to top of head	33.4	35.6
Seat to eye	28.7	30.5
Seat to shoulder	21.5	22.9
Seat to elbow	9.0	9.3
Floor to under knee	14.9	17.6
Floor to top of thigh	20.7	23.1
Floor to top of crossed knee	24.7	2.80
Lengths, inches		
Sitting (buttocks to front of knees)	23.4	25.1
Seat (buttocks to back of knees)	19.3	20.4
Lap (abdomen to front of knees)	13.3	15.2
Total arm (acromion to thumb tip)	25.0	27.0
Forearm (olecranon to thumb tip)	14.7	16.2
Widths, inches		
Maximum body	18.4	20.6
Shoulders	14.4	16.4
Upper body	17.8	19.7
Lower body (standing)	14.5	14.3
Lower body (sitting)	16.2	16.0
Elbows extended	33.6	37.7
Thickness, inches		
Maximum body	10.9	11.7
Lower body	10.7	10.9
Girth, inches		
Bust, Chest	36.6	37.4
Hip	39.0	
Bent at hips, inches		
Arms down, length	32.8	33.2
Reach from bent position (buttocks to thumb tips of arms extended)	45.0	48.2
One-knee kneel, inches	33.5	
Weight, lb	140.5	167.4

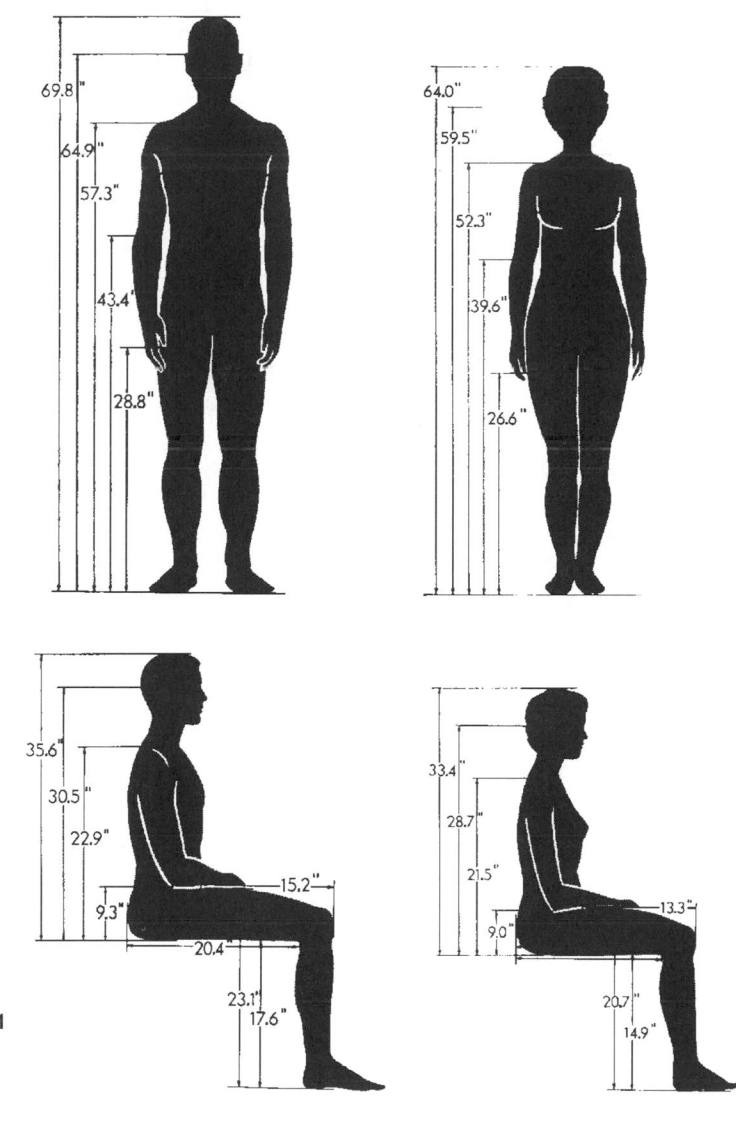

Fig. 1

Walking between
two high walls
(space adequate
for both men
and women)

Two people passing
(figure derived; twice the
space for one person to
walk between two high walls)

Walking between high
wall and 30" high
table (space adequate
for both men and
women)

Walking with elbows
extended (space ade-
quate for both men
and women)

Fig. 1

Kneeling on one
knee

Man bending at
a right angle

One person using
coat closet

42″

Two persons using coat closet in foyer
area with space for one person walking

60″   26″   72″   46″

Fig. 2   Coat closets.

RECOMMENDED CLEARANCES

The dimensions given for clearances are based upon European and North American studies of the measurements of human beings. The intention is to determine the smallest amount of space for comfortable circulation within a room. These dimensions are recommended as being minimum.

850
(2′-10″)

Fig. 3a   Limited access behind a chair.

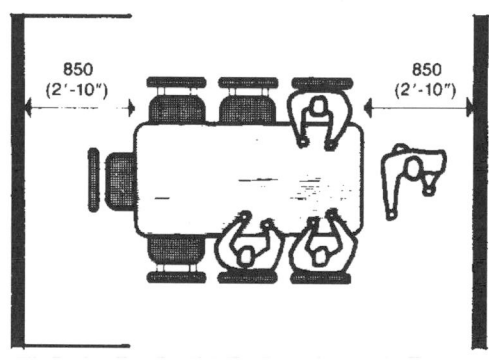

850
(2′-10″)   850
                  (2′-10″)

Fig. 3c   Access between a table and wall.

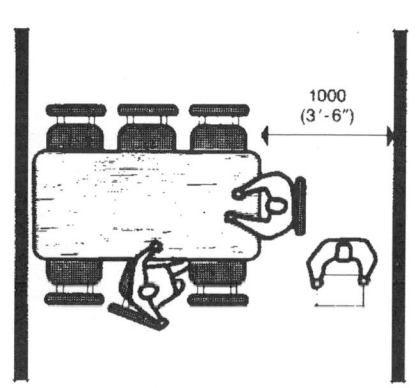

1000
(3′-6″)

Fig. 3b   Access behind a chair.

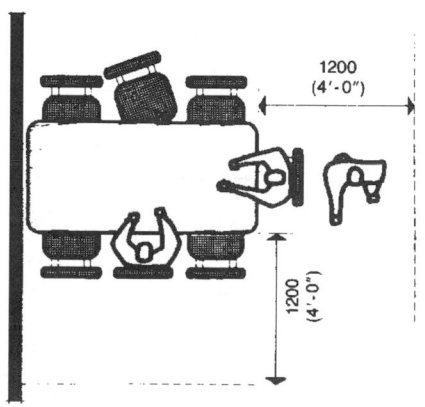

1200
(4′-0″)

1200
(4′-0″)

Fig. 3d   Access between a table and cabinets or appliances.

TABLE 1   Dining Area

Limited access behind a chair	850 mm (2 ft 10 in)
Access behind a chair	1000 mm (3 ft 6 in)
Access between a table and wall	850 mm (2 ft 10 in)
Access between a table and cabinets of appliances	1200 mm (4 ft 0 in)

ELEMENTAL ACTIVITIES

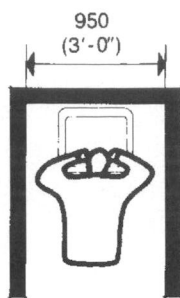

Fig. 4a   Space between walls for a handbasin.

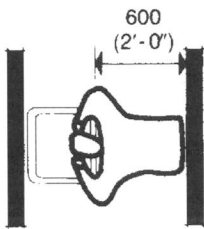

Fig. 4b   Space in front of a handbasin and wall.

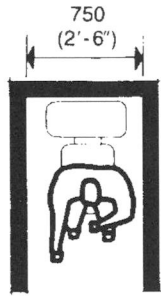

Fig. 4c   Space between walls for a toilet.

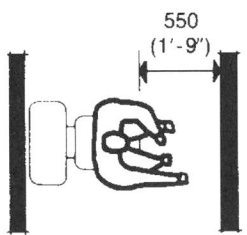

Fig. 4d   Space in front of a toilet and wall.

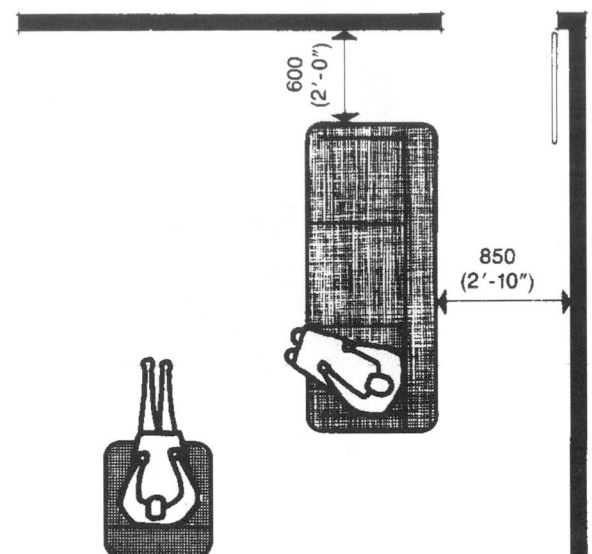

Fig. 5a   General access, 850 mm; limited access, 600 mm.

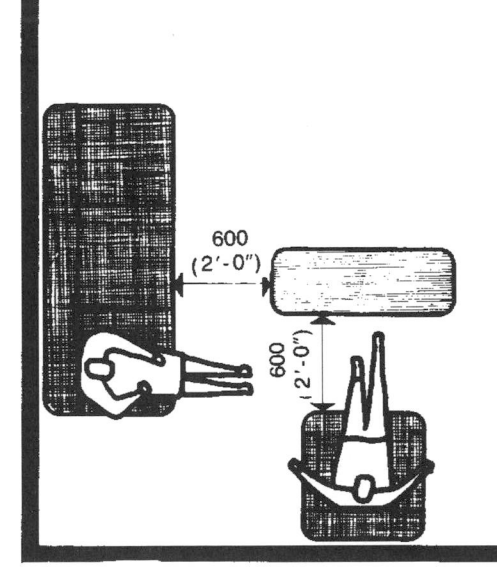

Fig. 5b   Limited access between a table and other furniture.

TABLE 2 Bathroom	
Space between walls for a washbasin	950 mm (3 ft 0 in)
Space in front of a washbasin and wall	600 mm (2 ft 0 in)
Space between walls for a toilet	750 mm (2 ft 6 in)
Space between front of a toilet and wall	550 mm (1 ft 9 in)

TABLE 3 Living Area	
General access	850 mm (2 ft 10 in)
Access between a table and other furniture	600 mm (2 ft 0 in)

The furniture sizes listed in this section have been determined from a study of moderately priced furniture on the market today. It is recommended that these dimensions be the minimum for furniture used in the design of rooms. Because the choice of furniture size and type will vary widely among families, sufficient floor space must be provided to allow for such variation.

The metric dimensions shown in the furniture illustrations approximate the mean dimensions; they are not necessarily those used in manufacture.

LIVING ROOM

Fig. 1*a*   Armchair, 800 by 850 mm (2 ft 8 In by 2 ft 10 in).

Fig. 1*b*   Chesterfield, 2000 by 800 mm (6 ft in by 2 ft 10 in).

Fig. 1*c*   End table, 650 by 450 mm (2 ft 2 in by 1 ft 6 in).

Fig. 1*d*   Coffee table, 1200 by 450 mm (4 ft 0 in by 1 ft 6 in).

Fig. 1*e*   Occasional chair, 700 by 750 mm (2 ft 4 in by 2 ft 6 in).

DINING AREA

Fig. 2*a*   Table, five to six persons, 1200 by 900 mm (4 ft 0 in by 3 ft 0 in).

Fig. 2*b*   Table, seven to eight persons, 1800 by 900 mm (6 ft 0 in by 3 ft 0 in).

Fig. 2*c*   Dining chair, 450 by 500 mm (1 ft 6 in by 1 ft 8 in).

Fig. 2*d*   Buffet, 1200 by 450 mm (4 ft 0 in by 1 ft 6 in).

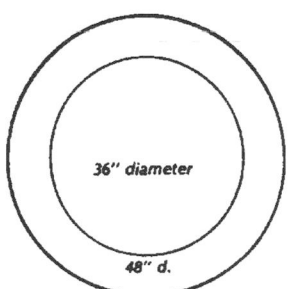

Fig. 2*e*   Tables, circular.

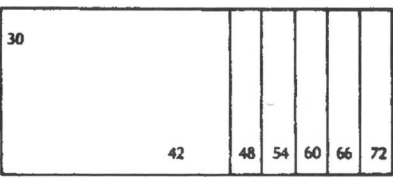

Fig. 2*f*   Tables, rectangular (dimensions in in.).

BEDROOM

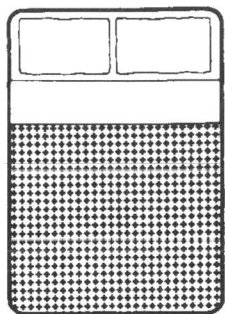

Fig. 3*a*   Double bed, 1350 by 2000 mm (4 ft 6 in by 6 ft 6 in).

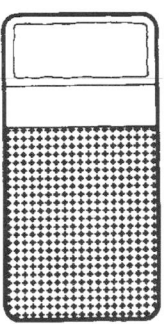

Fig. 3*b*   Single bed, 1000 by 2000 mm (3 ft 3 in by 6 ft 6 in).

Fig. 3*c*   Single dresser, 750 by 450 mm (2 ft 6 in by 1 ft 6 in).

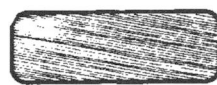

Fig. 3*d*   Double dresser, 1200 by 450 mm (4 ft 6 in by 1 ft 6 in).

Fig. 3*e*   Bedside table, 500 by 400 mm (1 ft 8 in by 1 ft 4 in).

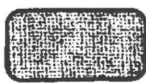

Fig. 3*f*   Work surface, 900 by 450 mm (3 ft 0 in by 1 ft 6 in).

FURNITURE SIZES

Couch
3'-0" x 6'-10"
Easy chair
2'-6" x 3'-0"
End table
1'-6" x 2'-6"
Television set
1'-4" x 2'-8"
Double bed
4'-6" x 6'-10"
Desk 1'-8" x 3'-6"
with chair
Crib
2'-6" x 4'-6"
Twin beds
3'-3" x 6'-10"
Table for two
2'-6" x 2'-6"
Table for four
2'-6" x 3'-2"
Chair
1'-6" x 1'-6"
Dresser
1'-6" x 3'-6"
or
1'-6" x 4'-4"
Dining table with chairs for six = 3'-4" x 4'-0"
for eight = 3'-4" x 6'-0" or 4'-0" x 4'-0"
60"
75"
85"
90"
100"

Fig. 4

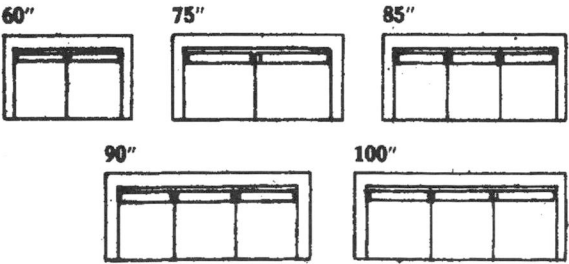

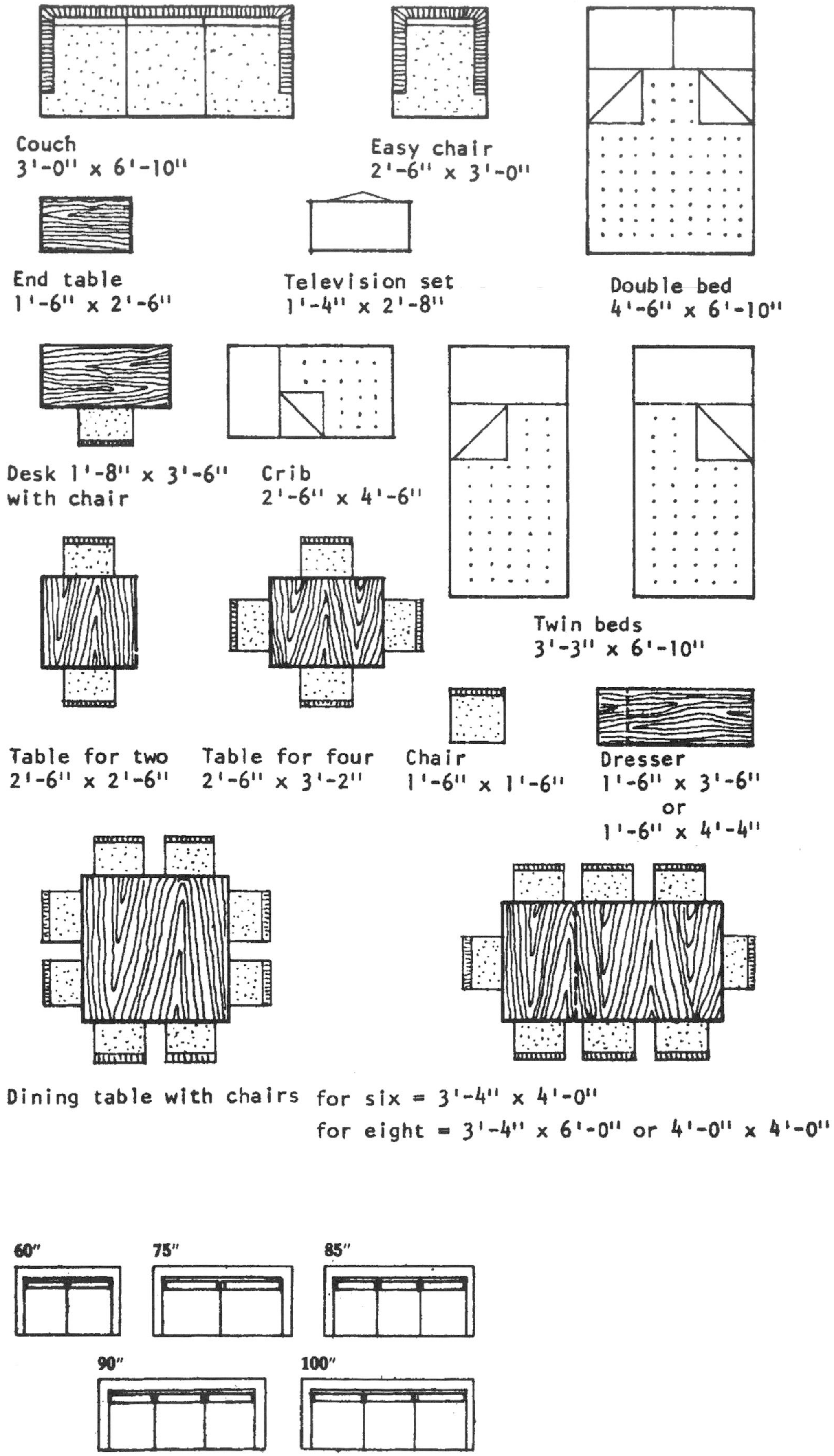

A full bathroom should be located on every floor within a dwelling unit on which bedrooms are located.

Bedrooms Where possible, bedrooms should all be located on one floor in a multifloor unit.

Access, stairs, and circulation One of the essential characteristics of the townhouse type of multiple dwelling unit is its ground orientation. Therefore all dwelling units shall have access from both sides of the unit. The secondary access point may be a sliding door.

Every effort should be made in the planning of units to avoid the use of rooms as primary circulation paths and to avoid excess floor area devoted to halls. Centralization of both vertical and horizontal circulation should be a design objective.

Halls and stairways should be designed to facilitate the easy movement of normal household furniture in and out of rooms.

Stairways within dwelling units should be located in close proximity to the primary entrance door of the unit. Basement stair entrances should be visually screened from living or dining spaces.

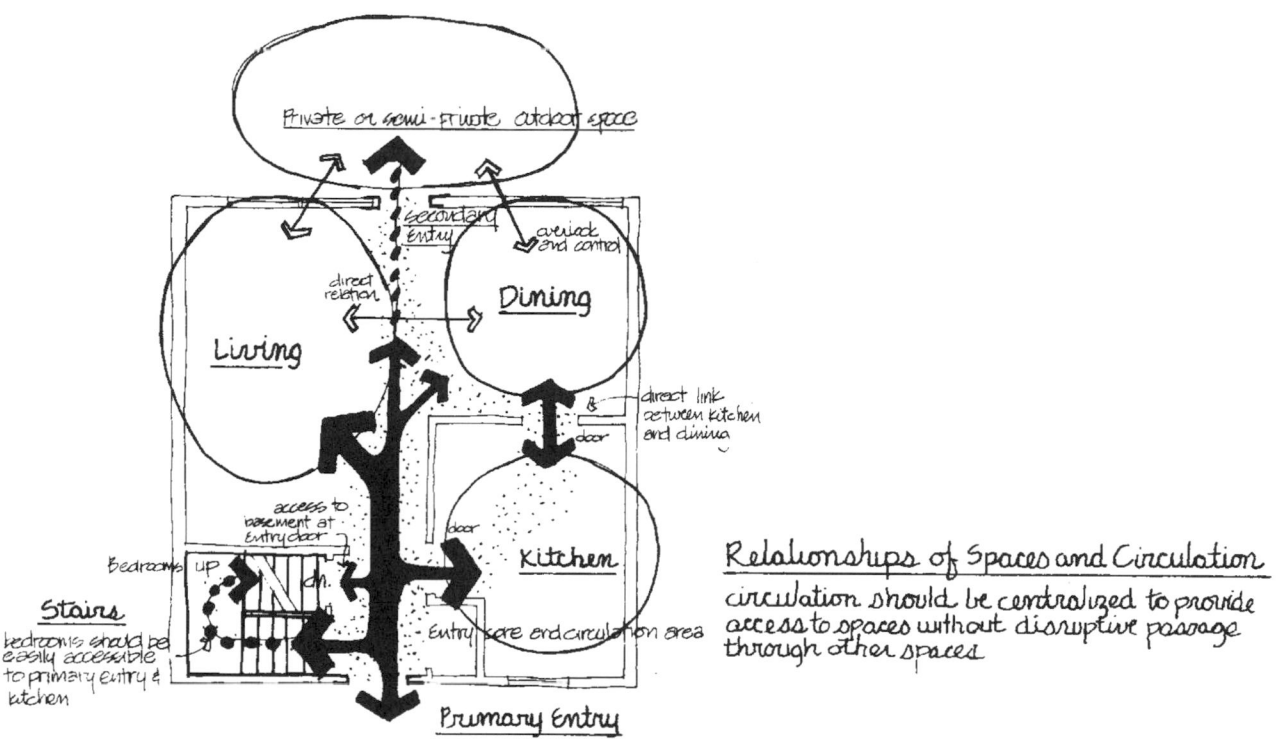

Fig. 1

FUNCTIONAL RELATIONSHIPS

CHARACTER AND FUNCTIONAL ORGANIZATION

Objective

The organization and character of individual dwelling units and residential buildings cannot be expressed by simple numerical standards, yet they significantly affect the quality of the living environment. Functional organization, privacy, visual appearance, etc., are issues which must be dealt with consciously and rationally if desirable housing is to be achieved. Therefore the following guidelines and performance criteria are set forth to assist the applicant in preparation of plans.

Functional Organization

The following should guide the organization of spaces within dwelling units.

Kitchens Kitchens should be located in proximity to the primary entrance door of the unit. It is not desirable to depend on the living room and/or dining space for access between entrance and kitchen.

Convenient access should be provided between the kitchen and private outdoor space.

If possible, access between bedrooms and the kitchen should be accomplished without passage through living and dining areas. Where this is not possible, circulation through living and dining areas should not interfere with normal use patterns within these areas.

Kitchens should be separated from other areas by doors.

The organization of kitchen cabinets, counters, and appliances should afford a logical sequence of food storage, preparation, serving, and clean-up activities which minimize steps and discontinuity of activity. Work and storage surfaces should be provided on both sides of the sink. The range and refrigerator should have work space on at least one side.

Dining spaces Dining space located in a combined living-dining room or in a separate dining room should be located adjacent and have direct access to the kitchen. Such dining space and its use should not impair normal circulation into, out of, or within the living room.

Dining space within kitchens should not impair normal circulation, food preparation, and clean-up activities.

Living spaces Living rooms should be accessible from main entrance doors and internal stairways without passage through kitchens, separate dining rooms, or other nonrequired living spaces within the dwelling unit. In a combined living-dining space, access to the living area should not require circuitous travel through the dining space. Either the living or the dining space, or both, must overlook and have access to the private outdoor space of the unit.

Bathrooms Bathrooms should be conveniently accessible from all rooms of the dwelling unit without entering a bedroom, except that in units where two full baths are provided one bath may open directly to the primary (master) bedroom. No bathroom should open directly off of the kitchen, living room, or dining space.

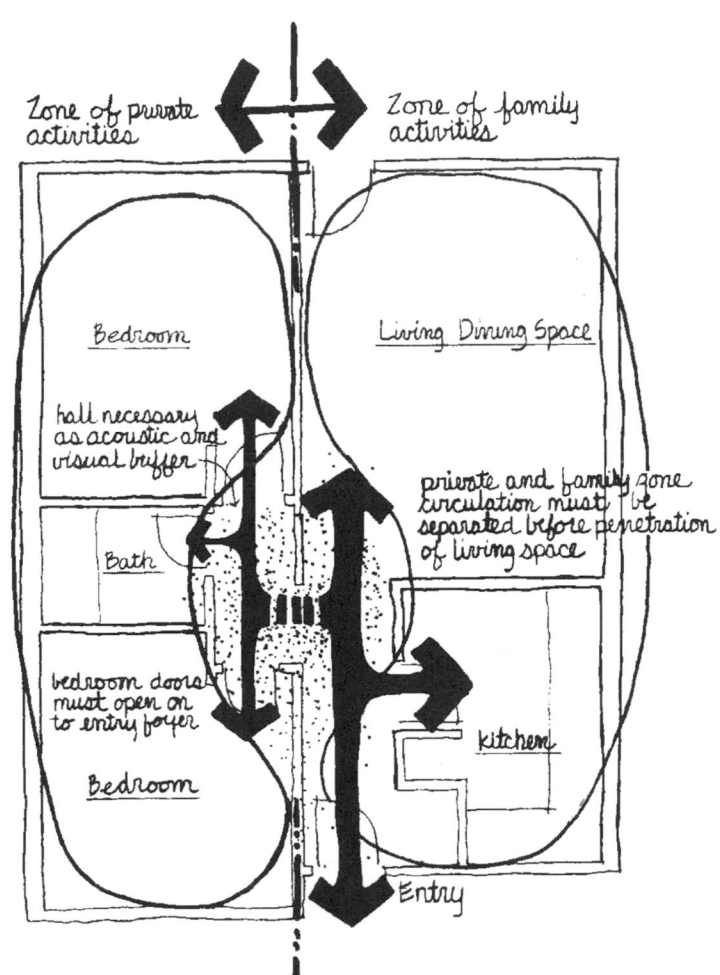

Zone of private activities

Zone of family activities

Bedroom

Living Dining Space

hall necessary as acoustic and visual buffer

private and family zone circulation must be separated before penetration of living space

Bath

bedroom doors must open on to entry foyer

Bedroom

kitchen

Entry

organization requirements of this approach extremely critical large units and must be satisfied conscientiously

Fig. 2 Horizontal zoning to achieve separation of activities.

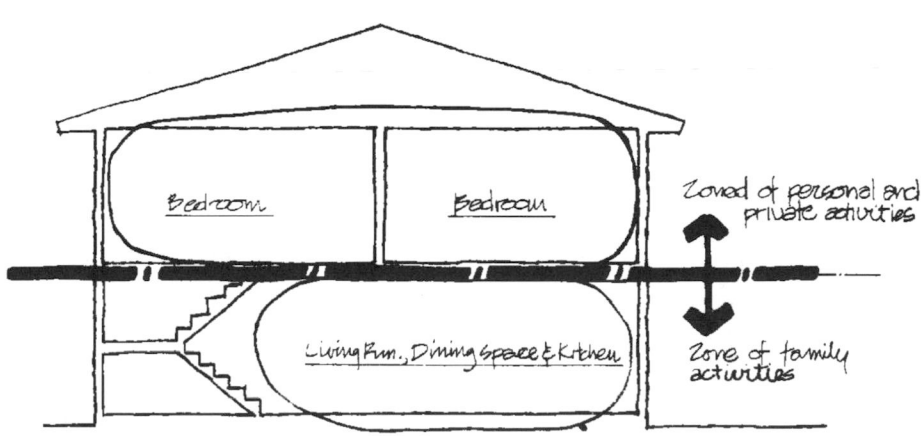

Bedroom

Bedroom

Zoned of personel and private activities

Living Rm., Dining space & Kitchen

Zone of family activities

good acoustic isolation of zones - especially desirable in large units without basements

Fig. 3 Vertical zoning to achieve separation of activities.

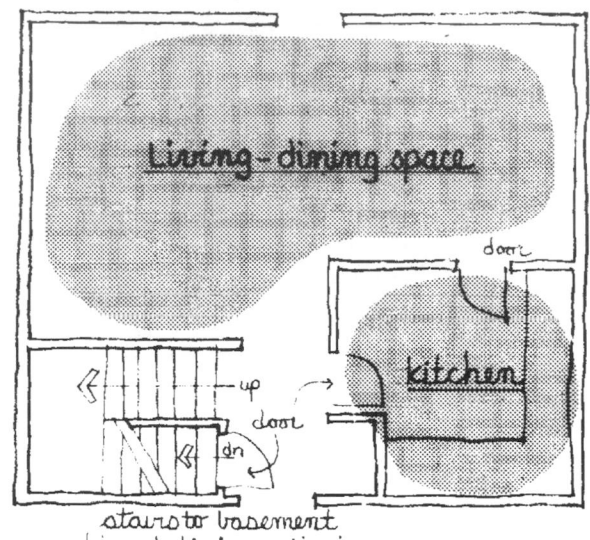

combined living-dining space

one large and one small activity space

satisfactory where basement is provided

unsatisfactory where no basement is provided, especially in large dwelling units

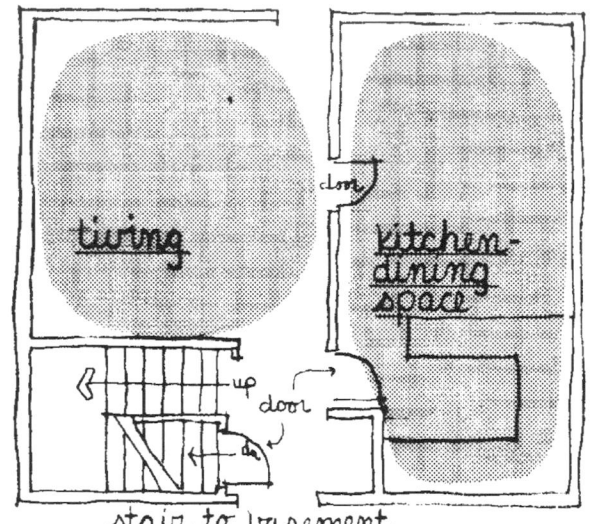

combined kitchen-dining space

two medium sized activity spaces

satisfactory where basement is provided

a desirable configuration for both large and small dwelling units where no basement is provided, if economically feasible in such cases, a larger than standard dining space should be provided

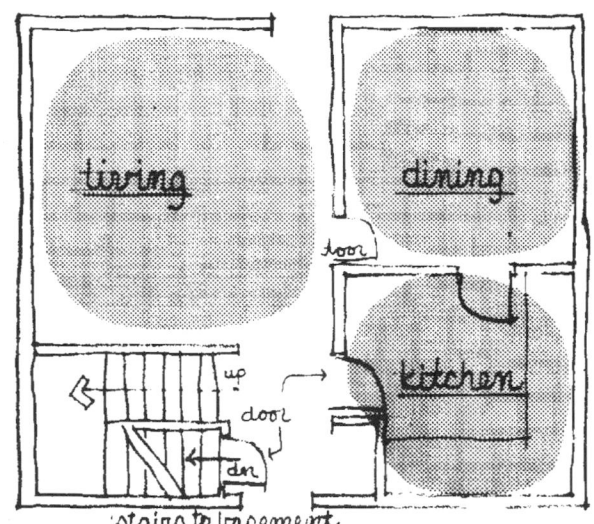

separate living, dining and kitchen spaces

three small activity spaces

configuration with three separable activity provide good first floor simultaneous use flexibility, however in dwelling units without basement no large activity area will be available, in such cases, where economically feasible, larger than standard living and dining spaces should be provided

Furnishability Requirements

The livability of the rooms within a dwelling unit cannot be assured simply by providing adequate floor area and room dimensions. The ability to successfully accommodate the furniture normally concomitant with a room's anticipated uses must also be assured if livability is to be achieved. Successful furnishability is achieved only when the furniture and activities of a room are accommodated while also providing free room circulation, access to furniture, allowances for door swings and windows, etc.

Furnishability, as a test of room adequacy, is a valuable tool for the housing designer. Dwelling units should demonstrate the capability of accommodating at least the furnishings listed below.

Bedrooms In primary bedrooms, there should be twin beds or a double bed, two small or one large dresser, and one chair. In a secondary bedroom for two persons there should be twin beds, two small or one large dresser, and one chair.

    Double bed, 4 ft 9 in × 6 ft 6 in
    Twin or single bed, 3 ft 3 in × 6 ft 6 in
    Small dresser, 1 ft 6 in × 2 ft 6 in
    Large dresser, 1 ft 6 in × 4 ft 6 in
    Chair, 1 ft 6 in × 1 ft 6 in

Living space

    Sofa, 6 ft 9 in × 2 ft 8 in
    Two easy chairs, 2 ft 6 in × 2 ft 6 in
    Desk or low table, 2 ft 0 in × 3 ft 4 in
    Other incidental furniture, 1 ft 10 in × 3 ft 0 in
In three- and four-bedroom units three easy chairs should be shown in living rooms.

Dining space Each dining space must contain sufficient space in a single location for an appropriate table and enough chairs to accommodate the maximum number of persons the unit will accommodate. Dining chairs should be at no less than 1 ft 6 in × 1 ft 6 in. The primary dining table should be not less than 3 ft 0 in in width, with an allowance of 2 ft 0 in lineal feet of table edge for each person to be seated. Only one person may be accommodated at each end for rectangular configurations. Secondary (kitchen) dining tables may be 2 ft 6 in wide.

The clearance from table edge to nearest wall or other obstruction should be as follows:

where circulation is intended and no seating is anticipated—2 ft 6 in; where seating is anticipated in the circulation passage—3 ft 6 in; where seating but no circulation is anticipated—3 ft 0 in except for secondary kitchen dining, where only one seat is anticipated—2 ft 0 in.

Private outdoor space Private outdoor space that may be connected with a dwelling unit should accommodate casual seating for four persons in one- and two-bedroom units and six persons in three- and four-bedroom units.

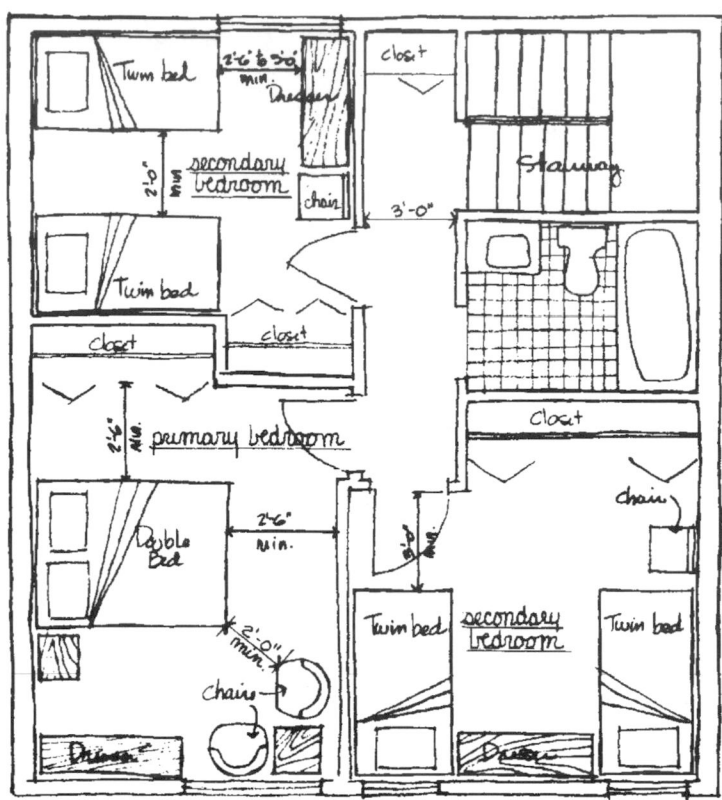

potential furniture
organization should
be considered when
locating windows
closets and doors

Fig. 5*a*   Furnishability—bedrooms.

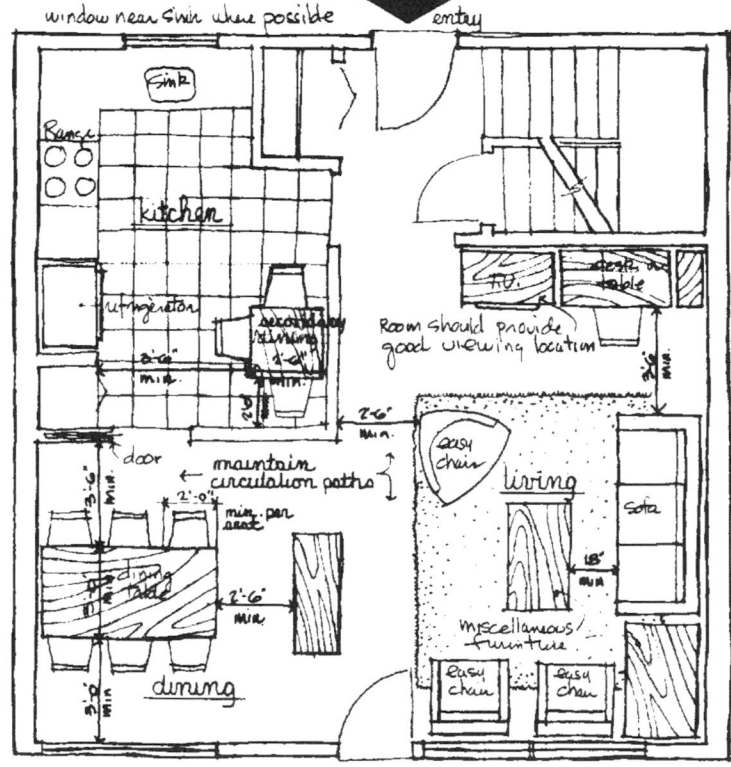

room configurations should
provide for reasonable furniture
groupings that are consistent
with room use, orientation, and
relationships

living rooms should provide
for conversation groups, tv viewing,
and miscellaneous activities

Fig. 5*b*   Furnishability—living room, dining room, kitchen.

GENERAL OBJECTIVES

Human-Centered Needs

The design and planning of living units should have a workable human-centered basis. Provision should be made for the essential needs of people for space, light, food, water, sleep, safety, sanitation, comfort, companionship, and periods of quietness. It is necessary that adequate housing quality be provided, yet reconciled with minimum cost by the efficient use of space.

Functional Considerations

Space needs should be determined by family size, the functions of daily living, and the normal possessions of the family. Living units should be planned to contain space sufficient to accommodate appropriate furniture or equipment for each habitable room. To demonstrate the furnishability, preliminary floor plans for each living unit design should show the appropriate furniture drawn to scale.

Room Relationships

The arrangement of the rooms should show a proper relation of one to the other, and provide reasonable privacy by:

1. Locating exterior openings in relation to exterior conditions
2. Having bathrooms accessible from bedrooms and other habitable rooms

Circulation Pattern

The circulation pattern throughout a living unit should function satisfactorily. Serious conflicts in the appropriate use of each room and its furniture and equipment should be avoided.

Access to Outdoors

Single-family houses and multiple living units at or near grade should have a convenient relationship to outdoor areas.

Indoor Space Needs

The indoor space needs for family recreation and self-service activities should be provided for (play space for children, minor home repairs, etc.)

Ceiling Heights

Ceiling heights should be such that the average person can move about comfortably, no difficulty is presented in the placement and use of furniture, and no unpleasant sensation is created by ceilings of insufficient height.

The minimum ceiling height of habitable and nonhabitable rooms should conform with the following:

Basement In nonhabitable basement rooms, the minimum clear ceiling height should be 6 ft 9 in. Structural beams, ducts, piping, and other such construction items in general should be installed with a minimum clearance above the floor of 6 ft 9 in, and in all cases should be installed in such a manner that they do not interfere with safe traffic and utilization of the nonhabitable spaces.

Habitable rooms For habitable rooms the minimum ceiling height should be 7 ft 6 in. Sloping ceiling should have a minimum average height of 7 ft 6 in and no height less than 7 ft 0 in. When necessary, beams, girders, or utilities such as ducts, pipes, or wiring installed as part of the ceiling and which cannot meet the 7-ft 6-in clearance should have a clearance height of not less than 7 ft 0 in.

Bathrooms, toilet compartments, utility rooms, and private halls In bathrooms, toilet compartments, utility rooms, private halls, etc., the clear height should be 7 ft.

Public halls In public halls the clear height should be 7 ft 8 in.

Garages Driving lanes should be 7-ft 6-in clear height. Parking areas should be 7-ft clear height. Truck unloading areas should be 10-ft clear height.

Other areas The clear ceiling height in any other area should be 7 ft 6 in.

Furnishability

The criterion for the amount of space provided in habitable rooms of a living unit is its furnishability. The test material for each room provides a list of furniture for which appropriate space must be provided.

Furniture Sizes

Furniture sizes given are standard sizes. However, it is well for the designer to keep in mind that families frequently have on hand or buy large and heavy furniture, particularly for the living room. In such cases, additional space is necessary to have planning arrangements function properly.

Built-in Furniture

In small rooms the use of built-in furniture can be advantageous as space savers. Properly designed built-in storage units can increase usable floor space and reduce the outlay for home furnishings. Where built-ins provide a functional equivalent to movable furniture as listed for the various areas of the living unit, their use can be considered an acceptable substitute.

Combination Rooms

The combination of more than one living function into a single space is a most common method of using space intensively and economically. Certain combinations and some limiting factors are given below.

Kitchen-Dining Area

A frequent and favorable planning arrangement in lower-income housing is a combined kitchen-dining area. This permits a wide use of the space not only for the kitchen-dining functions but for study and informal social activity for the entire family.

Definition of a Combined Room

For two adjacent spaces to be considered a combined room, the clear opening between the spaces should permit the common use of spaces for the expansion of the different functions. In general, the horizontal opening between combined spaces should be at least 8 ft.

Visual Separation between Areas

A combined living-dining-kitchen area should have the food preparation-cooking area screened from the living room-sitting area.

Limited Occupancy

The living unit without a separate bedroom generally provides more multiple use of space than any other combination. However, its acceptable use is limited to one or two persons.

Other Habitable Rooms

Apartment living units may contain an alcove or a separate room which may have multiple uses. It may be used as a den, a general-purpose family room, or additional sleeping space. Although space for no specific furniture is required for such a room, the floor area should be at least as large as a secondary, single-occupancy bedroom, and it should contain a clothes closet.

UNIT SIZE

Family units (units occupied by adults and children) should have no fewer than two bedrooms.

Individual privacy in a family is needed even if that family comprises one adult and one child.

Fig. 1

UNIT INTERIORS

A generous kitchen-dining area or a distinct dining area with a separate living room should be provided. A living room-dining space and a small working kitchen is not recommended.

As the unit size increases (i.e., more bedrooms), the kitchen should have more counter space and storage. Recognizing that in family living, the dining area is used for many activities (i.e., homework), provision should be made to allow these activities to occur with minimal conflict with living room activities.

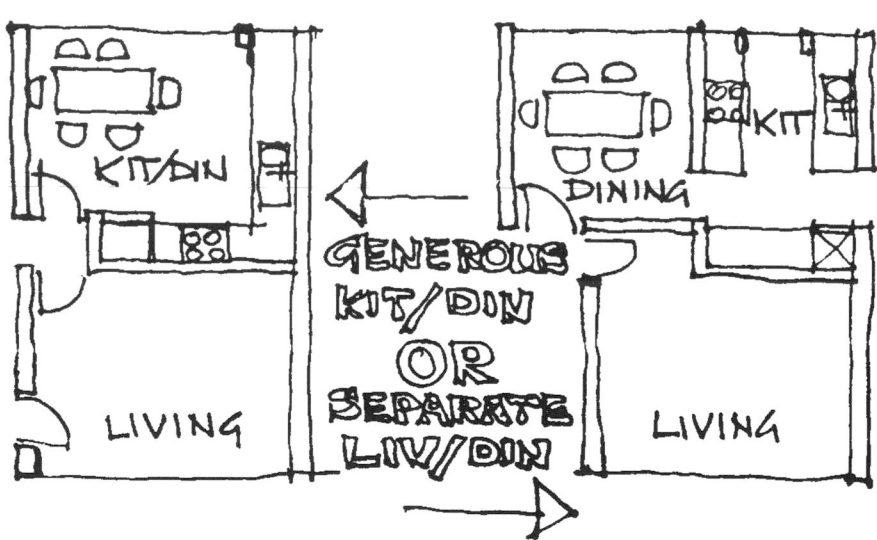

Fig. 2

TABLE 1 Typical Room Sizes (ft)

	Small	Average	Large
Basic Rooms			
Living room	12 × 18	16 × 20	22 × 28
Dining room	10 × 12	12 × 15	15 × 18
Kitchen	5 × 10	10 × 16	12 × 20
Utility room	6 × 7	6 × 10	8 × 12
Bedroom	10 × 10	12 × 12	14 × 16
Bathroom	5 × 7	7 × 9	9 × 12
Additional Rooms/Areas			
Halls	3′ wide	3′ 6″ wide	3′ 9″ wide
Area	10 × 20	20 × 20	22 × 25
Storage wall	6″ deep	12″ deep	18″ deep
Den	8 × 10	10 × 12	12 × 16
Family room	12 × 15	15 × 18	15 × 22
Wardrobe closet	2 × 4	2 × 8	2 × 15
One-rod walk-in closet	4 × 3	4 × 6	4 × 8
Two-rod walk-in closet	6 × 4	6 × 6	6 × 8
Porch	6 × 8	8 × 12	12 × 20
Entry	6 × 6	8 × 10	8 × 15
One-car garage	11 × 19	13 × 25	16 × 25
Two-car garage	20 × 20	22 × 22	25 × 25

ROOM DIMENSIONS

Living Area

The dimensions of a room are strong determinants of its furnishability. For example, in the living area one wall should be at least 14 ft long to accommodate the standard arrangement of a sofa, two end tables, and one corner easy chair. The more walls that are at least this length, the greater the flexibility in furniture arrangement. The minimum width of the room should be 11½ ft, based on furniture dimensions and access area. This approximates the recommended conversation distance of 10 ft measured from center points of sitting furniture. An additional 2 ft in width, for a total of 13.5 ft, is suggested as a maximum dimension for maintaining comfortable conversation and TV viewing distances.

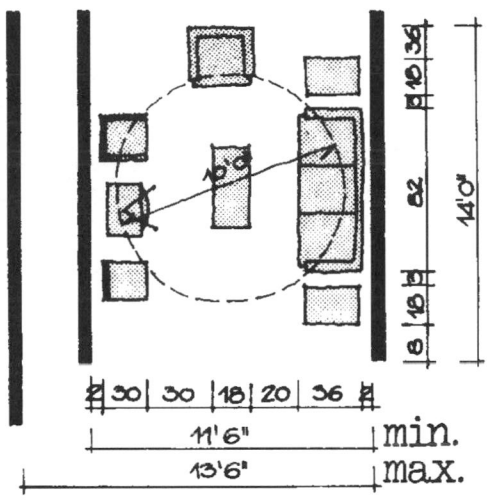

Fig. 1

Dining Area

A wall in or near the kitchen should be at least 8 ft long for the table, chairs, and access areas, with another wall nearby to accommodate the 42-in-long china cabinet. Enough floor area should be available for seating six to eight persons around a table.

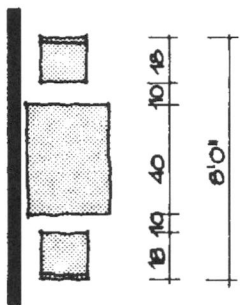

Fig. 2

Sleeping Area

In the single-occupancy sleeping area, one wall should be at least 9 ft long. For double occupancy, one wall should be a minimum of 11.5 ft in length to accommodate the standard furniture pieces and necessary access areas. Access to a closet area at either end of the room will require additional space. Minimum room widths are 10 ft if no bureaus are to be placed at the foot of the beds and 11.5+ ft if they are.

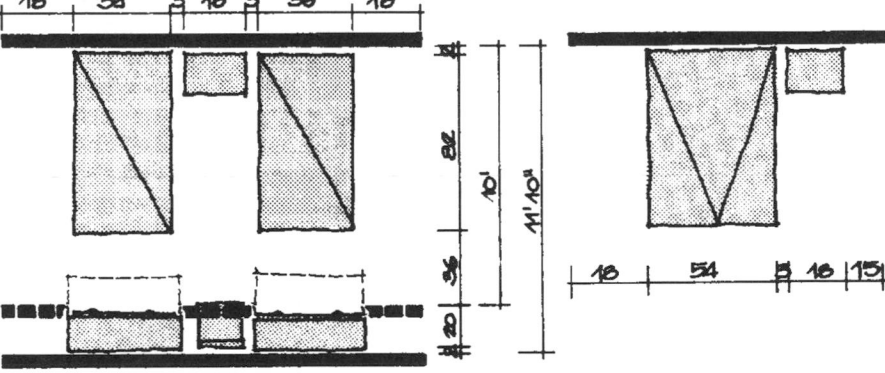

Fig. 3

FURNITURE REPLACEMENT

Entering

Most residents have a small table or set of shelves for displaying objects or setting packages on when returning to their apartments. A chair may also be used in the entryway for sitting down while putting on and taking off outer wear. This is typically placed near the coat closet where these items are stored.

Fig. 4

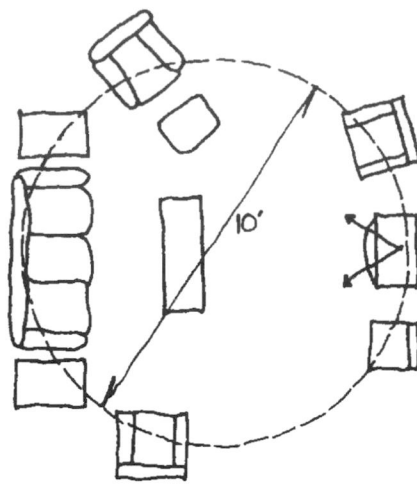

Fig. 5

Visiting

Conversation furniture typically includes a sofa with two end tables, coffee table, rocker, and one or more easy chairs arranged in a closed loop. Recommended distance between seats for comfortable conversation is 10 ft. If more seating is required, dining chairs or floor cushions are often used. When a snack or meal is served as part of the visit, the dining table is typically pulled away from the wall and chairs are arranged around it. For entertaining large numbers, some residents have folding tables and chairs which they keep stored away in closets when not in use. Some also have folding beds which they set up for overnight guests.

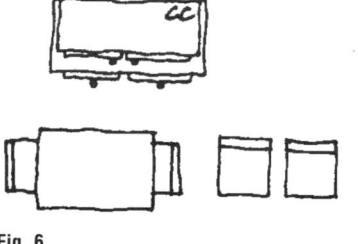

Fig. 6

Eating

To support this activity, most residents have a dining table and four side chairs as well as a china cabinet. The table is typically placed against a wall with two of the chairs while the other chairs are located in other areas in the apartment. For formal meals with visitors, this table is pulled away from the wall and chairs are placed around it. For many older residents, particularly women, the china cabinet is an important piece of furniture, as it typically symbolizes a previous role in the family as well as serving for the display and storage of dinnerware.

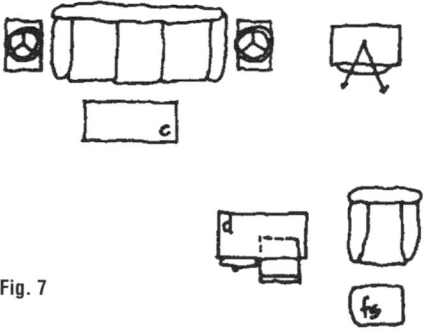

Fig. 7

Leisure Activities

All elderly residents have at least one TV which in modern times has replaced the fireplace as the focal point in the living room. It typically is placed against a wall directly across from the sofa. The sofa is almost always placed against a wall. Residents avoid putting it under a window where there is a potential problem with drafts on the back of the neck. Residents sit here to watch TV or in their "favorite" easy chair, which is usually placed at the end of the sofa and near a window to take advantage of natural light and outdoor views. This is also the favored setting for other leisure activities like reading, sewing, and watching outdoors. End and coffee tables accompany these primary seating areas to hold table lamps and materials associated with various other activities. Depending on the specific interests of the residents, there might also be a sewing machine, work surface, or storage and display piece for a particular hobby. Some residents might define a "communications area" with a desk and chair for telephoning and letterwriting.

Sleeping

For single residents, typical furniture in the sleeping and dressing area includes a double bed, two bureaus, a night stand, and a chair. Double occupants (who are most frequently married couples) are more likely to have twin beds rather than a double bed. Heads of beds are placed against walls, perpendicular to a window to permit a view of the outdoors while lying down. Placement of the head of a bed under a window is avoided because of problems with drafts. For ease in making the bed(s), residents prefer to place it so it is accessible from two sides and one end. A nightstand is located at the head of a bed (between twin beds) upon which is usually found a table lamp, clock, medicines, and other personal items. The two bureaus, along with a chair, are typically clustered near the closet(s) to form a convenient dressing area.

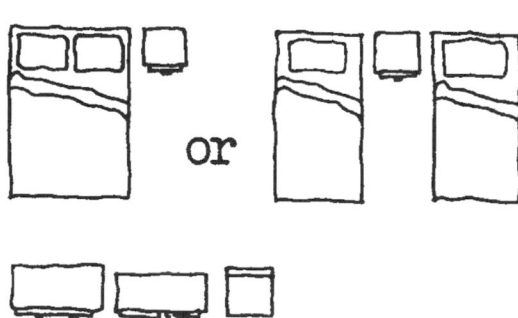

Fig. 8 or

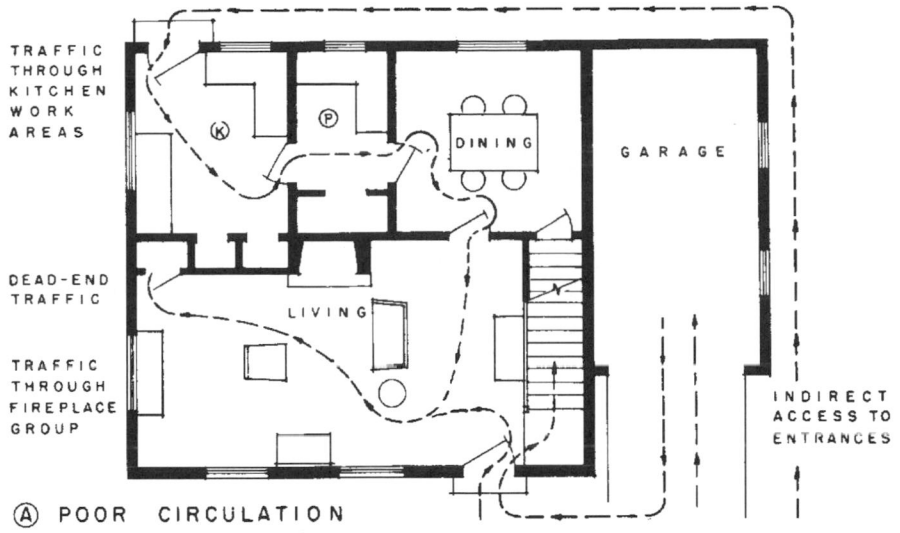

TRAFFIC THROUGH KITCHEN WORK AREAS

DEAD-END TRAFFIC

TRAFFIC THROUGH FIREPLACE GROUP

INDIRECT ACCESS TO ENTRANCES

Ⓐ POOR CIRCULATION

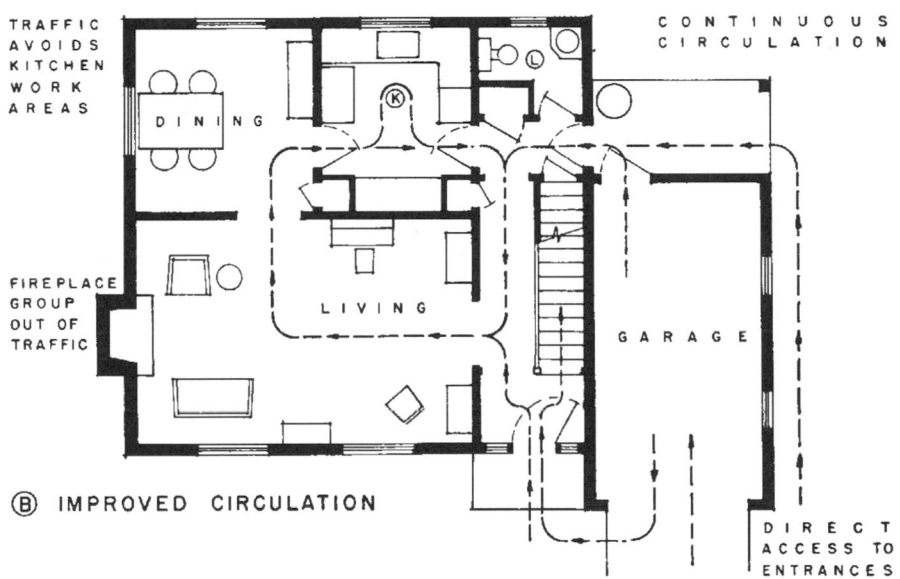

TRAFFIC AVOIDS KITCHEN WORK AREAS

FIREPLACE GROUP OUT OF TRAFFIC

CONTINUOUS CIRCULATION

Ⓑ IMPROVED CIRCULATION

DIRECT ACCESS TO ENTRANCES

Fig. 1 Paths of circulation. Dotted lines show most commonly used circulatory routes. The continuous path in B is more desirable than the dead-end and indirect circulation illustrated in A.

LIVING ROOMS

THE LIVING ROOM

Drawing room, parlor, minister's room—these are some of the various ancestors of today's living room, a formal space reserved for the most formal of guests. This is the place for serving afternoon tea, inspecting young courtiers, or hosting the card club (Fig. 1).

With the predominantly informal lifestyle of our contemporary society, there is some debate over how valid the living room really is today. Certainly there are very few households who confine their entertainment to the living room. Invariably, all guests will avoid spending too much time here if it is possible for them to drift to where the action is—the kitchen. Thus, one must ask why the living room has been retained over the years.

The answer lies partly in tradition, which, despite contemporary changes in lifestyle, is still very much a consumer value. For most households formal occasions have become more informal and more infrequent, but even the most casual homeowners enjoy having spaces that can rise to the special occasions.

Recognizing the diminishing use of the living room, many contemporary designers are reconfiguring it into a larger freestyle space that may include the balance of the ceremonial component. As with other transitional movements, this may be more prevalent in the southern climates where new communities devoted to casual, resort lifestyles favor a plan that combines the living, dining, and entry areas into one *grand room* or *great room*. In plans of relatively small square footage, this trend makes more sense than the traditional compartmentalization found in the northern climates. Yet even with this type of space, today's designs are incorporating more formal elements than those of 10 to 15 years ago (Figs. 2 and 3).

The formality of design in the living room also accommodates its role as the household museum or gallery. Here, the finest furniture, artwork, and heirlooms can be shown in their full splendor. They enhance the living room, and the room complements them. In family households, this complementary formality also discourages high-volume traffic—which encourages safe preservation of our treasures.

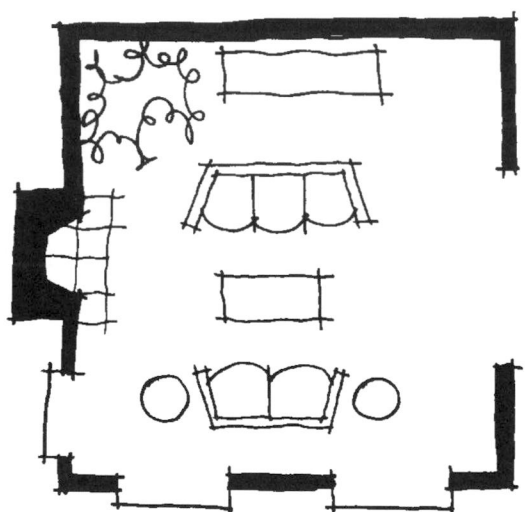

Fig. 1   Formal living room.

When the museum role is considered for the living room, several design objectives become apparent, whether for a separate room or for an area that is part of a great room plan. Like an art gallery, the living room should include *ample wall space* for display purposes. Door and window placement that creates interior symmetry and balanced dimensions promote a formal sense.

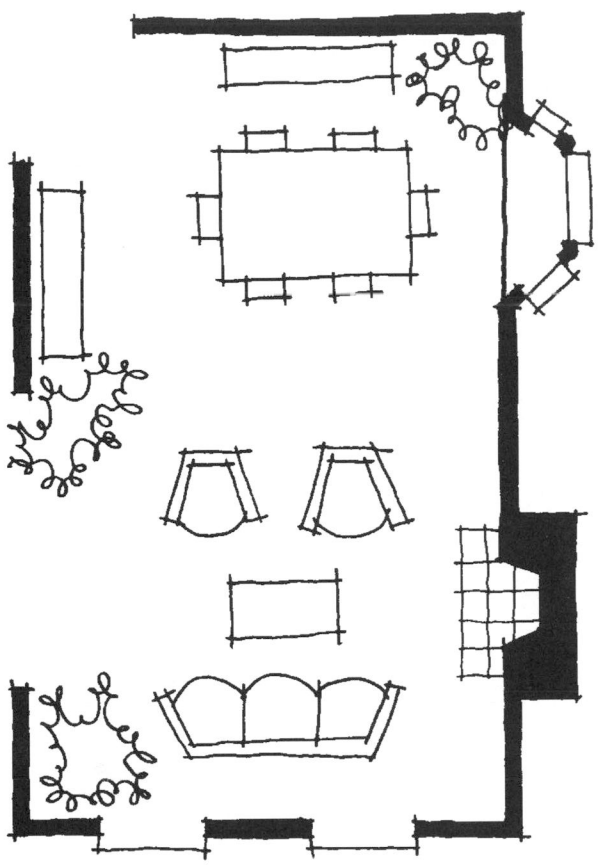

Fig. 2 Grand room (living and dining).

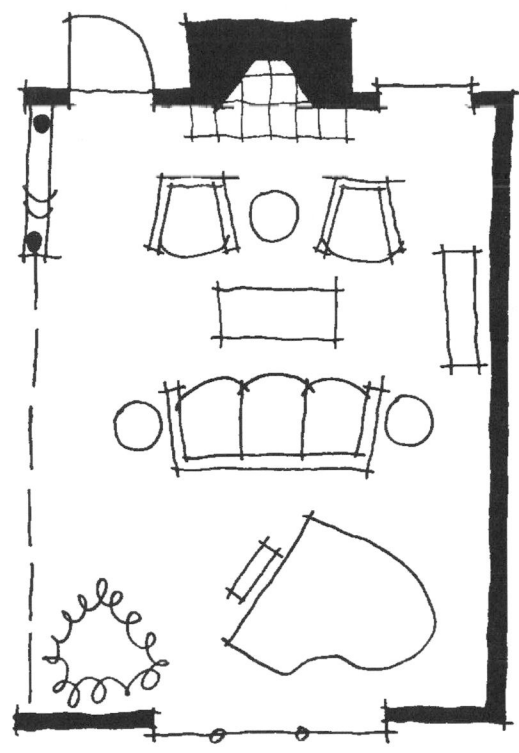

Fig. 3 Great room (living and family).

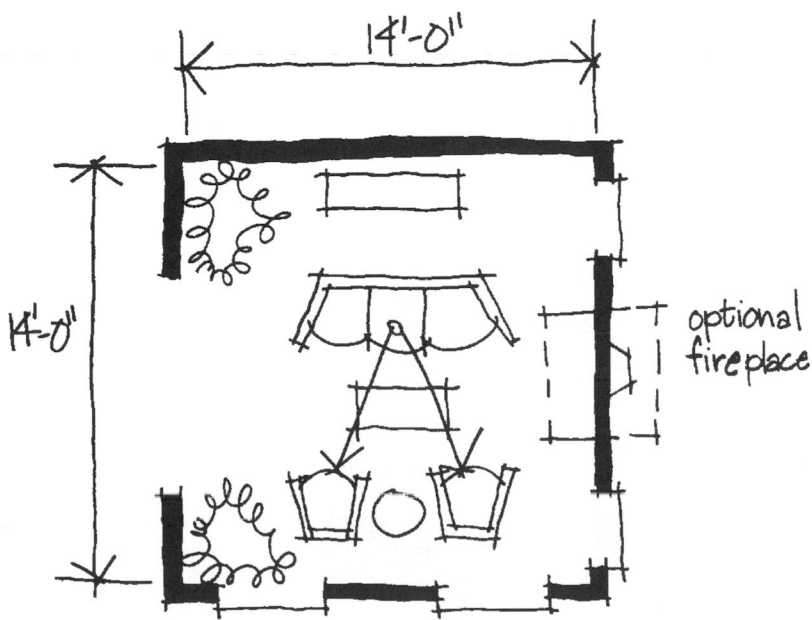

Fig. 4 Symmetrical living room.

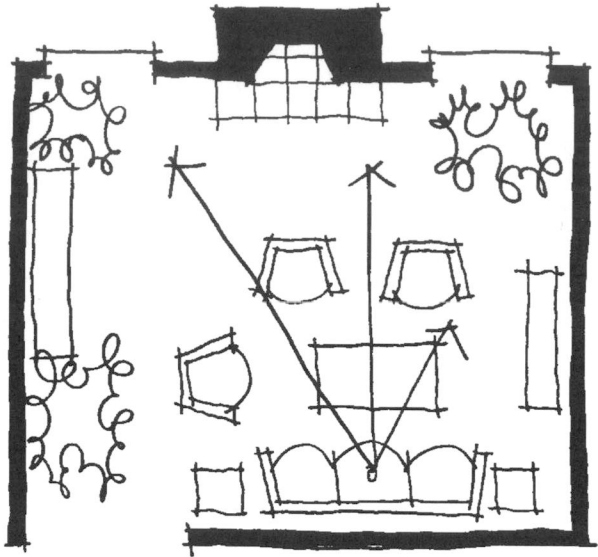

Fig. 5   Living room furniture placement.

The size of the living room is very dependent on market preferences and price limitations. In very small homes, the living room can be adequately accommodated in a 12- by 12-ft space, but a 14- by 14-ft space is a more acceptable minimum. Although some designs introduce volume ceilings into the living room, generally buyers will prefer more emphasis on materials and treatments, such as crown moldings, chair rails, and lavish wall veneers (Fig. 4).

Fireplaces for the living room are still popular but increasingly are offered as an option or are included as a secondary location to the fireplace in the family room. Location of the fireplace, windows, and openings to the living room must consider the primary requirements of accommodating furniture for seating, allowing comfort, and promoting conversation. Placement of windows for this element of the ceremonial component should include recognition that the people using this room will often be seated; lower, narrower windows afford views, as well as light to seating areas. Skylights and clerestory windows also add illumination and interest and are especially helpful to smaller living room plans.

Seating areas in the living room should be large enough to contain a sofa and side chairs, with a fireplace and/or specialty window as the focal point of the furniture arrangement. Balancing the room to include a sufficient amount of wall area, furnishing space, and windows is vitally important to the effective design of the living room. An overabundance of windows here will undoubtedly force the owners to partially cover a window with a sofa or chair. The most sensible solution designates one wall as a view wall, and one or two walls as furniture walls. A good view exposure should be assigned to the window wall, with the furniture wall left blank. The view wall may also include the fireplace, which provides the room with a focus for either day or evening uses (Fig. 5).

SPATIAL CHARACTERISTICS AND ARRANGEMENT

Living Area

Each living unit should contain space that is conducive to general family living activities, among which are entertaining, reading, writing, listening to music, watching television, relaxing, and frequently children's play. Unless specifically provided for elsewhere in the unit, appropriate space for these activities should be provided in the living area.

Living Furniture that can be accommodated in the living area should include the following items (sizes are minimums):

One couch, 3 ft 0 in by 6 ft 10 in
Two easy chairs, 2 ft 6 in by 3 ft 0 in
One desk, 1 ft 8 in by 3 ft 6 in
One desk chair, 1 ft 6 in by 1 ft 6 in
One television set, 1 ft 4 in by 2 ft 8 in
One table, 1 ft 6 in by 2 ft 6 in

Commentary Necessary planning considerations should include provision of adequate floor and wall space for furniture groupings, separation of trafficways from centers of activity, and ease of access to furniture and windows.

Circulation

Circulation through the living room should be as direct as possible, yet it should not interfere with the furniture placement.

Doors

The location of doors should fully consider the need for generous wall space for the placement of furniture.

Conversation Area

People gather or congregate during social activities in rather small groups. A desirable conversation distance is of relatively small size, approximately 10 ft in diameter.

Planning Considerations

Through traffic should be separated from activity centers.
   Openings should be located so as to give enough wall space for various furniture arrangements.
   Convenient access should be provided to doors, windows, electric outlets, thermostats, and supply grills.

Furniture Clearances

To assure adequate space for convenient use of furniture in the living area, not less than the following clearances should be observed: 60 in between facing seating, 24 in where circulation occurs between furniture, 30 in for use of desk, 36 in for main traffic, 60 in between television set and seating.
   Seating arranged around a 10-ft-diameter circle makes a comfortable grouping for conversation.

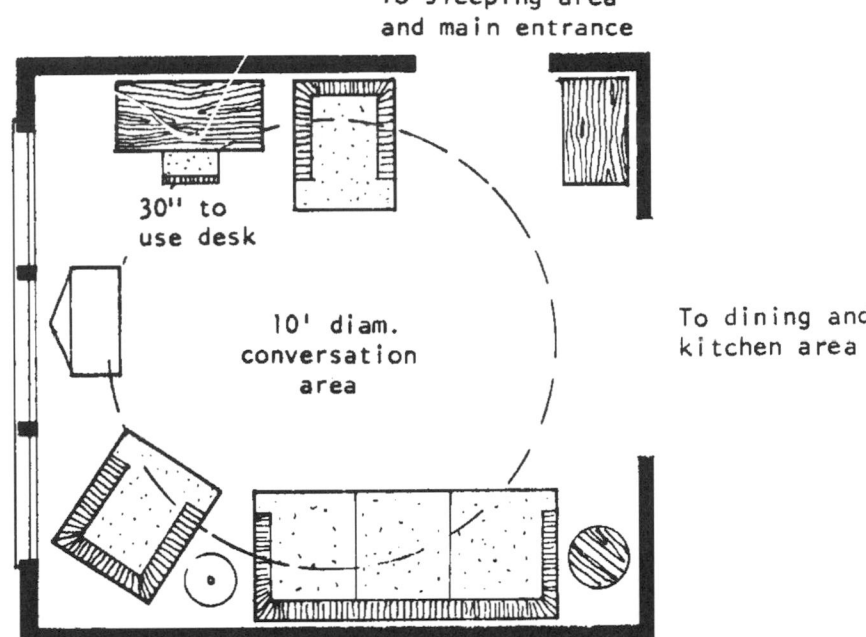

Fig. 6   Plan.

Fig. 7   Living rooms considered from the point of view of possible furniture groupings and interference by necessary circulation. Circles indicate conversational groups; dotted lines show circulation.

LIVING ROOMS

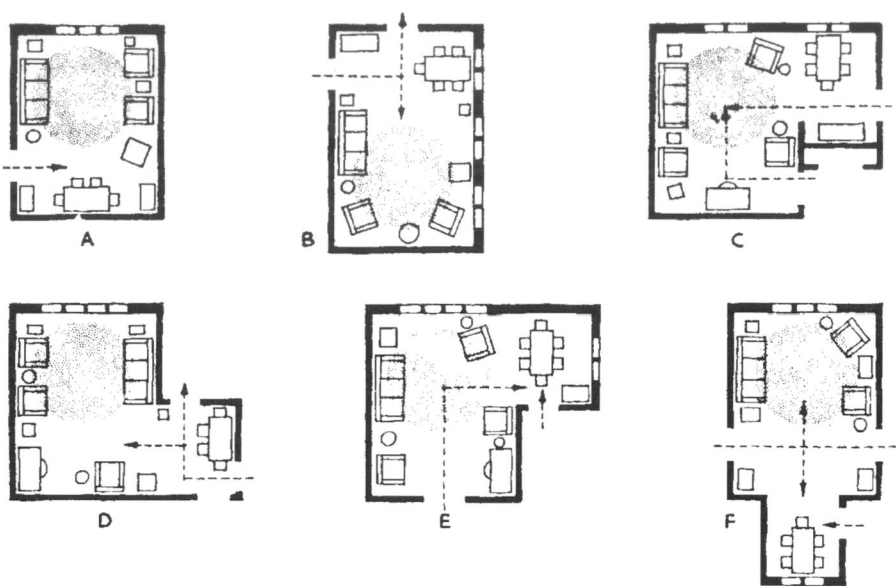

Fig. 8 A relationship similar to Fig. 7, complicated by adding constantly used dining space.

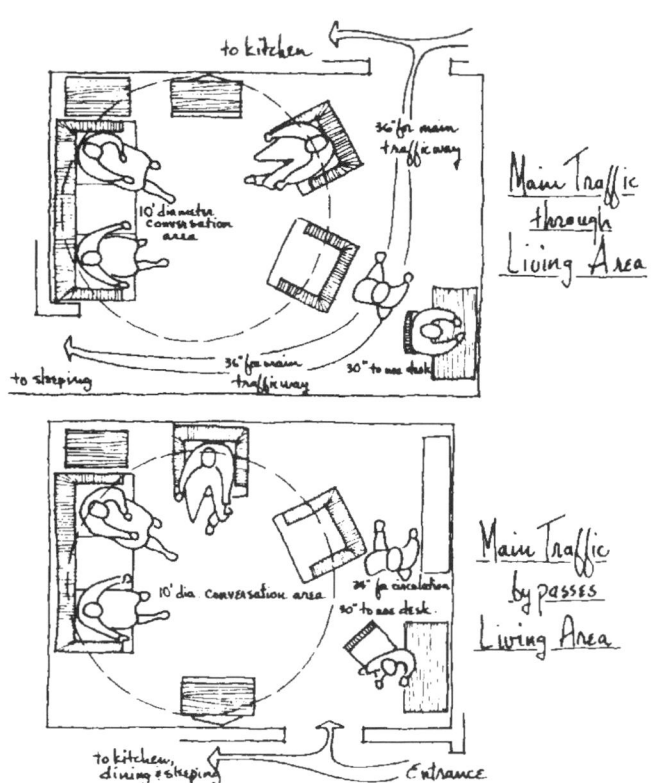

Fig. 9 Minimum clearances, circulation, and conversation areas for living rooms.

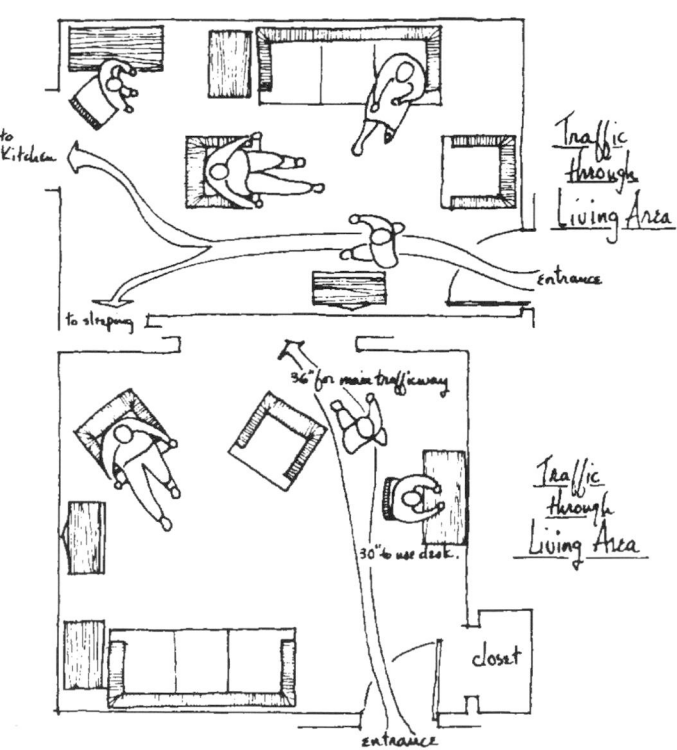

Fig. 10 Minimum clearances, circulation, and conversation areas for living rooms.

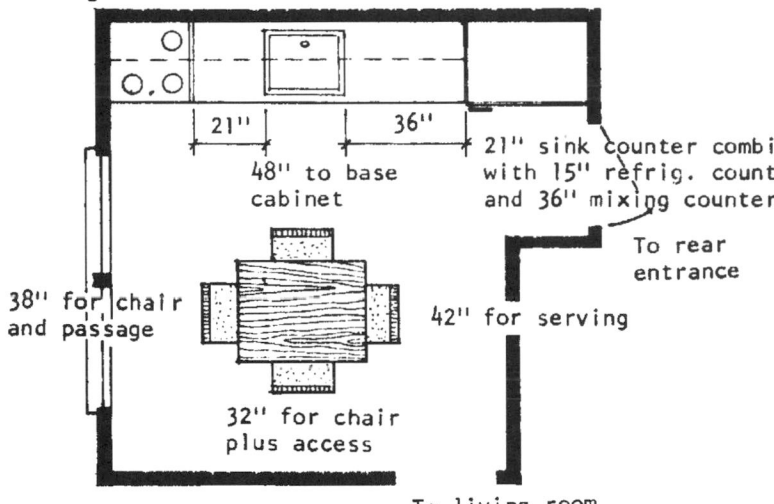

21" sink counter combined
with 21" range counter

21" 36"

48" to base
cabinet

21" sink counter combined
with 15" refrig. counter
and 36" mixing counter

To rear
entrance

38" for chair
and passage

42" for serving

32" for chair
plus access

To living room

COMBINED SPACES

A combination dining area-kitchen is preferred by some occupants of small houses and apartments. This arrangement minimizes housekeeping chores and provides space which can be used as the family's day-to-day meeting place.

Fig. 11   Combined dining area–kitchen, two-bedroom living unit.

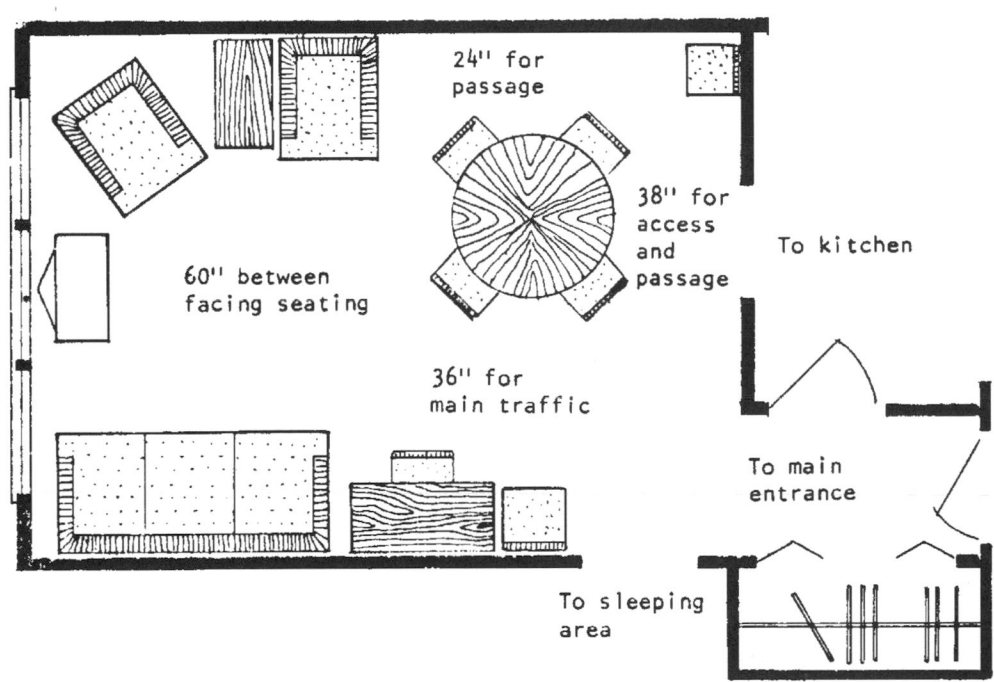

24" for
passage

38" for
access
and
passage

To kitchen

60" between
facing seating

36" for
main traffic

To main
entrance

To sleeping
area

Fig. 12   Combined living–dining room.

LIVING ROOMS

Often several compatible living functions can be combined advantageously in a single room. Some of the benefits of such arrangements are that less space is used but it is used more intensively, its functions can be changed, making it more flexible and serviceable space, it is adaptable to varied furniture arrangements, while visually it can be made more interesting and seem more generous than if the same functions were dispersed into separate rooms.

For adjacent spaces to be considered a combined room, the clear opening between them should permit common use of the spaces. This usually necessitates an opening of at least 8 ft.

A bed alcove with natural light and ventilation and which can be screened from the living area is desirable in a zero-bedroom living unit.

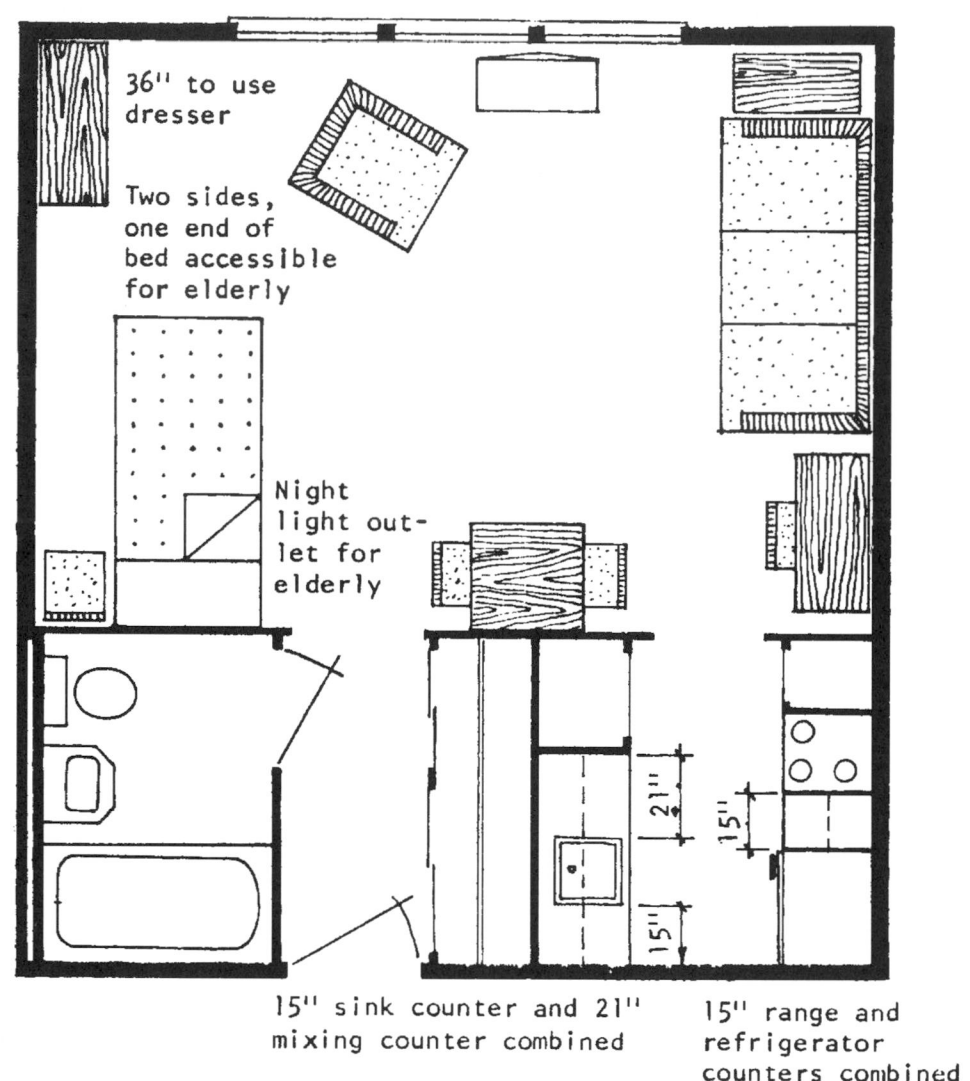

Fig. 13   Zero-bedroom living unit.

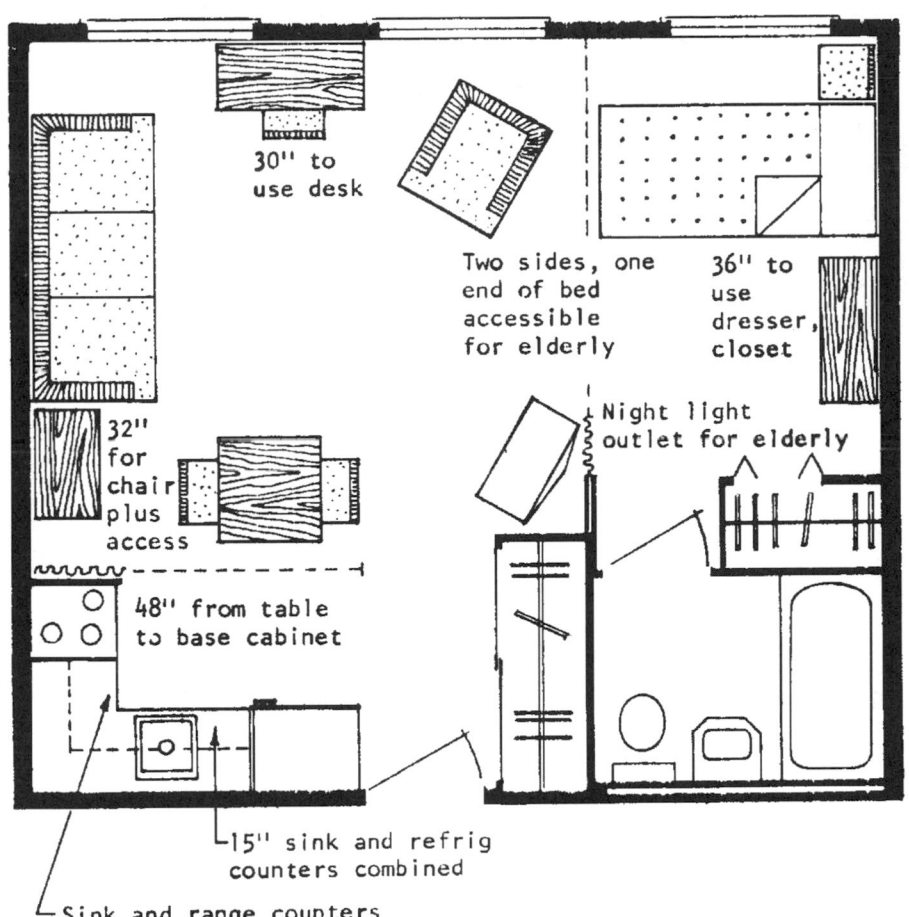

30" to
use desk

Two sides, one
end of bed
accessible
for elderly

36" to
use
dresser,
closet

32"
for
chair
plus
access

Night light
outlet for elderly

48" from table
to base cabinet

15" sink and refrig
counters combined

Sink and range counters
combined with 21" mixing
counter

Fig. 14 Zero-bedroom living unit with sleeping
alcove.

LIVING ROOMS

Typical Living Room Arrangements

Many different activities that must be accommodated in the living room complicate the space planning. These activities can be grouped under three headings: (1) social, (2) recreational, and (3) cultural. A thorough analysis of just what activities the family normally engages in is the first requisite for planning the space to provide both enough area and properly arranged area for each of the activities. The analysis must include primarily those activities that presumably will be engaged in simultaneously. Those that follow one another in point of time will permit dual use of both furniture and space.

As each of the activities usually demands its own quota of furniture and space for its pertinent paraphernalia, all must be enumerated and provided for in the planning.

The size of the living room should reflect the size of the dwelling unit and the economic status of the occupants. A living room for a three- or four-bedroom dwelling unit requires more space for its occupants than one for a one- or two-bedroom dwelling unit. Luxury units will necessarily need more space to accommodate more furnishings. In any case, the minimum living room with no dining facilities should be approximately 180 ft^2 but preferably around 200 ft^2. Figure 16a and b shows two living rooms with typical furniture groupings (no dining facilities).

Figure 16c shows a living room with one end used for dining. This area often is arranged in an L shape to achieve greater definition or privacy from the living activities. Dwelling units with three or more bedrooms should have separate dining rooms or clearly defined dining areas.

The range of living activities generally includes a conversation area (sitting area), relaxation area (books, TV, and music center), a work area (sewing machine, desk, and chair), and an entertainment area (bar, card table, terrace). Often it is the center of childplay if there is no space in the kitchen.

The minimum width of a living room should be 11 ft 0 in to 12 ft 0 in. The recommended width is 14 ft 0 in. There should be no through traffic in the living room. Preferably, the living room should be a dead-end space with all traffic handled at one end.

The major problem is to provide for the necessary flexibility in order to achieve the various activities. Separation and some degree of privacy are required. When a living room is combined with the dining area, the dining area should be offset into an alcove or be clearly identified as such.

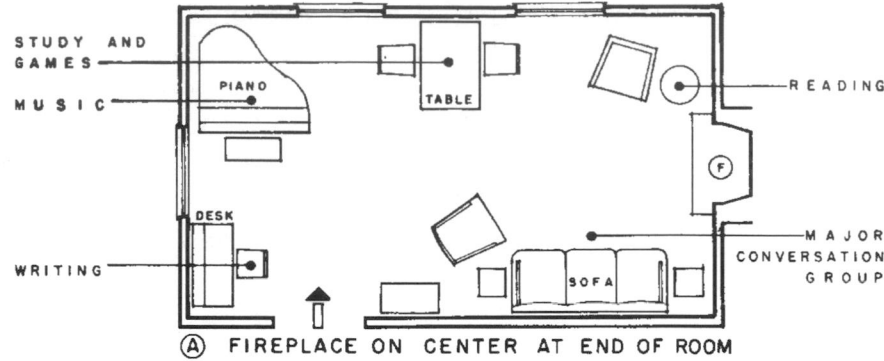

(A) FIREPLACE ON CENTER AT END OF ROOM

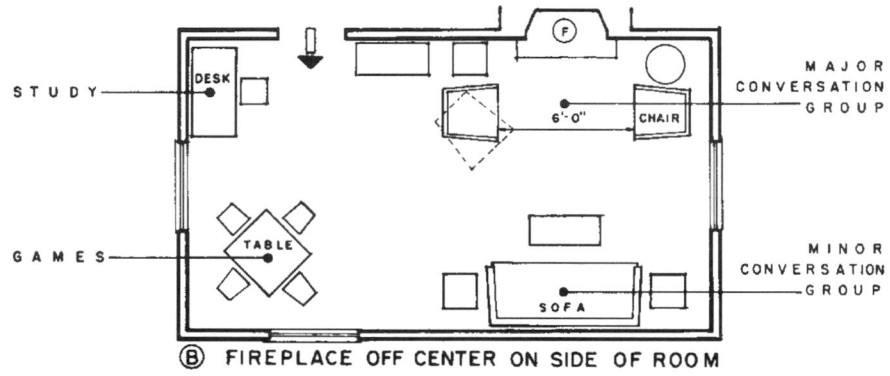

(B) FIREPLACE OFF CENTER ON SIDE OF ROOM

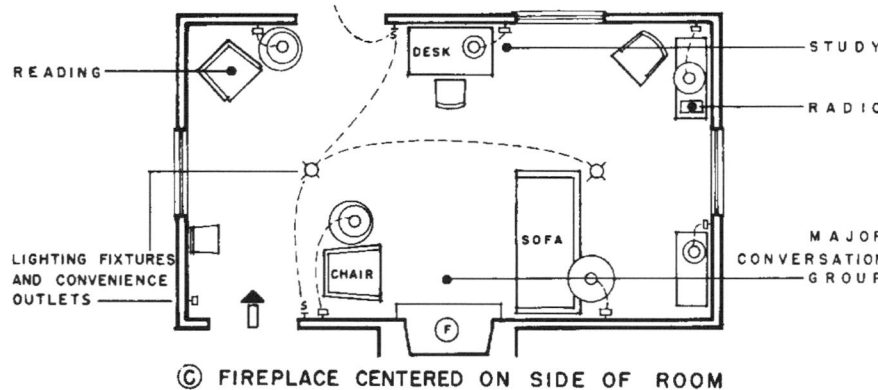

(C) FIREPLACE CENTERED ON SIDE OF ROOM

Fig. 15   These plans show how furniture groupings may be adapted to various types of recreational activities with the fireplace as the center of interest.

TABLE 1   Sizes of Living-Room Furniture

Sofas	2 ft 6 in to 3 ft 6 in deep by 6 to 7 ft long	Governor Winthrop	2 ft by 3 to 3 ft 8 in	Tables, circular:	
Love seats	2 ft 6 in to 3 ft deep by 3 ft 6 in to 4 ft 6 in long	Secretary	1 ft 6 in to 2 ft deep by 3 to 4 ft long	Lamp	2 ft diameter
				Coffee	3 ft diameter
Chairs		Highboy, lowboy	1 ft 6 in to 2 ft deep by 2 ft 6 in to 3 ft 6 in long	Drum	3 ft diameter
Club	2 ft 9 in by 3 ft 6 in			Piecrust	3 ft diameter
Wing	2 ft 6 in by 2 ft 9 in	Tables, rectangular		Pianos	
Bridge	1 ft 6 in by 1 ft 6 in	End	1 ft 3 in by 1 ft or 1 ft 8 in by 1 ft 8 in	Grand	4 ft 10 in to 5 ft wide by 5 to 9 ft long
Desks		Coffee	2 by 3 ft		
Flat-top	2 ft to 2 ft 6 in deep by 4 to 5 ft long	Bridge	2 ft 6 in by 2 ft 6 in	Upright	2 by 5 ft
		Console	1 ft 6 in by 3 ft		

A

for 1- and 2-bedroom apartments
12'6" x 16'0"
200 sf.

B

for 3-bedroom apartments
12'6" x 20'0"
250 sf.

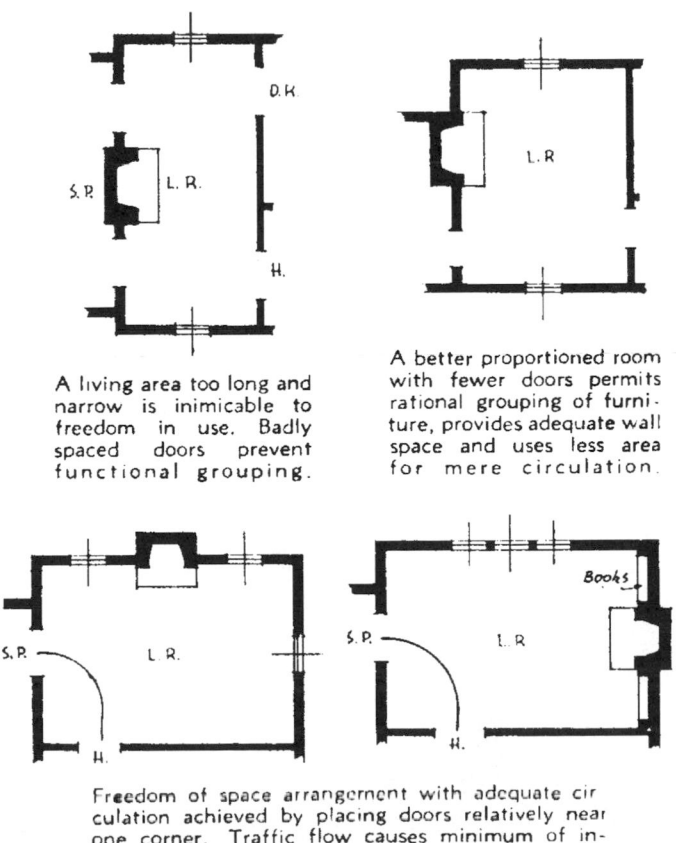

Fig. 16

C

for 3-bedroom apartments
12'6" x 22'0"
275 sf.

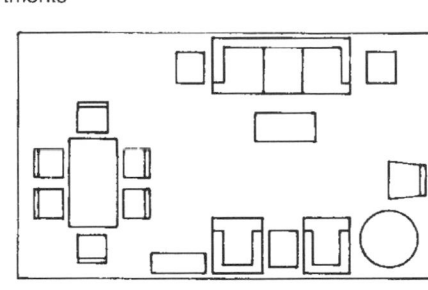

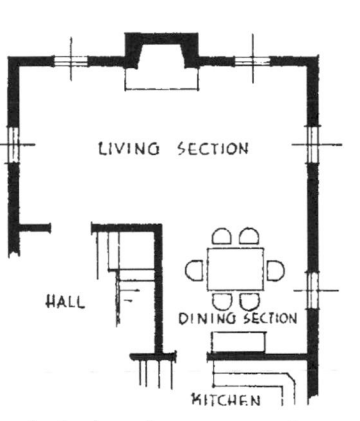

A living area too long and narrow is inimicable to freedom in use. Badly spaced doors prevent functional grouping.

A better proportioned room with fewer doors permits rational grouping of furniture, provides adequate wall space and uses less area for mere circulation.

An L-shaped room segregates dining from other activities and gains sense of space

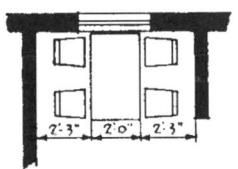

Minimum dimensions for informal dining area are 6 ft 6 in where freestanding chairs are to be used for better comfort and for ease of circulation

Freedom of space arrangement with adequate circulation achieved by placing doors relatively near one corner. Traffic flow causes minimum of interference with the groupings or activities. Location of fireplace depends on furniture groupings.

Fig. 17

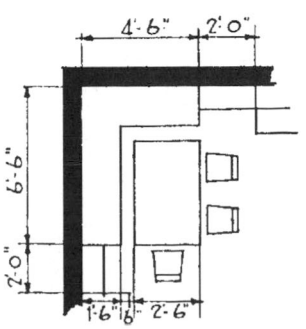

A banquette arrangement in a corner conserves space. It also makes a useful work or study area

Fig. 18

LIVING ROOMS

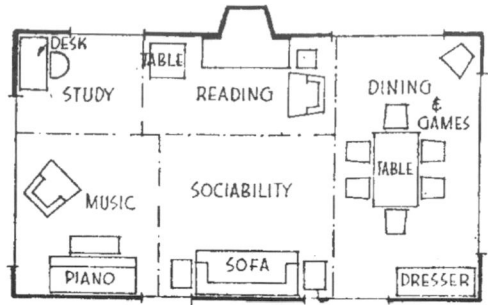

Use of part of living room for dining complicates planning, but increases sense of space. Dining table and chairs are useful for other purposes

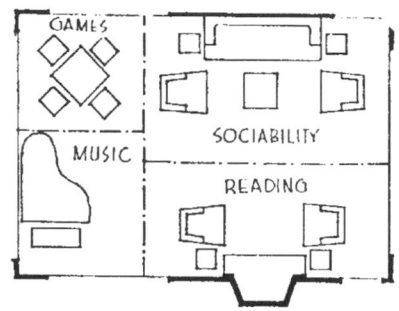

Freedom of interference in space-use can be attained by the diagrammatic study of the placing of furniture or equipment according to primary function. Window arrangement for light and view can follow.

Fig. 19

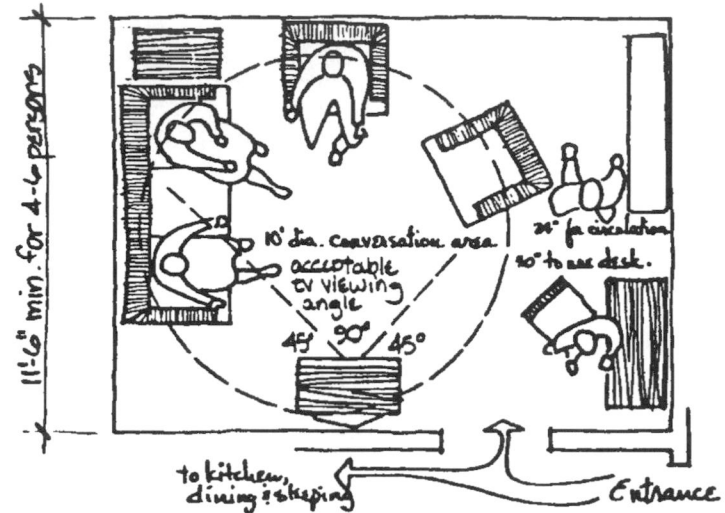

Fig. 20

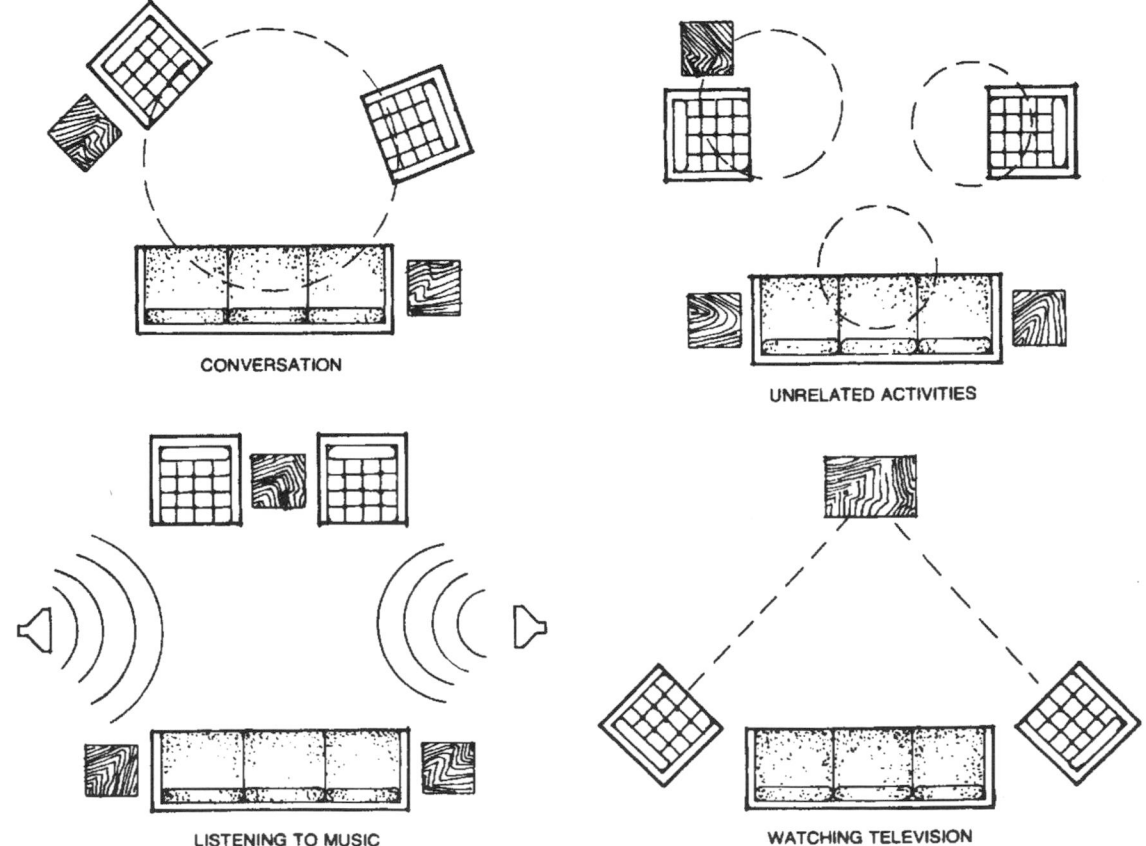

CONVERSATION

UNRELATED ACTIVITIES

LISTENING TO MUSIC

WATCHING TELEVISION

Fig. 21

THE DINING ROOM

For many of today's households, the dining room plays only a vestigial role in daily life. Fewer families gather regularly for evening meals, and most singles will bemoan how little time they spend preparing genuine evening meals. For all buyers, formal dining is also much less important. In response to this trend of increasingly casual lifestyles, some designers and builders have reduced the dining room to an area in the great room, even for large homes.

In recent years buyers have again begun to demand a "real" dining room, and today's new homes include either a dining room that is separate from the living room, or a more formal design treatment for the dining area of a great room floor plan. As with the living room, when the dining room is utilized, it is for special occasions of major significance. In fact, the infrequency of its use contributes to the significance of the meals shared here. Its importance is also heightened by the fact that it is the backdrop for the oldest and most enduring of social rituals—offering hospitality through the medium of food. Because of this context, the dining room is very much the epitome of the ceremonial component, and minimal "container" design will not suffice.

As with the living room, the dining room should emphasize grace and dignity through window placements, quality finishes, and materials. Because of the room's limited use, and because much use occurs during evening hours, provision for prime views to the outdoors is nice but not a mandatory requirement. And because most use is at night, the dining room is generally the only room in the house equipped with a fancy light fixture as standard equipment. If handsome views are easily available within the overall context of the floor plan, window placement should be low enough to afford views from a seated position.

Dimensions for the dining room should be no less than 12 by 12 ft, and a 14- by 14-ft space is far more comfortable. A rectangular configuration (12 by 14 ft) is also very acceptable and functional, as it can accommodate a rectangular dining room table.

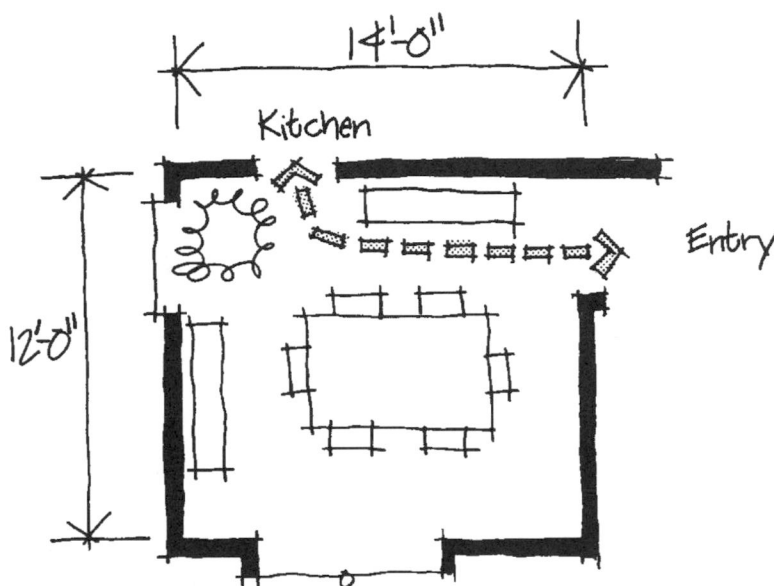

Fig. 1 Formal dining room clearances and circulation.

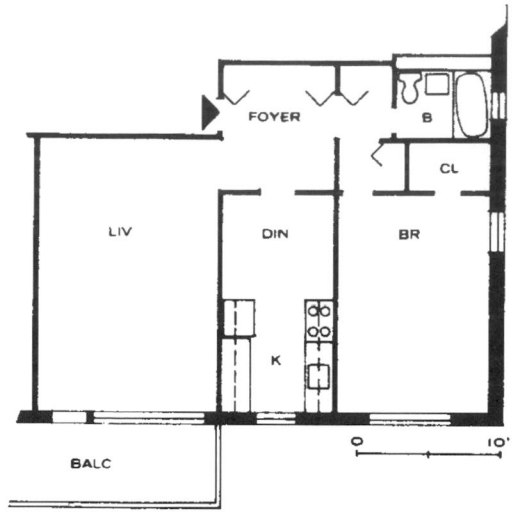

Fig. 2   Dining space as part of the kitchen.

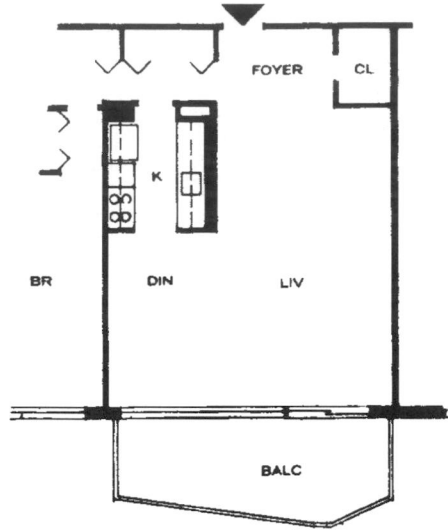

Fig. 3   Dining area—dining room.

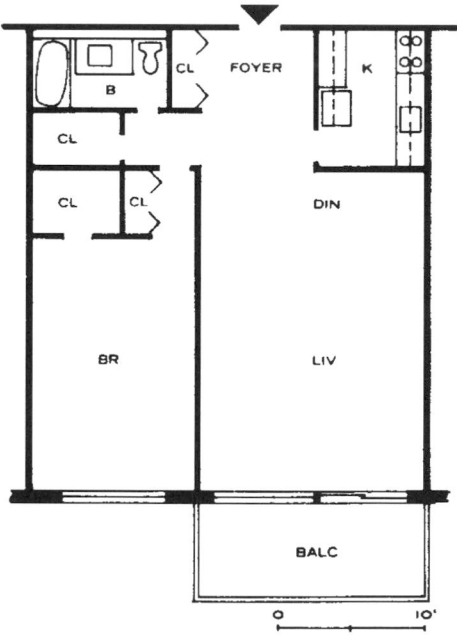

Fig. 4   Combined living room and dining area.

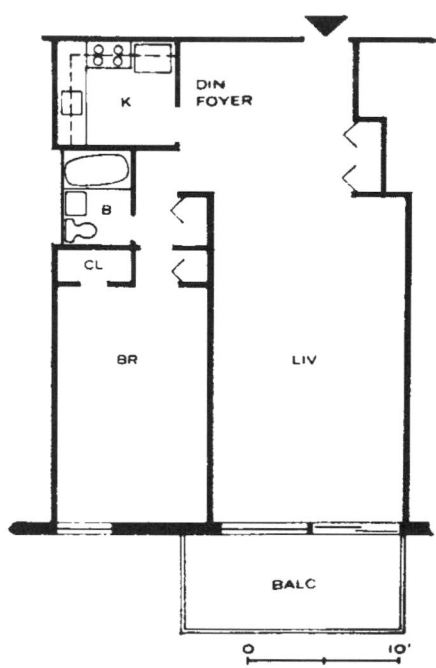

Fig. 5   Dining space in the foyer.

SPATIAL CHARACTERISTICS AND ARRANGEMENT

Requirement

Each living unit should contain space for the purpose of dining. This area may be combined with the living room or kitchen, or may be a separate room.

Criterion

The amount of space allocated to dining should be based on the number of persons to be served and the proper circulation space. Appropriate space should be provided for the storage of china and large dining articles either in the dining area itself or in the adjacent kitchen.

Space for accommodating the following sizes of tables and chairs in the dining area should be provided, according to the intended occupancy, as shown:

    1 or 2 persons: 2 ft 6 in by 2 ft 6 in
    4 persons: 2 ft 6 in by 3 ft 2 in
    6 persons: 3 ft 4 in by 4 ft 0 in or 4 ft 0 in round
    8 persons: 3 ft 4 in by 6 ft 0 in or 4 ft 0 in by 4 ft 0 in
    10 persons: 3 ft 4 in by 8 ft 0 in or 4 ft 0 in by 6 ft 0 in
    12 persons: 4 ft 0 in by 8 ft 0 in
    Dining chairs: 1 ft 6 in by 1 ft 6 in
    Buffet or storage unit: 1 ft 6 in by 3 ft 6 in

Commentary

Size of the individual eating space on the table should be based upon a frontage of 24 in and an area or approximately 2 ft². In addition, table space should be large enough to accommodate serving dishes.

Desirable room for seating is a clear 42 in all around the dining table. The following minimum clearances from the edge of the table should be provided: 32 in for chairs plus access thereto, 38 in for chairs plus access and passage, 42 in for serving from behind chair, 24 in for passage only, 48 in from table to base cabinet (in kitchen).

In sizing the separate dining room, provision should be made for circulation through the room in addition to space for dining.

The location of the dining area in the kitchen is desirable for small houses and small apartments. This preference appears to stem from two needs: (1) housekeeping advantages; (2) the dining table in the kitchen provides a meeting place for the entire family. Where only one dining location is feasible, locating the dining table in the living room is not recommended.

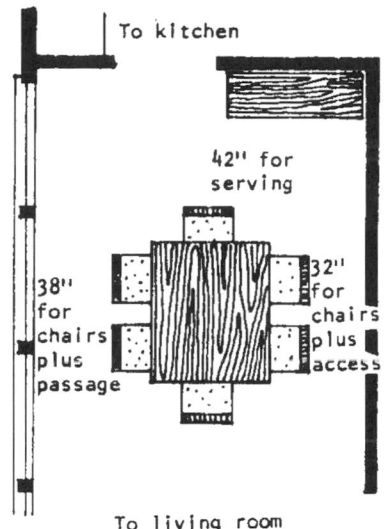

Fig. 6a Dining room, six persons, three-bedroom living room.

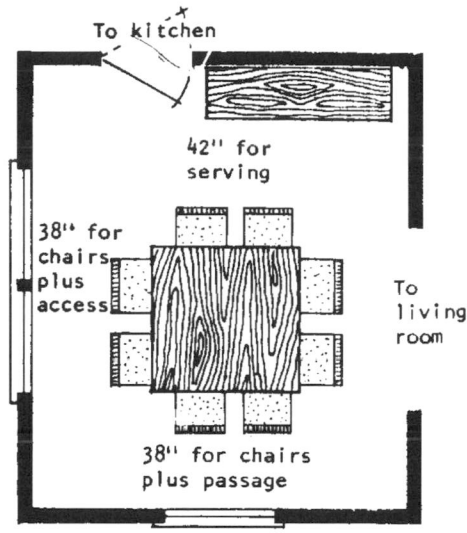

Fig. 6b Dining room, eight persons, four-bedroom living unit.

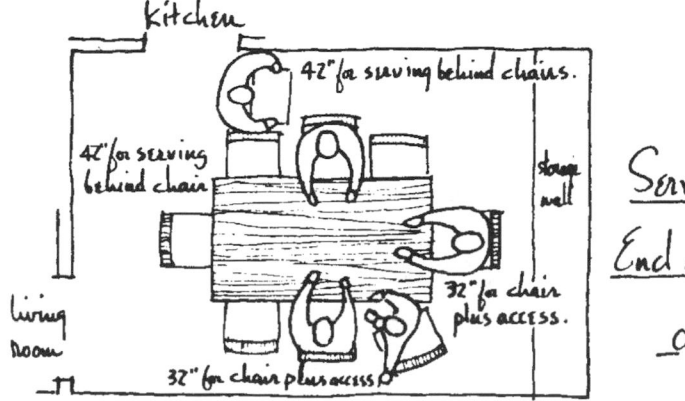

Fig. 7 Minimum clearances for dining areas.

Figure 8 illustrates a 36-in diameter round table seating four people and shows the necessary clearances around the perimeter. Four people cannot function comfortably around such a small table for anything but the lightest snacks. The individual place-setting zones are extremely restricted, and the shared access zone at the center is too small to accommodate much in the way of serving dishes, platters, or decorative elements. A 48-in clearance between the perimeter of the table and the wall or nearest physical obstruction is the minimal clearance necessary to allow circulation behind a seated person. A distance of 30 to 36 in between the table perimeter and the wall is the minimum clearance necessary to permit access to and adjustment of the chair.

The 48-in diameter table shown in Fig. 9, however, can function adequately for four people. The place-setting zone is reasonably sufficient to accommodate the various place-setting elements and provides generous elbow room as well. Although the central shared access zone is restricted, it provides far more space than the 36-in diameter table shown in Fig. 8. If used for light snacks or coffee, the table can seat five. The clearances for circulation are the same as for the 36-in table.

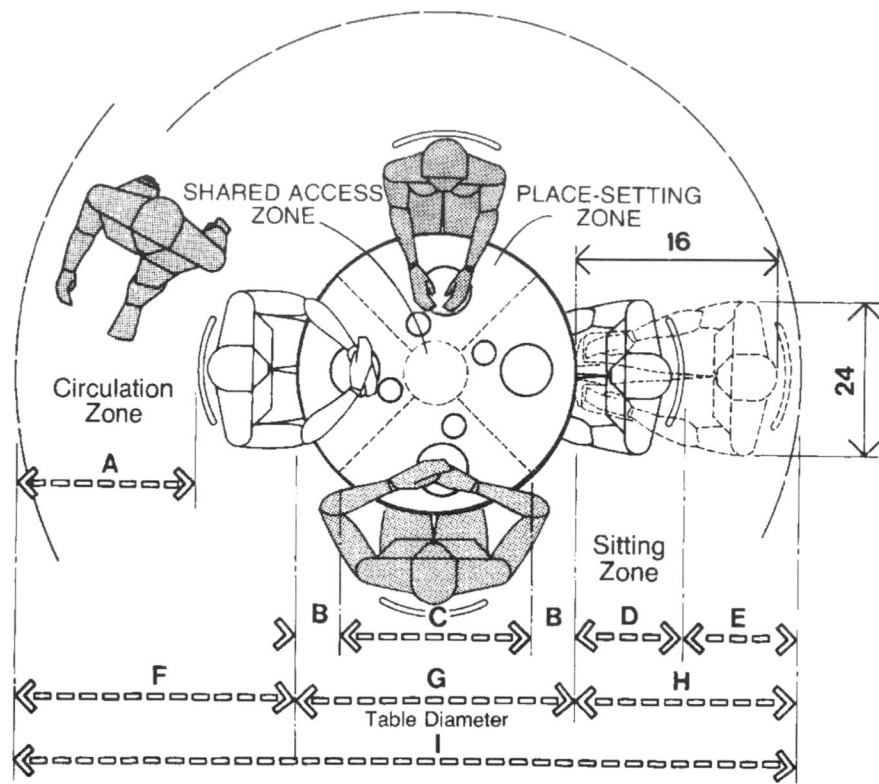

Fig. 8   A 36-in diameter breakfast/kitchen table for four. (See Table 1.)

Fig. 9   A 48-in diameter circular table for four—minimum scheme. (See Table 1.)

TABLE 1

	Inches
A	30 min.
B	6
C	24
D	18–24
E	12
F	48–54
G	36
H	30–36
I	114–126
J	84–96
K	48

Figure 10 applies the optimal incremental unit discussed on p. 296 to a rectangular table for formal dining for six. The table size shown is 54 by 96 in. This size will provide each person with an individual place-setting zone of 18 by 30 in and will allow a shared access zone at the center of the table with a depth of 18 in. The 30-in width provided for each person allows for elbow room.

Figure 11, in contrast, shows a minimal, square general purpose table for informal dining. Although the width and depth of the place-setting zones are the same as in the larger rectangular tables, their angular configuration reduces the area significantly as well as the area of the shared access zone. To allow clearance for the chair and head-on circulation behind the chair, a minimum distance of 48 in must be maintained between the edge of the table and the wall or nearest physical obstruction. A clearance of 36 to 42 in can be provided to allow restricted circulation. This will require a person to sidestep or the seated person to adjust the chair to allow passage.

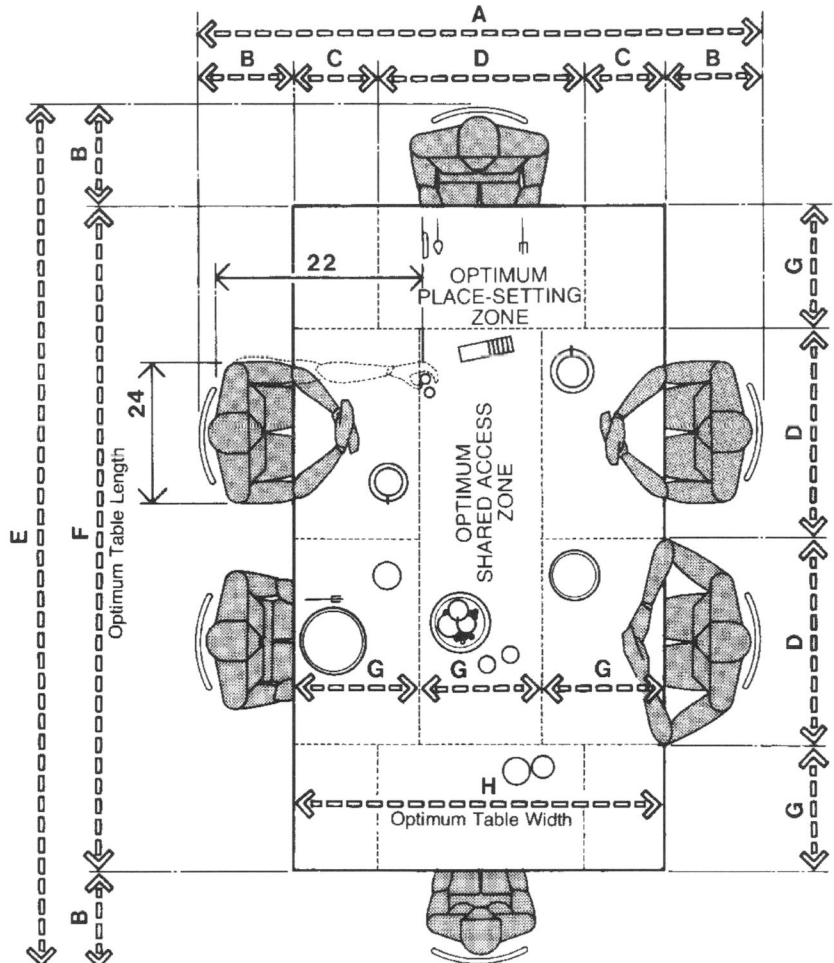

Fig. 10 Rectangular table with optimum length and width for dining for six. (See Table 2.)

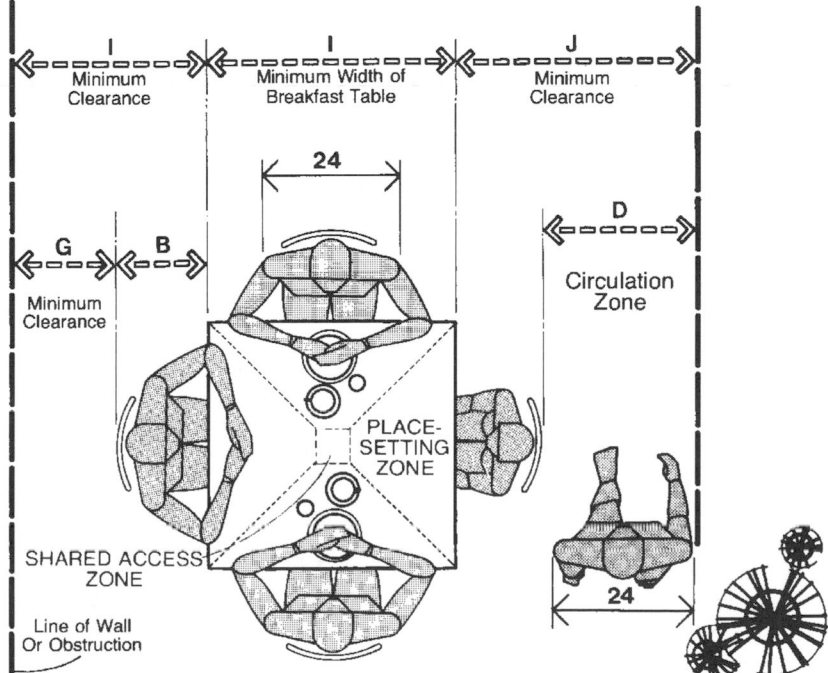

Fig. 11 Square table for four—minimal scheme. (See Table 2.)

TABLE 2

	Inches
A	96–102
B	18–24
C	12
D	30
E	132–144
F	96
G	18
H	54
I	36–42
J	48

DINING AREAS

Dining spaces for serving family meals to six persons are illustrated in Fig. 13. The size of table and space around the table for serving meals and clearing away dishes while people remain seated are shown for two seating arrangements—people on four sides and people on two opposite sides. The table space allowance for each individual is 24 in.

In Fig. 13, boldface indicates liberal dining area dimensions that provide for liberal clearances and tables of a larger size. Lightface indicates minimum dining area dimensions, providing for minimum clearances and minimum table size. Clearances allow only for straight-back, straight-leg chairs.

Dining areas in which company meals are to be served should provide at least the liberal allowances suggested for table and service space for family meals. At company meals more people may be served, more and larger serving dishes may be used, and individual covers may be more elaborate.

The dining areas shown do not provide space for storing guest china.

TABLE 3 Space Requirements for Individual Place Settings and Clearance around Table

Item	Minimum	Liberal
Space, inches, for individual place settings (cover):		
Width, side-to-side	24	29
Depth	12	15
Clearances, inches, table edge to wall:		
Getting up	24	30
Serving	30	36

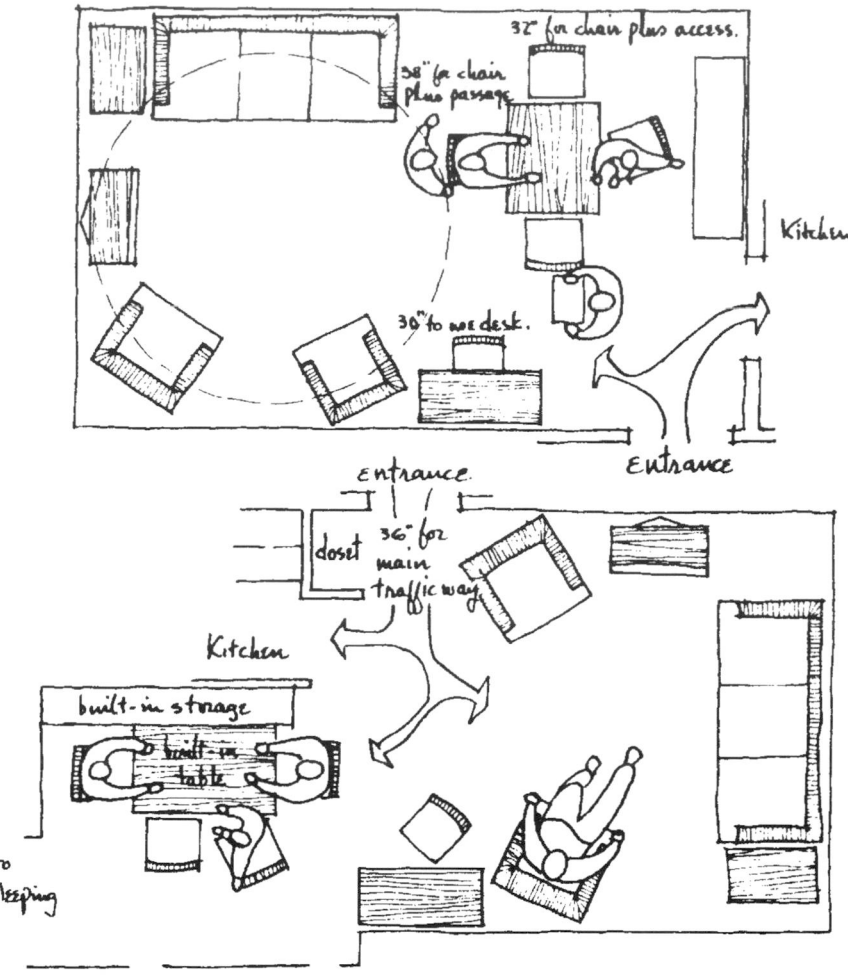

Fig. 12 Minimum clearances and circulation for combined living-dining areas.

SERVING TWO SIDES & ONE END

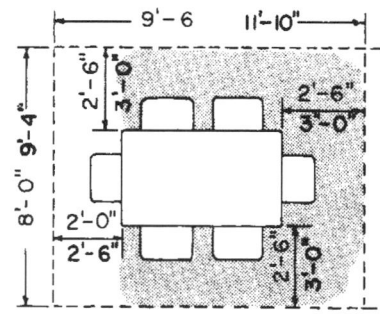

SERVING FOUR SIDES

SERVING TWO ENDS & ONE SIDE

SERVING ONE END & ONE SIDE

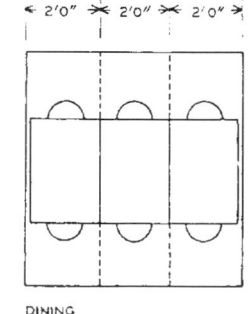

DINING

SERVING ONE SIDE

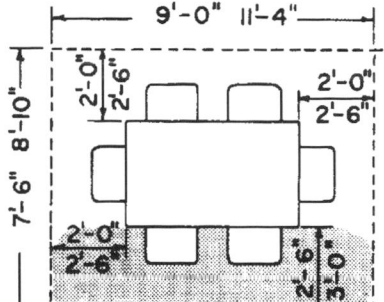

SERVING ONE END

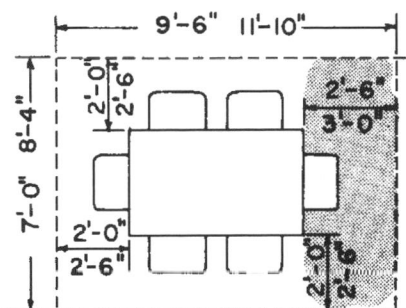

SERVING-ONE END

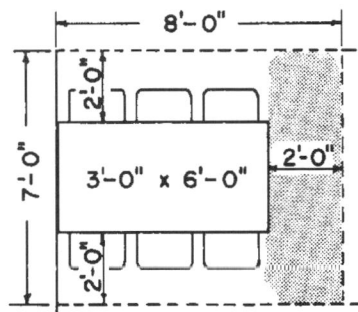

SERVING-ONE END & ONE SIDE

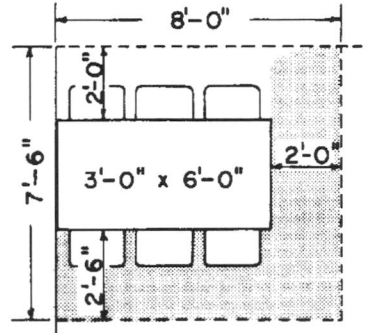

SERVING-TWO SIDES

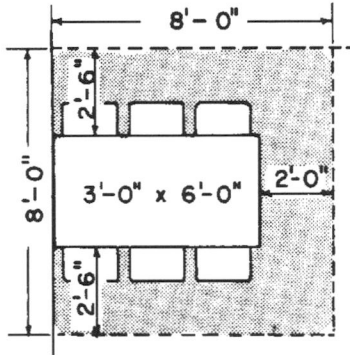

Table Sizes
Liberal — 3'-4" x 6'-4"

Minimum — 3'-0" x 5'-0"

 Space allowance
for serving

Fig. 14

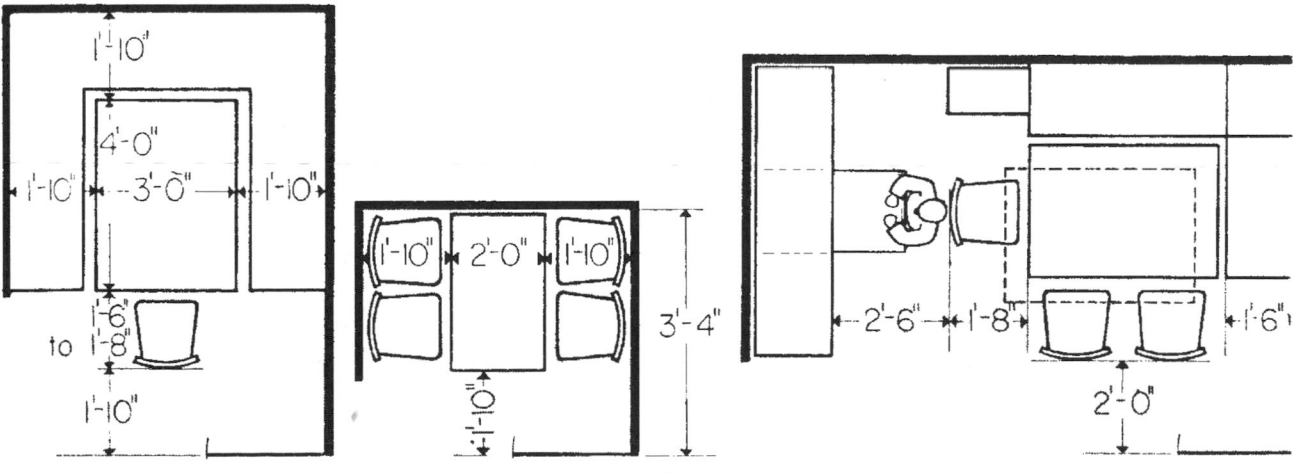

Fig. 15

DINING AREAS

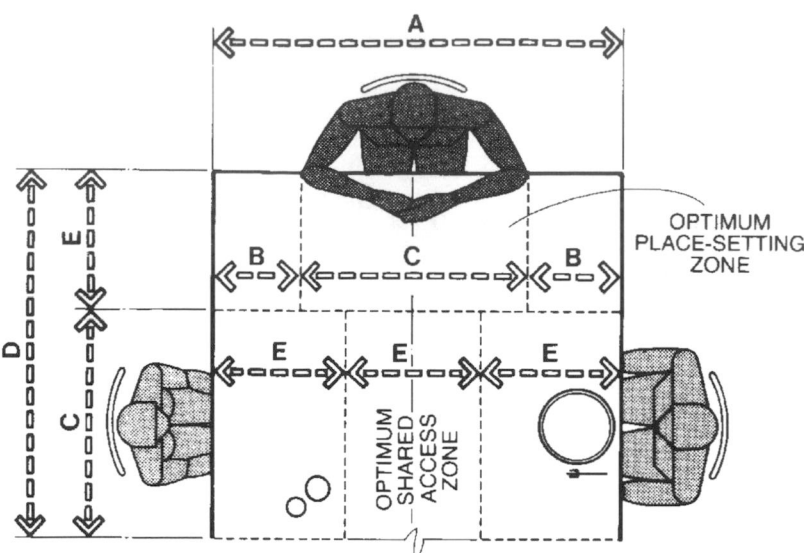

Fig. 16a   Optimum table width. (See Table 4.)

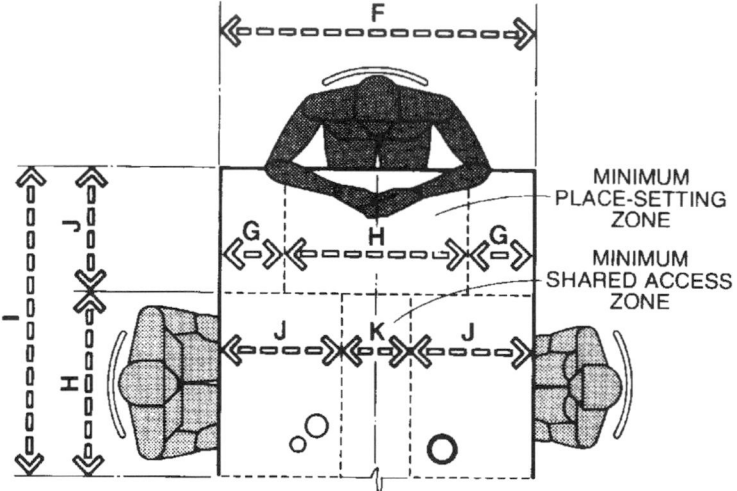

Fig. 16b   Minimum table width. (See Table 4.)

TABLE 4

	Inches
A	54
B	12
C	30
D	48
E	18
F	42
G	9
H	24
I	40
J	16
K	10

Tables and Chairs

Dining areas for eight persons with free-standing table 72 by 40 in, one armchair, and seven armless chairs (calculated on the basis of edging space on sides where there is not serving space, so that all persons can leave their seats without disturbing others).

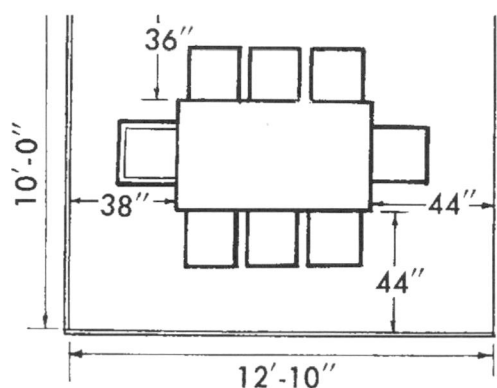

Serving space on one side and one end

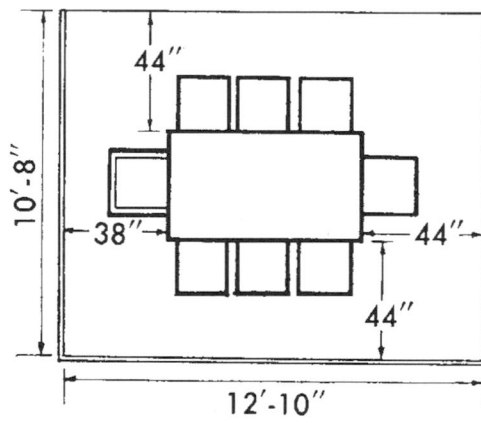

Serving space on two sides and one end

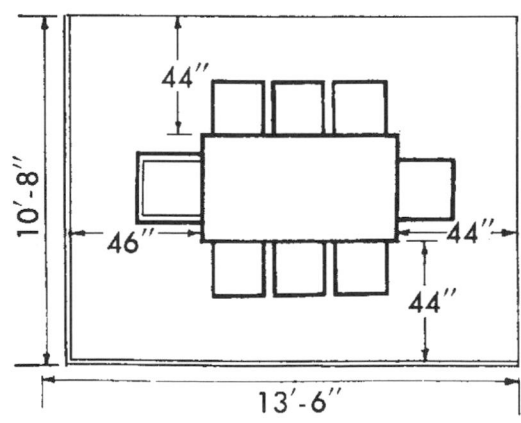

Serving space all around table

Fig. 17

Rising from table, armless
chair (armchair 2" more)

Foot extension, knees crossed,
not at table

Fig. 18

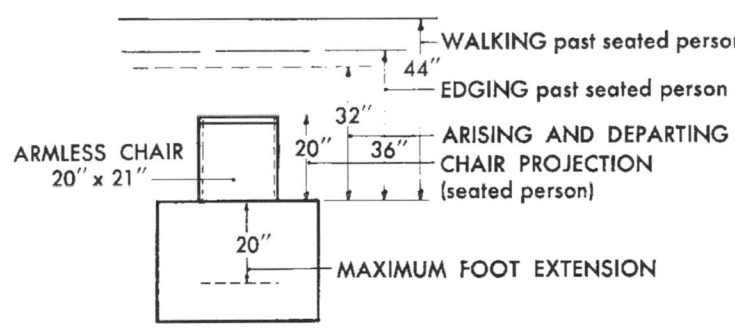

Armless chair in place at table

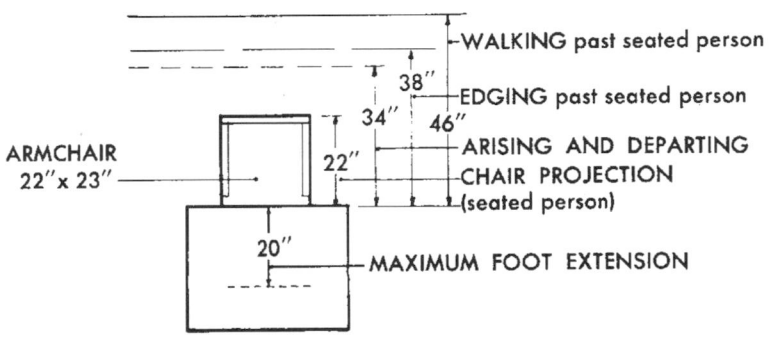

Armchair in place at table

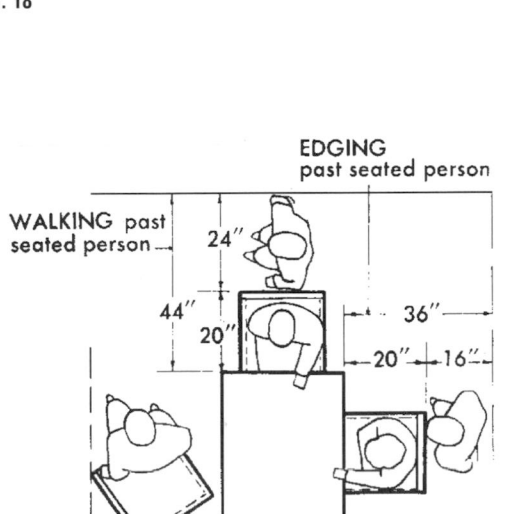

Using tables and chairs in free area

Fig. 19

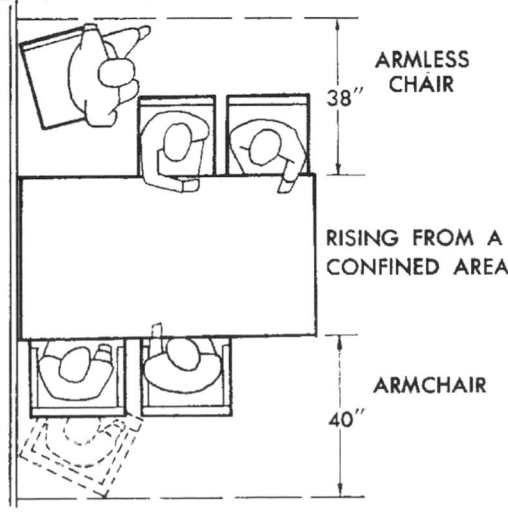

Using tables and chairs in confined area

ROUND TABLES

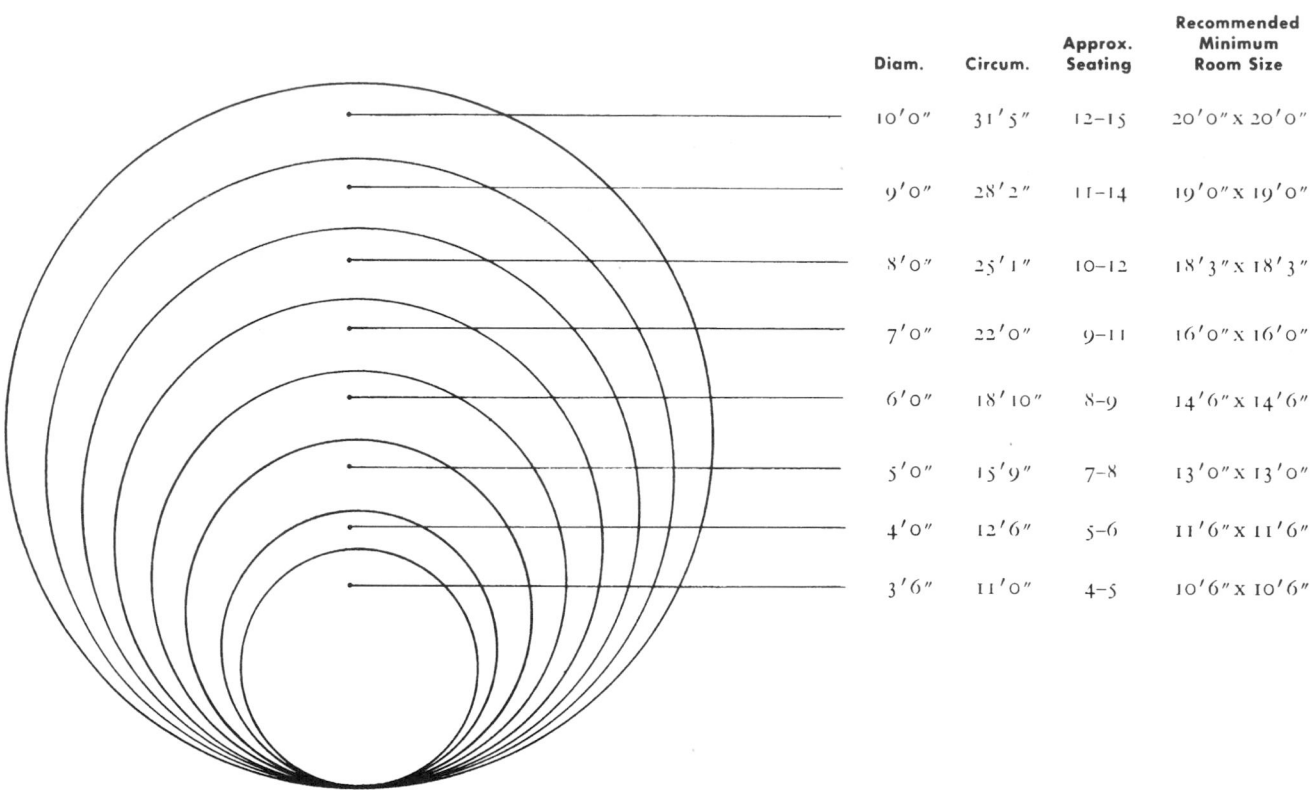

Diam.	Circum.	Approx. Seating	Recommended Minimum Room Size
10'0"	31'5"	12–15	20'0" x 20'0"
9'0"	28'2"	11–14	19'0" x 19'0"
8'0"	25'1"	10–12	18'3" x 18'3"
7'0"	22'0"	9–11	16'0" x 16'0"
6'0"	18'10"	8–9	14'6" x 14'6"
5'0"	15'9"	7–8	13'0" x 13'0"
4'0"	12'6"	5–6	11'6" x 11'6"
3'6"	11'0"	4–5	10'6" x 10'6"

SQUARE TABLES

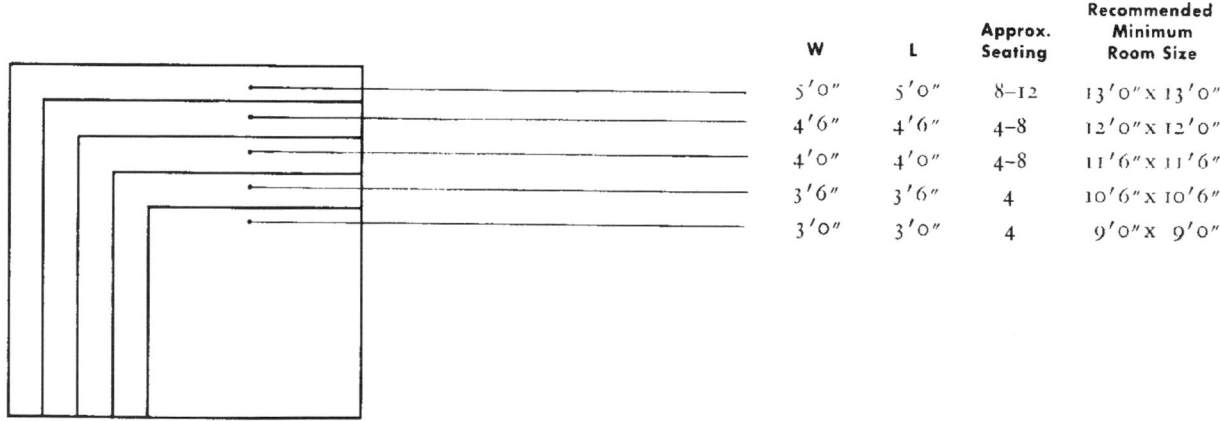

W	L	Approx. Seating	Recommended Minimum Room Size
5'0"	5'0"	8–12	13'0" x 13'0"
4'6"	4'6"	4–8	12'0" x 12'0"
4'0"	4'0"	4–8	11'6" x 11'6"
3'6"	3'6"	4	10'6" x 10'6"
3'0"	3'0"	4	9'0" x 9'0"

Fig. 20   Seating capacities for round, square, rectangular, and boat-shaped tables of various sizes and the recommended minimum room sizes to accommodate each.

RECTANGULAR TABLES

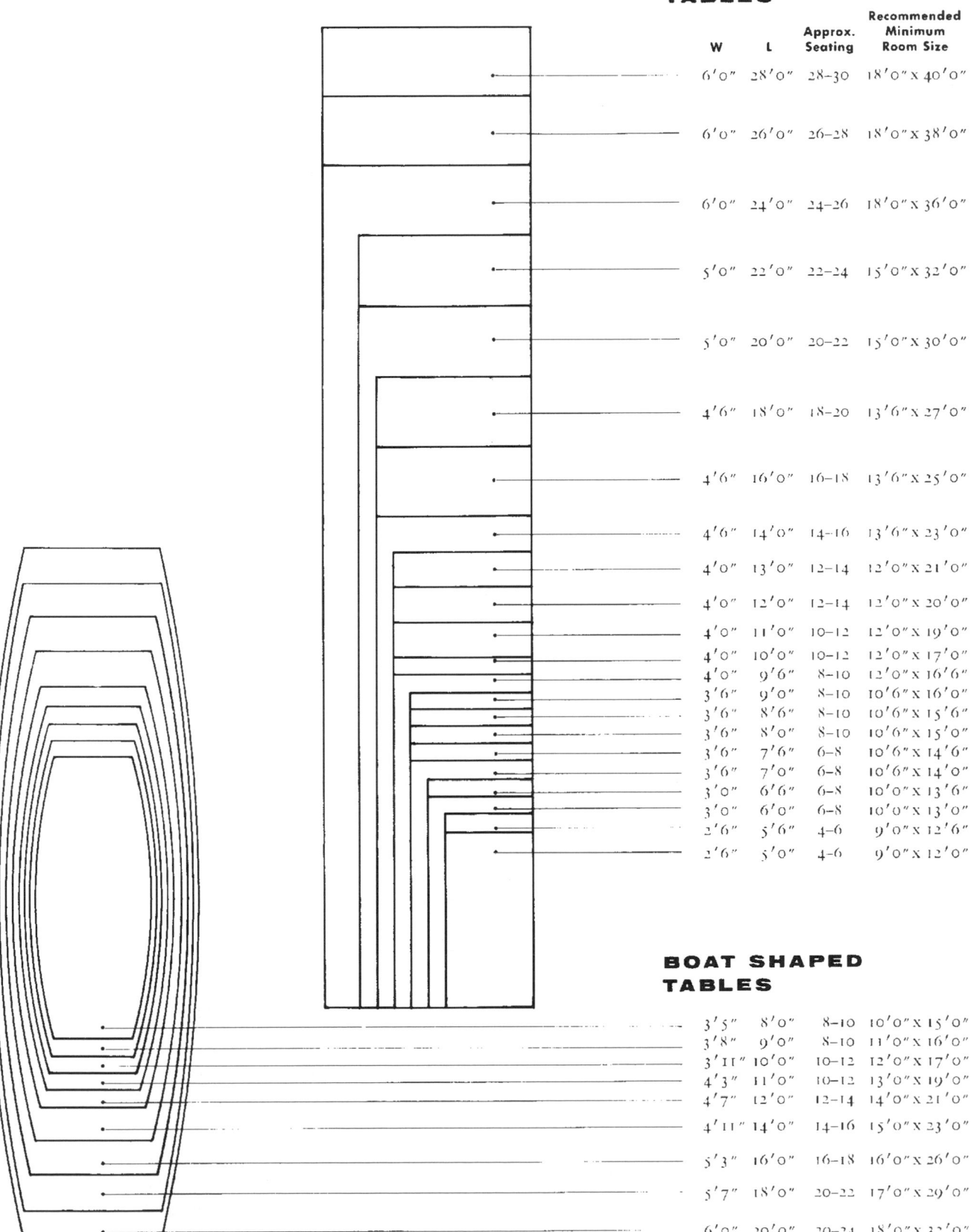

W	L	Approx. Seating	Recommended Minimum Room Size
6'0"	28'0"	28–30	18'0" x 40'0"
6'0"	26'0"	26–28	18'0" x 38'0"
6'0"	24'0"	24–26	18'0" x 36'0"
5'0"	22'0"	22–24	15'0" x 32'0"
5'0"	20'0"	20–22	15'0" x 30'0"
4'6"	18'0"	18–20	13'6" x 27'0"
4'6"	16'0"	16–18	13'6" x 25'0"
4'6"	14'0"	14–16	13'6" x 23'0"
4'0"	13'0"	12–14	12'0" x 21'0"
4'0"	12'0"	12–14	12'0" x 20'0"
4'0"	11'0"	10–12	12'0" x 19'0"
4'0"	10'0"	10–12	12'0" x 17'0"
4'0"	9'6"	8–10	12'0" x 16'6"
3'6"	9'0"	8–10	10'6" x 16'0"
3'6"	8'6"	8–10	10'6" x 15'6"
3'6"	8'0"	8–10	10'6" x 15'0"
3'6"	7'6"	6–8	10'6" x 14'6"
3'6"	7'0"	6–8	10'6" x 14'0"
3'0"	6'6"	6–8	10'0" x 13'6"
3'0"	6'0"	6–8	10'0" x 13'0"
2'6"	5'6"	4–6	9'0" x 12'6"
2'6"	5'0"	4–6	9'0" x 12'0"

BOAT SHAPED TABLES

W	L	Approx. Seating	Recommended Minimum Room Size
3'5"	8'0"	8–10	10'0" x 15'0"
3'8"	9'0"	8–10	11'0" x 16'0"
3'11"	10'0"	10–12	12'0" x 17'0"
4'3"	11'0"	10–12	13'0" x 19'0"
4'7"	12'0"	12–14	14'0" x 21'0"
4'11"	14'0"	14–16	15'0" x 23'0"
5'3"	16'0"	16–18	16'0" x 26'0"
5'7"	18'0"	20–22	17'0" x 29'0"
6'0"	20'0"	20–24	18'0" x 32'0"

Fig. 20 (*Continued*)

DINING AREAS

Typical Dining Areas and Furniture Grouping

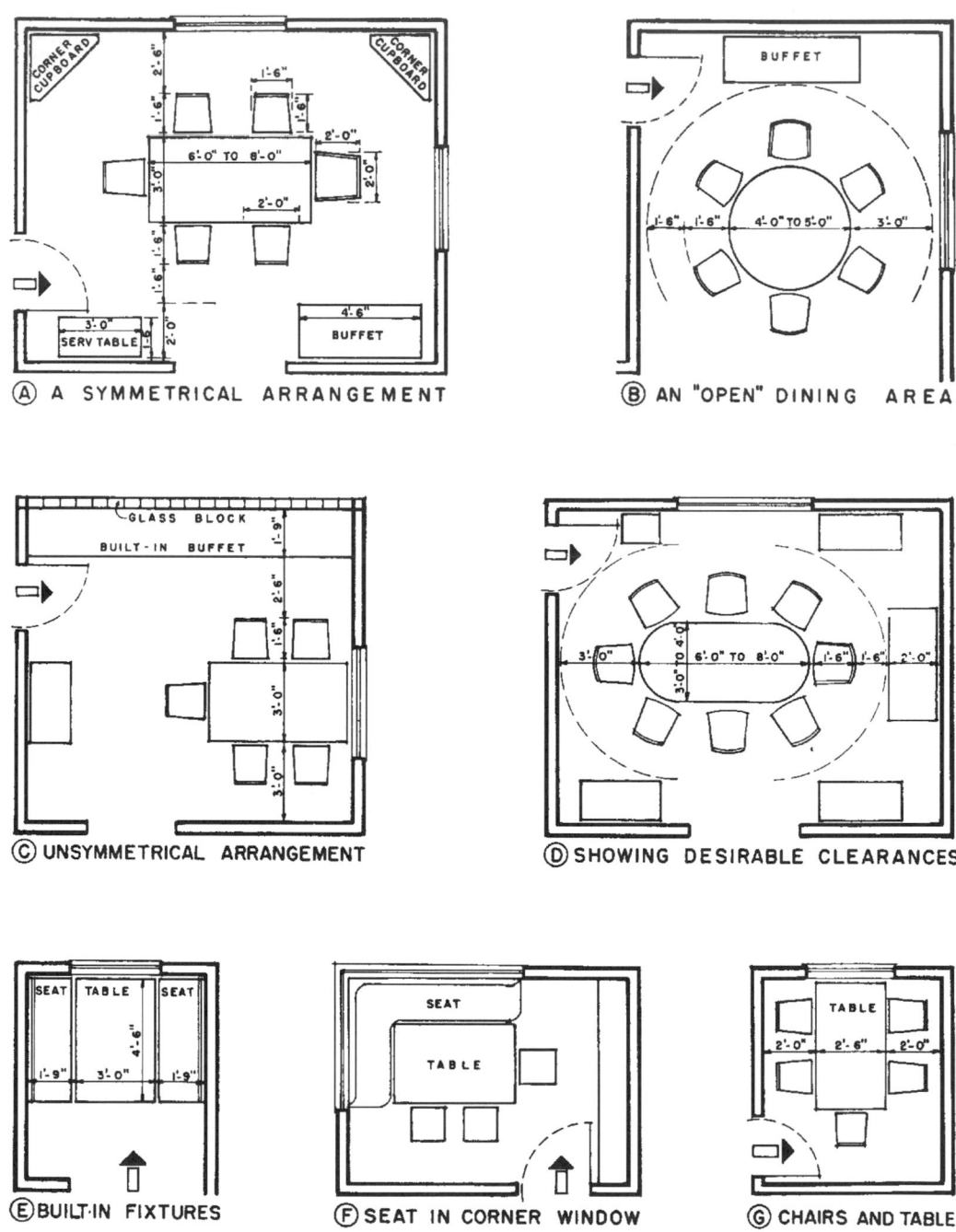

Ⓐ A SYMMETRICAL ARRANGEMENT

Ⓑ AN "OPEN" DINING AREA

Ⓒ UNSYMMETRICAL ARRANGEMENT

Ⓓ SHOWING DESIRABLE CLEARANCES

Ⓔ BUILT-IN FIXTURES

Ⓕ SEAT IN CORNER WINDOW

Ⓖ CHAIRS AND TABLE

Fig. 21 Plans showing various types of dining rooms and dining alcoves together with sizes and arrangements of furniture for formal and informal compositions.

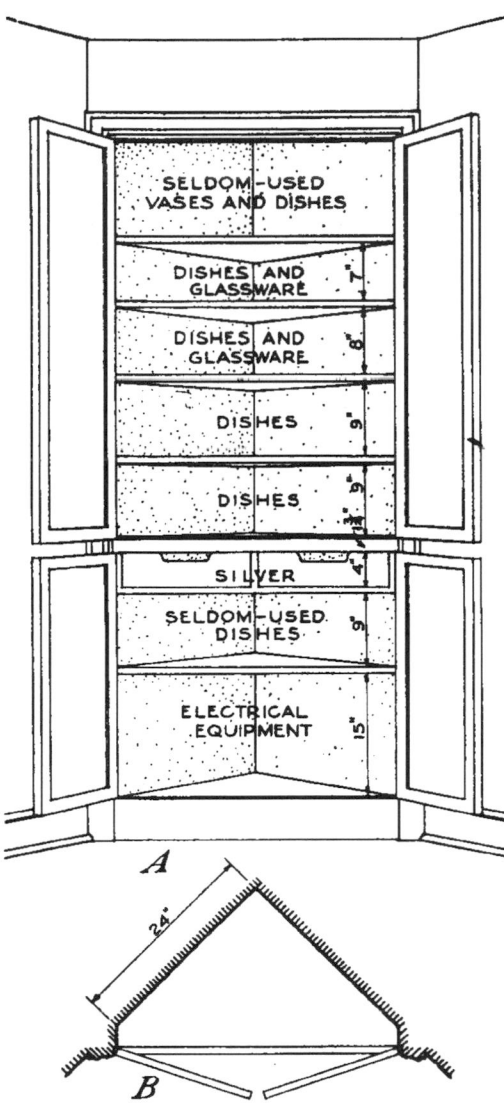

SELDOM-USED VASES AND DISHES

DISHES AND GLASSWARE

DISHES AND GLASSWARE

DISHES

DISHES

SILVER

SELDOM-USED DISHES

ELECTRICAL EQUIPMENT

A

B

Fig. 22 Corner cupboard: *A*, perspective *B*, plan.

Dining Area Corner Cupboard

For storing dishes and silver, especially the "best," used for entertaining, it is handy to have a closet or a cupboard in the dining room. This closet is also a good place to keep other articles used in serving meals.

It may be any depth, but the size shown in Fig. 22 will accommodate most articles that the average family uses in serving meals. Frequently cupboards are placed in two adjacent corners of the dining room.

Silverware should be kept in a separate drawer. Three inches is a satisfactory depth for the drawer unless more than 12 pieces are stored in a section. If the drawer has a separate section for forks, knives, and spoons, these sections should be at least 2½ in wide.

Shelves for china need to be at least 11 in wide. In estimating the distance between shelves, allow 1 in above stacks of plates, which are handled from the side, and 2 in above articles, such as cups, handled from the top. Table 4 may serve as a guide in planning shelf heights.

TABLE 5   Height Allowance for Dishes in Common Use

Article	Inches
Dinner plates, stack of 6	7
Soup plates, stack of 6	7
Salad plates, stack of 6	6
Sauce dishes, stack of 6	6
Bread-and-butter plates, stack of 6	5
Saucers, stack of 6	5
Cups, stack of 2	5
Platters, stack of 2	4
Sugar bowl and creamer	6
Covered vegetable dishes	6
Open vegetable dishes	5
Glasses	6
Sherbet glasses	6
Goblets	8

KITCHENS

SPATIAL CHARACTERISTICS AND ARRANGEMENT

Requirement

The basic activities in the kitchen consist of food preparation, serving, and cleanup after the meal. The kitchen design should permit efficient operation in the performance of these functions. In addition, storage space for staples, dinnerware, and utensils should be provided.

Location Criterion

The kitchen should be conveniently located near the dining area, living area, and utility area. The user should have easy access from the service entrance, or the only entrance if in an apartment. The kitchen should provide direct access to the dining area.

Space Criterion

The size of the kitchen should be determined by the number of bedrooms provided in the living unit. Work centers for the following equipment, cabinets, and space for their use should be provided:

1. Range space with base and wall cabinet at one side for serving and storage of utensils and staples.
2. Sink and base cabinet with counter space on each side for cleanup. Wall cabinets for storage of dinnerware.
3. Refrigerator space with counter space at latch side of the refrigerator door.
4. Mixing counter and base cabinet for electrical appliances and utensil storage. Wall cabinet for staple storage.

Light and Ventilation Criterion

Artificial ceiling light should be provided and ventilation should be provided by natural or mechanical means.

Fire Protection and Safety Criterion

Kitchen range should meet the provisions of the following:

1. Gas ranges, AGA "Directory of Approved Appliances and Listed Accessories."
2. Electric ranges, UL "Electric Appliances and Utilization Equipment List."

Commentary

Table 1 shows recommended minimum space standards for kitchen work centers. The spaces to be provided are sized according to the number of bedrooms in the living unit.

Height of shelving and counter tops

1. Maximum height of wall shelving 74 in. Height of counter tops should be 36 in.
2. Minimum clearance height between sink and wall cabinet 24 in; between base and wall cabinets 15-in clearance.

Recommended minimum edge distance

Equipment should be placed to allow for efficient operating room between it and any adjacent corner cabinet. At least 9 in from the edge of the sink and range and 16 in at the side of the refrigerator is recommended.

Circulation space

A minimum of 40 in should be provided between base cabinets or appliances opposite each other. This same minimum clearance applies when a wall, storage wall, or work table is opposite a base cabinet.

Traffic

Traffic in the kitchen should be limited to kitchen work only. Serving circulation to the dining area should be without any cross traffic.

KITCHEN ARRANGEMENTS

Kitchens and Kitchenettes

The most common types of kitchen arrangements are the (1) straight-line or galley, (2) parallel, (3) U type, and (4) L type. The galley-type arrangement is used for kitchenettes that require a minimum of equipment. The parallel and U arrangements are considered the most efficient in regard to movement and working relationships. The L is more common when used in a large kitchen, which makes possible an eating area.

Kitchen

The kitchen is an area where many different functions occur. These normally include:

1. Food preparation and cleanup
2. Food storage
3. Utensil and general storage
4. Eating
5. Laundry
6. Other, miscellaneous activities

The kitchen receives intensive use by most families, whether it is in a single-family house or a high-rise apartment.

Food Preparation

The food-preparation function has received a great deal of study and analysis. The sink, the range, and the refrigerator have been the traditional elements of food preparation and cleanup. Over the years, each of these elements has become increasingly more extensive. For example, from a simple sink it follows into a double sink, disposal units, and a dishwasher. The range has evolved into a wall oven, double oven, grille, barbecue pit, rotisserie, and infrared oven. The refrigerator has evolved into a large-sized refrigerator, separate freezer, and ice-making equipment.

The development of the sink disposal units and compactors has helped greatly in the cleanup of food preparation. The increasing use of a large battery of mechanical appliances, such as mixers, blenders, and slicers, is adding to the complexities of the kitchen.

This constant evolution of the food-preparation area of the kitchen is expected to continue and become more complicated.

Food Storage

Over the years, the type and packaging of foods available on the market have changed considerably. There has been a tremendous increase in prepackaged and semiprepared foods. All kinds of frozen foods are now available.

It can be reasonably anticipated that additional storage space for frozen foods will be needed. Sufficient storage for canned goods and other foods not requiring refrigeration is also needed.

KITCHEN AND DINING AREAS

It is a common misconception that a kitchen is only a work area. Considerable social activity is carried on there, not only among family members but also between residents and neighbors.

To create a feeling of space in a small unit, an open kitchen, dining, living area is frequently provided. In doing so, one sacrifices the possibility for several activities conveniently taking place concurrently with minimum infringement. Most families with small children prefer a large eat-in kitchen plus a multipurpose room which can be used alternatively for play or for formal dining. Alternatively, a kitchen-dining area with minimal separation (for example, an open counter) provides for both formal and informal use.

In apartment buildings, where there is greater pressure for young children to remain in the apartment, the added flexibility of this arrangement may be especially important. Younger children can play at the table under the supervision of the parent. Older children use the area as an evening study space.

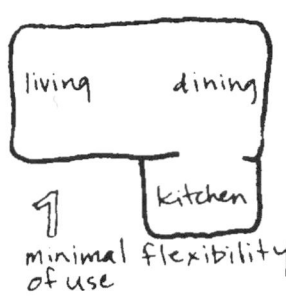

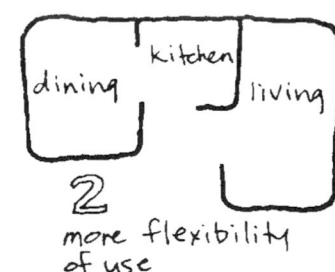

Fig. 1

1. Kitchens for larger units (i.e., more bedrooms) may not require more actual floor space than small ones but will require more counter space, a larger number of burners on a range, a larger oven, and more refrigerator and storage space.

2. Each center requires appropriate adjoining counter space and storage facilities. Dimensions on the following diagrams should be viewed as providing generous counter space.

A layout that provides a continuous sequence of sink-worktop-range has been shown to be the most functional. In this respect, the galley layout is in fact the least desirable of the alternatives.

L or U layouts provide a more flexible working space, especially when more than one person is using the kitchen. This fact may recommend the use of these layouts in family units.

Dimensions shown on the layouts are minimum requirements for two-bedroom units. Bracketed dimensions apply to three-bedroom, five-person units and larger.

It is recommended that the following widths of counter space be provided near the work centers:

Range	18 in
Refrigerator	36 in (door side)
Sink	18 in (each side)
Serve	36 in

Note: The in-line layout should have at least 3 ft 0 in clear space in front of cabinets.

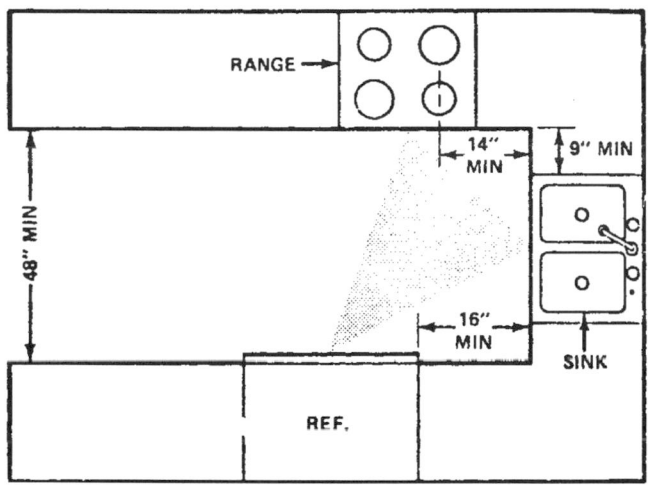

Fig. 2 Crowded kitchen floor plans create unsafe conditions. Inadequate space within the work triangle may result in painful injury if two people collide while preparing hot foods.

WORK TRIANGLE

The heart of the kitchen is the work triangle—formed by lines connecting the center fronts of the sink, range and refrigerator. The sum of the sides of the work triangle should not exceed 23 ft. As the major activity area in the kitchen this triangle should be out of the path of most traffic through the room to adjacent areas.

The four basic design categories of work triangles are straight wall (Fig. 3), U shape (Fig. 4), L shape (Fig. 5), and corridor (Fig. 6).

In planning a new kitchen, it is ideal for each of the kitchen centers to be a maximum of 12 to 22 ft apart.

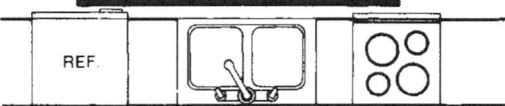

Fig. 3 Straight wall—Although this shape is not a "work triangle," it still works if the total distance between work areas is under 12 ft. This straight wall shape also lends itself well to the addition of a separate "island" cabinet.

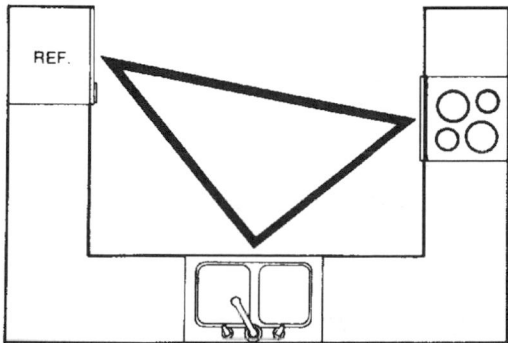

Fig. 4 U shape—This kitchen shape is most commonly seen adjacent to a dining room or family area, and is very popular in both small and large kitchens. One arm of the counter area can also be used as an island cabinet or as a peninsula which could serve as a snack bar.

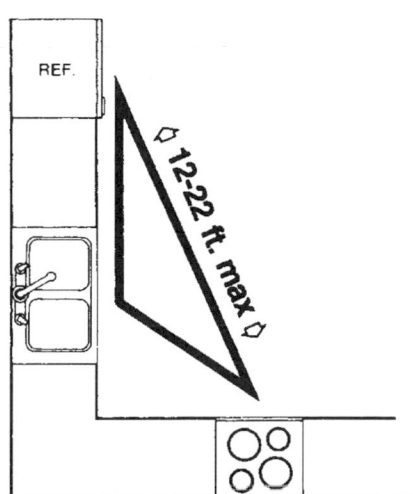

Fig. 5 L shape—This relationship between refrigerator, sink, and range makes an unbroken traffic flow possible. The two walls provide many options for both appliances and cabinetry. Ideal for large kitchens, this work triangle is perfect for adding an eating area or island cabinet.

KITCHENS

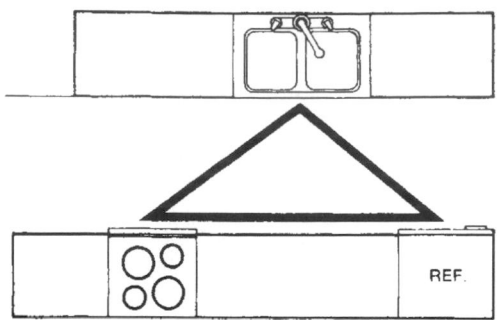

Fig. 6 Corridor—The perfect shape for small kitchens because it is the most efficient use of a work triangle. This corridor configuration works best, however, where there is a light amount of traffic.

TRAFFIC FLOW

Most traffic through a kitchen is bound for the backyard or an adjacent dining room or meal-serving area. A combination of hot dishes, appliances, kitchen utensils, and traffic flow can cause accidents. Good planning will keep traffic out of the work area and provide adequate clearance between fixtures and appliances. A minimum of 48 in should be provided between cabinets and appliances placed opposite each other. When such fixtures are placed at right angles to each other and separated by a passageway, they should be spaced a minimum of 30 in apart. In an "L" or "U" shaped kitchen, the minimum edge distance between an appliance and an adjacent corner should be 9 in from the edge of a sink, 16 in from the refrigerator, and 14 in from the center of the nearest range burner. These recommended distances allow people to pass each other safely while using the fixtures, carrying hot foods, and opening and closing cabinet and appliance doors and drawers.

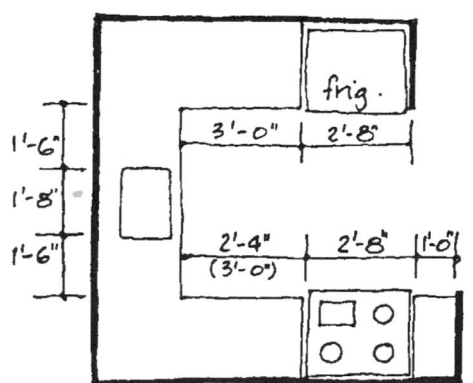

Fig. 7a U layout.

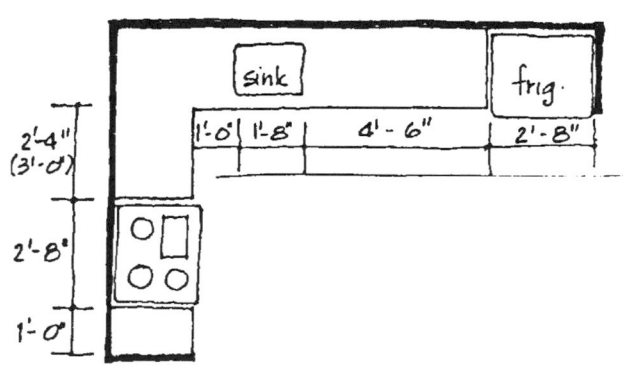

Fig. 7b L layout.

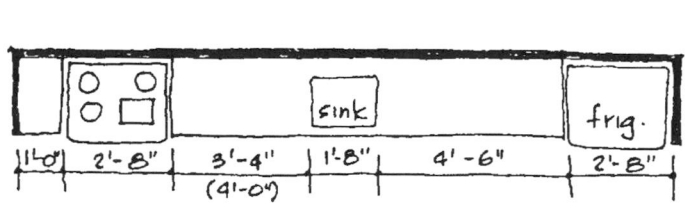

Fig. 7c In-line layout (not recommended).

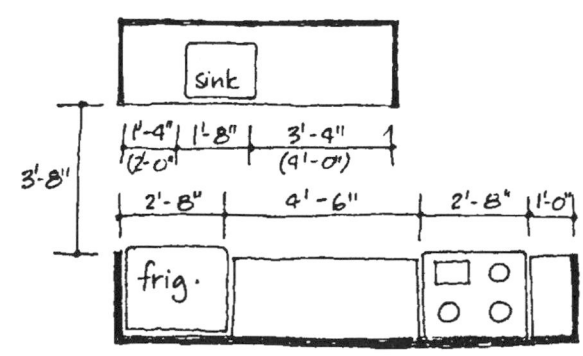

Fig. 7d Galley layout.

KITCHEN STORAGE

Each kitchen or kitchenette should have (1) accessible storage space for food and utensils, (2) sufficient space for the average kitchen accessories, (3) sufficient storage space for those items of household equipment normally used and for which storage is not elsewhere provided.

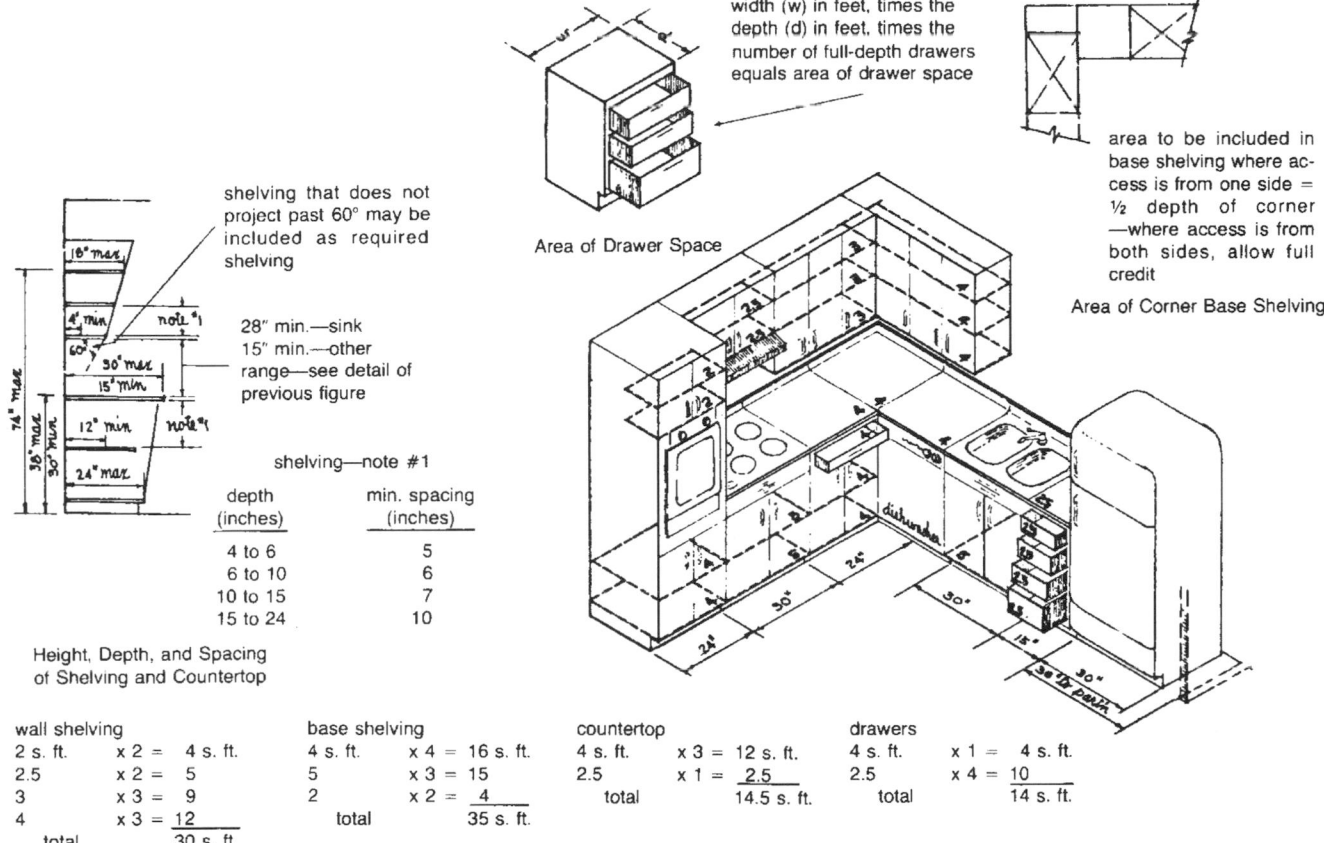

width (w) in feet, times the depth (d) in feet, times the number of full-depth drawers equals area of drawer space

Area of Drawer Space

area to be included in base shelving where access is from one side = ½ depth of corner —where access is from both sides, allow full credit

Area of Corner Base Shelving

shelving that does not project past 60° may be included as required shelving

28″ min.—sink
15″ min.—other range—see detail of previous figure

shelving—note #1

depth (inches)	min. spacing (inches)
4 to 6	5
6 to 10	6
10 to 15	7
15 to 24	10

Height, Depth, and Spacing of Shelving and Countertop

wall shelving			base shelving			countertop			drawers		
2 s. ft.	x 2 =	4 s. ft.	4 s. ft.	x 4 =	16 s. ft.	4 s. ft.	x 3 =	12 s. ft.	4 s. ft.	x 1 =	4 s. ft.
2.5	x 2 =	5	5	x 3 =	15	2.5	x 1 =	2.5	2.5	x 4 =	10
3	x 3 =	9	2	x 2 =	4	total		14.5 s. ft.	total		14 s. ft.
4	x 3 =	12	total		35 s. ft.						
total		30 s. ft.									

Fig. 8 Example: measurement of shelf and countertop areas.

KITCHENS

Utensil and General Storage

Space for utensils includes storage for dishes, pots and pans, utensils, and appliances. With the increased use of such electrical appliances, their storage becomes a significant problem. General storage requires space for linens, towels, and kitchen supplies. Included in this category are brooms, mops, and other cleaning equipment and supplies.

Laundry

The laundry activities are often assigned as part of the kitchen or directly related to it. The laundry function requires a clothes washer, dryer, sink, ironing board, and storage for both dirty and clean laundry.

1. An area occupied by sink basin(s) and by cooking units should not be included in countertop area.

2. Usable storage space in or under ranges, or under wall ovens, when provided in the form of shelving or drawers, may be included in the minimum shelf or drawer area.

3. The shelf area of revolving base shelves (lazy susan) may be counted as twice its actual area in determining required shelf area, provided the clear width of opening is at least 8½ in.

4. Drawer area in excess of the required area may be substituted for required base shelf area up to 25 percent of total shelf area.

5. At least 60 percent of required shelf space should be enclosed by cabinet doors.

TABLE 1 Minimum Kitchen Storage Required

	40 to 60 ft² Area—Kitchenette	
Item	0-bedroom living unit,* ft²	1-bedroom living unit,* ft²
Total shelving in wall and base cabinets	24	30
Shelving in either wall or base cabinets	10	12
Drawer area	4	5
Countertop area	5	6

	60 ft² Area and Over—Kitchen	
Item	1- and 2-bedroom living units, ft²	3- and 4-bedroom living units, ft²
Total shelving in wall and base cabinets	48	54
Shelving in either wall or base cabinets	18	20
Drawer area	8	10
Countertop area	10	12

*Kitchen unit assemblies serving the kitchen function and occupying less than 40 ft² area in 0-BR living units shall not be less than 5 ft in length and shall provide at least 12 ft² of total shelving in wall and base cabinets. Drawer and countertop space shall also be provided. No room count is allowable for this type facility.

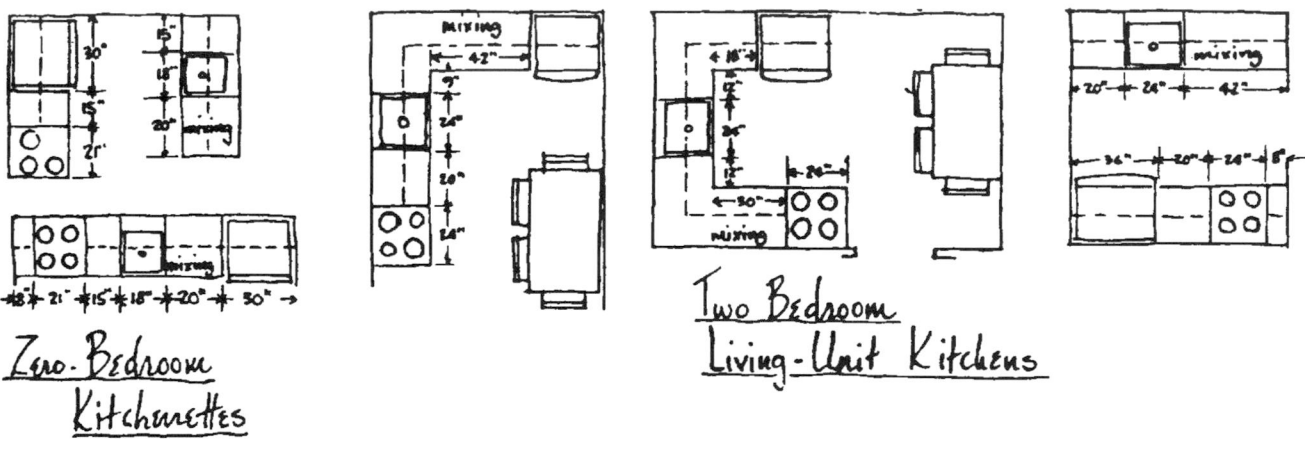

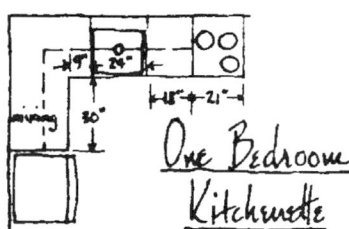

Fig. 9 Minimum frontages for various kitchens.

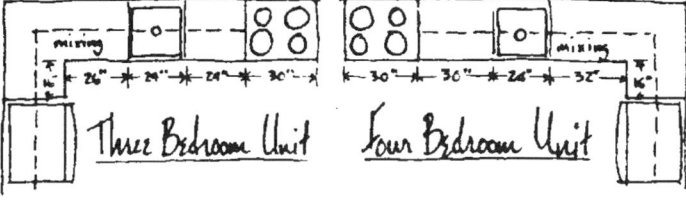

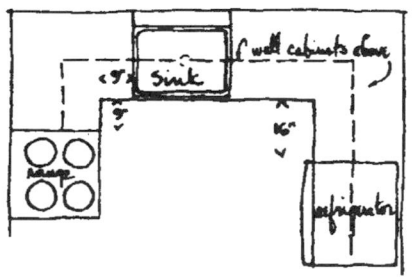

Fig. 10 Minimum edge distances.

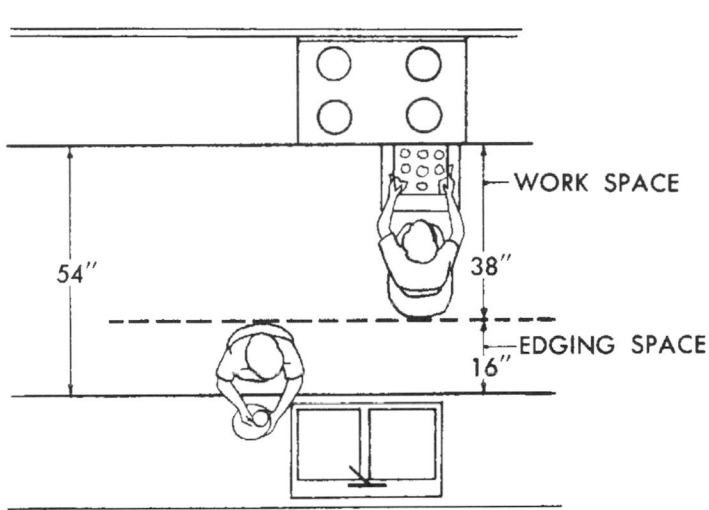

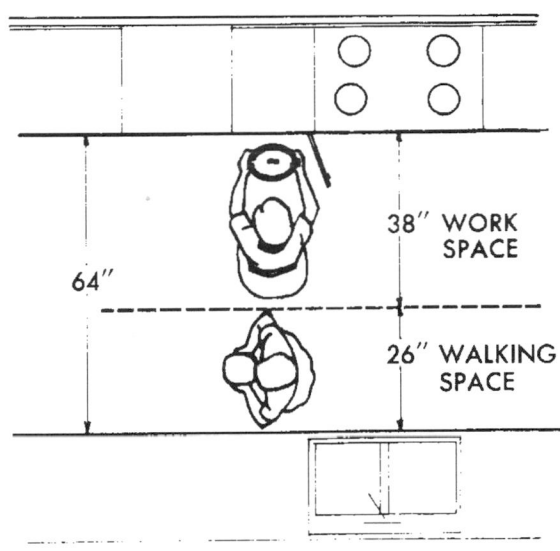

Fig. 11a Minimum space (allowing for edging) for two people working at cabinets and appliances opposite each other (except a front-opening dishwasher).

Fig. 11b Liberal space (allowing for walking) for two people working at cabinets and appliances opposite each other (except a front-opening dishwasher).

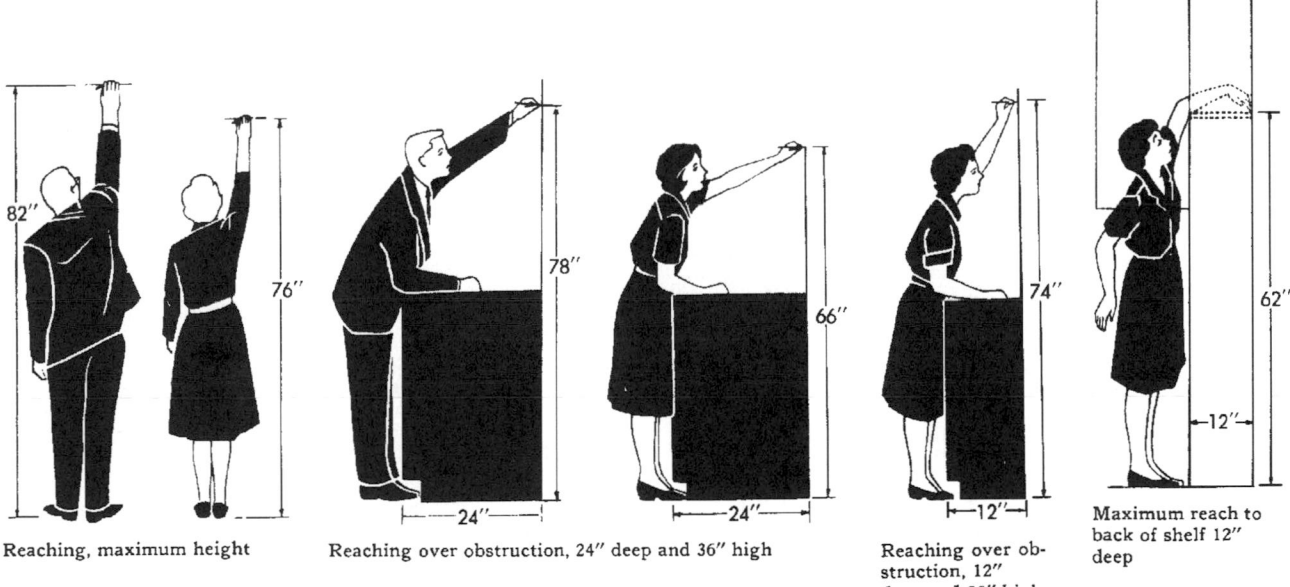

Reaching, maximum height

Reaching over obstruction, 24″ deep and 36″ high

Reaching over obstruction, 12″ deep and 36″ high

Maximum reach to back of shelf 12″ deep

Fig. 12

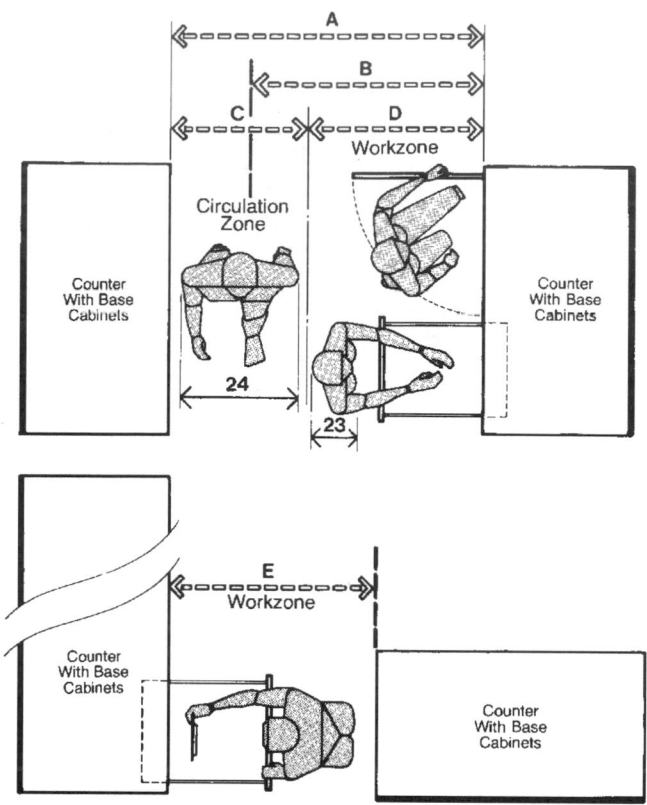

**COUNTER AND BASE CABINETS/
GENERAL CLEARANCE**

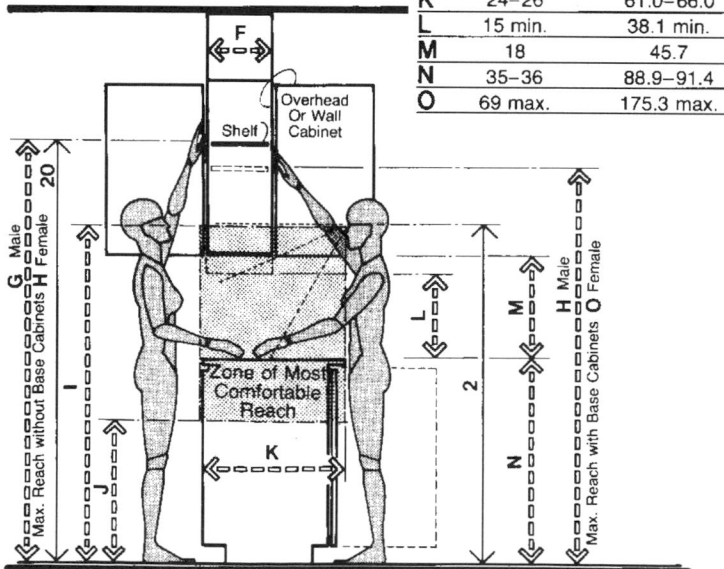

CABINET REACH COMPARISONS

Fig. 13

312

	in	cm
A	60–66	152.4–167.6
B	48 min.	121.9 min.
C	24–30	61.0–76.2
D	36	91.4
E	48	121.9
F	12–13	30.5–33.0
G	76 max.	193.0 max.
H	72 max.	182.9 max.
I	59	149.9
J	25.5	64.8
K	24–26	61.0–66.0
L	15 min.	38.1 min.
M	18	45.7
N	35–36	88.9–91.4
O	69 max.	175.3 max.

KITCHEN DIMENSIONS

This section gives recommended minimum counter space dimensions, comments on points to be considered in the layout, and illustrates various shapes of kitchens.

All dimensions given are measured along the front of base cabinets. The dimensions may be angled around a corner, provided that the minimum distances of equipment to base cabinet corners are observed. No dimensions should overlap with each other.

The illustrations show one arrangement of the "handing" of the kitchen—that is, the side of the sink on which the range is located. This may be varied according to design requirements and preferences.

Occupancy of the dwelling is indicated by code—for example, "5 and 6P" means "five and six persons."

All illustrations are diagrammatic; they do not represent detailed room plans.

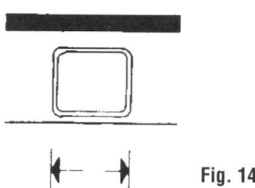

Fig. 14

TABLE 2   Minimum Width of Sink

1 and 2P	500 mm (1 ft 8 in)
3 and 4P	500 mm (1 ft 8 in)
5 and 6P	500 mm (1 ft 8 in)
7 and 8P	600 mm (2 ft 0 in)

A total clearance of 50 mm (2 in) is allowed (Table 3).

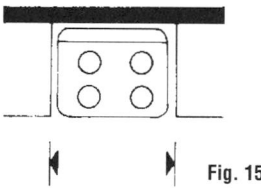

Fig. 15

TABLE 3   Minimum Space for Range, Including Clearances

1 and 2P	650 mm (2 ft 2 in)
3 and 4P	800 mm (2 ft 8 in)
5 and 6P	800 mm (2 ft 8 in)
7 and 8P	800 mm (2 ft 8 in)

A total clearance of 50 mm (2 in) is allowed. When the refrigerator is located at the end of base cabinets and is open to the room, only 25 mm (1 in) clearance between the refrigerator and cabinets is required (Table 4).

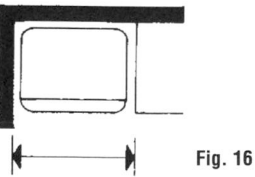

Fig. 16

TABLE 4   Minimum Space for Refrigerator, Including Clearances

1 and 2P	650 mm (2 ft 2 in)
3 and 4P	800 mm (2 ft 8 in)
5 and 6P	800 mm (2 ft 8 in)
7 and 8P	800 mm (2 ft 8 in)

This (Table 5) is the working space of the kitchen.

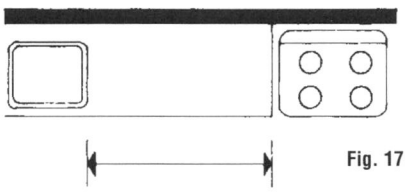

Fig. 17

TABLE 5   Minimum Uninterrupted Counter Space between Sink and Range

1P	600 mm (2 ft 0 in)
2P	750 mm (2 ft 6 in)
3 and 4P	1000 mm (3 ft 6 in)
5 and 6P	1200 mm (4 ft 0 in)
7 and 8P	1350 mm (4 ft 6 in)

The dimension in Table 6 is the minimum for the space. A small base cabinet with a drawer and lower door would be ideal. As this can be a costly item, the top and front only may be provided.

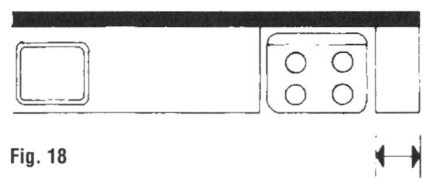

Fig. 18

TABLE 6   Minimum Counter Space at Other Side of Range

All sizes of dwellings (except for 1 and 2P)	300 mm (1 ft 0 in)

This space in Table 7 is for a drainer rack. In the case of a one- and two-person kitchen, this space is all that is required between the sink and the refrigerator. Kitchens for three to ten persons require a separate space adjacent to the refrigerator.

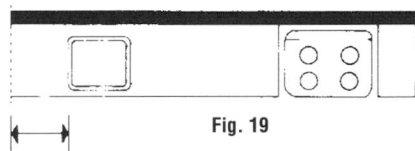

Fig. 19

TABLE 7   Minimum Counter Space at Other Side of Sink

1P	300 mm (1 ft 0 in)
2P	400 mm (1 ft 3 in)
3 and 4P	300 mm (1 ft 0 in)
5 and 6P	450 mm (1 ft 6 in)
7 and 8P	450 mm (1 ft 6 in)

The space in Table 8 is for placing items when using the refrigerator.

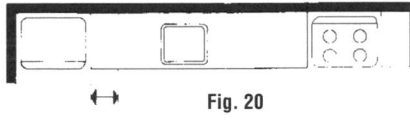

Fig. 20

TABLE 8   Minimum Space at One Side of Refrigerator

All sizes of dwellings (except for 1 and 2P)	300 mm (1 ft 0 in)

The range should not be located within 300 mm (1 ft 0 in) of a base cabinet corner.

This space is to provide sufficient clearance for opening drawers and doors.

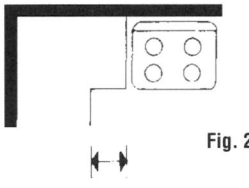

Fig. 21

The refrigerator should not be located within 450 mm (1 ft 6 in) of a base cabinet corner. This space is to permit access to the refrigerator when the latch is on the working side of the kitchen.

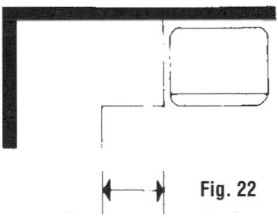

Fig. 22

The sink should not be located within 75 mm (3 in) of a base cabinet corner in one- and two-person dwellings. This space is to allow for framing around the sink if needed.

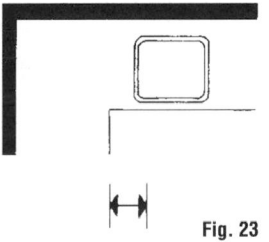

Fig. 23

For all other dwellings, the sink should not be located within 300 mm (1 ft 0 in) of a base cabinet corner. This minimum space is to allow access to the drainer rack.

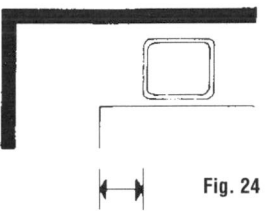

Fig. 24

The range should be located at least 600 mm (2 ft 0 in) from all windows. This dimension is to prevent the possibility of curtains touching the burners and catching fire.

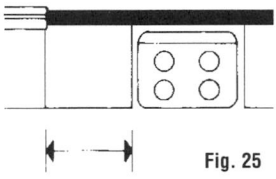

Fig. 25

Base cabinets, including counter top, should be 900 mm (3 ft 0 in) in height. The space between the base and wall cabinets should be approximately 400 mm (1 ft 4 in) in height. The top of all wall cabinets should be 2100 mm (7 ft 0 in) in height from the floor.

The latch or opening side of the refrigerator should be on the side nearest to the working area of the kitchen.

A pantry or full-height food and storage cupboard may be provided in addition to other cabinets and appliances. The depth should be a minimum of 280 mm (11 in) and should not exceed 400 mm (1 ft 4 in).

Wall cabinets should not be located over a range unless a fire-rated exhaust hood, the full width of the range, is provided.

At least one drawer should be provided in the base cabinets; it should be located adjacent to the sink, between the sink and the range. Other drawers in base cabinets are highly desirable.

A splashback, at the junction of the counter top and wall, should be provided of a minimum width of 100 mm (4 in).

The sizes given for counter space dimensions are the recommended minimum for kitchen design. If the sizes can be increased, consideration should be given to enlarging, first, the space between the sink and the range; and second, the space to one side of the refrigerator.

When provided in one- and two-person accommodation, the counter space dimensions for galley-shaped kitchens can be varied to locate the range and refrigerator against the same wall and against side walls, provided that there is a minimum counter space between them of 500 mm (1 ft 8 in). The sink should have a minimum space of 400 mm (1 ft 3 in) to one side.

In all kitchen layouts for one- and two-person accommodation, the range may be placed against a side wall.

Clear walkways at least 32 in wide should be provided at all entrances to the kitchen and 36 in of clearance should be provided for cabinet access. Note: If two counters flank the entry, this same minimum 32-in walkway space is recommended from the point of one counter to the opposite counter edge (Fig. 26).

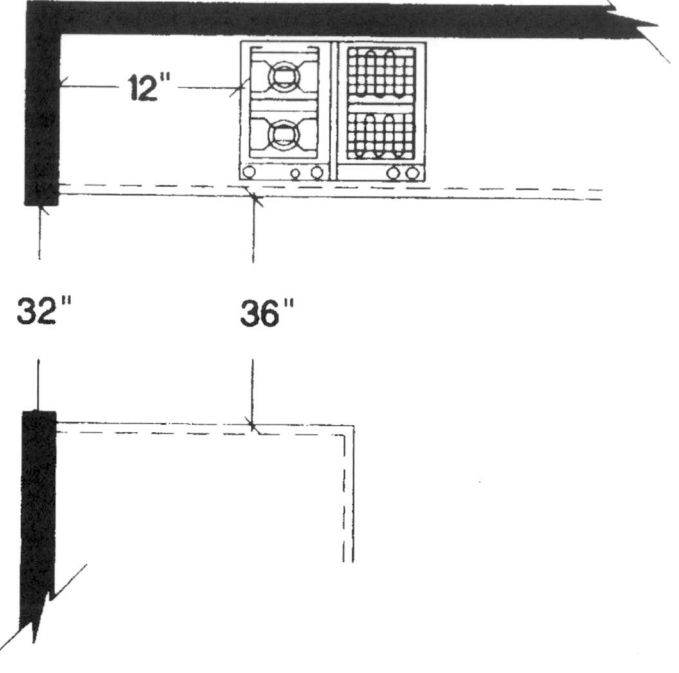

Fig. 26

Appliance doors and entries should not interfere with work center appliances. Note: In an island configuration, an appliance door on an island should not conflict with an appliance or cabinet door opposite it (Fig. 27).

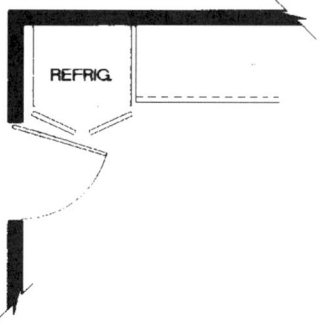

Fig. 27

Work aisles should be at least 42 in wide and passageways at least 36 in wide for a one-cook kitchen. Note: The minimum work aisle space increases 48 to 60 in from counter edge to counter edge when a kitchen is designed for more than one cook to work simultaneously in the space (Fig. 28).

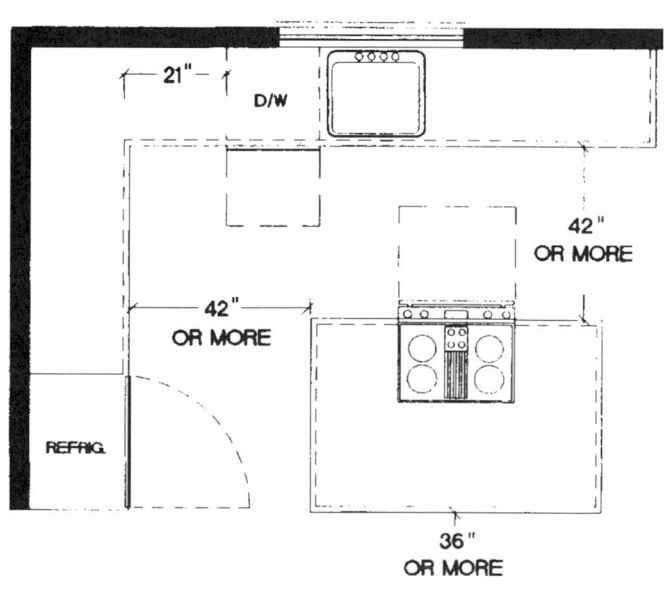

Fig. 28

Wall cabinet frontage: A kitchen under 150 sq ft should have at least 144 in of wall cabinet frontage, with cabinets at least 12 in deep and a minimum of 30 in high (Fig. 29*a*). Adjustable shelving must be installed over countertops. Diagonal or pie-cut wall cabinets count as a total of 24 in. Difficult-to-reach cabinets above the hood, oven, or refrigerator do not count unless specialized storage devices are installed inside to improve accessibility. If a cabinet has shelves, count twice the unit width for 12-in-deep tall cabinets and four times the unit width for 24-in-deep tall cabinets. A kitchen over 150 sq ft should have at least 186 in of wall cabinet frontage with cabinets that conform to the guidelines indicated above (Fig. 29*b*). Note: Tall cabinets can be counted as either base or wall cabinets, not both.

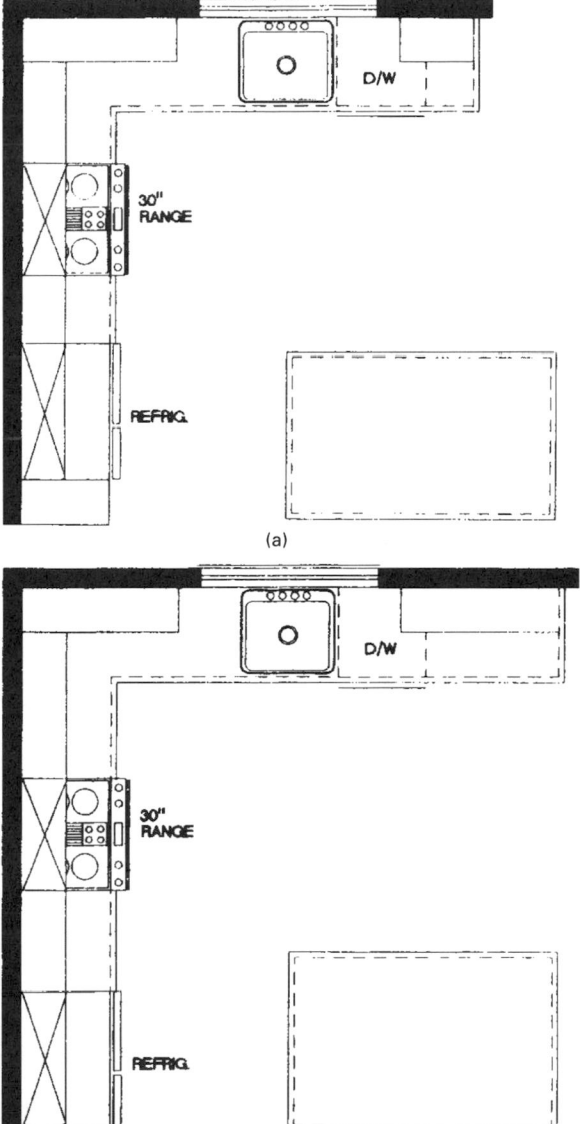

(a)

(b)

Fig. 29

Wall cabinet frontage should be at least 60 in, with cabinets at least 12 in deep and a minimum of 30 in high. These must be included within 72 in of the primary sink centerline (Fig. 30). Notes: A pantry cabinet can be substituted for the required wall cabinets if it is placed within 72 in of the sink centerline. Dish storage may be at the back of a peninsula near a table or in a hutch adjacent to a table.

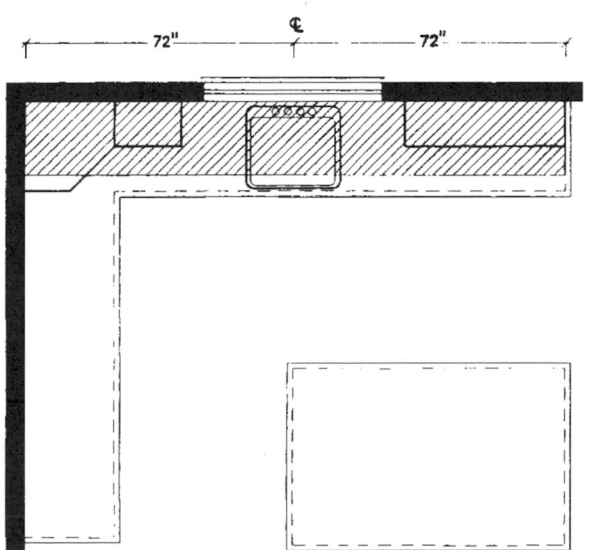

Fig. 30

KITCHENS

Base cabinet frontage: A kitchen under 150 sq ft should have at least 156 in of base cabinet frontage, with cabinets at least 21 in deep (Fig. 31*a*). Pie-cut/lazy Susan cabinets count as a total of 30 in. The first 24 in of a blind corner box do not count. A kitchen over 150 sq ft should have at least 192 in of base cabinet frontage that conform to the same guidelines (Fig. 31*b*).

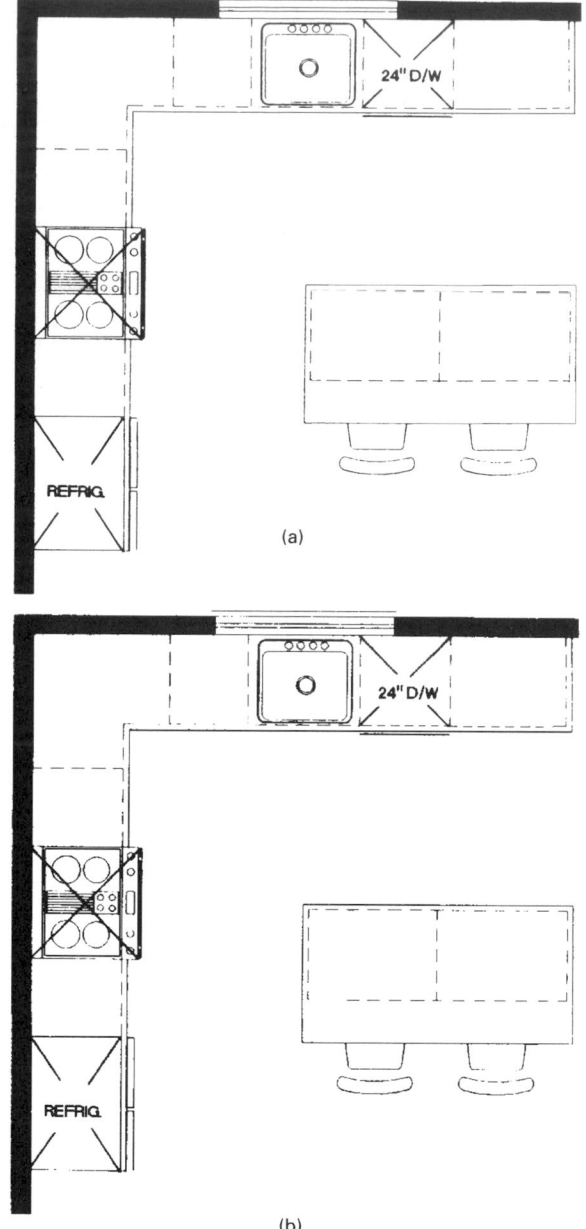

Fig. 31

Drawer frontage: A kitchen under 150 sq ft should have at least 120 in of drawer frontage or roll-out shelf frontage. A kitchen over 150 sq ft should have at least 165 in of drawer frontage or roll-out shelf frontage. Note: Measure cabinet width to determine frontage. Example: A 21-in-wide, three-drawer base would count as 63 in toward the drawer total and a 21-in-wide single-drawer base with one drawer and two sliding shelves would also count as 63 inches towards the total drawer measurement (3 by 21). Similarly, a 24-in-wide, three-drawer base or a 24-in-wide single-drawer base with one drawer and two sliding shelves would count as 72 in towards the drawer total (3 by 24) (Fig. 32).

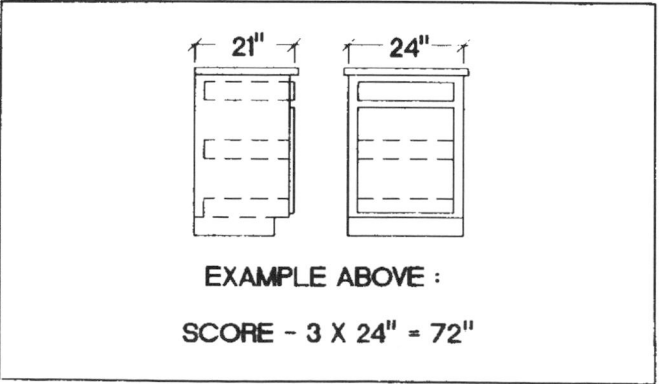

EXAMPLE ABOVE :

SCORE – 3 X 24" = 72"

Fig. 32

Five storage items at the minimum should be included in the kitchen to improve the accessibility and functionality of the plan (Fig. 33). These items include, but are not limited to, interior vertical dividers, specialized drawers, built-in bins/racks, swing-out pantries, or drawer/roll-out shelves greater than the 120-in minimum for small kitchens or the 165-in minimum for larger rooms.

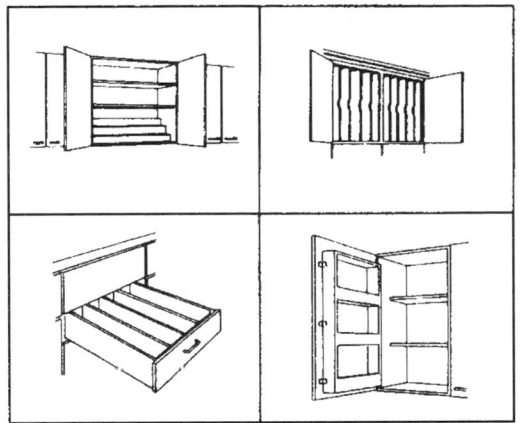

Fig. 33

Corner storage, for a kitchen with usable corner areas in the plan, should be included, with at least one functional corner storage unit (Fig. 34).

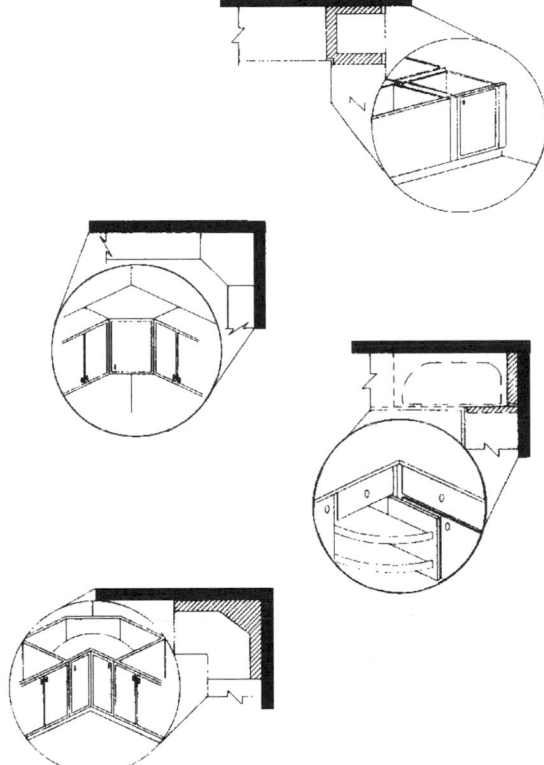

Fig. 34

Countertop clearance must be 15 to 18 in between the countertop and bottom of wall cabinets (Fig. 35).

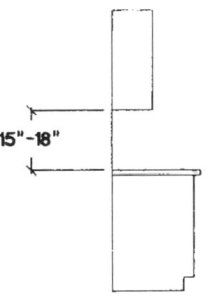

15"-18"

Fig. 35

KITCHENS

Countertop frontage should be at least 132 in of usable countertop frontage in a kitchen under 150 sq ft (Fig. 36a). A kitchen over 150 sq ft should have at least 198 in of usable countertop frontage (Fig. 36b). Notes: Counters must be 16 in deep to be counted; corner space does not count. If an appliance garage or storage cabinet extends to the counter, 16 in of clear space must be in front of this cabinet for the area to be counted.

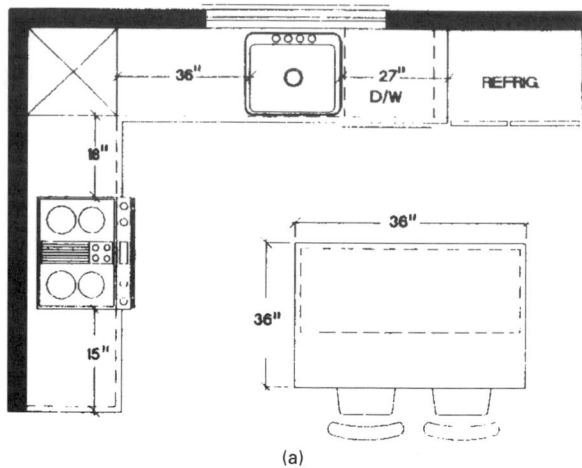

(a)

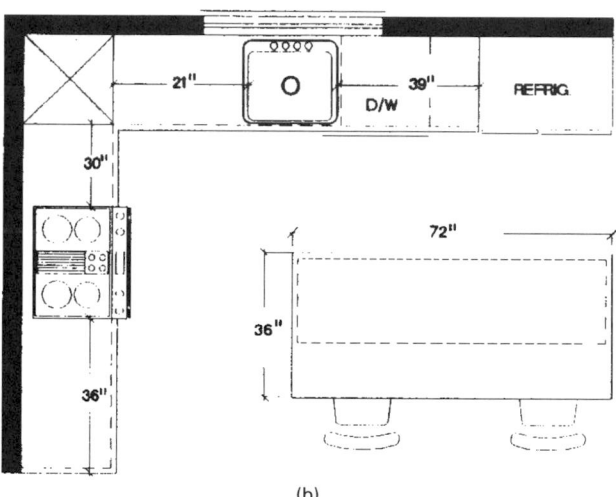

(b)

Fig. 36

Two primary work centers (the primary sink, refrigerator, and preparation or cooktop/range center) should not be separated by a full-height, full-depth tall tower, such as an oven cabinet, pantry or refrigerator (Fig. 37). Notes: A recessed corner wall oven is acceptable because the cook can move from the counter on one side of the oven to the counter on the opposite. The overall countertop frontage recommendation is formulated by identifying isolated centers of activities and the minimum and recommended counter space required in each center. If two centers are adjacent to one another, take the longest of the two counters and add 12 in to identify a new minimum.

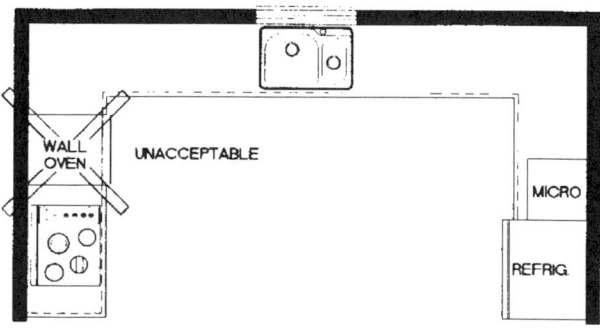

Fig. 37

Countertop frontage should be at least 24 in to one side of the sink, and 18 in on the other. (Measure only countertop frontage, not corner space.) The 18-in and 24-in counter sections may be continuous, or the total and two angle sections (Fig. 38). Notes: (a) If a kitchen has only one sink, it should be between or across from the cooking surface, preparation area, or refrigerator. (b) A corner sink should meet the same counter criteria on both sides. (c) A second sink should meet another set of standards. At least 3 in of landing space should be on one side and 18 in on the other. For a kitchen with two sinks, both should meet the planning criteria.

Two waste receptacles at minimum should be included in the plan: one for garbage and one for recyclables, or other recycling facilities should be planned.

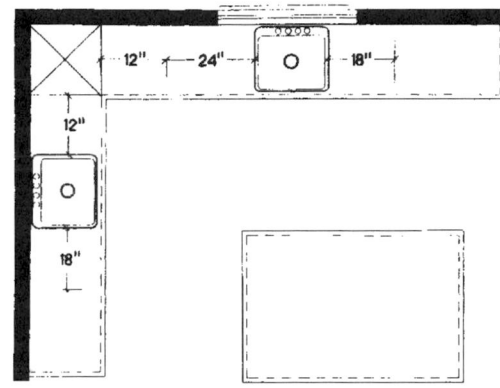

Fig. 38

KITCHENS

Counter space of at least 3 inches should be allowed from the edge of the sink to the inside corner of the countertop if more than 21 in of counter space is available on the return (Fig. 39). Note: Alternatively, there should be at least 18 in of counter space from the edge of the sink to the inside corner for the countertop if the return counter space is blocked by a full-height, full-depth cabinet or a tall appliance.

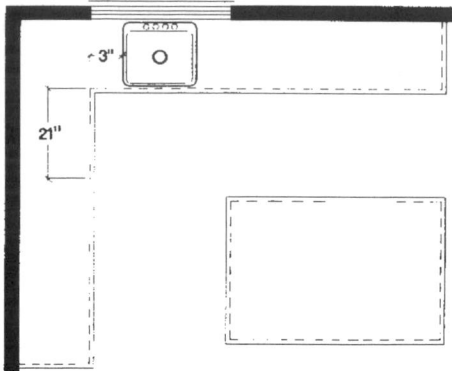

Fig. 39

Dishwasher placement requires 21 in of standing room between the edge of the appliance and adjacent counters, appliances, and/or cabinets that are placed at right angles to the dishwasher (Fig. 40). Note: Since the second cook or other family members may require access to the dishwasher, it should be accessible to more than one person in the kitchen.

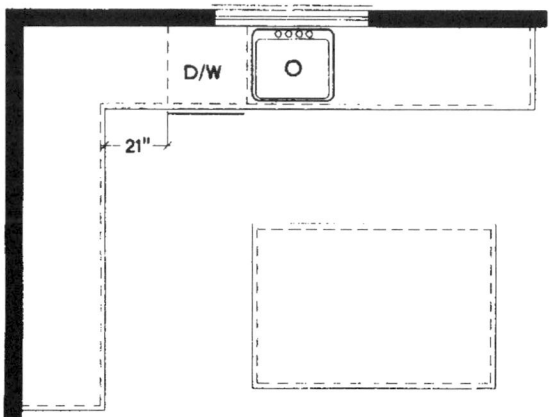

Fig. 40

The preparation center requires at least 36 in of continuous countertop. The preparation center should be immediately adjacent to a water source (Fig. 41). Notes: (a) The preparation center can be placed between the primary sink and the cooking surface, between the refrigerator and the primary sink, or adjacent to a secondary sink, island, or other cabinet section. (b) If more than one person works in the kitchen, each will need a 36-in preparation center of their own. If two people stand adjacent to one another, 72 in of space should be planned. Two people should never work at right angles to one another with their backs to each other. Try to orient people so that conversation can continue during the cooking and/or cleanup process.

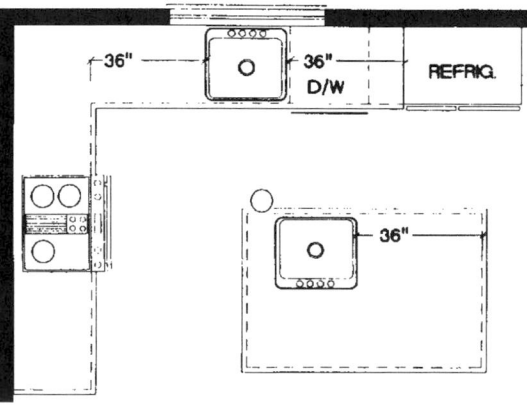

Fig. 41

Refrigerator placement should allow at least 15 in of counter space on the latch side of a standard unit, on either side of a side-by-side refrigerator, or at least 15 in of landing space no more than 48 in across from the refrigerator (Fig. 42). Notes: (a) Measure the 48-in walkway from the countertop adjacent to the refrigerator to the island countertop directly opposite. (b) Although not suggested, it is acceptable to place an oven adjacent to a refrigerator. For convenience, the refrigerator should be the appliance placed next to the available counter. If there is no safe area across from the oven, this arrangement may be reversed.

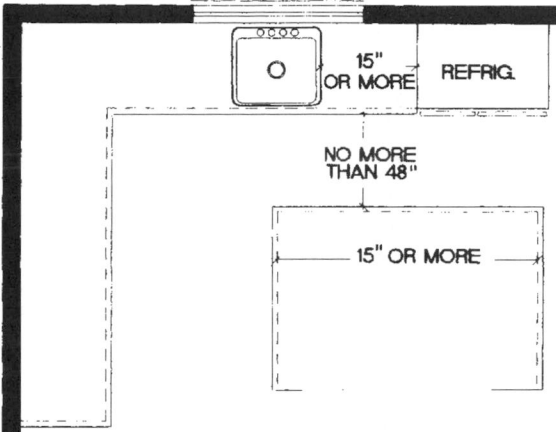

Fig. 42

Cooking surface requirements for an open-ended kitchen configuration (Fig. 43a) are at least 9 in of counter space required on one side and 15 in on the other. For an enclosed configuration (Fig. 43b), at least 3 in of clearance must be planned at an end wall protected by flame-retardant surfacing material, and 15 in must be allowed on the other side of the appliance.

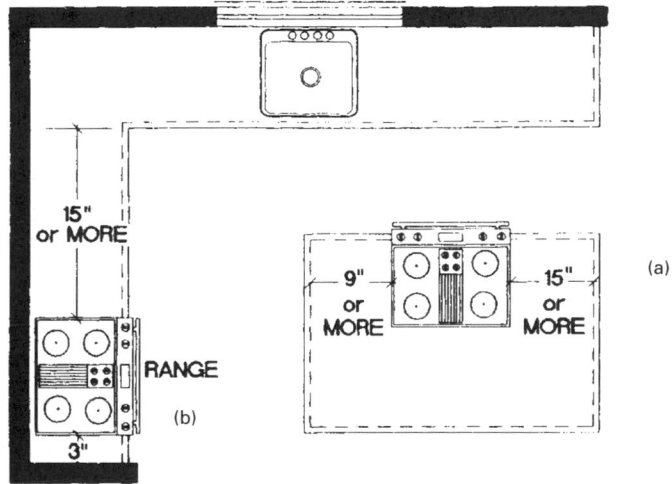

Fig. 43

The cooking surface should not be placed below an operable window unless the window is 3 in or more behind the appliance and more than 24 in above (Fig. 44).

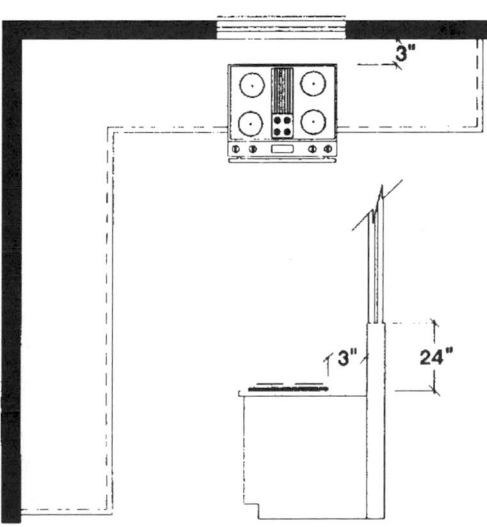

Fig. 44

KITCHENS

Oven placement requires at least 15 in of landing space next to or above the oven if the appliance door opens into a primary traffic pattern. A 15-in landing space no more than 48 in across from the oven is acceptable if the appliance does not open into a traffic area (Fig. 45). Note: Measure the 48-in walkway from the countertop adjacent to the refrigerator to the island countertop directly opposite.

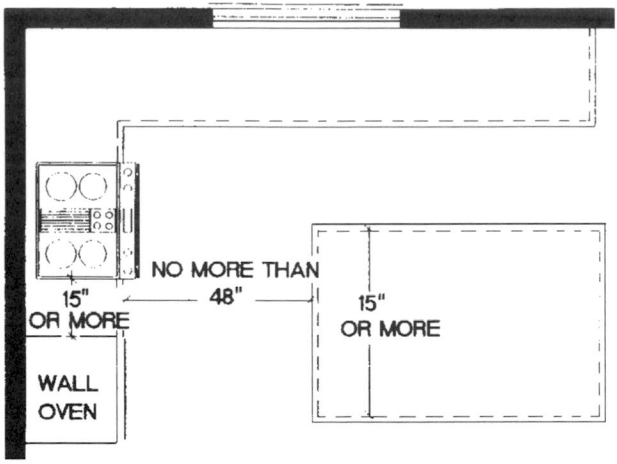

Fig. 45

A microwave oven should have at least 15 in of landing space above, below or adjacent to the appliance (Fig. 46).

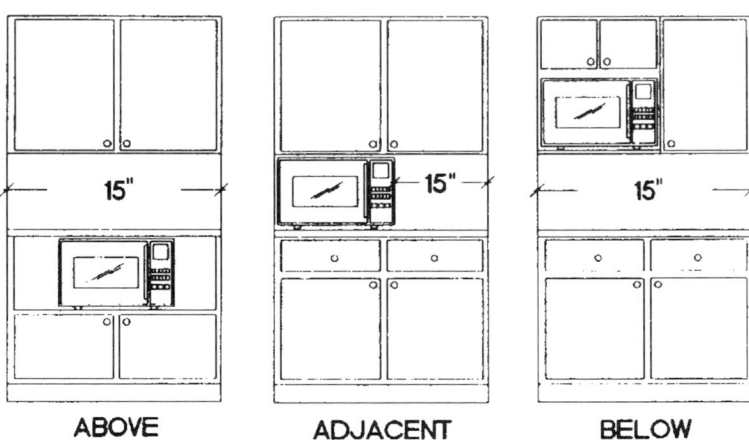

Fig. 46

Microwave oven placement should be such that the bottom of the appliance is placed between counter level and user eye level (36 to 54 in off the floor).

Ventilation is required for all major appliances used for surface cooking, with a fan rated at 150 cfm minimum (50 cfm times lineal footage of hood) (Fig. 47).

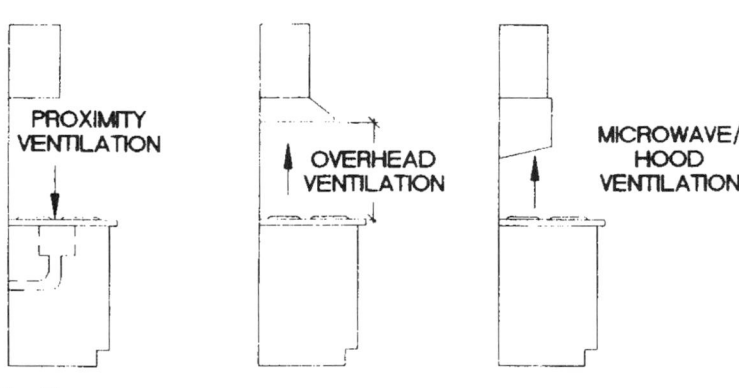

Fig. 47

The cooking surface must have at least 24 in of clearance between a protected surface above, or at least 30 in of clearance between the cooking surface and an unprotected surface above. A microwave-hood combination appliance may be lower than this 24-in dimension at the back wall (Fig. 48).

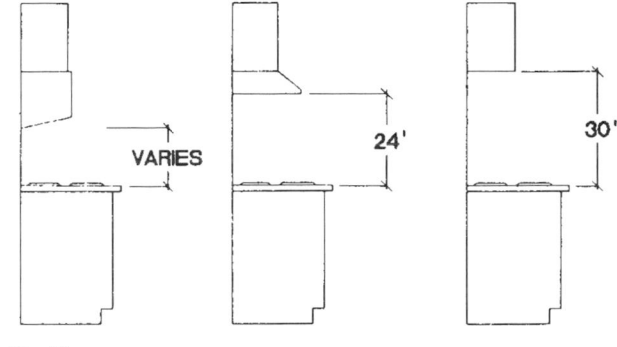

Fig. 48

The work triangle should be a total of 26 ft or less. The triangle is the shortest walking distance between the refrigerator, primary cooking surface, and primary food preparation sink. It is measured from the center front of each appliance (Fig. 49). The triangle should not intersect an island or peninsula cabinet by more than 12 in and no single leg should be shorter than 4 ft or longer than 9 ft. If more than one person cooks in the space, a triangle should be plotted for each cook. They may parallel one another, and the secondary triangle may coincide with one leg of the primary triangle, but they should not intersect one another. The secondary cook need not have as complete a work triangle as the primary cook. Note: A secondary triangle should include a cooking appliance and water source (either shared or separate); the refrigerator is typically shared.

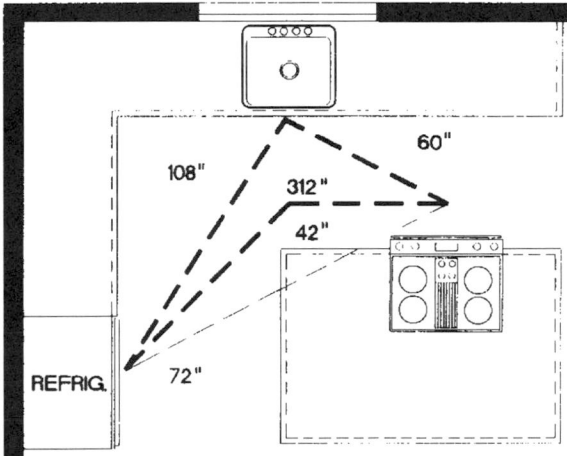

Fig. 49

Major traffic patterns should not cross through the work triangle connecting the primary work center (the primary sink, refrigerator, preparation area, cooktop/range center) (Fig. 50).

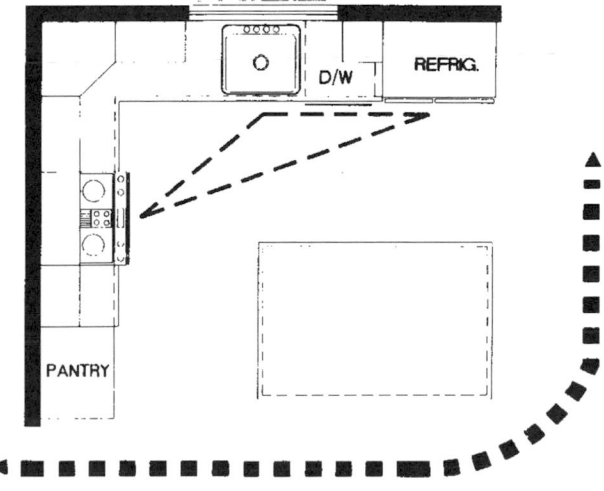

Fig. 50

KITCHENS

Counter/table space at least 24 in wide by 12 in deep should be planned for each seated diner (Fig. 51). The following knee-space clearance dimensions should be provided: 30-in-high table = 18-in knee-space clearance; 36-in-high counter = 15-in knee-space clearance; 42-in-high counter = 12-in knee-space clearance.

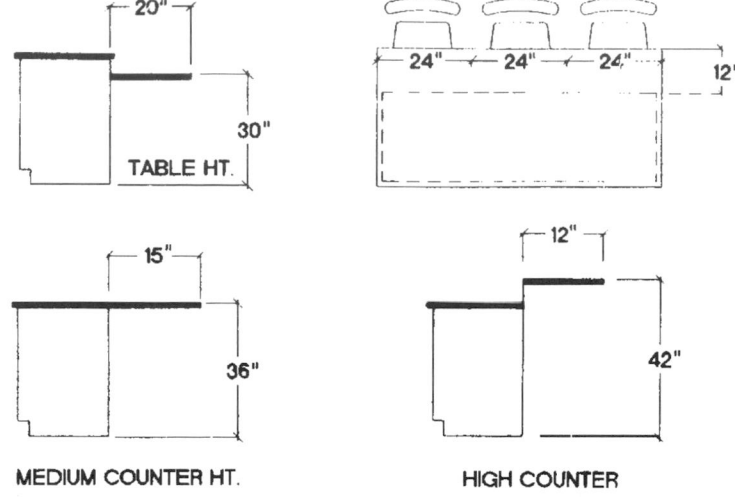

TABLE HT.

MEDIUM COUNTER HT.   HIGH COUNTER

Fig. 51

Walkway space must be at least 36 in from a counter/table to any wall or obstacle behind it if the area is to be used for passage behind a seated diner. At least 24 in of space from the counter/table to any wall or obstacle behind it should be planned if the area will not be used as a walk space (Fig. 52).

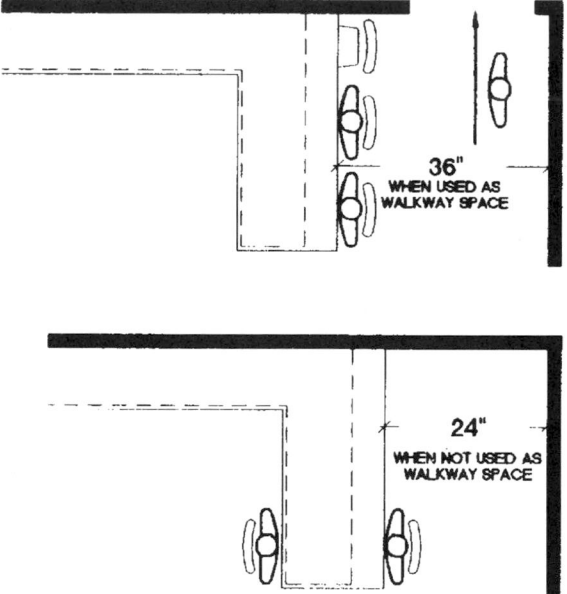

36"
WHEN USED AS
WALKWAY SPACE

24"
WHEN NOT USED AS
WALKWAY SPACE

Fig. 52

A window or skylight should equal at least 10 percent of the total square footage of the separate area or of a total living space that includes a kitchen.

Ground fault circuit interruptors should be specified on all receptacles that are within 6 ft of a water source in the kitchen. Also for safety reasons, a fire extinguisher should be located across from the cooktop, and smoke alarms should be included in the kitchen.

ALTERNATIVE KITCHEN SHAPES

Fig. 53a   In-line shape.

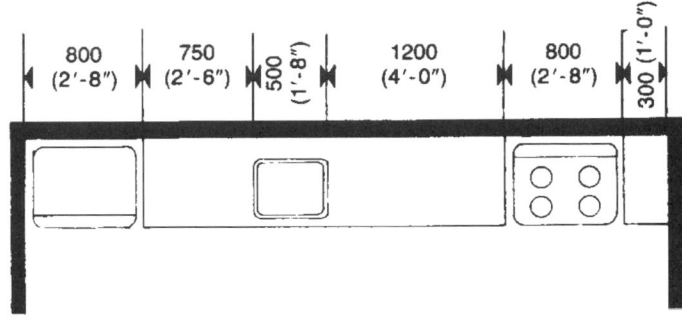

Fig. 53b   Five- and six-person in-line shape kitchen.

Fig. 54a   U shape.

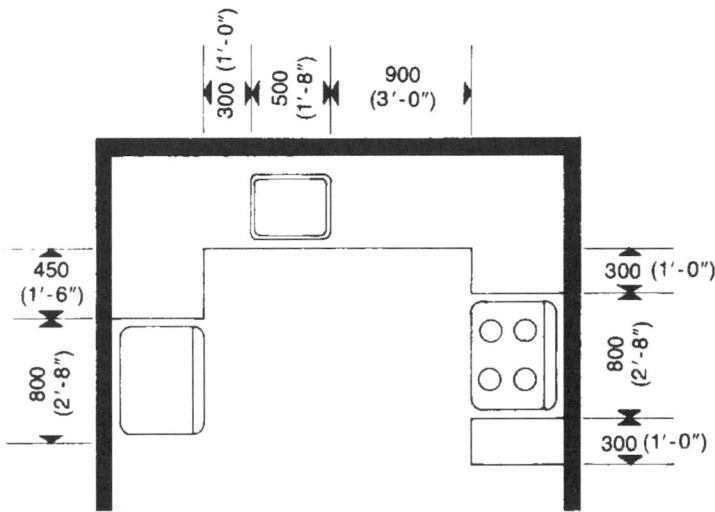

Fig. 54b   Five- and six-person U-shape kitchen.

Fig. 55a   L shape.

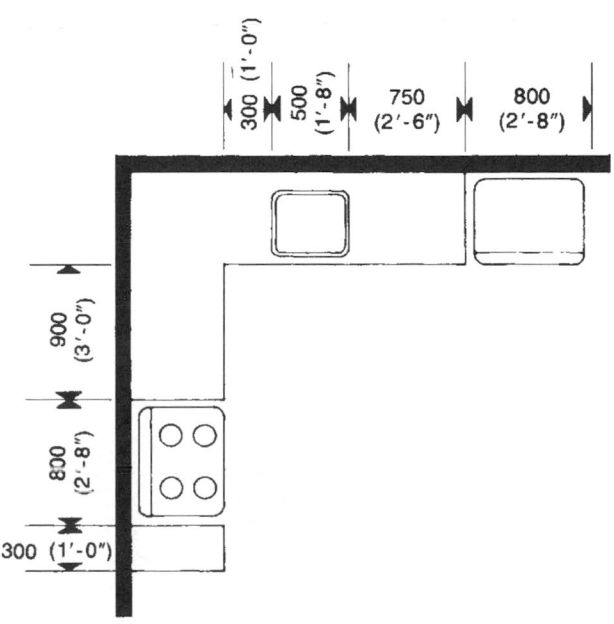

Fig. 55b   Five- and six-person L-shape kitchen.

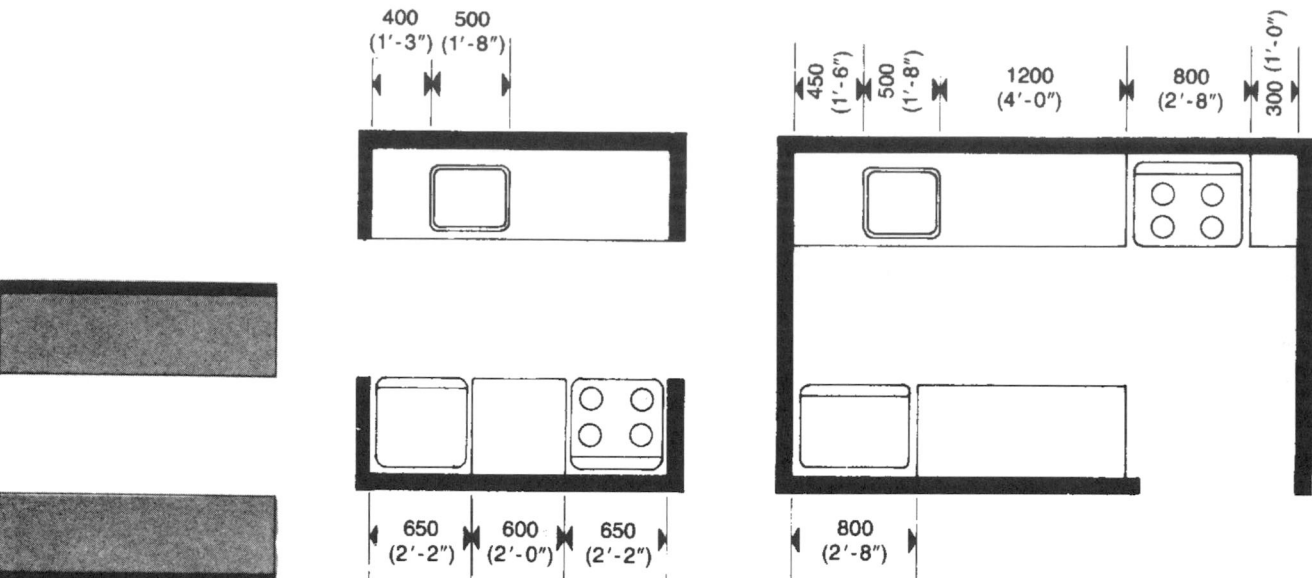

Fig. 56a Galley shape.

Fig. 56b One- and two-person galley-shape kitchen.

Fig. 56c Five- and six-person galley-shape kitchen.

KITCHEN PLANNING

Three work centers are involved: (1) storage and preparation, (2) preparation and cleaning, (3) cooking and serving.

Arrangement

As long as the operation sequence (from I to III) is preserved, organization of kitchen equipment may be varied to suit other requirements. Of the layout diagrams in Fig. 25, the U-shaped plan is most efficient in that work centers are most closely related and are entirely separated from general traffic areas.

Work Center Units and Sizes

Dimensions given are desirable minima to meet average small house requirements and to provide adequate working clearances.

I. Storage and preparation centers include a refrigerator, cabinets for other foods, and counter space. Average linear wall space of 4 ft 10 in to 5 ft 10 in should be allowed.

II. Preparation and cleaning require a sink, cabinets, and the greater part of the counter space. Sinks and special equipment vary greatly in size, but linear wall space of 5 ft 4 in to 6 ft 6 in should be allowed.

III. Cooking and serving centers contain the range, counter space, and storage cabinets. Average linear wall space required is 5 ft 0 in to 5 ft 10 in.

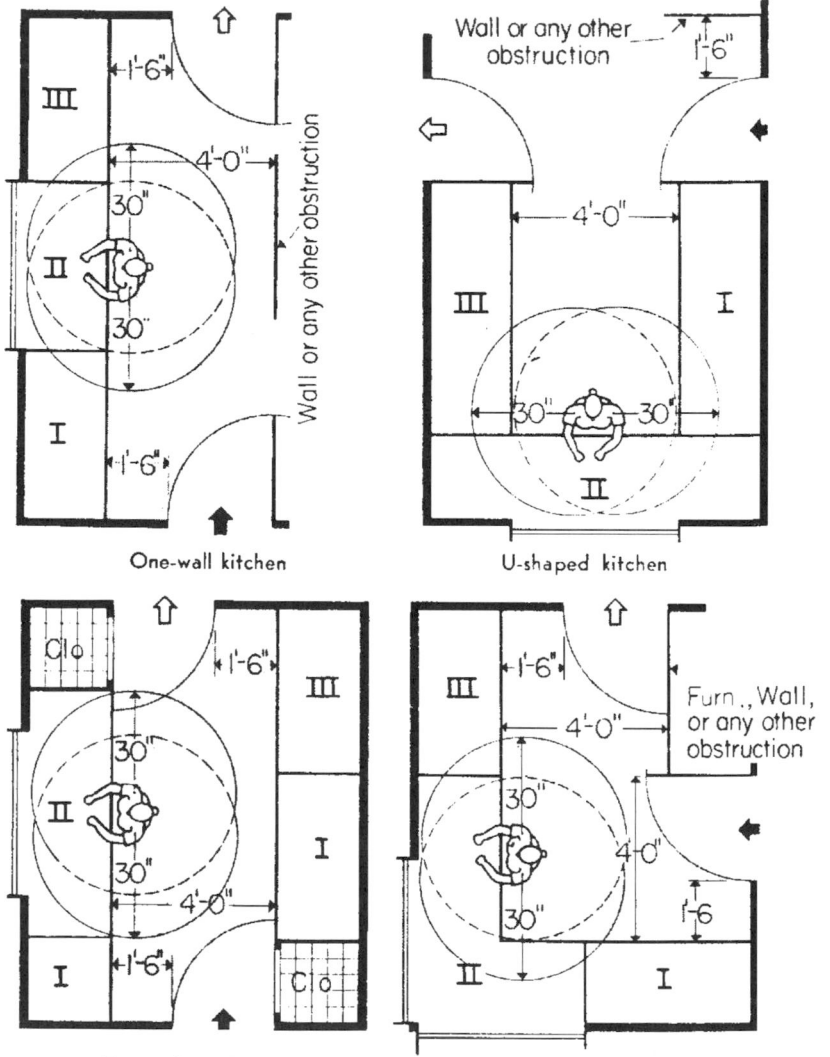

One-wall kitchen

U-shaped kitchen

Two-wall or Corridor

L-shaped kitchen

Cabinets

The size, number, and location of cabinets should be determined while the kitchen is being designed. Base cabinets and counter tops with wall cabinets above should not exceed 24 in in depth. This permits easy access to the rear of wall cabinets without standing on stools, or possibly dangerous makeshift climbing aids. For counter tops without wall cabinets above, a 30-in depth is the recommended maximum. The top shelves of cabinets should be placed no more than 72 in above the floor and wall cabinet depth should not exceed 12 in maximum.

Counters

Cabinet counters should be flush with installed appliances to provide a smooth, clear work area. For convenience and efficiency, a height of 36 in from the floor is recommended. Most manufacturers design appliances and fixtures for this level of operation. The trim used on counter top edges should be free of sharp edges, burrs and points.

Doors

Door swings which conflict with, or restrict the use of, appliances should be avoided. Doors should be installed to swing against the side or end of a cabinet or out of the kitchen. Where standard hinged doors would present a hazard, the use of sliding or folding doors is recommended.

Appliances

No matter how well designed, kitchen appliances are potentially hazardous when heating, grinding, cutting, shearing, mashing, or otherwise performing their necessary functions. They must be carefully located and properly installed. Here are some general rules.

■ Refrigerators should be installed where the interior is accessible to the work triangle and the door swing does not interfere with traffic flow or with other doors or drawers. For double-door refrigerator/freezer combinations the refrigerator door should swing away from the work triangle.

■ A minimum of 18 in of counter space on each side of ranges or 24 in on at least one side of separate oven units provides adjacent space for setting hot utensils, with room for their handles.

■ Choose a range with controls located at the front or side and eliminate storage areas above the burners. This reduces the risk of burns from spattering grease, boiling water or steam and the danger of garments catching fire when reaching over hot burners.

■ Range hoods are installed to remove heat, smoke, moisture, and odors and to provide a light source immediately above the cooking surface. A safely designed range hood should have rounded corners and rolled edges. It should be installed at such a height that the lower edge does not restrict the view of utensils on the rear heating elements. The optimum height for hood installations ranges between 56 and 60 in above the floor. The depth of the hood determines which height should be used for a particular hood. For example, a hood 17 in or less in depth should be placed no less than 56 in above the floor, while one 18 in or more in depth should be installed no less than 60 in above the floor.

■ Exhaust systems incorporated within the hood should mechanically direct their flow to the outside atmosphere and not into the attic or other overhead, unused space in the house.

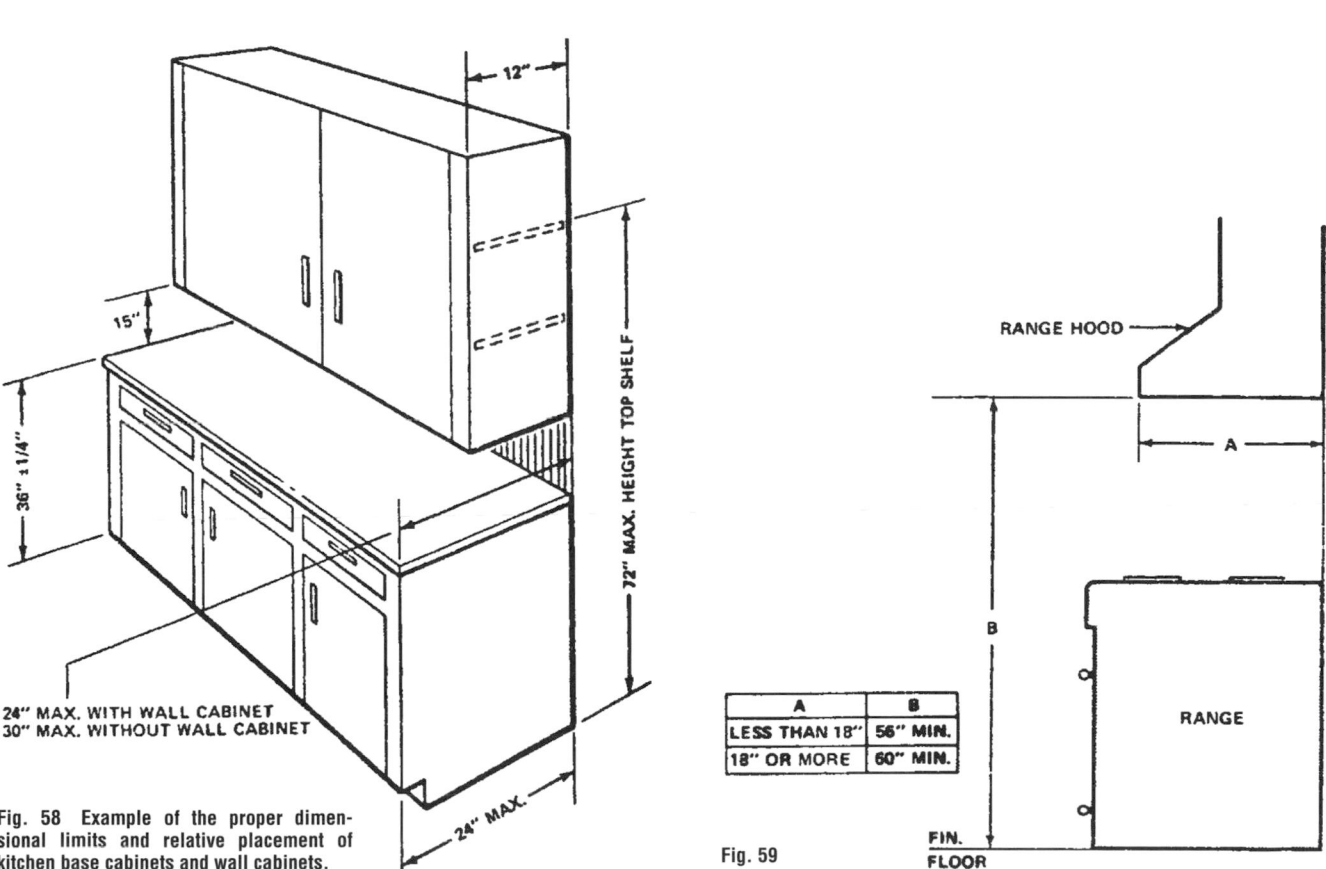

Fig. 58 Example of the proper dimensional limits and relative placement of kitchen base cabinets and wall cabinets.

Fig. 59

A	B
LESS THAN 18"	56" MIN.
18" OR MORE	60" MIN.

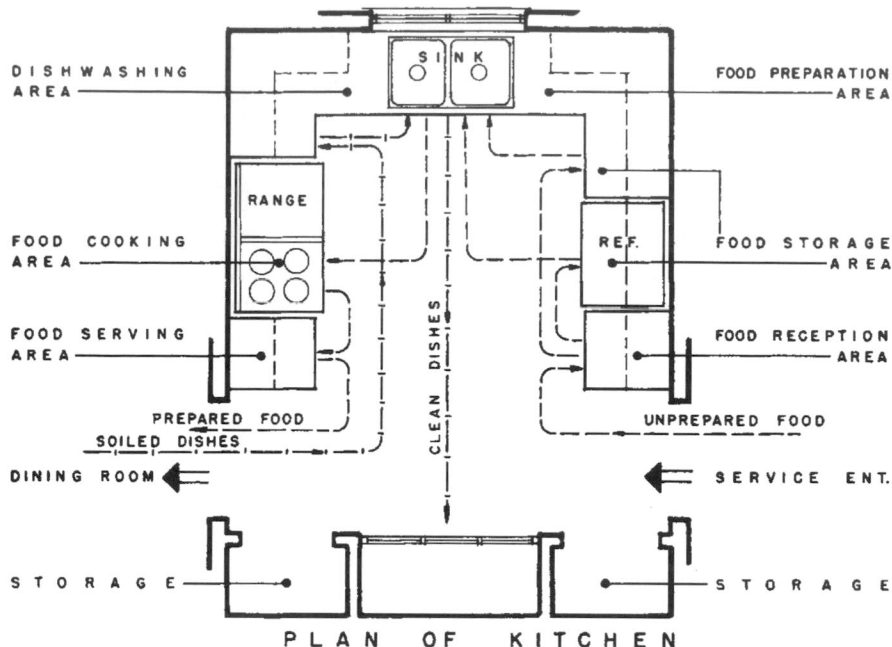

Fig. 60 Circulation of food through a kitchen. A U-shaped kitchen with a desirable arrangement of work areas according to use. Dotted lines show the routing of food through the kitchen to the dining room.

TABLE 9   Minimum Frontages for Work Centers, Inches

Work centers	Kitchenette	Kitchen			
	Zero bedroom	One bedroom	Two bedrooms	Three bedrooms	Four or more bedrooms
Sink	18	24	24	24	36
Counter and base cabinet at each side	15	18	20	24	30
Range	21	21	24	30	30
Counter and base cabinet at one side	15	18	20	24	30
Refrigerator (space)	30	30	36	36	36
Counter at latch side	15	15	15	15	18
Mixing (base and wall cabinet)	20	30	36	36	42

NOTES:

Work centers may be combined; the kitchenette multiple-use space should at least equal the largest frontage of any one of the work centers being combined; the kitchen multiple-use space should at least equal the largest frontage of any one of the work centers being combined plus 6 in.

Provide a drawer at each base cabinet.

Frontage may continue around a corner, except a space less than 12 in should not be counted.

A 72-in compact kitchen with 72-in wall cabinet may be substituted.

The frontages are based on typical cabinets. Base cabinet approximately 24 in deep by 36 in in height with one shelf and a drawer. Wall cabinet approximately 12 in deep by 30 in in height with two adjustable shelves.

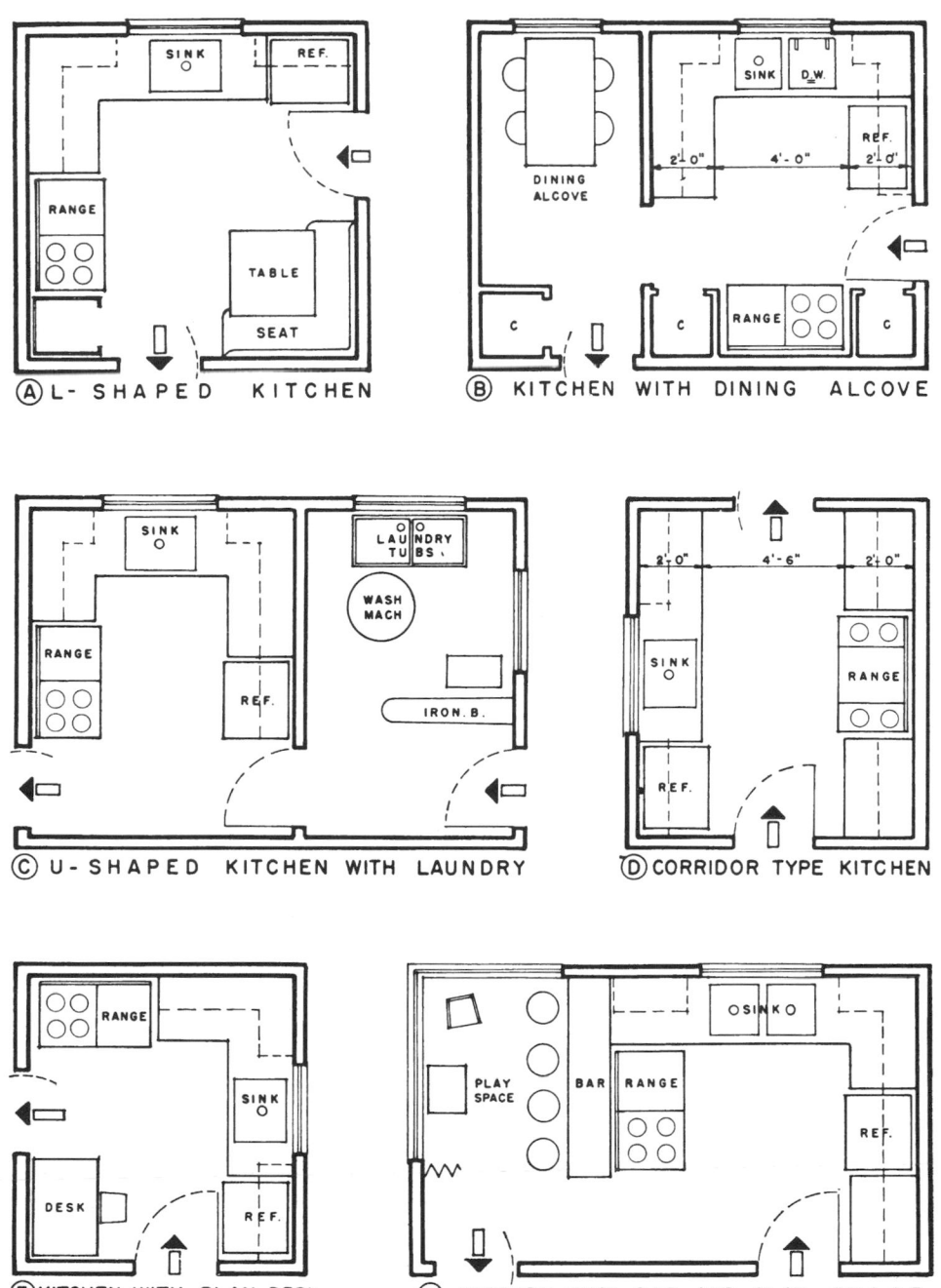

(A) L-SHAPED KITCHEN

(B) KITCHEN WITH DINING ALCOVE

(C) U-SHAPED KITCHEN WITH LAUNDRY

(D) CORRIDOR TYPE KITCHEN

(E) KITCHEN WITH PLAN DESK

(F) KITCHEN WITH BAR AND PLAY SPACE

Fig. 61 Various types of kitchen arrangements. Suggested plan types showing basic arrangements and also possible combinations of standard kitchen facilities with other activity areas and their equipment.

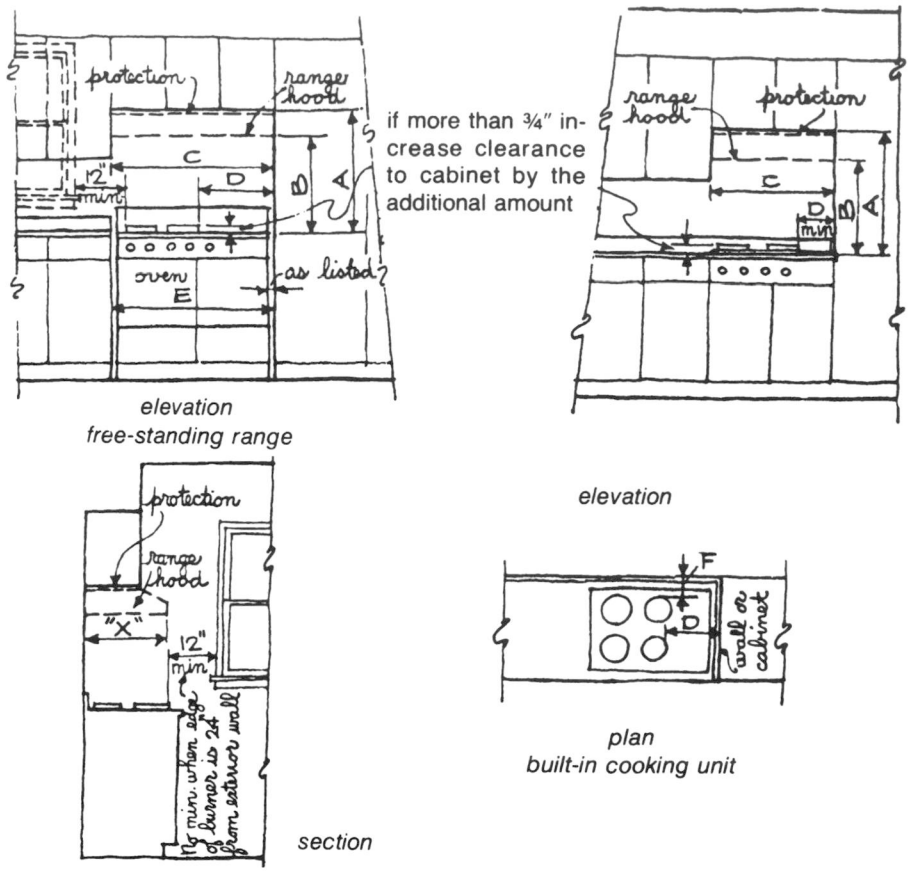

elevation
free-standing range

if more than ¾" increase clearance to cabinet by the additional amount

section

elevation

plan
built-in cooking unit

Fig. 62

CLEARANCES OVER COOKING RANGES

In Fig. 62, dimension A: 2 ft 6 in minimum clearance between the top of the range and the bottom of an unprotected wood or metal cabinet, or 2 ft 0 in minimum when the bottom of a wood or metal cabinet is protected.

Dimension B: 2 ft 0 in minimum when hood projection X is 18 in or more, or 1 ft 10 in min. when hood projection X is less than 18 in.

Dimension C: not less than width of range or cooking unit.

Dimension D: 10 in minimum when vertical side surface extends above countertops.

Dimension E: when range is not provided by builder, 40 in minimum.

Dimension F: Minimum clearance should be not less than 3 in.

Cabinet protection should be at least ¼ in asbestos millboard covered with not less than 28-gauge sheet metal (0.015 stainless steel, 0.024 aluminum, or 0.020 copper).

Clearance for D, E, or F should be not less than listed UL or AGA clearances.

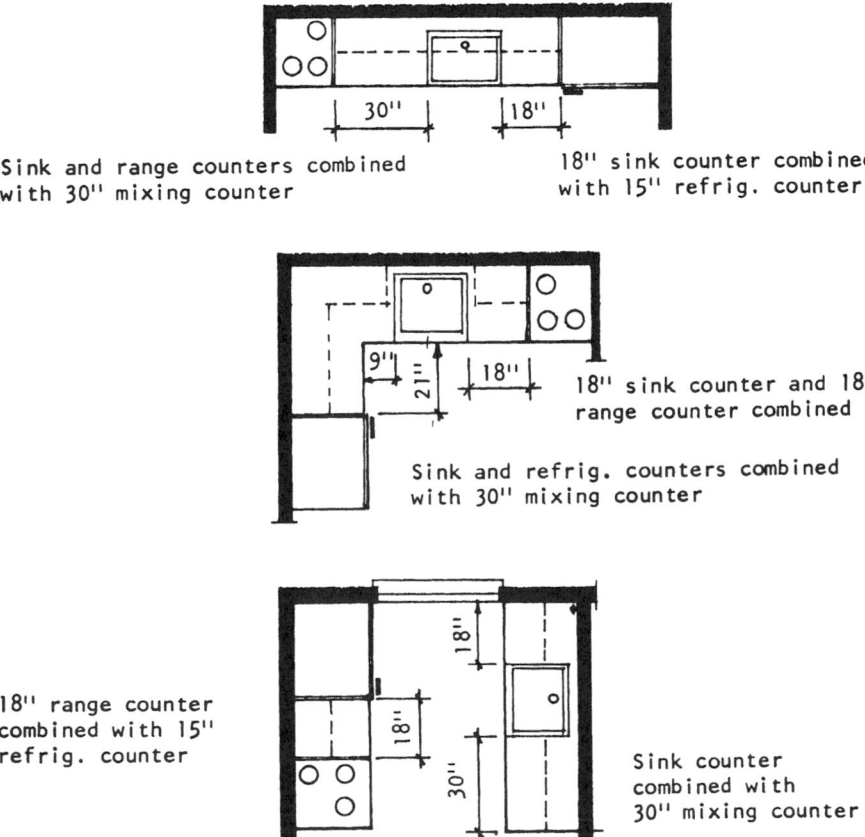

Sink and range counters combined with 30" mixing counter

18" sink counter combined with 15" refrig. counter

18" sink counter and 18" range counter combined

Sink and refrig. counters combined with 30" mixing counter

18" range counter combined with 15" refrig. counter

Sink counter combined with 30" mixing counter

Fig. 63 Kitchens for one-bedroom living units (with minimum storage, counter area, fixtures).

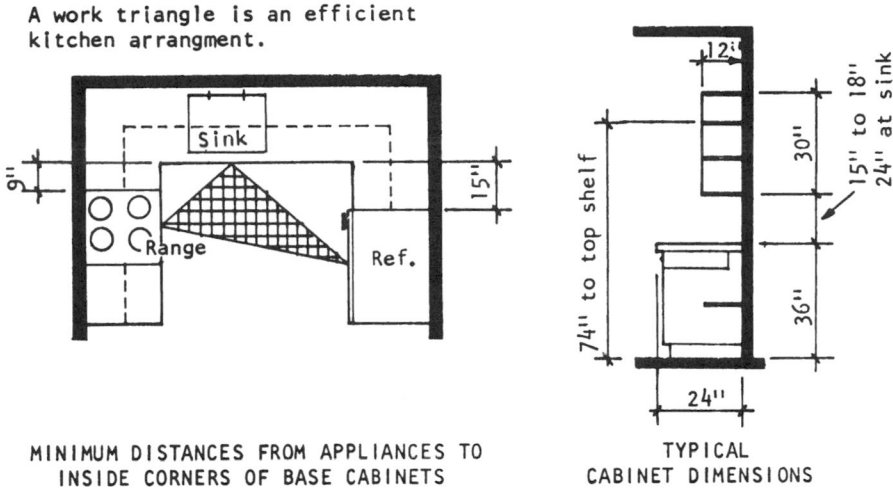

A work triangle is an efficient
kitchen arrangment.

MINIMUM DISTANCES FROM APPLIANCES TO
INSIDE CORNERS OF BASE CABINETS

TYPICAL
CABINET DIMENSIONS

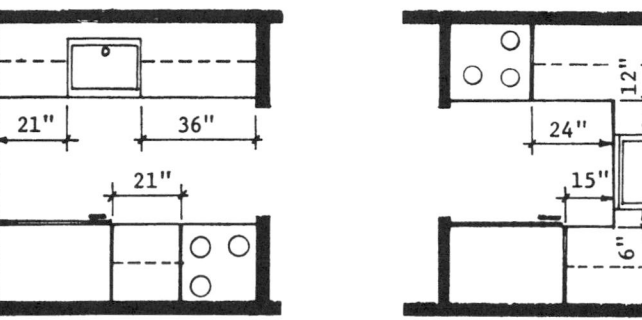

21" sink counter combined
with 36" mixing counter

Sink and range counters combined
with 36" mixing counter

21" range counter combined
with 15" refrig counter

21" sink counter combined
with 15" refrig counter

Fig. 64   Kitchens for two-bedroom living unit (with minimum storage, counter area, fixtures).

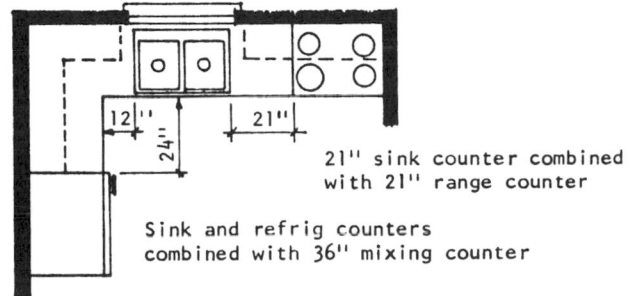

21" sink counter combined
with 21" range counter

Sink and refrig counters
combined with 36" mixing counter

Fig. 65   Kitchen for three-bedroom living unit (with minimum storage, counter area, fixtures).

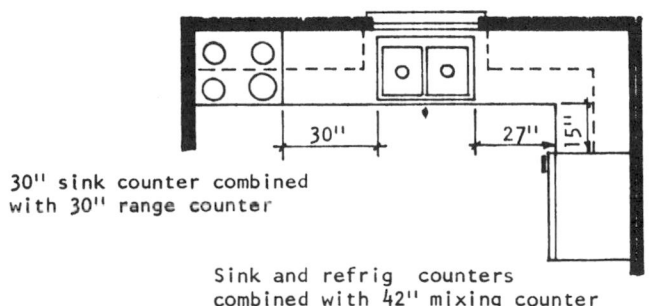

30" sink counter combined
with 30" range counter

Sink and refrig counters
combined with 42" mixing counter

Fig. 66   Kitchen for four-bedroom living unit (with minimum storage, counter area, fixtures).

331

MINIMUM SIZED ADAPTABLE KITCHEN OR KITCHENETTE

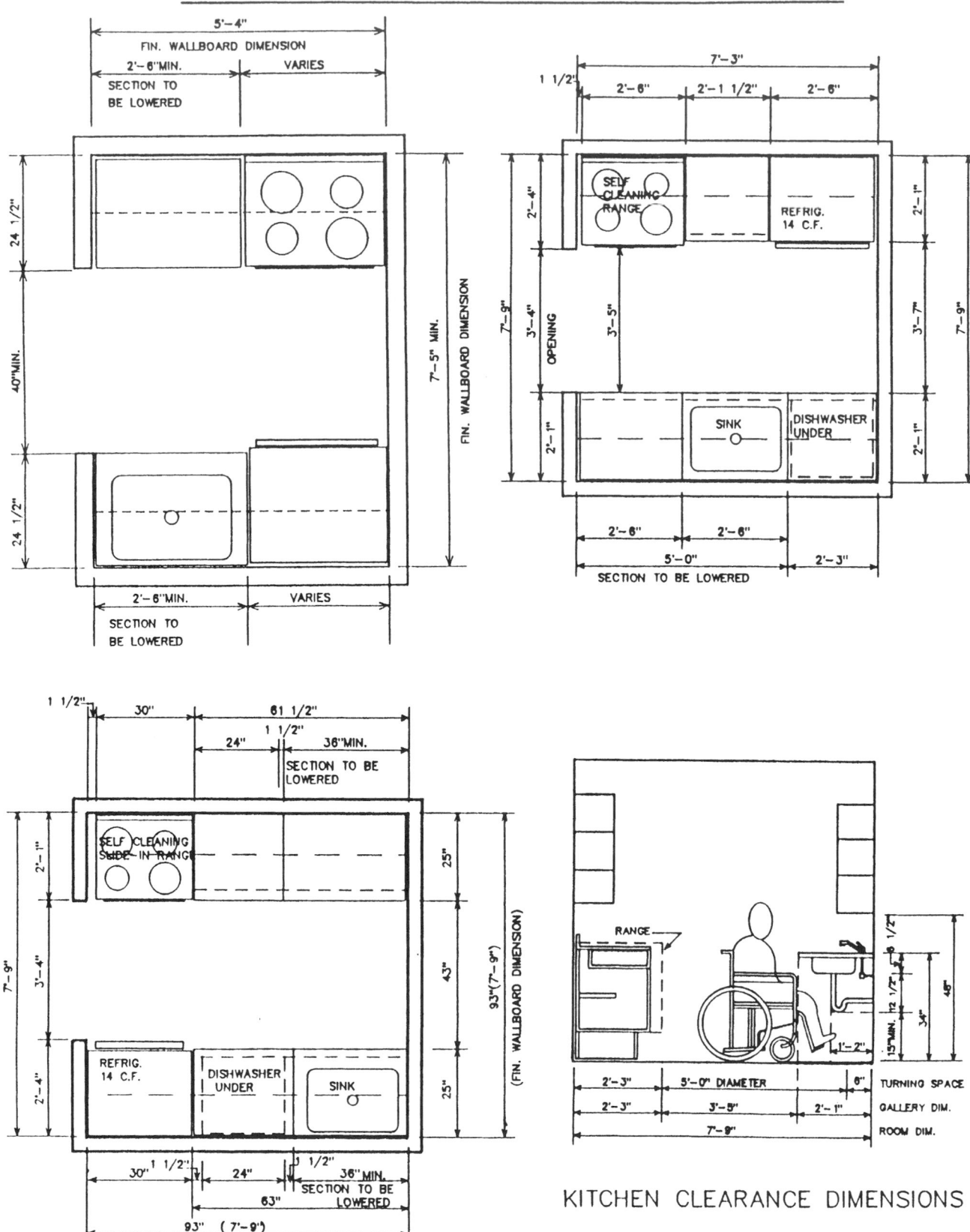

Fig. 67   Adaptable kitchen.

The L with an island

The U with an island

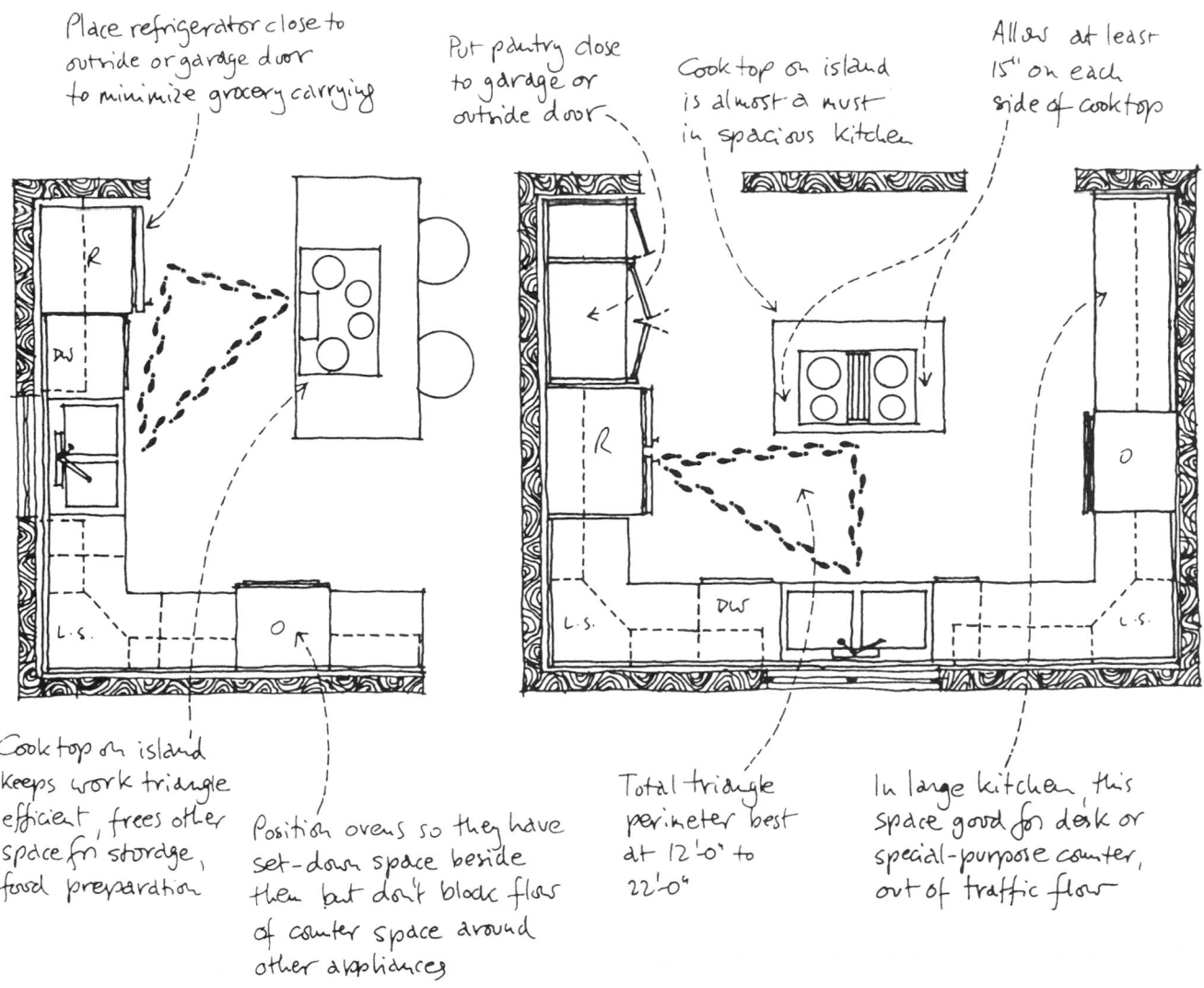

Place refrigerator close to outside or garage door to minimize grocery carrying

Put pantry close to garage or outside door

Cook top on island is almost a must in spacious kitchen

Allow at least 15" on each side of cooktop

Cook top on island keeps work triangle efficient, frees other space for storage, food preparation

Position ovens so they have set-down space beside them but don't block flow of counter space around other appliances

Total triangle perimeter best at 12'-0" to 22'-0"

In large kitchen this space good for desk or special-purpose counter, out of traffic flow

Fig. 68

The L-shaped kitchen

The U-shaped kitchen

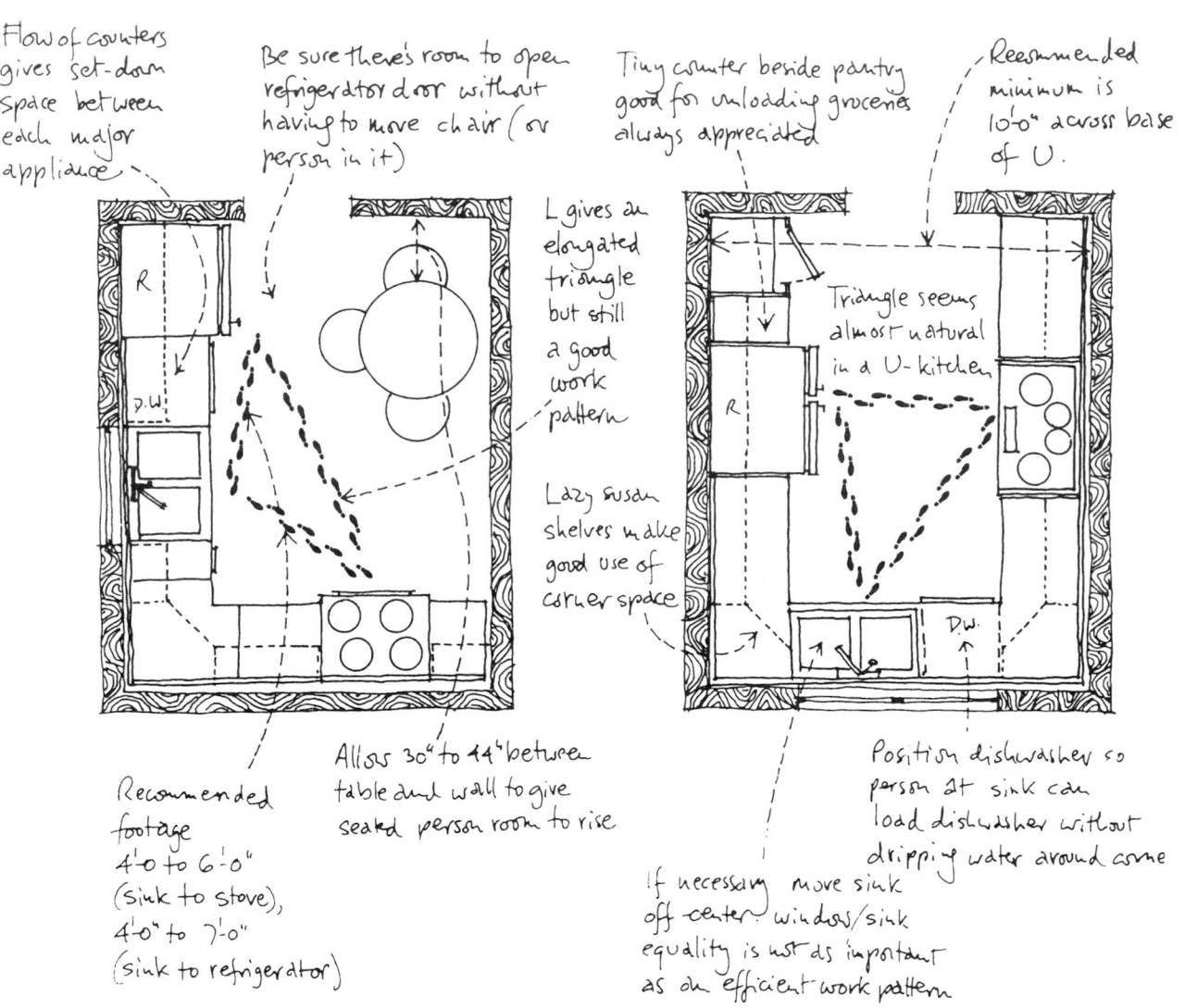

Flow of counters gives set-down space between each major appliance

Be sure there's room to open refrigerator door without having to move chair (or person in it)

Tiny counter beside pantry good for unloading groceries always appreciated

Recommended minimum is 10'-0" across base of U.

L gives an elongated triangle but still a good work pattern

Triangle seems almost natural in a U-kitchen

Lazy susan shelves make good use of corner space

Recommended footage
4'-0 to 6'-0"
(sink to stove),
4'-0" to 7'-0"
(sink to refrigerator)

Allow 30" to 44" between table and wall to give seated person room to rise

Position dishwasher so person at sink can load dishwasher without dripping water around some

If necessary move sink off center - window/sink equality is not as important as an efficient work pattern

Fig. 69

The one-wall kitchen

The corridor kitchen

At least 24" to right of sink for stacking dishes before washing; 36" much better, since this is prime mixing/preparation area

Double-basl sink especially helpful if there is no dishwasher

Recommended base cabinet minimum is 6'-0" plus under-sink area

Tiny counter to keep stove from wall; cabinet underneath a good spot for trays

Minimum counter to left of sink should be 18" for stacking dishes, giving set-down space beside stove

Give maximum counter space near sink for this is where most mixing and food preparation are done

Typical work triangle

Trash compactor-undercounter

Pantry or space for stacking washer and dryer

Allow minimum of 48"(60" is much preferable) for aisle space - to allow one person to pass behind another, with the dishwasher or oven door open

Lines of cabinet above

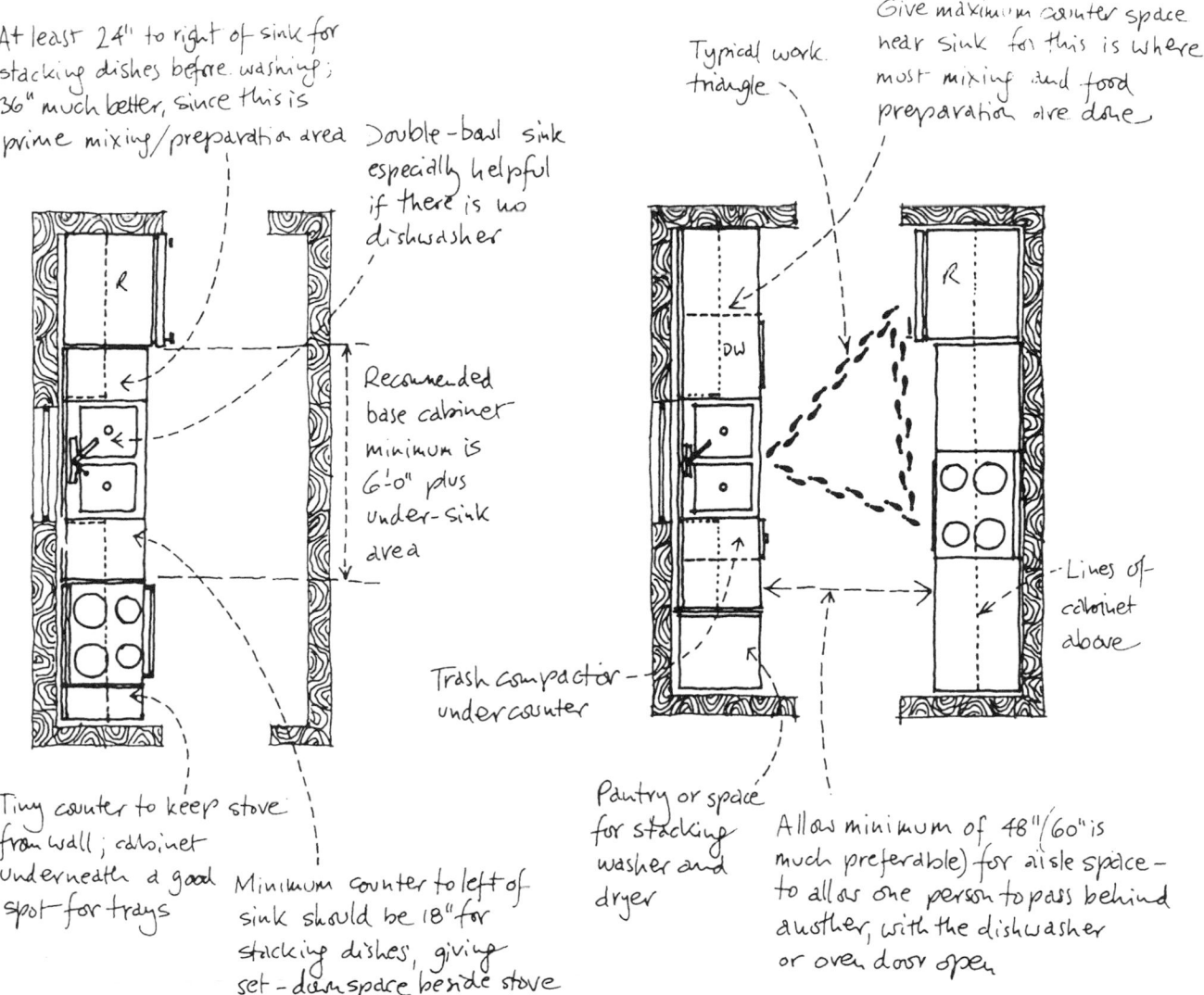

Fig. 70

KITCHEN/DINING AREAS

BREAKFAST AREA

The "breakfast" or informal dining area is part and parcel of the kitchen. A common historical example is the country kitchen, where the dining table is right inside the room. Contemporary plans have refined this definition into the breakfast area and have reduced the frequency of circulation through this space.

The breakfast area should generally accommodate a 48-in diameter table with four chairs. Typically, this translates to a 10- by 10-ft room. This is sufficient only if circulation through this space is not planned; otherwise, dimensions should increase to at least 10 by 12 ft (Fig. 1).

The breakfast space should be given high priority for orientation to natural light and views; despite the fact that this area will be used at various times, morning light should be given highest priority. Windows, skylights, or patio doors are all good sources of natural light; if possible, locate them so that they can provide maximum illumination throughout the day. Careful design will also extend this light or view to the adjoining kitchen (Fig. 2), particularly the sink workstation (Fig. 3).

Direct access to the outdoor deck/patio from the breakfast area is increasingly popular. Although French doors are more aesthetically pleasing, sliding glass doors may be functionally preferable, as they can remain open in warm weather with no impeding door swing. Direct access from the kitchen to the outdoor dining area is also popular. Locating a door in the heart of the kitchen work triangle will allow for easy transport of food to the outdoor table. If this is not possible, consider a pass-through window near the sink counter. (Figure 2.20)

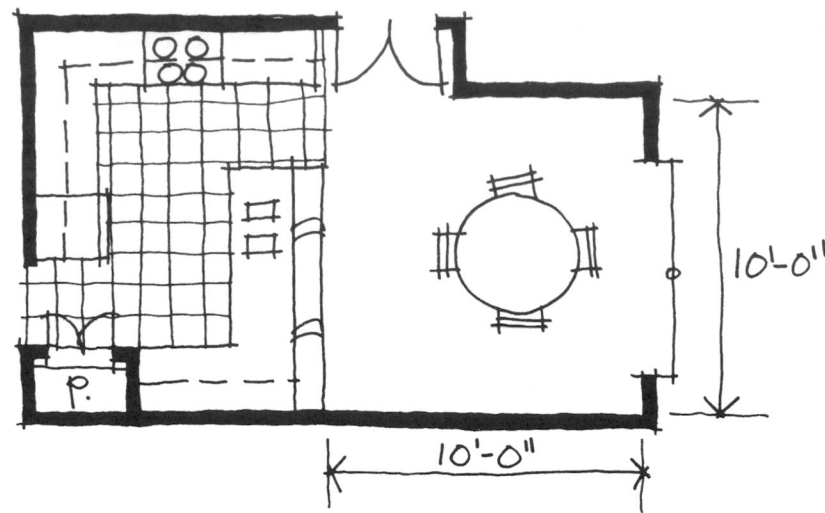

Fig. 1   Breakfast room clearances.

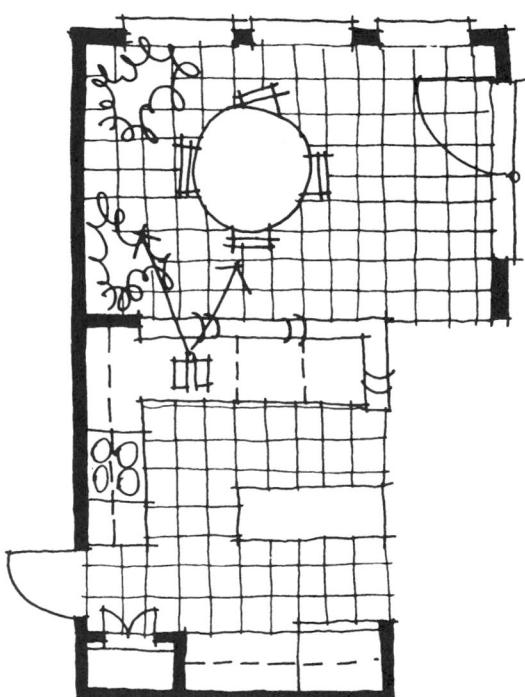

Fig. 2   Breakfast with windows shared by adjacent sink counter.

Fig. 3   View from sink blocked by upper cabinets (above) vs. area above sink left open (below).

THE LIVING-DINING-KITCHEN PLAN

The arrangement illustrated in Fig. 4 is the living-room-dining area-kitchen design that uses a movable wall. This layout and variations upon it have the principal advantage of offering the occupant the opportunity to use space in alternative ways at different times of the day and on different occasions. The arrangements tend to be more expensive than other less attractive layouts because of the cost of installing a movable wall. But it illustrates the idea of making the subdivision of space in the home as flexible as possible.

The movable wall between the dining area and the living room may be opened to provide a living room-dining area, and the kitchen may be closed off.

Alternatively (Fig. 5), the wall may be closed to provide a kitchen-dining area that is separated from the living room.

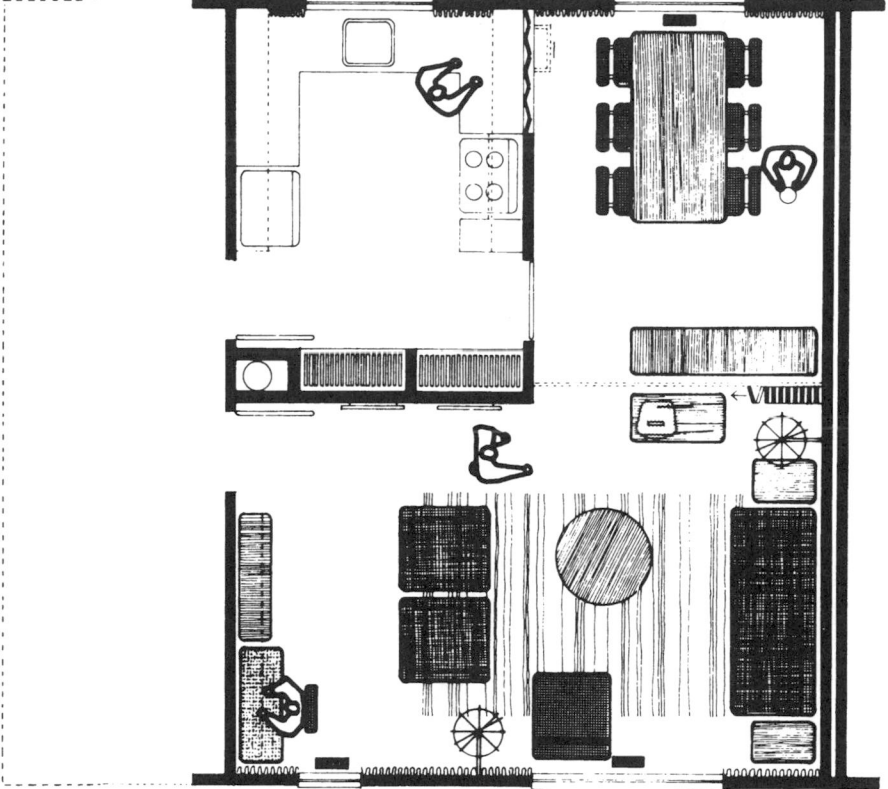

Fig. 4

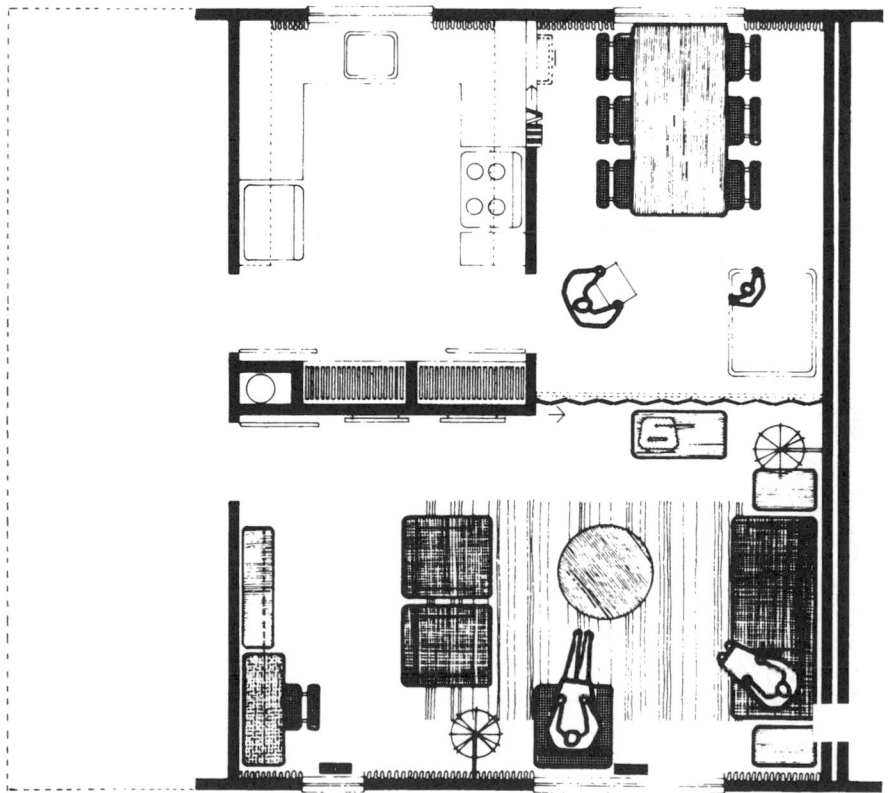

Fig. 5

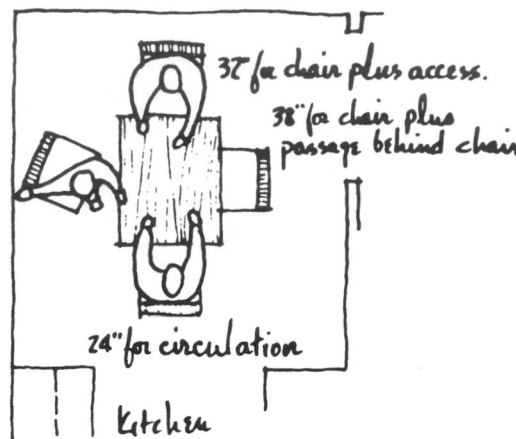

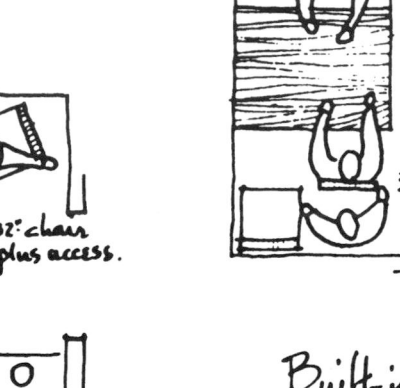

Eating

One of the primary functions of the kitchen has been to provide a place for informal or family eating. This is different from guest or formal dining in a separate dining room or area. The informal dining generally consists of breakfast, lunch, snacks, or just serving coffee to a neighbor. This eating area should be clearly defined as a separate functional area.

A frequent and desirable arrangement is the combined kitchen-dining area. Figures 6 through 32 show the various possible arrangements. Another arrangement is the kitchen-family room.

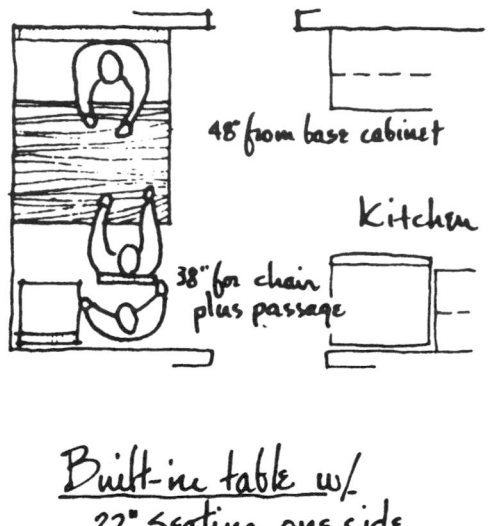

Fig. 6   Minimum clearances for dining area in kitchen.

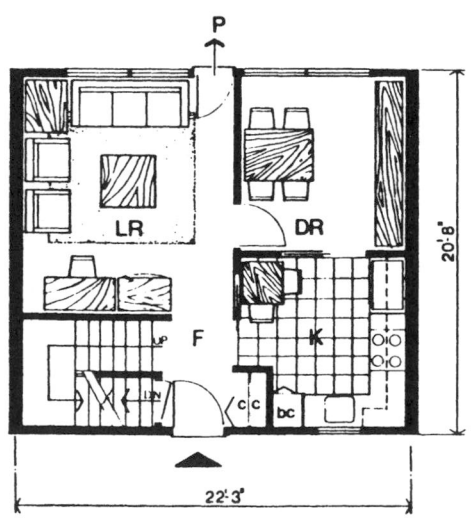

Fig. 7   Separate dining room.

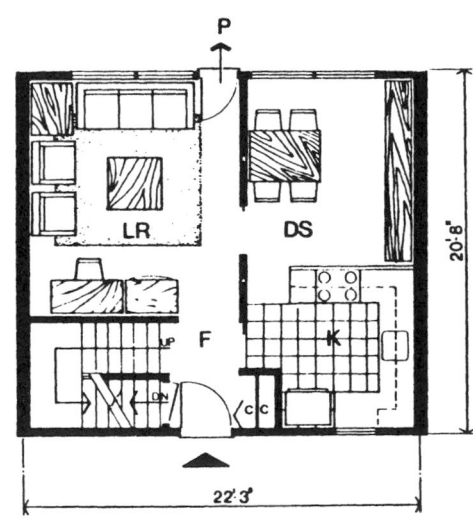

Fig. 8   Combined dining space and kitchen.

2 PERSONS

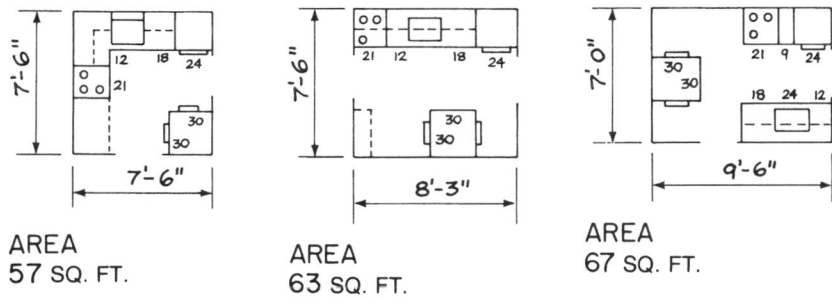

AREA
57 SQ. FT.

AREA
63 SQ. FT.

AREA
67 SQ. FT.

Fig. 9

4 PERSONS

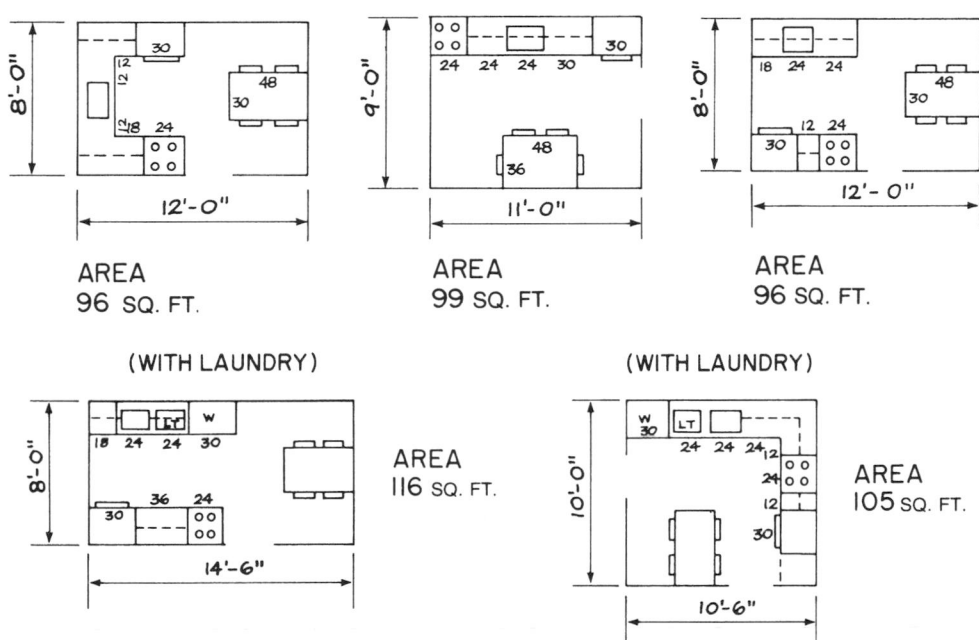

AREA
96 SQ. FT.

AREA
99 SQ. FT.

AREA
96 SQ. FT.

(WITH LAUNDRY)

(WITH LAUNDRY)

AREA
116 SQ. FT.

AREA
105 SQ. FT.

Fig. 10

6 PERSONS
(WITH LAUNDRY)

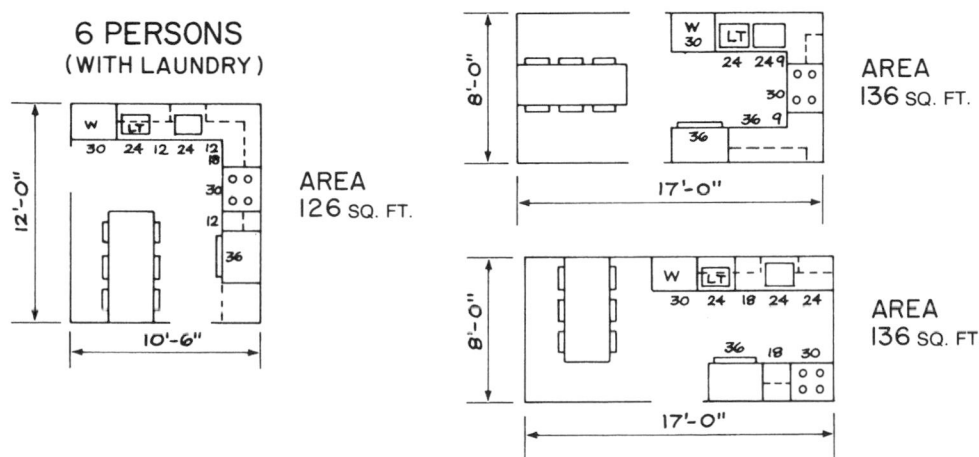

AREA
126 SQ. FT.

AREA
136 SQ. FT.

AREA
136 SQ. FT.

Fig. 11a

6 PERSONS

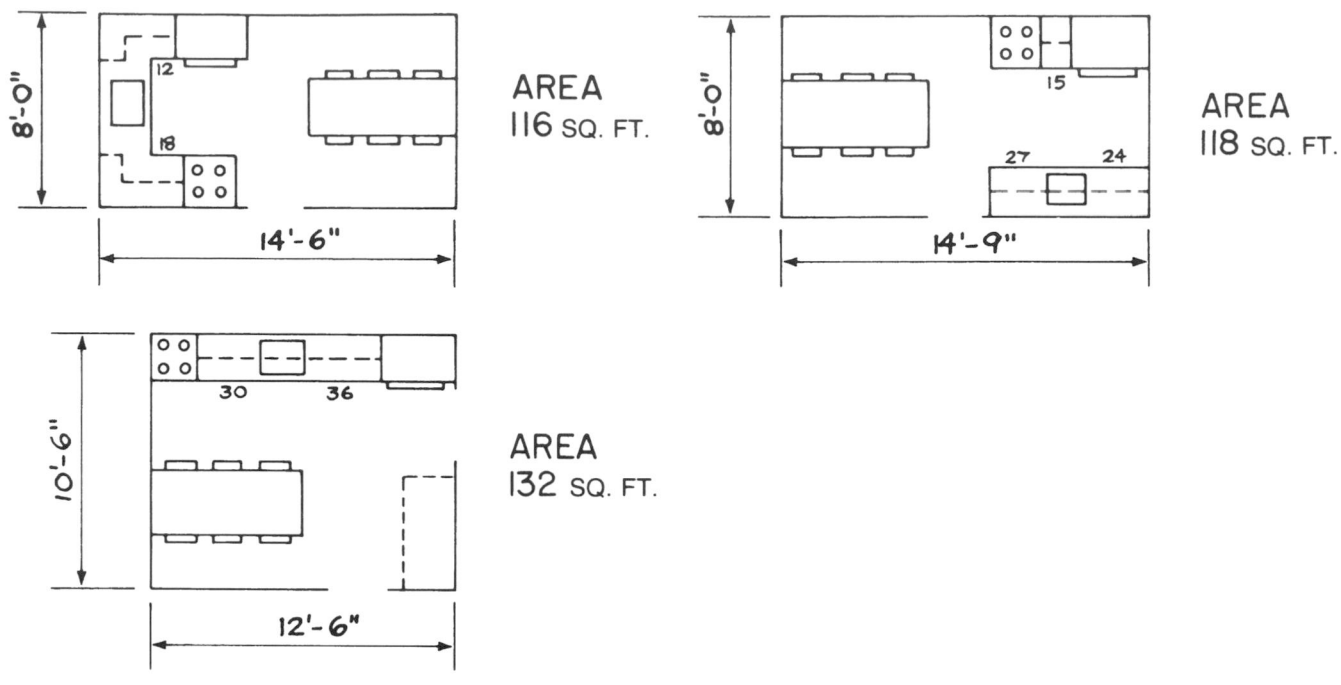

AREA
116 SQ. FT.

AREA
118 SQ. FT.

AREA
132 SQ. FT.

Fig. 11*b*

8 PERSONS

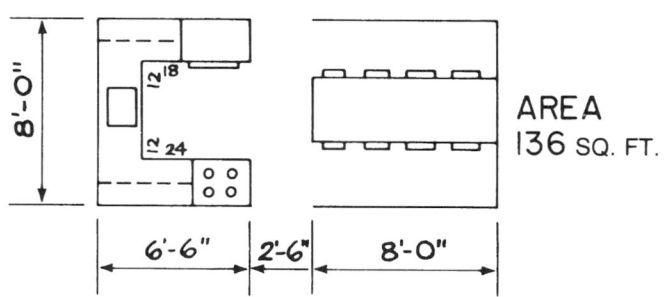

AREA
136 SQ. FT.

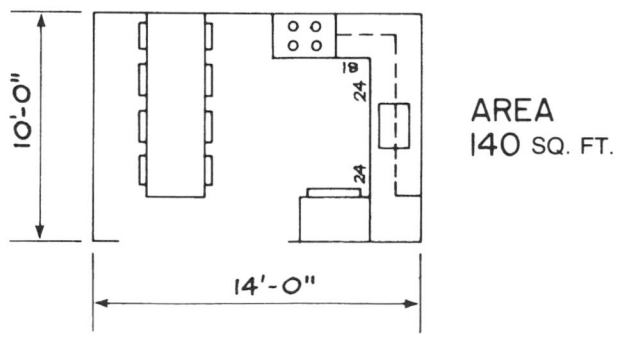

AREA
140 SQ. FT.

Fig. 12*a*

8 PERSONS
(WITH LAUNDRY)

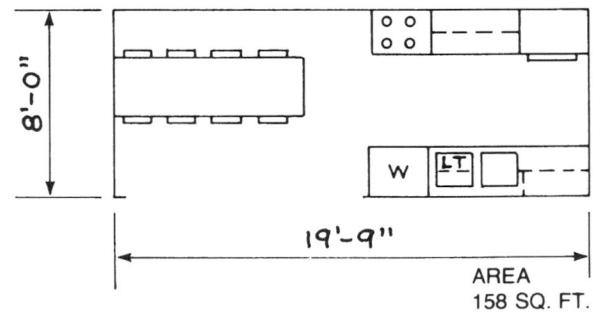

AREA
158 SQ. FT.

AREA
147 SQ. FT.

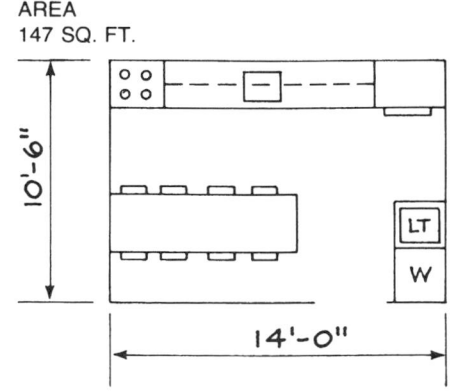

Fig. 12*b*

Kitchen/Breakfast Area

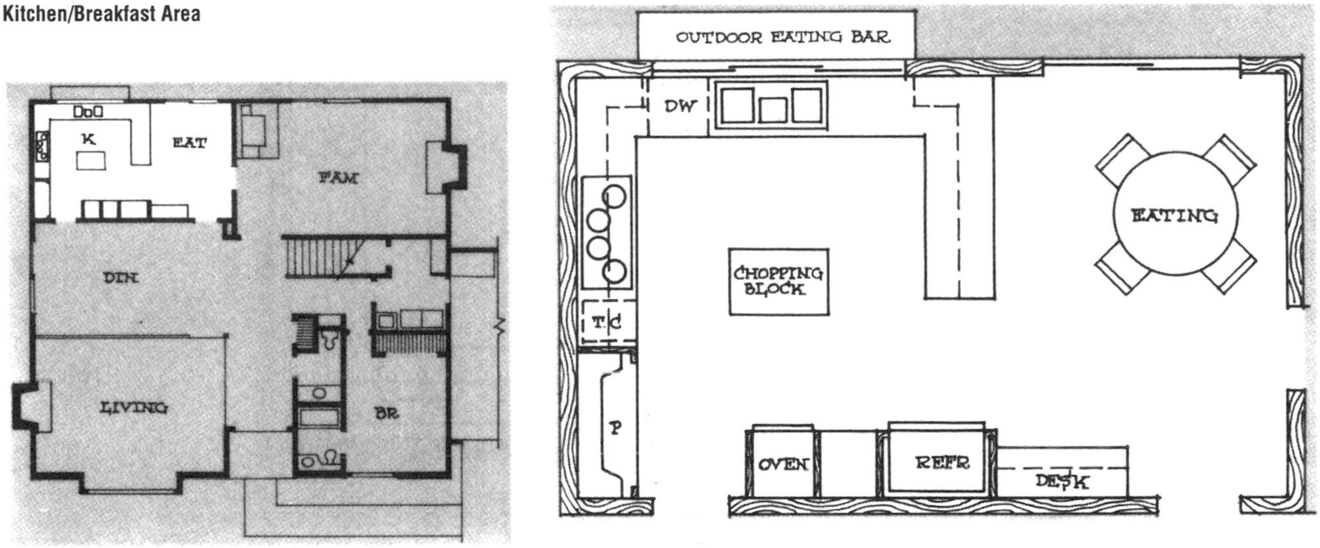

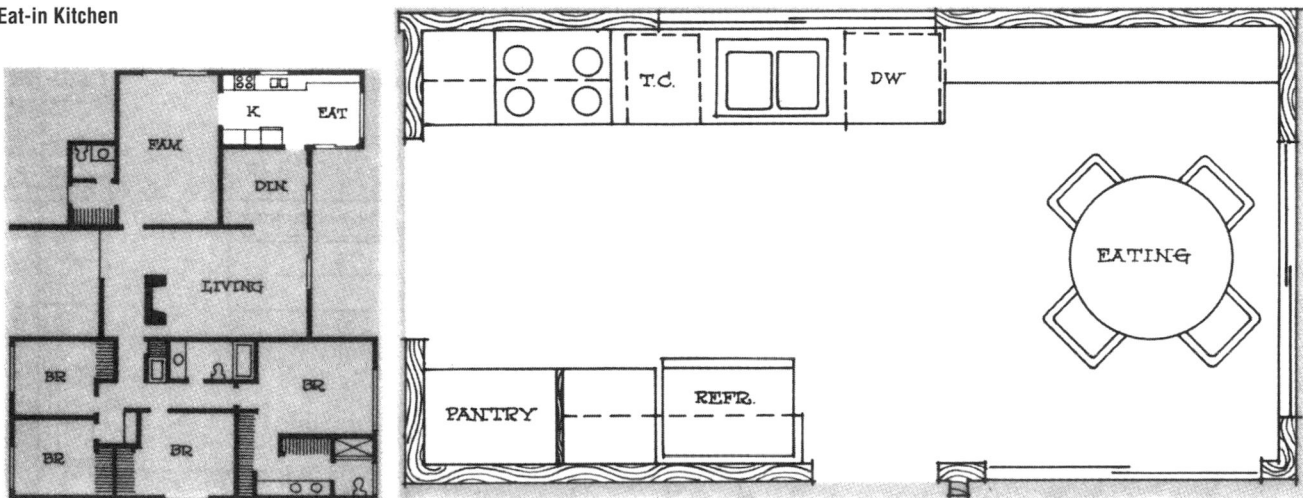

Fig. 13   Casiano Estates, Bel Air, California. Architect: Shapen Industries.

Eat-in Kitchen

Fig. 14   Anaheim Hills, Anaheim, California. Architect: Morris & Lohrbach.

Kitchen/Family Room

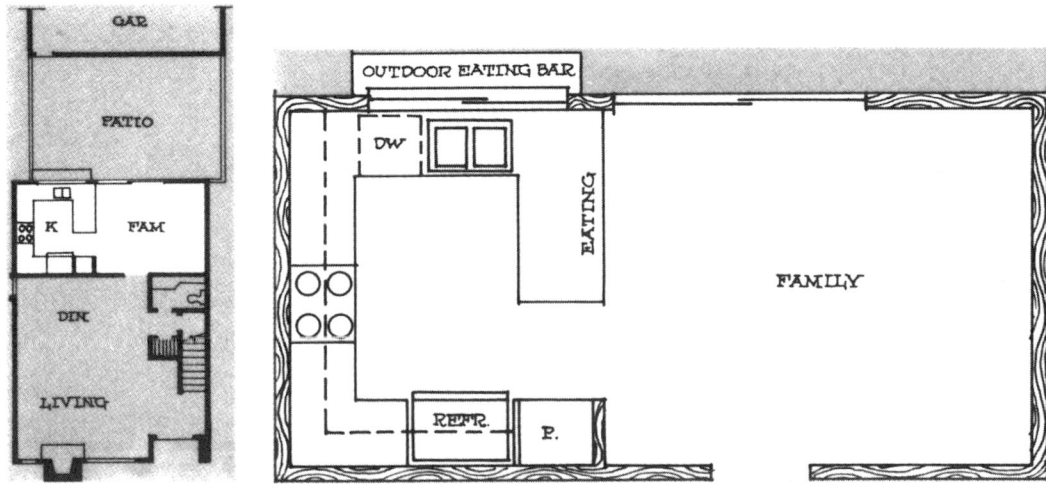

Fig. 15   Deerfield Town Homes, Irvine, California. Architect: Morris & Lohrbach.

Eat-in Kitchen

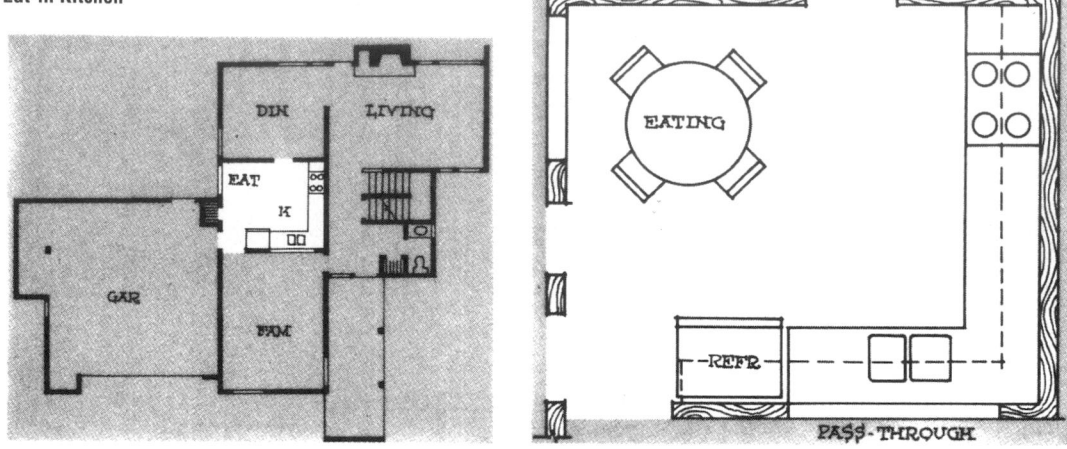

Fig. 16   Woodbridge Meadows, Montgomery County, Pennsylvania. Architect: Lynn Taylor.

Kitchen/Breakfast Room

Fig. 17   Anaheim Hills, Anaheim, California.
Architect: Morris & Lohrbach.

Indoor/Outdoor

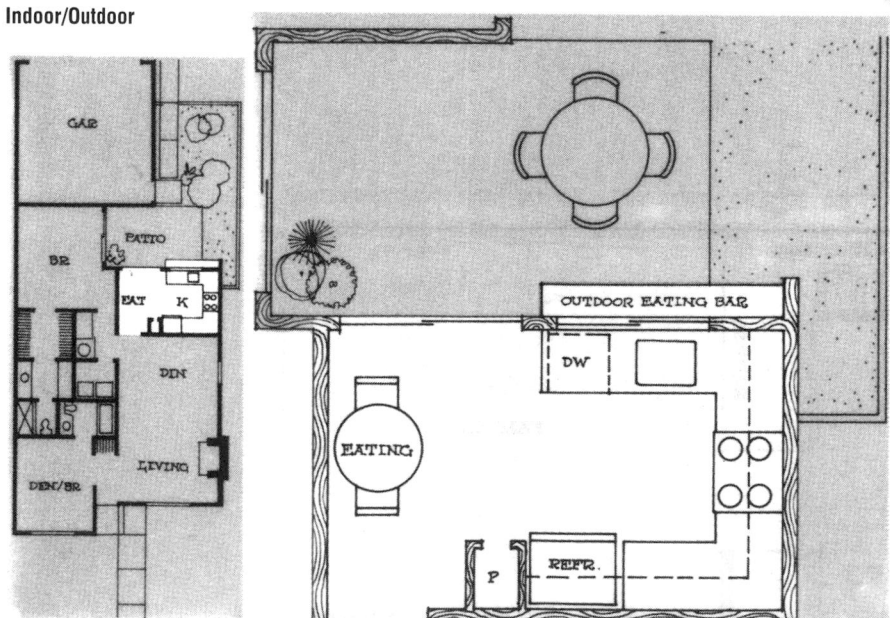

Fig. 18   Meadow Townhouses, Sacramento, California. Architects: Dreyfuss & Blackford.

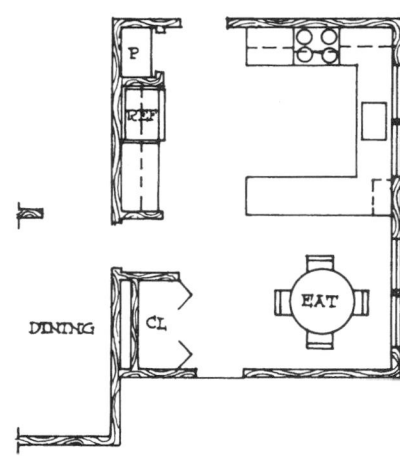

Fig. 19 Lyon Farm, Greenwich, Connecticut. Architect: SMS Arenitess.

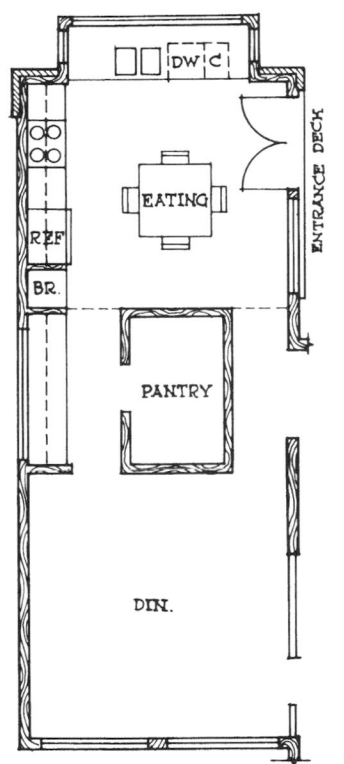

Fig. 20 Landen, Cincinnati, Ohio. Architect: Berkus Group.

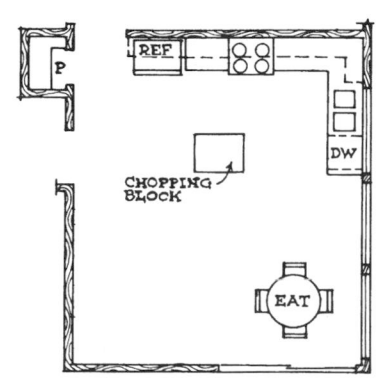

Fig. 21 Centennial Homes, Chino, California. Architect: L. C. Major & Associates.

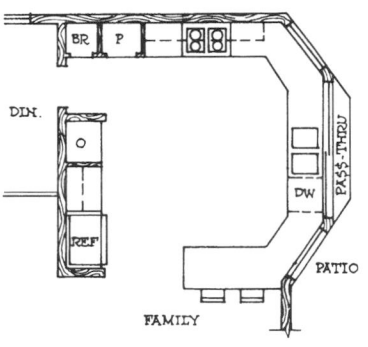

Fig. 22 Mission Viejo, California. Designer: Red Maltz Associates.

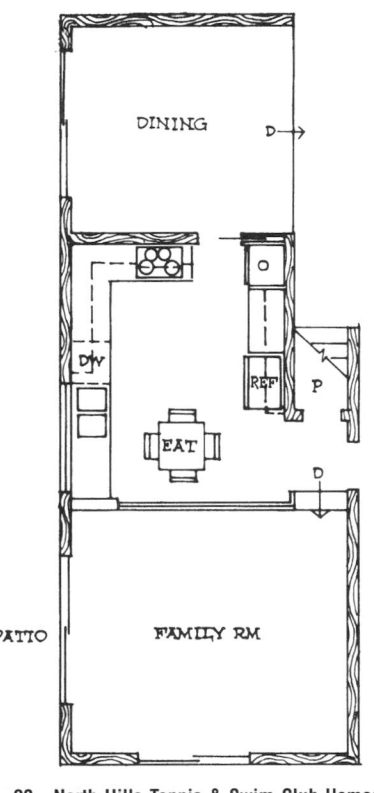

Fig. 23 North Hills Tennis & Swim Club Homes, Brea, California.

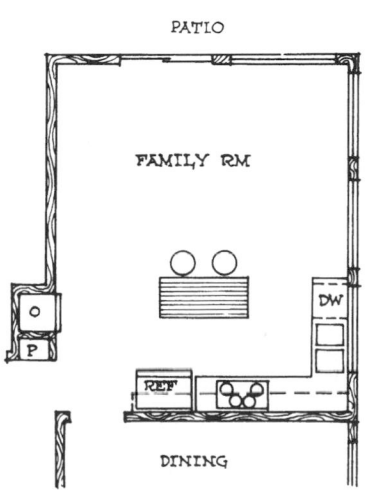

Fig. 24 Woodwalk Lake Forest, California. Architect: Frank Leslie Spangler.

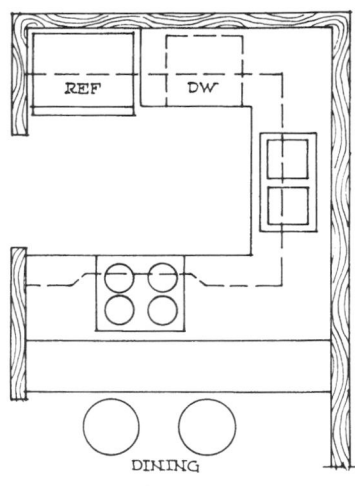

Fig. 25 Arroyo Santiago, Orange, California. Architect: Deck-Moffet & Associates.

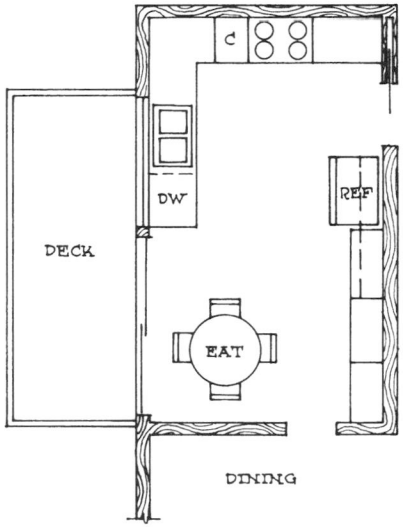

Fig. 26 Rancho San Joaquin, Irvine, California. Architect: Leitch/Klyotoki/Ben & Associates.

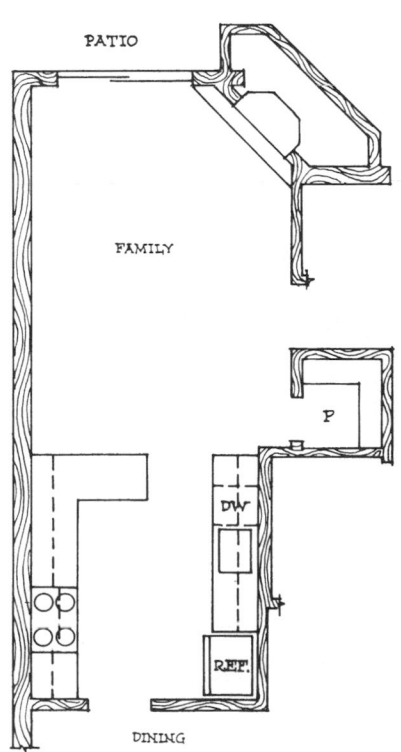

Fig. 27 Water's Edge, Columbia, Missouri. Designer: Bucher-Meyers & Associates.

343

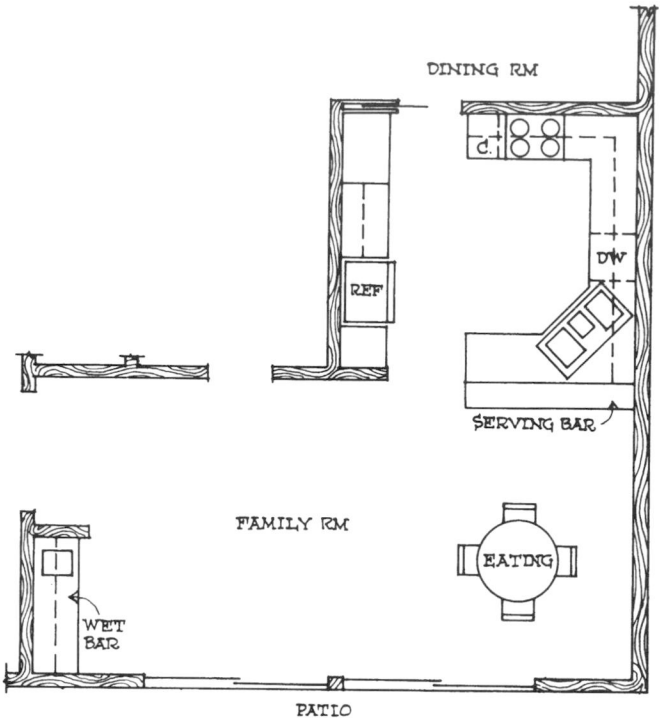

Fig. 28   Glenridge, Beverly Hills, California. Architect: McNarason, Nagy & Martin.

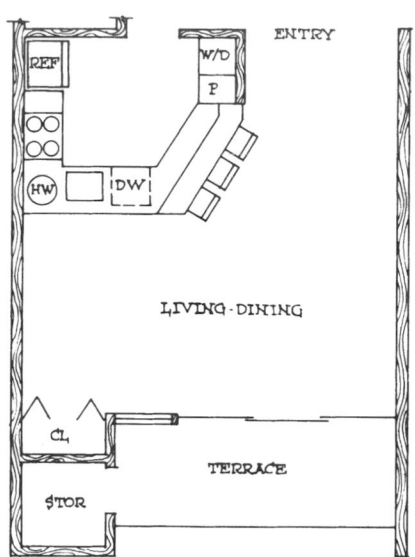

Fig. 29   Jupiter Ocean & Racquet Club, Jupiter, Florida. Architect: Schwab & Twitty.

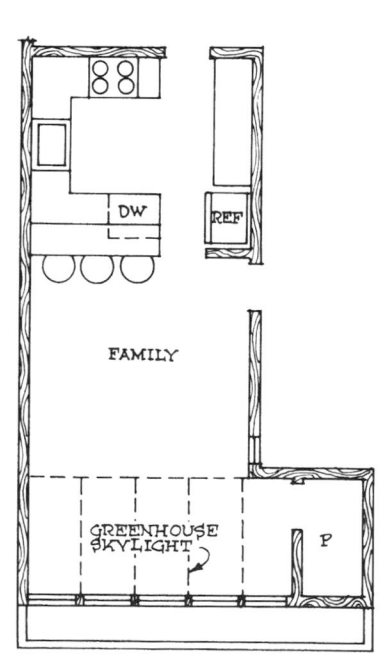

Fig. 30   The Islands, Foster City, California. Architect: Fisher-Friedman Associates.

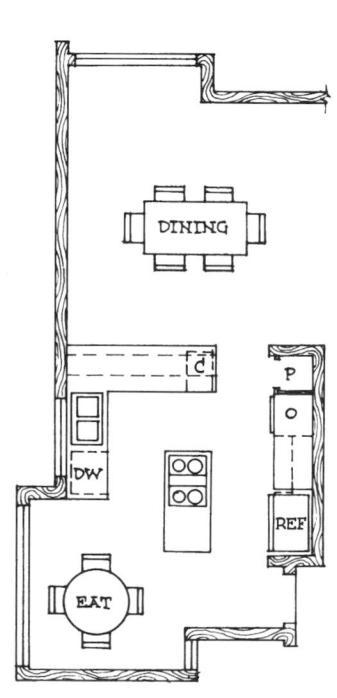

Fig. 31   Marina Strand Colonies, Los Angeles, California. Architect: Walter Richardson Associates.

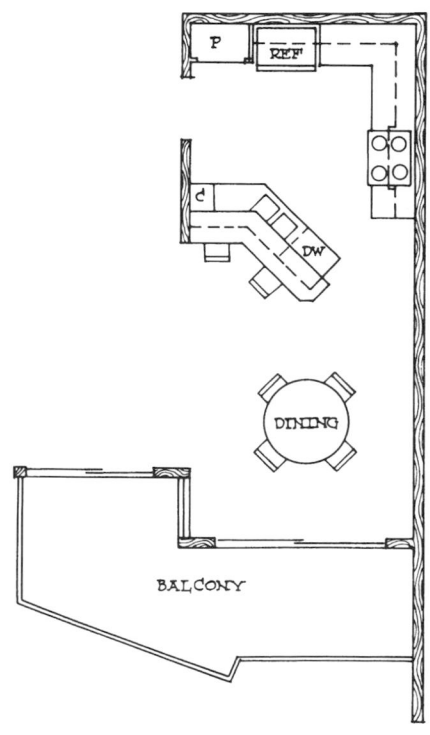

Fig. 32   Bay Vu, San Diego, California. Architect: Burkett & Wong.

APPLIANCES

Provided on the following pages are illustrations and approximate dimensions of various types of sinks, dishwashers, ranges, ovens, surface cooking units, refrigerators, freezers, and combination and package kitchen units. These dimensions may be used for preliminary planning purposes, but final selection of equipment and detailing of working drawings should always be based on specific manufacturers' data.

Door swings on refrigerators, ranges, cabinet sinks, and cabinet-type dishwashers should be checked against manufacturers' data after units have been tentatively selected. Side-hinged doors designed to swing from the left or right (such as those on refrigerators) can usually be obtained on order.

Dimensions have been drawn from the current catalogs of leading manufacturers of each type of equipment. Odd sizes and special and nonresidential equipment have been omitted. Dimensions are generally given only to the nearest ½ in, since dimensions of new models change slightly from year to year. The method of tabulation varies, depending on what is most appropriate to each type of equipment. Overall dimensions given for depth, width, and, where appropriate, height correspond to D, W, and H on the drawings. Inside dimensions of sink bowls are listed under the heading "Size of bowl" for the sections on stainless-steel sinks, porcelain-enamel sinks, stainless-steel sink tops, and porcelain-enamel sink tops.

STAINLESS-STEEL SINKS (FOR RECESSING INTO COUNTERTOPS)

Custom sizes and custom punching of fitting openings are available for most models. Some ledge-type sinks are available with fittings.

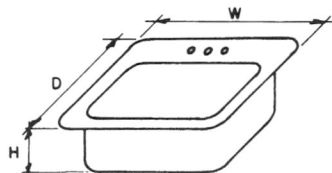

Fig. 1a Single-compartment ledge-type sink bowl. (See Table 1a.)

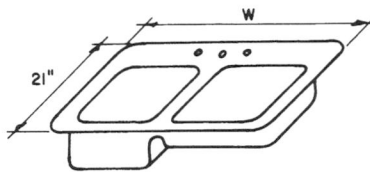

Fig. 1b Double-compartment ledge-type sink bowl. Right- or left-hand models available. (See Table 1b.)

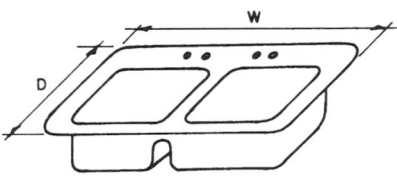

Fig. 1c Double-compartment ledge-type sink bowl. (See Table 1c.)

TABLE 1a

Size of bowl, in (D × W × H)	Depth, in	Width, in
16 × 14 × 3½	21	16
15 × 18 × 3½	21	21
16 × 20 × 3½	21	24
16 × 14 × 4	21	16
16 × 18 × 4	21	21
16 × 22 × 4	21	21
16 × 28 × 4	21	30
9 × 12 × 6	14	14
14 × 16 × 7	19	18
16 × 14 × 7	21	16
16 × 18 × 7	21	21
16 × 20 × 7	21	24
12 × 12 × 7½	17	14
16 × 12 × 7½	21	14
12 × 16 × 7½	16	21
16 × 16 × 7½	21	18
16 × 22 × 7½	21	24
16 × 28 × 7½	21	30
10 × 14 × 8	12	18
16 × 14 × 10½	12	16
15 × 18 × 11	21	21

TABLE 1b

Deep bowl, in (D × W × H)	Shallow bowl, in (D × W × H)	Width, in
16 × 14 × 7	16 × 14 × 3½	32
16 × 14 × 7	16 × 20 × 3½	37
16 × 14 × 7	16 × 28 × 4	46
16 × 16 × 7	16 × 22 × 4	42
16 × 22 × 7	16 × 22 × 4	48
16 × 14 × 10½	16 × 14 × 7	32

TABLE 1c

Size of bowl, in (D × W × H)	Depth, in	Width, in
12 × 9 × 6	16	21
12 × 12 × 7	12	28
12 × 16 × 7	17	36
16 × 12 × 7	21	28
14 × 16 × 7	21	36
16 × 14 × 7	21	32
16 × 16 × 7	21	36
16 × 18 × 7	21	42

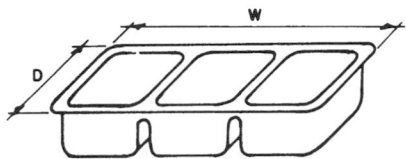

Fig. 1d Triple-compartment flat-rim sink bowl. (See Table 1d.)

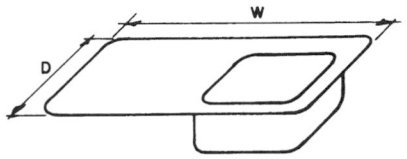

Fig. 1e Offset combination single ledge sink and drainboard. (See Table 1e.)

TABLE 1d

Size of bowl, in (D × W × H)	Depth, in	Width, in
12 × 9 × 7½	14	34
12 × 12 × 7½	14	42
16 × 12 × 7½	18	40
16 × 14 × 7½	18	46
16 × 16 × 7½	18	54
16 × 18 × 7½	18	60

TABLE 1e

Size of bowl, in (D × W × H)	Depth, in	Width, in
16 × 14 × 7½	21	42
16 × 18 × 7½	21	42

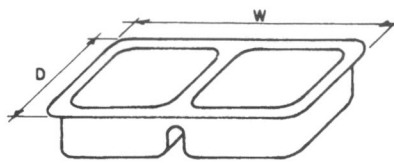

Fig. 1f Double-compartment flat-rim sink bowl. The rim is generally 1 to 1½ in on all sides. (See Table 1f.)

TABLE 1f

Size of bowl, in (D × W × H)	Depth, in	Width, in
12 × 12 × 7½	14	28
16 × 12 × 7½	18	28
16 × 14 × 7½	18	31
16 × 16 × 7½	28	36
16 × 22 × 7½	18	48
18 × 16 × 7½	20	36
18 × 18 × 7½	20	40
18 × 20 × 7½	20	44
18 × 25 × 7½	20	54

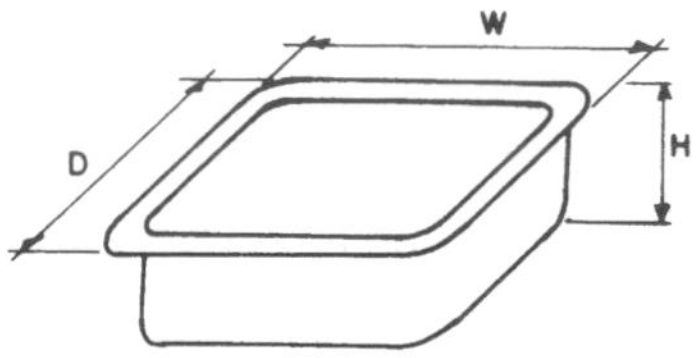

Fig. 1g Single-compartment flat-rim sink bowl. The rim is generally 1 to 1½ in on all sides. (See Table 1g.)

TABLE 1g

Depth, in	Width, in	Height, in
9	12	4
9	12	6
18	15	7
12	20	5
12	20	7
20	16	7
12	12	7½
12	16	7½
14	16	7½
16	16	7½
12	18	7½
16	18	7½
16	20	7½
16	22	7½
18	20	7½
18	25	7½
16	28	7½
18	18	7½
18	30	7½

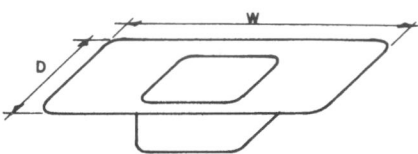

Fig. 1h Combination single ledge sink and drainboard. (See Table 1h.)

TABLE 1h

Size of bowl, in (D × W × H)	Depth, in	Width, in
6 × 14 × 7½	21	54
16 × 14 × 7½	21	60
16 × 18 × 7½	21	54
16 × 18 × 7½	21	60

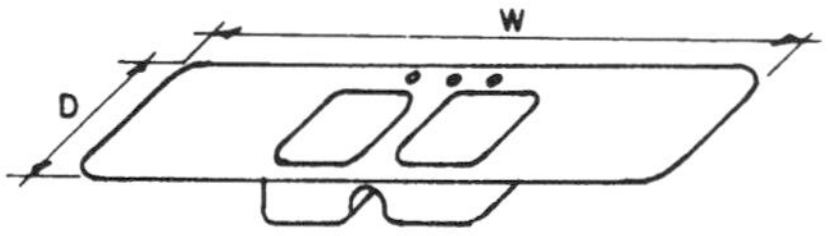

Fig. 1i Combination double ledge sink and drainboard. (See Table 1i.)

TABLE 1i

Size of bowl, in (D × W × H)	Depth, in	Width, in
16 × 14 × 7½	21	60
16 × 14 × 7½	21	72
16 × 18 × 7½	21	60
16 × 18 × 7½	21	72

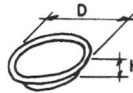

Fig. 1j Round sink bowls. The rim is generally 1 in. (See Table 1j.)

TABLE 1j

Inside diameter, in	Height, in
11	4
14	5½

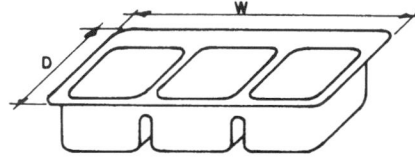

Fig. 1k Triple-compartment ledge-type sink bowl. (See Table 1k.)

TABLE 1k

Size of bowl, in (D × W × H)	Depth, in	Width, in
9 × 12 × 7½	14	42
12 × 9 × 7½	17	34
12 × 12 × 7½	17	42
16 × 12 × 7½	21	40
16 × 14 × 7½	21	46
16 × 16 × 7½	21	54
16 × 18 × 7½	21	60

PORCELAIN-ENAMEL SINKS (ON STEEL OR CAST-IRON BODIES FOR RECESSING INTO COUNTERTOPS)

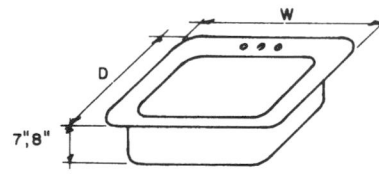

Fig. 2a Single-compartment ledge-type sink bowl. Some models are available in color. (See Table 2a.)

TABLE 2a

Size of bowl, in (D × W)	Depth, in	Width, in
16 × 22	21	24
16 × 28	21	30

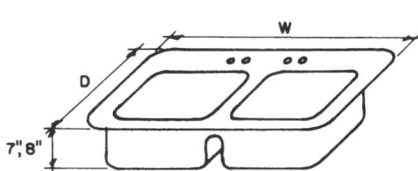

Fig. 2b Double-compartment ledge-type sink bowl. (See Table 2b.)

TABLE 2b

Size of bowl, in (D × W)	Depth, in	Width, in
15 × 13	20	30
16 × 14	21	32
16 × 19	21	42

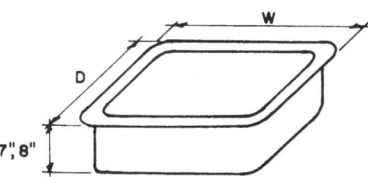

Fig. 2c Single-compartment flat-rim sink bowl. The rim is generally 1 to 1½ in on all sides. (See Table 2c.)

TABLE 2c

Depth, in	Width, in
12	12
16	20
16	24
18	24
18	30
20	24
20	30

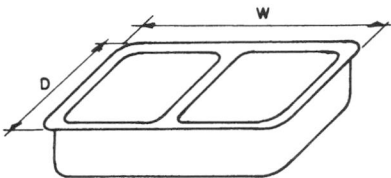

Fig. 2d Double-compartment flat-rim sink bowl. (See Table 2d.)

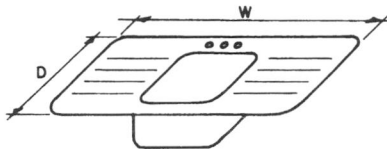

Fig. 2e Combination single ledge sink and drainboard. (See Table 2e.)

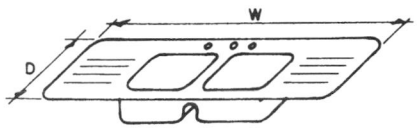

Fig. 2f Combination double ledge sink and drainboard. (See Table 2f.)

TABLE 2d

Size of bowl, in (D × W × H)	Depth, in	Width, in
18 × 14 × 8	20	30
18 × 14 × 8	20	32
18 × 19 × 8	20	42

TABLE 2e

Size of bowl, in (D × W × H)	Depth, in	Width, in
16 × 19 × 8	21	54

TABLE 2f

Size of bowl, in (D × W × H)	Depth, in	Width, in
16 × 14 × 8	21	60

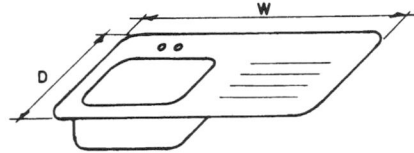

Fig. 2g Offset combination single ledge sink and drainboard. Right- or left-hand models are available. (See Table 2g.)

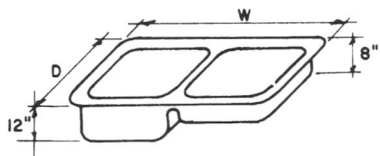

Fig. 2h Combination double-bowl flat-rim sink. (See Table 2h.)

TABLE 2g

Size of bowl, in (D × W × H)	Depth, in	Width, in
16 × 19 × 8	21	42

TABLE 2h

Size of bowl, in (D × W)	Depth, in	Width, in
18 × 18	21	42

STAINLESS-STEEL SINK TOPS

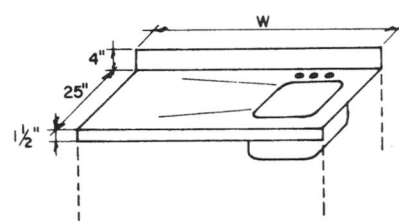

Fig. 3a Offset single-compartment sink top with drainboard. (See Table 3a.)

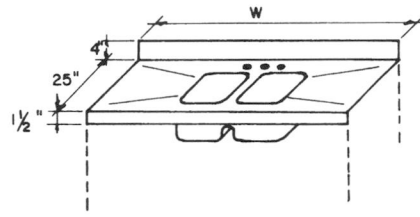

Fig. 3b Double-compartment sink top with drainboard. (See Table 3b.)

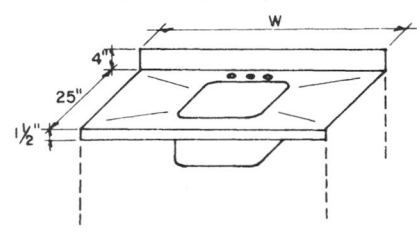

Fig. 3c Single-compartment sink top with drainboard. (See Table 3c.)

TABLE 3a

Size of bowl, in (D × W × H)	Width, in
16 × 12 × 7	39
16 × 16 × 7	42
16 × 18 × 7	48

TABLE 3b

Size of bowl, in (D × W × H)	Width, in
16 × 14 × 7	60, 66, 72
16 × 16 × 7	66, 72, 84, 96

TABLE 3c

Size of bowl, in (D × W × H)	Width, in
16 × 18 × 7	54
16 × 20 × 7	54
16 × 20 × 7	60
16 × 20 × 7	66
16 × 20 × 7	72
16 × 20 × 7	84
16 × 20 × 7	96
16 × 25 × 7	72

**PORCELAIN-ENAMEL SINK TOPS
(ON STEEL OR CAST-IRON BODIES)**

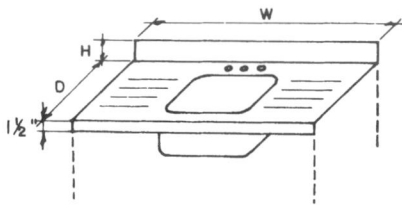

Fig. 4a Single-compartment sink top with drainboard. (See Table 4a.)

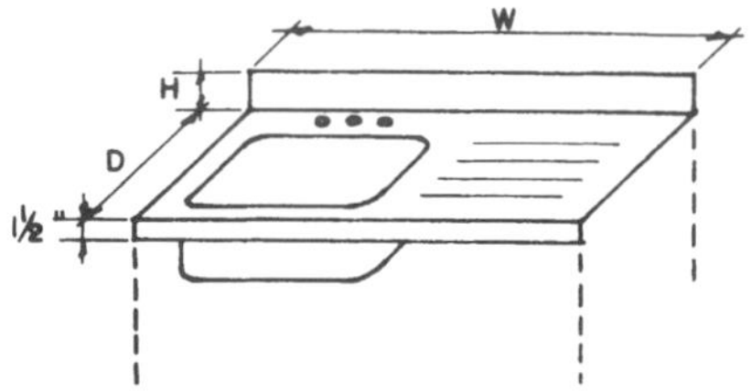

Fig. 4b Offset single-compartment sink top with drainboard. (See Table 4b.)

TABLE 4a

Size of bowl (D × W × H), in	Depth, in	Width, in	Height, in
15 × 18 × 8	22	54	4
15 × 20 × 8	22	60	4
18 × 20 × 8	25	54	4
18 × 20 × 8	25	60	4
18 × 22 × 6	22	60	8
20 × 16 × 6	25	54	3

TABLE 4b

Size of bowl (D × W × H), in	Depth, in	Width, in	Height, in
15 × 19 × 8	22	42	4
16 × 19 × 8	20	42	4
16 × 20 × 6	20	42	8
16 × 20 × 6	25	42	3
18 × 19 × 8	25	42	4

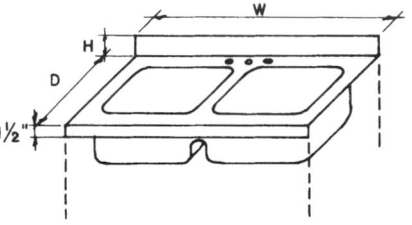

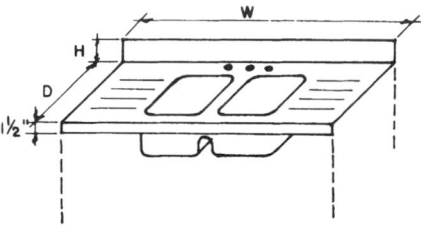

Fig. 4c Double-compartment sink top. (See Table 4c.)

Fig. 4d Double-compartment sink top with drainboard. (See Table 4d.)

TABLE 4c

Size of bowl (D × W × H), in	Depth, in	Width, in	Height, in
15 × 18 × 8	22	42	4
18 × 18 × 8	25	42	4

TABLE 4d

Size of bowl (D × W × H), in	Depth, in	Width, in	Height, in
18 × 14 × 8	25	60	4
18 × 16 × 8	25	72	4

CABINET SINKS (PORCELAIN ENAMEL ON STEEL BODIES)

All sinks are 36 in high; the backsplash varies from 3 to 4 in.

Some models are available with stainless-steel tops, plastic tops, or porcelain-enamel tops on cast-iron bodies. A variety of base cabinet arrangements is generally available.

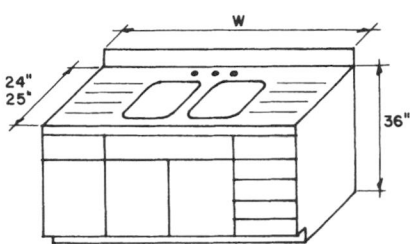

Fig. 5*a* Double-compartment sink with drainboard. This model is available in 60-, 66-, and 72-in widths.

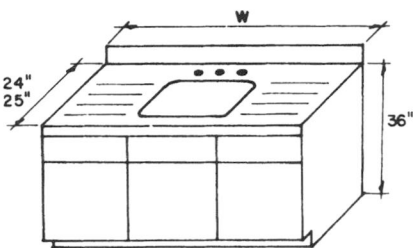

Fig. 5*b* Single-compartment sink with drainboard. This model is available in 54- and 60-in widths.

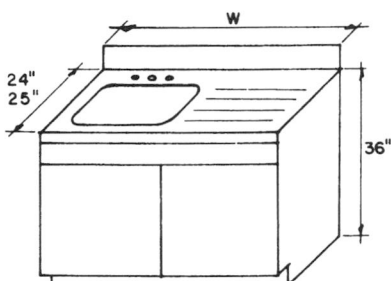

Fig. 5*c* Offset single-compartment sink with drainboard. Right- or left-hand models are available in 42-in widths.

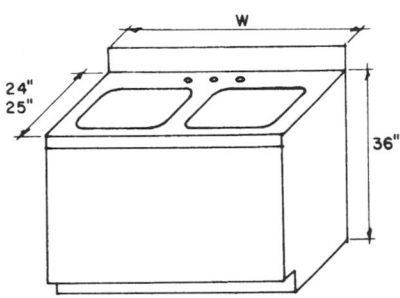

Fig. 5*d* Double-compartment sink. Some models have sliding drainboard; available in 36-, 42-, and 45-in widths.

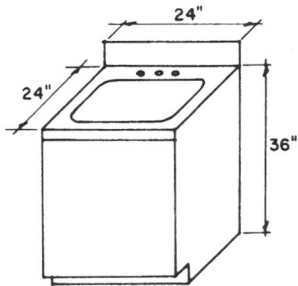

Fig. 5*e* Single-compartment sink.

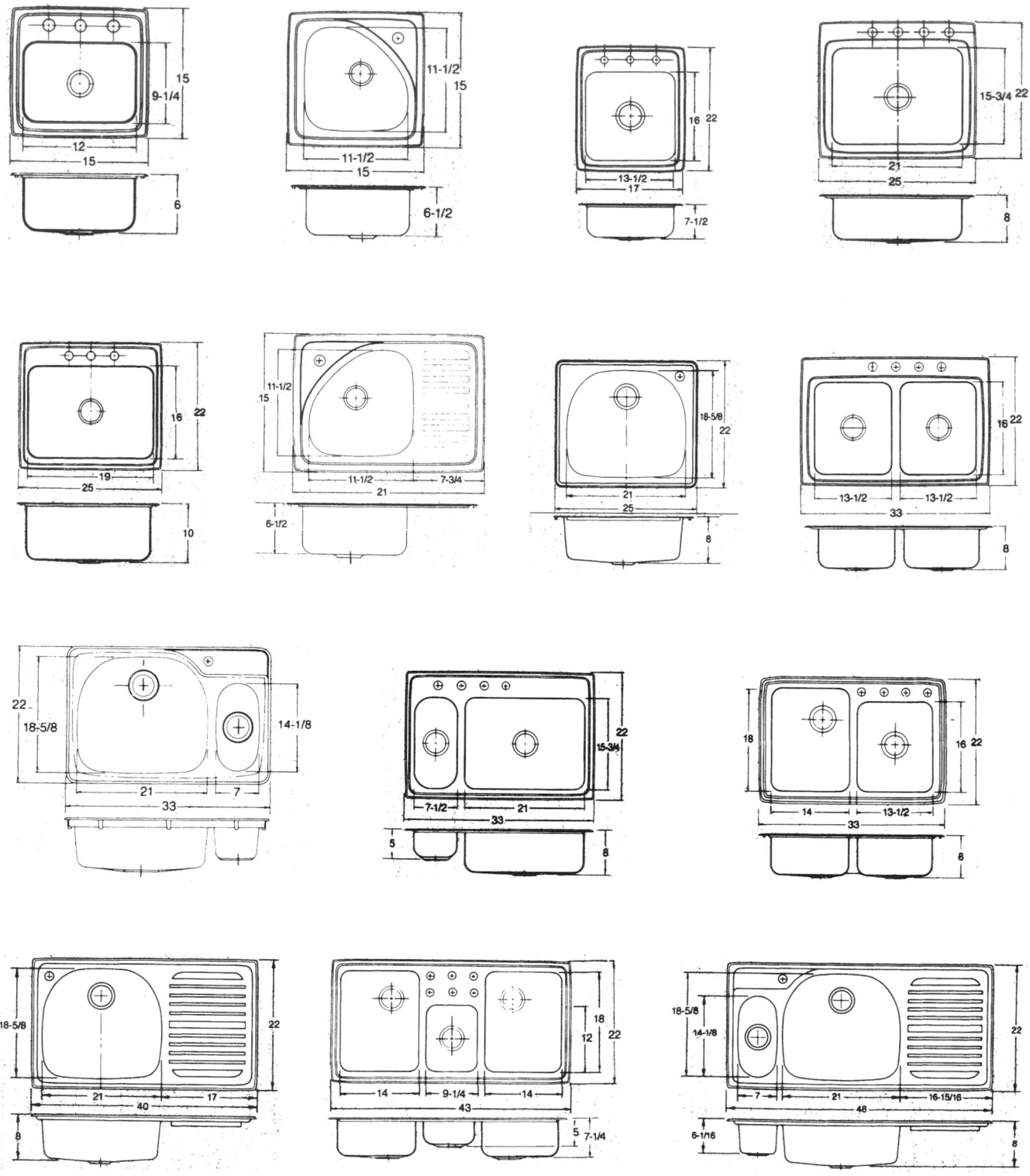

Fig. 6   Kitchen sinks.

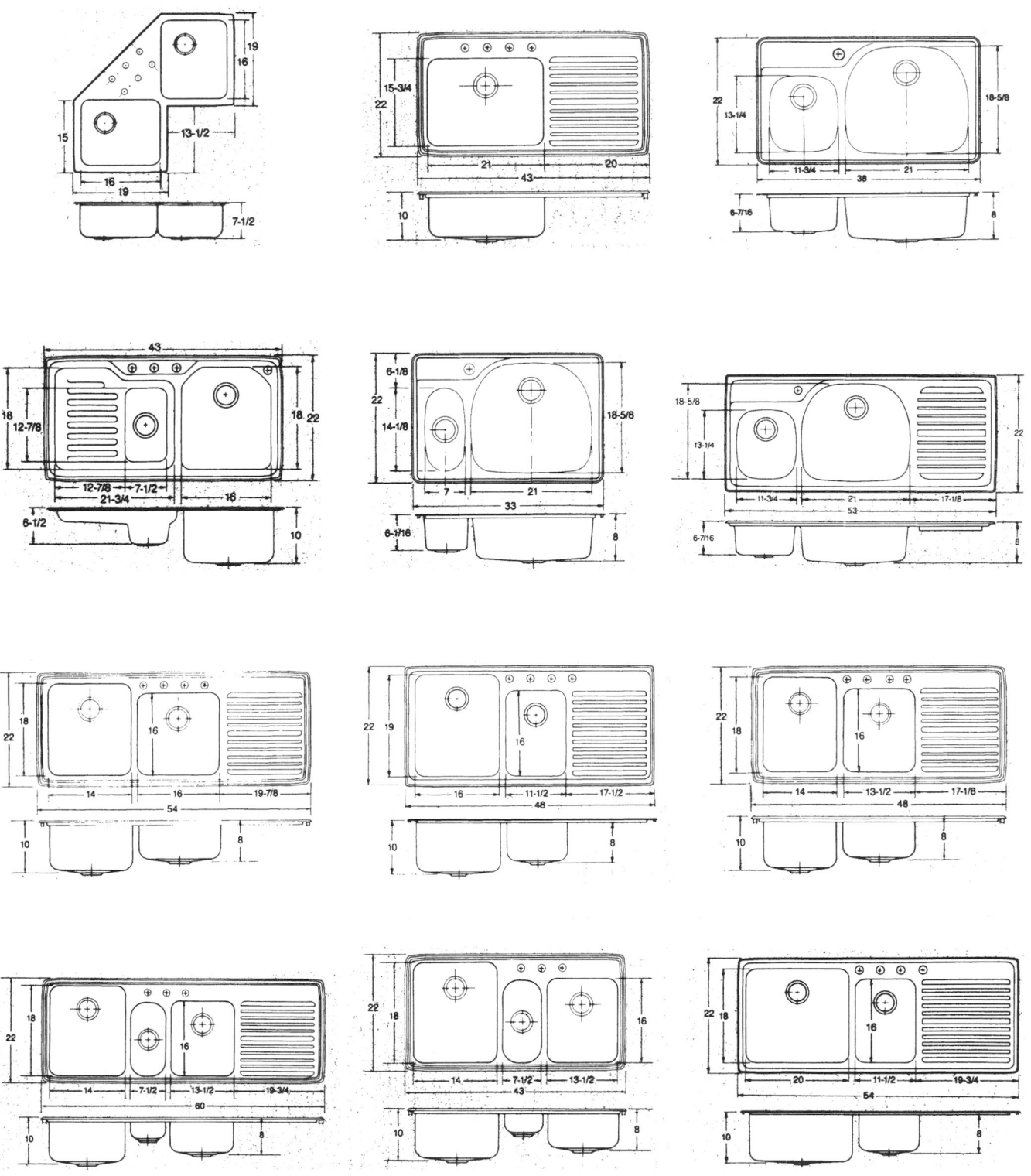

Fig. 6   (*Continued*)

GARBAGE DISPOSALS

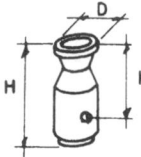

Fig. 7 Garbage disposers. The height of some models varies up to 12 in (*h*). (See Table 6.)

TABLE 6

Depth, in	Height, in	Height, in
7–9	13–16	6–8

COMPACTORS

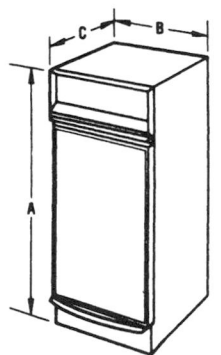

Fig. 8*a* A = ±34 in; B = 12, 15, or 24 in; C = 20 to 24 in.

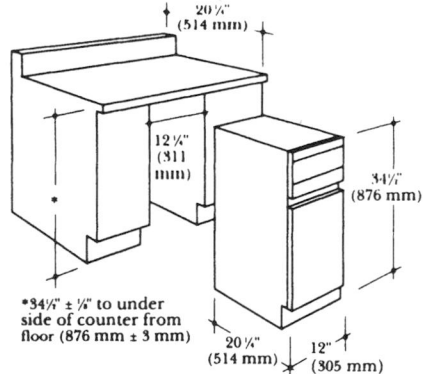

20¼"
(514 mm)

12¼"
(311 mm)

34½"
(876 mm)

*34½" ± ⅛" to under side of counter from floor (876 mm ± 3 mm)

20¼"
(514 mm)

12"
(305 mm)

Fig. 8*b*

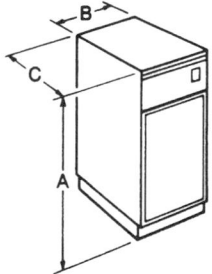

Fig. 8*c* Convertible compactors.

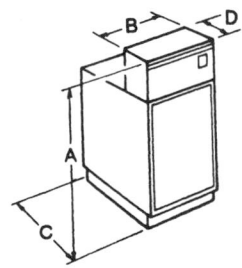

Fig. 8*d* Built-in compactor.

DISHWASHERS

Dishwashers may be either *built-in* or *portable;* the latter means the unit can be wheeled up to the kitchen sink and attached to the faucet with a hose. Waste flows out into the drain through another hose. Most models can convert to built-ins if you remodel or move, although some require the purchase of a *conversion kit.* Portable dishwashers tend to be somewhat more expensive than their built-in counterparts. The standard width of both types is 24 in, although several manufacturers now offer 18-in-wide machines. Another way to solve a space problem is with a 24-in unit that fits under the sink. It has a somewhat smaller capacity, since some of the upper level is sacrificed to the bowl of the sink.

Far from a luxury, not only does a dishwasher save time, it can save energy and water. A dishwasher uses 5.9 fewer gallons of water per load than it would take to do the same load by hand, according to a study done by Ohio State University. Today, *energy efficiency* and *quiet operation* are hot buttons in the dishwasher industry. The German firms of Gaggenau, Bosch, Miele, and A.E.G. and the Swedish firm Asko Asea have introduced high-quality—and generally high-priced—products to the American market, while American manufacturers have added numerous improvements to their product lines. Other important factors to consider when purchasing a dishwasher are *construction, performance, serviceability,* the *terms of the warranty,* and *convenience.* Finally, cost must be considered; however, the cost of an appliance only begins on the showroom floor. Once you figure in energy bills, water, and detergent, a machine with a higher price tag than another may actually wind up being the more economical choice. The yellow *energy-use hangtags* manufacturers must provide with each model give you the estimated annual energy cost and compare the model with those that are the least and most expensive to operate. But figures given do not reflect the cost of fuel in your locale, how often you use the machine, and which cycles you select, nor do these tags compute water usage and cost. Always locate your dishwasher away from the refrigerator/freezer as the moisture produced by the dishwasher will force your refrigerator to use more energy.

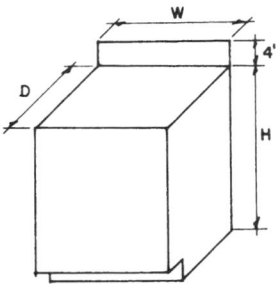

Fig. 9a Free-standing dishwasher. This model is 26 in deep, 24 in wide, 36 in high, and may be 42 or 46 in deep with the doors open. It fits standard base cabinets and is available with drop-down door or pull-out drawer for either front or top loading. Fronts are also available in a variety of materials, finishes, and colors to match standard kitchen cabinets.

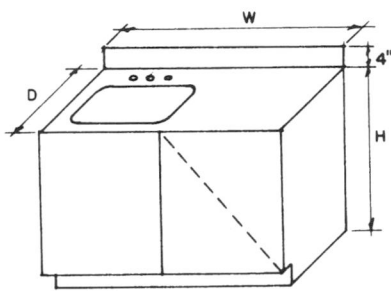

Fig. 9c Combination dishwasher and sink. This model is 25 in deep, 48 in wide, 36 in high, and 48 in deep with the doors open.

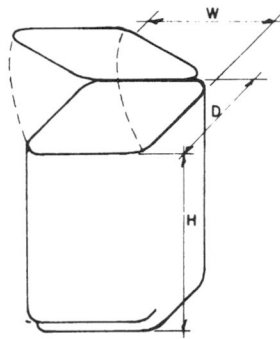

Fig. 9b Portable dishwasher. This model is generally the lift-top, top-loading type. (See Table 7.)

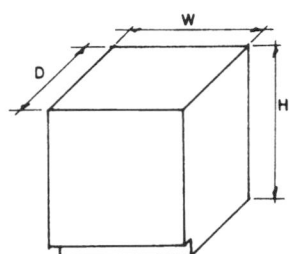

Fig. 9d Under-counter, built-in dishwasher. This model is 24 in deep, 24 or 30 in wide, 34 in high, and 45 in deep with the doors open. It is available with drop-down door or pull-out drawer for either front or top loading.

TABLE 7

Depth, in	Width, in	Height, in
25	22	33
25	22	35

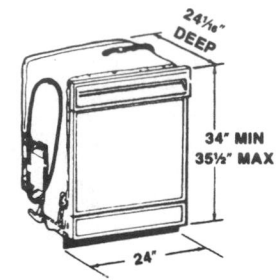

Fig. 10*a*   Built-in dishwasher.

Most dishwashers now come with *reversible door and access panels* that allow you to match the rest of your appliances, whether they are white, black, harvest gold, or almond. In more expensive models, a *trim kit* allows you to insert a ¼-in panel to match your wood-front cabinetry. Another popular look in pricier models is the European style of "white-on-white."

New features are the following:

■ Touch controls with readouts showing time left for cycle to complete.

■ Clips to hold down lightweight items that tend to flip over and fill with water.

■ Delay-start feature that programs cycle to begin later in the day or at night.

■ Extra insulation or sound-reducing materials to provide quieter operation.

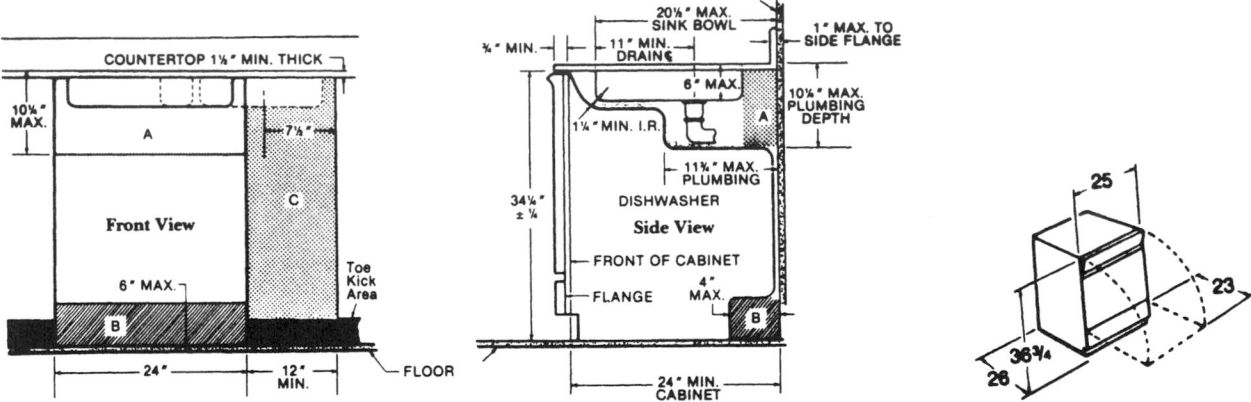

Fig. 10*b*   Undersink dishwasher.

Fig. 10*c*   Portable dishwasher.

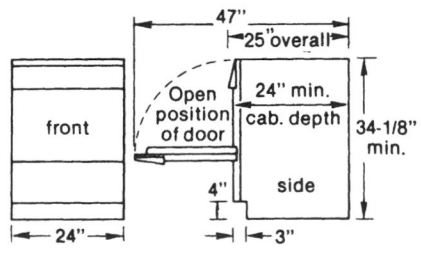

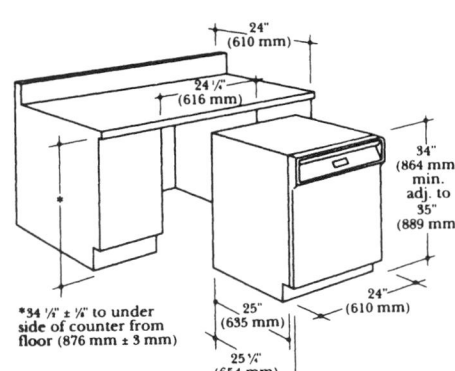

Fig. 11   Typical dishwasher opening dimensions.

**SURFACE COOKING UNITS
(COOKTOPS)—ELECTRIC AND GAS
(FOR BUILT-IN INSTALLATIONS**

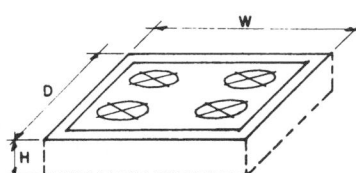

Fig. 12*a*   Electric-surface burners. If controls are remote, that is, in front of base cabinet or on the backsplash or alongside, additional space must be allowed. Some models are also available with a deep-well cooker in place of one surface element. (See Table 8a.)

TABLE 8a

No. of elements	Depth, in	Width, in	Height, in	Controls
2	21	14	5	Front
2	22	15	7	Front
2	21	14	8	Remote
2	22	14	5	Remote
2	22	17	6	Remote
4	19	31	5	Top, integral
4	21	27	5	Front
4	21	28	5	Top, integral
4	22	31	10	Remote
4	20	30	5	Top, integral
4	22	20	7	Remote
4	22	32	7	Top, integral
4	22	22	7	Remote
4	21	34	8	Top, integral

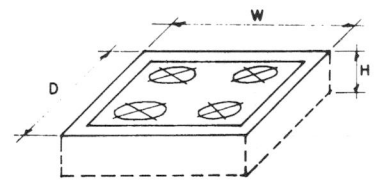

Fig. 12*b*   Gas surface burners. (See Table 8b.)

TABLE 8b

No. of burners	Depth, in	Width, in	Height, in	Controls
2	21	15	8	Front
2	22	15	8	Front
4	21	24	8	Front
4	22	24	8	Front

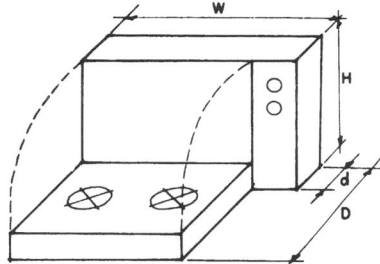

Fig. 12*c*   Fold-back surface burners. Some models include griddles. (See Table 8c.)

TABLE 8c

No. of elements	Depth, in	Depth d, in	Width, in	Height, in
2	18	6	24	14

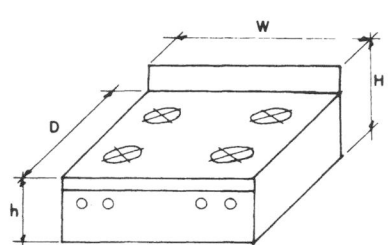

Fig. 12*d*   Stack-on surface burners. (See Table 8d.)

TABLE 8d

No. of burners	Depth, in	Width, in	Height, in	Height h, in
4	25	24	12	8

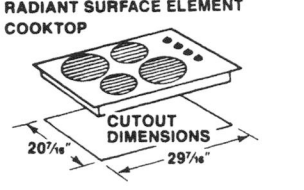

RADIANT SURFACE ELEMENT COOKTOP

20⁷⁄₁₆"   29⁷⁄₁₆"

OVERALL: 30" W x 21" D x 2¾" H

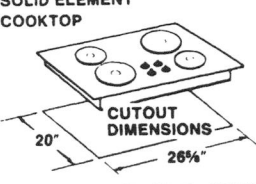

SOLID ELEMENT COOKTOP

20"   26⅝"

OVERALL: 30" W x 21" D x 2¾" H

ELECTRIC COOKTOP

21¹⁄₁₆"   33¹⁄₁₆"

OVERALL: 33⅝" W x 21⅝" D x 3¹⁄₁₆" H

ELECTRIC COOKTOP

19¹³⁄₁₆"   28⁵⁄₁₆"

OVERALL: 30" W x 21" D x 3¼" H

Fig. 13*a*   Electric cooktops.

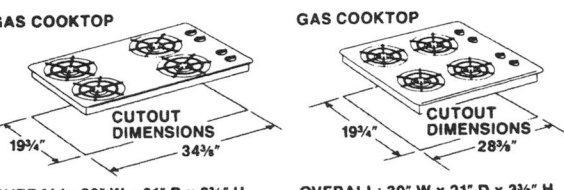

GAS COOKTOP

19¾"   34⅜"

OVERALL: 36" W x 21" D x 3⅜" H

GAS COOKTOP

19¾"   28⅜"

OVERALL: 30" W x 21" D x 3⅜" H

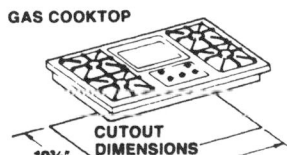

GAS COOKTOP

19¾"   34⅞"

OVERALL: 36" W x 21" D x 3⁹⁄₁₆" H

GAS COOKTOPS

19¹³⁄₁₆"   28⁹⁄₁₆"

OVERALL: 30" W x 21" D x 3⅜" H

Fig. 13*b*   Gas cooktops.

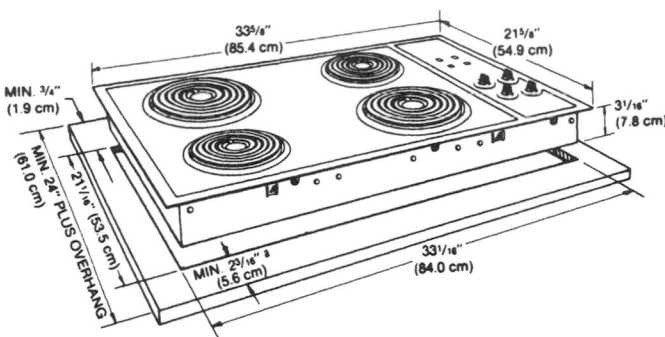

Fig. 14a   Conventional coil cooktop—electric updraft.

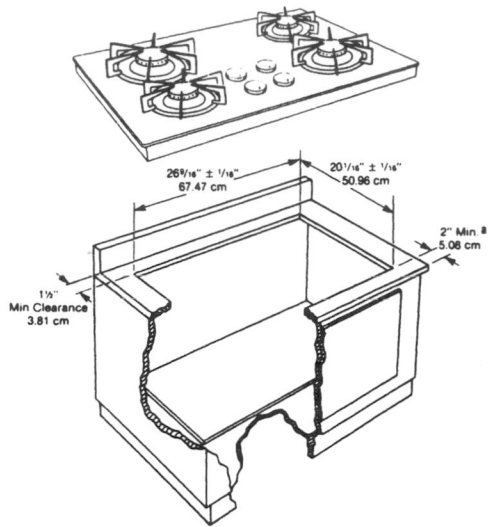

Fig. 14b   Gas cooktop—gas, updraft.

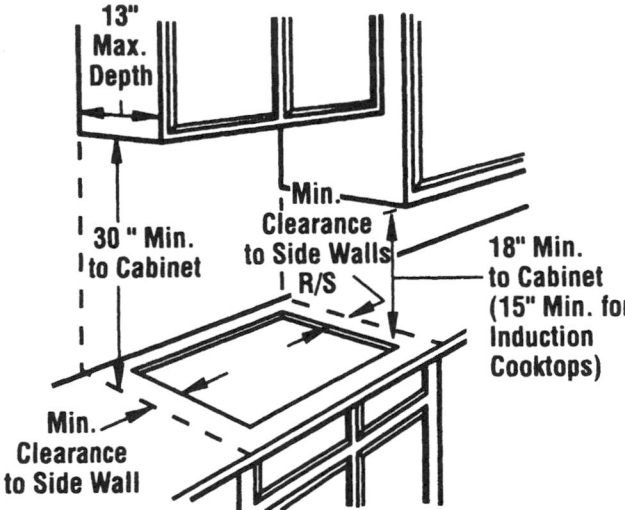

Fig. 15   Minimum clearances to adjacent construction.

OVENS AND RANGES

The choices available in cooking equipment today are enormous. The classic range, or stove, includes both burners and oven, or you can select separate oven and cooktop units. Ranges may be *free-standing, drop-in,* or *slide-in.* Separate cooktops and ovens are always *built-in.* Free-standing ranges look fine with either side exposed as both are painted or porcelain-enameled. The most common *width* is 30 in, but widths from 20 to 40 in are available. A space-efficient appliance includes a microwave or second conventional oven at eye level. For a built-in look, choose a slide-in or drop-in range. The former sit on the floor, the latter rest on a cabinet base. Both types generally have front-mounted controls and are 30 in wide, but several manufacturers make 27-in models and others offer models up to 48 in wide.

Wall ovens, which are always built-in, come in both single and double versions, and can be gas, electric, or a combination of electric and microwave. Widths vary between 24 and 30 in, although a few models do come in 36 in. Some models are wider inside and therefore have greater capacity. Be sure to measure the interior width of any model you are considering.

Increasingly popular are *convertible ranges* and cooktops. Available in both gas and electric, these essentially allow you to design your own appliance by interchanging *cooking modules* such as grills, coil elements, or smooth-top cooking units. At the high end of the market are professional models designed for home use with the look of commercial equipment.

Ease of *cleaning* is an important consideration in selecting a cooker, and manufacturers have responded to this need. On ranges and cooktops, in place of tough-to-clean coil electric heating elements, there are sleek *ceramic-glass surfaces* or *solid cast-iron* elements. Gas appliances offer *sealed burners* that make removing spills a cinch. Most models have even eliminated the joints and cracks between sections where grease and dirt can become lodged.

Electronic controls make it easier to use ranges and ovens, permitting the *programming* of different functions, such as delayed-start baking. Electronic devices also more closely control *temperature* and provide precise *information,* such as the exact temperature of the oven.

New features are as follows:

- Removable oven doors, range tops, storage drawers, and kickplates to make cleaning easier. Also, porcelain drip bowls for electric coil elements that can be cleaned in the dishwasher.
- Electronic controls with a safety lock feature.
- Delay-starting for cleaning cycles.
- Up to three oven racks and six rack positions.

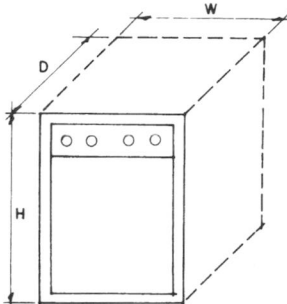

Fig. 16 Electric and gas wall oven. Microwave ovens are available in the same sizes and also as "combinations" with regular electric ovens. Some models are equipped with integral or separate broiler compartments. Some have single or double doors that open horizontally or vertically. Some doors have view panels. (See Table 9.)

TABLE 9

Depth, in	Width, in	Height, in
22	22	30
22	32	26
24	22	28
24	22	30
24	22	32
24	22	48
24	24	32
24	24	38
24	24	42
24	26	32

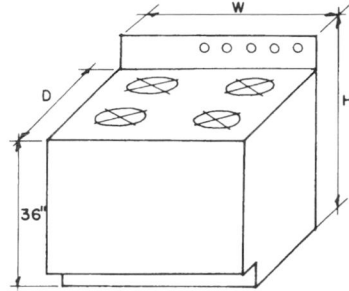

Fig. 17a Electric range. Some models have automatic controls and may include surface griddles, rotisseries, and single or double ovens. Twenty- and thirty-inch ranges generally require 10 to 14 kW; 30- and 40-in ranges require 10 to 18 kW. Forty-inch ranges generally have two ovens; 20- and 30-in ranges have one oven. (See Table 10a.)

TABLE 10a

No. of elements	Depth, in	Width, in	Height, in
4	24	20	45
4	24	42	48
4	25	20	40
4	25	30	44
4	25	30	45
4	25	36	45–49
4	25	40	48
4	26	21	40
4	26	30	44–47
4	26	30	48
4	26	39	43–50
4	26	40	48
4	27	21	42
4	27	30	45–49
4	28	30	48
4	28	40	44–48
4	29	40	48

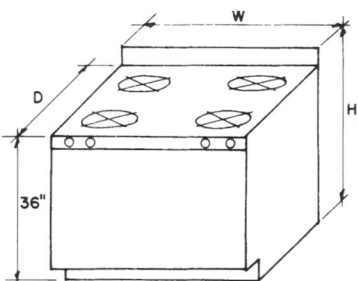

Fig. 17b Gas ranges. Forty-inch ranges generally have two ovens; 20- and 30-in ranges have one oven. (See Table 10b.)

TABLE 10b

No. of burners	Depth, in	Width, in	Height, in
4	24	20	45
4	25	30	45–49
4	25	36	45–49
4	25	42	47
4	26	30	43–45
4	28	31	44–49
4	28	41	44–49

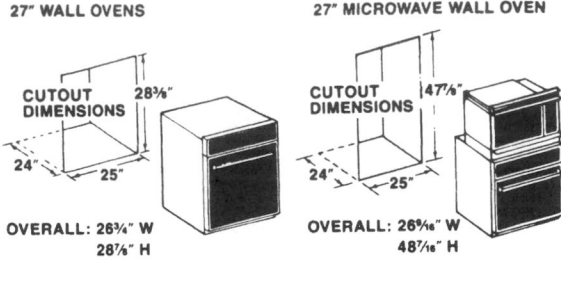

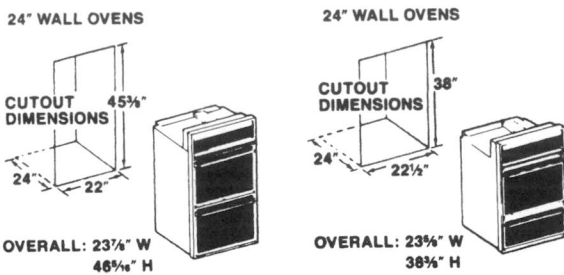

Fig. 19   Gas wall ovens.

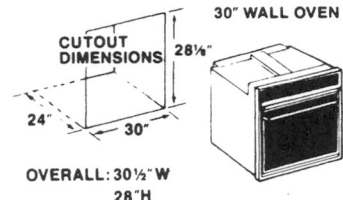

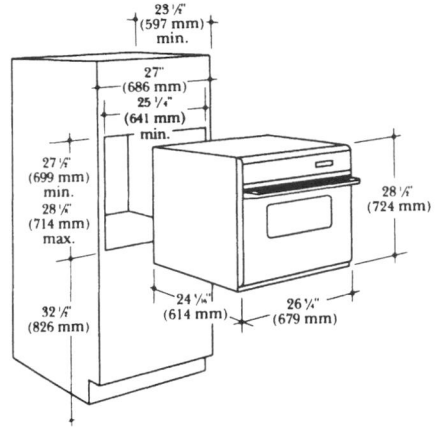

Fig. 20   Single oven.

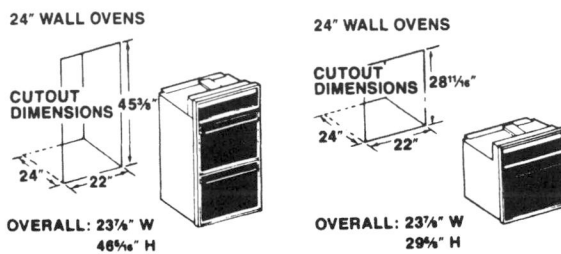

Fig. 18   Electric wall ovens.

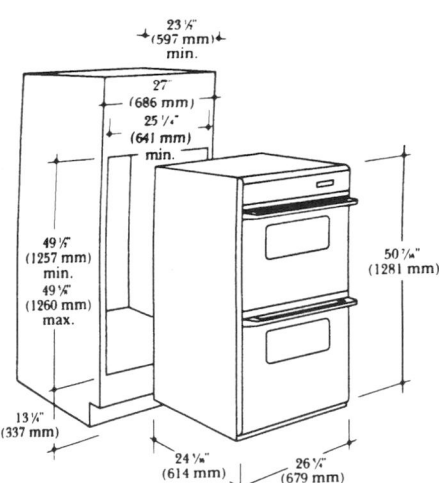

Fig. 21   Double oven.

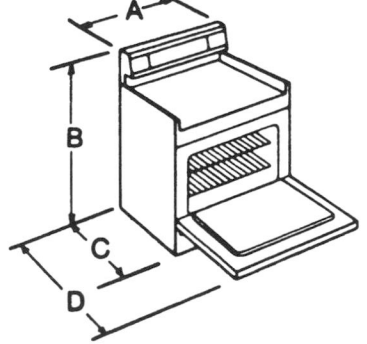

Fig. 22   Twenty-inch ranges. An economical compact range with convenient up-front controls, this model is ideal for apartments or homes with smaller kitchens.

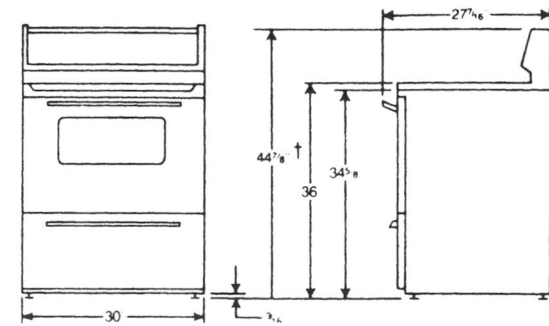

Fig. 23   Thirty-inch free-standing ranges.

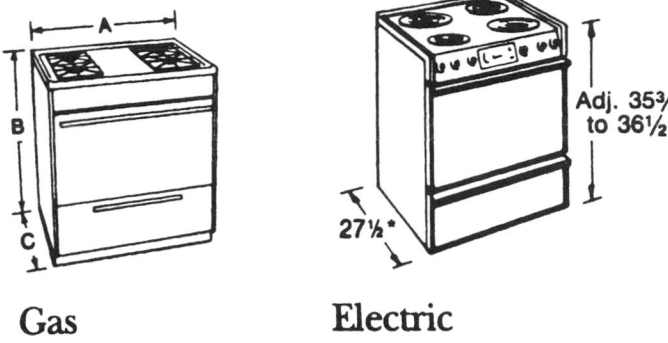

Gas   Electric

Fig. 24 Thirty-inch slide-in/drop-in ranges. For a clean, contemporary look, these popular ranges slide easily into place in kitchen cabinetry along the wall or in an island or peninsula installation.

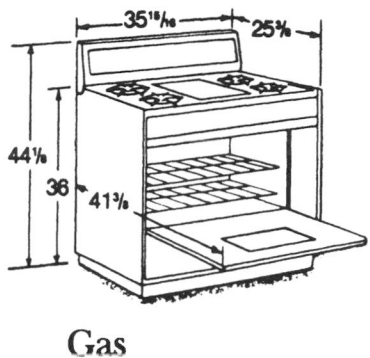

Gas

36″ Ranges

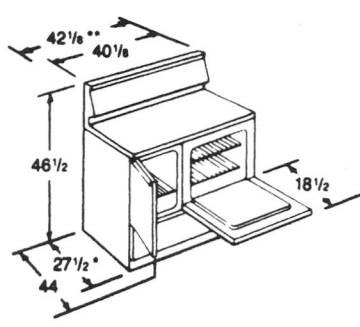

40″ Ranges

Fig. 25 Extra wide ranges offer expanded cooktop space, and a choice of standard or continuous cleaning oven plus a convenient side storage compartment.

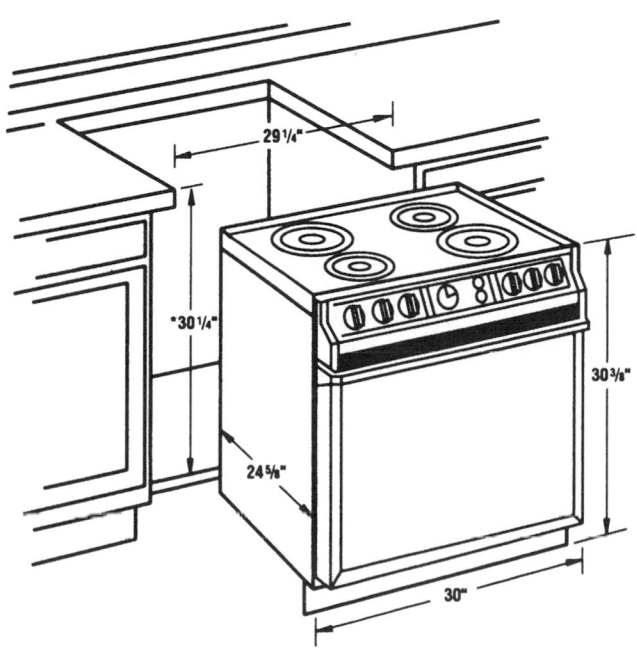

Fig. 26

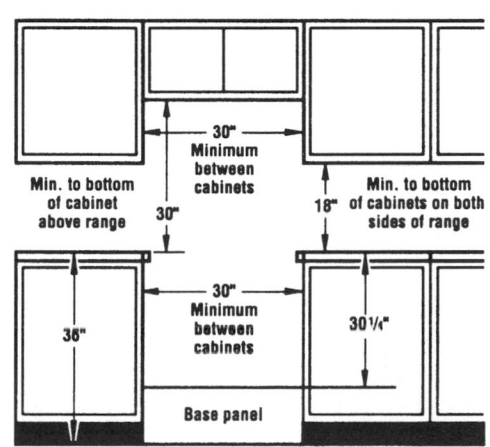

MICROWAVE OVENS

Microwave ovens use a *magnetron tube* to bombard food with tiny waves that agitate the food's water molecules, creating friction, which in turn produces heat. The heat goes directly into the food instead of into the surrounding air, meaning that the oven—and the kitchen—never gets hot.

Sizes range from 0.3 ft³ capacity to 1.6 ft³ units, and are known as *full-size, mid-size* (also known as family size), and *compact* units. *Wattage* also varies from 450 to 1000; the higher the wattage, the more powerful and faster the heating. Microwave ovens may be *free standing* or *built-in.* Many countertop models offer a kit with which to build the unit into the wall. Other models are designed to be installed above a range or under a cabinet and can be *vented outside.* Some models offer *combination convection* and/or *conventional thermal cooking,* allowing roasting and broiling results that a microwave alone cannot provide. A *browner* option gives the crispy brownness that adds appeal to many foods.

Turntable models help microwaves to penetrate food evenly for consistent results. Others use a *stirrer* that is mounted above the food and has a fan that moves the molecules of the food about. The most sophisticated models, known as *subsurface* or *circuwave,* actually cause the microwaves themselves to move around.

Microwave ovens offer from one to ten *power levels,* which correspond to temperature. For example, a microwave with 1000 W and ten power levels can operate at 100 percent of its power all the way down to 10 percent (or 100 W). More expensive machines have an array of *preprogrammed functions*—such as defrost, popcorn, pizza, beverage, and baked potato—that automatically set the time and power level. Top-of-the-line models have a *humidity sensor* or *probe* that reacts to the water in the food and adjusts the cooking time and power level appropriately.

Other special features include *multistage cooking,* which allows several sequential functions without having to reset the appliance after each one is finished. Look also for a *child lock, a minute-plus* feature, and *automatic reheat*—which keeps food at a certain temperature—and a *memory* feature that automatically repeats a certain sequence of cycles. Most models use *electronic touch-pad controls,* although a few of the budget models use *rotary dials.* Larger ovens often provide a metal rack to cook more than one dish at a time. A *glass door* allows you to see into the oven.

Virtually all microwave ovens carry a one-year parts-and-service *warranty.* All guarantee the magnetron tube for at least five years, others up to ten. Check whether you must bring in the appliance for servicing or whether the warranty includes in-home service, essential for over-the-range models.

Microwave placement should be such that the bottom of the appliance is placed between counter level and user eye level (36 to 54 in off the floor).

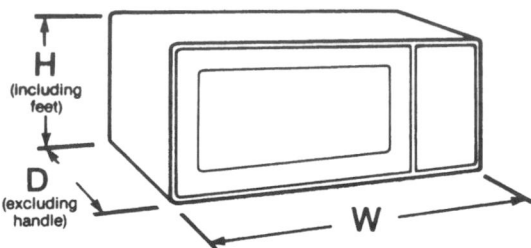

Fig. 27   Typical microwave oven dimensions: width—14 to 26 in; depth—12 to 18 in; height—10 to 16 in.

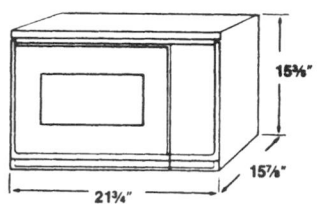

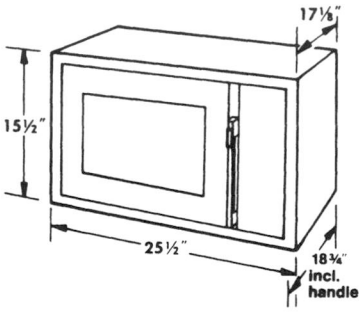

Fig. 28   Microwave oven dimensions.

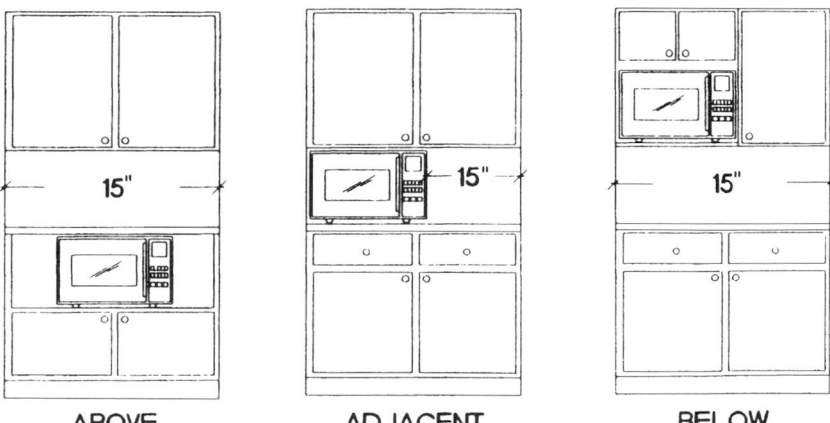

ABOVE   ADJACENT   BELOW

Fig. 29   A microwave oven should have at least 15 in of landing space above, below, or adjacent to the appliance.

REFRIGERATORS

Consider storage capacity first. A too-large model is cooling more space than necessary; a too-small model gets filled too full for adequate air circulation. Be sure to measure existing door and hallway clearances, and compare them with the outside dimensions of the model you are considering. Also check to see whether your plan requires a left- or right-hand opening door. Some models have reversible doors.

Single-door refrigerators have an overall capacity of 1.6 to 22 ft³. Most of these have a small freezer compartment, but some just make ice cubes. Almost all have a manual defrost, although a few larger ones have automatic defrost. *Combination refrigerator/freezers* include *top freezer units* and *bottom freezer units*. Overall capacity ranges from 2.0 to 25.1 ft³. With the exception of a few smaller models, combination refrigerator/freezers have either partial automatic or automatic defrost systems.

Side-by-side units have a capacity from 18.2 to 28.8 ft³, and all units have automatic defrost systems.

The built-in look: Some refrigerators are designed to be built in permanently for a flush look. These have a depth of 24 in, the same as most countertops. Standard models are 30 in deep. Trim kits for certain models allow fronts to be matched to cabinetry.

New features are the following:

- Adjustable gallon-container storage shelves on doors.
- On/off light illuminating through-the-door ice and water dispensers.
- Temperature-controlled beverage chillers.
- Optional crushed-ice setting in through-the-door ice dispensers.
- Quick-serve microwave storage dishes that fit snugly under refrigerator shelves.
- Easy-access compartment with pull-down door that doubles as a shelf for preparing snacks.
- Wine storage racks that provide ideal temperature and humidity levels.

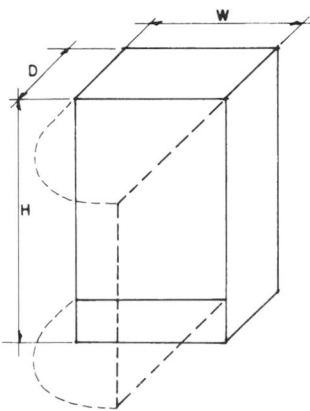

Fig. 30 Built-in refrigerator. This model may be installed under the counter, stacked, or side by side. (See Table 11a.)

TABLE 11a

Capacity, ft³	Depth, in	Width, in	Height, in
8	24	32*	33*
8	24	35	62*
12	24	35	56*

*Allowance must be made at the top or sides for an air grille.

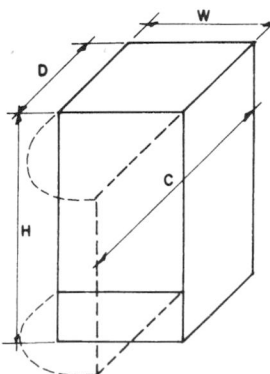

Fig. 30*b* Free-standing refrigerator. Some models can also be built in. Many of them include small freezing compartments. (See Table 11b.)

TABLE 11b

Approximate capacity, ft³	Depth, in	Width, in	Height, in	Door clearance C, in
4	27	24	34	50
8	28	24	58	49
8	28	24	56	48
8	31	24	56	
10	30	24	59	50
10	30	31	60	56
10	31	28	59	
11	28	28	64	53
12	25	36	50	
12	30	30	64	56
12	31	31	61	
13	28	32	64	57

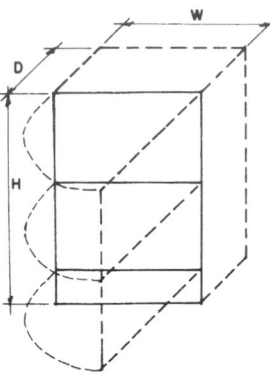

Fig. 31*a* Built-in refrigerator-freezer. This model is 24 in deep, 36 in wide, 74 in high, and has an approximate total capacity of 13 ft³.

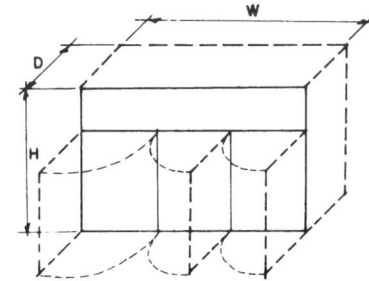

Fig. 31*b* Wall-hung or built-in refrigerator-freezer. (See Table 12a.)

TABLE 12a

Approximate total capacity, ft³	Depth, in	Width, in	Height, in	No. of doors
11	18	64	40	3
13	24	60	38	2

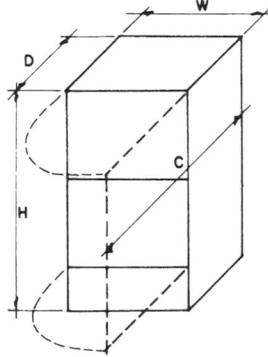

Fig. 31*c* Free-standing refrigerator-freezer. The freezer compartment may be above or below and may vary in size. Some models can also be built in. (See Table 12b.)

TABLE 12b

Approximate total capacity, ft³	Depth, in	Width, in	Height, in	Door clearance C, in
7	29	24	54	
8	32	24	56	50
10	29	28	64	53
10	29	32	59	
11	26	30	61	54
12	29	32	64	
12	29	30	64	55
12	32	30	59	56
12½	29	32	64	57
13	26	30	64	54
13	31	31	64	56
14	29	32	70	57
14	29	33	66	57
14	32	32	68	
15	26	30	71	54
16	32	32	66	
16	32	33	66	60

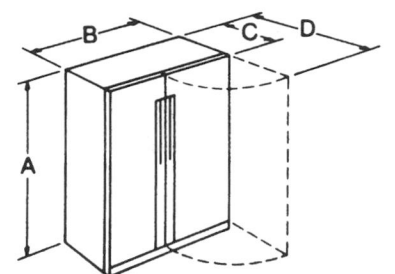

Fig. 32 Side-by-side refrigerator. A = 66 to 68 in; B = 30 to 38 in; C = 31 to 32 in; D = 46 to 50 in; E = 26 to 28 in.

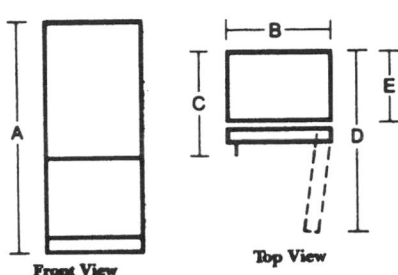

Fig. 33 Bottom-mount refrigerator. A = 64 to 68 in; B = 30 to 36 in; C = 31 to 32 in; D = 58 to 62 in; E = 26 to 28 in.

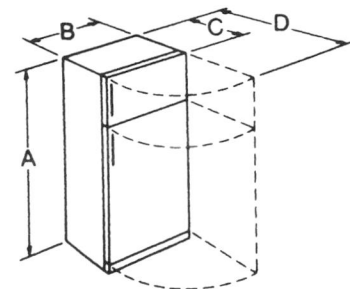

Fig. 34 Top-mount no-frost refrigerator. A = 64 to 68 in; B = 30 to 36 in; C = 31 to 33 in; D = 58 to 62 in; E = 26 to 28 in.

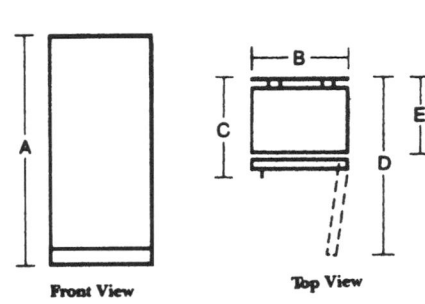

Fig. 35 Cycle and manual defrost refrigerator. A = 58 to 64 in; B = 24 to 30 in; C = 24 to 27 in; D = 44 to 52 in; E = 20 to 24 in.

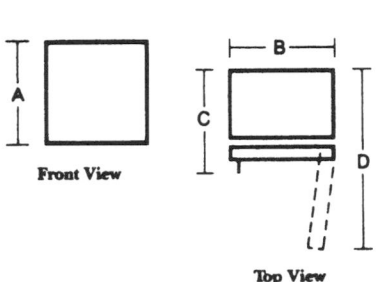

Fig. 36 Refrigerator/icemakers. A = 24 to 36 in; B = 18 to 24 in; C = 21 to 24 in; D = 36 to 48 in.

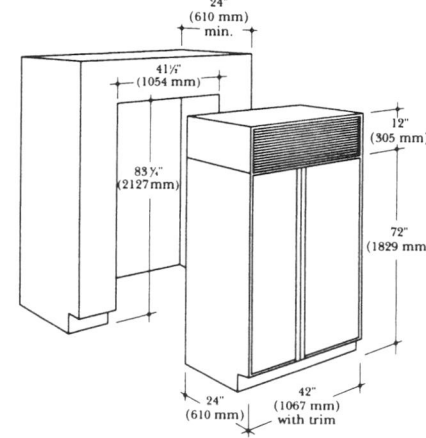

Fig. 37 Built-in refrigerator.

FREEZERS

If you're like most consumers, you'll keep your freezer for 15 years or more. As with refrigerators, *storage capacity* is probably your most important consideration. A too-large model could waste energy and offset the savings you can achieve by buying in bulk on sale or freezing garden produce and homemade dishes. A too-small model will require extra shopping trips or filling the unit too full for adequate *air circulation,* resulting in higher operating costs. The capacity of a freezer is measured in cubic feet of food storage space. Decide what capacity you need based upon family size, distance from stores, whether you have your own garden, how you prepare food, whether you freeze meals, and how often you entertain.

Freezers may be either *upright models*—with a capacity from 1.3 to 26.6 ft³—or *chest freezers*—with a capacity from 1.3 to 31.1 ft³. Ask your dealer for *specification sheets* on various models; these give outside dimensions, installation clearances, and information on features.

A key factor in deciding on what freezer size to purchase is the capacity and style of your refrigerator. Frozen foods should not be kept more than a few days in single-door refrigerators. In top freezer and side-by-side models, the freezer temperature is about 5°F, cold enough to store frozen food for several months. But for true long-term storage, nothing beats a separate freezer, which maintains a temperature of 0°F or less.

Also consider whether the freezer needs to be in the kitchen or whether it could just as well be located elsewhere.

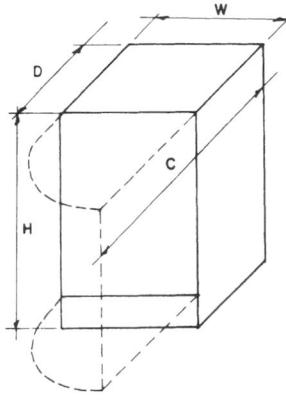

Fig. 38a Free-standing upright freezer. Some models can also be built in. (See Table 13a.)

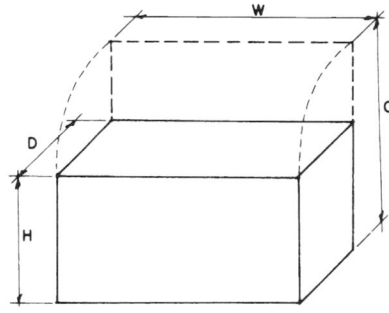

Fig. 38b Chest-type freezer. (See Table 13b.)

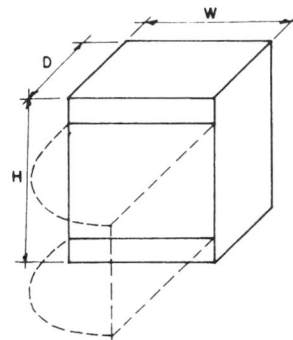

Fig. 38c Built-in freezer. This model can be built in under the counter, stacked with the refrigerator, or wall-hung. (See Table 13c.)

TABLE 13a

Capacity, ft³	Depth, in	Width, in	Height, in	Door clearance C, in
7	25	36	35	
9	29	28	57	52
11	25	36	50	
11	29	31	60	
12	30	32	59	57
12	31	31	62	
12	33	31	59	
13	26	30	64	
14	28	32	70	57
14	32	32	63	
14	30	32	65	57
16	33	34	65	
18	30	30	71	
18	34	32	68	
20	32	36	69	63
20	35	34	65	

TABLE 13b

Capacity, ft³	Depth, in	Width, in	Height, in	Door clearance C, in
4	24	33	36	
10	33	42	37	62
10	33	42	38	
12	30	54	36	59
15	33	55	38	62
15	33	55	38	
17	30	71	36	59
17	32	60	37	63
17	33	60	37	
20	33	71	38	62
20	33	60	37	
20	33	71	38	
21	32	71	37	63
26	33	84	37	

TABLE 13c

Capacity, ft³	Depth, in	Width, in	Height, in
6	24	36	42

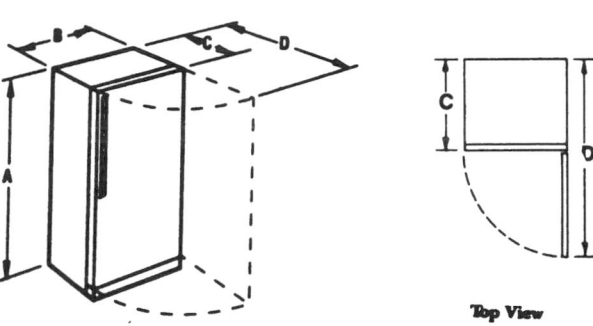

Fig. 39   Upright freezers. A = 24 to 32 in; B = 54 to 72 in; C = 24 to 30 in; D = 44 to 58 in.

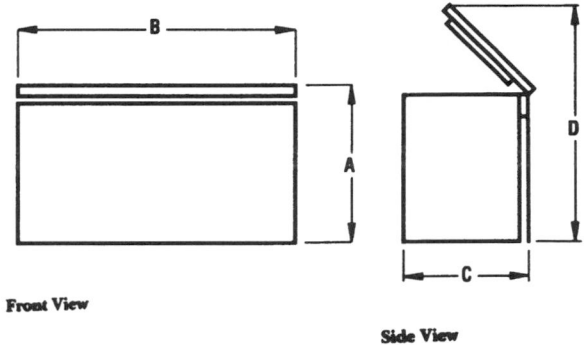

Fig. 40   Chest freezers. A = 30 to 74 in; B = 59 to 68 in; C = 34 to 36 in; D = 24 to 30 in.

COMBINATION UNITS AND APPLIANCE CENTERS

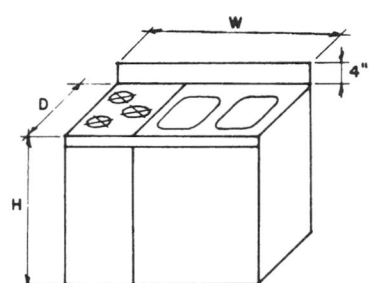

Fig. 41*a*   Combination unit. This model is available with electric or gas range. (See Table 14a.)

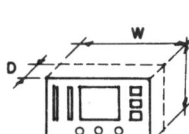

Fig. 41*b*   Automatic appliance center. This model may be built into walls or cabinets. Some models are equipped with receptacles, retractable cords, circuit breakers, clocks, and timers. (See Table 14b.)

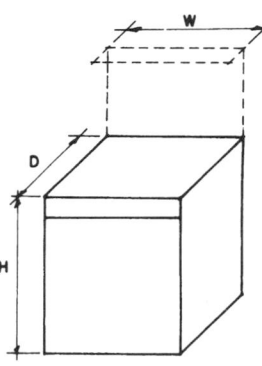

Fig. 41*c*   Executive-type unit. This model is available in various metal and wood finishes, as a free-standing unit, and with furniture styling. (See Table 14c.)

TABLE 14a

Depth, in	Width, in	Height, in	Features
25	24	36	2-burner range, refrigerator, single sink
25	24	36	4-burner range, refrigerator
25	29	36	Single sink, refrigerator
25	39	36	Single sink, refrigerator
25	48	36	Single sink, refrigerator
25	48	38	4-burner range, oven, broiler, refrigerator, single sink
28	28	36	3-burner range, refrigerator
28	28	36	3-burner range, refrigerator, single sink
28	42	36	3-burner range, oven, broiler, refrigerator, double sink

TABLE 14b

Depth, in	Width, in	Height, in
3	14	6
4	14	8
9½	12	12

TABLE 14c

Depth, in	Width, in	Height, in	Features
28	29	39	3-burner range, single sink, refrigerator
28	48	39	3-burner range, single sink, refrigerator, storage space

KITCHENS/APPLIANCES

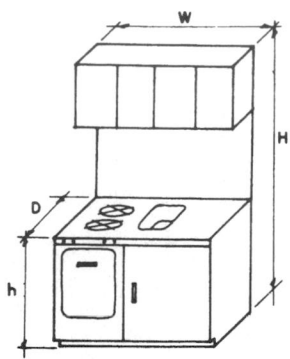

TABLE 14d

Depth, in	Width, in	Height, in	Height h, in	Features
25	39	81	39	2-burner range, refrigerator, single sink
25	48	81, 87	39	2-burner range, oven, broiler, refrigerator, single sink
25	60	87	36	3- to 4-burner range, oven, broiler, refrigerator, single sink
25	69	87	36	3- to 4-burner range, oven, broiler, refrigerator, single sink
25	72	87	36	4-burner range, oven, broiler,
25	63	87	36	refrigerator, single sink
25	48	87	36	3-burner range, oven, refrigerator,
25	51	87	36	single sink

Fig. 41d Pullman-type unit. This model is available with either gas or electric range, and as a free-standing or built-in unit. Garbage disposers are generally available. (See Table 14d.)

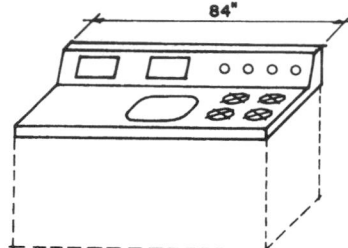

Fig. 41e Special utility package unit. This model fits over standard base cabinets and is available in a variety of counter materials. Its features include surface cooking units, griddles, appliance center, and sink. It can also be ordered in combination with under-counter ovens and dishwasher.

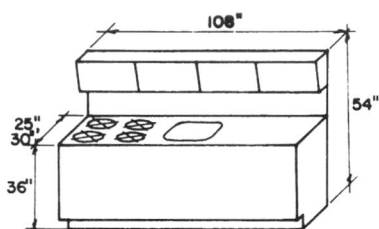

Fig. 41f Special utility package unit. Various combinations of equipment are available: surface cooking units, ovens, sink, appliance center, dishwasher, combination washer-dryer, garbage disposer, and storage units. This model may be free-standing or built-in.

COMPACT KITCHENS

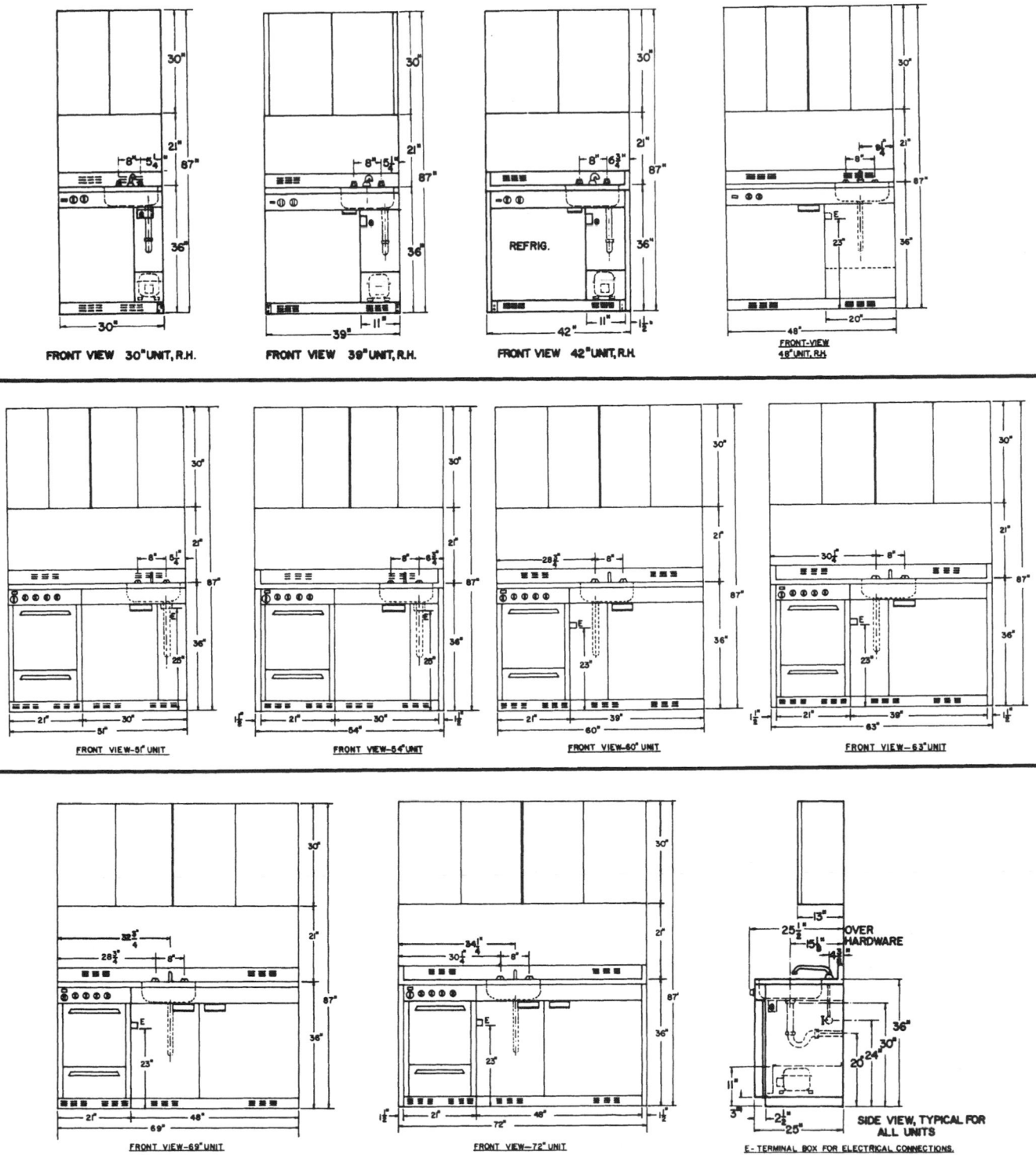

Fig. 42 Typical sizes and arrangements for compact kitchens.

KITCHEN STORAGE UNITS

CABINETS

Critical dimensions of the principal types of stock kitchen cabinets, of both metal and wood, are illustrated on the following pages. These dimensions should be used only for preliminary planning purposes, and specific manufacturers' data should be consulted before final selections and working drawings are made, since not all manufacturers make every cabinet in every size indicated.

The dimensions have been drawn from the current catalogs of the leading manufacturers of stock metal and wood cabinets. Special sizes, features, materials, and finishes are not individually discussed. Such items should always be investigated with the specific manufacturer.

Wood Cabinets

All base cabinets are 34½ in high without tops, 36 in high with tops, and 24 in deep. Wall cabinets are 13 in deep and vary in height.

A wide assortment of filler moldings and panels and side or end finishing panels are available for all cabinet sizes and types. Numerous special-purpose base cabinets are also available, such as mixer storage, tray storage, vegetable bins, breadbox drawers, linen storage, pull-out tables, and towel racks. In addition to such specially designed and equipped cabinets, many accessory items are available for installation in stock units.

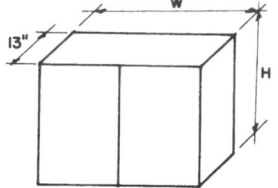

Fig. 1a   Wall cabinet. (See Table 1a.)

TABLE 1a

Height, in	Width, in							
	24	27	30	33	36	39	42	48
15			■	■	■	■		
16	■	■	■	■	■	■	■	
18			■	■	■	■	■	
22	■	■	■	■	■	■	■	
24		■	■	■	■	■	■	
30		■	■	■	■	■	■	
32	■	■	■	■	■	■	■	
33	■	■	■	■	■	■	■	
45	■	■	■	■	■	■	■	

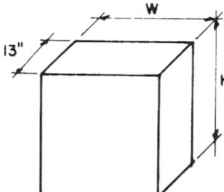

Fig. 1b   Wall cabinet. (See Table 1b.)

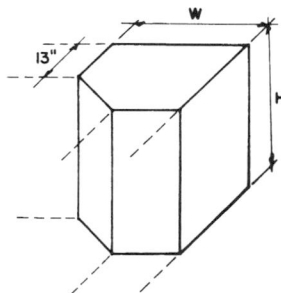

Fig. 1d   Wall cabinet (single-door corner unit). (See Table 1d.)

TABLE 1b

Height, in	Width, in				
	12	15	18	21	24
24				■	■
30	■	■	■	■	■
32	■	■	■	■	■
33	■	■	■	■	■
45	■	■	■	■	

Fig. 1c   Wall cabinet. (See Table 1c.)

TABLE 1c

Height, in	Width, in		
	42	45	48
16	■	■	■
22	■	■	■
24		■	■
32	■	■	■
33		■	■
45	■	■	■

TABLE 1d

Height, in	Width, in	
	24	27
24	■	■
32	■	■
33	■	■
45	■	■

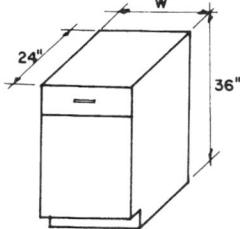

Fig. 1e   Base cabinet. This model is available in the following widths: 12, 15, 18, 21, and 24 in.

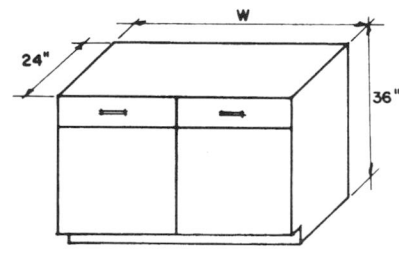

Fig. 1f   Base cabinet. This model is available in the following widths: 24, 27, 30, 33, 36, 39, 42, and 48 in.

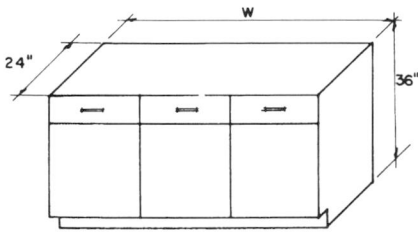

Fig. 1g   Base cabinet. This model is available in 42-, 45-, and 48-in widths.

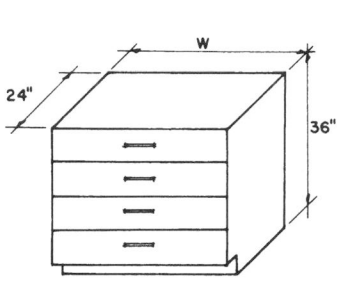

Fig. 1h Base cabinet with drawers. This model is available in 15-, 18-, 24-, and 30-in widths.

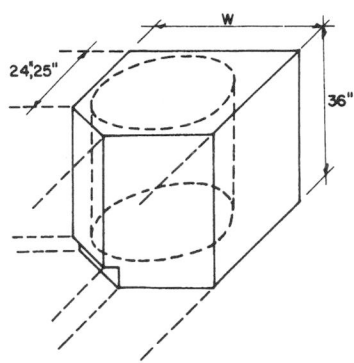

Fig. 1k Base cabinet (rotating corner unit). This model is available in 34- to 36-in widths.

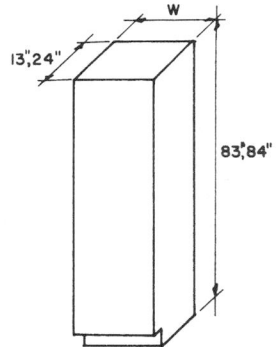

Fig. 1n Utility-storage cabinet. This model is available in the following widths: 15, 18, 21, 24, and 36 in.

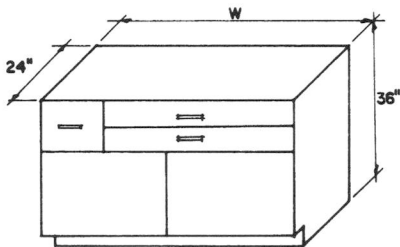

Fig. 1i Base cabinet with drawers. (See Table 1e.)

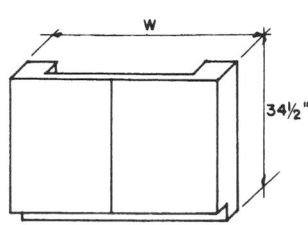

Fig. 1l Sink-front cabinets. The base is generally furnished, but not the cabinet floor. Some models are also available in a 28½-in height for low counters. (See Table 1f.)

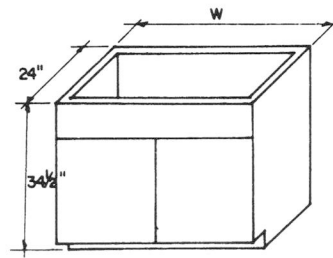

Fig. 1o Range cabinet (for built-in surface cooking units). Some models are 28 in high for use with "stack-on" cooking units. (See Table 1g.)

TABLE 1e

No. of drawers	Width, in
2	15, 18, 21, 24
3	27, 30, 33, 36, 39
4	42, 45, 48

TABLE 1f

Width, in
18, 21, 24, 27, 30, 33, 36
39, 42, 45, 48, 54

TABLE 1g

Width, in
18, 24, 27, 30, 36, 42, 48

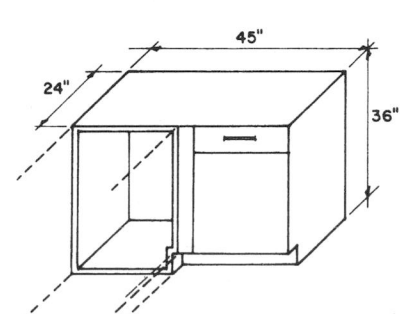

Fig. 1j Base cabinet (blind-corner unit).

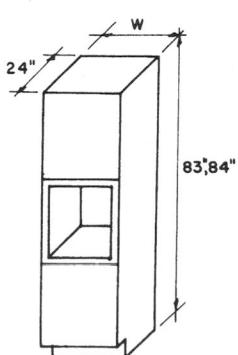

Fig. 1m Oven cabinet (for built-in wall ovens). This model is available in the following widths: 24, 27, 33, and 48 in.

KITCHEN STORAGE UNITS

Metal Cabinets

All base cabinets are 34½ in high without tops, and 36 in with tops. The total depth to the edge of the countertop is generally 25 in. The depth of the cabinet itself is generally 24 in plus or minus ½ in. Wall cabinets are 13 in deep and vary in height. Doors may be either right- or left-hand.

All manufacturers offer a wide variety of optional, special purpose accessories that can be fitted into any of the basic cabinets. A wide assortment of filler moldings and panels and side or end finishing panels are available for all cabinet sizes and types. Most manufacturers offer cabinets in several stock colors, and special colors can usually be matched on special order. Some models have wood doors. Base cabinets are available with or without tops.

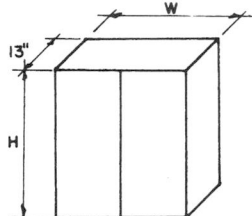

Fig. 2f Range-ventilator cabinet. Hoods are available in a variety of metals and finishes; however, some models have no projecting hood. (See Table 2f.)

TABLE 2f

Width, in	No. of doors
24	2
30	2
36	2
42	2–4

Fig. 2a Wall cabinet. (See Table 2a.)

TABLE 2a

Width, in	No. of doors
36	2

Fig. 2d Wall cabinet (island). This model has two doors and opens on two sides. (See Table 2d.)

TABLE 2d

Height, in	Width, in
18	21
18	24
18	30
18	36
24	24
30	12
30	15
30	18
30	21
30	24
30	27
30	30
30	36

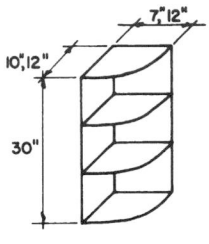

Fig. 2g Wall cabinet with open display shelves.

Fig. 2b Wall cabinet. (See Table 2b.)

TABLE 2b

Width, in	No. of doors
12	1
15	1
18	1–2
21	2
24	2
27	2
30	2
36	2
42	2
48	2
54	4

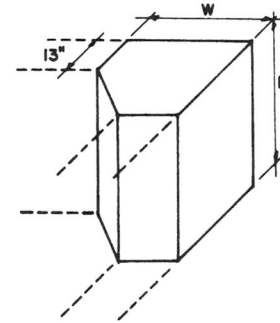

Fig. 2h Wall cabinet (single-door, corner-type). Some models are available with revolving shelves. (See Table 2g.)

TABLE 2g

Width, in	Height, in
21, 25, 26, 27	18, 30

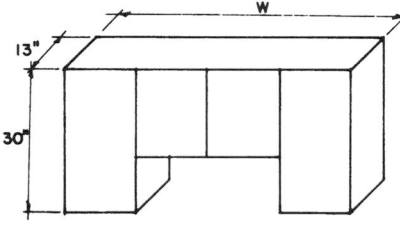

Fig. 2e Over-sink cabinet. (See Table 2e.)

TABLE 2e

Width, in	No. of doors
54	4
60	4

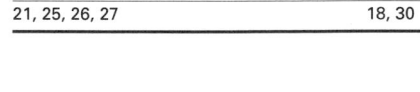

Fig. 2i Wall cabinet. This model has bypassing sliding doors and is available in the following widths: 21, 24, 30, 36, and 48 in.

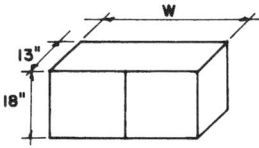

Fig. 2c Wall cabinet. (See Table 2c.)

TABLE 2c

Width, in	No. of doors
18	1
21	2
24	2
30	2
36	2
42	2–4

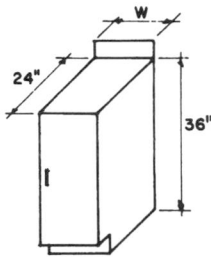

Fig. 2*j* Base cabinet. (See Table 2h.)

TABLE 2h

Width, in	No. of doors
9	1
12	1
24	2

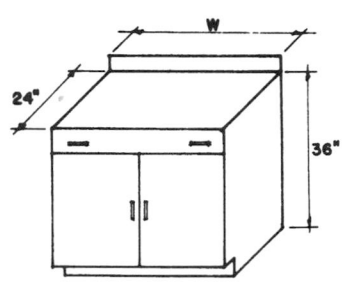

Fig. 2*k* Base cabinet. (See Table 2i.)

TABLE 2i

Width, in	No. of doors	No. of drawers
12	1	1
15	1	1–2
18	1	1
21	2	1–2
24	2	1–2
27	2	1–2
30	2	1–2
36	2	2
42	4	2

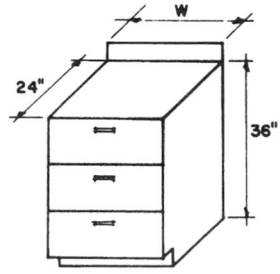

Fig. 2*l* Base cabinet with drawers. (See Table 2j.)

TABLE 2j

Width, in	No. of drawers
15	3–4
18	3–4
21	4
24	3–4
27	4

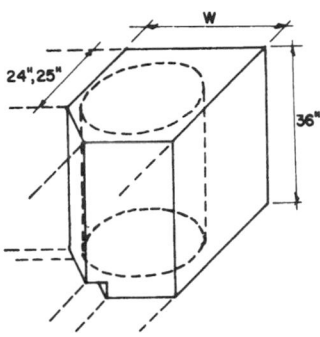

Fig. 2*m* Base cabinet (single-door, corner-unit). Most models have revolving shelves; available in 31-, 32-, and 33-in widths.

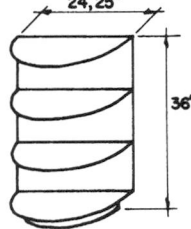

Fig. 2*n* Base cabinet (half-round, open-type).

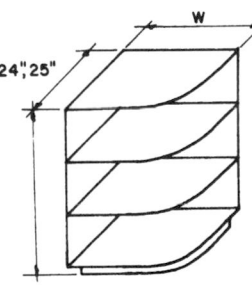

Fig. 2*o* Base cabinet (quarter-round, open-type). This model is available in 18-, and 24½-in widths.

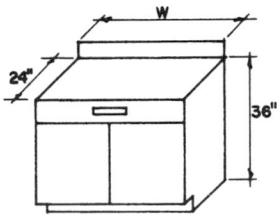

Fig. 2*p* Range base cabinet (for drop-in, surface cooking units). This model is available in 18-, 24-, 30-, and 36-in widths.

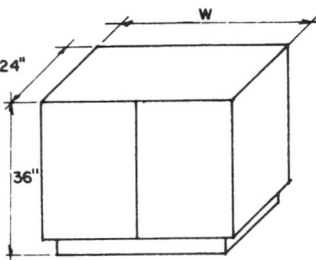

Fig. 2*q* Base cabinet (island). This model is available with single or double doors accessible on two sides. (See Table 2k.)

TABLE 2k

Width, in
12, 15, 18, 21, 24
27, 30, 36, 45

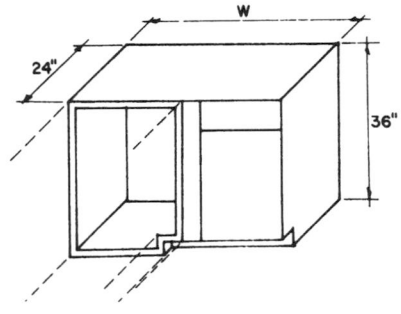

Fig. 2*r* Base cabinet (blind-corner unit). Right- or left-hand models are available; comes in 42- and 44-in widths.

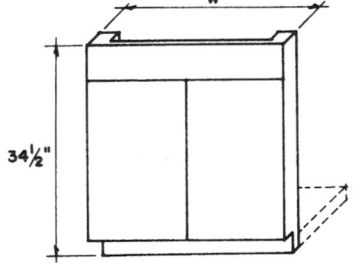

Fig. 2*s* Sink-front cabinet. This model is available in the following widths: 21, 24, 30, 36, and 42 in.

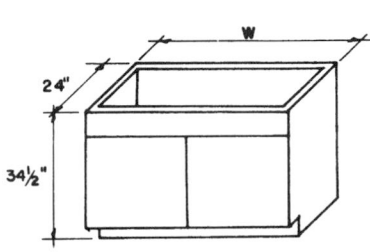

Fig. 2*t* Under-sink cabinet. (See Table 2l.)

TABLE 2l

Width, in
24, 30, 36, 42, 48
54, 60, 66, 72

KITCHEN STORAGE UNITS

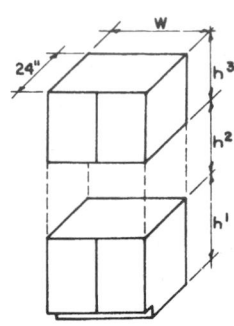

Fig. 2u Oven cabinet (for built-in wall oven). (See Table 2m.)

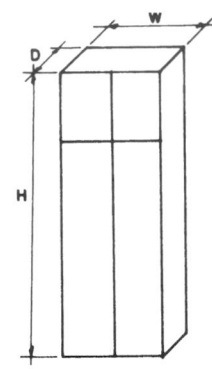

Fig. 2v Utility-storage cabinet. (See Table 2n.)

TABLE 2m

Width, in	Height h^1, in	Height h^2, in	Height h^3, in
24, 27	18, 22, 28	26, 34, 36, 40, 42	18, 24

TABLE 2n

Depth, in	Width, in	Height, in
13	21	80
24	24	84

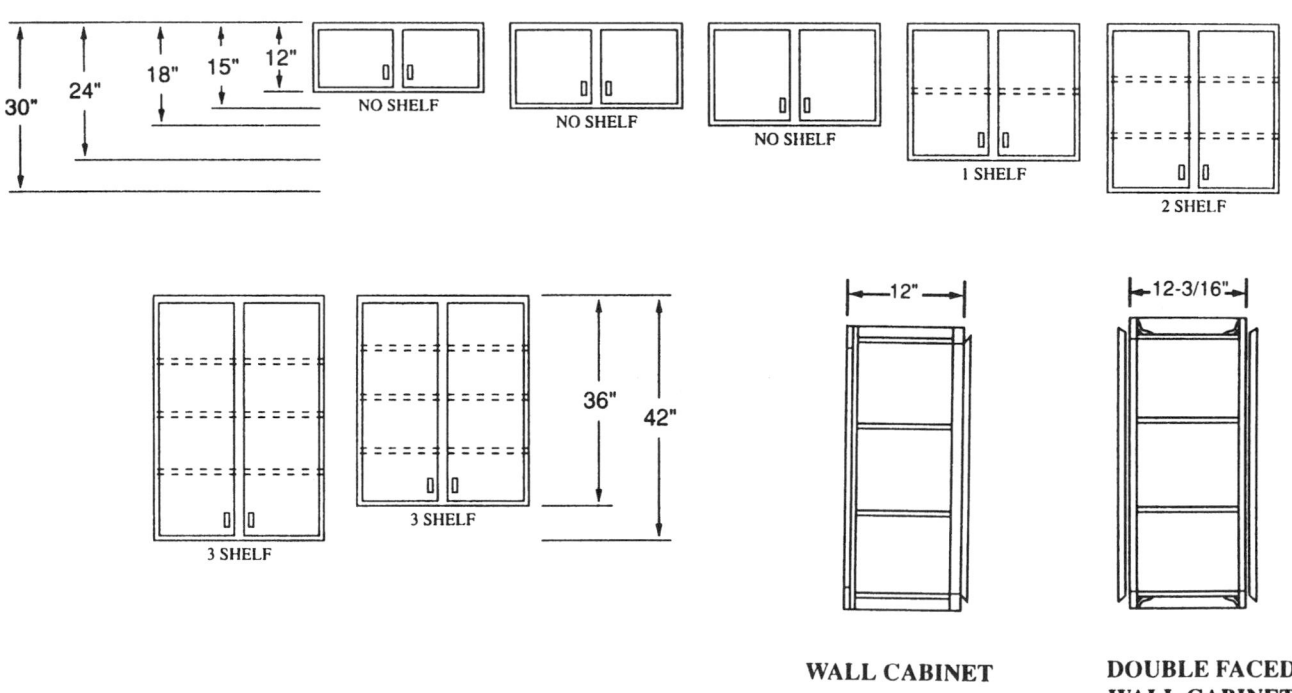

Fig. 3 Wall cabinet dimensions.

**12" High
Double Door**
• No shelf

Fig. 4   Wall cabinets: 12 in high.

**15" High
Double Door**
• For use over refrigerators
• No shelf

**15" High
24" Deep
Double Door**
• For use over refrigerators
• No shelf

Fig. 5   Wall cabinets: 15 in high.

**18" High
24" Deep
Single Door**
• Invertible for L or R hinging
• No shelf

**18" High
24" Deep
Double Door**
• No shelf

**18" High
Single Door**
• Not recommended for recessed or step-back" use over 24" deep utility or oven cabinets
• No shelf

**18" High
Double Door**
• Not recommended for recessed or "step-back" use over 24" deep utility or oven cabinets
• No shelf

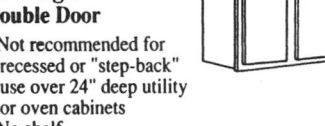

**18" High
Double Faced
Double Door**
• 2 doors on both sides
• No shelf

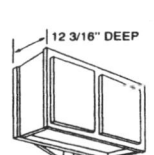

Fig. 6   Wall cabinets: 18 in high. (Cabinets that are 24 in deep are for use over pantry and utility and oven cabinets.)

**24" High
Single Door**
• 1 adjustable shelf

**24" High
Double Door**
• 1 adjustable shelf

**24" High
24" Deep Single Door†**
• 1 Adjustable shelf
• For use with 36" high wall cabinets
• Stacks on utility cabinets

**24" High
24" Deep
Double Door†**
• 1 adjustable shelf

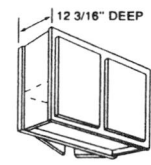

**24" High
Double Faced
Single Door**
• 1 adjustable shelf

**24" High
Double Faced
Double Door**
• 1 adjustable shelf
• 2 doors on both sides

**24" High
Blind Corner
Single Door**
• 1 adjustable shelf
• Blind panel with 2" filler included — shipped loose

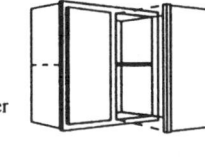

Fig. 7   Wall cabinets: 24 in high. (Cabinets that are 24 in deep are for use over pantry and utility and oven cabinets.)

**30" High
Single Door**
• 2 adjustable shelves
• Invertible for left or right hinging
• Specify L or R hinging for Arched doors

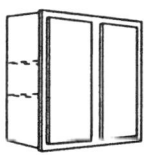

**30" High
Double Door**
• 2 adjustable shelves

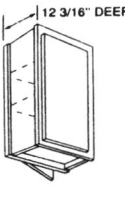

**30" High
Double Face
Single door**
• 2 adjustable shelves

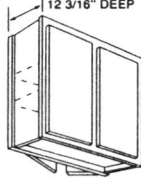

**30" High
Double Faced
Double Door**
• 2 adjustable shelves
• 2 doors on both sides

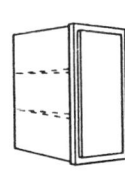

**30" High
24" Deep
Single Door**
• Use with 42" wall cabinets
• Stacks on top of utility cabinets
• 2 adjustable shelves

**30" High
24" Deep
Double Door**
• 2 adjustable shelves

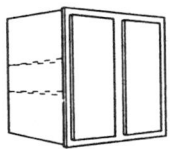

**30" High
90° Corner**
• Easy access corner storage
• 2 adjustable shelves

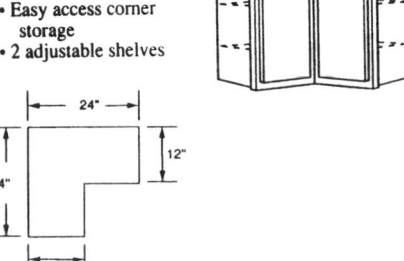

Fig. 8   Wall cabinets: 30 in high.

KITCHEN STORAGE UNITS

**30" High
45° Corner
Single Door**
• 2 adjustable shelves

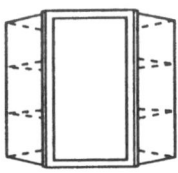
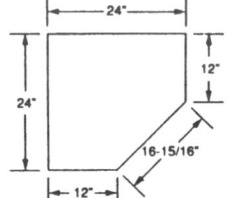

**30" High
Blind Corner
Single Door**
• 2 adjustable shelves

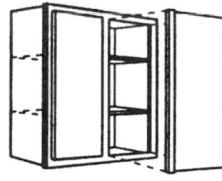
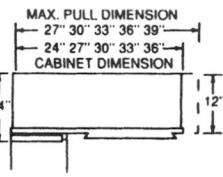

MAX. PULL DIMENSION
27" 30" 33" 36" 39"
24" 27" 30" 33" 36"
CABINET DIMENSION

Fig. 8 (*Continued*)

**36" High
Single Door†**
• 2 adjustable shelves

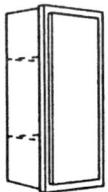

**36" High
Double Door†**
• 2 adjustable shelves

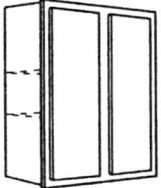

**36" High
45° Corner
Single Door†**
• 2 adjustable shelves

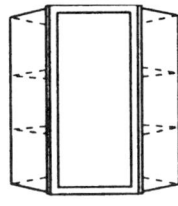
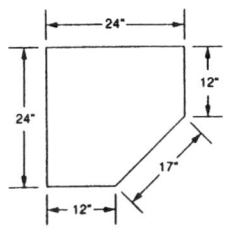

**36" High
Blind Corner
Single Door†**
• 2 adjustable shelves

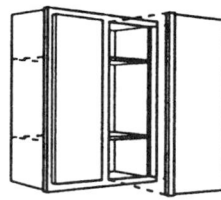

Fig. 9 Wall cabinets: 30 in high.

**42" High
Single Door**

• 3 adjustable shelves

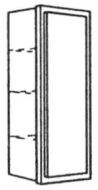

**42" High
Double Door**

• 3 adjustable shelves

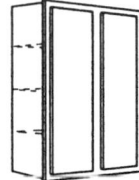

**42" High
45° Corner
Single Door**

• 3 adjustable shelves

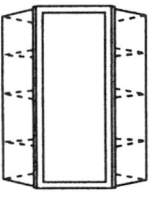
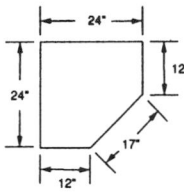

**42" High
90° Corner**

• Easy access corner
  storage
• 3 adjustable shelves

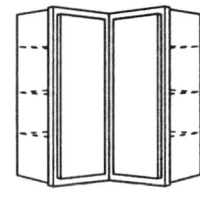
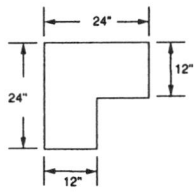

Fig. 10   Wall cabinets: 42 in high.

Single Door
• 1 door
• 1 drawer
• 1 roll-out tray

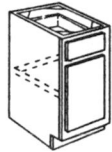

Double Door
• 2 door
• 2 drawer
• 2 roll-out trays

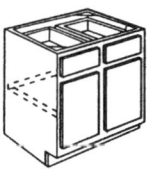

Single Drawer
• 2 doors
• 1 wide drawer with reinforced bottom
• 2 wide roll-out trays with
  reinforced bottoms

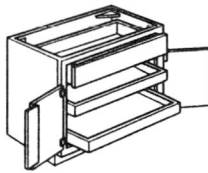

Three Drawer
• Two extra deep lower
  drawers for pots and
  pans storage
• One standard depth
  drawer

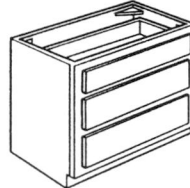

Four Drawer
• 1 deep bottom drawer
• 3 standard drawers

Tray Divider
• Full height door
• 2 vertical tray dividers

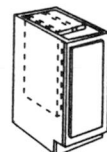

**Double Faced
Single Door**
• Drawer & roll-out tray
  operate from same side
• Door on each side hinged
  at the same end

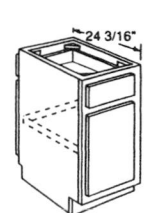

**Double Faced
Double Door**
• 2 doors on both sides
• 2 drawers & roll-out trays
  operate from same side

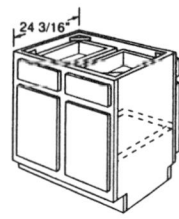

Fig. 11   Base cabinets.

Sink Base
• Double door

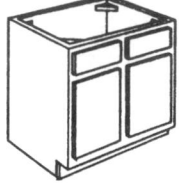

Sink Front Diagonal
• Single door

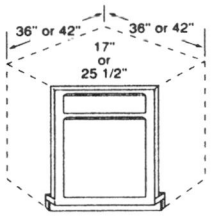

Sink Front 90° Corner
• Requires 40" of wall space in both directions

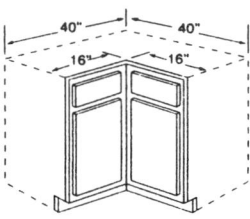

Fig. 12   Base cabinets: sink.

Base Corner Revolving
• Requires 36" wall space in both directions
• 2 piece bi-fold door
• 2-28" diameter pie-cut shelves with independent rotation

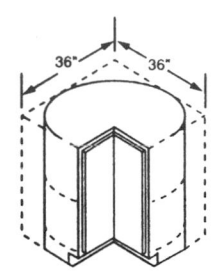

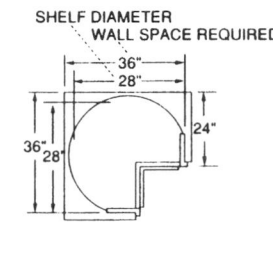

Base Corner
• 2 piece bi-fold door
• 1 fixed pie cut shelf
• Requires 36" wall space in both directions

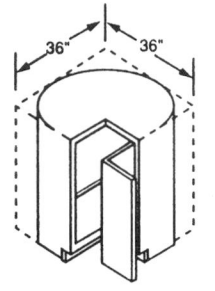

Fig. 13   Base cabinets: corners.

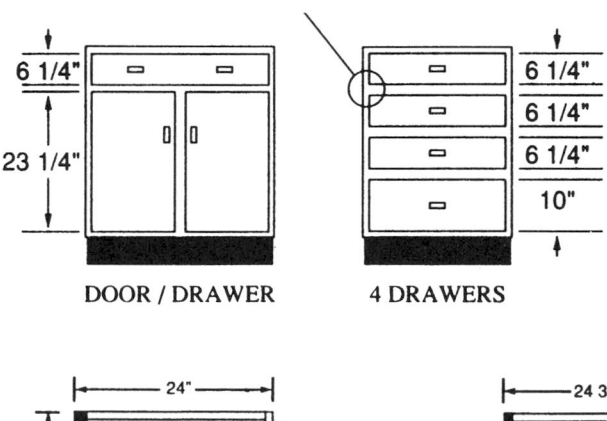

DOOR / DRAWER   4 DRAWERS

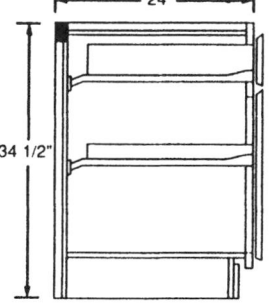

BASE CABINET

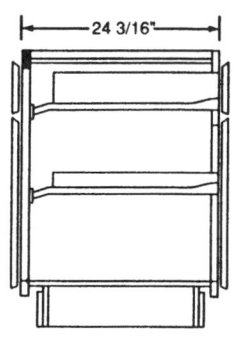

DOUBLE FACED
BASE CABINET

Fig. 14

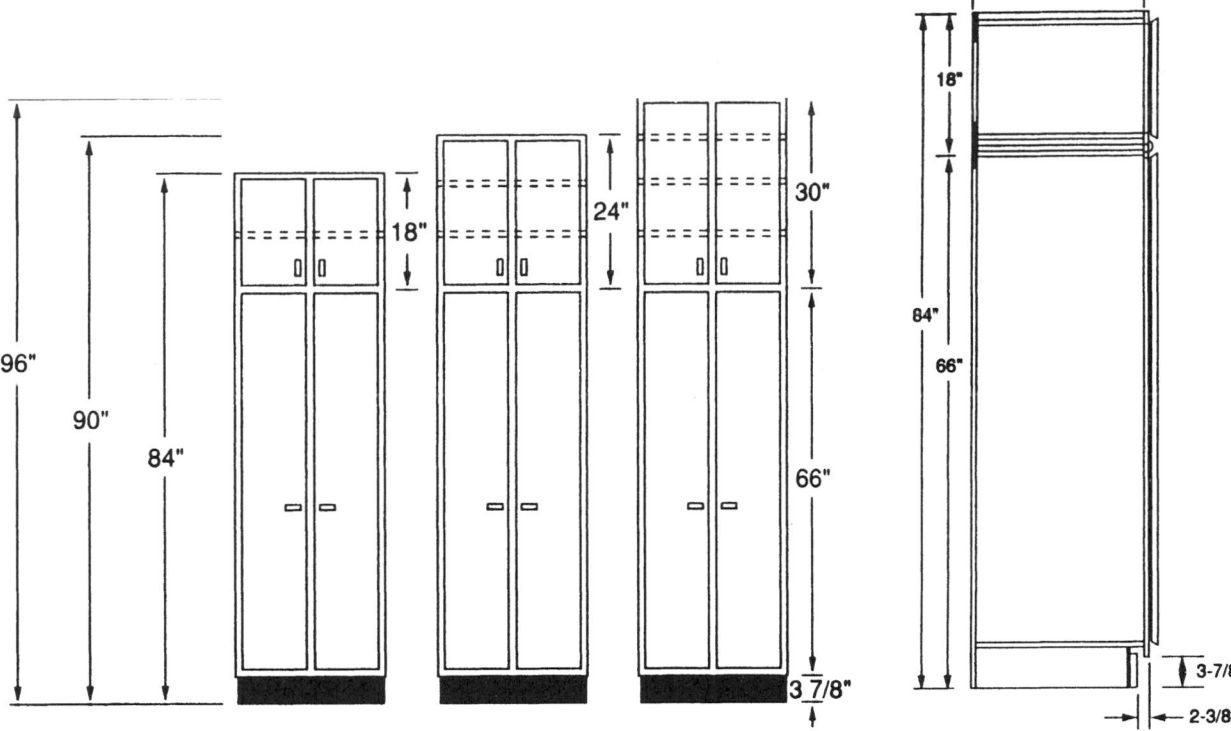

Fig. 15 Tall cabinets.

Fig. 16 Utility cabinet with wall cabinet.

KITCHEN STORAGE UNITS

TABLE 3   Items and Dimensions of Utensils, Equipment, and Supplies in Base Cabinets

Items* and storage units	Number stored		Unit space dimensions, in†		
	Limited	Liberal	Side to side	Front to back	Height
Mix utensils—B-3, 4, and 8					
Flour sifter	1	1	6½	9	7
3½-qt mixing bowl	1	1	12½	12½	6
2-qt mixing bowl	1	2	9½	9½	5½
Pint measure	1	1	4½	6½	5½
Cup measure	1	1	4	5	5
1-qt mixing bowl	1	1	7½	7½	5
Baking dish, 9½ in diameter	1	1	10	11½	4½
Baking dish, 10½ in long	0	1	11	12½	4½
Loaf pan	0	1	6	10½	3½
Biscuit pan	1	1	10	13½	3
Pie pans	1	2	10	10	2½
Cake pans	2	2	12	12	2½
Muffin pan	1	1	11	14	2½
Cookie sheet	0	1	12½	16	2
Pots and pans—B-1, 2, 3, and 5					
Double boiler	1	1	7½	12	10½
Coffee pot (6-cup)	1	1	6½	9	10
6-qt saucepot	0	2	10½	10½	9
3-qt saucepan	2	2	8½	15	8
Pressure saucepan	0	1	9	17	7½
4-qt saucepot	1	1	9	11	7½
2-qt saucepan	1	1	7½	14	7
1-qt saucepan	1	1	6½	13	6
Colander	1	1	11½	13	5
Frypans, pot lids—B-6 and 7					
10½-in frypan	1	1	11	17½	5½
9-in frypan	1	2	9½	16	5
6-in frypan	0	1	6	12	5
Pot lids	2	4	10½	10½	1
Tools for food preparation‡					
Mix					
Egg beater	1	1	4	12½	4
Cookie cutter	1	1	3	3	3½
Rolling pin	1	1	3	19	3½
Mixing and blending forks	4	8	3	12½	2½
Measuring spoons, sets	1	2	3	6	2½
Egg whisk	0	1	4	12½	2½
Knives and spatulas	2	6	3	14	2
Range					
Potato masher	1	1	3½	13	4½
Knives, forks, spoons	3	3	3½	13	3
Dishwashing utensils§					
Dishpans, nested	2	2	16½	18½	8
Dish drainer	1	2	14½	18½	6
Bread and pastries—B-12, 14					
Bread	2	2	5½	12	6
Cake, square	1	1	9½	9½	2½
Kitchen linens—B-12, 14					
Dishtowels	8	12	12	11	5(8)¶
Handtowels	8	12	12	10	5(8)
Aprons	4	6	11	10	5(4)
Dishcloths	6	12	8	8	4(6)
Paper napkins (box)	1	2	8	8	3½
Small tablecloths	0	1	10	14	3(2)
Medium tablecloths	1	2	10	19	2(2)
Potholders	4	8	7	7	2(6)
Produce—B-13 (nonrefrigerated fruits and vegetables)					
Potatoes, lb	10	10	9	11	8
Onions, lb	3	3	9	7	8
Fruit, lb	3	3	9	7½	5
Reserve staple supplies—B-13, 15					
Flour, 25-lb bag	1	1	5½	21	8
Cornmeal, 10-lb bag	1	1	4	13	6
Sugar, 10-lb bag	1	1	4½	12½	7

*Based on inventory data.
†The unit space dimension is the dimension of the item plus clearance for handling. Height provides for storing lids on utensils.
‡Drawers for tools are combined with storage for other items.
§Dishwashing supplies including dishpans and drainers are stored under sink.
¶Number in parentheses refers to number of items in stack for which unit space dimension is given.

TABLE 4 Items and Dimensions of Food Supplies Stored in Wall Cabinets

| Items* and storage units† | Number stored | | | Unit space dimensions,‡ in | | |
	Limited	Liberal	Capacity or size	Side to side	Front to back	Height
Mix supplies—S-1						
Canisters						
General-purpose flour	1	1	5 lb	8	8	9
Cornmeal	1	1	5 lb	8	8	9
Sugar	1	1	5 lb	8	8	9
Packages						
Pancake flour		1	2 lb	2½	6½	10½
Cake flour	1	1	2¾ lb	3	7	10½
Vinegar	1	1	1 qt	4	4	10
Powdered sugar	1	1	1 lb	2½	4	8½
Brown sugar	1	1	1 lb	2	4	8
Coconut	1	1	7 oz	2	4	8
Shortening	1	1	3 lb	5½	5½	8
Cornstarch		1	1 lb	2½	4	7½
Cocoa	1	1	1 lb	3	4	7½
Raisins	1	1	15 oz	2½	4½	7½
Syrup	1	1	5 lb	5½	5½	7½
Flavorings	3	5	6 in tall	1½	2½	7
Salt	1	2	1 lb 10 oz	4	4	7
Baking powder	1	1	1 lb	3½	3½	6½
Soda	1	1	1 lb	2½	4	6½
Package desserts	1	3	3⅝ oz	2	4	5½
Spice cans						
Medium	2	3	4½ in tall	2½	3½	5½
Small	3	3	3 in tall	1½	2½	4
Range supplies—S-2						
Oatmeal	1	1	3 lb	6	6	11
Macaroni	1	1	1 lb	2	5½	9
Grits	1	1	1 lb 8 oz	4½	4½	8
Tea	1	1	8 oz	2½	4½	7
Range supplies—S-2						
Rice	1	1	1 lb	2½	4	6½
Spaghetti	1	1	1 lb	2½	11½	6
Coffee	1	1	1 lb	5½	5½	4
Sink supplies—S-3						
Lentils and peas	1	1	2 lb	3½	5	9½
Dry beans	1	1	2 lb	3½	5	8½
Prunes	1	1	2 lb	3	5	8
Canned food	6	8	No. 2	4	4	5½
Serve supplies—S-4						
Prepared cereals	1	2	11 in tall	3	8	14
Cookies	1	1	1 lb	3	6½	11½
Mayonnaise		1	1 qt	4	4	8
Crackers	1	2	1 lb	4½	10½	6½
Peanut butter	1	1	1 lb 4 oz	3	3	6½
Mayonnaise	1		1 pt	3½	3½	6
Jam and pickles	6	3	1 pt	3½	3½	2

*Based on inventory data.

†The storage unit indicated for each grouping of foods will accommodate only those foods in the specified group.

‡The unit space dimension includes dimensions of package plus allowance of ½ in on width, depth, and diagonal height of package to permit handling in storage.

KITCHEN STORAGE UNITS

TABLE 5 Everyday Dishes—Kinds and Number of Dishes Stored and Unit Space Dimensions

	Limited quantity			Liberal quantity			Unit space dimensions, in		
Item	Pieces	Stacks	Weight of dishes*	Pieces	Stacks	Weight of dishes†	Side to side†	Front to back†	Vertical clearance
Dinnerware									
Cups	8	4	2 lb 8 oz	12	6	3 lb 12 oz	4½	5½	6
Cereal dishes	8	2	4 lb	8	2	4 lb	7½	7½	5
Dinner plates	8	1	10 lb	12	1	15 lb	11	11	4½
Salad or pie plates	8	1	3 lb 8 oz	12	1	5 lb 4 oz	9	9	4½
Sauce dishes	8	1	4 lb	12	1	6 lb	5½	5½	6
Saucers	8	1	3 lb	12	1	4 lb 8 oz	7½	7½	4½
Soup bowls				8	2	4 lb			
Glassware‡ and pitchers									
Iced tea glasses				8		4 lb 8 oz			
Juice glasses	8		1 lb	8		1 lb	3	3	5
Pitcher									
Large	1		2 lb 8 oz	2		5 lb	7½	10½	10
Medium	1		1 lb	1		1 lb	7	8	10
Water glasses	8		2 lb	8		2 lb	3½	3½	6
Serving dishes									
Bowels									
Oval	2	1	2 lb	3	2	3 lb	13½	9½	9½
Round	2	1	2 lb 8 oz	4	2	5 lb	9½	9½	7½
Creamer			6 oz	1	1	6 oz	5	7	5
Gravy boat				1	1	12 oz	6	10½	5½
Jelly-relish dishes	2	1	1 lb 2 oz	2	1	1 lb 2 oz	7½	7½	2
Platter									
Large	1	1	5 lb	1	1	5 lb	16½	13	2½
Medium	1	1	2 lb	2	1 or 2	4 lb	14	11	2½
Small				1	1	1 lb	12	9	2½
Serving plates				1	1 or 2	4 lb	11	11	4½
Sugar	1	1	9 oz	1	1	9 oz	5½	6½	5½
Tray, medium				1	1	1 lb	15½	11½	3
Refrigerator dishes, set of 4	1 set	1		1 set	1		8	8½	7

*Medium-weight dishes.

†One-half inch added to side-to-side and front-to-back measurement of item or stack and ½ to ¾ in to height to permit safe handling. For stacked items, clearance is sufficient to remove single item from stack.

‡Glasses placed three rows to a shelf instead of stacking.

TABLE 6   List of Kitchen Utensils

Utensil	Storage utensil alone, inches L × W × H or diameter	Dimensions with handle, inches L × W × H
1. Top-of-range cooking		
Regular use—saucepans		
1 sauce pot, 6-qt, straight sides, 2 handles	10 × 5	12½
1 cover	10¼ × 2	
1 saucepan, 4-qt, straight sides	8¾ × 4¾	14¾
1 cover	9 × 2	
1 stew kettle, 4-qt, flared sides, bale handle and	9½ × 5½	11
1 side handle		
1 cover	9 × 2	
1 saucepan, 2-qt, straight sides	7½ × 3½	13¾
1 cover	7½ × 1½	
1 saucepan, 1½-qt detachable handle		15½
1 saucepan, 1-qt	5½ × 3	10¼
1 double boiler, stacked	6½ × 9	11
Upper section, 2-qt	6 × 5¾	11
Lower section, 2½-qt	6½ × 5½	11
1 cover	6 × 1¾	
Occasional use		
1 large kettle, rounded sides, bale handle	12¼ × 8½	14
1 cover	12 × ½	
Skillets		
1 double skillet, stacked, sloping sides	10¼ × 4¾	17
Top section	10 × 2¼	17
Lower section	10¼ × 2½	17
1 open double-lipped skillet	8¾ × 2	12½
1 griddle	10½ × 2½	16½
1 teapot, 6-cup	6½ × 5	8½
1 coffeepot, drip, 6-cup	5 × 8	8 × 8
2. Oven cooking		
Casseroles and bake pans		
1 casserole, 2-qt	8¾ × 3	10 × 3
1 cover	8¾ × 2½	
1 casserole, 1½-qt	8½ × 2¾	9½ × 2¾
1 pie pan cover	8½ × 1½	
1 set of 8 custard cups, stacked in 4's	4 × 3	
1 open roaster	14 × 10 × 2¼	
With trivet	12 × 8 × 2	
1 meat loaf pan	12¾ × 4 × 3	
1 quick break pan	9 × 5 × 3	10 × 5 × 3
Cake, cookie, muffin pans		
2 layer cake pans, round	9½ × 1½	
1 cake pan, square	9 × 9 × 2½	
1 angel cake pan, with tube center	10 × 5	
1 cake cooler rack	17 × 9 × ¾	
2 muffin pans, 9-hole	9 × 9 × 1½	
2 cookie sheets	15½ × 12 × ½	
Pie pans		
1 large	10 × 1¼	
1 medium	8½ × 1¼	
1 casserole top	8½ × 1¼	

KITCHEN STORAGE UNITS

TABLE 6   List of Kitchen Utensils (*Continued*)

Utensil	Storage utensil alone, inches L × W × H or diameter	Dimensions with handle, inches L × W × H
3. Food preparation—preliminary: Paring, dicing, grinding, grating, juicing, rolling, and cutting		
1 paring knife, short	3 in blade	7 in overall
1 paring knife, long	4 in blade	7¾ in overall
1 meat slicer	8½ in	13½ in
1 French pattern knife	8¼ in	13½ in
1 apple corer	7 in	
1 potato parer, floating blade	6½ in	
1 grater, all-purpose	4 × 3 × 9	
1 grinder, 3 blades	11 × 9 × 4	2-in diameter blade
1 fruit juicer, glass, or extractor, metal	8¼ × 6 × 3	
1 can and bottle opener	6½ × 2½ × 1½	
1 corkscrew		
1 scissors	7 × 2½	
1 sharpener, steel or carbon	15½	
1 hardwood cutting board	12 × 10 × ¾	
1 pastry board	20 × 24 × ½	
1 biscuit cutter	2½ × 2	
1 cookie cutter	3 × 1½	
1 rolling pin	10 × 2¼	17 in with handle
4. Food preparation—Measuring, mixing, beating, stirring		
1 qt measure, glass, flared sides	5½ × 5¼	7½ in overall
1 pt measure, glass, flared sides	4½ × 4¼	6 in overall
1 cup, dry, glass	3¼ × 3	4¼ in overall
1 cup, liquid, glass, heatproof	3½ × 3½	5 in overall
1 set of 4 nested cups, stacked in 4—overall	4 × 3¼	
Individual:		
1-cup size	3¼ × 2½	4
7½ i-cup size	3 × 1¾	3¾
⅓-cup size	2¾ × 1⅛	3½
¼-cup size	2½ × 1	3¼
2 sets measuring spoons	3¾ × 1¾ × 1½	
1 pastry blender	5½ × 47½ × 1½	
1 wooden or large bowl spoon	14 × 2¼ × 1½	
1 perforated spoon	1½ × 2½ × 2	
1 sifter	5 × 6	8
1 rotary beater	12½ × 5¼ × 3½	
1 one-egg beater	13 × 2½ × 2	
1 mixing bowl, 4–47½ i qt	12 × 5½	
1 mixing bowl, 2–27½ i qt	10 × 4	
1 mixing bowl, 1–17½ i qt	8½ × 3½	
1 mixing bowl, 1 qt	7 × 3	
5. Food preparation—final spreading, leveling, loosening, turning, mashing, testing		
1 griddle cake turner	13 × 3¼ × 1¾	
1 spatula, long narrow	10 × 1	
3 teaspoons (individual)	6 × 1¼	
3 tablespoons	8 × 1¾	
1 fork, short	7½ × 1	
1 fork, long	15 × 1	
1 lifter, hot foods	7¾ × 2¾	
1 ladle	13 × 4	
1 masher	10 × 4 × 3	

TABLE 6 List of Kitchen Utensils (*Continued*)

Utensil	Storage utensil alone, inches L × W × H or diameter	Dimensions with handle, inches L × W × H
6. Cleanup		
1 brush, soft	8½ × 3 × 1½	
1 brush, stiff	8 × 4 × 3	
1 colander	10 × 4½	13 in overall
1 strainer, fine mesh, large	6¼ × 3¼	12¼ in overall
1 strainer, tea	3 × 1¼	7½ in overall
1 funnel	4 × 4	
1 plate scraper	8½ × 3½	
1 dishpan	17⅓ × 10¾ × 5⅓	
1 dish drain	19 × 12 × 3½	
1 garbage unit	13 × 10½	
1 trash basket	16 × 15	
7. Food storage		
1 bread box	14 × 11 × 11	
1 cake or cookie box	12 × 12 × 7	
Empty jars and casseroles		
Wax paper and cartons	12 × 2½ × 2½	
8. Food service—4 trays, service and utility	17 × 13	
9. Electric appliances—suggested 4 pieces		
Toaster		
Waffle iron		
Mixer		
Coffee maker		
10. Optional		
Cannister sets for holding supplies		
Scoops for flour, sugar		
Dredge for flour, sugar		
Utility pan for vegetable washing		
Refrigerator storage dishes		
Pyrex platter for fish cooking		
Ring molds		
Individual molds for salad and desserts		
Steamer molds for bread and puddings		
Thermometers—meat, candy, oven, refrigerator, room		
Turkey roaster		
Cookie press		
Pastry tube		
Clock and timers		
Picnic supplies		
Lunch box supplies		
Outdoor cooking supplies		
Thermos bottles		
Ice cream scoop		
11. Tools, household		
1 pliers		
1 hammer		
1 screwdriver, large		
1 screwdriver, small		
12. Cart on wheels		

BEDROOMS

SPATIAL CHARACTERISTICS AND ARRANGEMENT

Requirement

Each dwelling unit should have space(s) allocated to sleeping and such related activities as dressing and personal care, and study or reading. Sufficient space should also be provided for clothes storage and housekeeping in the bedroom areas.

Criterion

Although the required area in a primary bedroom varies depending upon the given room layout, adequate space should be provided to permit comfortable use of essential furniture and circulation as follows:

Two twin beds, 3 ft 3 in by 6 ft 10 in
One dresser, 1 ft 6 in by 4 ft 4 in
One chair, 1 ft 6 in by 1 ft 6 in
One crib, 2 ft 6 in by 4 ft 6 in
One table, 1 ft 6 in by 2 ft 6 in for sewing or other work

A secondary double-occupancy bedroom should have space to facilitate circulation and accommodate the following sized furniture:

Two twin beds, 3 ft 3 in by 6 ft 10 in, or one double bed, 4 ft 6 in by 6 ft 10 in
One or two dressers, 1 ft 6 in by 3 ft 6 in
One chair, 1 ft 6 in by 1 ft 6 in
One desk, 1 ft 8 in or 3 ft 6 in, or storage chest for toys

A secondary single-occupancy bedroom should have space to facilitate circulation and accommodate the following sized furniture:

One twin bed, 3 ft 3 in by 6 ft 10 in
One dresser, 1 ft 6 in by 3 ft 6 in
One chair, 1 ft 6 in by 1 ft 6 in
One desk, 1 ft 8 in by 3 ft 6 in

Dormitory-type (extra large) sleeping spaces to serve three or four persons may be provided in living units containing at least three other bedrooms. The design and area should accommodate the required number of beds (not bunks) and other necessary furniture and have appropriate closet space.

Each bedroom should have proper provisions for natural light and ventilation.

Visual and auditory privacy is needed for all bedrooms.

Commentary

1. For reasonable access to and use of bedroom furniture and equipment, the following minimum clearances should be observed:

42 in at one side or foot of bed, for dressing
12-in clearance for least-used side of double bed or pair of twin beds. The least-used side of a single bed may be placed against wall

6-in clearance from side of bed to side to dresser or chest
36-in clearance in front of dresser, closet, or chest of drawers
24-in clearance for major circulation path (door to closet, etc.)
22-in clearance on one side of twin bed

2. It should not be necessary to move double beds in order to make them up.

3. Bedrooms should be of sufficient size to permit an alternate arrangement of furniture if at all possible.

4. Primary bedrooms should have at least one uninterrupted wall space of at least 10 ft. There should also be space provided for private working or resting separate from dressing spaces.

Location

1. The bedroom or sleeping area of the living unit should be located away from the living and working areas for privacy.

2. In the analysis of the bedroom area, there should be a regard for the makeup of the family. With preschool children, it is convenient if the master bedroom is located close to children's bedrooms. With teenage children, separation of the master bedroom from other bedrooms wherever possible to reduce noise is desirable.

3. The location of doors, windows, and closets should be planned to allow the best placement of the bed and other furniture.

4. Placement of the closet so it is next to the door into the bedroom minimizes the use of wall space.

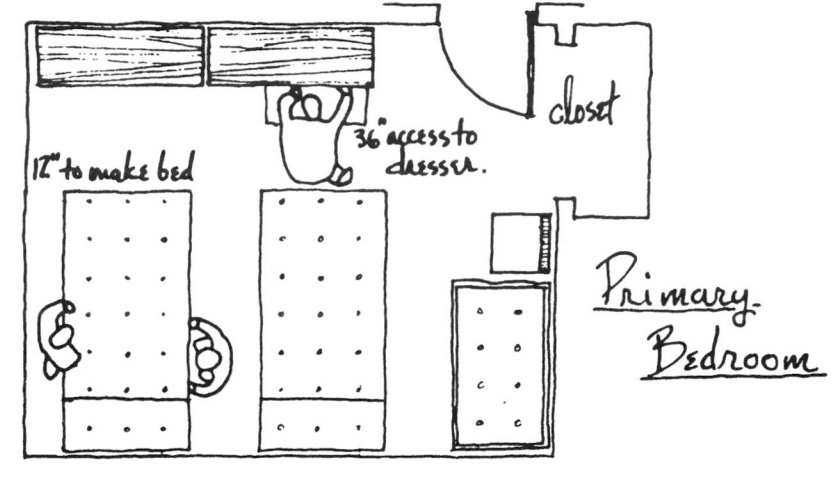

Fig. 1   Minimum clearances in primary bedrooms.

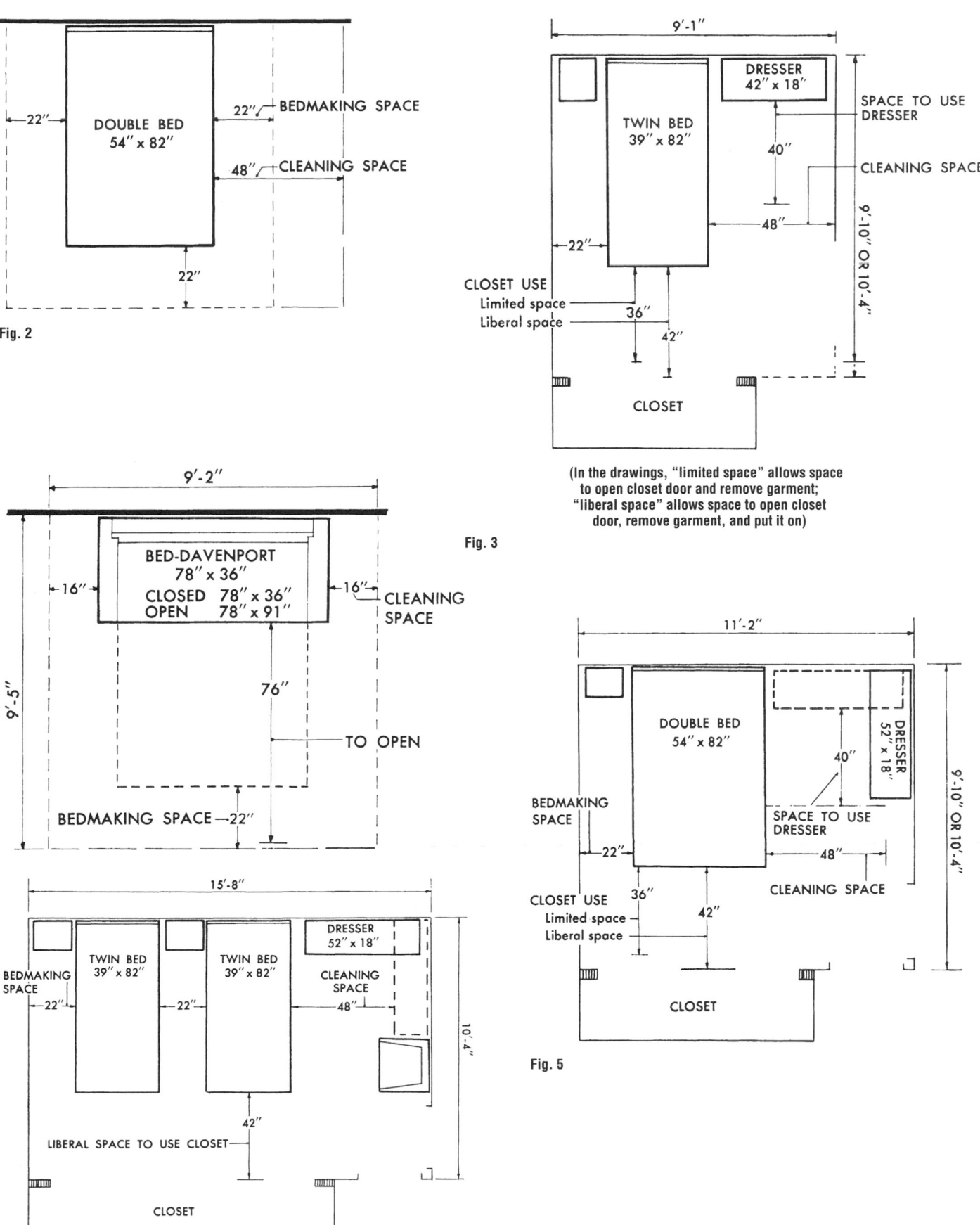

Fig. 2

DOUBLE BED 54" x 82"
22" — BEDMAKING SPACE
22"
48" — CLEANING SPACE
22"

Fig. 3

9'-1"
DRESSER 42" x 18"
SPACE TO USE DRESSER
TWIN BED 39" x 82"
40"
CLEANING SPACE
48"
9'-10" OR 10'-4"
22"
CLOSET USE
Limited space
Liberal space
36"
42"
CLOSET

(In the drawings, "limited space" allows space to open closet door and remove garment; "liberal space" allows space to open closet door, remove garment, and put it on)

9'-2"
BED-DAVENPORT 78" x 36"
CLOSED   78" x 36"
OPEN   78" x 91"
16"
16" — CLEANING SPACE
9'-5"
76"
TO OPEN
BEDMAKING SPACE → 22"

15'-8"
BEDMAKING SPACE
22"
TWIN BED 39" x 82"
22"
TWIN BED 39" x 82"
DRESSER 52" x 18"
CLEANING SPACE
48"
10'-4"
42"
LIBERAL SPACE TO USE CLOSET
CLOSET

Fig. 4

11'-2"
DOUBLE BED 54" x 82"
DRESSER 52" x 18"
40"
BEDMAKING SPACE
22"
SPACE TO USE DRESSER
48"
9'-10" OR 10'-4"
CLEANING SPACE
CLOSET USE
Limited space
Liberal space
36"
42"
CLOSET

Fig. 5

FURNITURE CLEARANCES

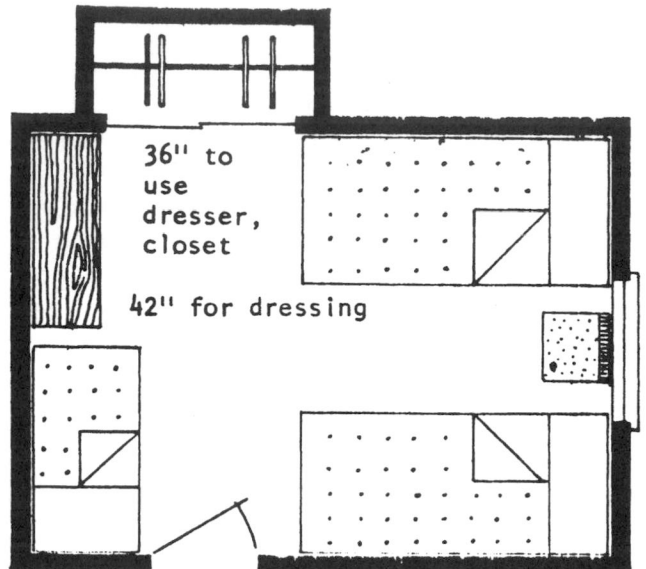

36" to use dresser, closet

42" for dressing

Fig. 6a   Primary bedroom.

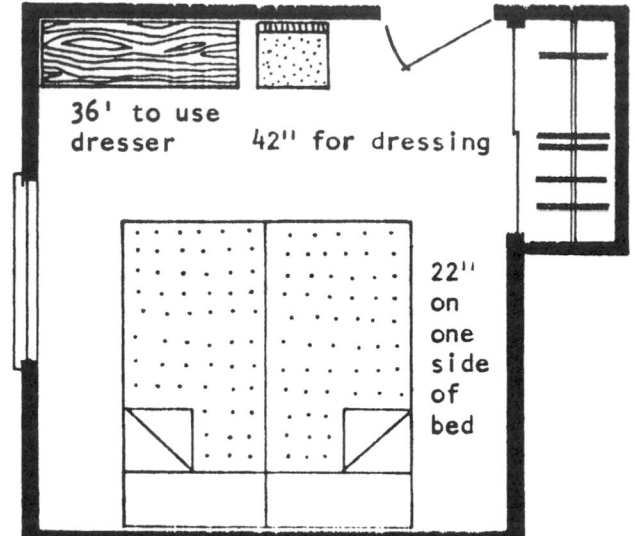

36' to use dresser   42" for dressing

22" on one side of bed

Fig. 6b   Primary bedroom without crib.

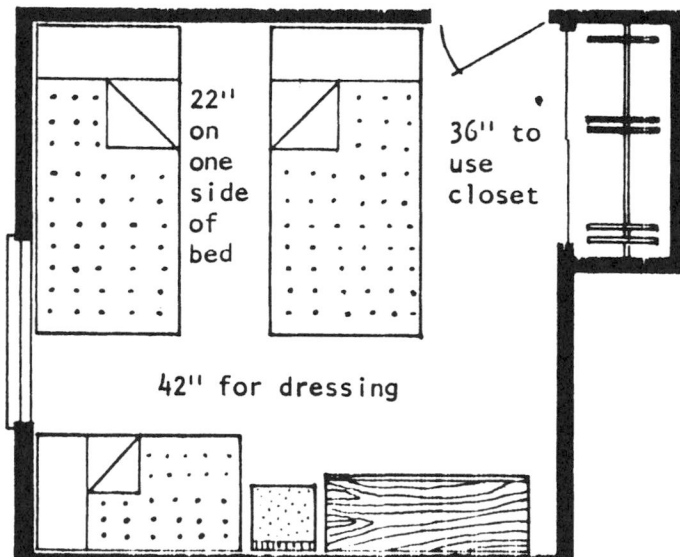

22" on one side of bed

36" to use closet

42" for dressing

Fig. 6c   Primary bedroom.

The location of doors and windows should permit alternate furniture arrangements.

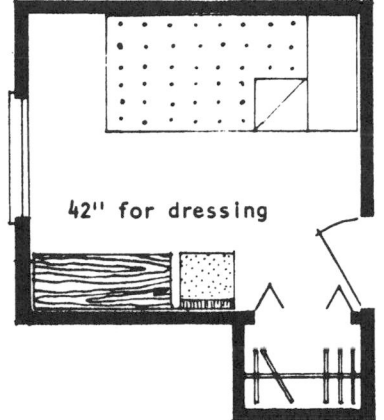

42" for dressing

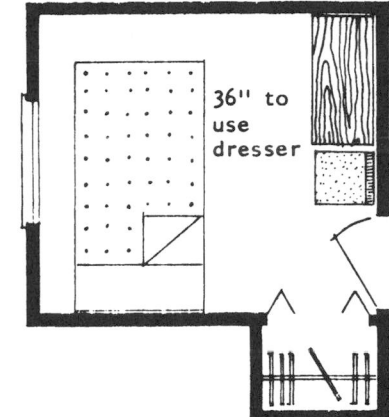

36" to use dresser

Fig. 7   Single-occupancy bedroom.

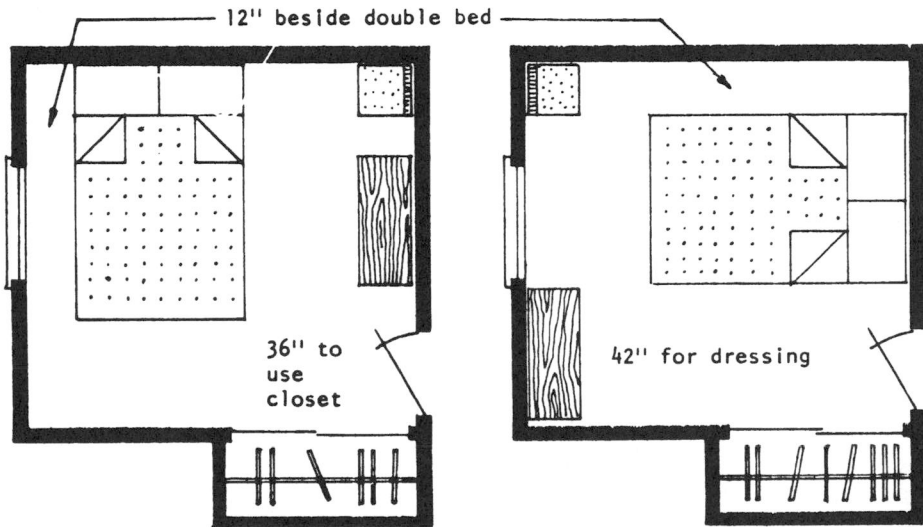

12" beside double bed

36" to use closet

42" for dressing

Fig. 8   Double-occupancy bedroom.

Where at least two other sleeping spaces are provided, a dormitory is sometimes preferred by larger families.

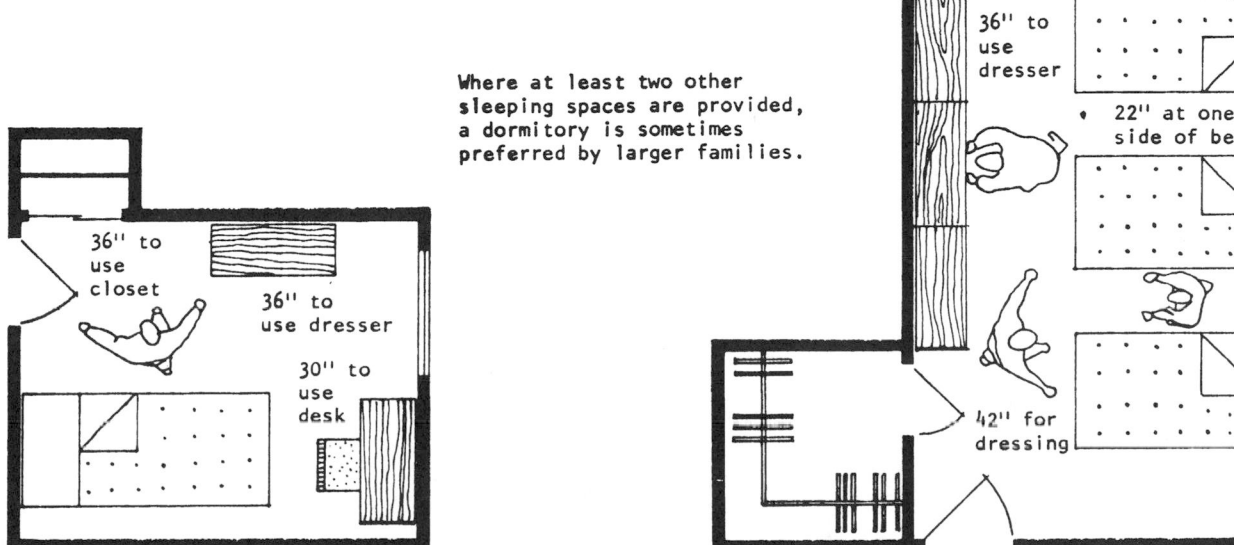

36" to use closet

36" to use dresser

30" to use desk

36" to use dresser

22" at one side of bed

42" for dressing

Fig. 9   Single-occupancy bedroom for elderly.

Fig. 10   Dormitory bedroom.

BEDROOMS

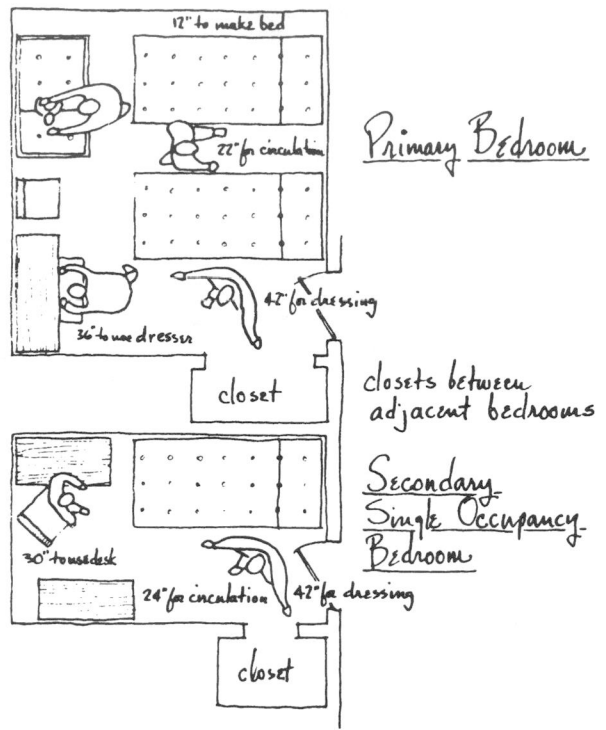

Fig. 11 Minimum clearances in primary and secondary bedrooms.

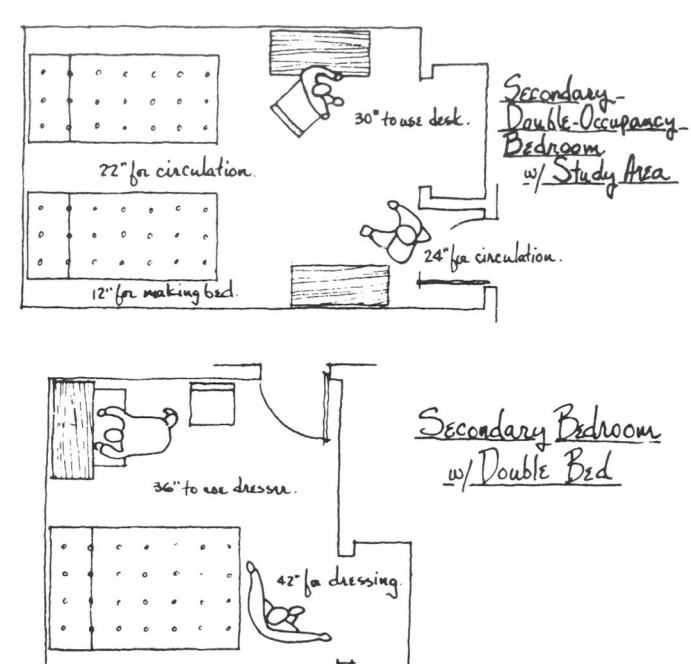

Fig. 12 Minimum clearances in secondary bedrooms.

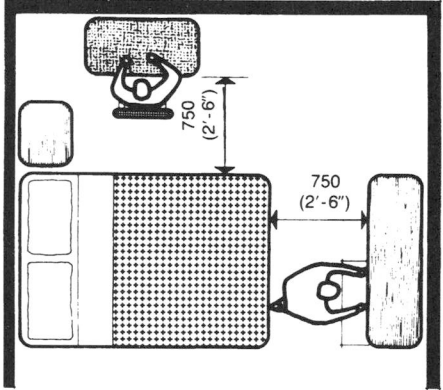

Fig. 13a Access between a bed and furniture.

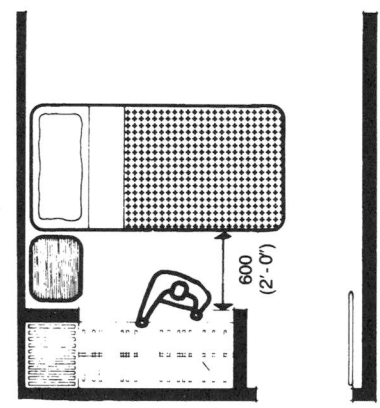

Fig. 13b Access between a bed and closet.

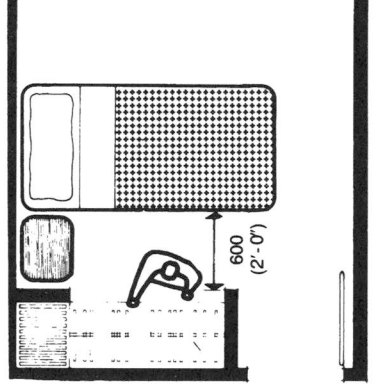

Fig. 13c Bed-making space.

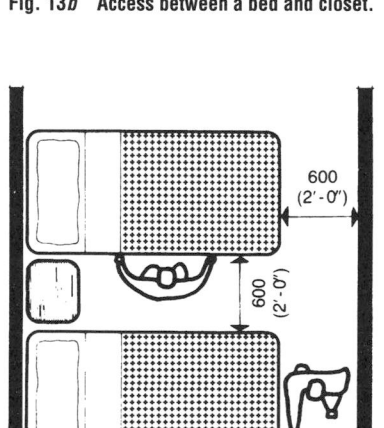

Fig. 13d Access between a bed and wall and access between beds.

TABLE 1	Bedroom
Access between a bed and furniture	750 mm (2 ft 6 in)
Access between a bed and closet	600 mm (2 ft 0 in)
Bed-making space	400 mm (1 ft 3 in)
Access between a bed and wall	600 mm (2 ft 0 in)
Space between beds	600 mm (2 ft 0 in)

ROOM PLANS

The room plans presented in this section are designed to illustrate various techniques for providing choices in the use of living space in the home. It must be remembered that a reduction in room size inevitably limits the options for alternative arrangements of furniture; but it should always be possible to use more than one layout in any room designed to the minimum requirements of residential standards.

The Bedroom

The bedrooms illustrated show the approximate size and shape of rooms that will permit at least three variations in furniture layouts. In each case, the location of the closet, door, hot air heating register, and window should be noted.

Parents' bedroom This bedroom has an approximate size of 3200 by 3350 mm (10 ft 6 in by 11 ft 0 in), i.e., 10.7 m² (115.5 ft²). The size of this bedroom will only just permit the use of twin beds if placed together.

One-person bedroom The bedroom in Fig. 15 has an approximate size of 2800 by 2900 mm (9 ft 3 in by 9 ft 6 in), i.e., 8.1 m² (87.9 ft²). The window location in the corner of the room is acceptable for this kind of layout. If required, the window could be moved to the center of the room in arrangement C.

This would mean that it would partially overlap the bed. This size and shape of room permits a variety of layouts without interfering with the closet, door, or heating register.

Two-person bedroom The bedroom in Fig. 16 has an approximate size of 3200 by 4400 mm (10 ft 6 in by 14 ft 6 in), i.e., 14.0 m² (152.2 ft²).

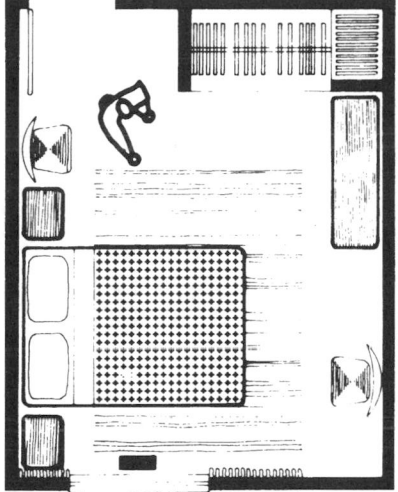
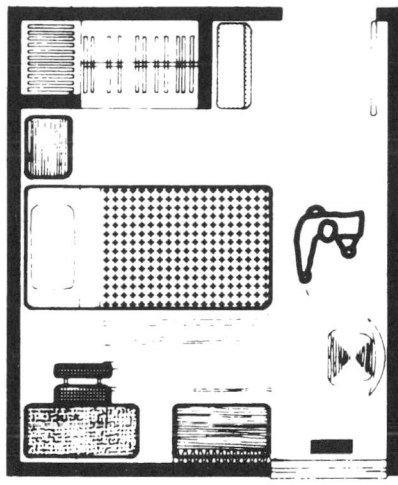

Fig. 14*a* This layout provides good circulation and dressing space.

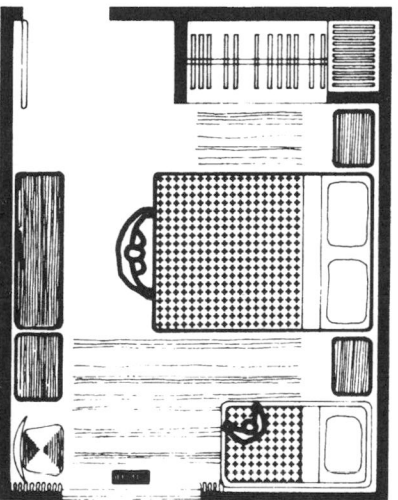
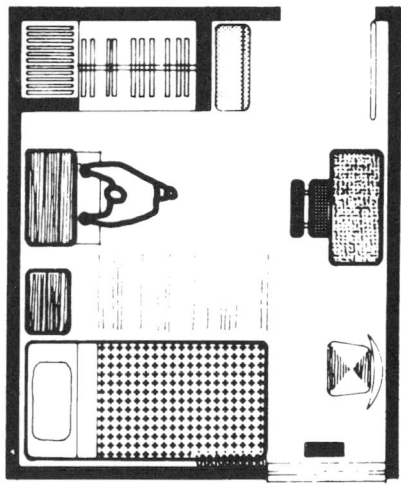

Fig. 14*b* This arrangement can accommodate a crib and extra pieces of furniture.

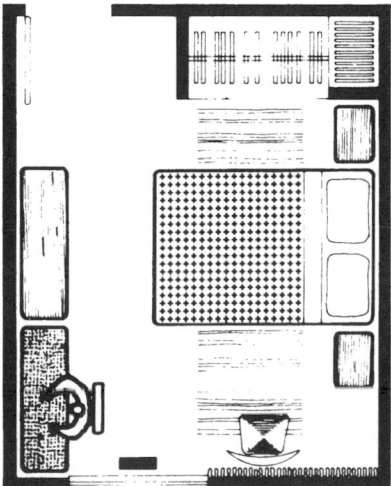
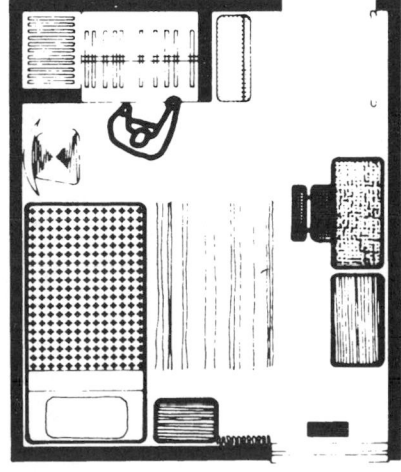

Fig. 14*c* Occasionally a small writing desk is required in this bedroom. Another piece of furniture, such as an armchair, could be substituted for the desk.

BEDROOMS

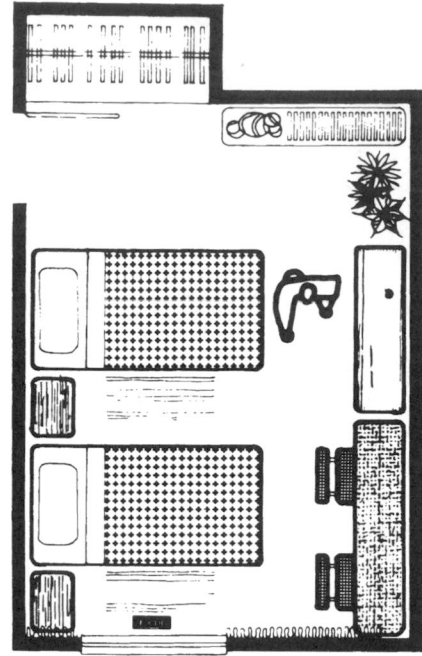

Fig. 15a This is a conventional arrangement and allows for one large extra piece of furniture, such as a bookcase.

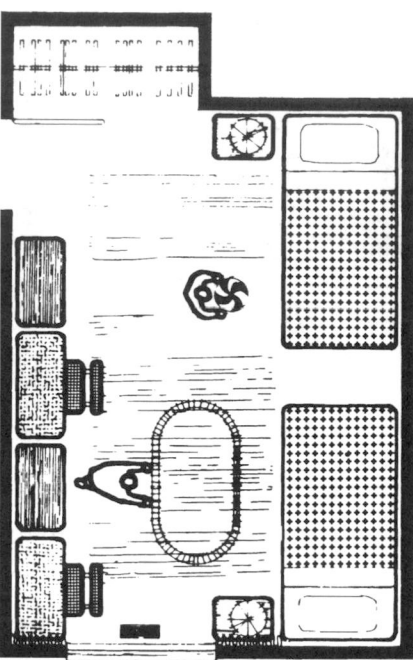

Fig. 15b Placing beds against the wall gives children a maximum amount of floor space for playing with toys.

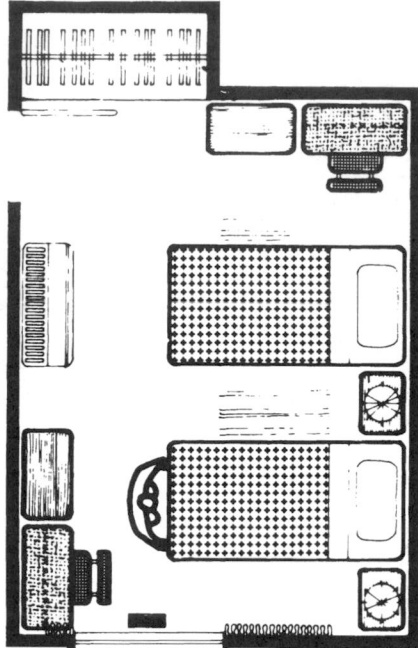

Fig. 15c This layout permits each occupant to control his or her own individual space and is therefore particularly desirable for older children.

TABLE 2   Sizes of Bedroom Furniture

Beds	
Single	3 ft × 6 ft 6 in
Twin	3 ft 3 in × 6 ft 6 in
Three-quarter	4 ft × 6 ft 6 in
Double	4 ft 6 in × 6 ft 6 in
Bed tables	1 to 2 ft square
Dressers	1 ft 6 in to 2 ft deep × 3 to 4 ft wide
Chests	1 ft 6 in to 2 ft deep × 2 ft 6 in to 4 ft wide
Dressing tables	1 ft 6 in to 1 ft 10 in deep × 3 to 4 ft wide
Bench	1 ft 6 in × 2 ft
Chairs	
Boudoir	2 ft 6 in × 2 ft 6 in
Side	1 ft 6 in × 1 ft 6 in

**TYPICAL BEDROOM
ARRANGEMENTS**

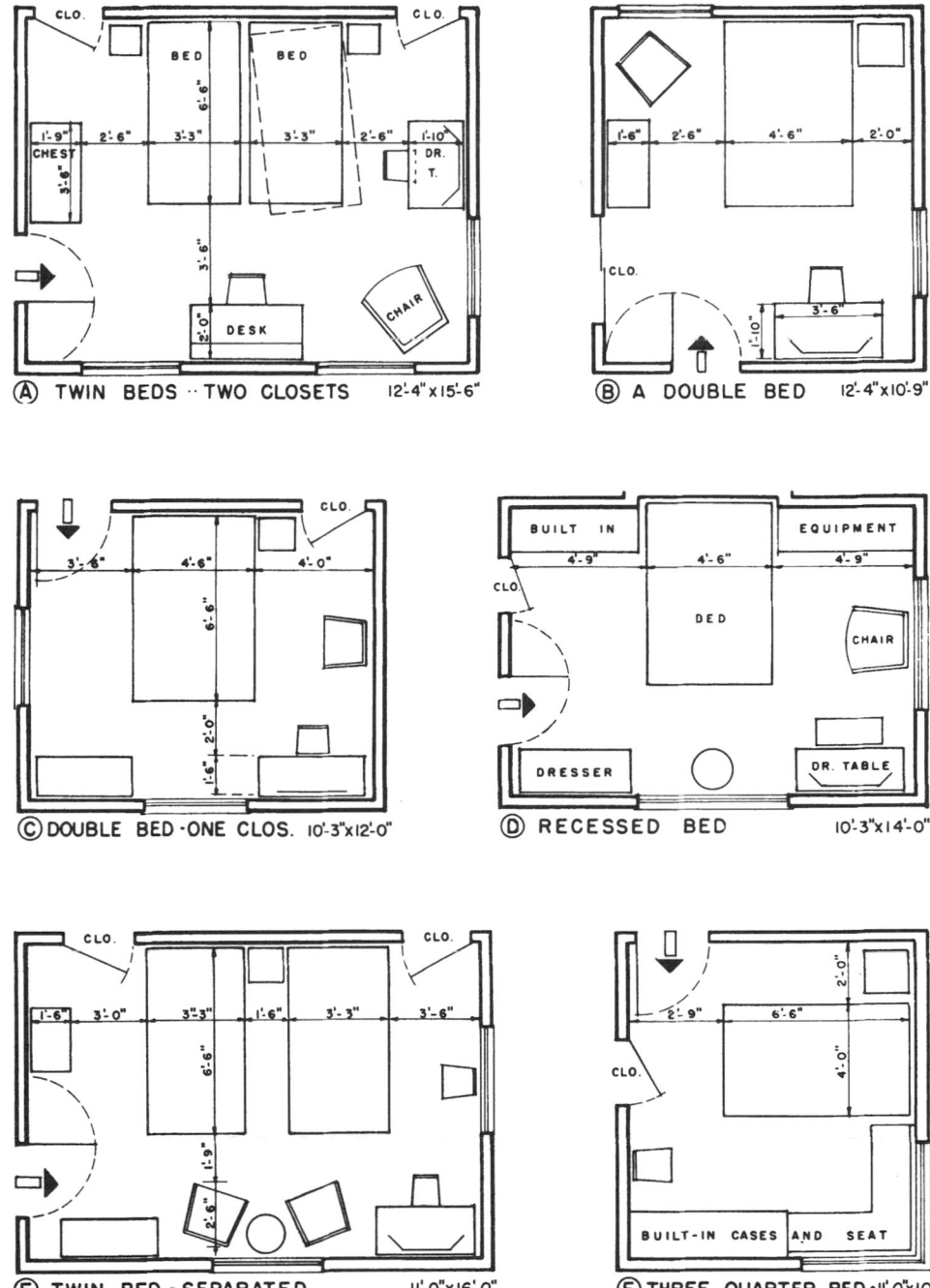

Fig. 16   Various ways in which bedrooms may be designed around different pieces of furniture, or furniture may be related to walls, doors, and windows.

BEDROOMS

THE MASTER SUITE

The master suite's role within a contemporary home is one of isolated retreat for the heads of household. Although we generally think of the master suite as having two spaces—bedroom and bathroom—the ideal configuration of a well-defined master suite contains five distinct areas: sleeping, sitting, dressing, closet, and bath. A careful balance of function, perceived value, and emotion dictates their relative sizes and relationship to one another (Fig. 17).

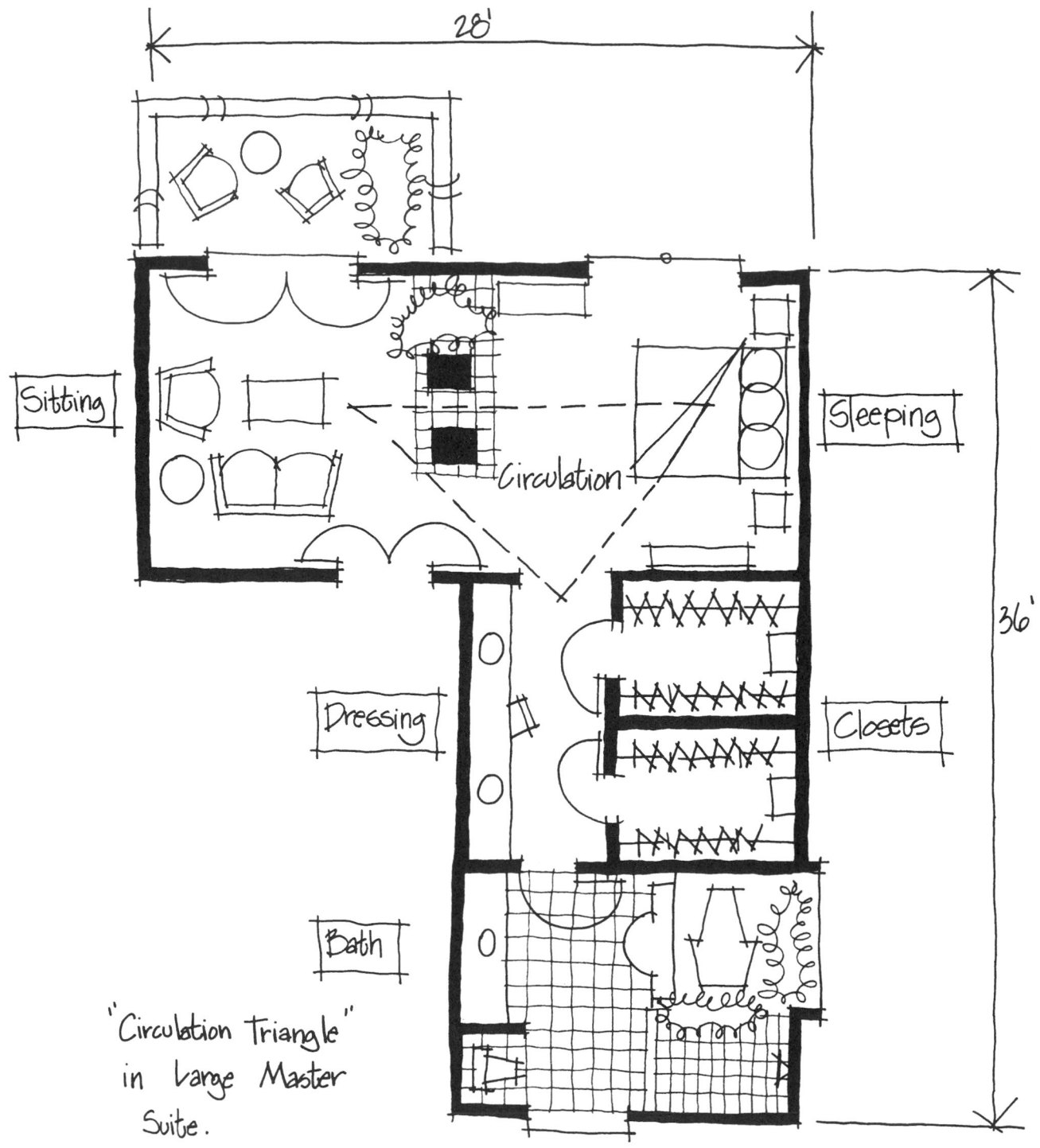

Fig. 17 Typical master suite.

Sleeping Area

Of the five areas in the master suite, the most hours are spent in the sleeping area, but the location of the sleeping area may be subordinate to view considerations for the sitting area. The sleeping area may be one area of the home where darkness is an asset.

The sleeping area should accommodate a king-size bed with 30 in of clearance around the perimeter. The bedroom configuration should be designed to accommodate a bed on at least one wall, with an option for an alternative location if possible. If the bed wall is an exterior wall, window placement should recognize headboard location. Besides being aesthetically undesirable, a window blocked by the headboard or other bed framing is also unreachable for operation, which diminishes comfort and privacy (Fig. 18).

If the bed is to be located along an interior wall, use of the adjoining space must be considered. As with any bedroom, the master suite should not share common walls with such active, noisy rooms as the laundry or bath.

Although the bed is the major furnishing to be accommodated, additional standard items to consider when dimensions are established include two night stands, one for either side of the bed, and two dressers. A location for a television should also be anticipated, one which accommodates viewing both from the reclining position of the bed and the sitting area (Fig. 19).

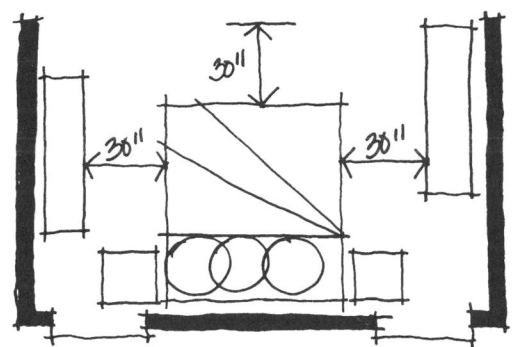

Fig. 18   Sleeping area clearance.

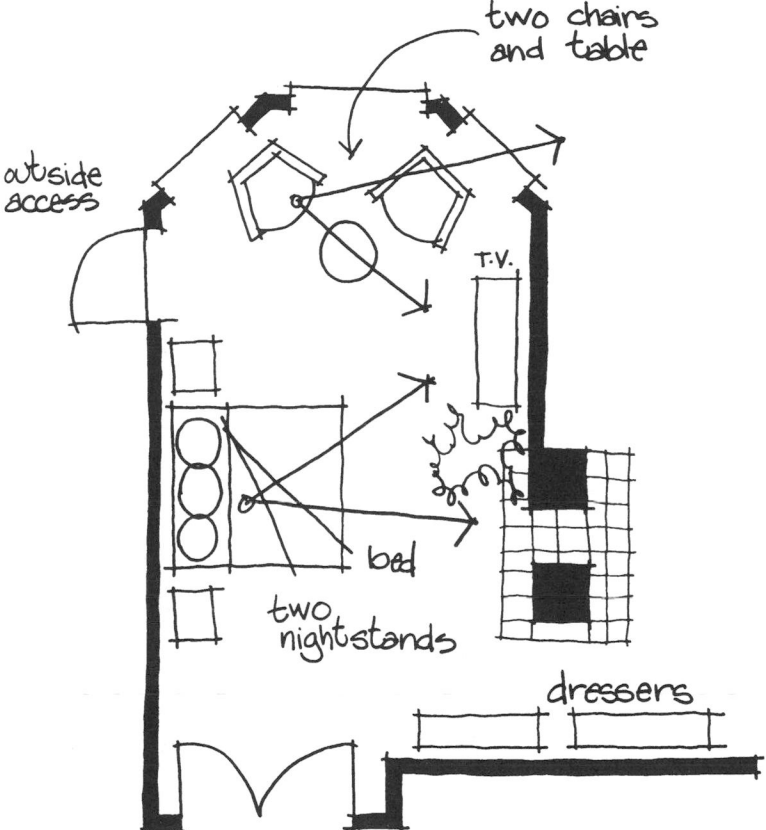

Fig. 19   Master suite with furniture placement.

BEDROOMS

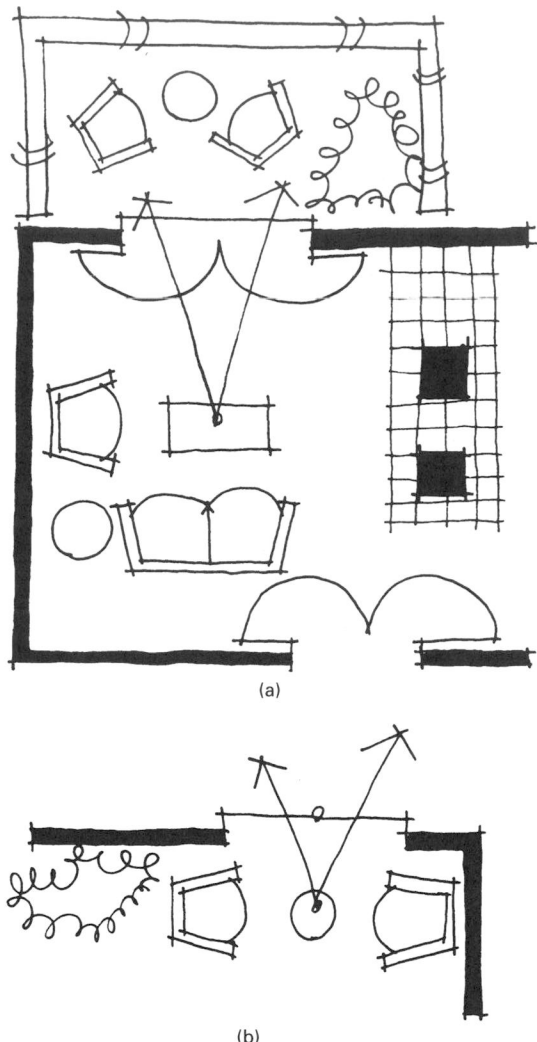

(a)

(b)

Fig. 20 (*a*) Large sitting area; (*b*) small sitting area.

Sitting Area

The sitting area is the other space that should be provided for in the bedroom portion of the master suite. Even when seldom used, a sitting area expands the master bedroom, adding graciousness to the entire suite. In smaller master suites the sitting area may simply be a location for two chairs, while in larger suites it may be a well-defined niche or even a separate room. In the latter case, the sitting area should accommodate at least several chairs and possibly a sofa for reading, relaxing, or conversing (Fig. 20).

The sitting area should receive primary consideration for views, access to a balcony or patio, and possibly a fireplace. Natural light is a vital ingredient of the design here, and since pleasant views will most often be enjoyed while one is sitting or lounging, lower sills on these windows are desirable (Fig. 21).

In some cases, the master suite may be designed with a sitting room that can function easily as a separate library or guest bedroom. If this room has direct access to both the master suite and hall, this option can be easily exercised by the homebuyer before or after purchase (Fig. 22).

Fig. 21 Windows that allow view while seated.

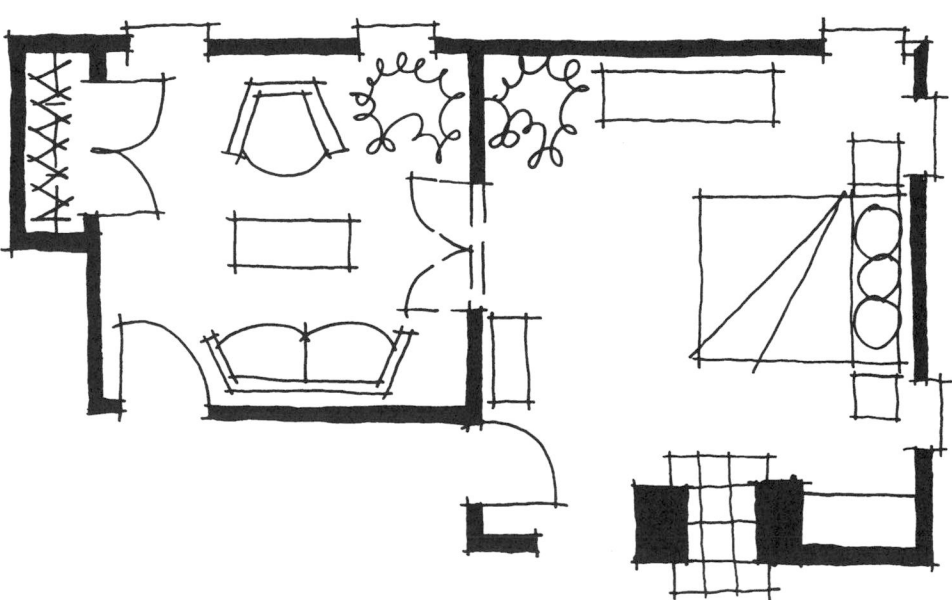

Fig. 22 Sitting room can double as extra bedroom.

Sitting alcoves can be separated from the sleeping area by discreet room dividers. Two-sided fireplaces are increasingly popular, along with partial-height walls, planters, and steps. Changes in ceiling height or wall materials can also reinforce the definition of this alcove (Fig. 23).

A view of the sitting area at the entry to the master suite makes an appealing first impression that may include a picturesque balcony or fireplace. Alternatively, master suite layouts should keep the sleeping area out of immediate view (Fig. 24).

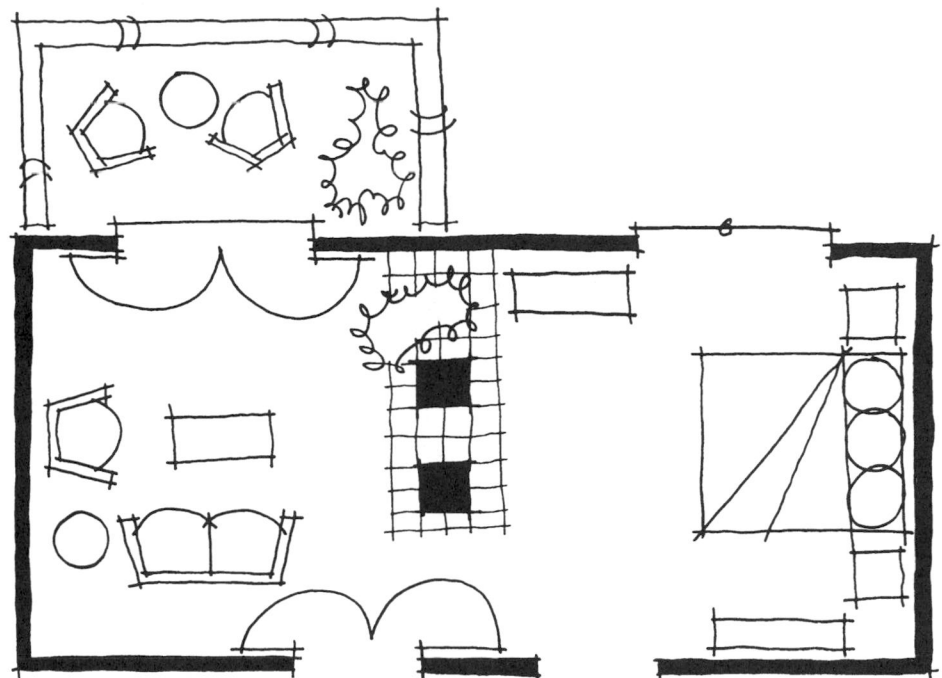

Fig. 23   Sitting area and sleeping area separated by "through" fireplace.

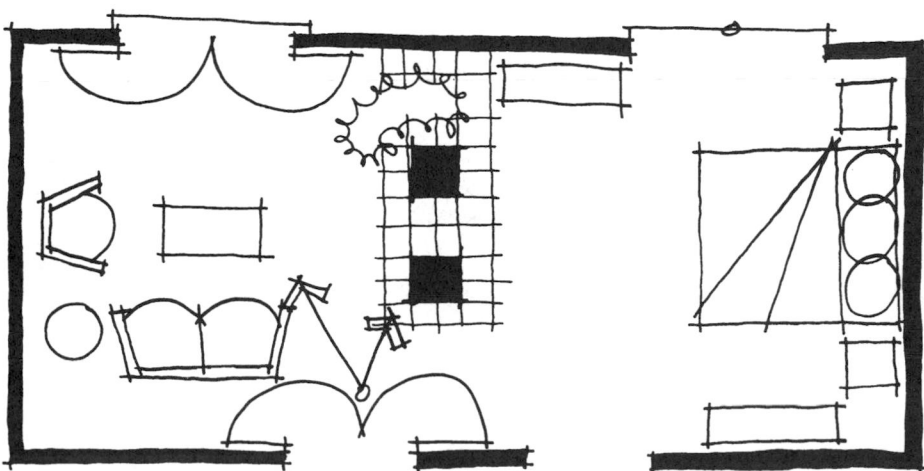

Fig. 24   Initial view impact in master suite.

BEDROOMS

Closet

Tucked out of sight from the rest of the suite, but of critical importance, is the closet. Small or poorly planned closets in the master suite can be the kiss of death for a sales program. If closets do not correspond to the spacious dimensions of a large master suite, they can make the entire suite seem poorly designed.

In the master suite, *walk-in closets* are preferred hands down. Linear closets should be used only if square footage is truly at a premium. In larger models, separate closets are highly marketable, reflecting the natural desire of people to have their own space.

Closet dimensions should reflect the true maximum configuration for storage of clothes. Rod and shelf length is the measurement of performance. Probably the optimal shape is the rectangle, with rods and shelves on both sides and circulation down the middle. The addition of a third perpendicular rod and shelf may often look good on paper, but in reality, may not produce any additional storage. With this configuration, the closet door should swing out, so as not to block any of the interior rods or shelves (Fig. 25).

The minimum width of a double-loaded walk-in closet is 6 ft. This allows for a rod and shelf along both sides with a minimum aisle space between. Wishful thinking on paper may seek to reduce this, but a reduction will create difficult accessibility. If the 6-ft width cannot be achieved, an L-shaped rod and shelf configuration is the next most desirable. In this case the door can swing in toward the interior blank wall of the closet (Fig. 26).

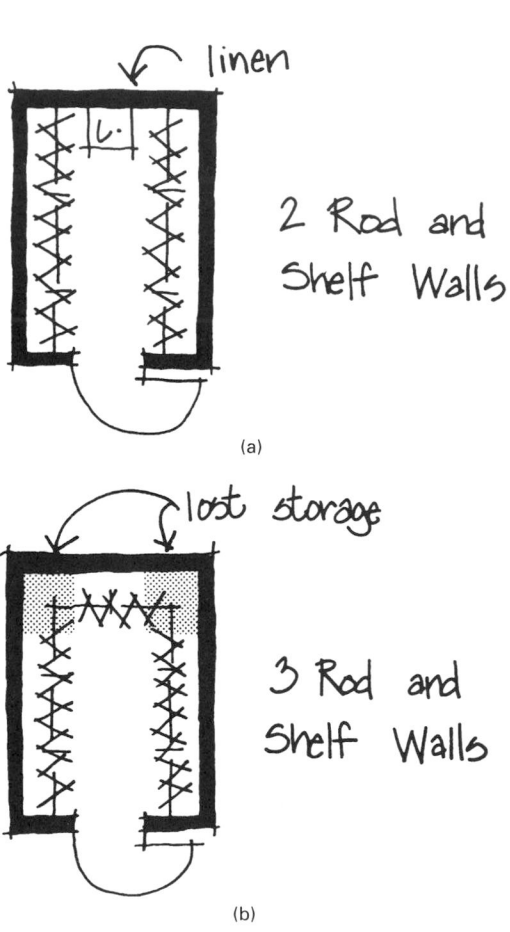

Fig. 25   (*a*) Two-rod walk-in closet vs. (*b*) three-rod walk-in closet.

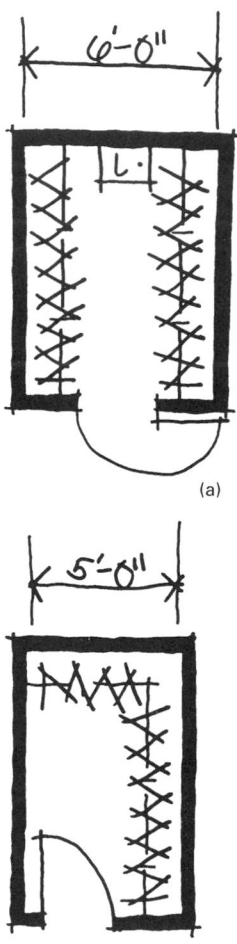

Fig. 26   (*a*) Six-foot walk-in closet vs. (*b*) five-foot L-shaped closet.

Dressing Area

Adjacent to the closet and the bath, but slightly out of view from the balance of the suite, is the dressing area. This is the locale for selection and donning of clothing, for primping and preparing to meet the world.

The focus of the dressing area is the *vanity*. This may include one or two washbowls, preferably with an integral place to sit. A large mirror should be located above the vanity, and the entire space should incorporate generous lighting. Natural light is a marketing plus in the dressing area, with skylights as a popular option.

Vanity lengths of 6 ft are desirable as a minimum. A shorter vanity should include only one washbowl, as two will reduce valuable counter space. To accommodate two washbowls and seating, the vanity should include 8 to 12 linear feet. Drawers below and medicine cabinets at either side wall of the vanity will provide for necessary storage of toiletries (Fig. 27).

The dressing area should have a generous *linen closet*, either under the vanity or along a wall near the closet. In smaller homes, the linen closet may be located within the closet itself. In larger homes, separate linen closets for each person are emerging as popular extras for the master suite (Fig. 30).

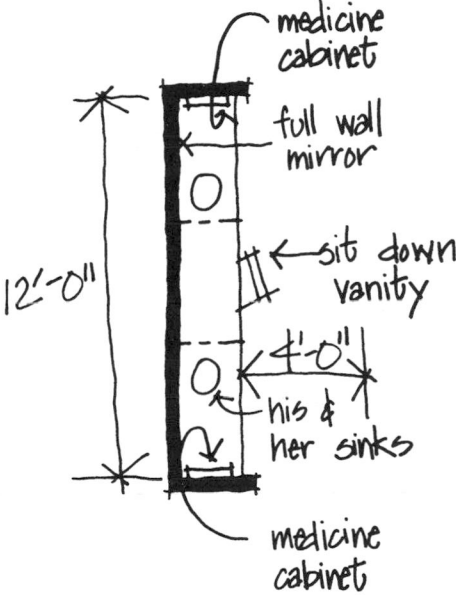

Fig. 27   Vanity features, sizes, and clearances.

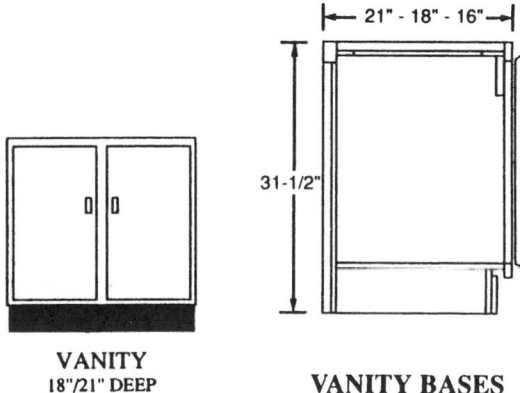

VANITY
18"/21" DEEP

VANITY BASES

Fig. 28

Vanity Bowl Unit 16" Mini Depth
• 1 full height door

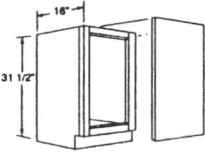

Vanity Bowl Unit 16" Mini Depth
• 2 full height doors

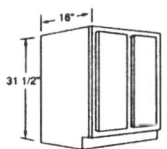

Vanity Bowl Unit 18" Space Saver Depth
• 2 full height doors

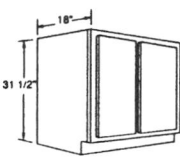

Vanity Bowl Unit 21" Standard Depth
• Full height door

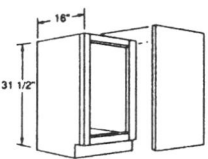

Vanity Bowl Unit 21" Standard Depth
• 2 full height doors

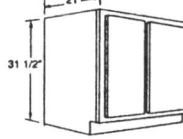

Fig. 29

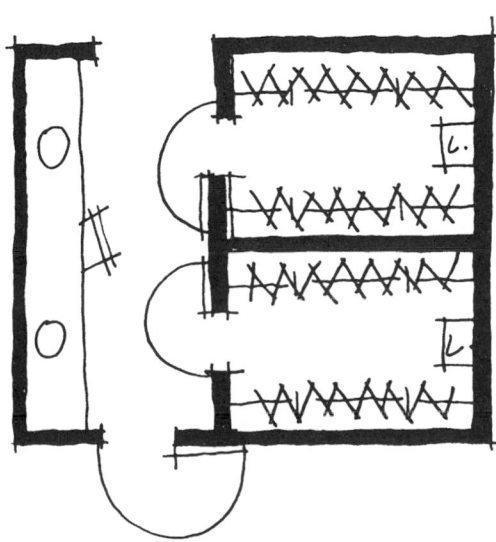

Fig. 30   His and her walk-in closets and sinks.

BEDROOMS

Bath Area

The celebrated master bath in contemporary homes is on a par with the kitchen relative to design significance. As with the kitchen, function is important, but fantasy and the desire for luxury are even more significant. Regardless of its location within overall layout, the master bath has become the centerpiece of a typical master suite bathroom.

Current emphasis on the master bath can be attributed to a variety of trends, including an increased concern for physical health and well-being. Public bathing has for centuries been regarded as a healthy and uplifting process as reflected in the great public baths of the Roman empire and in contemporary public baths in Japan. One sees remnants of these values in today's health spas and in the all-important *garden tub* of the master bathroom.

Indeed, the garden or oversized tub is the focal point of the master bath. Often sunken or elevated, the tub has become a domestic icon, framed with accent walls of decorative tile and illuminated by an adjacent window or skylight above. Sometimes called a spa tub, this fixture often includes optional water jets to create a whirlpool effect (Fig. 31).

Although the *shower* will probably be used as the facility for daily washing, the bath area is more than symbolic. A luxurious tub is highly desirable for soaking and relaxing, for daydreaming, and for the fantasy or reality of a romantic *tête à tête*.

Although in smaller bathrooms the garden tub must double as a shower, a common alternative is to separate the shower enclosure. Generally, the shower is used daily, and its design is evolving away from the standard 3- by 3-ft bay, with shower stalls increasing to 4 by 6 ft or larger, with generous glass doors and low curb steps for easy access. Because the garden tub is used less frequently, it is given locational preference as a display item (Fig. 32).

The third essential item in the master bath is the *toilet*. Although the garden tub is the centerpiece and the shower can easily be left on display, the toilet should be discreetly tucked away from view. If space permits, compartmentalizing the toilet provides for total privacy—universally appreciated in any market.

If the toilet cannot be compartmentalized, it should at least be screened by partial-height walls in a remote location within the bath, away from the entry view. If compartmentalization is possible, care should be taken to avoid making the enclosure claustrophobic. Here a small window for ventilation may be one of the home's most appreciated openings.

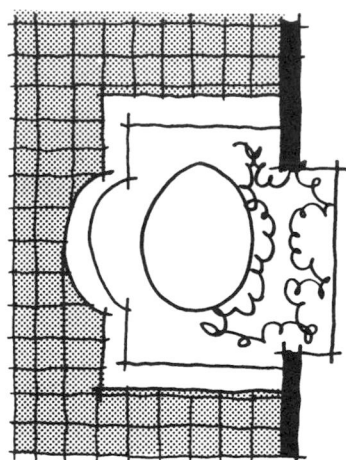

Fig. 31 Garden tub.

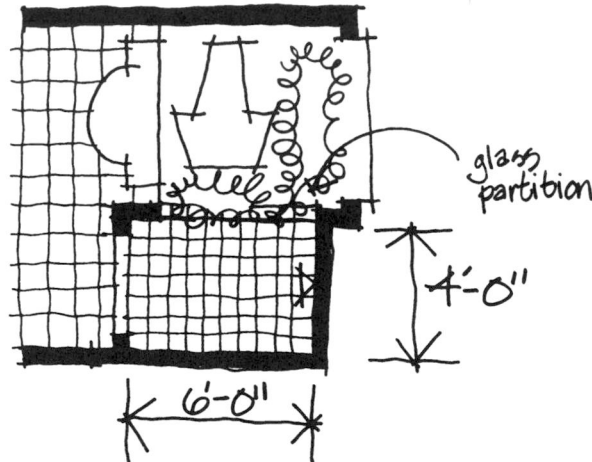

Fig. 32 Walk-in shower separate from garden tub.

Each dwelling unit should have a bathroom with enough area to accommodate a lavatory, a water closet, and a bathtub or shower. Arrangement for fixtures should provide for comfortable use of each fixture and permit at least a 90° door swing unless sliding doors are used.

The bathroom should be convenient to the bedroom zone, and accessible from the living and work areas. Linen storage should be accessible from the bathroom but not located within the bathroom.

Each complete bathroom should be provided with the following:

1. Grab bar and soap dish at bathtub
2. Toilet paper holder at water closet
3. Soap dish at lavatory (may be integral with lavatory)
4. Towel bar
5. Mirror and medicine cabinet or equivalent enclosed shelf space
6. In all cases where shower head is installed, a shower rod or shower door

Each half bath should be provided with items 2, 3, 4, and 5 as listed above.

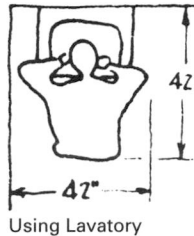

Using Lavatory

Drying Child at Tub

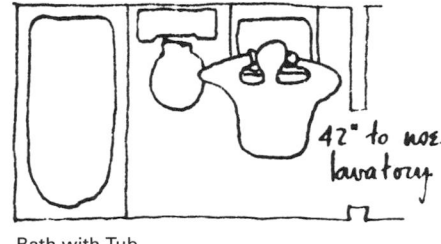

Bath with Tub

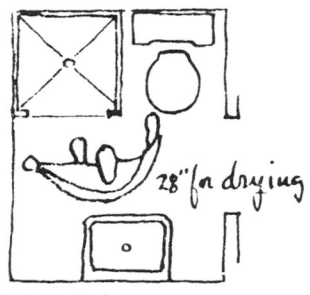

Bath with Shower

Bath with Water Closet
in Separate Compartment

Fig. 1   Minimum clearances for bathrooms.

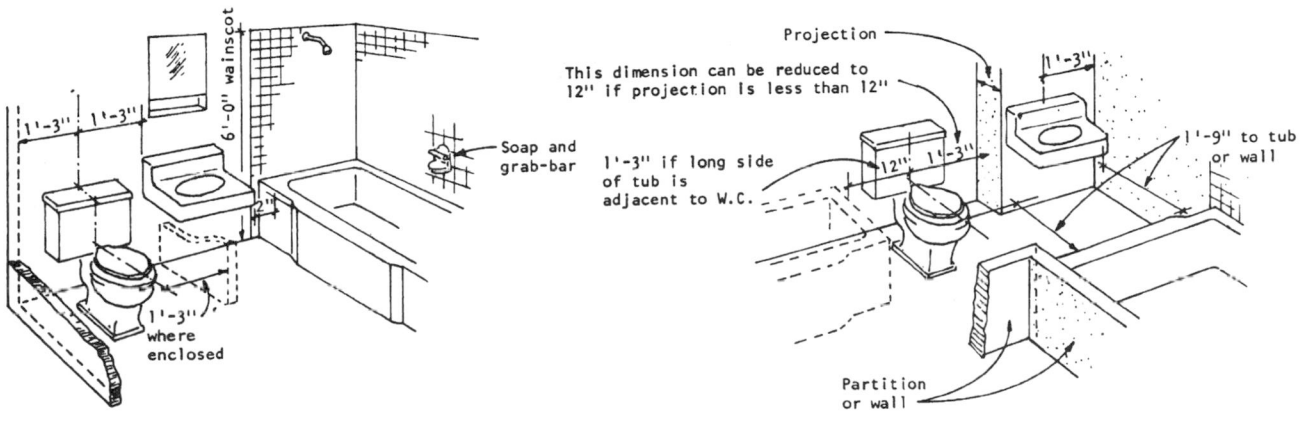

Fig. 2

BATHROOMS

CLEARANCES FOR BATHROOMS

The standard 5- by 7-ft bathroom found in many housing units is grossly inadequate in serving the sporadic heavy use and the variety of demands found in most families. Toilets should be separated from washing facilities.

For units designed to accommodate five people or more, a second toilet should be provided.

For dwellings on two levels, the second toilet ideally should not be on the same level as the bathroom.

5'-0"
5'-0"
tub
seat
W.C.
basin 1
basin 2
storage

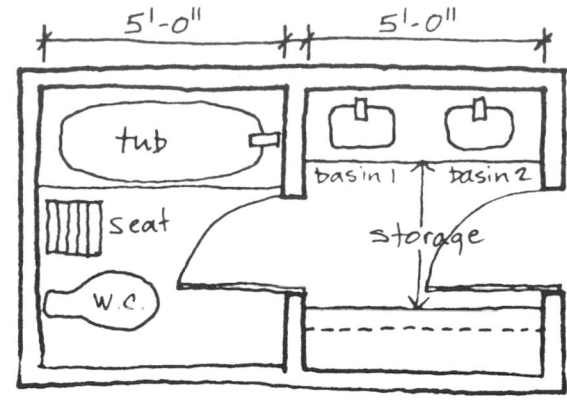

Fig. 3

BASIC BATHROOM LAYOUTS

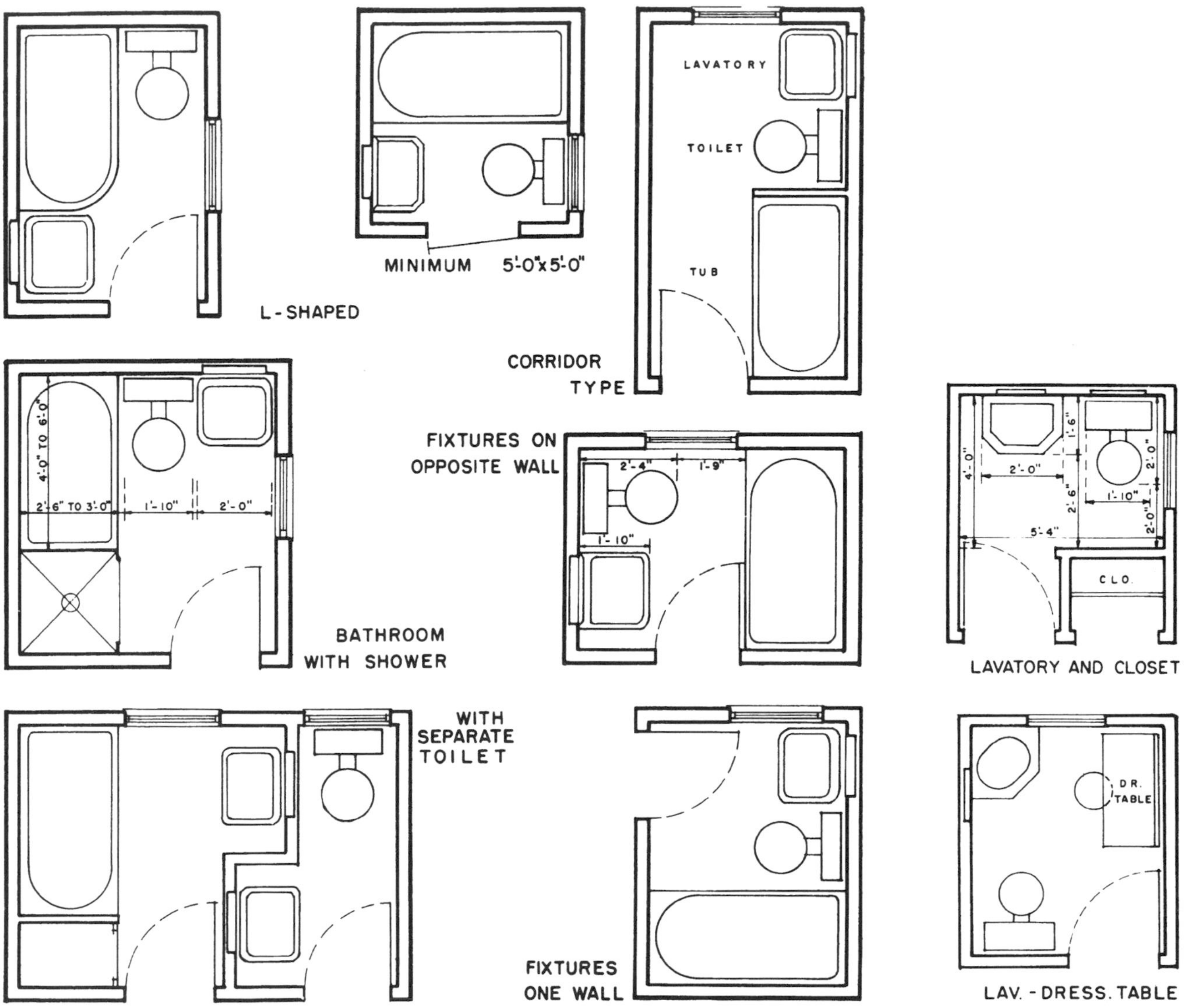

L-SHAPED
MINIMUM 5'-0"x5'-0"
LAVATORY
TOILET
TUB
CORRIDOR TYPE
4'-0" TO 6'-0"
2'-6" TO 3'-0"
1'-10"
2'-0"
BATHROOM WITH SHOWER
FIXTURES ON OPPOSITE WALL
2'-4"
1'-9"
1'-10"
4'-0"
1'-6"
2'-0"
2'-6"
1'-10"
2'-0"
5'-4"
CLO.
LAVATORY AND CLOSET
WITH SEPARATE TOILET
FIXTURES ONE WALL
DR. TABLE
LAV. - DRESS. TABLE

Fig. 4 Suggestions for arrangements of fixtures in bathrooms of varying sizes and shapes. Attention should be given to dimensions of fixtures and to the clear space in front of them.

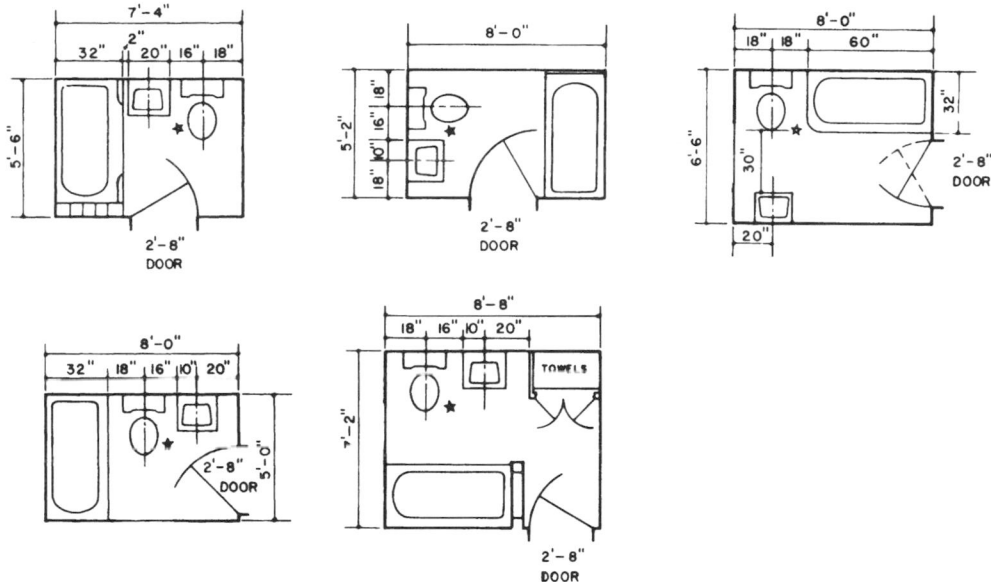

Fig. 5   Family bathrooms.

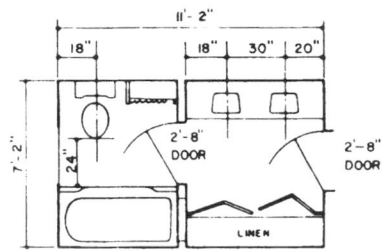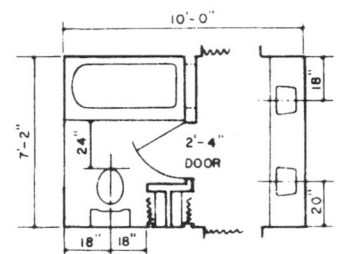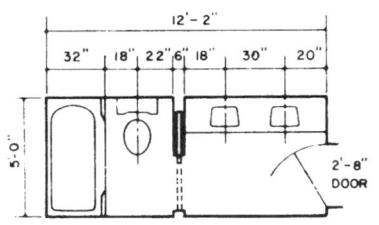

Fig. 6   Two lavatories, one tub, and one toilet.

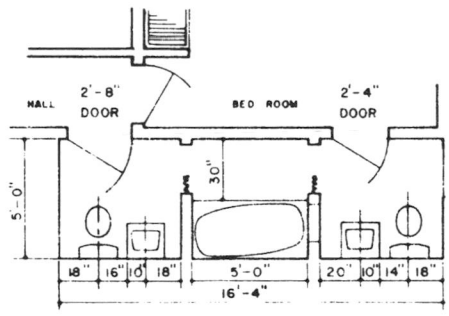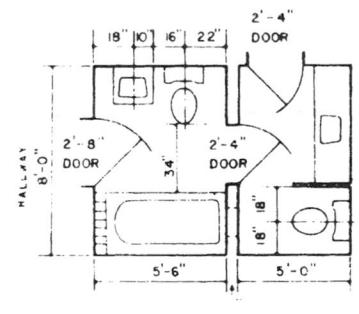

Fig. 7   Two lavatories, two toilets, and one tub.

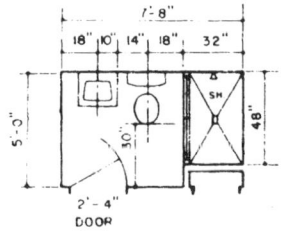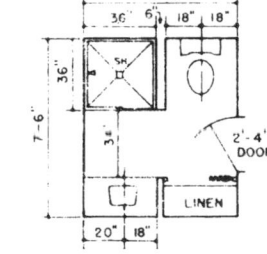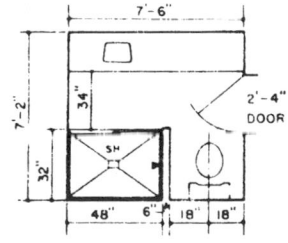

Fig. 8   Bathrooms with shower stalls.

BATHROOMS

POWDER ROOMS

Traditionally, the powder room has been extremely small, most often no larger than 3 by 6 ft. In most markets, a 5- by 5-ft space is better, as it can accommodate a vanity and toilet side by side. Popular current options in the powder room include pedestal vanities, or "banjo" vanity tops that partially cover the toilet (Fig. 9).

The powder room may be paired with a guest suite. When this is the case, design may include either a shower/tub combination or a shower, and the inclusion of a window and compartmentalization of the toilet become more important (Fig. 10).

Design of the powder room should also include careful attention to sound mitigation. Once inside the powder room, guests do not want to feel as if their every move can be heard through paper-thin walls. As with all bathroom locations, such buffer spaces as stairs and closets should surround the powder room to help muffle noise (Fig. 11).

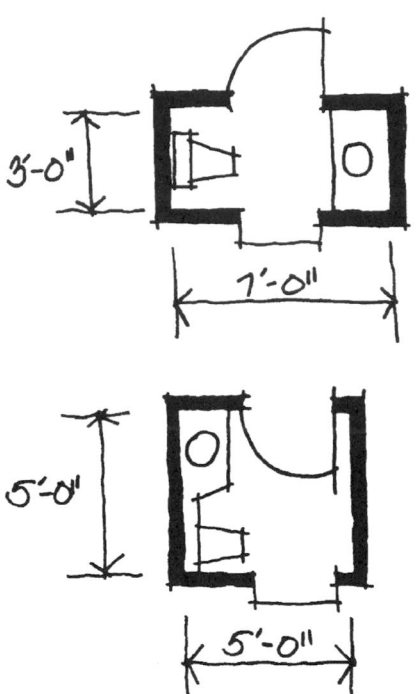

Fig. 9   Rectangular powder room vs. square powder room.

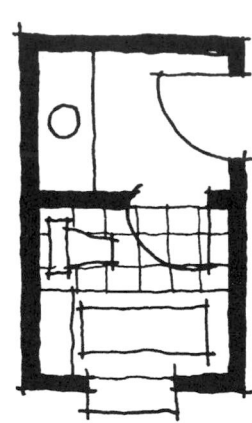

Fig. 10   Compartmentalized powder room with tub.

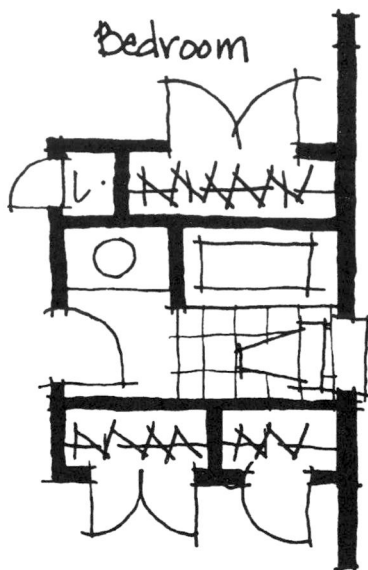

Fig. 11   Closets as noise buffers around bathroom.

LAVATORIES

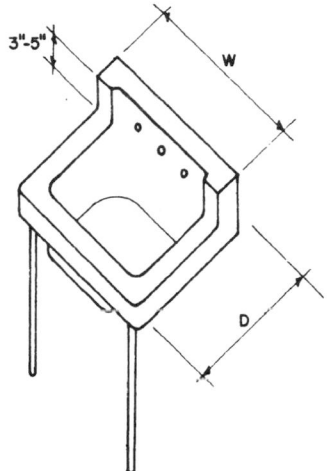

Fig. 1a Shelf-back lavatory.

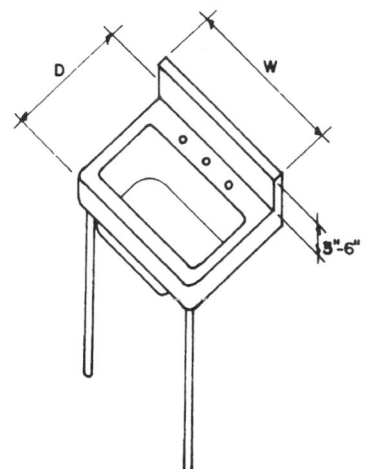

Fig. 1b Ledge-back lavatory.

Fig. 1c Plain-back lavatory.

TABLE 1a

Width, in	Depth, in	Material
18	14	Vitreous china
18	15	Vitreous china
19	17	Vitreous china, steel, cast iron
20	14	Vitreous china
20	18	Vitreous china, steel
22	14	Vitreous china
22	18	Vitreous china, steel, cast iron
24	20	Vitreous china
26	14	Vitreous china
26	22	Vitreous china

TABLE 1b

Width, in	Depth, in	Material
18	15	Vitreous china
19	17	Vitreous china, steel
20	18	Vitreous china, steel
22	18	Vitreous china, steel
24	20	Vitreous china
27	21	Vitreous china

TABLE 1c

Width, in	Depth, in	Material
18	14	Vitreous china
19	17	Cast iron
20	14	Vitreous china
20	18	Vitreous china, cast iron
22	19	Cast iron
24	18	Cast iron
24	20	Vitreous china
26	14	Vitreous china

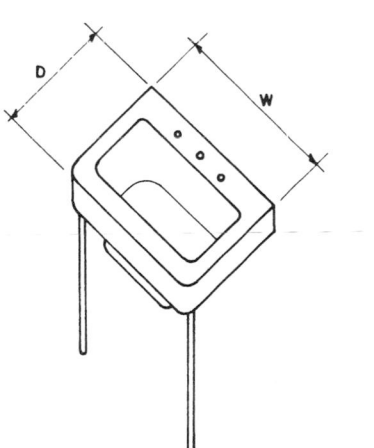

Fig. 1d Flat-slab lavatory.

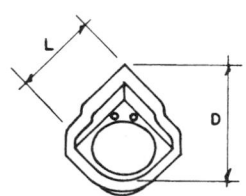

Fig. 1e Corner lavatory.

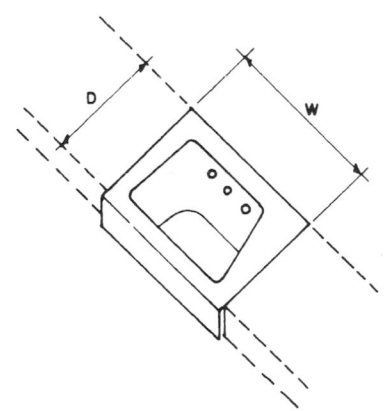

Fig. 1f Open-front built-in lavatory.

TABLE 1d

Width, in	Depth, in	Material
20	18	Vitreous china
24	18	Vitreous china
24	20	Vitreous china, steel, cast iron
27	22	Vitreous china
28	20	Vitreous china
32	18	Vitreous china
32	24	Vitreous china
36	18	Vitreous china
36	22	Vitreous china
42	18	Vitreous china

TABLE 1e

Depth, in	Length, in	Material
15	12	Vitreous china
16	14	Vitreous china
21	17	Vitreous china, cast iron

TABLE 1f

Width, in	Depth, in	Material
20	18	Vitreous china
22	18	Vitreous china
24	20	Vitreous china
26	22	Vitreous china
27	21	Vitreous china
30	22	Vitreous china

BATHROOM FIXTURES

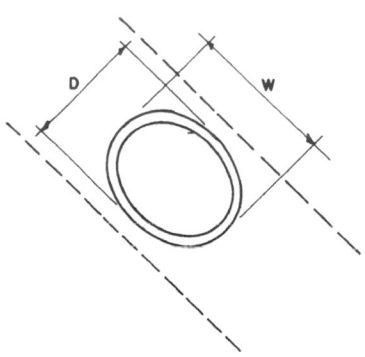

Fig. 1g   Round or oval built-in lavatory.

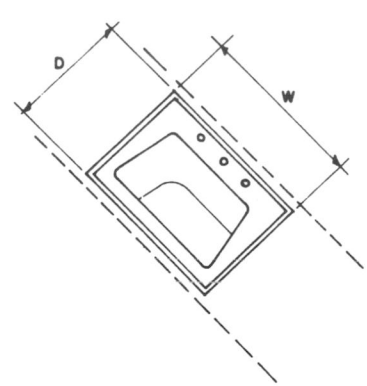

Fig. 1h   Flat-rim built-in lavatory.

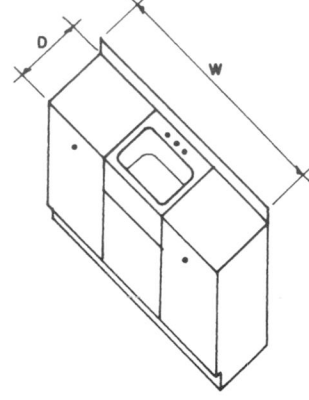

Fig. 1i   Lavatory-dressing table. Some models are available with two bowls. Bowls are available in vitreous china, steel, and cast iron. Cabinets are generally available in steel.

TABLE 1g

Width, in	Depth, in	Material
19	16	Vitreous china
14	14	Vitreous china
15	15	Vitreous china

TABLE 1h

Width, in	Depth, in	Material
16	14	Vitreous china
19	18	Steel
20	17	Cast iron
20	18	Vitreous china, steel, cast iron
20	19	Vitreous china
21	17	Vitreous china, steel
22	18	Vitreous china
24	18	Vitreous china
24	21	Vitreous china
28	20	Vitreous china

TABLE 1i

Width, in	Depth, in	Size of bowl, in
24	22	20 × 18
30	18	16 × 13
36	22	20 × 18
43	18	20 × 18
43	22	20 × 18
47	18	24 × 18
59	18	20 × 28
63	18	24 × 18

TOILETS

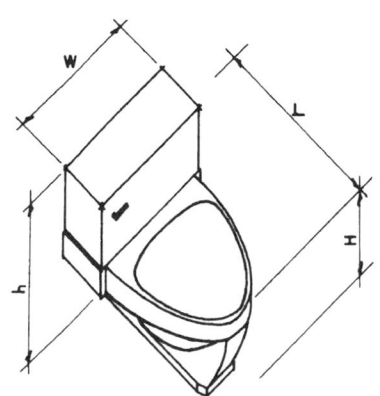

Fig. 2a   One-piece or close-coupled toilet. The styles available are whirlpool jet, reverse trap, or washdown siphon. This model is available only in vitreous china.

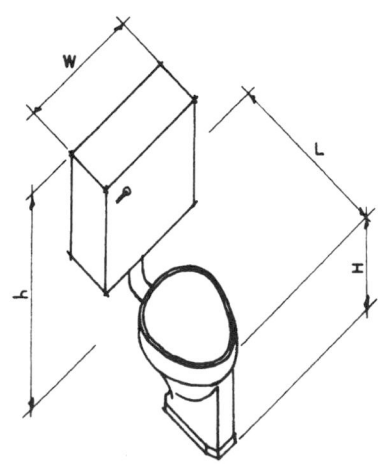

Fig. 2b   Flush-tank toilet. The styles available are washdown or reverse trap. This model is available only in vitreous china.

BIDET

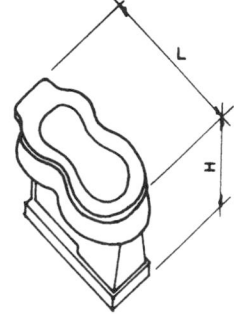

Fig. 3   Bidet. This model is available only in vitreous china.

TABLE 2a

Width, in	Length, in	Height, in	Height h, in
21	28	14	19
23	30	14	29
23	28	14	29
24	26	14½	30
23	29	14	24
22	27	14	24
20	29	14½	30

TABLE 2b

Width, in	Length, in	Height, in	Height h, in
22	27	14¾	36
20	27	15	36

TABLE 3

Length, in	Height, in
26	14½
25	15
24	15

SHOWERS

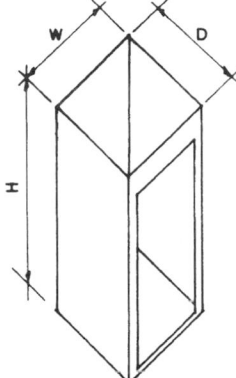

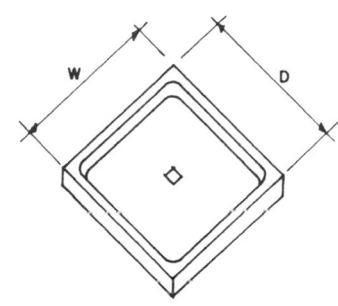

Fig. 4a Shower stalls. Prefabricated metal shower stalls are available with or without integral receptor base. Corner and built-in models are also available. All glass enclosures are available in approximately the same stock sizes. Special sizes are generally available on order.

Fig. 4b Shower receptors. Preformed and precast shower receptors are available in steel and terrazzo and, on special order, in marble and slate. Special sizes and shapes can generally be obtained on special order.

TABLE 4a

Width, in	Depth, in	Height, in
32	32	81
36	36	81
40	40	81

TABLE 4b

Width, in	Depth, in
32	32
36	36
36	42
40	40
42	44

BATHTUBS

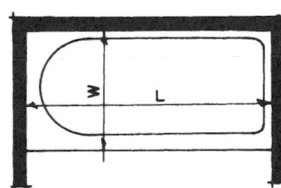

Fig. 5a Recessed bathtub.

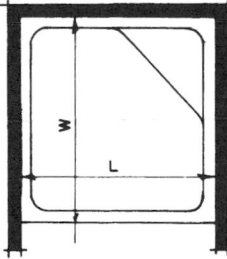

Fig. 5b Square recessed bathtub.

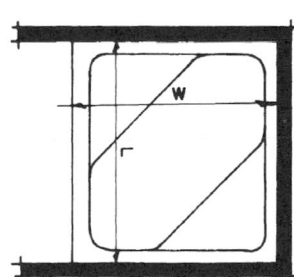

Fig. 5c Square recessed bathtub.

TABLE 5a

Width, in	Length, in	Height, in	Material
30	60	16	Cast iron
28½	60	14	Cast iron
29	54, 60	16	Steel
32	60	16	Steel
31	54, 60	16	Steel

TABLE 5b

Width, in	Length, in	Height, in	Material
38	39	12	Cast iron
51	48	16	Cast iron

TABLE 5c

Width, in	Length, in	Height, in	Material
48	50	16	Cast iron

BATHROOM FIXTURES

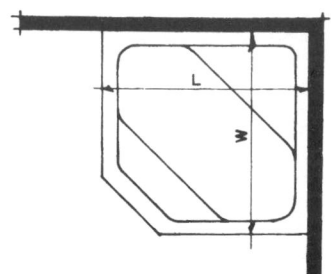

Fig. 5d   Square corner bathtub.

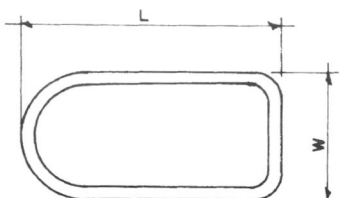

Fig. 5e   Free-standing bathtub.

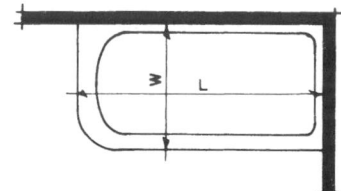

Fig. 5f   Corner bathtub.

TABLE 5d

Width, in	Length, in	Height, in	Material
50	50	16	Cast iron

TABLE 5e

Width, in	Length, in	Height, in	Material
30	54, 60	23	Cast iron
30	60	20	Cast iron

TABLE 5f

Width, in	Length, in	Height, in	Material
31	54, 60	16	Cast iron
32	60	16	Steel
30	60, 66	16	Cast iron
31	60	16	Steel
31	60	15	Steel

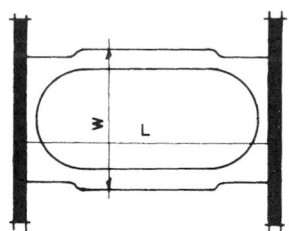

Fig. 5g   Recessed two-way bathtub.

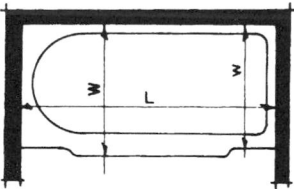

Fig. 5h   Recessed wide-ledge bathtub. Stock glass enclosures are available.

TABLE 5g

Width, in	Length, in	Height, in	Material
35	60	15	Steel

TABLE 5h

Width, in	Width w, in	Length, in	Height, in	Material
33	31	54, 60, 66	16	Cast iron
30	25	60	16	Cast iron
32	29	60	16	Steel
33	30	64	17	Cast iron
32	30	48, 54, 60, 66, 72	16	Cast iron
29	28	60	14	Cast iron
31	29	54, 60	16	Steel
30	28	60	14	Cast iron
33	31	54, 60, 66	16	Cast iron
32	31	60	14	Cast iron

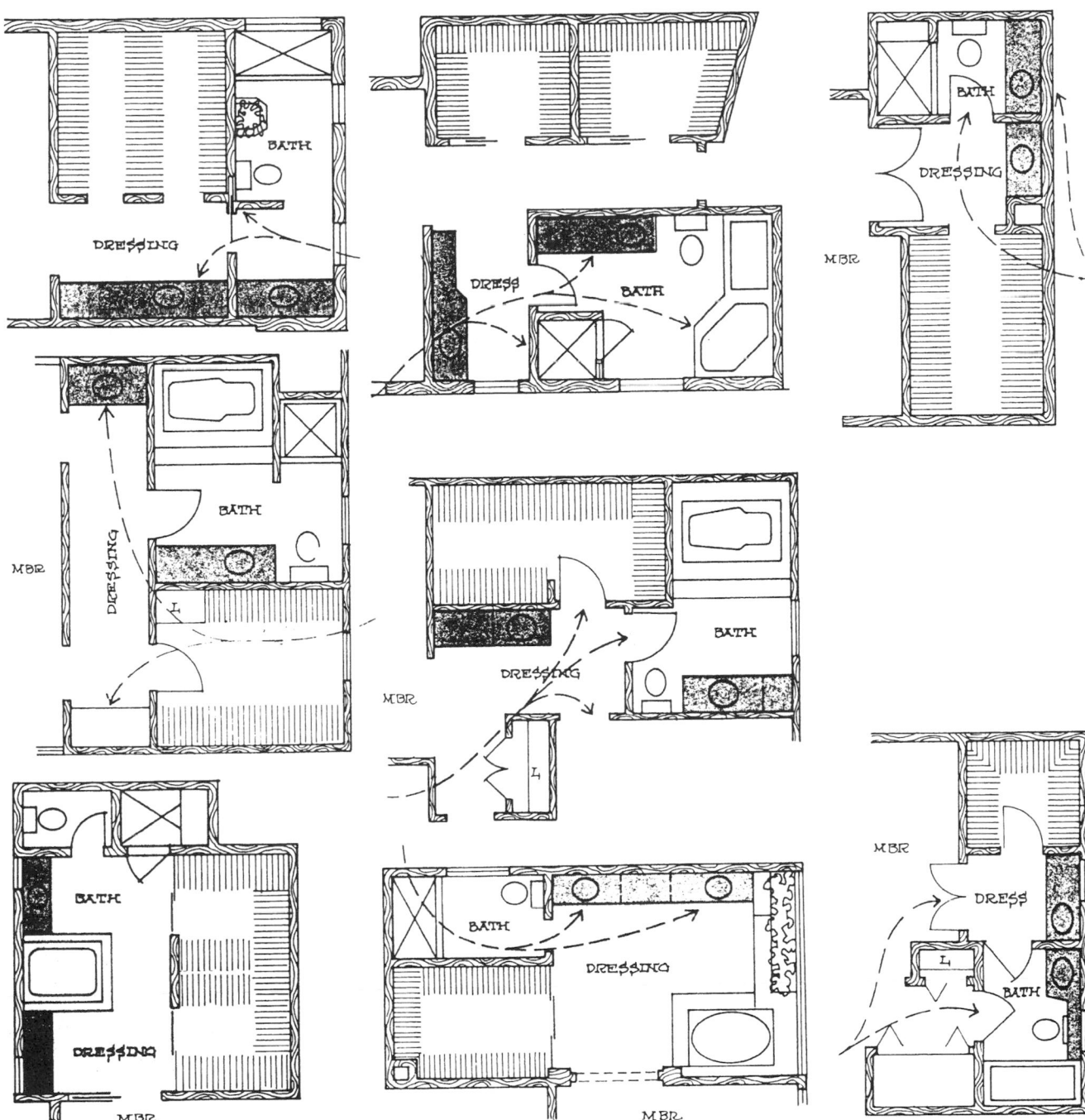

Fig. 6

BATHROOM FIXTURES

PRESTIGE 6032

PRESTIGE 6634

PRESTIGE 7236

FUTURA SERIES

FUTURA 6042

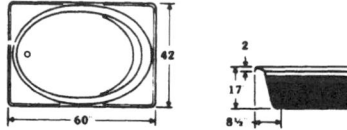

FUTURA 7242

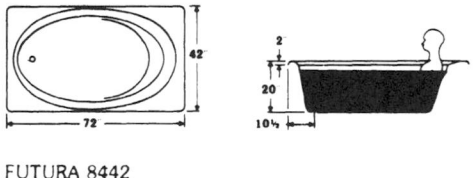

FUTURA 8442

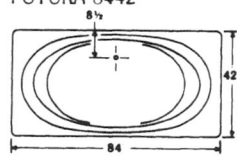

CLASSIC

ELITE

CONCORDE

PLAZA

ÉTOILE

KYOTO

CONFIDENCE

Style	Size	Water Capacity* (Approx.)	Motor Size	Jets
PRESTIGE 6032	60x32x19	50 gal.	¾ h.p.	6
PRESTIGE 6634	66x34x19	54 gal.	1 h.p.	6
PRESTIGE 7236	72x36x19	57 gal.	1 h.p.	6
FUTURA 6042	60x42x17	46 gal.	1 h.p.	6
FUTURA 7242	72x42x20	70 gal.	1 h.p.	8
FUTURA 8442	84x42x20	80 gal.	1 h.p.	8
CLASSIC	60x40x32	95 gal.	1½ h.p.	8
ELITE	84x66x20	122 gal.	1½ h.p.	8
CONCORDE	70x43x18	82 gal.	1 h.p.	8
PLAZA	74x47x19	95 gal.	1 h.p.	8
ÉTOILE	69(dia.)x17	95 gal.	1½ h.p.	8
KYOTO	68x68x22	140 gal	1½ h.p.	8
CONFIDENCE	75x59x19	95 gal.	1½ h.p	10

*All water gallonage measured to 2" below overflow
Note: All dimensions are approximate

Fig. 7

WHIRLPOOL BATHS

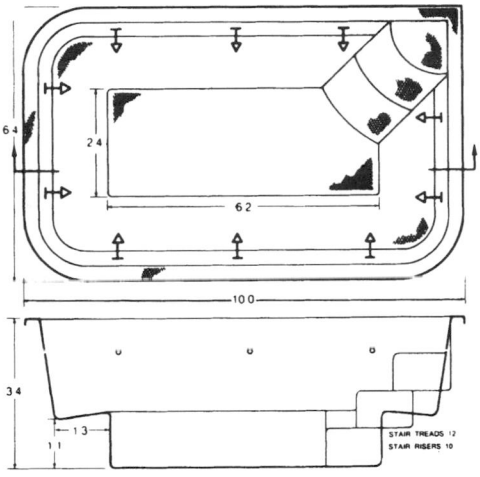

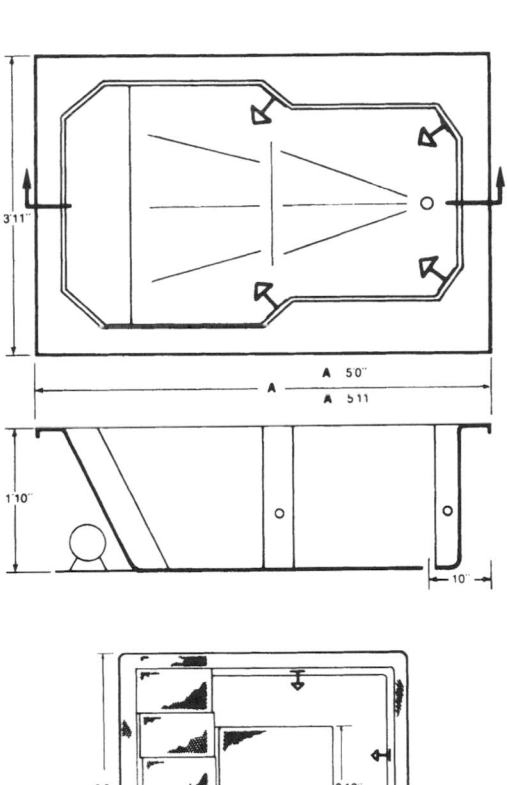

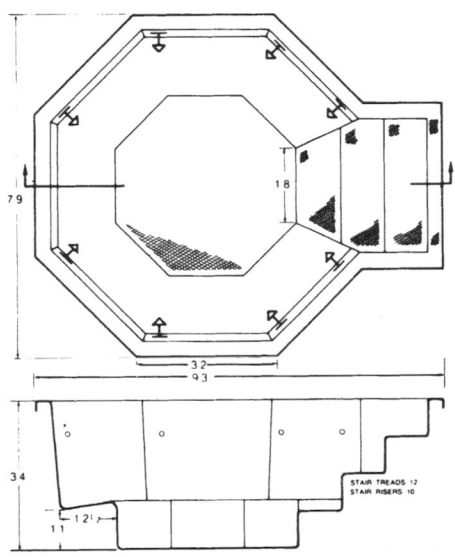

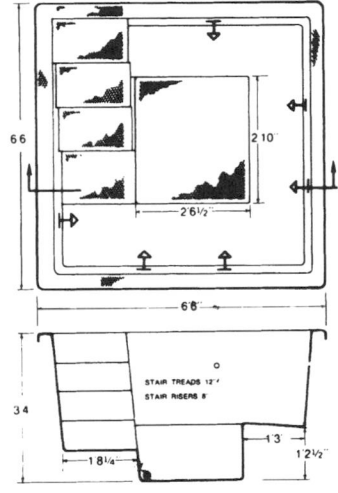

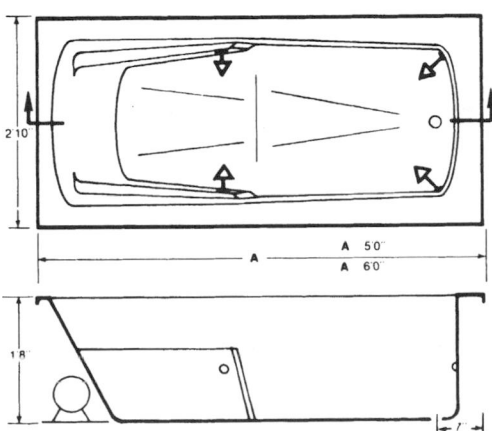

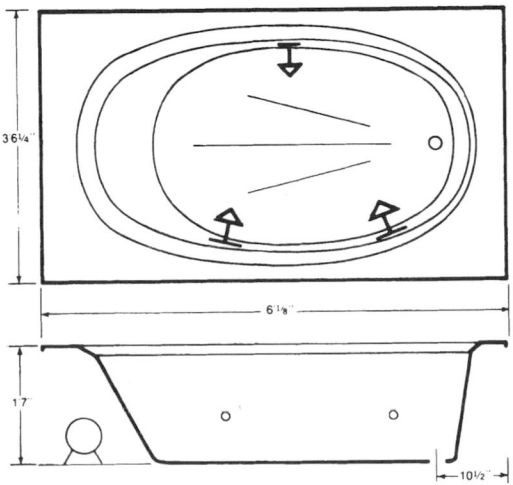

Fig. 8

FAMILY ROOMS

In smaller homes, the family room may double as the living room and be called a "great room." Under either name, this is the place household members gather in a relaxed manner, to "let their hair down," and share time together.

The family room role has gradually evolved into what may also be called the *media center*. Regardless of one's views about the positive or negative effects of the media, one cannot deny that television is an immensely popular appliance for most families. The average American spends up to 30 hours a week in front of it, and home designers must ensure that room configurations can provide for it (Fig. 1).

The television is in fact only one part of the media center. Many family room designs provide an entertainment wall that also holds stereo equipment, storage for tapes and records, and a video recorder. With the popularity of home video, the family room is now an in-home movie theater, which provides another excellent reason to make this area accessible to the home snack bar—the kitchen (Fig. 2).

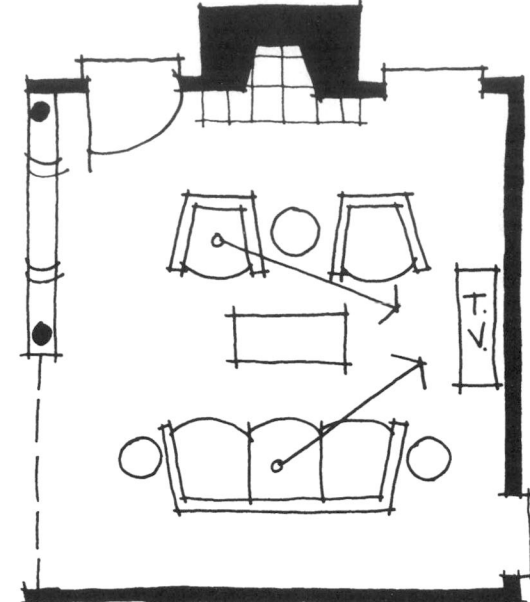

Fig. 1   Family room with space for television viewing.

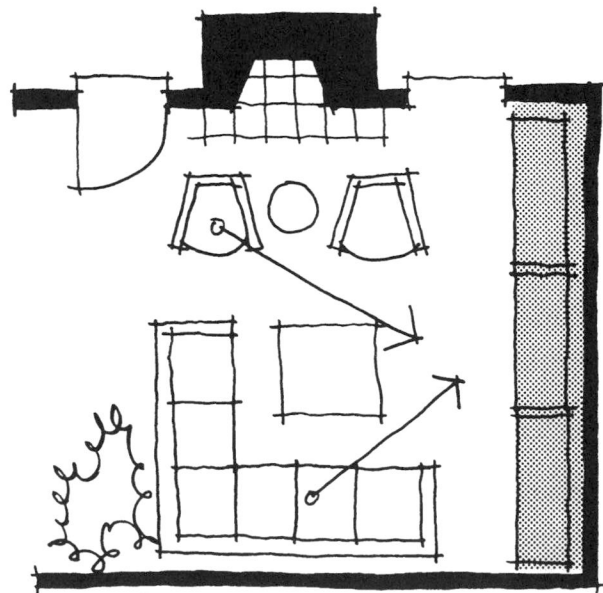

Fig. 2   Family room with media wall.

A *fireplace* is also integral to the family room layout. Although households seldom use the fireplace, this amenity is symbolically important, retaining at least a vestigial role as the true "hearth" and favorite place for intimate gatherings. Surveys indicate that buyers would rather have the fireplace in the family room than in any other room in the house.

In addition to the television/media wall and fireplace, the family room must also accommodate group seating, within view of both media center and fireplace. Furnishing plans may include an L-shaped sectional couch or a sofa with two or three side chairs, and the layout should also allow room for a coffee table and side tables.

Because of the activity in the community component, space flow between the three areas that form it is extremely important. Occupants of the conversation area should be able to view both the TV and the fireplace. They should also enjoy outdoor views and access, as well as be able to converse with people in the kitchen or breakfast area (Fig. 3).

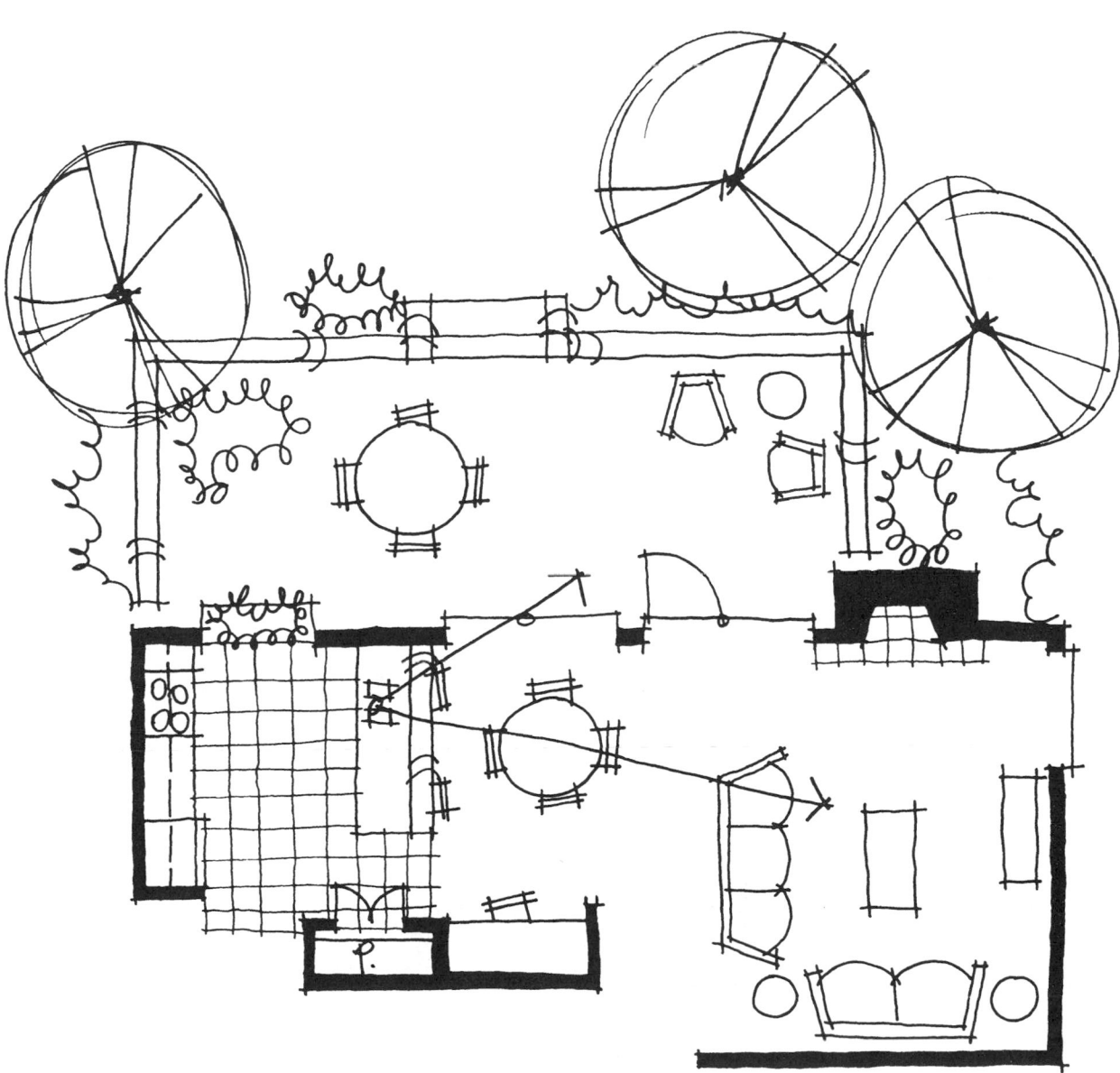

Fig. 3   View of family room from kitchen and breakfast area.

RELATIONSHIP OF HUMAN DIMENSIONS

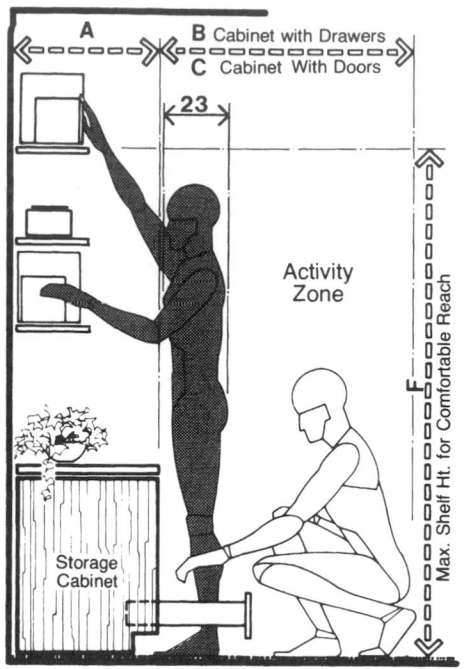

Fig. 1   Wall unit—access by male. (See Table 1.)

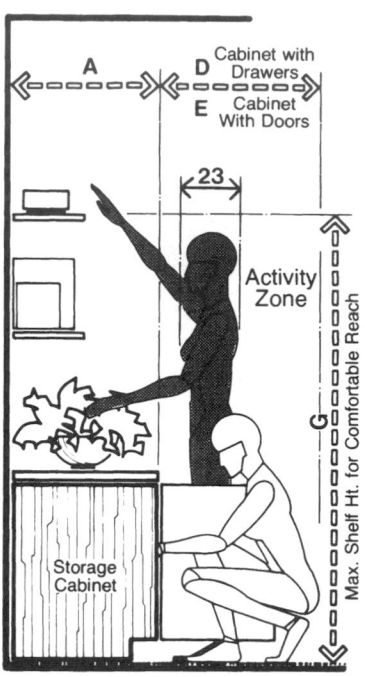

Fig. 2   Wall unit—access by female. (See Table 1.)

Figures 1 to 4 illustrate the relationship of human dimension and accessibility to low and high storage or furniture usually associated with living spaces. The configuration of the furniture is not intended as a realistic illustration of any specific element of furniture, but rather as a general representation of furniture types normally found in a living space. In situations where the user is not a known entity, either in terms of sex or body size, the body size data of the smaller person should govern. In the event the user is known, dimensions more appropriate to that body size should be used where practical. It should be noted that for each sex two dimensions are shown on the drawing. In each case the lower figure is based on 5th percentile body size data and the larger on 95th percentile data.

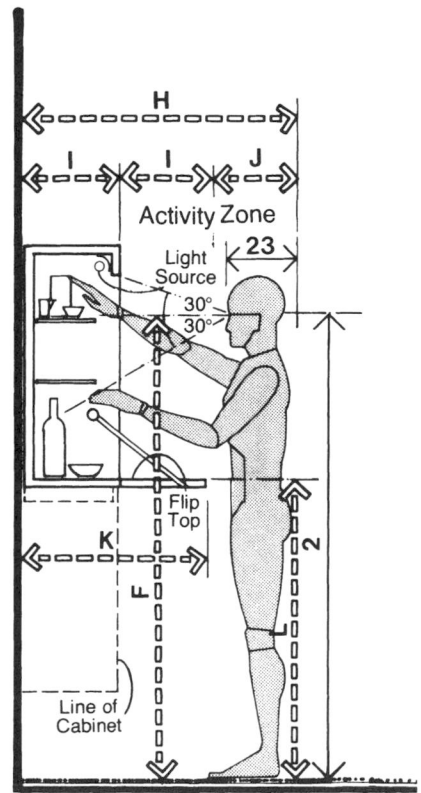

Fig. 3   Wall-mounted bar unit—access by male. (See Table 1.)

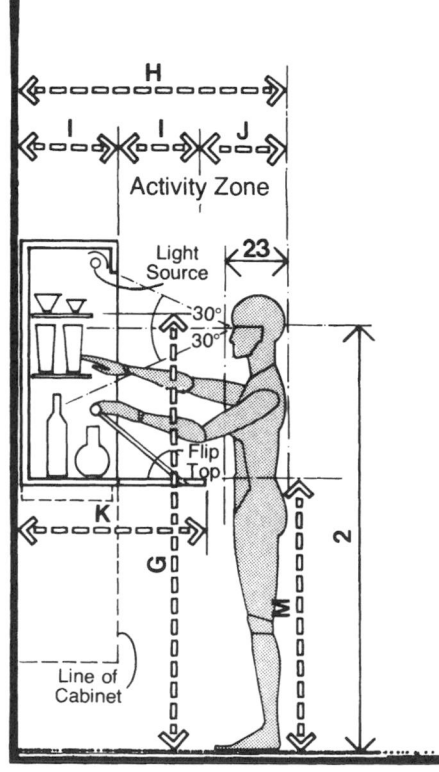

Fig. 4   Wall-mounted bar unit—access by female. (See Table 1.)

TABLE 1

	Inches
A	18–24
B	48–58
C	36–40
D	46–52
E	30–36
F	72
G	69
H	42–50
I	12–16
J	18
K	24–32
L	39–42
M	36–39

EXERCISE AREAS

Figures 1 and 2 show typical exercise equipment available on the market. Figure 1 typifies the classic exercise bicycle and shows some of the clearances required in a commercial installation. Figure 2 is representative of the many weight-lifting devices currently in use. The front and side views indicate some of the overall dimensions as well as the relationship of the human body to the equipment. Dimensions and general configuration vary with model and manufacturer, but the information shown can be used for making preliminary design assumptions.

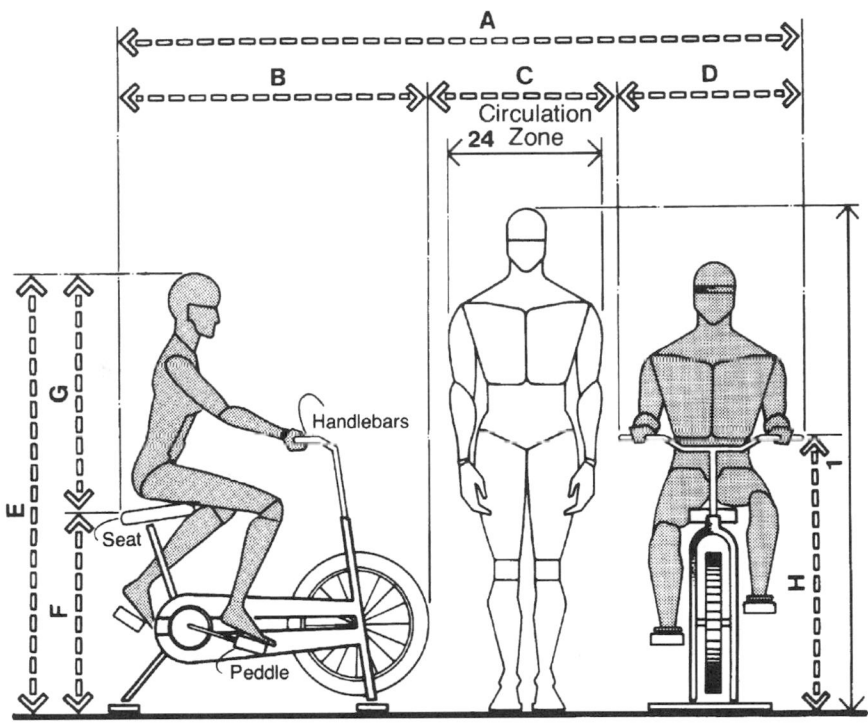

Fig. 1   Exercise bicycle. (See Table 1.)

TABLE 1

	Inches
A	83–104
B	35–48
C	30
D	18–26
E	55–68
F	25–30
G	30–38
H	46

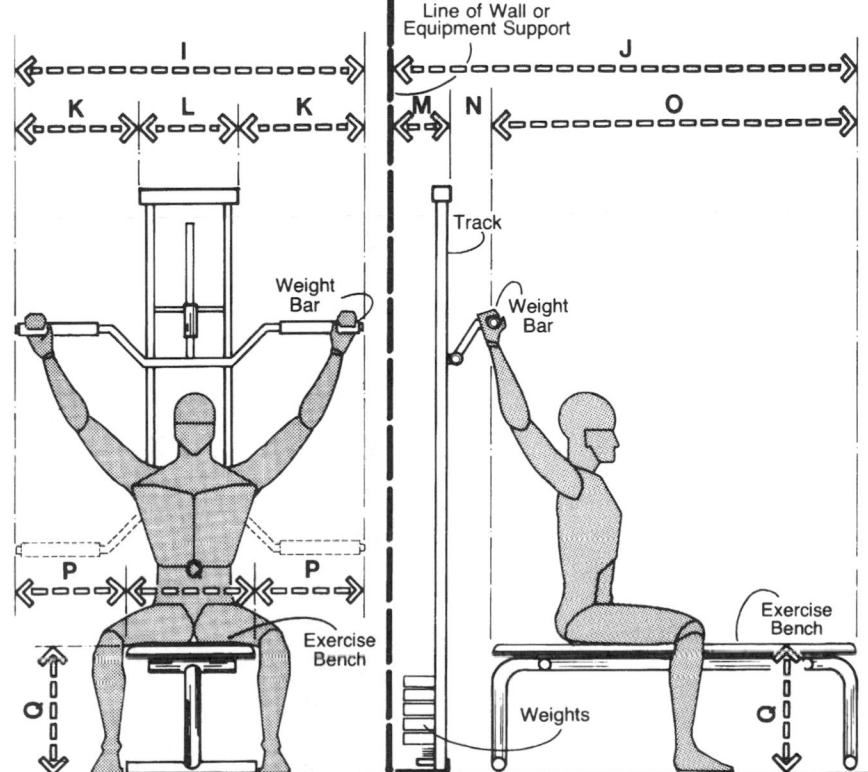

Fig. 2   Wall-mounted latissimus power lift unit. (See Table 2.)

TABLE 2

	Inches
I	36–48
J	58–76
K	12–18
L	12
M	6–12
N	4–10
O	48–54
P	9–14
Q	18–20

EXERCISE AREAS

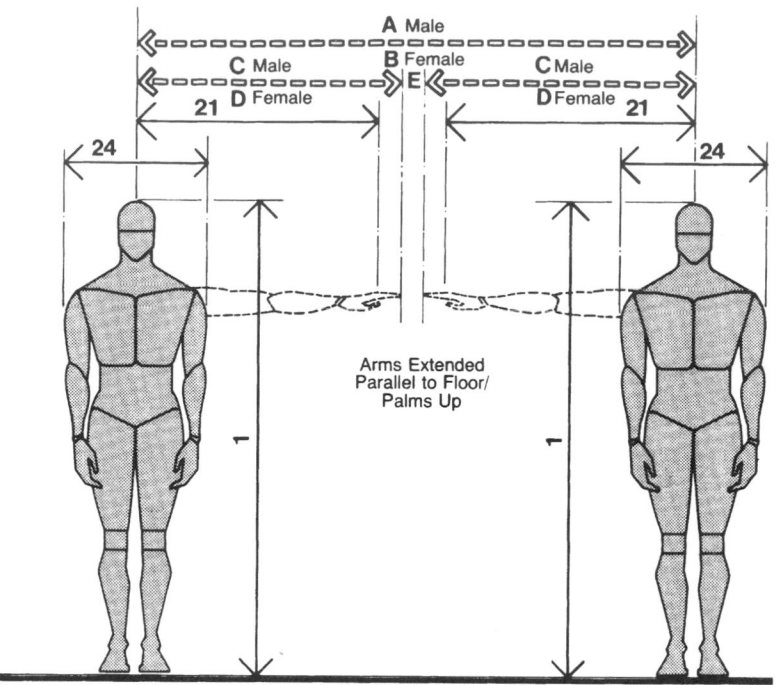

Fig. 3 Minimum exercise clearance requirements. (See Table 3.)

Figure 3 should be helpful in establishing minimum center spacing for standing exercises in place. The drawing is not intended as a standard, but rather as a base of reference for preliminary design assumptions. The nature of the particular exercise and the intensity of body movements involved should all be taken into consideration.

Certain exercises require significant head room. Dance and similar activities, for example, require considerable clearance to avoid accidents. Figure 4 shows only two such possibilities. There are, obviously, many vaiations. Tables 3 and 4 should provide the necessary data with which to establish clearances appropriate to those variations.

TABLE 3

	Inches
A	65–80
B	61–88
C	31–37
D	29–41
E	3–6

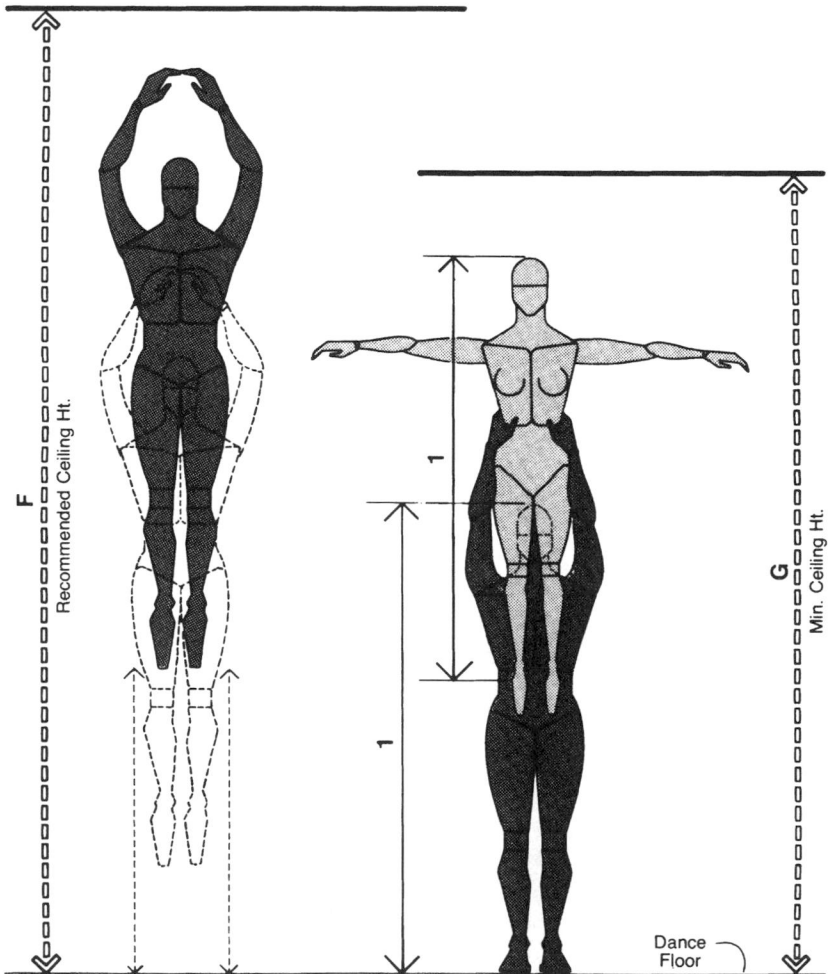

Fig. 4 Dance and exercise practice rooms—ceiling height requirements. (See Table 4.)

TABLE 4

	Inches
F	144
G	120

Figure 5 indicates in side and front view the clearances required by the human body while engaged in sit-up exercises. Although it is recommended that in establishing clearances, the person of larger body size be used as a model, the ranges shown reflect small and large male and female data. The 5th and 95th percentile vertical grip reach measurements were used as the basis of the dimensions, with an allowance to compensate for the fact that the anthropometric measurement does not quite extend to the tip of the fingers. The authors suggest that even if the design is intended for a particular population of smaller body size, the larger measurements be used. The largest clearance required would be for the large male, and is shown as 91.5 in.

Figure 6 provides the designer with the dimensional information necessary to establish basic spacing for an exercise class.

Figure 7 shows the clearance required for push-up exercises. Stature would be the most useful anthropometric measurement to consider.

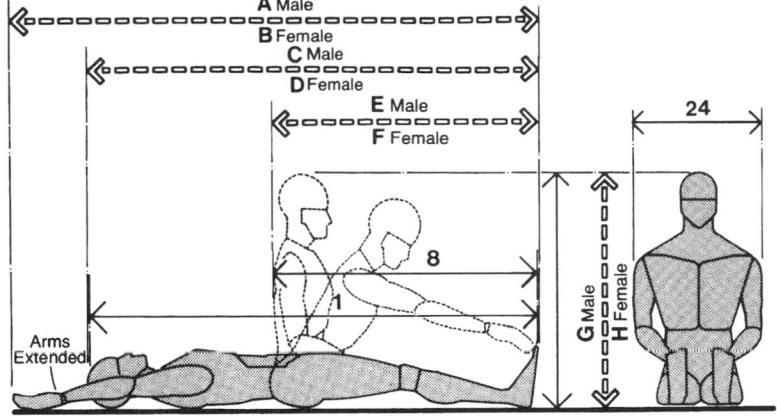

Fig. 5   Sit-up floor exercise. (See Table 5.)

TABLE 5

	Inches
A	80–91.5
B	75–87
C	65–74
D	60–69
E	32–37
F	27–37
G	33.2–38.0
H	30.9–35.7

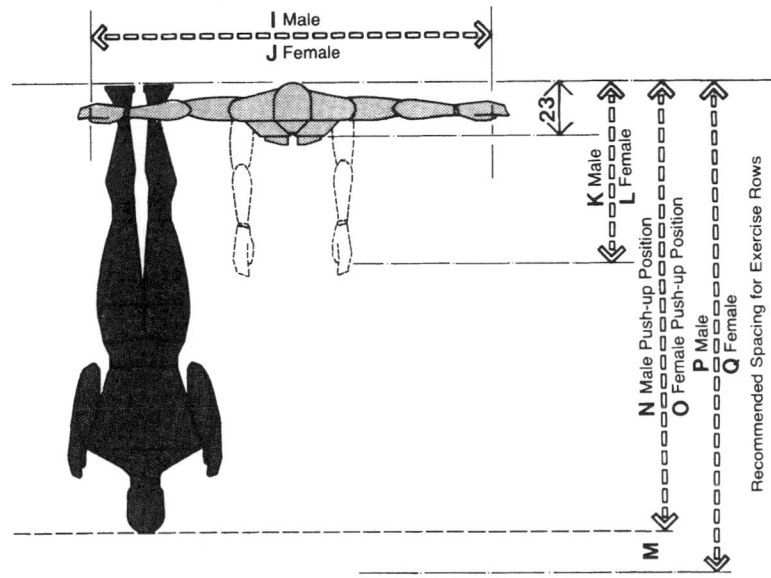

Fig. 6   Basic spacing for exercise class. (See Table 6.)

TABLE 6

	Inches
I	58–68
J	54–76
K	29.7–35.0
L	26.0–31.7
M	6–12
N	63–73
O	61–67
P	79–85
Q	73–79
R	23–38
S	10–16

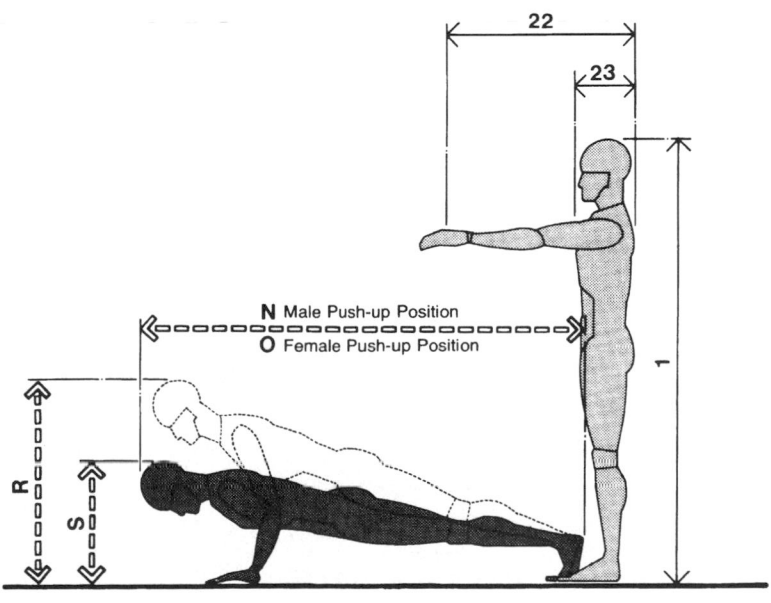

Fig. 7   Space requirements for basic push-up position. (See Table 6.)

EXERCISE AREAS

The sauna is essentially a thermal bath using dry heat, unlike the low heat and high humidity of the steam bath. Although there are many complete prefabricated models on the market, the heater units can be purchased separately. It is therefore relatively simple to custom design an individual installation.

Figure 8 illustrates some of the critical dimensions involved. Two possible ceiling heights are indicated. The alternate height will allow more comfortable access to the second tier bench, while the normal height will permit installation within the conventional 96-in ceiling limitations of most residential interior spaces.

Figure 9 shows a section through a typical locker room. The restricted circulation zone shown at the right would require either the seated or the standing person to move out of the way to avoid body contact. The circulation zone at the left would allow more comfortable passage without body contact.

TABLE 7

	Inches
A	108
B	24
C	84
D	36–40
E	44–48
F	12–14
G	18–20
H	78 min.

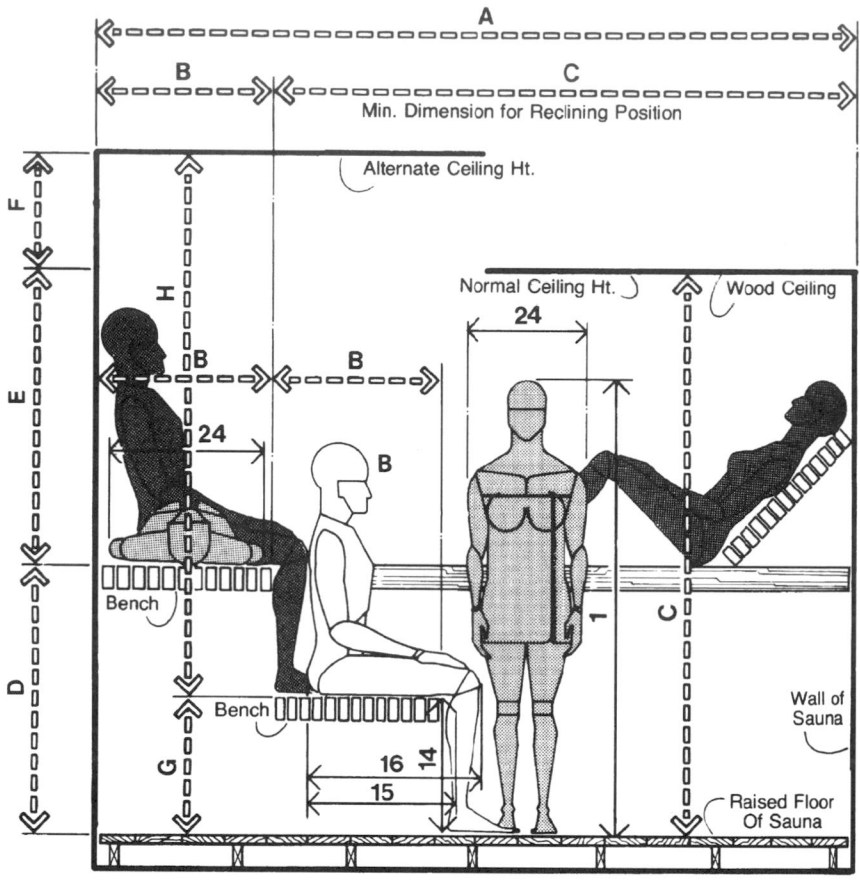

Fig. 8 Section through sauna room. (See Table 7.)

TABLE 8

	Inches
I	56–64
J	12–15
K	42–48
L	12–18
M	30
N	14–16
O	4–6
P	14–17
Q	60–72

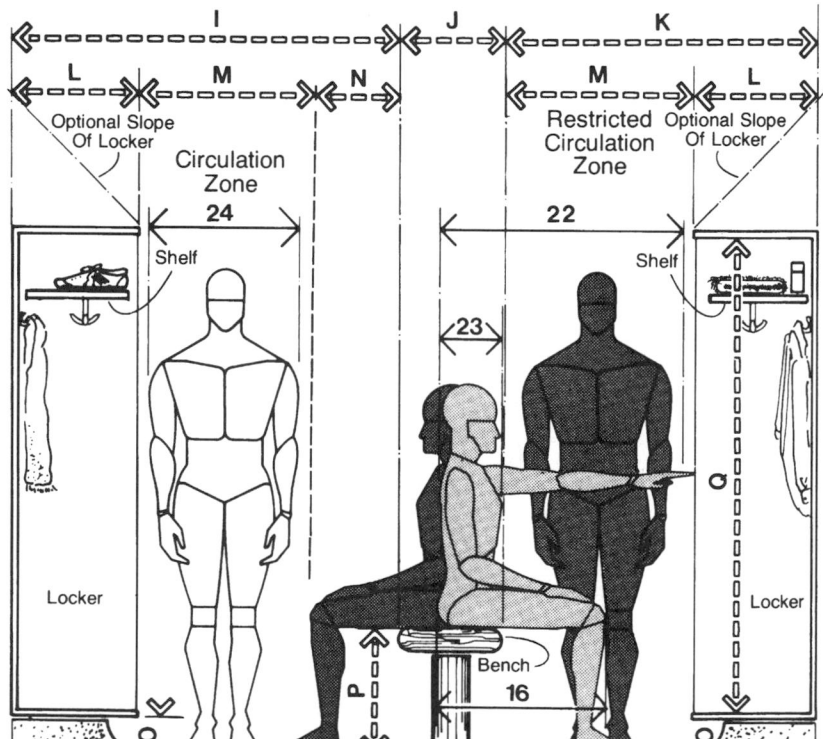

Fig. 9 Locker room. (See Table 8.)

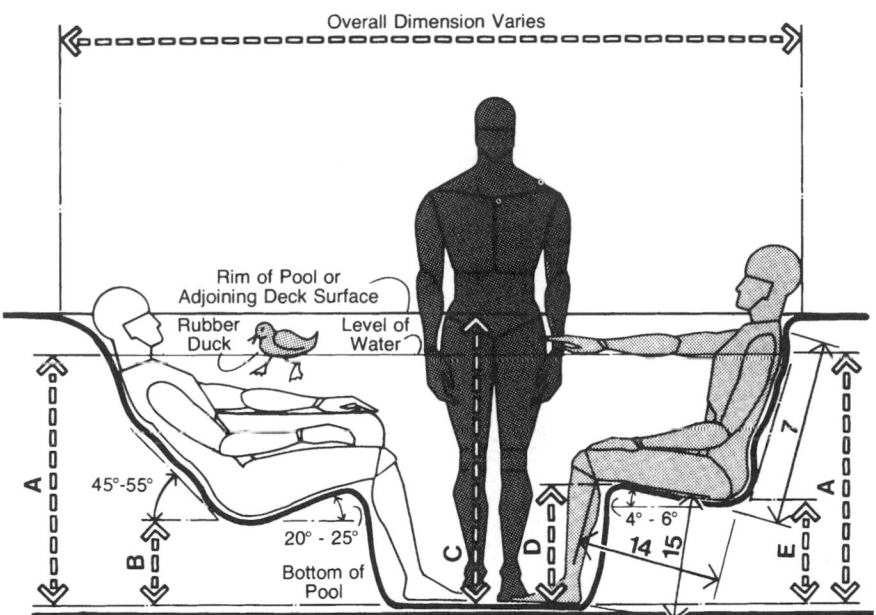

Most hydrotherapy pools provide turbulent hot water massage. Some models, such as the ones shown in Fig. 10 have been anthropometrically contoured to provide proper support for the back, particularly in the lumbar region. The pools are manufactured in a variety of profiles to accommodate different body positions. The height of the pools is between 33 and 38 in. The lengths and widths vary with the model.

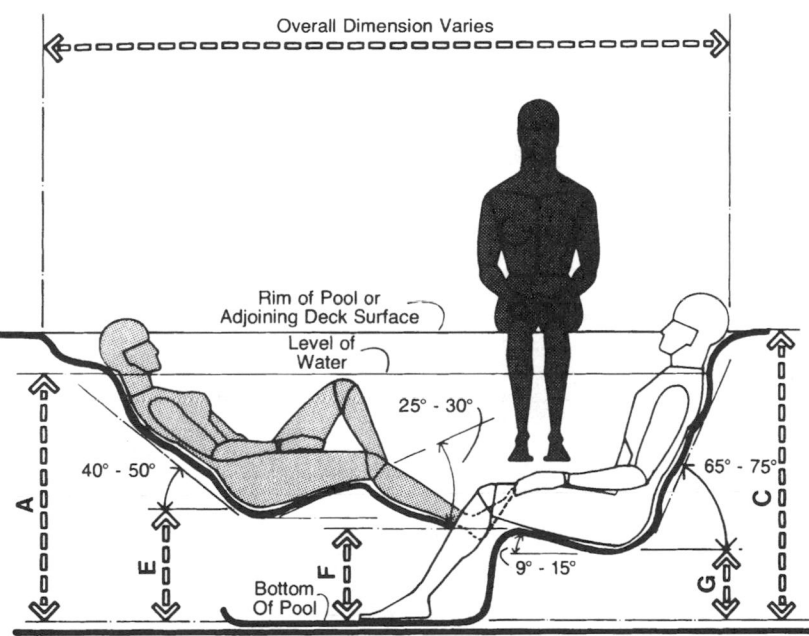

Fig. 10   Anthropometrically contoured hydrotherapy pools. (See Table 9.)

TABLE 9

	Inches
A	33–38
B	9–12
C	38–44
D	13–16
E	12–15
F	11–14
G	8–11

INDOOR PLAY AREAS

Space standards for indoor play areas for preschool children are presented in tables and illustrations. Any play area should provide furnishings and activity space. As distinguished from a playroom, a play center is an organized play space in a room used primarily for other purposes. When there are two or more play centers in a given house, the major play center is the one that permits moderately active play and a variety of activities; the minor center is one that permits quiet play only and a limited variety of activities.

Usually only one play center can be provided in houses with no living center other than a bedroom-living room. In houses with one or two social rooms—such as living or dining room, den, or family room—a major play center and one or two minor centers can be incorporated. In larger houses with more social rooms, one room can usually be used exclusively for play and minor play centers can be planned in other areas of the house.

Location of centers will vary with the needs of the family as well as the plan of the house. For single or major play centers the living room in most homes is a suitable location; minor centers can usually be incorporated in work and sleeping areas of the house. Location of play centers may be different in winter and summer.

Only one-fourth of the floor space allocated to the play center should be occupied by toy storage devices, child's table and chairs, or large play equipment. To permit free movement during play, the width of the center should be not less than 3 ft 6 in. Play centers of 20 feet^2 are satisfactory as minor centers; those of 35 ft^2 are large enough for comfortable play for only one child.

TABLE 1   Total Floor Space for Preschool Children's Play by Number of Social Rooms in the House

Number of social rooms in the house	Floor space, ft^2		
	Minimum	Moderate	Liberal
None	35	50	
One	35	50	70
Two		70	90
Three or Four		90	170+

TABLE 2   Suggested Size of Separate Play Areas by Location

Location in house	Floor space, ft^2		
	Minimum	Moderate	Liberal
Bedroom-living room	35*	50	
Kitchen-utility area (including utility porches)	20†	35*	50
Family social room (den or similar room)	20†	50	70
Child's bedroom	20†	50	70

*Appropriate for one child.
†Minor play center for one child. Smallest floor dimension should be not less than 3 ft 6 in.

Fig. 1

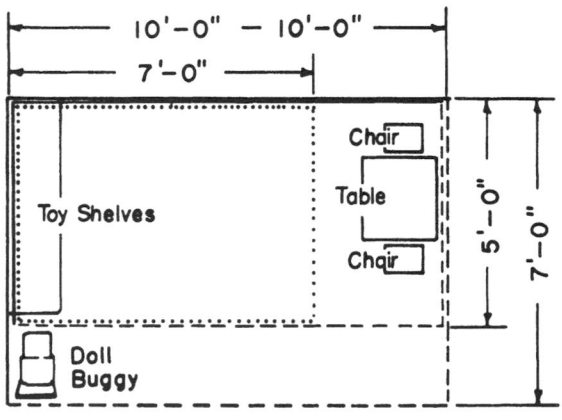

············· Minimum One Child
========= Minimum Two Children
■■■■■■■ Moderate Two Children

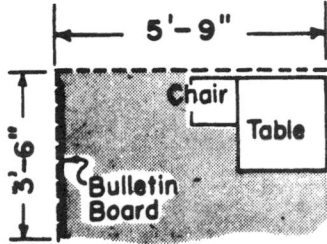

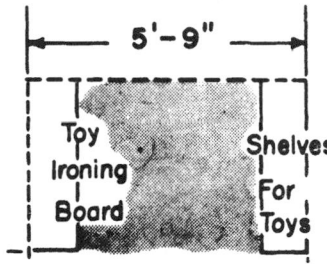

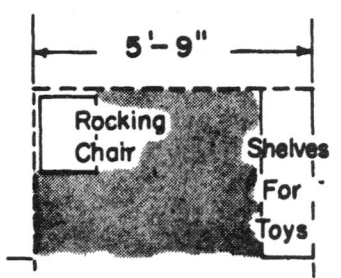

Fig. 2

TABLE 3 Dimensions of Typical Play Equipment, Furniture, and Toys Requiring Floor Space

Item	Length, in	Width, in	Height, in
Play equipment and furniture			
Bulletin board	31½	½	18
Car	37	17	22
Chalkboard (wall)	24	½	18
	36	½	24
	48	½	36
	72	½	48
Chairs, straight	10½	8½	10[1]
	13	11	12[1]
	12½	12½	14[1]
Easel	18	18	36
	24	22	44½
Record player	12	10	4½
Table	17	24	21
	18	24	21
	26½	26½	20½
Tricycle	27½	20½	22
	31½	20½	28
Wagon	34	15½	14
Toys			
Barn	25	9	12
Baking table	23½	13	32
Doll furniture			
Bed	22½	12½	19
	24½	14½	23
Buggy	25	15½	28½
	18	9	25
Chair	24	12	28
	9½	8	10½*
House	9	9	10½
	19½	9	15½
Ironing table	27	10½	17
Piano	30	8	23
Piano bench	20	13½	22
Railroad station	12	8	9½
Service station	21	10	6½
Trucks	26	15	8
	17	8½	8½
	21	5½	7½
	22	6½	7½
	32	9	15

*Height of chair seat.

RECREATIONAL ACTIVITIES

Indoor recreational activities invariably require definite spaces for equipment and clearances for using it. Not all games occupy floor areas indicated as necessary for those diagramed on this page. But if interiors are planned to accommodate large units of equipment such as that required for table tennis, and provide necessary playing clearances, spaces will be adequate for many other uses as well.

Dimensions of game equipment and floor areas required for its use are both subject to variation. Sizes noted here are comfortable averages, not absolute minima.

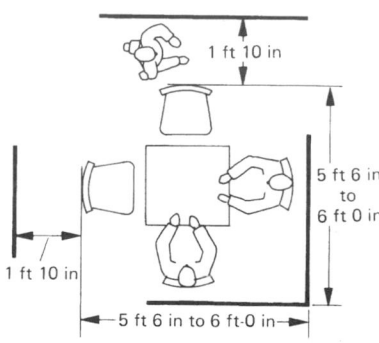

Clearances for playing bridge

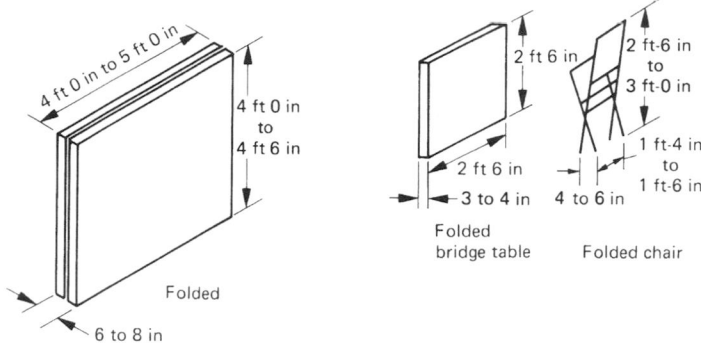

Folded

Folded bridge table   Folded chair

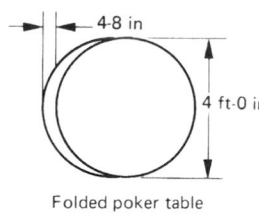

Folded poker table

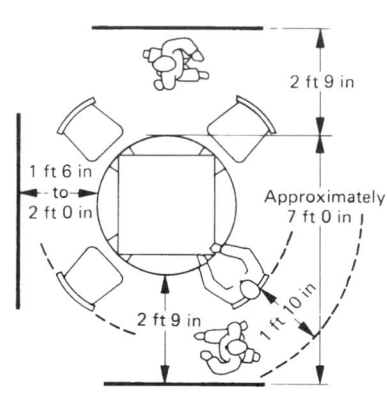

Clearances for playing poker

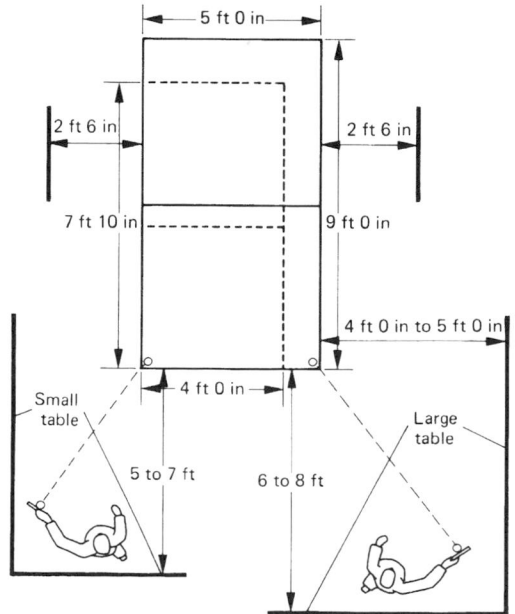

Small table

Large table

Left: clearances for table tennis.
Right: same for pool or billards.

Standard table sizes
are: 3 ft x 6 ft
3 ft 6 in x   7 ft 0 in;
4 ft 0 in x   8 ft 0 in;
4 ft 8 in x   8 ft 6 in;
5 ft 0 in x   9 ft 2 in;
5 ft 6 in x 10 ft 2 in;
6 ft 8 in x 12 ft 8 in.

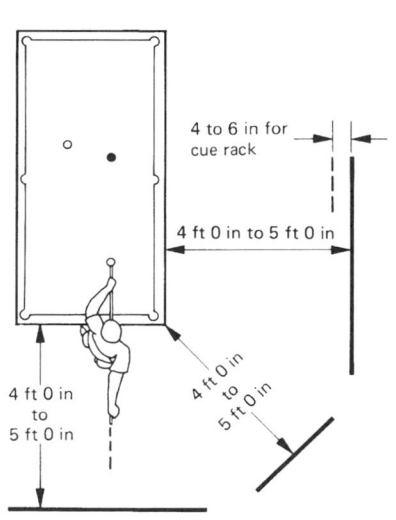

Fig. 1

Figure 2 indicates side clearance requirements for a table tennis installation within a residential environment: 48 in is the absolute minimum, while 72 in is preferred. Figure 3 indicates the clearances required at either end of the table. In a close-up position, the player usually functions within 24 to 36 in of the edge of the table. An overall clearance between the edge of the table and the wall or nearest physical obstruction—between 84 and 120 in—is suggested. The smaller figure should be regarded as an absolute minimum, and the larger figure as the preferred clearance. The latter, however, may be difficult to provide in terms of the room size required. The extent of clearance is a function of the size of the players and the intensity and skill with which the game is played. What must be considered is not only the space required for low-key volleying but the space required, for example, to chase a strategically placed ball, return it, decelerate, and ultimately stop, all in enough time to avoid colliding into the wall at the rear or side of the playing area.

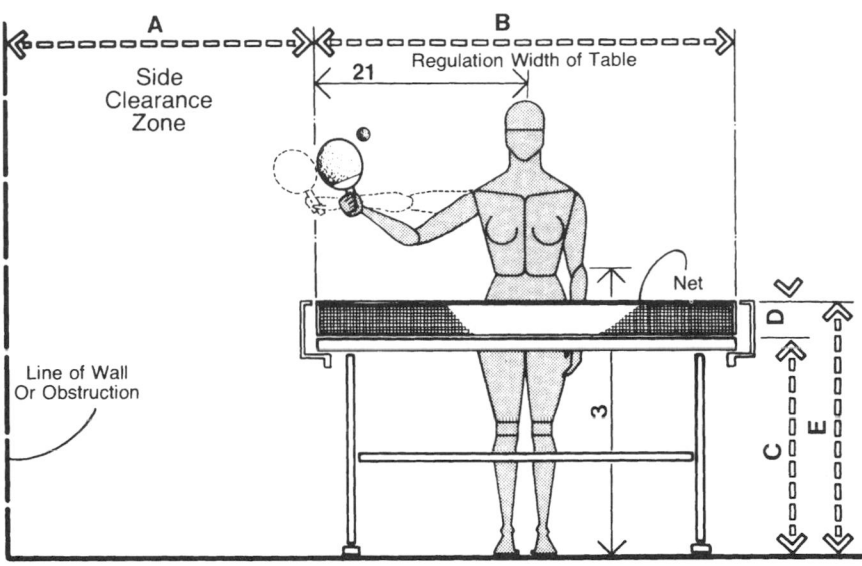

Fig. 2 Residential table tennis side clearance requirements. (See Table 1.)

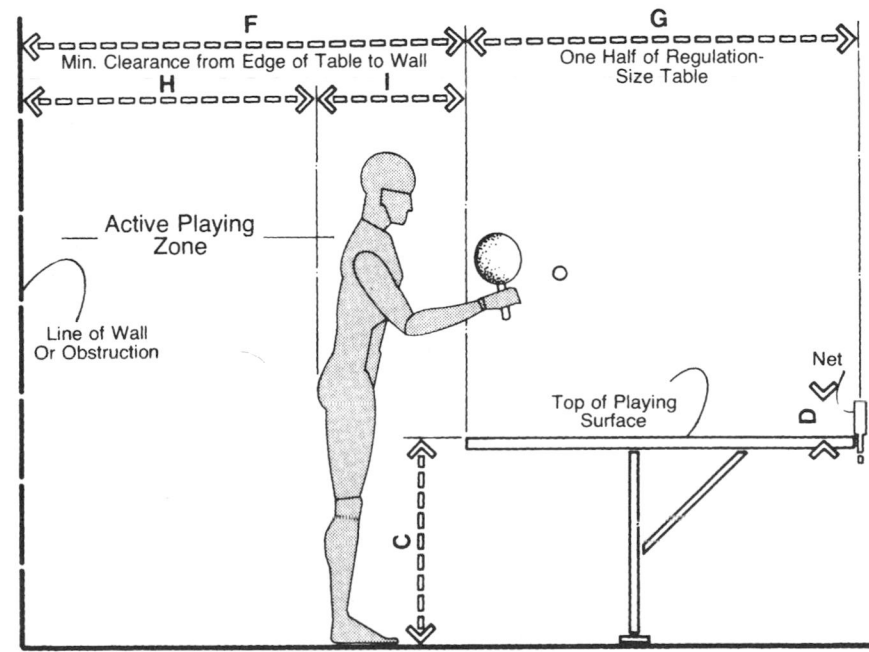

Fig. 3 Residential table tennis requirements for end of table clearances. (See Table 1.)

TABLE 1

	Inches
A	48–72
B	60
C	30
D	6
E	36
F	84–132
G	54
H	60–96
I	24–36

RECREATIONAL ACTIVITIES

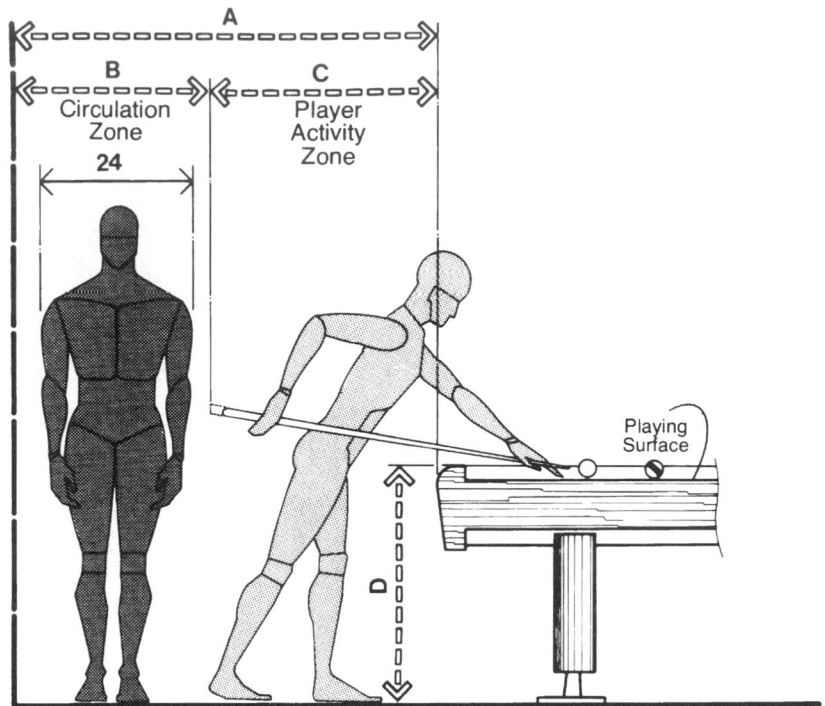

Figure 4 indicates the clearance required from the edge of a pool or billiard table to the wall or nearest physical obstruction. A clearance of 60 to 72 in is suggested to allow the possibility for some circulation behind the active player. The activity zone shown applies for most shots. In some instances, due to the nature of the play, the stance of the player and the length of the cue stick, there may be some intrusion into the circulation zone.

Fig. 4   Billiard and pool table requirements.

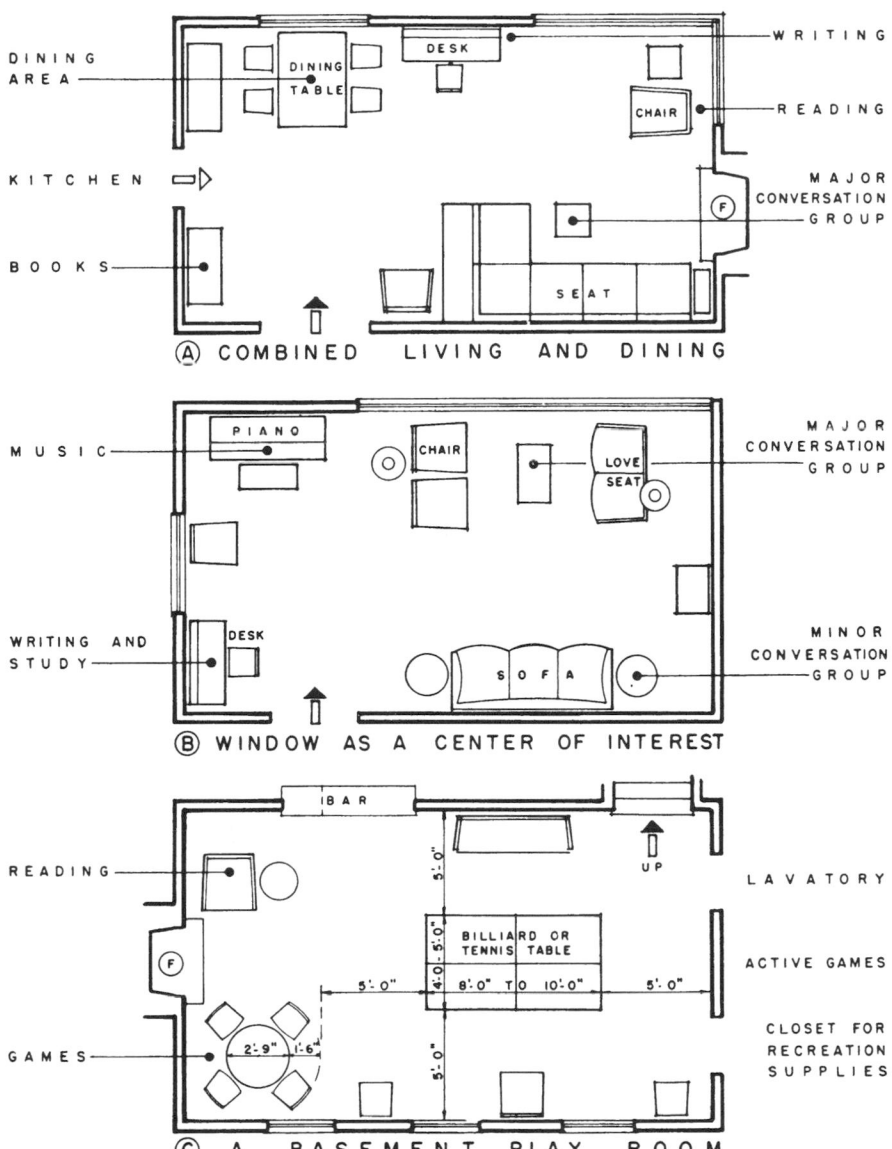

Fig. 1   The position and proportions of walls and openings may determine the arrangement of groups of furniture in the general living area, as suggested in these plans.

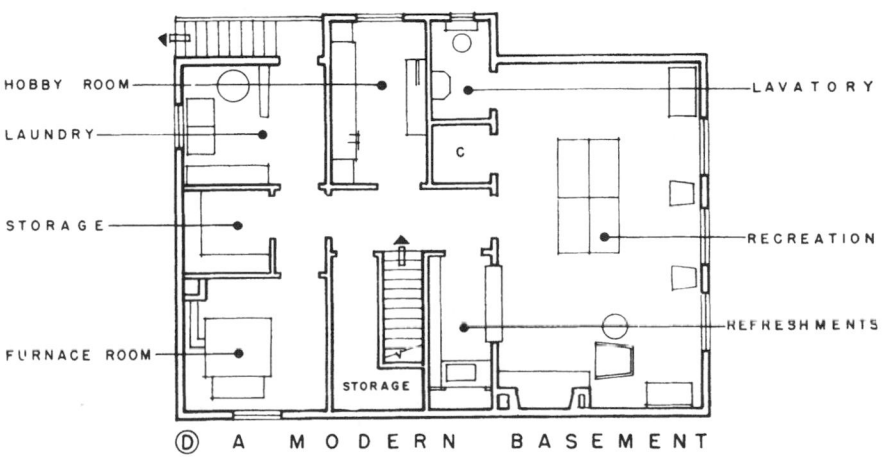

Fig. 2   Basement.

OUTDOOR LIVING SPACE

Space for living, dining, or just relaxing need not be confined to the interior of a dwelling unit. Full advantage of outdoor living areas should be taken, especially where the climate permits it for more than a few months of the year. By properly designing and integrating the indoor-outdoor areas, the entire living space can be greatly enhanced at a minimum of cost. The most common methods used are:

Balcony Usually designed as an extension of an apartment unit in a high-rise building. In the past, most have been constructed for limited use and inadequately sheltered. Since the balcony is above the ground and in close proximity to other balconies, privacy and protection are most essential. Also, because of the construction cost, balconies have tended to be minimal in size.

Terrace Generally related to the extension of the first floor at grade level of the living-dining area. Because of the simplicity of the slab on grade, terraces have tended to be more expansive than balconies. Terraces related to multi-purpose rooms or even bedrooms are not uncommon.

Patios Patios are similar to terraces and always located on the lowest levels on grade.

Enclosed porch The enclosed porch is not truly an outdoor space because it usually has a roof overhead and is partially enclosed. However, it does serve as a transitional or in-between area of the indoors and the outdoors.

All outdoor spaces should have ample space to function properly, be adequately screened and sheltered for privacy, and be functionally integrated with the adjacent living area.

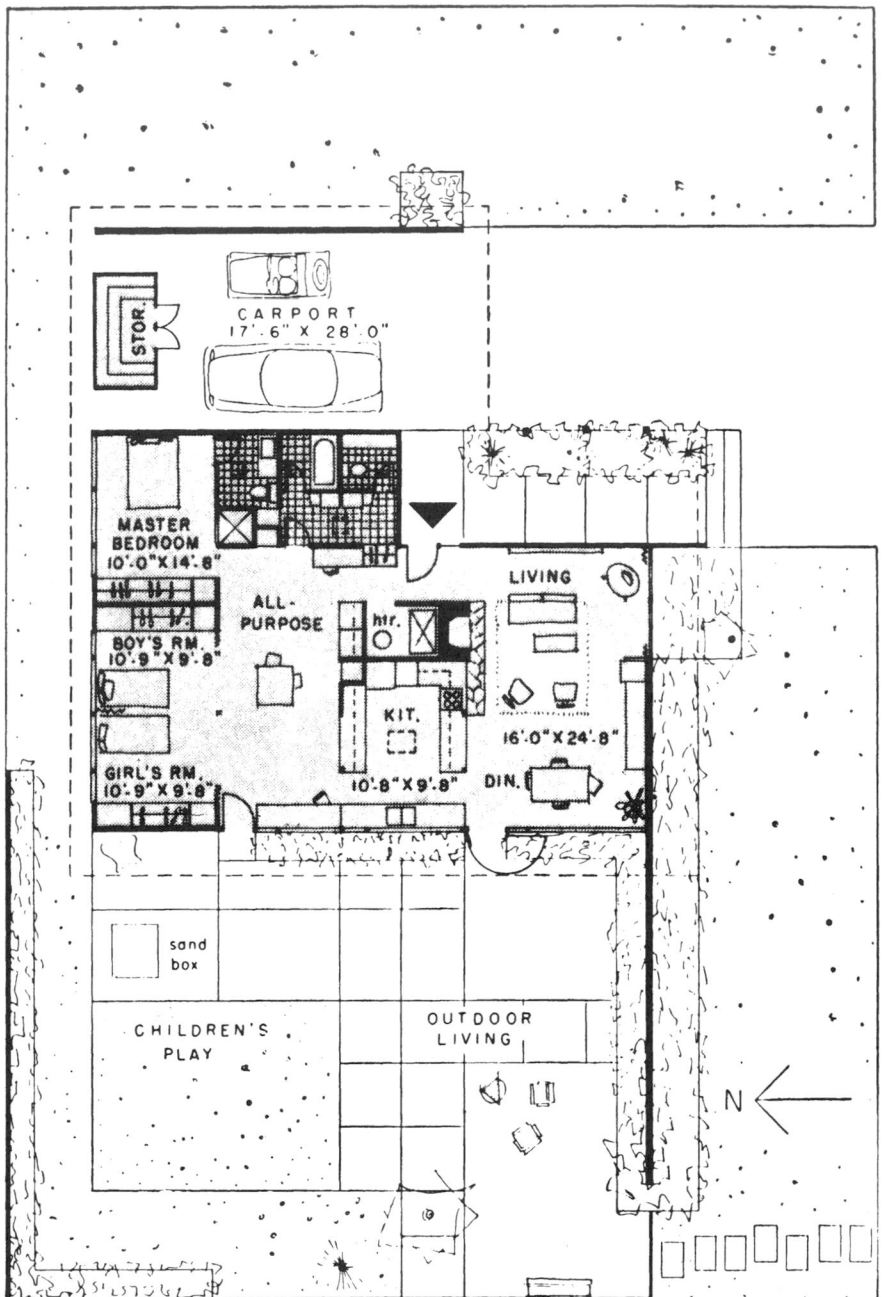

Fig. 1

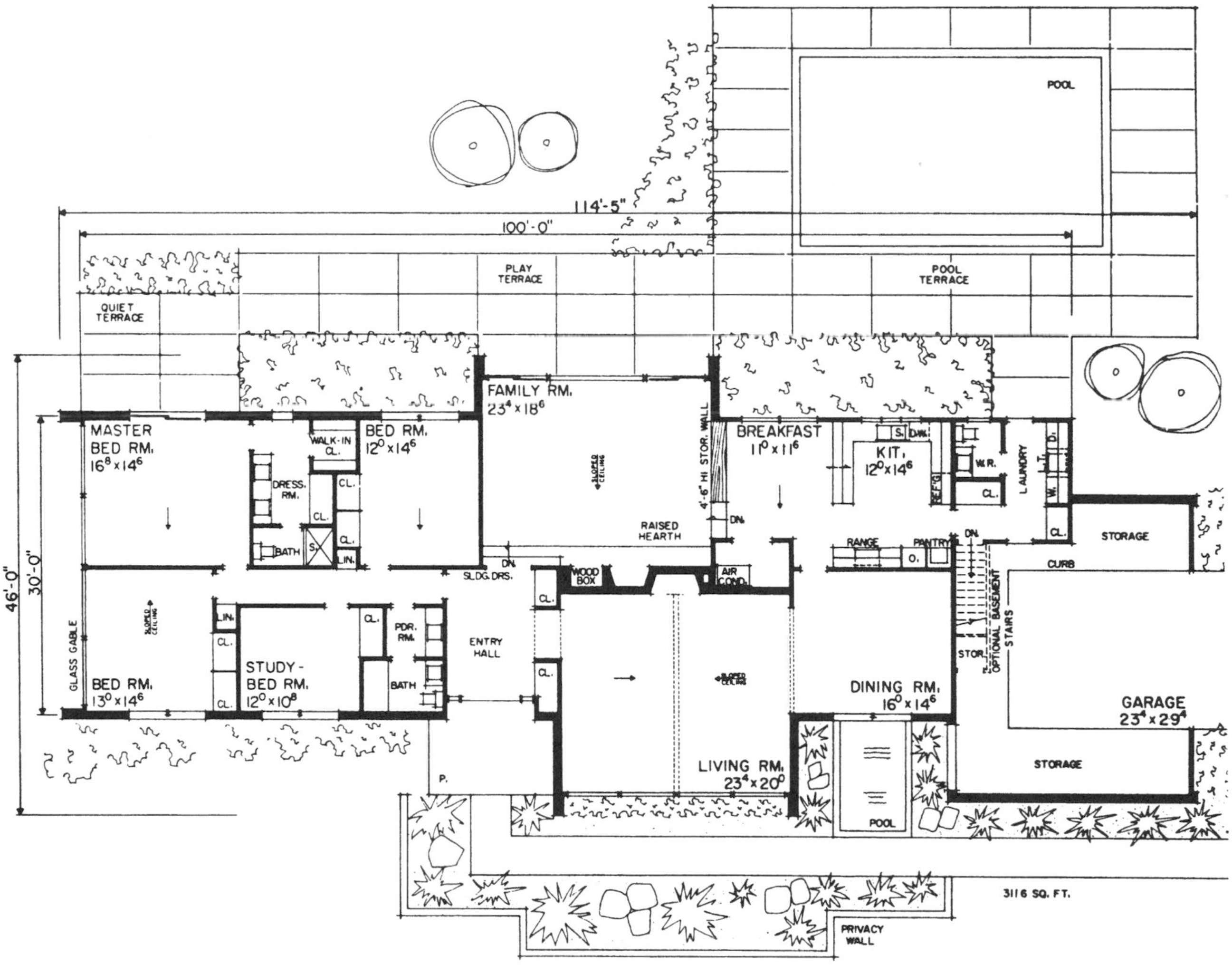

Fig. 2   Relationship of indoor and outdoor areas.

Fig. 3   Community space—outdoors and upstairs.

Fig. 4   Private terrace.

OUTDOOR LIVING SPACE

Balconies and Patios

1. Make the private outdoor space large enough to allow a range of family activities to occur there.

 a. Main balconies should be designed as an outdoor room (approximately 8 by 12 ft). Balconies should be a minimum of 90 ft, tend to a square shape, and should be provided with some screening for privacy.

 b. Patio (ground-level outdoor space) should, if possible, be between 400 and 800 ft^2; 400 ft^2 is a minimum requirement.

2. Shelter the private outdoor space against rain and wind; provide good drainage. On balconies, partial recessing helps.

3. Orient the private outdoor space to receive sun at least part of the day.

4. Planter boxes provide opportunities for personal expression and assist in reducing visual intrusion.

5. Arrange and locate private outdoor space to make it as private and as secure against intrusion as possible.

In developments where all households are likely to be families with children the off-grade private open space might overlook the central court or children's play area, thus facilitating surveillance of children's play.

Fig. 5

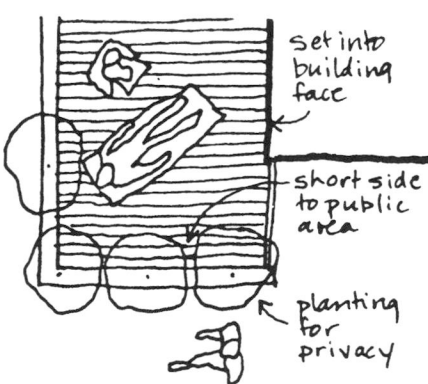

Fig. 6

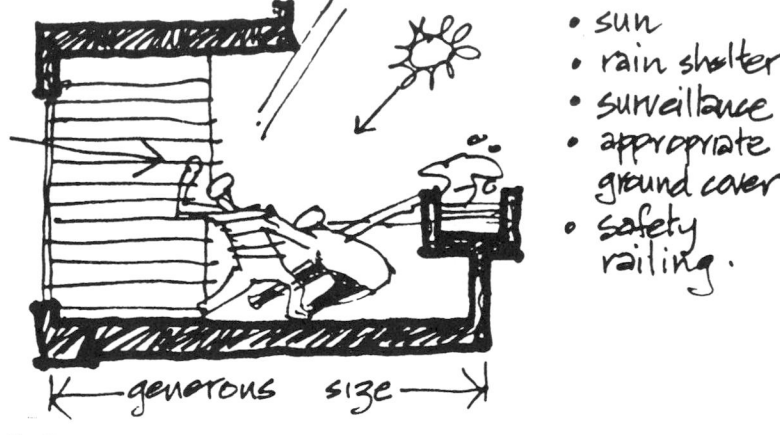

Fig. 7

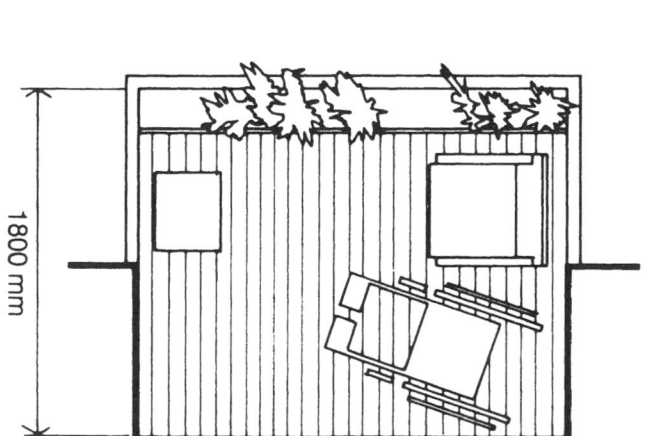

Fig. 8   Plan of balcony.

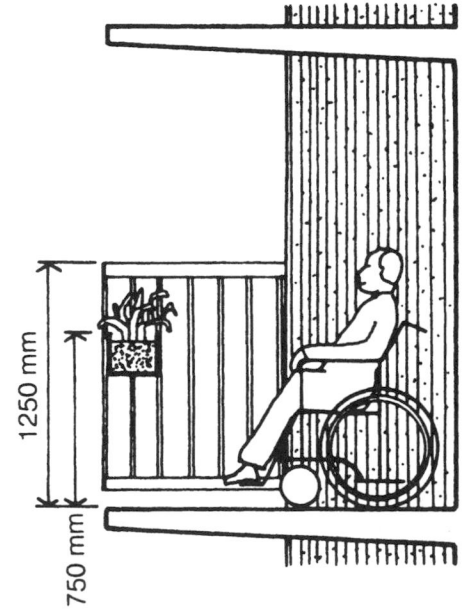

Fig. 9   Section through balcony.

LAUNDRY

Facilities should be provided for household laundry needs.

Where central laundry facilities are not provided, each living unit should contain at least a laundry tray, or a combination sink and tray fixture, or the installation of water and waste piping and space for a clothes-washing machine.

Artificial light should be provided. Clothes dryers, where provided, should be adequately vented.

The laundry facility may be located in the kitchen or family room, but its location in a utility room or bathroom is preferable in order to provide more noise control and improved sanitation.

Laundry space allowances:

One laundry tray, 24-in frontage by 20 in

One combination sink and tray, 42-in frontage by 21 in

One clothes washer, 30-in frontage by 30 in

One clothes dryer, 30-in frontage by 30 in

CLOTHES-CARE CENTERS

Second only to meal preparation in importance, in the average house, is the care of clothing and household linens. This care includes washing, drying, ironing, mending, and construction.

It seems logical to combine the sewing center and laundry centers to make a complete clothes-care center. Such a center encourages mending in relation to laundry, as well as pressing in relation to both sewing and laundry. The table used for laundry sorting and folding can double as a cutting surface.

Figure 2 is an example of a clothes-care center that meets space requirements of both activities. A free-standing ironing table (board) is easier to adapt to room arrangements than a built-in one.

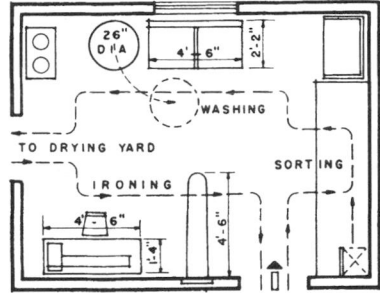

Fig. 1a Laundry—showing routines.

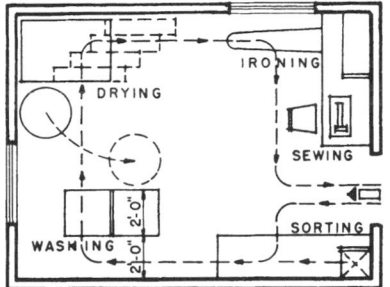

Fig. 1b Laundry—free-standing tubs.

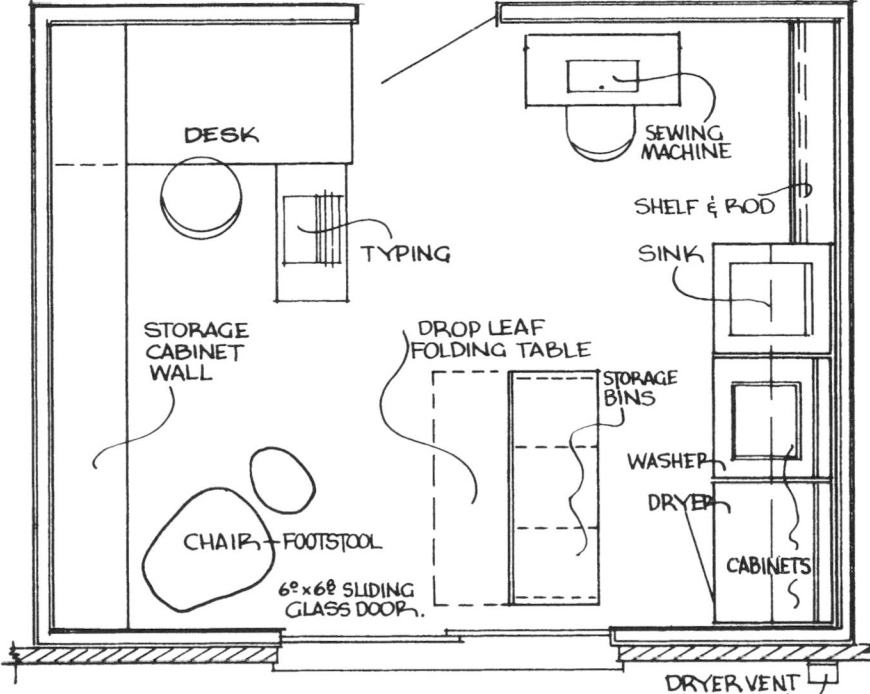

Fig. 1c

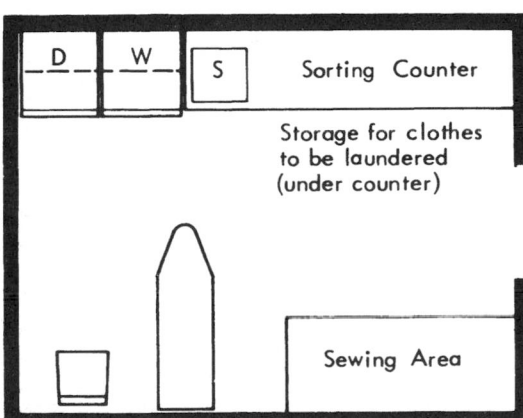

Fig. 2   Floor plan of a clothes-care center.

HOME LAUNDRY AREAS

LAUNDRY AREA

Location

Changes in laundry equipment, new developments in fabrics, changes in standards of cleanliness, and available space all influence where the laundry center will be located. Other considerations are venting, noise, and possible breakdowns of the equipment.

Some house plans will show a central location for the laundry; others require a location at the end of the house. While a central location reduces carrying laundry and lots of steps, this may be secondary to some other features of the house. See Figs. 3 and 4.

There are very few hard and fast rules about where a laundry center should be located. Although many laundry areas are still located in the basement, some are moving upstairs into fashionable settings. A bathroom or bedroom area is ideal, since most dirty laundry accumulates in these locations. In warmer climates the laundry is often located in carports, patios, or breezeways. And locating the laundry in the kitchen makes it easy for everyone to help out with both laundry and kitchen duties.

In looking for the best laundry location, think in terms of saving time and energy.

First-floor utility and/or mudroom A utility room, right off the kitchen, is an excellent location. It can share plumbing facilities with the kitchen. Sewing and mending also might be located here. This location for the laundry area eliminates stair climbing and allows for messiness. This area is usually adjacent to the kitchen so it is possible to supervise activities in both areas at the same time.

Direct access to the outside is important if the area is to serve as mudroom, cleanup area, or garden work area. If space permits, a sink can be used by children for washing up when they come in from play. Muddy or dirty play clothes can be changed here and left for laundering.

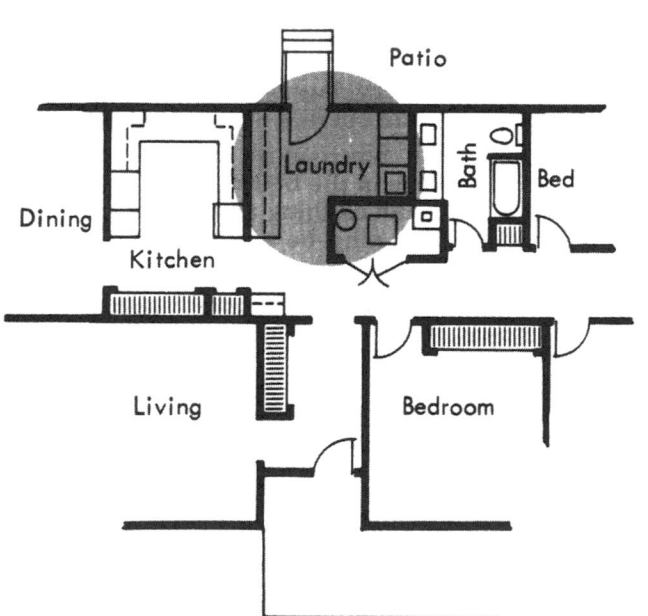

Fig. 3   Floor plan of central laundry location.

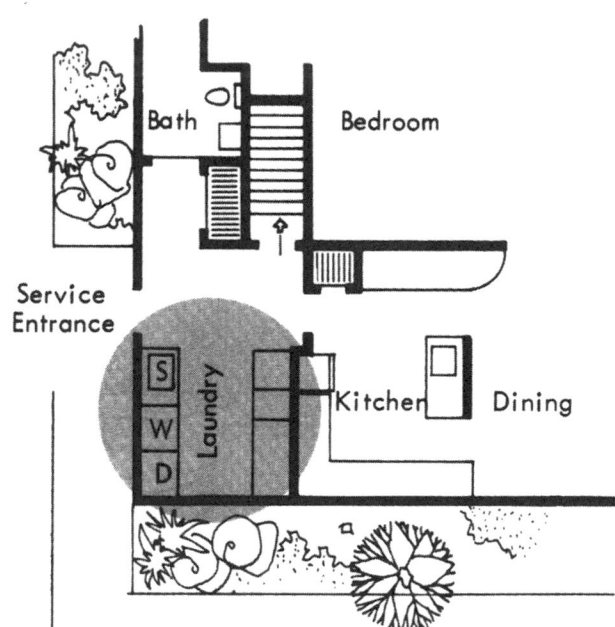

Fig. 4   Floor plan of end laundry location.

Kitchen Putting the laundry center in or near the kitchen concentrates the location of two busy work areas of the home. Having the washer and dryer in or near the kitchen area helps in dovetailing household tasks. Also, plumbing lines are close and there is water available for diluting laundry products and pretreating, if necessary. If a wall or corner of the kitchen is used for the laundry center, some type of partition or peninsula should separate the laundry and food preparation areas for sanitation reasons. The laundry area needs its own work space and storage.

If you plan to dry laundry outside, convenient access to the outside is important.

Bedroom/bath area The bathroom is a logical accumulation point for soiled clothing and linens, plumbing lines are close and the existing sink can be used for diluting laundry products and for any pretreating that might be necessary. In addition, the floor, wall and counter surfaces are the same type as recommended for a laundry—moisture-resistant, durable and easy to clean.

If space is available in an older home or can be planned into a new or remodeled home, the bedroom hallway is an excellent location. It's at the hub of the bedrooms and bath where most laundry is collected and bathroom plumbing is nearby. The hallway itself can provide the floor space needed for working in the center. Pass-through storage directly to the rooms behind the laundry center is quite feasible.

The laundry center could be in a corner of a bedroom or along one wall that backs up to a bathroom. This permits utilization of existing plumbing, thus, cutting down appliance installation costs. A spare bedroom could be turned into a combination laundry-sewing-hobby and general-activity room. The area can be attractively hidden from view with sliding or folding doors that blend with the room decor.

But laundry equipment does not operate silently; so you may not want it next to your bedrooms. A separate room is highly desirable, especially if you develop a complete clothes-care center.

Basement Because of less demand on basement space, the laundry center is frequently placed there. Other advantages are the removal of the noise from the rest of the house and proximity to the source of the hot-water supply and other plumbing lines.

However, a basement location for a laundry center is not recommended, unless no other choice is feasible. Studies show that an upstairs location can reduce footsteps by as much as one-half. Carrying heavy loads up and down stairs adds to the inconvenience. Also, it is difficult to combine laundering with other household activities.

For a basement location, a clothes chute from the upper floors helps eliminate the extra labor of carrying the clothes.

Fig. 5 Kitchen.

Fig. 6 Bedroom/bath.

HOME LAUNDRY AREAS

Family room laundry center This room frequently provides pleasant surroundings for laundry facilities because of the other activities that take place here. It often has a wall or corner that can be utilized for a laundry center. A sink could serve a dual-purpose for both the laundry work and entertainment purposes. Depending upon the floor plan of the house, you should keep in mind these disadvantages of a family room laundry: transporting the laundry from the point of accumulation to the laundry center may involve considerable walking and one or more flights of steps, and plumbing lines may not be readily available. On the other hand, a family room might make the task more pleasant and encourage some volunteer help.

Patio, breezeway, carport, or garage In warm climates, one of these locations may be your best choice for a laundry center, particularly if it's located near the bedroom/bath or kitchen areas of the home.

Keep in mind that a laundry center needs hot and cold water lines, a drain, electricity and, with a gas dryer, a gas connection.

Other A hall, closet, or pantry location usually is not satisfactory, because available space is not adequate to develop a complete and efficient laundry center.

Space Requirements

An efficient laundry area should accommodate the following related activities for clothing and soiled linens:

Collection and storage
Sorting
Pretreating
Hand laundry
Machine washing
Drip drying
Machine drying
Folding and sprinkling clothes
Ironing
Temporary storage for cleaned and ironed clothes

Preparation area needs: sink, sorting table or area, containers for dirty clothes and storage area for supplies and equipment used in pretreating.

Sorting may be done on a table or counter at a comfortable height. A counter 6 ft long will provide adequate space to sort a 32-lb wash (Fig. 9). Three to six bins, baskets, or boxes can also provide a convenient way of sorting clothes as they accumulate.

Fig. 7   Family room.

Fig. 8   Patio, breezeway, carport, or garage.

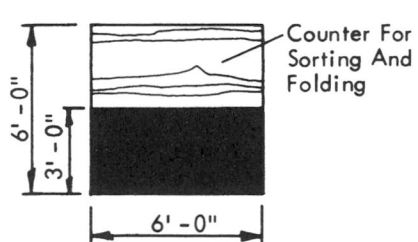

Fig. 9   6- by 6-ft desirable space for sorting and folding clothes.

Counter For Sorting And Folding

6' - 0"
3' - 0"
6' - 0"

8' - 0"   8' - 0"

Fig. 10   Nonautomatic washer and rinse tubs.

Machine washing area needs Washer, detergent, storage, water conditioners, fabric softeners, bleaches, and measurers.

The type of washer determines the space requirement of this area. Automatic washers may vary from 24 to 30 in in width. An area approximately 8 by 8 ft is needed for an efficient arrangement of a nonautomatic washer and rinse tubs and space for the worker to move around (Fig. 10). The location of drain openings and agitator controls influences the actual arrangement.

Top-opening washers require space above them for opening the door and loading clothes. Front-opening washers require space for the door to open, and for the worker to bend. About 36 in should be available in front of the washer. If there is a passageway behind the worker, then 24 in more is desirable.

Drying area needs Dryer, folding table, rods for hanging, laundry cart, clothes line, clothes pins.

Dryers are 24 to 30 in wide. If space is limited, stack models and the combination washer-dryer are available. Provide a minimum of 42 in in front of the dryer to work without being cramped.

For convenience in handling laundry, place the dryer adjacent to the washer, either side by side (Fig. 12) or at right angles with the doors opening away from each other (Fig. 13). Place the dryer where it can be easily vented.

Space is needed for hanging no-iron garments as they are removed from the dryer.

Table space of 36 by 60 in is also needed for folding laundry (Fig. 9).

Ironing area needs Ironing board, iron, facilities for stacking and hanging clothes, dampening equipment, and pressed clothes, and storage for spray starch, fabric finishes, etc.

To provide adequate space for using an ironing board, approximately 4 ft 9 in by 6 ft of floor space is needed. Thirty inches is needed on the working side of the board, 6 in on the other side, and 18 in at the point of the board (Fig. 14).

Containers for ironed and unironed clothing should be within easy reach of the worker.

Arrangement

The arrangement of equipment, work surfaces, and storage areas should permit the worker to progress in a logical sequence. This will reduce the time and energy necessary to perform the tasks. Consider elevating a front-opening washer or dryer, or combination, 1 ft to allow easier use.

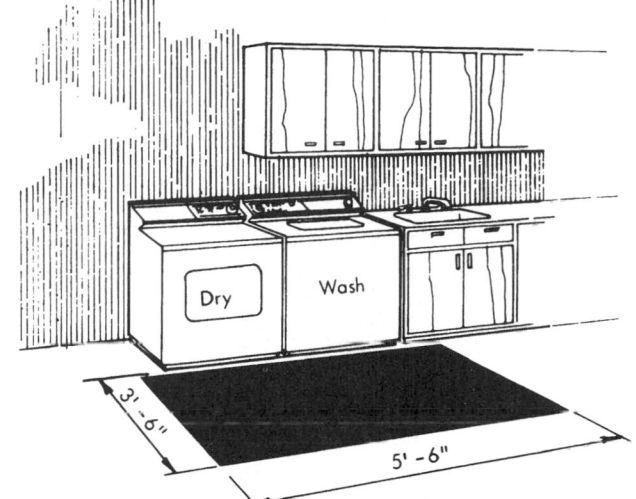

Fig. 12   Side-by-side washer-dryer location.

Fig. 13   Right-angle washer-dryer location.

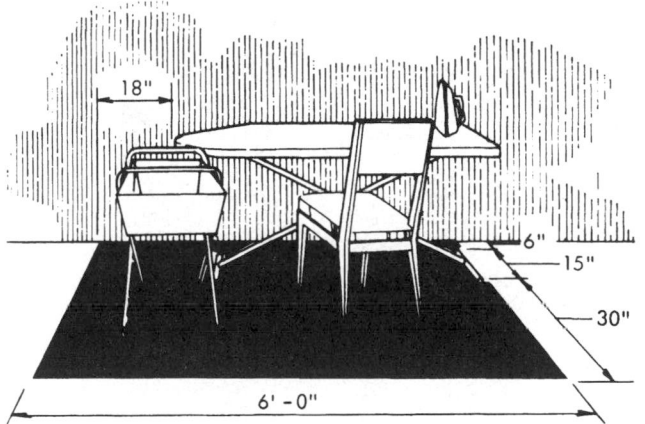

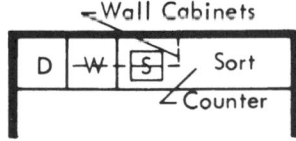

Fig. 11

Fig. 14   Ironing space requirements.

HOME LAUNDRY AREAS

LAUNDRY ACTIVITIES

Home laundry includes the processes from sorting through ironing of clothes and household linens, including pretreating, washing, drying, and sprinkling.

General Planning Suggestions

1. It is desirable to plan space for specific laundry processes.
2. Moistureproof surfaces are needed for pretreating and sprinkling of clothes.
3. Drying areas should be accessible for use under all climatic conditions.
4. To control moisture in the room, some types of dryers should be located to permit venting to the outside of the house.
5. Adequate storage for washing equipment and supplies should be located near the place of first use.
6. Facilities for hanging drip-dry garments after washing should be provided.
7. In locating the washing equipment consideration should be given to convenience of interrelated household activities, distances from the source of soiled clothes and the drying areas, and the isolation of clutter.

Figure 17 illustrates arrangements of laundry equipment. Space needed by a single worker in front of equipment or between equipment placed opposite is indicated. Overall dimensions of areas will vary with type and size of equipment selected. No allowance has been made between the back of equipment and the wall for electrical, plumbing, and dryer vent connections. The space required will depend on the type of installation used.

Counter space is provided for sorting and folding three washer loads of clothes. The space under the counters has been used for bins, one for soiled clothing and the other for dry, clean articles that require further treatment before use or storage. Additional counter space can be provided by the tops of the dryer and washer, depending upon the type selected.

A tall storage cabinet for laundry supplies is shown in each arrangement. In this cabinet, an ironing board, iron, mops, and buckets (needed for cleaning the laundry area) may also be stored.

TABLE 1   Space Requirements for Washer-Dryer Arrangements

Type and size of equipment	Auxiliary equipment	Work area	Total floor area, in	
			Width	Depth
Stacked arrangement: washer, 31 × 26 in; dryer, 31 × 26 in	Basket, 19-in diameter	43 × 37	43	63
Angle arrangement: washer, 26 × 26; dryer, 31 × 26 in	Basket, 19-in diameter	36 × 59	62	76
Straight-line arrangement: washer, 26 × 26 in; dryer, 31 × 26 in	Basket, 19-in diameter	36 × 66	62	66

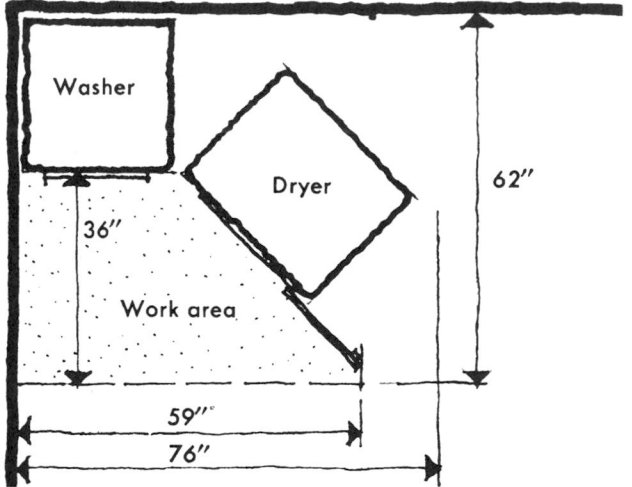

Fig. 15a   Angle arrangement.

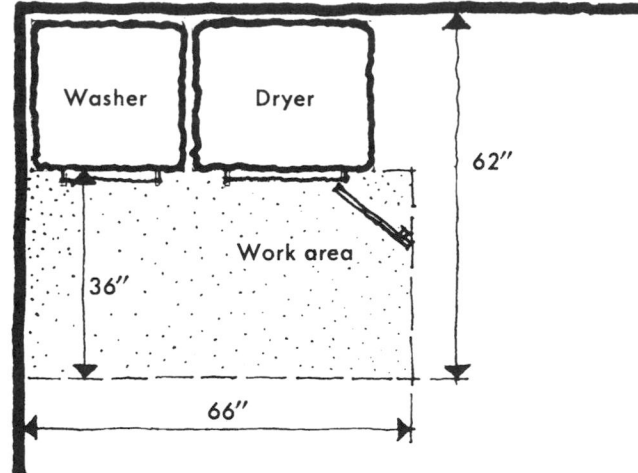

Fig. 15b   Conventional arrangement.

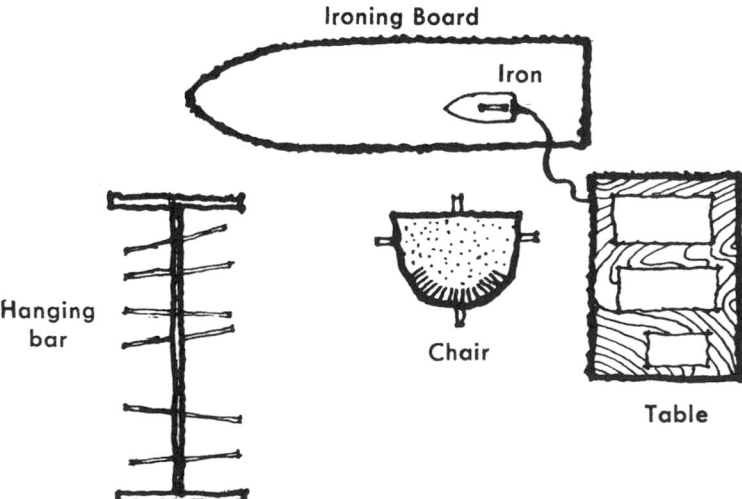

Fig. 16   Arrangement of ironing equipment based on flow of work.

TABLE 2 Space Requirements for Ironing at an Ironing Board

Measurement	Adequate dimensions, in	
	Depth	Length
Clearance for working		
At front of board	32	
At back of board	8	
At tapered end of board		18
Area		
Standard ironing board	15	54
Board plus working space	55	72

TABLE 3 Counter Requirements for Laundry Activities

Activity	Counter width,* in	
	Limited	Liberal
Sorting soiled clothing loads	24	46
Folding clean, dry articles	20	30

*Counter depth of 25 in.

TABLE 4 Key to Units in Fig. 17

Symbol	Unit	Approx. Width, in	Approx. Depth,* in
W	Washer	30	27
D	Dryer	30	27
T	Tub	20	26
St.	Storage	24	25
S.Cl.	Soiled clothes	25	25
C.Cl.	Clean clothes	21	25

*Depth, front to back.

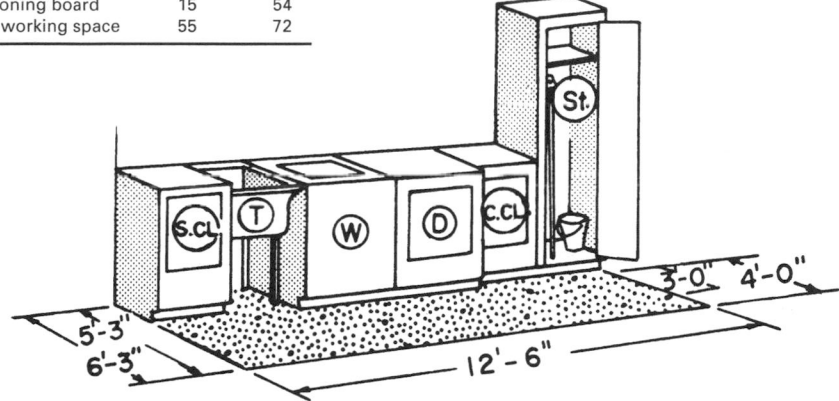

Fig. 17a One wall.

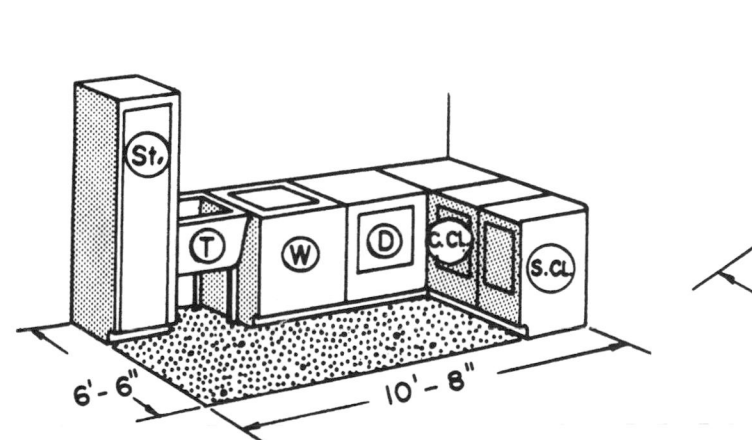

Fig. 17b L shaped.

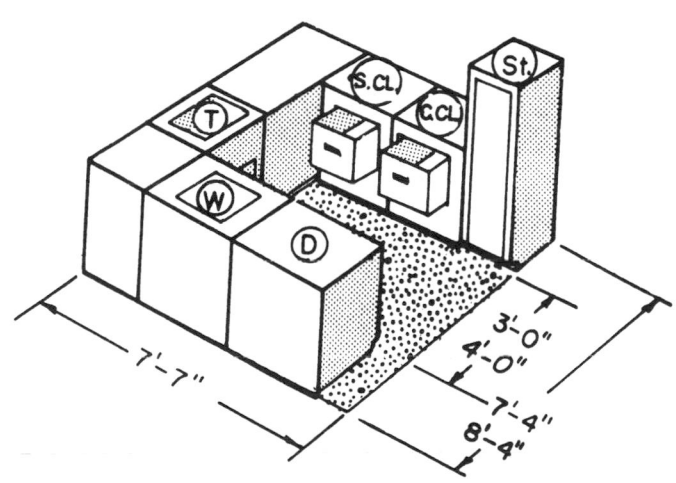

Fig. 17c Parallel wall.

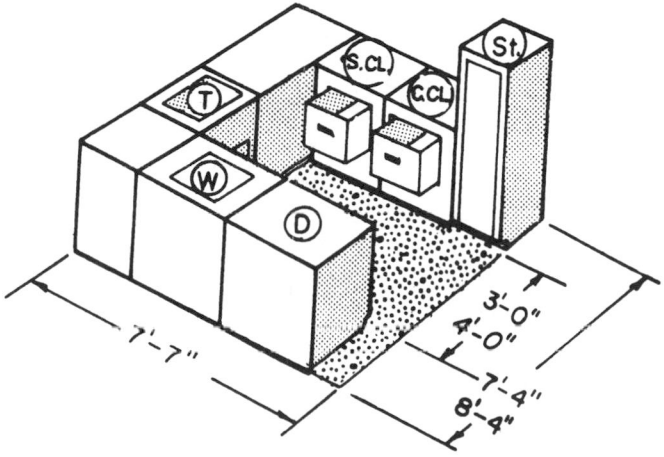

Fig. 17d U shaped.

ELEMENTS OF A LAUNDRY CENTER

When laying out the floor plan, consider each step in the laundering process and eliminate all unnecessary motions. Try to arrange equipment, as nearly as possible, in the order that the laundry steps are performed.

Each of these steps requires some kind of facility—either storage or work space—and the more of these facilities you can factor into your laundry center, the more efficient it will be.

Storage for Accumulated Laundry

If at all possible, storage for soiled laundry should consist of several smaller containers, rather than one large hamper. If clothing is sorted as it accumulates and is put with other items that will be laundered in the same manner, you don't have to re-sort every time you want to wash a load of clothes. Family members can help by dropping their laundry into the appropriate containers.

Soiled laundry should be sorted by color, degree of soil, and construction or type of fabric. The number of containers needed for adequate presorted storage depends upon the type of laundry that accumulates. The loads described below offer guidelines for proper sorting.

1. White household linens, underwear, shirts, handkerchiefs, towels, and similar cottons. (White nylons should be washed separately because they pick up color from other fabrics.)

2. Light colored dresses, blouses, sportswear, children's clothes, and light colored items of the type listed in 1 above.

3. Bright and deep colored garments and household items for which color retention is the most important factor.

4. Knits, acrylics, polyesters, nylons, and garments labeled *permanent press.*

5. Delicately constructed garments, lace-trimmed articles, lingerie, and articles made from sheer fabrics.

Transportation of the Laundry

Be sure to include some method of moving the clothes from one part of the house to another, such as a clothes chute, if laundry must go from one floor of the home to another, or some type of rolling cart. Some laundry carts provide compartments for presorting, as well as transportation of soiled laundry.

Storage for Laundry Products

How much and what type of storage you need depends upon the variety of laundry supplies that you use and the size of packages you purchase. In addition to detergents or soaps, bleaches, fabric softener, and other regularly used laundry products, it is a good idea to plan for the storage of stain-removal supplies or special pretreatment products.

Overhead storage is preferable because it eliminates stooping and is best for keeping supplies out of the reach of small children.

Counter Workspace

You will need counter workspace for activities that are preliminary to the actual laundering process: sorting the load (if not already sorted), examining garments for any that need special attention before laundering, space for removing spots and stains, mending small rips or tears, or pretreating garments with areas of heavy soil. If you have no other choice, you can use the tops of the appliances as a work area.

For folding and sorting freshly dried items, you can use the same workspace discussed earlier for treating garments before laundering.

If it is not always convenient to return clean, folded clothes and linens immediately to the places where they are stored, a simple shelf arrangement over the washer and dryer could be used for the temporary storage of clean clothes. In a more elaborate laundry room, separate bins or cupboards could be used for this purpose. One practical idea is to equip a cabinet or cupboard with removable bins or trays, each one labeled with a family member's name. Clothes are sorted accordingly. Family members are then responsible for returning their clothes to their own rooms.

If you have space, it is a convenience to have a small sink in the laundry for dilution of laundry aids and for spot and stain removal or other pretreating, if needed. The need for such a sink is minimized if you select a washer with up-to-date features, such as a prewash or soak cycle, bleach dispenser, and fabric softener dispenser.

Hanging Area

As soon as the dryer stops, articles should be removed and folded or hung on hangers promptly. This is especially important with nylon, acrylic, polyester fabrics, and garments with a permanent press finish that will become wrinkled if allowed to remain in the bottom of the dryer drum after tumbling stops.

To get maximum benefit from the many garments that shed their wrinkles in the dryer, you will want to provide a place to hang permanent press and other no-iron garments as they are removed from the dryer. Space permitting, a full-length hanging closet is ideal. If a separate closet is not feasible, a portable garment rack, clothes pole, extension rod, or even a wall hook will suffice.

Space for Ironing

Depending upon the wardrobe and upon laundering procedures, it may not be necessary for you to allow space for this activity.

Permanent press and other no-iron garments may, by now, have reduced the ironing load to a few occasional items. Still, when ironing is required, it is generally most convenient to provide for this activity in the laundry center.

For an ironing area in the laundry center, plan on space at least 5 ft 10 in wide and 4 ft 3 in deep to accommodate an ironing board, chair, and laundry cart or basket.

Space to Mend and Sew

Small holes or openings in garment seams should be mended before laundering to keep them from getting bigger. A small sewing box, with the necessary thread, needles, and scissors should be kept on a shelf or in a drawer in the laundry center.

If you do a lot of sewing, however, it makes sense to incorporate this activity into the laundry center. In addition to the sewing machine, you'll want a pattern file, a large flat surface that can be used as a cutting table, and drawers, shelves, or bins for garments being constructed or fabrics waiting to be cut. The hanging closet can be used for partially constructed garments and the ironing equipment will be close at hand for pressing during construction.

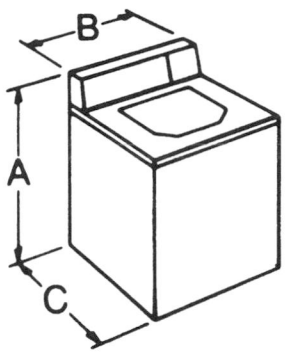

Fig. 18   Washer dimensions. A = 42 to 44 in; B = 24 to 30 in; C = 25 to 27 in.

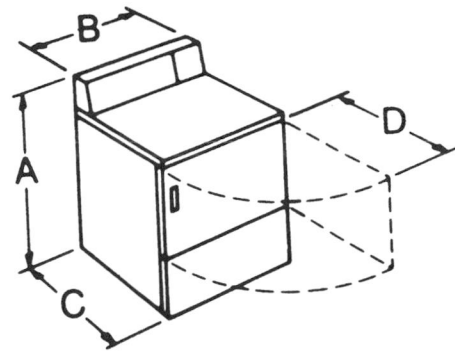

Fig. 19   Dryer dimensions. A = 42 to 44 in; B = 27 to 31 in; C = 25 to 27 in; D = 26 to 28 in.

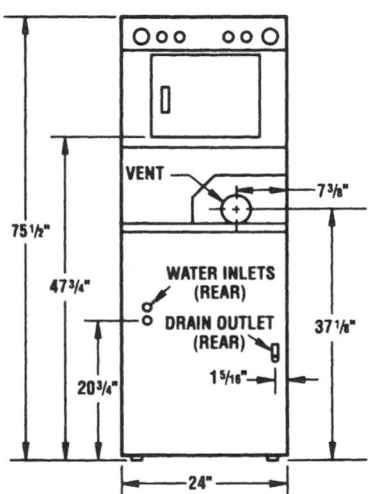

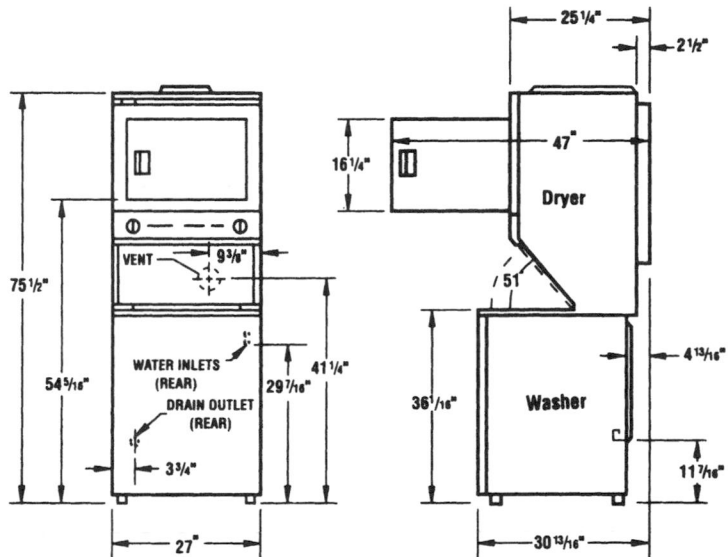

Fig. 20   Typical stacked units.

Elements of the Dwelling Unit

HOME LAUNDRY AREAS

TABLE 5

Processes	Equipment	Supplies
Collection of soiled linens Bring clothes from place of soiling to place of washing.	Laundry bags for each person in his or her room. Hampers in the hall, the bathroom, in or near the kitchen. Chutes from bedroom areas to laundry areas on lower levels. Built-in bins in the bathroom, utility room, or halls.	
Sorting Sort into loads according to white, colorfast and noncolorfast, heat sensitive fabrics, slight or heavy soil.	Built-in bins provide presorting. A size 13 by 12 by 22 will accommodate a typical load.* Work counter 6 ft by 2 ft 6 in adequate for 32 pounds of laundry.† This amount of clothes equals four loads. A file of hang tags from garments giving washing instructions.	
Pretreating Stain removal. Wash out water soluble stains with clean hot or cold water. Use chemical solvents for other stains. Treat much soiled parts with concentrated soap or syndets.	Work counter with moisture proof finish 20 by 60 in. Tops of washer and dryer may be a part of this. A small sink is desirable. Storage space in wall cabinets. Tea kettle and heating unit.	Dry cleaning fluid. Bleaches. Soap or synthetic detergents. Scouring powder.
Mending Darn, patch, repair seams of anything needing it, except unpleasantly soiled clothing.	Needles, scissors, sewing machine. Ironing board and iron.	Threads of suitable colors, iron-on patches.
Washing Wash in washer and treat as required during last rinse.	Hot and cold water supply. Automatic washer with outlet to sewerline, or Wringer or nonautomatic washer, 2 to 3 portable tubs, or 2 stationary tubs. Sticks for lifting hot clothes, or combination washer-dryer.	Water softener. Soaps or synthetic detergents. Starches. Clothes softeners. Bluing.
Drying *With a dryer.* Place directly into dryer, if adjacent space is available. Hang up wash-and-wear clothes and shape as soon as from the dryer. *By line hanging.* Remove to drying yard or indoor hanging space. Hang. Remove from line and fold. *Drip-dry.* Some garments cannot be satisfactorily dried in dryer, or spun dry. They should be line hung.	Clothes dryer set to the *right* of the automatic washer, or stacked over the washer. Hangers and hooks near the dryer. Clothes basket. Laundry cart. Clothes line 124 ft long. Floor drain, laundry sink, with rod over it.	Clothes pins. Pan. Clothes hangers. Pant stretchers, hose stretchers.
Sprinkling and folding Lay clothes on a work surface. Sprinkle. Fold and set away to cure. Many clothes require no sprinkling, and may be ironed slightly damp from the line or dryer. Others require no ironing.	Work surface, may be the top to washer or dryer. Source of water. Sprinkler bottle, or brush and small pan. Plastic bag. Clothes basket.	
Ironing Ironing set up near the dryer for touchups of wash-and-wears. In other areas a setup for standing or sitting to iron other garments is needed. Shake and smooth out, iron. Hang garments to air after ironing. Fold linens to convenient size for storage and use.	Dry and steam irons, sleeve board. Ironing board adjustable from 23 to 39 in high, or ironer. Hanging racks. Work surface. Clothes basket at suitable height.	Paraffin. Distilled water. Sponge.
Store		

*Bernice Strawn, "New Laundry Layout," *New Homes Guide,* Vol. 46 (Summer, Fall, 1961), p. 121.
†Cecile Sinden, and Kathleen Johnston, *Space for Home Laundering* (Agricultural Experiment Station, Bulletin No. 658; Pennsylvania State University, College of Home Economics, Research Publication No. 162 [Pennsylvania State University, 1959]), p. 7.

General Planning Suggestions

1. An area especially planned for sewing, convenient to other activity areas, is desirable.

2. Most houses need storage space for sewing materials and equipment. The amount and kind of storage required varies according to the quantity and frequency of sewing.

3. A minimum sewing area should include the machine, auxiliary work surfaces, a chair that permits freedom of motion, and storage arrangements. The work surface for layout and cutting may be outside the area for sewing machine operations and serve multiple purposes.

4. Consideration should be given to work surfaces at comfortable heights for the varying activities of sewing.

5. Lighting should be adequate for the activity.

TABLE 1   Dimensions of Area for Layout and Cutting Garments

Measurement	Dimensions, in	
	Minimum	Adequate
Working surface		
Length	56	72
Width		
Table, free-standing	28	36
Table obstructed on one side	28	32
Height	34–40 (range)	36 (median)
Clearance for worker	18	24

TABLE 2   Dimensions of Fitting Space

Use of space	Minimum	Adequate
Viewing in mirror		
Mirror dimensions, in		
Width	16	18
Length	42	60
Top to floor	70	72
Clearance in front of mirror, ft		
Width	3	4
Length	6–8	10
Clearance while fitting self, ft	6 × 4	
Clearance while being fitted, ft	8½ × 4	
Fitting garment on dress form, ft	5 × 4	7 × 6

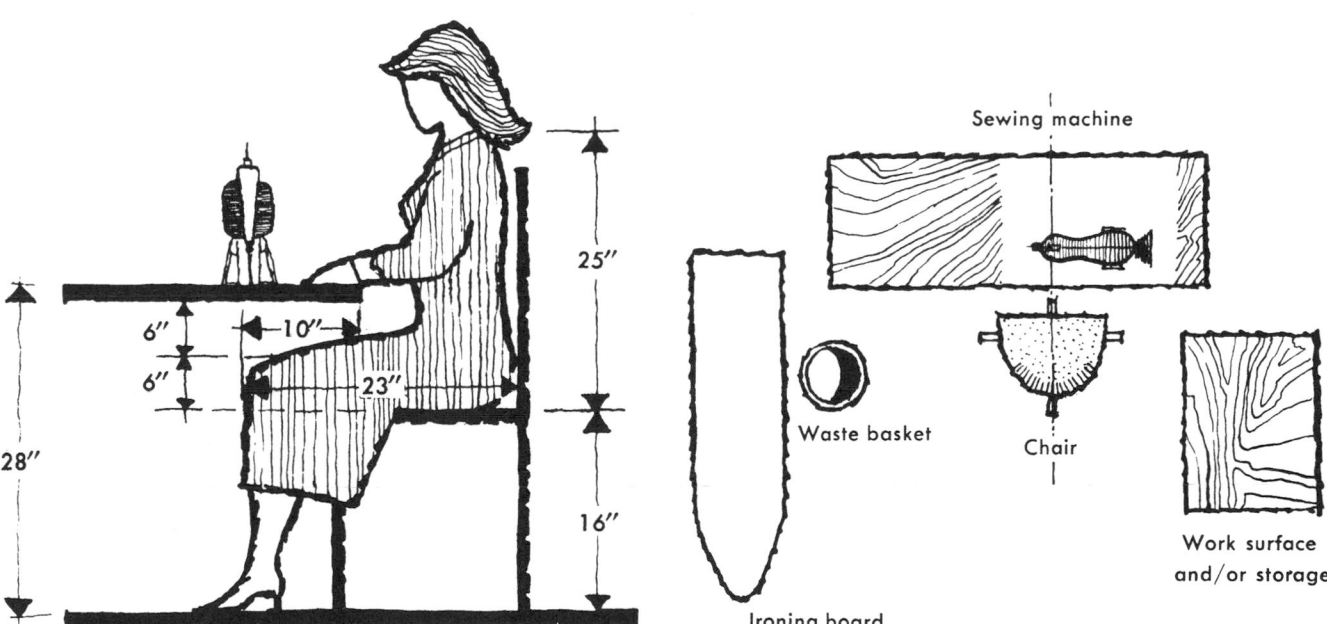

Fig. 1a   Mean heights and clearances for sewing machine use.

Fig. 1b   Arrangement of sewing equipment based on flow of work.

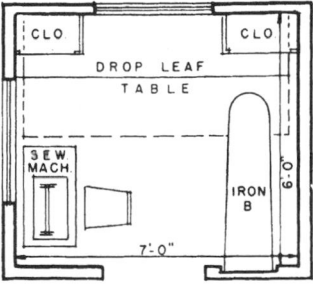

Fig. 1c   Sewing room.

WORK AND CRAFTS CENTERS

For standing work height (Fig. 1) the height of the elbows above the floor (elbow height) should be considered. If considerable muscular force is required, the distance from the elbow to the top of the bench should be clearly greater. If minimal physical force is involved, a distance between the elbow and the bench top of between 3.5 and 6 in should be adequate. For preliminary design assumptions, a height of 34 to 36 in would be reasonable.

In regard to bench heights for seated work (Fig. 2), 24 to 29 in can be used for preliminary design assumptions. The limitations of human reach must also be taken into account in locating overhead tool storage.

Figure 3 indicates some of the critical dimensions related to an arts and crafts center for children ranging in age from 6 to 11 years. The critical anthropometric consideration is in making the design responsive to the body size of the child as well as the adult. A teacher forced to bend to the surface of tables scaled down to the body size of a child would suffer fatigue and backache in a short time. Adjustability in both chair and table, however, can reconcile the needs of differing requirements.

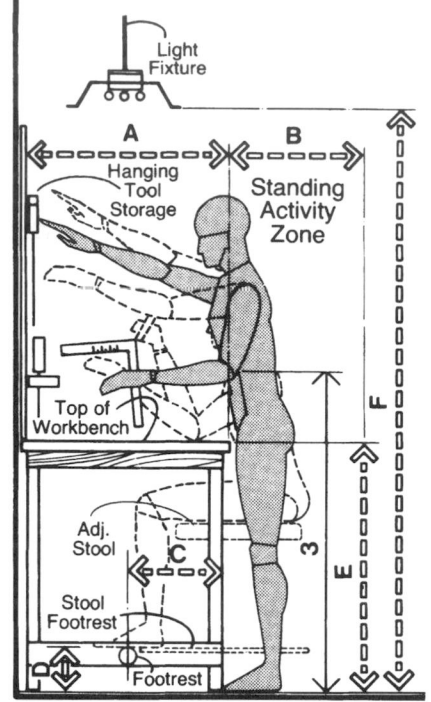

Fig. 1   High workbench. (See Table 1.)

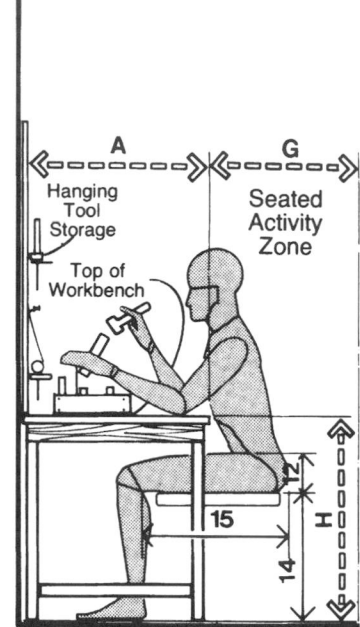

Fig. 2   Low workbench. (See Table 1.)

TABLE 1

	Inches
A	18–36
B	18
C	6–9
D	7–9
E	34–36
F	84
G	18–24
H	29–30
I	65
J	36
K	30
L	15
M	21
N	24
O	22–27
P	29
Q	34
R	33
S	26
T	16

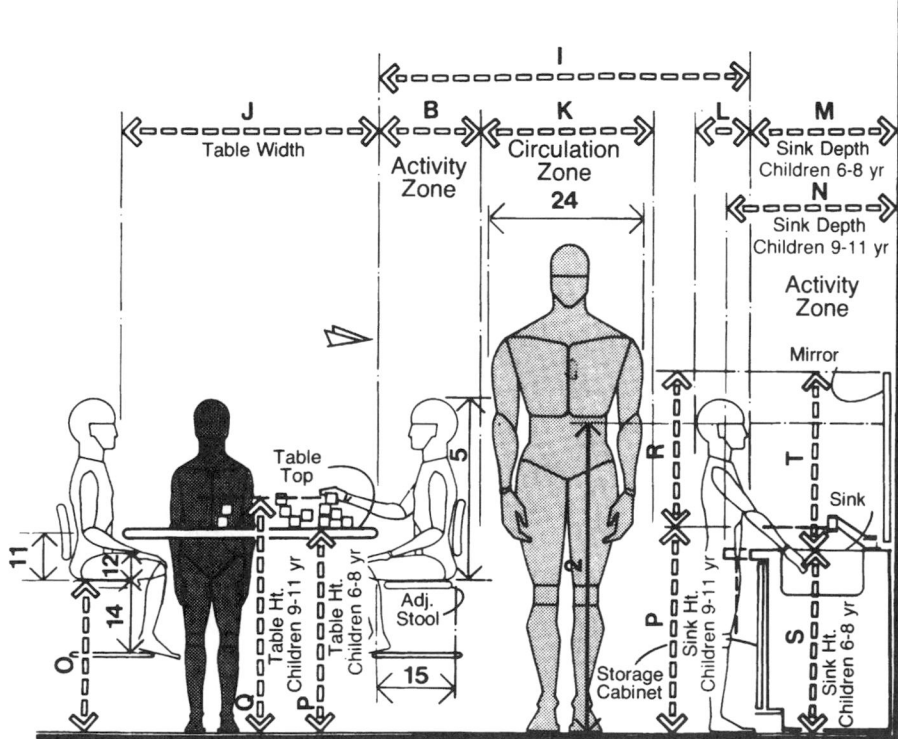

Fig. 3   Child's art and craft center. (See Table 1.)

Most artists have individual preferences regarding the arrangement of their particular studio or workplace. In regard to human dimension and the artist's interface with his or her space, the factors to consider also vary greatly. Techniques, media, style, and process all impact on the anthropometric requirements. Figure 4 therefore, should not be taken too literally. It is not intended to illustrate in detail a specific plan that will necessarily be responsive to the personal needs of all artists. It is intended simply to illustrate some of the components of the space. The anthropometric considerations involved must be examined with respect to the individual artist and the specific activities involved.

There are, however, some basic considerations that apply in most situations. Vertical reach from a standing and sitting position is helpful in locating shelving for art supplies. Side and forward arm reach measurements can be useful in locating various components of the space, relative to each other and the artist, in the most efficient manner possible. The eye height of a seated and standing person can be used to determine the location of visual displays and reference materials above the floor. Elbow height can be extremely helpful in establishing the height of a utility table. The text related to workbenches on the preceding page is also applicable to the artist's utility or prep table.

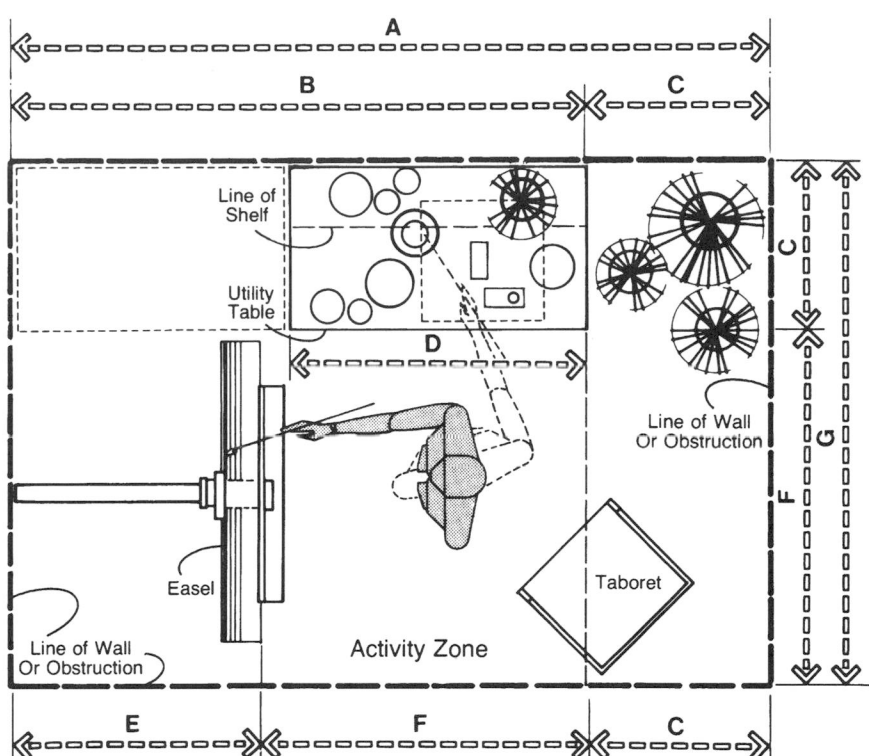

Fig. 4   Painting facilities. (See Table 2.)

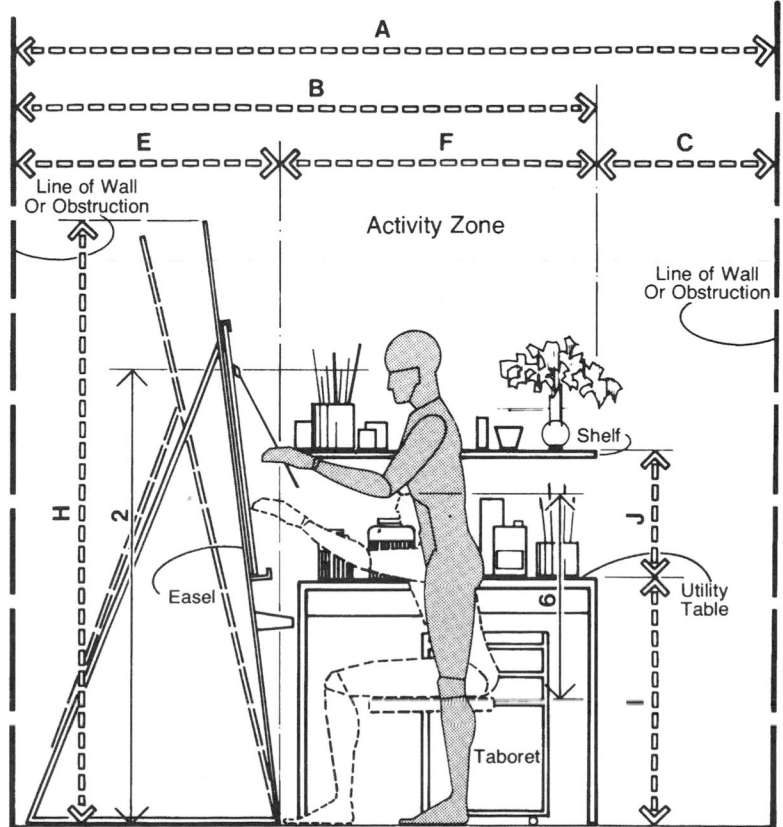

Fig. 5   Vertical view of painting facilities. (See Table 2.)

TABLE 2

	Inches
A	108
B	84
C	24
D	42
E	36
F	48
G	72
H	72–86
I	30–36
J	18

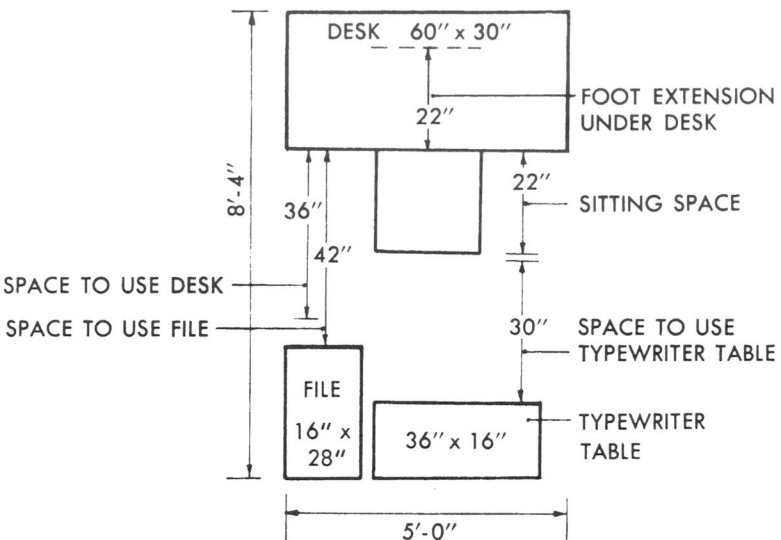

Parallel arrangement of office equipment

Fig. 1

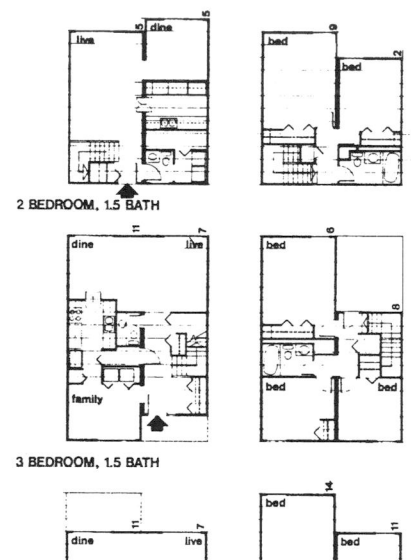

2 BEDROOM, 1.5 BATH

3 BEDROOM, 1.5 BATH

Fig. 2

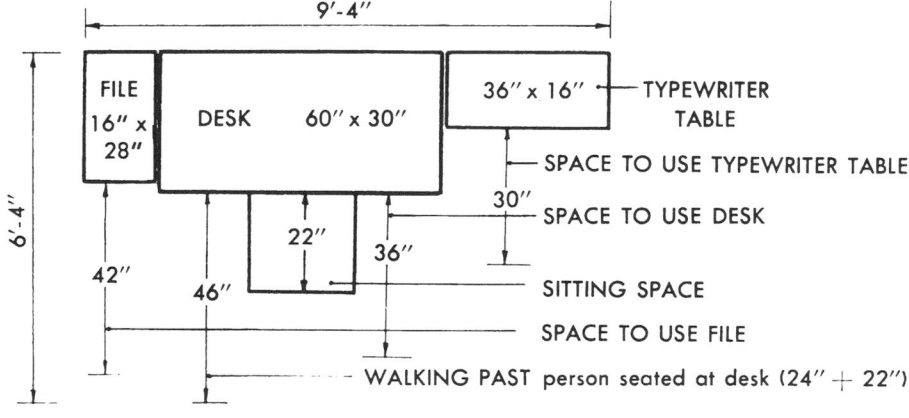

One-wall arrangement of office equipment

Fig. 3

The combined sitting/work zone shown in Fig. 4 permits the male or female user to rotate 180° for ease of access to a lateral file drawer in the rear. If the minimum clearance shown is not met, access to the file drawer is inhibited, and more awkward body motions or positions for file access are required. A minimum overall dimension to accommodate such a workstation should not be less than 96 in.

In addition to providing appropriate clearance for seat rotation and access to files, it is important to consider the circulation zone requirement for passage behind the seat at the typical workstation (see Fig. 5). The edge of this zone should take into account the movement of the chair within the chair clearance zone to avoid obstruction of any people circulating behind it. Minimal recommended clearance to allow for that circulation is predicated upon the maximum clothed body breadth of the larger user. Accordingly, this minimum dimension, allowing for the circulation of only one person, should not be less than 30 in. Based upon this minimum dimension and allowing for the requirements of the worktask and chair clearance zones, the overall distance from the front of the worksurface to the line of wall or obstruction should fall between 94 and 114 in.

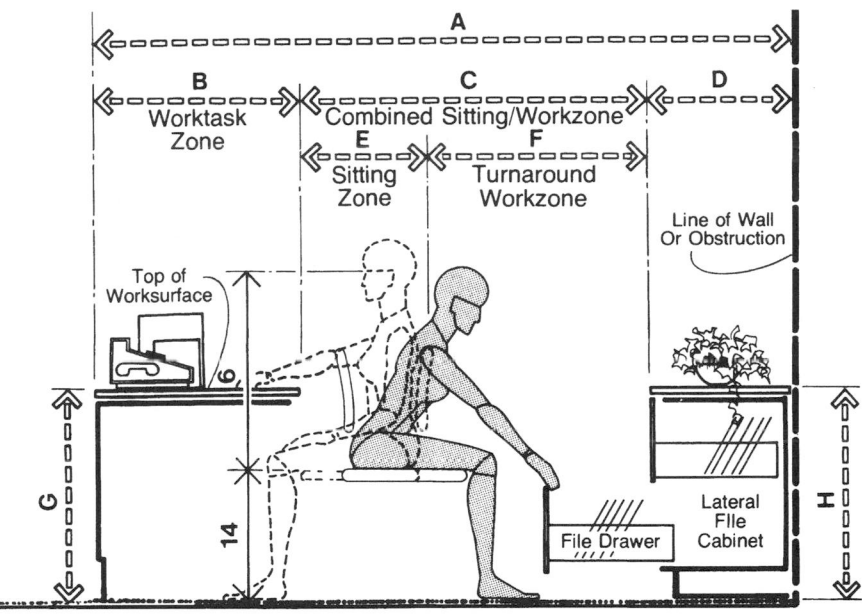

Fig. 4   Workstation with back lateral file storage. (See Table 1.)

TABLE 1

	Inches
A	96–128
B	30–36
C	48–68
D	18–22
E	18–24
F	30–44
G	29–30
H	28–30
I	90–102
J	30
K	12
L	7.5 min.
M	15–18

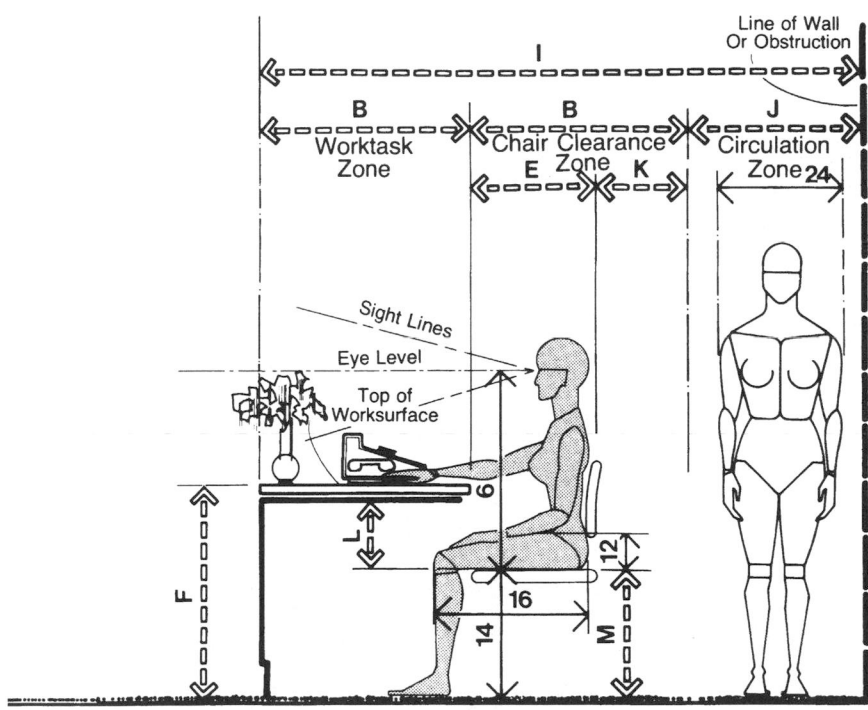

Fig. 5   Clearance and circulation zone behind workstation seat. (See Table 1.)

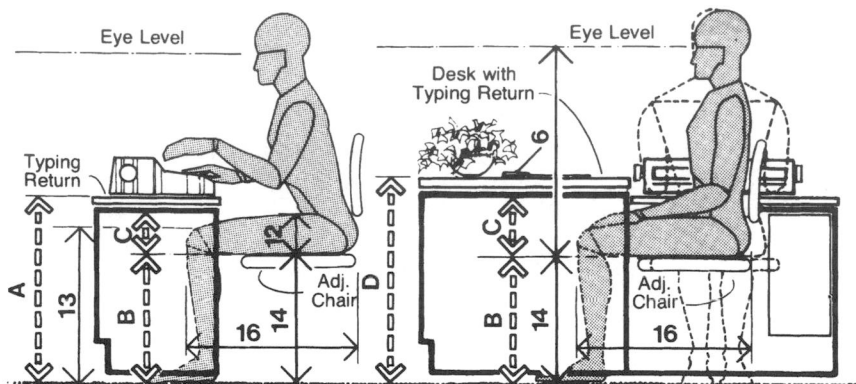

Fig. 6 Typing return and desk—male user. (See Table 2.)

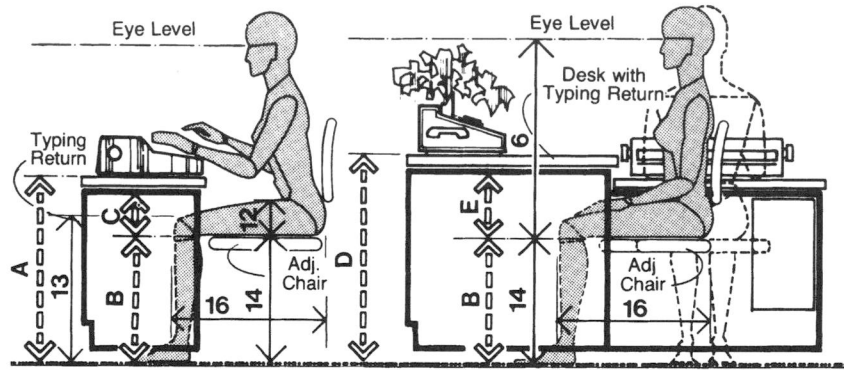

Fig. 7 Typing return and desk—female user. (See Table 2.)

The two elevations in Figs. 6 and 7 illustrate the major anthropometric considerations for the seated male and female user at both workstation and typing return. What should be noted is the seat height of the chair (a function of popliteal height) and its relationship to the specific task. When the worksurface is lowered to accommodate a specialized function, as in the case of the typing return, special attention must be given to the requirements for thigh clearance. Most standard office typing returns have been geared to the anthropometric requirements of the female user. The popliteal height and thigh clearance requirements of the larger male user may not be readily met.

The plan in Fig. 8 shows the typical workstation expanded into the basic U-shaped configuration. The work/activity zone dimension range is shown as 46 to 58 in; additional space is needed to allow for drawer extension of the lateral file. Not only does it provide more storage, the lateral file unit is generally the same height as that of the worksurface and is often utilized as a supplementary worksurface. The distance between this unit and that of the primary worksurface must be sufficient to allow for movement and rotation of the chair.

TABLE 2

	Inches
A	26–27
B	14–20
C	7.5 min.
D	29–30
E	7 min.

TABLE 3

	Inches
F	18–24
G	46–58
H	30–36
I	42–50
J	18–22
K	60–72
L	76–94
M	94–118

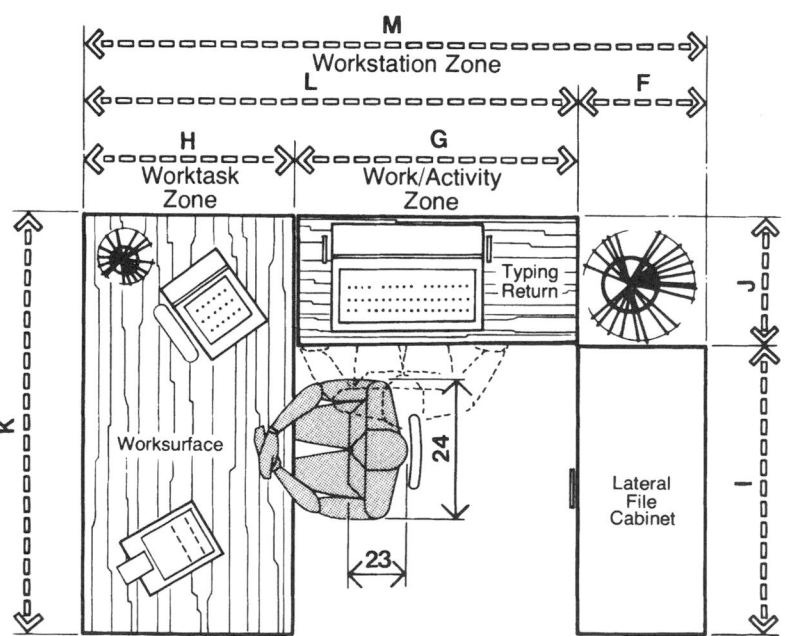

Fig. 8 Basic U-shaped workstation. (See Table 3.)

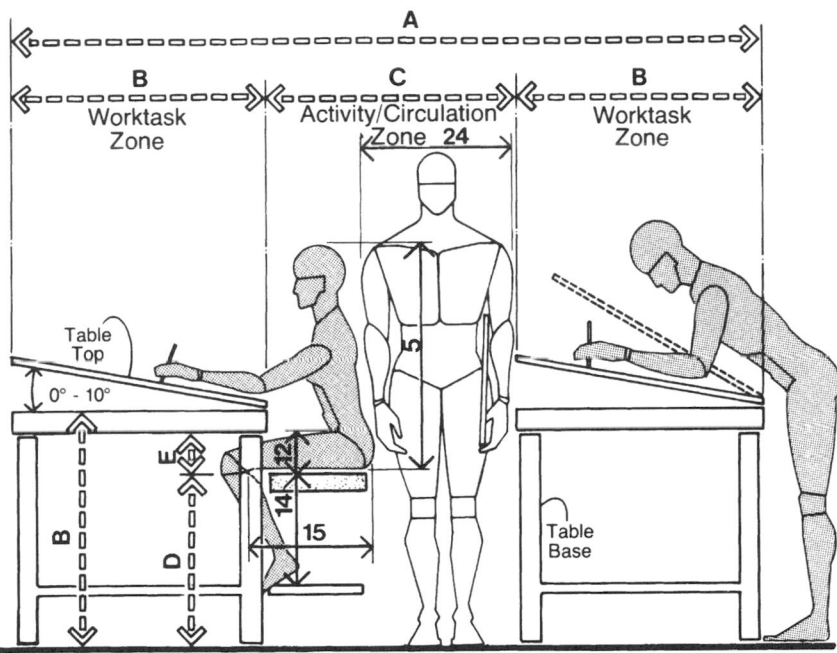

Fig. 9 Clearance between drafting tables. (See Table 4.)

Workplaces for drafting and related types of activities for general group use or instructional purposes can be arranged on the basis of individual drafting tables, as shown in Fig. 9, or as cubicles or workstations, as indicated in Fig. 10. Figure 9 shows the clearances involved between tables as well as the clearances necessary for proper interface between the seated and standing person and the table. A table height of 36 in, as opposed to regular desk height, will permit use of the table from both a seated and a standing position. Proper minimum clearance between the top of the seat surface and the underside of the table, as shown, is essential. An adjustable stool can be extremely helpful in compensating for variability in body size. Provisions for a footrest are also a critical consideration. Because of the height of the table, the distance of the seat above the floor will invariably be higher than normal and exceed the popliteal height of most, if not all, intended users. This will cause the feet to dangle above the floor, resulting not only in a lack of proper body stability but pressure on the underside of the thigh just behind the knee. This pressure will cause irritation of the tissue involved and impede blood circulation, resulting in considerable discomfort. The lack of body stability will require compensatory muscular force to maintain equilibrium, resulting in additional discomfort and pain.

TABLE 4

	Inches
A	108–120
B	36
C	36–48
D	21–27.5
E	7.5
F	48–60
G	36–60
H	30
I	12
J	54–60
K	27–30

Fig. 10 Drafting cubicle. (See Table 4.)

Requirement

Sufficient closets and storage space should be provided for living and housekeeping within each living unit. All closets and storage spaces should be appropriately located in relation to their principal uses.

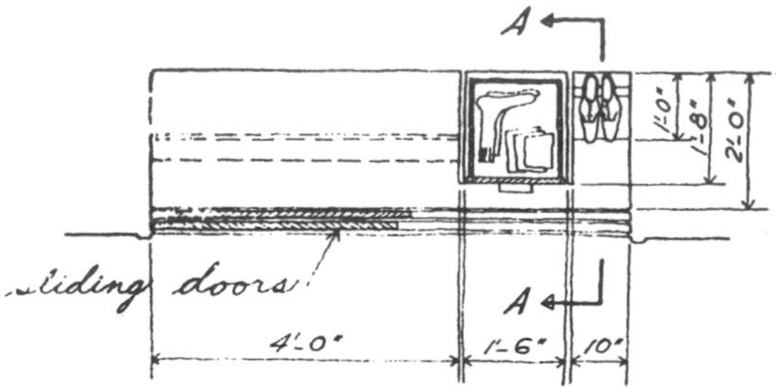

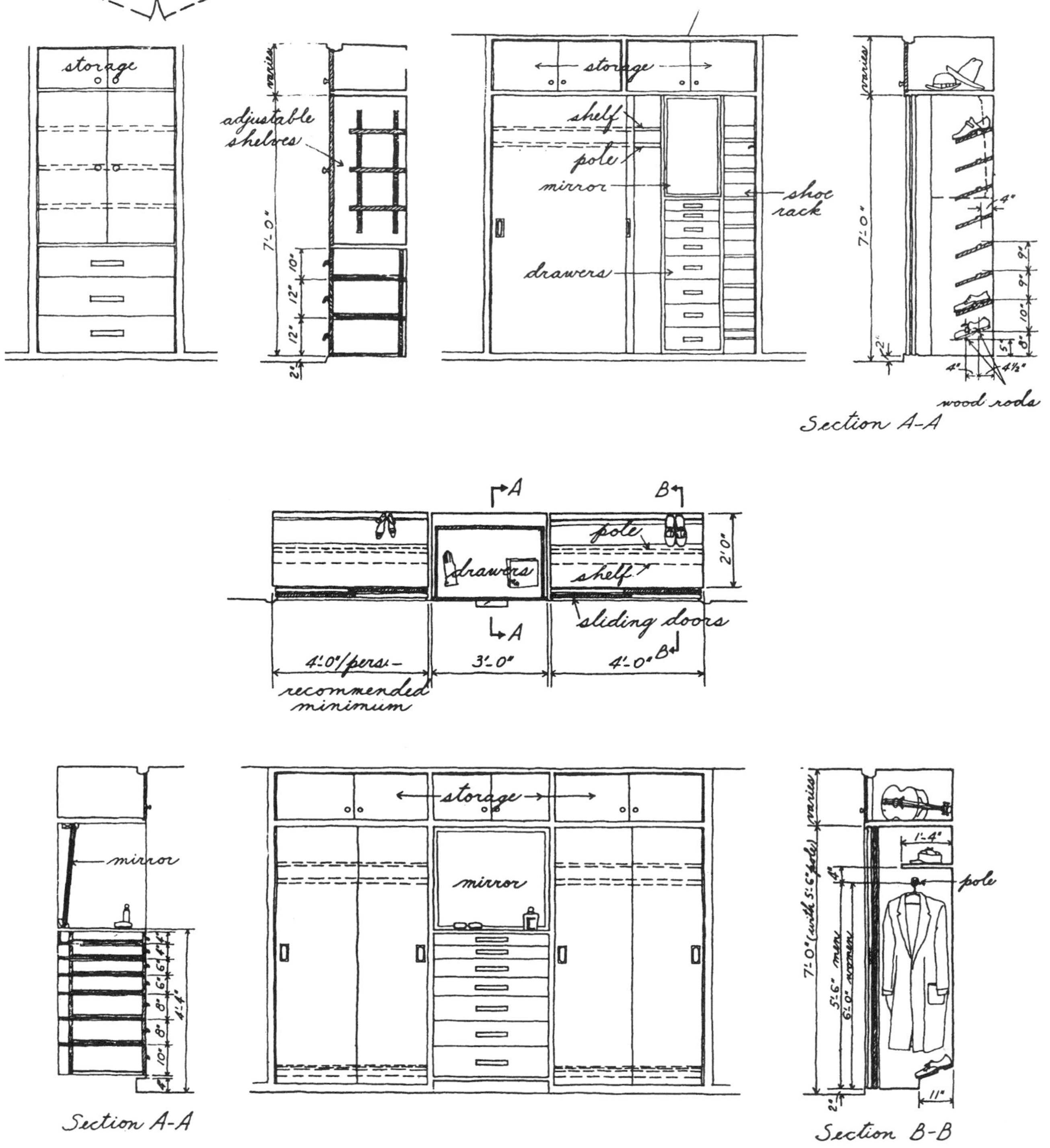

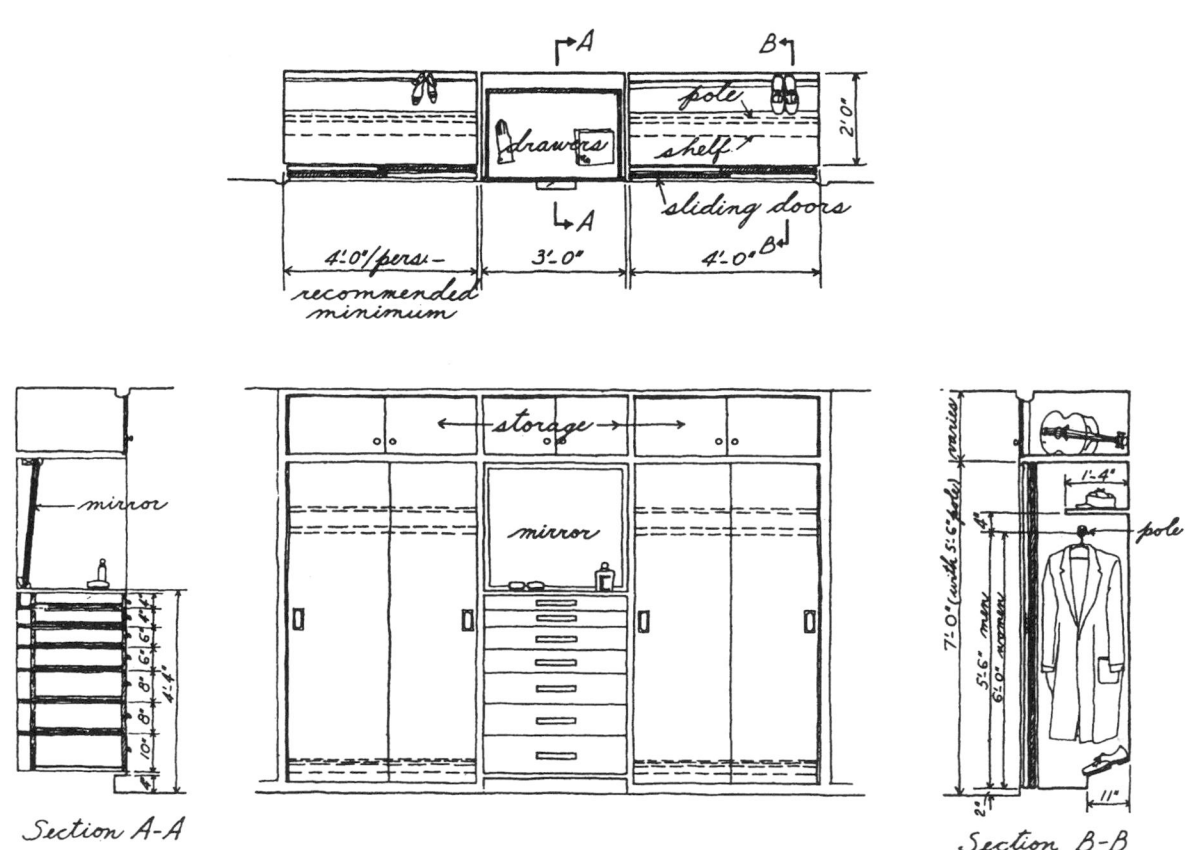

Section A-A

Section A-A

Section B-B

Fig. 1

Criterion

Bedroom closets Each bedroom should have at least one closet for clothes hanging and storage which meets or exceeds the following:

Depth: 2 ft, clear

Length: Primary bedroom, 5 lin ft, clear

Double secondary bedroom, 5 lin ft, clear

Single secondary bedroom, 3 lin ft, clear

Each bedroom closet should have a shelf and hanging pole.

Coat closet Each living unit should have an appropriately sized coat closet near the entrance of the unit.

Linen closet Provide a suitable linen closet for each living unit which is accessible to the bedroom area. Shelves should be provided but drawers may be substituted for a portion of the shelf space.

General storage Usable general storage space should be provided for each living unit in at least the following amounts:

1. Storage located entirely within the living unit:

Zero and one bedroom, 80 ft³

Two bedroom, 112 ft³

Three bedroom, 144 ft³

Four bedroom, 176 ft³

2. Where at least a minimum of one-third of the storage space is located within the unit, provide the following total storage:

Zero and one bedroom, 120 ft³

Two bedroom, 168 ft³

Three bedroom, 216 ft³

Four bedroom, 264 ft³

Interior general storage may be located in bedroom closets when these closets are larger than the required space.

3. Ground storage space should provide for such items as mowers, prams, and ladders.

Commentary

1. In bedrooms, the placement of the closet so that it is next to the door into the bedroom minimizes the use of valuable wall space for furniture.

2. Closets should be used between bedrooms and the living-working zones, and between individual bedrooms wherever possible, to enhance privacy.

Unit Storage

The square footage provided for storage may meet a minimum standard, but if it is broken up into areas too small for storing large pieces of equipment, or located inappropriately, storage may still be considered inadequate by users.

Households with children require a greater amount and variety of space for storage. Infants and young children introduce a considerable amount of bulky equipment such as cribs, bathinettes, strollers, and wheeltoys into the household. These items have to be stored close to the areas where they are used if full convenience and utility are to be realized.

Locate storage space so that items are near the activity area where they are used.

Children's bedrooms should have storage for clothes and toys.

TABLE 1 Minimum Storage Allowance

Dwelling type	Minimum storage, net ft²
Two- to three-person, two-bedroom	35
Four-person, two- and three-bedroom	45
Five-person, three- and four-bedroom	55
Six-person, three- and four-bedroom	65

Secure communal storage for bicycles should be provided within easy access of the dwelling (i.e., at grade).

The storage areas in Table 1 have been found to be very minimal. Occupants of three-bedroom units identified a need for more square feet of storage distributed between entry, 10 ft²; living, 6 ft²; kitchen, 54 ft²; master bedroom, 12 ft²; child's room, 8 ft²; at grade (bicycles, etc.), 18 ft².

Dimensions indicate minimum practical clearances for the storage of articles, commonly used, not their actual sizes. Diagrams and dimensions refer primarily to storage requirements of adults and require adjustment for children. Articles stored on shelves higher than 6 ft 8 in are difficult to reach and should be considered in dead storage. Similarly, 2 ft 6 in is the usual limit of human reach; shelves are in most cases inconvenient if they are deeper than 1 ft 3 in.

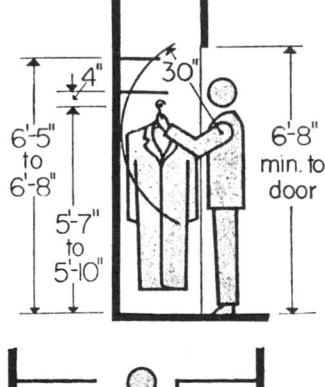

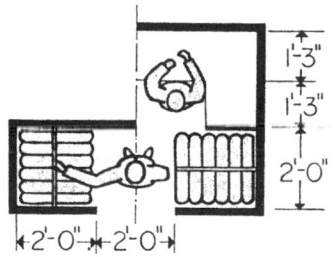

Fig. 2

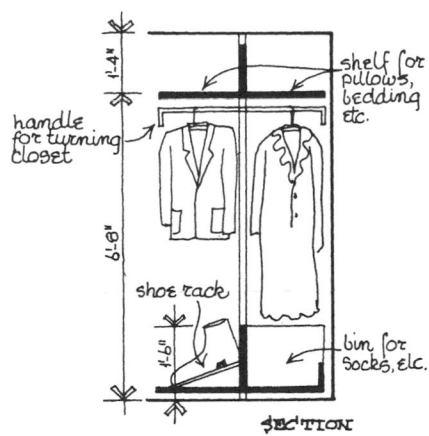

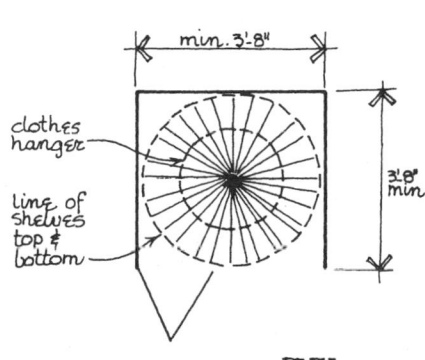

Fig. 3 Revolving hanger storage.

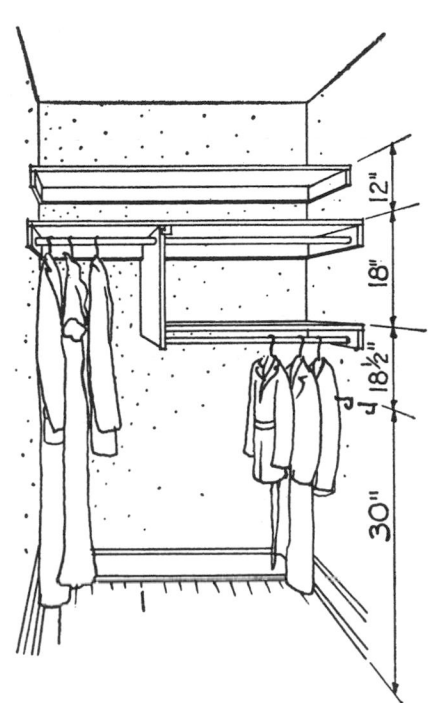

Fig. 4 Coat closet for children and adults.

CLOSETS AND STORAGE

Rods take care of practically all clothing on hangers. Usually this will include all dresses, except those for infants, all skirts, blouses, trousers, and coats. Table 2 shows the space to allow on the rod for different types of garments. This table also shows how much space to allow from wall to rod and from floor to rod, depending on the width and length of garments hanging from it. If there is a shelf above the rod, a minimum of 2½ in should be allowed between the top of the rod and the bottom of the shelf.

TABLE 2 Rod Allowance for Garments and Location of Rod with respect to Wall and Floor

Garments	Space allowance on rod, in	Desirable minimum distance from wall to rod center, in	Minimum distance from floor to top of rod, in
Adults'			
Skirts	2	12	45
Jackets	3	12	45
Shirts	1½	12	45
Suits	2	12	45
Trousers	3	12	45
Dresses	1½	12	63
Overcoats	4	12	63
Coats with fur collar	3–6	12	63
Coats without fur collar	2–5	12	63
Evening gowns	2	12	72
Garments stored in mothproof bags	3	12	72
Children's, 6 to 12 years		10	45
Children's, 3 to 5 years		8	30

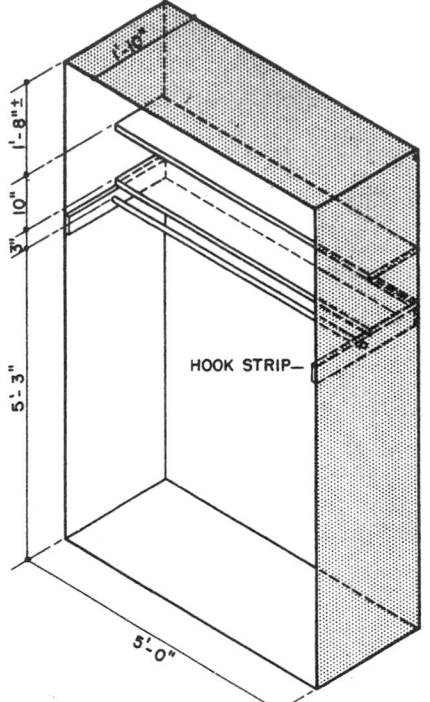

Fig. 5 Typical sizes and arrangement for clothes closet.

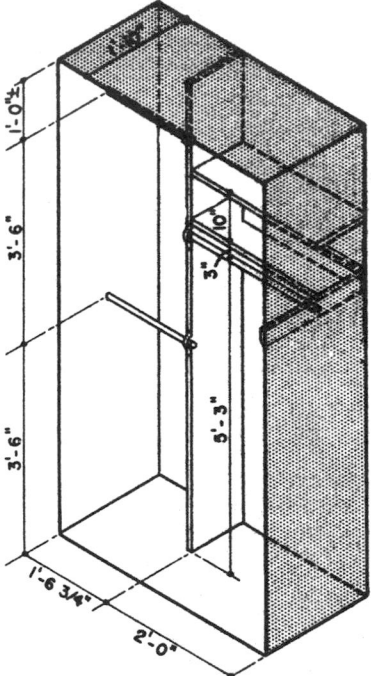

Fig. 6 Hooks.

Fig. 7 Vertical rack for shoes.

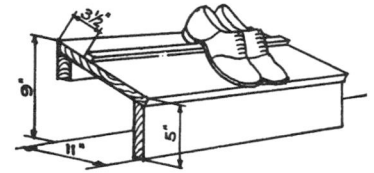

Fig. 8 Tilted shelf for shoes.

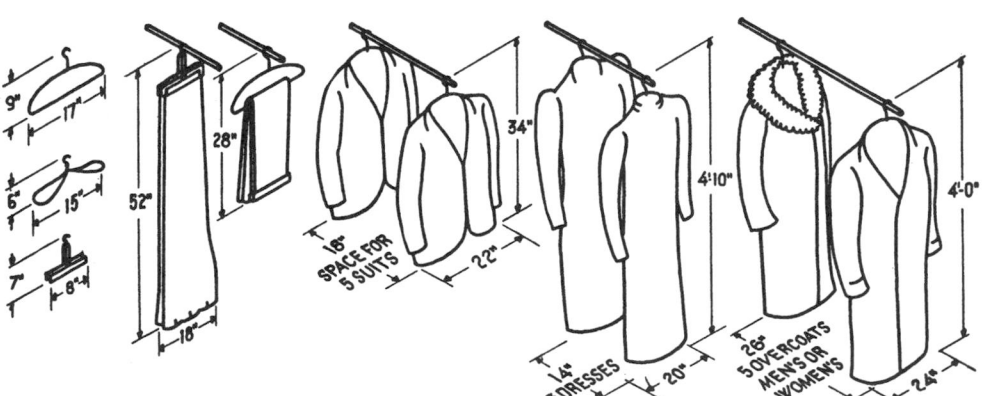

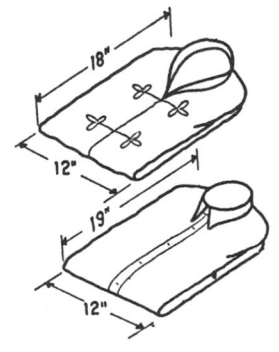

Fig. 9

Bedroom Closets

In the closet shown in Fig. 10 a rod provides 24 to 27 in of hanging space. Two feet of rod length is about the minimum to allow for each person. The tops of both the rod and the hooks on the closet door are 63 in from the floor, a good height for the garments of most adults. Just above the rod, 65½ in from the floor, is a shelf 18 in from front to back. The second shelf above the rod is narrower and may be omitted if height is limited.

On the right-hand side is a section of shelves and drawers, 18 in wide. The two lowest shelves for shoes are spaced 7 in apart. The two drawers for ties, handkerchiefs, and toilet articles are 4 in deep. Four movable shelves for folded articles are spaced 9 in apart and have guards on the front to keep articles from falling off.

One of the closet doors is a handy place to put a full-length mirror. Which door is the better one for the mirror depends upon the way the room is arranged and the lighting conditions. The top of a full-length mirror for the use of adults should be no less than 5 ft 11 in, from the floor. To accommodate children as well as adults, the bottom of the mirror must be no more than 14 in from the floor.

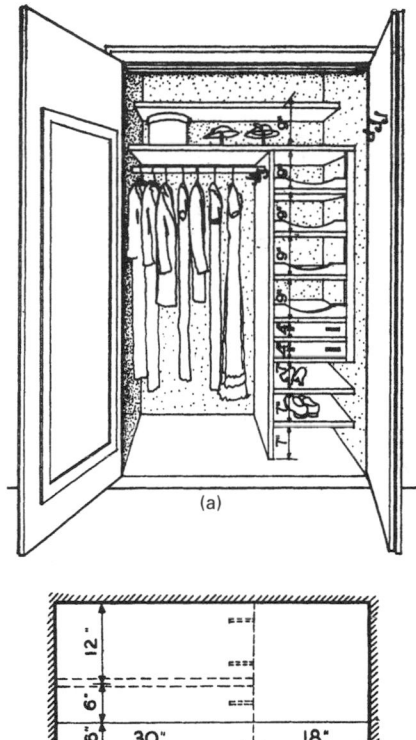
(a)

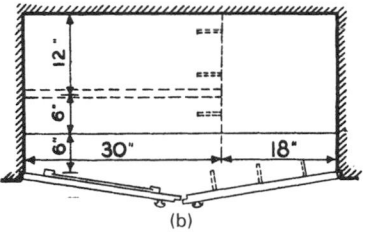
(b)

Fig. 10 Bedroom closet designed for one person: (*a*) perspective; (*b*) plan.

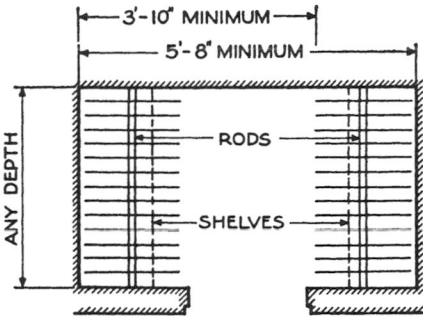

Fig. 11 Bedroom closet.

Fig. 12 Two-rod walk-in closet.

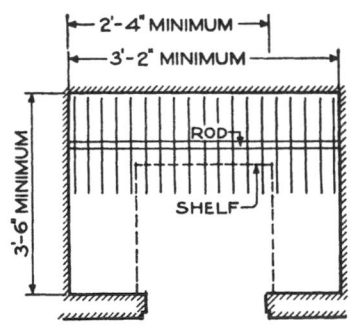

Fig. 13 One-rod walk-in closet.

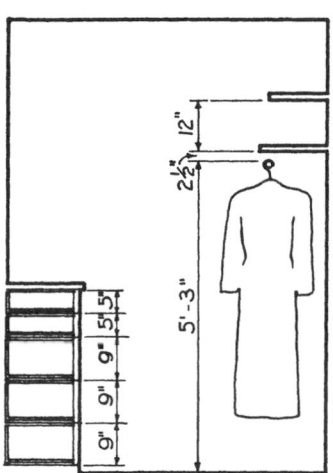

Fig. 14

CLOSETS AND STORAGE

1'-4" to 1'-6"
2'-0" to 2'-2"
1'-0"
A
10" to 12"
1'-0"
B
2'-0"
1'-4" to 1'-8"
2'-0" to 2'-4"
1'-0"
D
9" to 12"
1'-0"
E
1'-10" to 2'-0"

7" to 9"
1'-0"
C
9" to 12"

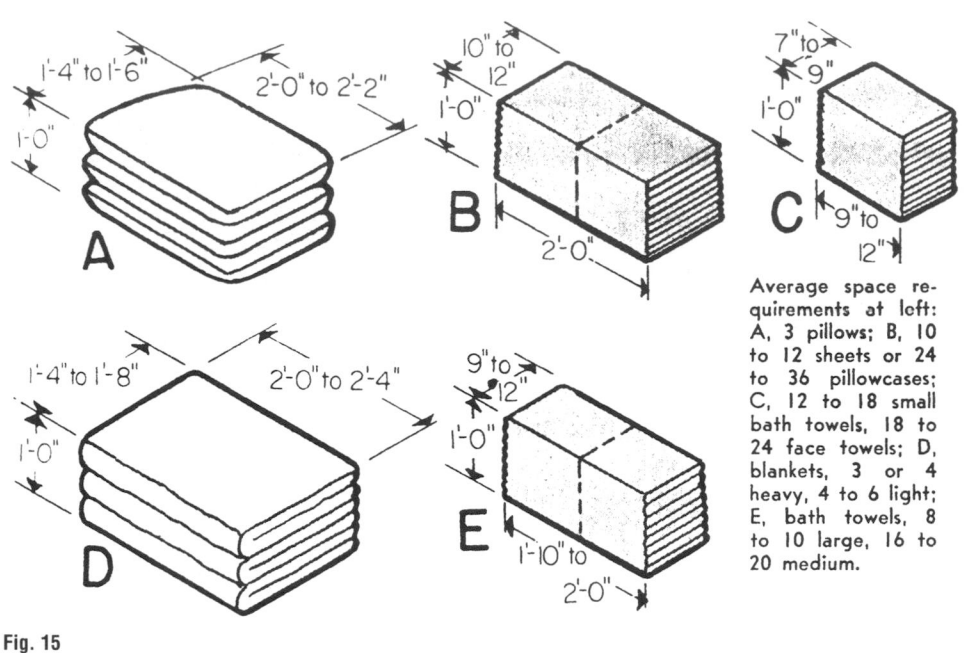

Average space requirements at left: A, 3 pillows; B, 10 to 12 sheets or 24 to 36 pillowcases; C, 12 to 18 small bath towels, 18 to 24 face towels; D, blankets, 3 or 4 heavy, 4 to 6 light; E, bath towels, 8 to 10 large, 16 to 20 medium.

Fig. 15

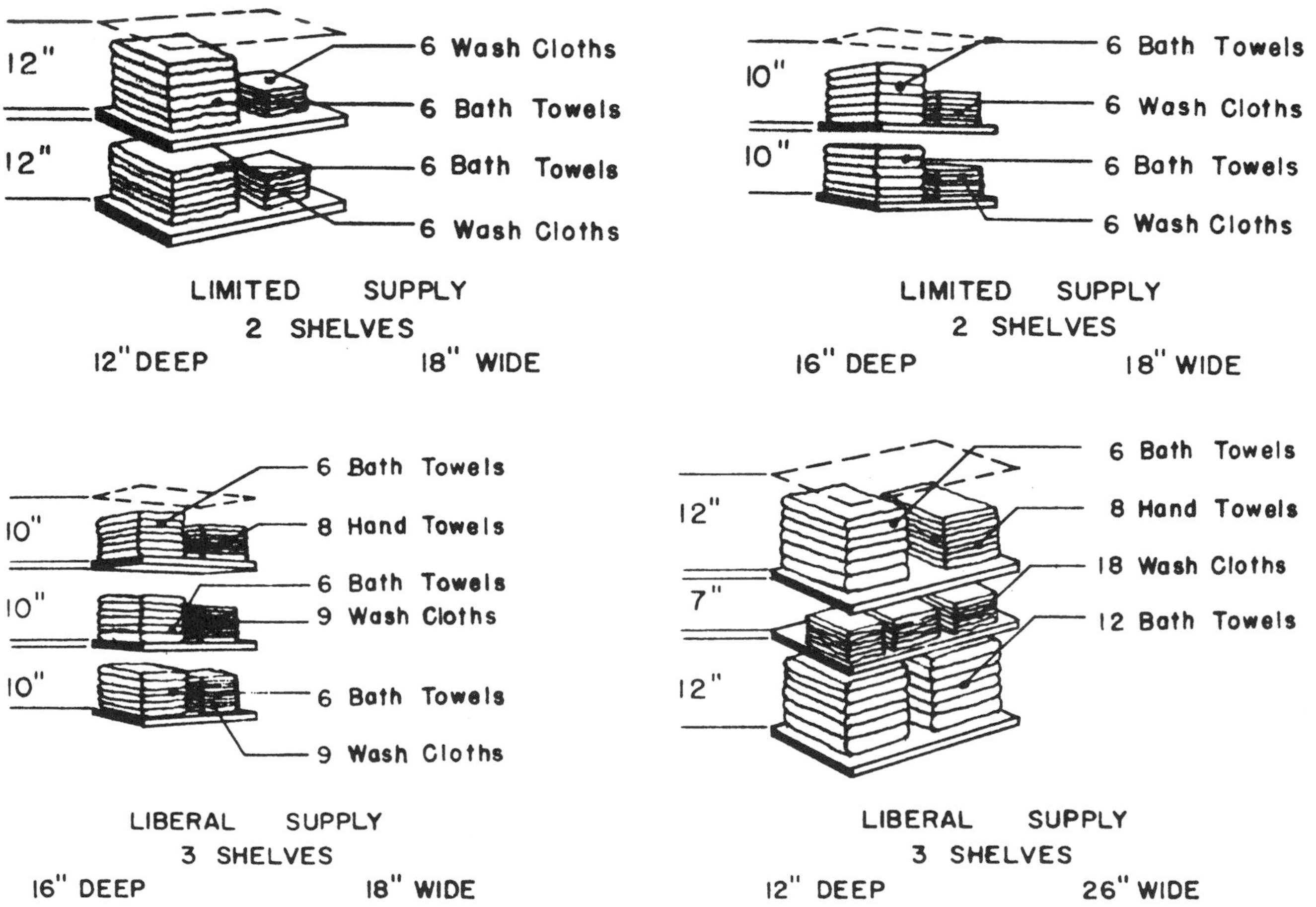

12"
12"
6 Wash Cloths
6 Bath Towels
6 Bath Towels
6 Wash Cloths
LIMITED SUPPLY
2 SHELVES
12" DEEP
18" WIDE
10"
10"
6 Bath Towels
6 Wash Cloths
6 Bath Towels
6 Wash Cloths
LIMITED SUPPLY
2 SHELVES
16" DEEP
18" WIDE
10"
10"
10"
6 Bath Towels
8 Hand Towels
6 Bath Towels
9 Wash Cloths
6 Bath Towels
9 Wash Cloths
LIBERAL SUPPLY
3 SHELVES
16" DEEP
18" WIDE
12"
7"
12"
6 Bath Towels
8 Hand Towels
18 Wash Cloths
12 Bath Towels
LIBERAL SUPPLY
3 SHELVES
12" DEEP
26" WIDE

Fig. 16 **Suggested shelf storage for liberal supply of bath linens.**

Linen Storage

The linen closets shown in Fig. 18 a recommend arrangement of shelves for storing various linens and bedding for the average family. A good average size is 36 in wide and 24 in deep.

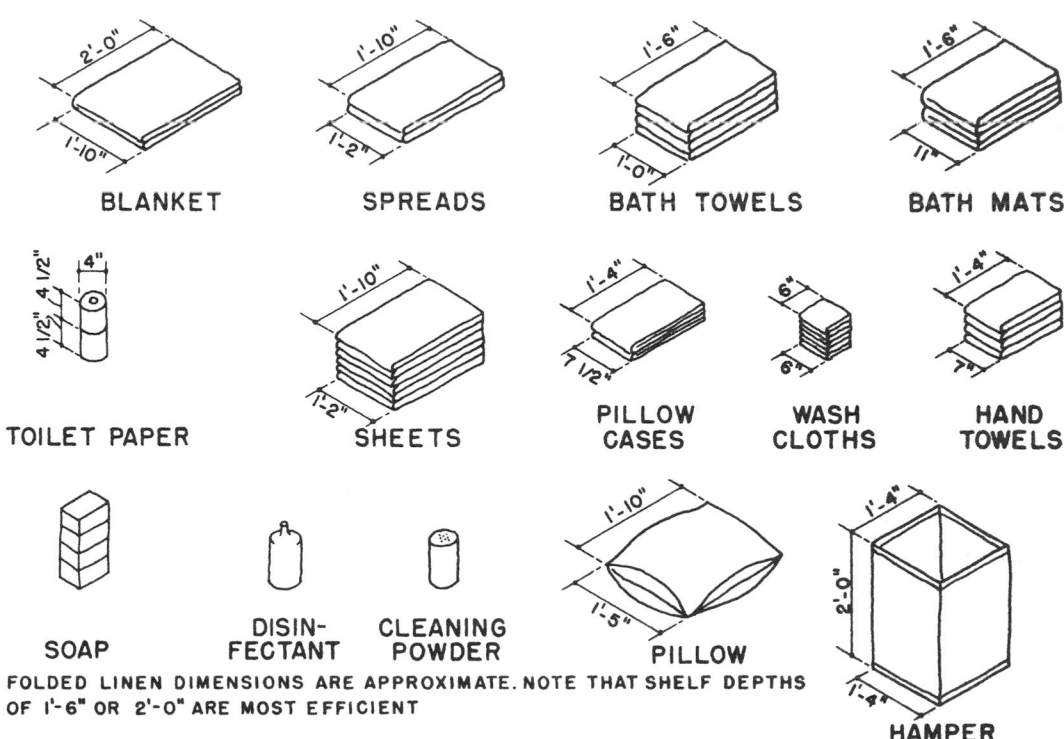

BLANKET   SPREADS   BATH TOWELS   BATH MATS

TOILET PAPER   SHEETS   PILLOW CASES   WASH CLOTHS   HAND TOWELS

SOAP   DISINFECTANT   CLEANING POWDER   PILLOW   HAMPER

FOLDED LINEN DIMENSIONS ARE APPROXIMATE. NOTE THAT SHELF DEPTHS OF 1'-6" OR 2'-0" ARE MOST EFFICIENT

Fig. 17   Typical storage items for linen closets.

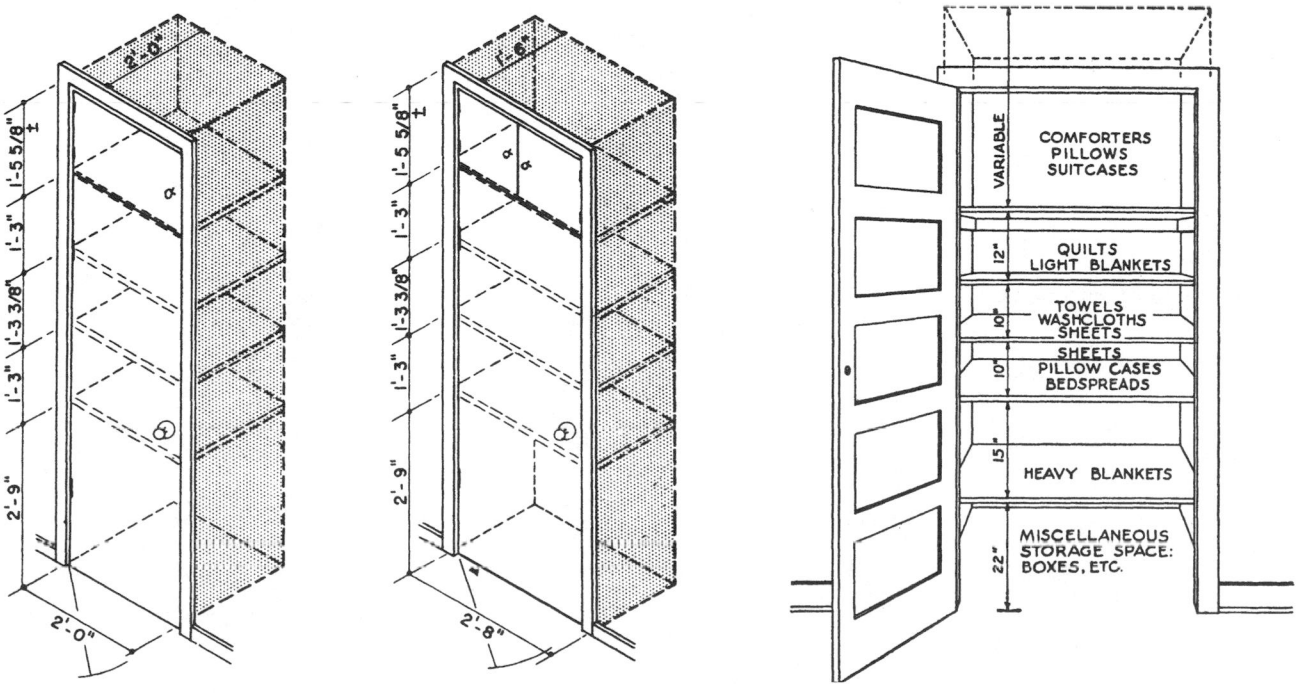

Fig. 18   Typical sizes of linen closets.

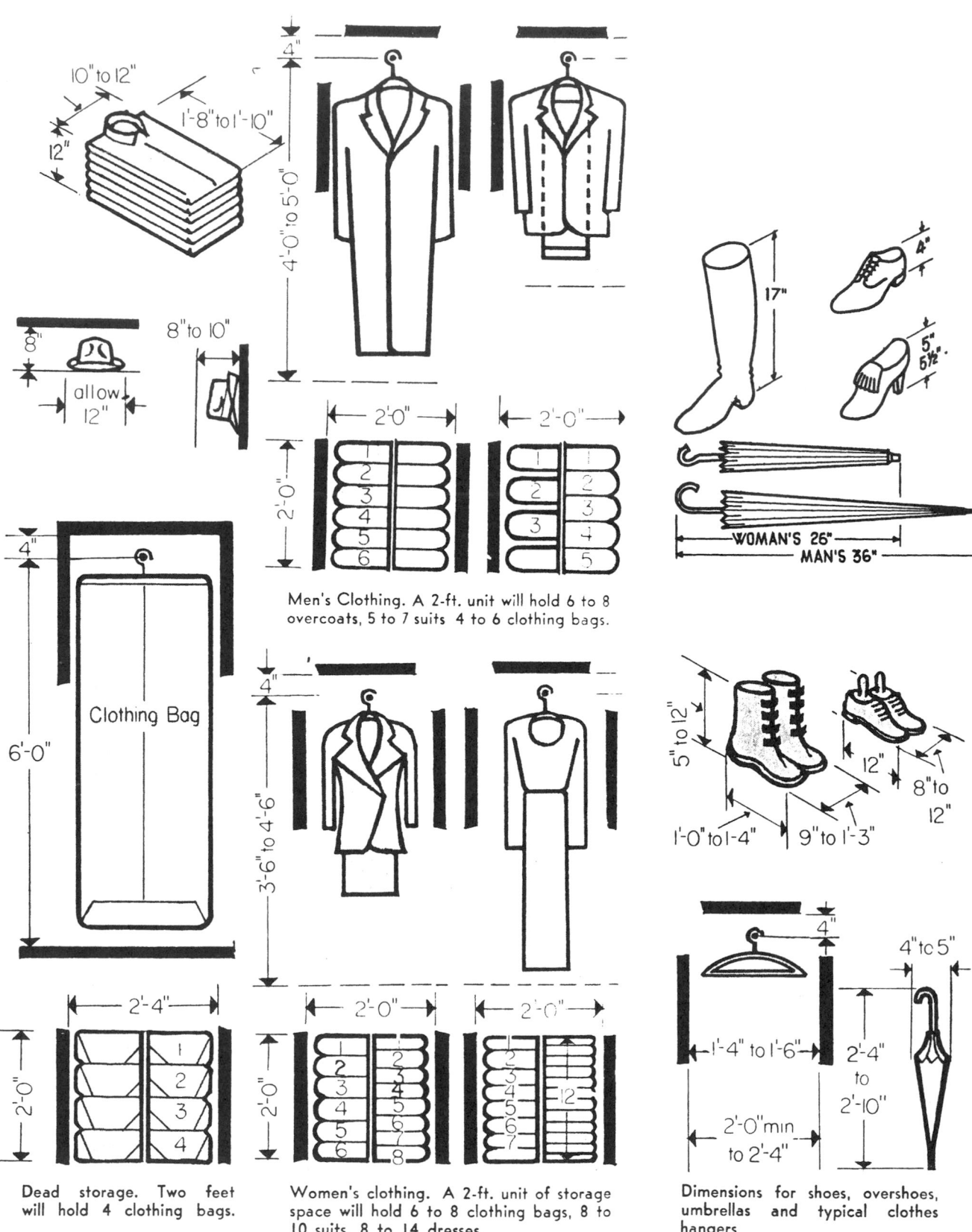

Men's Clothing. A 2-ft. unit will hold 6 to 8 overcoats, 5 to 7 suits 4 to 6 clothing bags.

Dead storage. Two feet will hold 4 clothing bags.

Women's clothing. A 2-ft. unit of storage space will hold 6 to 8 clothing bags, 8 to 10 suits, 8 to 14 dresses.

Dimensions for shoes, overshoes, umbrellas and typical clothes hangers

Fig. 19

Storage Arrangements

TABLE 3 Storage of Clothing

Ample space per garment on rod	
Suits	3 in
Trousers	2 in
Overcoats	4 in
Dresses	2 in
Skirts	2 in
Coats, heavy	5 in
Distance between floor and rod	
Adults' topcoats and suits	5 ft to 5 ft 3 in
Short coats, shirts, skirts	4 ft
Evening gowns	6 ft
Children's garments	3 to 4 ft
Width of garments on hangers	
Suits and coats	1 ft 8 in to 1 ft 10 in
Dresses and blouses	1 ft 6 in to 1 ft 8 in

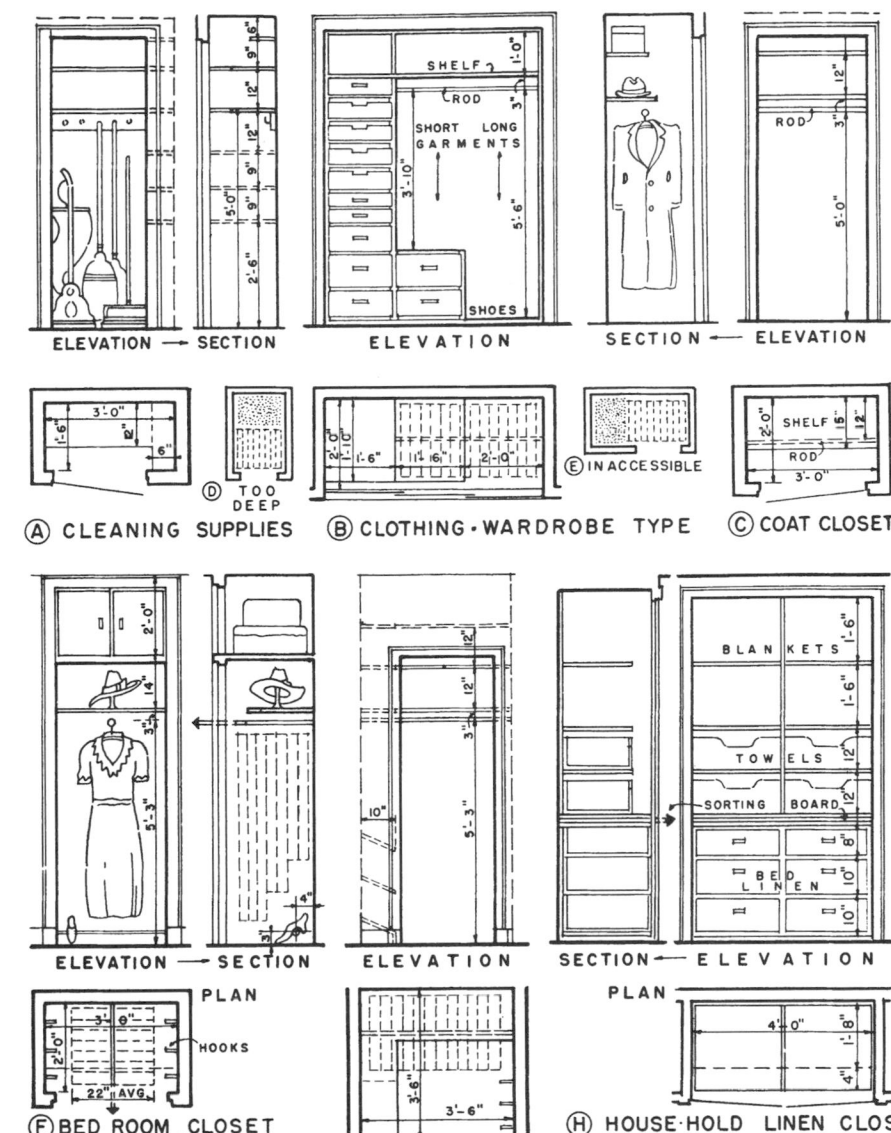

Fig. 20 Storage requirements vary greatly with families of different sizes and interests. These drawings suggest only a few of the many types of necessary closets.

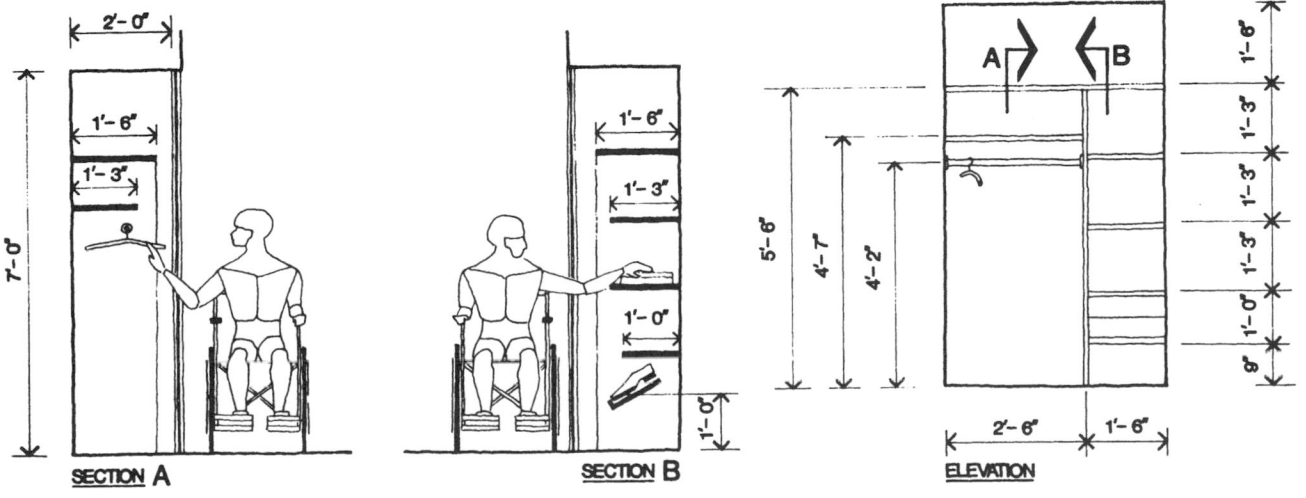

Fig. 21 Handicapped access to closets.

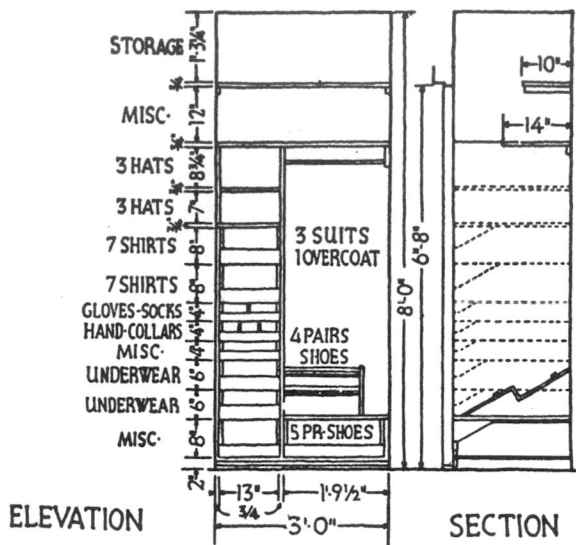

Fig. 22   Closet arrangement for a man.

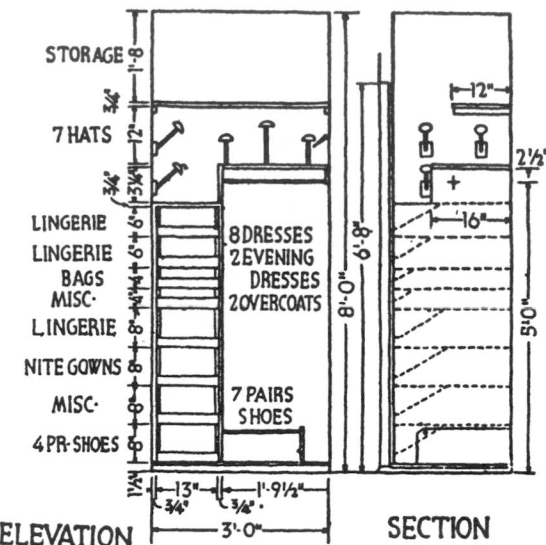

Fig. 23   Closet arrangement for a woman.

When planning the garage, adequate space should be allowed for all intended uses. A good rule to follow is to allow a minimum of 2 ft 6 in clearance space to the front of work benches, storage cabinets, laundry facilities, and on both sides of the automobile.

Commodious storage, a hobby room, and other special-purpose space can be handy to the carport and a private patio on the street side of a townhouse.

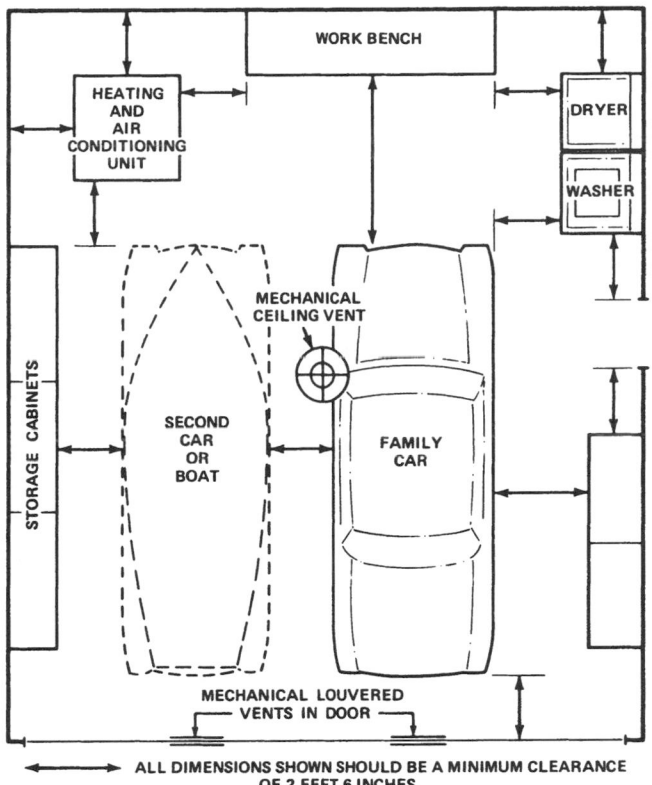

ALL DIMENSIONS SHOWN SHOULD BE A MINIMUM CLEARANCE OF 2 FEET, 6 INCHES.

Fig. 1

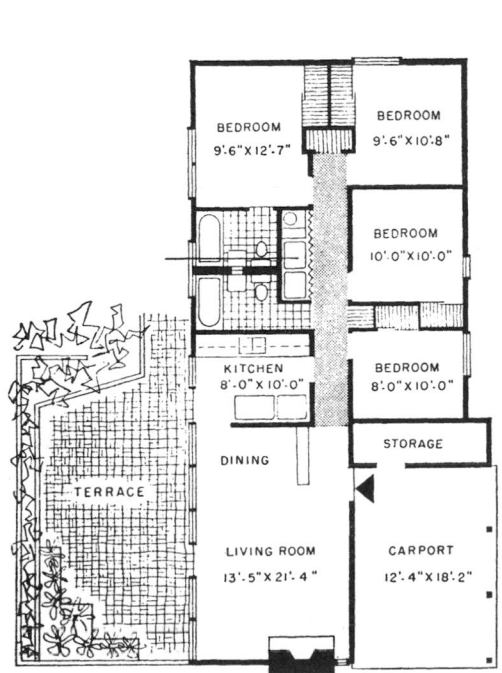

Fig. 2 Carport for one car.

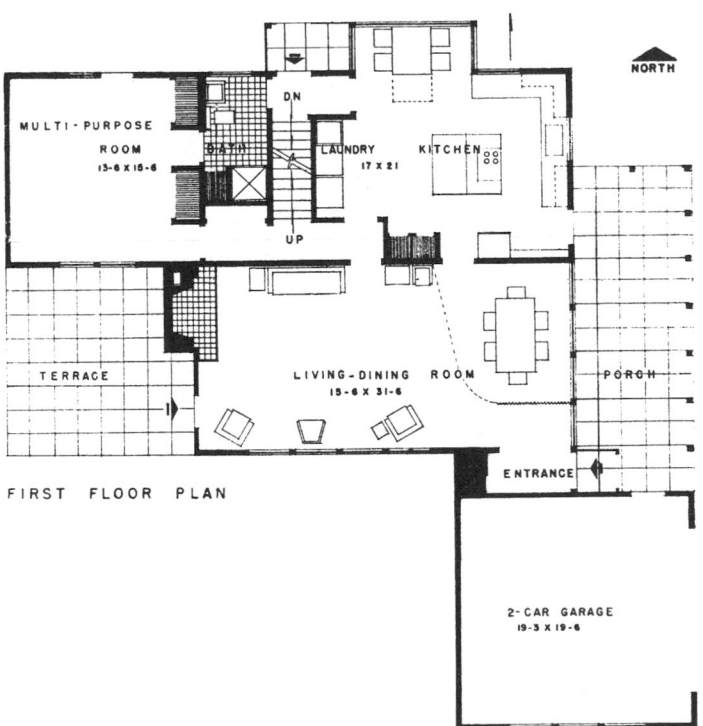

FIRST FLOOR PLAN

Fig. 3 Two-car attached garage.

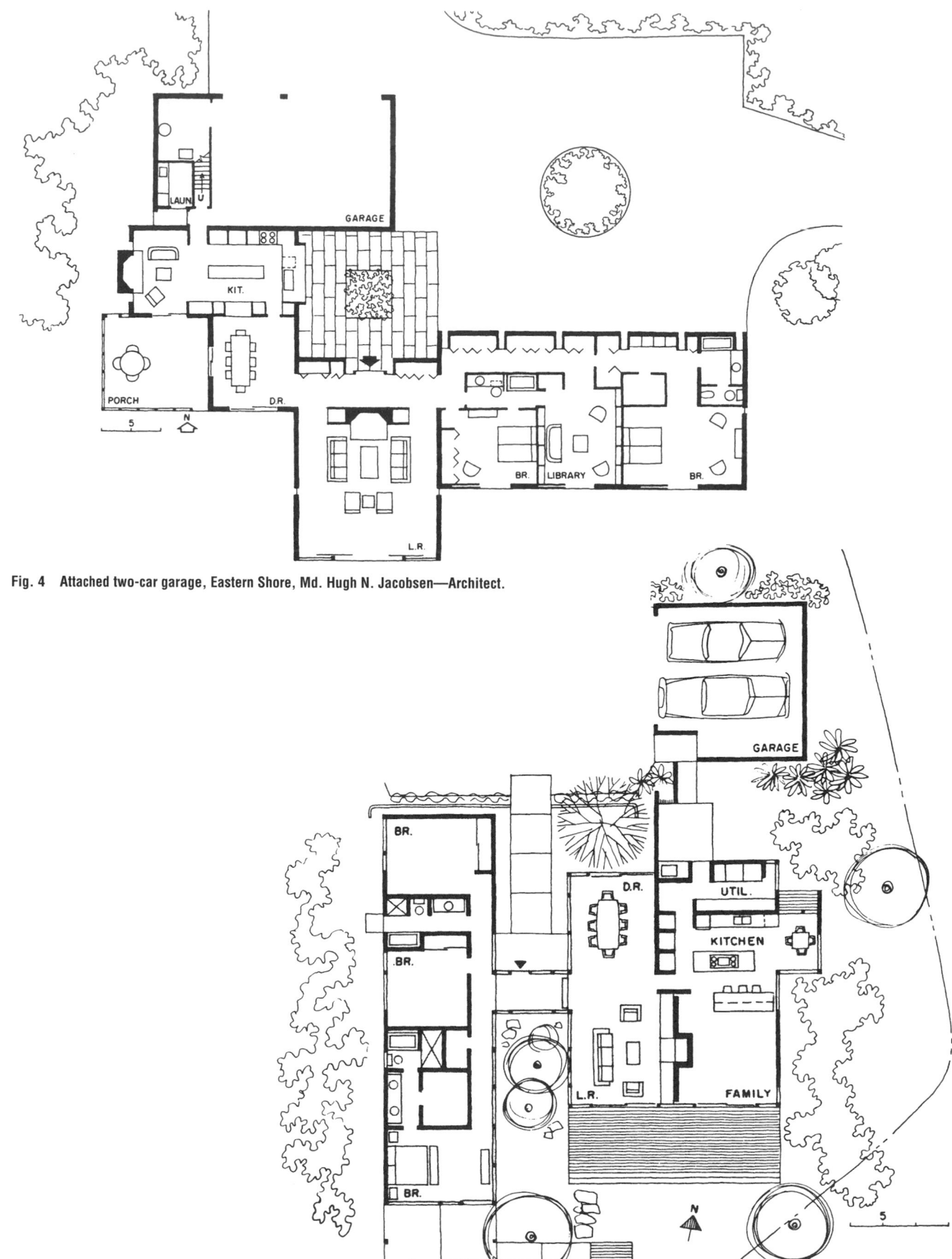

Fig. 4   Attached two-car garage, Eastern Shore, Md. Hugh N. Jacobsen—Architect.

Fig. 5   Detached 2-car garage, W. Los Angeles, F. Dorman/Muselle—Architects.

By establishing a plan theory based on user needs for dwelling units in general, a design framework has been created. The internal design is based on a theory of usage "zones" and "rules." Four "zones" have been identified:

Private Quiet, personal spaces (bathrooms, bedrooms)

Meeting Less quiet, communal spaces (living, family, and dining rooms)

Transition Connecting spaces (stairs, corridors, entries)

Service Support spaces (kitchens, storage rooms, mechanical spaces, utility room)

The approximate location and relationships of these "zones" are determined by "rules." Some of the "rules" which evolved during the planning study can be summarized as follows:

■ Ground-level living areas should be oriented to the opposite side, since most units (townhouse or garden apartment) will have parking on the entry side of the building.

■ Entries should not be directly into living spaces but should be into a circulation space from which various parts of the unit can be reached.

■ Townhouse plans should allow all bedrooms to be on the second floor, and stairs to be accessible from entry space.

■ Family rooms, when included, should have access to the kitchen.

■ Plans should allow for the inclusion of eat-in kitchens and/or separate dining room arrangements.

■ Living rooms should have the capability of expanding beyond the limits of a 12-ft module; therefore, living rooms and dining rooms should ideally be located adjacent to each other or in an L-shaped configuration.

TOWN HOUSES

The plans in Fig. 2 reflect space-planning rules established by HUD's Operation Breakthrough guide criteria, which specify the use of space rather than abstract size requirements. Use of the criteria has allowed somewhat more space-planning flexibility, yet final design plans appear to be compatible with most FHA and model code standards.

TABLE 1   Town Houses

Two-bedroom	1036 ft²
Three-bedroom, 1.5 baths	1248 ft²
Three-bedroom, 2.5 baths	1348 ft²
Four-bedroom	1536 ft²

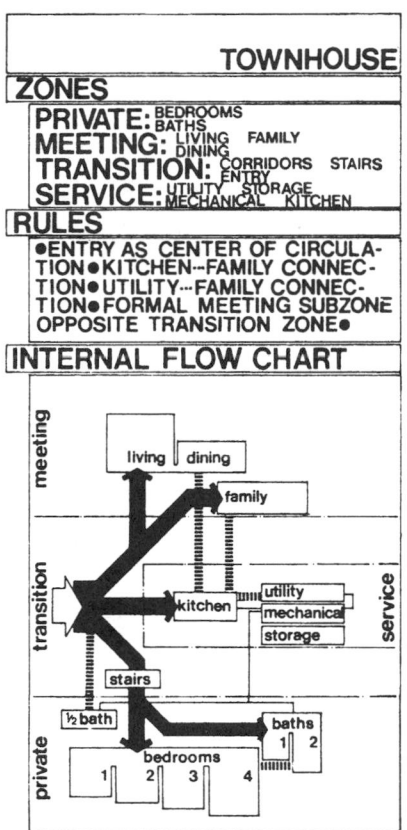

Fig. 1

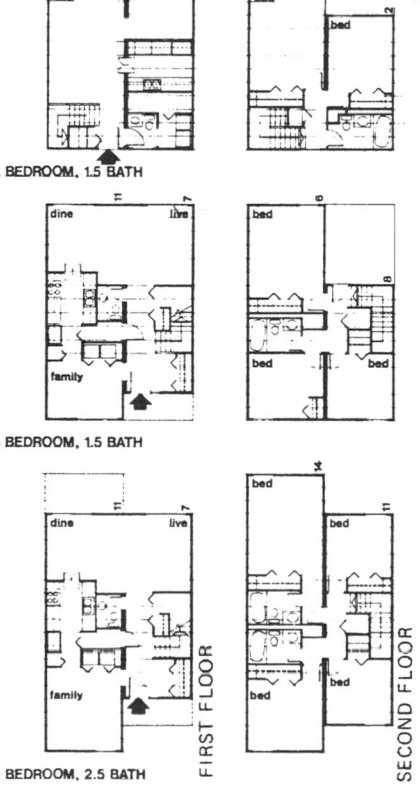

2 BEDROOM, 1.5 BATH

3 BEDROOM, 1.5 BATH

4 BEDROOM, 2.5 BATH

FIRST FLOOR   SECOND FLOOR

Fig. 2

6

Types of
Single-Family Houses

The types of buildings used for housing range from detached single-family dwellings to high-rise apartment houses. Often a housing development will consist of only one type of housing, but in relatively large developments a mix or range will occur. Typical combinations would be single-family and garden apartments, or town houses and high-rise apartments.

The actual types are dictated by planning considerations such as densities, type of occupancy, economics, and community housing needs.

Housing types can be divided into several categories:

1. Single-family detached—those located on a separate and independent lot. This type is generally owner-occupied.

2. Two-family and town houses—those that have two or more housing units in a group. This includes duplexes, quadruplexes, row houses, and town houses. Most frequently this type of housing is located on separate lots and is also owner-occupied.

3. Garden apartments—clusters of apartments with higher densities, generally rental units.

4. Low- and high-rise apartments—those types of housing with the greatest densities.

No one type of housing is superior to any other. Each has an appropriate time and place for its use. In built-up urban areas, high-rise apartments are generally the most appropriate, while in outlying areas, detached housing is generally more successful.

On the following pages, a review of all types of housing is given. Apartments, or multiple dwellings, obviously are more complicated structures and require greater regulation, by building codes and lending institutions.

ONE-FAMILY HOUSES

Detached Single-Family House

The single-family detached house is the most common type of housing in the United States. It is characterized by being a completely independent structure and housing one family. The type of house can usually be described as a ranch, high ranch, split-level, or two-story.

The single-family detached house is generally considered to be the best type of housing for families with growing children. Only the one-family house provides full use of private outdoor facilities. Another distinct advantage is the freedom to make normal amounts of noise without disturbing the neighbors. Also, the close proximity to open space, grass, and trees is considered desirable and healthful.

This dwelling occupies its own structure from ground to roof, is separated from other dwellings by yards or other open space, and is designed for occupancy by one family. The property lines and density of development are influenced by zoning and subdivision regulations which specify both minimum lot area and dimensions, and features such as tile fields which are required for individual septic tanks.

Type of construction can be of frame, brick veneer, solid masonry, or stucco. The most common type is frame construction.

Lot sizes vary tremendously and reflect the cost of land, zoning requirements, and the general character of the neighborhood. Prior to World War II, lot sizes tended to be small, some even as small as 20 by 100 ft. Since then, lot sizes have increased substantially. Modest lot sizes range from 50 by 100 ft to ½-acre lots. Lots over an acre can be considered large. Lots up to 100 ft front normally have depths of 100 ft. Lots with frontage greater than 100 ft usually have depths greater than 100 ft.

A garage is often included, which is either attached or detached. Generally, older houses have detached garages while the new houses have attached garages.

Ranch

The ranch-type house is the traditional one-story house. All activities, cooking, dining, living, and sleeping are on one level close to the ground. The house may or may not have a cellar, which is generally used for storage or minor activities. Older houses had high-pitched roofs for expansion. The newer houses have low-pitched roofs without provisions for expansion. This is the simplest type of construction.

High Ranch

The high ranch is similar to a ranch except that the main level is raised out of the ground, allowing light and air into the basement. This lower level is then utilized as additional living space. One of the kitchen-dining-living areas can be located there, or the space can be used for additional bedrooms. The major advantage of this type of house over the traditional ranch is the utilization of the lower level for living purposes rather than storage or incidental use.

Split-Level

The split-level house separates the living activities into three levels. The kitchen-dining-living is the main level close to the ground. The sleeping level is located one-half level above the main level. The garage-recreation room-utility level is one-half level below the main level. The main advantage is the partial separation of activities and greater privacy. Disadvantages are the up and down stair movement and more complicated construction.

Two-Story

The two-story house is characteristic of most older houses. The lower level contains the kitchen-dining-living areas. The upper floor contains the sleeping areas. This type of house most often has a cellar for storage. The main advantage is the complete separation of living and sleeping activities for maximum privacy. The major disadvantage is the up and down stair movement. Construction is more complicated than in the ranch-type house. Also, there is less lot coverage than the other types.

A garage is most often included, which is either attached or detached. Generally, older houses have detached garages while the newer houses have attached garages.

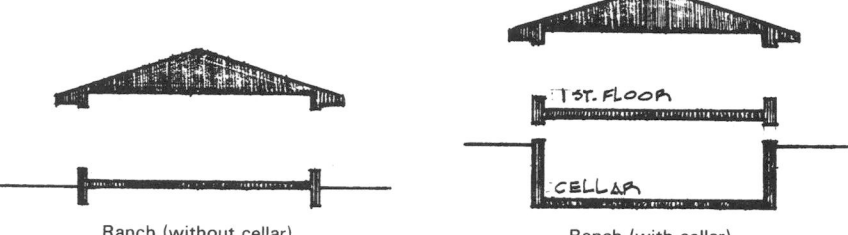

Ranch (without cellar)

Ranch (with cellar)

Fig. 1

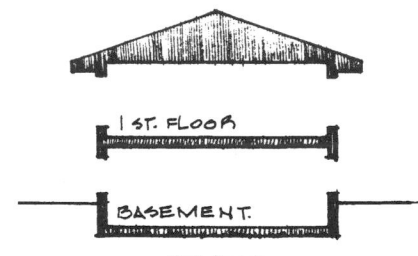

1 ST. FLOOR

BASEMENT

High Ranch

Fig. 2

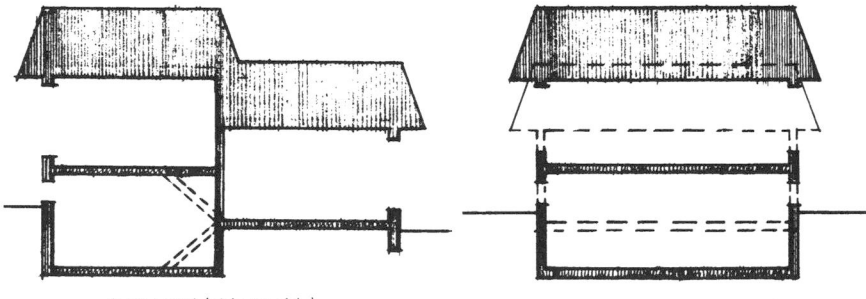

Split Level (side to side)

Split Level (front to back)

Fig. 3

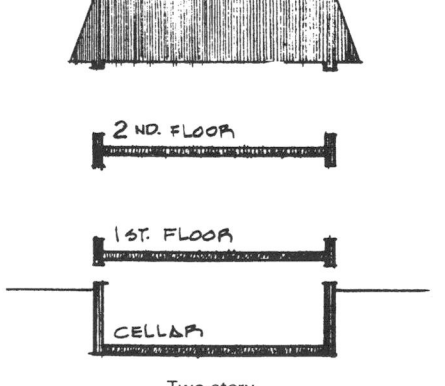

2 ND. FLOOR

1 ST. FLOOR

CELLAR

Two story

Fig. 4

Earth-Covered Homes

The site is important for earth-covered homes. Homes covered with earth, either totally or partially, are gaining wider acceptance. When properly used, earth is a strong, durable, and weather-resistant material which can be used to create an energy-efficient home.

The site's natural contours should be kept intact as much as possible and slope gently to the south. This allows any exposed windows to face the sun, and facilitates natural and controlled drainage of the roof.

There are three basic concepts for earth-covered homes. The first, called "elevational," includes an exposed wall on the south side for the access and collection of sunlight. This concept is well suited for construction on sloping terrain and hillsides.

Another concept, called "atrium," has the earth-covered home surrounding a central open court area or atrium. Sunlight and access are provided through the central area. This idea is appropriate for a flat site or the top of a hill or knoll.

A third idea, called "penetrational," totally covers the house with earth except for skylights and access doors punched through the roof.

The two most important considerations in an earth-covered home are structure and waterproofing. The structure must be very strong to hold the weight and pressure of the earth on the top and sides. It may be constructed with concrete, steel, and even wood.

The earth-covered home must be properly waterproofed. Waterproofing and internal air circulation should be carefully determined by a professional designer.

Earth-covered homes can provide a comfortable living environment with little energy consumption. This is possible by utilizing the heat of the surrounding soil and minimizing exposure to winds. Soil temperature below the frost line fluctuates very little during the year. Thus, the home requires a little additional heat in winter and little or no cooling in summer.

Additional heat needed in winter can easily and inexpensively be provided by the sun. The combination underground-solar design concept is an excellent approach to energy efficiency.

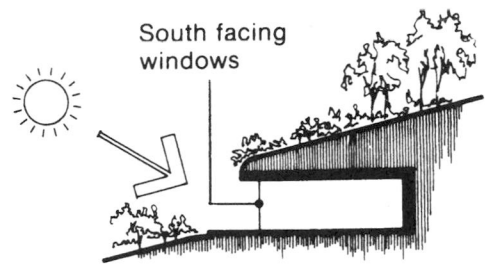

Fig. 5   Elevational earth-covered home.

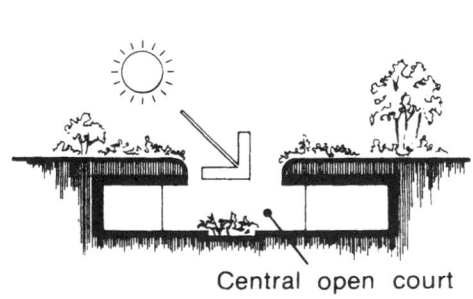

Fig. 6   Atrium earth-covered home.

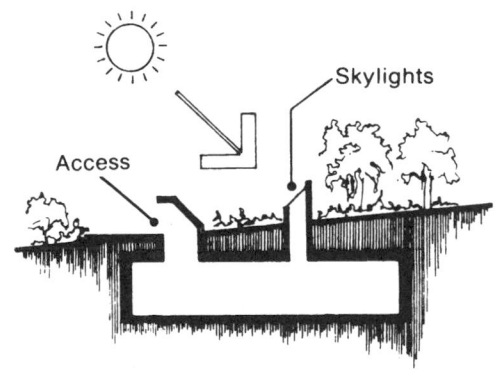

Fig. 7   Penetrational earth-covered home.

earth covered walls only

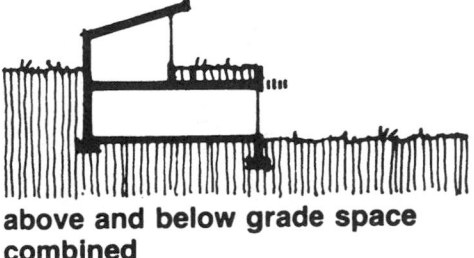

above and below grade space combined

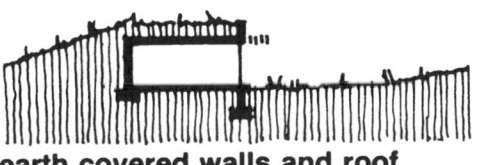

earth covered walls and roof partially recessed (bermed)

Fig. 8   Various relationships to the surface.

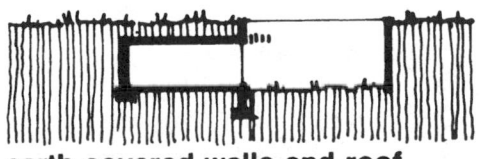

earth covered walls and roof fully recessed

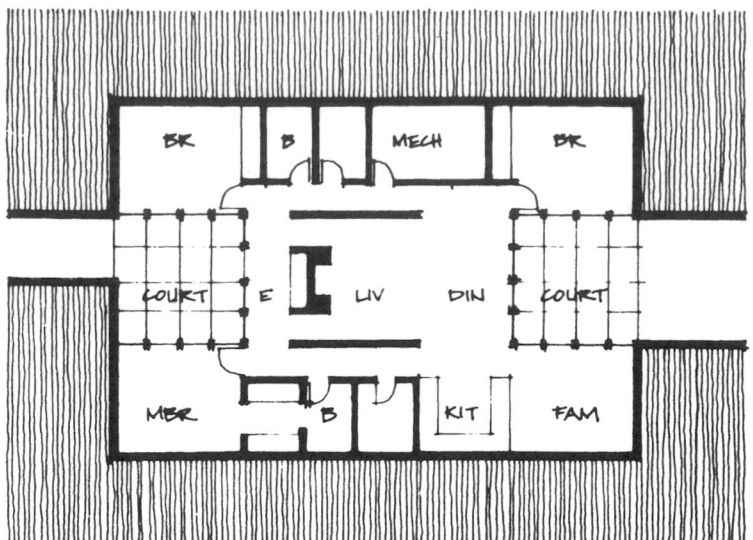

floor plan

Fig. 9*a*   Atrium design.

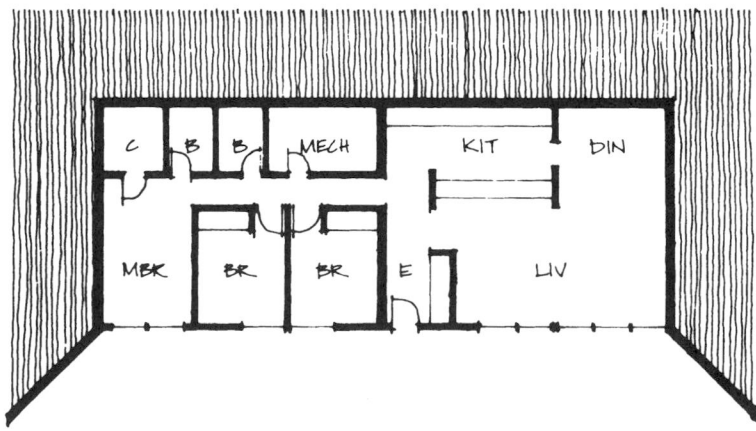

floor plan

Fig. 9*b*   Elevational design.

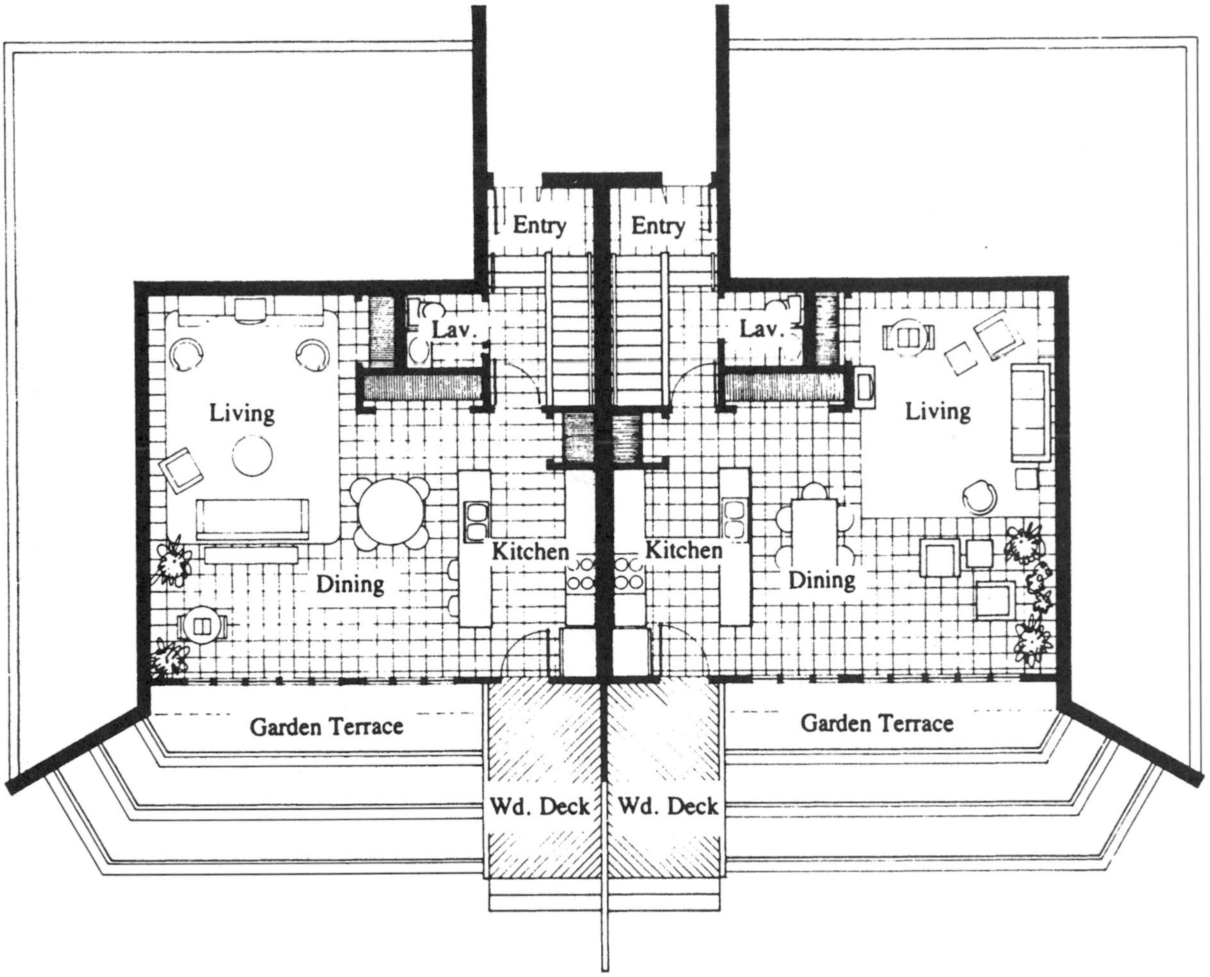

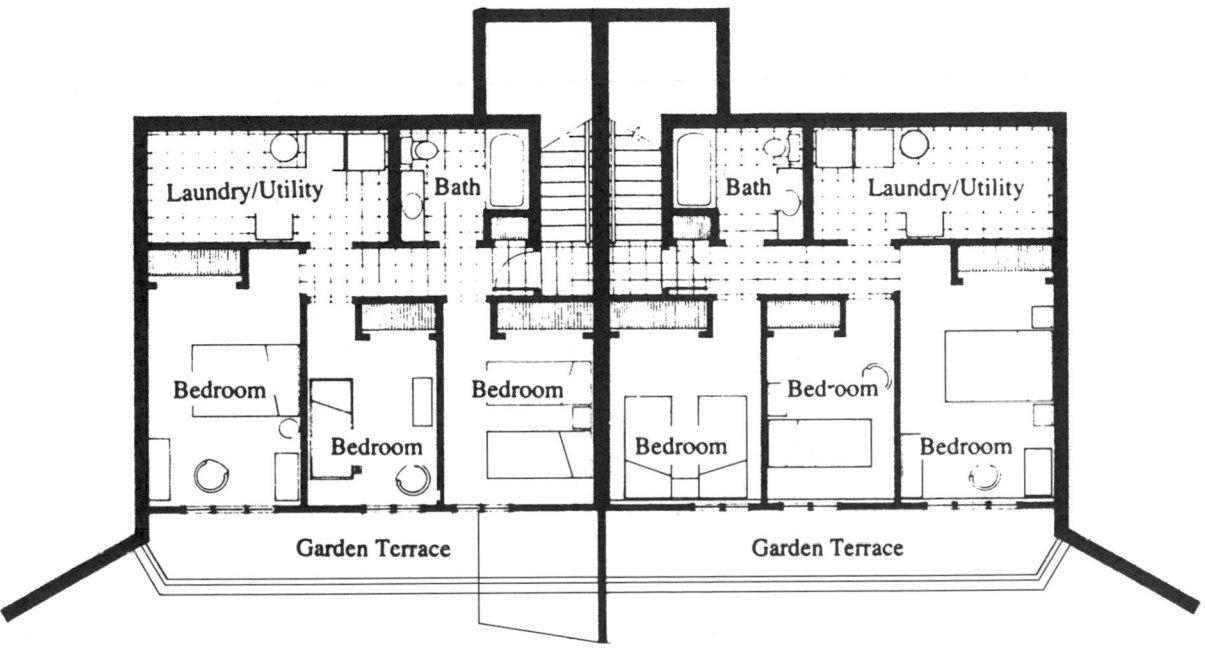

Fig. 10a Upper level.

Fig. 10b Lower level.

TYPICAL HOUSE PLANS

SIMPLE RECTANGLE

The rectangle plan is the simplest and one of the most common plan types. Most minimal or economical houses utilize this kind of plan because it encloses greater floor area per exterior wall length than other plans. Its simplicity also results in uncomplicated framing. The plan is compact, which results in a minimum of circulation space. A garage or carport is generally located alongside the kitchen or the front of the house.

Because of its compactness, there is a minimum of separation between the living and sleeping activities, thereby lessening the amount of privacy. This type of house is usually referred to as a "ranch" type.

A variation on the simple rectangle plan is the offset rectangle where the living area is pushed forward.

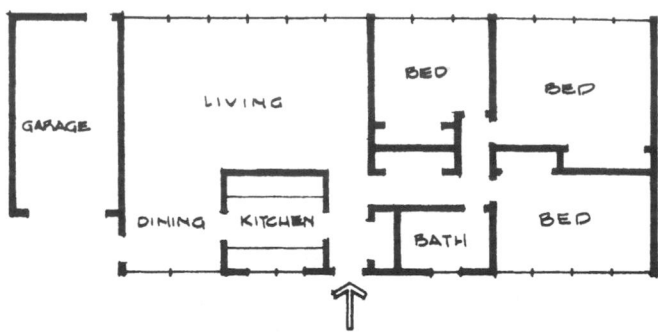

Fig. 1   The simple rectangle plan.

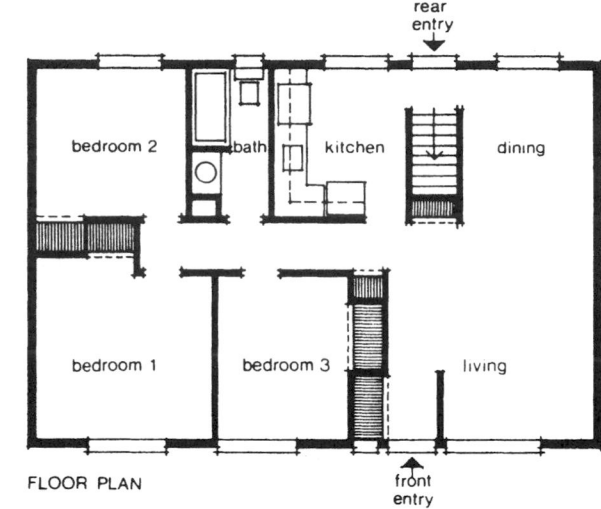

Fig. 3

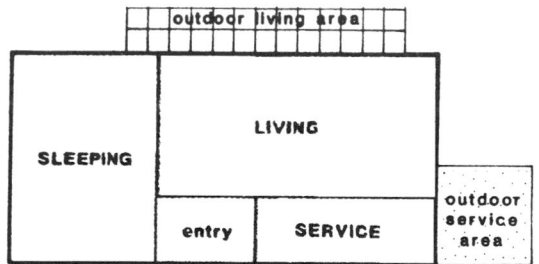

Fig. 2   Rectangular plan.

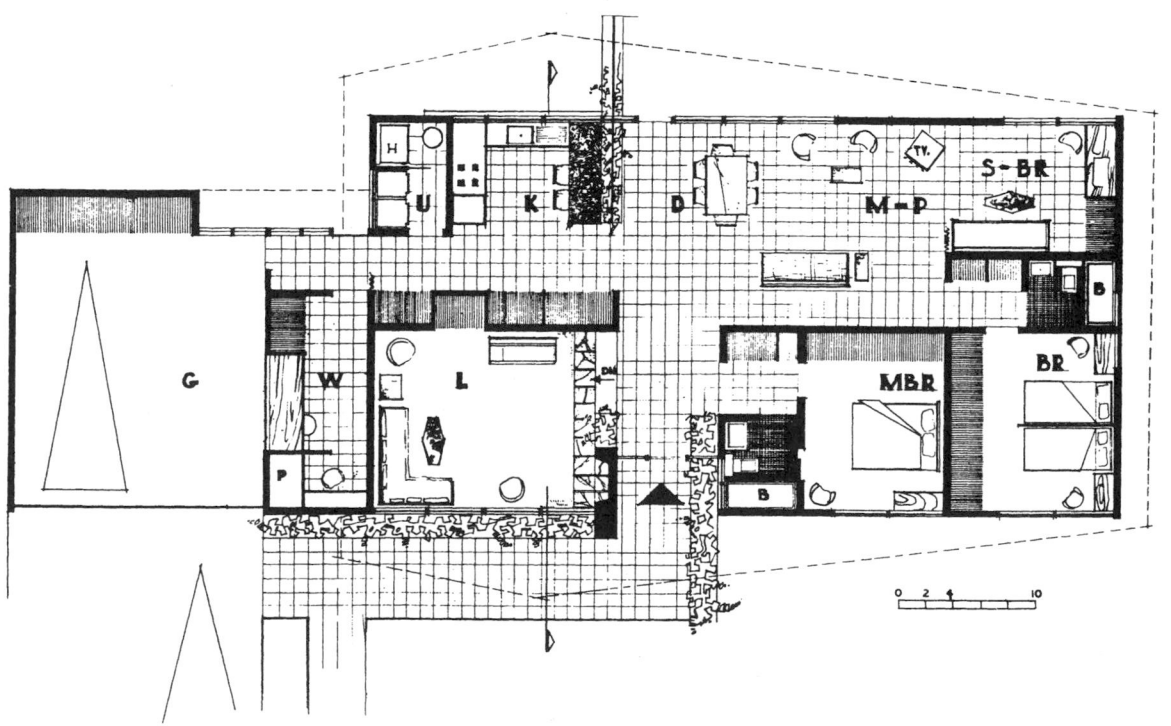

Fig. 4   Donald E. Evenson—Architect.

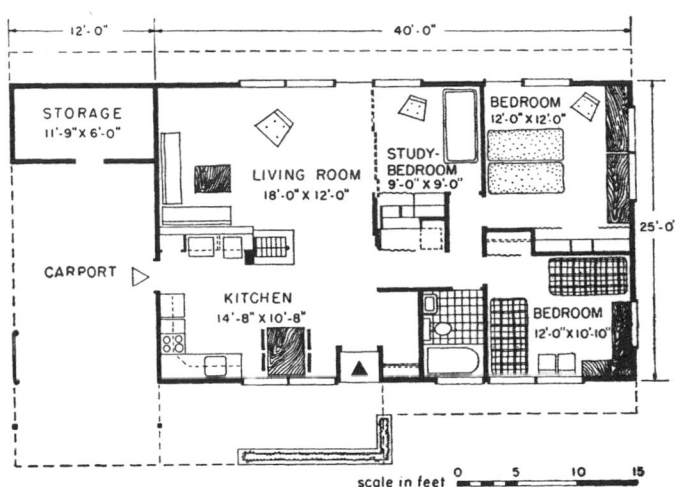

STORAGE
11'-9" x 6'-0"

LIVING ROOM
18'-0" X 12'-0"

BEDROOM
12'-0" X 12'-0"

STUDY-
BEDROOM
9'-0" X 9'-0"

25'-0'

CARPORT

KITCHEN
14'-8" X 10'-8"

BEDROOM
12'-0" X 10'-10"

12'-0"

40'-0"

scale in feet   0   5   10   15

Fig. 5

TYPICAL HOUSE PLANS

THE IN-LINE PLAN

The in-line plan is an excellent solution for many unusual site conditions. On a narrow lot it allows access to side patios and outdoor areas; on steep hillsides it allows the maximum economy of construction and land usage. It can have good circulation (at the expense of a long corridor) and the same good orientation for all the rooms.

The plan may be adapted to a two-story house, where it helps to concentrate circulation and utilities, while retaining the advantage of providing the best orientation for both floors.

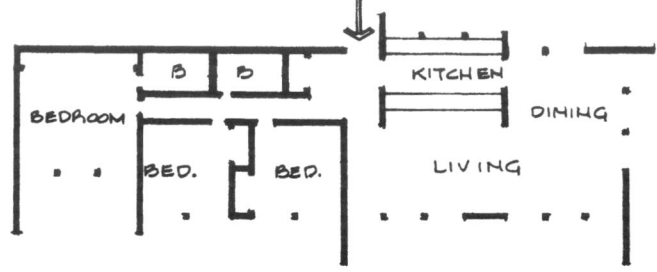

Fig. 6   The in-line plan.

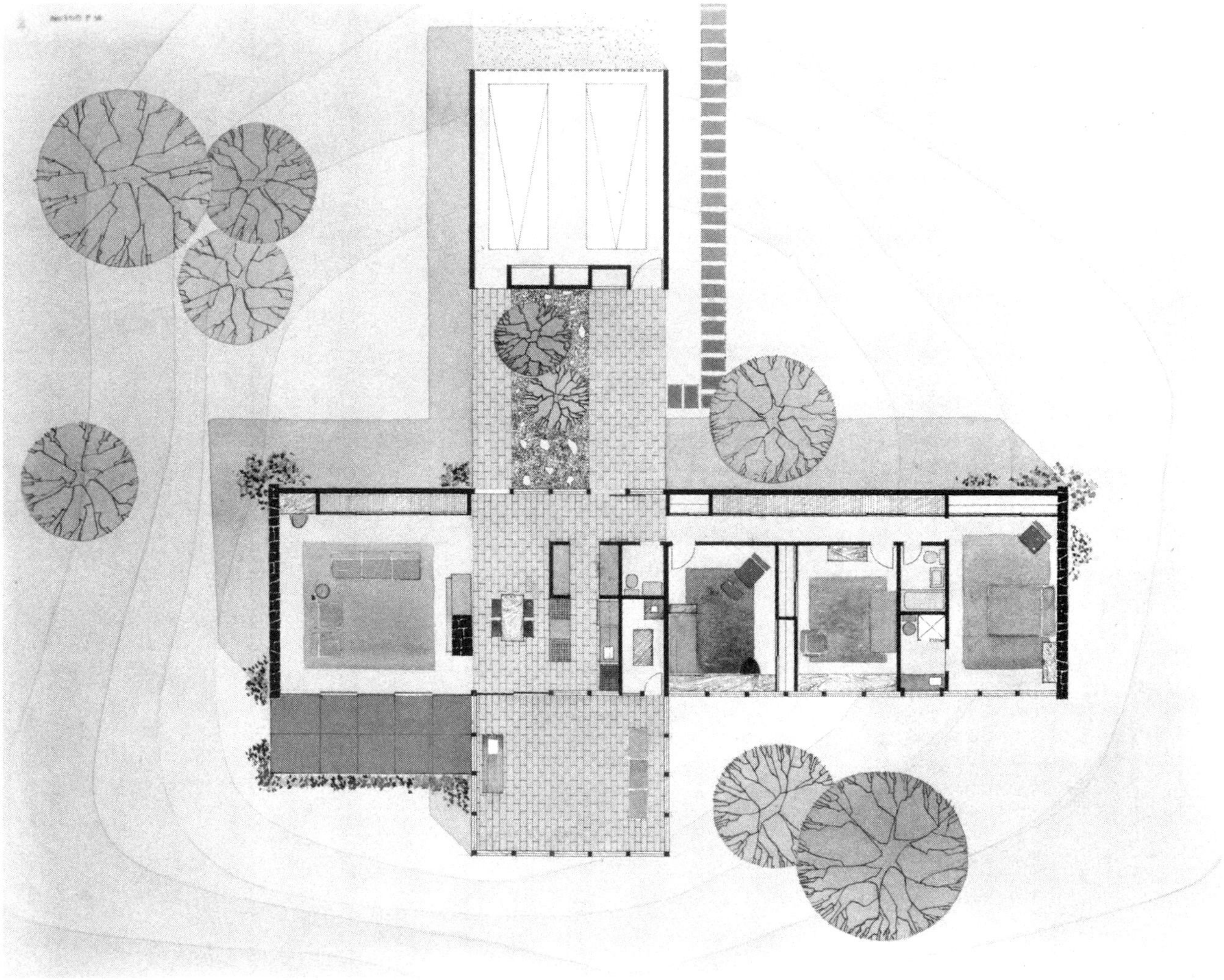

Fig. 7

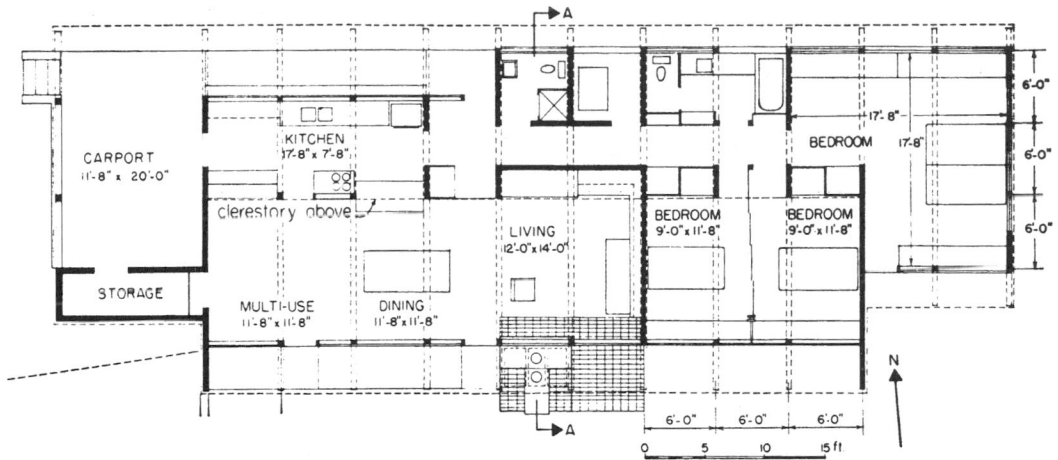

Fig. 8   Raleigh, N.C. George Matsumoto—Architect.

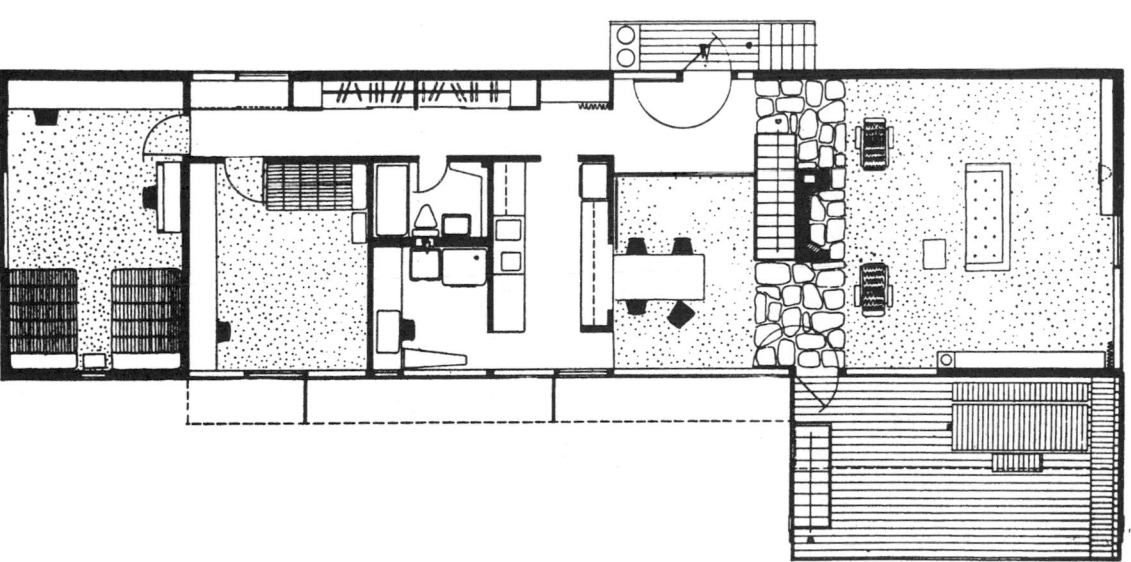

Fig. 9   Breuer house, New Canaan, Conn.

TYPICAL HOUSE PLANS

T OR L PLAN

The T plan is the placement of the living and sleeping areas at right angles to each other. By such juxtaposition, excellent separation and privacy of the two functions is achieved. It may also be possible to achieve better orientation for both functions since they are relatively independent of each other.

The internal circulation is compact and access to all rooms direct if the entrance is located at the junction of the two wings.

This plan type is best on a flat site. If the site slopes, it is possible to locate the garage, recreation, and utility areas under one of the wings.

A variation of the T plan is the L plan. This occurs when the living area is located at the top or bottom of the sleeping wing instead of at the center.

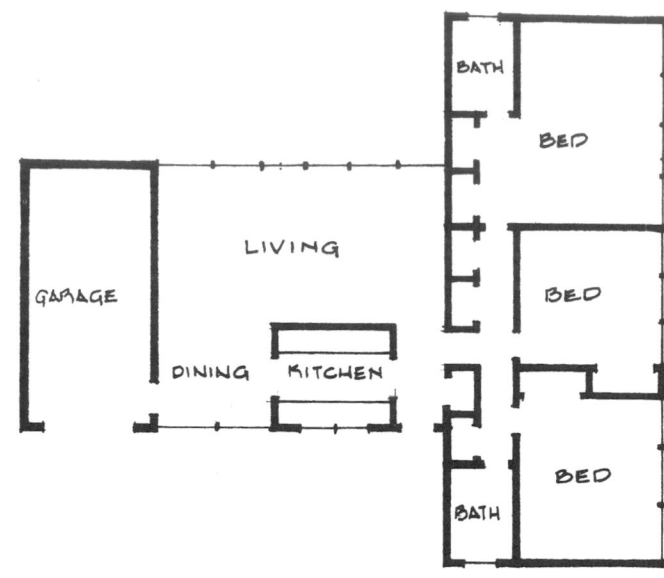

Fig. 10   T plan.

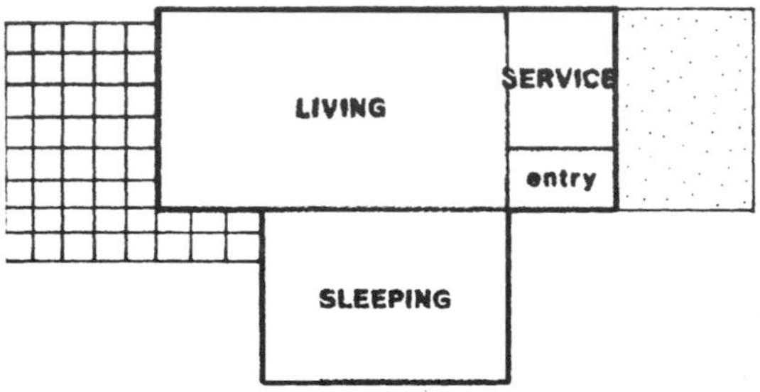

Fig. 11   T plan.

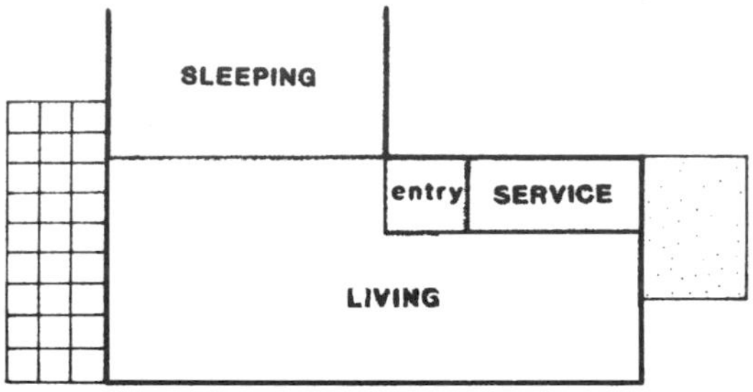

Fig. 12   L plan.

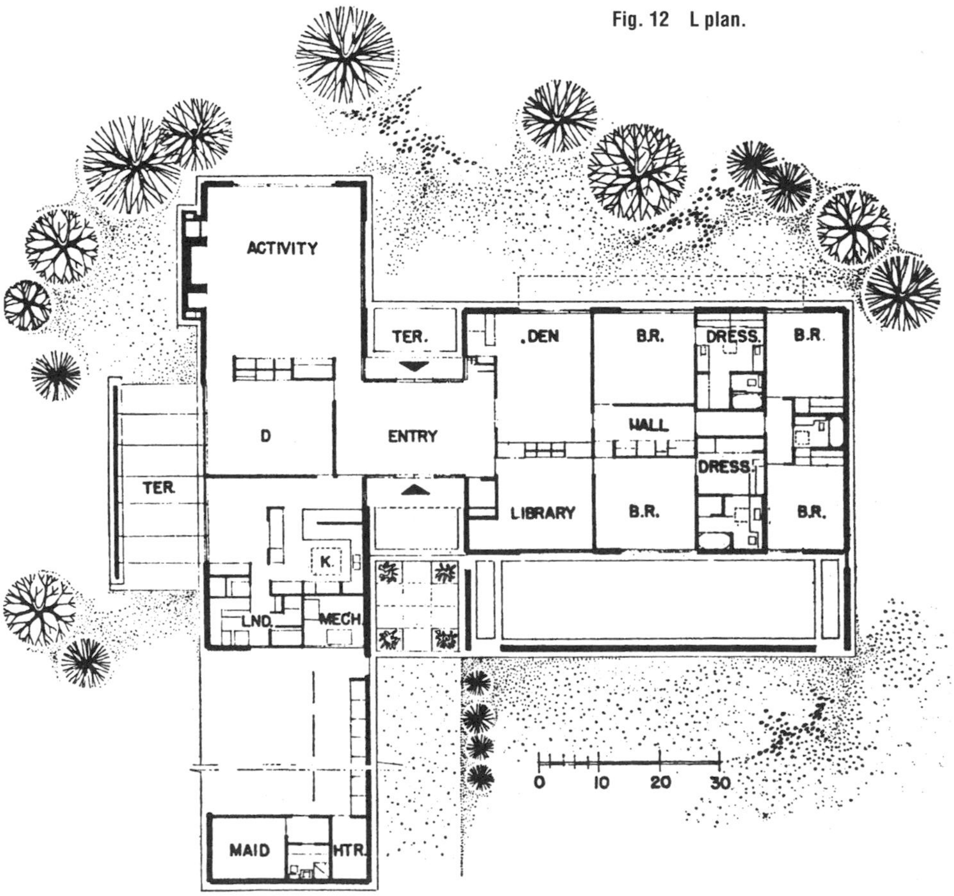

Fig. 13   Bolton and Barnstone—Architects.

H AND U PLANS

H and U plans divide living and sleeping units into separate sections. This layout is especially applicable to a utility core concept in which the kitchen becomes part of the connecting link. Excellent separation of activities is achieved, and useful patios afford shelter and privacy. In addition, each room can receive cross ventilation. The chief disadvantage of these types is in the long perimeter walls (almost 50 percent more than the same space in a simple rectangle), resulting in higher construction cost as well as increased expense of heating and air conditioning.

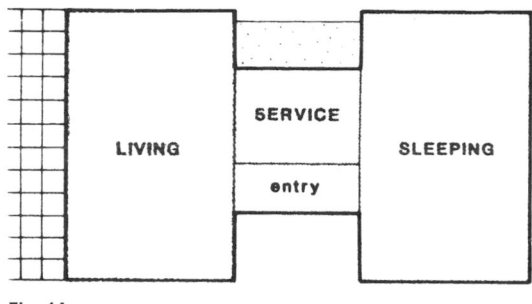

Fig. 14

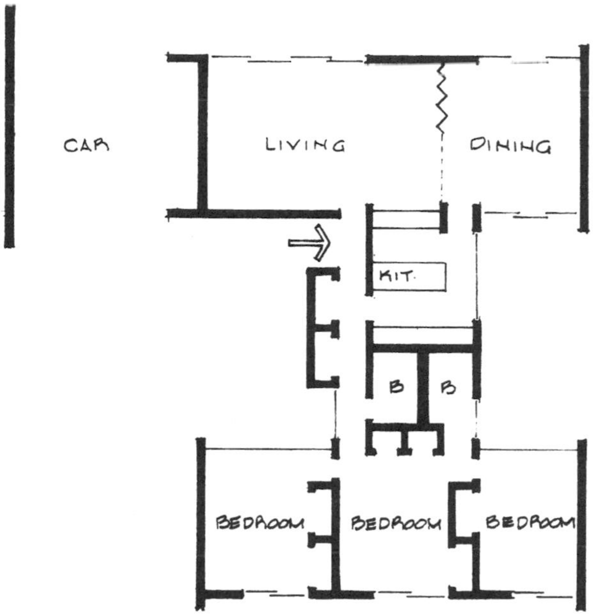

Fig. 15   H plan.

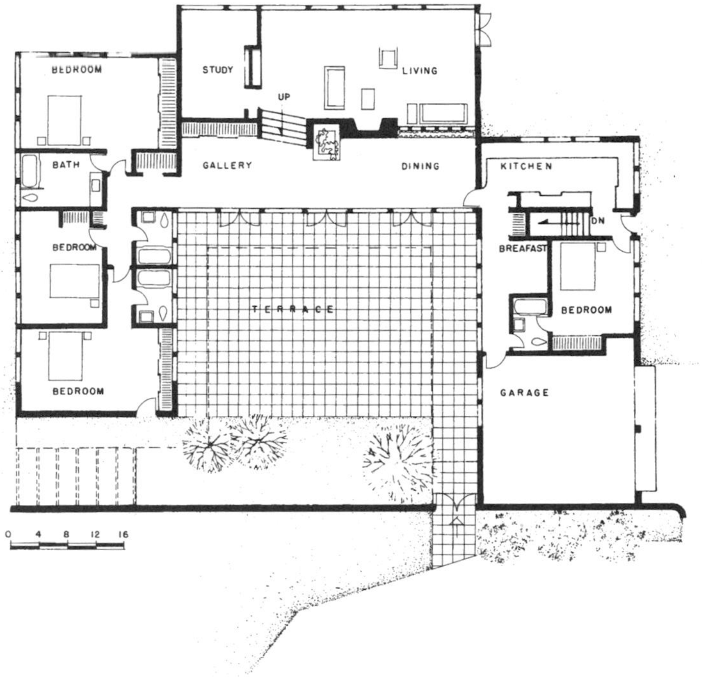

Fig. 16   Maduro House, Great Neck, N.Y. Edward D. Stone—Architect.

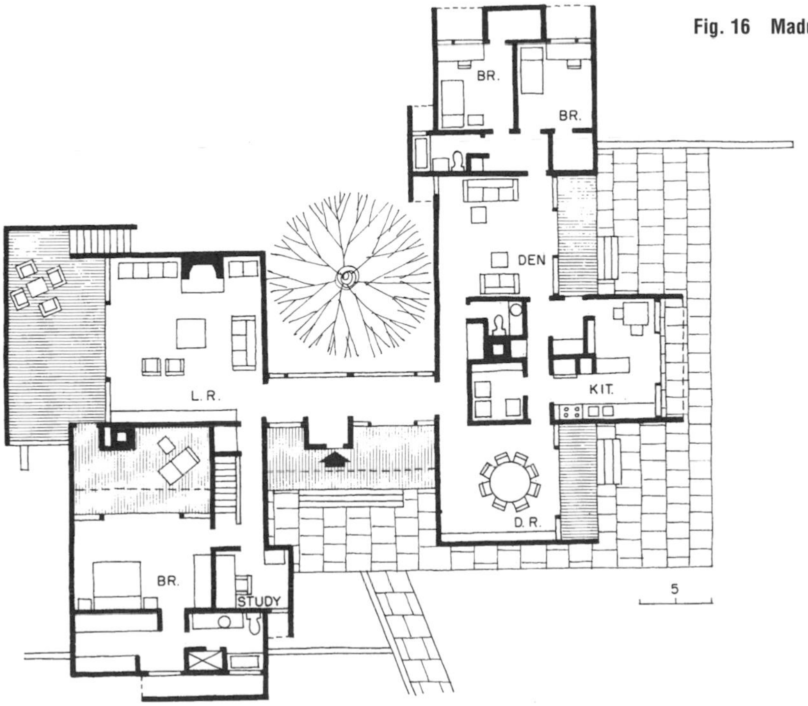

Fig. 17   Marcel Breuer—Architect.

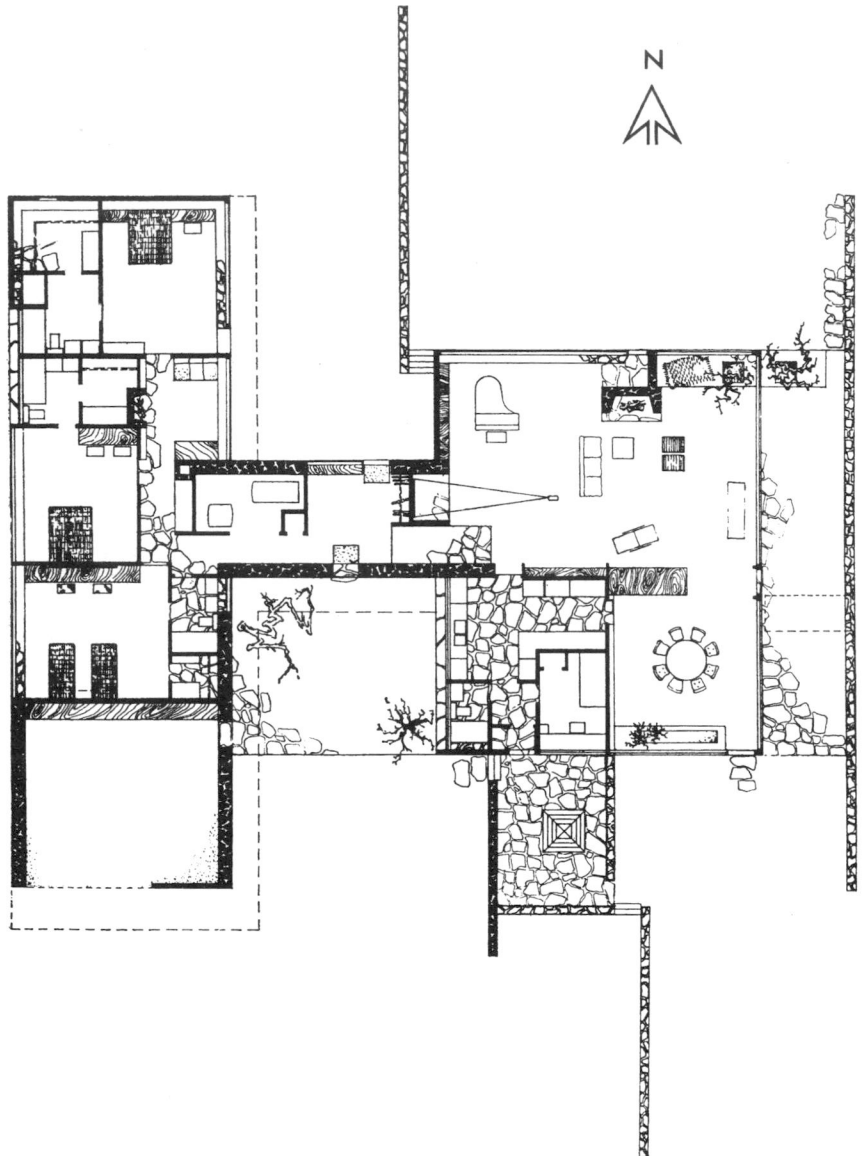

Fig. 18   Robinson house, Williamstown, Mass. Marcel Breuer—Architect.

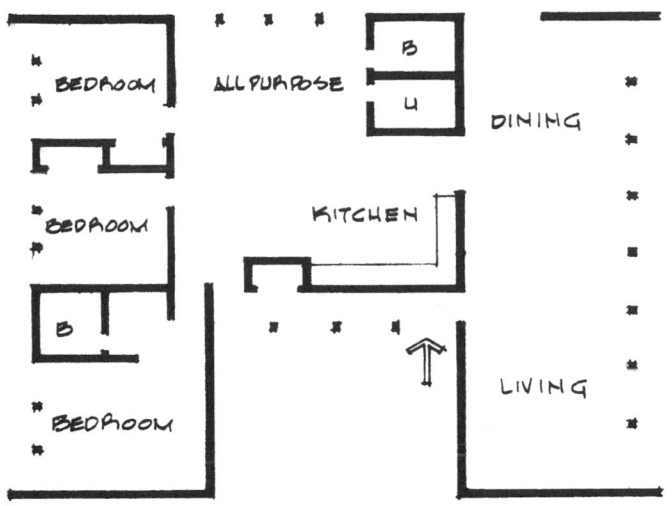

BEDROOM

ALL PURPOSE

B
U

DINING

BEDROOM

KITCHEN

B

BEDROOM

LIVING

Fig. 19   U plan.

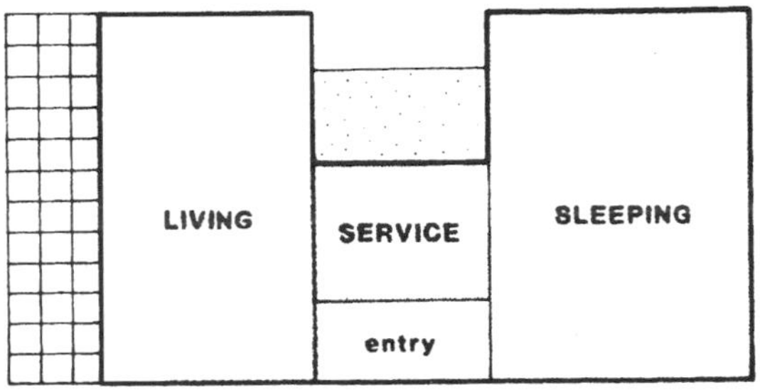

LIVING

SERVICE

SLEEPING

entry

Fig. 20   U plan.

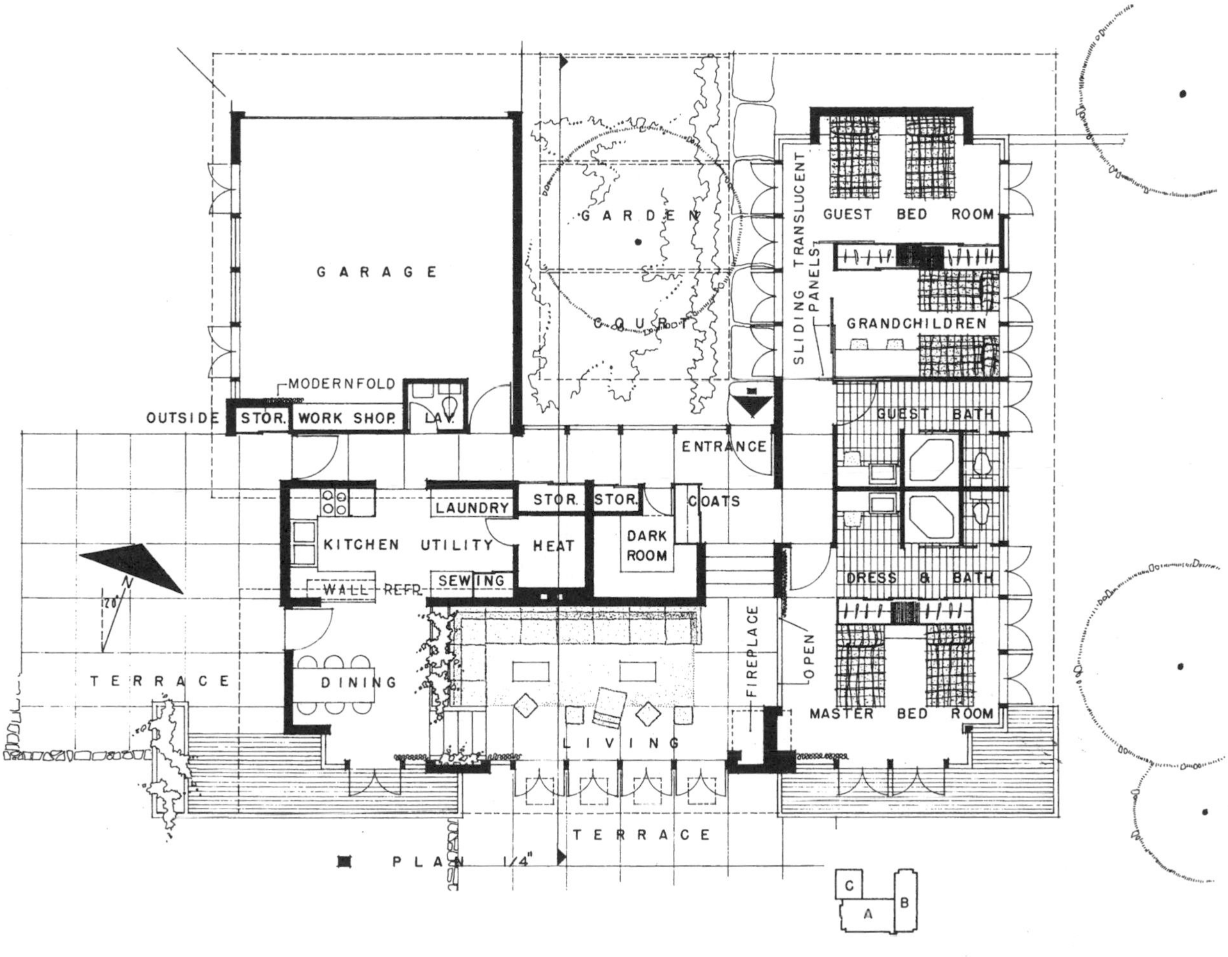

GARAGE

GARDEN

COURT

SLIDING TRANSLUCENT PANELS

GUEST   BED   ROOM

GRANDCHILDREN

GUEST   BATH

OUTSIDE   STOR.   WORK SHOP.   LAV.
MODERNFOLD

ENTRANCE

KITCHEN   UTILITY
LAUNDRY   STOR.   STOR.   COATS
HEAT   DARK ROOM
WALL   REFR.   SEWING

DRESS   &   BATH

N
20

TERRACE

DINING

FIREPLACE   OPEN

LIVING

MASTER   BED   ROOM

TERRACE

PLAN   1/4"

C
A   B

LIVING   AREA   "A"   951 2/3 □'
SLEEPING   AREA   "B"   808 2/3 □'
GARAGE   1/2   AREA   "C"   238   □'
TOTAL   1998 1/3 □'

Fig. 21

TYPICAL HOUSE PLANS

UTILITY-CORE PLAN

The rectangular utility-core plan has several advantages. The house may be almost square and very compact, with a good concentration of utilities. In addition, the core acts as a buffer between the sleeping and living zones. The problems of this plan include the difficulty of properly relating the kitchen, garage, and main entrances, and the excessive circulation space that is often required. This can be helped by opening up the exterior walls and actually using the lot as circulation and access in areas of mild climate.

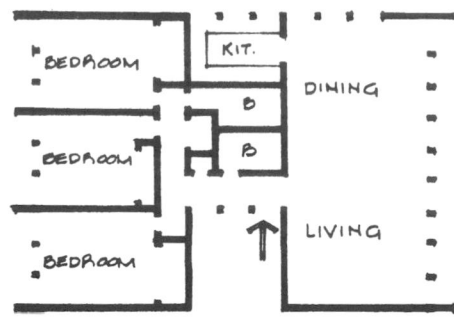

Fig. 22   The utility-core plan.

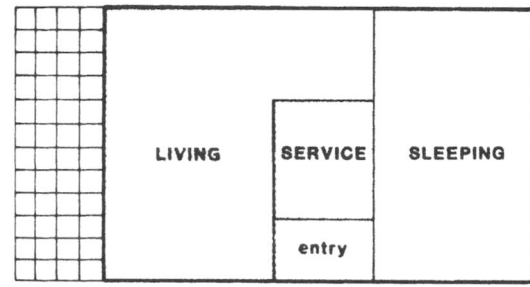

Fig. 23   Service-core plan.

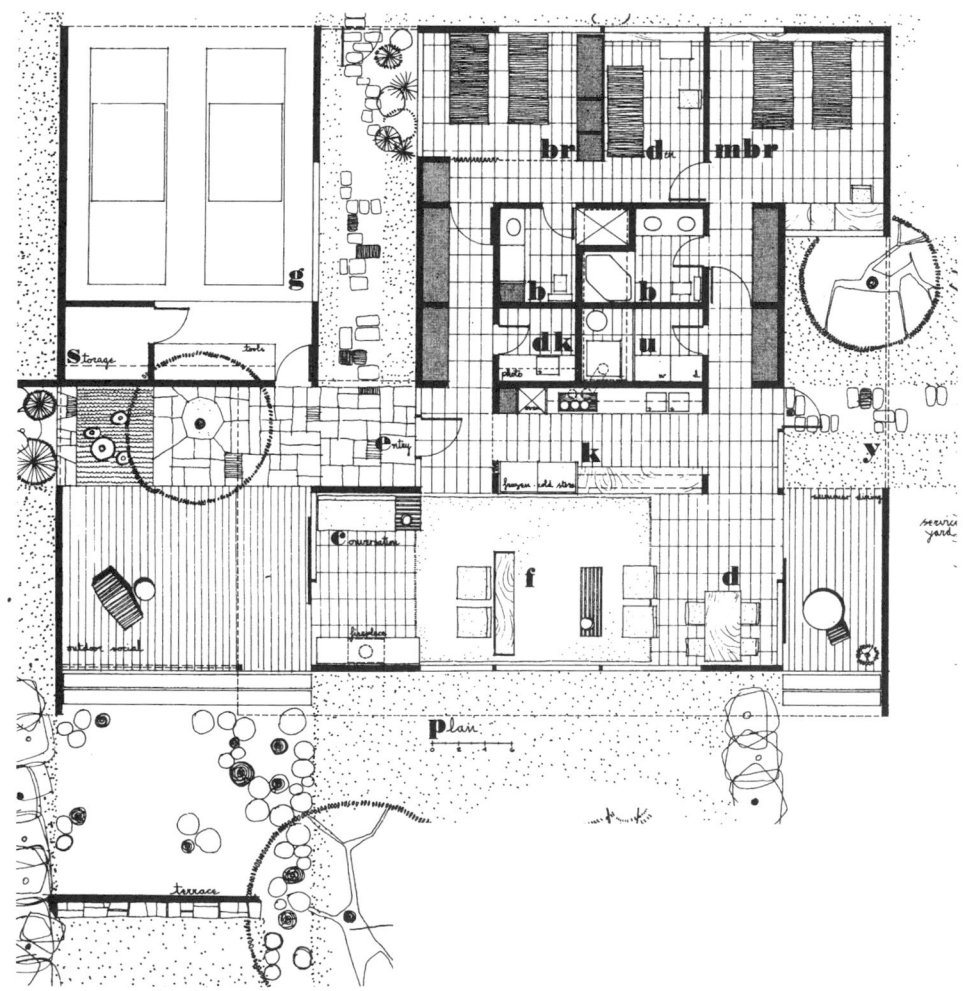

Fig. 24   Charlotte, N.C. Frederick F. Sadri—Architect.

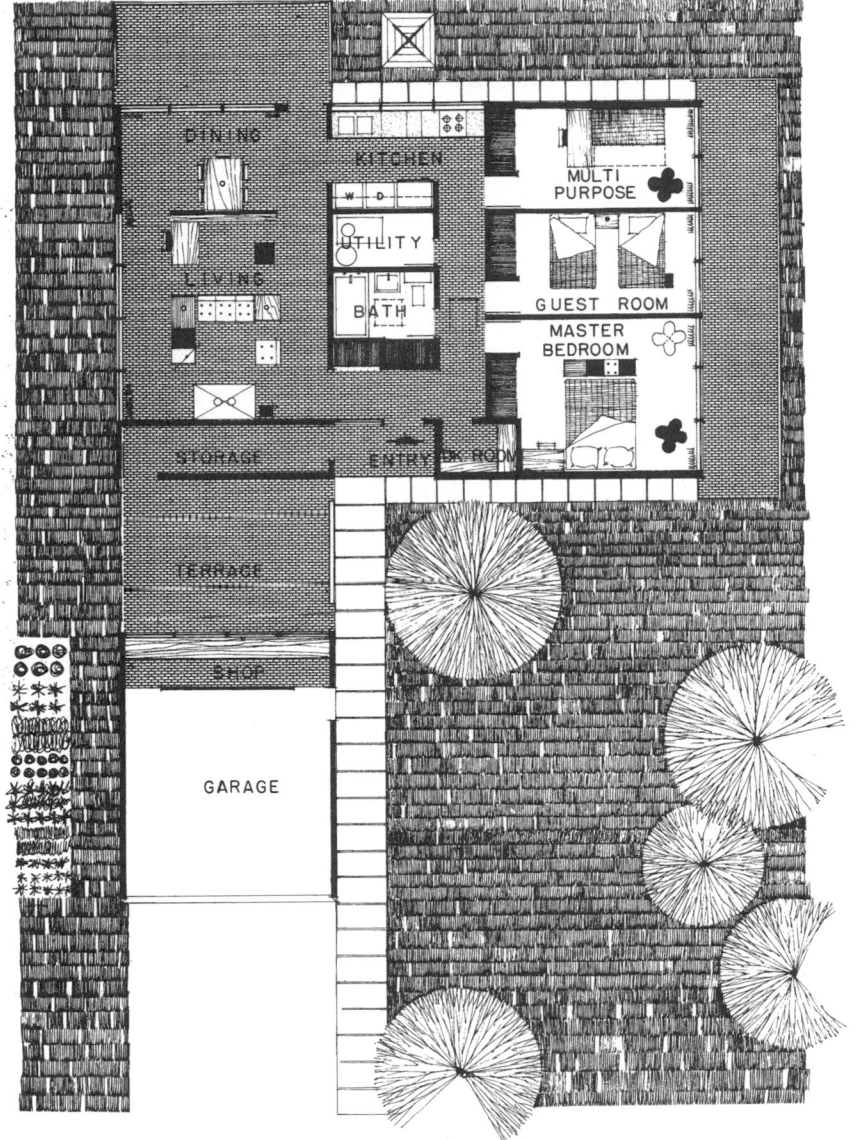

DINING

KITCHEN

MULTI PURPOSE

W D

UTILITY

GUEST ROOM

BATH

MASTER BEDROOM

LIVING

STORAGE

ENTRY

TERRACE

SHOP

GARAGE

Fig. 25   Floor plan. Gus W. Kostopulos, University of Illinois—Architect.

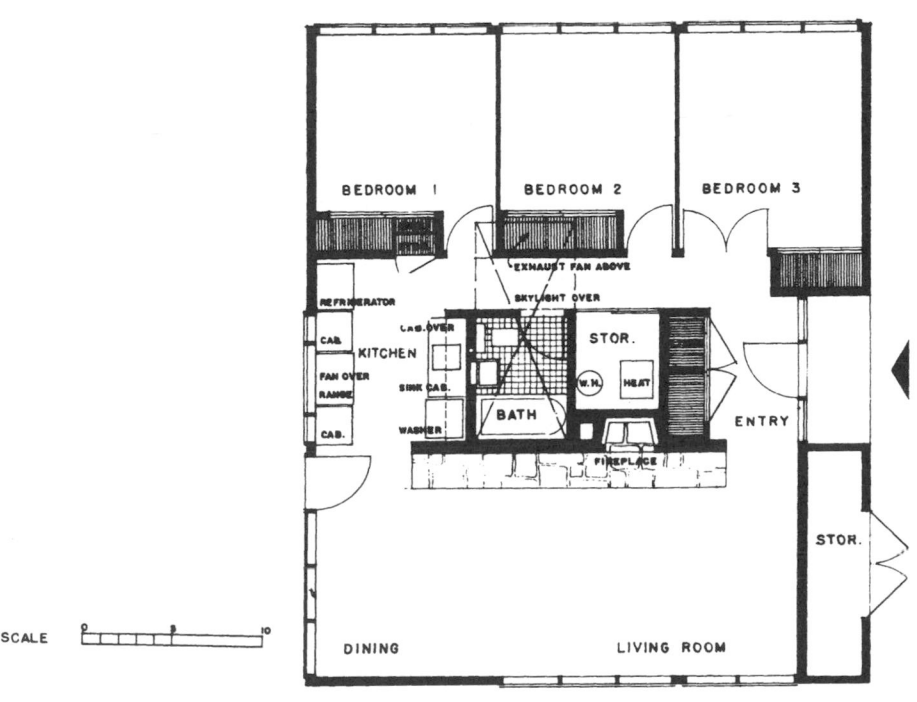

BEDROOM 1

BEDROOM 2

BEDROOM 3

EXHAUST FAN ABOVE

SKYLIGHT OVER

REFRIGERATOR

CAB. OVER

STOR.

CAB.

KITCHEN

FAN OVER RANGE

SINK CAB.

W. H.   HEAT

CAB.

WASHER

BATH

ENTRY

FIREPLACE

STOR.

DINING

LIVING ROOM

SCALE   0   5   10

Fig. 26

TYPICAL HOUSE PLANS

THE SPLIT-LEVEL PLAN

The split-level plan produces a maximum of total interior area for a house of small overall size, and its separate levels can give greater privacy and interest in each area. It is very adaptable to sloping lots, and helps to solve the problems of deep foundations in northern climates. However, it may require a somewhat complicated framing system, and is difficult to relate to outdoor areas without special terracing or grading.

The split-level house is a multilevel dwelling unit consisting of either three or four levels separated by one-half floor heights, but all connected by a single stair. The most common type is the three-level design, which has the main living area (living room, kitchen, and dining area) on the middle level. The upper level, one-half flight up, contains the sleeping area (bedrooms and bath), and the lower level, one-half flight down, contains the utility, recreation, work areas, or a garage. Sometimes a level below the lower level is introduced as a cellar.

The split-level house was originally designed to accommodate or take advantage of sloping terrain. Frequently, because of advantages of the split-levels, this type of house is placed on flat sites. The result is either an excessive amount of grading and retaining walls, or an awkward relationship of house to site.

The main advantage is the separation of the living, sleeping, and recreation-utility areas, yet they are more accessible than in a traditional two-story house. One drawback to the split-level is that it is more complicated and costly to build than a one- or two-story house.

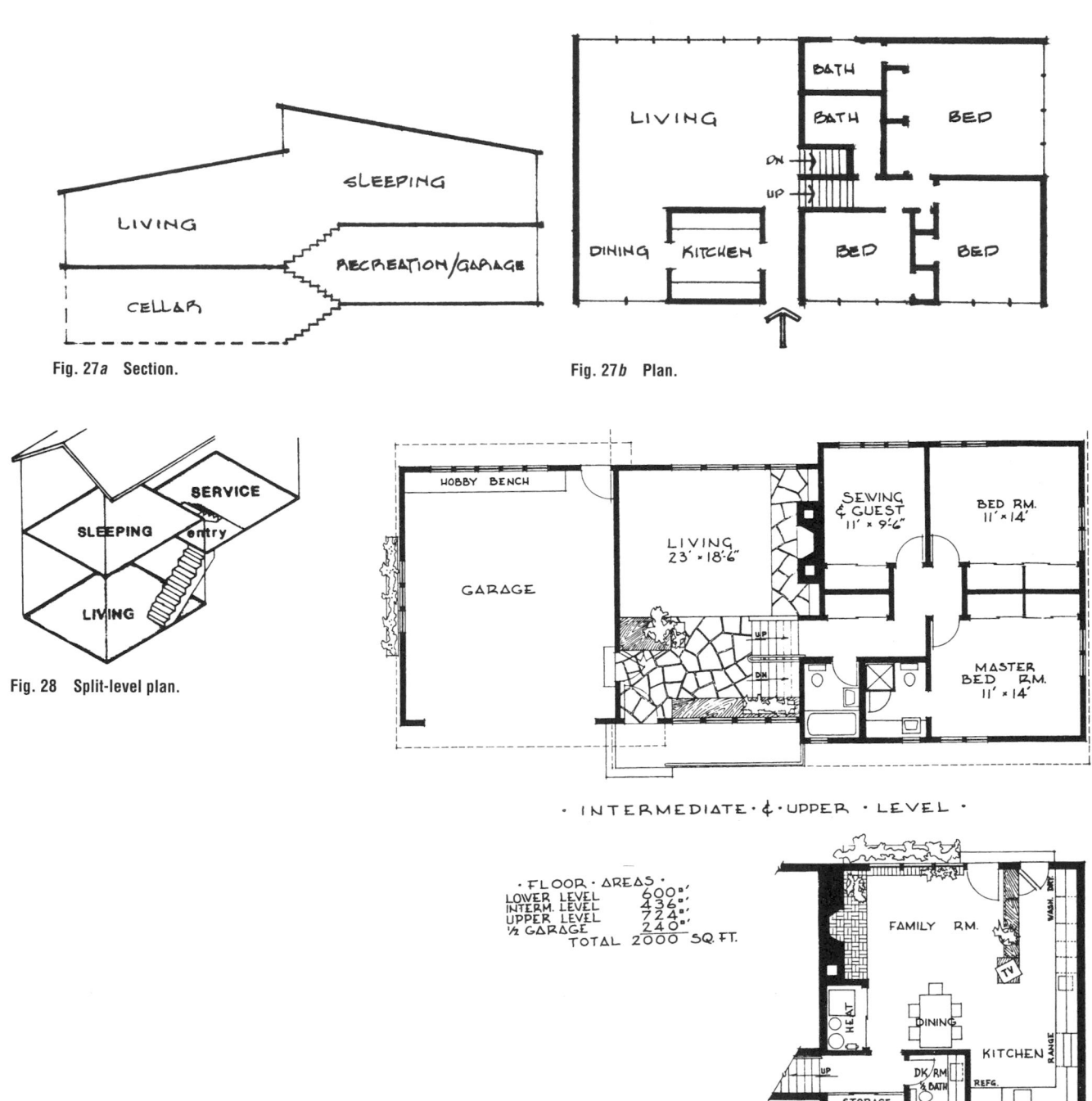

Fig. 27a   Section.

Fig. 27b   Plan.

Fig. 28   Split-level plan.

· FLOOR · AREAS ·
LOWER LEVEL   600
INTERM. LEVEL   436
UPPER LEVEL   724
½ GARAGE   240
          TOTAL 2000 SQ. FT.

· INTERMEDIATE · & · UPPER · LEVEL ·

· LOWER · LEVEL ·
SCALE 1'2'3'4'5'

Fig. 29

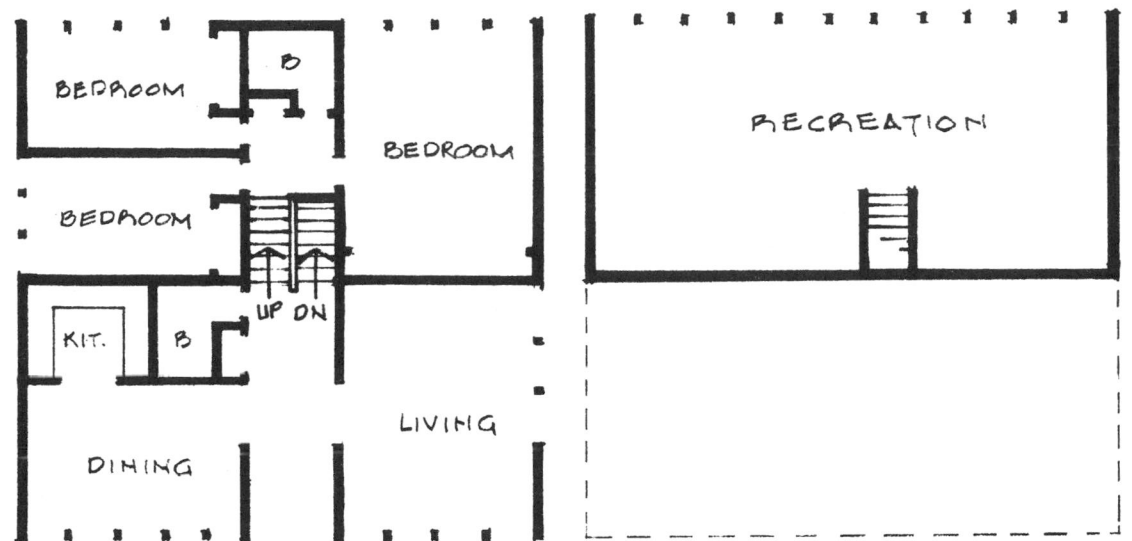

Fig. 30*a*   Entry and sleeping level, split-level plan.

Fig. 30*b*   Recreational level (under sleeping level).

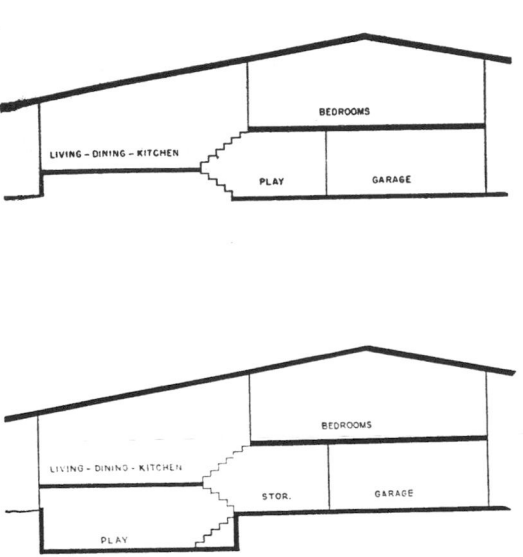

Fig. 31

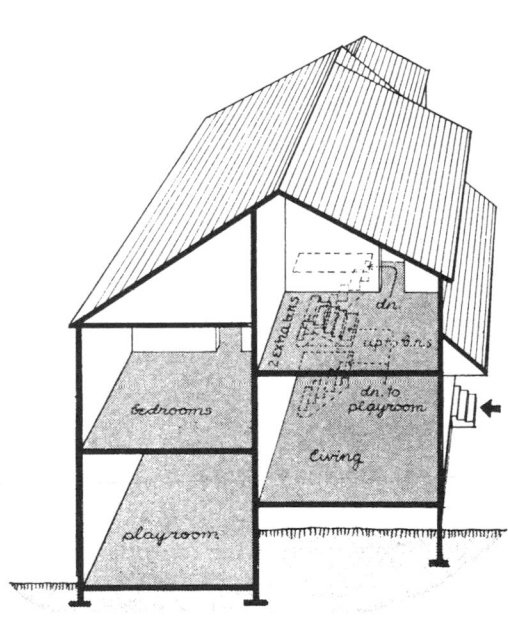

Fig. 32*a*   Front-to-back split-level.

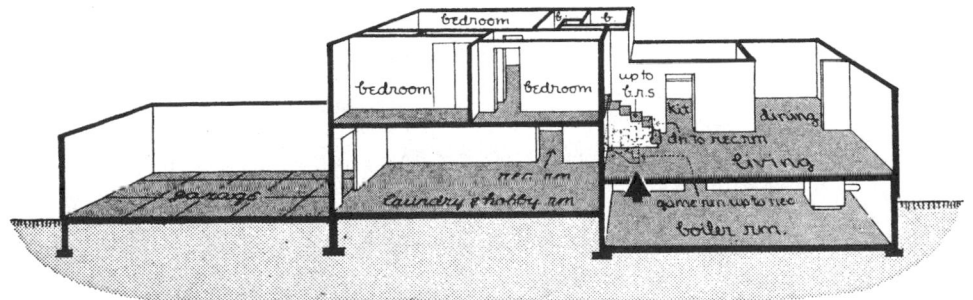

Fig. 32*b*   Typical side-by-side split-level with garage at side rather than under house.

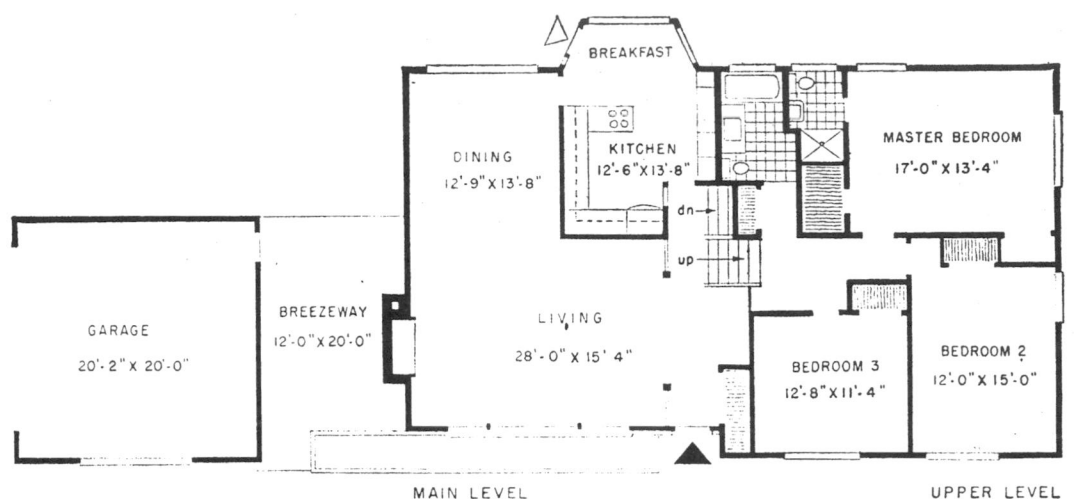

MAIN LEVEL

UPPER LEVEL

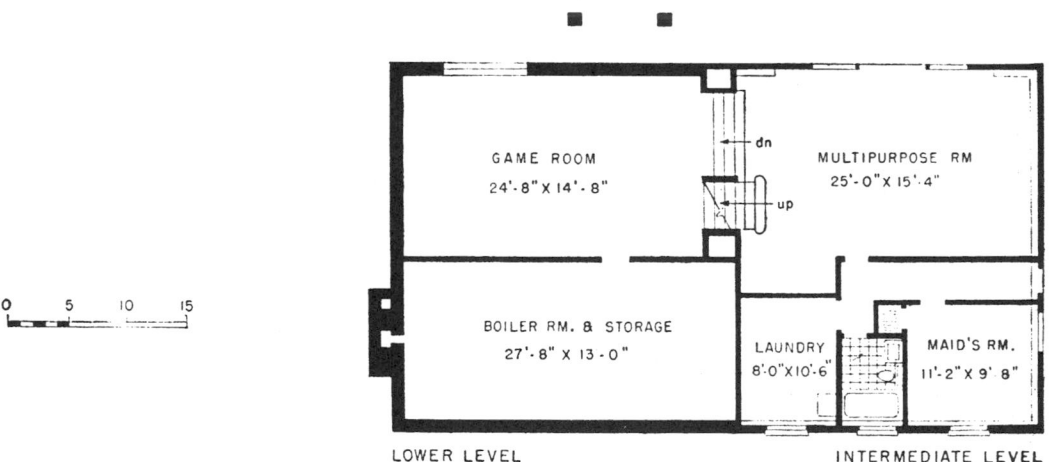

LOWER LEVEL

INTERMEDIATE LEVEL

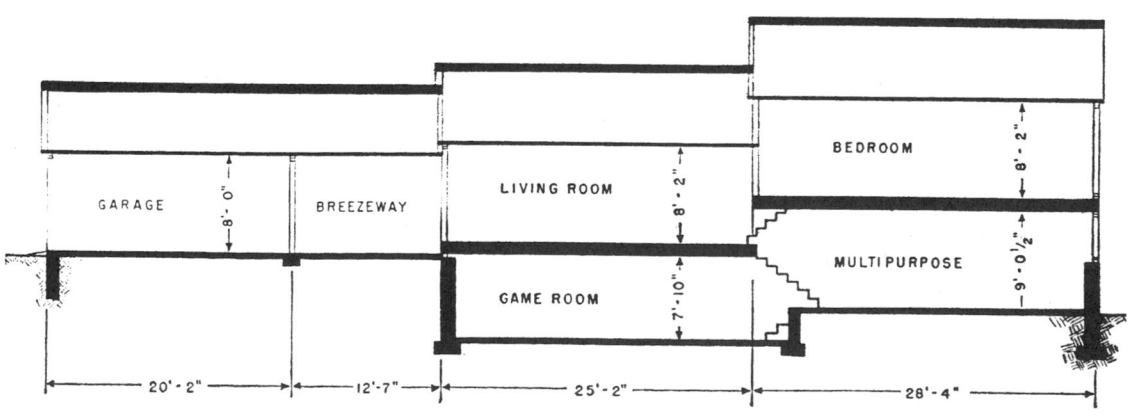

Fig. 33

TWO-STORY

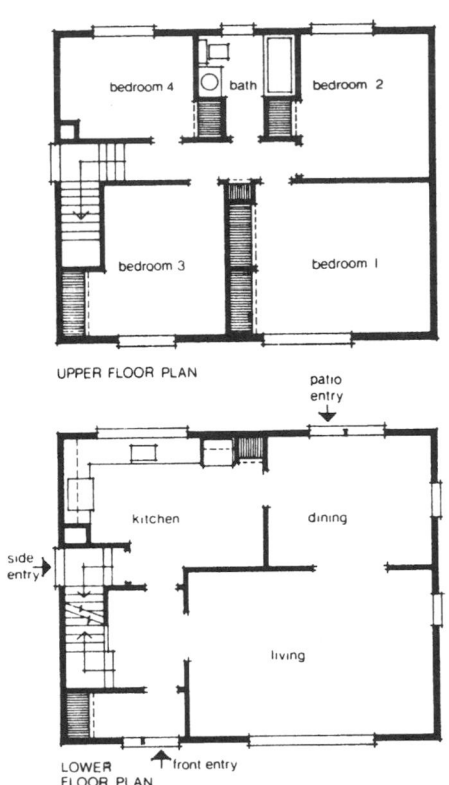

Fig. 34   Four-bedroom two-story house.

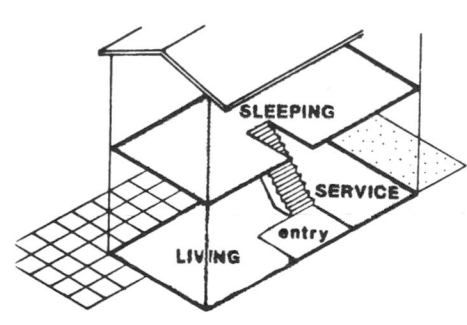

Fig. 36   Two-story plan.

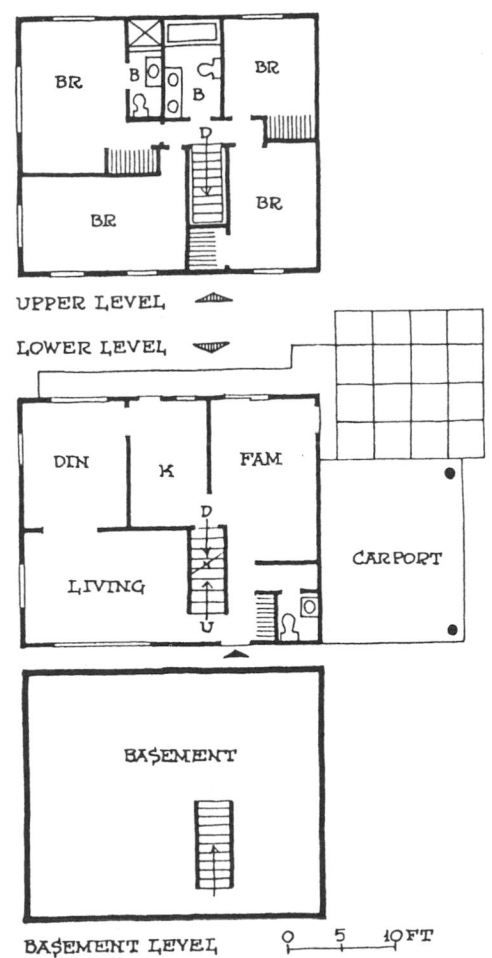

Fig. 35   Characteristic single-family home: two-story and basement, frame, with 1695 sq ft of living area.

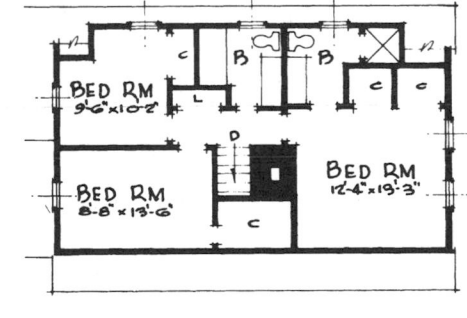

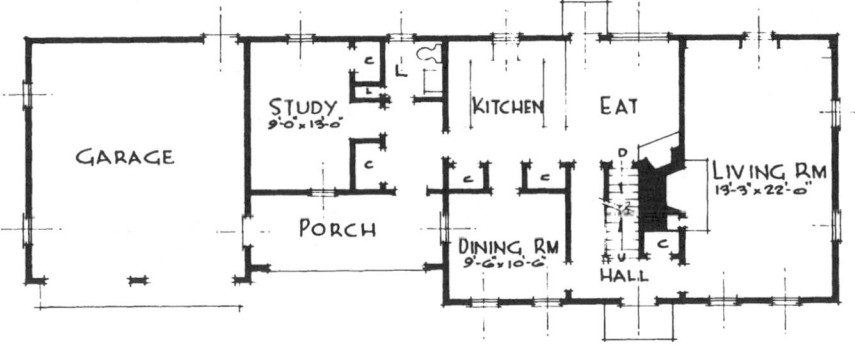

Fig. 37

COURT OR ATRIUM HOUSE

This type of house is a single-family, one-story dwelling unit and is described as either a court, garden-court, atrium, or patio house. The common element is an open landscaped courtyard partially or completely surrounded by living areas. The major source of light and air is through the open garden courtyard. Sometimes it is attached, as in a row house, or clustered as in a checkerboard pattern, but most frequently it is a detached structure. Historically, this type of house dates back several thousand years to Egyptian, Greek, and Roman houses. The open court is a variation of the Greek peristyle and the Roman atrium. All the living areas opened out into the atrium, creating a secluded indoor-outdoor space. This inward-directed house provides maximum privacy and livability. A large degree of integration of the house with the landscaping can be achieved. When used as an attached house, it makes maximum use of the lot and generally can be located on a much narrower lot than a conventional detached house. When enclosed by high walls or parts of the indoor living space, the house completely shuts off the outside world and assures greater protection from intruders. A garage may be incorporated within its enclosing walls, or a common parking area can be provided at the ends of the groupings or clusters.

The density of this type of housing, depending on size of units and site development, can be described as medium-density and generally will be similar to town house or row house densities. An approximate range would be from 12 units per acre for large court houses to 18 units for small ones.

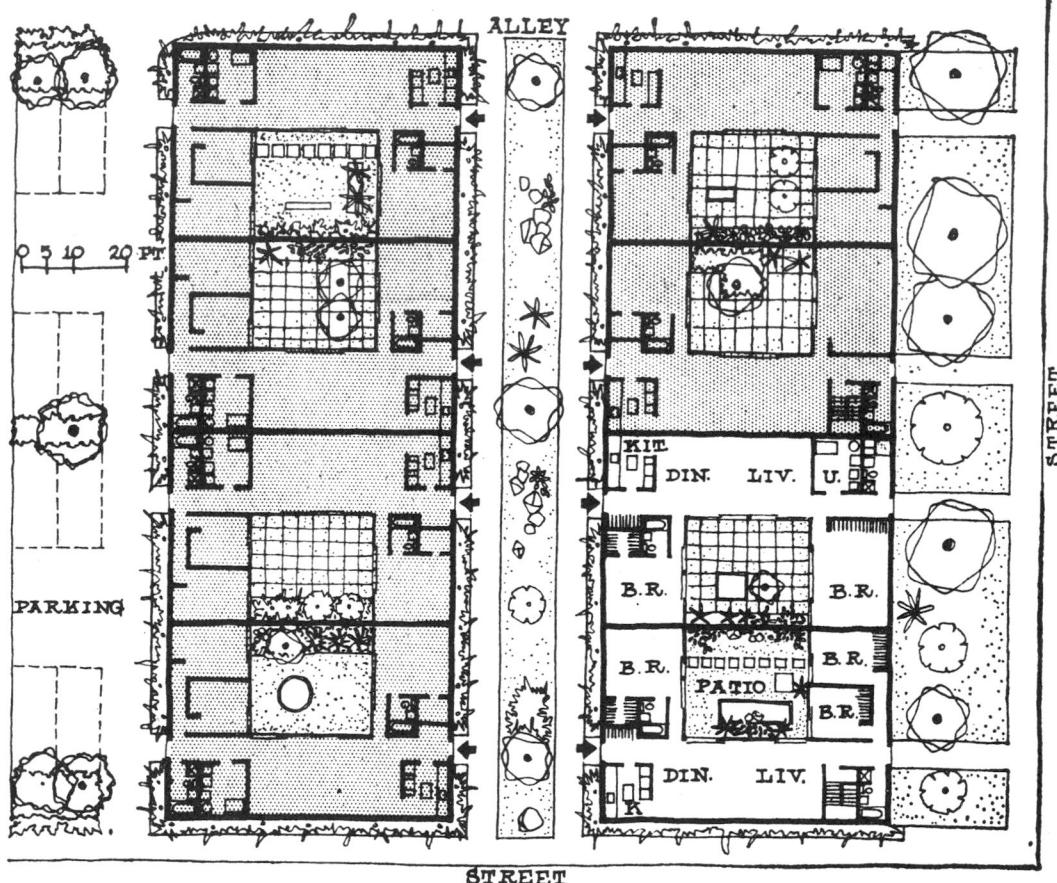

Room arrangements are varied to meet individual needs. They range from a one-bedroom-and-study layout to three-bedroom plans. Driveway, left, through the site from street to alley reduces congestion in the parking area.

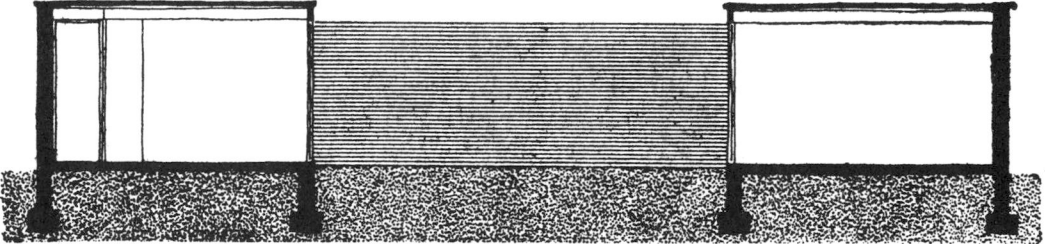

Cross-section of the structure is through on-slab house. Foundation walls of basement units are carried to greater depth. Non-bearing interior walls permitted different room arrangements within identical areas.

Fig. 38

478

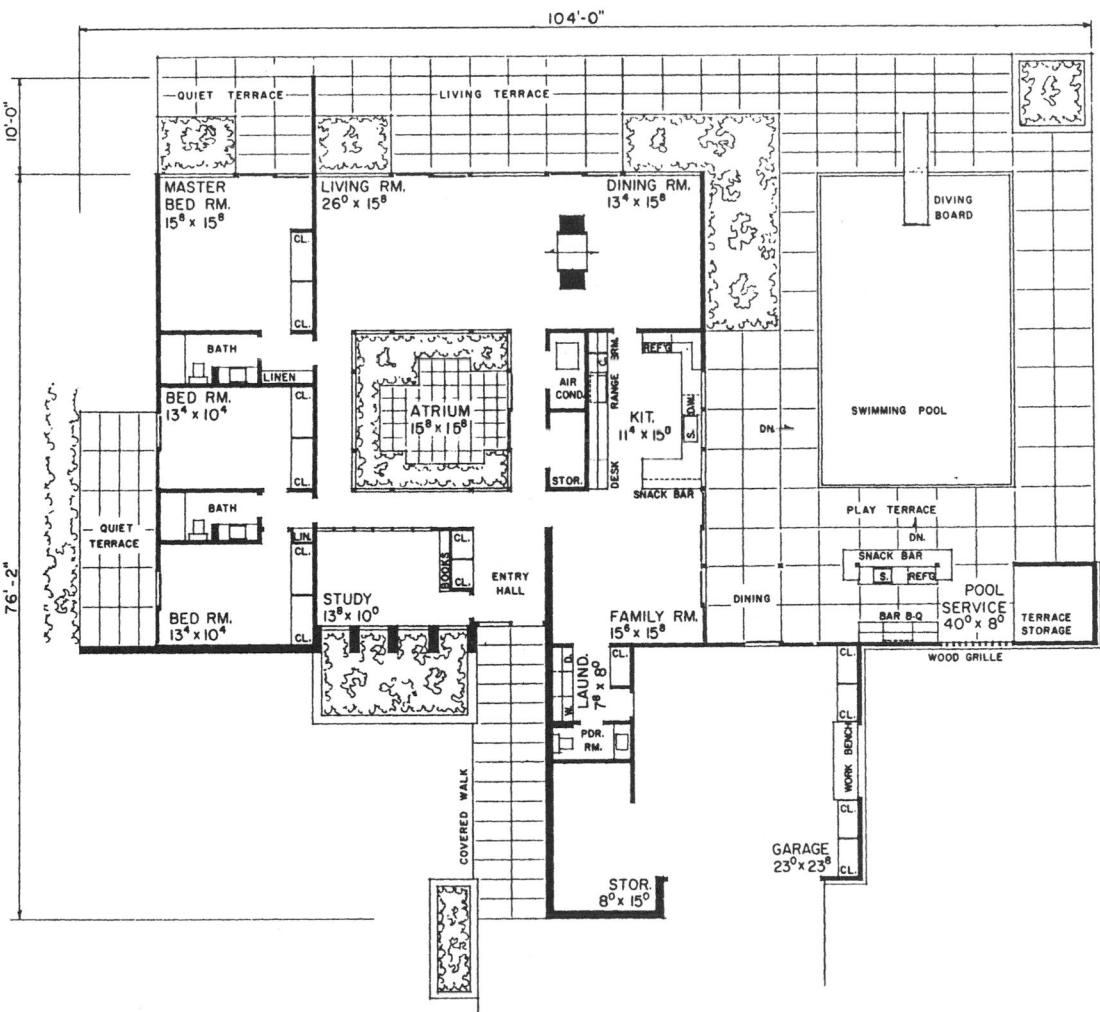

Fig. 39   House plan totaling 2569 ft²—excluding atrium.

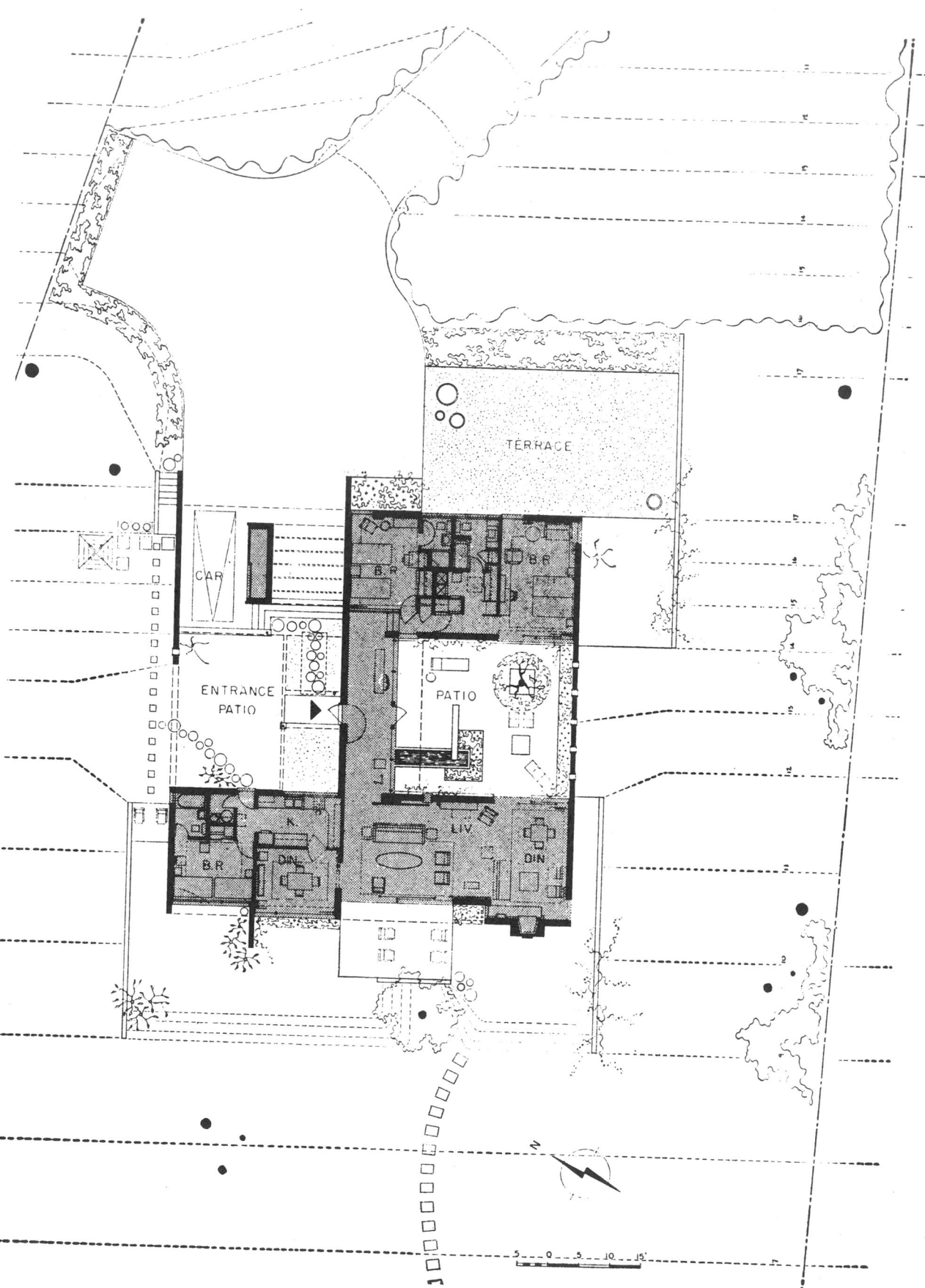

CAR

TERRACE

B.R

B.R

ENTRANCE
PATIO

PATIO

B.R

K

DIN

LIV

DIN

Fig. 40

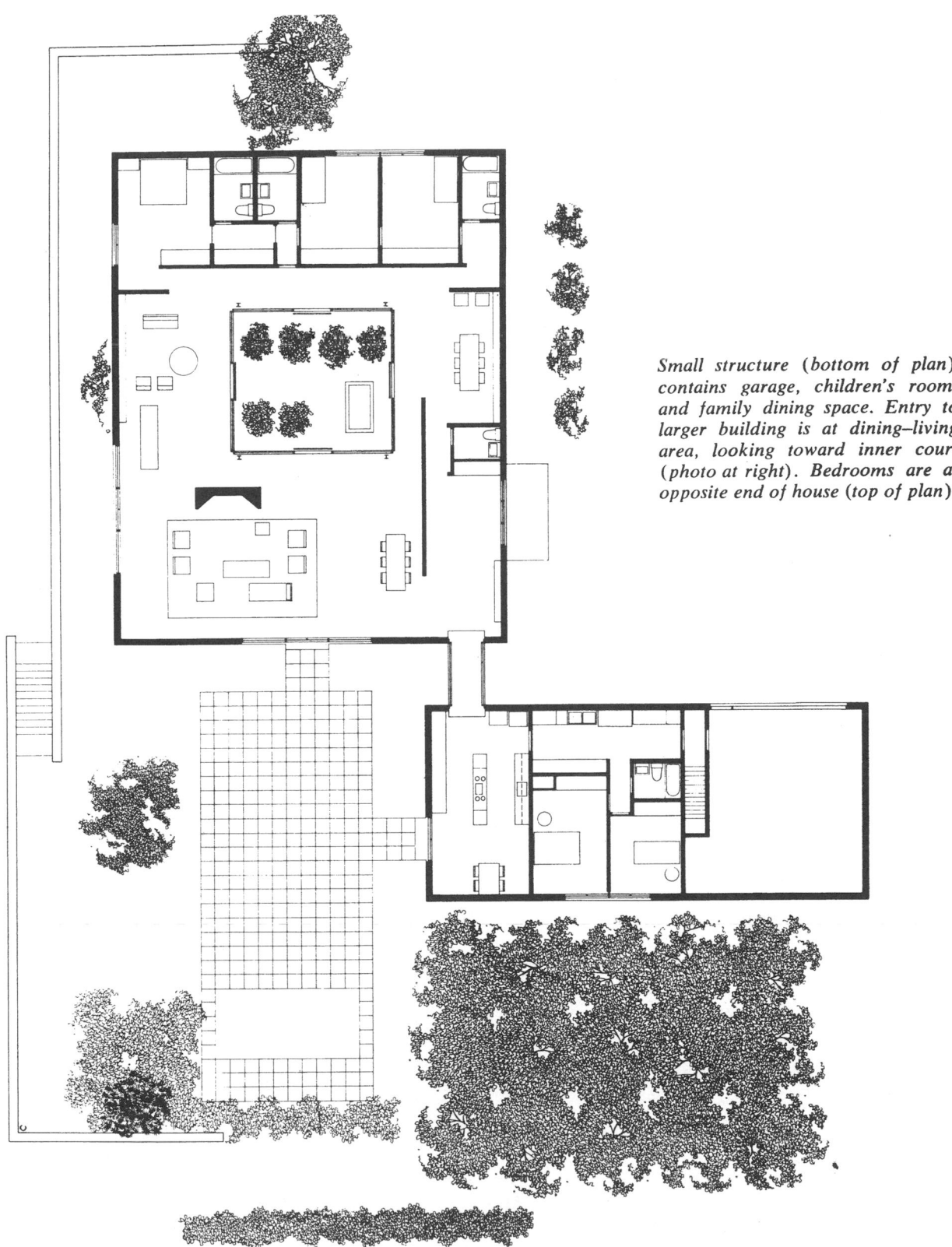

Small structure (bottom of plan) contains garage, children's room, and family dining space. Entry to larger building is at dining–living area, looking toward inner court (photo at right). Bedrooms are at opposite end of house (top of plan).

Fig. 41   Philip Johnson—Architect.

TYPICAL HOUSE PLANS

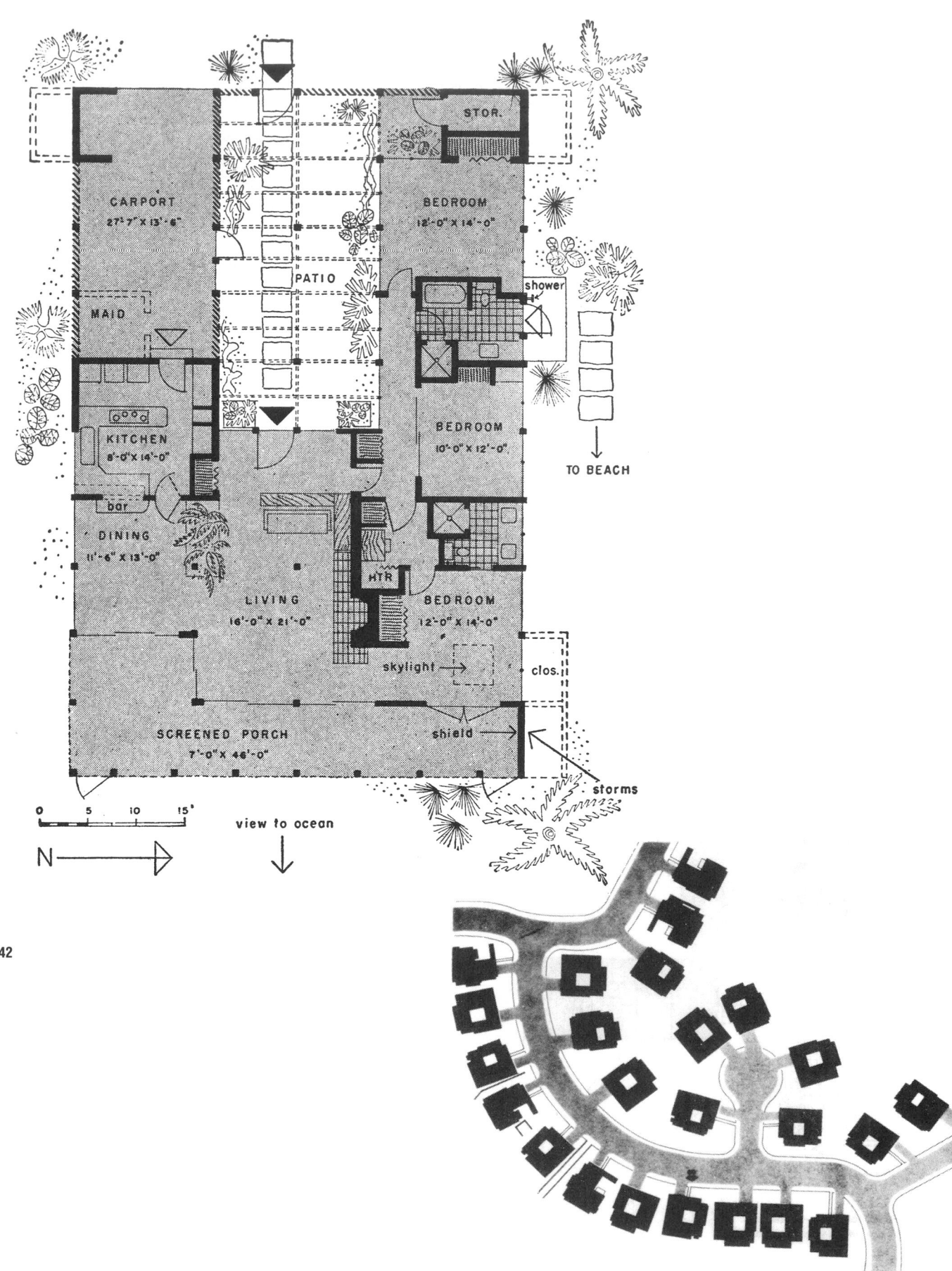

CARPORT
27'-7" X 13'-6"

STOR.

BEDROOM
12'-0" X 14'-0"

PATIO

MAID

shower

KITCHEN
8'-0" X 14'-0"

BEDROOM
10'-0" X 12'-0"

bar

TO BEACH

DINING
11'-6" X 13'-0"

LIVING
16'-0" X 21'-0"

HTR

BEDROOM
12'-0" X 14'-0"

skylight →

clos.

SCREENED PORCH
7'-0" X 46'-0"

shield →

storms

0   5   10   15'

view to ocean

N →

Fig. 42

Fig. 43

Fig. 44 Attached houses with courts. Genesee River Houses, Rochester, N.Y. Conklin & Rossant—Architects.

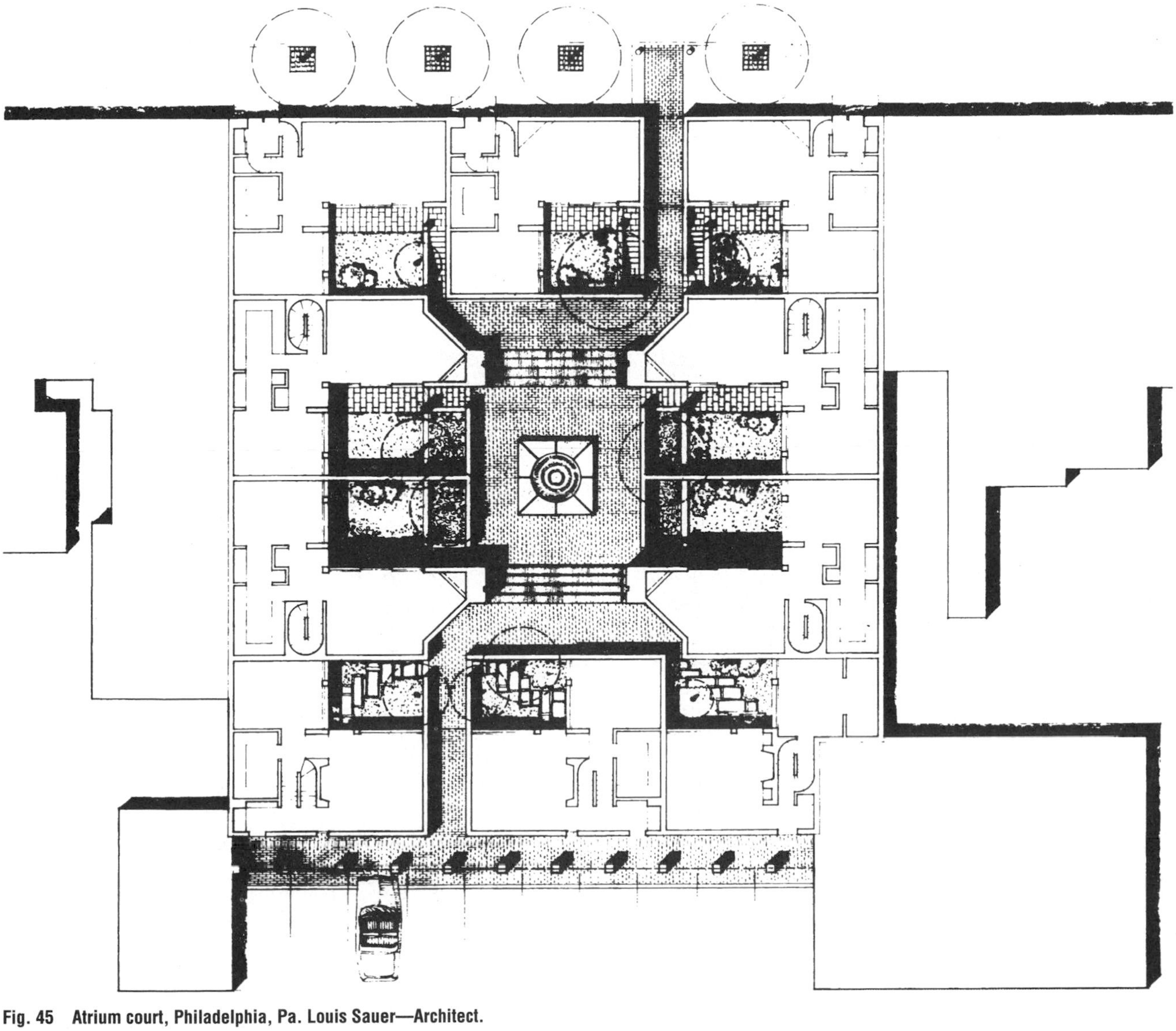

Fig. 45 Atrium court, Philadelphia, Pa. Louis Sauer—Architect.

TYPICAL HOUSE PLANS

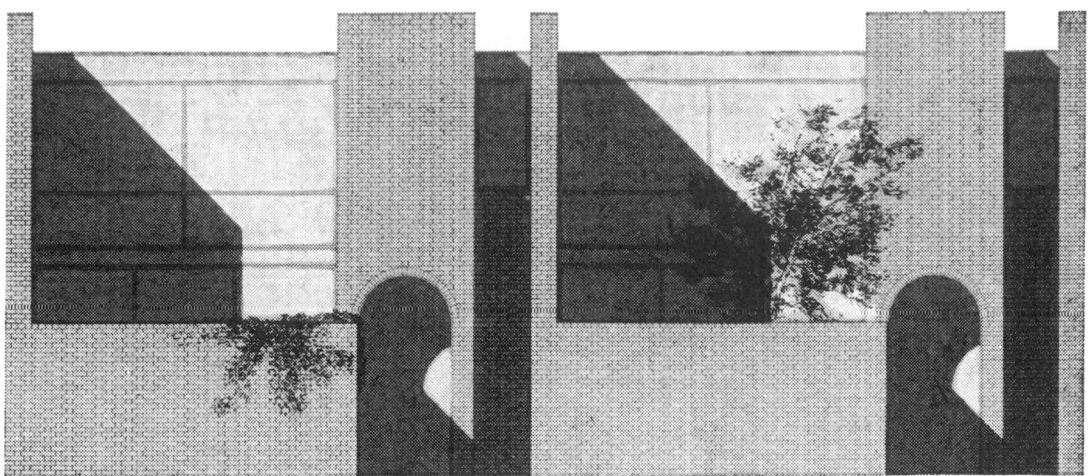

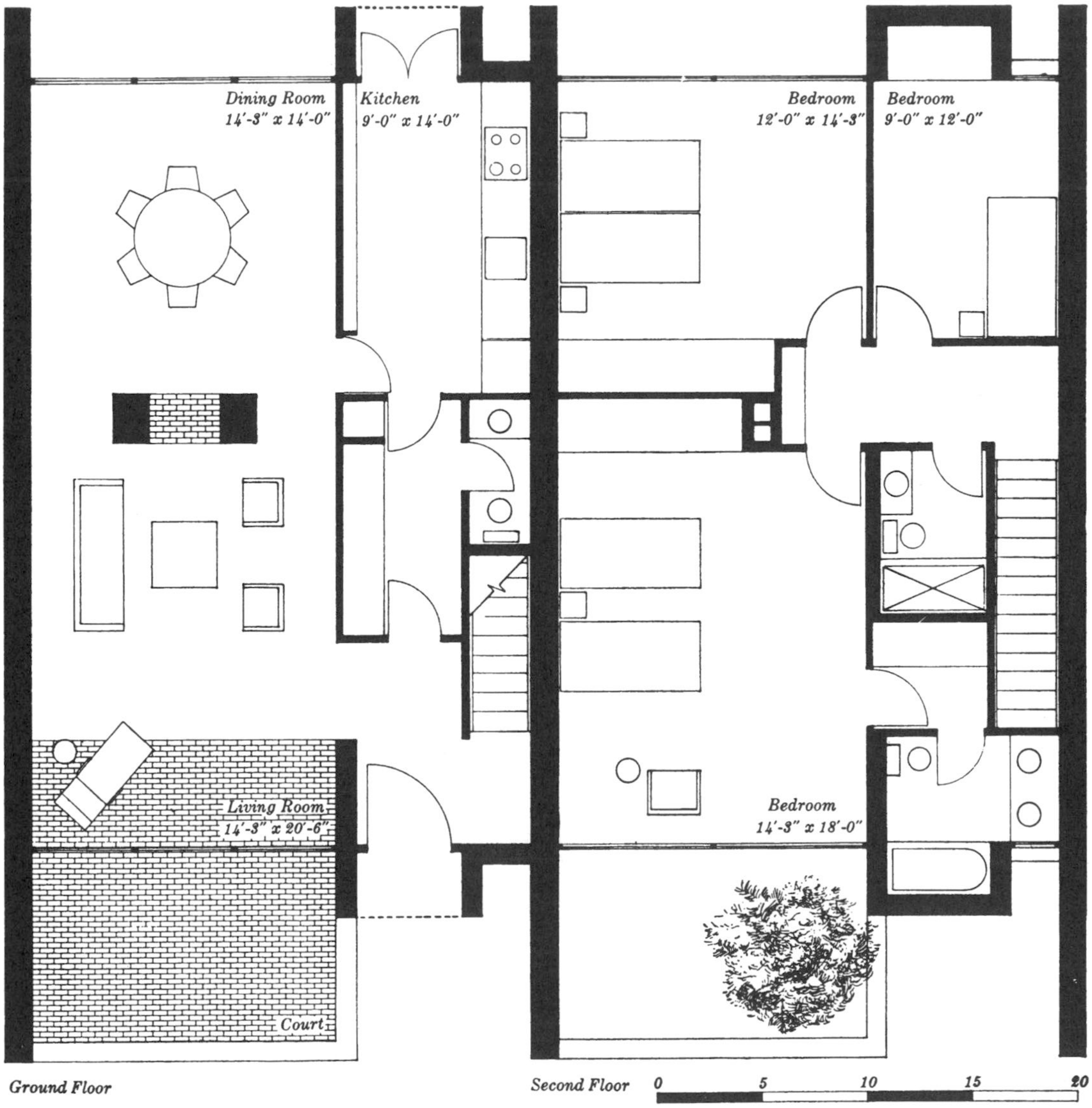
Dining Room
14'-3" x 14'-0"
Kitchen
9'-0" x 14'-0"
Living Room
14'-3" x 20'-6"
Court
Bedroom
12'-0" x 14'-3"
Bedroom
9'-0" x 12'-0"
Bedroom
14'-3" x 18'-0"
Ground Floor
Second Floor
0
5
10
15
20

Fig. 46 Washington Sq. Urban Renewal Area, Philadelphia, Pa. I.M. Pei & Associates—Architect.

ZERO-LOT-LINE MODELS

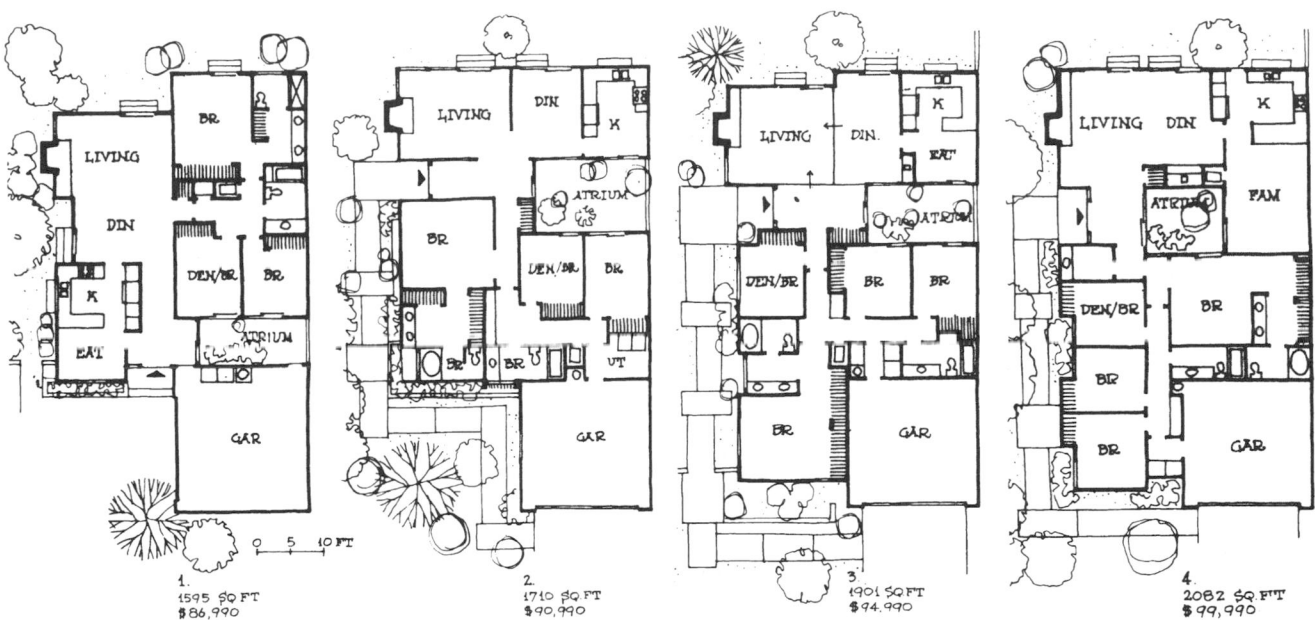

1.
1595 SQ FT
$86,990

2.
1710 SQ FT
$90,990

3.
1901 SQ FT
$94,990

4.
2082 SQ FT
$99,990

Fig. 1   Crow Canyon, Contra Costa County, California. Morris Lohrbach—Architect.

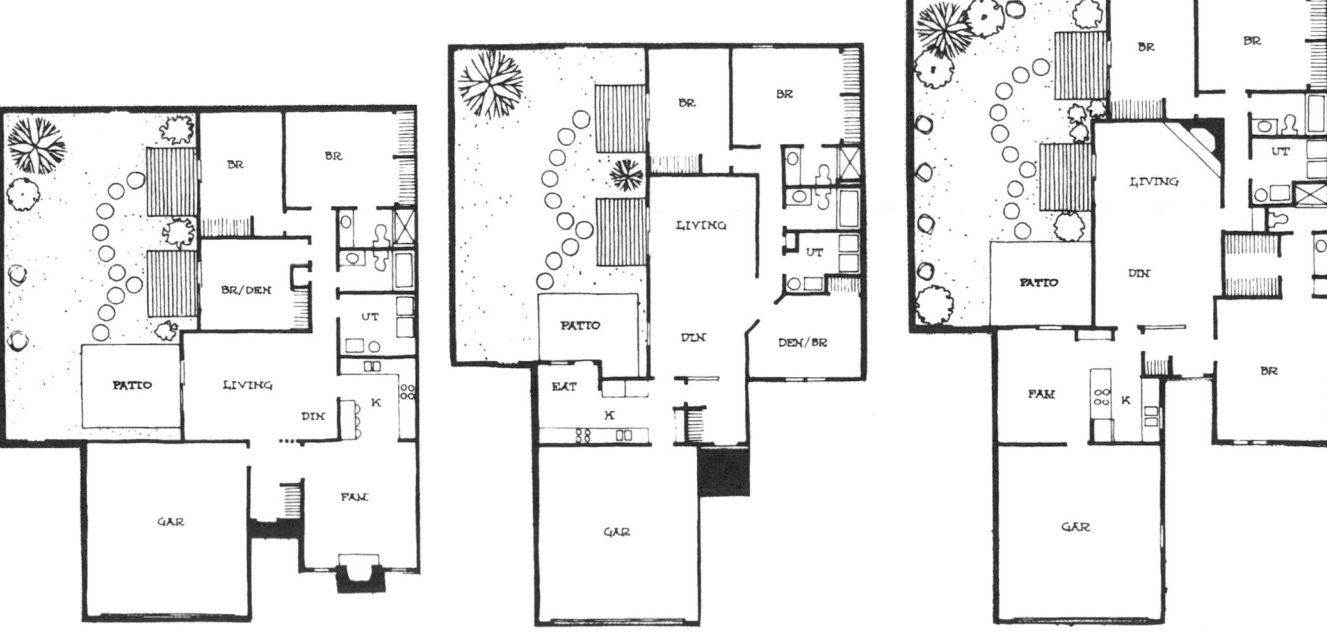

Fig. 2   1400 ft². Terre Haute, Ind.   **Fig. 3   1200 ft². Terre Haute, Ind.**   **Fig. 4   1600 ft². Terre Haute, Ind.**

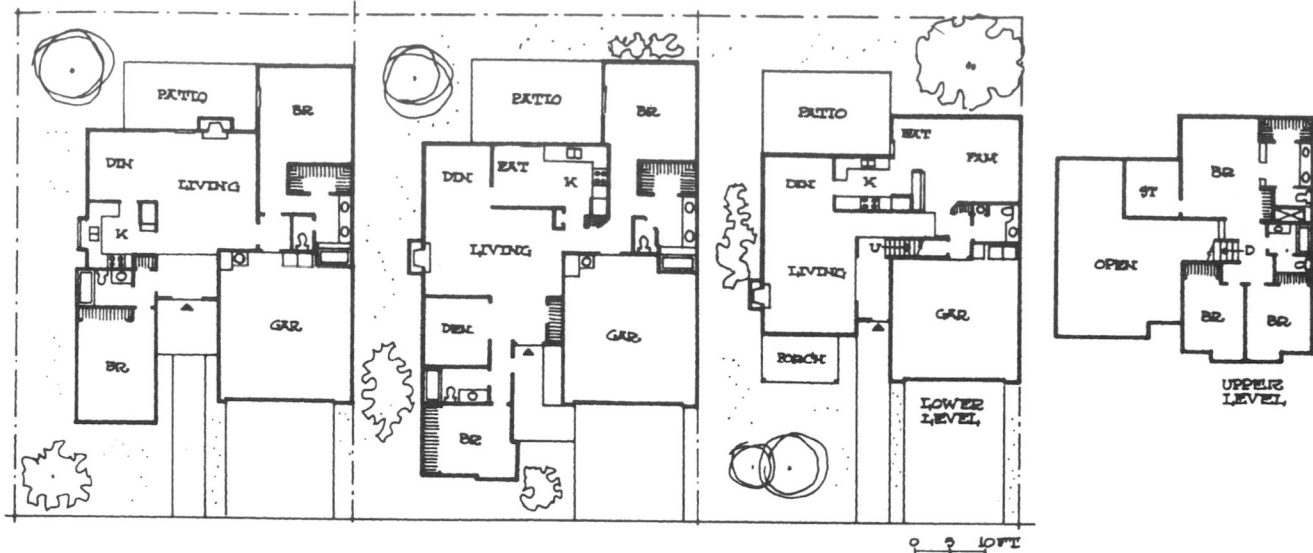

Fig. 5

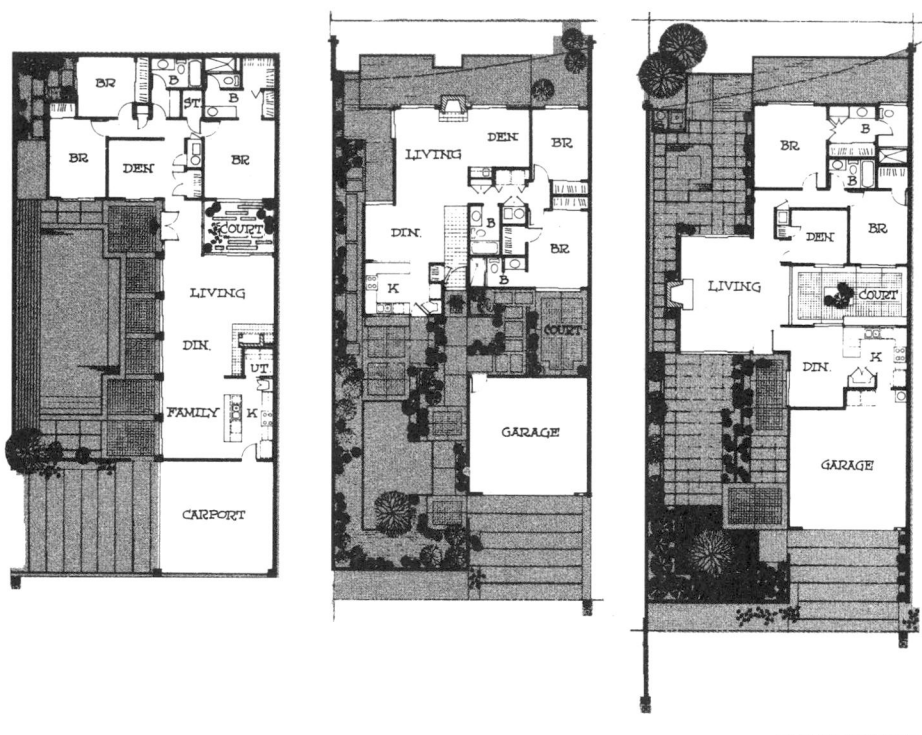

Fig. 6 Typical plans for zero-lot-line sitting of a patio house on a 45 ft by 90 ft lot. Plans show a variety of layouts available in one- and two-story patio homes. All lots allow for the construction of a pool within the walled garden.

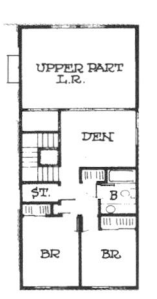

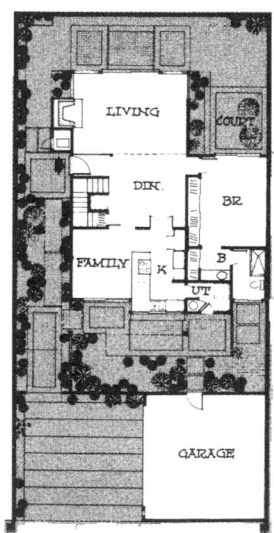

DUPLEXES

The term *duplex* refers to a single structure consisting of two separate dwelling units. It may be a two-story walk-up building in which one dwelling unit is situated over another, with access to the upper apartment by means of a private staircase. It may also be a structure having two dwelling units located side by side, with the individual units on one or more levels. (*Duplex* is also used to describe an apartment of two floors in a multistory building.)

The majority of duplexes are found in older sections of cities and in the inner ring of suburbia. Exceptions are in communities with a tradition of two-family dwellings, such as Philadelphia, and where local zoning specifically refers to the duplex as the most intensive residential type allowed, usually in transition or buffer areas between single-family and multifamily housing districts. Duplexes are no longer as popular as they once were. They have been eclipsed by other housing—single-family houses in the suburbs, and large new apartment complexes with many conveniences on the same site.

From *duplex* has grown a whole family of -plexes—*triplex, fourplex, eightplex,* and so on. The prefix identifies the number of dwelling units in the building, which is usually two stories high. A variation on the two-story fourplex, with two units each per floor, is the quatrefoil, which is a one-story square building with a unit in each corner providing each apartment with at least two exposures. The converted single-family structure can also be classified as a -*plex*. Sometimes houses are built with eventual conversion in mind. The separate quarters can become a source of extra income or may be intended for some elderly member of the family. When local building and zoning ordinances permit, the original structure may remain intact internally and the second unit may be added on the back or side.

DUPLEX (SEMIATTACHED)

The semiattached house is an independent lot that is attached on one side to a similar dwelling on an adjacent lot. The attachment is made along a common or "party" wall, which is jointly owned. The main advantage of this type of construction is the economy achieved in the construction of the party wall. Because one side yard is eliminated, it is also possible to build on a narrower lot than if it were a detached dwelling. This type of dwelling can be used for either one or two families. Usually this type of dwelling is two stories high, but it can be one story also.

As a one-family dwelling, the living room, kitchen, and dining areas are on the first floor while the sleeping areas are on the second floor. This type is usually owner-occupied.

As a two-family dwelling, each floor has a separate unit with its own independent entrance. The owner usually lives in the lower unit and rents the upper unit. Garages are either detached or incorporated into the cellar, or lowest level of the structure.

The term *duplex* refers to a single structure consisting of two separate dwelling units. It may be a two-story walk-up building in which one dwelling unit is situated over another, with access to the upper apartment by means of a private staircase. It may also be a structure having two dwelling units located side by side, with the individual units on one or more levels.

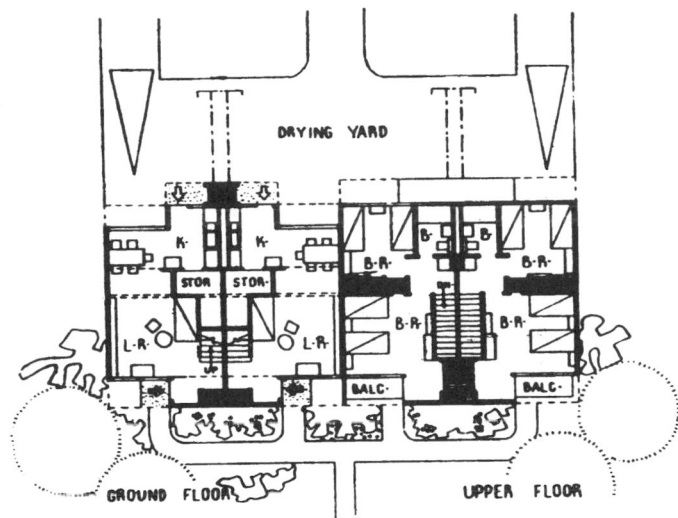

Fig. 1

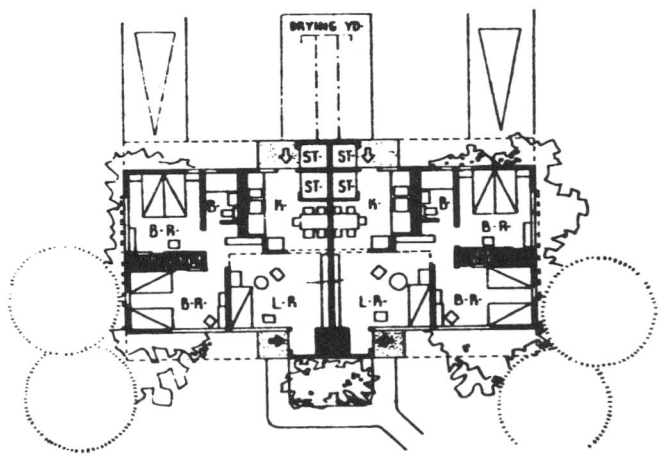

Fig. 2

SITE PLAN

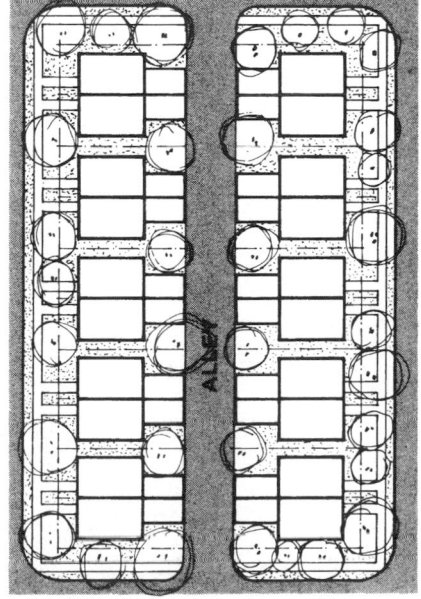

CONVENTIONAL LAYOUT
20 UNITS

Fig. 3

SEMIATTACHED (DUPLEXES)

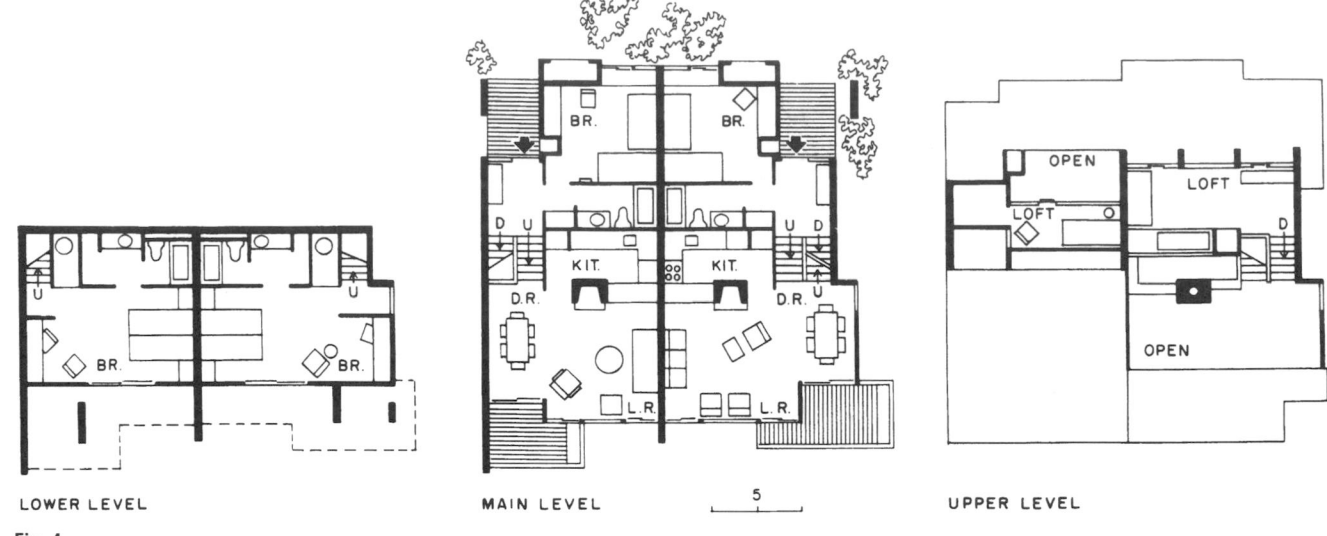

LOWER LEVEL

Fig. 4

MAIN LEVEL

5

UPPER LEVEL

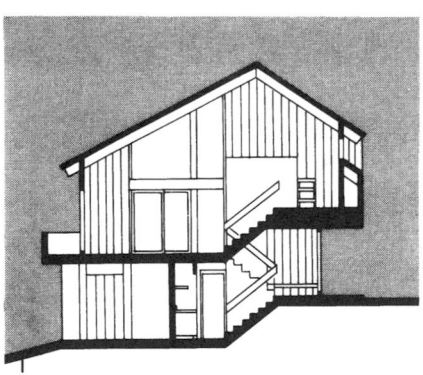

Fig. 5

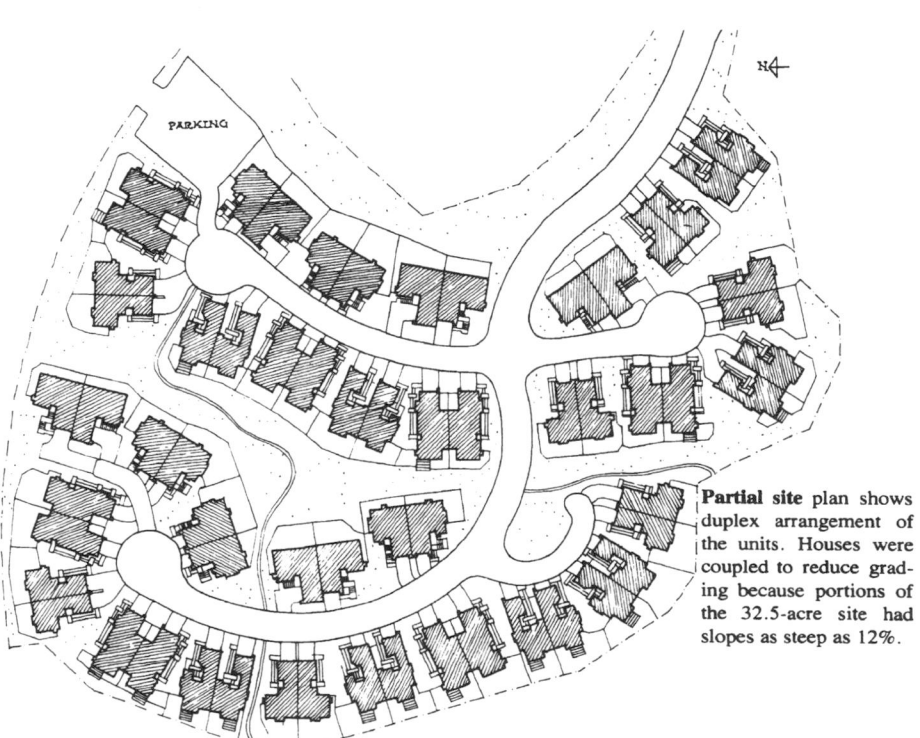

Partial site plan shows duplex arrangement of the units. Houses were coupled to reduce grading because portions of the 32.5-acre site had slopes as steep as 12%.

Fig. 6 Turtle Rock Glen, Irvine, Calif. Richardson, Nagy, Martin—Architects.

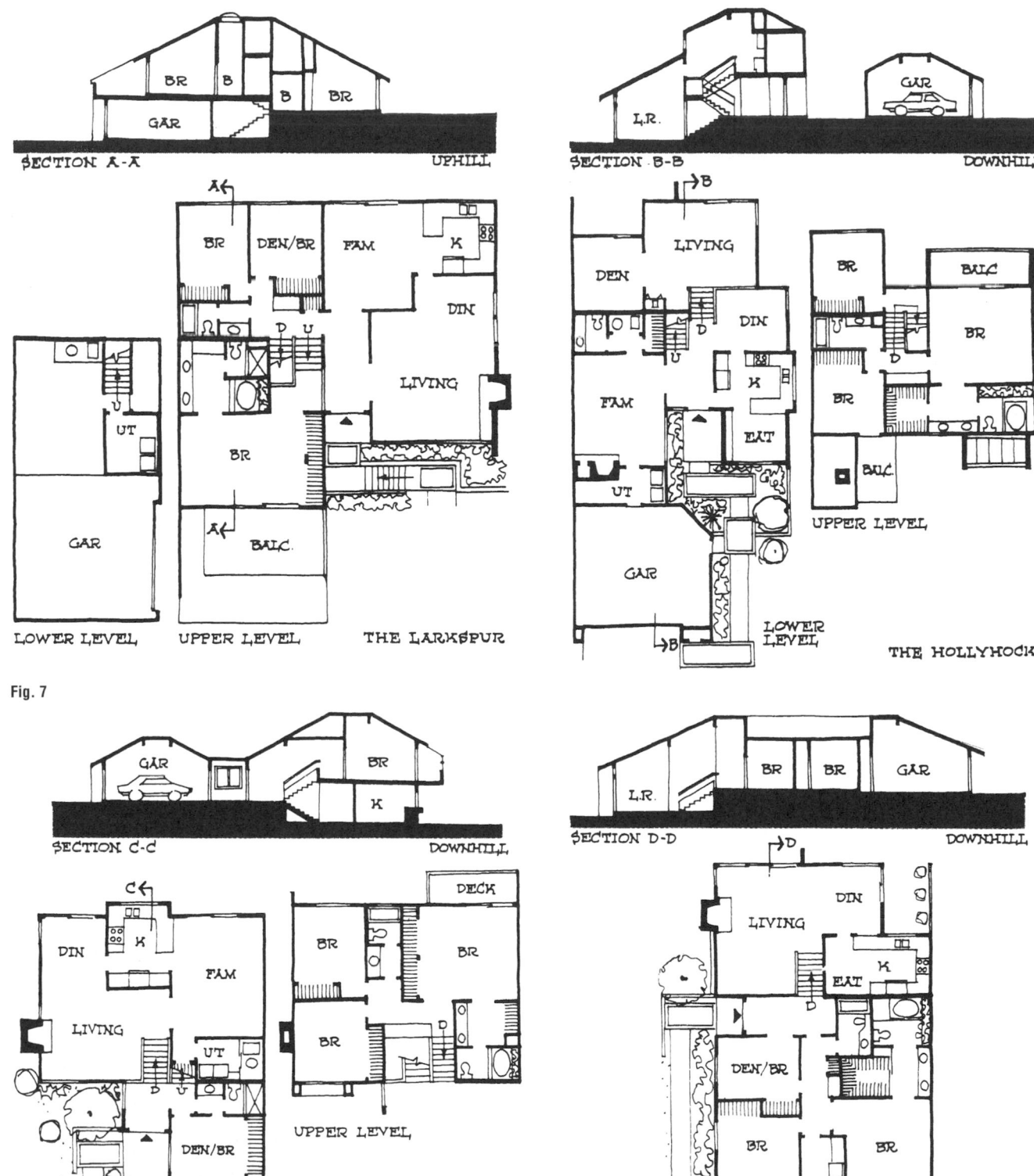

Fig. 7

Fig. 8   Turtle Rock Glen, Irvine, Calif. Richardson, Nagy, Martin—Architects.

Fig. 9   Turtle Rock Glen, Irvine, Calif. Richardson, Nagy, Martin—Architects.

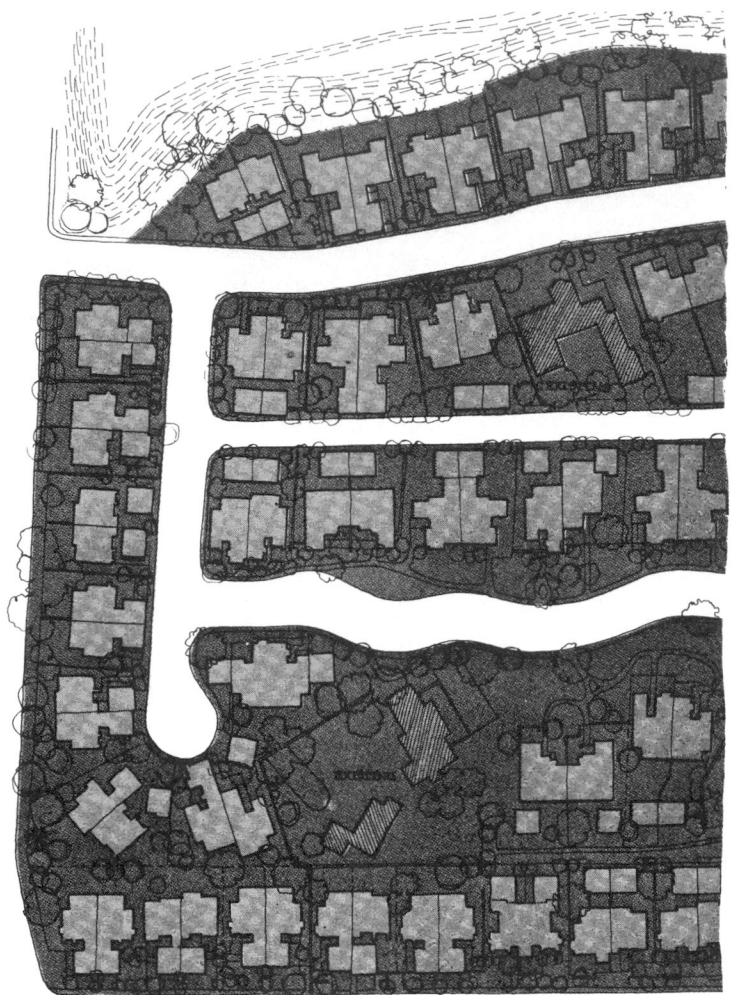

Fig. 10 Crown Pointe project in Long Beach, California, uses walled front yards, varied roof pitches, and variously coupled plans to disguise duplex siting. Site plan for 6.3-acre first phase arranges 42 duplex units around four manor houses that are still occupied (shaded). New units are owned fee-simple. Homeowner association maintains streets.

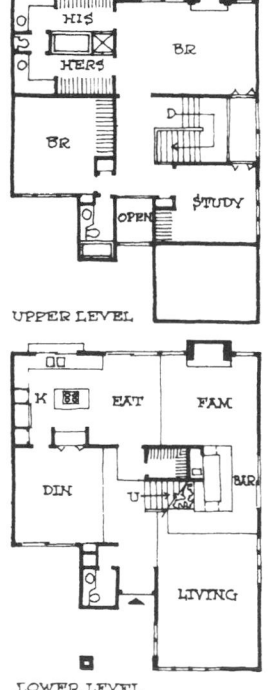

Fig. 11

Fig. 12 Five floor plans feature conversation areas on first level.

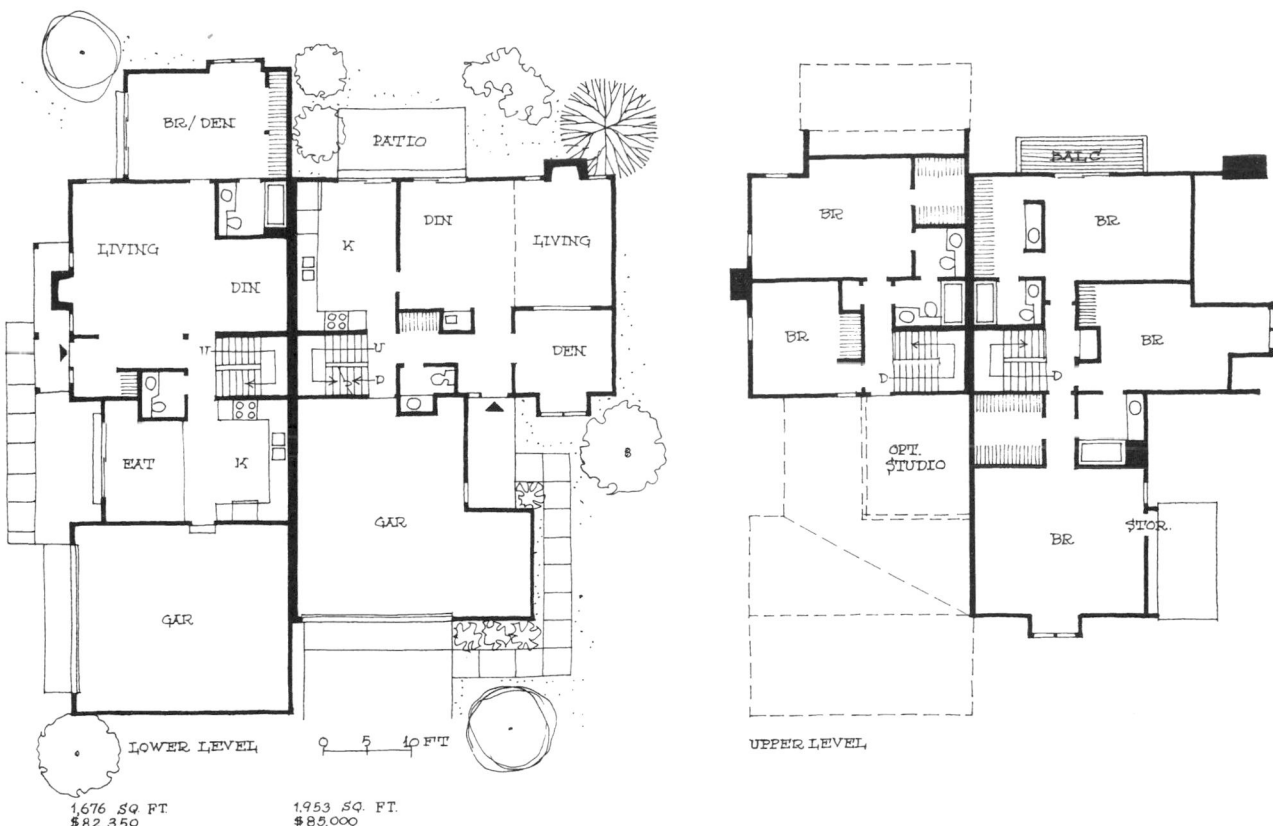

Fig. 13   Manor Houses, St. Charles, Ill.

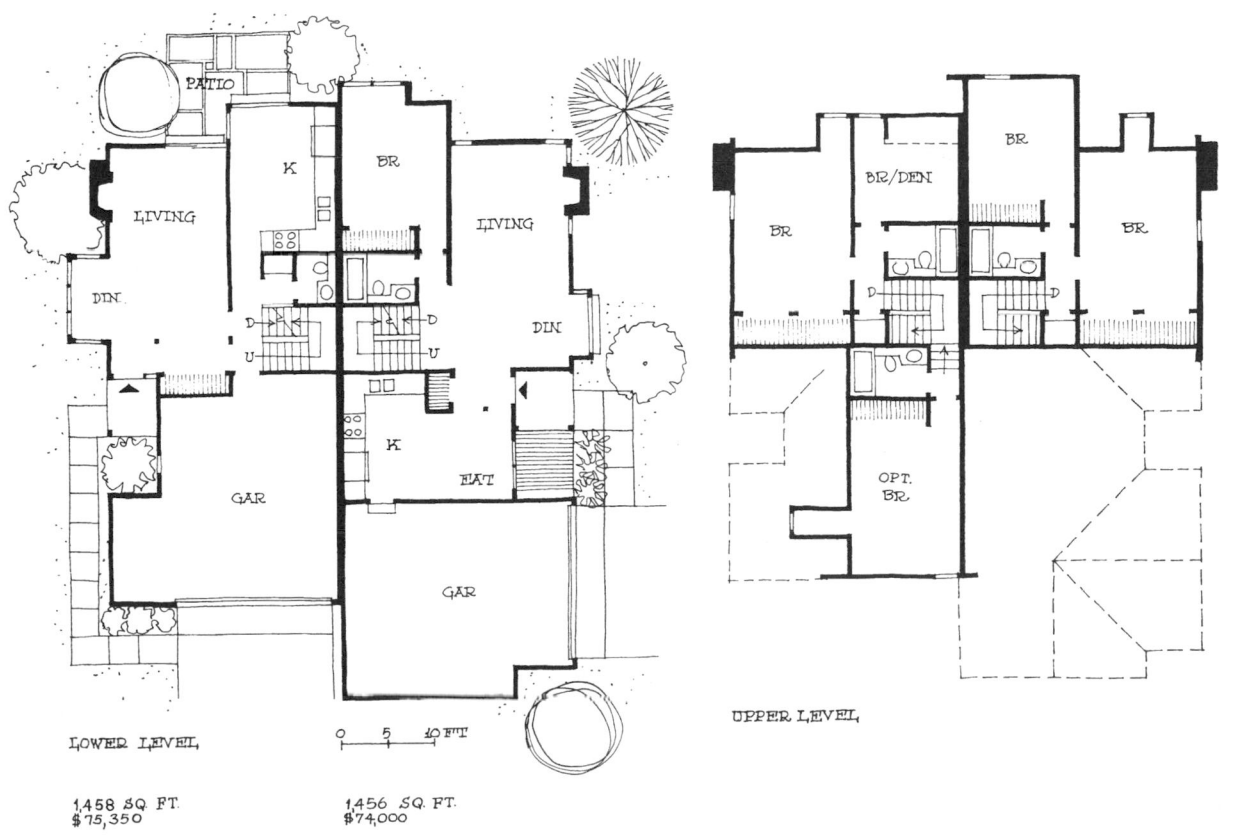

Fig. 14   Manor Houses, St. Charles, Ill.

SEMIATTACHED (DUPLEXES)

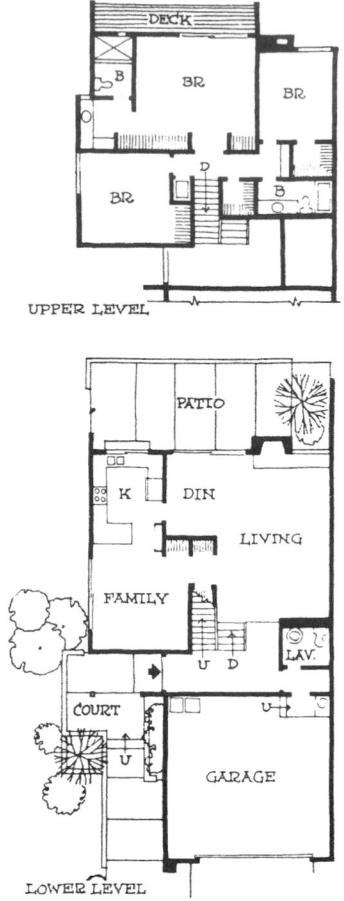

Fig. 15   Two-story unit; three bedrooms.

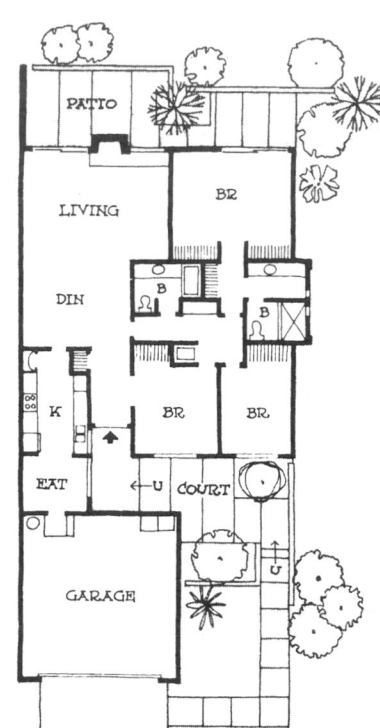

Fig. 16   Single-story unit; three-bedrooms, Mount La Jolla, Calif. Walter Richardson and Associates—Architects.

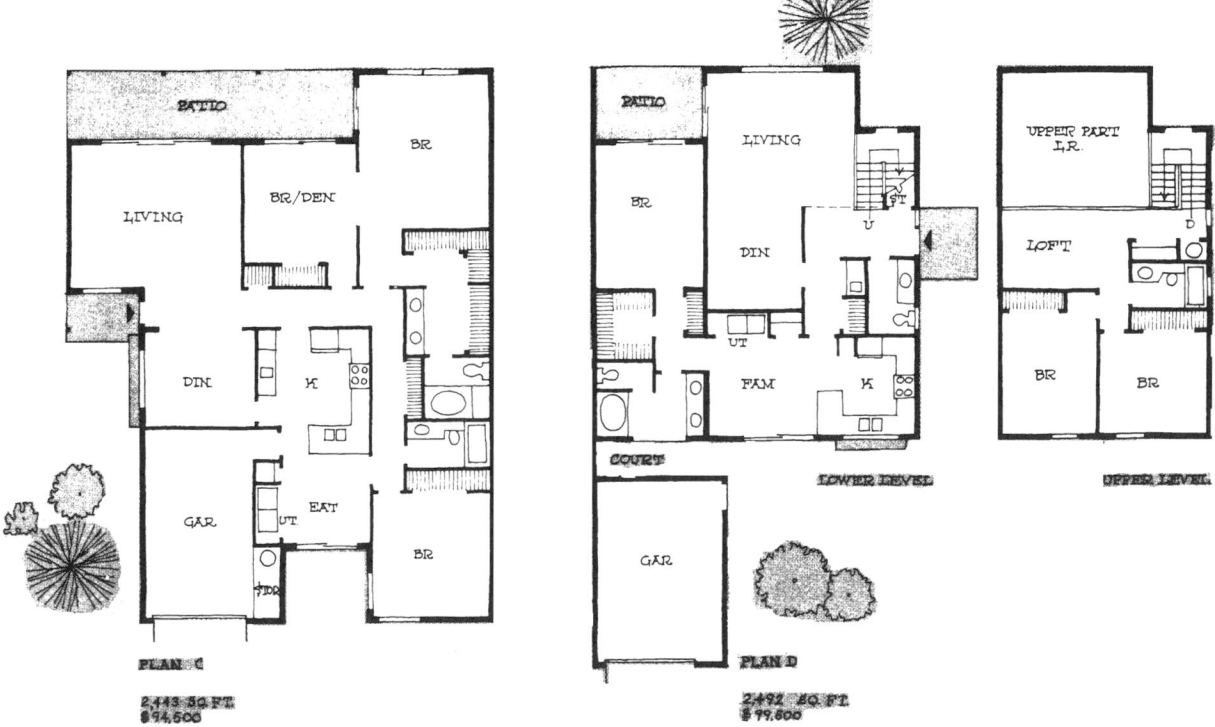

Fig. 17   Attached single-family (duplex). Twelve Oaks, North Palm Beach, Florida. Dudley, Omura & Larry Winkler—Architects.

QUADRUPLEX/FOURPLEX

The consolidation of four single-family houses into one structure utilizing common walls need not destroy the privacy of the individual dwellings. It permits greater use of the total site for outdoor living. The advantages of such an arrangement are shown on the accompanying plan. Concrete wall construction provides for better sound insulation and the back-to-back arrangement of plumbing cores. Further economies will accrue in reduced fuel costs and maintenance of shared driveways and foot-paths. All utility services are combined and economically run underground rather than overhead. The elimination of rear and side yards allows maximum use of the site, while careful orientation and screening assures privacy for each residence. Automobile parking is centralized to serve the four units and thus requires less area. The amount of open land between these "quadruplexes" is greater than in many present subdivisions and can be assigned either to community functions or more dwelling units, developing higher densities without overcrowding. Abundant landscaping and the use of varied setbacks would greatly enhance the total appearance.

Adjacent to the common walls are the kitchen, bathrooms, stair hall, and power core with utility room, none of which requires a window. The bedrooms, dining and living areas, and family room have outside exposure, and in some cases they open to terraces. Figure 4 shows how entries, as well as patios, of units in the same building are isolated from each other.

Fourplexes are spotted throughout the 9.6-acre site (Fig. 5). The floor plans in Fig. 6 show the configuration of the four-unit buildings, and how much privacy for entrances and outdoor living spaces this kind of design provides. All units have open lofts overlooking two-story-high living rooms and two full bedroom suites.

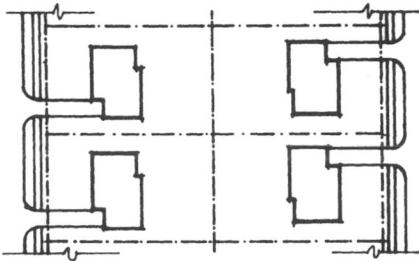

Fig. 1   Conventional layout.

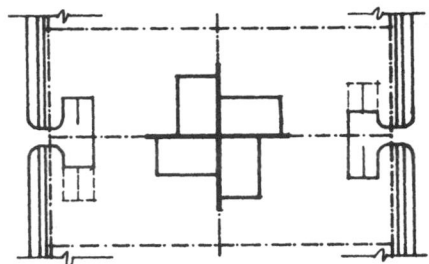

Fig. 2   Suggested layout.

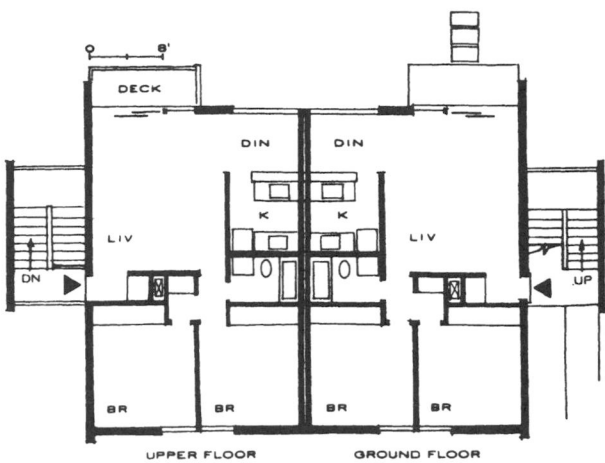

Fig. 3   Two-bedroom unit.

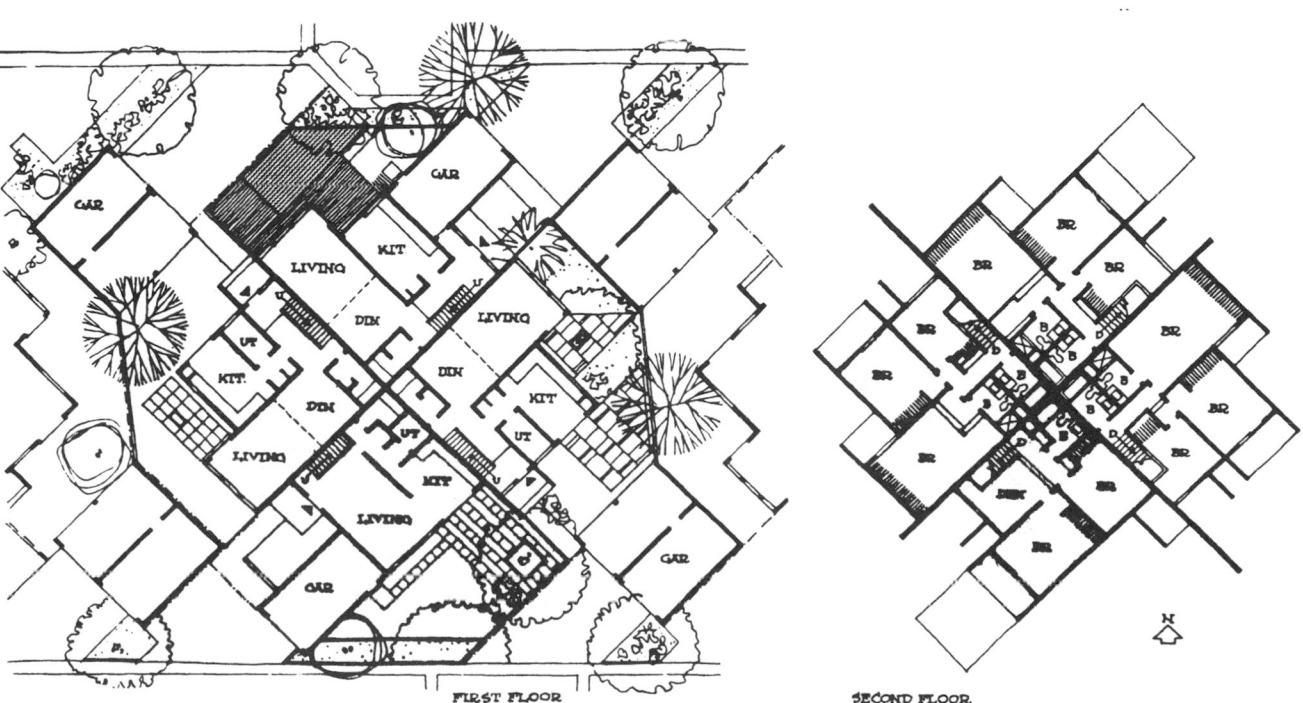

Fig. 4

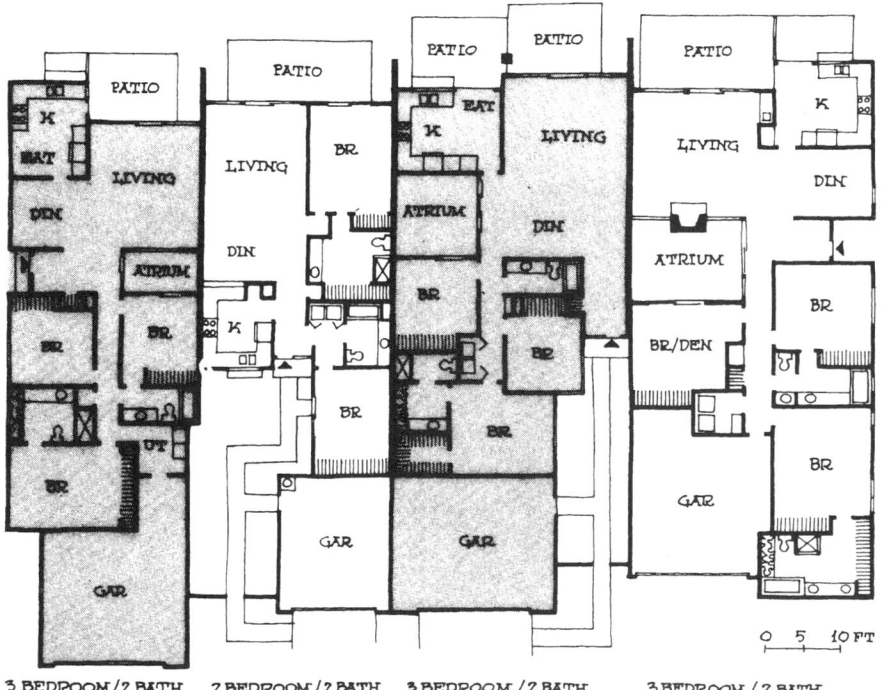

Fig. 5   Tennis Village, Sun River, Oregon. Paschall & Associates—Architects.

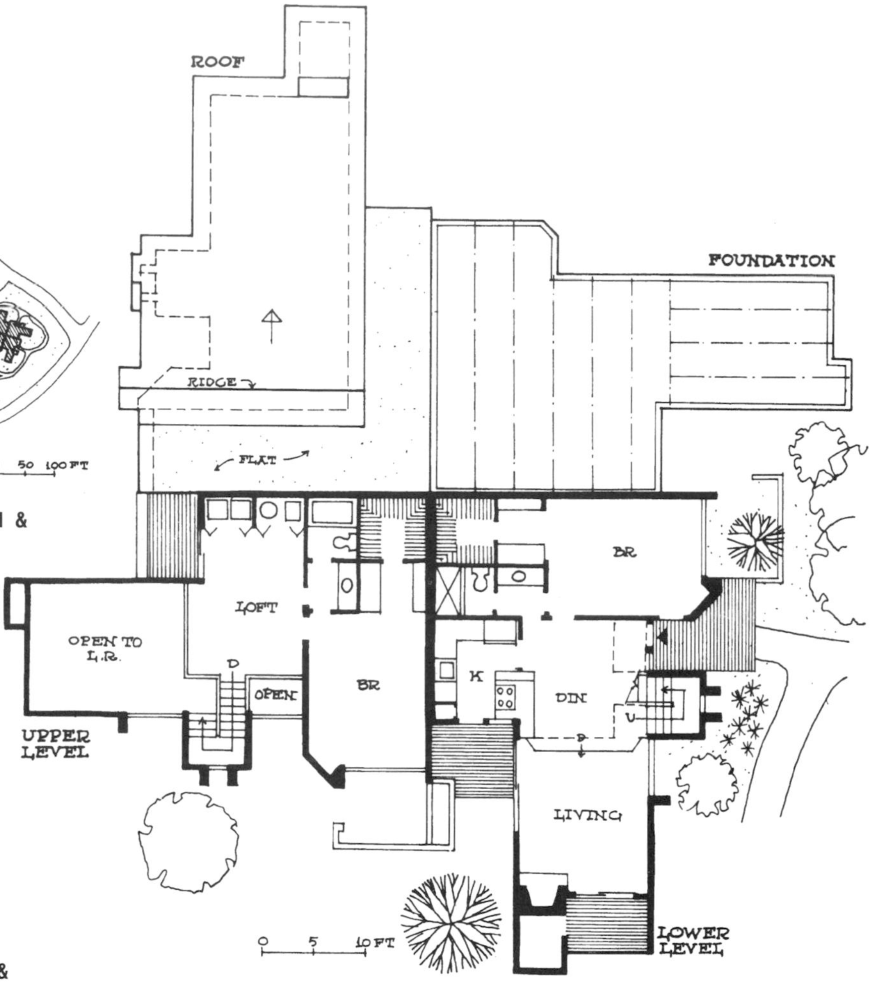

Fig. 6   Tennis Village, Sun River, Oregon. Paschall & Associates—Architects.

3 BEDROOM / 2 BATH   2 BEDROOM / 2 BATH   3 BEDROOM / 2 BATH   3 BEDROOM / 2 BATH

Fig. 7   Rancho Las Palmas, Palm Springs, Calif.

Fig. 8a

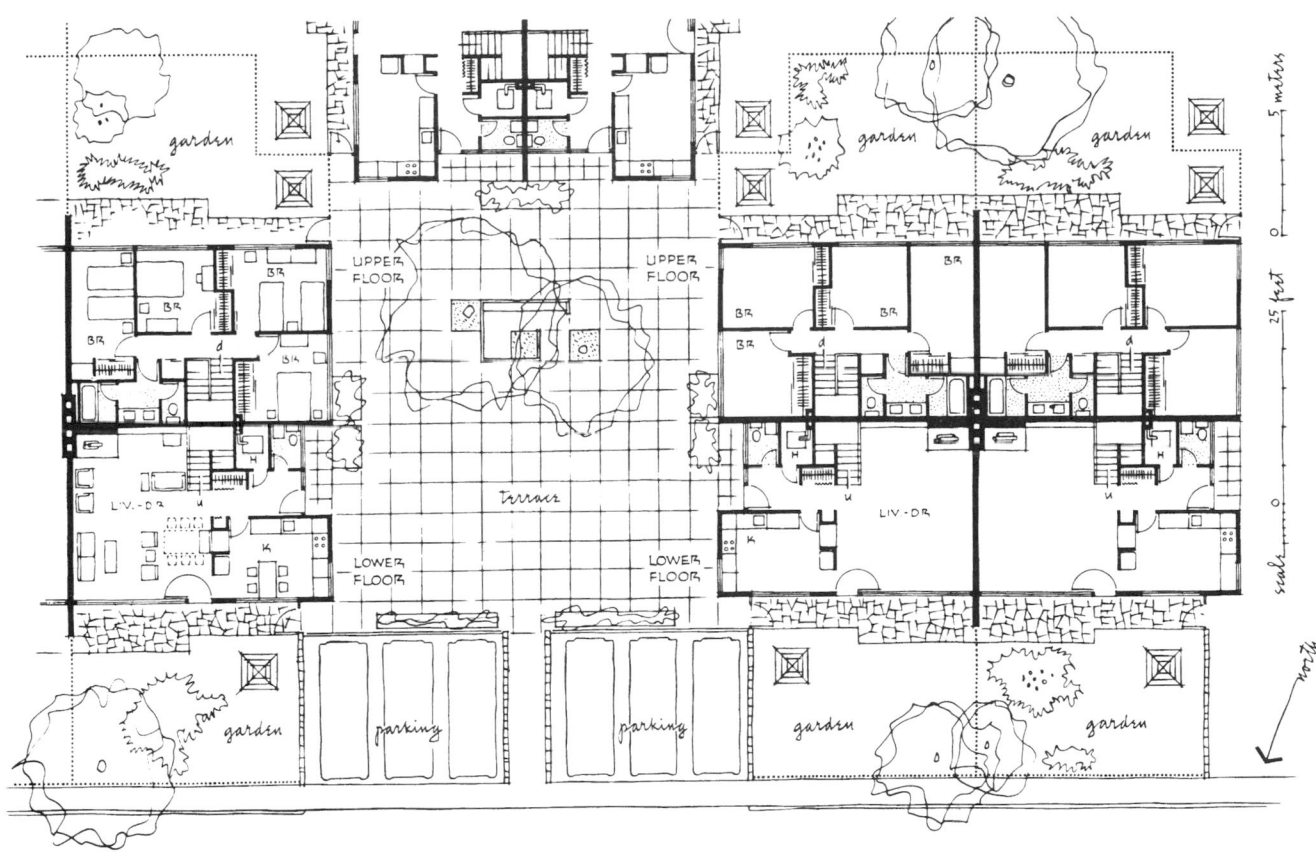

Fig. 8*b*

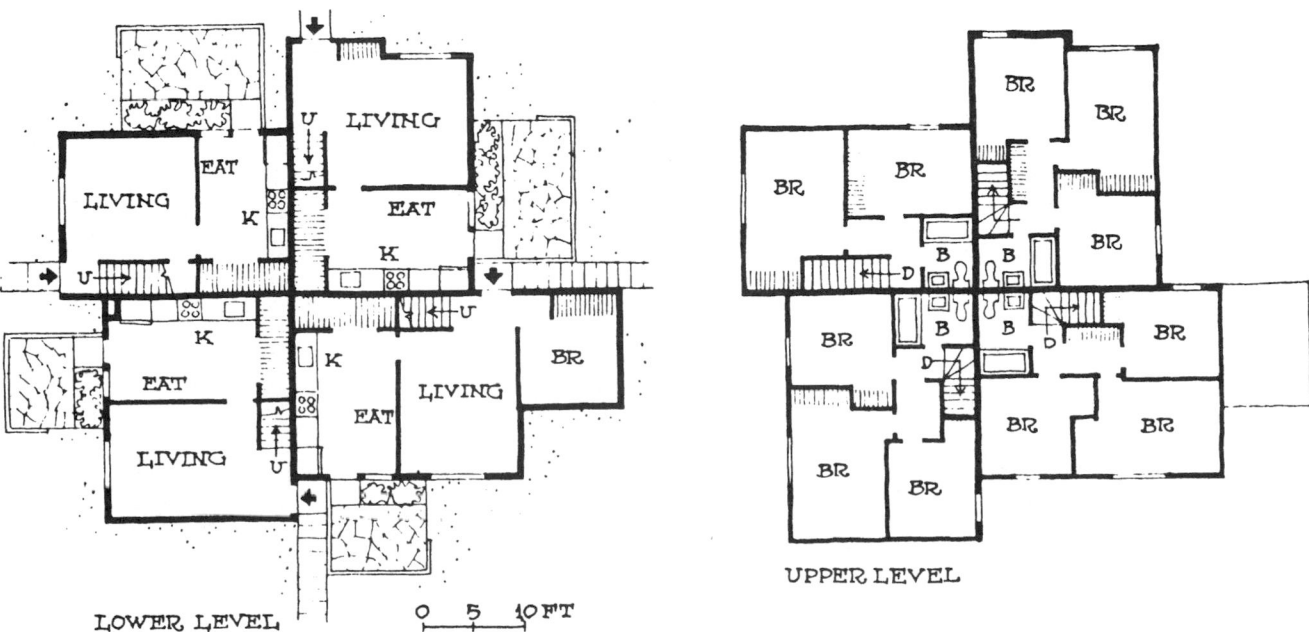

LOWER LEVEL

UPPER LEVEL

0 5 10 FT

Fig. 9 Randles Estate, Cleveland, Ohio. Whitley–Whitley—Architects.

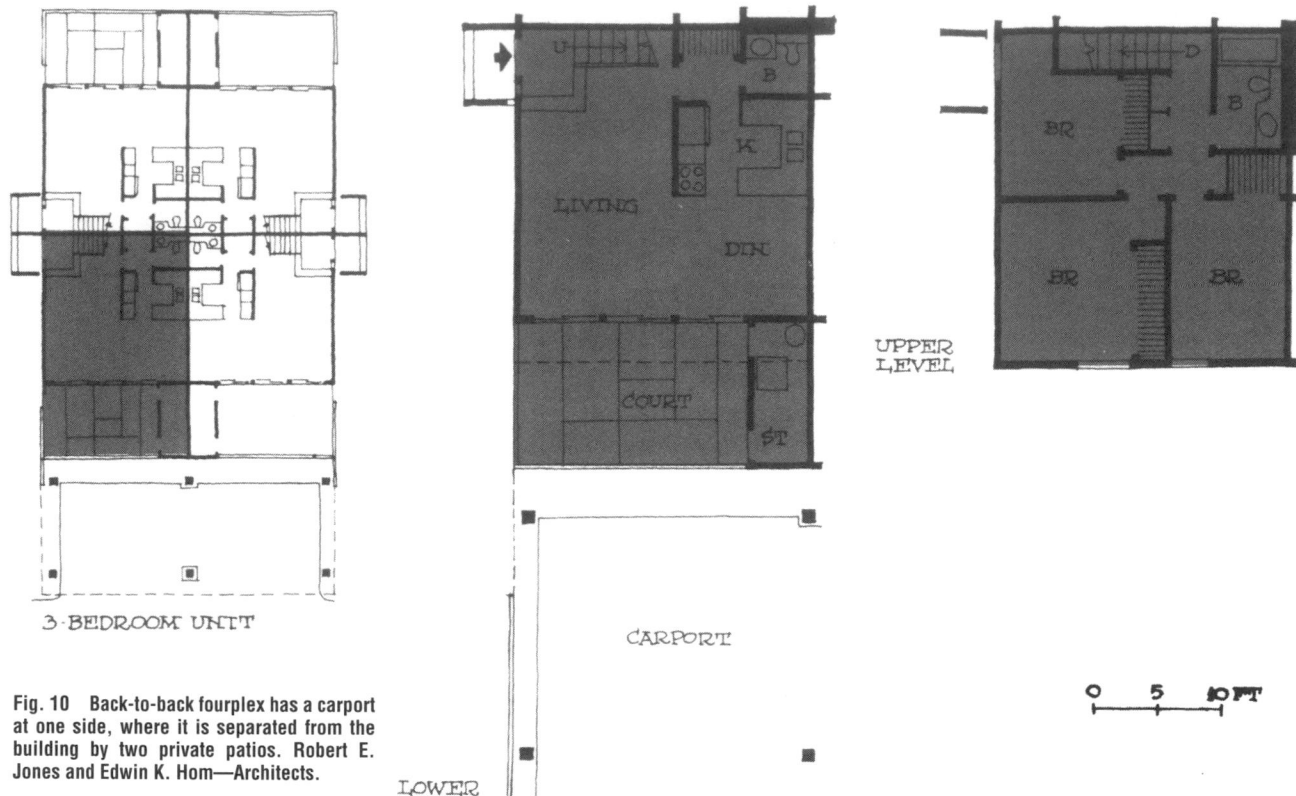

3·BEDROOM UNIT

Fig. 10 Back-to-back fourplex has a carport at one side, where it is separated from the building by two private patios. Robert E. Jones and Edwin K. Hom—Architects.

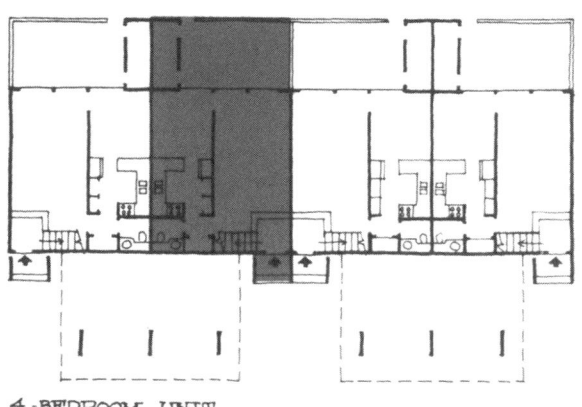

4·BEDROOM UNIT

Fig. 11 In-line fourplex has rear patios with outdoor storage and front carports beneath master bedrooms and balconies. Robert E. Jones and Edwin K. Hom—Architects.

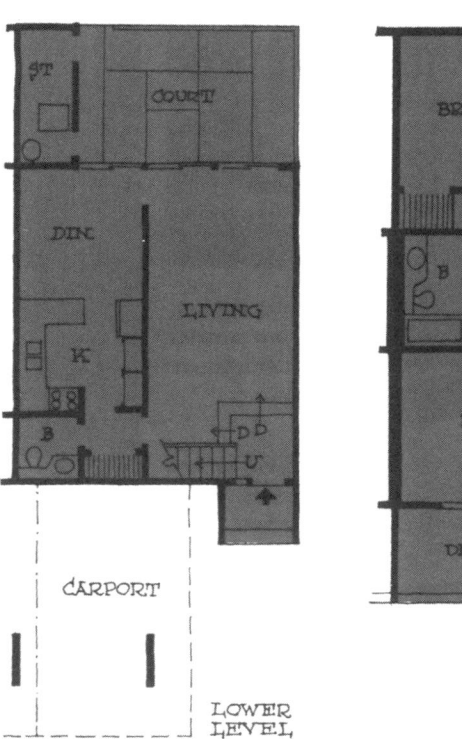

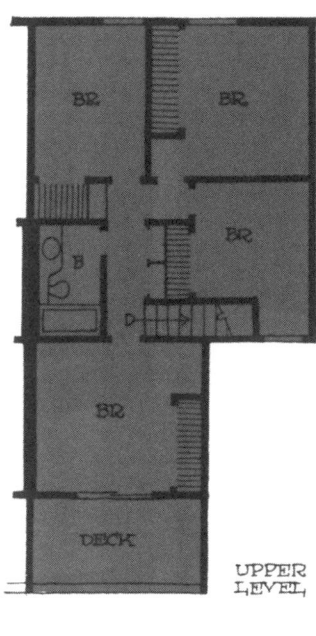

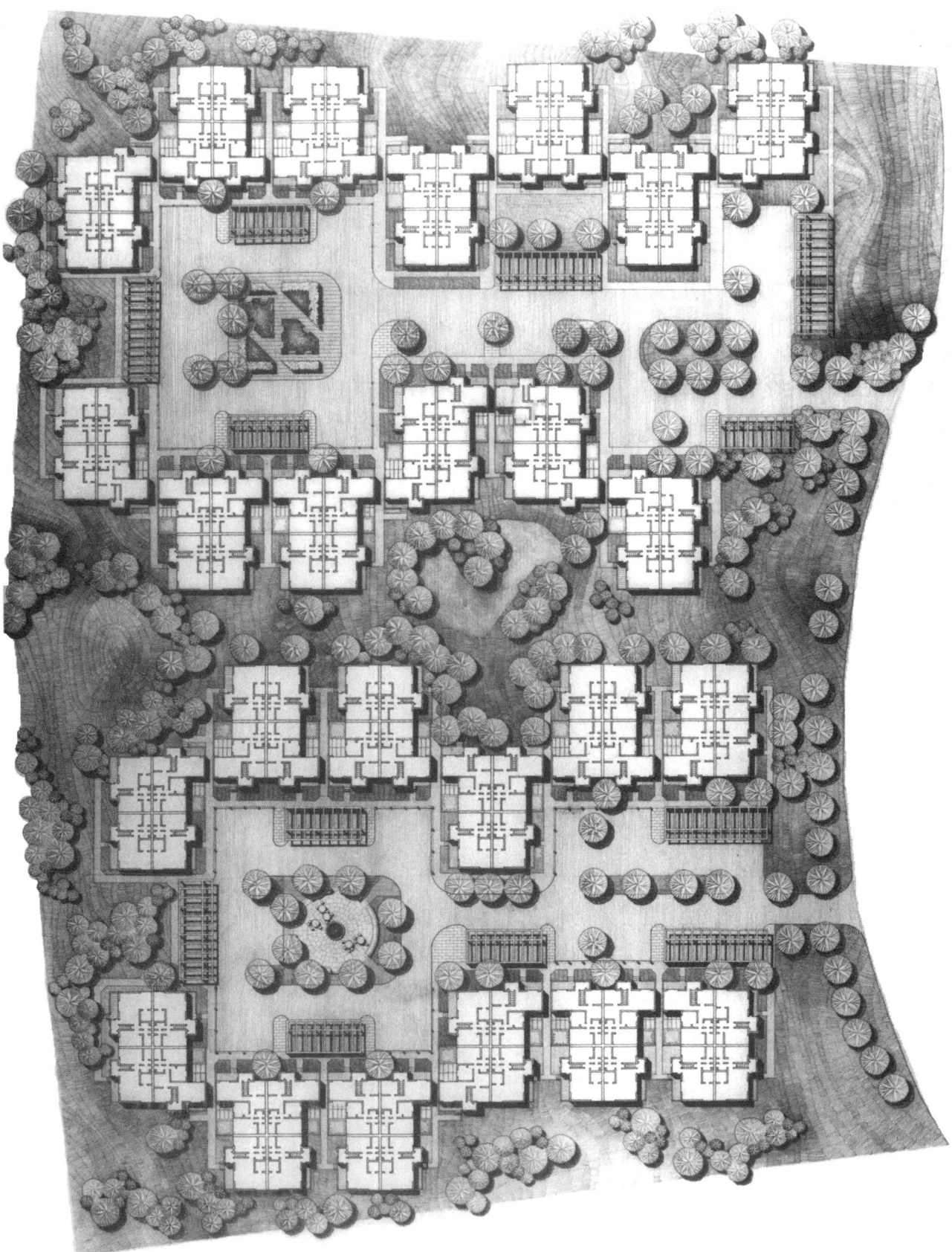

Fig. 12 Wilton Condominiums, New Castle, Delaware. Louis Sauer Associates—Architects.

SEVENPLEX

Seven two-car garages form a right angle within which sit four back-to-back town houses (units C and D, Fig. 2). These are separated from the garages by covered walkways and landscaped patios. The three remaining units are flats built on top of the garages (units A and B). The second bedrooms of the B units extend over the walkway below to provide covered entries to the B units.

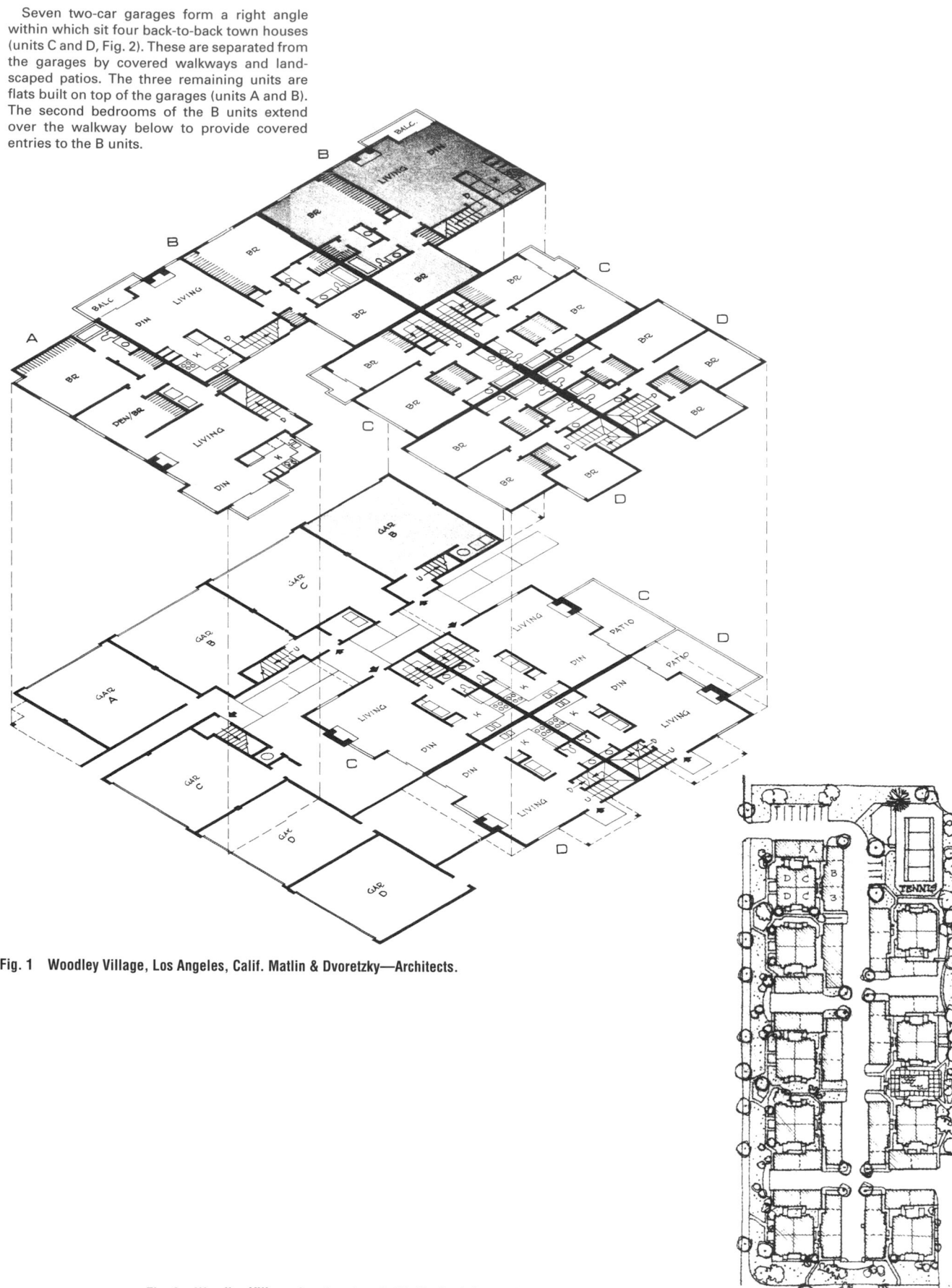

Fig. 1 Woodley Village, Los Angeles, Calif. Matlin & Dvoretzky—Architects.

Fig. 2 Woodley Village, Los Angeles, Calif. Matlin & Dvoretzky—Architects.

ROW HOUSES

The row house is a standard dwelling type in many American cities. It is characterized by great economy in the use of land, moderate construction cost, and low maintenance and operating costs. It affords each family its own home and the opportunity of developing a plot of land for its own use. The economy of a row house plan obviously derives from the length of the rows, although the per-unit saving diminishes with the length of the building. The economic advantage of a long building results in part from the elimination of end walls, but there are also savings in land utilities and walks.

Row houses are not adaptable to steep contours except by heavy cutting, ordinarily, buildings should tend to parallel the contours, since longitudinal slopes can only be accommodated by costly breaks of floor level.

In the organization of a row-house plan an effort should be made to avoid movements of traffic in any considerable volume parallel to the rows. If the movement of vehicles is restricted to streets lying at the ends of the buildings, the space between the buildings, adjacent to the living units, is left free for the safe and comfortable use of the occupants.

Row houses may be assembled in many patterns. In general these fall into two types—court plans and parallel row plans. Each of these has advantages. A court layout may attain spaciousness of effect and aesthetically satisfying enclosed areas. When the rows are predominantly parallel, the distance from row to row will be somewhat less than the average width of a court, but the longitudinal views will be longer. It is obvious that when a particular orientation to sun or prevailing wind is strongly favored, such orientation must result in predominantly parallel rows. The parallel row plan usually facilitates a simple and practical servicing scheme in which all units are handled uniformly. A court plan can seldom be arranged without streets being parallel to some of the buildings.

Two-story row houses can economically accommodate families requiring two- and three-bedroom units. Units with one bedroom or less are best handled as two-story flats or as one-story rows.

Eighteenth and nineteenth century forerunners of contemporary row housing still exist both in Europe (the crescents at Bath and the town houses of London and Paris, for example) and in the United States (in such cities as Philadelphia, New York, Boston, and Washington).

Row housing can be defined as a line of dwelling units, attached at the side or rear by means of common walls, comprising an architectural whole. Each dwelling unit occupies the internal space from the ground to the roof. Depending upon local usage, this housing type is variously called town, terrace, group, chain, or attached housing. Although they can range from one to three and even four stories in height, ordinarily row houses are two stories. This is the densest type of housing that permits direct access to private outdoor space at grade for every dwelling unit. Some maisonettes built on sloping sites also have this characteristic where the upper unit has land on one side of the building and the ground-floor unit on the other side. These are sometimes called *stacked row houses.*

Historically, these houses were built as individual dwelling units constructed along the lot lines of individual land parcels. This in part helps to explain the contemporary practice of treating the facades of row units differently to give the impression that each dwelling in the row was constructed separately.

Until fairly recently, row housing was found primarily in central city locations. The spread of cities, however, and the parallel increase in multifamily housing construction have resulted in a scattering of row housing throughout metropolitan areas. Like the duplex, it often serves as a transition between detached dwellings and apartment buildings. Location influences the size of yards. In outlying areas, where land costs are lower than in central city areas, the siting of row housing is like that of walk-up buildings. Both try to duplicate the large front yards of expensive single-family units. In central areas the yards are much smaller and even nonexistent. Attached housing is often built right up to the front lot line, again consistent with the site planning of neighboring construction.

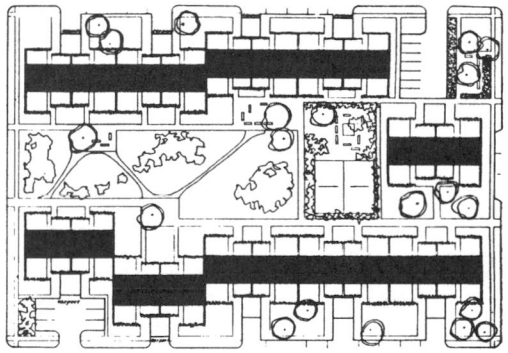

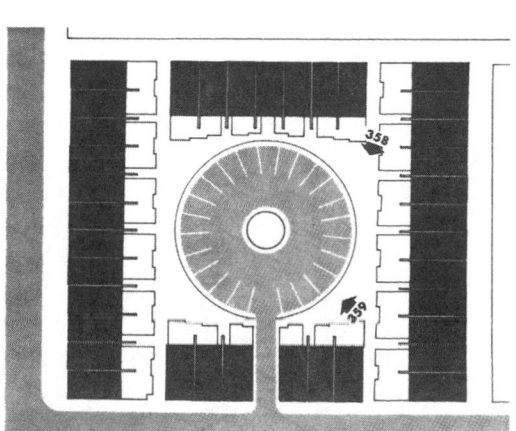

Fig. 1 Washington Square East Townhouses, Philadelphia, Pa. I. M. Pei—Architect.

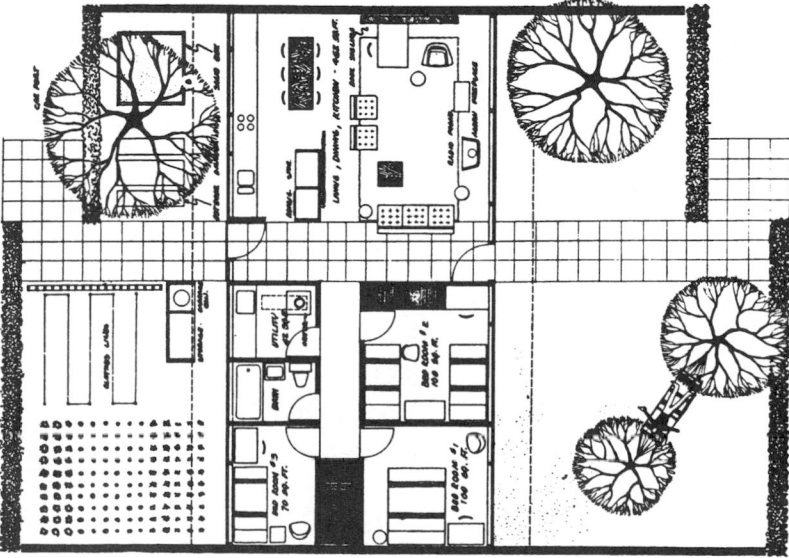

Fig. 2 Edward C. Weren and Willo von Moltke—Architects.

ROW HOUSES

The 20-ft Row House

The 20-ft row house has a gross floor area of 1000 ft². Living-dining areas are combined into one room, 12 by 19 ft. This room should face south. The dining area has direct communication with the kitchen by means of a pass, and can be screened off from the living area with a curtain, bookshelves, a permanent plywood screen, or other media.

A large coat closet, 2 ft by 6 ft 6 in, separates the stairhall from the living room. Its height can be held to 6 ft if an effect of greater spaciousness for the whole living area is desired. The space under the stairs is used for storage.

In addition to accommodating standard equipment, the kitchen provides space for a heater, water heater, and washing machine. If individual heat is planned, duct work is reduced to a minimum. If central heat is provided, the kitchen will gain 3 more feet of counter and cabinet space. On the second floor are two bedrooms, the bathroom (tub on opposite side from the window), and a small dressing alcove with a storage closet. Bedroom window sills are high so that furniture can be placed under them. All plumbing is concentrated in one wall. One fuel services the heater and hot water heater. Hot water lines are short. The outdoor terrace, linking garden to living room, can be used in complete privacy.

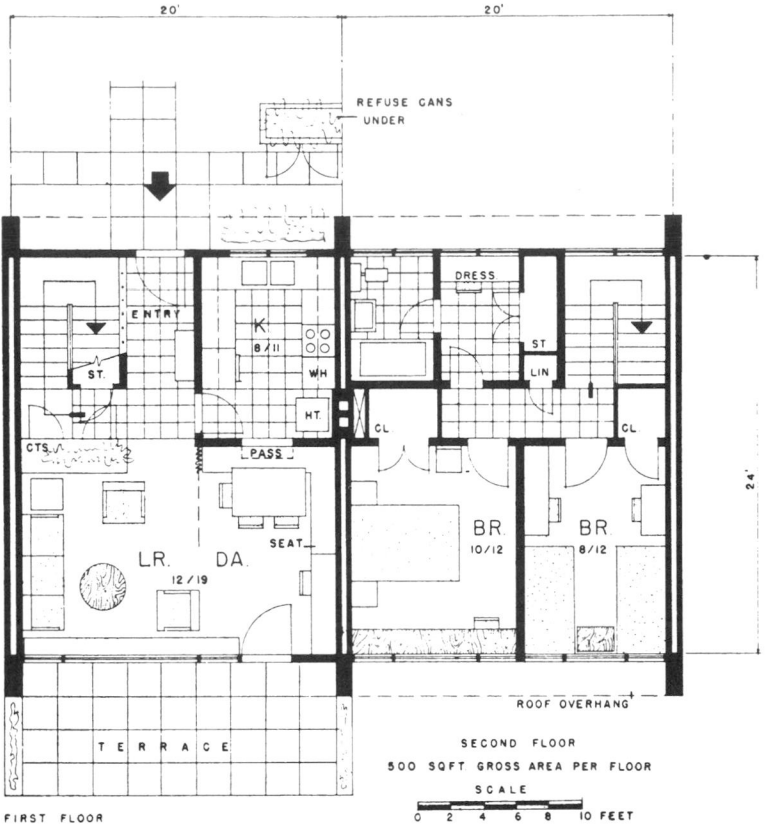

Fig. 3   The 20-ft row house.

The 25-ft Row House

The 25-ft row house has a gross floor area of 1250 ft². Living and dining are combined into one spacious room facing the garden site and should have south orientation. Two of the three bedrooms on the second floor will then have south orientation, also. Storage closets are ample and include a large storage space off the entry, as well as a smaller one accessible from the outside, for tools and deliveries. Mechanical installations are similar to those of the 20-ft row house.

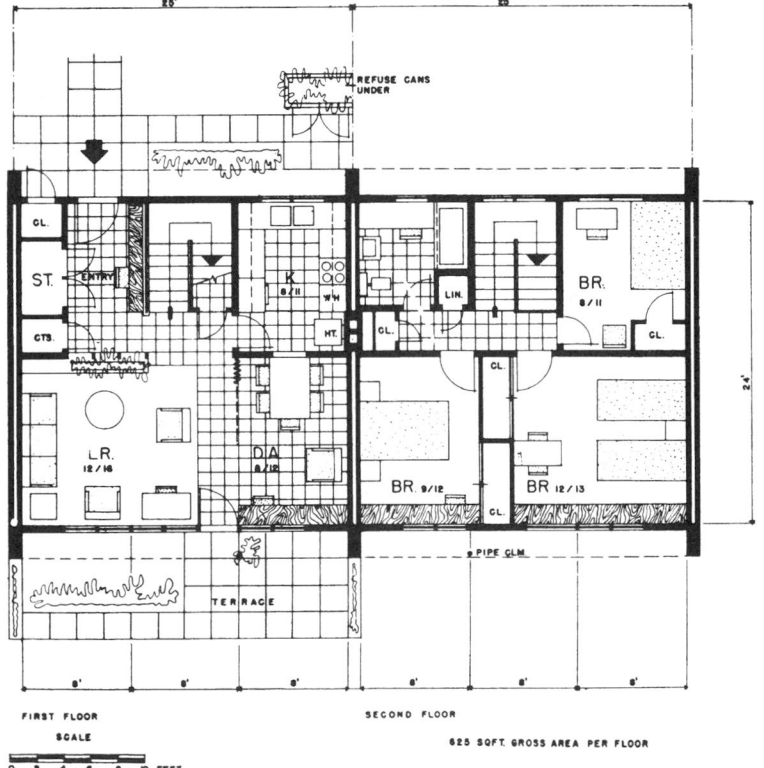

Fig. 4   The 25-ft row house.

The 30-ft Row House

The 30-ft row house has a gross floor area of 1500 ft². The basic arrangement of rooms is similar to the preceding row-house types, except for the addition of a first-floor lavatory off the entry. Three of the four bedrooms upstairs face the same direction as the living-dining combination downstairs. This direction should generally be south.

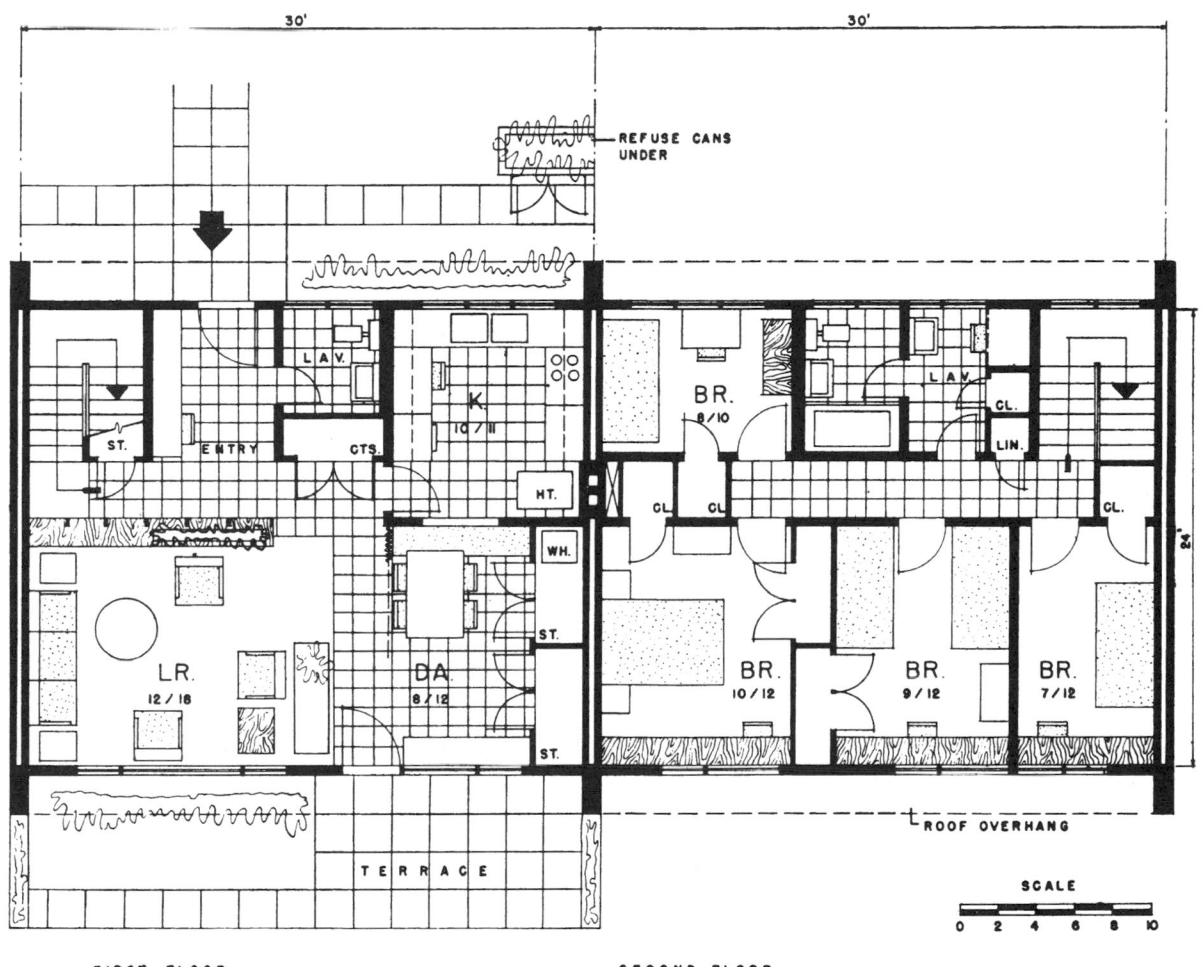

FIRST FLOOR   SECOND FLOOR

Fig. 5   The 30-ft row house.

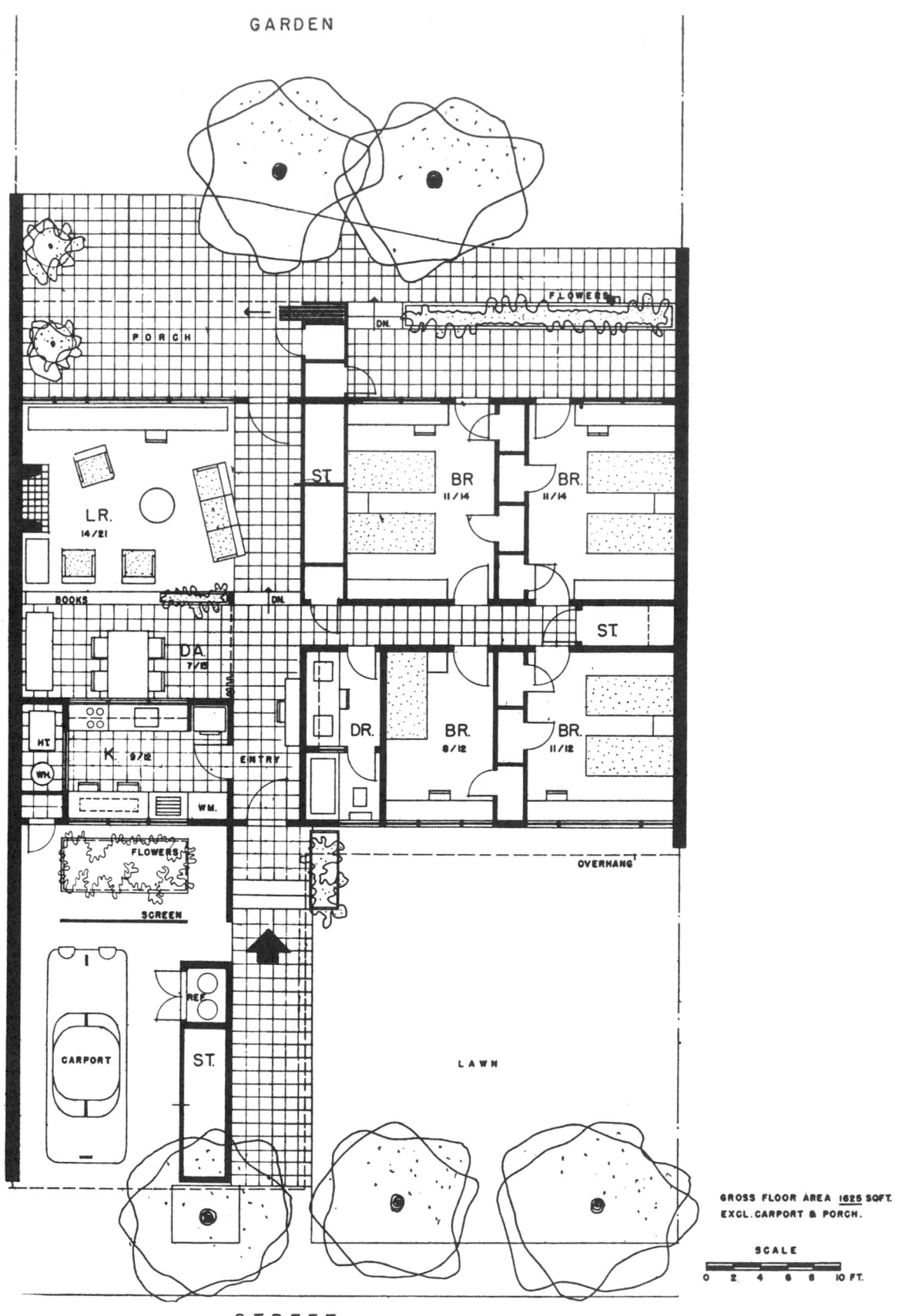

GARDEN

PORCH

FLOWERS

DN.

ST.

BR.
11/14

BR.
11/14

LR.
14/21

BOOKS

DN.

ST.

DA.
7/15

DR.

BR.
8/12

BR.
11/12

K. 9/12

HT

WH

WM.

ENTRY

FLOWERS

OVERHANG

SCREEN

LAWN

CARPORT

REF.

ST.

GROSS FLOOR AREA 1625 SQFT.
EXCL. CARPORT & PORCH.

SCALE
0 2 4 6 8 10 FT.

STREET

Fig. 6

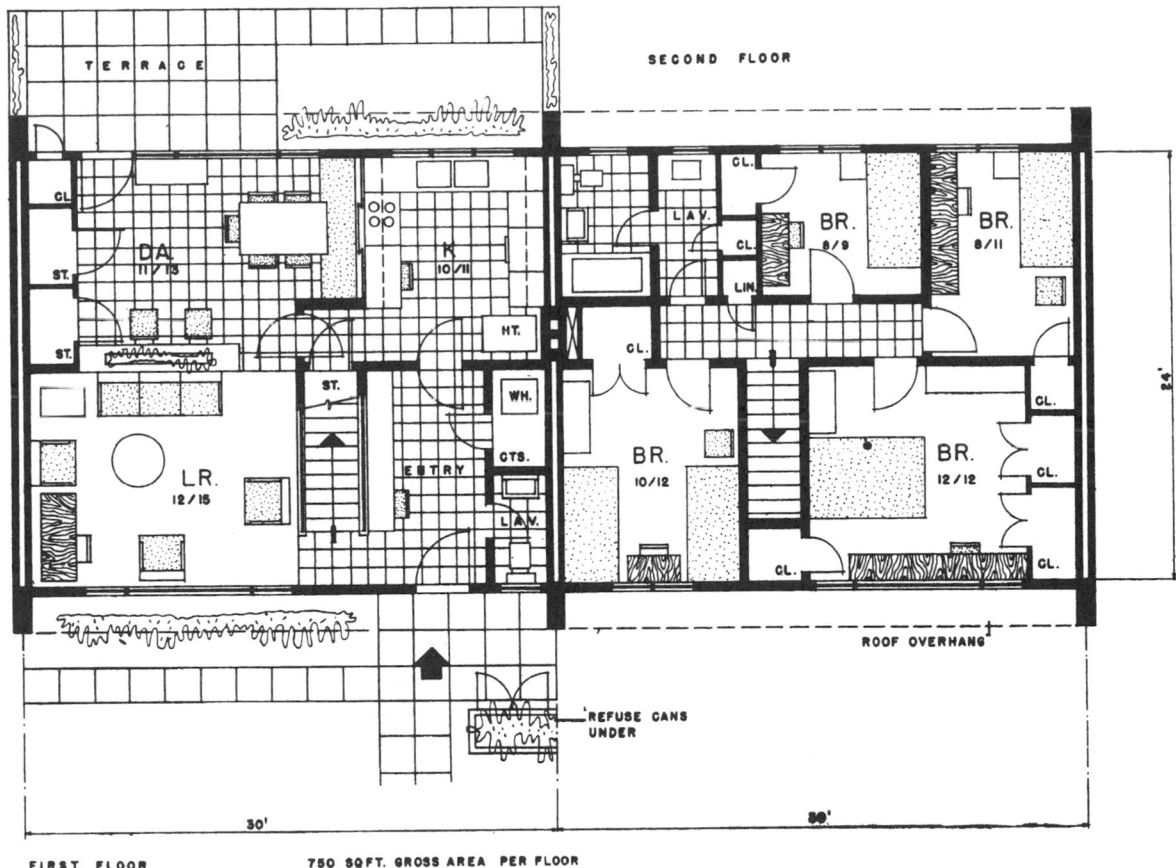

Fig. 7   The 30-ft row house.

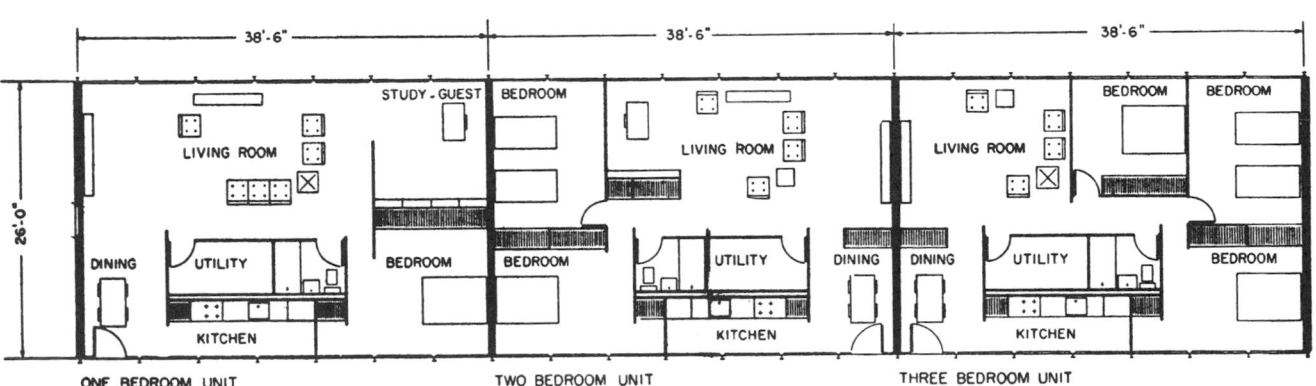

Fig. 8   Mies van der Rohe row house.

The 40-ft Row House

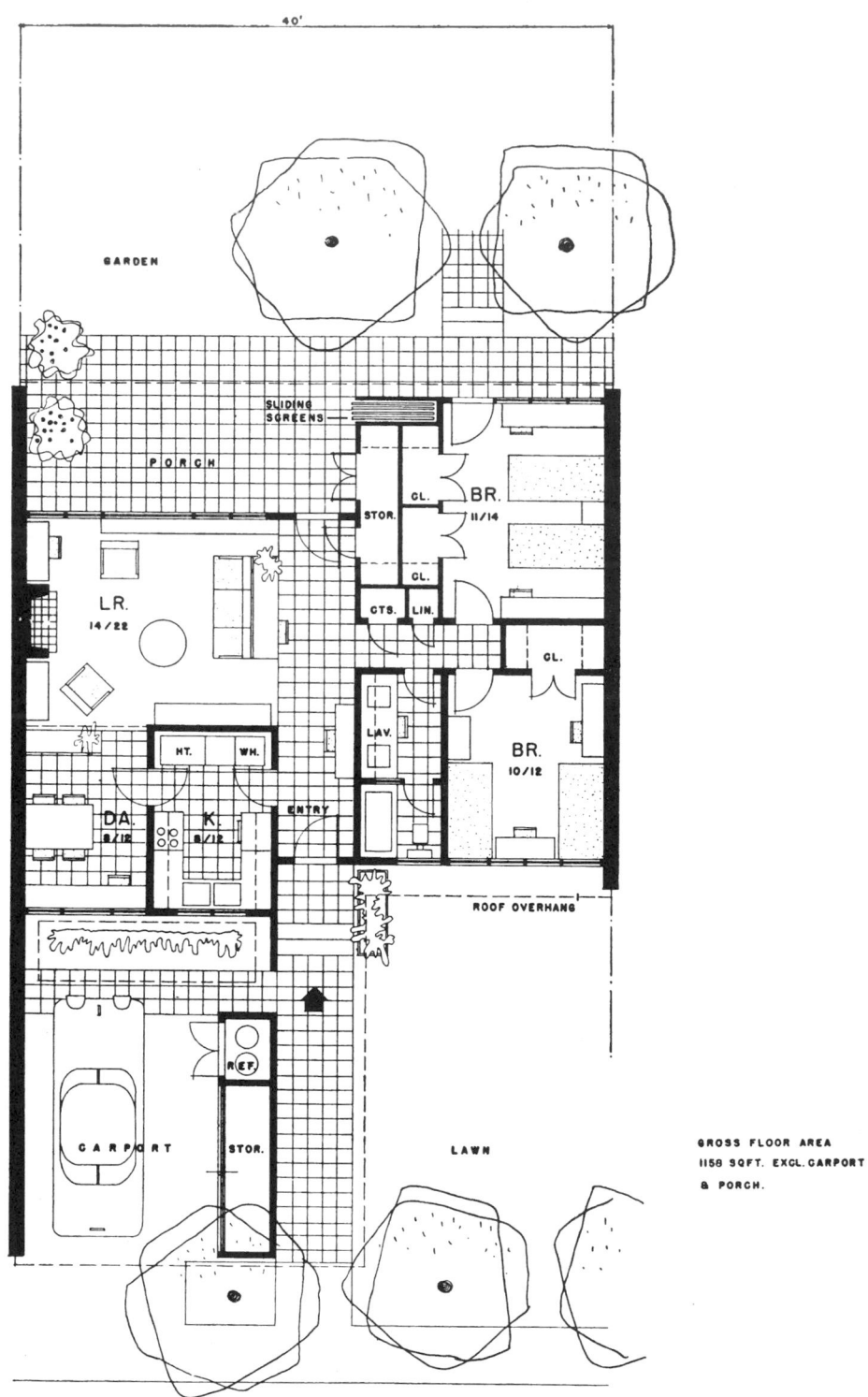

Fig. 9   The 40-ft row house on one floor.

One-Story Row Houses

Even though the apartments in Fig. 10 lie in long ribbons there is privacy from neighbors, created by extending partywalls and offsetting adjacent apartments. A full 800 sq ft of lawn is available to each family (and they maintain it) plus individual clothes-drying facilities, screened by cinder block fences.

Control of outdoor play is easy. A parent looking out through the big windows can see a wide play space at one glance. There is through ventilation in every unit, and the plans provide plenty of closets and big enough bedrooms for play space on rainy days.

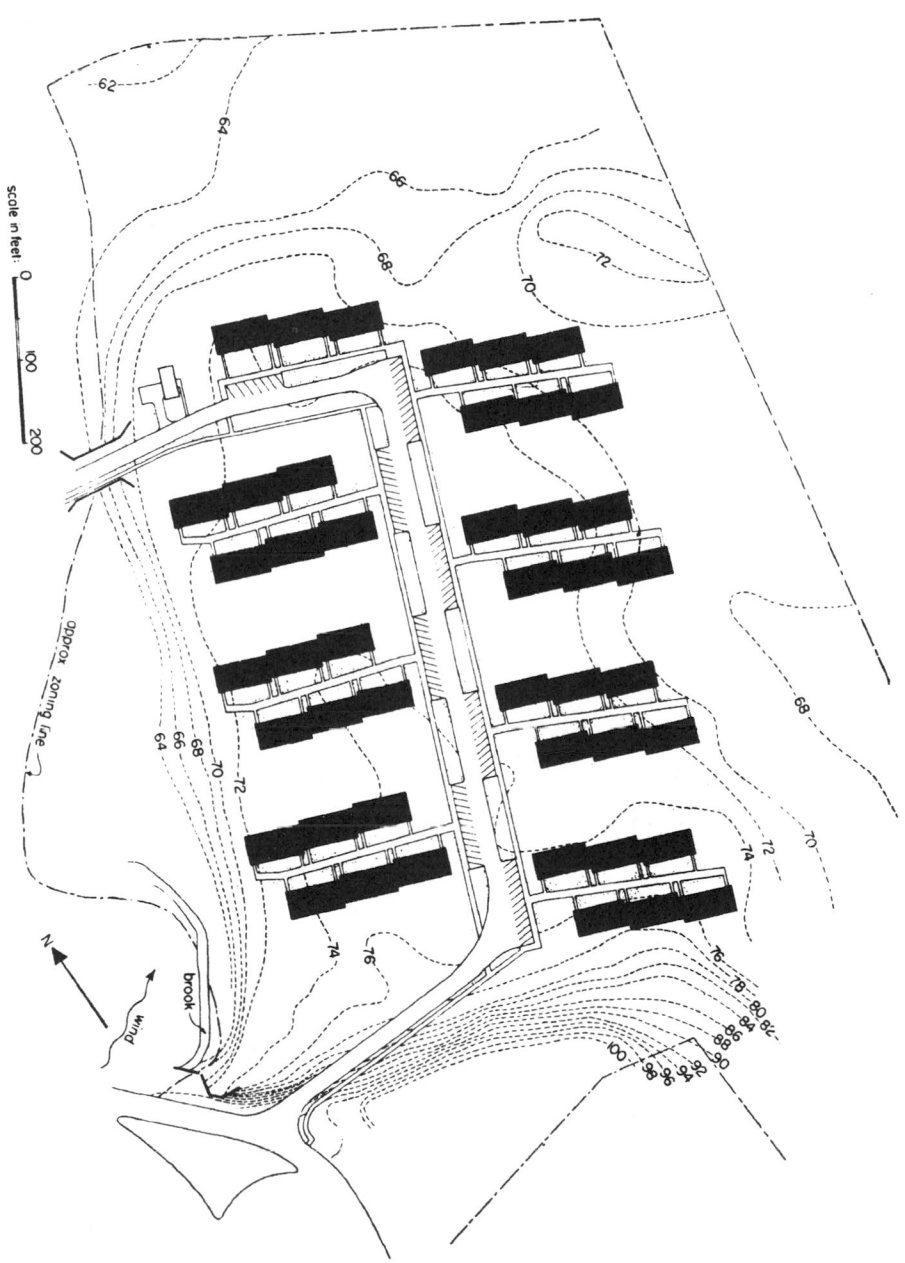

Fig. 10   Site plan. Row houses for Wellesley Housing Authority, Wellesley, Mass. Hugh Stubbins, Jr.—Architect.

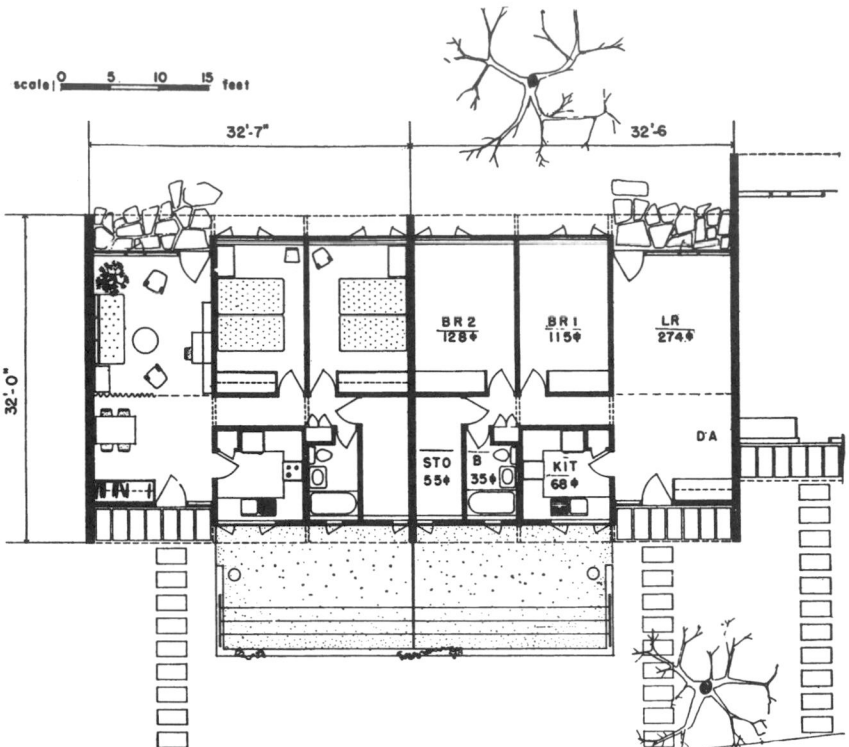

Fig. 11*a*   Wellesley, Mass. Hugh Stubbins, Jr.—Architect.

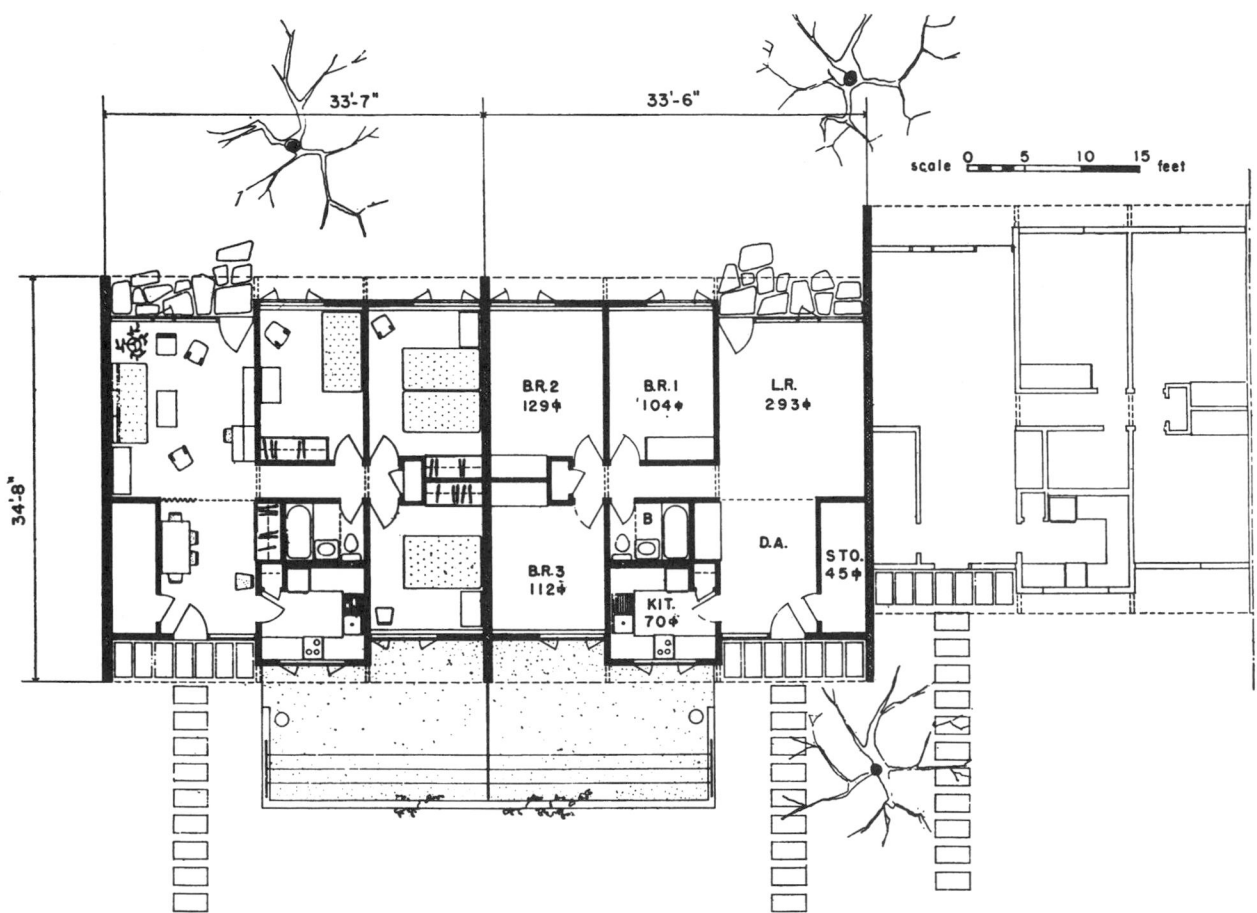

Fig. 11*b*   Wellesley, Mass. Hugh Stubbins, Jr.—Architect.

TOWN HOUSES

The town house, which has become popular in recent years, is similar to the old row house. It is an independent dwelling on an independent lot, which is attached on both sides to a similar dwelling on both adjacent lots. The attachment is made along two common or party walls that are jointly owned. The chief advantage of this type of construction is the economy of the party walls. Also, because no side yards are required, it is possible to build on a relatively narrow lot. Old row houses were built on lots as narrow as 16 and 18 ft wide. However, this is not recommended. Lot widths should be a minimum of 20 ft, but preferably wider.

Town houses are usually one-family dwellings with the living room, kitchen, and dining area on the lower level and the sleeping areas on the upper level. Row houses may contain two dwelling units, one above the other.

A built-in garage is desirable if it can be reasonably incorporated within the house. Alternate solutions are a parking space either in front of the house or in a group parking area close by.

The town house has long been advocated for rental housing for urban families with children as a good compromise between the desirability of a detached single-family house and the economic necessity of multifamily units. It is decidedly preferable from the viewpoint of the tenants because of greater livability. The results of surveys in both public and private housing indicate that families want to have direct access to the house, an individual yard or garden, and a place for small children to play close to the house where they can easily be supervised. These are features that the town house can provide.

From the management point of view, town house projects can be designed for maximum tenant maintenance of land area. They can also be designed for either individual heating installations or a central heating plant. Individual heating installations, though of higher operating cost to the tenant, result in lower maintenance cost to the management.

Private Garden

Privacy is an important factor in town house design. All house types show, therefore, an extension of the party walls beyond the face of the building on either side. Sitting out terraces on the garden side are separated by wing walls, approximately 6 ft long and 6 ft high. These wing walls do not have to be of masonry material, although preferably they should be of a permanent rather than of a temporary nature.

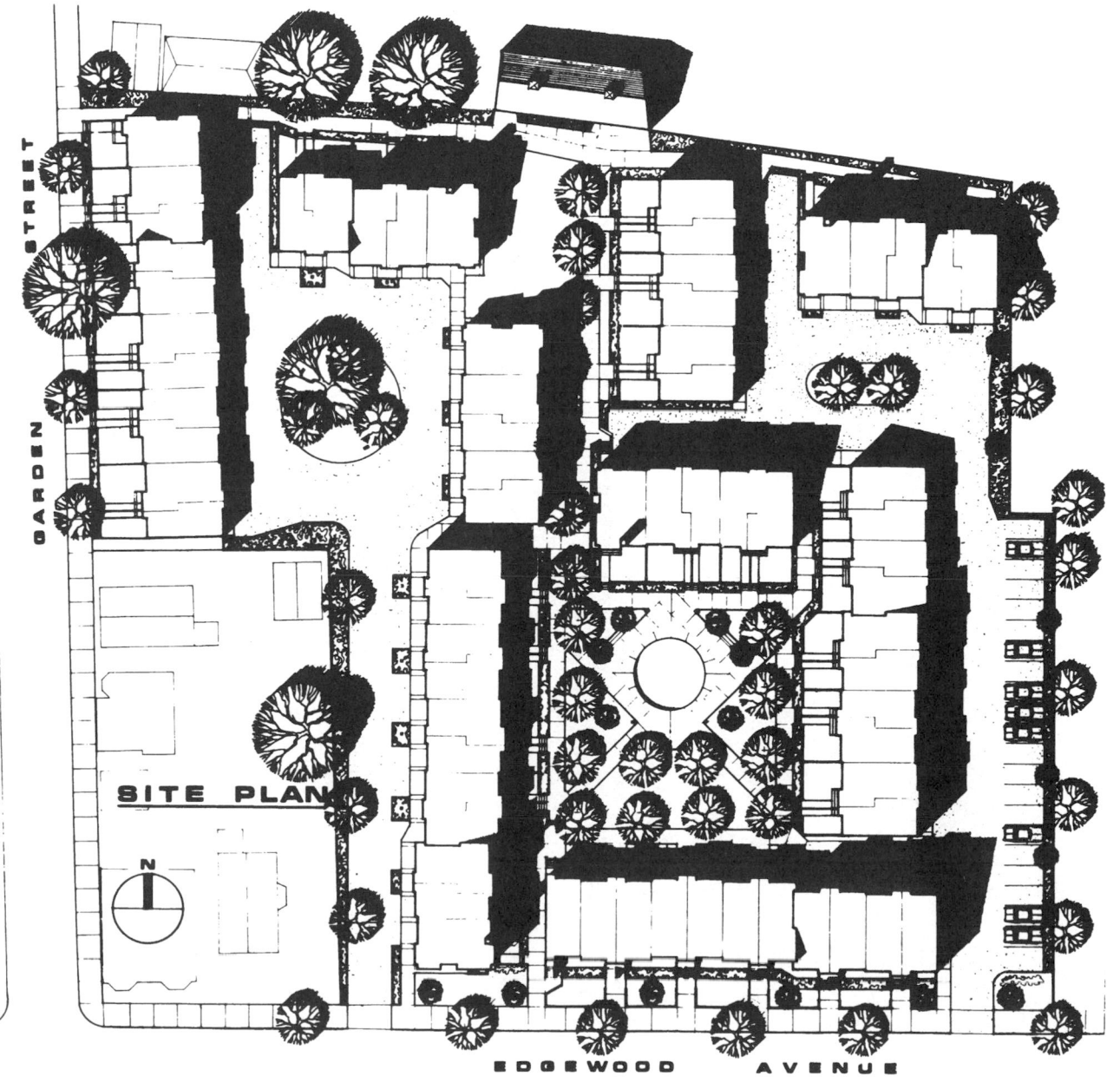

Fig. 1

TOWN HOUSES

Public Access

Another arrangement that ensures more privacy is the concentration of services from the front. The problem of refuse collection is solved by means of a masonry enclosure, 3 ft wide, 4 ft high, and 10 ft long for two living units. Access doors to the enclosure are from the side facing the building, away from street view. A flower box built into the top of the enclosure makes the appearance pleasing and attractive to the passerby. A hose-bib connection facilitates cleaning and reduces odors to a minimum.

An entry space for each living unit presents another privacy feature and is absolutely necessary for service-from-the-front planning. The conventional direct entrance from the street into the living room reduces privacy and is the cause of annoyance to many people.

Recently in urban areas, the town house has emerged as a popular type of dwelling. The town house is usually one-family and owner-occupied.

The height is most frequently two stories, and construction is brick or brick veneer.

The two-level town house is probably the most common, with the living activities on the first floor and the sleeping activities on the second floor. A powder room is generally provided on the lower level to minimize up and down traffic. In more elaborate developments, a second bathroom can be provided on the second floor.

The work area (kitchen) on the first floor is best located near the entrance and on the service side. This enables the living room and dining area to be oriented toward the terrace or open green areas. The bedrooms can easily be provided with a balcony.

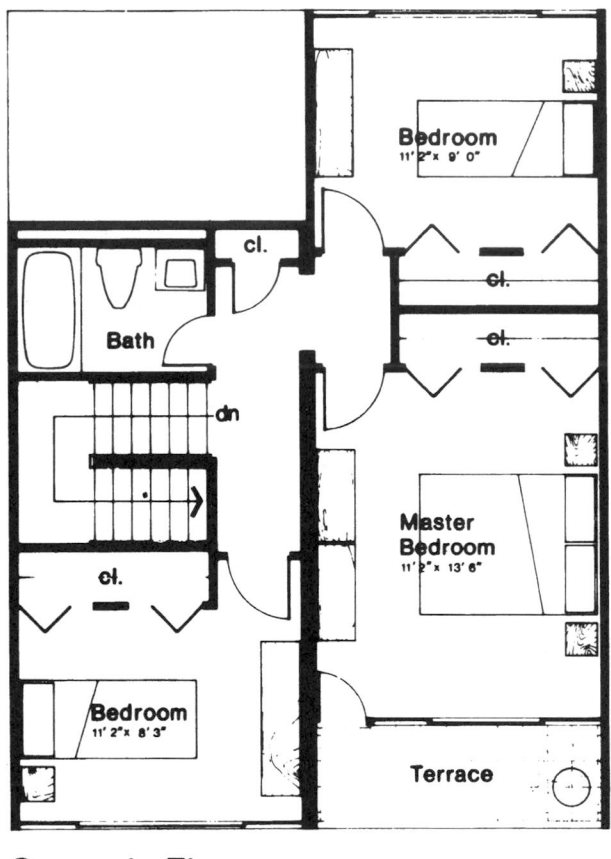

Second Floor

Fig. 2

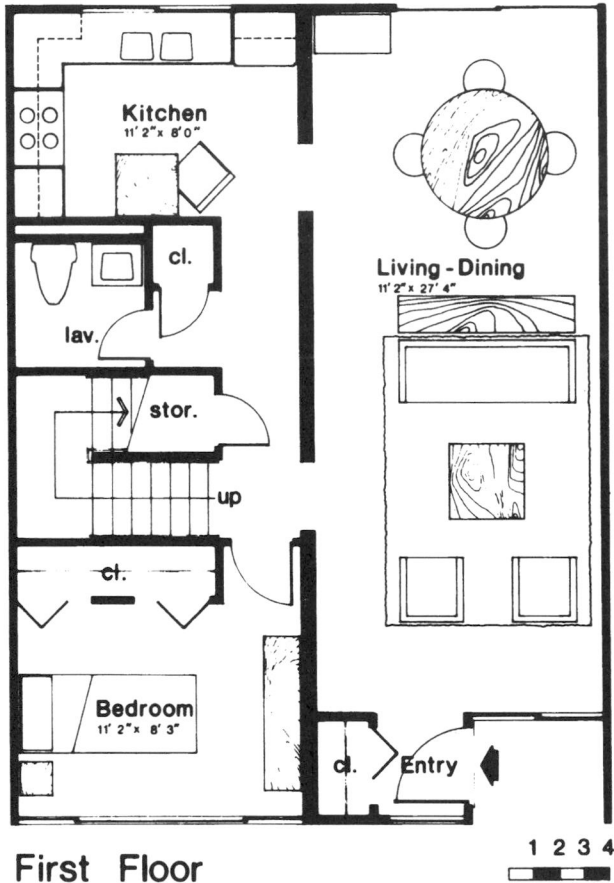

First Floor

1 2 3 4

Three-Level

In this development the town houses, which are designed on three levels, were built on a steeply sloping site. The houses are sited in line along the existing contours. The land lends itself ide- ally to the provision of a walk-out basement from the family room. The main entrance to the house is on the opposite side on the first floor.

The basic house type is three bedrooms. Alternate schemes are shown for the two-bed- room and three-bedroom plan.

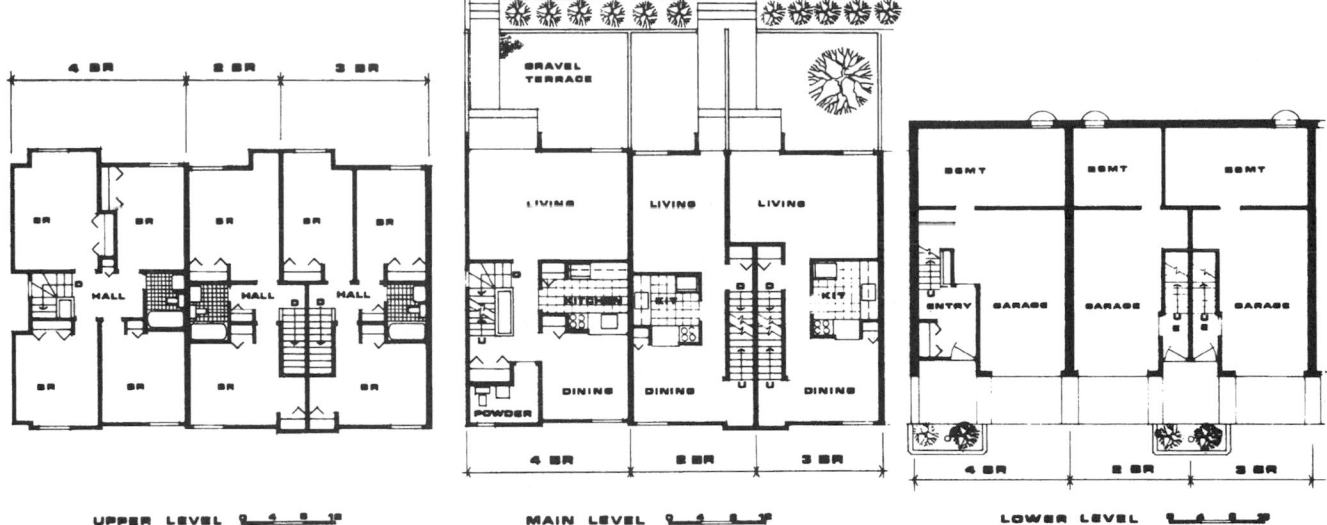

Fig. 3 Wycliffe Hill, Toronto, Canada.

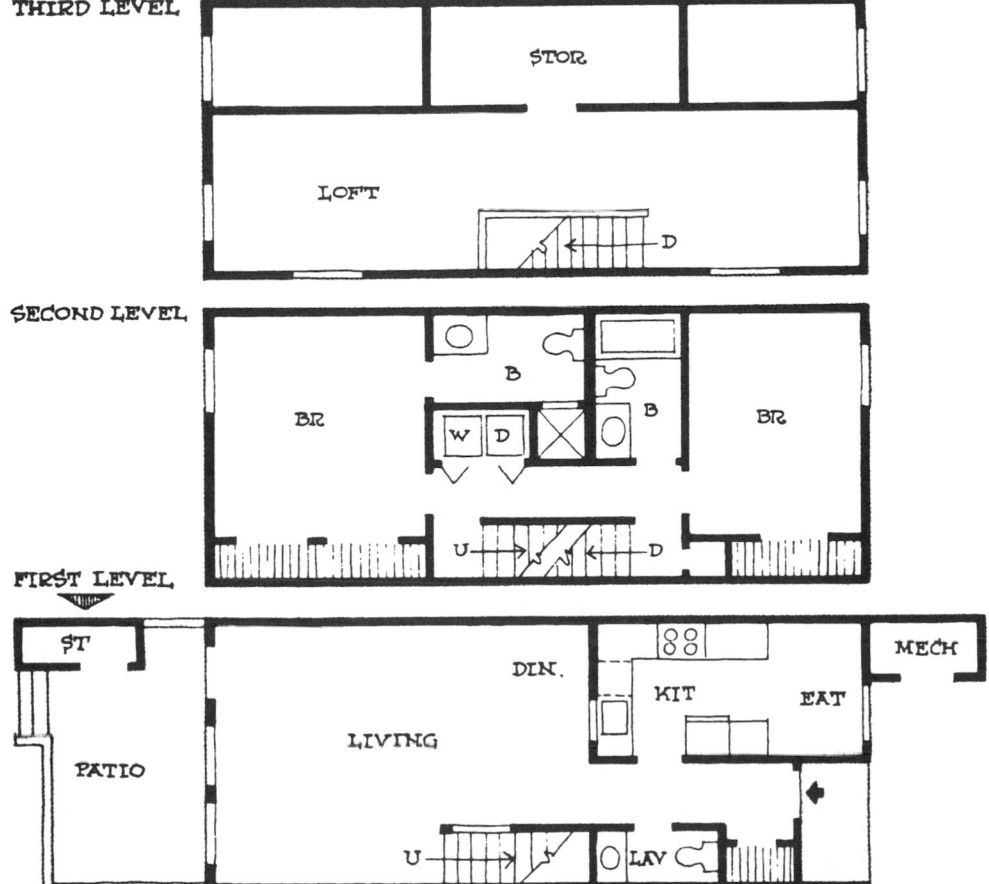

Fig. 4 Two-bedroom and loft unit.

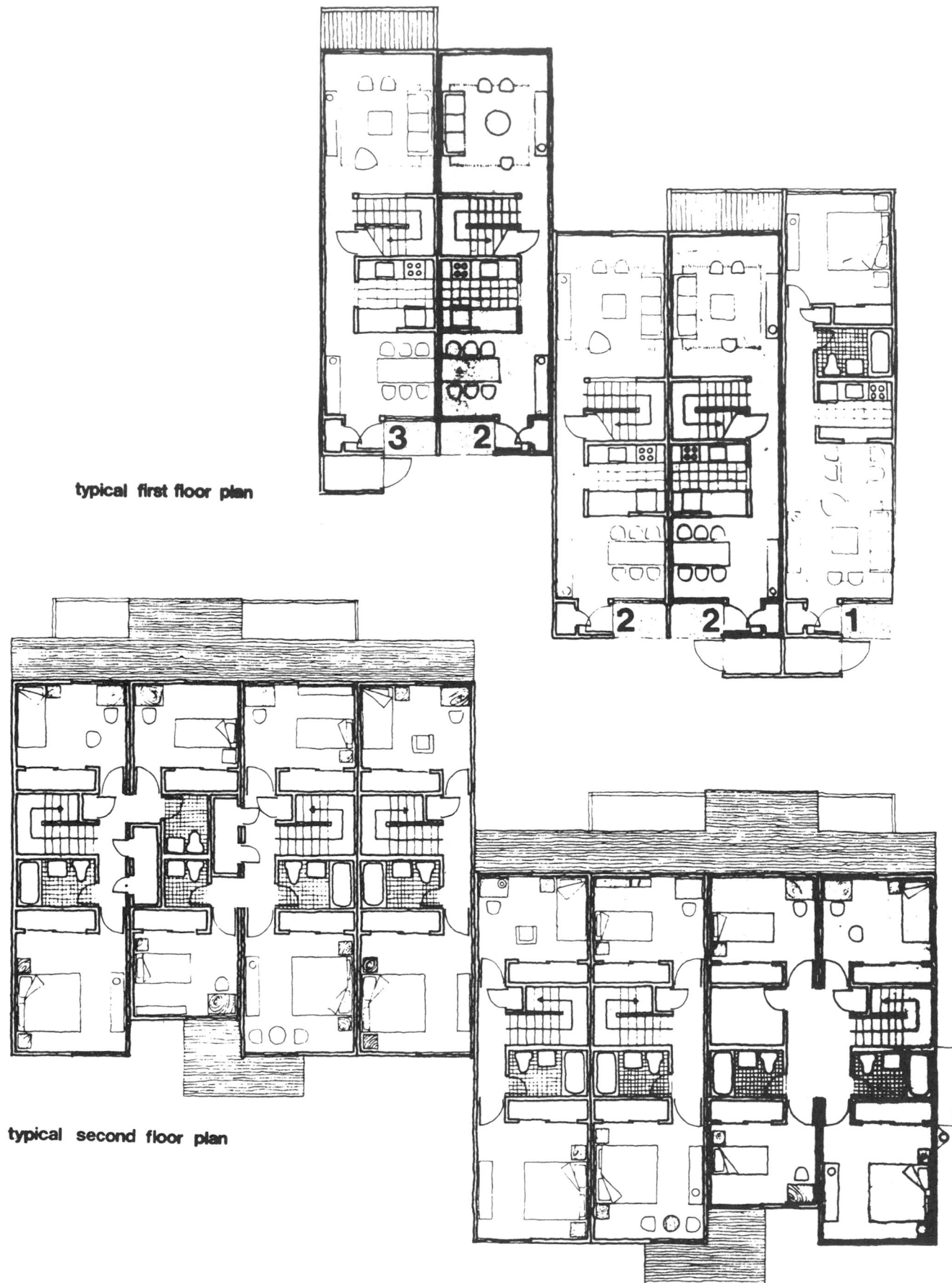

typical first floor plan

typical second floor plan

Fig. 5   College Hill Housing, Middletown, N.Y. Todd/Pokorny—Architects and Planners.

Clusters

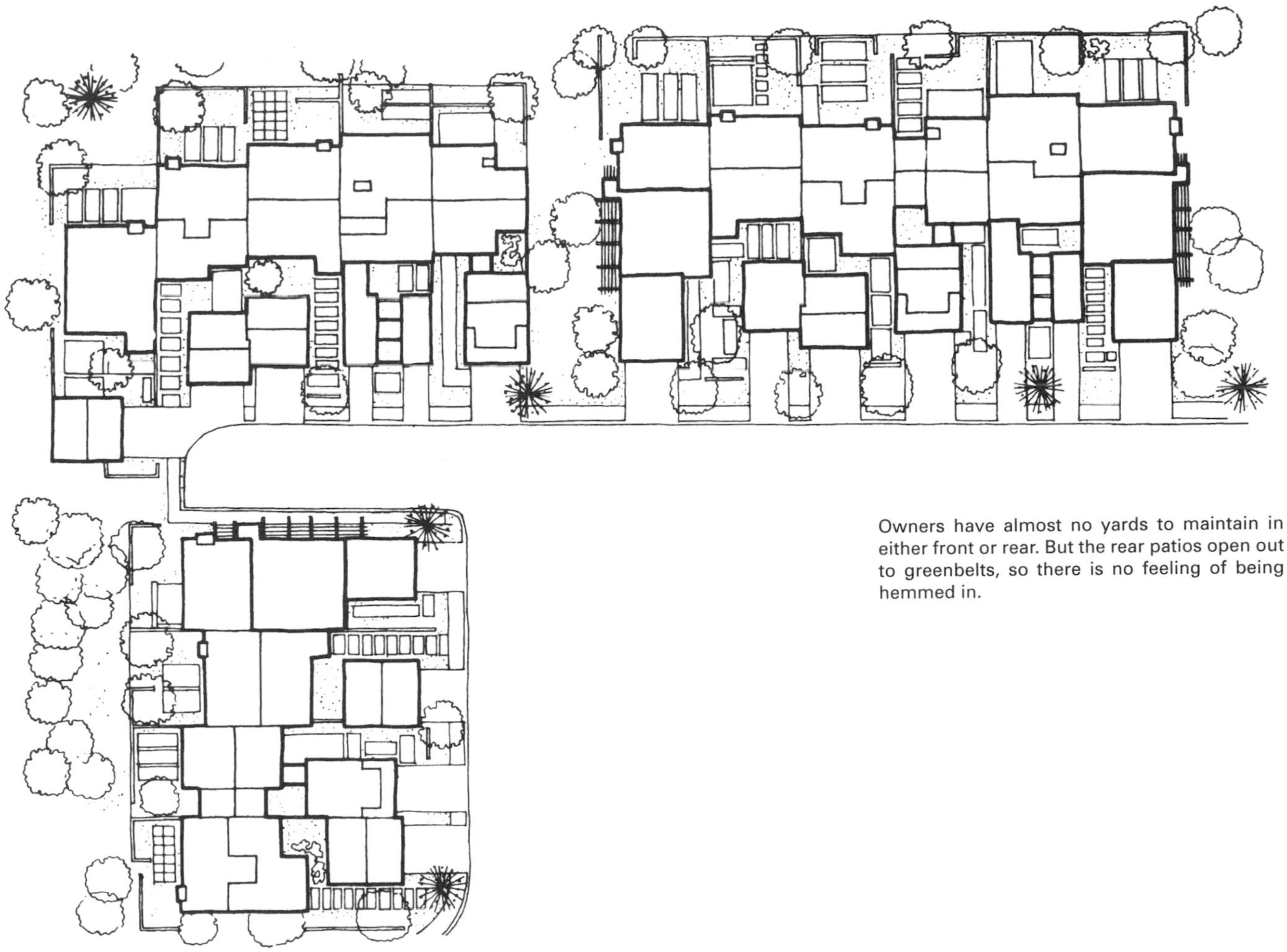

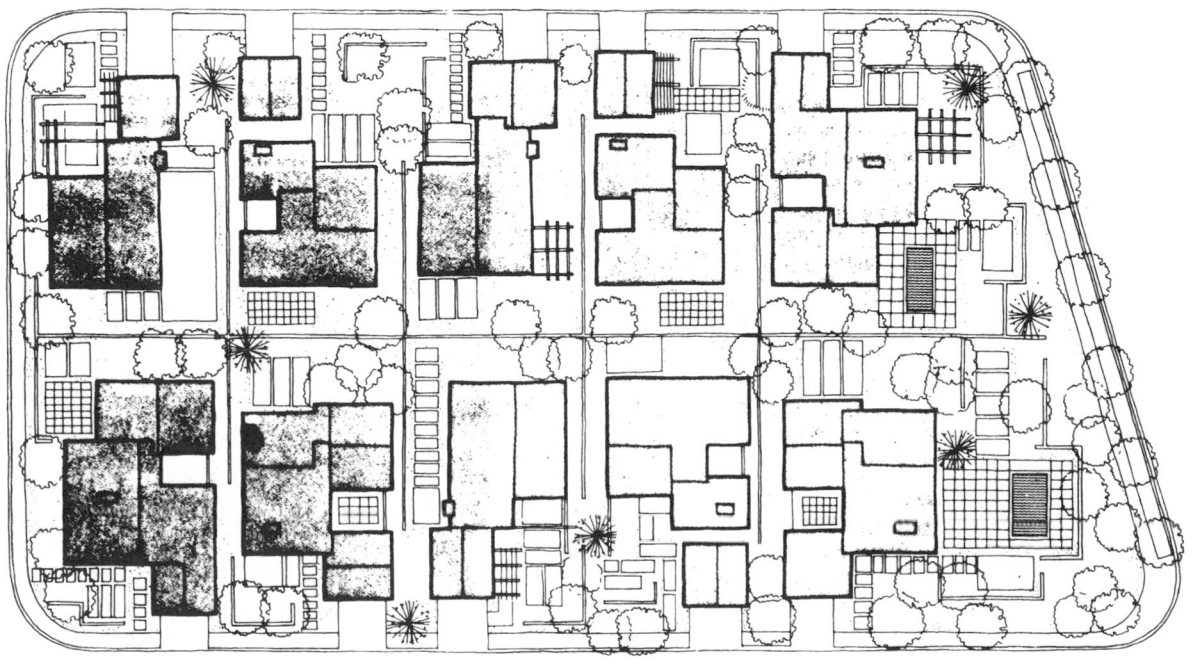

Owners have almost no yards to maintain in either front or rear. But the rear patios open out to greenbelts, so there is no feeling of being hemmed in.

Fig. 6

Fig. 7

Formal town house has a large entrance foyer and a separate dining room, with a small family room in the front. Second floor has three bedrooms and two baths.

Informal town house is exactly the same as the one at left, but lower floor is divided into just two big areas with kitchen opening to the rear patio.

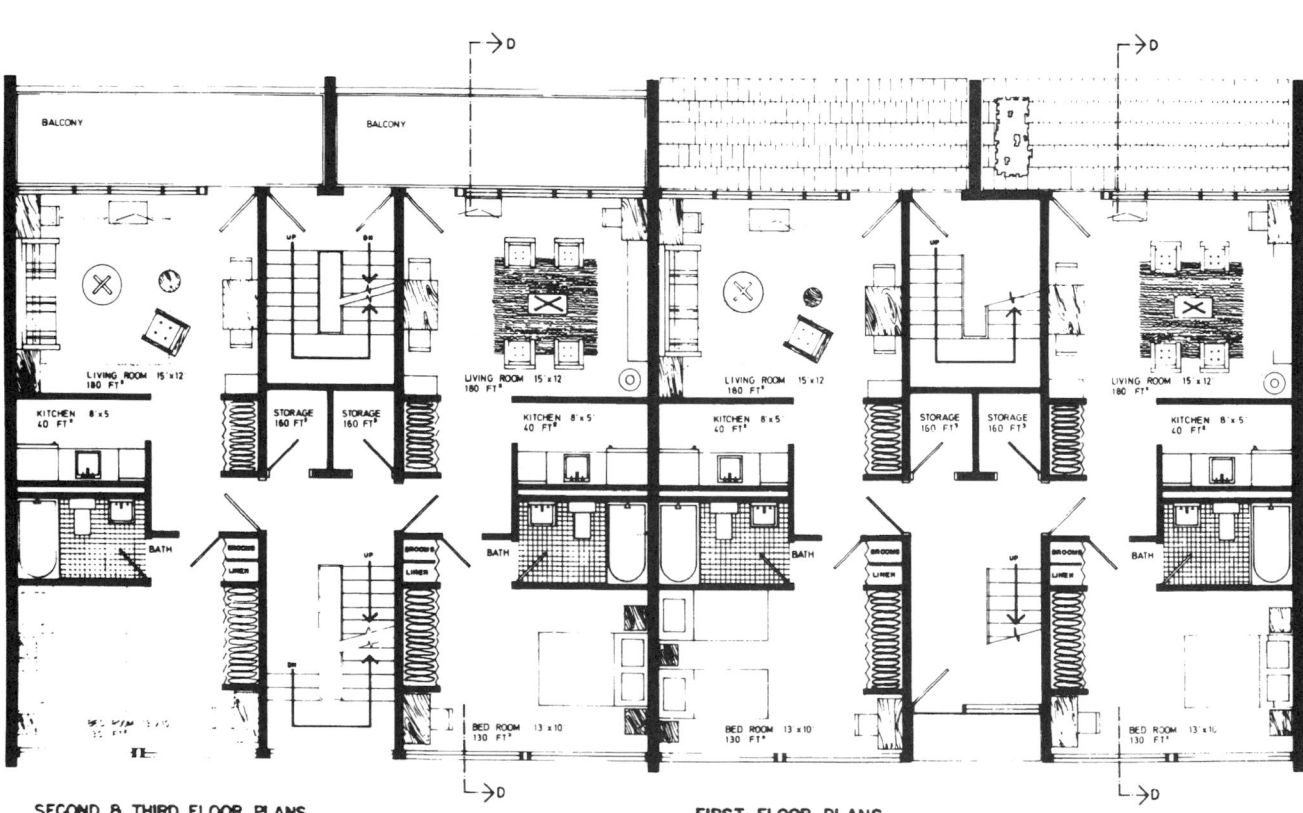

SECOND & THIRD FLOOR PLANS

FIRST FLOOR PLANS

Fig. 8   Charlotte Area Project, Rochester, N.Y. Northrup, Kaelber, & Kopf—Architects.

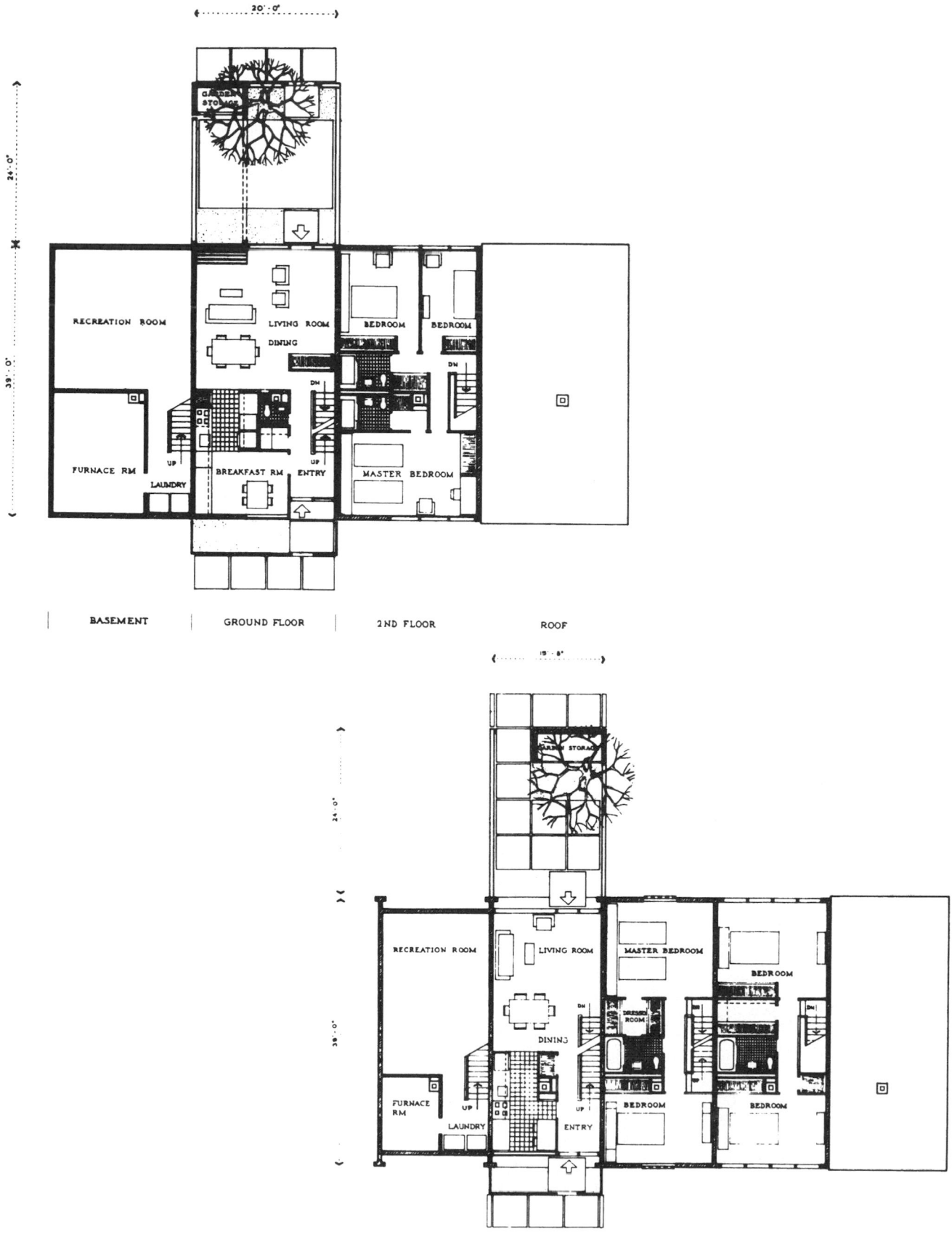

BASEMENT   GROUND FLOOR   2ND FLOOR   ROOF

BASEMENT   GROUND FLOOR   2ND FLOOR   3RD FLOOR   ROOF

Fig. 9

First floor.

All dimensions approximate.

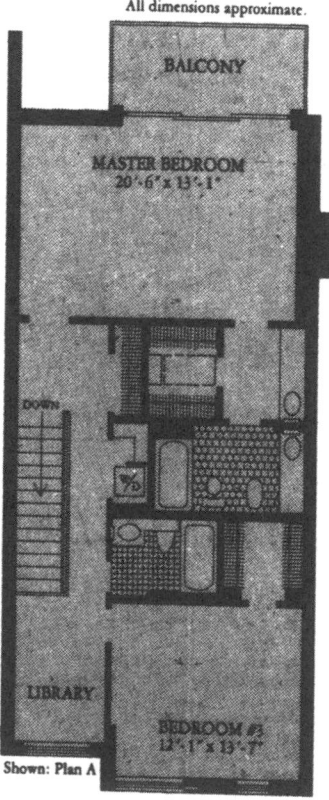

Shown: Plan A

Second floor.

Fig. 10   Water's Edge at Rye, N.Y.

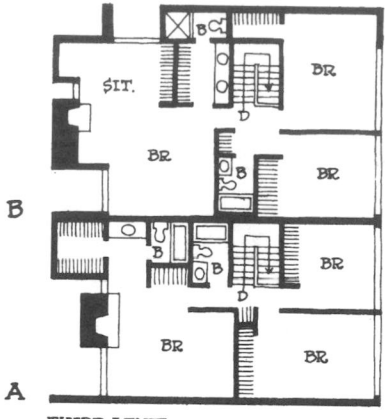

THIRD LEVEL

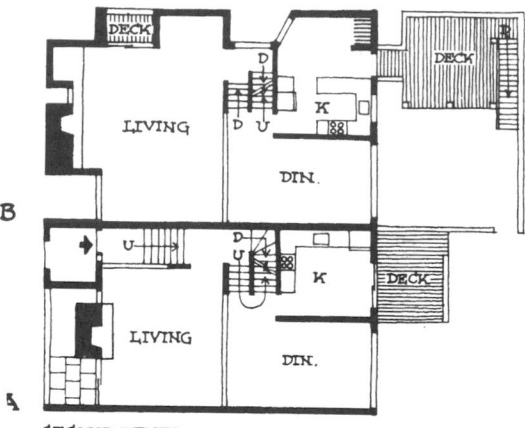

SECOND LEVEL

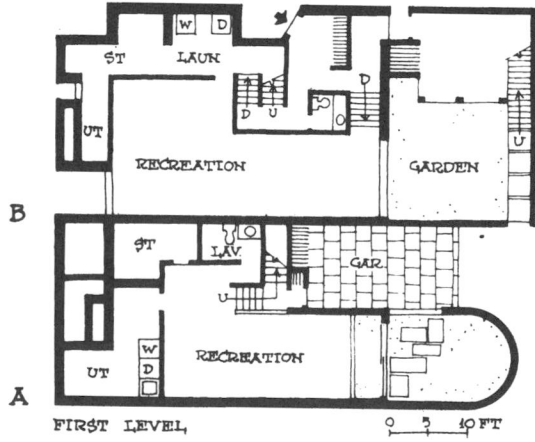

FIRST LEVEL

Fig. 11

Condominiums

Five of six available floor plans are shown in Fig. 113 as they are grouped within town house rows. Most popular, the three-bedroom with family room, is second plan from the right. Square footage ranges from 1247 to 1753.

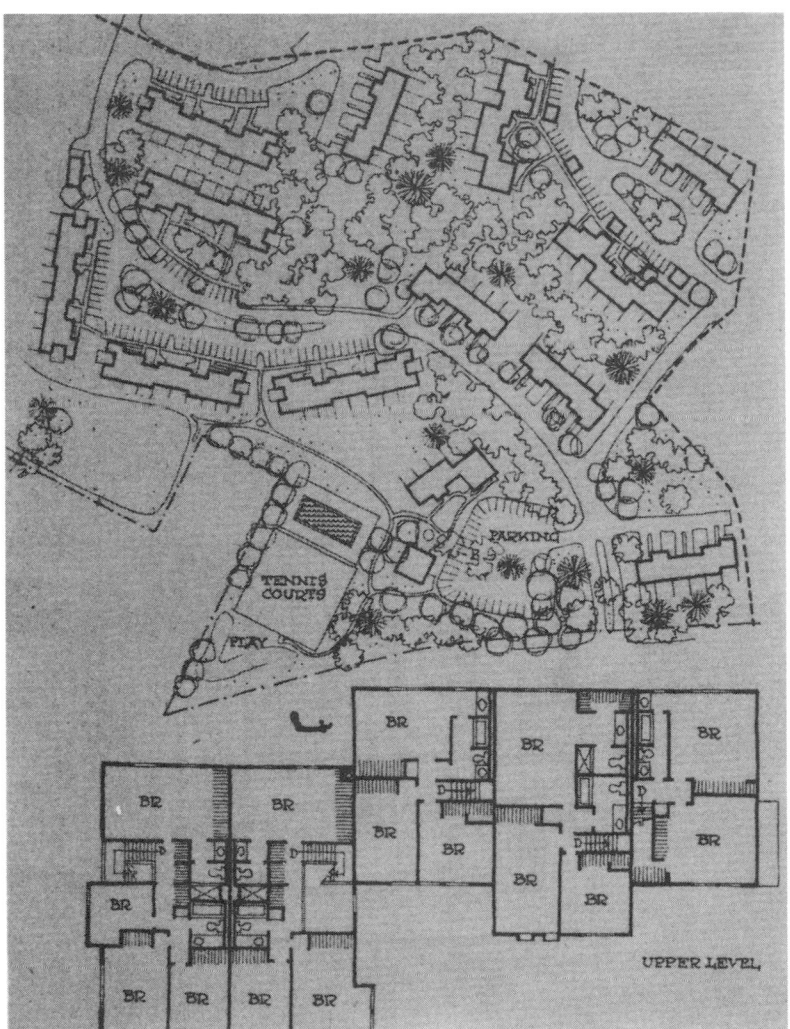

Fig. 12   Wren's Cross, Chamlee, Ga.

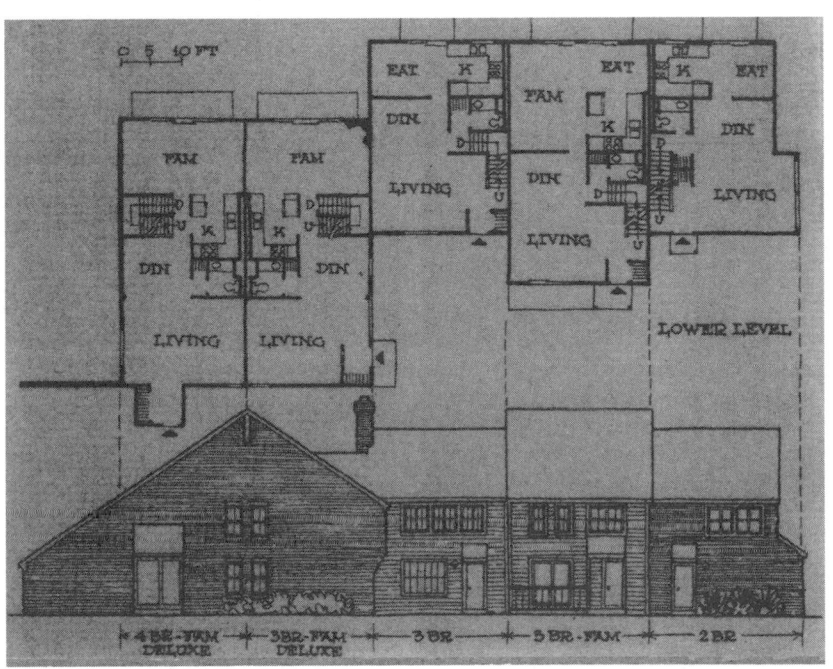

Fig. 13   Wren's Cross, Chamlee, Ga.

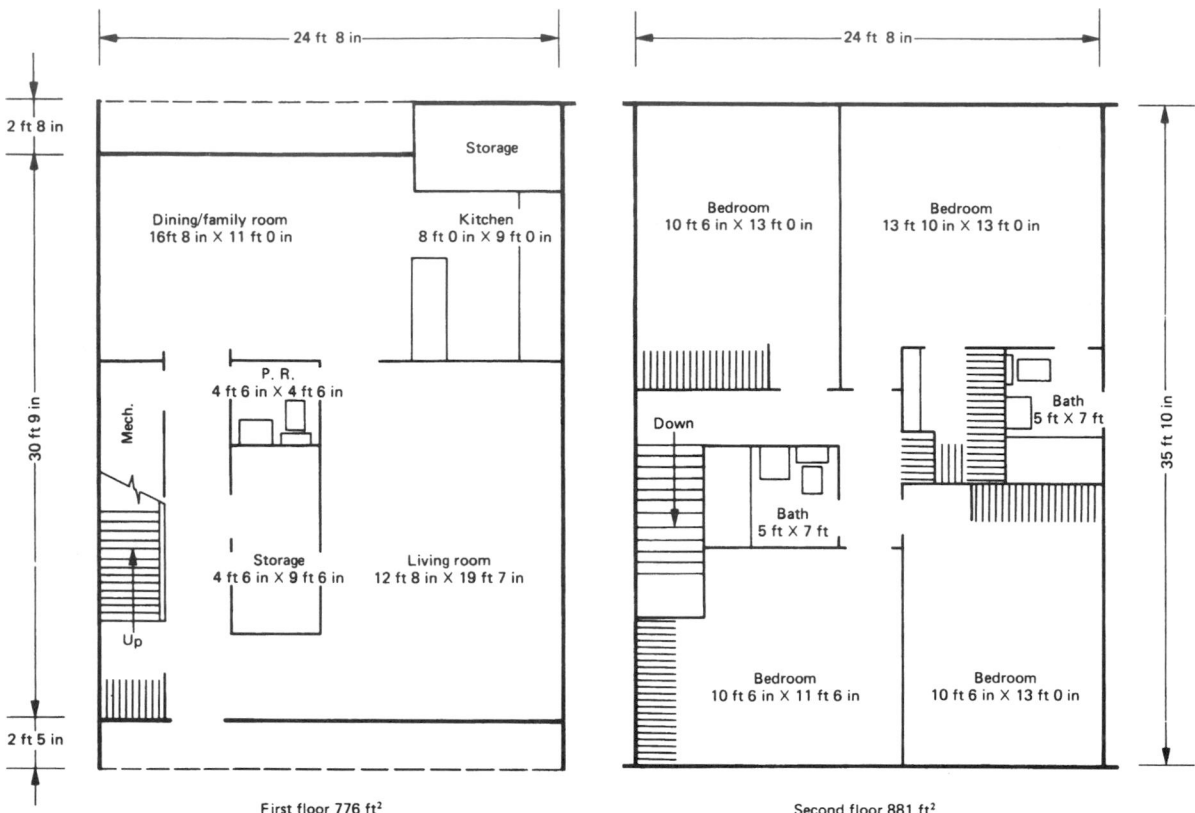

Fig. 14   Four-bedroom town house (1657 ft²). New Jersey Housing Finance Agency.

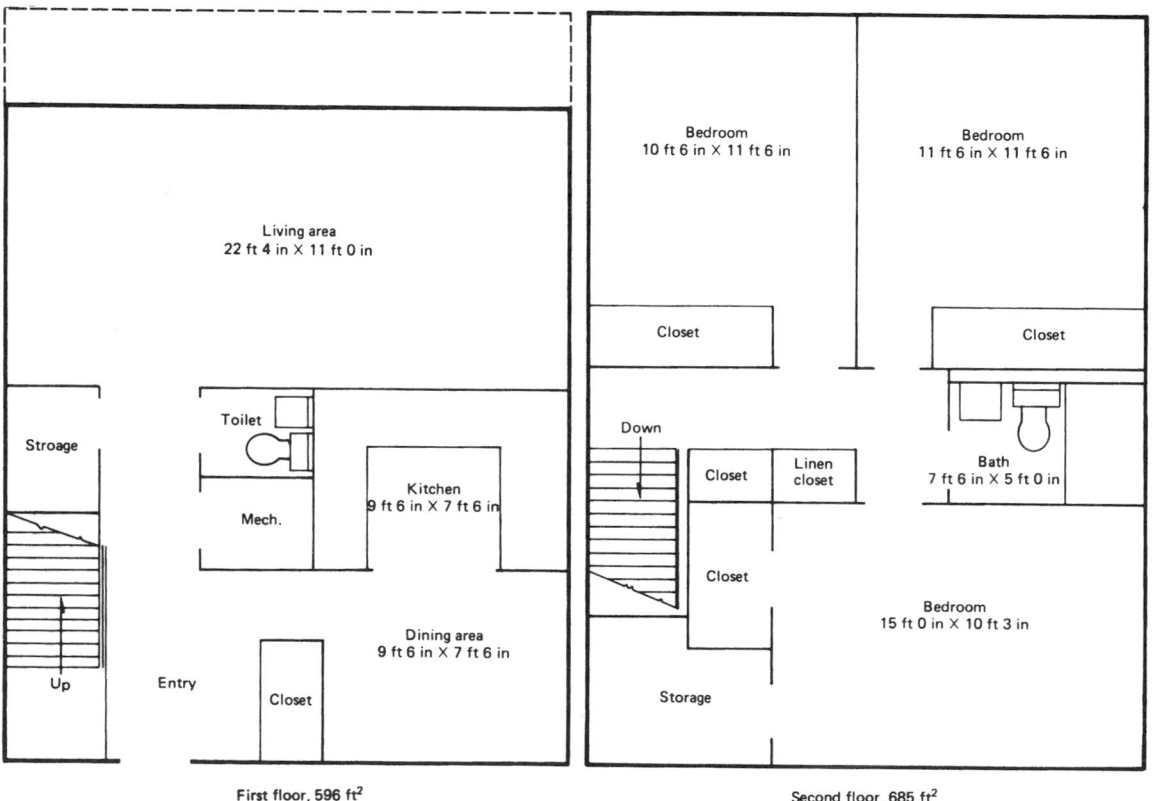

Fig. 15   Three-bedroom town house (1281 ft² total). New Jersey Housing Finance Agency.

Three-, Four-, Five-, and Six-Bedroom Units

Two units are placed back to back, and each has its front entry court set off from the public walkway by a planting bed. Each unit has its own on-grade entrance, giving tenants a sense of proprietorship rare in apartments.

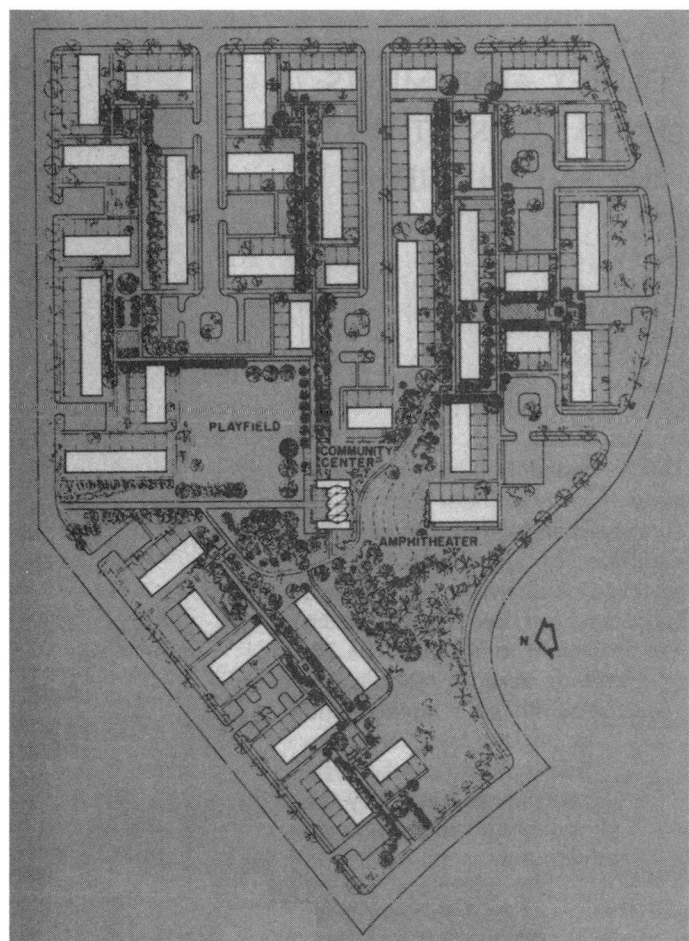

Fig. 16   Site plan. Eastgate Gardens, Washington, D.C. Chapman & Miller—Architects.

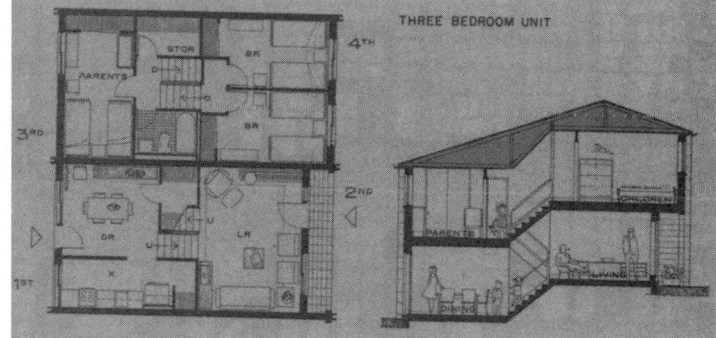

Fig. 17   Eastgate Gardens, Washington, D.C. Chapman & Miller—Architects.

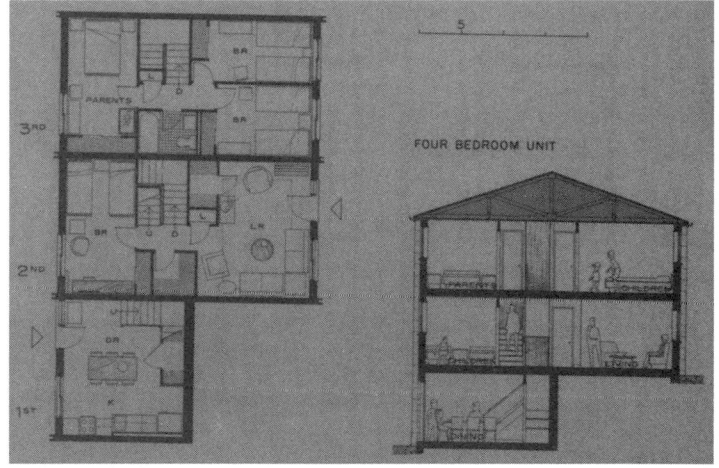

Fig. 18

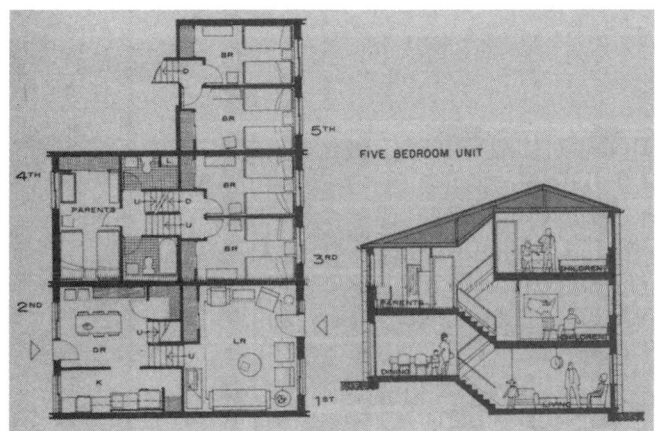

Fig. 19

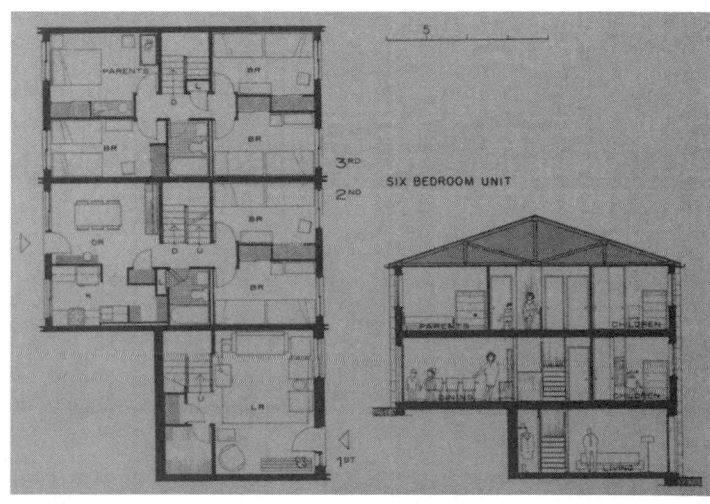

Fig. 20

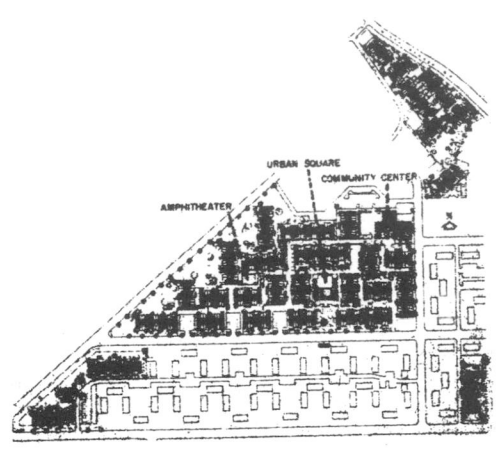

Fig. 21*a* Highland Extension, Washington, D.C. Brown, Wright, Mano—Architects.

Fig. 21*b* Highland Extension, Washington, D.C. Brown, Wright, Mano—Architects.

FOUR BEDROOM ELEVATION

THREE BEDROOM ELEVATION

Fig. 22*a* Highland Extension, Washington, D.C. Brown, Wright, Mano—Architects.

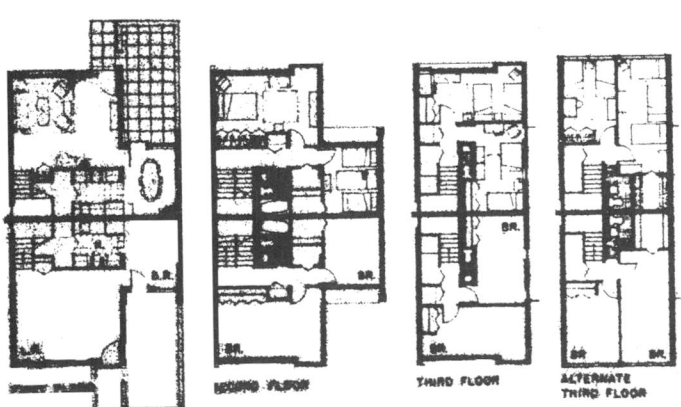

THIRD FLOOR

ALTERNATE THIRD FLOOR

Fig. 22*b* Highland Extension, Washington, D.C. Brown, Wright, Mano—Architects.

Attached Units (Two, Three, and Four Units)

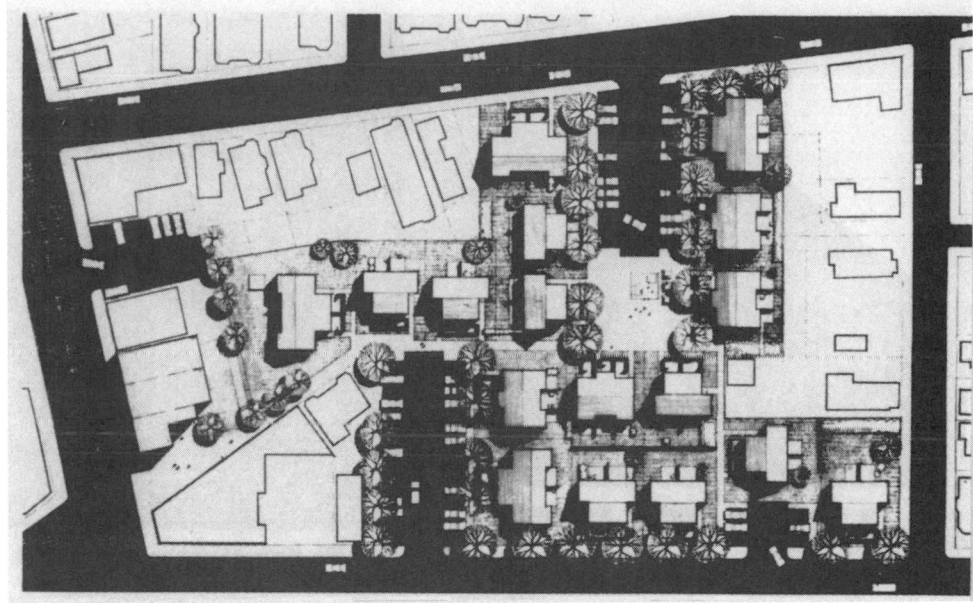

Fig. 23   Wellington-Harrington Urban Renewal Area, Cambridge, Mass. Huygens and Tappé—Architects.

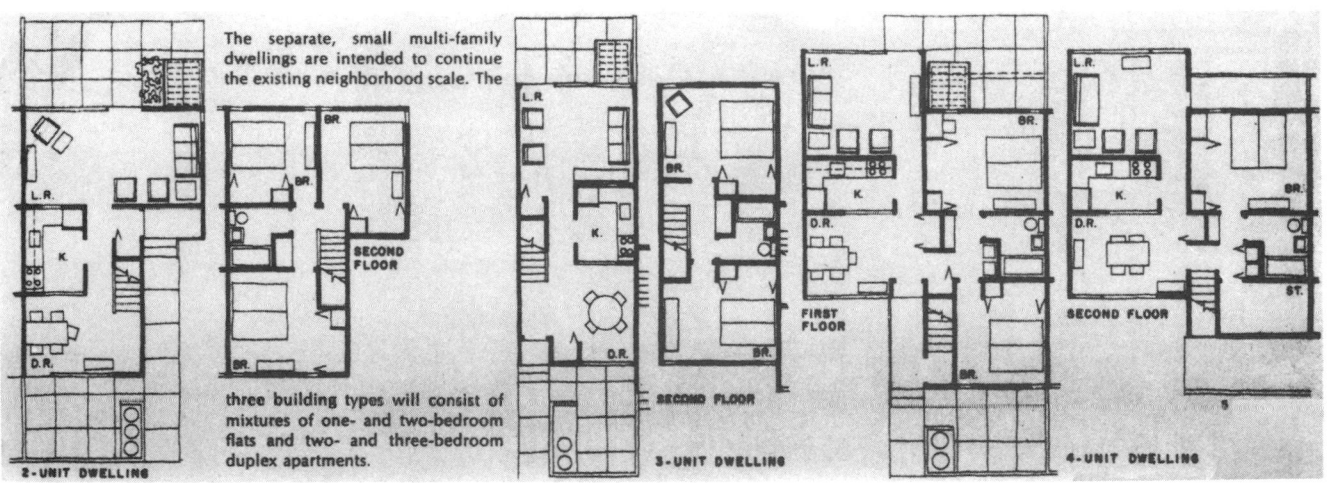

The separate, small multi-family dwellings are intended to continue the existing neighborhood scale. The three building types will consist of mixtures of one- and two-bedroom flats and two- and three-bedroom duplex apartments.

Fig. 24

Fig. 25

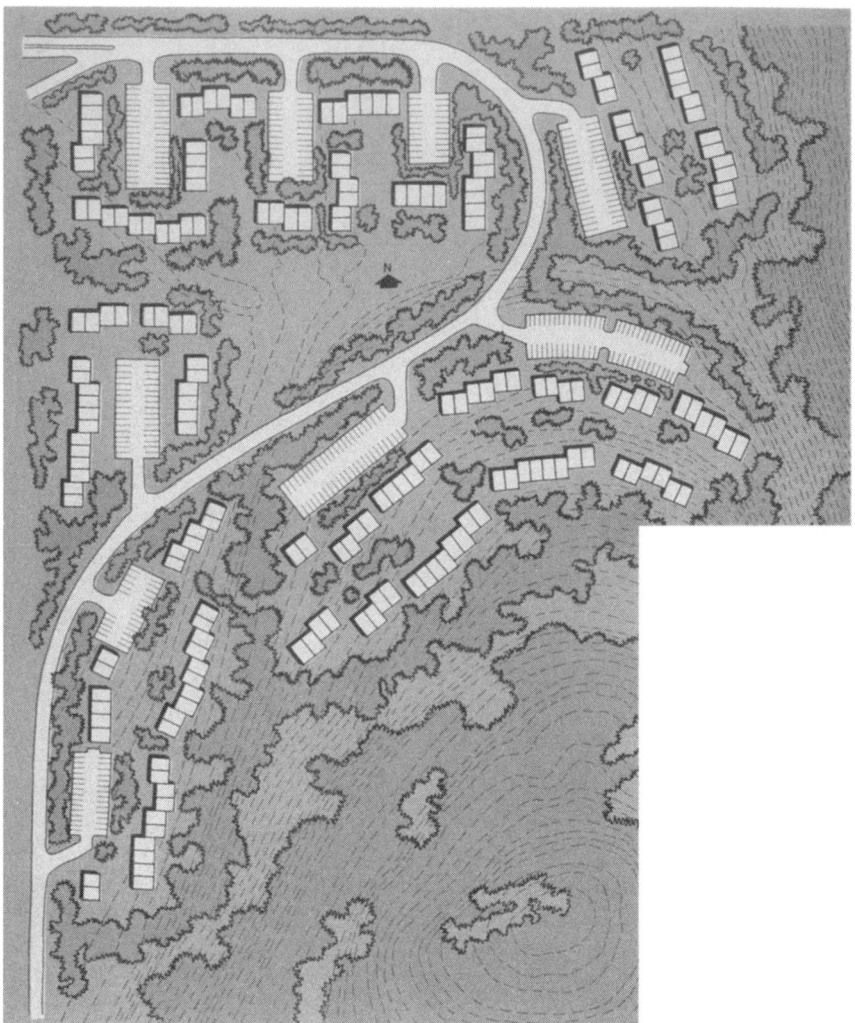

Fig. 26   Site plan. Ely Park Housing, Binghamton, N.Y. The Architects Collaborative—Architects.

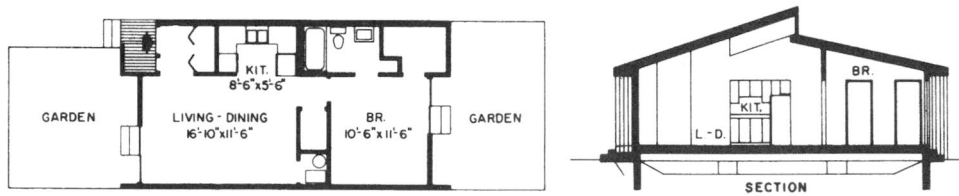

Fig. 27   One-bedroom design.

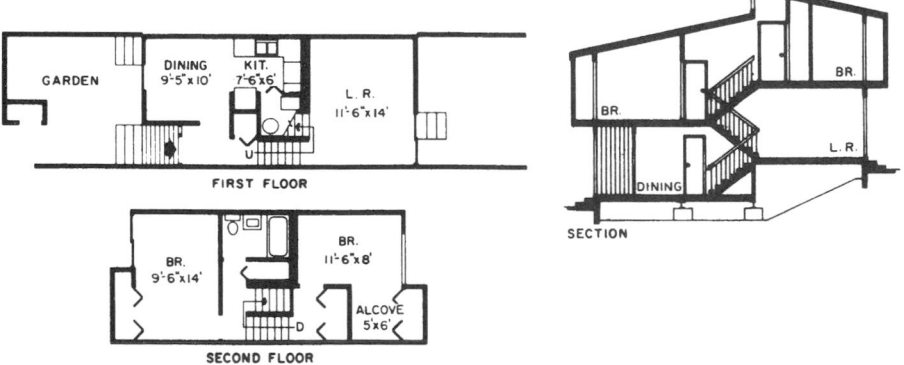

Fig. 28   Two-bedroom design.

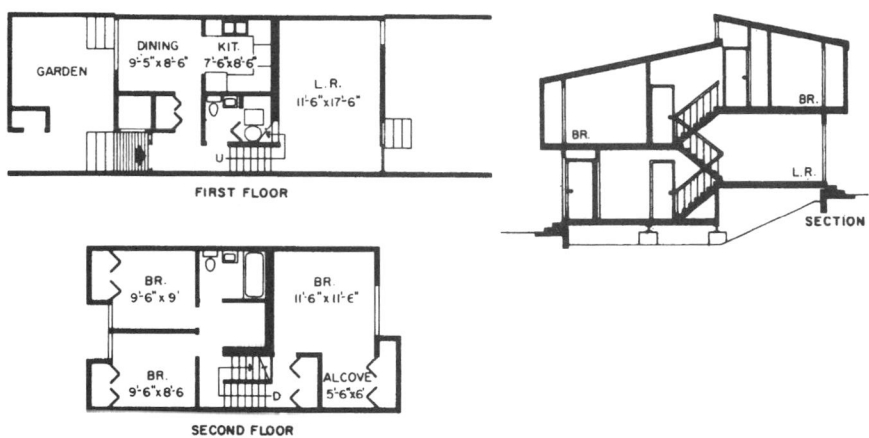

Fig. 29   Three-bedroom design.

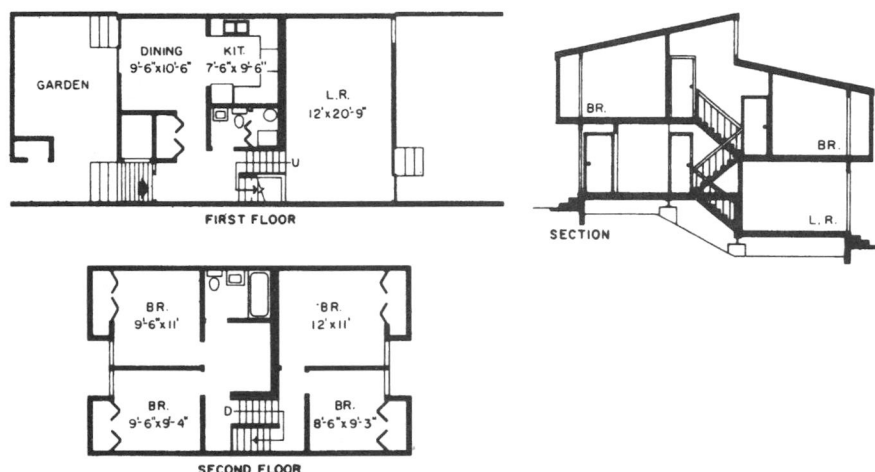

Fig. 30   Four-bedroom design. Eastover Gates, Charlotte, N.C. Wolf Associates—Architect.

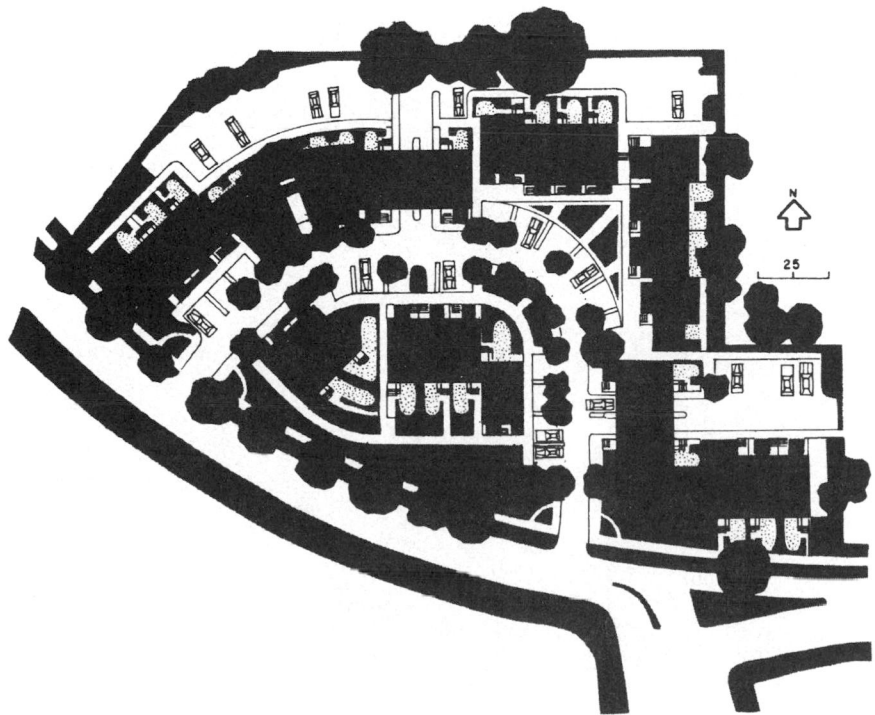

Fig. 31   Eastover Gates, Charlotte, N.C. Wolf Associates—Architect.

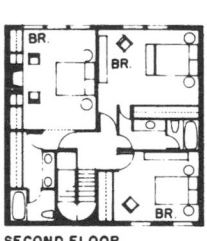

SECOND FLOOR

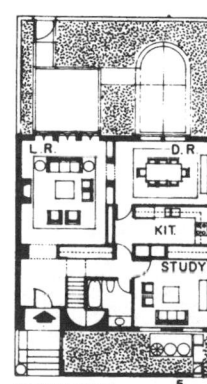

FIRST FLOOR

Fig. 32   Eastover Gates.

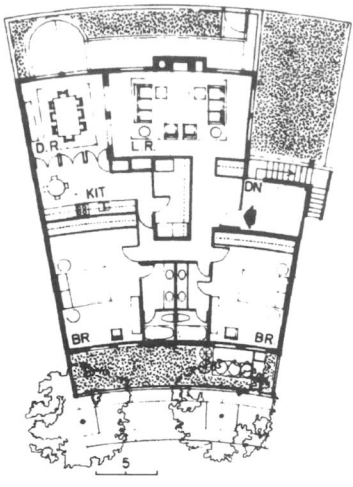

Fig. 33   Eastover Gates, Charlotte, N.C. Wolf Associates—Architect.

Fig. 34   Eastover Gates, Charlotte, N.C. Wolf Associates—Architect.

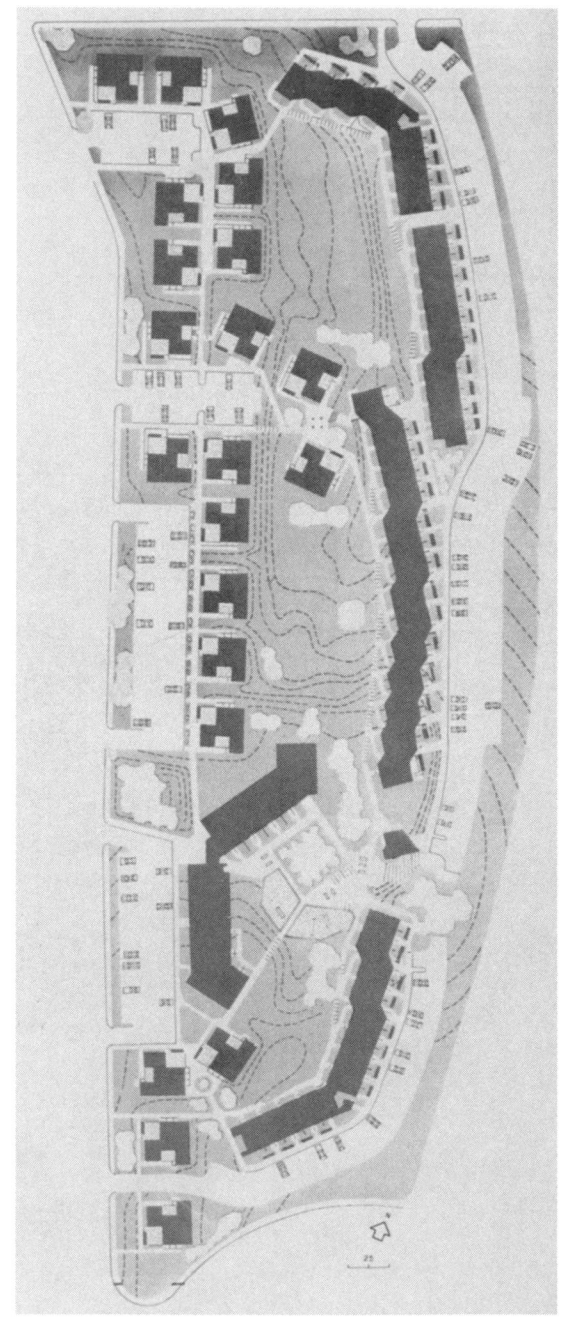

Fig. 35   Whitman Village, Huntington, N.Y. MLTV/Moore-Turnbull—Architects.

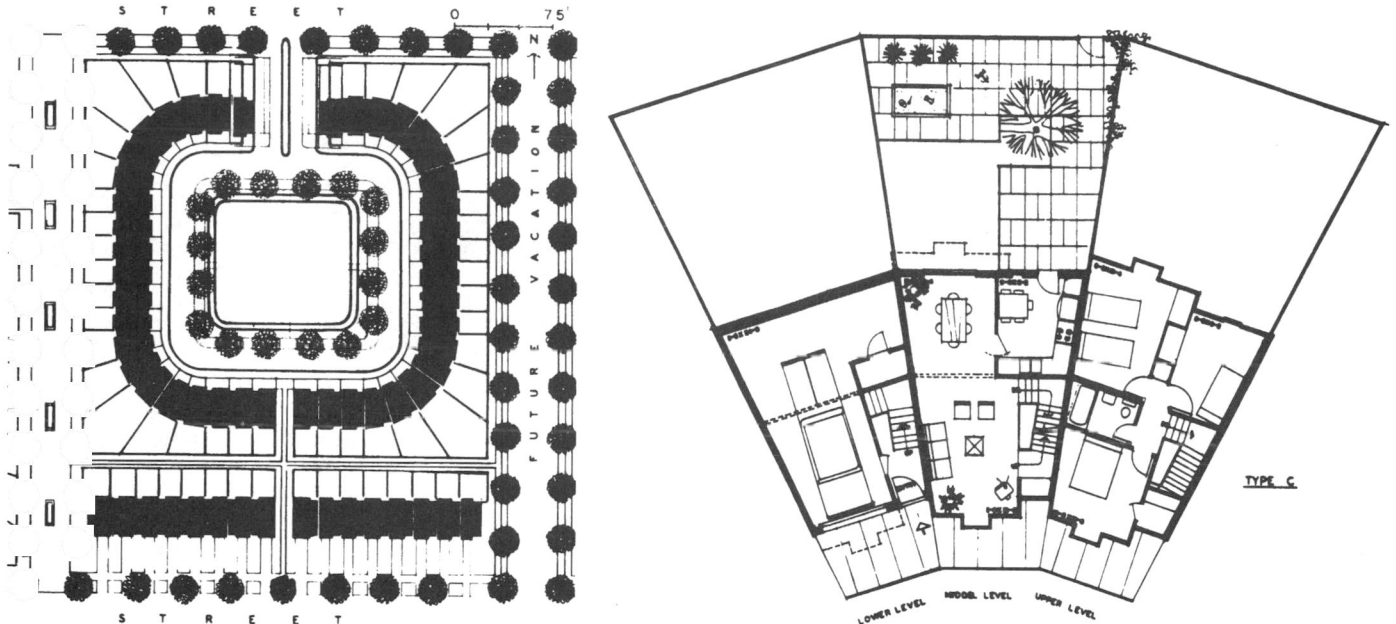

Fig. 36a Plymouth Hill, Milwaukee, Wis. Don M. Hisaka—Architect.

Fig. 36b Plymouth Hill, Milwaukee, Wis. Don M. Hisaka—Architect.

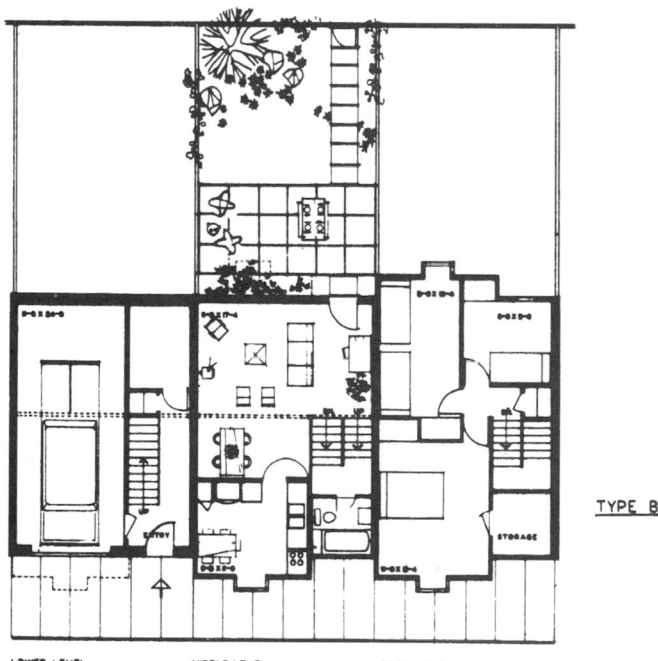

Fig. 36c Plymouth Hill, Milwaukee, Wis. Don M. Hisaka—Architect.

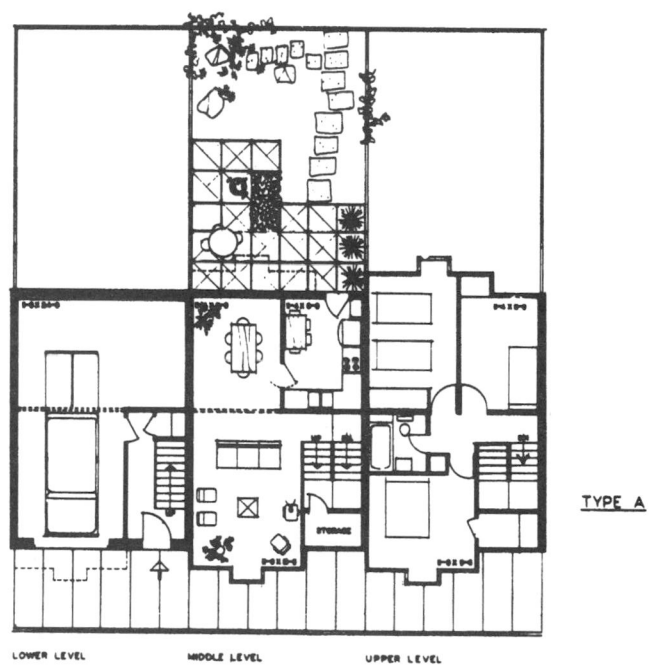

Fig. 36d Plymouth Hill, Milwaukee, Wis. Don M. Hisaka—Architect.

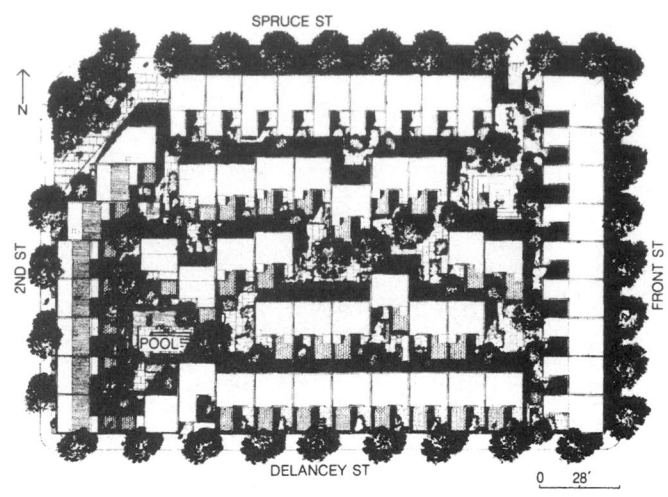

Fig. 37 Site plan. Well-protected inner courtyard can be entered only at three secured places. Penn's Landing, Philadelphia, Pa. Louis Sauer—Architect.

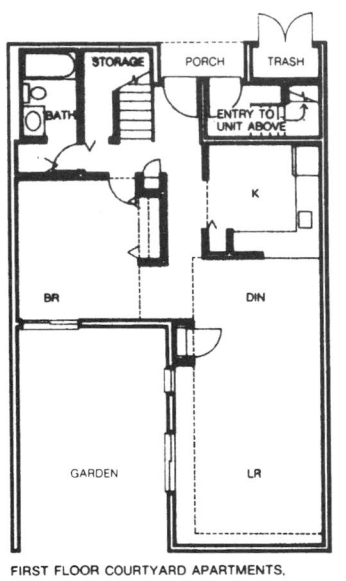

FIRST FLOOR COURTYARD APARTMENTS,
2 BR DUPLEX OVER 2 BR GARDEN DUPLEX

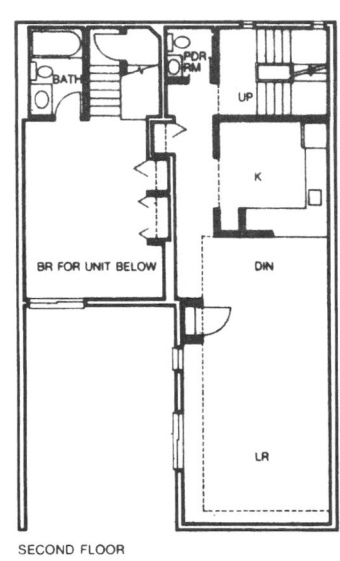

SECOND FLOOR

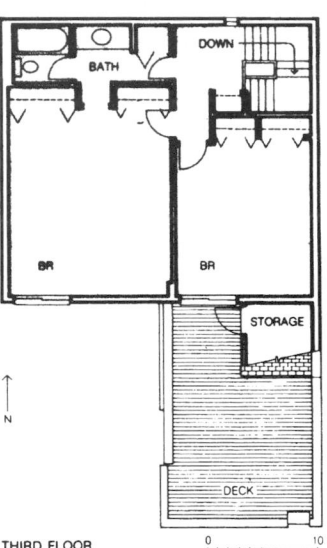

THIRD FLOOR

Fig. 38 Penn's Landing, Philadelphia, Pa. Louis Sauer—Architect.

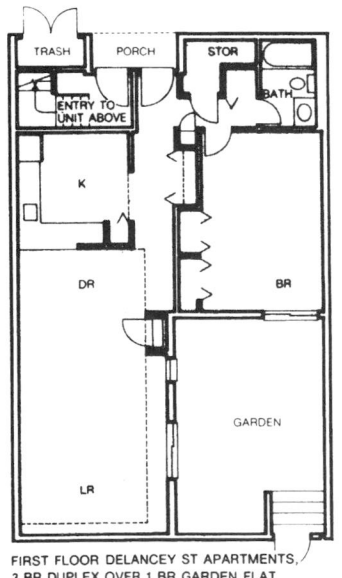

FIRST FLOOR DELANCEY ST APARTMENTS,
3 BR DUPLEX OVER 1 BR GARDEN FLAT

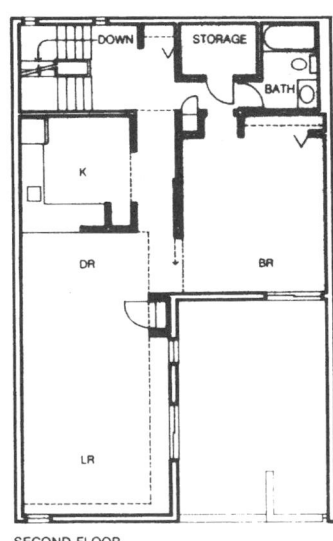

SECOND FLOOR

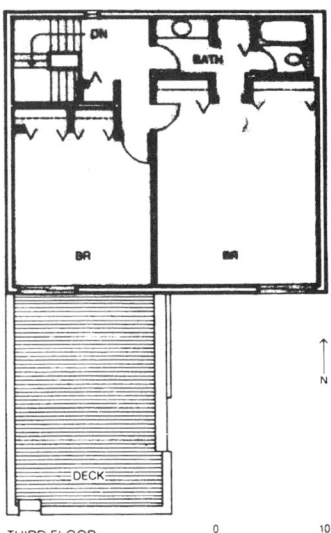

THIRD FLOOR

Fig. 39 Penn's Landing, Philadelphia, Pa. Louis Sauer—Architect.

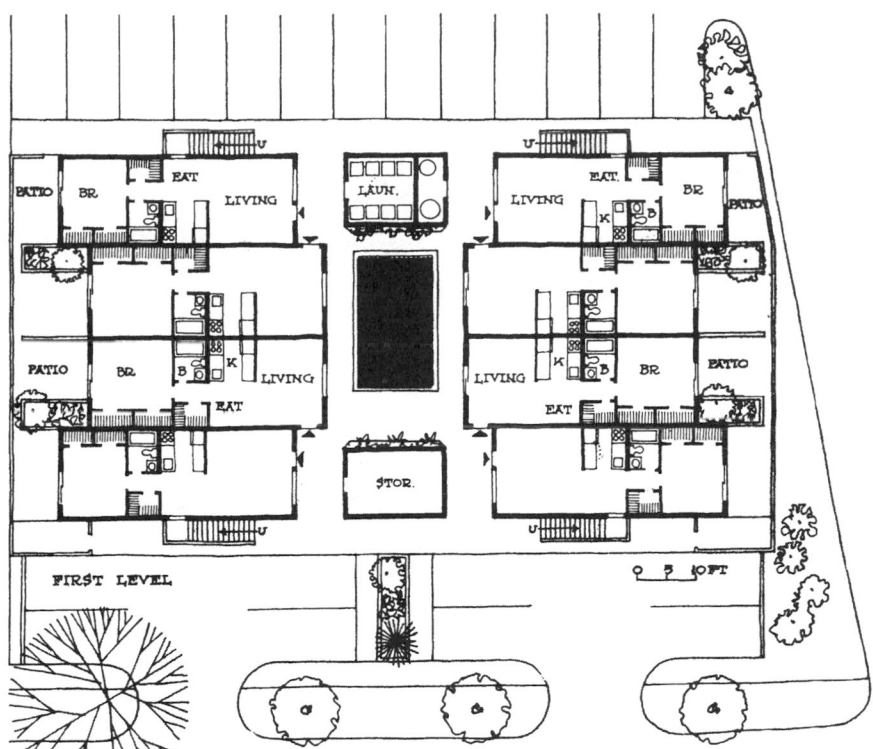

Fig. 40 Baton Rouge, Louisiana. B. & K. Claus—Architects.

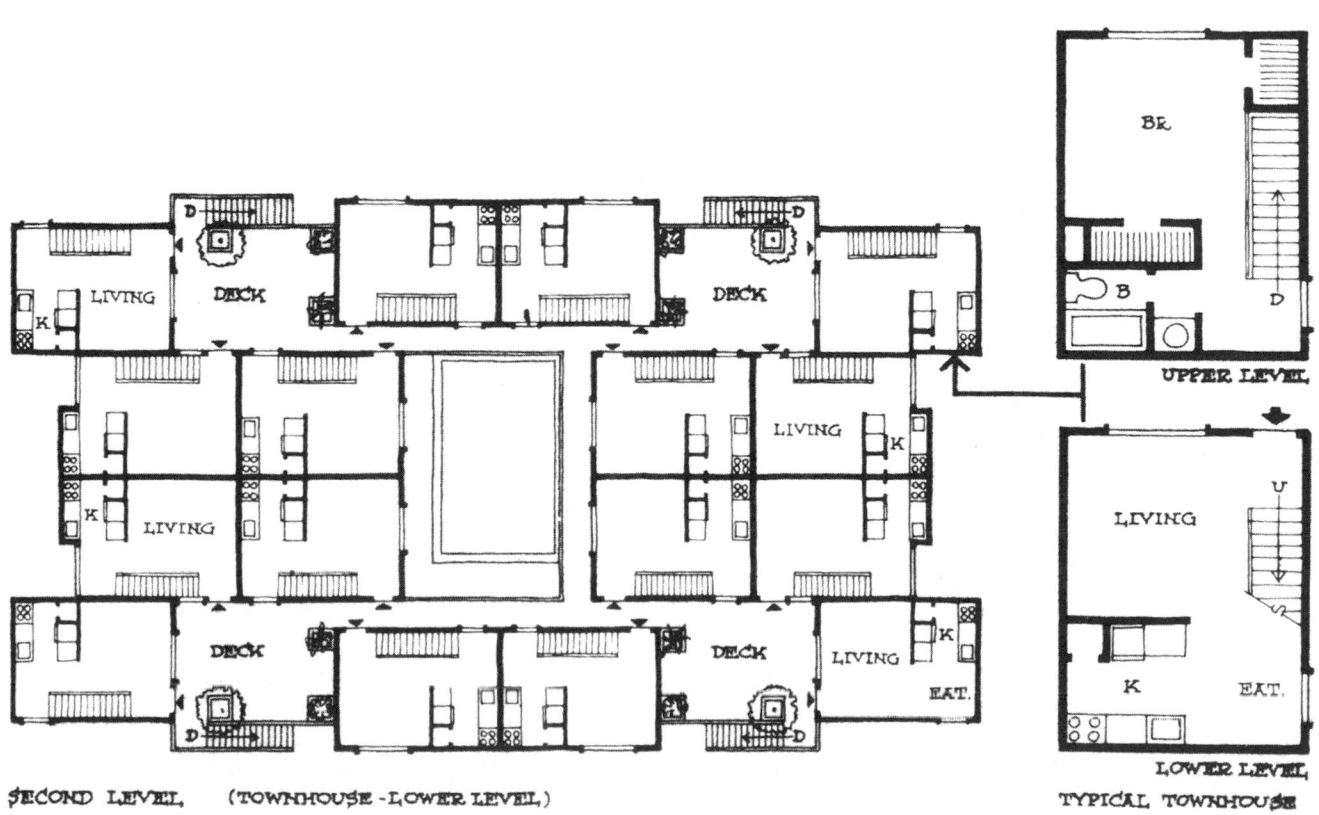

SECOND LEVEL. (TOWNHOUSE - LOWER LEVEL.)

TYPICAL TOWNHOUSE

UPPER LEVEL.

LOWER LEVEL.

Fig. 41

Types of Single-Family Houses

TOWN HOUSES

Three-story town-house units (section and plans in Fig. 42a) have access to grade at both the first and second levels. Most units (site plan, Fig. 42b) face inward on private streets; the higher, three-story units are located along a main thoroughfare, to create a buffer zone.

Fig. 42a Warren Gardens, Roxbury, Mass. Hugh Stubbins & Associates, Ashley Myer & Associates—Architects.

SECTION

STOR

KIT/DINING

LOWER LEVEL

LIVING

BR   BR

MAIN LEVEL

BR

UPPER LEVEL

0   5   10 FT

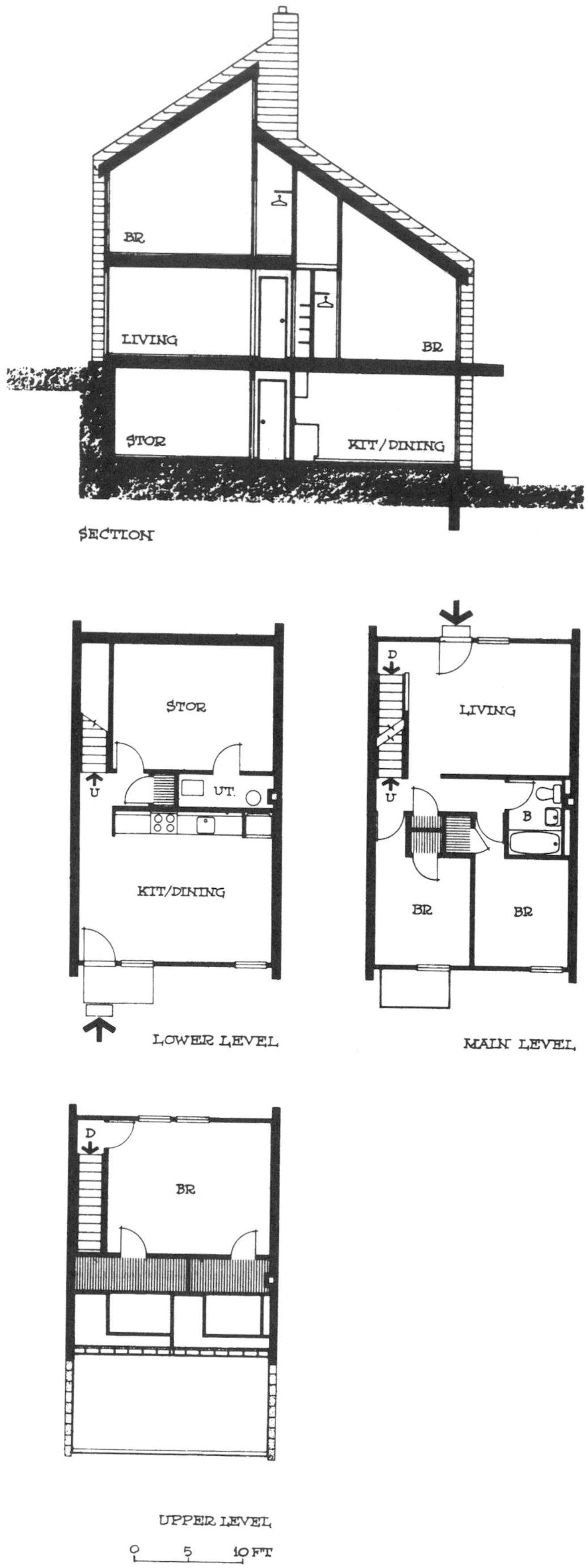

Fig. 42b Warren Gardens, Roxbury, Mass. Hugh Stubbins & Associates, Ashley Myer & Associates—Architects.

Wedge-shaped units follow curved site topography. Piggyback arrangement puts one-story efficiency apartments on the lowest level, with two-story units above. Unit size is 440 to 1425 ft².

Variable density of town houses is shown in site plan of 94 units. Overall density is 9.2 per acre, but siting is tightest along prime waterfront and loosest away from water, where most common area and all manmade amenities are located. Only identical units are joined; project's four plans are so different that mixing them would have meant enormous siting and design problems.

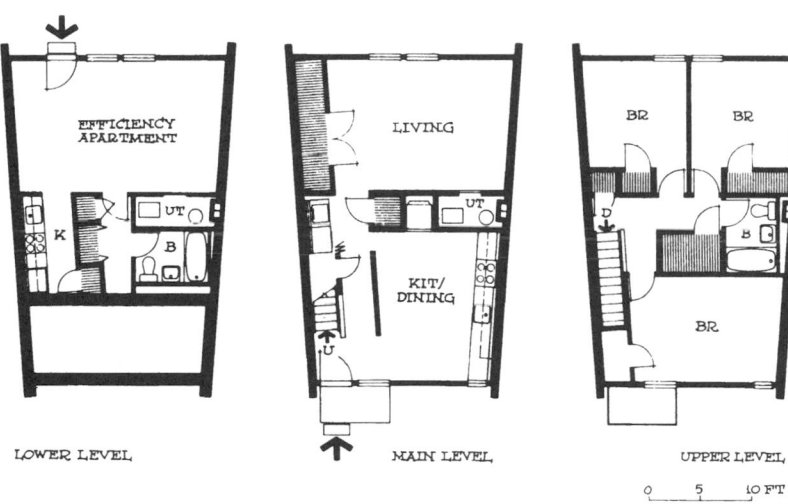

Fig. 43a

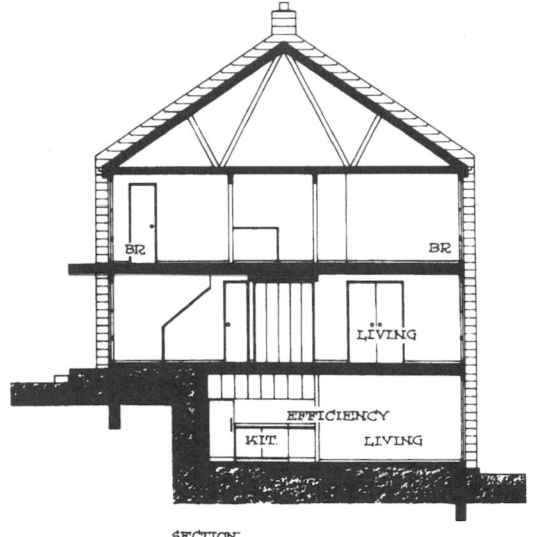

Fig. 43b

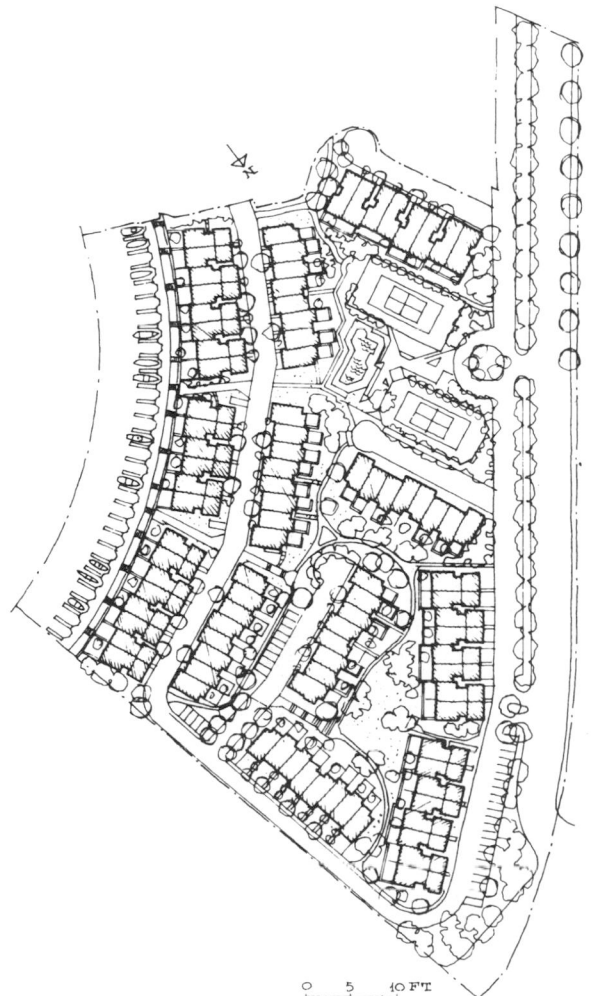

Fig. 44   Seabridge, Huntington Harbor, The Landau Partnership—Architects.

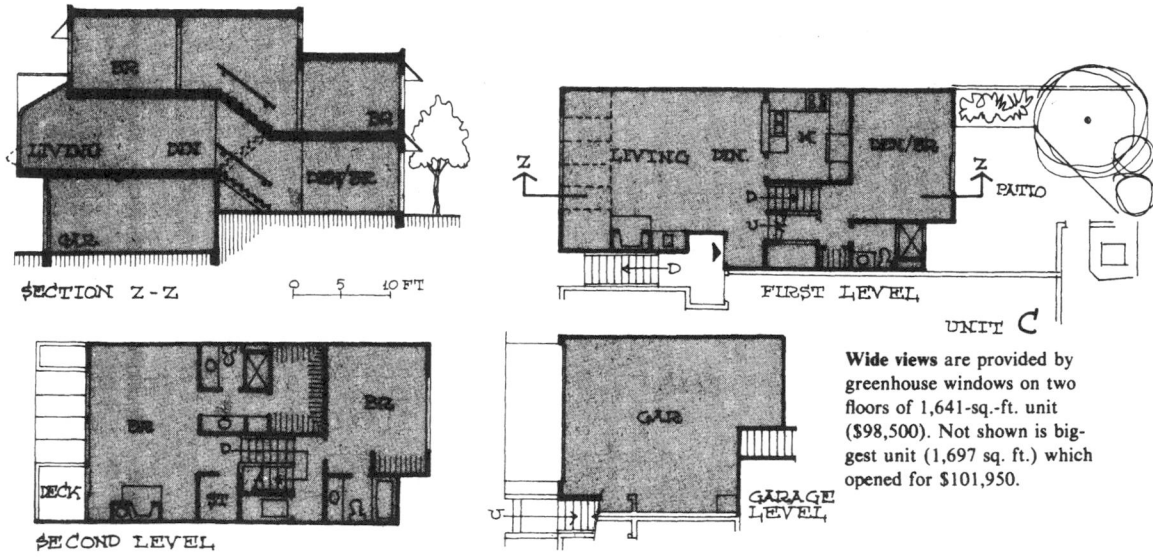

SECTION X-X

SECTION Y-Y

SECOND LEVEL

SECOND LEVEL

FIRST LEVEL

UNIT A   **Waterfront unit** opens to three outdoor areas. Roof over master-suite deck is cut back to give widest possible water views. Unit has 1,620 sq. ft. and opened for $113,400 plus cost of boat slip.

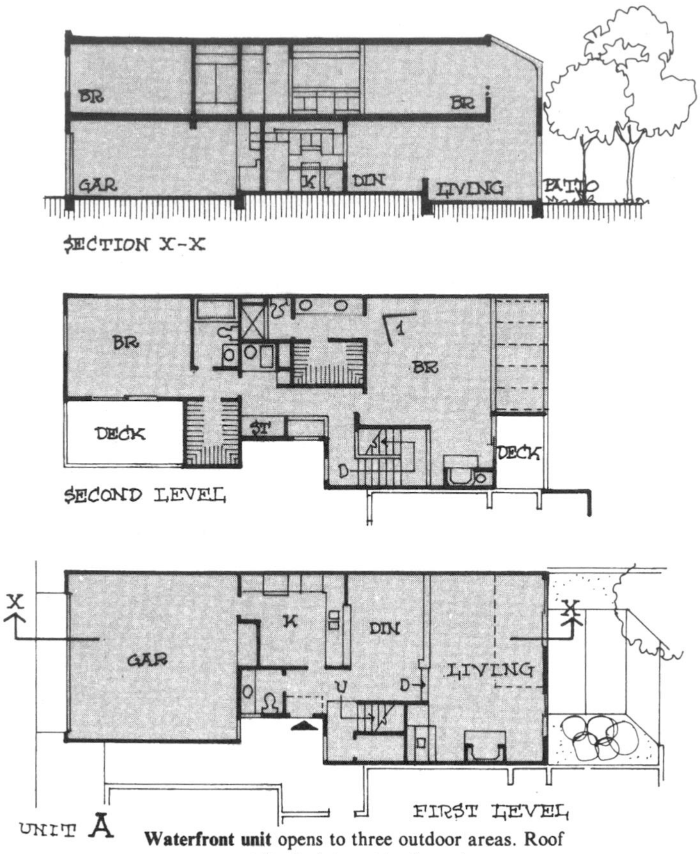

Fig. 45 Seabridge, Huntington Harbor, The Landau Partnership—Architects.

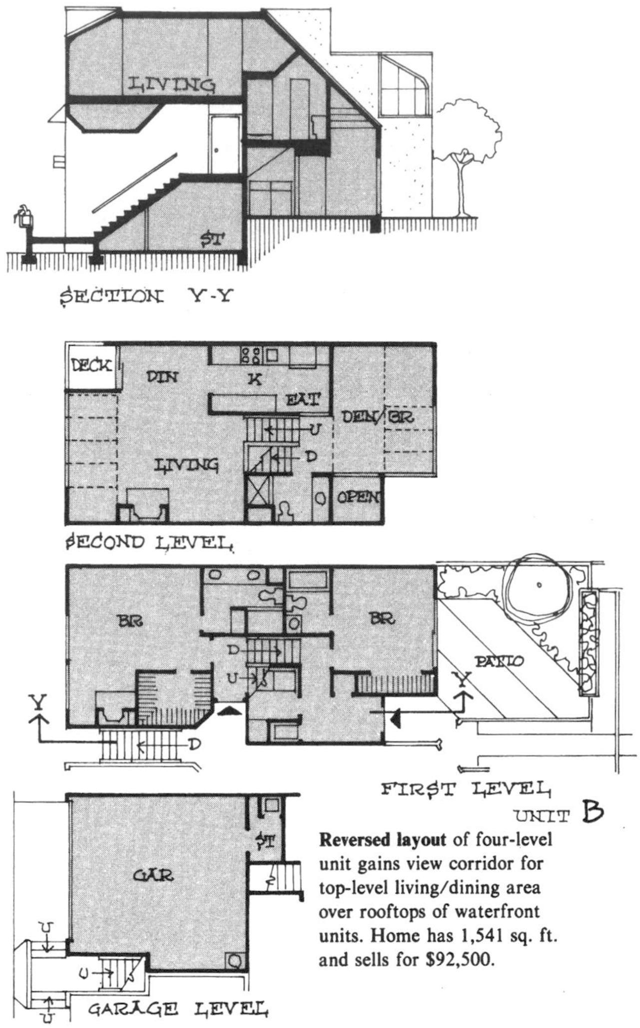

FIRST LEVEL
UNIT B

GARAGE LEVEL

UNIT B   **Reversed layout** of four-level unit gains view corridor for top-level living/dining area over rooftops of waterfront units. Home has 1,541 sq. ft. and sells for $92,500.

Fig. 46 Seabridge, Huntington Harbor, The Landau Partnership—Architects.

SECTION Z-Z   0   5   10 FT

FIRST LEVEL

UNIT C

SECOND LEVEL

GARAGE LEVEL

Wide views are provided by greenhouse windows on two floors of 1,641-sq.-ft. unit ($98,500). Not shown is biggest unit (1,697 sq. ft.) which opened for $101,950.

Fig. 47 Seabridge, Huntington Harbor, The Landau Partnership—Architects.

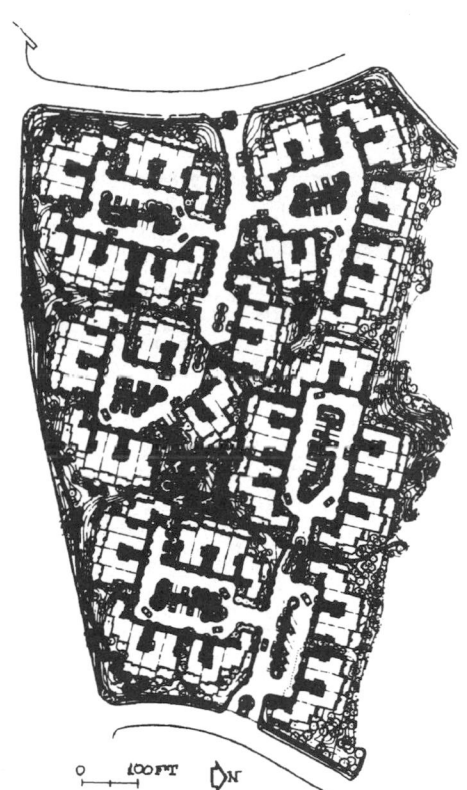

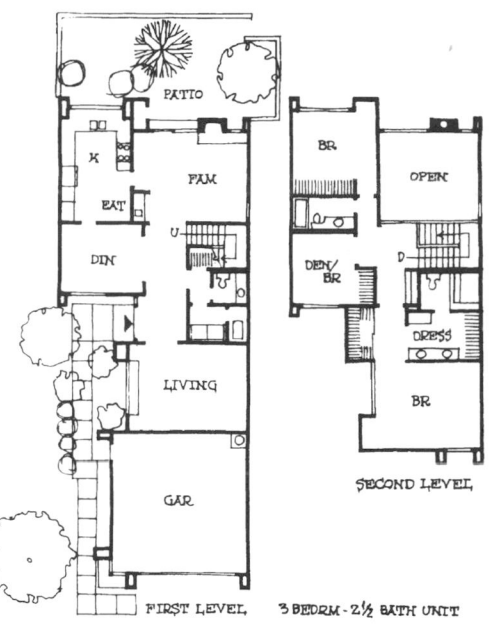

Fig. 48*b* 2200 ft². University Park Town Houses, Irvine, Calif.
Richardson, Nagy, Martin—Architects.

Fig. 48*a* University Park Town Houses, Irvine, Calif.
Richardson, Nagy, Martin—Architects.

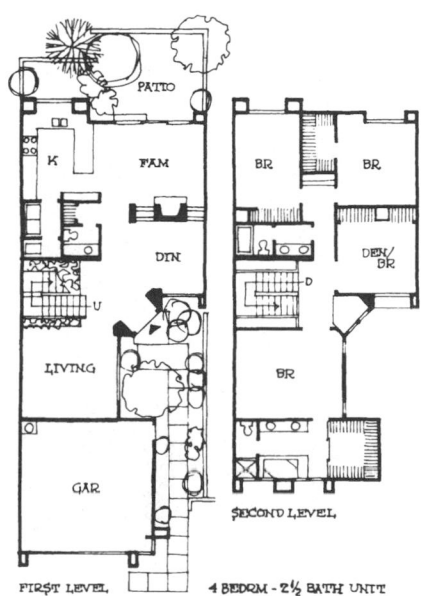

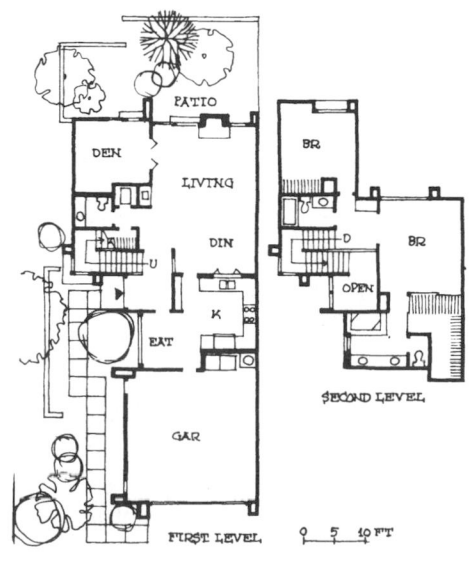

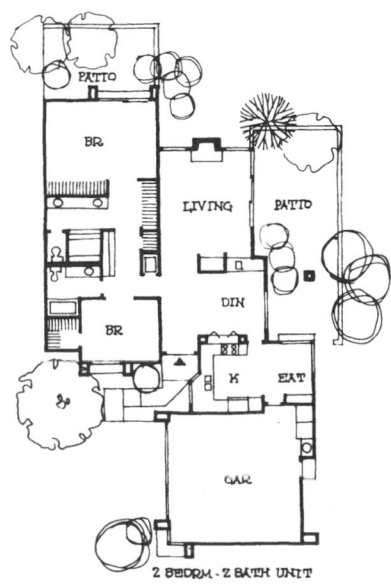

Fig. 48*c* 2550 ft². University Park Town Houses,
Irvine, Calif. Richardson, Nagy, Martin—Architects.

Fig. 48*d* 1750 ft². University Park Town Houses,
Irvine, Calif. Richardson, Nagy, Martin—Architects.

Fig. 48*e* 1488 ft². University Park Town
Houses, Irvine, Calif. Richardson, Nagy,
Martin—Architects.

529

TOWN HOUSES

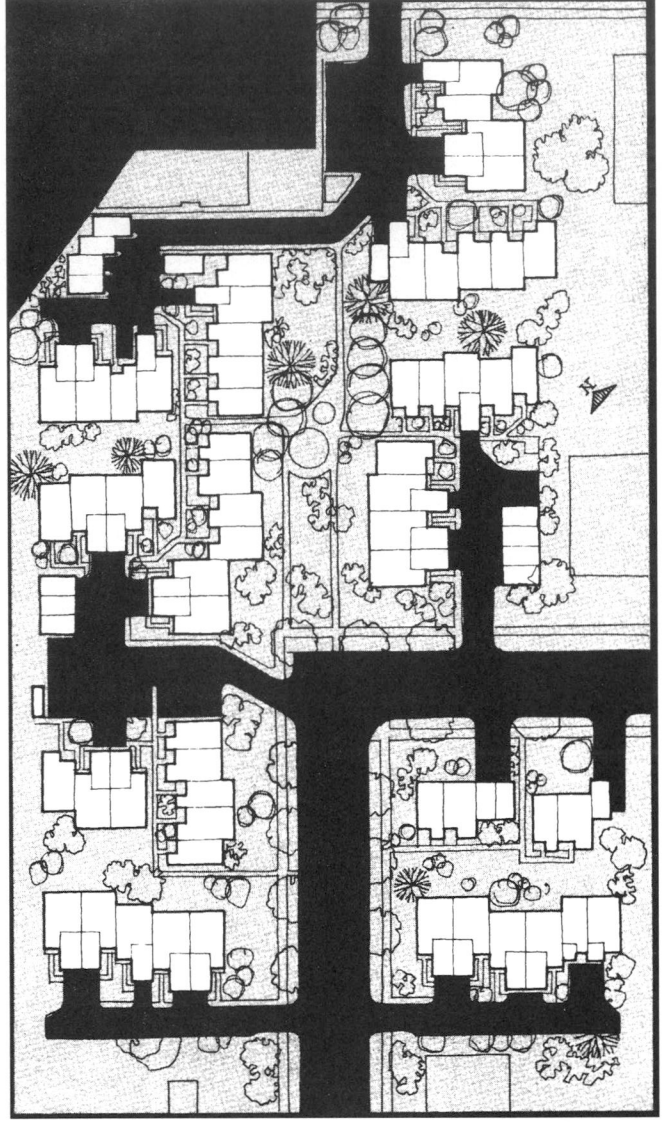

High-density plan—87 units per acre—puts 189 condo town houses (plans in Fig. 48) and flats (plans in Fig. 50) on 2.17-acre site. Town-house buildings are four stories; apartment buildings are seven and nine stories. Central plaza is paved, landscaped, and waterscaped. Parking is underground.

Apartment plans offer wide variety of room layouts and shapes. Most units open to private outdoor areas—patios at ground level, concrete decks on second floor, balconies at other levels. All units on top three floors have fireplaces. Six of nine available plans are keyed to building plans.

Fig. 49a Town houses, Minneapolis, Minn. Miller, Hanson, Westerbeck, Inc.—Architects.

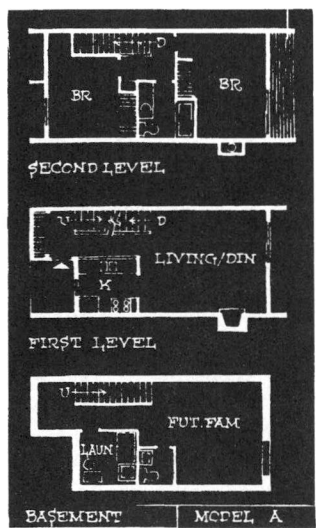

Fig. 49b Town houses, Minneapolis, Minn. Miller, Hanson, Westerbeck, Inc.—Architects.

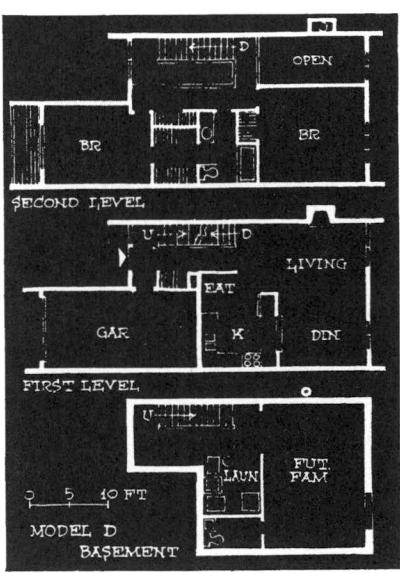

Fig. 49c Town houses, Minneapolis, Minn. Miller, Hanson, Westerbeck, Inc.—Architects.

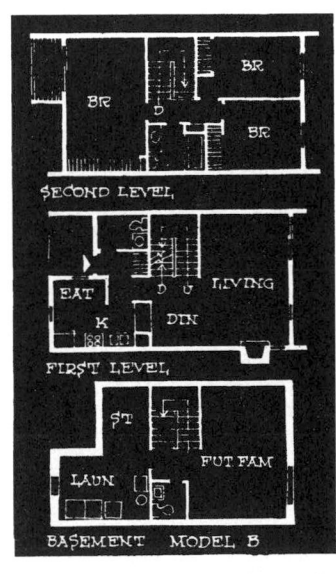

Fig. 49d Town houses, Minneapolis, Minn. Miller, Hanson, Westerbeck, Inc.—Architects.

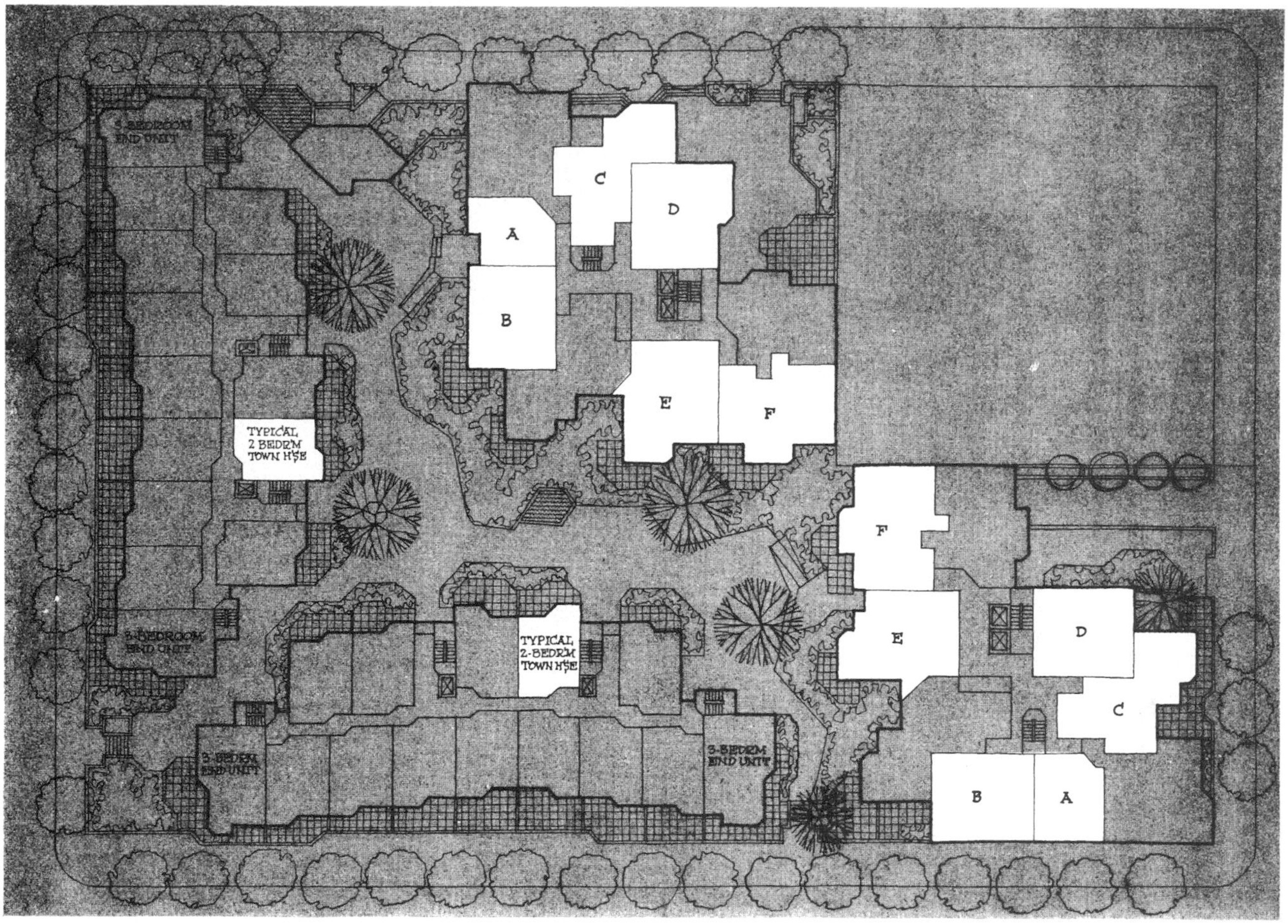

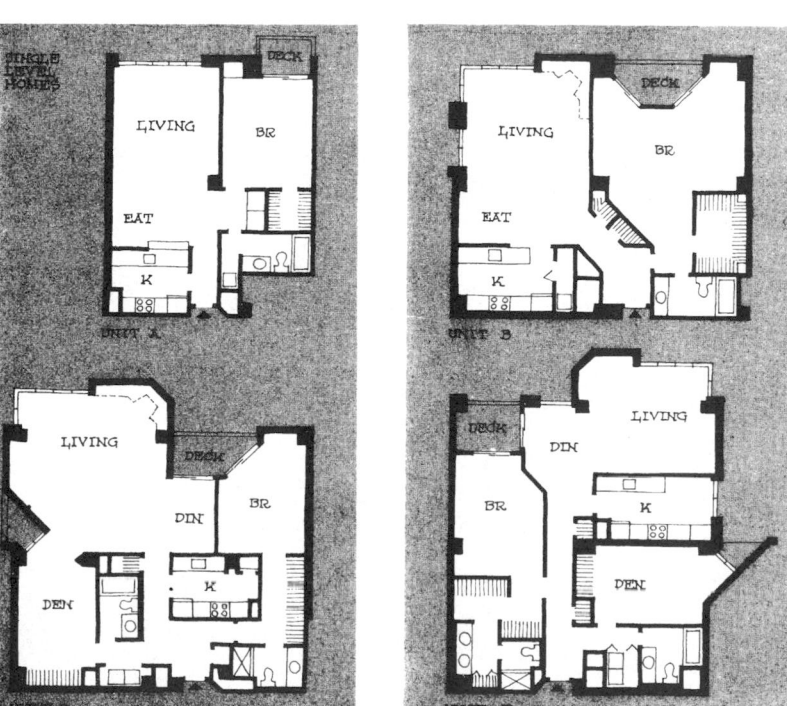

Fig. 50 Telegraph Landing, San Francisco, Calif. Bull, Field, Volkmann & Stockwell—Architects.

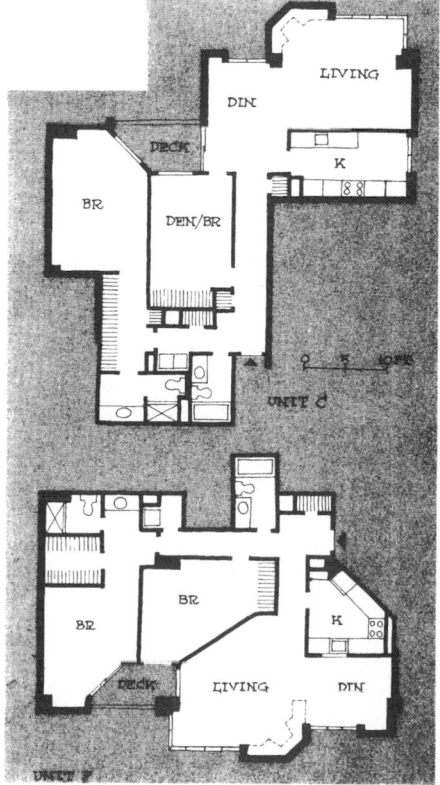

Fig. 51 Telegraph Landing, San Francisco, Calif. Bull, Field, Volkmann & Stockwell—Architects.

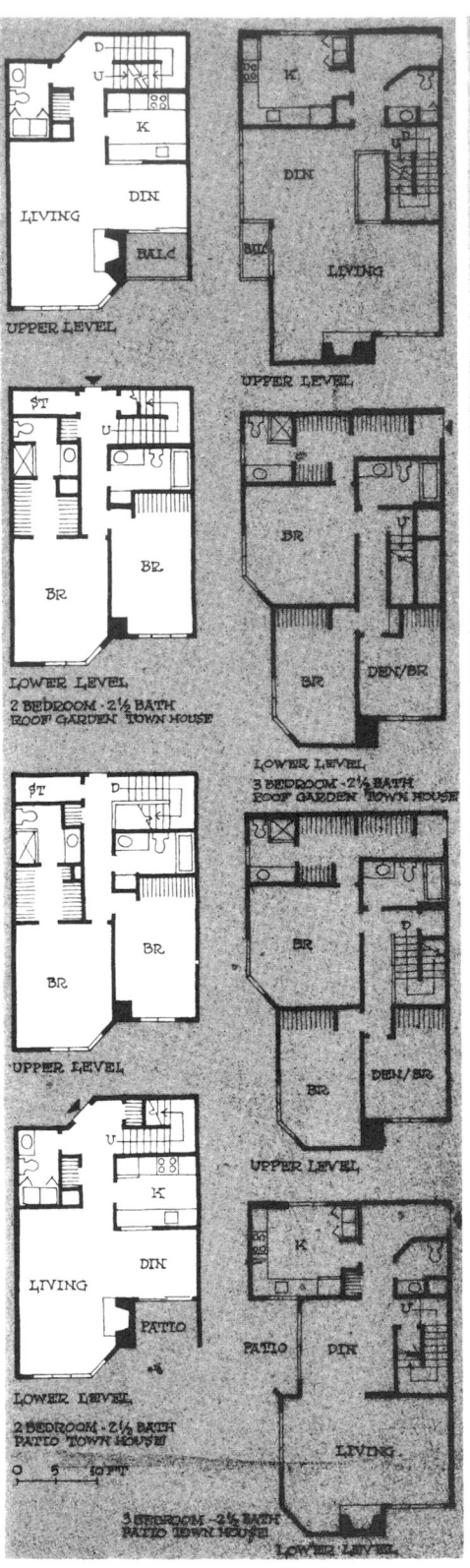

Fig. 52 Hancock Square, Los Angeles, Calif. Richardson, Nagy, Martin—Architects.

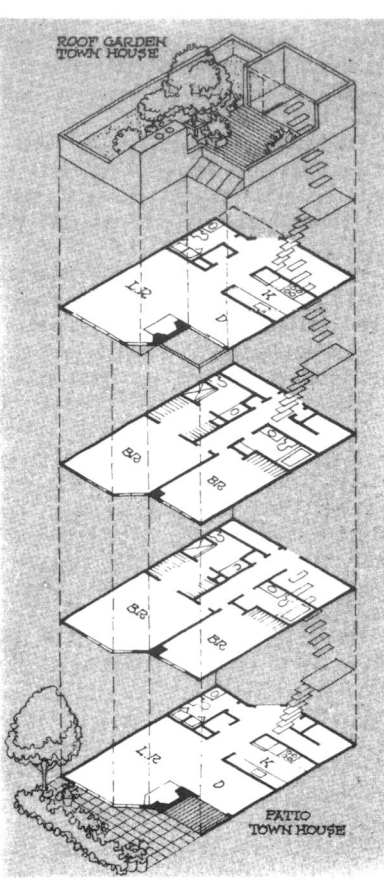

Two-level town houses are stacked two high in both of the project's four-story buildings. Each building contains 28 townhouses—24 with two bedrooms and four end units with three bedrooms. Bedrooms are on the upper level of each bottom unit and on the lower level of each top unit (see Fig. 52). This arrangement reduces potential noise problems by isolating the bedrooms from the active living areas of upstairs or downstairs neighbors. It also gives active-living levels direct access to private outdoor areas—an enclosed patio off the bottom unit and a roof garden above the top unit. The dining room of the top unit also opens to a balcony.

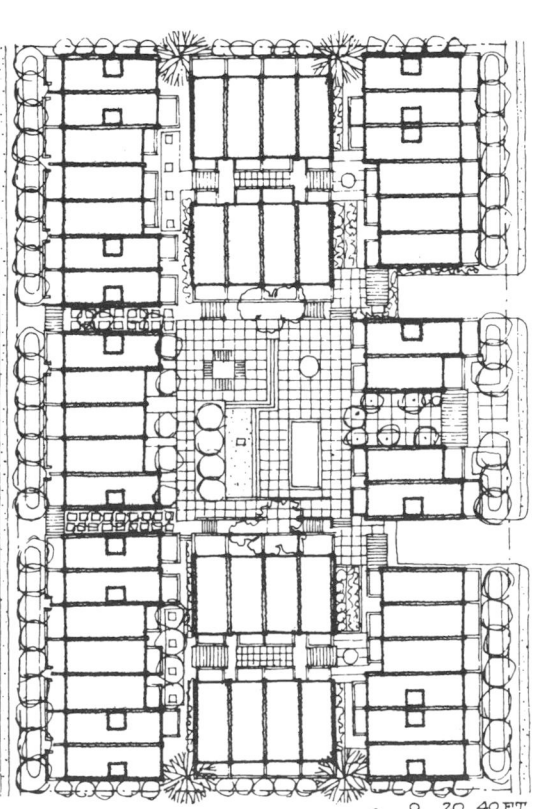

Fig. 53a Site plan. Hancock Square, Los Angeles, Calif. Richardson, Nagy, Martin—Architects.

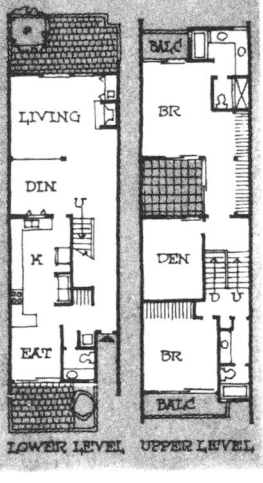

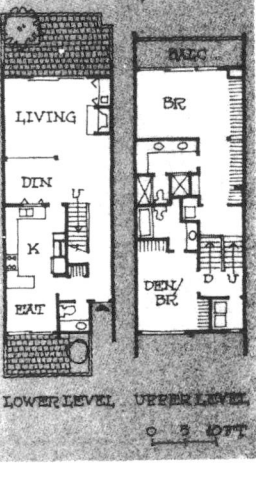

Fig. 53b Typical units. Hancock Square, Los Angeles, Calif. Richardson, Nagy, Martin—Architects.

Town-house clusters comprise five units sited so as to provide complete privacy for each owner. Two-story privacy walls isolate the patios and balconies of adjoining units.

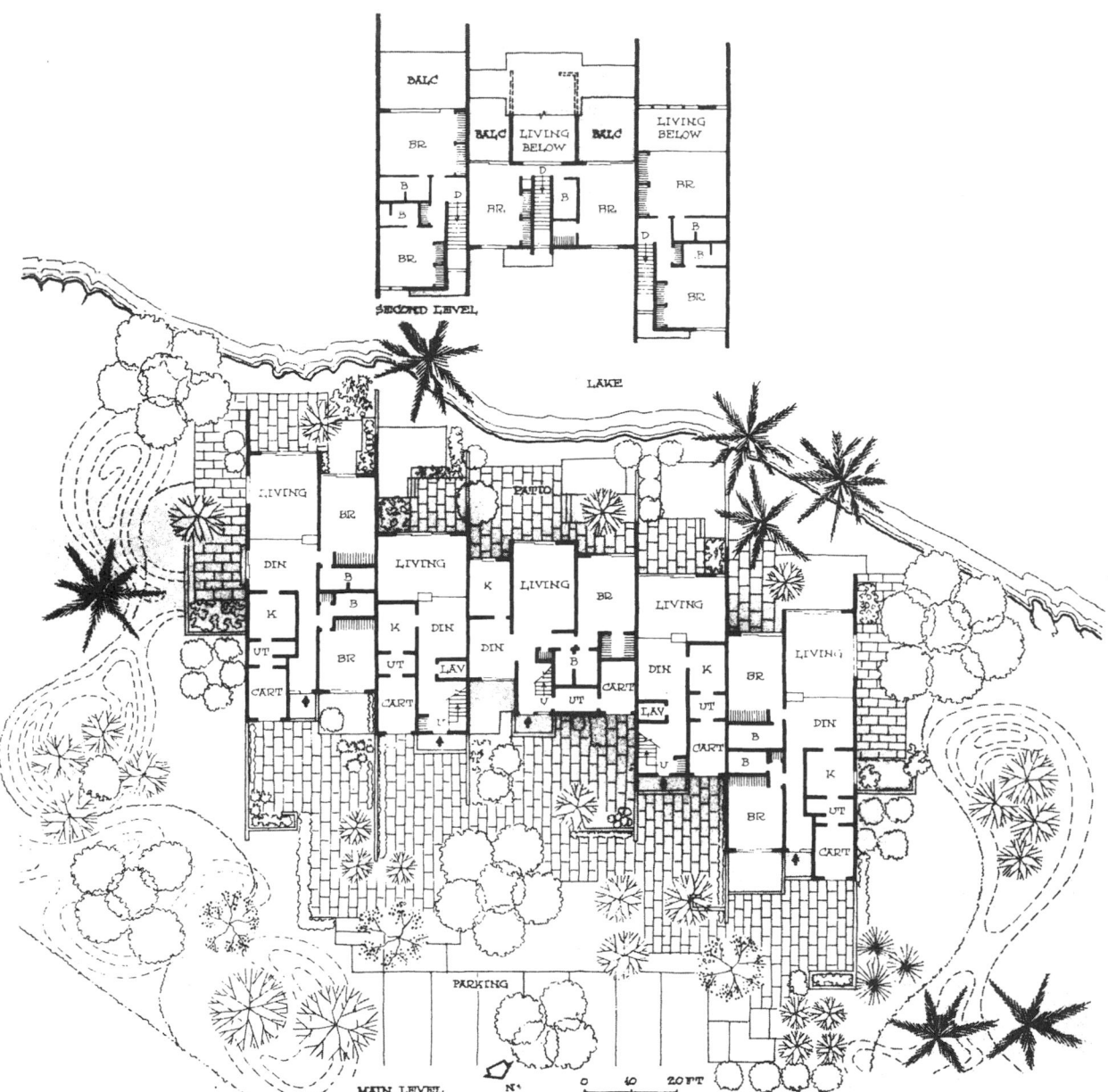

Fig. 54 Ocean Reef Club, North Key Largo, Fla. Hatcher, Zeiger, Gunn & Associates—Architects.

TOWN HOUSES

Waterfront Development

The site plan in Fig. 55a shows how a large portion of the 9-acre site was turned into a big lake, with the town houses sited on islands and promontories.

The two plans in Fig. 55b include an 829-ft² efficiency with a loft and a 1135 ft², two-bedroom unit, of which there were only 18.

The site is 24 acres of orchard land bounded by a commuter rail line to the north, a major highway to the east, and a group of single-family houses to the south. The township has proud Revolutionary associations and so was sensitive to the potential impact of this 220-unit cluster on the community. The architects therefore worked especially closely with the township to accommodate their wishes in matters of siting and development.

Paved and built-upon areas were arranged to preserve the existing landscape wherever possible. Finish materials and building forms were designed to respond to the region's historical character. Two entrances have been provided to reduce on-site vehicular traffic and apartment units form cul-de-sac clusters off the main loop road. The original farmhouse was retained to give the new community a firm historical centerpiece.

The orientation of the apartments is away from the railroad and the main highway and toward a brook that runs through the site.

The two-story town houses in Fig. 58 are skillfully arranged at relatively high land-use intensity. Each unit has its own carport and two fenced patios. Though space for recreation building and pool is minimal, good site planning, well-designed auxiliary open areas, and effective planting provide a high degree of livability and visual appeal. The plan provides 11.0 living units per gross acre with 1570 ft² units.

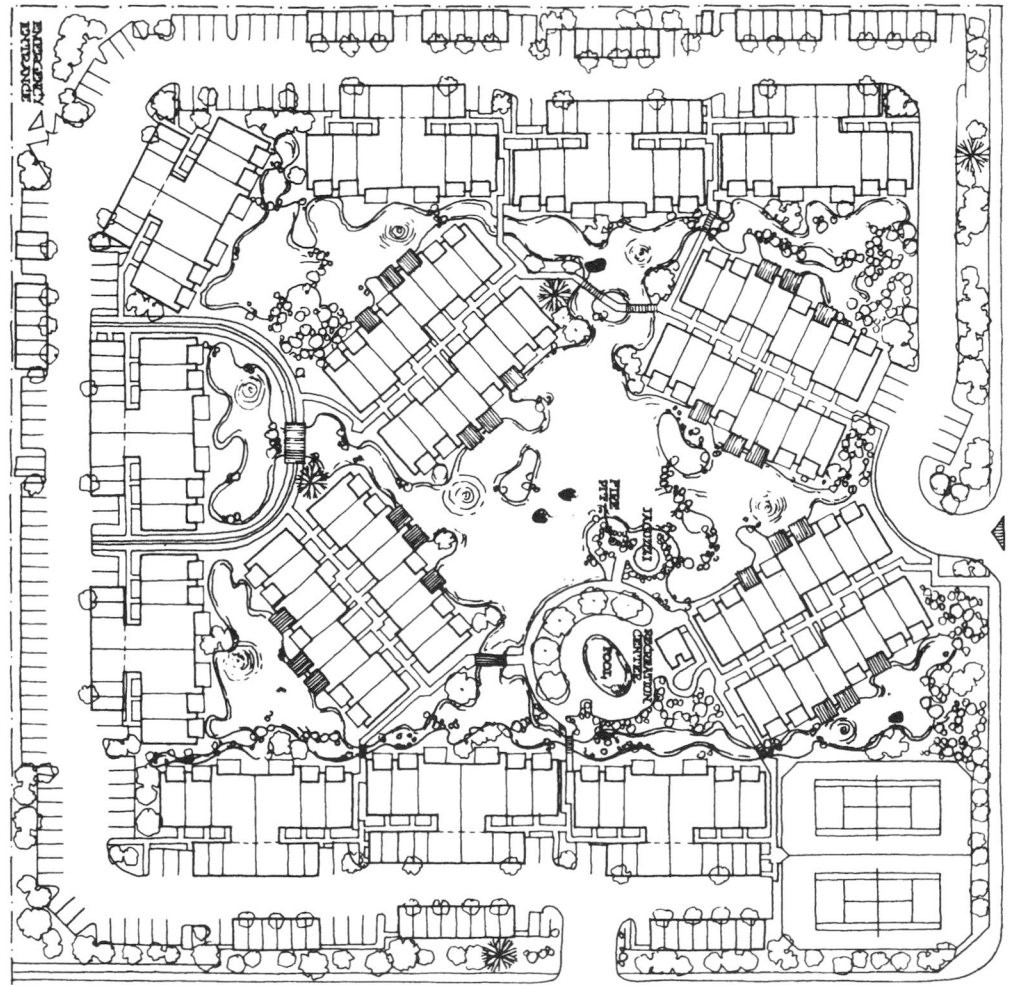

Fig. 55a   Site plan. Orange Lakes, Orange, California. Carl McLarand & Associates—Architects.

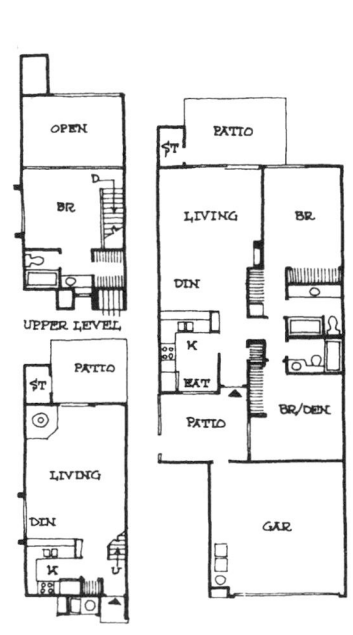

Fig. 55b Two floor plans. Orange Lakes, Orange, Calif. Carl McLarand & Associates—Architects.

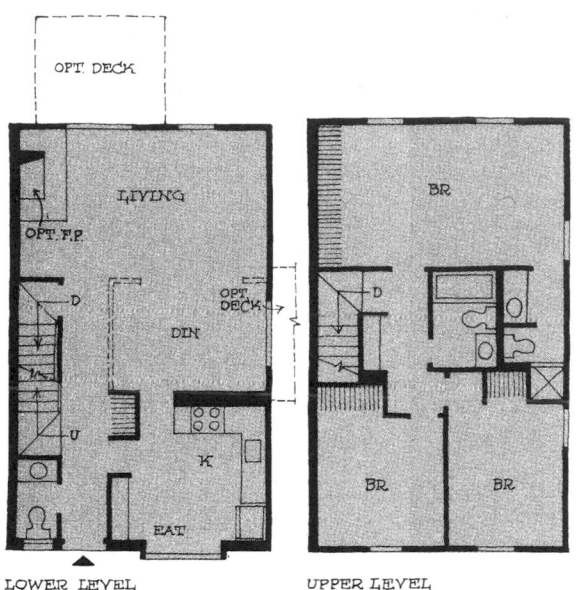

Fig. 56a Front-kitchen model. Montpelier Oaks, Laurel, Md. Victor Smolen & Associates—Architects.

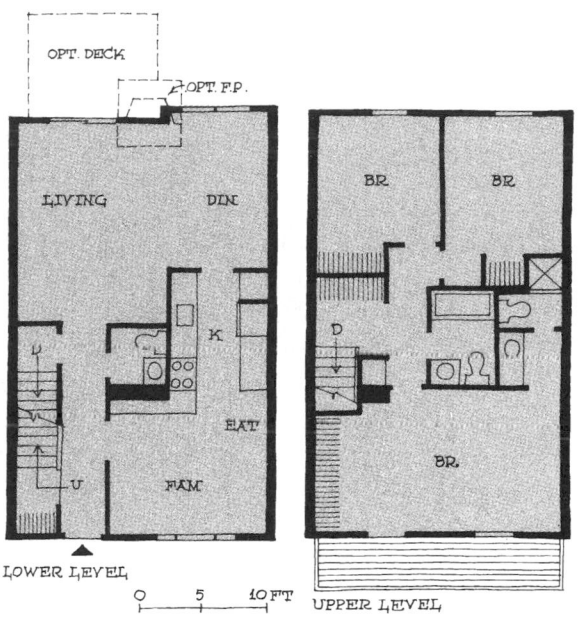

Fig. 56b Center-kitchen model. Montpelier Oaks, Laurel, Md. Victor Smolen & Associates—Architects.

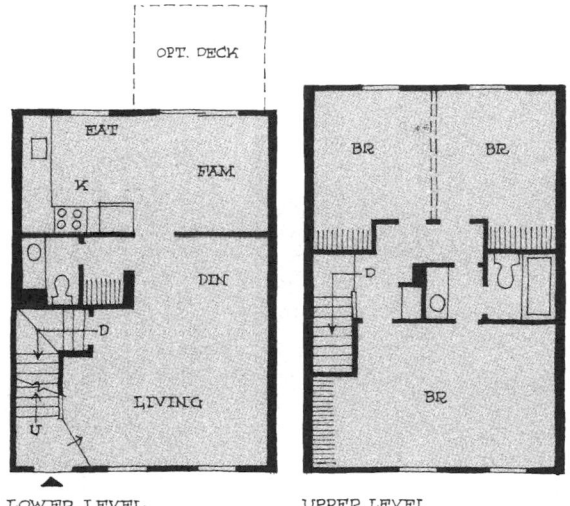

Fig. 56c Rear-kitchen model. Montpelier Oaks, Laurel, Md. Victor Smolen & Associates—Architects.

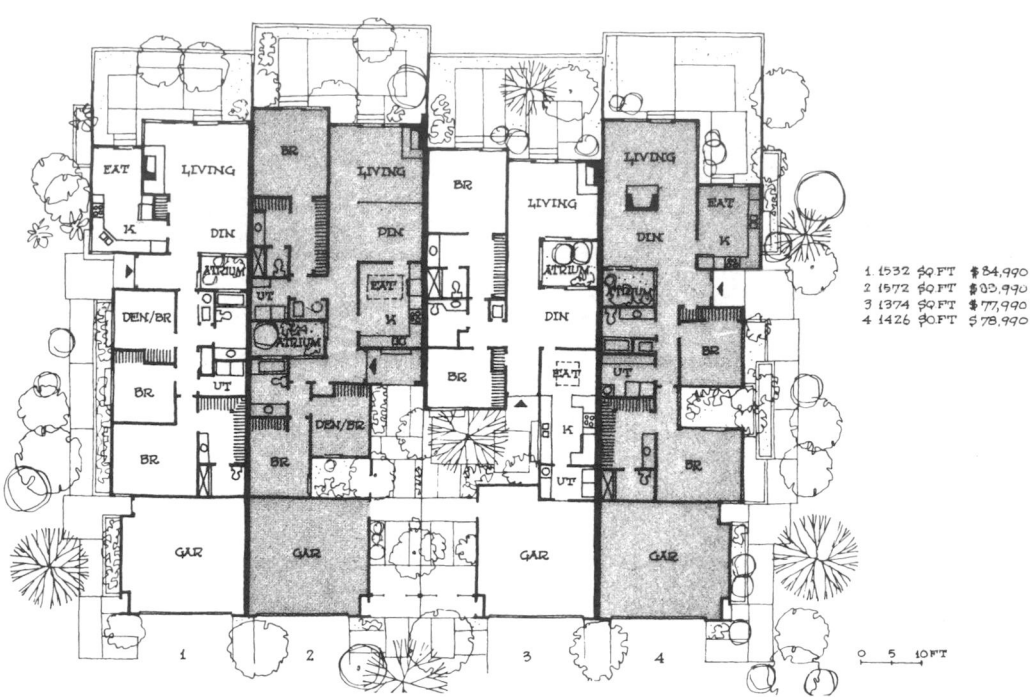

1. 1532 SQ FT $84,990
2. 1572 SQ FT $83,990
3. 1374 SQ FT $77,990
4. 1426 SQ FT $78,990

Fig. 57 Crow Canyon, Contra Costa County, Calif. Morris Lohrbach—Architect.

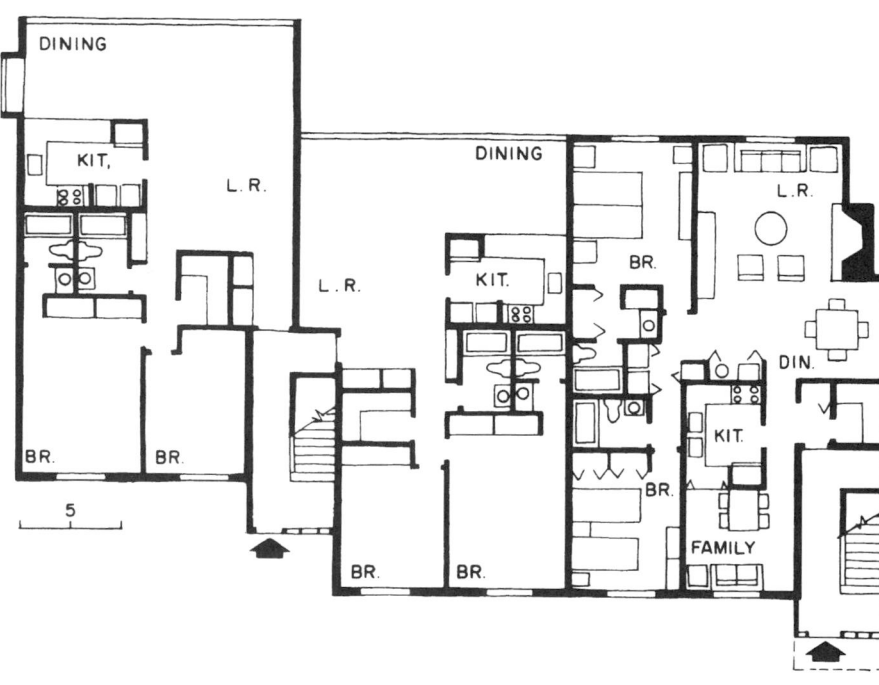

Fig. 58b Typical plans. Concord Greene Apartments.

Fig. 58a Site plan. Concord Greene Apartments.

Fig. 58c Section. Concord Greene Apartments. Concord, Mass. Huygens & Tappé—Architects.

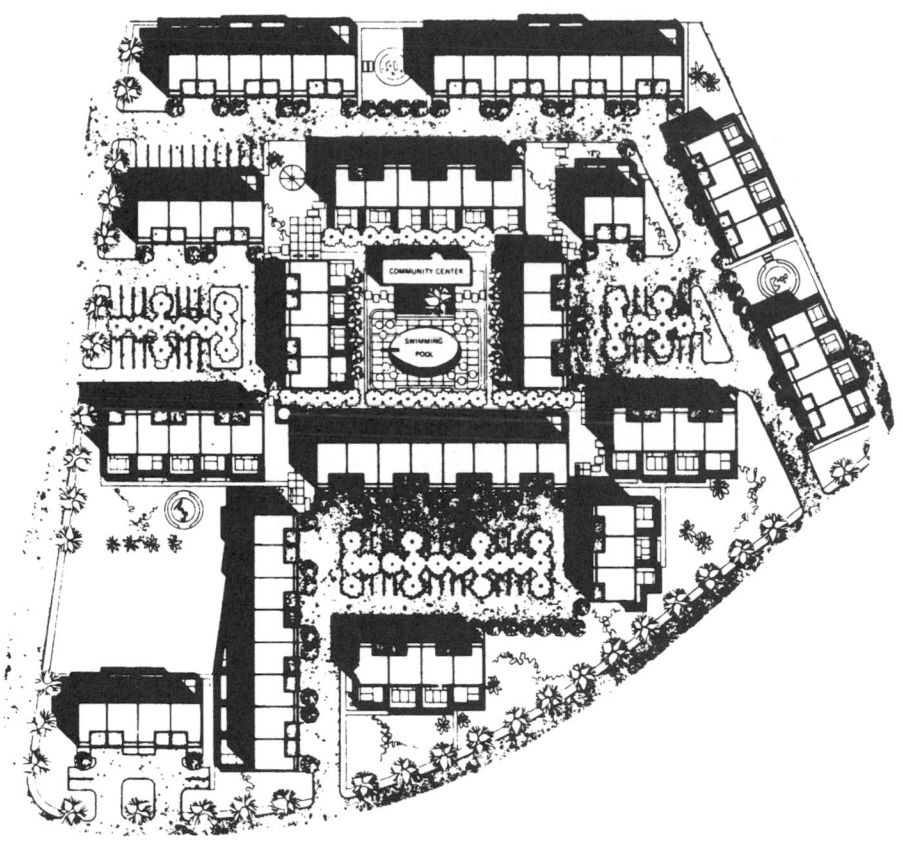

Fig. 59

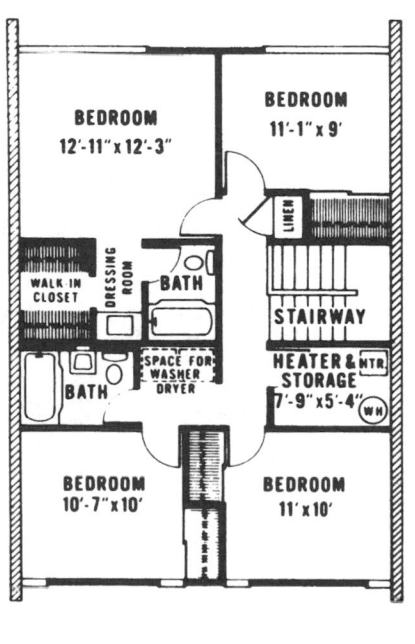

PATIO

PARLOUR & DINING
24'-4"x10', x12'-3"

SPACE
FOR
REFRIG

RANGE

GARB
DISP

POWDER
ROOM

STAIRWAY

OVEN

KITCHEN

DISHWR

FOYER

STORAGE

REFUSE

STORAGE

MULTI-PURPOSE
10'-8"x11'-4"

CARPORT

PATIO

FIRST FLOOR

BEDROOM
12'-11" x 12'-3"

BEDROOM
11'-1" x 9'

LINEN

WALK-IN
CLOSET

DRESSING
ROOM

BATH

STAIRWAY

BATH

SPACE FOR
WASHER
DRYER

HEATER & MTR.
STORAGE
7'-9"x5'-4"

WH

BEDROOM
10'-7" x 10'

BEDROOM
11' x 10'

SECOND FLOOR

Fig. 60

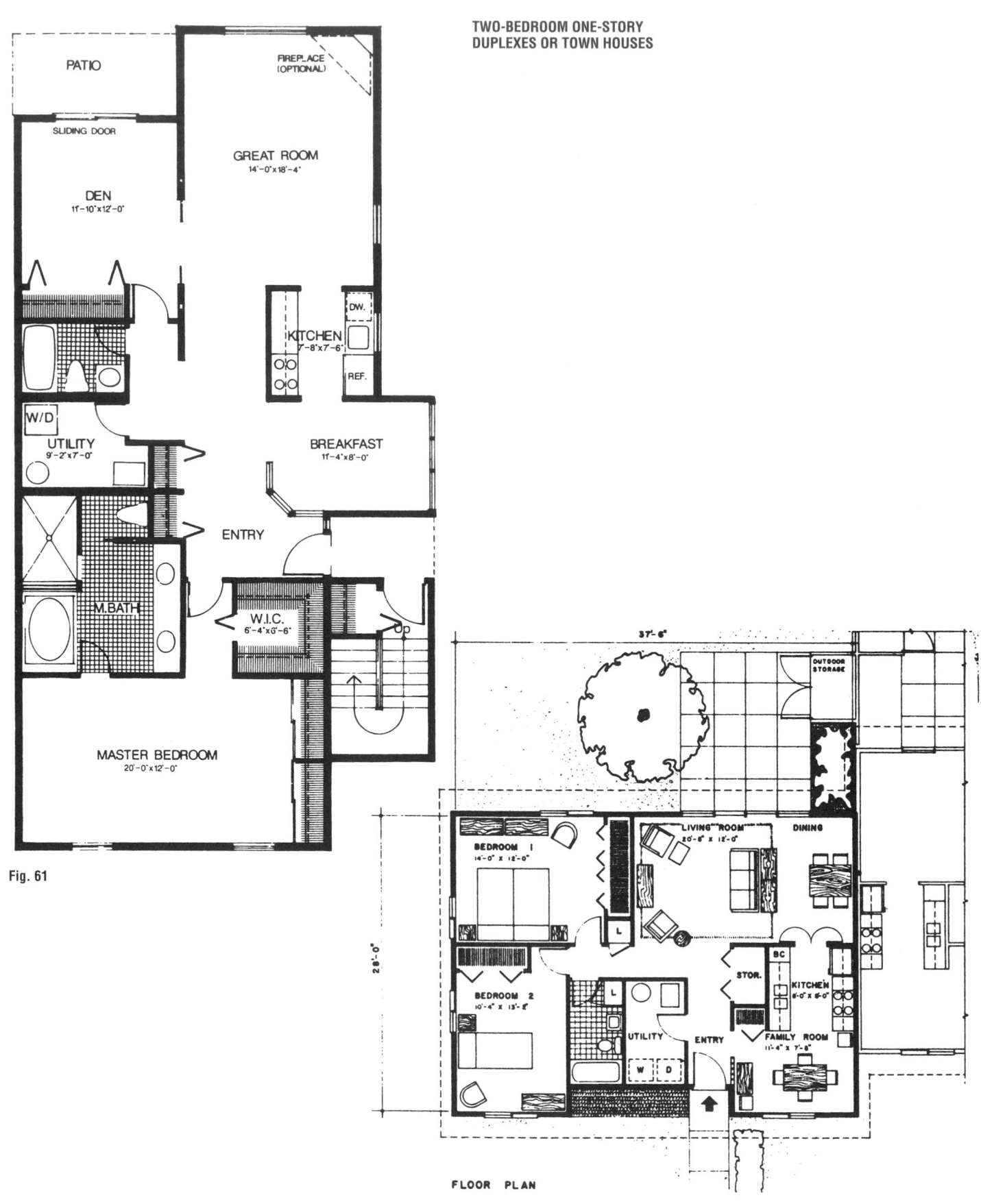

PATIO

FIREPLACE
(OPTIONAL)

SLIDING DOOR

GREAT ROOM
14'-0"x18'-4"

DEN
11'-10"x12'-0"

W/D

UTILITY
9'-2"x7'-0"

KITCHEN
7'-8"x7'-6"

DW.

REF.

BREAKFAST
11'-4"x8'-0"

ENTRY

M.BATH

W.I.C.
6'-4"x6'-6"

Up

MASTER BEDROOM
20'-0"x12'-0"

Fig. 61

TWO-BEDROOM ONE-STORY
DUPLEXES OR TOWN HOUSES

37'-6"

OUTDOOR
STORAGE

28'-0"

BEDROOM 1
14'-0" X 12'-0"

LIVING ROOM
20'-8" X 12'-0"

DINING

BEDROOM 2
10'-4" X 13'-8"

L

L

STOR.

BC

KITCHEN
8'-0" X 8'-0"

UTILITY

ENTRY

FAMILY ROOM
11'-4" X 7'-8"

W   D

FLOOR PLAN

GROSS AREA   1065 SQ. FT.
NET AREA   930 SQ. FT.

Fig. 62

**TWO-BEDROOM TWO-STORY
TOWN HOUSES**

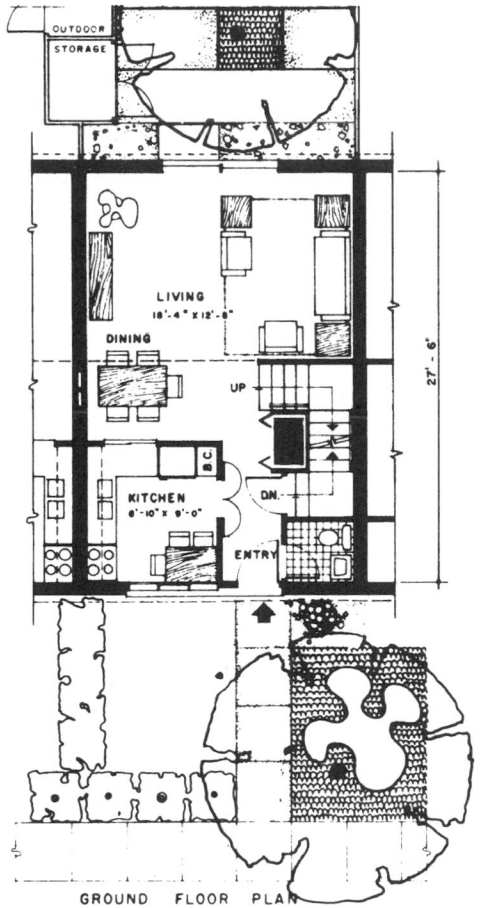

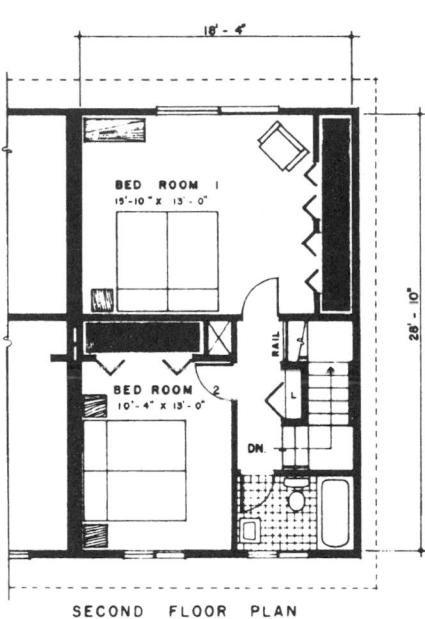

SECOND FLOOR PLAN

GROUND FLOOR PLAN

Fig. 63

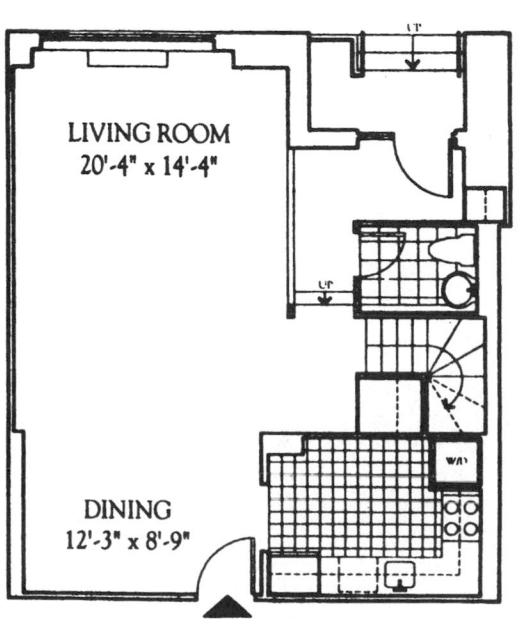

LIVING ROOM
20'-4" x 14'-4"

DINING
12'-3" x 8'-9"

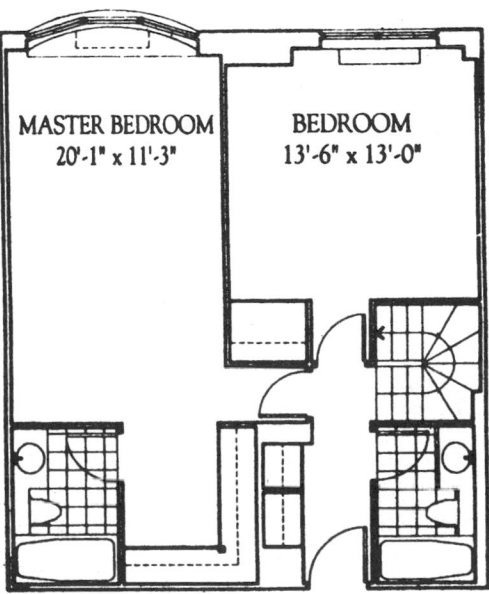

MASTER BEDROOM
20'-1" x 11'-3"

BEDROOM
13'-6" x 13'-0"

Fig. 64 2250 Broadway, New York, N.Y.

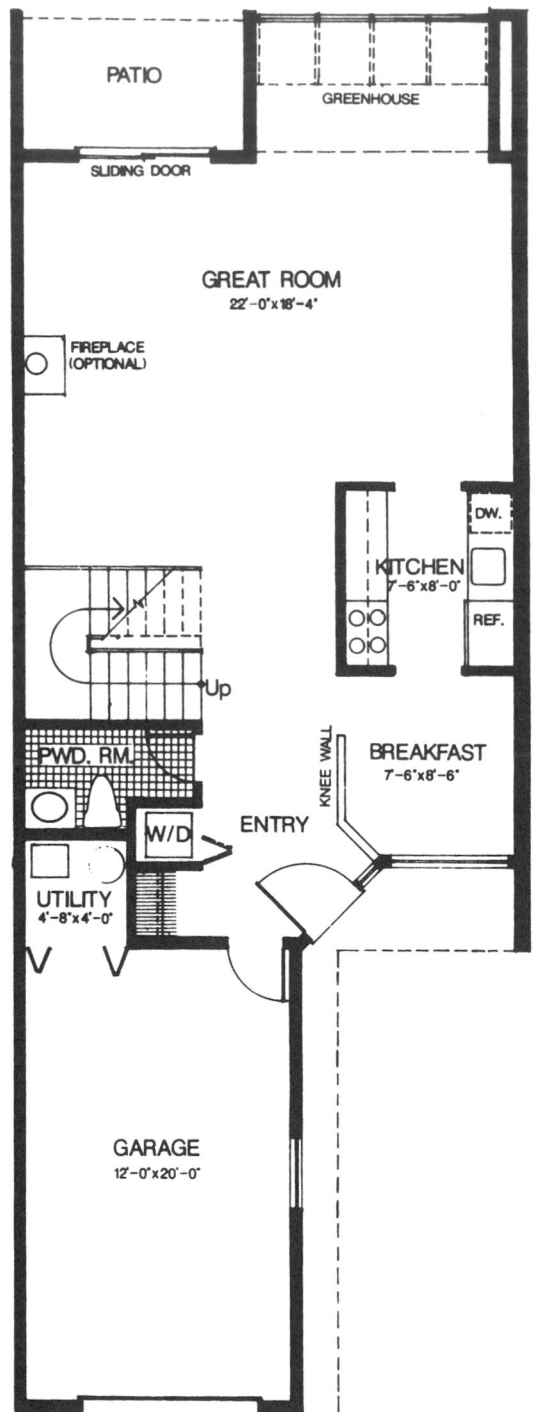

LOWER LEVEL

Fig. 65

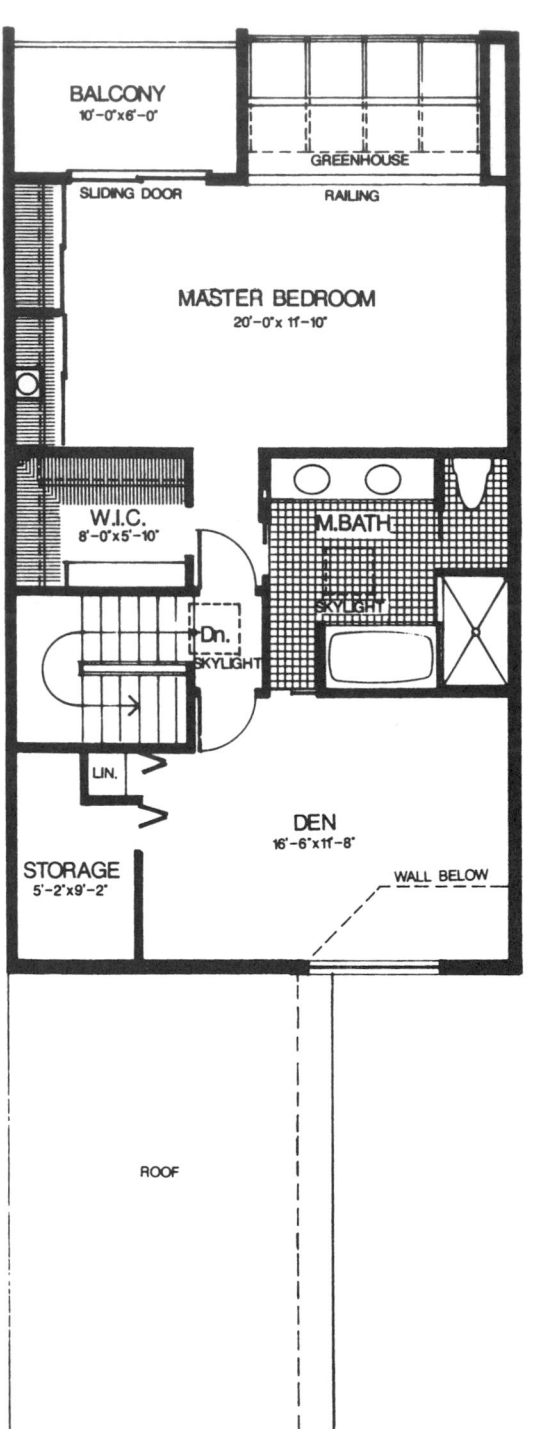

UPPER LEVEL

**THREE-BEDROOM TWO-STORY
TOWN HOUSES**

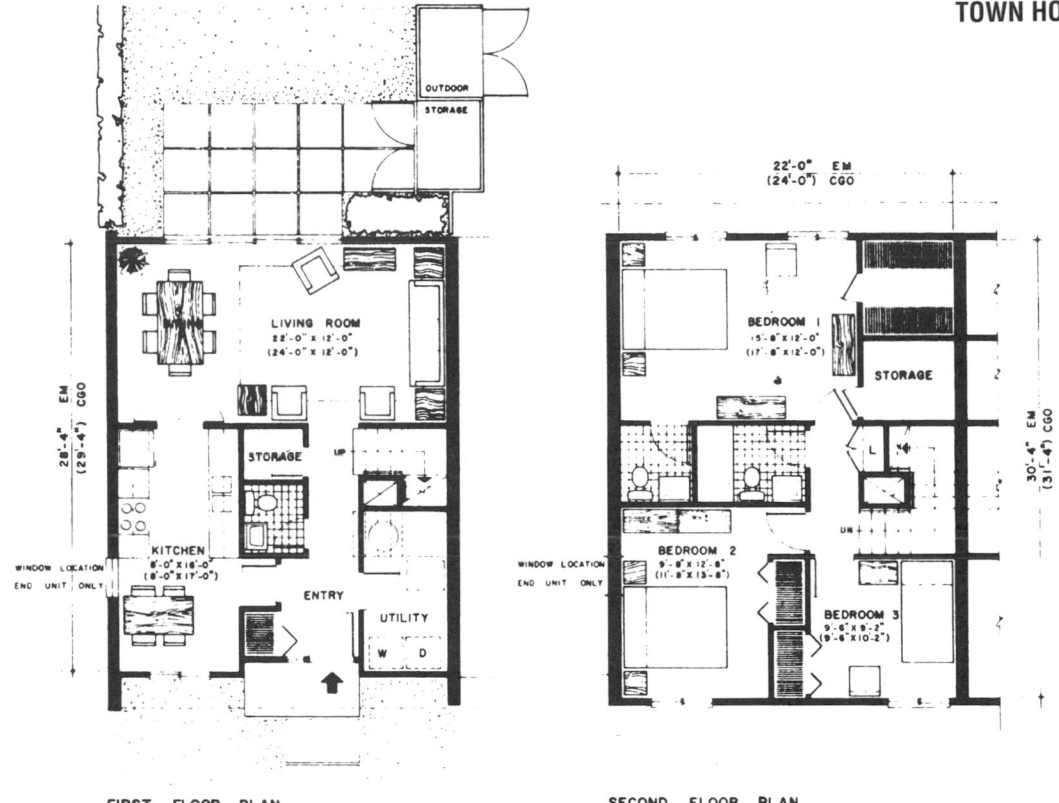

FIRST FLOOR PLAN   SECOND FLOOR PLAN

Fig. 66

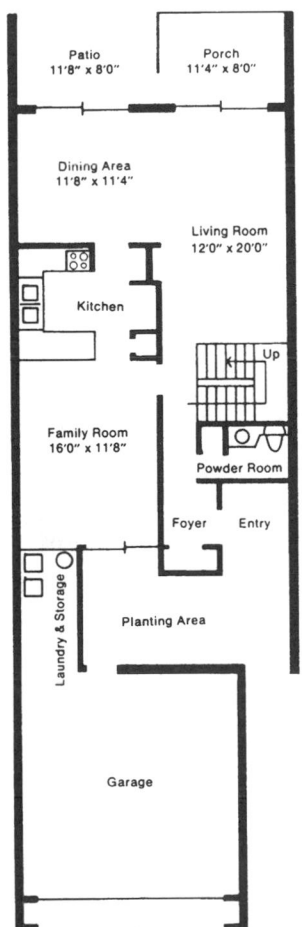

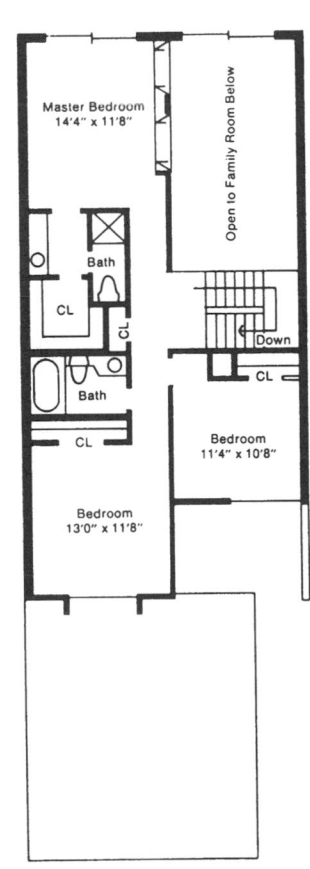

Fig. 67

TOWN HOUSES

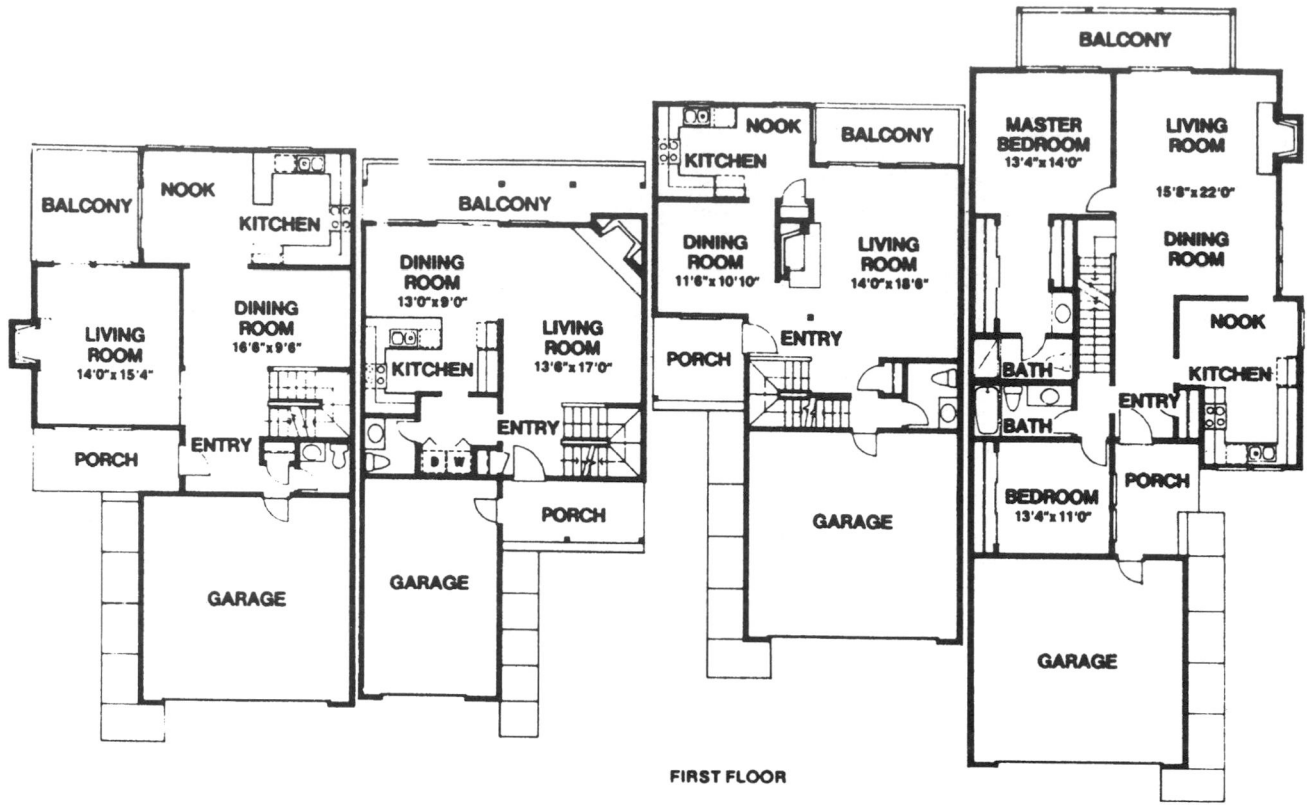

FIRST FLOOR

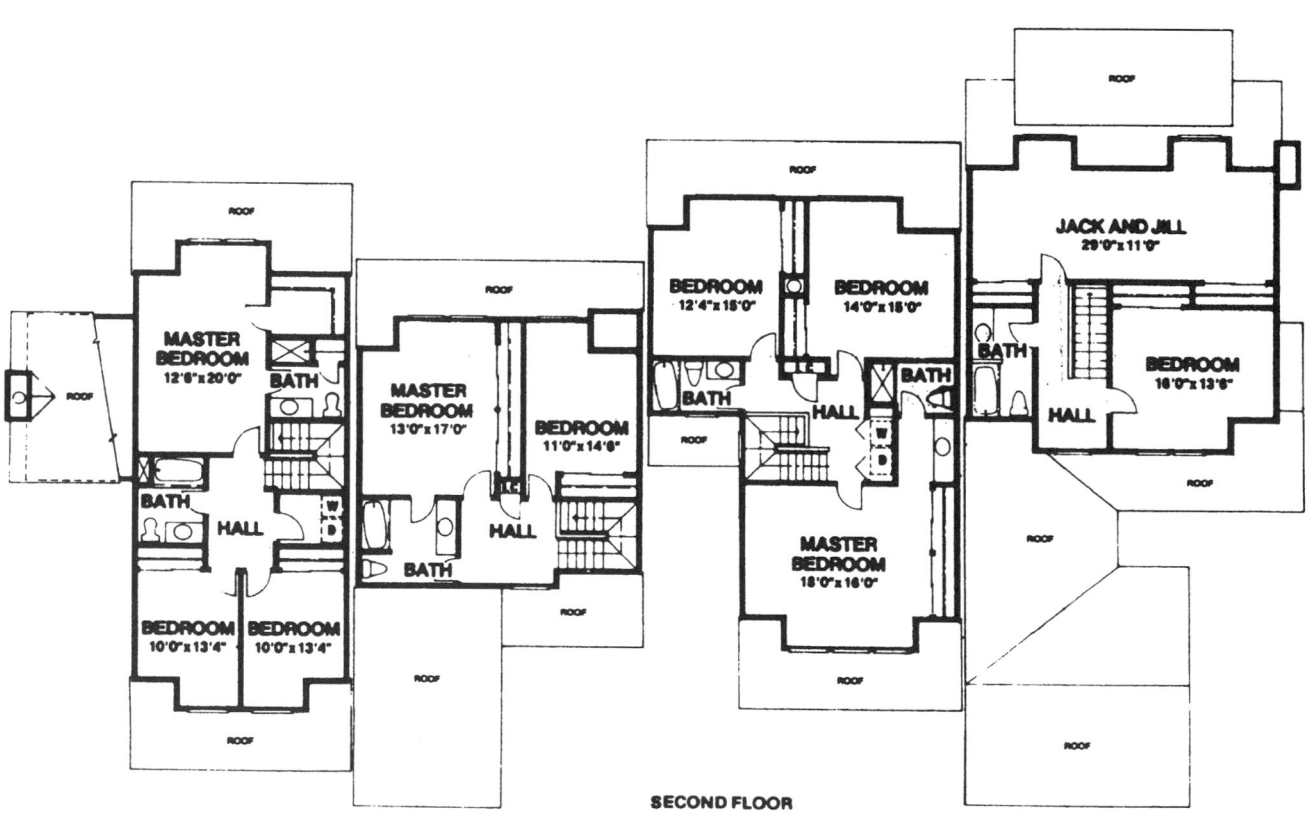

SECOND FLOOR

Fig. 68

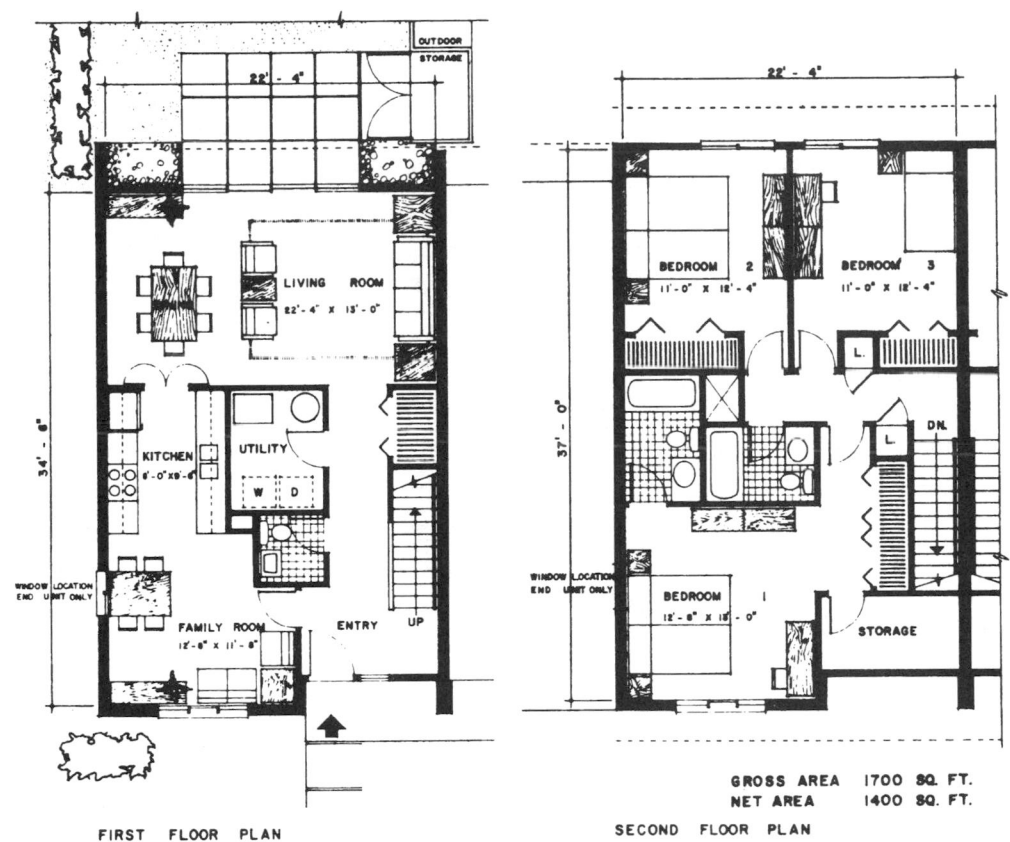

FIRST FLOOR PLAN

SECOND FLOOR PLAN

GROSS AREA 1700 SQ. FT.
NET AREA 1400 SQ. FT.

Fig. 69

FOUR-BEDROOM TWO-STORY TOWN HOUSE

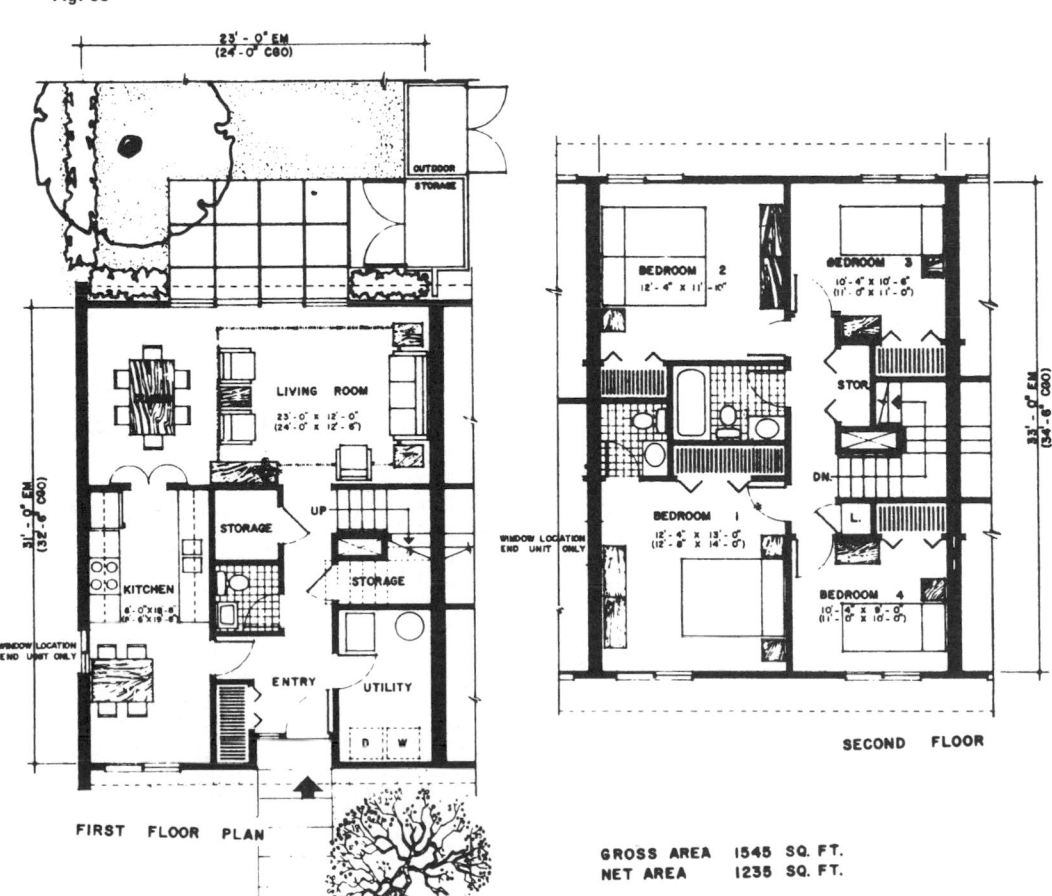

FIRST FLOOR PLAN

SECOND FLOOR

GROSS AREA 1545 SQ. FT.
NET AREA 1235 SQ. FT.

Fig. 70

MOBILE HOMES

The mobile home is considered a special category of single-family housing. The mobile home is perhaps the most economical form of single-family housing built today. It combines economy of indoor and outdoor space and construction costs and, in addition, is not bound by the same standards other single-family housing is.

Since it is often designed for and occupied by families with few or no children, it is smaller than conventional housing. By using less land per house site and by clustering common facilities—laundries, indoor and outdoor recreation, etc.—mobile home parks can be built at densities well above those for other detached dwellings. Finally, low costs are possible by using construction methods and materials which the codes do not permit for other types of housing.

According to the mobile home industry, there are three basic types of mobile home parks: the housing-oriented park, the service-oriented park, and the resort-oriented park. These are usually owned and managed by private companies or private citizens who lease individual lots for parking a single unit. However, the latest trend is toward mobile home subdivisions where individual lots are sold, not leased.

The housing-oriented park is not designed for a specific clientele, although its occupants are generally young married couples. Of the three types of parks, this is the only one marked by any degree of mobility. But the ever-increasing size of the units and the resultant cost of transporting them from one location to another inhibit frequent moves. The service-oriented parks are primarily designed for retired persons, although some parks have areas set aside for young families with children. The resort parks are almost exclusively for adults, generally retired. These parks generally operate on a system similar to that of a country club with memberships for park residents. In the past many were open only on a seasonal basis, but today year-long occupancy is common.

In the past, most mobile homes had rather undesirable locations. They were, and still are in most instances, excluded from residential districts by local zoning ordinances. The newer large parks are being developed at outlying locations, beyond the limits of restrictive zoning, where sizable parcels of land are available at prices developers are willing to pay. Better planned, more attractive units and park sites should make the mobile home next to the railroad tracks or the town dump an anachronism.

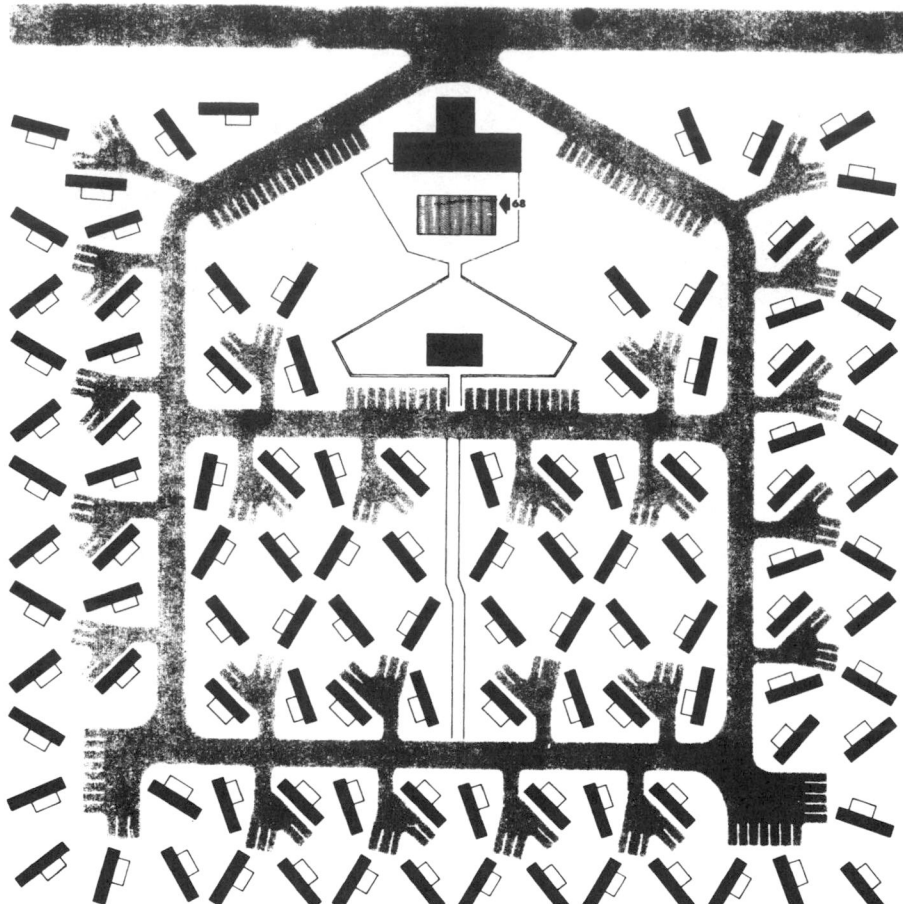

Fig. 1

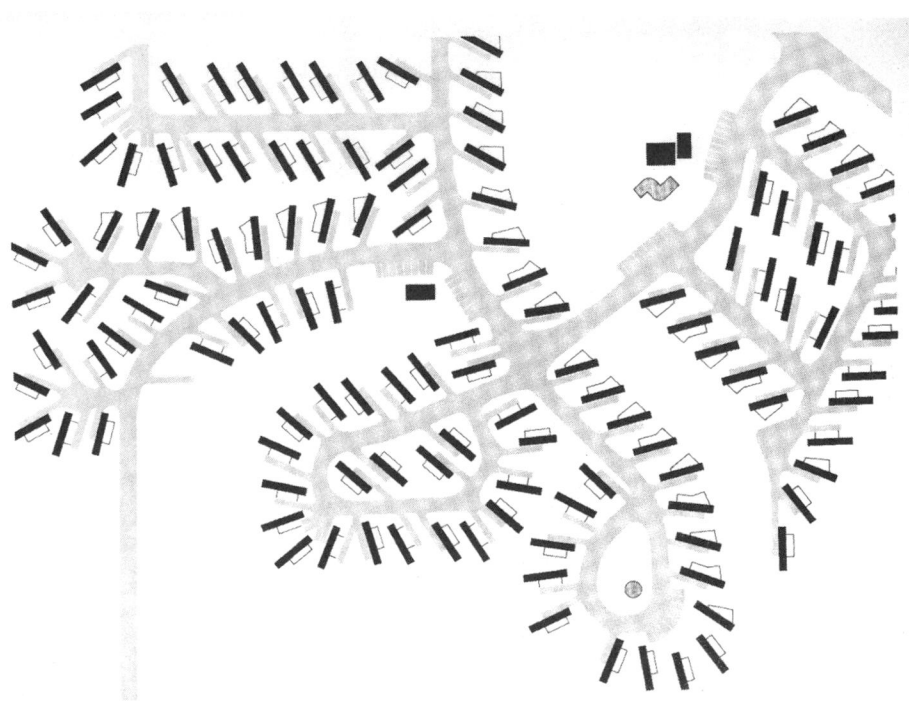

Fig. 2

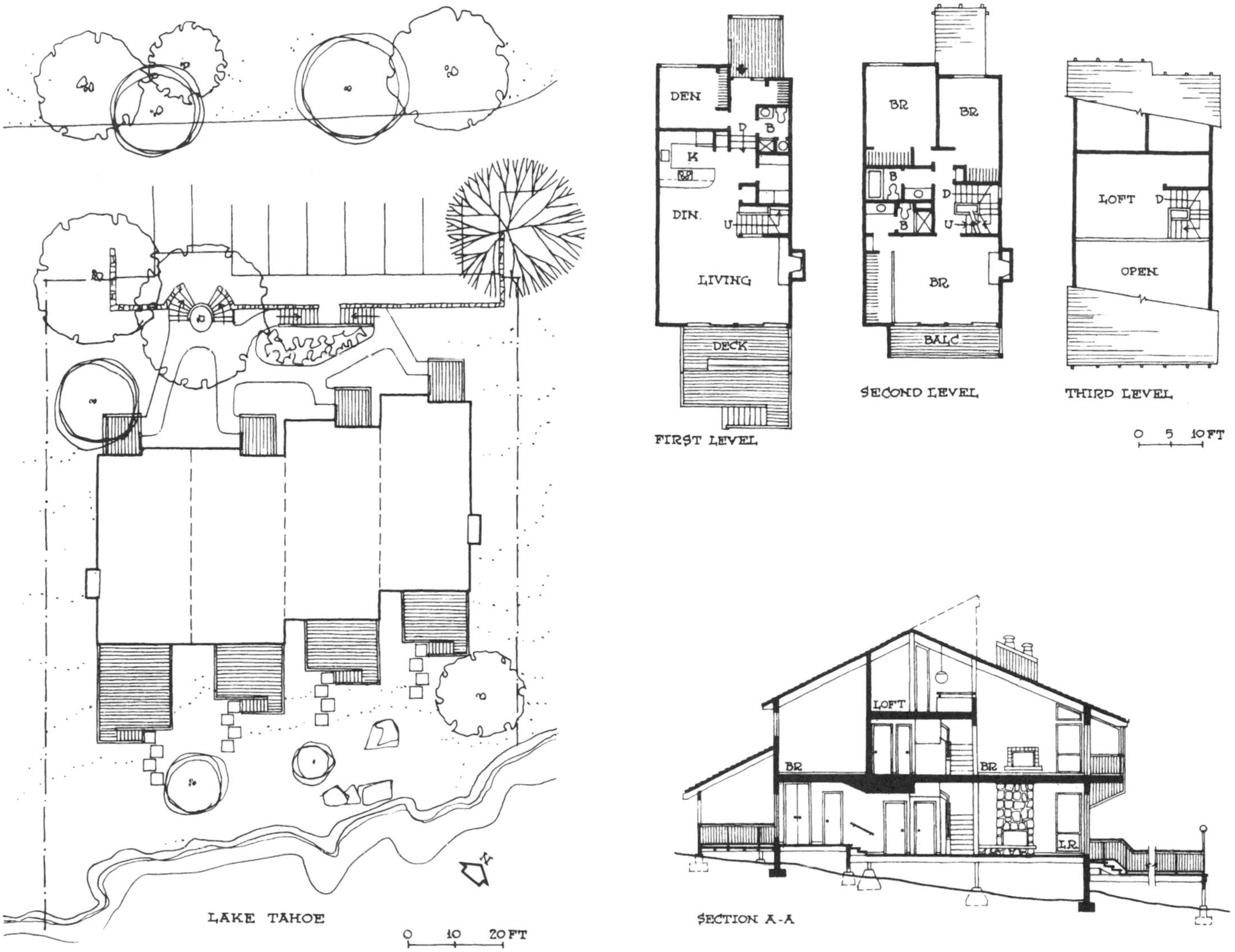

DEN

K

DIN.

B

LIVING

DECK

FIRST LEVEL

BR

BR

B

D

BR

BALC

SECOND LEVEL

LOFT

D

OPEN

THIRD LEVEL

0 5 10 FT

LAKE TAHOE

0 10 20 FT

Fig. 1 Lake Tahoe, Nevada, Wisser/Olin—Architects.

LOFT

BR

BR

L.R.

SECTION A-A

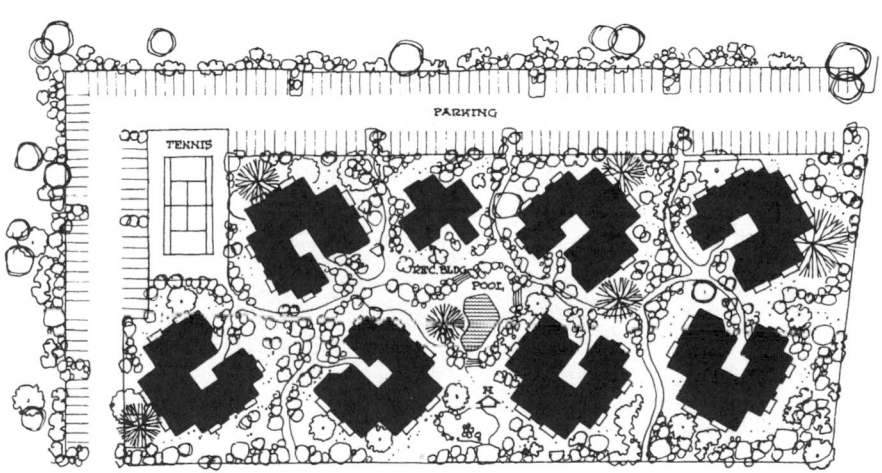

PARKING

TENNIS

REC. BLDG.

POOL

**Fig. 2a Sherwin Villas, Monmouth, Calif. Duplanty/
Huffaker—Architects.**

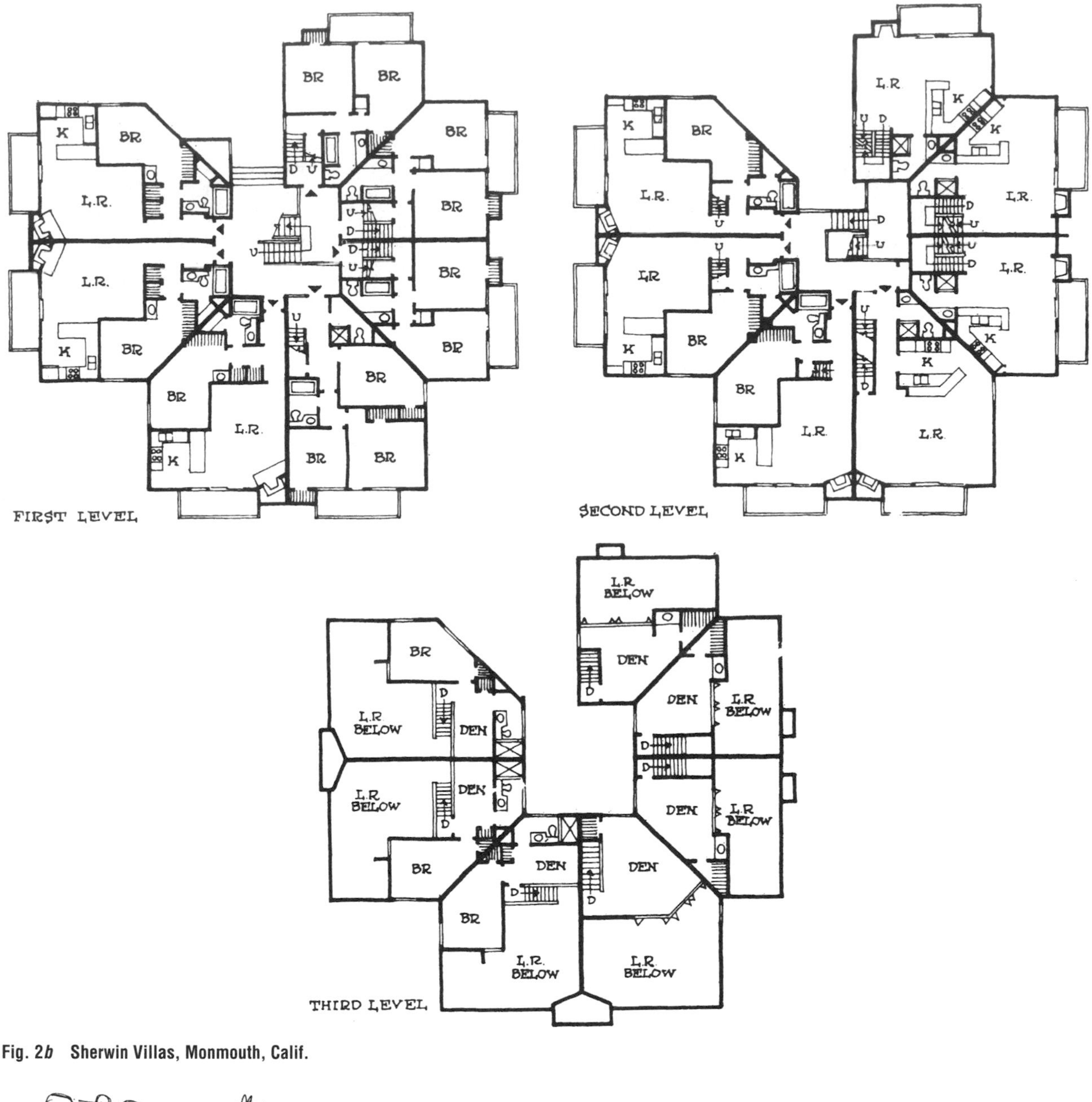

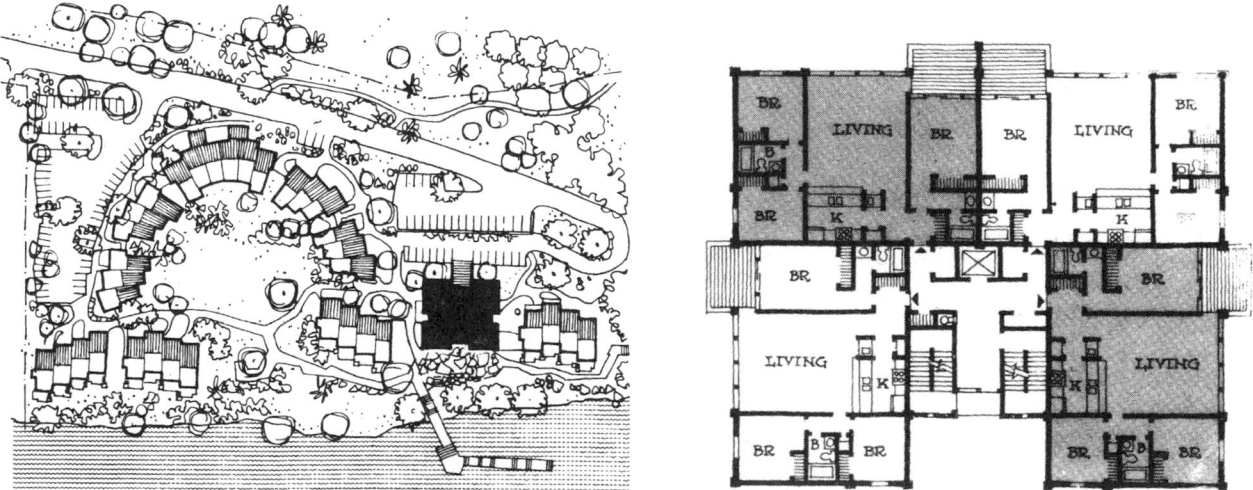

FIRST LEVEL
SECOND LEVEL
THIRD LEVEL
BR
K
L.R.
DEN
L.R. BELOW

Fig. 2*b* Sherwin Villas, Monmouth, Calif.

BR
LIVING
K

Fig. 3 Brockway Springs, Lake Tahoe, Nev. Bull, Field, Vollmann, & Stockwell—Architects.

FARMSTEAD PLANNING ZONES

Zones help the planner organize each activity relative to all other activities, and to allow for each planning factor for each activity. Each zone is 100 ft wide—less space may lead to crowding, and wider zones are often desirable. The areas of the zones as shown are: zone 1 = 0.7 acres; zone 2 = 1.6 acres; zone 3 = 2.3 acres; zone 4 = 2.8 acres. The first three zones include most basic buildings and equipment and use about 4.6 acres.

For a farmstead with a family living area, place the house at the center of the planning zones. For a farmstead without a house, the farm court is usually the center, because vehicles, materials, and labor tend to work from the court. These 100-ft bands are activity zones, and they help locate major activity areas, help preserve a desirable family living environment, and encourage spreading the farmstead out, leaving space for present operations and future expansion. See Figs. 1 and 2.

Zone 1: Family living
Lawns, recreation space, flower and vegetable gardens, and guest parking are close to the house. Protect zone 1 from noise, odor, and dust as much as possible.

Zone 2: Machinery center
Shop, storage, and related services that are relatively quiet, dry, and odor-free are in zone 2. Consider screening the center from family view.

Much of the driveway and farm court may be in zone 2. Put fuel and chemical storage toward the outer edge—near the machinery, but removing odors, fire danger, and some hazard to children perhaps 200 ft from the house.

Zone 3: Grain, feed, and some livestock
These areas cause dust, noise, traffic, and odor and are therefore moved another zone farther from the house.

Grain and feed handling and processing require electric power and good vehicle access. But keep heavy equipment, large dryers, and fire hazards away from the house. Zone 3 is a compromise.

Small animal units may also be in zone 3; that is, small animals or a small number of animals may not seriously degrade family living. A livestock unit close to the house is convenient for active management of maternity and nursery units or for care of pet or hobby animals.

Zone 4: Major livestock facilities
A large unit, whether confined to a building or on drylot, creates demand for adequate space, drainage, waste management, access, loading facilities, feed distribution, and other services. It also creates noise, dust, traffic, and odors. Space for expansion is usually important. Locate major livestock production in zone 4 or beyond.

Moving away from the old farmstead is frequently the most economical, as well as satisfactory, way to solve major expansion problems.

Zone planning applies to both cash grain and livestock farms. A grain farm can become a livestock farm and vice versa; so allow for both grain and livestock in your master design, to protect future growth, efficiency, and sale value. Adjusting to changes in health, labor supply, or economics can be difficult unless space is available for expanded and new facilities.

FARMSTEAD AND MAIN ROAD

The illustrations in Fig. 2 show some of the problems encountered in designing a farmstead plan. Prevailing winds are assumed from the northwest or west in winter, and from the northwest, southwest, and southeast in summer.

Figure 2a shows space between buildings, an adequate court, and a good windbreak for windy climates. But, the house is southeast of the livestock; so some winter winds will carry odors to the living area. If possible, plan the house farther west or southwest and the livestock center farther northeast.

In Fig. 2b a straight drive would permit north and northwest winds to blow directly toward the court. The layout can be improved by exchanging the house and machinery centers and using the alternate drive location shown.

In general, study prevailing wind directions. Position the house so that fewest winds blow toward it from the rest of the farmstead during the times of the year when dust, noise, odors, and insects are problems. Using the zones as described, locate the other activity areas. Location within the farmstead area is also determined by many other factors, such as drainage, electric and water lines, sewage system, and topography.

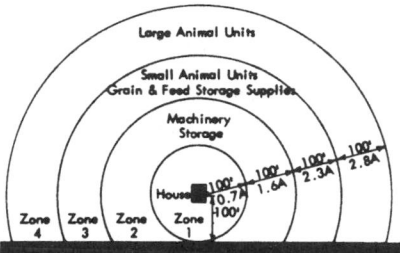

1a. Four planning zones.
If the road is busy, or if a tree windbreak is between the house and the road, set the house back further than 100'.

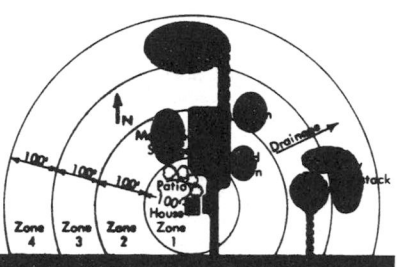

1b. Example: Livestock enterprise north of the road.
Major centers: living, livestock. Secondary centers: machinery, grain. One driveway serves all centers; a separate drive could serve a new large livestock unit.
The living area can be screened from other areas, yet it is convenient for family use, visitors, and observation. Leave space near all areas for expansion.

2a. Farmstead west of the road.
Some winter winds come from the NW. Locate the house as far west, and the livestock area as far north, as practical.

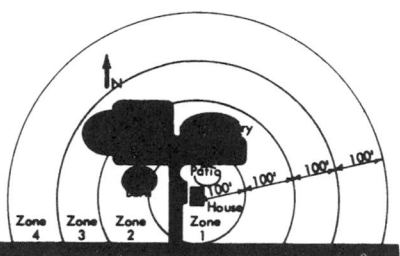

1c. Example: Grain enterprise north of the road.
Major centers: living, grain. Locate machinery and supply areas for convenience and accessibility.

Fig. 1

2c. Farmstead north of the road.
A good relation between house, windbreak, livestock center and main road is easy with this layout.

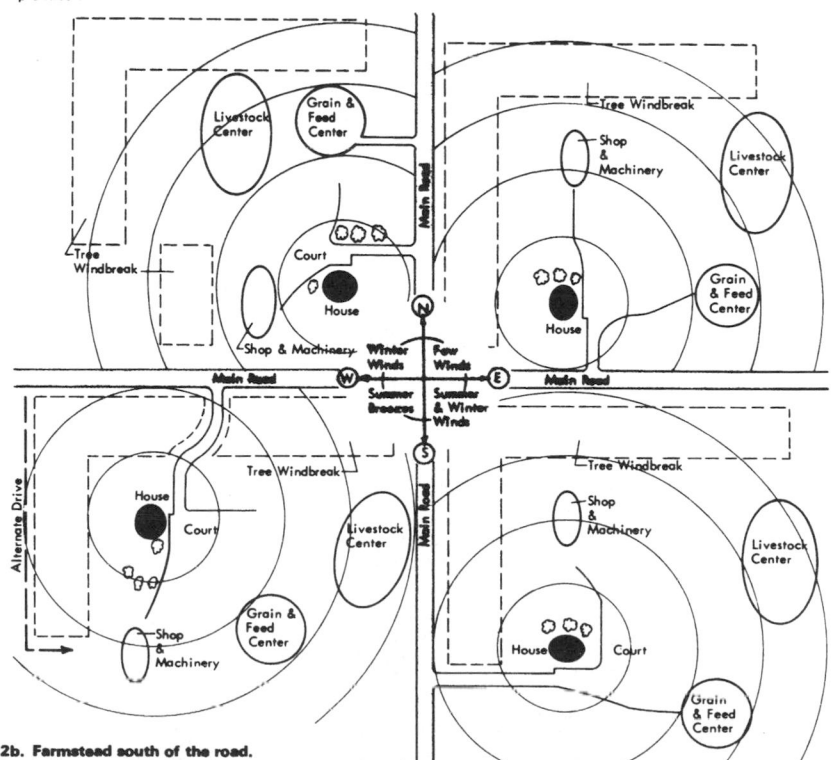

2b. Farmstead south of the road.
Note that a curved drive avoids a straight cut through the windbreak. Moving the house further south and the livestock area NE is desirable. An alternate drive location makes a good layout if the house and machine center can be reversed.

2d. Farmstead east of the road.
As in 2c, good layout is easy, assuming drainage and other factors permit this arrangement.

Fig. 2

FARMHOUSE PLANNING

Family Type I

The most distinguishing feature of family life for this group is the presence of infants and very young children. Children under 6, for instance, are reflected primarily in playing and resting. These activities, not surprisingly, are the two foremost simultaneous activities occurring in the house as a whole. These and the other activities involving young children (special and frequent feedings, and general child care) should be provided for in such a way that supervision can be maintained and dangers to the children minimized.

A successful house for this family type, therefore, must accommodate the needs and activities of young children and also provide for ease in communication between parents and children. Specifically, any space devoted to play should be easily supervised by the mother while she is preparing food or busy at other household activities.

The attention required by the children leaves comparatively little time to the mother for sewing or house care. (These tend to become more prominent as the children grow older and more self-sufficient.) Food preparation activities, which take up a larger proportion of the family's time in later stages, are now less time-consuming for the family as a whole than playing, resting, eating, and care of family members. Eating takes up a considerable proportion of time at this stage, and is one of the most frequently occurring activities, probably because young children are both fed more often and take a longer time to eat than adults.

The leisure time available to the parents is primarily spent in informal entertaining and watching television. Activities such as music and hobbies tend to develop in later years. Reading also assumes more importance in the later stages of the cycle. Most of the evenings for type I families are spent at home because of the children, and entertainment of guests is simple. It is at this stage that little time is spent outside the home by the wife and the children. This contrasts significantly with the families in their later stages, when, although the house is then used for more varied activities and by more people, much of the time of the family members is spent outside the house.

Family Type II

The essential difference between the families of this type and those of type I is that some of the children are away at school for part of the day. Also there are generally more children in the family.

It is in this stage that the roughest and most active play goes on, and at any time there are likely to be several age groups involved in different play activities. Friends from school are frequently in the house during the afternoons. Because of this and because of the general desire on the part of both children and adults to work and enjoy their leisure in undisturbed surroundings, a separate play area is recommended. The ideal location for this play area would be on the first floor, near the food preparation center and farm entry. This would provide the possibility of easy supervision when required.

Although the older children were found to spend much less time resting than those under 6 years of age, the activity was still the second most frequent one for the house as a whole, and quiet places for children to nap are still necessary. Almost all the resting at this stage, however, took place in bedrooms in contrast with

family type I, where an appreciable amount of resting also took place in the living area. This suggests that with a greater number of people and more active children in the house there is a need for a relatively isolated and quiet area.

There are a number of other space implications concerning families of this type. Since the children generally require less care as they grow older, the mother has more time available to her for household activities. Sitting and talking is now a major activity for these families. Hobbies and reading are also significant for these families. There tends to be more leisure

engaged in by the family as a unit, since some of the children are now old enough to participate in more leisure activities.

Activities of the children also tend to become diversified as they grow older. There is some studying done by the children attending school, and some participating in "chores" both inside and outside the house.

Snacks may tend to be scattered throughout the day for these larger, more complex families. There may also be guests and friends from school frequently eating with the family.

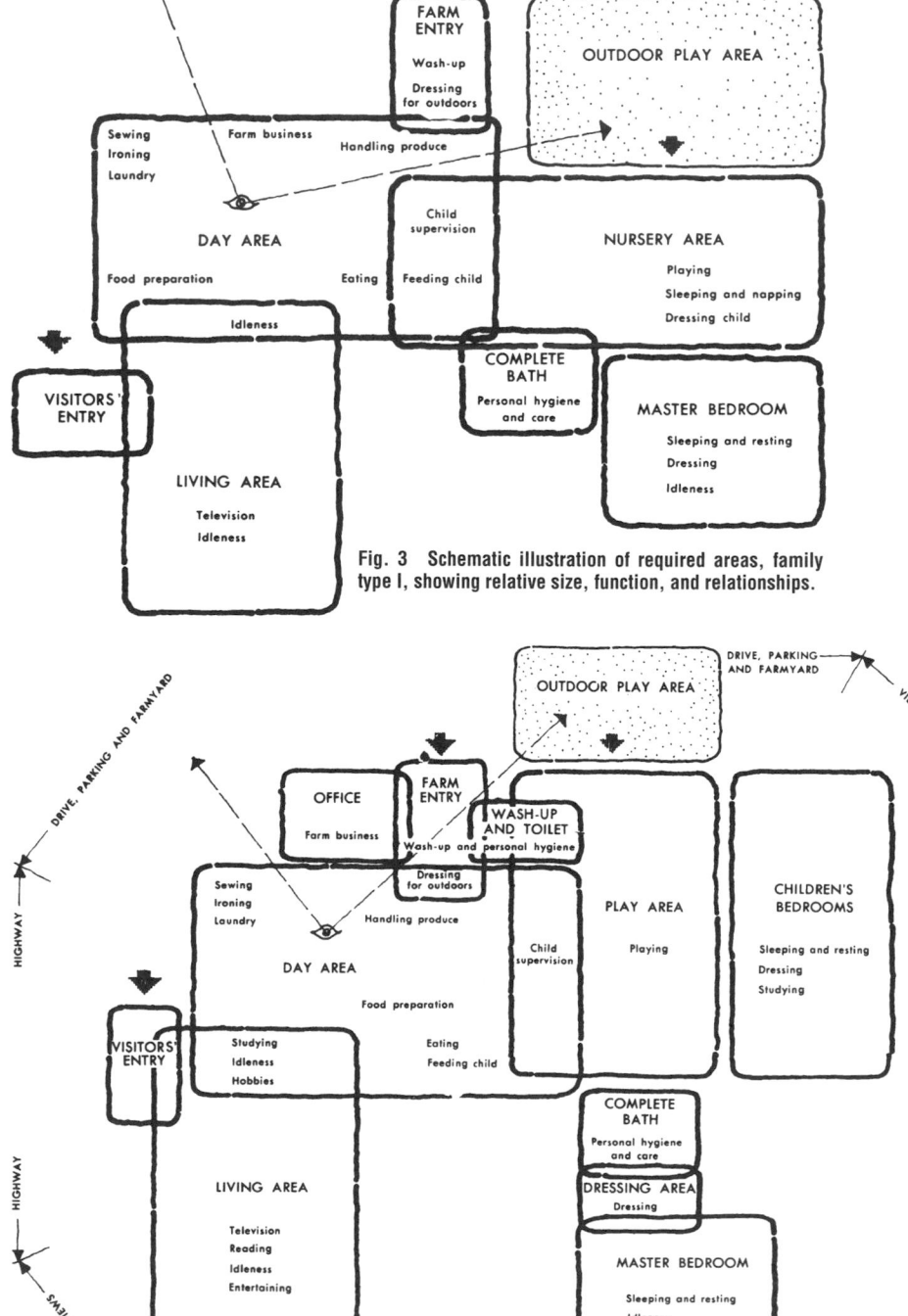

Fig. 3   Schematic illustration of required areas, family type I, showing relative size, function, and relationships.

Fig. 4   Schematic illustration of required areas, family type II, showing relative size, function, and relationships.

Family Type III

There is a distinct change in the atmosphere of the houses of this family type, compared with those of the two types already discussed. Mainly this is brought about by the lessened frequency of children's play activities. In the kitchen, for example, more time is spent on food preparation than by families of types I and II. A larger proportion of time in the living area is spent on leisure than by families of type I (but less than by those of type II). Business and study activities in the dining area consume more time than for any other family type.

Although the children in this family type may be still relatively young or they may be in their teens, in either case they are characterized by a growing maturity and diversity of interests. While the family tends to do some things together, the older children will begin to spend more time in their own rooms, or in areas where they can be alone with their friends. This clearly indicates the need for separate bedrooms adequate for a range of activities. There is also a need for two distinct leisure areas so that adults and children can entertain their friends simultaneously and in relative privacy.

The children are developing a variety of interests and consequently engaging in a greater range of activities. Their homework keeps them busy after school and their participation in all farm and home "chores" increases. The teenage girls spend somewhat more time on cooking, sewing, laundering, and ironing than younger girls.

With the children off at school for most of the day, the house also tends to be used less extensively throughout the day but more intensively during certain portions of the day. The homemaker's principal activities are changed from those of child care and supervision to those of increased food preparation, ironing, business, and leisure.

Family Type IV

This phase of the family cycle is characterized by more modest space needs than the earlier phases. The general pattern of activities tends to be more limited in scope.

The necessary activities such as eating, sleeping, personal hygiene, and dressing should be accommodated in the easiest possible manner. There will be a need for facilities for passive entertainment such as a radio or television set. A quiet, warm, and comfortable leisure area for talking or watching the activities of others characterizes the needs of this area. An important part of the life of the elderly person is spent in reading and reminiscing, so that the appropriate storage space for books, pictures, and other items should be provided.

Particular attention in these homes should be given to safety features as they relate to older persons. Strength and speed as well as sensory perception and ability to judge physical relationships are somewhat lessened as a person grows older. Stairways, changes in floor level, and slippery surfaces should be carefully avoided. From the point of view of easy housekeeping, the house should be a compact one. Furthermore, two elderly people alone in a house need to be within easy talking distance of one another. A one-floor arrangement is therefore indicated for this family type, quite apart from the obvious hazards and waste of energy caused by stairs.

Research shows that one of the problems concerning planning for families at this stage is the fact that there may be as many as three or four substages, each of which has its own special problems. As the family unit decreases in size, there is obviously far less conflict in the use of space and also less disturbance caused by the activity of other people. For example, when the family first enters this stage, the couple itself may still be fairly active and independent; in fact, there may still be a grown child living at home. At the next substage, the couple has aged but both husband and spouse are still active in farming. Following this, the husband and wife may still be living, but they probably have become inactive in farming, or one spouse may now be living alone and evidencing a steadily diminishing ability for activity.

Fig. 5 Schematic illustration of required areas, family type III, showing relative size, function, and relationships.

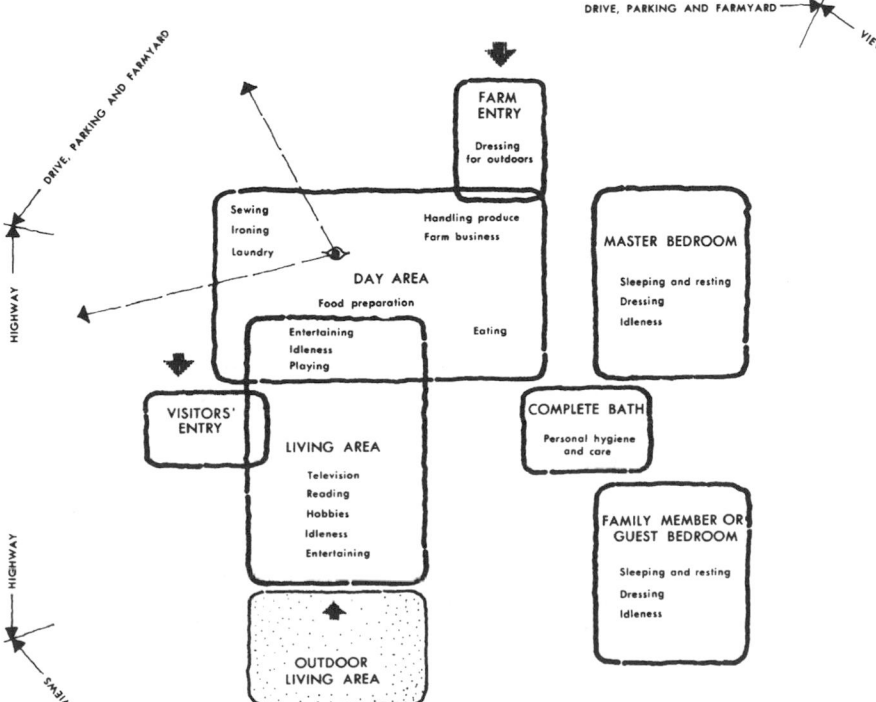

Fig. 6 Schematic illustration of required areas, family type IV, showing relative size, function, and relationships.

GARAGES

LOCATION OF GARAGES

Garages for one- and two-family houses are usually attached directly to the house or connected to it by a covered passage. If attached, most building codes require 1-h fire resistance for the wall between house and garage. The one opening permitted in this wall must be protected by a self-closing door having a 1-h fire rating and a sill 6 in above the garage floor. If the house extends above the garage, the ceiling must also have a 1-h fire rating.

Detached garage

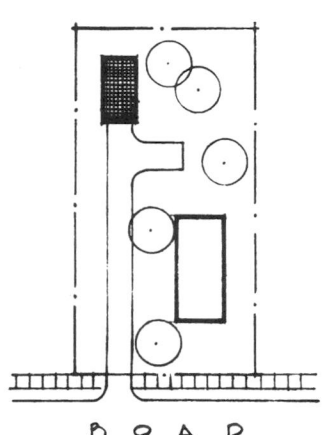

Fig. 1   Detached garage.

Carport or semi-attached

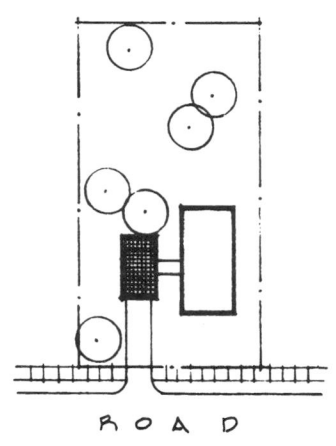

Fig. 2   Carport or semi-attached garage.

Attached garage

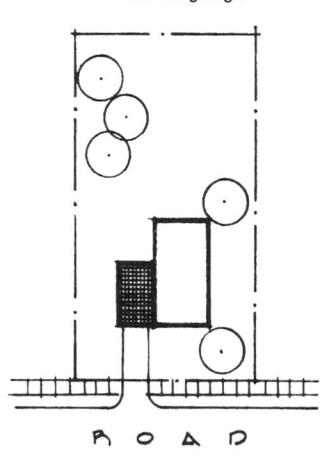

Fig. 3   Attached garage.

Garage within unit

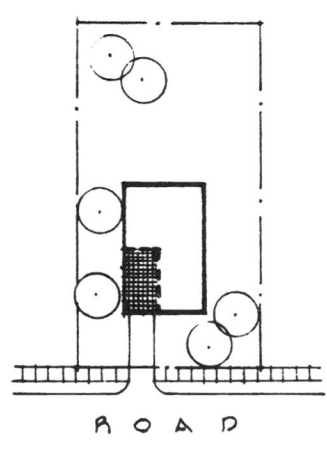

Fig. 4   Garage within unit.

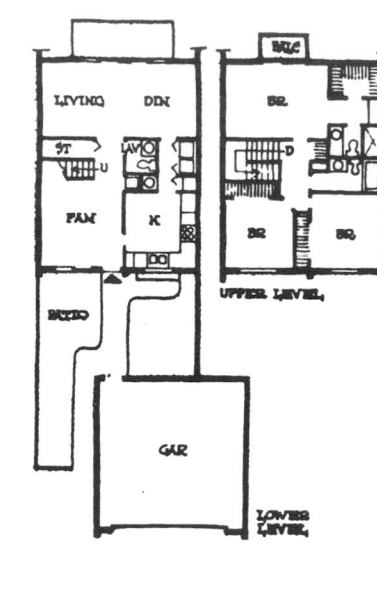

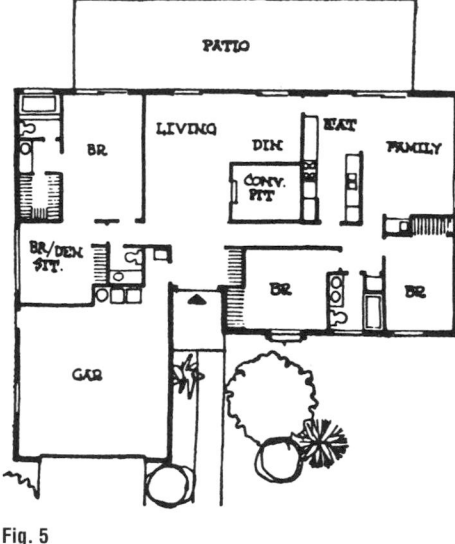

Fig. 5

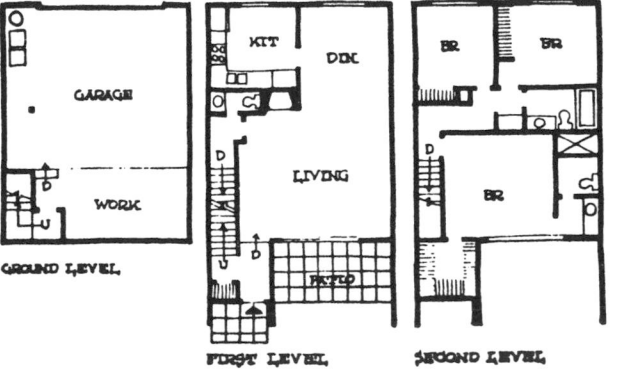

Fig. 6

SIZE OF GARAGES

The minimum size for a one-car garage is 10 by 20 ft, inside dimensions (Fig. 7a). This permits access to one side of the car only. The addition of 2 ft in both directions, or 12 by 22 ft, is recommended for comfortable access to both sides of a large car (Fig. 7b). For two cars, inside dimensions of 22 by 22 ft are recommended (Fig. 7c). These dimensions are for garage use only; if work or storage space is to be included, the dimensions must be increased accordingly. Generally speaking, the garage is not the best location for these spaces.

CARPORTS

Carports require approximately the same roof area as garages. Supporting posts need not be located at the outer edges of the roof provided they are spaced so that there is no interference with car door openings (Fig. 7d).

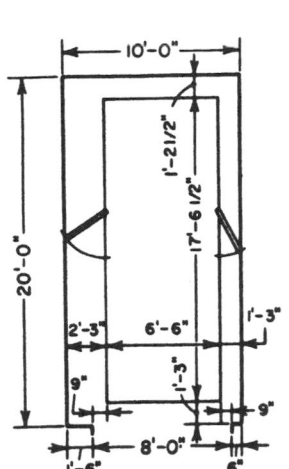

Fig. 7a Minimum one-car garage with average medium-sized car (9-ft door recommended; 8-ft door, if used, should be offset as shown).

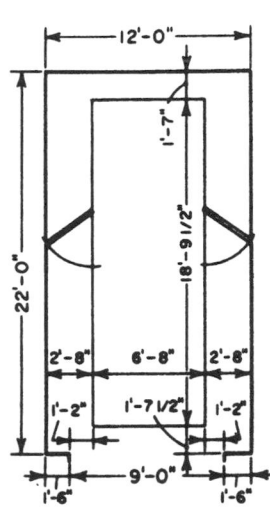

Fig. 7b Adequate one-car garage with average large car.

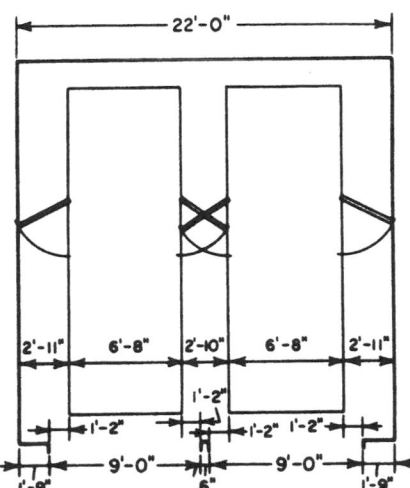

Fig. 7c Adequate two-car garage with average large cars.

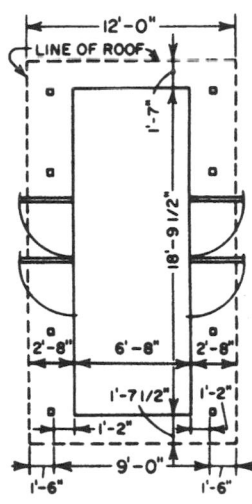

Fig. 7d Adequate one-car carport with average large car.

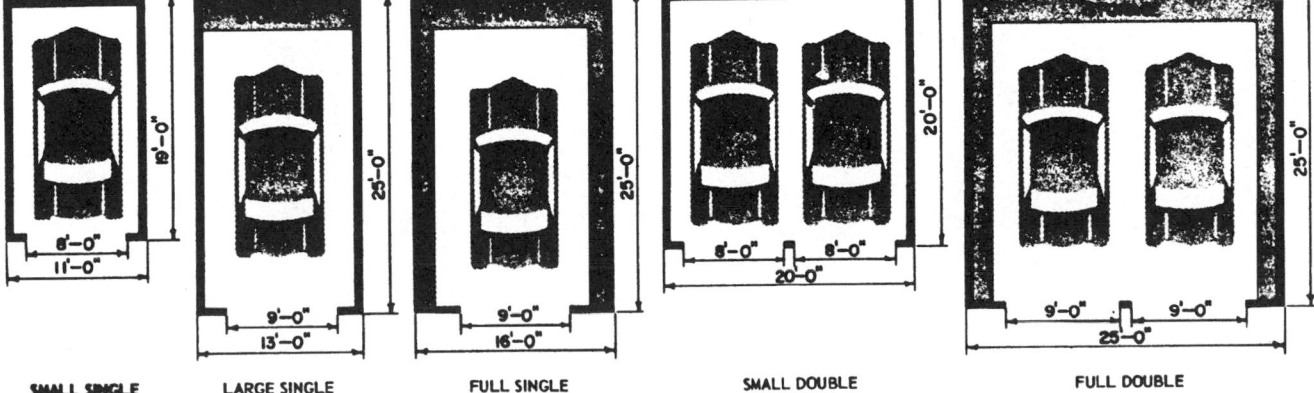

SMALL SINGLE   LARGE SINGLE   FULL SINGLE   SMALL DOUBLE   FULL DOUBLE

Fig. 8 Garage sizes.

Types of Apartments

INTRODUCTION

Most aspects of apartment unit design today reflect the need to comply with basic minimum standards. Such standards are enforced by building and housing codes, and by administrative requirements set by governmental agencies that approve housing proposals. Generally, these standards deal with minimal floor areas, access, privacy, ventilation, sanitation, food preparation, safety, utilities, and construction.

These requirements provide adequate space for typical family activities, e.g., relaxation, sleeping, preparation of meals, eating, and maintaining sanitary conditions and storage.

It is recognized that all of the criteria relative to selection of apartment types cannot be met completely on any one development. Since many factors in the problem may be in some degree opposed to a number of the others, a well-balanced compromise is often the best solution that can be obtained. For instance, dwelling types that naturally lend themselves to the physical nature of the site may not be well suited to the special needs of the tenants; types that are best fitted to the general economic level of the tenants may not be in accord with the zoning regulations in effect, or may be incongruous with the existing neighborhood pattern or its trend.

Many conflicts between the several factors might be cited. Moreover, there are no rules whereby any single criterion can be met with assurance that the right answer has been found.

All the factors listed, however, have proved significant and worthy of full consideration in this phase of project development. Hasty decisions or rigid adherence to preconceived ideas as to types of housing not only may jeopardize the success of a project but may work an injury to the orderly and proper development of the community of which it is a part.

Food Preparation

A food preparation center is required that will allow normal food preparation and serving activities for the number of occupants of the living unit. For the smaller units, efficiency and one-bedroom, a kitchenette would be adequate; for the larger units a full kitchen is required.

The food preparation center must be equipped with at least a sink, range, refrigerator, and full complement of storage space including wall and base cabinets, counters, and broom closet. Optional equipment may include a dishwasher, washing machine, clothes dryer, and a separate freezer.

Bathrooms

Every living unit must have at least one private bathroom equipped with a water closet, lavatory and bathtub. Larger units with two or three bedrooms should have an additional compartment equipped with a water closet and lavatory. Other requirements include a medicine cabinet, a clothes hamper, and other accessories.

Apartment Layout and Privacy

All parts of the dwelling unit should be so arranged that each may function properly without interfering with any other. Specifically, the arrangement of rooms should be such that no person need walk through one room to get to another room. To walk through a bedroom or a living room to get to another bedroom is unacceptable. The bathroom must be accessible to all bedrooms without entering any other room intended for sleeping purposes. Also, the bathroom must not open directly off the kitchen, living room, dining area, or bedroom.

TABLE 1

Only access from	To	Shall not be through
Habitable room*	Bathroom	Bedroom
Habitable room	Habitable room	Bedroom
Habitable room†	Habitable room	Bathroom
Bedroom‡	Bathroom	Another bedroom
Bedroom	Bathroom	Habitable room

*In one-bedroom living units only, access to the bathroom from the living room may be through the bedroom.

†A required bathroom opening directly into a kitchen is not acceptable.

‡An only bathroom shall not be located on a separate floor (full story height) from all bedrooms of a living unit. Even exterior spaces such as balconies or porches must be provided with privacy. This can easily be achieved by physical separation or adequate visual screening from each other.

Table 1 shows the room arrangements that are not acceptable by FHA standards.

Storage

Every dwelling must be provided with adequate closet and storage facilities for a variety of uses. This should include provision for a normal amount of frequently used personal and household items. The following types of locations of closets are considered to be minimal.

The FHA requires the following minimum amounts of general storage if it is located entirely within the living unit. If storage is provided outside the unit, the total volume must be increased by approximately 100 percent.

Storage Outside Living Unit

Each development or building requires central storage spaces for individual tenant storage. This area would be for baby buggies and wheeled toys. Adequate protection and ready accessibility is essential. These needs require management control and security with locks and other devices. If space is located in basement areas, ramps or elevators are required.

Coat Closet

Each living unit requires one coat closet near the front entrance of the unit, preferably off the foyer. The length should be sufficient to hold all the outer garments of the family.

Linen Closet

A linen closet is needed to store all kinds of linens that are used daily. The best location is usually considered to be near the bedrooms or near the bathroom. The greater the number of bedrooms, the larger the linen closet should be.

Storage within Living Unit

Each living unit requires at least one separate closet for general storage purposes. It should be located in a conveniently accessible place within the unit. Sometimes it can be combined with clothes closets or coat closets.

Light and Ventilation

Each living unit must be provided with an adequate amount of light and ventilation. The purpose of this requirement is to prevent the accumulation of unpleasant odors and provide adequate fresh air for healthful living.

The generally accepted standard for natural light is that each habitable room must be provided with a minimum window area of at least 10 percent of the floor area of the room.

For natural ventilation, each habitable room must be provided with a minimum ventilation area of at least 5 percent of the floor area of the

TABLE 2   Schedule of General Storage

0- and 1-bedroom	100 ft^3
2-bedroom	140 ft^3
3-bedroom	180 ft^3
4-bedroom	200 ft^3

room. In most cases this is achieved by having half of the window area open to get the required ventilation. Only in kitchens or bathrooms may mechanical ventilation be substituted for natural ventilation. For bathrooms, mechanical ventilation is generally considered to be more effective because a positive movement of air is achieved. For kitchens, both natural and mechanical ventilation are used. For obvious reasons, mechanical ventilation does remove cooking odors much more effectively than natural ventilation, and is preferable in most cases.

The ideal arrangement of rooms in a living unit for natural ventilation is to achieve through ventilation; that is, air enters at one end of the unit and exits at the other end. If the building is properly oriented toward the prevailing breezes, through ventilation can easily be obtained with the proper floor layout.

The next best arrangement of rooms is to have the living unit face in two different directions at a minimum of 90°. This will achieve cross ventilation through the living unit. In recent years, with the installation of air conditioning in most luxury and middle-income housing, the need to depend on natural ventilation has lessened. Also, the poor quality of our air in urban areas reinforces the desire to utilize mechanical ventilation or air conditioning rather than natural ventilation.

Utilities

1. *Water.* All living units must have hot and cold running water.

2. *Electricity.* All living units must have enough current capacity in the electrical system to allow the use of normal electric lights and appliances.

3. *TV.* Provisions must be made for the installation of a master antenna, amplifier, and distribution system to an outlet in each apartment.

4. *Heat.* A system of heating must be provided that is adequate for the maintenance of a temperature of 70° when the outside temperature is 0°.

5. *Garbage and trash removal.* An adequate system for the storage, removal, or disposal of trash and garbage must be provided. This includes temporary storage within easy reach of the living unit, a method of collection, and incineration or compaction.

The balance of this chapter is devoted to a description and illustration of apartment floor plans ranging from the efficiency unit to the triplex layout. In addition, specific bathroom and kitchen arrangements are depicted.

GENERAL CONSIDERATIONS AND STANDARDS

GENERAL CONSIDERATIONS AND STANDARDS

The minimum requirements ensure sufficient space to allow the placement of normal furniture in all the rooms. In addition, reasonable allowances are made for circulation between and around the furniture and to have access to all drawers, cabinets, and work spaces.

These standards are defined as (1) minimum room dimensions and (2) minimum floor areas. The most widely used standards are Federal Housing Administration standards, which are given below. Many states, cities, and housing authorities have adopted variations on these standards. Table 1 indicates the minimum room sizes and dimensions for individual rooms. Table 2 indicates the minimum room sizes and dimensions for combined spaces. To ensure adequate volume for each space, the FHA requirement for ceiling height is 8 ft, which is considered to be the absolute minimum for all habitable rooms. Halls, public corridors, toilets, and storage areas may have slightly less height.

A safety requirement is that the distance of travel within a living unit and the door of any room leading to the doorway of an exit corridor must not exceed 50 ft.

Floor Areas

Total net floor areas per dwelling unit, as shown in Table 1, are calculated at the principal floor level and measured between the inner finished faces of exterior walls in detached dwellings, and to the centerline of partitions.

Light and Ventilation

1. *Method of measurement.* Measurements in this section are based on distance between finished floor surface and ceiling surface (if finished) and between finished wall or partition surface.

2. *Minimum ceiling heights.* (a) Basements: 7 ft clear under joists; for basement dwelling units, 7 ft 6 in. (b) For all dwelling units on floors above basement, 7 ft 6 in.

3. *Living space in basements.* In rooms used for living, sleeping, or eating, or dwelling units, the finished floor should be not more than 2 ft 6 in below the outside finished grade at required windows.

4. *Habitable rooms.* Provide light and ventilation in rooms used for living, sleeping, eating, and cooking, as indicated below. In computing the floor area of rooms with sloping ceilings, the area with less than 5 ft of headroom should not be included.

   a. Total glass area. Not less than 10 percent of floor area of room.

   b. Ventilating area. Not less than 4 percent of floor area of room.

   c. If windows open on covered porches and terraces, or are in rooms any portion of which is more than 18 ft from a window, the glass area should be not less than 15 percent of the floor area of the room.

   d. Unless separately lighted and ventilated by windows that provide the required glass and ventilating area, include any alcove adjoining a habitable room as part of that room in computing required glass and ventilating area.

   e. An alcove may receive light and ventilation from the window of an adjoining habitable room only when the common wall between the alcove and the habitable room contains an opening, the area of which is not less than 80 percent of the area of the entire wall on the alcove side.

5. *Bathrooms.* Provide light and ventilation in bathrooms.

TABLE 1 Minimum Floor Space Required for Household Activities, Furniture, Equipment, and Storage, in Square Feet

	Number of persons					
	1	2	3	4	5	6
For basic activities						
Sleeping and dressing	74	148	222	296	370	444
Personal cleanliness and sanitation	35	35	35	70	70	70
Food preparation and preservation	8	76	97	97	118	118
Food service and dining	53	70	91	105	119	146
Recreation and self-improvement	125	164	221	286	357	383
Extrafamilial association	17	17	34	34	51	51
Housekeeping	48	91	110	127	146	149
Care of the infant or the ill		124	124	124	124	124
Circulation between areas	20	20	35	35	45	45
Operation of utilities		20	20	20	20	20
Total basic dwelling unit area	380	765	989	1159	1450	1550
For other activities						
Laundry	36	48	65	80	96	112
Household maintenance		42	42	42	42	42
Circulation, two-story		32	32	32	32	32
Total with other activities	416	887	1128	1313	1590	1736

SOURCE: "Planning a home for occupancy," Standards for Healthful Housing, Public Adm. Service, 1950, American Public Health Assoc.

TABLE 2

	Elderly		Nonelderly					
	Efficiency 0-bedroom	1-bedroom	1-bedroom	2-bedroom	3-bedroom	4-bedroom	5-bedroom	6-bedroom
Maximum Allowable Dwelling-Unit Areas*								
Occupancy (persons)	1	2	2	2	6	8	10	12
Room count	3	3½	3½	4½	5½	6½	7½	8½
Maximum area within perimeter walls, gross ft²	400	525	550	720	900	1120	1320	1540
Guide to Maximum Space Areas (Not Mandatory)								
Living room minimum dimension 10 ft 6 in			145	155	160	165	170	175
Living room–dining room combination	120	120	170	185	205	220	230	240
Kitchen	40	40	50	60	75	90	100	110
Kitchen–dining room combination			70	90	110	130	150	170
Guest coat closet		4	4	6	8	10	12	14
Linen closet	2	3	4	5	6	7	8	9
Kitchen work top (countertop)	4	4	4	6	8	9	9	9
Kitchen shelving	20	20	30	36	42	54	60	60
General storage (20% should be near kitchen)	10	20	25	30	35	40	45	50
Bedroom 1 minimum width 8 ft 6 in†	65	120	125	125	125	125	125	125
Bedroom 1 closet	6	8	10	10	10	10	10	10
Bedroom 2†				100	100	100	100	100
Bedroom 2 closet				8	8	8	8	8
Bedroom 3†					90	100	100	100
Bedroom 3 closet					8	8	8	8
Additional bedrooms						90	90	90

*These areas do not include stairs and stair landings inside unit, general storage and circulation outside unit, public facilities (stair, elevators, etc.), or space for heating equipment.

For heating equipment, add 15 ft² for equipment operated by tenant; add 30 ft² for heater room for gas equipment, add 45 ft² for heater room for coal or oil equipment.

†All bedrooms 100 ft² or larger shall accommodate twin beds.

SOURCE: "Low-Rent Housing Manual"—221.1, Housing Assistance Adm., Dept. of Housing and Urban Development, Sept. 1967.

SPACE STANDARDS

MAXIMUM AREAS

The total floor area of a dwelling unit, measured between the linear finish of enclosing exterior walls and between partitions separating units, should not exceed the following:

These areas do not include stairs and stair landings inside the unit, general storage and circulation outside the unit, or public facilities (stairs, elevators, etc.).

Occupancy (persons)	1	2	4	6	8	10	12
Description	Efficiency	1-bedroom	2-bedroom	3-bedroom	4-bedroom	5-bedroom	6-bedroom
Room count	3	3½	4½	5½	6½	7½	8½
Area, ft²	360	550	720	900	1120	1320	1540

FURNISHABILITY REQUIREMENTS*

A dwelling unit must contain space so planned as to accommodate the following furniture, facilities, and equipment, and permit free circulation with due allowance for heating devices, door swings, accessibility to electric outlets, etc. Such furnishability should be demonstrated on the dwelling plans.

1. Living space:
   Couch, 3 ft 0 in by 6 ft 9 in
   Large chairs, 2 ft 6 in by 3 ft 0 in 1 for one-person unit, 2 for all others
   Desk, 2 ft by 2 ft by 3 ft 4 in None required for efficiency unit
   TV, 1 ft 4 in by 2 ft 8 in
2. Dining space:
   Table, one or two persons, 2 ft 6 in by 2 ft 6 in
   Table, four persons, 2 ft 6 in by 3 ft 2 in
   Table, six persons, 3 ft 4 in by 4 ft 0 in
   Table, eight persons, 3 ft 4 in by 6 ft 0 in or 4 ft 0 in by 4 ft 0 in
   Table, 10 persons, 3 ft 4 in by 8 ft 0 in or 4 ft 0 in by 6 ft 0 in
   Table, 12 persons, 4 ft 0 in by 8 ft 0 in
   Chairs, 1 ft 6 in by 1 ft 6 in
3. Sleeping spaces (per two persons):
   Twin beds, 3 ft 6 in by 6 ft 9 in or double bed, 4 ft 6 in by 6 ft 9 in (single bed for efficiency unit)
   Dresser, 1 ft 10 in by 3 ft 6 in (one for efficiency unit)
   Chair, 1 ft 6 in by 1 ft 6 in
   Crib, 2 ft 4 in by 4 ft 5 in (for first bedroom of family unit)

*"Low-Rent Housing Manual"—207.1 Dept of Housing and Urban Development Housing Assistance Adm.

TABLE 1 Minimum Room Sizes and Allowable Room Count for Separate Rooms

Name of space*	Room count	Minimum area, ft²					Least dimension
		LU with 0 BR	LU with 1 BR	LU with 2 BR	LU with 3 BR	LU with 4 BR	
LR	1	NA	160	160	170	180	12 ft 0 in
DR or DA	1	NA	100	100	110	120	8 ft 4 in
K	1	NA	60	60	70	80	5 ft 4 in
Kitchenette	½	40	40	NA	NA	NA	3 ft 6 in
BR (primary)	1	NA	120	120	120	120	9 ft 4 in
BR (secondary)	1	NA	NA	80	80	80	8 ft 0 in
Total area, BRs		NA	120	200	280	380	
OHR	1	NA	80	80	80		808 ft 0 in
Bathroom	½						
Half bathroom	¼	NA					
Foyer	¼	25	25	25	25	25	4 ft 0 in
Balcony or porch	¼	70	70	70	70	70	6 ft 0 in
Terrace	¼	120	120	120	120	120	8 ft 0 in

*Abbreviations:
LU = living unit
LR = living room
DR = dining room
DA = dining area
K = kitchen
K'nette = Kitchenette
NA = not applicable
BR = bedroom
OHR = other habitable room
0 BR = LU with no separate bedroom
SOURCE: Adapted from Federal Housing Adm., *Minimum Property Standards.*

TABLE 2 Minimum Room Sizes and Allowable Room Count for Combined Spaces

Combined space*†	Combined room count	Minimum area, ft²				
		LU with 0 BR	LU with 1 BR	LU with 2 BR	LU with 3 BR	LU with 4 BR
LR-DA	1½	NA	200	200	220	230
LR-DA (DR size)	2	NA	240	240	260	270
LR-DA-BR	2	240	NA	NA	NA	NA
LR-DA-K	2½	NA	260	270	290	310
LR-BR	1½	190	NA	NA	NA	NA
K-DA†	1½	100	110	110	120	140
K-DA (DR size)†	2	NA	150	150	160	180
K'nette-DA	1	80	80	NA	NA	NA

*Abbreviations:
LU = living unit
LR = living room
DR = dining room
DA = dining area
K = kitchen
K'nette = kitchenette
NA = not applicable
BR = bedroom
OHR = other habitable room
0 BR = LU with no separate bedroom
†For two adjacent spaces to be considered a combined space, the clear horizontal opening between spaces should be at least 8 ft wide, except for a combined K-DA. For 1½-room count of K-DA, the least dimension should be 6 ft 0 in and the clear opening at least 4 ft 0 in. For 2-room count of K-DA, the least dimension should be 10 ft 0 in and the clear opening at least 6 ft 0 in. All requirements having to do with light and ventilation should be complied with.
SOURCE: Adapted from Federal Housing Adm., *Minimum Property Standards.*

TABLE 1 Summary of Apartment Types

	Efficiency	One-bedroom	Two-bedroom
General characteristics	Minimal apartment with cooking and bathroom facilities. Desired by people who will occupy premises for short periods of time or only for sleeping. Meals will be simple and few.	Smallest apartment having separate living and sleeping accommodations. Contains adequate cooking facilities and dining areas.	Apartment having adequate space for family living. Contains full cooking and bathing facilities. May have separate dining room. Requires storage space for many activities.
Elements	Combined living, sleeping, dining area; minimal cooking facilities; bathroom; minimal storage space.	Combined living-dining area, kitchenette, separate bedroom, bathroom.	Living room, dining room, kitchen, master bedroom, bedroom, bathroom, small outdoor area.
Toilet facilities	Full bathroom plus storage and dressing	One bathroom (three fixtures)	One bathroom (four fixtures)
Site	Minimum area permitted—150–300 ft^2	400–600 ft^2	600–800 ft^2
Number of occupants	One or two persons	Two persons	Three or four persons
Type of occupancy	Single persons Young couples Elderly Temporary personnel or transients	Young couples Elderly Families with one child	Family with 1 or 2 children Couple with older or younger relatives
Planning considerations	Generates the least number of children per dwelling unit. Fastest turnover in occupancy. Minimum requirements for community facilities.	Generates a small number of children per dwelling unit. Provides flexibility for variety of dwelling unit types in a community. Relatively stable type of occupancy.	Adequate school facilities required for children. Adequate facilities for recreation required. Generally stable type of occupancy.
Parking requirements	One car per dwelling unit. Elderly people may require less.	One car per dwelling unit.	1¼–1½ cars per dwelling unit.

Three-bedroom	Four-bedroom	Five-bedroom
Apartment for average-sized family.	Apartment for large-sized family or occupancy with different generations living in the same household.	Apartment for extremely large family or with several generations living in the same household.
Living room, separate dining room, kitchen, master bedroom, two smaller bedrooms, bathrooms, outdoor area.	Living room, separate dining room, kitchen, family room, master bedroom, three smaller bedrooms, bathrooms, outdoor area.	Living room, separate dining room, kitchen, family room, master bedroom, four smaller bedrooms, bathrooms, outdoor area.
1½ or 2 bathrooms	Two bathrooms	2½–3 bathrooms
800–1100 ft^2	1100–1500 ft^2	Minimum area 1400 ft^2
Four to six persons	Six to eight persons	Eight to twelve persons
Family with two to four children Family with elderly persons or relatives	Family with three to six children Family with elderly persons or relatives	Family with six or more children Family with elderly persons or relatives
Generates a variety of school needs and recreational facilities.	Generates a variety of school needs and recreational facilities.	Generates a variety of school needs and recreational facilities. May require facilities for the elderly.
1½–2 cars per dwelling unit.	Two cars per dwelling unit.	Two to three cars per dwelling unit.

SIMPLEX APARTMENT/FLAT

SIMPLEX APARTMENT/FLAT

The simplex is an apartment that has all its rooms on one level. The size may range from an efficiency up to a multibedroom unit. The simplex apartment is the most common type of apartment because it is the simplest and most economical to build. Because both the living and sleeping activities occur on the same level, circulation is simplified. The close proximity of the two activities, however, may be disturbing if they are not properly zoned.

It may be located either in a high-rise apartment building or in a garden apartment develop-ment. A major criticism of the simplex has been the excessive amount of floor area required in corridors or stairs in order to gain access to the apartment. For example, in a central-corridor scheme, the length of public corridor is approxi-mately equal to the length of the apartment. The configuration or shape of the apartment will be determined by the type of building in which it will be located. With a central-corridor scheme, the apartment shape will be long and narrow to obtain maximum perimeter. With an open-cor-ridor scheme, the shape would tend to be deep and narrow, while with a tower scheme, the shape tends to be square.

A flat is another way of describing a simplex apartment. It is a common term used in England and some parts of the United States. Generally, the flat refers to a small simplex apartment such as an efficiency or one-bedroom unit. Another use of the term was to describe apartments in two-story buildings, with one flat above the other. The access hall or stairs for the upper flat is often incorporated into the living unit and is maintained by the tenant. The flats are usually grouped continuously in one building.

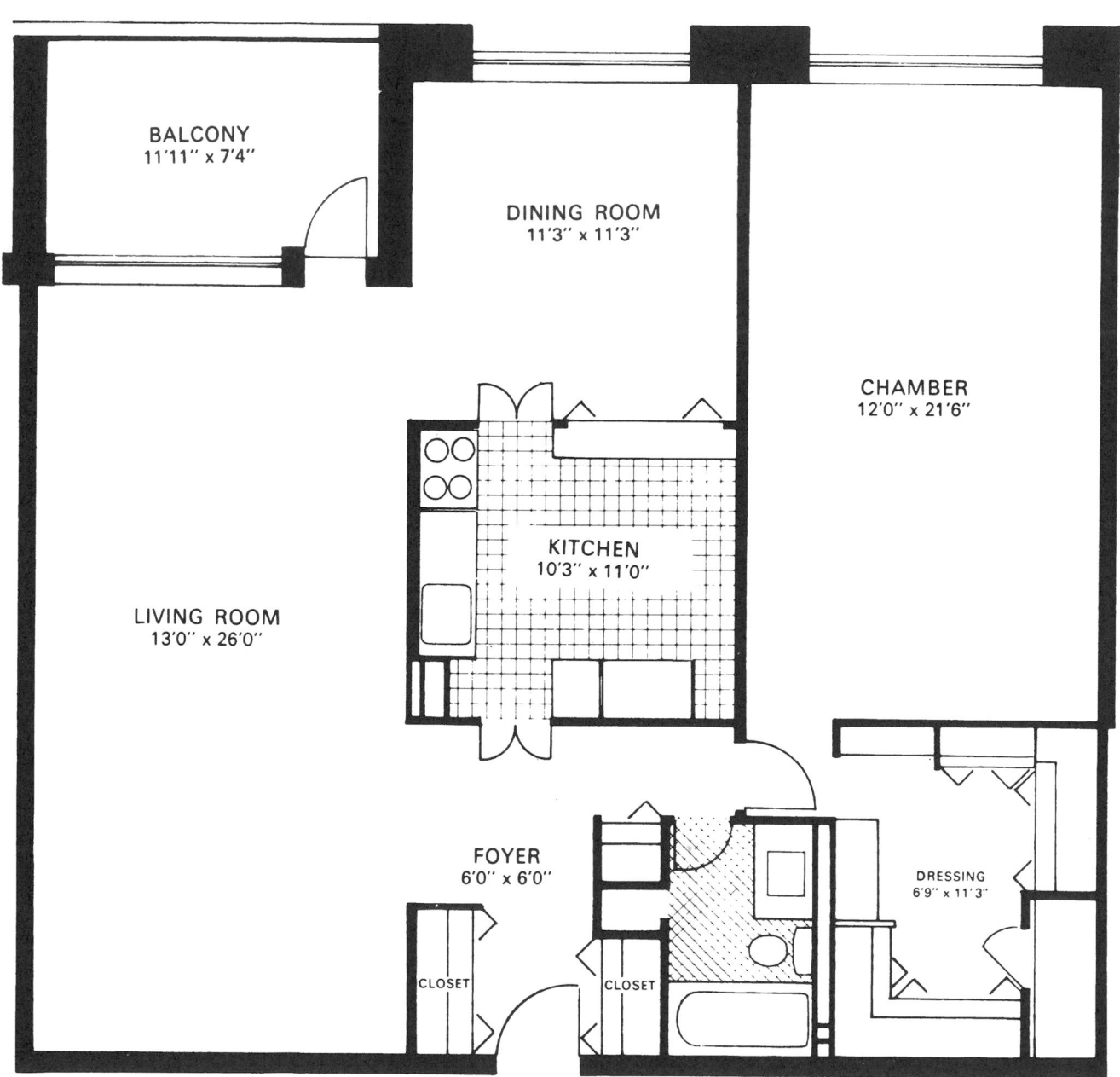

Fig. 1

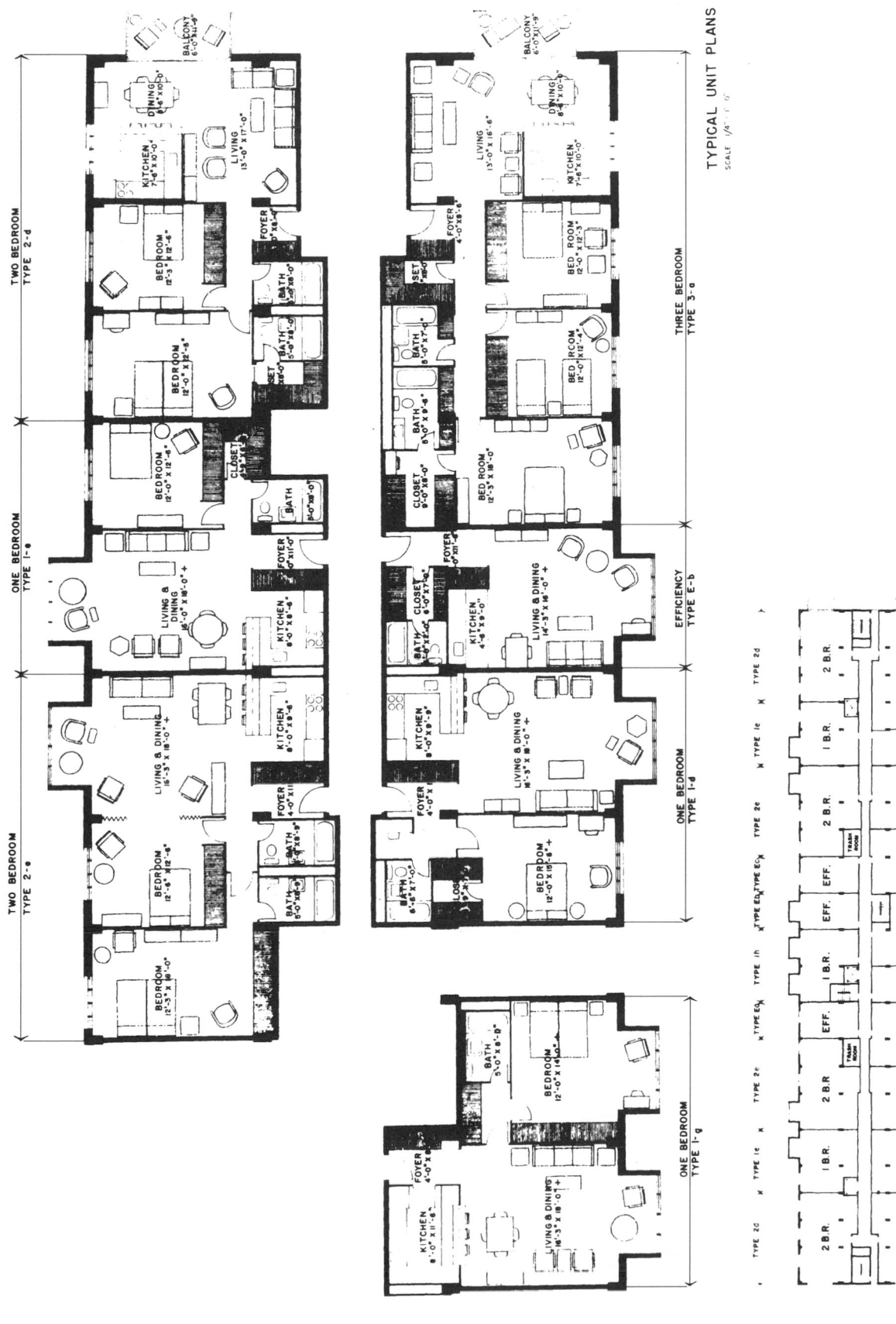

TYPICAL UNIT PLANS
SCALE 1/4"=1'-0"

TWO BEDROOM TYPE 2-d

ONE BEDROOM TYPE 1-e

TWO BEDROOM TYPE 2-e

THREE BEDROOM TYPE 3-a

EFFICIENCY TYPE E-b

ONE BEDROOM TYPE 1-d

ONE BEDROOM TYPE 1-g

TYPICAL FLOOR PLAN
SCALE 1/6"=1'-0"

Fig. 2   Victor Gruen Associates—Architects.

DUPLEX APARTMENT

DUPLEX APARTMENT

The typical duplex apartment is located on two levels with the living room, kitchen, and dining area on one level and the sleeping area on the other level connected by an interior stair. The illustrations shown are for a typical duplex apartment with an open-corridor building, and one utilized with a center-corridor building type.

The major economic advantage of the duplex apartment is the elimination of a corridor and elevator doors on every other floor. However, this saving is partially offset by the need for an interior stair for each apartment. From a livability standpoint, the main advantage is the separation of the living activities from the sleeping activities. This separation approximates the relationship in a typical two-story, one-family house. It provides greater privacy and feeling of more space.

The duplex apartment, whether located in a center-corridor scheme or open-corridor scheme, has the added advantage of through ventilation for the upper level and two exposures, which permits better building orientation. With an open-corridor plan, both levels can have through ventilation. The major disadvantage for the duplex is the need for the interior stair. For handicapped people or for the elderly, this may be a severe problem. Another advantage, which is less tangible, is the fact that a duplex apartment generally has more prestige or value than an equal simplex apartment. This can be seen in the development over the past years of greater numbers of duplex apartments in middle-income and luxury-type buildings.

A variation of the traditional duplex is the use of two levels separated by only one-half flight of stairs. This arrangement reduces the separation of the two levels and reduces the number of stairs a person needs to go up and down.

Also the interrelationship of levels allows for more imaginative and exciting use of space.

This kind of separation of levels, however, tends to be more expensive than conventional construction.

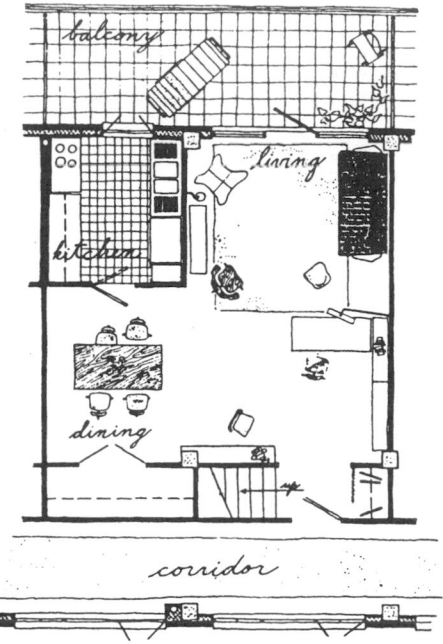

Fig. 1a Lower level. Fig. 1b Upper level.

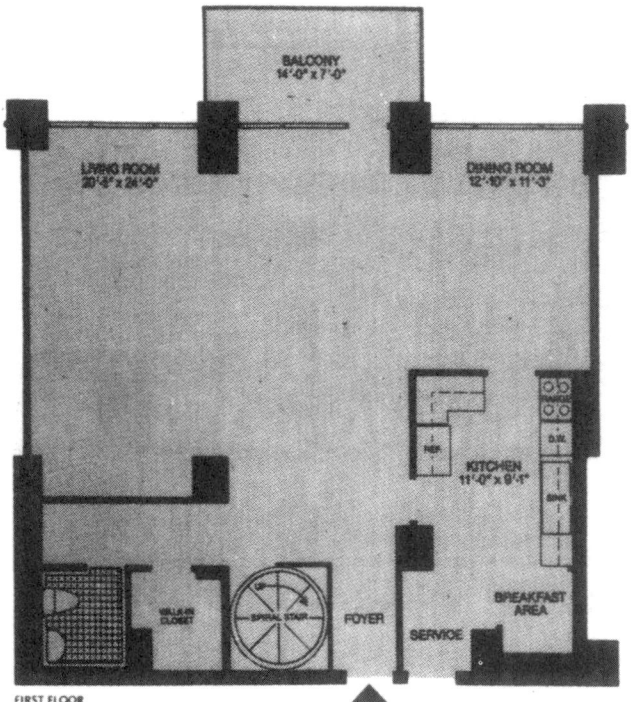

FIRST FLOOR

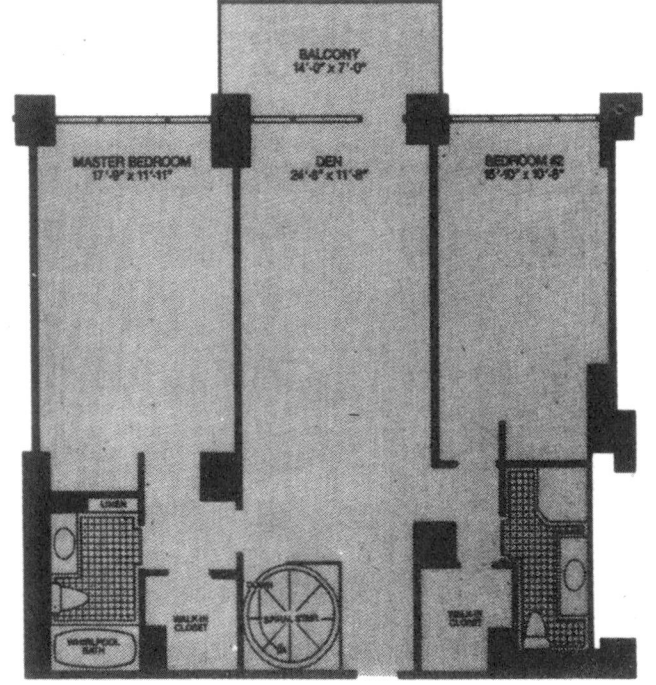

SECOND FLOOR

Fig. 2 Century Tower, Fort Lee, N.J.

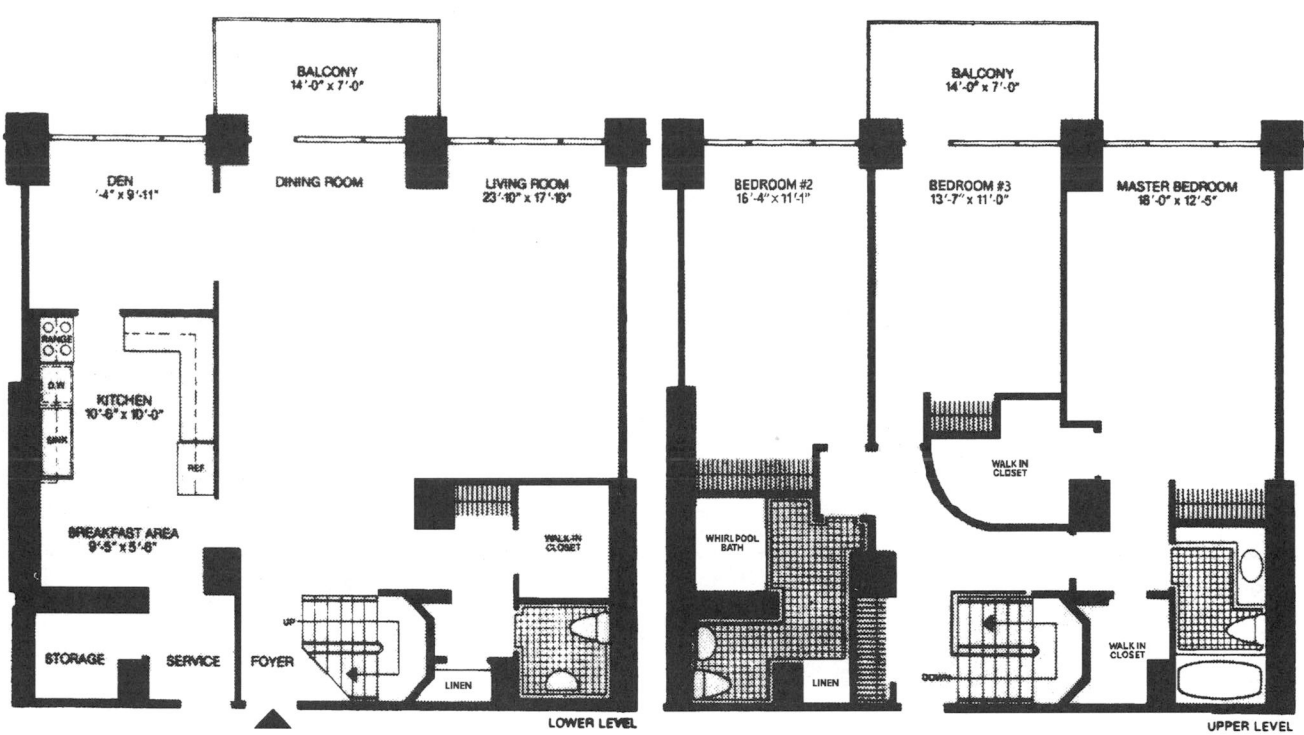

Fig. 3   Century Tower, Fort Lee, N.J.

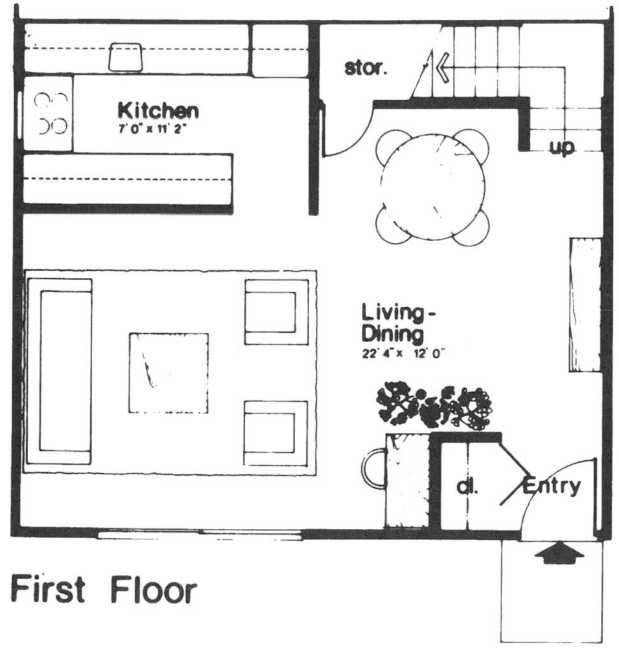

First Floor

Second Floor

Fig. 4

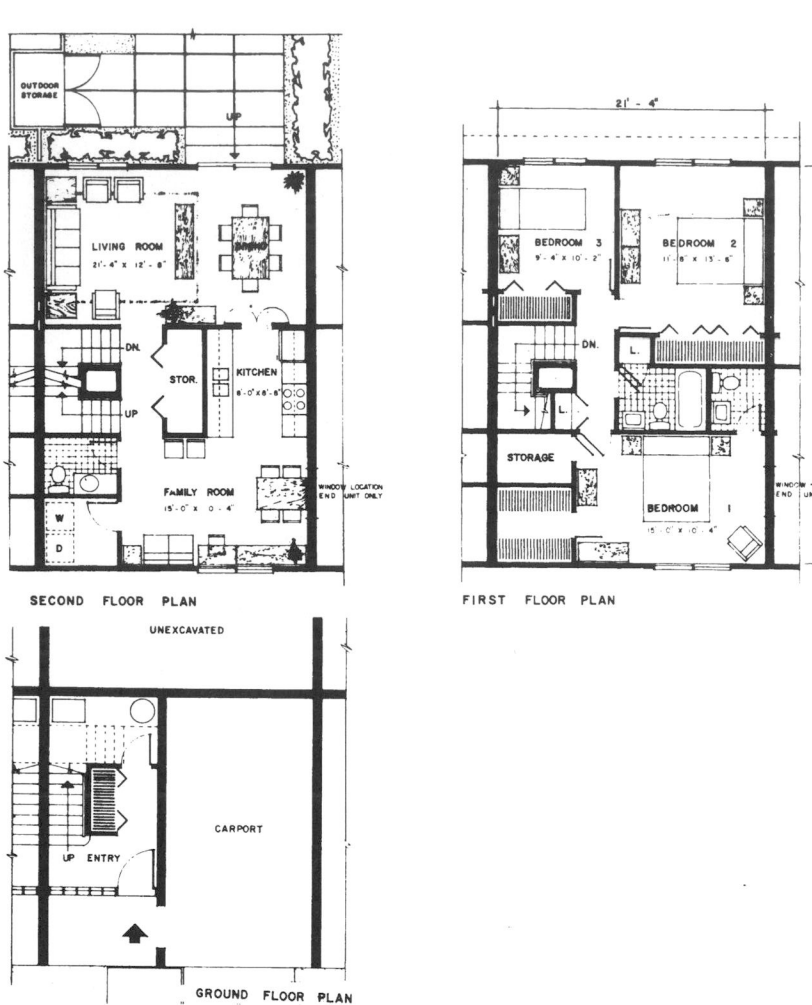

Fig. 5

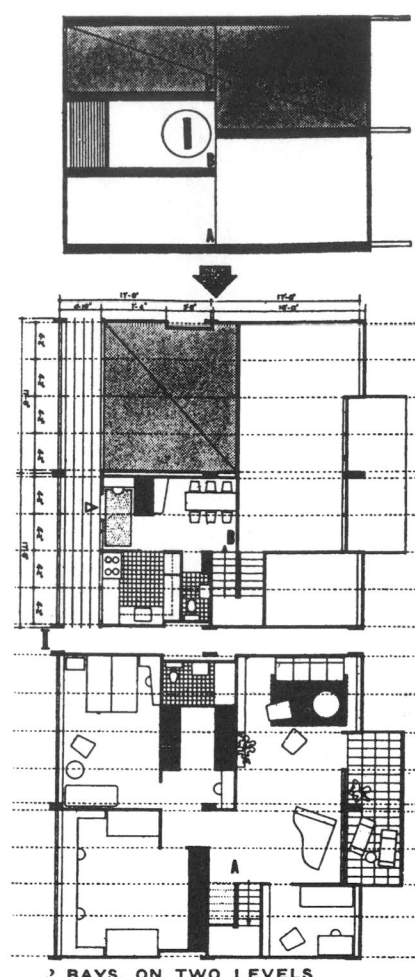

Fig. 6

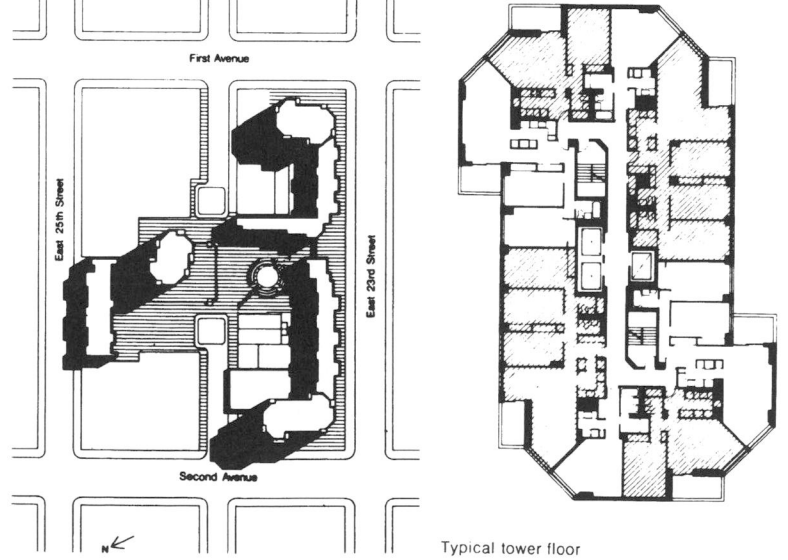

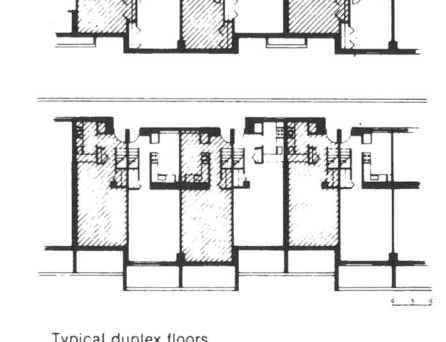

Fig. 7   East Midtown Plaza, New York, N.Y. Davis, Brody and Associates—Architects.

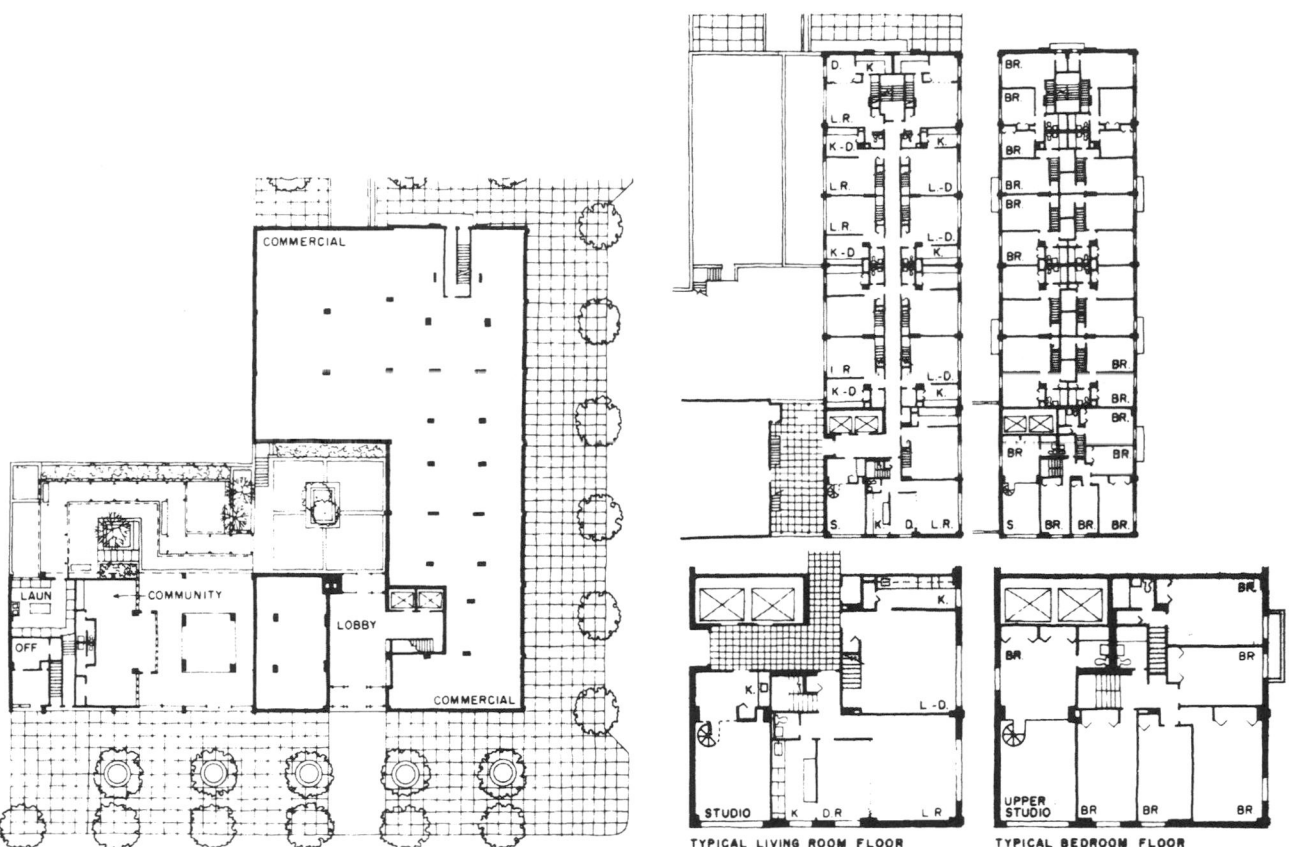

Fig. 8 Turin House, New York, N.Y. Holden, Yang, Raemsch, & Corser—Architects.

TYPICAL LIVING ROOM FLOOR

TYPICAL BEDROOM FLOOR

In Fig. 8 between the legs of the "L" are open breezeways, each two stories high. Besides two-story artists' studios, there are a variety of one- to four-bedroom duplex apartments. Public halls and elevator stops are required only on every other floor where entrances are located.

DUPLEX APARTMENT

For a densely forested site near the beach of Hilton Head Island, architects Stoller/Glasser and Marquis Associates developed an extraordinarily inventive scheme that puts 308 units on 20.6 acres—while leaving intact the forest atmosphere. Seemingly complex, the basic plan is quite simple: in each building, two units are stacked. (See Fig. 9.)

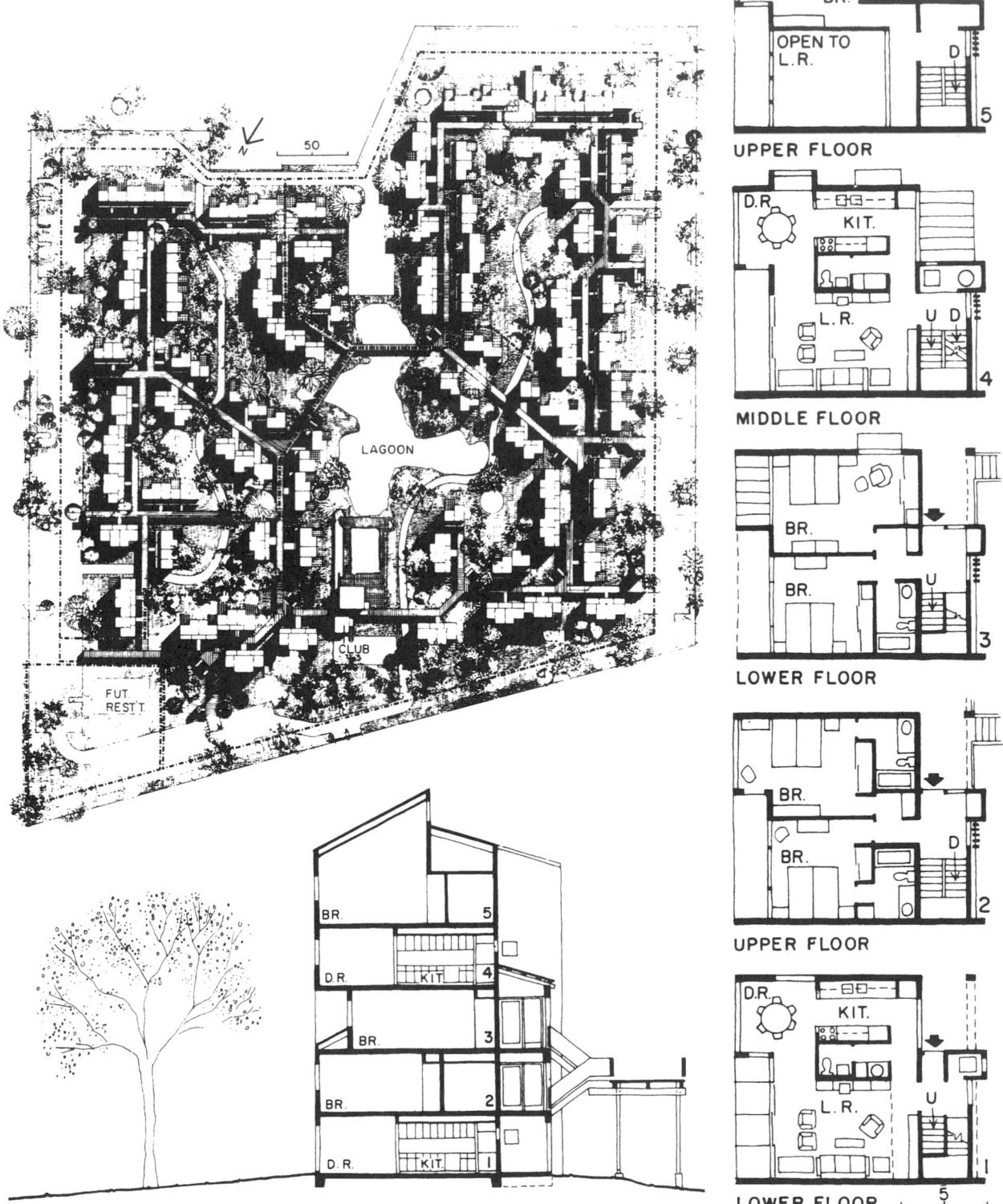

Fig. 9 The Treetops, Hilton Head, S.C. Stoller/Glasser and Marquis Associates—Architects.

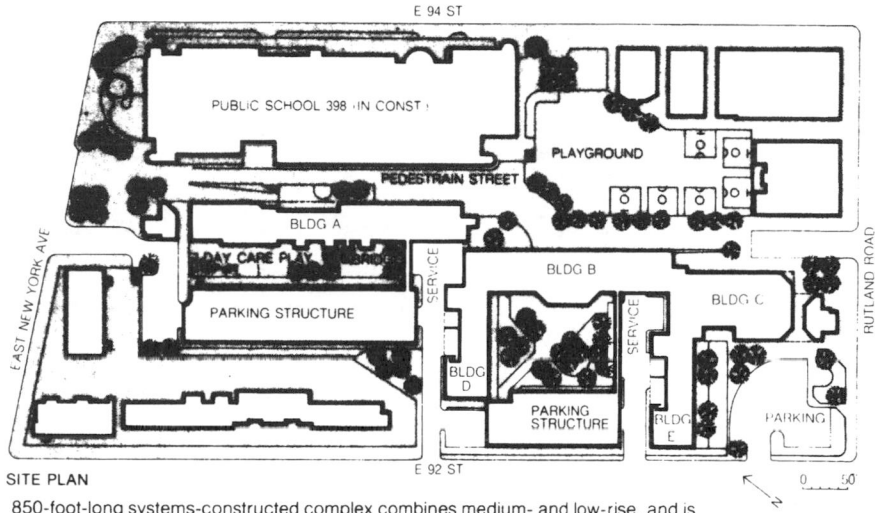

SITE PLAN

850-foot-long systems-constructed complex combines medium- and low-rise, and is staggered across two-city-block site where existing older houses were retained.

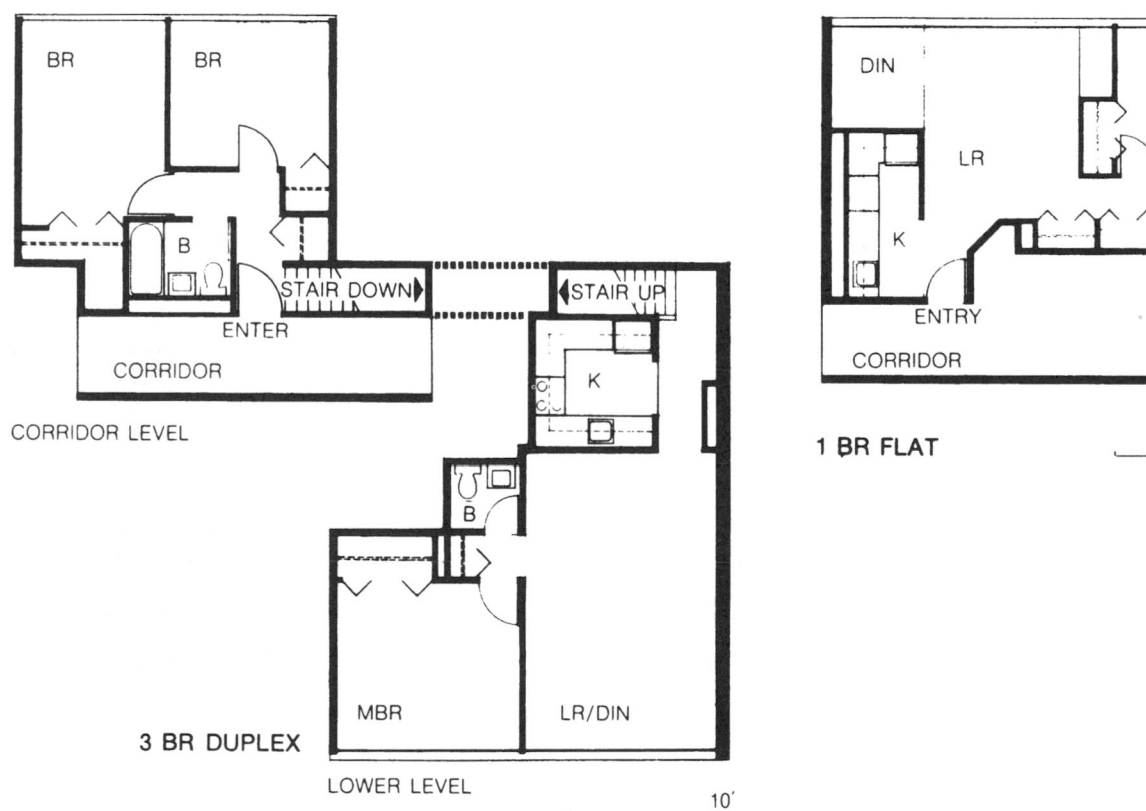

Fig. 10   Rutland Road Houses, Brooklyn, N.Y. Stull Associates, Inc.—Architects.

DUPLEX APARTMENT

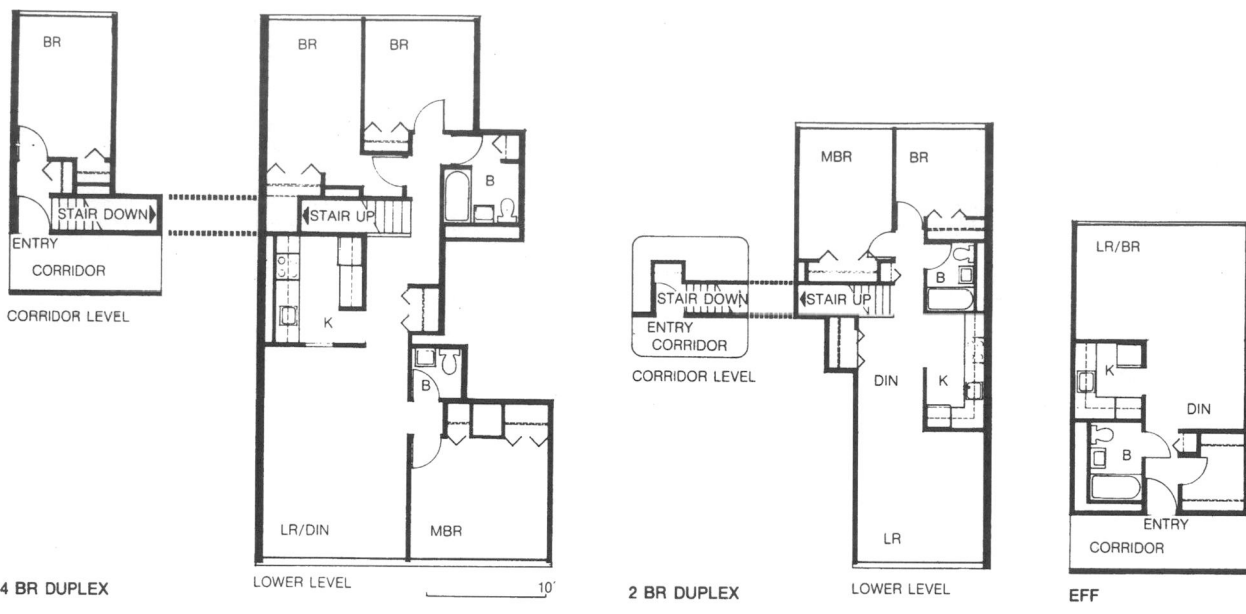

4 BR DUPLEX LOWER LEVEL 2 BR DUPLEX LOWER LEVEL EFF

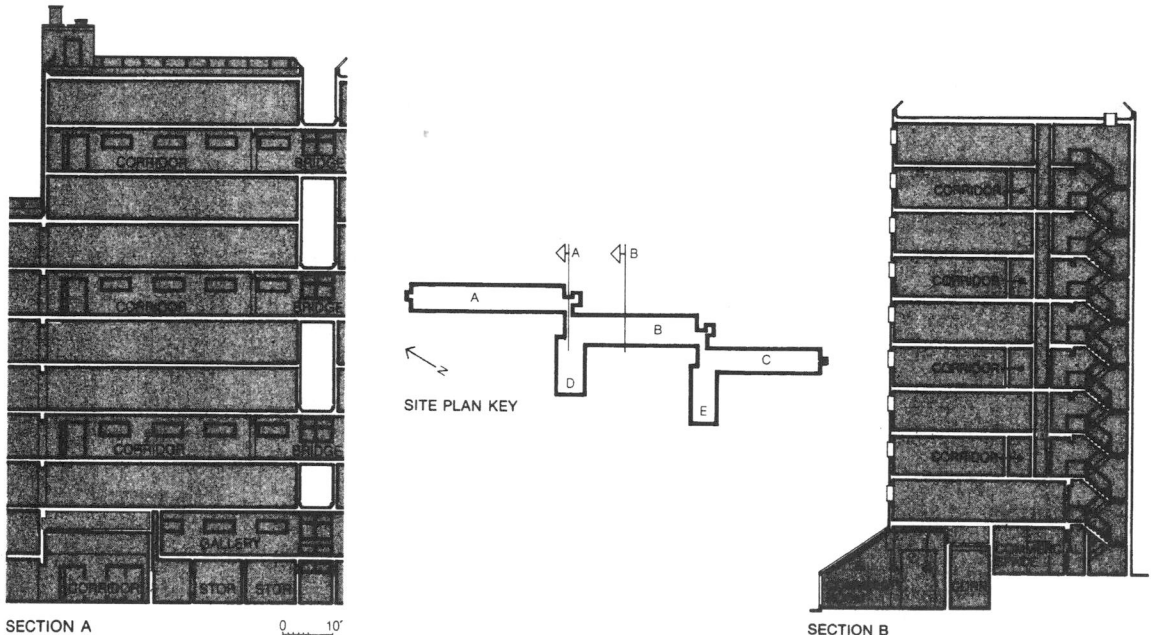

SECTION A SITE PLAN KEY SECTION B

Fig. 11 Rutland Road Houses, Brooklyn, N.Y. Stull Associates, Inc.—Architects.

The intricate idea in Fig. 12 works clearly and well: Living and dining facilities of each duplex are on the lower level; bedrooms are on the next higher level. The scheme steps vertically up the building, so that bedrooms of one duplex sit next to the living rooms of the next. Connecting stairways are tucked neatly into vacant cubage in the fire stair, packing it full.

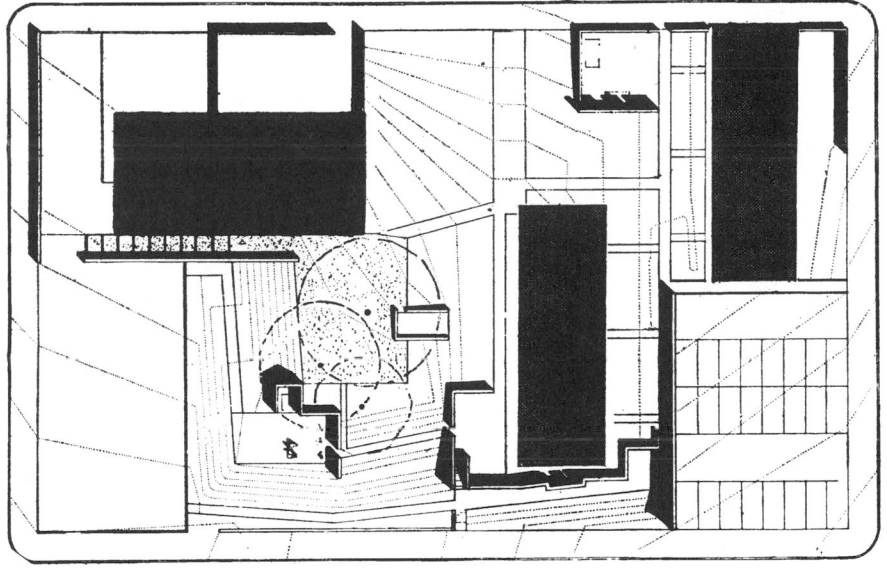

Fig. 12 Mill Creek Apartments, Philadelphia, Pa. Kahn, McAllister, Braik, & Day—Architects.

TRIPLEX APARTMENT

TRIPLEX APARTMENT

The triplex apartment is, as its name implies, located on three different levels. This type of apartment is restricted to the most luxurious high-rise apartment buildings or to the three-level town houses. The kitchen, dining, and living rooms are generally located on the lower levels while the bedrooms are on the upper levels. The separation of activities creates greater privacy and livability. In high-rise buildings, corridors on the upper levels may be omitted, but interior stairs must be provided. To justify the use of stairs, the apartment must be relatively large, consisting of three bedrooms or more. Such apartments can only be supported in private luxury-type buildings. Even then triplex apartments are not extensively used.

A variation of the triplex utilizing three half levels of floors, similar to a split-level house, is shown. A compact unit and interesting space relationships are possible with this approach. However, the most common use of this type of housing unit is with the town house. In this arrangement, the living area may be at the lowest level and the bedrooms on the upper two levels. An alternate arrangement is to have the living activities on the middle level, sleeping on the top level, and a garage, family recreation room, or den on the lowest level.

Essentially, this type of housing unit approximates the traditional three-story, single-family house.

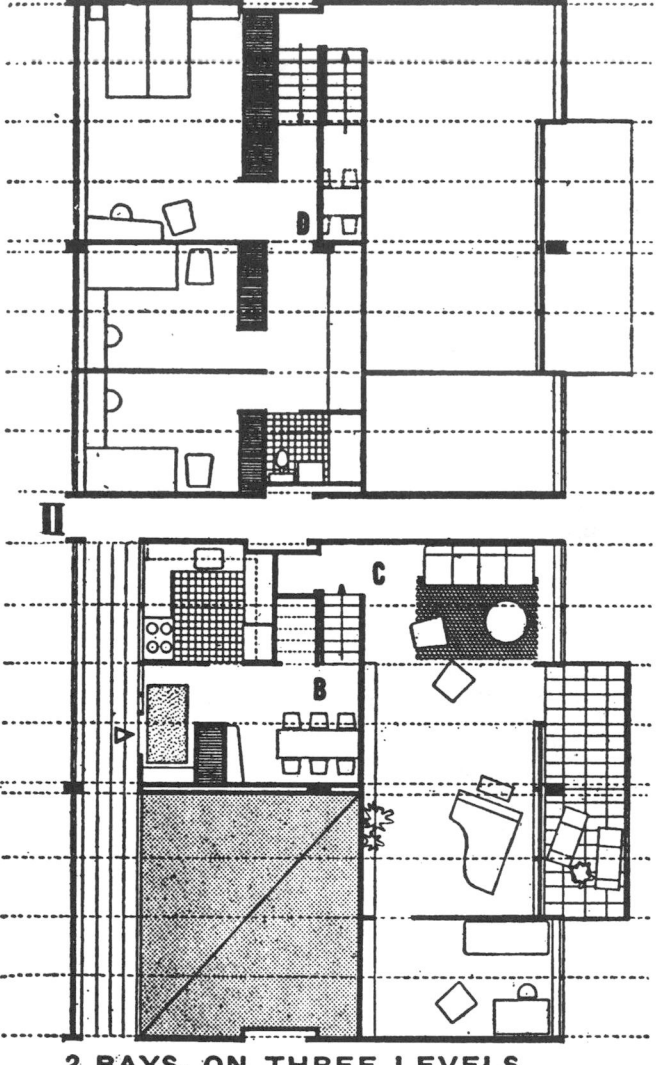

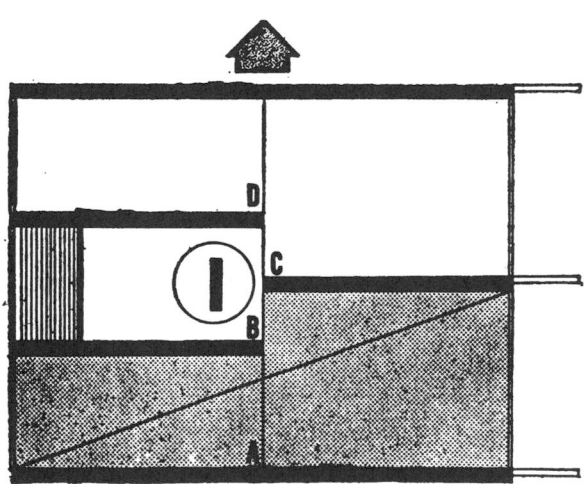

Fig. 1

2 BAYS, ON THREE LEVELS

EFFICIENCY APARTMENT

Elements

The efficiency apartment consists mainly of one large room combining living, eating, and sleeping activities. An alcove may be provided for a kitchenette with minimum facilities. It is also provided with a full bathroom.

Design

The essential design feature is the flexibility of the main space to be used alternatively for living, sleeping, and dining. Most often, a convertible sofa bed is used to achieve this flexibility, and the entry foyer can be used as a dining area. A critical problem is storage of clothes and a dressing area, which is often less than minimum.

Size

The size of an efficiency can range from approximately 200 to 500 ft². The FHA minimum requirement for an efficiency apartment is approximately 300 ft². One way to add to the spaciousness of the apartment is to provide a terrace with an all-glass wall.

Type of Occupancy

Efficiency apartments are generally occupied by single persons, young married couples without children, or elderly couples. Maximum number of occupants should be two persons.

Planning Implications

Because of the type of occupancy, no or few children are generated from this size apartment. This can easily avoid the need for new schools or the addition to an existing one. Single persons and young married couples are a relatively mobile group because of changes in status or arrival of children. As a result, this type of occupancy has more turnover than older-family units have. The elderly tend to be stable in family size.

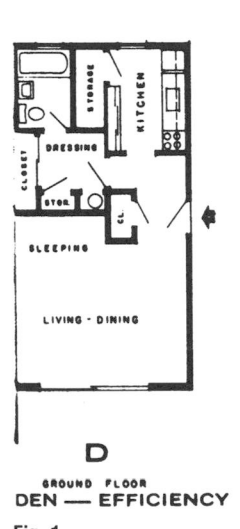

D

GROUND FLOOR
DEN — EFFICIENCY

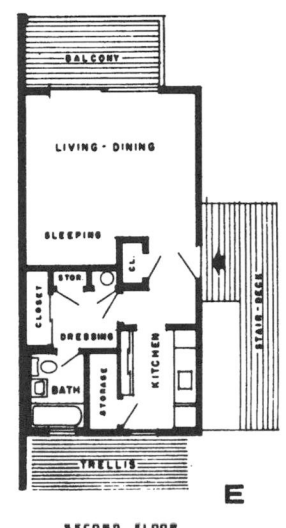

E

SECOND FLOOR
BALCONY — EFFICIENCY

Fig. 1

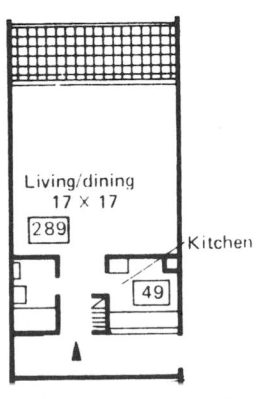

Fig. 2   Total area, 416 ft².

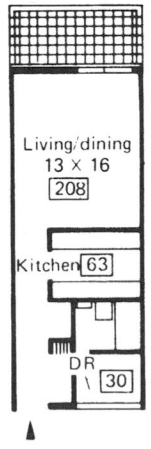

Fig. 3   Total area, 432 ft².

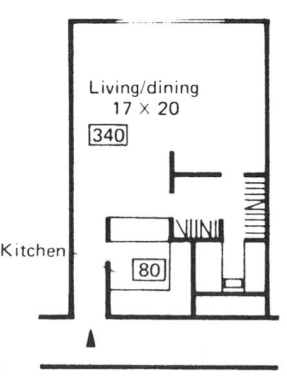

Fig. 4   Total area, 600 ft².

EFFICIENCY APARTMENT

Studio

When the efficiency unit increases substantially in size beyond the minimum size, it usually is referred to as a "studio" apartment. This type of unit occupies as much space as a one-bedroom apartment or more. However, it still is arranged as an efficiency apartment but with a feeling of openness and spaciousness. This type of apartment is usually restricted to luxury-type housing and has a special appeal for professional use such as for artists and photographers.

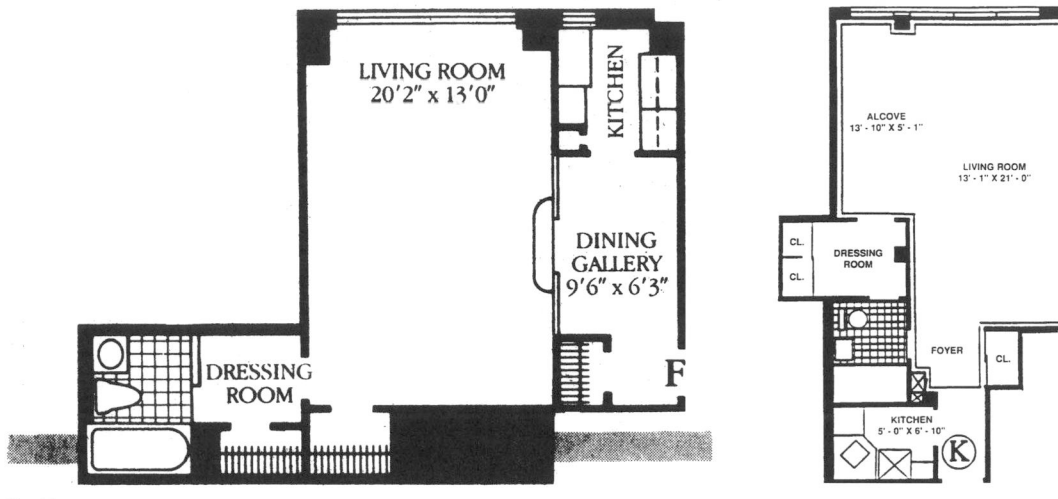

Fig. 5

Fig. 6

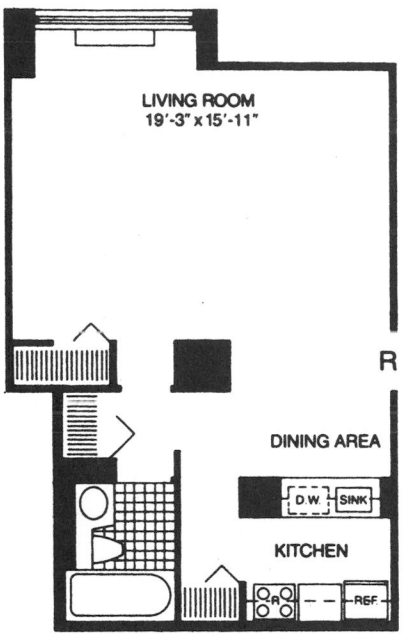

Fig. 7 One-bathroom studio apartment. Liberty Court, Battery Park City, New York, N.Y.

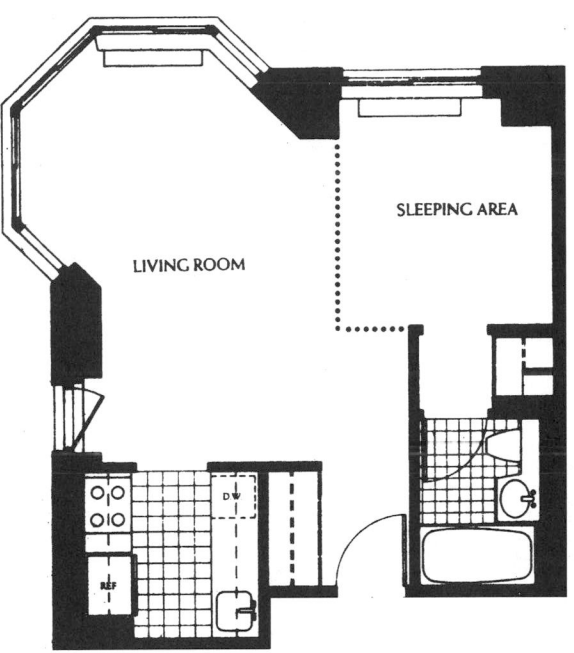

Fig. 9 Tribeca Tower, New York, N.Y.

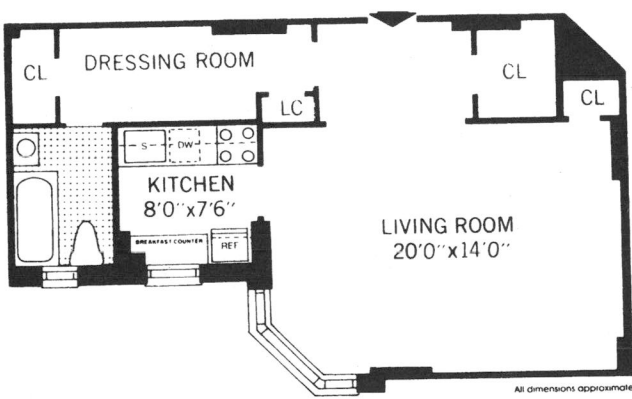

Fig. 8 Castle Village, New York, N.Y.

Fig. 10 32 Gramercy Park South, New York, N.Y.

EFFICIENCY APARTMENT

lanai
living—sleeping
dining
foyer
patio
kitchen
6'-9"
13'-0"
10'-6"
5'-9"
36'-0"
15'-0"
lanai
sleeping
living
kitchen
foyer
entrance patio
6'-9"
10'-0"
10'-8"
5'-4"
4'-9"
37'-4"
18'-0"

EFFICIENCY APARTMENT **JUMBO EFFICIENCY APARTMENT**

Fig. 11 Queen Emma Plaza, Honolulu, Hawaii. I. M. Pei & Partners—Architects.

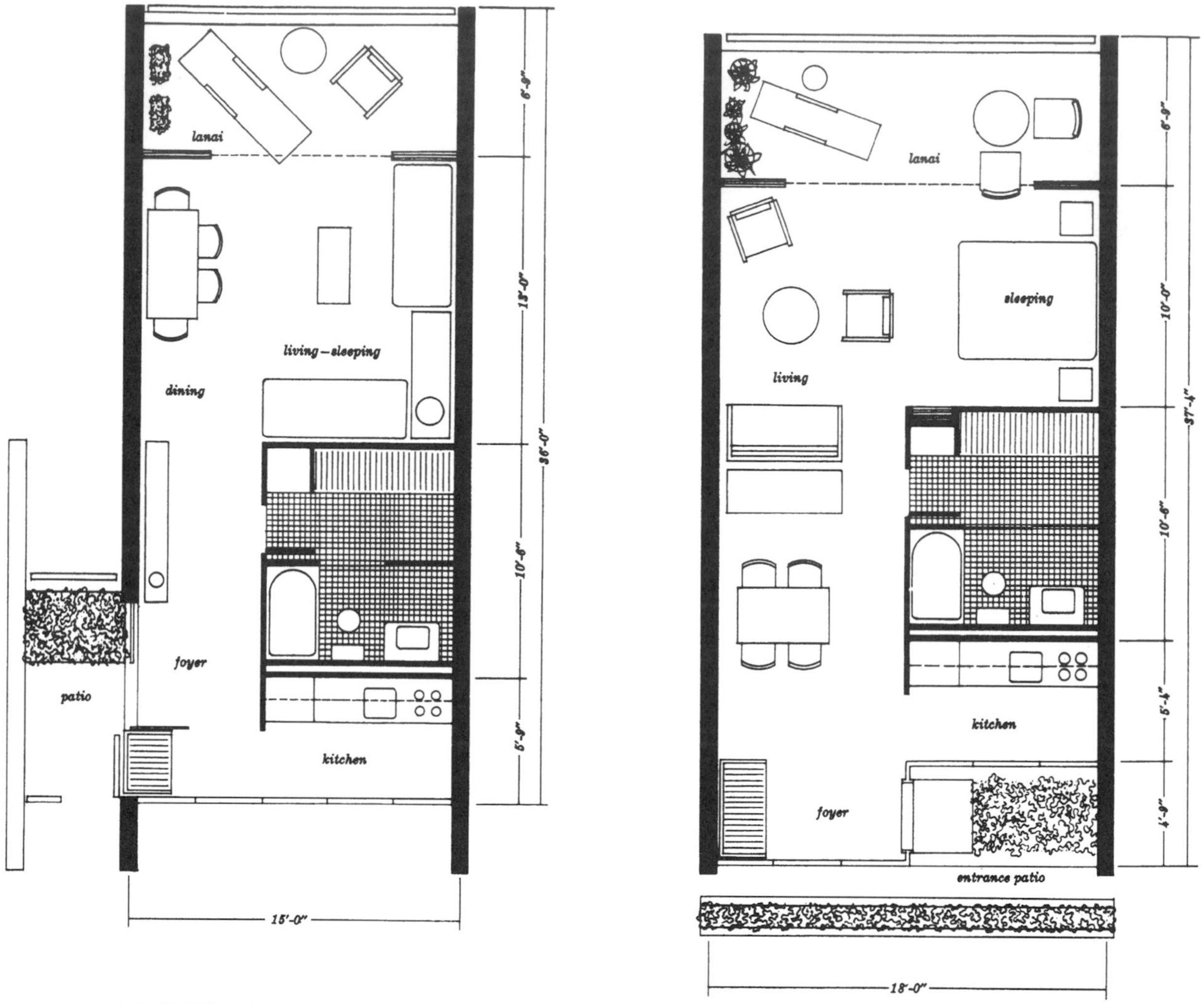

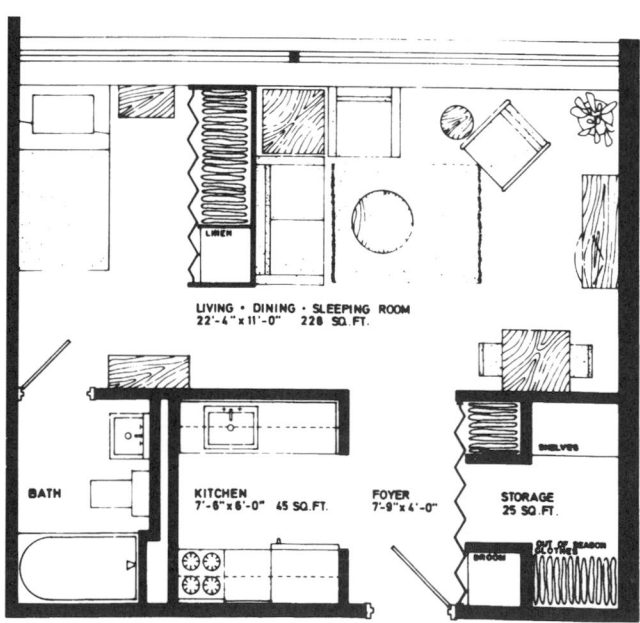

LINEN
LIVING · DINING · SLEEPING ROOM
22'-4" x 11'-0" 228 SQ. FT.
SHELVES
BATH
KITCHEN
7'-6" x 6'-0" 45 SQ. FT.
FOYER
7'-9" x 4'-0"
STORAGE
25 SQ. FT.
BROOM
OUT OF SEASON CLOTHES

Fig. 12 Charlotte Area Project, Rochester, N.Y. Northrup, Kaelber & Kopf—Architects.

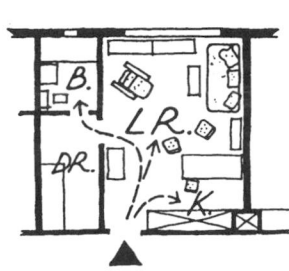

Fig. 13a Efficiency apartment with kitchenette unit.

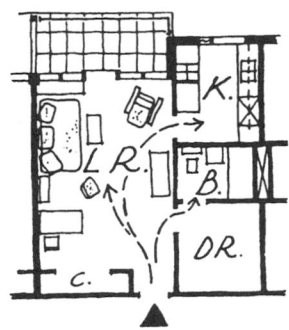

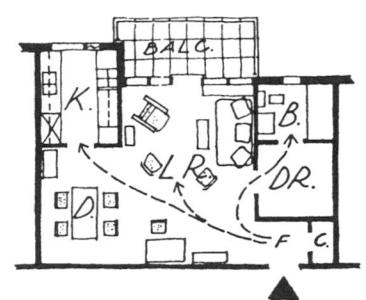

Fig. 13b Efficiency apartment with separate kitchen.

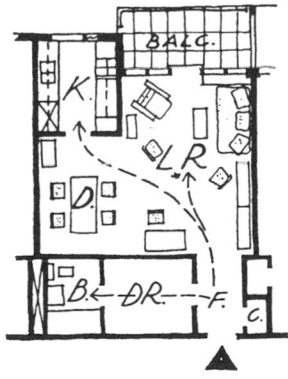

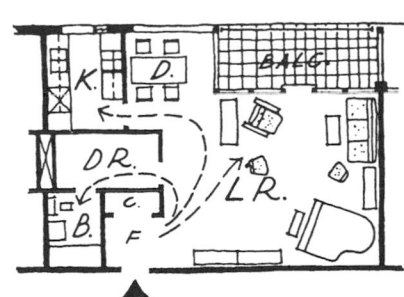

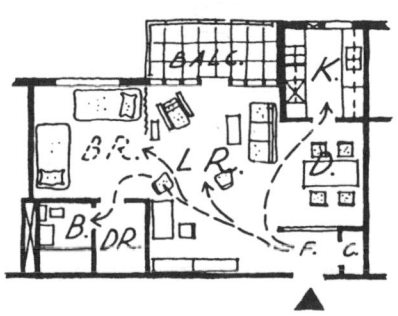

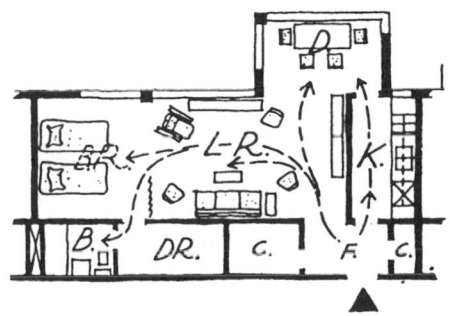

Fig. 13c Studio apartment with separate kitchen and sleeping area.

ONE-BEDROOM APARTMENT

ONE-BEDROOM APARTMENT

Elements

The one-bedroom consists of (1) a living-dining room (a separate dining room is very rare), (2) a kitchen area, (3) a bedroom, (4) a bathroom, (5) an outdoor terrace, which is optional.

Design

The main object of a one-bedroom apartment is its compactness. A full range of activities is anticipated within a minimal area. The foyer is frequently used as a dining space. The kitchen is often minimal.

Size

The size of a one-bedroom apartment can range from 400 to 600 ft². The FHA minimum requirement for a one-bedroom apartment is approximately 500 ft². The addition of an outdoor terrace will add to the spaciousness of the apartment.

Type of Occupancy

One-bedroom apartments are occupied by two or three persons. This could include a wide range of individuals, such as young married couples with or without a child, elderly persons, or unrelated single persons sharing an apartment. Whether the occupancy is stable or has a quick turnover will be directly related to the type of occupancy.

Planning Implications

The one-bedroom apartment can be expected to yield a greater number of children than the efficiency. Also, this type of apartment can be used as a transition between home ownership and relocation to a different community by elderly persons.

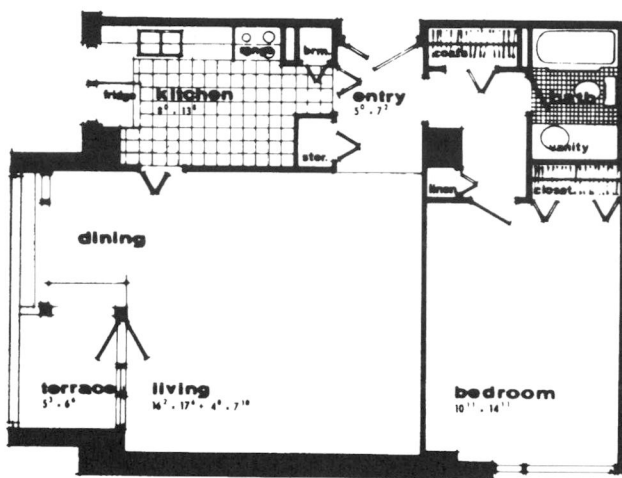

Fig. 1

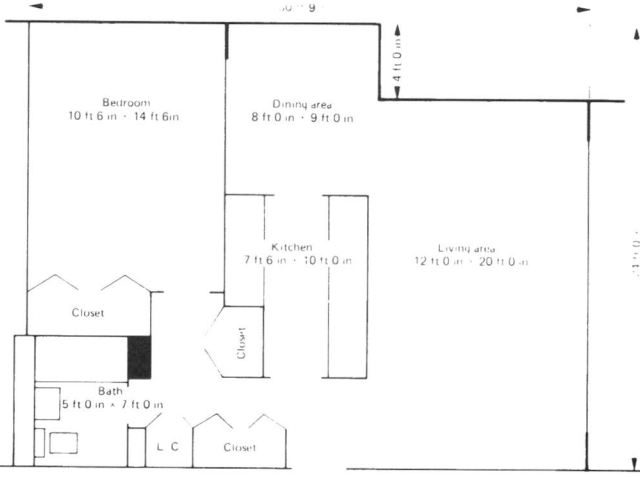

Fig. 2   One-bedroom family apartment, 690 ft². N.J. Housing Finance Agency.

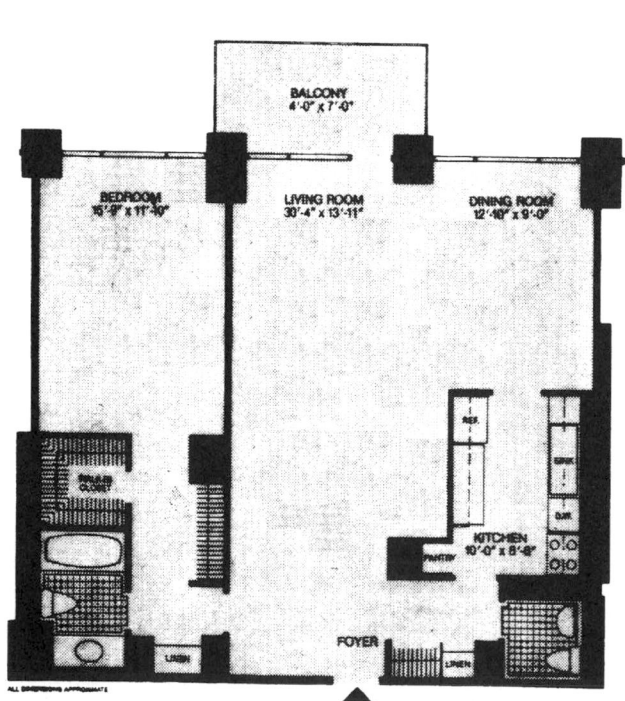

Fig. 3   Century Tower, Fort Lee, N.J.

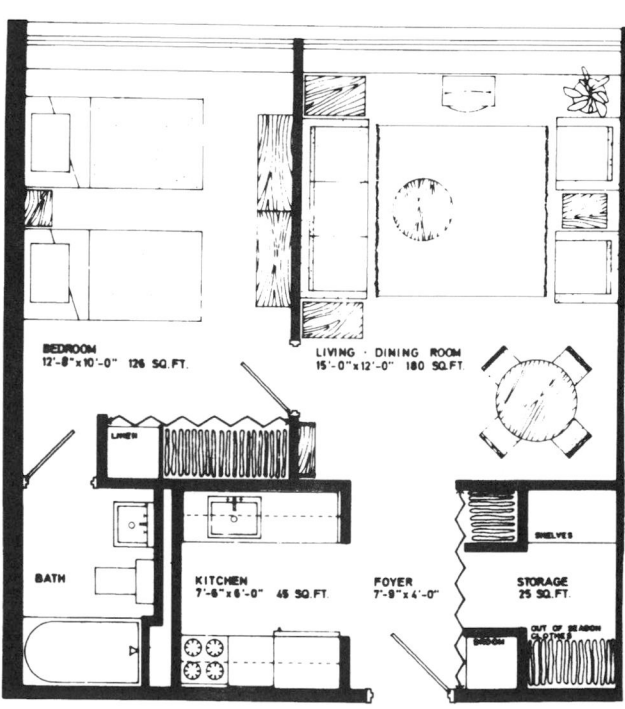

TYPICAL ONE BEDROOM APARTMENT

GROSS AREA 575 SQ. FT.

Fig. 4

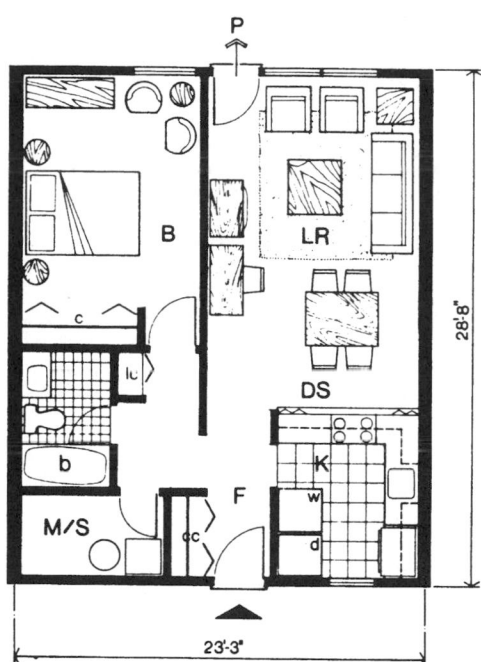

Fig. 5

In the one-bedroom apartment shown in Fig. 5, the gross area is 620 ft^2.

Features	Dimensions, ft	Area, ft^2
Living room	12×18.5	222
Dining space		
Kitchen	7.5×8	60
Bedroom	10×13	130
Bathroom	5×7.5	38

The unit in Fig. 5 has standard room sizes, good entrance privacy and protection, efficient organization, internal privacy for bath and bedroom, and good circulation paths minimize disruption of activities. The orientation of the unit is toward the private outdoor space away from main entrance.

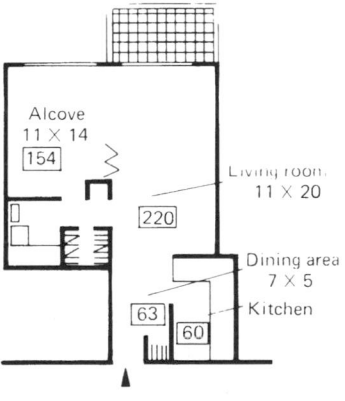

Fig. 6   Total area, 563 ft^2.

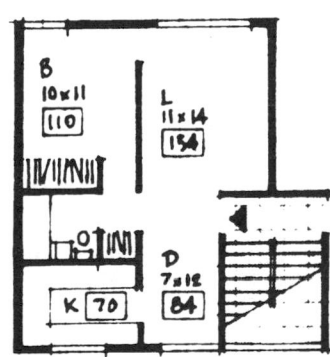

Fig. 7   Total area, 522 ft^2.

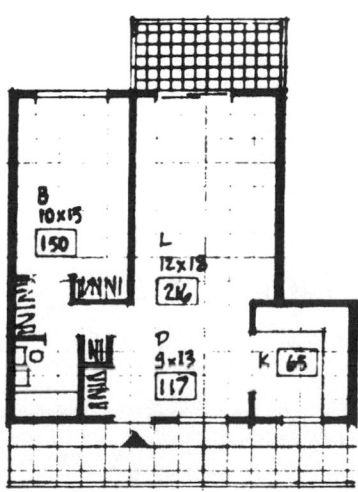

Fig. 8   Total area, 650 ft^2.

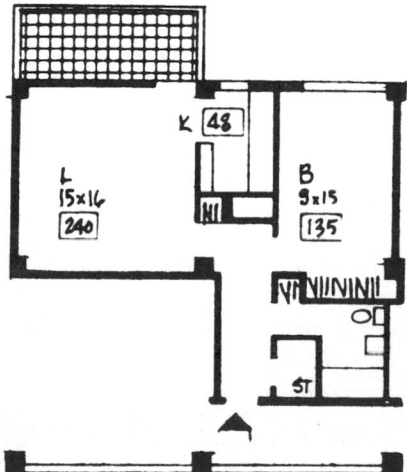

Fig. 9   Total area, 600 ft^2.

ONE-BEDROOM APARTMENT

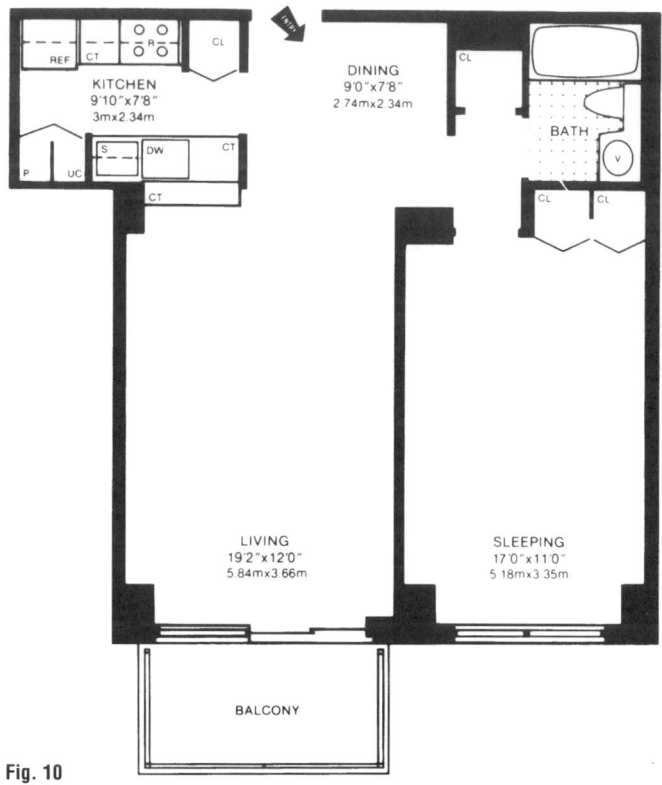

KITCHEN
9'10"x7'8"
3mx2.34m

DINING
9'0"x7'8"
2.74mx2.34m

BATH

LIVING
19'2"x12'0"
5.84mx3.66m

SLEEPING
17'0"x11'0"
5.18mx3.35m

BALCONY

Fig. 10

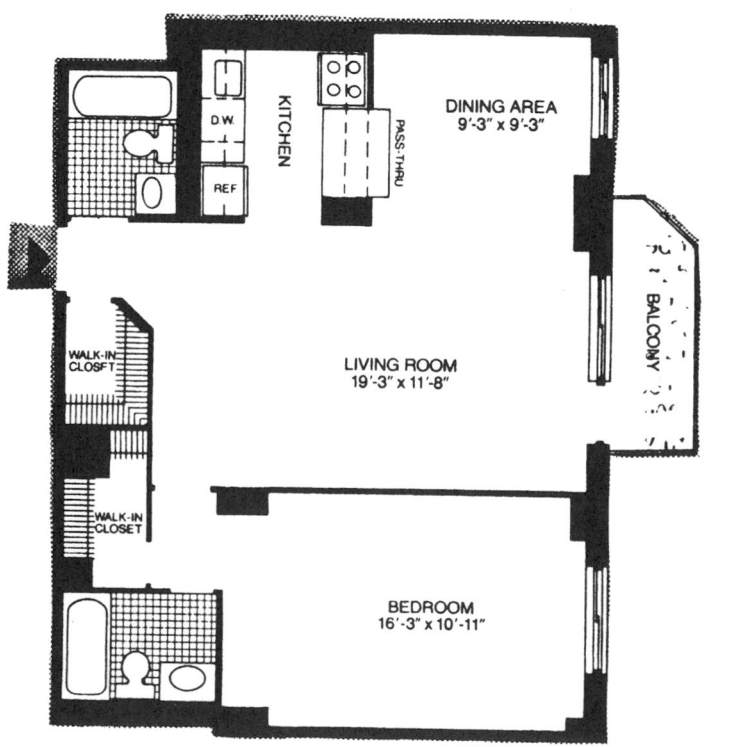

DINING AREA
9'-3" x 9'-3"

KITCHEN

PASS-THRU

WALK-IN CLOSET

LIVING ROOM
19'-3" x 11'-8"

BALCONY

WALK-IN CLOSET

BEDROOM
16'-3" x 10'-11"

Fig. 11

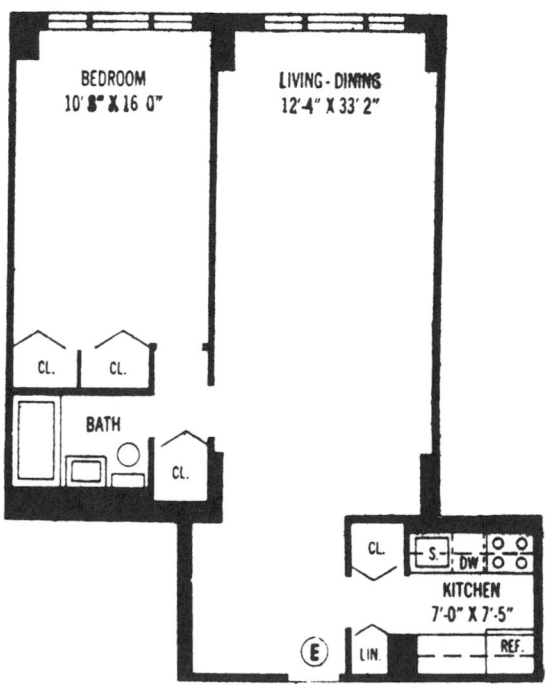

BEDROOM
10' 8" X 16' 0"

LIVING - DINING
12'-4" X 33' 2"

CL.   CL.

BATH

CL.

CL.

KITCHEN
7'-0" X 7'-5"

LIN.

REF.

Fig. 12   Parker 86th, New York, N.Y.

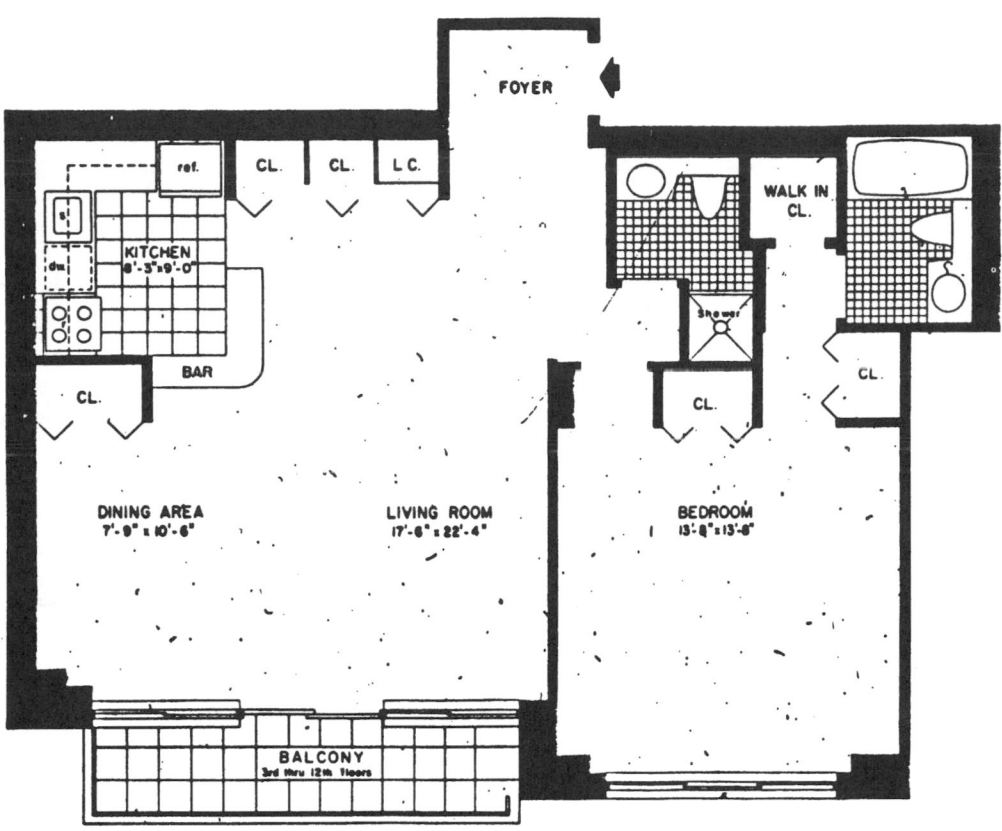

Fig. 13   Carriage House Condominium, 510 E. 80th St., New York, N.Y.

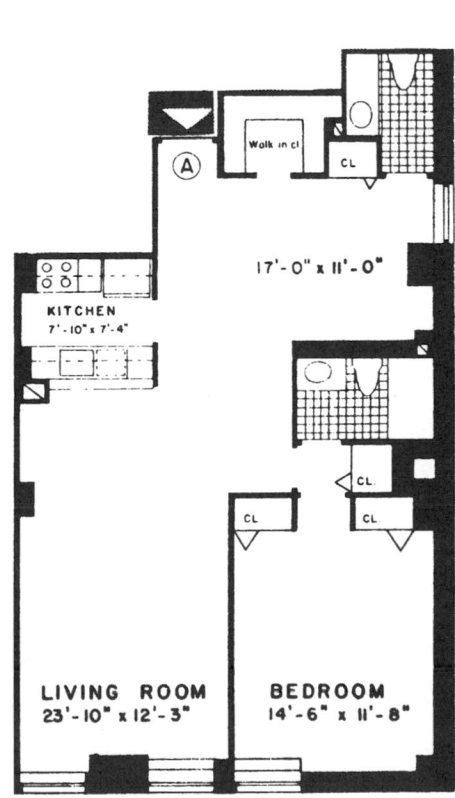

Fig. 14   Morgan House, 153 E. 87th St., New York, N.Y.

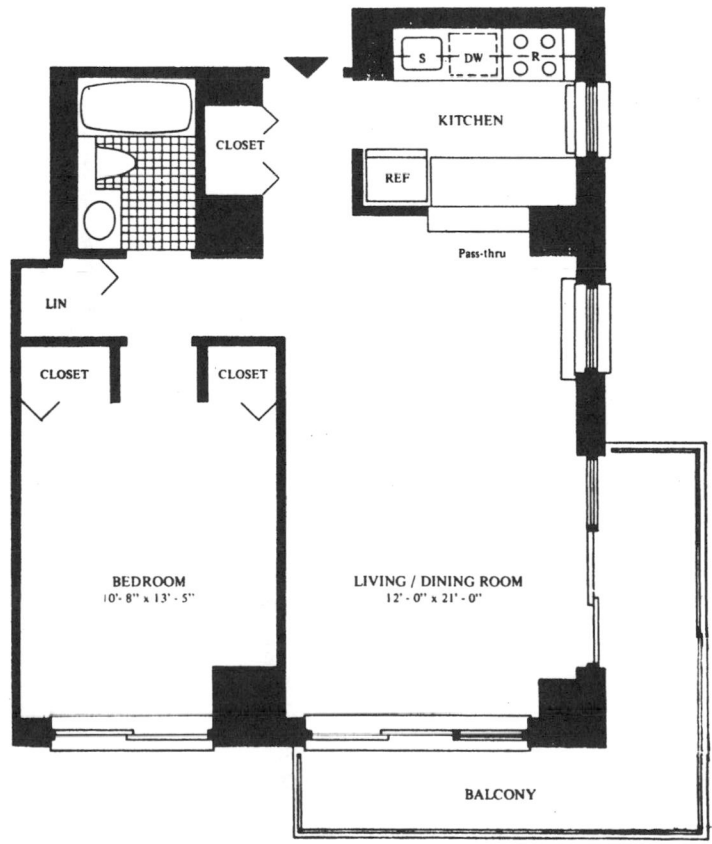

Fig. 15   The Monarch, 260 E. 84th St., New York, N.Y.

ONE-BEDROOM APARTMENT

LIVING ROOM
23'-6" x 13'-5"

BALCONY

BEDROOM
15'-6" x 12'-2"

walk in
closet

DINING ROOM/DEN
14'-5" x 11'-10"

GALLERY
9'-8" x 12'-11"

KITCHEN

LIVING ROOM
29'-2" x 13'-4"

BALCONY

DINING

BEDROOM
15'-3" x 12'-1"

GALLERY
10'-0" x 9'-6"

KITCHEN

Fig. 16   The Corinthian Condominiums, New York, N.Y.

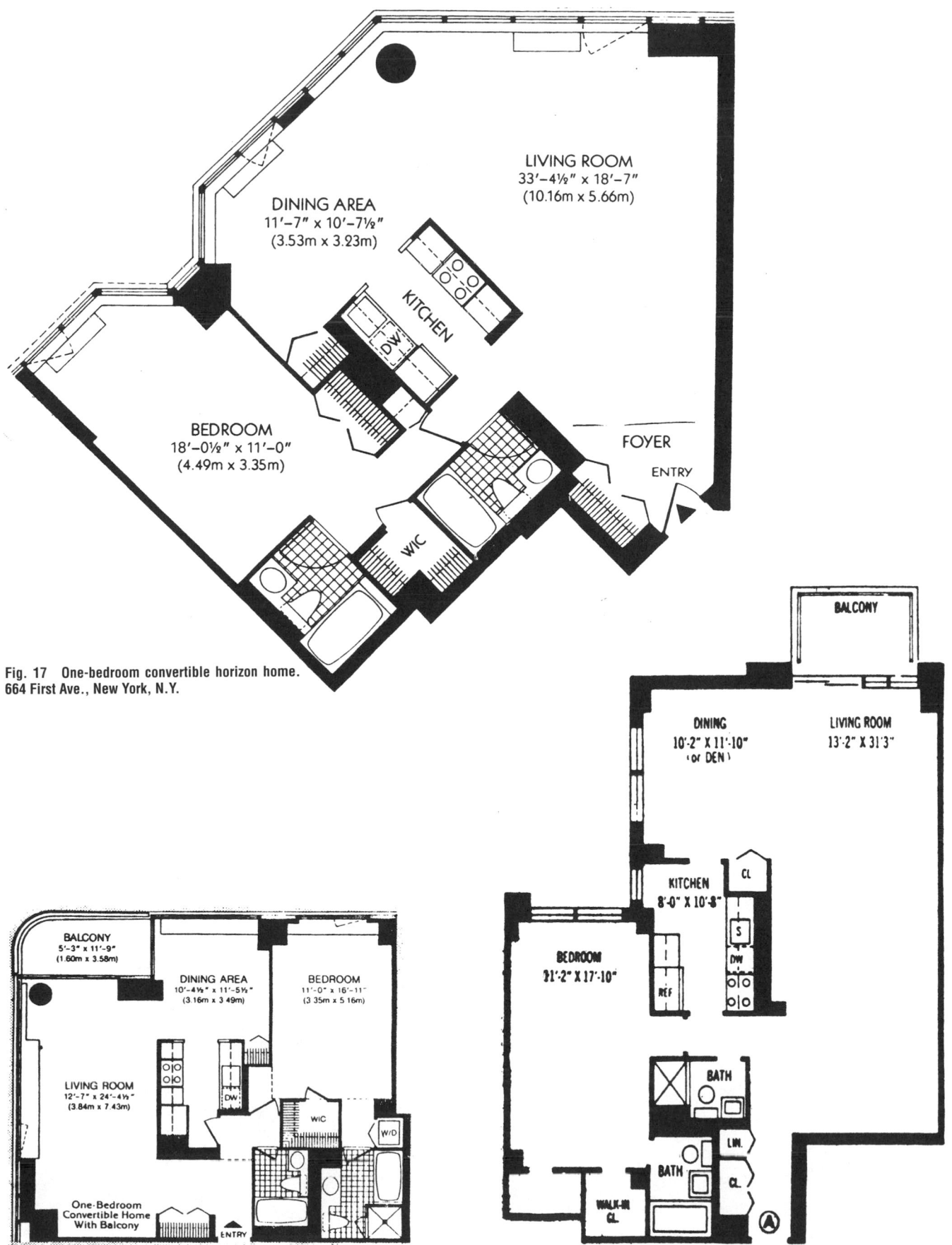

LIVING ROOM
33'-4½" x 18'-7"
(10.16m x 5.66m)

DINING AREA
11'-7" x 10'-7½"
(3.53m x 3.23m)

KITCHEN

DW

BEDROOM
18'-0½" x 11'-0"
(4.49m x 3.35m)

FOYER

ENTRY

WIC

Fig. 17 One-bedroom convertible horizon home.
664 First Ave., New York, N.Y.

BALCONY

DINING
10'-2" X 11'-10"
(or DEN)

LIVING ROOM
13'-2" X 31'3"

KITCHEN
8'-0" X 10'-8"

CL

S

DW

REF

BEDROOM
21'-2" X 17'-10"

BATH

BATH

LIN.

CL.

WALK-IN
CL.

Ⓐ

BALCONY
5'-3" x 11'-9"
(1.60m x 3.58m)

DINING AREA
10'-4½" x 11'-5½"
(3.16m x 3.49m)

BEDROOM
11'-0" x 16'-11"
(3.35m x 5.16m)

LIVING ROOM
12'-7" x 24'-4½"
(3.84m x 7.43m)

DW

WiC

W/D

One-Bedroom
Convertible Home
With Balcony

ENTRY

Fig. 18 The Promenade. 530 E. 76th St., New York, N.Y.

Fig. 19 Parker 86th, New York, N.Y.

ONE-BEDROOM APARTMENT

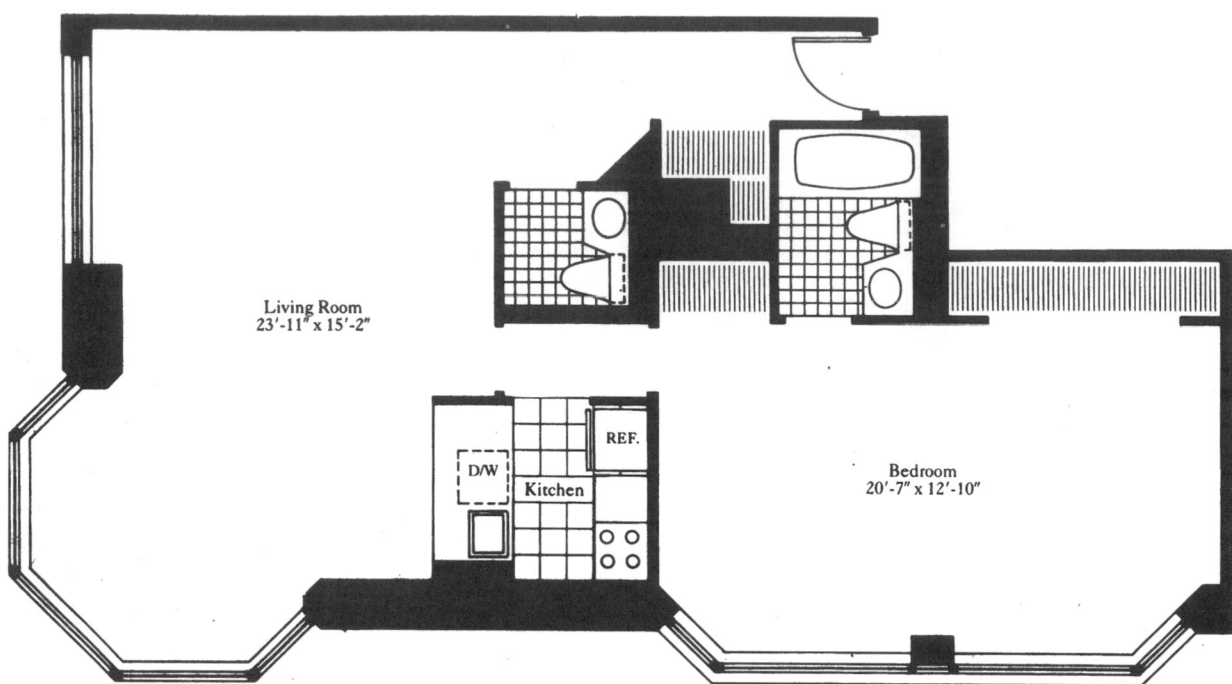

Fig. 20   200 E. 65th St., New York, N.Y.

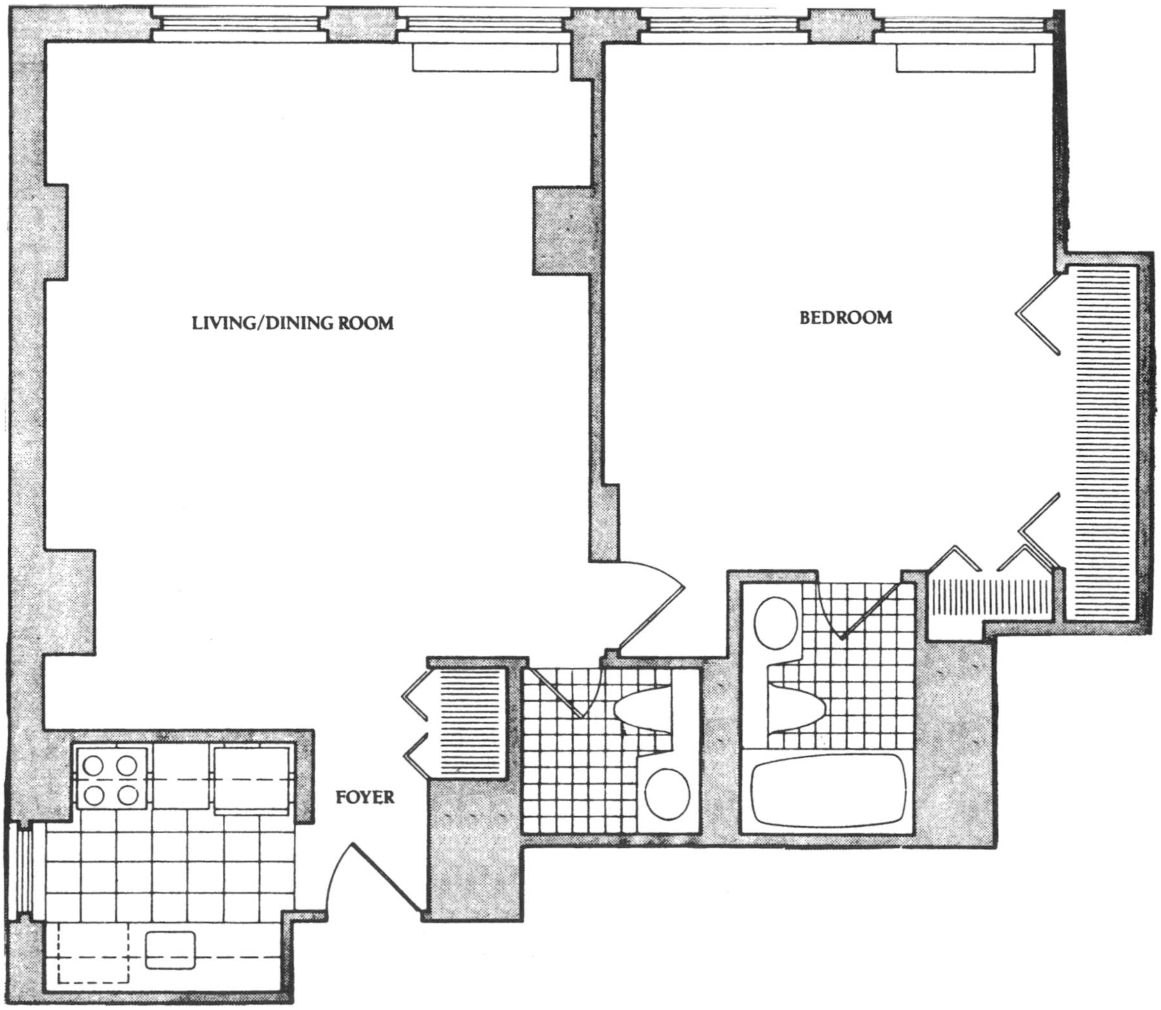

Fig. 21   255 W. 85th St., New York, N.Y.

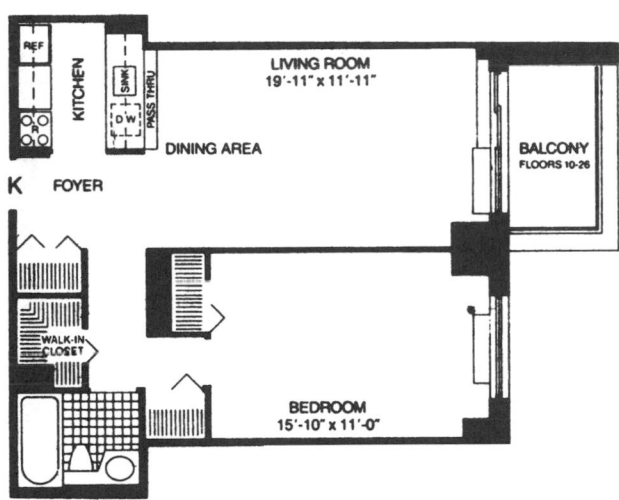

Fig. 22 One-bedroom one-bath apartment. Liberty Court, Battery Park City, New York, N.Y.

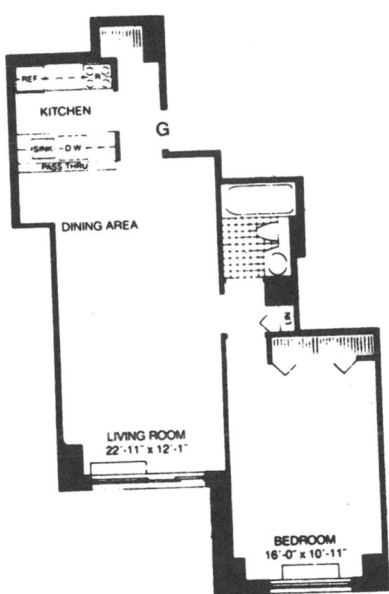

Fig. 23

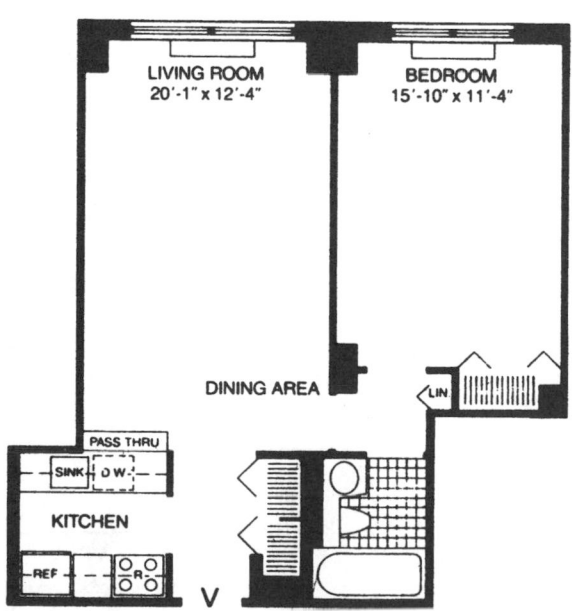

Fig. 24 One-bedroom one-bath apartment. Liberty Court, Battery Park City, New York, N.Y.

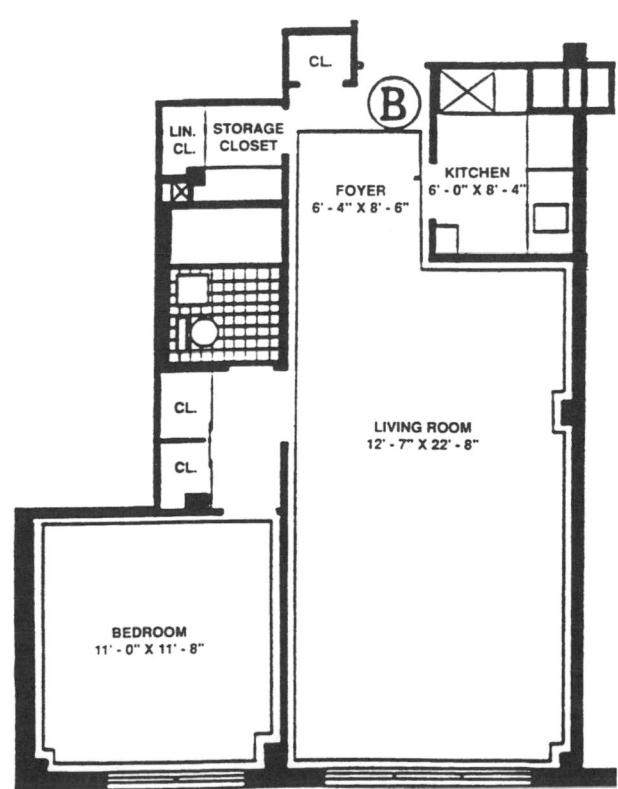

Fig. 25 32 Gramercy Park South, New York, N.Y.

ONE-BEDROOM APARTMENT

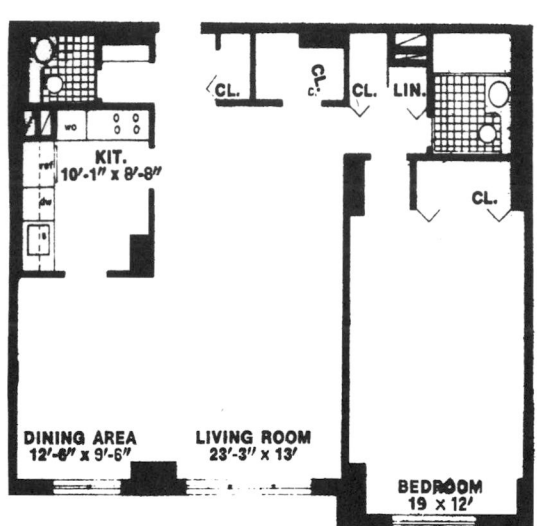

Fig. 26   The Century, Riverdale, N.Y.C.

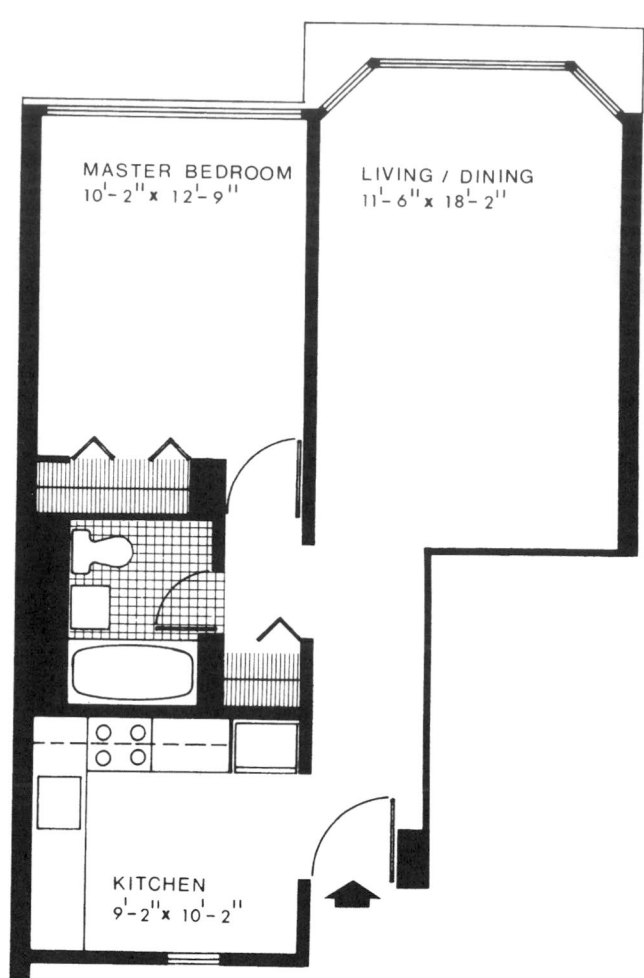

Fig. 27   Eastwood, Roosevelt Island, N.Y.C.

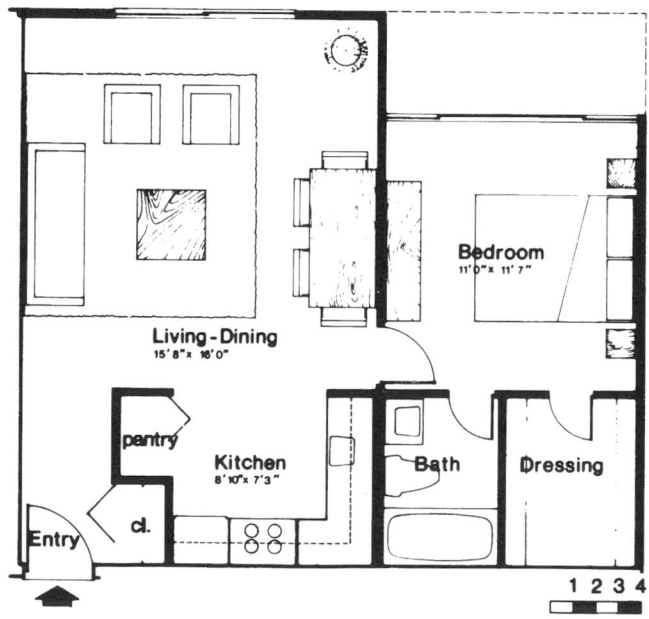

Fig. 28

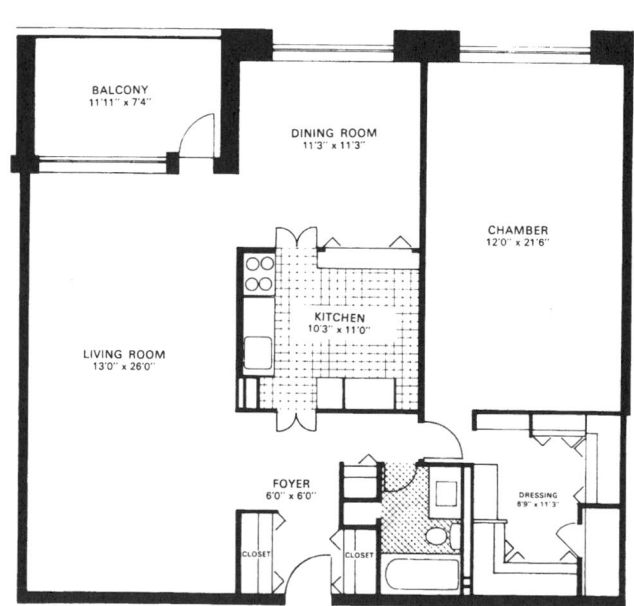

Fig. 29

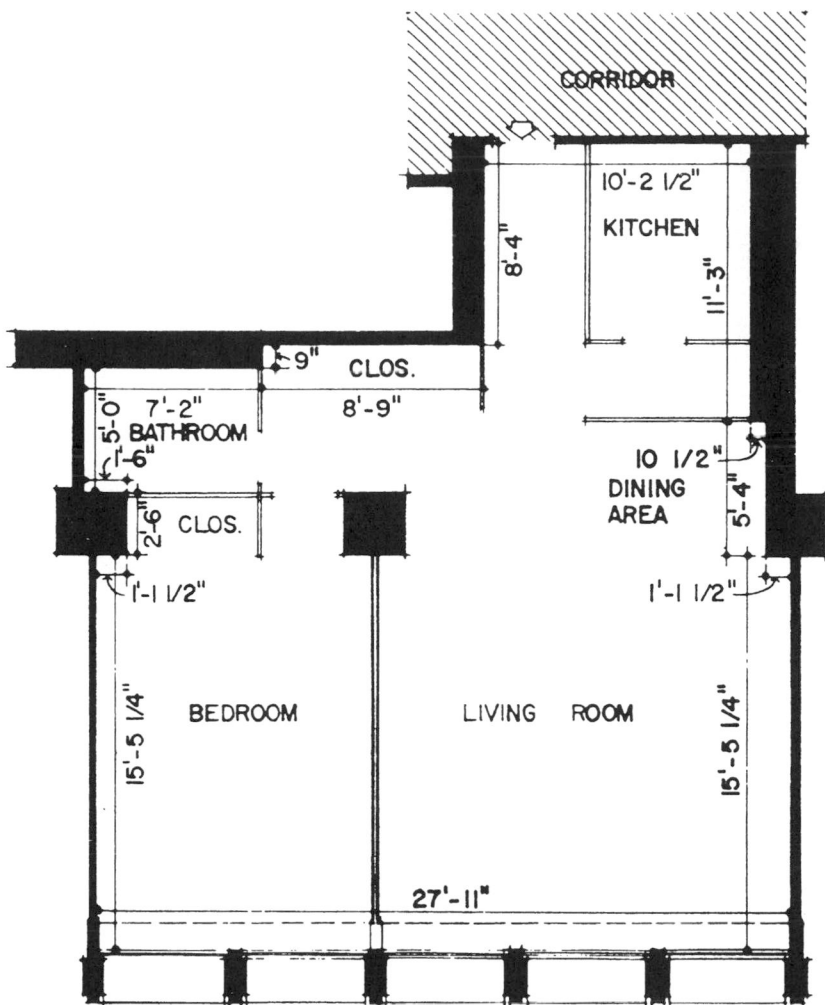

Fig. 30   Kips Bay Plaza, New York, N.Y.

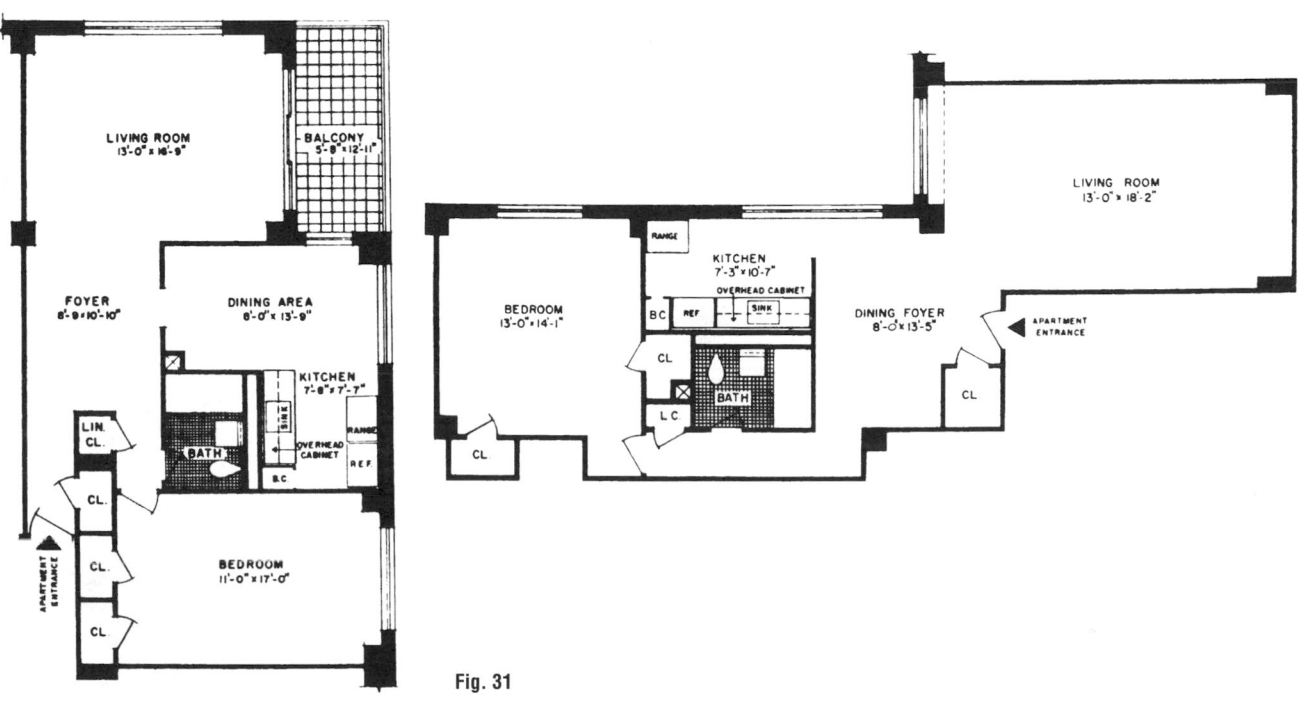

Fig. 31

ONE-BEDROOM APARTMENT

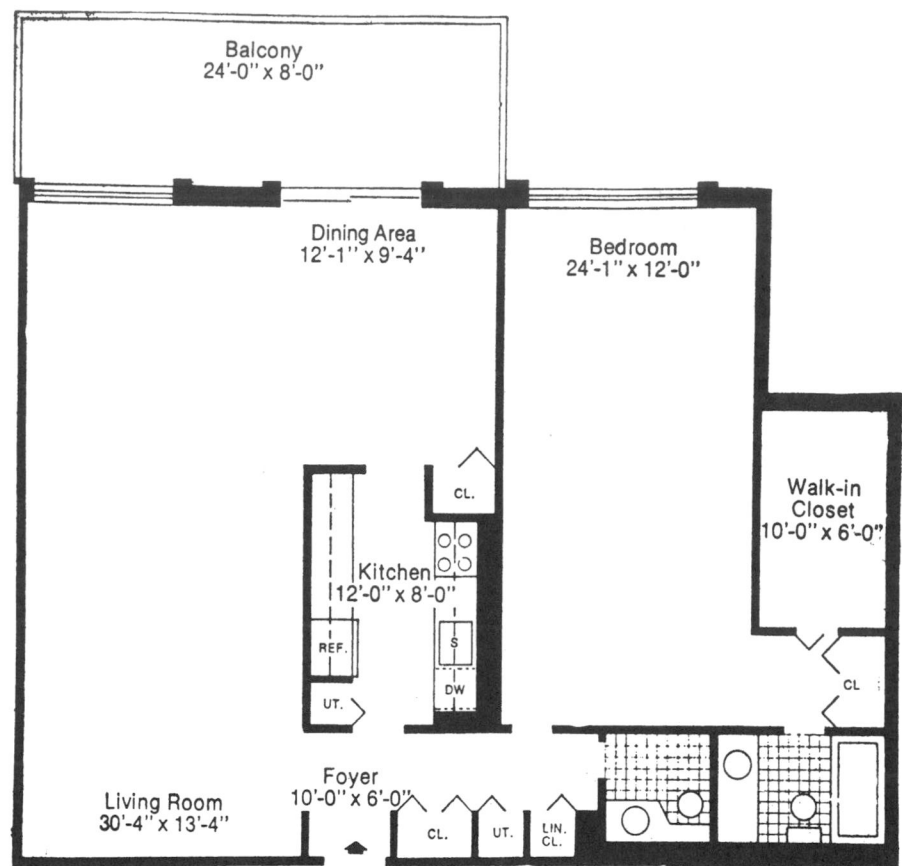

Fig. 32   The Parke Imperial, North Bergen, N.J.

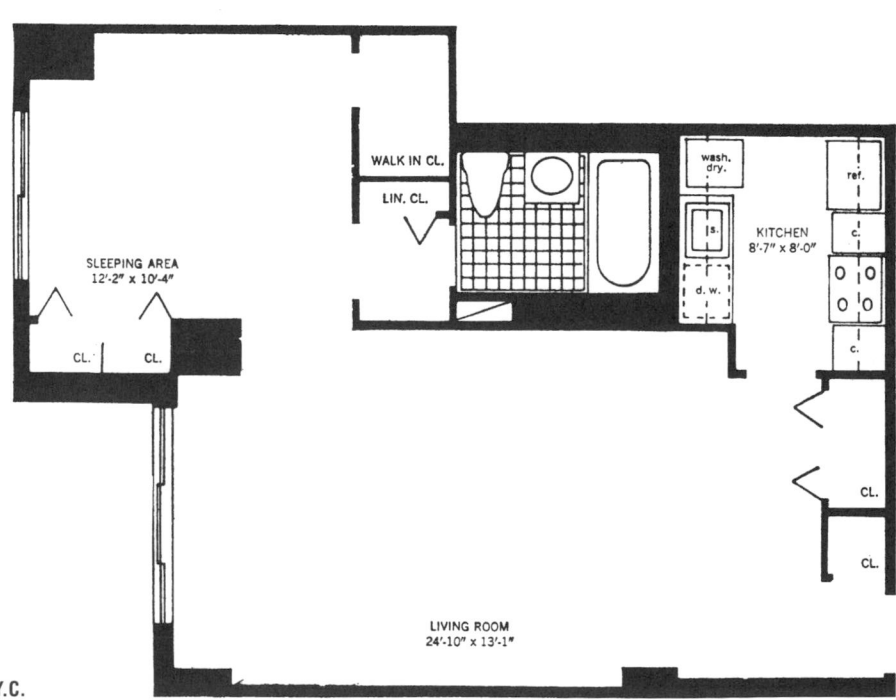

Fig. 33   North Shore Towers, Queens, N.Y.C.

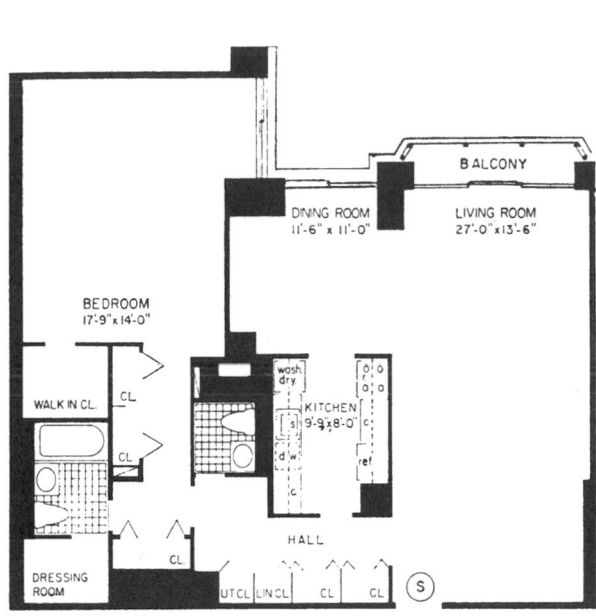

Fig. 34   North Shore Towers, Queens, N.Y.C.

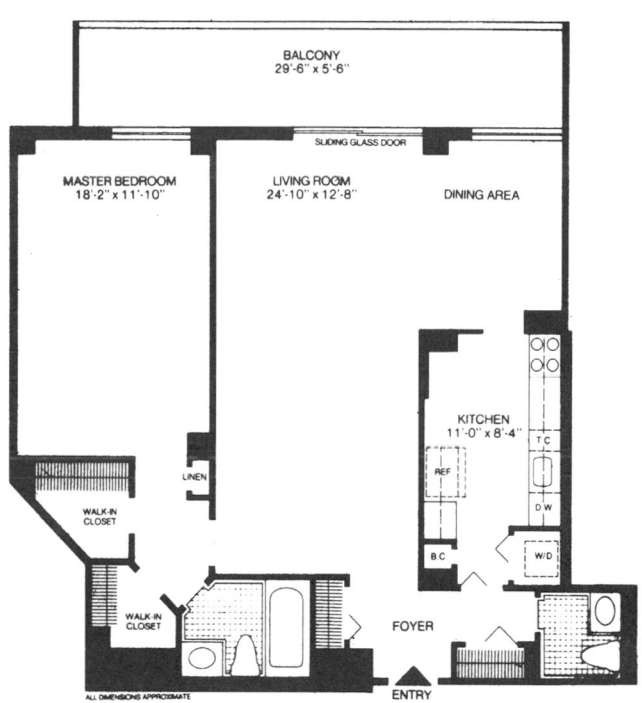

Fig. 35   Channel Club Tower, Monmouth Beach, N.J.

Fig. 36   Century Tower, Fort Lee, N.J.

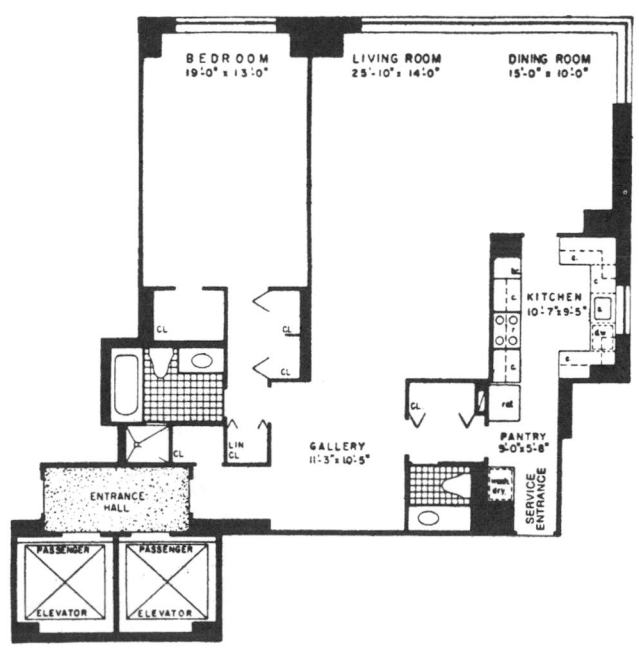

Fig. 37   The Sovereign, New York, N.Y.

ONE-BEDROOM APARTMENT

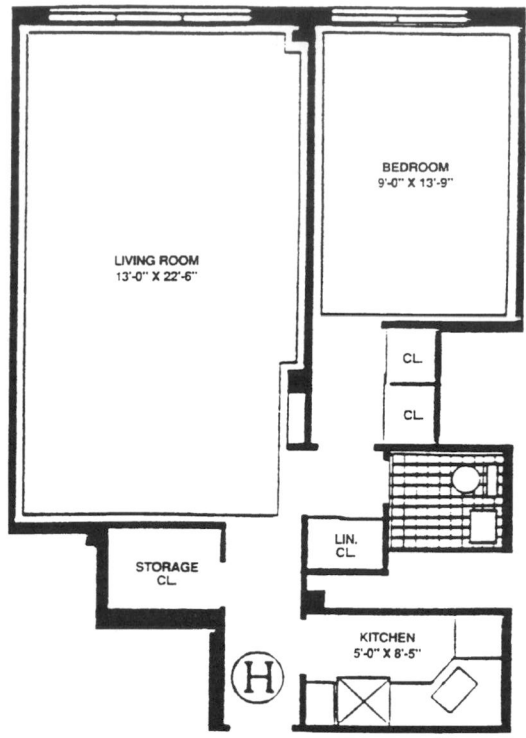

Fig. 38   32 Gramercy Park South, New York, N.Y.

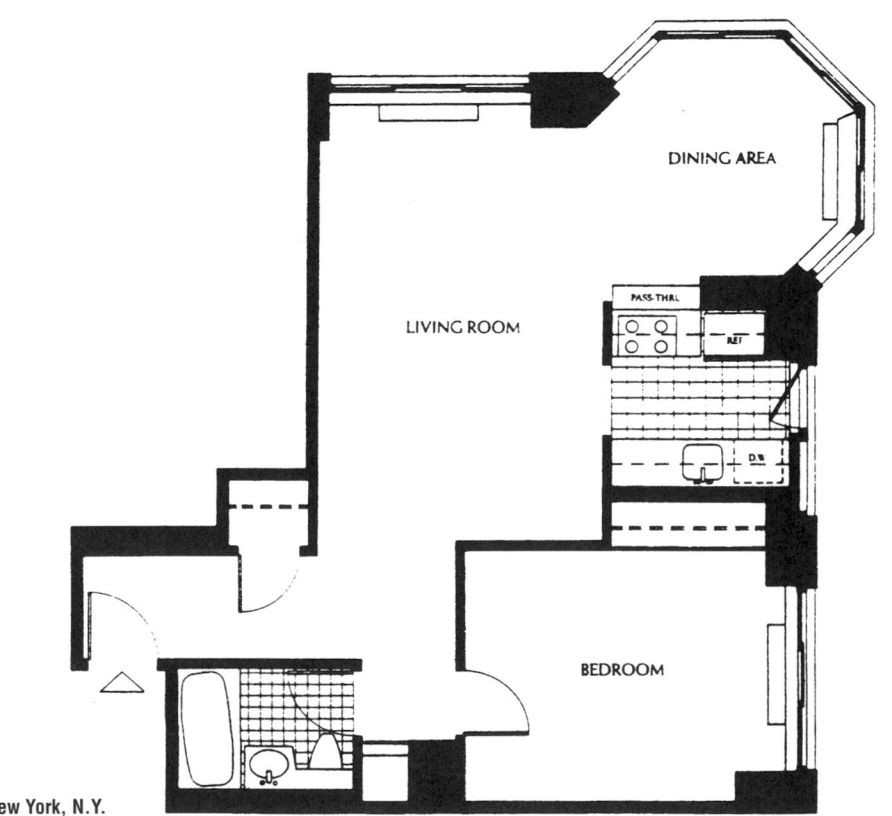

Fig. 39   Tribeca Tower, New York, N.Y.

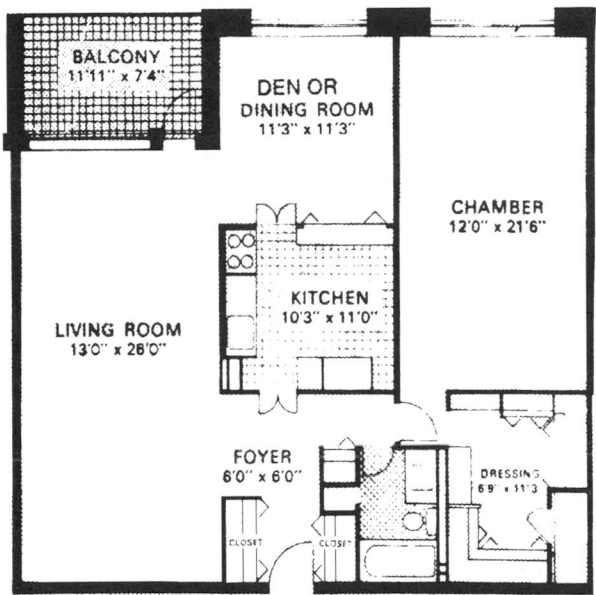

Fig. 40

1 BR

HARBOR VILLAGE AT PAERDEGAT

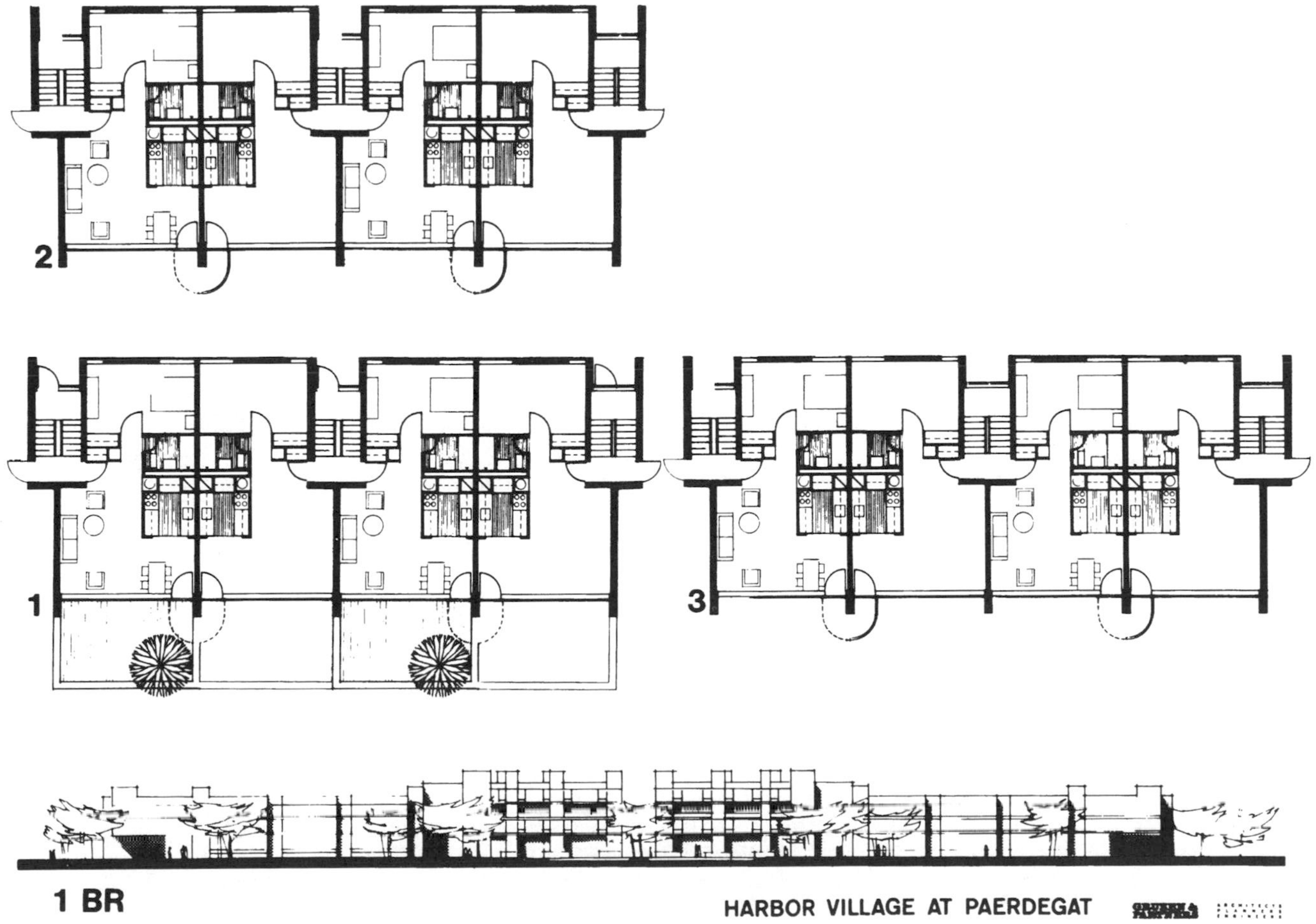

Fig. 41 Harbor Village at Paerdegat. Gruzen & Partners—Architects.

TWO-BEDROOM APARTMENT

TWO-BEDROOM APARTMENT

Elements

The two-bedroom apartment consists of two bedrooms, living room, dining area (usually part of the living room), full kitchen, bathroom, and possibly an outdoor terrace. In luxury apartments, an additional half bath, consisting of a water closet and washbasin, may be included.

Design

The two-bedroom apartment is considered the average size for a typical family with one or two children. The range of family activities is anticipated within the dwelling unit. The arrangement of rooms should be such as to permit a reasonable separation of living activities (kitchen, dining, living) from sleeping activities.

Size

The size of a two-bedroom apartment will range from 500 to 1000 ft². The FHA minimum requirement for such an apartment is approximately 650 ft². An outdoor terrace is important to add to the livability of the apartment.

Type of Occupancy

As stated, the average occupancy of two-bedroom apartments will be a family with one or two children. Often a third child or an older relative would be included. Number of occupants will be a minimum of three persons and maximum of four persons.

Planning Implications

Because of the type of occupancy, the two-bedroom apartment will yield one or two children of school age. Also, as an average family, they will require a full range of municipal services.

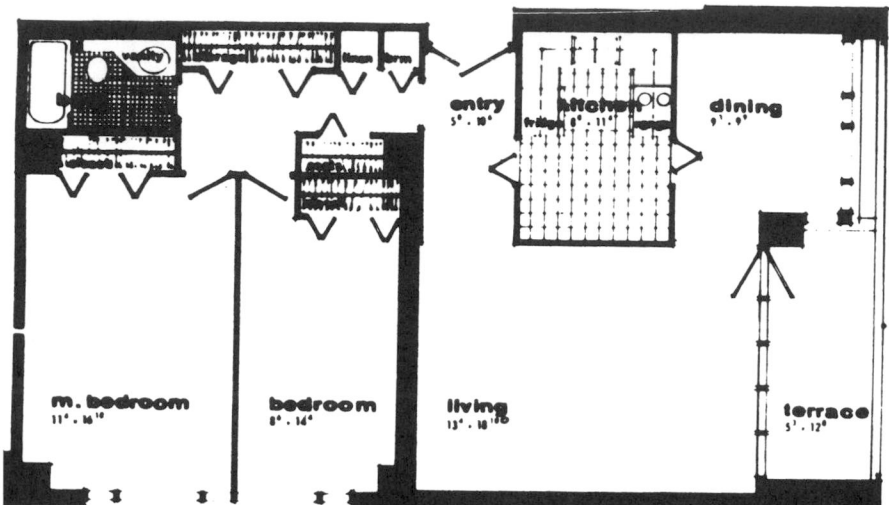

Fig. 1

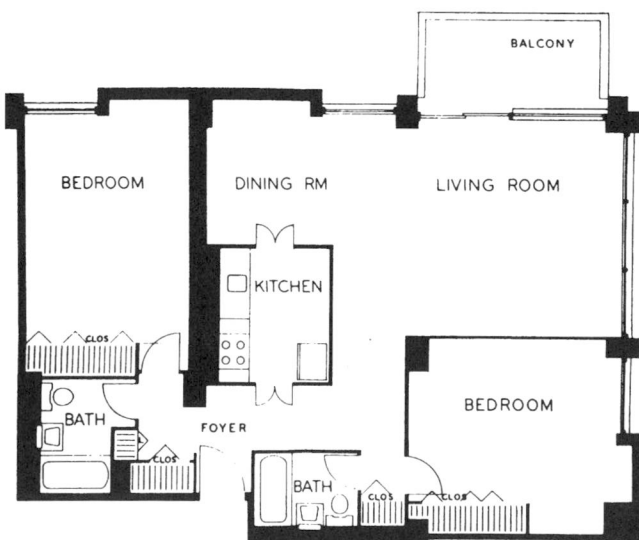

Fig. 2 Carl Sandburg Village, near Chicago. Cordwell & Partners—Architects.

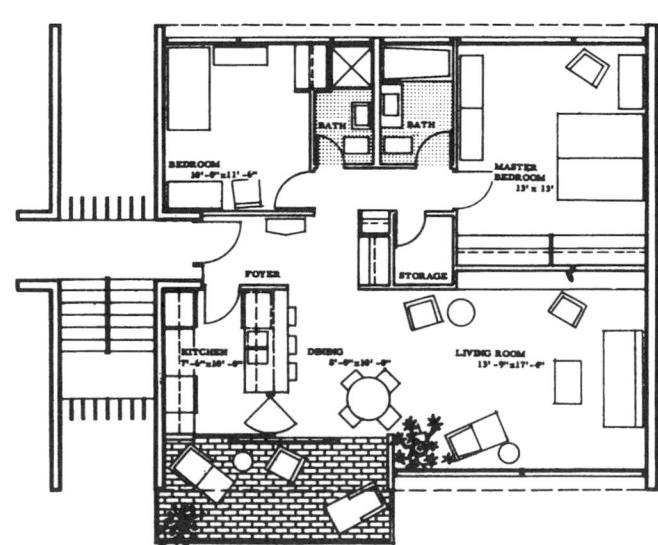

Fig. 3 Two-bedroom simplex plan.

TWO-BEDROOM APARTMENT

In the two-bedroom apartment shown in Fig. 4, the gross area is 920 ft².

Features	Dimensions, ft	Area, ft²
Living space	12 × 16.5	240
Dining space		
Kitchen	9 × 10	82
Bedroom 1	13.25 × 10	132
Bedroom 2	11 × 11	121
Bathroom	5 × 8	40

The unit in Fig. 4 has standard room sizes, good entrance privacy, and efficient plan organization. The internal privacy for bedrooms is good, and the circulation paths are organized to minimize disruption of activities. The furnace, hot water heater, and laundry facilities are in the basement. The unit has a primary orientation toward private outdoor space and combined space for living and dining areas.

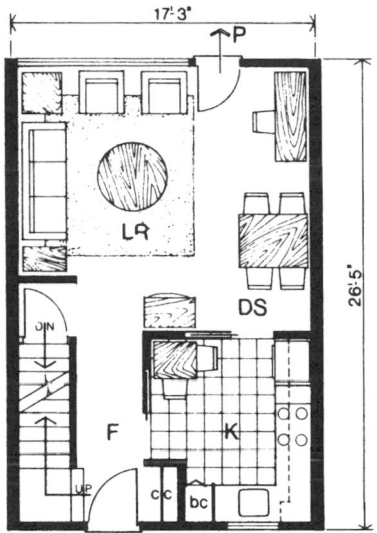

FIRST FLOOR

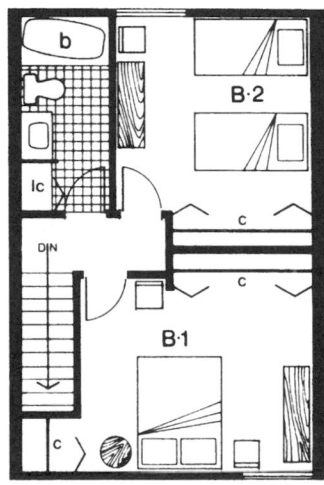

SECOND FLOOR

Fig. 4

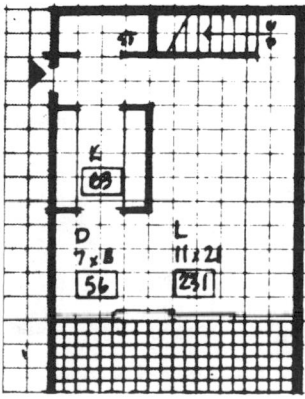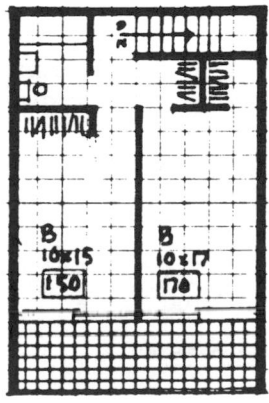

Fig. 5   Total area, 945 ft².

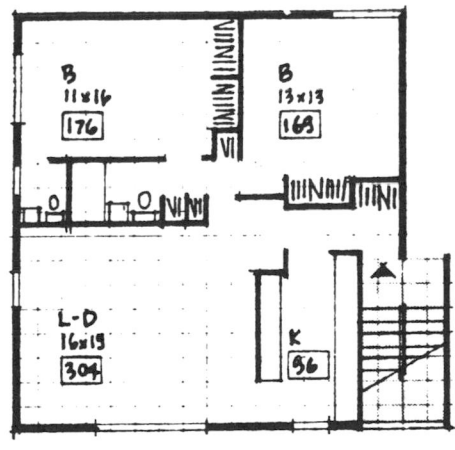

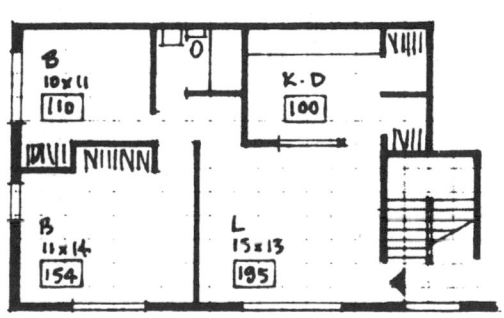

Fig. 6   Total area, 966 ft².

TWO-BEDROOM APARTMENT

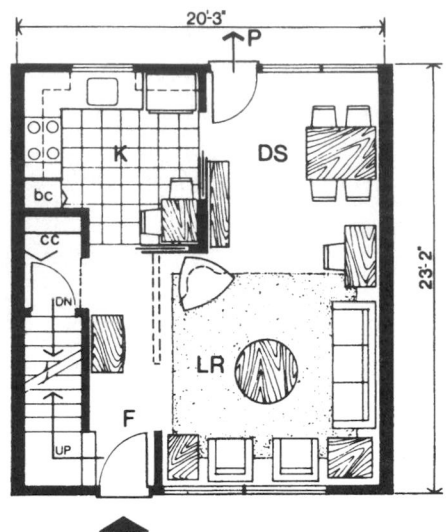

FIRST FLOOR

In the two-bedroom apartment shown in Fig. 7, the gross area is 940 ft^2.

Features	Dimensions, ft	Area, ft^2
Living space	16 × 12.5	295
Dining space	10 × 9.5	
Kitchen	9.5 × 9.5	90
Bedroom 1	10 × 13	130
Bedroom 2	9.5 × 14.5	133
Bathroom	6.5 × 7.6	50

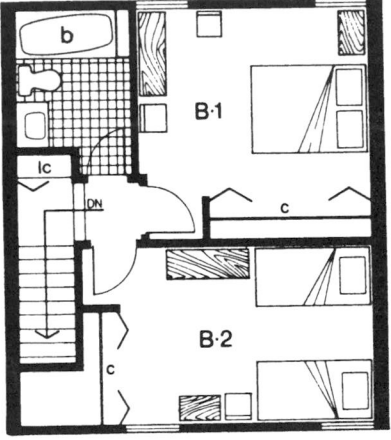

SECOND FLOOR

Fig. 7

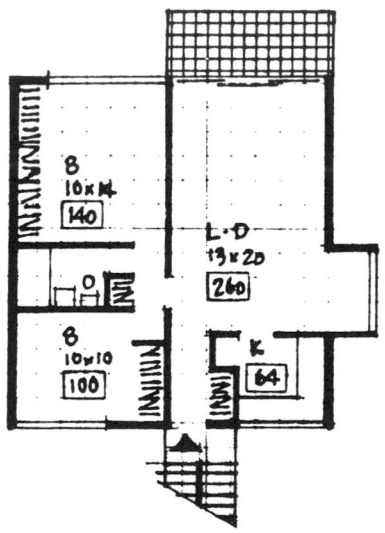

Fig. 8   Total area, 628 ft^2.

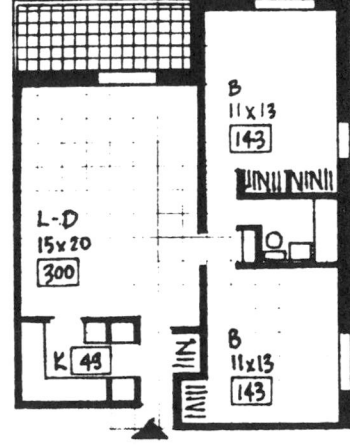

Fig. 9   Total area, 830 ft^2.

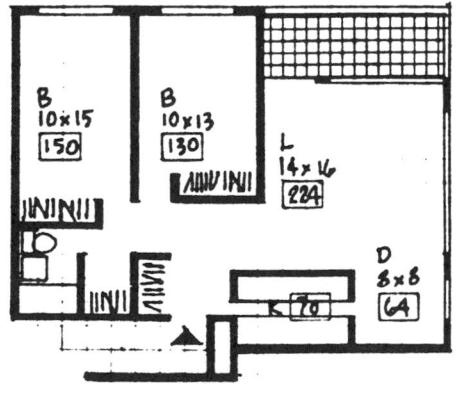

Fig. 10   Total area, 870 ft^2.

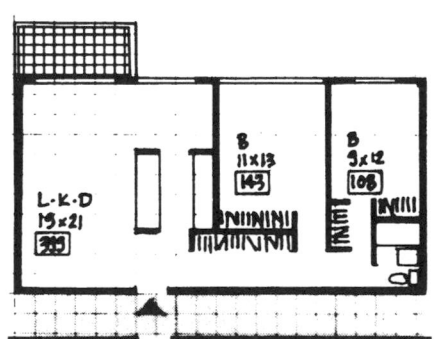

Fig. 11   Total area, 800 ft^2.

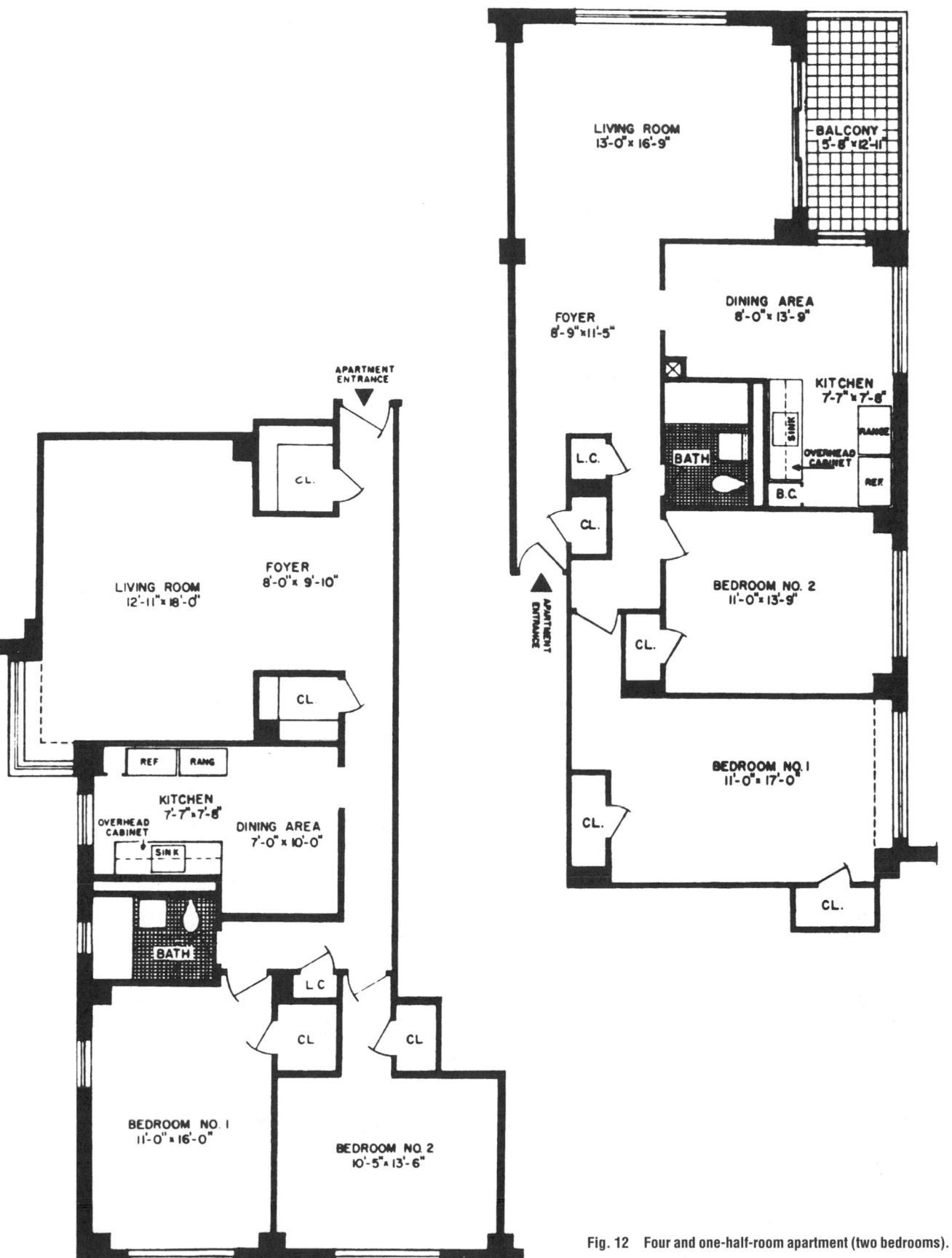

Fig. 12 Four and one-half-room apartment (two bedrooms).

TWO-BEDROOM APARTMENT

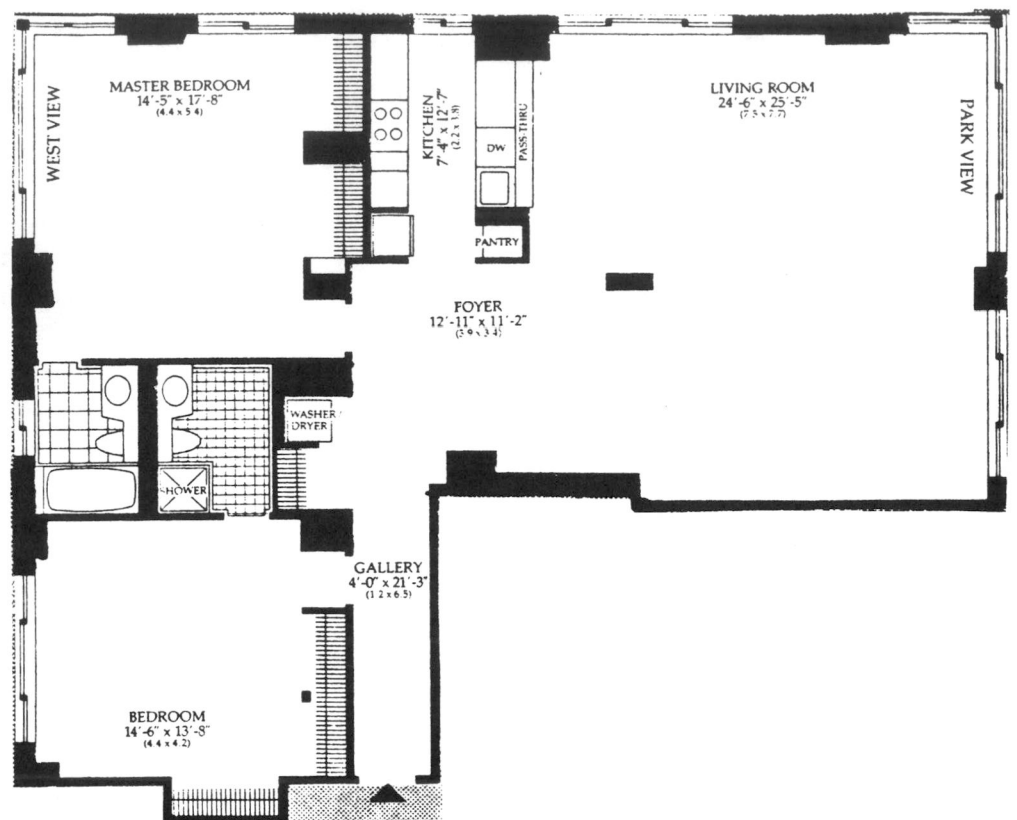

Fig. 13   Park Bevedere, 79th St. and Columbus Ave., New York, N.Y.

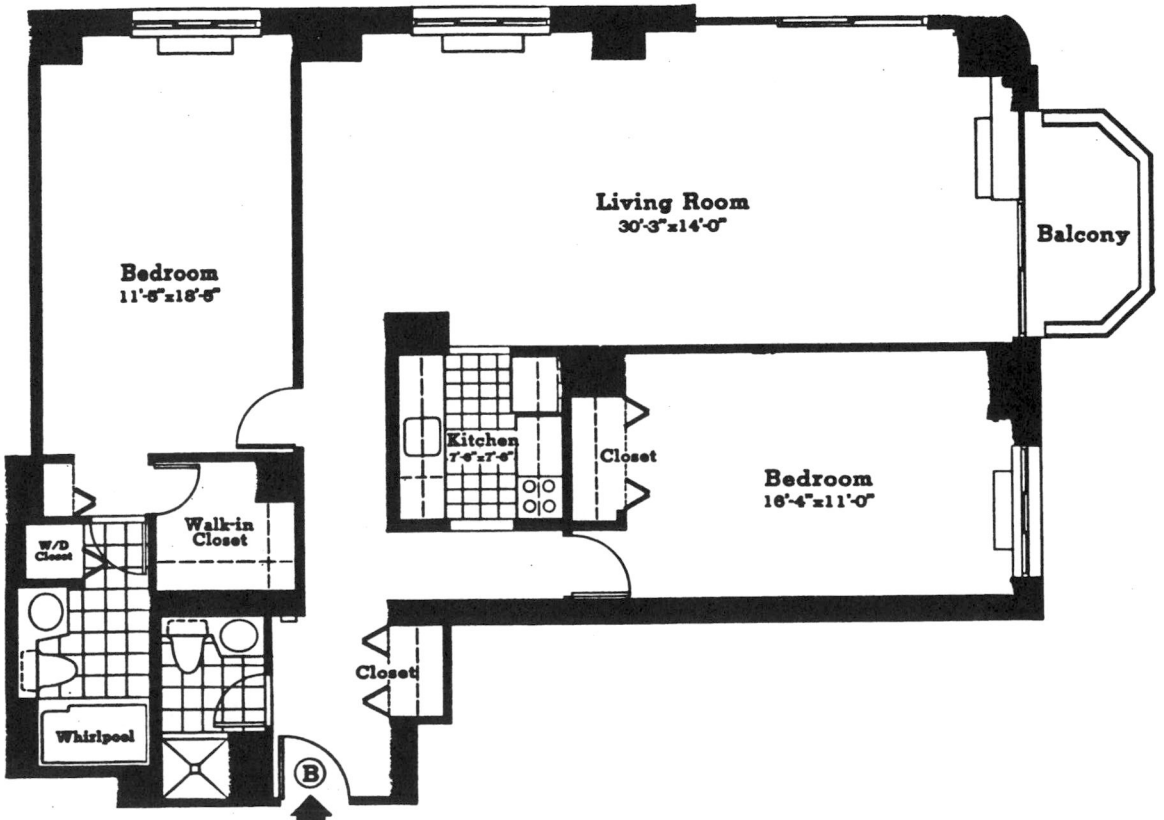

Fig. 14   108 5th Ave., New York, N.Y.

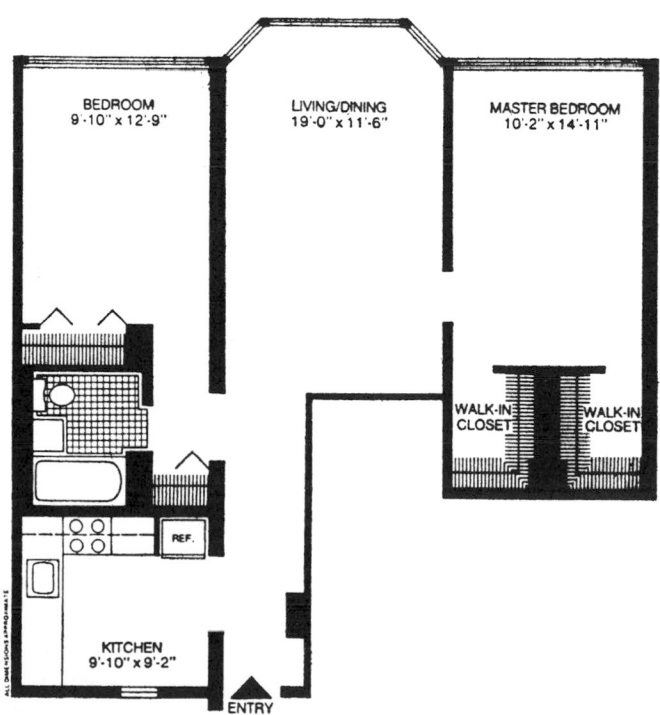

Fig. 15

LIVING ROOM
24'-9" x 16'-10"

fireplace

BEDROOM
18'-6" x 11'-3"

walk in cl.
5'-9" x 4'-9"

BEDROOM
17'-1" x 10'-11"

cl.

cl.

lin.

cl.   cl.

cl.   cl.

DINING
16'-7" x 10'-4"

pass-thru

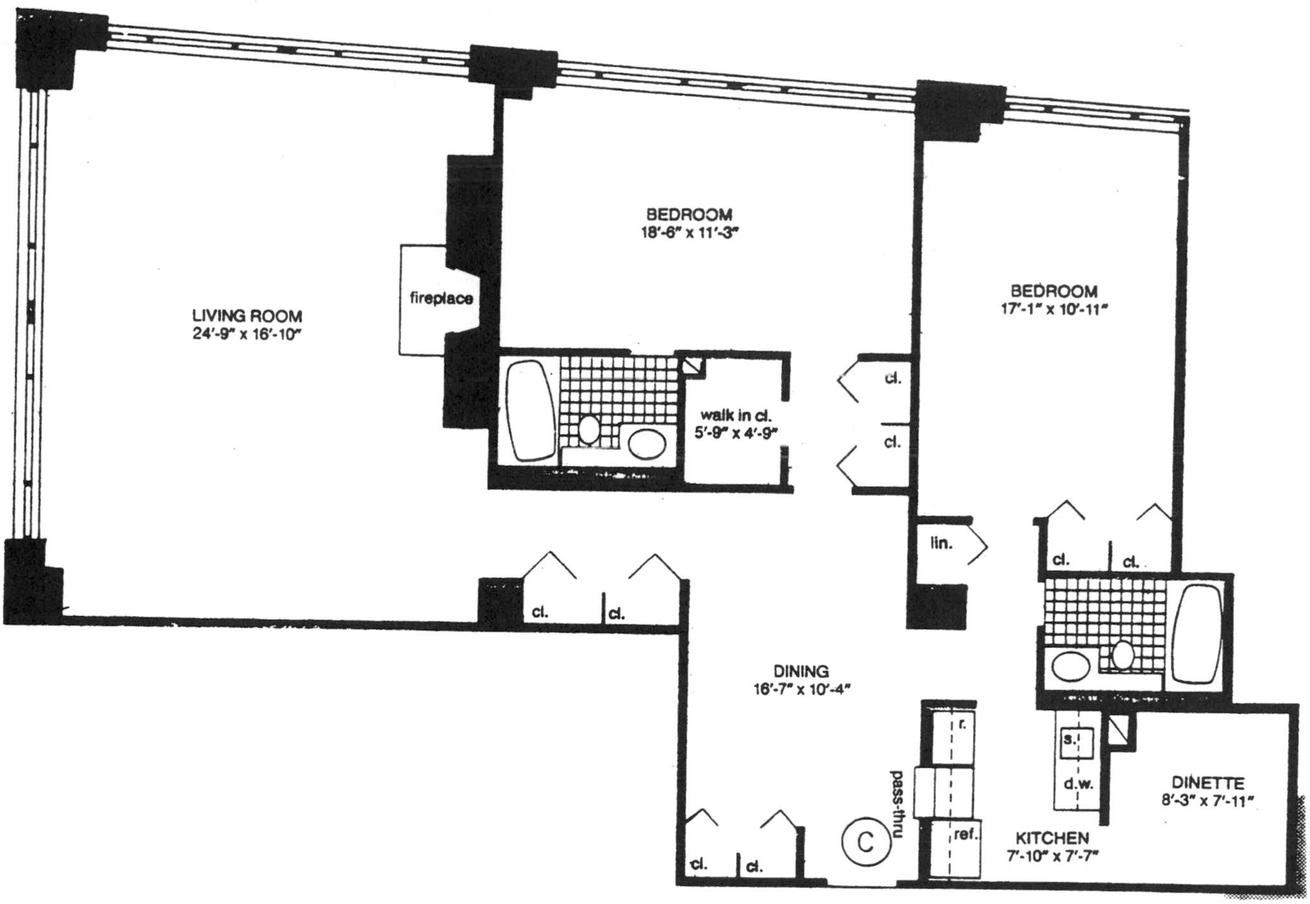

C

r.

ref.

s.

d.w.

KITCHEN
7'-10" x 7'-7"

DINETTE
8'-3" x 7'-11"

cl.   cl.

BEDROOM
9'-10" x 12'-9"

LIVING/DINING
19'-0" x 11'-6"

MASTER BEDROOM
10'-2" x 14'-11"

WALK-IN
CLOSET

WALK-IN
CLOSET

REF.

KITCHEN
9'-10" x 9'-2"

ALL DIMENSIONS APPROXIMATE

ENTRY

Fig. 16   Eastwood, Roosevelt Island, N.Y.C.

TWO-BEDROOM APARTMENT

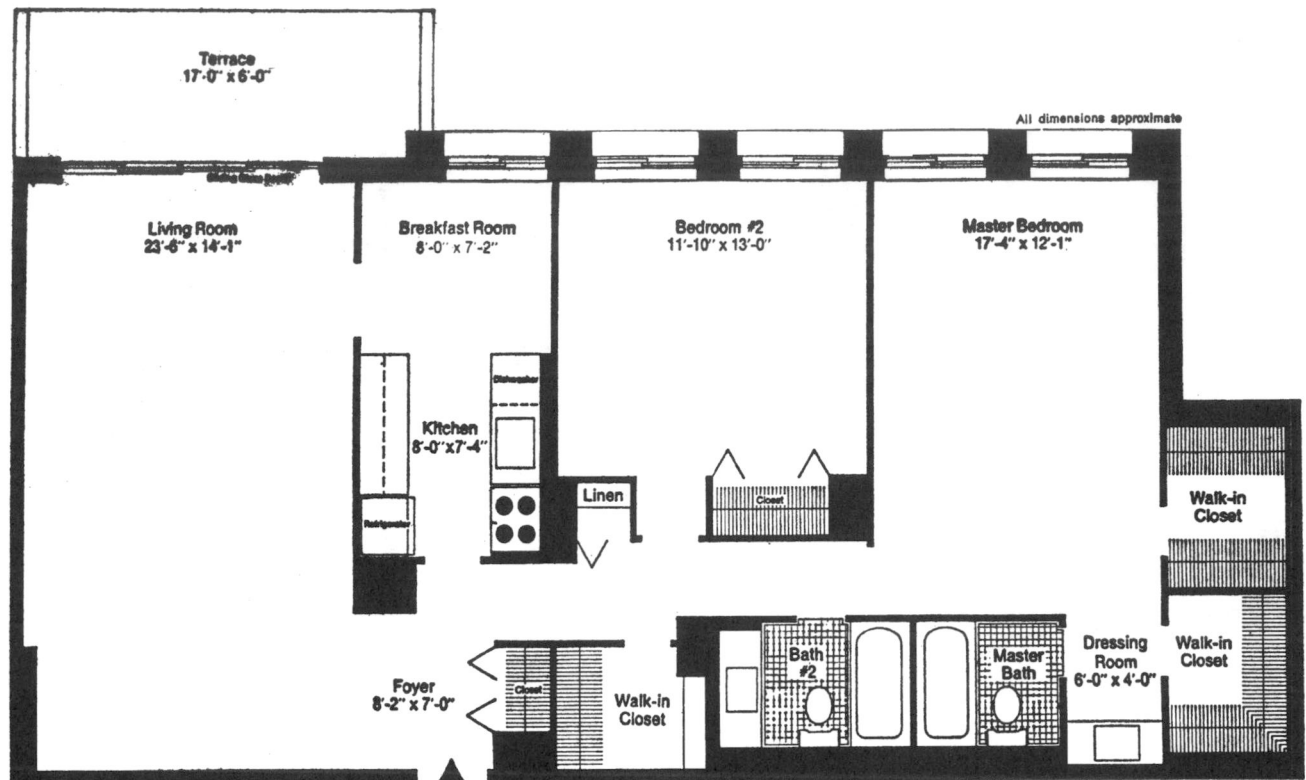

Terrace
17'-0" x 6'-0"

All dimensions approximate

Living Room
23'-6" x 14'-1"

Breakfast Room
8'-0" x 7'-2"

Bedroom #2
11'-10" x 13'-0"

Master Bedroom
17'-4" x 12'-1"

Kitchen
8'-0" x 7'-4"

Dishwasher

Refrigerator

Linen

Closet

Walk-in
Closet

Walk-in
Closet

Foyer
8'-2" x 7'-0"

Closet

Walk-in
Closet

Bath
#2

Master
Bath

Dressing
Room
6'-0" x 4'-0"

Fig. 17   Winston Towers, Fort Lee, N.J.

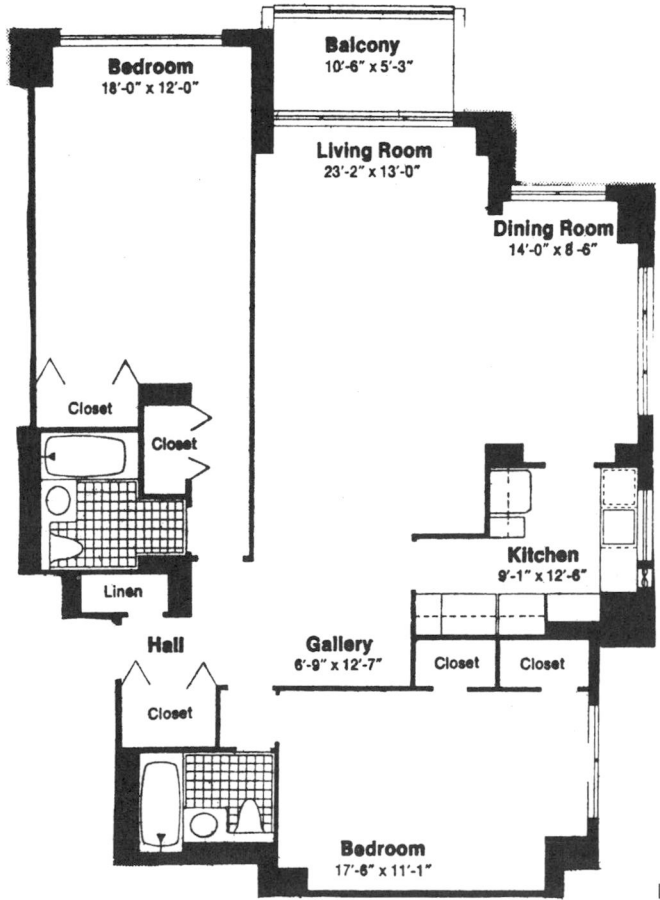

Bedroom
18'-0" x 12'-0"

Balcony
10'-6" x 5'-3"

Living Room
23'-2" x 13'-0"

Dining Room
14'-0" x 8'-6"

Closet

Closet

Linen

Kitchen
9'-1" x 12'-6"

Hall

Closet

Gallery
6'-9" x 12'-7"

Closet

Closet

Bedroom
17'-6" x 11'-1"

Fig. 18   Park Regis, New York, N.Y.

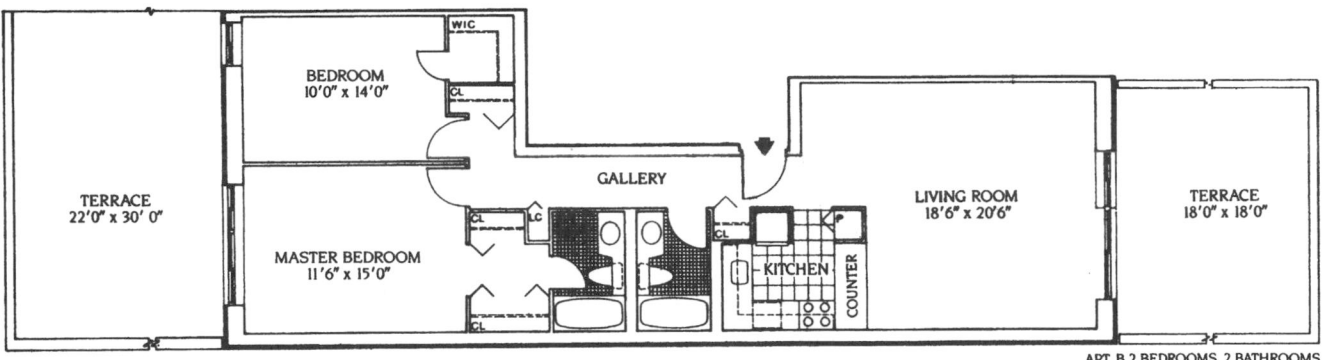

Fig. 19   The Atrium at Chelsea, New York, N.Y.

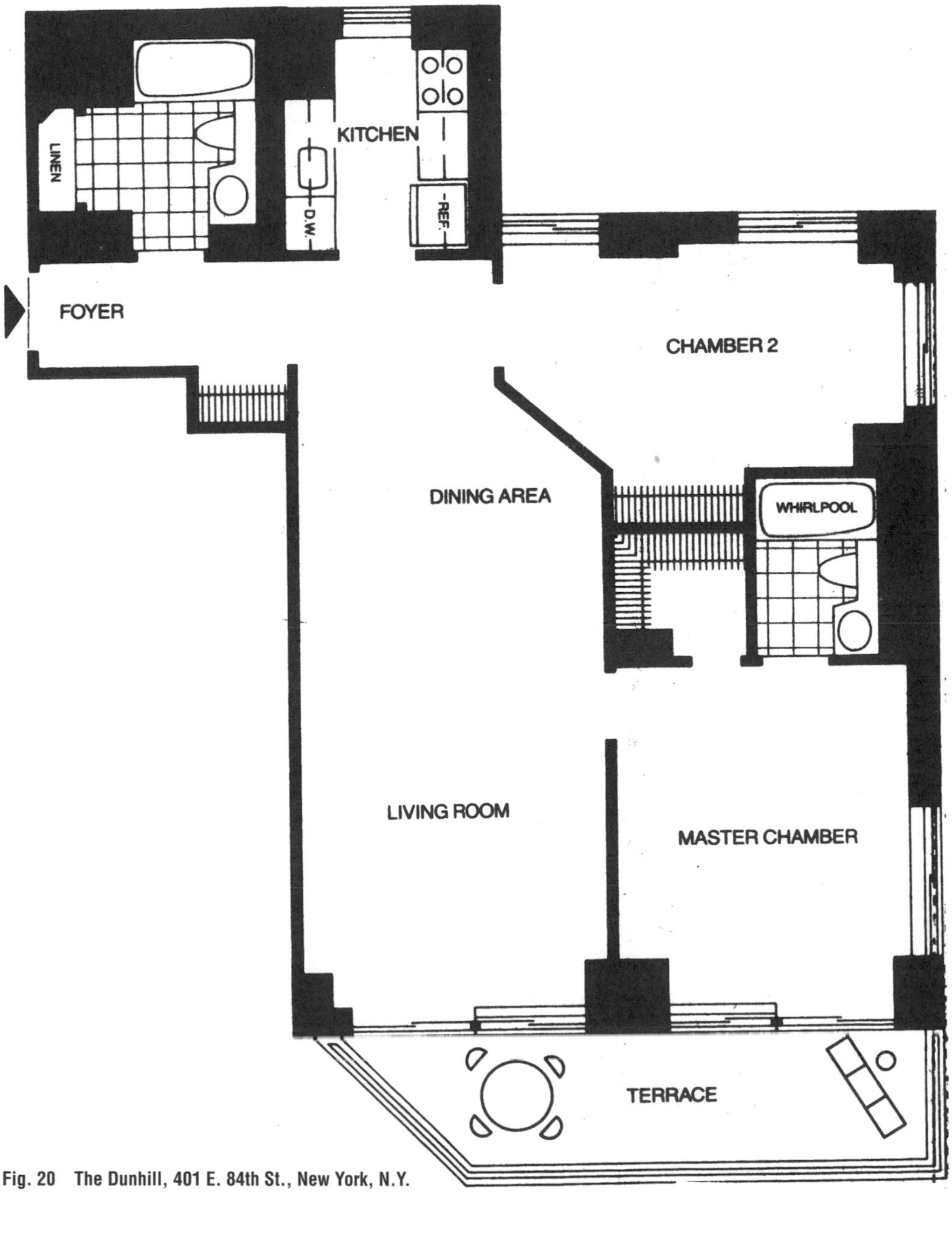

APT. B 2 BEDROOMS, 2 BATHROOMS.

Fig. 20   The Dunhill, 401 E. 84th St., New York, N.Y.

TWO-BEDROOM APARTMENT

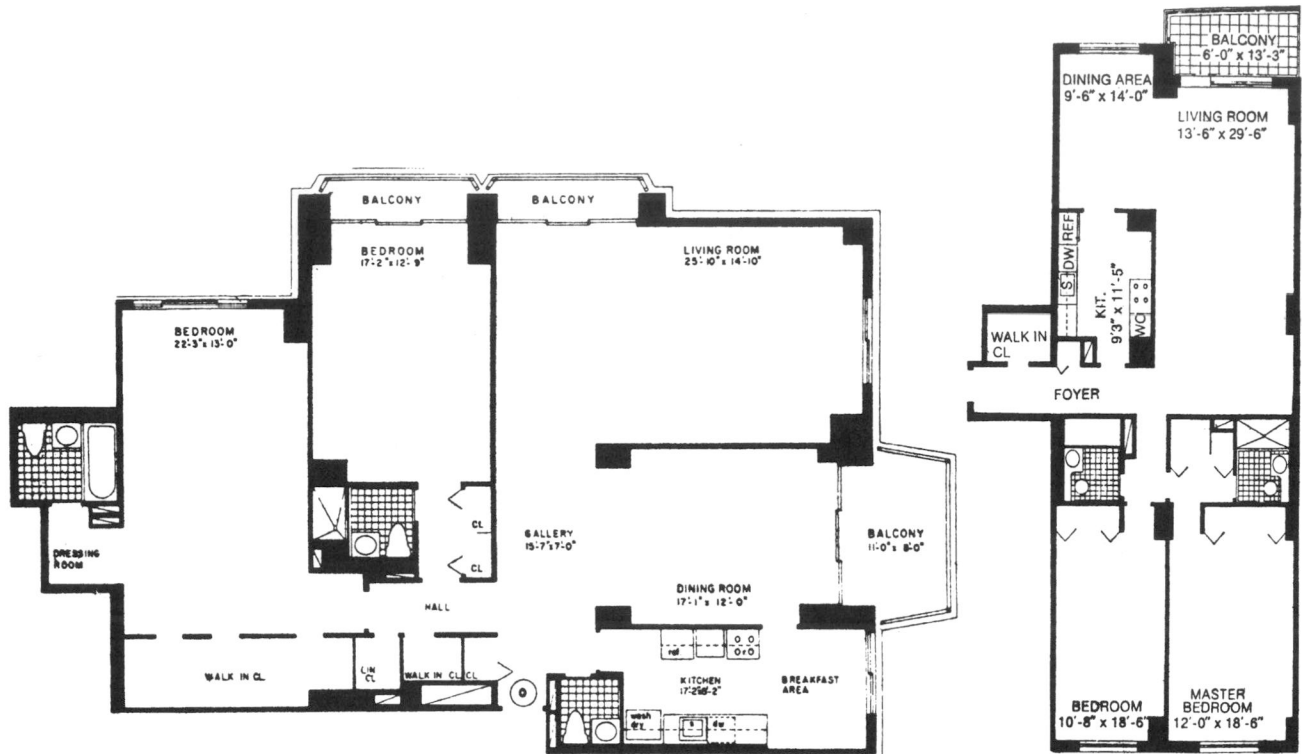

Fig. 21   North Shore Towers, Queens, N.Y.C.

Fig. 22   The Century, Riverdale, N.Y.C.

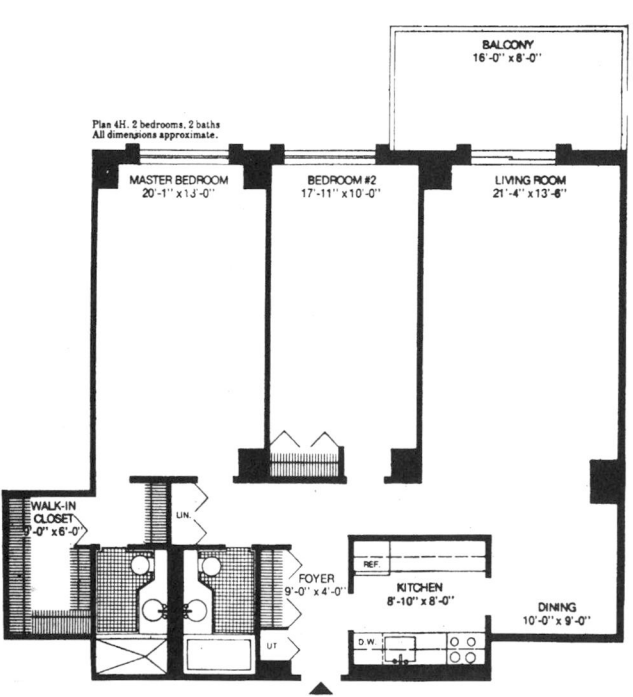

Fig. 23   The Parker Imperial, North Bergen, N.J.

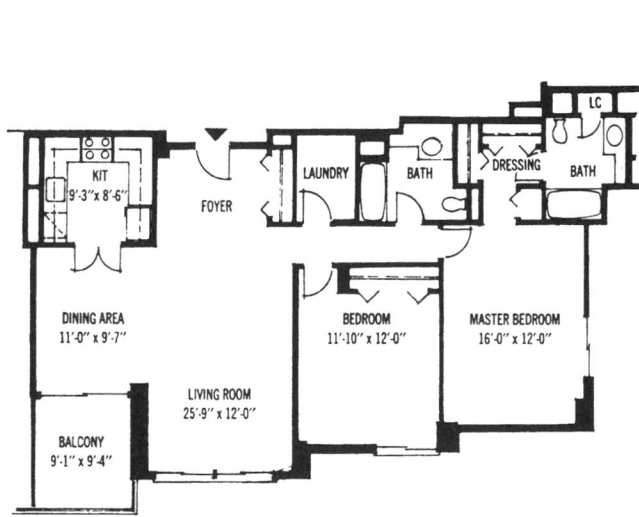

Fig. 24   Eastpoint, Highlands, N.J.

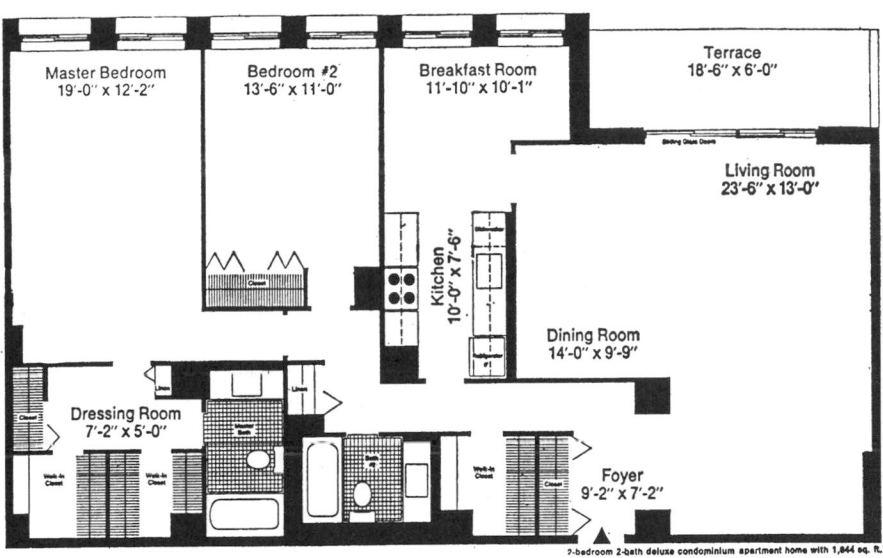

Fig. 25   Winston Towers, Fort Lee, N.J.

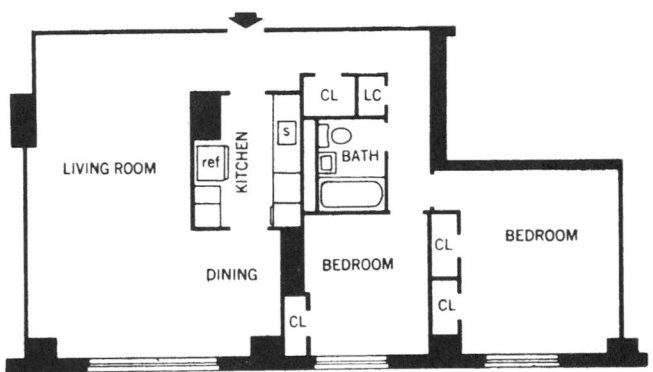

Fig. 26

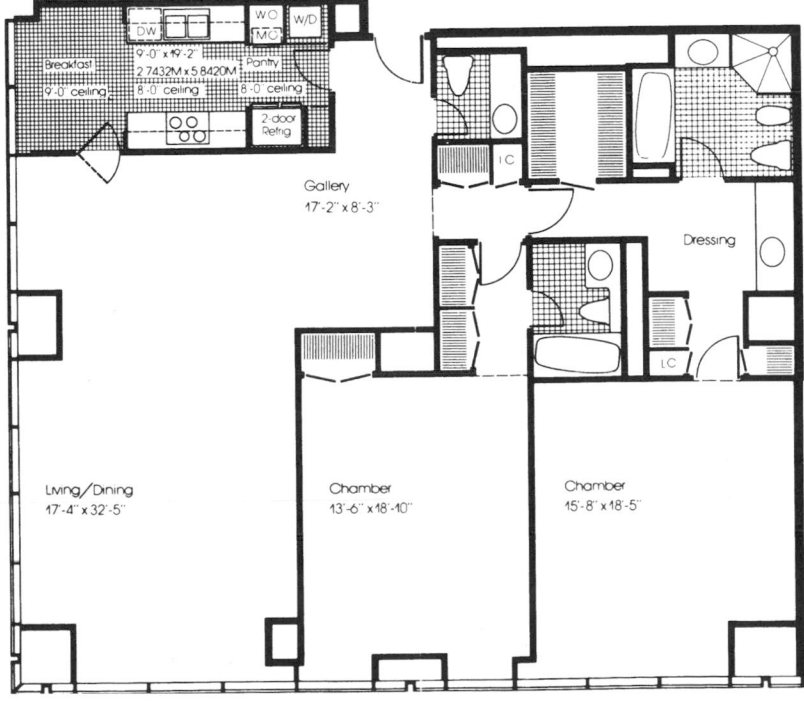

Fig. 27   Olympic Tower, New York, N.Y. Skidmore, Owings, & Merrill—Architects.

TWO-BEDROOM APARTMENT

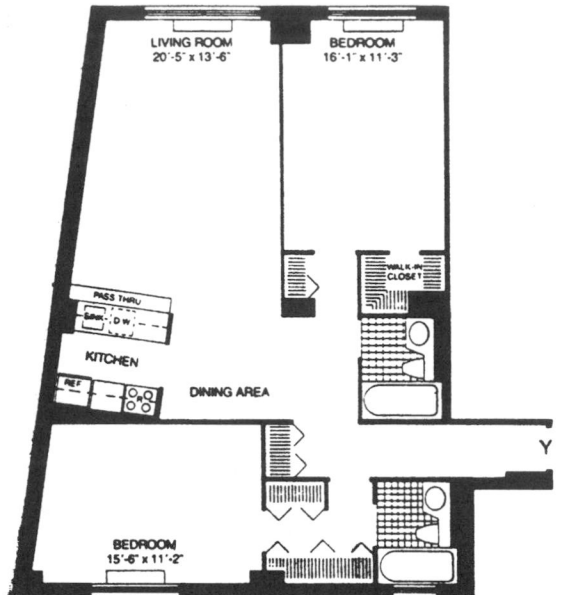

Fig. 28   Two-bedroom two-bath apartment. Liberty Court, Battery Park, New York, N.Y.

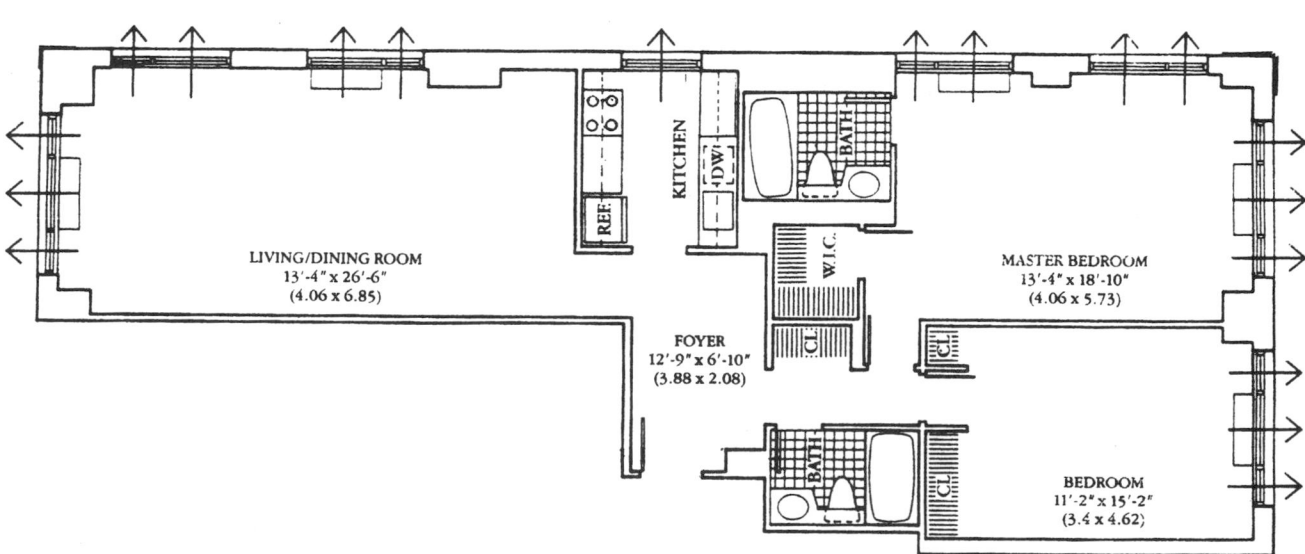

Fig. 29   Two-bedroom two-bath apartment. Zeckendorf Towers, New York, N.Y.

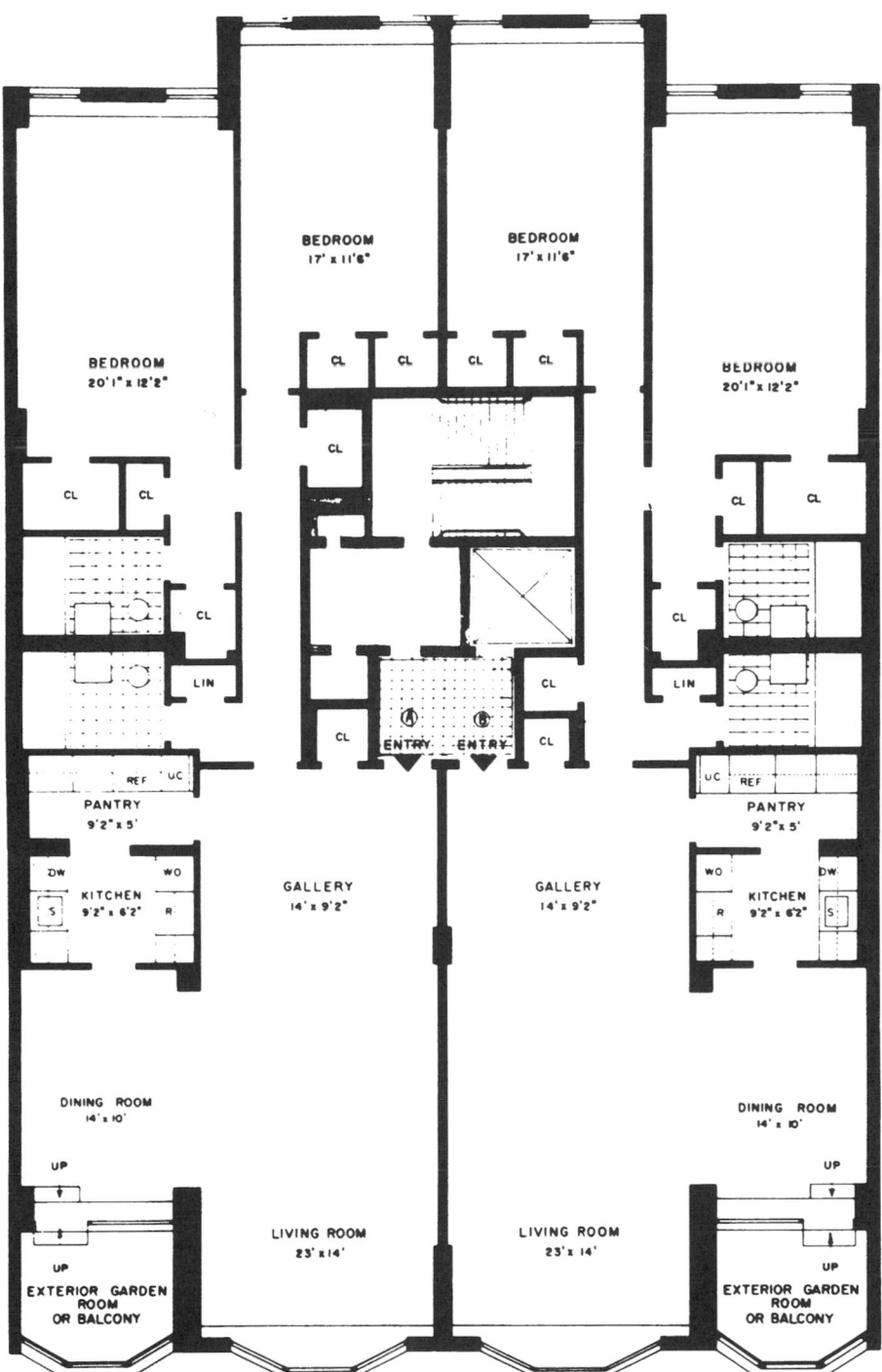

Fig. 30 Butterfield House, New York, N.Y. Mayer, Whittlesey and Glass—Architects.

TWO-BEDROOM APARTMENT

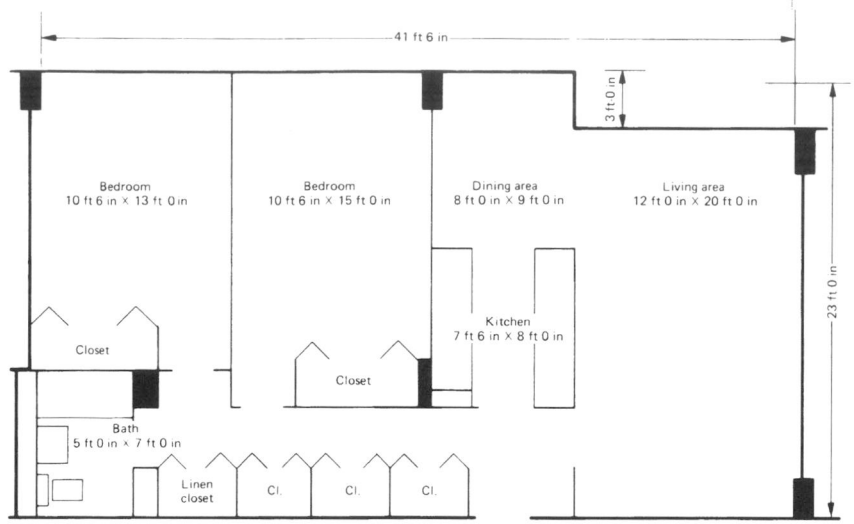

Fig. 31 Two-bedroom family apartment, 918.5 ft². N.J. Housing Finance Agency.

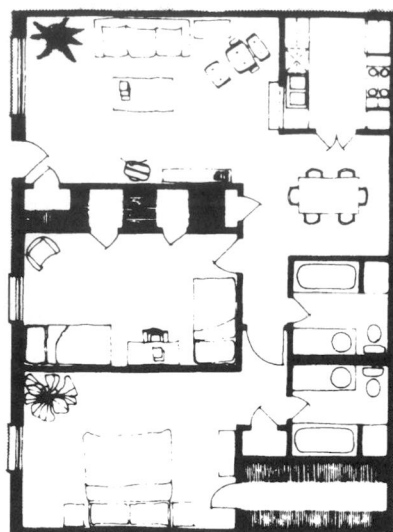

two bedroom/two bath

Fig. 32

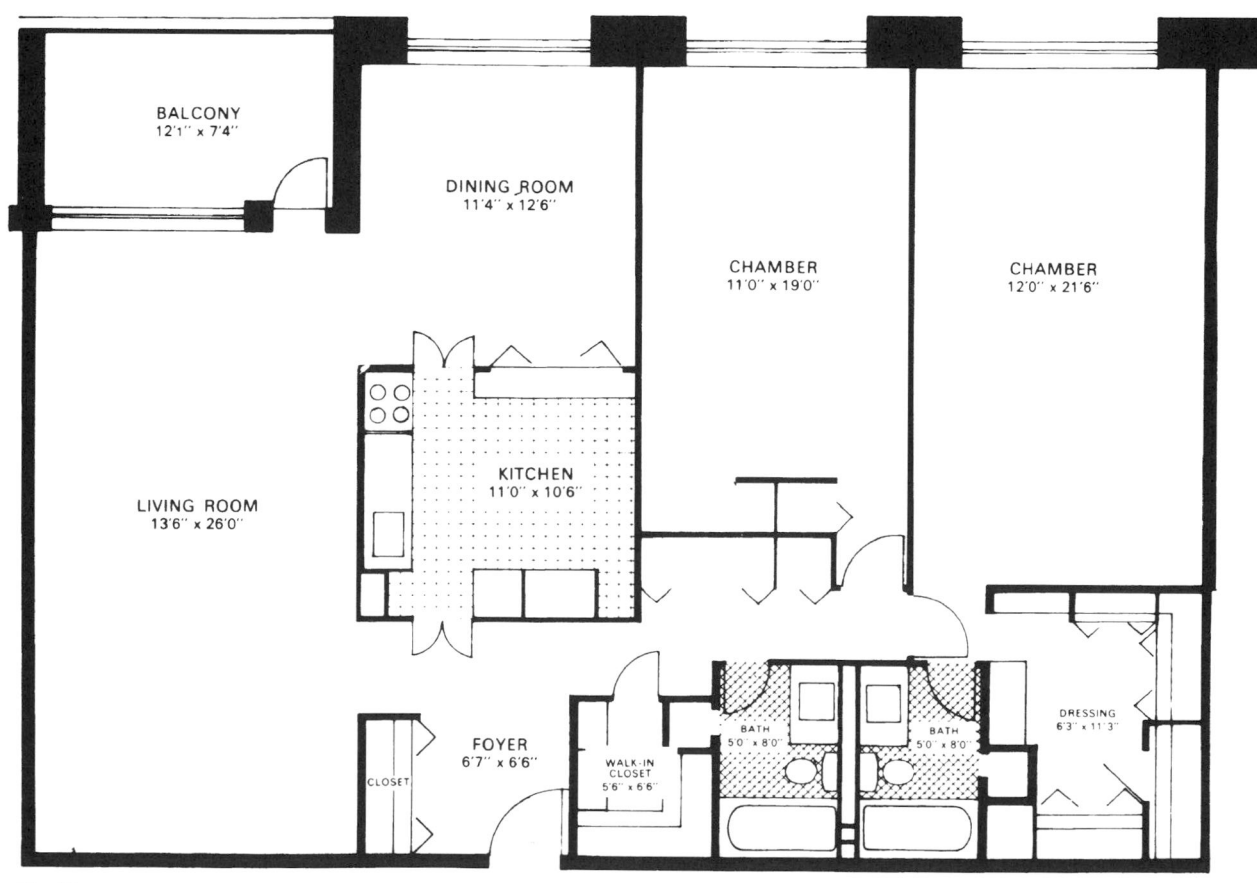

Fig. 33

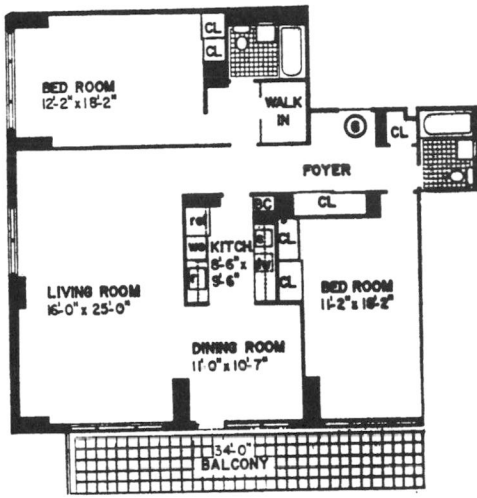

Fig. 34   Riviera Towers, West New York, N.J.

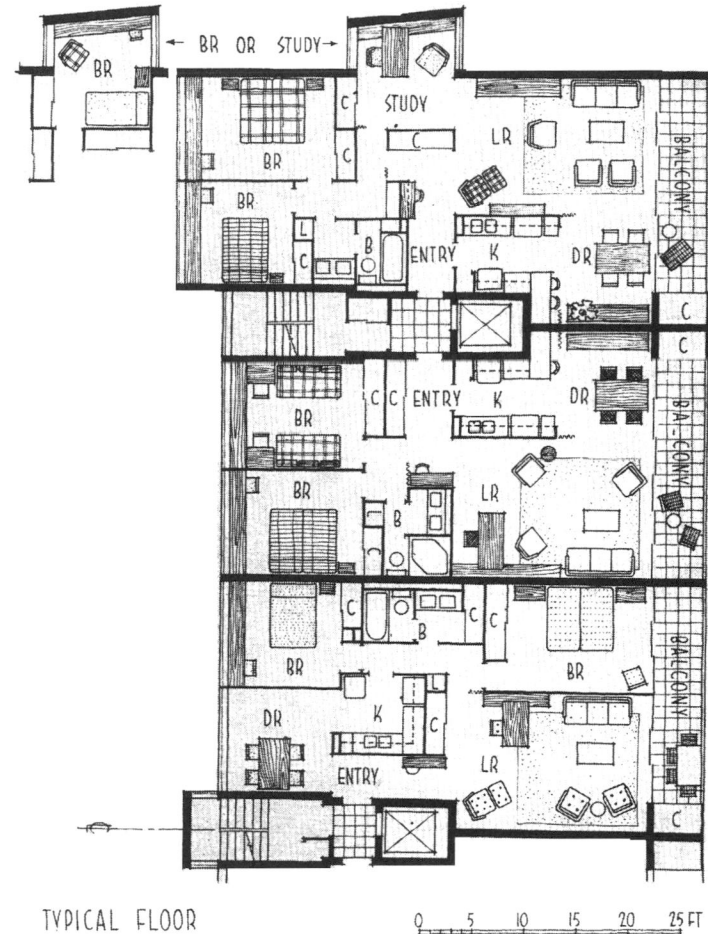

TYPICAL FLOOR

Fig. 35   River Ledge Garden Apts., Hastings-on-Hudson, N.Y. Raymond & Radd and Holsman, Holsman, Klekamp, & Taylor—Architects.

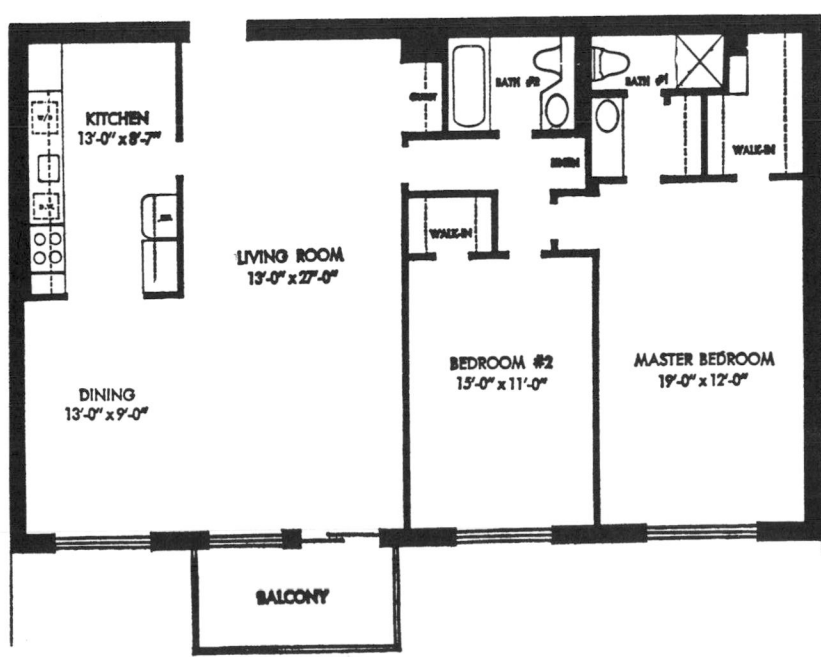

Fig. 36   Hampshire House, White Plains, N.Y.

TWO-BEDROOM APARTMENT

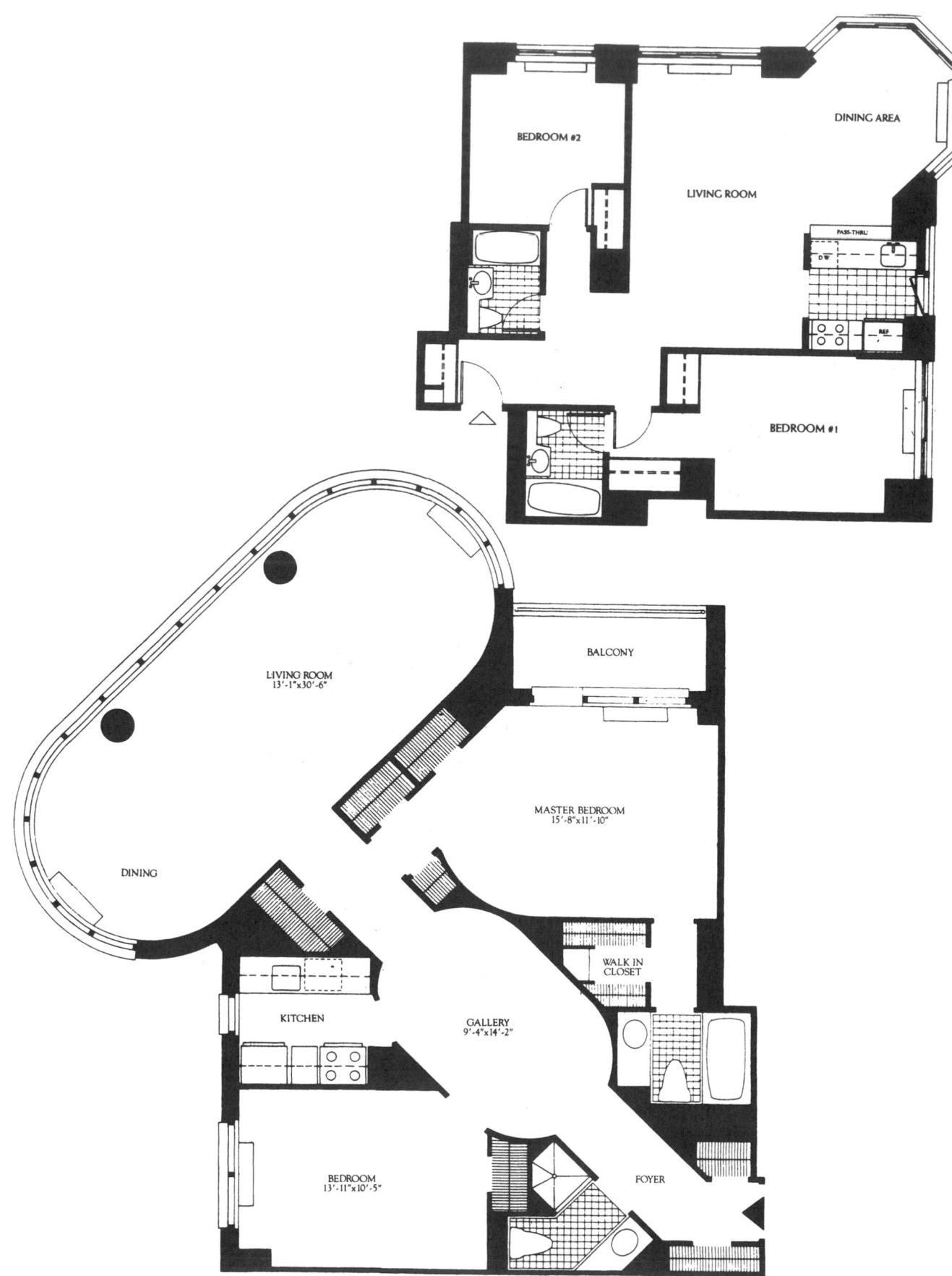

BEDROOM #2

DINING AREA

LIVING ROOM

PASS-THRU

D/W

REF.

BEDROOM #1

LIVING ROOM
13'-1"x30'-6"

BALCONY

MASTER BEDROOM
15'-8"x11'-10"

DINING

WALK IN
CLOSET

KITCHEN

GALLERY
9'-4"x14'-2"

BEDROOM
13'-11"x10'-5"

FOYER

Fig. 37   Tribeca Tower, New York, N.Y.

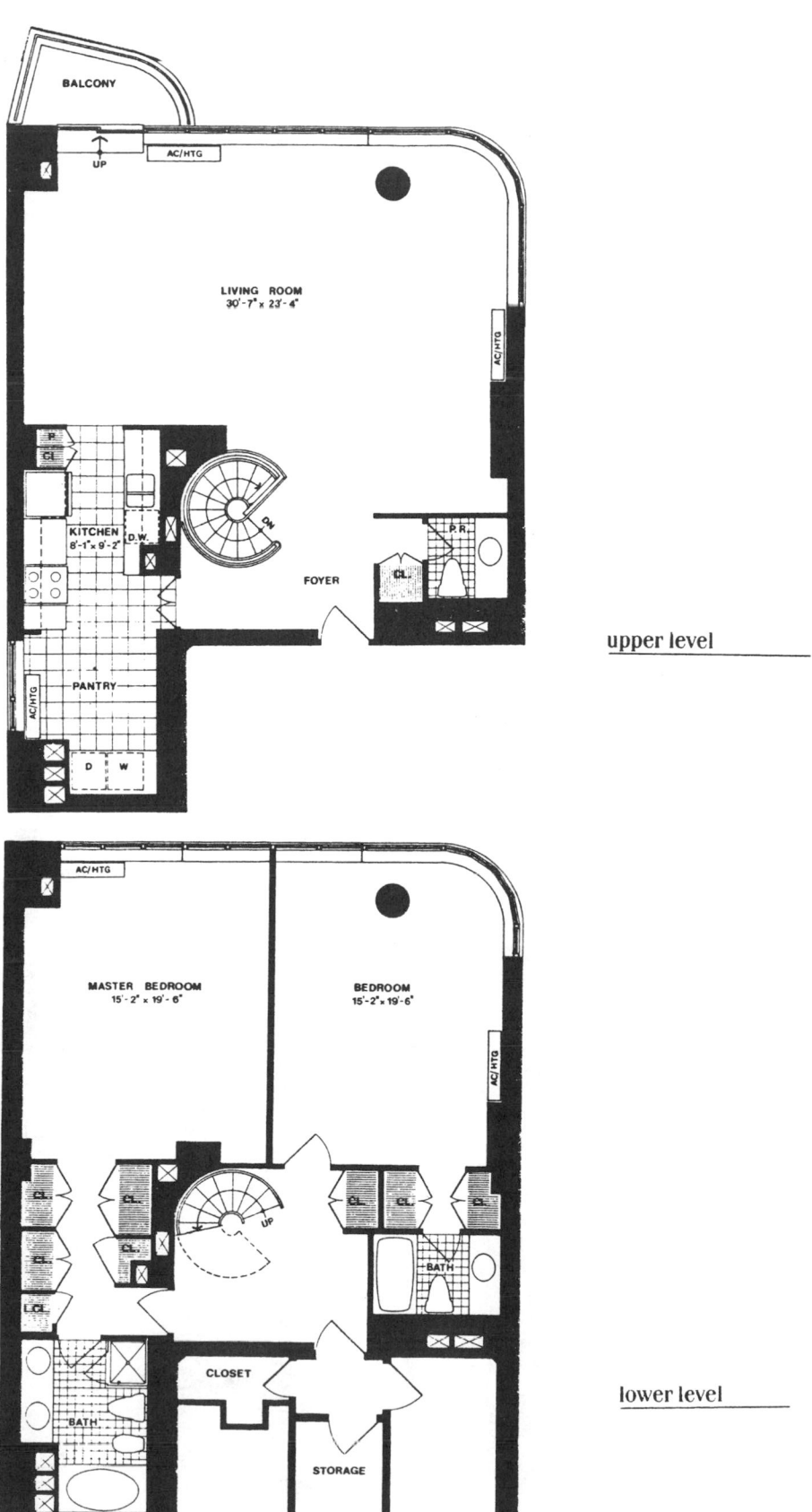

upper level

lower level

Fig. 38   Morgan Court duplex condominium, 211 Madison Ave., New York, N.Y.

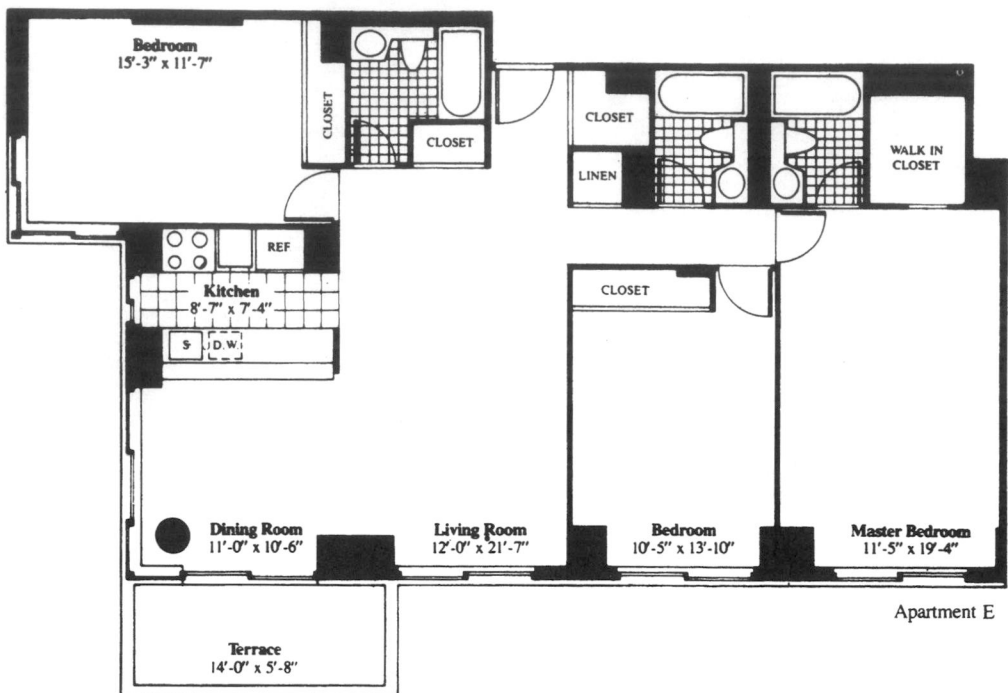

Fig. 39   The Columbia, New York, N.Y.

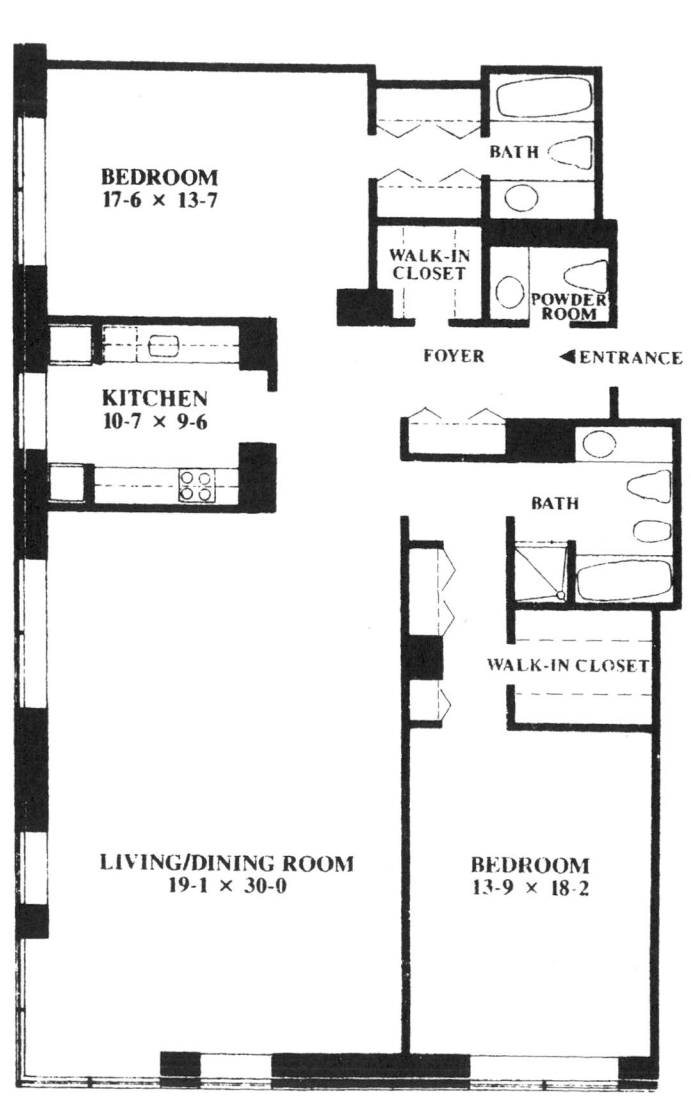

Fig. 40   Museum Tower, New York, N.Y.

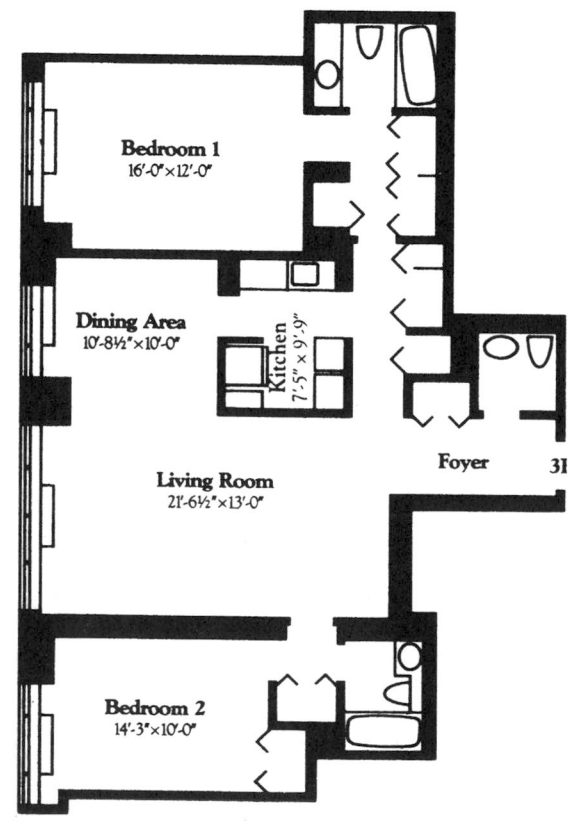

Fig. 41   Astor Terrace, 245 E. 93rd St., New York, N.Y.

1st floor   1st floor

2nd floor   2nd floor

Fig. 42   Lavanburg Community, New York, N.Y. Conklin & Rossant—Architects.

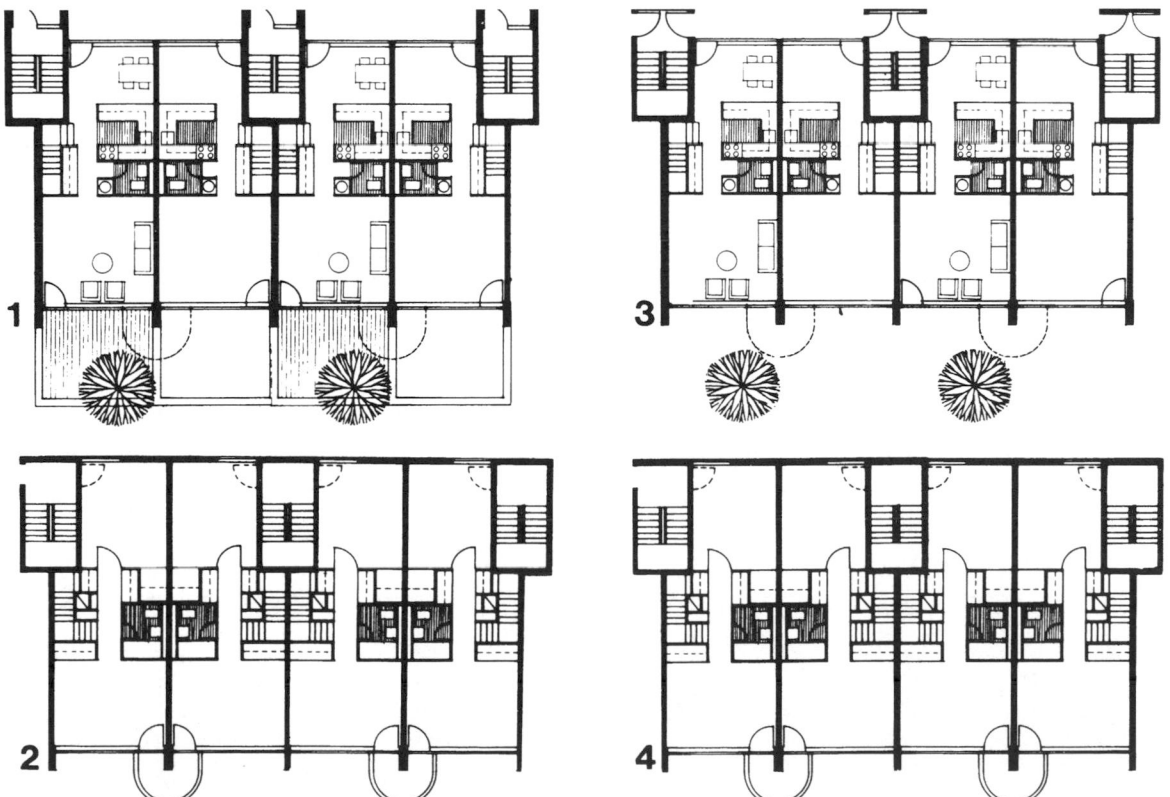

1   3

2   4

Fig. 43   Harbor Village at Paerdegat.

THREE-BEDROOM APARTMENT

THREE-BEDROOM APARTMENT

Elements

The three-bedroom apartment consists of three bedrooms, living room, dining area, full kitchen, one to two bathrooms, and an outdoor terrace. Luxury apartments will have two bathrooms, while public housing may still have only one.

Design

The three-bedroom apartment is generally considered for large families with three or more children. A larger living and dining area is necessary for the larger family. Consideration should be given for the greater privacy for each member of the family.

Size

The size of the three-bedroom apartment will range from 600 to 1200 ft². The FHA minimum requirement for such an apartment is approximately 800 ft². An outdoor terrace is important to add to the livability of the apartment.

Type of Occupancy

Occupancy is expected to be for a large family with three or more children. Frequently, however, family size may be increased by other family members, such as in-laws or grandparents, rather than more children.

Planning Implications

The additional children will definitely add to the school population. Elderly persons will also require additional services.

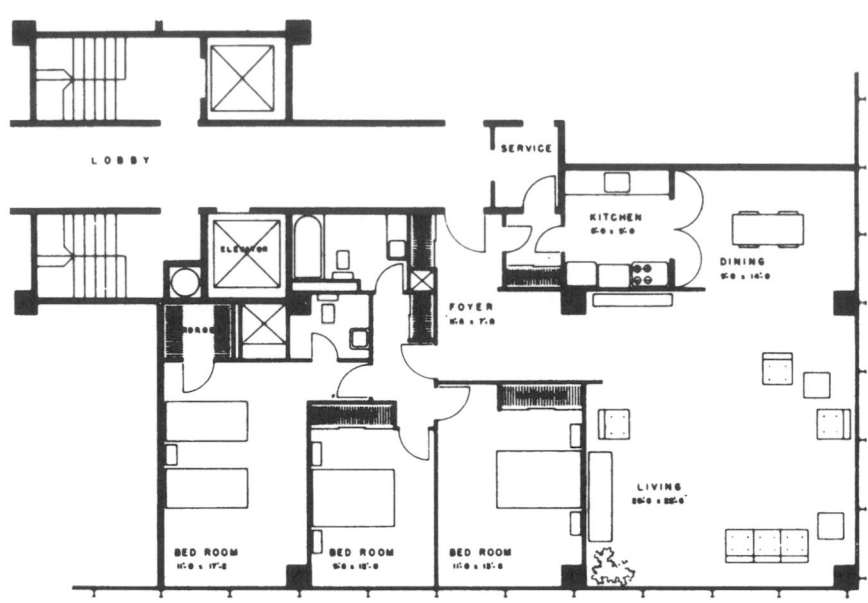

Fig. 1   Lake Shore Drive Apartments, Chicago, Ill. Mies van der Rohe—Architect.

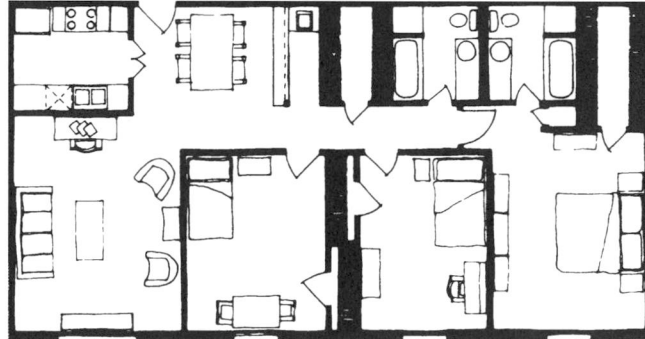

three bedroom/two bath

Fig. 2

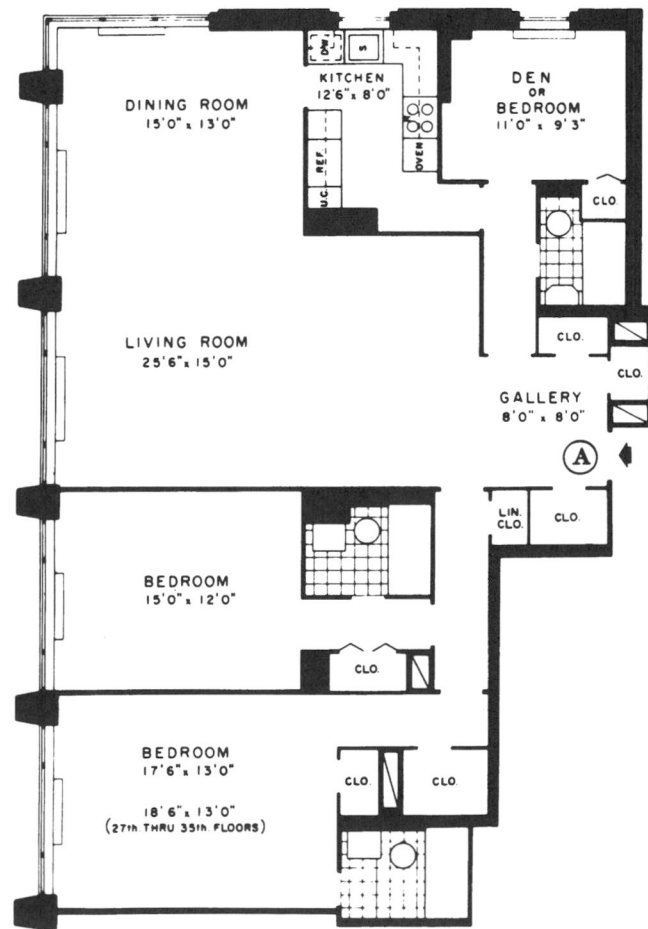

Fig. 3   Tower East, New York, N.Y. E. Roth & Sons—Architects.

THREE-BEDROOM APARTMENT

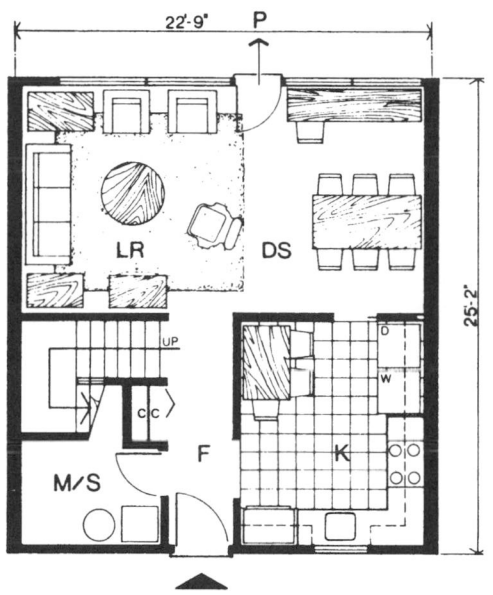

FIRST FLOOR

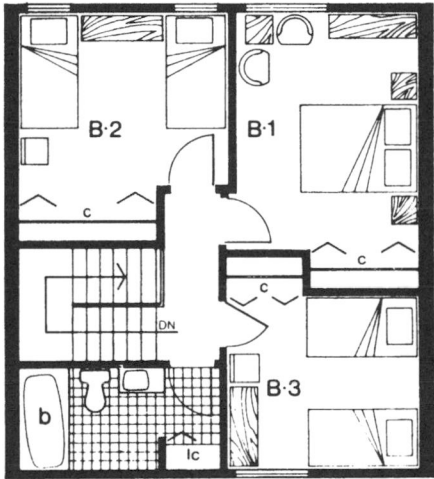

SECOND FLOOR

Fig. 4

The three-bedroom apartment shown in Fig. 4 has a gross area of 1050 ft².

Features	Dimensions, ft	Area, ft²
Living room	12 × 14.5	270
Dining space	9.5 × 10	
Kitchen	11 × 9.5	90
Bedroom 1	10 × 13	130
Bedroom 2	11.5 × 9.5	110

The unit in Fig. 4 has standard room sizes, good entrance privacy and protection, efficient activity organization, kitchen and living-dining area are separable thus providing two living floor activity spaces, circulation is centralized and will not disrupt activities. Good bedroom privacy. Unit is oriented toward private outdoor space.

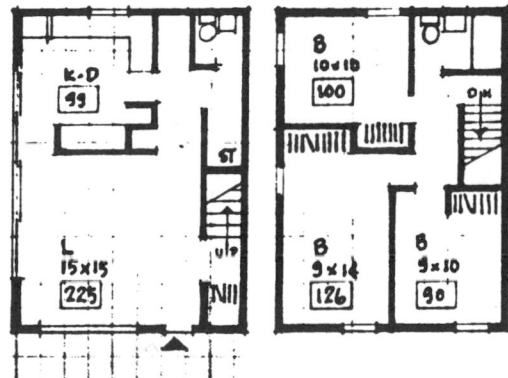

Fig. 5   Total area, 935 ft².

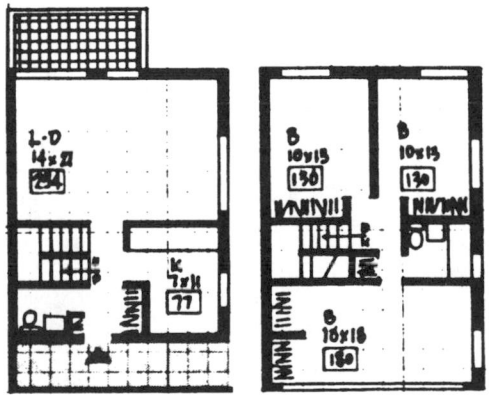

Fig. 6   Total area, 1134 ft².

THREE-BEDROOM APARTMENT

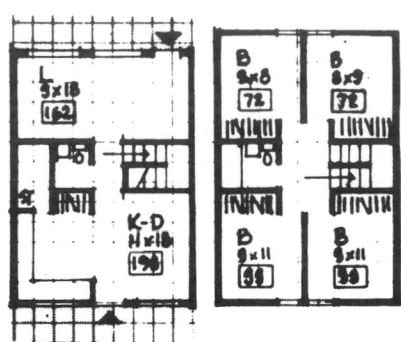

Fig. 7   Total area, 950 ft².

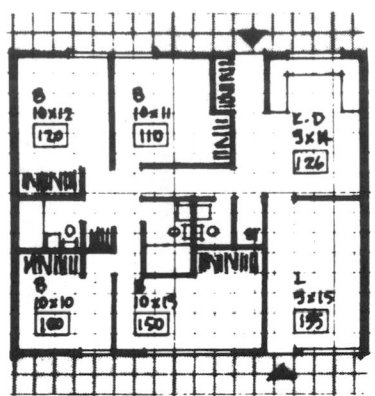

Fig. 8   Total area, 900 ft².

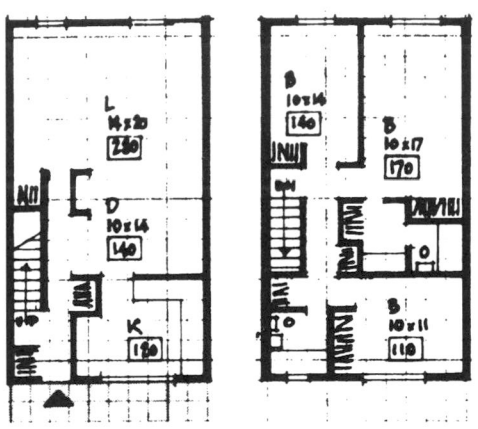

Fig. 9   Total area, 1400 ft².

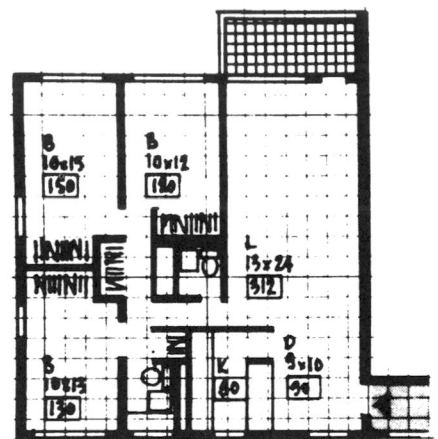

Fig. 10   Total area, 1120 ft².

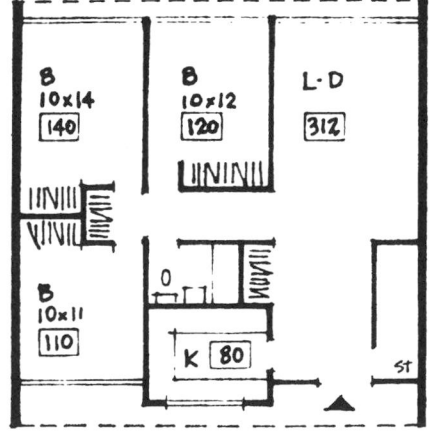

Fig. 11   Total area, 1020 ft².

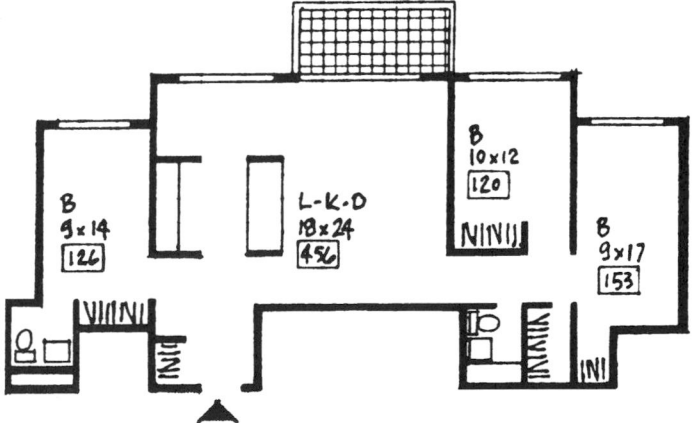

Fig. 12   Total area, 1105 ft².

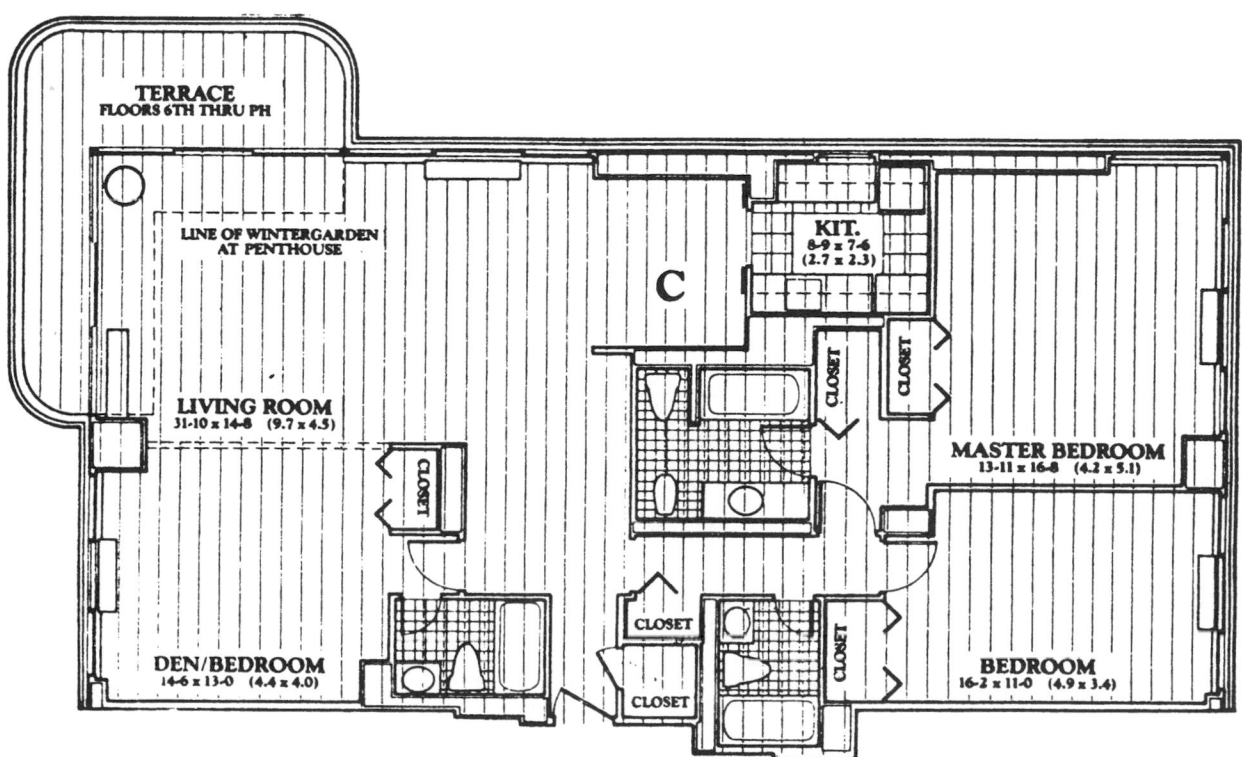

THREE-BEDROOM APARTMENT

Dining Room
15'-6" x 14'-6"
Kitchen
13'-0" x 12'-6"
Living Room
22'-0" x 21'-0"
Master Bedroom
22'-0" x 14'-0"
Gallery
14'-6" x 14'-0"
Bedroom/Study
14'-0" x 13'-0"
W/D
Bedroom
17'-6" x 11'-6"
6 Rooms
2447 Square Feet
3 Bedrooms
3½ Baths

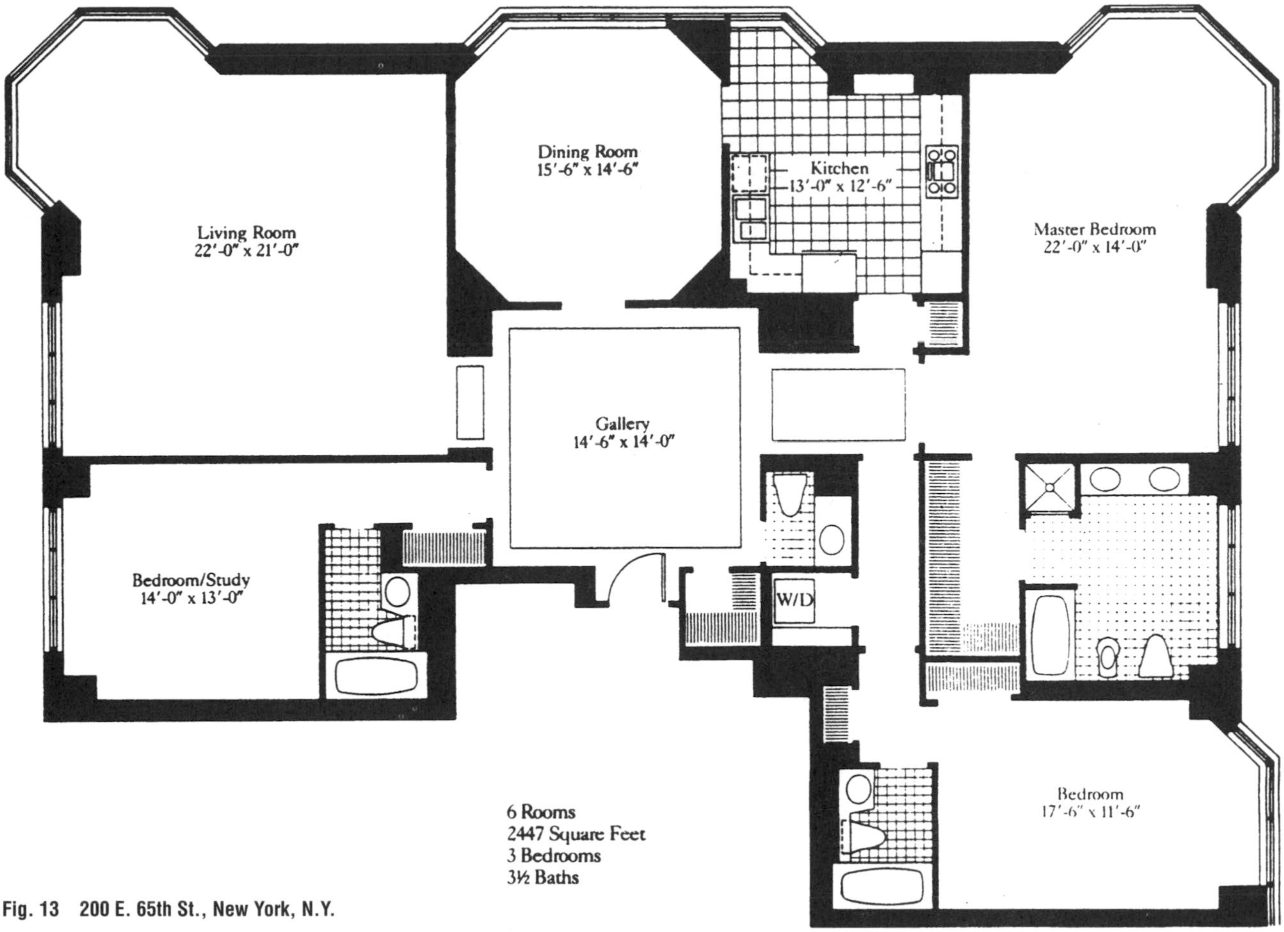

Fig. 13 200 E. 65th St., New York, N.Y.

TERRACE
FLOORS 6TH THRU PH
LINE OF WINTERGARDEN
AT PENTHOUSE
KIT.
8-9 x 7-6
(2.7 x 2.3)
C
LIVING ROOM
31-10 x 14-8 (9.7 x 4.5)
CLOSET
CLOSET
CLOSET
MASTER BEDROOM
13-11 x 16-8 (4.2 x 5.1)
CLOSET
CLOSET
CLOSET
DEN/BEDROOM
14-6 x 13-0 (4.4 x 4.0)
BEDROOM
16-2 x 11-0 (4.9 x 3.4)

Fig. 14 Trump Plaza, 167 E. 61st St., New York, N.Y.

THREE-BEDROOM APARTMENT

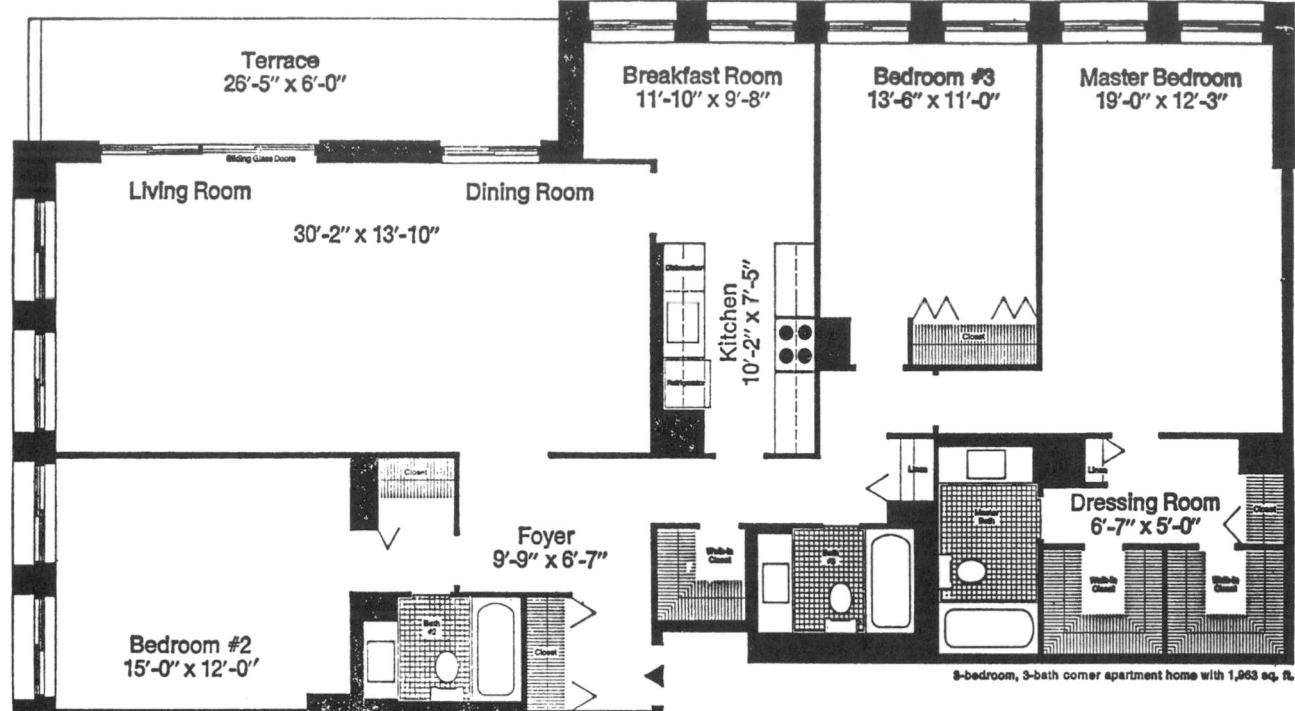

Fig. 15   Winston Towers, Fort Lee, N.J.

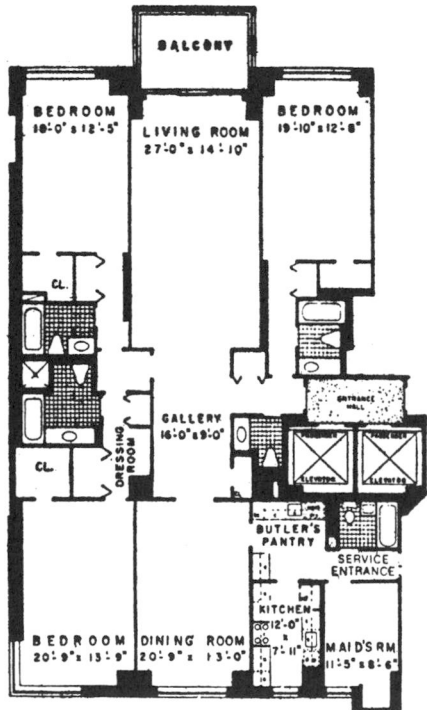

Fig. 16   The Sovereign, New York, N.Y.

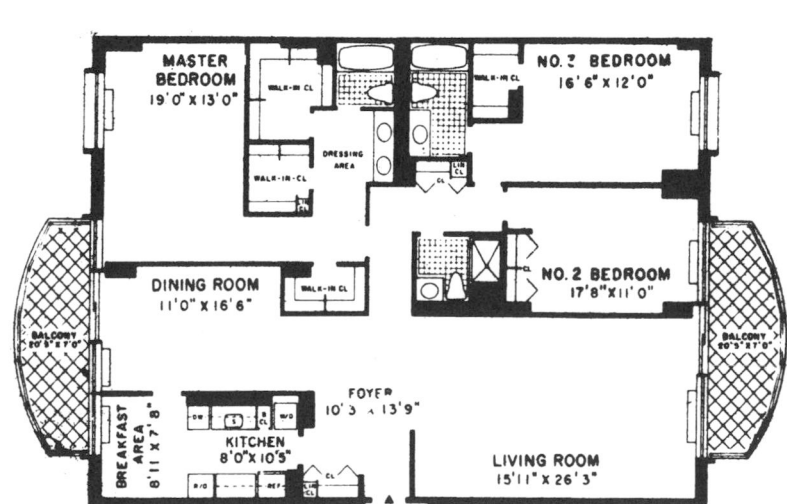

Fig. 17   Claridge House, Verona, N.J.

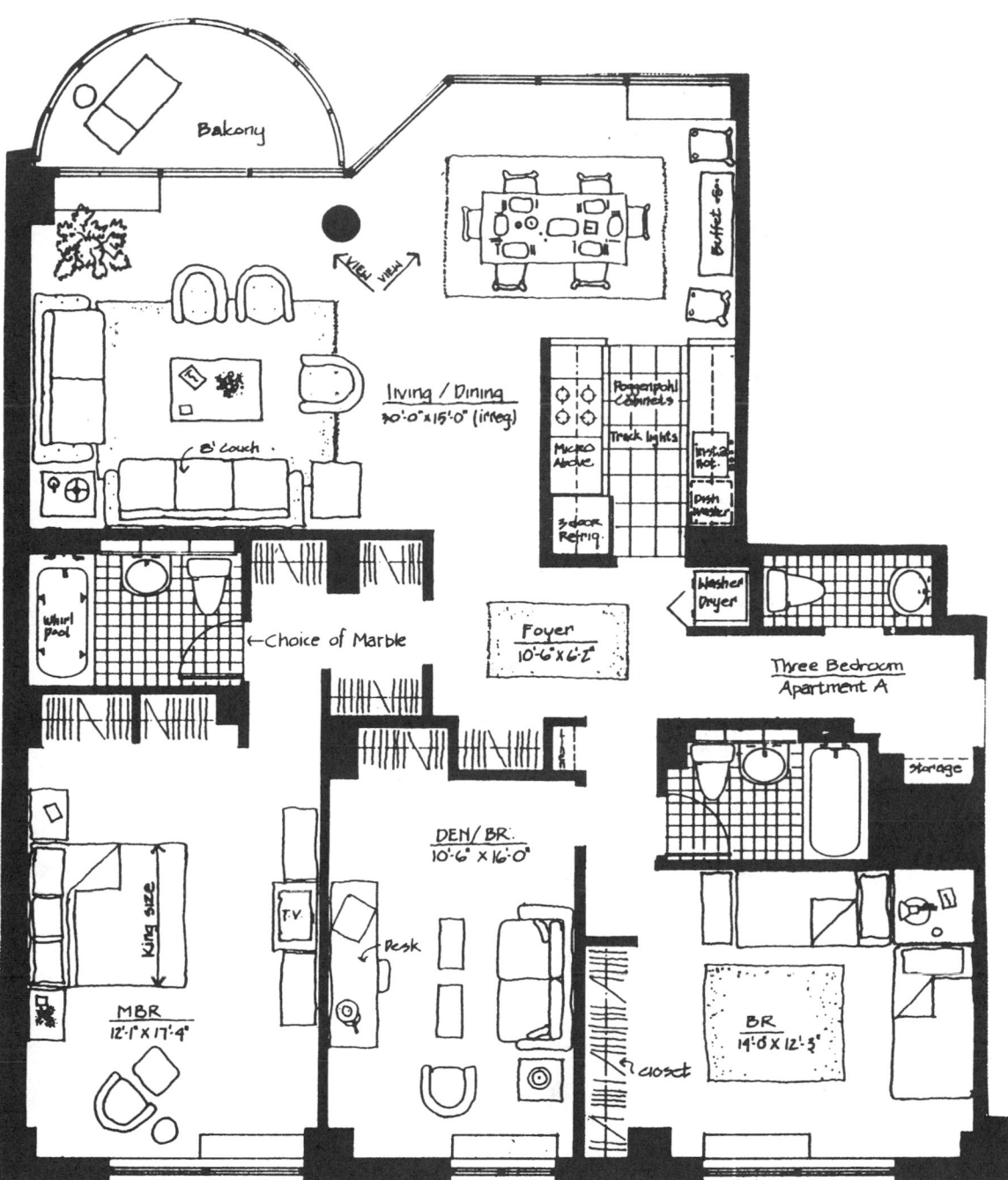

Balcony

living / Dining
30'-0" x 15'-0" (irreg.)

8' couch

← Choice of Marble

Poggenpohl
Cabinets

Track lights

Micro
Above

3-door
Refrig.

Inst. hot.

Dish washer

Washer
Dryer

Foyer
10'-6" x 6'-2"

Three Bedroom
Apartment A

storage

Whirl pool

King size

MBR
12'-1" x 17'-4"

T.V

Desk

DEN / BR.
10'-6" x 16'-0"

closet

BR
14'-0 x 12'-5"

Fig. 18   La Rive, Riverdale, N.Y.C.

THREE-BEDROOM APARTMENT

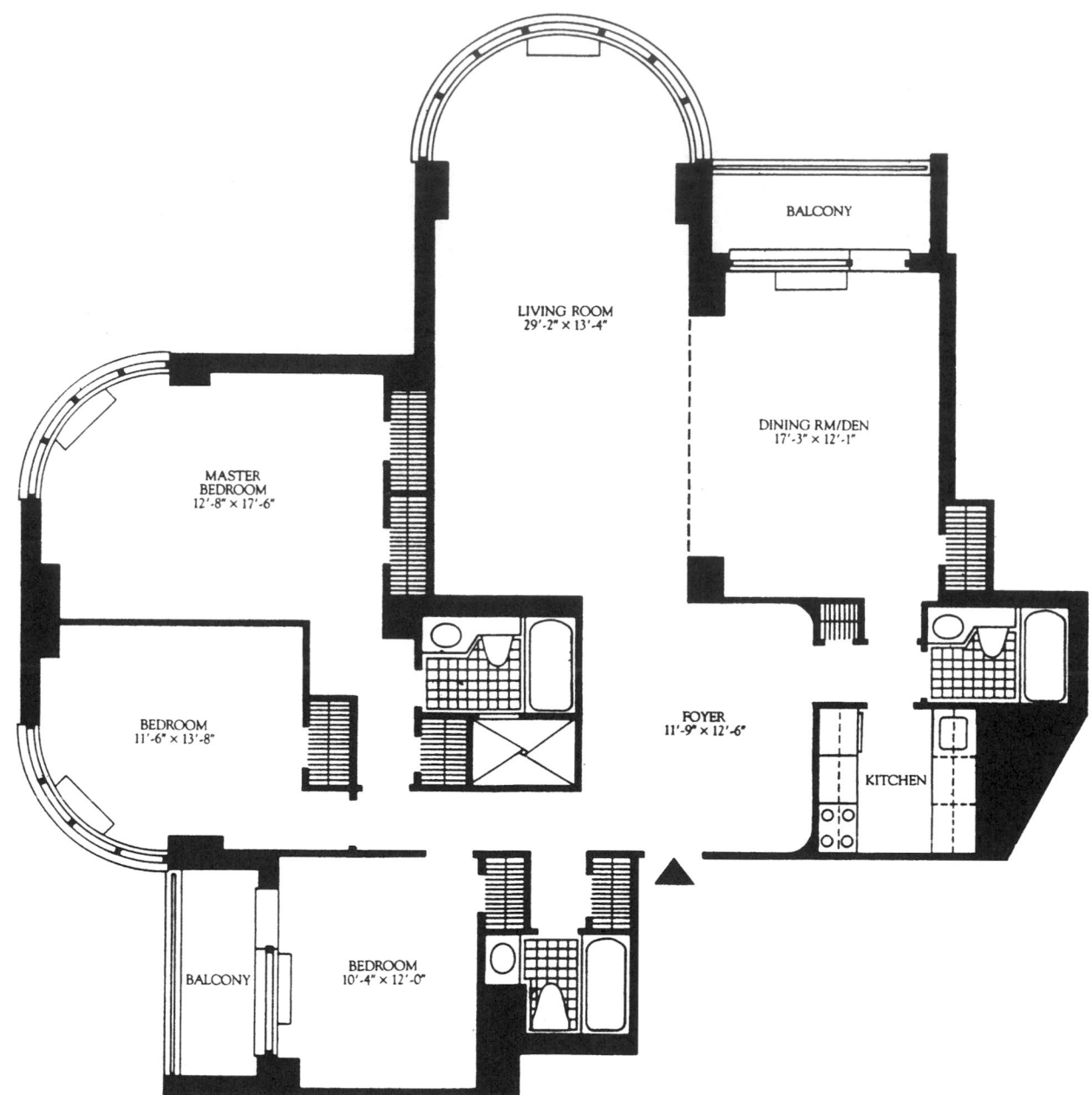

BALCONY

LIVING ROOM
29'-2" × 13'-4"

DINING RM/DEN
17'-3" × 12'-1"

MASTER
BEDROOM
12'-8" × 17'-6"

BEDROOM
11'-6" × 13'-8"

FOYER
11'-9" × 12'-6"

KITCHEN

BALCONY

BEDROOM
10'-4" × 12'-0"

Fig. 19   The Corinthian Condominiums, New York, N.Y.

Maid's Room
11'-0" x 10'-0"

W.

D.

Kitchen
16'-0" x 12'-6"

Bedroom
17'-6" x 11'-6"

Dining Room
17'-0" x 14'-0"

Bedroom
15'-0" x 11'-0"

Wet Bar

Gallery
16'-0" x 13'-0"

Living Room
22'-0" x 21'-0"

Master Bedroom
22'-6" x 13'-6"

Library
15'-6" x 14'-0"

All measurements are approximate.

8 Rooms
3022 Square Feet
3 Bedrooms
3½ Baths
Maid's Room and Bath

Fig. 20   200 E. 65th St., New York, N.Y.

THREE-BEDROOM APARTMENT

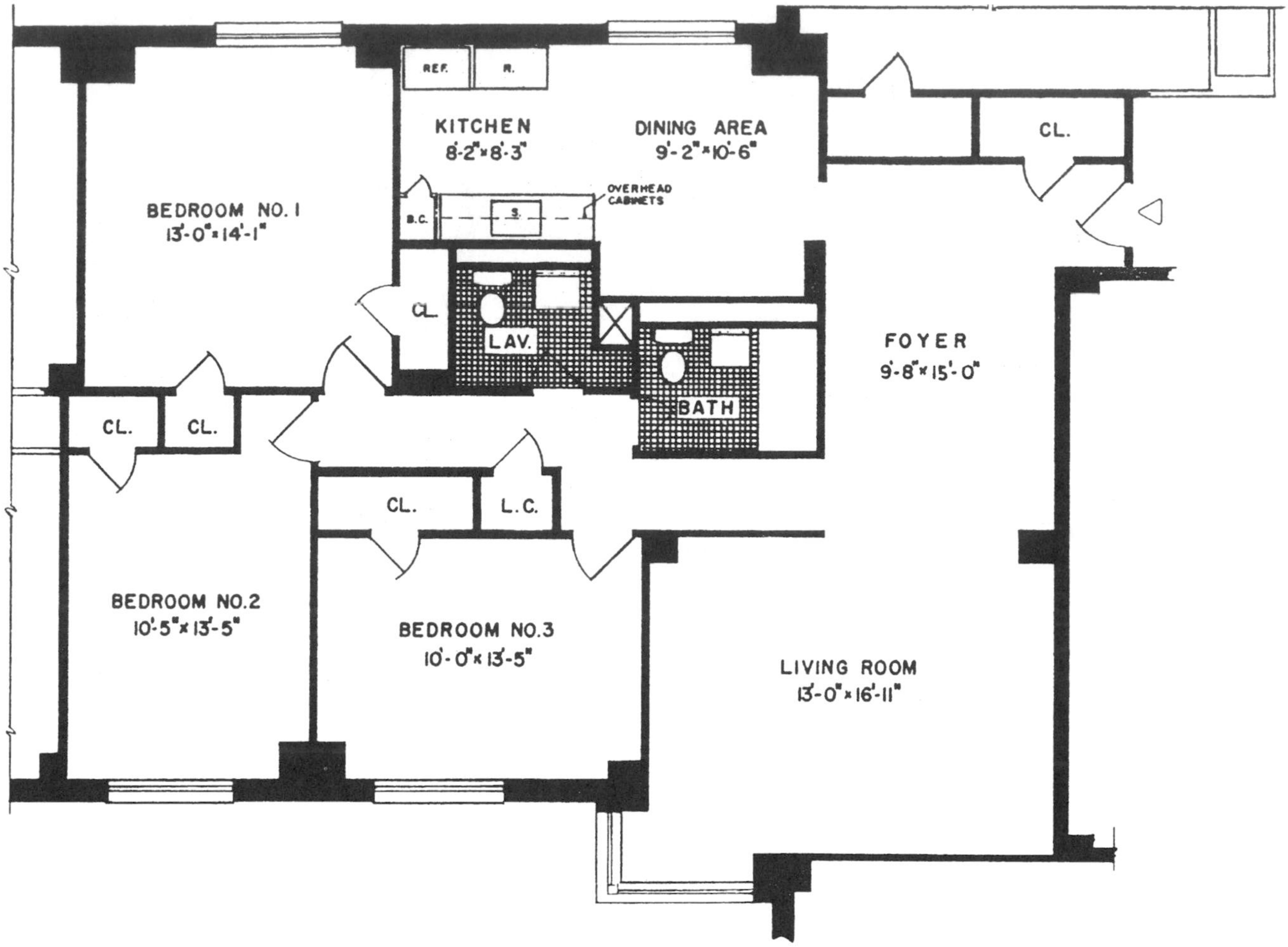

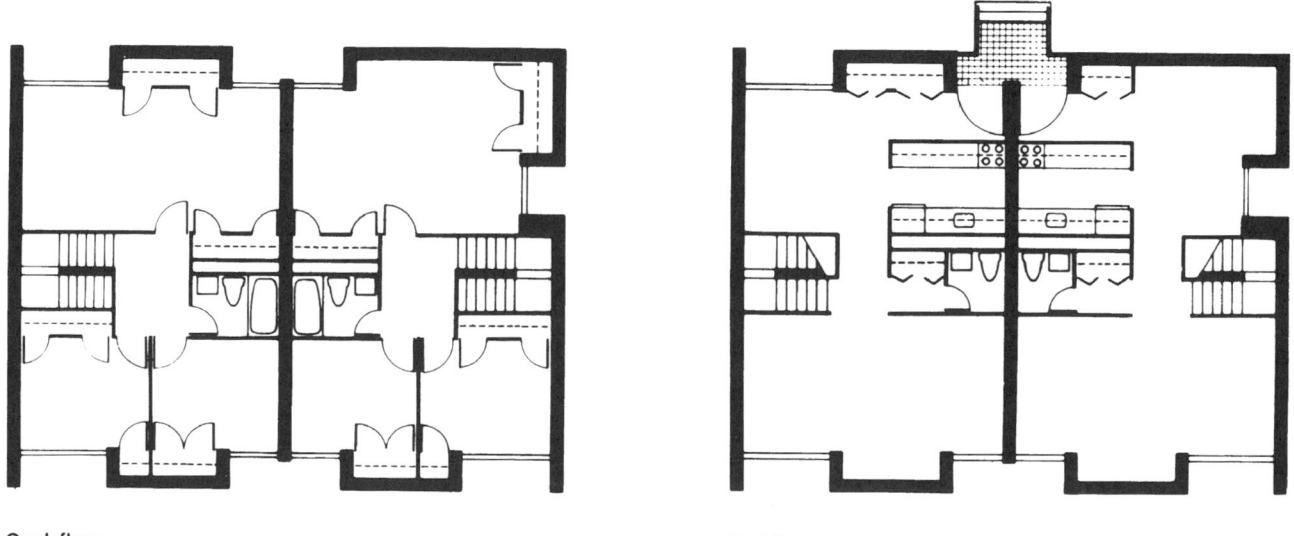

Fig. 21

KITCHEN
8'-2"x 8'-3"

DINING AREA
9'-2"x 10'-6"

REF.   R.

B.C.

OVERHEAD CABINETS

BEDROOM NO.1
13'-0"x 14'-1"

CL.

LAV.

BATH

CL.

FOYER
9'-8"x 15'-0"

CL.

CL.   CL.

CL.   L.C.

BEDROOM NO.2
10'-5"x 13'-5"

BEDROOM NO.3
10'-0"x 13'-5"

LIVING ROOM
13'-0"x 16'-11"

2nd floor

1st floor

Fig. 22   Lavanburg Community, New York, N.Y. Conklin & Rossant—Architects.

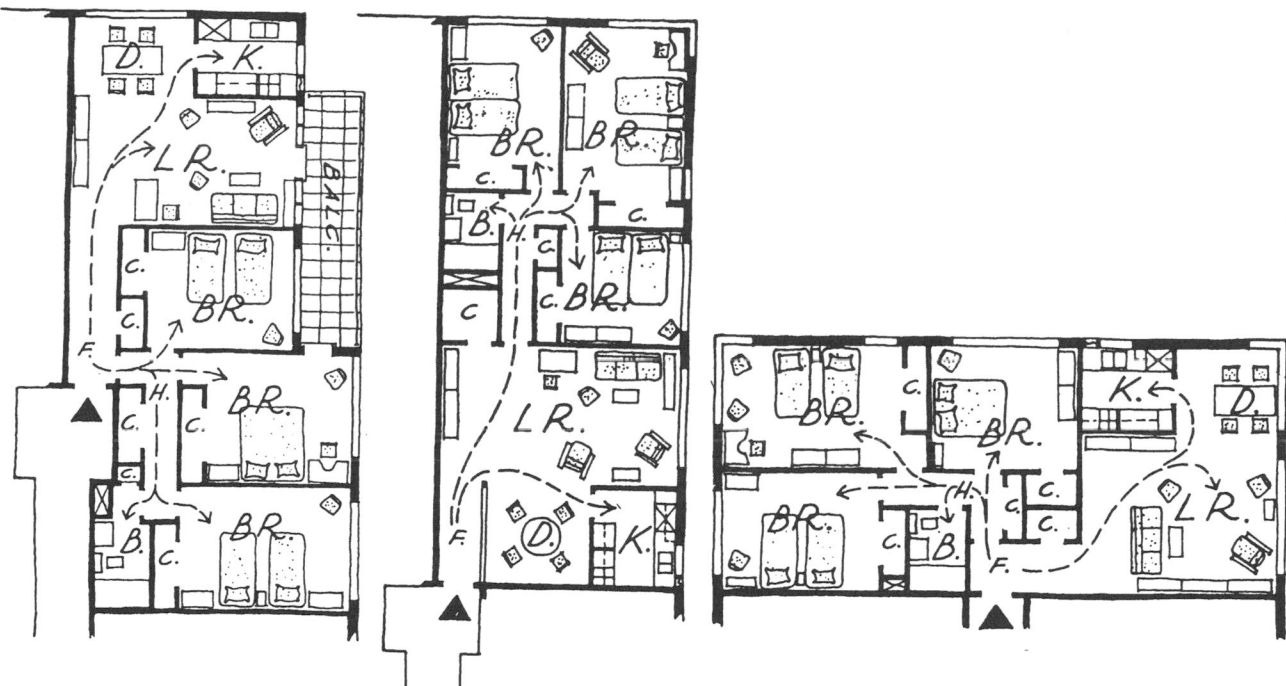

Fig. 23

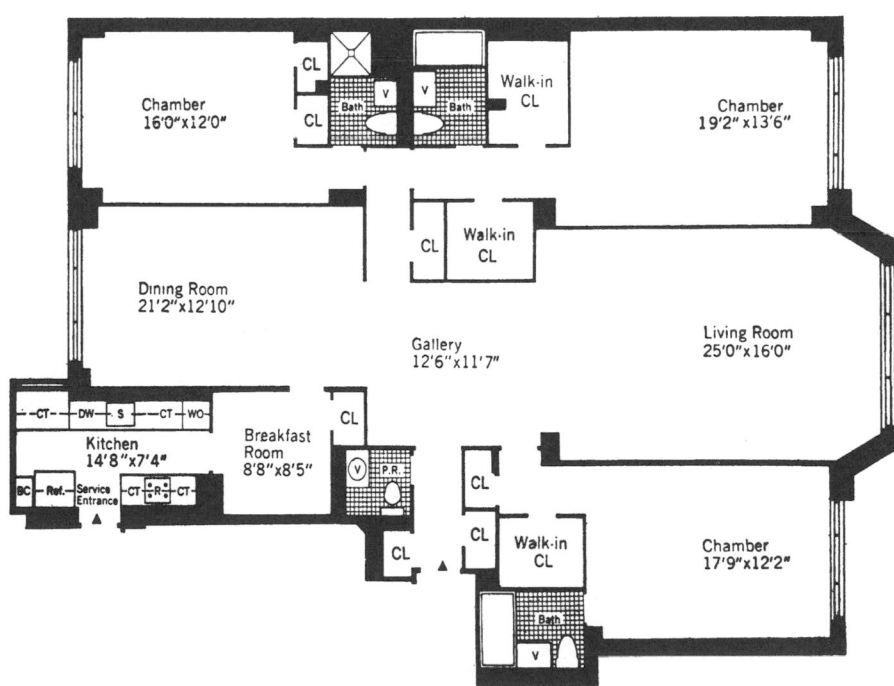

Fig. 24   60 East End Ave., New York, N.Y.

THREE-BEDROOM APARTMENT

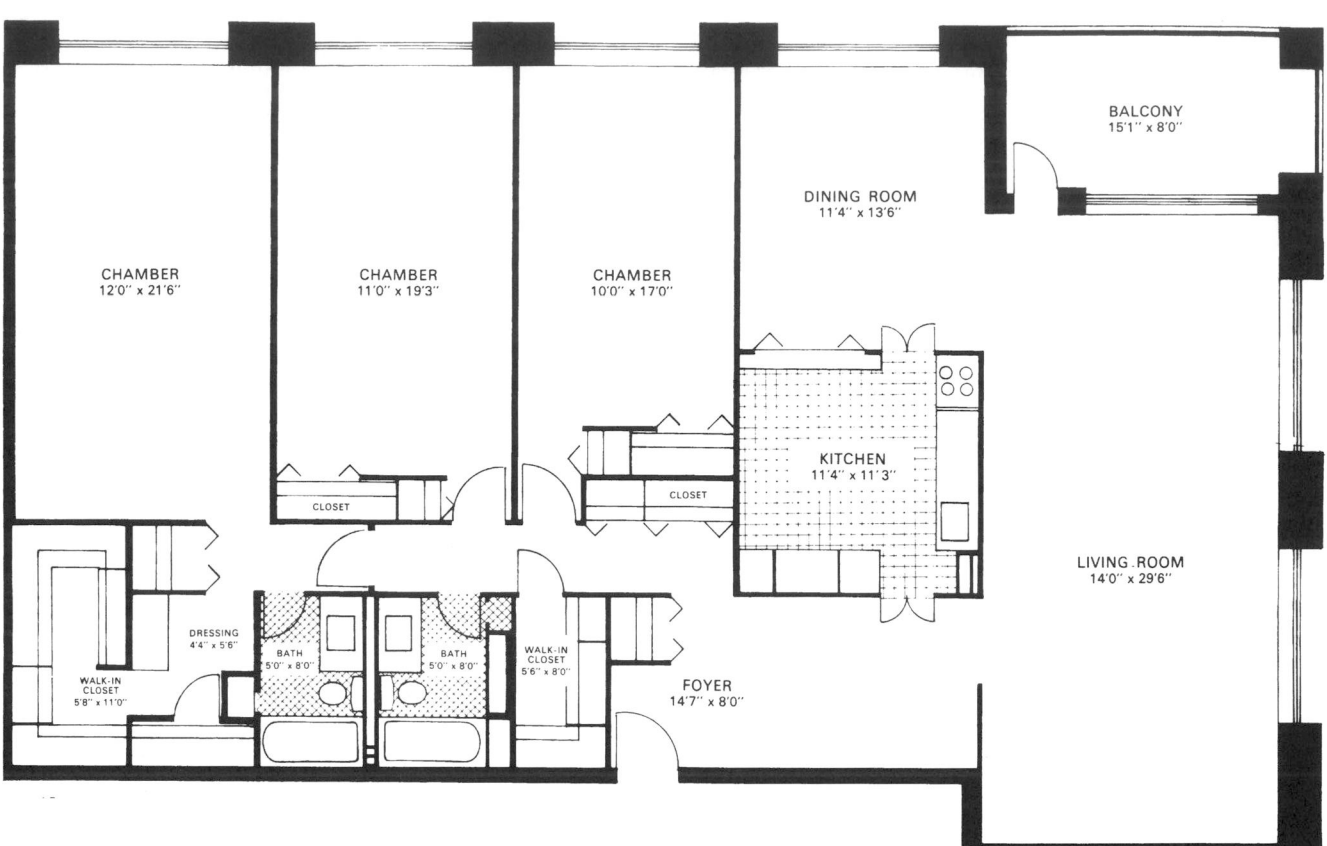

CHAMBER
12'0" x 21'6"

CHAMBER
11'0" x 19'3"

CHAMBER
10'0" x 17'0"

DINING ROOM
11'4" x 13'6"

BALCONY
15'1" x 8'0"

KITCHEN
11'4" x 11'3"

CLOSET

CLOSET

LIVING ROOM
14'0" x 29'6"

DRESSING
4'4" x 5'6"

BATH
5'0" x 8'0"

BATH
5'0" x 8'0"

WALK-IN
CLOSET
5'6" x 8'0"

WALK-IN
CLOSET
5'8" x 11'0"

FOYER
14'7" x 8'0"

Fig. 25

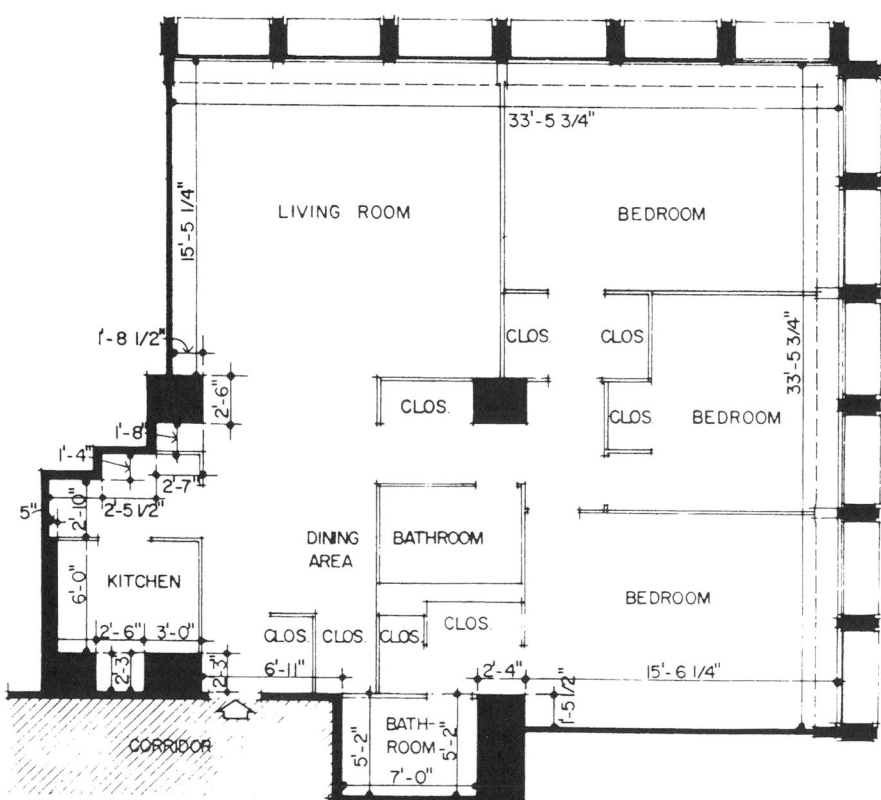

33'-5 3/4"

15'-5 1/4"

LIVING ROOM

BEDROOM

33'-5 3/4"

1'-8 1/2"

2'-6"

1'-8"

1'-4"

2'-7"

2'-5 1/2"

5"

2'-0"

6'-0"

KITCHEN

CLOS.

CLOS.

CLOS.

CLOS.

BEDROOM

DINING
AREA

BATHROOM

BEDROOM

2'-6"

3'-0"

CLOS. CLOS. CLOS.

CLOS.

2'-3"

2'-3"

6'-11"

2'-4"

15'-6 1/4"

1'-5 1/2"

CORRIDOR

5'-2"

BATH-
ROOM
7'-0"

5'-2"

Fig. 26   Kips Bay Plaza, New York, N.Y. I. M. Pei & Partners—Architects.

618

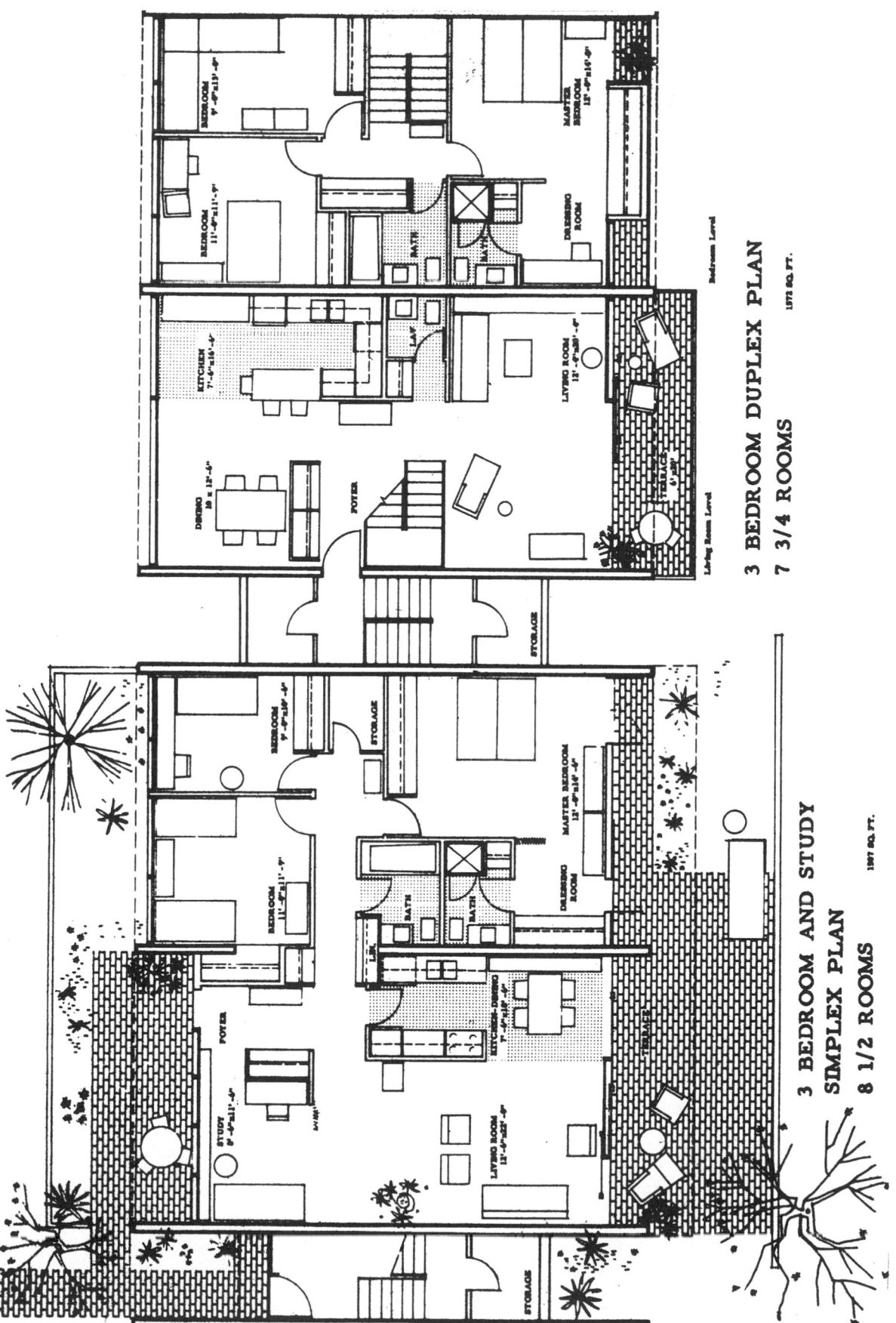

3 BEDROOM DUPLEX PLAN
7 3/4 ROOMS

3 BEDROOM AND STUDY
SIMPLEX PLAN
8 1/2 ROOMS

Fig. 27

THREE-BEDROOM APARTMENT

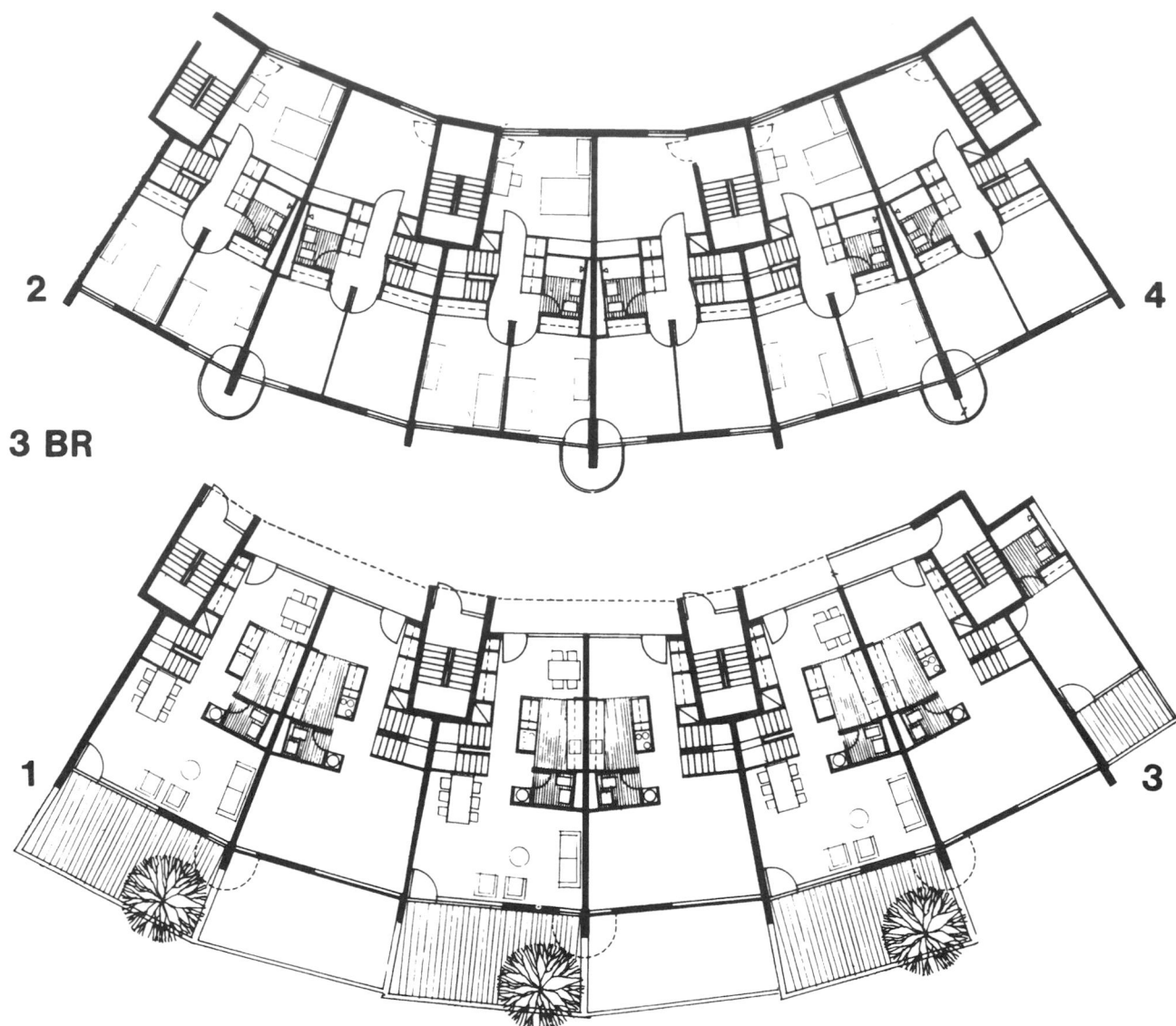

3 BR

Fig. 28   Harbor Village at Paerdegat, Gruzen & Partners—Architects.

FOUR-BEDROOM APARTMENT

Elements

The four-bedroom apartment consists of four bedrooms, living room, dining room, full kitchen, and two bathrooms. In addition to a terrace off the living room, an additional terrace may be provided off the master bedroom. A large amount of storage space is essential.

Design

The four-bedroom apartment is considered to be a large apartment and is not very common. Since more occupants are expected with a wider spread of age differences, more living space is required for a greater number of activities. This arrangement of rooms should be

such as to provide maximum privacy for each one grouping. A separate dining room should be provided. Long corridors to the bedrooms should be avoided.

Size

The size of the four-bedroom apartment will range from 1100 to 1500 ft². The FHA minimum requirement for such an apartment is approximately 1200 ft². A larger percentage of floor area will be utilized for circulation than in the smaller apartments.

Type of Occupancy

The number of occupants can range from five to eight persons. The most typical would be a normal family with three to six children. The

minimum occupancy would be one child per bedroom plus the parents in the master bedroom. The maximum would be two children per bedroom. Another fairly common type of occupancy would be occupancy by three generations, that is, grandparents, parents, and children. As the life expectancy increases, some other people are sharing housing accommodations with their children or other relatives.

Planning Implications

Because of this type of occupancy, large numbers of school-age children could easily result. This will require additional space in the schools and related facilities. If older persons are part of the household, additional facilities may be required in the form of medical, social, and leisure-time activities.

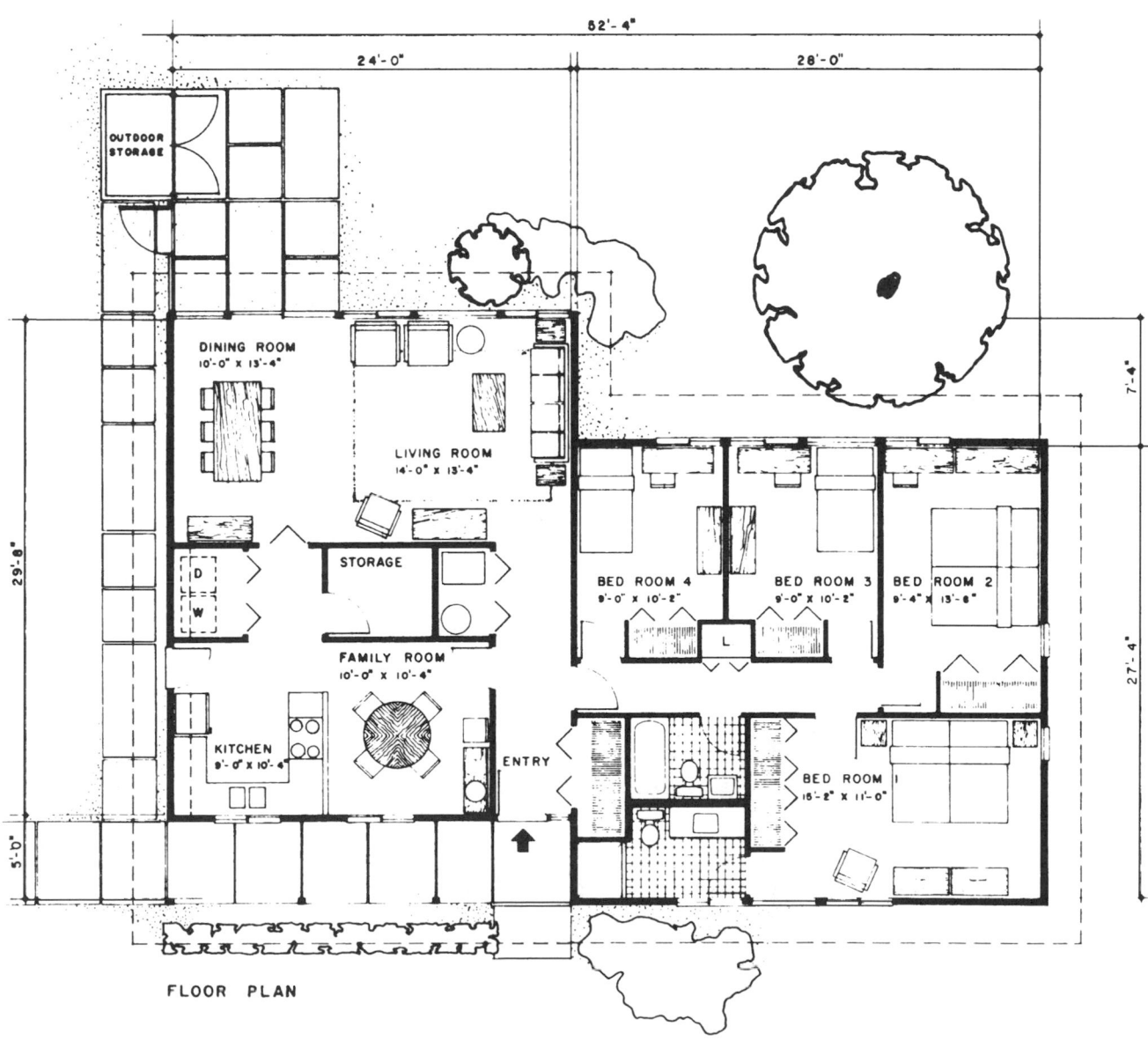

FLOOR PLAN

Fig. 1 Gross area = 1560 ft²; net area = 1400 ft².

FOUR-BEDROOM APARTMENT

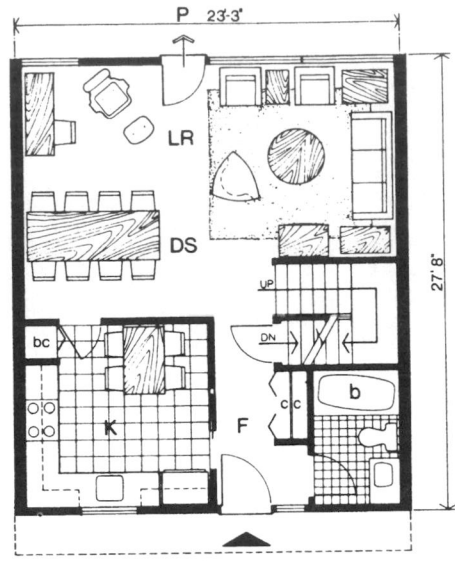

FIRST FLOOR

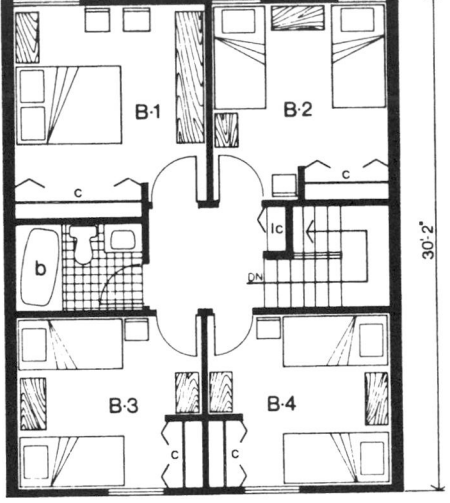

SECOND FLOOR

Fig. 2

The four-bedroom apartment shown in Fig. 2 has a gross area of 1350 ft².

Features	Dimensions, ft	Area, ft²
Living space	15.5 × 22.5	320
Dining space		
Kitchen	11 × 11	121
Bedroom 1	11.5 × 11	125
Bedroom 2	10.5 × 10.5	110
Bedroom 3	9 × 10.5	95
Bedroom 4	9 × 10.5	95
Bathroom 1	5 × 7.5	38
Bathroom 2	5 × 8	40

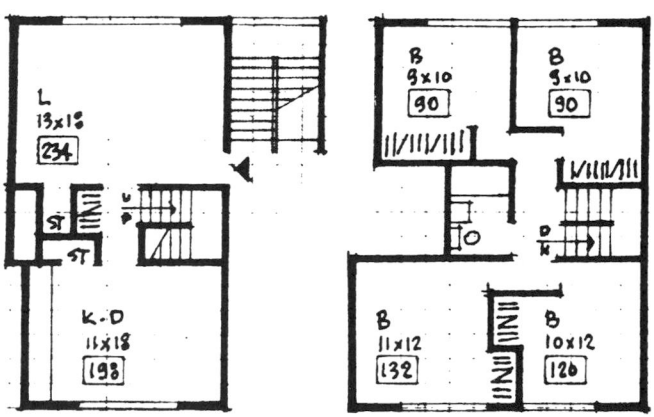

Fig. 3   Total area, 1180 ft².

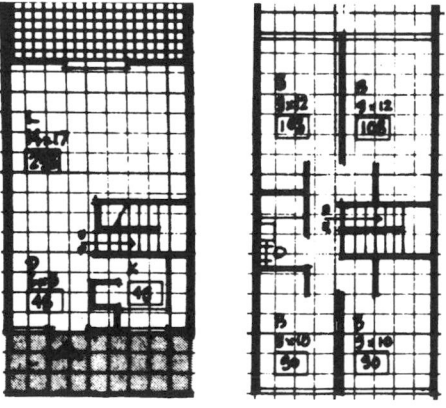

Fig. 4   Total area, 1080 ft².

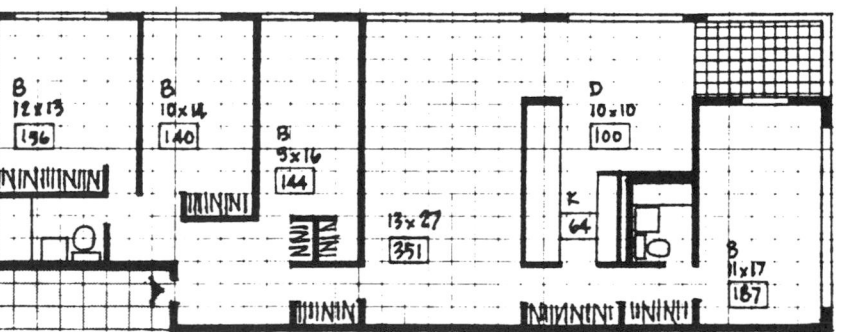

Fig. 5   Total area, 1655 ft².

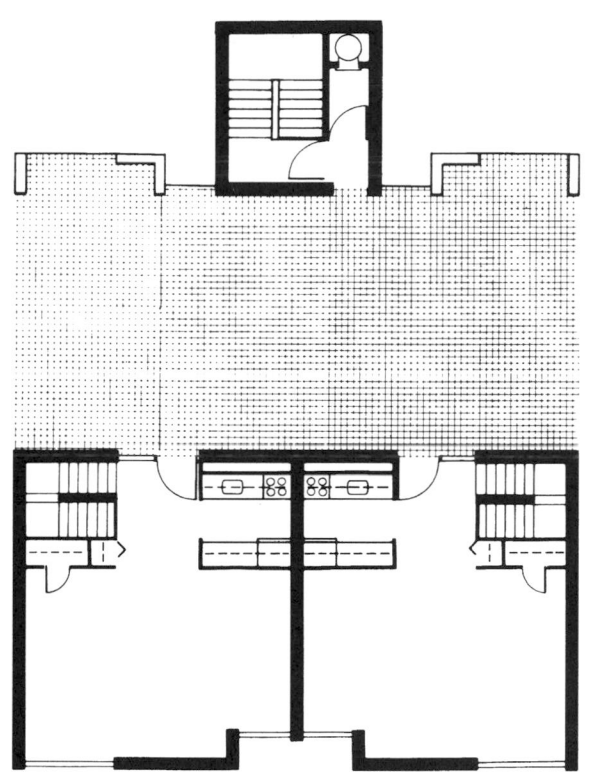

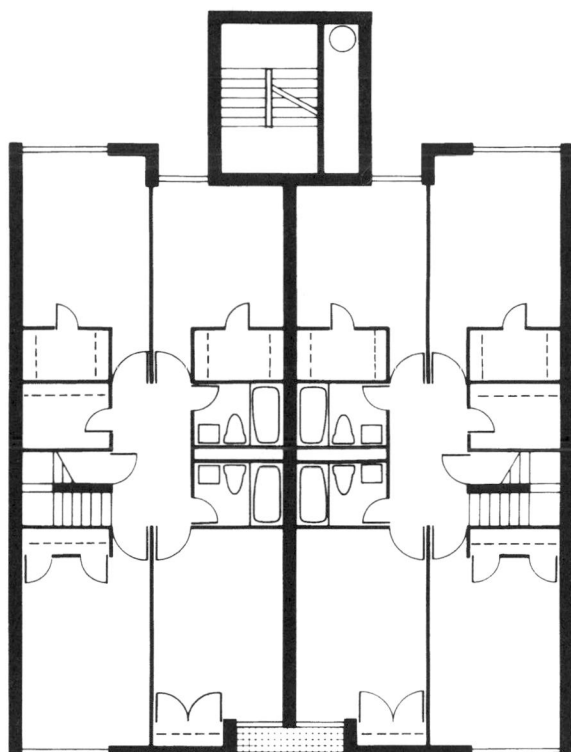

Fig. 6   Lavanburg Community, New York, N.Y. Conklin & Rossant—Architects.

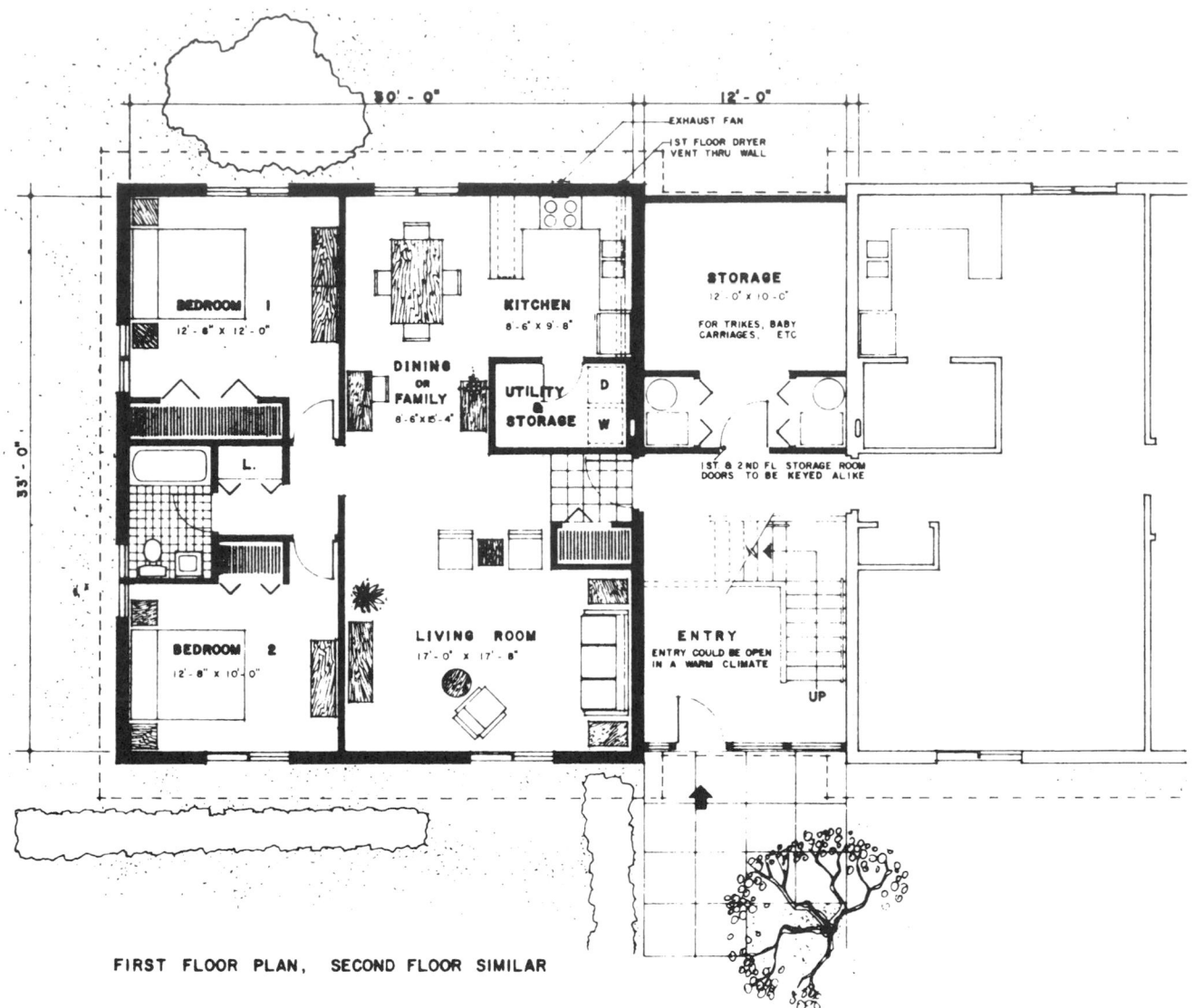

FIRST FLOOR PLAN, SECOND FLOOR SIMILAR

Fig. 1

Building plans show how the ground-floor flats and piggybacked two-story units fall into line within the curved, three-story structures. Staggering the back-to-back units provides spaces for private entries and also allows for maximum separation between most of the mid-story balconies. Where those balconies do abut, they are separated by wing walls.

Two-story units—with one exception—occupy the second and third floors of the buildings. The exception: lower-floor end units at the rear of the buildings (plan E), which are included to keep the buildings in scale. The two-story units (plans C and D) are available with one or two bedrooms. All have two-story spaces in the living rooms. The master bedroom in plan C is open to this two-story space; so is the single bedroom in plan D.

Ground-floor flats are either efficiency units (plan A) or one-bedroom layouts (plan B). In the efficiency plan, the sleeping space is large enough to be separated visually from the dining area by a screen or a folding partition. Patios for ground-floor units are screened for privacy on three sides by wooden fencing. A planning feature worth noting: Although kitchens in the project are windowless, all have a pass-through above the sink so they can gain light from the living room.

For Marcus Garvey Park Village in Brooklyn, two types of mews units and four types of street units were designed; the most common of each unit type is shown. In actual application the arrangement of units was modified to accommodate site characteristics, which forced a tighter arrangement. Minimal outdoor play area is compensated for by a park adjacent to site.

The four main units of the prototype consist of the street unit, the mews unit, the mews itself, and the public stoop in relation to the inset parking. Plans show that each mews unit contains two upper and two lower duplexes, giving a total of three three-bedroom apartments and one four-bedroom apartment. The street unit contains a two- and a three-bedroom duplex on the lower floors, two one-bedroom apartments and two two-bedroom apartments on the upper floor. For purposes of security, all apartments have private or semiprivate entrances at street level.

Located on two islands in a large, man-made lagoon near the southern tip of San Francisco Bay, this condominium development provides a fascinating geometrical configuration in which all apartments have a view to the water. Excellent site planning and design amenities in individual units create one of the most pleasant new housing developments in the area.

Each of the five buildings shown in the site plan contains twin, mirror layouts that were designed for one of the three specified markets: first-floor for retirees; second floor for entertainment-oriented, mature couples; penthouse for growing families. Dotted line in penthouse plans shows area above which the loft (left) is located. Units contain from 1170 to 1610 ft^2.

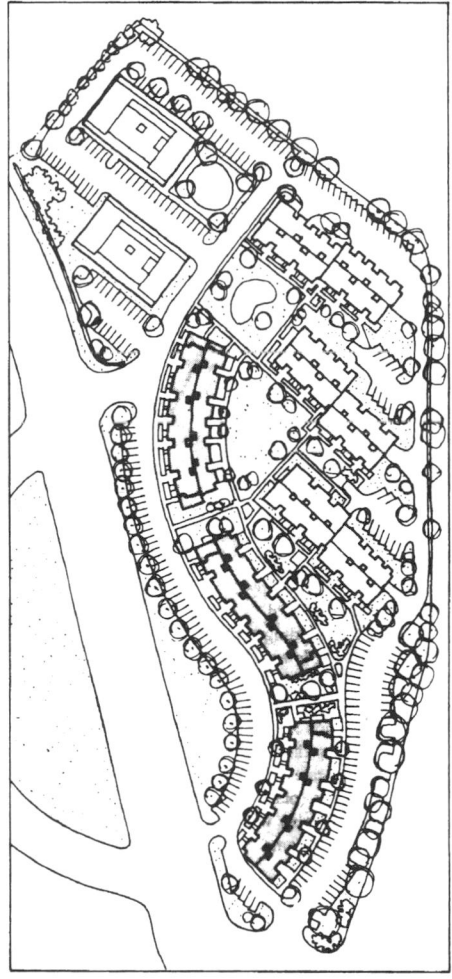

Fig. 1 The Commons, Greenwich, Conn. SMA—Architects.

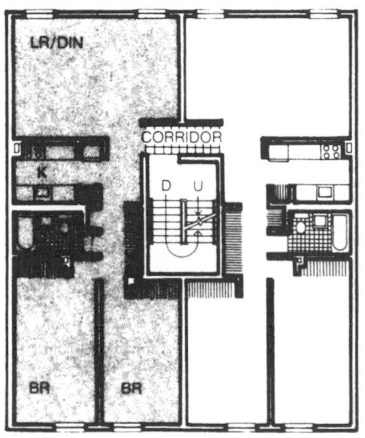
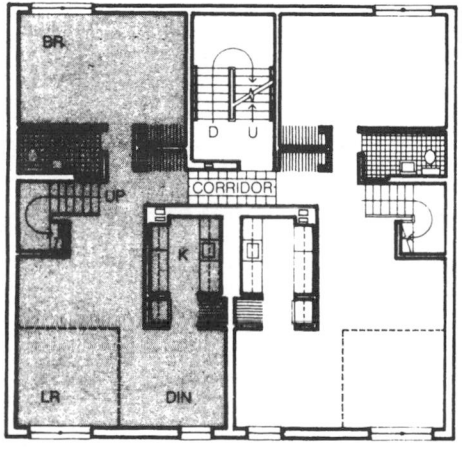
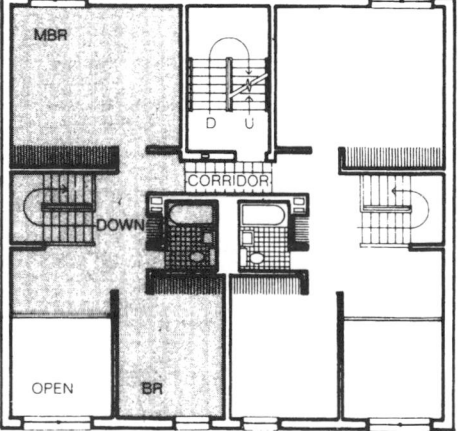

Fig. 2 West Village Houses, New York, N.Y. The Perkins & Wills Partnership—Architects.

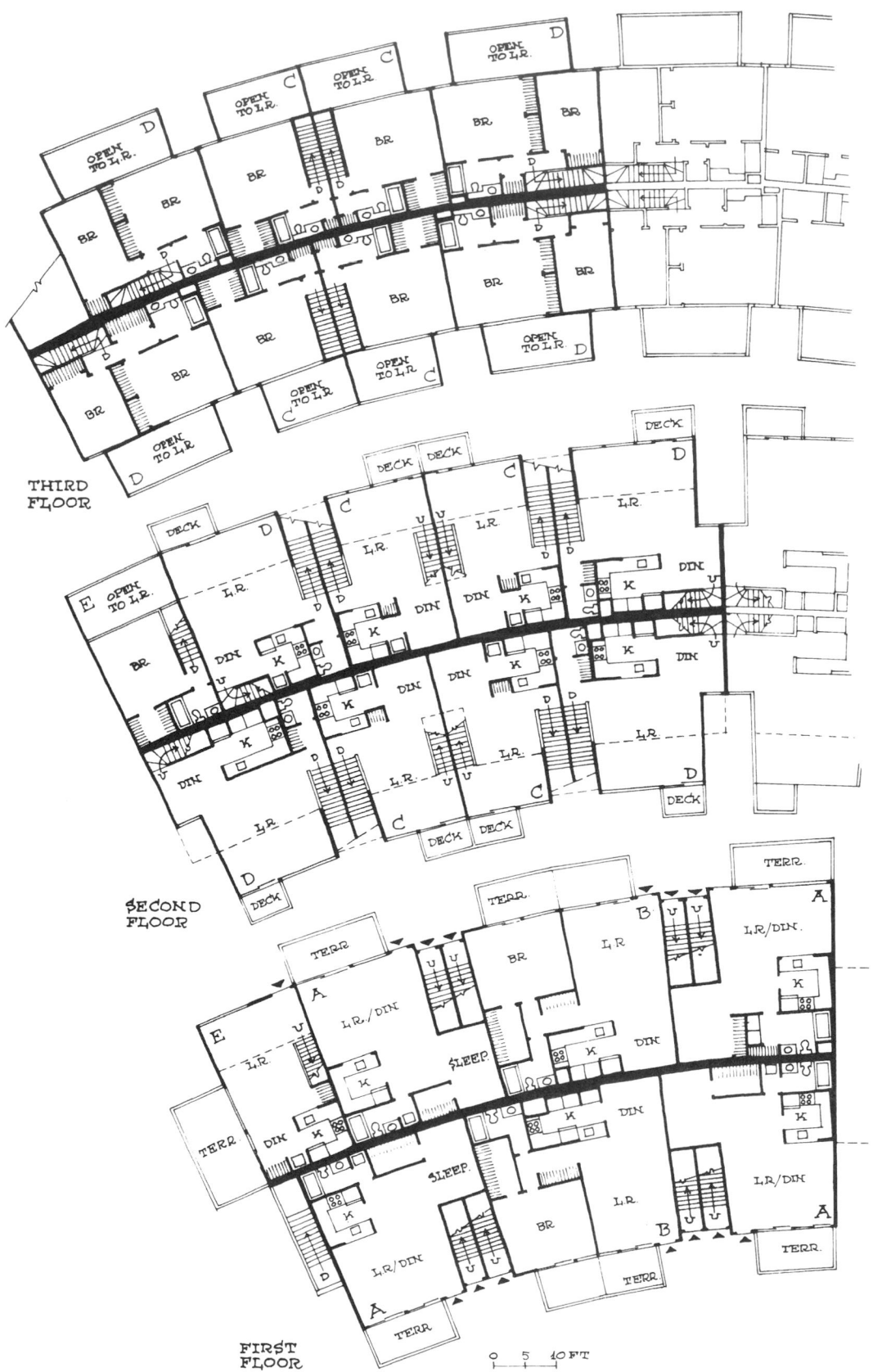

THIRD FLOOR

SECOND FLOOR

FIRST FLOOR

0 5 10 FT

Fig. 3 The Commons, Greenwich, Conn. SMA—Architects.

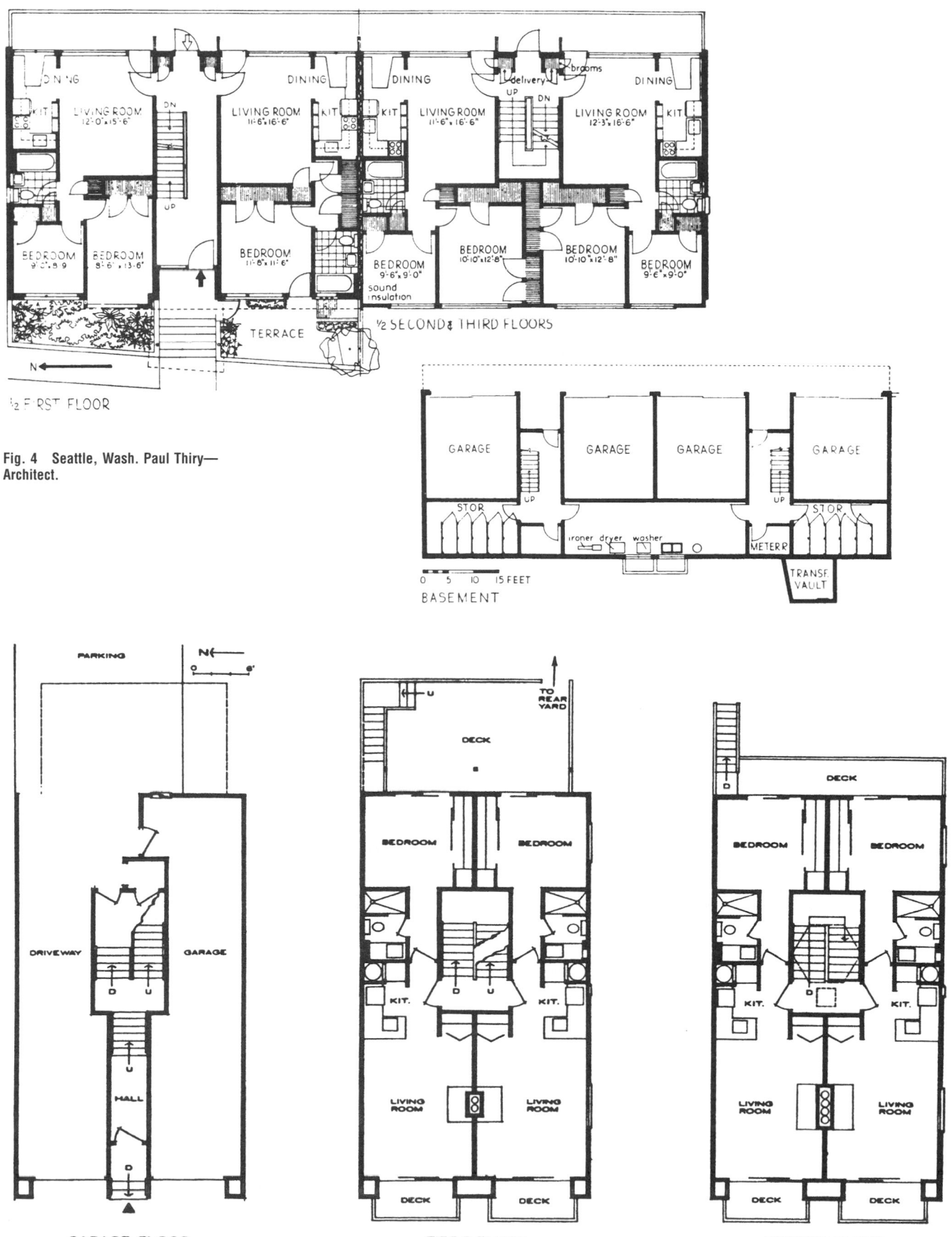

Fig. 4   Seattle, Wash. Paul Thiry—Architect.

Fig. 5   Jonathan Buckley & Associates—Architects.

LOW-RISE, WALK-UP

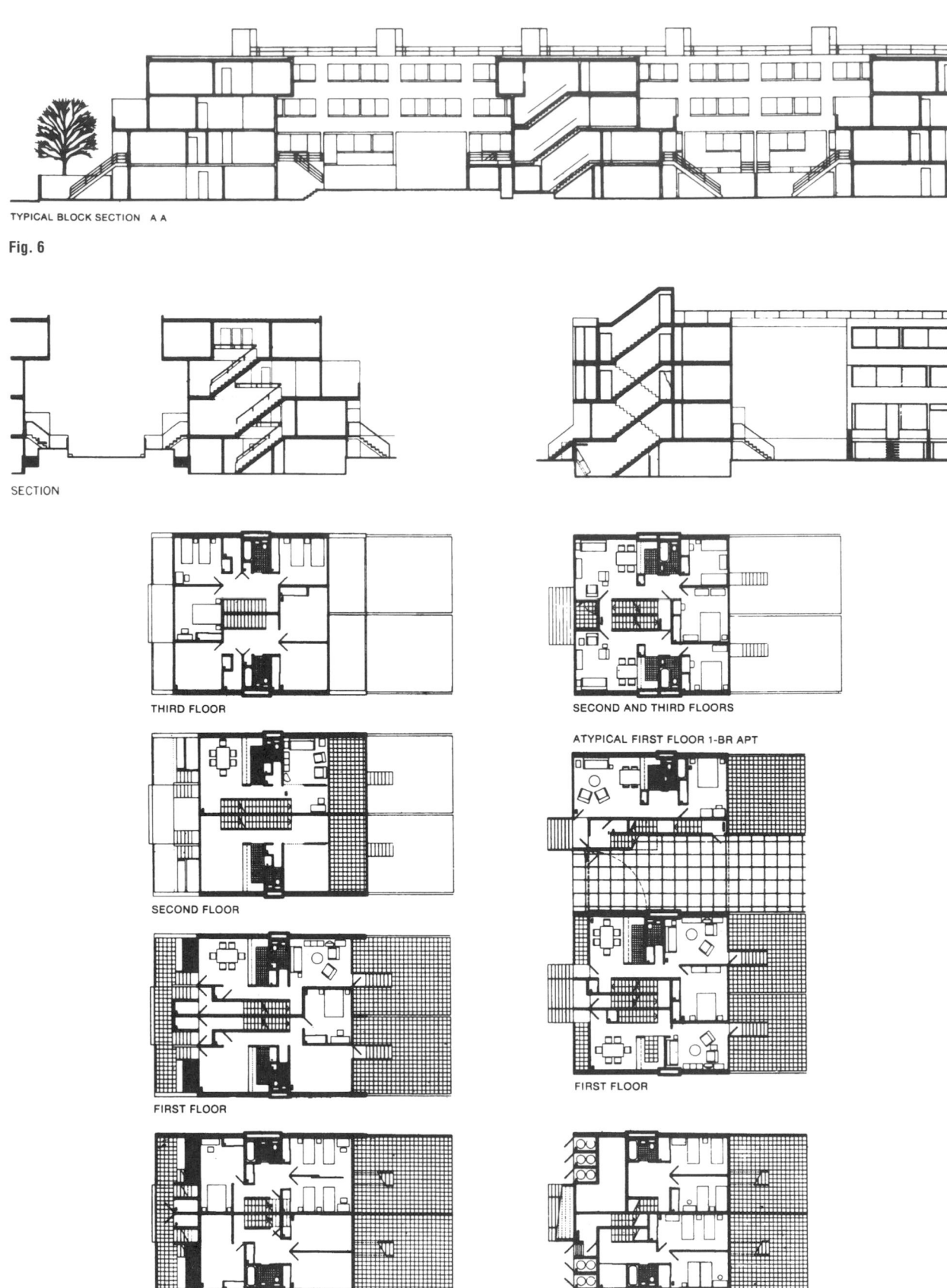

TYPICAL BLOCK SECTION A A

Fig. 6

SECTION

THIRD FLOOR

SECOND AND THIRD FLOORS

ATYPICAL FIRST FLOOR 1-BR APT

SECOND FLOOR

FIRST FLOOR

FIRST FLOOR

GROUND FLOOR

GROUND FLOOR

Fig. 7

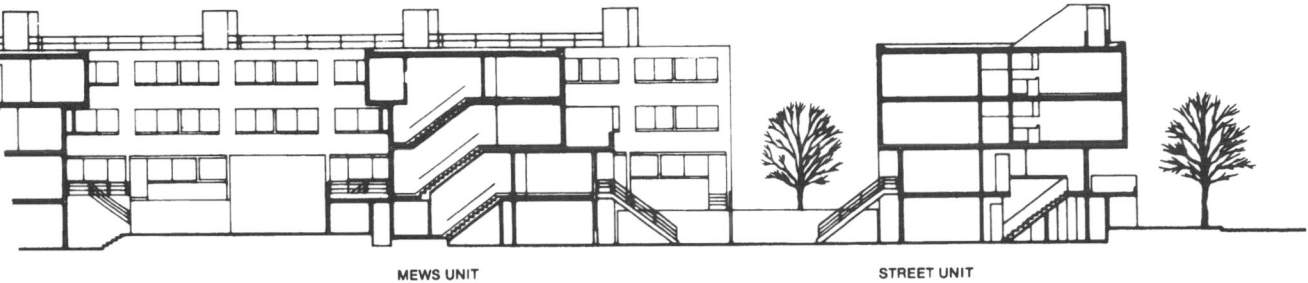

MEWS UNIT   STREET UNIT

Fig. 6 (*Continued*)

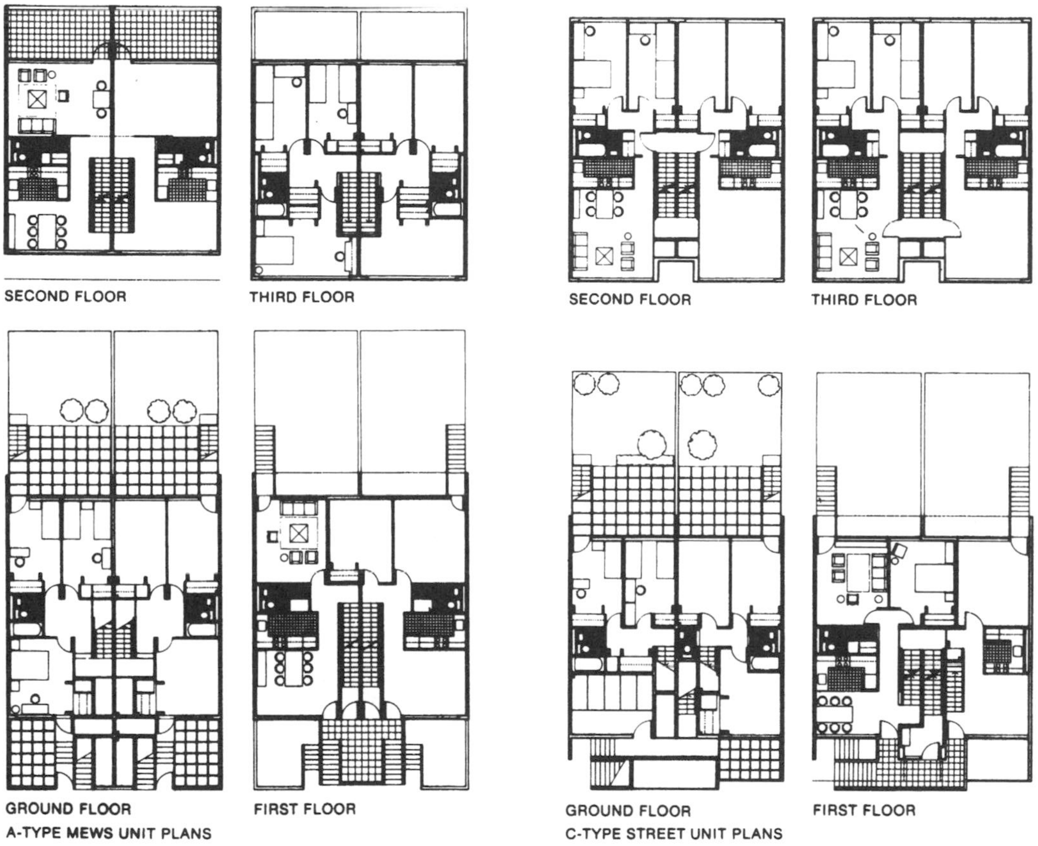

SECOND FLOOR   THIRD FLOOR   SECOND FLOOR   THIRD FLOOR

GROUND FLOOR   FIRST FLOOR   GROUND FLOOR   FIRST FLOOR

A-TYPE MEWS UNIT PLANS   C-TYPE STREET UNIT PLANS

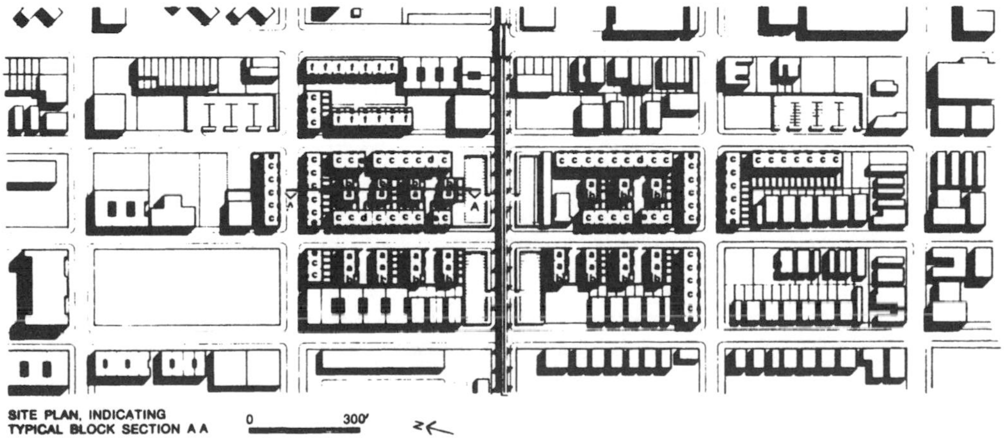

SITE PLAN, INDICATING
TYPICAL BLOCK SECTION A A   0   300'

Fig. 8 Marcus Garvey Park Village, Brooklyn, N.Y. The Institute for Architecture and Urban Studies—Architects.

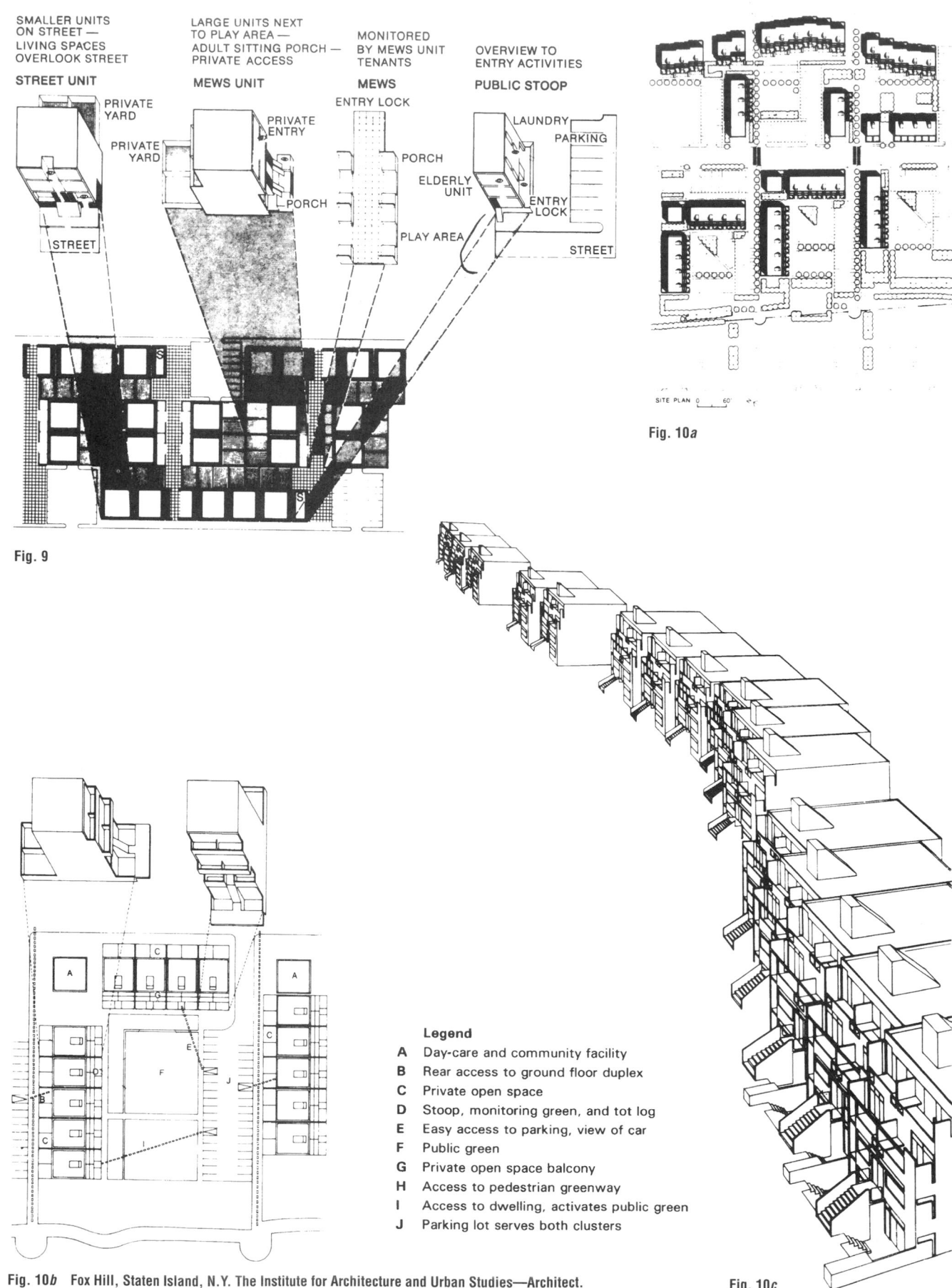

STREET UNIT
SMALLER UNITS ON STREET — LIVING SPACES OVERLOOK STREET

PRIVATE YARD

PRIVATE YARD

STREET

MEWS UNIT
LARGE UNITS NEXT TO PLAY AREA — ADULT SITTING PORCH — PRIVATE ACCESS

PRIVATE ENTRY

PORCH

MEWS
MONITORED BY MEWS UNIT TENANTS

ENTRY LOCK

PORCH

PLAY AREA

PUBLIC STOOP
OVERVIEW TO ENTRY ACTIVITIES

LAUNDRY

PARKING

ELDERLY UNIT

ENTRY LOCK

STREET

Fig. 9

SITE PLAN 0   60'

Fig. 10a

Legend

A Day-care and community facility
B Rear access to ground floor duplex
C Private open space
D Stoop, monitoring green, and tot log
E Easy access to parking, view of car
F Public green
G Private open space balcony
H Access to pedestrian greenway
I Access to dwelling, activates public green
J Parking lot serves both clusters

Fig. 10b   Fox Hill, Staten Island, N.Y. The Institute for Architecture and Urban Studies—Architect.

Fig. 10c

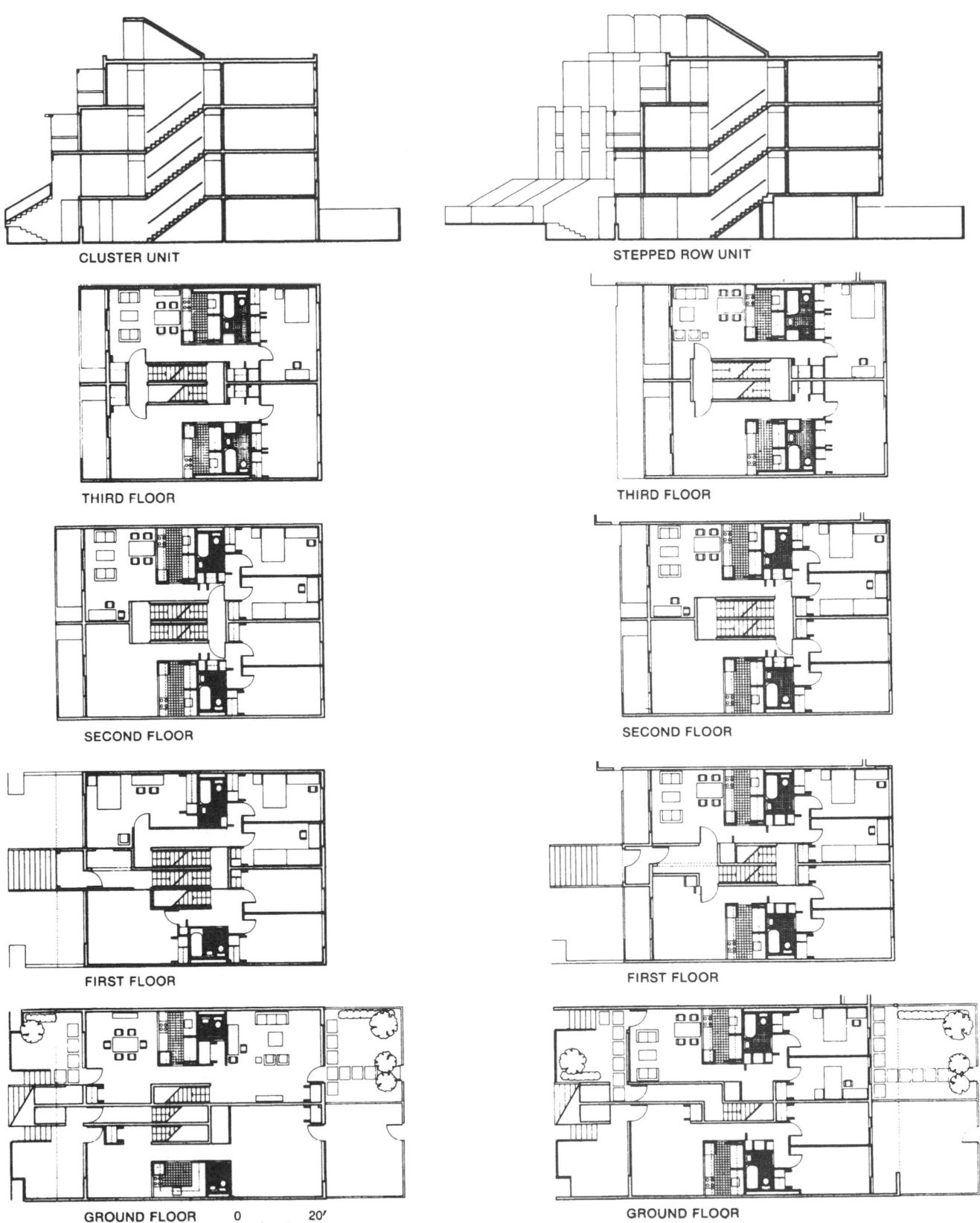

CLUSTER UNIT

STEPPED ROW UNIT

THIRD FLOOR

THIRD FLOOR

SECOND FLOOR

SECOND FLOOR

FIRST FLOOR

FIRST FLOOR

GROUND FLOOR   0   20'

GROUND FLOOR

Fig. 11   Fox Hill, Staten Island, N.Y. The Institute for Architecture and Urban Studies—Architect.

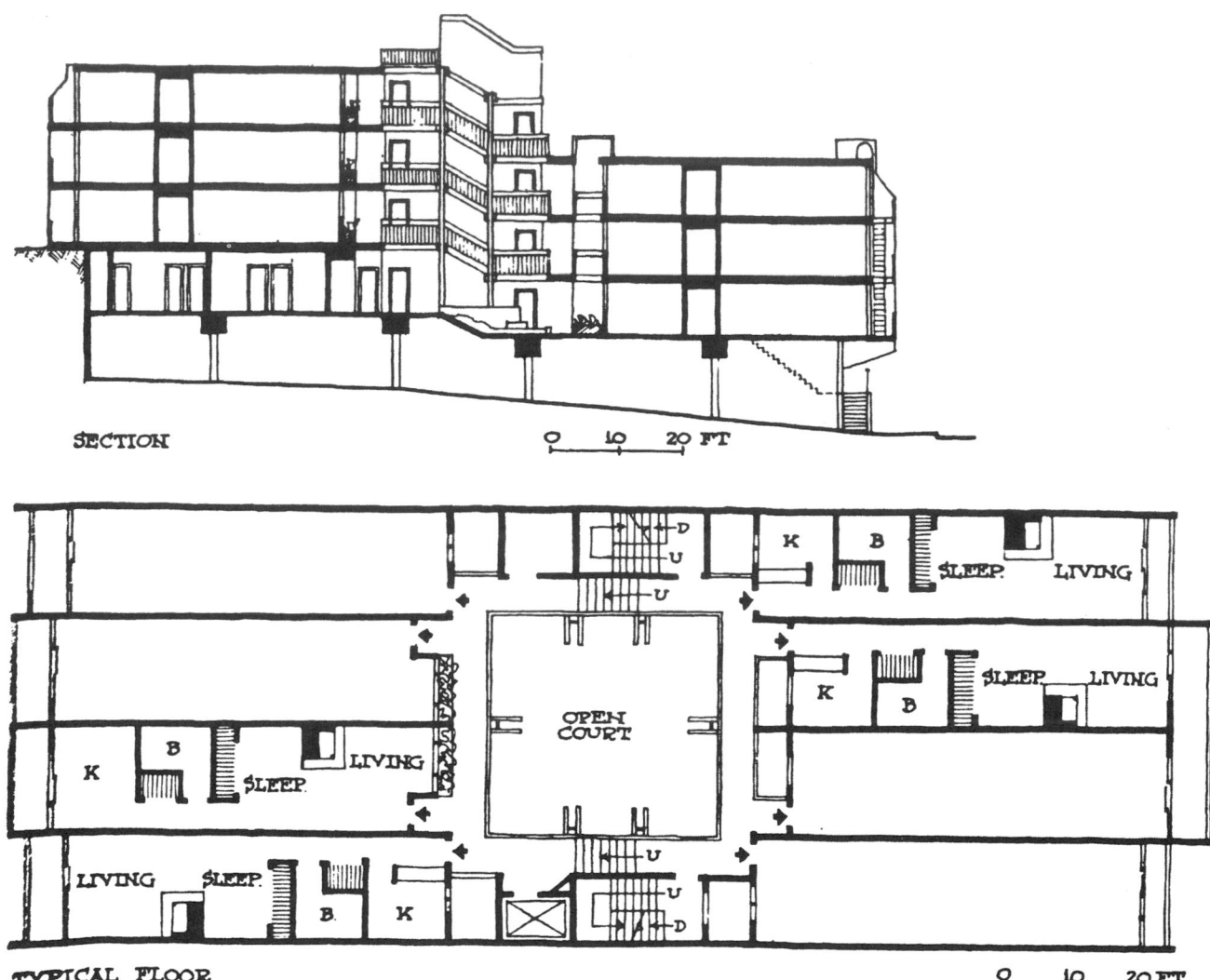

SECTION   0 10 20 FT

TYPICAL FLOOR   0 10 20 FT

Fig. 12

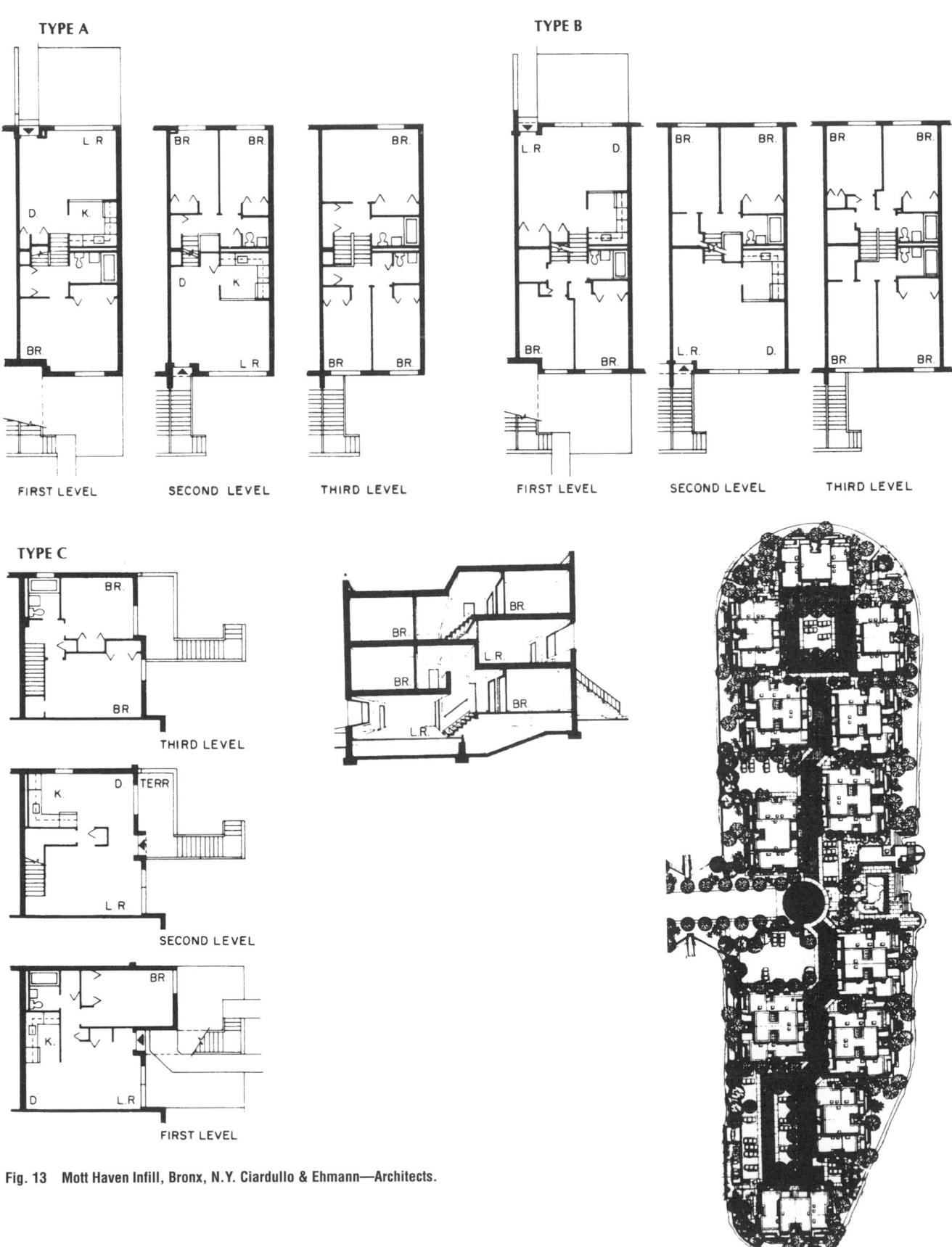

TYPE A

FIRST LEVEL   SECOND LEVEL   THIRD LEVEL

TYPE B

FIRST LEVEL   SECOND LEVEL   THIRD LEVEL

TYPE C

THIRD LEVEL

SECOND LEVEL

FIRST LEVEL

Fig. 13   Mott Haven Infill, Bronx, N.Y. Ciardullo & Ehmann—Architects.

Fig. 14   The Islands, San Francisco, Calif.
Fisher-Friedman Associates—Architects.

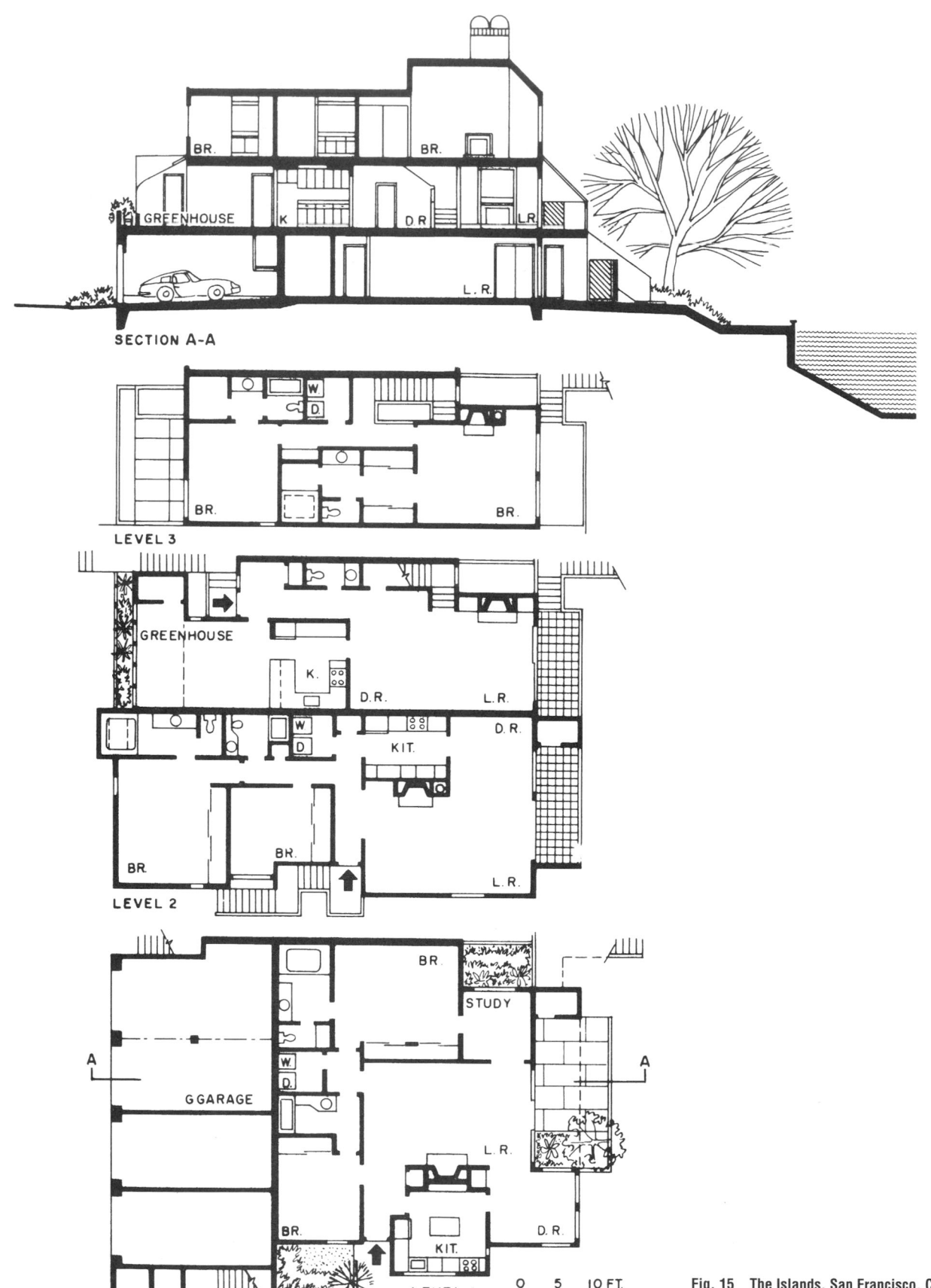

SECTION A-A

LEVEL 3

LEVEL 2

LEVEL 1

0 5 10 FT.

Fig. 15   The Islands, San Francisco, Calif. Fisher-Friedman Associates—Architects.

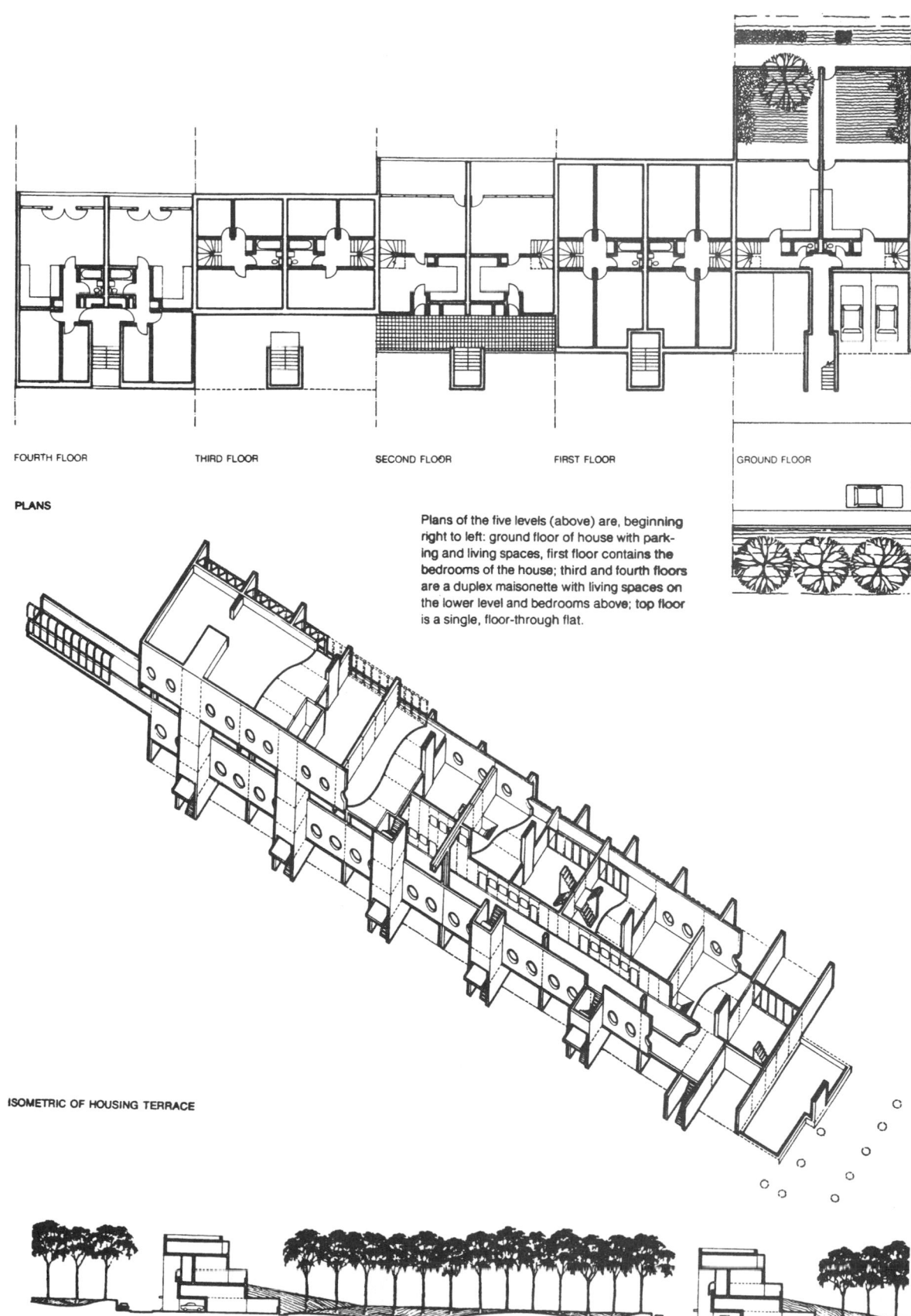

FOURTH FLOOR   THIRD FLOOR   SECOND FLOOR   FIRST FLOOR   GROUND FLOOR

PLANS

Plans of the five levels (above) are, beginning right to left: ground floor of house with parking and living spaces, first floor contains the bedrooms of the house; third and fourth floors are a duplex maisonette with living spaces on the lower level and bedrooms above; top floor is a single, floor-through flat.

ISOMETRIC OF HOUSING TERRACE

Fig. 16   Town Center Housing, Runcorn, England.

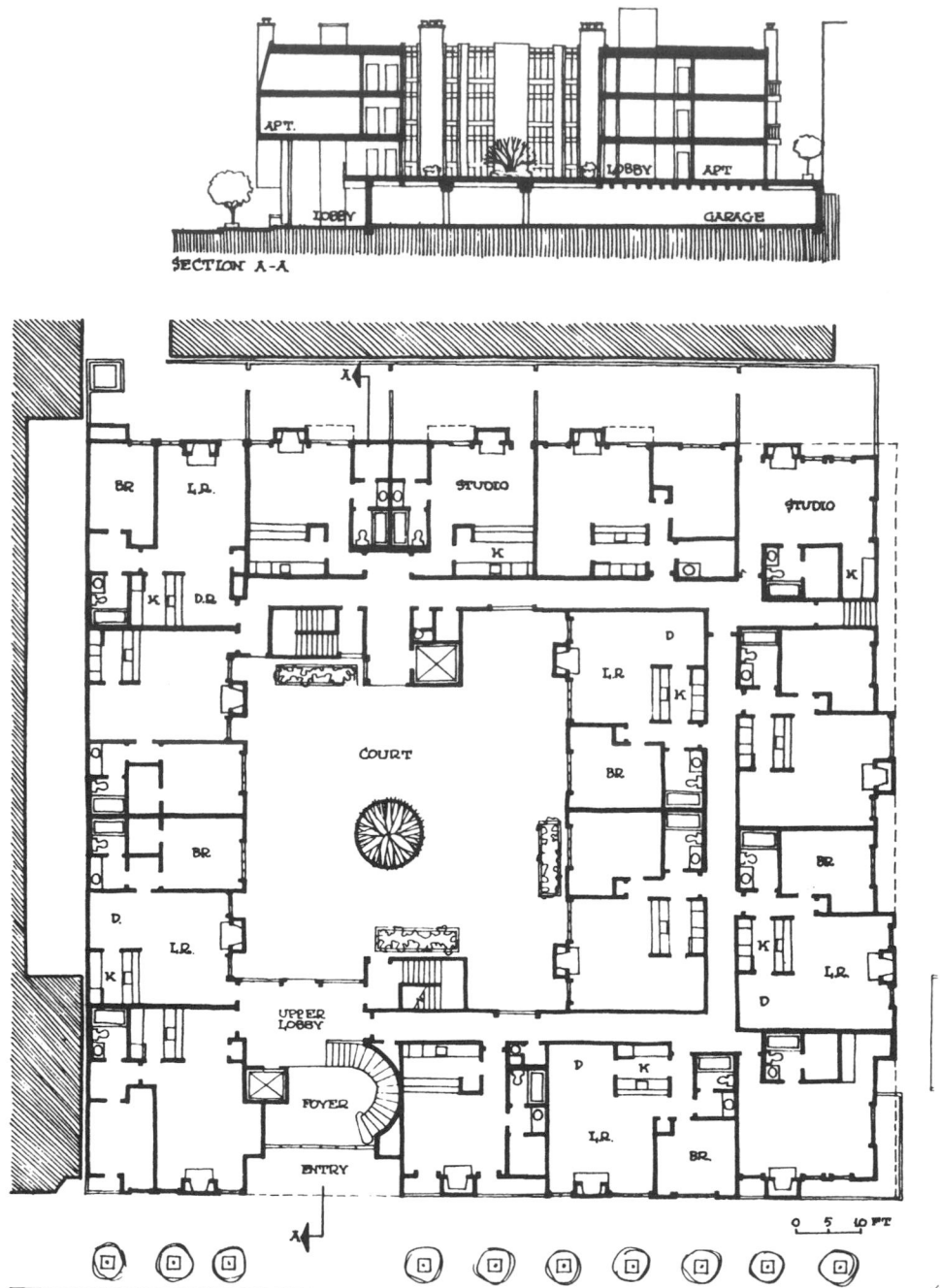

Fig. 17   This building contains nine efficiency units, 34 one-bedroom apartments, and 2 two-bedroom units. Willis and Associates—Architects.

MULTISTORY APARTMENT BUILDINGS—ELEVATOR

Of all the housing types this is the easiest to identify and to define. Often known as "high-rise," to distinguish it from lower walk-up build-ings, a multistory apartment building is over three floors in height with one or more apart-ments per floor. In the United States such resi-dential structures are almost always equipped with at least one elevator, either by regulation or by common practice. Beyond the three-story level, the number of floors is determined by construction techniques, elevator manufacture, and local zoning ordinances. Traditionally, tall buildings were found only in the central por-tions of large cities. However, land shortages, high housing costs, and a widening acceptance of apartment living have changed this picture. Such buildings are frequently situated along major waterways and other locations offering outstanding features or views (no doubt reflect-ing high land value as well) or adjoining major regional shopping facilities and in the centers of new, planned communities. Many buildings accommodate nonresidential as well as resi-dential uses in the same structure or on the same parcel of land. Parking facilities are planned to serve both.

Multistory apartments vary in shape as well as in height. The most common shapes are: the slab (whose form derives from the internal, central, double-loaded corridor or the single-loaded corridor), the tower or "point block" (characterized by a central circulation core from which a limited number of apartments open at each floor), and the multiwing building (with a combination slab and point internal circulation arrangement).

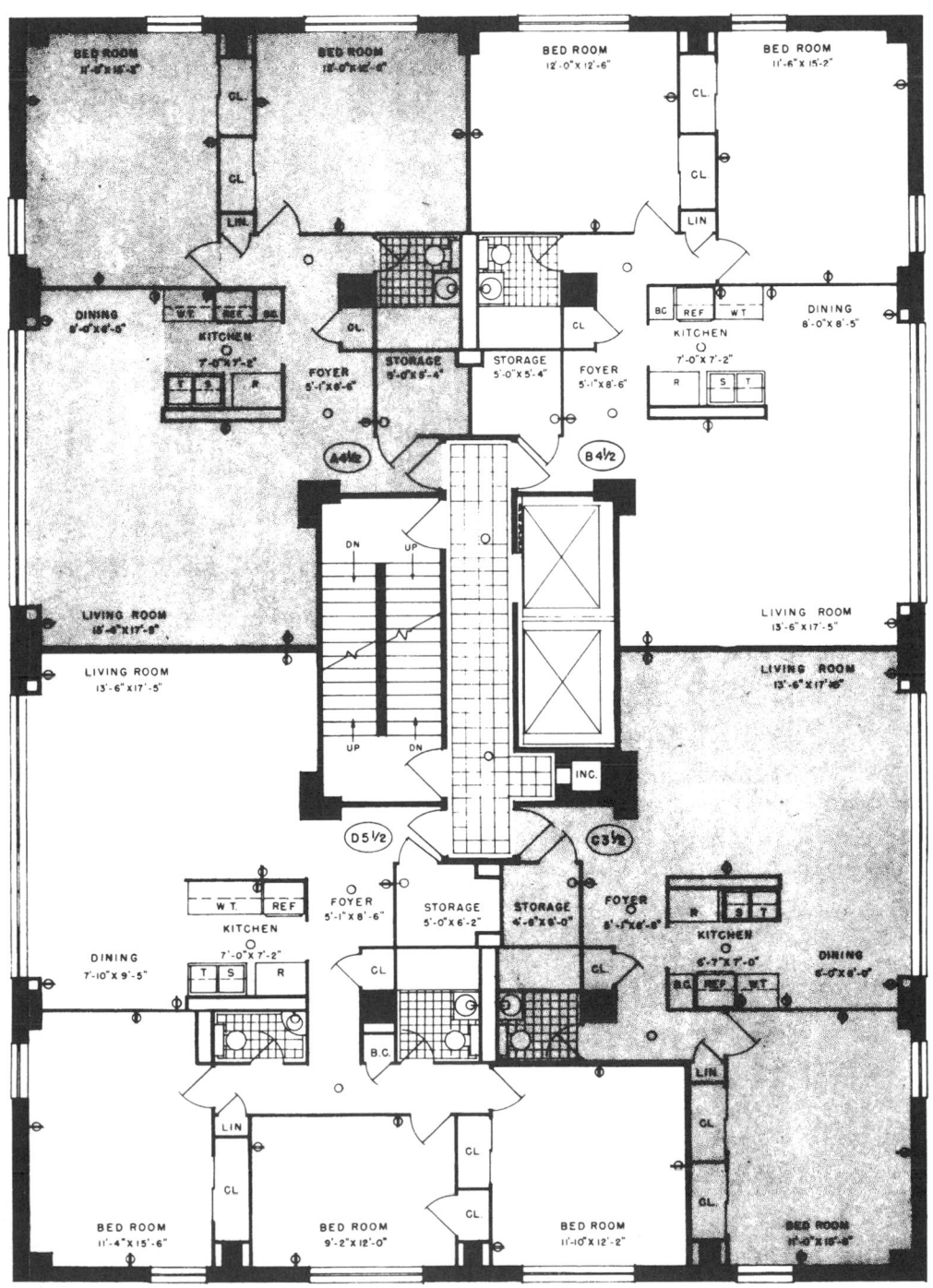

Fig. 1   Typical floor plan—14-story. Kingsview Apartments, Brooklyn, N.Y.

CENTER-CORRIDOR PLAN

The center-corridor scheme, or interior-corridor scheme or double-loaded corridor, is the most common type of floor plan. It is characterized by having apartments on both sides of the corridor, thereby making it economical. The corridor can easily be extended, allowing a maximum number of apartments per floor and per elevator/stair core. In fact, the building length can be extended indefinitely, provided adequate exit stairs are located in proper intervals to meet local code requirements. The total width of the building is limited to the depth of two apartments plus the corridor, or approximately 50 to 80 ft. The result of this is a narrow width and long length, making a building configuration called a "slab" as opposed to a "tower."

One drawback is that it does not provide for through or cross ventilation except for the four corner apartments. Another limitation is orientation of the building. If the building is facing south, half of the apartments will have good orientation while the other half will be facing north, which is not desirable. The compromise orientation is east-west, which is acceptable but not considered the best. Also, all apartments, except the corners, have only one exposure. One planning consideration for this type of building is the placement of these slablike buildings on the site. If they are too long or are placed parallel to each other, such buildings tend to visually create a "Chinese wall" effect.

With respect to orientation, in high-rise housing developments, the largest apartments should ideally go in the corners for two-way orientation.

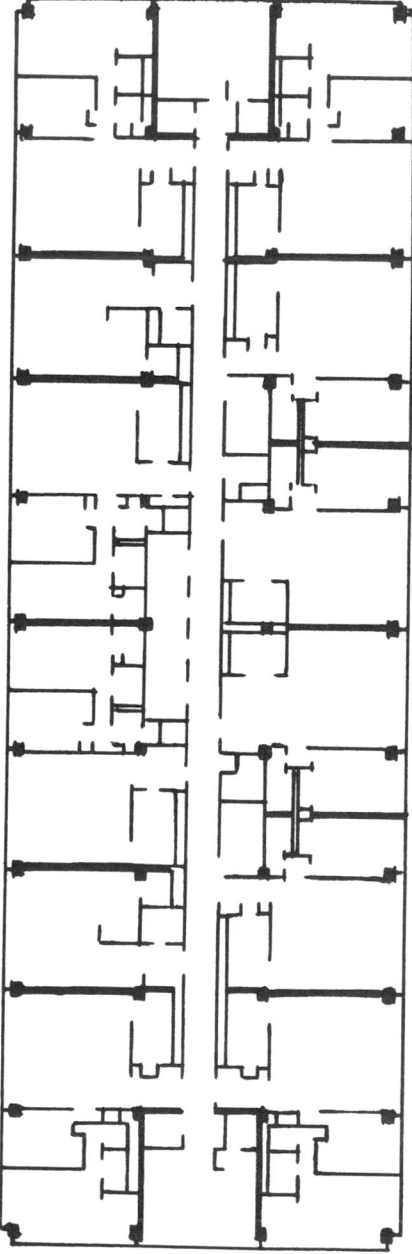

Fig. 1

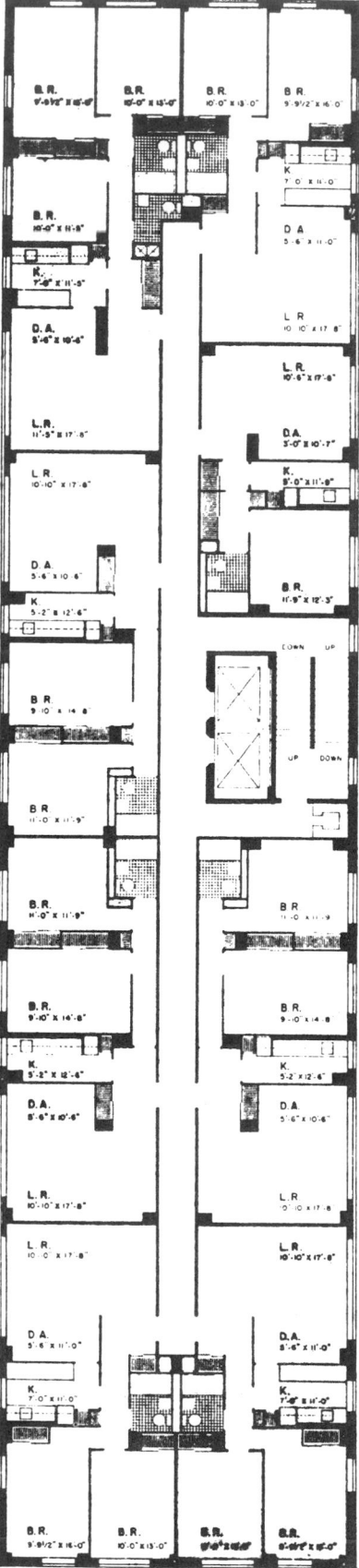

Fig. 2 Typical floor plan.

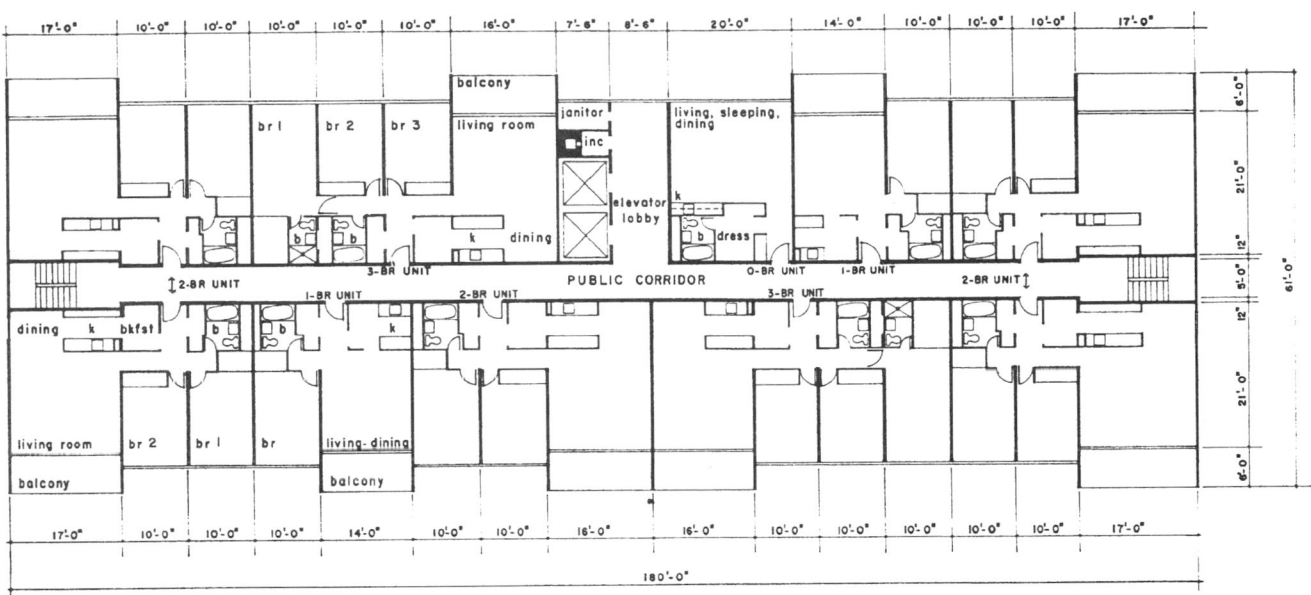

Fig. 3   Double-loaded corridor scheme.

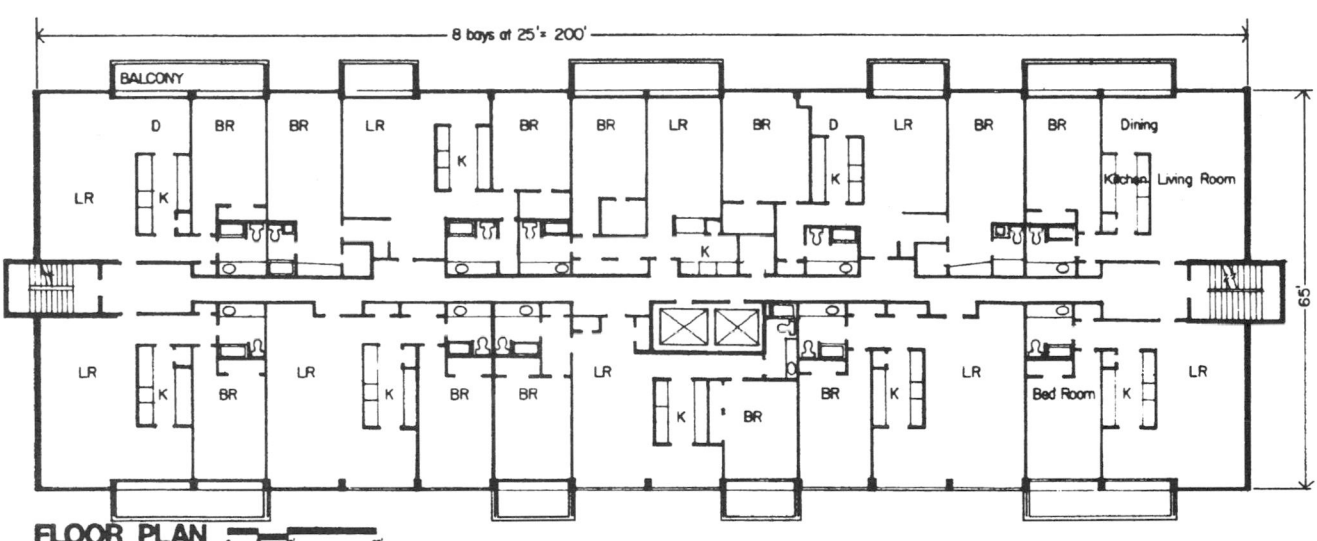

Fig. 4

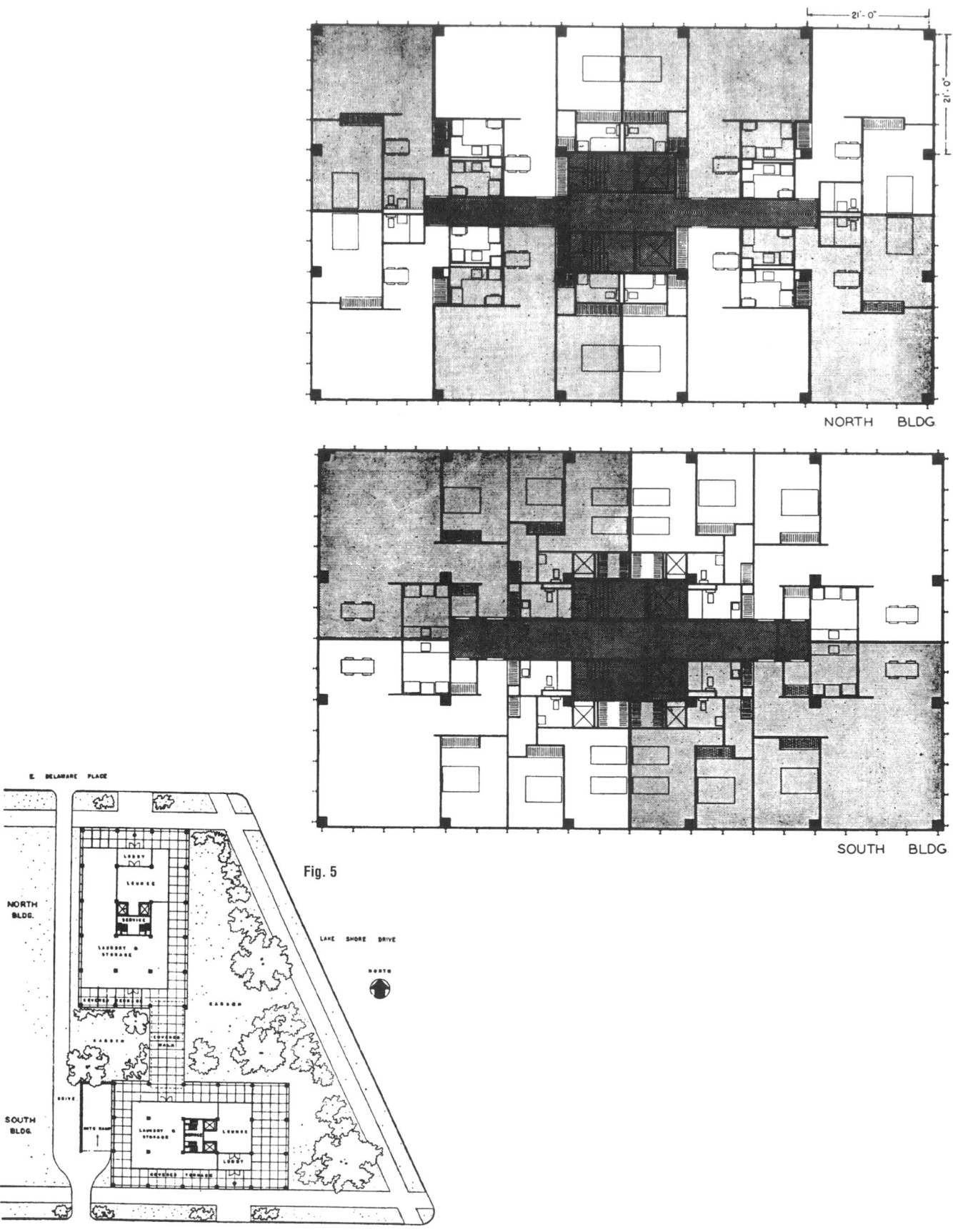

NORTH   BLDG.

SOUTH   BLDG.

Fig. 5

Fig. 6

Fig. 7a Two studio apartments. Lakeshore Drive Apartments, Chicago, Ill. Mies van der Rohe—Architect.

Fig. 7b Three-bedroom apartment. Lake Shore Drive Apartments. Chicago, Ill. Mies van der Rohe—Architect.

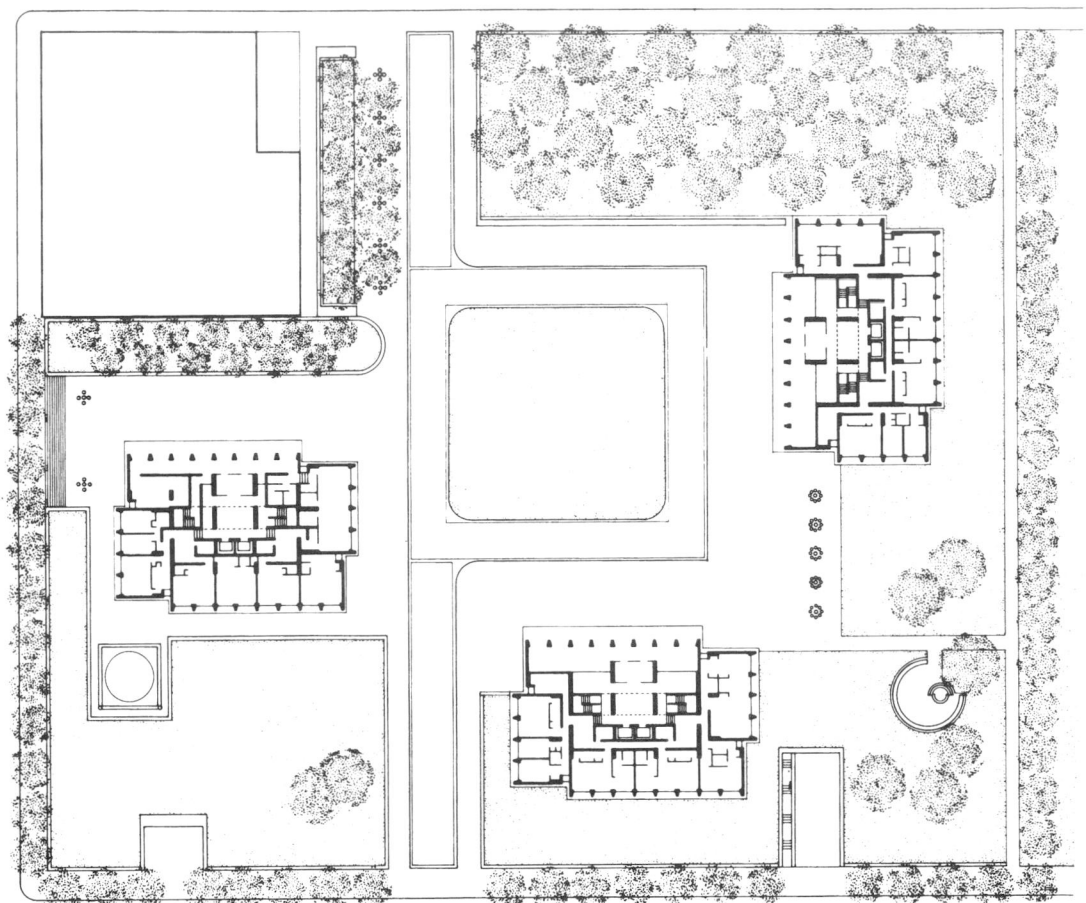

Fig. 8*a* Site plan.

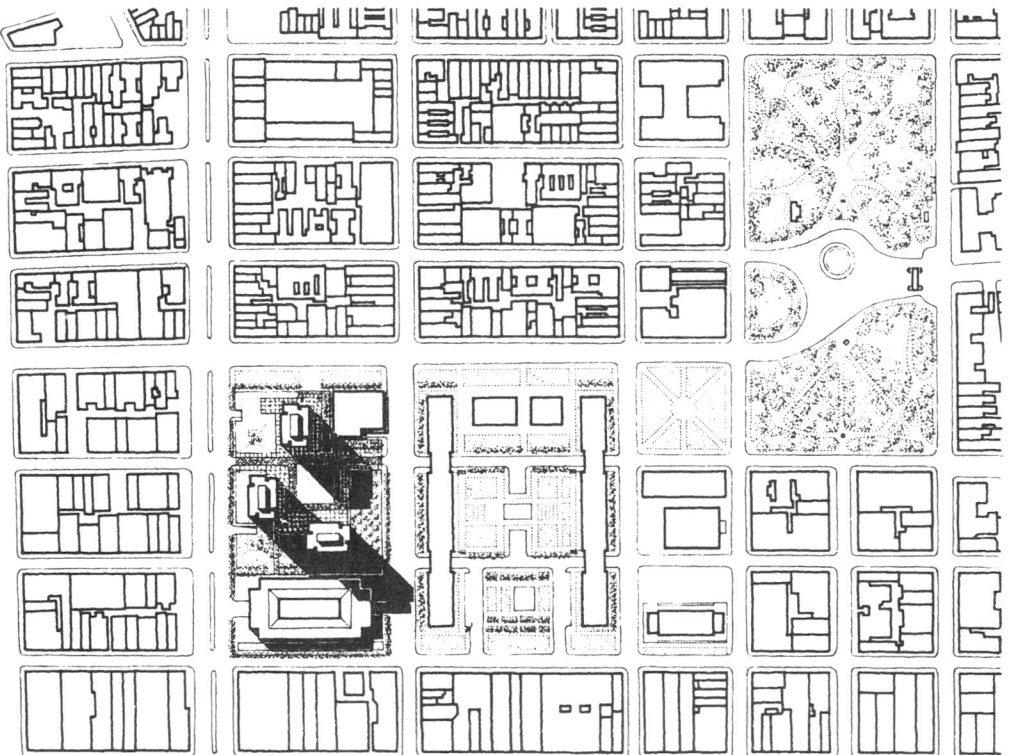

Fig. 8*b* Location plan. University Plaza, N.Y. University. I. M. Pei & Partners—Architects.

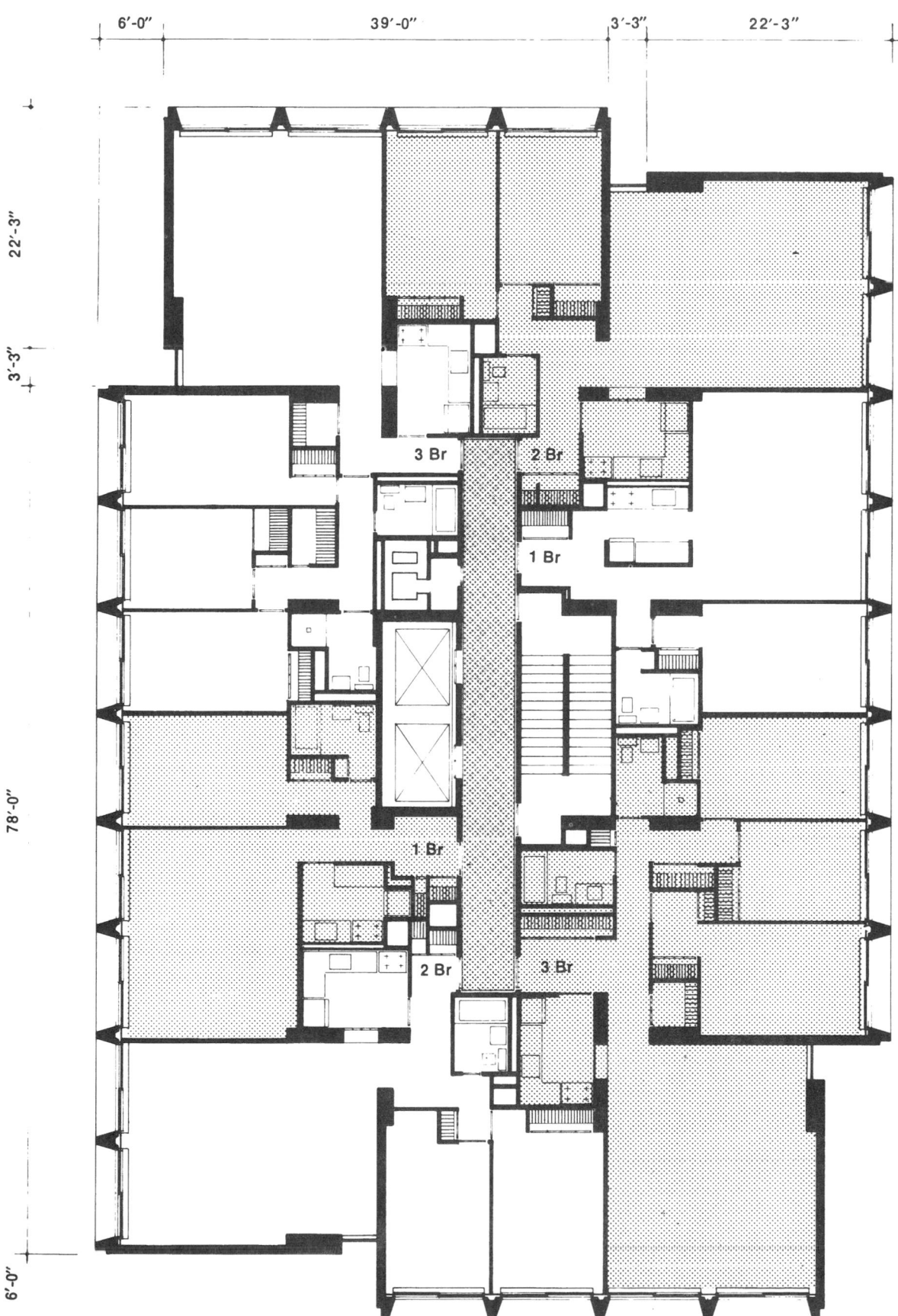

6'-0"   39'-0"   3'-3"   22'-3"

22'-3"

3'-3"

78'-0"

6'-0"

3 Br

2 Br

1 Br

1 Br

2 Br

3 Br

Fig. 9   Typical floor plan. University Plaza, N.Y. University. I. M. Pei & Partners—Architects.

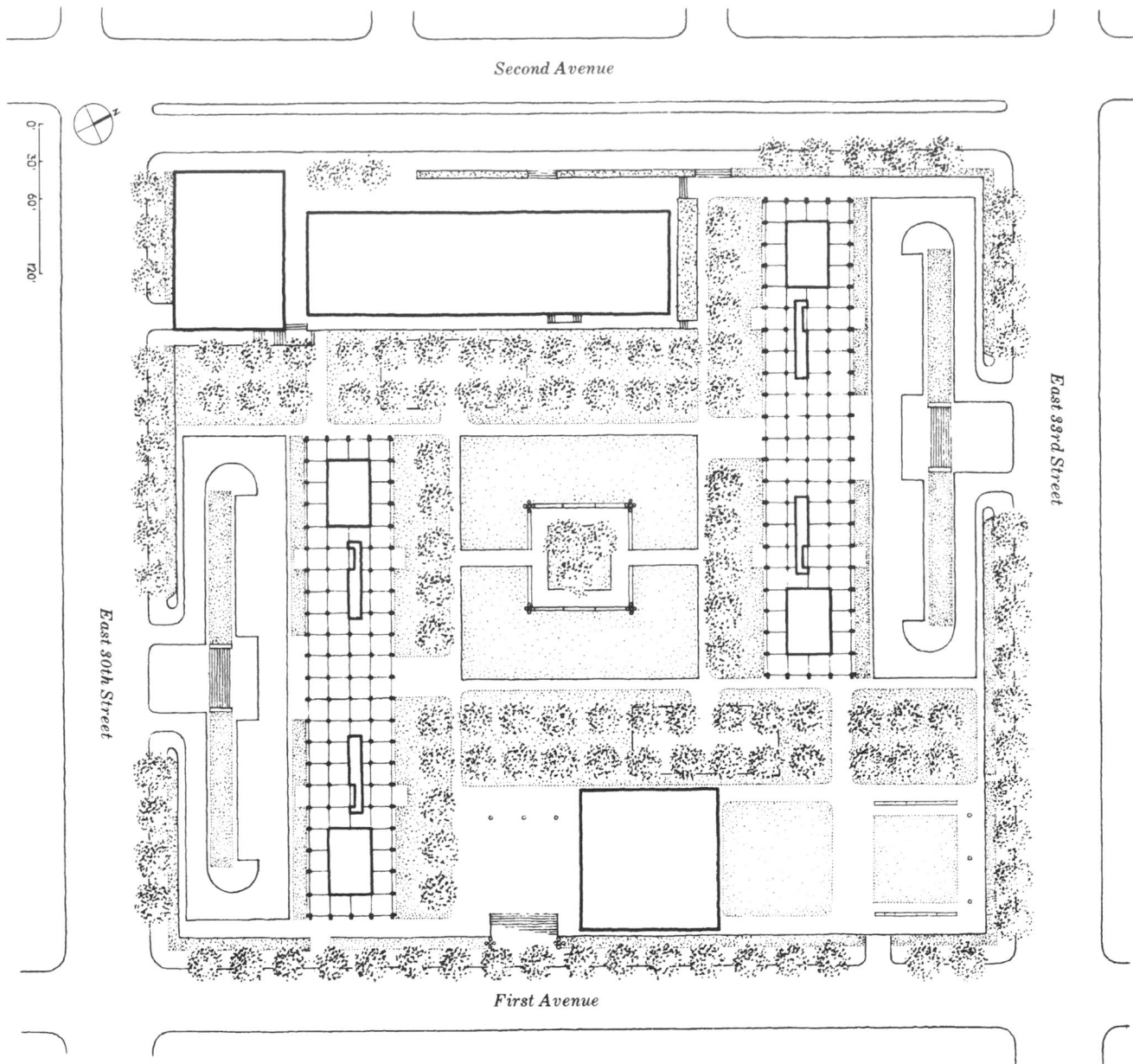

Fig. 10   Site plan. Kips Bay Plaza, New York City. I. M. Pei & Partners—Architects.

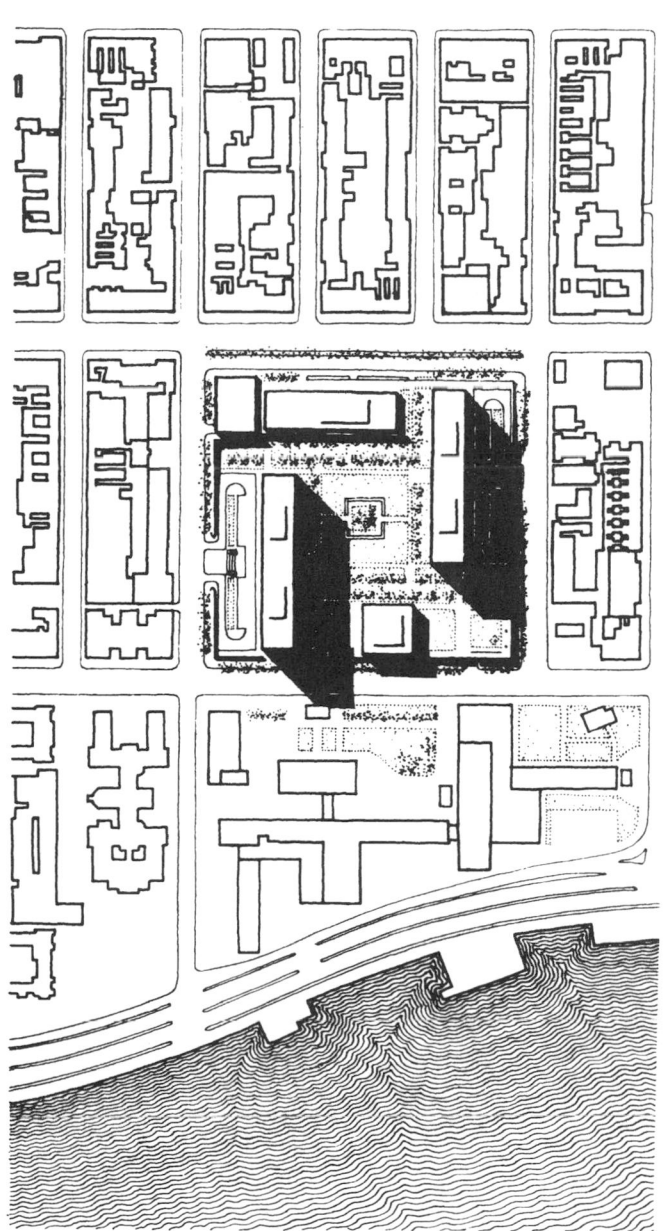

Fig. 11a Location plan. Kips Bay Plaza, New York City. I. M. Pei & Partners—Architects.

Fig. 11b Typical floor plan. Kips Bay Plaza, New York City. I. M. Pei & Partners—Architects.

CENTER-CORRIDOR PLAN

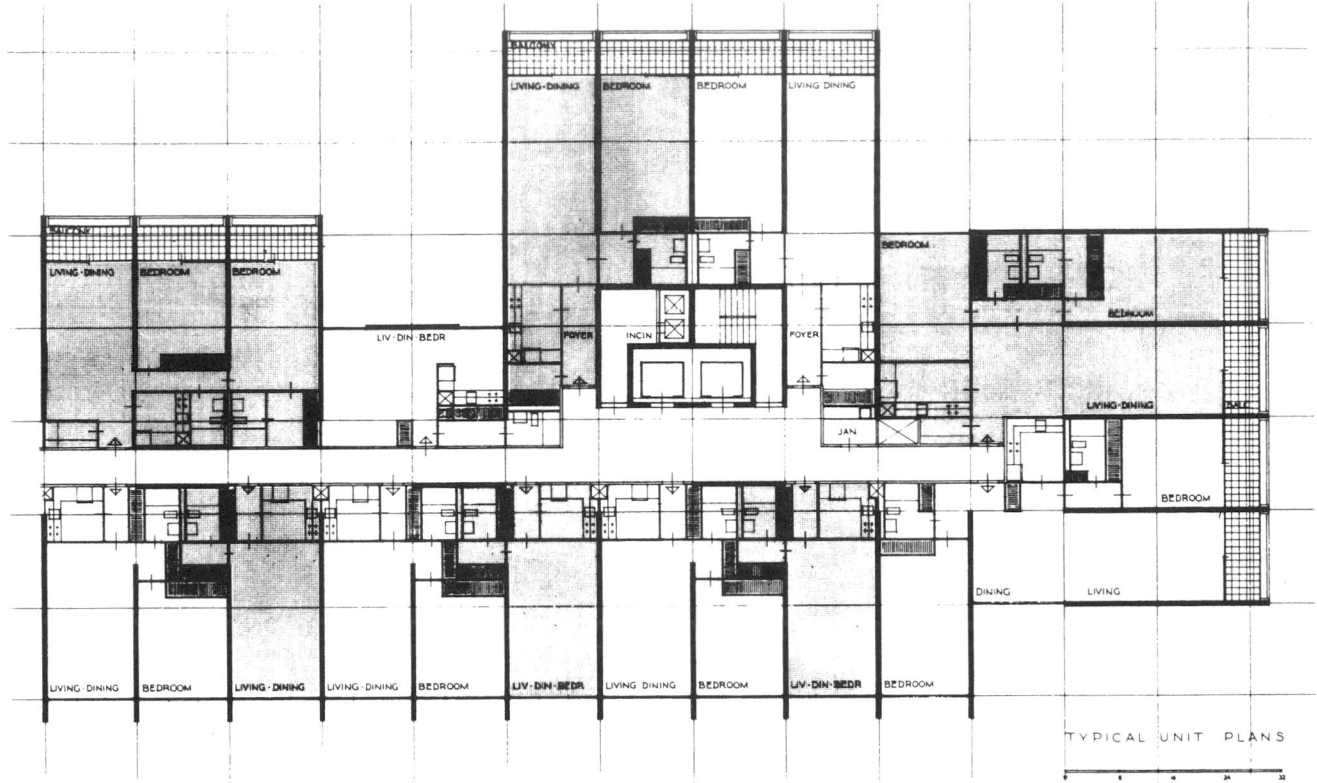

TYPICAL UNIT PLANS

Fig. 12   Lower Hill, Pittsburgh, Pa. I. M. Pei & Partners—Architects.

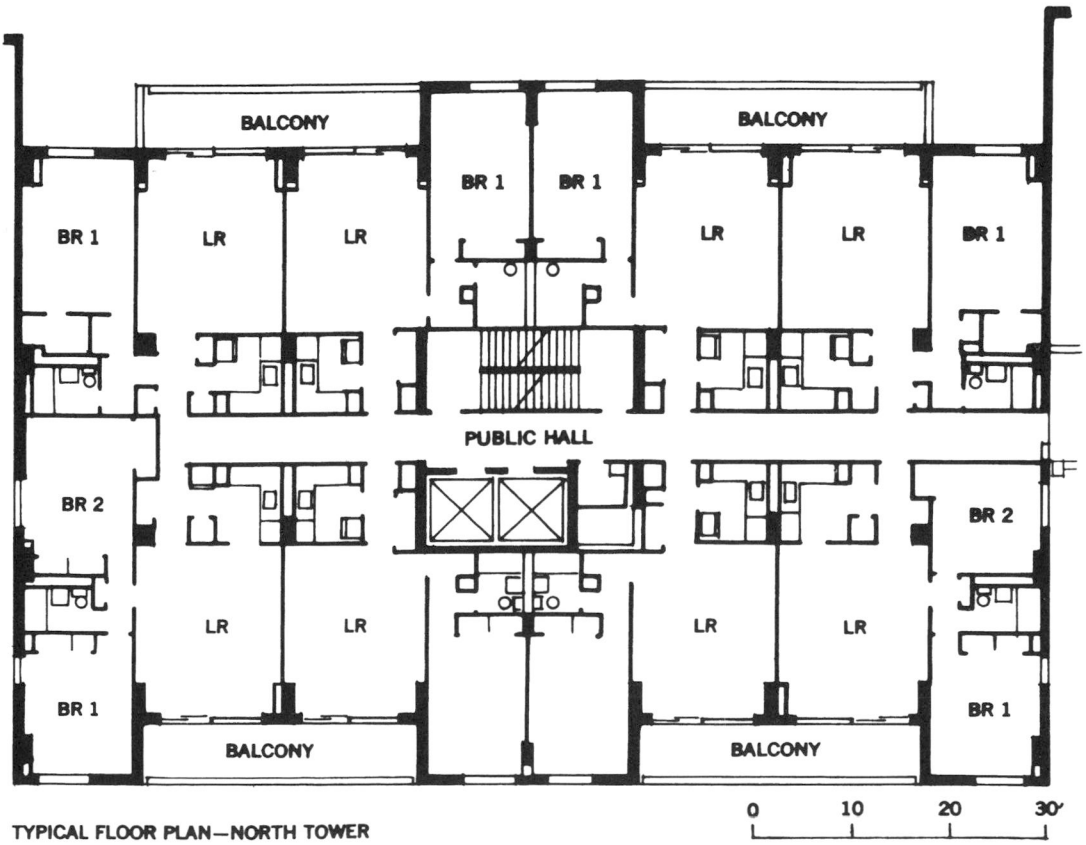

TYPICAL FLOOR PLAN—NORTH TOWER

Fig. 13   Riverbend, New York City. Davis, Brody & Associates—Architects.

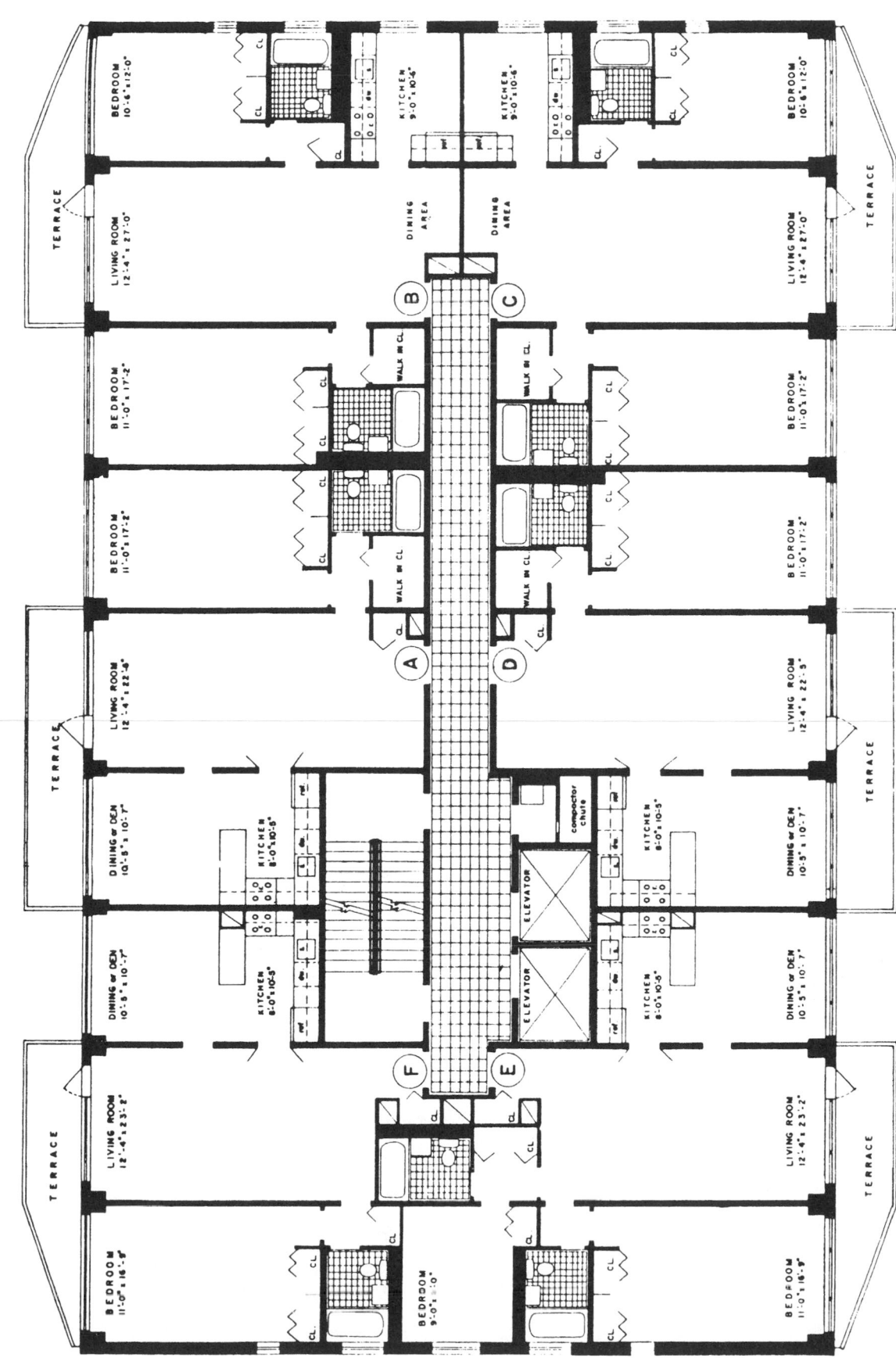

Fig. 14

679

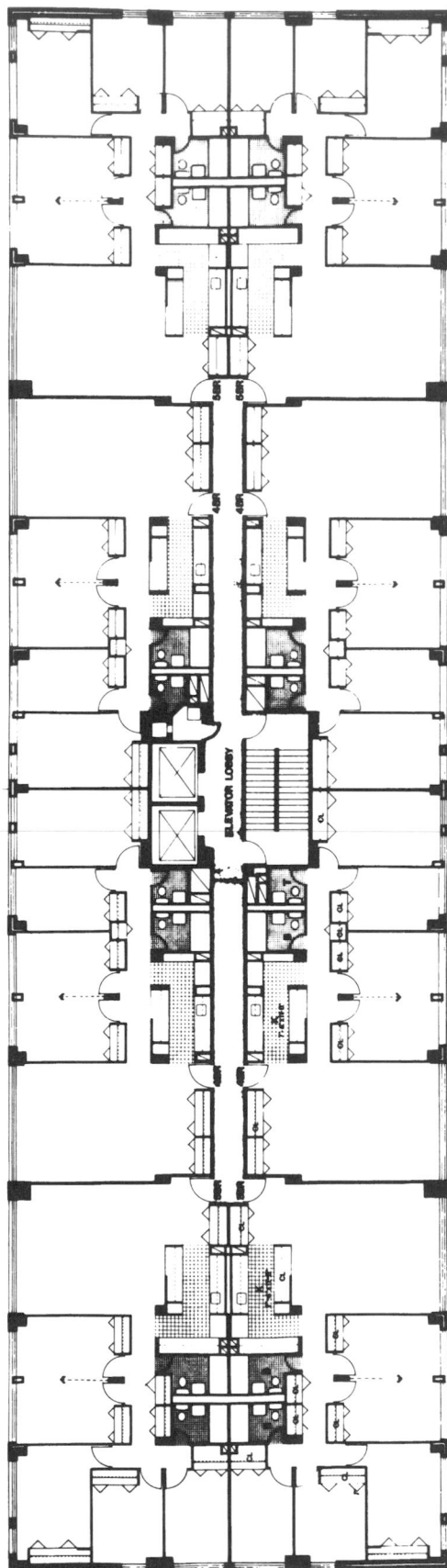

Fig. 15a   A. Schomburg Plaza, New York City. Gruzen & Partners—Architects.

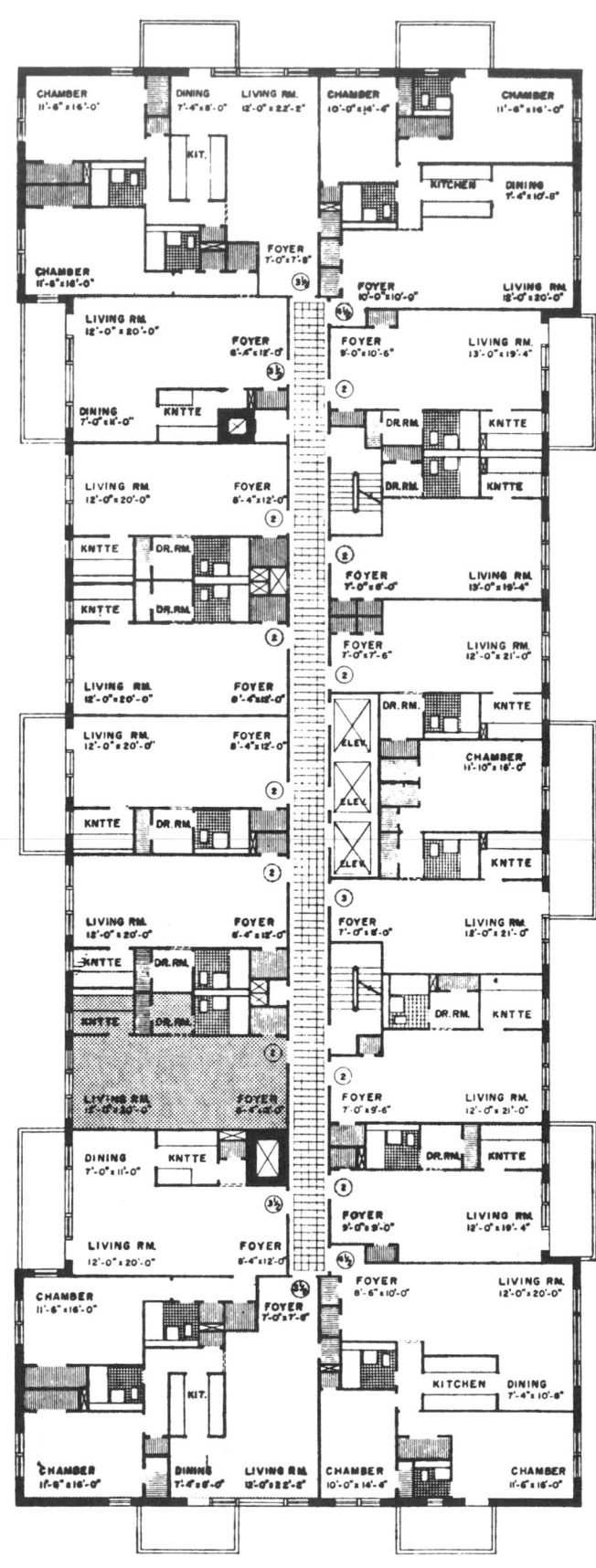

TYPICAL FLOOR PLAN

Fig. 15b

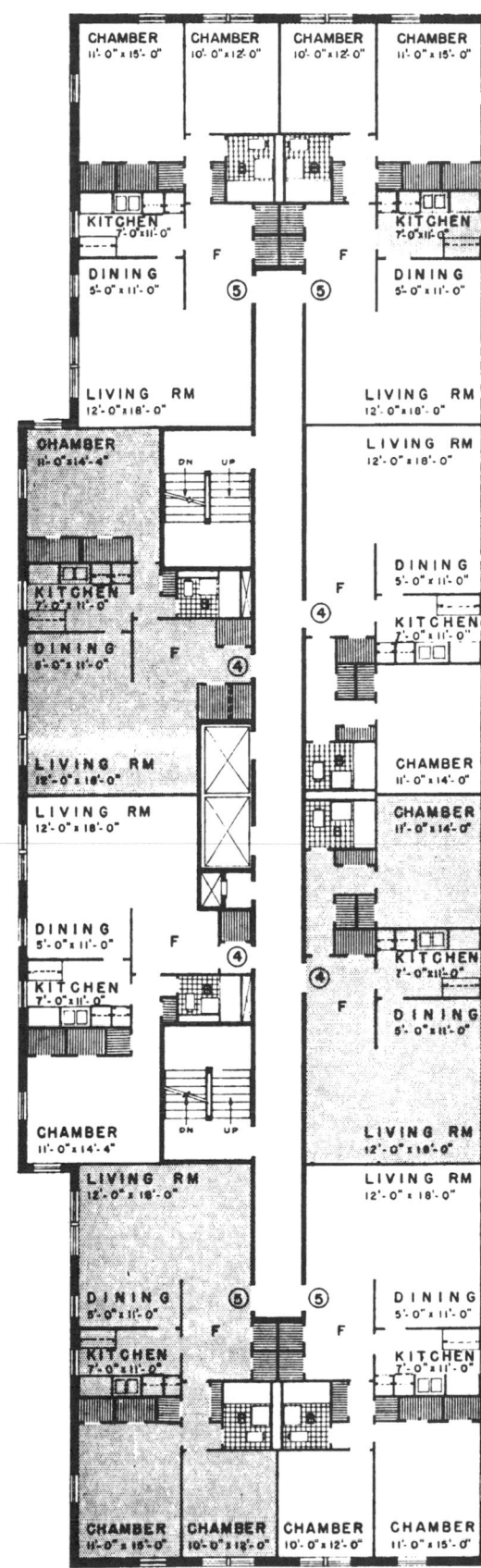

TYPICAL FLOOR PLAN

Fig. 15c

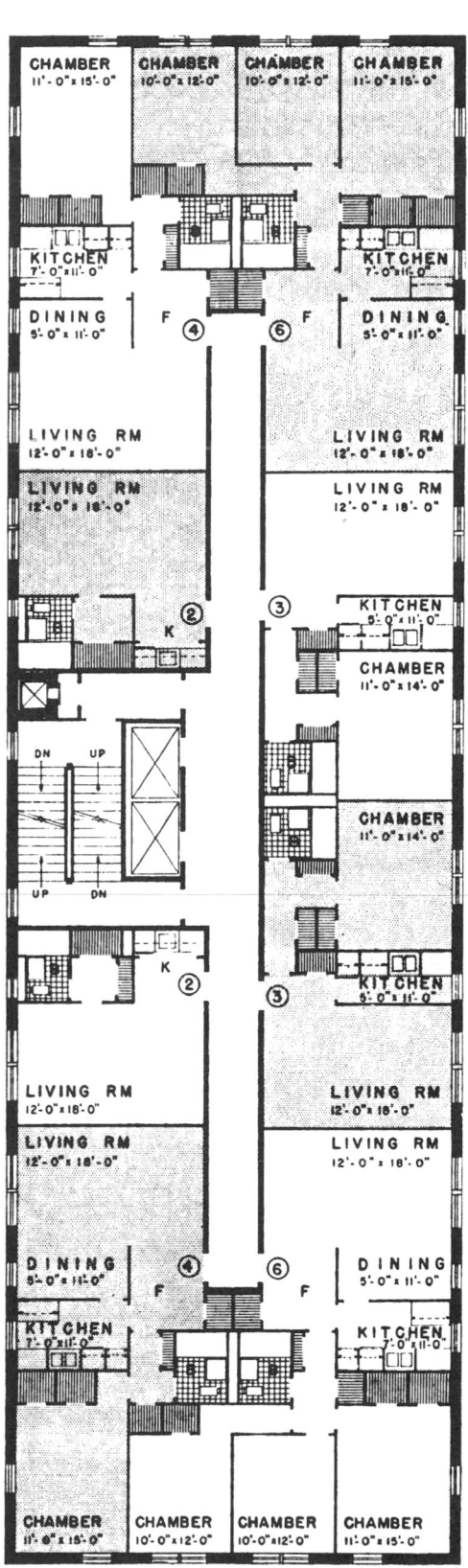

TYPICAL FLOOR PLAN

Fig. 15d

CENTER-CORRIDOR PLAN

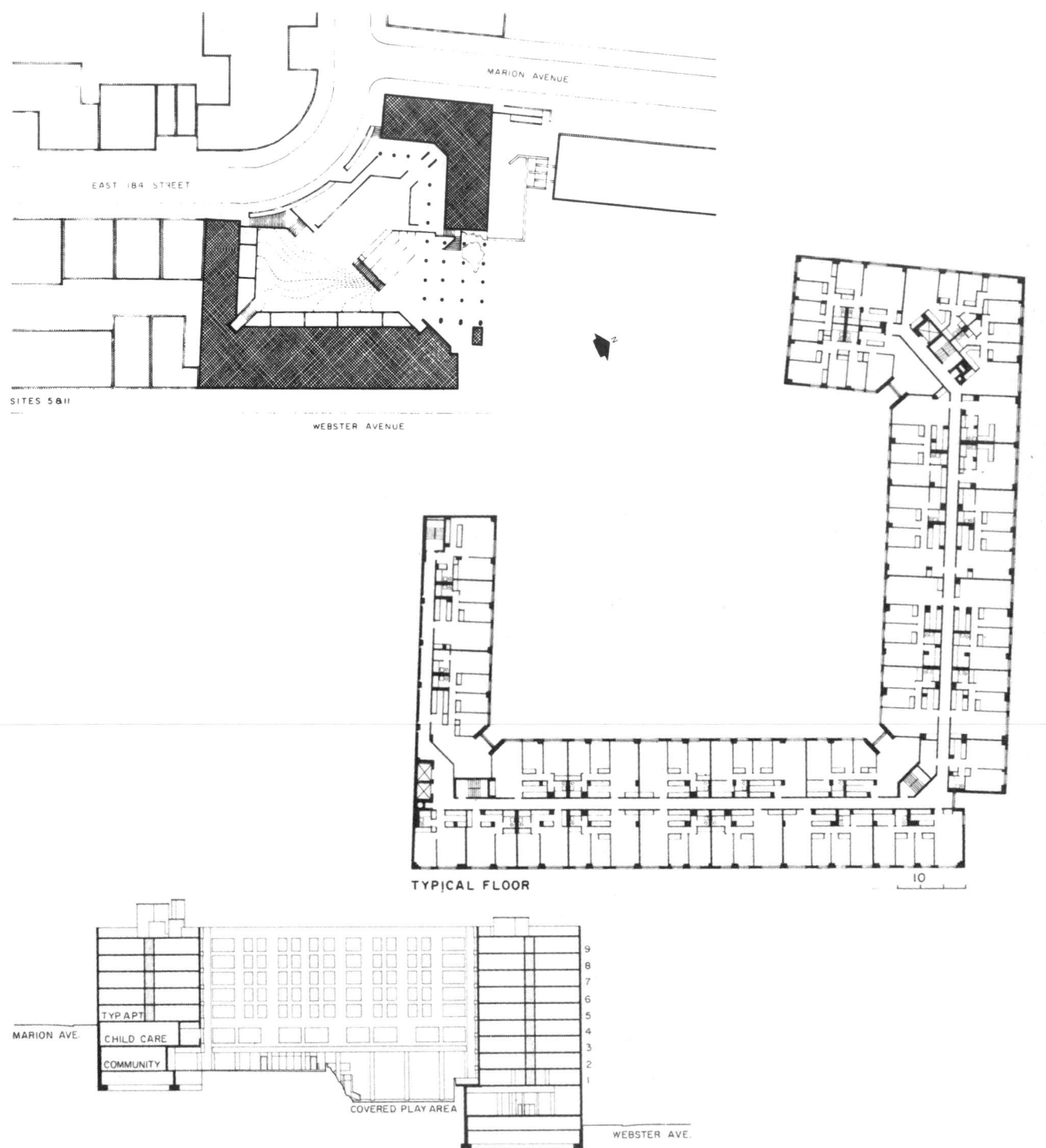

Fig. 16 Twin Parks Northwest, Bronx, N.Y. Prentice, Chan, Ohihausen—Architects.

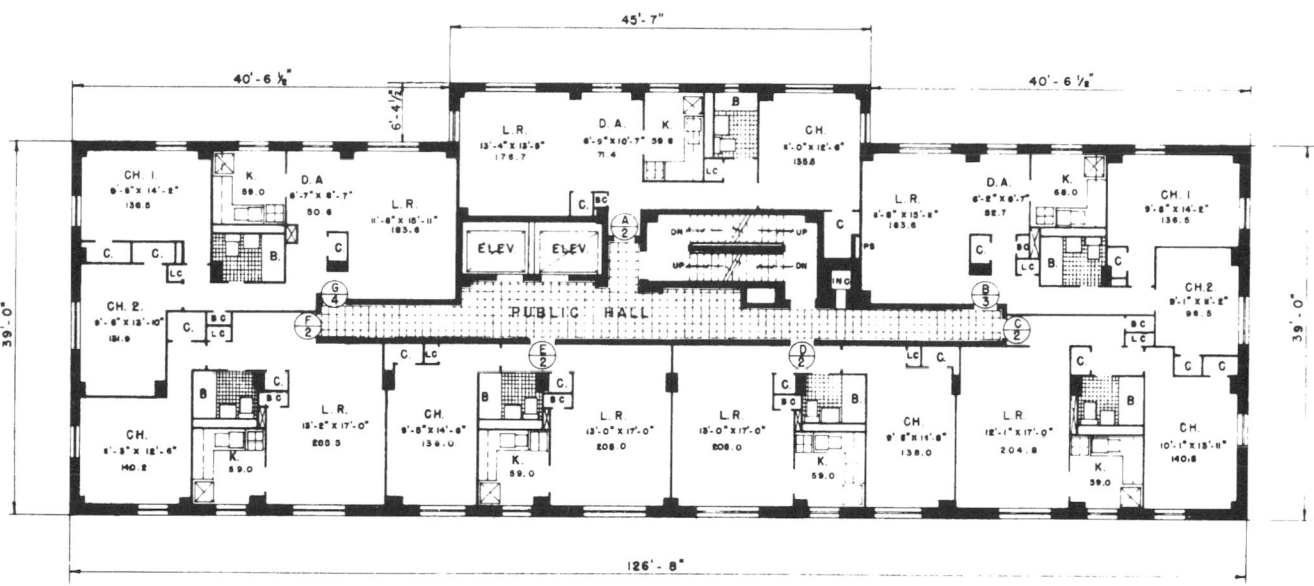

Fig. 17   Public Housing Authority, New York City.

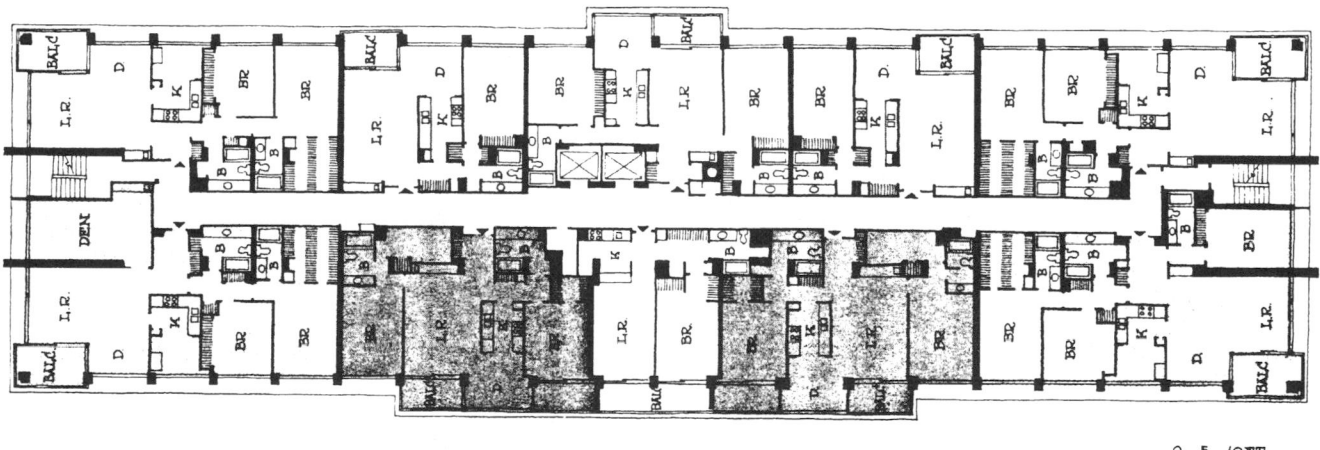

Fig. 18   Coronado Shores, Coronado, Calif. Krisel, Shapiro & Associates—Architects.

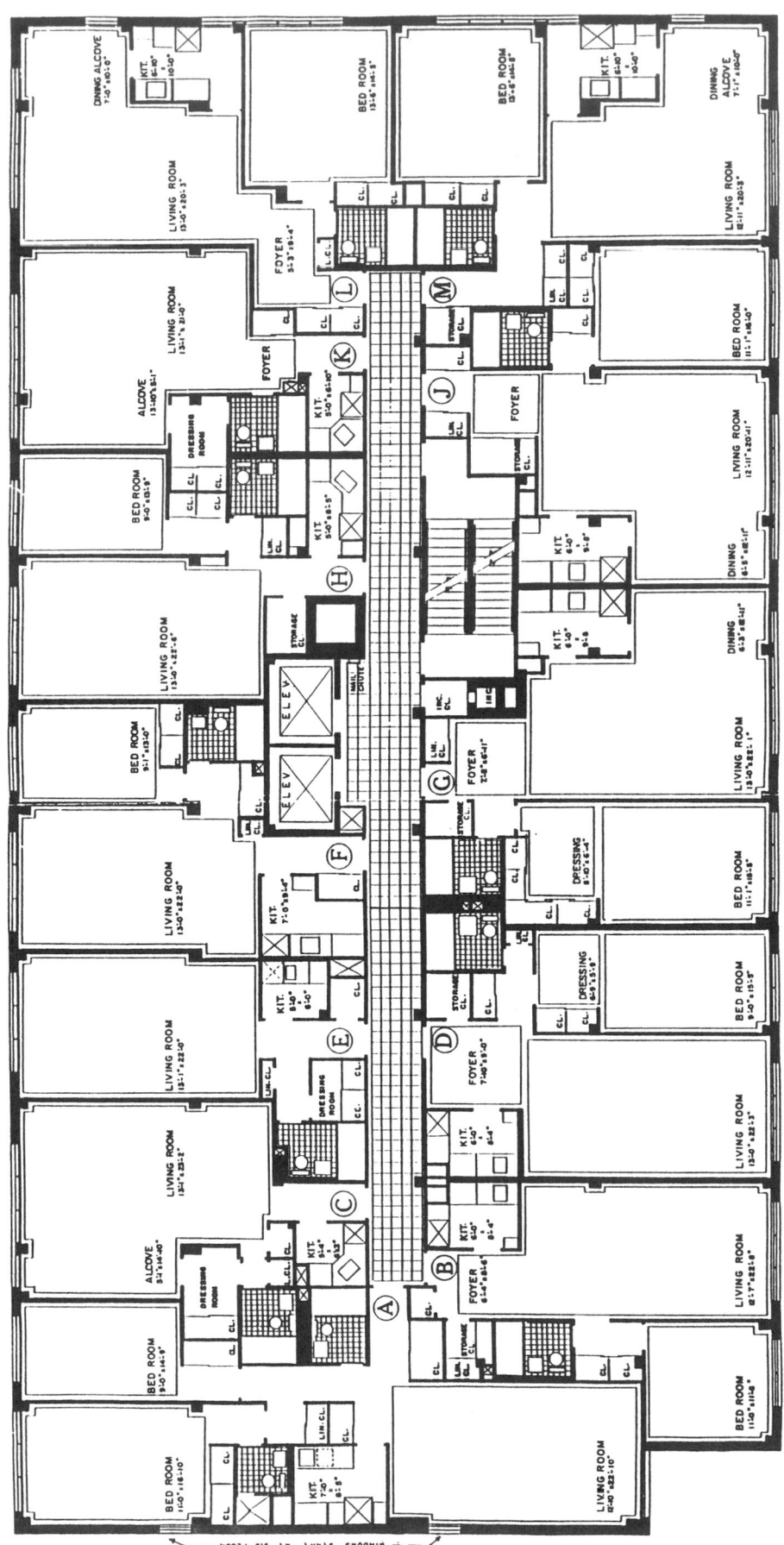

Fig. 19 Typical floor plan, 32 Gramercy Park South, New York, N.Y.

CENTER-CORRIDOR PLAN, OFFSET

Floor-through split-level apartments, in which living and sleeping areas are separated by a half-level change in elevation, are a radical departure from conventional high-rise housing design. For what is believed to be the first time in any New York City high-rise structure, public corridors and elevator stops do not serve every building level. Rather, one corridor—and elevator stop—serves two and one-half floors, saving 60 percent of the public corridor space for redistribution into the apartments.

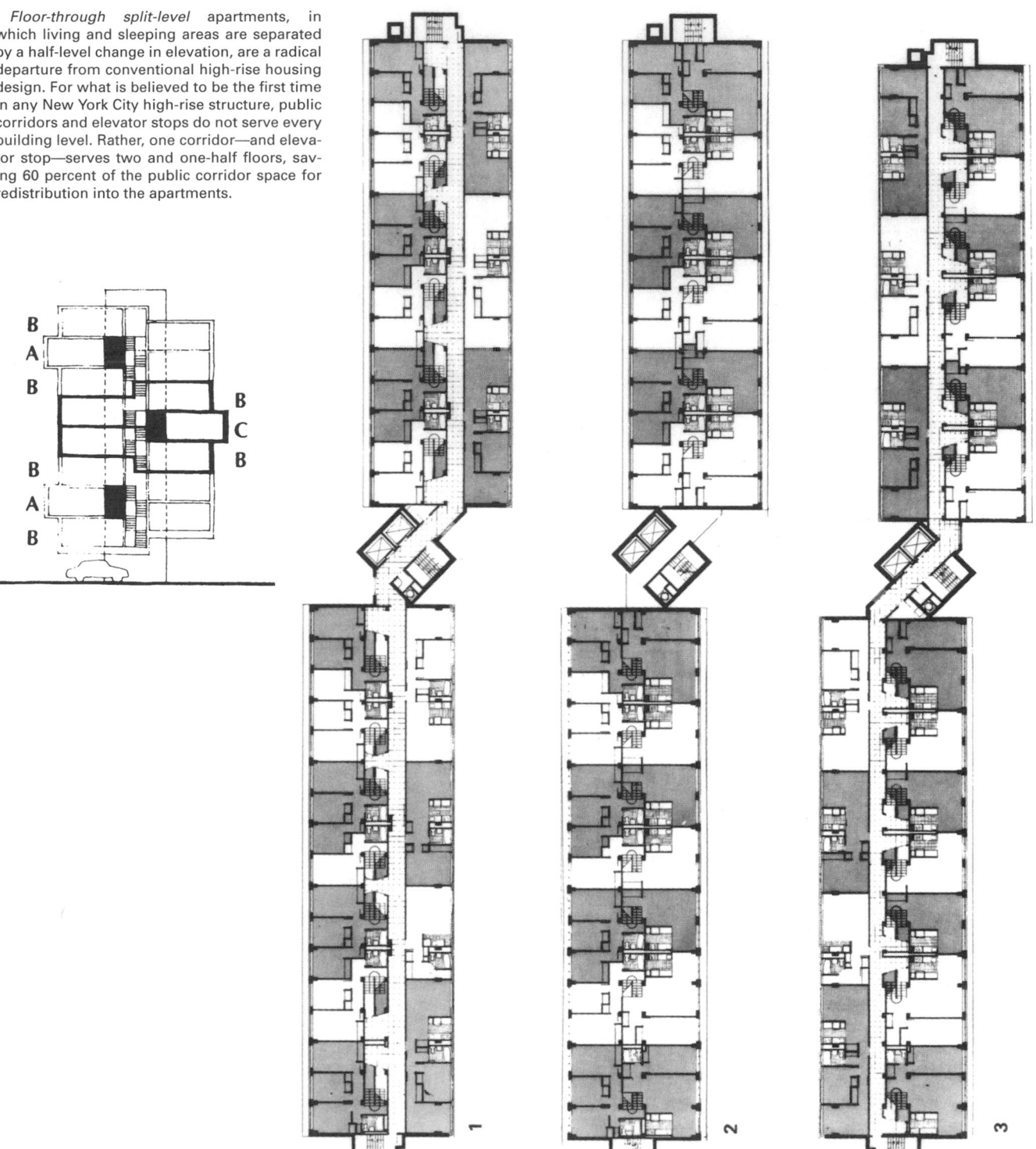

Fig. 1   Twin Parks Northwest, Bronx, N.Y. Giovanni Pasanella—Architect.

CENTER-CORRIDOR PLAN, OFFSET

TYPICAL UPPER FLOOR

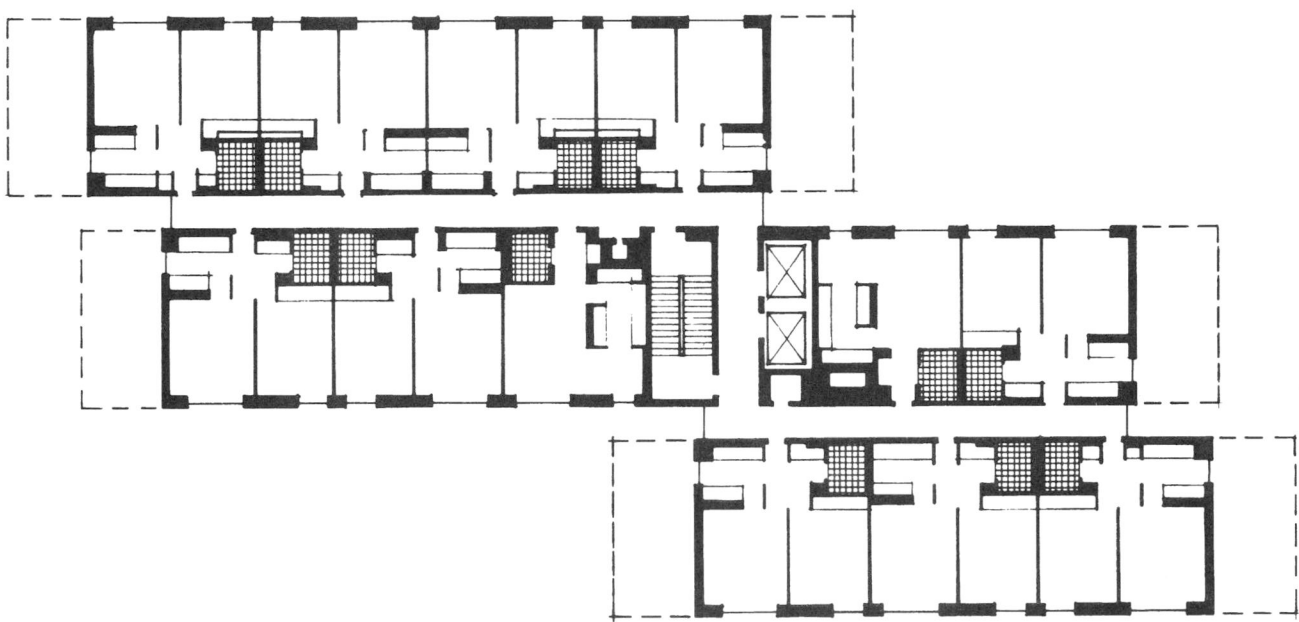

Fig. 2   Davis, Brody & Associates—Architects.

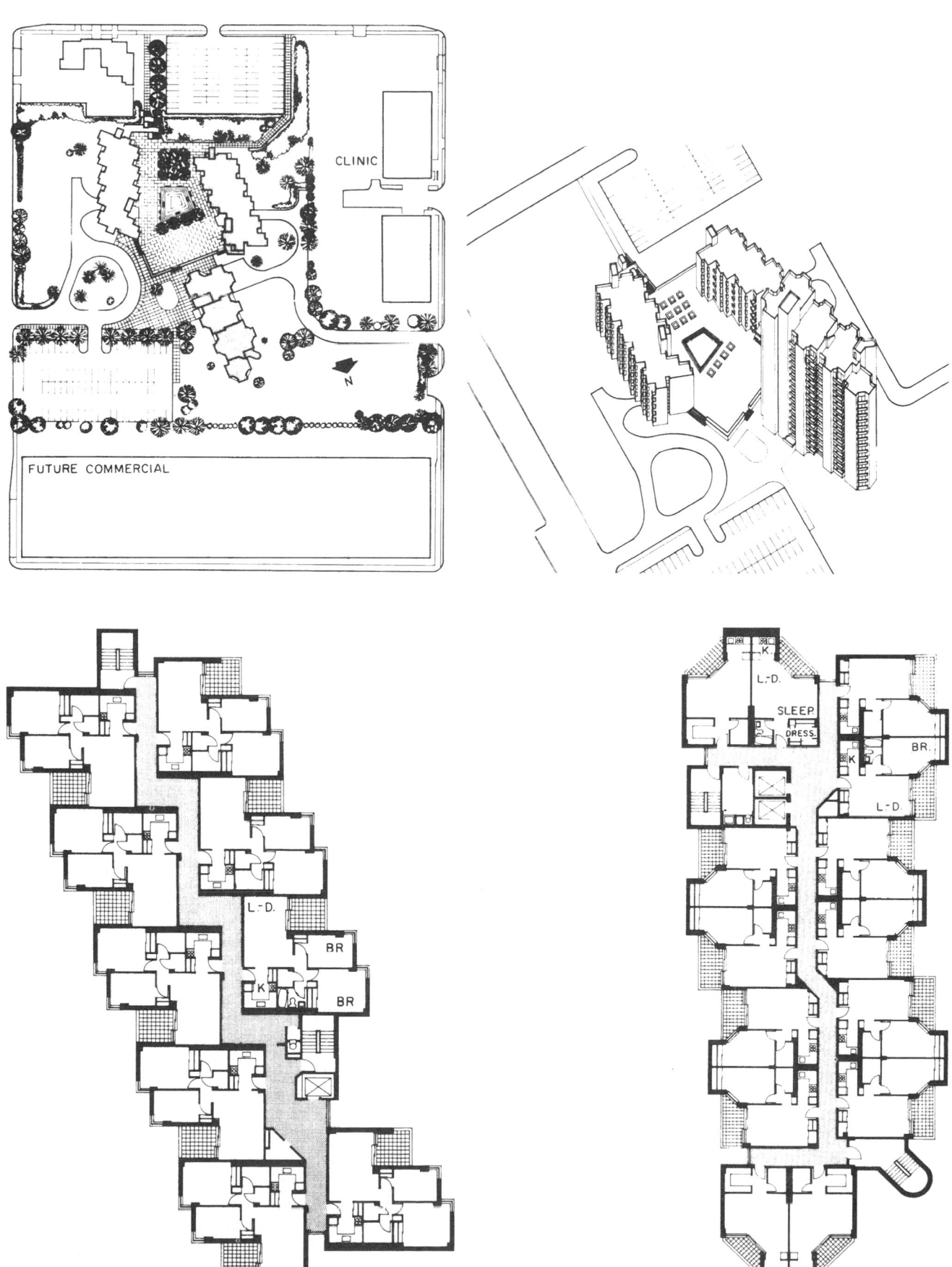

CLINIC

FUTURE COMMERCIAL

N

L.-D.

BR

K

BR

LOW RISE

K

L.-D.

SLEEP.

DRESS.

BR.

L-D.

HIGH RISE

Fig. 3   Kennedy Plaza, Utica, N.Y. U. Franzen & Associates—Architects.

CENTER-CORRIDOR PLAN, OFFSET

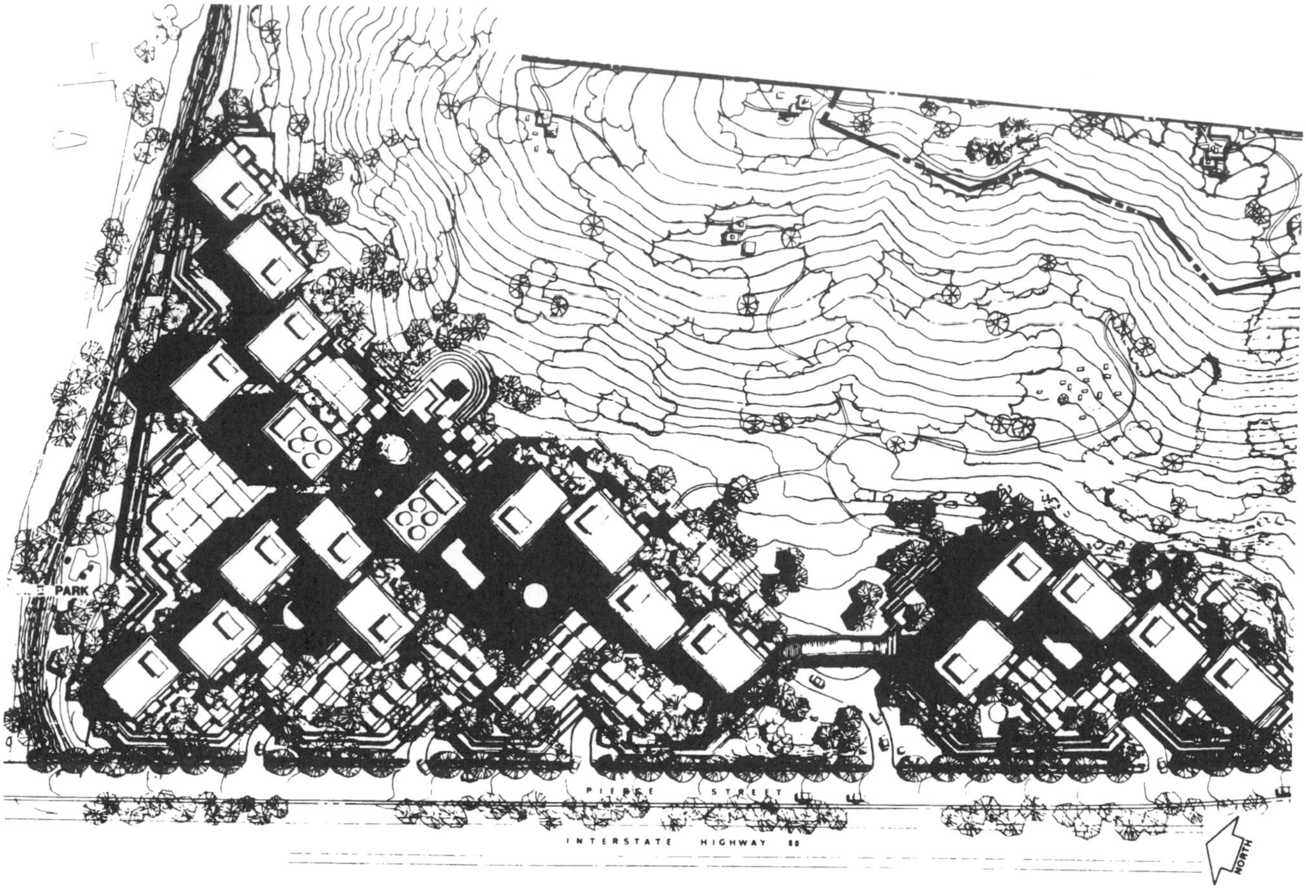

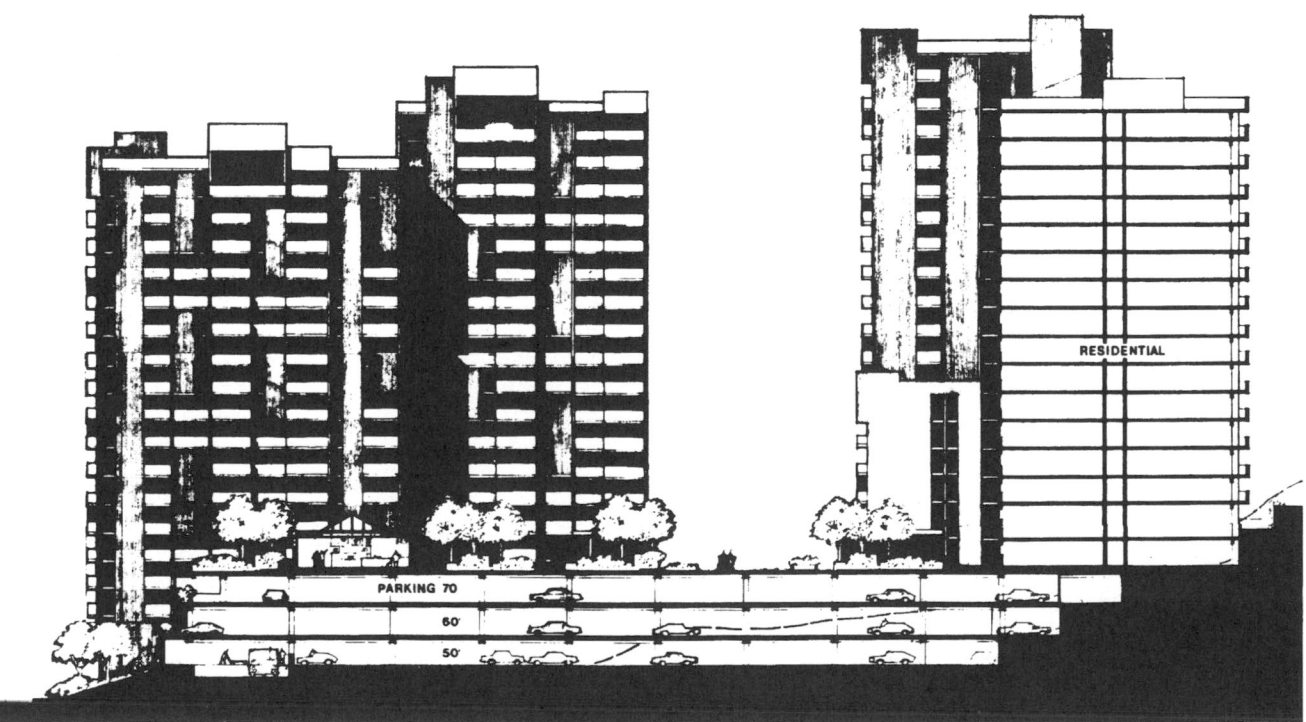

Fig. 4a   Site plan. Gateview at Albany Hill, Albany, N.Y. Hallenbeck, Chamorro & Associates—Architects.

Fig. 4b   Section. Gateview at Albany Hill, Albany, N.Y. Hallenbeck, Chamorro & Associates—Architects.

LOW-RISE, WALK-UP

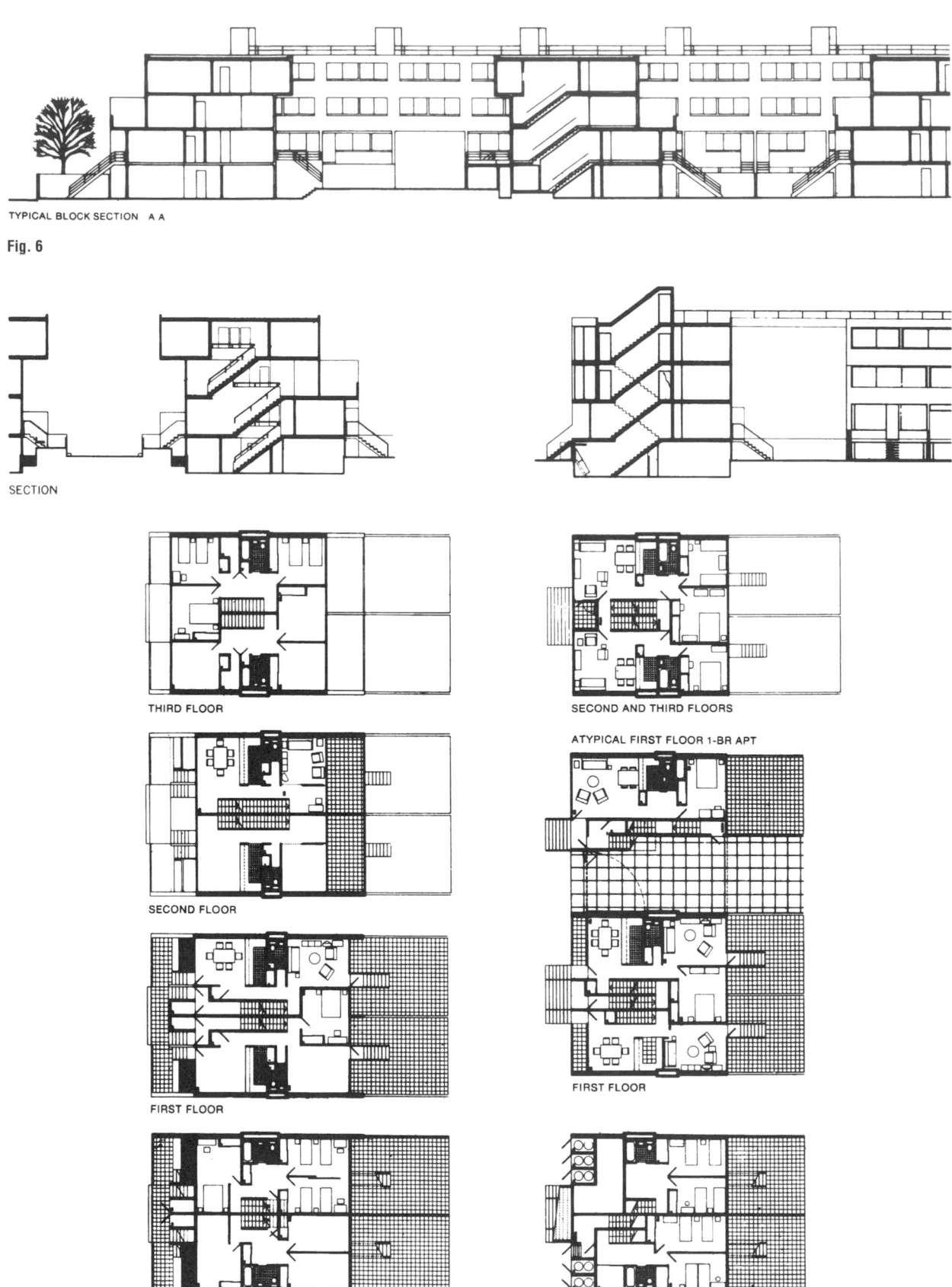

TYPICAL BLOCK SECTION A A

Fig. 6

SECTION

THIRD FLOOR

SECOND AND THIRD FLOORS

SECOND FLOOR

ATYPICAL FIRST FLOOR 1-BR APT

FIRST FLOOR

FIRST FLOOR

GROUND FLOOR

GROUND FLOOR

0   20'

Fig. 7

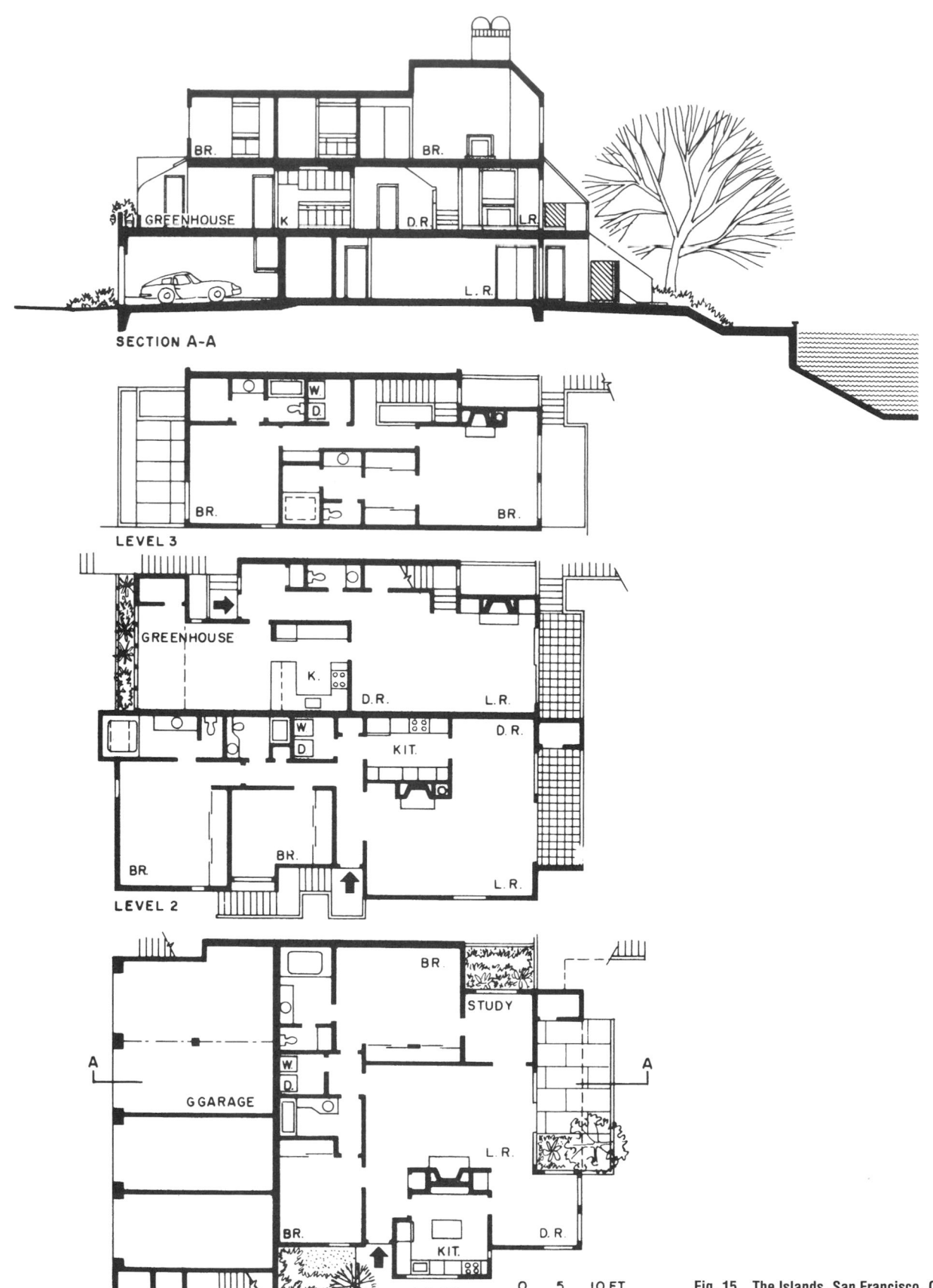

SECTION A-A

LEVEL 3

GREENHOUSE

LEVEL 2

LEVEL 1

Fig. 15   The Islands, San Francisco, Calif. Fisher-Friedman Associates—Architects.

0   5   10 FT.

LOW-RISE, WALK-UP

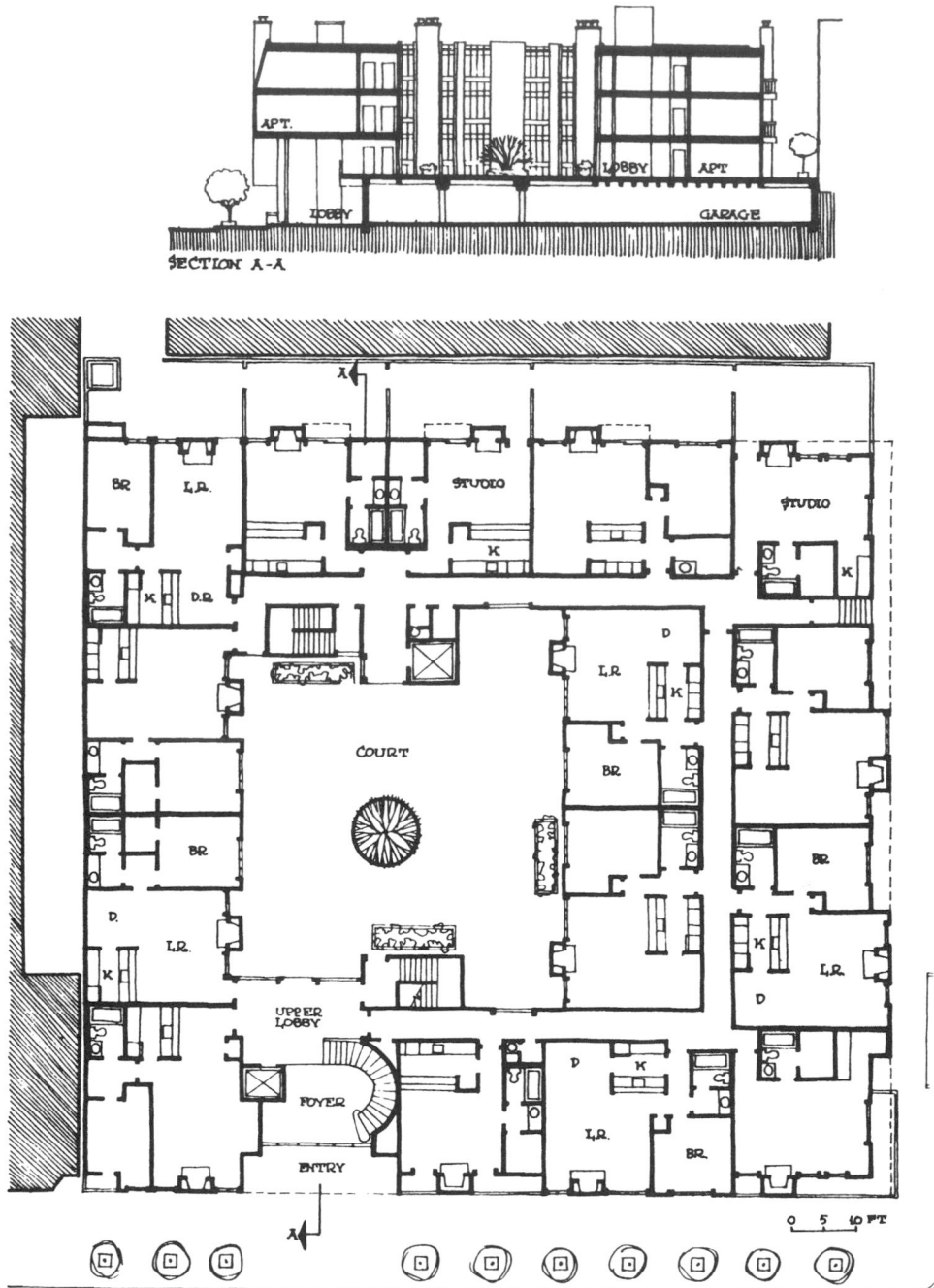

Fig. 17 This building contains nine efficiency units, 34 one-bedroom apartments, and 2 two-bedroom units. Willis and Associates—Architects.

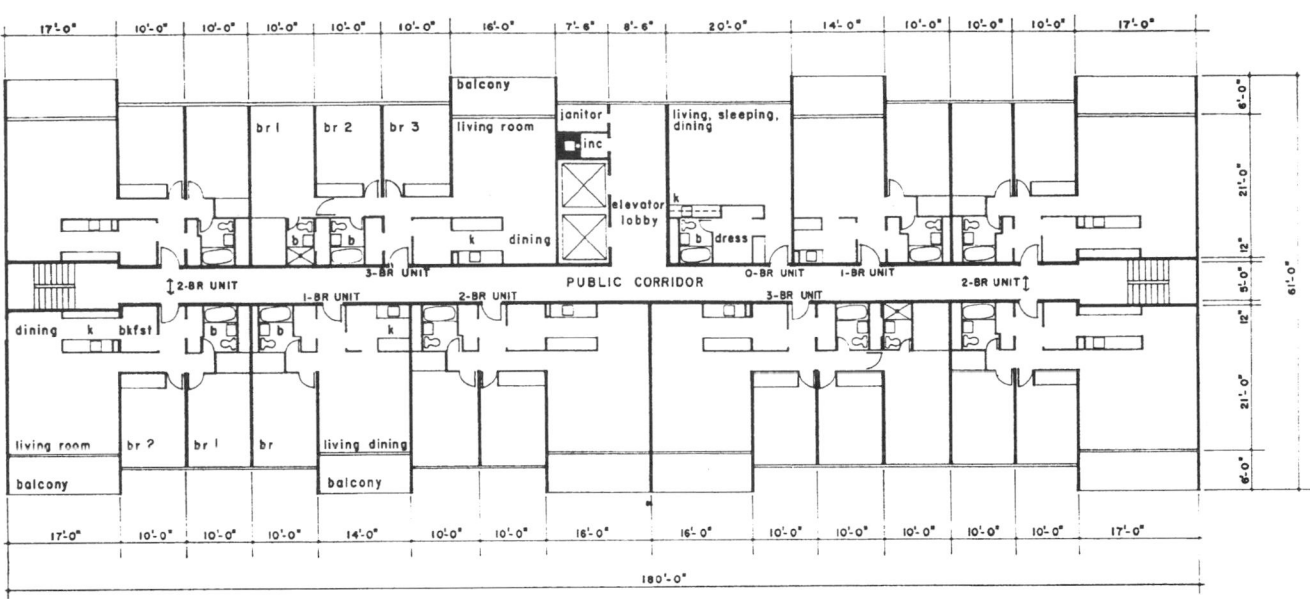

Fig. 3   Double-loaded corridor scheme.

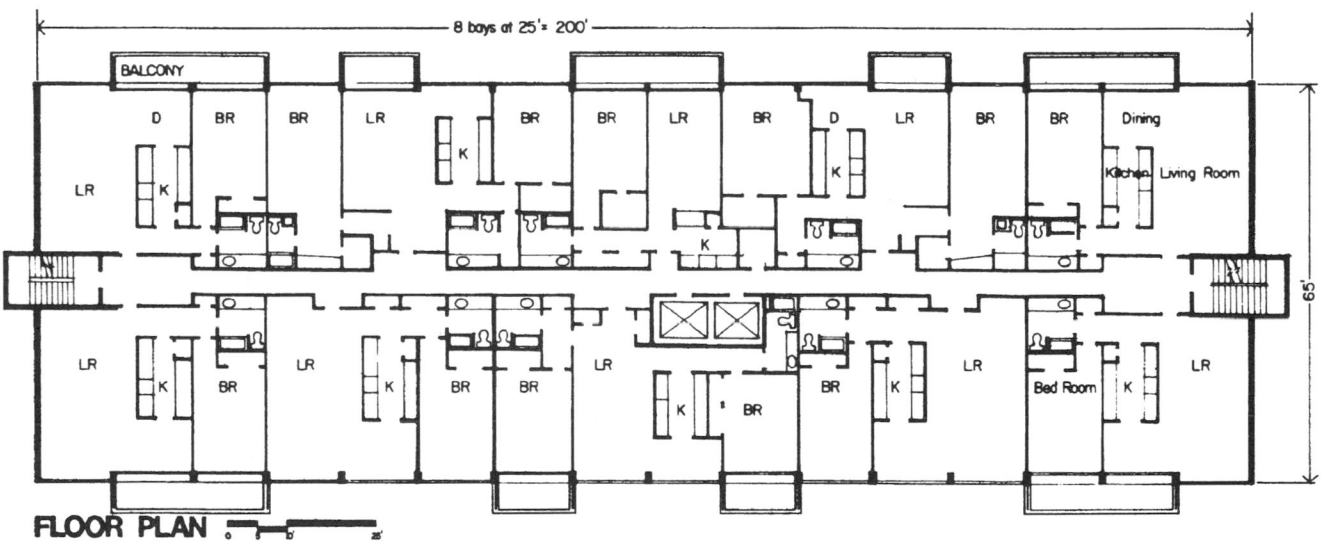

FLOOR PLAN

Fig. 4

CENTER-CORRIDOR PLAN

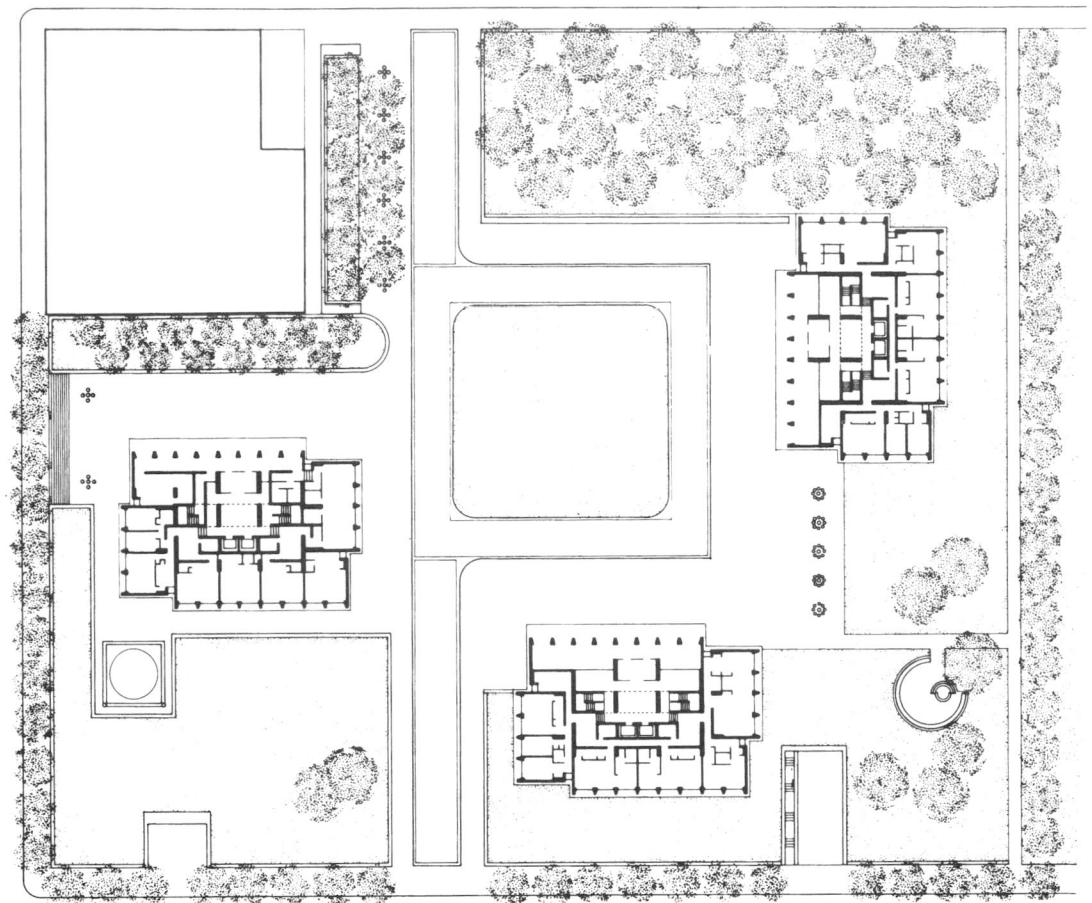

Fig. 8a Site plan.

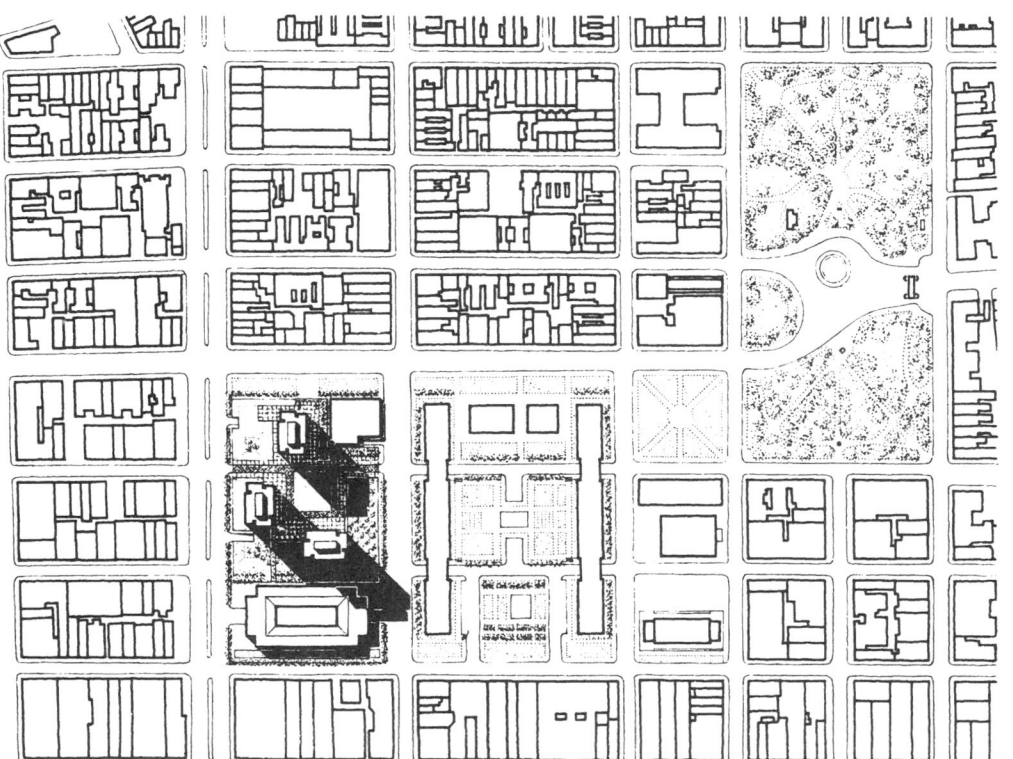

Fig. 8b Location plan. University Plaza, N.Y. University. I. M. Pei & Partners—Architects.

CENTER-CORRIDOR PLAN

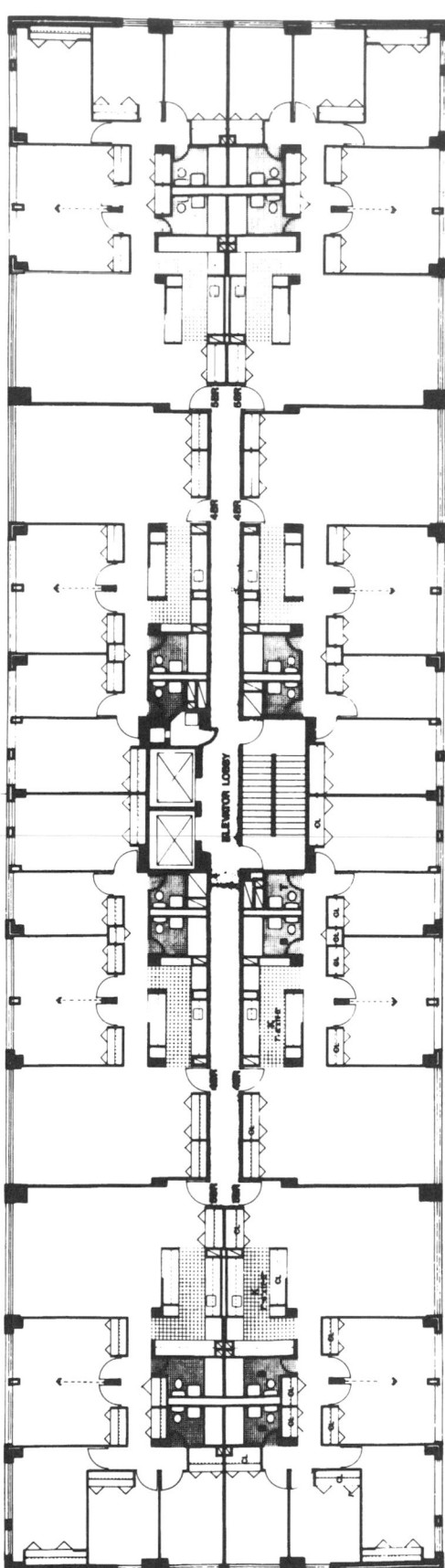

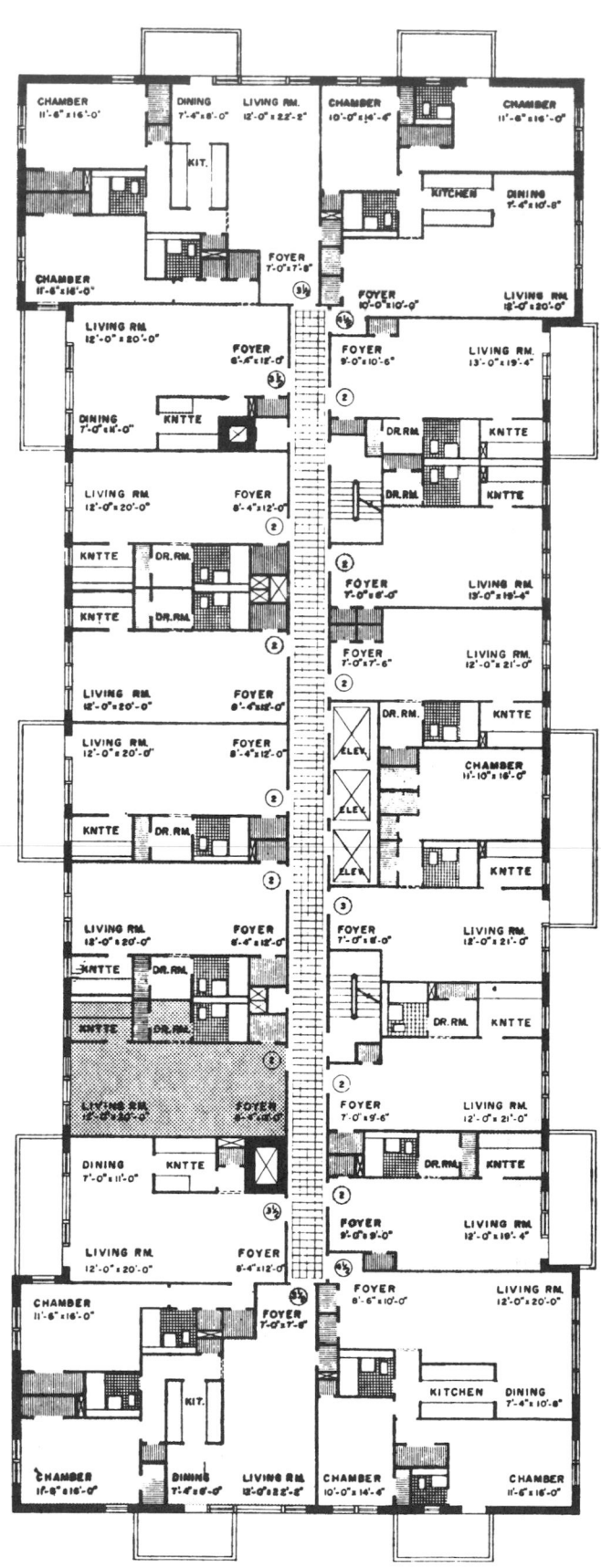

TYPICAL FLOOR PLAN

Fig. 15a   A. Schomburg Plaza, New York City. Gruzen & Partners—Architects.

Fig. 15b

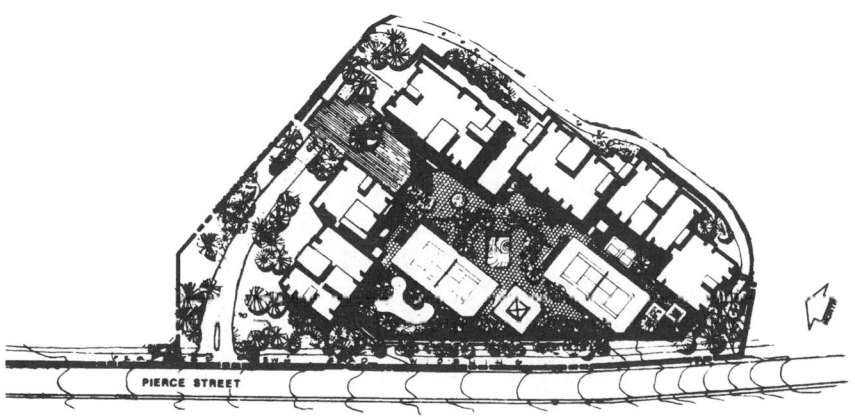

ONE BEDROOM UNIT

TWO BEDROOM UNIT

5

17 FLOORS

15 FLOORS

13 FLOORS

6 FLOORS

15 FLOORS

13 FLOORS

11 FLOORS

N

TOWER PLAN

20

PIERCE STREET

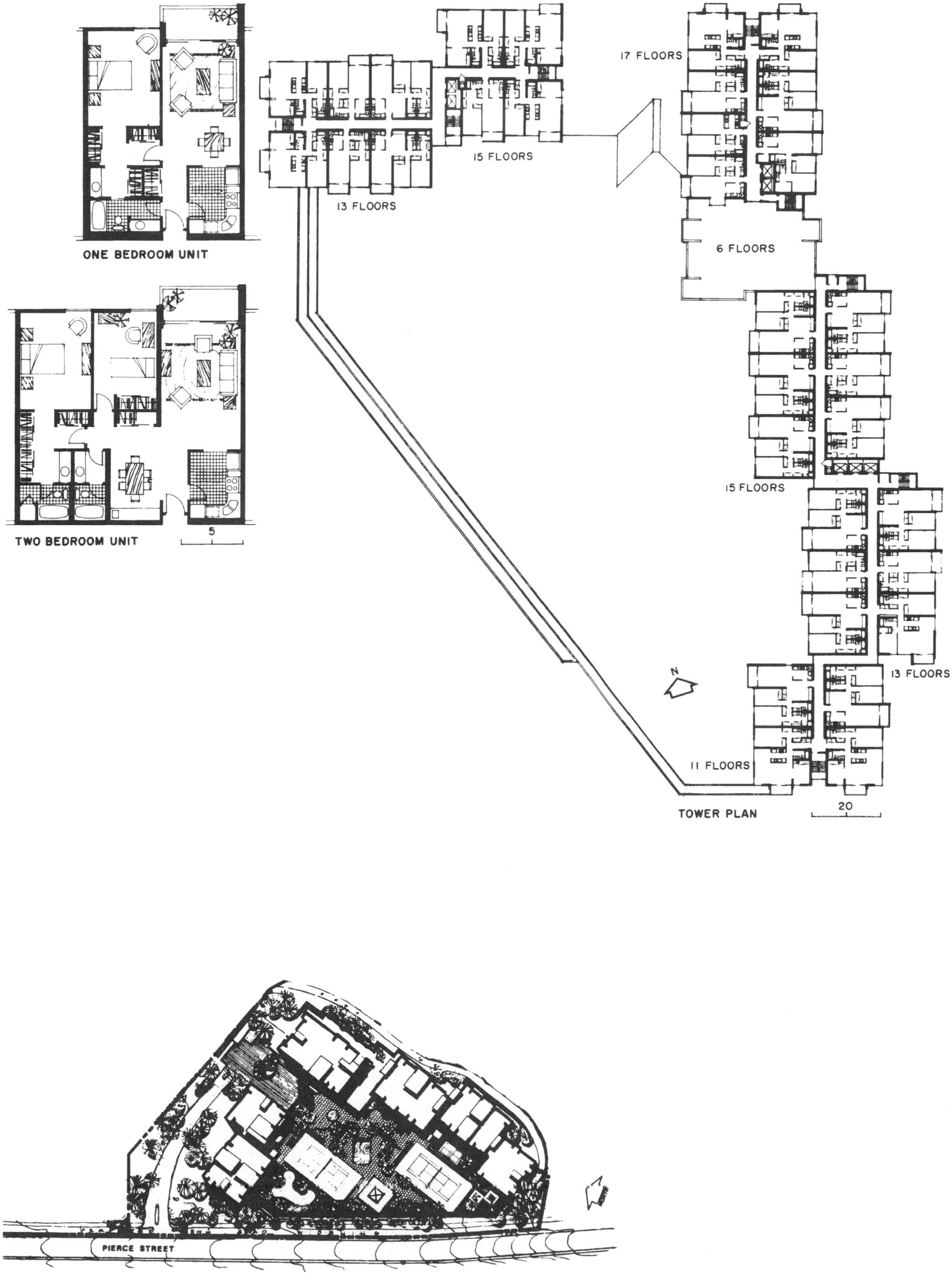

Fig. 5 Gateview at Albany Hill, Albany, N.Y. Hallenbeck, Chamorro & Associates—Architects.

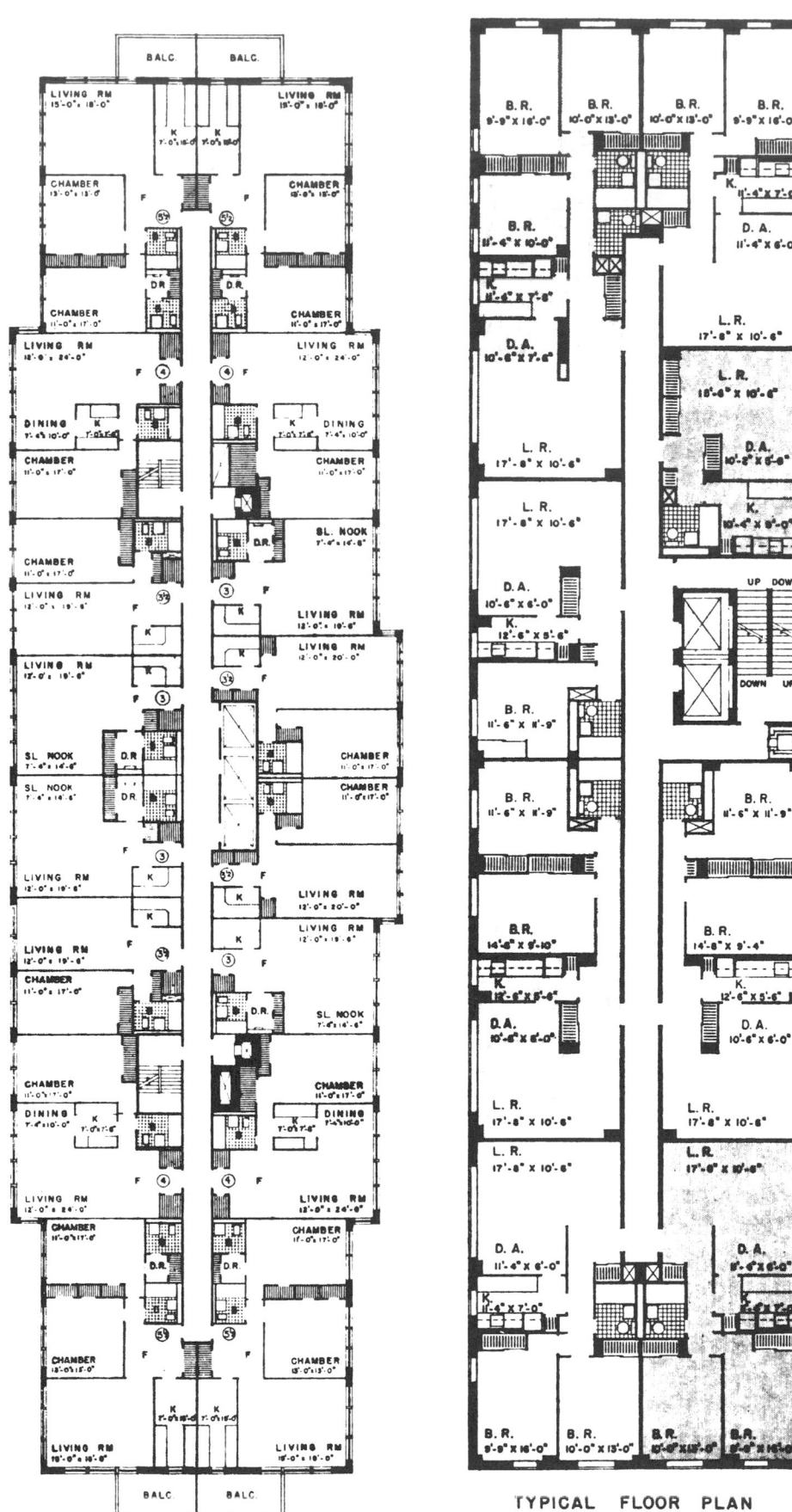

Fig. 6a

Fig. 6b

TYPICAL FLOOR PLAN

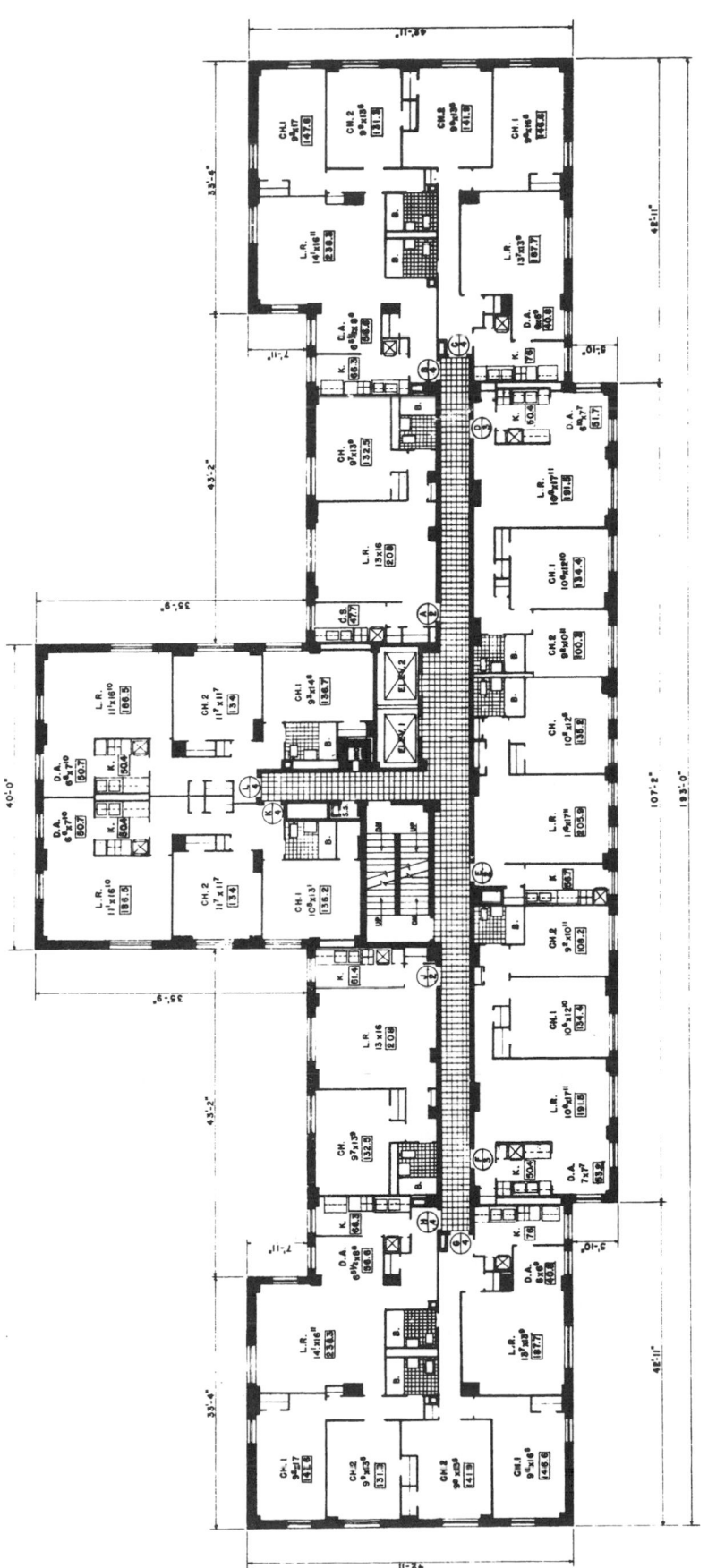

Fig. 6c

OPEN-CORRIDOR PLAN

OPEN-CORRIDOR PLAN

The open-corridor scheme consists of an exterior corridor serving a single line of apartments. The characteristic shape of this type of building is long and thin. The actual width of the building is limited to the corridor and depth of one apartment. Because of this arrangement, every apartment has through ventilation and two exposures. All rooms, including baths and kitchens, can have natural light and ventilation if desired. The open corridor provides literally a "sidewalk-in-the-sky." If additional outdoor space is provided to the open corridor, a "front yard" can be created.

The disadvantages are the long corridors and distances from the elevator to an apartment, and the possible loss of privacy by the movement of people in front of one's apartment.

A variation of the open-corridor scheme is the single-loaded corridor. Essentially it is the same as the open corridor except that it is enclosed from the elements.

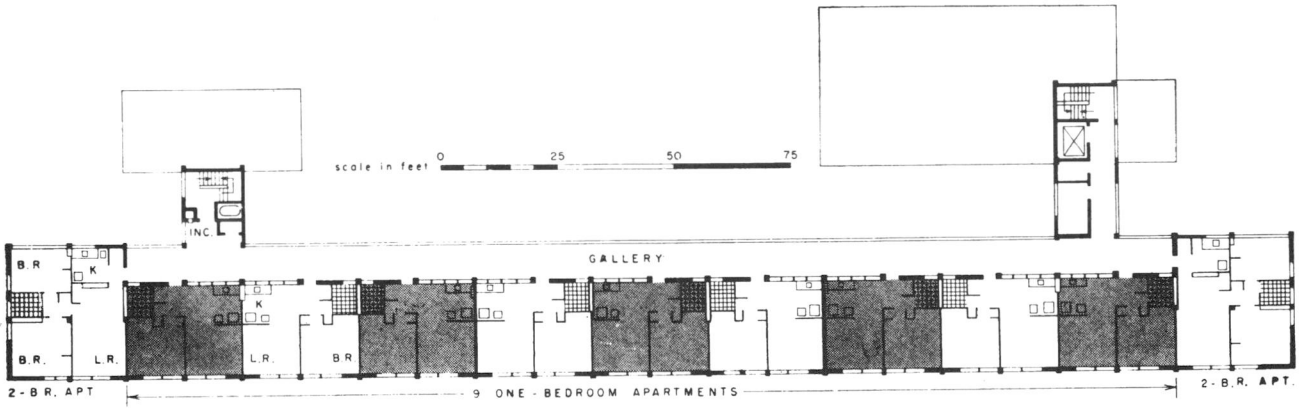

Fig. 1 Gallery sweeps 225 ft (depth, 8 ft). These two seven-story slabs have low adjacent laundries and free-standing elevator and stairway stacks. Gallery exposure is northwest. Total number of apartments, 148. Archer Courts, Chicago, Ill. E. F. Quinn & Associates—Architects.

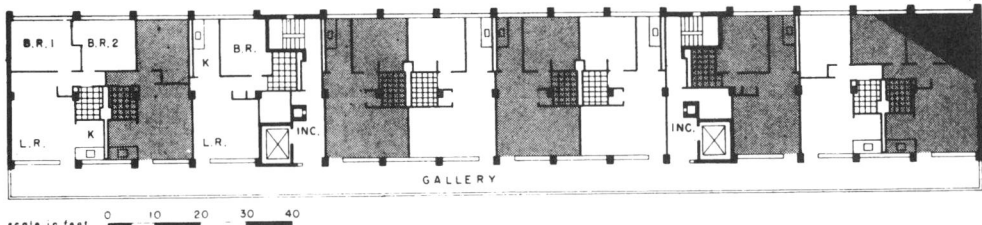

Fig. 2 South gallery is a unique feature. Eight feet deep, the gallery will admit winter sun warmth to apartments through obscure fronts, but galleries above will shade the summer sun. Total number of apartments in this 14-story apartment, 130. Prairie Avenue Courts. George Fred Keck & William Keck—Architects.

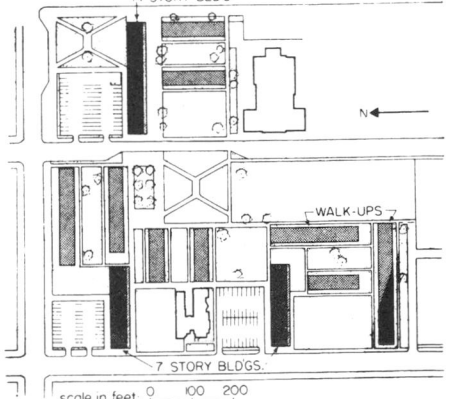

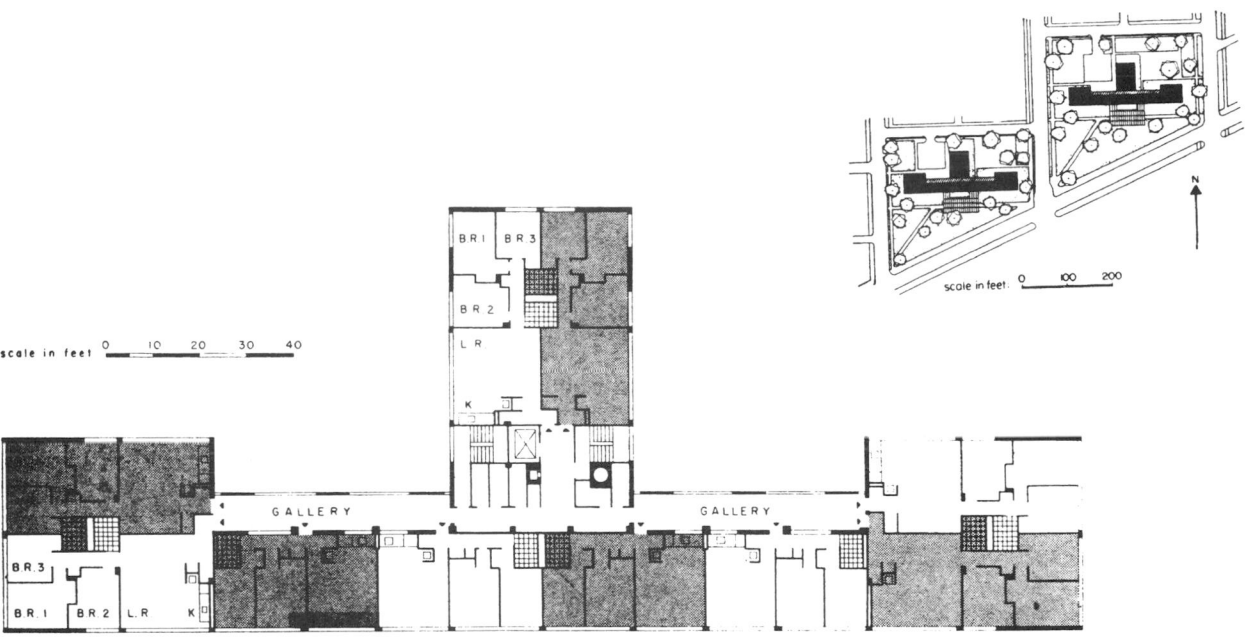

OPEN-CORRIDOR PLAN

scale in feet
0 100 200
N
scale in feet
0 10 20 30 40
BR1
BR3
BR2
LR
K
GALLERY
GALLERY
BR3
BR1
BR2
LR
K

Fig. 3 Gallery (8 ft deep, north exposure) is split in center by another apartment wing. This gallery is not entirely open, alternating bright-colored tile squares with mesh fencing. Total number of apartments, 136. Ogden Courts, Chicago, Ill. Skidmore Owings & Merrill—Architects.

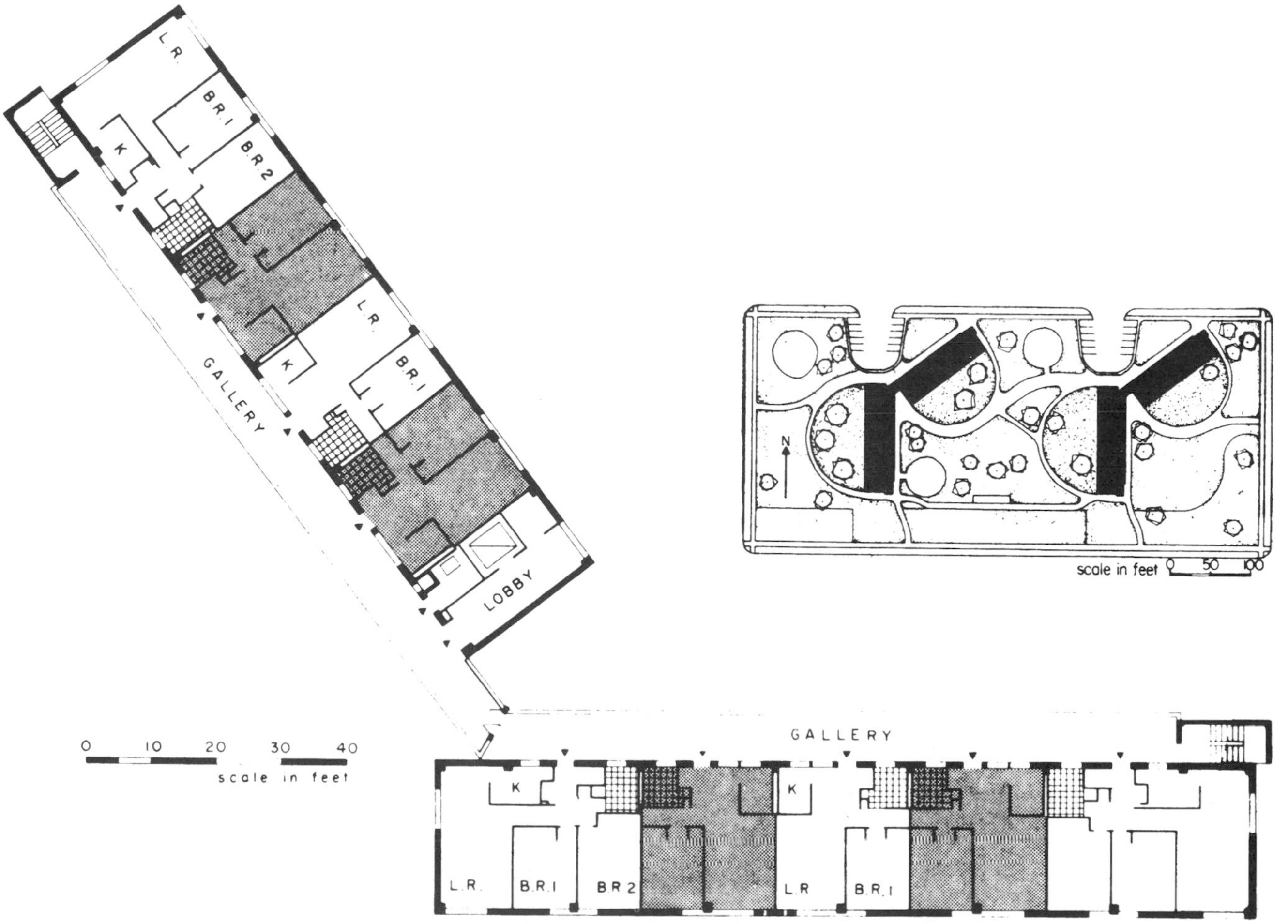

LR
BR1
K
BR2
LR
K
BR1
GALLERY
LOBBY
N
scale in feet 0 50 100
0 10 20 30 40
scale in feet
GALLERY
K
K
LR
BR1
BR2
LR
BR1

Fig. 4 Canted two-wing apartment is served by same galleries (8 ft deep) bridged from one wing to the other. Gallery exposures are east and northwest. Elevator is at inner end of one wing. Total number of apartments in two of these buildings, 126. Loomis Courts Loewenburg & Loewenburg—Architects and Engineers; Weese & van der Meulen—Associate Architects.

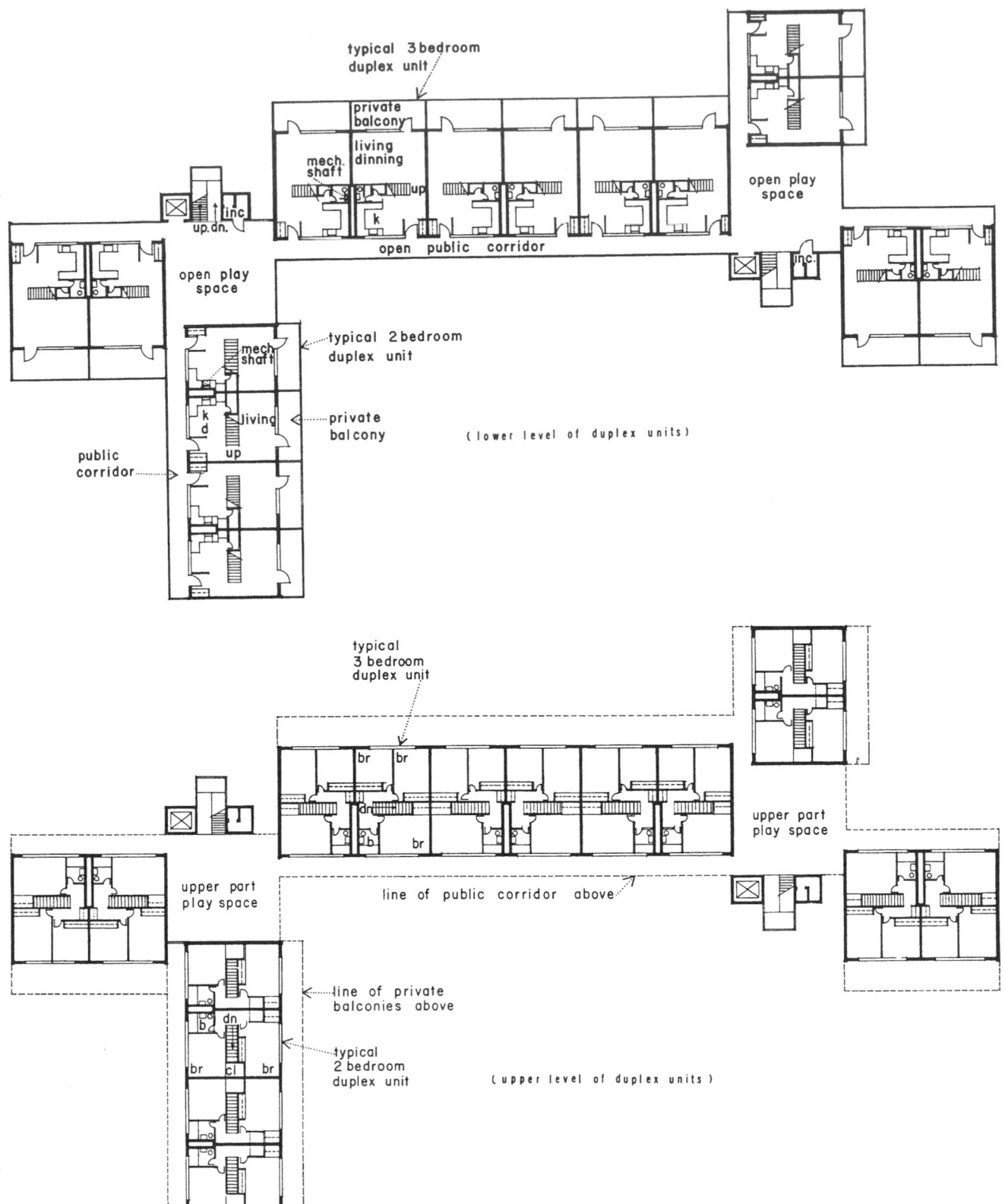

Fig. 5 Duplex scheme: This plan type offers floor-through duplex apartments, and exterior play spaces on each entry floor.

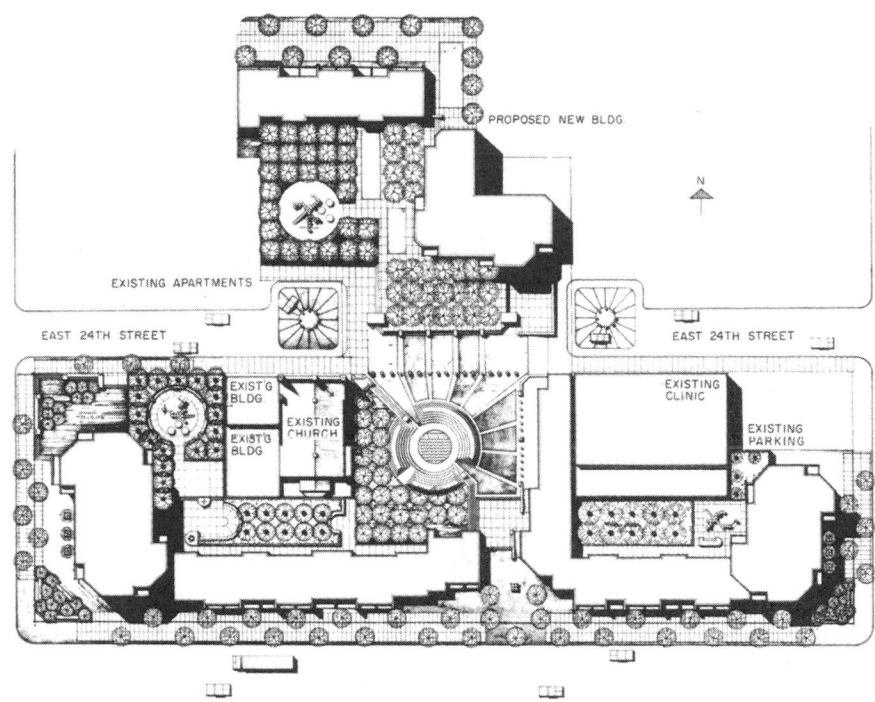

Fig. 6   East Midtown Plaza, New York City. Davis, Brody & Associates—Architects.

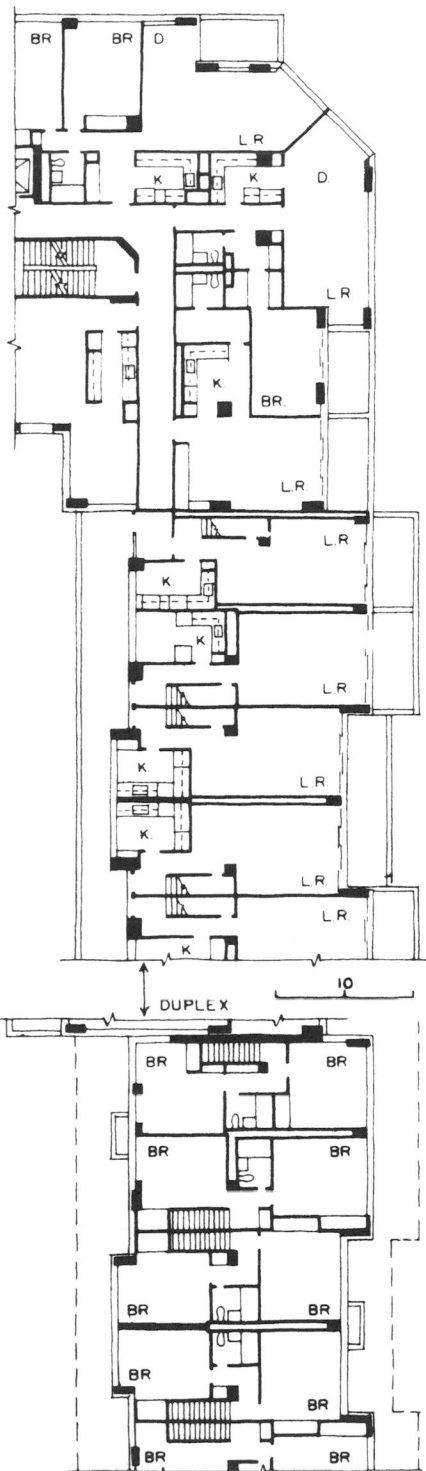

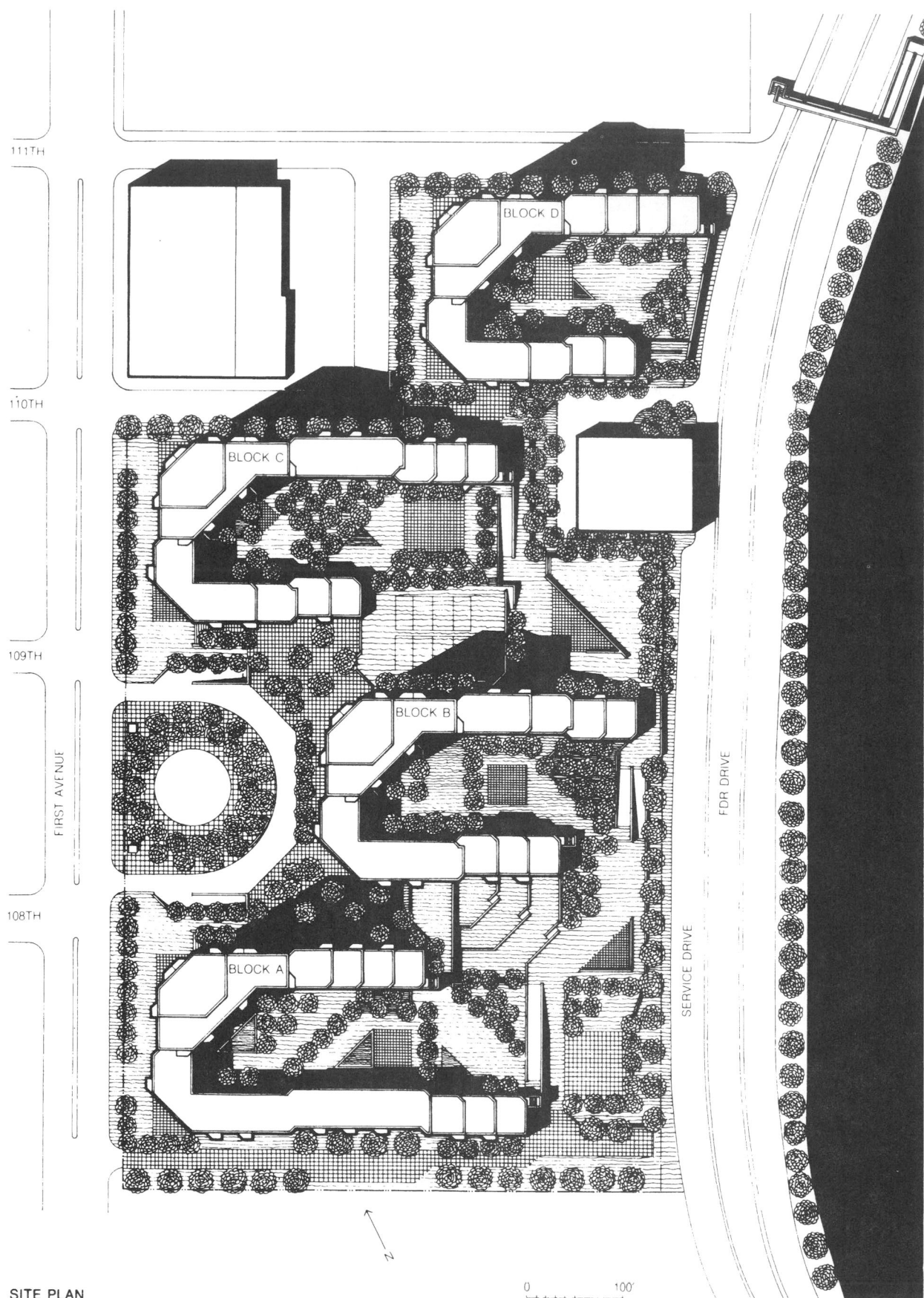

SITE PLAN

Fig. 7   1199 Plaza Cooperative Housing, New York City. The Hodne/Stageberg Partners—Architects.

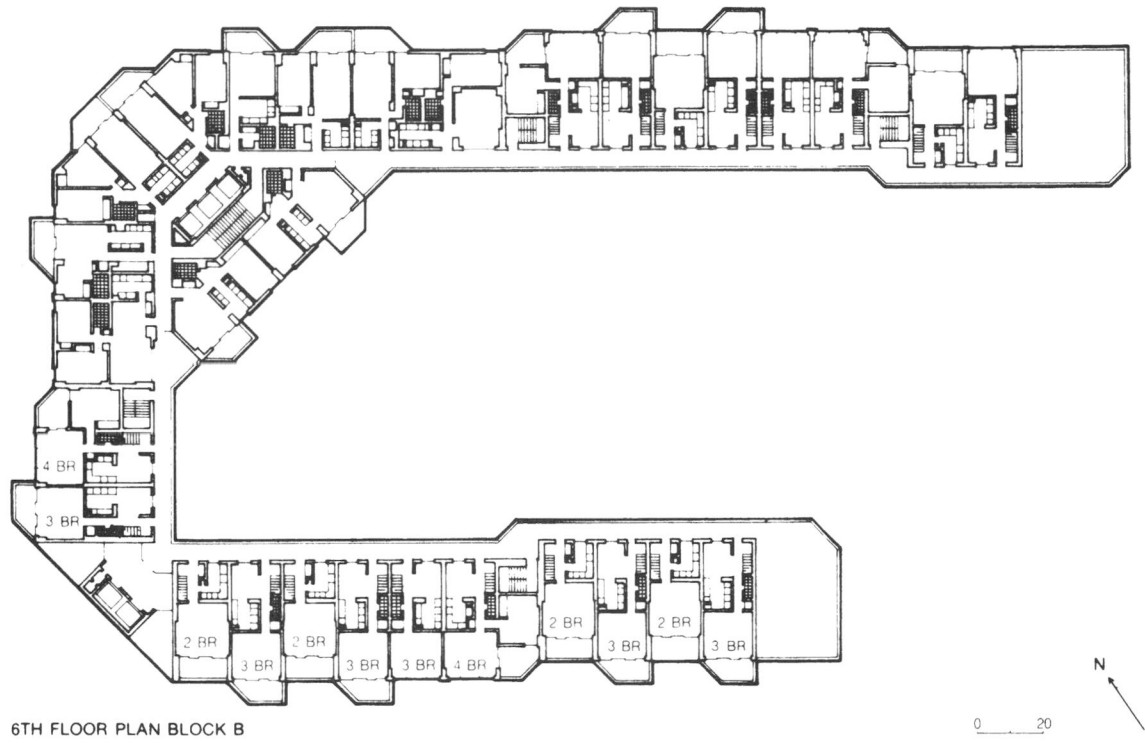

6TH FLOOR PLAN BLOCK B

Fig. 8a 1199 Plaza Cooperative Housing, New York City. The Hodne/Stageberg Partners—Architects.

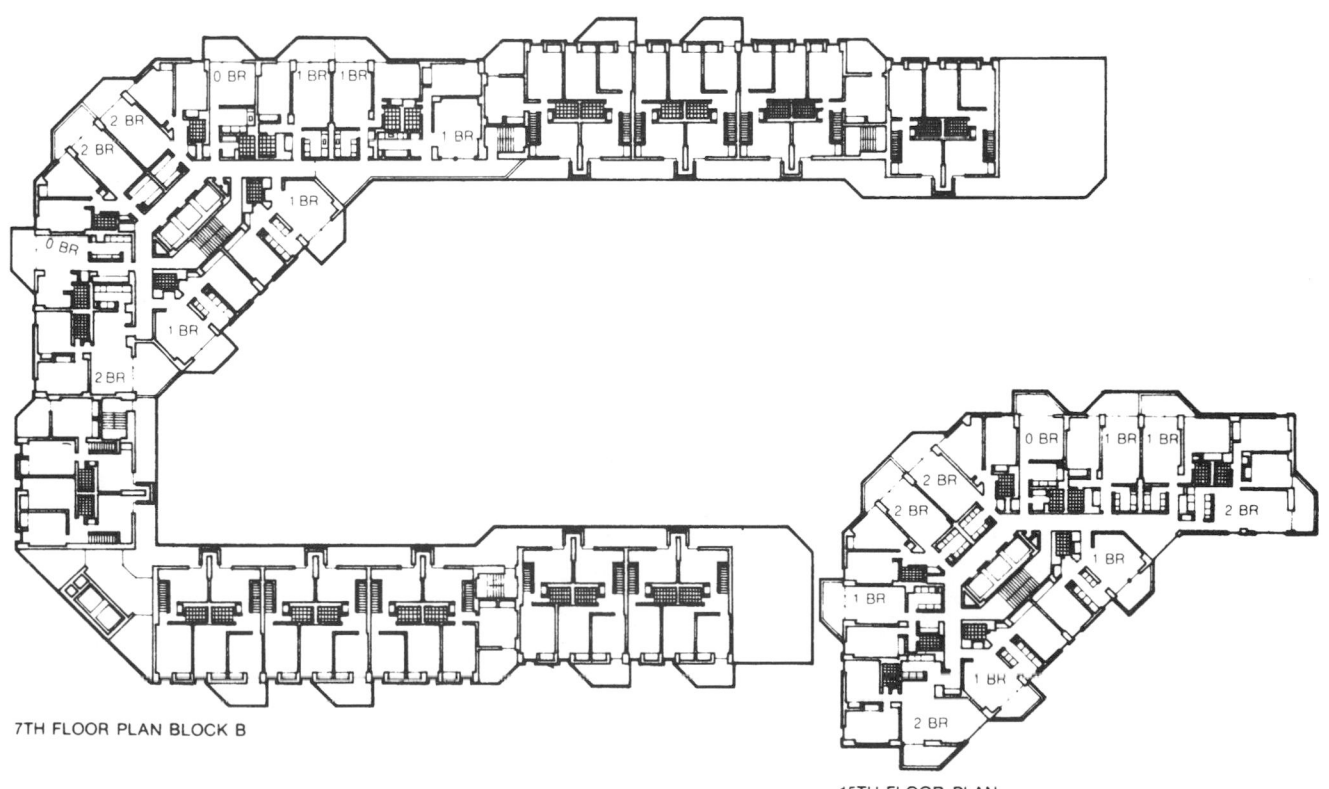

7TH FLOOR PLAN BLOCK B

15TH FLOOR PLAN

Fig. 8b 1199 Plaza Cooperative Housing, New York City. The Hodne/Stageberg Partners—Architects.

OPEN-CORRIDOR PLAN

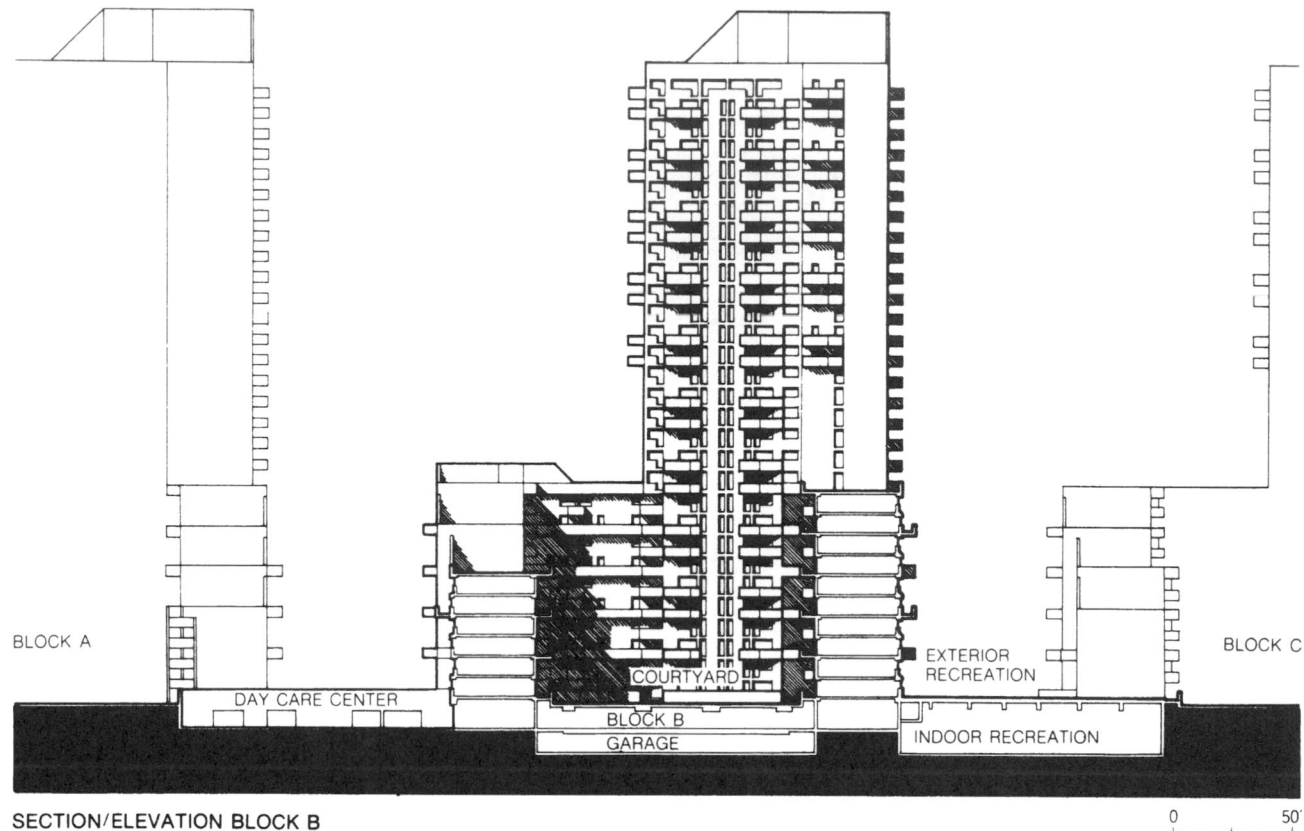

BLOCK A

DAY CARE CENTER

COURTYARD

EXTERIOR
RECREATION

BLOCK B

GARAGE

INDOOR RECREATION

BLOCK C

SECTION/ELEVATION BLOCK B

0   50'

Fig. 8c   1199 Plaza Cooperative Housing, New York City. The Hodne/Stageberg Partners—Architects.

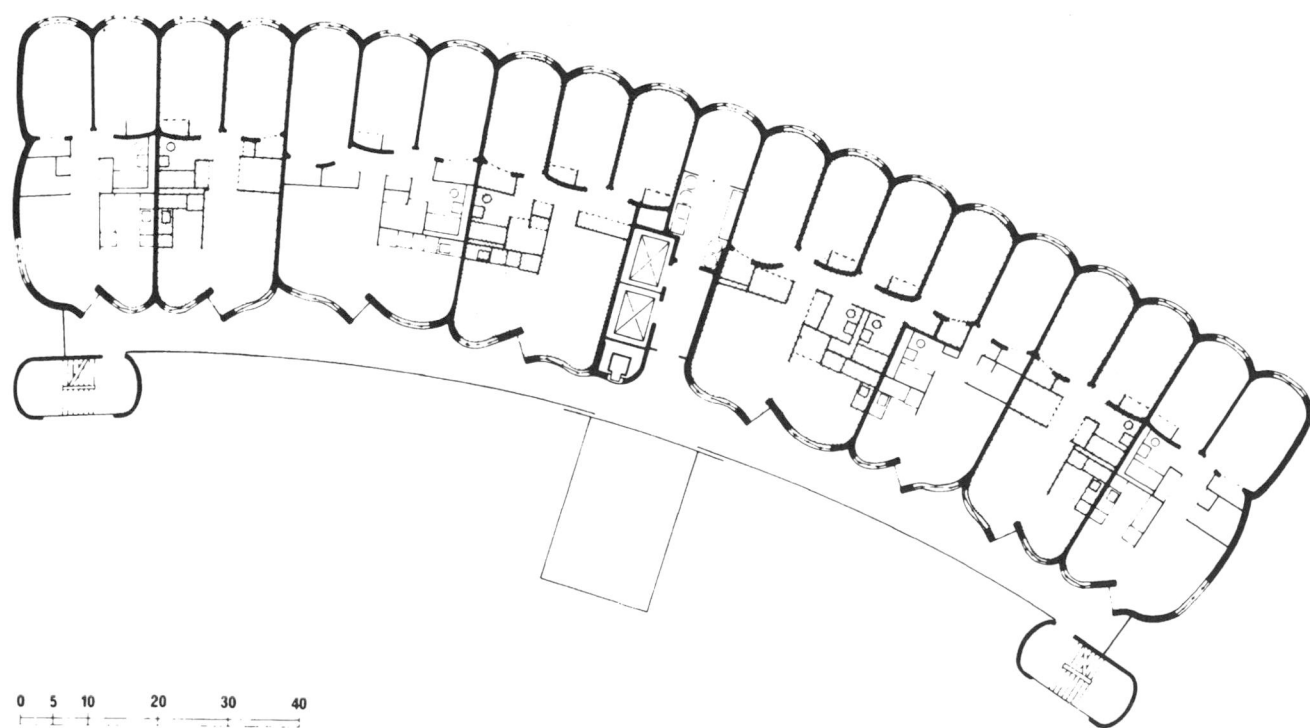

0   5   10   20   30   40

Fig. 9   Hilliaro Center, Chicago, Ill. Bertrand Goldberg—Architect.

SKIP-STOP PLAN

A variation of the center-corridor scheme is the so-called *skip-stop* plan. In this arrangement, the elevator stops only on alternate floors and eliminates completely the public corridor on the nonstop floors. The objective is to eliminate nonliving floor area and reduce the number of elevator doors, thereby reducing construction costs and making the building more efficient.

By using duplex apartments, through ventilation can be achieved on the noncorridor level. Also, apartments will have two exposures, which will simplify the problem of orientation.

The main objection to the skip-stop scheme is that each apartment is required to have an interior stair to the alternate level. This may create serious problems for elderly and handicapped people. Another problem is providing two means of egress from each floor, which most building codes require. Exit through common balconies is sometimes accepted as meeting this requirement. A more extreme variation of the skip-stop is to have the elevator stop on every third level. People would walk either up or down to the level of their apartment.

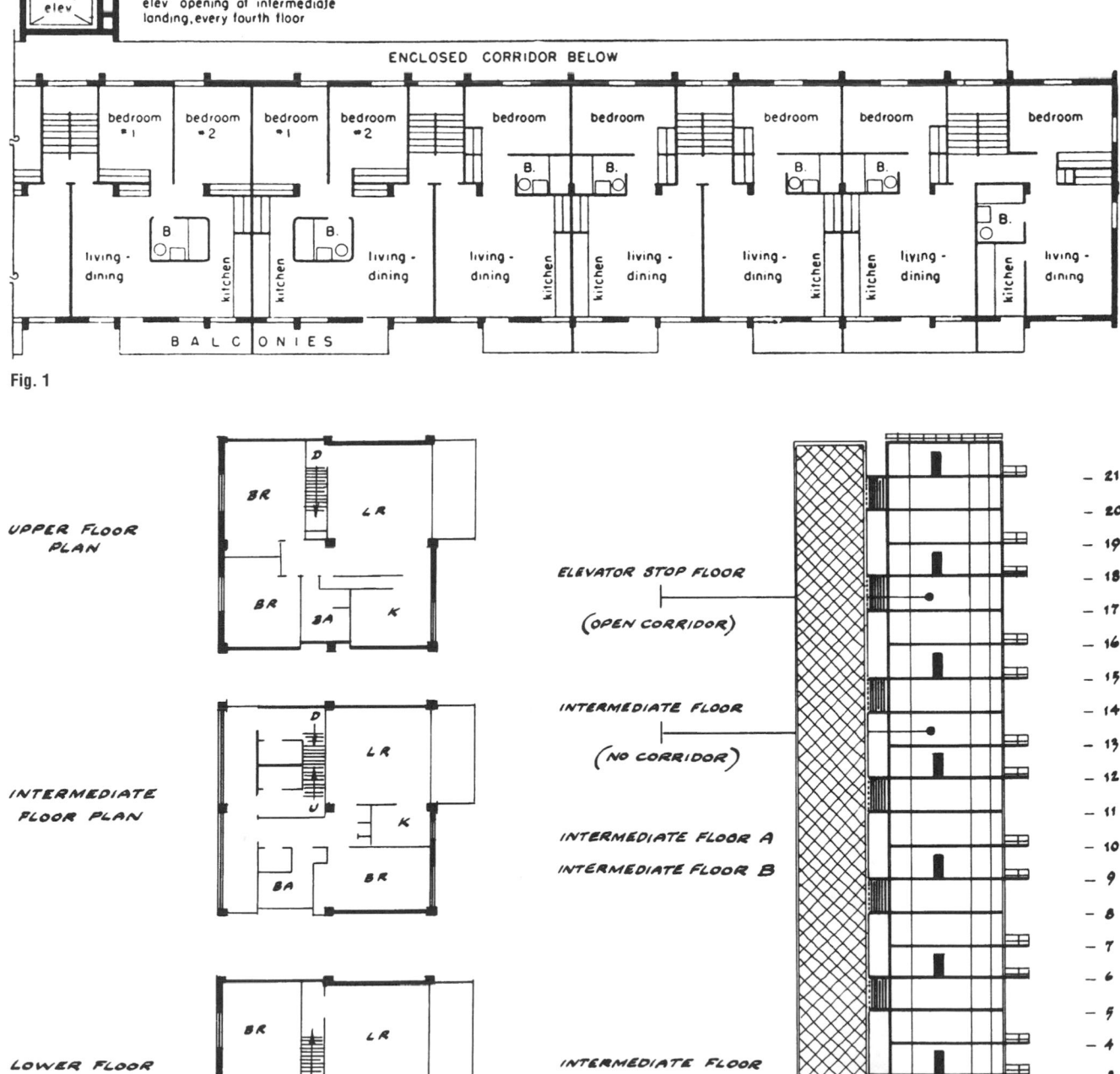

Fig. 1

Fig. 2

Fig. 3

SKIP-STOP PLAN

SKIP-STOP APARTMENTS

A skip-floor corridor scheme has four components:

1. Elevators stop only at intermittent floors, where there are long access corridors.

2. All occupants leave elevator on the corridor floor. Those living on noncorridor floors use auxiliary private stairways, going up or down.

3. Noncorridor floors extend from exterior wall to exterior wall, have privacy and through ventilation.

4. Secondary means of egress must be provided.

Fig. 4

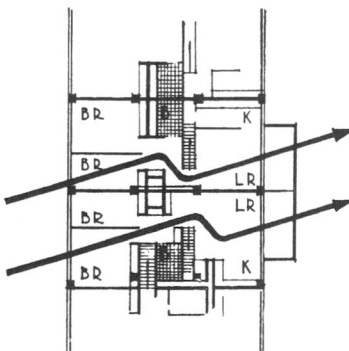

Fig. 5

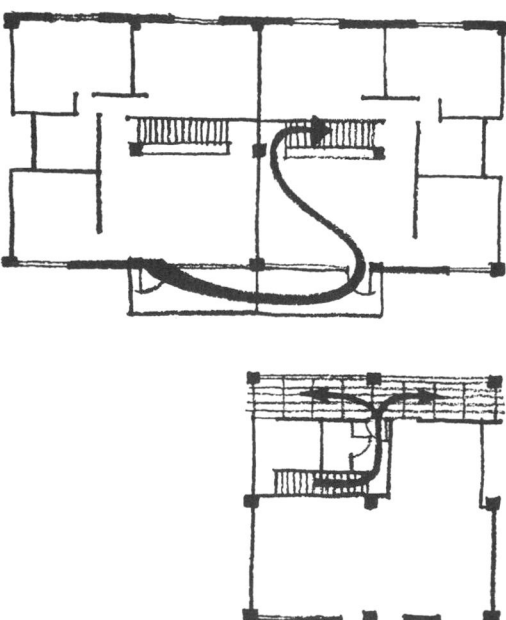

Fig. 6

TWO BR APT

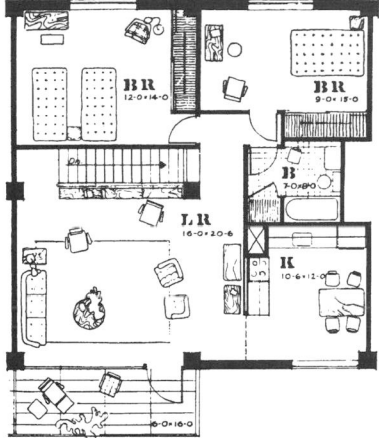

FL. ABOVE

ONE BR APT

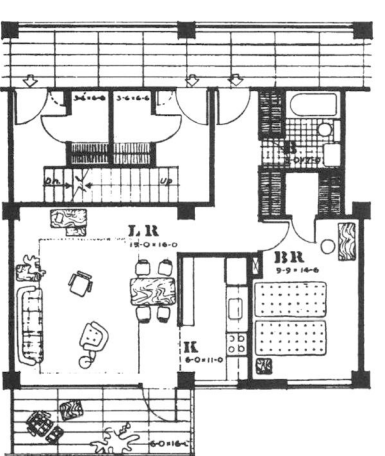

CORRIDOR FL.

TWO BR APT

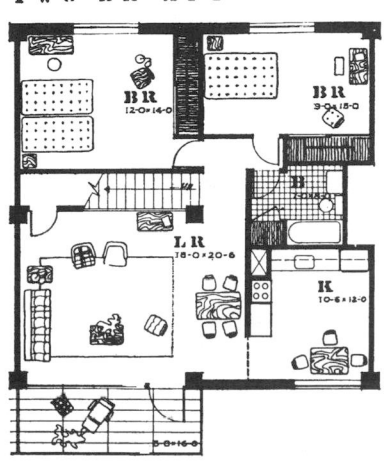

FL. BELOW

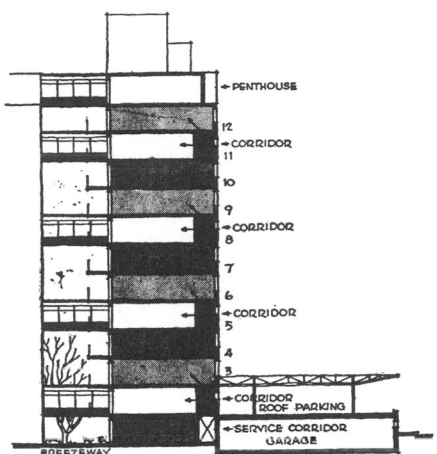

TWO BR APT

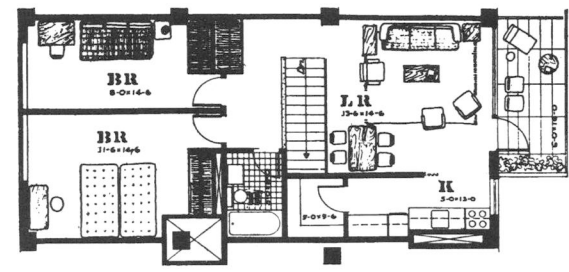

FL. ABOVE

STUDIO ONE BR APT

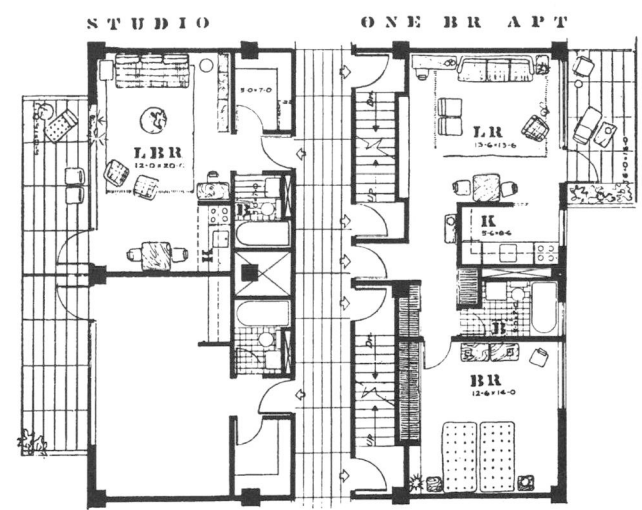

CORRIDOR FL.

TWO BR APT

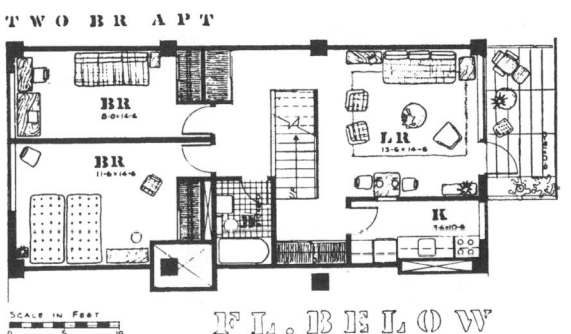

FL. BELOW

Fig. 7 Eastgate Apts., Cambridge, Mass.

SKIP-STOP PLAN

In the duplex apartments of this design, you need walk only one half flight up or down between bedrooms and living rooms. But this is also a skip-stop apartment, an unusually subtle one, so the other one half flight of stairs which places all entrances on *stop* floors instead of *skip* floors must be accounted for.

Solution: the duplex apartments actually are entered on *half* levels on half flight above or below the regular *stop* floor planes (*see section*). The *skipped* floors are filled by the rest of the duplexes while straightforward single level "efficiency" apartments fill out the rest of the *stop* floors.

This is luxury and economy too. The contradiction of the duplex within an elevator building constitutes much of its considerable lure to city dwellers, who want "a house within the apartment." Economies in elevator service and hall space justify the complications in this apartment.

The plans and the section in Fig. 9 indicate the complexity of the program for this tower, with zero- to five-bedroom apartments arranged as duplexes. One out of every three floors has no public corridor.

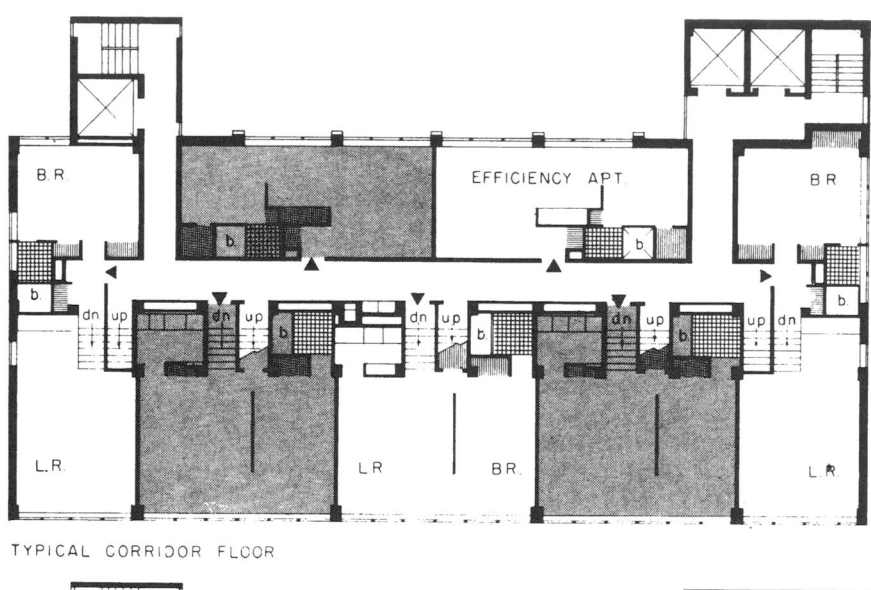

TYPICAL CORRIDOR FLOOR

TYPICAL FLOOR - TWO LEVEL APT'S

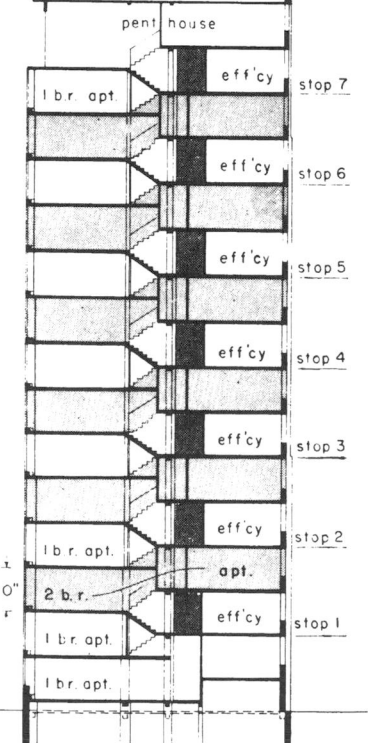

Fig. 8   Apartment House, Boston, Mass. Glaser & Gray—Architects.

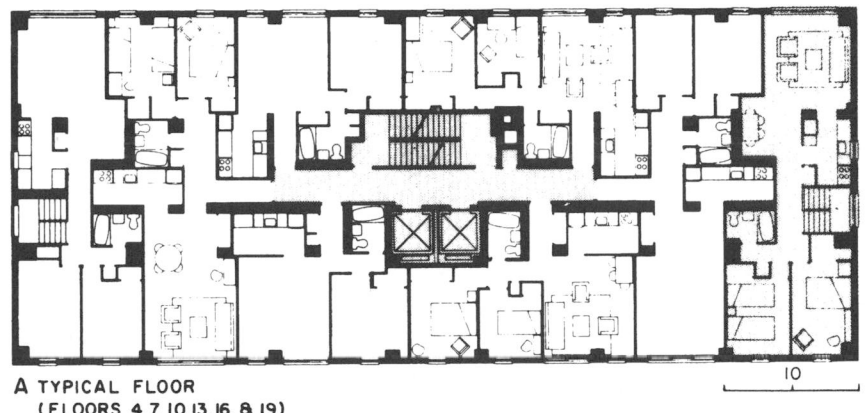

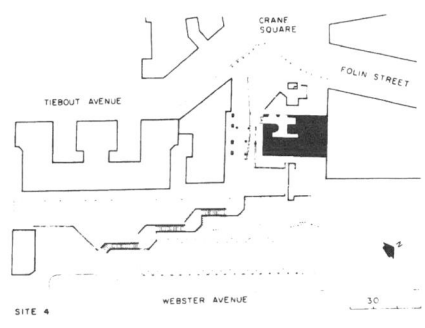

A TYPICAL FLOOR
(FLOORS 4,7,10,13,16,& 19)

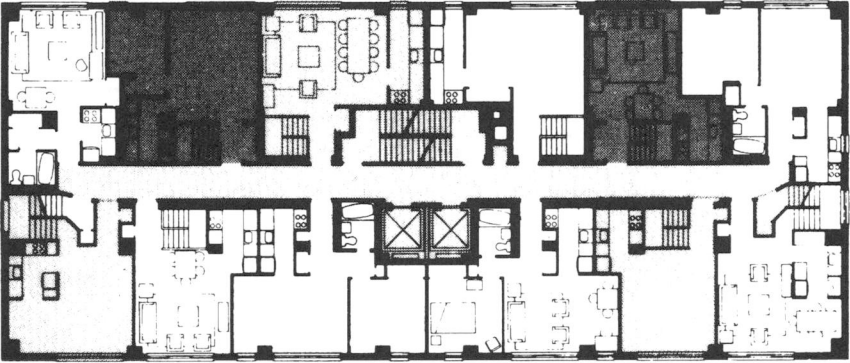

B TYPICAL FLOOR
(FLOORS 3,6,9,12,15,& 18)

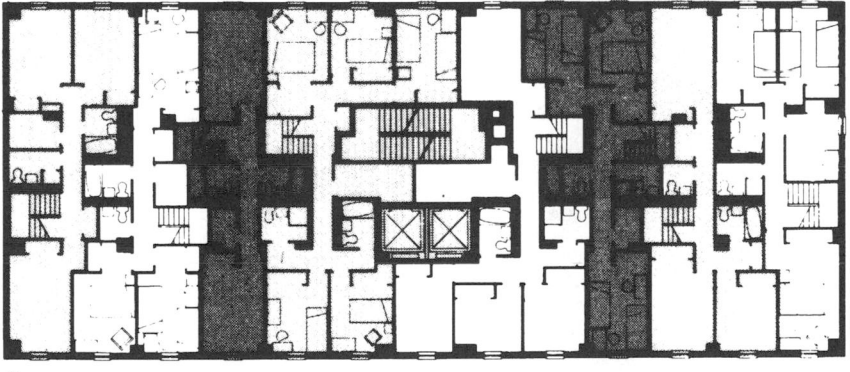

C TYPICAL FLOOR
(FLOORS 2,5,8,11,14,& 17)

☐ SINGLE LEVEL APTS.

▨ DUPLEX APTS.

Fig. 9 Twin Parks Northwest, Bronx, N.Y. Prentice, Chan, Ohihausen—Architects.

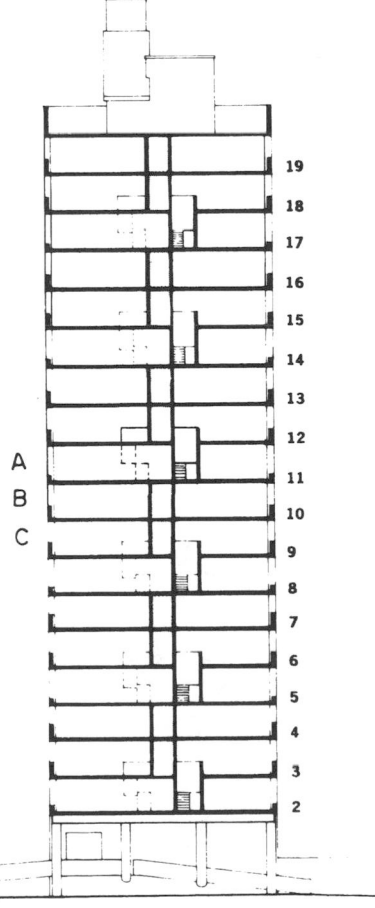

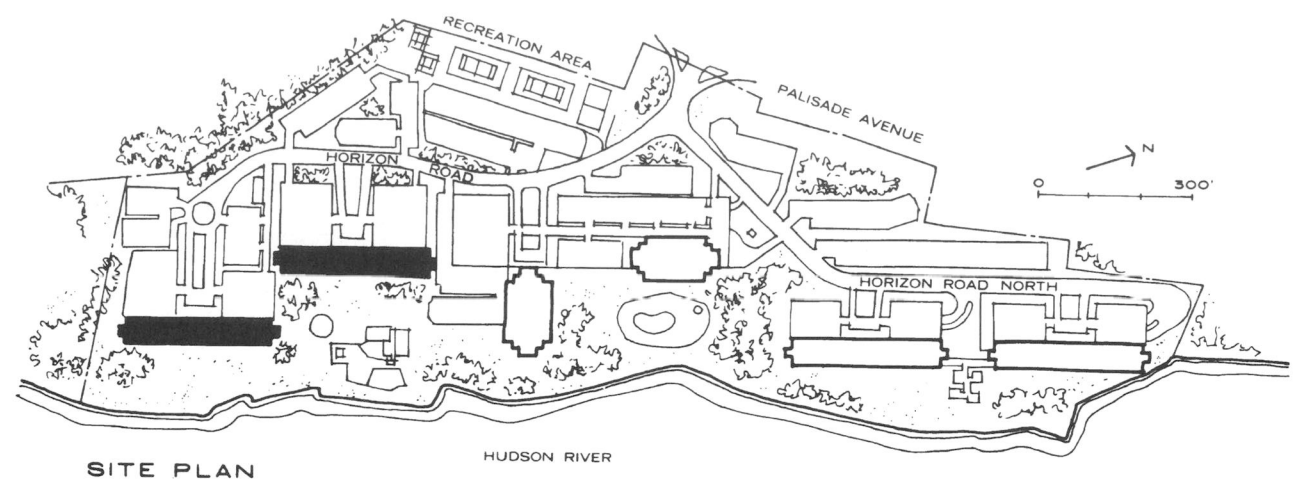

SITE PLAN   HUDSON RIVER

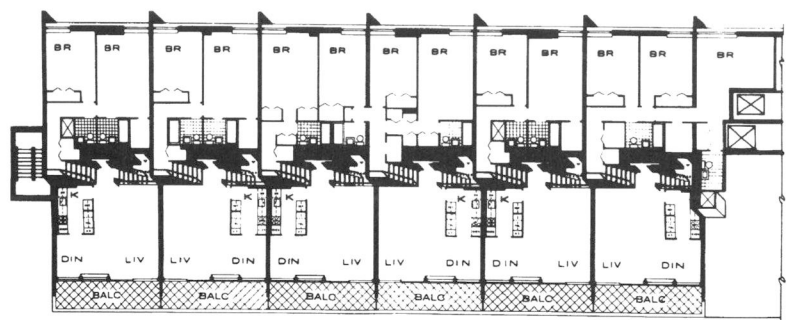

PART TYPICAL FLOOR PLAN

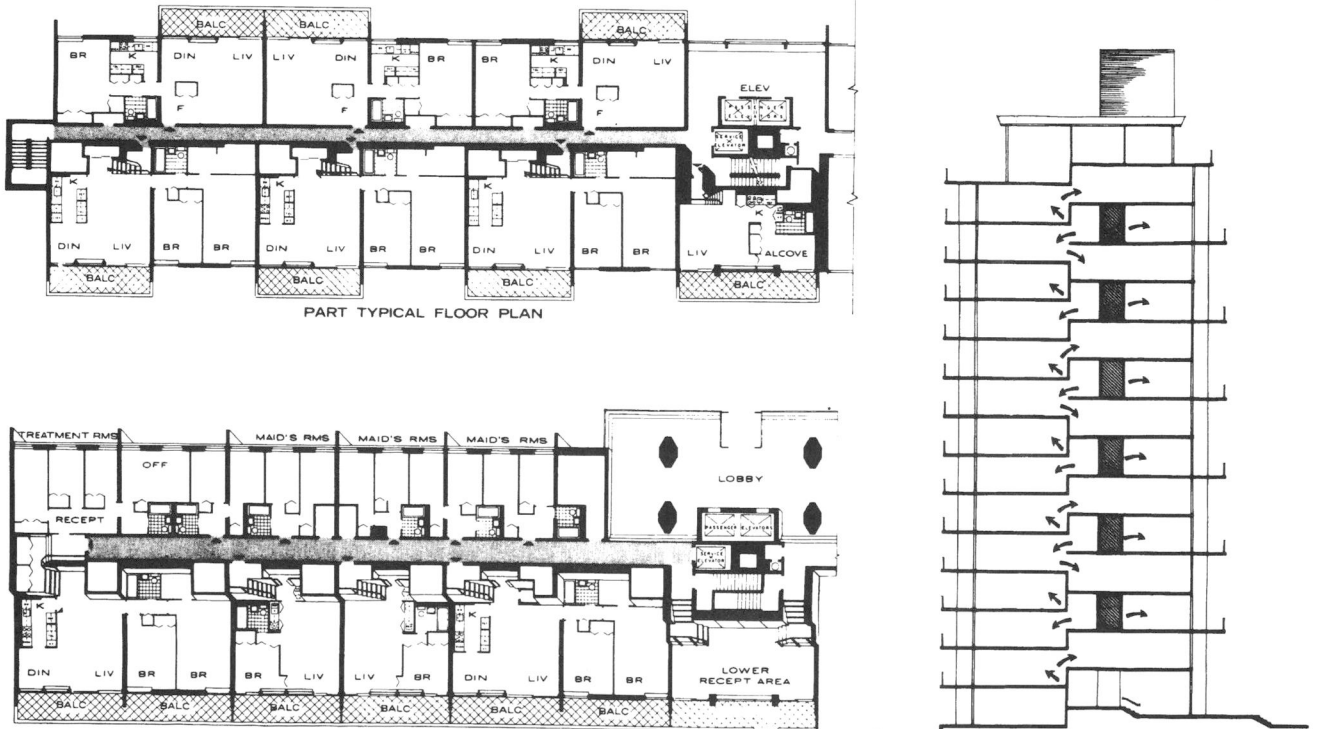

PART TYPICAL FLOOR PLAN

PART GROUND-FLOOR PLAN

DIAGRAMMATIC SECTION

Fig. 10 Horizon House, Fort Lee, N.J. Kelley & Gruzen—Architects.

Le Corbusier's 20-story apartment building in Marseilles has partly through ventilation and partly not. Corridors are every third floor. All apartments are duplexed and have interior baths and kitchens. There are five apartment types in the building's schedule, but only one of these is "cross" duplexed to a single exposure. The master bedroom is on the gallery above the living room and faces the brise-soleil and balcony system, which occurs on east, west, and south exposures.

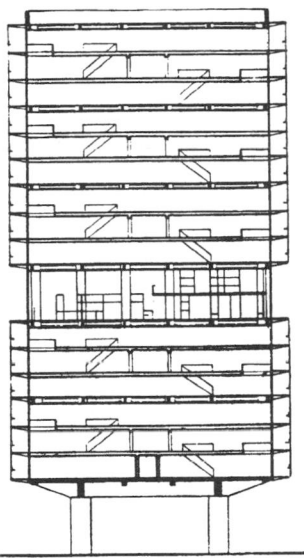

Fig. 11a Apartment house, Marseilles, France. Le Corbusier—Architect. Cross section through building.

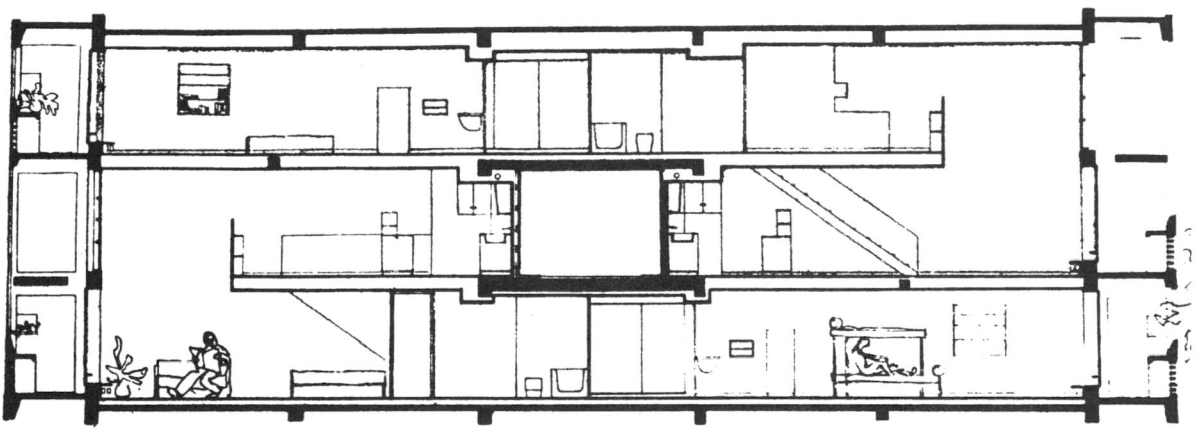

Fig. 11b Apartment house, Marseilles, France. Le Corbusier—Architect. Section through typical unit.

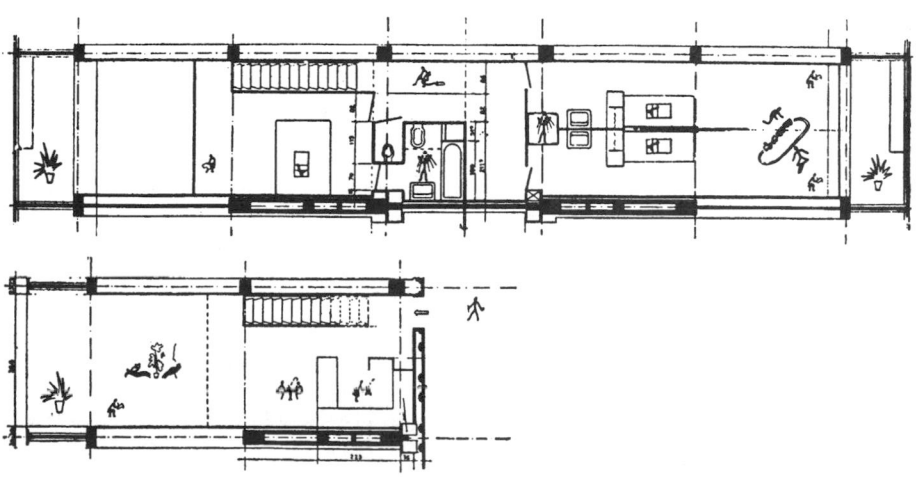

Fig. 11c Apartment house, Marseilles, France. Le Corbusier—Architect. Plan of lower level.

Fig. 11d Apartment house, Marseilles, France. Le Corbusier—Architect. Plan of mezzanine/bedroom level.

SKIP-STOP PLAN

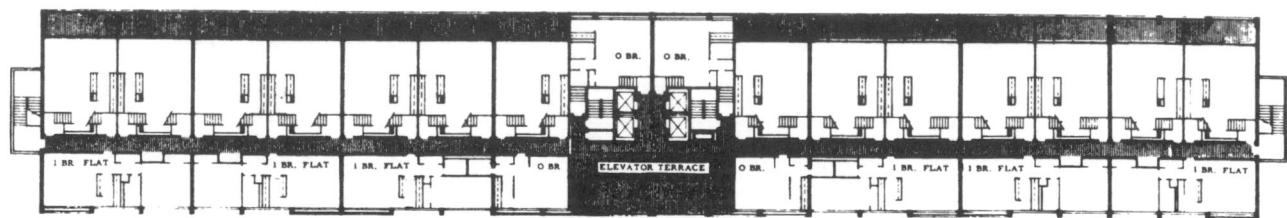

TYPICAL CORRIDOR FLOOR PLAN

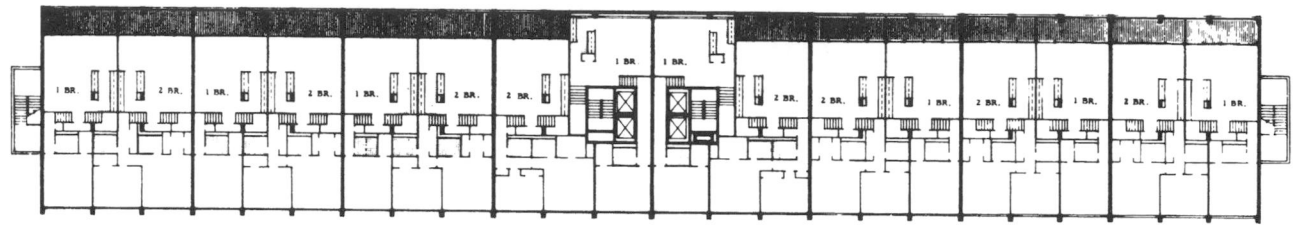

TYPICAL SPLIT LEVEL FLOOR PLAN

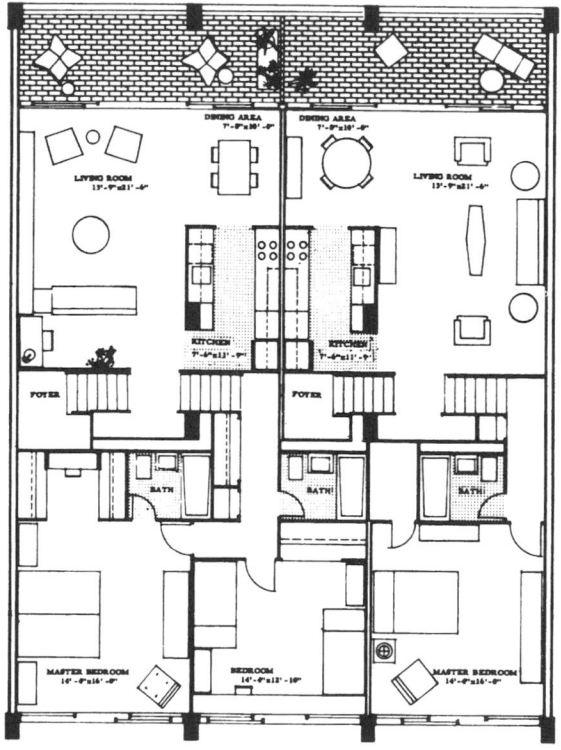

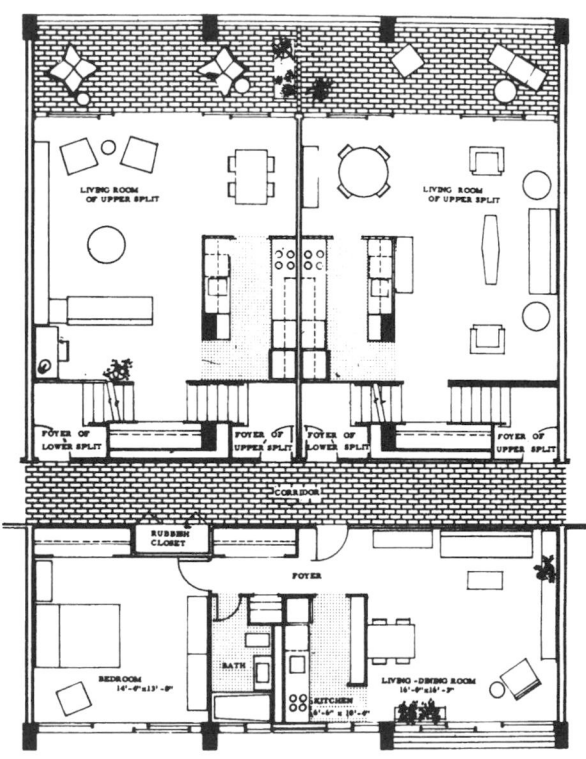

LOWER SPLIT LEVEL
2 BEDROOM UNIT

LOWER SPLIT LEVEL
1 BEDROOM UNIT

1 BEDROOM FLAT
4 3/4 ROOMS

Fig. 12

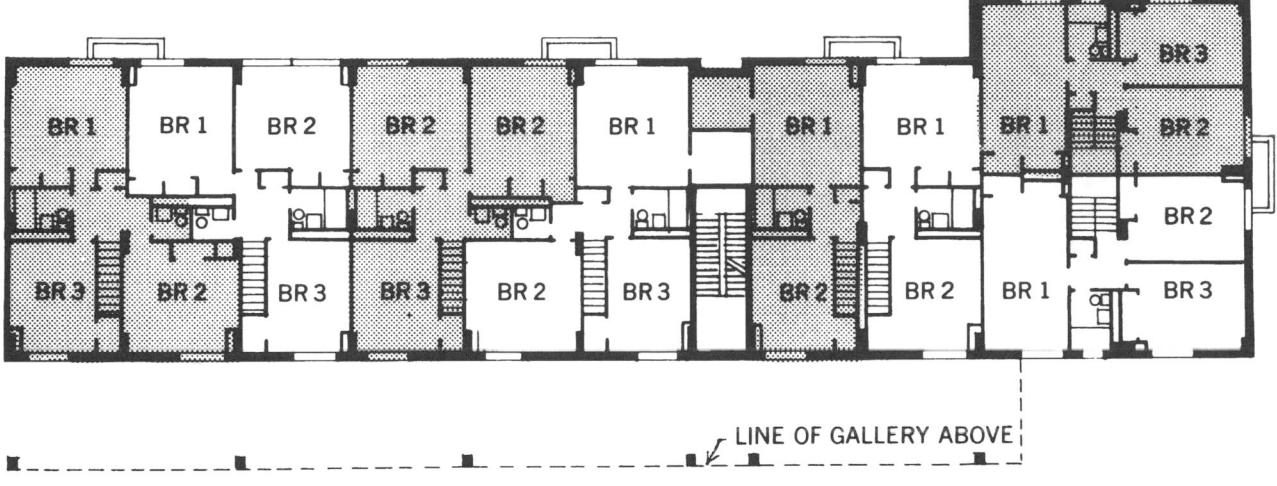

LINE OF GALLERY ABOVE

UPPER LEVEL

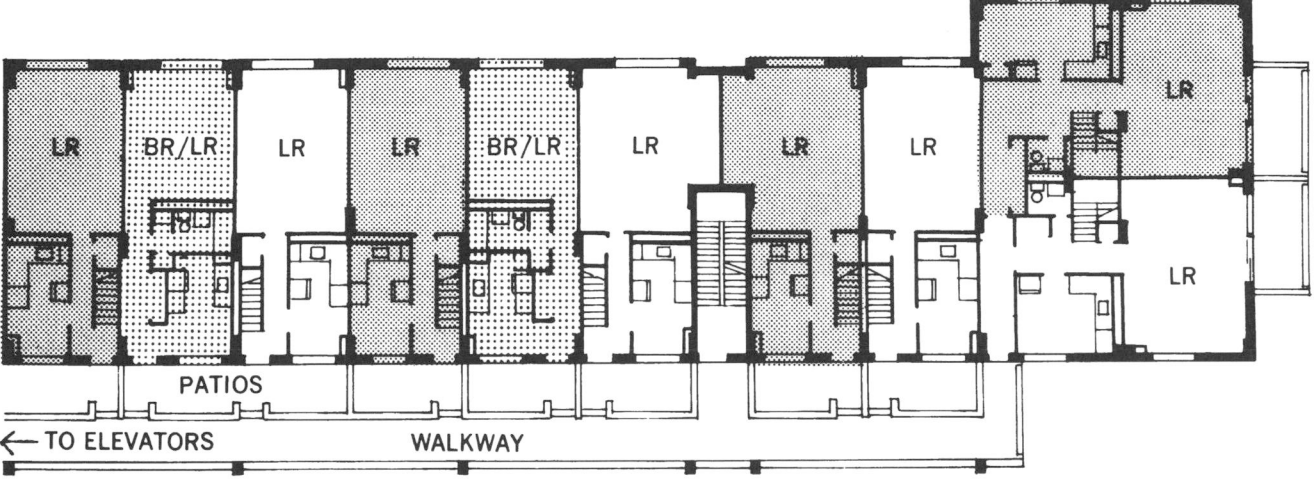

PATIOS

← TO ELEVATORS   WALKWAY

LOWER LEVEL

Fig. 1

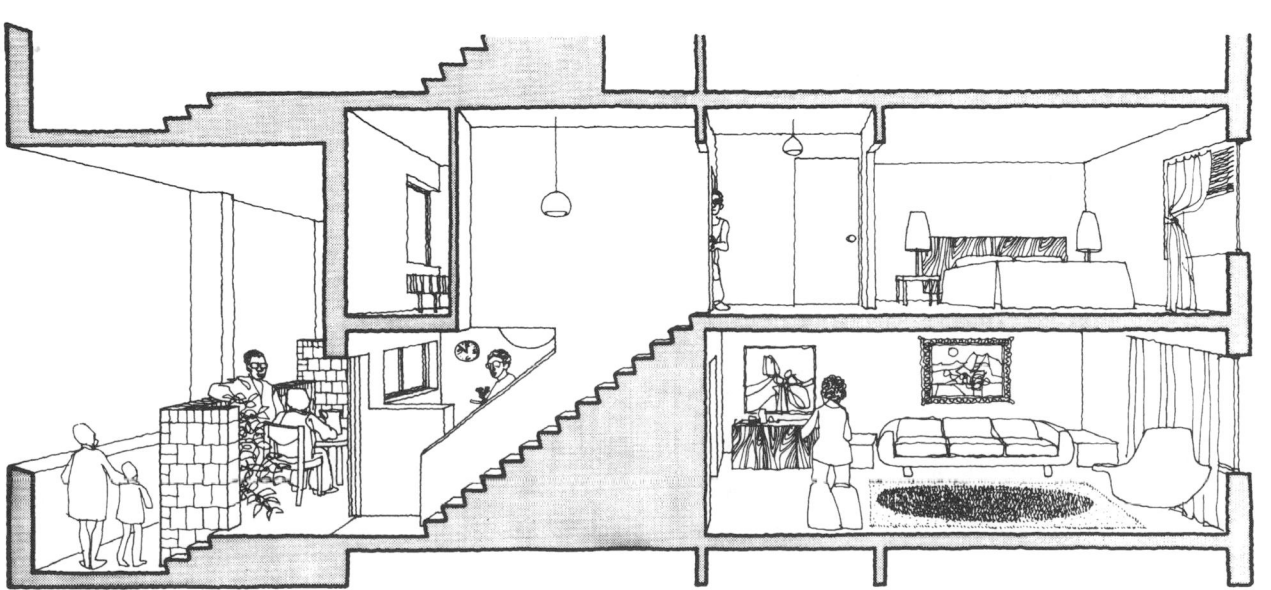

Fig. 2   Riverbend, New York City. Davis, Brody & Associates—Architect.

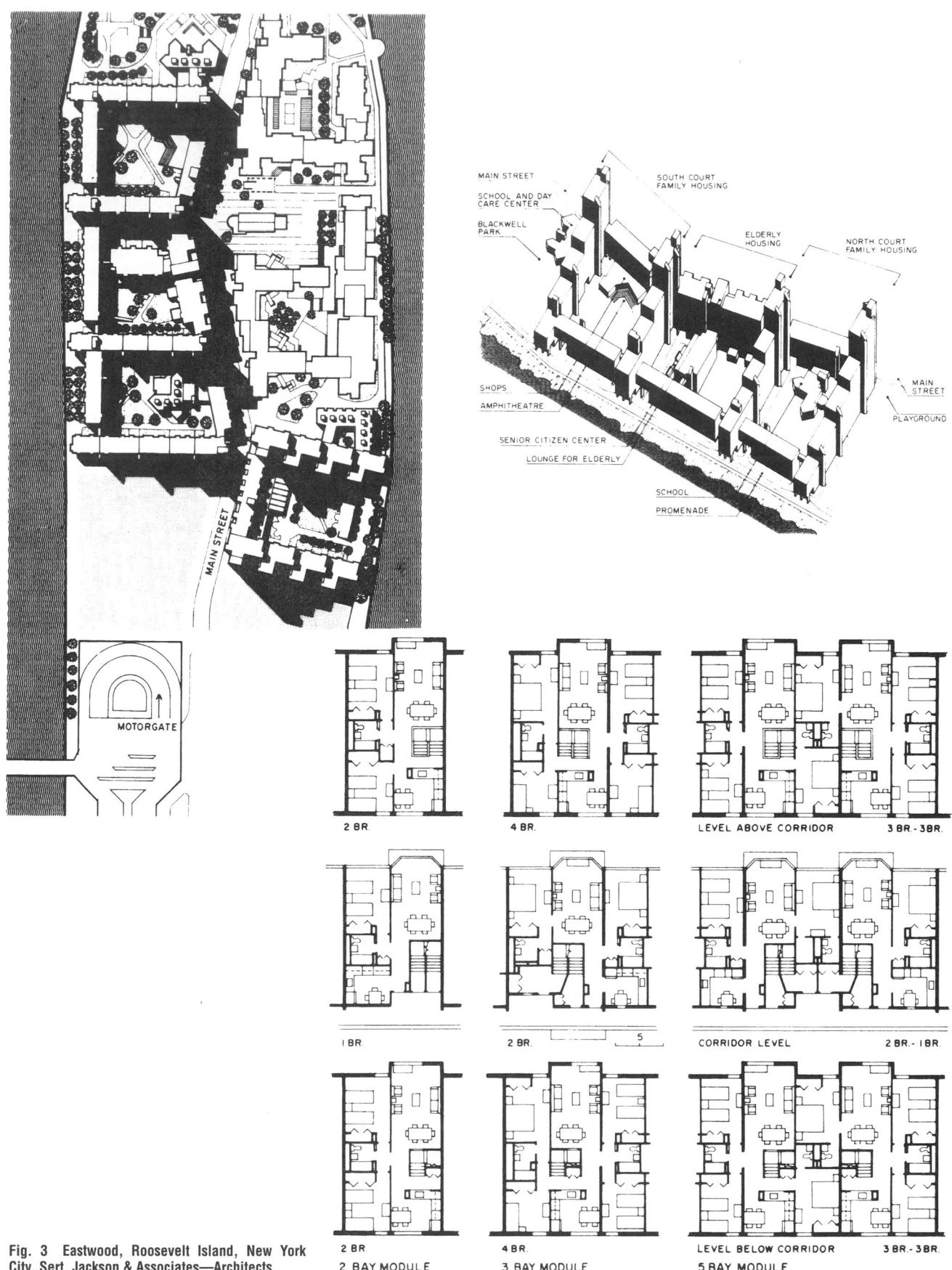

Fig. 3   Eastwood, Roosevelt Island, New York City. Sert, Jackson & Associates—Architects.

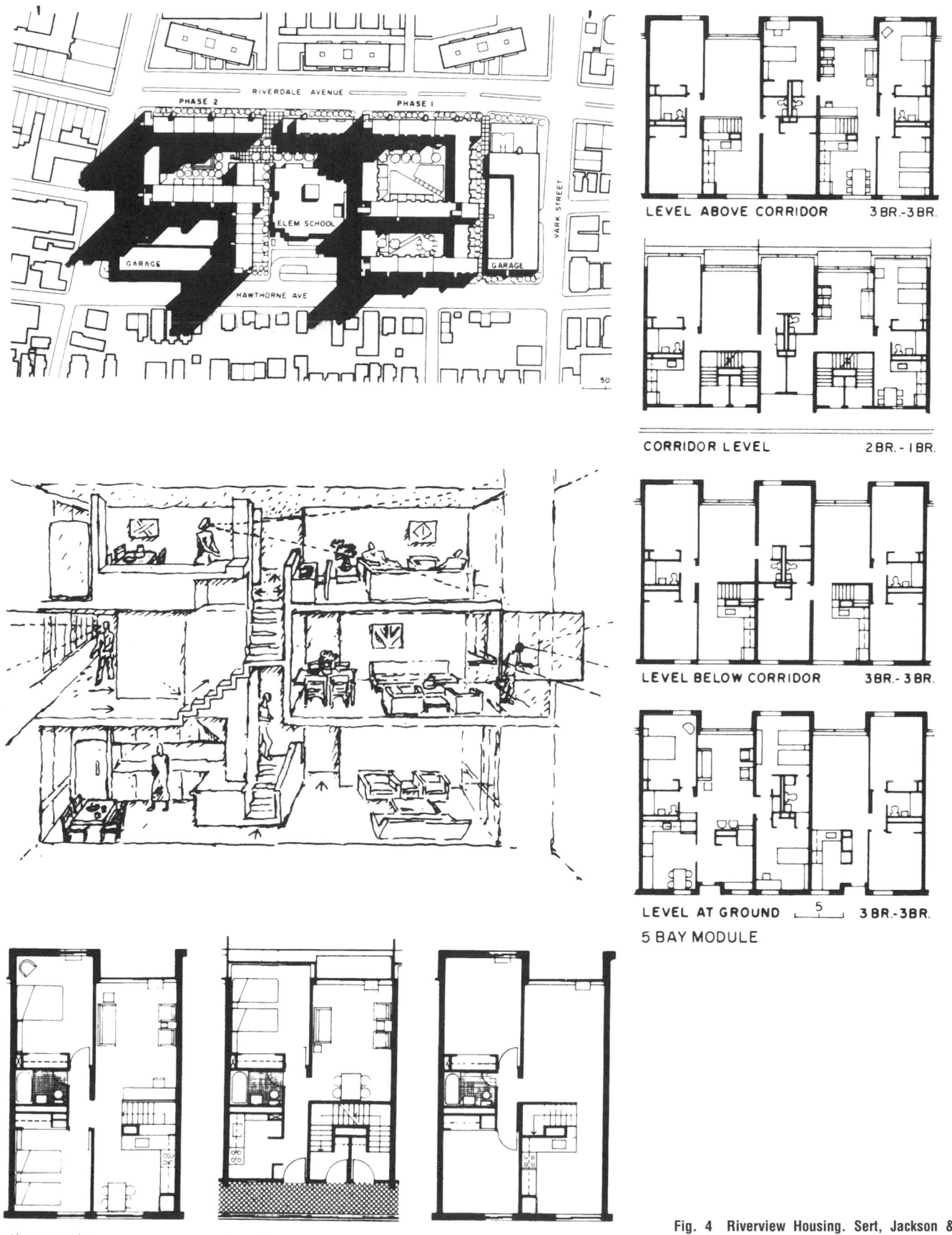

LEVEL ABOVE CORRIDOR   3 BR.-3 BR.

CORRIDOR LEVEL   2 BR.-1 BR.

LEVEL BELOW CORRIDOR   3 BR.-3 BR.

LEVEL AT GROUND   5   3 BR.-3 BR.

5 BAY MODULE

Above corridor

At corridor

Below corridor

Fig. 4 Riverview Housing. Sert, Jackson & Associates—Architects.

TOWER PLAN

The tower scheme consists of a central core with apartments wrapped compactly around it. The overall plan configuration is, or approximates, a square. This provides maximum floor area with minimum exterior perimeter. The usual number of apartments per floor is limited to four or six. The tower scheme has a number of advantages and one serious disadvantage. The most significant advantage is the reduction of lengthy and expensive public corridors. The tower scheme provides cross ventilation and two exposures for each apartment, which enhances its attractiveness and livability. The tower scheme is also advantageous in site planning. The square plan results in a greater feeling of openness than a slab building. It is also easier to situate on an irregular site or a site with topographical difficulties.

The main disadvantage of the tower scheme is the small number of apartments per core. If a typical floor is limited to only four apartments for each elevator and stair core, this becomes inefficient. Most often the same core can serve up to eight or ten apartments per floor. This

drawback has usually restricted the tower to middle-income or luxury development.

Another minor disadvantage is that one side of the building usually is facing north, which is not ideal. However, since each apartment has two exposures, this is not critical.

The project consists of three 28-story buildings containing a total of 738 apartments covering only 23 percent of the site. The remainder of the land is used for landscaping, sitting, and play areas. Multilevel garages placed between the buildings contain more than adequate parking facilities and further develop level areas for recreation on this steeply sloping site. A swimming pool and children's wading pool are provided. Magnificent views of the New York skyline and surrounding Long Island are obtained from the upper stories (see Fig. 15).

The apartment towers, planned around a central core, contain 10 units per floor for a total of 738 families in three buildings. The layout is typical for each floor except at the top 10 stories, where a two-bedroom and a one-bedroom unit are combined to form a three-bedroom unit and an efficiency.

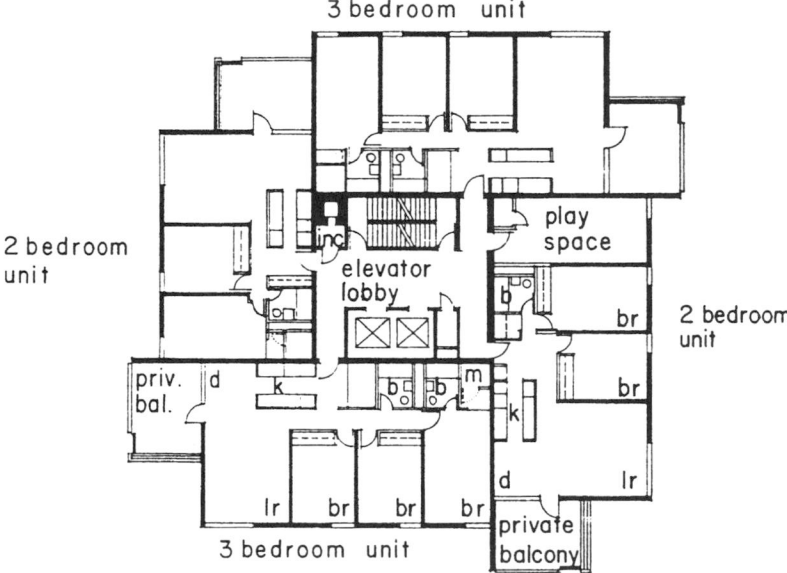

Fig. 1 Tower scheme—typical floor plan. This scheme offers the amenity of corner rooms, cross ventilation, and short corridors, but is less efficient than the double-loaded corridor plan.

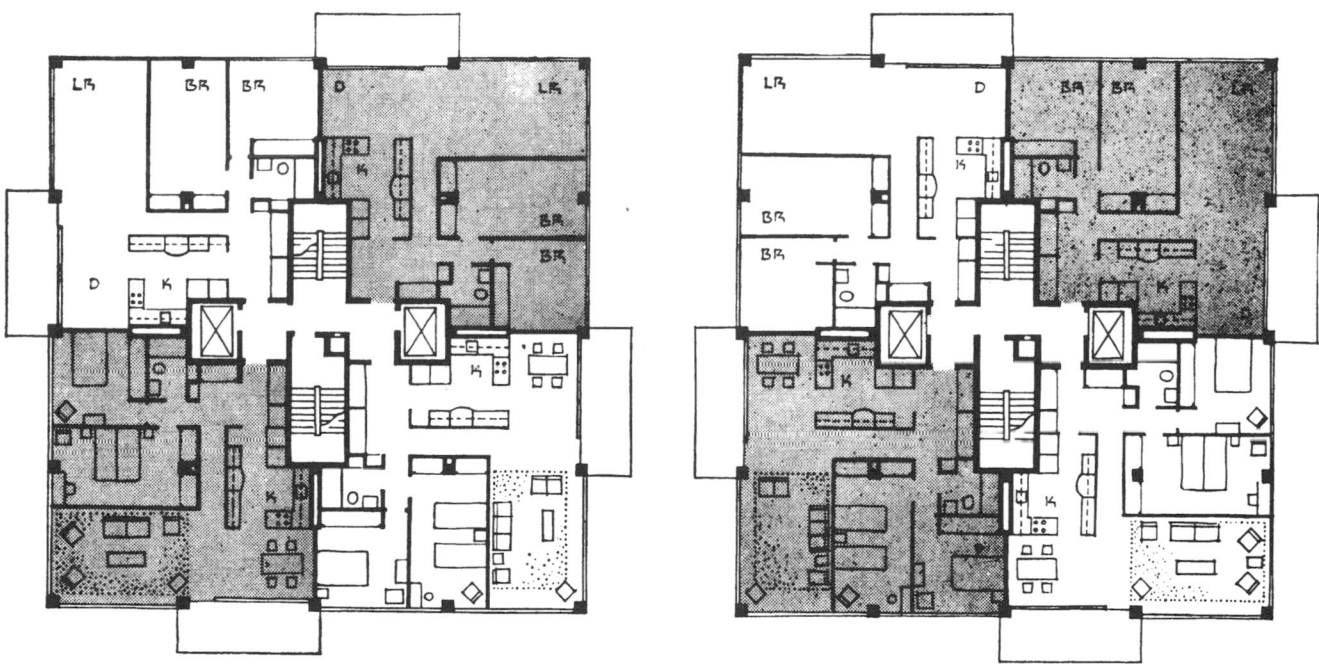

Fig. 2   Typical alternating upper floors.

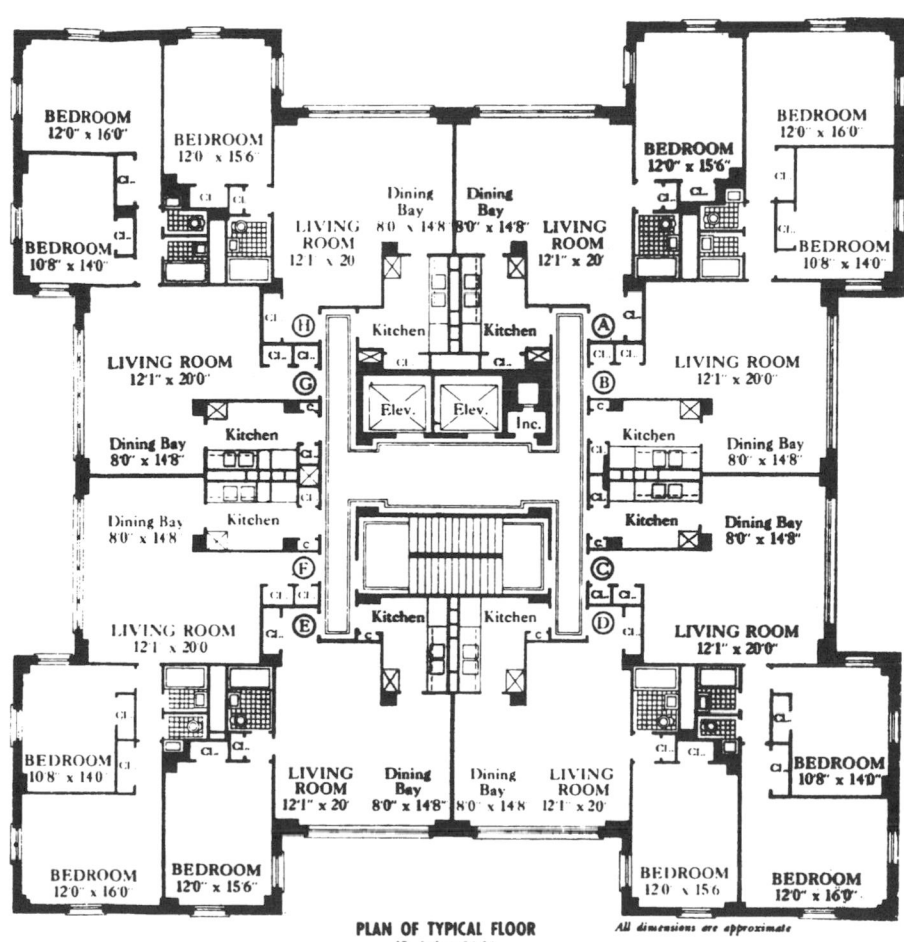

PLAN OF TYPICAL FLOOR
(2nd thru 16th)

All dimensions are approximate

Fig. 3   Fordham Hill, Bronx, N.Y. Schultze & Associates—Architects.

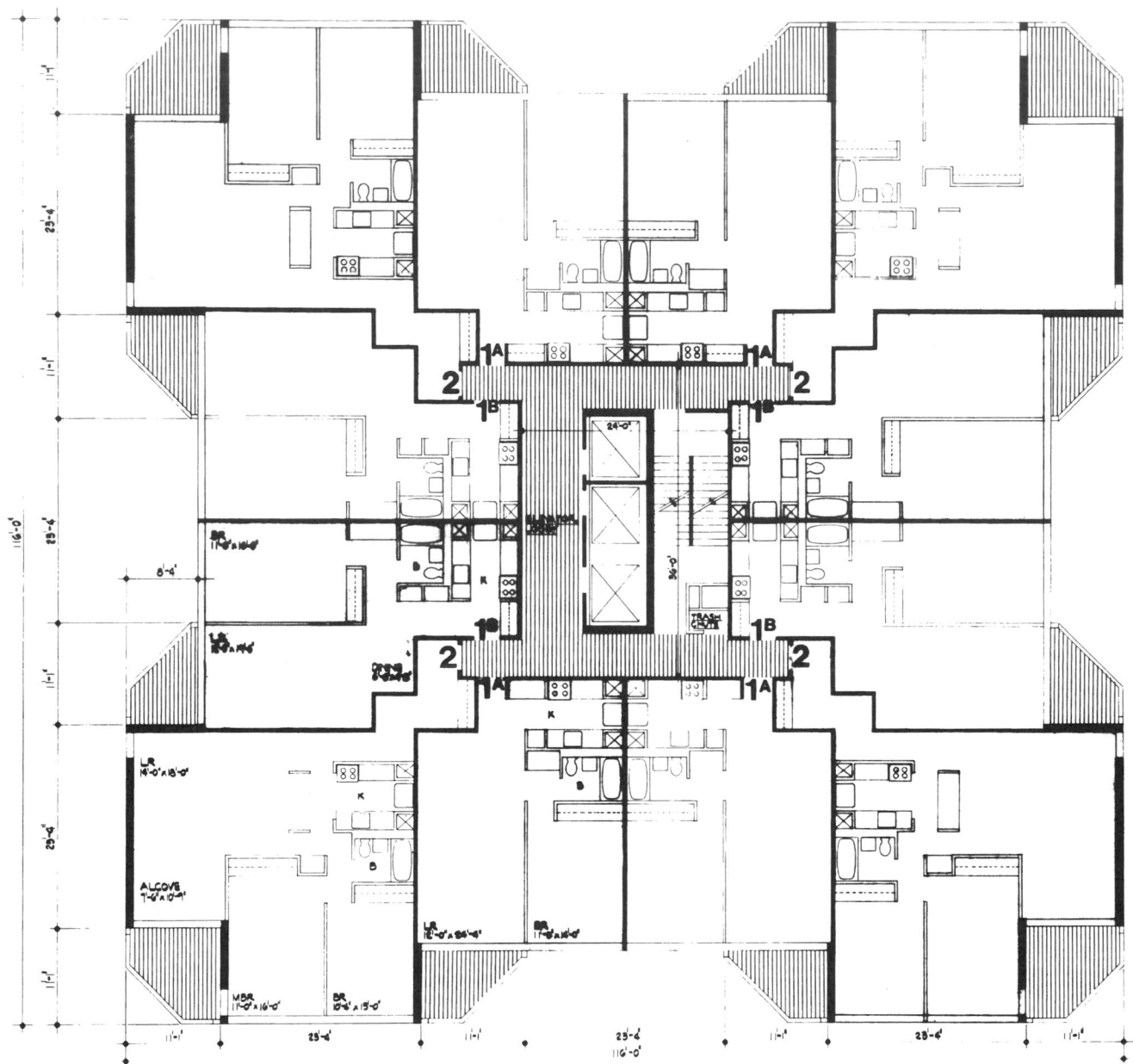

Fig. 4   Seven Pines, Yonkers, N.Y. Gruzen & Partners—Architects.

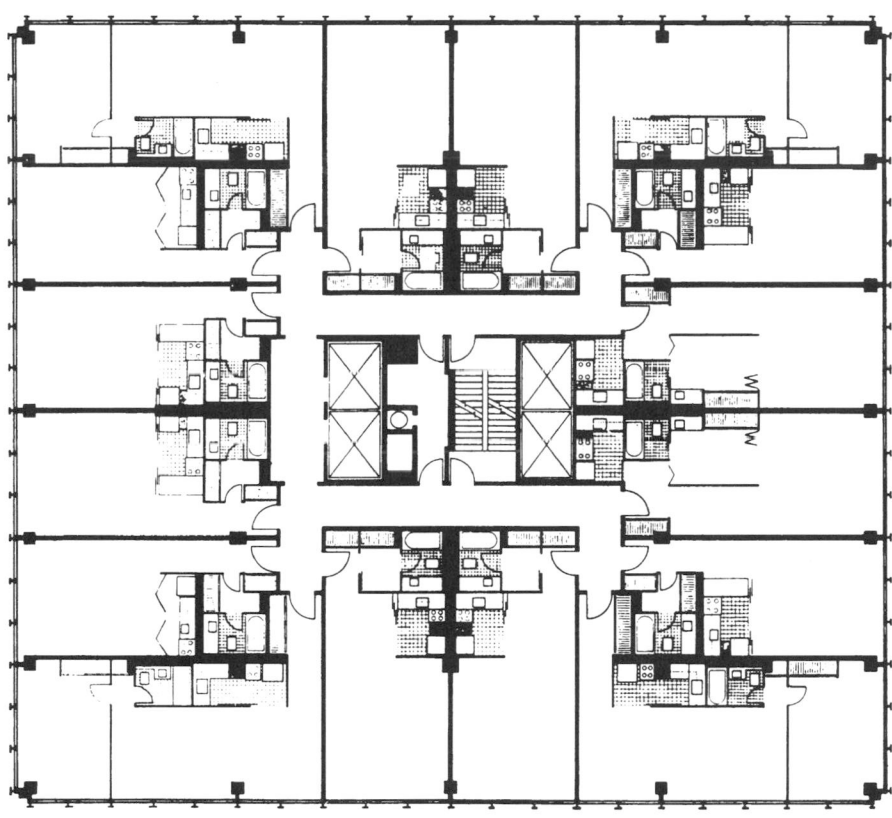

Fig. 5 Typical floor plan. Lake View, Chicago, Ill. Mies van der Rohe—Architect.

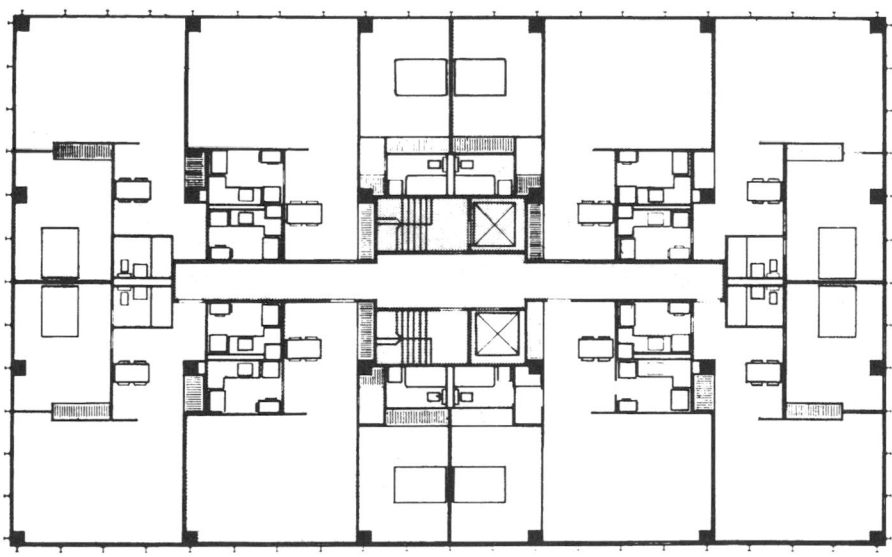

Fig. 6 Typical floor plan. Lake Shore Drive, Chicago, Ill. Mies van der Rohe—Architect.

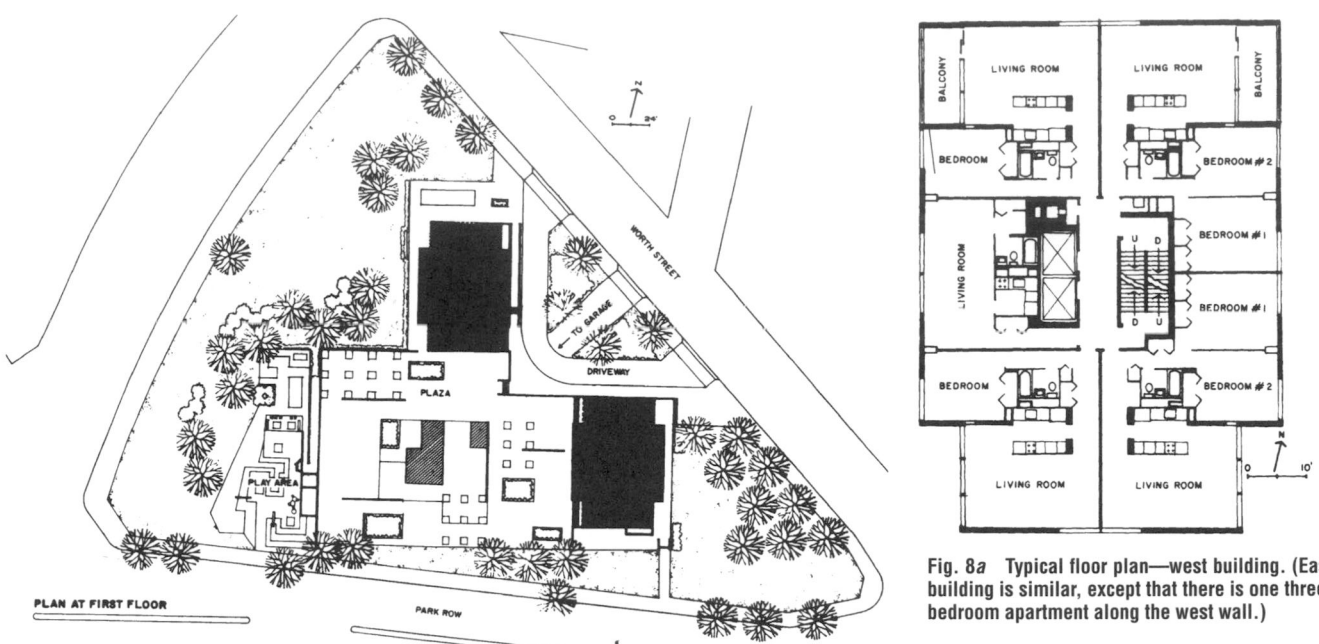

TOWER PLAN

WORTH STREET
TO GARAGE
DRIVEWAY
PLAZA
PLAY AREA
PLAN AT FIRST FLOOR
PARK ROW

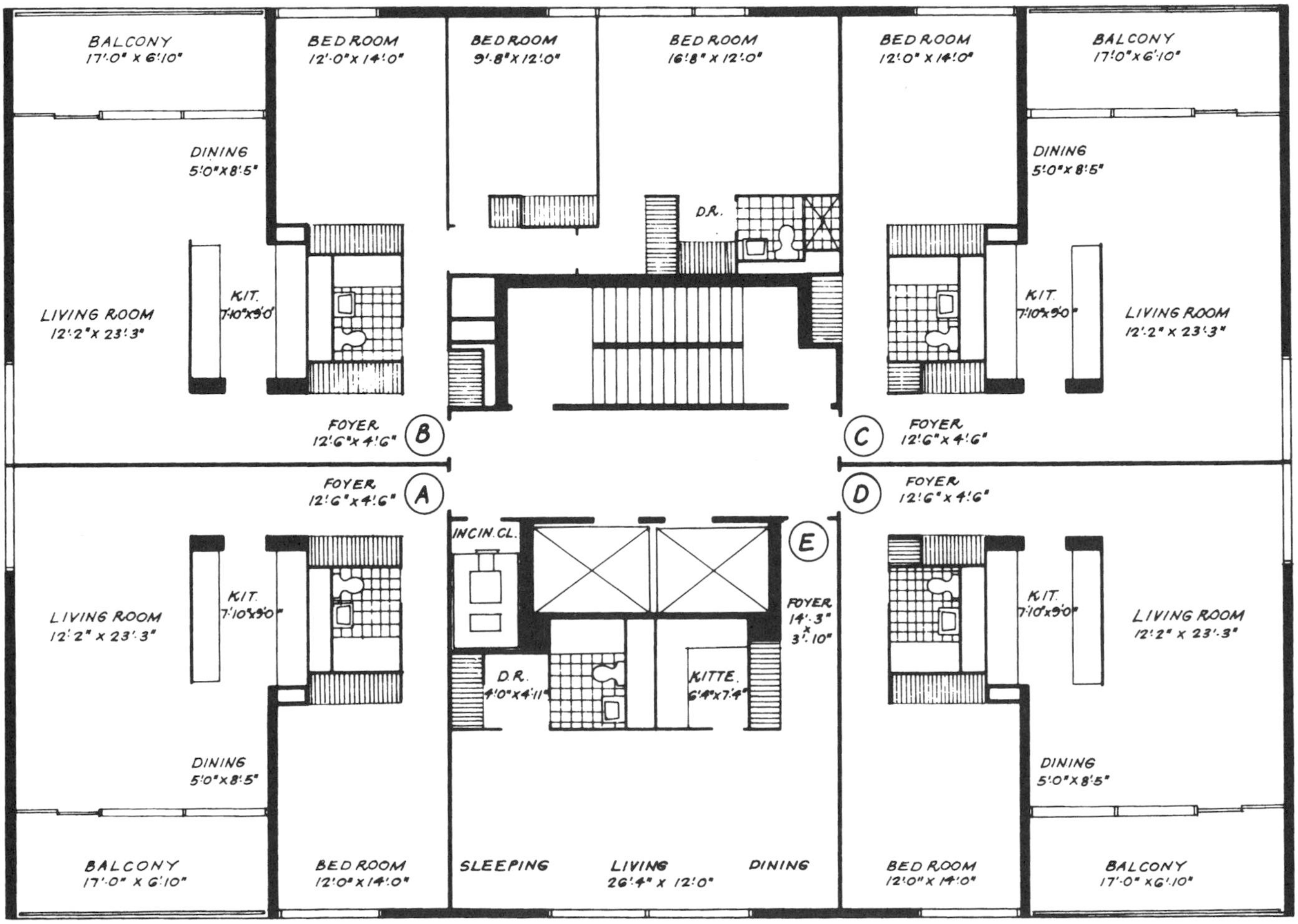

Fig. 7 Chatham Towers, New York City. Kelley & Gruzen—Architects.

BALCONY
LIVING ROOM
LIVING ROOM
BALCONY
BEDROOM
BEDROOM #2
BEDROOM #1
LIVING ROOM
BEDROOM #1
BEDROOM
BEDROOM #2
LIVING ROOM
LIVING ROOM

Fig. 8a Typical floor plan—west building. (East building is similar, except that there is one three-bedroom apartment along the west wall.)

BALCONY 17'-0" x 6'-10"
BED ROOM 12'-0" x 14'-0"
BED ROOM 9'-8" x 12'-0"
BED ROOM 16'-8" x 12'-0"
BED ROOM 12'-0" x 14'-0"
BALCONY 17'-0" x 6'-10"
DINING 5'-0" x 8'-5"
D.R.
DINING 5'-0" x 8'-5"
KIT. 7'-10" x 9'-0"
LIVING ROOM 12'-2" x 23'-3"
KIT. 7'-10" x 9'-0"
LIVING ROOM 12'-2" x 23'-3"
FOYER 12'-6" x 4'-6"
B
C
FOYER 12'-6" x 4'-6"
FOYER 12'-6" x 4'-6"
A
D
FOYER 12'-6" x 4'-6"
INCIN. CL.
E
LIVING ROOM 12'-2" x 23'-3"
KIT. 7'-10" x 9'-0"
FOYER 14'-3" x 3'-10"
KIT. 7'-10" x 9'-0"
LIVING ROOM 12'-2" x 23'-3"
D.R. 4'-0" x 4'-11"
KITTE. 6'-4" x 7'-4"
DINING 5'-0" x 8'-5"
DINING 5'-0" x 8'-5"
BALCONY 17'-0" x 6'-10"
BED ROOM 12'-0" x 14'-0"
SLEEPING
LIVING 26'-4" x 12'-0"
DINING
BED ROOM 12'-0" x 14'-0"
BALCONY 17'-0" x 6'-10"

Fig. 8b Plan, Tower No. 2.

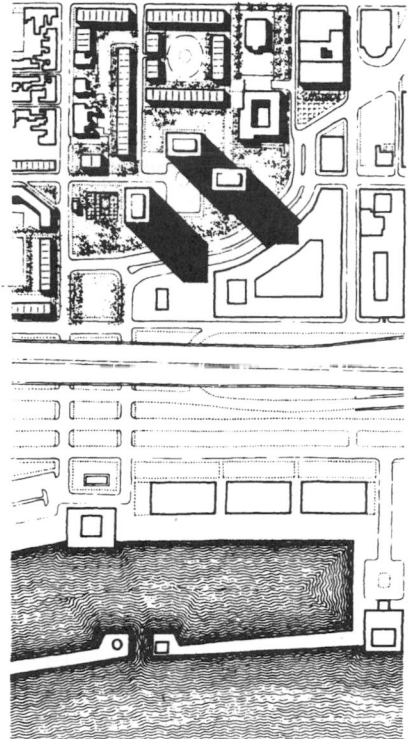

Fig. 9

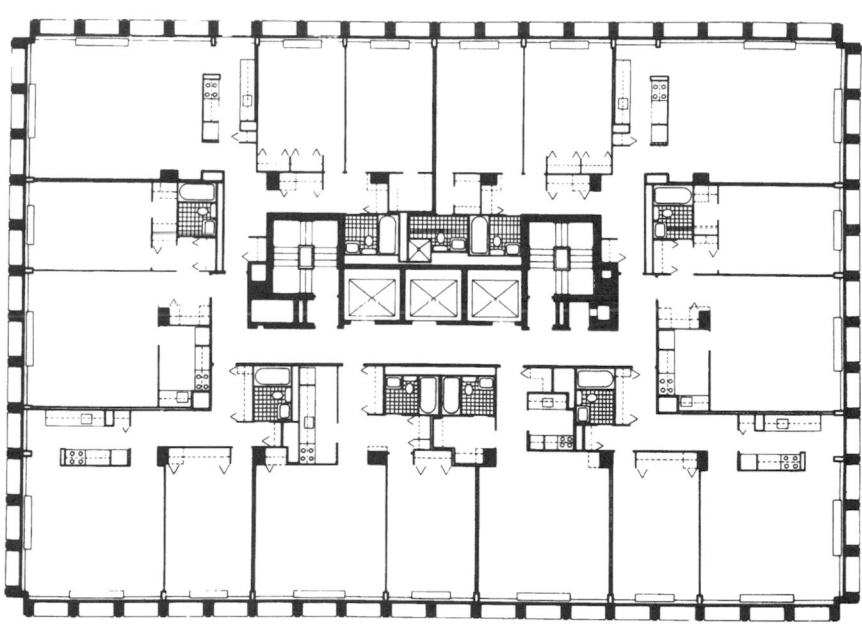

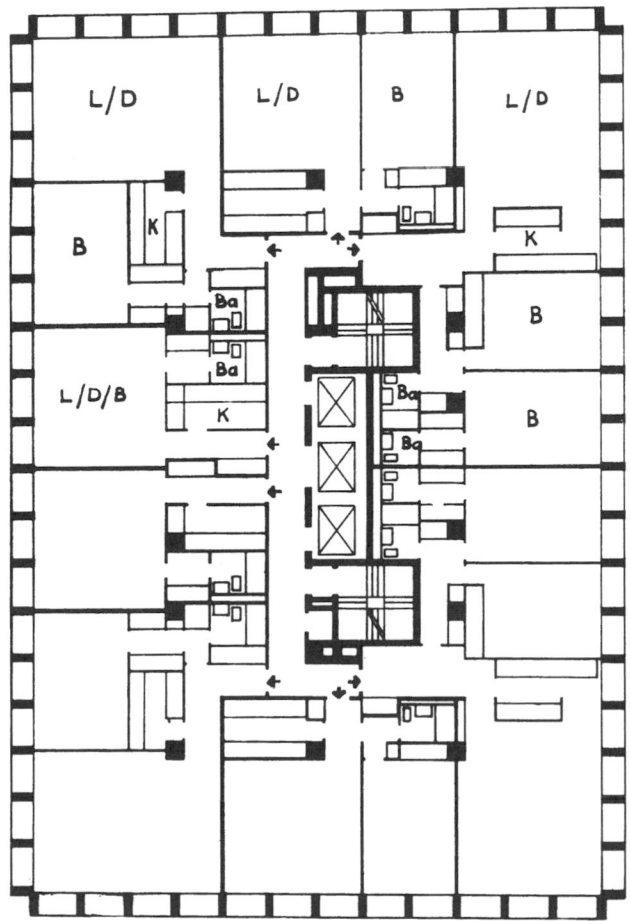

Fig. 10   Typical floor plan. Washington Sq. East, Philadelphia, Pa. I. M. Pei & Partners—Architects.

TOWER PLAN

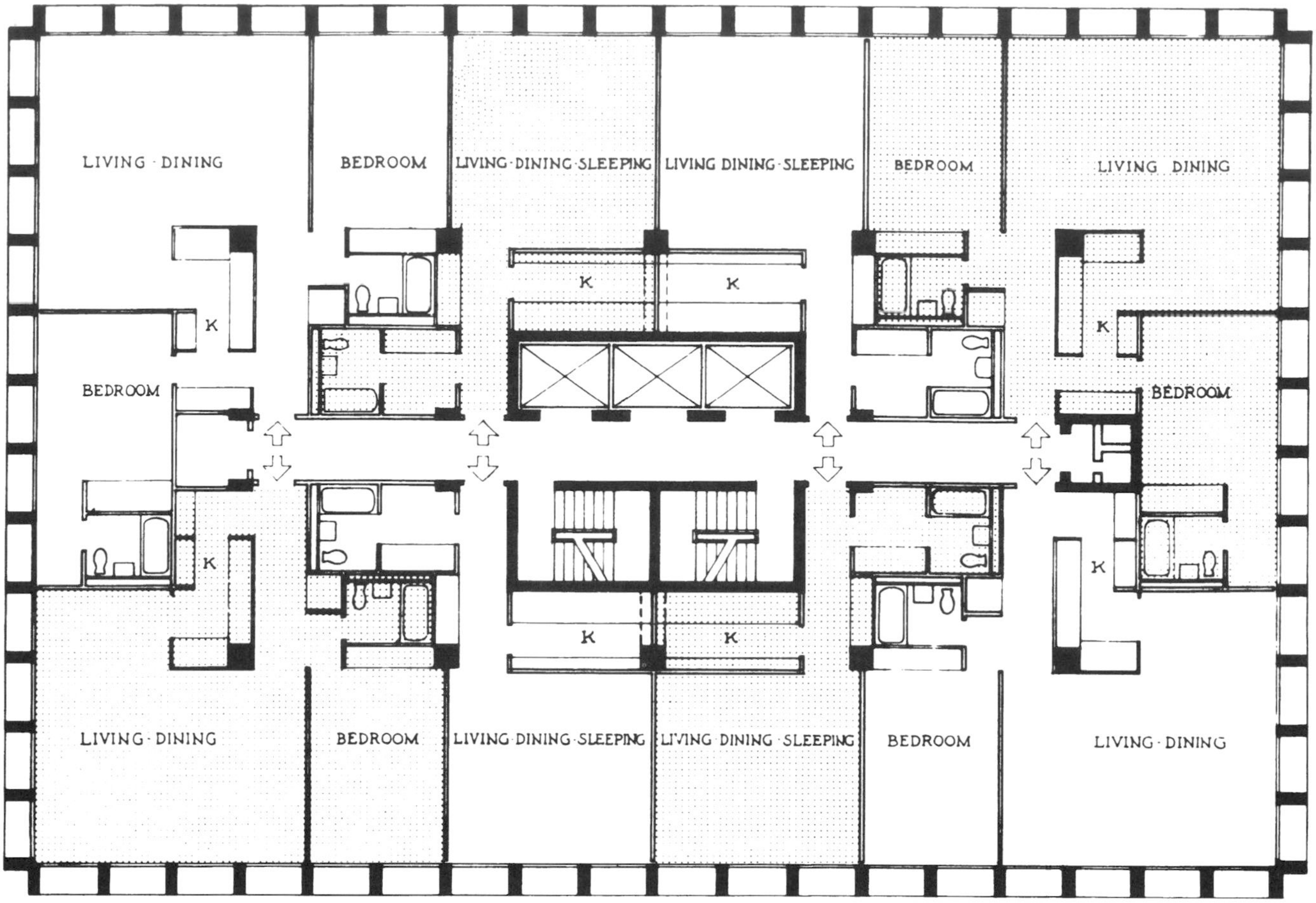

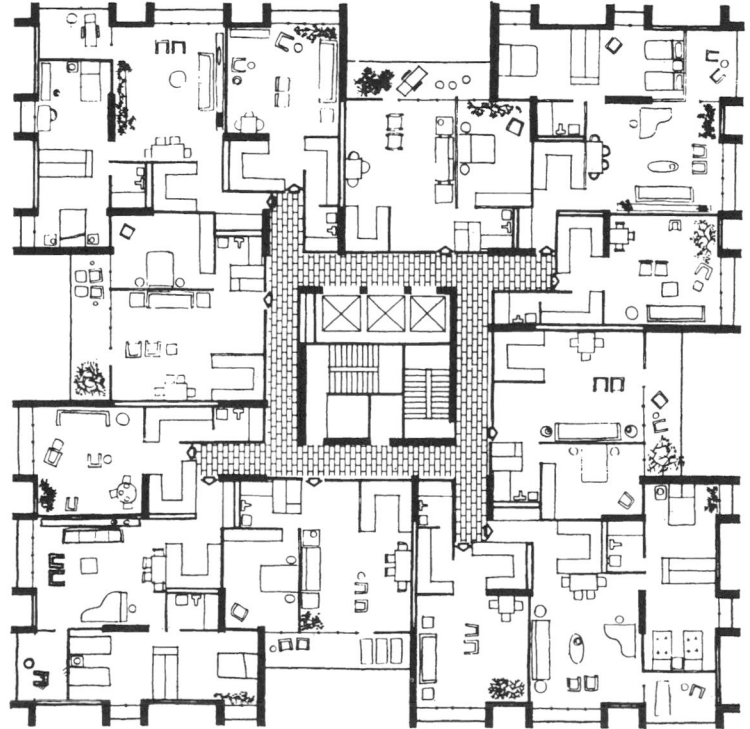

Fig. 11 Society Hill, Philadelphia, Pa. I. M. Pei & Partners—Architects.

Fig. 12 St. Louis Hills, St. Louis, Mo. Hellmuth, Obata & Kassabaum—Architects.

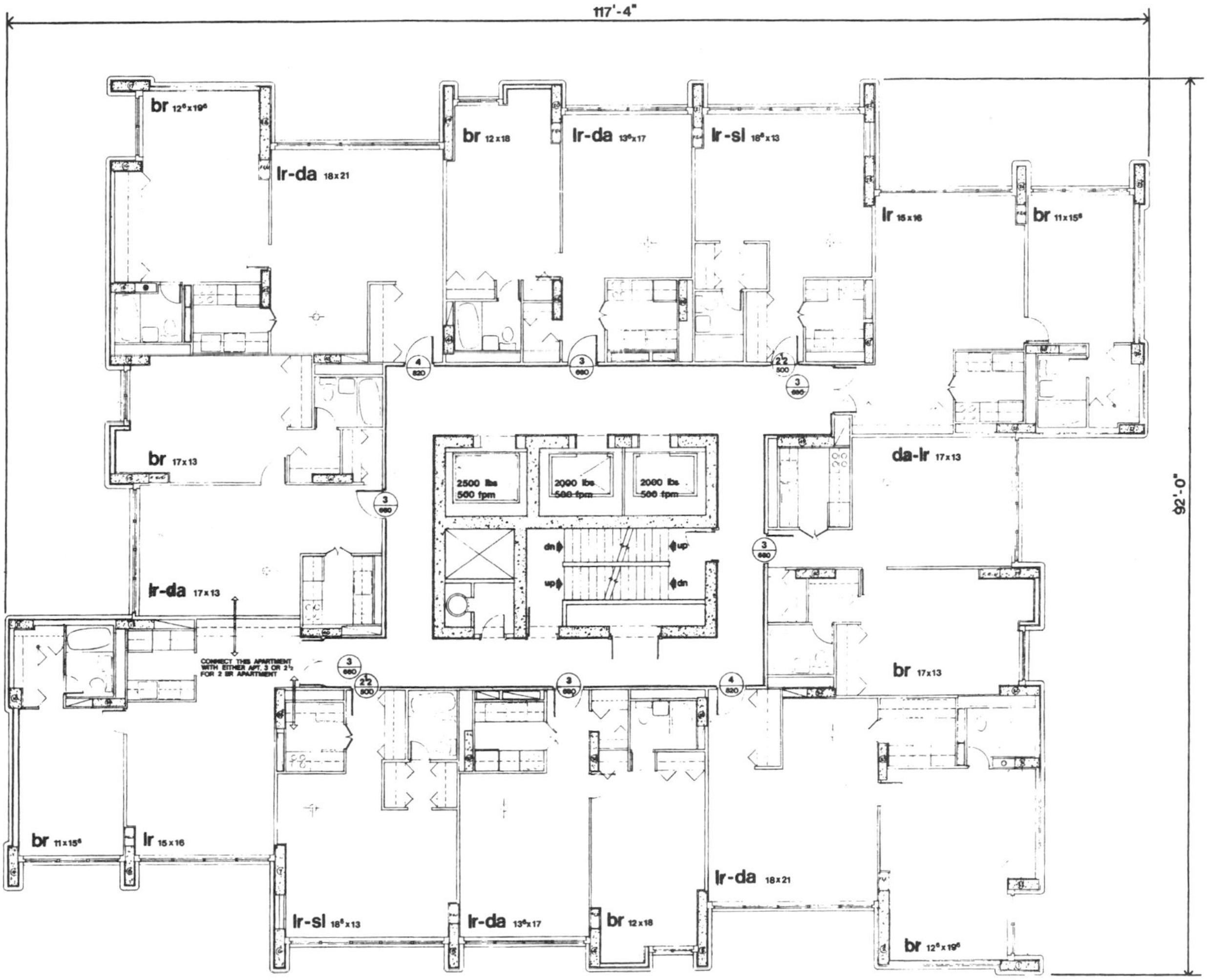

117'-4"

92'-0"

br 12⁶x19⁶

lr-da 18x21

br 12x18

lr-da 13⁶x17

lr-sl 16⁶x13

lr 15x16

br 11x15⁶

br 17x13

da-lr 17x13

2500 lbs 500 fpm

2000 lbs 500 fpm

2000 lbs 500 fpm

dn

up

up

dn

lr-da 17x13

br 17x13

CONNECT THIS APARTMENT WITH EITHER APT. 3 OR 2½ FOR 2 BR APARTMENT

br 11x15⁶

lr 15x16

lr-da 18x21

lr-sl 16⁶x13

lr-da 13⁶x17

br 12x18

br 12⁶x19⁶

Fig. 13   Conklin & Rossant—Architects.

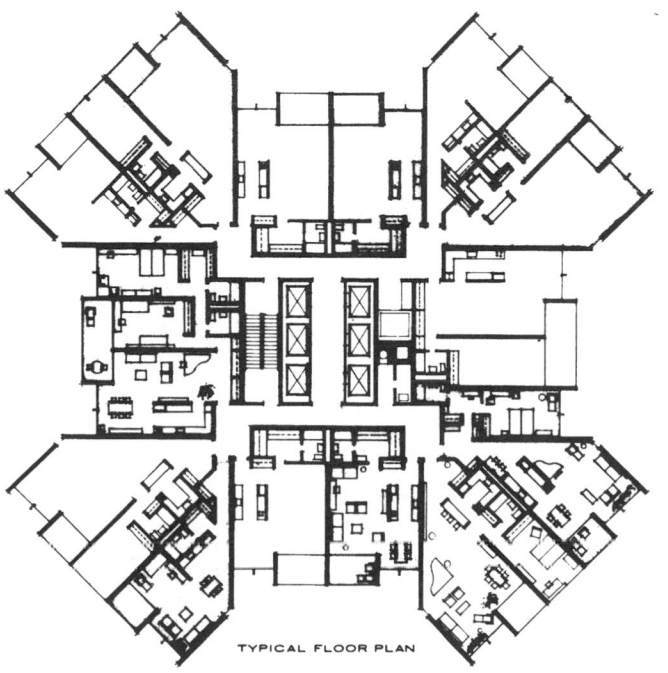

TYPICAL FLOOR PLAN

Fig. 14   Harbour House, St. George, Staten Island, N.Y. Kelley & Gruzen—Architects.

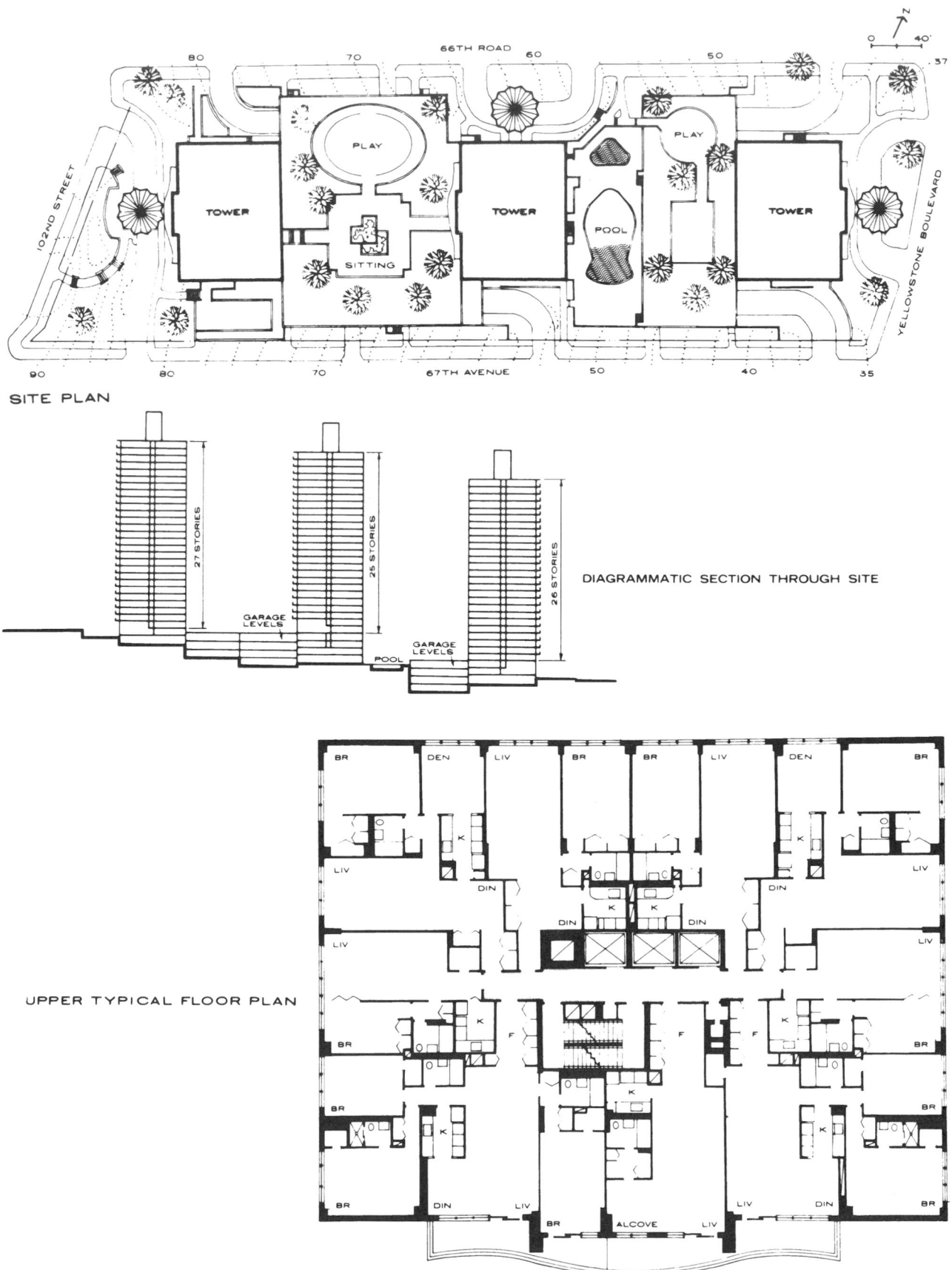

SITE PLAN

DIAGRAMMATIC SECTION THROUGH SITE

UPPER TYPICAL FLOOR PLAN

Fig. 15 Birchwood Towers, Forest Hills, New York City. Paul & Jarmul—Architects.

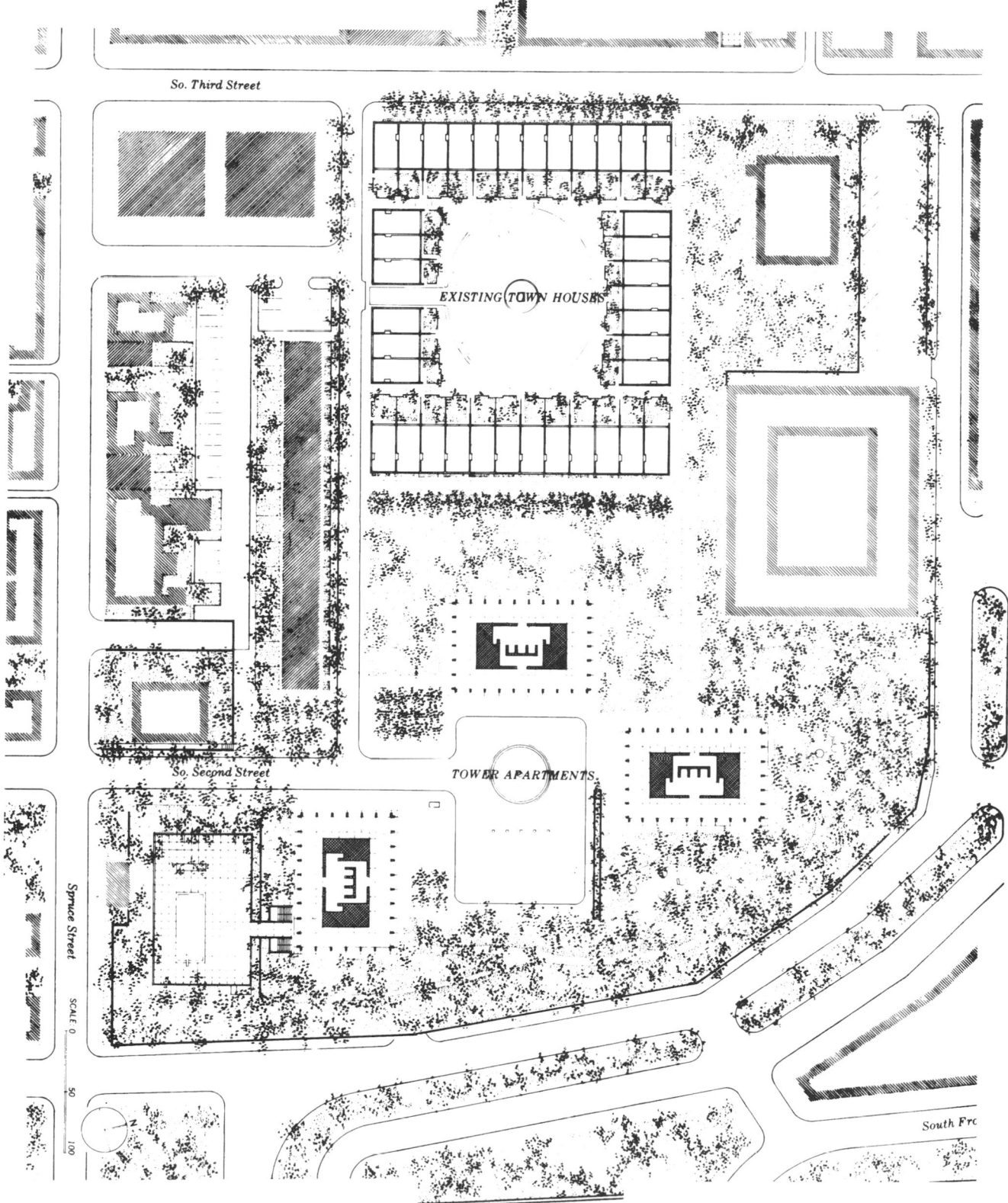

Fig. 16   Washington Square East (Society Hill), Philadelphia, Pa. I. M. Pei & Partners—Architects.

TOWER PLAN

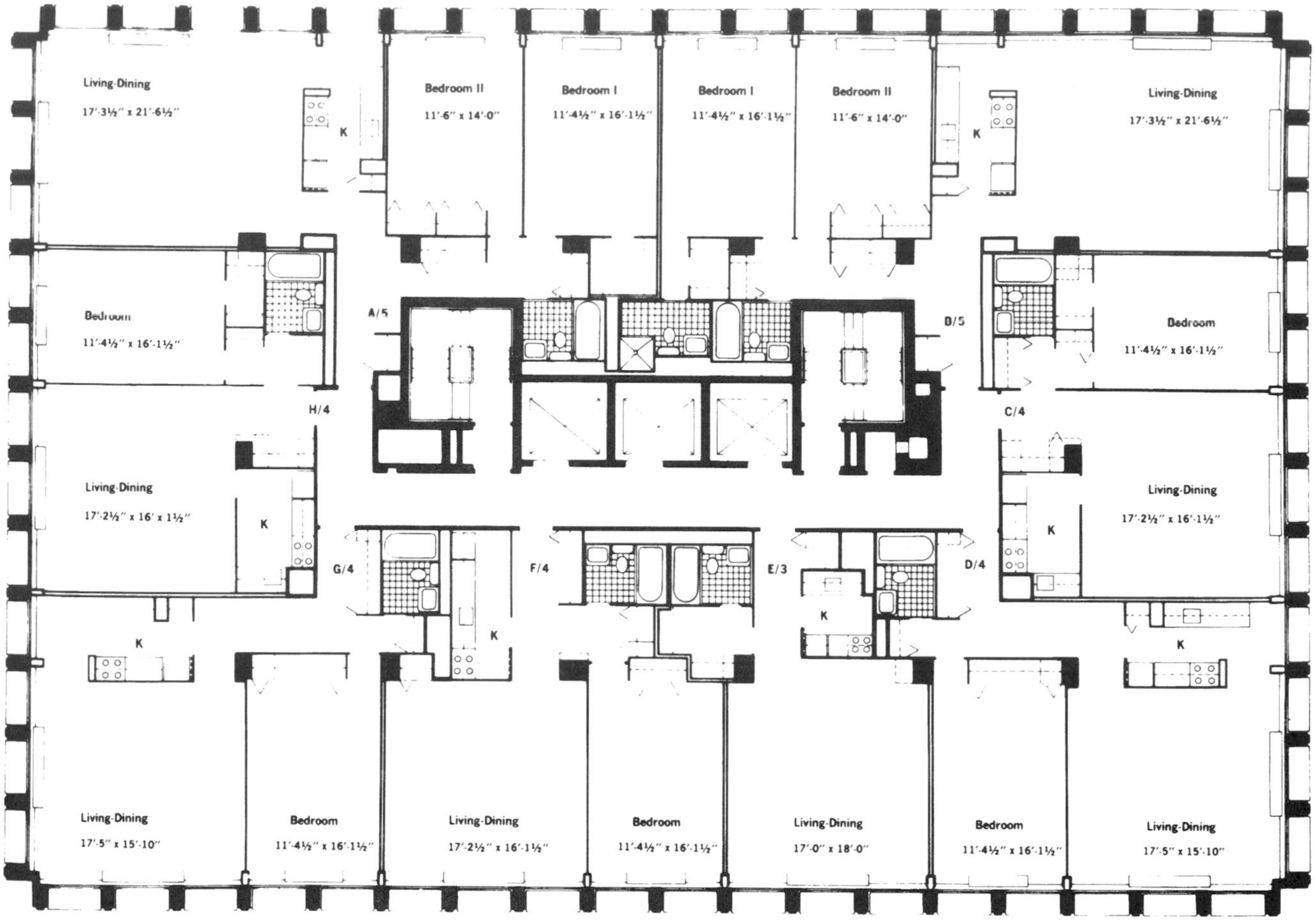

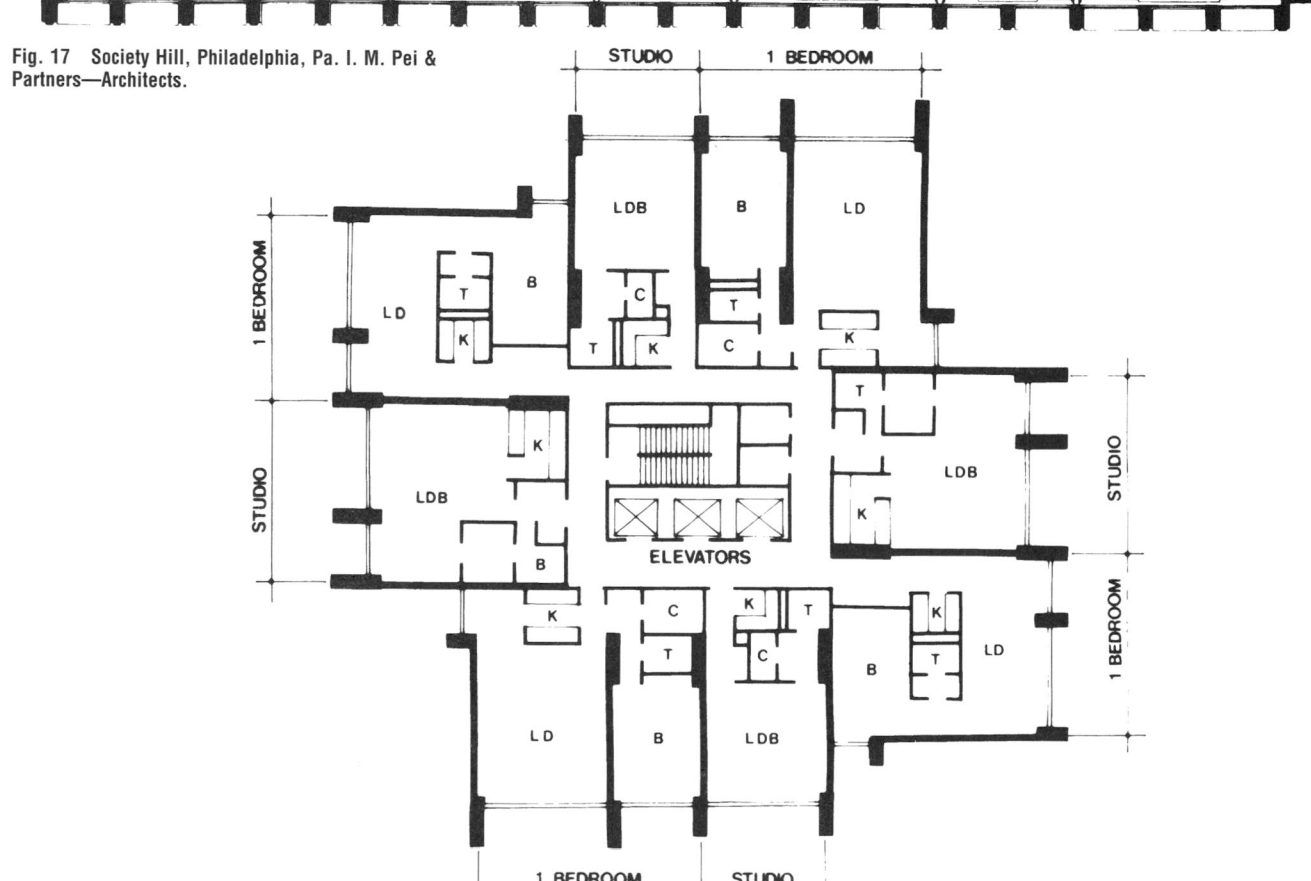

Fig. 17 Society Hill, Philadelphia, Pa. I. M. Pei & Partners—Architects.

Fig. 18 Charles Tower. Conklin & Rossant—Architects.

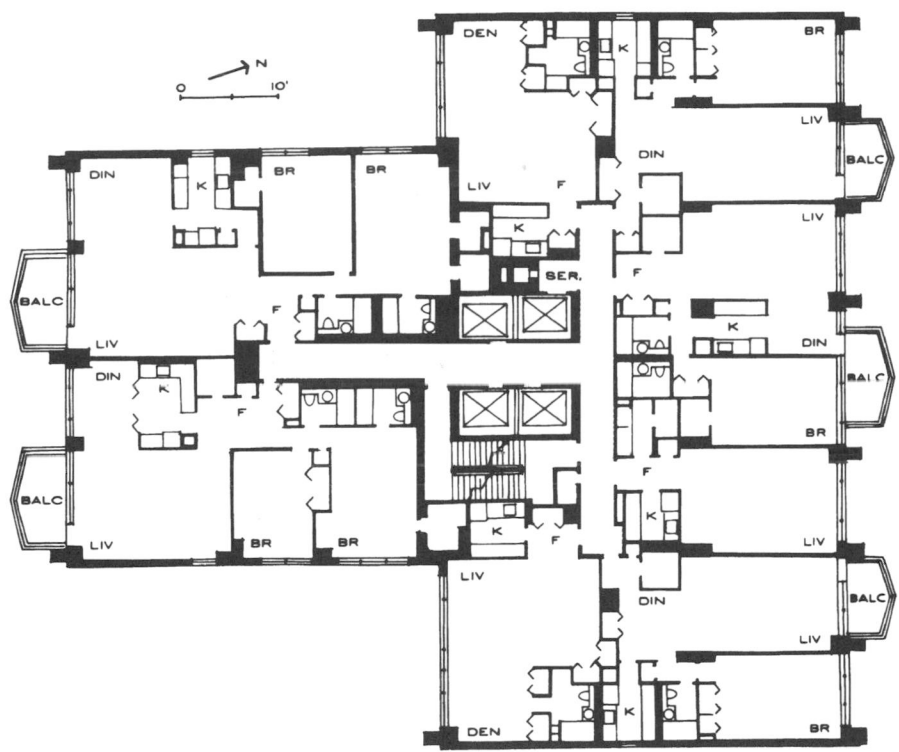

TYPICAL FLOOR PLAN-4TH THROUGH 21ST FLOOR

Fig. 19   Plaza Towers, New York City. Paul & Jarmul—Architects.

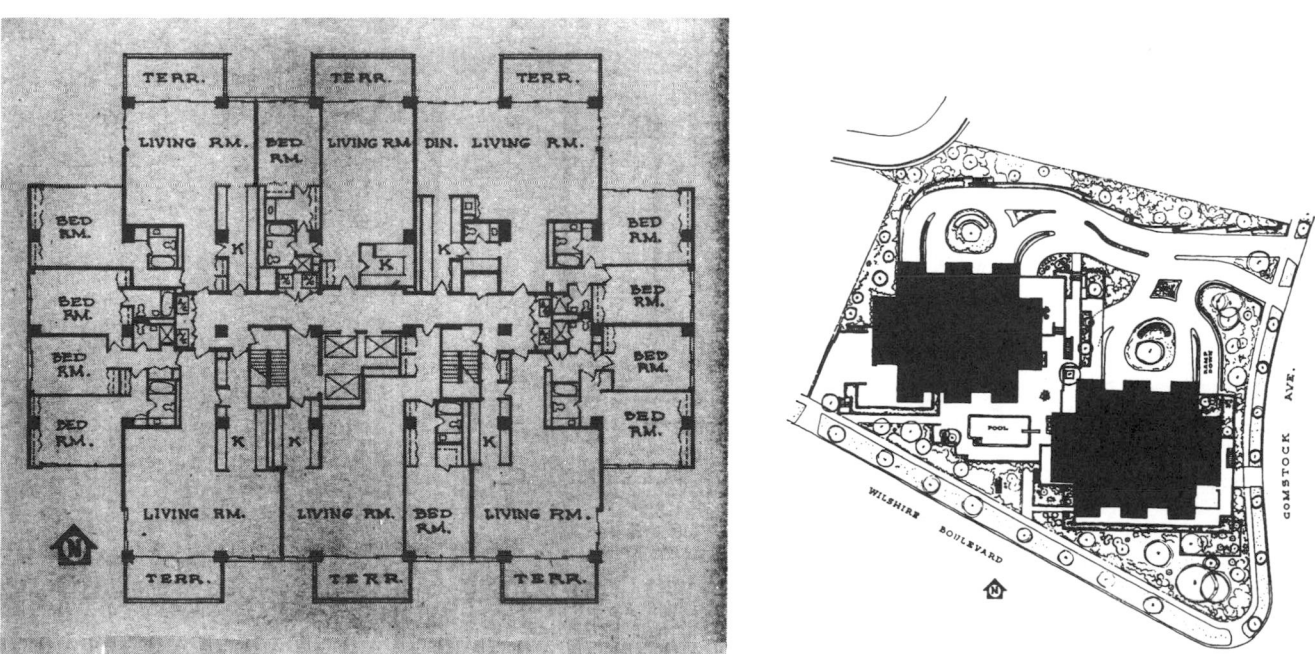

Fig. 20   Wilshire Comstock Apartments. Los Angeles, Calif. Victor Gruzen Associates—Architects.

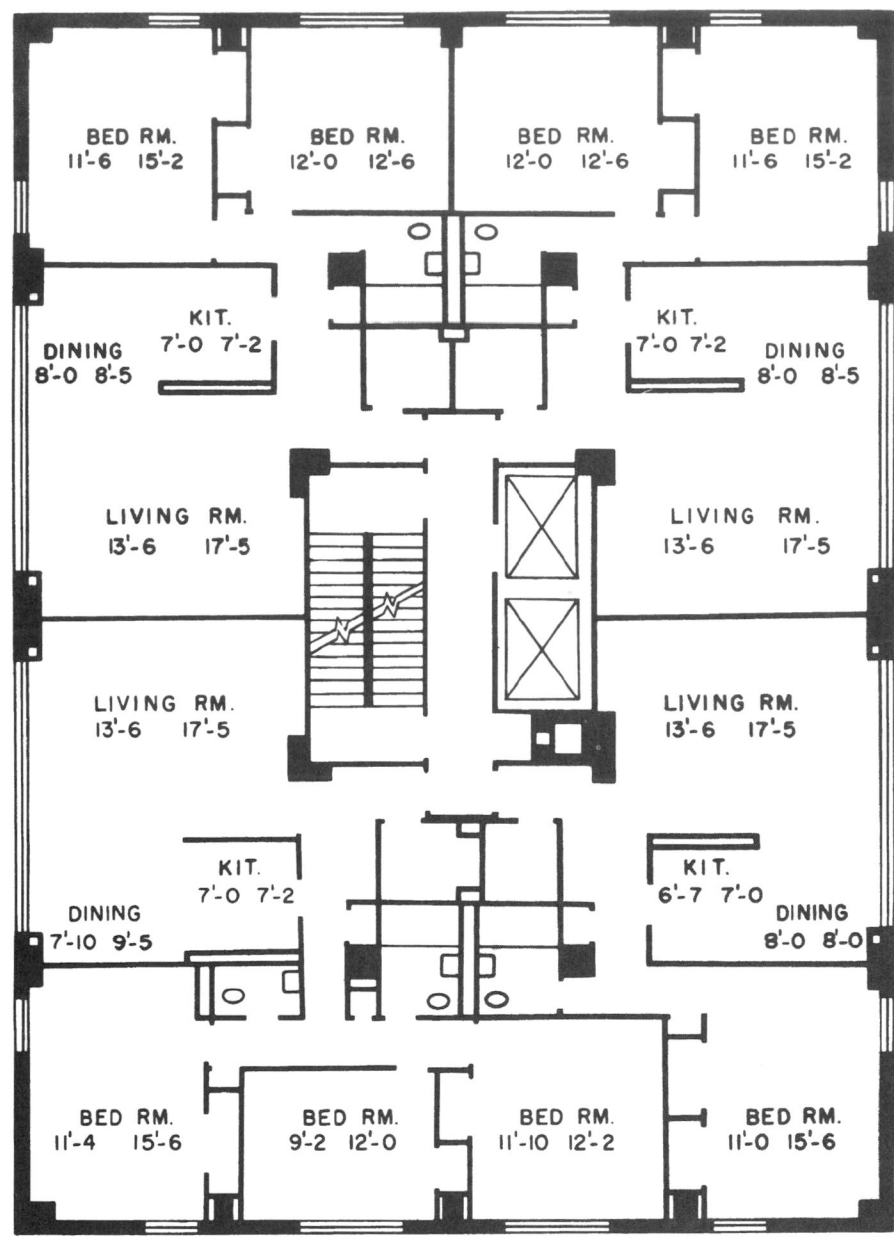

Fig. 21 Typical floor plan of tower apartment building.

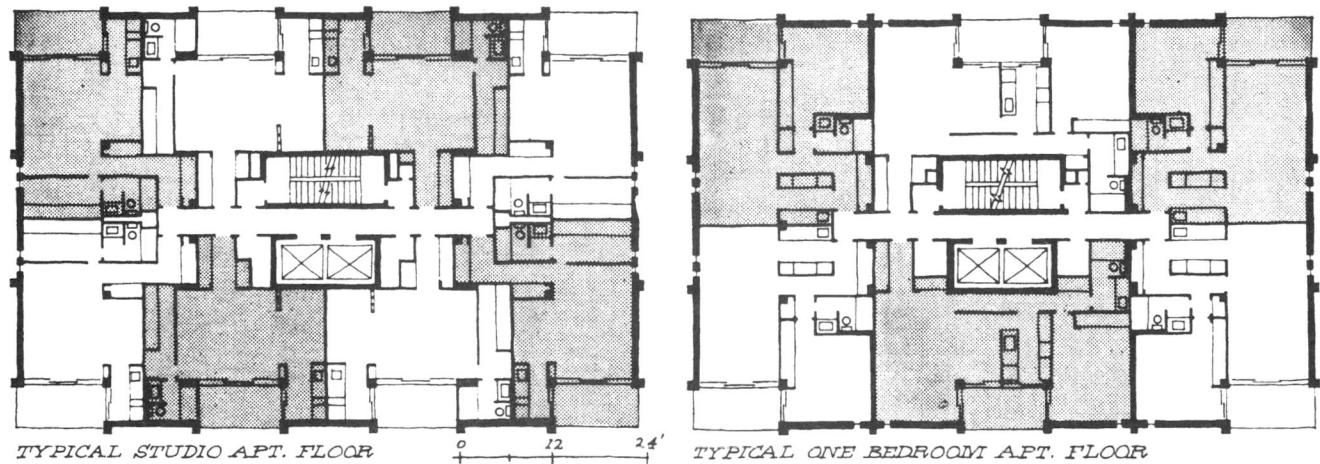

TYPICAL STUDIO APT. FLOOR

TYPICAL ONE BEDROOM APT. FLOOR

Fig. 22 Eugenie Lane Apartments, Chicago, Ill. Harry Weese & Associates—Architects.

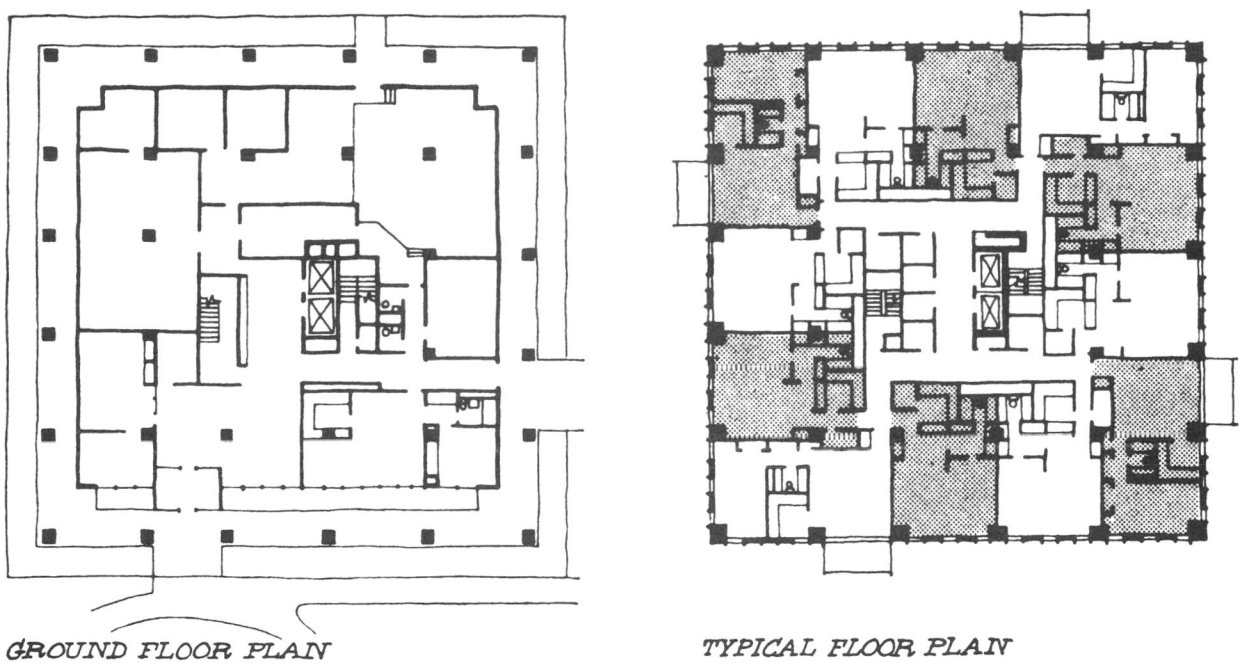

GROUND FLOOR PLAN

TYPICAL FLOOR PLAN

Fig. 23   Charles Bank Apartments, Boston, Mass. Hugh Stubbins & Associates—Architects.

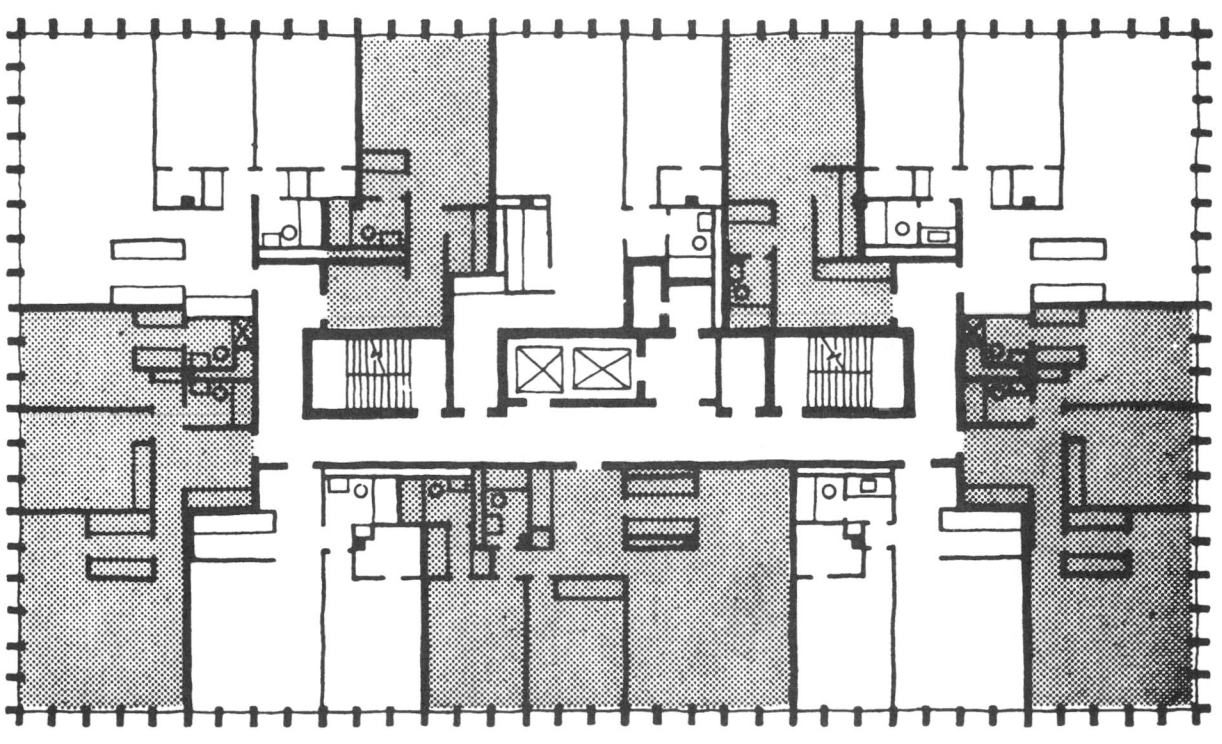

TYPICAL FLOOR PLAN

Fig. 24   Dearborn Towers, Dearborn, Mich. King & Lewis, Inc.—Architects.

EXPANDED TOWER PLAN

The expanded tower scheme is similar to the tower scheme except that it attempts to alleviate its main objection. As indicated, the tower scheme usually has only four apartments per floor, thus making it inefficient in its relationship to the maximum utilization of the elevator/ stair core. By expanding or enlarging the floor area, more apartments can be incorporated into the typical floor. However, when this is done, additional public corridor is created and some apartments lose their cross ventilation and two exposures. Also, some apartments will be forced to be completely oriented toward the north. The main advantage is the cost reduction in providing a more efficient elevator/ stair core.

High-rise housing set in a restricted site in a park is being developed along the Harlem River overlooking Manhattan. The plan includes 1650 units in two towers with new school and community facilities included. Setting is a 17-acre park.

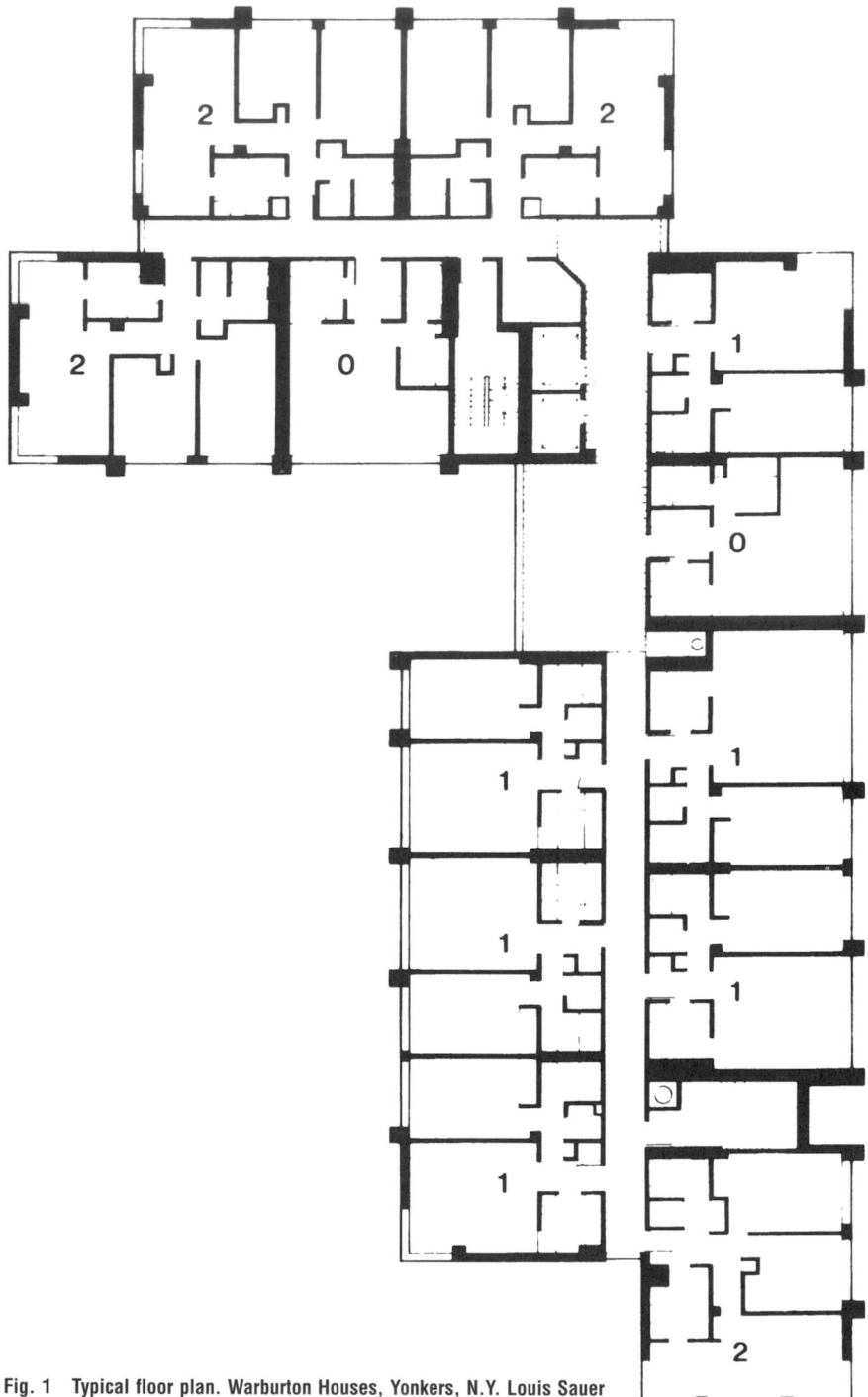

Fig. 1 Typical floor plan. Warburton Houses, Yonkers, N.Y. Louis Sauer Associates—Architects.

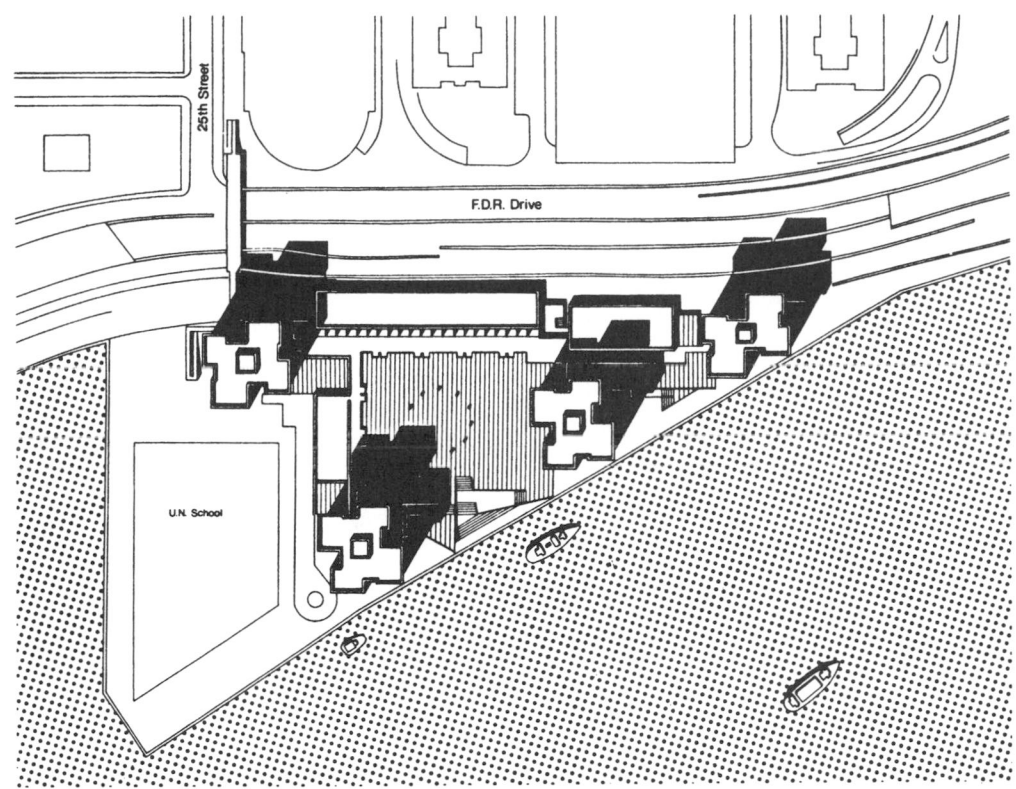

Fig. 2   Waterside, New York City. Davis, Brody & Associates—Architects.

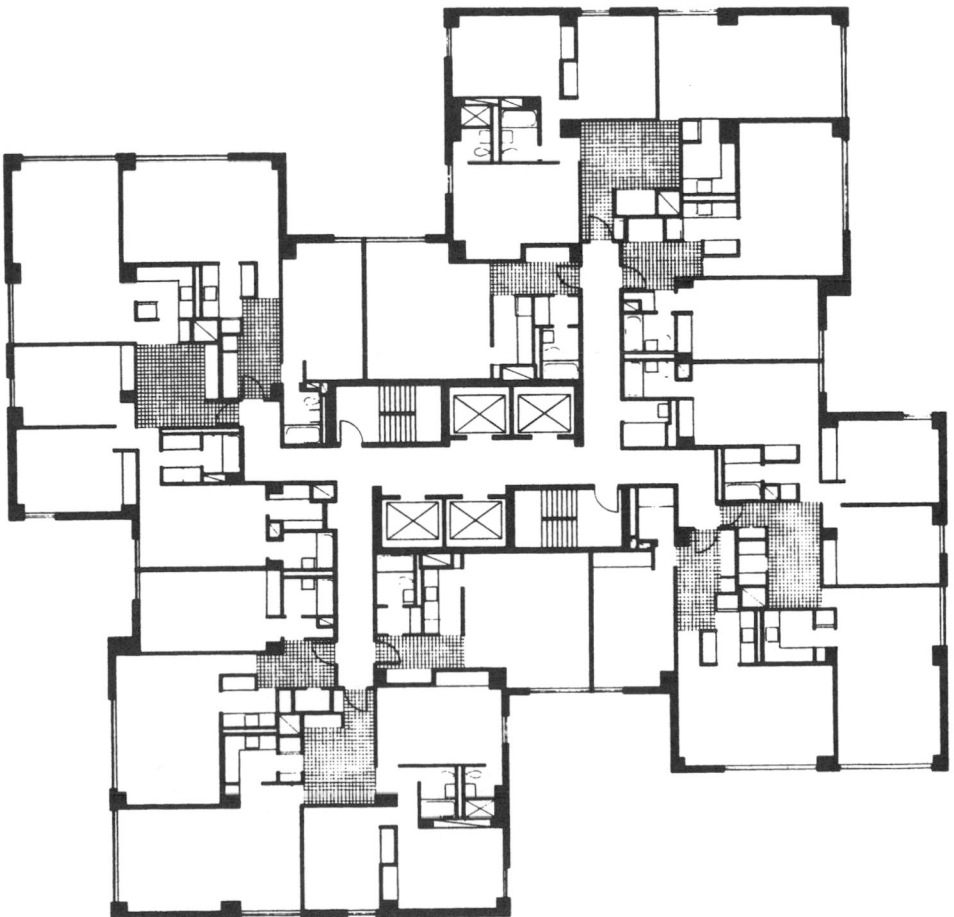

Fig. 3   Typical floor plan. Waterside, New York City. Davis, Brody & Associates—Architects.

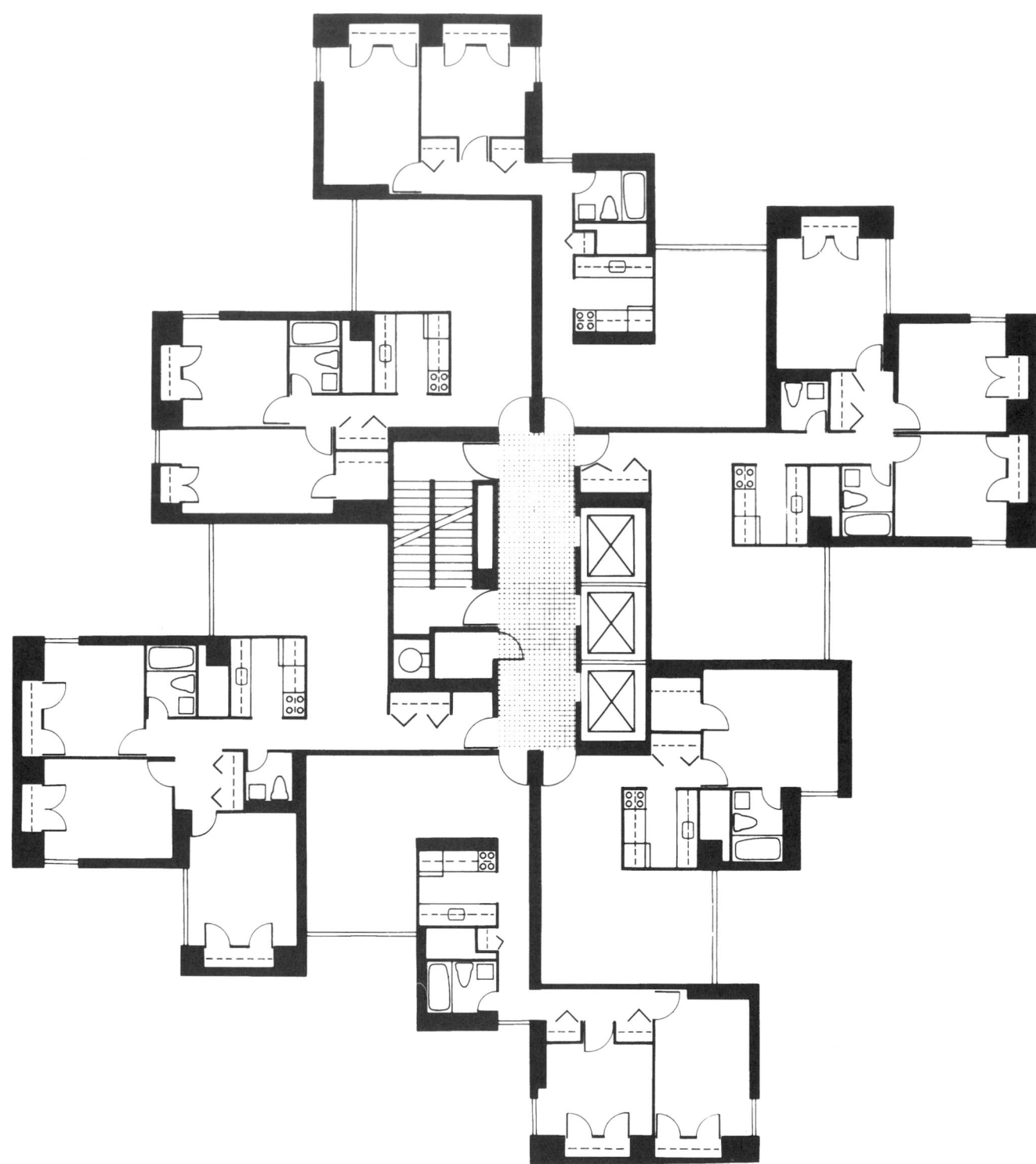

Fig. 4   Lavanburg Community, New York City. Conklin & Rossant—Architects.

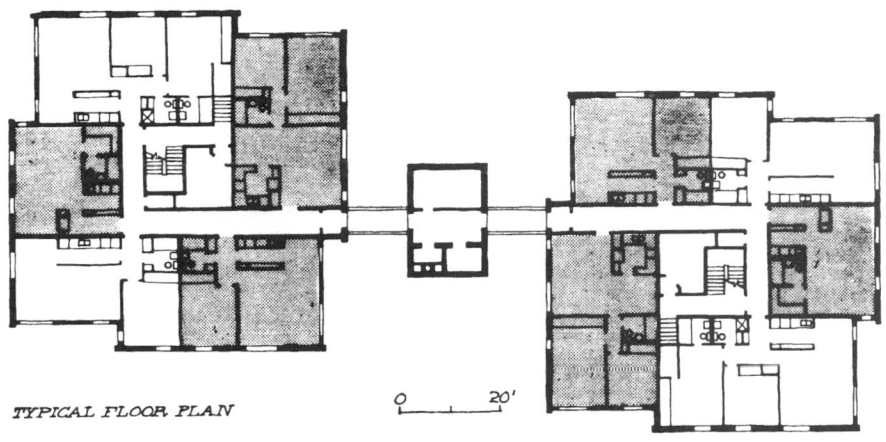

TYPICAL FLOOR PLAN

0   20'

Fig. 5 Sampson Plaza Apartments, Madison, Wis. Harry Weese—Architect.

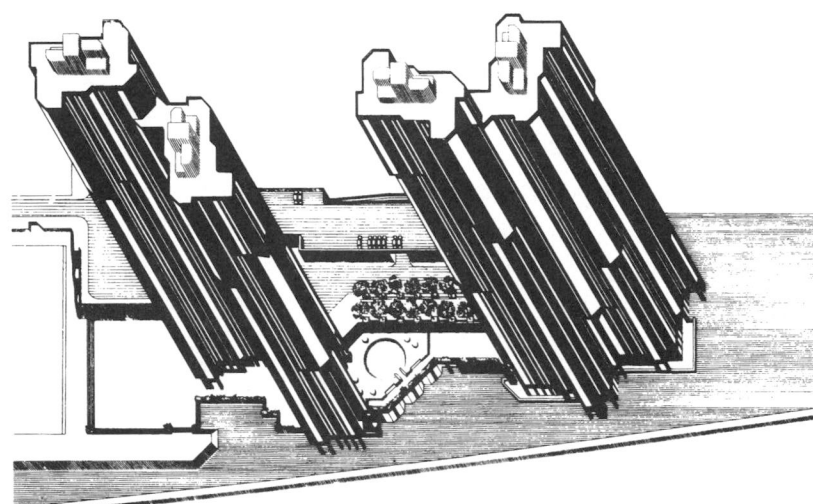

Fig. 6a River Park Towers, Bronx, N.Y. Davis, Brody & Associates—Architects.

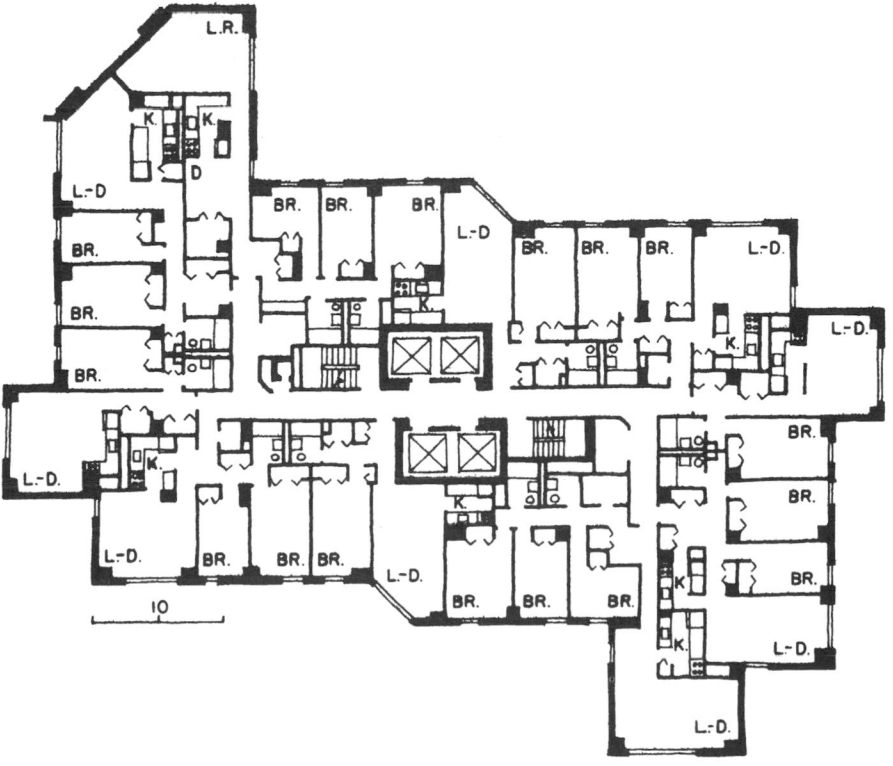

Fig. 6b Typical floor plan. River Park Tower, Bronx, N.Y. Davis, Brody & Associates—Architects.

727

THREE-WING PLAN (Y PLAN)

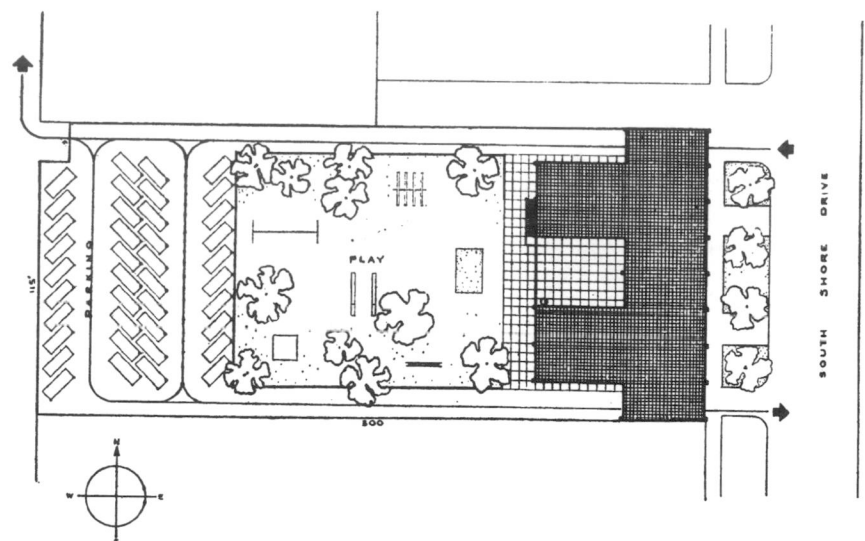

Fig. 1 Promontory Apartments, Chicago, Ill. Mies van der Rohe—Architect.

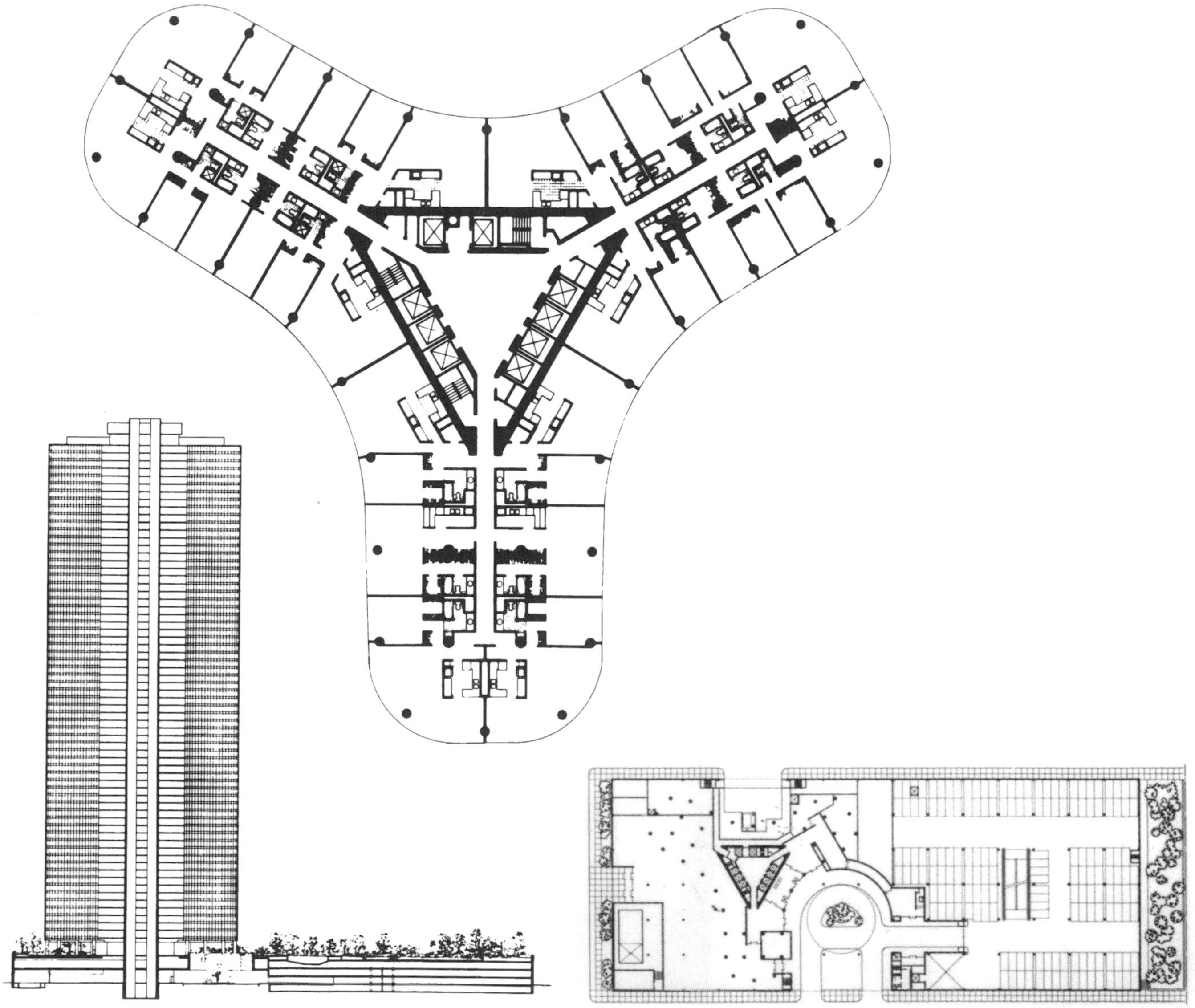

Fig. 2

THREE-WING PLAN (Y PLAN)

The project in Fig. 3 will consist of two 14-story reinforced concrete Y-shaped towers, each containing 42 apartments on a landscaped plaza. In characteristic Puerto Rican fashion, the entrance is kept completely open and the building is raised above the plaza by columns which support bearing walls. The exterior concrete is left exposed.

The Y shape was chosen to provide uninterrupted vistas toward the east from the apartment balconies. Repetitive elements of the plan enabled rapid construction utilizing a climbing crane placed on the elevator shaft.

Open parking is provided at the sides and rear of the site and is slightly depressed to keep it below sight level from the plaza.

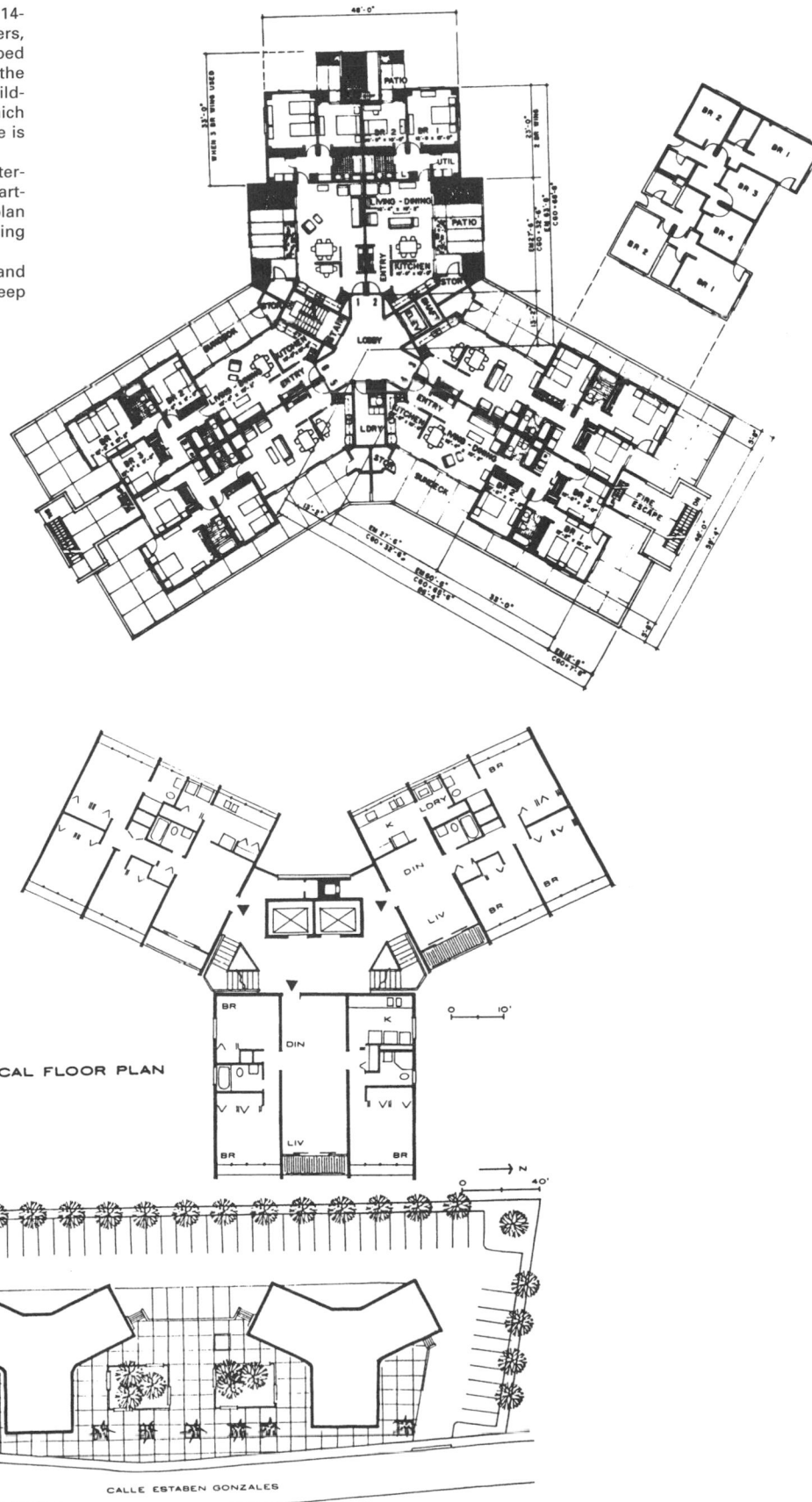

TYPICAL FLOOR PLAN

CALLE ESTABEN GONZALES

Fig. 3

CROSS PLAN

CROSS PLAN

The cross scheme has four equal wings extending out from a central service core. The most common apartment arrangement is to have two apartments per wing for a total of eight. The amount of corridor space is compact and access to the apartments is directly off the elevator. Since each wing has two apartments, each apartment has some cross ventilation at the bend of the apartment and two exposures. Because of the configuration, the typical floor has an extensive perimeter, which permits practically all rooms to have windows. Each apartment is situated at a 90° angle with the opposite apartment, which is acceptable.

One major difficulty with the cross plan is orientation. No matter which direction the building is facing, some apartments will have unacceptable orientations, while others will have good ones. For example, if the building is placed on a north-south axis, two apartments will face each of the four compass points.

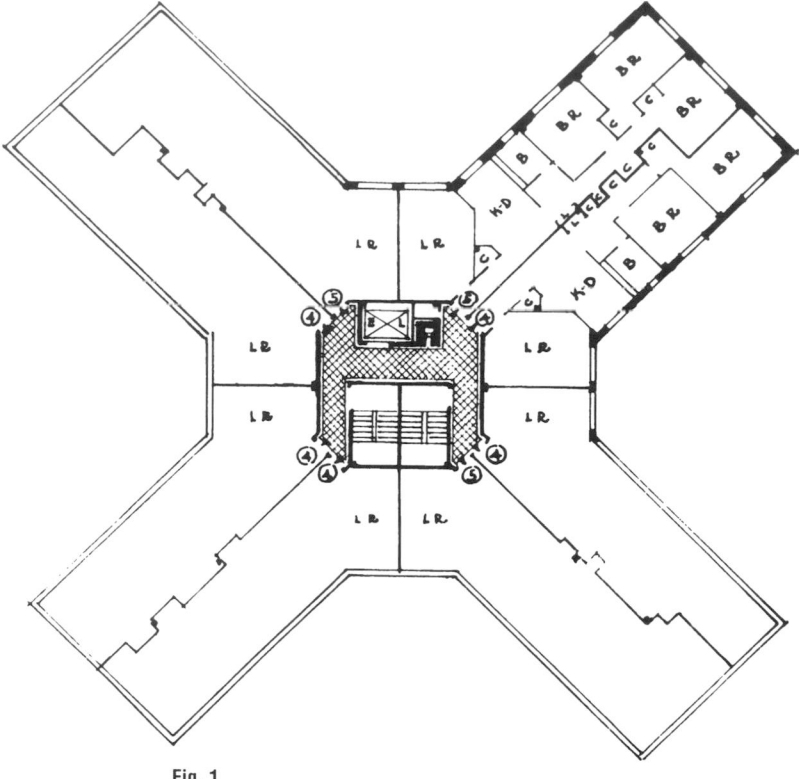

Fig. 1

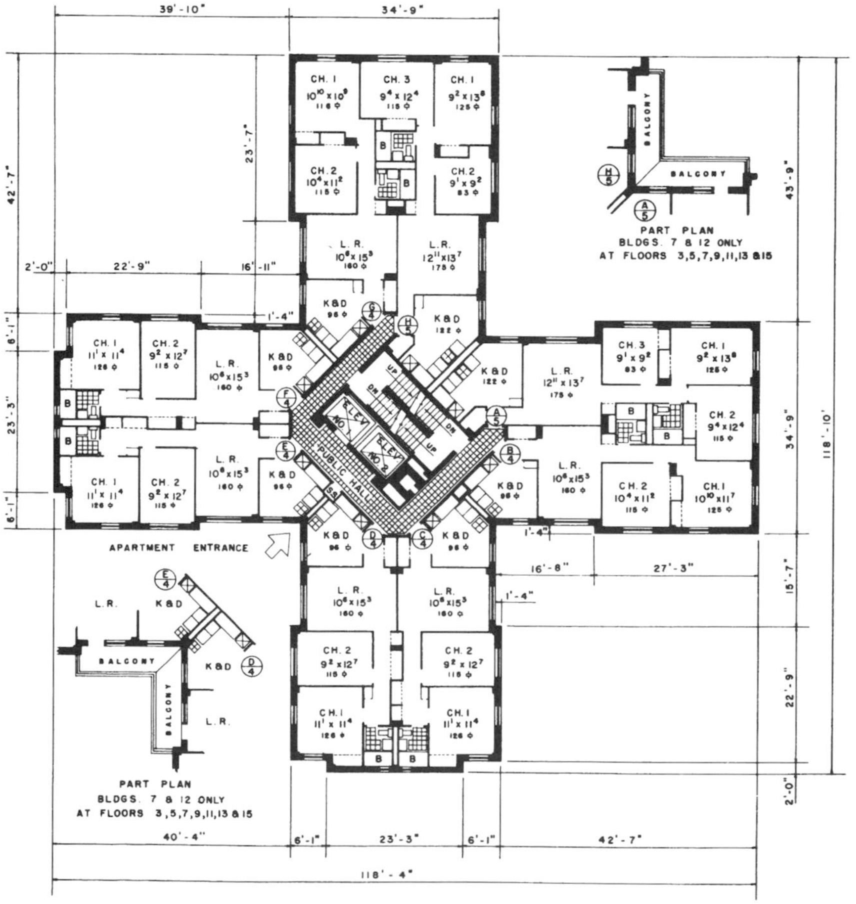

Fig. 2   Carver Houses, New York City Housing Authority.

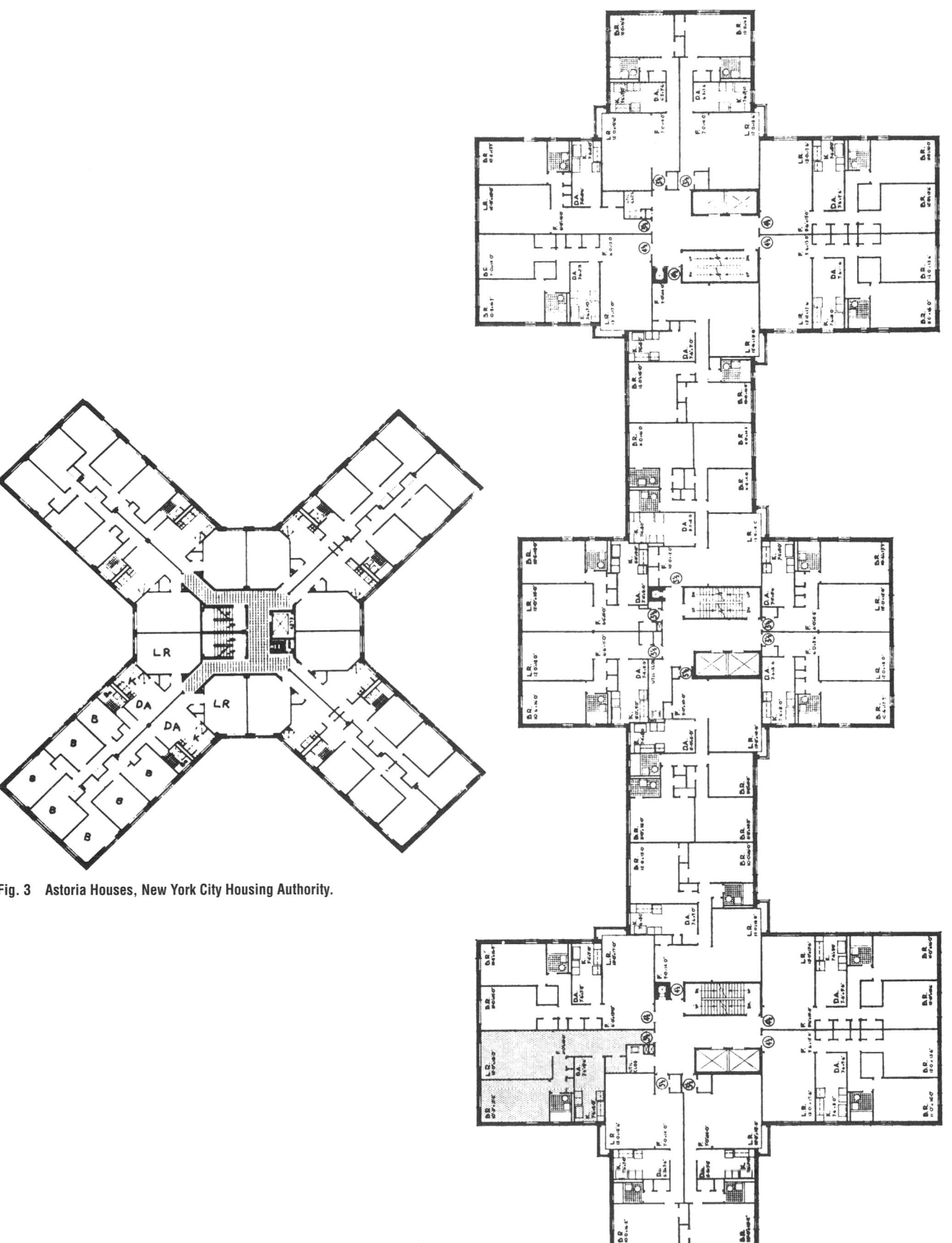

Fig. 3   Astoria Houses, New York City Housing Authority.

Fig. 4   Typical floor plan.

CROSS PLAN

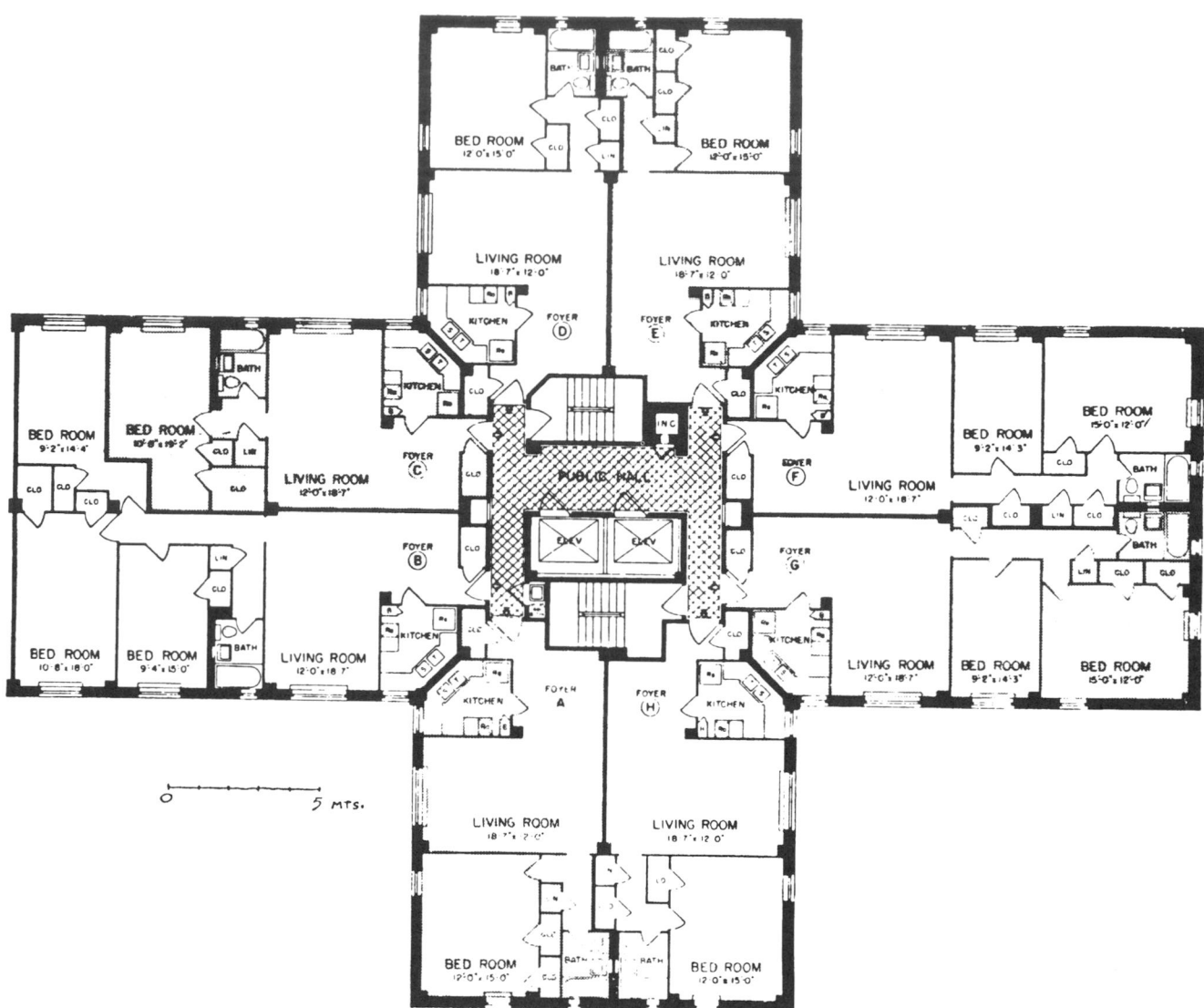

Fig. 5   New York City Housing Authority.

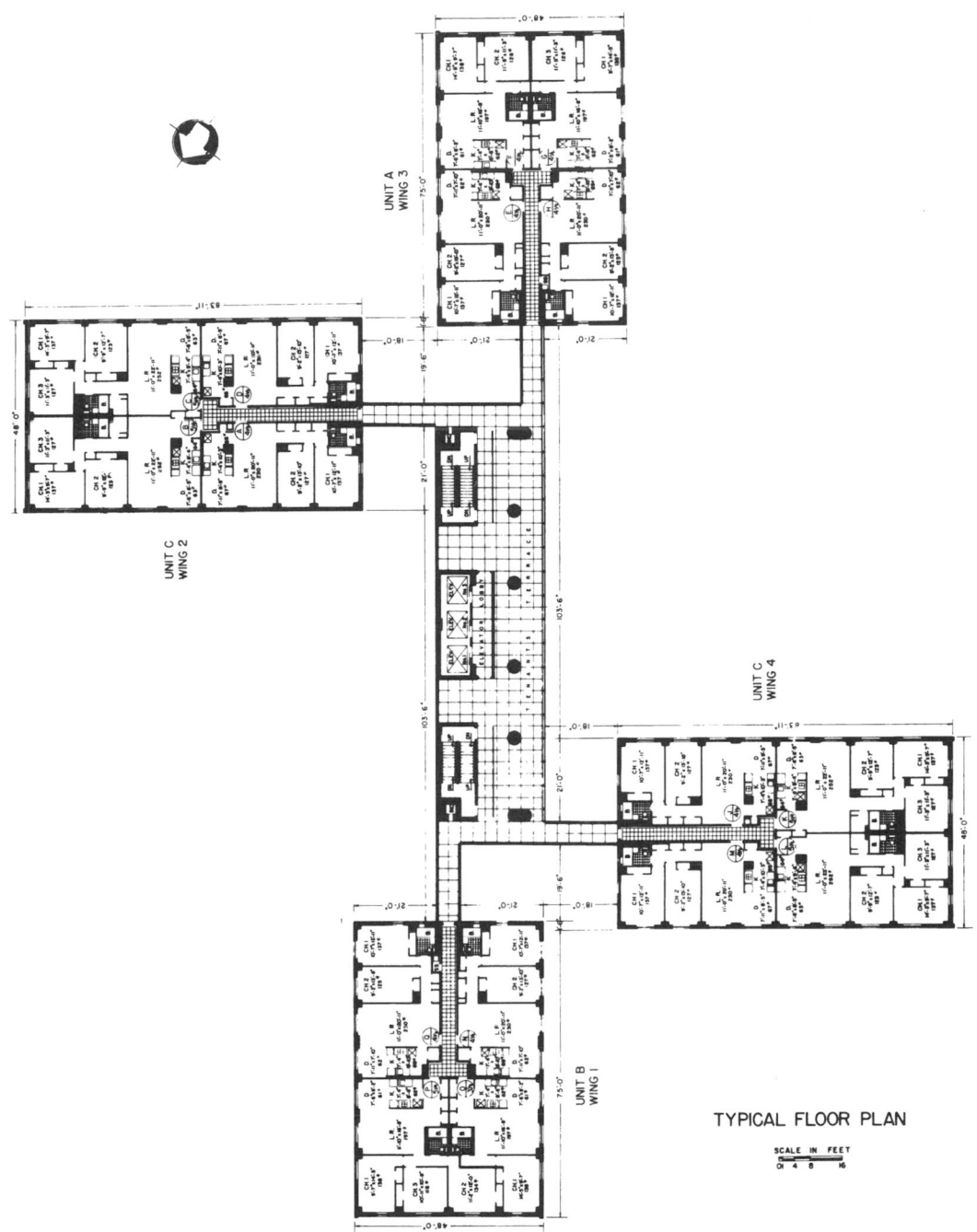

UNIT A
WING 3

UNIT C
WING 2

UNIT C
WING 4

UNIT B
WING I

Fig. 1

TYPICAL FLOOR PLAN

SCALE IN FEET

FIVE-WING PLAN

FIVE-WING PLAN

The five-wing radial scheme is similar to the cross scheme except that it has an additional wing. Access to the apartments is direct from the service core. Each wing has two apartments, making a total of 10 apartments per floor. The motivating factor for this scheme is the larger number of apartments per elevator/stair core.

Each apartment has some cross ventilation at the end of the unit. Also, two exposures are achieved. The length of building perimeter is more extensive and complicated than the simple cross plan. The cost savings in providing 10 apartments per service core can easily be offset by the complications in the perimeter construction. The orientation limitations are even more severe than in the cross plan. No matter which direction the building is facing, more apartments will have unacceptable or poor orientations. More critical, however, is the fact that each wing is only 72° to each other. This results in apartments that can partially face each other and thereby reduces privacy.

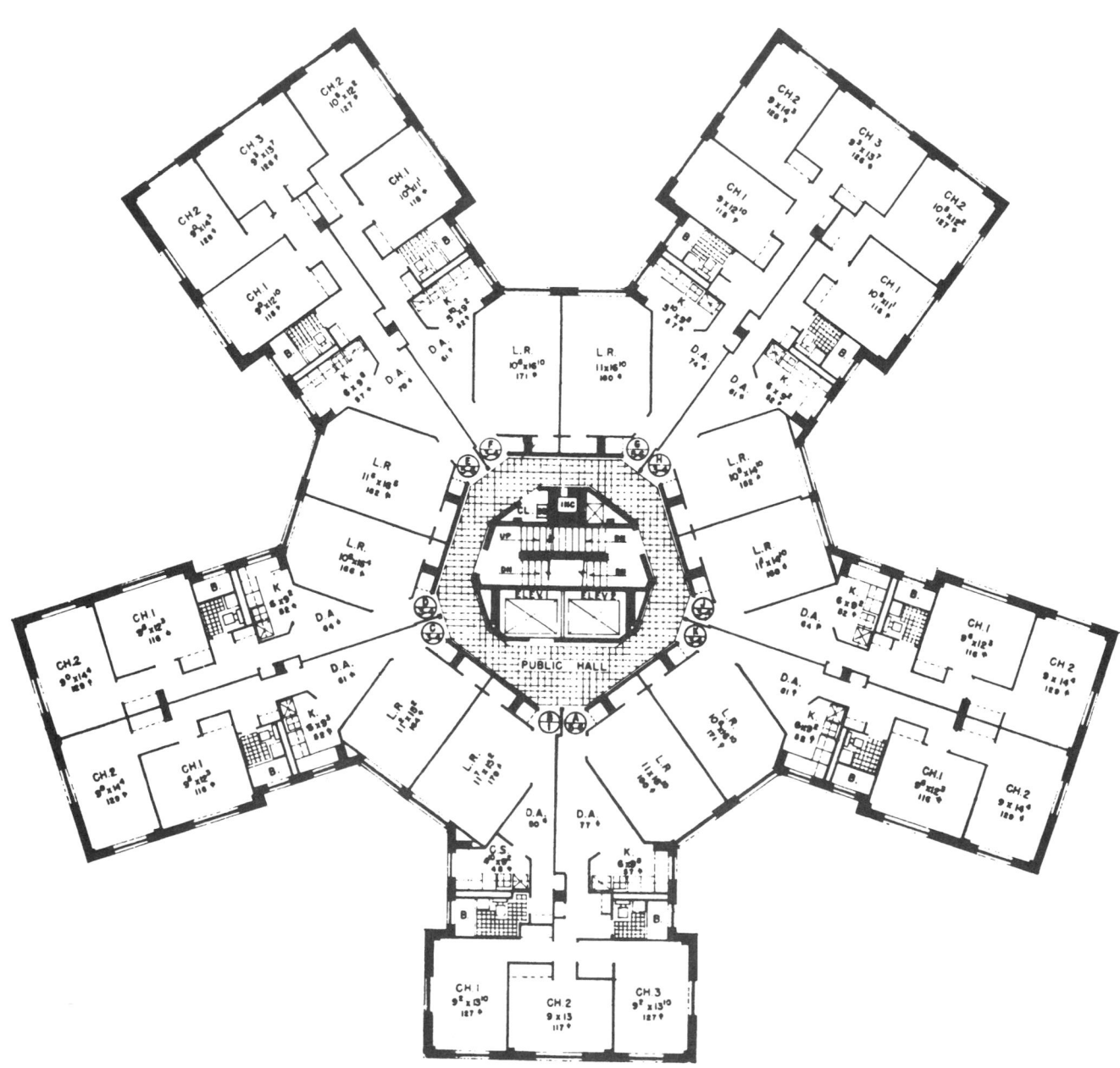

Fig. 1   Albany Houses, Brooklyn, N.Y. Fellheimer, Wagner & Vollmer—Architects.

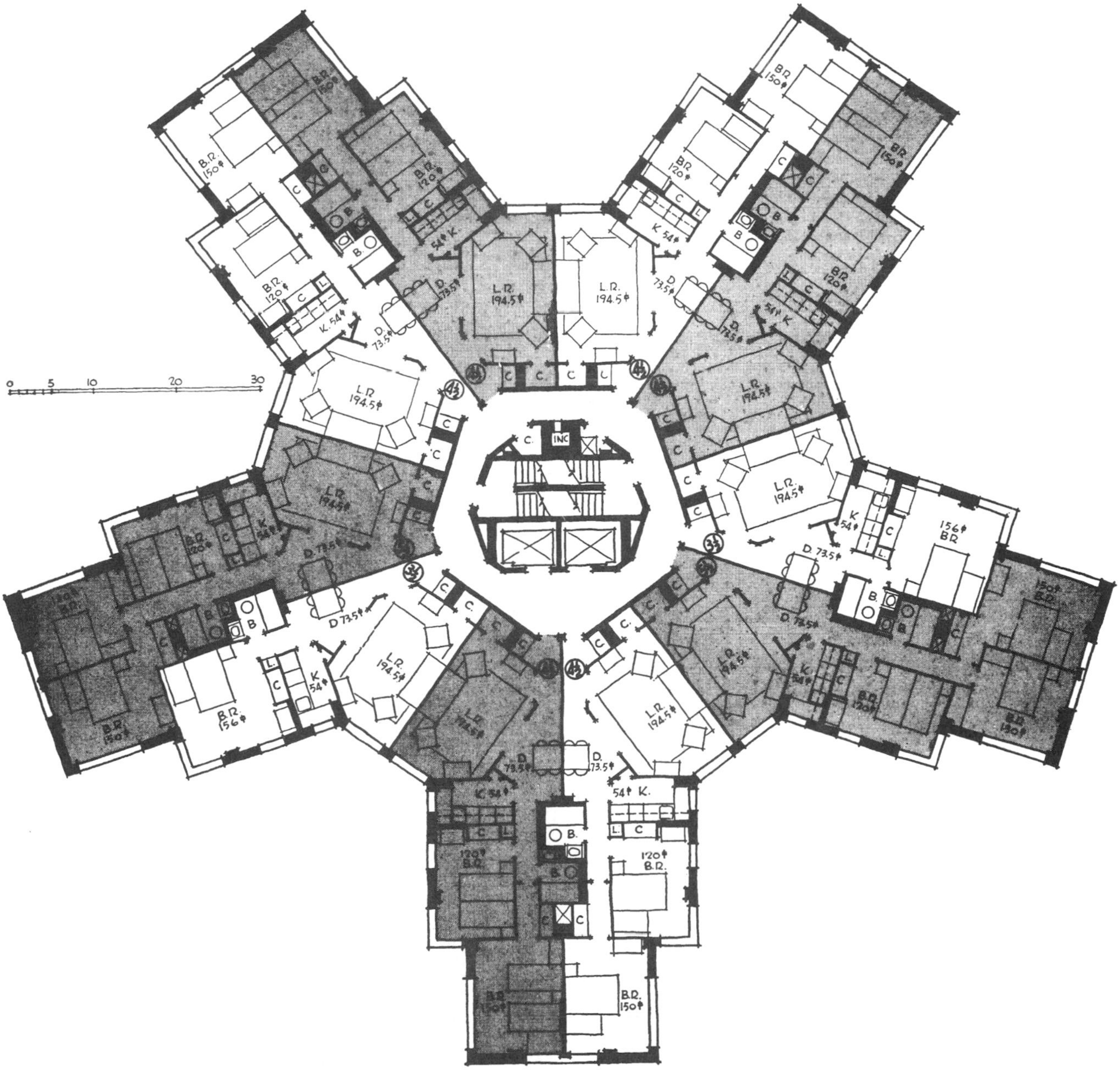

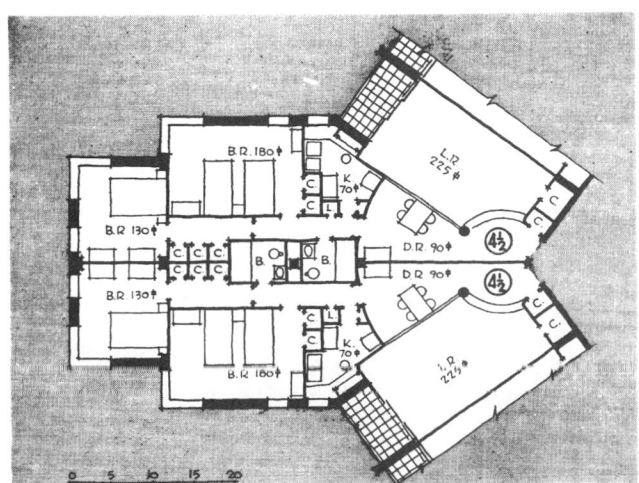

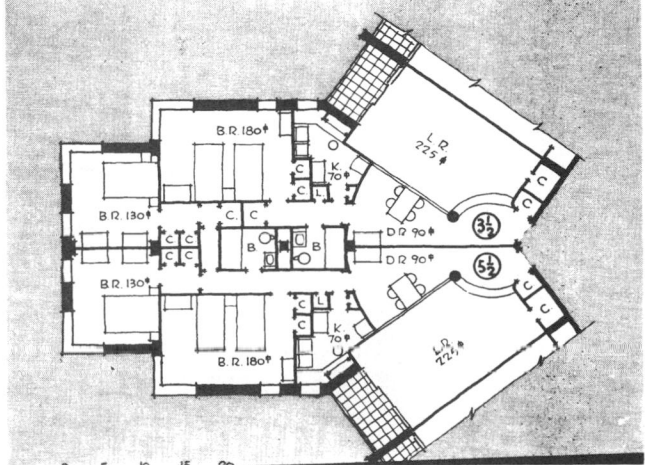

Fig. 2   C. Vollmer, Fellheimer & Wagner—Architects.

CIRCULAR PLAN

The circular scheme is similar to the tower. It contains a central service core with apartments wrapped around it. The number of apartments possible depends upon the size of the apartments and the diameter of the building. Access to the apartments is directly off a corridor at the center.

The major advantage is that, as in any circle, the plan encompasses the largest floor area with the minimum amount of perimeter. Also, the service core area is minimal and located at the center, which is undesirable as living space.

A drawback of this plan is in orientation. One side is always facing north while the other sides have the full range of orientations. Through or cross ventilation is not possible.

Like the tower plan, the circular plan is advantageous in site planning. It creates a greater feeling of openness and spaciousness than the slab. It is also easier to situate on irregular sites or those with topographical difficulties.

A variation of the circular plan is the spiral. The spiral scheme has a continuous corridor from bottom to top.

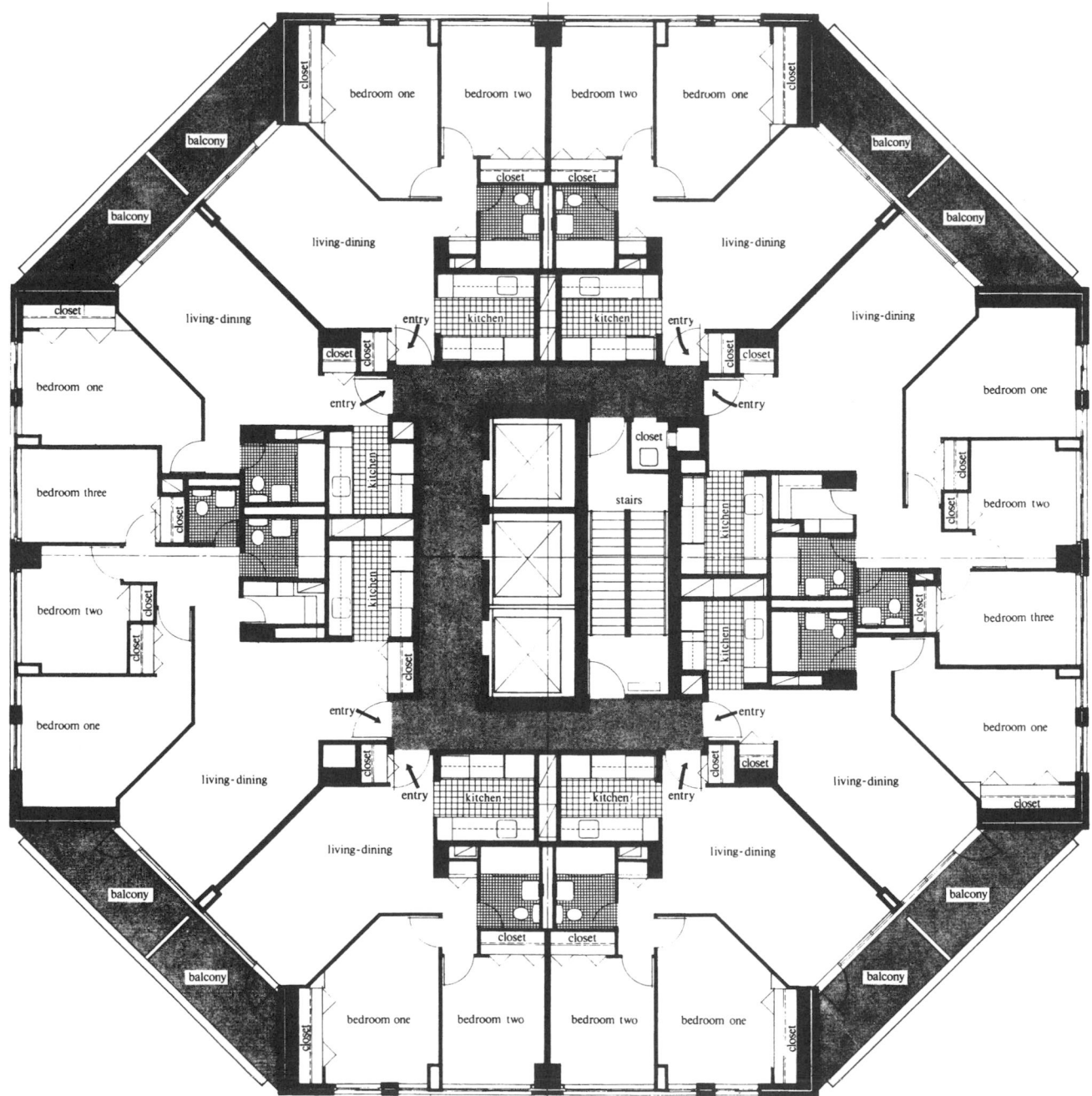

Fig. 1   Typical tower floor plan. Frawley Plaza, New York City. Gruzen & Partners and Castro-Blanco, Piscioneri & Feder—Architects.

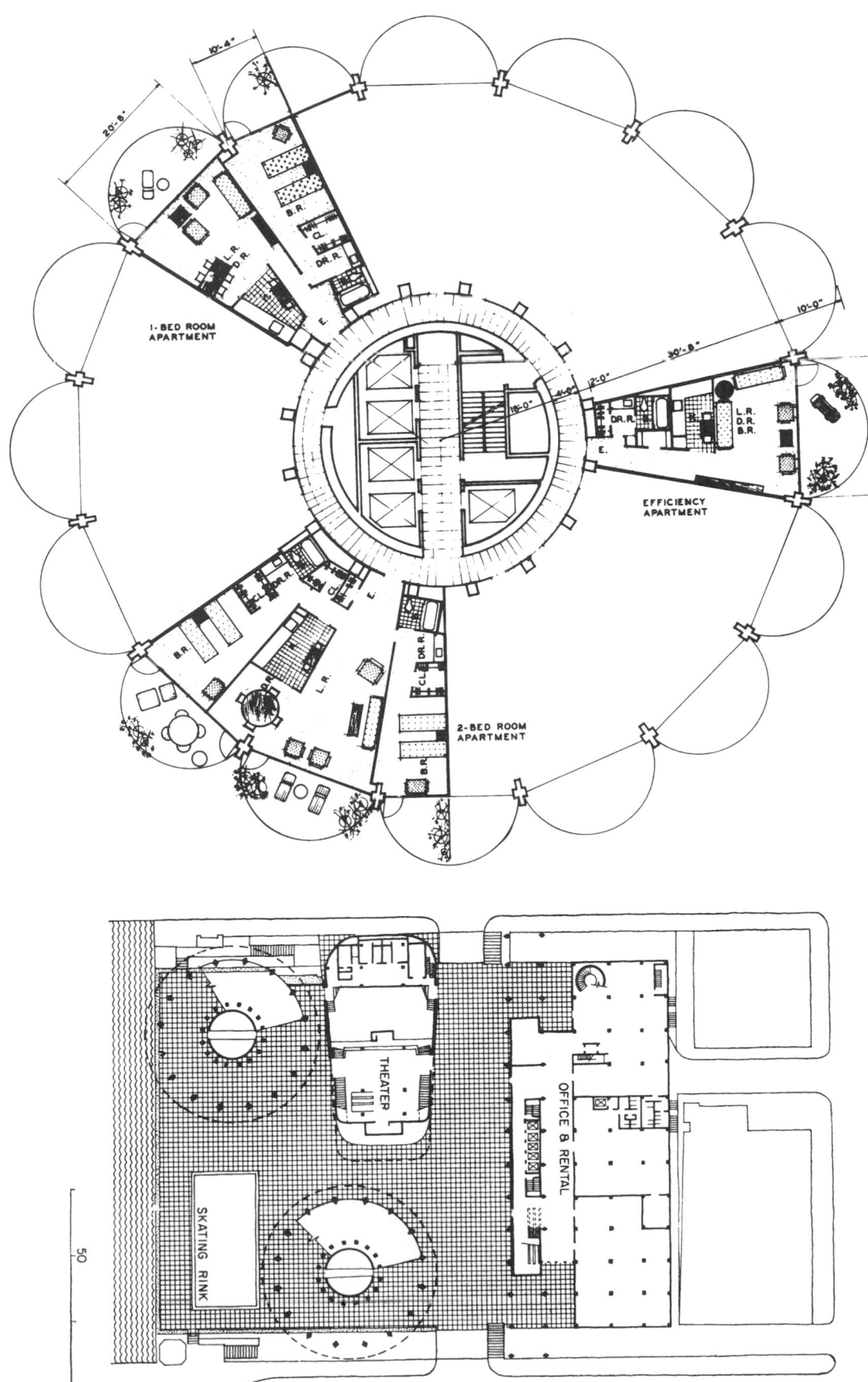

Fig. 2   Marina City, Chicago, III. Bertrand Goldberg—Architect.

CIRCULAR PLAN

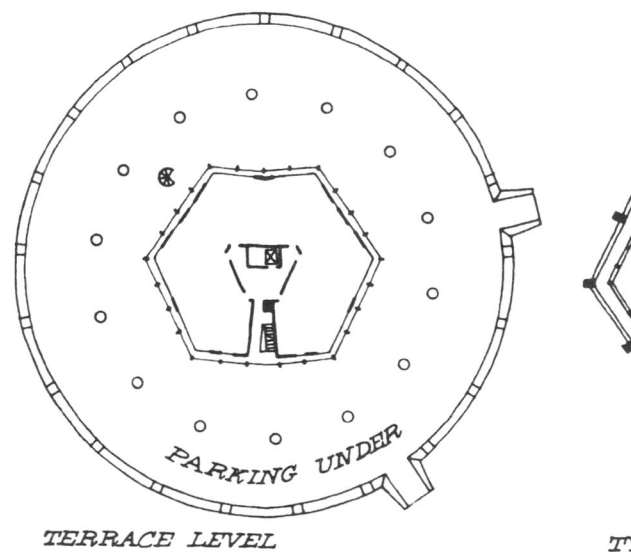

0 5 10 20 30 40

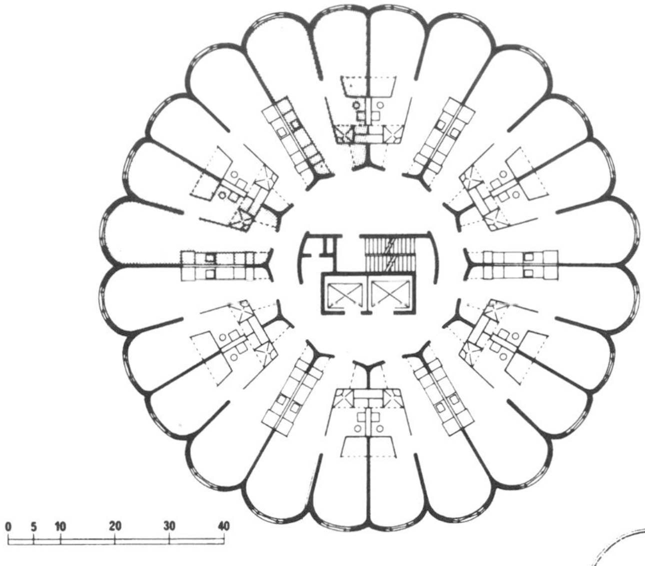

Fig. 3 Hilliard Center, Chicago, Ill. Bertrand Goldberg—Architect.

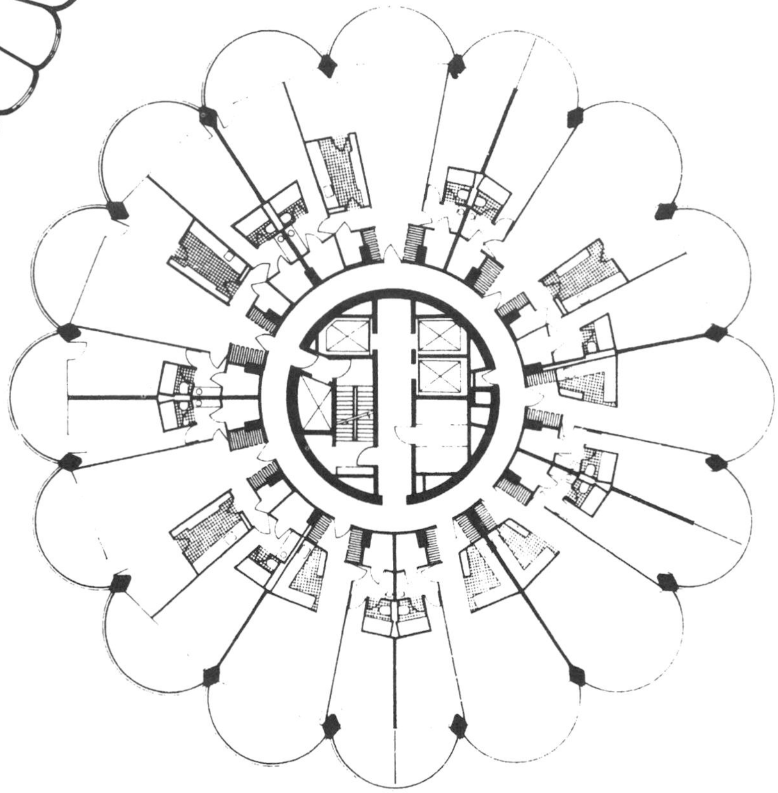

Fig. 4 Marina City, Chicago. Bertrand Goldberg—Architect.

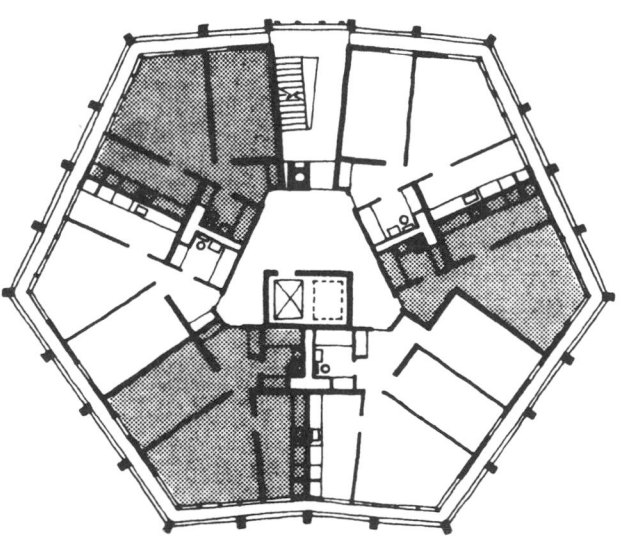

PARKING UNDER
TERRACE LEVEL
TYPICAL FLOOR PLAN

Fig. 5 Point Royal High Flats. Bracknell, England—Architects.

The spiral plan has four revolutionary features:

1. It is circular instead of rectangular, with six rings adding up to a 102-ft diameter. The three inner rings are the utility core—first the elevators and fire stairs at the center, then a circular corridor, then a 2½-ft shaft ring for ducts and pipes. The three outer rings are for tenant living—first a 10-ft utility ring for interior kitchens and toilets backed up to the pipe shaft, then a 25-ft depth of living and sleeping space with continuous windows, and then a ring of big balconies 8 ft wide and 40 ft long.

2. The rentable part of the circle (i.e., the three outer rings) is divided into eight wedges or pie slices, each containing 800 to 820 ft² plus balcony. Tenants can rent as many wedges as they want and subdivide each into two or three rooms plus kitchen, bath, and storage space to suit their needs. If they want more wedges, they can add one or two. If they need fewer, they can turn one or two back to the landlord.

3. The wedges are arranged in a spiral, with each wedge half a floor (5 ft) higher or lower than the wedges on either side. This gives the duplex apartment's advantages of privacy without the duplex's sacrifice of space for long stairs and duplicate corridors.

4. The whole building would be supported without columns by eight radiating fins of prestressed concrete which would do double duty as soundproof dividing walls between the eight wedges. The sections of these fins will be precast on the ground and lifted into position, as will also the inverted U-shaped sections at the apexes of the triangles. The fins will leave an opening for a door from wedge to wedge almost 25 ft back from the windows.

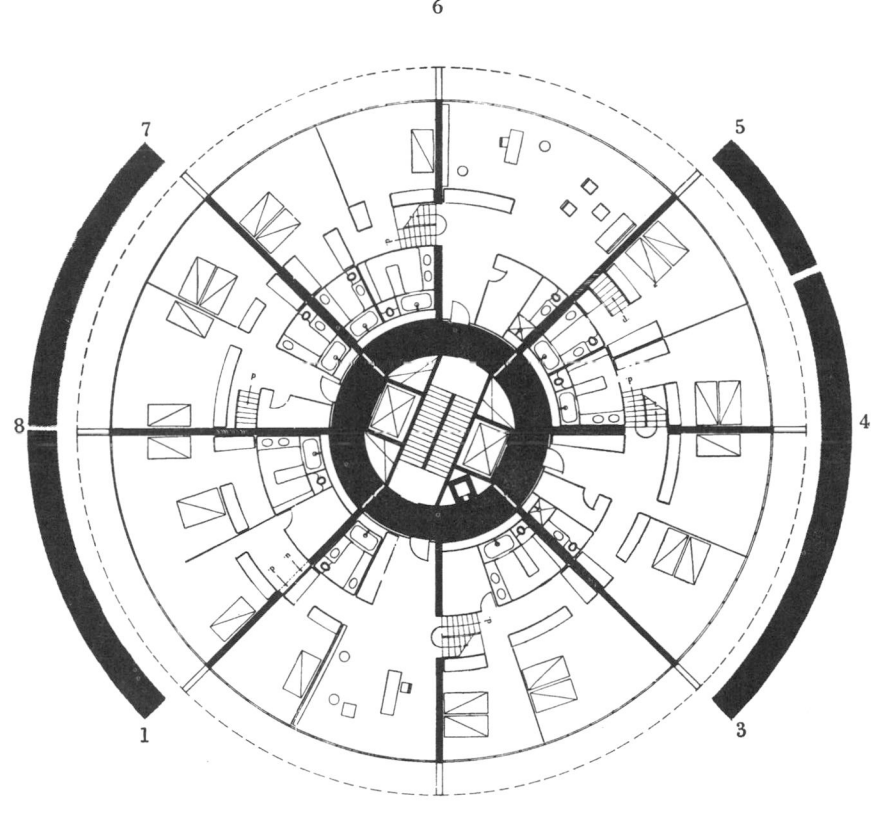

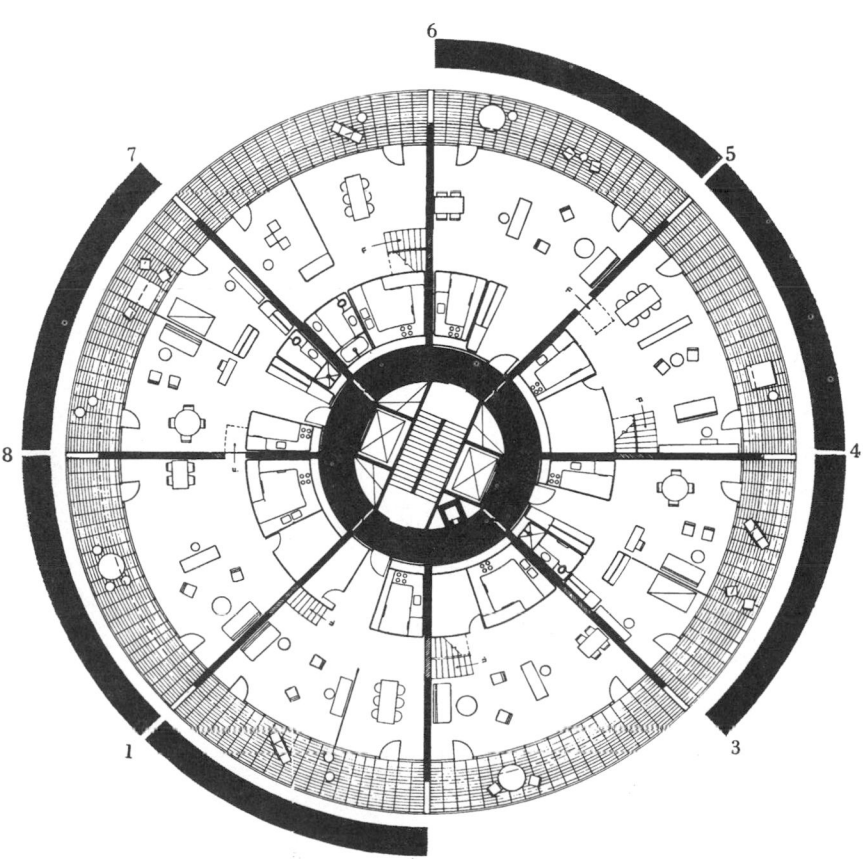

Fig. 1   Apartment helix, New York City. I. M. Pei—Architect.

FREE-FORM PLAN

The Watergate development consists of an office building, a residential hotel, and two additional cooperative apartment houses for a total of 1600 units. The site is a 10-acre triangle overlooking the Potomac River in the Foggy Bottom section of Washington and is adjacent to the John F. Kennedy Cultural Center and Rock Creek Park (see Fig. 3).

The unusual curvilinear design creates an exciting and dramatic effect. The form was developed to produce a unique type of structure for multiple use in an urban area while still maintaining a distinct residential character. The curved forms also permit efficient use of the triangular site and exploit the surrounding views.

All the buildings rest on a platform extending three levels below street grade. The platform contains the commercial facilities and parking for 1200 cars. The open space above, developed for the exclusive use of the pedestrian, contains garden areas, fountains, and swimming pools. Other amenities include a health club, an indoor pool, and restaurants.

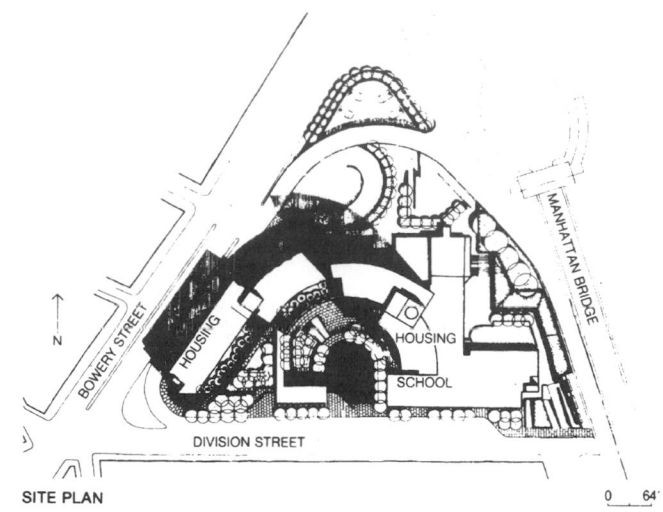

SITE PLAN   0   64'

Fig. 1

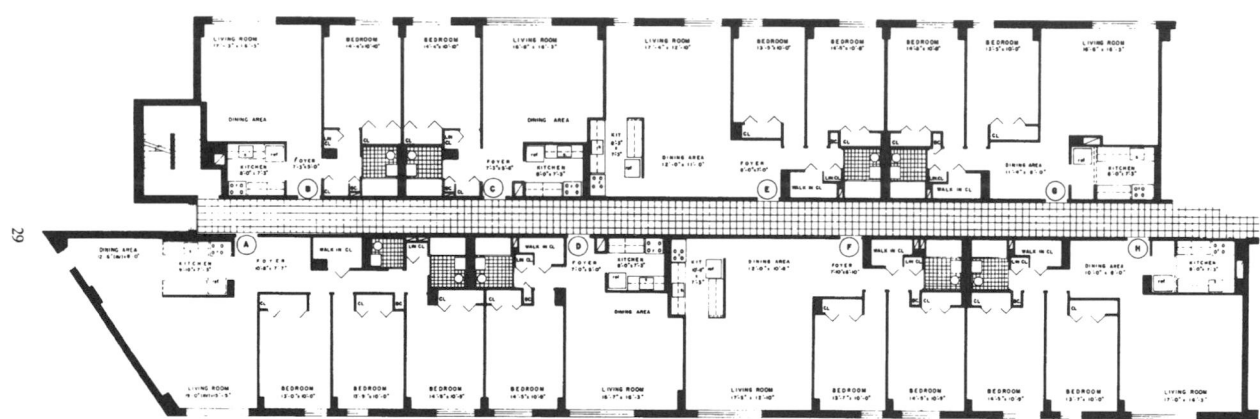

CONFUCIUS PLAZA- Typical Floor Plan
SECTION A

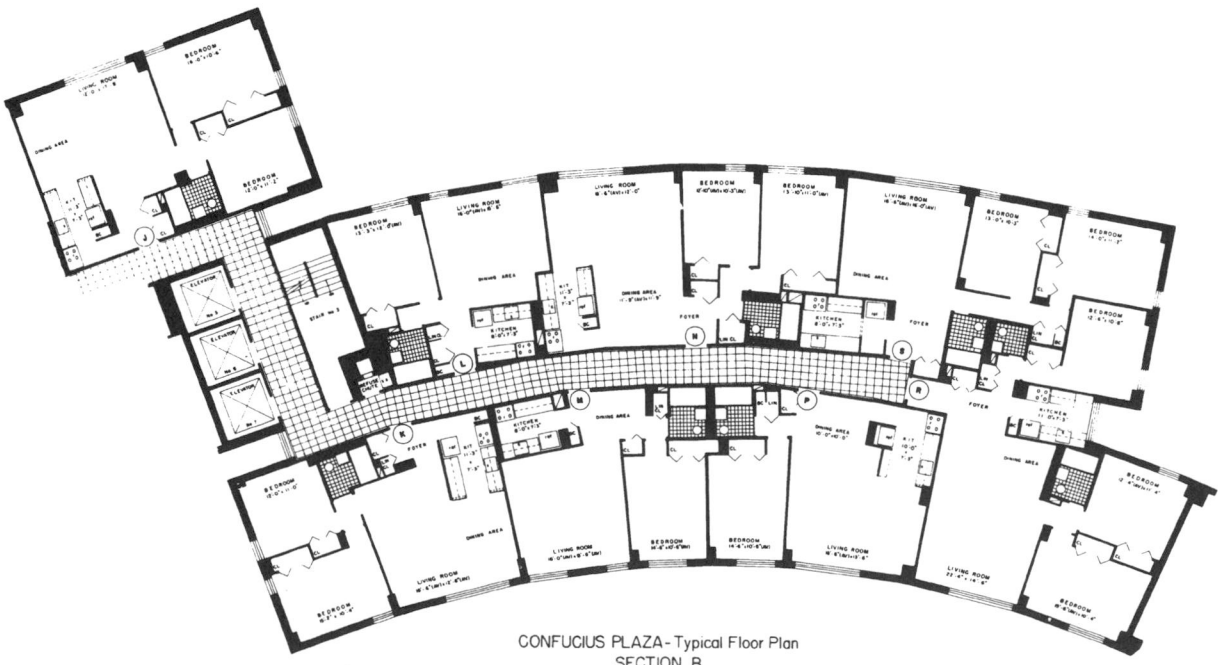

CONFUCIUS PLAZA- Typical Floor Plan
SECTION B

Fig. 2   Confucius Plaza, New York City. Horowitz & Chun—Architects.

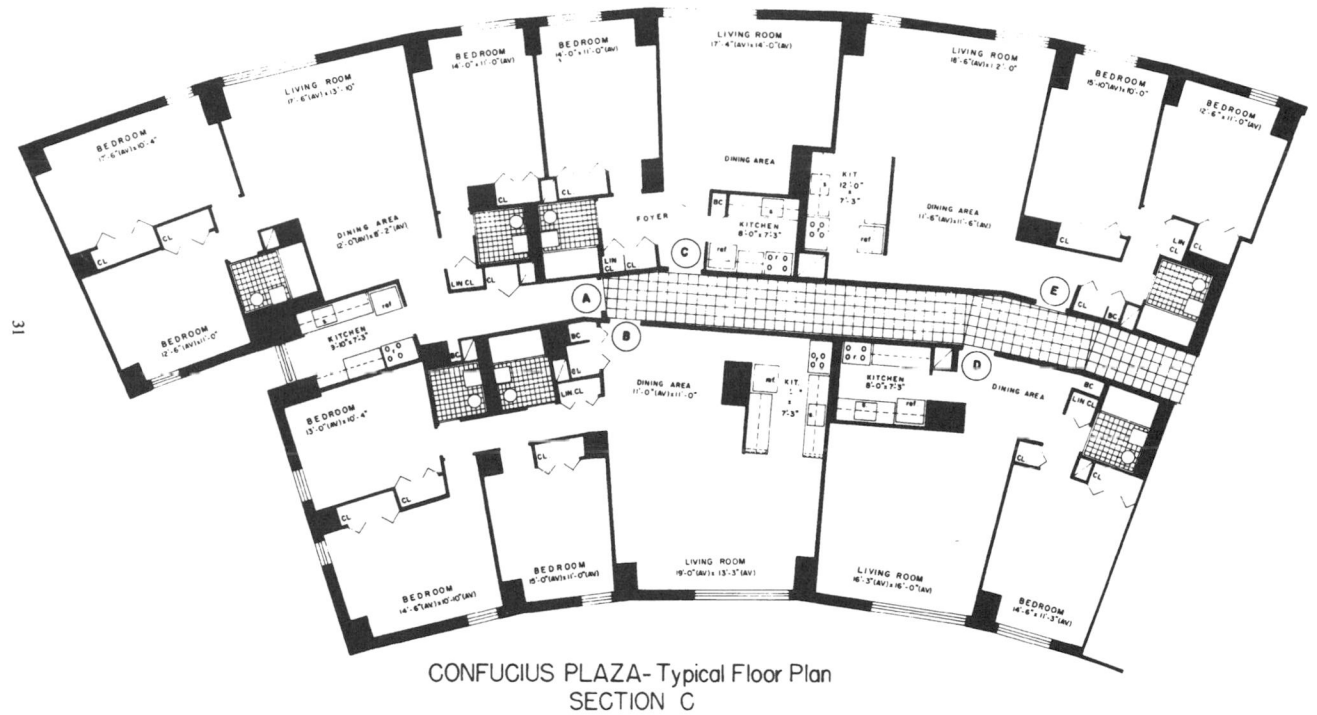

CONFUCIUS PLAZA- Typical Floor Plan
SECTION C

CONFUCIUS PLAZA- Typical Floor Plan
SECTION D

Fig. 2 (*Continued*)

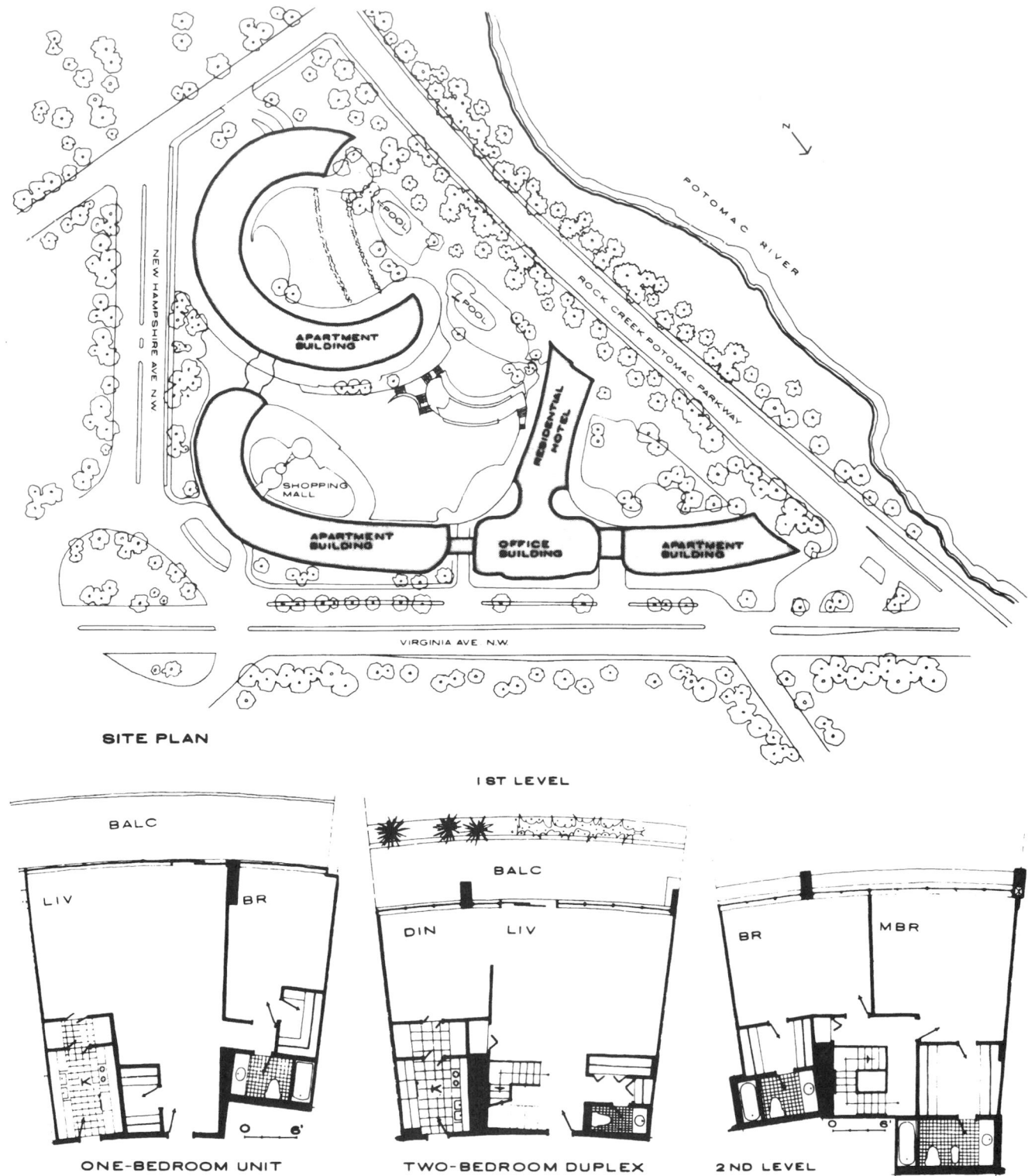

SITE PLAN

ONE-BEDROOM UNIT

TWO-BEDROOM DUPLEX

1 ST LEVEL

2 ND LEVEL

Fig. 3   Watergate Apartments, Washington, D.C. Luigi Moretti, Fischer, & Elmore—Architects.

TERRACE PLAN

The terrace plan is generally a single-loaded arrangement with each succeeding floor set back from the floor below. This creates an extensive terrace area that can be utilized as an extension of the adjacent apartment. The building is usually oriented toward a view or the sun.

As a single-loaded corridor scheme it has the same characteristics. It can have through ventilation and a good orientation for all or most of its apartments. The terrace area is expensive to build, but its cost can be offset by the added livability it creates. One major difficulty is the vertical alignment of stairs, elevators, and utility lines. In the best of schemes,

these elements would be in direct vertical alignment. With a terrace scheme, this is difficult to achieve and therefore creates serious problems.

The ideal siting for this type of building is along a steeply sloping terrain where it would naturally take advantage of the physical conditions.

Fig. 1a

TERRACE PLAN

Fig. 1*b*

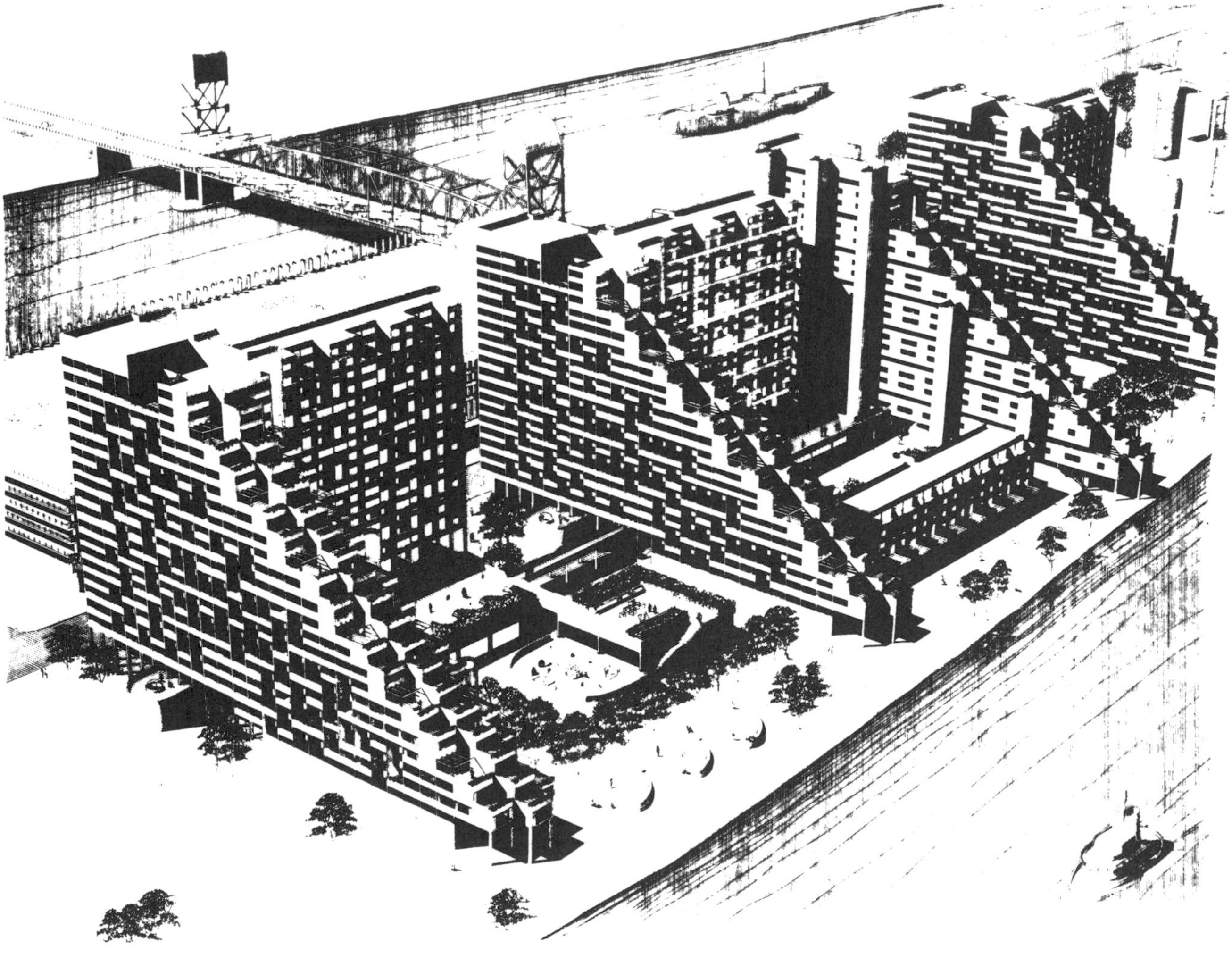

Fig. 1*c*

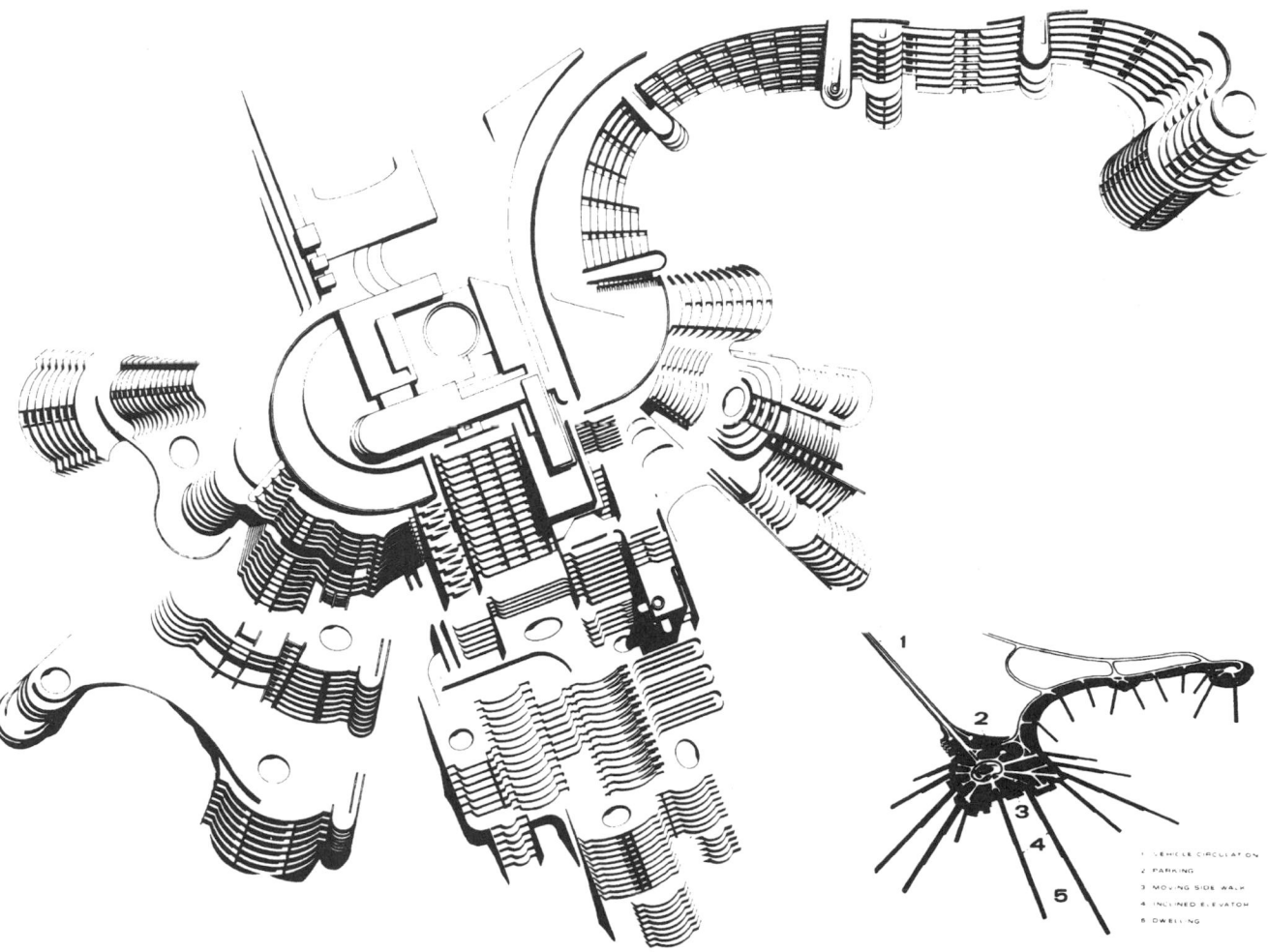

Fig. 2 Urban Nucleus, Sunset Mountain Park, Santa Monica Mountains, Los Angeles, Calif. Daniel, Mann, Johnson & Mendenholl—Architects.

TERRACE PLAN

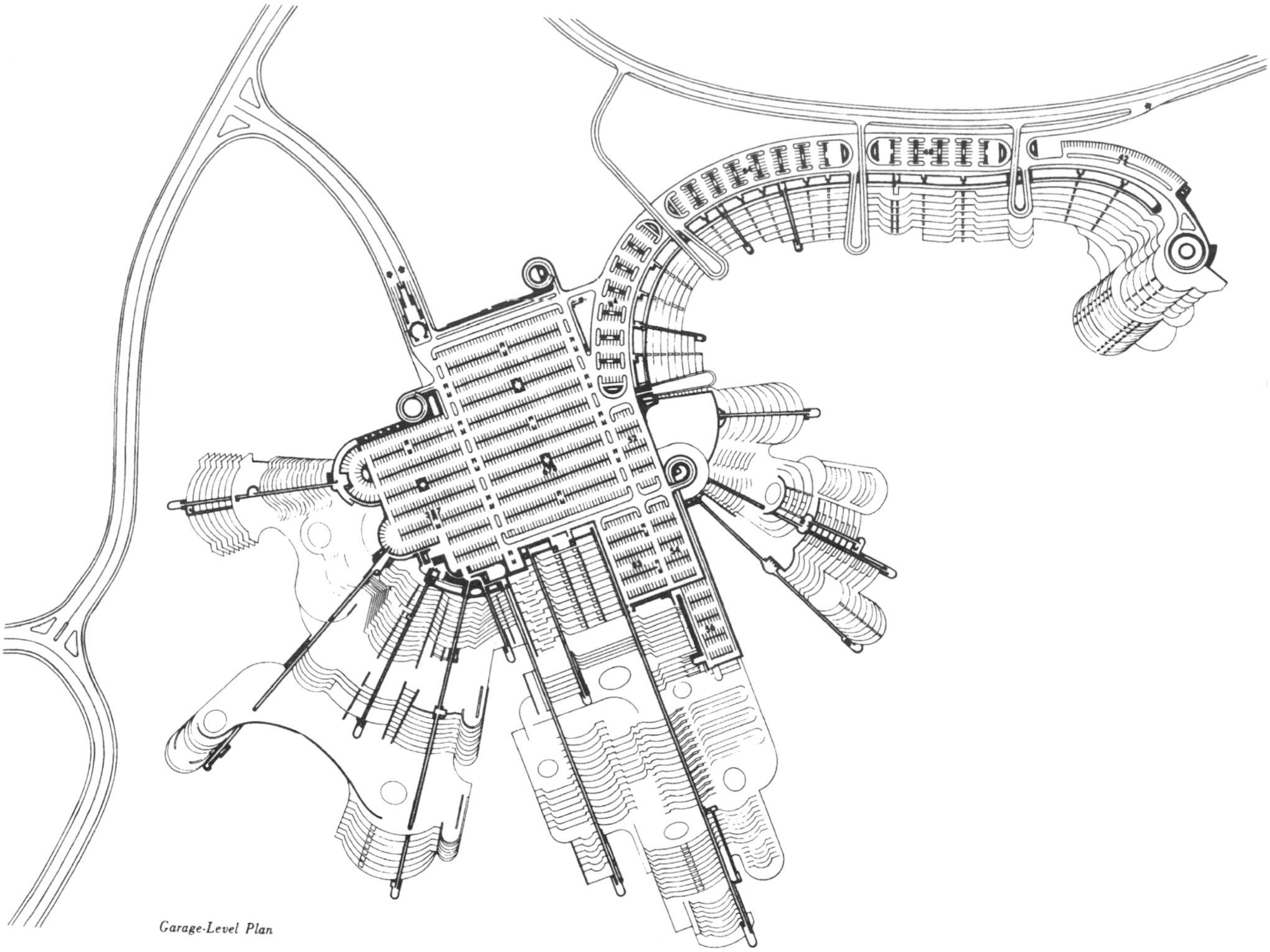

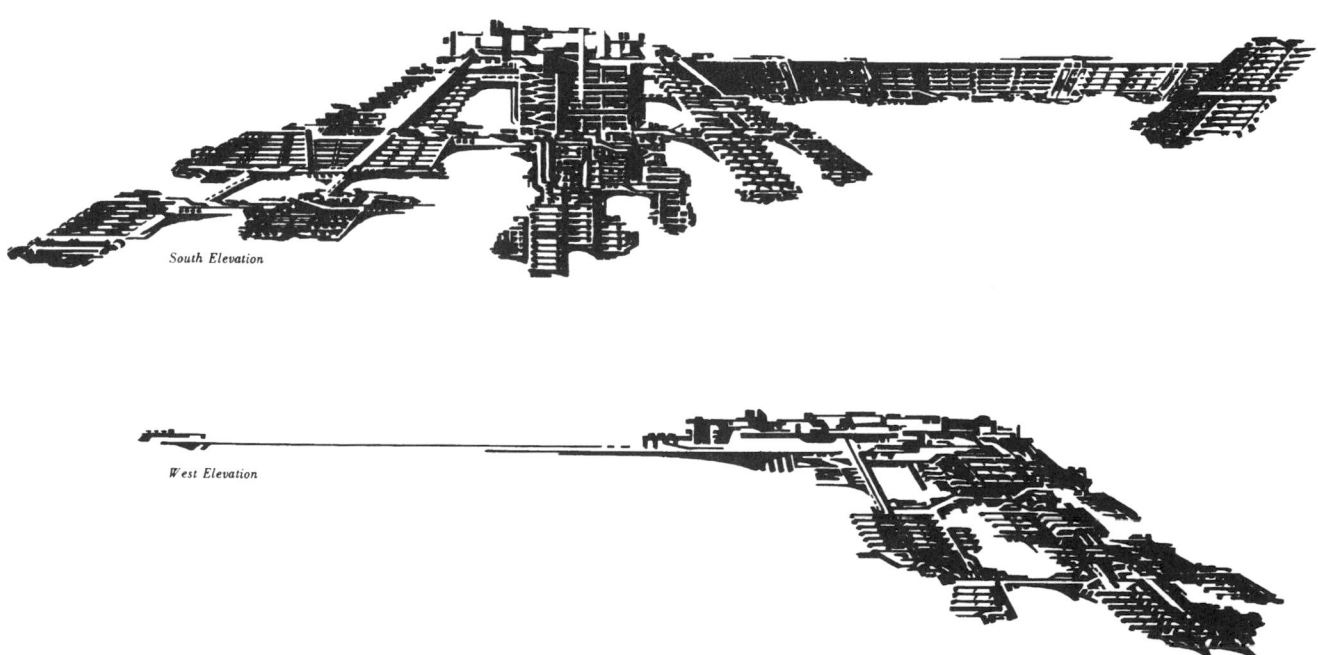

Garage-Level Plan

Fig. 3 Urban Nucleus, Sunset Mountain Park, Santa Monica Mountains, Los Angeles, Calif. Daniel, Mann, Johnson & Mendenholl—Architects.

South Elevation

West Elevation

Fig. 4 Urban Nucleus, Sunset Mountain Park, Santa Monica Mountains, Los Angeles, Calif. Daniel, Mann, Johnson & Mendenholl—Architects.

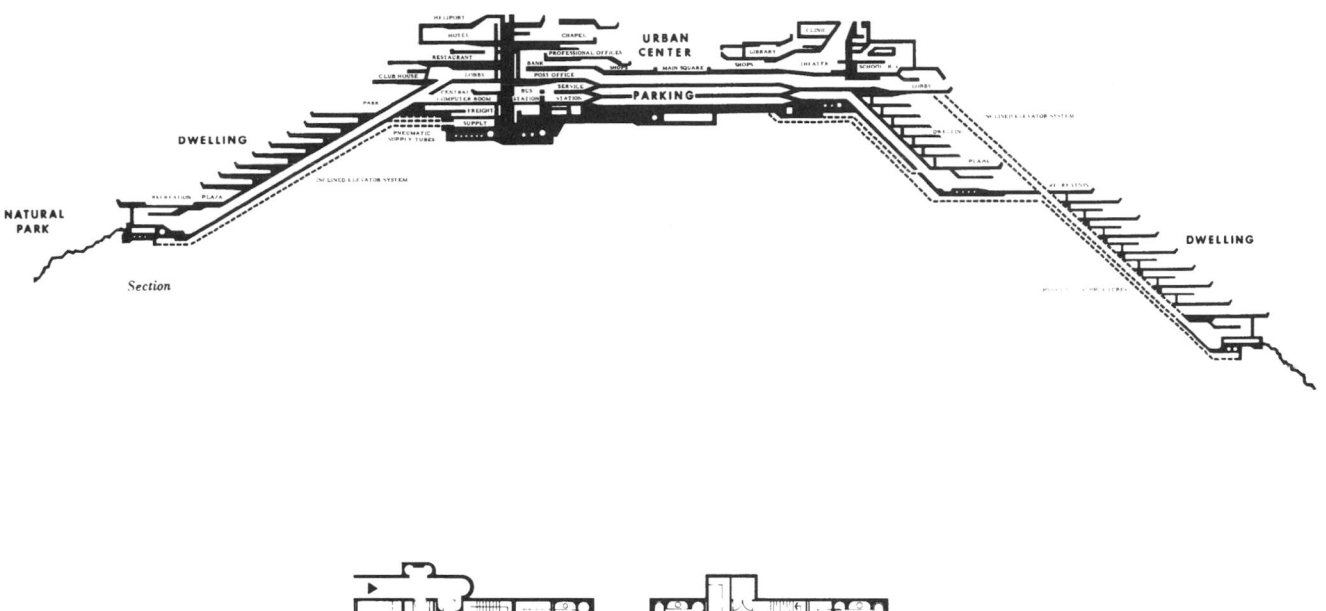

Section

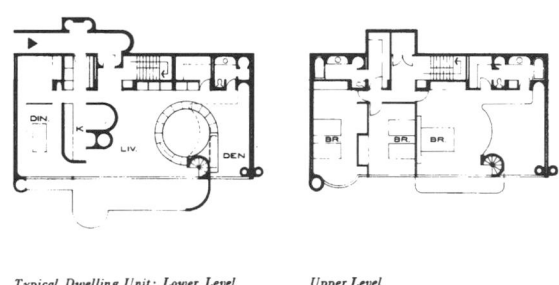

Typical Dwelling Unit: Lower Level   *Upper Level*

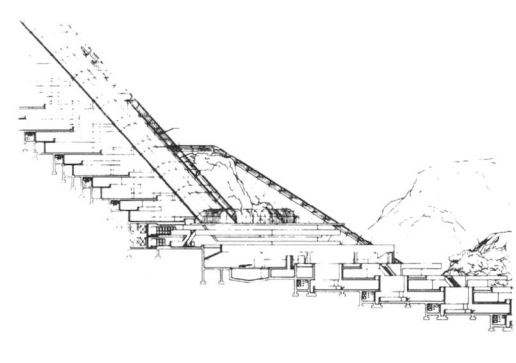

Partial Section

Fig. 5   Urban Nucleus, Sunset Mountain Park, Santa Monica Mountains, Los Angeles, Calif. Daniel, Mann, Johnson & Mendenholl—Architects.

TERRACE PLAN

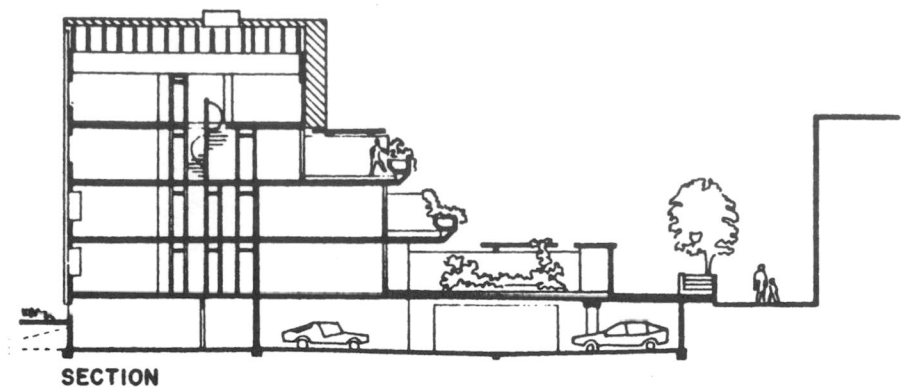

SECTION

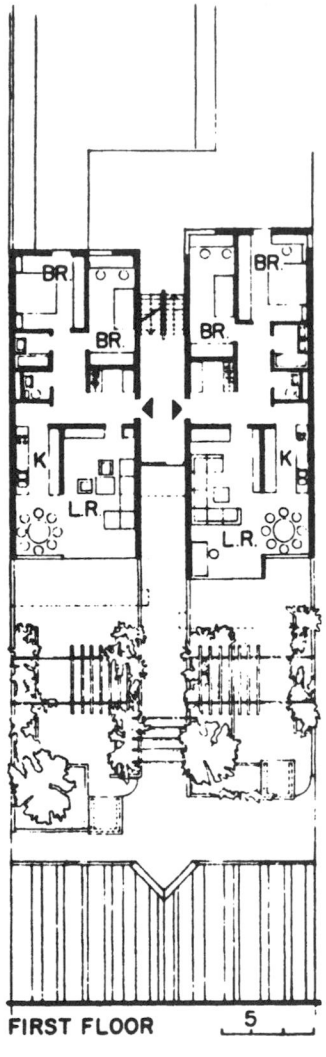

FIRST FLOOR 5

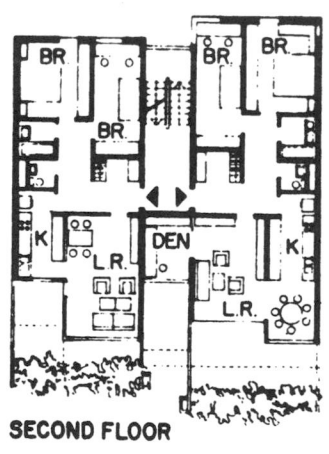

SECOND FLOOR

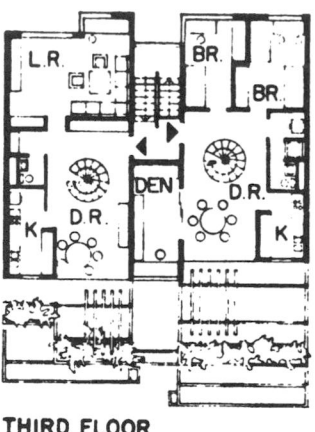

THIRD FLOOR

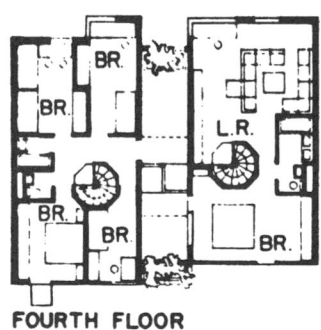

FOURTH FLOOR

Fig. 6

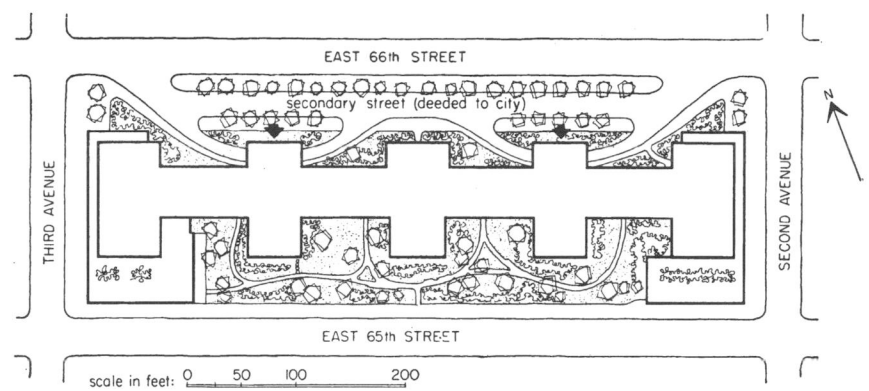

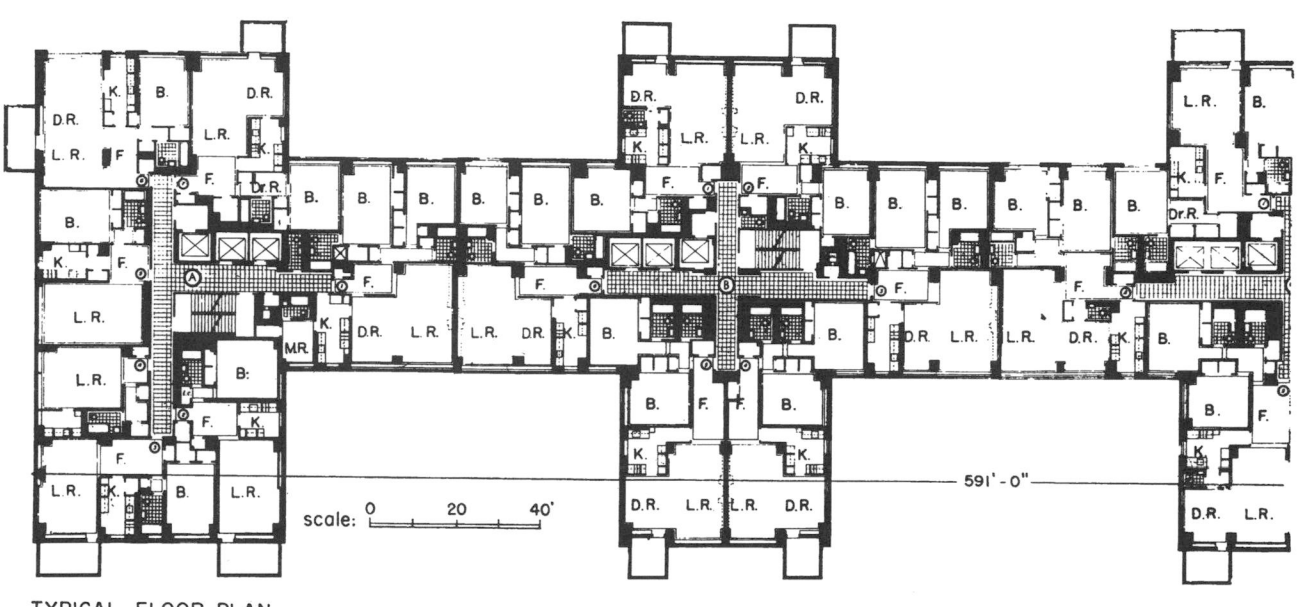

TYPICAL FLOOR PLAN

scale: 0   20   40'

591'-0"

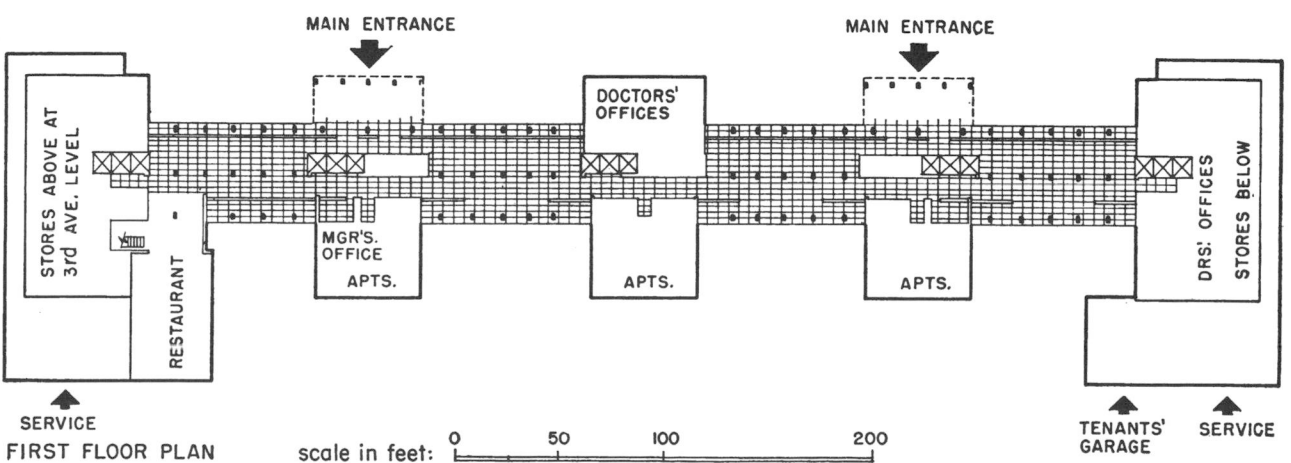

MAIN ENTRANCE

MAIN ENTRANCE

FIRST FLOOR PLAN   scale in feet: 0   50   100   200

SERVICE

TENANTS' GARAGE   SERVICE

Fig. 1   Manhattan House, New York City. Mayer, Whittlesey & Skidmore, Owings & Merrill—Architects.

LUXURY APARTMENT HOUSES

Fig. 2   Manhattan House, New York City. Mayer, Whittlesey & Skidmore, Owings & Merrill—Architects.

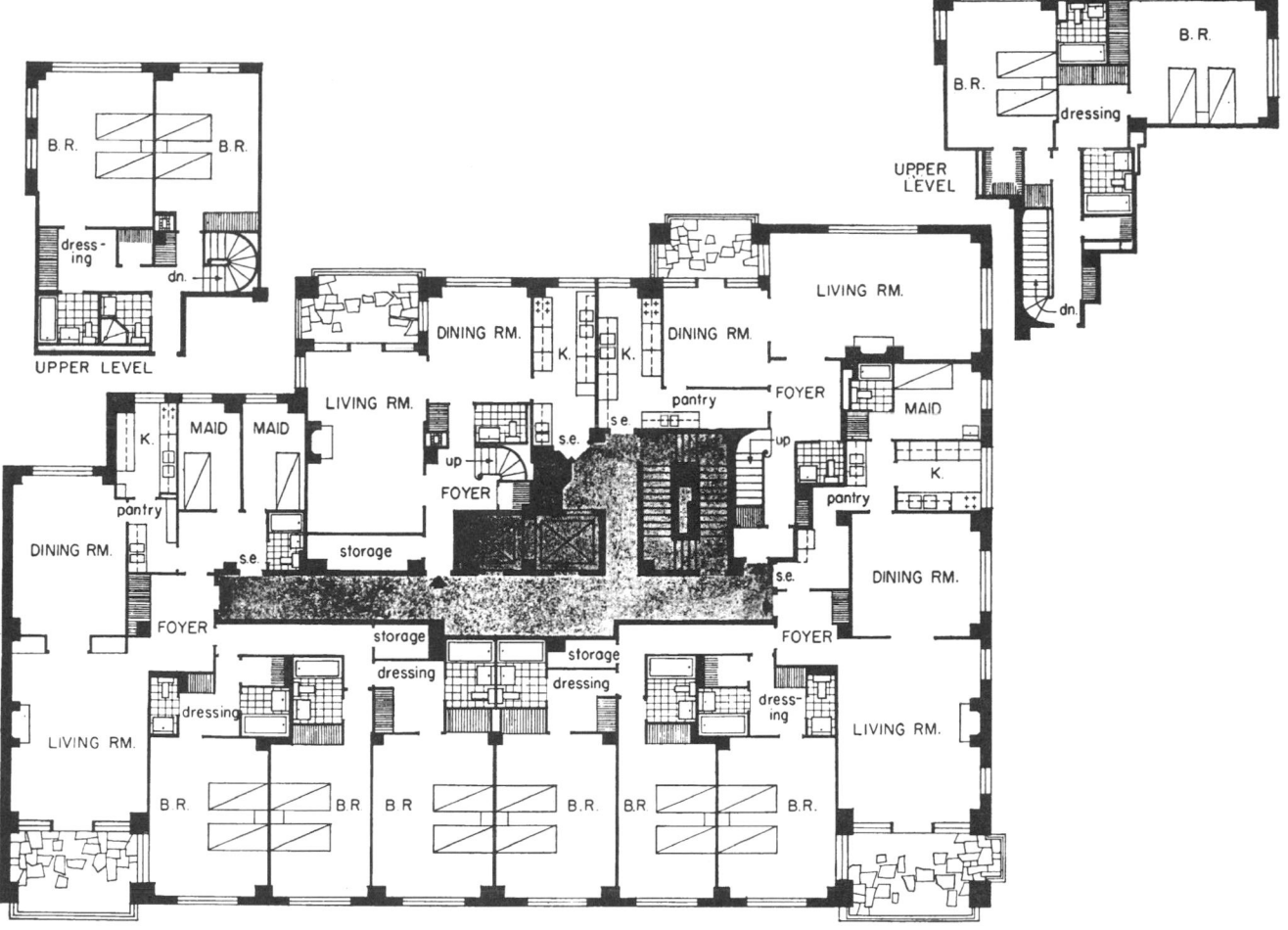

Fig. 3   Apartment House, New York City. L. Shultze—Architect.

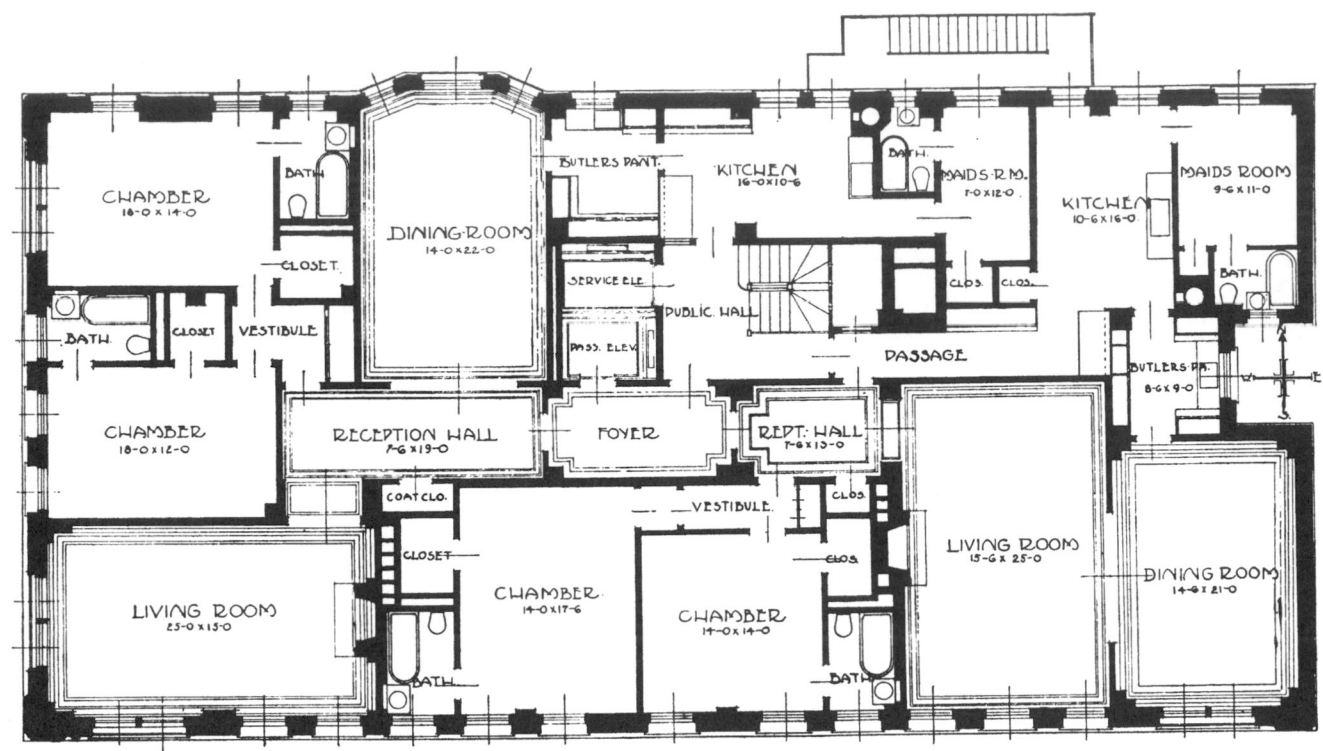

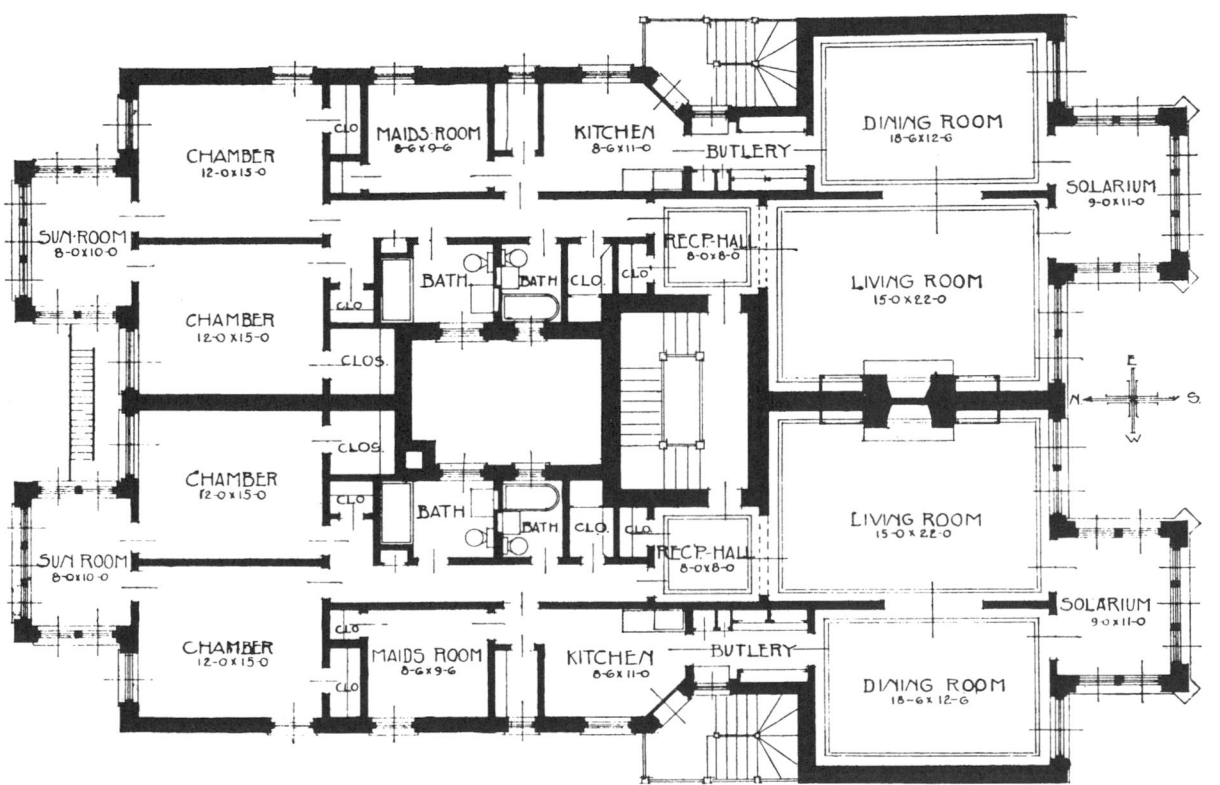

Fig. 4   Typical traditional floor plans, Chicago, Ill.

Fig. 5 630 Park Ave., New York City. This building has twelve stories.

Fig. 6   903 Park Ave., New York City. This building has 17 stories.

8 ROOMS, 3 BATHS, 10 CLOSETS

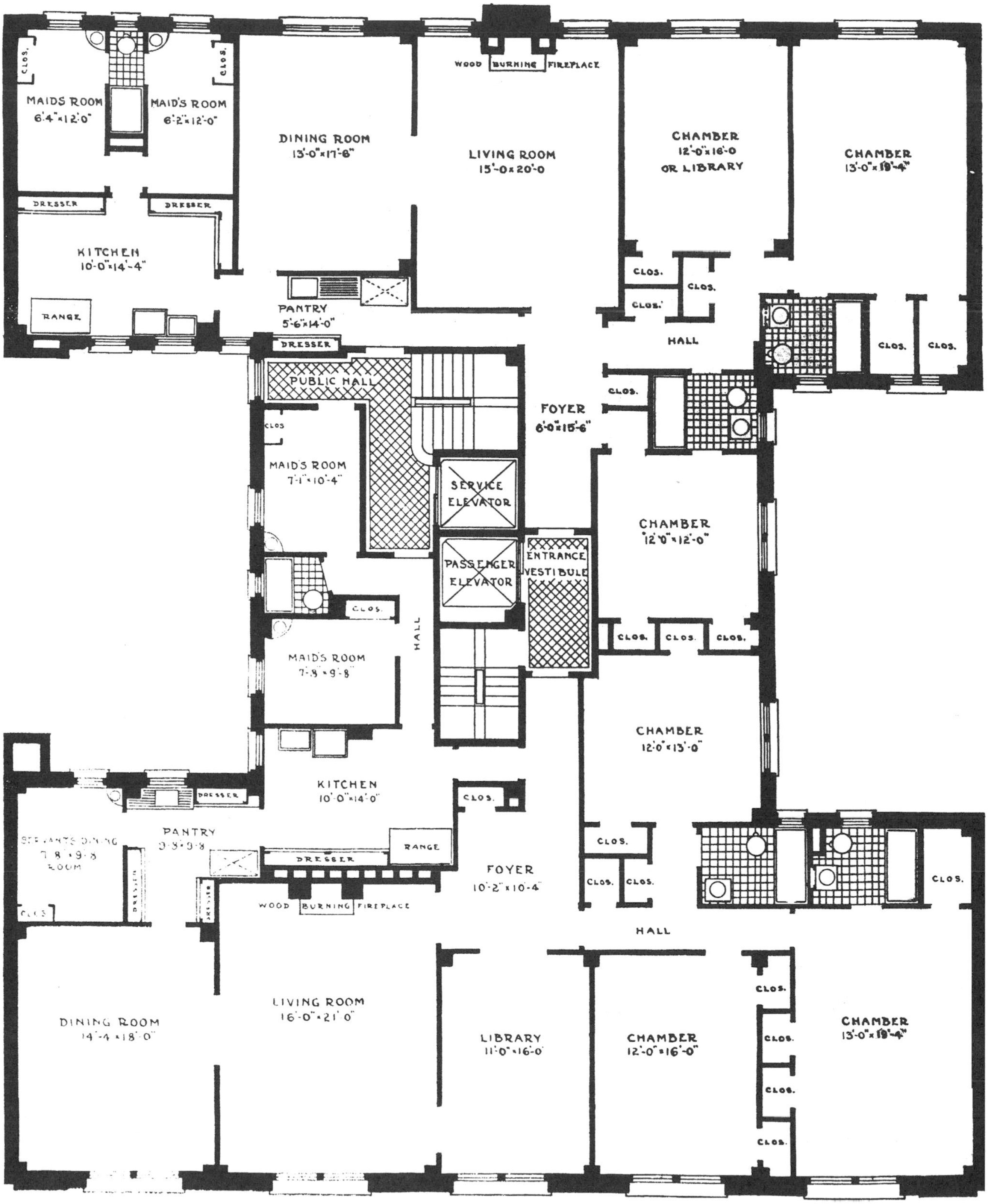

CLOS.

MAID'S ROOM
6'-4"×12'-0"

MAID'S ROOM
6'-2"×12'-0"

CLOS.

DRESSER

DRESSER

DINING ROOM
13'-0"×17'-6"

WOOD BURNING FIREPLACE

LIVING ROOM
15'-0"×20'-0

CHAMBER
12'-0"×16'-0
OR LIBRARY

CHAMBER
13'-0"×19'-4"

KITCHEN
10'-0"×14'-4"

RANGE

PANTRY
5'-6"×14'-0"

DRESSER

CLOS.

CLOS.

CLOS.

HALL

CLOS.

CLOS.

CLOS.

PUBLIC HALL

CLOS

MAID'S ROOM
7'-1"×10'-4"

SERVICE
ELEVATOR

FOYER
6'-0"×15'-6"

CLOS.

CLOS.

CHAMBER
12'-0"×12'-0"

CLOS.

PASSENGER
ELEVATOR

ENTRANCE
VESTIBULE

MAID'S ROOM
7'-8"×9'-8

HALL

CLOS.

CLOS.

CLOS.

CHAMBER
12'-0"×13'-0"

KITCHEN
10'-0"×14'-0"

CLOS.

CLOS.

PANTRY
9'-8"×9'-8

RANGE

SERVANTS DINING
7'-8"×9'-8
ROOM

DRESSER

FOYER
10'-2"×10'-4

CLOS.

CLOS.

CLOS.

CLOS.

CLOS.

WOOD BURNING FIREPLACE

HALL

DINING ROOM
14'-4"×18'-0"

LIVING ROOM
16'-0"×21'-0"

LIBRARY
11'-0"×16'-0

CHAMBER
12'-0"×16'-0"

CLOS.

CLOS.

CLOS.

CHAMBER
13'-0"×19'-4"

CLOS.

10 ROOMS, 3 BATHS, 13 CLOSETS

Fig. 7   112 E. 74th St., New York City. This building has nine stories.

COMBINATION OF HOUSING TYPES

The design of the following dwelling unit and cluster plans was based on data gathered from a review of current housing design practice, a review of the limited information available that relates social patterns directly to physical design, and a survey of several public housing projects in the Boston area.

The urban design method was to develop systematically a range of dwelling unit plans, combine them into buildings and clusters, and group them in progressively larger and more complex levels of community. The result is a fully dimensioned, illustrative housing study ranging from the private dwelling to the aggregate of dwellings—the city.

The plans are examples of the type of development that could take place in the city. The actual housing design and grouping are matters for careful architectural study. The urban design study has established the probable limits within which architectural development could take place as well as providing an expression of urban design intent. The summary illustrates a few representative examples selected from two dozen building types applied in the development of the demonstration plans.

The primary considerations in the distribution of population throughout the city as a whole, and within the several residential areas, have included proximity to transportation facilities, to commercial activity, to open space, to education facilities, and to the quality of the specific sites for containing people. The result is a range of net residential density from 83 to 282 per acre. The gross density, including the recreation areas, is about 40 per acre.

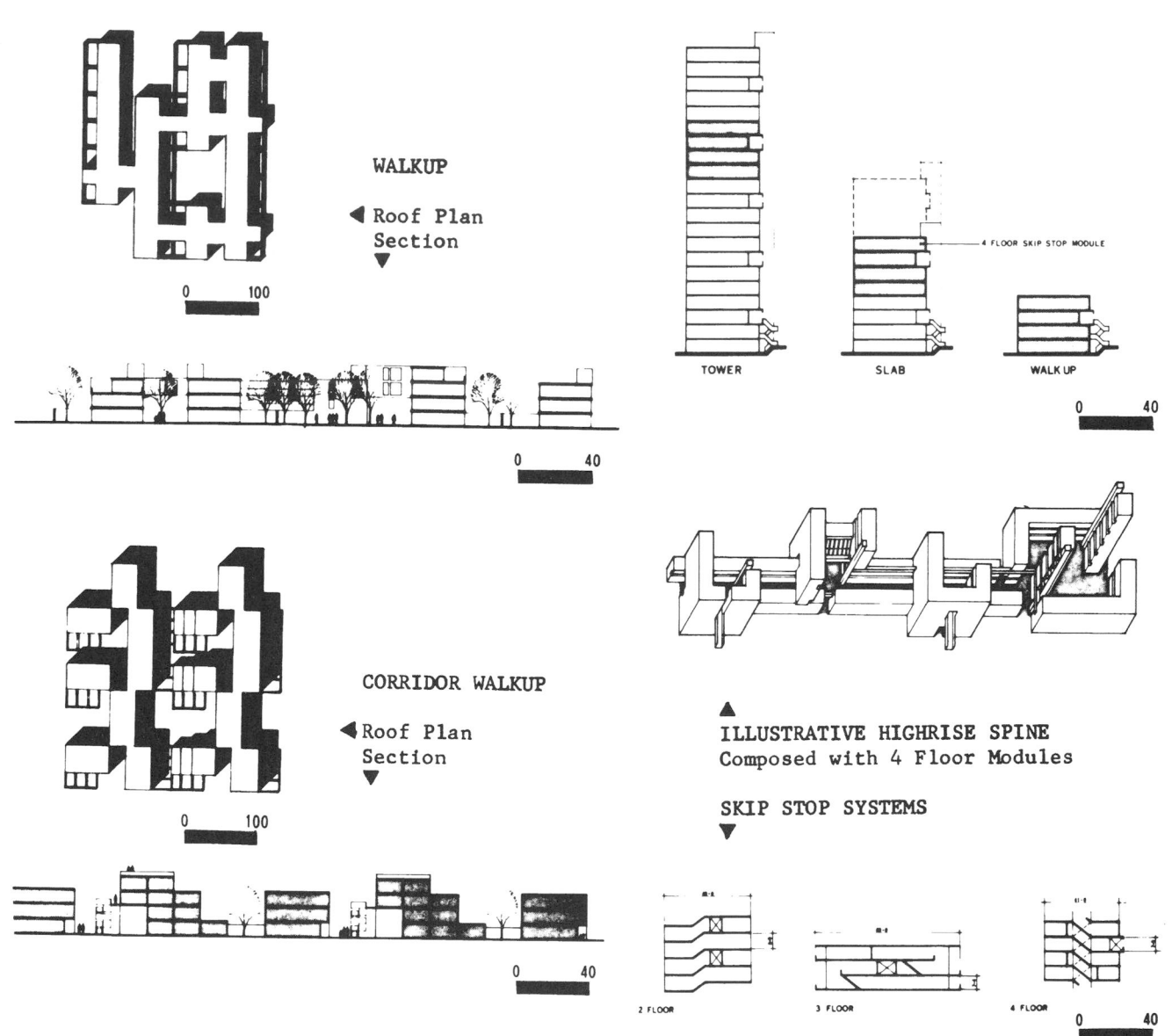

Fig. 1 New communities: one alternative, Graduate School of Design, Harvard University, Cambridge, Mass., 1968.

Comparison of Three Bedroom Units from
various housing types studied

COURTHOUSE

Net Unit Living Area	722	SF
Gross Unit Area	1016	SF
Efficiency	71	%

ROWHOUSE

Net Unit Living Area	697	SF
Gross Unit Area	1030	SF
Efficiency	68	%

WALKUP

Net Unit Living Area	698	SF
Gross Unit Area	966	SF
Efficiency	73	%

CORRIDOR WALKUP

Net Unit Living Area	811	SF
Gross Unit Area	1024	SF
Efficiency	79	%

SKIP STOP SYSTEM

Net Unit Living Area	781	SF
Gross Unit Area	1070	SF
Efficiency	73	%

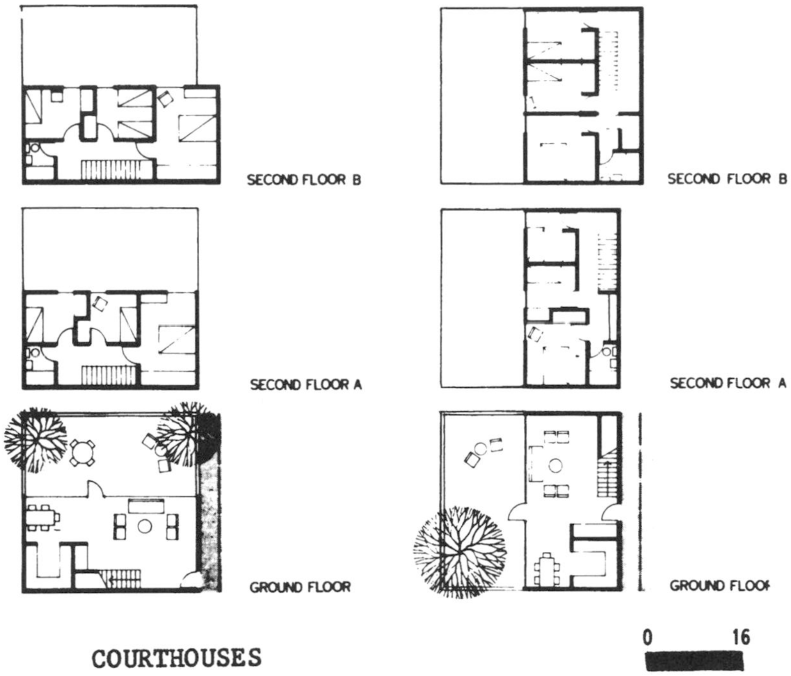

COURTHOUSES

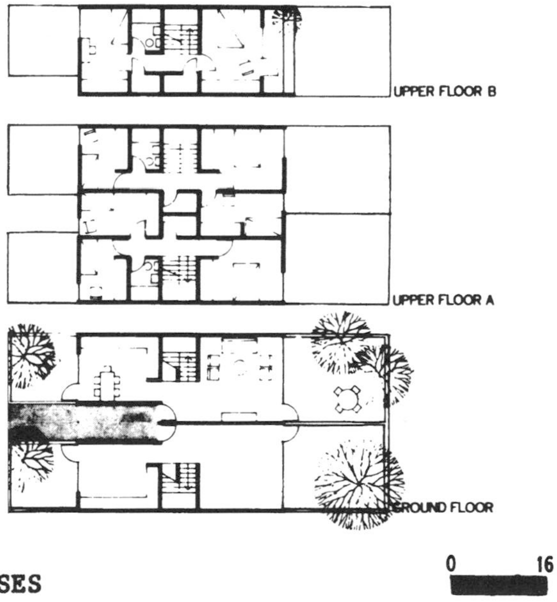

ROWHOUSES

Fig. 2

COMBINATION OF HOUSING TYPES

Comparison of typical clusters of housing types studied

COURTHOUSE CLUSTER
Dwelling Units per Acre	40
Building Coverage	50 %
Floor-Area Ratio	.91

ROW HOUSE CLUSTER

Dwelling Units per Acre	34
Building Coverage	43.5 %
Floor-Area Ratio	.81

WALK-UP CLUSTER

Dwelling Units per Acre	69
Building Coverage	49 %
Floor-Area Ratio	1.63

CORRIDOR WALK-UP CLUSTER

Dwelling Units per Acre	73
Building Coverage	41.6 %
Floor-Area Ratio	1.66

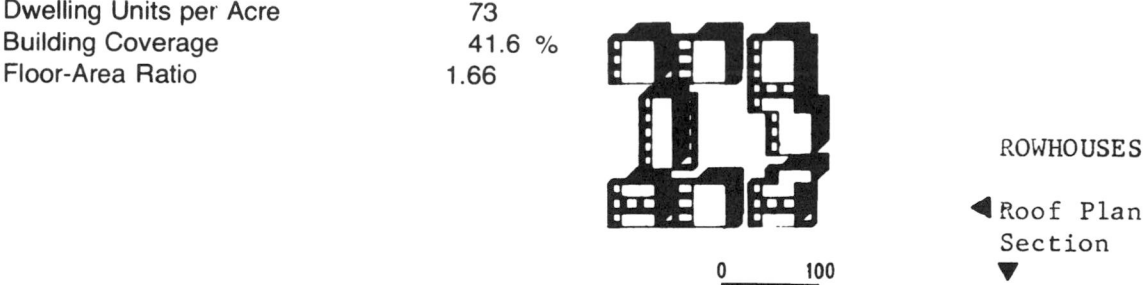

COURTHOUSES

◀ Roof Plan
Section
▼

0   100

ROWHOUSES

◀ Roof Plan
Section
▼

0   100

0   40

Fig. 3

The site of this project consists of an L-shaped corner parcel dropping sharply, 40 ft in some cases, to the Jamaicaway. Perkins Street to the south provides the only vehicular access. To disturb the land the least, to provide open space for recreational use, and to create a park-like setting which would integrate the project with adjacent Jamaicaway Pond Park, 282 apartments were placed in a 30-story tower and seven 2-story town houses. The town houses, which contain three-bedroom units, are strategically placed to continue the low residential character along Perkins Street onto the site.

The plan of the tower provides studio, one-bedroom, and two-bedroom units with several duplex apartments placed at the upper floors. Some of the amenities include a health club, a sauna bath, a valet room, a community room, a swimming pool, a tennis court, a putting green, a convenience shop, and a mezzanine coffee shop. All apartments are air-conditioned from a central system.

Parking for 270 cars is provided in a three-level garage, formed by a continuous ramp, set into the side of the hill. The roof of the garage is landscaped and used for children's play, a sitting area, a putting green, and a tennis court.

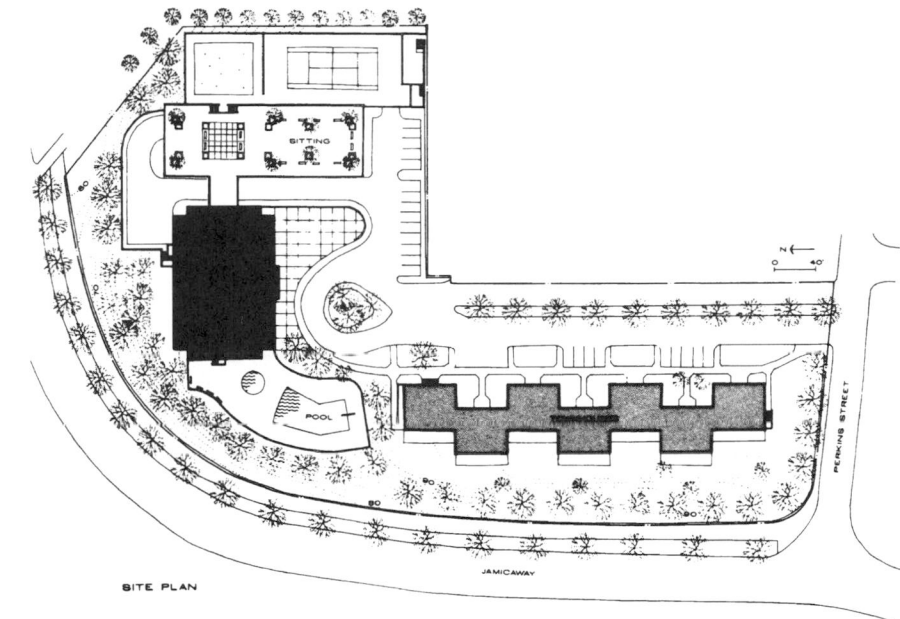

SITE PLAN

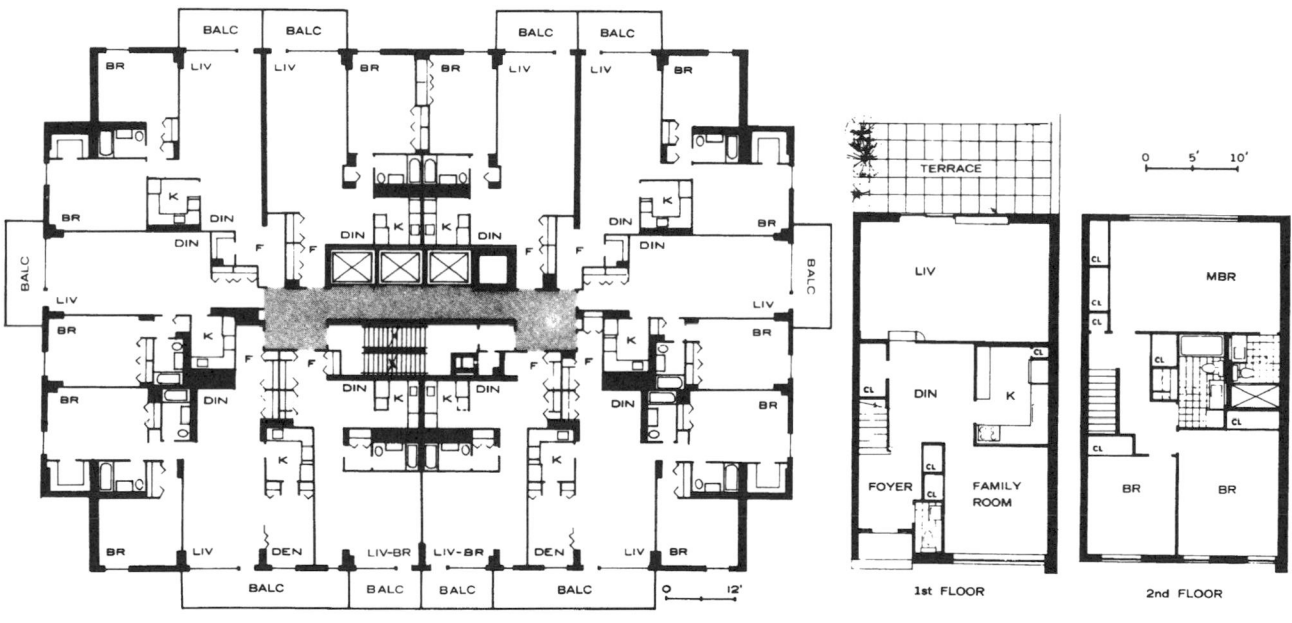

TOWER FLOOR PLAN – 13TH TO 27TH FLOOR

TOWNHOUSE PLAN

Fig. 4   Jamaicaway Tower and Town Houses, Boston, Mass. Paul & Jarmul—Architects.

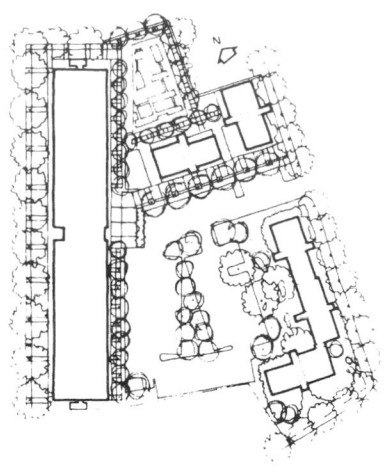

TYPICAL 2-BR UNIT   3-BR UNIT   4-BR UNIT

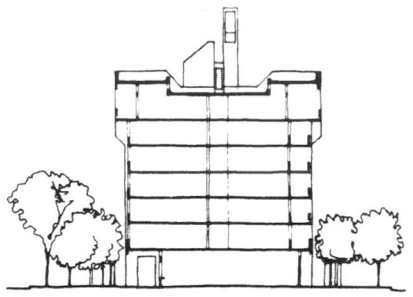

In the seven-story building, the 4-ft width within each "module" formed by the bearing walls contains one two-bedroom apartment and half (either the living room or the bedroom) of a one-bedroom unit.

The three-story walk-ups completing the complex have a total of 228 units and are based on the same construction system. The site plan provides one-to-one parking for the walk-ups, 30 percent parking for the seven-story building.

GROUND FLOOR   TYPICAL FLOOR   SEVENTH FLOOR

Fig. 5 Roxse Housing, Boston, Mass. The Architects Collaborative—Architects.

The community building is at the heart of the project, which will eventually comprise 775 units of housing, 40,000 ft² of medical office space and 6000 ft² of neighborhood shops and recreational facilities. Thirty percent of the units are low income, 65 percent are for moderate incomes, and 15 percent are for market income.

The town houses are constructed over a 1274-car underground garage of precast concrete. Available to all the residents in the high-rise, mid-rise, and town houses are dishwashers, disposals, a swimming pool, tennis and basketball courts, six play areas for small children, a formal plaza with a fountain and extensive landscaping (see Fig. 7).

The mid-rise building steps up from four stories to thirteen.

The high-rise building is 27 stories high.

Fig. 6 Mission Park, Boston, Mass. John Sharratt Associates—Architects.

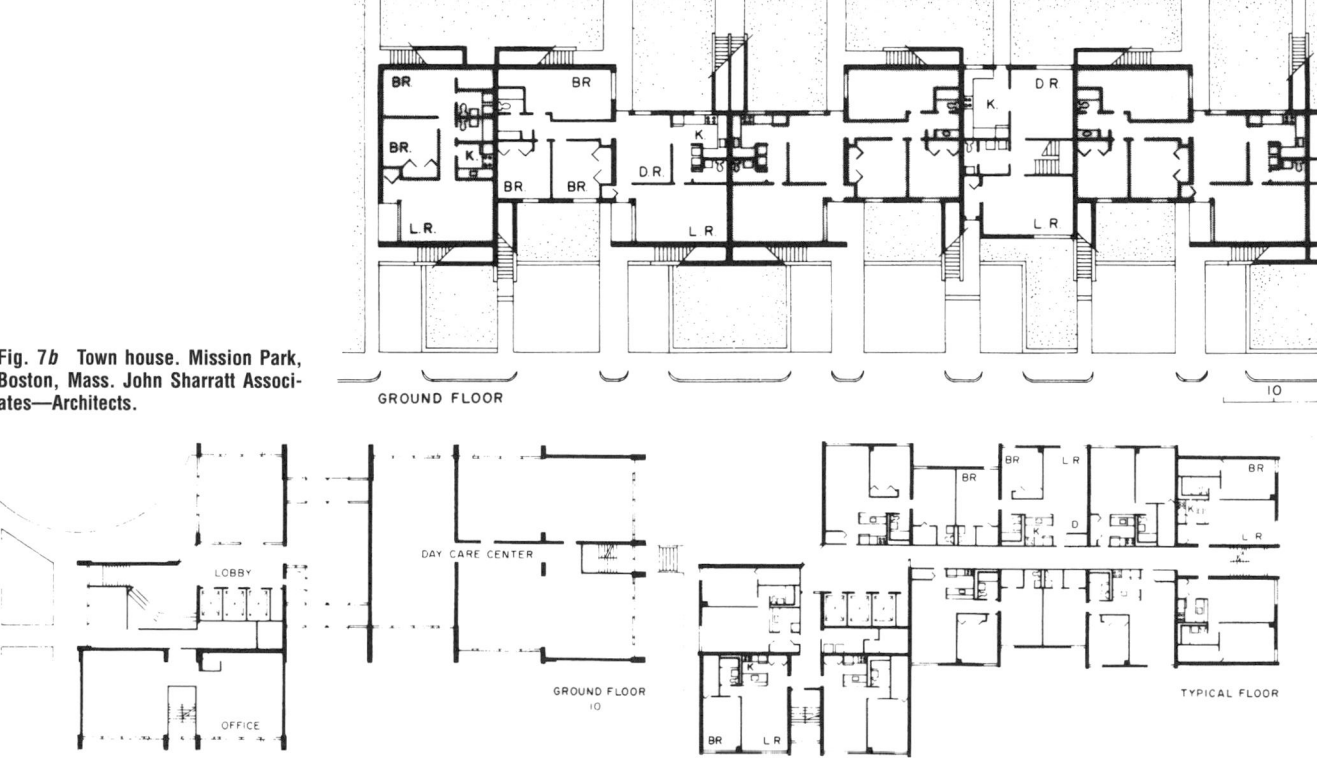

COMBINATION OF HOUSING TYPES

BR
LR
K
D
LAUN
COMMUNITY SPACE
TYPICAL FLOOR
7TH FLOOR
LOBBY
COMMUNITY SPACE
GROUND FLOOR
10

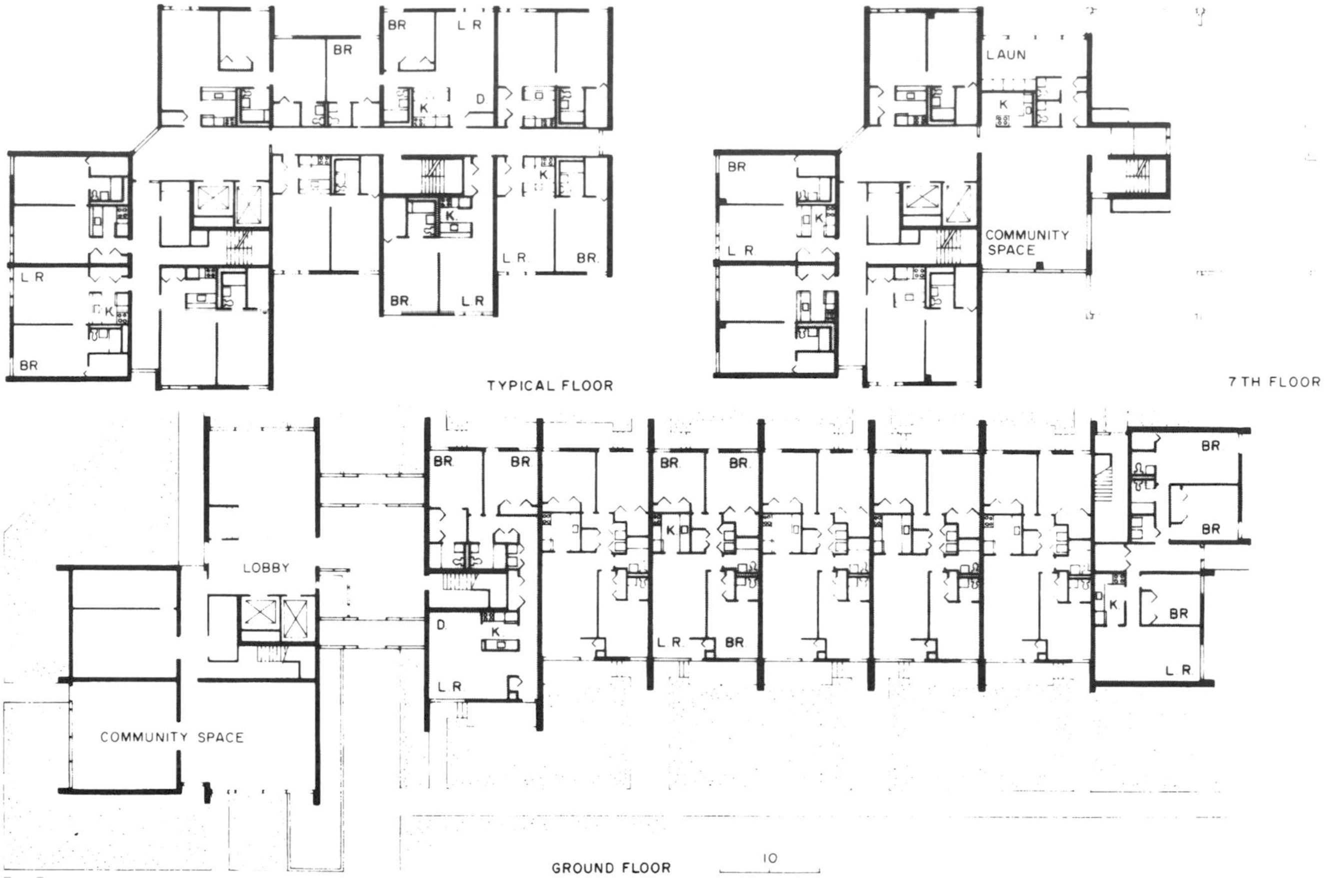

Fig. 7*a* Mid-rise building. Mission Park, Boston, Mass. John Sharratt Associates—Architects.

BR
K
DR
LR
GROUND FLOOR
10

Fig. 7*b* Town house. Mission Park, Boston, Mass. John Sharratt Associates—Architects.

LOBBY
OFFICE
DAY CARE CENTER
GROUND FLOOR
10
BR
LR
K
D
TYPICAL FLOOR

Fig. 8 High-rise building.

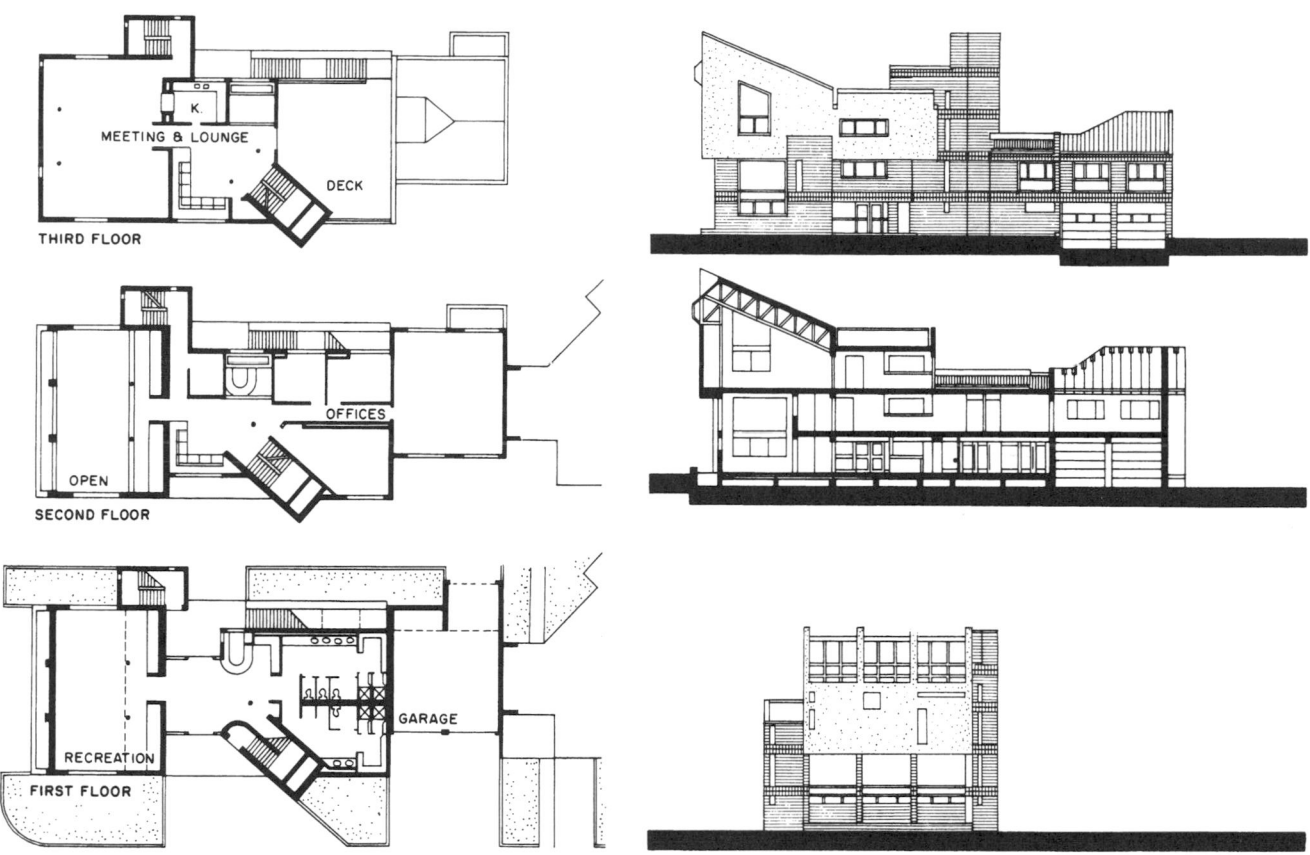

THIRD FLOOR

MEETING & LOUNGE

K.

DECK

SECOND FLOOR

OPEN

OFFICES

FIRST FLOOR

RECREATION

GARAGE

Fig. 9 The community center. Mission Park, Boston, Mass. John Sharratt Associates—Architects.

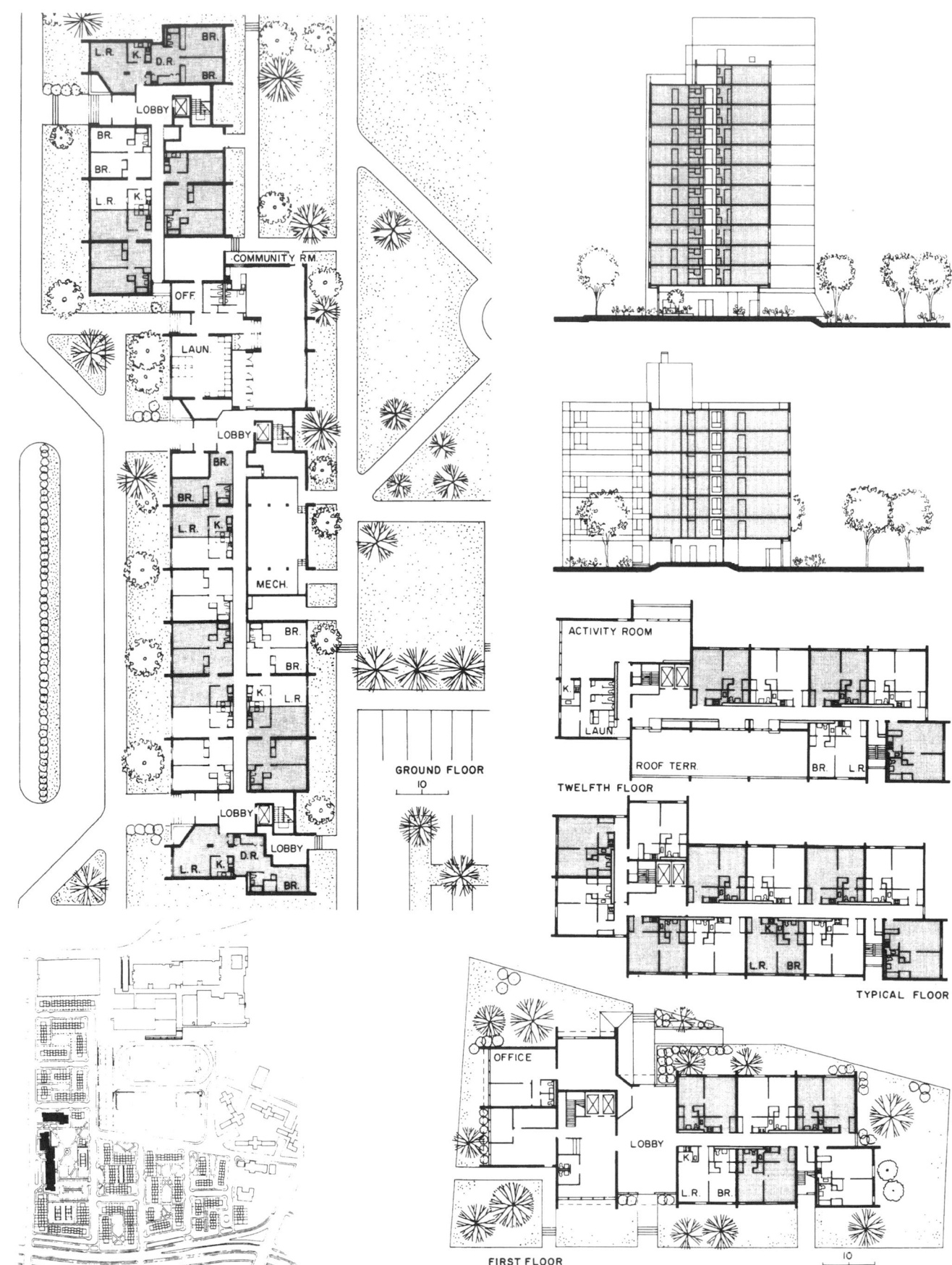

Fig. 10   Haynes House, Smith House, Boston, Mass. John Sharratt Associates—Architects.

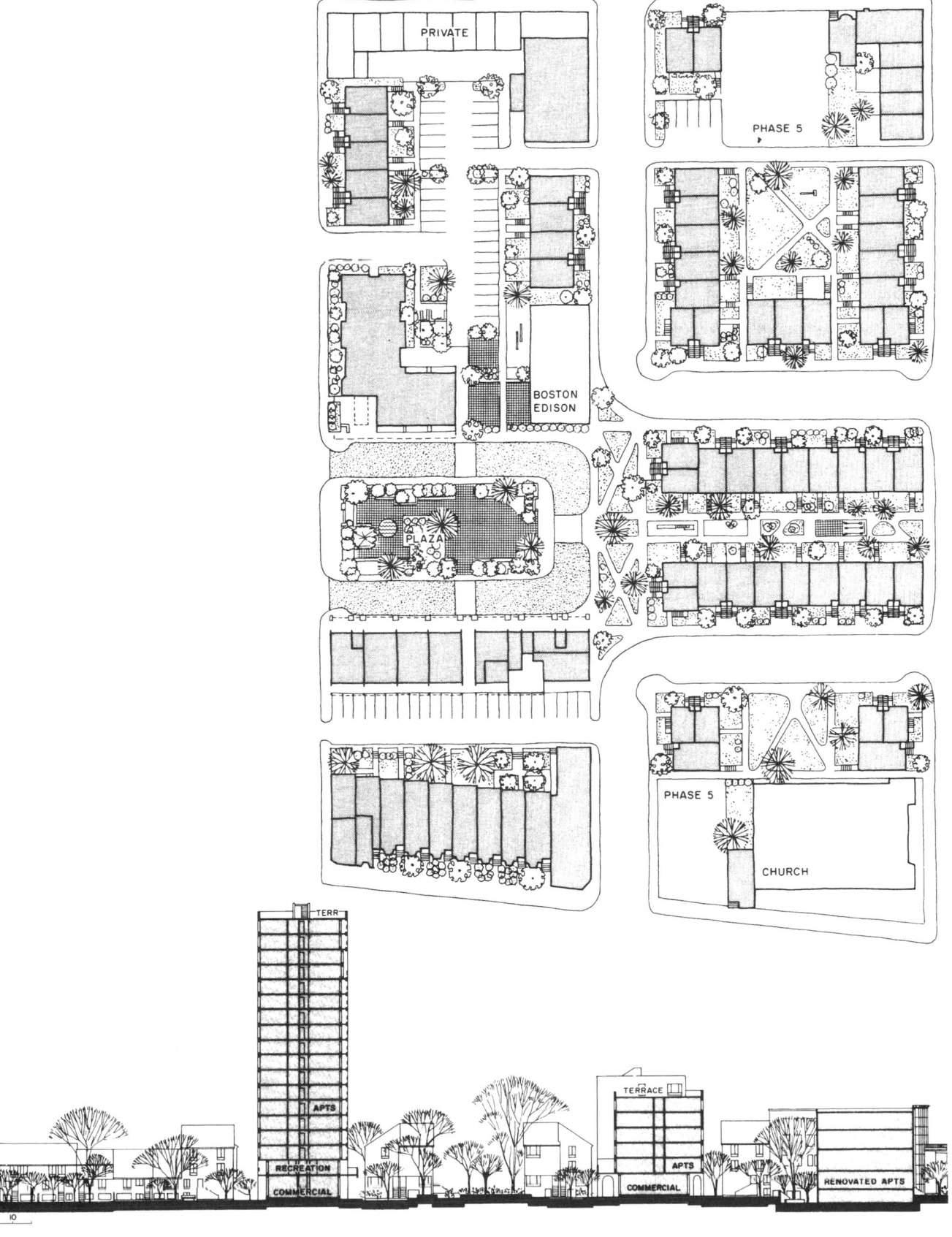

Fig. 11 Haynes House, Smith House, Town Houses of Madison Park, Boston, Mass. John Sharratt Associates—Architects.

NINETEENTH FLOOR

TYPICAL FLOOR

SECOND FLOOR

HEALTH & SOCIAL SERVICES

COMMUNITY RM.

FIRST FLOOR

GROUND FLOOR

FIRST FLOOR

SECOND FLOOR

THIRD FLOOR

Fig. 12   Haynes House, Smith House, Town Houses of Madison Park, Boston, Mass. John Sharratt Associates—Architects.

The through-type apartments created by the exterior circulation scheme have two-way views and through ventilation. In the high-rise approximately half the apartments have balconies, which are distributed freely over the facade.

The plan of the Fairview Heights housing development in Ithaca, New York, consists essentially of two protected groupings of row houses, separated by a high-rise apartment building which is also the visual dominant of the scheme. The high-rise unit is reached by a driveway that splits around its base and is the only automobile access to the site. Parking is provided under the large building at ground level, and along the driveway, otherwise only on the periphery of the 5.3-acre plot. The row houses thus face inward to protected and pleasantly landscaped areas that center on unusual and attractive play areas designed by the architects. The larger of these areas is enclosed by a group of 32 houses (in five buildings) at the lower end of the plot—which slopes 1 ft in 15. The smaller area—for 10 houses in two units—looks out over a space which ends against the steep slope of the adjacent lot (see Fig. 14).

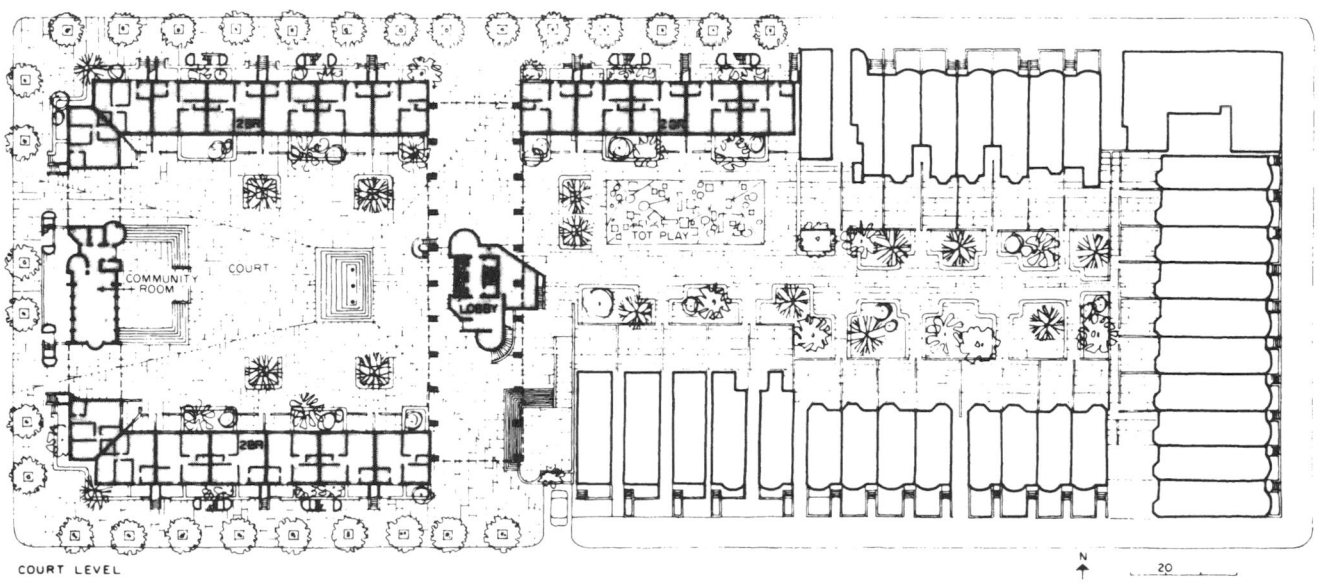

COURT LEVEL

N
20

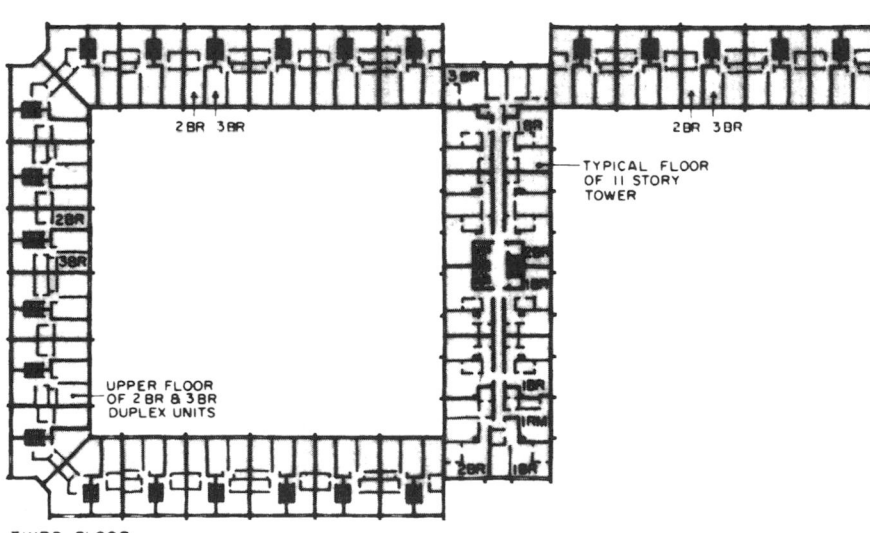

THIRD FLOOR

Fig. 13 Crown Gardens, New York City. Richard D. Kaplan—Architect.

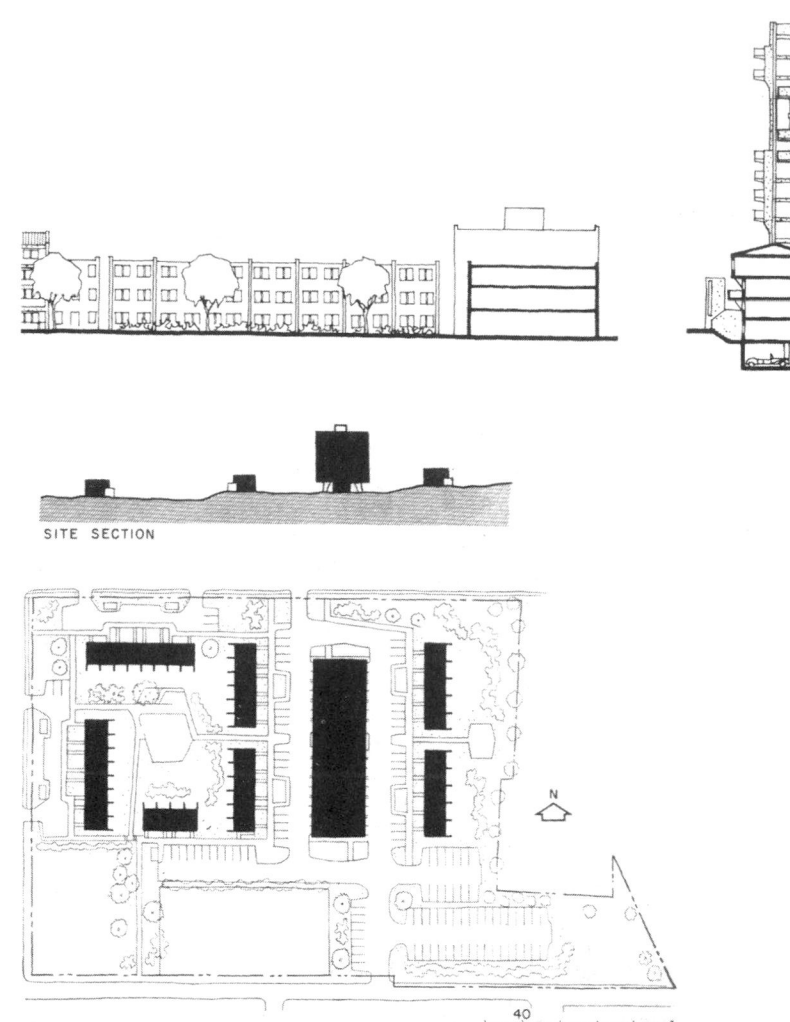

SITE SECTION

40

N

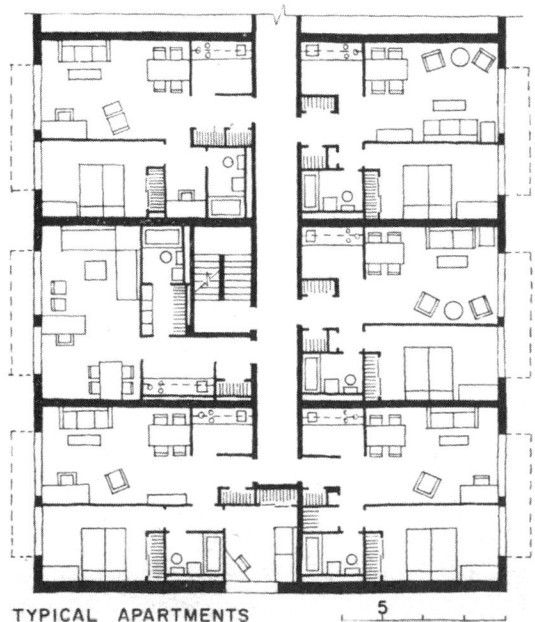

TYPICAL APARTMENTS   5

Fig. 13 (*Continued*)

Two heavily wooded sites in the hills over-looking Ithaca are developed with a long five-story apartment building at the top, and one-story atrium units with private gardens sloping down the hill to take advantage of the views. The plan includes 300 units on 4.8 acres.

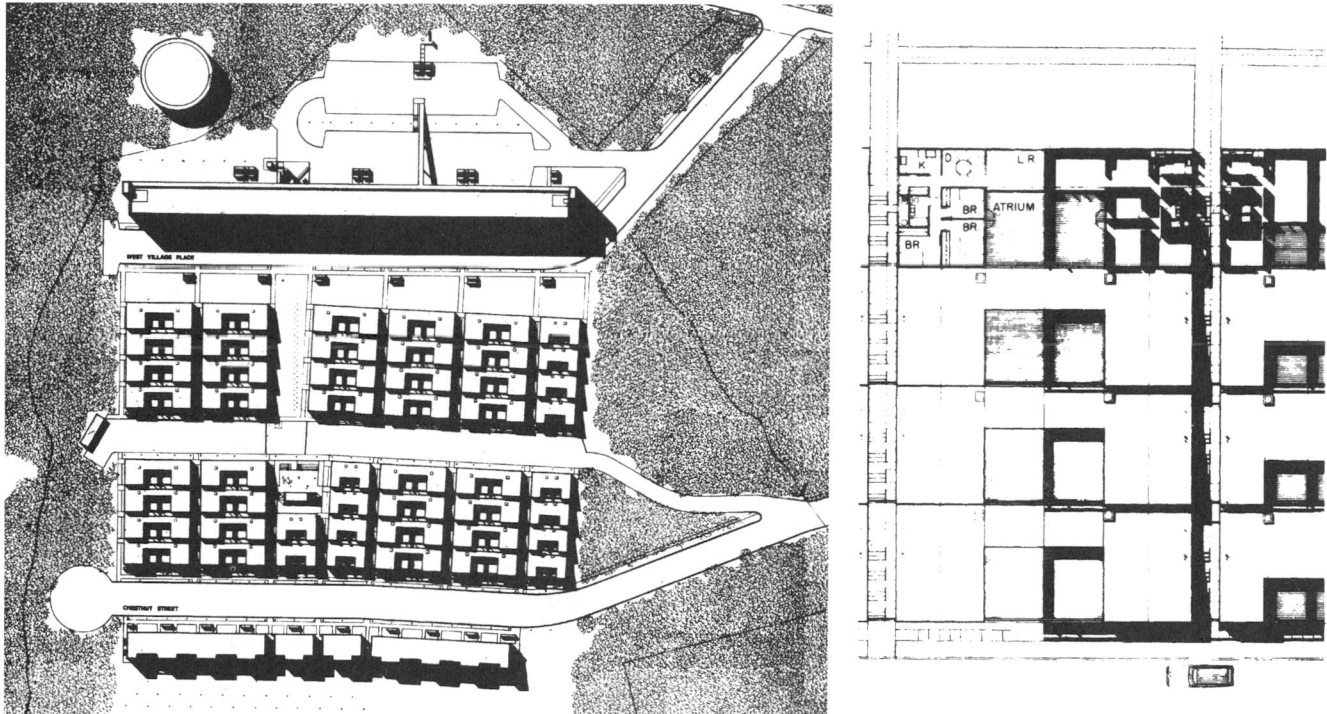

Fig. 14

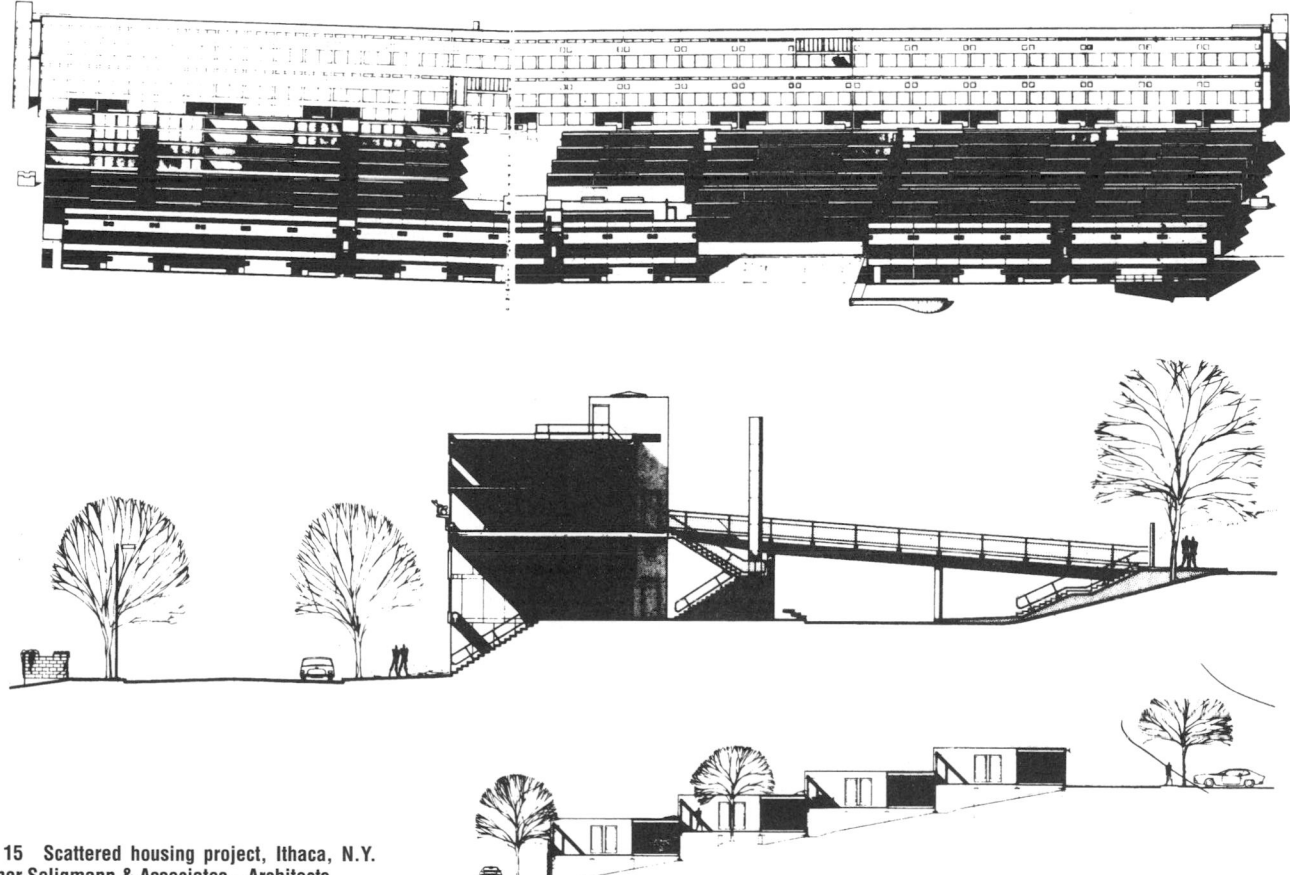

Fig. 15   Scattered housing project, Ithaca, N.Y. Werner Seligmann & Associates—Architects.

COMBINATION OF HOUSING TYPES

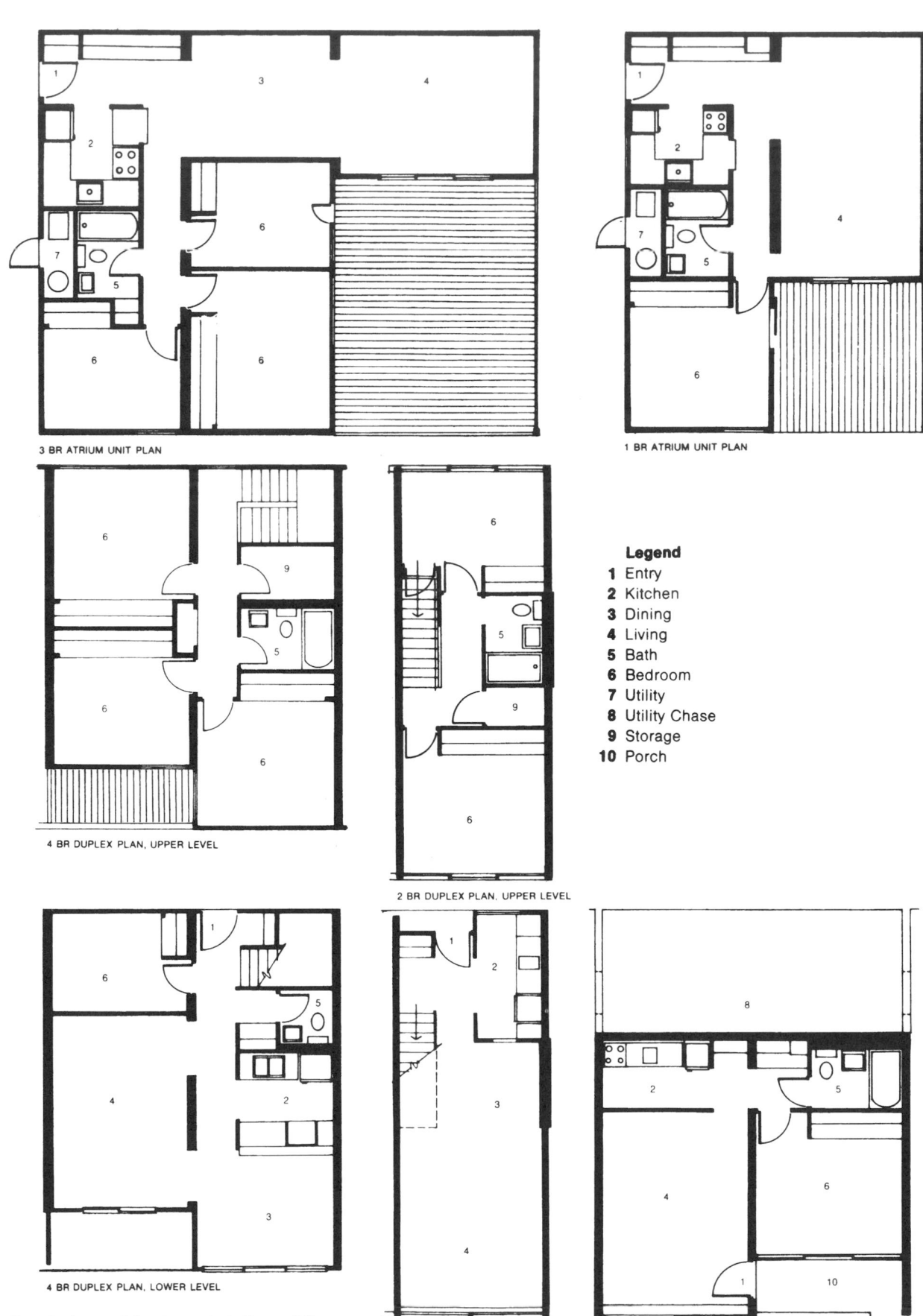

3 BR ATRIUM UNIT PLAN

1 BR ATRIUM UNIT PLAN

4 BR DUPLEX PLAN, UPPER LEVEL

2 BR DUPLEX PLAN, UPPER LEVEL

Legend
1 Entry
2 Kitchen
3 Dining
4 Living
5 Bath
6 Bedroom
7 Utility
8 Utility Chase
9 Storage
10 Porch

4 BR DUPLEX PLAN, LOWER LEVEL

2 BR DUPEX PLAN, LOWER LEVEL

1 BR LOWER-LEVEL UNIT PLAN

Fig. 16 Scattered housing project, Ithaca, N.Y.
Werner Seligmann & Associates—Architects.

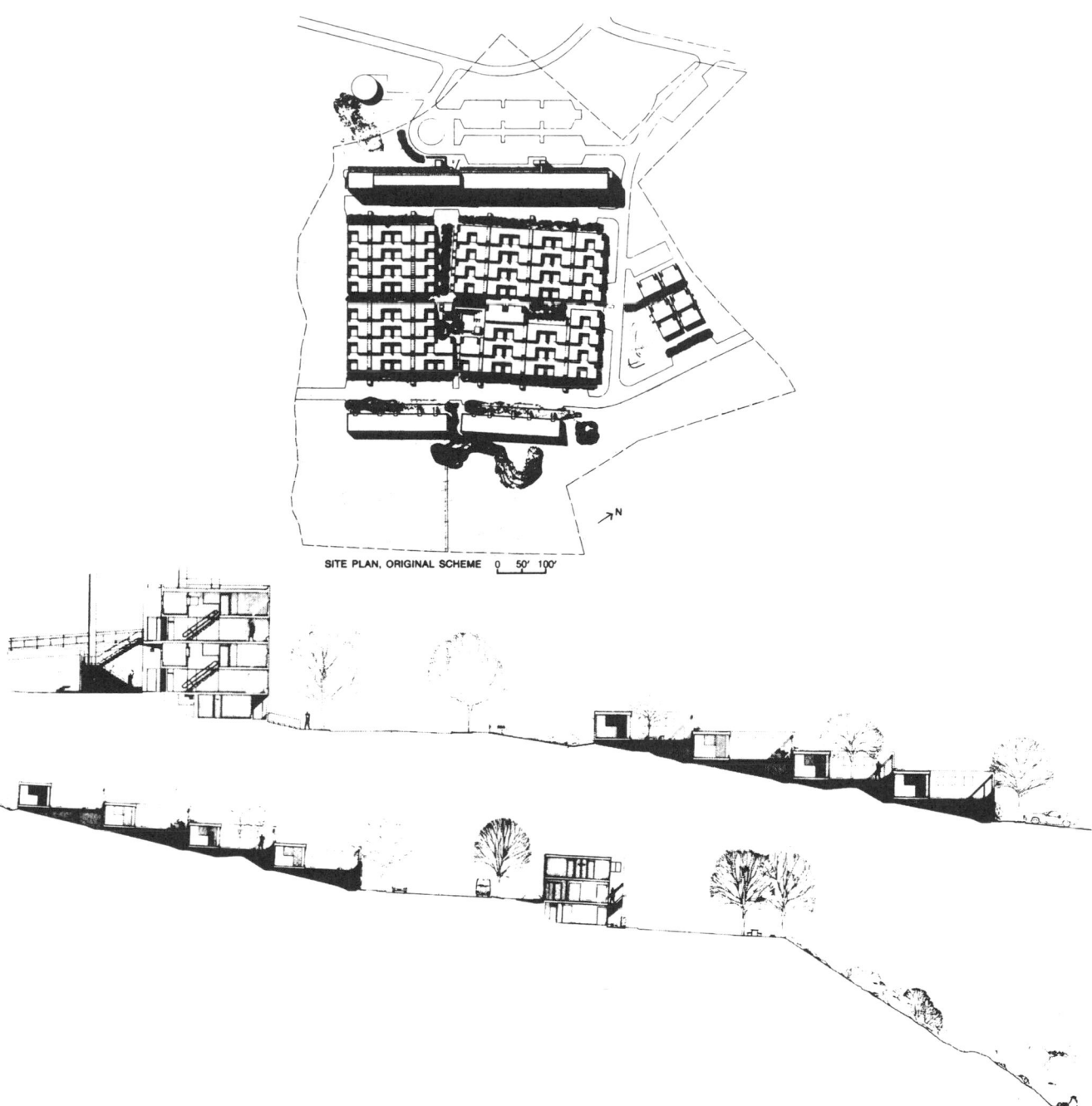

SITE PLAN, ORIGINAL SCHEME   0   50'   100'

Fig. 17   Scattered housing project, Ithaca, N.Y. Werner Seligmann & Associates—Architects.

Apartment Building Amenities

The external design is based on categories or *zones* of usage.

Private Personal or family areas (dwelling units, rear yards, balconies)

Meeting Communal, public areas (open space, parks, etc.)

Transition Connecting, spaces (entry courts, stairs)

Service Support functions (roads, paths, utilities)

The *rules* that determine the location and interrelationships are functions of particular localities and therefore not explicitly definable. As a general rule, however, a linear movement (service to transition to private or meeting) is made.

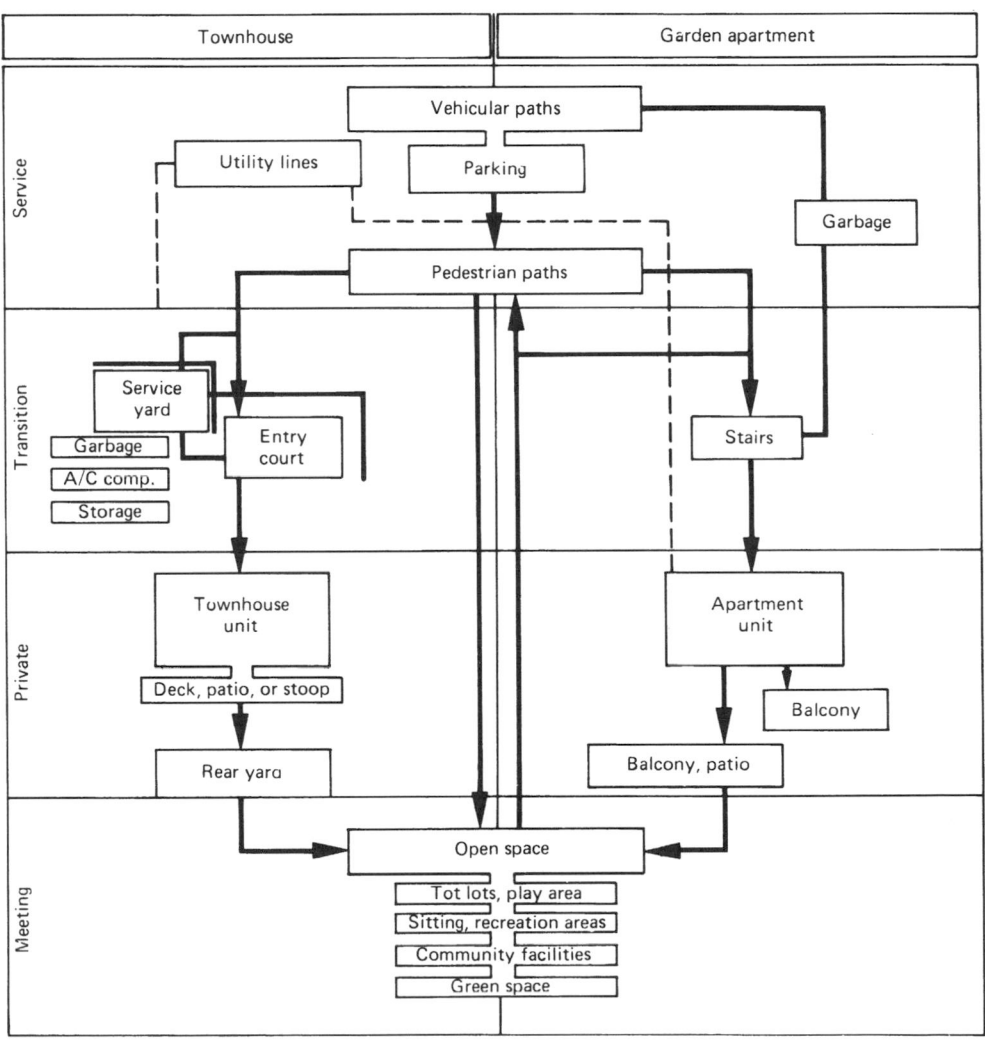

Fig. 1

OPEN SPACE AND RECREATION

The recreational activities occurring within the project are directly controlled by the open space concept. The size, shape, method of containment and ground plane treatment set the stage for most outdoor activities. The following categories of spaces should be incorporated in the site plan.

Spatial classification
1. Private or semiprivate spaces
2. Activity spaces
3. Social spaces
4. Neighborhood spaces

Open space There should be a smooth, flowing open space network. In some cases, it may be possible to take advantage of an existing grove of trees or other natural amenities on the site. This is desirable. Great benefits in livability can be achieved through an imaginative and logical combination of open spaces.

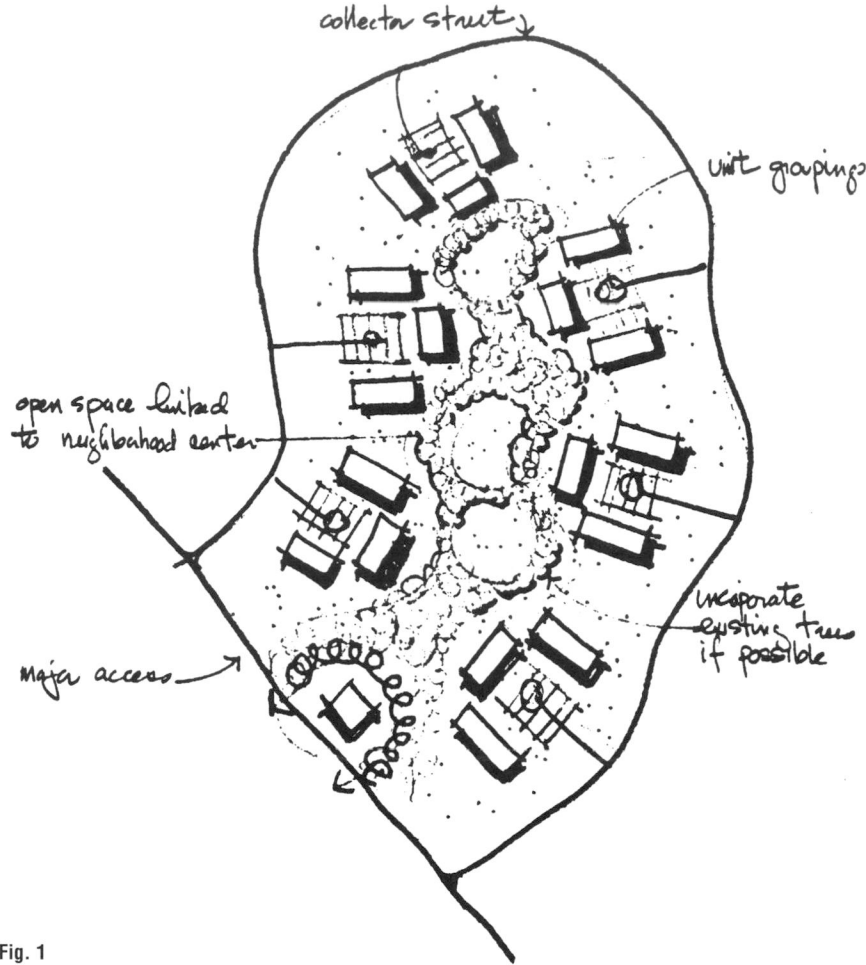

Fig. 1

Private or semiprivate space Each family is entitled to its own outdoor space. The space shall be a minimum of 12 ft deep and extend the width of the unit. Complete enclosure is desirable. However, wooden fences or freestanding walls should be located to either side of the space. The ground plane should include a 100 ft² paved patio. The remaining area may be lawn or ground cover. The end of the space opposite the face of the building may be contained with a planting screen. Building offsets help create interest in the building facades, while at the same time helping to enclose private spaces.

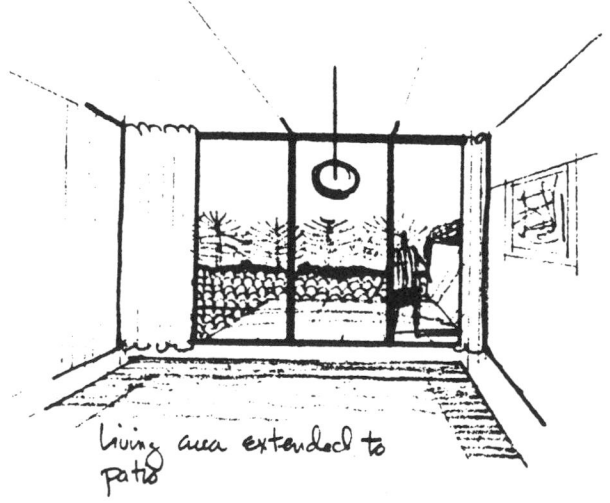

Fig. 2

Activity spaces Generous expanses of lawn area are desirable and will provide a setting for a virtually limitless number of informal games and other activities. (See Fig. 3.) Simultaneously, a feeling of openness is experienced. These areas may be visually accessible to the general public, but are only to be used by the project's residents. In general, minimum dimensions for these spaces should be 40 by 90 ft. They should be varied in size and shape and should be woven into the overall open space concept.

The individual play courts dispersed throughout the project should be located away from private or semiprivate spaces. Planting, grading, or architectural elements may also be used to buffer these two zones. There should be one play court for every 100 bedrooms. Each of the play courts must have a minimum of five separate activities. A partial list of typical activities is provided in Fig. 4.

Social spaces Opportunity for social contact should be encouraged but not forced. Social contact occurs in entrance courts, play courts, and other places within the overall walkway system. These spaces might include benches and canopy trees for shade and interest. This provides the occupant with a place to relax and converse with neighbors.

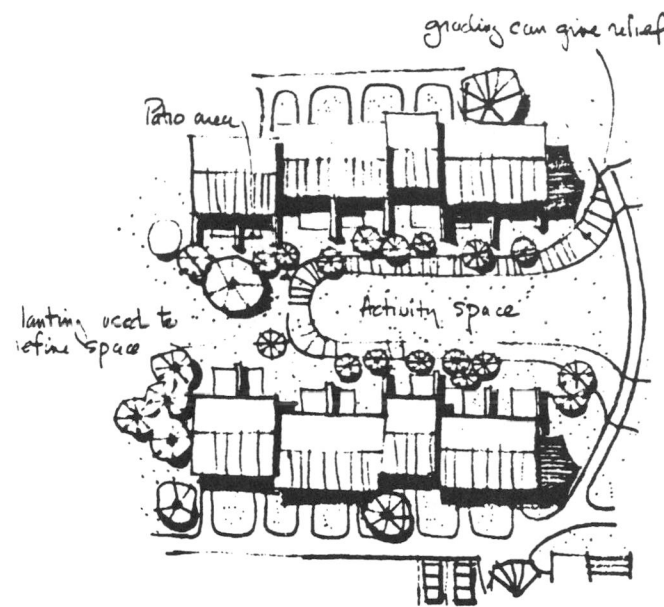

Fig. 3

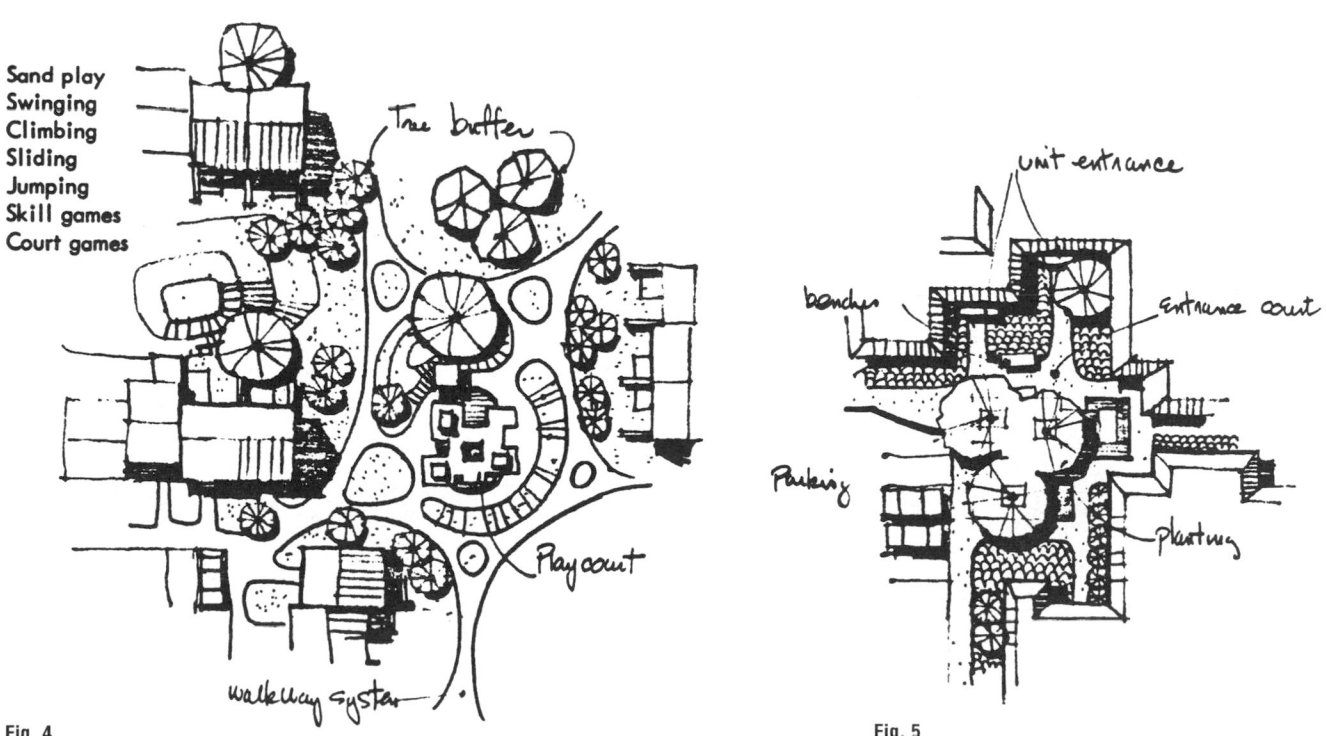

Sand play
Swinging
Climbing
Sliding
Jumping
Skill games
Court games

Fig. 4

Fig. 5

OPEN SPACE AND RECREATION

Neighborhood spaces These areas include outdoor facilities clustered about the neighborhood center and the various play courts located throughout the development. The major emphasis in the development of outdoor facilities included at the neighborhood center should be to provide a spectrum of activities which can accommodate all age groups. The facilities and activities should be arranged in a manner which keeps user conflict to a minimum.

In determining the types and quantities of facilities to be provided, the surrounding off site facilities must be considered. For example, baseball fields, basketball courts, etc., of an adjacent public playground or school may reduce the need for such activities on the site. Instead of duplicating these facilities, money might better be spent for expanding the variety of facilities. Conversely, a lack of adjacent recreational facilities may increase the demand and number of facilities required on a site.

Recreational facilities may include court games, shuffle board, open space for field games, swimming pool, etc. Facilities and spaces of this type are strong factors in creating a sense of neighborhood or community cohesiveness for the development's residents.

Fig. 6

Outdoor Recreation

The outdoor recreation area should provide for a wide range of possible activities. Flexibility is a major consideration in this area.

Accessibility This area should be protected and separate from adjacent properties and roads. It should be easily accessible to the building.

Orientation The area should get some sun.

Furnishability The area should be designed to utilize portable furniture. Effort should be made to provide multiuse recreation potential.

Spatial characteristics The activities which can occur within a recreation area are divided into two categories, active and passive activities. It is important that a visual stimulus or attraction be a part of the passive activity area. One possibility is to organize passive and active areas so that the passive areas become a place to watch the active areas. There must be enough separation, however, so that active areas do not overpower the passive areas. A range of passive areas going from a quiet nook to a sideline seat at a volleyball court might be a solution to this problem. The outdoor recreation area should have human scale.

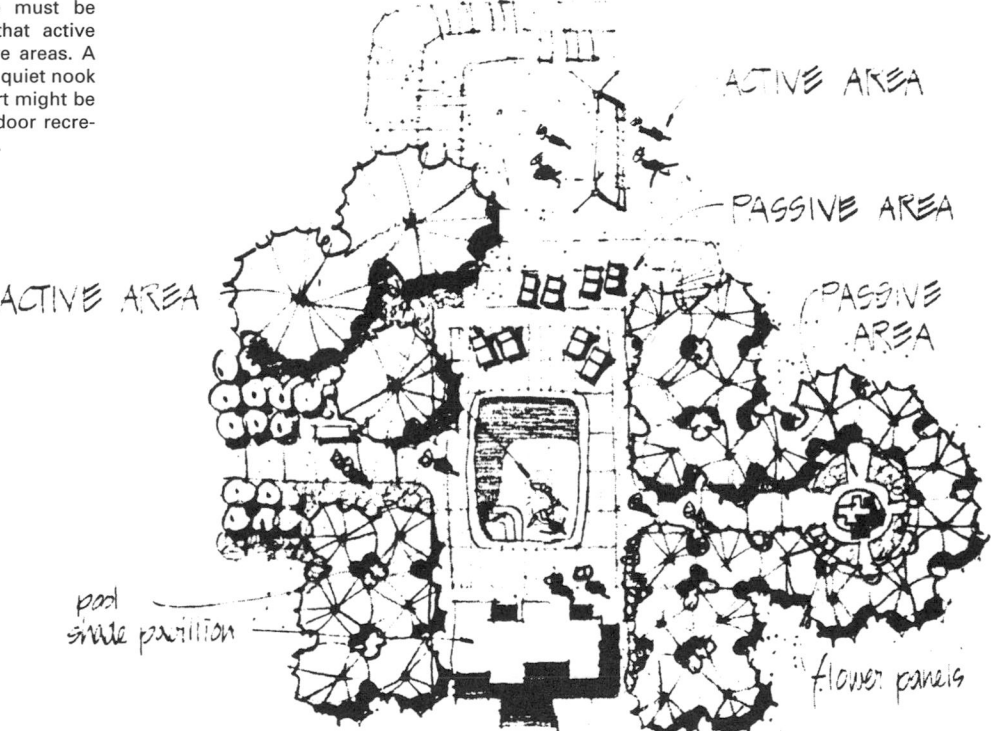

Fig. 7   Passive and active area relationships for a typical high rise—low rise.

INDOOR/OUTDOOR FACILITIES FOR HIGH-DENSITY LIVING

TABLE 1 Indoor Facilities for High-Density Housing

Facility or area	Space/design standards	Location and description
Indoor pool, 1 per 200 units	60 × 30 ft with 3- to 5-ft depth; no diving board	Adjacent to exercise rooms, sauna, day nursery, sun deck, and patio; may be operated by public or private agency
Sauna and exercise area, 1 per 200 units	Occupancy space for 24 adults with three-tiered sauna to give temperature variation	Near shower rooms; may include coin-operated exercise machines and be staffed on a part-time basis
Games room according to expressed interest	Minimum 20 × 30 ft with good storage space; space as flexible and adaptable as possible for multiuse	Near laundry area; may include fireplace, bar facilities and sitting space for socializing; indoor play space for parties and movies
Handball or squash courts, 1 per 200 units	Two courts within 50 × 50 ft × 20 ft high	Near shower area; could be matted for karate, judo, and wrestling; could be used for table tennis
Workshop and auto bay	Minimum two-car capacity with lockable cupboards; ample electrical outlets and work benches	Adjacent to parking area; away from incompatible areas; tools could be rented
Craft room according to expressed need	Minimum area of 20 × 30 ft; adequate sinks, electrical outlets, and good lighting; blackboard and work tables should be included	Near laundry area; space should be as adaptable as possible for multiuse

TABLE 2 Outdoor Facilities for High-Density Housing

Facility or area	Space/design standards	Location and description
Preschool play area, 1 per building	800 to 4000 ft^2; can be part of a larger open space area and provide sitting area for adults; should be visually pleasant	Adjacent to laundry room and within view of balconies; space should be protected from dominance by older age groups; equipment scaled to preschool size
Open area, 1 per building	Minimum space of 150 × 200 ft within a level, well-drained grassed area	Within a few minutes walking distance; equipped for active needs of all age groups
Hard-surfaced area, 1 per building	Minimum of 40 × 60 ft	May be part of the visitor parking area equipped for organized and unorganized sports and games, i.e., badminton, volleyball, floor hockey
Passive area, 1 per building	Minimum of ¼ acre with mixed sun and shaded areas, a natural environment with trees, shrubs, and flowers	Adjacent to children's playground equipped for sun bathing, barbecuing, and such games as horseshoes, shuffleboard, and croquet
Garden plots, 1 per building	Area will vary according to interest	May be adjacent to building or part of a community garden area
Trails and linkage system	Walkways, bikeways that link the site to other open space and public areas	Develop wherever possible to provide access to nearby school yards or park areas

TENNIS

Recommended area Ground space is 7200 ft² minimum.

Size and dimension Playing court is 36 by 78 ft plus at least 12 ft clearance on both sides or between courts in battery and 21 ft clearance on each end.

Orientation Orientation of long axis is to be north-south.

PADDLE TENNIS

Recommended area Ground space is 3200 ft² minimum to edge of pavement.

Size and dimension Playing court is 20 ft 0 in by 30 ft 0 in plus a 15-ft minimum space on each end and a 10-ft minimum space on each side or between courts in battery.

Orientation Preferred orientation is for the long axis to be north-south.

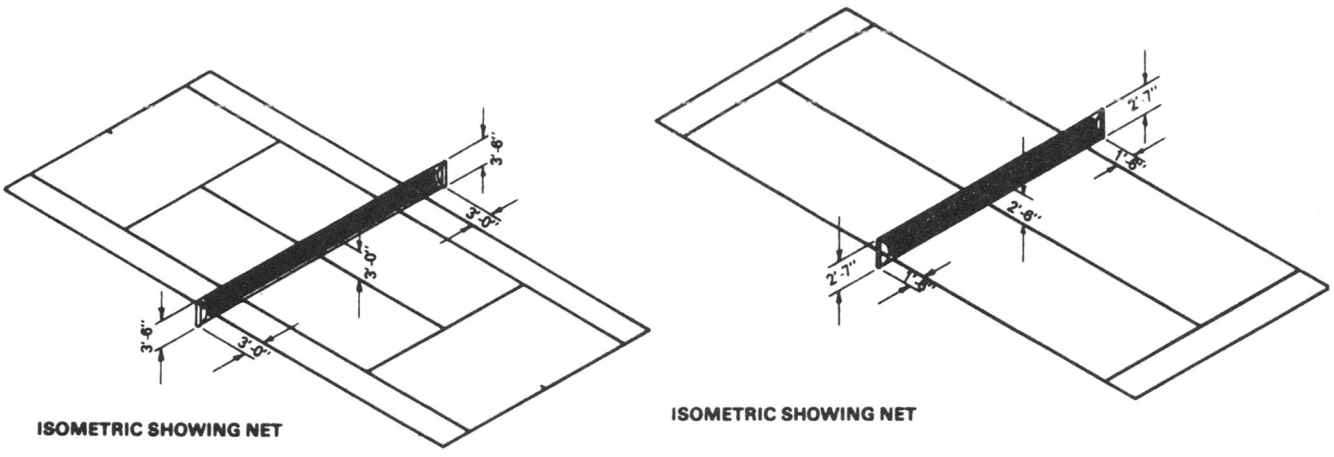

ISOMETRIC SHOWING NET

ISOMETRIC SHOWING NET

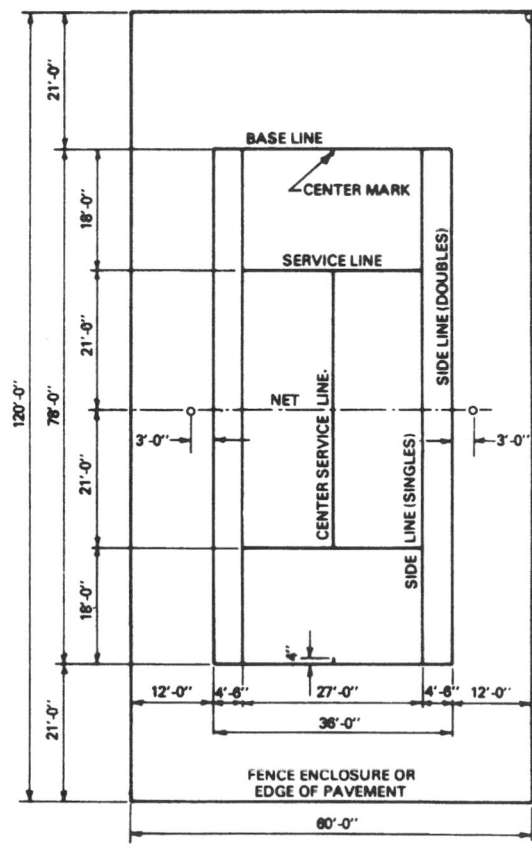

COURT LAYOUT

Fig. 1

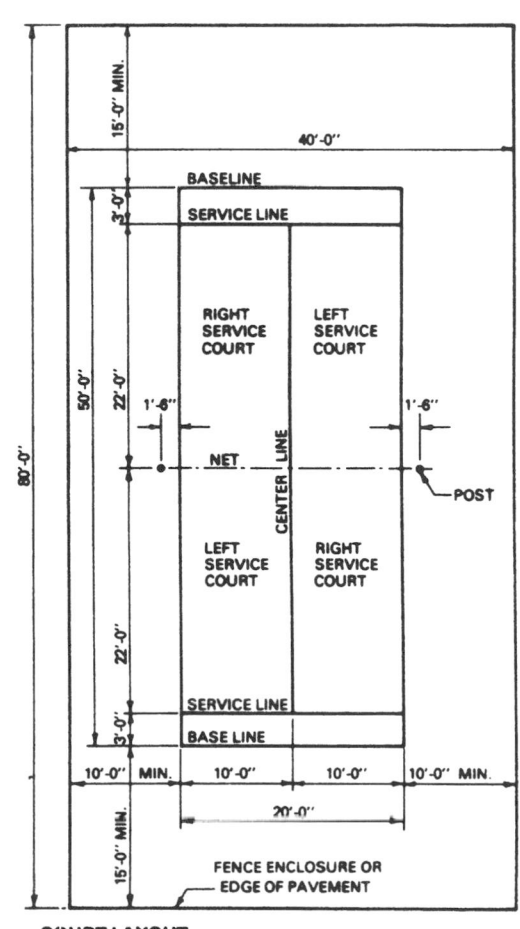

COURT LAYOUT

Fig. 2

DECK TENNIS/PLATFORM TENNIS

DECK TENNIS

Recommended area Ground space is 1300 ft² including clear space.

Size and dimension Singles court is 12 ft 0 in by 40 ft 0 in. Doubles court is 18 ft 0 in by 40 ft 0 in. Additional paved area at least 40 ft 0 in on sides and 5 ft 0 in on ends is recommended.

Orientation Preferred orientation is for the long axis to be north-south.

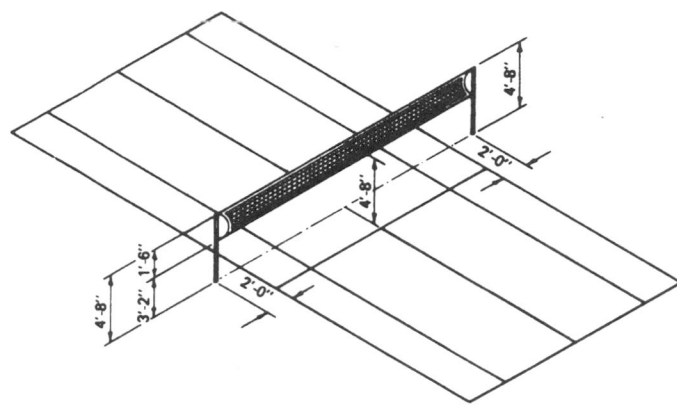

ISOMETRIC SHOWING NET

Fig. 1

PLATFORM TENNIS

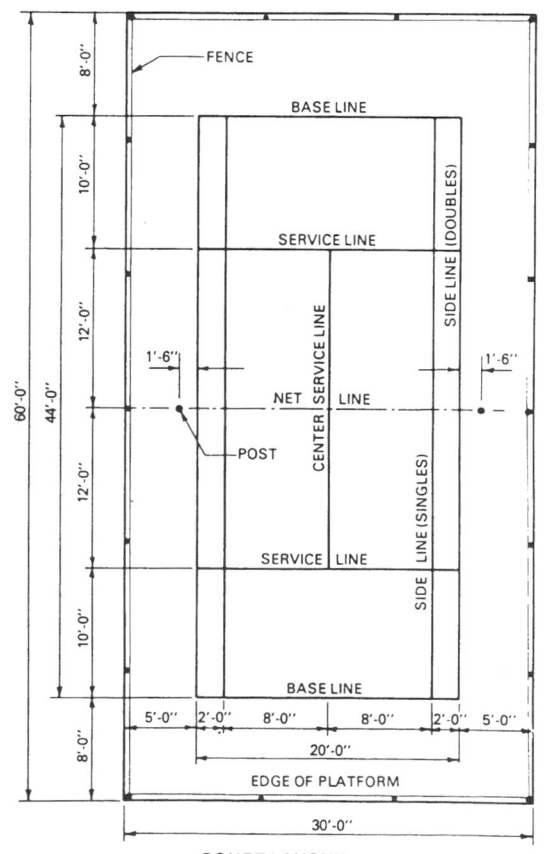

COURT LAYOUT

Fig. 2

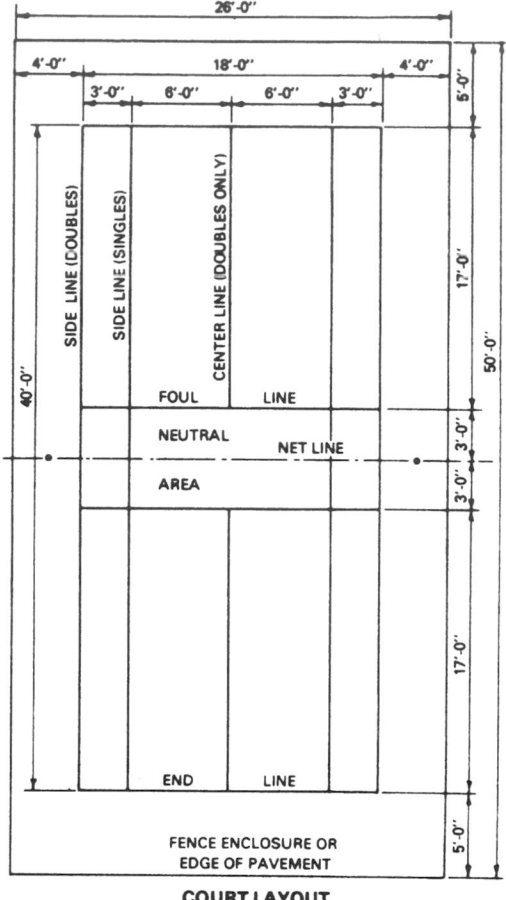

COURT LAYOUT

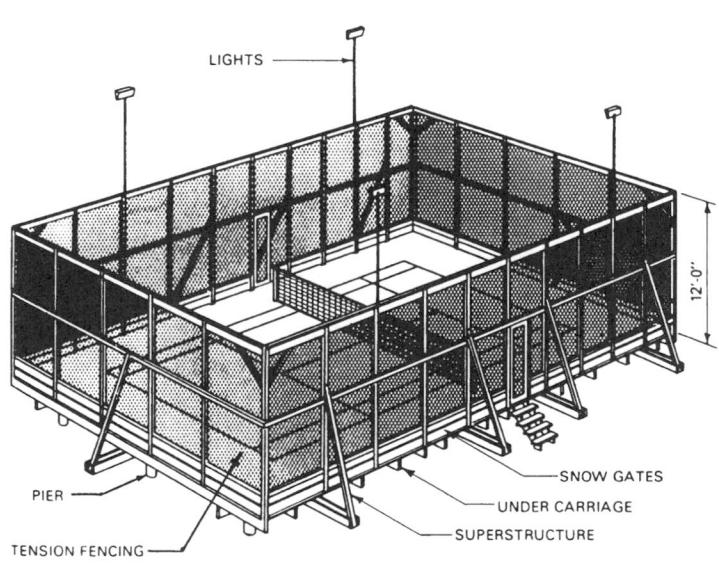

ISOMETRIC SHOWING FENCE (TYPICAL WOOD CONSTRUCTION)

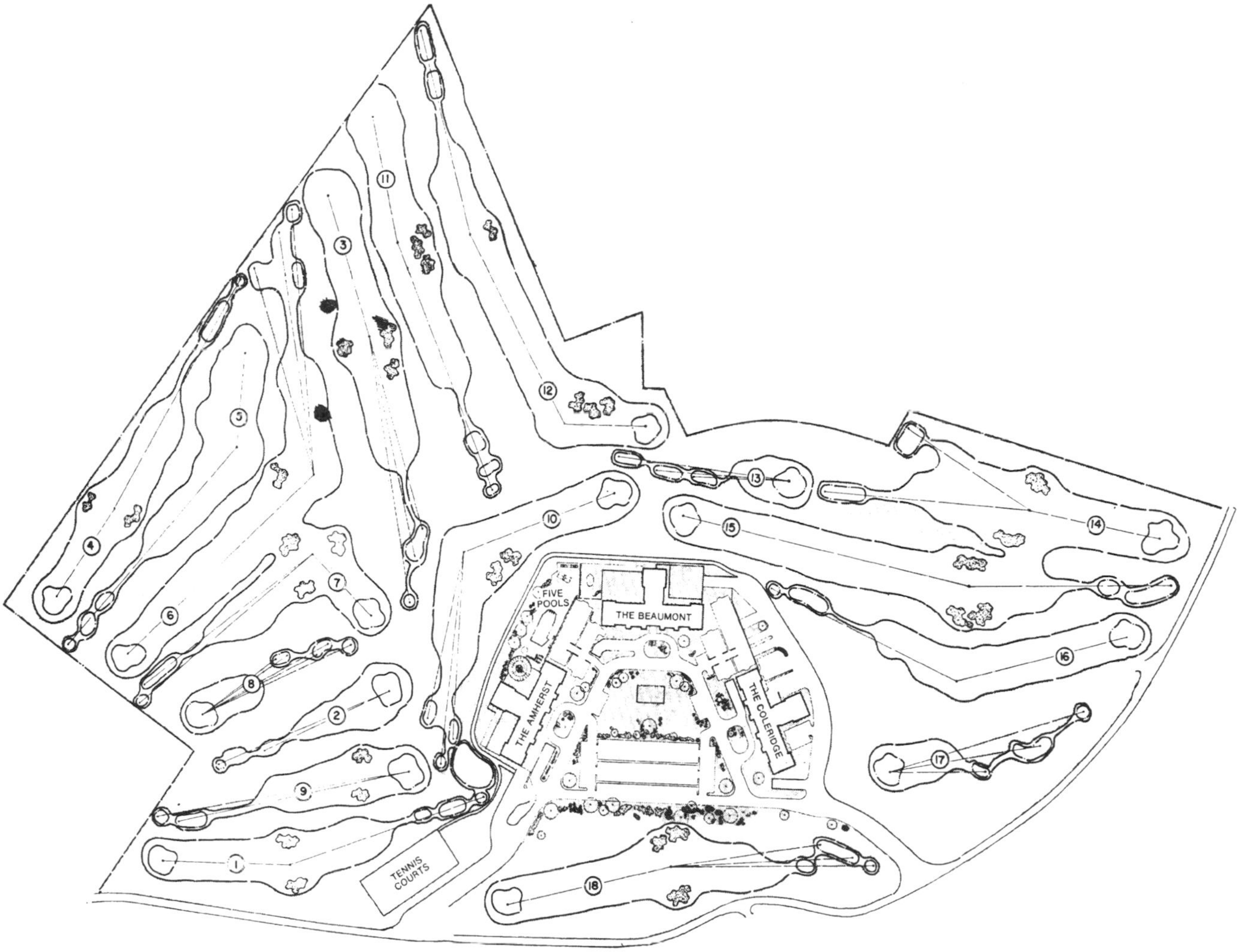

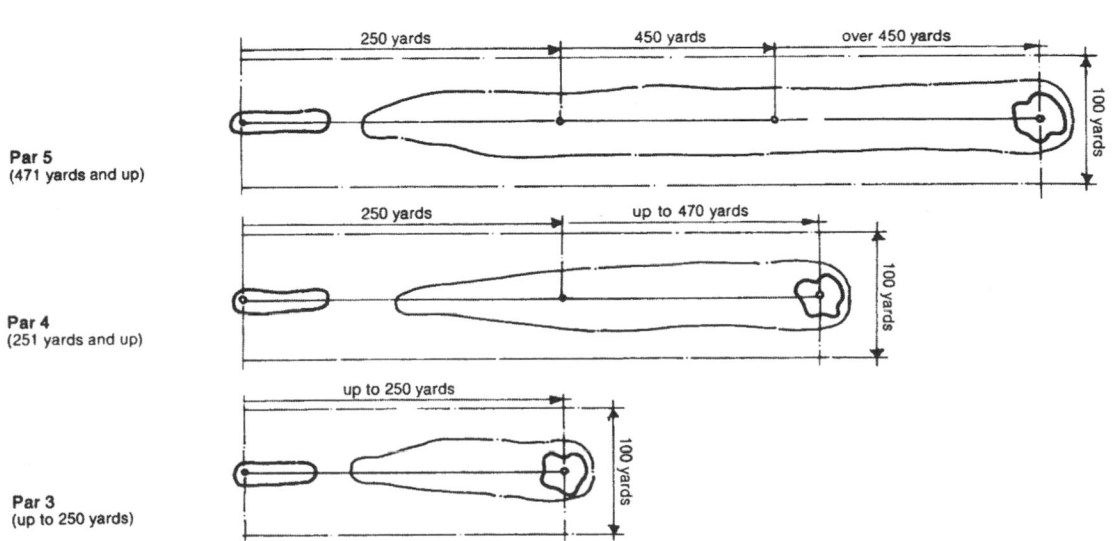

Fig. 1

Fig. 2 Different types of golf holes in relation to par.

RECREATIONAL FACILITIES

Preschoolers

Preschoolers require continual visual supervision. Therefore, they are mostly confined to playing right outside the kitchen door. Tot lots should be located where parents can observe the children. Since it is not feasible to build numerous tot lots, they should be located within clusters of units. No preschoolers should cross a major street or parking area to reach a tot lot. Parents prefer that tot lots be visible from the kitchen, where they spend much of their time.

These play areas should include sand and water play areas and simple climbing devices. The ground surface should be cushioned and nonabrasive, using materials such as pea gravel. The area should be shaded and protected by plantings and screening from wind, older children's activities, passersby, parking, and other potentially dangerous uses. Comfortable, tree-shaded sitting areas should be provided for parents.

TOT LOTS SHOULD HAVE SAND & WATER PLAY, SIMPLE CLIMBING STRUCTURES, BENCHES & SHADE AREAS & SCREENING

Fig. 1

Children Ages 6 to 13

Older children require larger play spaces and like to play without supervision, some distance from their homes. These play areas should be equipped with challenging and complex equipment. Drinking fountains should be provided. Since these children are generally quite active and noisy, their play areas should be removed from dwelling areas and separated by dense plantings and/or earth mounts.

PLAY AREAS FOR GRADE SCHOOL CHILDREN HAVE COMPLEX & CHALLENGING EQUIPMENT AND LOCATED AWAY FROM DWELLING UNITS

Fig. 2

Teenagers

Teenagers have two distinct recreation needs. They like informal gathering places where they can socialize or "watch the action" and be away from their homes. They also engage in very active group games that require a hard court or a field (e.g., basketball, baseball, or football). Areas for both kinds of teenage activities should be together, away from the dwelling area. Drinking fountains should be provided near active play areas. Where the project site is small but open space is available nearby, efforts should be made to utilize it for teenage recreation, or nearby community facilities may satisfy this requirement.

TEENAGERS LIKE COURT GAMES & SOCIALIZING AREAS THAT ARE SEPARATED FROM THEIR DWELLING UNITS

Fig. 3

Elderly

Sitting areas for socializing and passive activity are important to elderly residents. They should be located so as to afford a view of activity in and around the project, separated from intrusion and noise and well shaded. They should be convenient to the dwelling units of most elderly residents.

Equipment will include comfortable benches with backs, fixed chairs and tables for card games and other table games, and possibly, provision of such semiactive game courts as shuffleboard, horseshoes, croquet, or roque.

Community Building

Every project of over 50 units should have indoor community space useful to residents of all age groups. The building should be large enough to accommodate two or more activities without conflict at one time, and floor space should lend itself to a variety of uses. Space should be useful for meetings, indoor sports and games, hobbies, study and library, and service agency activities (including social and health services such as disease screening and diagnostic clinics).

Uses of community facilities are determined by what services the surrounding community offers, what other facilities are nearby and accessible, and what the tenants want. In rare instances, where adequate facilities exist nearby, this money might be better spent on other project improvements. A full survey of residents and careful needs analysis are necessary in planning community facilities and programming their use.

Programming of community building facilities cannot be limited to very time- and space-

THE ELDERLY NEED A SITTING AREA THAT HAS SHADE TREES, BENCHES WITH BACKS & LIGHTING. THESE AREAS SHOULD PROVIDE A GOOD VIEW OF ACTIVITY YET BE FAR AWAY FROM SUCH ACTIVITY.

Fig. 4

consuming uses for special-interest groups. Facility spaces must be flexible and available to different groups at different times of the day. For example, young children may use them after school; a social meeting of parents or an elderly group may occur in the early afternoon; and teenagers may use the space for evening social activities.

abstract structures allow the child's imagination to supply the details

sculptured forms in a continuous play-scape

Fig. 5

PLAYGROUND AND PLAYGROUND EQUIPMENT

Parks and playgrounds are a major factor in establishing good environmental quality for a community. Appropriate public and/or private facilities should be available for both active and passive recreation in or near all residential developments.

In one- and two-family dwelling developments often ample recreation space is provided within the area of the private lot. In high-density situations where individual lots approach minimum sizes or where multifamily buildings are used, the best service to recreation space needs can be provided by the development of common use areas.

ACTIVE RECREATION AREA

Active recreation space design should include appropriate locations for sports such as baseball, football, basketball, volleyball, and tennis. Playgrounds should be considered for grammar-school-aged children, and tot lots are needed by preschoolers. Elderly residents enjoy activities such as shuffleboard or horseshoes.

Fig. 1

TABLE 1

Type of apparatus	Dimensions of apparatus		Approximate use space requirements, ft	Space, ft²
	Length, ft	Height, ft		
Balance beam	12	.5	6 × 20	120
Climbing structure (average)	10	10	20 × 20	400
Climbing tree or ladder	5 (diam.)	12	12 × 12	144
Giant stride	—	12	30 × 30	900
Horizontal bar (single)	6	5.5–7.5	12 × 20	240
Horizontal bar (double)	11	5.5–7.5	18 × 20	360
Horizontal ladder	12–16	6.5–7.5	8 × 30	240
Junglegym (junior)	6.5	7	12 × 15	180
Junglegym (medium)	10	10.5	20 × 20	400
Merry-go-round	10 (diam.)	3.5	22 × 22	484
Sand box	6 × 10 (min.)	1	12 × 16	192
See-saws (set of 4)	12	2	20 × 20	400
Slide	16	8	12 × 30	360
Slide, gang	16	8	25 × 40	1,000
Slide, kindergarten	8	4.5	8 × 16	128
Slide, racer	16	8	20 × 30	600
Swings (set of 3)	15 at top	12	25 × 35	875
Swings (set of 6)	30 at top	12	25 × 50	1,250
Swings (set of 4)	18 at top	10	20 × 30	600
Swings, chair				
(set of 3)	10 at top	8	16 × 20	320
(set of 6)	20 at top	8	16 × 30	480
Traveling rings (in line)	36 at top	12	20 × 60	1,200
Traveling rings (circular)	10 (diam.)	12	25 × 25	625

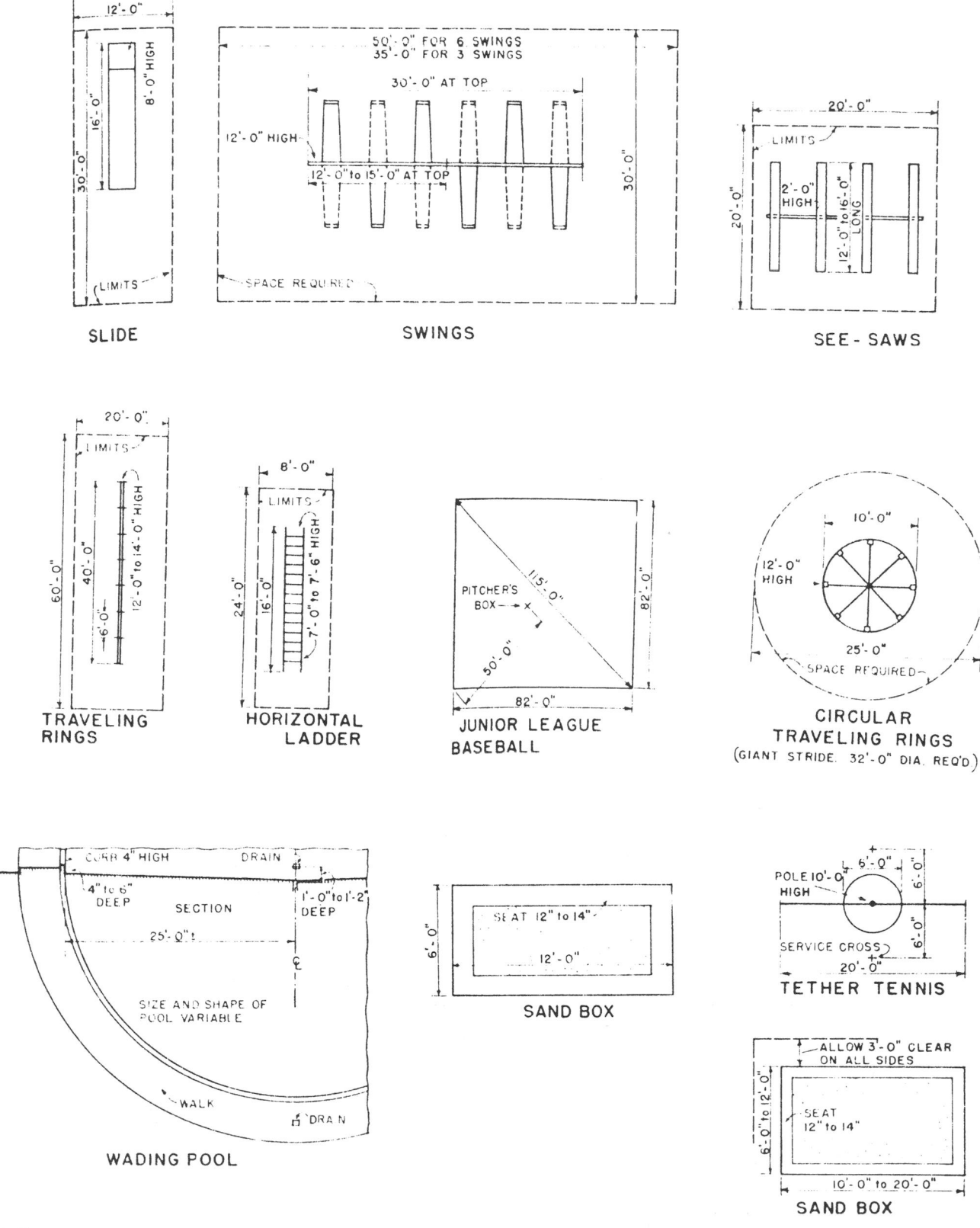

SLIDE

SWINGS

SEE-SAWS

TRAVELING RINGS

HORIZONTAL LADDER

JUNIOR LEAGUE BASEBALL

CIRCULAR TRAVELING RINGS
(GIANT STRIDE, 32'-0" DIA. REQ'D)

WADING POOL

SAND BOX

TETHER TENNIS

SAND BOX

Fig. 2

PLAYGROUND AND PLAYGROUND EQUIPMENT

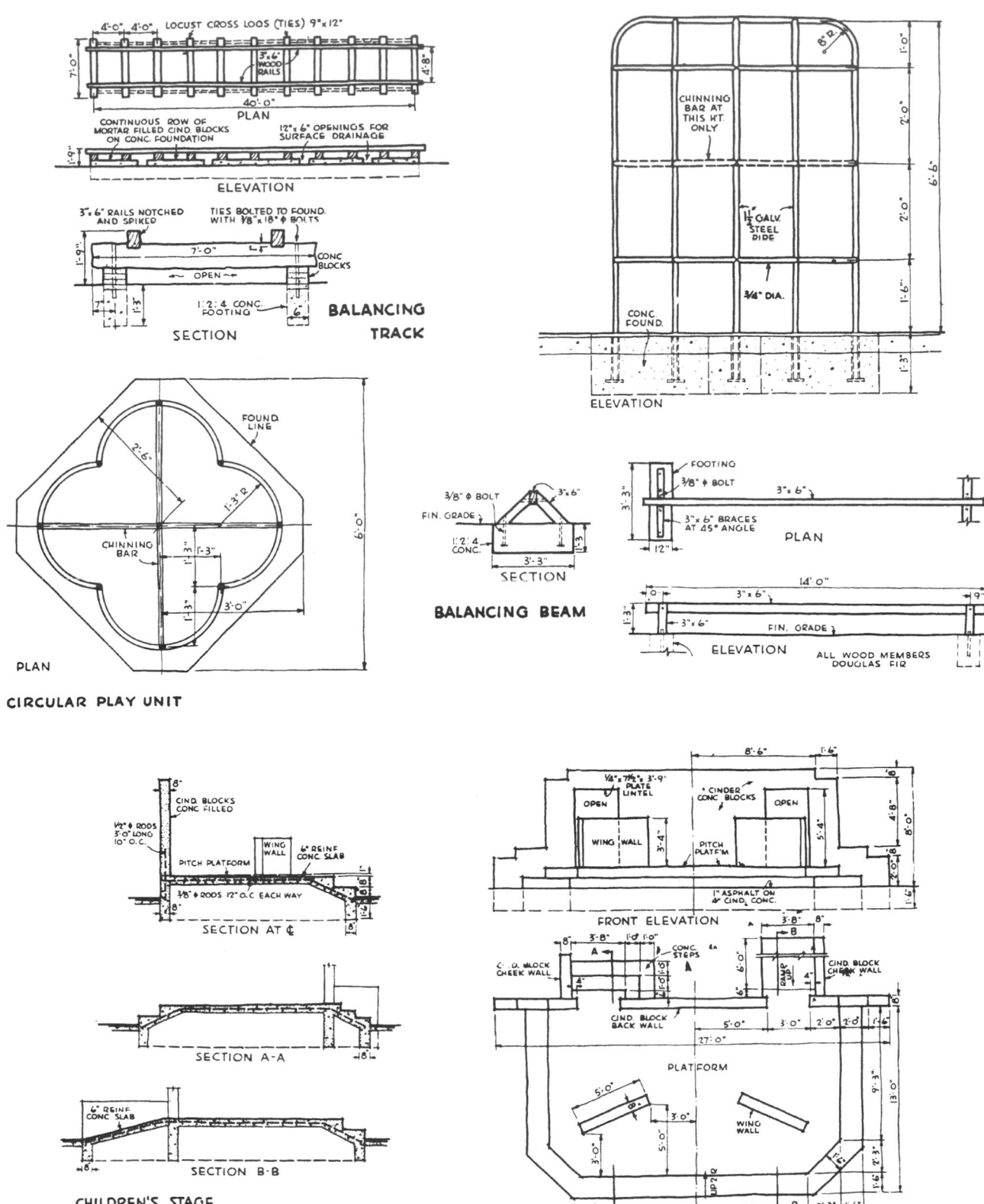

Fig. 3

USE ZONES FOR EQUIPMENT

The use zone for each piece of equipment is made up of two parts:

1. *Fall zone:* an area under and around the equipment where protective surfacing is required.

2. *No-encroachment zone:* an additional area beyond the fall zone where children using the equipment can be expected to move about and should have no encroaching obstacles.

With the exception of spring-rocking equipment, equipment under 24 in in height, and the zone between adjacent swings, the fall zones of adjacent pieces of equipment should not overlap. However, adjacent pieces of equipment may share a single no-encroachment zone.

Regardless of the type of equipment, the use zone should be free of obstacles that children could run into or fall on top of and thus be injured. For example, there should not be any vertical posts or other objects protruding from the ground onto which a child may fall.

Recommendations for the Fall Zone

Stationary equipment The fall zone should extend a minimum of 6 ft in all directions from the perimeter of the equipment.

Slides The fall zone in front of the access and to the sides of a slide shall extend a minimum of 6 ft from the perimeter of the equipment. *Note:* This does not apply to embankment slides.

The fall zone in front of the exit of a slide shall extend a minimum distance of 6 ft from the end of the slide chute or for a distance of $H + 4$ ft, whichever is the greater. H is the height of the slide platform and the $H + 4$ ft measurement is made from a point on the slide chute where the gradient has been reduced to 5° from the horizontal (see Fig. 4).

Single-axis swings Because children may deliberately attempt to exit from a single-axis swing while it is in motion, the fall zone in front of and behind the swing should be greater than to the sides of such a swing. It is recommended that the fall zone extend to the front and rear of a single axis swing a minimum distance of 2 times the height of the pivot point above the surfacing material measured from a point directly beneath the pivot on the supporting structure (see Fig. 15). The fall zone to the sides of a single-axis swing should follow the general recommendation and extend a minimum of 6 ft from the perimeter of the swing structure in accordance with the general recommendation for fall zones. This 6-ft zone may overlap that of an adjacent swing structure.

Multiaxis swings The fall zone should extend in any direction from a point directly beneath the pivot point for a minimum distance of 6 ft + the length of the suspending members (see Fig. 16). In addition, the fall zone shall extend a minimum of 6 ft from the perimeter of the supporting structure. This 6-ft zone may overlap that of an adjacent swing structure.

Merry-Go-Rounds The fall zone should extend 6 ft beyond the perimeter of the platform.

Spring-rocking equipment The fall zone should extend a minimum of 6 ft from the "at rest" perimeter of the equipment but adjacent spring rockers with a maximum seat height of 24 in may share the same fall zone.

Composite equipment The above recommendations for individual pieces of equipment should be used as a guide in establishing the fall zones around pieces of composite playground equipment.

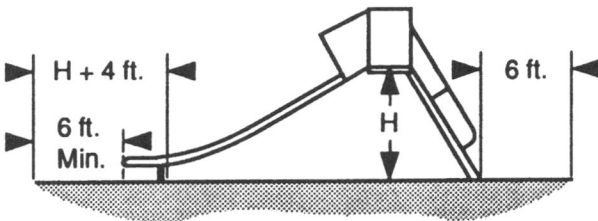

Denotes Fall Zone with Protective Surfacing

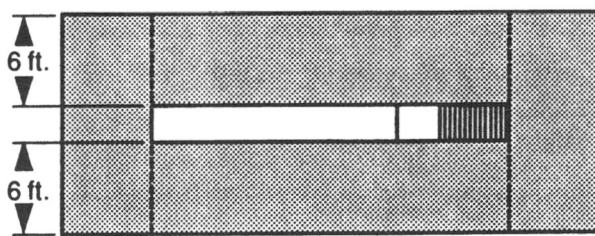

Fig. 4   Fall zone for slides.

PLAYGROUND AND PLAYGROUND EQUIPMENT

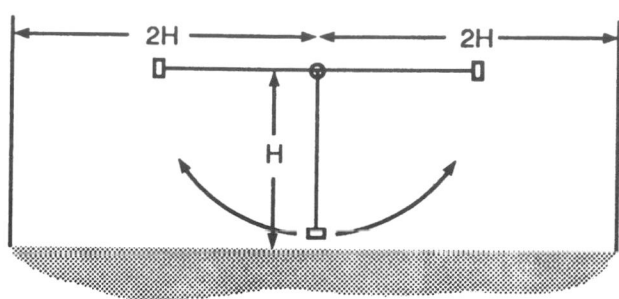

▨ Denotes Fall Zone with Protective Surfacing

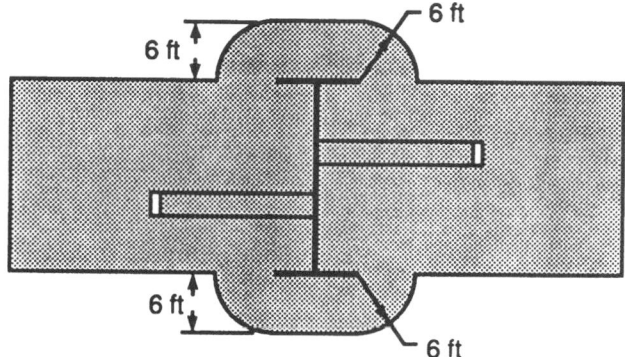

Fig. 5   Fall zone for single axis swings.

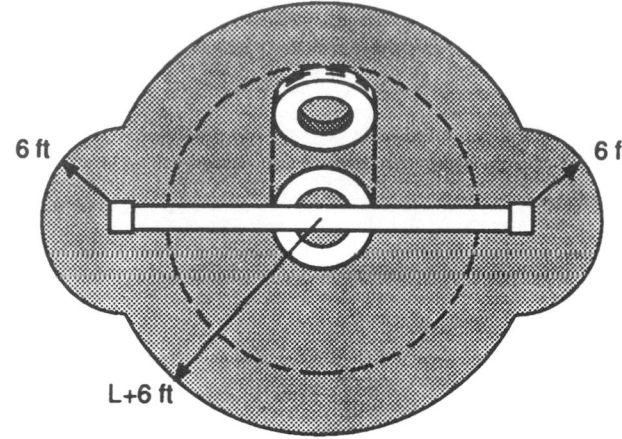

▨ Denotes Fall Zone with Protective Surfacing

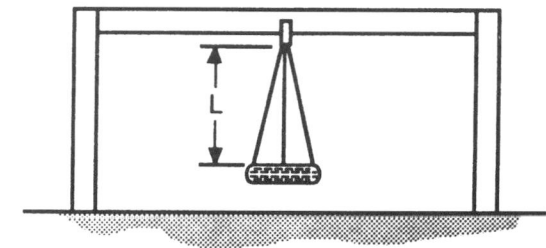

Fig. 6   Fall zone for multi axis tire swings.

Recommendations for the No-Encroachment Zone

No specific dimensions can be recommended for the no-encroachment zone around individual pieces of playground equipment. These dimensions will vary according to the types of adjacent pieces of equipment and their orientation with respect to one another.

For example, the recommended fall zone at the side of both a slide and a swing is 6 ft. Since fall zones should not overlap, a slide could be placed with its side no closer than 12 ft to the side of a swing. Therefore, there may be no need to add an additional no-encroachment zone. Conversely, it would not be desirable to have a slide exit facing the front or rear of single axis swing.

No-encroachment zones extending beyond the fall zones are recommended for moving equipment or equipment from which the child is in motion as he or she exits. This allows more space for children to regain their balance upon exiting the equipment and also provides added protection against other children running into a moving part.

For a single-axis swing, it is recommended that there be a barrier beyond the fall zone in front of the swing if it is located in a playground facing other pieces of equipment.

Fundamentals of Play Areas

1. ACCESSABLE to all groups.
2. SAFETY in choice of elements
3. INTERESTING facilities & apparatus
4. CHALLANGING & INNOVATIVE in the design and layout.

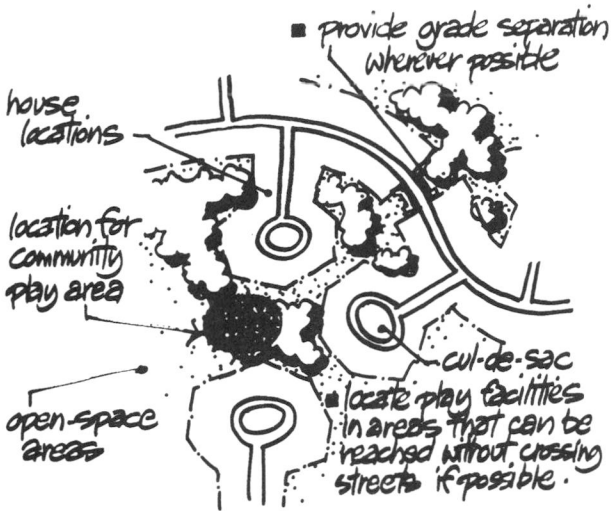

- provide grade separation wherever possible

house locations

location for community play area

open-space areas

cul-de-sac

- locate play facilities in areas that can be reached without crossing streets if possible.

1. Community Access to Play

- connect playground paved walkways to community sidewalk system.

- logical organization of play facilities as to type, age, etc.

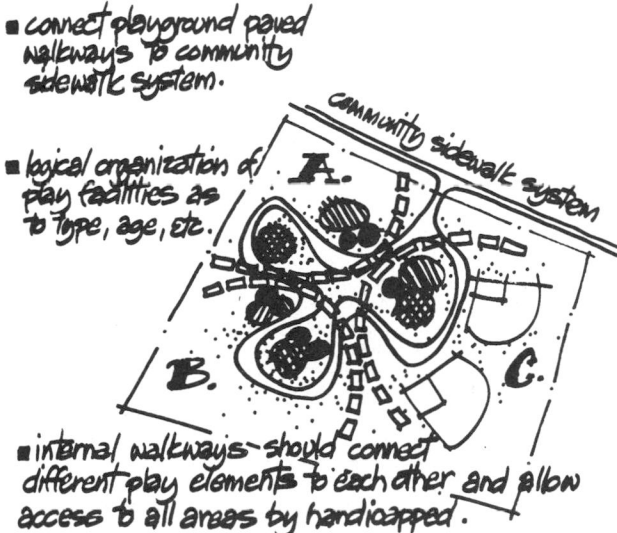

community sidewalk system

- internal walkways should connect different play elements to each other and allow access to all areas by handicapped.

2. Internal Walkways

Fig. 7

HANDICAPPED ACCESS

In general, play can be grouped into two categories: (1) defined play and (2) creative play.

Defined play refers to the channeling of play activities into certain prescribed directions. For instance, swings and slides define the child's play within the limits of their function. Although children do many creative things on swings and slides, they are primarily outgrowths of the basic functions of swinging and sliding. On the other hand, creative play primarily arises from the child's imagination. The play element is somewhat amorphic and therefore undefined. A child in a sand area creates sand castles, mountains, rivers, roads, and a plethora of other fantasies straight from his mind. Likewise, free-form sculpture, random climbing blocks, or simply open areas of lawn act as springboards for the imagination.

There seems to be a current trend in which designers heavily specify creative play apparatus for playgrounds, sometimes to the exclusion of defined apparatus. This trend does not well serve children since it does not account for the child who is unable to play creatively.

There are, for instance, thousands of children in this country alone who, handicapped by severe mental and emotional problems, are only able to achieve satisfying play through the use of defined apparatus. Likewise, an imaginative child may quickly lose interest in traditional play equipment whereas a creative apparatus may hold his attention. Therefore, the designer should strive to create a playground that will provide a rich and wide ranging set of both defined and creative experiences. (See Fig. 7.)

Just as designers have been designing the environment for the "normal" adult, so have playgrounds been designed for the "normal" child. Unfortunately, the child who is physically handicapped usually has restricted motor development, and as a consequence of limited movement, does not see the world and himself or herself in the same way as a normal individual would.

By designing play situations in which a disabled child can manipulate his environment as much as possible by himself, regardless of the extent of his disability, the child can have motor experiences comparable to those of normal children. These experiences give a child a broader range of perceptual sophistication and thus a fuller and more normal base for academic growth and self appreciation.

1. Slides & Climbing Areas

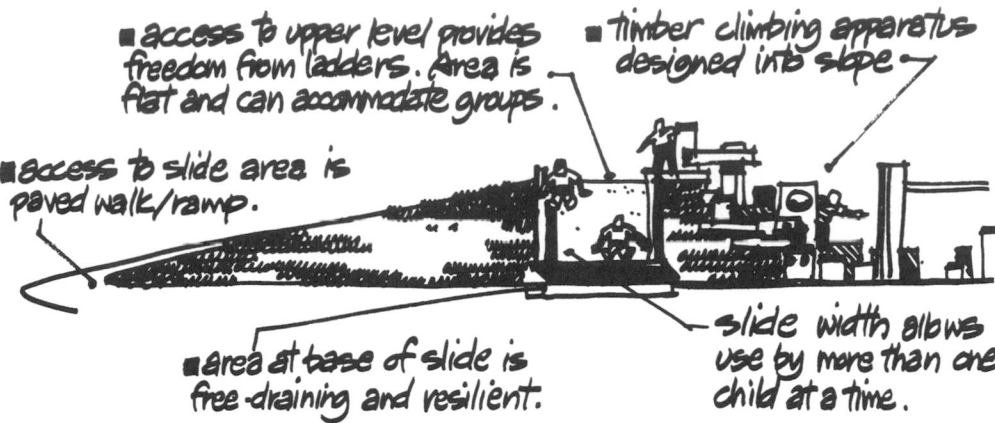

- access to upper level provides freedom from ladders. Area is flat and can accommodate groups.
- timber climbing apparatus designed into slope
- access to slide area is paved walk/ramp.
- area at base of slide is free-draining and resilient.
- slide width allows use by more than one child at a time.

2. Elevated Sand-Tables

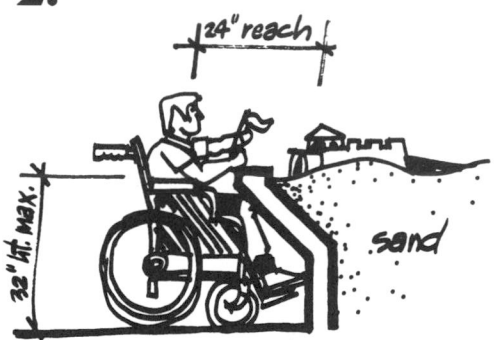

24" reach

32" ht. max.

sand

- elevated area containing sand or water provides access for those in wheelchairs. Flat area is useful for toy cars, crafts, etc.

3. Basketball Hoops

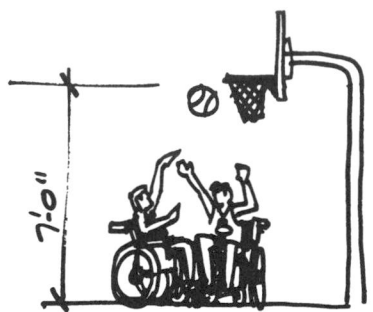

7'-0"

- basketball hoops lowered to 7'-0" from standard 10'-0" ht. allow those in wheelchairs and young children to enjoy the game.

Fig. 8

The following criteria are given for consideration in enhancing the use of play facilities both from the standpoint of serving more people and of making the facility safer.

1. A playground should be easily accessible from the adjacent community over hard surface paths, with ramps placed where necessary.

2. Access within the playground should include a system of hard surface paths. Not only does this improve mobility for the handicapped but can double as a tricycle path.

3. The play area should be reasonably organized in order that a child who is blind may learn how to locate equipment as he or she enters and moves about the grounds.

4. Apparatus able to accommodate a greater diversity of children does not need to be drastically altered from those now in use. Rather, they must be placed and modified in such a way as to make them both more safe and accessible. Sharp edges, splinters or poorly designed appurtenances should be eliminated.

5. Playgrounds that are accessible to handicapped children require a certain amount of adult supervision. The amount of supervision varies, depending on the type of handicap the child has, the type of equipment present, and the number of handicapped children using the facility. This may mean that in certain cases, parents will have to accompany their child in order that they may supervise his or her play. In other cases, a single attendant may be sufficient.

6. A series of small vignettes have been prepared to illustrate some of the many recreational devices that can be incorporated into play grounds and can be used by most handicapped children. (See Figs. 8 and 9.)

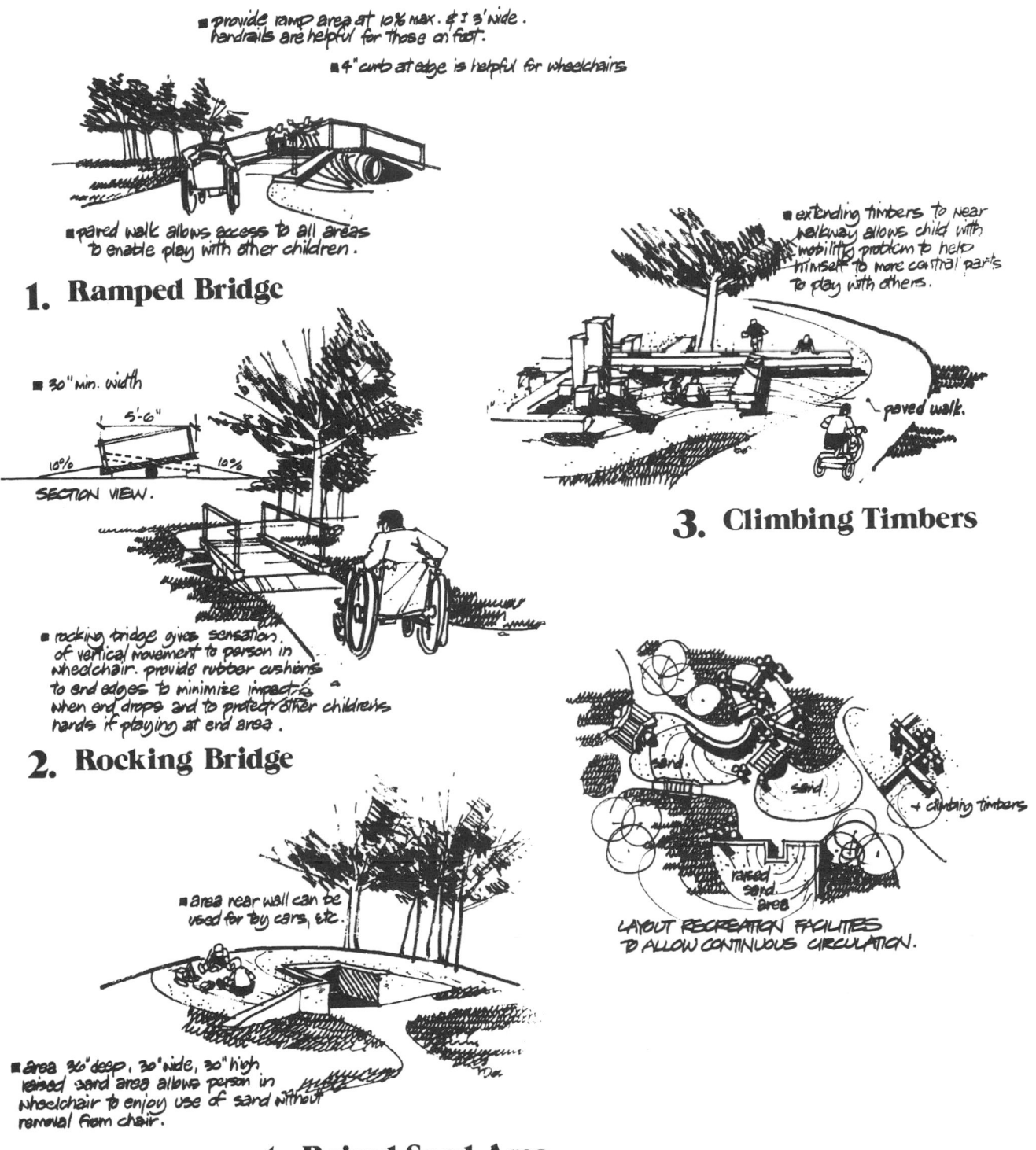

■ provide ramp area at 10% max. & 3' wide.
handrails are helpful for those on foot.

■ 4" curb at edge is helpful for wheelchairs

■ paved walk allows access to all areas
to enable play with other children.

1. Ramped Bridge

■ 30" min. width

5'-6"

10%   10%

SECTION VIEW.

■ rocking bridge gives sensation
of vertical movement to person in
wheelchair. provide rubber cushions
to end edges to minimize impacts
when end drops and to protect other children's
hands if playing at end area.

2. Rocking Bridge

■ extending timbers to near
walkway allows child with
mobility problem to help
himself to more central parts
to play with others.

~ paved walk.

3. Climbing Timbers

■ area near wall can be
used for toy cars, etc.

+ climbing timbers

sand

sand

raised
sand
area

LAYOUT RECREATION FACILITIES
TO ALLOW CONTINUOUS CIRCULATION.

■ area 36" deep, 30" wide, 30" high
raised sand area allows person in
wheelchair to enjoy use of sand without
removal from chair.

4. Raised Sand Area

Fig. 9

Fig. 1

Fig. 2

UNSUPERVISED AND SUPERVISED PLAY AREAS

PLAYGROUND ADJACENT TO HOUSING DEVELOPMENT

1. Children's play area (junior), ages 6 to 8:
   Junior exercise units
   Play logs
   Wood dodgers
   Pipe tunnels
   Log piles
   Removable shower
2. Children's play area (intermediate), ages 9 to 12:
   Pipe frame exercise units
   Wood dodgers
   Basketball backstops (single)
   Roller skating areas, ice skating
3. Children's play area (senior), ages 13 to 15 (if no public playground adjacent):
   Basketball courts
   Roller skating areas
   Softball fields
   Ice skating rink[1]
4. Older children's playground, ages 16 to 18 (if no public play fields adjacent):
   Softball fields
   Basketball courts
   Football fields[1]
   Soccer fields[1]
   Ice skating rink[1]
   Baseball fields[1]

PLAY AREAS

Playgrounds and play areas should be located for minimum noise nuisance to apartment dwellers and should not be placed in courts or across main walkways. All play area entrances shall be located to discourage through circulation. Play areas should be enclosed by chain-link fence and paved with a bituminous surface (except as noted). They should generally be surrounded outside by trees and occasional shrubbery.

HOUSING PROJECT SITE

1. Children's center play areas, ages 3 to 5:
   Junior exercise units
   Play pyramids
   Pipe tunnels
   Log piles
2. Community center, ages 12 and over:
   Paddle tennis
   Table tennis
   Shuffleboard
   Volleyball courts[1]
   Tables and benches
   Croquet (or roque)[1]
   Badminton courts[1]
   Areas for movies and dancing[1]

PROJECT PLAYGROUND ADJACENT TO HOUSING PROJECT

1. Children's project playground (junior), ages 3 to 7:
   Kindergarten slides
   Kindergarten swings
   Sandpits
   Seesaws
   Wading pools
2. Children's project playground (senior), ages 8 to 12:
   Play slides
   Play swings
   Pipe frame exercise units
   Wading pools
   Paddle tennis
   Roller skating areas

3. Older children's project playground, ages 13 to 18:
   Basketball courts
   Volleyball courts
   Handball courts
   Shuffleboard
   Horseshoes
   Paddle tennis
   Table tennis
   Roller skating areas
   Softball fields
   Baseball fields[1]
   Football fields[1]
   Soccer fields[1]
   Track[1]
4. Adult project playground, ages 19 and over:
   Basketball courts
   Volleyball courts
   Handball courts
   Paddle tennis
   Badminton courts
   Tennis courts[1]
   Baseball fields[1]
   Softball fields
   Football fields[1]
   Soccer fields[1]
   Track[1]
   Area for dancing and movies
   Boccie courts (where played)
   Bowling greens (where played)
   Cricket fields (where played)[1]
   Croquet (or roque)
   Horseshoes
   Shuffleboard
   Ice skating rinks[1]

[1]For large projects.

TABLE 1 Characteristics of High-Density Living

Common characteristic	Effect on recreation experience	Possible solution
Little or no private front/back yard space	Restricted activities, such as outdoor play for children, family-oriented activities, socializing with friends, barbecues, and gardening	Provide adjacent open space for children's play and improve accessibility to local parks with walkways; plan community garden plots
Little or no basement area	Limits indoor active play, hobbies, furniture refinishing, building or tinkering jobs	Apartment basements or ground floor space can be adapted for play areas with adjacent outdoor space; indoor hobby work space with locked storage areas
Little storage space	No space for storing bulky outdoor recreation equipment, i.e., canoes/boats, camping gear, skis, bicycles; this limits related activities	Arrange for rented space in building or adjacent to parking area; include added storage areas in building plan
Restricted space for entertaining	Entertaining at home is limited by space and noise restrictions	Provide party rooms that can be booked for private use in a soundproof area; use as play areas for children during day if it has access to outdoor play area and is adequate for this purpose
Restricted outside view	Visual access to pleasant outdoor areas is limited at certain levels or from some units; creates supervisory problems for children's play	Locate play areas within view of balconies or kitchen windows; restrict lower levels to families with children; plan units to have visual access to outdoors; provide comfortable seating for parents in play areas where supervision from unit is impossible
No front porch or veranda	Restricts sitting outdoors, chatting with neighbors, and outdoor play for small children	Increase balcony size and provide sheltered small attractive outdoor sitting areas adjacent to units

PASSIVE RECREATION AREAS

Major areas for passive recreation may include a plaza, a landscaped park, or a public square. Minor areas for passive recreation should vary according to the type of projects, from a sun deck, balcony, or roof garden to a paved area overlooking a pleasant view, or a shaded sitting area along a walk.

PASSIVE RECREATION AREA

Natural Areas

Natural site features such as wooded areas or stream valleys are highly desirable assets and should be preserved when appropriate for recreation.

Fig. 1

RECREATION FOR THE ADULT: Chess and Checkers

leon brand

Fig. 2

BUILDING ROOF DECK

This deck is located at the very top of a building. Access to it is usually by elevator and a connecting hall, although this may not be so when there are other communal facilities at deck level, such as a party room, pool, sauna or laundry rooms. In these cases, the roof deck has a horizontal as well as a vertical relationship with the building.

Advantages

■ Its position and means of access make it more private and secure than other types of decks.
  ■ It is exposed to the sun throughout the day.
  ■ The wind speed, though high, is constant.
  ■ It is away from street noise.

Disadvantages

■ This type of deck usually has fewer access points than others.
  ■ Surveillance from other parts of the building is not possible.
  ■ It is usually higher than other types of decks and may cause vertigo.
  ■ Wind speed is generally higher than on lower types of decks.

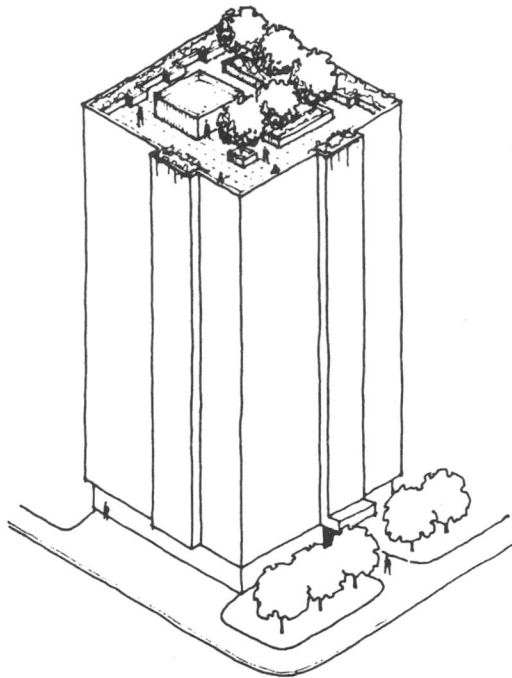

Fig. 1   Building roof deck.

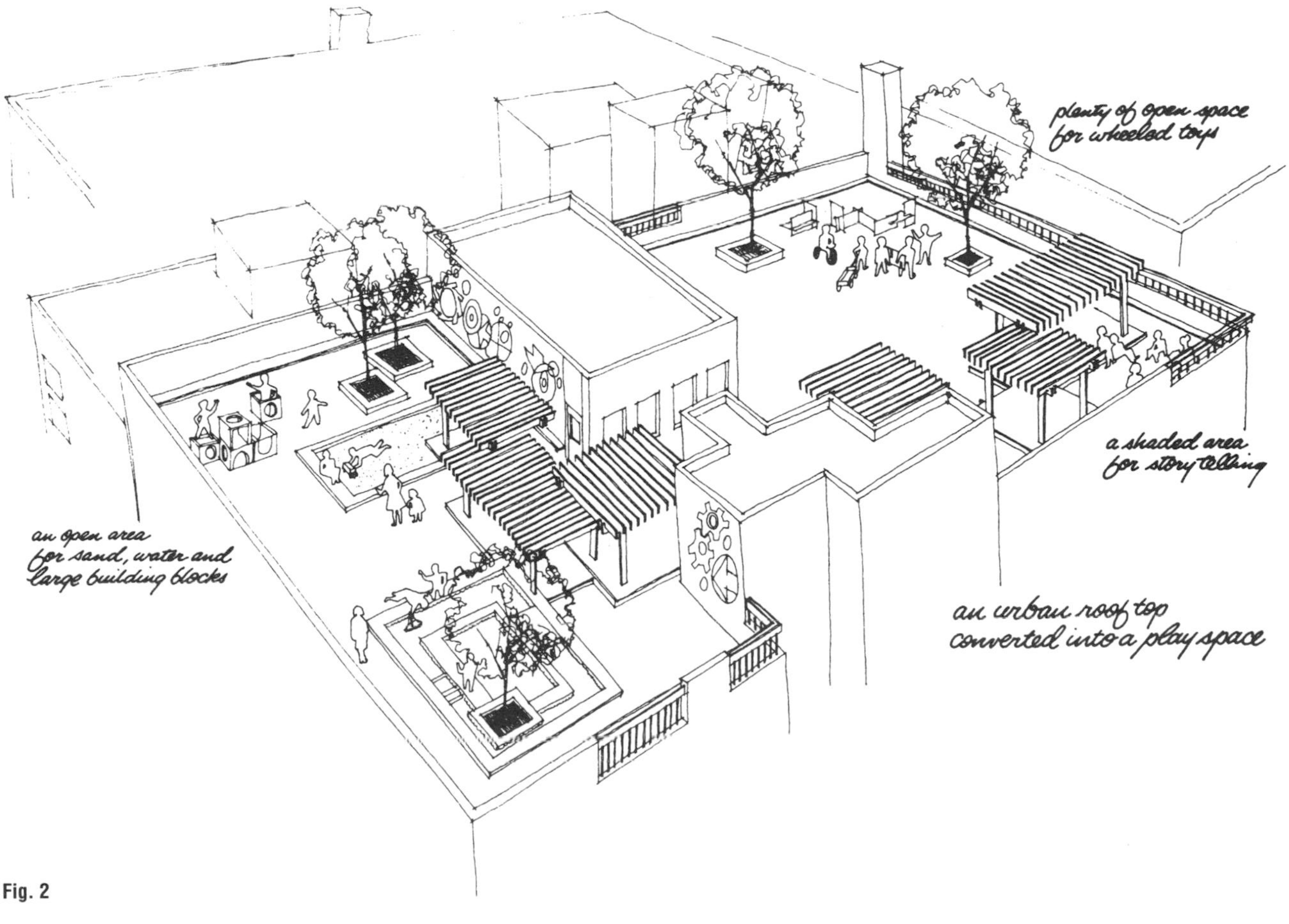

plenty of open space for wheeled toys

a shaded area for story telling

an urban roof top converted into a play space

an open area for sand, water and large building blocks

Fig. 2

BUILDING ROOF DECK

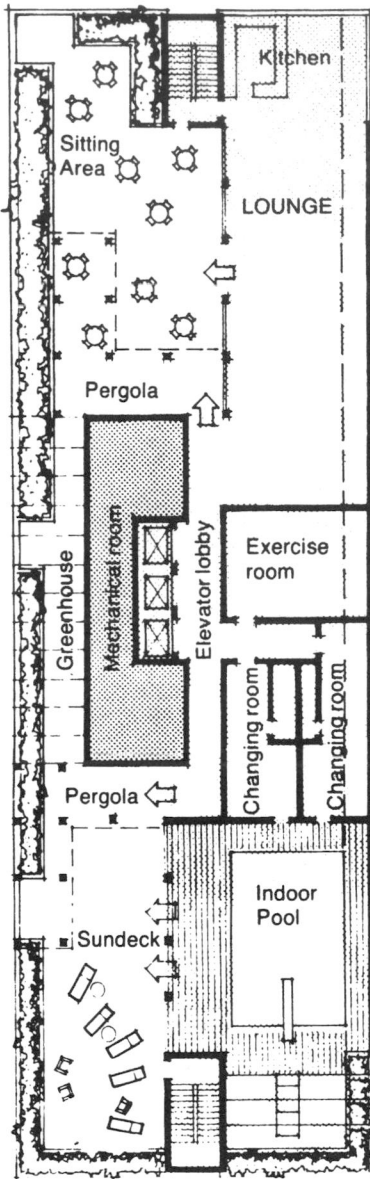

Fig. 3   Building roof deck.

Activity Types

The dominant activities are sunbathing, swimming, viewing, sitting out, and playing board games.

Particular Layout Features

Entries   The roof deck has three entry points. From the lobby the users can have direct access to the viewing zone; they can also have access to the sunbathing zone from the swimming pool area, and they have access to the sitting area from the lounge.

Sunbathing zone   Part of the roof deck is immediately adjacent to an interior swimming pool. This provides an opportunity for a cycle of swimming and drying off.

Sitting area and greenhouse   Another part of the deck is immediately adjacent to a lounge where people may sit or socialize. A greenhouse is adjacent to the lounge.

Viewing zones   The two above-mentioned areas and the greenhouse have viewing stations which overlook the deck as well as the surrounding locale.

Space-defining components   High parapet walls, planters, and trellises provide protection from wind and sun.

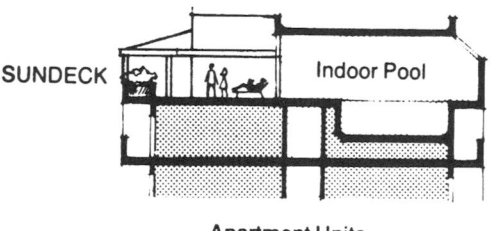

Fig. 4

The toddlers' room plan shows carpet in all areas except those near the sink cabinets, the small secure wading pool, and the toilet. The room is divided into quiet and active areas. The quiet area is furnished with thick rugs and daybeds with bright covers. The furniture in Fig. 1 should be bright, lightweight, and child-sized.

Fig. 1 Toddlers' room.

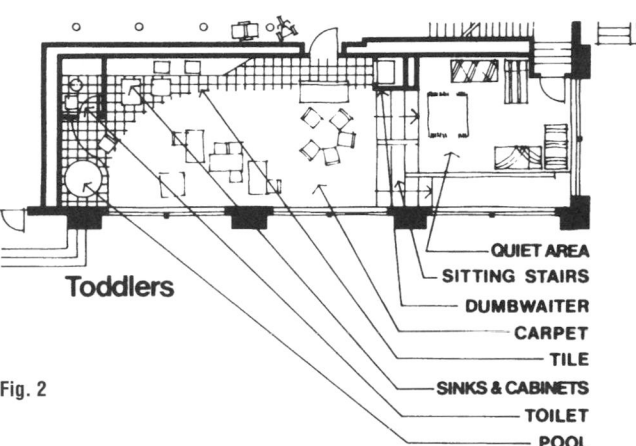

Toddlers

Fig. 2

- QUIET AREA
- SITTING STAIRS
- DUMBWAITER
- CARPET
- TILE
- SINKS & CABINETS
- TOILET
- POOL

Fig. 3

KINDERGARTEN

The kindergarten plan is one of four kindergarten areas, each of which is connected by shared supply spaces with adjacent toilets. These smaller service spaces have low ceilings, and playloft spaces have been provided above them. They are shown in the background of Fig. 1. The toilets and storage areas are tiled. All other areas, including the playloft and stairs, are carpeted. The kindergartens are furnished with light, child-sized tables and chairs, and with movable display panels and chalkboards.

In the prekindergarten plan the carpeted active play area is located near the entry door, with light movable tables, chairs, and display boards nearby. Wide, carpeted sitting stairs lead up 2 ft to a quiet area, which can be separated from the active area by a sliding partition. Bathroom floors and those near the sink cabinets are tiled.

Fig. 1 Kindergarten.

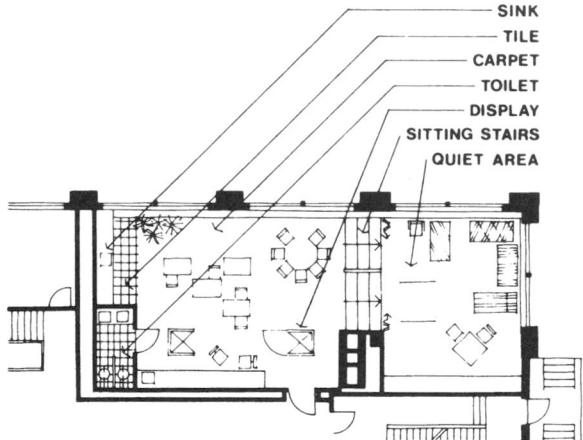

Fig. 2 Prekindergarten.

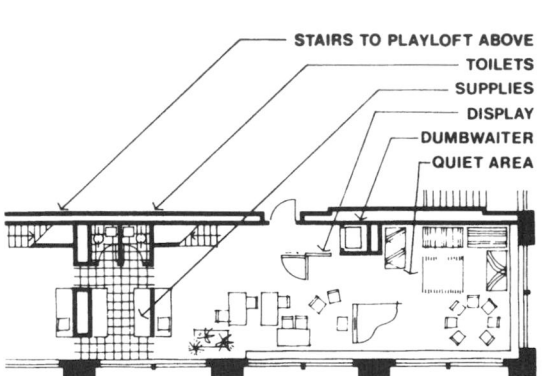

Fig. 3 Kindergarten.

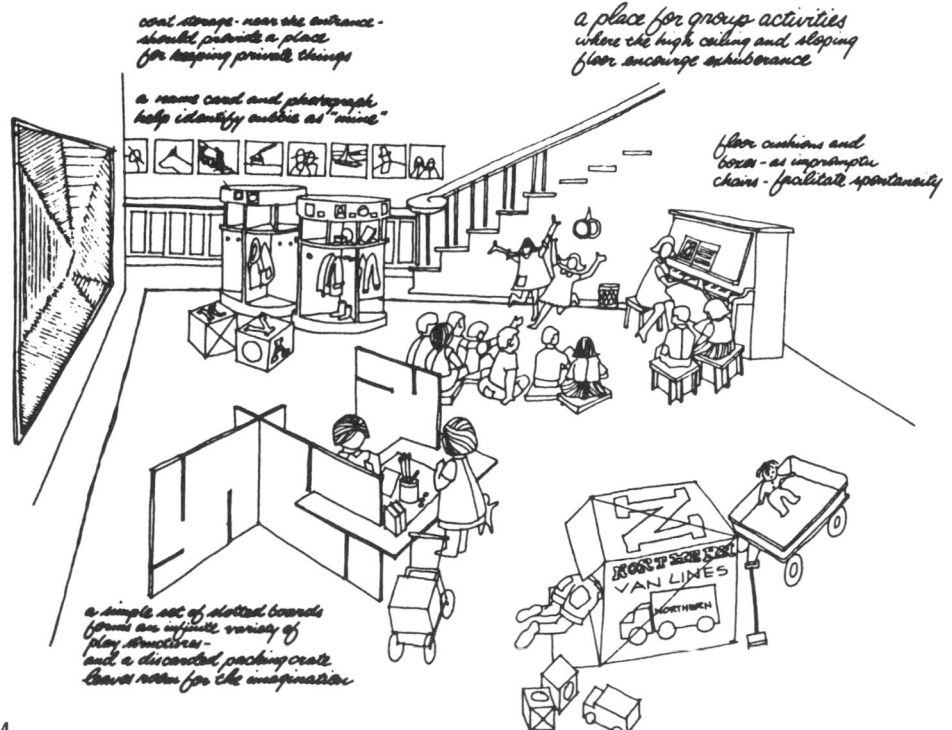

Fig. 4

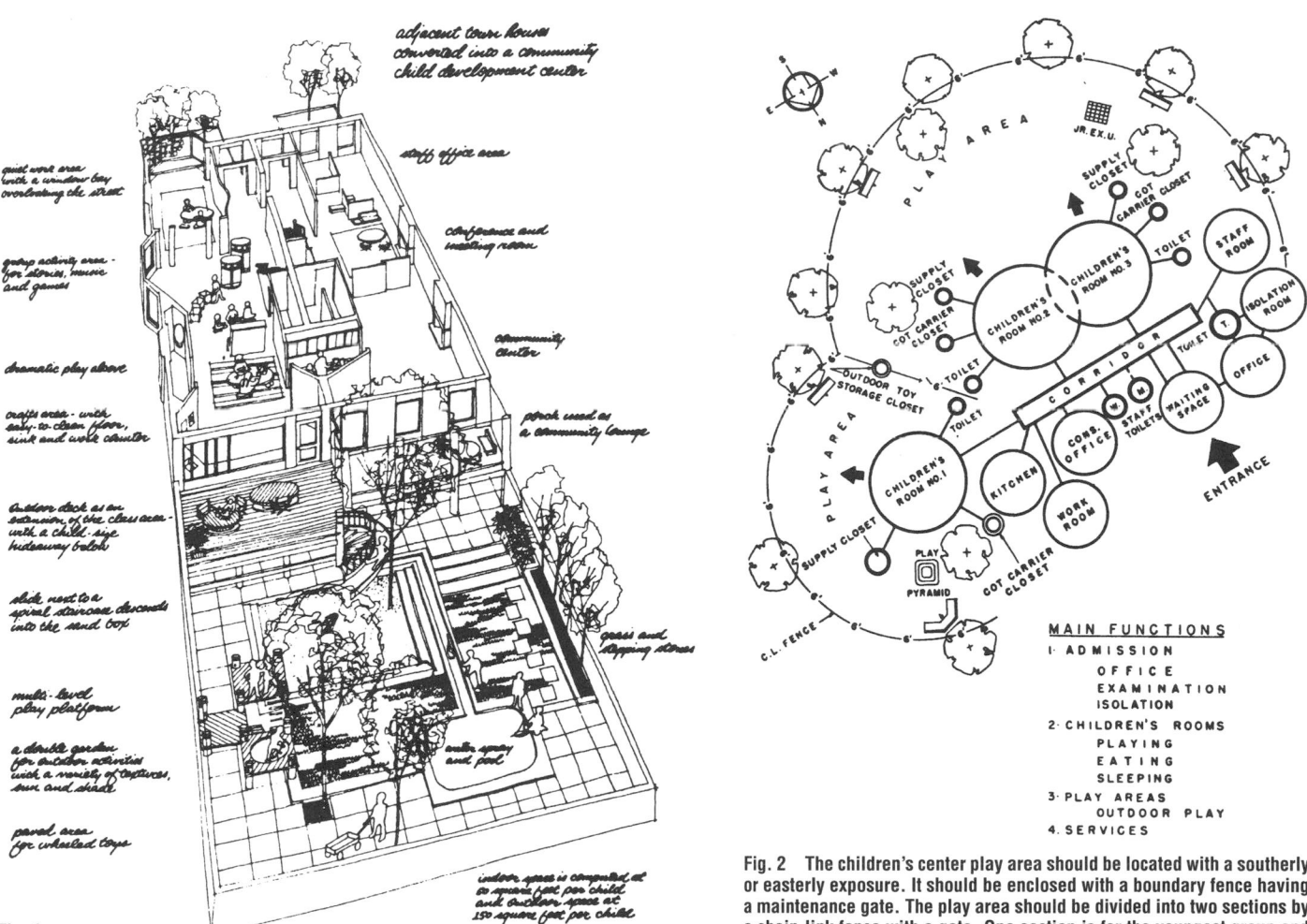

Fig. 1

Fig. 1 labels:

- quiet work area with a window bay overlooking the street
- group activity area for stories, music and games
- dramatic play above
- crafts area - with easy-to-clean floor, sink and work counter
- outdoor deck as an extension of the class area with a child-size hideaway below
- slide next to a spiral staircase descends into the sand box
- multi-level play platform
- a double garden for outdoor activities with a variety of textures, sun and shade
- paved area for wheeled toys
- adjacent town houses converted into a community child development center
- staff office area
- conference and meeting room
- community center
- porch used as a community lounge
- grass and stepping stones
- water spray and pool
- indoor area is computed at 50 square feet per child and outdoor space at 150 square feet per child

MAIN FUNCTIONS

1. ADMISSION
   OFFICE
   EXAMINATION
   ISOLATION
2. CHILDREN'S ROOMS
   PLAYING
   EATING
   SLEEPING
3. PLAY AREAS
   OUTDOOR PLAY
4. SERVICES

Fig. 2 The children's center play area should be located with a southerly or easterly exposure. It should be enclosed with a boundary fence having a maintenance gate. The play area should be divided into two sections by a chain-link fence with a gate. One section is for the youngest group and should be without apparatus except for a play pyramid. The other has a climbing unit, another piece of equipment, and if required, a digging area.

Fig. 3

- wheeled vehicle storage combined with a play shelter
- a shaded terrace... as an extension of the indoors
- easily made equipment that invites creativity and provides opportunities for problem solving

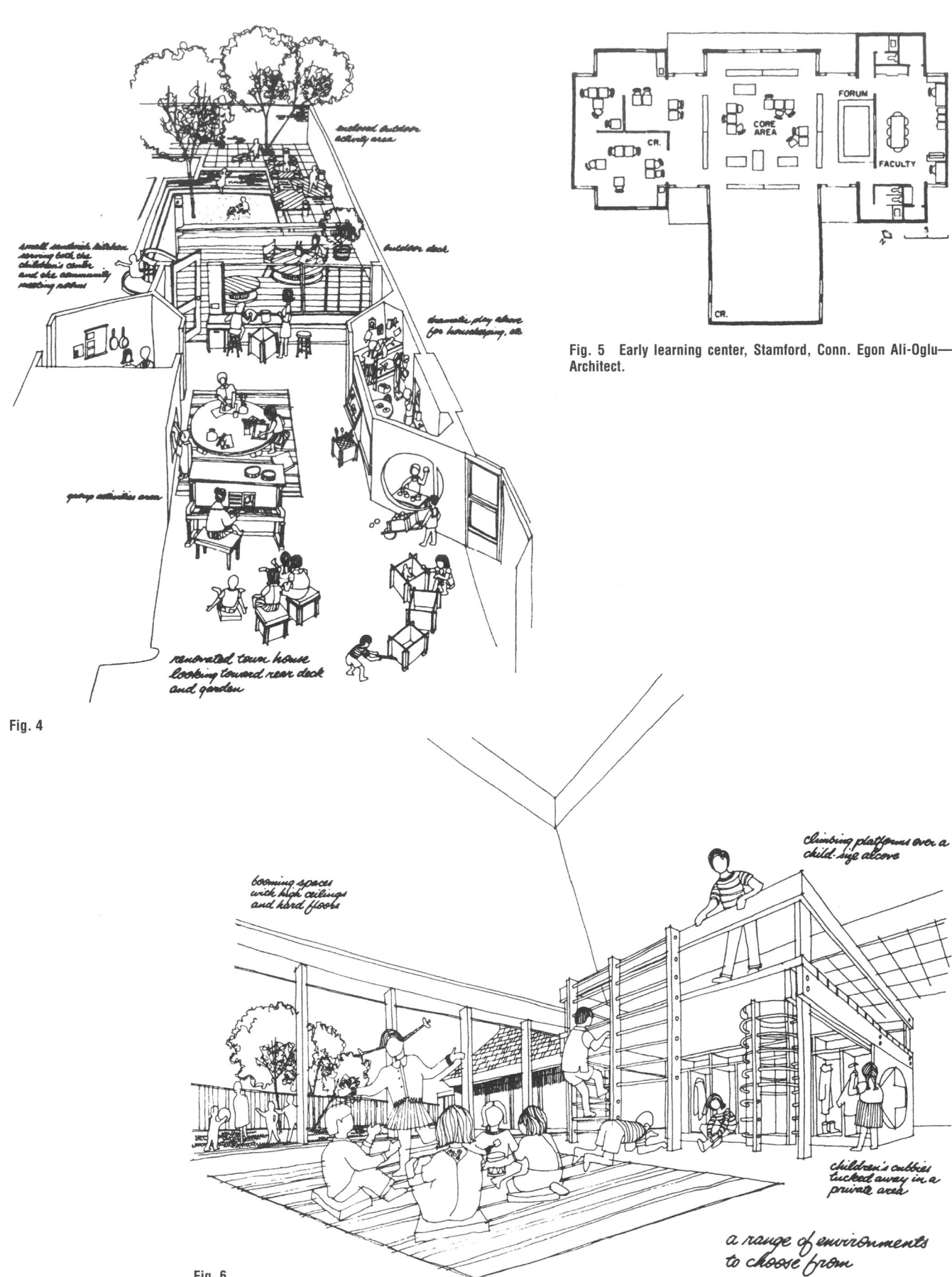

Fig. 4

enclosed outdoor activity area

small sandwich kitchen serving both the children's center and the community meeting rooms

builders deck

dramatic play alcove for housekeeping, etc.

group activities area

renovated town house looking toward rear deck and garden

Fig. 5 Early learning center, Stamford, Conn. Egon Ali-Oglu—Architect.

FORUM

CORE AREA

FACULTY

CR.

CR.

looming spaces with high ceilings and hard floors

climbing platforms over a child-size alcove

children's cubbies tucked away in a private area

a range of environments to choose from

Fig. 6

parts of the outdoors set aside for quieter activities

an outdoor garden away from active play

a work counter and sink close to indoors, on a hard surface

a well-drained area for water-play

Fig. 7

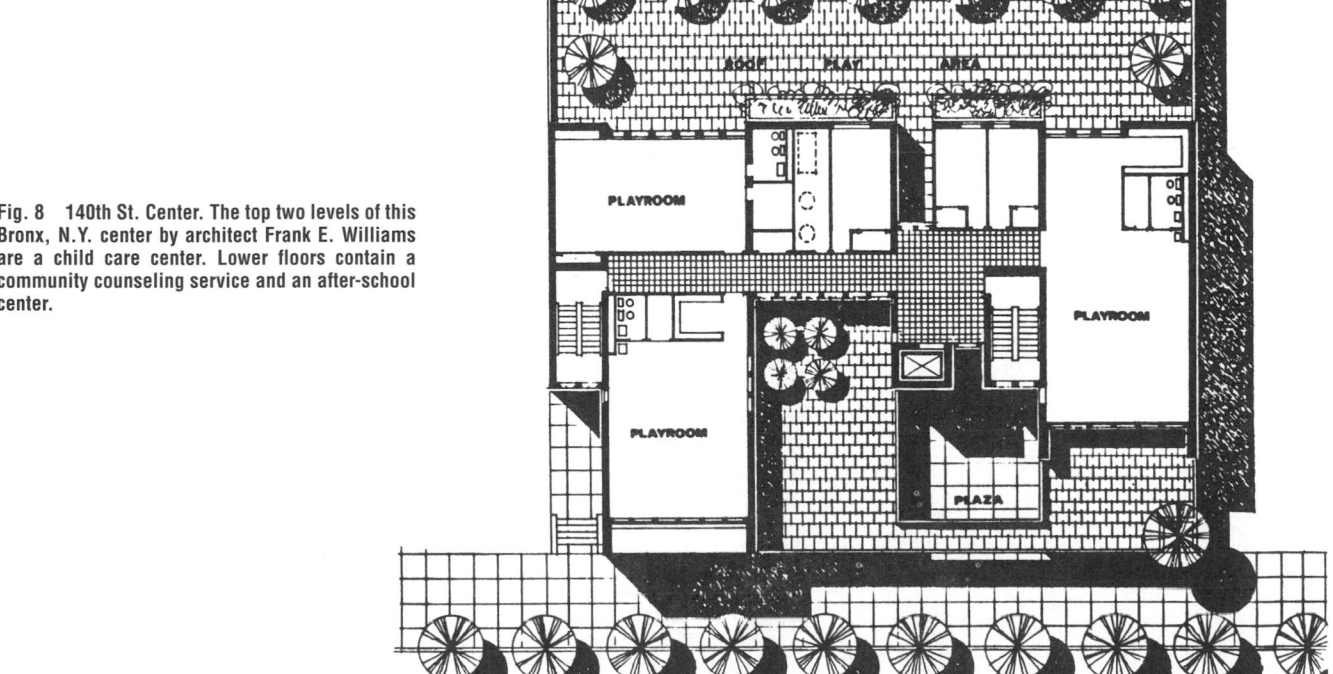

Fig. 8   140th St. Center. The top two levels of this Bronx, N.Y. center by architect Frank E. Williams are a child care center. Lower floors contain a community counseling service and an after-school center.

PLAYROOM

PLAYROOM

PLAYROOM

PLAZA

COMMUNITY SPACES

A community room should be located so that its entrance and interior can be observed from nearby public space, apartments, buildings, or streets.

Community spaces might include a large activity room with adjacent kitchen, a smaller room for meetings or office use, men's and women's toilets, appropriate storage, and a janitor's closet. Social and recreational facilities should reflect the preferences of the anticipated residents. Multiple-purpose use of common areas should be encouraged.

In projects for the elderly and handicapped, suitable and adequate facilities for social and recreational activities are especially important. In such facilities, it may be desirable to provide a craft room with sink, work counter, and storage.

Community facilities in nonelderly projects should not exceed the following:

Number of dwellings served	Net area (area within the finished interior surfaces excluding circulation)
Under 50	25 ft^2 per dwelling
50–99	1250 ft^2 plus 20 ft^2 for every dwelling over 50
100 or more	2250 ft^2 plus 15 ft^2 for every dwelling over 100

Number of bedrooms served	Net area (area within the finished interior surfaces excluding circulation)
Under 100	8 ft^2 per bedroom
100 and over	800 ft^2 plus 4 ft^2 for every bedroom over 100

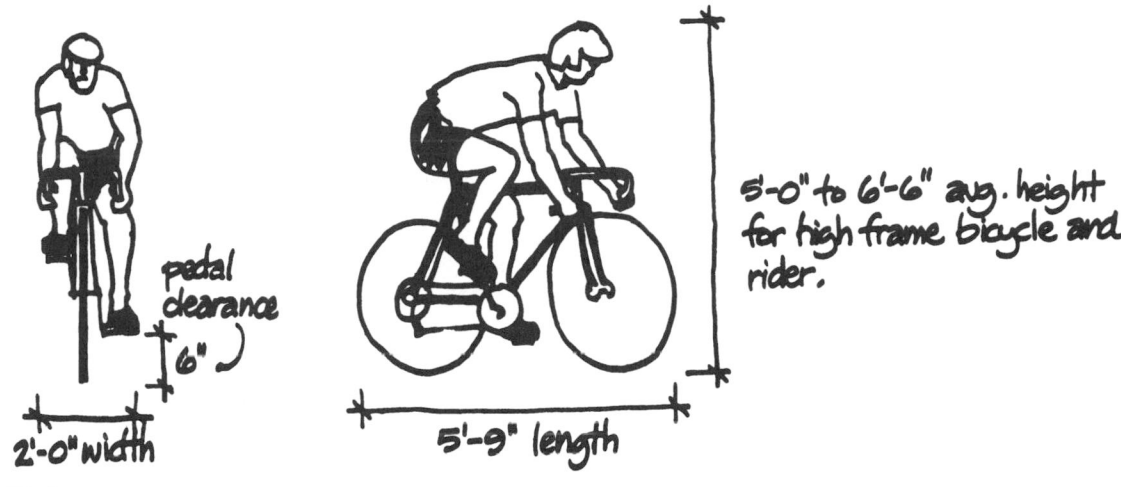

5'-0" to 6'-6" avg. height for high frame bicycle and rider.

pedal clearance 6"

2'-0" width

5'-9" length

Fig. 1

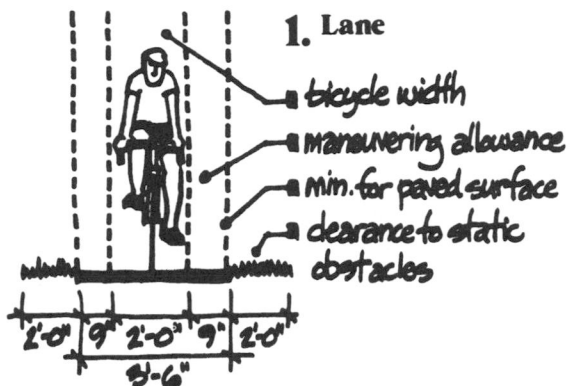

1. Lane

- bicycle width
- maneuvering allowance
- min. for paved surface
- clearance to static obstacles

2'-0" 9" 2'-0" 9" 2'-0"

5'-6"

2. Lanes

- min. recommended paved surface

6'-6"

3. Lanes

- min. recommended paved surface

11'-0"

Fig. 2

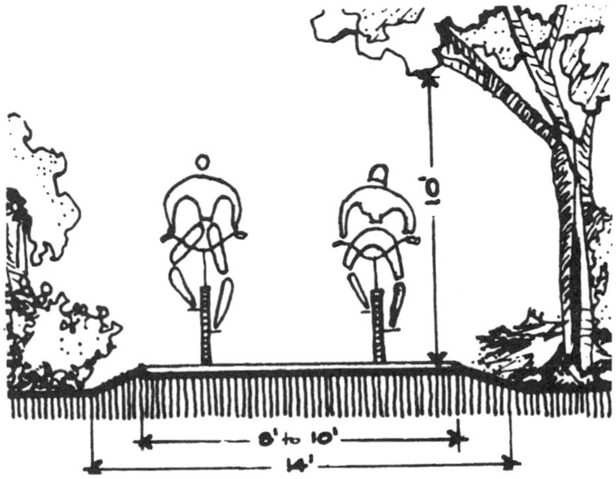

Fig. 3   Bikeways should have a width of 8 to 10 ft (two-way) or 6 ft (one-way). The grade should be a maximum of 5 percent for lengths up to 1000 ft and 15 percent for very short distances. Clearance should be 10 ft. The surface should be hard, preferably asphalt.

8' to 10'

14'

2'-0"
min. to drop in grade

1'-6"
min. to raised curb

1'-6"
recommended pavement extension
min. to soft shoulder

2'-0"
min. to static obstacles

Fig. 4   General dimensional requirements of bikeways.

RESIDENTIAL SWIMMING POOLS

DEFINITIONS

Residential pool A residential pool is any constructed pool, permanent or portable, which is intended for noncommercial use as a swimming pool by not over three owner families and their guests, and which is over 24 in in depth and has either (1) a surface area exceeding 250 ft² or (2) a volume over 3250 gal.

Residential pools are further classified into types as an indication of the suitability of a pool for use with diving equipment.

Type I: Any residential pool where the installation of diving equipment is prohibited.

Type II through type V: Residential pool suitable for the installation of diving equipment also classified by type, provided further that the diving equipment classified at a higher type; i.e., type III, may not be used on a pool of a lesser type, i.e., type II.

Permanently installed swimming pool One that is constructed in the ground, on the ground, or in a building in such a manner that the pool cannot be readily disassembled for storage.

Nonpermanently installed swimming pool One that is so constructed that it may be readily disassembled for storage and reassembled to its original integrity.

Vertical Vertical is defined as not exceeding an 11 percent (1 ft horizontally for each 5 ft vertically) slope from plumb.

Water line The water line referred to in these standards is established in one of the following ways:

1. The water line falls in the midpoint of the operating range of the skimmers.

2. On pools with overflow systems, the water line is established by the height of the overflow rim.

In-ground swimming pools Any pool whose sides rest in partial or full contact with the earth.

On-ground swimming pool Any pool whose sides rest fully above the surrounding earth.

Nonswimming area Any portion of a pool wherein water depths, offset ledges, or similar would prevent normal swimming activities.

Public pool A public pool is any pool, other than a residential pool, which is intended to be used collectively by numbers of persons for swimming or bathing and is operated by any person (owner, lessee, operator, licensee, or concessionaire), regardless of whether a fee is charged for such use. Public pools are listed in the following categories:

Class A: Competition pool—any pool intended for use for competitive events such as Olympic, AAU, and NCAA.

Class B: Public pool—any pool intended for public recreational use.

Class C: Semipublic pool—any pool operated solely for and in conjunction with lodgings such as hotels, motels, apartments, and condominiums.

Class D: Special-purpose pool—any pool operated as a treatment, water therapy, or nonrecreational function.

Dimensional Design

No limits are specified for shape of swimming pools except that consideration should be given to shape from the standpoint of safety and the recirculation of the swimming pool water.

Water depths at the shallow end of the swimming area should be 2 ft 9 in minimum and 3 ft 6 in maximum, except for special-purpose pools. More shallow depths may be used in the nonswimming area.

Walls in the shallow portion of the pool should be vertical from the water line for a minimum of 2 ft 3 in, from which point a tangent radius or vertical section can be used to join the wall section to the floor.

The slope of the floor from the shallow end wall toward the deep end should not exceed 1 ft in 7 ft to the point of first slope change.

The point of the first slope change is the point at which the floor slope exceeds 1 ft in 7 ft and is at least 6 ft from the shallow end wall.

The slope of the floor from the point of first slope change to a water depth of 5 ft 6 in should not exceed 1 ft in 2 ft 6 in.

In water depths over 5 ft 6 in the slope of the floor should not exceed 1 ft in 1 ft.

If the point of first slope change from the shallow end to the deep end occurs in water depths less than 4 ft 6 in, a permanently attached safety line, supported by buoys, should be affixed to the sidewalls at a point at least 1 ft on the shallow side of the point of first slope change.

All slopes should be uniform.

Dimensions of length and depth are taken at the longitudinal cross section of the pool corresponding with the centerline of the diving board; dimensions of minimum width provide one-half of this dimension as the minimum clearance from the centerline of the diving board to each side of the pool.

Pools intended for use with diving equipment should have the minimum dimensions of width, length, and water depth according to Fig. 1. The slope of the walls in the deep end should be as outlined in Fig. 1.

Minimum Dimensions for Residential Pools with Diving Equipment

TABLE 1

Pool type	Radius to be used	Radius to be drawn tangent to plumb deep end wall at the following depth
II	4 ft	3 ft
III	4 ft 6 in	3 ft
IV	5 ft	3 ft
V	4 ft 6 in	4 ft

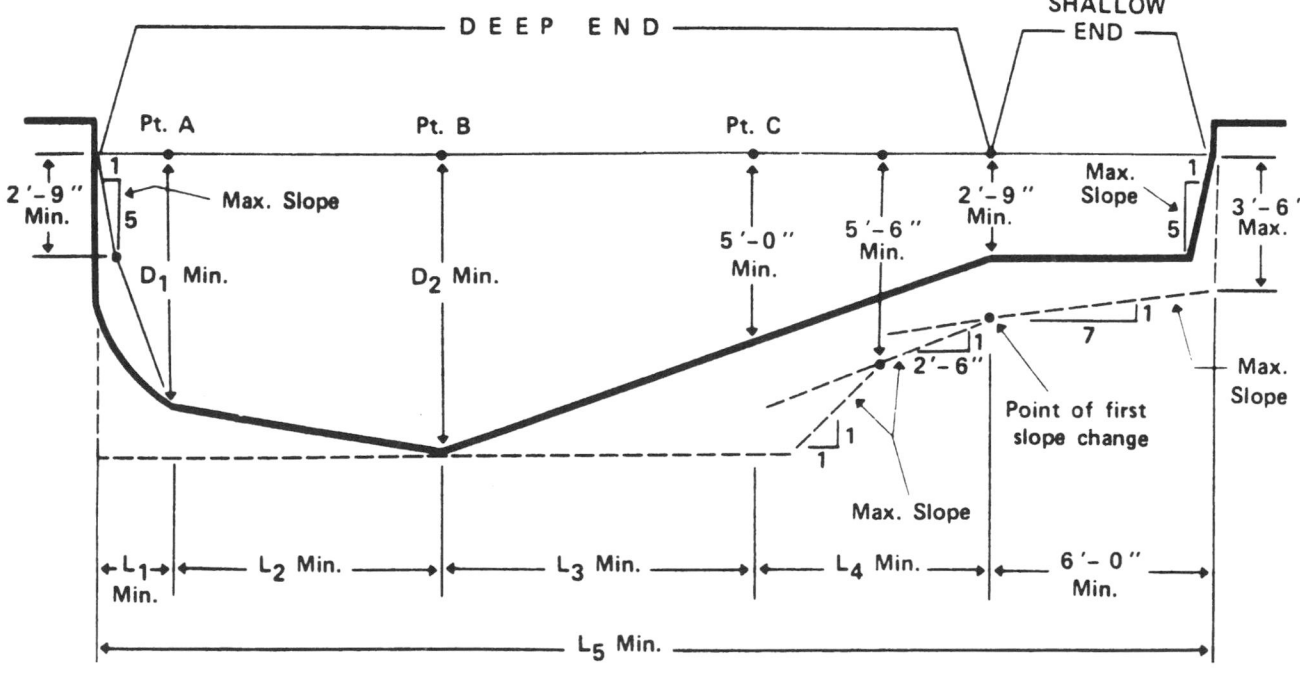

Fig. 1

TABLE 2

Pool type	Minimum dimension							Minimum width of pool		
	D_1*	D_2	L_1†	L_2	L_3	L_4	L_5	Point A	Point B	Point C
I	No minimum dimension									
II	6 ft 0 in	7 ft 6 in	1 ft 6 in	7 ft 0 in	7 ft 0 in	6 ft 6 in	28 ft 0 in	12 ft 0 in‡	15 ft 0 in‡	12 ft 0 in
III	6 ft 10 in	8 ft 0 in	2 ft 0 in	7 ft 6 in	8 ft 0 in	6 ft 6 in	30 ft 0 in	12 ft 0 in	15 ft 0 in	12 ft 0 in
IV	7 ft 8 in	8 ft 6 in	2 ft 6 in	8 ft 0 in	10 ft 0 in	6 ft 6 in	33 ft 0 in	15 ft 0 in	18 ft 0 in	15 ft 0 in
V	8 ft 0 in	9 ft 0 in	3 ft 0 in	9 ft 0 in	11 ft 6 in	6 ft 6 in	36 ft 0 in	15 ft 0 in	18 ft 0 in	15 ft 0 in

*The point where D_1 strikes the floor may be smoothed with a radius as noted in Table 1. The innermost lines of the resultant drawing represent the minimum allowable underwater longitudinal section.

†Point A is a base reference point where depth D_1 intersects the water surface and from which point all other dimensional points may be derived. It should be specifically noted that the L_1 distance from point A to the deep end wall is a minimum distance.

‡On a pool type II the width dimension may be reduced to 10 ft at point A and 12 ft at point B provided the following minimum conditions are met:

The cross sections at points A and B show that the walls in these cross sections are plumb for the first 3 ft below the water level.

The walls in the cross section at point B beneath the plumb 3-ft portion should be curved to a depth of 7 ft 3 in with a tangent radius of 4 ft 6 in, from which point a slope of 1 ft in 3 ft should be used to a depth of 8 ft.

Diving equipment with a 6-ft board length should be used.

Any other configuration that creates a water envelope in excess of that outlined here should be permitted within the classification of a type II pool.

Cross-sectional dimensions of pools designed for the installation of a diving board should have as their minimum dimensions the widths specified at points A, B, and C as outlined above and as located in Fig. 2.

Plan View (Minimum Water Surface Area of Minimum Diving Water Envelope)

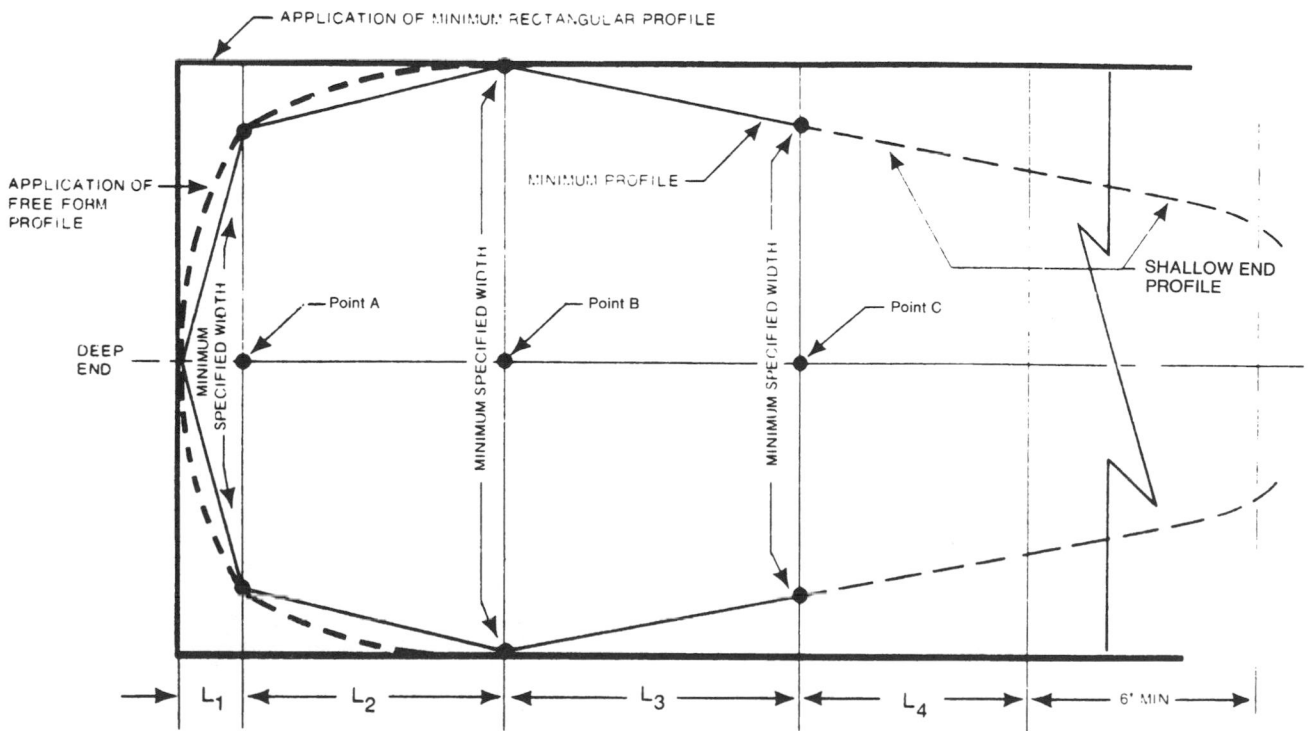

Fig. 2

RESIDENTIAL SWIMMING POOLS

Minimum Allowable Underwater Cross Section at Point *B*

The minimum allowable underwater cross section at point *B* should be established as the minimum profile represented by the superimposure of case A and case B as outlined below. Any other configuration that creates a water envelope in excess of the superimposure of case A and case B should be permitted.

Superimposure of case A and B (Fig. 3) Minimum dimensions to be in accordance with those outlined below.

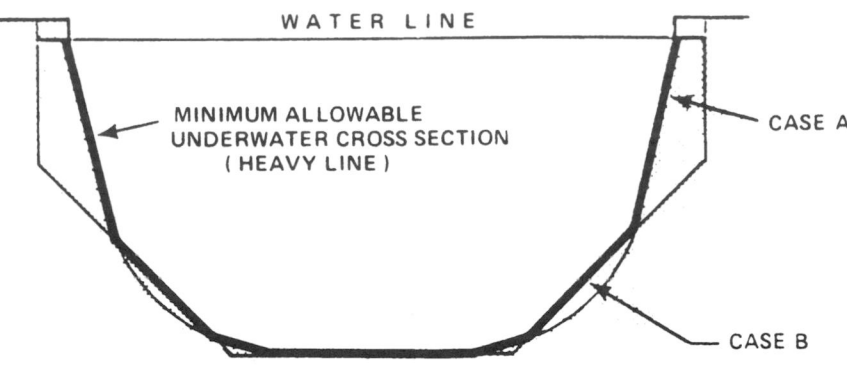

Fig. 3

Case A (Fig. 4) Walls should be vertical for a distance of not less than 2 ft 9 in below the water level, below which the wall should be curved to the bottom with a tangent radius not greater than the difference between the depth and 2 ft 9 in.

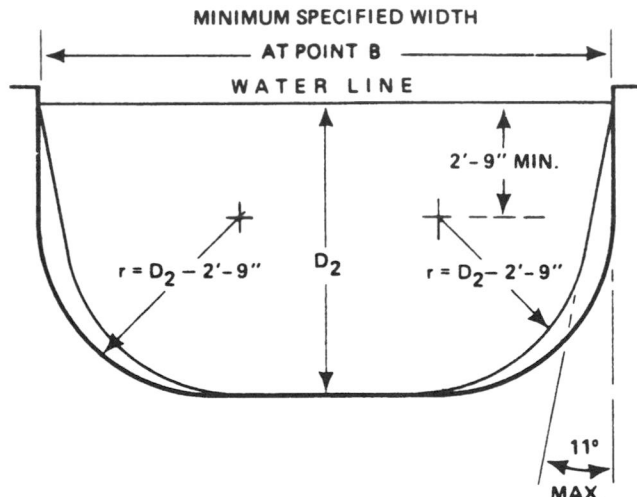

Fig. 4

Case A (Fig. 5) Walls should be plumb for a distance of not less than 3 ft, below which a 4 ft 6 in tangent radius should extend to a depth of 7 ft 3 in, from which point a slope of 1 ft in 3 ft should extend to a depth of 8 ft.

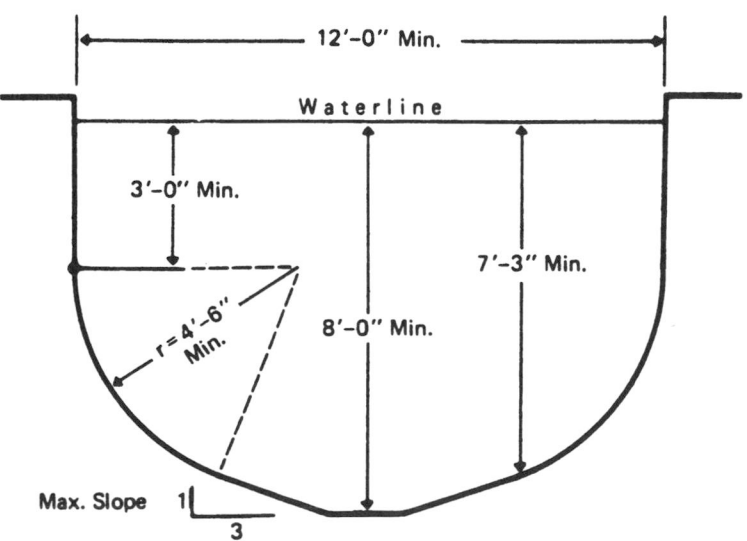

Fig. 5

Case B (Fig. 6) Walls should be plumb for a distance of not less than 2 ft 9 in below the water level, below which the walls should slope uniformly to intersect the bottom at a point conforming to the minimum geometry called for in Table 3.

The minimum allowable underwater cross section at point A should be a uniform intersection or blending of the minimum longitudinal section called for and the minimum cross section called for at point B.

Diving equipment Diving equipment should be designed for swimming pool use and be installed in accordance with Table 4.

Diving equipment should be of a nonslip design.

Minimum unobstructed headroom from the top of the diving equipment should be provided for diving in accordance with Table 5 unless greater dimensions are called for by the board manufacturer.

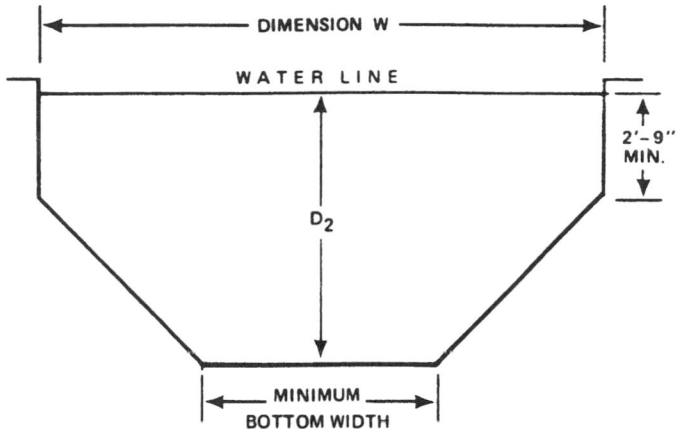

Fig. 6

TABLE 3

	Pool types			
	II	III	IV	V
Minimum depth specified at point B (D2)	7 ft 6 in	8 ft 0 in	8 ft 6 in	9 ft 0 in
Case A minimum width specified at point B	15 ft	15 ft	18 ft	18 ft
Case B minimum width (dimension W)*	16 ft	16 ft	18 ft	18 ft
Minimum bottom width	7 ft	7 ft	10 ft	10 ft

*Use dimension W in conjunction with 2-ft 9-in minimum plumb wall to establish case B profile.

TABLE 4

Pool type	Maximum diving board length, ft	Maximum jump board length, ft	Maximum board height over water, m (in)
I	No diving equipment permitted		
II	8	6	½ (20)
III	10	8	⅔ (26)
IV	12	8	¾ (30)
V	12	8	1 (40)

TABLE 5

Board type	Headroom, ft
II	12
III	12
IV	13
V	14

Slide equipment Slide equipment should be installed in accordance with Fig. 7.

The A dimension should be the height of the slide above the deck or ground on which the slide rests.

The B dimension is the distance to the minimum depth of water that should be maintained. This distance should be equal to 1½ times the height of the slide (B = 1.5A).

The C dimension is the minimum distance to the water that should be maintained in front of the end of the slide. This distance should be equal to 2 times the height of the slide (C = 2A).

The D dimension is the minimum depth of water. Since the required minimum depth of water is being studied, no standard is now available.

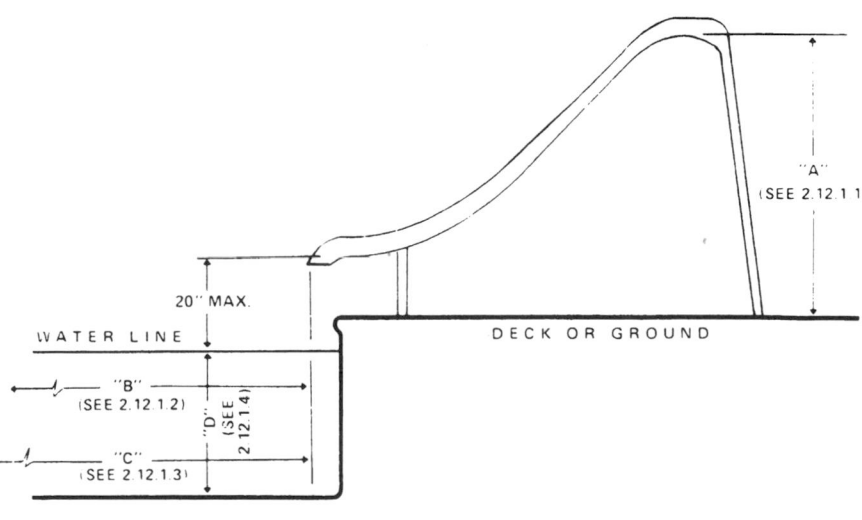

Fig. 7

RESIDENTIAL SWIMMING POOLS

Position of Slide at Edge of Pool

Slides should be positioned at the edge of the pool so that any and all water flowing off the end of the slide runway drops into the pool.

All slides should be positioned so that the centerline of the slide does not intersect the centerline of the diving board for a minimum distance of 7 ft from point A.

All slides should be positioned so that the slide user cannot hit the edge of the pool, diving board, or other equipment at any point with arms extended. The minimum distance from the centerline of the slide runway to the edge of the pool or diving board should be 3 ft 6 in at a point 2 ft 6 in minimum from the end of the slide.

All slides should be firmly anchored.

Slides may be straight or curved. Centerlines of the straight portion of the slide ends are referenced in Fig. 8.

Materials of Construction

Swimming pools and all their appurtenances should be constructed of materials which are nontoxic to human beings, which can withstand the design stresses, which will provide a watertight structure with a smooth and easily cleaned surface without cracks or joints, excluding structural joints, or to which a smooth, easily cleaned surface finish is applied or attached.

Deck Equipment (Steps, Ladders, Stairs, Diving Boards, and Platforms)

Steps or a ladder or other approved means should be provided at the shallow end of the pool if the vertical distance from the bottom of the pool to the deck or the top of the wall is over 2 ft. If a pool is over 40 ft long, a second step or ladder or other approved means should be provided to serve the deep or opposite end.

A ladder or steps or other approved device should be provided at the deep end of the pool if the depth of water is over 5 ft. If the pool is over 30 ft wide in the deep end, such steps or ladders should be installed at each side.

All step treads should have a minimum unobstructed width of 10 in and a minimum surface area of 240 in². Risers at the centerline of the steps should have a maximum uniform height of 12 in. Where treads do not terminate at a pool wall, they should be protected by a handrail or grabrail. Treads may be narrowed to 8 in unobstructed, provided a suitable handrail or grabrail is installed. Seats may be provided as part of the steps.

Pool ladders should be corrosion-resistant and should be equipped with nonslip treads. All ladders should be so designed as to provide a handhold and should be securely installed. There should be a clearance of not more than 6 in or less than 3 in between any ladder and the pool wall.

If steps are inserted in the walls or if step holes are provided, they should be arranged to drain into the pool to prevent the accumulation of dirt. Step holes should have a minimum tread of 5 in and a minimum width of 12 in. They should have a grabrail at the top of both sides.

Supports, platforms, and steps for diving boards should be of substantial construction and of sufficient structural strength to carry the anticipated loads safely. Steps should be of corrosion-resistant material, easily cleanable, and of nonslip design. All diving stands higher than 21 in measured from the deck to the top butt end of the board should be provided with step(s).

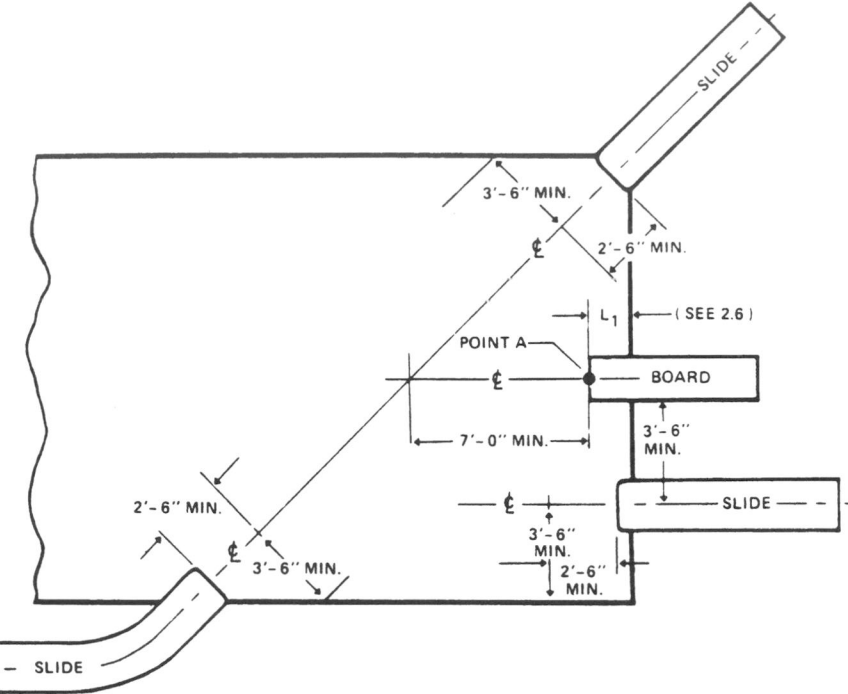

Fig. 8

Safety Features

When a "permanently attached" safety line is called for, it should be supported by buoys and installed approximately at water level and at the location specified so that one end of it cannot be removed without the use of tools. Further, in lieu of providing a safety line meeting the above "permanent" description, a removable safety line may be installed provided that there are easily readable warnings firmly affixed and located near the anchors for the safety line, indicating that use of the swimming pool without the safety line in place may be hazardous.

All residential pools should be provided with a suitable handhold around their perimeter in areas where depths exceed 3 ft 6 in. Handholds should be provided no farther apart than 4 ft and may consist of any one or a combination of the following:

1. Coping, ledge, or deck that is not higher than 12 in above the water line.
2. Ladders, steps, or seat ledges.
3. A rope or railing placed at or not over 12 in above the water line fastened to the wall.

Outdoor swimming pools should be protected by a fence, wall, building, enclosure, or solid wall of durable material (of which the pool itself may be constructed) or any combination of barriers. Artificial barriers should be constructed so as to afford no external handholds or footholds, of materials which are impenetrable by toddlers, at least 4 ft in height so that a toddler cannot grasp its top by jumping or reaching, and equipped with a self-closing and positive self-latching closure mechanism at a height above the reach of toddlers and provided with hardware for permanent locking.

FACTORS AFFECTING POOL DESIGN

Size of Pools

One of the major criteria for pool design is the need for at least one competitive course as a design requirement for new swimming pools. Private apartment-house pools, smaller semipublic facilities, and combination display-fountain swimming pools are the noted exceptions.

Pool lengths are determined by AAU and NCAA specifications. After length requirements are established for the design program, the water area can be studied with respect to other program requirements. A minimum area will automatically be derived from the number and width of lanes used (7-ft lane recommended). Sometimes the rectangular pool is all that can be permitted. Obviously, the other program activities must be tailored to such a basic, competitive pool.

When diving or recreational swimming and group water activities are programmed, distinct pool areas have to be designated and designed for these purposes. Code regulations and previous experience dictate use factors that minimize the safety hazard inherent in all aquatic programs.

Pool dimensions for multiple usage are usually determined by a combination of the several pool areas required, superimposed or joined, for the best architectural and engineering results. The T and L designs have been modified in endless procession. The I and H have evolved. Fans, boomerangs and trapezoids have been designed.

In many situations, pool size is regulated by the occupancy limit. For years, an accepted factor for pool use has been 25 ft² of pool area per capita. More recently, 15 ft² per capita has been shown to be adequate and safe for shallow-water recreational swimming, while 175 ft² per capita for instructional diving and 45 ft² per capita for swimming lessons are minimum areas.

Recent studies have shown that the total complex area has to be considered in determining required pool size and safety limitations. Health and municipal agencies, recognizing the problem of area overcrowding, have established design criteria regulating maximum occupancy of combined deck-and-pool areas.

Minimum recommended occupancy design factors are shown in Table 6. Minimum deck widths have also been established.

Final determination of the *safe bathing load,* using recommendations in Table 6, can be established by considering the supervision available, the method of instruction, and the nature of the activity being conducted.

Water Depth

Water depths are determined by the various age groups using the pool. For educational purposes, a pool should be divided with lifelines, separating the water areas into deep (over 5 ft 0 in), intermediate (3 ft 9 in to 5 ft 0 in), and shallow (2 ft 10 in to 3 ft 9 in). Familiarization pools, or so-called "wading" pools, are kept separate from the main pool. They are less than 24 in in depth and are designed for separate activity and supervision.

Shape of Pools

The rectangular pool is generally satisfactory for all-purpose instructional use and is most frequently designed for indoors, where available space and enclosure cost are limiting factors. Fan-shaped pools provide the services of the rectangular pool plus added areas of shallow water which meet the needs for instruction and accommodate both swimmers and beginners.

L-shaped pools provide the standard rectangle, with an additional section at one end. The advantage of separation of activities is important.

T-shaped pools feature a bay projecting from one side and are used as alternates to Ls in larger pool sizes. The added section is always smaller than the main pool, and usually constitutes the deep-water area. Separating deep-water from shallow-water programs is a safety improvement.

Z-shaped pools permit a very-shallow-water section in addition to the deep-water appendage. Again, the separate-activity section features are apparent.

TYPES OF POOLS BY DESIGN

The *rectangular swimming pool* was probably the original pool shape, and a great many rectangular pools are still being built. Some of the first pools were 50 by 100 ft, or even 100 by 200 ft. Many indoor pools were 30 by 60 ft, and, in some few cases, 35 by 75 ft. The rectangular pool of 45 by 75 ft is quite acceptable today both indoors and outdoors, since it includes the 25-yd short course for competitive swimming. In some cases, a pool 60 ft wide—or preferably 75 ft (25-yd short course)—and 50 m long (164 ft ½ in) is also quite acceptable.

The rectangular pool has a comparatively low construction cost and is easy to supervise. However, in small pools there is a low percentage of shallow water, and only one competitive activity can be operated at a time. The rectangular pool is a traditional-type pool, but it is not as dramatic or as interesting as some of the "free-form" pools which can easily include the proper competitive-swimming lengths, make possible larger deck areas, offer an interesting shape which fits into the landscape, and give variety and interest to the activities.

The *T-shaped pool* was probably one of the first pools to break away from the rectangular shape, and it is today a fine shape for a pool. The top of the T can be 45 ft by 164 ft ½ in, preferably 60 ft by 164 ft ½ in, and sometimes 75

ft by 164 ft ½ in, thereby providing both the short and long courses and an entire area of so-called "shallow water" which is 3 ft 3 in to 5 ft 0 in in depth.

The lower part of the T should be reserved for the diving area and should be separated from the top of the T by a float line. This is probably one of the most popular pool designs, although the *L-shaped pool* includes the same features, except that the deep area is placed at one end rather than in the middle.

The *Z-shaped pool* is also extremely functional, having the competitive-swimming lengths and general swimming area in the center of the Z, the diving area offset from the main part on one side, and the extremely shallow area (extending to 2 ft) off the other side of the general swimming area. In this shape, the swimming area could be the 25-yd or 50-m distance. In some cases, the Z-shaped pool can include both the 25-yd and 50-m lengths, keeping in mind, of course, that recreational swimming is the prime objective, but the objectives of a good recreational area and competitive swimming lengths can be combined in such a pool. The most obvious advantages in the T-, L-, and Z-shaped pools are the separation of divers and swimmers, tending to eliminate the danger of collisions; and the provision of more shallow-water area. These shapes do, however, present some supervision problems.

The *multiple-pool* idea is an extremely good one, especially in instances where funds are not available to build all the swimming facilities desired at one time. The program can be phased so that only one pool is built at a time.

In a country club where there is need to spread out the use of the pool area and have pools for different groups—adults and juniors— a competitive pool, a diving pool, and a play pool (for the small children) will be found most advantageous. The adult pool has been conceived to provide a special area for adults only, and this is particularly desirable in a country club setting where alcoholic beverages are served around a pool.

The *junior pool* is a pool ranging from 2 to 3 ft in depth, where swimming instruction can be conducted for young people—ages up to 10.

The *tots' play pool,* more commonly known as the wading pool, has a water depth of 0.0 to 15 or 18 in. It is designed for the comfort and convenience of youngsters below the age of 6. The more complete tots' play pool includes a 6 to 10-ft walk surrounding the pool, and the entire area is enclosed with a barrier, such as a 3-ft chain-link fence (knuckle-finish top and bottom) or a masonry sitting wall for the parents or others accompanying the small tots.

The separate diving pool has many advantages. Since there is usually a conflict between swimmers and divers, the safety feature is a good justification for the separation of the main pool from the diving pool.

There are advantages, certainly, to the separate pools, safety being the principal one. However, the separate pools cost more, require a greater number of lifeguards, require more extensive mechanical and water-treatment systems, and occupy more space.

Free-form pools are ordinarily restricted to resort pools and residential pools. This shape is not favored for the public or institutional-type pool.

The *spray pool* is popular in many communities. In some instances, the water goes to waste, and in other locations, the spray pool is combined with a wading area. Spray pools of 30 to 40 ft diameter have been designed as multiple-use areas so that they can also serve as dance and roller-skating areas. In the majority of cases, the pool with a wading area has proved more popular than the spray pool.

GENERAL PLANNING CONSIDERATIONS

Adequate area is needed around a pool, not only to dignify its setting and location but to serve as a buffer from nearby streets and residences. Space must also be provided for parking and for other recreational areas and facilities.

A pool should not be located in a low spot. Water from the surrounding area will drain into the pool, and unless precautions are taken for proper drainage around the edge of the pool, considerable water will penetrate the area under the deck and floor and be a source of constant annoyance and engineering problems.

A pool should not be located in a grove of trees. Leaves fall into the pool and keep it dirty, clog up filters and the hair and lint catcher, and in many ways keep a pool in an unsatisfactory condition. Trees also keep the sun away from the pool, and to be successful, the pool area must have sunshine. The pool should be so located that buildings and trees to the west of the pool are at such a distance that they will not shade the pool in the late afternoon.

Location in relation to streets is most important. A pool must be near main traffic arteries for good circulation and for accessibility, but the pool itself should not be too near a busy street. The dirt from the street will blow into the pool and give considerable trouble in the filtration system. If possible, the pool should be set back 200 to 300 ft or more from the street.

TABLE 6   Minimum Recommended Occupancy Design Factors

Activity	Indoor pools	Outdoor pools
Shallow-water area (under 5 ft 0 in), ft^2 per capita		
Recreational swimming	14	15
Advanced swimming instruction	20	25
Beginning swimming instruction	40	45
Deep-water area (over 5 ft 0 in), ft^2 per capita		
Recreational swimming	20	25
Advanced swimming	25	30
Diving (based on area within 30 ft of deep-end diving wall)	175	200
Minimum walk width,* ft	6	12
Sum of walk dimensions,* ft on either side of pool length or width, should not be less than	18	30

*Walk dimensions should be horizontal clear deck width, not including any portion of the coping or interior gutter sections.

COMPETITION { INDOOR COURSE: 25 YDS. X 45' RECOMMENDED MINIMUM.
OUTDOOR COURSE: 50 METERS X 60' RECOMMENDED MINIMUM.

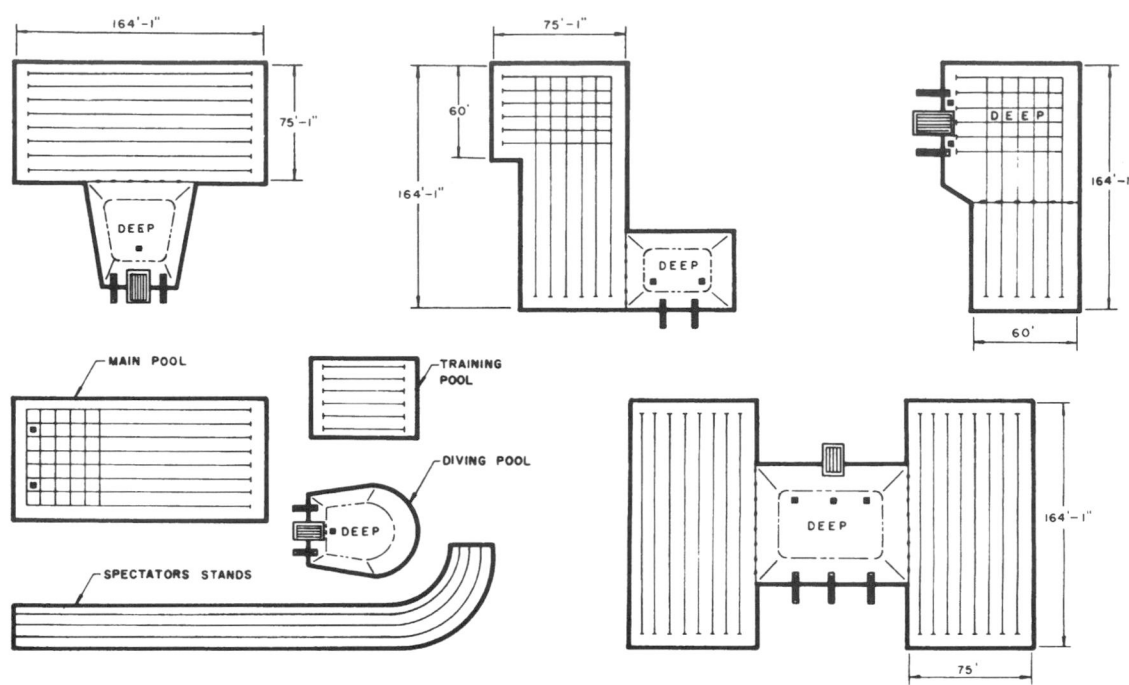

FREE-FORM (PRIVATE HOME, CLUB, MOTEL, ETC.)

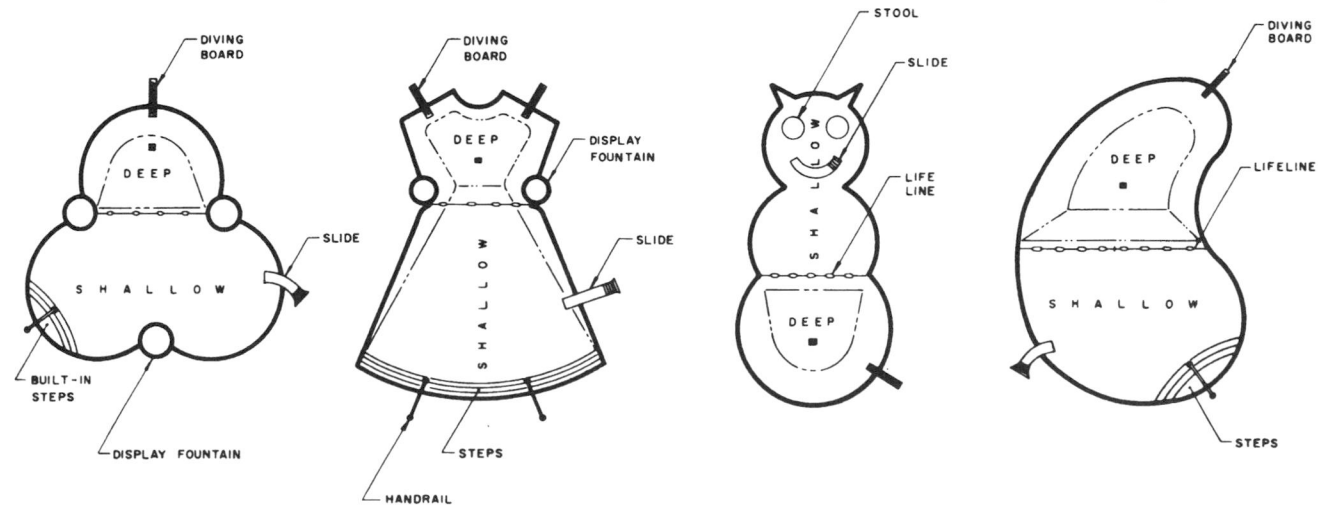

Fig. 9   Prepared by Milton Costello, Consulting Engineer.

Recreation facilities supplementing a swimming pool can consist of recreation buildings or community centers; softball and baseball diamonds; a football field; pitch-and-putt, par-3, or regulation golf courses; a multiple-use paved area; an area for playground equipment for both small children and older groups; and parking areas. Court games can also be included—tennis, croquet, shuffleboard, badminton, handball, horseshoes, paddle tennis, table tennis, deck tennis, roller skating, and volleyball.

WADING, JUNIOR, AND SHALLOW-WATER INSTRUCTIONAL-TYPE POOLS

Wading pools for small children vary from the very simple square or rectangular pool to a series of pools with sprays and water cascading from one to the other. Depth of the water varies from 8 to 24 in. The pool should be enclosed by a fence and should be separately controlled for the safety of the small children.

The pool should also have a nonskid bottom to prevent slipping on the part of the children. It is not considered economical to wall off one section of a swimming pool to serve as a wading pool. The reason for this is that a wading pool costs only about 30 percent of the unit cost of a regular swimming pool. The separate wading pool or tots' play pool provides a separate and safer place for the young children.

In a typical design utilizing three small circular wading pools, water flows from the center pool into the two outer ones and is then pumped back into the upper pool. These pools can be elevated above the deck, and the pool coping or curb will serve as a seat for parents and others watching over the small children.

The tots' play pool or wading pool should be separated from the swimming-pool area by a 3-ft chain-link fence or low sitting wall. This barrier will prevent the children from straying into the deeper-water area and also adds to the safety of the smaller children in that it keeps the older children from running through the wading pool.

A pool for the smaller children—not simply a wading pool or spray pool, but rather a play pool—can utilize a free form with some innovations to provide play opportunities, and will be more attractive (even to older children) than the conventional round or rectangular wading pool. One such play pool, designed as two squares lying parallel to each other, is joined together at a center point by a passageway approximately 5 ft wide.

Popular additions to these tots' play pools are sculptured concrete figures—usually a frog, seal, whale, turtle, baby elephant, or others with sprays—on which the children play and slide. These figures create an interesting environment for the youngsters and give play value to the pools.

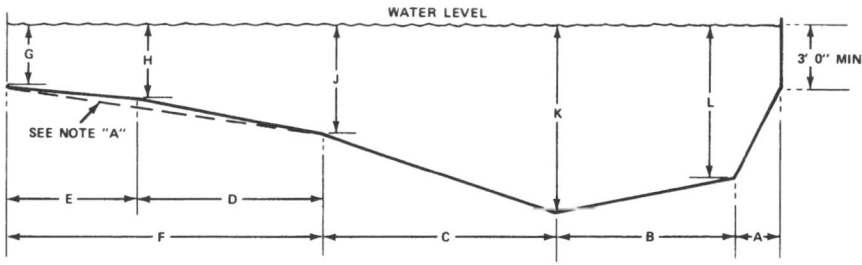

TYPICAL SECTION

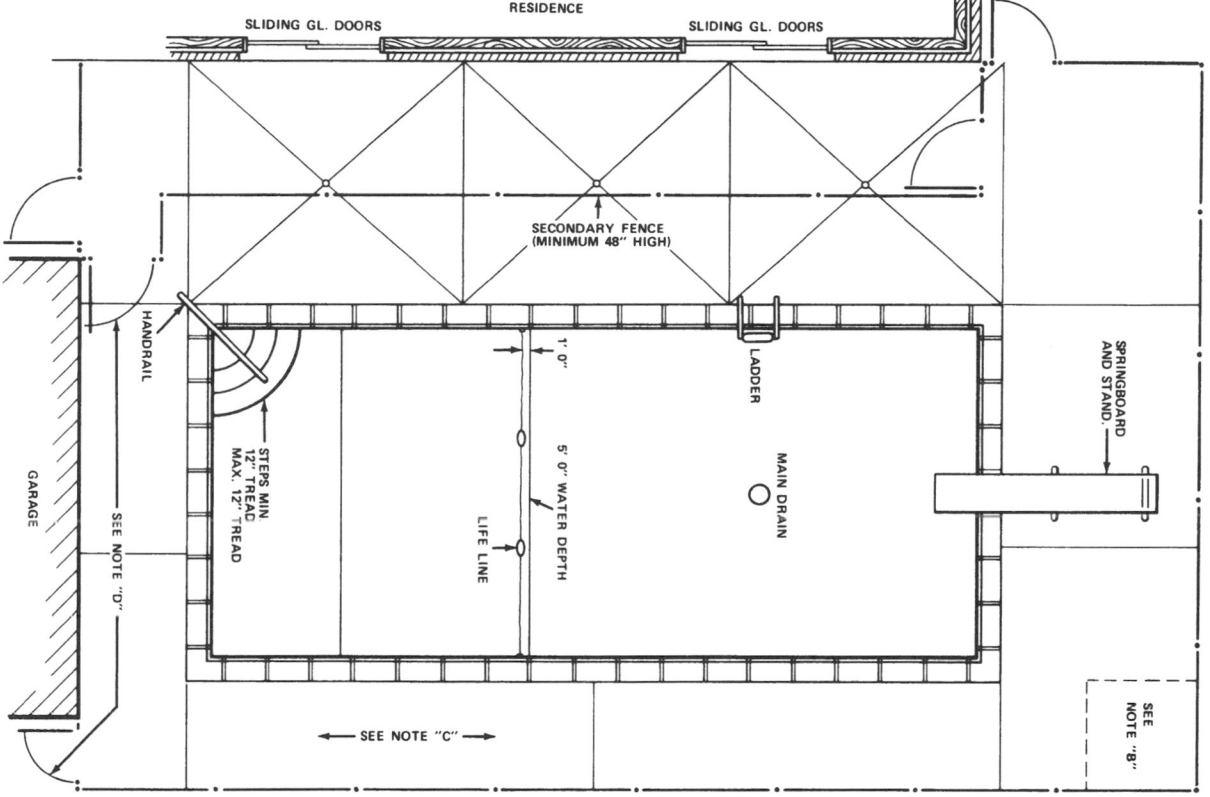

Fig. 10

TABLE 7

RECOMMENDED SWIMMING POOL DIMENSIONS														
SIZE	A	B	C	D	E	F	G	H	J	K	L	Length of Springboard	Overhang	Height of Diving Board Stand
12 x 28	1' 6''	7' 0''	7' 0''	6' 6''	6' 0''		2' 6''	3' 0''	4' 6''	6' 6''	5' 0''	0	0	None
12 x 30	1' 6''	7' 0''	9' 0''	6' 6''	6' 0''		2' 6''	3' 0''	4' 6''	6' 6''	5' 0''	0	0	None
12 x 32	1' 6''	7' 0''	9' 0''	8' 6''	6' 0''		3' 0''	3' 6''	5' 0''	7' 6''	6' 0''	8' 0''	1' 6''	Deck Level
15 x 30	1' 6''	7' 0''	8' 6''	7' 0''	6' 0''		3' 0''	3' 6''	5' 0''	7' 6''	6' 0''	8' 0''	1' 6''	Deck Level
15 x 32	1' 6''	7' 0''	9' 0''	8' 6''	6' 0''		3' 0''	3' 6''	5' 0''	7' 6''	6' 0''	8' 0''	1' 6''	Deck Level
15 x 35	2' 0''	8' 0''	10' 6''	8' 6''	6' 0''		3' 0''	3' 6''	5' 0''	8' 6''	7' 0''	10' 0''	2' 0''	Deck Level to 12''
16 x 35	2' 0''	8' 0''	10' 6''	8' 6''	6' 0''		3' 0''	3' 6''	5' 0''	8' 6''	7' 0''	10' 0''	2' 0''	Deck Level to 12''
16 x 40	2' 0''	8' 0''	10' 6''	13' 6''	6' 0''		3' 0''	3' 6''	5' 0''	8' 6''	7' 0''	10' 0''	2' 6''	12'' to 18''
18 x 30	2' 0''	8' 0''	12' 0''	9' 6''	6' 0''		3' 0''	3' 6''	5' 0''	8' 6''	7' 0''	10' 0''	3' 0''	12'' to 18''
18 x 40	3' 0''	9' 0''	11' 6''	-	-	16' 6''	3' 0''	See Note A	5' 0''	9' 0''	7' 6''	12' 0''	3' 0''	12'' to 39''
20 x 40	3' 0''	9' 0''	11' 6''	-	-	16' 6''	3' 0''	See Note A	5' 0''	9' 0''	7' 6''	12' 0''	3' 0''	12'' to 39''

NOTE "A"   FLOOR TO SLOPE FROM 3' 0" DEEP TO A UNIFORM SLOPE TO THE 5' 0" DEPTH.
NOTE "B"   PROVIDE APPROX. 30 SQ. FT. OF FLOOR SPACE FOR FILTRATION EQUIPMENT (PREFERABLY INSIDE BUILDING).
NOTE "C"   SLOPE WALKS AWAY FROM POOL AT 1/4" PER 1' 0". PROVIDE DRAINS AS REQUIRED.
NOTE "D"   PROVIDE SELF-CLOSING, SELF-LATCHING GATES. CAPABLE OF BEING LOCKED.

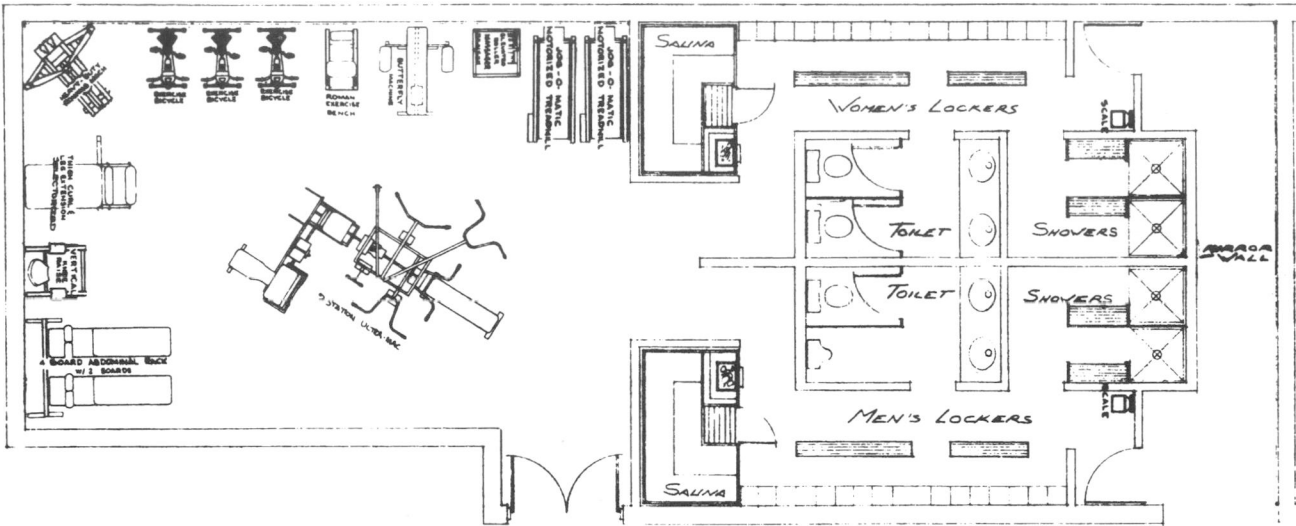

Fig. 1

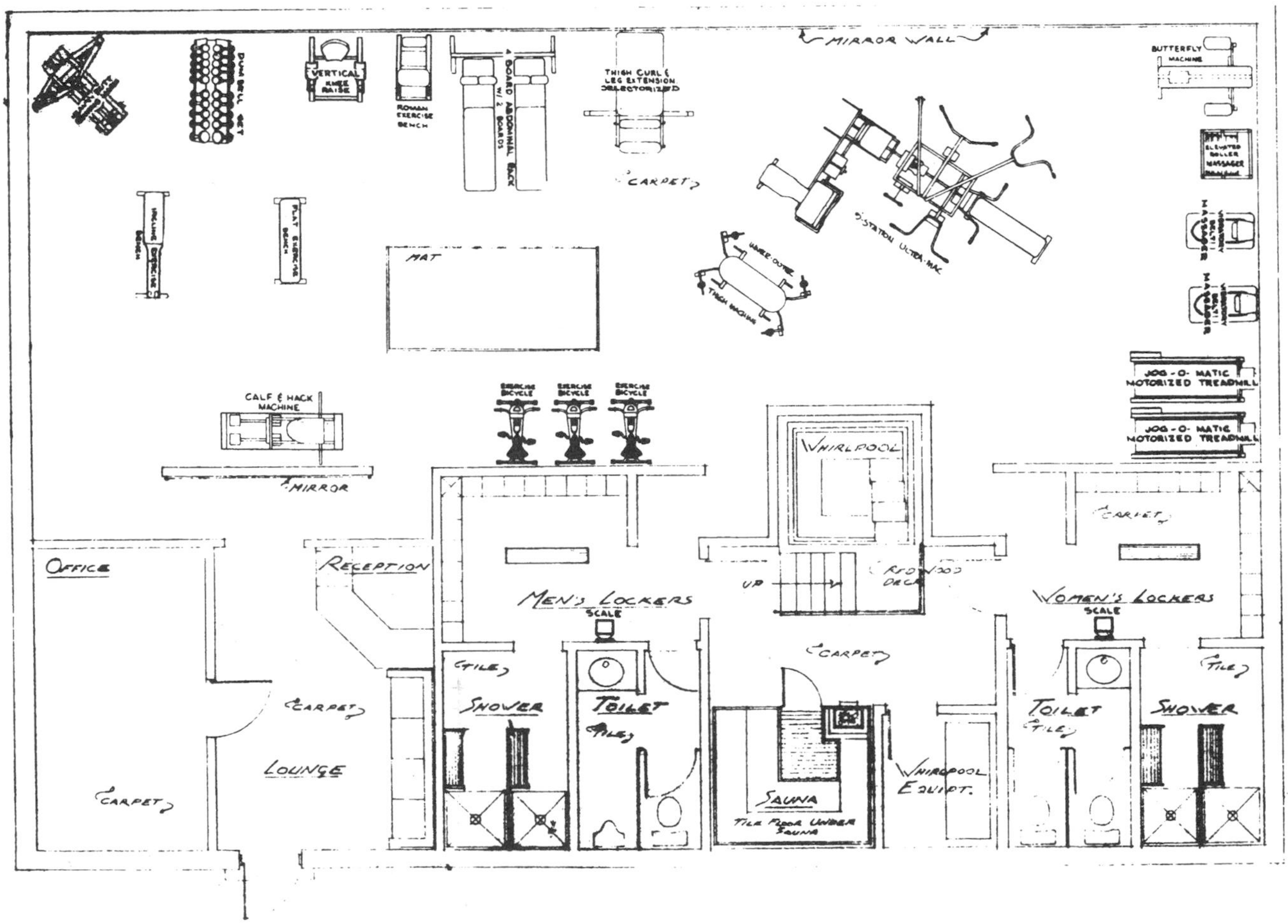

Fig. 2

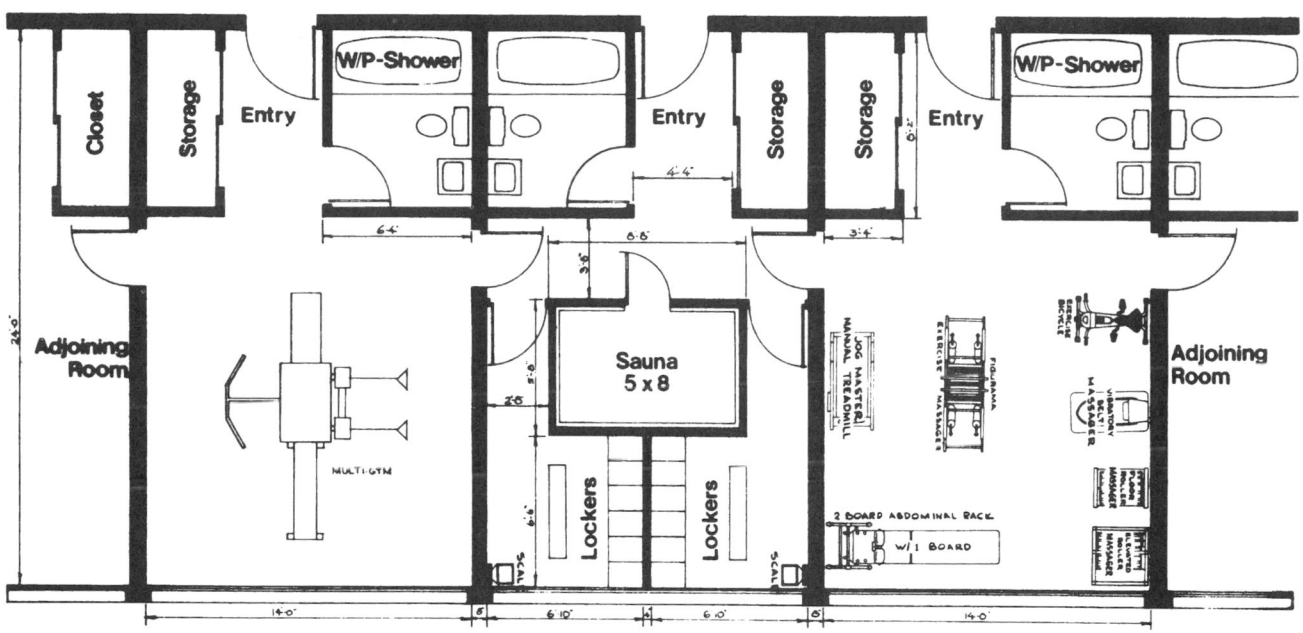

Fig. 3   Deluxe alternate facility.

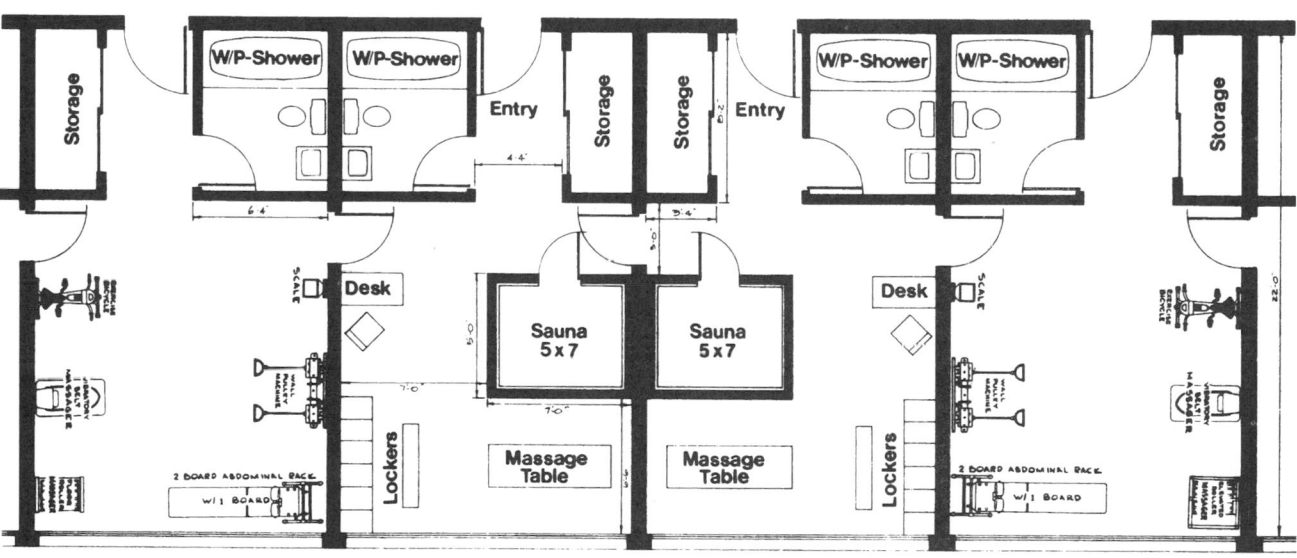

Fig. 4

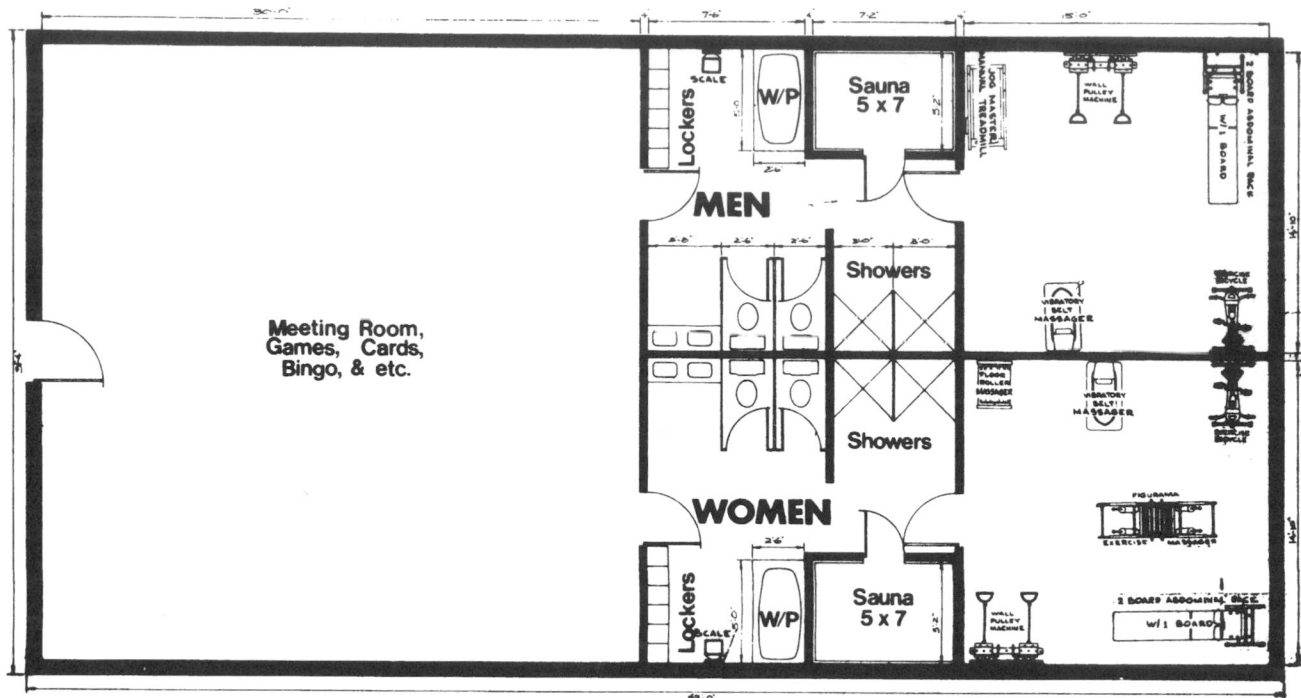

Fig. 5.   Standard dual facility.

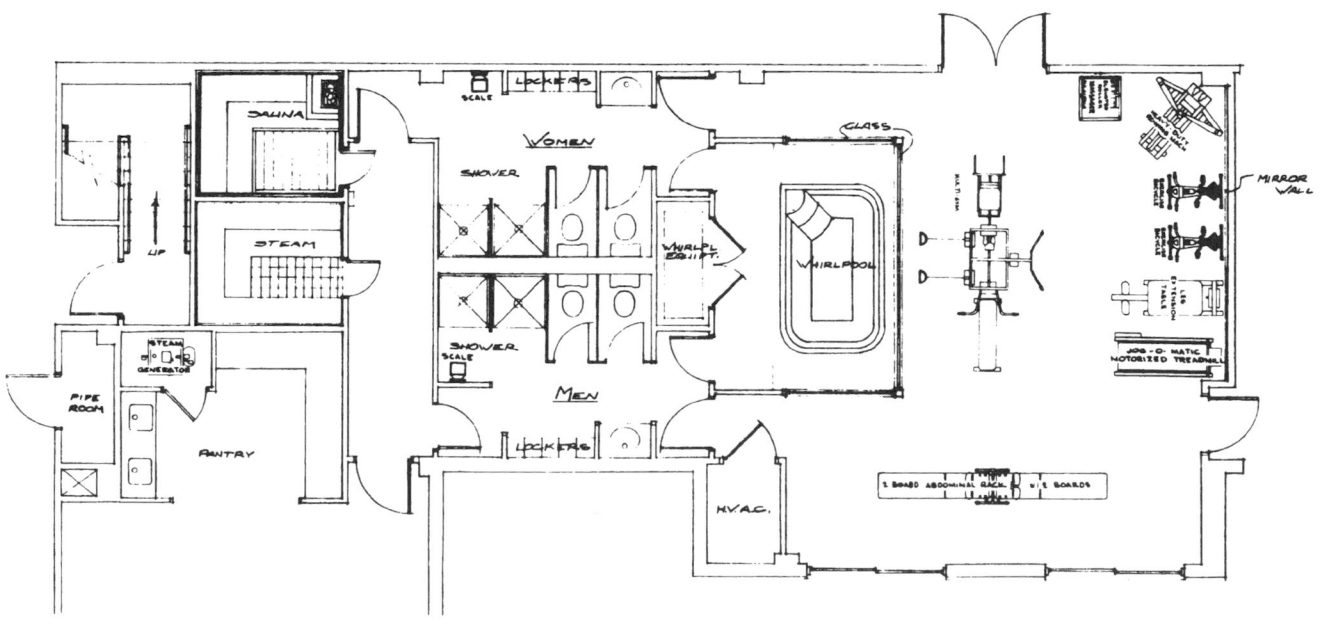

Fig. 6

TYPICAL ROOM LAYOUTS

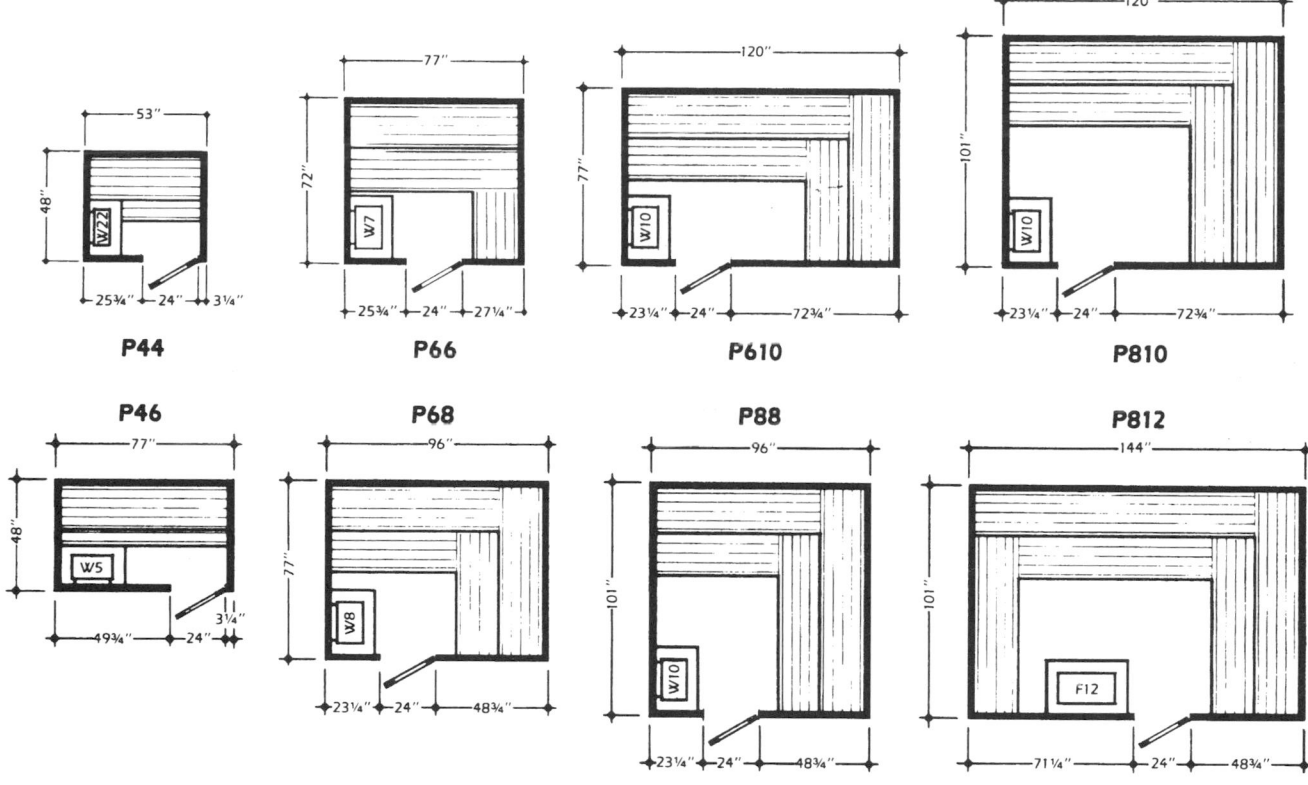

P44

P66

P610

P810

P46

P68

P88

P812

Fig. 1

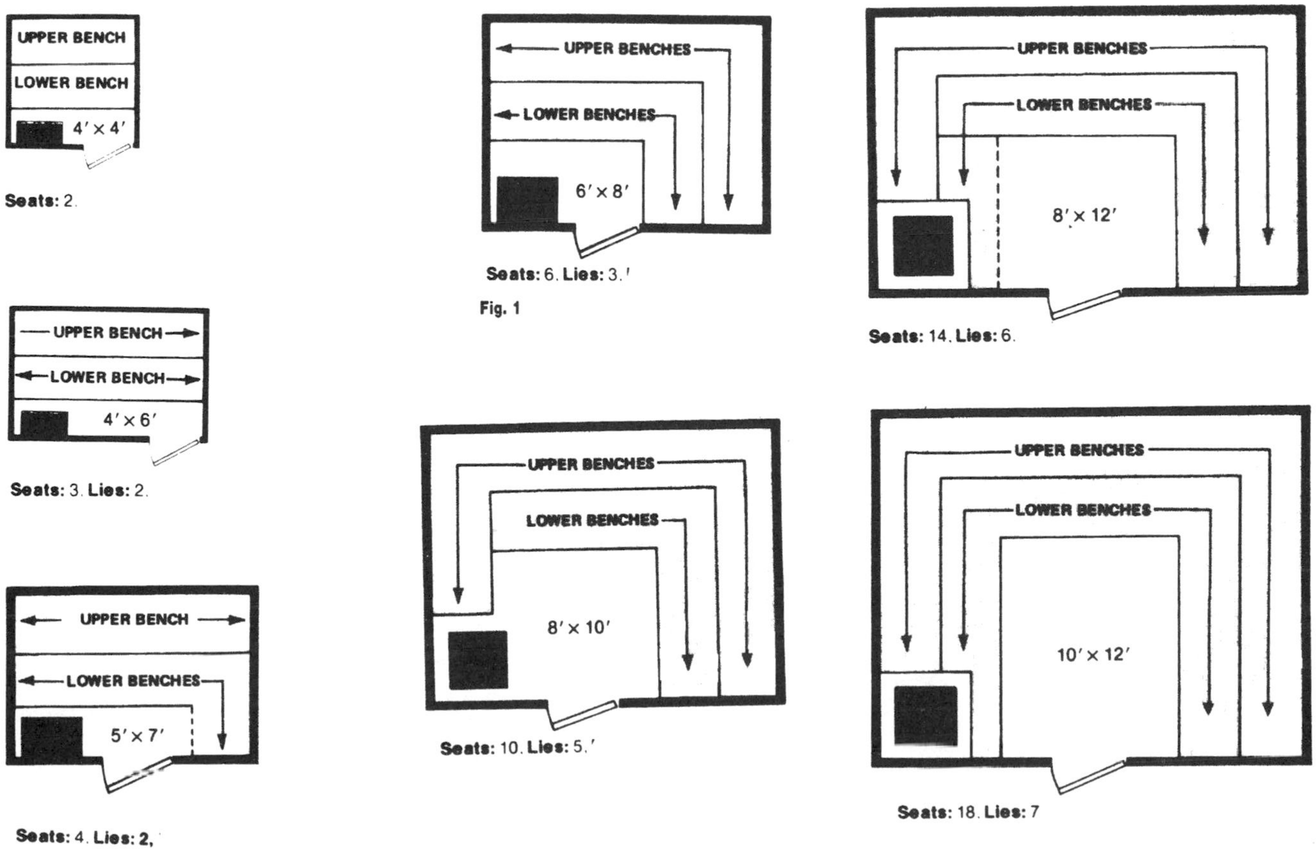

UPPER BENCH

LOWER BENCH

4′ × 4′

Seats: 2.

UPPER BENCH

LOWER BENCH

4′ × 6′

Seats: 3. **Lies:** 2.

UPPER BENCH

LOWER BENCHES

5′ × 7′

Seats: 4. **Lies:** 2,

Fig. 2

UPPER BENCHES

LOWER BENCHES

6′ × 8′

Seats: 6. **Lies:** 3.′

Fig. 1

UPPER BENCHES

LOWER BENCHES

8′ × 10′

Seats: 10. **Lies:** 5,′

UPPER BENCHES

LOWER BENCHES

8′ × 12′

Seats: 14. **Lies:** 6.

UPPER BENCHES

LOWER BENCHES

10′ × 12′

Seats: 18. **Lies:** 7

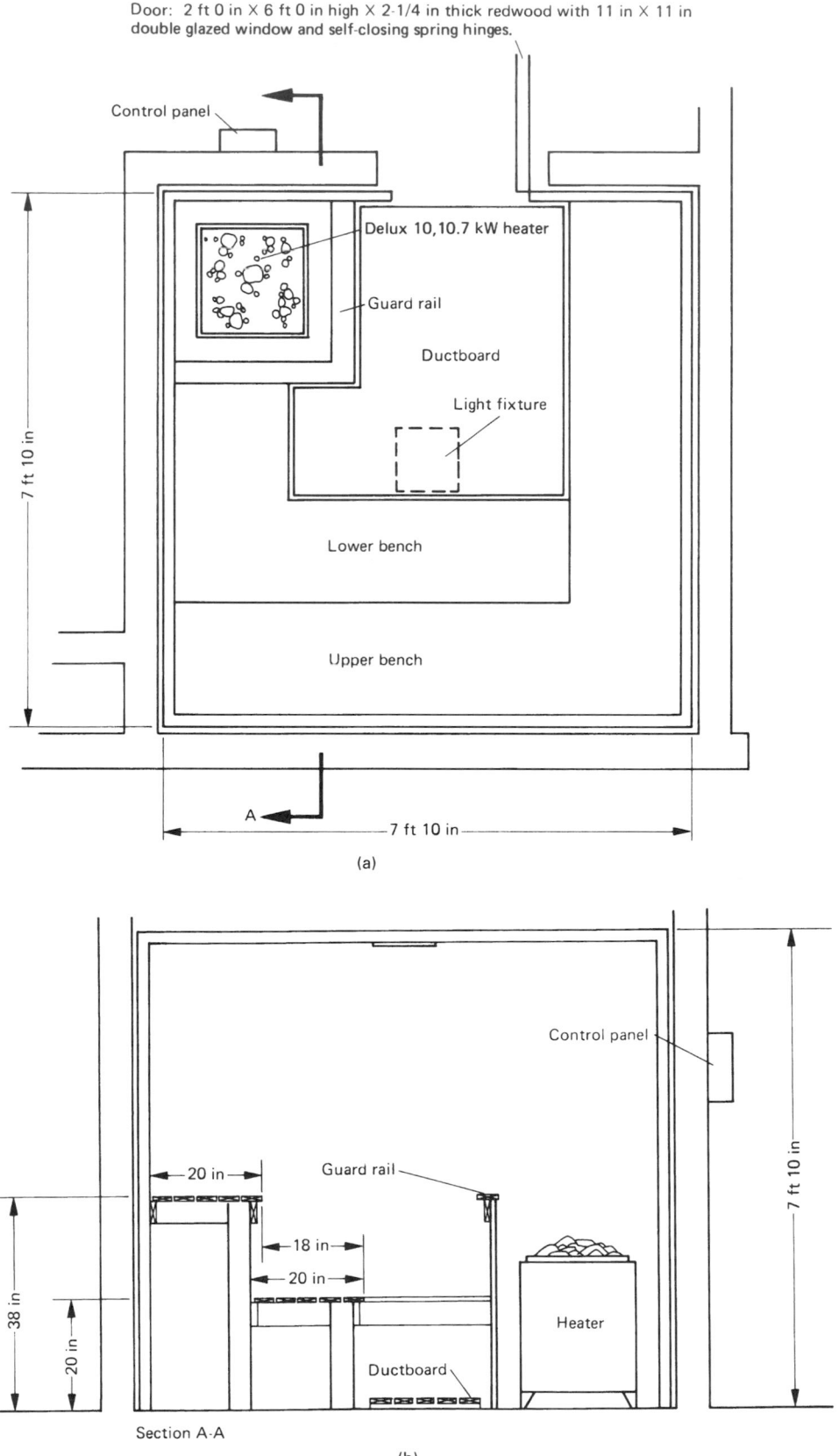

Door: 2 ft 0 in × 6 ft 0 in high × 2-1/4 in thick redwood with 11 in × 11 in double glazed window and self-closing spring hinges.

Control panel

Delux 10,10.7 kW heater

Guard rail

Ductboard

Light fixture

7 ft 10 in

Lower bench

Upper bench

A

7 ft 10 in

(a)

Control panel

20 in

Guard rail

18 in

20 in

7 ft 10 in

38 in

20 in

Heater

Ductboard

Section A-A

(b)

Fig. 3

BUILDING ENTRANCE/EXIT

It should be remembered that, because first impressions tend to be lasting, it is very important that the experience related to entering or leaving a residential building be pleasant for residents and visitors. The psychological response to building entry/exit points will either enhance or destroy the sense of residential appropriateness. All buildings shall have at least one main entry/exit point as well as additional entry/exit points as dictated by the nature of the circulation system and/or the relative location of interior and exterior facilities. In these guidelines, the main building entry/exit function shall consist of the following activities and/or spaces:

1. *Front porch:* outdoor sitting, viewing, and waiting area.

2. *Foyer:* to serve as an air lock against the weather and the point where visitors are identified (in low-density developments this may be only a sheltered outdoor area)

3. *Lobby:* primary circulation node to provide access to vertical circulation and public functions.

4. *Lounge:* flexible space to sustain observation, waiting, visiting, chance meeting, and communal activities such as bake sales, activity sign-up, etc. (this may not be required in low-rise, low-density developments)

Secondary building entry/exit points shall have at least a foyer as a weather protection.

Accessibility The main entry/exit point of each building in a development should occur at the focal point of both the internal and external pedestrian circulation systems, and should function as the transition zone between them. This point should be directly accessible to a vehicular drop-off/pick-up point. It should have continuous overhead weather protection from building to driveway, and should conform to applicable codes concerning the physically handicapped.

The normal arrival or departure sequence of events will best be accommodated by providing direct physical and visual accessibility (no intervening space or activity) between functional subcomponents in the following order:

1. Front porch
2. Foyer
3. Lobby
4. Horizontal and vertical circulation elements

The lobby is the focal point of this sequence and should provide direct physical and visual accessibility to the following other activities:

1. Management office
2. Mail and package rooms
3. Lounge

The physical and visual relationship between the lobby and common facilities may be indirect to the extent that access is through an intervening circulation element. This access should, however, be as simple as possible and should not include passage through other activity areas. Visual and physical access between entry/exit spaces and residential areas should

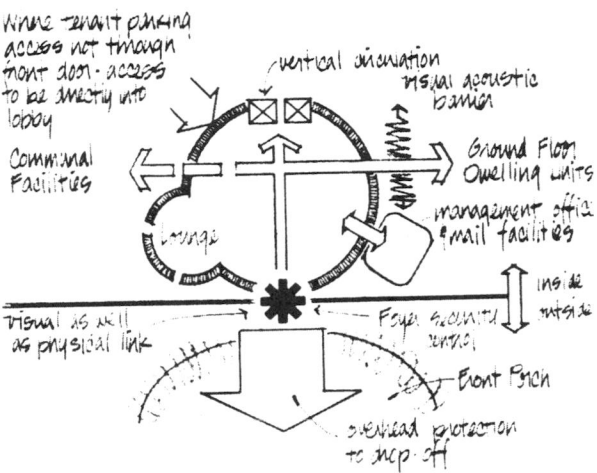

Fig. 1 Building entry/exit.

BUILDING ENTRANCE/EXIT AND LOBBY

always be through intervening circulation elements. Where residential areas are located on the same floor as the entry/exit the circulation elements should provide positive visual and audio screening to ensure privacy.

Where resident parking is not directly accessible from the main building entry/exit, a secondary entry/exit shall directly connect the lobby and the parking area.

The visual and physical access between the lobby and the lounge must be such that access between the entry/exit and other activities does not require circulation through the lounge. At the same time it must be possible to observe the traffic in and out of the main entry/exit from the lounge. The main entry/exit point should be prominently visible from building approaches.

Security Main and secondary entry doors shall be locked. The door shall be controlled and opened by a buzzer/call system located in each dwelling unit and in the management office.

All entry/exit points shall provide a full view of the outside from behind the locked door in lobbies. The view shall cover a sufficient field of vision to assure the resident or visitor that there is no danger before leaving the security of the building. Such a view shall extend unobstructed all the way to potential destinations such as the parking lot, the public street in urban situations, and the vehicular pick-up point.

Orientation The entry/exit spaces (especially the lounge) should be oriented to receive some sunlight each day. Northern orientations should be avoided where possible. The lounge and its windows should be organized to capture views of surrounding street life.

Furnishability Lounges should be provided with comfortable seating for at least ten people in developments of 100 units or more; in very large developments several lounge areas may be a desirable solution. The lobby should provide adequate wall and floor area for displays, bulletin boards, information signs, bake sales, and the like without interference with circulation. The space should be such that alternative furniture arrangements and furniture arrangements which provide both private conversation areas and less private viewing areas are possible. The use of the space for a sale or sign-up activity should not totally negate its other purposes.

Spatial characteristics The foyer and the lobby should be bright, welcoming, and easy to maintain. They should project an image of spaciousness and residential activity and engender pride in residents. The spatial volume of the lobby, principally its height, should be greater than the normal residential ceiling height. It should have sufficient floor area to accommodate both circulation and people standing still in chance meeting or decision making.

The lounge should be a more intimate space having a more nearly residential ceiling height. It should not be a room with doors, but rather should be a contained space with a free opening to the lobby and surrounding circulation elements. Individual or small groups of people should feel comfortable in the lounge. Its spatial composition and furniture arrangement should provide both very private areas and less private transition areas to the very public lobby areas.

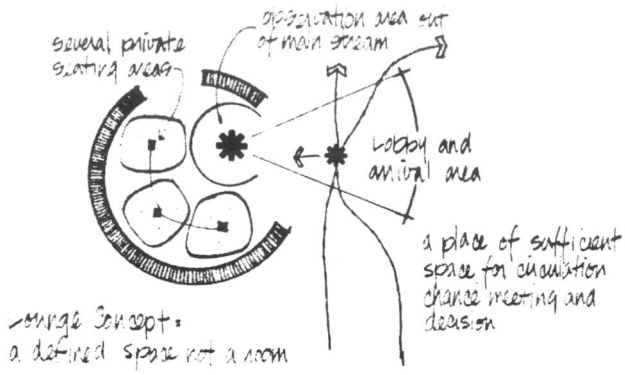

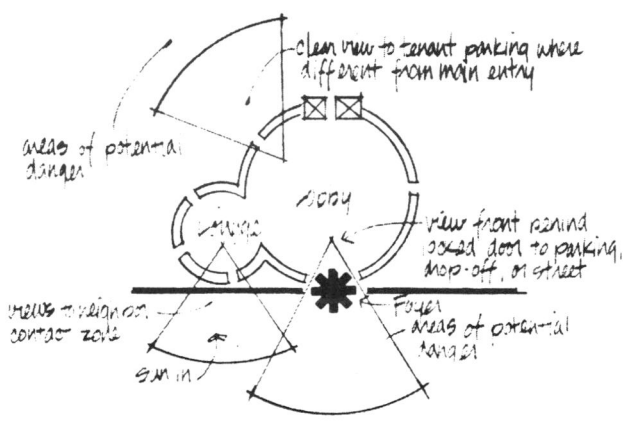

Fig. 2

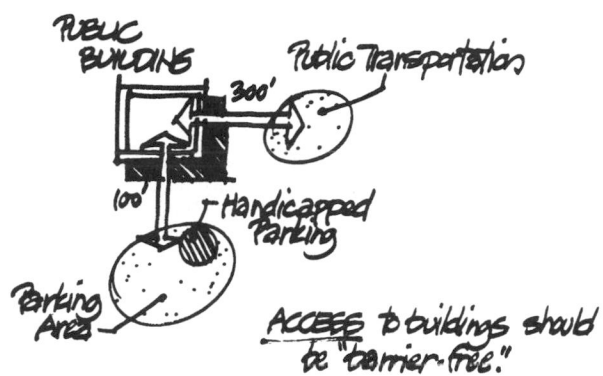

Fig. 3

PUBLIC LOBBIES SHOULD CONTAIN THE FOLLOWING ITEMS IN AREAS ACCESSIBLE TO THE HANDICAPPED AS CLOSE TO DOORWAYS AS POSSIBLE *:

- PUBLIC TELEPHONES
- REST ROOM FACILITIES
- DRINKING FOUNTAINS
- WAITING AREA WITH APPROPRIATE SEATING
- INFORMATION AND DIRECTIONAL SIGNAGE
- ELEVATORS, ESCALATORS, ETC.

*NOTE: ALL FACILITIES SHOULD BE FUNCTIONALLY USEABLE BY HANDICAPPED INDIVIDUALS.

- ENTRANCE PLAZAS REQUIRE 10'-0" LENGTH FROM DOORWAY TO CHANGE IN GRADE (STAIRS)
- PROVIDE 5'-0" MIN. CLEAR SURFACE AT ALL LANDINGS FOR BOTH STAIRS AND RAMPS.
- RAMPS: MAX. % SLOPE = 8.33%
  MAX. LENGTH/RAMP = 30'-0"
  MIN. ONEWAY WIDTH = 3'-0"
  MIN. TWO-WAY WIDTH = 6'-0"
- PROVIDE ADEQUATE RAILINGS, HANDLES, CURBS, AT ALL STAIR AND RAMP LOCATIONS. SEE "STAIRS, RAMPS, AND HANDRAILS."
- PROVIDE SIGNAGE SHOWING POINTS OF ACCESS FOR HANDICAPPED PEOPLE.

Doorways at Entrances

- MIN. 32" CLEAR OPENING.
- NO GRADE CHANGE AT THRESHOLD.
- HORIZONTAL THROW-BARS ARE RECOMMENDED OVER KNOBS, LATCHES, VERTICAL HANDLES, ETC.
- RECOMMENDED FORCE REQUIRED TO OPEN IS 5 lbs. TO 8 lbs.
- PROVIDE 18" SET BACK FROM NEAREST OBSTACLE (WALL, EDGE OF PAVEMENT, ETC.) SEE "GATES & DOORWAYS."
- PROVIDE AUTOMATIC DOOR AT HEAVILY USED LOCATIONS.

AUTOMATIC DOOR TRUNDLE

RAMP

- PROVIDE 5 FOOTCANDLES LIGHTING AT ALL ENTRANCES.
- OVERHEAD CANOPY PROTECTS PEDESTRIANS DURING INCLEMENT WEATHER.
- MAX. GRADE CHANGE BETWEEN STAIRWAY LANDINGS = 6'-0"
- SEE "RAMPS, STAIRS, AND HANDRAILS," FOR DETAILS OF TREADS AND RISERS.
- PROVIDE MIN. 5 FOOTCANDLES OF ALL WALKWAYS, RAMPS, STAIRWAYS, SEE "LIGHTING CONSIDERATIONS."

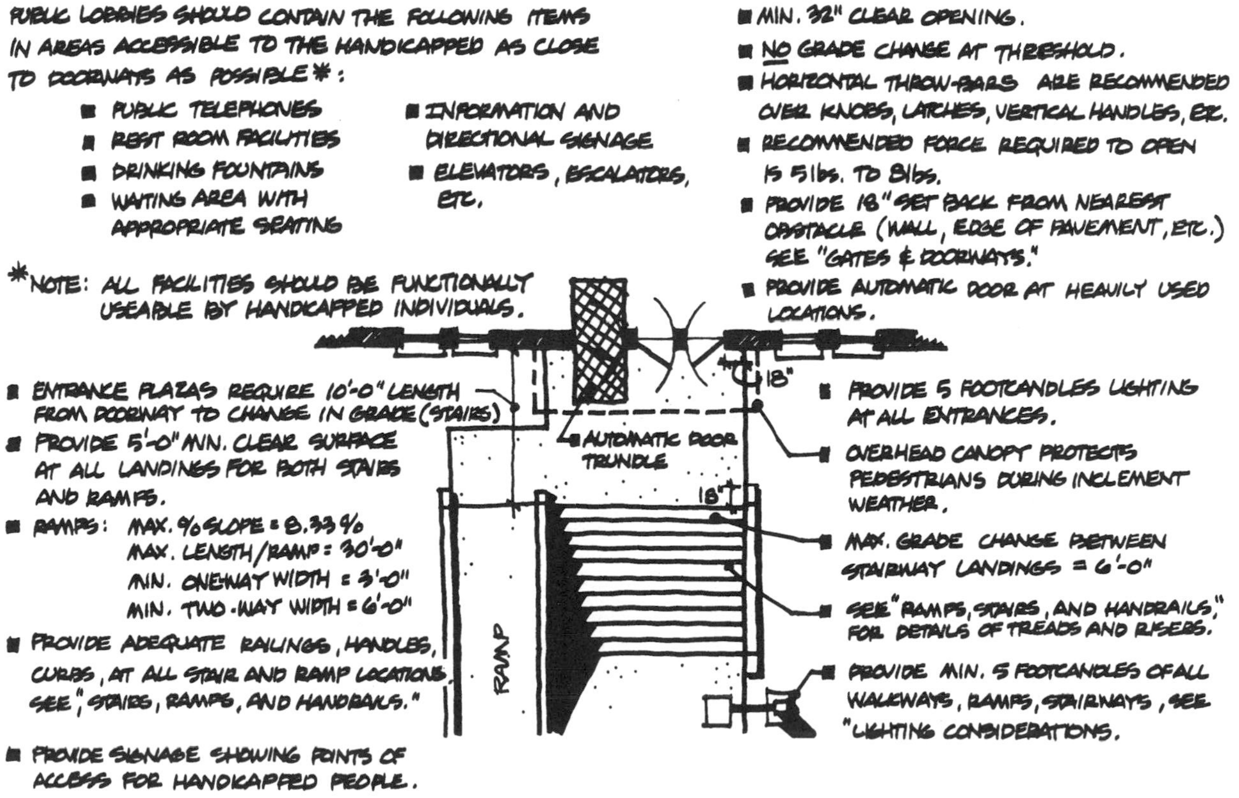

Fig. 4

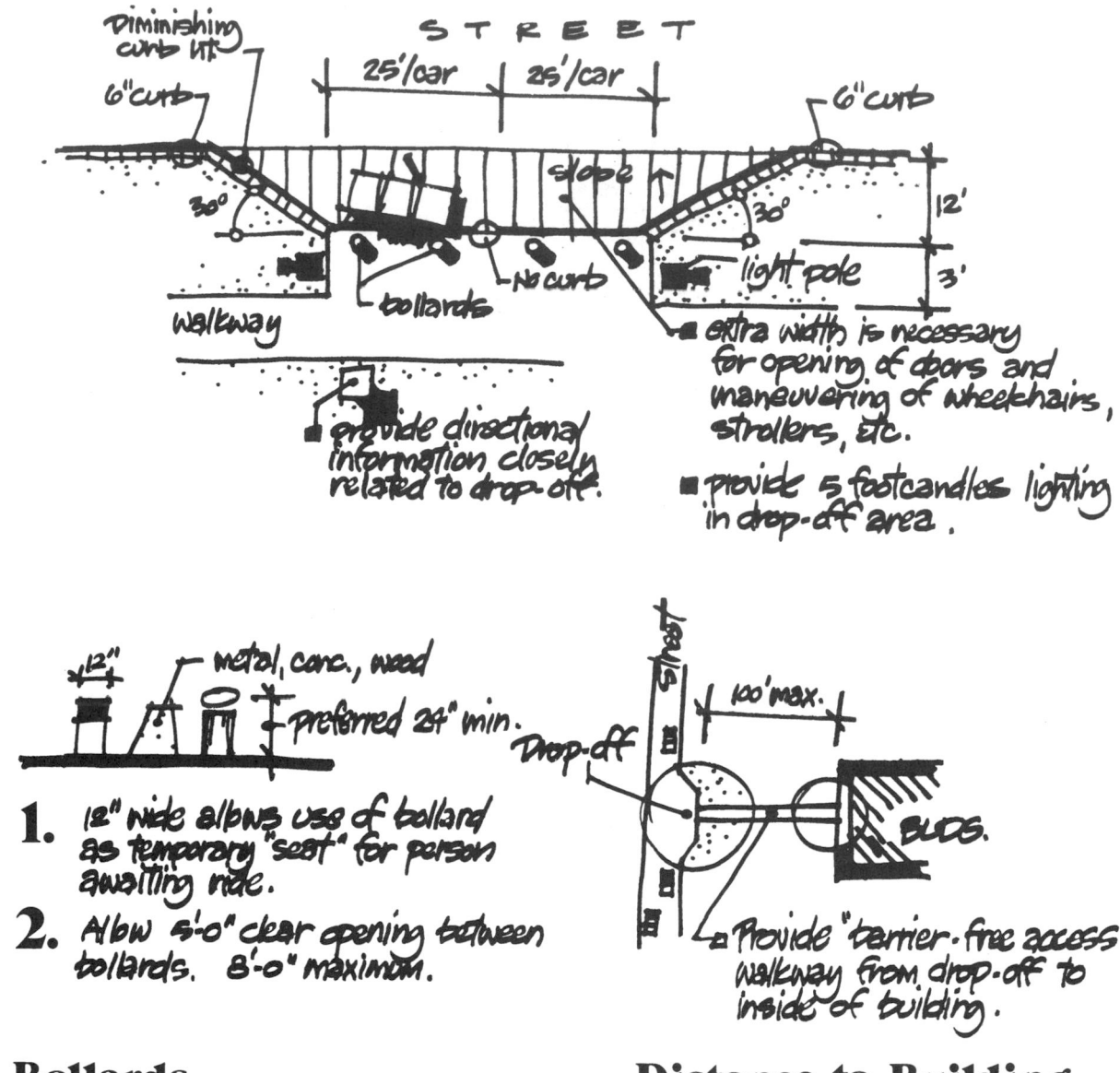

Bollards

Distance to Building

Fig. 5

Drop-Off Zones

Drop-off zones are beneficial for letting off and picking up people who are laden with packages, have children in strollers, or are physically restricted in some way.

The designer should consider the following items:

1. The width of the drop-off zone should be a minimum of 12 ft wide to allow the car doors to be fully opened for ease of access.

2. The length of the zone should accommodate at least two cars, allowing 25 ft for each, and should have gradual access to the main road.

3. Where the zone is at the same grade as the adjacent walk, bollards or some other suitable device should be used to separate the two functions. Where a curb exists and cannot be removed, one small 1:6 ramp per car should be provided to make the grade change.

4. Signage should be provided to identify the drop-off zone and limit its defined use to a "pick-up–drop-off" function.

Bollards

1. Bollards are useful as traffic control devices as they allow for pedestrian access while halting vehicular access. They should be spaced a minimum of 3 ft apart to allow a wheelchair to pass.

2. Bollards can be useful for seats if they are at least 12 in wide, and between 18 in to 24 in high.

3. Bollards should be painted in a contrasting color to the paving they rest on and should be well illuminated at night to minimize the risk of a person inadvertently walking into them.

Bus Stop Shelters

a.) allow views of oncoming busses
b.) bus route information
c.) shelter from elements
d.) 5 footcandle lighting
e.) transparent sides for visability & safety.
f.) provide space for wheel-chairs.

■ allow space for strollers, canes, wheelchairs, etc.

■ space large enough for people in groups.

■ shaded sitting provides greater comfort for extended wait.

3'-6" minimum

BUS

Fig. 6

Considerations for Waiting Areas

Waiting areas for mass transit are perhaps the most common of all exterior waiting areas. Due to the large amount of time spent waiting for buses and trains, it is important that these areas be physically accommodating for all people.

When designing exterior waiting areas, the following items should be considered:

1. The waiting area should be large enough to accommodate the average number of people normally using it comfortably.

2. Seating should be provided for the average number of daily users, with space also allotted to park wheelchairs, strollers, and other wheeled vehicles.

3. Where possible, an overhead shelter or canopy should be used to minimize the effects of the weather. Care should be taken to locate vertical support posts out of the paths of pedestrians either using or passing near the shelter. If the shelter is enclosed, adequate space must be allotted for easy movement into and through it.

4. Make sure that waiting area designs allow passengers to see approaching vehicles before they arrive at the stop. This courtesy allows all passengers time to adequately prepare themselves for boarding and as a result, shorten loading times for vehicles and reduce embarrassing situations for handicapped individuals.

5. Loading areas should be designed so that circulation from the waiting area is uncomplicated and over paved surfaces. The loading area itself should not have a curb that must be climbed. If a curb cannot be avoided, a 1:6 ramp will be necessary.

MANAGEMENT OFFICE

MANAGEMENT OFFICE

There shall be a space or spaces provided in each development for visitor reception, rental activities, and development administration.

Accessibility The management office should be located with direct physical, visual, and auditory accessibility to the main building entrance and lobby, to common facilities when they are on the main floor, and to the mail/package room. Accessibility to residential areas should be through an intervening circulation element. Direct visual access is here taken to mean visual control.

Security The management office should control the building entrance and lobby to discourage unwanted guests and add to the sense of security of the residents. This should be accomplished in an unobtrusive manner so as to avoid the feeling that all those who pass by are under surveillance.

Orientation The management office should have natural light and also overlook critical areas of the site whenever possible such as parking lots, entrance drives, etc.

Furnishability This shall be determined for each development, but it should at least include facilities for a manager and secretary, that is, two desks with chairs, filing cabinets, and several chairs for visitors.

Spatial characteristics The management office should accommodate its intended functions without becoming officious or overbearing in character. It should probably have normal residential ceiling height; and it should be enclosed with glass as much as possible, both to enhance surveillance and to achieve an open, welcoming quality.

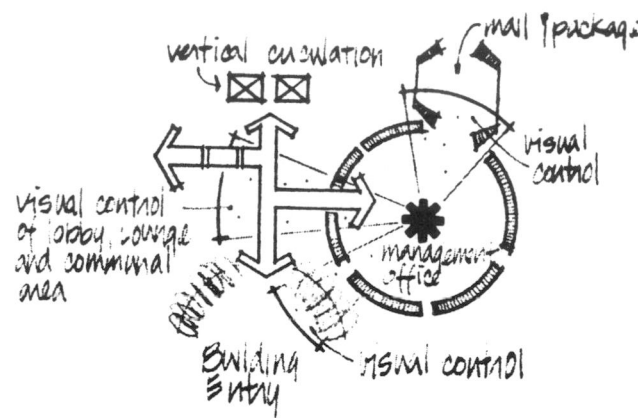

Fig. 1 Management office.

MAIL AND PACKAGE DELIVERY

A facility for the delivery of mail and packages to tenants must be provided in all developments. In developments consisting of a large single building, a separate mail room should be provided.

Accessibility These facilities should have direct physical accessibility from the entrance lobby and from horizontal and vertical circulation elements. They should also be directly accessible to the service entrance of the building for convenience of delivery. Visual and audio accessibility to the lounge should be indirect (screened). Visual and audio access to other areas will be determined by other considerations.

Spatial Characteristics In large developments, care should be taken to avoid the institutional appearance which results from a single large bank of mailboxes. Perhaps an island with mailboxes on several sides or a U-shaped configuration offers the opportunity to present small groups of mailboxes to which an individual resident can relate. It should be possible to reach this room without being forced into extensive public contact or unwanted interpersonal confrontations.

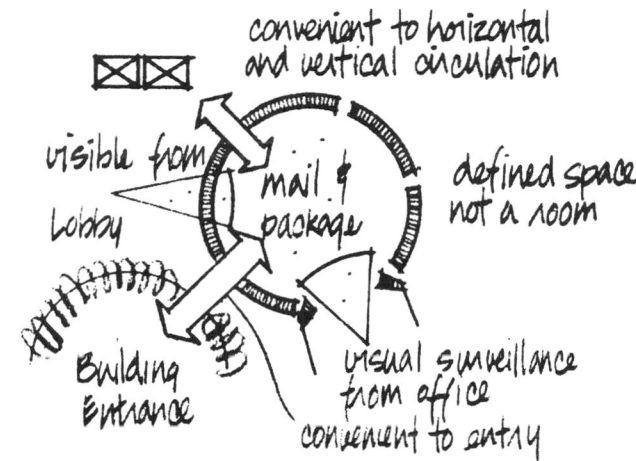

Fig. 1   Mail and package room.

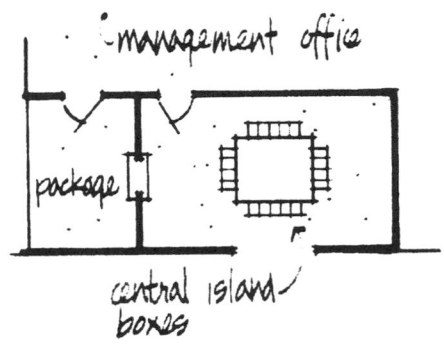

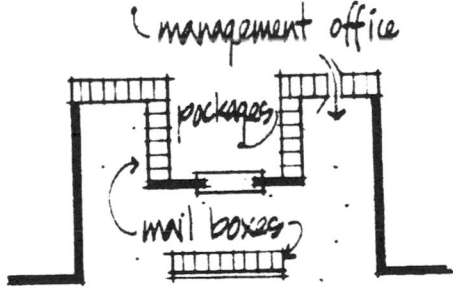

Fig. 2   Examples of mailbox organization.

VEHICULAR STORAGE

In terms of space demands and appearance, parking lots present one of the most difficult problems for the site planner. Problems increase proportionally as the density of housing increases. Streets around housing sites cannot be expected to handle overflow parking as they once did; adequate on-site parking facilities have become a necessity.

The amount of space required for parking is determined by many factors. The most significant ones which influence the ratio of space per dwelling unit are occupant characteristics (total population, age, family status) and location of the site with respect to public transportation, shopping and recreational facilities, and employment centers. Housing for the elderly generally requires less parking space for each dwelling unit than housing occupied by younger families. More than a one-to-one ratio is likely to be needed at sites where there are families with many teen-age children or where apartments are shared by bachelors. This is especially true of multifamily sites in outlying areas. Here isolation from community facilities, particularly schools and shopping, and poor public transportation service result in greater and greater demands for parking space. Regulations applying to on-site parking show a lower ratio for central city locations than they do for outlying sites. This reduced ratio reflects the fact that people living at high-density urban sites are often less dependent upon the automobile. In fact, some families deliberately choose central locations just to avoid using a car for everyday activities. Presumably they have easy access to many community facilities and to some form of public transportation. The higher cost of land in town also acts to hold down the parking per dwelling unit ratio.

In addition to resident parking, many multifamily housing projects allot on-site spaces for guest parking. The optimum number of such spaces is obviously not predictable for all sites, and it, too, is influenced by site location, occupant income, and family characteristics. Parking spaces for service vehicles are also sometimes provided at multifamily housing sites (usually at rear entrances of the buildings).

There are no rigid rules to be followed in the location of parking facilities vis-à-vis dwelling units. The quality of the pedestrian route from the parked car to the door of the dwelling is often more significant than proximity.

A route that offers no overhead protection, is indirect, or is inconvenient is not very satisfactory. Parking areas directly in front of a dwelling may also be unsatisfactory because of exhaust fumes, bright headlights, and engine noise, or because cars may have to back across major pedestrian routes to leave the site.

There is a correlation between density and the ease with which vehicular storage is accomplished. At housing sites with either high or low densities, providing for vehicular storage seems to cause fewer problems, at least visually, than it does at sites with a middle-density range. At both high and low densities, it is often economically feasible to build garages, with the result that the car is completely eliminated from sight. At low densities these garages are often attached to individual dwelling units; at high densities there may be underground garages with direct access into the dwelling structures or even separate parking buildings. Between these two extremes of density, problems arise in trying to achieve a balance between convenience and nuisance. Middle-density housing sites tend to appear completely dominated by the automobile. Buildings are surrounded by seas of paved parking areas.

There are at least two remedies, both of which have drawbacks. First, large, unpleasant-looking parking lots could be replaced by a number of smaller lots scattered throughout a site. However, these might be inconvenient if located too far from the buildings they serve. Second, parking facilities could be built either above or below grade, but these would probably be quite costly for the developers. Screening is an inexpensive remedy which has been used effectively at a range of sites. This technique has a twofold advantage—it makes possible a separation between pedestrian and vehicular traffic and it hides cars from view. It improves both the appearance and the usability of a site. Aesthetic and safety features are both important in the planning of parking facilities,

regardless of site density. Designs which create blind corners at the edge of a parking lot or along a roadway and which do not clearly divide pedestrian routes from parking areas endanger drivers and pedestrians alike.

For several reasons, small parking lots are preferable to large ones: they are easier to screen, safer, and depending upon their location, more convenient. Large lots are more dangerous because their long straight lanes tend to produce high speeds.

Other problems associated with large lots are as follows: they require sidewalks and dividing strips to prevent uncontrolled pedestrian cross circulation; for aesthetic reasons, they require careful landscaping to minimize their size; and they encourage unwanted uses. It is possible, for example, that children might use them as playgrounds if on-site playgrounds were not available. Allocating separate areas for auto maintenance and repair might also discourage the use of parking lots for other than parking purposes.

The open parking lot, though the most common, is not necessarily the best type of storage facility for a particular site. There are additional choices—carports, garages either on grade, underground, or in multistory structures—which are dependent upon the size, location, and type of housing project, the number and density of the units, and topographic and subsoil conditions.

Protected car-storage areas, attached to the dwelling unit if possible, are preferred by the majority of automobile owners, particularly in northern climates. At central city sites the solution is frequently a base of parking, above or below ground, upon which the building sits. At outlying sites the car is often parked behind and/or under a row of housing units. Hilly sites permit terracing or parking lots, partly below and partly above ground. Cluster developments facilitate a grouping of garages or carports conveniently near entrances to a number of single units. Often the most outstanding aspect of many plans is the ingenuity with which site planners have accommodated huge numbers of cars without destroying the usability of open space.

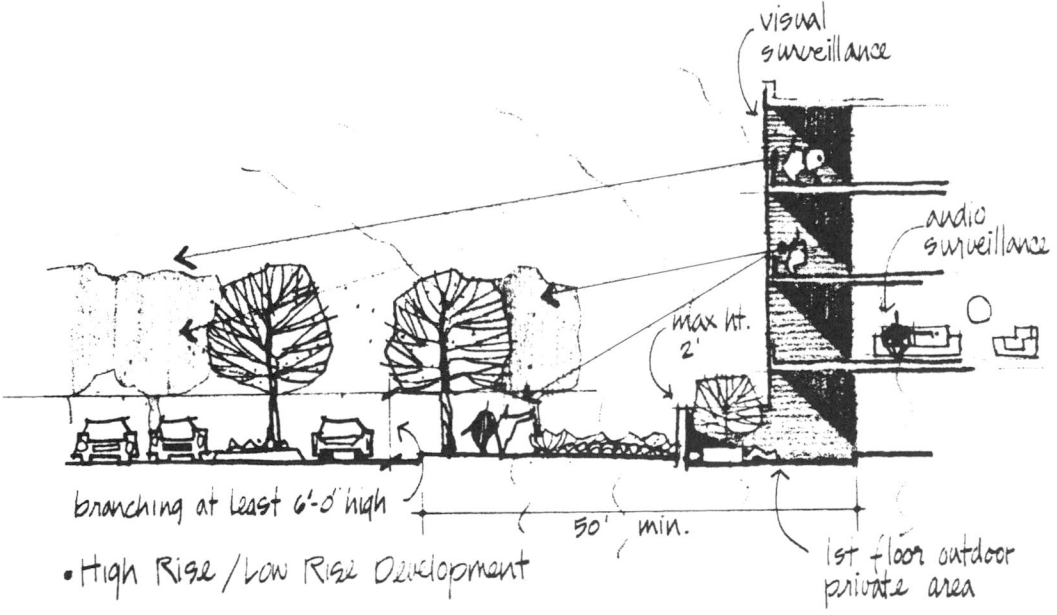

Fig. 1   Typical parking area/living-area relationship.

VEHICLE DIMENSIONS

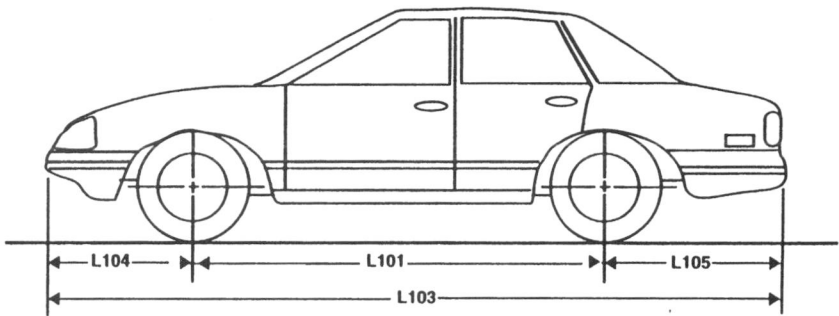

L101 – Wheelbase (WB). The dimension measured longitudinally between front and rear wheel centerlines. In case of dual rear axles, the dimensions shall be to the midpoint of the centerlines of the rear wheels.

L103 – Vehicle length. The maximum dimension measured longitudinally between the foremost point and the rearmost point on the vehicle, including bumper, bumper guards, tow hooks and/or rub strips, if standard equipment.

L104 – Overhang – front. The dimension measured longitudinally from the centerline of the front wheels to the foremost point on the vehicle, including bumper, bumper guards, tow hooks and/or rub strips, if standard equipment.

L105 – Overhang – rear. The dimension measured longitudinally from the centerline of the rear wheels; or in the case of dual axles, the dimension shall be midpoint of the centerlines of the rear wheels, to the rearmost point on the vehicle, including rear bumpers, bumper guards, tow hooks and rub strips, if standard equipment.

Fig. 2   Exterior length dimensions.

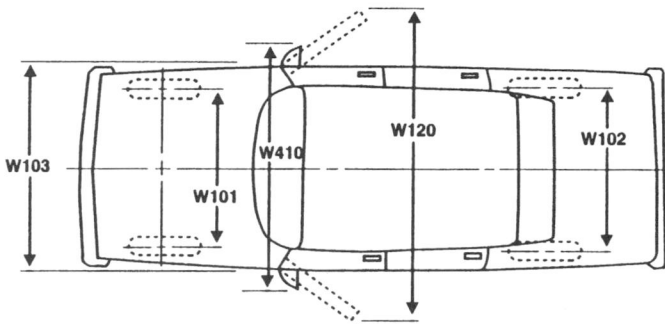

W101 – Tread – front. The dimension measured between the tire centerlines at the ground.

W102 – Tread – rear. The dimension measured between the tire centerlines at the ground. In case of dual wheels, the dimension will be measured to the centerline of tire and wheel assemblies.

W103 – Vehicle width. The maximum dimension measured between the widest point on the vehicle, excluding exterior mirrors, flexible mud flaps, marker lamps, but including bumpers, moldings, sheet metal protrusions or dual wheels, if standard equipment.

W120 – Vehicle width – front doors open. The dimension measured between the widest point on the front doors in maximum hold-open position.

W410 – Outside mirror width. The dimension between the widest point on the outside mirrors. The standard right and left mirror adjusted for normal driving will be shown unless otherwise noted. When only one outside mirror is standard, the dimension will be to the zero "Y" plane.

Fig. 3   Exterior width dimensions.

Wall to Wall Turning Diameter is the diameter of the smallest circle which will enclose the outermost points of projection of the vehicle while executing its sharpest practicable turn. This is equal to the minimum turning circle plus twice the radial overhang beyond the turning radius.

Curb to Curb Turning Diameter is the diameter of the smallest circle within which the vehicle will clear a curb 6 in. high, while the vehicle is executing its sharpest practicable turn. This is equal to the turning circle plus twice the horizontal distance from the center of tire contact with the road to the arc subtended by a chord drawn between the points of intersection of the outermost projection of the tire shoulder on a horizontal plane 6 in. above the surface on which the tire rests.

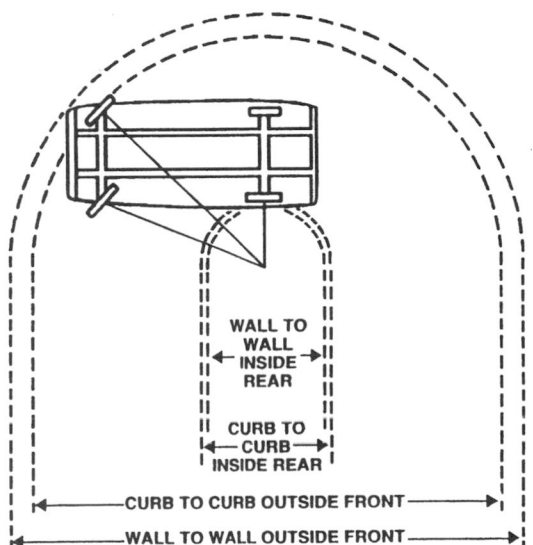

Fig. 4   Turning diameter dimension.

PARKING, GENERAL

TABLE 1   Overall Dimensions for 1993 Vehicles (Figures in Inches)

Models	L101 Wheelbase	L103 Overall Length	W103 Overall Width	W120 Overall Width Doors Open	H101 Overall Height	H156 Minimum Running Ground Clearance
Chrysler Motors						
Chrysler New Yorker 5th Avenue, Chrysler Imperial	109.6	198.6–203.0	68.9	137.0	55.1–55.3	5.0
Chrysler New Yorker Salon, Dodge Dynasty	104.5	192.0–193.6	68.9	137.9	53.6	5.0
Dodge Spirit, Plymouth Acclaim, Chrysler LeBaron Sedan	103.5	181.2–182.7	68.1	135.1	53.5–53.7	4.6
Dodge Stealth	97.2	178.9–180.3	72.4	157.5	49.1–49.3	4.1–4.4
Dodge Daytona, Daytona ES, IROC, IROC R/T	97.2	179.0–179.8	69.3	154.9	50.3–50.6	4.7
Plymouth Sundance, Dodge Shadow	97.2	171.9	67.3	135.1–160.4	53.0–54.3	4.6
Chrysler LeBaron Coupe/Convertible	100.5	184.8	69.2	156.1	51.2–52.4	4.6 4.7
Chrysler Concorde, Dodge Intrepid, Eagle Vision	113.0	201.6–202.8	74.4	144.6	55.8–56.3	5.9
Plymouth Voyager, Dodge Caravan, Chrysler Town & Country	112.3–119.3	178.1–192.8	72.0	142.3	65.9–66.1	5.6–5.7
Plymouth/Dodge Colt, Eagle Summit, 2-door/4-door	96.1–98.4	171.1–174.0	66.1		51.4	5.9
Plymouth Colt Vista, Eagle Summit Wagon FWD/AWD	99.2	168.5	66.7	137.4	62.1–62.6	5.7
Eagle Talon, Plymouth Laser	97.2	172.4–172.8	66.7–66.9	150.4	51.4–52.0	6.2
Viper	96.2	175.1	75.7		44.0	5.0
Ford Motor Company						
Crown Victoria/Grand Marquis	114.4	212.4	77.8	142.9	56.8	6.1
Taurus/Sable	106.0	192.0–193.1	71.2	140.6	54.1–55.5	5.4–5.8
Thunderbird/Cougar	113.0	198.7–199.9	72.7	173.6	52.5–53.0	5.4–5.7
Tempo/Topaz	99.9	176.7–177	68.3	150.8	52.9	5.3
Probe	102.9	178.9	69.8	154.5	51.8	5.2
Mustang	100.5	179.6	68.3	153.5	52.1	4.5
Escort	98.4	170.0–171.3	66.7	154.3	52.5–53.6	5.1
Festiva	90.2	140.5	63.2	147.2	55.3	5.39
Lincoln Town Car	117.4	218.9	76.9	144.4	56.7	6.0
Lincoln Continental	109	205.1	72.7	142.5	55.5	5.9
Lincoln Mark VIII	113.0	206.9	74.6	165.0	53.6	5.7
Tracer	98.4	170.9–171.3	66.7	154.3	52.7–53.6	5.1
Capri	94.7	166.1	64.6	145.3	50.2	4.9
General Motors Corporation						
Oldsmobile Cutlass Ciera, Buick Century	104.9	194.4–189.1	69.5–69.4	149.3–130.6	54.5–53.7	6.3–5.7
Chevrolet Caprice Classic, Buick Roadmaster/Estate	115.9	217.6–214.1	79.8–77.5	141.3	60.9–55.7	6.2–4.8
Oldsmobile '98 Regency/Touring Sedan, Buick Electra Park Avenue, Cadillac De Ville	113.7–110.8	206.3–203.3	74.9–73.3	164.6–136.4	55.1–54.8	5.9–5.1
Cadillac Fleetwood	121.5	225.1	78.0	141.3	57.1	5.5
Buick Riviera, Cadillac Eldorado	108.0–86.6	202.2–142.5	75.5–64.2	158.6–135.8	65.6–52.9	7.9–5.4
Chevrolet Camaro, Pontiac Firebird	101.0	195.6–193.2	74.5–74.1	165.3–165.2	52.0–51.3	4.4
Pontiac Bonneville, Oldsmobile '88 Royale, Buick LeSabre	110.8	201.1–199.5	74.9–74.1	140.4–140.1	55.7	5.4–5.3
Chevrolet Cavalier, Pontiac Sunbird	101.3–86.6	182.3–142.5	66.3–64.2	147.2–127.8	65.6–51.9	7.9–5.5
Cadillac Seville	111.0	204.1	74.2	144.3	54.5	5.6
Chevrolet Corsica, Beretta	103.4	187.3–183.5	68.2–67.9	153.6–130.6	54.0–52.7	5.7–5.4
Chevrolet Geo Metro	93.1–89.2	151.4–147.4	62.7–62.0	141.3–128.0	53.5–52.0	6.1
Pontiac Grand AM, Oldsmobile Achieva, Buick Skylark	103.4	189.1–186.9	67.5	150.7–127.9	53.2–53.1	5.7
Chevrolet Geo Storm	96.5	164.0	66.7	152.9	51.1	5.2
Chevrolet Geo Prizm	97.0	173.0	66.3	132.4	53.3	4.7
Pontiac LeMans	99.2	177.0–167.9	65.7–65.5	146.4–129.9	53.7–53.5	5.3–5.2
Chevrolet Lumina APV, Pontiac Trans Sport, Oldsmobile Silhouette	109.8	194.5–194.2	74.6–73.9	132.9–132.4	65.7	7.4
Cadillac Allante	99.4	178.7	73.4	154.7	51.5	5.7
Chevrolet Lumina, Pontiac Grand Prix, Oldsmobile Cutlass Supreme, Buick Regal	107.5	198.3–193.6	72.5–71.0	162.8–.0	54.8–52.8	5.9
Chevrolet Corvette	96.2	178.5	70.7	145.9	47.3–46.3	4.2–3.6
Saturn	102.4–99.2	176.3–175.8	67.6–65.9	150.2–139.1	53.7–50.6	5.7–5.6

Parking bays can be both directly off the street and within property limits. Parking bays within the property lines offer a solution where parking is not permitted on a public way. Parking bays directly off the street, as illustrated in Fig. 1A, are not only more convenient for the tenants but are less expensive to construct and more economical to maintain. These should be used, however, only on minor streets. The illustrations are for streets with two-way traffic.

Minimum dimensions are for 90°, 60°, 45° parking in basements with and without columns. Where 60° and 45° parking is necessary, one-way traffic should be planned.

The diagrams and dimensions in Figs. 1 to 3 are for typical use where tenants park their own cars. Where cars are parked by a paid attendant in a large garage, 8-ft stalls and back-in parking for 60° and 90° may be used with dimensions as in Table 1.

TABLE 1

	45° Drive-in	60° Back-in	90° Back-in
Aisle width	12 ft 8 in	17 ft 4 in	22 ft 0 in
Stall depth perpendicular to aisle	17 ft 2 in	18 ft 10 in	18 ft 0 in
Unit parking depth	47 ft 0 in	55 ft 0 in	58 ft 0 in
Stall width parallel to aisle	11 ft 4 in	9 ft 3 in	8 ft 0 in

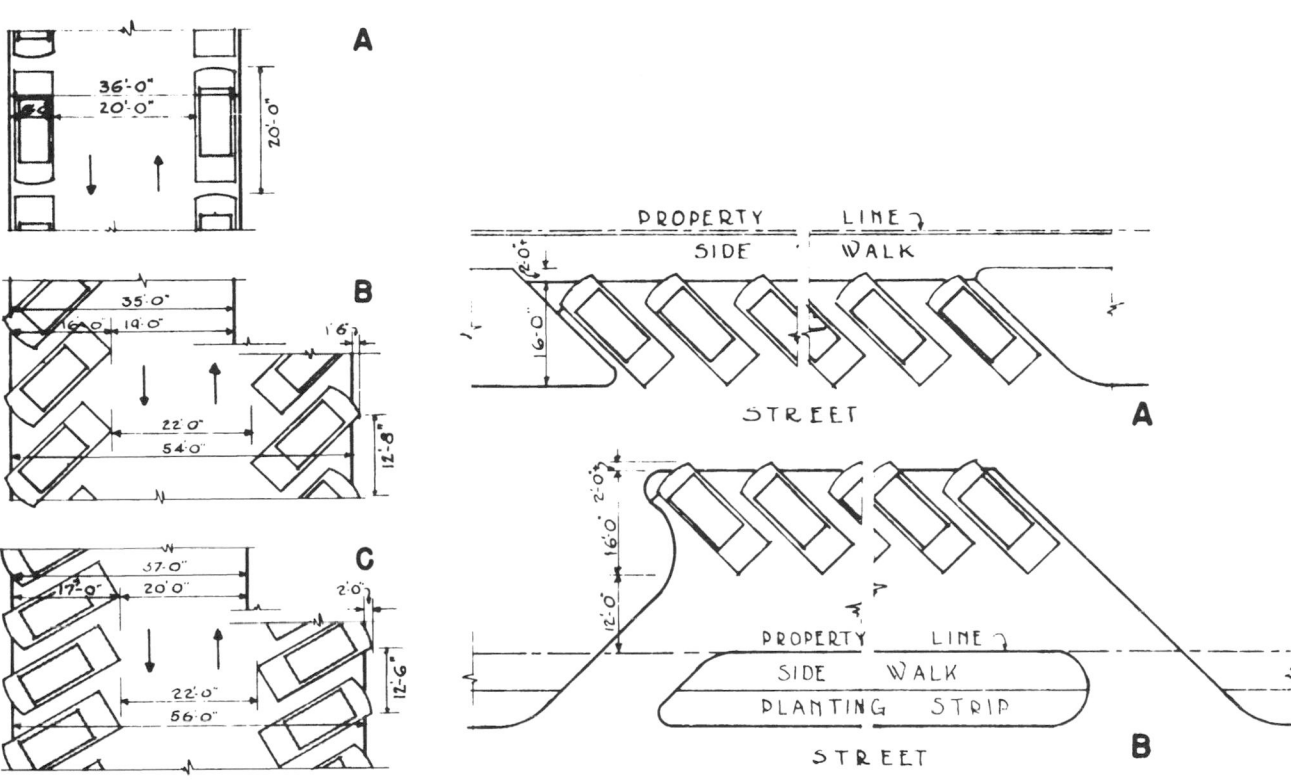

Fig. 1

Fig. 2

PARKING WITH COLUMNS

PARKING WITHOUT COLUMNS

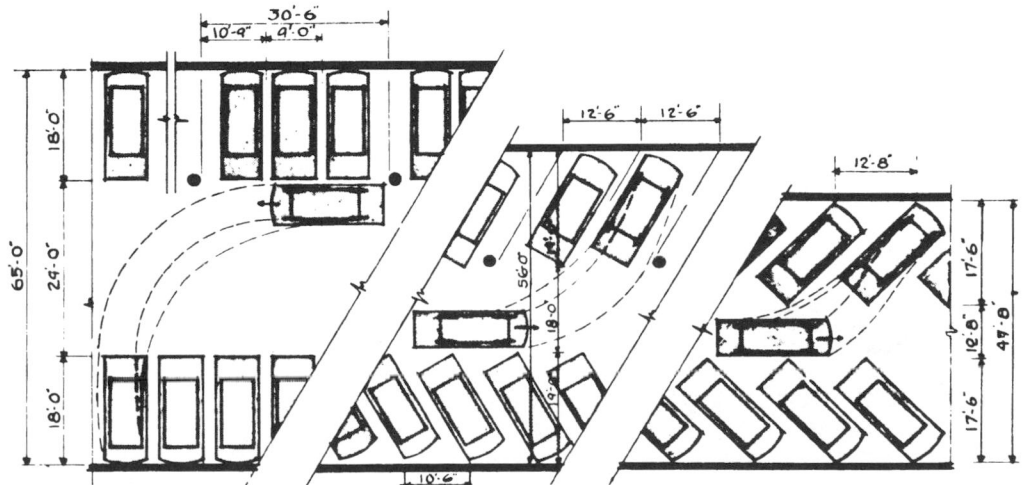

Fig. 3

PARKING AREAS AND GARAGES

Occupant car parking should always be located so that all homesites are conveniently served. The best accommodation is a garage or carport attached to the dwelling, or separated only by a court or patio. Uncovered parking spaces on the lot are also acceptable. Of course, the two-car family prefers having both cars on the home property.

Where the market does not require the parking space on the individual lot, parking for the home owners may be in nearby parking courts, carports, or uncovered car spaces. Each home owner should be assigned one or more specific spaces which are conveniently located, usually not more than 100 ft from the home. Parking courts should be closed at one end, or the driveway alignment sharply broken at the exit, to assure slow car speed.

Guest parking space on individual properties is desirable. Even with on-lot guest car parking, however, some off-lot guest parking is needed for occasional overflow crowds as well as for the use of service vehicles. Such parking should be located near all homesites to assure maximum use and convenience, for example, within 150 ft of any dwelling. It may be located in parking courts, along street curbs, or in bays along the streets. Often more car storage should be provided than indicated by the ratios in the FHA Land-use Intensity Standards, as the people being served may have more cars than the indicated minimum.

Skillful design of parking courts is very important as wide open stretches of pavement or massive parking structures reduce the visual appeal of a project. Parking court design should include areas for planting. The right shrubs are needed for screening, and suitable trees to shade the paved surface and to reduce glare. Good planting not only reduces the unsightly but adds visual appeal.

GARAGES

Ramp Breakover Angle

The ramp breakover angle is the measure of ability of the car to break over a steep ramp, either climbing or descending, without scraping (see Fig. 1). The Society of Automotive Engineers calls for a minimum of 10° as a design standard. A number of models have not met this standard in recent years. The average for all groups has remained relatively constant during the past four decades despite appreciable vehicle height reductions.

The ramp breakover angle influence can be altered through use of design techniques. Transitional blends at the top and bottom of ramps composed of two or more break points can multiply the ramp steepness, with workable break angles, beyond the normal capacities of car or driver. In existing structures these problems are overcome by building a pad of asphalt or concrete on each side of the break point. In this manner cars having a low breakover angle can negotiate potential critical points without scraping.

Long wheelbase cars combined with low center clearance are most susceptible to inadequate breakover angles.

Angle of Departure

A reasonable minimum value is necessary to reduce the incidence of tailpipe and rear-bumper dragging. The standard calls for a minimum of 10°, violated only in the 1957–1959 period. Most cars are substantially above 10°. The most critical condition is at driveways where the apron is steep, or a combination of excessive crown to gutter and apron slope.

Angle of Approach

The standard developed in 1960 by the Society of Automotive Engineers calls for a minimum value of 15°. (See Fig. 2.)

Ramp Slopes

The maximum ramp slope should be 20 percent. For slopes over 10 percent, a transition at least 8 ft long should be provided at each end of the ramp at one half the slope of the ramp itself. (See Fig. 3.)

Driveway Exits

A ramped driveway exit rising up to a public sidewalk must have a transition section that is almost level (maximum slope: 5 percent) before intersecting the sidewalk to prevent the hood of the car from obscuring the driver's view of pedestrians on the walk. This transition should be 16 ft long. (See Fig. 4a.)

Property line walls should also be regulated so as not to interfere with the driver's view of pedestrians on a public sidewalk. Wherever an exit driveway is parallel and adjacent to a property line wall which extends all the way to a sidewalk, the edge of the driveway should be physically established, by curb or railing, at least 6 ft from that wall. For each foot that the wall is held back from the sidewalk, the required distance between driveway and wall may be reduced by one foot. (See Fig. 4b.)

Fig. 1 Ramp breakover angle.

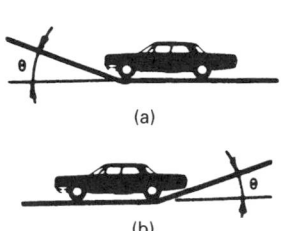

Fig. 2 (a) Angle of approach. (b) Angle of departure.

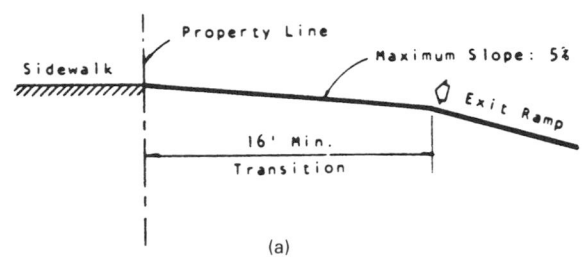

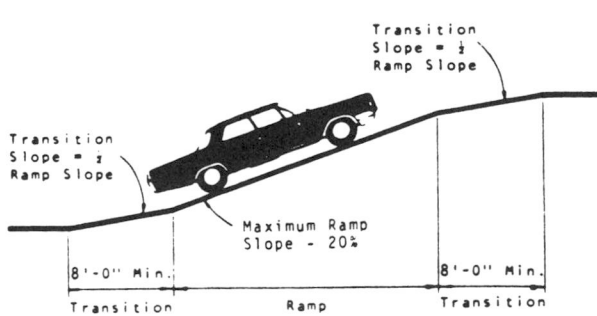

Fig. 3 Ramp slopes. (Transitions are required only if ramp slope exceeds 10 percent.)

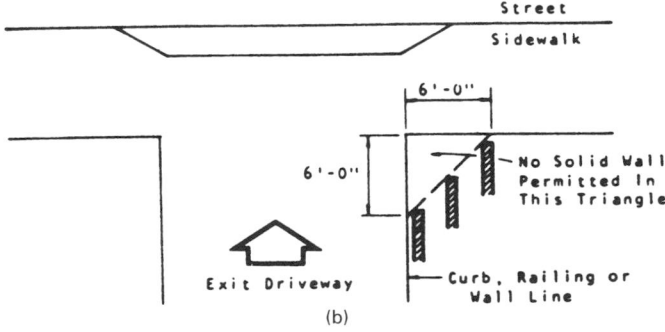

Fig. 4 Driveway exits.

PARKING AREAS

Figure 5 is an on-site parking area with single entrance and double 90° open parking. In order to enter the end spaces nearest the entrance in one motion, extra width must be maintained for a distance of 35 ft as indicated. A double compound for cars parked at 90° under a shed roof requires a total width of 65 ft. Posts supporting the roof should be spaced so as not to interfere with maneuvering (see Fig. 5). It must be assumed that amateur drivers will park in this type of compound; therefore, generous maneuvering space should be provided to allow for backing out in one motion. Figure 17 shows an on-site parking area with single-entrance and single-90° parking without enclosing walls.

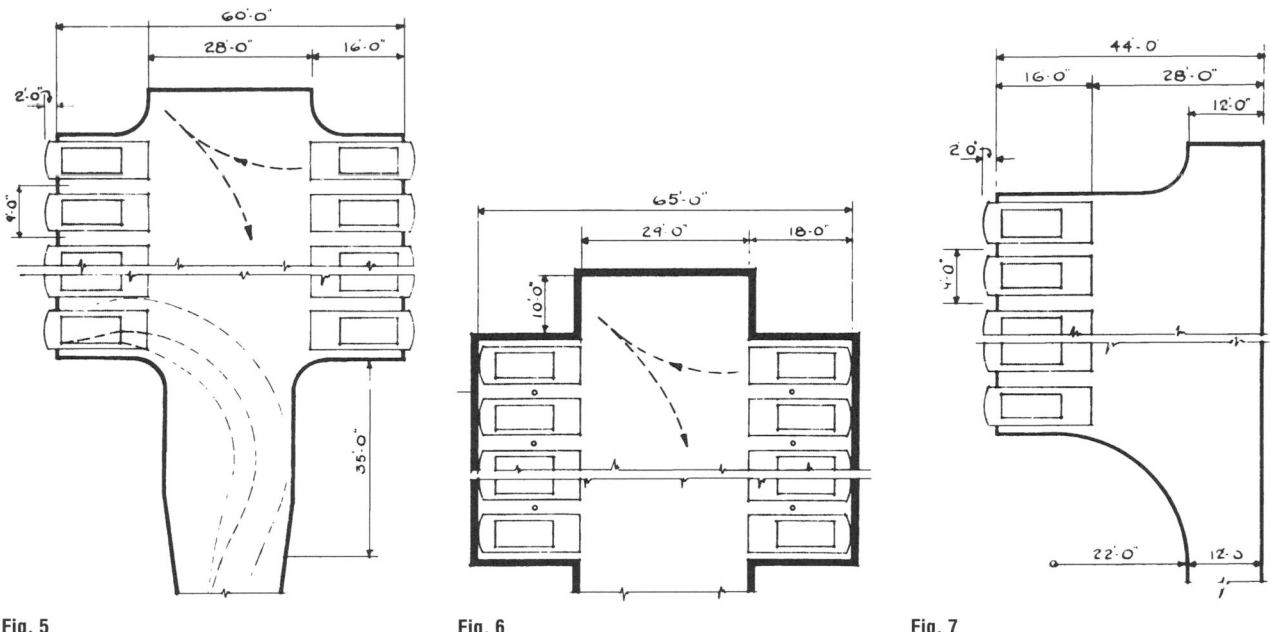

Fig. 5   Fig. 6   Fig. 7

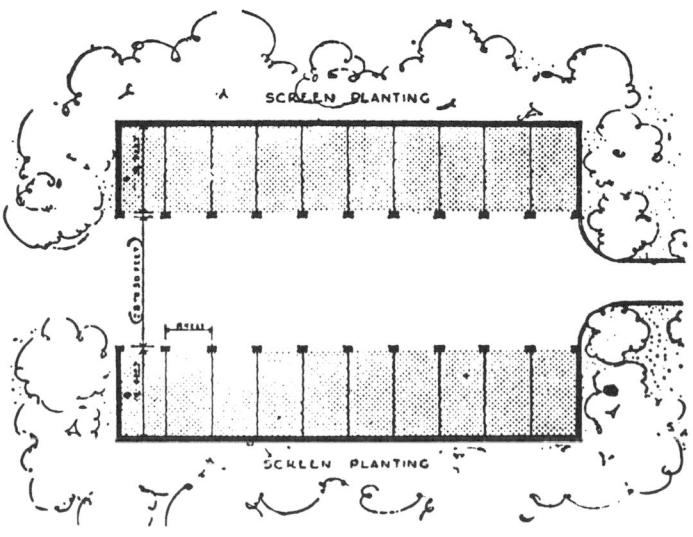

Fig. 8   Garage compound showing enclosure wall and screen planting.

PARKING AREAS AND GARAGES

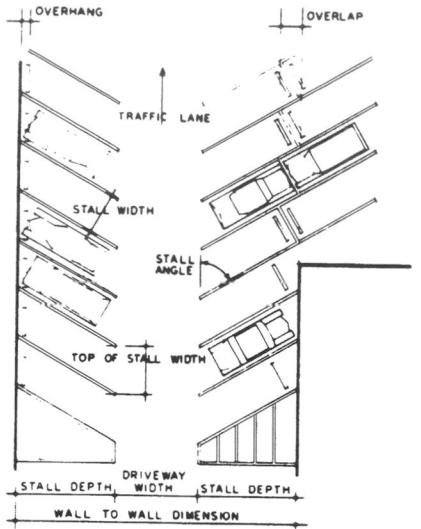

Fig. 9 Definitions.

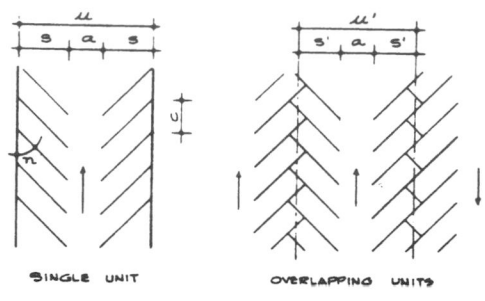

n	s	a	c	u	s'	u'
90°	19'0"	24'0"	9'0"	62'0"	19'0"	62'0"
60°	21'0"	18'0"	10'5"	60'0"	18'9"	55'6"
45°	19'10"	13'0"	12'9"	52'8"	16'7"	46'2"

Fig. 10 Parking layout dimensions.

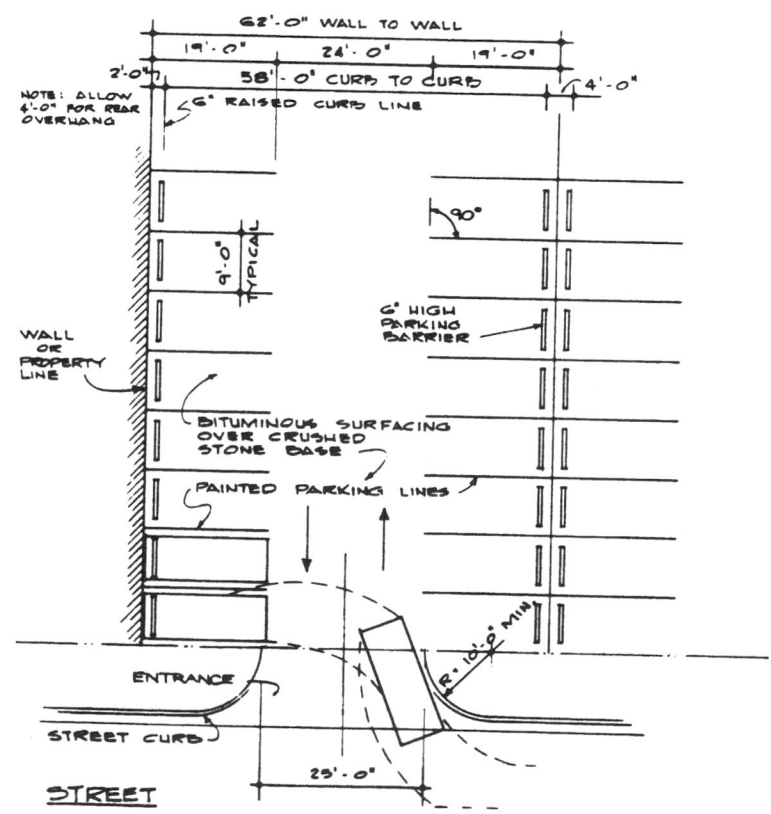

Fig. 11 Parking plan—90° parking.

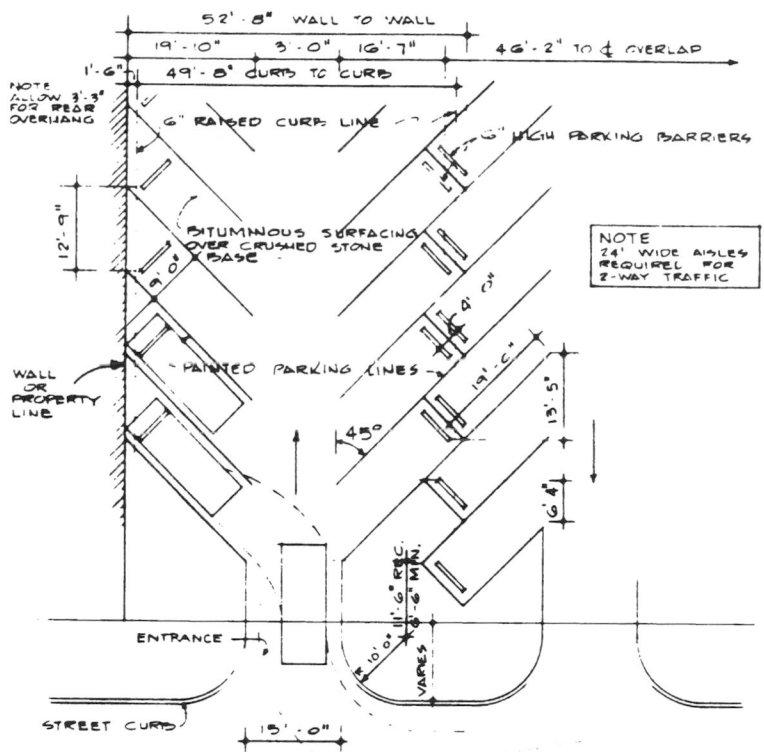

Fig. 12 Parking plan—45° parking.

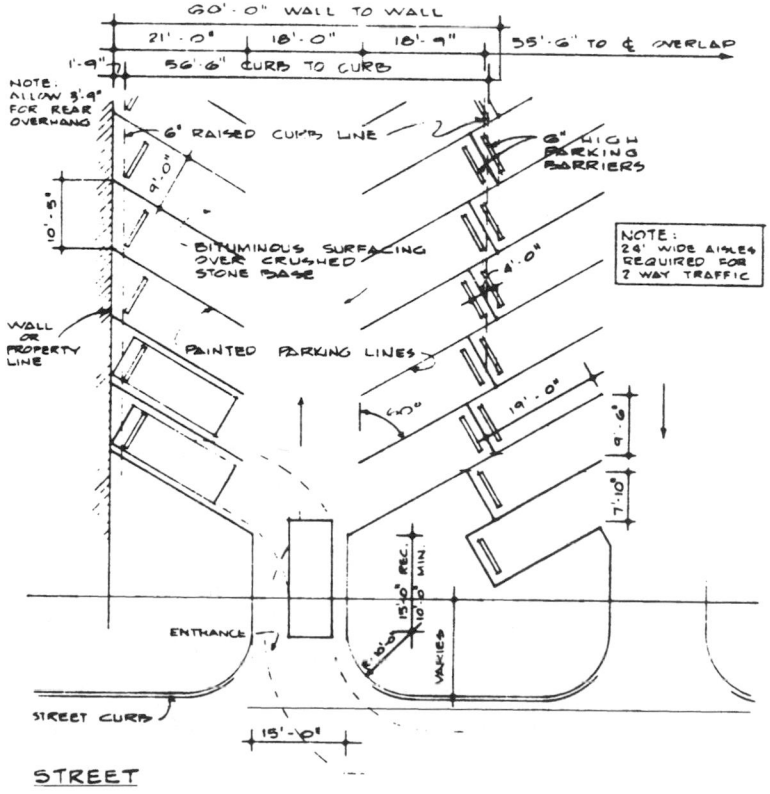

STREET

Fig. 13 Parking plan—60° parking.

PARKING AREAS AND GARAGES

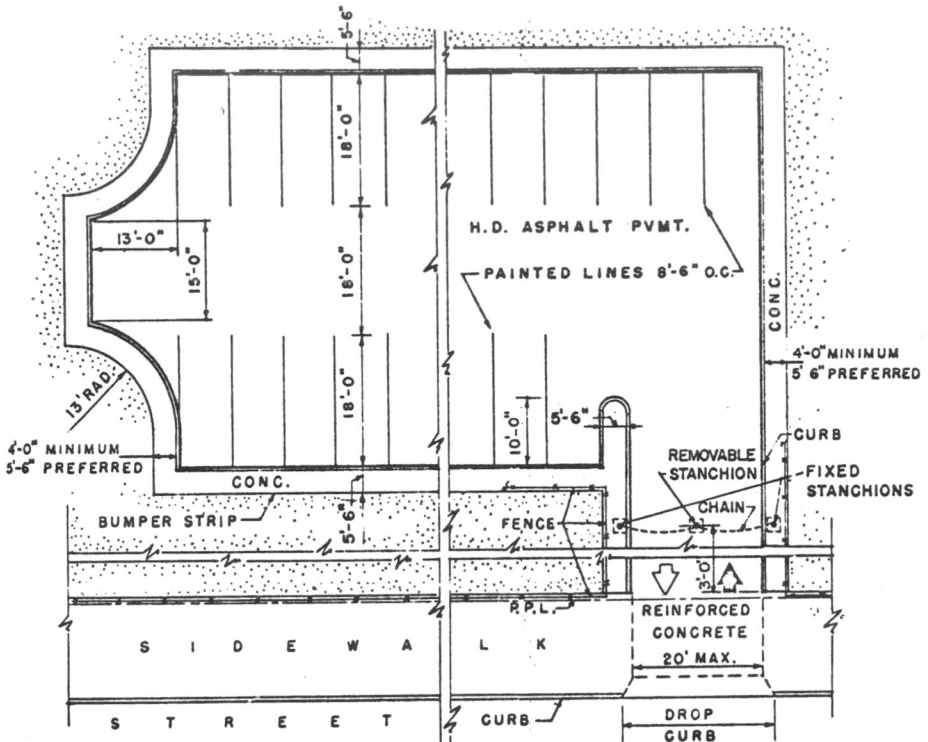

Fig. 14   Parking area—parallel to street, one entry/exit.

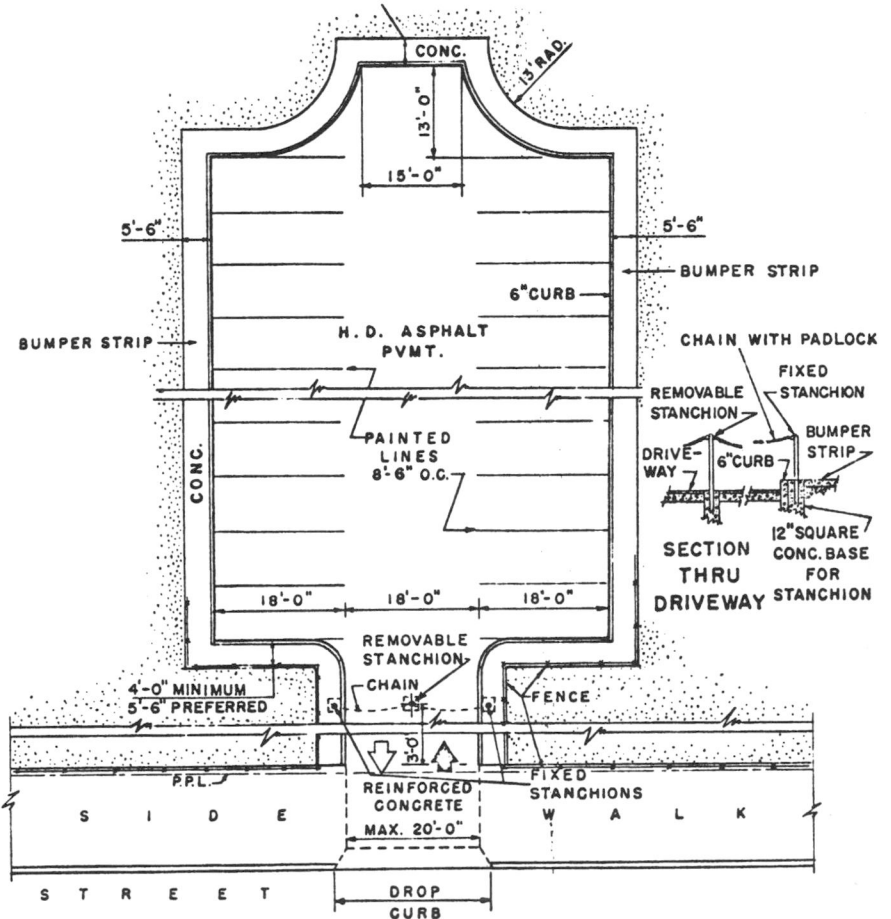

Fig. 15   Parking area at a right angle to street.

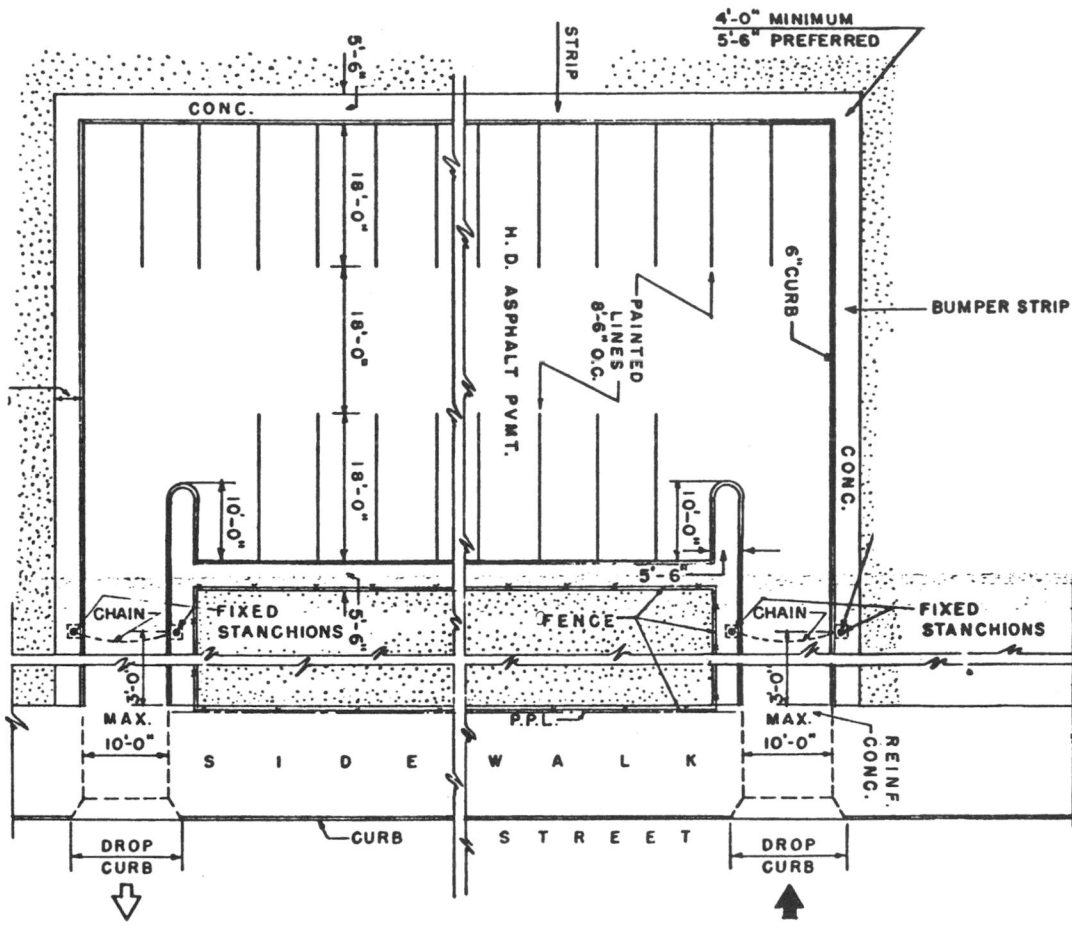

Fig. 16   Parking area—parallel to street, with two entry/exit lanes.

CENTRAL LAUNDRY ROOM

The laundry room is an essential part of any multifamily building. Not only does it provide a place to wash clothes but also it is often the only community area within a building. Providing laundries designed for convenience, accessibility, safety, and maintainability will make any project more attractive no matter what the economic status of the occupants. For low-income occupants it takes the place of the village store. For middle-income occupants it may replace the small town community center. For high-rent units it is no less welcome by maids or others actually doing the laundry. For every segment of the market, laundries can be planned so that they serve as a gathering place—a place to work—a place to communicate—a place to get to know your neighbors.

Such facilities shall provide for the mechanical washing and drying of clothes as well as soaking and sorting of clothes.

LOCATION

The primary design consideration in locating laundries should be the convenience of the tenants. The laundry room should be near elevators or other means of vertical transportation or access. Laundries should not be located in secluded areas of buildings or projects; instead, they should be near main traffic patterns. The location of the laundry room or rooms will depend upon the size and type of project and the number and composition of families.

Housing that is in townhouse form shall have laundry hookups provided within the dwelling unit. For dwelling units in apartment buildings, laundry facilities may be contained in the dwelling unit or they may be centralized at one or more convenient locations to serve groups of dwelling units.

For most projects tenants' convenience is achieved best by a centrally located laundry facility with washers and dryers concentrated in one room. Sometimes in high-rise apartments, the single, centralized laundry room is replaced by laundry rooms on each floor. However, there are disadvantages in this type of location:

1. When the machines are in use or out of order, the tenant is required to go to another floor. This may cause resentment by tenants who find an "outsider" using "their" equipment.

2. Sufficient equipment often is not available if tenants wash more than one load at a time.

3. Additional supervision and maintenance by the building management is required when laundry rooms are on many floors.

4. Less security is provided for the tenant who is more likely to be alone.

5. Some efficiency may be lost because it is more difficult to vent the hot air expelled from the dryers satisfactorily.

6. Costs of installing and servicing the equipment in a multifloor operation are higher.

Where dwelling units are in groups of buildings, an attempt should be made to provide a covered passageway to the laundry room.

Recommendations

1. Use one or more laundry rooms and locate them centrally.

2. Locate the laundry room so that one wall is an outside wall.

DESIGN

Laundries are often danger spots; therefore, special attention must be paid to design in order to eliminate screened or pocketed areas. To assure that all laundry areas are visible from the corridor, corridor walls should be of glass or contain glass panels. For small laundries a door of approved glass may afford the necessary visibility. Placing the change vendor and detergent dispenser where they are visible from the corridor makes theft more difficult. The room arrangement should not include areas that are totally screened from view from the hall or entry door.

Orientation

Laundry rooms should be provided with natural light and views of the outdoors. It is also desirable to achieve a close visual link between the laundry and natural focal points in the building circulation system.

Other design considerations relate to laundry equipment and its arrangement. There must be adequate space for the equipment, for working, for passage, and for equipment servicing.

Tables of adequate size for folding laundry should be provided as well as sufficient numbers of washers and dryers.

A large table (at least 3 by 5 ft) should be provided for folding clothes. An ironing board and extra convenience outlets are desirable. A water closet and lavatory basin in an appropriate room should be provided adjacent to laundries wherever possible. Seating should be provided both for waiting and to accommodate the natural social intercourse resulting from the gathering of people.

The total number of pieces of equipment will determine the size of the laundry. The typical washer is 27 in wide and 28¼ in deep; the average single-load dryer is 34 in wide and 28¼ in deep; the average double-load dryer is 34 in wide, 37½ in deep, and 72 in high.

Laundry rooms shall be designed with adequate clearance around the equipment for use and maintenance. Work space for using the laundry folding table shall be provided such that circulation is not impaired.

The social functions of a laundry room can best be accommodated by room design which provides a natural lounge area.

Recommendations

1. Determine the size of the laundry room by allowing a minimum of 25 ft² per machine.

2. Do not provide raised concrete bases for top-loading equipment. However, some front-load washers do require a raised base.

3. Show the floor as concrete having a trowled finish and sloped toward the floor drain. At least one floor drain should be provided in each laundry.

4. Provide a minimum clearance of 8 ft between floor and overhead pipes.

5. Washtubs are not required. However, if they are installed, provide adequate space for them and for the tenants using them. Such space should be in addition to that recommended for each machine.

6. Locate dryers on outside walls, since long ducts increase the installation cost and are less effective for proper venting.

7. If fewer than five pieces of laundry equipment are installed, show them placed side by side.

8. Provide sufficient lighting, preferably fluorescent.

MACHINES

The type and number of washers, dryers, and other equipment required will depend upon the family composition of the occupants, the number of apartments to be served, and whether or not commercial laundry facilities are available in the particular locality.

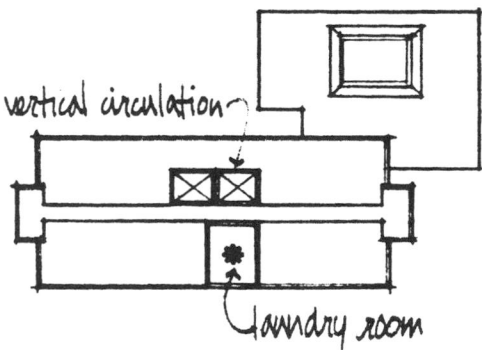

Fig. 1 Laundry room accessibility.

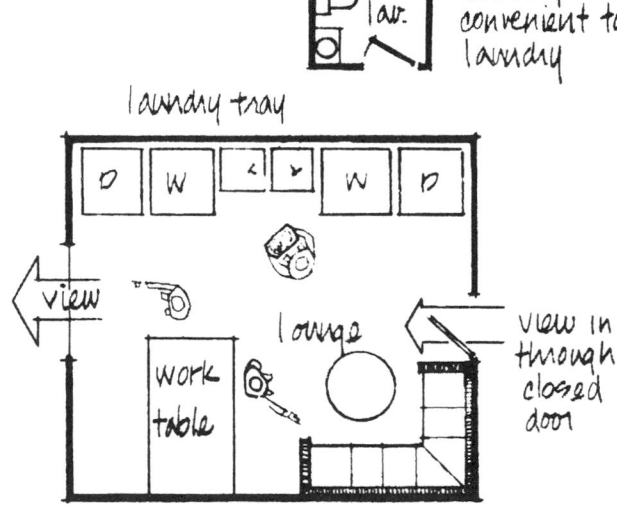

Fig. 2 Example of laundry room orientation.

The most important single factor in determining the number of machines required is the nature of the building occupancy. To be considered are the presence or absence of children, the age of the occupants, their status, and the type of laundry service required.

Sufficient numbers of machines should be provided to assure the occupants of available equipment at all times when the laundry room is open and to allow for washers or dryers temporarily unavailable because of repairs or breakdowns.

In order to approximate the equipment needed, the following should be considered as a guide. Washing machines normally operate on a 30-min cycle. A washing machine could therefore theoretically do 24 washes in a 12-h day and 144 washes in a 6-day week. Dryers generally run for 10-min periods. However, in order to dry the load contained in a washing machine, a single-load dryer usually requires three periods of 10 min.

Washers and dryers should be of the domestic size rather than the large capacity commercial size.

Recommendations

The following guides are recommended for multifamily dwellings:

Washers

1. Lower and middle rental levels (predominantly young families)—one washer per 20 bedrooms.

2. Lower and middle rental levels (predominantly efficiency and one-bedroom apartments)—one washer per 20 apartments.

3. Luxury apartments—one washer per 25 apartments.

4. Apartments for senior citizens—one washer per 40 apartments.

Dryers One single-load dryer per washer or one double-load dryer for every two washers.

Scheme I in Fig. 3 shows a laundry conveniently located. It is adjacent to the elevators. There are toilets nearby. Notice that all areas are visible through the glass which forms the corridor wall. This affords greater protection for occupants. The change vendor and detergent dispenser are next to the glass. This makes theft more difficult.

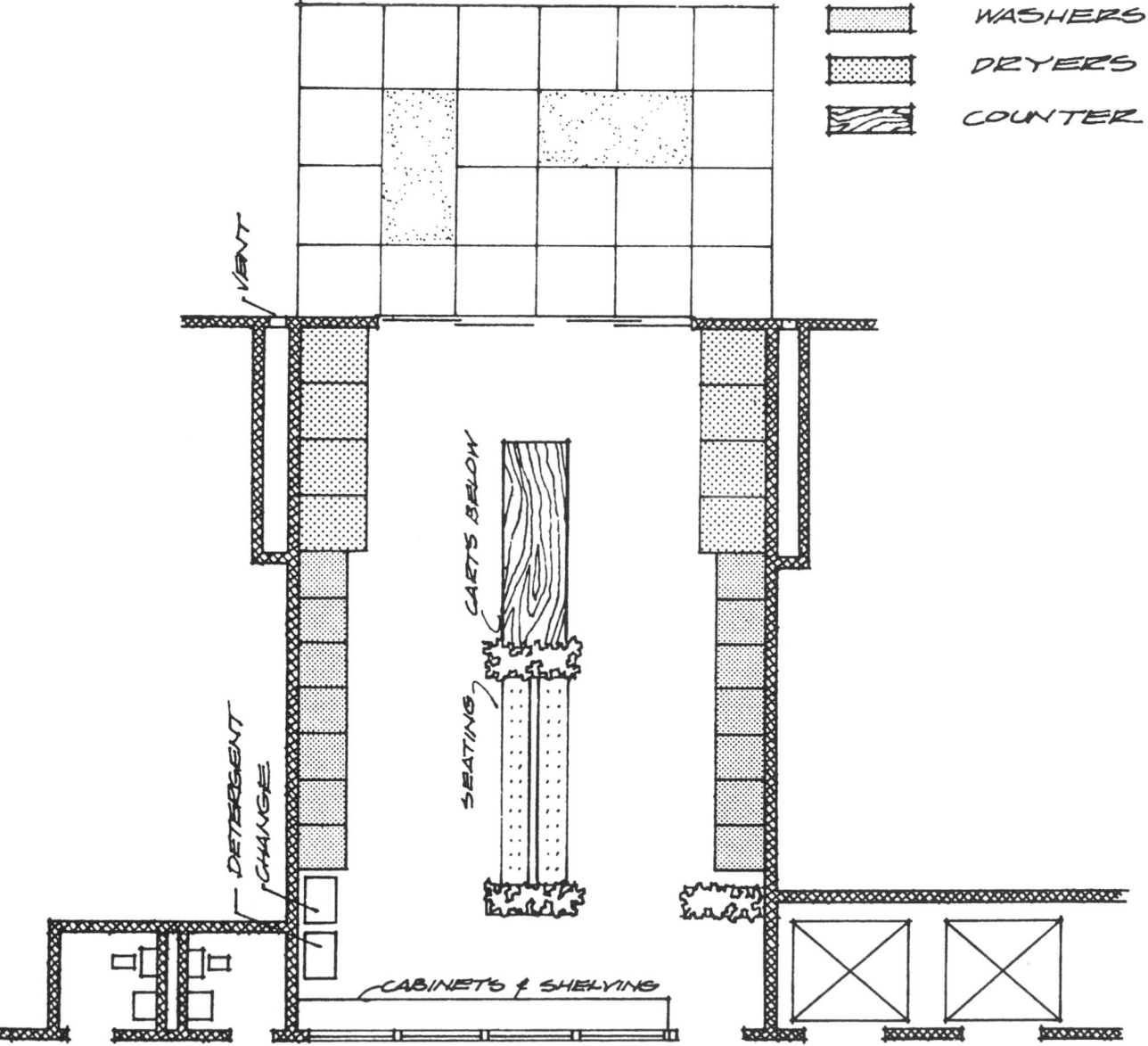

Fig. 3

CENTRAL LAUNDRY ROOM

The arrangement of the laundry, the planters, and the cushioned bench makes it attractive. Light and ventilation are more than adequate. The full-glass, bypassing doors make the attractive patio a supervised child play area in good weather.

There is a washer for every 20 bedrooms. In luxury apartments, this ratio can be decreased to one washer for every 25 apartments. One double-load dryer has been provided for every two washers. Note that the dryers are next to the exterior wall. This arrangement is required for proper venting. Long horizontal runs in vents collect lint and create a fire hazard. The central location within the building groups the equipment, provides a larger selection, and reduces installation and maintenance costs. The table for folding clothes is readily accessible and convenient to both washers and dryers.

Scheme II in Fig. 4 provides an adjoining room separated from the laundry by a glass-paneled partition. This room provides additional seating capacity. It also functions as a close-supervision play area for use in inclement weather. Note that all areas are visible from the laundry for maternal supervision and all areas, except one very small portion of the added room, are visible from the corridor. Here again, safety is included as a design factor.

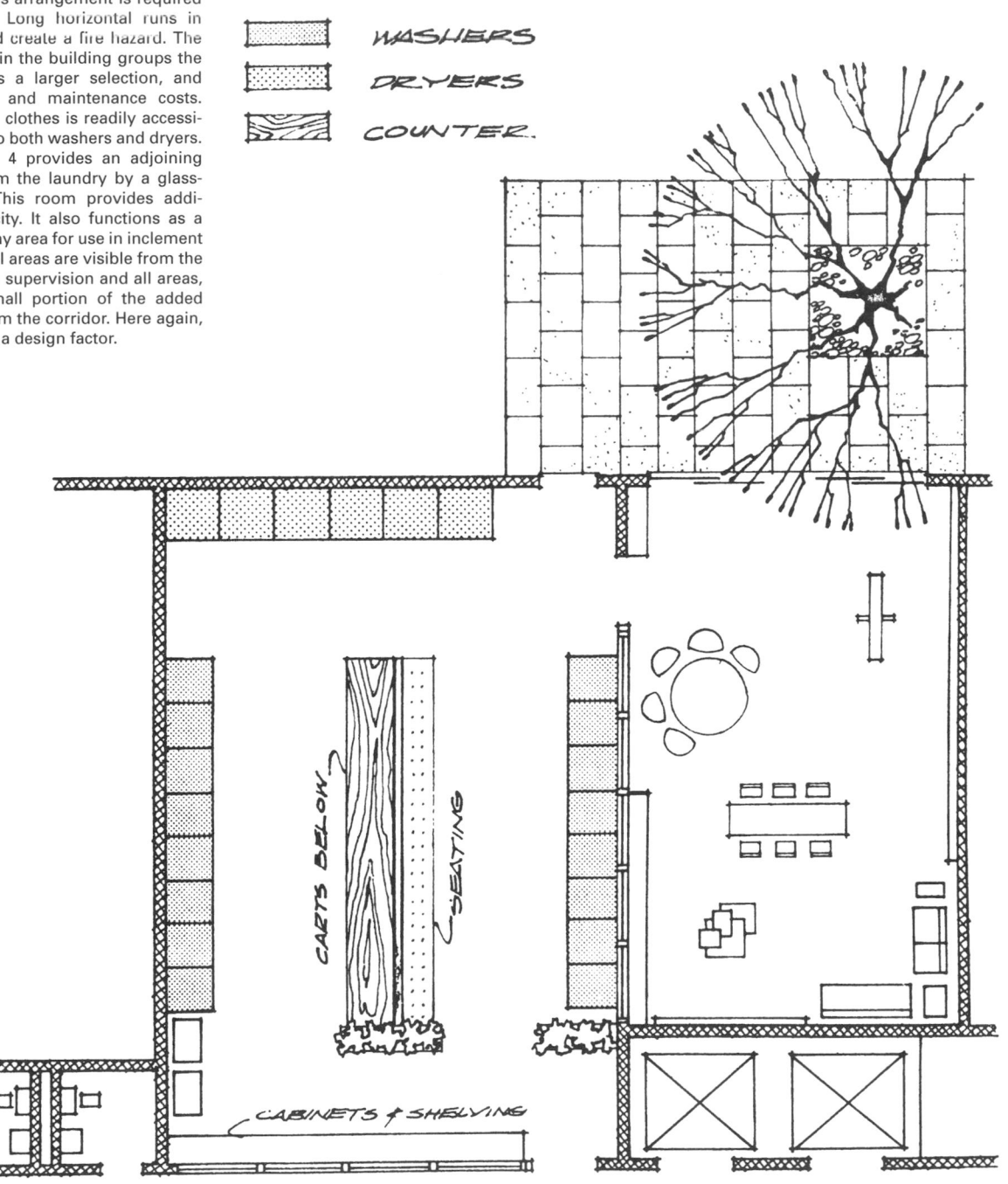

WASHERS

DRYERS

COUNTER

CARTS BELOW

SEATING

CABINETS & SHELVING

Fig. 4

Scheme III in Fig. 5 shows a more luxurious design arrangement. The number of washers and dryers has been reduced. The total work space and seating space in relation to the number of washers is now quite generous. There is additional furniture. The furniture is arranged to provide a conversation grouping which is warm and inviting. This space has the atmosphere we mentioned earlier—a place to communicate—a place to get to know your neighbor.

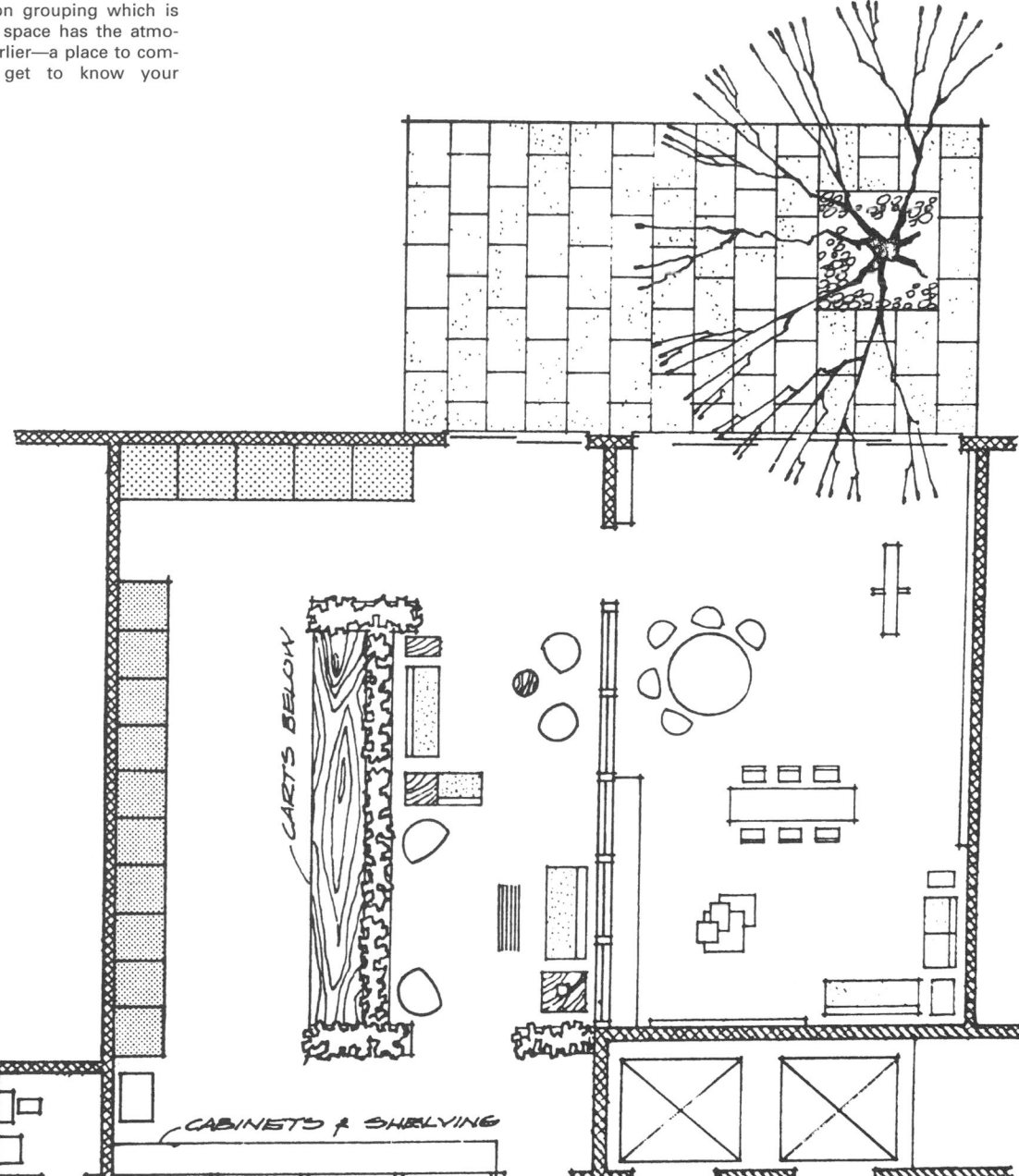

CARTS BELOW

CABINETS & SHELVING

▨	WASHERS
▨	DRYERS
▨	COUNTER.

Fig. 5

CENTRAL LAUNDRY ROOM

GENERAL ROOM INFORMATION

For most facilities, the central laundry will probably be best from the residents' standpoints. More equipment will be available for their immediate use and large tumbler dryers will undoubtedly be used for faster drying and a higher rate of user turnover. From a design standpoint too, the central room offers some advantages. Foremost would be the central location of utility and venting facilities for dryers.

Where it is not possible to utilize one central room, it might be best to locate a laundry room on every second or every third floor so as to have more equipment in the room than if some equipment were located on every floor. This arrangement would also tend to produce some savings in utilities and venting facilities.

In any case, if at all possible in the laundry room design, make provisions for future expansion. As time passes, more equipment may be required for the same building, and as improvements in equipment occur and new products may be developed, tenants may demand more laundry facilities than originally necessary.

LOW-RISE APARTMENT

Garden apartments with low density can be served well with the typical plan in Fig. 6 for a multibedroom low-rise building. Three washers and three dryers serve a 20-unit apartment.

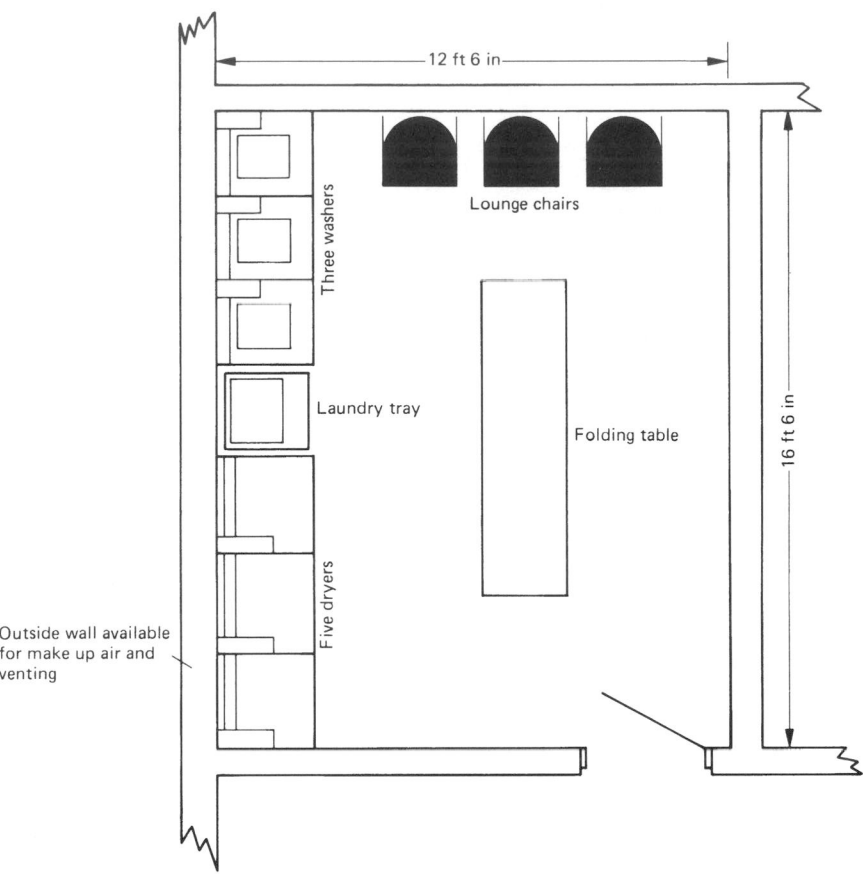

Fig. 6   Typical 20-unit multiple-bedroom, low-rise apartment building.

HIGH-RISE APARTMENT

The plan in Fig. 7 is for a typical 400-unit high-rise apartment building. The plan suggests 20 washers, 10 drying tumblers, and 1 washer extractor. This plan presupposes one central laundry room for the entire building. Some very tall apartment buildings will space smaller laundry rooms throughout the living area of the building.

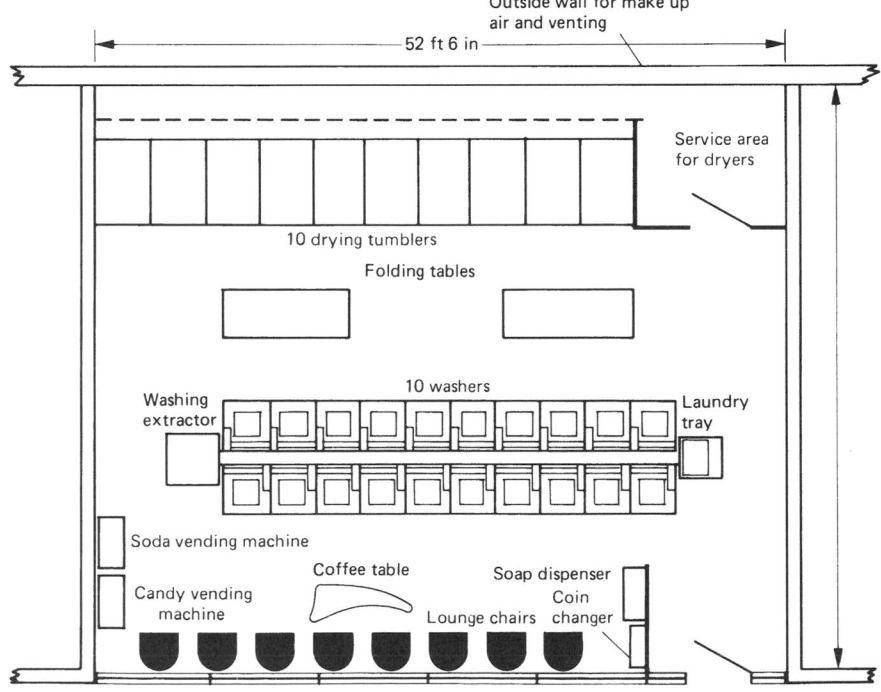

Fig. 7   Typical 400-unit, high-rise apartment building.

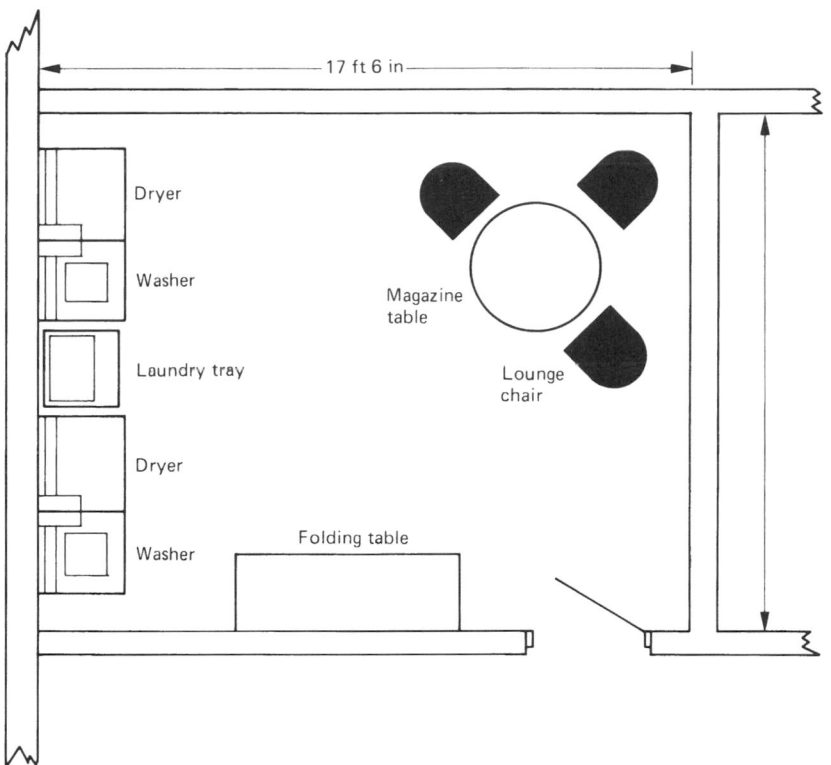

Fig. 8   Typical 18-unit-per-floor, high-rise housing for the elderly.

SENIOR CITIZENS RESIDENCE

Convenient laundry facilities are important to senior citizens. The typical plan in Fig. 8 shows two pairs of washers and dryers serving 18 residential units in a multistory home for the elderly. Additional room here is provided for seating and lounge facilities.

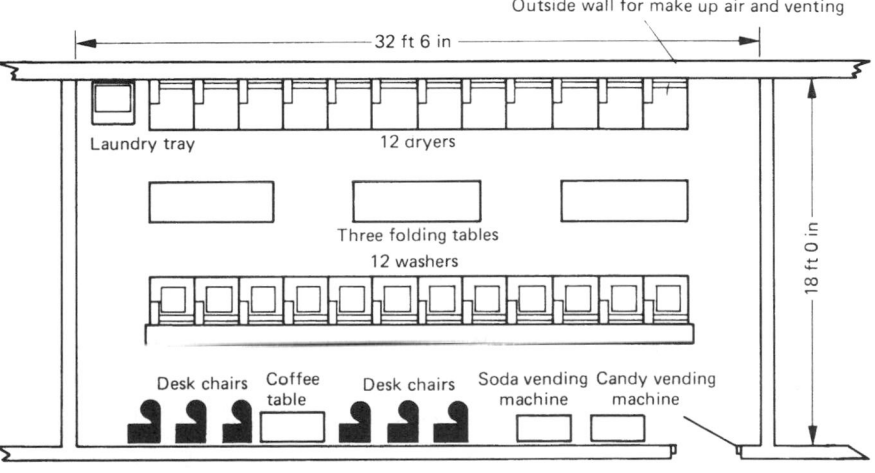

Fig. 9   Typical 500-unit student dormitory.

COLLEGE DORMITORY

This typical laundry facility in Fig. 9 for a large 500-student residence features 12 washers and 12 dryers.

CENTRAL LAUNDRY ROOM

LAUNDRY ROOM LAYOUT/HOUSING PROJECT

Common laundries can range in size and completeness from a single large and elaborate facility to a washer and a dryer in a closet on each floor of a multistory building. A single central laundry permits the maximum amount of equipment, including large fast-working tumbler dryers, to be available in one place. It requires a minimum of supervision and maintenance because equipment, utilities, and ventilators are concentrated in one place. In a large common laundry a cheery clublike atmosphere can be created and space can be allocated to an adult seating area and sometimes to a children's play area.

In spread-out projects it may be more convenient for occupants to have several smaller laundry rooms located nearer the living units. This is also true in high-rise projects where laundry rooms can be situated on every floor or every second floor.

Convenient laundry facilities are especially important to the elderly. Elderly occupants can use more seating (socializing) space but need fewer machines because they are less active and tend to have smaller wardrobes.

DESIGN CONSIDERATIONS

Laundries should be located conveniently near elevators and toilets.

A troweled-finish concrete slab makes a durable and economical floor. At least one floor drain should be provided.

All overhead obstructions should be 8 ft clear of the floor.

Use of punchcard-type machines eliminates the need for change machines and discourages thievery.

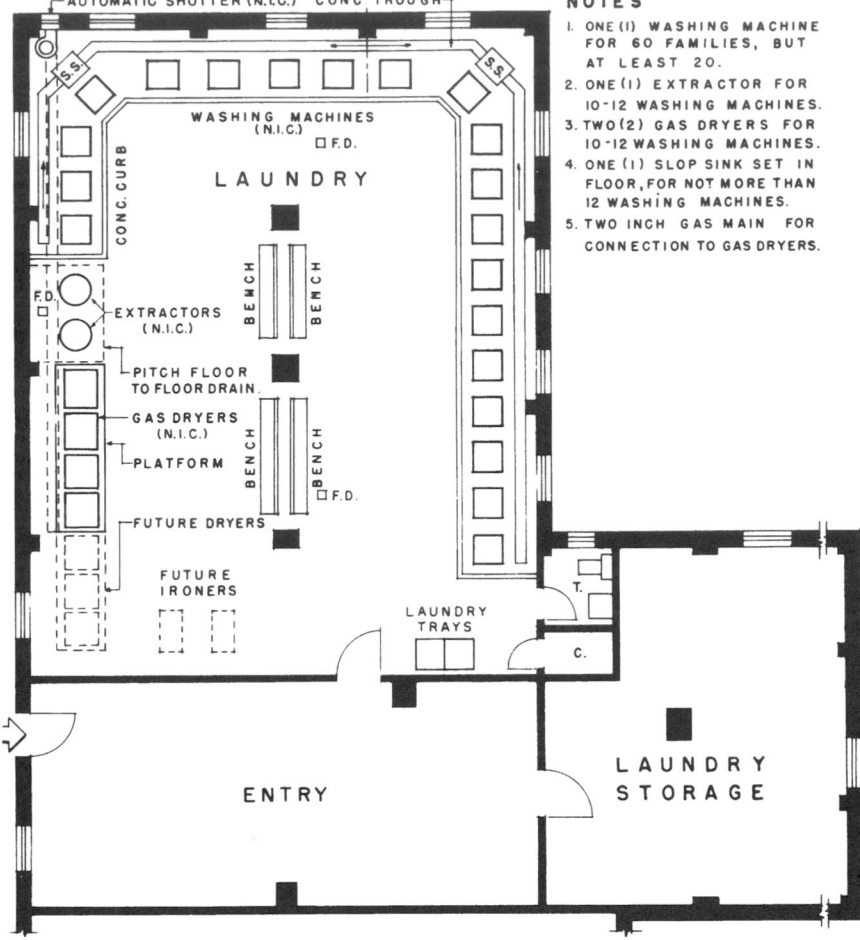

NOTES

1. ONE (1) WASHING MACHINE FOR 60 FAMILIES, BUT AT LEAST 20.
2. ONE (1) EXTRACTOR FOR 10-12 WASHING MACHINES.
3. TWO (2) GAS DRYERS FOR 10-12 WASHING MACHINES.
4. ONE (1) SLOP SINK SET IN FLOOR, FOR NOT MORE THAN 12 WASHING MACHINES.
5. TWO INCH GAS MAIN FOR CONNECTION TO GAS DRYERS.

Fig. 10

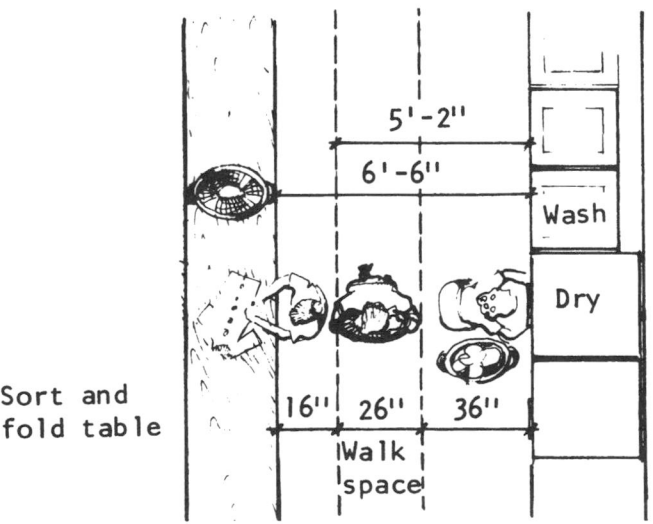

Fig. 11 Clearances for central laundries.

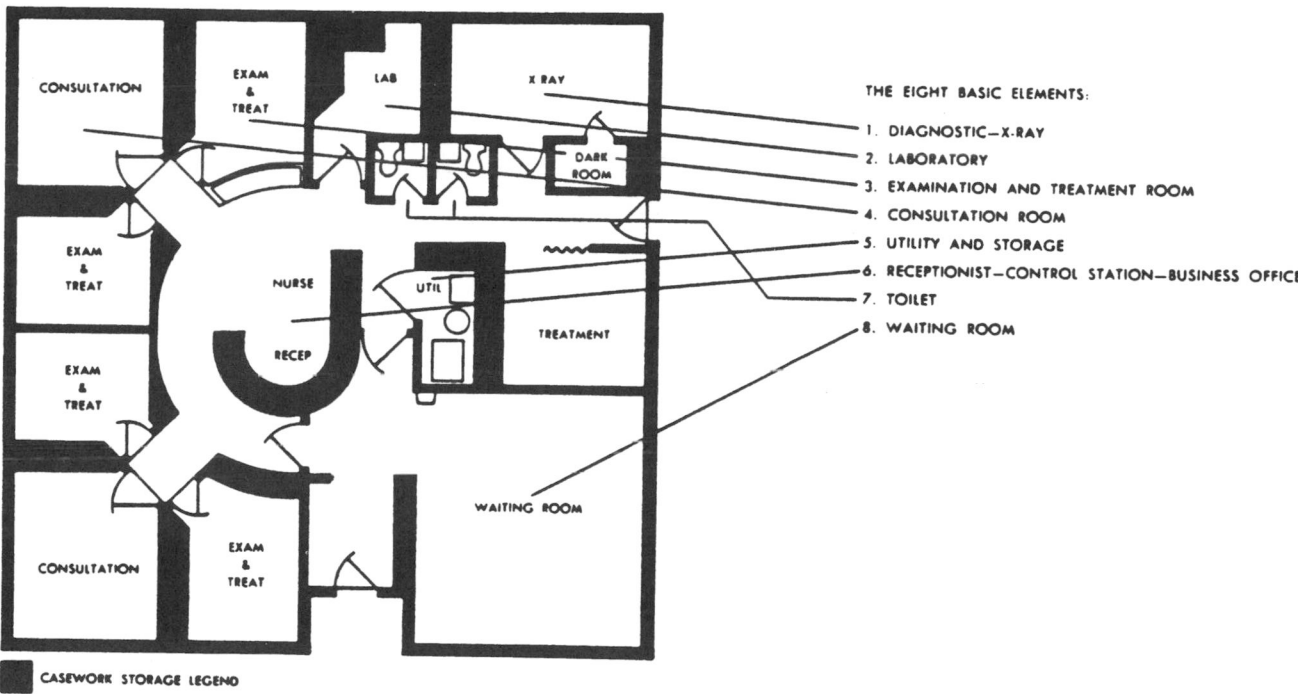

THE EIGHT BASIC ELEMENTS:

1. DIAGNOSTIC—X-RAY
2. LABORATORY
3. EXAMINATION AND TREATMENT ROOM
4. CONSULTATION ROOM
5. UTILITY AND STORAGE
6. RECEPTIONIST—CONTROL STATION—BUSINESS OFFICE
7. TOILET
8. WAITING ROOM

CASEWORK STORAGE LEGEND

Fig. 1

A medical practice facility can have no fixed, ideal plan. First, no two individuals or groups of individuals think alike or work alike. Second, the physical and geographical limitations which characterize a medical practice facility, whether for a new building, a remodeled building, or rental space, do not permit the adoption of any single plan. Each facility must be custom-made to express the individuality and to satisfy the working habits of those who will use it.

The eight basic elements found in nearly all medical offices can be thought of as the *building blocks*. While these eight elements may change in size and shape depending on methods of operation, they are always integrated in the medical practice facility, or their counterpart is conveniently available (Fig. 1).

The eight basic elements of a medical office are as follows:

1. Receptionist—control station—business office
2. Waiting room
3. Consultation room
4. Examination and treatment room
5. Laboratory including EKG and BMR
6. X-ray
7. Utility and service areas
8. Toilet

ADMINISTRATION AREAS

Adequate space should be provided for patient reception, waiting, and admitting procedures. The business manager's office, business records office, patient billing area, and telephone and toilet facilities should be located close to the waiting space. The extent of the administrative service in a central location will depend on the organization of the facility. Since the patient's initial impression of the clinic is formed in the reception area, cheerful and attractive spaces and furnishings are desirable. Good reading light and comfortable seating are needed. In large facilities, the administrative area may also provide for a doctors' lounge, conference room, library, isolation of patients suspected of being infectious, and where applicable, a waiting area for small children.

Space is needed for centralized medical records. This area is usually located within the administrative area or adjoining it. Medical records for all patients of the clinic are kept here.

In larger facilities, work space for a medical record librarian should also be provided. It may also be necessary to provide space and equipment for microfilming medical records.

PATIENT EXAMINATION AND TREATMENT AREAS

Most group practice facilities, which provide for practitioners such as the internist, the surgeon, the obstetrician-gynecologist, the pediatrician, and the orthopedist, require certain similar areas: consultation rooms, examination rooms, treatment rooms, and recovery rooms. Various combinations of these rooms are often desirable, such as consultation-examination room or a treatment-recovery room.

A typical consultation room should be large enough for a full-sized physician's desk and comfortable chairs for two or three people. Space for bookcases and one or more file cabinets should also be provided.

Adjacent to the consultation room and usually accessible directly from it should be an examination room. An examination room may be required on each side of one consultation room. The examination room should be large enough for a full-sized adjustable examining table, an instrument supply cabinet, a small instrument table, a small desk or wall-hung writing surface for the physician, and one or two straight chairs. A good examining light should also be provided. A small curtained-off space in a corner of the room for patients to undress is desirable, and a patient toilet should be easily accessible.

A recovery room is needed in most clinics. It may also be used for tests and treatments such as injections and EKGs. Space should be provided for a bed, equipment, an instrument and supply cabinet, and a chair.

Surgery Clinic

In a clinic where minor surgery is contemplated, a small surgery room with a small dressing room adjoining should be provided. This room requires an operating table instead of an examining table, portable examining light, and instrument table.

Dental Examination

Dental examination and diagnosis may require one or more operatories. While a fully equipped operatory, such as one described below, is not always considered necessary, it is sometimes desirable in order to provide greater flexibility in use. Taking x-rays is part of the examination and diagnosis procedure and requires that appropriate radiology equipment be located adjacent to or as close as possible to the operating room. For a small dental facility, x-rays may be taken in the dental operatory, but particular care must be given to locating the equipment to provide proper shielding from gamma and x-rays. In larger facilities, a separate radiology section as described below under Dental X-Ray Service should be used.

MEDICAL PRACTICE OFFICE FACILITIES

Dental Operatories

A dental operatory varies in size from a cubicle of 7 by 10 ft to a room of 10 by 14 ft. Modern dentistry is performed with the patient in a supine position and the dentist and chairside assistant in a seated position. Many dentists who operate from a seated position prefer the mobile dental unit to the older fixed floor-mounted unit. Another type of dental unit is designed to be mounted in the cabinetry behind the head of the patient. While not having the flexibility of the mobile unit, the cabinet-mounted unit can be located in a position that is more accessible to both the dentist and the assistant than the fixed-floor mounted unit. The method of operation and the selection of equipment will strongly influence the design of the operatory. With the exception of oral surgery, the operatories of all the other dental specialties, such as pedodontics, periodontics, and orthodontics, are basically the same.

The oral surgery operatory can be the same size as other operatories. Operatories designed specifically for oral surgery often do not have dental units like those described. A portable dental engine is used instead. Scrub-up sinks and provisions for sterilization and sterile supplies can be within the area but are usually in an alcove adjacent to the oral surgery operatories. One or more small recovery rooms with accessible toilet facilities should be provided in the oral surgery area.

SUPPORTING SERVICE AREAS

Laboratory and X-Ray Services

Some type of laboratory work will generally be done in all but the very smallest group practice facilities. In the very large facilities all, or almost all, of the laboratory work will be done on the premises.

In many facilities, space for routine fracture and chest x-ray work must be provided. The x-ray facilities may include both diagnostic x-ray and x-ray therapy. A separate waiting area nearby, particularly for x-ray therapy, would be desirable. Besides the x-ray room, the diagnostic x-ray and fluoroscopy space should include a consultation-viewing room for the radiologist, and a darkroom. Films should be filed in an adjoining room or in the central medical records room. Toilet and dressing rooms should be arranged so that the patient need not leave the immediate area of the x-ray room. One x-ray room could be used for the chest work, gastrointestinal work, and routine radiography, depending on the caseload. Gastrointestinal work is time-consuming; therefore, patients need waiting space near the x-ray office. Therapeutic x-ray providing facilities for superficial types of treatment should include convenient dressing space, a toilet room, and at least one recovery room. Care must be taken in placing the x-ray equipment in the room. In the case of diagnostic x-ray, the chest cassette should be on an outside wall, and if possible, the tube should be installed so that it does not point in the direction of persons working in

adjacent rooms. For appropriate shielding to reduce the effects of gamma and x-rays, refer to National Bureau of Standards Handbooks 73 and 76.

Physical Therapy Service

Some clinics will include physical therapy as a part of their total services. Such a service generally includes some hydrotherapy—arm baths, leg baths, and perhaps Hubbard tanks; and alcoves for massages, electrotherapy, and diathermy. A large room with various types of exercise equipment may also be included in the clinic.

Dental X-Ray Service

X-ray cubicles or rooms should be a minimum of 8 by 10 ft and should have radiation protection equivalent to $\frac{1}{16}$ in lead to a height of 7 ft 0 in. Controls are usually outside the areas beneath a lead-glass observation window. A lavatory should be provided within the room. A darkroom should be located as close as possible to the x-ray cubicle or room. A combination transfer box and dryer, installed in the partition separating the darkroom from another room or alcove in which the dental films could be mounted and examined in white light, would facilitate operations.

Dental Laboratory

While many dental group practice clinics will have the major portion of their prosthetic laboratory work done by private dental laboratories, all clinics will wish to have some laboratory facilities available. The full range of work can be divided into five main categories: (1) crown and bridge, (2) full dentures, (3) partial dentures, (4) ceramics, and (5) orthodontics. Each category requires specialized equipment which in turn determines the work area and overall size of the laboratory.

Pharmacy Service

Some group practice clinics may find it advantageous to arrange for the inclusion of a pharmacy. The need for one and its size depend upon the number and types of medical disciplines it will serve.

Even a minimum pharmacy will need storage space for a large reserve stock of many items. Smaller quantities will be kept in drug cabinets, which should adjoin a compounding and dispensing counter. Several hundred chemicals and pharmaceuticals in various forms and strengths must be stored in sectional cabinets, ready for immediate use. A refrigerator with drawers will be needed for keeping biologicals and thermolabile preparations.

AREAS FOR OTHER SPECIALISTS

Specialized examination and treatment rooms may be required for special services. These are usually provided on the basis of one room to each consultation room.

Otolaryngology Service (Ear, Nose, and Throat)

The examination and treatment room for the otolaryngologist should be approximately the same size as a typical examination room. In this room, instead of an examining table, an adjustable examining and treatment chair with a cuspidor connected to the water supply and drain should be provided. This specialty needs a special ENT treatment cabinet with suction and compressed-air equipment. Provision for undressing other than to hang outer garments is not necessary. One room should be sound isolated from street noises for audiometry if a speech and hearing service room is not a part of the complex.

Speech and Hearing Service

A speech and hearing clinic may also be included in a group practice facility, particularly in a very large one. The basic unit consists of a control room and a test room which are sound-isolated with an observation window between them. The testing equipment is located in the control room, which should accommodate two persons; a room 8 by 10 ft is suggested. The test room where the patient sits is usually 10 by 10 ft. The typical consultation room is also used for hearing and speech therapy and should accommodate four persons and the required instruments.

Ophthalmology Service

The ophthalmologist will need a consultation room and an examination room. The size and number of the eye examination rooms depend on the patient load. As a minimum, one examination room would be required to house the refracting equipment and a small desk. Additional equipment the ophthalmologist uses for the examination could be in this same room. In larger clinics it is frequently desirable to have a separate room for eye refraction, perimeter, and corrective treatment equipment. Patient load may justify a separate room for measuring eye tension, although this procedure can be performed in the eye therapy room. A small examination room with a treatment and examining chair or table is all that is required for this test.

Optometry Service

Except for surgery, the space needs of the optometrist would be similar to those of the ophthalmologist. If optician services are included, work space and equipment may be provided here for filling prescriptions for corrective lenses and for frames. A waiting room and space for fitting frames may be needed.

Psychiatry Service

This service may be found in large group practice facilities. In some facilities, separate exits may be desirable. The consultation room is the focal point of the service. A larger room may be required for conference or group therapy. A more extensive program may include psychologists, social workers, or both, each of whom would also need a typical consultation room.

COMMON STORAGE SPACES

Each housing project should have:

1. Tenant storage room or rooms within about 500 ft of the farthest building to be served. Allow approximately 3 ft^2 per family. Tenant storage rooms usually are planned in basements. They may not be used for passage.

2. Perambulator storage rooms. Provide perambulator space for 75 percent of the dwelling units, allowing 8 ft^2 plus passageway of 4 ft^2 for each perambulator. Each building should have a perambulator room, preferably on the entrance floor. If this is not practicable, the room should be in the basement reached by ramp conveniently near the main entrance.

MANAGEMENT AND MAINTENANCE SPACES

In determining the extent and character of management and maintenance space to be provided, due recognition should be given to the project's type and size, its proximity to other developments, the kind of construction, nature of equipment and utilities, and special local problems. The areas for management and maintenance space (measured between the inner faces of enclosing and dividing walls) listed in Table 1 should not be exceeded. Management and maintenance space areas should not be combined in applying this table.

When the number of units falls between the above, interpolate between the areas shown. For more than 500 units, the additional management and maintenance space to be provided should be determined in consultation with the HUD field office. If management or maintenance space is required for less than 50 units, the area for management space should be not more than 6½ ft^2 per unit and not less than 80 ft^2 total; the area for the maintenance space should be not more than 8 ft^2 per unit and not less than 100 ft^2 total.

Management space might include a public waiting area separated by a counter from a rent-collection or general office, a private office, staff toilet, and supply storage. Where more than 100 units are served, one or more small interview rooms, a record vault, janitor's closet, and an additional staff toilet room should be provided. Office and public spaces should be air-conditioned.

Maintenance space could include a heated shop area with workbench, service sink, tool storage, and toilet. A separate paint storage room may be advisable in larger projects.

TABLE 1

Number of dwelling units	Management space, ft^2	Maintenance space, ft^2
50	325	400
100	500	800
200	775	1400
300	1000	1900
400	1200	2300
500	1375	2650

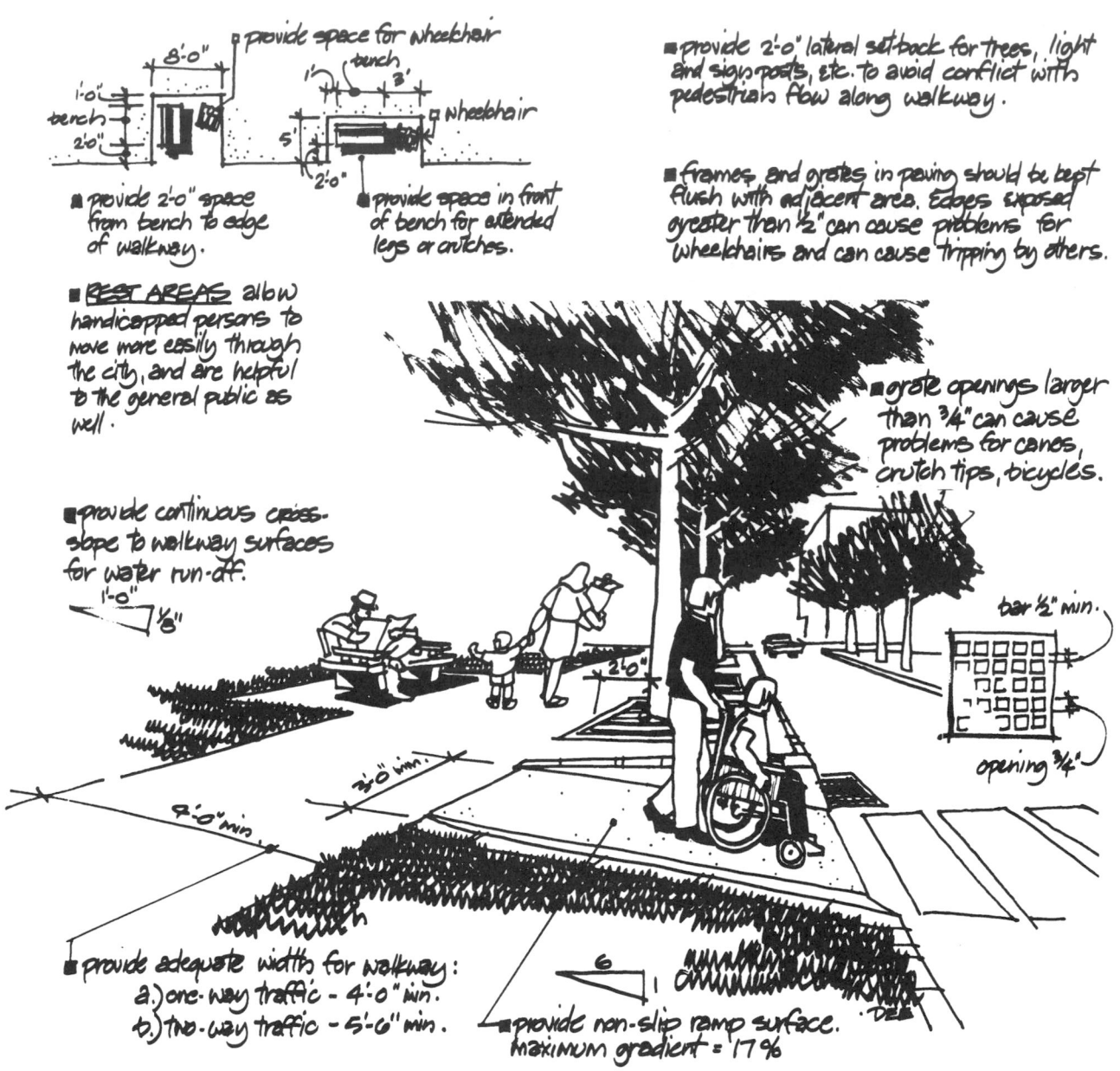

Fig. 1 Walks—general dimensions.

Walks should be designed to allow the greatest diversity of people to move safely, independently, and unhindered through the exterior environment.

Items to consider in the design or modification of walk systems are as follows.

1. *Surfaces:* Walks should possess stability and firmness, be relatively smooth in texture, and have a non-slip surface. The use of expansion building contraction joints should be minimized, and their size should be as small as possible, preferably under ½ in in width. (The chart in Fig. 2 shows some different types and

characteristics of materials when used as walkway surfaces).

2. *Rest areas:* Occasional rest areas off the traveled path are enjoyable and helpful for all pedestrians, and especially for those with handicaps that make walking long distances exhausting.

3. *Gradients:* Pedestrian paths with gradients under 5 percent are considered walks. Walks with gradients in excess of 5 percent are considered ramps and have special design requirements. (Also see "Ramps") Routes with gradients up to 5 percent can be negotiated independently by the average wheelchair user, but sustained

grades of 4 percent and 5 percent should have short (5 ft) level areas approximately every 100 ft to allow a chairbound person using the walk to stop and rest. Gradients up to 3 percent are preferable where their use is practical.

4. *Lighting:* Lighting along walkways should vary from ½ to 5 ft candles, depending on the intensity of pedestrian use, hazards present, and relative need for personal safety.

5. *Maintenance:* Proper maintenance of walks is imperative. Where they are deteriorating, repairs should be made to eliminate any conditions that may cause injury.

Surfaces for Walkways

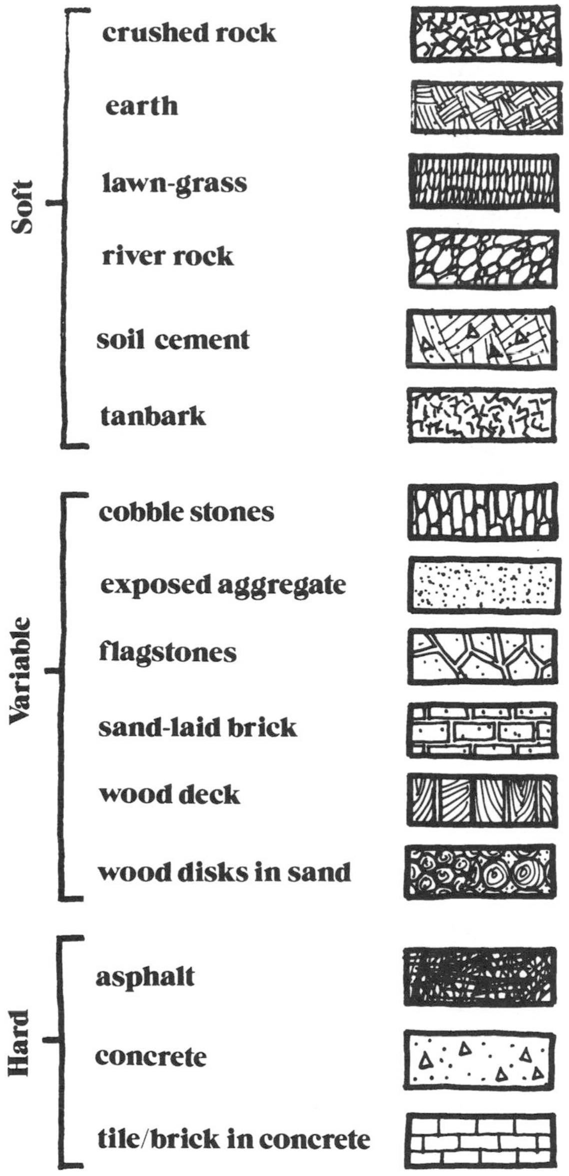

Soft

- crushed rock
- earth
- lawn-grass
- river rock
- soil cement
- tanbark

Variable

- cobble stones
- exposed aggregate
- flagstones
- sand-laid brick
- wood deck
- wood disks in sand

Hard

- asphalt
- concrete
- tile/brick in concrete

Fig. 2

Comments

Soft Surface Characteristics

- IRREGULAR AND SOFT SURFACES MAKE WALKING EXTREMELY DIFFICULT FOR PEOPLE WITH MOBILITY HANDICAPS.
- POOR SURFACES FOR WHEELCHAIRS AND OTHER SMALL-WHEELED VEHICLES.
- THE BLIND HAVE DIFFICULTY WITH ORIENTATION.
- SURFACES ARE SUSCEPTIBLE TO EROSION.
- SURFACES WILL WITHSTAND ONLY LIGHT TRAFFIC.
- SURFACES ARE USEFUL FOR AREAS WHERE LIGHT PEDESTRIAN TRAFFIC WILL NEED A MODERATELY FIRM SURFACE, I.E. RECREATION AREAS, PARKS, NATURE AREAS, ETC.
- HIGH MAINTENANCE REQUIREMENTS, LOW INSTALLATION COSTS.

Variable Surface Characteristics

- IRREGULAR SURFACES AND WIDE JOINTS MAKE WALKING EXTREMELY DIFFICULT FOR PEOPLE WITH MOBILITY HANDICAPS.
- JOINTS EASILY TRAP CRUTCH AND CANE TIPS, HEELS, NARROW WHEELS; JOINTS SHOULD BE FILLED AND NO WIDER THAN ½".
- IRREGULAR SURFACES MAKE MOVEMENT DIFFICULT FOR WHEELCHAIRS AND OTHER SMALL-WHEELED VEHICLES.
- ICE AND SNOW CAN BE A PROBLEM BY DAMAGING THE SURFACE OR BEING DIFFICULT TO REMOVE.
- MODERATE MAINTENANCE REQUIREMENTS, MODERATE TO HIGH INSTALLATION COSTS.

Hard Surface Characteristics

- FIRM AND REGULAR SURFACES FOR WALKING AND MOVING WHEELED VEHICLES.
- JOINTS ARE KEPT TO A MINIMUM, LESS THAN ½" WIDE AND FILLED.
- ICE AND SNOW REMOVAL POSSIBLE WITHOUT EXTENSIVE DAMAGE TO SURFACES.
- HIGH INSTALLATION COSTS, LOWEST MAINTENANCE COSTS.

6. *Curb ramps:* Changes in grade from street to sidewalk and from sidewalk to building entrances create most problems for people with physical handicaps. To facilitate movement over low barriers, a curb ramp should be installed. Surfaces should be nonslip, but not currugated as the grooves may fill with water, freeze, and cause the ramp to become slippery.

7. *Drainage structures:* Improperly designed, constructed, or installed drainage structures may be hazardous to people who must move over them. They should be placed flush with the surface on which they occur and grates having narrow parallel bars or patterns with openings larger than ¾ in should not be used. Grates should likewise be kept clean so as not to lessen the efficiency of the overall storm system. Obviously, a surface buildup of water, especially in the winter, may present a hazard. For this reason, drainage structures should not be located between a curb ramp and the corner of a street or immediately downgrade from a curb ramp.

8. *Dimensions:* Walkway widths vary according to the amount and type of traffic using them. Walks should be a minimum of 4 ft wide, with 5 ft 6 in (6 ft preferred) being the minimum width for moderate two-way traffic.

9. *Wheel stops:* Wheel stops are necessary where wheeled vehicles may roll into a hazardous area. They should be 2 in to 3 in high, 6 in wide, and should have breaks in them every 5 to 10 ft to allow for water drainage off of the walk.

TRASH REMOVAL AND SERVICE

Each development should be equipped with or serviced by an approved means of trash collection and disposal in addition to the garbage disposal in the kitchen sink of each dwelling unit.

In apartment developments of three stories or more, an approved means of garbage collection will include central deposit points on each floor occupied by tenants. At the present time the only approved system is based on central trash compaction. In this system trash chutes are located at a central point on each floor which vent into a central trash compactor in a room on the first floor. The system design and the pick-up schedule should be scaled to accommodate at least 2 pounds of trash per occupant per day at 7.5 pounds per cubic foot before compaction. The trash compactor shall feed a dumpster or other similar transfer device which shall be stored within the building. Doors of sufficient width shall be provided to allow free access to the outside so that the trash container may be moved to a pick-up vehicle.

At this time, incinerators or the transport of raw garbage and trash in the building for conventional pick-up are unacceptable solutions.

An acceptable system will fulfill the following specifications:

1. A substantial round or rectangular vertical metal chute, not less than 24 in nominal diameter of aluminized or stainless steel with not less than #16 U.S. gauge wall thickness

2. Fire-protected vertical shaft for chute installation with sprinklers in the chute at not less than every other floor and at the top of the building

3. Bottom-hinged, safety-type, quiet self-closing hopper door of class B-UL fire rating on each floor, with nominal dimensions not less than 15 by 18 in and not greater than 18 by 18 in

4. Hopper doors on each floor located in a room or enclosure of at least 20 ft² in area separated from the other parts of the building by wall, floor, and ceiling assemblies of not less than 1-h fire rating with the top of the hopper no greater than 3 ft 6 in above the floor

5. No projections of any kind inside the free-fall area of the chute below the top hopper door, and all joints ground smooth or overlapped to avoid resistance to normal trash flow

6. Remote control washing and disinfectant sprays at top of chute

7. Chute that discharges directly into a system which automatically compacts and loads the material into containers for removal from the building

8. System should be capable of withstanding impact of dense, heavy articles without loss of function

9. Trash compaction room of 1-h fire-rated construction, protected by sprinklers, equipped with rodent protection, readily cleanable, and equipped with grate-protected floor drains

10. Bottom of trash chute shall be equipped with a normally open, self-closing fire safety door

Since compactor devices are to be used inside the buildings, loading docks will not be necessary. However, a service drive must connect the service/compactor area door with the vehicular circulation system. This drive shall be a minimum of 12 ft wide and paved to withstand use by heavy trucks. Radii and maneuvering space should permit easy access.

Fig. 1 Typical trash chute facilities for each floor.

TERRITORIALITY

The site (building, open space, circulation) should be organized to achieve a clear hierarchy of territorial order.

Each of the four realms (private, semiprivate, semipublic, and public) should be physically identifiable so that residents can know their degree of control as well as their responsibilities for the upkeep and surveillance of their environment.

PRIVACY AND TERRITORIALITY

Privacy refers to the ability of people living within their own dwellings to carry on personal, family, and other social activities under conditions which are free from intervention or observation by others. It also refers to the ability, when desired, to be free of the sights, sounds, and other stimuli of persons outside the dwelling unit. The difference between privacy and isolation is control over situational conditions and the right to relinquish privacy when desired.

Whatever the form of tenure, the degree of security and satisfaction obtained and the responsibility and caring manifested by residents can be related to the degree of real or perceived control over their environment. Territorial needs of residents require a clear indication of what areas are "ours" or "theirs." When spheres of control are clearly defined, individual and shared territorial responsibilities become more evident.

1. The public realm: streets, sidewalk, public areas to which each person has equal right, freedom and comfort in use; privacy is achieved through anonymity; no private responsibility for maintenance.

2. The semipublic realm: indoor and outdoor areas shared by residents; includes circulation, access, shared facilities, and open areas; requires clear physical statement of transition from public realm to limited shared area for security and growth of community identity; group responsibility.

3. The semiprivate realm: transition space between semipublic areas (courts, hallways) and unit (including private open space): requires physical articulation but not necessarily barrier or screen; ignoring this realm may overexpose residents to public scrutiny; individual responsibility.

4. The private realm: indoor and outdoor private space of the home; the private outdoor space should be clearly defined—barriers will prevent nonmembers of the household from trespassing; individual responsibility.

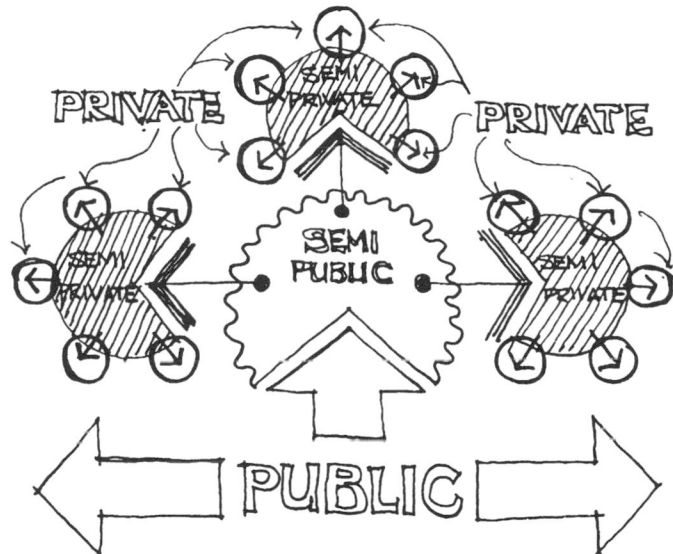

Fig. 1

Fig. 2

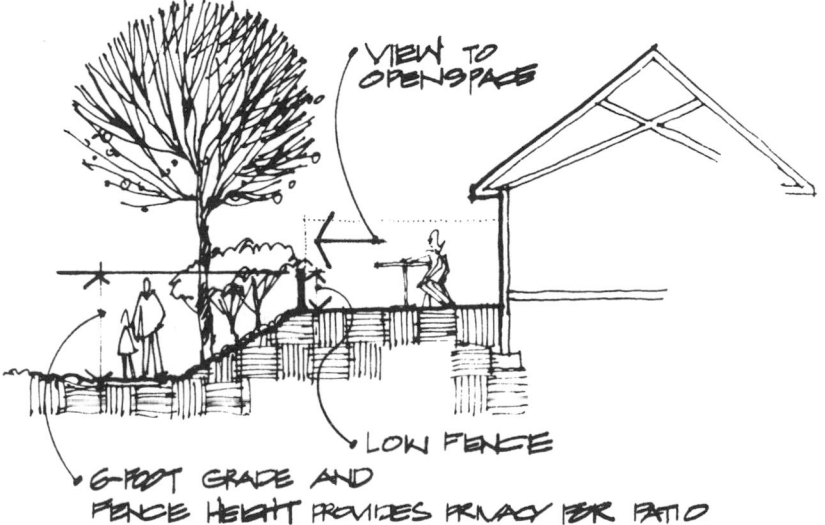

Fig. 3

TERRITORIALITY

Checklist

1. Use transitional space to make people aware of the act of leaving one territory to enter another. This is important to establishing a sense of control and a definition of responsibility.

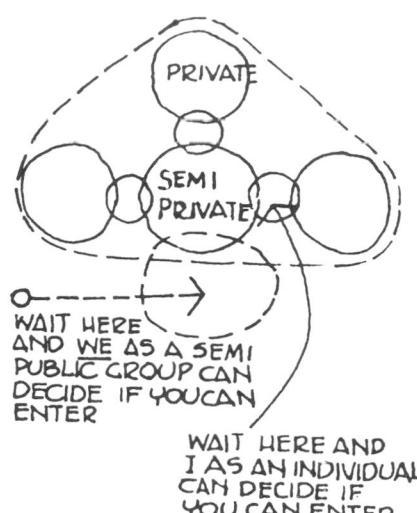

Fig. 4

2. Associate spaces that residents are to be responsible for with their private territories. Ensure that a clear understanding of shared responsibilities is built into the physical organization.

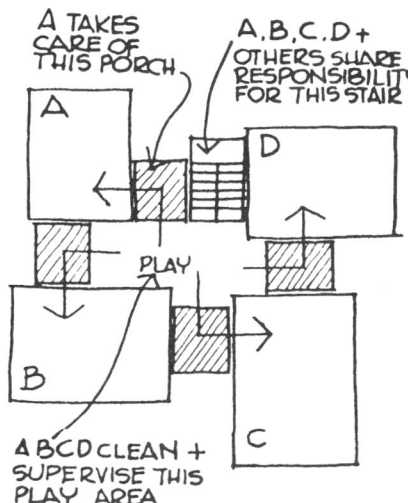

Fig. 5

3. Unit windows and doors should be separated from public or semipublic areas, e.g.:
   By semiprivate zone
   By visual screening
   By "passive" outdoor landscaped area

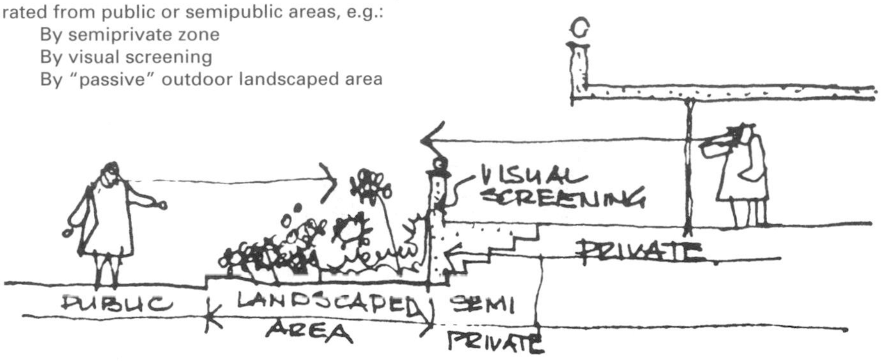

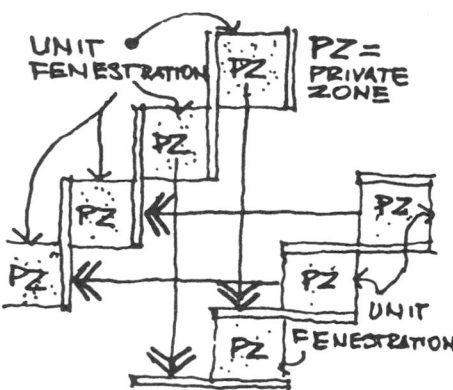

Fig. 6

4. Where space separation alone is insufficient, physical screening devices should be used, e.g.:
   Where two buildings face one another
   Where balconies or patios overlook
   Where front doors are close together

5. Buildings should be articulated to reduce opportunities for households to overlook each other's private areas.

Fig. 7a

6. Sound-compatible areas within or between dwellings should be adjoining to assist in effective noise privacy. Sound travels more easily down and sideways than up. Sounds made by people are more disturbing than sounds made by equipment (washers).

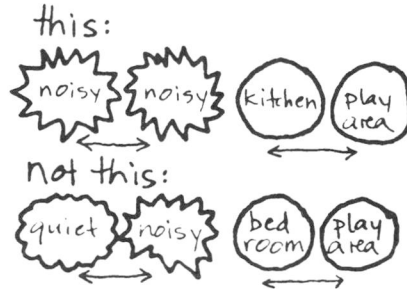

Fig. 7b

Fig. 8

Special Types of
Housing

INTRODUCTION

Physically disabled people not only need accessible places of employment and public buildings—they also need places to live. But they have found it difficult to find suitable housing. The trend toward deinstitutionalization of physically and mentally disabled people also has been severely impeded by the shortage of accessible housing.

The conventional policy toward the provision of accessible housing has been to allocate a certain percentage of units within a building or complex to the handicapped, usually 5 to 10 percent. HUD standards and those of other government agencies require that these units have additional space and design features: larger bathrooms, larger kitchens, wider doorways, grab bars, etc.

Accessible dwelling units have been provided within federally subsidized housing for many years. In the private sector a few states have required a percentage of accessible units within any rental housing or multifamily housing projects. These units are sometimes reserved for disabled people and held off the market until qualified tenants appear. Although this assures that they will actually be used for their intended purpose, there have been reports that, particularly in market-rate housing, they often remain vacant for a long time. This results in reduced profit for developers or financial liabilities for nonprofit sponsors and thus resistance toward broader policies on accessible housing.

If accessible dwelling units are not reserved, they will often be rented to ablebodied people. This leads to no increase in the availability of accessible housing, a poor use of resources, and a false perception of low demand for such units. Many disabled people do not know that this housing exists, and thus, without outreach, potential demand does not become effective demand.

Most accessible dwelling units have been built within projects for the elderly. Although there is clearly a need for these units, they do not consider the needs of younger disabled people. Similarly, within the last 10 years, there have been a number of subsidized housing projects built solely for the disabled. However, many disabled people reject such segregated housing because it isolates them from the broader community. Housing for disabled people is not often constructed for families, yet disabled people who are parents or parents of disabled children want to live among other families.

Two other problems with the conventional policy on accessible housing are that not all people who live in one dwelling may be disabled and not all disabled people have the same needs. The design of dwelling units cannot satisfy the requirements of both nonambulant and ambulant users adequately without a degree of flexibility. Needs vary even among people with the same disability. The cost of designing for the lowest common denominator can be very high. Developers ask why they have to install features that will not be used and in fact may reduce the unit's marketability.

Finally, in housing for the elderly, research has demonstrated that the incidence of disability is very high—much higher than the usual percentage of accessible units. With advancing age older people in such housing are likely to become more disabled, but the lack of vacant accessible units in their building can force them to relocate, become institutionalized, or be dependent on home-delivered services.

TYPES OF ACCESSIBLE HOUSING

When independent housing for disabled people was first initiated, specific features were recommended to provide accessibility and convenience in use by residents. Usually, the needs of wheelchair users were considered the "lowest common denominator." However, experience and research have shown that disabled people are quite varied in their needs and abilities. Even among wheelchair users, there is a great range in degree of disability. Some wheelchair users can stand to transfer, others cannot, some have good balance, others do not. Because of these differences, not all disabled people need all the features of accessibility. Moreover, some people need more physical accommodations than others.

In response to the differences in accessibility needs, the concept of *adaptability* has developed. Adaptability means design features that allow a dwelling unit to be modified to meet the needs of residents with a wide range of disabilities.

Using the adaptability concept, there can be two basic types of accessible housing. The first is called *adapted* housing. Such housing is intended to be occupied by disabled people from the initiation of building occupancy. Thus it is provided with all the necessary supports for independent living required by disabled residents. Adapted housing may be designed specifically as an alternative to institutionalization or as a means of deinstitutionalizing people such as in a group home for mentally retarded people. Usually such projects will house only disabled residents. When this is the intent of a project, the residents are likely to be much more seriously disabled than those who will live in accessible dwelling units which are only a small proportion of the total number of units in a building. The minimum requirements of accessibility should be enhanced in adapted housing, particularly in buildings housing only disabled people. Adaptability is provided by the *flexibility* to accommodate the needs of people with a wide range of disabling conditions.

The second type of accessible housing is called *adaptable* housing. Adaptable housing is used when it is not known who will live in a building before it is designed—either what kind of disabilities residents will have or whether there will be any disabled residents at all. It is designed to be basically accessible, in terms of gaining access to the building and dwelling unit and adaptable at a later date for use by people with any disability or degree of disability. This type of housing is also usable, with no inconvenience, by people who are not disabled. Unlike *adapted* housing, it would not look any different from conventional dwelling units and would not necessarily require any additional square footage beyond typical minimum sizes normally found in low-cost housing. Adaptable housing will not provide the same degree of convenience for very severely disabled people as adapted housing; however, it will be minimally accessible and usable enough for independent living.

Adaptable housing, of course, has a much lower or initial cost than adapted housing. However, the adaptable approach should not be used as a way to save money when it is clear that disabled people with serious disabilities will always be residents of the building.

In housing for older people, and, in particular, projects where a range of services such as housekeeping, meals, some health services, transportation, and social services are provided, some features of adapted housing should be provided initially, even though the full range of flexibility found in adapted units might not exist.

DESIGN ISSUES

The design of accessible housing includes the design of the site and site amenities, the dwelling unit, and common spaces used by residents of buildings or complexes in which accessible dwellings are located.

On the site, issues of barrier-free design include parking, getting from parking to the walkway system, using the walkways, and using site amenities. In common spaces of buildings, accessibility issues include using entrances, using mailboxes, laundries, elevators, and negotiating interior circulation to dwelling units. Within the dwelling unit, barrier-free design encompasses use of the kitchen and bathroom, circulation, doorways, access to storage, and alarm systems.

For design issues where circulation is the main concern, the use of wheelchairs becomes the basis for design recommendations. Generally, if sites and common interior spaces are accessible to wheelchair users, they will also be accessible and usable by people with other disabilities. In dwelling units, circulation clearances in both adaptable and adapted housing must accommodate wheelchairs from the start. Most other features in adaptable housing can be provided when necessary as long as the basic built elements have been designed to accept them.

PARKING

Severely disabled people depend heavily on personal automobile transportation for mobility, if they can afford it. There is a trend today among wheelchair users toward use of vans with automatic lifts. This equipment requires slightly more space when transferring in and out of the vehicle than an automobile. Lifts are usually side-mounted and many require space in front of the lift platform for maneuvering. Total required space for convenient use is 13 ft, although 12 ft can be used as a minimum assuming that some space within the adjacent parking bay can be used for maneuvering room. In adapted housing, spaces reserved for disabled drivers should be designed on the assumption that vans with lifts will be used. For adaptable housing, local requirements for parking are sufficient. Most local building codes require one and a half to two parking spaces per unit in multifamily housing. This allotted space is sufficient to serve a vehicle used by a person with a disability. If there are no specifically assigned handicapped parking places in a lot, one can be restriped and reserved as a 13-ft-wide space, or one-half of a space can be allocated as an access aisle next to a reserved space.

Signs reserving parking spaces for disabled drivers are essential. If there are only a few such spaces in a lot, every space should have a sign. One sign for a group of spaces is satisfactory where 50 percent or more of the spaces are accessible. Enforcement of reserved parking is critical to its usefulness; thus signs must be positioned so that they are not obscured by vehicles when they are parked in the space. In climates that have snow, signs are not effective during winter when they are lettered onto the pavement. In other climates, such signs are satisfactory but must be located in front of the parking space to be seen.

Special Types of Housing

BARRIER-FREE HOUSING

If accessible parking spaces are not on the same grade level as walkways leading to the dwelling units, then curb ramps should be provided. Two spaces can share one curb ramp if it is located in a transfer zone common to both spaces. Such an arrangement is acceptable and economical; however, it requires one driver to back into the space. Thus it cannot be used with anything but 90° parking layouts. In adaptable housing, curb ramps should be installed initially, even if reserved spaces are not created at first, since the cost of providing curb ramps is negligible within the scope of total site development costs but substantial if done at a later date. In adaptable housing, only a small percentage of spaces would need curb ramps. They should be carefully located, with a restriping plan in mind so that when reserved spaces are created the ramps will not be blocked by parked cars.

If apartments are specifically designated for handicapped people, the parking spaces serving the designated units should be as close as possible to the unit itself.

WALKS

To be accessible to disabled people, walks should be kept to a maximum running slope of 5 percent (1:20) and a cross slope of 2 percent (1:50) or less. If a section of walk must be steeper, it should be designed as a ramp. Many wheelchair users cannot negotiate the maximum slope allowed by barrier-free design building codes. In adapted housing, it is especially important that site circulation, even if there are no stairs, is still usable by severely disabled people without assistance. This assures independence in the immediate living environment. Thus, ramp slopes should be no steeper than 1:16 and preferably 1:20 or less.

SITE AMENITIES

Someone who rents or buys a unit in a multifamily complex is entitled to the full range of opportunities available to all residents. This means that wherever adaptable or adapted dwelling units are provided, at least one of each type of facility available for common use should be accessible to disabled people. This includes swimming pools, playgrounds, picnic areas, or other recreation and service facilities.

Children's play areas, for example, provide social interaction both for the children and for parents who accompany their young children while they play. Whenever adaptable or adapted housing is provided in a project, at least one play area should be accessible to the handicapped.

All accessible dwelling units in a housing facility should be connected by at least one accessible walkway to all the accessible facilities on the site. On most sites it is relatively easy to achieve full accessibility to all common use facilities. On some sites, the terrain may make it difficult to provide access to certain buildings and certain site facilities.

Steeply sloping sites are not always impossible to make accessible. In fact, sometimes the topography itself allows a greater degree of accessibility. For instance, walk-up apartments running with the contours can have entries directly on grade on two different levels. If topography does make it difficult to achieve access to all buildings in a complex, the main facilities used by the entire group of residences should be made accessible as well as the pathways leading to them. Other facilities, such as individual tot lots, that serve individual buildings or groups of dwelling units can be made

accessible on a selected basis. Accessible dwelling units can be clustered together on those parts of the site to provide more convenient use of central facilities. In some cases, recreation trails are planned to provide a challenging and physically demanding experience; these will inherently be inaccessible. Perhaps only a small segment of the trail can be made accessible; at least the disabled residents will be able to use part of it.

ENTRIES

In adapted housing, all common facilities in a building should be accessible; however, in adaptable housing, it is sufficient that only those common facilities serving the site as a whole and groups of dwelling units containing accessible units are accessible as a minimum.

Entries to buildings with accessible dwelling units should be accessible. Sometimes, site conditions allow on-grade access to first-floor dwelling units in walk-up buildings. While this is desirable, common facilities such as mailboxes and laundry and garbage rooms are often related to shared stair towers or entrances. Even in walk-up buildings, accessible dwelling units should have access to all facilities shared in common by all residents. Thus, if ground-floor accessible units do not share stair towers and building entrances, there should still be a convenient accessible route of travel from the entrances of accessible units to these common facilities.

It is preferable in walk-up buildings that the entrances to as many units as possible (e.g., all grade-level units) be made accessible so that disabled residents can visit with their neighbors and provide mutual assistance if necessary.

It is important that the entrances to accessible dwelling units be similar to those of other units. Entrance location and orientation with respect to other entrances can influence friendship and security in housing. If the entrances to accessible dwellings are isolated from those of their neighbors—for example, on another side of the building—it may be a barrier to friendship formation and reduce security for their residents.

COMMON AMENITIES

Mailboxes, garbage chutes, and storage rooms serving accessible dwellings should be designed to be usable by disabled people. In adapted housing, particularly, if all residents have disabilities, special attention should be given to ensure that all mailboxes are within reach and can be opened easily by people with grasping and reaching impairments. Locating bulk storage within the dwelling unit itself is preferable to common storage rooms in adapted housing, since many severely disabled people will have extra prosthetic equipment such as wheelchairs and crutches. They need to have these items readily available.

CIRCULATION

Within buildings, ramps are not feasible for level changes greater than a few feet because they take up so much space. Multistory housing with accessible units above or below grade more than 3 or 4 ft must be served by elevators. Elevators must be large enough to accommodate wheelchairs. Sizing of elevators includes consideration of many factors including number of stories, total number of units, and number of units per floor. While in adaptable housing, elevators with platform dimensions large enough to carry one wheelchair user with-

out any other passengers are satisfactory, in adapted housing with a large proportion of wheelchair users, elevators may have to be much larger than in conventional applications to keep waiting time within satisfactory limits. Elevators should be usable by blind people as well as people with ambulatory impairments.

Doors and hallways must have adequate clearances to allow use by people in wheelchairs. There should be an accessible path of travel from all accessible dwelling units to all accessible entries and amenities. In adapted housing, more than minimum clearance is required, since residents will often be severely limited in strength and mobility. Moreover, in interior spaces, two wheelchair users might have to pass each other quite frequently.

Standardization in the design of dwelling units is always an effective way to keep costs low. In walk-up apartments, it may sometimes prove cost-effective to make not only ground-level units adaptable but also units that are reached by stairways; they can all be designed identically. There are many disabled people including those who are deaf and blind and others with limitations of walking who can use stairs; so there is no reason to restrict allocation of adaptable dwelling units to only those on the ground level. Adapted units, of course, should only be on accessible floors.

CIRCULATION FOR WHEELCHAIRS

An accessible dwelling unit must have at least one path of travel without stairs from the main entry of the unit to at least the following rooms or spaces: kitchen, dining, bedroom, bathroom, living room, and storage. In housing for families with children, both a bedroom for parents and a bedroom for small children should be accessible. If there is only one accessible bathroom, that must be a full bathroom. It is preferable that all spaces in the dwelling unit be accessible to handicapped people, but in the interest of making as many dwelling units accessible as possible, there can be spaces other than those listed above which are not accessible.

In terms of circulation, accessibility (for both adaptable and adapted units) means that halls and doors are wide enough and have enough maneuvering room for passage by a person in a wheelchair and that there are no stairs along the path of travel to required accessible spaces. Clearances in hallways can be provided by careful design without additional space being added to conventional dwelling unit floor plans. More space may be necessary in front of doorways. However, the overall square footage of dwelling units can be kept within generally minimum size limits. In adapted units, it is advisable to increase minimum clearances for convenience and to avoid damage to walls and corners caused by wheelchairs.

Where flooring materials change, there is often a change in floor surface height. These changes should not be greater than ½ in, and edges should be beveled. It has been found that most wheelchair users can negotiate abrupt changes in height of ½ in and sometimes more. However, floor surface changes in dwelling units are often at doorways where this abrupt change in height can be extremely difficult to manage while manipulating the door. Adapted units should not have any change in floor surface heights or thresholds.

Because of the generally tight spaces in dwelling unit plans, careful attention must be given to the clearances at doors. Doorways must have at least a 32-in clear opening. Maneuvering room is necessary in front of doors if wheelchair users are to be able to reach door openers and pass through easily. Entry doors to apartments are sometimes set back in alcoves. Such doors can be extremely difficult for wheelchair users to open unless there is a clearance at the latch side of the door within the alcove. Often this latch side clearance can be created by changing the swing direction of the door or relocating it in the room plan. Bathroom doors should always open out so that if someone is injured or falls behind the door it is still possible to open the door without hurting the individual. A side benefit of the out-swinging bathroom door is that the space within the bathroom can be kept to a minimum because there is less space needed on the push side of the door than on the pull side and the door swing will not intrude upon clearances at bathroom fixtures.

Sliding doors and, in particular, pocket doors are not recommended except in situations such as mobile homes, where they might be necessary to provide accessibility within very rigid modular constraints. Sliding doors have a tendency to come off their tracks, and if someone falls against them from the inside of a room, it may be extremely difficult to open the door. If sliding or pocket doors are used, very reliable products should be specified.

Door openers should be levers or other types of opening hardware that do not require tight grasping or twisting of the wrist to operate. In adaptable housing, such equipment is not necessary on doors inside dwelling units, since those doors are typically left open for most of the day and night. Moreover, inexpensive adapters are available that can be attached to doorknobs. Entry doors to adaptable dwelling units should, however, have accessible openers. In adapted housing, all door openers should be the accessible type, since residents are much more likely to have impaired grip and difficulty using their hands in general.

KITCHENS

The kitchen is one area of the dwelling unit that should be designed for future modification depending on the needs of the individual who moves in. This is true even if it is known that disabled people will be living in the dwelling unit from the start, since people with different kinds or levels of disability have different needs.

In adaptable housing, the kitchen cabinets at the sink and the mix center area of the counter should be designed so the cabinet fronts and base can be removed to provide a clear area underneath for access by a person using a wheelchair or someone who sits while working. The countertops of these two areas should be set initially 36 in high but designed so that they can be lowered to at least 32 and 28 in. Although some accessibility codes have established a 34-in height for kitchens in "wheelchair units," this height is not really appropriate. There is no acceptable compromise height for a person working in a kitchen in a seated position. People standing need a counter height of between 35 and 36 in. When sitting down, the optimum height for most kitchen work is as close to knee height as possible. A 1½-in-thick counter positioned at about 28 in is about as low as it can be and allow the knees of a tall person to fit underneath. Some wheelchairs have detachable arms, others have a desk-type arm which allows a closer positioning to a low counter, and still others have arms that cannot be detached. When the counter is lowered to a height of 28 in, wheelchairs with fixed arms

must be positioned with the armrests behind the edge of the counter. Many wheelchair users having chairs with such arms would rather work at a counter height that is higher than optimum for working but high enough so that their wheelchair arms can slip underneath the counter itself. Thus, the 28-in height provides a low work surface and clearance for knees, while the 32-in height provides clearance for the arms of a wheelchair. Ideally, more variability in counter height between 28 and 36 in should be provided to account for differences in stature.

In adaptable kitchens, the counter fronts of the cabinets at the sink and mix center can look conventional in appearance and simply detach when accessibility is needed. The counter itself should be seamed so that if it has to be lowered, it can simply be relocated without being cut. Usually the backsplash will have to be deeper than normal to accommodate the lower counter position. Moreover, the base cabinet at either end of the adjustable counter must have enclosed sides. All these features can be provided using standard kitchen cabinet parts with the exception that the counter top itself may have to be reinforced structurally if it spans 60 in. There are several methods that can be used successfully to provide adjustable support of the countertop; these include metal L-shaped brackets that are bolted into the back wall, ledger strips at each side of the adjoining cabinet and a back wall, and construction of cabinets in sectionalized units so that bottom portions can be removed and heights added.

In the kitchens of dwelling units that are initially intended for disabled people, the base cabinets can be omitted and replaced by a full-height pantry cabinet. In adaptable dwellings it is preferable to provide a closet or other storage area conveniently located to the kitchen to make up for the cabinet space lost when base cabinets are removed. There are many simple ways to improve the accessibility of storage cabinets in kitchens for disabled people. The inside of the cabinet doors can be provided with storage racks so that when the door is opened equipment and supplies are easily accessible without reaching into the cabinet itself. Lazy Susan shelves, sliding shelves in base cabinets, and the addition of extra drawers are also useful.

In adaptable dwelling units, over-the-counter wall cabinets can be used, as in conventional kitchens. The bottom shelf of such cabinets, however, should be mounted no higher than 48 in from the floor. In dwelling units where it is known that disabled people will be residing from the start, wall cabinets can be omitted in favor of a single shelf within 48 in of the floor running above the counter area. Again, the space lost from wall cabinets can be made up by the addition of full-height storage units.

Conventional ovens and ranges are very difficult for disabled people to clean. Thus, in adaptable dwelling units the ovens should be self-cleaning or a separate cook-top and wall-hung oven should be provided. Self-cleaning ovens are initially more expensive than others, but energy savings through the increased insulation provided result in a long-term saving in operation costs. Wall-mounted ovens are easier for nonambulatory or semiambulatory people to use, since they do not require bending. They should have a place to rest heavy items while transferring them in and out of the oven. This may be the oven door itself, an adjoining counter, or a pull-out board under a side-opening oven.

The conventional minimum clearance of 42 in between base cabinets is sufficient to provide wheelchair maneuverability in the kitchen, but

only when base cabinet fronts are removed. Some codes require a 60-in turning radius in kitchens. While such additional space is helpful, it is not necessary for basic access because the space under the open counters can be used for maneuvering. In a U-shaped kitchen, however, a 60-in clearance between base cabinets is needed to provide enough space for access to the mix center, sink, and refrigerator.

Having a space next to the oven or refrigerator where a wheelchair can be pulled in is extremely advantageous in making the oven easier to use. Kitchens should be planned so that the accessible counter space is next to either the oven or the refrigerator, preferably both. If base cabinets have a large toe space where they meet the floor, extra maneuvering room for wheelchairs can be obtained.

BATHROOMS

Bathrooms are another area of the dwelling unit where adjustability to the specific needs of individuals must be taken into account, even in adapted housing. Accessible bathrooms can be provided within the constraints of a 5- by 8-ft bathroom, but only if the fixtures and doorway are located so that adequate clearances and maneuvering room for wheelchairs are available. The most important consideration in space planning of a minimal-sized accessible bathroom is that the water closet be located parallel and next to an uninterrupted wall so that a grab bar of adequate length can be installed on it. Since the tub or shower must be located on the opposite wall, the door to such bathrooms must be on the long side. Minimum-sized bathrooms are not recommended for adapted housing. More space is necessary in case a person requires assistance, for extra prosthetic devices that may be used, and generally for greater convenience on a daily basis. The height of water closets in adaptable dwelling units can be the conventional 15½ or 16 in to the top of the seat. Where it is known that disabled people will be living in the units and in housing for the elderly, a 17- to 19-in height should be used with a wall-hung fixture.

Either tubs or showers can be provided in adaptable bathrooms. Where it is known that many severely disabled people will live in a building—for example, in service-supported housing for elderly—shower stalls should be provided in all units. Although many disabled people can use bathtubs if they have seats, hand-held shower spray units, and grab bars, there are quite a few who cannot use the tub at all. Since soaking in a bathtub is a good therapy for many individuals, a tub room should be provided somewhere in the building whenever the dwelling units do not have bathrooms.

There are two types of accessible shower stalls. The small size of a 3- by 3-ft stall makes it easy for individuals to maintain their balance and to catch themselves on the opposite side if they start to fall. The other type is a 5-ft-long shower stall that takes the same amount of space as a bathtub. This stall has no advantage over the 3-ft stall unless it has no curb. Without a curb, the area within the stall provides additional maneuvering room for people who use wheelchairs and makes a minimum-sized bathroom much easier for them to use. The absence of a curb in a shower stall allows a wheelchair to be pulled into the stall while a person transfers from the wheelchair to a shower seat or a standard chair. Shower stalls of any type should not have curbs in adapted housing.

Some accessibility codes require a 4- or 5-ft-square shower stall. These sizes provide no advantage over the stalls described above. Although they may be accessible, they take more room in a bathroom than necessary and require custom-made stalls.

The 3-ft shower stall should have a folding seat installed. A seat that is in a fixed open position is not appropriate because a fixed seat that is adequately sized in a stall as small as this would be a barrier to an ambulant resident using the shower. A 5-ft-long stall has enough space to accommodate a fixed chair or bench.

If bathtubs are provided, they should have seats; a ledge at the back of the tub can be used as a seat, or portable seats can be used. The seat must be designed and attached so that it will not move as the person transfers into the tub.

Many people cannot use shower stalls or tubs unless they have grab bars. People use grab bars to maintain balance as they step or transfer into the tub and for support as they lower themselves down or pull themselves up. The best location of grab bars varies considerably from person to person. Generally they are needed on each side of tubs or showers. Many people need assistance from grab bars at water closets. Side bars must be located so that people can pull themselves forward. This means that the bar must project beyond the front of the toilet or the water closet. Other people find a bar at the back of the water closet useful. Adaptable dwelling units do not need grab bars initially; they can be installed according to individual requirements. However, reinforcement in the walls must be provided so that grab bars can be attached securely wherever they might be needed. The illustrations show where reinforcement is necessary to accommodate the needs of most individuals.

In adapted dwelling units, horizontal grab bars should be installed initially. Vertical bars and diagonal bars do not provide as much safety if a person should start to fall. A single set of horizontal bars at toilets and shower stalls is sufficient. At bathtubs, however, bars should be provided at both ends and two bars should be provided along the side of the tub. Grab bars are not needed at the seat side of a 36-in shower stall. In fact, they are a barrier to sitting on the seat. All bathtubs and shower stalls should be equipped with a hand-held shower spray.

ELECTRICAL CONTROLS

If multifamily housing has fire alarms, adaptable dwelling units should be equipped with connections and wiring necessary for the installation of an emergency alarm system. For deaf people this system should be visual or visual and vibratory. Equipment is now available that connects to the auditory emergency alarm and activates portable flashing units when the alarm rings. If a portable system is not utilized, the wiring for the emergency visual alarm system must be connected to the emergency power supply. If a deaf person should occupy an adaptable dwelling unit, the emergency light can be connected so that when the fire alarm rings, the light will flash in the individual's apartment. In apartments where it is known that deaf people will live, a visual indicator

should also be provided to substitute for doorbells. Vibratory systems are available that utilize a pad that can be placed under pillows or carpets at places where residents sit or sleep. Such systems will not be sufficient if there is not also a flashing light. Strobe lights are most effective for attracting attention.

In adaptable housing and housing designed specifically for disabled people, electrical switches, controls, and thermostats should be located at a height of 48 in and wall outlets should be located no lower than 15 in from the floor, on center.

BUILDING REGULATIONS

There is great variation in building regulations for barrier-free housing. State and federal laws and regulations vary considerably in the extent of accessibility required, from no requirements in some cases to extensive requirements, even in privately sponsored housing, in others. The individual recommendations presented here may be either less stringent than or more stringent than the regulations governing a particular project. These recommendations present a range of solutions, taking into consideration the variety of needs among disabled people and design objectives of housing that are related to accessibility.

The designer using this material is urged to check it against the regulations that apply to any specific project. The minimum conditions of accessibility here are based on ANSI A117.1 (1980), *Making Buildings and Facilities Accessible to and Usable by Physically Handicapped People*. The recommendations for other than minimum conditions go beyond ANSI A117.1 based on available findings from research and practical experience.

SUMMARY OF DESIGN CRITERIA

The features described above for adaptable housing can be summarized as follows:

1. Basic space clearances and doorway design necessary to maneuver a wheelchair through the dwelling unit.

2. Kitchen cabinetry that can be modified to provide space clearances for wheelchair and appropriate counter and shelf heights for people working in a sitting position or who use wheelchairs.

3. Structural reinforcement in bathroom walls for future installation of grab bars.

4. Electrical controls located within reach of people who use wheelchairs.

5. Wiring and necessary connections for future installation of visual emergency alarms for deaf people.

Units adapted fully for disabled people should have the same provisions because there may be a need to adjust and modify the unit to fit specific needs. In addition, they should have grab bars already installed and, if a deaf person is to use the unit, should have the visual emergency light installed. Base cabinets and wall cabinets above counters can be omitted in fully adapted units, but it is necessary to substitute full-height cabinets or pantries. In addition, space clearances should be larger for more convenience in use by severely disabled people.

MINIMUM DESIGN CRITERIA

Parking

1. Parking ratio of 1.5 to 2.0 spaces per unit.

2. Parking spaces reserved for disabled drivers: 13-ft-wide (accommodates a van with side lift) or 5-ft-wide access aisle next to 8- or 9-ft-wide space.

Shared spaces

1. Facilities used by all residents in common: all should be accessible or at least one of each type provided.

2. Facilities serving group of dwelling units: accessible if one or more units is accessible.

Circulation routes

1. Exterior: at least one accessible route from each accessible dwelling to each accessible site facility.

2. Interior: at least one accessible route from entry to accessible dwellings to all shared spaces (laundry, mailboxes).

3. Elevators: all accessible.

4. At least one accessible entry to each accessible dwelling unit.

Circulation in dwelling units

1. No stairs to reach: kitchen, one full bath, living room, dining room, the bedroom in one-bedroom apartments and two bedrooms or sleeping spaces if dwelling units have two or more bedrooms, any equipment necessary for routine maintenance, private outdoor spaces, carport, or garages.

2. Doors: 32-in clear width, lever handle on entry doors; clearances in front to allow wheelchair access, floor surface changes and thresholds less than ½ in high and beveled.

3. Halls: 36-in width minimum.

4. Room dimensions: enough space for 36-in clearance at sides of bed and in front of closet, bureau.

5. Kitchen: 42 in between base cabinets, 60 in needed in U-shaped kitchens.

Kitchens

1. Removable base and cabinet fronts at sink and mix center.

2. Sink and mix center: height of counter adjustable to 36, 32, and 28 in. from floor to top of counter; 30-in clearance width under counters.

3. Wall cabinets above counters mounted at 48 in high to top of bottom shelf.

4. Plumbing: rough-in for sink low enough to attach drain when sink is mounted at 28 in.

5. Counter: 2-in-thick maximum, including supporting structure.

6. Ovens: self-cleaning or wall-mounted; if wall-mounted, locate next to mix center.

7. Refrigerator/freezers: self-defrosting or vertical side-by-side type or have 50 percent of freezer space below 54 in.

8. Cook tops: if provided, insulate and protect underside to prevent burns or scrapes.

9. Closet or other storage area conveniently located to make up for lost cabinet space if base cabinets are removed.

10. Adapted kitchen: eliminate wall-hung cabinets and base cabinets under sink and mix

center and replace with full-height cabinet or pantry; provide shelves above counters at 48 in from floor.

Bathrooms

1. Water closets located at 18 in on center to side wall; normal height; structural reinforcement at side wall and rear wall (see illustrations); 30- by 48-in minimum clearance in front of fixture.

2. Lavatory: 32 in high, 30- by 48-in floor clearance (lavatory may project up to 19 in into clearance); clear space underneath or removable vanity cabinet.

3. Shower stall preferred over tub; 36- by 36-in or 30- by 60-in size; folding seat in 36-in size but not necessary in 60-in size; hand-held shower spray; structural reinforcement for grab bars (see illustrations), controls mounted on wall opposite seat in 36-in stall and to one side of back wall in 60-in stall; 36- by 48-in minimum floor space, 4-in curb maximum in 36-in stall, no curb in 60-in stall.

4. Tub: provide seat; structural reinforcement for grab bars (see illustrations); controls mounted near entry side of head wall; 30- by 60-in minimum clear floor space; hand-held shower spray.

5. Mirror: 40-in maximum from floor to bottom edge.

6. Faucets and other plumbing controls: single-lever type.

7. Adapted bathroom: install grab bars at shower or tub and at water closet; toilet height 17 to 19 in to top of seat in housing specifically for disabled or elderly people.

Electrical system

1. Outlets: 15-in-high minimum.

2. Visual emergency alarm: wiring and connections to install flasher unit, connected to emergency power supply and fire alarm control.

3. Thermostats: 48-in maximum mounting height.

NOTE: This section on barrier-free housing is included here courtesy of Edward Steinfeld, Associate Professor, Department of Architecture, State University of New York at Buffalo.

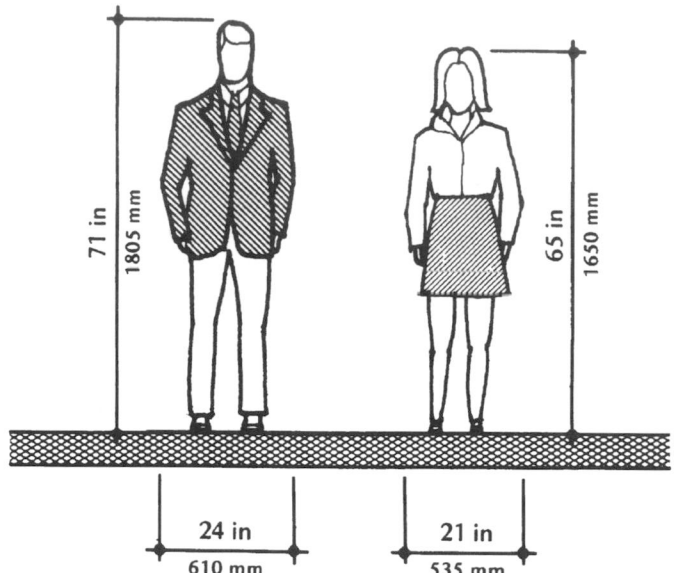

Able-Bodied Man and Woman

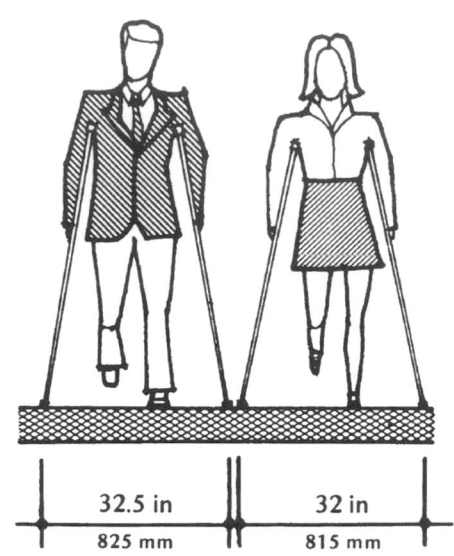

Man and Woman on Crutches

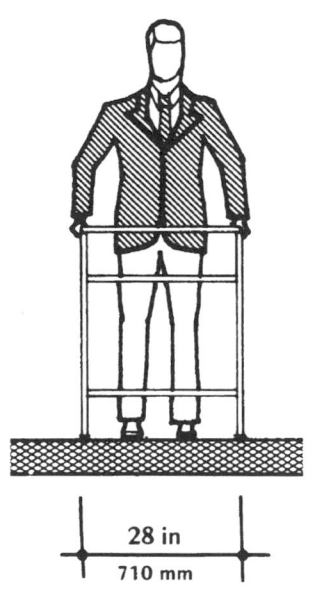

Man with Walking Aid

Fig. 1

Visually-Impaired Man with Guide Dog

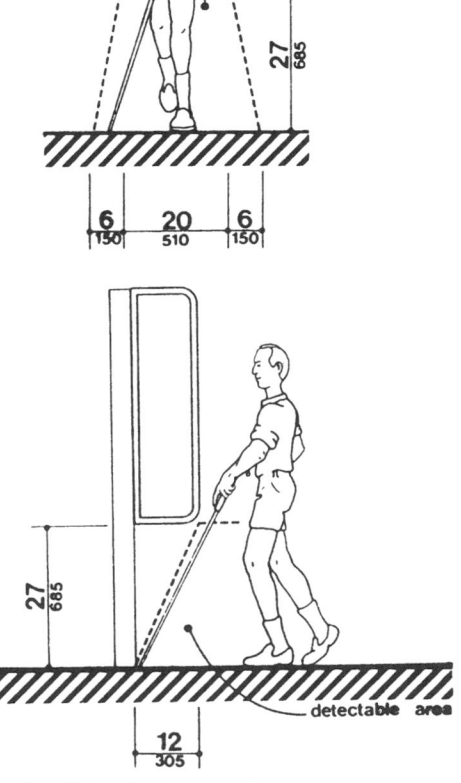

Fig. 2   Visually impaired person with long cane.

WHEELCHAIR DIMENSIONS

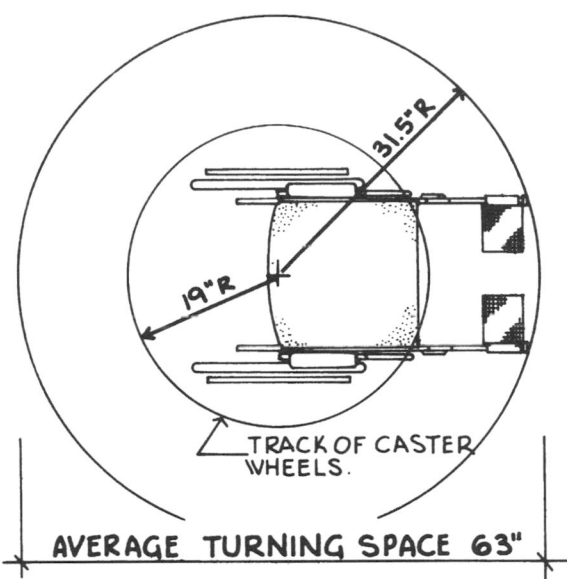

AVERAGE TURNING SPACE 63"

PIVOT POINT AT CENTER

USUAL TURNING METHOD -
MOVING ONE WHEEL FORWARD
& THE OTHER BACKWARD TO PIVOT
ABOUT CENTER.

Fig. 1   Turning radii of wheelchair.

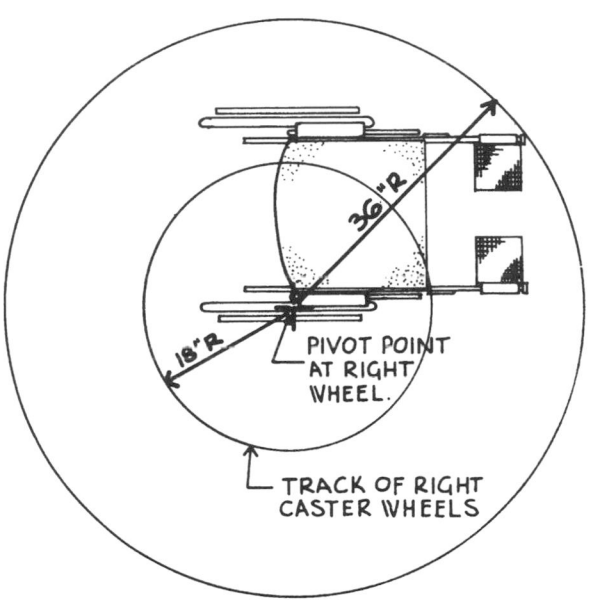

PIVOT POINT AT ONE WHEEL

ALTERNATE TURNING METHOD-
LOCKING ONE WHEEL & TURNING
THE OTHER.

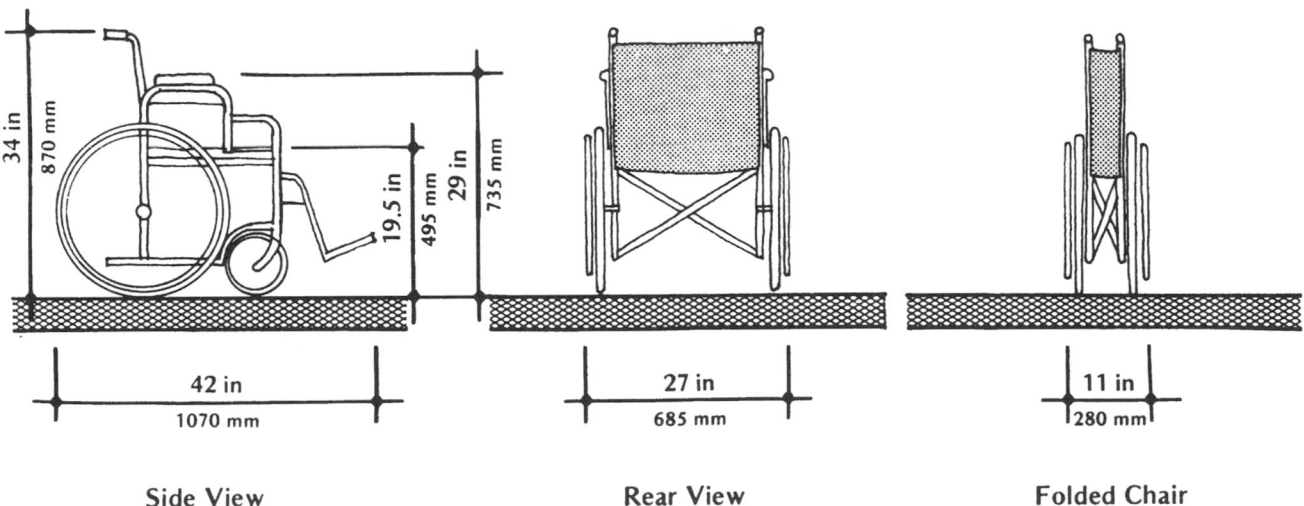

Side View	Rear View	Folded Chair

Fig. 2

Average Adult Chair Measurements (Nonmotorized)

Height: Arms from floor	29 in	(735 mm)
Height: Seat from floor	19.5 in	(495 mm)
Footrest: Min. extension	14.5 in	(365 mm)
Footrest: Max. extension	20.75 in	(525 mm)
Wheel diameter—front	8 or 5 in	(205 or 127 mm)
Wheel diameter—rear	24 in	(610 mm)

CLEARANCES

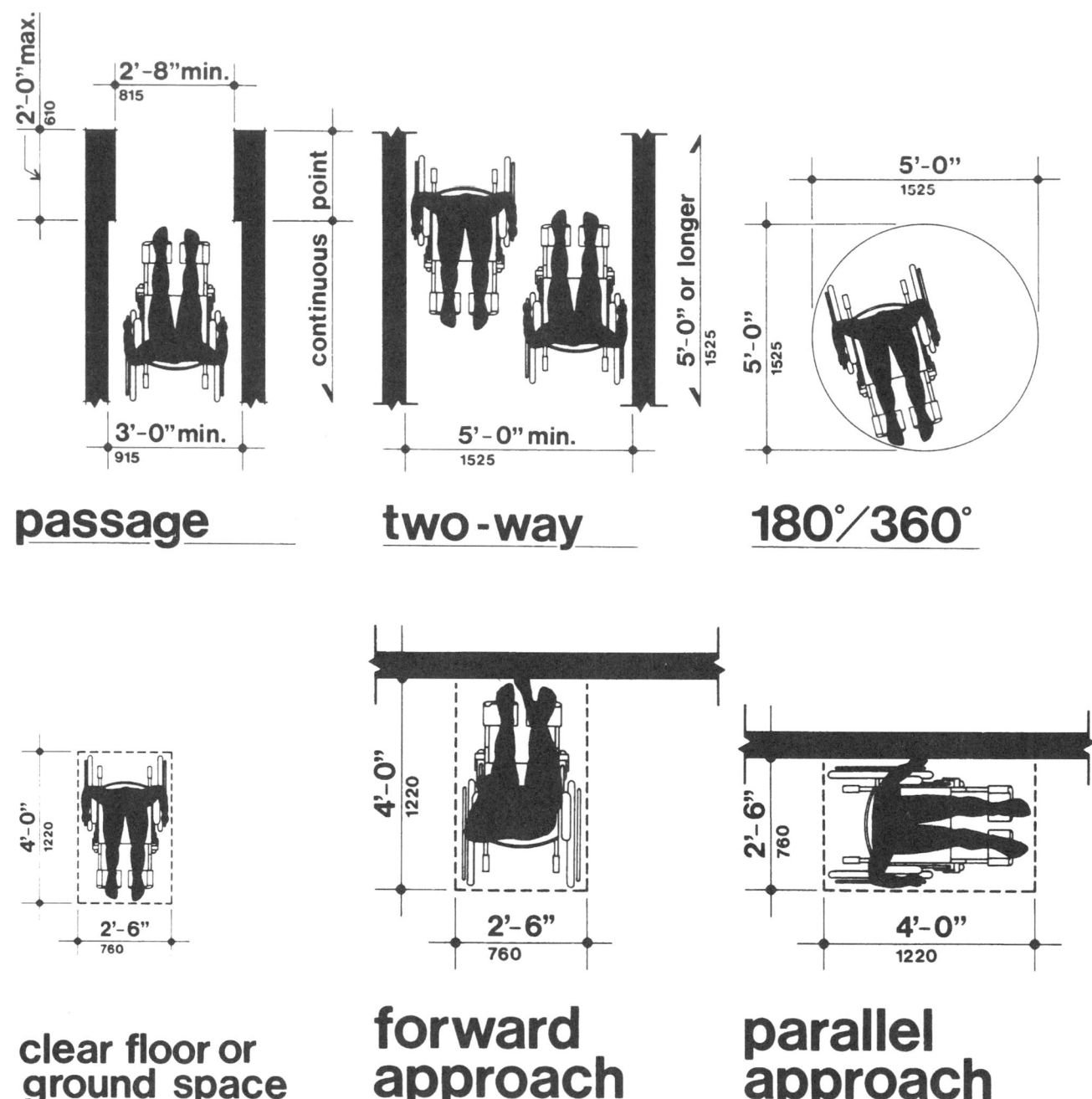

Fig. 3

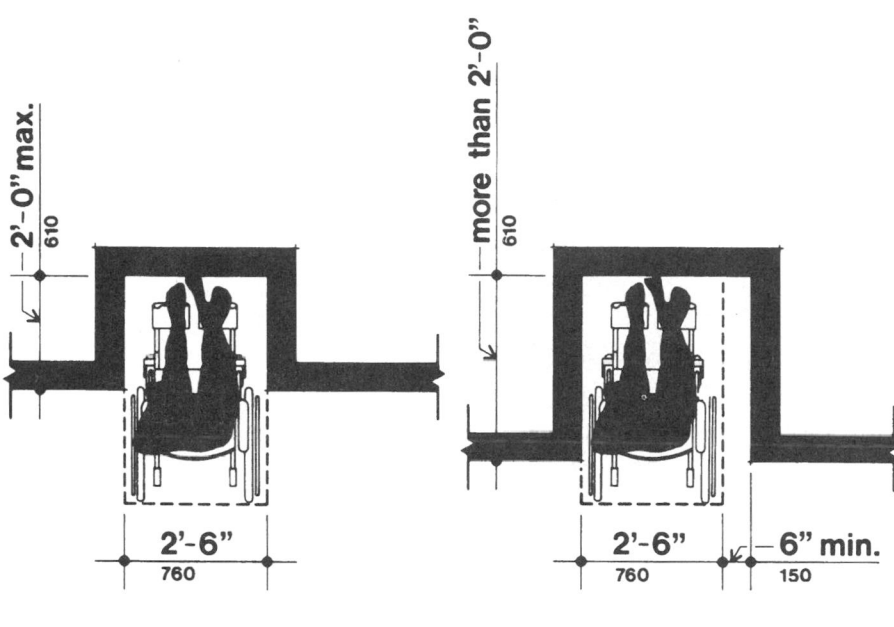

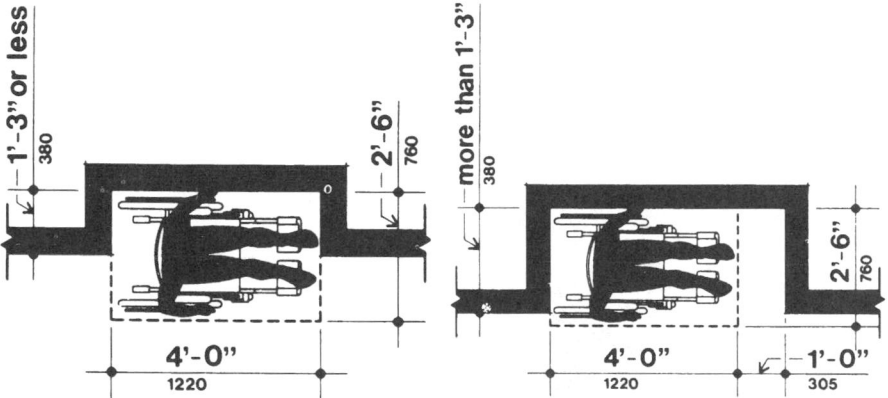

alcove

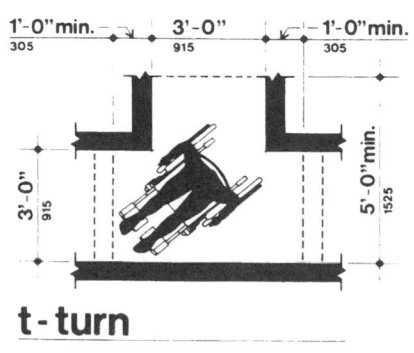

t-turn

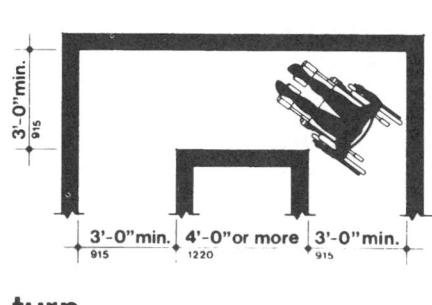

turn
maneuvering clearances

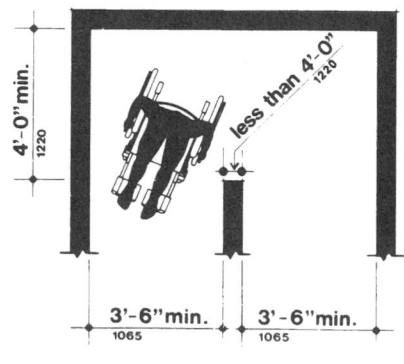

turn
maneuvering clearances

Fig. 4

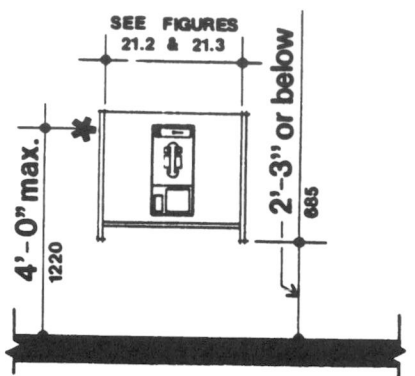

#HEIGHT TO HIGHEST OPERABLE PARTS WHICH ARE ESSENTIAL TO THE BASIC OPERATION OF THE TELEPHONE

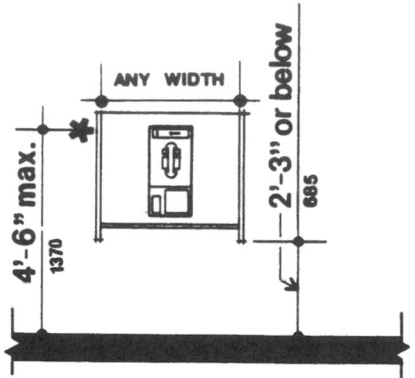

#HEIGHT TO HIGHEST OPERABLE PARTS WHICH ARE ESSENTIAL TO THE BASIC OPERATION OF THE TELEPHONE

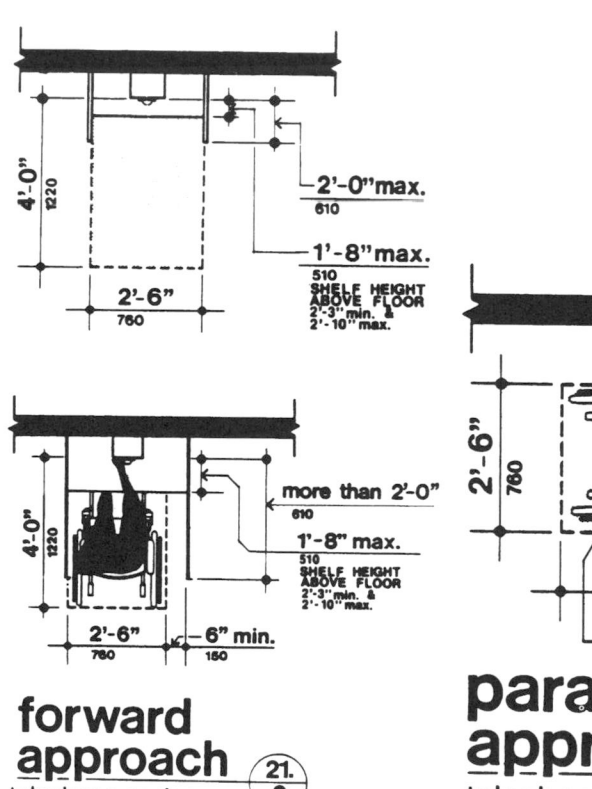

forward
approach
telephone enclosures (21. / 2)

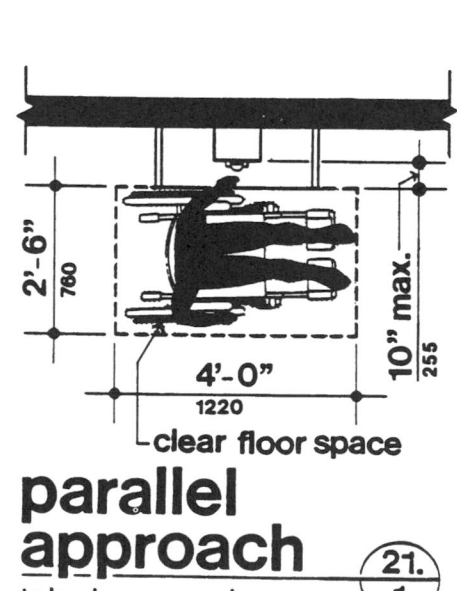

clear floor space

parallel
approach
telephone enclosures (21. / 1)

Fig. 5

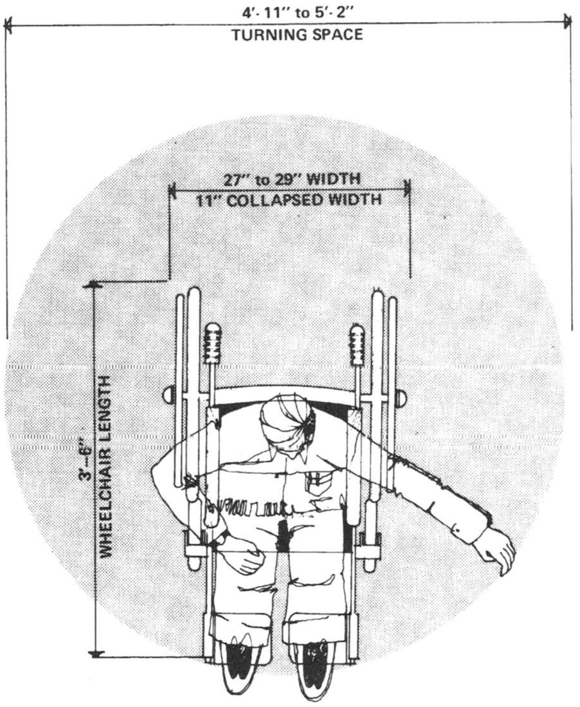

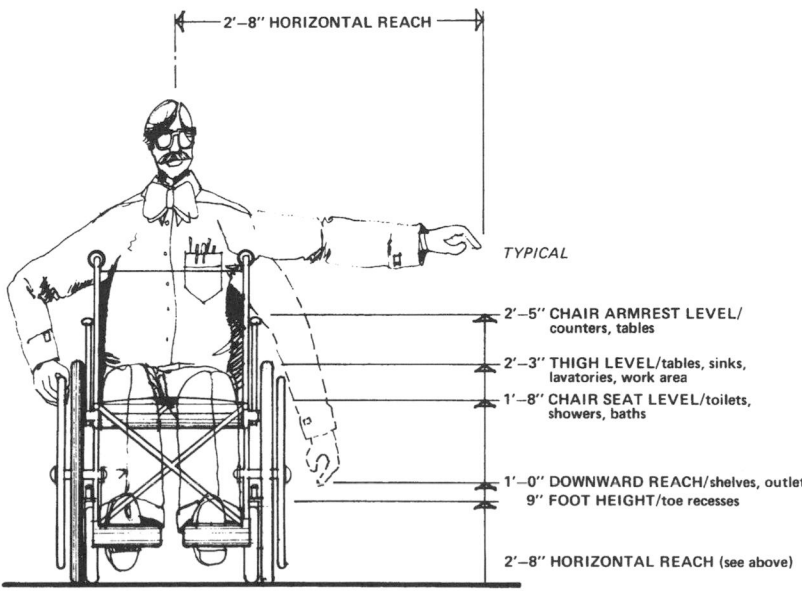

4'-11" to 5'-2"
TURNING SPACE

27" to 29" WIDTH
11" COLLAPSED WIDTH

WHEELCHAIR LENGTH
3'-6"

Fig. 1a   Wheelchair dimensions.

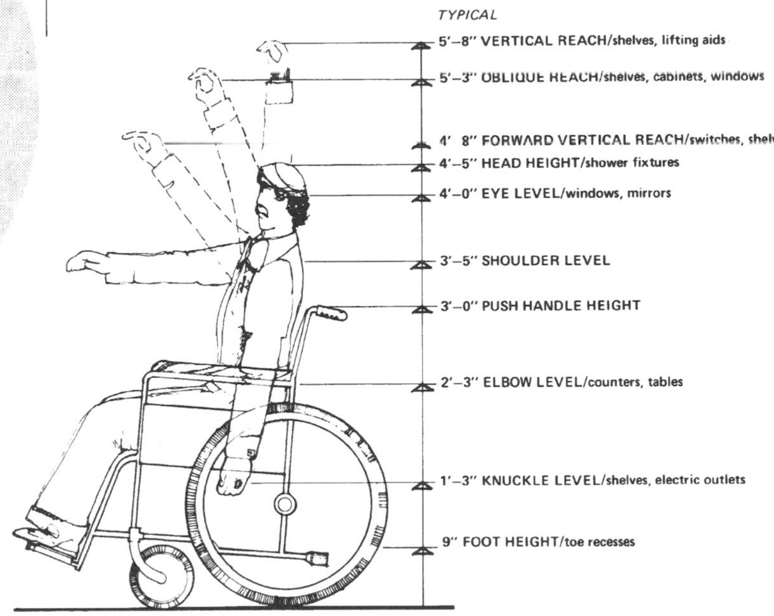

TYPICAL

5'-8" VERTICAL REACH/shelves, lifting aids

5'-3" OBLIQUE REACH/shelves, cabinets, windows

4' 8" FORWARD VERTICAL REACH/switches, shelves
4'-5" HEAD HEIGHT/shower fixtures
4'-0" EYE LEVEL/windows, mirrors

3'-5" SHOULDER LEVEL

3'-0" PUSH HANDLE HEIGHT

2'-3" ELBOW LEVEL/counters, tables

1'-3" KNUCKLE LEVEL/shelves, electric outlets

9" FOOT HEIGHT/toe recesses

Fig. 1b   Typical dimensions.

2'-8" HORIZONTAL REACH

TYPICAL

2'-5" CHAIR ARMREST LEVEL/
counters, tables

2'-3" THIGH LEVEL/tables, sinks,
lavatories, work area
1'-8" CHAIR SEAT LEVEL/toilets,
showers, baths

1'-0" DOWNWARD REACH/shelves, outlets
9" FOOT HEIGHT/toe recesses

2'-8" HORIZONTAL REACH (see above)

Fig. 1c   Typical dimensions

A clothes hanger rod at a height between 3 ft 6 in and 4 ft 0 in is adequate for most clothing and within easy reach of the wheelchair user. The maximum rod height is 4 ft 6 in. Adjustable hanger rods are often provided.

Shelves should be mounted at a maximum height of 4 ft 6 in. Shelf depths should not exceed 1 ft 4 in.

Where sliding doors are used, floor-mounted tracks or guides must not pose an obstruction to the wheelchair user.

The minimum width of all doors and openings is 3 ft 0 in. An opening of 2 ft 8 in is accept-able only in renovation work where 3-ft 0-in openings are impossible.

All doors should allow a clear area of at least 1 ft 6 in in width adjacent to the door on the side opposite the hinges. The clear area must be provided on both sides of the door. This allows the wheelchair user to back the wheelchair to open the door.

A pull handle on the trailing side of the door will enable the user to pull the door closed as he or she passes through.

Doors to bathrooms or similar confined spaces should swing out. In-swinging doors

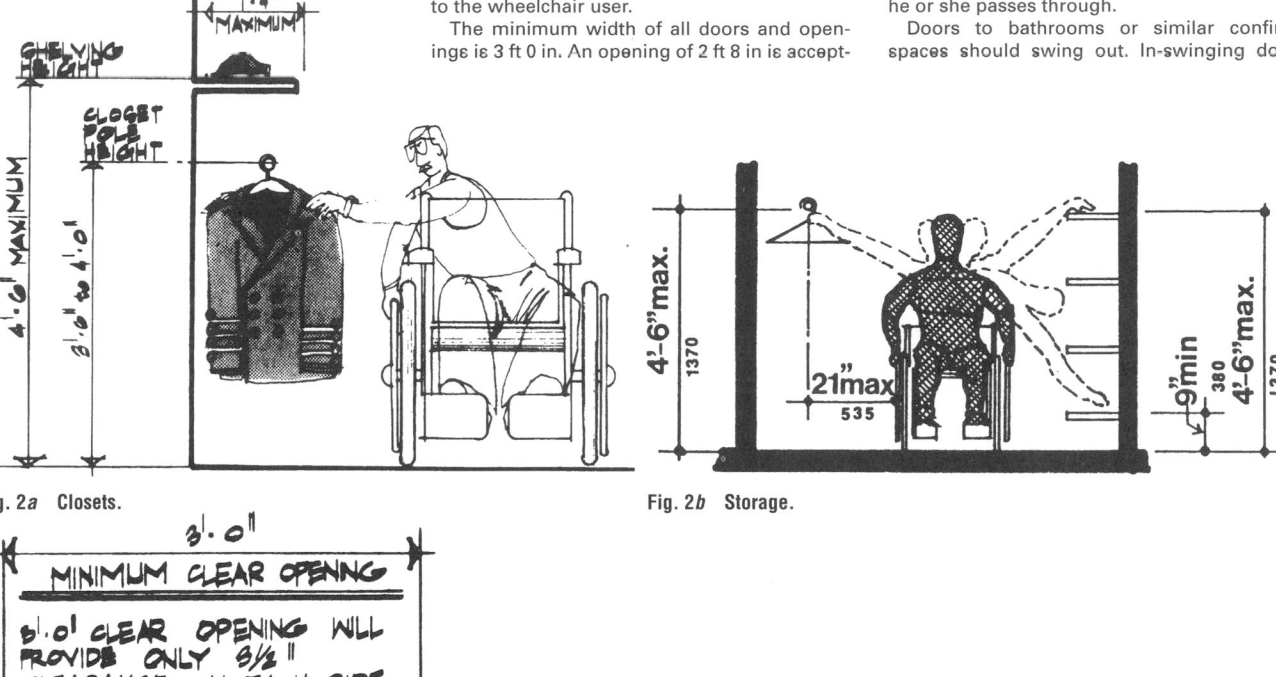

Fig. 2a Closets.

Fig. 2b Storage.

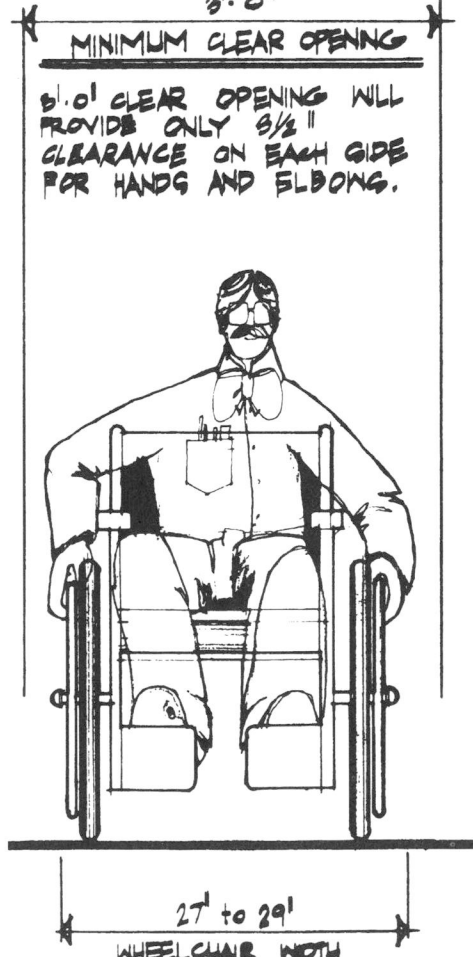

Fig. 3a Door width.

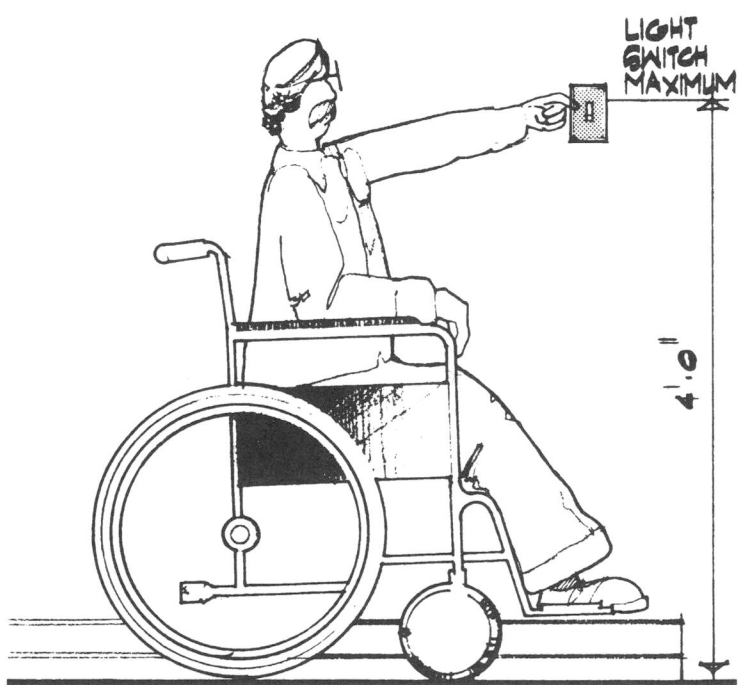

Fig. 3b Wall switches.

pose a potential danger should the wheelchair user fall and block the door. Alternatives are sliding doors or "breakaway" hardware.

All switches should operate with a single positive action. Individuals may prefer rocker to pushbutton-type switches to normal toggle switches.

Switches and thermostats should be placed at maximum height of 4 ft 0 in. The preferred height is 3 ft 0 in to 3 ft 9 in.

Switches adjacent to doors should be horizontally aligned with door handles to aid the user in locating the switch. At certain locations, light switches with locator lights are a convenience.

All switches and outlets must be conveniently located and easily accessible to the wheelchair user.

Wall-mounted outlets should be located at a maximum height of 4 ft 0 in and a minimum

height of 1 ft 6 in.

Wall-mounted telephones should be at a maximum height of 4 ft 0 in. A height of 2 ft 9 in to 3 ft 3 in is preferred. Do not mount telephones above counters which restrict access.

Telephone extensions or plug-in jack outlets should be provided at critical locations such as bedrooms and bathrooms.

Many individuals find pushbutton telephones easiest to operate.

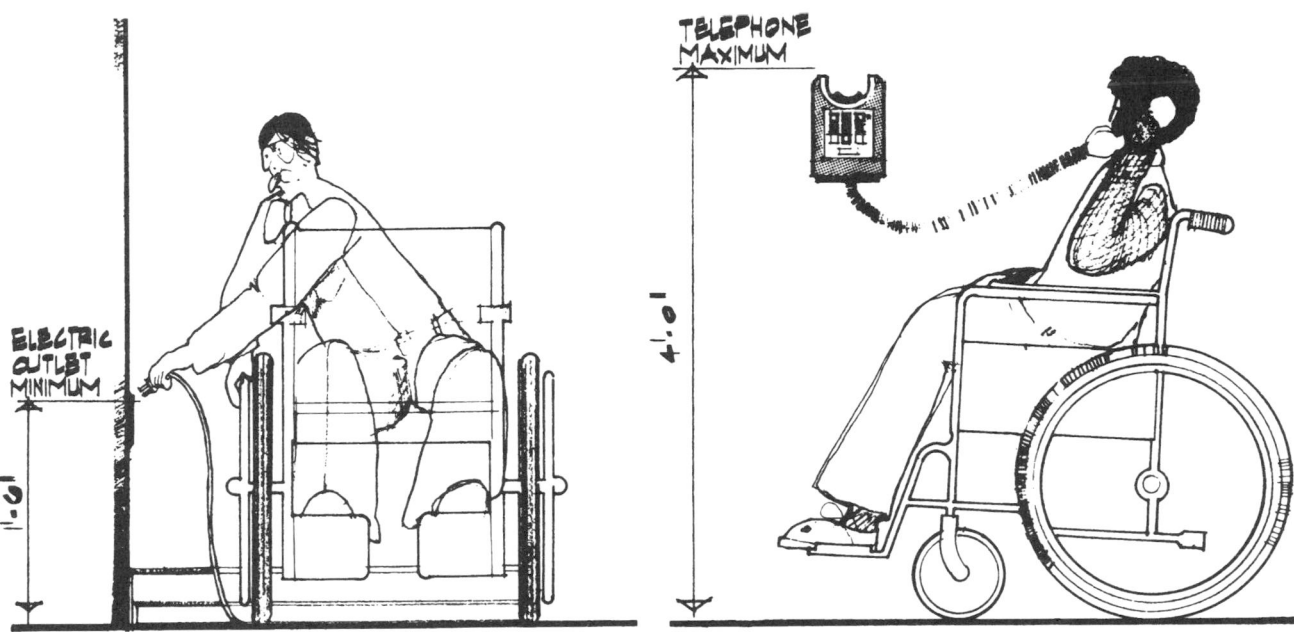

Fig. 4a Electrical outlets.

Fig. 4b Telephones.

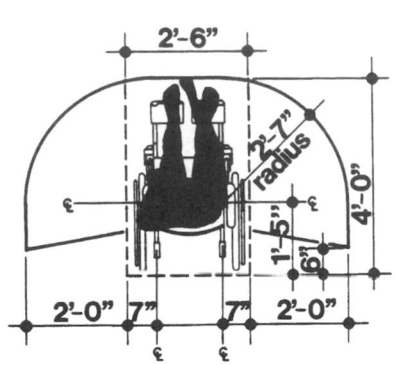

reach range
objects 4'-0" high max.

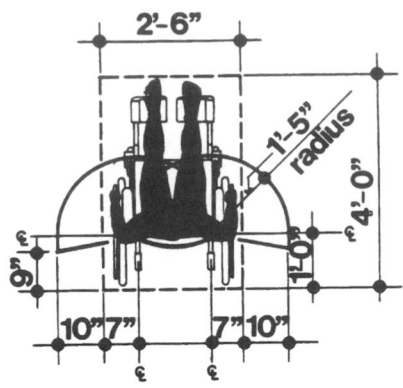

reach range
objects 4'-6" high max.

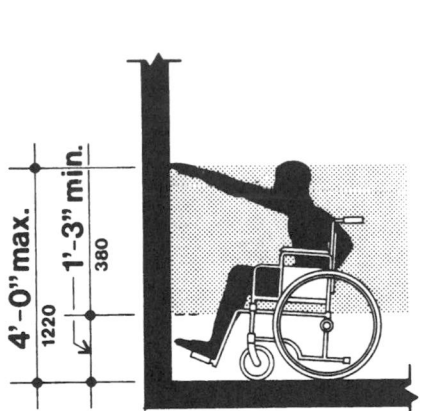

forward reach

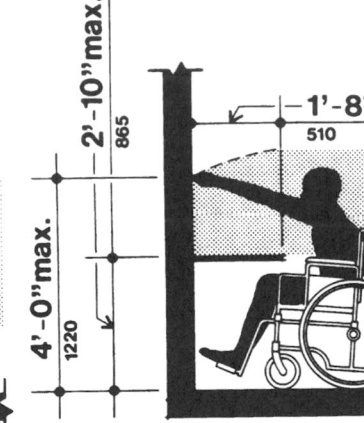

reach over obstacle
forward approach

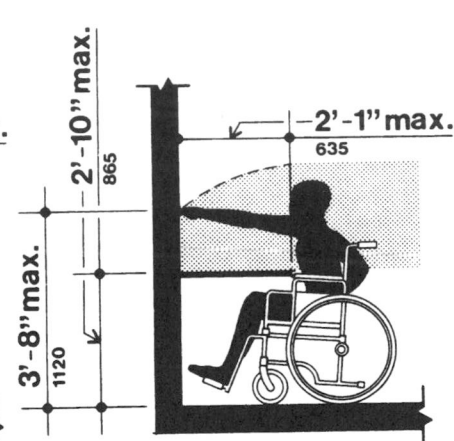

reach over obstacle
forward approach

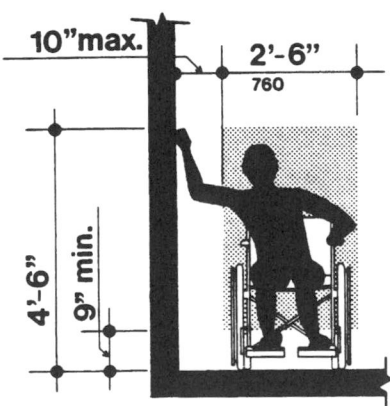

side reach
parallel approach

Fig. 5

TOILETS AND URINALS

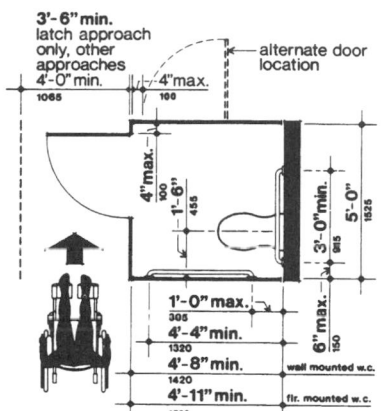

standard stall
(left-hand approach)

clear floor space
(right-hand approach)

door may swing in if additional clear
floor space is provided as shown

alternate stall - 3' wide

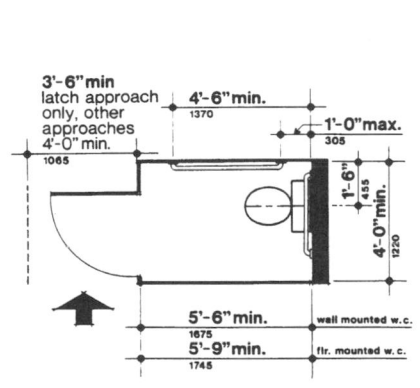

alternate stall - 4' wide (15. 8)

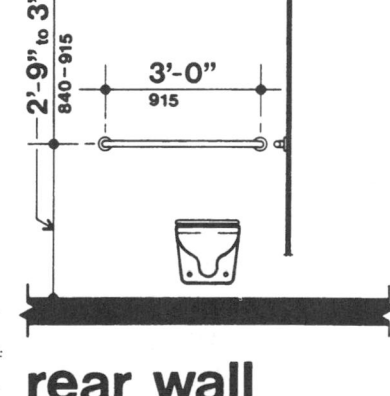

rear wall elevation
standard stall

side wall
standard stall (15. 11)

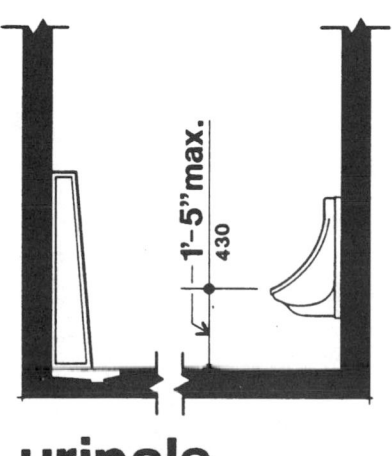

urinals

urinal shields

Fig. 6

DRINKING FOUNTAINS

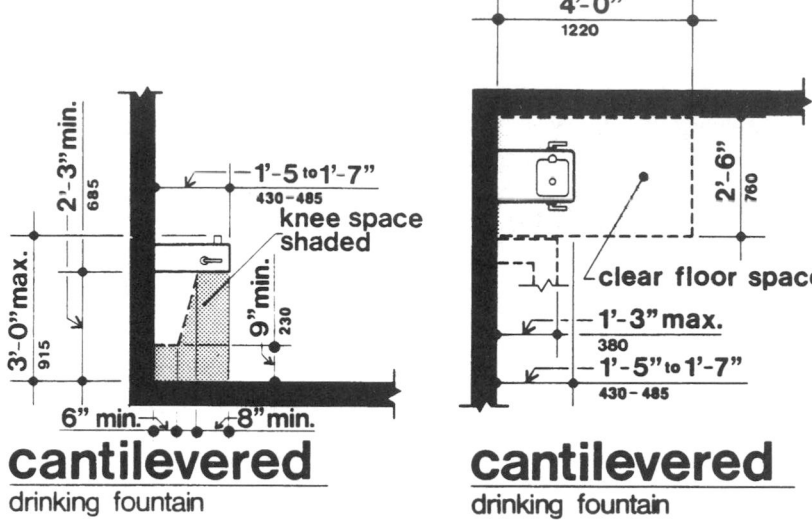

cantilevered
drinking fountain

cantilevered
drinking fountain

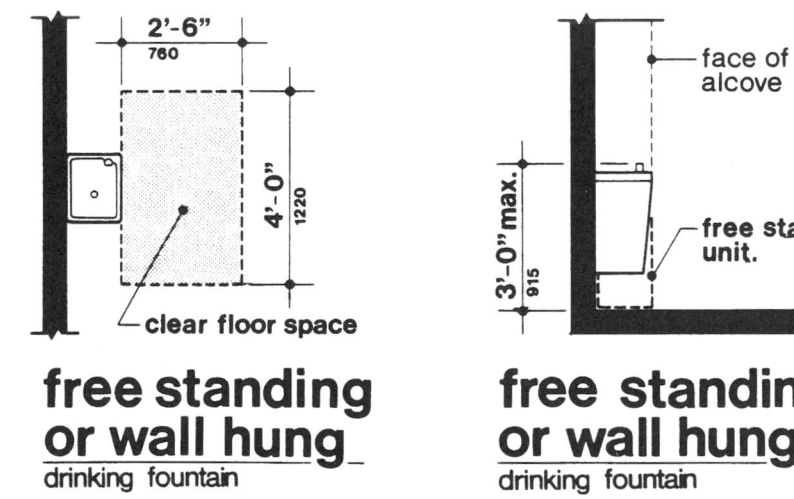

**free standing
or wall hung**
drinking fountain

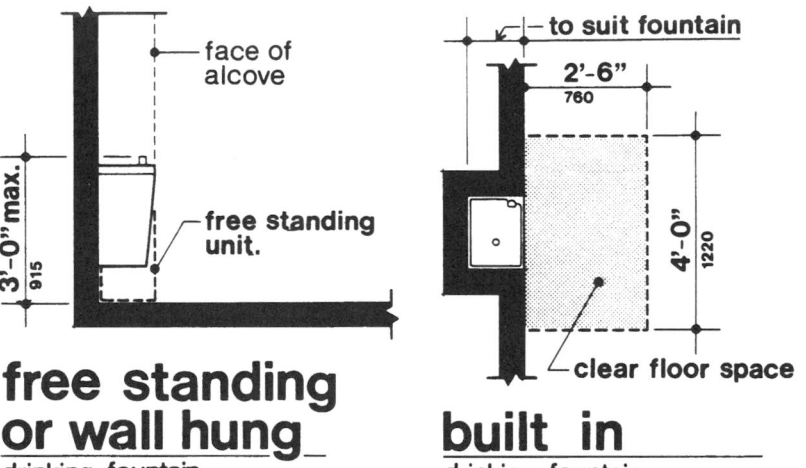

**free standing
or wall hung**
drinking fountain

built in
drinking fountain

Fig. 7

ELEVATORS

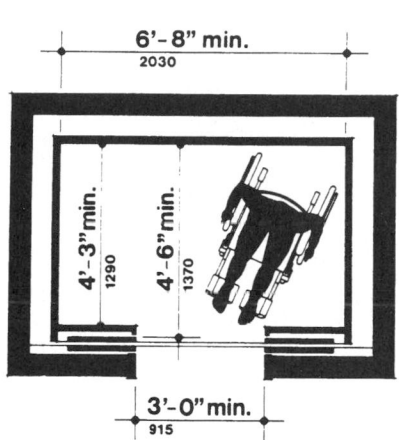

**elevator car
center opening**

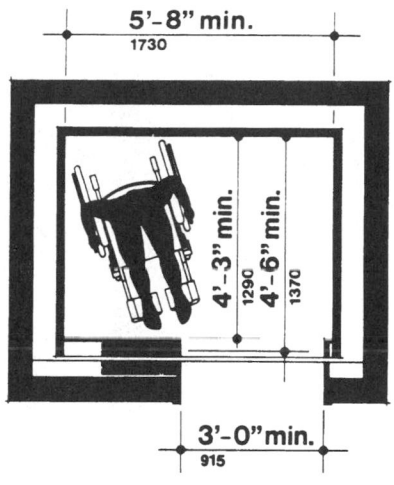

**elevator car
side opening**

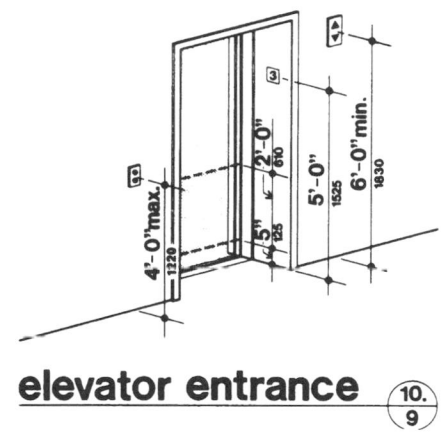

elevator entrance (10. 9)

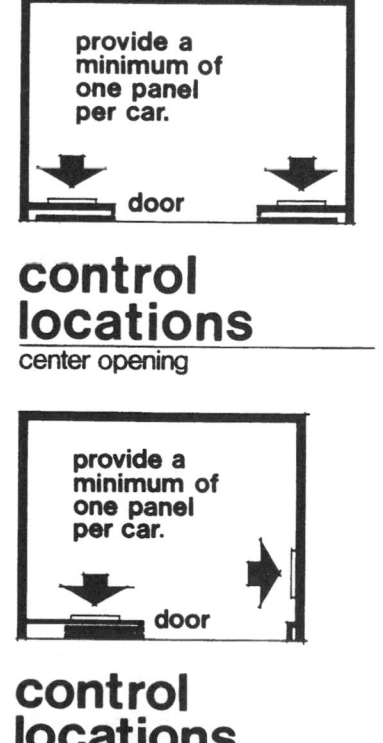

**control
locations**
center opening

**control
locations**
side opening

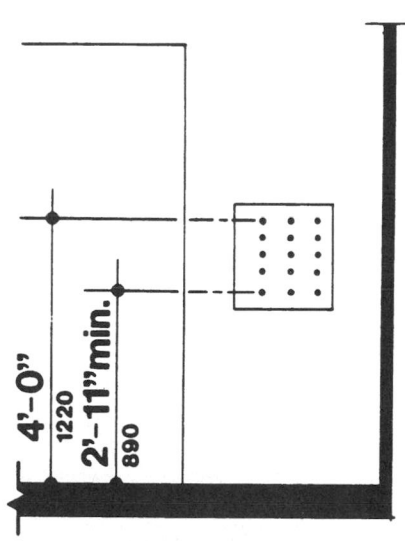

**control
panel**

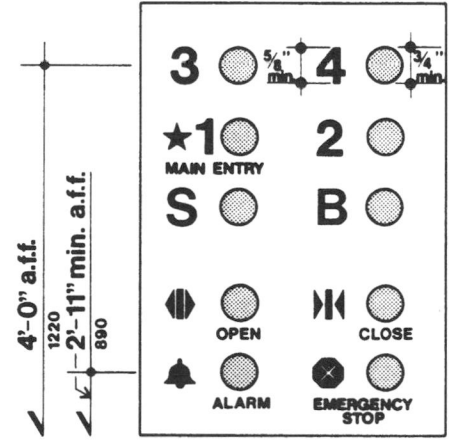

**elevator
control panel**

Fig. 8

STAIRS

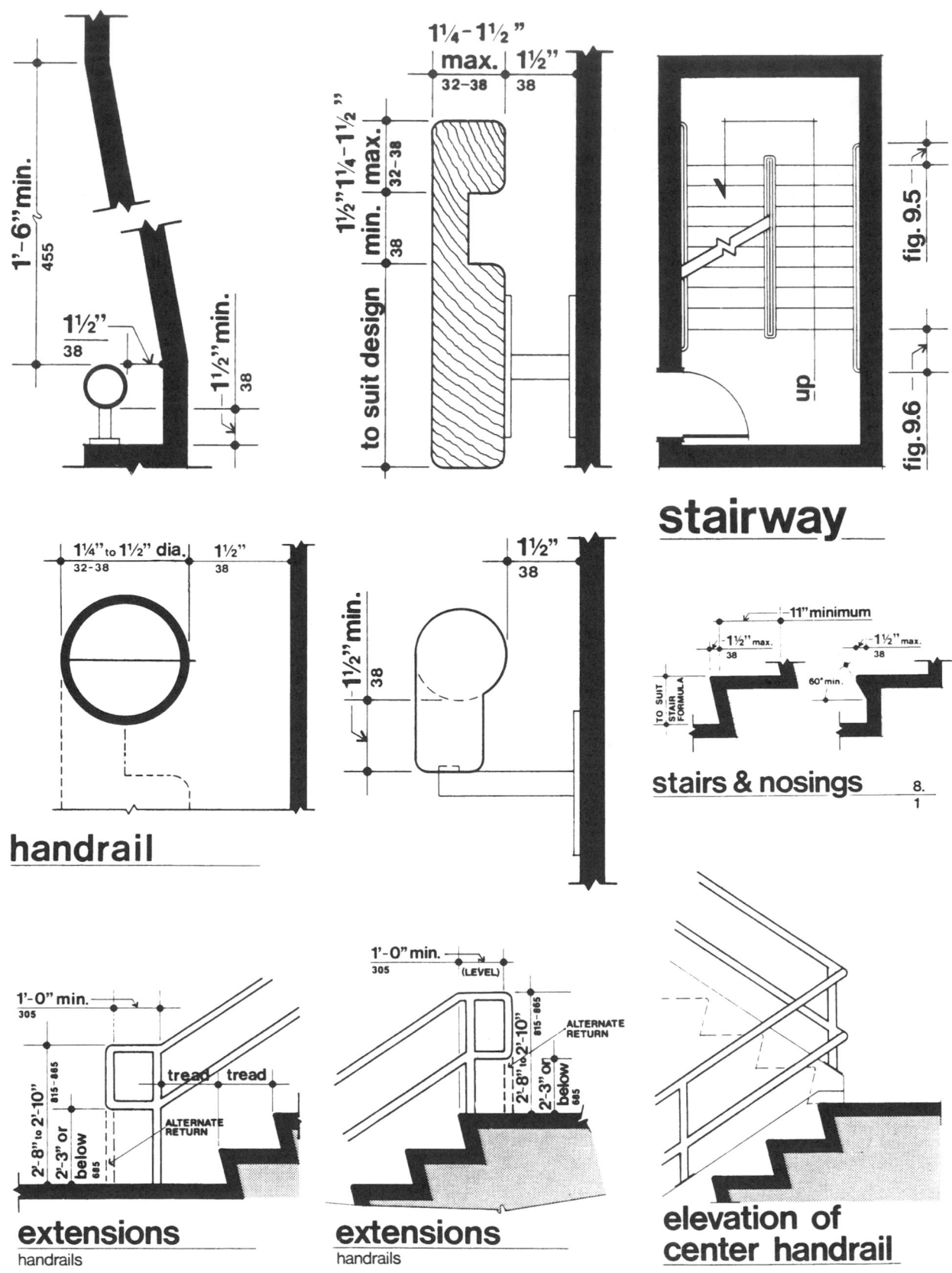

stairway

handrail

stairs & nosings

8.
1

extensions
handrails

extensions
handrails

elevation of center handrail

Fig. 9

CONVENIENCE CONTROLS

Protruding objects No protruding object should reduce the clear width of an accessible route or maneuvering space below the minimum required.

Objects mounted with their leading edge at or below 2 ft 3 in (685 mm) above the finished floor may protrude any amount (see Fig. 11).

Objects 2 ft 0 in (610 mm) long or less that are fixed to wall surfaces should not project into accessible routes more than 4 in (100 mm) if mounted with their leading edges between 2 ft 3 in and 6 ft 8 in (685 and 2030 mm) (nominal dimension) above finished floor.

Free-standing objects mounted on posts or pylons may overhang the circulation path in the direction(s) of approach a maximum of 1 ft 0 in to 6 ft 8 in (685 to 2030 mm) above ground or finished floor surface.

Objects fixed to wall surfaces may project more than 4 in (100 mm) if mounted with the lower extreme of their leading edge at or below 2 ft 3 in (685 mm) above the finished floor. These objects should not project into the required minimum clear width (see Fig. 12).

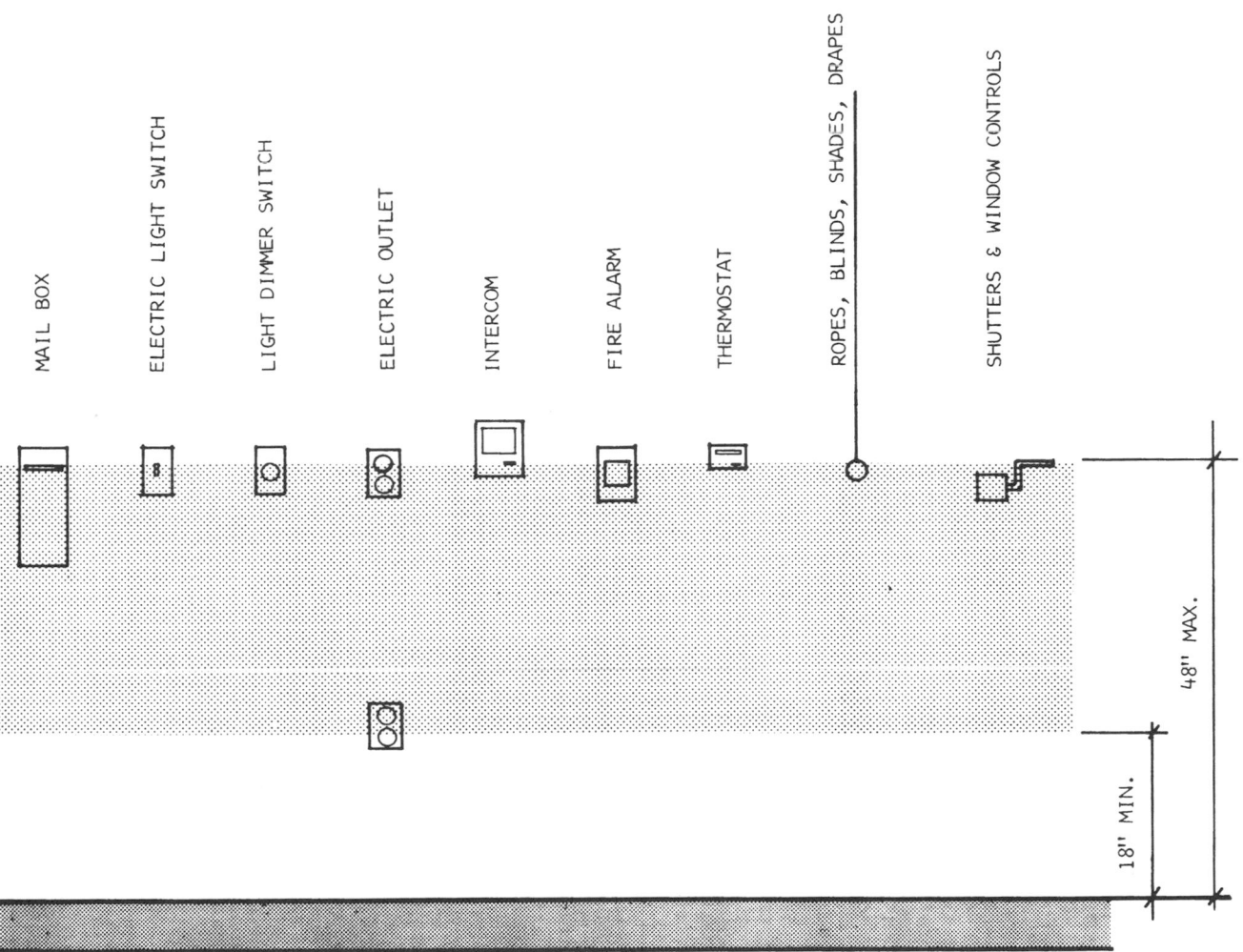

Fig. 10   Convenience controls.

WALKWAY CLEARANCES

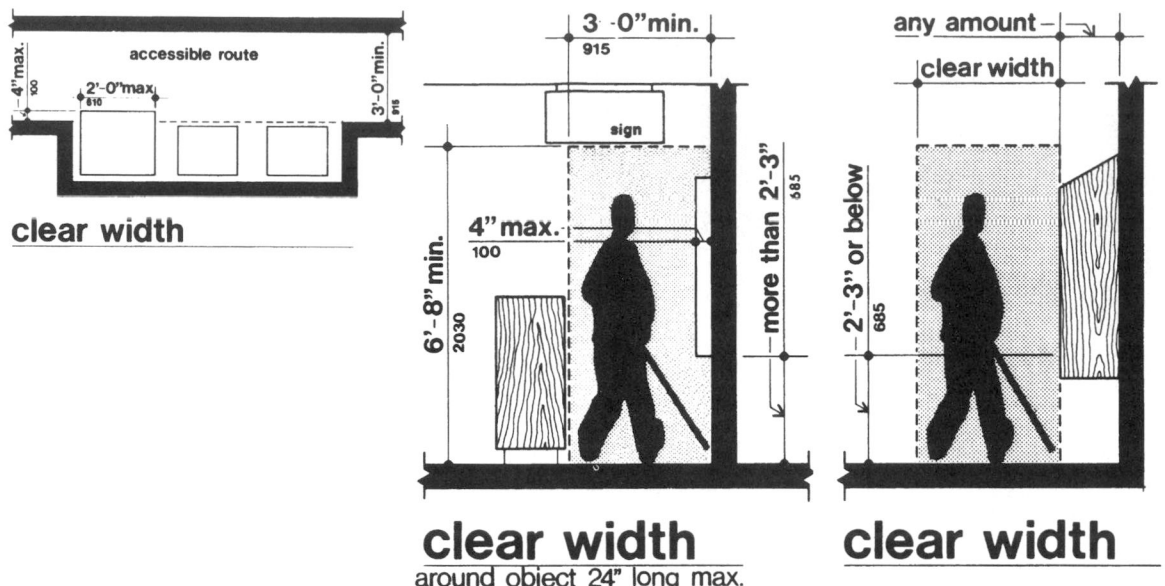

clear width

clear width
around object 24" long max.

clear width

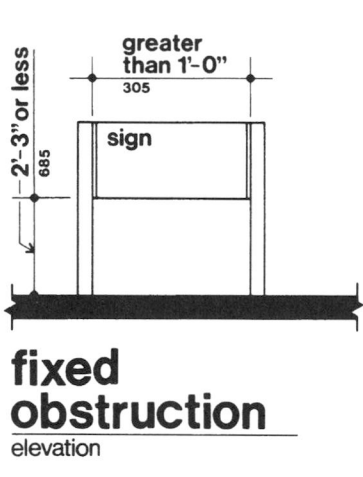

fixed obstruction
elevation

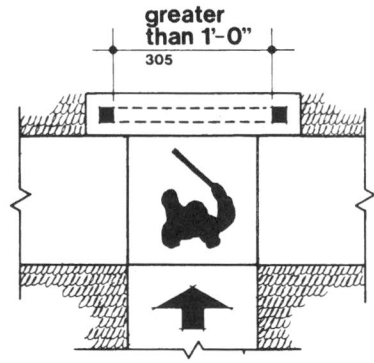

fixed obstruction
plan view

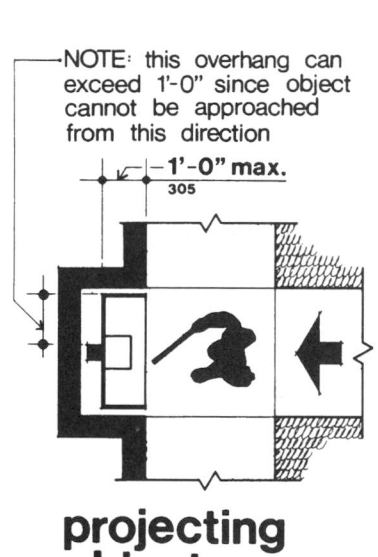

projecting object
plan view

projecting object
elevation

Fig. 11

GENERAL PLANNING

All multifamily apartment and condominium housing projects should be designed to be as barrier-free as possible, whether publicly or privately financed. If a few common sense design guidelines are followed, all units can be made accessible to most persons with mobility limitations or visual or auditory impairments. In addition, 10 percent (or at least one) of each unit type in a project should be especially designed to accommodate a person using a wheelchair. These accessible units should not be segregated. Most of these suggestions also apply to the design or modification of single-family homes for increased accessibility.

ALL UNITS

The following design recommendations apply to *all* living units in any project and are aimed at increasing accessibility for all.

Unit Entrances

Unit entrances should be unobstructed and direct. There should be no step at the doorsill, and entrance doors should be protected from extreme weather. Entrance doorways should be well lighted. Large, raised numerals are recommended for unit identification.

Hallways

The amount of hallway space in a unit should be minimized but should be at least 40 in (1015 mm) wide. Adequate lighting is essential, and three-way switches at the hallway entry and near the master bedroom are recommended.

Doors

Spatial arrangements which emphasize open areas, enclosing only bedrooms and bathrooms with doors, are recommended. All doors should provide a minimum clear width opening of 32 in (815 mm).

Living-Dining Areas

Living-dining areas should be planned for flexibility in use and furniture arrangement. Diagonal traffic paths through living rooms should be avoided. Room sizes and shapes can be found in other planning sources, such as "HUD Minimum Property Standards."[1]

Kitchens

Kitchens should be planned to maximize reachable storage space. Modern ventilation equipment makes closed-in kitchens unnecessary; pass-through openings can contribute to the suggested open plan. Appliances should be placed for work efficiency and convenience. The highest storage shelf should not be more than 56 in (1425 mm) above the floor.

Bathrooms

Ease of access and convenience for cleaning should be top priorities in bathroom design. The door should open outward to facilitate access and movement within the room. Sufficient space should be allowed and reinforcing should be provided so that grab bars can be added, as required, around water closets, tubs,

[1]HUD Minimum Property Standards, U.S. Department of Housing and Urban Development, Washington, D.C.

and shower. It is recommended that a complete bathroom include both a bathtub and a shower stall.

Conveniences and Controls

Electrical outlets should be at least 21 in (535 mm) above the floor. Wall switches should be no higher than 48 in (1220 mm). Illuminated "decorative" rocker switches are strongly recommended. Thermostats, window and drapery operating controls, wall telephones, etc., can all be located for access and use by everyone. Drapery and window blind controls should not be located in corners.

Closets

Adjustable shelving is most easily used by everyone. Shelf heights between 15 in (380 mm) and 63 in (1600 mm) above the floor are recommended. Coat closets with adjustable hanging rods and shelves are best. Closet doors should be either sliding or bifold types.

TOTALLY ACCESSIBLE UNITS

All multifamily housing projects should contain at least 10 percent (minimum—one) of the total number of units designed for total accessibility for all. The criteria below are indicative of the residential requirements of people using wheelchairs, and should be added to the suggestions in the last section for total accessibility.

Entry/Exit

The entry/exit should be level and should provide a minimum clear width of 32 in (815 mm).

Circulation

Hallways and other circulation spaces should be designed with wheelchair maneuvering in mind. An open space at least 60 in (1525 mm) by 72 in (1830 mm) should be provided at some point in the hallway to allow a person in a wheelchair sufficient space to change direction.

Travel routes from the kitchen to the dining area should be generous to allow space for serving meals.

Kitchens

The kitchen in a unit designed for a person using a wheelchair should be a finely honed complex of specialized work centers—much like a laboratory—with associated storage areas. The following recommendations will result in kitchen designs that accommodate the needs of persons in wheelchairs. (See Figs. 1 to 3.)

Mixing and preparation center

1. Recommended counter height is 34 in (870 mm) with at least 30 in (760 mm) of clear space below the countertop. Minimum knee space 30 in (760 mm) wide is suggested.
2. Mixing and preparation should take place adjacent to the refrigerator and sink, with nearby space for storage of mixing bowls, cutting board, utensils, pots and pans, electric mixer, etc. If overhead cabinets are provided, at least 18 in (460 mm) clear between countertop and bottom of cabinets, to accommodate canisters and small electric appliances, is necessary.

Cooking center

1. Countertop range units, installed 34 in (870 mm) from the floor, are recommended for

use by persons in wheelchairs. A clear space 30 in (760 mm) high and 30 in (760 mm) wide should be maintained.
2. Countertop range controls should be located at the front of the cooking surface to avoid the need to reach past hot heating elements.
3. Open counter space—perhaps the mixing and preparation center—should be located adjacent to the countertop range.
4. Wall-mounted ovens, installed with the bottom at the same level as the countertop range, can be used by most persons in wheelchairs. The oven should have a side-hinged door, with the opening side facing counter space.
5. Access to pots and pans and utensil storage should be nearby and convenient.

Clean-up center

1. Kitchen sinks should be mounted 34 in (870 mm) from the floor. Since 30 in (760 mm) of vertical clear space is required under the sink, a wide, shallow sink is indicated. The drain should be located in a rear corner, to place the garbage disposer out of the required knee space. A single-lever mixing valve and faucet, mounted to the *side* of the sink—within easy reach—is recommended.
2. Under-sink piping should be plastic, if possible. If exposed metal piping is used, both hot-water supply and drainpipes must be insulated to prevent burns.
3. Automatic dishwashers are recommended for convenience. When furnished, dishwashers should be located immediately adjacent to the sink, on the drain/disposer side.

Storage

1. Food items which do not require refrigeration can best be stored on adjustable shelving, installed between 15 in (380 mm) and 48 in (1220 mm) above the floor. Several sliding and pantry-type cabinets have been devised for use by persons in wheelchairs.
2. The best refrigerator/freezer unit for accessible kitchens is a "side-by-side" model. These units have narrow doors which do not require much clearance, and are convenient because the frozen food compartment is neither above nor below the refrigerator compartment.
3. Storage for dishes, glasses, etc., should be convenient to both dishwasher and dining room. One suggestion is a cabinet with doors on both sides, serving as a divider between kitchen and dining area.
4. Pots, pans, and cooking utensils should be stored close to the range and oven, as noted above.
5. Standard kitchen base cabinets, with swinging doors and interior shelves, do not provide adequate storage for persons using wheelchairs, because of their reaching limitations. One recommended under-counter storage system is a combination of standard drawers and pull-out storage bins or racks below. These racks or bins should be on rollers or glides so that they operate with minimum effort. Corner base cabinets should be equipped with lazy Susan revolving storage shelves.

Bedrooms

At least one bedroom in each living unit should have sufficient room for twin beds as well as adequate wheelchair maneuvering space. Easy access to the bathroom from the primary bedroom is essential.

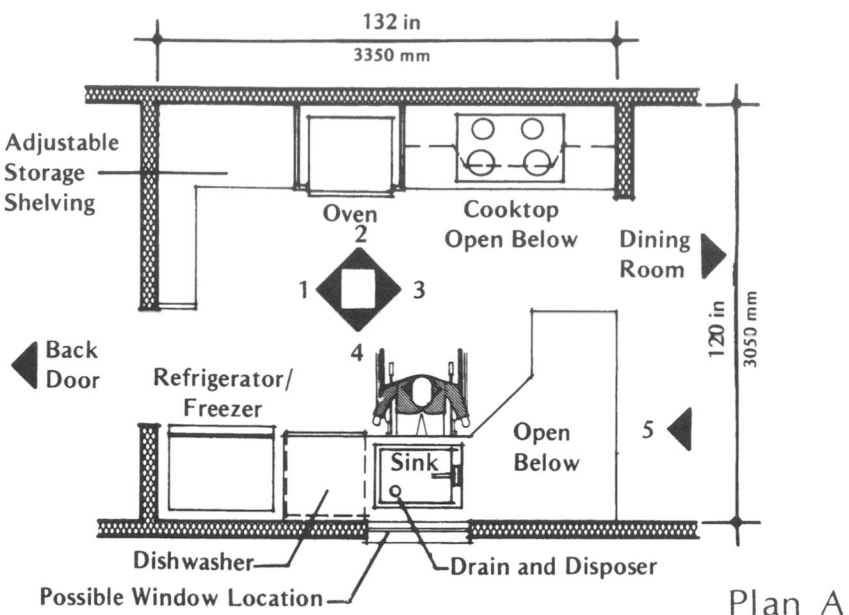

Fig. 1

Plan A

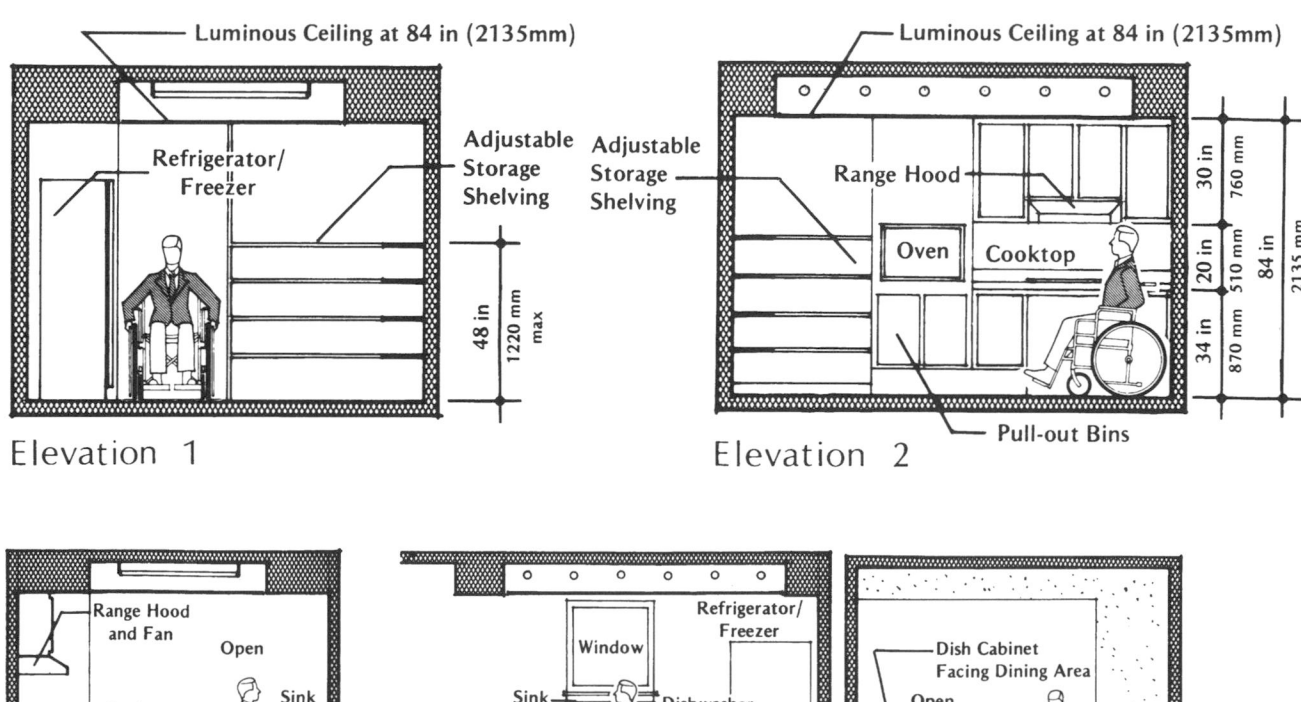

Elevation 1

Elevation 2

Elevation 3

Elevation 4

Elevation 5

Fig. 2   Toe space at all base cabinets 8 in (205 mm) high and 6 in (150 mm) deep.

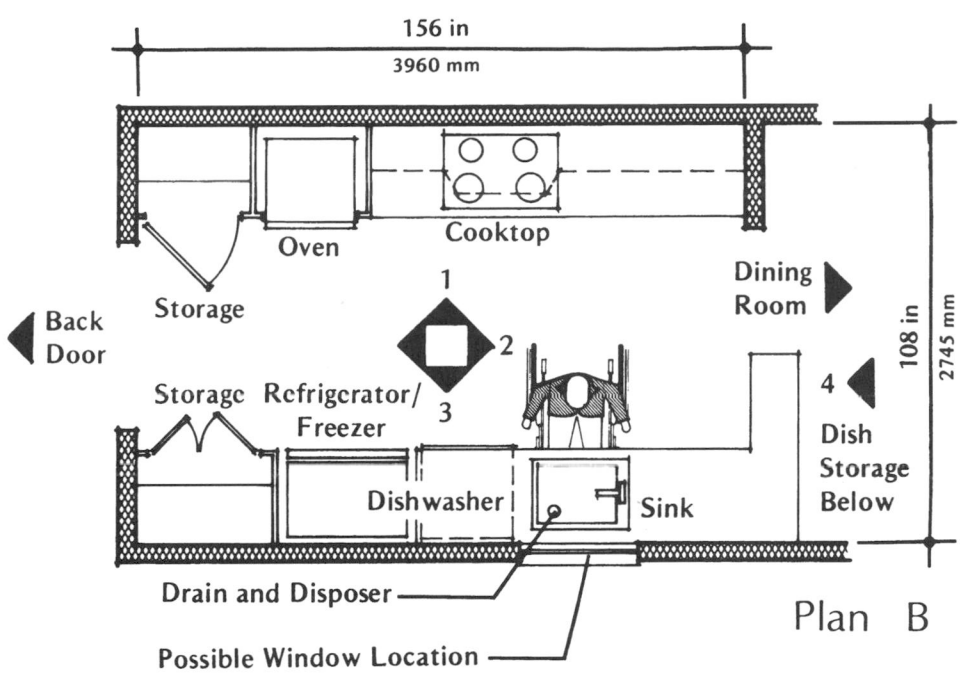

Plan B

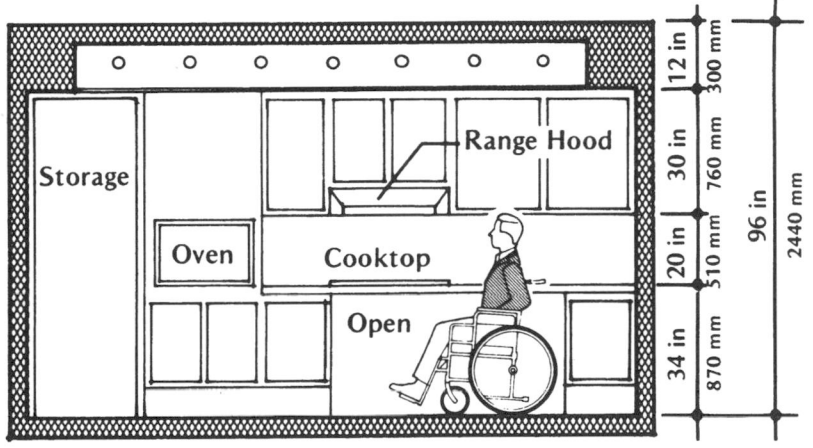

Elevation 1

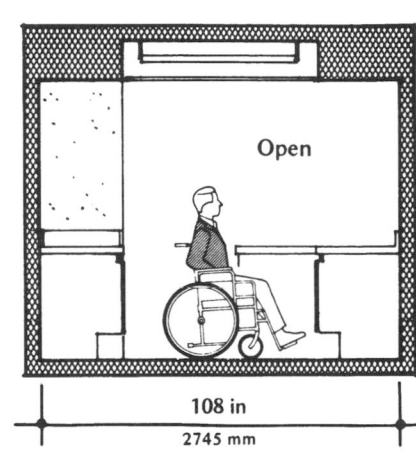

Elevation 2

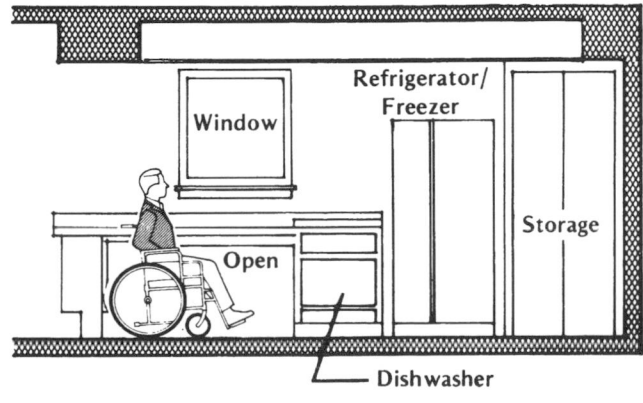

Elevation 3

Elevation 4

Fig. 3

HANDICAPPED HOUSING—GENERAL PLANNING

Bathrooms

Fully accessible bathrooms should include grab bars at the water closet, tub, and shower. Lavatories should be mounted at 34 in (870 mm), with a minimum clear knee space below the fixture 30 in (760 mm) deep and 34 in (870 mm) wide. Lavatory drain and supply piping should be either enclosed or insulated.

Adequate wheelchair maneuvering space—for entering, turning, and using the fixtures—is essential. Full-length mirrors are suggested. Medicine cabinets, towel bars, light switches, and electrical outlets should be within easy reach.

Storage

The amount of storage space required by disabled persons varies little from that needed by ablebodied people. Clothes closets with adjustable hanging rods and linen closets with adjustable shelving are recommended.

Windows

Whenever possible, single- or double-hung window sash, which require raising and lowering operations, should be avoided, since many disabled persons have difficulty using them. Hopper-type windows project inward and may create a hazard. Casement- and awning-type windows, while safe, present operational difficulties for some persons. *Horizontal rolling* windows require the least operating effort and are the safest.

Window sill heights between 28 in (710 mm) and 32 in (815 mm), which allow views for persons using wheelchairs, are recommended. Operating hardware for windows, draperies, shades, blinds, shutters, etc., should be easy to use and conveniently located. Window shades should have pulls that are long enough so that the ring is no higher than 48 in (1220 mm) above the floor when the shade is fully open.

Doors

Horizontal rolling doors or bifold doors are recommended for residential closets because they are more convenient for persons in wheelchairs than hinged swinging doors.

Floors

Floor finishes should have low maintenance requirements and carpets should be low-pile to accommodate wheelchair traffic. Color selection for resilient flooring should account for the possibility of marks left by rubber tires.

Lighting

Lighting should be controlled by accessible wall switches. Some wall outlets should be switched so that table or floor lamps can be easily used.

Telephones

Telephones are essential tools for disabled persons. Wall telephones should be mounted no higher than 48 in (1220 mm) above the floor. A bedside telephone should be provided, in case of emergency. Phones which provide a visual signal and adjustable volume controls are available for use by persons with hearing impairments.

Signals

The design of living units for disabled persons should include a full study of the various communication and warning device needs of the occupants. Provisions for *visual indicators* in telephones, doorbells, and smoke detectors are extremely important. In addition, attention should be given to the need for an individually tailored signaling system to summon assistance for disabled occupants who live alone.

Eliminate dead ends. These three words are the key to a wheelchair house principle that goes a step beyond the conventional open floor plan. It offers the ultimate in freedom of motion to the chairborne person while providing practical convenience to the rest of the family, especially to the homemaker, by cutting mileage out of the daily chores.

We call this principle the *circulating house.* It describes *any* floor plan for a one-level house designed so that one can circle the interior and visit all the most actively used rooms without backtracking. Each room that is part of the circuit must be approachable from either of two directions. Though the concept does not necessarily include bedrooms and bathrooms, even they could be included if desired.

Many wheelchair homes have been designed for open circulation because the concept is so practical and logical. Where a tract developer is constructing a number of homes in an area, one of the plans may be convertible to a circulating house by simply opening one new passageway, provided, of course, that all other requirements for accessibility are met.

A hallway in a wheelchair home should not be less than 4½ ft wide for comfortable maneuvering at closets and doorways and for easy, about-face turns. But with building costs skyrocketing from year to year, halls are an expensive luxury and, in most budgets, must be kept to a minimum. A hall leading from the entranceway, however, can be one end of the circulating plan and will serve both functional and aesthetic purposes in the overall design.

The chairborne resident should not face the frustration of being barred from any part of the house; all rooms should be accessible, even those that might be visited only rarely. Often the difference in welcome or no welcome is made by the furnishings and their placement in a small room. In a child's room, for example, situating the bed against a far wall will leave space for a visiting wheelchair, while the same bed placed in the center of the room would make visits difficult or impossible. In small rooms the windows and closets should be planned with wheelchair access in mind; this is especially important if the chairborne person is the homemaker and a parent of small children.

Three demonstration house plans are illustrated. These point up a few of the many possible specific features of a wheelchair-conscious design. Each of the three belongs to a member of the Paralyzed Veterans of America, Inc., and has been proved in service over years.

SITE CONSIDERATIONS

Since wheelchairs must be operated on relatively level hard surfaces, it is important that outdoor spaces such as patios, decks, or gardens be properly developed. For example, flower beds in raised planters may be provided for the wheelchair user with an interest in gardening. Patios or decks should be appropriately sheltered from wind and sun, and adequate lighting should be provided to allow nighttime use of such areas. Paved, hard-surfaced walks should be provided between all appropriate points.

A sloped site may often be adapted by careful design and regrading to match the elevation of the driveway and the residence. A walk or ramp must connect the driveway and the front door. The portion of the driveway adjoining this walk must be level and of adequate width and length to allow maneuvering room around the car. The private automobile is important for the independence of the wheelchair user. Garage or parking facilities

must be provided in conjunction with any type of residence.

It is recommended that specially adapted homes be single-story designs. For the wheelchair user, it is mandatory that all essential facilities be located on one level. Where level changes between rooms cannot be avoided, ramps must be provided. Individual rooms should be generous in size to provide increased circulation space. Unnecessary doors or partitions should be omitted to allow maximum freedom of movement.

Interior finishes should be carefully selected for ease of cleaning and maintenance. Special consideration should be given to floor finishes. Bathroom and shower floors must be of a nonslip material. Low-pile, high-density carpet may be installed in any appropriate location. In addition to its aesthetic qualities, carpet greatly reduces sound transmission and serves to cushion accidental falls. Loose weave or shag rugs, however, make travel difficult for ambulants and wheelchair users. If carpet pads are necessary, the pads should be thin and firm. All

carpets must be well fitted and properly secured to the floor.

Many individuals are especially vulnerable to cold and drafts. It is therefore important that the residence be well insulated and adequately weatherstripped at all doors and windows. In some instances, it may be desirable to provide zone-controlled heating to allow the user independent temperature control for the master bedroom and bath. Bathroom heat may be further supplemented by a radiant-heat lamp.

The bedroom should allow at least one clear area for maneuvering with a minimum diameter of 5 ft 0 in. Ideally, such an area should be provided in front of all bedroom closets.

A clear area with a minimum width of 3 ft 0 in must be provided on at least one side of the bed. This space will allow the user to position the wheelchair for transfer to the bed. A similar clear area is desirable on the opposite side of the bed to allow the user to make up the bed. A passageway 4 ft 0 in in width should be provided between the end of the bed and the opposite wall.

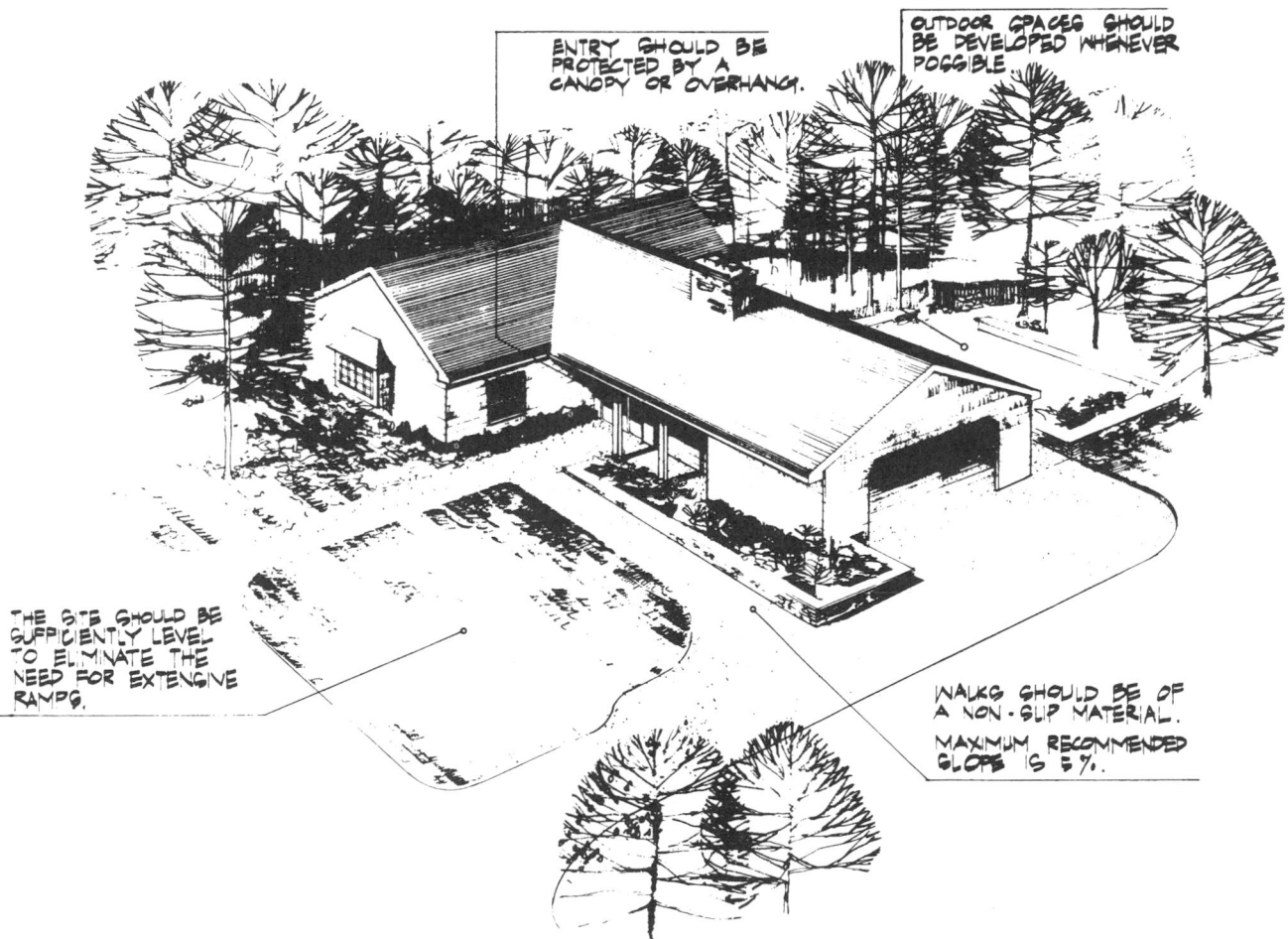

Fig. 4   Site considerations.

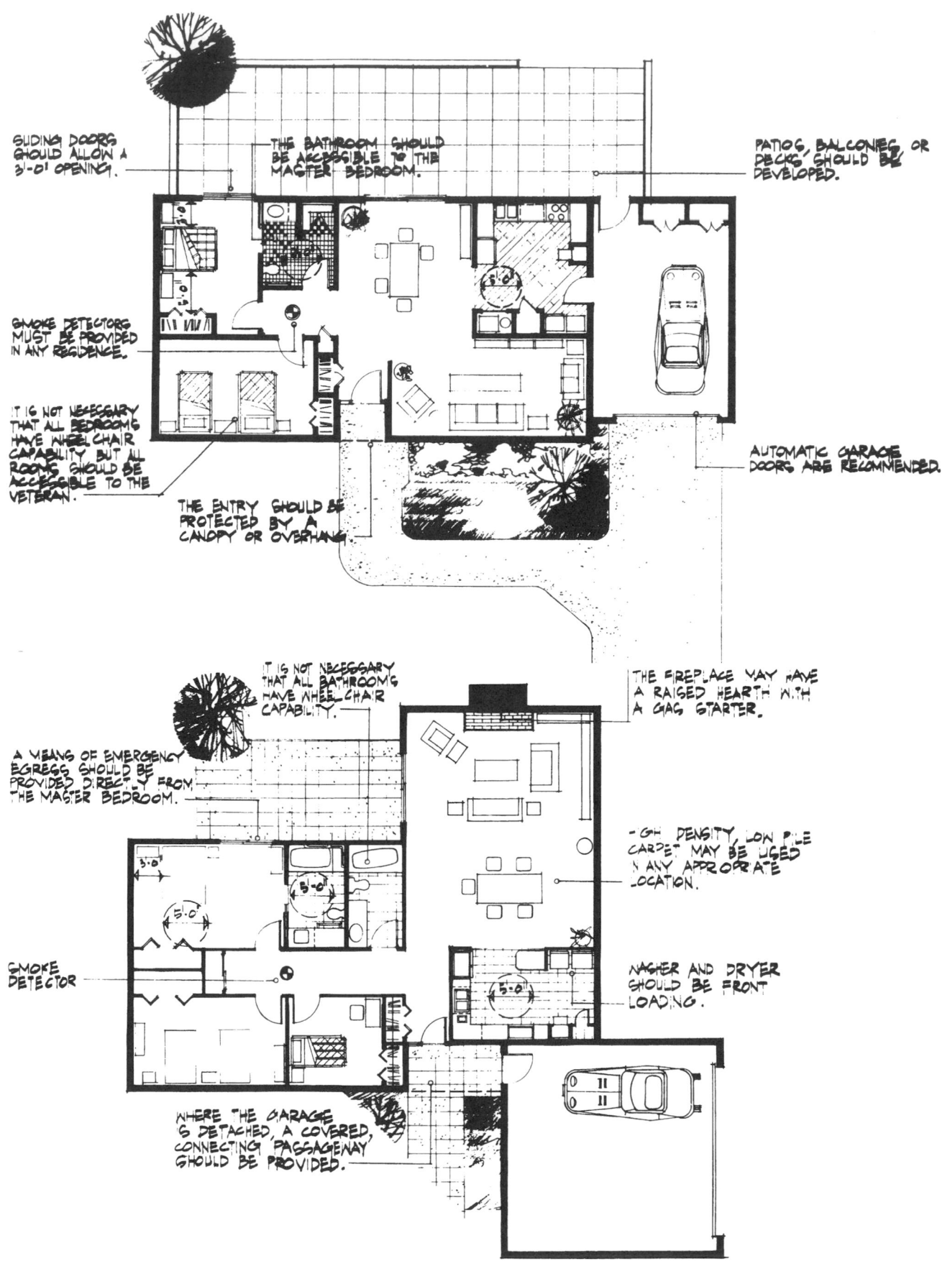

Fig. 5 Typical plan.

ADAPTABLE DWELLING UNITS

1. *Definition:* An adaptable dwelling unit is initially accessible to disabled people in terms of entry and circulation and is adaptable, through minor renovations and additions, to use as a residence by one or more severely disabled people.

Housing designed to be accessible to and usable by physically disabled people has typically been constructed with all the necessary equipment and design features already in place. This approach has serious limitations in both cost and marketability. A definition of accessible housing that is based upon the provision of adaptable dwelling units can overcome these limitations. An adaptable dwelling unit would have the features described below.

2. *Circulation:* At least one path of travel without stairs should be provided from the main entry of the unit to all rooms and spaces necessary for cooking, eating, sleeping, personal hygiene, storage, leisure, and child rearing.

3. *Kitchens:* Kitchens should allow sufficient space for maneuvering wheelchairs, provide storage units within reach and alternate working heights at counter tops and fixtures.

4. *Bathrooms:* Bathrooms should have enough space for maneuvering wheelchairs, provisions for future installations of supports needed to transfer to and from toilets, bathtubs, and showers, and fixtures at comfortable heights for use.

5. *Controls, operable hardware, and telephones:* All controls, operable hardware, and telephones should be located within reach of a person who uses a wheelchair.

6. *Closets and bulk storage:* Bulk storage areas and 50 percent of all storage volume in the dwelling unit should be within reach of a wheelchair user.

7. *Alarm systems:* If audible emergency alarm systems are provided, the unit should be equipped with all necessary wiring and connections to install a visual emergency alarm system.

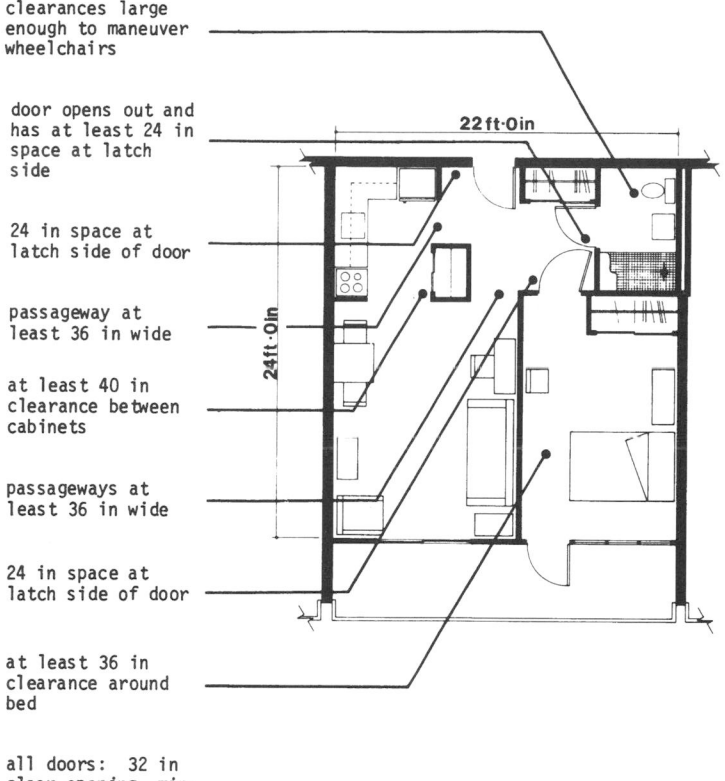

clearances large enough to maneuver wheelchairs

door opens out and has at least 24 in space at latch side

24 in space at latch side of door

passageway at least 36 in wide

at least 40 in clearance between cabinets

passageways at least 36 in wide

24 in space at latch side of door

at least 36 in clearance around bed

all doors: 32 in clear opening, minimum

Fig. 1   Single-bedroom apartment—adaptable dwelling unit.

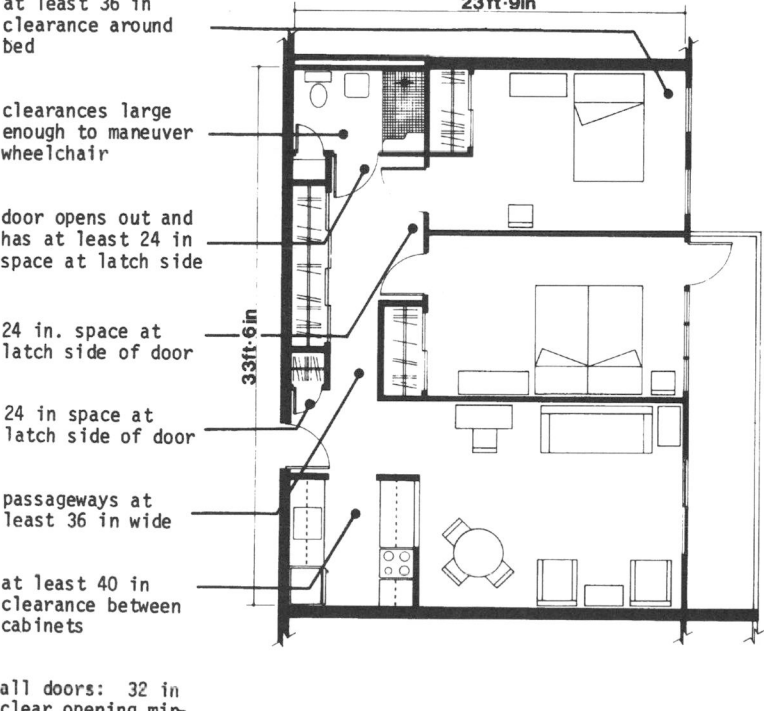

at least 36 in clearance around bed

clearances large enough to maneuver wheelchair

door opens out and has at least 24 in space at latch side

24 in. space at latch side of door

24 in space at latch side of door

passageways at least 36 in wide

at least 40 in clearance between cabinets

all doors: 32 in clear opening minimum

Fig. 2   Two-bedroom apartment—adaptable dwelling unit.

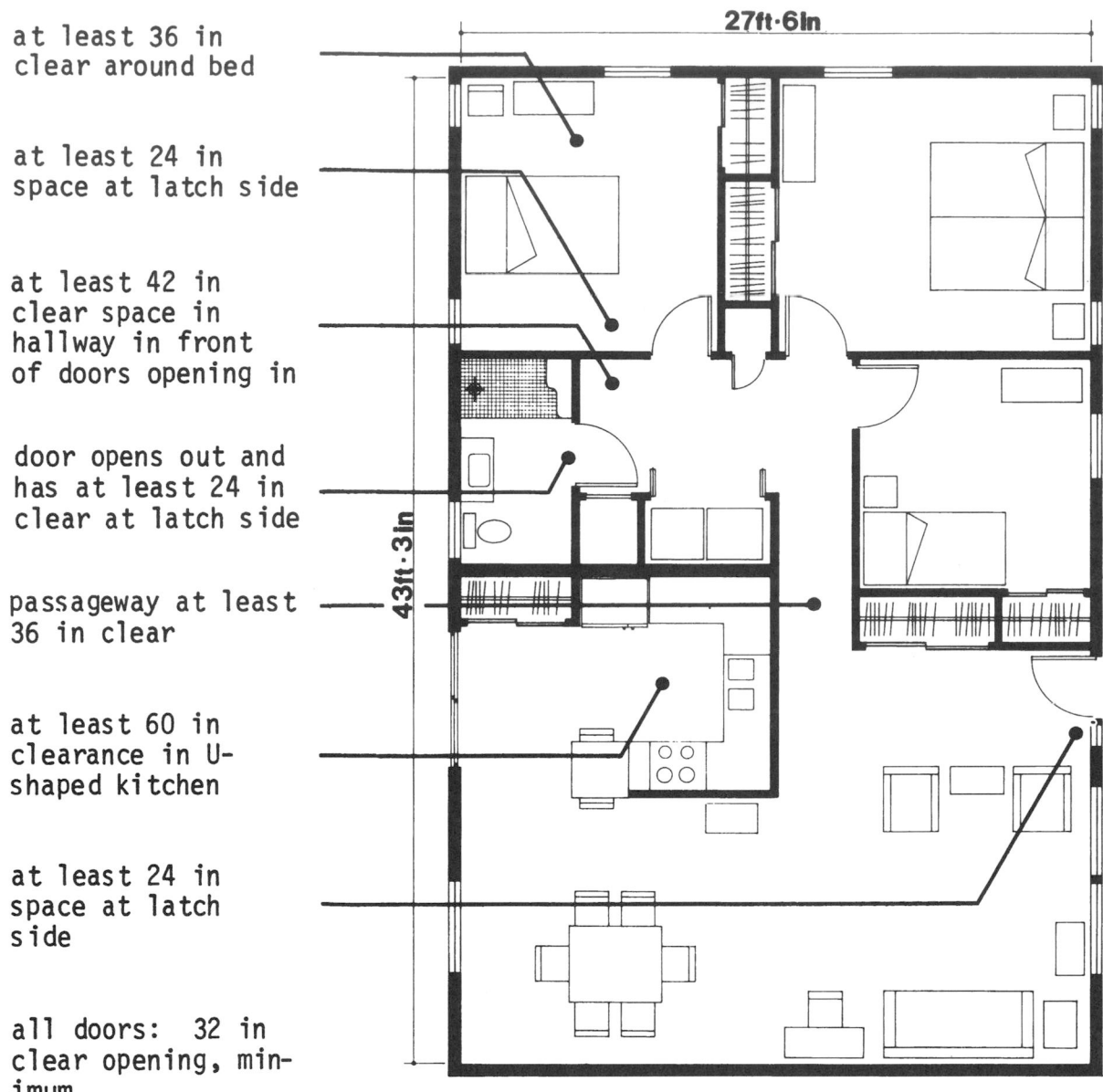

at least 36 in
clear around bed

at least 24 in
space at latch side

at least 42 in
clear space in
hallway in front
of doors opening in

door opens out and
has at least 24 in
clear at latch side

passageway at least
36 in clear

at least 60 in
clearance in U-
shaped kitchen

at least 24 in
space at latch
side

all doors: 32 in
clear opening, min-
imum

27ft·6In

43ft·3In

Fig. 3   Single-family home—adaptable dwelling unit.

KITCHEN

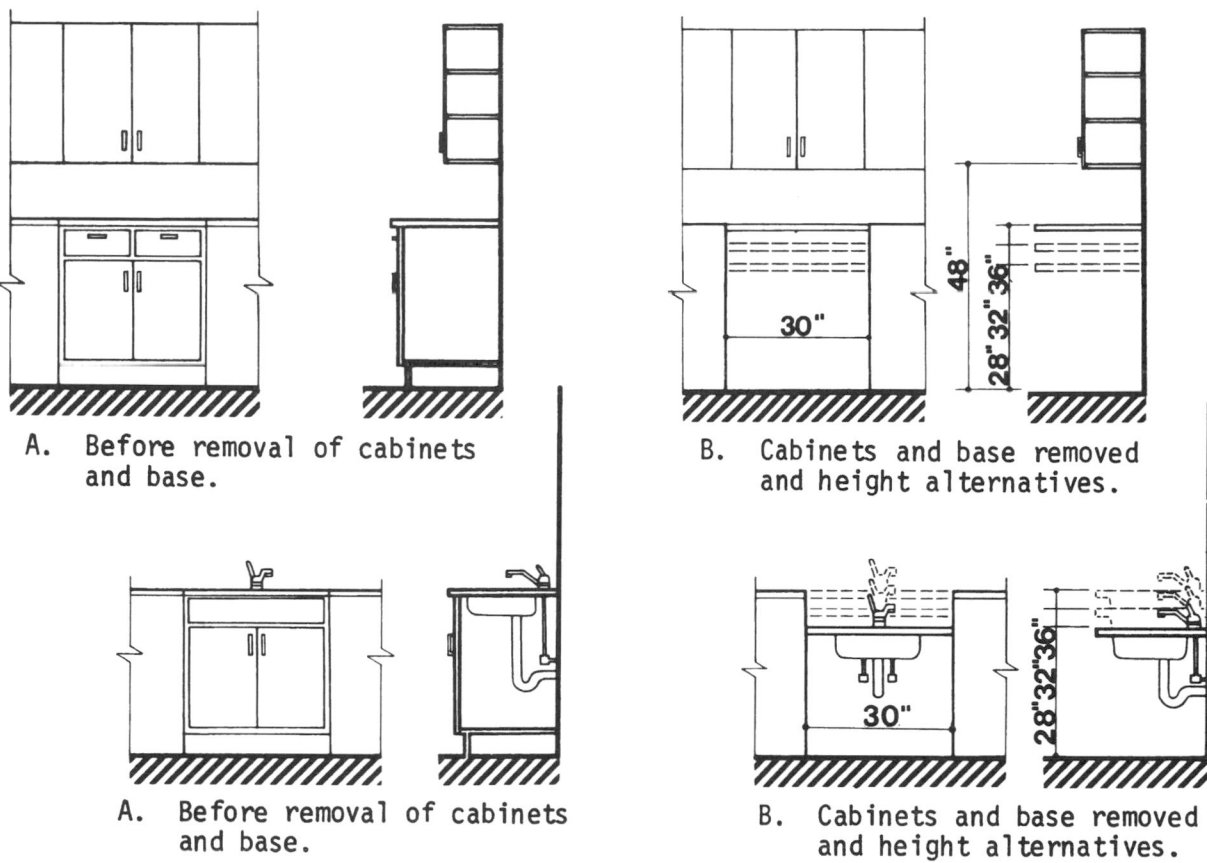

A. Before removal of cabinets and base.

B. Cabinets and base removed and height alternatives.

A. Before removal of cabinets and base.

B. Cabinets and base removed and height alternatives.

Fig. 4a   Adaptable counters and sinks.

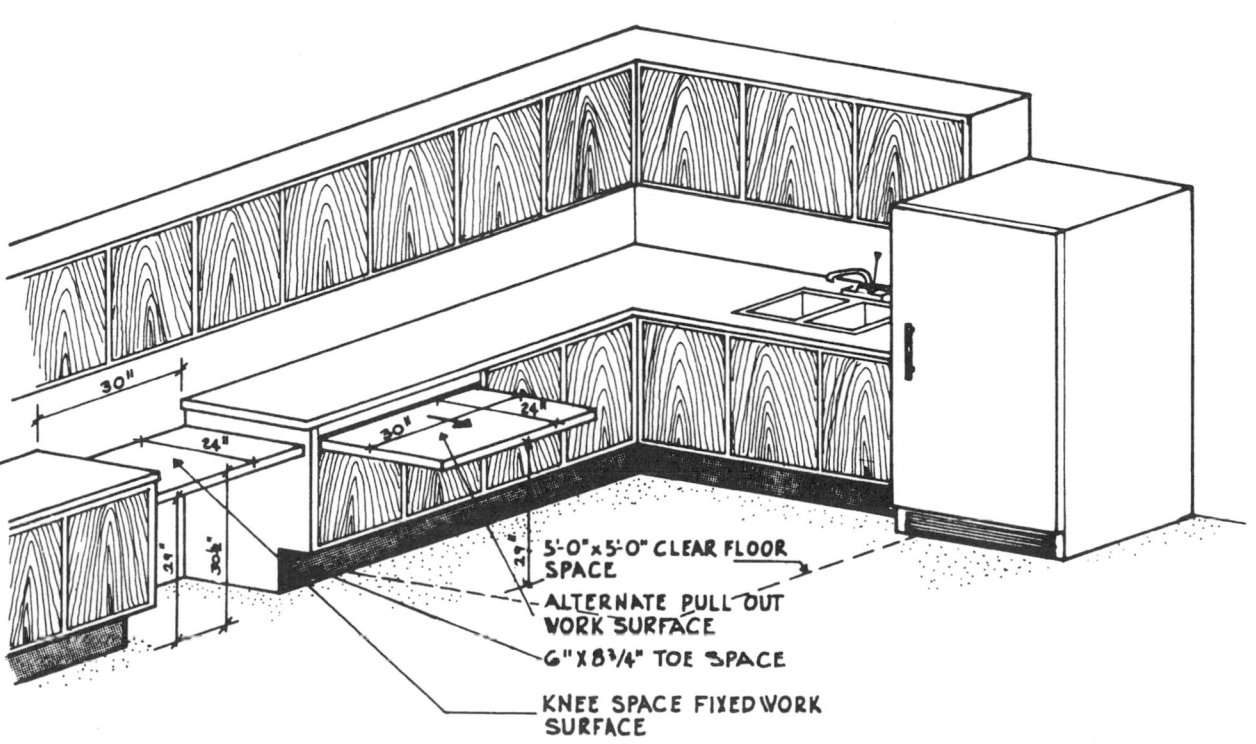

5'-0" x 5'-0" CLEAR FLOOR SPACE

ALTERNATE PULL OUT WORK SURFACE

6" X 8¾" TOE SPACE

KNEE SPACE FIXED WORK SURFACE

Fig. 4b   Adaptable kitchens.

ADAPTABLE DWELLING UNITS

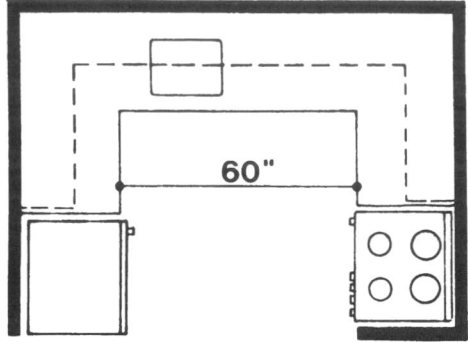

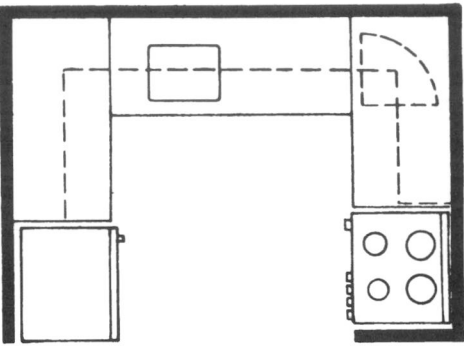

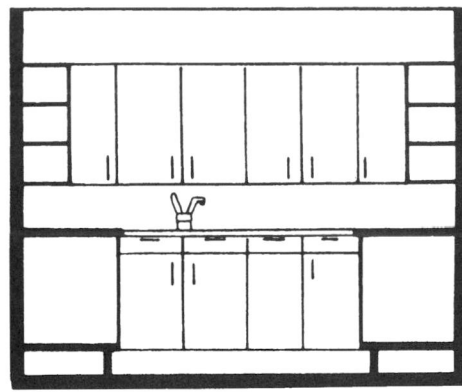

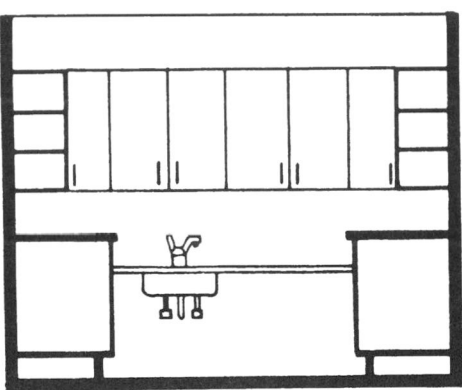

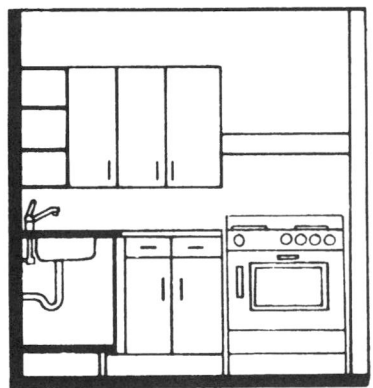

Accessible; before re-
moval of cabinets and
base.

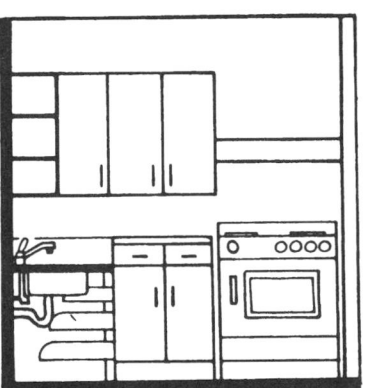

Cabinets and base re-
moved, counter height
lowered.

Fig. 5 Example of adaptable kitchen—U-shaped plan.

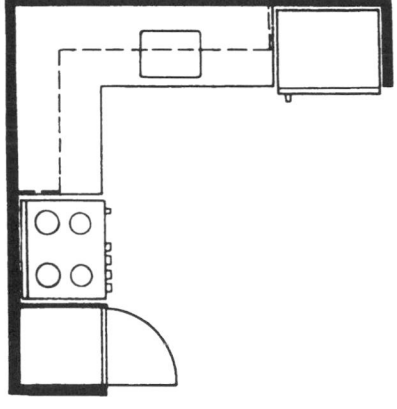
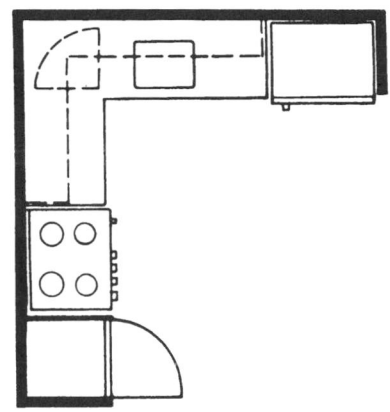

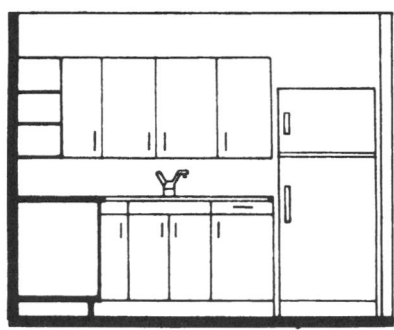
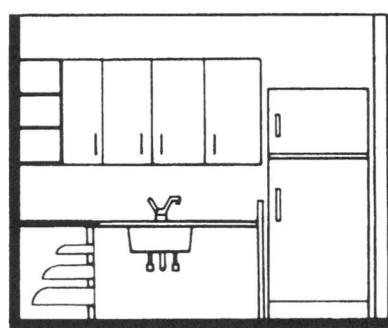

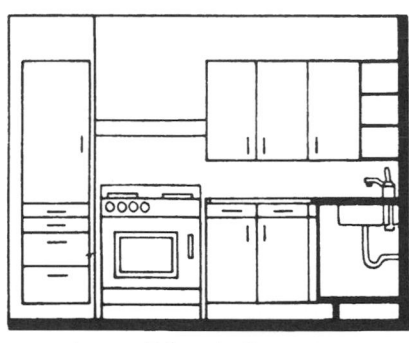
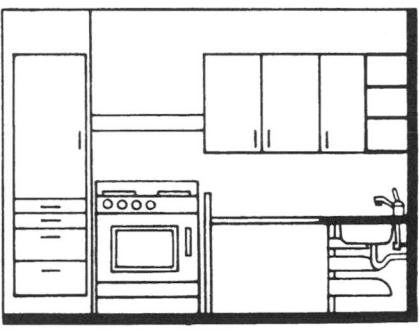

Accessible; before re-
moval of cabinets and
base.

Cabinets and base re-
moved, counter height
lowered.

Fig. 6 Example of adaptable kitchen—L-shaped plan.

ADAPTABLE DWELLING UNITS

BATHROOM

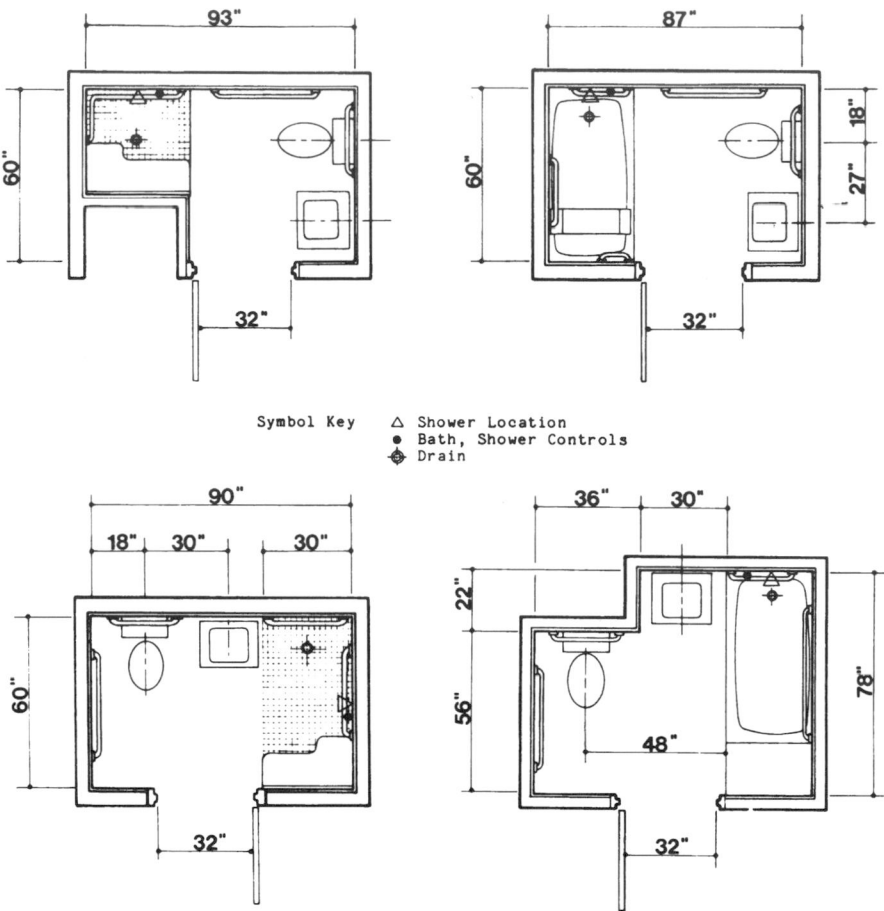

Symbol Key △ Shower Location
            ● Bath, Shower Controls
            ⊕ Drain

Fig. 7   Minimum-sized adaptable bathrooms.

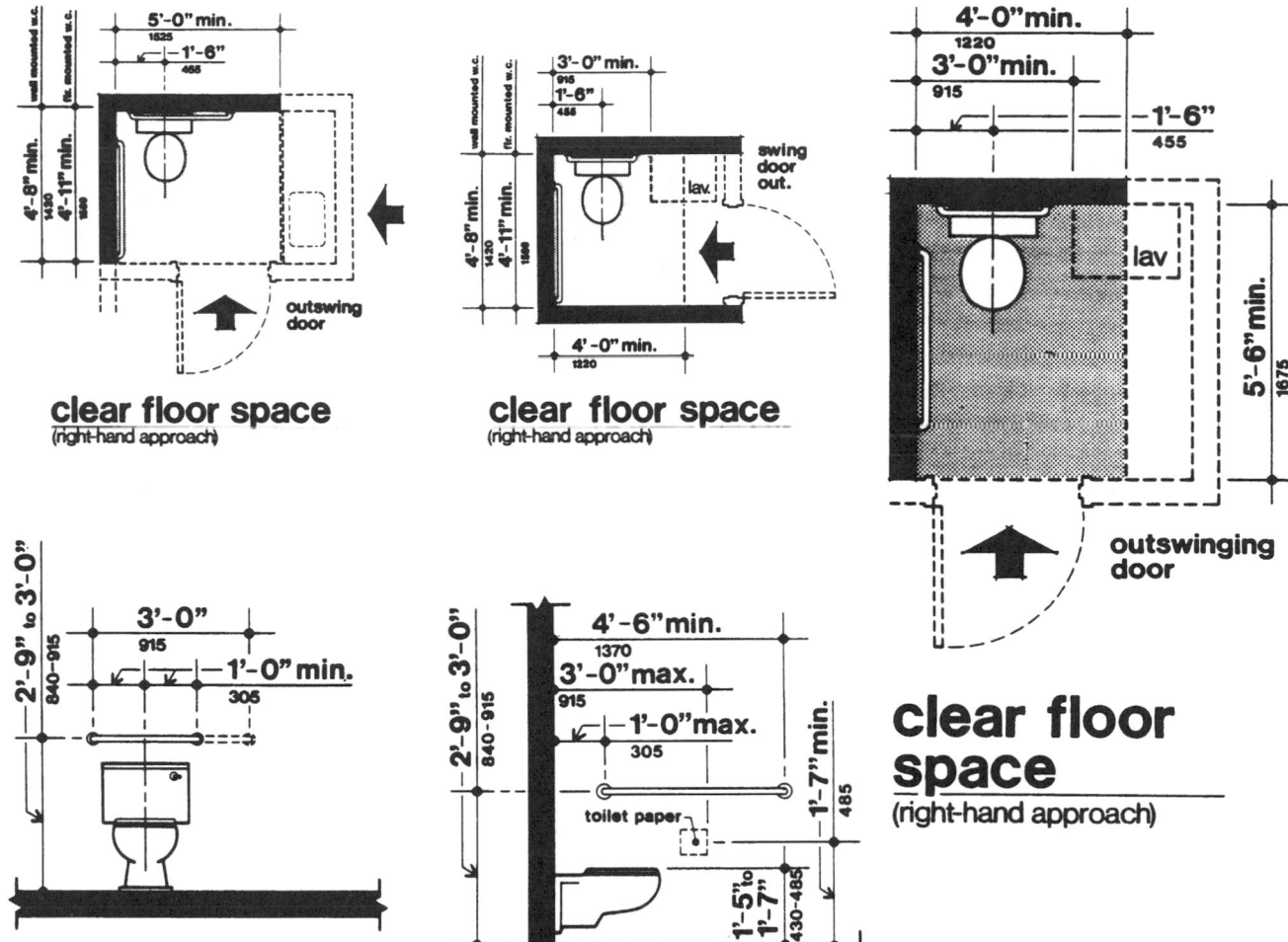

clear floor space
(right-hand approach)

clear floor space
(right-hand approach)

clear floor space
(right-hand approach)

rear wall elevation
without stall
or alternate stalls

side wall
without stall
or alternate stalls

Fig. 8

SHOWER

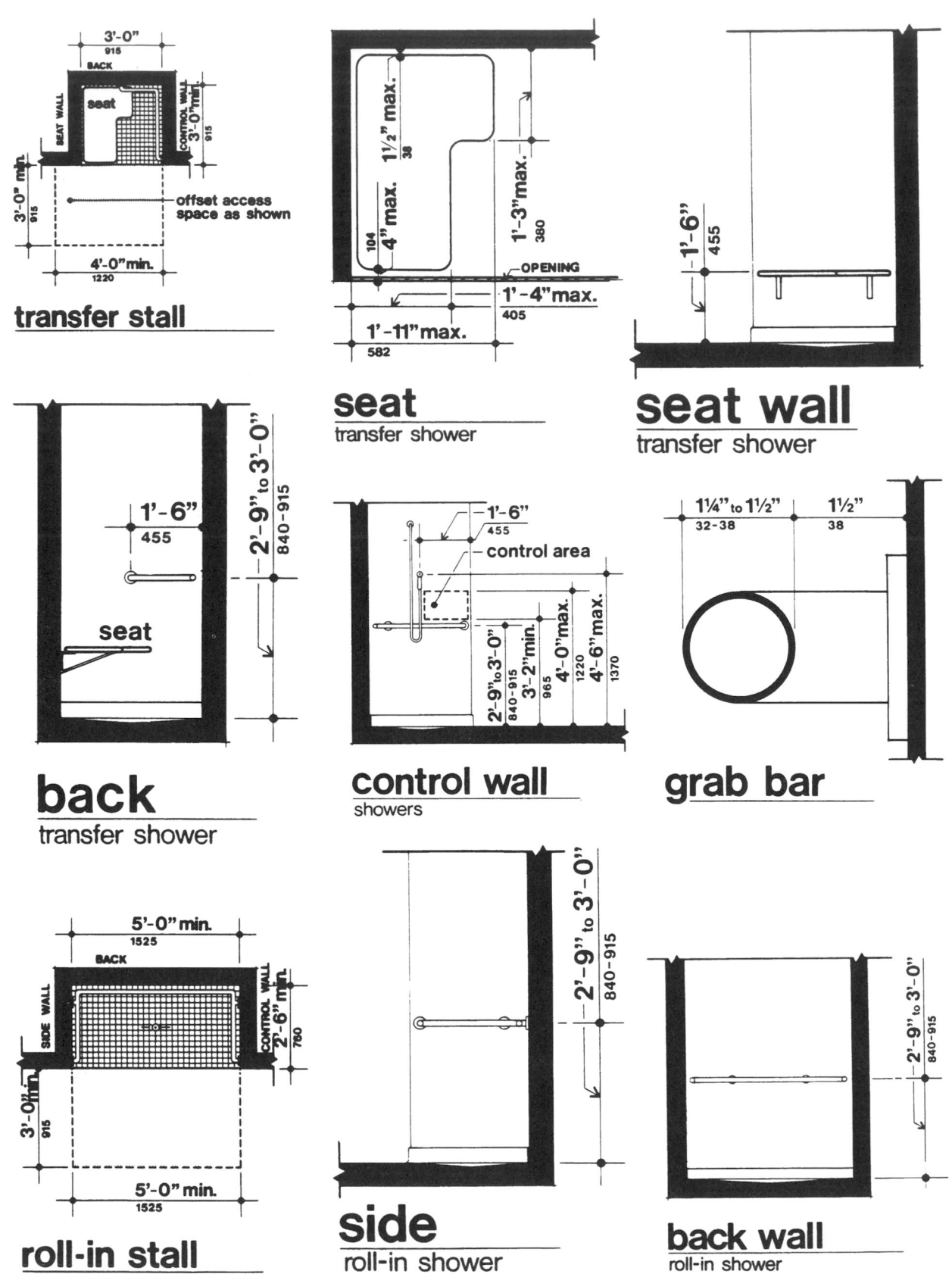

transfer stall

seat
transfer shower

seat wall
transfer shower

back
transfer shower

control wall
showers

grab bar

roll-in stall

side
roll-in shower

back wall
roll-in shower

Fig. 9

BATHTUB

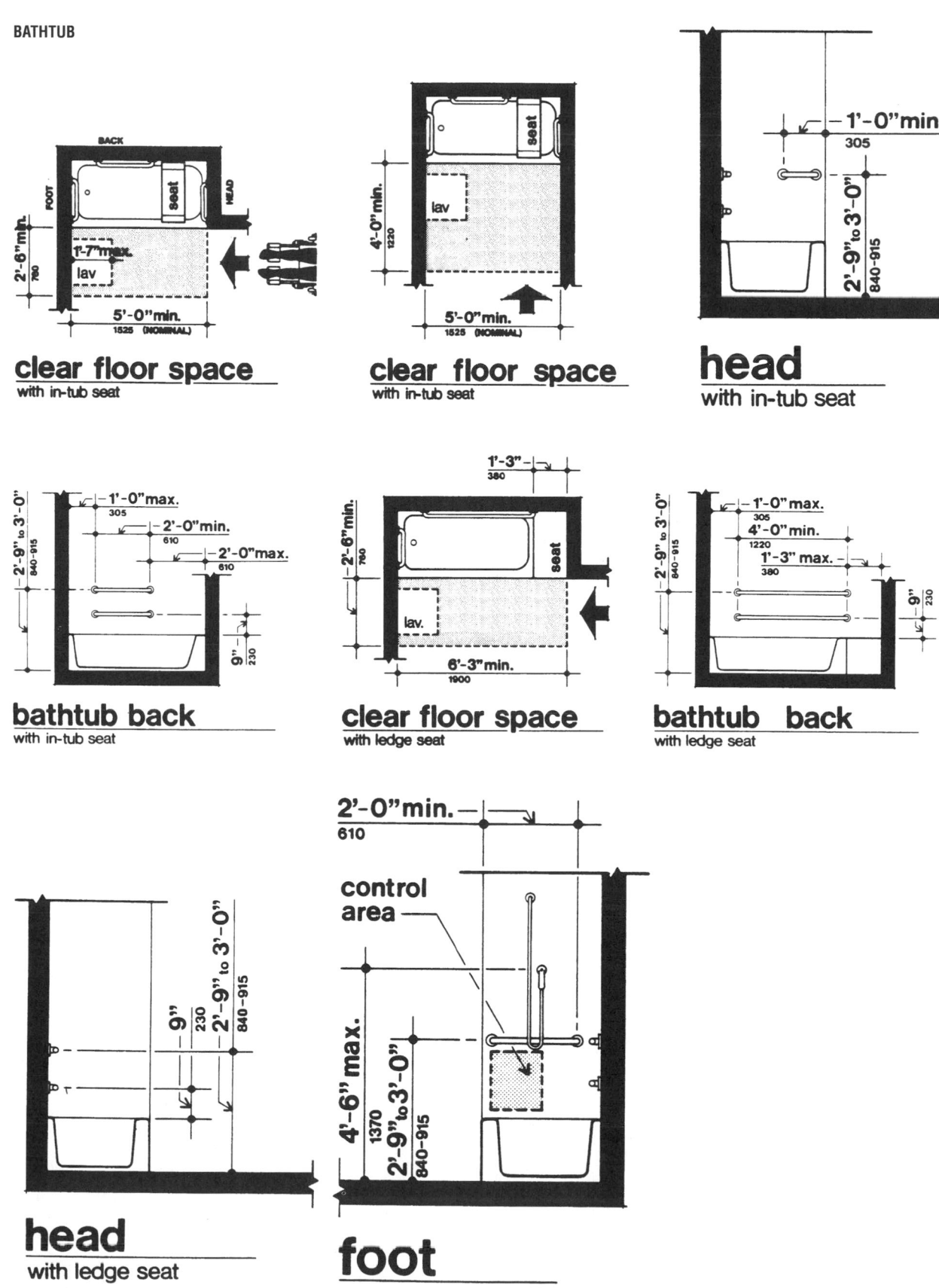

clear floor space
with in-tub seat

clear floor space
with in-tub seat

head
with in-tub seat

bathtub back
with in-tub seat

clear floor space
with ledge seat

bathtub back
with ledge seat

head
with ledge seat

foot

Fig. 10

HANDICAPPED HOUSING—ONE-FAMILY HOUSE

DEMONSTRATION HOUSE NUMBER ONE

This house, built around a central plumbing core, permits total wheelchair circulation. Except for the bathrooms, every room is accessible from more than one direction. Though not large (1500 ft² exclusive of the basement), the house has an entrance foyer and hallway wide enough for turnaround wheelchair maneuvering. A two-car carport shields the front entrance, which is approached by a ramp graded slightly upward.

The spacious living room and the owner's wheelchair bedroom look out onto a fully accessible cantilevered balcony and a woodland view, the living room through a glass wall and double glass doors, the bedroom through wide windows and one door. A large bathroom adjoins the bedroom.

The children's bedrooms, which adjoin each other, are a large playroom area by day, divided by an accordion-type folding door for privacy at night. The kitchen is fully accessible, with all its facilities at reachable levels.

The basement is finished to include a guest bedroom and bath and a playroom. These rooms and the back patio are accessible by elevator from the main floor.

Note the unique emergency exit inside a closet in the main bedroom (far from the front entrance). It is a bottom-hinged push-out door that becomes a drawbridge wide enough for a wheelchair, as is the closet entrance itself.

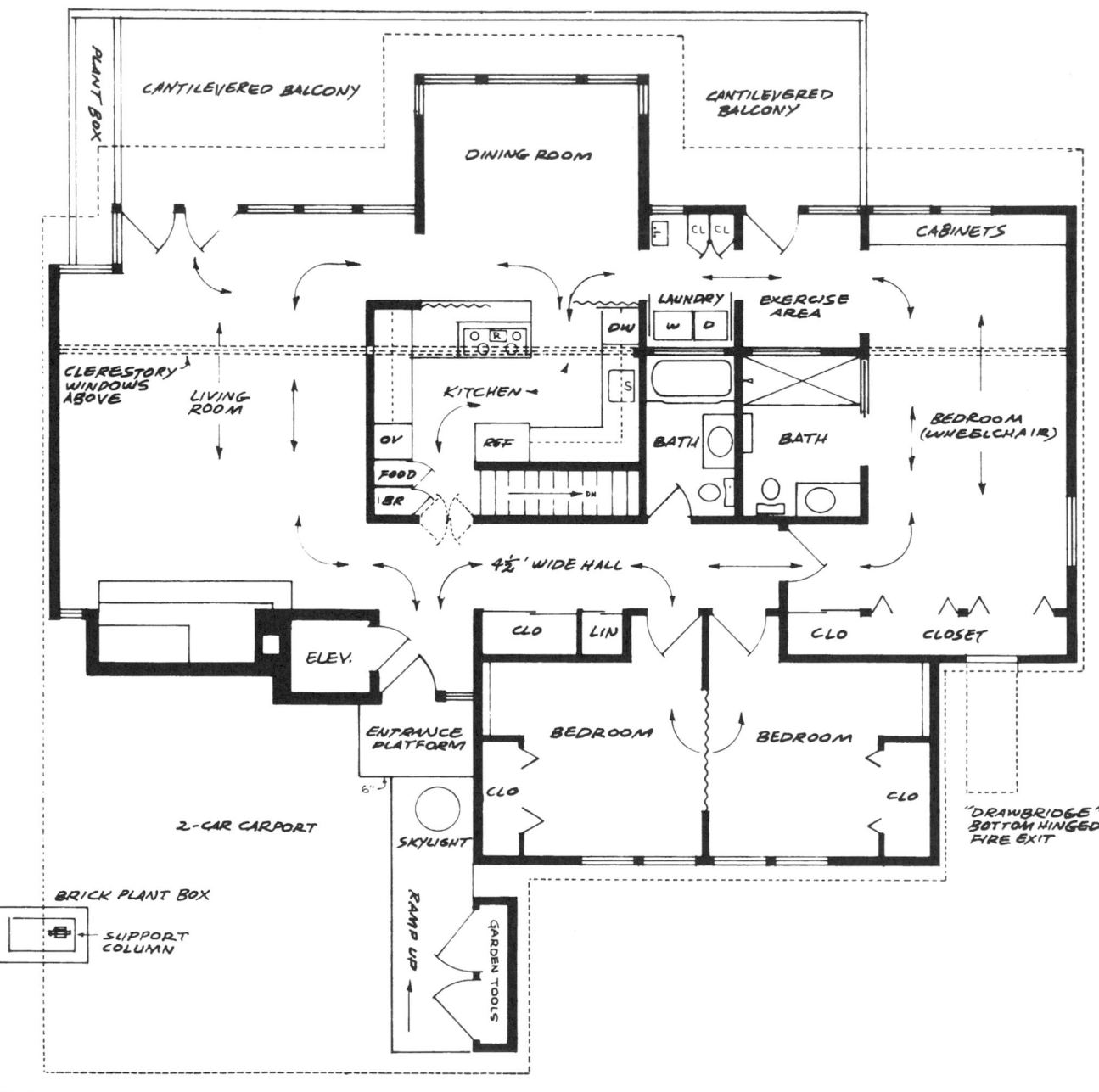

Fig. 1

DEMONSTRATION HOUSE NUMBER TWO

This house was adapted for a chairborne home-maker from a builder's plan. Excellent two-way movement is provided in the family and social activities area. Not even the crowded conditions of a social gathering would impair too much the mobility of the wheelchair resident.

To overcome the fact that the house floor is about a foot and a half above ground level, a gently rising ramp approaches the entrance under a sheltering roof overhang. Another ramp in the garage leads to the inside doorway, and a third one in the rear gives access to the backyard and its outdoor facilities, including a sunken swimming pool.

Adaptations in the kitchen include counter tops 32 in high for comfortable wheelchair use and a slideway for direct transfer of filled pots from the sink to the smooth ceramic surface of the electric cooktop. A wheeled cart serves to bridge the gap between these facilities and the refrigerator-freezer and built-in wall oven. The cleaning accessories—washer, dryer, broom closet, and others—and food storage are conveniently close to the kitchen area.

A wide hallway off the foyer makes all parts of the bedroom wing accessible. A screened Dutch-style (two-part) door in the exercise room serves as a window and ventilator when the upper half is open. Fully opened as a door, it is an emergency fire exit.

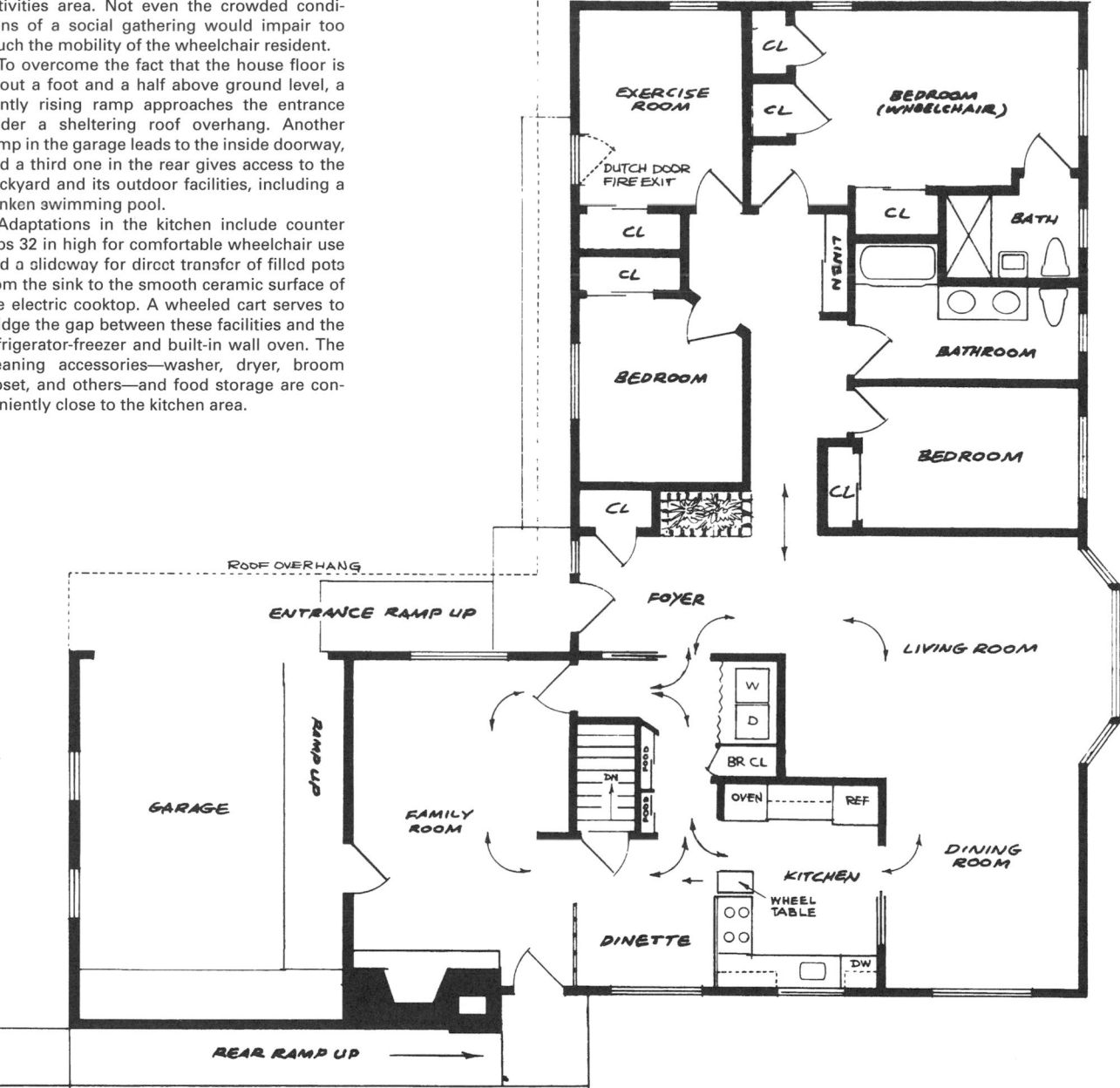

Fig. 2

HANDICAPPED HOUSING—ONE-FAMILY HOUSE

DEMONSTRATION HOUSE NUMBER THREE

Despite its modest size, this house gives a spacious appearance and easy wheelchair mobility. The open floor plan provides two-way circulation in the family and social activities areas. In addition to the two bedrooms, a library-den is convertible to a guest bedroom, which can be closed off for privacy by a folding door-wall. A side-hung door leading to the hall then serves for access. When the door-wall is open, the den adds a feeling of lavish size to the living area.

The two large bathrooms, each with enough space for complete turns, are both designed for wheelchair convenience. They include a wheel-in shower in one and a tub with a built-in transfer seat at one end in the other. Both rooms have wheel-under wash basins sunk into countertops, with wall-size mirrors above and supply cabinets to one side, under the counters.

The large kitchen is an important part of the open circulating floor plan. It has plenty of space for a snack table. This room was not designed as a wheelchair-work-kitchen for the chairborne resident; so counters are at standard 36-in height.

A custom-built, noncommercial elevator gives access to the basement.

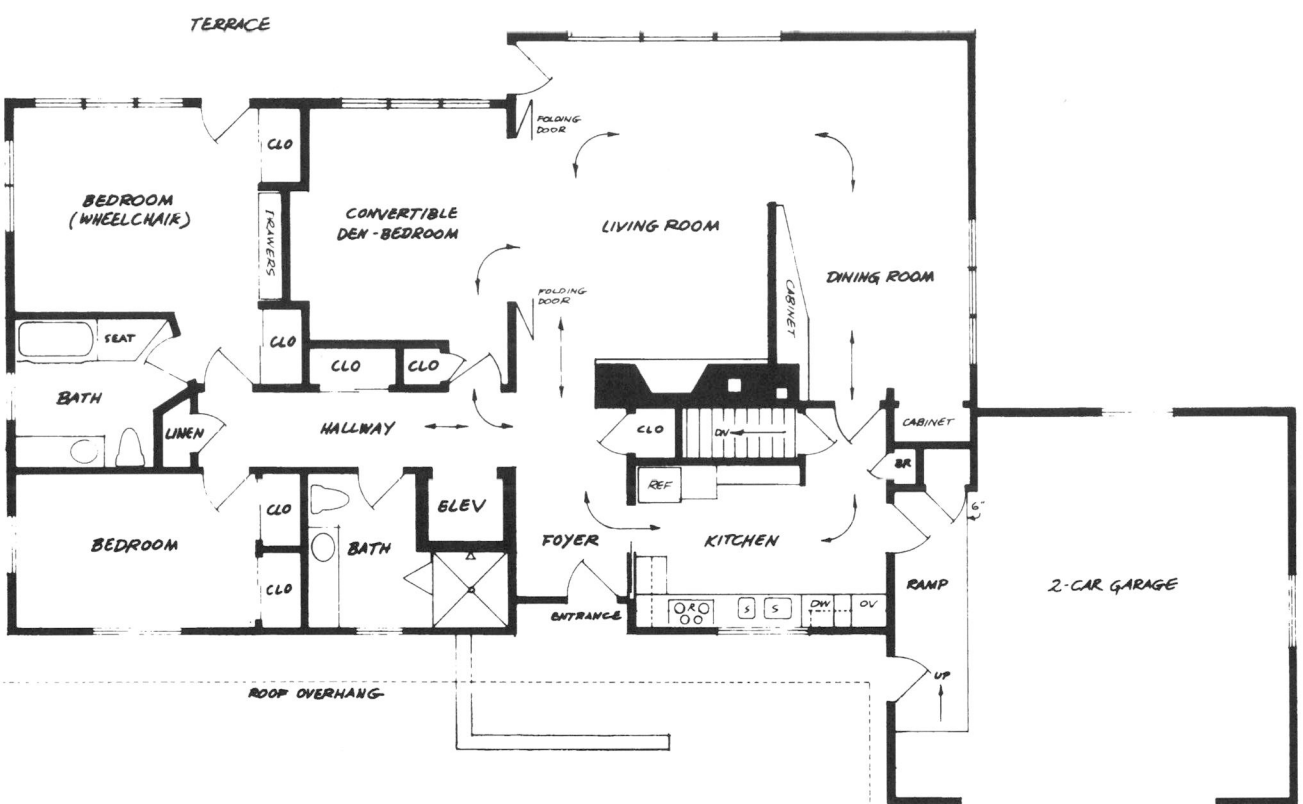

Fig. 3

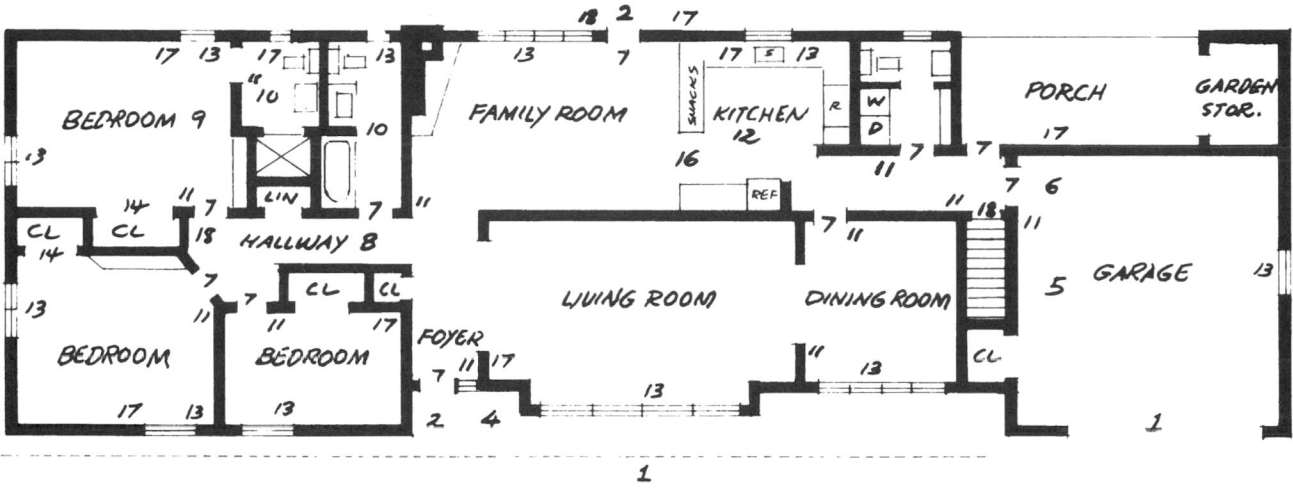

Fig. 4

Here's a checklist of the minimum requirements for a house to be built or remodeled for functional wheelchair living:

1. *Site:* Level, with sufficient grade for drainage.

2. *Access:* No steps, front or rear, inside or out. If ramping is necessary, the slope must not exceed 1 in of rise for every 12 in of length. If a basement is planned, a deeper foundation may be necessary to lower the house to an acceptable level. Find out about the groundwater level and possible seepage problems in the area.

3. *Ramps* must be at least 42 in wide, made of permanent, fireproof material, with a surface to prevent slipping when wet.

4. *Ramp approaches* to front and rear doorways must have level platforms at least 5 ft square. Entrance platform and interior floor must be at identical levels; no step up or down.

5. *Garage or carport* must be wide enough to allow maneuvering of the wheelchair for transfer to and from the car. It is important to consider in planning whether the chairborne resident habitually uses the passenger's or the driver's side of the car for transfers, because in most cases a change is virtually impossible.

6. *Entrance to the house* should be possible directly from within the garage or carport. If not, a pathway from the car shelter to the house entrance should be roofed over and sheltered from the weather.

7. *All doorways,* inside and out, should be 36 in wide. In the event that this width for some *interior* doors may not be feasible, the minimum allowable width is 32 in. The use of special "full-clear-opening" hinges for such doors is recommended.

8. The *hallway* should be no less than 42 in wide. This will permit a 90° turn by the wheelchair. However, a 48-in wide hallway is a much more practical minimum for the average wheelchair, and serious effort should be directed to modifying a house plan to meet this minimum.

9. The *bedroom* to be occupied by the chairborne resident must be adequately large for wheelchair maneuvering. Give thought to the potential placement of furniture and any special equipment such as a transfer lift. Direct and easy access to a bathroom is essential.

10. The *bathroom* is vital in the activities of everyday living for the chairborne person. It must be planned for comfortable access to and use of each fixture. Check the pointers concerning showers, washbasins, and so on in the bathroom. Consider the possibility of taking space from an adjoining room, if necessary, to enlarge the wheelchair bathroom.

11. *Electric wall switches* to be installed 36 in above the floor; outlets 18 in high, all at comfortable reach levels from wheelchair. Plan for multiple switch control as needed, delayed-action switches, remote-control systems. Service entrance panel (with master switch and fuses or circuit breakers) must be reachable from the chair.

12. If the chairborne person is the homemaker, plan the *kitchen* accordingly, with lowered base cabinets and other features for comfortable, efficient functioning. This may be the most costly modification in the house plan and must be given very careful study.

13. *Windows* should be operable from the wheelchair. If double-hung windows are in the original plan, arrange for a change to casement or awning type, with built-in inside screening. Consider the advantages of insulating glass.

14. Lowered hanger rods in *clothes closets.*

15. Discuss and consider choices available for low-maintenance exterior and interior surfaces.

16. Can the house plan be opened to improve traffic circulation? A simple change may make a world of difference to easier living!

17. Consider a built-in *intercom system* with call boxes in appropriate locations in the house.

18. Install *UL-approved smoke-detector alarms* and plan an emergency *fire exit* that can be used by the chairborne resident as well as others in the house.

Ideally, a house designed for wheelchair accessibility is on a level site, with no more slope than necessary for effective drainage. The approaches to the entrance door and other walkways are at least 42 in wide, and any gradient no more than 3 percent (1 in of rise for every 30 in of length).

When the ideal is not available, however, and the pitch is greater than 1 in 30, one or more ramps may be required. The slope of a ramp must not exceed 1 in in 12 in of length. A ramp must be constructed of concrete or other permanent, weatherproof material and provided with a nonskid surface.

Level platforms for resting should be included on a ramp that:

1. Is excessively long (to provide a breather for a person in a wheelchair with limited arm strength).

2. Has a sharp change in direction.

3. Approaches a door. To allow ample room for opening, the platform before a door should be at least 5 by 5 ft. It should extend alongside the door at least 1 ft—preferably 18 in—on the side next to the door handle, to permit positioning a wheelchair so that an outward-swinging door can be opened without backing up.

Either a curb or railing, or both, should be run along the edges of a ramp to prevent a wheelchair from going over accidentally, especially where a steep drop-off is involved. A handrail should be 29 to 32 in high, depending on the arm reach of the chairborne individual, who can use a railing to assist mobility, which is an advantage it has over a curb alone. The handrail should extend slightly beyond the ramp at top and bottom, but care must be taken that the extension is not a hazard to passersby.

THE KITCHEN

In a wheelchair house the kitchen, with its expensive appliances and custom cabinetry, may well be the costliest room of all. Careful planning is essential, if only to avoid having to redo it all because of errors that place hurdles in the way of effective use by a handicapped homemaker.

The wheelchair resident may not expect to be involved with kitchen activities. In that case, its design can be geared toward wheelchair accessibility for meals and social, rather than working, purposes—with perhaps some concession to the possibility of occasional meal preparation. When use will probably be limited to reaching for a glass of water or other casual refreshment, the kitchen can be designed for the ablebodied worker (using standard rather than custom-built cabinetry, for example) with some minor wheelchair-oriented modifications.

The suggestions that follow, however, are aimed at creating a functional, attractive, and cheerful kitchen for the chairborne homemaker—the person who will use it many hours every day for preparing, serving, and cleaning up after meals. This section provides general guidelines for planning and design of a wheelchair kitchen; more detailed development of these suggestions may be found in *The Wheelchair in the Kitchen,* published by the Paralyzed Veterans of America.

GENERAL

- Accessibility should be easy. Doorways should have at least 34-in clear openings. Direct, level entrance from an outside delivery door and the car shelter is advisable.
- No doorsills that might be bumped over with a load of hot food.
- Adequate space for the chair to maneuver, including complete turns in any direction.
- Floor must be absolutely level so braking is not constantly needed at work centers to keep chair from rolling.
- A triangular work-center plan: refrigerator to sink to range, with adequate counter space next to each.
- Natural ventilation and an exhaust fan to remove airborne cooking grease, steam, and heat.
- Pleasant, well-lit working conditions; cheerful color and overall decor to reflect the personality of the homemaker.

WORK SURFACES

- Countertops, sink, and range at comfortable working height for the seated person; that is, no more than 32 (preferably 31) in instead of the standard 36 in. Counter surfaces continuous and at one level to form a freeway on which filled pots and other burdens can be slid from place to place instead of carried.
- Mixing work center: 27 in high, with knee space beneath. Can be pullout shelf, drop-leaf table, built-in or perhaps roll-out mixing center.
- Adequate electrical outlets for small appliances, at front of base cabinets, especially near mixing center.
- Roll-about serving table (may be built-in for storage under a counter) to serve an entire meal safely in one trip and for help in after-meal clean-up. When an under-counter roll-out table is out of its place, the counter above it can serve as a mixing center, with plentiful knee space below.

REFRIGERATOR-FREEZER

- Plan only a two-door model, not a one-door with a freezer section as a compartment inside. Freezer section should be large, and capable of maintaining 0°F to keep foods safely frozen solid, in order to permit maximum storage and reduce shopping trips.
- Both refrigerator and freezer compartments should be frost-free to eliminate difficult and time-consuming defrosting chore.
- A freezer compartment at top should have side-opening door to facilitate reaching in. Avoid too-high top freezer, making food in rear inaccessible. Consider a side-by-side (French-door) type refrigerator-freezer combination, or pullout-drawer-type freezer at bottom.
- Look for slide-out or rotating shelves. Or install shallow turntables, available from most houseware stores, on each shelf to bring food stored at rear within reach.
- For special convenience: automatic ice cube maker inside freezer or chilled water and ice cube dispensers on outside of door of freezer compartment, available in some side-by-side models (requires plumbing connection).
- Provide counter space on open-door side of refrigerator for easy transfer of food.

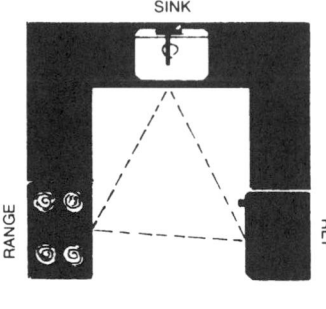

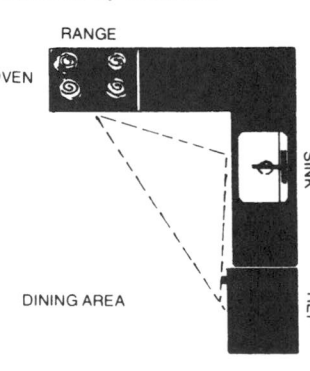

Left, U-shaped kitchen, triangular work center plan, continuous counters. Locate wall oven, storage, on extension of one or both legs. L-shape, below, also offers continuous counters. Good basic plan for many variations.

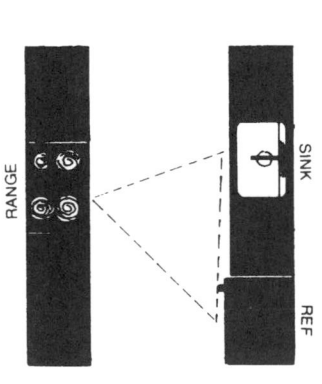

Corridor, left, needs 5-foot turning space between counters. Bridge them with roll-table or tray. Below, one-wall or studio plan is least efficient for wheelchair use.

Fig. 1

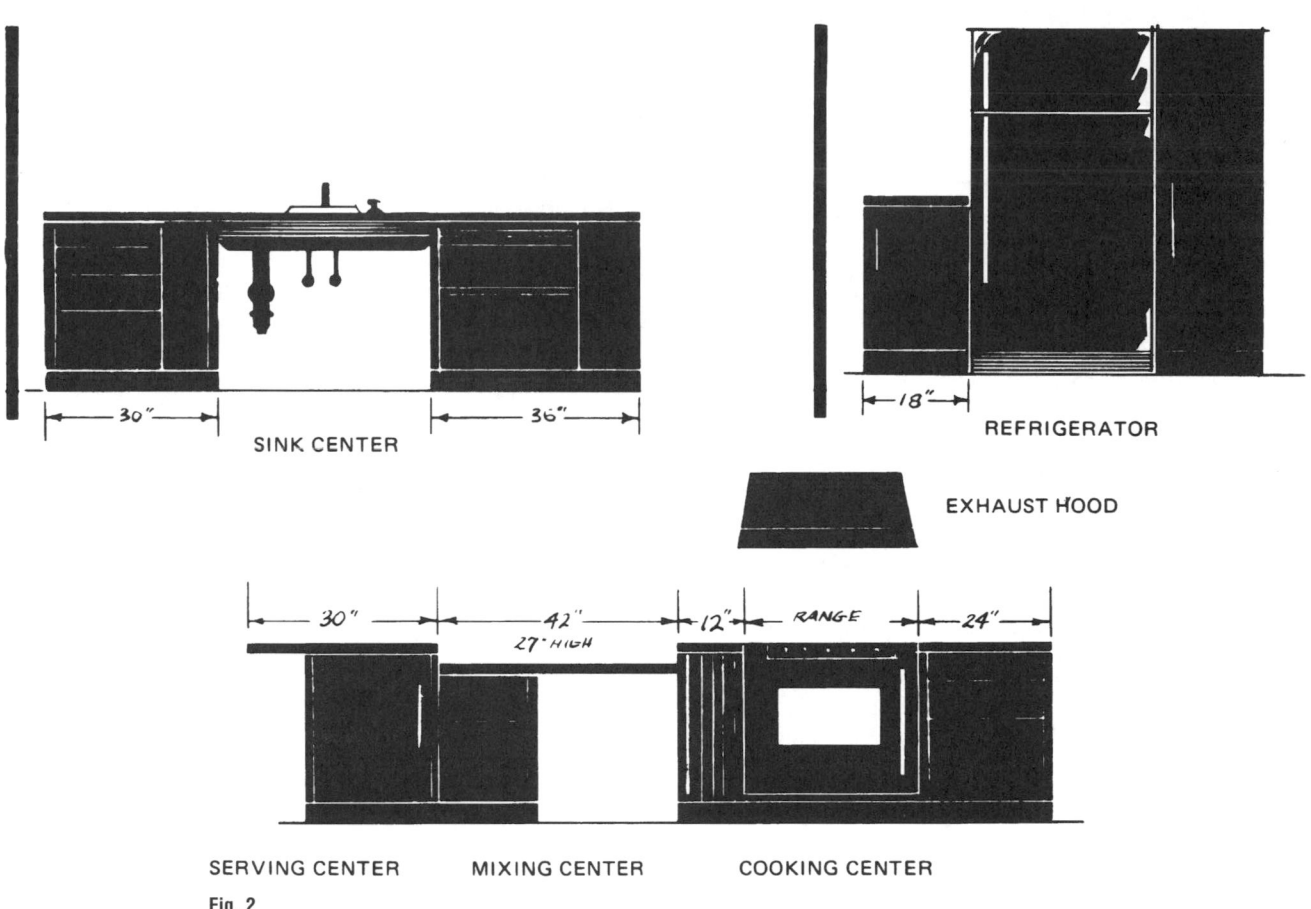

30" · SINK CENTER · 36"

18" · REFRIGERATOR

EXHAUST HOOD

30" · 42" 27" HIGH · 12" · RANGE · 24"

SERVING CENTER · MIXING CENTER · COOKING CENTER

Fig. 2

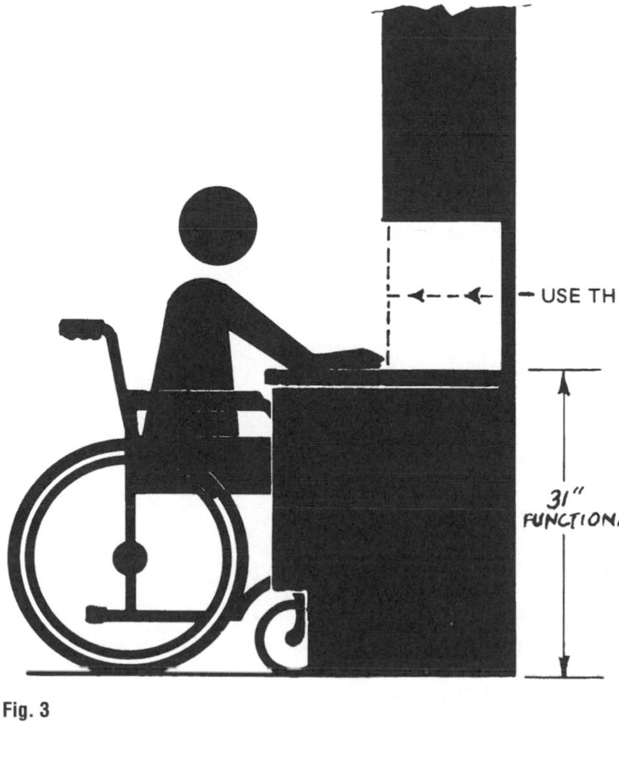

— USE THIS SPACE FOR STORAGE

31" FUNCTIONAL

Fig. 3

Better posture, less tiring work at a 31-inch-high (about) surface, especially with good knee space. Above: Suggested countertop lengths near sink, refrigerator, serving, mixing, range; with knee space.

Countertop dimensions The ideal unbroken U or L shape is not always possible because doors and windows may interfere; but in any shape the countertops should conform to these minimum specifications (established by the Small Homes Council of the University of Illinois):

- The refrigerator should have about 1½ ft of countertop next to the door-opening side. Location should be close to the outer door where grocery and other deliveries are made.
- The sink should have 3 ft on the right and 2½ ft on the left, for use in food preparation and dishwashing.
- The range should have 2 ft of heat-resistant countertop to its left—preferably—or right. In addition, the SHC recommends a "mixing center"—3½ ft of counter that may be next to any of the three appliances. Also a "serving center," at least 2 ft of countertop, preferably next to the range, with next to the refrigerator a second choice. In either case it should be as close as possible to the dining area.

SINK

- Should be easy-to-clean stainless steel, no more than 5 in deep, sunk into the 31- or 32-in high counter; 5-in depth is easier to reach into and provides ample knee and lap space beneath (facilitated by desk-type recessed armrests on the wheelchair). Double sink, if space permits—but not needed if a dishwasher is planned.
- Underside of sink, hot-water pipes, and drainpipes insulated to prevent burns.
- Drain or garbage disposal unit at rear, preferably a rear corner, to keep knee space open.
- Retractable spray hose has many valuable uses, especially filling pots on the counter, eliminating need to lift filled pot out of sink.
- Easily handled water controls; preferably a one-handle faucet.

DISHWASHER

- Front-loading, under-the-counter unit installed close to sink—a must in the chairborne homemaker's kitchen. Can save hours of tedious work daily; does a much better job than hand washing.
- Because of height of all such units now available, will probably have to be at end of continuous slide-along counter surface, perhaps next to refrigerator.

COOKING CENTER

- Gas or electric surface cooktop unit set into wheelchair work-height counter is preferable, with knee space beneath.
- Controls in front for easy access without reaching over hot burners and utensils. Second choice: controls on side. (Pushbutton-type controls available in electric ranges may be an advantage for some.)
- Similarly, staggered burner arrangement for safe access to pots on rear burners.
- *Oven:* wall-type, installed at height convenient to user, preferably with most-used shelf on level with adjacent counter. Heat-resistant counter surface or pullout shelf alongside for placing hot utensils removed from oven. Side-opening door for easy reach, easy cleaning. Self-cleaning type is advisable. (Consider microwave oven, which greatly reduces cooking and clean-up time.)

KITCHEN STORAGE

- All utensils, foods and condiments, cleaning supplies, and so on, within easy reach, preferably at points of their first use.
- All base cabinets custom-built to the required height for wheelchair use. Provide toe space for closer approach. Use smooth-gliding drawer hardware, tap catches for feathertouch door openers.
- Wall cabinets standard, but lowered to place items on lower shelves within reach.
- Use space between wall cabinet and rear of countertop for easy access to items in frequent use.
- Pegboard on wall space for miscellaneous items.
- Heavy items and breakables stored low. Use reachers for light packages and plastic items stored above.
- Broom closet with roll-about cleaning caddy.

THE LAUNDRY

For the chairborne homemaker, having laundry facilities close to the kitchen is a practical way of coordinating several time-consuming activities with a minimum of movement from place to place.

The basic necessities are an efficient automatic washer, a dryer, storage space for supplies, a good lightweight steam iron, and an adjustable ironing board that can be set at a comfortable seated-work height of about 29 in. The board should also be capable of changing to a standing-work height for the convenience of others who may wish to use it.

Soiled laundry storage in the same area is a convenience, but if space is at a premium it may be found closer to the bedrooms and bathrooms where most soiled linen originates. Rollabout hampers are then needed for easy carting of materials to and from the laundry machines.

Look for a front-loading drum type of washer, which may be difficult to find but is available. It is more accessible than the usual top loader; this is too high and its tub too deep for the average wheelchair homemaker.

Most automatic dryers are designed as front loaders. Select a model with controls up front and within reach.

If the laundry area has adequate space, consider the possible value of an electric roller type of ironer. But do not overlook that boon of modern textile science "permanent press." This invention has taken the labor of ironing out of a large percentage of our everyday clothing, as well as bed and table linen.

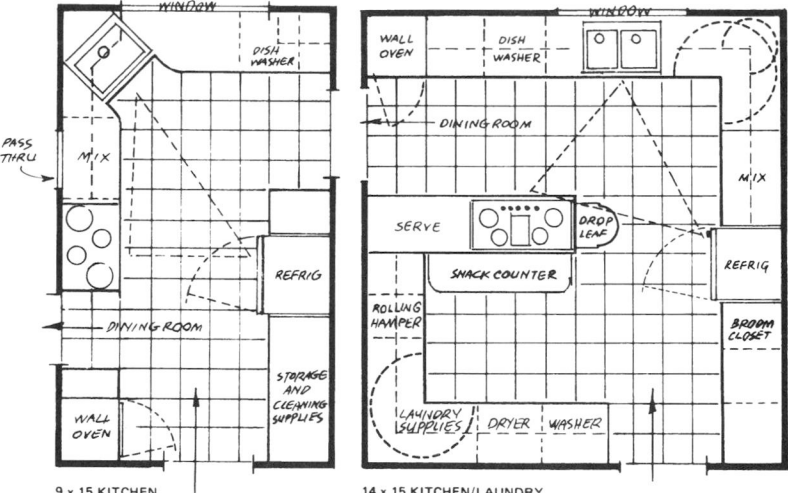

9 x 15 KITCHEN

Fig. 4

14 x 15 KITCHEN/LAUNDRY

Fig. 5

Fig. 6

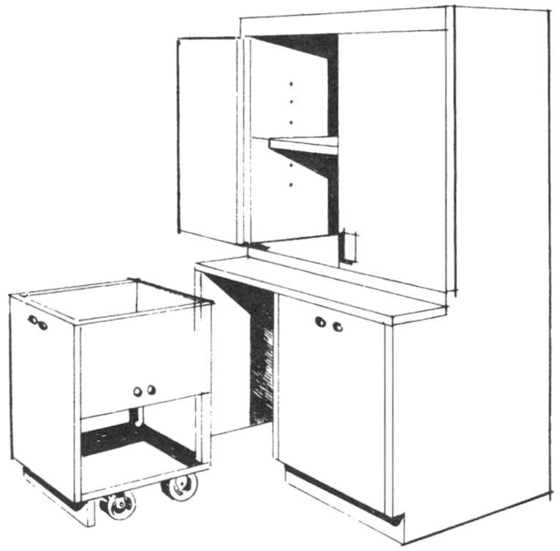

Fig. 7

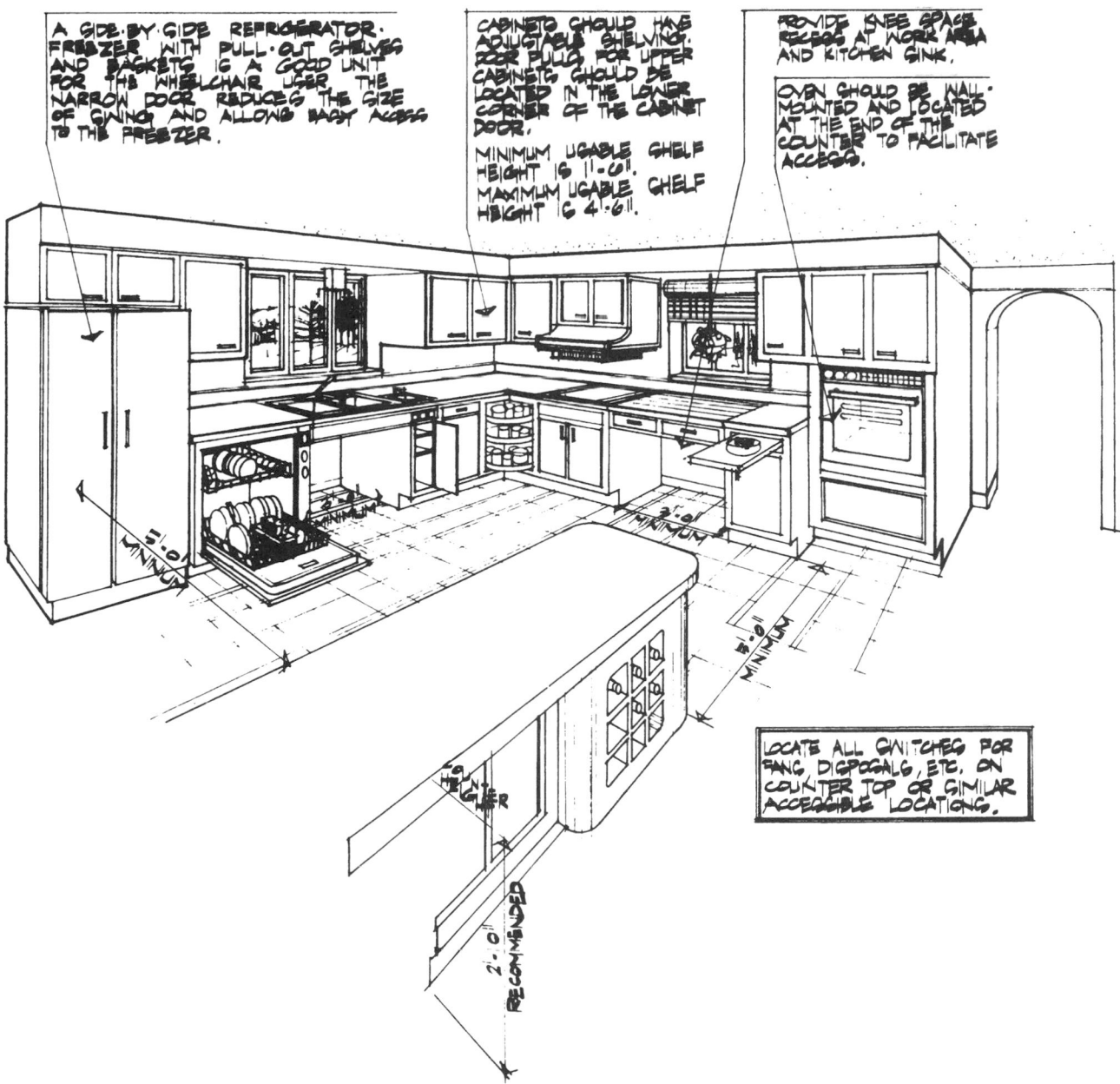

A SIDE-BY-SIDE REFRIGERATOR-FREEZER WITH PULL-OUT SHELVES AND BASKETS IS A GOOD UNIT FOR THE WHEELCHAIR USER. THE NARROW DOOR REDUCES THE SIZE OF SWING AND ALLOWS EASY ACCESS TO THE FREEZER.

CABINETS SHOULD HAVE ADJUSTABLE SHELVING. DOOR PULLS FOR UPPER CABINETS SHOULD BE LOCATED IN THE LOWER CORNER OF THE CABINET DOOR.

MINIMUM USABLE SHELF HEIGHT IS 1'-0". MAXIMUM USABLE SHELF HEIGHT IS 4'-6".

PROVIDE KNEE SPACE RECESS AT WORK AREA AND KITCHEN SINK.

OVEN SHOULD BE WALL-MOUNTED AND LOCATED AT THE END OF THE COUNTER TO FACILITATE ACCESS.

LOCATE ALL SWITCHES FOR FANS, DISPOSALS, ETC. ON COUNTER TOP OR SIMILAR ACCESSIBLE LOCATIONS.

Fig. 8   Kitchen arrangements.

HANDICAPPED HOUSING—KITCHEN

KITCHEN WORK AREAS

Two types of work area counters are illustrated. The first type (Fig. 9) provides a recess which is only high enough to accommodate the individual's legs and knees. Wheelchair armrests (unless they are removed) limit the user's access to the counter. The comfortable forward reach is approximately 1 ft 9 in (from the front of the counter) and therefore, any counter space deeper than 1 ft 9 in is unusable. In many instances, this solution is acceptable; however, the designer should be aware of the limitations. Such a workspace should have a minimum height of 2 ft 3 in.

The second type of counter (Fig. 10) provides a recess which will allow the wheelchair user to approach the countertop more closely. To accommodate the wheelchair armrests, the recess should have a minimum height of 2 ft 6 in. Such a recess should have a minimum depth of 2 ft 0 in to prevent the wheelchair footrests from limiting the individual's approach. Pullout

lapboards at a suitable height also provide convenient workspace.

The kitchen workspace is often a convenient location for a telephone extension.

The kitchen sink should be a maximum height of 2 ft 10 in. Controls should be lever-type and located no farther than 1 ft 9 in from the edge of the counter.

Sinks should be no deeper than 5 in. A level counter area 2 to 3 in in front of the sink should be provided for arm support.

A pullout spray attachment is useful for rinsing dishes, filling pots, or cleaning the sink itself.

A knee recess must be provided below the kitchen sink with a minimum height of 2 ft 3 in and a minimum width of 3 ft 0 in. The drain should be located in the back of the sink to allow the maximum possible knee space.

Insulation must be provided around any source of heat. Insulation should be provided for the sink and all supply and drainpipes, as well as dishwasher connections.

Disposals are a great convenience and should be provided whenever possible. The designer must take special care, however, that the disposal motor is enclosed in a manner that will protect the individual from shocks or burns. Disposal installation will normally limit the knee space below the sink. This can be minimized by locating the drain at the back of the sink and offset to one side. A more satisfactory solution is the installation of a separate disposal sink to one side of the knee recess.

The cooktop should be mounted at a maximum height of 2 ft 10 in to allow the seated individual to monitor food while it is cooking. Even at 2 ft 10 in, food on the back burners may be difficult to see. The Canadian publication, *Housing the Handicapped,* suggests the installation of a mirror set above the cooktop mounted at an angle. The cleaning problem associated with such a mirror may, however, limit its convenience. The cooktop should have flush burners or a ceramic surface to reduce the possibility of spills. The cooktop should be flush with the adjoining counters.

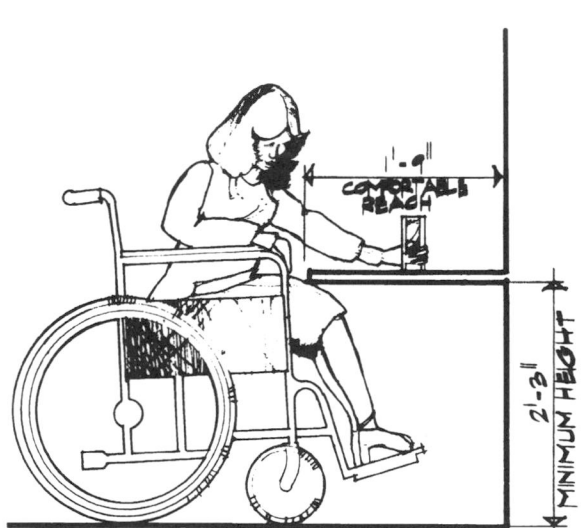

Fig. 9   Knee space clearance.

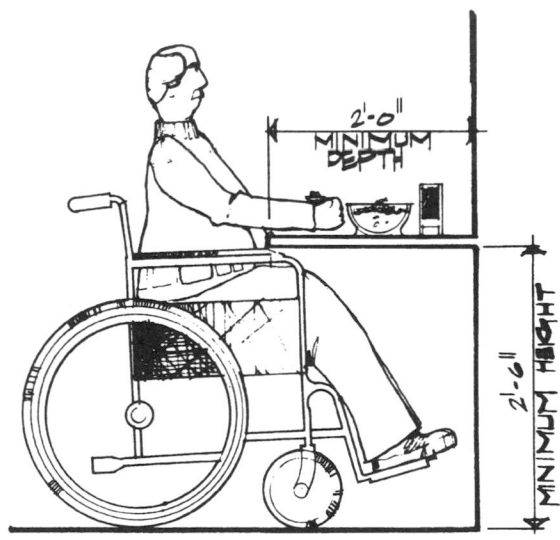

Fig. 10   Armrest clearance.

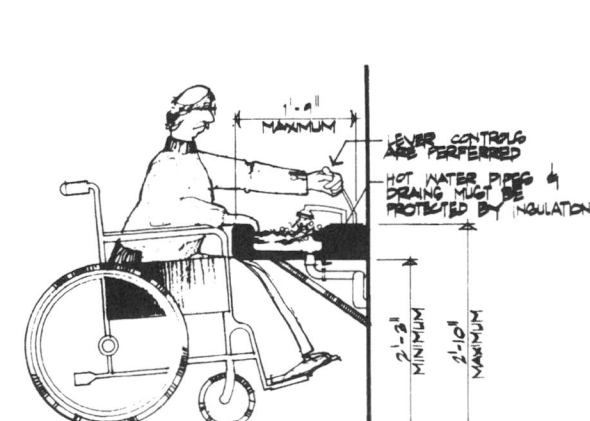

Fig. 11   Sink with knee space.

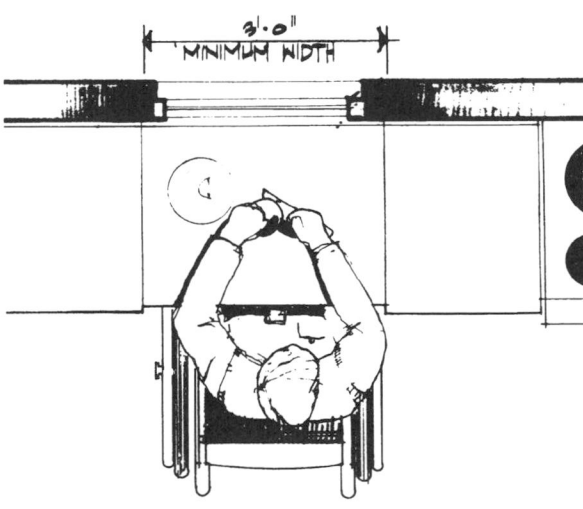

Fig. 12   Knee recess work area.

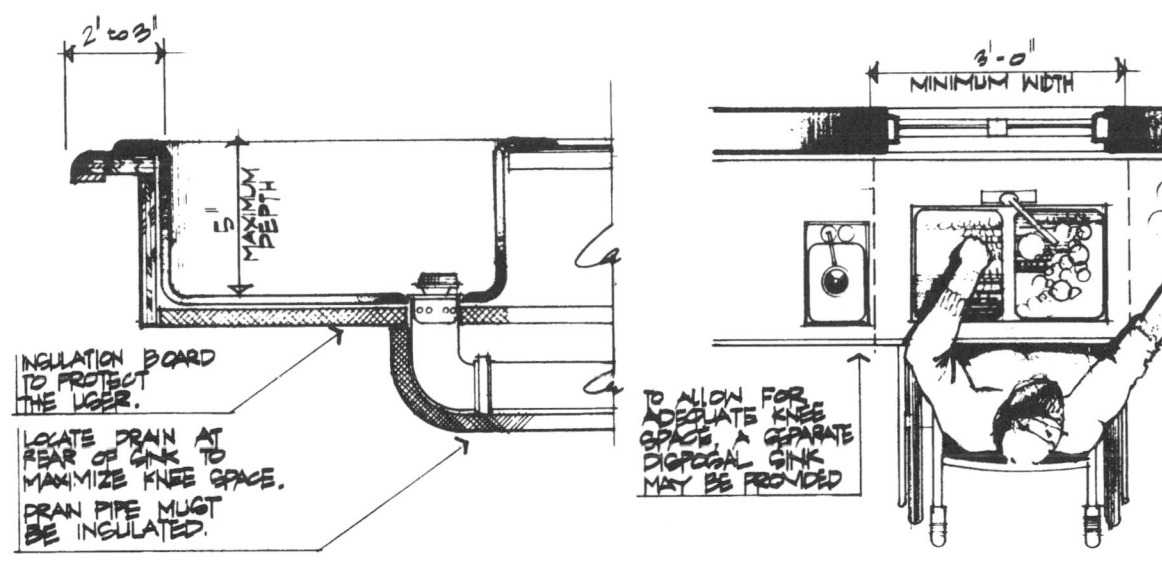

INSULATION BOARD TO PROTECT THE USER.

LOCATE DRAIN AT REAR OF SINK TO MAXIMIZE KNEE SPACE.

DRAIN PIPE MUST BE INSULATED.

2' to 3'

5" MAXIMUM DEPTH

Fig. 13 Sink.

3'-0" MINIMUM WIDTH

TO ALLOW FOR ADEQUATE KNEE SPACE, A SEPARATE DISPOSAL SINK MAY BE PROVIDED

Fig. 14 Disposal sink.

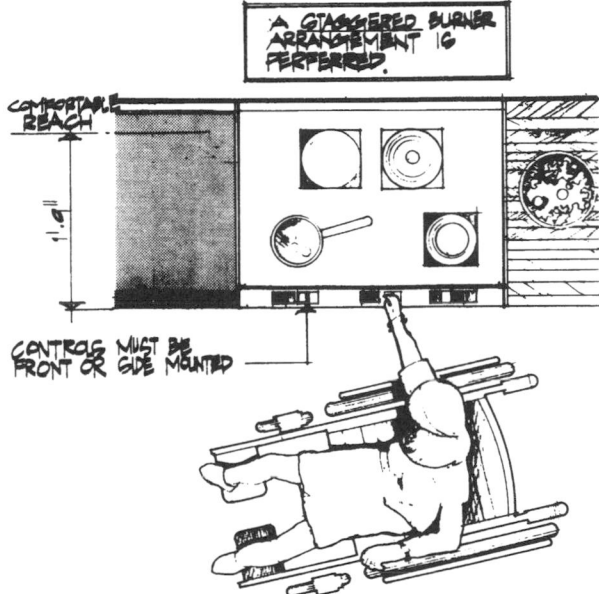

A STAGGERED BURNER ARRANGEMENT IS PREFERRED.

COMFORTABLE REACH

11.9"

CONTROLS MUST BE FRONT OR SIDE MOUNTED

Fig. 15 Counter-mounted cooktop.

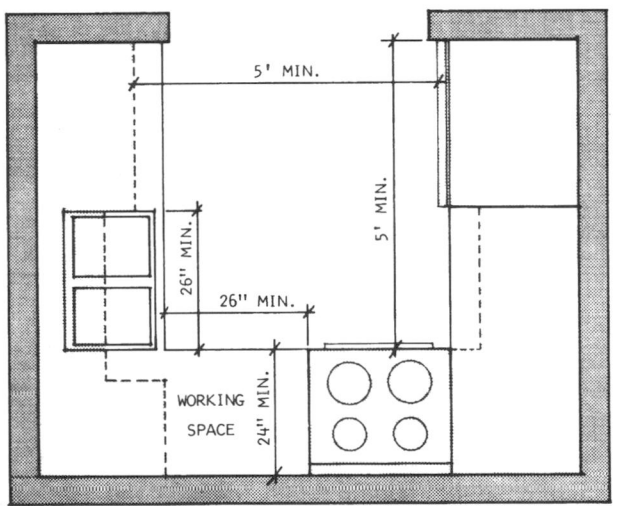

5' MIN.

5' MIN.

26" MIN.

26" MIN.

24" MIN.

WORKING SPACE

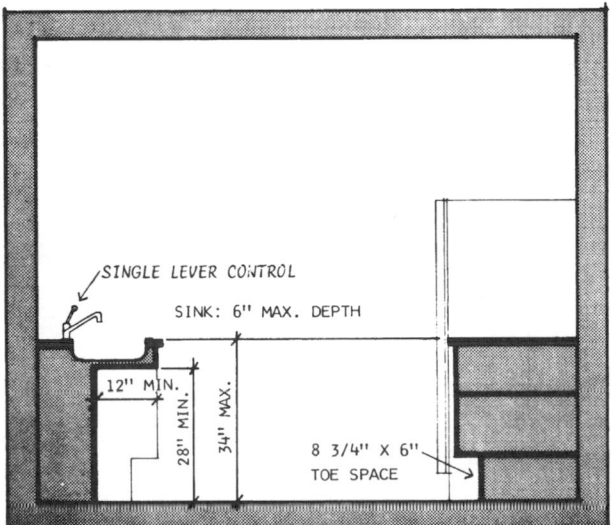

SINGLE LEVER CONTROL

SINK: 6" MAX. DEPTH

12" MIN.

28" MIN.

34" MAX.

8 3/4" X 6" TOE SPACE

Fig. 16

HANDICAPPED HOUSING—KITCHEN

WORK SURFACES WITH KNEE
SPACE

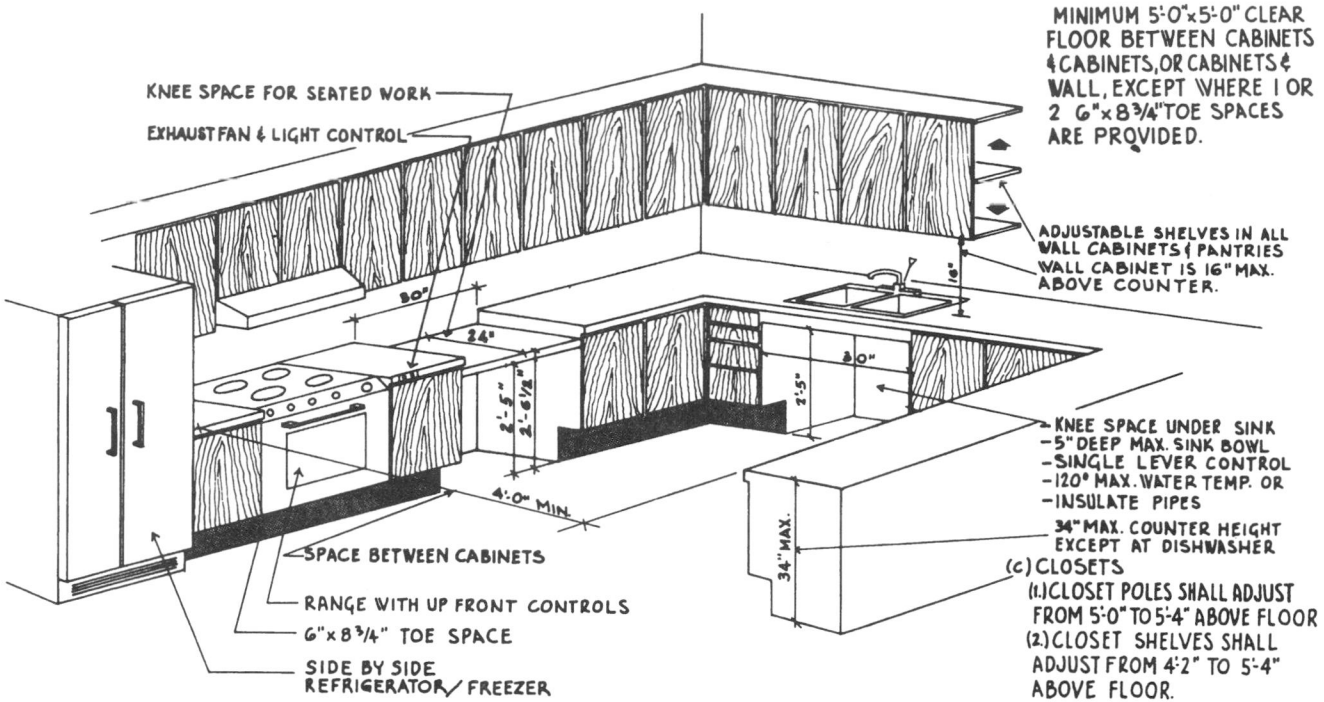

KNEE SPACE FOR SEATED WORK

EXHAUST FAN & LIGHT CONTROL

MINIMUM 5'-0" x 5'-0" CLEAR
FLOOR BETWEEN CABINETS
& CABINETS, OR CABINETS &
WALL, EXCEPT WHERE I OR
2 6" x 8¾" TOE SPACES
ARE PROVIDED.

ADJUSTABLE SHELVES IN ALL
WALL CABINETS & PANTRIES
WALL CABINET IS 16" MAX.
ABOVE COUNTER.

KNEE SPACE UNDER SINK
—5" DEEP MAX. SINK BOWL
—SINGLE LEVER CONTROL
—120° MAX. WATER TEMP. OR
—INSULATE PIPES
34" MAX. COUNTER HEIGHT
EXCEPT AT DISHWASHER
(c) CLOSETS
(1.) CLOSET POLES SHALL ADJUST
FROM 5'-0" TO 5'-4" ABOVE FLOOR
(2.) CLOSET SHELVES SHALL
ADJUST FROM 4'-2" TO 5'-4"
ABOVE FLOOR.

SPACE BETWEEN CABINETS

RANGE WITH UP FRONT CONTROLS

6" x 8¾" TOE SPACE

SIDE BY SIDE
REFRIGERATOR / FREEZER

4'-0" MIN.

Fig. 17

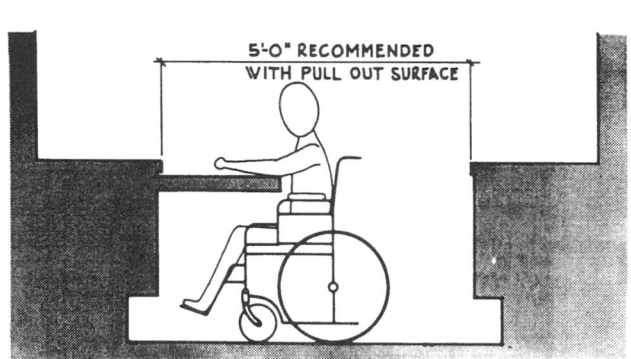

5'-0" RECOMMENDED
WITH PULL OUT SURFACE

Fig. 18a   Pull-out.

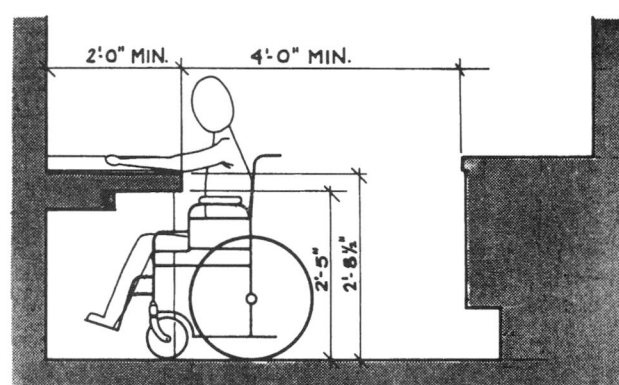

2'-0" MIN.   4'-0" MIN.

2'-5"
2'-8½"

Fig. 18b   Built-in.

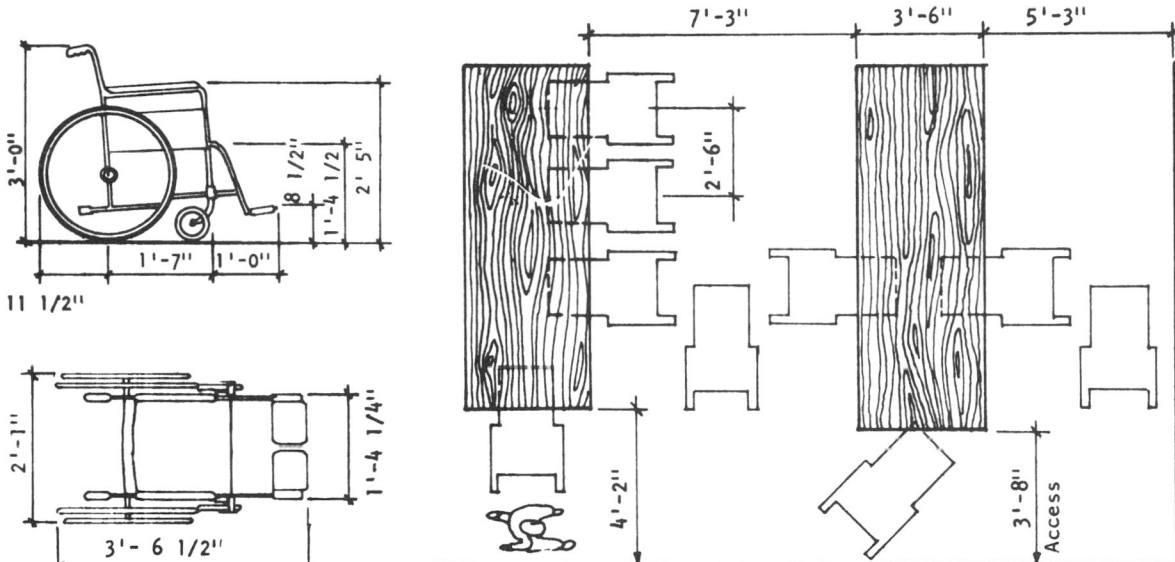

Fig. 1   The wheelchair.

Fig. 2   Clearances for central dining—wheelchair users.

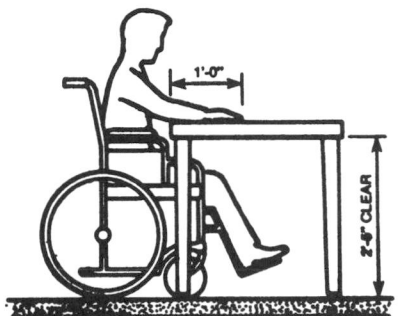

Fig. 3   Clearance under table.

Seating for wheelchair users should be on at least 2-ft 6-in centers. Tables should be 3 ft 6 in wide if chair users are to face each other. Wider tables are not recommended because of chair users' restricted reaching ability.

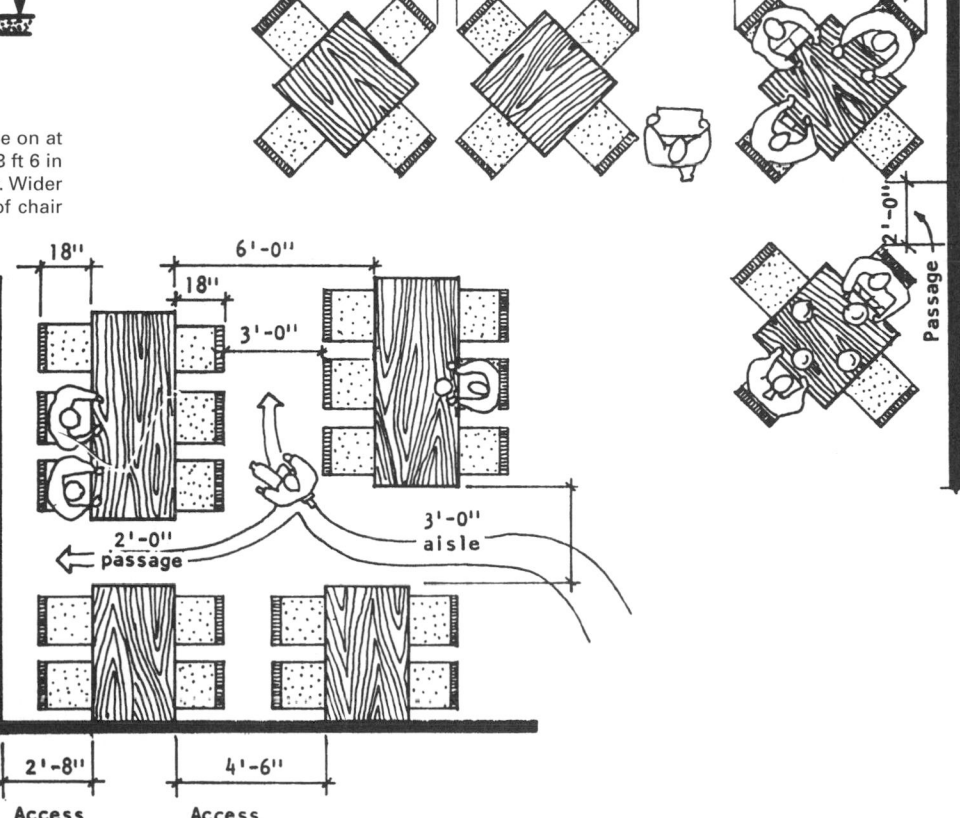

Fig. 4   Clearances for central dining.

HANDICAPPED HOUSING—OTHER FEATURES AND ROOMS

STORAGE

Household storage space, for the chairborne, is essentially closet space. Attic and basement storage, as well as most shelf space higher than a seated person can easily reach, is usually physically off limits. A certain amount of household storage involves seasonal items and things for which there is only occasional use. Such things can be relegated to the unreachable spaces, provided that they can be placed there and retrieved by others in the family as required. It must be considered and accepted that most wheelchair people who live in one-family homes cannot do so alone.

Things in active use regularly and daily, in every area of the house, must be within comfortable reach of the wheelchair resident if he or she is to function independently without frustration.

Planning wheelchair-accessible storage calls for ingenuity. It involves, first, appropriate location of the storage space. Then, doors and drawers must open and close with little effort, not act as barriers to the reaching hand. Here the secret lies in selection of appropriate hardware of the finest quality.

Examples are kitchen cabinet doors, without door pulls, that open by a light tap on the door surface and close just as effortlessly. Also, drawers that open at a finger touch, no matter how heavily laden, and close when gently nudged, literally pulling away and closing themselves 2 or 3 in before complete closure. Also, clothes closets with smoothly moving sliding doors and low, reachable hanger poles. A wheelchair does not need maneuvering around a sliding door as it does for a side-hung door. But if a side-hung door must be used, it should be capable of opening a full 180° to permit unobstructed access for the chair.

The popular walk-in type of closet is particularly well adapted to wheelchair use for clothing and other storage needs. There should be adequate space, however, for reasonable wheelchair maneuvering at the clothes racks or shelves, as well as easy access and egress. A standard side-hung door may be most appropriate for this use.

For storage of packaged food, cleaning supplies or cooking utensils, a broom-closet-sized cupboard with a vertical center pole around which circular shelves revolve can present items directly to the chairborne person's reach at heights that would make them unreachable at the rear of ordinary fixed shelves.

FLOORS AND FLOOR COVERINGS

For a wheelchair, wooden or tile flooring is preferable to carpeting. The chair moves easily on firm flooring but is much more difficult to propel over carpet pile. Smooth and well-main-

tained woods or tiles are attractive; they can be kept so by the occasional application of a non-skid polish.

If carpeting must be used, a very low, dense pile should be chosen, in carpet of a tough fiber such as nylon, of the kind generally used in commercial establishments. Avoid deep piles and shags. A test should be made before buying by rolling a wheelchair over a large sample piece in the store. If the pile drags at the wheels or mats down and shows tracks, it is not suitable. The indoor-outdoor type of carpeting might do—provided it passes an actual wheelchair test.

When laid, carpeting should be tightly stretched in a wall-to-wall installation. Any exposed edge may present an obstruction to a chair.

Scatter rugs and larger area rugs should not be used unless they can be firmly anchored to the floor. This may be possible with a good grade of double-faced adhesive tape. Otherwise, loose rugs are likely to be clutched at, pulled out of place, and lumped up by the swiveling action of the wheelchair's small casters—a source of continual irritation and frustration to the chairborne individual.

THE BEDROOM

The size requirements of a bedroom may vary considerably, depending on the furnishings that must go into it. A single bed, either twin size (39 by 75 in) or hospital-type, can fit into a room that is 14 by 11½ ft. This will allow adequate wheelchair mobility and transfer, provided, of course, that the room is not cluttered with excess furniture.

If two twin beds are involved, they may be spaced apart so that a wheelchair homemaker can get around and between them in making them up and cleaning. For this, the minimum space required is considerably greater, as much as 20 by 1½ ft. Pushing the beds together to make a close-to-king-size (78 in wide) unit can reduce the needed room length to about 17 ft. These are minimums; as always, the luxury of extra space wherever possible makes for more comfortable wheelchair living.

If possible, avoid positioning the head of a bed against an outside wall, unless the wall is exceptionally well insulated. *Never* place the head of a bed against a window.

Arrange furniture so as not to create an obstacle course for the wheelchair, especially if it must sometimes move around in the dark.

Bedroom planning should include direct and clear passageway to an adequate bathroom, preferably within the privacy of the bedroom.

The bedroom should have adequate and accessible electrical outlets for all appliances likely to be used there. These might include such

easily overlooked things as the charger for an electric wheelchair or an electrically operated adjustable bed, as well as a vaporizer, television, radio, phonograph, and other personal items.

Selecting the appropriate spot for an intercom may need thought; also, the possible need for remote-control switches at bedside for certain house lights. Then there is the right place for the telephone and possibly for a citizen's-band base station transceiver. Organizing any such needs into a comprehensive package is easier when the planning is done in advance, along with the positioning of the bed.

EXERCISE AREA

The health and sense of well-being of a chairborne individual are improved by physical activity and exercise, whether self-motivated and administered or applied with the aid of a physical therapist. The person with a strong upper body can achieve a large part of his or her daily exercise while seated in the wheelchair itself, partly by the normal activities of daily living, and partly by simple calisthenics such as push-ups from the seat, weight lifting, bending, flexing, twisting, and isometrics.

Additional setting-up exercises can be done on a bed, either alone or with the aid of a physical therapist. The individual with limitations of upper-body strength must have the assistance of a physical therapist or other trained aide.

Where possible, an exercise area set aside exclusively for the purpose is recommended for the person who needs or desires the use of specialized, space-consuming apparatus for arm development, weight-bearing (standing), or other such activities. A spare room at least 10 ft by 10 ft will serve the purpose, if properly planned.

Wall-hung upper-body and arm exercisers take almost no wall space but do require floor space in front and to the sides, for the chair and for wide arm movements. Parallel bars must be approachable by the chair, a maneuver that in itself takes over 4 ft; so the length of the bars can hardly be more than 5 to 6 ft. A floor mat may be attached to a hinged board, which can be folded up to a wall and out of the way when not in use.

Planning a room for convenient use will encourage the urge to exercise. In addition, such a room may be the ideal place to store such other frequently used items as a shower chair or an electric wheelchair.

WORK AREA

The work area countertop should be no higher than 2 ft 10 in, and a recess must be provided below. This recess must have a minimum width of 3 ft 0 in.

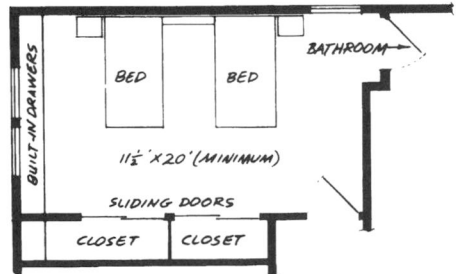

Fig. 1   Twin beds spaced apart for wheelchair access between. Keep furniture out of wheelchair path.

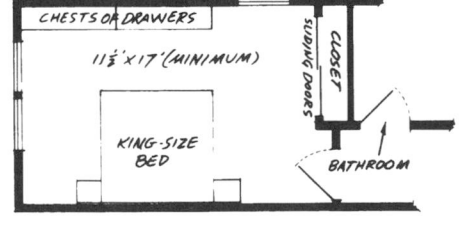

Fig. 2   Twin beds together, or king-sized bed, need less space.

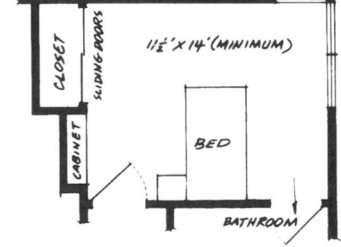

Fig. 3   Single bed, or hospital type, needs least space.

THE BATHROOM

Chiefly for reasons of economy, no doubt, it is a long-standing American practice to make bathrooms small, crowded, and inconvenient even for people who are fully ambulant. Such a room may be totally unusable for a chairborne person.

Designing an adequate bathroom needs careful and thoughtful planning. Any barrier to its convenient use can be an emotional and health problem to the wheelchair resident, as well as creating unnecessary dependence on others. The bathroom should be *more* than adequate. A room of minimum size, carefully planned to admit a wheelchair and serve its occupant's needs, might do. But generous size combined with well-selected and properly placed facilities will create an atmosphere of comfort and convenience in a psychologically sensitive area; this can give dividends in morale and self-sufficiency.

Size in itself is probably the most beneficial feature of a wheelchair bathroom. It should be sufficient for easy entering, turning, and maneuvering of the chair. Ideally, the chairborne individual should be able to wheel in, close the door, approach any of the facilities with ease, use them comfortably, and turn to wheel out without backing up.

Accessibility is the first consideration in planning the use of the space allotted. The door should be approachable directly, without obstructions and preferably without need for a turn. A 34-in clear opening is recommended, but it cannot be less than 32 in.

If space does not permit closing an inward-opening door with the chair inside, the door can be reversed to open outward. Such a door should have a handhold close to the hinge for pulling it shut from inside. For safety, an outward-opening door may be preferable even if the space is adequate. Should an emergency arise needing outside aid, an inward-opening door might be blocked by the chair itself from being opened from the outside.

The bathroom door might be the double-hung type, opening in either direction (again, a handhold near the hinge is desirable); or the accordion type; or a folding one. A sliding door is probably the most efficient. The doorsill should be shallow or absent completely.

Facilities. The purpose of a bathroom is, of course, to make possible two things: elimination of body wastes and maintenance of body cleanliness. The first requires an accessible toilet commode. The second calls for a wash basin, plus a shower or tub or both—possibly in a two-in-one combination.

Wheelchair users of the facilities may have preferences determined by the nature of their disability and the strength of their arms and upper body. Adaptive equipment can be installed, such as support bars, an overhead trapeze to enable them to lift themselves by one or both arms, a bathtub lift device, or a built-in seat outside of and at one end of the tub.

The tub's upper edge should be flat, not rounded. Removable wheelchair armrests and in some instances removable footrests may facilitate movement from chair to tub.

Shower. For the person who prefers a shower, a stall model may be more practical than the shower in the tub.

A roll-in stall shower, about 4 by 4 ft (inside dimensions) is ideal. A special wheelchair may be used—perhaps an old one adapted by lacquering or painting to make it reasonably rustproof, with seat and back of canvas or other waterproof material and perhaps with a cutout seat to permit cleansing of the lower body. A hand-held telephone-type shower fixture on a flexible hose greatly eases showering.

Instead of the usual high sill at the stall entrance to prevent overflow, it should be very low or absent, replaced with a slight grade toward the drain. An extra-long curtain that rests 2 or 3 in on the floor will help control splashing and overflow.

A shower stall too small to accept the wheelchair can be used if the minimum inside dimensions are 3 by 3 feet. A smooth, softly padded bench may be situated inside the stall. Space in front of the stall should be adequate for a head-on approach by the wheelchair with footrests removed to facilitate transfer to the seat inside. Or a side-of-chair transfer with an armrest removed may work. A sturdy grab bar inside the stall will aid in the transfer and in balance while showering.

Special commode-like shower chairs to fit into the smaller shower stall are available but may require the help of an attendant.

Where a tub already exists, it may be converted to a wheelchair shower by removing it and using its space and plumbing connections to install a shower fixture and floor drain at one end of the vacated area. The floor, when water-proofed and tiled, should slope slightly toward the drain. The room wall should be tiled several feet high on three sides of the area. A low tiled partition (1 or 1½ ft high) can be built on the fourth side, extending about halfway out from the wall at the shower-drain end. The other half should be left open for a roll-in entranceway. Curtains or sliding plastic walls above the partition and extra-long curtains over the entranceway will control splashing.

Toilet. The toilet should have maneuvering space beside it for sideways movement from wheelchair to seat. Again, removable chair arms and perhaps footrests may be helpful. A toilet placed close enough to the wash basin will permit anal cleaning with water after toilet use. Placed alongside the tub, the toilet seat may be used as a bench for transfer to the tub and for drying after the bath. Strong, well-secured grab bars, tailored to the needs of the wheelchair person, may be advisable.

Basin. The wash basin should be wall-hung, without a pedestal or corner legs, unless the legs are far enough apart and the underside of the basin high enough to permit a wheelchair underneath. An excellent solution is a countertop vanity into which the basin has been recessed. Proper knee space beneath should be tailored to the individual.

Mirror. A large mirror, installed over the basin low enough for comfortable use at wheelchair level, is important for ease in grooming.

Controls. Equip the wash basin, shower, and bathtub with easily handled water controls. The wash-basin faucet, for example, might be the one-lever type, with which temperature is set by movement from side to side, volume by up or down. This permits setting the desired temperature before the volume is turned on.

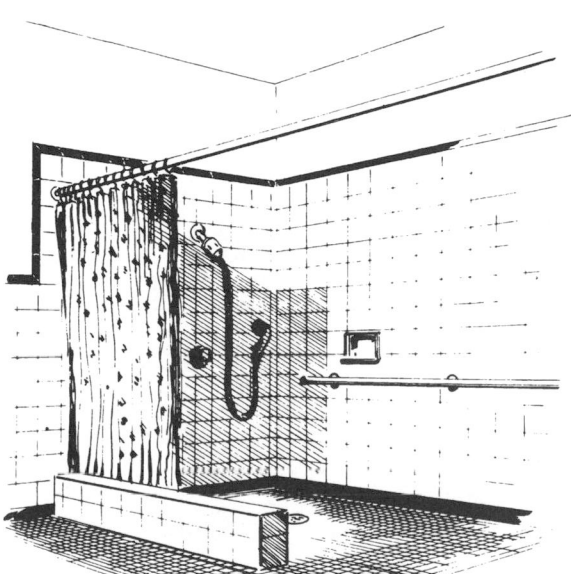

Fig. 1 Tub space converted to wheelchair shower.

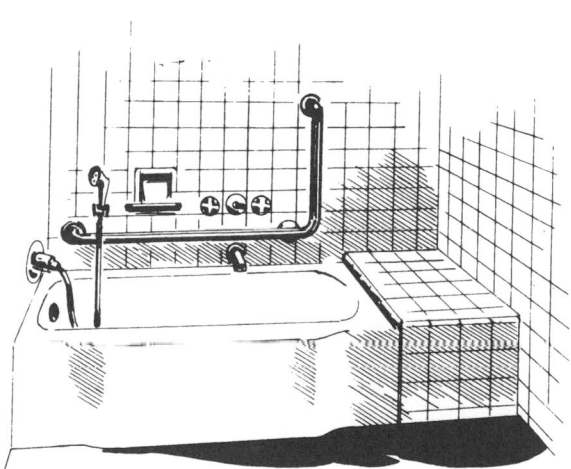

Fig. 2 Seat built outside tub, with grab bar and reachable water controls, to facilitate transfer from and to wheelchair.

Safety. Both shower head and bathtub faucet must be equipped with a protective temperature-sensitive device that prevents excessively hot water from reaching the bather.

Protect exposed hot water and steam pipes in the bathroom with asbestos or other insulation. Contact with heat by body areas that are without sensation can do serious damage. Even when they feel it, handicapped people may not be able to recoil quickly from a hot surface.

Never install electric outlets or switches where they could be reached by a person in the bathtub or shower.

Other conveniences. Cabinet space should be adequate, within reach, and well planned, taking into account how often and when the stored items are used. For example, one cabinet should be reachable from a seated position on the toilet. Also, a hook on the wall between the wash basin and toilet may make possible the unaided use of enema equipment.

Another convenience is an auxiliary heating device, installed in the wall or ceiling and controllable by a switch at the room entrance.

For someone who requires assistance in using the bathroom facilities, space must be provided for the attendant to maneuver. It should not be necessary to remove the wheelchair from the room to enable the attendant to assist with an enema, use a bathtub transfer lift, or give other services.

Finally, avoid clutter in the bathroom. Keep aids and equipment for both the wheelchair person and others to a necessary minimum. Also hold down the materials customarily stored in the bathroom. Make the room an attractive place as well as a convenient one.

1. Doors should have a 32-in clear opening and swing out or slide.

2. 6 ft 0 in required between walls except at end of tub wall.

3. Where 5- by 5-ft clear floor space is not provided, a wall-hung lavatory must be used.

4. Wall adjacent to toilets and tubs should be capable of supporting handrails which can support a 250-lb load.

5. Where provided, at least one mirror should be placed above lavatory no higher than 40 in above floor.

6. Where provided, at least one towel rack should be not more than 40 in above floor.

7. Maximum water temperature should not exceed 120°, or exposed hot-water lines and drains should be fully insulated.

Showers for the handicapped should have:

1. Nonslip floor.

2. Seat 19 in above floor and placed opposite controls and grab bar.

3. Grab rail opposite seat and around back wall.

4. Single-lever water controls.

5. Flexible hose with hand-held shower head.

6. Soap tray no more than 40 in above floor.

Cabinets should have minimum toe space 6 in in depth and 8¾ in in height.

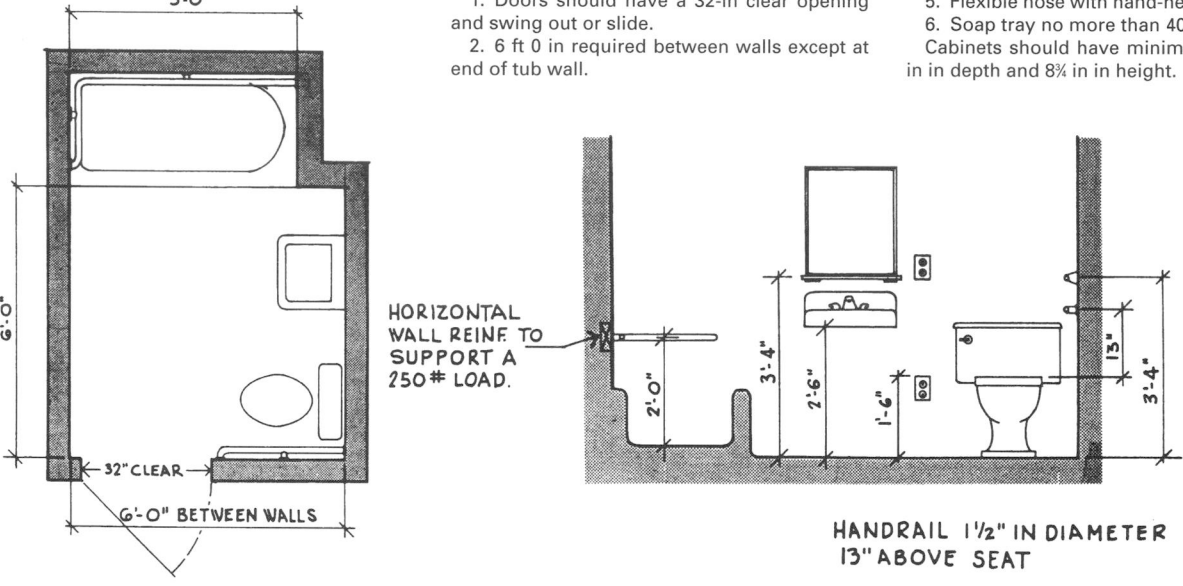

Fig. 3

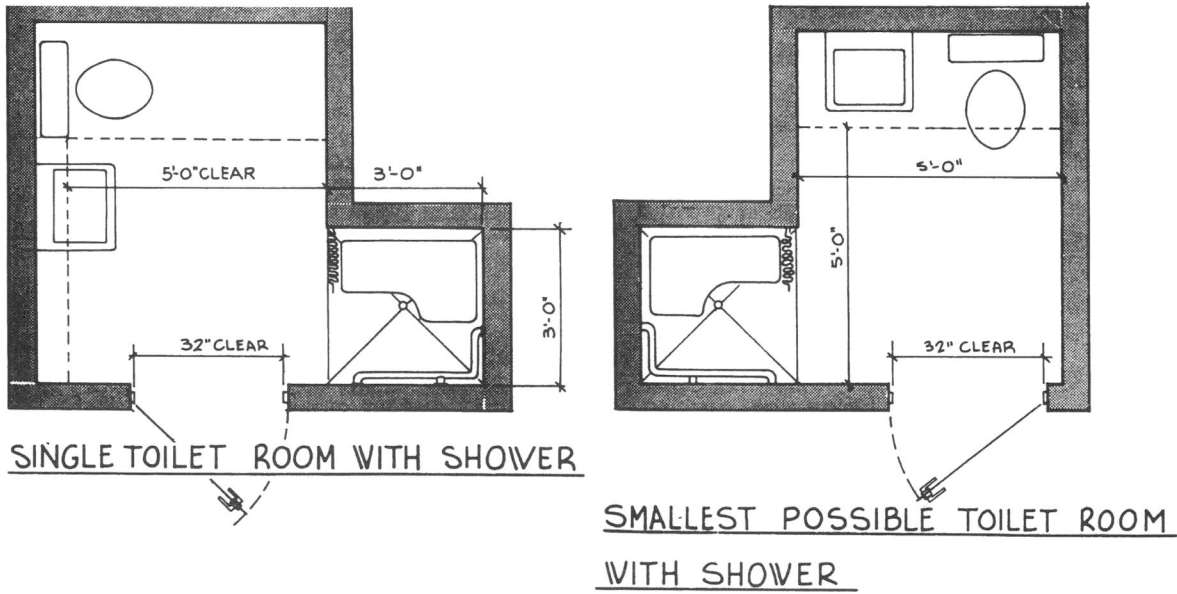

SINGLE TOILET ROOM WITH SHOWER

SMALLEST POSSIBLE TOILET ROOM WITH SHOWER

Fig. 4

1. 5- by 5-ft clear floor space may have one or two 6-in toe spaces.

2. Doors to toilet rooms must have 32-in. clear openings.

Toilet seat must be 16½ to 20 in above the floor.

In individual toilet rooms, at least one chrome or stainless-steel handrail 52 in long, 1½ in diameter, must be wall-mounted 13 in above the seat.

3. Lavatories must be mounted with bottom of apron 30 in minimum above floor and rim 34 in maximum above floor.

4. Maximum water temperature must not exceed 120°, or exposed hot-water lines and drains must be fully insulated.

5. Mirrors and shelves should not be more than 40 in above the floor.

6. Towel racks and dispensers should not be more than 40 in above floor.

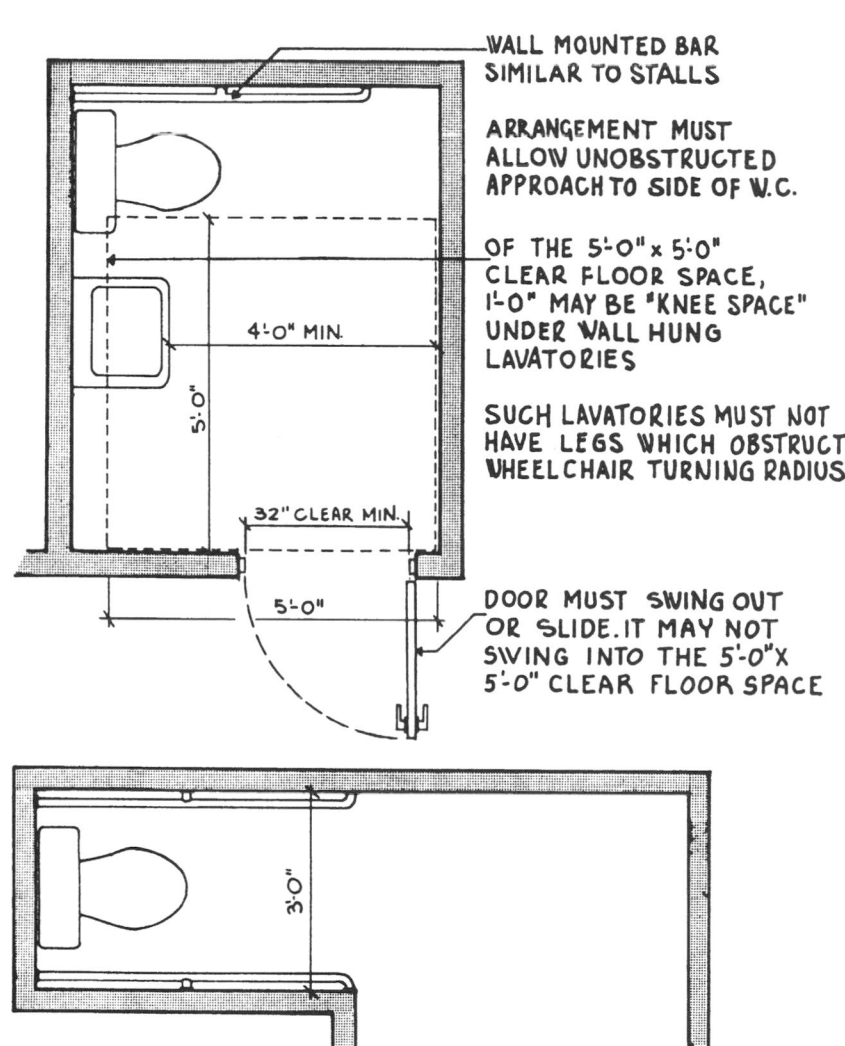

WALL MOUNTED BAR SIMILAR TO STALLS

ARRANGEMENT MUST ALLOW UNOBSTRUCTED APPROACH TO SIDE OF W.C.

OF THE 5'-0" x 5'-0" CLEAR FLOOR SPACE, 1'-0" MAY BE "KNEE SPACE" UNDER WALL HUNG LAVATORIES

SUCH LAVATORIES MUST NOT HAVE LEGS WHICH OBSTRUCT WHEELCHAIR TURNING RADIUS.

DOOR MUST SWING OUT OR SLIDE. IT MAY NOT SWING INTO THE 5'-0" X 5'-0" CLEAR FLOOR SPACE

COMPARTMENTED TOILET ROOM SIMILAR TO STALL REQUIRES BOTH HANDRAILS SINCE THERE IS NOT ENOUGH ROOM FOR SIDE APPROACH.

Fig. 5

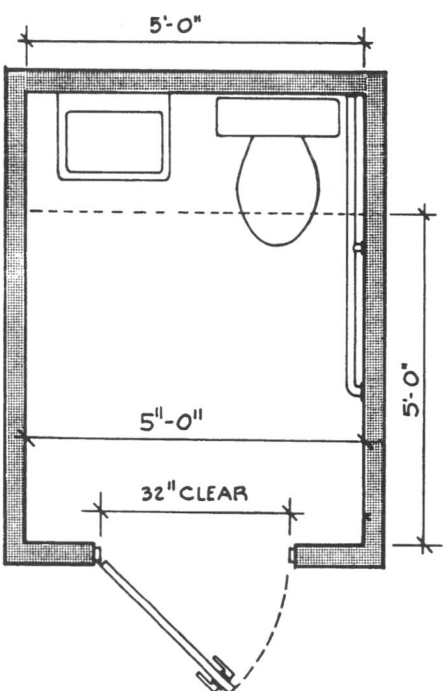

SMALLEST POSSIBLE TOILET ROOM WITH 5'-0" x 5'-0" CLEAR FLOOR SPACE. PART OF CLEAR FLOOR SPACE MAY BE UNDER TOILET IF TOILET HAS A DEEPLY RECESSED BASE TO PROVIDE CLEARANCE AS INDICATED ON PAGE 43.

FOR SIDE TRANSFER TOILETS A 3'-0" GRAB BAR IS ACCEPTABLE WHEN MOUNTED TO EXTEND 1'-6" BEYOND THE FRONT EDGE OF W. C.

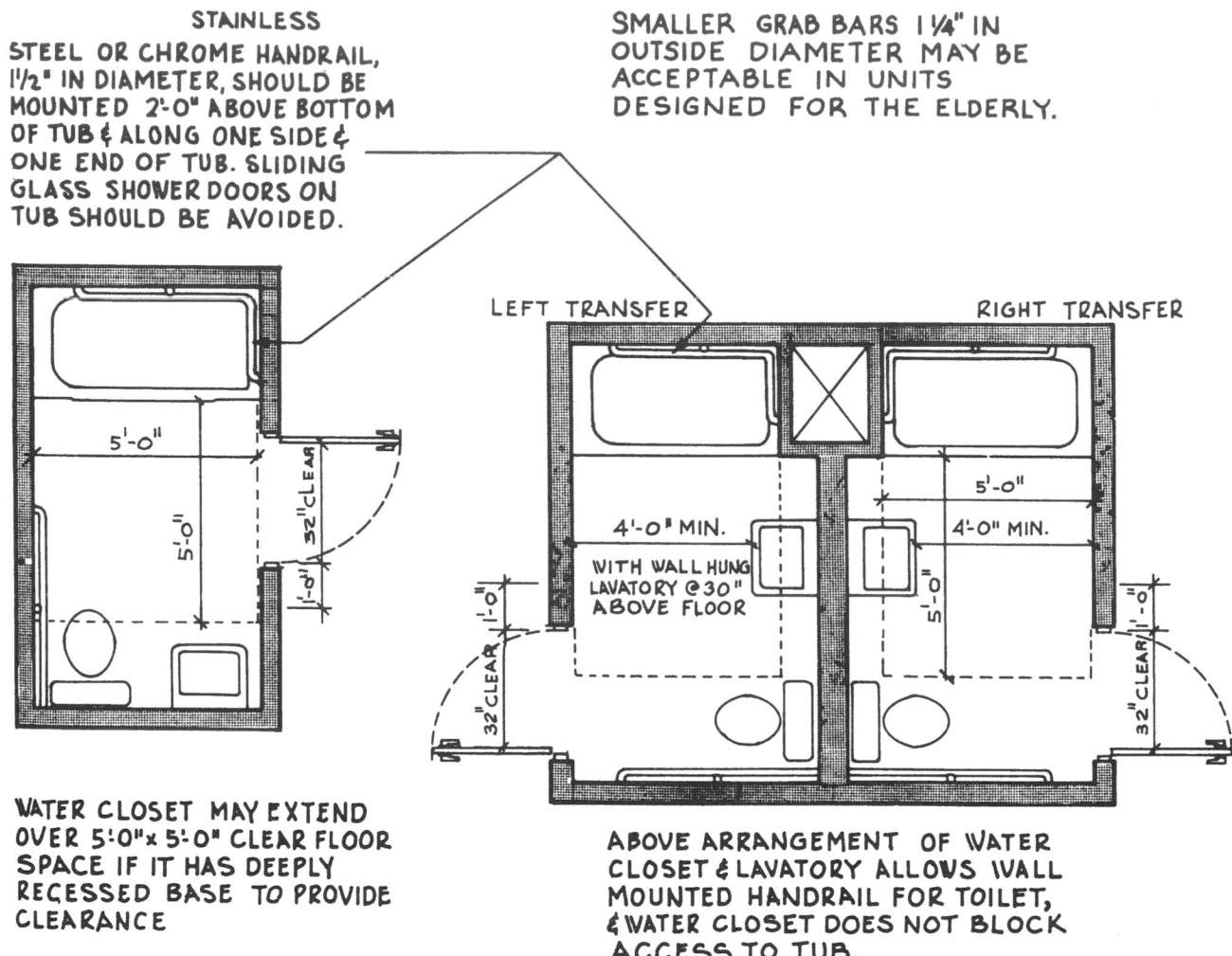

STAINLESS
STEEL OR CHROME HANDRAIL,
1½" IN DIAMETER, SHOULD BE
MOUNTED 2'-0" ABOVE BOTTOM
OF TUB & ALONG ONE SIDE &
ONE END OF TUB. SLIDING
GLASS SHOWER DOORS ON
TUB SHOULD BE AVOIDED.

SMALLER GRAB BARS 1¼" IN
OUTSIDE DIAMETER MAY BE
ACCEPTABLE IN UNITS
DESIGNED FOR THE ELDERLY.

LEFT TRANSFER   RIGHT TRANSFER

5'-0"

5'-0"

32" CLEAR

1'-0"

4'-0" MIN.

WITH WALL HUNG
LAVATORY @ 30"
ABOVE FLOOR

5'-0"

4'-0" MIN.

5'-0"

32" CLEAR 1'-0"

32" CLEAR 1'-0"

WATER CLOSET MAY EXTEND
OVER 5'-0" x 5'-0" CLEAR FLOOR
SPACE IF IT HAS DEEPLY
RECESSED BASE TO PROVIDE
CLEARANCE

ABOVE ARRANGEMENT OF WATER
CLOSET & LAVATORY ALLOWS WALL
MOUNTED HANDRAIL FOR TOILET,
& WATER CLOSET DOES NOT BLOCK
ACCESS TO TUB.

Fig. 6

EACH BATHROOM SHALL
MEET THE REQUIREMENTS
ILLUSTRATED HERE.

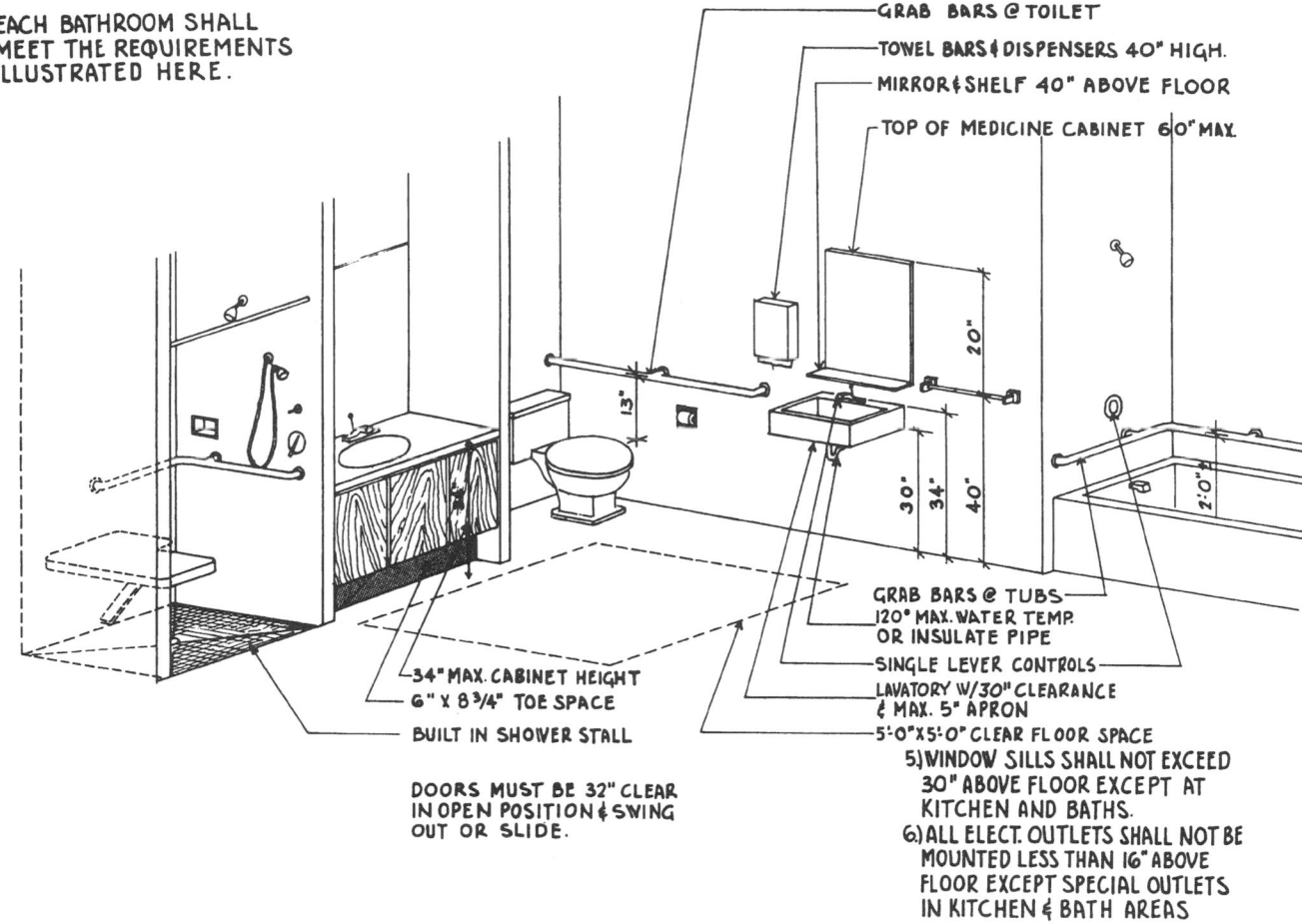

GRAB BARS @ TOILET

TOWEL BARS & DISPENSERS 40" HIGH.

MIRROR & SHELF 40" ABOVE FLOOR

TOP OF MEDICINE CABINET 6'0" MAX.

13"

20"

30"
34"
40"

2'-0"

34" MAX. CABINET HEIGHT
6" X 8¾" TOE SPACE

BUILT IN SHOWER STALL

DOORS MUST BE 32" CLEAR
IN OPEN POSITION & SWING
OUT OR SLIDE.

GRAB BARS @ TUBS
120° MAX. WATER TEMP.
OR INSULATE PIPE

SINGLE LEVER CONTROLS

LAVATORY W/30" CLEARANCE
& MAX. 5" APRON

5'-0"X5'-0" CLEAR FLOOR SPACE

5) WINDOW SILLS SHALL NOT EXCEED
30" ABOVE FLOOR EXCEPT AT
KITCHEN AND BATHS.

6) ALL ELECT. OUTLETS SHALL NOT BE
MOUNTED LESS THAN 16" ABOVE
FLOOR EXCEPT SPECIAL OUTLETS
IN KITCHEN & BATH AREAS

Fig. 7

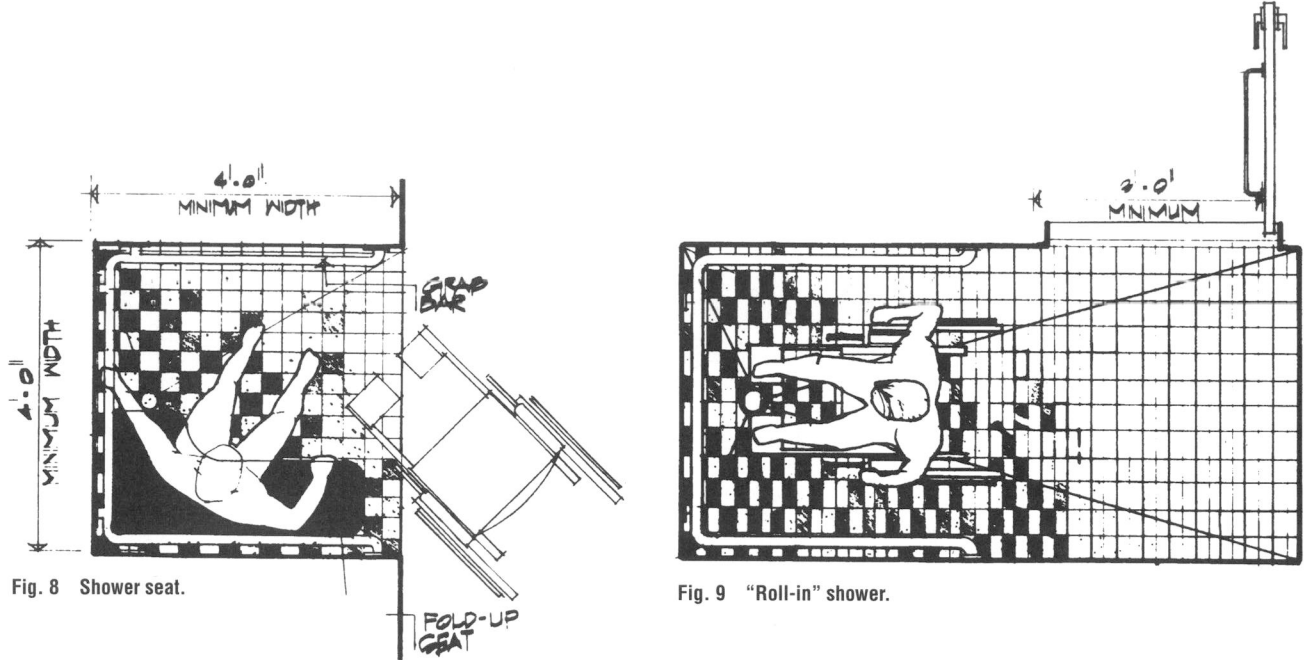

Fig. 8   Shower seat.

Fig. 9   "Roll-in" shower.

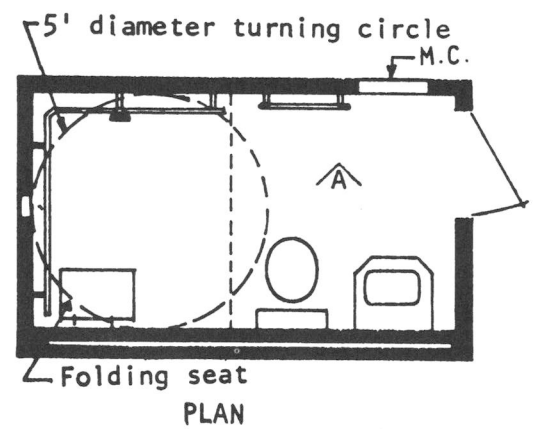

PLAN

BATHROOM FOR WHEELCHAIR USER

The outswinging bathroom door has a flush threshold.

Space is provided for both frontal and lateral transfer from wheelchair to water closet.

At least 6 in is left between lavatory and toilet to accommodate a toilet chair or toilet-mounted side grab bars as per tenant preference.

Standard-height toilet seat (1 ft 3 in) permits use of toilet chair (1 ft 5 in).

Lavatory with 4-in-deep undercoated bowl and lever-type faucet, drain at side or rear, any exposed hot-water piping to be well insulated, and front edge capable of withstanding 250-lb load, mounting height 2 ft 10 in.

Tilting mirror over lavatory.

Recessed medicine cabinet with unbreakable shelves is located near the lavatory, mounting height, 5 ft to top shelf.

All grab bars are 1½ in outside diameter with 1½-in clearance at wall.

Shower with vertical and horizontal grab bars mounted 3 ft above floor, recessed soap dish and water regulator mounted at 3 ft 6 in, regulator accessible from inside and outside the shower, folding seat, no curb, floor sloped to drain.

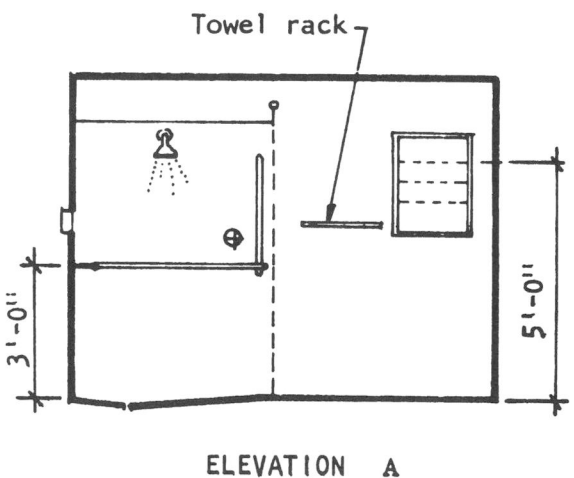

ELEVATION   A

Fig. 10

In housing for the elderly and handicapped, the units suitable for wheelchair users often can be placed advantageously on the ground floor.

For details of the kitchen shown in Fig. 1, see Fig. 3.

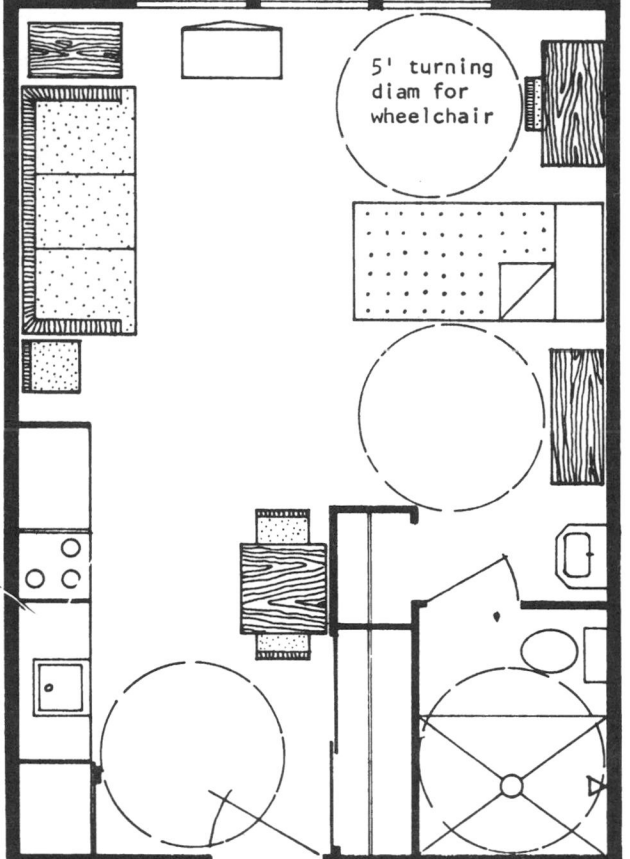

Fig. 1 Zero-bedroom living unit for wheelchair unit.

Omission of an easy chair is acceptable to give more space for occupant's wheelchair.

For details of this bathroom, see Fig. 10 on p. 908.

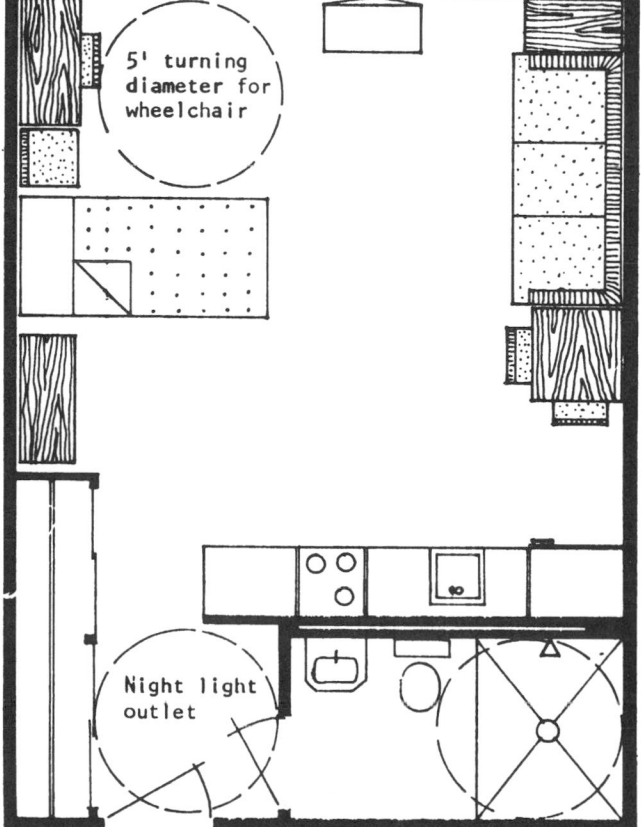

Fig. 2 Zero-bedroom living unit for wheelchair user.

HANDICAPPED HOUSING—COMBINED SPACES

Design Features of the Zero-Bedroom Living Unit

Storage cabinet with adjustable shelves, 18 in deep plus 6-in-deep shelves on door, 30 in wide, 48 in to top shelf.

Work counter, 34 in high—usable by person standing or in wheelchair, 29-in-high knee space to clear wheelchair arms.

Sink with 4-in-deep undercoated bowl and single-lever faucet, drain at side or rear. Any exposed hot-water piping to be well insulated.

Eight-inch-deep wall shelf mounted 12 in above work counter.

Standard range with controls at front.

Standard refrigerator.

Outsized cabinet door and drawer pulls.

Not shown are lap or chop boards which rest on a person's lap or on the arms of a wheelchair. If two boards are available, one should have an 8-in-diameter cutout to receive a mixing bowl. For a plan of the kitchen see Fig. 3.

Floor-mounted or wall-hung water closets should generally have a seat height of 1 ft 8 in. For some individuals, slight adjustments in height may be required to facilitate transfer to or from the wheelchair.

A horizontal grab bar at a height of 2 ft 9 in should be provided adjacent to the water closet to aid the individual to transfer to or from a wheelchair. If an individual prefers the "front transfer," grab bars should be provided on both sides.

Grab bars should be 1½ in in diameter, and anchorage must be adequate to support the individual's weight. Typical grab bar locations are illustrated; however, individual needs must be considered. As a general rule, horizontal grab bars are used for pushing up while vertical grab bars are used for pulling up.

Lavatories should be mounted at a maximum height of 2 ft 10 in. Lavatory basins should have as shallow a profile as possible to maximize knee space below. Lavatories should project approximately 2 ft 3 in from the wall.

All exposed water supply and drain lines must be insulated to prevent burns and scrapes.

The water spigot should be a minimum of 4 in clear of any rear obstruction and at least 4 in above the lavatory rim to allow for ease of rinsing. Lever-type temperature controls are recommended.

To maximize knee space below the lavatory, P traps in drain lines should be offset horizontally, or traps may be located in wall with an access panel.

Shelving and medicine cabinets should be carefully located within easy reach (see Figs. 4 and 7).

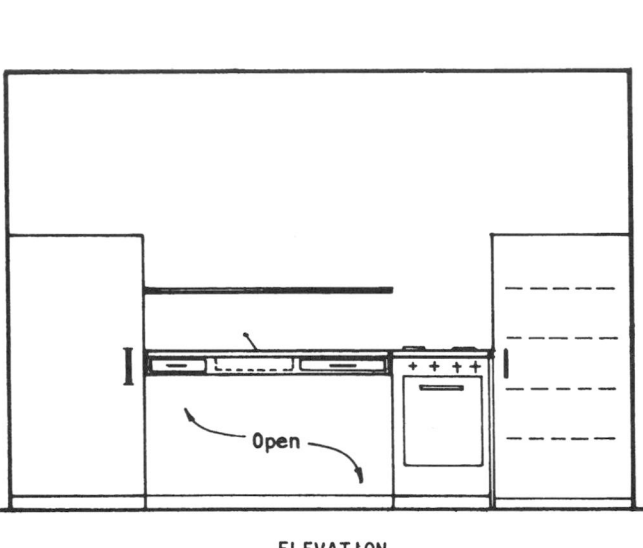

ELEVATION

Fig. 3

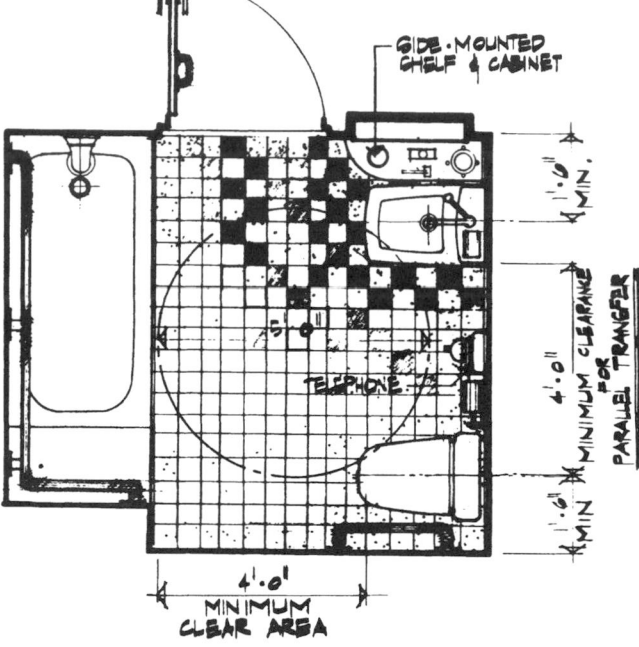

Fig. 4   Typical bathroom arrangement.

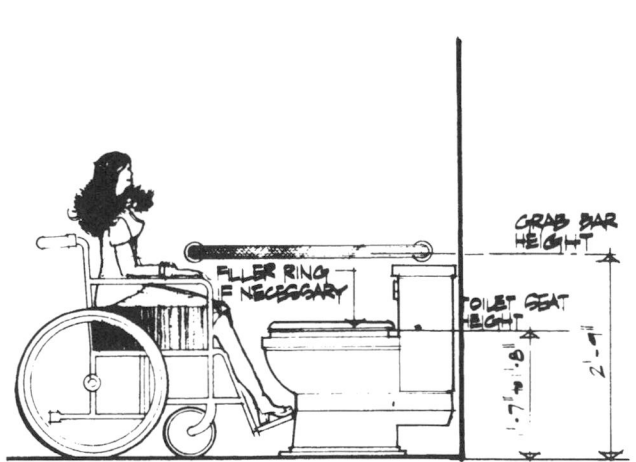

Fig. 5   Floor-mounted water closet.

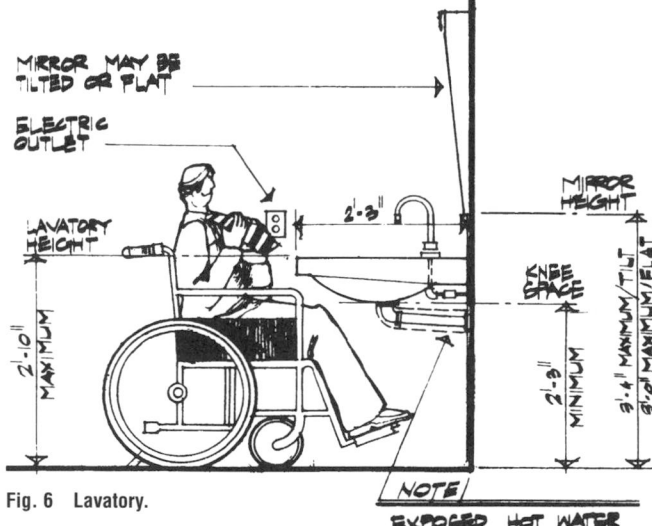

Fig. 6   Lavatory.

Mirrors should be tilted or lowered to accommodate the individual in a seated position. The bottom edge of a flat mirror should be no higher than 3 ft 0 in.

A platform or seat at the end of the bathtub aids the wheelchair user in transferring to and from the bathtub. This platform should be at the same level as the bath rim and be of the same width as the bathtub. The design must provide a clear area beside the platform to position a wheelchair for transfer. Bathtub height should ideally be set at 1 ft 8 in.

Grab bars or a suspended hoist or stirrup grip attached to the ceiling should be provided to aid the transfer. Grab bars along one side of the tub provide support during bathing.

Bathtub controls must be easily accessible before and during immersion. Some individuals may prefer side-mounted controls and remote-control drain operation.

The most flexible solution is the bathtub and shower combination. This may be used by a person sitting in the bath, seated on the platform, or standing in the bath. A hand-held shower head should be provided and all controls must be carefully located to be accessible from the platform or the bathtub. Thermostatic controls must be provided to protect against a sudden change in water temperature.

The minimum size of any wheelchair shower is 4 ft 0 in by 4 ft 0 in. The minimum shower opening is 3 ft 6 in except where a door is provided for a "roll-in" shower.

Showers should not have curbs or thresholds which impede wheelchair access.

Shower floors (as well as bathroom floors) should be of a nonslip material.

Thermostatic controls must be installed to protect the user from sudden changes in water temperature. Many people find lever-handle temperature controls easiest to operate. All controls must be easily accessible to the shower occupant.

A bench seat may be incorporated in the shower. This seat may be hinged to fold up against the wall when not in use. The seat should be mounted at a height of approximately 1 ft 8 in. Grab bars or a suspended stirrup grip should be provided to aid in transfer to the seat. Grab bars also provide support during showering.

Where shower seats are incorporated, they should be positioned consistent with the user's preference for left or right transfer.

All showers should be equipped with a flexible hose and hand-held shower head. The hand-held shower head should be stored within easy reach of the shower occupant.

Curtains are normally provided to help contain water within the shower. Small "wing walls" may be incorporated if they do not restrict access or impede transfer. Doors are sometimes used in roll-in showers and may incorporate a rubber sweep strip to prevent the escape of water. Shower doors must meet all the requirements of other interior doors.

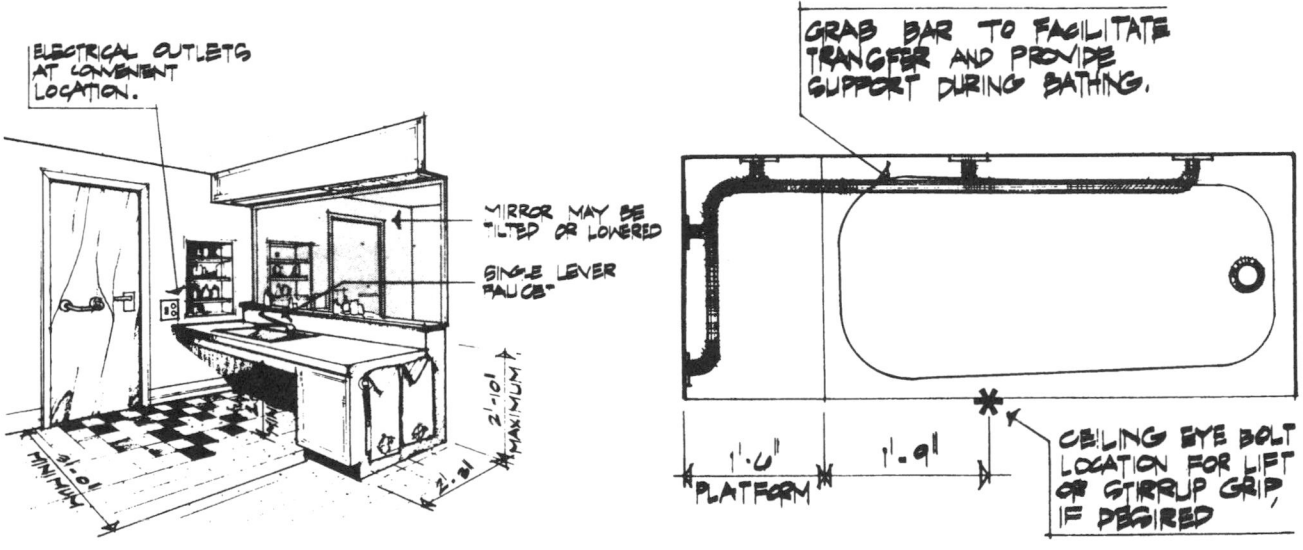

Fig. 7 Vanity.

Fig. 8 Bathtub.

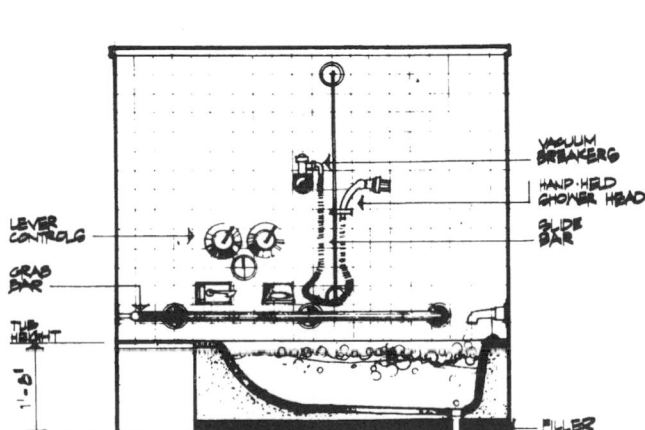

Fig. 9 Combination bathtub/shower.

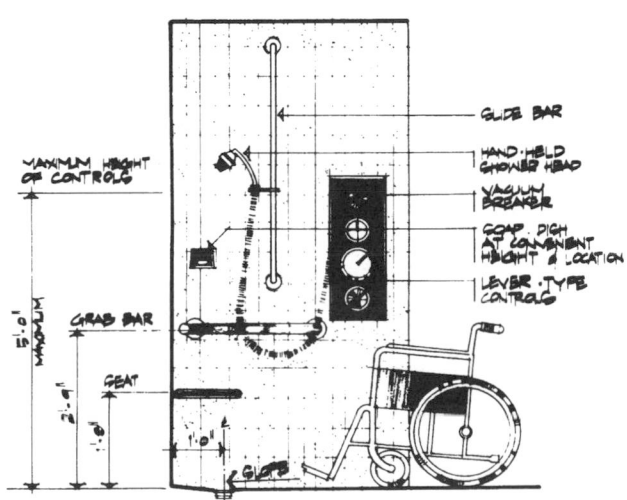

Fig. 10 Shower.

HANDICAPPED HOUSING—CORRIDORS AND RAMPS

WALKWAYS AND CORRIDOR
WIDTHS

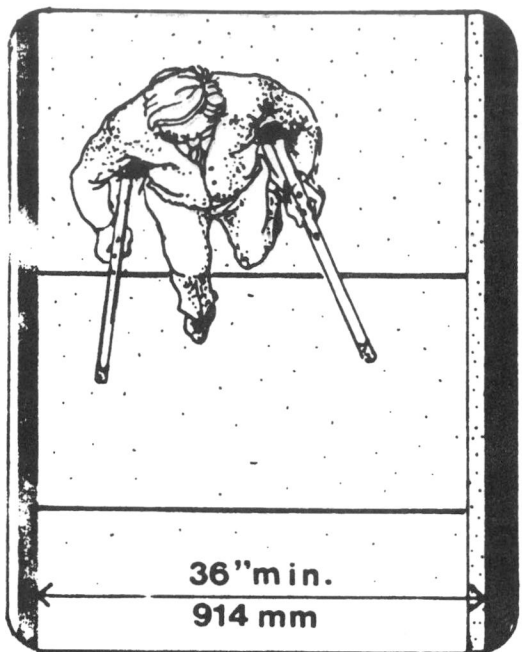

Fig. 1a   One-way traffic.

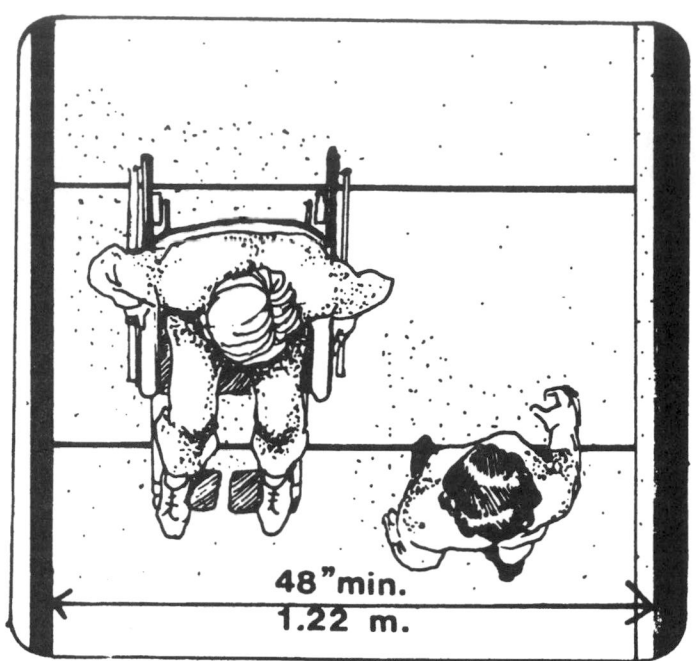

Fig. 1b   Two-way traffic.

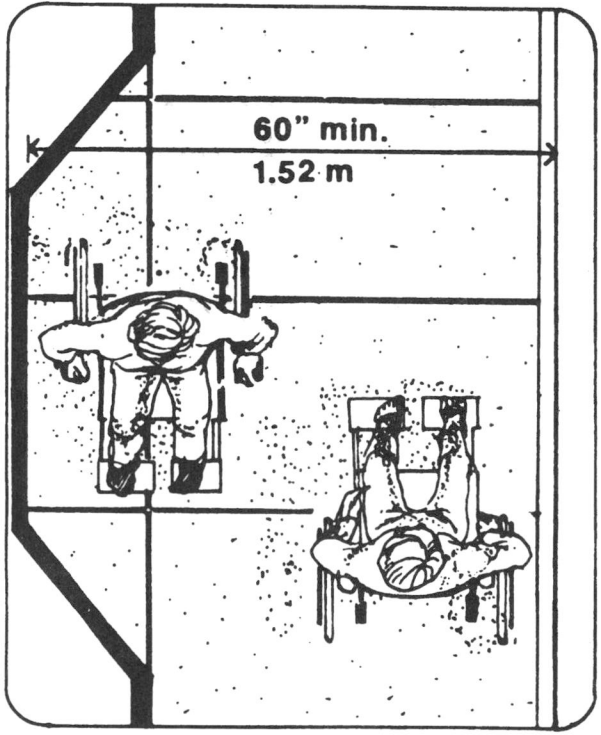

Fig. 1c   Passing areas.

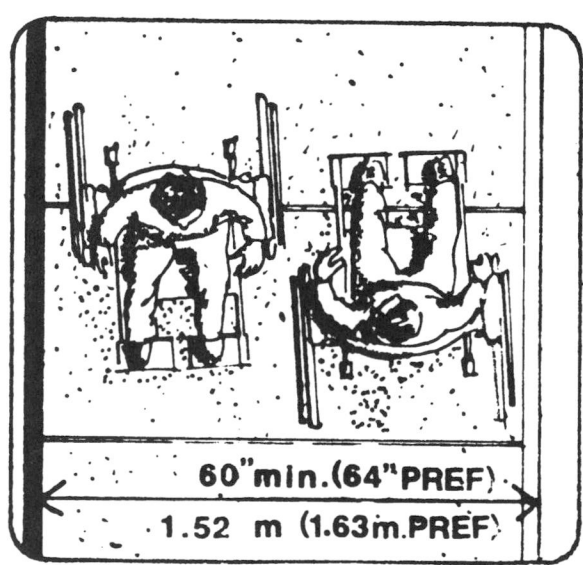

Fig. 1d   Recommended minimum widths for passing.

CIRCULATION

To get through a doorway: The clear opening (the measured width of the actual door opening less the 2 in taken up by the thickness of the door itself, standing wide open) must, as a minimum, be 27 in for a head-on approach.

A livable clear opening could be 30 in, for a head-on or slightly oblique approach. The best clear opening is 34 in (that is, a 36-in actual door opening). The 2 in taken up by the thickness of the door can be retrieved, of course, by removing the door.

■ Remove doorsills for a smoother passage. This is essential if oblique entry is necessary, especially for a homemaker with weak arms and hands. Eight-inch casters make nonremovable sills easier to cross than do 5-in wheelchair casters.

■ Common door widths in an apartment house are apartment entrance, 36 in; room doors, 32 in; bathroom, 22 to 27 in (all actual door openings).

■ Two-way swinging doors give less trouble to users who cannot easily pull a door toward them. Very light swing action hinges or "gravity" hinges (which push a door closed by its own weight) are best. Caution: A two-way door must have a window at wheelchair height to make visible traffic coming from the other direction.

To get through a hallway: A 36-in-wide hallway allows a wheelchair to move forward or to back up, but not to make a complete reverse-direction turn. Four feet (48 in) allows such a turn, with some back-and-forth maneuvering. For an easy continuous-movement reverse-direction turn, 4 ft 6 in is needed.

Four and a half feet also allows easy oblique turns into doorways as narrow as 27-in clear opening.

A wheelchair with footrests will not be able to turn from a 36-in-wide hallway into a 27-in doorway, clear opening, without assistance. For such turning, the doorway should be at least 30 in, preferably 34-in clear opening.

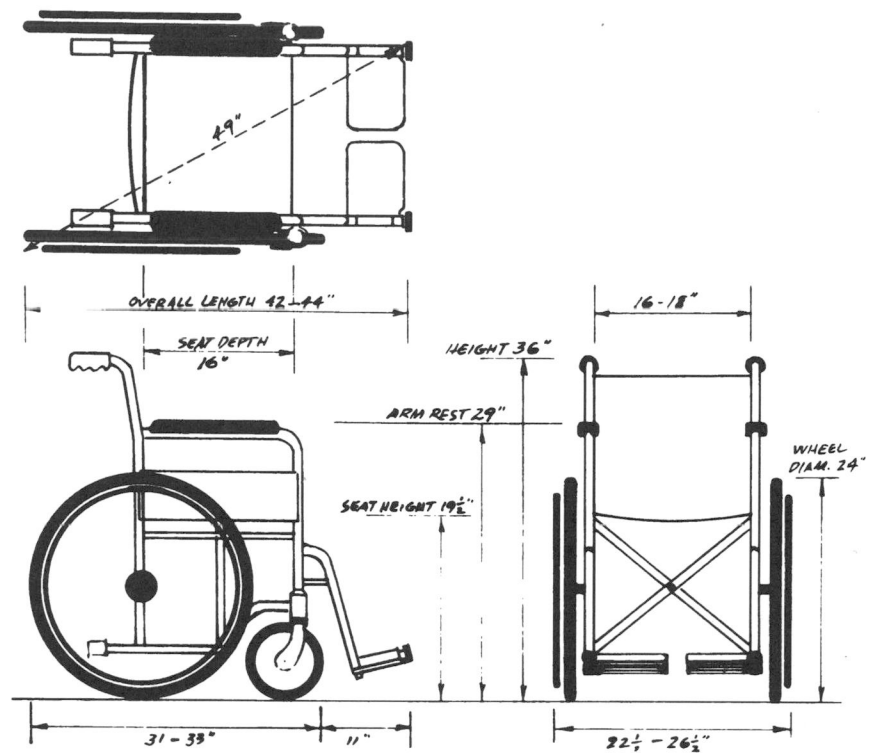

Fig. 2

A smaller chair can make the turn in a 36-inch hallway. One that's extra wide or long cannot make it.

When the door stands wide open at right angle to the opening, it reduces it by about two inches.

Fig. 3

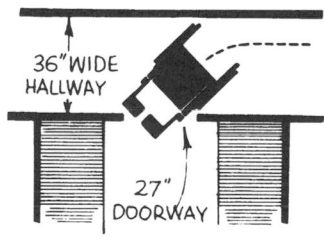

No! Without extra help, a wheelchair cannot turn into a 27-inch doorway from a 36-inch-wide hall.

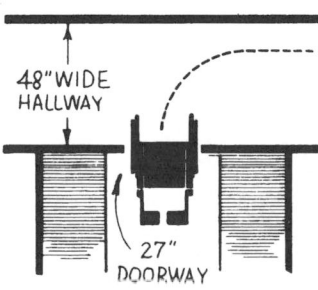

A right-angle turn approach, made possible by a 48-inch hall, makes a 27-inch doorway also negotiable.

Fig. 4

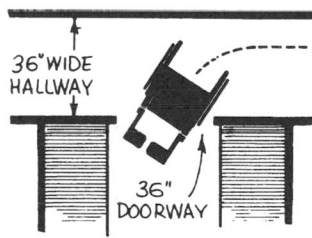

Yes! Widen the doorway to 36 inches and the chair can make it. Removing the doorsill helps, too.

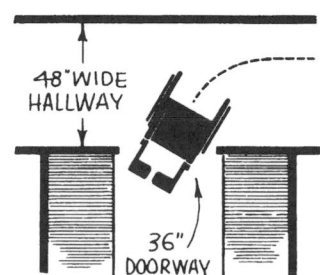

Best of all: a 48-inch hall and 36-inch doorway. Without a sill, it is ideal for weak arms, poor balance.

ENTRANCE APPROACHES

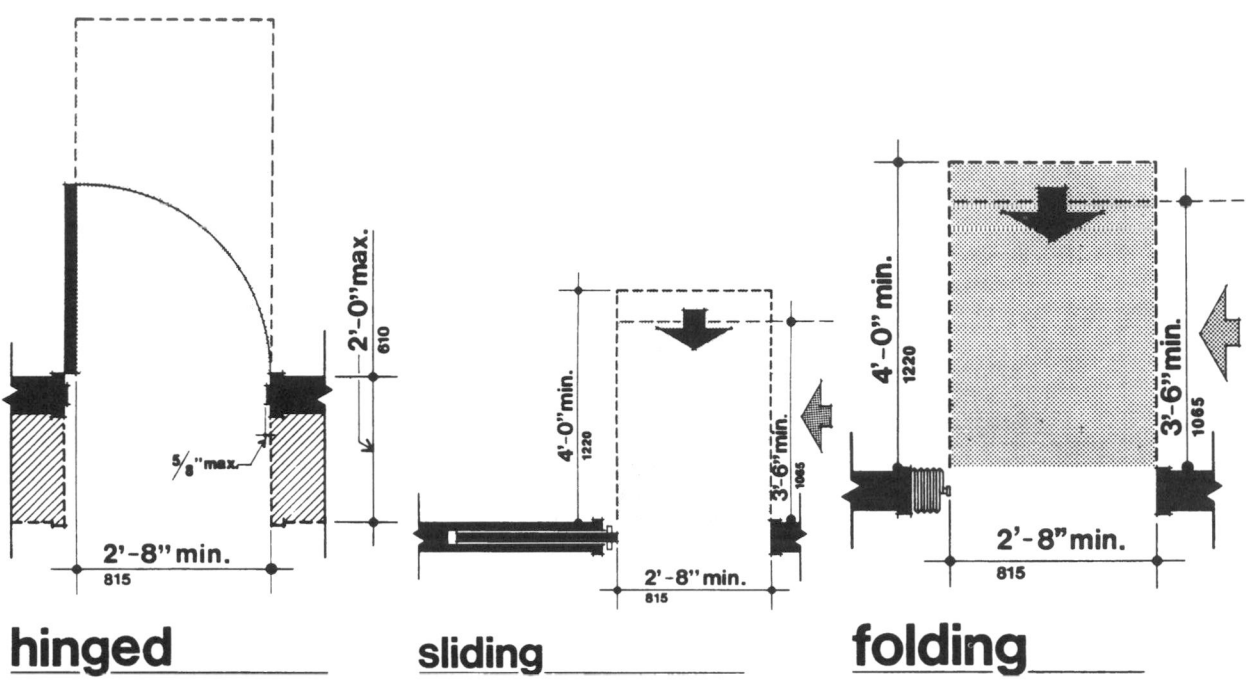

hinged   sliding   folding

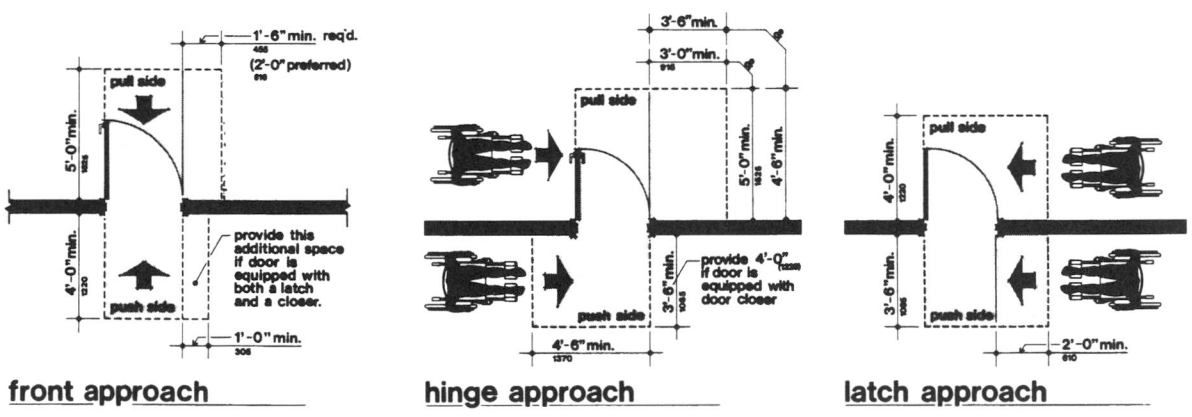

front approach   **hinge approach**   **latch approach**

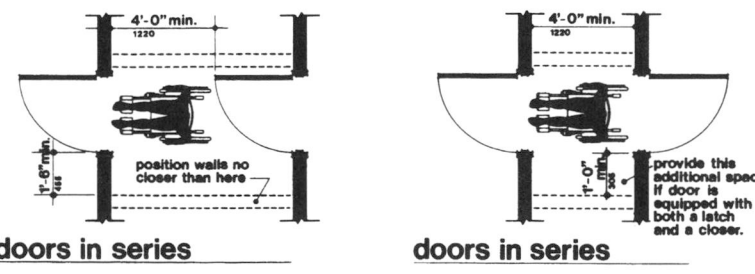

doors in series   **doors in series**

Fig. 5

914

RAMPS

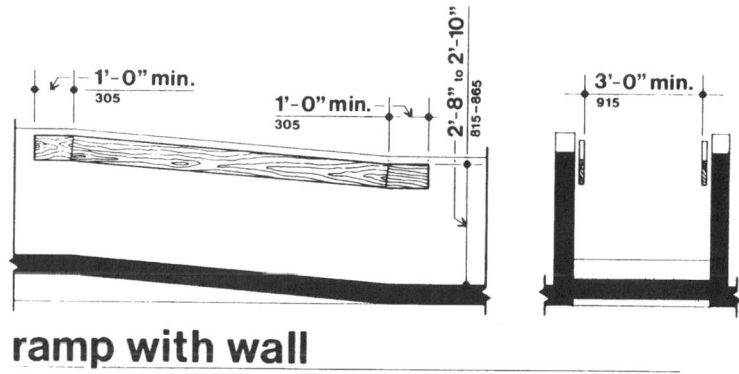

ramp with wall

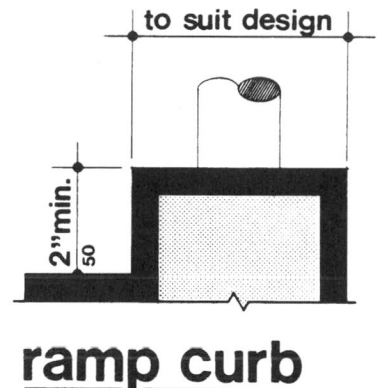

ramp curb

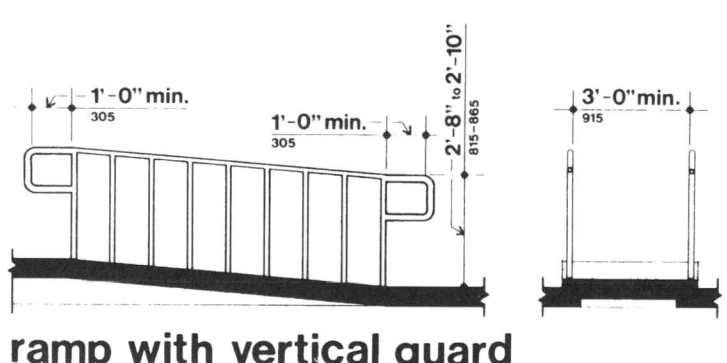

ramp with vertical guard

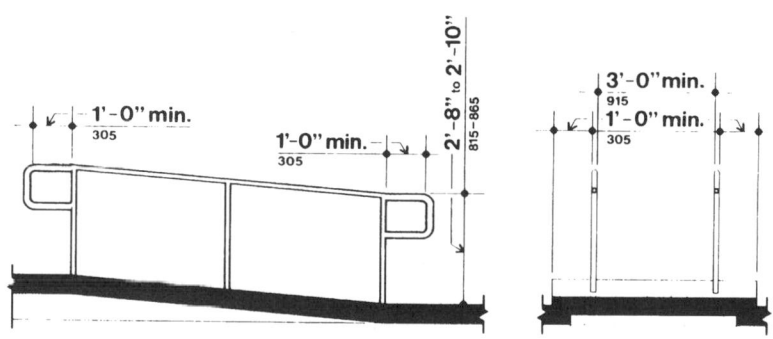

ramp with extended edge

Fig. 6

HANDICAPPED HOUSING—CORRIDORS AND RAMPS

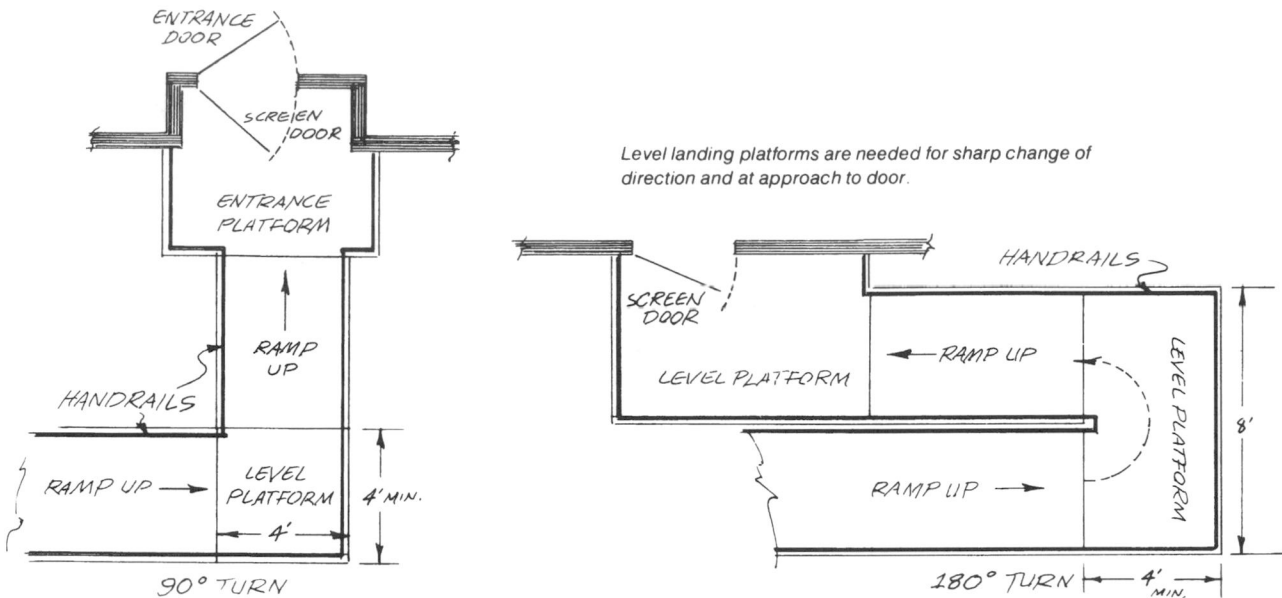

Level landing platforms are needed for sharp change of direction and at approach to door.

Fig. 7

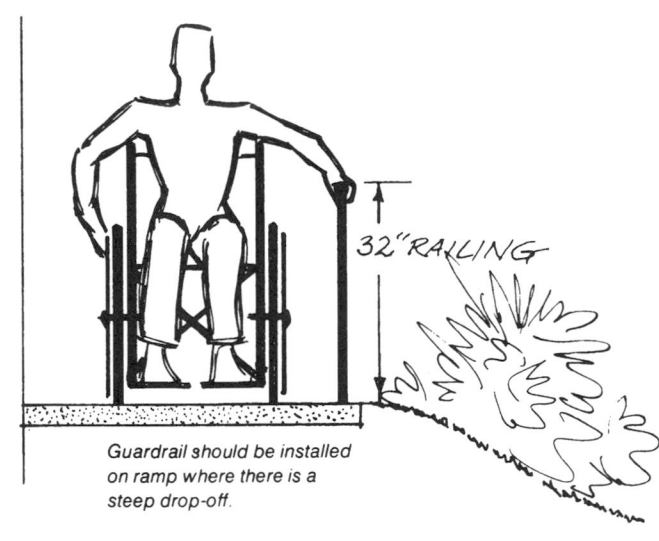

Guardrail should be installed on ramp where there is a steep drop-off.

Fig. 8

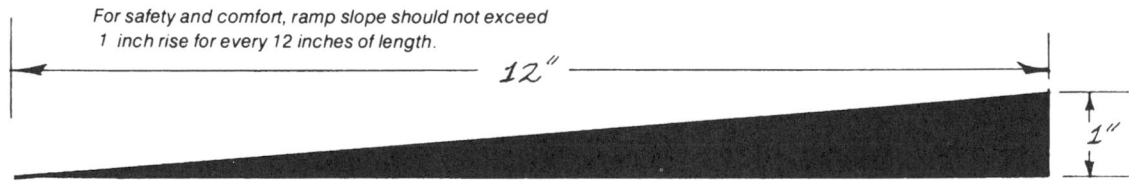

For safety and comfort, ramp slope should not exceed 1 inch rise for every 12 inches of length.

Fig. 9

CURBS AND RAMPS

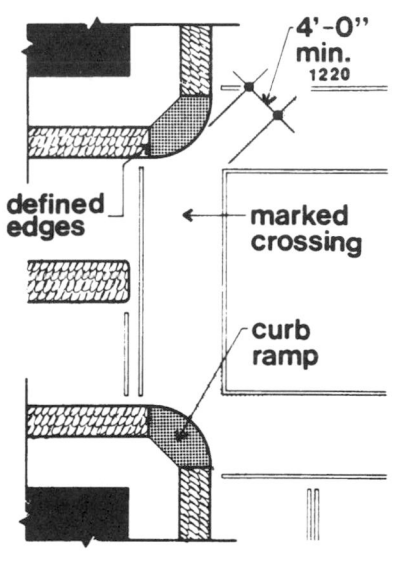

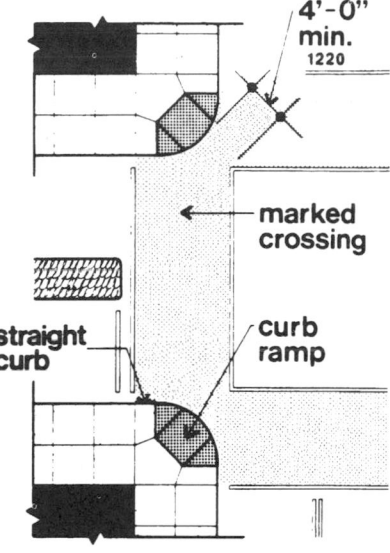

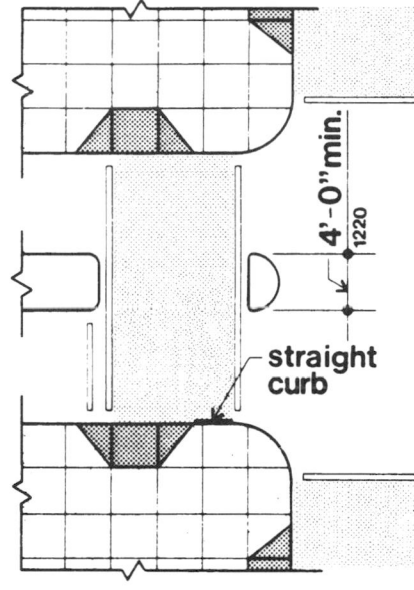

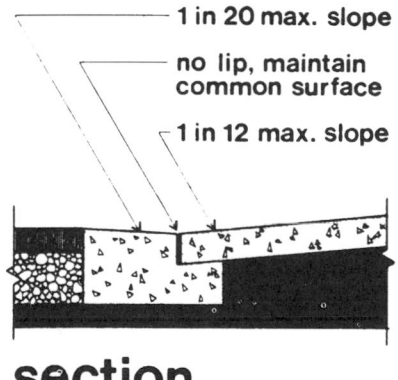

Fig. 10

Diagonal or corner-type curb ramps having returned curbs or well-defined edges should have such edges parallel to the direction of pedestrian flow. Diagonal or corner-type curb ramps having flared sides should have at least a 2-ft 0-in (610-mm)-long segment of straight curb located on each side of the curb ramp and within marked crossings.

Curb ramp discharge (bottom) should be to a 4-ft 0-in (1220-mm) minimum deep clear space. If the marked crossings are provided, locate bottom discharge entirely within marked crossings.

Locate curb ramps to prevent blockage of discharge areas by parked vehicles.

Cut any islands through flush with street surfaces or ramp each side to permit crossing. Provide 4-ft 0-in (1220-mm)-long rest area.

Slopes and rise Provide the least practical slope for any ramp or curb ramp subject to the following maximums:

New construction requirements:

1. Maximum running slope should not exceed 1:12 (8.3 percent).

2. Maximum rise for any run should not exceed 2 ft 6 in (760 mm).

Curb ramps should comply with the following requirements:

1. Provide flared sides if a circulation path crosses any part of the ramp or curb ramp not protected by handrails or guardrails; flared slope should not exceed 1:10 where a 4-ft 0-in (1220-mm) landing is provided at the top of the curb ramp. If less than 4 ft 0 in (1220 mm) is provided, the flared slope should not exceed 1:12. Where pedestrians will not normally walk across a ramp, returned curbs may be used.

2. Locate built-up curb ramps so that they do not project into vehicular traffic.

Maximum slopes of adjoining gutters, road surface immediately adjacent to the curb ramp, or accessible routes should not exceed 1:20.

Existing construction requirements:

If space limitations prevent compliance with standards, slopes and rises listed may be used.

Width Ramps and curb ramps should have a minimum clear width of 3 ft 0 in (915 mm) exclusive of edge protection or flared sides.

HANDICAPPED HOUSING—CORRIDORS AND RAMPS

TABLE 1 Maximum Rise and Projection, New Construction

Slope	Maximum rise		Maximum projection	
	in	mm	ft	m
1:12 to <1:16	30	760	30	9
1:16 to <1:20	30	760	40	12

TABLE 2 Maximum Rise and Projection, Alterations to Existing Construction

Slope	Maximum rise		Maximum projection	
	in	mm	ft	m
1:10 to 1:8	3	75	2	0.6
1:12 to 1:10	6	150	5	1.5

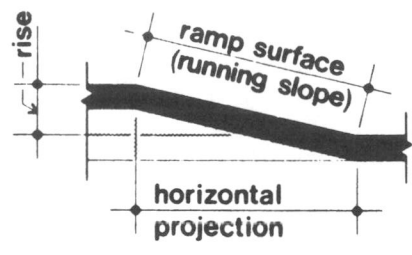

ramp slope

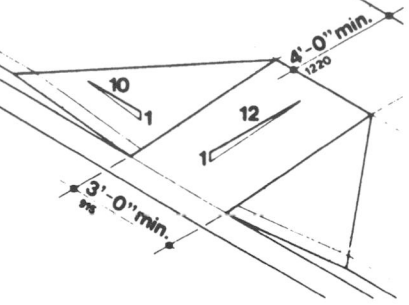

curb ramp

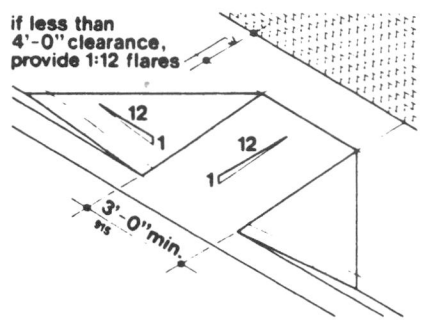

curb ramp

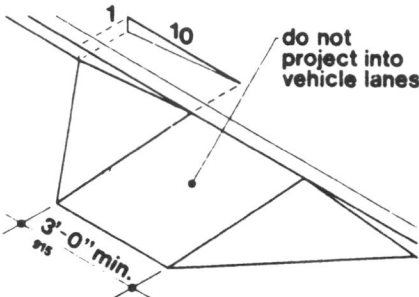

curb ramp

curb ramp

Fig. 11

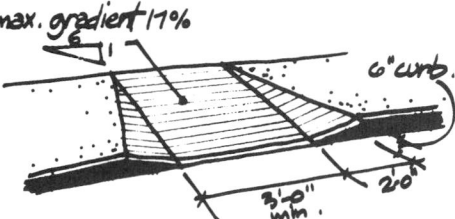

■ avoid "lip" greater than ½" wherever ramp meets adjacent paving at top or bottom.

max. gradient 17%

Flared Ramp

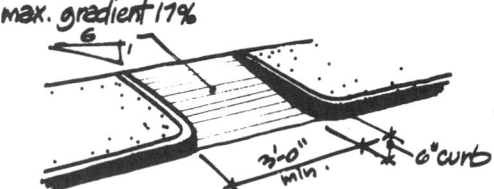

■ corrugated lines in ramps should be avoided since they can hold water in freezing weather and become icy.

max. gradient 17%

Ramp With Continuous Curb

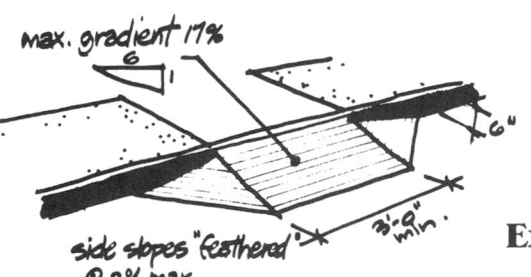

max. gradient 17%

side slopes "feathered" @ 8% max.

■ use of this type often interferes with curb-side storm drainage & snow plowing.

Extended Ramp

Fig. 12

SPACE REQUIREMENTS FOR RAMPS

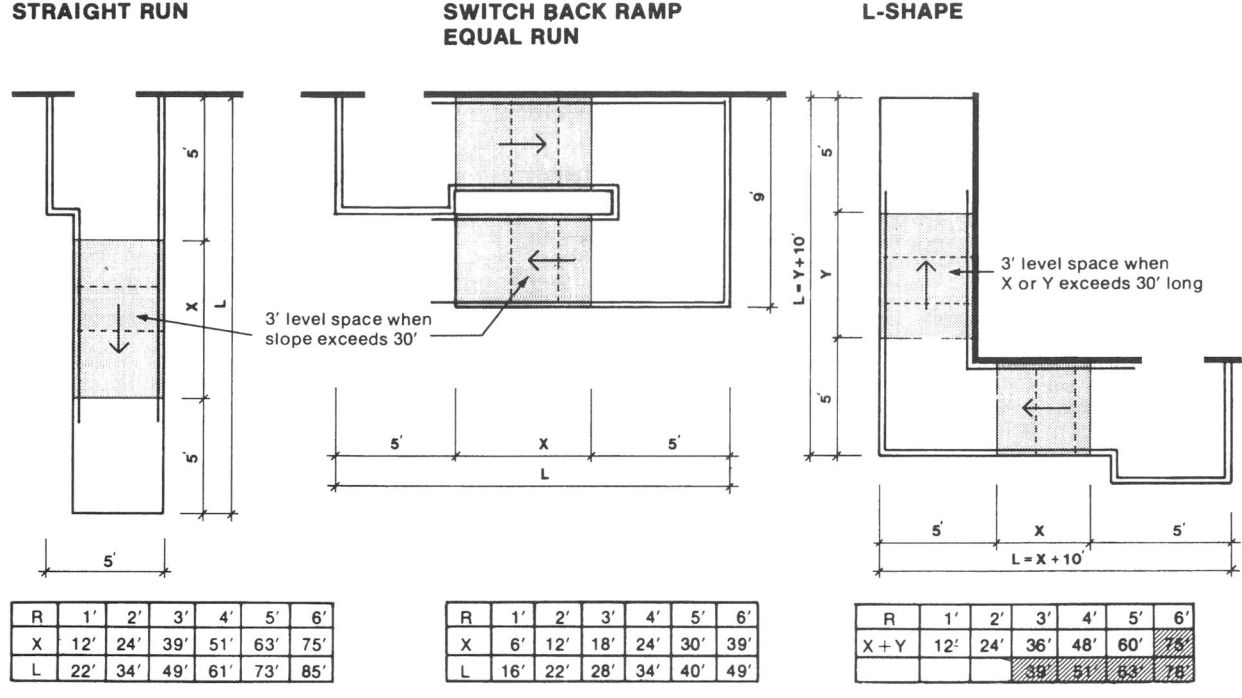

STRAIGHT RUN

SWITCH BACK RAMP EQUAL RUN

L-SHAPE

3' level space when slope exceeds 30'

3' level space when X or Y exceeds 30' long

R	1'	2'	3'	4'	5'	6'
X	12'	24'	39'	51'	63'	75'
L	22'	34'	49'	61'	73'	85'

R	1'	2'	3'	4'	5'	6'
X	6'	12'	18'	24'	30'	39'
L	16'	22'	28'	34'	40'	49'

R	1'	2'	3'	4'	5'	6'
X + Y	12'	24'	36'	48'	60'	75'
			39'	51'	63'	78'

Note: Tables assume flat sites and 1 in 12 slopes.
Sites which slope may require longer or shorter ramps depending on direction of ramp and slope of site. Ramps should be oriented to minimize their length.

Wherever possible the length of the slope of ramps should be evenly divided.

L = total length of ramp
R = rise
X and Y = length of the slope

Whenever either X or Y exceeds 30' add 3' for rest area.

Fig. 13

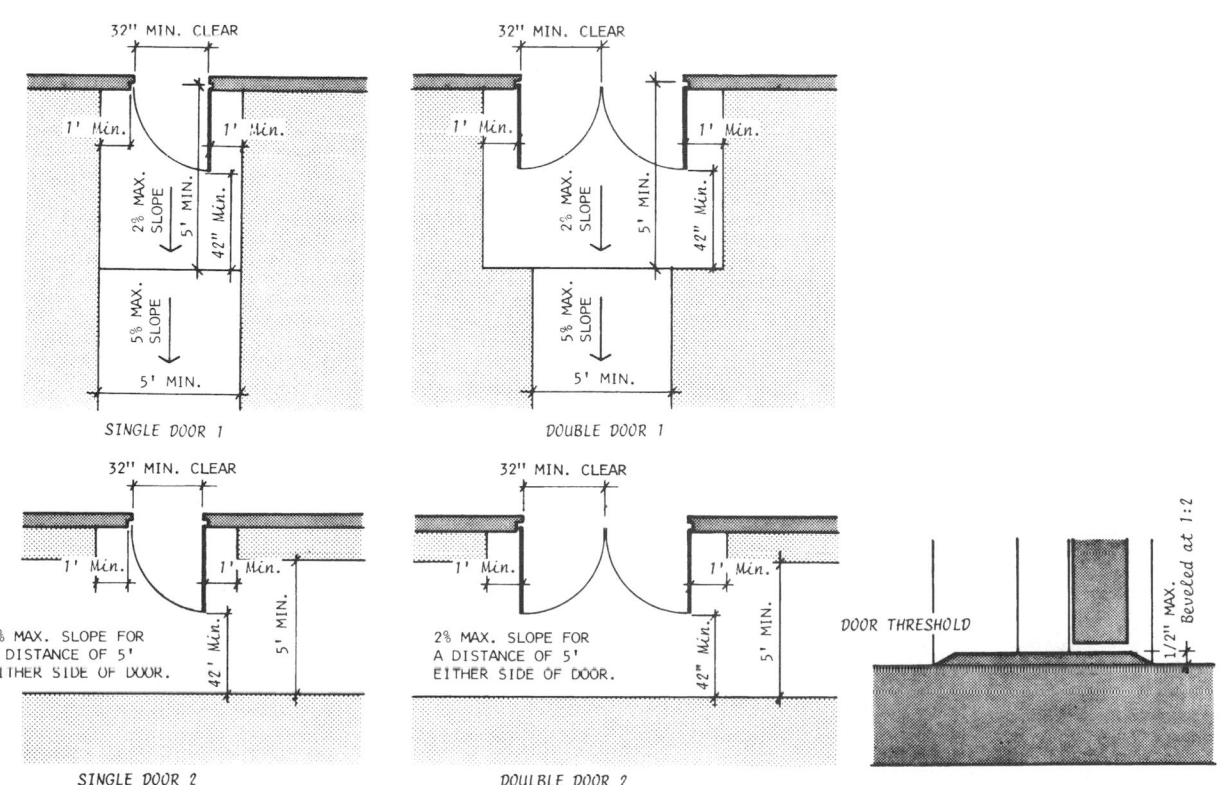

SINGLE DOOR 1

DOUBLE DOOR 1

SINGLE DOOR 2

DOUBLE DOOR 2

DOOR THRESHOLD

Fig. 14

HANDICAPPED PARKING AND GARAGES

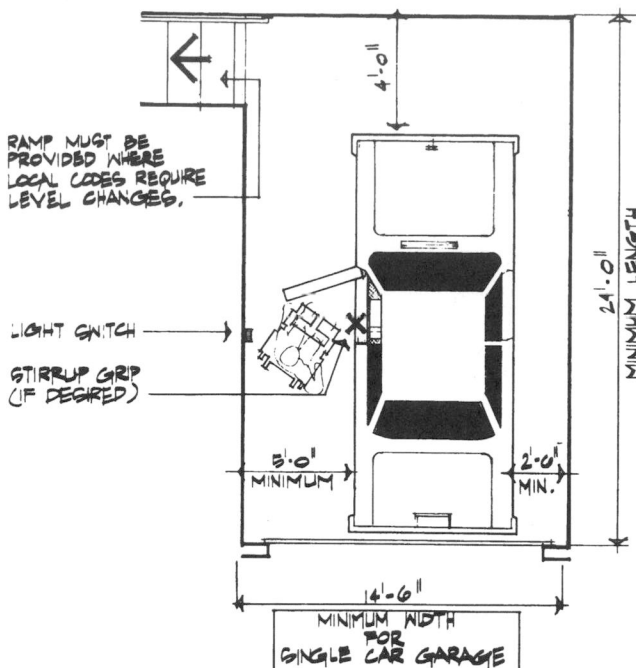

Fig. 1 Garage or carport.

Parking spaces should have a minimum width of 13 ft 6 in.

Garages or carports must allow a clear area with a minimum width of 5 ft 0 in in on at least one side of the car. To accommodate this area, a single-car garage should have a minimum width of 14 ft 6 in. A passageway 4 ft 0 in wide should be provided in front of or behind the automobile. A garage or carport should therefore have a minimum length of 24 ft 0 in.

Width of Parking Spaces

A parking space for people who use wheelchairs and walking aids should be at least 13 ft (3.9 m) wide. If two adjacent spaces are to be designated for the handicapped, then the total width of both spaces together should be 21 ft (6.4 m) if a 5-ft (1.52-m) wide access aisle separates its two spaces.

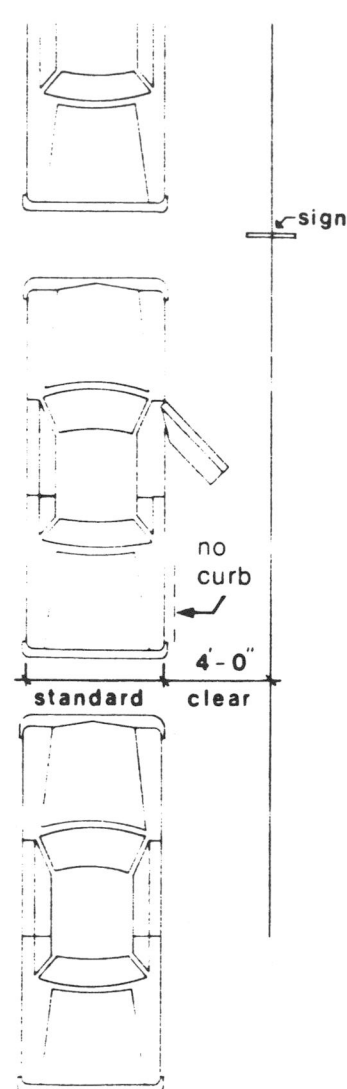

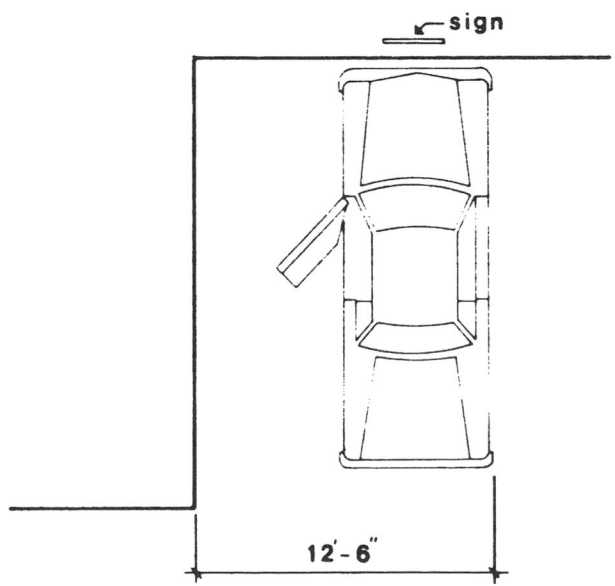

Parking spaces for handicapped persons must be 12'-6" wide to allow opening of doors fully for loading and unloading wheelchairs, and must have above ground sign ✳ designating the space for the handicapped only.

Fig. 2

Parallel parking may also be used, provided at least 4'-0" clear space to side is available *at same level as the parking space.* Some such spaces should be on the driver's side and some on the passenger side.

Fig. 3

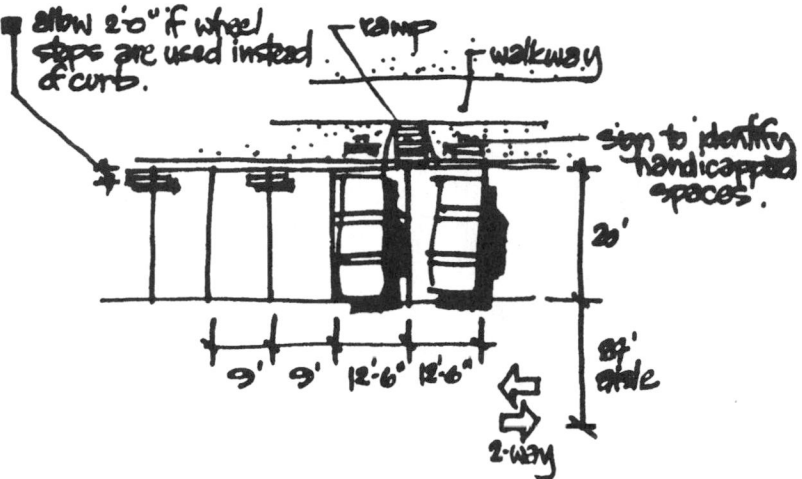

Fig. 4 90° parking.

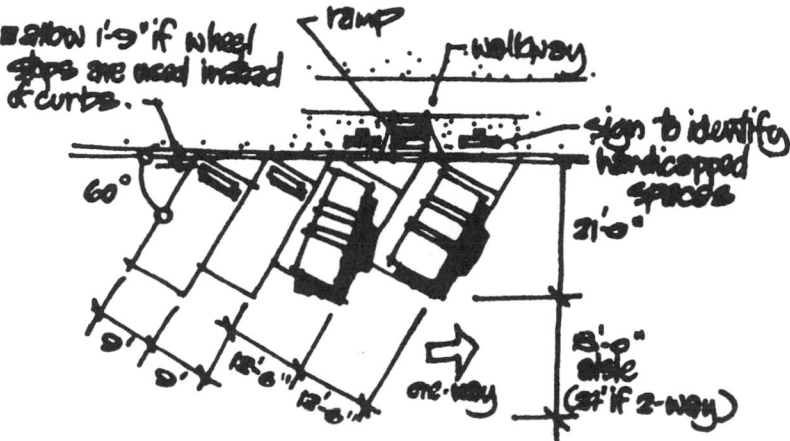

Fig. 5 60° parking.

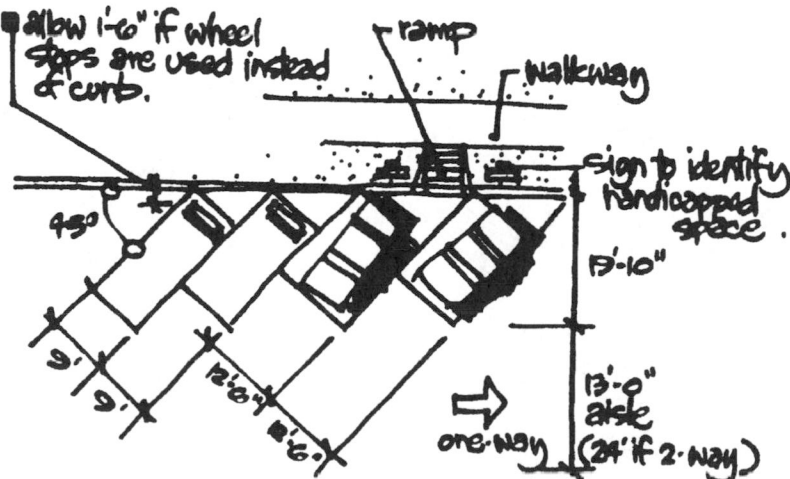

Fig. 6 45° parking.

HANDICAPPED PARKING AND GARAGES

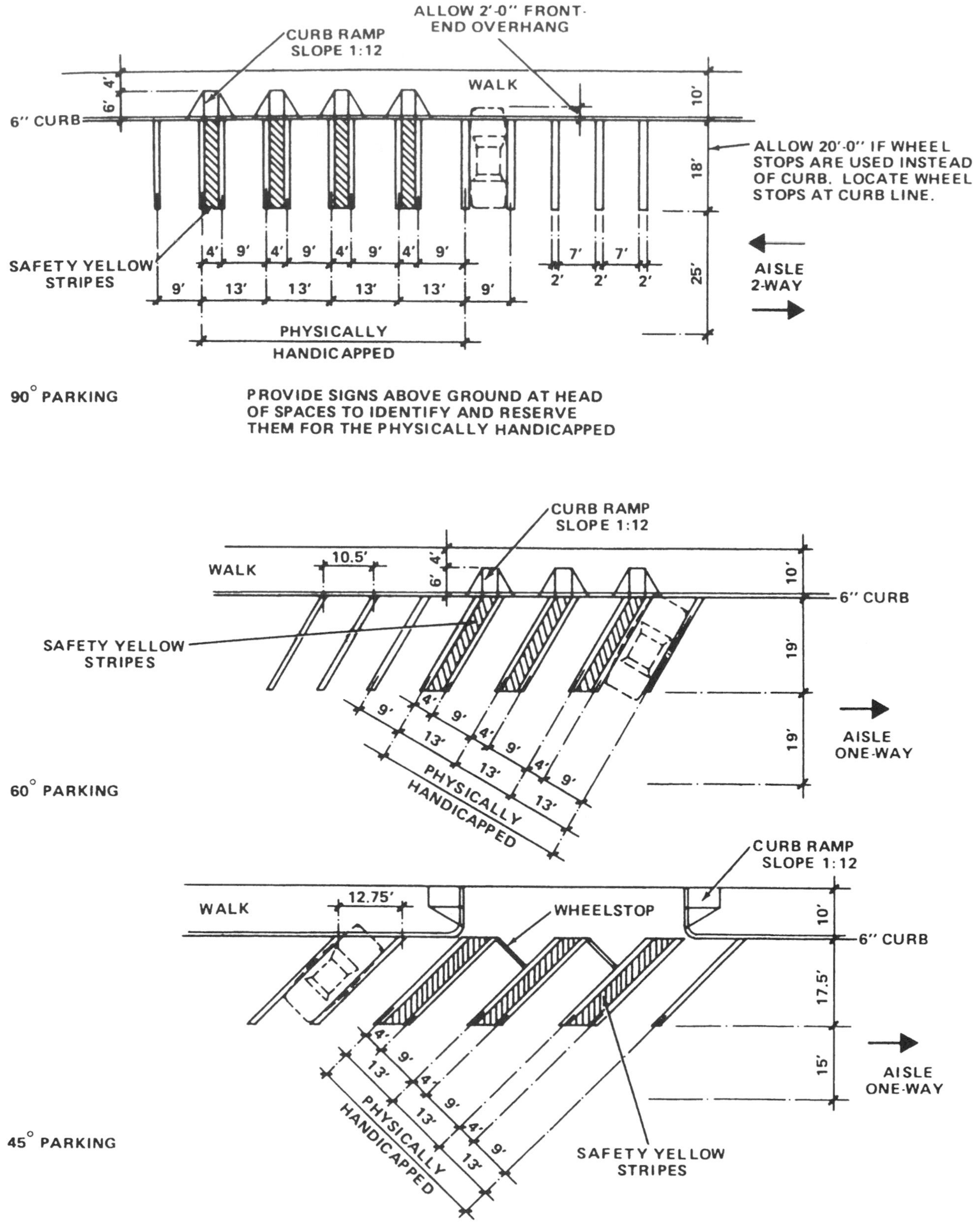

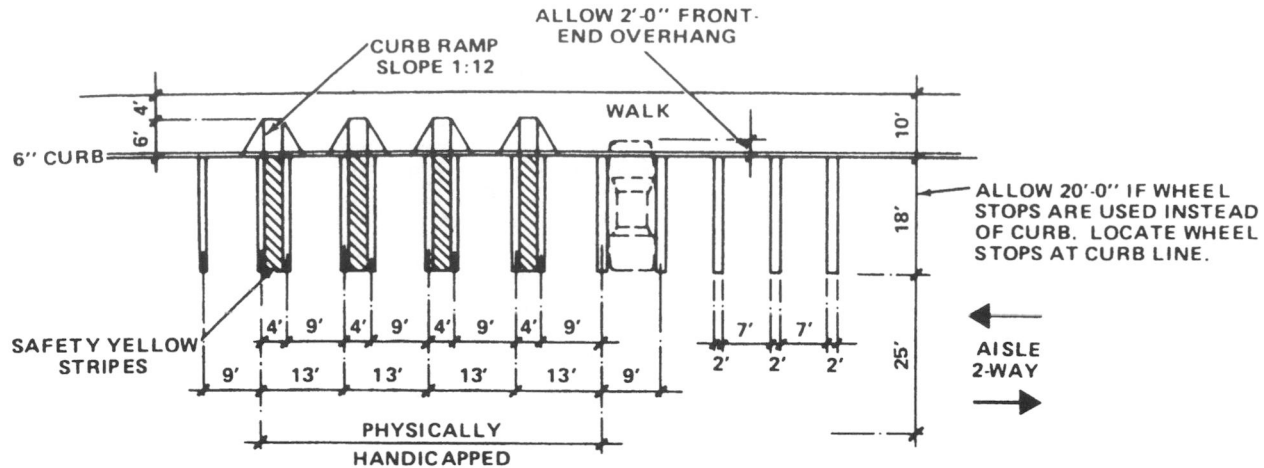

90° PARKING

PROVIDE SIGNS ABOVE GROUND AT HEAD
OF SPACES TO IDENTIFY AND RESERVE
THEM FOR THE PHYSICALLY HANDICAPPED

60° PARKING

45° PARKING

Fig. 7   Parking spaces for use by the handicapped.

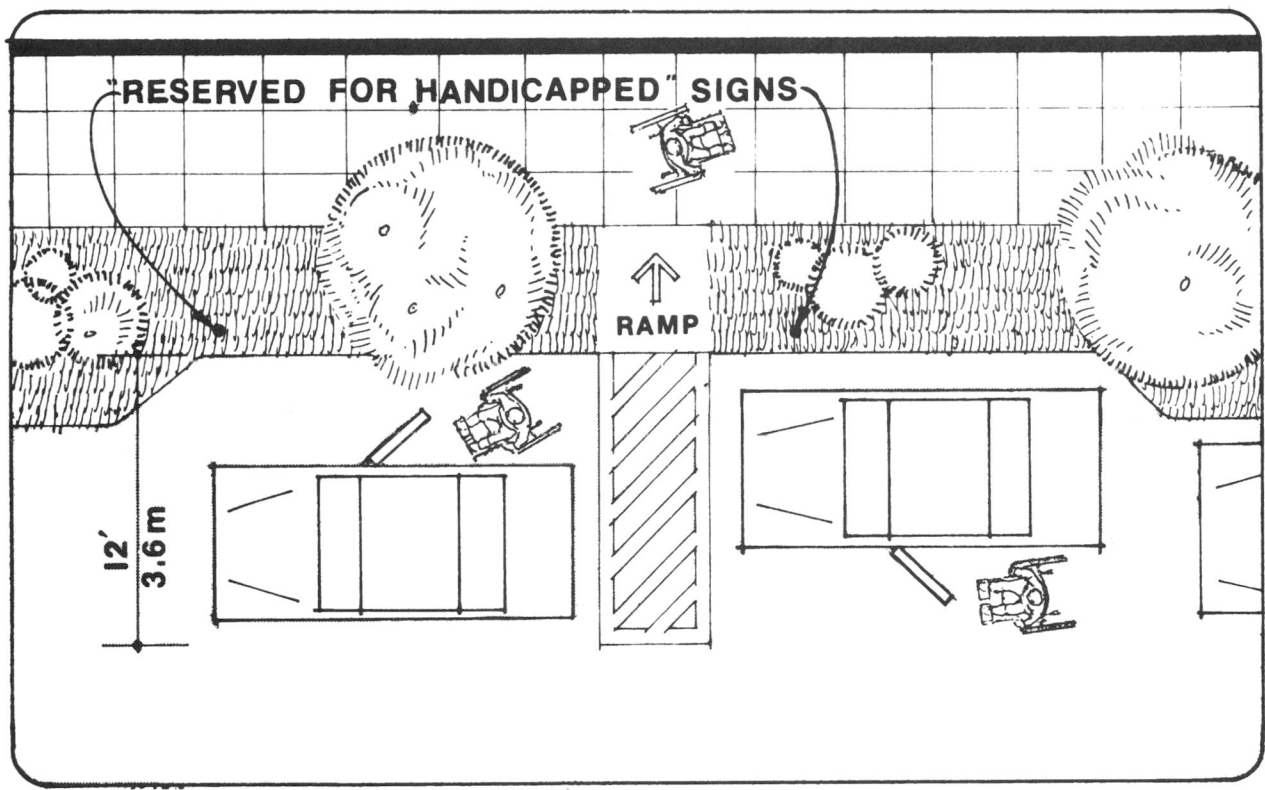

Fig. 8 Parallel parking for handicapped.

Where parallel parking spaces for the handicapped are provided, they must be placed adjacent to a walk system with a hard surface that is accessible from the space. The parking space including the access aisle must be at least 12 ft (3.6 m) wide.

Accessible parking spaces Provide accessible parking spaces that:

1. Are at least 8 ft 0 in (2440 mm) wide.
2. Have an adjacent access aisle at least 5 ft 0 in (1525 mm) wide.

Exception: If accessible parking spaces for vans designed for handicapped persons' vans are provided, each should have an adjacent access aisle at least 8 ft 0 in (2440 mm) wide.

Passenger loading zones Provide accessible passenger loading zones that:

1. Have an access aisle at least 5 ft 0 in (1525 mm) wide by 20 ft 0 in (6 m) long adjacent, parallel, and level with the vehicle standing space.
2. Have vehicle standing spaces and access aisles with surface slopes not exceeding 1:48 (¼ in/ft) in all directions.

Vertical clearance Provide minimum vertical clearances of 9 ft 6 in at accessible passenger loading zones and along vehicle access routes to such areas from site entrances. Minimum vertical clearances of 9 ft 6 in (3.45 m) at accessible van parking spaces should be provided.

TABLE 1

Total parking in lot	Required minimum number of accessible spaces
1–25	1
26–50	2
51–75	3
76–100	4
101–150	5
151–200	6
201–300	7
301–400	8
401–500	9
501–1000	2% of total
Over 1000	20 plus 1 for each 100 over 1000

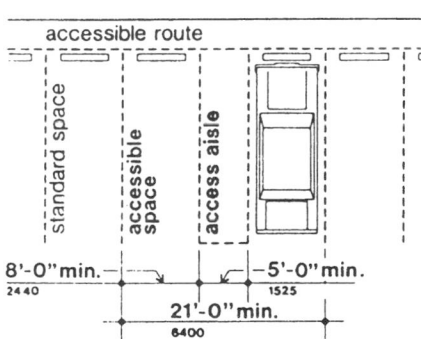

accessible parking
two spaces sharing one aisle

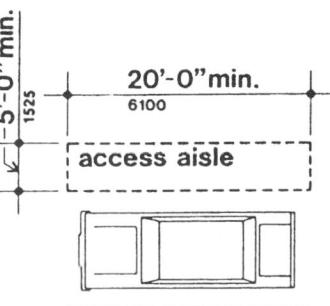

unloading zone

Fig. 9

CRITERIA FOR LOCATING HOUSING FOR THE ELDERLY

Neighborhood Requirements

Housing for the elderly should be in a neighborhood which provides the opportunity for independence in an atmosphere of comfort and security. The elderly are reluctant to go out when living in or near unsafe neighborhoods because of their fear of being assaulted and robbed while shopping, waiting for buses, or even walking for pleasure. Because of their relative helplessness, the elderly are easy prey for thieves, and their physical as well as monetary losses are relatively very great as the result of such encounters.

Housing for the elderly should be in areas suitable for residential use. While this principle may seem quite simple, it is not unusual for projects for the aged to be proposed for those leftover pieces of land which have no obvious use. In contrast to other residential land uses, accessibility to schools and playgrounds is not an important factor in the location of housing for the elderly.

The aged should be given the opportunity to live where there is activity. The elderly will find greater safety on the sidewalks if other members of the community are moving about in the course of their own affairs. Moreover, many elderly persons feel most happy where they can watch or be keenly aware of human activity, whether it be shopping or cars passing in a busy street. When housing is in residential neighborhoods near shopping, churches, and parks, elderly people have reported satisfaction with the location because, in being surrounded by the activities of that community, they can feel a part of it.

The elderly show a decided preference for residing in a relatively flat topographical area, and housing projects should be located accordingly. Even the presence of low hills, either on the site or near their homes, can be an unreasonable physical strain and therefore unpopular. Climbing of any sort can be greatly disliked and dangerous and should be avoided in the physical design of the project itself.

Facility Relationship Requirements

Because of their physical inabilities and often fixed low incomes, the elderly rely upon public transportation and their own walking capacity to move them about. Very few elderly in housing projects own or operate automobiles. When making a trip in excess of several blocks, they use public transit if convenient. For shorter journeys the elderly will generally walk. These are circumstances which strongly determine the proper distance relationship of service facilities to the homes of the aged.

Accessibility to facilities that are used regularly or are critical when needed is an important location determinant. A brief descriptive listing of the facilities is presented to describe the kind of services required to satisfy the physical and psychological needs of the elderly. Of the eleven facilities listed, the first six are considered as essential to the overall health and continued well-being of the elderly.

1. *Supermarket or food retailer*. A preference has been shown for supermarkets over grocery stores because of lower prices, better-quality products, and broader selection of items.

2. *Public transit stop*. This facility is frequently indispensable to the elderly for attaining accessibility to some important facilities and to other communities.

3. *Place of worship*. Although several studies have cited this facility as filling the purpose of a social center, its fundamental function is to serve both as a religious center and as a psychological support in a time of anxiety concerning death.

4. *Medical facilities*. These facilities include a physician's office, clinic, or hospital.

5. *Drugstore*. A source of supplies necessary to maintain medical care.

6. *Laundry*. The facility is frequently installed on the dwelling premises.

7. *Beauty Parlor or Barber Shop*. A frequently used facility which provides support for individual self-esteem.

8. *Social center*. The facility acts as a reference point and as an aid to communication among the elderly in the community. It frequently provides the elderly with the means for reentry into and involvement in society.

9. *Bank*. It may provide the elderly with the only convenient means for cashing their social security checks.

10. *Restaurant*. Its use is discouraged by general infirmity, by the high cost of eating out, or by special dietary needs.

11. *Department store*. Serves primarily as an entertainment facility in which shopping for pleasure may take place.

Other facilities, which are apparently of far less importance to the living patterns of the elderly, include the library, post office, news-tobacco store, movie house, bar, and a variety store.

The maximum reasonable distance to a facility, unlike its overall importance, is closely related to its frequency of use. A reasonable and comfortable walk for the elderly has been determined to be two to three blocks for most services frequently used and no more than ¼ to ½ mile for most other needs.

Table 1 sets forth recommended standards for the relationship of service facilities to housing for the aged. Its content is taken from several sources but very heavily draws upon analysis of interviews with the elderly themselves and with the managers of projects in which they live.

Two vitally important facilities, the house of worship and the medical facility, should always be available but need not be located within the critical walking distance if direct and convenient public transit is available to the elderly when going to and returning from them.

Social service facilities are being recognized as equal in importance to those which provide basic necessities. To many elderly, adult education, counseling, information and referral services, monetary assistance, health services, and cultural, recreational, and social activities may mean the difference between enjoying life and only staying alive. Sometimes limited collections of such services are provided in housing projects for the elderly.

It is recommended that the relationship to a social service center be carefully considered when locating elderly housing within a neighborhood. If an independent center exists or is likely to develop, housing should be located as close as possible. If a center is to be built within the project, thought should be given to its expansion beyond the needs of the project and to placement of the project such that its facilities can serve the elderly of the entire community.

Parking Requirements

The location of housing will be affected by the availability of adequate land area for both structures and necessary parking. Despite the needs of visitors and staff, housing for the elderly has unusually small parking requirements because few elderly residents own and operate automobiles. Accordingly, the following is a set of informal standards, based upon limited surveys and upon parking provision/usage data for public and private housing for the elderly.

1. For housing not subject to the following exceptions, parking spaces numbering at least 30 percent of the total number of dwelling units should be provided.

2. Where service by public transportation is very poor, parking spaces numbering more than 30 percent, but less than 50 percent, of the total number of dwelling units should be provided.

3. For housing located in or easily accessible to the central city or located in or adjacent to regional shopping centers, parking spaces numbering at least 15 percent of the total number of dwelling units should be provided.

4. For housing intended for moderate- to low-income occupants (other than public housing) parking spaces numbering at least 15 percent of the number of dwelling units should be provided.

5. For public housing for the elderly, parking spaces numbering at least 10 percent of the total number of dwelling units should be provided.

TABLE 1   Facility Location Standards

Facility	Importance	Critical distance*	Optimum distance†
Supermarket or grocery	Vital	2 blocks	1 block
Public transit stop	Vital	1 block	On site
House of worship	Vital	½ mile	2 blocks
Medical facilities	Vital	½ mile	On site
Drugstore	Vital	3 blocks	1 block
Laundry	Vital	2 blocks	On site
Beauty parlor or barber shop	Useful, but not essential	½ mile	2 blocks
Social center	Useful, but not essential	½ mile	On site
Bank	Useful, but not essential	½ mile	2 blocks
Restaurant	Useful, but not essential	½ mile	2 blocks
Department store	Useful, but not essential	½ mile	3 blocks

*Critical distance refers to the limit of comfortable walking ability by the elderly.
†Optimum distance refers to the most desirable for fulfilling the needs of the elderly.

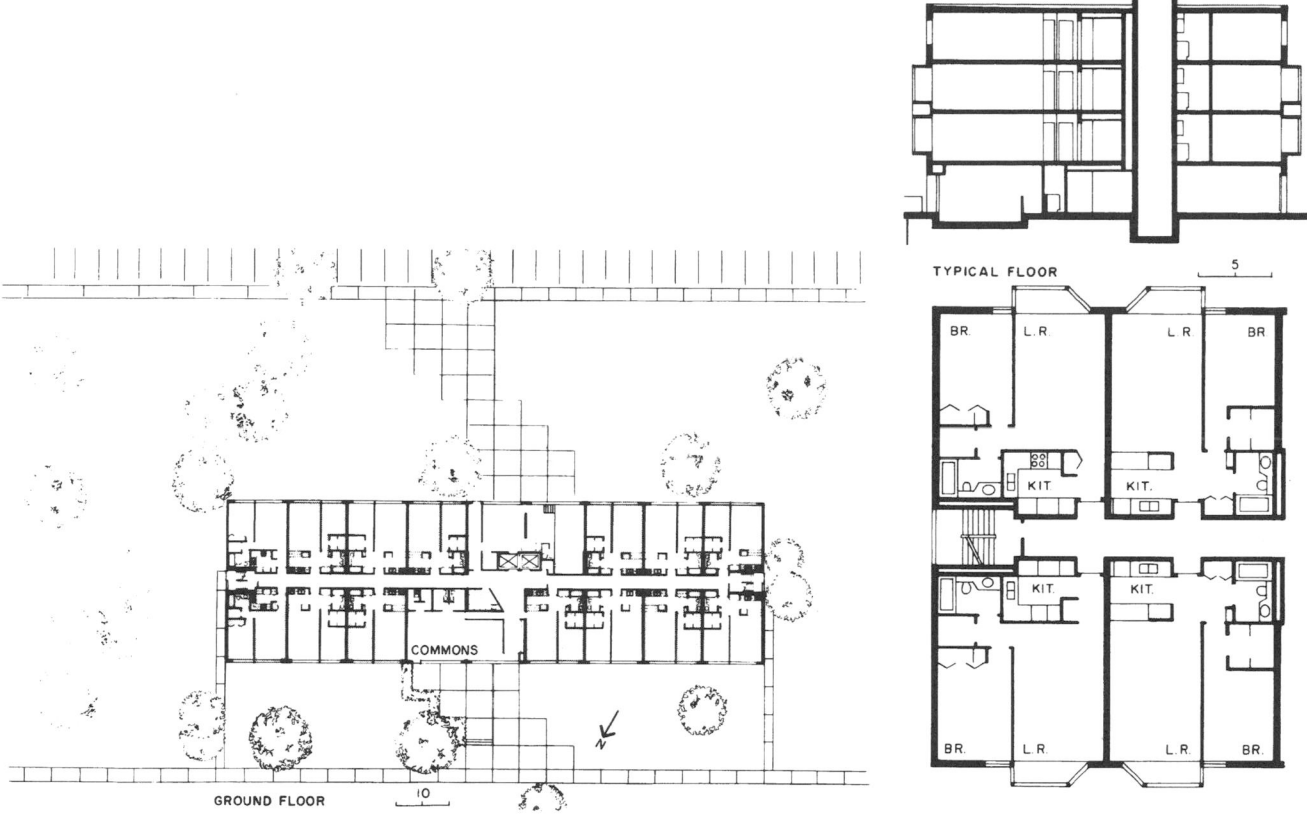

TYPICAL FLOOR

GROUND FLOOR

COMMONS

Fig. 1a

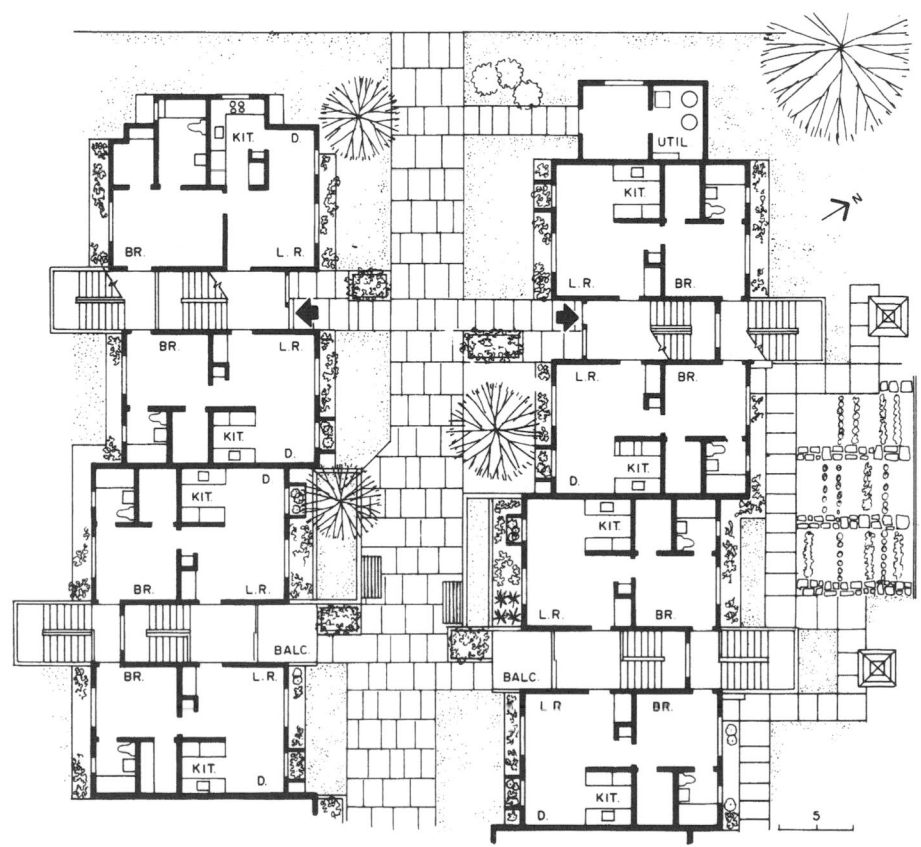

Fig. 1b

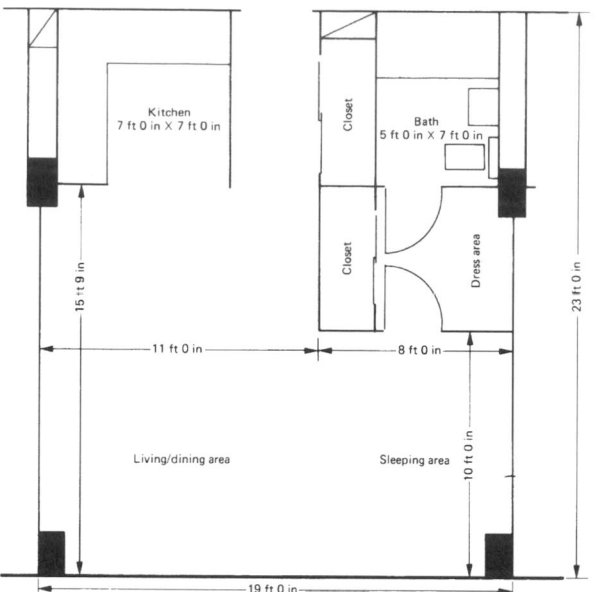

Fig. 1c Efficiency apartment for the elderly, 437 ft². New Jersey Housing Finance Agency.

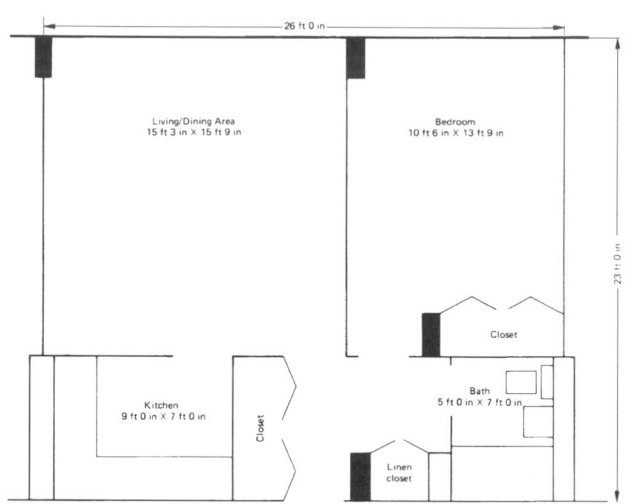

Fig. 2 One-bedroom apartment for the elderly, 598 ft². New Jersey Housing Finance Agency.

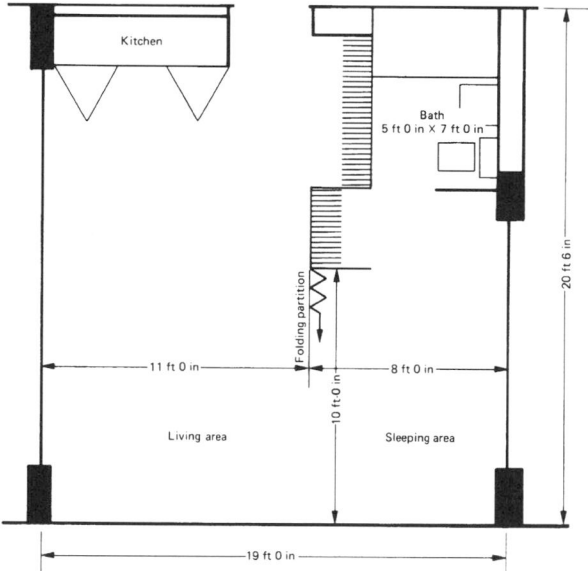

Fig. 3 Efficiency apartment for the elderly with central dining facilities, 390 ft². New Jersey Housing Finance Agency.

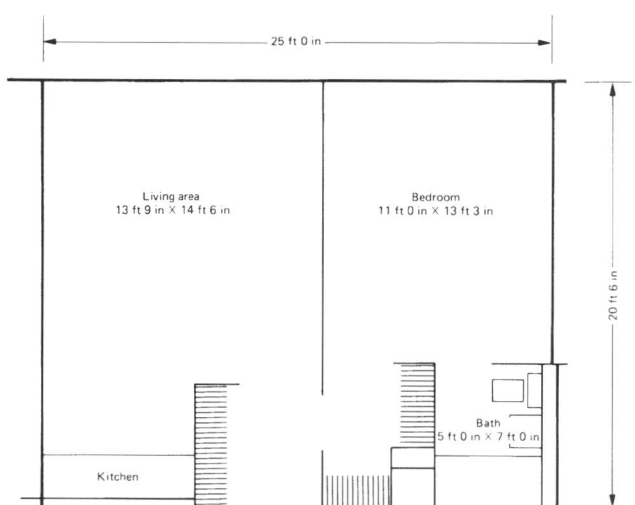

Fig. 4 One-bedroom apartment for the elderly with central dining facilities, 513 ft². New Jersey Housing Finance Agency.

TYPICAL UNIT PLANS

The one-bedroom unit shown in Fig. 5 is a typical senior citizen unit modified to meet the standards for occupancy by a disabled person. The original senior citizen unit is shown in Fig. 6 for comparative purposes.

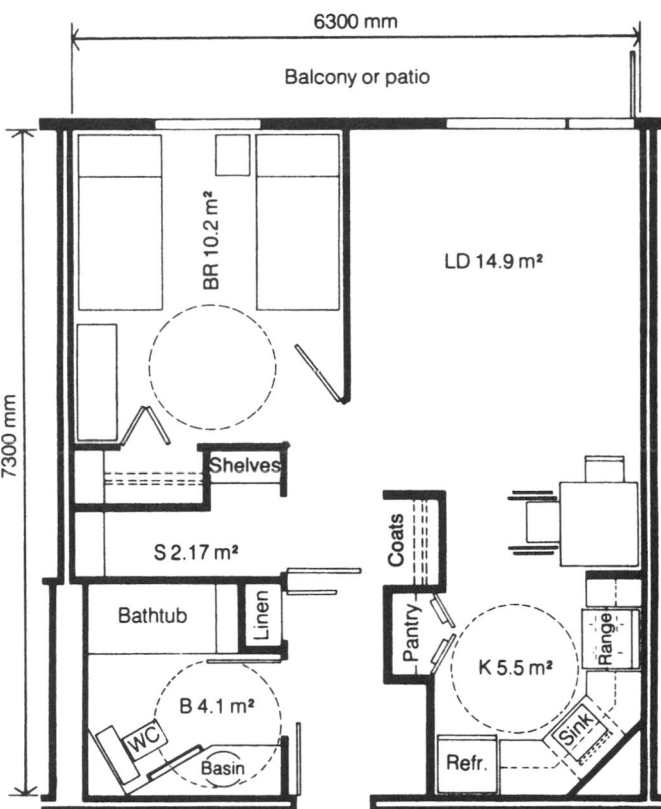

Fig. 5 Barrier-free one-bedroom unit. Total net area: 46 m².

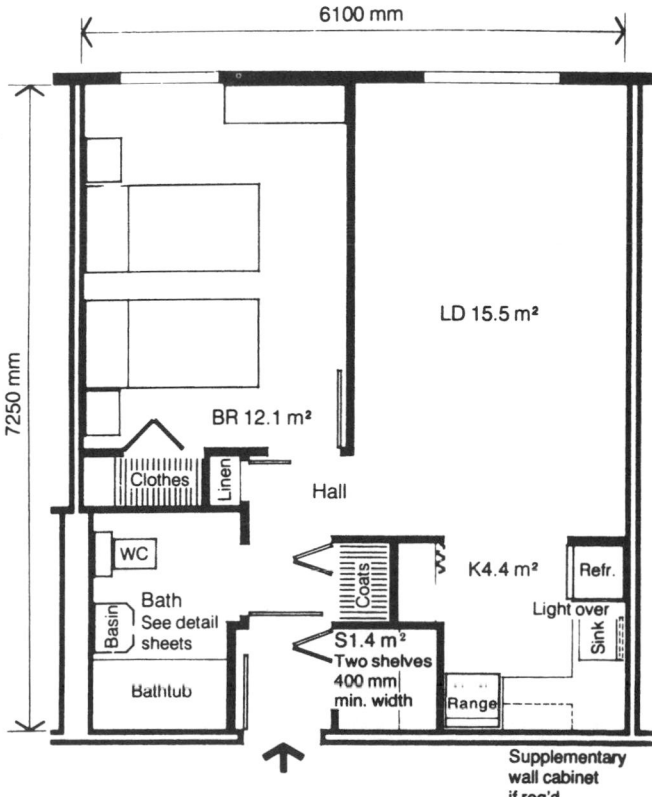

Fig. 6 Senior citizen's one-bedroom unit. Total net area: 44.25 m².

APARTMENTS FOR THE ELDERLY

Bathroom door is outswinging with flush threshold.

A diagonal grab bar is mounted forward of the toilet on the side wall. All grab bars are 1½ in outside diameter by 2 ft 0 in long and 1½ in clear of wall.

The lavatory has lever-type faucets and a front edge capable of withstanding a 250-lb load.

The medicine cabinet is recessed with unbreakable shelves.

The bathtub has a vertical grab bar near the faucets to facilitate vertical entry and another running diagonally across the center of the back wall to aid in rising and in showering. If a shower is provided, a second soap dish might be mounted at 4 ft 6 in to obviate some stooping.

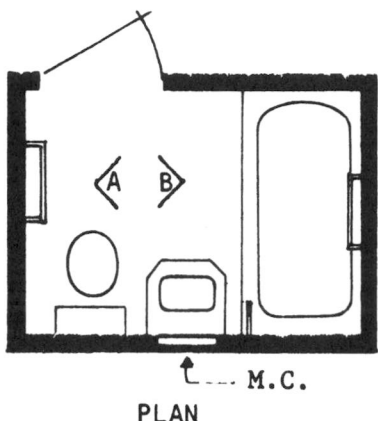

PLAN

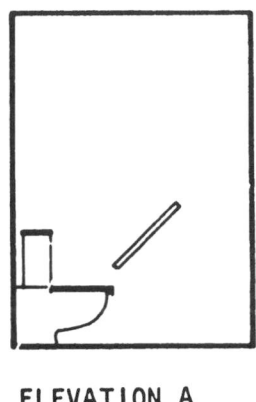

ELEVATION A

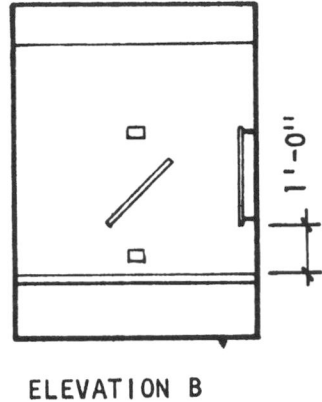

ELEVATION B

Fig. 7   Edward Steinfeld, Associate Professor, Department of Architecture, SUNY at Buffalo.

Congregate housing, an emerging concept of assisted group living for older persons, is intended to serve the needs of frail, but not ill, elderly who require a supportive environment in order to cope with the demands of daily living. Its major goal is to maintain independence in a residential setting and prevent unnecessary or premature institutionalization which often results in excessive costs to the individual and to society.

The term "institution" has many connotations, most of them unpleasant. Its definition varies from state to state, depending upon specific code requirements. It is certainly intended that congregate housing not be classified as an institution or provide an institutional environment. Nevertheless, it is intended that the sponsors and managers of such housing provide more than room and board. In order to foster the necessary supportive environment, sponsors and managers must assume responsibility for the general welfare of residents. In some states this may serve to classify the facility technically as an institution and may require licensing.

The architectural program described reflects current trends in housing and services for the elderly. It has been developed by a number of individuals in the field of aging from a wide variety of backgrounds. It is intended to provide a framework for financial planning and a pro forma operational and development budget. The basic principles should apply in any locale, but modifications will be needed to account for specific local conditions.

In general, an architectural program has three basic components—planning and design directives, a space program, and space relationship diagrams. A discussion of the first two is presented here, but in view of the influence of local site requirements, it is inappropriate to develop space relationship diagrams here. The planning and design directives provide the framework for the architect and the sponsor in relating location, form, and design to the physical, social, and psychological needs of residents. The space program consists of a list of all spaces in the building and a tabulation of the percentage of the total area attributable to specific spaces.

The model described here is a facility of 100 apartments that will house approximately 125 elderly, both single persons and couples. Undoubtedly the largest number of occupants will be widows. Nearly all residents will have some disability that has encouraged them to seek out a supportive environment, including the inability, physically or financially, to maintain a private home, a need for social involvement, difficulty in preparing adequate meals, and, in some cases, restricted mobility. A few may be handicapped to the degree that they need the use of a wheelchair. Five percent of the apartments are designed to accommodate the handicapped.

The philosophy of design, operation, and services is one of selective intervention to encourage independence and, at the same time, offer physical and social supports as needed, but not before they are needed. Design details such as the speed of automatic closing doors, apartment furnishability, bathroom doors that open out, and the legibility of graphics play an important role in fostering independence and should not be underestimated.

No recommendation has been made for developing this model as either a high-rise or a low-rise. This decision can only be made on the basis of such factors as local preferences and experience of the elderly, construction costs, and the availability and cost of land. Although site selection criteria are not included as part of the architectural program described here, the location of congregate housing is of such importance that an initial digression must be made to discuss this subject.

SELECTING A CONGREGATE HOUSING SITE

According to an old adage, the three most important factors in the purchase of real estate are location, location, and location. This is especially true for the older person. As we age, our mobility tends to decrease and travel often is restricted to the surrounding neighborhood. For many elderly, the neighborhood is the life support system that provides most of the opportunities for recreation and enjoyment. Site selection must be contingent on the quality and supportiveness of the neighborhood. The more supportive it is, the fewer expensive services the sponsor may have to provide on site.

The general criteria that apply to the location of congregate housing are the same as those applicable to the location of all housing for the elderly. However, in congregate housing, emphasis is placed on easy access to neighborhood services or on making sure that transportation to them is provided. There are daily needs that arise for which no services are available, despite the fact that residents can obtain meals, housekeeping, and some personal services in the project. Such a minor activity as going to the drugstore or shopping center may become the most important event of the day, offering sights and sounds to talk about later. This outing helps break through the isolation from life and from others that residents may feel due to frailty or impairment.

The first consideration, then, is to choose a site where things happen, where residents can take part in life or, at least, observe some activity. It is an error—proved false over and over, but still persisting in some circles—to assume that the proper environment for the elderly is a quiet, peaceful setting amid the beauties of nature. For some elderly this may be accurate. But for most, we have learned that, as persons age and take part less frequently in community life, their desire to observe activity increases. Observing events can substitute, in part, for more active participation in them. It thus enables residents to feel they are still part of the human community.

A second factor in site location is neighborhood safety. Residents will be even more vulnerable to street crime or the fear of it than those in housing for fully independent living. Unsafe streets and the fear of assault will surely result in limited excursions into the neighborhood and probably none at all after the sun goes down.

A third and final factor in site location is the availability of emergency care resources. This not only will relieve the need for costly in-house clinical services but also will provide access, when needed, to more comprehensive services than can possibly be offered in a limited in-house health service program.

In summary, a good location in an active area close to service facilities and transportation makes it easier to guarantee that services will be available on a regular basis at minimum cost.

PLANNING AND DESIGN DIRECTIVES

As stated previously, one of the basic components of an architectural program is a list of planning and design directives to relate the facility to its resident population. Each directive that follows is explained further by a brief rationale and by an example of how it can be implemented in a congregate housing model.

1. The operational and physical characteristics of the housing environment should encourage independence.

   Rationale: Self-esteem and well-being are closely related to independence. A high degree of self-reliance permits residents to provide for themselves, rather than be provided for, and, in turn, helps reduce operating expenses.

   Example: Include a kitchenette in each apartment to complement the central dining service.

2. The physical design and services should maximize the options open to the individual in daily living.

   Rationale: Aging is a process of attempting to avoid closing out life's options which creates stress. By having options, the older person is better able to maintain well-being and physical and mental health. Offering options will attract a diversity of residents and foster a variety of interests and participation among them.

   Example: Allow for flexibility in service utilization as economy of operation permits.

3. Architectural design, especially in the dwelling units, should permit and encourage individuals to project their personalities onto the spaces by changing them to suit themselves.

   Rationale: Most individuals, if given the means and the opportunity when moving into a new apartment, will decorate or modify it to suit their tastes. This alteration of one's personal space is one of the few options possible and should not be denied.

   Example: Permit residents some selection in colors and draperies for their apartment.

4. Management policy, service programs, and physical spaces should permit and encourage the continuation of previous roles and lifestyles.

   Rationale: Involuntary change in roles or lifestyles can cause severe stress. Avoiding this reduces the institutional ambience for both the resident and the potential resident.

   Example: Select a site as accessible as possible to community activities and services.

5. The design of physical spaces should encourage social interaction among residents, between residents and staff, and between residents and the community.

   Rationale: A fundamental prerequisite for sound health is social connectedness. Because many persons attracted to age-congregate settings have often lost many previous social ties and roles, new role formations are desirable.

   Example: Locate service, activity, and recreation spaces in main patterns of pedestrian movement. Relate resident laundry rooms to other activities.

6. The institutional characteristics of building design, service programs, and management policy should be minimized.

   Rationale: Overinstitutionalization can induce premature dependence and reduce the motivation of relatively competent older people to do things for themselves. The institutional environment attracts residents who are more dependent, and it will be difficult to maintain a balanced age and competence distribution among the resident population.

   Example: Keep operational rules and regulations governing daily living and the use of

CONGREGATE HOUSING

services and spaces to a minimum. Minimize fluorescent lighting.

7. The architectural features and the management of the project should permit the resident a reasonable degree of security.

Rationale: Older people are very concerned with their personal security, and this will greatly affect their behavior and degree of involvement. The elderly have been singled out as prime targets in many areas for attack, robbery, and fraudulent practices.

Example: Locate emergency call system outlets in the bedroom and bathroom of each dwelling unit with a 24-hour contact point.

8. The congregate housing facility should include outreach programs and services, serving the community at large as well as residents of the project.

Rationale: Interaction between the community and residents will be enhanced by this practice. The "island approach" of serving "one's own" is no longer an economical or socially viable concept.

Example: On-site kitchen facilities can prepare meals for sale and delivery to neighborhood elderly.

9. The management should provide information and referral services that will make a continuum of social and health services readily available to residents.

Rationale: Information and referral service has been shown to be one of the most critical needs of older persons. Relocation to secure an increased level of service is often unnecessary, expensive, and may cause severe stress.

Example: Form associations or affiliations with local information and referral agencies. Provide on-site space for use by visiting service providers.

10. The building design and service programs should be planned to accommodate changes in the resident population.

Rationale: The service needs of residents will change as they age and gradually become more infirm and dependent. The neighborhood can change and influence the utilization of a building's spaces and services.

Example: Plan and design all public areas as well as 5 percent of the apartments in accord with general standards of design for handicapped persons.

11. Meal services should be provided within the housing project but should be voluntary to the extent possible, within the requirements of economic operation.

Rationale: Requiring a resident to take and pay for (in effect, involuntarily) three standard meals a day will overinstitutionalize the setting. Only the very dependent will sacrifice the freedom to prepare their own food at least occasionally. Eating is a social event, and the opportunity to prepare one's food facilitates small group gatherings, reciprocal dinner invitations, and similar socially beneficial activities.

Example: An extended brunch can be scheduled instead of breakfast and lunch. Require that only a minimum number of meals be included in the basic rent structure.

12. The design, layout, and architectural details of furnishings and equipment, particularly in the dwelling unit, should be planned to reduce hazards and induce a feeling of competence on behalf of residents.

Rationale: Self-esteem and well-being are very closely related to environmental competence. Individual self-maintenance

is enhanced by needed prosthetic devices and appropriate design.

Example: Items such as storage and shelves, closets, cabinets, ovens, and refrigerators should be designed and located to maximize their accessibility.

13. The physical surroundings should enhance the residents' orientation and increase their ability to negotiate their environment.

Rationale: Environmental competence and satisfaction are closely related. If people become easily confused by the settings around them, they will likely avoid those areas, reducing the utilization and enjoyment of the total facility.

Example: Provide visual cues through the use of colors, decorative planting, and other coding mechanisms to differentiate one floor from another in a high-rise.

14. Activity spaces of all types, i.e., recreation, lounge, and lobby, should have individual identifies but he conceived and interrelated as a series of interacting spaces.

Rationale: Spaces having an identity assist the individual in determining how to behave or what role to play and make participation less stressful. Certain interrelationships greatly increase utilization; as distance between activities increases, utilization often declines.

Example: A lounge adjoining the dining room.

15. The design of outdoor spaces should be given as much consideration as indoor spaces.

Rationale: Many older people are outdoor-oriented, and are quite mobile. They should be encouraged to remain so for as long as possible. The use of outdoor spaces can considerably expand the opportunities for varied activities, recreation, and social interaction, and thus contribute to the maintenance of sound health.

Example: Partial shelter, outdoor lighting, sun shading, and heating can be used to increase utilization for longer periods during the day and night as well as during seasonal changes.

16. Special consideration should be given to the design of heating, ventilation, and air-conditioning systems for the entire complex, especially in the dwelling units.

Rationale: Older people are more sensitive than younger people to fluctuations in temperature and drafts. Elderly residents spend a substantial portion of each day in their apartments.

Example: Provide each apartment with an individually controlled air-conditioning unit.

THE SPACE PROGRAM: SPECIAL CONSIDERATION

Another basic component of an architectural program is the space program which lists all spaces in the building and accounts for the percentage of the total area attributable to specific spaces. Table 1 shows special considerations for each department listed in the space program prepared for the congregate housing model discussed here.

Residential Facilities

It is recognized that an elderly resident, like nearly everyone else, prefers to select from various types of units. But in view of economic considerations, it is felt that there would be basic economy and clarity in limiting this particular model to one-bedroom units. Efficiencies are, in general, not preferred by older people and do

not provide flexibility for use by two persons. Two-bedroom units, though desirable in some cases, are nearly always those in least demand (this naturally depends upon the income range of the persons being served) and therefore have also been omitted. Five percent of the one-bedroom units have been programed for handicapped persons and should have design features required to accommodate a wheelchair.

Administration and Service Areas

The secretary's office and waiting area are the focal point for general information and serve a variety of functions such as rent collection and bookkeeping.

The multiuse office is intended for such purposes as counseling and, in particular, as a space where service personnel from the community (e.g., social service counselors) can meet with residents.

The health maintenance counseling room does not imply that there are in-house medical staff. This is an area where a visiting physician, nurse, or counselor can schedule meetings with residents on certain days during the week.

The duplication and work room is intended for use by both staff and residents. It is an area in which, for example, a photocopy machine could be located for use in making copies of an in-house newsletter.

Dietary Spaces

The proposed food service for this model serves two meals a day: a buffet breakfast or brunch from 8 to 11 a.m. and dinner at approximately 6 p.m. An area should be set aside for residents to obtain snacks at any time during the day or early evening.

Housekeeping

In addition to general maintenance and cleaning of public spaces, the housekeeping department is programmed to clean residents' rooms once a week.

Maintenance and Engineering

The boiler and mechanical room areas must, by the very nature of the model, be quite arbitrary. They will very considerably depending upon the region, climate, system, and type of building construction.

Common Use Facilities

The activity rooms should be developed in such a way that particular types of activities can be screened off from others to give identity to a specific activity and to provide some visual and sound isolation. It would be well to have these different areas open off one main open-access corridor.

The waiting and lounge area should definitely adjoin the dining room so that persons arriving early for a meal can become involved with others who are waiting for the meal to start. It should be a comfortable area for them to sit after the meal as well.

Parking

Parking needs and code requirements will vary greatly from city to city and have been the subject of much debate in the field of housing for the elderly. A final determinant of the number of spaces required for parking must therefore be arbitrary in a model program. Nevertheless, five parking spaces should be sized to accommodate handicapped drivers and should be conveniently located.

TABLE 1 Space Program—List of Spaces

Department and space	Number	ft²	% total ft²	ft³/unit
Residential facilities				
One-bedroom apartments, 600 ft²	95	57,000		
One-bedroom apartments (for handicapped), 750 ft²	5	3,750		
Manager's apartment	1	900		
Total residential facilities		61,650	65.7	616.5
Administration and service areas				
Manager's office	1	230		
Business office, secretary and waiting	1	150		
Multiuse office	1	150		
Health maintenance counseling	1	150		
Duplication and work room	1	150		
Total administration and service		830	0.8	8.3
Dietary				
Kitchen including storage and office	1	2,000		
Total dietary		2,000	2.1	20.0
Housekeeping				
Housekeeping office	1	100		
Housekeeping storage	1	200		
Janitor's closets 20 ft²	4	80		
Total housekeeping		380	0.4	3.8
Maintenance and engineering				
Boiler and mechanical room to include				
Transformer room				
Electric equipment room				
Elevator equipment room	1	1,200		
Repair shop	1	200		
Receiving and loading	1	400		
General storage	1	300		
Trash and compactor room	1	150		
Total maintenance and engineering		2,250	2.3	22.5
Staff facilities				
Men's toilet and locker	1	100		
Women's toilet and locker	1	150		
Employees' lounge	1	150		
Total staff facilities		400	0.4	4.0
Common use facilities				
Activity rooms, 400 ft²	2	800		
Kitchenette adjoining activity rooms	1	20		
Storage and sound equipment (adjoining activity rooms)	1	200		
Resident dining room—seat 120	1	1,800		
Waiting, lounge, and lobby area	1	1,200		
Craft room	1	400		
Coat room	1	50		
Women's toilet and powder room	1	120		
Men's toilet	1	120		
Mail room	1	200		
Men only room	1	400		
Reception	1	150		
Laundry rooms for residents, 100 ft²	2	200		
Public telephone, 10 ft²	2	20		
Total common use facilities		5,680	6.0	56.8
Total net areas		73,190	77.9	731.9
Partitions and circulation		20,644	22.0	206.4
Project total		93,834		938.34
Parking, 400 ft²	30	12,000		

Additional Spaces

The following spaces not listed in the space program presented here should be considered for inclusion in a congregate housing project if the budget permits and if the size or scope of the project is expanded:

Assistant administrator's office
Conference room
Social service counseling area
Garage
Snack bar
Game room
Library
Classroom
Balconies and patios
Concession and gift shop
Staff toilet in administration area
Barber shop
Beauty shop
Workshop
Chapel
Activity director's office
Greenhouse
Fixed-seat auditorium

CORE SUPPORTIVE SERVICES

Congregate housing for the elderly should be more than bricks and mortar. The housing sponsor must enhance the quality of life for residents as well as add to their years of independence by ensuring the availability of special and essential supportive services: a food service, housekeeping, transportation/escort services, personal counseling, and social and recreational services. These *core* services are directed toward a level of intervention appropriate for residents typically found in congregate housing settings. They should be viewed as enabling devices which compensate for activities and functions that individuals may not be able to perform independently. Such services are directed toward maintaining individuals in their place of residence, stimulating their capacity for social interaction and communication, strengthening their community ties, and ensuring access to previously inaccessible resources. These services should be incorporated into any congregate housing project and should be provided on-site by the sponsor with qualified staff. The design of the facility should allow adequate space for on-site service delivery.

Whenever possible, the delivery of services should be arranged in conjunction with available community resources. The formation of working relationships with agencies, such as health and welfare councils, health care programs, and area agencies on aging, should be encouraged. Coordination activities might include joint planning, information sharing, and agreements for joint funding and operation of programs and for reimbursements (third party payments) for services.

Although the supportive services which are essential to congregate housing are intended to assist the resident population, they should be extended to elderly persons in the community whenever possible. In so doing, costs to residents of the project can be reduced and broader community support for the facility will be generated. Similarly, on-site supportive services should be developed in such a way as to complement community resources rather than duplicate or overlap with them.

Determination of which approach to follow (e.g., coordination with existing services, development of core services from scratch, or contracting for them from other agencies) depends upon a number of factors, chief among them being the availability of services, access to these resources, and the financial capacities of the sponsor, the resident, and the community. Of particular importance is the availability of third party reimbursement payments for social and health maintenance services. In essence, the decision regarding which approach to take in providing core services can only be made on a case-by-case basis after careful examination of resident needs, the facility, and the community and its service delivery system. That is, the modalities for delivering core services in congregate housing will vary, depending upon the specific situation.

The following are brief descriptions of the core services. There are other supportive services, e.g., health and welfare counseling, and information and referral, that will undoubtedly be utilized also as the occasion dictates.

Food Service

The food service offers nutritionally balanced meals, at reasonable costs, in a central dining room. This is a major necessity because residents, especially single persons, may not always desire to cook for themselves, because some may have difficulty with meal preparation due to the infirmities of advancing age, and because the service provides opportunities for social interaction.

Although there are a variety of approaches for food service delivery, the model proposed here requires that each resident purchase a minimum amount which can be flexibly used according to his or her preference for different meals.

Housekeeping Service

This service, delivered to residents in their own units, includes assistance with housecleaning, window cleaning, washing, and ironing. These services are not included in the basic fee structure but should be available (on a fee basis) to residents who require temporary assistance because of illness, injury, or some other cause.

Although housekeeping services do not encompass home health care (i.e., assistance with bathing, dressing, and medications for those who may be ill or find it difficult to leave home to reach local services), arrangements for this assistance can be made by the service staff of the housing sponsor with a local visiting nurse program or a home health services agency.

Transportation/Personal Escort Services

Transportation and/or personal escort services should be provided by the sponsor for residents who require help with access to needed services, that is, when residents are unable to provide such transportation, when it is extremely inconvenient for them to do so, or when their mobility is impaired.

Some residents will require escort services in addition to transportation. The sponsor should furnish this service or contract with an agency to provide it. Escort service refers to an activity which is designed to assist those who are physically handicapped and require some personal assistance and special modes of transportation (e.g., barrier-free). This type of service entails more than the provision of transportation; it is equally a companion service.

Personal Counseling and Emotional Support Services

Older people often need personal contacts and emotional support services because they are separated from families or are widowed. These services, unless developed and provided on an organized basis, will be of little if any value. Therefore, the sponsor should arrange to make them available for delivery by trained personnel with experience and understanding of elderly needs and problems.

A number of programs can be orchestrated to provide emotionally supportive services. For example, friendly visitors can be organized to make regular visits to each resident. This social interaction is often mutually rewarding to both the older person and the volunteer. Another programmatic example is the "buddy system" whereby residents are organized to look after one another on a daily basis. Any irregularities (e.g., illness, no response, etc.) are reported immediately. These activities are relatively simple to organize; however, they must be carefully developed in order to sustain their effectiveness over time.

Social and Recreational Services

Social and recreational programs should be organized by trained personnel. Such programs are intended to keep residents involved in the activities they enjoy. Among those often offered in congregate housing are lending libraries, cards, crafts, discussion groups, field trips, clubs, and the like. These are intended to stimulate social interaction through personal contacts in a supportive atmosphere. While the sponsor should provide full- or part-time staff to assist with program development and implementation, residents should be involved not only as participants but also as initiators.

SPACE REQUIREMENTS

In considering accommodation for single young people, an initial breakdown of an individual's requirements will be necessary. The younger, more mobile single person may be in less need of a permanent home, often treating furnished accommodation as a *base* rather than *home,* which for many single people still exists on weekends back at the parental home. Having at least one room used primarily as a bedroom as their own domain, sharing facilities, kitchens, bathrooms, living space, can prove important both as a social and as a financial alternative to self-contained accommodation. However, the latter type of provision may better suit older, more settled single people, perhaps in their thirties, as a permanent home.

Single Self-Contained Accommodation

This concept is easily understood, implying an independent dwelling for one person containing exclusive facilities for normal daily life, i.e., a kitchen, bathroom, or toilet, living area, sleeping area, and storage space. It can be a small detached house or bungalow, or more usually forms a part of a complex of similar units.

Shared Accommodation

This term can cover a multitude of tenure patterns, from shared *dormitory* bedrooms with adjacent communal washing facilities, to small groups of single people sharing existing housing. For the latter each member of a group will have a bedroom but will share domestic cooking, living, and sanitary facilities.

An individual "private" room for each person for use primarily as a bedroom should be considered as a basic requirement. Shared bedrooms where provided in the past have proved unpopular, as they do not allow a personal retreat for each occupant.

Shared accommodation has attracted a number of terms, the most usual being *multiple occupation* or *cluster flat.* The former term is based on the normal definition of a dwelling occupied by more than one household, or family. The term "cluster flat" probably derives from the pattern of individuals' rooms grouped around a common kitchen, bathroom, and living space, as with the parallel of family housing. Inclusion of cooking and washing facilities within an individual's room will reduce the cluster effect, producing instead semi-independent units commonly referred to as *bedsitting rooms.* (This term is imprecise, as it is also sometimes used to imply only a bed/living space without the serviced amenities of a cooker and sink.)

ROOM TYPES AND DEFINITIONS

With family housing, each room within the dwelling can be clearly defined—bedroom, living room, kitchen, etc. The different tenure arrangements with shared use can blur these categories, room functions being adjusted to suit the individual circumstances. In an attempt to determine a range of room notations, which can apply to both new buildings and adaptation of existing housing, the different room permutations possible are outlined in the illustration. Related to space and activity function, individual situations may have factors of overriding significance, e.g., circulation pattern of the existing dwelling, availability and proximity of services and drainage, aspect and natural light-

ing, and these factors will have to be balanced accordingly.

As the upper range of room functions does not include an individual bathroom (or shower room) with a toilet, these will rely on shared use of facilities elsewhere in the dwelling. The greater the increase in washing and cooking facilities within the larger room, the less the need for shared facilities except for the fundamental bathroom.

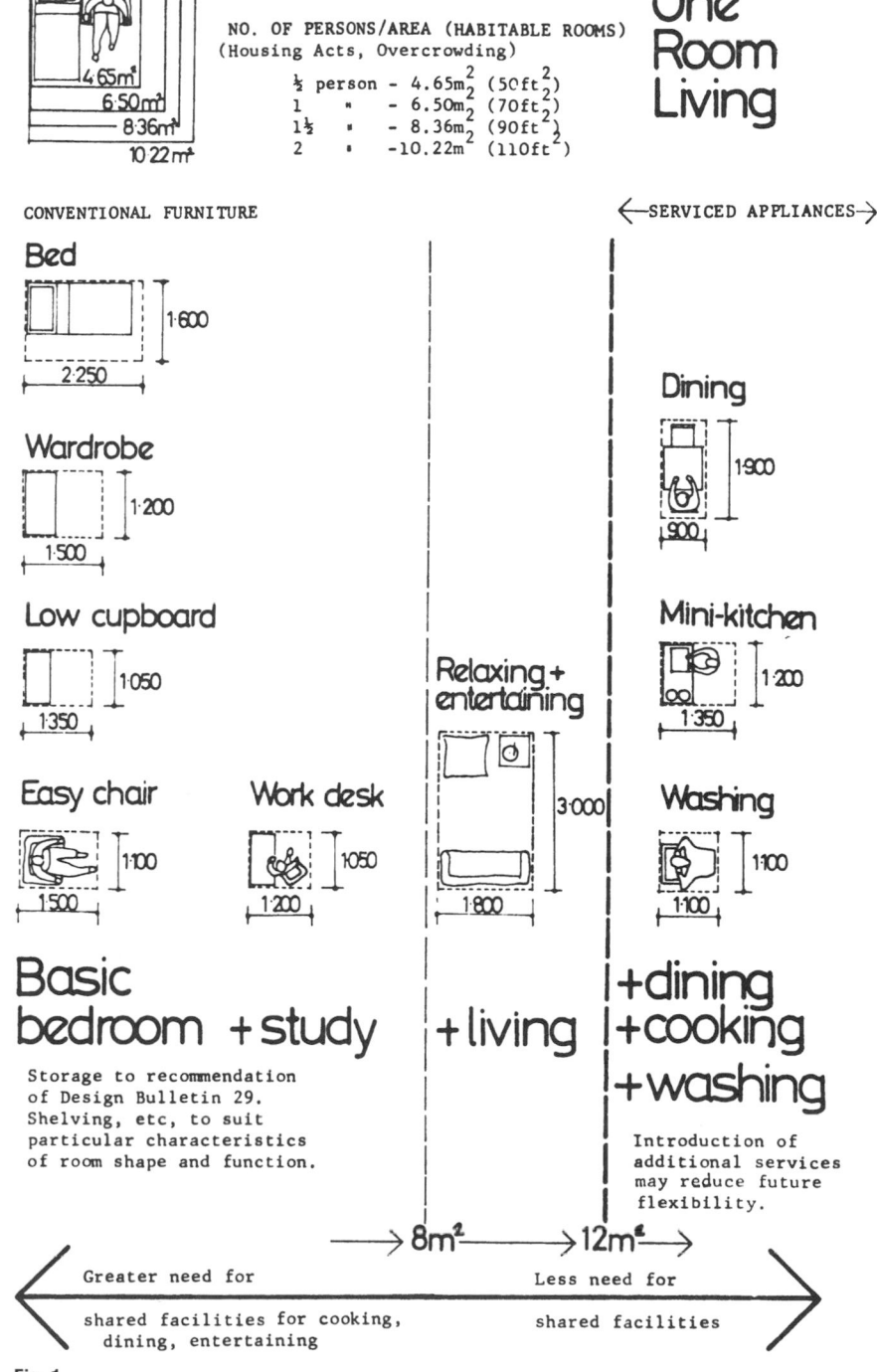

Fig. 1

HOUSING FOR SINGLE YOUNG PEOPLE

ROOM	NOTATION	DESCRIPTION	
CABIN BEDROOM	CB	Very small room relying on fixed or built in furniture and storage, little prospect for re-arrangement.	SMALLEST ROOM
BEDROOM	B	Small room, conventional furniture, single bed, storage, wardrobe, chair. Primarily for sleeping.	
STUDY/BEDROOM	SB	As B, with inclusion of desk, chair, book storage, shelves, lighting to enable use for dual function.	
STUDY/BEDROOM (SERVICED)	SB(S)	As SB, with wash basin, h&c water supply, drainage.	SHARED ACCOM- MODATION
LIVING/STUDY/ BEDROOM	LSB	As SB, but additional area used for entertaining.	
LIVING/STUDY/ BEDROOM (SERVICED)	LSB(S)	As LSB, inclusion of wash basin, necessary h&c water supply, drainage.	
DINING/LIVING STUDY/BEDROOM (SERVICED)	DLSB(S)	As LSB(S), inclusion of cooker/ grill, food storage, refrigerator, washing-up sink, h&c water supply.	LARGEST ROOM
SMALL FLAT (SELF-CONTAINED)	SF(SC)	Living/study/bedroom with sep. individual kitchen, bathroom and WC, storage and entrance lobby.	
MEDIUM FLAT (SELF-CONTAINED)	MF(SC)	As SF(SC) above	SELF- CONTAINED
LARGE FLAT (SELF-CONTAINED)	LF(SC)	Separate bedroom, living room, kitchen, bathroom and WC, storage and entrance lobby. (Conforms to Parker-Morris standards.)	

Fig. 2   A notation of living spaces used as private space.

Shared Kitchens/Living Space

In relation to existing housing, the "shared" household of single young people would appear to work better where individuals' rooms do not have exclusive washing and cooking facilities. For these cases, the nucleus of the dwelling is the shared kitchen/living area rather than individuals' rooms in a similar manner to a family house.

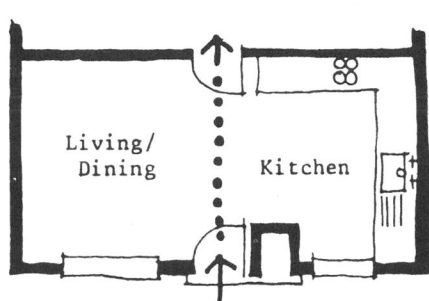

1. 'Farmhouse' kitchen; living, dining, kitchen and main circulation.

2. Kitchen, dining living areas separate from main circulation.

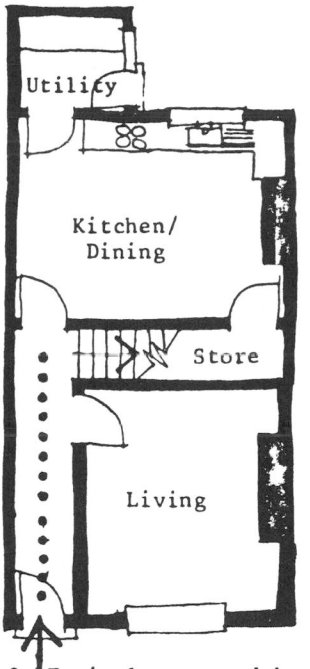

3. Typical terraced house separate living area, kitchen as secondary circulation.

Fig. 3 Typical current provision for main communal and shared space.

Shared Bathrooms

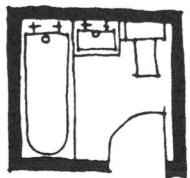

1-2 persons

3 persons

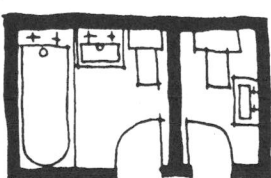

4-5 persons

Fig. 4

Self-contained, One Person

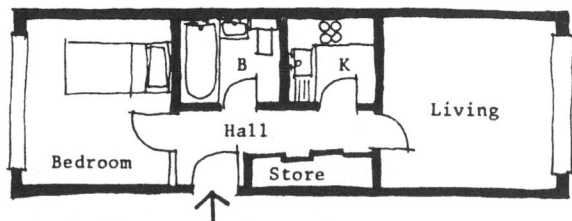

Fig. 5 Note that a typical variant of this arrangement will have a single bed/living room in place of the two separate spaces shown here.

Semi-independent, Four Persons

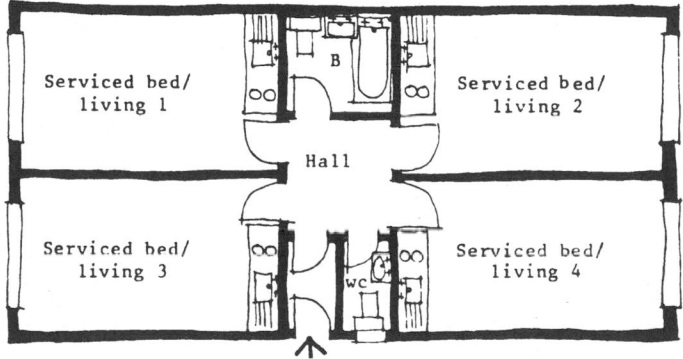

Fig. 6

935

Shared Accommodation, Four Persons

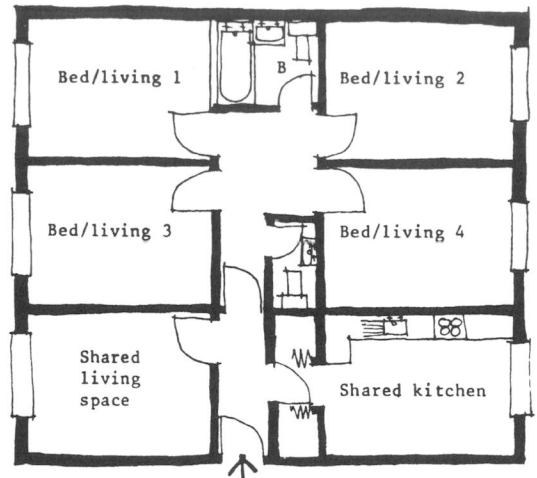

Fig. 7 Variations of single-person tenure; illustrated by notational dwelling arrangement.

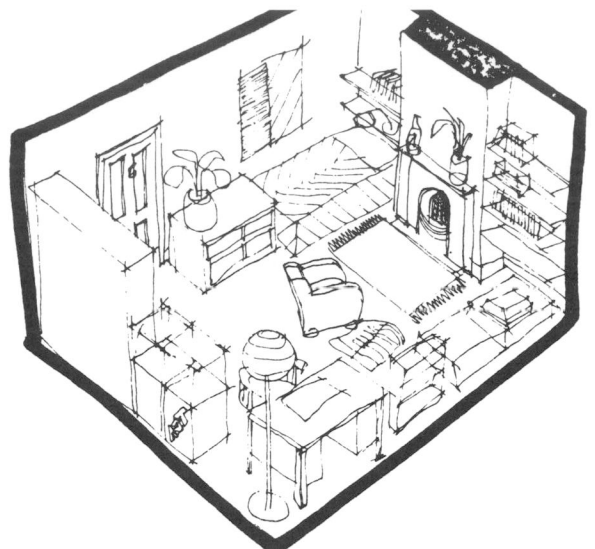

Fig. 8 Typical room used as private space, 15 m².

PRIVATE SPACE/SHARED SPACE/GROUP SIZE

Projects that involve shared accommodation within a dwelling have two main components of space allocation, private space and shared space. Of these, private space can be considered in its most basic form as a bedroom/retreat with some personal storage, whereas shared space is the total of the communal living and amenity areas (and circulation) contained within the shared dwelling.

Arrangements of private space depend on the number of people within the sharing group related to the building form. With purpose-built housing for single young people, the group size may be determined for social reasons with relatively small sharing units or by design constraints with provision of set group size for existing ratios of persons/sanitary appliances. Older housing has the physical limitations of the existing building envelope with varied room sizes and arrangement as the main criterion.

Suitable combinations of private spaces and shared spaces should reflect the respective functions; the nucleus of the dwelling should be the shared living/dining area, and the relatively noisy activities should be separated from the private spaces horizontally by story level or vertically by circulation "buffer" zones. As such, the shared space component can be further subdivided into living areas, serviced areas (kitchen, bathroom, utility), and circulation spaces (halls, stairs).

PRIVATE SPACE COMPONENTS

From the components of one room living, a relationship can be established between activity requirements and suitable room size for private space within a shared dwelling. Given the premise of using conventional furniture within these rooms, different arrangements and living patterns can be established; for the exception of a minimal cabin bedroom, it may be possible to rely on ingenuity with dual-purpose furniture and fittings. A room used primarily as a bedroom will require little more than the zones of space surrounding the items of furniture; such an arrange-

ment will, however, rely on living space in addition to the confines of the private space.

Rooms in new family housing described as "single bedrooms" imply room sizes that accommodate little more than a single bed. The needs of single young people may require a larger size of single room if it is to serve as a bed/living room or as a study bedroom. These requirements can, however, be satisfactorily accommodated in rooms often labeled as "double bedrooms" in new housing (to cost yardstick standards) or even more capaciously in the larger rooms common with older housing.

The introduction of serviced amenities, initially a washbasin or sink, or perhaps later more complex cooking and food-preparation facilities can produce a further degree of independence for the one-room unit. Such a move toward self-containment does though tend to ignore the social structure engendered within shared dwellings. If the kitchen/living spaces are no longer the shared nucleus for semi-independent one-room units (leaving only the bathroom with toilet as the communal spaces), there is less and less opportunity for social integration within the dwelling.

SHARED KITCHENS/LIVING SPACE

As with the term "cluster flat" applied to a shared dwelling, the "farmhouse kitchen" has been the description for living/kitchen areas in shared use by single people. This space forms a nucleus for the shared dwelling, where cooking and eating together provides the greatest opportunity for social integration. In legislative terms, a "household" is an organization that shares at least one meal daily, and there should be space and facilities for all the group, with the inevitable guests, to share meals at the same time.

Existing housing tends to have rather small kitchen areas with larger living/dining spaces leading off. This can be a more effective arrangement for social contact with a sharing group, having the advantage of isolating cooking, food preparation, and washing up. The proportions of private space/shared space should be suitably balanced; those arrangements with limited private space can be offset, with an increase in

shared space made available. This can be arranged by having a separate shared living space away from the bustle of the kitchen/dining area. In terms of the normal ground-floor room arrangement of older housing, a front lounge can be a "quiet" shared living space, the back living/dining room being linked to the rear kitchen as the busy social core of the dwelling.

SHARED BATHROOMS

The inclusion of bathrooms within many older houses has been related to the demand for increasing amenity standards. Many bathrooms are converted from previous bedrooms and boxrooms, others form purpose-built extensions. The basic bath and washbasin with hot and cold water supply, together with a toilet are normally grouped in one "bathroom," although it is advantageous both for a sharing group of single people and the family situation to have a toilet in a separate compartment. A second toilet, together with a washbasin, will meet the requirements for a sharing group of up to six single people in new-built accommodations, although the plan form and proximity to "bedrooms" will determine whether such an arrangement is viable in the adaptation of existing housing. In general, a bathroom and toilet should not be more than one floor distant from a bedroom; hence a ground-floor bathroom will be inadequate for a three-story dwelling with "bedrooms" on the upper floor.

The "bedsitter" arrangement common in traditional multiple occupation often has each single room used as a bedroom, living room, and kitchen with a sink and water supply used for both personal washing and washing dishes. Use of bath and toilet common to all the units is the only shared amenity. Although this arrangement is common, the insertion of a water supply and drainage to each "bedroom" is often difficult and relatively expensive to incorporate in existing housing and may restrict any plans for a reversion back to family housing. As with current recommendations for new-purpose-built accommodation, extra washbasins can instead be added to the shared bathroom facility rather than to individual rooms.

A typical room is 300 ft².

All rooms have televisions and microwave ovens. Communal spaces include a first-floor reading room. Key to Figs. 1 and 2 is as follows:

1. Garage entry
2. Café
3. Recreation
4. Front desk
5. Deck
6. Lobby
7. Reading room
8. Courtyard
9. Typical unit
10. Live/work space
11. Existing building
12. Future park

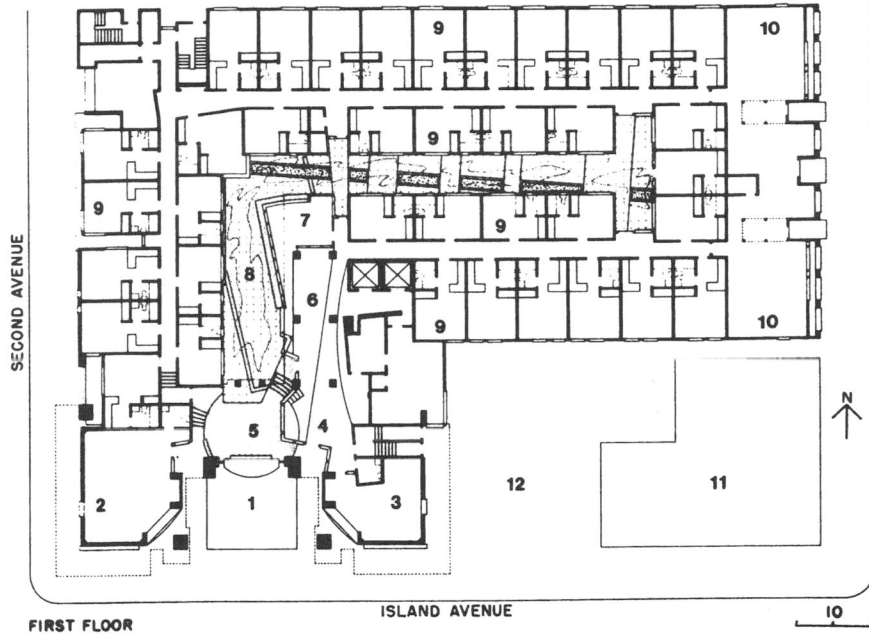

FIRST FLOOR

Fig. 1 202 Island Inn, San Diego, Calif. Rob Wellington Quigley—Architect.

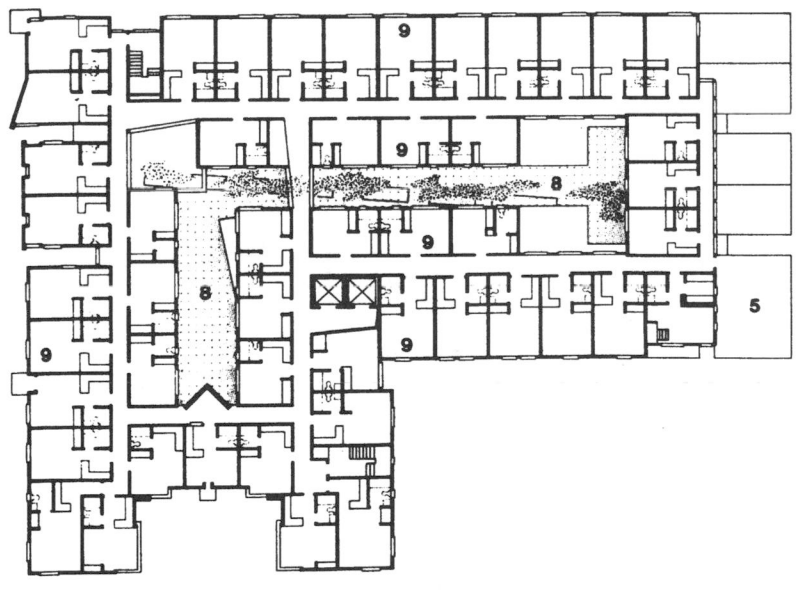

SECOND FLOOR

Fig. 2 202 Island Inn, San Diego, Calif. Rob Wellington Quigley—Architect.

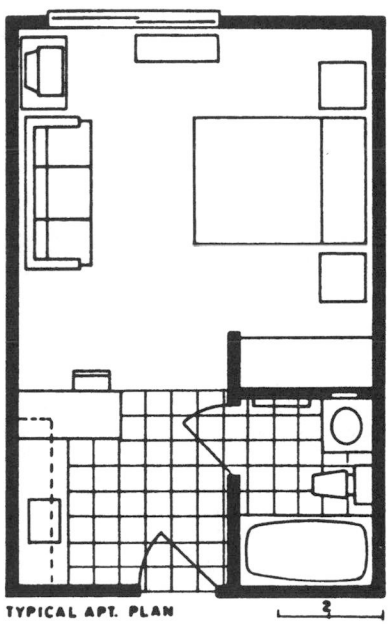

TYPICAL APT. PLAN

Fig. 3

TYPICAL PLAN

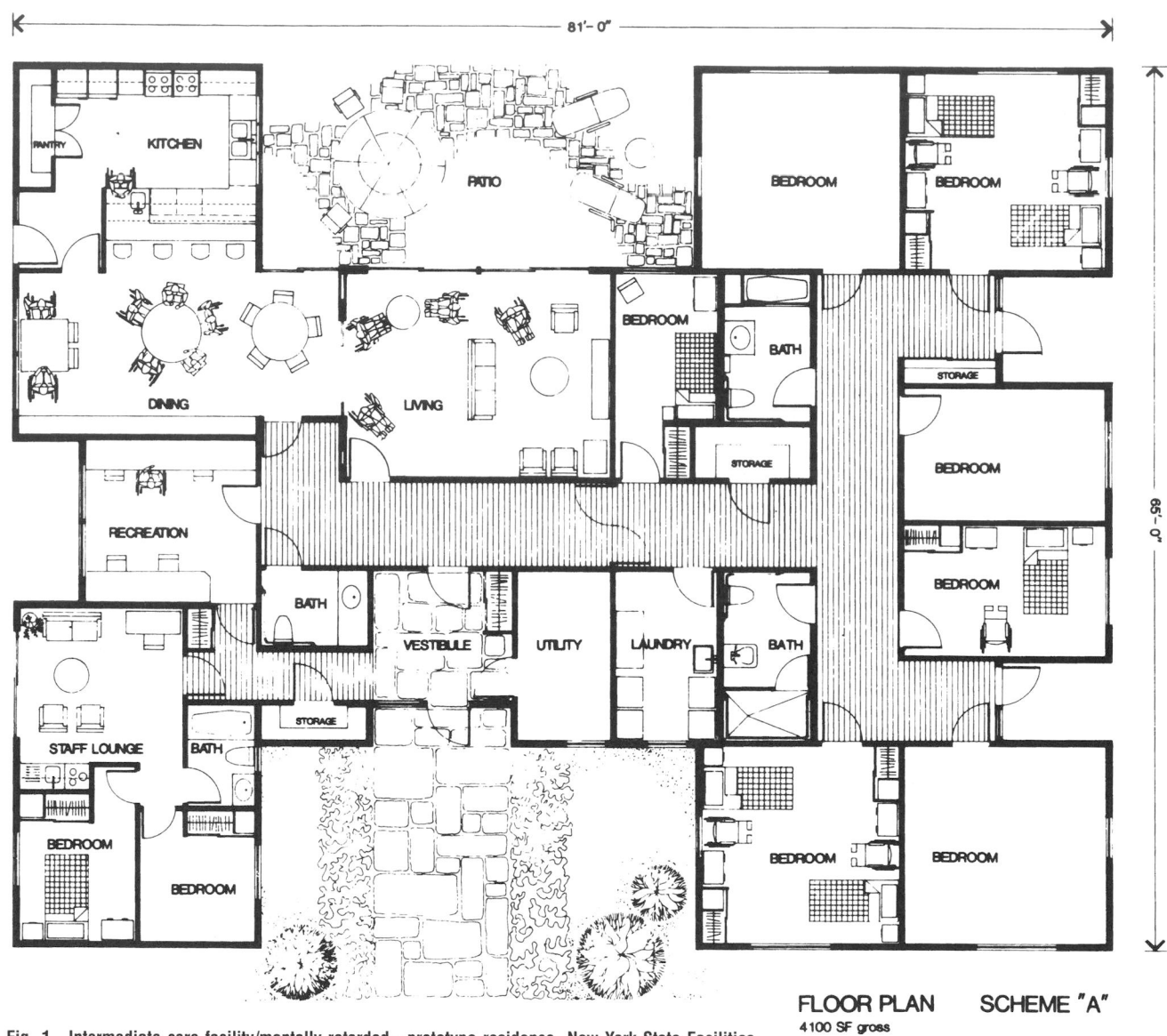

FLOOR PLAN   SCHEME "A"
4100 SF gross

Fig. 1 Intermediate care facility/mentally retarded—prototype residence, New York State Facilities Development Corporation, OMRDO. Rudolph Horowitz—Architect.

LIVING ROOM

Layout to foster group discussions. It is desirable to locate living in close proximity and relationship to the dining room so that all residents and some visitors might be accommodated at one time.

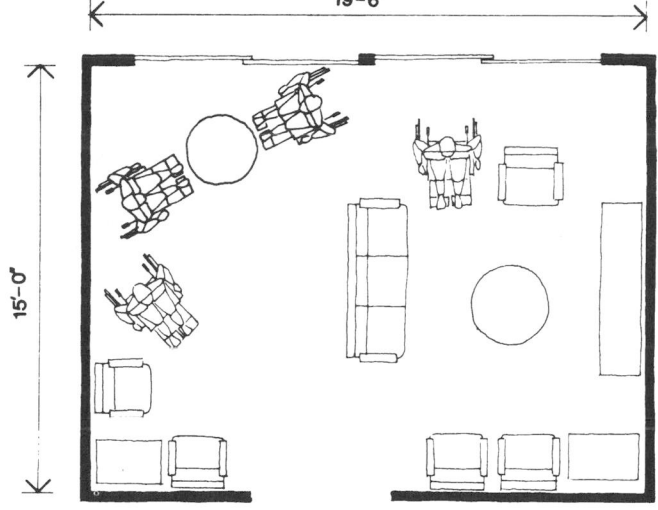

Fig. 2   Living room, 293 ft².

DINING ROOM

Layout to seat 14 persons, at two or more tables. Wall storage unit for dishes, games, etc.

KITCHEN

A residential-type kitchen. One working position with a small sink should be provided for training a person seated in a wheelchair. An open eating counter should be provided between kitchen and dining. Kitchen should be laid out in such a way as to control access to it when desired.

PANTRY

Small lockable storage room or large closet with open shelving, accessible to persons in wheelchairs.

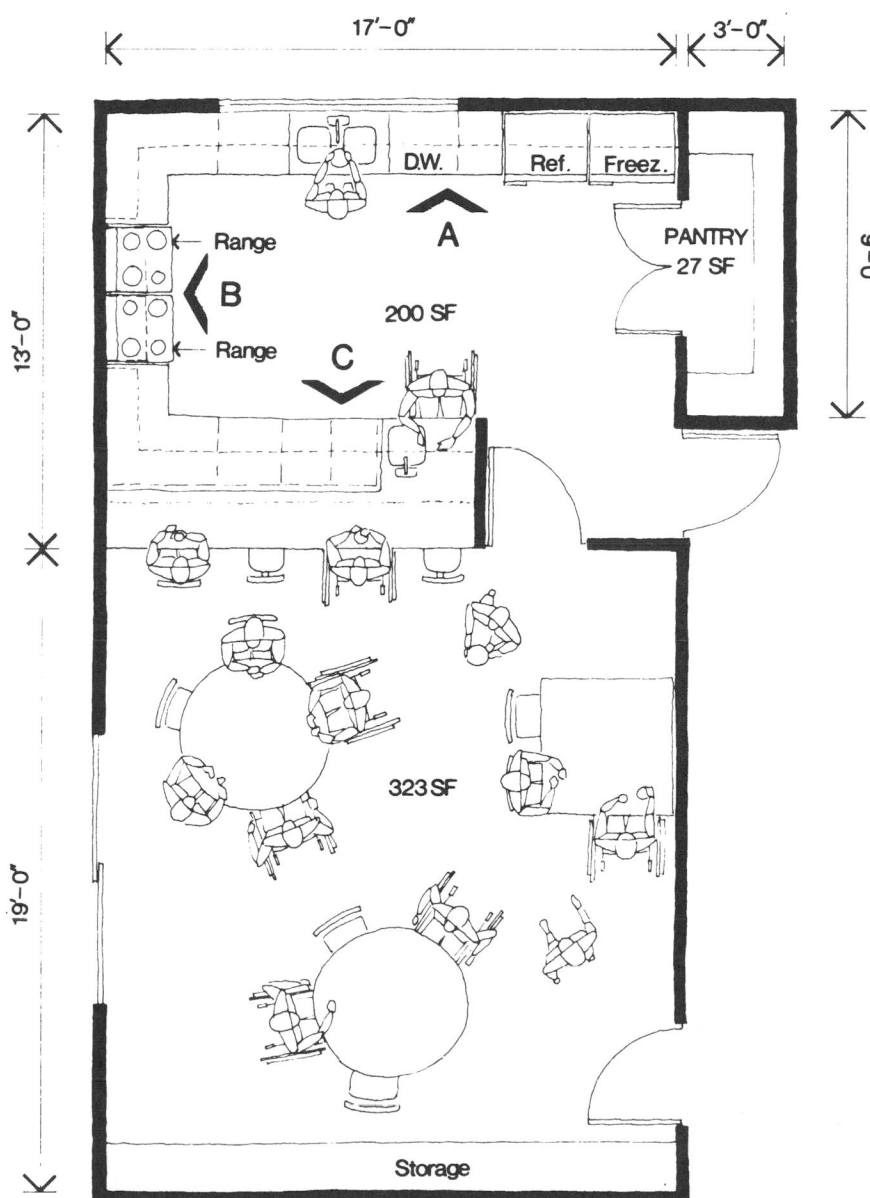

Fig. 3

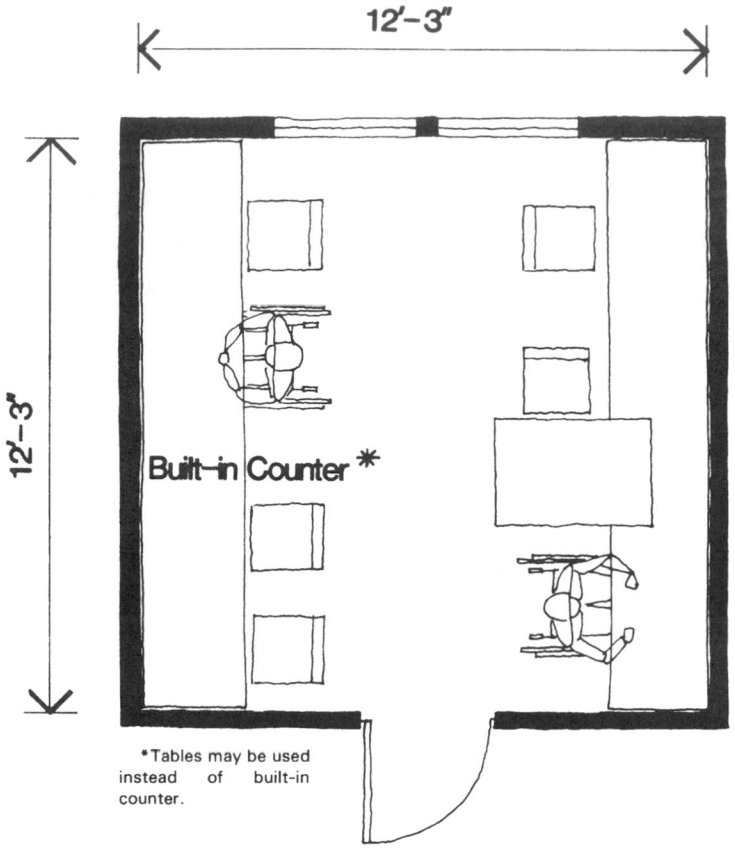

12'-3"

12'-3"

Built-in Counter *

*Tables may be used instead of built-in counter.

RECREATION ROOM

Provide room in vicinity of living room and dining. Layout should preclude through traffic. This room could be used as rest area for overnight, on-duty staff.

Fig. 4   Recreation room, 150 ft².

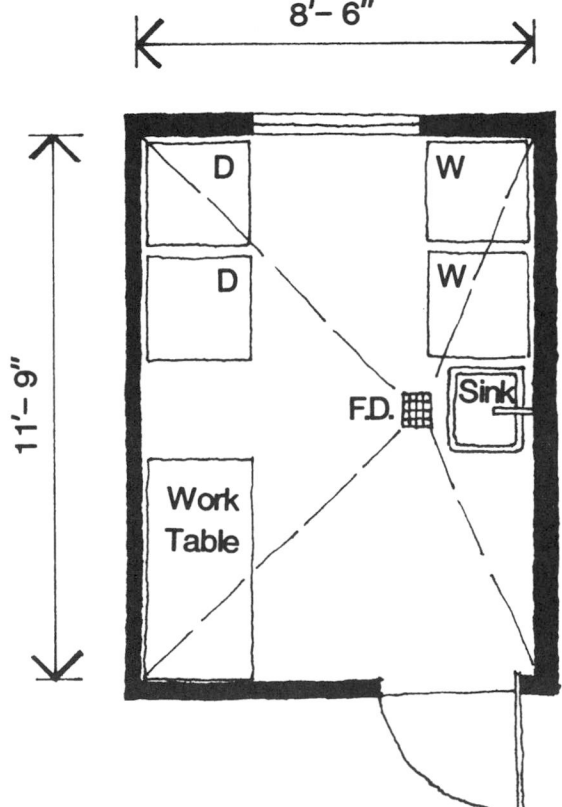

8'-6"

11'-9"

D   W

D   W

F.D.   Sink

Work Table

LAUNDRY

The laundry should be accessible to wheelchairs. Provide two washers and two heavy-duty dryers, and a laundry tub. Provide space for work tables. Locate away from bedrooms for acoustic reasons. Provide floor drain.

Fig. 5   Laundry, 100 ft².

SINGLE BEDROOM, AMBULATORY

DOUBLE BEDROOM, AMBULATORY

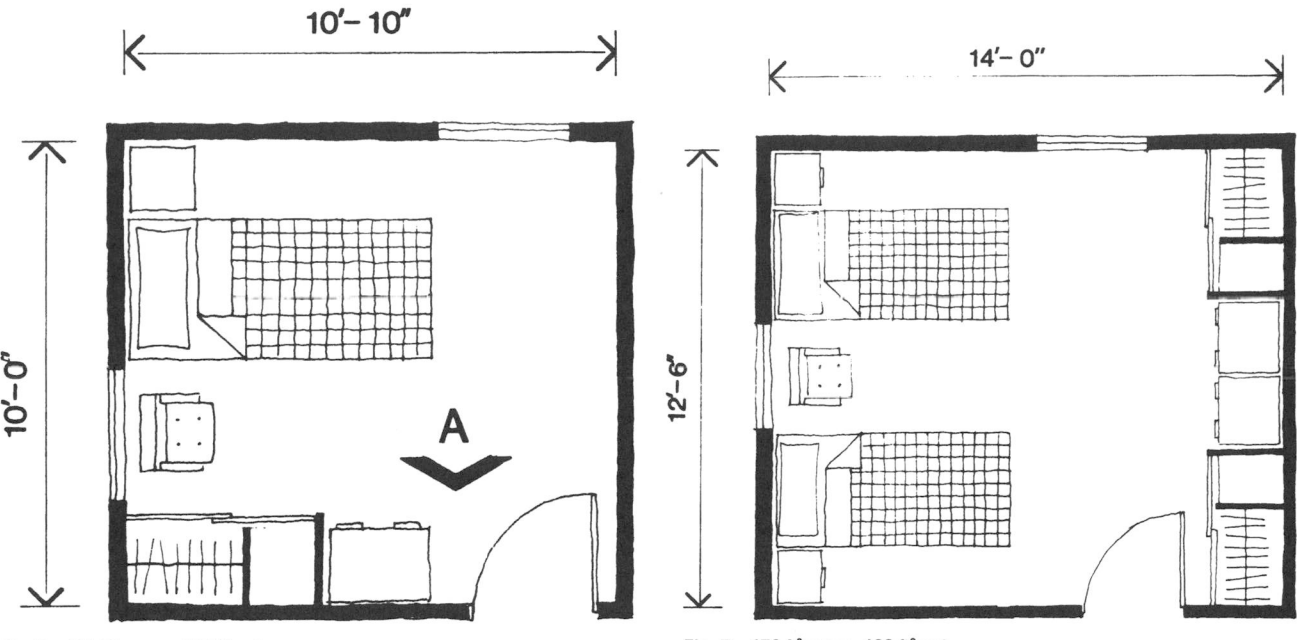

Fig. 6   100 ft² gross, 100 ft² net.

Fig. 7   176 ft² gross, 160 ft² net.

**SINGLE BEDROOM,
NONAMBULATORY**

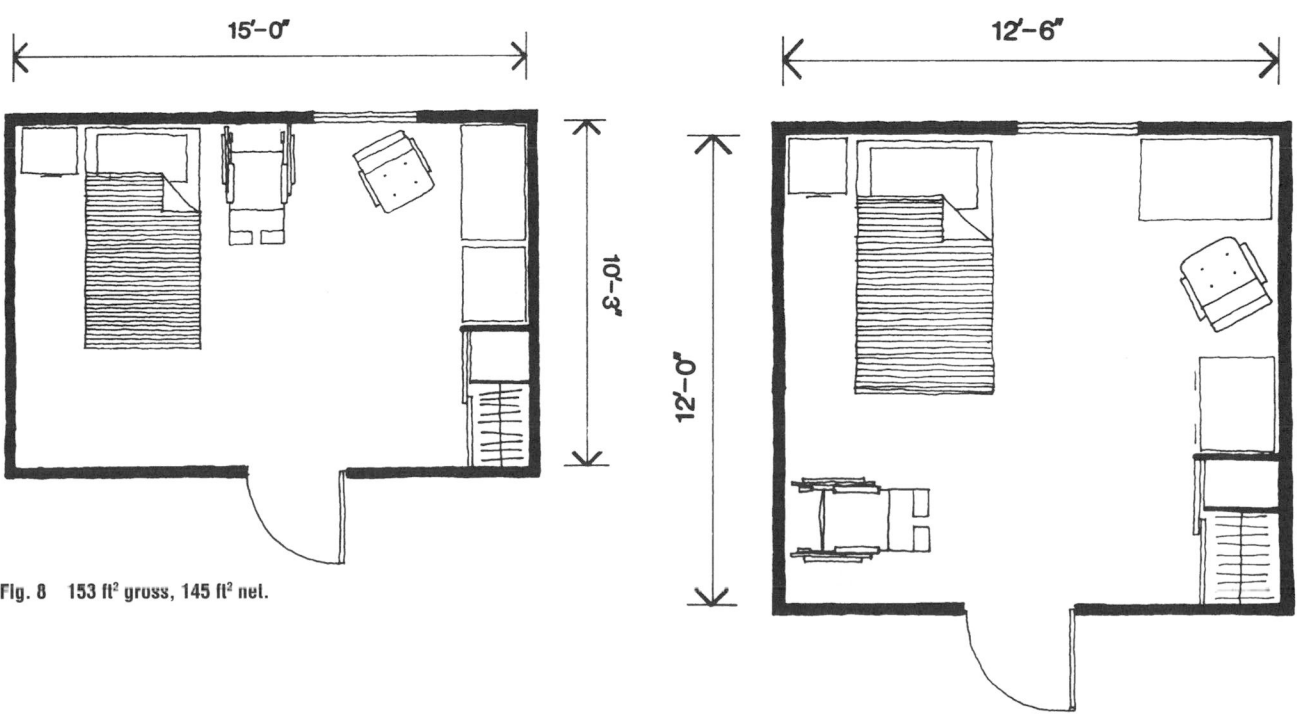

Fig. 8   153 ft² gross, 145 ft² net.

Fig. 9   150 ft² gross, 142 ft² net.

COMMUNITY RESIDENCE/MENTALLY RETARDED

DOUBLE BEDROOM,
NONAMBULATORY

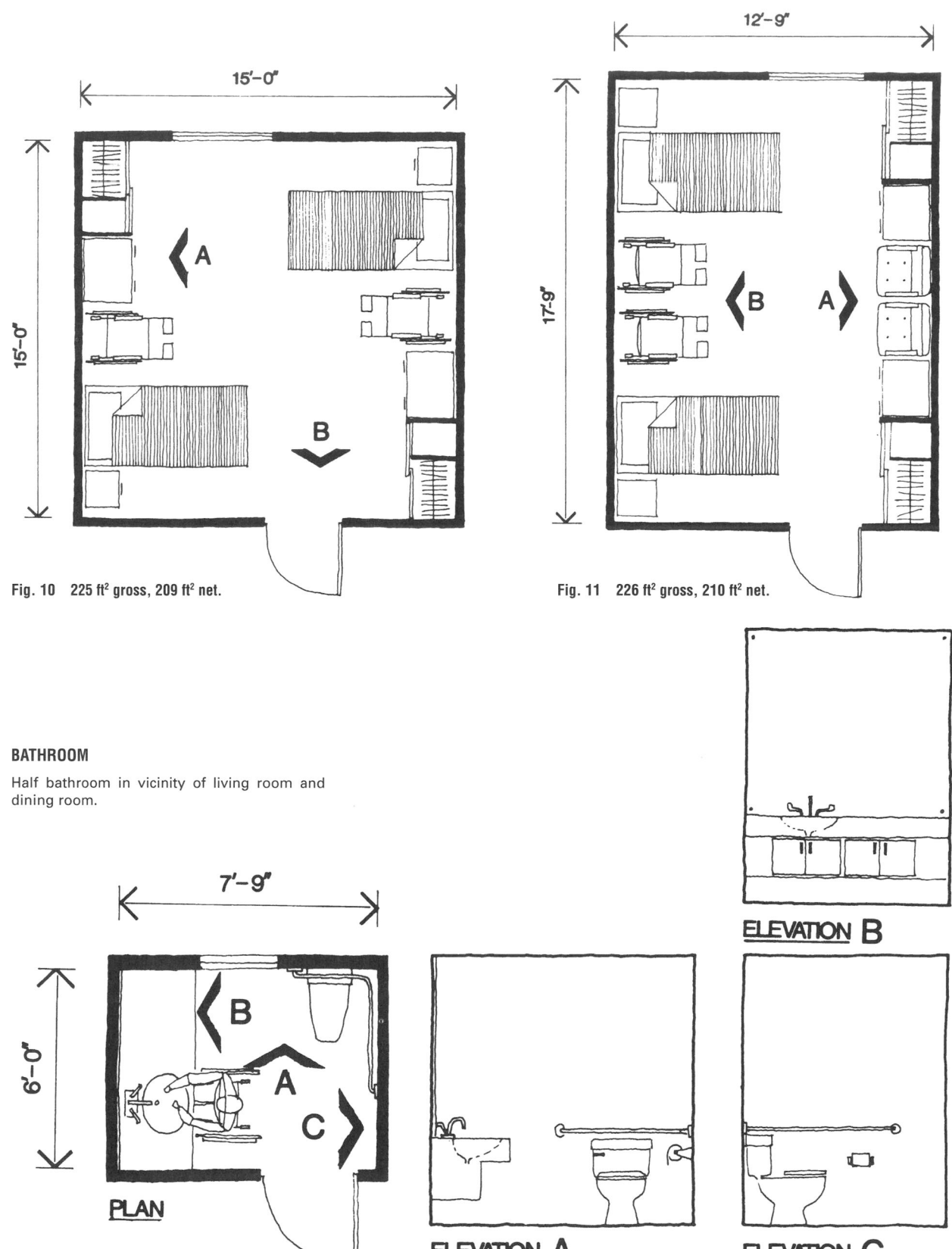

Fig. 10   225 ft² gross, 209 ft² net.

Fig. 11   226 ft² gross, 210 ft² net.

BATHROOM

Half bathroom in vicinity of living room and
dining room.

ELEVATION B

PLAN

ELEVATION A

ELEVATION C

Fig. 12   Bathroom, 47 ft².

BATHROOM

One bathroom with roll-in shower.

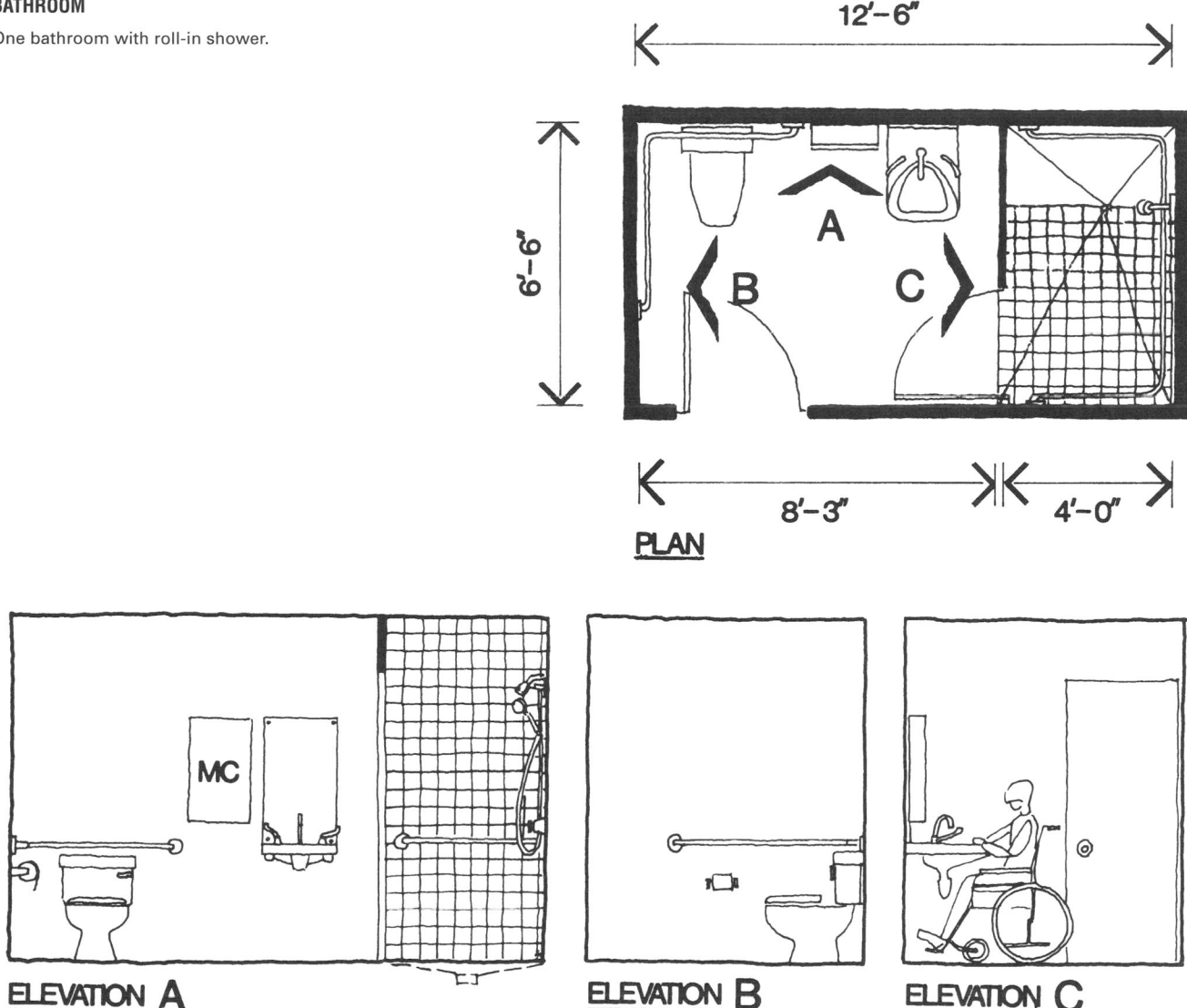

PLAN

ELEVATION A

ELEVATION B

ELEVATION C

Fig. 13 Bathroom, 81 ft².

BATHROOM

One bathroom with tub with grab bars.

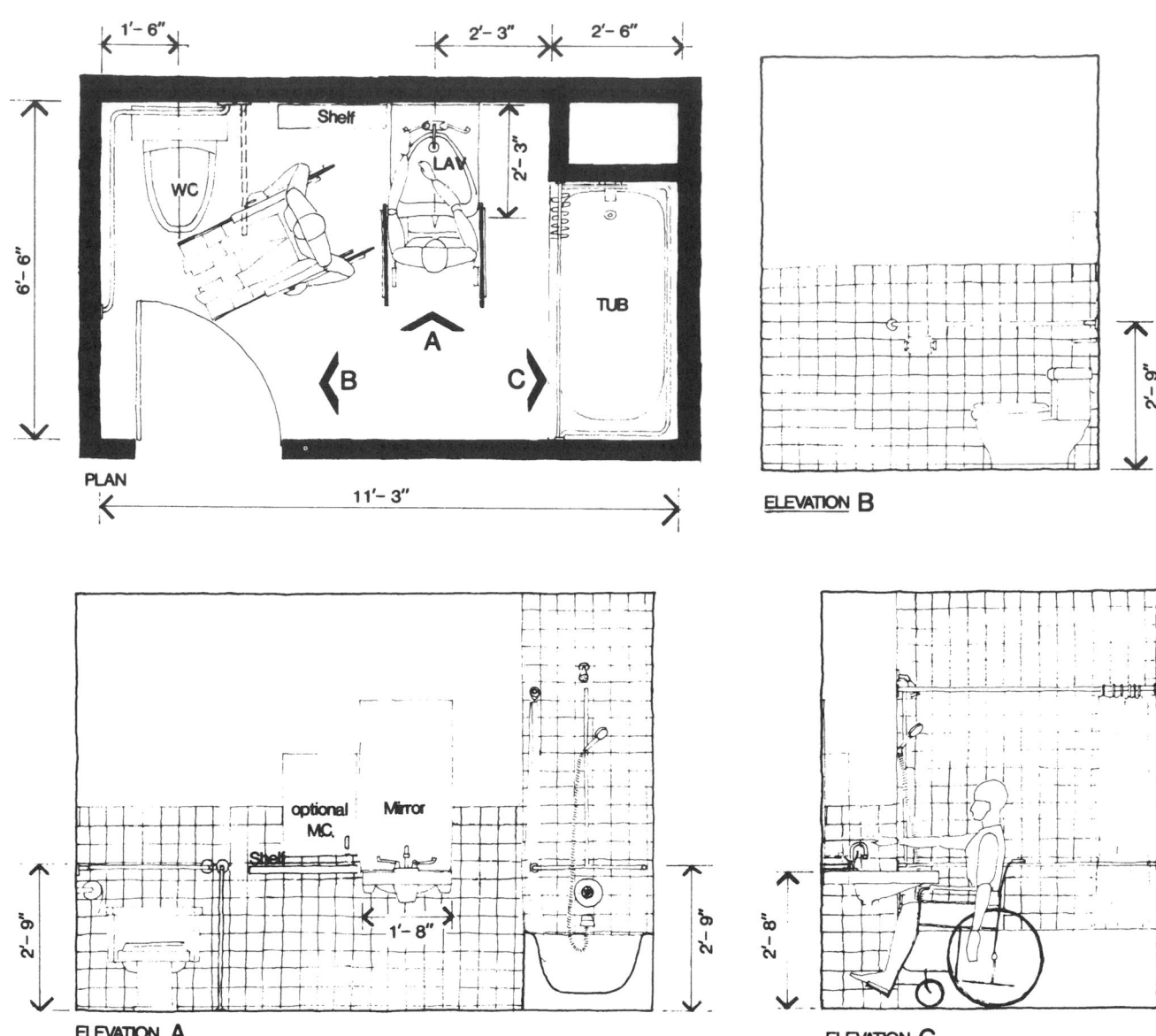

Fig. 14   Bathroom, 69 ft².

STAFF AREAS

Staff lounge with access to outdoors is to be used as lounge and office; provide space for desk and file cabinet. Provide a unit kitchenette in lounge; provide two bedrooms, and a bathroom.

In programs where there is no live-in staff, this staff area can be used as an additional program space or a training apartment. Alternatively, it could be left out of the program.

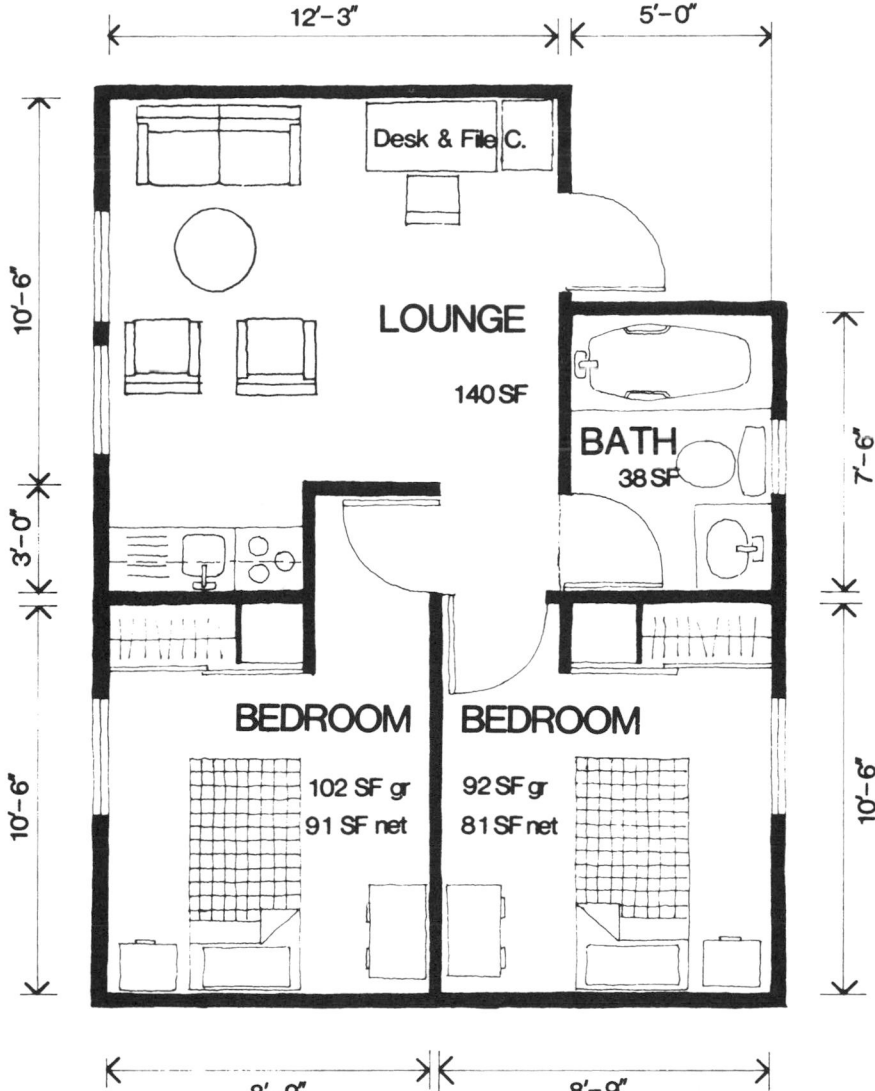

Fig. 15 Staff residence.

HOMELESS CENTER

The homeless center shown in Fig. 1, located in San Francisco's South of Market district, was a vacant, reinforced-concrete frame warehouse, used as a shelter even as construction took place. Generous space is provided at the check-in area, so that lines do not form in front of the building, greatly lessening the impact of the building on the neighborhood and responding to concerns expressed during planning meetings for the project. Inside, spaces are arranged along sight lines, to give staff members visual control over the goings-on in the center, which feeds and sleeps 200 men.

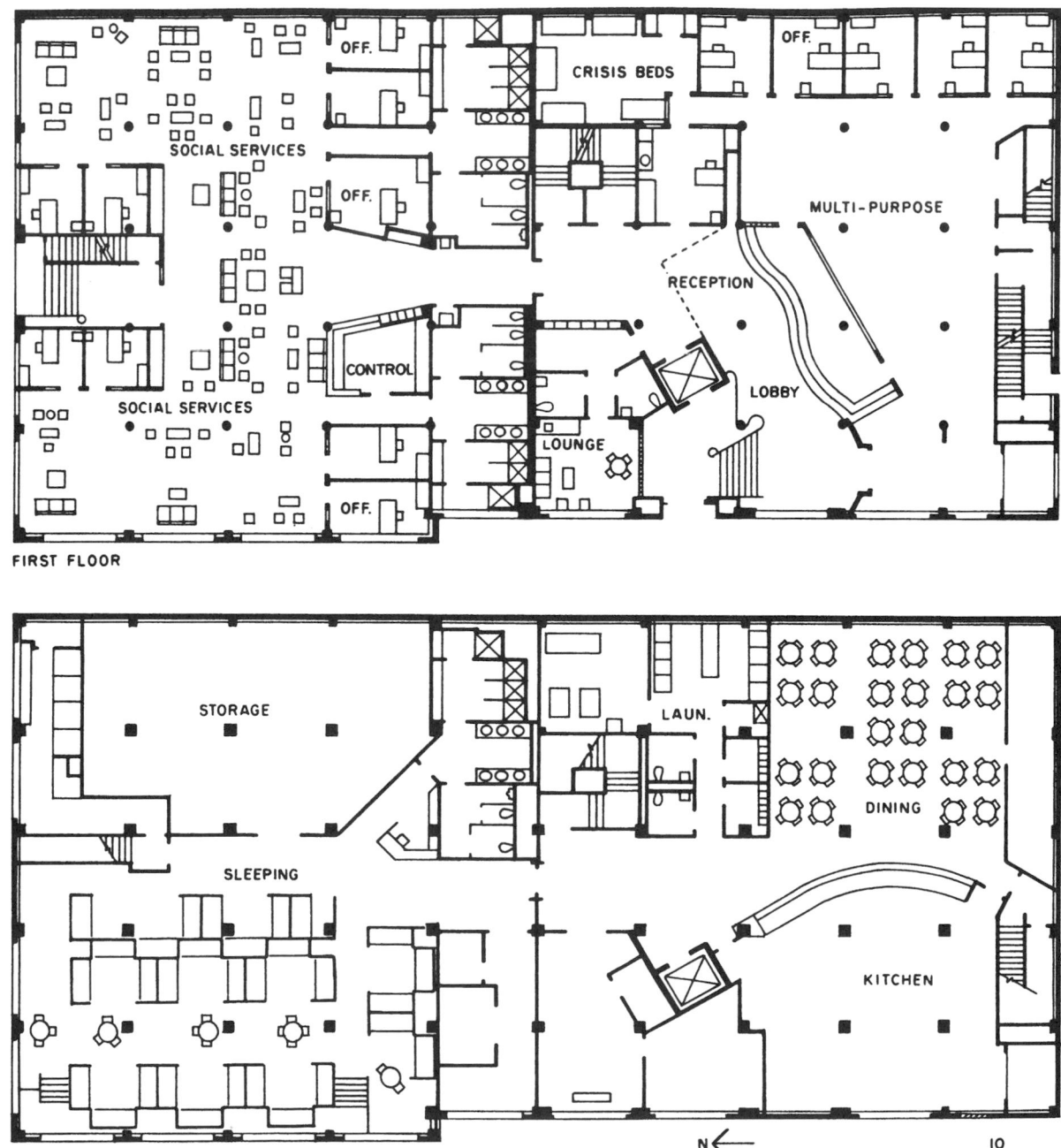

FIRST FLOOR

BASEMENT

Fig. 1 Multiservice homeless center, San Francisco, Calif. Asian Neighborhood Design—Architect.

The ground floor is used for check-in, social services, and the drop-in center, a librarylike space where both men and women may spend the day. The basement contains a sleeping area, dining room, commercial kitchen, laundry, and kennel, the last for pet owners who might otherwise balk at taking advantage of services if that meant leaving their pets unattended outside. The mezzanine houses counseling and social-service offices, and the second floor is devoted to sleeping areas and shower facilities.

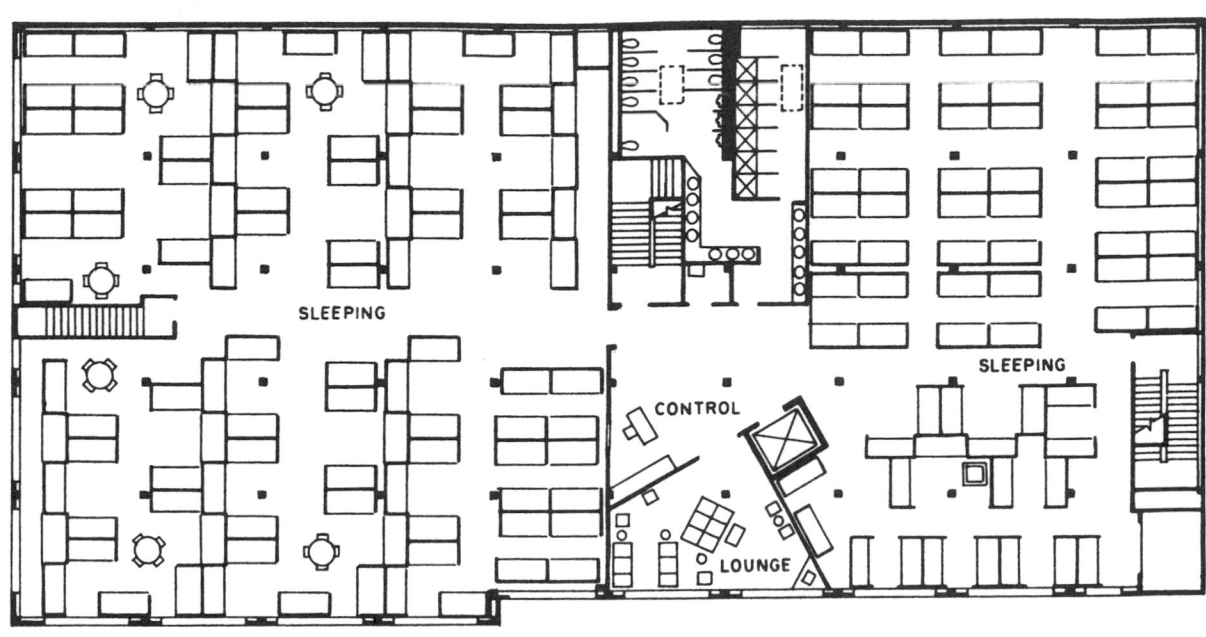

SECOND FLOOR

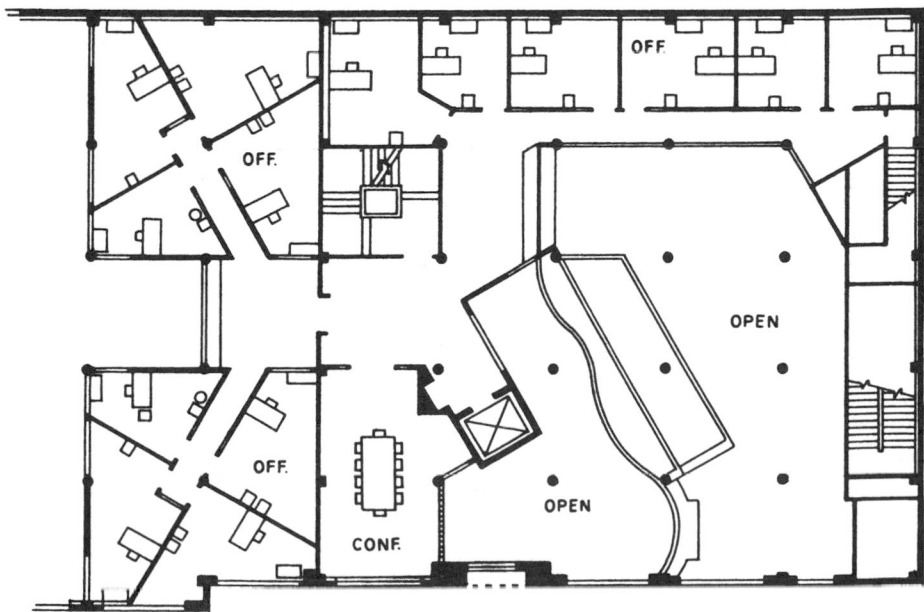

MEZZANINE

Fig. 1 (*Continued*)

ARRANGING THE SPACES

Types of floor plans

Most of the many possible plan shapes for residence halls are based on the principle that a short rectangle is most economical because it reduces the length of exterior walls, and hence the cost of the walls themselves and the amount of heat loss through them. The traditional pattern is an open or closed quadrangle, but wings arranged in T, H, or X plans are also common. All these arrangements may increase the problem of exterior noise, and long wings usually have the added disadvantage of long corridors. However, wing plans make it possible to place common-use and service facilities at the crossing or core of the wings, sometimes in a separate central service core linked to the bedroom wings by covered or enclosed passages. This kind of plan separates units with different structural requirements, isolates noise sources from study and sleeping areas, and provides an opportunity to vary architectural treatment.

Carrying the core principle to its logical conclusion, several colleges have explored the possibility of circular plans with service and common-use areas in the center and student rooms along the periphery so that every student has an outside room isolated from noise sources, and space for circulation is cut to a minimum.

From the point of view of economy, it may be practical to plan common-use facilities so that they can serve several purposes. Their location should be related to the flow of student traffic. Exterior and interior common-use spaces should be readily accessible from student rooms without creating noise or otherwise interfering with the use of other facilities.

Forming student groups

One of the objectives of college housing is the formation of effective student communities. From the college's point of view, properly constituted social groups can aid individual adjustment, provide training in getting along with others, and exert a desirable influence on behavior.

Unfortunately, the factors that contribute to the development of student groups are largely a matter of speculation, although physical proximity and frequency of contact within normal traffic patterns certainly help to determine which students are likely to share common activities.

While the optimum size of residential living groups is not definitely established, the experience of housing officers indicates that they should be small enough for each member to know the others as more than casual acquaintances. At many institutions groups of from 35 to 50 students—either men or women—have been regarded as suitable, but there are tentative indications that groups of 8 to 20 tend to develop into better working communities, and some colleges have planned their housing units for student groups within this size range.

Student Room Arrangements

Student rooms themselves can be arranged in a variety of plans, although the most common

arrangements merely string a series of boxlike rooms together in a pattern all too often reminiscent of a cell block.

The *traditional vertical house plan* is no longer generally used for two principal reasons: (1) fire safety regulations that require access to two stairwells for emergency exit, and (2) the increased cost of housekeeping. The plan does, however, reduce space for circulation by eliminating through corridors, and it creates small, readily identifiable living groups. To retain these advantages while overcoming the drawbacks, the vertical house plan has occasionally been modified by connecting two houses with doors that are closed to normal traffic but provide access for housekeeping and serve as emergency exits.

The *corridor plan* is the most common and the least satisfactory way to arrange student rooms. The most frequent version is the double-loaded corridor with rooms opening off either side as in a conventional hotel.

This arrangement, however, poses perennial noise and conduct problems that have given rise to such solutions as the offset corridor—which simply provides for a turn or a jog to interrupt sight and sound about midway in the structure—and the single-loaded corridor.

The single-loaded arrangement has several variations. At one university the corridor has been widened to include informal lounge and meeting space, with student rooms on one side and a glass window wall on the other. This plan has some of the advantages of a suite arrangement, but it juxtaposes several different types of building use in a way that could interfere with all of them. At many colleges, especially in the southern states, balcony-type exterior corridors are used to simplify plans, permit through ventilation, and reduce the cost of constructing and maintaining corridors and of providing a second fire exit.

Still another variation consists of widening the building and placing service facilities in an island in the center of each floor. As a result, there are two corridors per floor, with bathrooms and other service facilities on the inner side and student rooms on the outer side. This arrangement has the advantage of reserving all exterior wall space for student rooms. Furthermore, the service island acts as a sound barrier between corridors and helps to subdivide the floor.

The *suite plan* combines a common study or sitting room with one or more connecting bedrooms and, usually, a private bath. These spaces have been arranged in various ways: study and relaxation in one room, with sleeping and storage in others; study and dressing in one room with sleeping and social activity in others; or several standard single or double rooms opening on a common study. The major value of the suite plan is the opportunity it affords for closer student association and the freedom it gives students for using the various spaces as they wish. Where suites are arranged in a vertical house plan, special effort may be needed to bring the smaller groups together.

Variations on the Theme

With the addition of kitchens or kitchenettes, suites become *apartments*. This type of housing unit is rarely assigned to single undergraduates because of such drawbacks as difficulty of supervision, lack of common lounges and meeting rooms, and undue fragmentation of living groups. However, if these problems are recognized and overcome during the early stages of planning, it might be advantageous to construct apartment units which could be assigned to married students, to faculty members, or to single graduates or undergraduates, depending on policy and demand. In this case the apartment project might require some common-use space, as well as physical arrangements for somewhat closer supervision.

For married students and faculty, apartment projects are usually patterned more or less after commercial practice, with units arranged in vertical house plans or along corridors. In most cases, basic furnishings are provided by the institution, largely to reduce property damage due to moving in and out. Sometimes a few unfurnished units are available, or extra bedrooms are unfurnished.

Cooperative housing, the undergraduate version of apartment living, is believed to be more important than the relatively few examples and small number of students involved would seem to indicate. Such houses offer a unique opportunity to put into practice many of the theories of desirable group size and organization discussed earlier. But their main advantage is minimum individual living expense due to shared responsibility for housekeeping and for food preparation.

However, cooperatives should not be labeled merely as facilities for needy students. Sometimes membership in cooperatives, with its accompanying increase in both freedom and responsibility, is considered an honor, as in the case of scholarship houses. In other cases, groups are formed around mutual interests. Coops may be supervised by church groups or private organizations as well as by the colleges.

Coeducational housing, another departure from more usual housing arrangements, also has a number of advantages. Sharing of public areas eliminates duplication of facilities and contributes savings in construction costs. With proper design, there is greater flexibility in reallocating space to meet changing demands. Joint participation in educational programs and social activities seems to lead to more mature relationships between men and women.

Coeducational housing often groups student rooms for men and women in separate buildings, with public rooms such as lounges, libraries, and dining rooms in a central structure for joint use. In other versions, common-use rooms in the separate residence halls may be open to both men and women, or a single coeducational building may be divided vertically or horizontally into separate living sections for men and women.

The Vertical House Revisited

Proposed Housing, Washington University In this version of the vertical house plan (Fig. 2), four suites for six students each (two single bedrooms and two doubles plus living room-study and bath) are arranged around a vertical service-circulation core. As a result, student groups are small and flexible, and corridor area is virtually nonexistent, but the problem of housekeeping access is neatly sidestepped. The two exits required for fire safety are provided by separating the stairways in the core.

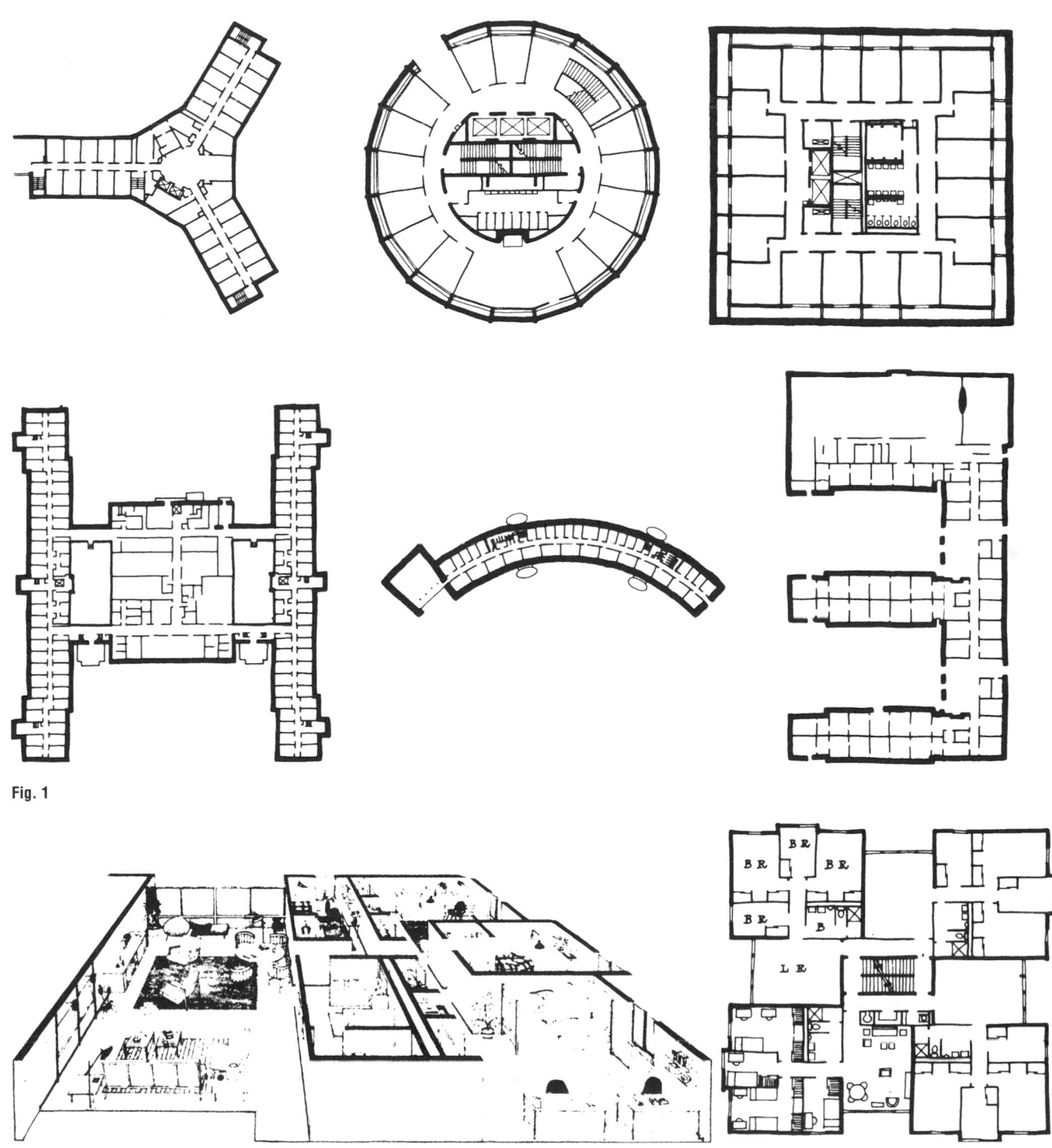

Fig. 1

Fig. 2

The Corridor Plus

Residence Hall, Thompson Point Development, Southern Illinois University Because it arranges student rooms along one side instead of two, the single-loaded corridor cuts in half the number of potential noise sources. In the example shown in Fig. 3, the passage was widened to double as a lounge, although such a use could cancel out the acoustic advantage of the single-loaded corridor. The window wall breaks the visual monotony and makes the narrow lounge seem more spacious.

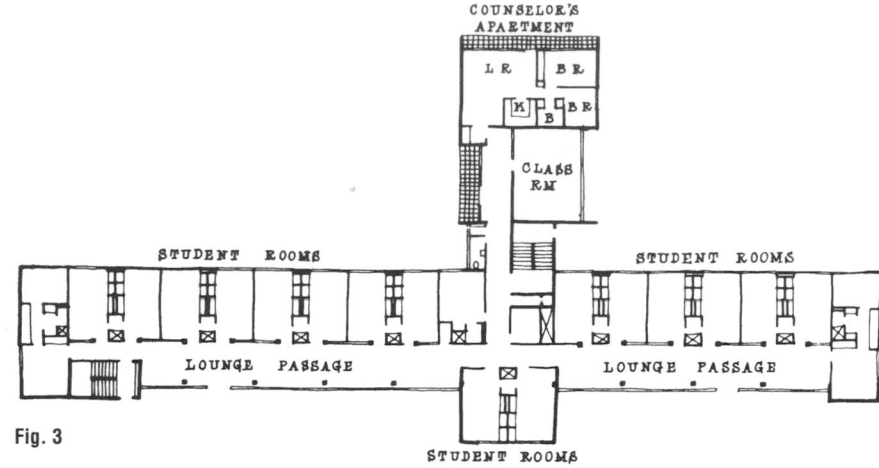

Fig. 3

The Corridor Minus

Hume Hall, University of Florida If uninterrupted, the typical double-loaded corridor can look like a tunnel and sound like bedlam. Hence devices like the jog corridor, which reduces the distance sight and sound must travel. The offset is also a convenient location for common spaces. (See Fig. 4.)

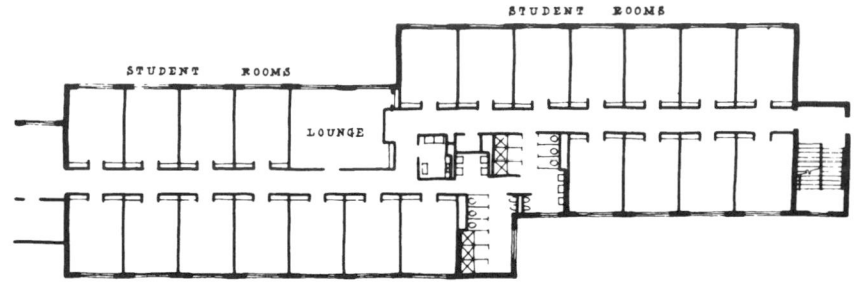

Fig. 4

Suites for Sleep and Study

Suites, which arrange student rooms around a common space, usually provide for some separation of use.

Tate Hall, Central Michigan College In the suite plan in Fig. 5 two double bedrooms flank a study.

Bragaw Hall, North Carolina State College In the plan in Fig. 6 students in four double study-bedrooms share a bath. Sound isolation of each room is provided by the wardrobes and hallway. Access to the suite is from a balcony-corridor.

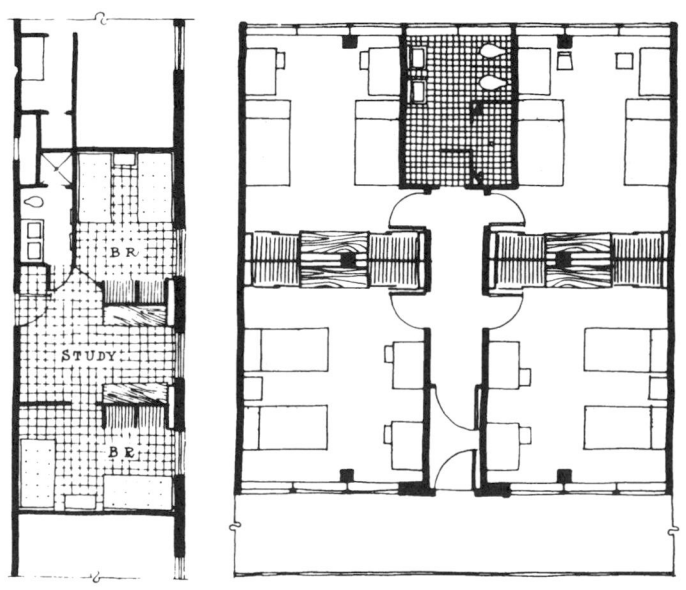

Fig. 5   Fig. 6

Suite Living and Single Loading
Residence Hall, Christian Brothers College
Stacking beds on top of one another, bunk-style, made it possible to borrow space from the study-bedrooms in this scheme and add it to a living room, thus creating living-bedroom suites off a single-loaded, window-walled corridor. The four-person suites are paired, with a connecting bath between each pair. (See Fig. 7.)

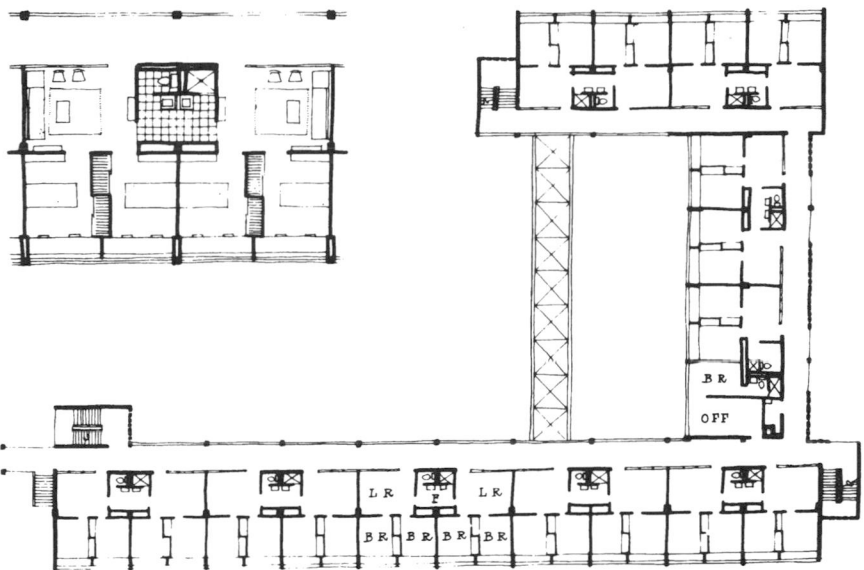

Fig. 7

Bailey Hall, Illinois Institute of Technology
This nine-story structure is one of four similar high-rise apartment buildings in a newly developed residential area that also includes a shopping center, a chapel, recreational facilities, and parking lots. Bailey Hall's 88 efficiency, one-bedroom and two-bedroom apartments are assigned to faculty and staff as well as to married students. (See Fig. 8.)

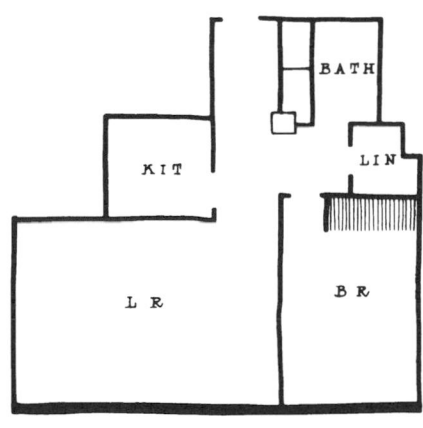

Fig. 8

Dexter M. Ferry, J. Cooperative House, Vassar College This example of cooperative housing is the home of 27 sophomores, juniors, and seniors chosen on the basis of scholarship, citizenship, and, to some extent, financial need. The students who live in the house are responsible for its operation, sharing expenses as well as cooking and other household tasks. As a result, each student's expenses are reduced.

The clean horizontal lines of the house give it a residential quality very expressive of the activities within. The students are assigned to double study-bedrooms on the second floor, but the community life of the group centers about the downstairs living room, dining room, terraces, and kitchen. The ground floor, which criss-crosses the second floor, also contains the chaperone's quarters. (See Figs. 9 and 10.)

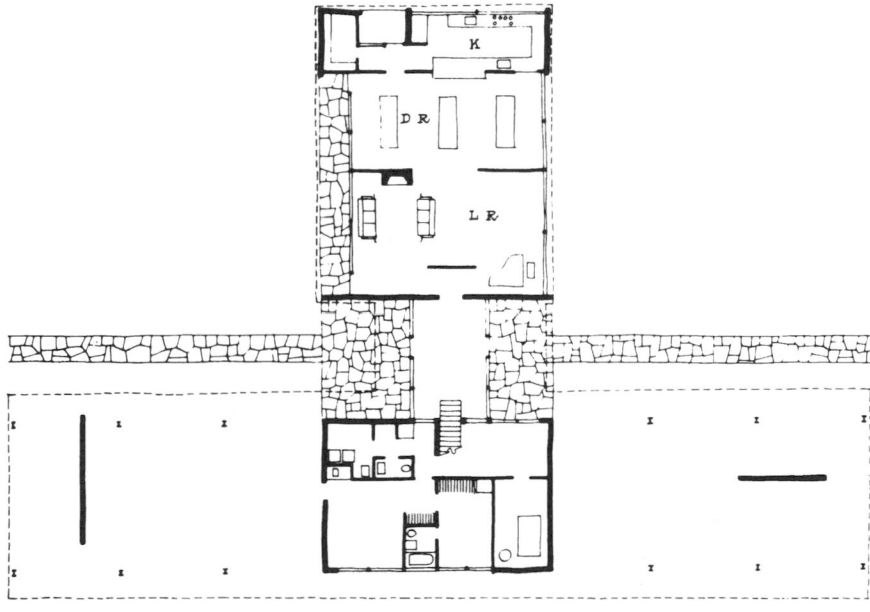

Fig. 9

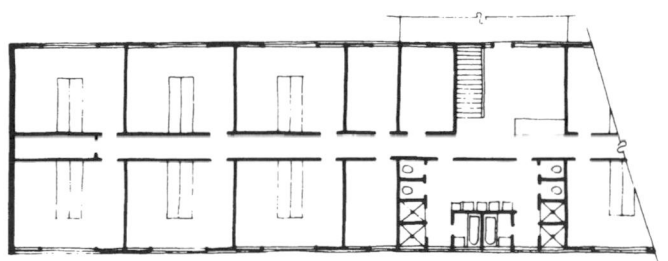

Fig. 10

Apartments

A variety of apartment types offers options for the various sorts of communal living that are programmed. Duplexes occur on the lower two floors of each straight portion of the building, and simplexes are above. As the apartments are mainly reached by vertical access only, the effect on the courtside is to be that of a series of linked houses, although the walls on the opposite sides will have a unified character.

The spine of the complex is a pedestrian road from which students will directly enter their apartments by means of open stair towers. This road passes through a variety of spaces which alternately recognize the project's semirural location and the need for a central focus. The focus is created by the tight relationship of the buildings to create a contained plaza that will be highly usable in the spring and fall. A multilevel court arrangement will provide access to the varied ground-floor levels and visual interest, and the intermediate stairs will provide seating.

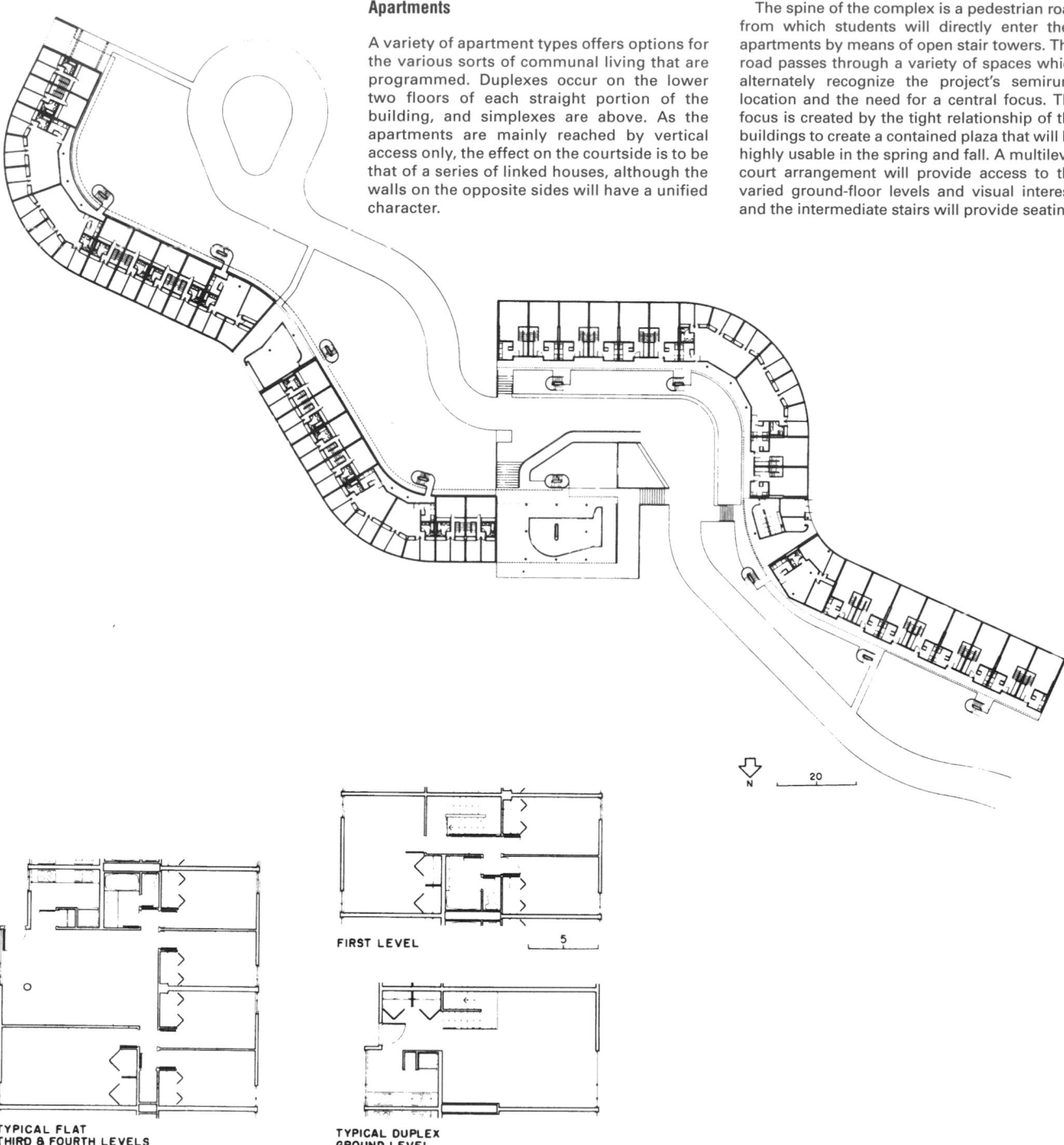

N   20

**TYPICAL FLAT
THIRD & FOURTH LEVELS**

FIRST LEVEL   5

**TYPICAL DUPLEX
GROUND LEVEL**

Fig. 11

Married Student Apartments

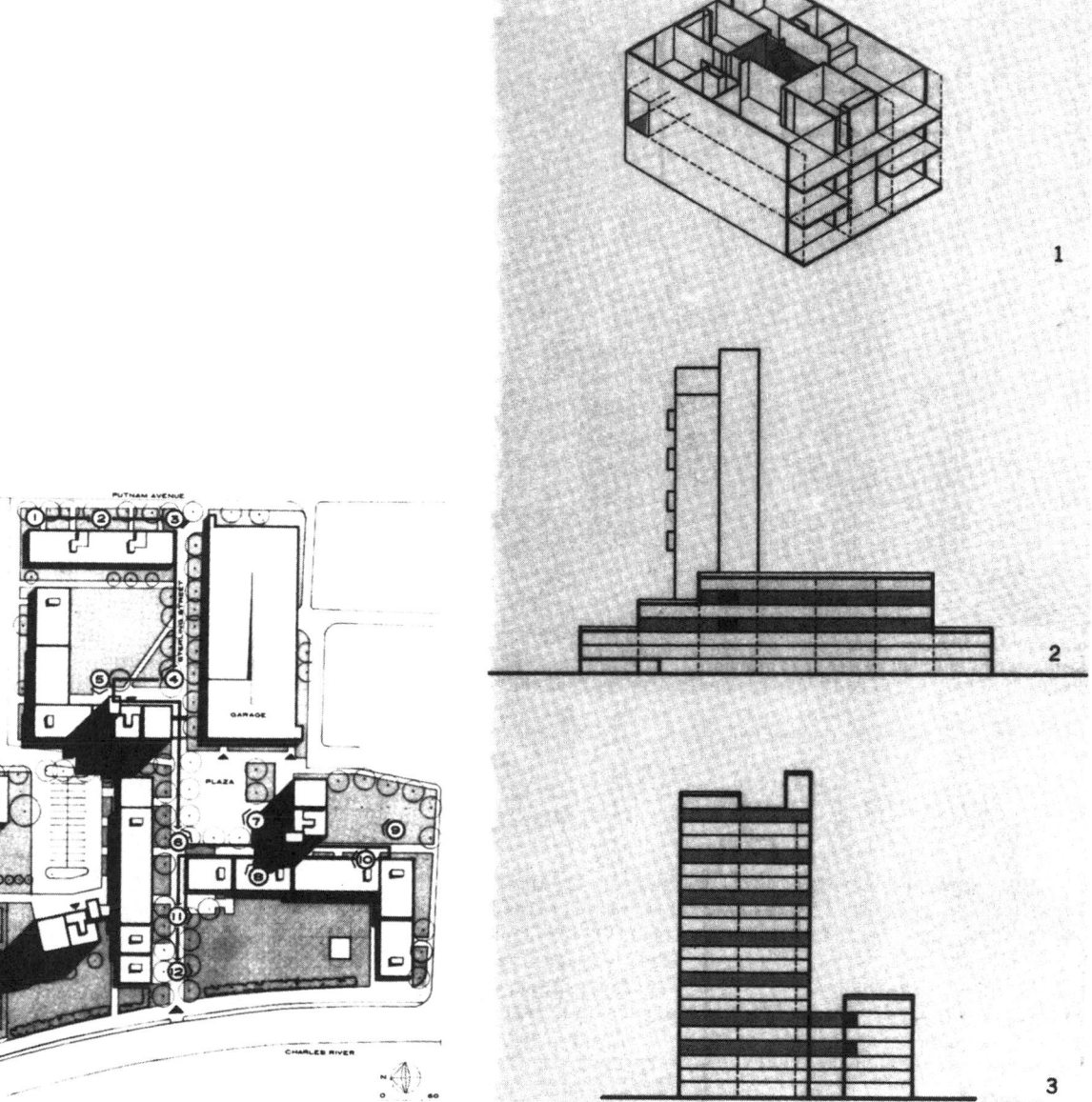

Fig. 12

STUDENT HOUSING

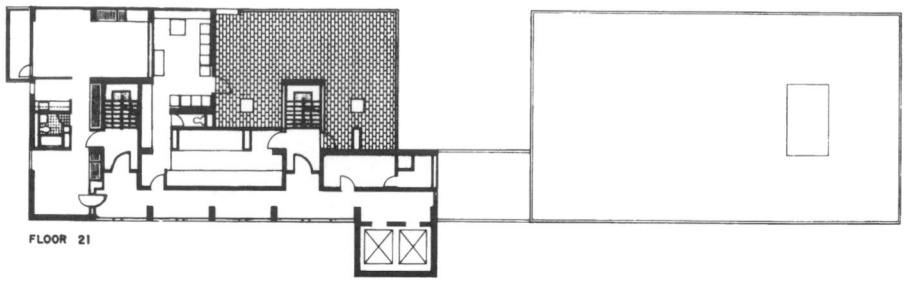

FLOOR 21

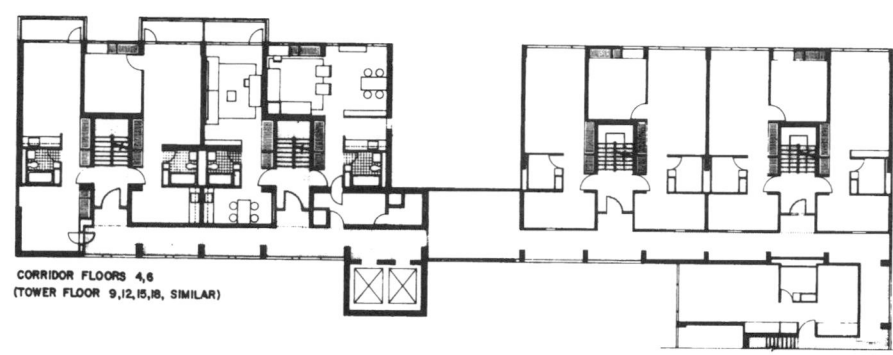

CORRIDOR FLOORS 4,6
(TOWER FLOOR 9,12,15,18, SIMILAR)

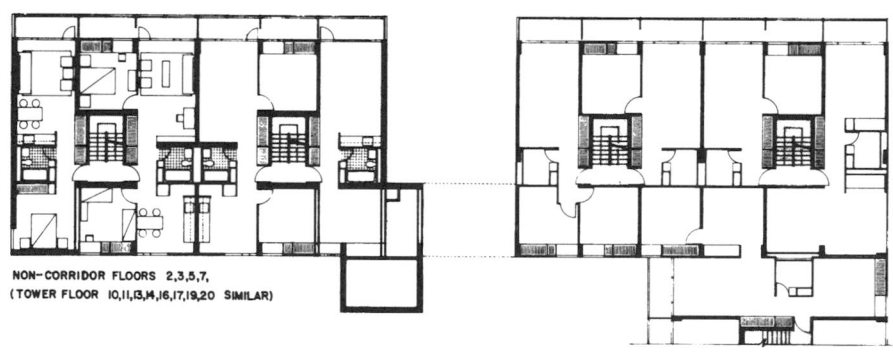

NON-CORRIDOR FLOORS 2,3,5,7,
(TOWER FLOOR 10,11,13,14,16,17,19,20 SIMILAR)

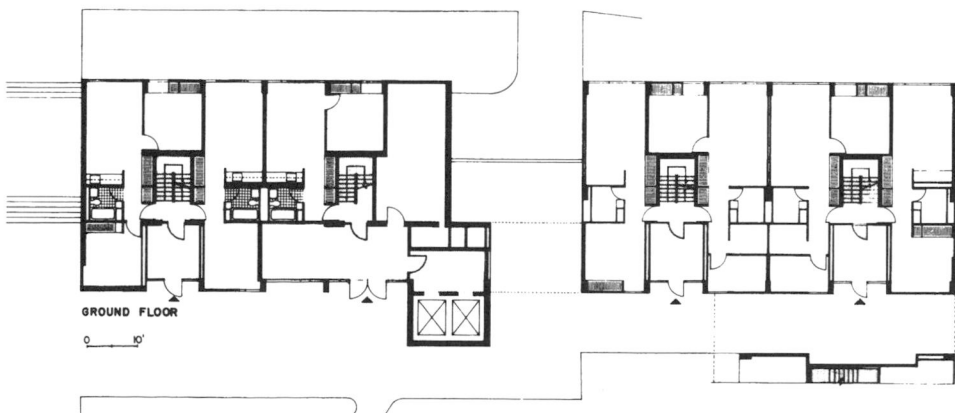

GROUND FLOOR

0   10'

Fig. 13

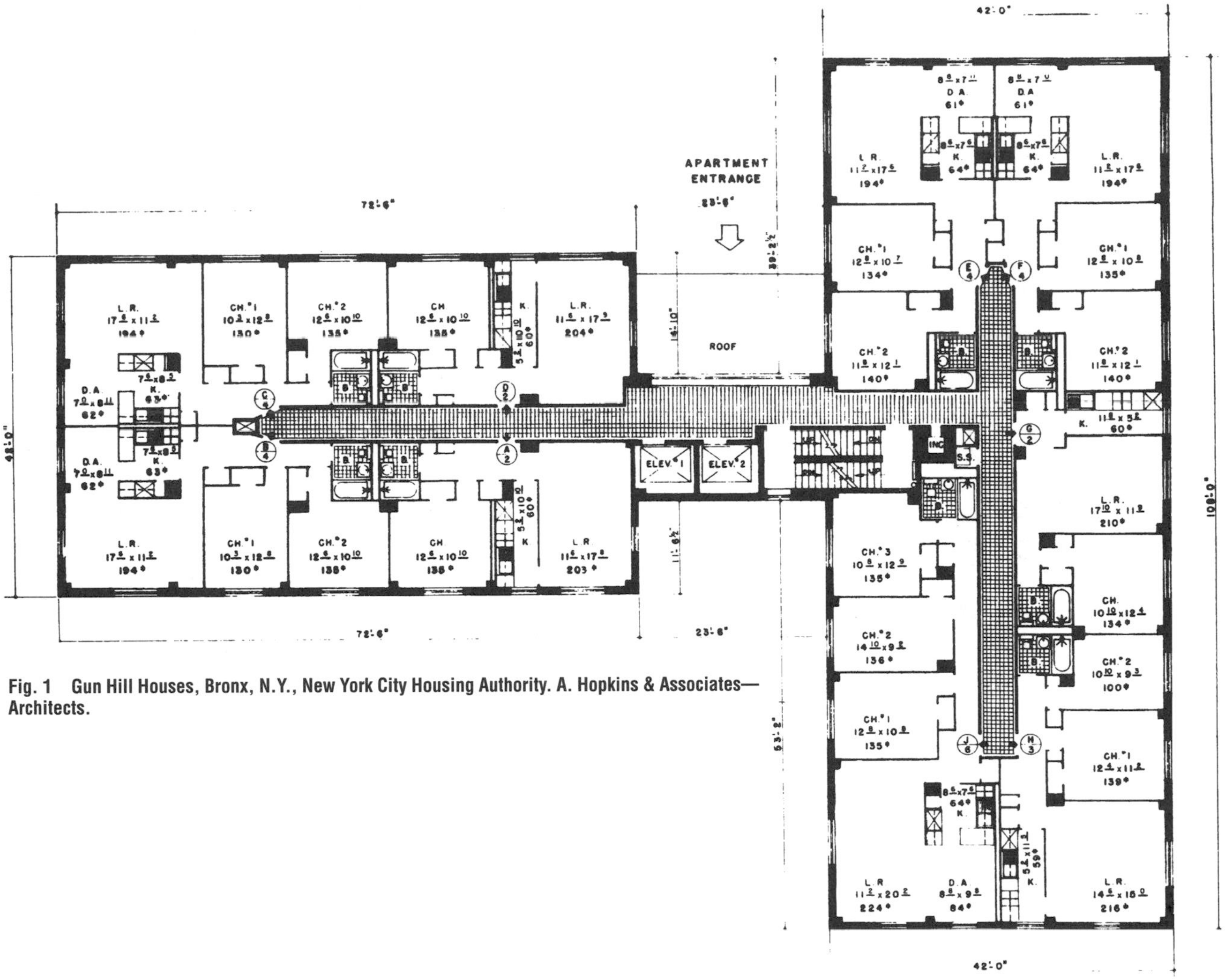

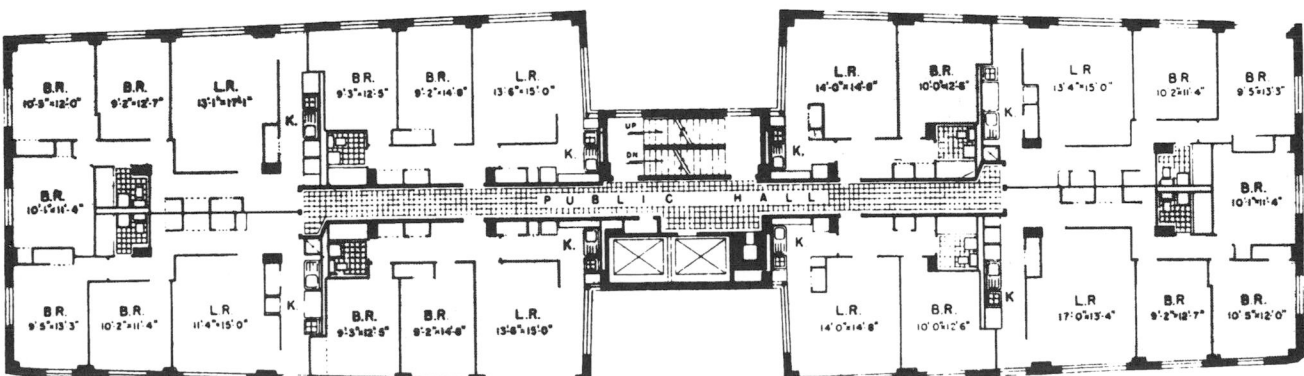

Fig. 1 Gun Hill Houses, Bronx, N.Y., New York City Housing Authority. A. Hopkins & Associates—Architects.

Fig. 2 George Washington Houses, New York City Housing Authority.

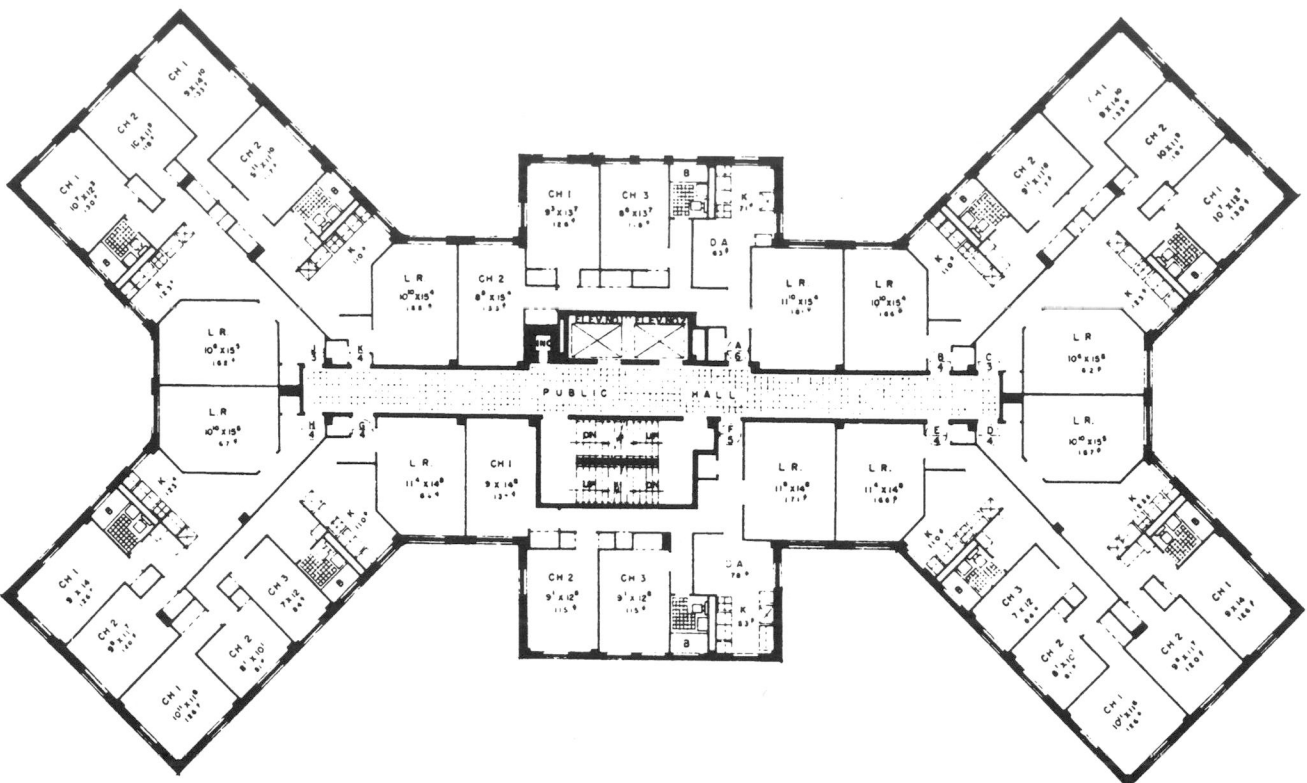

Fig. 3   Bronx River Houses, New York City Housing Authority.

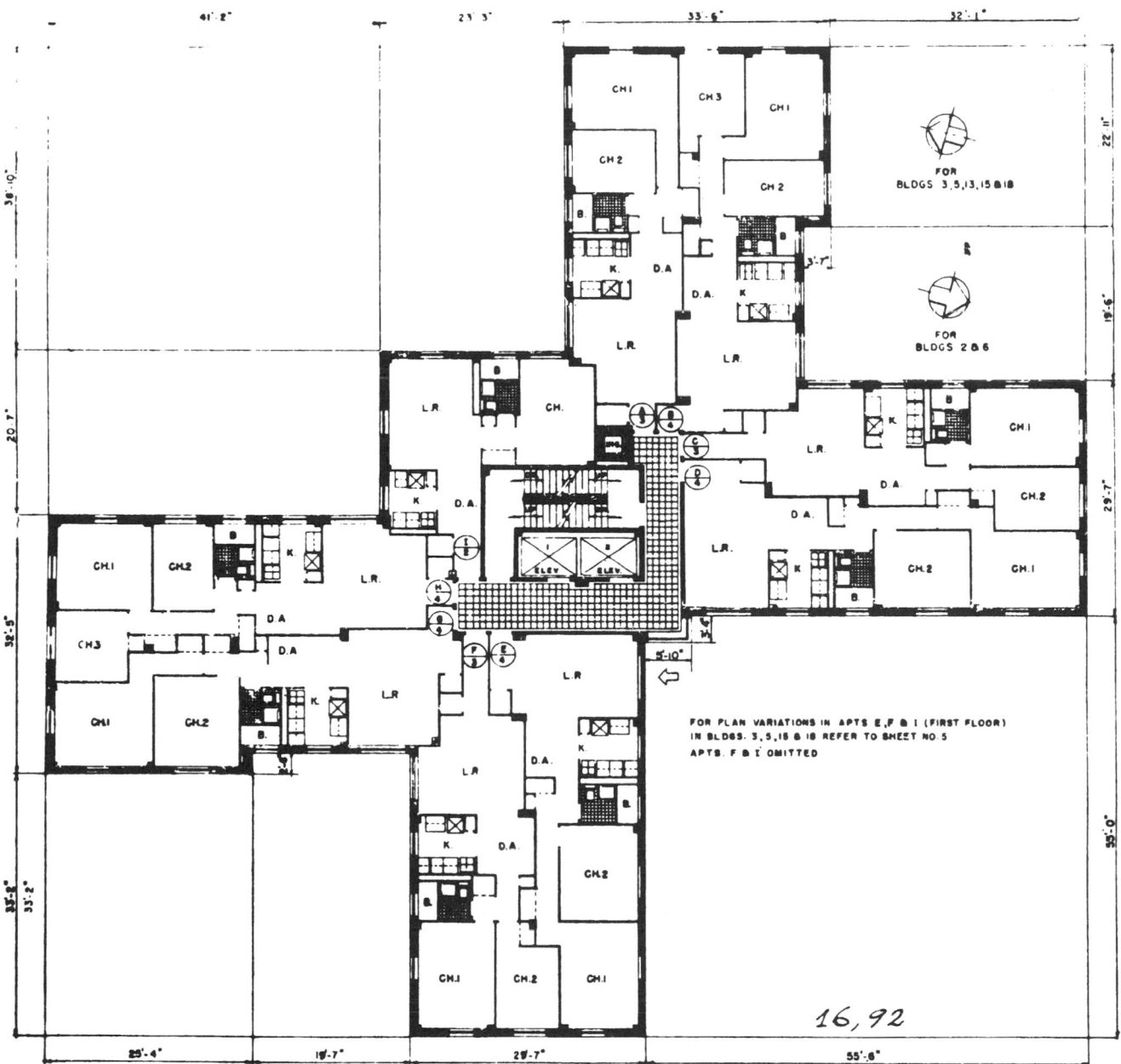

Fig. 4   Pelham Parkway Houses, New York City Housing Authority.

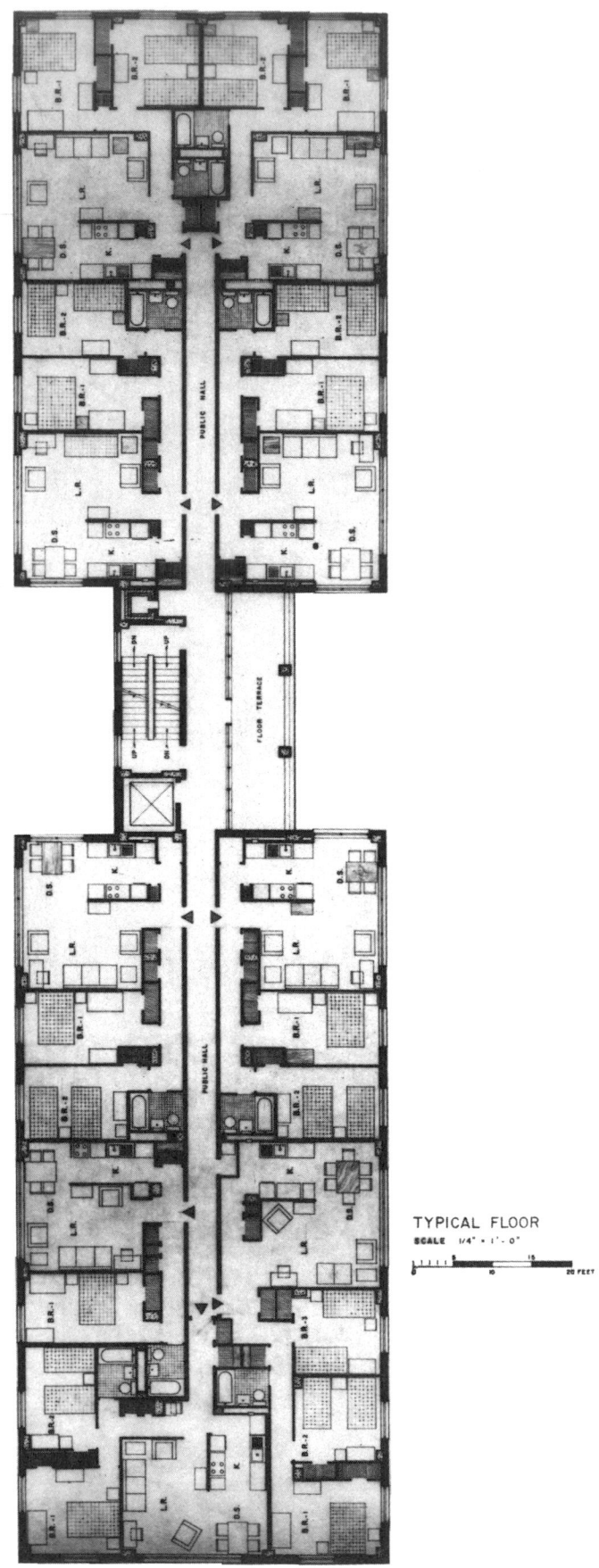

TYPICAL FLOOR
SCALE 1/4" = 1'-0"

Fig. 5   Bayview Houses. Katz-Waisman-Blumenkranz-Stein-Weber—Architects.

THE PHYSICAL PLANT

The location of the site, the planting, the traffic routes, the kinds of buildings and their arrangement, the play areas for children, and the recreation areas for adults are all important.

Site

A well-drained, sunny site with a gentle slope, and oriented so that prevailing winds can be a help rather than a hindrance, has proved most successful.

Good drainage is of paramount importance. The area should not be subject to flooding either by streams or by excessive surface run-off.

Proximity to Highway

Accessibility to public highways is important so that traffic to and from work will be convenient and not too time-consuming.

Entrance Driveway

There should be only one driveway into the camp area for successful camp operation. A camp manager's building, facing the driveway and set back far enough from the drive to provide space for growers' trucks, implements the assignment of workers in the morning and their check-in at the end of the day.

Building arrangement

Studies have shown that there are definite minimum and maximum desirable distances between certain types of structures. This is par-

ticularly true with regard to the sanitary facilities. No living unit should be more than 150 ft from toilet facilities and bath houses. In camps where meals are prepared in a central structure, this building should be at least 100 ft from the toilets and bath houses.

Child-Care and Play Areas

The care and supervision of small children can be accomplished best if definite areas are set aside for these activities. Desirable locations and amounts of space needed for each should be considered from the start of the planning.

Play areas for older, more active children require more space than the child-care center.

A central location is recommended for the child-care center so that parents may reach it easily when leaving children in the morning or calling for them after work.

Dining area

If meals are served to workers, the dining hall should be near the manager's office so that the morning pickup of the workers is not delayed.

CAMP LAYOUT

Eighteen living units are arranged in three groups of six each. (See Fig. 1.) All units are

within a 150 ft radius of a wash and bath house. The number of people who could be housed depends on the size and type of structure used. By staggering the units as shown, good ventilation, some privacy, and outdoor space for resting are assured. Clothes-drying lines placed as indicated provide convenient hanging space for the families' washing.

Play areas are located between each two groups of living quarters so that children can reach them without having to cross the main drive.

Clumps of trees between the groups of quarters provide shade and serve as natural divisions of space. The trees and shrubs just inside the property line also provide shade as well as screening. The plantings bordering the drive where it enters the property should be low to reduce traffic hazards at that point.

The child-care center is near the central eating space, and the latter is adjacent to the camp manager's post. The toilet next to the manager's office is convenient to the truck loading area, the recreation area, and the eating center.

Adequate space is provided for a number of trucks. This facilitates the loading and unloading of workers and minimizes traffic problems.

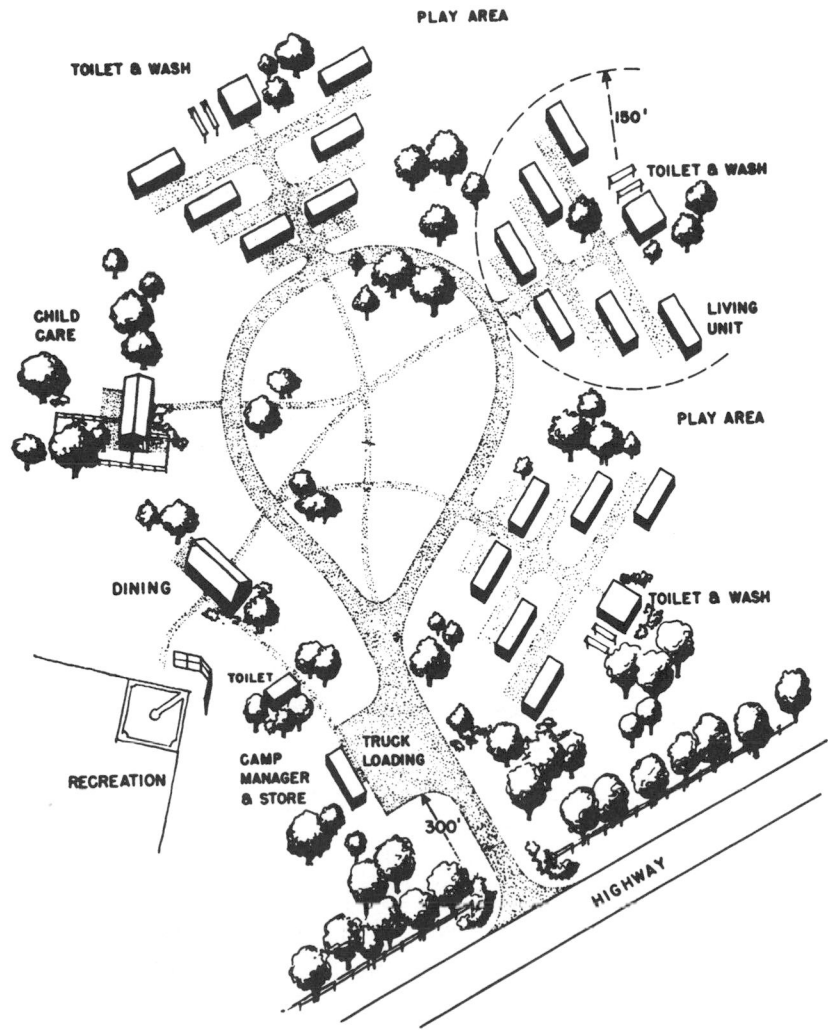

Fig. 1 A suggested camp layout.

HOUSING FOR MIGRANT FARM WORKERS

BARRACKS-TYPE LIVING QUARTERS

The simplest type of migrant housing provides space for sleeping only and is often called *barracks*. Space is usually more or less open, as shown in Fig. 2, and is best suited for groups of one sex. No provision for cooking or eating is made.

The narrow, rectangular shape of the building offers two distinct advantages. No trussing or post supports are needed for the roof, and adequate cross ventilation is assured.

The building is 16 ft wide and approximately 50 ft long. Ten double bunk beds, arranged as shown, provide uncrowded sleeping space for 20 people. The beds are located to leave three open floor areas where chairs and tables may be placed.

Although only eight storage areas are provided, they are large enough to give the necessary hang and shelf space. Definite assignment of these areas to individuals will prevent confusion and avoid friction among the occupants.

The storage areas are placed to subdivide the floor space and thereby assure some semiprivacy.

Another arrangement for this same building is shown in Fig. 3. The addition of two complete partitions and one partial partition provides additional wall space, thereby making it possible to install 12 double bunk beds. The capacity is increased to 24 persons. If the partitions across the building are about 6½ ft high, air can circulate over the top of them. No changes have been made in size or location of doors and windows. Free floor space is maintained in each of these sections.

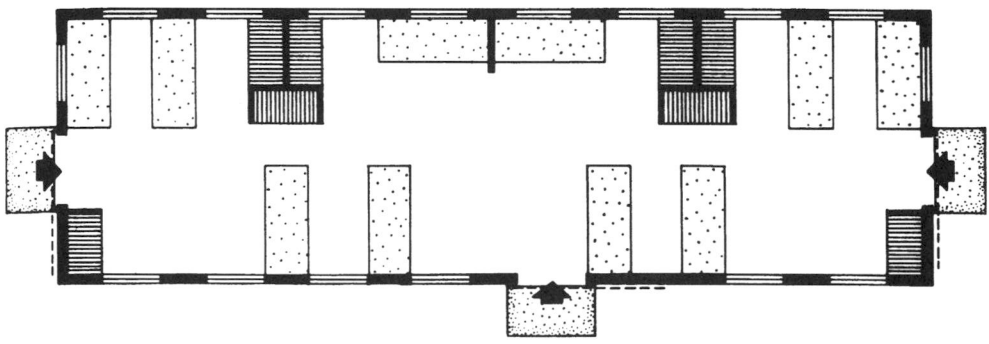

Fig. 2   Barracks-type quarters for 20 people.

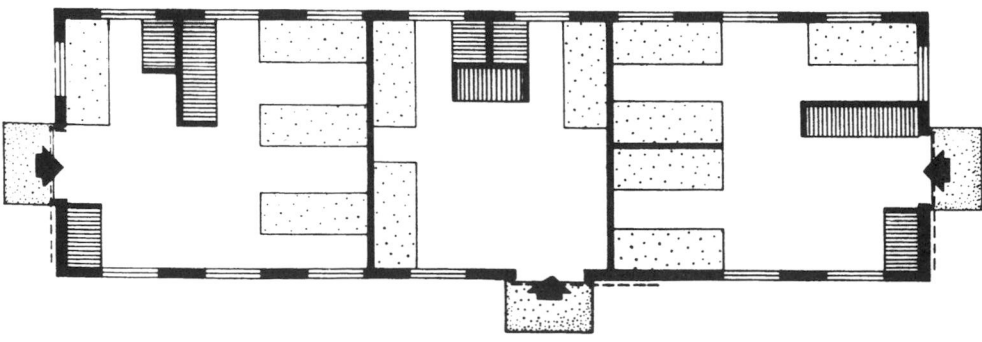

Fig. 3   Barracks-type quarters for 24 people.

FAMILY LIVING UNITS

Figure 4 shows another arrangement of space, within the same building shell, that is suitable for four small families. Three full-height interior partitions are needed. These subdivide the space so that each end section will house a family of six, and each center section provide room for a family of four.

The storage areas are ample and separate the parents' space from that occupied by the children in all but one section. In the latter case a partial wall is used for separation. Free floor space also is allowed although it is not so great as in the preceding arrangements.

Another method of subdividing the space in this structure Is shown in Fig. 5. Three units for larger families are obtained by using two full-height partitions and some partial partitions. Each unit will accommodate a family of eight.

Again, free floor space and satisfactory storage spaces are provided.

The parents' bed in each section of both arrangements has been placed where the outside door can be seen. It is not advisable to locate the children's quarters where they can be entered easily.

Since none of these arrangements allows for cooking or eating, these facilities would have to be provided elsewhere.

Migrants who travel in family groups prefer to cook and eat their meals in their own quarters, rather than in a central place. If at all possible, new construction should be designed to include kitchen space.

The two arrangements shown in Figs. 6 and 7 provide sleeping and food preparation space for 16 and 18 people, respectively.

The common kitchen in Fig. 6 is large enough for two stoves, two sinks, and two tables. In this way, both families may prepare and eat meals at the same time, if desired. However, since the seating space is somewhat restricted, the families may prefer to eat at different times. An advance agreement about eating time will be necessary to avoid difficulties.

Provision for separate preparation of food eliminates long delays and prevents confusion or disagreements concerning each family's supplies.

The smaller sleeping space to the right of the kitchen accommodates a family of six. The area to the left of the kitchen can house two related

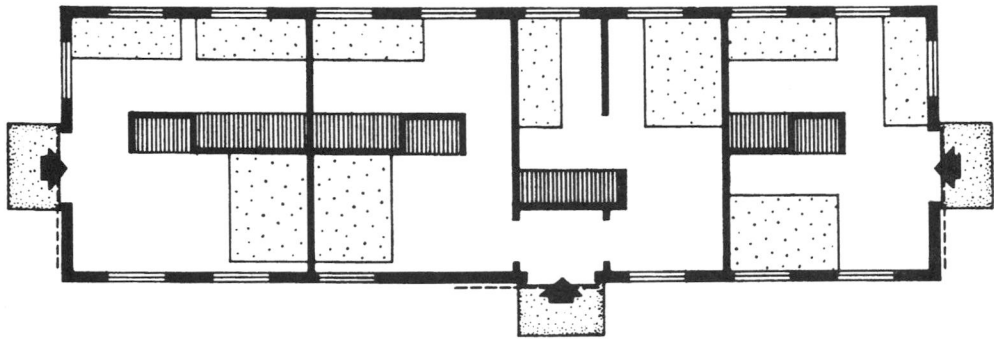

Fig. 4 Quarters for four families—two of 4 and two of 6 members—a total of 20 persons. Sleeping and dressing only.

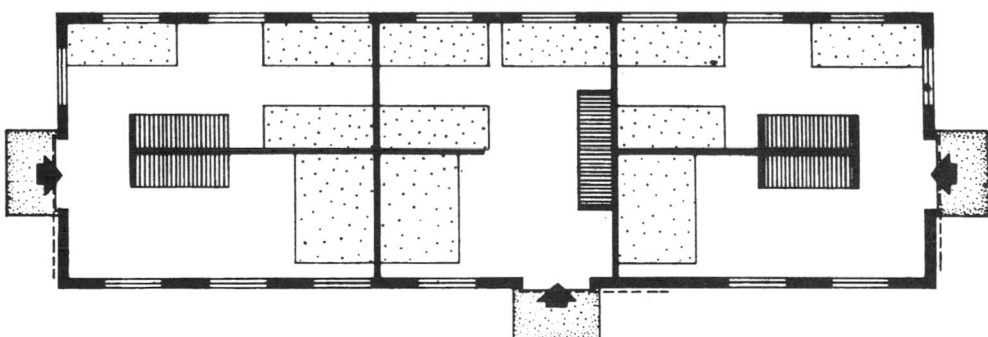

Fig. 5 Quarters for three families of 8 members each—a total of 24 persons. Sleeping and dressing only.

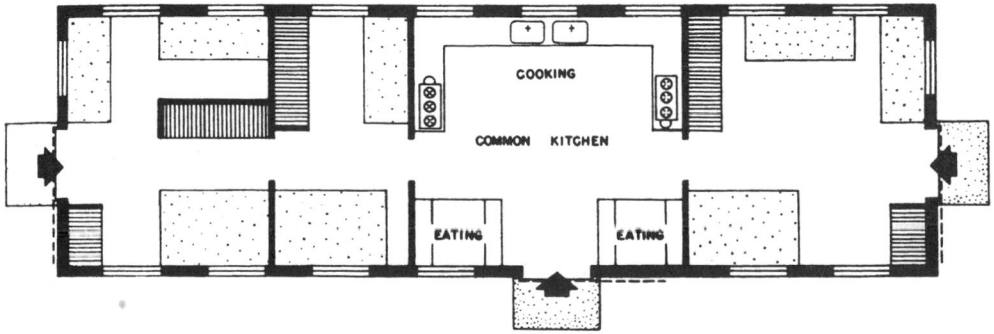

Fig. 6 Quarters for two or three families—a total of 16 people. The common kitchen, containing duplicate equipment, to be shared.

HOUSING FOR MIGRANT FARM WORKERS

families—one of four and one of six, or one family of 10. In the latter case, the substitution of a double bunk bed for one of the double beds might be desirable.

In Fig. 7 the kitchen space is smaller and contains only one table. A definite agreement about eating times is essential in this case. Duplicate preparation facilities are recommended, however, even though the counter space for each must be reduced.

This smaller kitchen leaves more space elsewhere for sleeping. Consequently the building's capacity is increased to 18.

Figure 8 shows another arrangement that provides sleeping space for three families, totaling 18 persons, and a cooking and eating area for each family. There are two sections for families of four members and one section for a family of 10.

Quarters for camp managers, other camp staff, or work group leaders can be provided by making a few alterations in the basic shell of the structure used for preceding arrangements.

One possible arrangement is shown in Fig. 9. This arrangement provides space for six people, facilities for cooking and eating, three storage areas, and a bathroom containing a lavatory, a toilet, and a shower. Ample free floor space is left for chairs and small tables.

Another basic building, approximately 11 ft wide and 61 ft long, is shown in Fig. 10. A permanent wall through the center divides it into two sections. Each section provides sleeping space plus a cooking and eating area for a family of six. Adequate storage is available, and its location ensures the needed privacy.

Figure 11 shows workers' living quarters that include cooking and eating space. This structure would accommodate four families of six members each, or a total of 24 people.

Figure 12 shows the same arrangement as it can be built for use by the camp manager, other camp staff personnel, and/or work group leaders. The center section has been lengthened to provide bathrooms for each living unit.

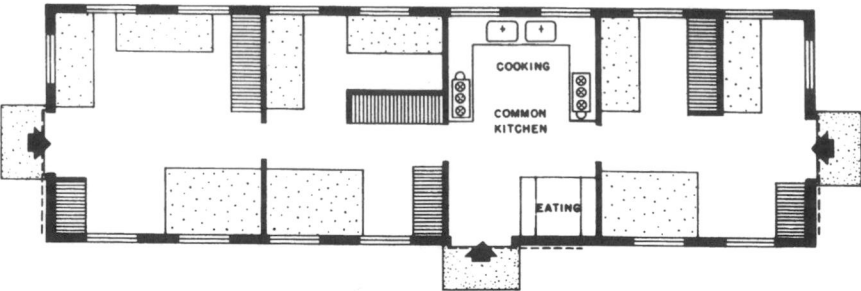

Fig. 7 Quarters for two or three families—a total of 18 persons. The common kitchen, having two stoves, two sinks but only one table, to be shared.

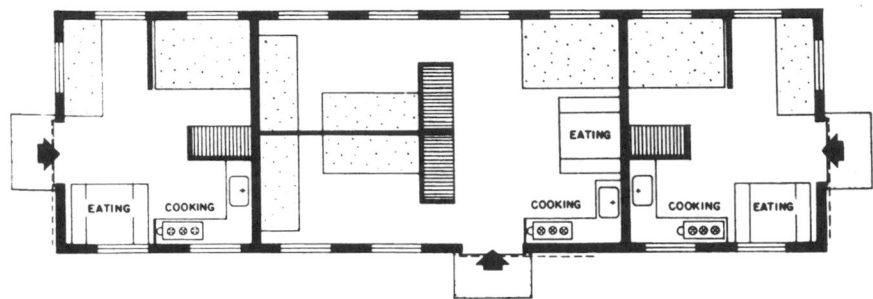

Fig. 8 Quarters for three families—one of 4, one of 6, and one of 8 members—a total of 18 persons. Cooking and eating space for each family.

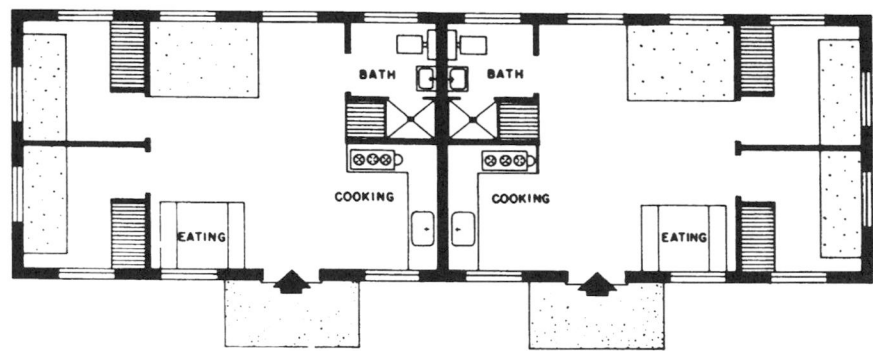

Fig. 9 Quarters for camp manager, camp staff, or work group leaders—two families of 6 members each. Separate cooking, eating, and bath facilities.

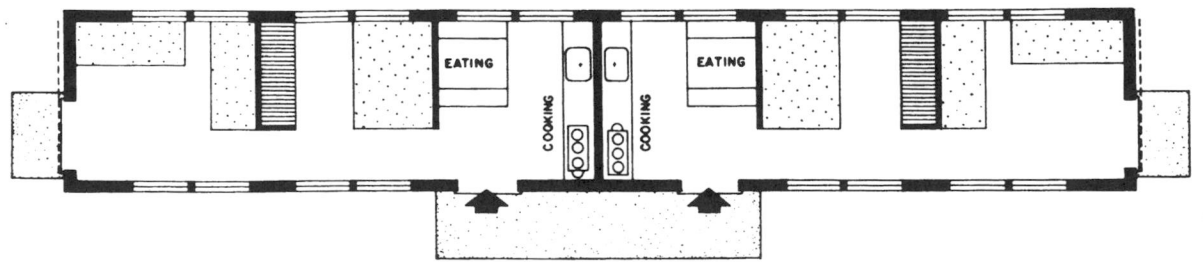

Fig. 10 Quarters for two families of 6 members each—a total of 12 persons. Separate cooking and eating space for each family.

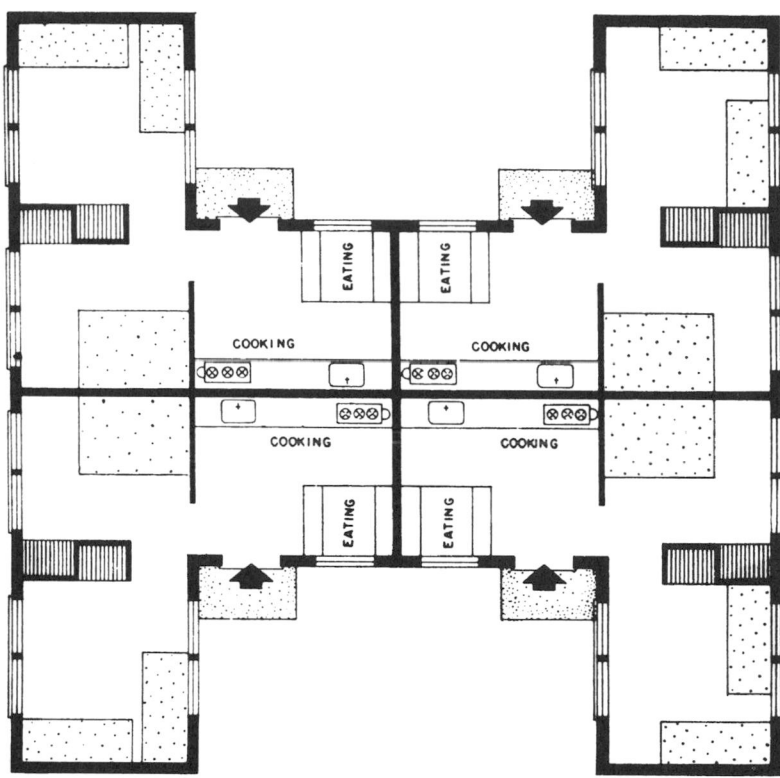

Fig. 11   An H-shaped building providing space for four families of 6 members each—a total of 24 persons. Cooking and eating space for each family.

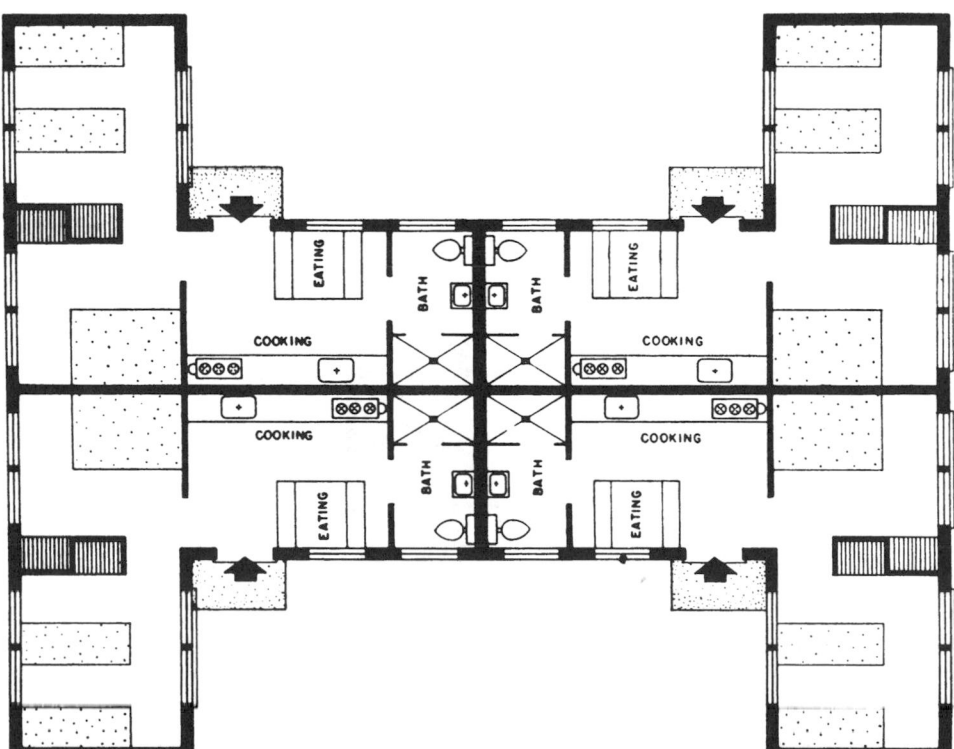

Fig. 12   A variation of the plan shown in Fig. 11 suitable for quarters used by camp manager, camp staff, and work group leaders. Individual bath facilities are included.

HOUSING FOR MIGRANT FARM WORKERS

BATH AND WASH HOUSES

Adequate toilet, bath, and laundry facilities are essential in every labor camp. They are important for good sanitation and increase the workers' morale. By combining all these facilities in one building, the problem of supplying sufficient hot water is simplified and the cost kept within practical limits.

Disposal of sewage from such buildings must be in accordance with Public Health Department requirements.

Three suggested designs are shown in Figs. 13 to 15. The 22- by 32-ft building in Fig. 13 contains four pairs of laundry trays; three showers, four toilets, and a long industrial-type wash sink in the women's section; a common shower room with adjacent dressing space, four toilets, a trough-type urinal, and an industrial-type wash sink in the men's section; and a heater room.

Entrance to the men's section is in one end of the building; entrance to the women's section is in the other end. The door to the laundry room is located in the side of the building. Access to the heater room is through the laundry room. The bathing and toilet facilities are placed so that only the wash sinks are visible through the entrances. The shower partitions in the women's section are extended to provide dressing space with each shower.

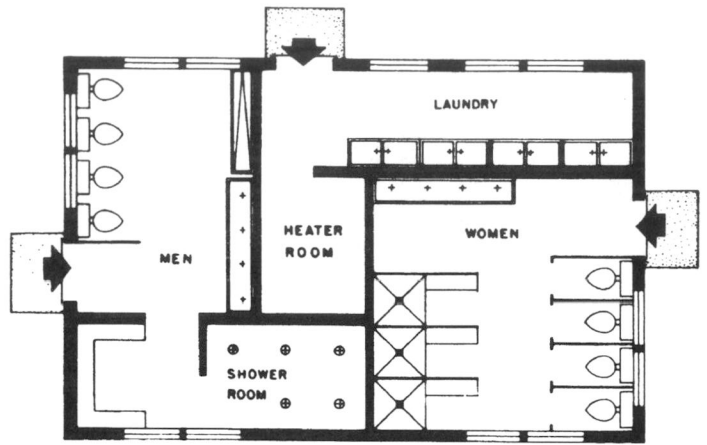

Fig. 13   A 22- by 32-ft combination bath and wash house.

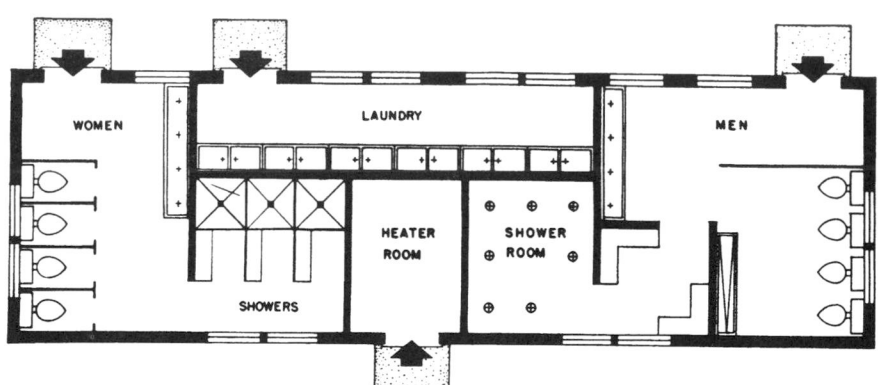

Fig. 14   A 16- by 52-ft combination bath and wash house.

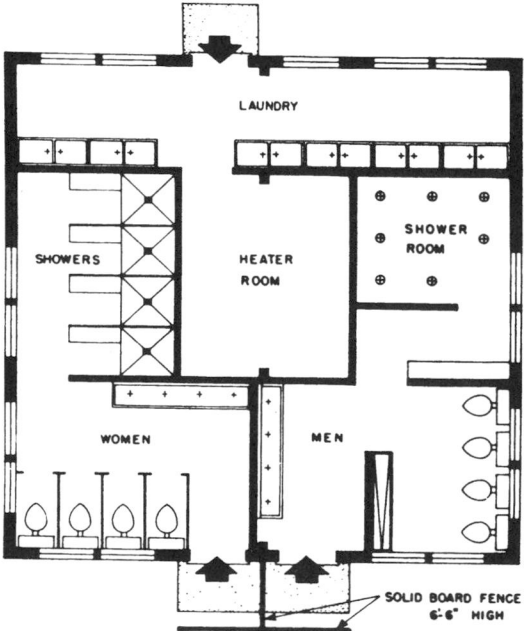

Fig. 15   A 30- by 30-ft combination bath and wash house.

CENTRAL COOK HOUSE

If cooking and eating facilities are not included in the family living structures nor meals served to the workers by the management, a central building where meals may be prepared is necessary.

The design in Fig. 16 is for a 24- by 35-ft building. Food centers, if used for both cooking and eating, should be planned on the basis of from 18 to 20 ft² of floor space per person.

CHILD-CARE CENTERS

Many migrant families have children of preschool age. Provisions for care of these children permit all of the adult family members to work. Certain definite facilities are needed if this care is to be adequate. Space and equipment must be provided so that the children may play, eat, and sleep. Toilet and washing facilities are also needed. The amount of space and the number of facilities depend on the size of the camp.

Space Needs and Space Arrangement

The kind and amount of space needed depends upon the number of children to be cared for. A small center, for 20 children, would require an infants' room, a playroom, a kitchen, a bathroom, and a room for the staff. Space for the staff is necessitated by the long days required during harvest season. When harvest work is at its peak, children are often brought to the center at 6:30 a.m. and not called for until late evening.

When the playroom must be used for eating, sleeping, and playing by both the toddlers and the older group, a minimum of 30 ft² of floor space is needed per child. If other rooms are provided for sleeping, the space may be reduced to 20 ft² per child. In each case this space is in addition to that required for halls, bathrooms, kitchens, and stationary pieces of furniture.

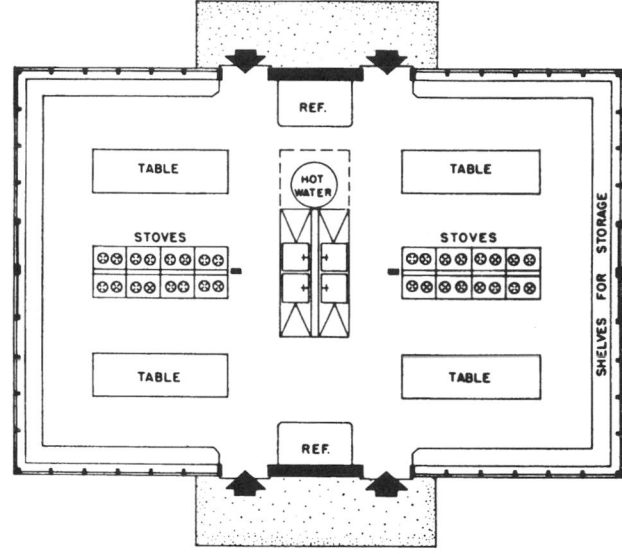

Fig. 16   A central cook house containing facilities for the preparation and eating of meals, the washing of dishes, and the storage of food and utensils.

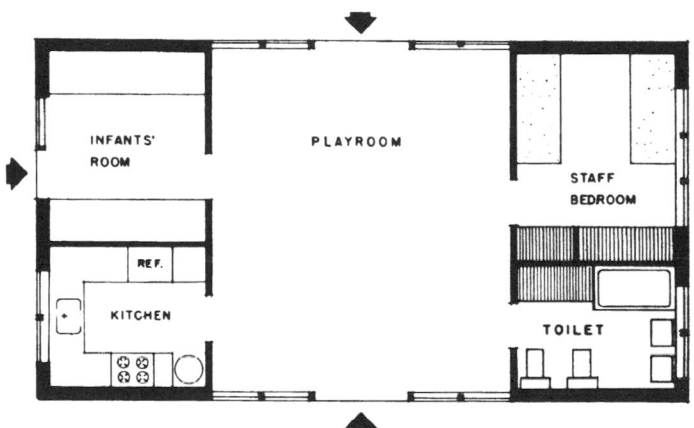

Fig. 17   A 20- by 37-ft child-care center suitable for 20 children.

HOUSING FOR MIGRANT FARM WORKERS

Children are served several meals a day. The kitchen must be large enough so that meals can be prepared easily. It must have a stove, a sink, and a refrigerator, in addition to ample counter and storage space. Food is purchased in large quantities—usually in amounts sufficient for a month. Therefore plenty of storage space is needed.

The bathroom is used by the children and the staff. The fixture ratio is one toilet and lavatory to not more than 15 children. A more desirable ratio is one toilet and lavatory for every 10 children. One bathtub is sufficient as the infants are bathed in their own room. The tub should be installed on a raised platform for the convenience of the staff when bathing the children.

The size of the staff room depends upon the number of persons on the staff. Space is needed for single beds and clothes storage.

The floor plan of a 20-child unit is shown in Fig. 17. Space is provided for an infants' room,

a kitchen, a general purpose playroom, a staff room for two, and a bathroom. Hang space and built-in storage are provided in the staff room and in the bathroom. The storage needed in the playroom can be obtained with boxes and movable shelves.

The 40-child unit shown in Fig. 18 has two playrooms—one for toddlers and one for older children—and a storage and medical room in addition to an infants' room, a kitchen, a staff room, and the bathroom. The storage and medical room can be used as an isolation room if necessary. All of the rooms are proportionately larger than in the 20-child unit.

Easy access to the out-of-doors from both the infants' room and the playroom is important. On sunny days the infants are taken out-of-doors to a protected area. The toddlers and older children like to play out-of-doors part of the time and sometimes are fed there when the weather permits.

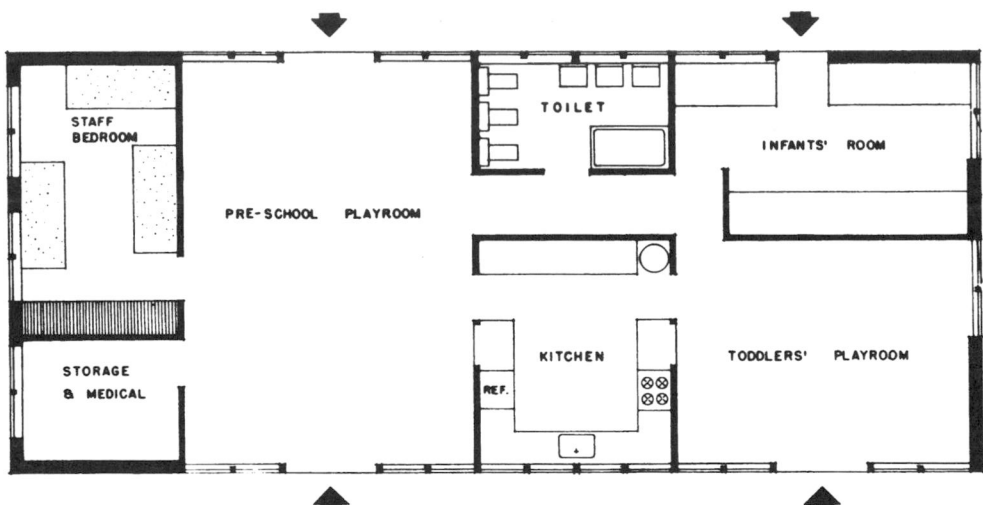

Fig. 18   A 24- by 56-ft child-care center suitable for 40 children.

Mobile Homes
and Parks

SITE PLANNING

Location and Area of Site

Sites selected for mobile home parks should be well drained and free from topographical or geographical hindrances or other conditions unfavorable to a proper residential environment.

Sites should not be located near swamps, marshes, or other breeding places for insects and rodents, or heavy industrial zones with objectionable odors or noise. The site should have good natural drainage, or a storm-drainage system should be provided. Drainage from the park should not endanger any water supply. The site should be graded to eliminate depressions and provide a uniform ground surface. Steep slopes should be graded as much as possible to minimize the hazards they present.

Because mobile homes are for residential use, they should preferably be placed in residentially zoned districts rather than in commercial or industrial districts. Wherever possible, mobile home parks should be so located as to be accessible to public water and sewage systems.

The area of a mobile home park must be sufficient to accommodate (1) the desired number of mobile home lots (it is recommended that a minimum of 50 mobile home lots be constructed in any new park to obtain a better-designed, more economical plan), (2) parking areas for motor vehicles, (3) access roads and walkways, and (4) recreation facilities. Additional area must be provided for management buildings, service buildings, or other structures to be included.

Experience has shown that mobile home park designs should be based on local conditions since neither the repetitious application of one lot design nor any arbitrary conglomeration of various modules will result in good planning. Local conditions that might affect a particular site plan include the size, shape, and topography of the site and surrounding area; land costs; local codes and ordinances; the use of adjacent property; the availability of water supply and sewage-disposal facilities; and the needs of any special groups in the park, such as the elderly.

The best designs are those that make effective use of existing topographical conditions and properly blend in an adequate amount of open area, recreation area, and other common-use area. See illustrations for possible modules and arrangements of mobile homes.

Before land is acquired for a mobile home park, the health authority should be consulted regarding the compliance of the proposed site with existing health regulations. Other local agencies, such as zoning or planning commissions, should also be consulted.

Roads and Parking Facilities

All roads in mobile home parks should provide for convenient vehicular circulation. Pavements should be of adequate widths to accommodate anticipated parking and traffic loads.

Entrance streets that connect the internal streets of a mobile home park to a public street or road should have a minimum width of 34 ft if parking is permitted along both sides or 27 ft if parking is permitted on only one side. If parking is not allowed, the width can be reduced to 24 ft provided the entrance street is more than 100 ft long and does not provide access to abutting mobile home lots within the first 100 ft.

All typical internal streets should have a minimum width of 24 ft. However, the width of minor internal streets can be reduced to 18 ft if parking is prohibited on both sides. Minor streets are (1) two-way streets that are less than

500 ft in length and serve less than 25 mobile homes and (2) one-way streets, of any length, that provide access to abutting mobile home lots on one side only. Cul-de-sacs should be limited in length to 1000 ft and should be provided with a surface turning circle at least 60 ft in diameter.

The proper design of street intersections is an important safety consideration. Within 100 ft of intersections, streets should be at approximately right angles. Street intersections should be at least 150 ft apart and the intersection of more than two streets at one point should be avoided.

Street grades should not be excessive, especially at intersections. It is suggested that grades be less than 8 percent whenever possible; however, short runs of up to 12 percent can be used if necessary.

All streets should be provided with a smooth, hard, and dense surface that is properly drained and durable under normal use and weather conditions.

Off-street parking, in the form of parking bays or individual parking spaces on each lot, should be provided to reduce traffic hazards and improve the appearance of the mobile home park. Parking space should be provided in sufficient number to obtain a ratio of at least five spaces per every four mobile home lots in order to accommodate two-car tenants and guests. Every parking space should be designed and located so as to be convenient for use and should be within 100 ft of the mobile home it is to serve.

Walkways

All mobile home parks should be provided with walkways where pedestrian traffic is expected to be concentrated, such as around recreation, management, or service areas and between individual mobile homes. It is recommended that these common walks be at least 3½ ft wide.

Walks should be provided on each individual lot to connect the mobile home with a common walk, street, or paved surface. Such walks should be at least 2 ft wide.

Mobile Home Lots

Every mobile home lot should contain at least 2500 ft^2 of area to accommodate modern mobile homes and their appurtenances and to assure adequate clearances between mobile homes and other structures. Many of the mobile homes presently being manufactured are between 50 and 60 ft long and 10 and 12 ft wide. Some are as large as 70 ft long and 24 ft wide. These larger units require correspondingly larger lots. Lot sizes of 3000 ft^2 and more are frequently used to accommodate the larger mobile homes and provide more privacy to residents. Some other advantages of larger lots are that they facilitate later changes in design, such as the addition of carports or other accessory structures to mobile homes, and they also provide assurance against premature obsolescence of the mobile home park. All lots within any mobile home park should not be the same size and shape if different-sized mobile homes are to be accommodated and if effective use is to be made of the available space.

It is generally agreed that small lots contribute to overcrowding and create an undesirable appearance, especially when used to accommodate the larger mobile homes. A practical program to eliminate undersized lots should be developed by the local governmental agency having authority that is agreeable to all organizations concerned, including mobile home park operators and owners, the local health authority, and other involved groups.

Once adopted, such a program should be enforced to assure that all mobile home lots not meeting established minimum space requirements will be eliminated.

There should be a clearance of at least 15 ft between adjacent mobile homes and between mobile homes and other structures except that mobile homes placed end to end need a clearance of only 10 ft when opposing rear walls are staggered. Mobile homes should be at least 25 ft from any park property line abutting upon a public street or highway, 15 ft from all other park property lines, and 10 ft from any area such as a park street, a common parking area, or a common walkway. When determining clearances, any accessory structure that has a horizontal area exceeding 25 ft^2, located within 10 ft of a window on a mobile home, should be considered as part of the mobile home if the accessory structure has an opaque top or roof higher than the window.

If driveways are provided for individual mobile home lots, they should be at least 8 ft wide, with an individual 2 ft added if they also serve as walks. The on-lot parking space served by the driveway should have dimensions of 9 ft wide by 20 ft long.

It may be desirable to provide storage facilities for each lot in order to discourage the storing of objects under mobile homes. Many mobile homes presently built do not contain ample space for storing equipment such as rakes, shovels, garden hose, lawn chairs, and other similar items. The storage of such items under a mobile home is undesirable since they can provide a potential harborage for rodents, snakes, insects, and other pests.

Recreation Areas

Mobile home parks that accommodate 25 or more mobile homes should be provided with at least one easily accessible recreation area. When several different age groups are to be provided for, it may be desirable to have two or more separate areas to serve the varied interests.

For safety reasons, recreation areas should always be located where they are free of traffic hazards. It may also be desirable to provide some sort of buffer zone around the area such as trees, bushes, or other vegetative growth. A recreation area can be located adjacent to recreation or service buildings, if provided for efficient construction, use, and maintenance of both the area and the structure.

Recreation areas should be provided in a ratio of at least 100 ft^2 of space per each mobile home lot. However, many planners will provide more recreational space than the minimum, depending on the availability of recreational facilities in the neighborhood of the mobile home park. Swimming pools, recreation buildings, and child play areas can be considered as fulfilling part of the total requirement for recreational area. Each outdoor recreation area should contain at least 2500 ft^2 of area to assure adequate space for all activities.

Swimming pools should be constructed and operated in accordance with all applicable state and local requirements and regulations.

Service Buildings and Other Structures

Every mobile home park should be provided with a service building containing emergency sanitary facilities consisting of at least one lavatory and one flush toilet for each sex per each 100 mobile home lots. Where feasible, the consolidation of sanitary, laundry, management, and other service facilities in a single building

SITE PLANNING

and location is recommended if the single location will adequately serve all mobile home lots. Consolidation is preferable for efficient construction, use, and maintenance of all facilities.

Service Areas

Where areas for the outdoor drying of clothes are necessary, it has been found that approximately 2500 ft² per 100 mobile home lots is adequate with rotated use. It may be desirable to locate the drying yard near the service or laundry building, if provided, and as far as possible from roadways or traveled areas. It has been found practical to provide clothes-drying facilities on the individual mobile home lots provided that drying units are standardized and are properly located and installed. Where the clothes-drying facilities are permitted on the individual mobile home lots, it is suggested that they be provided as part of the basic facilities to assure that the same type of unit, located in the same general area of each mobile home lot, is used throughout the park. Umbrella-type lines in permanent sockets are recommended. The use of individual drying facilities also requires that the owner or operator develop and enforce rules that permit clothes drying only on the facilities provided. All clothes-drying areas, whether centrally located near the service building or on the individual mobile home lot, should be adequately screened from view so as not to detract from the appearance of the mobile home park or be objectionable to residents on adjacent property.

If desired, car-wash and other general-purpose facilities can be provided as a service to residents of the mobile home park. Any such facilities should be properly constructed and preferably screened from view.

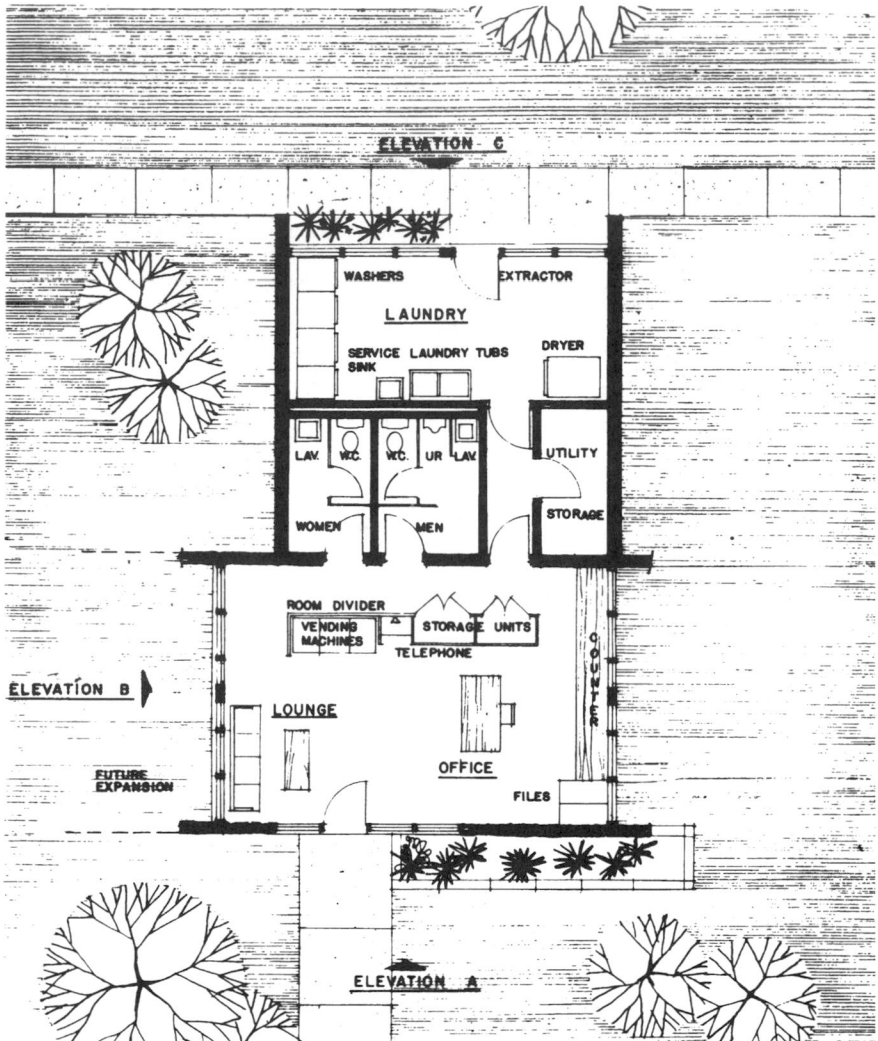

Fig. 1 Floor plan.

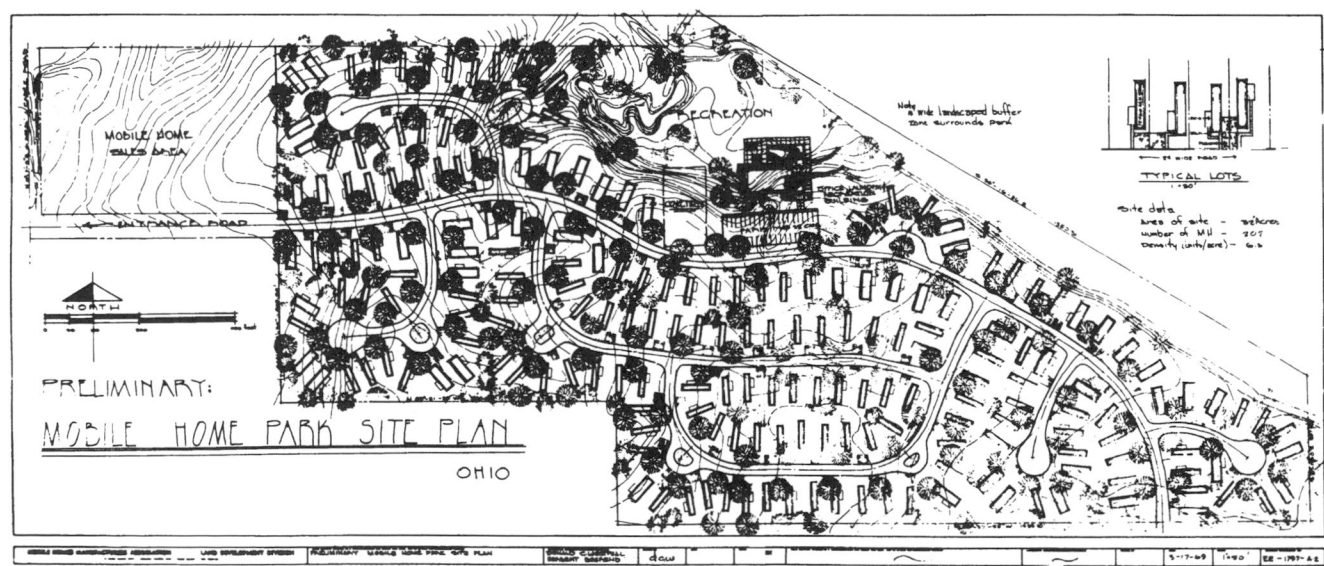

Fig. 2

INDIVIDUAL MOBILE HOME LOTS
(SITES, PARKS, AND SUBDIVISIONS)

Meeting Basic Requirements

The mobile home lot is the land area, large or small, upon which the home is placed and which provides space for all the belongings and activities of its occupant.

Required Functional Areas

The individual mobile home lot consists of six component areas which reflect the basic functions of the mobile home site—pad, parking, entrance, outdoor living, utility corridor, and storage. The arrangement of these six functional components of the lot is somewhat variable, but typically looks like Fig. 1.

The only component of this arrangement that is fixed is the utility side of the home, which is always on the right side when facing the mobile home hitch. The other areas are variable, depending primarily upon the lot size and unit orientation on the lot.

Each mobile home lot is usually required, by ordinance, to provide the pad, parking area, outdoor living, and storage areas. Existing standards vary but typically define a minimum lot size and minimum yard areas which does not give the flexibility of lot size required because of the highly variable size of homes. A minimum distance between homes and setbacks allows this lot size flexibility.

Convenience in Relationship of Use Areas

The arrangement of the six functional component areas of the lot should be determined by the floor plan of the home, the characteristics of each site, and the logical and convenient relationship of on-lot space. The sequence in which residents use the component areas should be reflected in the lot arrangement. Occupants arrive at home in a car, make way from the car to the door, and go in. They live in the house and occasionally use the yard area as an extension of living space. Logically, the parking pad

should be between the outside parking area and door. The outdoor living area should be adjacent to the home and near one of the two entrances.

Accessibility and Barrier-Free Access

The mobile home lot should be accessible for handicapped people. With alterations in traditional design and setup, barriers can be removed, making the mobile home entrance accessible to people with limited mobility.

Present site barriers to handicapped people relate mainly to home access. Floor elevations are typically 17 to 33 in higher than the surrounding grade, thus requiring an entry stair to reach the door level. Accessibility for the handicapped can be achieved by eliminating steep grades and steps, and by providing wide walks and ramps.

Walks Walks from the parking area to the main entrance should be at least 42 in wide to allow for a person on crutches or wheelchair to move freely. If the slope of the walk is greater than 5 percent, a handrail should be provided and the slope of the walk should not exceed 8.33 percent. Walks should have a continuous smooth common surface not interrupted by steps or abrupt changes in level greater than ½ in in height. Where walks meet parking areas, they should blend to a common level by means of a ramp or curb cut where necessary. Curb cuts or ramps should have a textured nonslip surface such as a broom-finish concrete.

Ramps Ramps from walk or parking surface to door level are not normally feasible because of the height of floor level above grade and the limitations on wheelchair climbing ability. Ramps should not be greater than 8.33 percent in slope or less than 42 in in clear width.

Ramp lengths depend upon the height of the floor above grade, generally 17 to 33 in, and would have to be 17 to 33 ft in length. In addition, an 8-ft entrance landing would have to be provided at door level to allow the door to

swing past the landing area if it opens to the outside.

The most viable solution to providing barrier-free access to the mobile home entry levels is to lower the home to grade by placing the home over an area excavated to accommodate the wheels, frame, and axles of the home. Approximately 6 in of height should remain above grade for ventilation of the home, and at least 18 in in areas where under-unit utility connections are made. The 6-in change in grade between walk and entry level can be handled by providing a short ramp at up to 8.33 percent grade.

CREATING PRIVACY

Mobile home sites should provide a private outdoor living area suitable for eating, entertainment, and relaxation. This area should be reasonably spacious, private, and as free as possible from the visual and noise intrusion of neighboring areas.

Lot Size and Unit Orientation

The outdoor spaces of mobile home lots are defined by the mobile home itself, its floor plan, and its relationship to other mobile homes, roads, adjacent structures, and physical features. Each of the spaces around the mobile home may accommodate one or more functions (parking, entrance, outdoor storage building, or patio) which are normally part of every lot.

The size and arrangement of the exterior spaces of mobile home lots are variable. Consequently, the location of parking and outdoor living areas may vary from one lot to the next, depending mainly upon lot size. They may consist of the arrangement of functional components unique to the mobile home park, or they may be similar to those of the typical subdivision lot with site-built homes.

Small-lot and medium-lot developments (5000 ft² or less) typically consist of homes positioned perpendicular to the street. Outdoor living space occupies the front yard segment from the midpart to within 30 ft of the street.

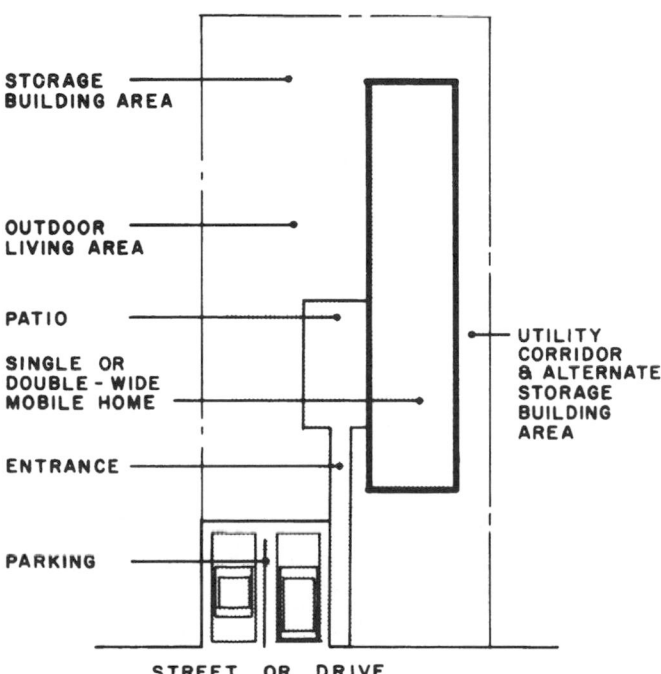

Fig. 1

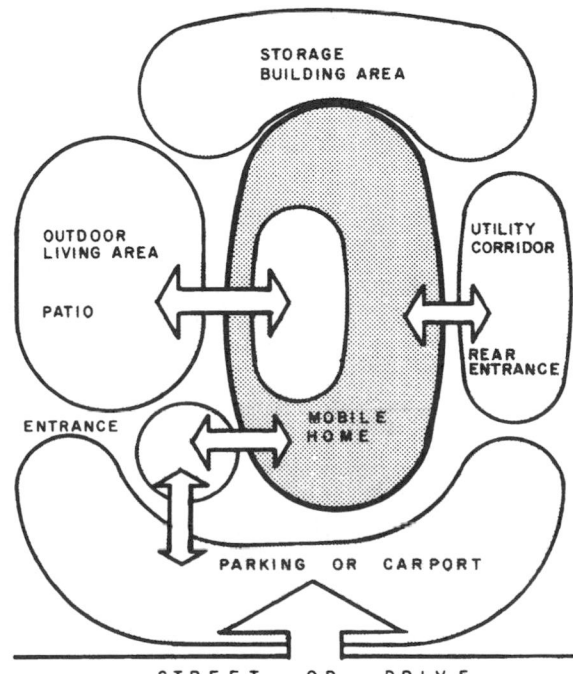

Fig. 2

MOBILE HOME LOTS

Private living space on the front "main entrance" or "side" of a mobile home is a deviation from the traditional single-family home where rear yards are reserved for private activities and the front yards are the main entrances. In the mobile home lot, this front and rear yard function is combined and is normally found on the side of the unit. This "side" or front yard location, unique to mobile housing, is caused partially by the long and narrow design of the unit with no significant rear entrance. The front "side" location is also due to siting practices where the unit is usually placed perpendicular to, rather than parallel to, the street. This front yard living space is usually small, and is bordered by elements which are not totally compatible with its use as a private living area. On the side nearest the street, it abuts the driveway and parked cars; on the other side, the utility hook-ups and windows of the adjacent mobile home limit attractiveness and privacy.

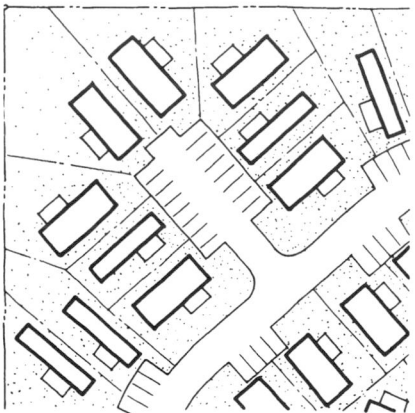

Fig. 3

Larger lots allow more flexibility in the placement of homes and result in the exterior spaces being less rigidly defined.

Even with the same perpendicular arrangement to the road where there is more distance between the patio and the neighboring utility connections, electric meter and windows, the patio becomes a more private and livable area.

Large lots, those greater than 5000 ft² in area, allow the home to be placed parallel to the street in conventional single-family subdivision fashion. On these large lots, the private living space moves to the traditional rear yard location. However, the design characteristics of the typical mobile home unit regarding a rear entrance do not enable the yard location to function as well as it does in the case of on-site-built homes. Privacy can be maximized with this arrangement for mobile homes, an essential in mobile home living where the indoor space is limited.

Other more innovative lot arrangements, such as homes clustered around a parking court or open space area, are sometimes used in small- to medium-sized lot mobile home subdivisions. These arrangements are usually employed where the double loading of a through street is not possible or where lots are desired in an isolated corner of the site. Although patios still occupy the front yard, these arrangements sometimes result in more usable exterior spaces without the intrusion of parking and adjacent utility corridors on outdoor living areas.

Height of Surrounding Structures

On most mobile home lots where distances between homes are small, the height of the neighboring home affects the spaciousness and privacy of the patio area. A higher structure blocks and dominates views. Screening the view of adjacent mobile homes will lessen the feeling of closeness arising from unit height. The effect of height on an enclosed space is an important design variable, because mobile homes are normally placed 1½ to 2 ft higher than necessary. There is no reason why the floor of the home has to be 2 to 2½ ft above grade after the home is in place. The only reason for this height is to provide room for the wheels and axles upon which the home is transported to the site and for sanitary utility connections. The home can be lowered to the ground by a method called low profiling, as long as clearance of 18 in under the frame is provided to allow hookup of the sanitary sewer line under the home.

Low-profile grading of the mobile home pad involves creating an 18 in depression in the pad area for the full length of the home. The home is then rolled into place and piers are constructed to support the frame. Wheels and axles may be removed and used again for transporting other homes. The 18-in lowering of the home allows about 6 to 8 in for ventilation of the underfloor area, which is necessary to prevent moisture buildup and mildew damage. This 6- to 8-in space is then skirted to conceal the underside of the home and seal off the crawl space while maintaining adequate ventilation. Provisions for adequate drainage should be made depending upon groundwater, drainage, and soil conditions.

Sloping Sites

Mobile homes, by nature, are fairly adaptable to moderately sloping sites. Their long narrow shape requires that they normally be placed parallel to the contours but also requires roads to run perpendicular to the contours if road frontage is to be minimized.

The placement of homes on a slope can affect the spaciousness and privacy of each lot. It is desirable if homes are adjusted so that the patio area faces downslope to take advantage of views and to provide visual separation of units.

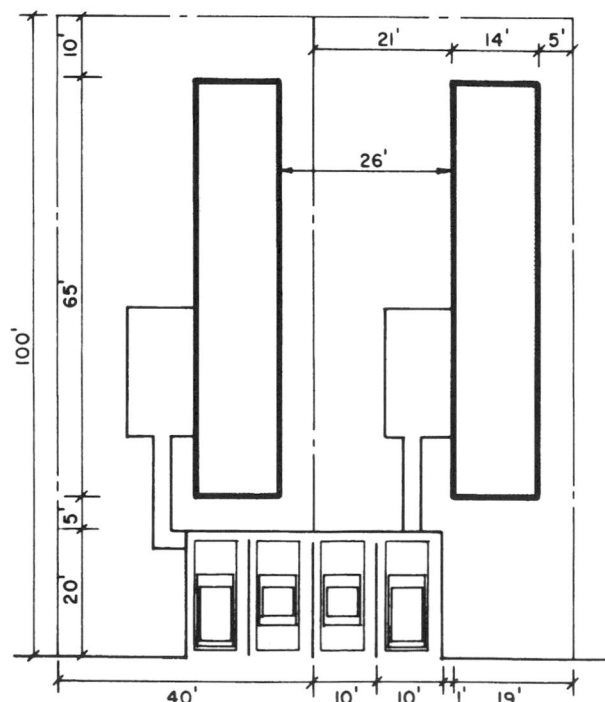

Fig. 4 4000-ft² lot.

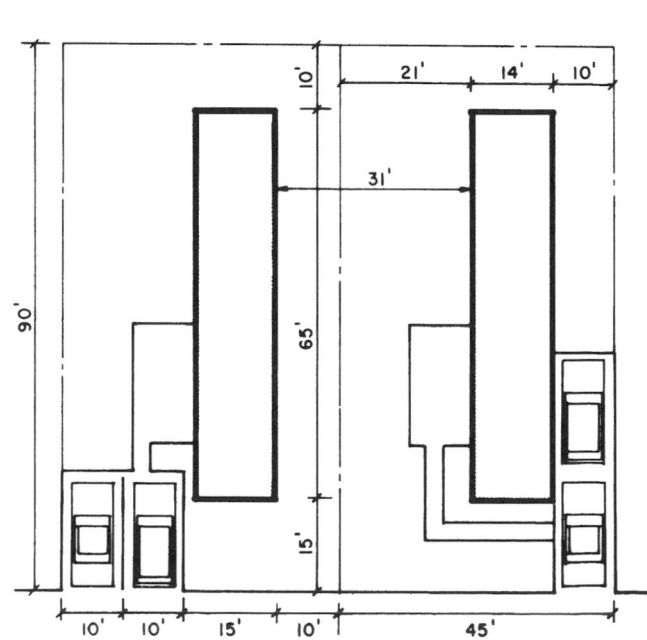

Fig.5 4050-ft² lot.

Screen Plantings and Natural Vegetation

Where lot sizes and spaces between mobile homes are small, screen plantings can create as much privacy as is found in larger lot developments. Evergreen trees or shrubs can be planted along lot lines or close to the patio area to visually separate neighboring homes. Evergreen trees such as pines, firs, or spruces are best suited to lots with considerable room for mature growth.

Of course, planting trees for screen purposes requires a compromise of space for mature growth and massing for immediate effect. Usually evergreen trees should be planted at about 10 to 15 ft apart and staggered to create a reasonably dense screen, although some may eventually have to be removed.

Smaller lots can use other screen plantings like hemlock or arborvitae, tall shrubs like privet, tall hedge, or broadleafed evergreen shrubs to create the screen effect and require less space.

All screen plantings should be massed in clusters of one or two plant types rather than scattered about the site.

Screen Fences

Screen fences can also create privacy in the patio area but if not properly executed are more likely than plantings to detract from the general appearance of the lot and neighborhood. Fences offer the advantage of taking up little space and thus being adaptable to small lots while creating a complete visual barrier between homes. Fences should not be over 6 ft high and should be used in small sections placed close to the patio area. Their direct exposure to the street should be minimized because they can become a rather imposing structure if overdone. L-shaped sections of fence work very well. Fences should be constructed to create a simple surface texture. The pattern of shadow created on the fence surface is all the ornamentation that it needs. Colors may be very subdued. The simple cedar picket or redwood basketweave fences have been successful in residential use. Board-and-batten or louvered fences are also good designs. Small shrubbery can be used in conjunction with fences to soften their typical hard lines and edges.

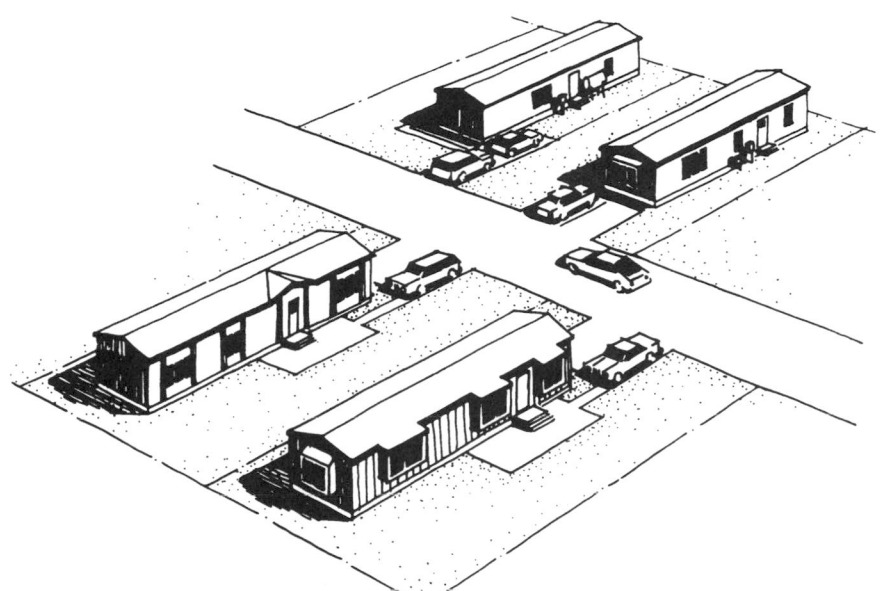

Fig. 6

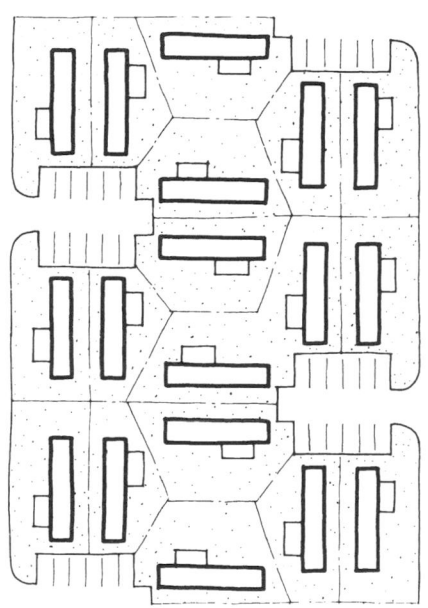

Fig. 7

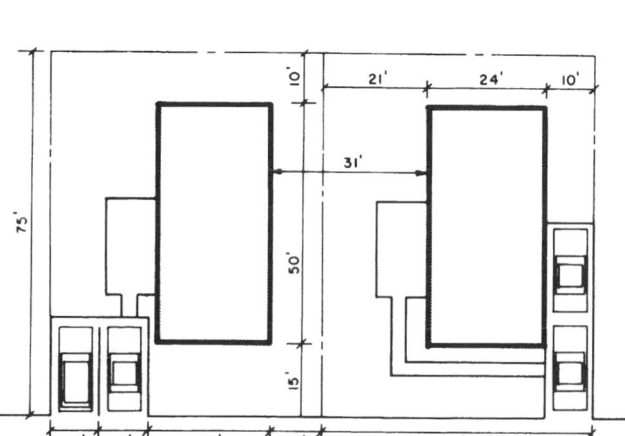

Fig. 8   4125-ft² lot.

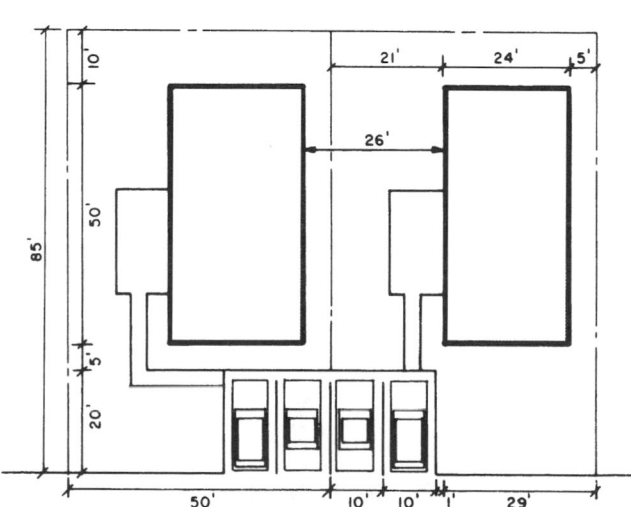

Fig. 9   4250-ft² lot.

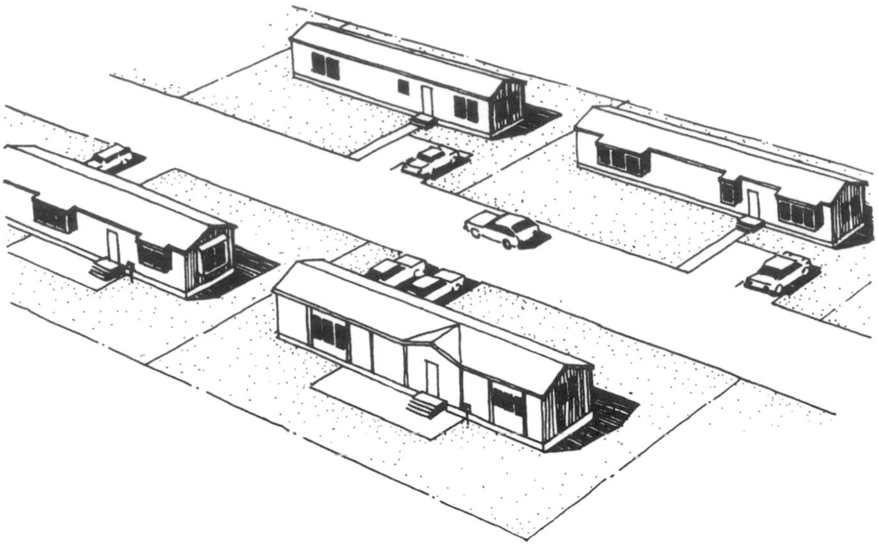

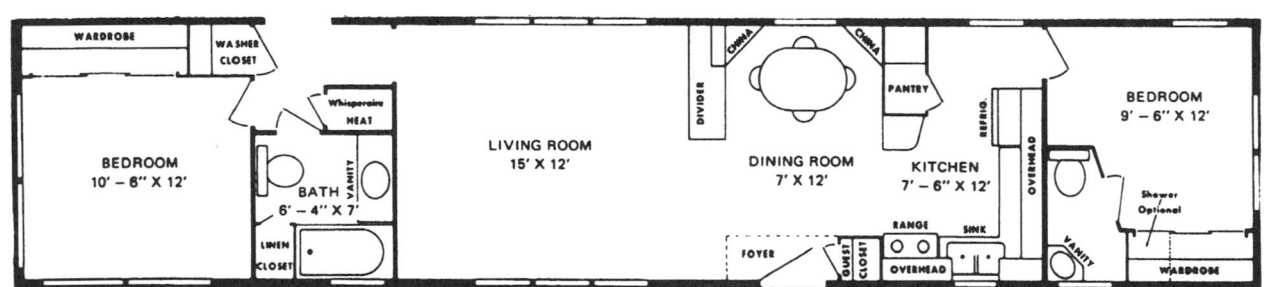

Fig. 10

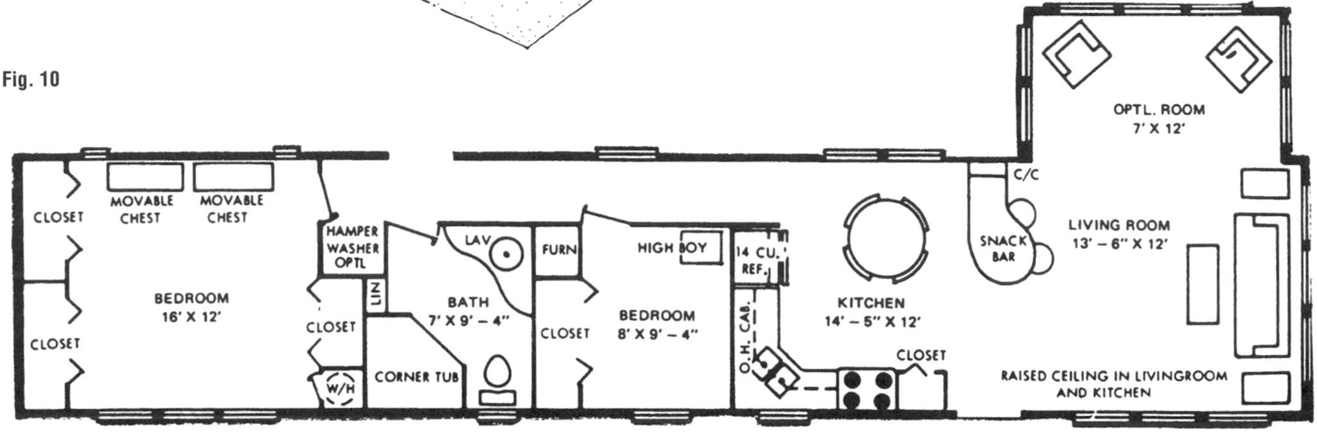

Fig. 11*a* 12-wide expandable—59 by 12 ft, 792 ft², two bedrooms, one bath, expandable living room.

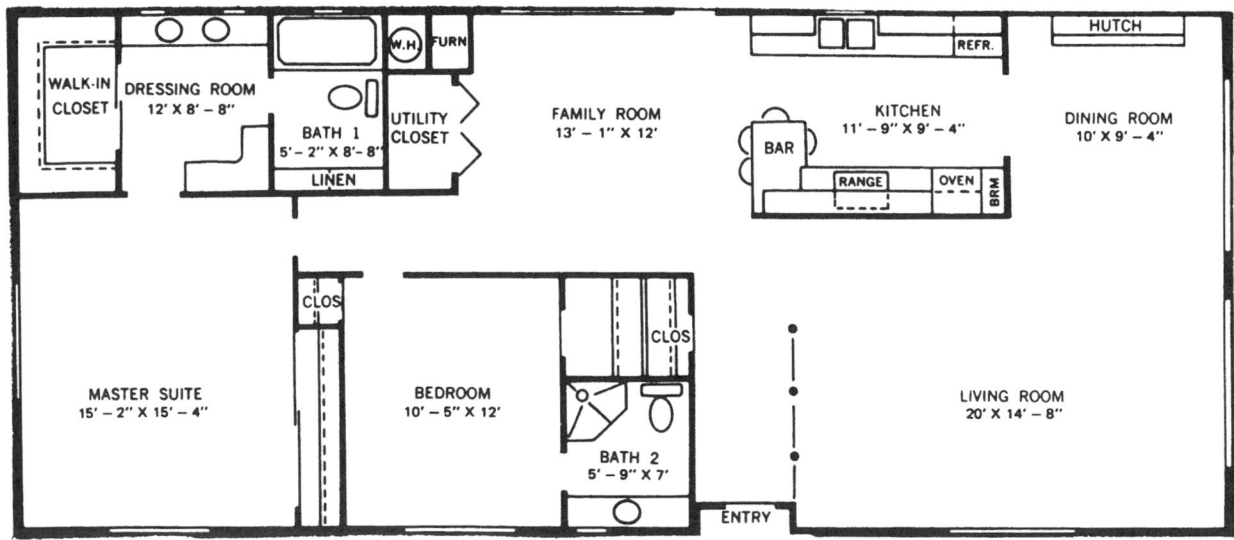

Fig. 11*b* 12-wide—56 by 12 ft, 672 ft², two bedrooms, two baths.

Fig. 11*c* Double-wide—24 by 56 ft, 1344 ft², two bedrooms, family room, two baths, utility room.

TYPICAL MODULES AND ARRANGEMENTS

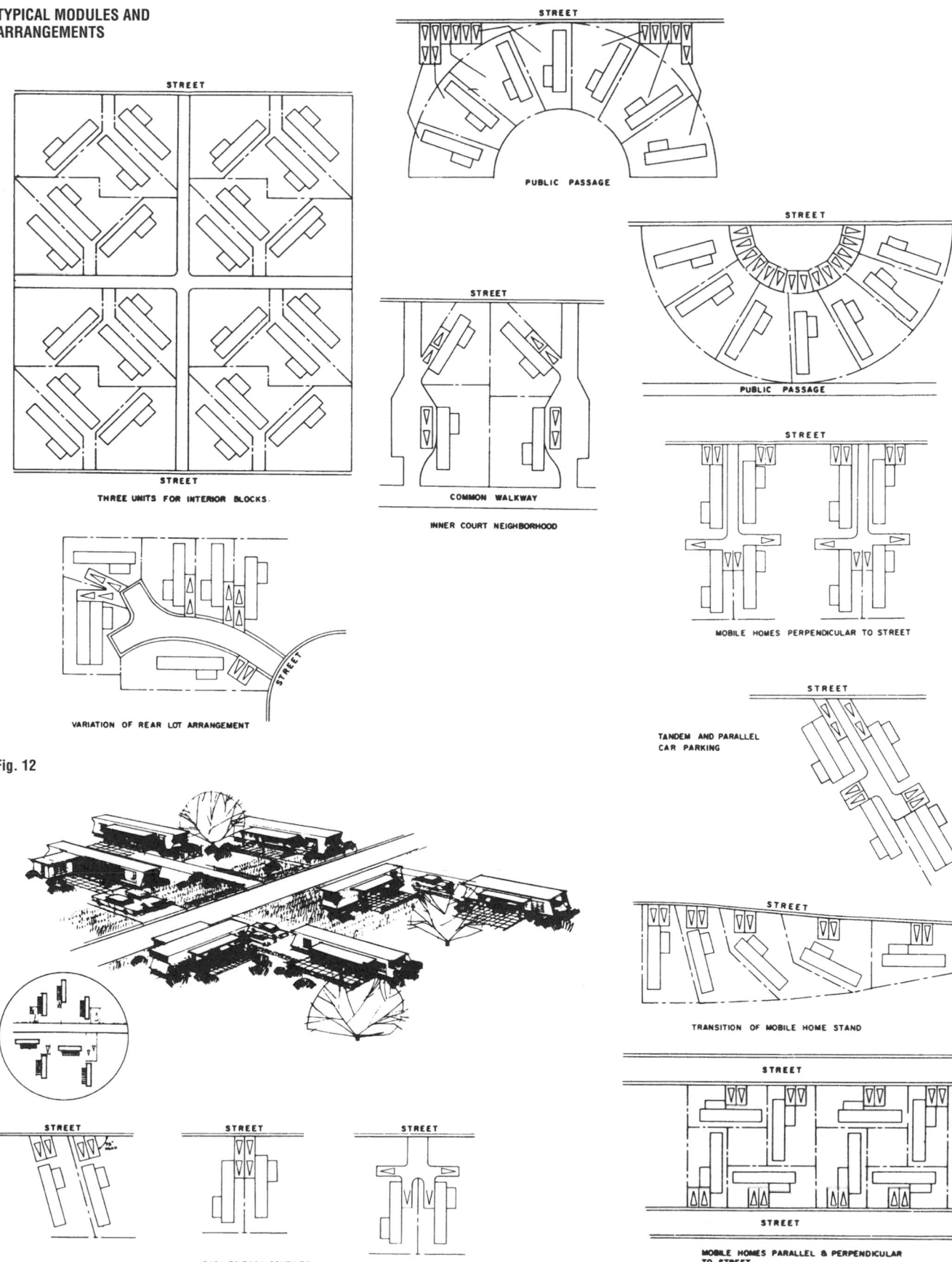

STREET

PUBLIC PASSAGE

STREET

THREE UNITS FOR INTERIOR BLOCKS.

STREET

COMMON WALKWAY

INNER COURT NEIGHBORHOOD

STREET

PUBLIC PASSAGE

STREET

MOBILE HOMES PERPENDICULAR TO STREET

STREET

TANDEM AND PARALLEL CAR PARKING

VARIATION OF REAR LOT ARRANGEMENT

Fig. 12

STREET

TRANSITION OF MOBILE HOME STAND

STREET

STREET

STREET

STREET

PARALLEL CAR PARKING

BACK TO BACK COMBINED UTILITY CORE & CAR PARKING

FOR SUB DIVISION

STREET

MOBILE HOMES PARALLEL & PERPENDICULAR TO STREET.

Fig. 13

MOBILE HOME LOTS

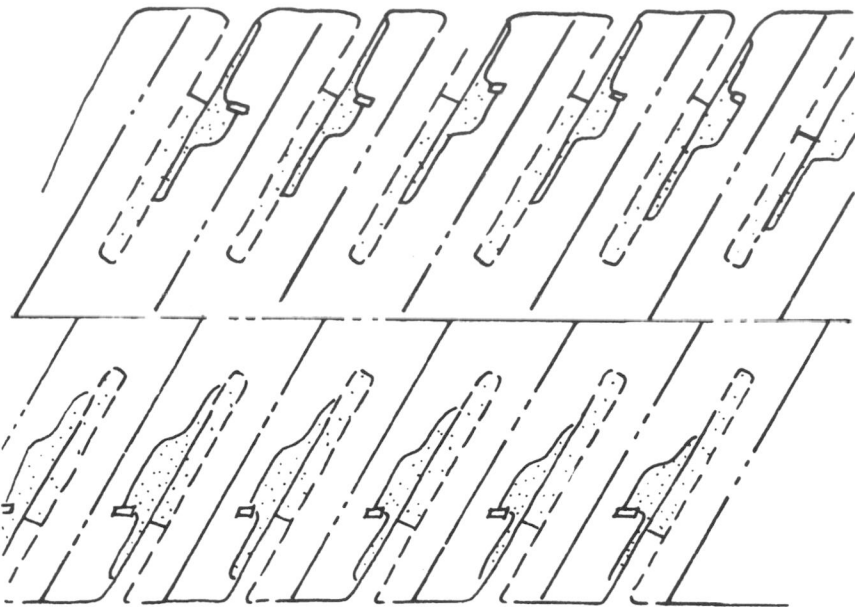

Fig. 14

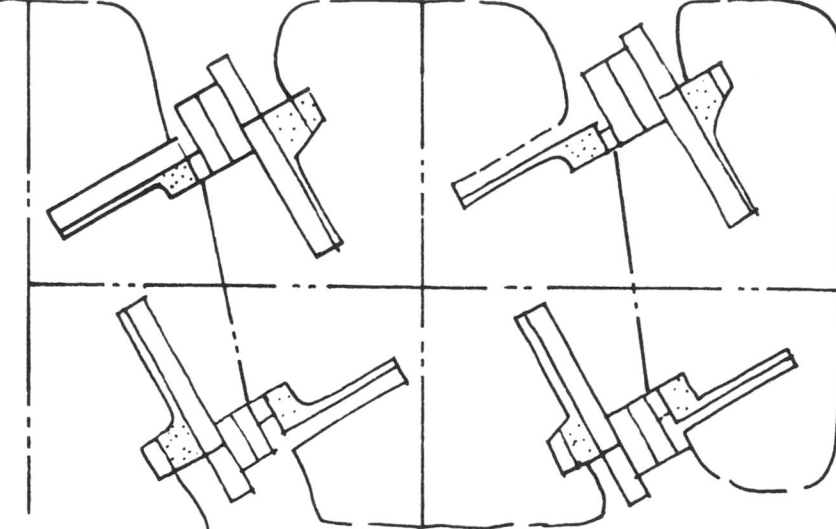

Fig. 15

Parallel Arrangement—Zoning

The site should be in a residential zone if mobile home courts are permitted or in a heavier zone provided the site is not subject to unhealthful or adverse influences.

If unzoned, the location chosen should be such that the mobile home court will not be subject to unhealthful or adverse influences and will not itself adversely affect adjacent neighborhoods.

Community Facilities

The site should be accessible to schools, churches, and shopping facilities as for other residential uses.

The site should be within reasonable commuting distance of employment.

Perpendicular Arrangement—Lot Area

Lot areas should be 3000 ft^2 or larger for each mobile home, with possibly a few lots somewhat smaller.

OPTIMIZING ATTRACTIVENESS OF HOME AND LOT

Optimum Orientation

Views of mobile homes from the street are most attractive when units are placed parallel or angled toward the street. This arrangement creates the same relationship of home to street found in most on-site-built housing developments.

In some instances, various groupings of homes have a positive effect, typical of some of the newer, creative land development concepts.

Low Profile

The mobile home is an unusually long, narrow structure compared with typical on-site-built homes. It is usually placed on a long, narrow lot. The height of the home tends to accentuate this appearance because the single-wide home

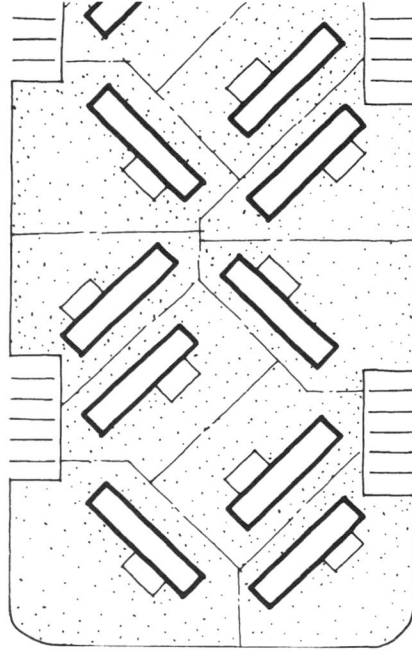

Fig. 16

is almost as high as it is wide, when resting 2 to 3 ft above grade.

Lowering the home to the ground helps improve its appearance by making it look wider. This is especially effective in groupings of mobile homes. The appearance of both single- and double-wide homes is equally improved by low profiling, and the double-wide with this treatment really appears like a traditional ranch home.

Exterior Finish of Homes

The exterior treatment of mobile homes varies from manufacturer to manufacturer. Two recent trends in mobile home exteriors seem to offer an especially pleasing appearance. Some manufacturers are offering wood siding, the textures and colors of which create a warm appearance and are easily blended with the natural features of the lot. Many homes, especially double-wides, are offered with a shingled and peaked roof which makes the home look more like a traditional on-site-built ranch house.

Landscaping

Landscaping of the mobile home lot is the key to unifying the home and lot. Landscaping frames the home, defines the yard, and gives the home an image of permanence in relation to the site.

Landscaping may include street trees, on-lot shade trees, screen plantings, shrubs, lawns, and ornamental plantings. The most important landscape elements are those which visually define private outdoor living space on the site. Deciduous shade and evergreen screen tree plantings provide the most dramatic impact on the visual character and livability of the mobile home lot. They define and limit exterior space by creating private spaces on the lot.

Buildings are usually the dominant vertical elements which define exterior spaces for human use in most residential developments. The density of mobile home developments with their one-story, long, narrow homes doesn't normally result in adequate exterior spaces. Without major landscaping to diminish the visual impact of many units or to further articulate the community and private space in mobile home developments, exterior spaces are monotonous. Trees provide an opportunity to define and enhance yards and living spaces within the mobile home development and between individual lots.

Trees for planting in mobile home developments should be selected with two considerations in mind. First, they should be space definers but not significant space users, and paved parking surface may be desirable. Screening of the parking area can be accomplished by leaving a small area between the patio surface and parking pavement for a planting bed. A few upright shrubs or yews, or a tall hedge, decrease the view of parked cars from the patio area.

Fig. 17

CONTROLLING CLIMATE

Shade Trees on South in Warm Climate

Plantings can help moderate climate, thus reducing energy demand and improving the livability of mobile homes. In hot or temperate climates, it is desirable to plant deciduous shade trees on the southern side of the home. Deciduous trees provide shade during the summer months yet allow sun to penetrate to the home during the colder months of the year when branches are bare.

Fig. 18   Summer.

Fig. 19   Winter.

Windbreak on North or Northwest in Cool Climate

In cool climates, it is desirable to plant evergreen windbreaks on the northern or northwestern exposures of mobile homes to deflect cold winter winds around mobile home sites. Plantings will reduce wind for a horizontal distance equal to 25 times their height.

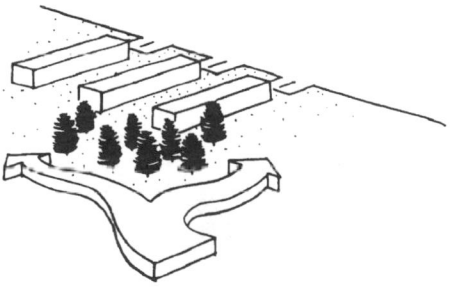

Fig. 20

MOBILE HOME LOTS

Optimum Orientation of Living Areas

It is desirable to orient mobile home units, when possible, to take advantage of the characteristics of their respective climatic regions.

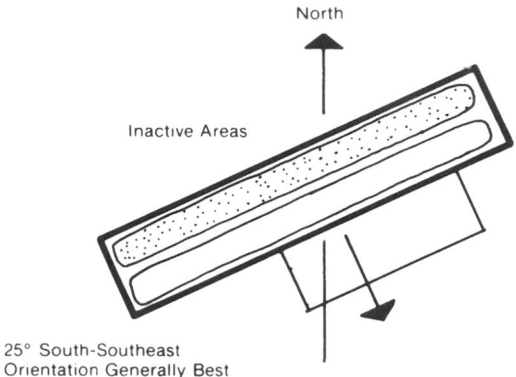

25° South-Southeast Orientation Generally Best

Fig. 21a Hot, arid regions.

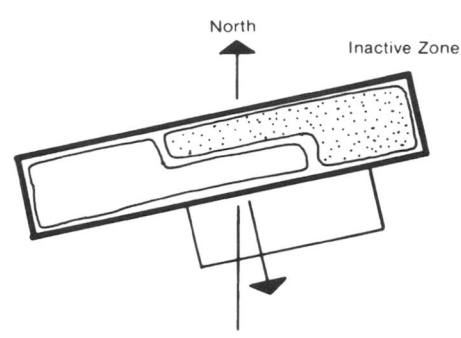

12° South-Southeast Orientation Generally Best

Fig. 21b Hot, humid regions.

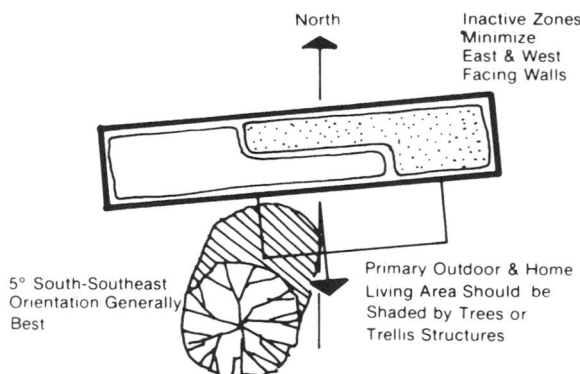

5° South-Southeast Orientation Generally Best

Primary Outdoor & Home Living Area Should be Shaded by Trees or Trellis Structures

Fig. 21c Temperate regions.

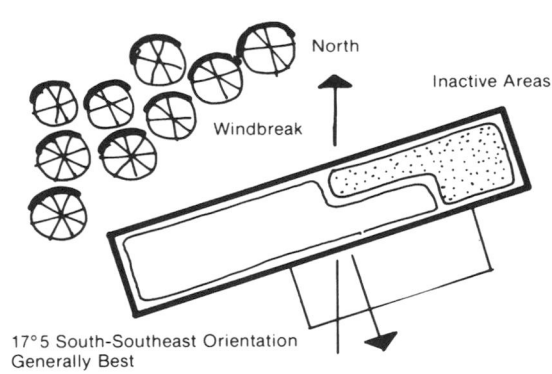

17°5 South-Southeast Orientation Generally Best

Fig. 21d Cool regions.

Control Air Circulation—Capture Breezes

Screen plantings or fences can also be positioned to direct and channel daily or seasonal breezes in hot climates.

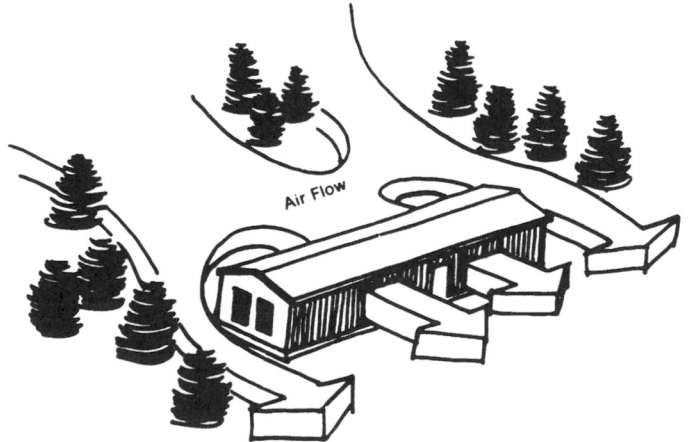

Fig. 22

GROUP MOBILE HOME SETTING (PARKS AND SUBDIVISIONS)

Meeting Basic Requirements

Clustering of mobile homes Mobile home group settings in both rental parks and subdivisions are mobile home sites clustered together to utilize organized support systems including streets, sewer, water, electric, and fuel services. The mobile home grouping benefits from sharing the costs of support facilities. Also, lower land costs may result because smaller lots are feasible when public sewer and water systems are used.

Organized support systems—utilities Mobile home park development standards usually regulate the basic design of utility distribution systems. In subdivisions, utilities may be located beneath the street pavement, outside of the pavement area but within the right-of-way or in easements on individual lots. Location beneath the street pavement does not require additional right-of-way width, but the pavement must be disrupted each time a utility must be serviced at locations other than access holes. Utility easements can sometimes be used to advantage within the mobile home subdivision. Locations along the rear lot lines usually do not require additional utility lengths. In this approach, the easement area is not wasted because it is included in the tax-revenue-producing deeded lot, and street pavement is not disrupted for servicing.

In mobile home developments, it is generally more economical to run utility mains under the mobile homes, thereby reducing the cost of providing laterals to each home. Costs can sometimes be reduced up to 30 percent where utilities are under the homes, but utilities handled this way can normally not be dedicated to a municipality for ownership and maintenance. Utilities located beneath the units usually require a condominium ownership of land with agreements for the maintenance of the subsurface improvements. Utilities, or even roads, do not have to be dedicated to a municipality when condominium ownership is involved and therefore may not have to meet standards required of dedicated streets or utilities in conventional single-family subdivisions.

Subsurface utility lines, while not visible, still require that land be cleared for their installation. Therefore, the amount of disturbance to existing natural features determines the extent to which utility line locations will adversely affect a site.

Streets

Streets are provided for safe, convenient access to individual mobile home sites and to community facilities. Street design for mobile home developments is based upon a few general engineering principles which ensure reasonable traffic and pedestrian safety.

Street alignments should be based upon sight distance and probable roadway speeds using computation methods endorsed by the Institute of Traffic Engineers. Generally, a minimum practical curve radius in residential areas is 100 ft, with 30 ft acceptable on minor streets. Street alignment at intersections is especially critical. The preferred angle of street intersections is 90°; for safety purposes, streets should never intersect at angles less than 80°. When two streets intersect the same street, they should either form a through intersection or be offset by at least 100 ft.

Street gradients affect the visual character, safety, and accessibility of the mobile home development. Generally speaking, grades between 2 and 7 percent are the most desirable; and a minimum of 0.5 percent is necessary on all curbed streets to prevent pooling of water. If gradients must be less than the minimum in very flat areas, special subgrade compaction and street construction controls are necessary. Streets of less than 2 percent grade are visually perceived as flat. Moderate slopes of 2 to 7 percent usually result in a more interesting streetscape and encourage more imagination in the siting of homes.

Streets should generally not exceed 12 percent grade, but on minor streets, grades of up to 15 percent are acceptable. Where steep road gradients are unavoidable, care must still be taken to flatten grades at intersection areas; gradients within 100 ft of intersections should not exceed 10 percent, with 4 to 6 percent preferable in snow or ice areas for a distance of 50 ft.

Circulation layout determines the accessibility of the mobile home site within each development. In properly designed residential neighborhoods without through traffic, travel distances from residences to collector streets are short, actual traffic speeds are low, lane capacity is not a controlling design factor, and inconvenience or short delay is a minor consideration.

In conventional single-family residential neighborhoods, traffic speed should be slow, approximately 25 mi/h. In mobile home areas where density is higher, speed should not exceed 15 mi/h. Momentary delays to allow other traffic to pass around parked cars is acceptable, and it is customary to drive slowly to avoid children and pets.

Pavement widths should be determined by considering probable peak-traffic volume, parking needs, and limitations imposed by sight distance, climate, terrain, and maintenance requirements. It is senseless for streets to be wider than absolutely necessary; excessive widths only increase development costs which are passed on to the lot renter or owner. Also, from an ecological point of view, avoiding excessively wide streets means less impervious surface, which results in less storm-water runoff. The special problem of delivering mobile homes to lots is not a major consideration in determining street widths. Spacing between the mobile homes and the grades from street to lot are of much greater concern. Movement of mobile homes from their original placement on a lot is uncommon. When homes are to be moved through narrow streets, notice may be given to remove parked cars from the street.

Where streets also serve as pedestrian walks, they should be built with a 2 ft wider cartway than otherwise required. All entrance streets and other collector streets with guest parking should be 28 to 30 ft wide; this provides for two moving lanes and a parking lane on one side. Collector streets without parking should be 24 to 26 ft wide. Minor streets with parking on one side should be 26 ft wide; and local streets, courts, plazas, and cul-de-sacs with no parking should be 20 ft wide. A 20-ft-wide pavement is the minimum width which generally offers year-round utility and convenience where snow and ice control is necessary.

One-way streets may be allowed at 11-ft widths in the following situations: (1) adequate off-street parking is assured; (2) the climate is mild, and snow and ice control problems cannot be foreseen; (3) total loop length will not exceed about 500 ft; (4) no more than about 25 dwelling units are served; (5) adequate longitudinal sight distances can be provided; and (6) vehicle speeds may be reasonably expected in the 10 to 15 mi/h range. A 16-ft-wide pavement may be a practical loop street alternative in difficult terrain where cross-pavement ground slopes are severe, where vehicle speeds will not exceed 10 mi/h, and where other above-outlined considerations can be met. Under the various conditions outlined, the 16-ft-wide pavement can be functionally effective, but will result in a higher level of resident inconvenience than a wider pavement. Sixteen feet cannot be considered a desirable pavement width but must be conceded to be acceptable under certain conditions, where necessary to avoid destruction of natural features.

Street rights-of-way are a consideration unique to the mobile homes subdivision development. Streets within rental parks simply are not dedicated to the municipality, since the entire park property is owned by the park developer-operator. In order to achieve maximum density, a park developer will usually retain ownership of roads and therefore does not have to meet local standards for roads or reserve a wide right-of-way.

Within the mobile home subdivision, the streets and street rights-of-way are dedicated or retained in the ownership of the homeowners' association.

When roads are to be dedicated, they have to meet the same standards applied to all roads within residential areas. This generally means that a 50-ft right-of-way is required. The resulting arrangement of space, small lots, and wide rights-of-way is comparatively wasteful of space. Space is still lacking where it is most needed (in the outdoor living area of individual homes) because lot sizes are still usually small. On the other hand, an excess of space is provided where it is needed least. The distance between homes on opposite sides of streets is sometimes 70 to 80 ft or a more generous spacing than is needed in a small lot development.

Street rights-of-way must be adequate to provide required street pavements, sidewalks, drainage facilities, and utilities as needed, when they are placed in the rights-of-way. Right-of-way widths are too frequently fixed uniformly by local ordinances, regardless of the actual space required to accommodate necessary improvements. Excessively wide rights-of-way waste land and result in avoidable maintenance costs to the municipality; a community realizes no tax revenue from street rights-of-way. This land would be better devoted to individual building sites rather than public right-of-way.

Sidewalks along the road edge in suburban residential areas are being provided less frequently than in the past, and the amount of pedestrian use varies. Placement of sidewalks immediately adjacent to the road is not really a safe location unless curbs are provided. More elaborate developments have interior pedestrian paths linking logical origins and destinations such as clusters of individual homes to community facilities or to convenience commercial areas. Paths or sidewalks other than these are not necessary in low-traffic-movement areas.

Drainage facilities may include either grassed swales or curb gutters and subsurface storm drainage structures. Where roadside drainage swales are used, they normally require a right-of-way at least 10 ft wider than the pavement width. Thus if a 28-ft pavement is used and swales are located on both sides, a total right-of-way of 48 ft would be required. If streets are curbed, there may be no justifiable reason for right-of-way widths to be much wider than roadway pavements.

Dead-end streets must be provided with turnaround areas. Turnarounds in most conventional single-family subdivisions are cul-de-sac streets with a 75- to 80-ft-diameter paved area. It is fairly common in mobile home parks to eliminate the turning circle for streets with fewer than 25 homes, substituting a T or Y turnaround incorporated into a parking lot cluster. T and Y turnarounds should utilize an 18-ft minimum radius on all turns. The residential dead-end turnaround is basically for automobile use, but larger vehicles must sometimes be accommodated. Residential streets will also be used, in decreasing order of frequency, by refuse collectors, delivery trucks, snow plows, moving vans, and fire trucks. Experience has shown that circular paved turning areas of 75 to 80 ft in diameter function very well.

Curbs along residential streets are usually justified for three reasons: (1) preventing the roadway pavement from breaking down, (2) controlling traffic from encroaching beyond paved surfaces, and (3) concentrating and channelizing storm-water runoff. Valid arguments have also been made against the use of curbs, making a clear-cut answer difficult. Proponents of curbless streets regard curbs as both a needless expense and an ecologically unsound practice which disrupts natural surface drainage and requires expensive storm drainage facilities. Such facilities frequently concentrate storm water and produce water velocities which necessitate the collection of water in pipes and storm-water systems. These conditions tend to minimize the amount of water infiltrating the ground and can cause a considerable amount of fast-moving water to leave the site and cause off-site flooding and erosion. Swales, an alternative to curbs, can collect water where velocity is slowed and allow it to be absorbed into the ground. Proponents of curbless streets also question the reliability of curbs as a safety measure, citing how easy it is for a vehicle to strike or go over a curb and out of control.

The decision to use curbs should largely be based on how effectively storm water can be removed from the site without causing harmful on- or off-site impacts. The feasibility of using swales should be explored as a preferred alternative over the use of curbs during the early planning phase. If curbs are used, the rolled curb may be more desirable because of the numerous crossovers required for frequent on-lot parking areas, yet its effectiveness in controlling traffic may be less. If curbs are not used, pavements can be prevented from unraveling by using a thickened-edge pavement, extending the base course beyond the paving surface by 6 to 8 in, or using anchored steel edging flush with the pavement surface.

Community Facilities and Open-Space Systems

Regulations for community facilities and open-space systems typically require that at least 8 percent of the gross site area be devoted to recreational facilities and that a community building, storm shelter, laundry and drying facilities, toilets and a management office be provided. Depending on the size of the development, however, all these facilities may not be desirable or necessary.

Accessibility and barrier-free access All community facilities should be designed to eliminate barriers to handicapped users. Basic provisions which should be incorporated into site design include five major areas:

Walks should be 42 in wide. If the slope of a walk is greater than a 1-in rise in a 20-in run, a handrail should be provided and the slope of a walk should not exceed a 1-in rise in a 12-in run. Walks should have a continuous common surface, uninterrupted by steps of abrupt changes in level greater than ½ in. Where walks cross driveways or parking lots, they should blend to a common level by means of curb cuts, ramps, or other means. Curb cuts should have a textured nonslip surface (such as broom-finish concrete). Walks should be provided with a level area no less than 5 by 5 ft where they terminate at doors; in no case should such walks extend less than 11 ft beyond the side from which the door opens.

Ramps should not have a slope greater than 1 ft in 12 ft and should be no less than 4 ft in clear width. If the ramp slope is greater than 5 percent and there is no drop-off, one handrail should be provided; where a ramp drops off on one or both sides, handrails should be required on both sides of the ramp. Handrails should be 32 in high measured from the surface of the ramp and should extend 1 ft beyond the top and bottom of the ramp or turn at right angles. The ramp should have a nonslip surface. Each ramp should have a level platform at the top which is at least 5 by 5 ft, and this platform should extend at least 1 ft on the side from which a door opens. Each ramp should have at least 5 ft of straight level clearance at the bottom. Continuous ramps should have 3 ft minimum long intermediate-level platforms at 30-ft intervals for purposes of rest and safety and should have level platforms wherever they turn, which should be at least as wide as the ramp and 5 ft long (deep).

Doors and doorways, exterior and interior, should have a clear opening of no less than 32 in when the door is open and should be operable by a single effort with one hand.

Outside stairs should not have abrupt (square) nosing; a 1-in rounded nosing is desirable. Stairs should have at least one continuous handrail 32 in high from the tread at the face of the riser. The handrail should extend at least 18 in beyond the top step and beyond the bottom step or should be turned at right angles. Care should be taken that the extension of the handrails is not in itself a hazard, and the extensions should be made on the side of a continuing wall where available.

Other facilities in common indoor and outdoor facilities available for the convenience of physically handicapped persons include:

- Control devices for light, power, heat, ventilation windows, draperies, doors, and similar devices
- Elevators
- Kitchen arrangements
- Swimming pool facilities
- Telephone
- Toilet rooms and toilet fixtures (including showers)
- Water fountains

Optimizing Attractiveness of Group Setting

Variation in placement of homes In many mobile home developments, homes are arranged perpendicular to streets to reduce street frontage and minimize costs. This causes long narrow corridors lined with the trailer-hitch sides of mobile homes, showing mobile homes at their worst. Homes angled toward or parallel to the street are always more attractive because entrance facades are toward the street.

Curvilinear street alignments are especially desirable over grid alignments because they decrease the inherent monotony or sameness of mobile home developments. Curved streets provide an ever-changing angle of view and the opportunity to see beyond the facades of mobile homes. Curves do not have to be sharp or long to be effective. Even gentle serpentine alignments help overcome the monotony created by rows of mobile homes.

Other more innovative lot arrangements, such as houses clustered around a parking court or open space area, are sometimes used in small to medium lot developments. They are usually used where the double loading of a through street is not possible, or in isolated corners of the development. These clusters are generally more attractive than rigid, perpendicular arrangements of homes, since they provide some sites which are parallel to the street.

Group configurations suited to specific site shapes and harmonious with topography should be used wherever possible to create siting variety and interest. Homes placed other than perpendicular to the street help create individual home and occupant identity—a desirable design and psychological asset.

Topography Development of land parcels with some topographic variation, either uniform slopes or rolling land, is aesthetically desirable for the mobile home development. The advantage of the moderately sloping site is that its development encourages clever unit siting and design solutions, and increased visual separation of units. Mobile home parks can be developed most easily on sites with 1 to 5 percent slopes. They can also be sited on slopes of 5 to 8 percent with particular care given to the placement of each unit, but with some increased earthwork costs. Slopes of 8 to 12 percent require special attention and substantially increase costs of grading and benching for the placement of streets and homes. Slopes greater than 12 percent are difficult to develop and require intense grading operations, resulting in greatly increased costs.

The unique shape of mobile homes requires a special relationship of streets and lots to topography. Because homes are long and narrow, they are turned parallel to the slope where grades are above 3 to 4 percent to avoid excessive change in grade over the length of the home. With homes turned parallel to the slope, streets normally have to run perpendicular to the slope to service this arrangement of homes. Overall, the relationship of homes and streets to the topography makes economical development of mobile home sites practical only on slopes of less than 10 percent without excessive grading.

Capacity of site to visually absorb development The density which you perceive in a mobile home development is determined not only by the number of homes per acre but also by the capacity of the site to visually absorb a certain number of units. In other words, the topography and vegetation of a site affect the number of homes in your field of view at any one time.

Rolling sites are especially desirable in this respect. They are visually closed, in that they do not expose the entire site to view from any one vantage point; the result of a site which is visually broken into smaller areas is that only a few homes are visible at one time.

Vegetation density also affects the capacity of the site to absorb development. The more vegetation which remains on the site after development, the fewer mobile homes will be visible from any vantage point. The preservation of existing trees and other vegetation, while desirable, is difficult because of the densities at which mobile home parks are typically developed. Mobile home pads, roads, and the network of utility locations in high-density developments leave little undisturbed land. Many times only hedgerows, groupings of trees at selected locations or along stream banks, and tree stands on steep sloping areas unsuitable for development are preserved.

Where large lots are involved or where particular care has been taken in locating underground utilities, existing trees may be retained among the mobile homes.

Utility placement beneath streets, with laterals run to units as in conventional subdivision design, results in higher development costs but makes possible the preservation of tree stands

between homes. The difficulty associated with preservation of existing trees and vegetation in a high-density development establishes the need for significant landscaping to supplement or replace any natural features retained in the mobile home development.

Fig. 23 Mobile home park sketch plan.

Landscaping

Landscaping of the group mobile home setting in both park and subdivision is an extremely important determinant of overall appearance. Major plantings can create visual absorption capacity and reduce the impact of extensive development by eliminating the feeling of high density and giving the development an image of permanence. Street trees should be planted along all streets or on individual lots to create a street-tree effect. Street trees within street rights-of-way or on individual lots should be planted at a substantial size (2½ to 3 in caliper) which provides a good balance of immediate effect and cost. Street trees should be planted about 40 to 50 ft apart. If an even spacing is used, they may also be clustered at intervals along the road to create an alternating sequence of open and planted areas along the drive.

Street trees should be selected to assure that they will not create a hazard or nuisance to the streets, walks, or lots. Species of trees to be avoided are shallow-rooted or weak-wooded ones, those that shed fruits and large seed pods, trees with dangerous thorns, and especially those attractive to insects.

Coordination of Signs, Lighting, and Other Elements

Mobile home developments can sometimes benefit from a group identity established by repetition of a design element. For example, a development containing all homes with wooden siding, or peaked, shingled roofs, will benefit from having attractive homes which harmonize throughout the development.

Other design elements such as signs, lighting systems, and group mail receptacles can be tied into a coordinated design scheme by using similar materials, colors, or designs throughout the development. For example, a total wooden scheme for site furniture including all entrance and directional signs, street and walkway light standards, guardrails, mail receptacles, and benches can help give the development a desirable overall identity and help unify the total development design.

Caution should be exercised in creating a "design theme" because the repetition of a bad design element is worse than no design coordination. Occasionally, mobile home developers will develop a rather weak common design feature which ties the development together with a

group identity. The unifying elements should be subtle, not gaudy, and should retain a dignity worthy of a residential area.

Miscellaneous

Unique natural features such as lake frontage, ponds, streams, and rock outcrops provide interest and opportunities for extraordinary design solutions. Their value in creating attractive living environments is often demonstrated by their portrayal in the promotional material advertising a particular mobile home development. These features can be beautiful, positive landscape elements, but only if they are properly improved and maintained. These types of natural features can easily become safety liabilities and nuisances if not properly located, designed, and cared for.

Curbs, in addition to facilitating surface water drainage, can affect the appearance of mobile home developments; they give an image of quality and permanence to the streets, portray a clean, crisp edge, and help maintain nice lawns. If curb is low height, rolled curbs are the most practical.

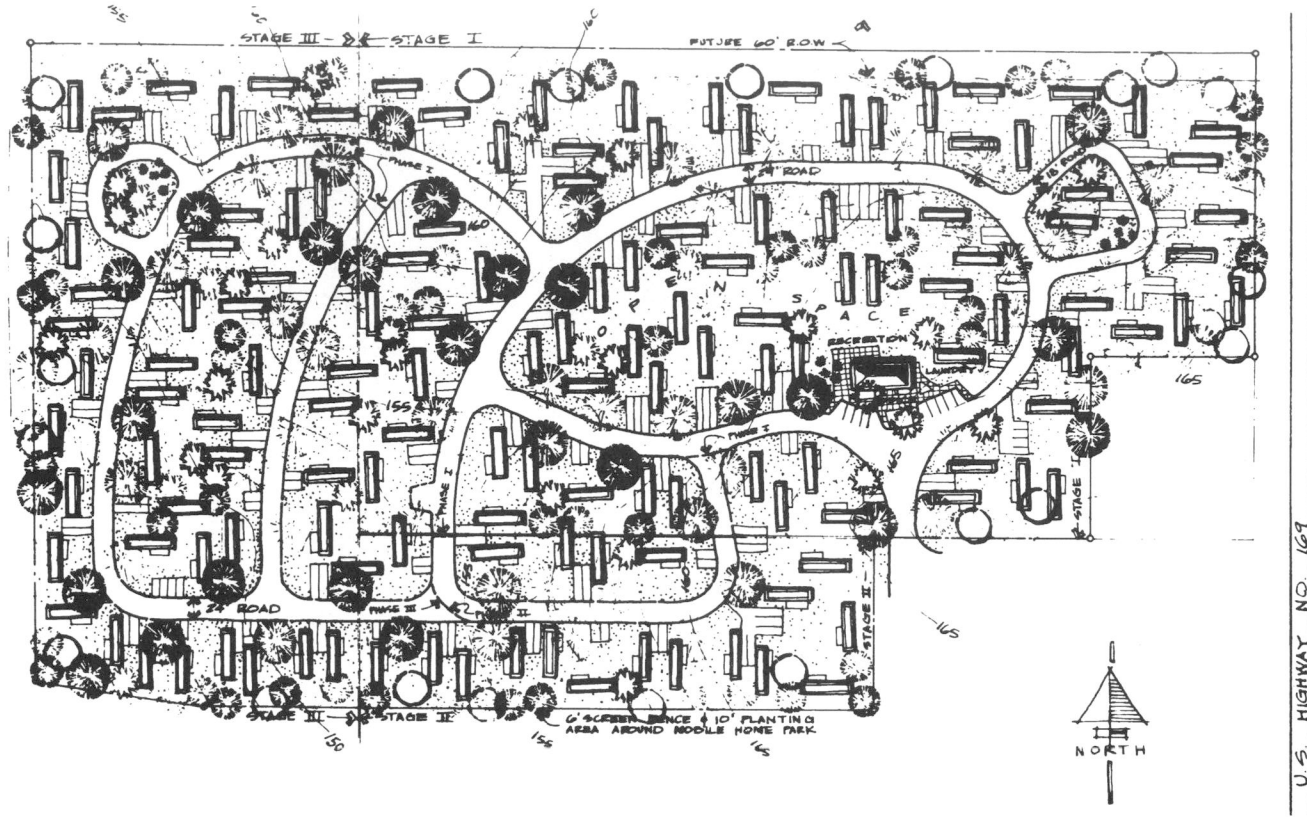

Fig. 24   Mobile home park sketch plan.

Lawns are the lowest-cost, most maintenance-free and effective ground cover for large areas and provide the most acceptable setting for the mobile home. Lawn quality has a great impact on the attractiveness of the lot. Developments where lawns are well maintained are much more orderly and clean looking than those where lawns and weeds are unmowed or only sparse lawn is evident.

Organizing Uses

Use area separation Community facilities should be separated from living areas but be accessible to all residents. Additional spacing between active community facilities and living areas should be provided; and landscape buffers, earth mounds, or screens should be considered in preparing the plan.

Entrances Entrance drives should focus on community buildings, common facilities, or natural areas rather than on living areas and should establish a theme for a residential living environment.

Convenience of community facilities Community facilities should be pedestrian-oriented facilities, separated from streets and parking areas and centrally located within the development to facilitate accessibility, unless unique natural features dictate another location.

Separate off-street parking areas should be provided for the community facilities in relationship to size of park; normally not more then one parking space for each 10 dwelling units is

required in order to encourage pedestrian orientation. Development of the common open space should be planned so that natural amenities of the site can be preserved and accentuated. Natural site drainage patterns can dictate the paths of lineal open space features or areas which may not be logically suited for the siting of homes.

The most intensive community facility area should have the greatest separation from residential living areas. Where outstanding natural features such as streams, trees, rock outcroppings, and topographic variety can be incorporated into the open space, they should become an important part of the development plan. Mature trees in community facility areas greatly enhance the character of these common areas and the total mobile home development.

Pedestrian pathways serve an important function in mobile home developments. Since the role of community facilities and open space may be emphasized, a safe, convenient pedestrian link from private lots is desirable. Paths should be located to connect logical points of origin and destination. Individual lots should have access to paths leading to recreation facilities, convenience commercial areas, laundry facilities, bus stops, and community buildings. Paths should be separated from streets, where possible, by locating them within a lineal open-space corridor to the rear of homes or within a planted buffer strip between the walk and street where parking is in courts. If this is not possible it may be more desirable to locate walks ahead of the parking where it is immediately adjacent to the street rather than along the street.

Circulation—hierarchy of streets Streets within mobile home developments should be grouped into four functional categories (see Fig. 25): (1) courts, places, cul-de-sacs, (2) local streets, (3) subcollectors, and (4) collectors.

Courts, places, or cul-de-sacs are very minor residential streets, the primary purpose of which is to serve individual lots and provide access to local or higher forms of streets. A place may be a dead-end, cul-de-sac street, or court with no through traffic and with limited on-street parking. Local streets are generally short and may have cul-de-sacs, courts, or occasionally two or three branching places. Usually no through traffic should exist between two streets of a higher classification. The purpose of a local street is to connect traffic to and from dwelling units to subcollectors.

Subcollectors provide access to local streets and courts, places, or cul-de-sacs and conduct this traffic to an activity center or to a collector street. A subcollector may be a loop street connecting one collector or outside arterial street at two points, or conducting traffic between collector streets or arterial streets.

Collectors are the principal traffic arteries within residential areas and carry fairly high traffic volumes. They function to conduct traffic to or between major arterial streets outside the residential areas. A well-planned neighborhood maximizes the number of mobile homes located on either local streets or courts, places, and cul-de-sacs where there is no through traffic between streets of higher classification. Homes having direct access to subcollectors

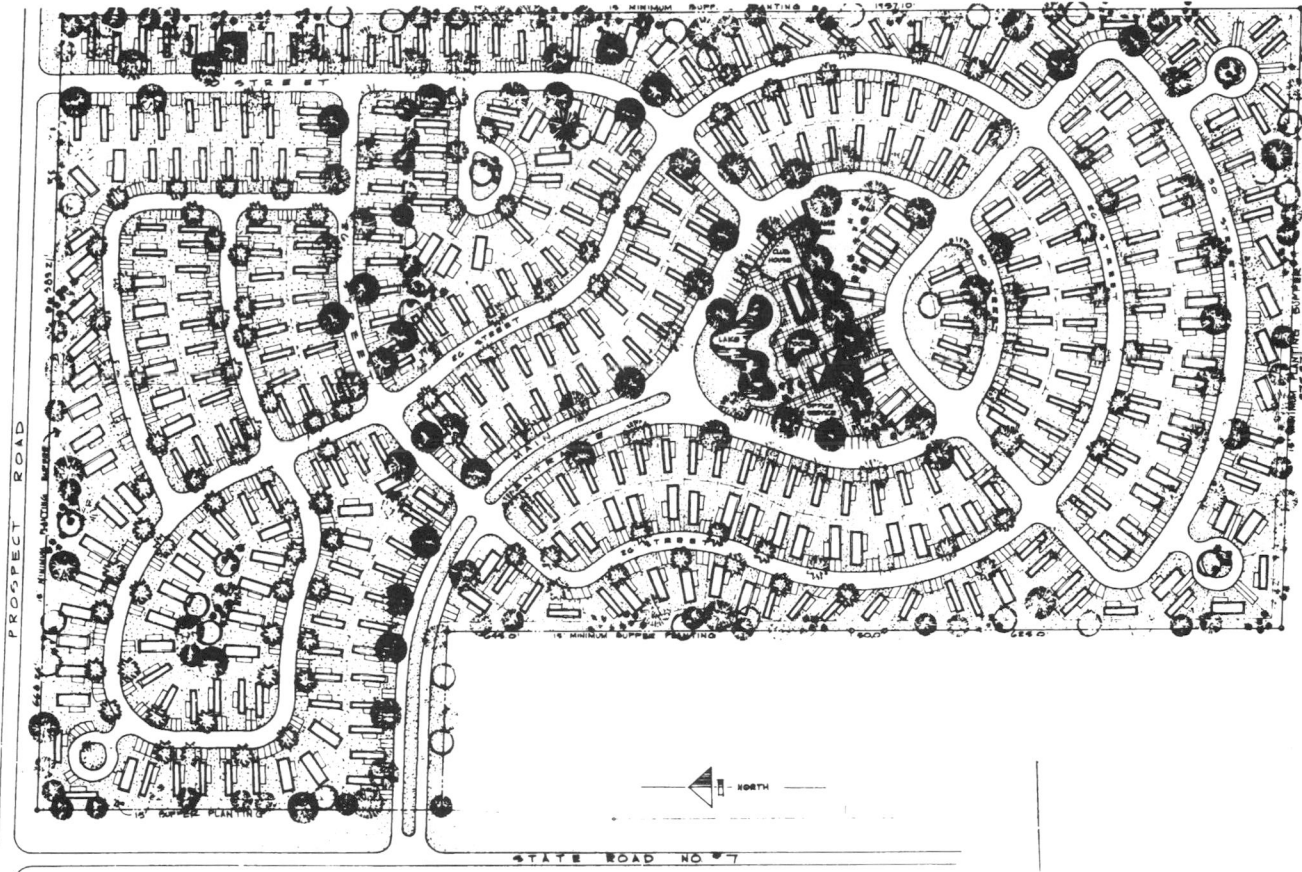

Fig. 25 Mobile home park sketch plan.

and collectors are allowable, but should be minimized. Local streets and courts are safe and desirable places to live; living areas are dominant and traffic movement is subordinate.

Only a single moving traffic lane is necessary on local streets or courts, while subcollectors and collectors should have two moving lanes. Frontage of living areas on entrance roads and collector streets should be minimized.

Providing Community Facilities

The need for community facilities is related to the density of the development. Community facilities are especially important in small lot developments where private outdoor space is limited; they are somewhat less critical where lots are large enough to allow many activities in individual yards. At higher densities, community open space can compensate for small private exterior living space.

Tot lots and areas for children to play away from the mobile homes are especially necessary to minimize disturbance of the individual residents' outdoor living areas. In developments where lots are greater than 10,000 ft^2, there is less dependence upon community space; but playground and park areas for large-scale activities are desirable as in any residential area.

Community areas should have a parklike atmosphere compatible with residential living environments. Community buildings and structures should also be designed in a manner compatible with a residential living environment rather than a commercial development.

Public outdoor open space commonly consists of two types: "structured" and "unstruc-

tured" facilities. Structured facilities include formal playgrounds, golf courses, shuffleboard courts, tennis courts, swimming pools, and related facilities. Structured facilities are normally developed in a complex with a community building. Equally as important are the unstructured public open spaces, which can be as simple as open grass areas for spontaneous team games and other activities. The type of community facilities necessary for any mobile home development is determined by the occupants to be served. A family-oriented development may require more extensive outdoor open space for active recreation, whereas a retirement community may require less space but a greater variation of activity areas. Community facilities commonly include such things as swimming pools, community buildings, vehicle storage areas, pedestrian paths, tot lots, and court games.

Community buildings A community facility which is common to most new mobile home parks is the community building. It contains more than one activity and serves more than one function. Uses commonly built into a community building include laundry facilities, meeting rooms, recreation rooms, and in the case of family-oriented parks, day-care centers. The building is normally constructed as part of a complex including structured outdoor recreation facilities, such as swimming pools and limited off-street parking. Community buildings should be designed with a residential character harmonious with the mobile home development. When constructed and managed properly, the community building can be a major asset to the mobile home environment. The key

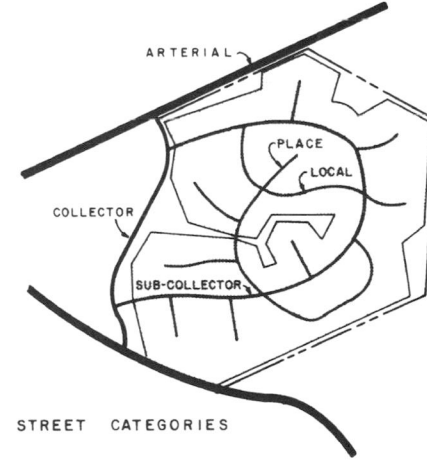

STREET CATEGORIES

Fig. 26

word becomes "management," for after the mobile home development is established and the community building is constructed, it is the responsibility of the homeowners' association in subdivisions, and the operator in parks, to maintain the structure and operate the activity programs.

A secondary, but very important, function of the community building is that of storm shelter for mobile home residents. In areas of the country where dangerous storms might occur, a structurally adequate community building of ample capacity must be provided for the residents' safety.

The Mobile Home Manufacturers Association recommends that approximately 10 to 15 ft^2 of floor area per mobile home unit should be provided.

Common vehicular storage area Much of the clutter and disarray in mobile home parks is due to the lack of a defined storage area for seldom used vehicles or recreational vehicles. Provisions for storage of these vehicles should be included in the mobile home development, especially where lots are small. In many mobile home parks, residents have more leisure time than their conventional housing counterparts; recreational equipment, snowmobiles, boats, and travel trailers are sometimes abundant. Recreational vehicles generally take up too much space to be stored on each individual home site. Common areas accessible to all residents of the development are necessary to store such vehicles or equipment. The storage area should be separated from the living areas of the site and should be a gravel or hard surface area enclosed by a security fence and adequately screened from sight. At least one storage space should be provided for every 19 mobile homes.

Swimming pools Swimming pools do much to enhance the image of a mobile home community. In fact, most high-quality mobile home parks include a swimming pool or some equivalent structured recreational facility. The generally isolated location of mobile home developments suggests that such a facility is desirable, especially under certain climatic conditions and for specific segments of the mobile home market. Swimming pools are usually located near a community building and other structured facilities, and should be designed to accommodate the anticipated usership without undue crowding. An estimate of participation rate during typical summer weekends provides the basis for determining an appropriate pool size. This rate of participation varies with the expected population characteristics of the development. Approximately one-quarter of the persons at the pool will be in the water at any one time, and the pool should be designed to provide 10 to 15 ft^2 of water surface for each wader and 30 ft^2 for each swimmer. Deck area equal to or larger than the pool surface area should be provided. Most participants also desire a large, fenced-in turf area of equal size for sunbathing.

A general rule of thumb for estimating required pool area is to provide 3 ft^2 of pool surface for each mobile home lot. (This standard assumes two potential participants per home, 20 percent participation rate, 25 percent of actual participants in pool at any one time, and 30 ft^2 of surface per swimmer.)

Tot lots and playgrounds Tot lots are small playgrounds consisting of several pieces of play apparatus, swings, or climbing equipment provided especially for use by young children. They should be located close to the homes which

they serve or within the community recreation area where they can be easily observed and supervised. Ideally, a small tot lot could be established for each grouping of homesites so that children could use them without crossing collector streets in the development. Tot lots also work well, when located adjacent to adult recreation areas so that children may be observed by adults using other facilities.

Playgrounds are somewhat larger in scale than tot lots and are normally oriented to elementary-school-aged children. They should have safe apparatus which provide opportunities for children to use a variety of motor skills. Such equipment can include tire or other flexible seat swings, seesaw with tire safety stops, climbing arches or other apparatus on "soft" surface, and splinter-free climbing blocks.

Court games Basketball and tennis courts are popular facilities for adult recreation. They can often be incorporated into a centralized recreation clubhouse complex where they are easily accessible via streets and pedestrian paths.

Both facilities require much space, serve a limited number of people at any one time, and can benefit from night lighting which increases the number of people which can be served.

General court games
■ Provide a variety of facilities to serve various age groups including:

   Basketball courts (hard surface)—50 users per half court, daily capacity

   Volleyball (in lawn area)—72 users per court, daily capacity

   Shuffleboard (hard surface)—20 users per court, daily capacity

■ Lighting for night use of court areas is desirable and will increase daily capacity by 20 to 30 percent.

Tennis courts
■ Provide a fenced, low-maintenance, all-weather (hard-surface) court.

■ General capacity is 20 participants per day per court.

■ Lighting for night use is desirable and will increase capacity by 40 percent.

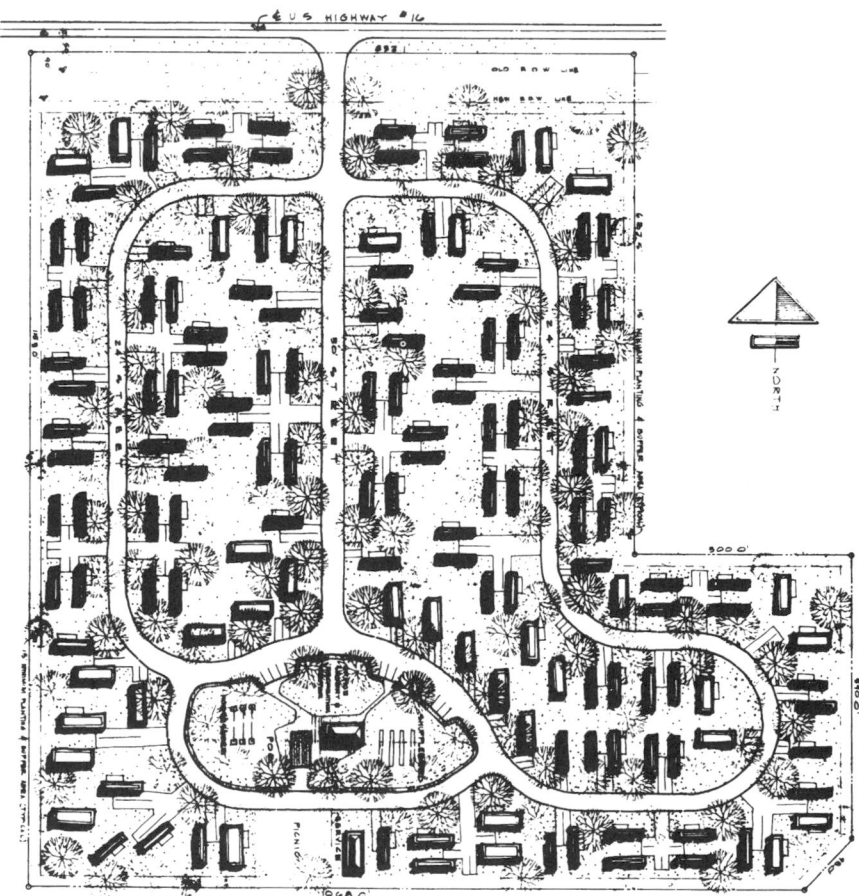

Fig. 27

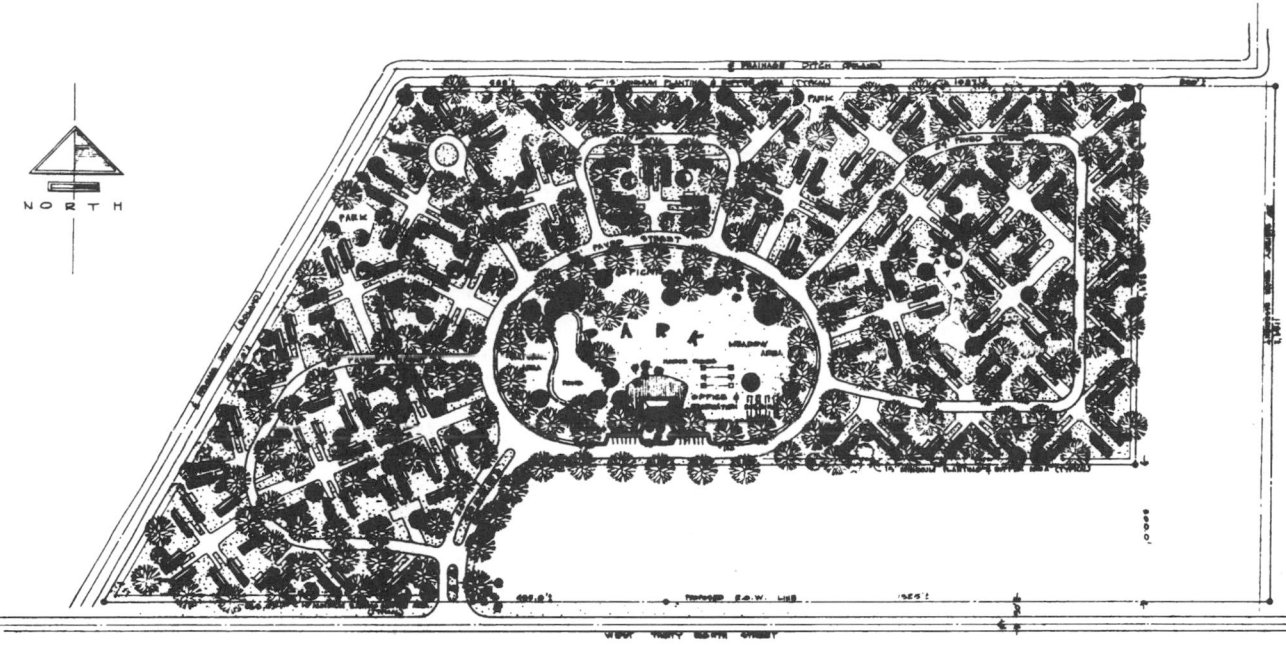

Fig. 28   Mobile home park sketch plan.

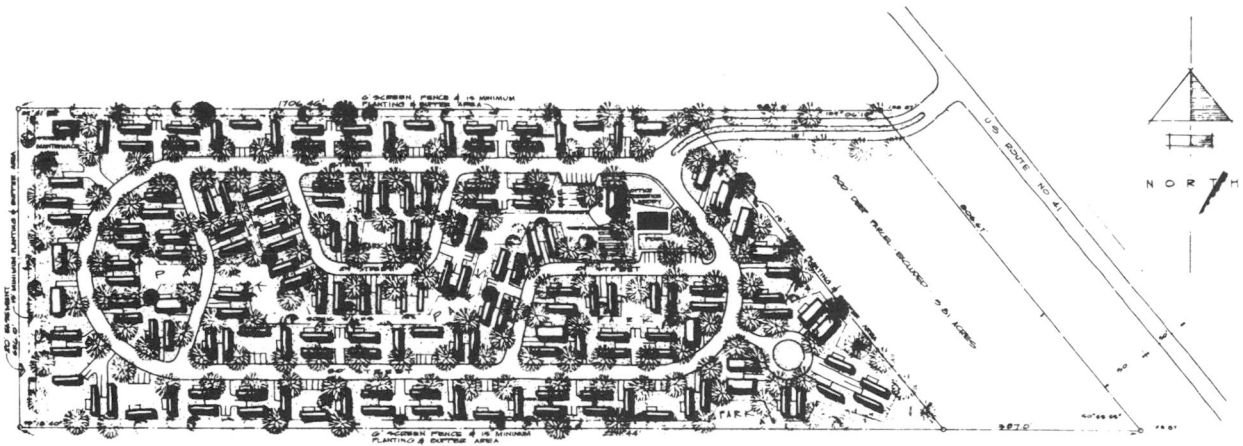

Fig. 29   Mobile home park sketch plan.

MOBILE HOME STAND

General

A stand should be provided on every mobile home lot to accommodate the mobile home and its attached accessory structures. The stand should provide an adequate foundation and anchoring facilities to secure the mobile home against any accidental movement.

The stand should not heave, shift, or settle unevenly under the weight of the mobile home as a result of any frost action, poor drainage, vibration, or other such forces.

Because the mobile homes now produced vary in size and shape, the stands should be individually designed to fit the dimensions of the mobile homes that will be accommodated. Consideration should also be given to the fact that many mobile home owners may later want to add carports or other accessory structures. If future additions are anticipated, the stand should be so located on the mobile home lot that the proper clearances can be maintained between the mobile home and other structures.

Patios are frequently constructed as an integral part of the mobile home stand. The patio area provides useful outdoor living space for mobile home occupants and can also be utilized for future additions to or expansions of the mobile home. It is recommended that patios have a minimum size of about 180 ft². Often the construction of the patio is delayed until after the mobile home is placed in order to best fit the patio to the design of the mobile home.

Location of Service Connections

Individual connections should be provided at each mobile home stand for water, sewerage, electricity, telephone, gas, and other services.

WATER SUPPLY

General

An adequate safe supply of water under pressure should be provided to each mobile home

lot. The source and distribution system should be satisfactorily constructed and approved by the health authority having jurisdiction.

Sewage Treatment

If possible, the sewerage system of a mobile home park should be connected to a public sewerage system. If public sewers are not available within a reasonable distance of the mobile home park, adequate treatment facilities must be installed to dispose of the sewage. Where the sewerage system of the mobile park is not connected to a public sewer, any proposed sewage disposal facilities should be approved by the health authority prior to construction. The effluent from such treatment facilities should not be discharged into any body of water without approval.

DIMENSIONS OF MOBILE HOME STAND

To accommodate modern mobile homes, the stand where the home is placed should be 10 by 50 ft.

From the mobile home stand to the stand line on the opposite side of street should be 60 ft minimum.

To a common parking area, roadway, or walk, 20 ft is typical, 10 ft minimum.

The distance to a public highway or major street should be 50 ft with protective screening.

To the other boundary of the mobile home court the distance should be 25 ft if adjoining uses are compatible; otherwise, 50 ft with protective screening.

The sum of the side yards at the entry side and nonentry side of the mobile home stand should be 32 ft minimum with at least 15 ft on the entry side and 5 ft on nonentry side.

REFUSE HANDLING

General

Public health problems are often associated with improper storage, collection, and disposal of refuse. There are significant relationships between the incidence of certain diseases in humans and animals and improper refuse handling. It is also common knowledge that many hazards and nuisances, such as fire, smoke, odors, and unsightliness, are created by poor refuse handling practices. Experience has shown that application of the basic principles of sanitation to refuse handling results in substantial reductions in insects, rodents, and related health problems.

Notes The design criterion for mobile home stands is 14 by 60 ft. (Most states allow 14-ft-wide movement for mobile homes.) Most double-wide units are 24 ft wide.

Use landing mats or similar means for easy and accurate placement of mobile homes on mobile home stands.

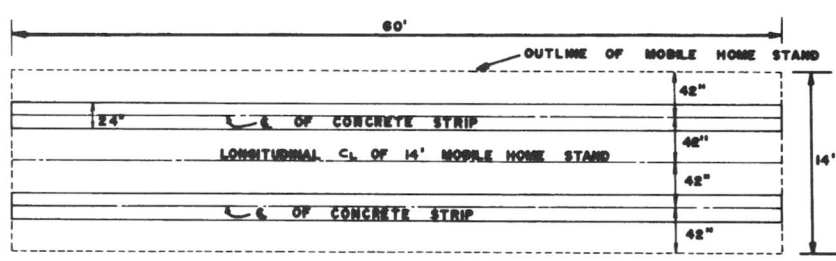

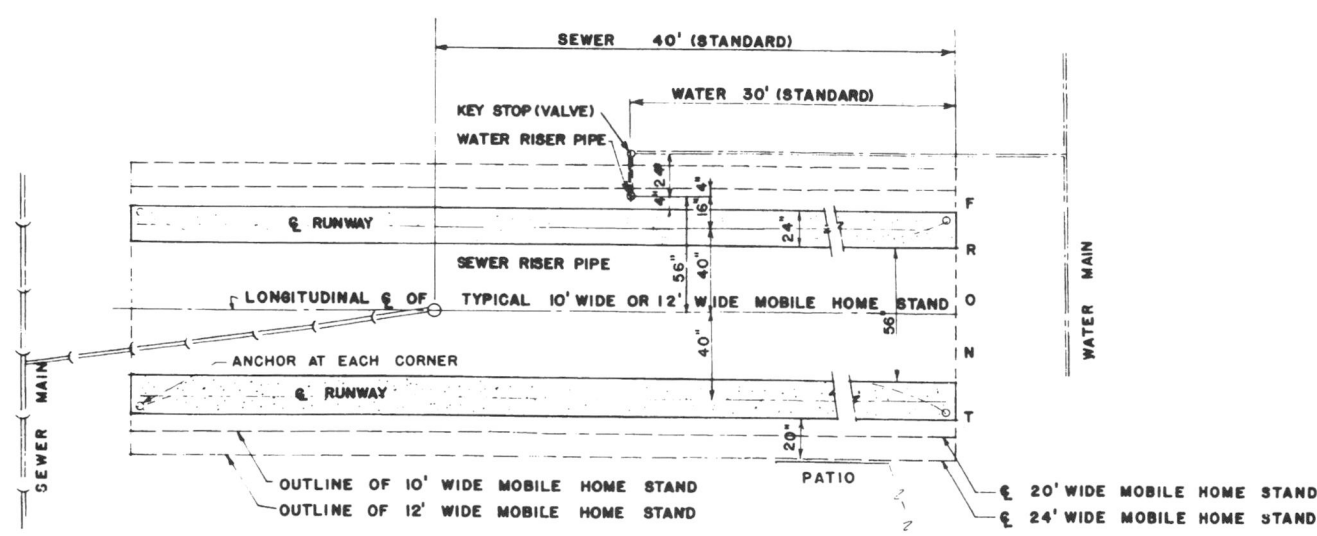

Fig. 1   Location of water and sewer riser pipes.

Site considerations and placement of the mobile home on the site must suit the buyer's needs.

A fully accessible mobile home placed on a site so that it is difficult or impossible for a disabled person to enter defeats the goal of independent living.

Offered here are some observations and recommendations for ways of placing the mobile home on the site to best accommodate the disabled user.

Mobile homes resting upon their wheels have floor levels approximately 3 ft 0 into 3 ft 6 in (91.5 to 107 cm) above ground. If they are raised and placed upon pier supports so the wheels and/or axles are clear of the ground, their floor levels may be from 3 ft 6 in to 4 ft 0 in (107 to 122 cm) or more above ground. The wheelchair user will require a ramp or other means of gaining access to the mobile home.

The maximum slope for a ramp for severely disabled wheelchair users should not exceed 1 in 20—1 unit of rise for every 20 units of run. At this slope a 4 ft 0 in (122 cm) floor height would require an 80 ft 0 in (24 m) long ramp. Moderately disabled wheelchair users may be able to negotiate a ramp slope of 1 in 15. This slope would require a 60 ft 0 in (18 m) long ramp.

Clearly it would help to reduce the distance from ground to floor level and thereby shorten the length of the required ramp. Careful planning and site positioning of the mobile home may eliminate the need for ramps altogether. The problem of access to the mobile home has several solutions.

HILLSIDE PLACEMENT

Horizontal floor-level access to the unit is most desirable for everyone and could be accomplished on a hillside site by placing the unit parallel to the hill and constructing a bridge from the unit to the ground level on the uphill side. Another method would be building a retaining wall and using earth fill.

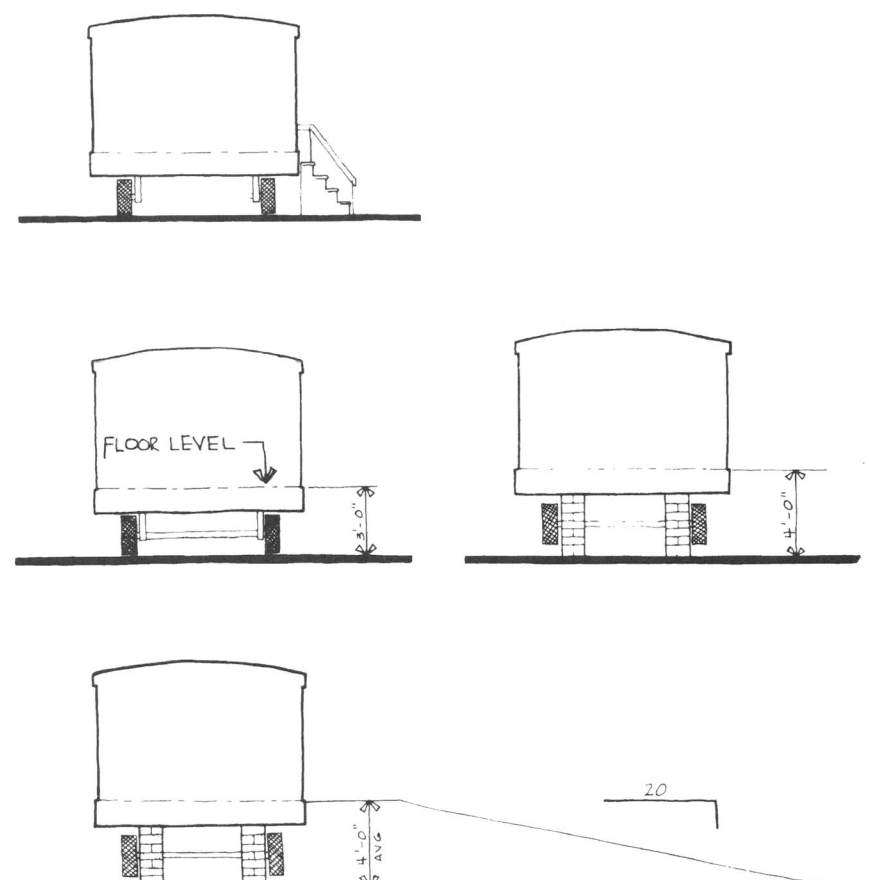

Fig. 1

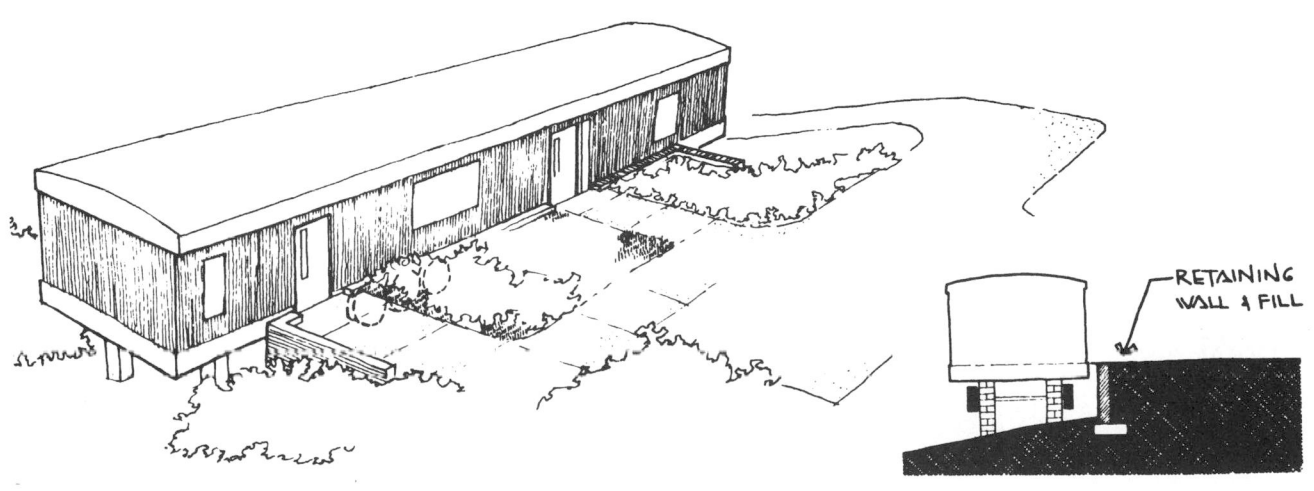

Fig. 2   Earth fill.

HANDICAPPED HOUSING

Fig. 3 Bridges.

EXCAVATION

On flat sites a depression might be excavated to accommodate the wheels and axles of the unit. If this solution is employed, a storm drain must be installed to prevent flooding of the pit, and careful attention must be paid to the elevation of plumbing drains and sewer system at the location to ensure gravity flow of sewage.

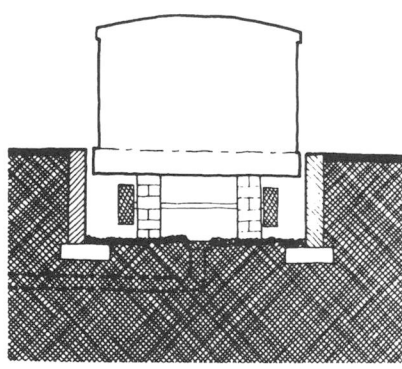

Fig. 4

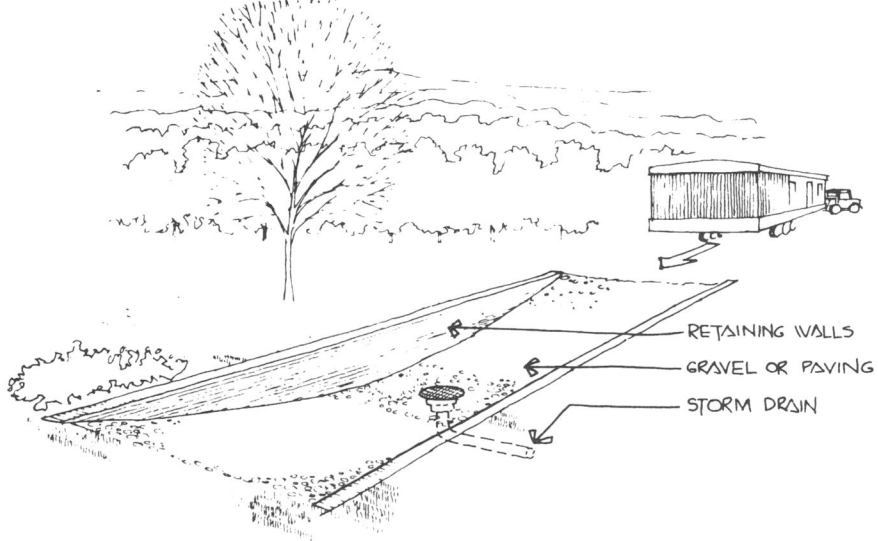

RETAINING WALLS
GRAVEL OR PAVING
STORM DRAIN

Fig. 5

This solution has the added advantage of providing ground-level access to all doors and emergency exits. Also placing the unit in the ground would eliminate the need for foundation enclosure on two or three sides and would help offset costs of the additional site work.

In areas where no excavation is possible, and where hillside sites are not available, ramps may be necessary.

Ramps should meet the following criteria:
1. Slopes must not exceed 1 in 15.
2. Ramps should be at least 4 ft 0 in (122 cm) wide.
3. Handrails on both sides must be 2 ft 8 in (81 cm) high and have rails which are easily grasped.
4. Handrails should extend horizontally at least 1 ft 0 in (30.5 cm) beyond the top and bottom of the ramp.
5. There should be large flat platforms at the top and a large flat paved area at the bottom of ramps.
6. Long ramps should be broken with flat platform rest areas at 30-ft (9-m) intervals.
7. Ramps may turn or switch back, and they should have full-width level platforms wherever they change direction.

8. Ramps and platforms may be built of any durable weather-resistant material including wood, concrete, or brick. Ramp surfaces should be of a slip-resistant material or texture such as epoxy-sand, carborundum grit, or brushed concrete, and in cold climates the ramp surface should be protected from snow buildup or have built-in snow-melting devices.

It is recommended that steps be used in conjunction with ramps for the convenience of able-bodied people and those who find ramps difficult to negotiate.

LIFTS

Small electrically powered mechanical wheelchair lifts may also be used to provide floor-level access to mobile homes where none of the above methods is feasible.

Lifts of this type are commercially available and are designed for permanent outdoor installation. They require the construction of a large level landing platform at the door and an electric circuit for power. They are therefore not usable in the event of power failure. The lifts can be key-operated to prevent misuse.

Some disabled people who may not be able to use a ramp independently may be able to use a wheelchair lift.

CIRCULATION

For wheelchair circulation, corridors, halls, and all doors must be a minimum of 3 ft 0 on (91.5 cm) clear width. Most rooms have a diameter of at least 5 ft 0 in (152.5 cm) in which a person in a wheelchair can make a 36-° turn.

KITCHENS

Food Flow

Some severely disabled people are unable to pick up a bowl or pot full of food and carry it across a room while at the same time maneuvering their crutches, wheelchair, or other support device. The kitchen in the demonstration unit is arranged in an extended U plan. The sequence of the work areas and appliances is designed to facilitate food preparation. All counters and surfaces provide a continuous, common level surface from food-preparation area to the table. Food containers and other itmes can remain on the counter surface while being moved from place to place and onto the table.

Fig. 6

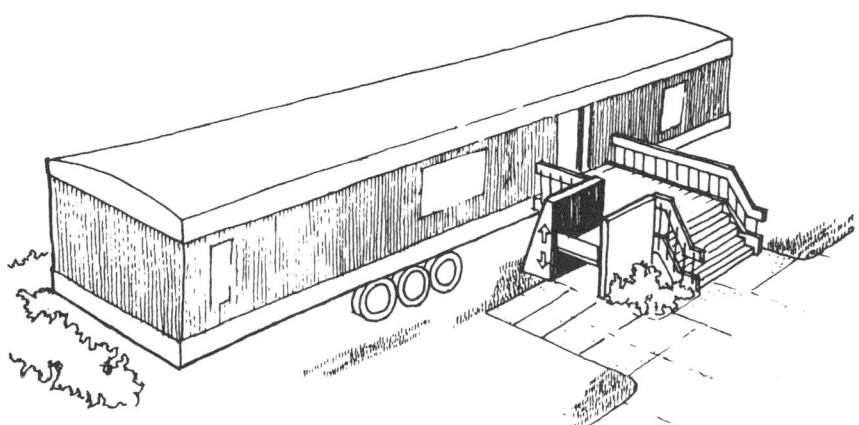

Fig. 7

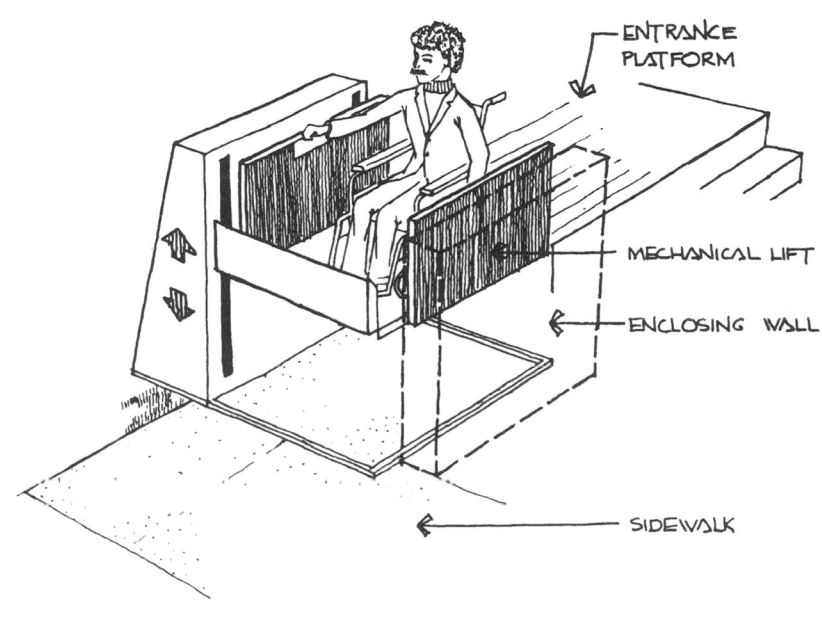

Fig. 8

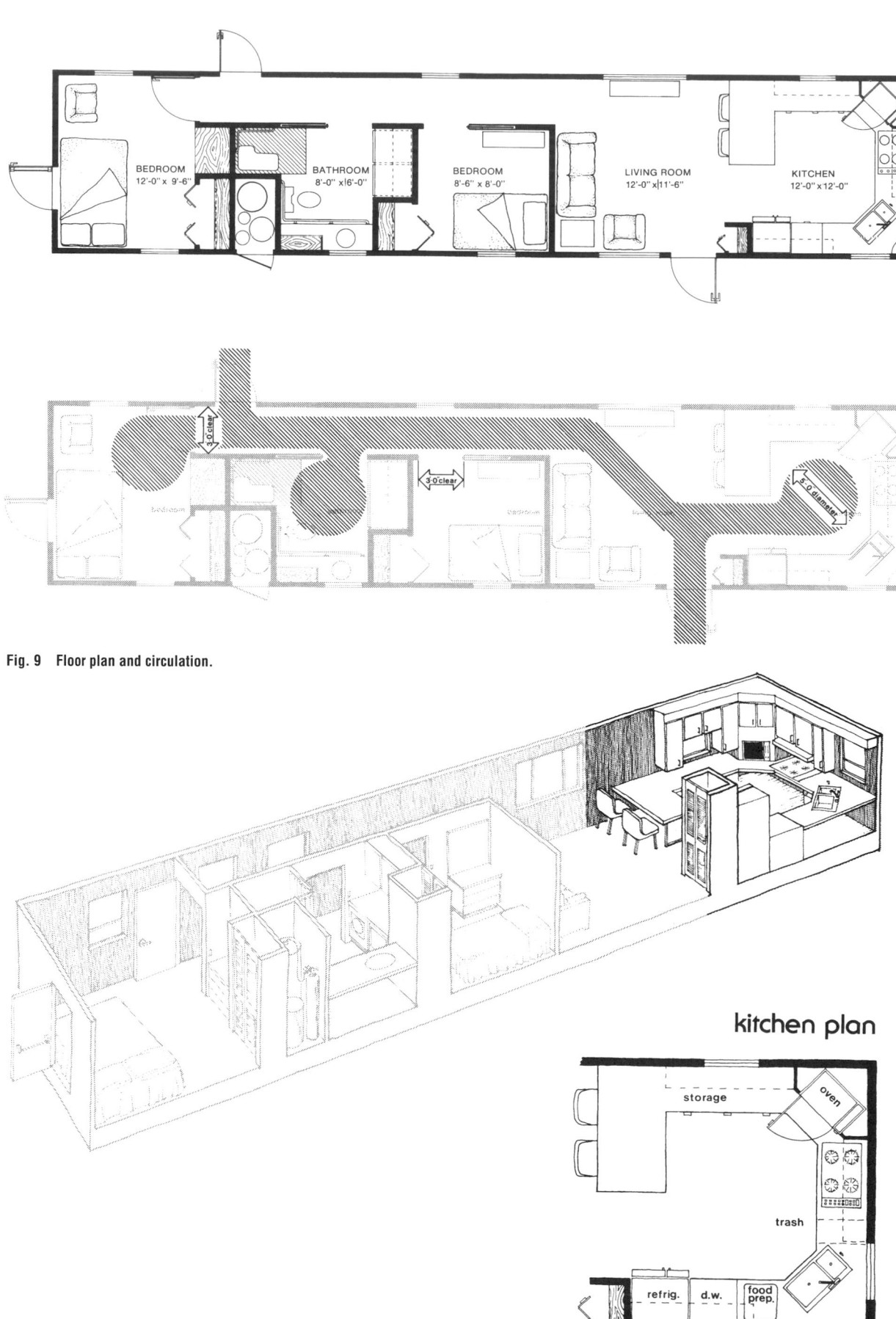

BEDROOM
12'-0" x 9'-6"

BATHROOM
8'-0" x 6'-0"

BEDROOM
8'-6" x 8'-0"

LIVING ROOM
12'-0" x 11'-6"

KITCHEN
12'-0" x 12'-0"

3'-0"clear

3'-0"clear

5'-0 diameter

Fig. 9 Floor plan and circulation.

kitchen plan

storage

oven

trash

refrig.   d.w.   food prep.

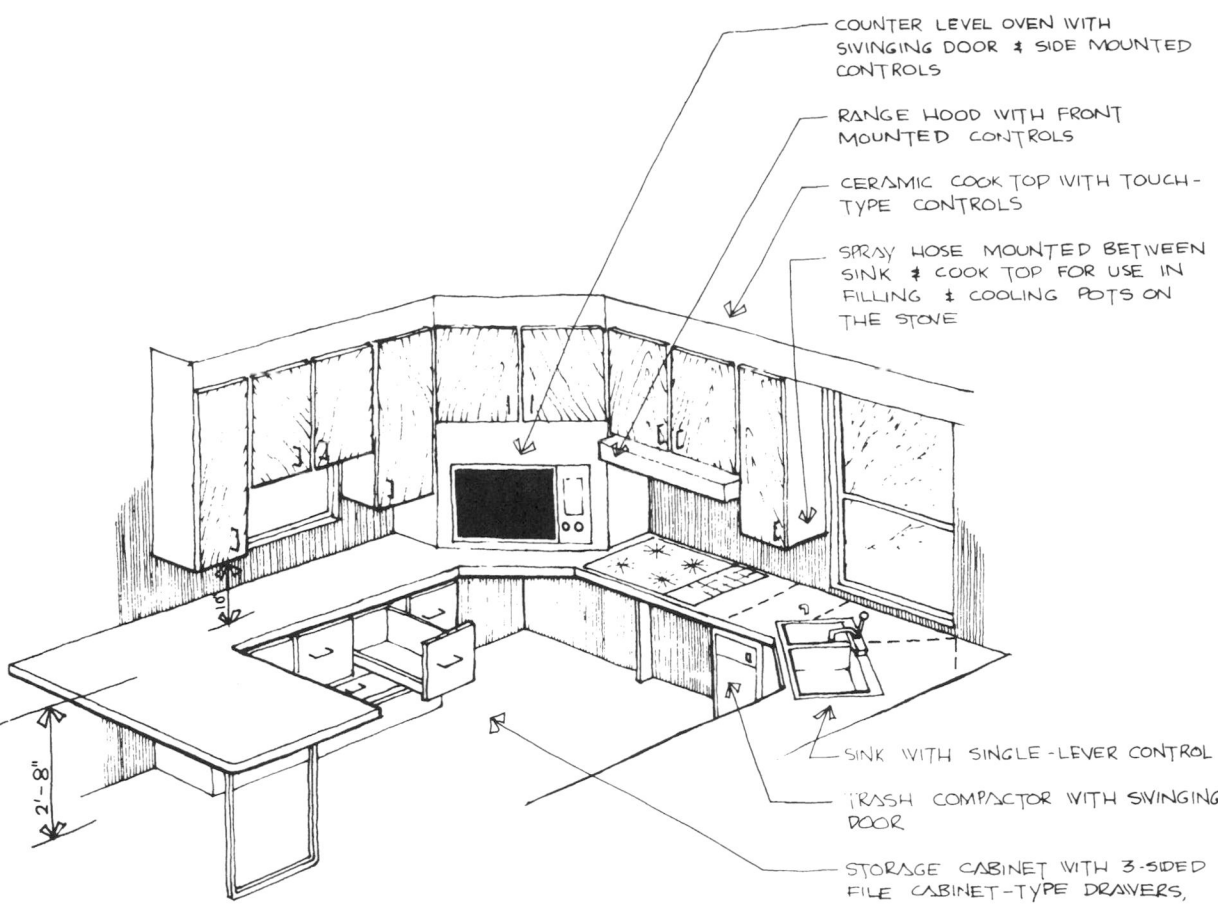

COUNTER LEVEL OVEN WITH SWINGING DOOR & SIDE MOUNTED CONTROLS

RANGE HOOD WITH FRONT MOUNTED CONTROLS

CERAMIC COOK TOP WITH TOUCH-TYPE CONTROLS

SPRAY HOSE MOUNTED BETWEEN SINK & COOK TOP FOR USE IN FILLING & COOLING POTS ON THE STOVE

2'-8"

SINK WITH SINGLE-LEVER CONTROL

TRASH COMPACTOR WITH SWINGING DOOR

STORAGE CABINET WITH 3-SIDED FILE CABINET-TYPE DRAWERS,

Fig. 11 Kitchen interior.

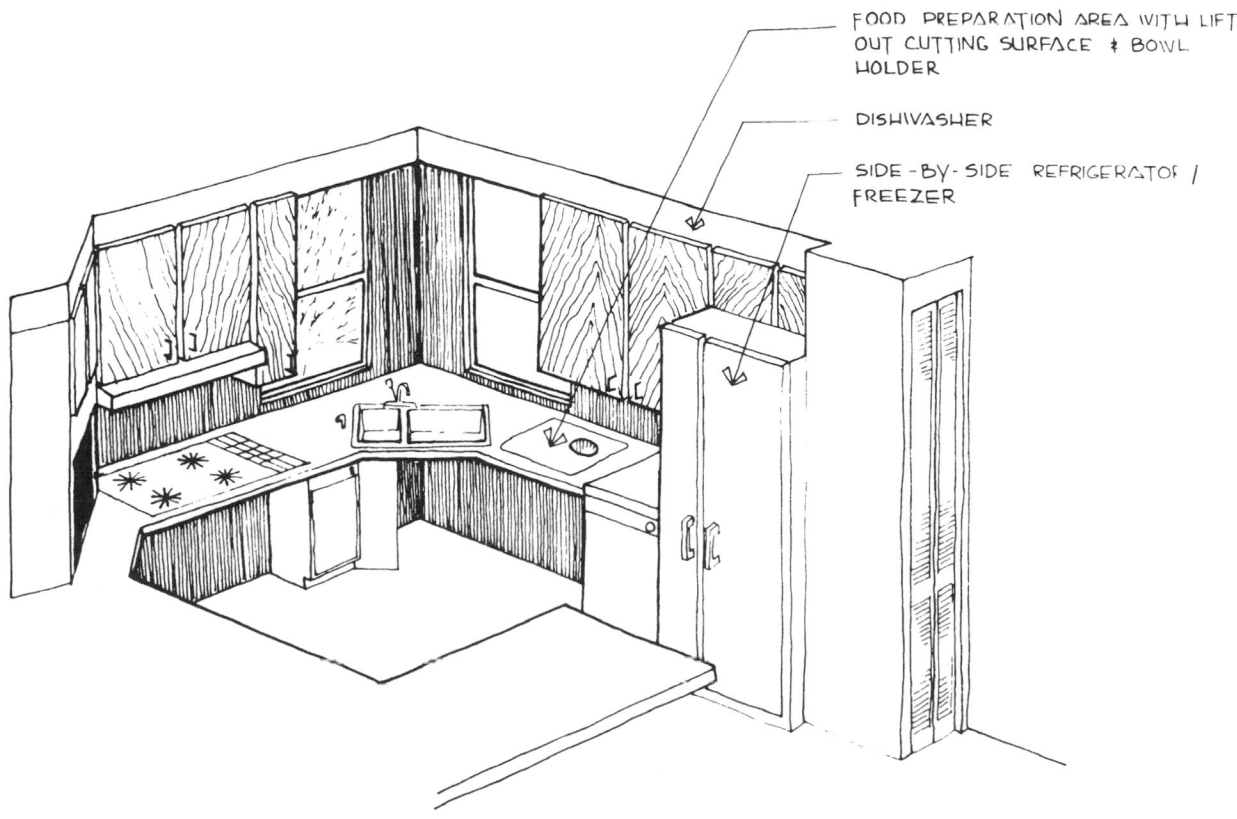

FOOD PREPARATION AREA WITH LIFT OUT CUTTING SURFACE & BOWL HOLDER

DISHWASHER

SIDE-BY-SIDE REFRIGERATOR / FREEZER

Fig. 12 Kitchen interior.

HANDICAPPED HOUSING

Some base cabinet space must be left open to provide necessary knee space below counters to make it possible for wheelchair users to maneuver close to the counters while moving containers from point to point.

Refrigerator/Freezer

The refrigerator/freezer that best suits most people is a vertical side-by-side type which provides freezer and refrigerator spaces at all reach levels.

The refrigerator is placed so that doors can swing back 180° to allow close parallel approach and to provide easy access to door racks by people in wheelchairs.

Sink

Corner position The sink is located in the corner so that more counter space is within easy reach of a person in a fixed position in front of the sink. This location reduces required movements and facilitates food flow from preparation area to stove. The knee space required for seated people uses corner space usually allocated to inefficient storage.

Spray hose The spray hose is located between the sink and cook top where it can be used for rinsing at the sink and for filling pots while they are on the stove. This is necessary for people with limited arm or hand use, who either may not be able to move pots full of water or for whom moving pots of boiling water may be especially hazardous. Cold water added to boiling pots will lower the temperature to a safe level before they are moved to the sink for draining.

Food Preparation/Dishwasher

A food preparation/work area with knee space for seated people is provided between the sink and dishwasher. This location plces the sink within easy reach of food-preparation area and also allows easy transfer of dishes from sink to dishwasher.

A standard front-opening dishwasher with roll-out baskets and front controls is adequate. The food-preparation area has a lift-out cutting board which can be replaced with a bowl holder for those people who have difficulty holding bowls in place while mixing. The bowl holder consists of a ¾-in (2-cm) plywood panel with openings to fit bowls or plans of various sizes and shapes. A foam-rubber strip at the lip of each opening prevents bowls from spinning during mixing and makes one-hand mixing possible. Additional panels can be easily fabricated to meet specific user needs.

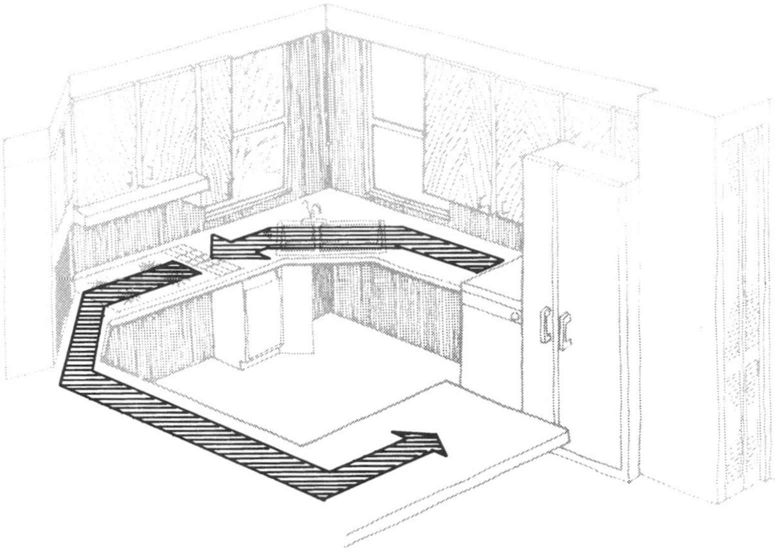

Fig. 13 Flood flow diagram.

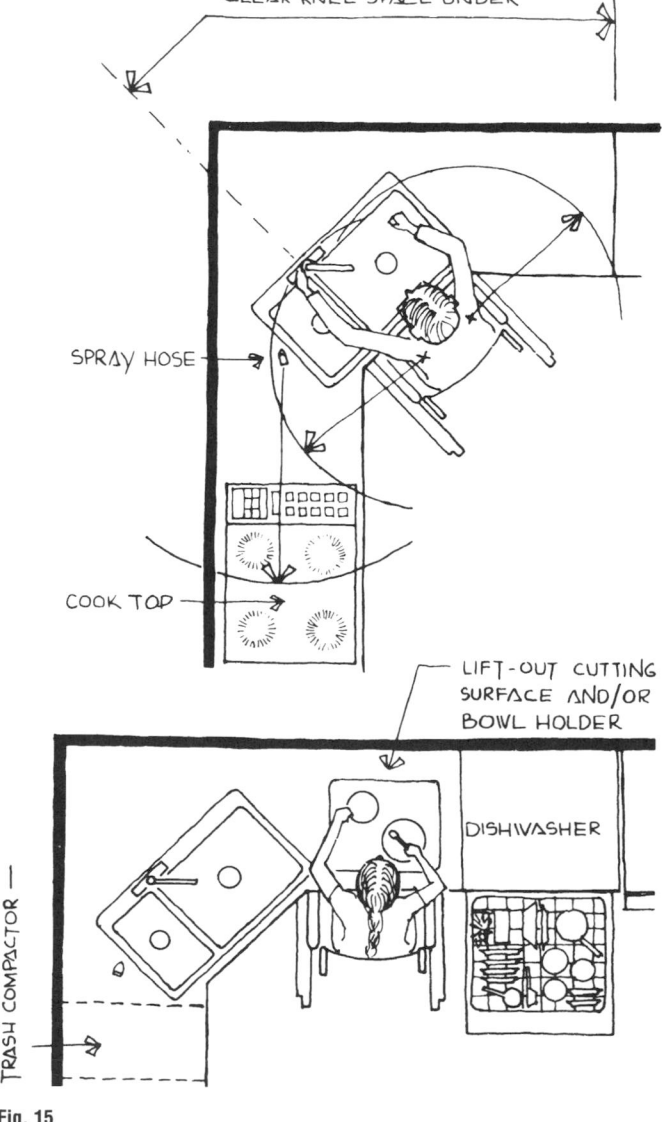

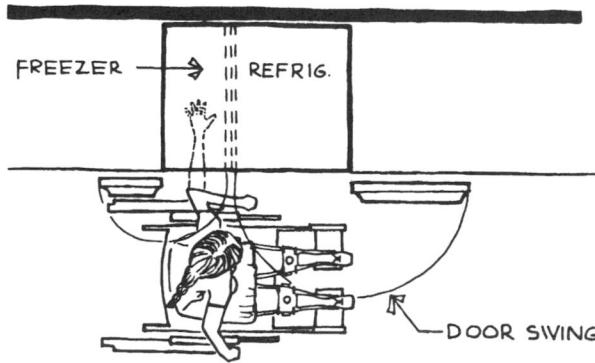

Fig. 14

Fig. 15

Waste Removal

Disposal unit A standard electric disposal unit is placed in the smaller side of a double sink. The sink is offset, and the disposal unit and plumbing are enclosed in a cabinet to prevent interference with wheelchair knee space.

Trash compactor The compactor included in Fig. 16 is a standard model with a swinging door. Many severely disabled people have difficulty carrying trash or garbage to outdoor containers. A trash compactor can reduce the bulk and the need for frequent trips. However, it can also make the trash bundle heavier so that a severely disabled person may require assistance or may need to empty the compactor before the bundle becomes too heavy to handle.

Cook Top

A cook top is used instead of a conventional stove to allow knee space under the unit. This allows severely disabled people who otherwise would not be able to use a stove to approach close enough to operate controls and handle pans. This arrangement exposes the disabled person to the hazard of spilling hot liquids in the lap, but it may be the only way some people can use the stove. Great caution must be exercise by the user.

Because of the increased probability of spills with raised burners, a smooth ceramic cook top unit with electronic touch controls to one side was installed which has no raised burners or knobs and provides a smooth continuous surface upon which cooking utensils can be easily maneuvered.

The location of burners on cooking units is important. People—especially seated people—should never have to reach over or between front burners to reach those in the rear. Burners located in positions similar to those shown in Fig. 17 would be safest.

Oven

Corner position The oven is located in the corner to make use of the under-counter space for knee clearance and to make it possible to move

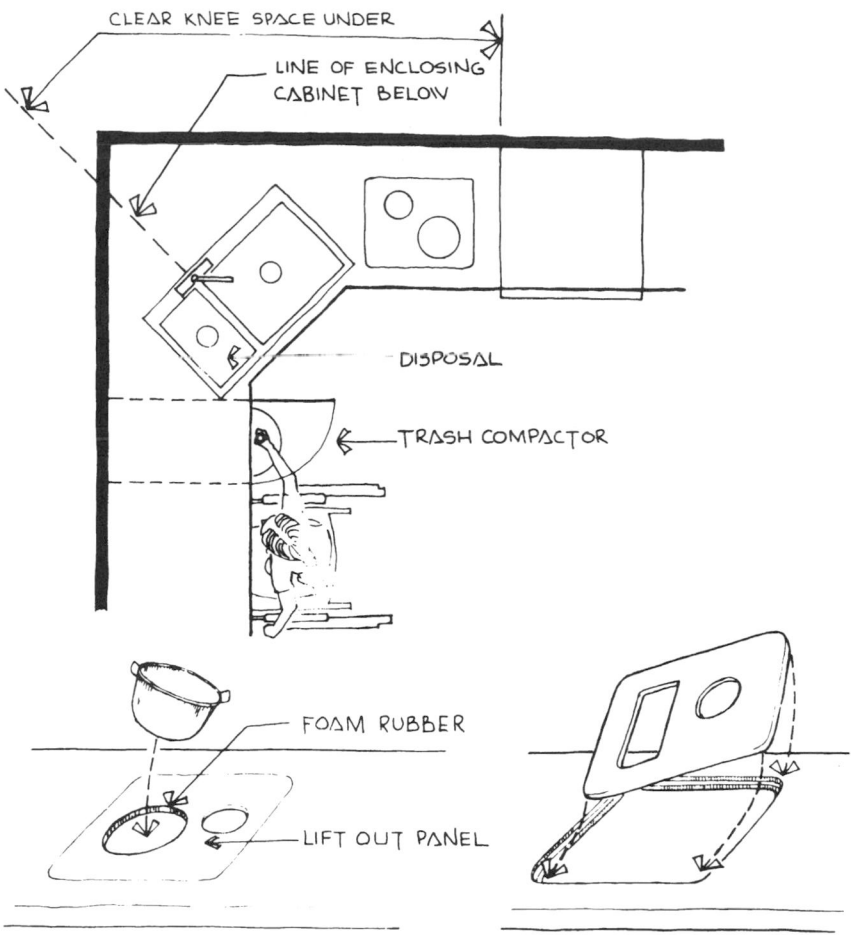

Fig. 16

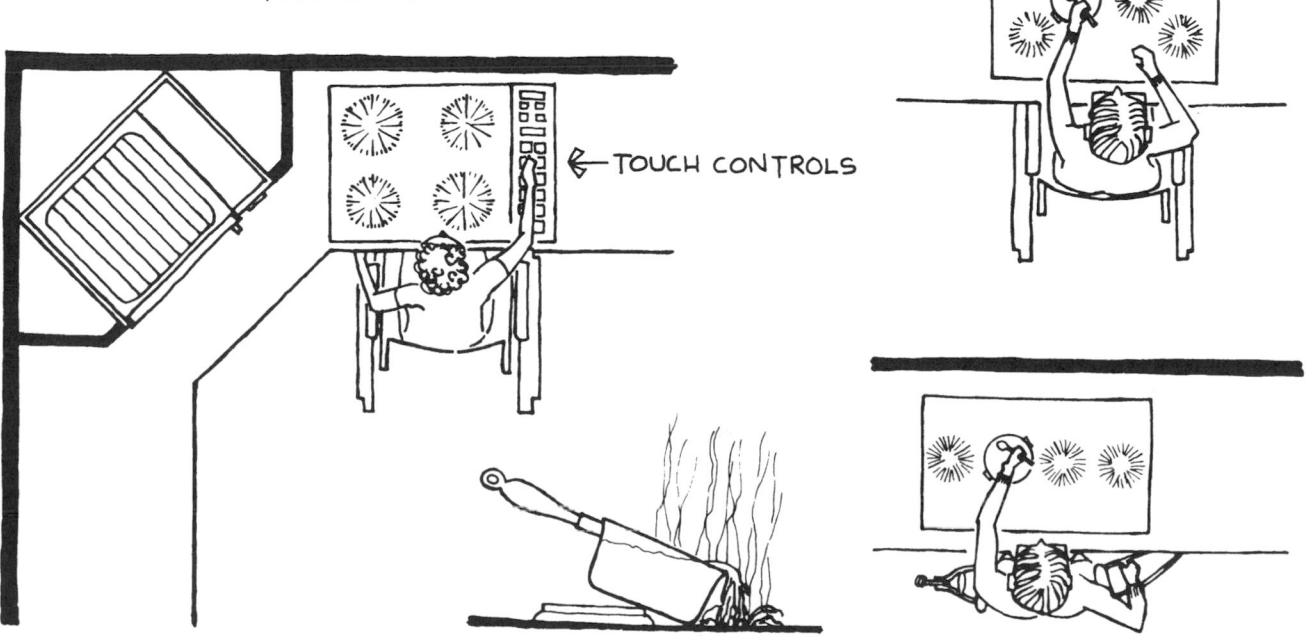

Fig. 17 Burner location.

HANDICAPPED HOUSING

a dish from the counter into the oven, and vice-versa, while remaining in a fixed position in front of the oven. The oven has a swinging door and low, side-mounted controls.

Continuous counter The counter top continues around the corner forming a 10-in (25.5-cm) shelf in front of the oven. Hot dishes may be moved from the oven to the shelf without lifting or carrying and slide along the counter to the serving area.

Storage

Drawers Many seated disabled people cannot adequately use over-counter cabinets. In the demonstration unit a food and utensil base storage unit of approximately 40 ft^3 (1.13 m^3) is included. The storage cabinet has four standard utensil drawers and three 1-ft 0-in (30.5-cm)-deep full-extension "file cabinet" type drawers. Each "file cabinet" drawer has one low side which makes it possible for a seated person to reach the contents from the side and eliminates difficult over-the-side reach.

The deep drawers can be fitted with racks for dishes and pots or used for dry food storage. Additional drawers are provided for large flat objects such as place mats and table cloths.

All drawers in the storage unit are equipped with full-extension, ball-bearing drawer slides to allow the drawers to come fully out of the cabinet for easy access to items in the back.

The storage cabinet has a large toe space to accommodate wheelchair foot rests and a protective metal bumper where contact with chair tubing occurs.

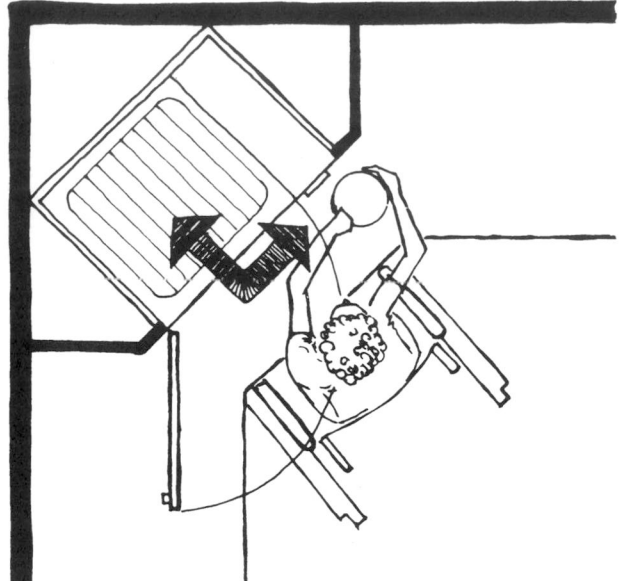

Fig. 18

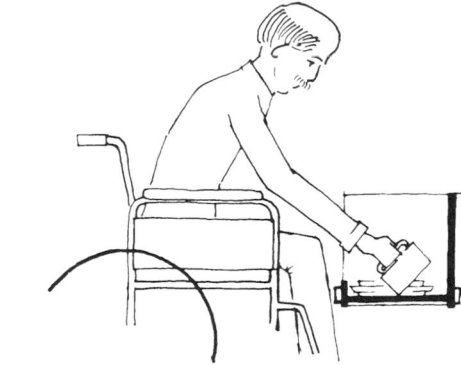

Fig. 19

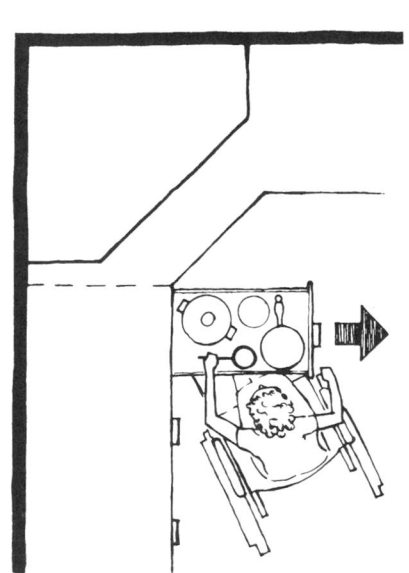

Fig. 20

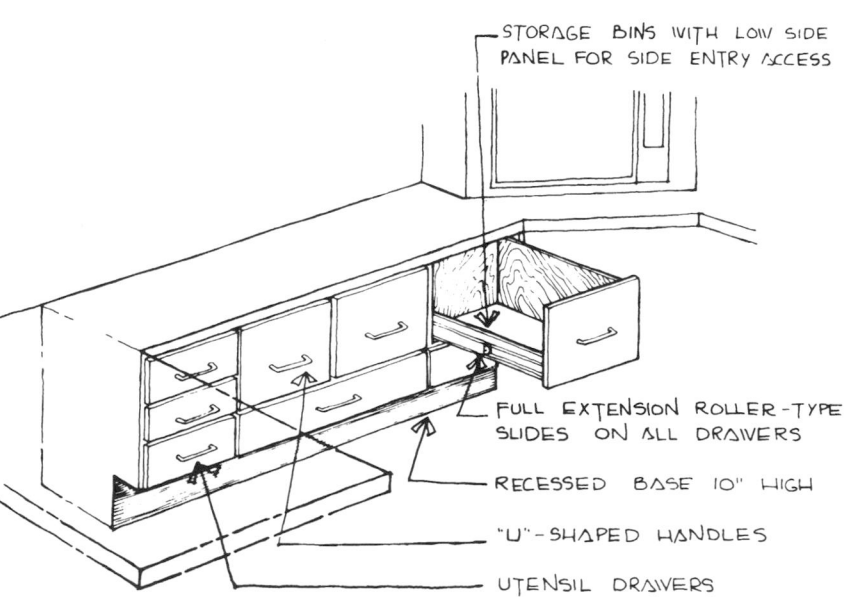

STORAGE BINS WITH LOW SIDE PANEL FOR SIDE ENTRY ACCESS

FULL EXTENSION ROLLER-TYPE SLIDES ON ALL DRAWERS

RECESSED BASE 10" HIGH

"U"-SHAPED HANDLES

UTENSIL DRAWERS

Fig. 21 Storage cabinet.

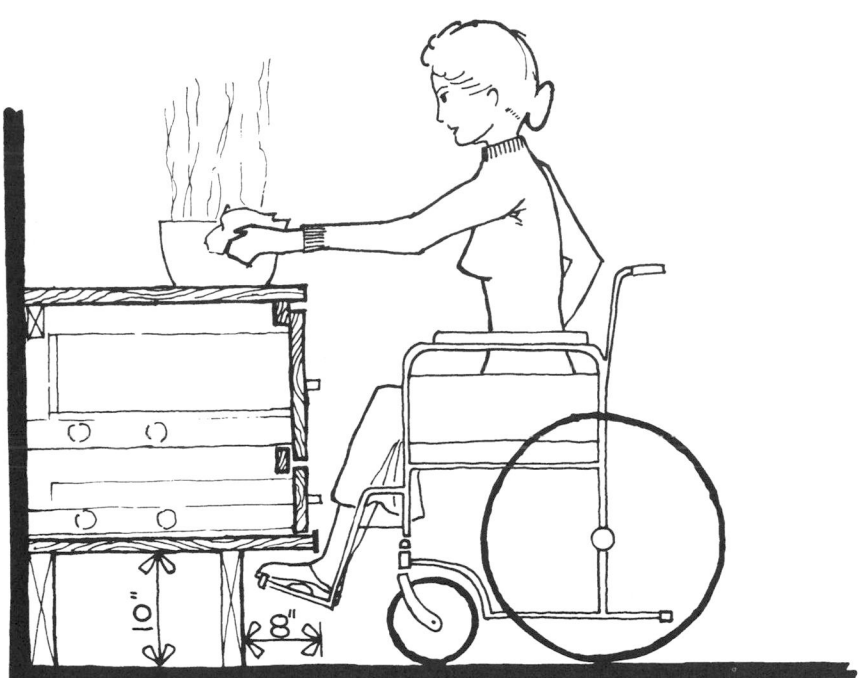

Fig. 22   Toe space.

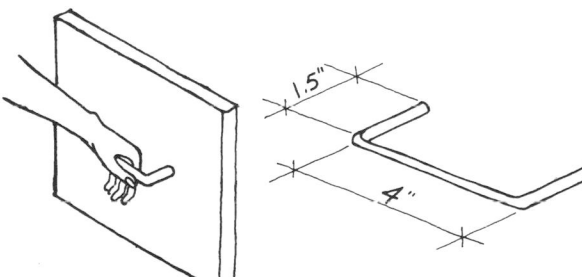

Fig. 23

Wall cabinets   Wall cabinets for storage are mounted with the bottom shelf 10 in (30.5 cm) above the counter top and are equipped with large U-shaped handles mounted as close as possible to the bottom of doors. This is done so that handles will be within reach of a person whose elbow is resting on the counter. This location makes it possible for people who cannot reach high and forward to use the cabinets and places handles within easy reach of standing people.

Handles   Some disabled people who have loss of use or limited use of their hands cannot grasp knobs or small handles to pull doors and drawers open. If the handles are large enough to put a hand or wrist through, most people can use them.

The handles on all cabinets are U-shaped with ½-in (3.8-cm) clearance and 4-in (10-cm) width.

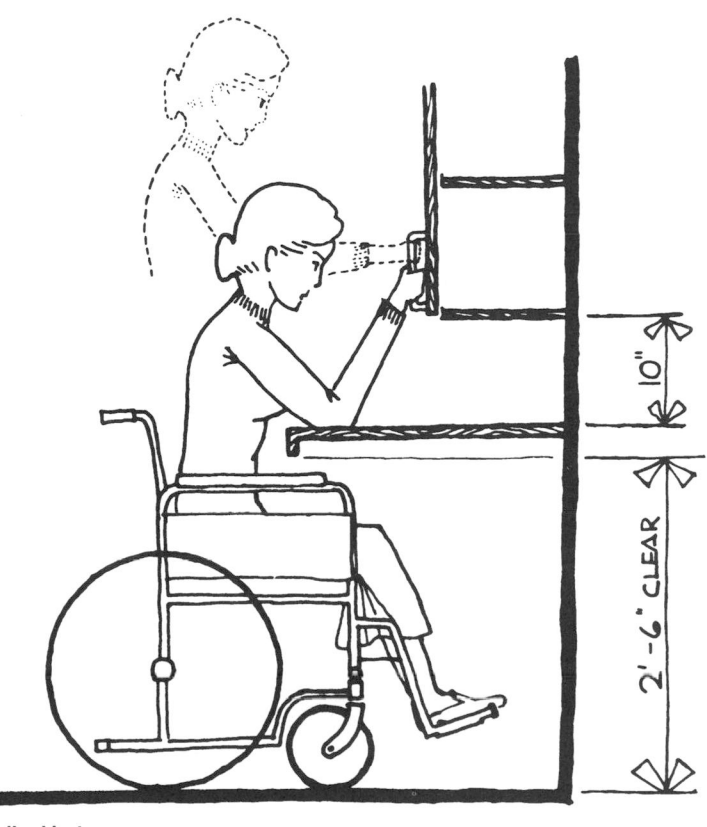

Fig. 24   Wall cabinet.

BATHROOMS

Plan

People with different disabilities require different types of bathing facilities—a tub or shower. The installation of one or the other determines the bathroom plan. The bathroom has been designed so that the plan with the tub may be fitted into the same space allocated to the shower-equipped room.

Some disabilities, such as hemiplegia, may require a bathing arrangement with support on the left or the right depending on the person's functioning side. The bathroom plans may be flipped within the same space to provide either left or right approach.

Washer and Dryer

Washer and dryer units were placed in the bath area of the demonstration unit for user convenience and to maintain centralized plumbing. Front-loading units are best for wheelchair users. They should have doors which swing fully out of the way, i.e., 180°, and the units should be located so that walls and other projections do not interfere with close parallel approach and people in wheelchairs can reach easily into the machine.

Washers and dryers with controls up front are more convenient for everyone. Standing people, both disabled and ablebodied, may prefer top-loading machines, and options of this type should be considered.

Shower

Wheelchair users may roll their chairs into the shower stall or may transfer onto a seat, or some may stand with support. Since most disabled people may safely use a shower stall, a 3-ft 0-in-wide by 4-ft 6-in-deep (91.5- by 107-cm) stall with a wood slat floor, level bottom, and fold-up seat is shown in Fig. 27.

Shower head Hand-held shower heads are best for disabled people who may need to remain seated when taking a shower. These shower heads are attached by a clip to a vertical slide bar on the shower wall and can be used in this position, or the head may be removed from the bar and held in the hand.

Controls Single-lever-operated thermostatic controls are safer and can be operated with one hand or by people who cannot use their hands well.

Grab bar Grab bars in the configuration shown are necessary for many disabled people and are a safety feature for all. Such bars should be 1¼ to 1½ in (3.2 to 3.8 cm) clearance from the wall, and should support at least 250 lb (113.4 kg). Because of differing disabilities, people often need these bars in different locations. The entire wall area of the shower in the demonstration unit is reinforced with ¾-in (2-cm) plywood so that bars can be securely fastened in any location.

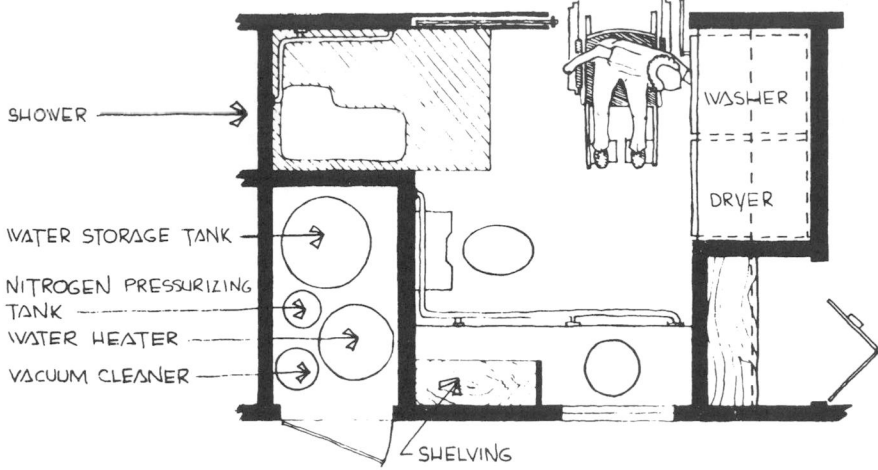

Fig. 25 Bathroom plan.

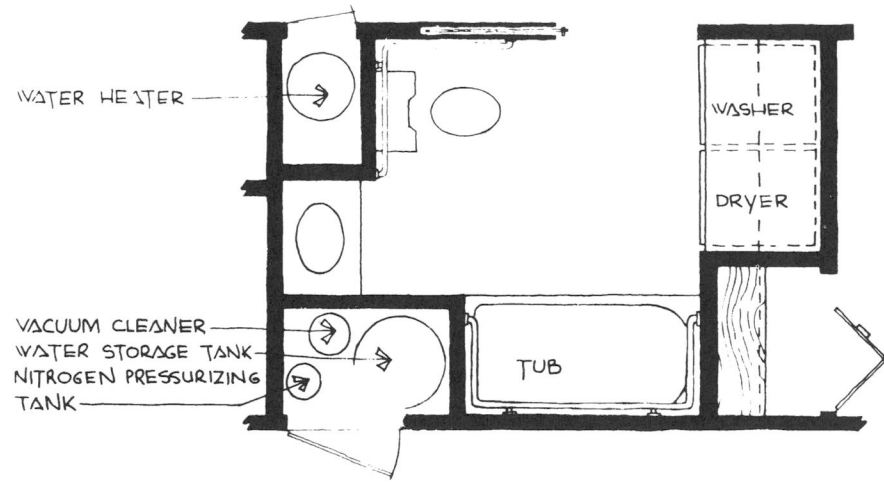

Fig. 26 Alternate plan.

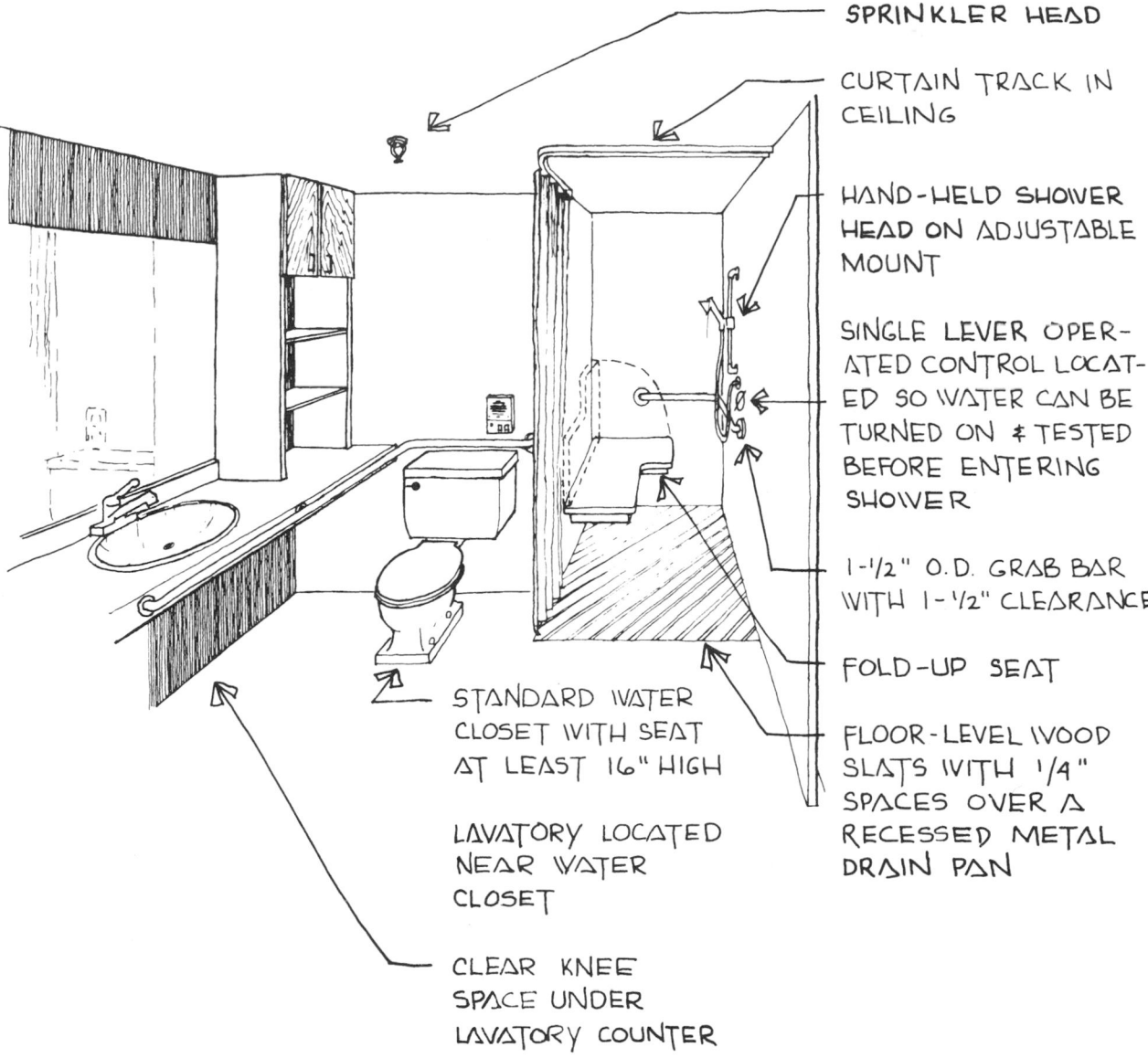

SPRINKLER HEAD

CURTAIN TRACK IN CEILING

HAND-HELD SHOWER HEAD ON ADJUSTABLE MOUNT

SINGLE LEVER OPERATED CONTROL LOCATED SO WATER CAN BE TURNED ON & TESTED BEFORE ENTERING SHOWER

1-1/2" O.D. GRAB BAR WITH 1-1/2" CLEARANCE

FOLD-UP SEAT

FLOOR-LEVEL WOOD SLATS WITH 1/4" SPACES OVER A RECESSED METAL DRAIN PAN

STANDARD WATER CLOSET WITH SEAT AT LEAST 16" HIGH

LAVATORY LOCATED NEAR WATER CLOSET

CLEAR KNEE SPACE UNDER LAVATORY COUNTER

Fig. 27

HANDICAPPED HOUSING

Seat The seat in the shower stall is an essential item for those disabled people who must sit to bathe. The L-shaped seat comes fully to the front of the stall so that a wheelchair person can pull up to the side and easily transfer onto the long end of the L. Thed short end of the L is in the corner and provides a wide surface which allows an unstable person to sit diagonally in the corner where side walls provide lateral support to prevent falling to one side.

The seat is positioned opposite the controls and the 3-ft 0-in (91.5-cm) width of the stall puts the controls within the reach of a person on the seat.

Some disabled people must have a padded seat. The shower seat can be constructed of wood, metal, or plastic, and could be upholstered with waterproof fabric or have a removable cushion which attaches securely.

The seat can be hinged to swing up into a vertical position when the shower is to be used by a standing person or for wheelchair roll-in bathing.

Floor The shower floor must be flat and at the same level as the bathroom floor. It must not have a curb which would block wheelchair access and be hazardous for unstable people to step over.

These requirements were met in Fig. 28 by installing a recessed metal shower pan into the floor and adding a 1- by 2-in redwood slatted floor on wood sleepers. The salts have ¼-in (0.6-cm) spaces between to allow water to drain into the pan below. This quick removal of water allows the shower floor to remain at the same level as the bathroom floor and eliminates the need for a curb to retain water.

Lavatory

The lavatory in Fig. 29 is a standard residential type with shallow bowl, single-lever faucet, and rear-mounted drain. A clear knee space of 2 ft 8 in (81 cm) is provided under the lavatory counter, and plumbing is routed backward toward the all so as not to interfere with wheelchair maneuvering. A bathtub bend can be used in place of the standard tailpiece for this purpose. Pipes are also insulated and padded to prevent burn or abrasion should someone bump the pipes.

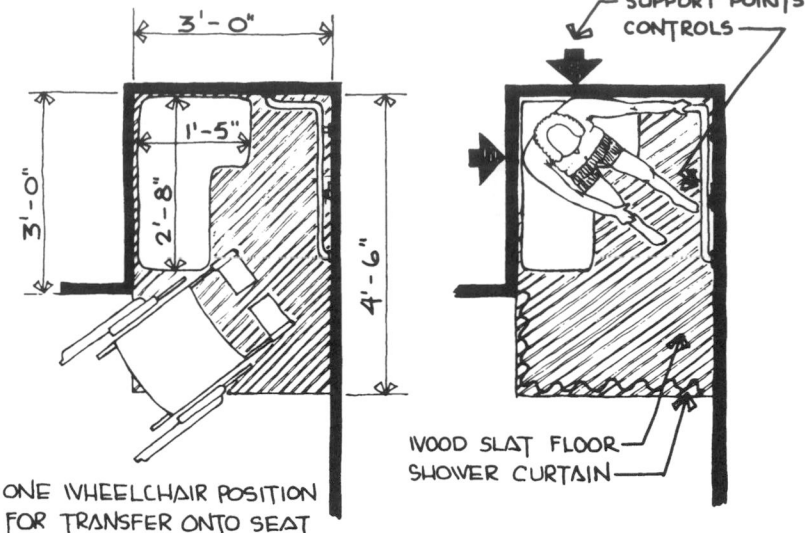

ONE WHEELCHAIR POSITION FOR TRANSFER ONTO SEAT

SUPPORT POINTS
CONTROLS
WOOD SLAT FLOOR
SHOWER CURTAIN

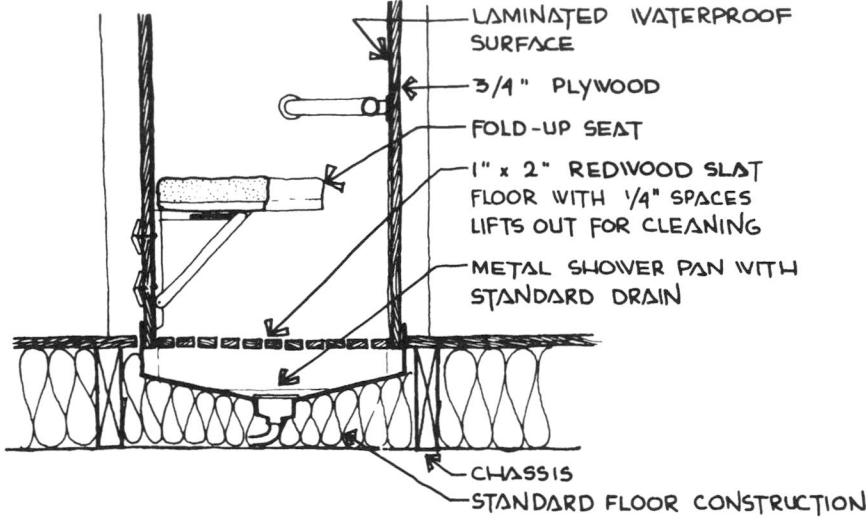

LAMINATED WATERPROOF SURFACE
3/4" PLYWOOD
FOLD-UP SEAT
1" x 2" REDWOOD SLAT FLOOR WITH ¼" SPACES LIFTS OUT FOR CLEANING
METAL SHOWER PAN WITH STANDARD DRAIN
CHASSIS
STANDARD FLOOR CONSTRUCTION

Fig. 28   Section through shower floor.

Toilet

The toilet should be a standard fixture with a seat height of at least 16 in (40.5 cm) and no more than 18 in (45.5 cm). The fixture is placed where it can be approached from the front or to one side, and sufficient space is provided to maneuver wheelchairs and walkers while approaching the fixture. Some disabled people need to be able to use the lavatory while seated on the toilet; therefore, the lavatory is placed next to the toilet.

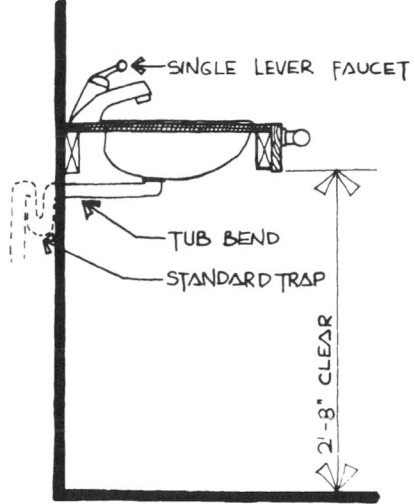

Fig. 29   Section through lavatory.

Storage

Linen and supply storage is provided on open shelving near the lavatory and water closet. Open shelving is preferred by many people, but both open and closed storage spaces should be provided. All storage should be within reach of both seated and standing people.

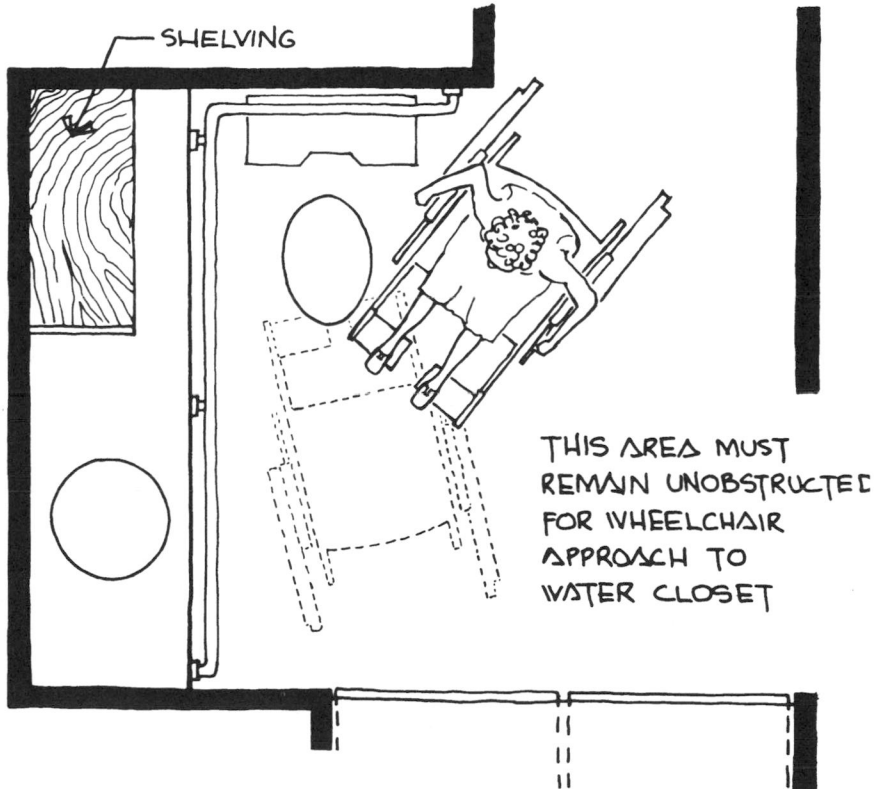

Fig. 30   Clear floor space.

DISASTER HOUSING

SITE DESIGN

The following recommendations and criteria should be observed in designing the site.

Overall Goal and Summary

A site must be designed and developed as quickly as possible after a disaster.

For most efficient development, the design should be reduced to its basic components, and special development or operating costs should be avoided. Development of imaginative layouts and amenities cannot be permitted to delay implementation. For efficient site design, the planner should understand the mobile home unit to be used and should analyze the mobile home unit, how it relates to other units and to open spaces between units.

The 12- by 60-ft mobile home unit which is currently stockpiled will continue to be used in disasters. Sites shown here have been designed to accommodate this unit.

Two prototypical lots have been developed to meet the criteria of density, privacy, and cost-effectiveness. If the two prototypes are used exclusively, emergency situations can be handled quickly.

No matter what size unit is used, a spatial relationship exists between units in the design of a site. A 30-ft space between units is efficient and yet creates a sense of privacy. This space acts as a buffer between units and provides appropriate density. The lot prototypes incorporate these elements.

The space just outside a unit is private and semiprivate. Private space, the area beneath the bedroom windows, is passive, undefined activity; the semiprivate space, viewed from the unit's entry, living, dining, and kitchen areas, is for specific activities such as parking and recreation.

Planners should know the type and size of unit that will be provided so they can design a site plan which will ensure privacy, provide private outdoor yards, and orient units appropriately.

These elements are integral parts of site design and are illustrated in the prototypical layouts which follow.

Lot prototype A Lot prototype A in Fig. 1 is a simple and efficient layout of lots in which units are parallel to each other on adjacent lots. Each lot is 112 by 42 ft. The important characteristics of prototype A are:

Mobile home units are placed parallel to the contours of the site.

A 25-ft setback between the unit and the on-site access road provides a private parking space, direct vehicular access to the front door of each unit, and paved access to the lot to position the mobile home. As a result, a larger private yard is provided adjacent to each unit.

The overall lot depth of 112 ft extends to the centerline of the 24-ft-wide access road. That is, 12 ft of the 24-ft-wide access road right-of-way is provided in each prototype so the placement of two lots side by side will provide the total required roadway width.

The lot dimensions provide at least 30 ft between units, enough to maintain privacy. If the recommended dimensions are exceeded, the length of roadway and utility lines, plus total development costs, will be increased. Therefore, dimensions should be carefully observed.

The windows of the private or bedroom areas of each unit face the windowless wall of an adjacent unit. This feature allows units to be placed only 30 ft apart.

The schemes in Fig. 2 illustrate efficient arrangements of prototype A. A utility line for essential utilities—sewer, water, and electric service—is aligned above or below ground along rear lot lines. The servicing of units on either side or double loading the utility line is the most efficient utility layout. Gravel across roadwayas are provided at the perimeter of the grouping to avoid conflicts with the installation and maintenance of utilities.

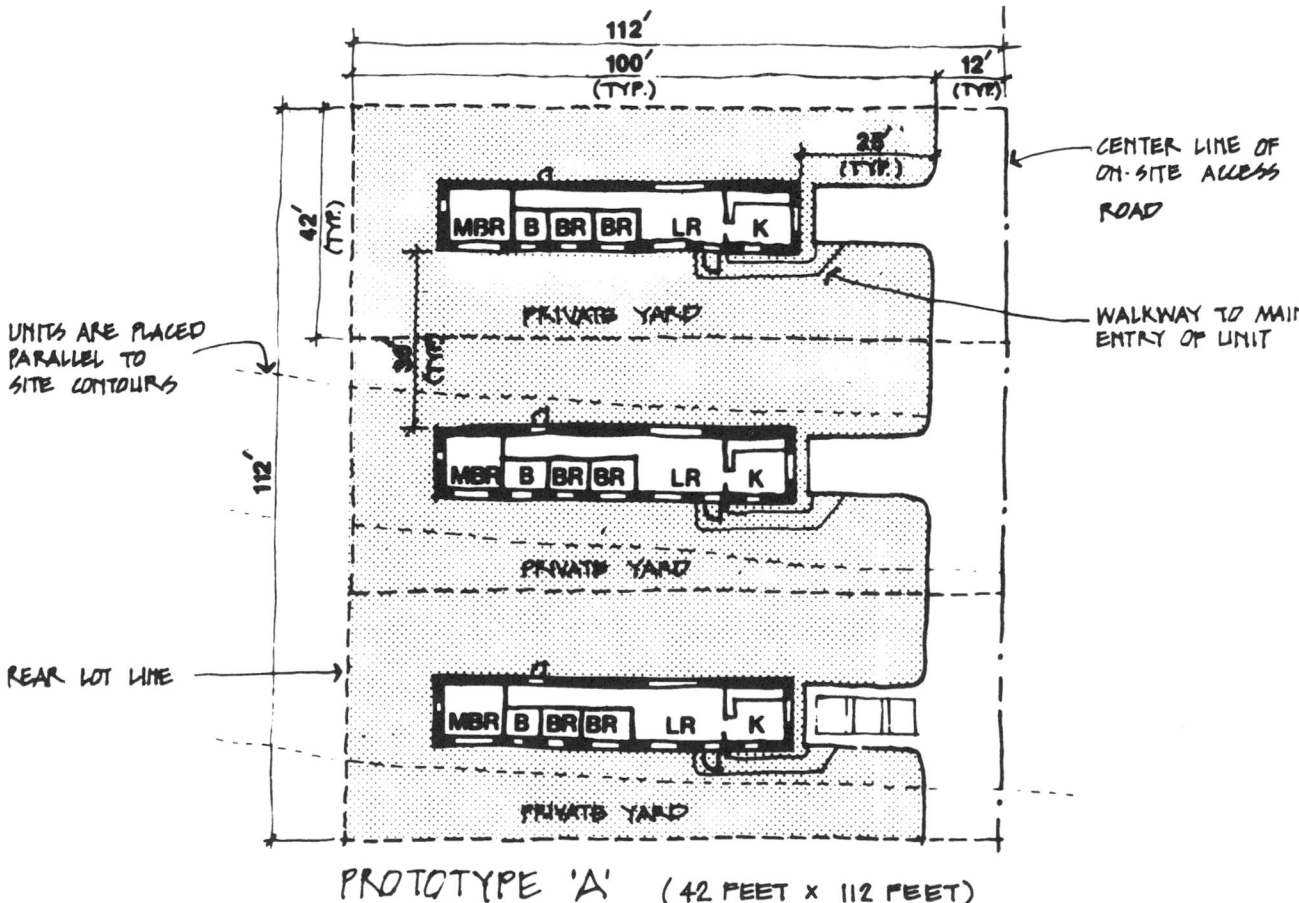

PROTOTYPE 'A' (42 FEET x 112 FEET)

NOTE: THE 12 FOOT x 65 FOOT MOBILE HOME SHOWN ABOVE IS ONE TYPE OF UNIT WHICH MAY BE USED AT TEMPORARY DISASTER GROUP SITES. INTERIOR LAYOUTS AND OVERALL UNIT DIMENSIONS MAY VARY.

Fig. 1 Lot prototype A.

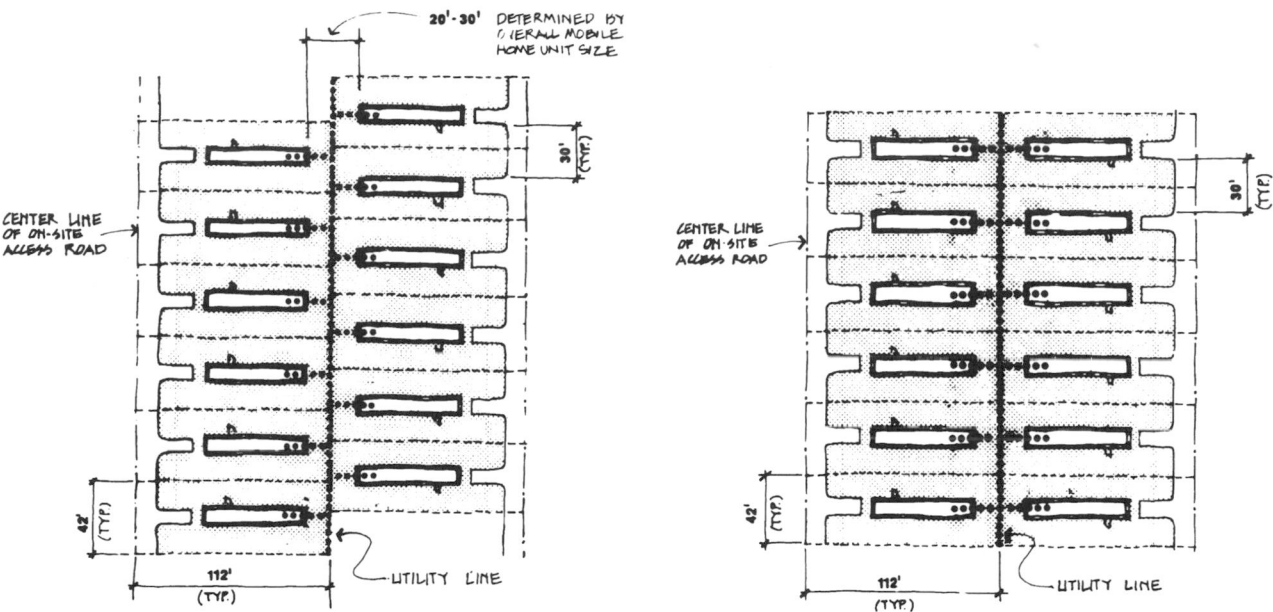

Fig. 2 Layout variations for lot prototype A.

Lot prototype B   The scheme in Fig. 3 illustrates a second lot size, prototype B. Each lot is 80 by 58 ft. Units are placed at a 45° angle to the access road. Prototype B differs from prototype A in these important respects:

Prototype B is more appropriate when the lot layout is limited and grading problems can be reduced if units are placed parallel to contour lines. This prototype is also valuable when sun, wind, or natural site features make it desirable to place units on an angle.

Installation is simplified, since a truck delivering a unit need not negotiate a 90° turn to place the unit on its lot.

Placing units at an angle gives the planner greater flexibility in designing an imaginative layout while maintaining desirable efficiency.

The variations illustrated in Fig. 4 for prototype B contain the same features as the prototype A groupings. Units are placed two abreast, 30 ft apart, and are provided with a doubleloaded utility line and a perimeter access road.

GUIDELINES FOR SITE PLANNING

If the criteria for site selection are observed, difficulties in site planning should be eliminated or reduced. However, the planner should keep in mind the following guidelines.

Sites of 100 units or more are usually more difficult to develop, require more amenities, and demand greater attention to lessen monotony and dreariness.

Small sites, averaging 25 units or 4 acres, have been used most often, regardless of the size of the disaster. Such sites are simplest to design. However, in the past, sites selected were sometimes larger than necessary to house the displaced families. In such cases it is tempting to decrease density by placing units farther apart. This should be avoided, since it may lead to less efficient and costlier site plans. The criteria for overall site planning, which follow, should be strictly observed to produce the most efficient site plans. Undeveloped area can be used for recreation or to increase setbacks.

In any case, the site plan should be designed so units are grouped in clusters of 25 to achieve an efficient and economical plan and a sense of community.

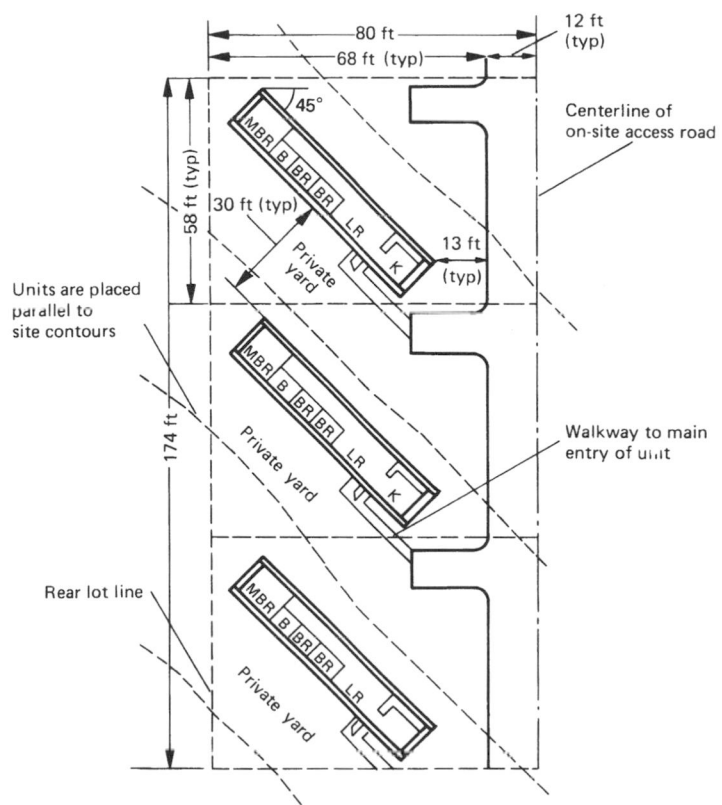

Prototype B (58 ft × 80 ft)

Note: The 12-ft × 65-ft mobile home shown above is one type of unit which may be used at temporary disaster group sites. Interior layouts and overall unit dimensions may vary.

Fig. 3   Lot prototype B.

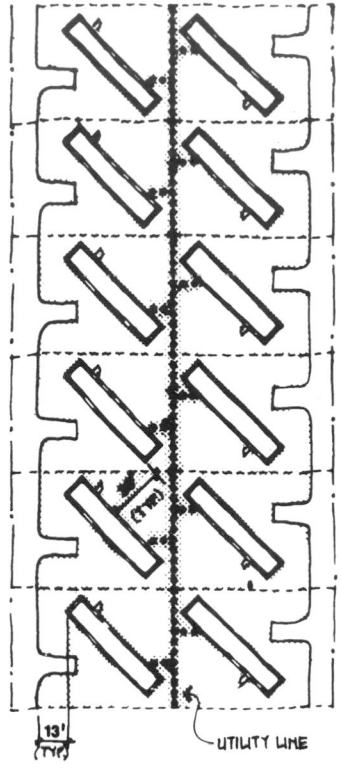

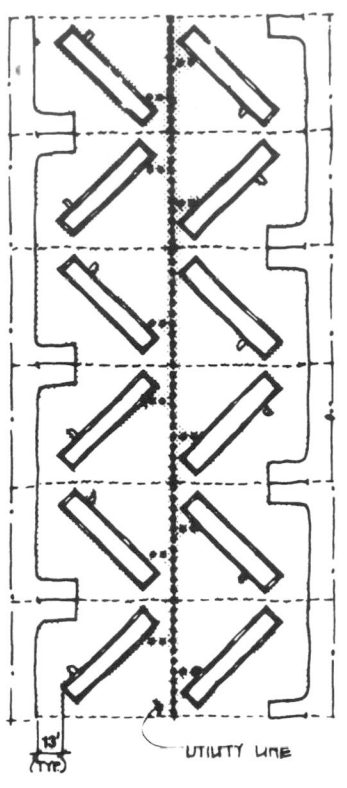

EXAMPLES OF ACCEPTABLE SITE PLANS

To apply the recommended prototypes to an actual site efficiently, the planner should refer to the following layout variations.

Three site shapes are shown—a narrow rectangle, a square, and a triangle. Each contains 8.26 acres and can accommodate at least 50 mobile homes. Each site has 2 to 8 percent slope and no unusual features. Each plan incorporates double-loaded utility lines, perimeter roadways, and centrally located community areas.

A narrow rectangular site may dictate the use of one prototype throughout, as shown in Fig. 5.

Both site plans require the same length of utility and are equally acceptable in terms of density and development costs. However, the site plan in Fig. 5a is less expensive to develop, since 25 percent less access roadway is required. This site plan is therefore more desirable in terms of cost-effectiveness.

The site shown in Fig. 5b, however, provides a larger community space. If necessary, additional mobile homes could be added. This plan also incorporates the use of prototype B, allowing units to be placed at a 45° angle to the access road. In this way, natural site features such as sunlight, trees, and land contours are respected. The use of prototype B also reduces monotony.

The square sites shown in Figs. 6 and 7 each provide an equal number of units with desirable density and acceptable efficiency. However, the site plan shown in Fig. 7, containing lot prototypes A and B, uses the given site area less efficiently and at a higher development cost.

The site plan shown in Fig. 7 also requires more utility line and roadway. Furthermore, the plan in Fig. 7 provides minimal community space, while the plan in Fig. 6 provides a large community space adjacent to half the units.

Although both schemes are acceptable, the site plan in Fig. 6 is more desirable in terms of cost-effectiveness because of greater efficiency and lower development costs.

The triangular site shown in Fig. 8 is often difficult to develop. The triangular site plan demonstrates that an efficient layout can be achieved, with densities and lengths of roadways and utility lines which compare favorably with rectangular sites.

Cul-de-sacs, or turnarounds, are provided. Vehicular turnarounds should be considered whenever unnecessary roadway length can be eliminated or where dead-end streets are called for.

All these plans achieve a desirable level of efficiency. The planner should therefore consider various layouts as guidelines, not as fixed rules.

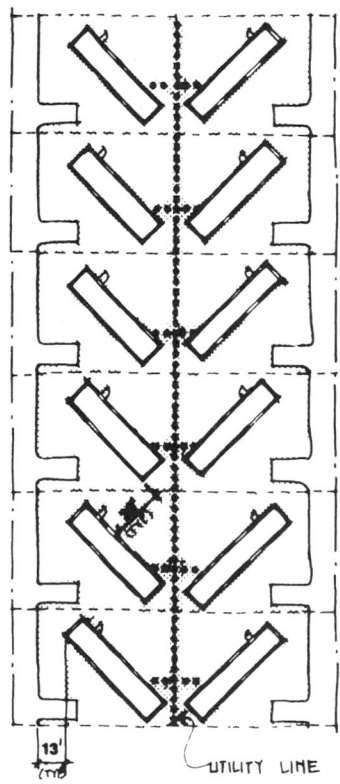

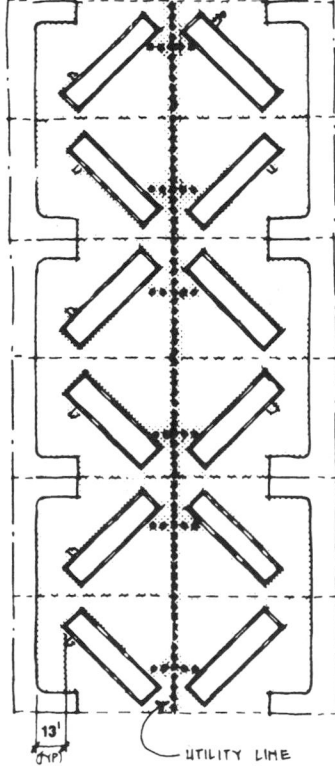

Fig. 4   Layout variations for lot prototype B.

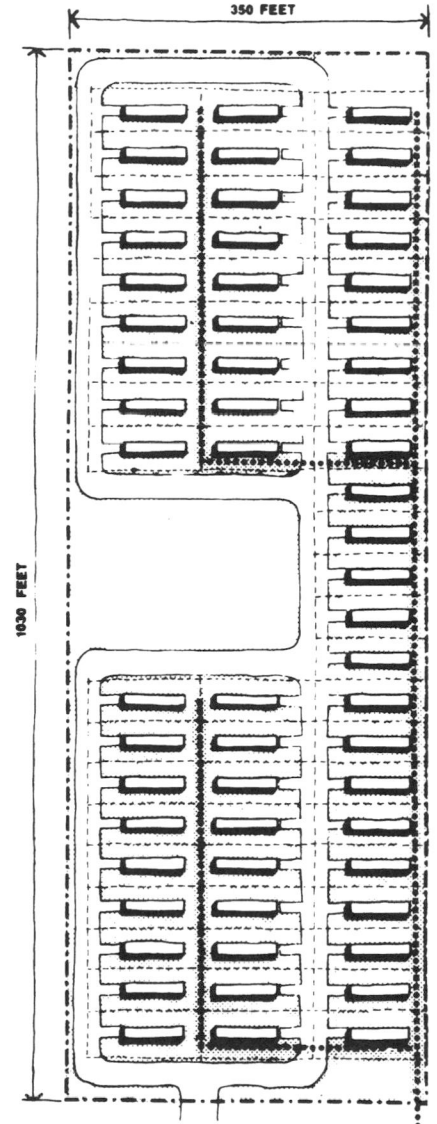

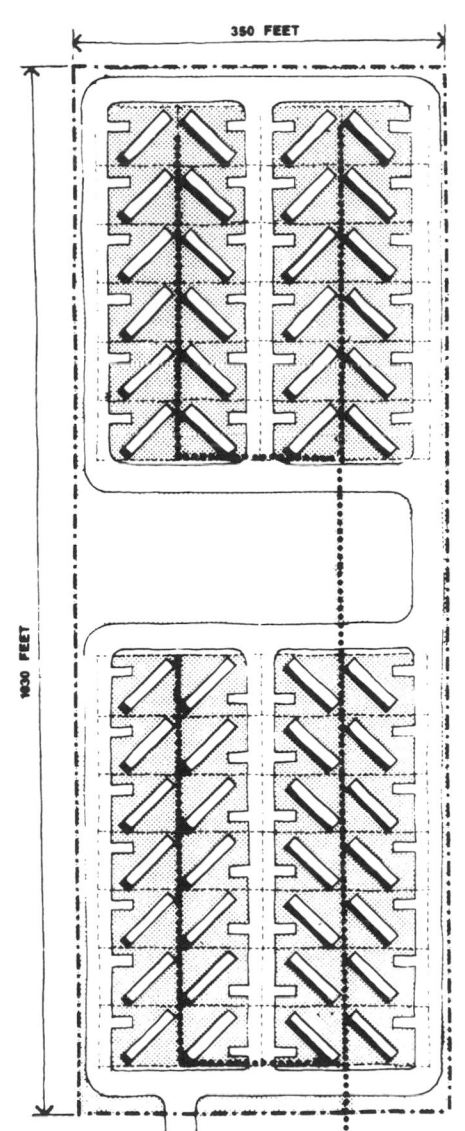

	PER UNIT	TOTAL
LINEAR FT. OF ONSITE ROADS:	45	2680
LINEAR FT. OF ONSITE UTILITY LINES:	37	2230
COMMUNAL RECREATION FACILITIES (DEFINED SPACES):	735 SQ. FT.	43380 SQ. FT.
COMMUNAL OPEN AREAS (UNDEFINED SPACES):	660 SQ. FT.	38420 SQ. FT.
INDIVIDUAL LOT AREAS (INCLUDING 50% ROADWAY):	4704 SQ. FT.	277536 SQ. FT.
PARKING SPACES PER UNIT:	1	59

GROSS SITE AREA	8.26 ACRES	
MOBILE HOMES ON SITE	59 UNITS	
DENSITY	7.14 UNITS/ACRE	

MOBILE HOME UNIT	
UTILITIES	
SITE BOUNDARY	
PARKING	

	PER UNIT	TOTAL
LINEAR FT. OF ONSITE ROADS:	73	3810
LINEAR FT. OF ONSITE UTILITY LINES:	39	2040
COMMUNAL RECREATION FACILITIES (DEFINED SPACES):	1066 SQ. FT.	58660 SQ. FT.
COMMUNAL OPEN AREAS (UNDEFINED SPACES):	1275 SQ. FT.	66310 SQ. FT.
INDIVIDUAL LOT AREAS (INCLUDING 50% ROADWAY):	4640 SQ. FT.	241280 SQ. FT.
PARKING SPACES PER UNIT:	1	52

GROSS SITE AREA	8.26 ACRES	
MOBILE HOMES ON SITE	52 UNITS	
DENSITY	6.3 UNITS/ACRE	

MOBILE HOME UNIT	
UTILITIES	
SITE BOUNDARY	
PARKING	

Fig. 5 Site plan for rectangular site.

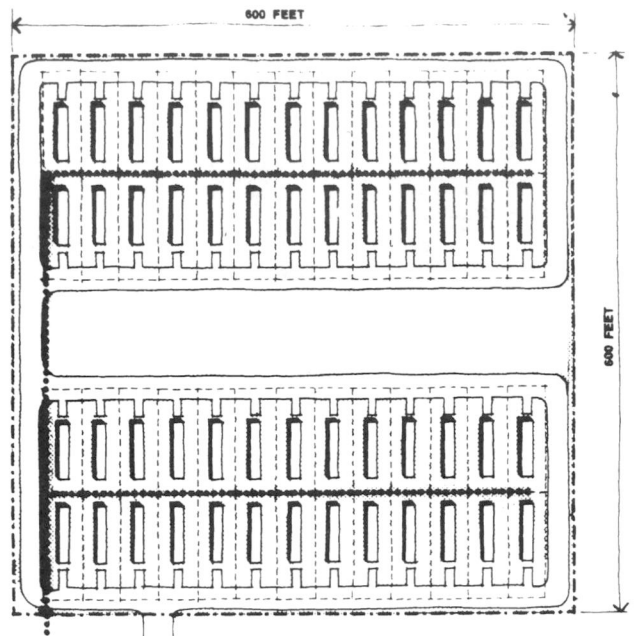

	PER UNIT		TOTAL
LINEAR FT. OF ONSITE ROADS:	62		3200
LINEAR FT. OF ONSITE UTILITY LINES:	29		1520
COMMUNAL RECREATION FACILITIES (DEFINED SPACES):	1033	SQ. FT.	53740 SQ. FT.
COMMUNAL OPEN AREAS (UNDEFINED SPACES):	1162	SQ. FT.	60940 SQ. FT.
INDIVIDUAL LOT AREAS (INCLUDING 50' ROADWAY):	4704	SQ. FT.	244608 SQ. FT.
PARKING SPACES PER UNIT:	1		52

GROSS SITE AREA: 8.26 ACRES
MOBILE HOMES ON SITE: 52 UNITS
DENSITY: 6.3 UNITS/ACRE

MOBILE HOME UNIT
UTILITIES
SITE BOUNDARY
PARKING

0 50 100 150

Fig. 6   Site plan for square site.

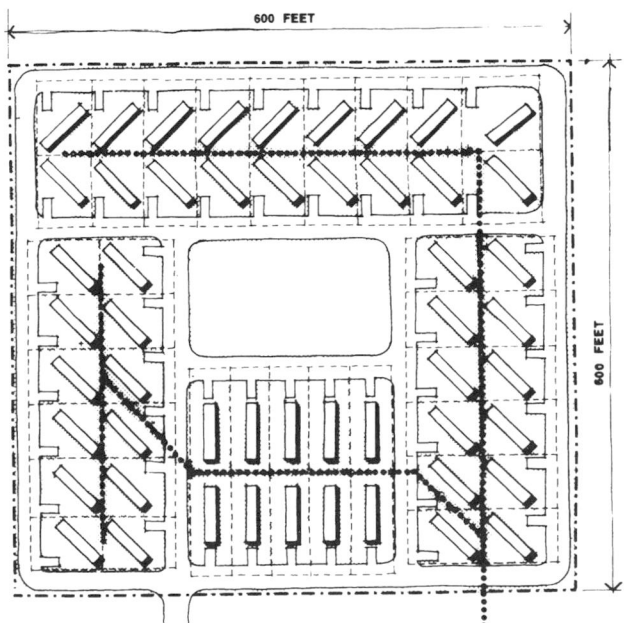

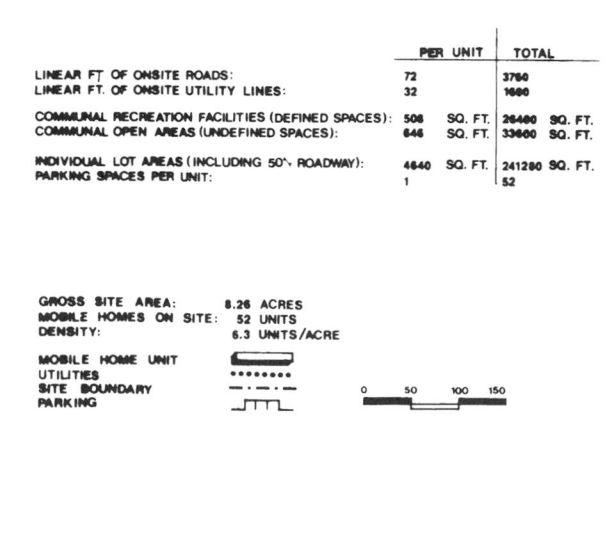

	PER UNIT		TOTAL
LINEAR FT. OF ONSITE ROADS:	72		3760
LINEAR FT. OF ONSITE UTILITY LINES:	32		1680
COMMUNAL RECREATION FACILITIES (DEFINED SPACES):	508	SQ. FT.	26400 SQ. FT.
COMMUNAL OPEN AREAS (UNDEFINED SPACES):	646	SQ. FT.	33600 SQ. FT.
INDIVIDUAL LOT AREAS (INCLUDING 50' ROADWAY):	4640	SQ. FT.	241280 SQ. FT.
PARKING SPACES PER UNIT:	1		52

GROSS SITE AREA: 8.26 ACRES
MOBILE HOMES ON SITE: 52 UNITS
DENSITY: 6.3 UNITS/ACRE

MOBILE HOME UNIT
UTILITIES
SITE BOUNDARY
PARKING

0 50 100 150

Fig. 7   Site plan for square site.

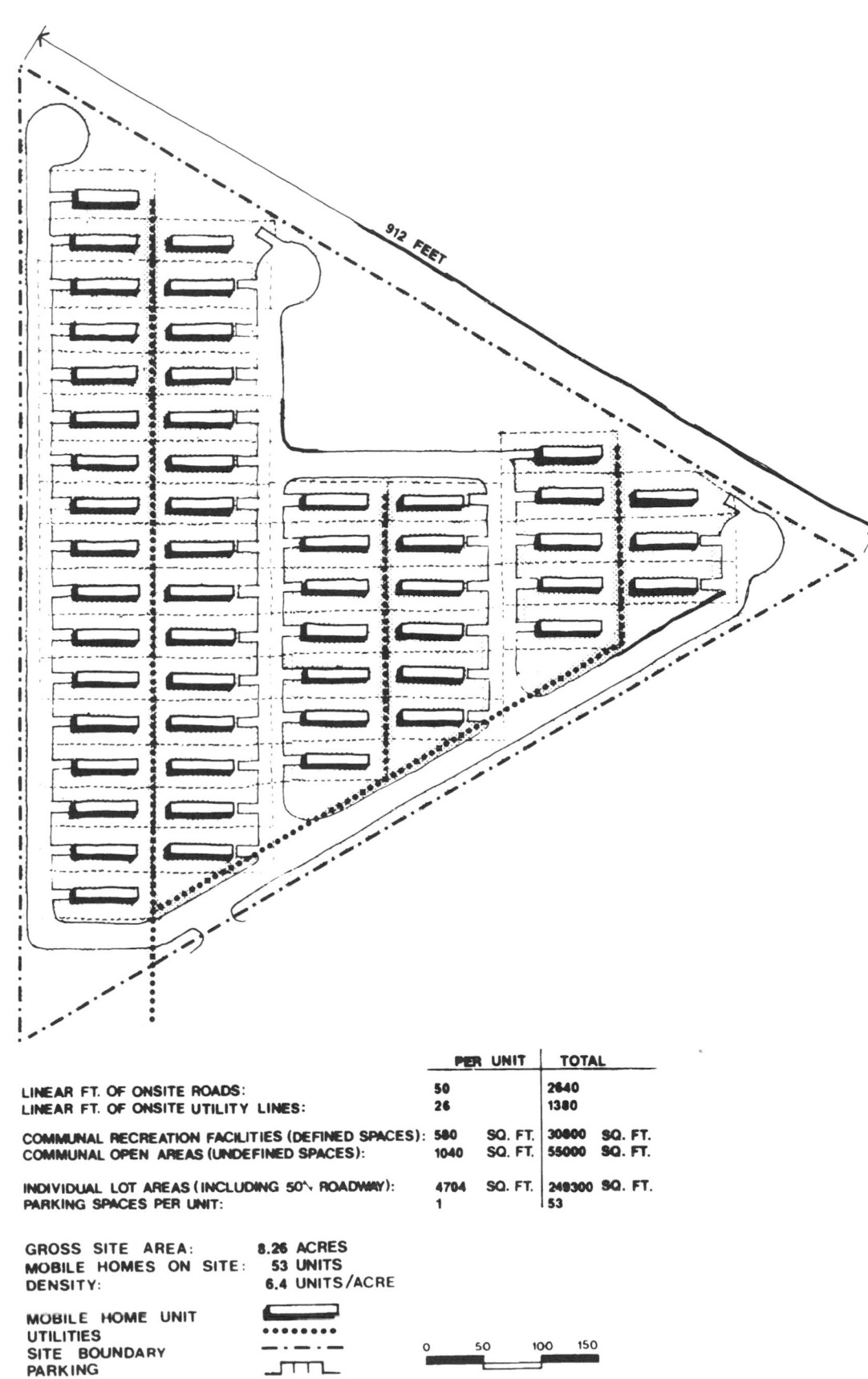

912 FEET

	PER UNIT	TOTAL
LINEAR FT. OF ONSITE ROADS:	50	2640
LINEAR FT. OF ONSITE UTILITY LINES:	26	1380
COMMUNAL RECREATION FACILITIES (DEFINED SPACES):	580 SQ. FT.	30800 SQ. FT.
COMMUNAL OPEN AREAS (UNDEFINED SPACES):	1040 SQ. FT.	55000 SQ. FT.
INDIVIDUAL LOT AREAS (INCLUDING 50' ROADWAY):	4704 SQ. FT.	249300 SQ. FT.
PARKING SPACES PER UNIT:	1	53

GROSS SITE AREA: 8.26 ACRES
MOBILE HOMES ON SITE: 53 UNITS
DENSITY: 6.4 UNITS/ACRE

MOBILE HOME UNIT
UTILITIES
SITE BOUNDARY
PARKING

0 50 100 150

Fig. 8 Site plan for triangular site.

DISASTER HOUSING

TABLE 1 Site Selection Criteria Checklist

	Locational Criteria		
	At least	But not more than	Desirable
Required regional roadway clearance for delivery of mobile home	12-ft width, 13½-ft height, 15-ft turning radius at right-angle turns	14-ft width, 14-ft height, 20-ft turning radius at right-angle turns	Adequate roadway clearance and turning radius for the transportation and delivery of mobile home units to the site; refer to the National Policy Statement of the American Association of State Highway and Transportation Officials (AASHTO), July 5, 1973.
Distance to original neighborhood	Not applicable	4–5 miles	Adjacent to or as near as possible to original neighborhood in effort to reestablish normal or routine living patterns.
Distance to shopping	Within walking distance	4–5 miles	Walking distance of 10 min (½ mile) or less; the usual impaired mobility of the disaster victims elevates the importance of locating group sites near shopping facilities.
Distance to employment	Same as predisaster	45 min	30 min or less, or same as predisaster.
Distance to schools	Same as predisaster	30-min bus ride to predisaster	Predisaster schools should be attended in order to avoid further psychological disorientation, preserve the community fabric, and reestablish normal daily routine; if necessary, busing children to their predisaster schools should be instituted.
Distance to medical facilities	Access to disaster relief station	4–5 miles	Access to medical services or facility either near or on-site; the need for medical and dental attention will be determined by the scope of the disaster and the characteristics of the dislocated population (children, elderly).
Distance to existing recreational facilities	Within walking distance	Walking distance of 15 min (¾ mile)	Walking distance of 10 min (½ mile) or less; the need for recreational facilities will be determined by the characteristics of the population (children, elderly) and the intent to provide some temporary on-site facilities.
Distance to social, cultural, and other amenities	Same as predisaster	45 min	30 min or less, or same as predisaster; proximity to such amenities is not essential but is important in combating the disruptive social and psychological effects of a disaster.
Distance to existing public transportation	Within walking distance	Walking distance of 10 min (½ mile)	Walking distance of 10 min (½ mile) or less to existing public transportation; the need for additional or temporary public transportation will be determined by the group site size, population (low income, elderly), the availability of private transportation, and the scope of the disaster.
Adjacent land use	Not applicable	Not applicable	Heavy industrial, air, and/or noise pollution should be avoided where possible.
Local codes and ordinances	Not applicable	Not applicable	Local restrictions should be waived by local authorities in a disaster situation so as to accommodate temporary group housing design and development.
Site information	Not applicable	Not applicable	Accurate site surveys and existing utility information should be available; the lack of site information will cause delay in design and development.

TABLE 1 Site Selection Criteria Checklist (*Continued*)

Physical Site Features			
	At least	But not more than	Desirable
Potential net density	5 units/acre	10 units/acre	6–8 units/acre; past experience and site development studies have shown that a density of 6–8 units/acre is most desirable for efficient economical development.
Size and number of units	Not applicable	Not applicable	25 units on 4–5 acres; potential capacity of any site can be estimated by multiplying the total site acreage by 6, which is the recommended unit density per acre; that is, a 10-acre site can accommodate 60 mobile home units. If physical site and utility constraints prohibit the use of a single site, several smaller sites should be considered which may be in closer proximity to original neighborhood(s).
Shape	Not applicable	Not applicable	Square or rectilinear; for 50 units or less, the units are most efficiently placed on a square or rectilinear site; shape is less important for larger sites.
Slope	2% slope	8% slope; 12% slope for short runs may be acceptable	Uniform ground surface with a site gradient between 2 and 8%; a too uniform site with too little slope could result in inadequate site drainage.
Heavily wooded areas	Not applicable	Not applicable	Should be avoided since the sites are to be vacated in a short period of time and the use of the land is temporary.
Floodplains	Not applicable	Not applicable	Should be avoided unless no alternative is available; refer to the Flood Disaster Prevention Act, PL.93.234.
Falling rock zones, swampy areas or other hazards	Not applicable	Not applicable	Should be avoided because of the potential development difficulties and dangers related to child play.
Soil conditions for bearing of mobile unit	Not applicable	Not applicable	High water table or swampy sites should be avoided. Sandy soils are preferable.
Soil conditions for proper site drainage	Not applicable	Not applicable	Sandy soils are most preferable and clay soils are least preferable for site drainage.
Availability and Requirements for Utilities			
Water	Potential for portable water-supply systems	Not applicable	Existing local water line, adjacent to site, with an adequate pressure to serve the total number of units required; adequate pressure can be determined by multiplying the total number of units by 280 gal/unit/day.
Sewer	Potential for portable sewage disposal systems	Not applicable	Existing local sewer line, adjacent to site, with an adequate capacity to serve the total number of units required; adequate capacity can be determined by multiplying the total number of units by 250 gal/unit/day.
Electric	An existing electrical power source in a remote location from the site from which electrical lines can be extended	Not applicable	Existing electrical utility line which can provide 120/220-V, single-phase, three-wire, 60-Hz service.
Gas	Not applicable	Not applicable	Existing gas lines are not appropriate for temporary disaster housing; bottled gas should be used exclusively, if required; bulk liquid propane may also be used.

Rehabilitation, Conversions, and Historic Preservation

INTRODUCTION

Although specifically developed to assist property owners eligible to receive Historic Preservation Loans and for local officials responsible for the community development block grant program of the Housing and Community Development Act of 1974, these guidelines will help any property owner or local official in formulating plans for the rehabilitation, preservation, and continued use of old buildings, neighborhoods, and commercial areas. They consist of eight principles that should be kept in mind when planning new construction or rehabilitation projects. The checklist suggests specific actions to be considered or avoided to ensure that the distinguishing qualities of buildings or neighborhood environments will not be damaged by new work. In addition, whenever possible, advice should be sought from qualified professionals, including architects, architectural historians, and planners, who are skilled in the preservation, restoration, and rehabilitation of old buildings and neighborhoods.

When the buildings or areas being considered for rehabilitation are listed or eligible for listing in the National Register of Historic Places, property owners and local officials responsible for the work should, as a first step, contact the appropriate state historic preservation officer, in addition to consulting with experienced professionals. Where comprehensive surveys (to identify properties eligible for National Register listing) have not yet been completed in a project area, the undertaking of such surveys should be discussed with appropriate local officials and with the state historic preservation officer.

GUIDELINES

1. Every reasonable effort should be made to provide a compatible use for buildings which will require minimum alteration to the building and its environment.

2. Rehabilitation work should not destroy the distinguishing qualities or character of the property and its environment. The removal or alteration of any historic material or architectural features should be held to the minimum, consistent with the proposed use.

3. Deteriorated architectural features should be repaired rather than replaced, wherever possible. In the event replacement is necessary, the new material should match the material being replaced in composition, design, color, texture, and other visual qualities. Repair or replacement of missing architectural features should be based on accurate duplications of original features, substantiated by physical or pictorial evidence rather than on conjectural designs or the availability of different architectural features from other buildings.

4. Distinctive stylistic features or examples of skilled craftsmanship which characterize older structures and often predate the mass production of building materials should be treated with sensitivity.

5. Many changes to buildings and environments which have taken place in the course of time are evidence of the history of the building and the neighborhood. These changes may have developed significance in their own right,

and this significance should be recognized and respected.

6. All buildings should be recognized as products of their own time. Alterations to create an appearance inconsistent with the actual character of the building should be discouraged.

7. Contemporary design for new buildings in old neighborhoods and additions to existing buildings or landscaping should not be discouraged if such design is compatible with the size, scale, color, material, and character of the neighborhood, building, or its environment.

8. Wherever possible, new additions or alterations to buildings should be done in such a manner that if they were to be removed in the future, the essential form and integrity of the original building would be unimpaired.

GUIDELINES CHECKLIST

The Environment

In new construction, retain distinctive features of the neighborhood's existing architecture, such as the distinguishing size, scale, mass, color, materials, and details, including roofs, porches, and stairways, that give a neighborhood its special character.

Use new plant materials, fencing, walkways, and street lights, signs, and benches that are compatible with the character of the neighborhood in size, scale, material, and color.

Retain existing landscape features such as parks, gardens, street lights, signs, benches, walkways, streets, alleys, and building setbacks that have traditionally linked buildings to their environment.

Existing Buildings: Lot

Inspect the lot carefully to locate and identify plants, trees, fencing, walkways, outbuildings, and other elements that might be an important part of the property's history and development.

Retain plants, trees, fencing, walkways, and street lights, signs, and benches that reflect the property's history and development. Base decisions for new work on actual knowledge of the past appearance of the property found in photographs, drawings, newspapers, and tax records. If changes are made, they should be carefully evaluated in light of the past appearance of the site.

Existing Buildings: Exterior Features

Masonry buildings Retain original masonry and mortar, whenever possible, without the application of any surface treatment.

Duplicate old mortar in composition, color, and textures.

Duplicate old mortar in joint size, method of application, and joint profile.

Repair stucco with a stucco mixture duplicating the original as closely as possible in appearance and texture.

Clean masonry only when necessary to halt deterioration and always with the gentlest method possible, such as low-pressure water and soft natural-bristle brushes.

Repair or replace, where necessary, deteriorated material with new material that duplicates the old as closely as possible.

Replace missing architectural features, such as cornices, brackets, railings, and shutters.

Retain the original or early color and texture of masonry surfaces, wherever possible. Brick or stone surfaces may have been painted or whitewashed for practical and aesthetic reasons.

Frame buildings Retain original material, whenever possible.

Repair or replace, where necessary, deteriorated material with new material that duplicates the old as closely as possible.

Roofs Preserve the original roof shape.

Retain the original roofing material, whenever possible.

Replace deteriorated roof coverings with new material that matches the old in composition, size, shape, color, and texture.

Preserve or replace, where necessary, all architectural features which give the roof its essential character, such as dormer windows, cupolas, cornices, brackets, chimneys, cresting, and weather vanes.

Place television antennae and mechanical equipment, such as air conditioners, in an inconspicuous location.

Windows and doors Retain existing window and door openings including window sash, glass, lintels, sills, architraves, shutters, doors, pediments, hoods, steps, and all hardware.

Respect the stylistic period or periods a building represents. If replacement of window sash or doors is necessary, the replacement should duplicate the material, design, and hardware of the older window sash or door.

Porches and steps Retain porches and steps which are appropriate to the building and its development. Porches or additions reflecting later architectural styles are often important to the building's historical integrity and, wherever possible, should be retained.

Repair or replace, where necessary, deteriorated material with new material that duplicates the old as closely as possible.

Repair or replace, where necessary, deteriorated material with new material that duplicates the old as closely as possible.

Existing Buildings: Exterior Finishes

Discover and retain original paint colors, or repaint with colors based on the original to illustrate the distinctive character of the property.

Existing Buildings: Interior Features

Retain original material, architectural features, and hardware, whenever possible, such as stairs, handrails, balusters, mantelpieces, cornices, chair rails, baseboards, paneling, doors and doorways, wallpaper, lighting fixtures, locks, and door knobs.

Repair or replace, where necessary, deteriorated material with new material that duplicates the old as closely as possible.

Retain original plaster, whenever possible.

Discover and retain original paint colors, wallpapers, and other decorative motifs or, where necessary, replace them with colors, wallpapers, or decorative motifs based on the original.

REHABILITATION GUIDELINES

Existing Buildings: Plan and Function

Use a building for its intended purposes.

Find an adaptive use, when necessary, which is compatible with the plan, structure, and appearance of the building.

Retain the basic plan of a building, whenever possible.

New Construction

Make new additions and new buildings compatible in scale, building materials, and texture.

Design new work to be compatible in materials, size, scale, color, and texture with the earlier building and the neighborhood.

Use contemporary designs compatible with the character and mood of the building or the neighborhood.

Mechanical Services in Existing Buildings: Heating, Electrical, and Plumbing

Install necessary building services in areas and spaces that will require the least possible alteration to the plan, materials, and appearance of the building.

Install the vertical runs of ducts, pipes, and cables in closets, service rooms, and wall cavities.

Select mechanical systems that best suit the building.

Rewire early lighting fixtures.

Have exterior electrical and telephone cables installed underground.

Safety and Code Requirements

Comply with code requirements in such a manner that the essential character of a building is preserved intact.

Investigate variances for historic properties under local codes.

Install adequate fire-prevention equipment in a manner that does minimal damage to the appearance or fabric of a property.

Provide access for the handicapped without damaging the essential character of a property.

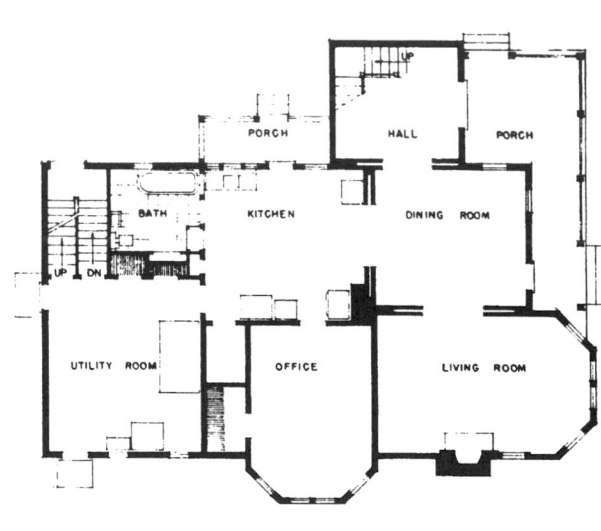

FIRST FLOOR PLAN (BEFORE)

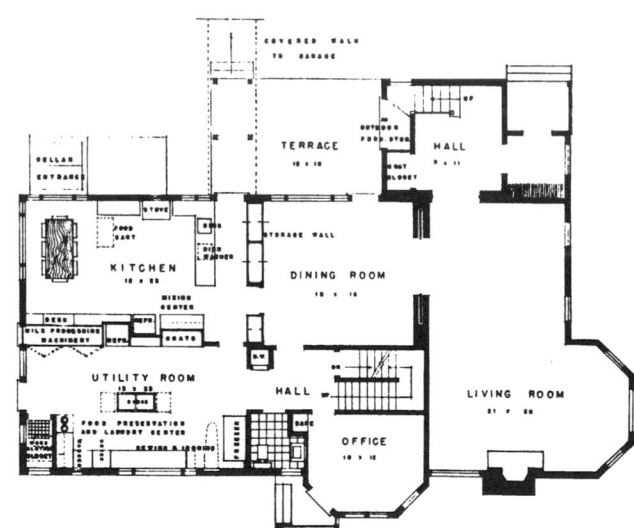

FIRST FLOOR PLAN (AFTER)

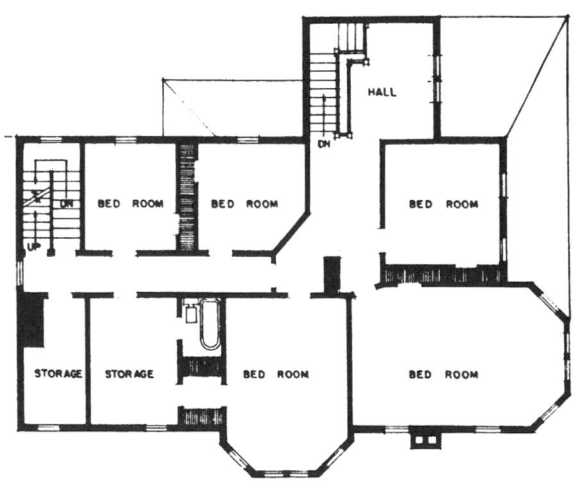

SECOND FLOOR PLAN (BEFORE)

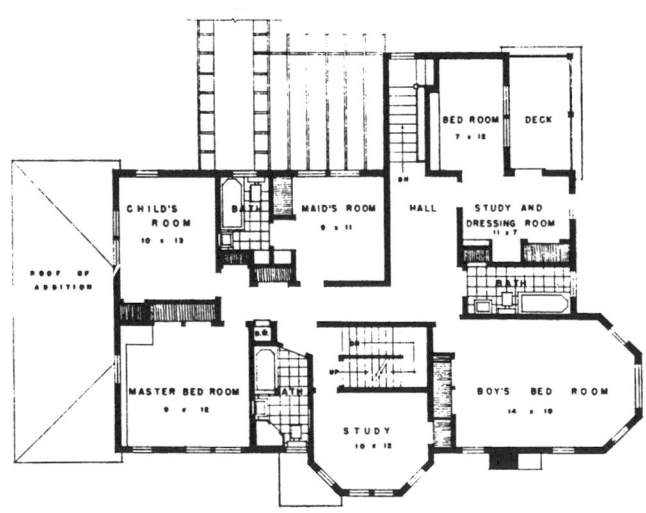

SECOND FLOOR PLAN (AFTER)

Fig. 1

BROWNSTONE REHABILITATION

Floor plans of the existing buildings, and plans of the three grades of remodeling (minimum, intermediate, and extensive) are presented. All three grades showed the best results by having two small apartments per floor, particularly above the garden and first floor levels, rather than single floor-through apartments. The additional rent more than justifies the cost of extra bathrooms and kitchens.

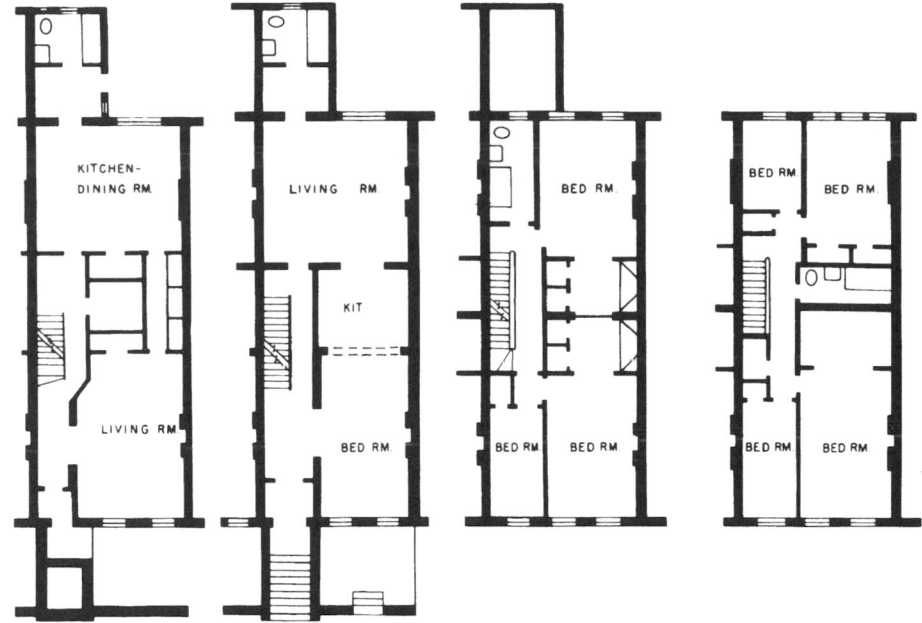

Fig. 1   Brownstones before rehabilitation.

Minimum Rehabilitation

In the scheme in Fig. 2, every effort was made to avoid structural changes and moving of partitions. The old stoop and cornice remain; electric and heating facilities are to be augmented rather than replaced; plaster and floors are to be patched. However, in regard to plumbing, it was found that it would be cheaper to rough-in pipes, back to back, in a central location, rather than to replace old lines in isolated spots—a factor which also improved the room layouts.

The results of these alterations are eight apartments per building, each with bath and kitchenette mechanically ventilated by ducts in a shaft rising from the basement up through the roof. Five of the apartments have separate small bedrooms; three have only one room. The apartments shown are entered from the street, but a building entrance from the rear could be included.

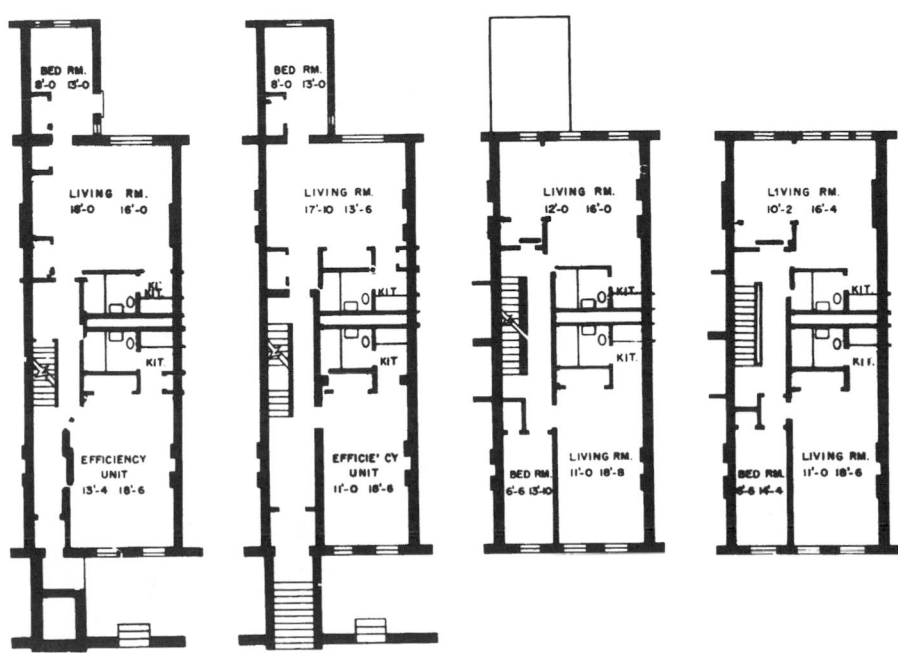

Fig. 2   Minimum rehabilitation of single brownstone.

BROWNSTONE REHABILITATION

Intermediate Rehabilitation

Intermediate rehabilitation (Fig. 3) involves removal of the stoop and resurfacing of the front facade; shifting of some of the interior partitions; improving items of finish such as tile in bathrooms; laying new floors and extensive new plaster; and installing completely new wiring, a new oil burner, and other elements. The floor plan is quite similar to the minimum grade, with two dwelling units per floor, except that a 3½-rental-room apartment is provided on the garden level.

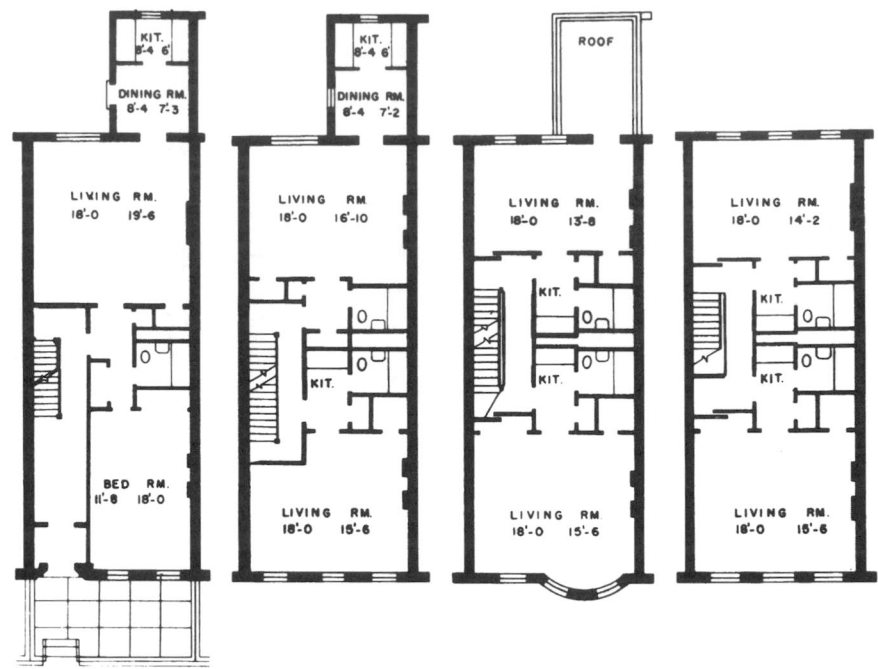

Fig. 3   Intermediate rehabilitation of single brownstone.

Extensive Rehabilitation

Extensive rehabilitation (Figs. 4 and 5) involves gutting the existing structure, removing all interior partitions, and leaving only exterior walls and subflooring. The plans call for new plastered partitions, new ceiling plaster, finished flooring, heating, plumbing, and electrical work, the removal of the stoop and cornice, and replacement of the old interior stairs.

Architectural studies were made of a single building and also of two adjoining buildings, with a new central staircase serving both structures. Alternate solutions for street and garden entrances are shown for the latter. In the plan for two buildings, a 3½-room apartment and two smaller apartments would be provided on the garden level, with four apartments on each of the other floors.

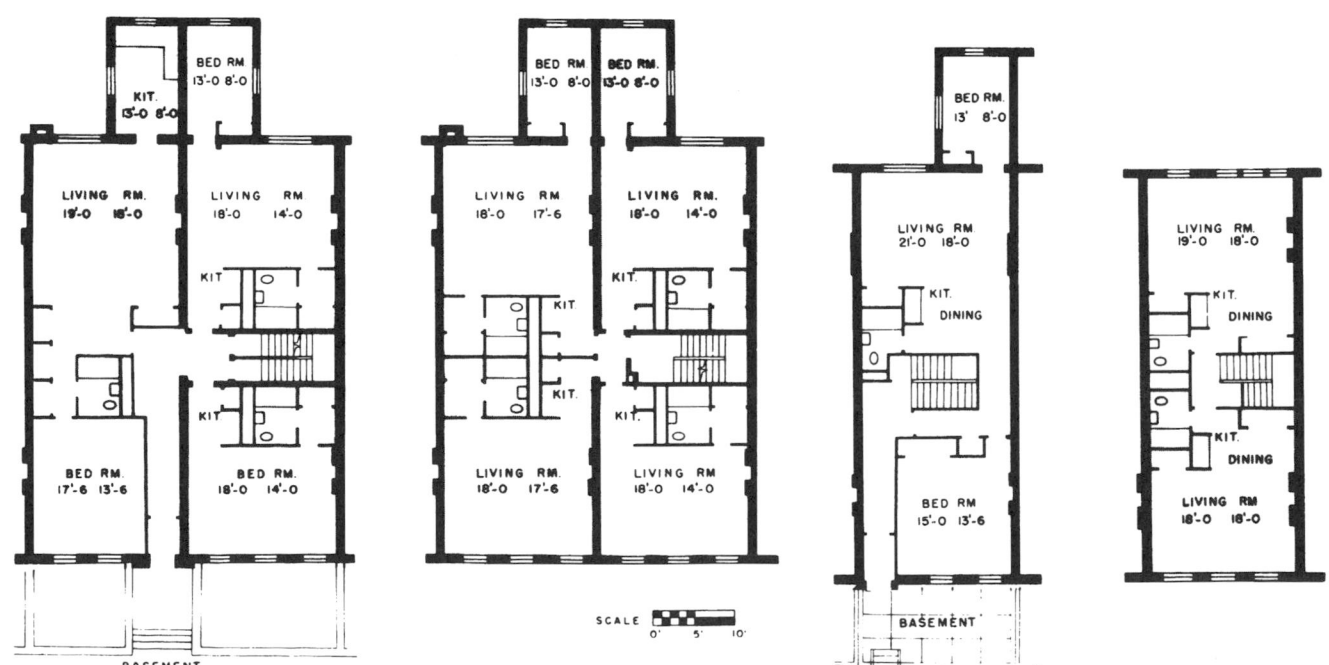

Fig. 4   Extensive rehabilitation of two adjacent brownstones, entrance from street.

Fig. 5   Extensive brownstone rehabilitation with floor-through apartment.

Desirability of Brownstone Rehabilitation

Whether or not any of the rehabilitation studies described above are advisable on a large scale depends on two considerations. The first is whether the rehabilitated building is satisfactory for living, that is, whether it has enough space, light, and air. Is it attractive both within and without, in terms of outlook and appearance? Another consideration is whether the costs of alteration can be carried by prospective rents under available financing terms. In other words, will the returns to builders, entrepreneurs or owner-occupiers be sufficient to induce them to undertake remodeling?

All these rehabilitation proposals are judged to be acceptable with respect to the quality of the resultant living space. The brownstone as a building type was originally intended for single-family use and is therefore seldom over four or five stories high. The density per acre even after remodeling will be low compared with a new multistoried apartment building. Forty percent of the lot is generally available for rear gardens, held individually or in common, which form one of the most desirable aspects of city living. The streets on which the buildings face are local streets, have less noise from traffic, and can be made more attractive with trees. These low buildings along the side streets form a pleasing contrast to the proposed tall buildings along the avenues. Retaining the brownstones will result in a desirable combination of high and low structures. The sense of privacy in the smaller buildings and their more human scale are added attractions for many prospective tenants. It should also be noted that unless the brownstones are rehabilitated they will probably be replaced, for economic reasons, with tall apartment structures. While such replacement might result in lower ground coverage, it would also mean a higher and undesirable overall density. Thus the choice is not between new row houses of a better design and the proposed remodeling, but between some existing brownstones rehabilitated and mixed with high-rise buildings on the one hand, and an area completely covered with high-rise buildings on the other.

Fig. 6 Extensive rehabilitation of two adjacent brownstones, entrance from gardens.

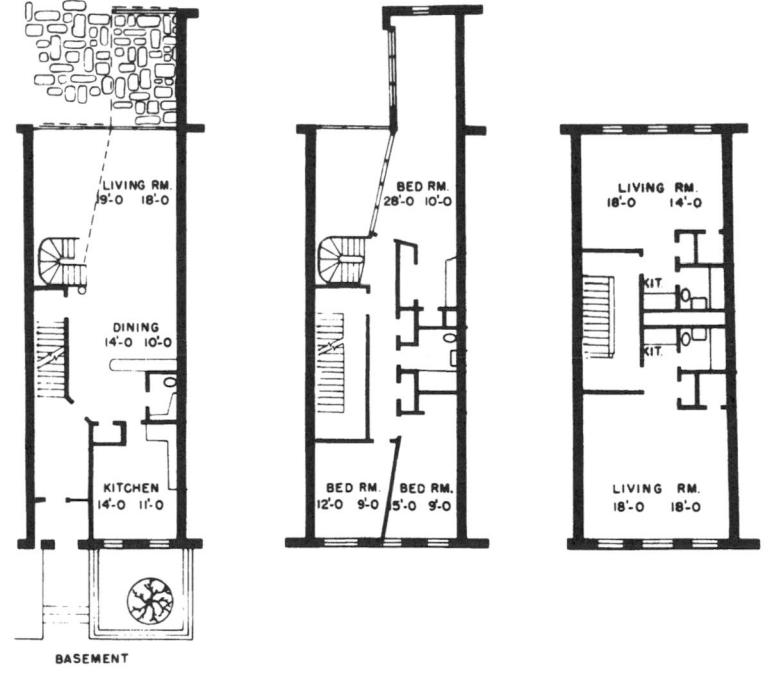

Fig. 7 Extensive brownstone rehabilitation with duplex apartment.

BROWNSTONE REHABILITATION

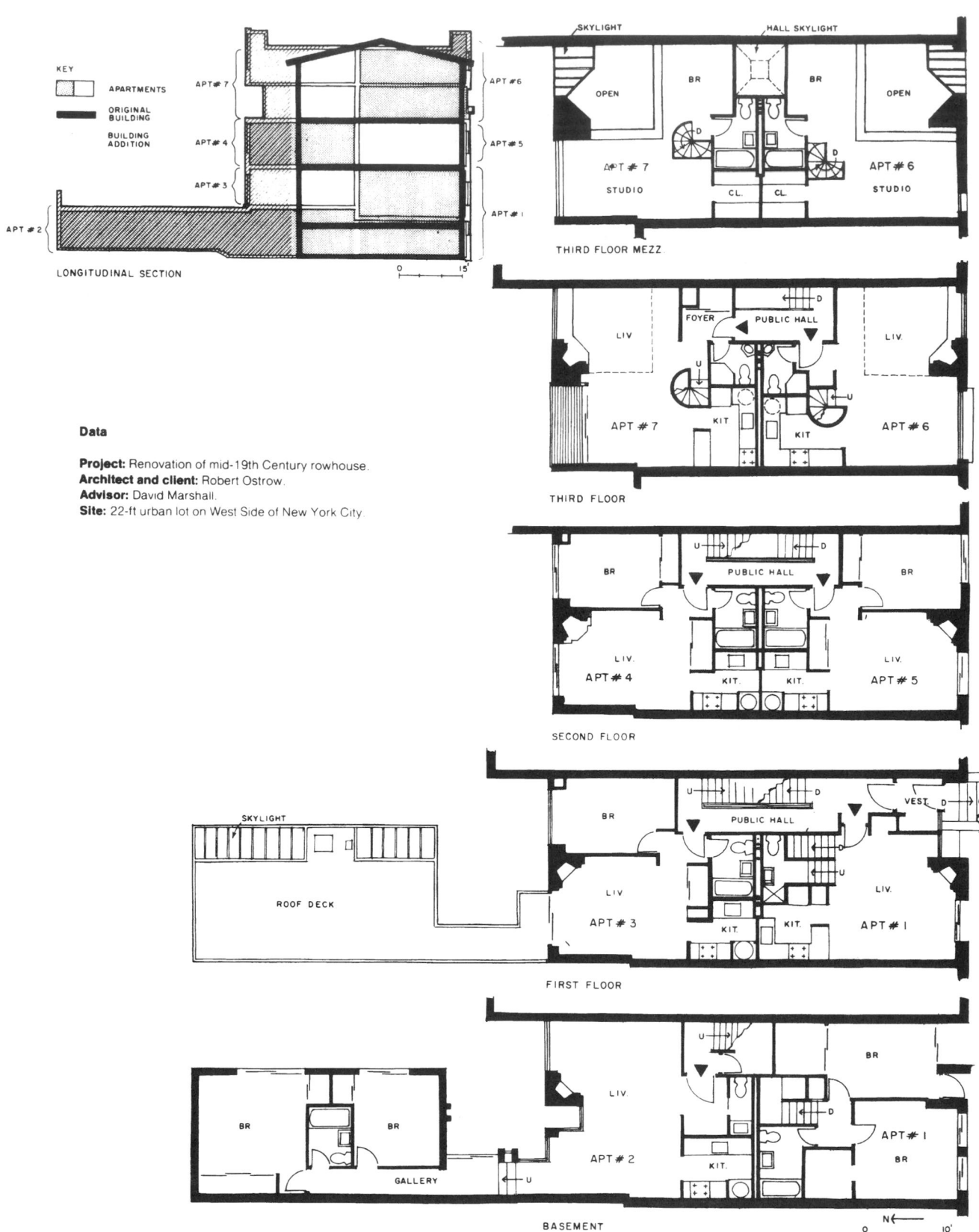

KEY

APARTMENTS

ORIGINAL BUILDING

BUILDING ADDITION

LONGITUDINAL SECTION

Data

Project: Renovation of mid-19th Century rowhouse.
Architect and client: Robert Ostrow.
Advisor: David Marshall.
Site: 22-ft urban lot on West Side of New York City.

THIRD FLOOR MEZZ.

THIRD FLOOR

SECOND FLOOR

FIRST FLOOR

BASEMENT

Fig. 8

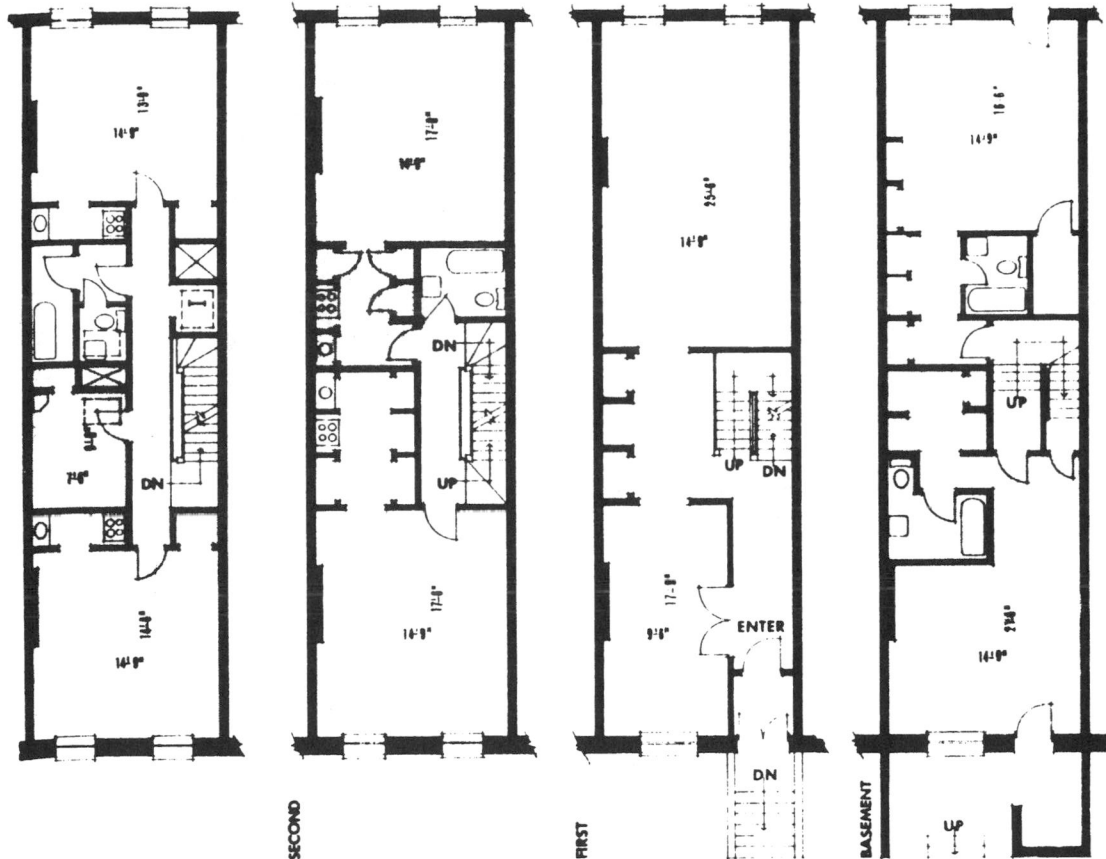

Fig. 9a   Before rehabilitation.

Fig. 9b   Standard rehabilitation.

Fig. 10 Extensive rehabilitation.

REHABILITATION OF OLD-LAW TENEMENTS

A similar analysis was conducted to test the feasibility of rehabilitating old-law tenements. In contrast to the brownstone, which is essentially a Victorian variation on the Georgian town house, the five-story walk-up old-law tenement is not basically a good building type. Land coverage is excessive—in many cases 80 percent of the lot. Many of its small rooms look into outer "courts" not more than 4 ft wide and over 40 ft deep. Little air and almost no light and privacy are the rule. The street facades do not lend themselves to economical facelifting. Nevertheless, most of these buildings are structurally sound.

Five old-law tenements were selected for study as representative of the great majority of the old-law tenements in New York City generally. The front of these tenements faces on a residentially developed thoroughfare; the rear overlooks a group of lower brownstone buildings, which permit some sun, light, and air to penetrate the narrow air shafts. Five tenements were chosen because the layout of this type of building requires alteration in groups to achieve a satisfactory plan.

The same technique was used in analyzing this building type as in the case of the brownstones. Three degrees of rehabilitation—minimum, intermediate, and extensive—were developed for these five buildings which are shown in Fig. 1 together with a plan of the existing structures.

Minimum Rehabilitation

This scheme in Fig. 2 involves the use of existing space where possible, little change in partitions, and generally the type of finish proposed for the "minimum" brownstone alteration. As a floor-through six-room apartment would require excessive rents for this type of walk-up structure, the minimum scheme provides for breaking up the existing six-room flats into two apartments, one facing the street, the other the rear yard and air shaft. The old staircase would be kept in its present location, and no elevator or incinerator would be provided. The basic defects of deep narrow airshafts and small rooms would remain.

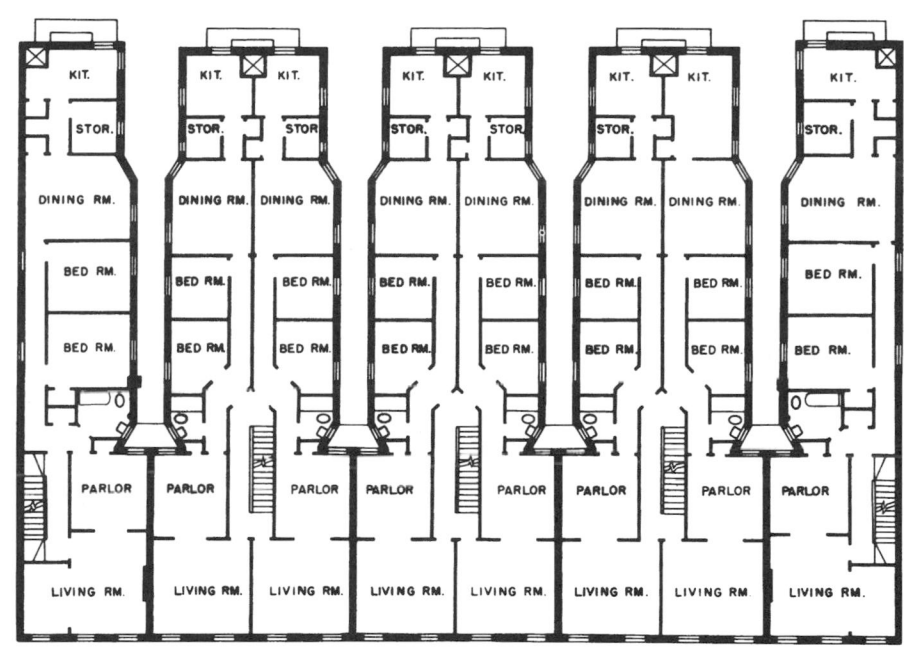

Fig. 1   Before rehabilitation.

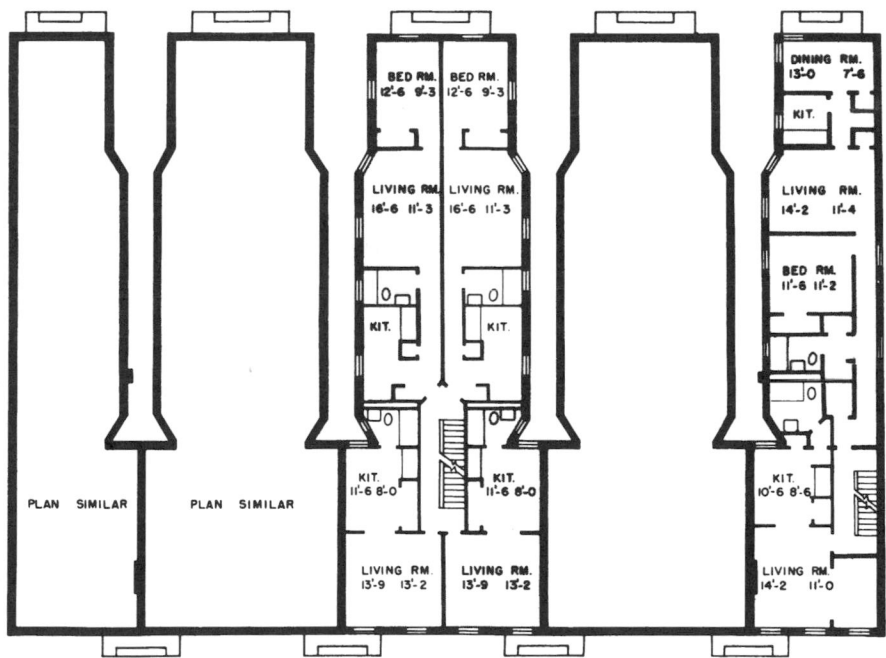

Fig. 2   Minimum rehabilitation.

REHABILITATION OF OLD-LAW TENEMENTS

Intermediate Rehabilitation

The scheme in Fig. 3 would eliminate the deep narrow airshafts by the partial demolition of the rear wings of alternate buildings to provide better light, air, and outlook for all rear apartments. Here again the old staircase remains and there is no elevator. As may be seen in floor plans, the apartment layouts have been improved and the specifications call for better renovation than proposed in the minimum scheme.

Rather than the "railroad" layout of the original tenement, which was kept in the minimum scheme, the Intermediate plan makes a single apartment out of each rear wing. Thus where the rear apartment is not located in one of the end buildings, it has windows on three sides, opening onto generously proportioned courts. All rooms are good-sized and are entered from an internal hall and foyer. The large apartment between the rear wings has through ventilation with an equally good layout. Even the smallest apartments have through ventilation.

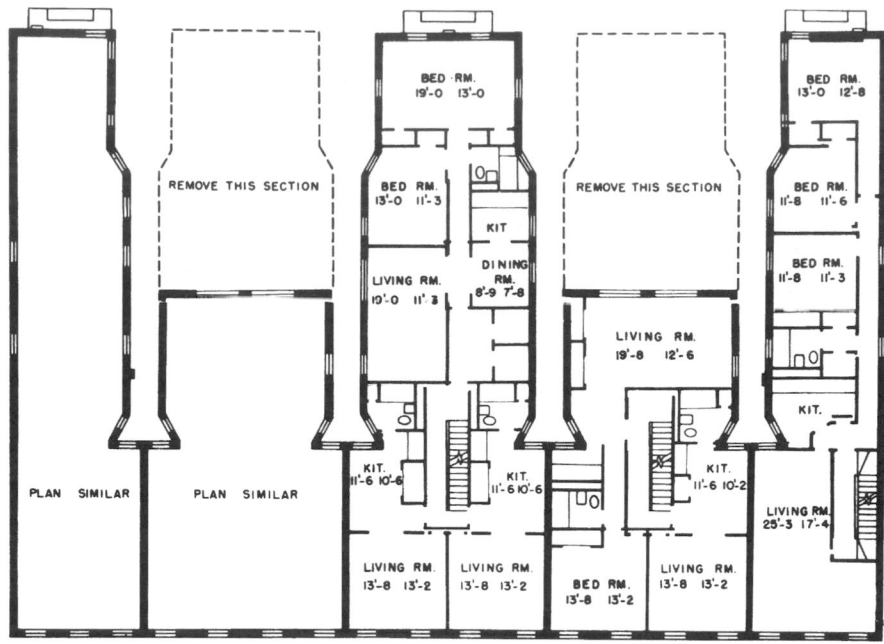

Fig. 3   Intermediate rehabilitation.

Extensive Rehabilitation

The scheme in Fig. 4 goes the whole way toward complete modernization. All five buildings are connected by a common complex of hall, stair, elevator, and incinerator. Alternative wings would be demolished, most partitions would be new, and specifications would be similar to those used in the maximum brownstone rehabilitation scheme. As shown, the resulting apartments are quite satisfactory.

Fig. 4   Extensive rehabilitation.

Fig. 5*a* Before rehabilitation.

Fig. 5*b* Standard rehabilitation.

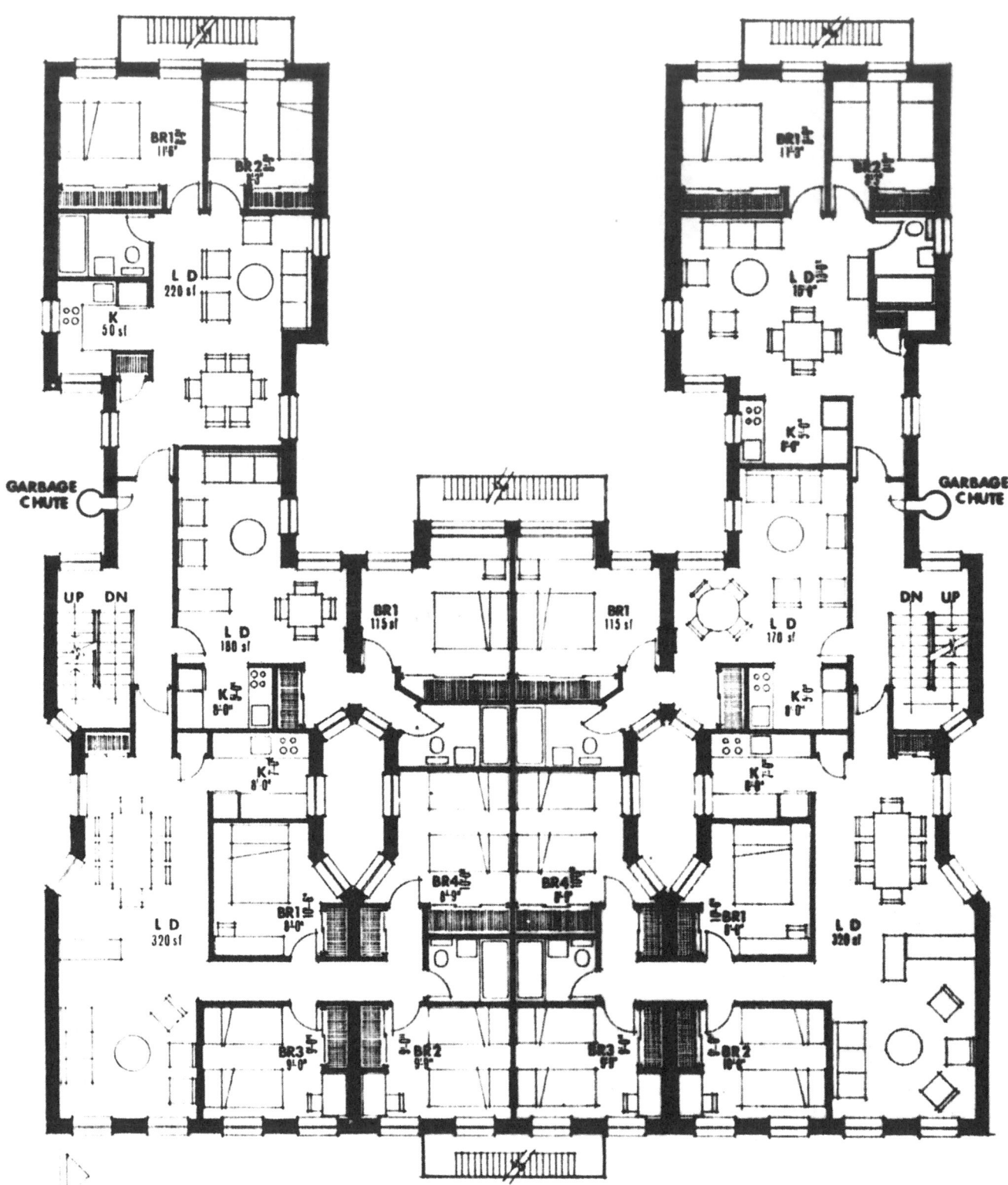

Fig. 6   Extensive rehabilitation.

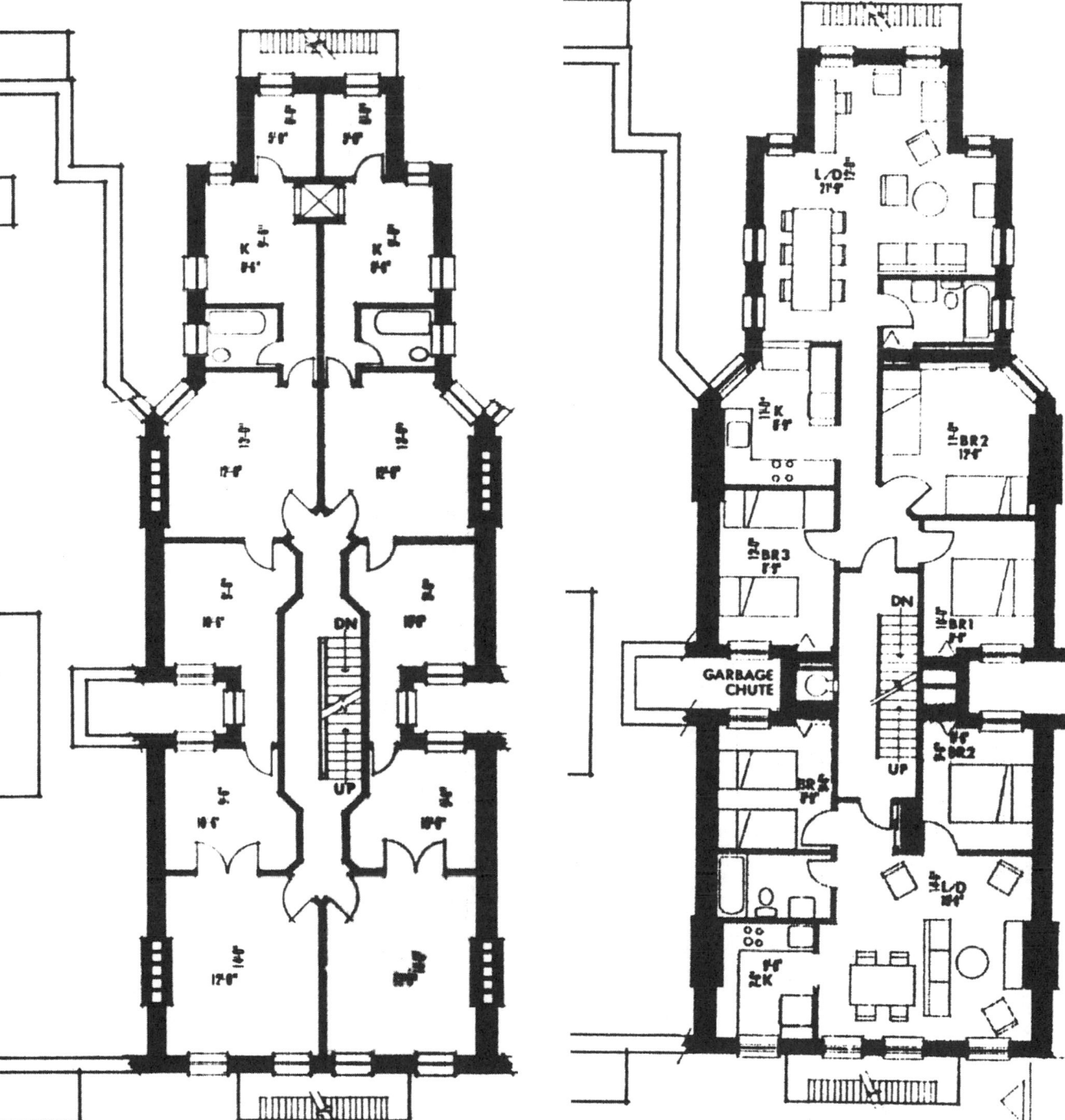

Fig. 7*a* Before rehabilitation.

Fig. 7*b* Standard rehabilitation.

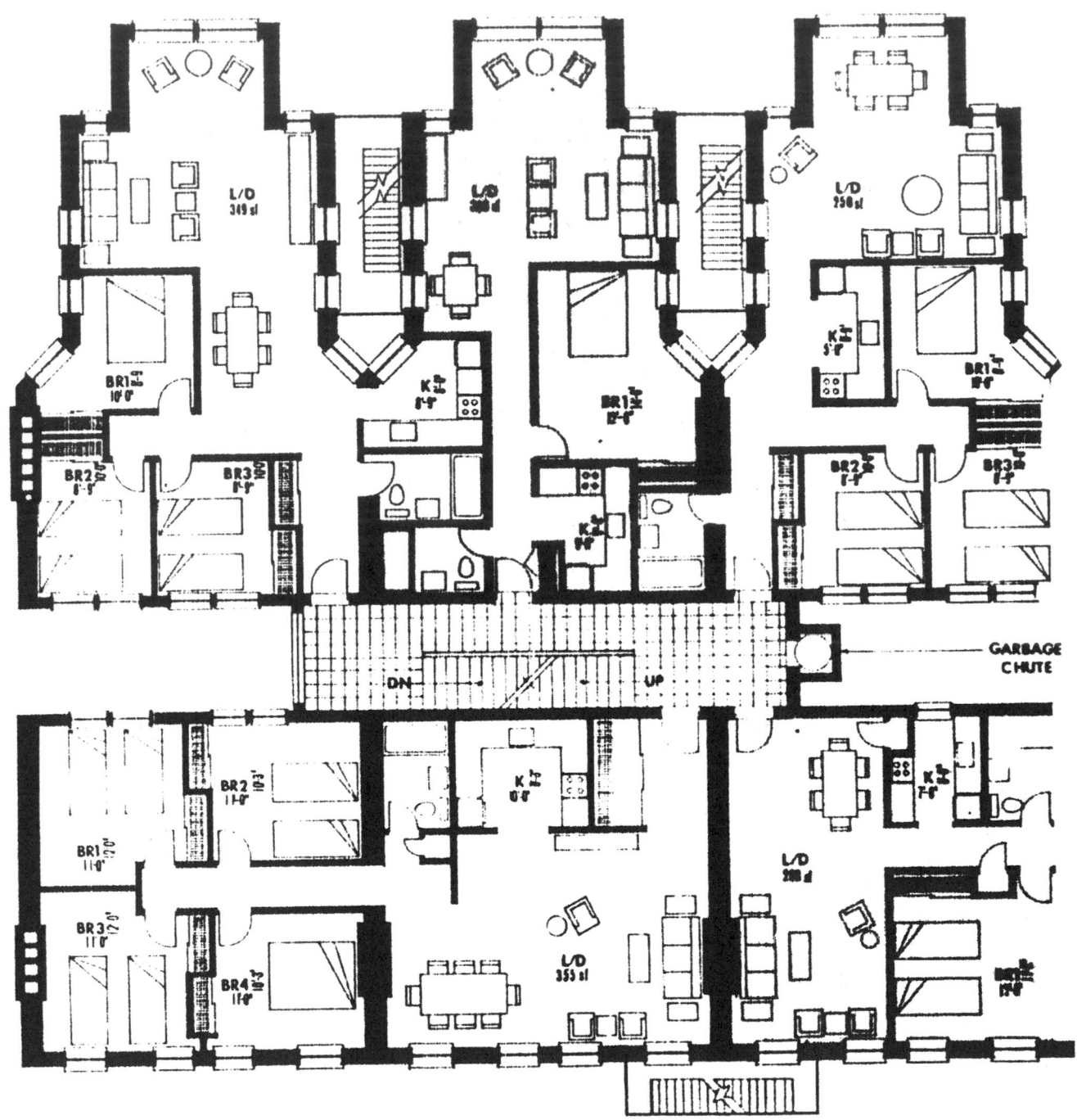

Fig. 8   Extensive rehabilitation.

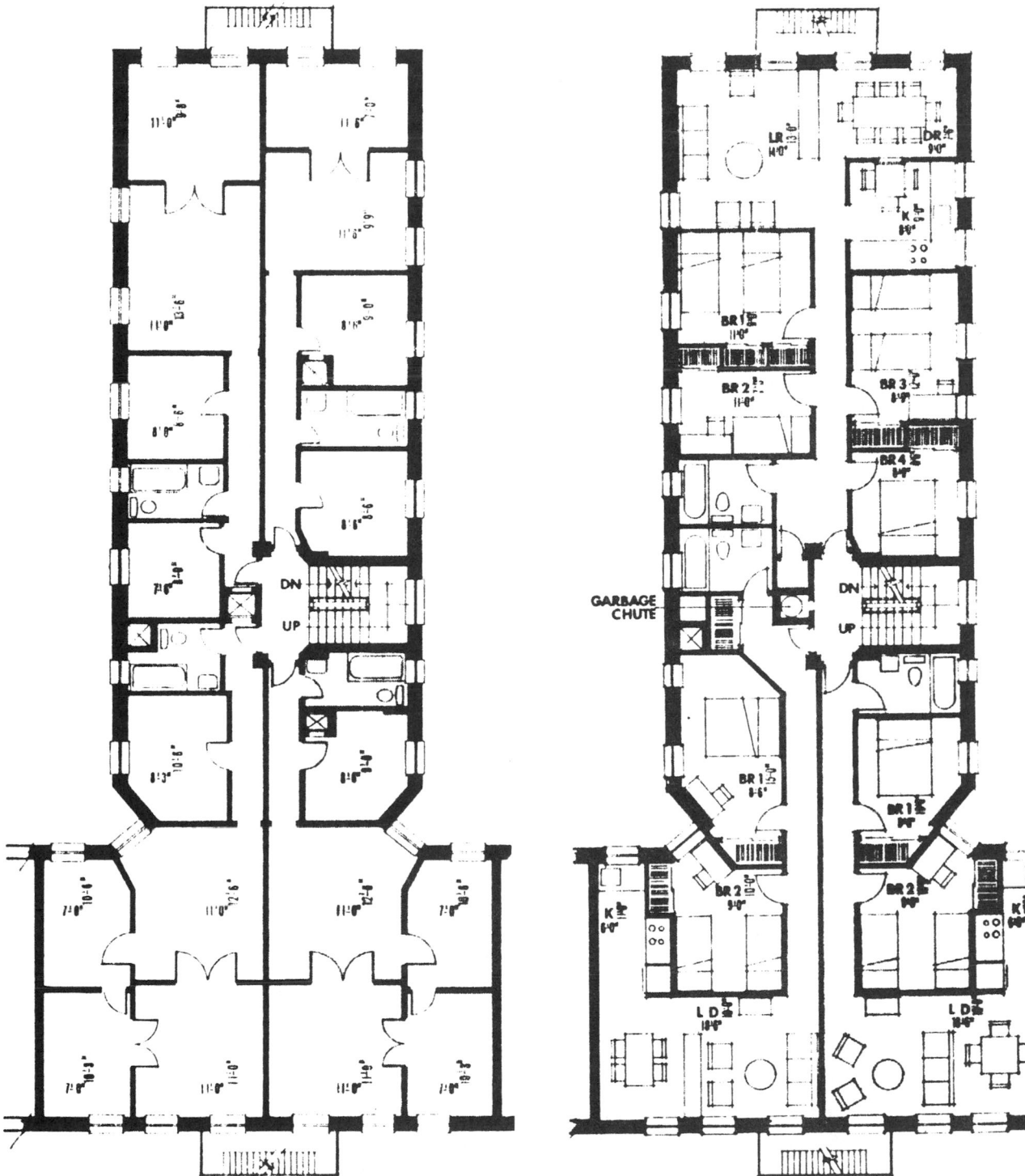

Fig. 1*a* Before rehabilitation.

Fig. 1*b* Standard rehabilitation.

REHABILITATION OF NEW-LAW TENEMENTS

Jose de Diego Beekman houses in the Bronx were originally T-shaped tenement structures built under the "new law" tenement legislation of 1901. New bridges, shaded in Fig. 2, connect what originally were two separate buildings so that one elevator could serve both.

The H-shaped plan, another type of early tenement structure, is part of the same project.

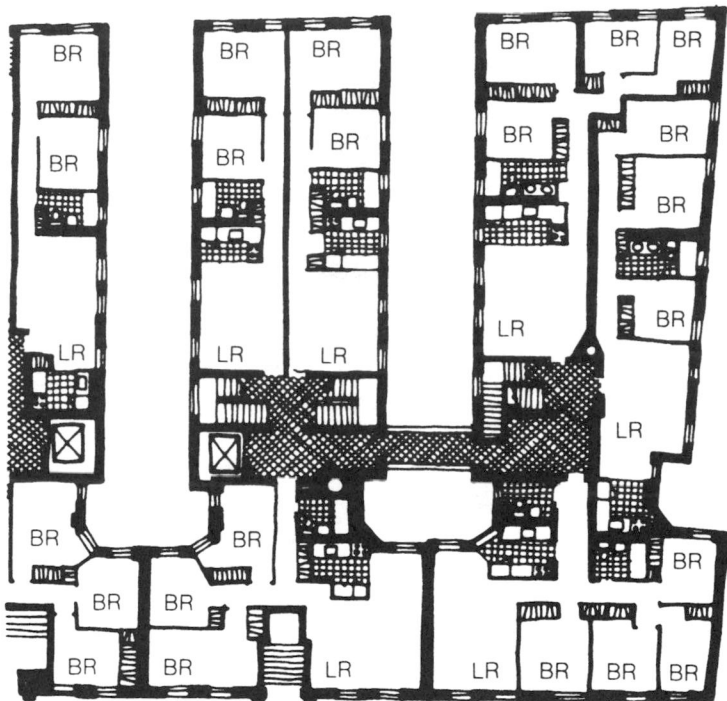

Fig. 2   Beyer, Blinder, Belle—Architects.

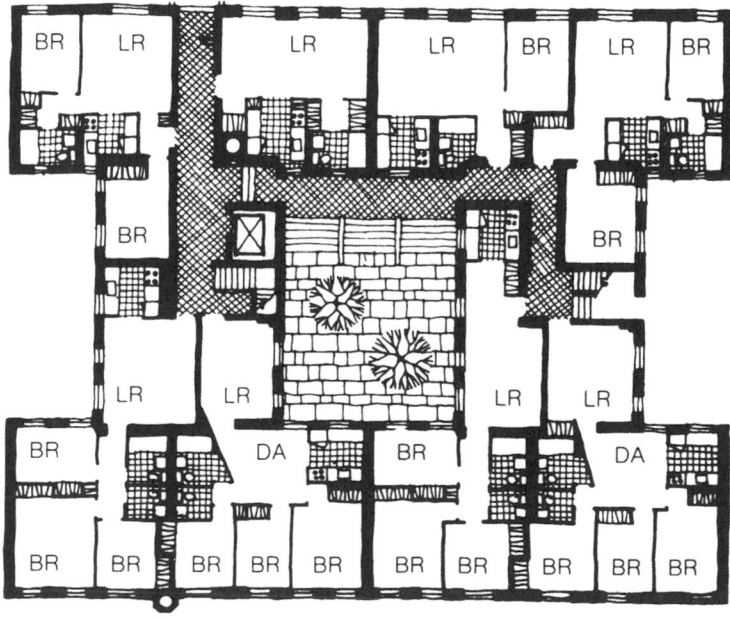

Fig. 3   Beyer, Blinder, Belle—Architects.

The building in Fig. 1 contains 121 apartments ranging from duplexes and triplexes with lofts and skylights to one-bedroom apartments.

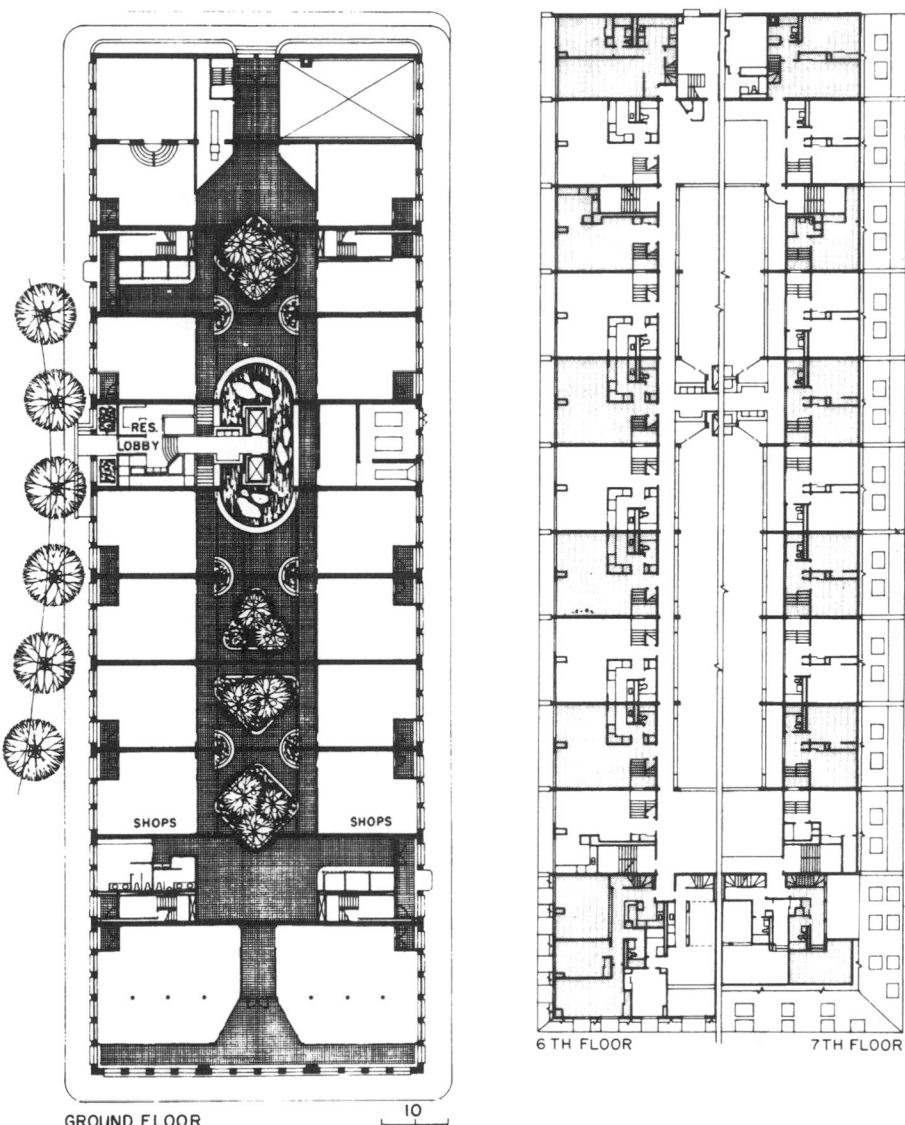

GROUND FLOOR

6TH FLOOR   7TH FLOOR

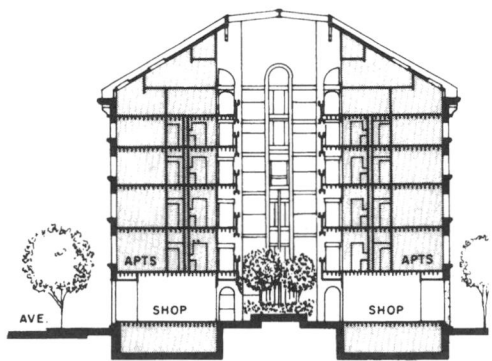

Fig. 1   Mercantile Wharf Building, Boston, Mass. Mercantile Associates—Owner. John Sharratt Associates, Inc.—Architects.

Westbeth Artists Housing is one of the largest and most complex rehabilitation projects in the country. Designed by Richard Meier, the renovation has turned a square block of old buildings in Greenwich Village (formerly used as laboratories and warehouses) into some 384 new apartments for working artists and their families.

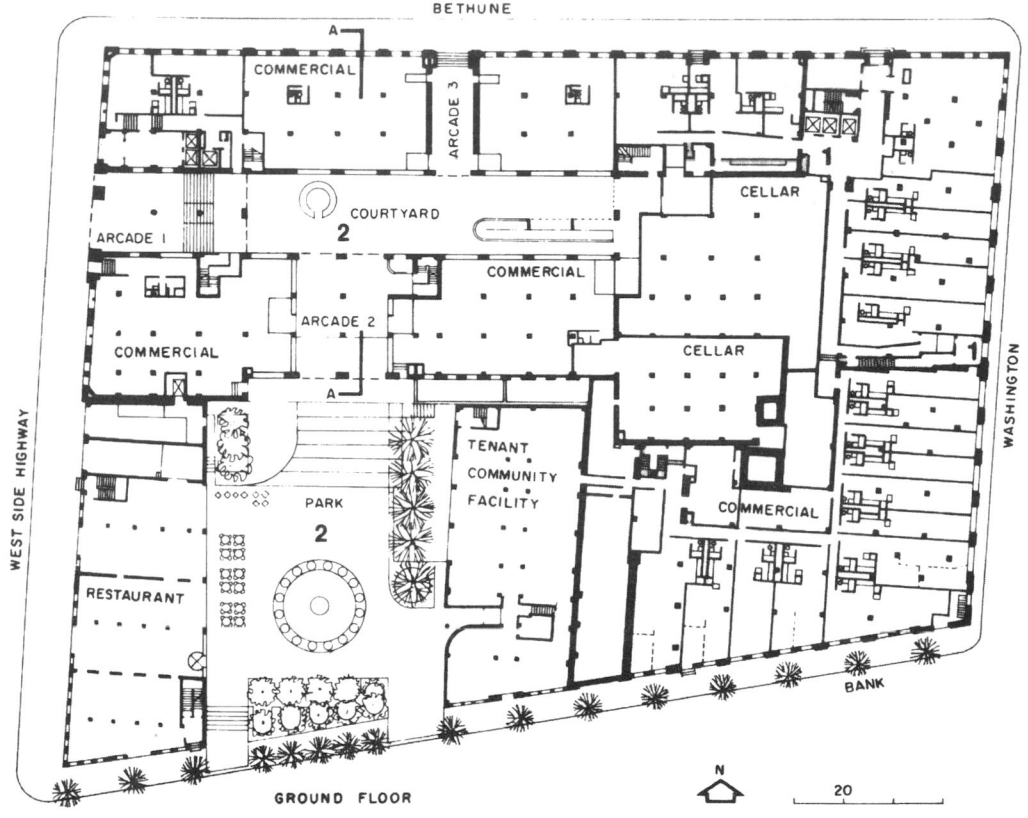

Fig. 1   Westbeth Artists Housing, Greenwich Village, New York City. Richard Meier—Architect.

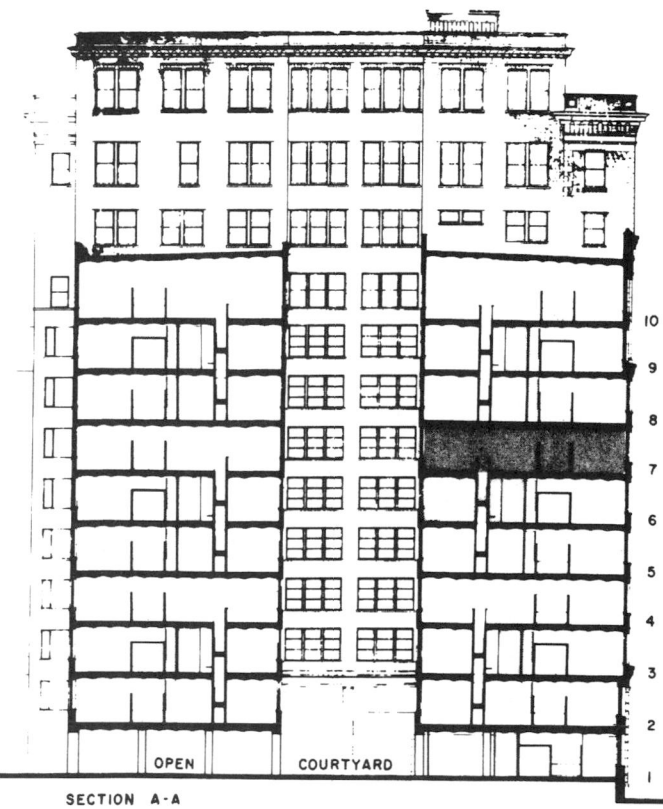

Fig. 2   Westbeth Artists Housing, Greenwich Village, New York City. Richard Meier—Architect.

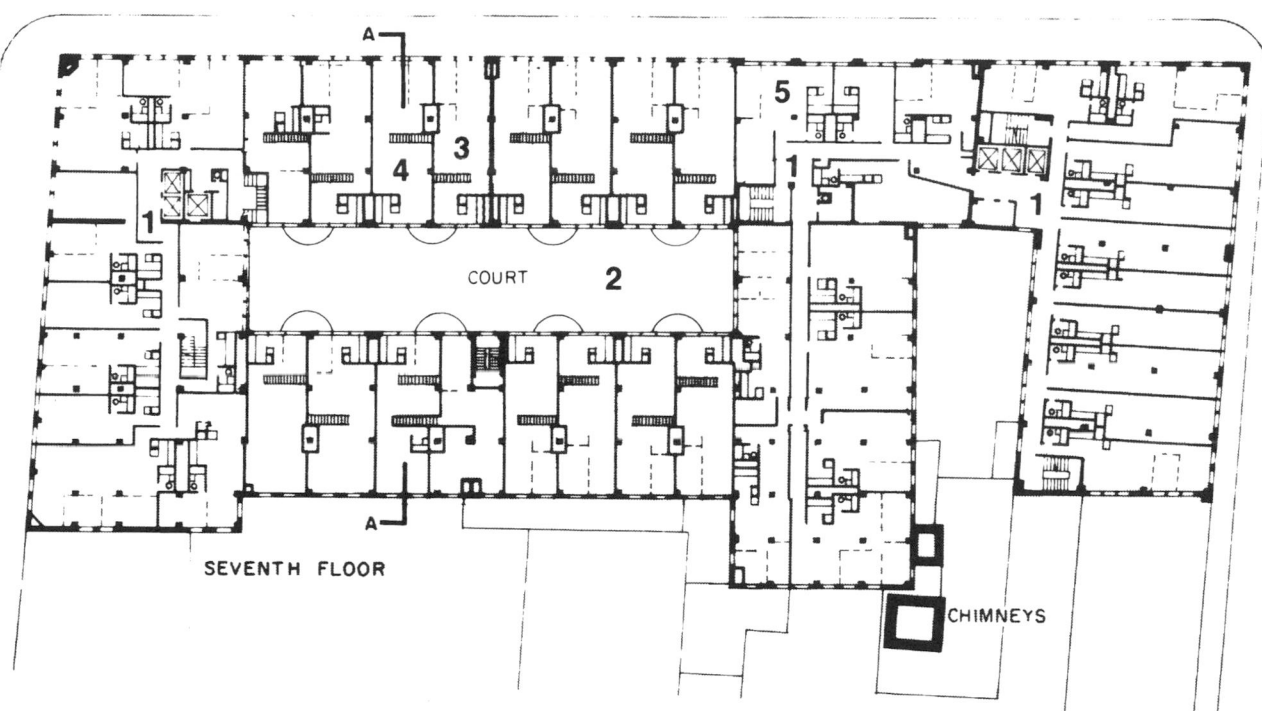

SEVENTH FLOOR

COURT

CHIMNEYS

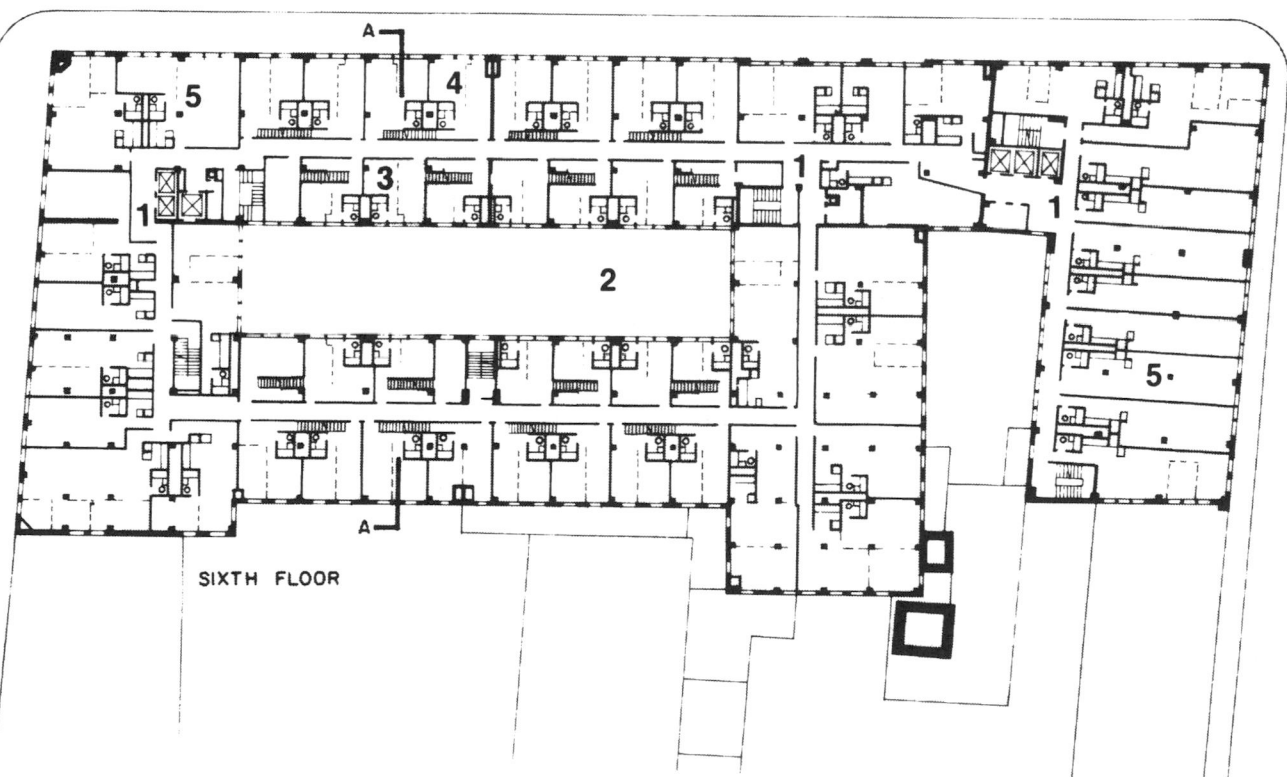

SIXTH FLOOR

Fig. 3   Westbeth Artists Housing, Greenwich Village, New York City. Richard Meier—Architect.

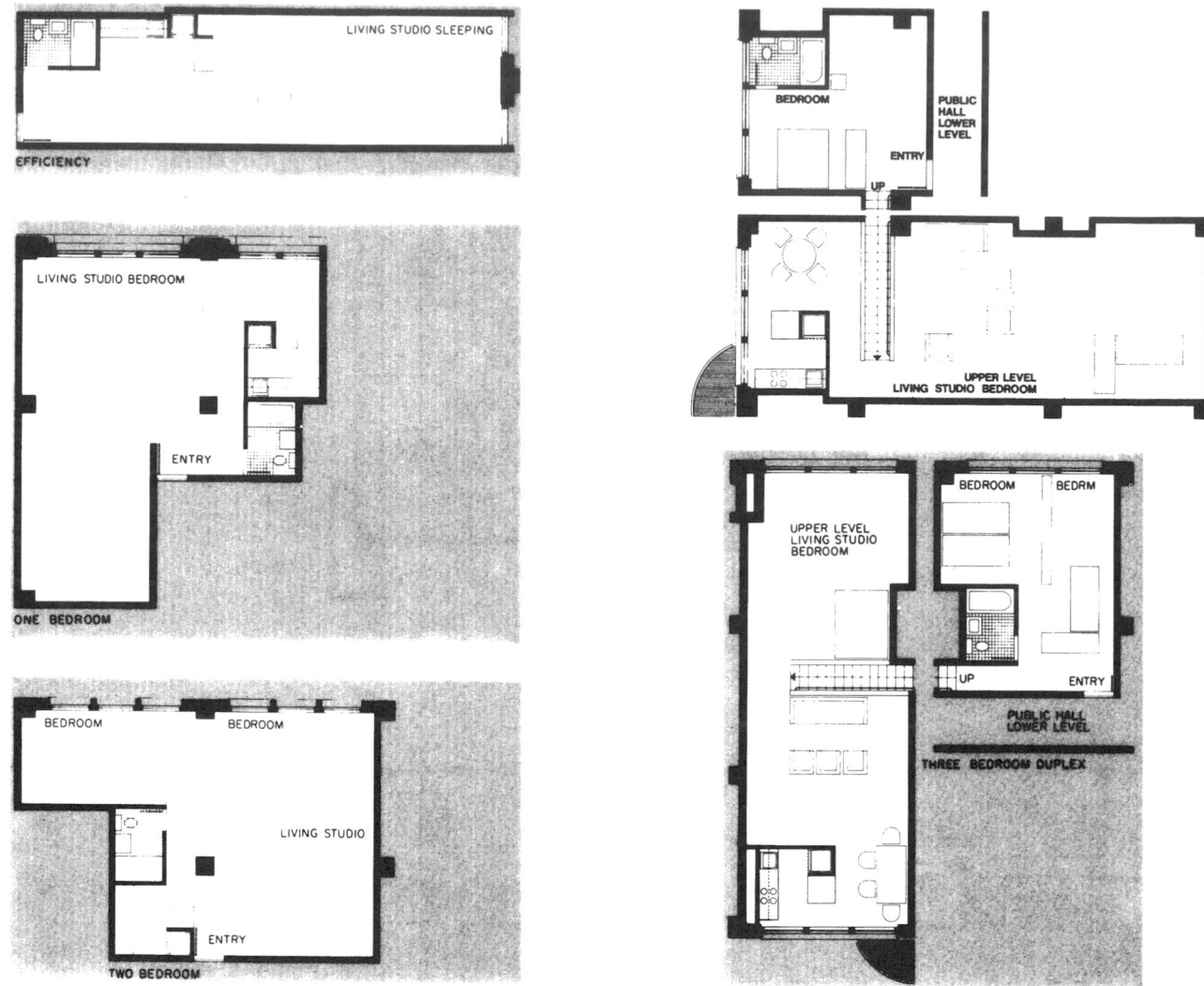

Fig. 4 Westbeth Artists Housing, Greenwich Village, New York City. Richard Meier—Architect.

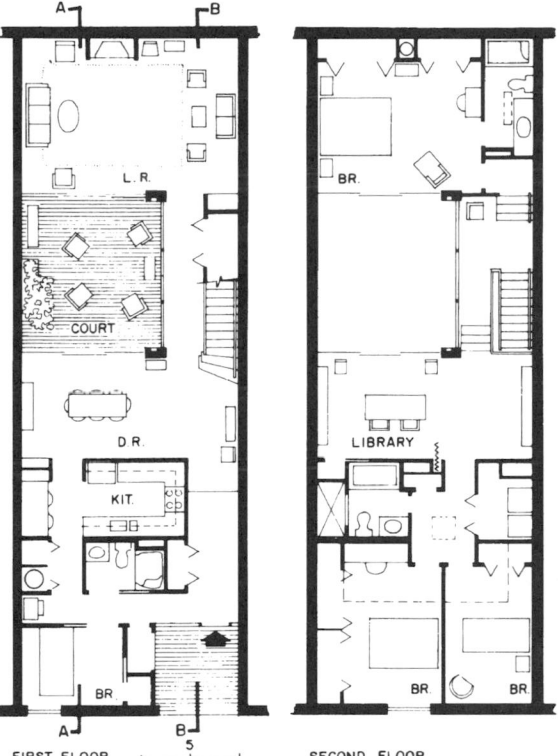

FIRST FLOOR   SECOND FLOOR

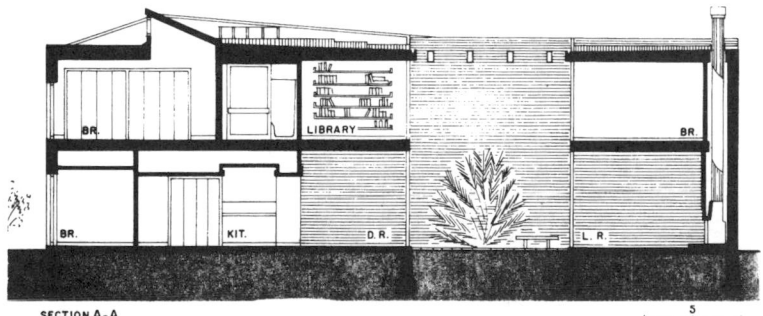

SECTION A-A

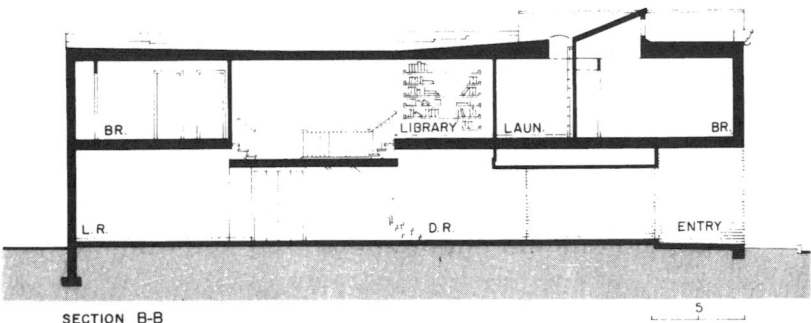

SECTION B-B

Fig. 5 Townhouse on Beacon Hill, Boston, Mass. Childs Bertman Tseckares Associates, Inc.—Architects.

The 19th century stable on Boston's Beacon Hill (Fig. 5), remodeled as a house, preserves a scale and character which is important in that historic district. But it also provides a place to live in town within walking distance of the owner's place of business.

In remodeling the old stable, some restrictions were imposed which determined the end result in unusually pleasant ways. The façade could not be changed because the building is in a designated historic district, and the side and rear walls precluded any new windows. The handsome courtyard was a natural and delightful solution to light and air for otherwise inside rooms. The rooms which surround the court are glass-walled, floor to ceiling, and the height of the principal rooms on the first floor was increased for added spaciousness and light.

The courtyard is a tradition in the part of Boston where this house is located, and its use here proved compatible with the owner's wishes. Its enclosed space acts as an additional room and is enlivened by a curtain and many plants. The court is the source of daylight for the principal rooms on both floors. In other parts of the house, colored clerestory windows, skylights and light shafts bring in natural light.

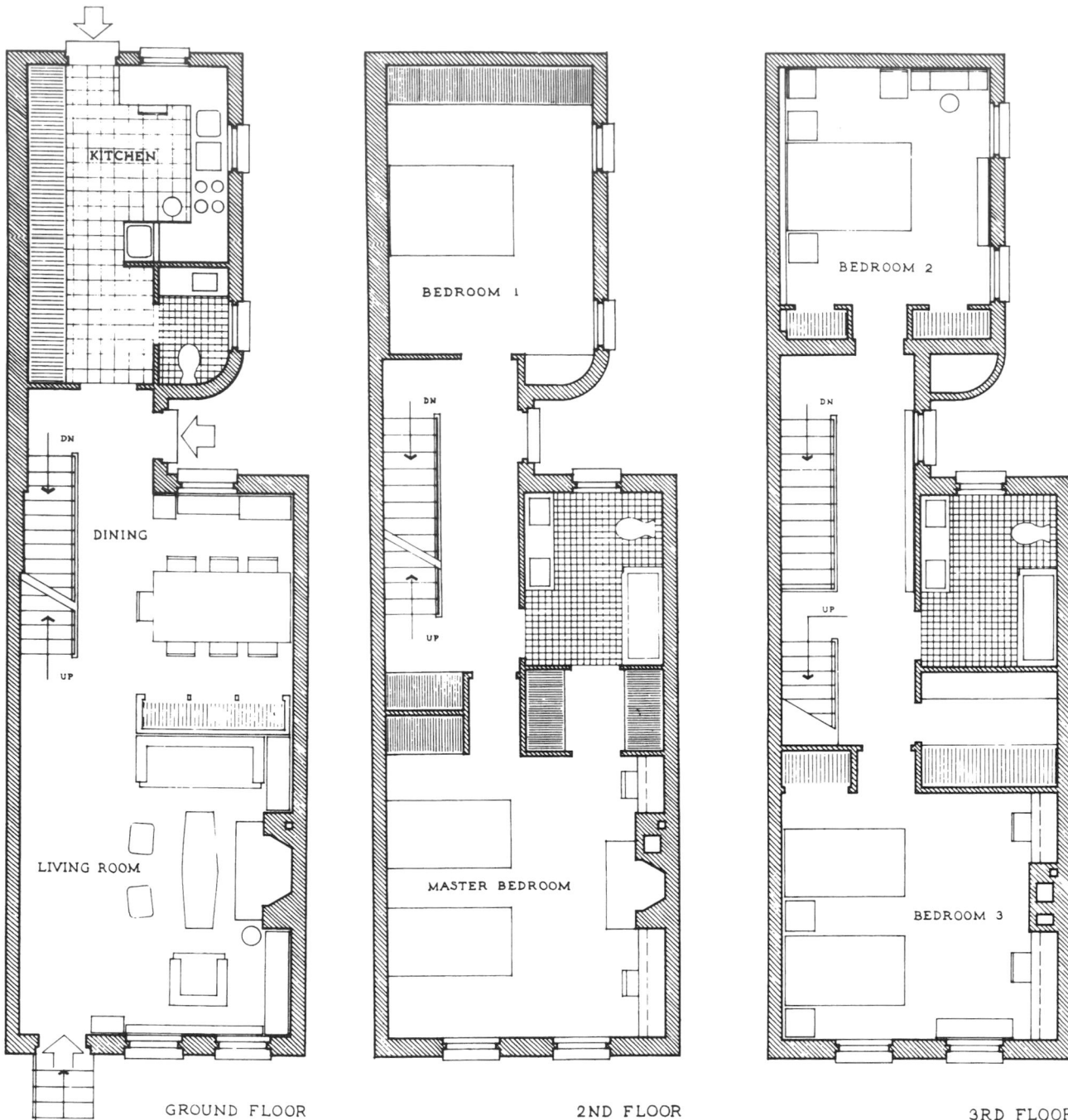

GROUND FLOOR   2ND FLOOR   3RD FLOOR

Fig. 6a Society Hill, Philadelphia, Pa. I. M. Pei—Architect.

223 SPRUCE STREET: A REHABILITATED TOWN HOUSE

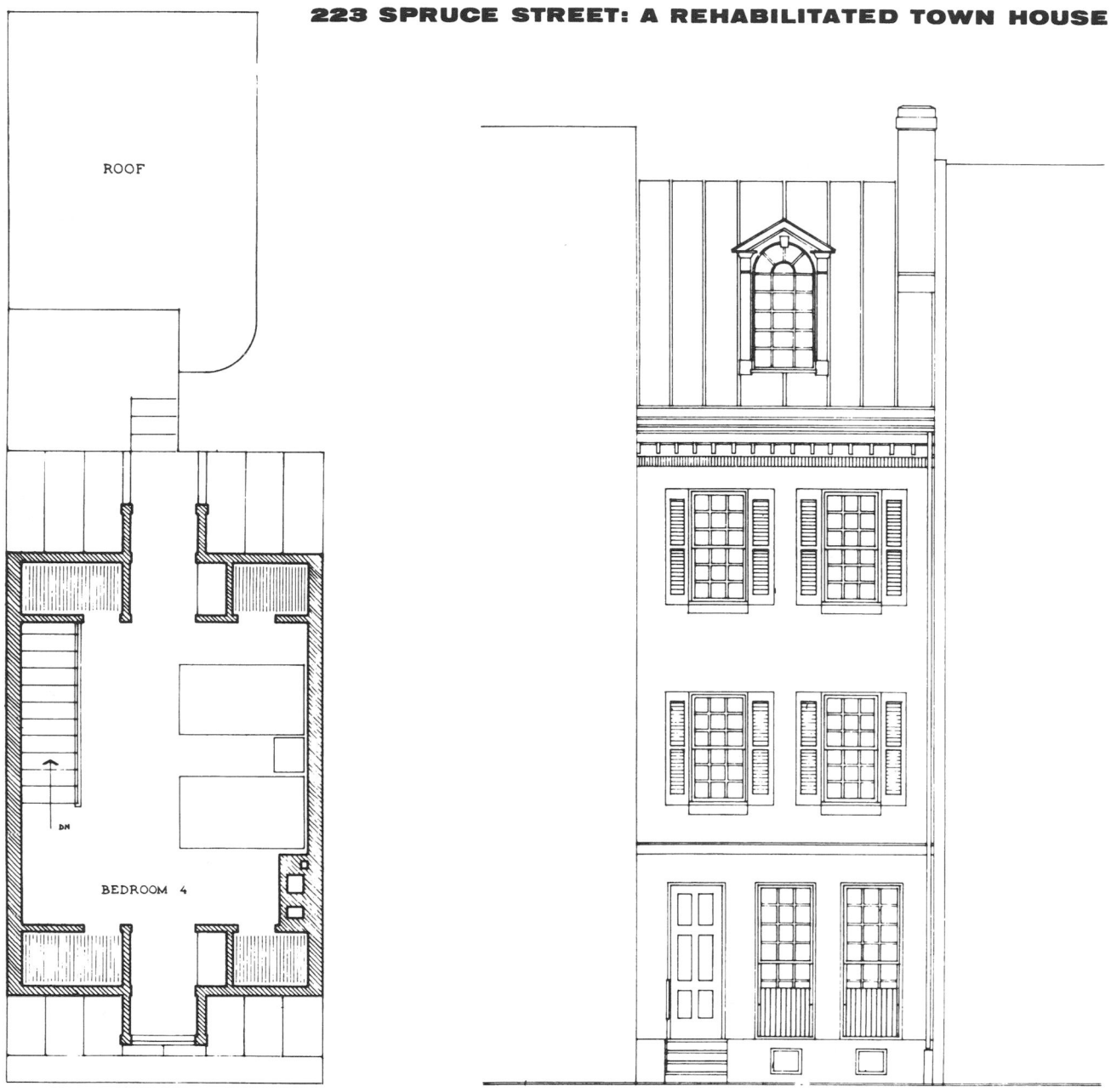

Fig. 6b   Society Hill, Philadelphia, Pa. I. M. Pei—Architect.

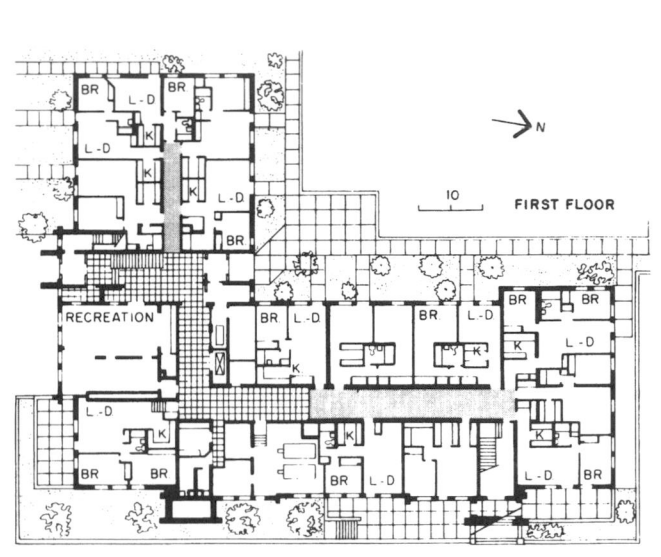

FIRST FLOOR

FOURTH FLOOR

SECOND AND THIRD FLOORS

Fig. 7   Anderson Notter Finegold—Architects.

The retail space in Fig. 8 includes 17 bays which open to outside streets and interior court at left.

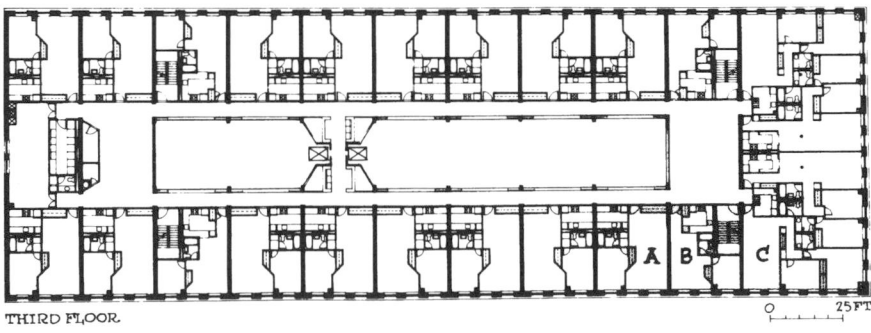

THIRD FLOOR

Fig. 8a   Mercantile Wharf, Boston, Mass. John Sharrett Associates, Inc.—Architects.

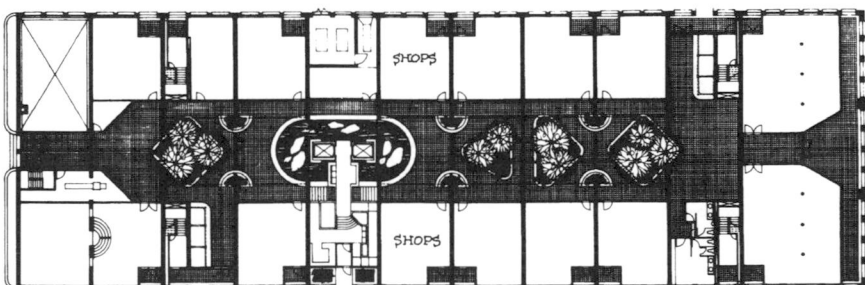

Fig. 8b   Mercantile Wharf, Boston, Mass. John Sharrett Associates, Inc.—Architects.

An interior court (Fig. 9) was carved out of the center of the building. It is topped by skylight and surrounded by ground-level shops and upper-level apartment corridors that are reached by two elevators.

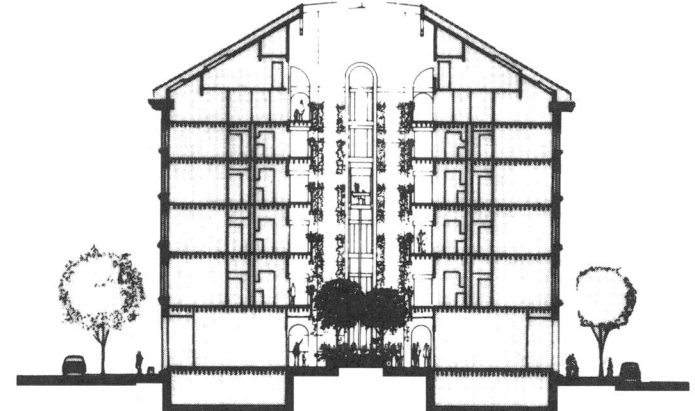

Fig. 9   Mercantile Wharf, Boston, Mass. John Sharrett Associates, Inc.—Architects.

Apartment space includes 14 different plans—eight flats on the second through fifth floors, and six multilevel units on the top floor. The smallest flat (B) is 535 ft²; the largest flat (C) is 985 ft²; flat A is 690 ft².

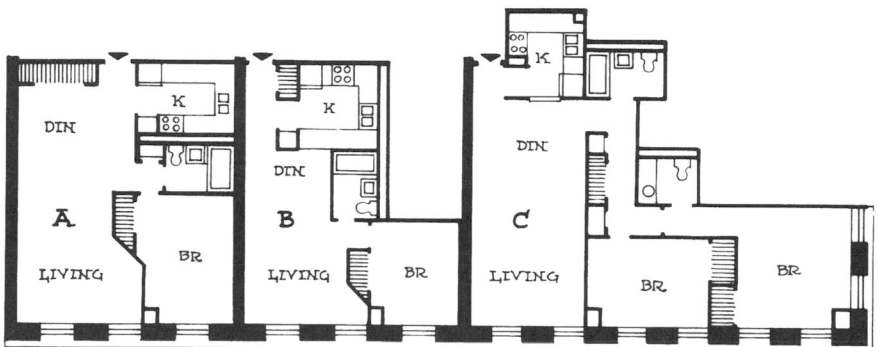

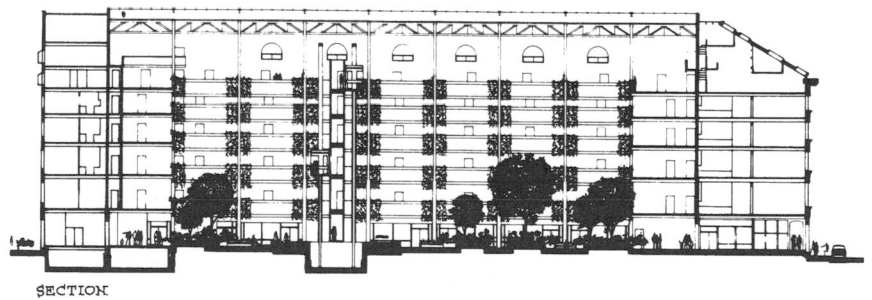

Fig. 10

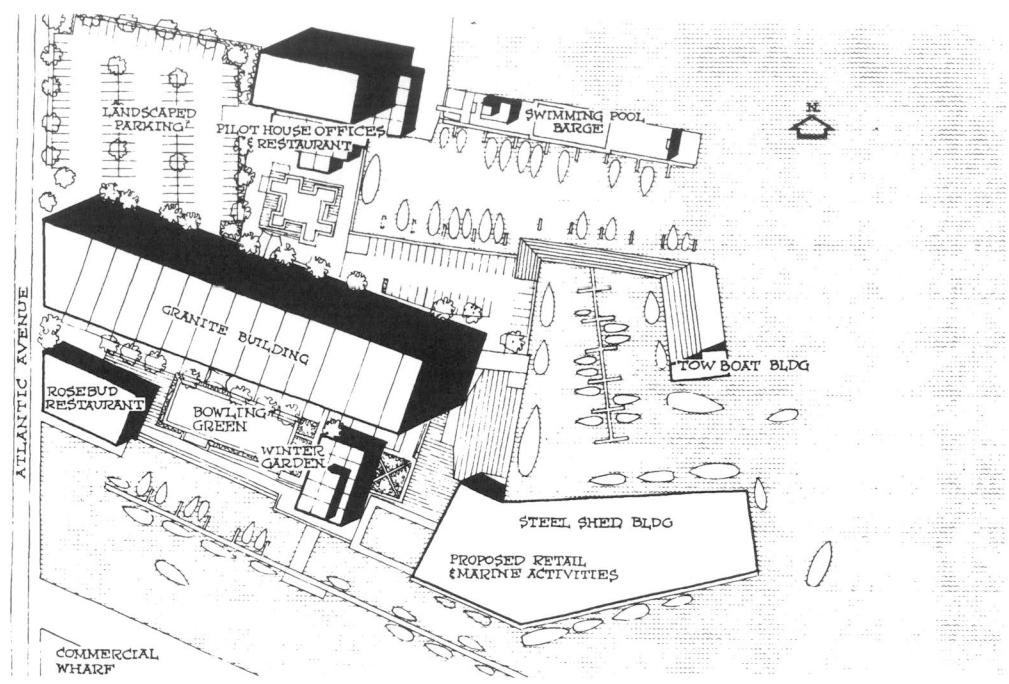

Fig. 11a Site plan. Lewis Wharf, Boston, Mass. Carl Koch—Architect.

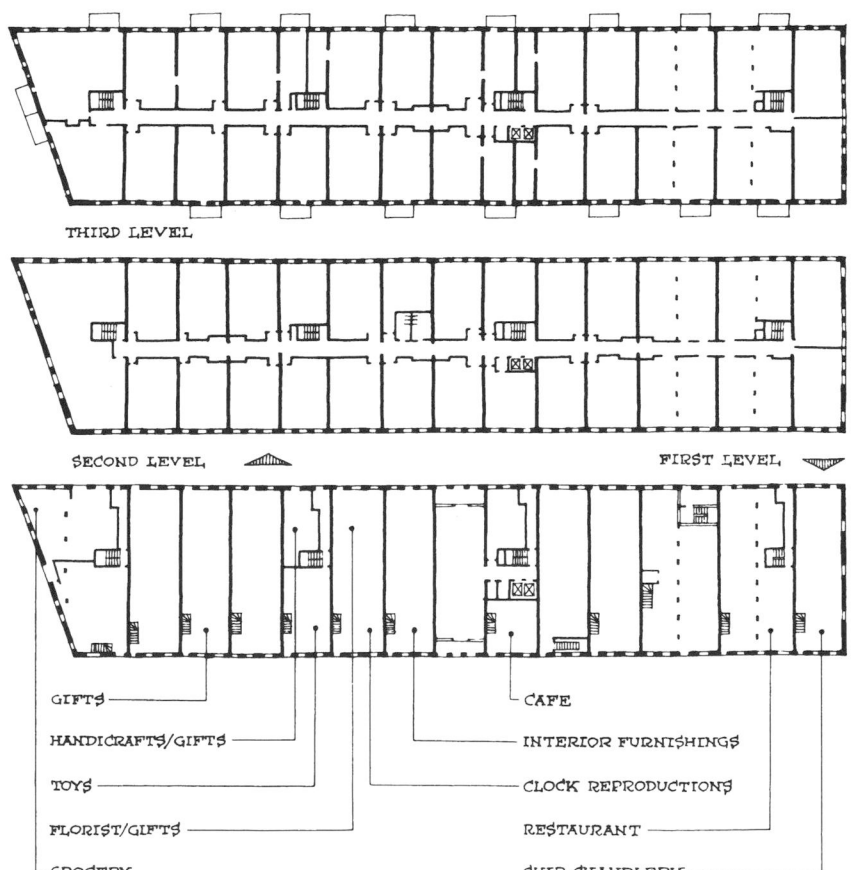

Fig. 11b Floor plan. Lewis Wharf, Boston, Mass. Carl Koch—Architect.

The two plans in Fig. 12 are typical of most apartments on the third through the sixth floors.

1-BEDROOM-1 BATH UNIT

2-BEDROOM-2 BATH UNIT

3-BEDROOM-2 BATH UNIT

2-BEDROOM-1 BATH UNIT

Fig. 12

Typical floor plans (Fig. 13) include balconies for both one- and two-bedroom units. Existing concrete columns were incorporated into the design of each room. Special plans for the handicapped, such as the one shown far right, provide extra space in bathrooms and hallways, as well as special amenities, such as oversized baths.

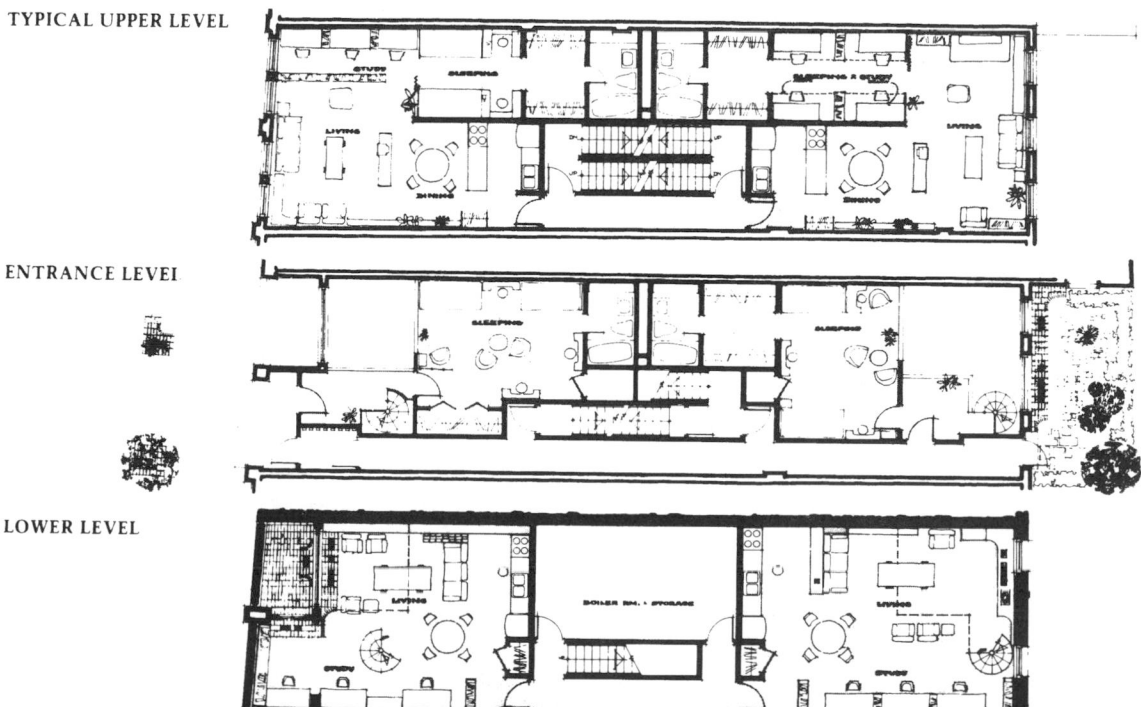

TYPICAL UPPER LEVEL

ENTRANCE LEVEL

LOWER LEVEL

Fig. 13a

LOFTS

The long corridor on each floor in Fig. 13*b* is broken into four zones by two triangular storage areas and a laundry room.

Rooms were designed around existing stairwells and elevator shafts; so a total of 22 floor plans were needed—13 one-bedroom, 8 two-bedroom and 1 studio (located by stairwell at left in plan.) Plans range in size from 478 to 1002 ft².

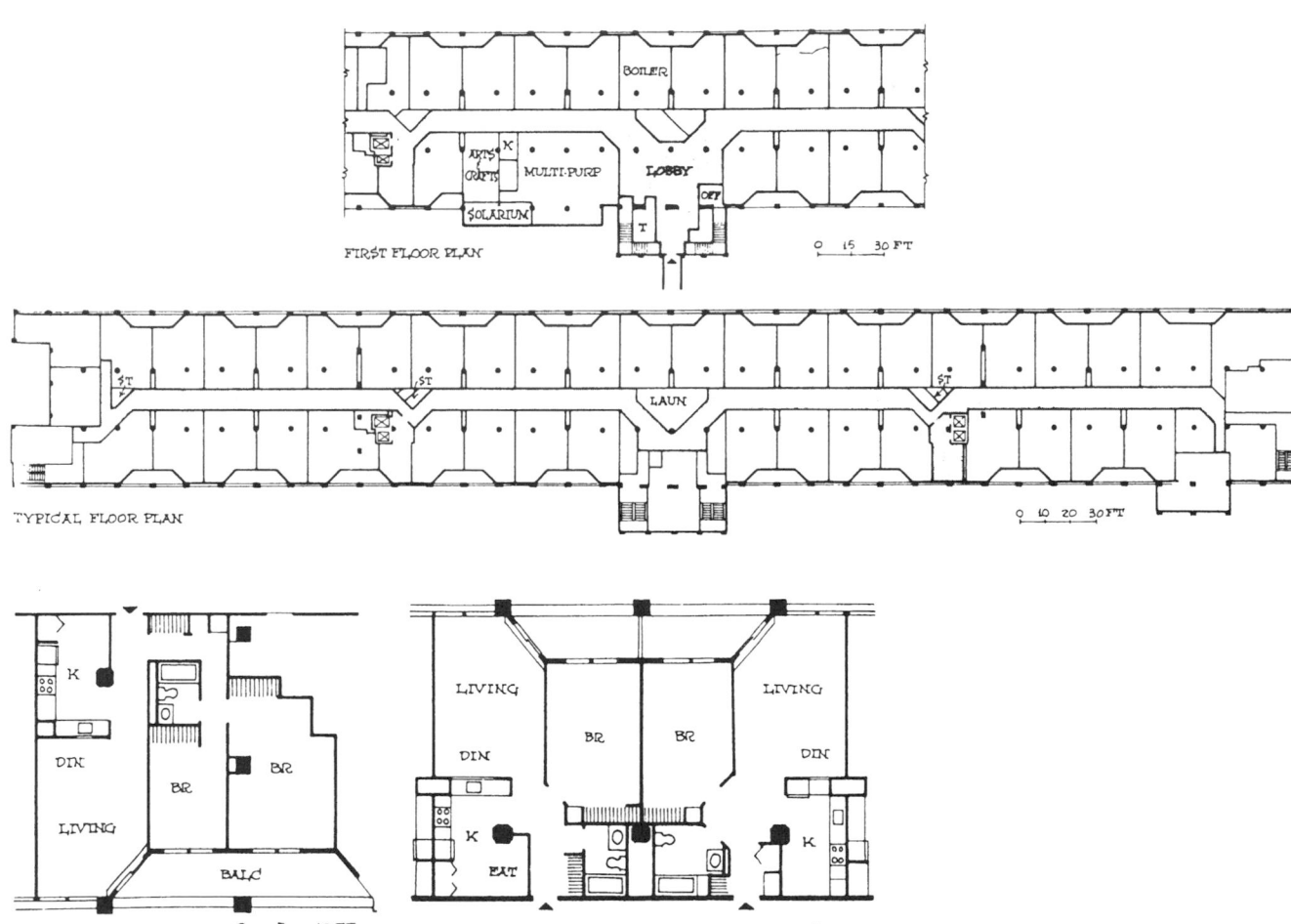

Fig. 13*b* **Keystone Apartments, Quincy Apartments. V. Strekalovsky—Architect.**

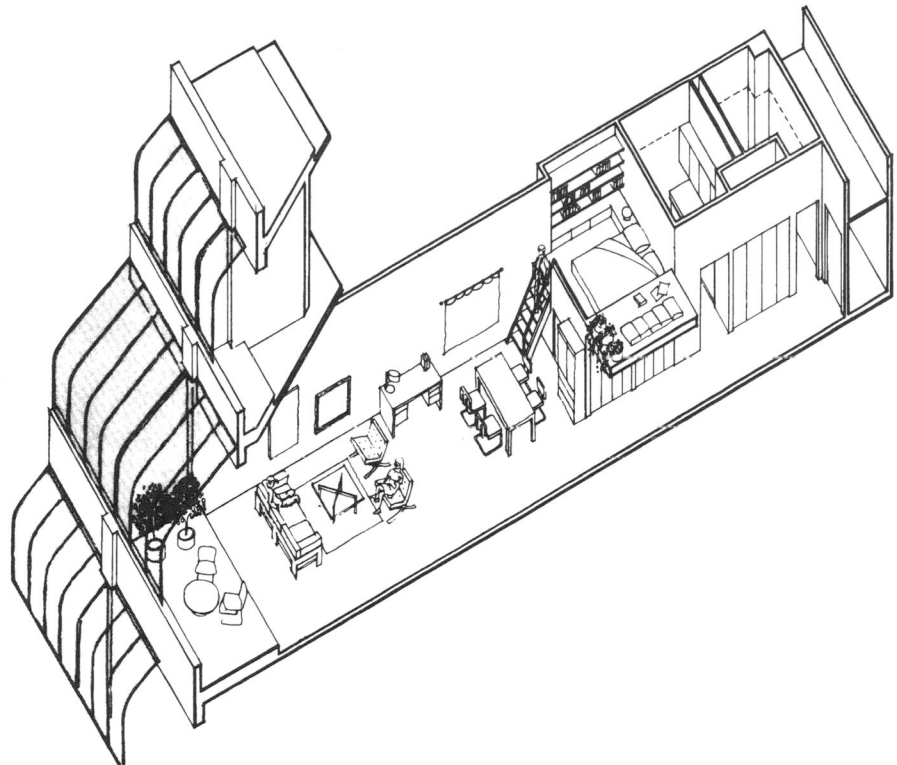

Fig. 14*a* Turtle Bay Towers, New York, N.Y. Bernard Rothzeld & Partners—Architects.

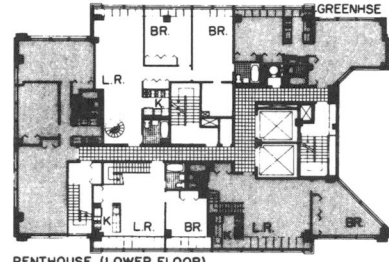

PENTHOUSE (LOWER FLOOR)

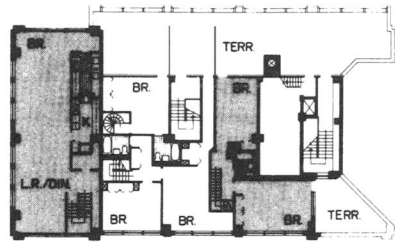

PENTHOUSE (UPPER FLOOR)

Fig. 14*b* Turtle Bay Towers, New York, N.Y. Bernard Rothzeld & Partners—Architects.

Lofts were included in the design of many apartments, especially studios, for they provide a spatial variety to the predominantly linear units. There are 341 apartments on the ½-acre site, with configurations varying from studios to "town houses" on the upper floors.

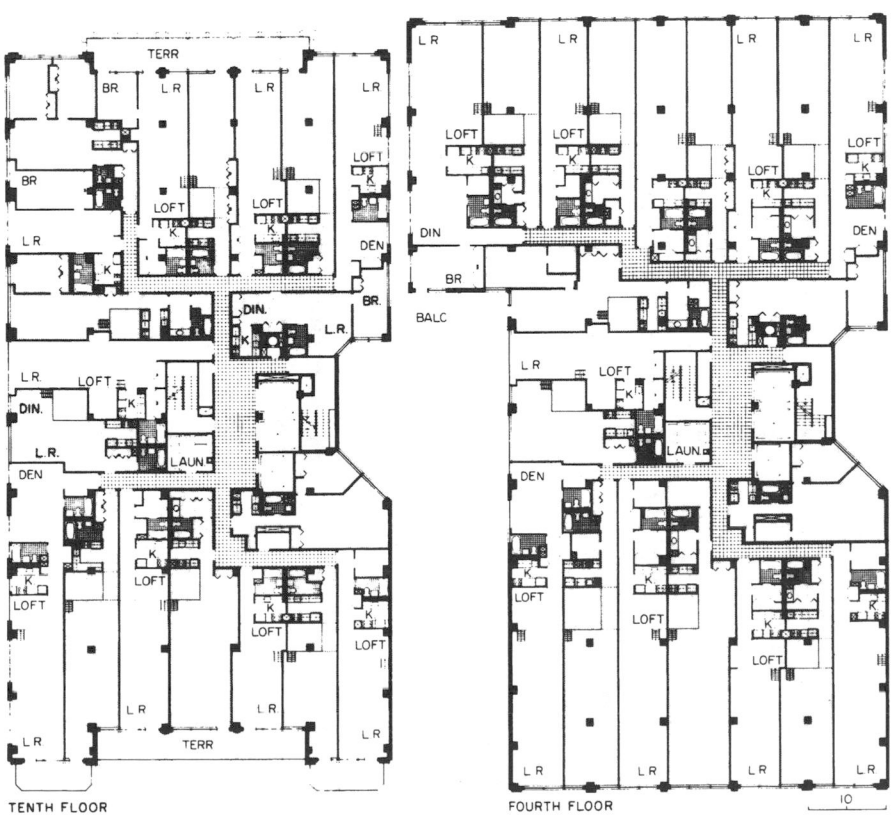

TENTH FLOOR

FOURTH FLOOR

Fig. 14*c* Turtle Bay Towers, New York, N.Y. Bernard Rothzeld & Partners—Architects.

FACTORY

Remnants of the former tannery in Fig. 15 have been saved and reused to provide *historic continuity* to the complex and site. Old tannery vats and drying wheels are now used as planters, including in the parking areas. Some foundations or walls are other than functional, lending an element of surprise or delight while retaining a special quality to the old structure. The pond was also cleaned-up to make a pleasant spot.

Community facilities at the tannery, including the management's office, laundry, and meeting rooms, are housed in an old mansion on the site. Because of varying structural and dimensional conditions, interior features such as high ceilings, heavy timber beams, and exposed cast-iron details provide attractive living and meeting spaces.

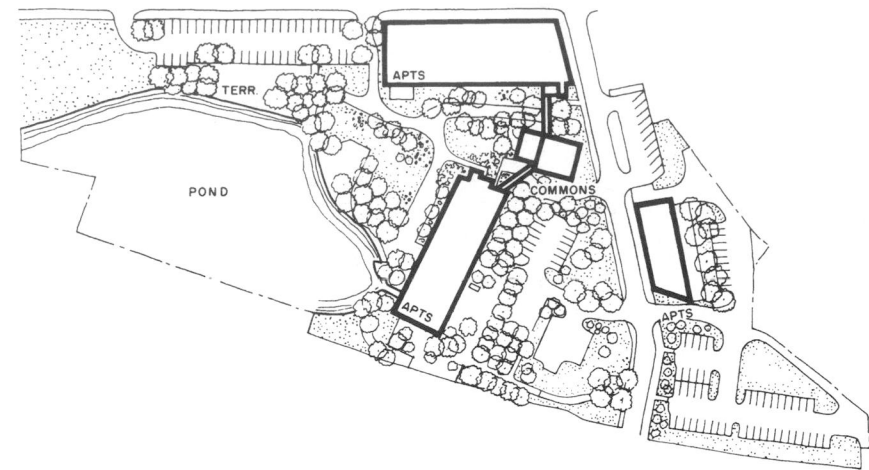

Fig. 15a   The Tannery, Peabody, Mass. Anderson Notter Associates, Inc.—Architects.

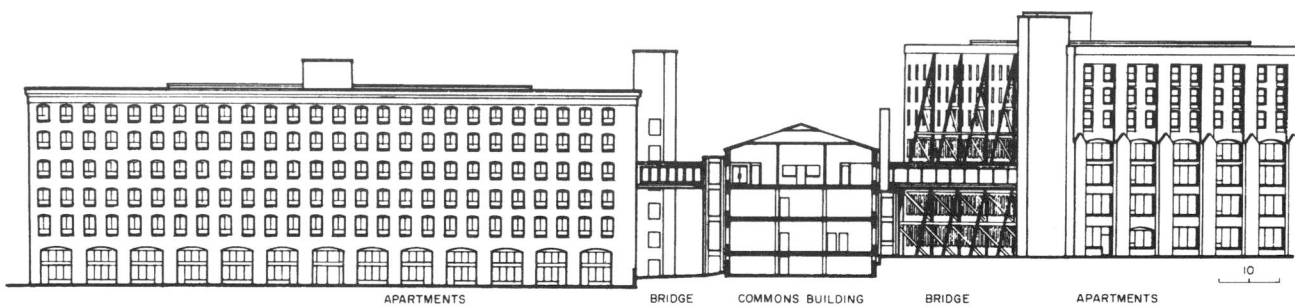

Fig. 15b   The Tannery, Peabody, Mass. Anderson Notter Associates, Inc.—Architects.

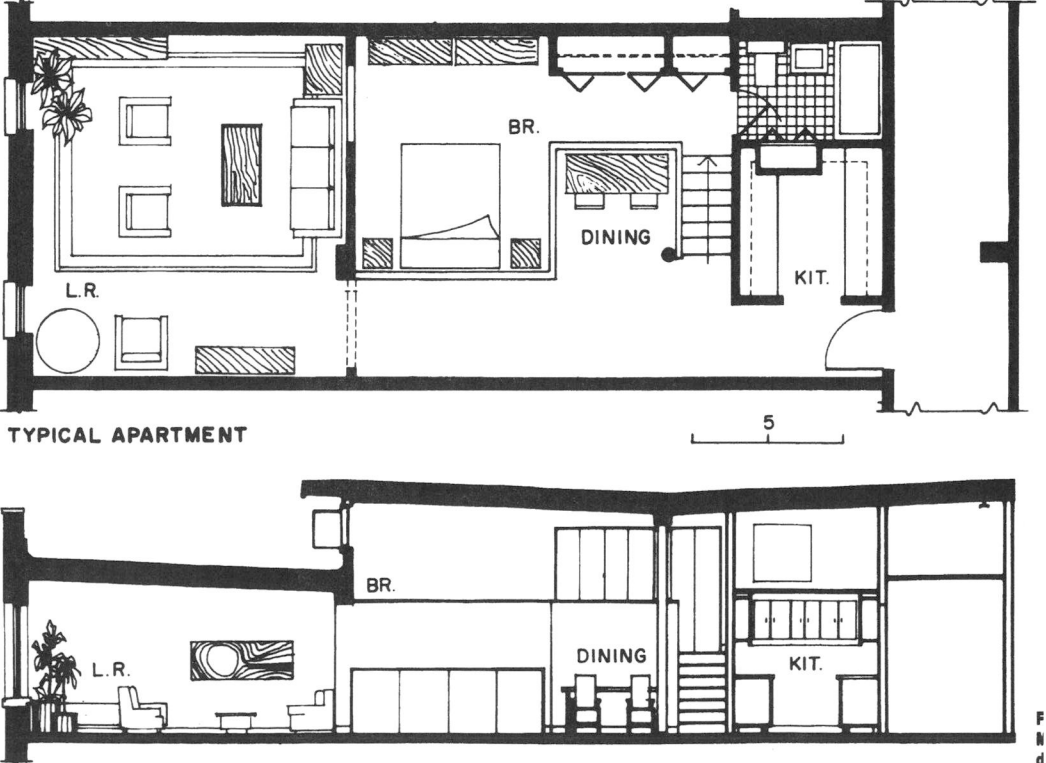

Fig. 15c   The Tannery, Peabody, Mass. Anderson Notter Associates, Inc.—Architects.

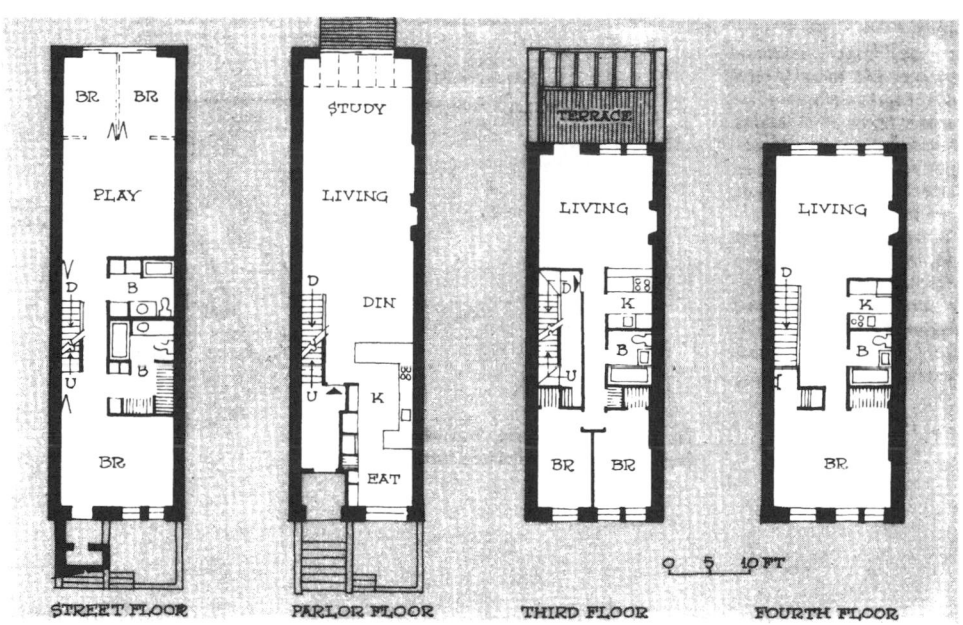

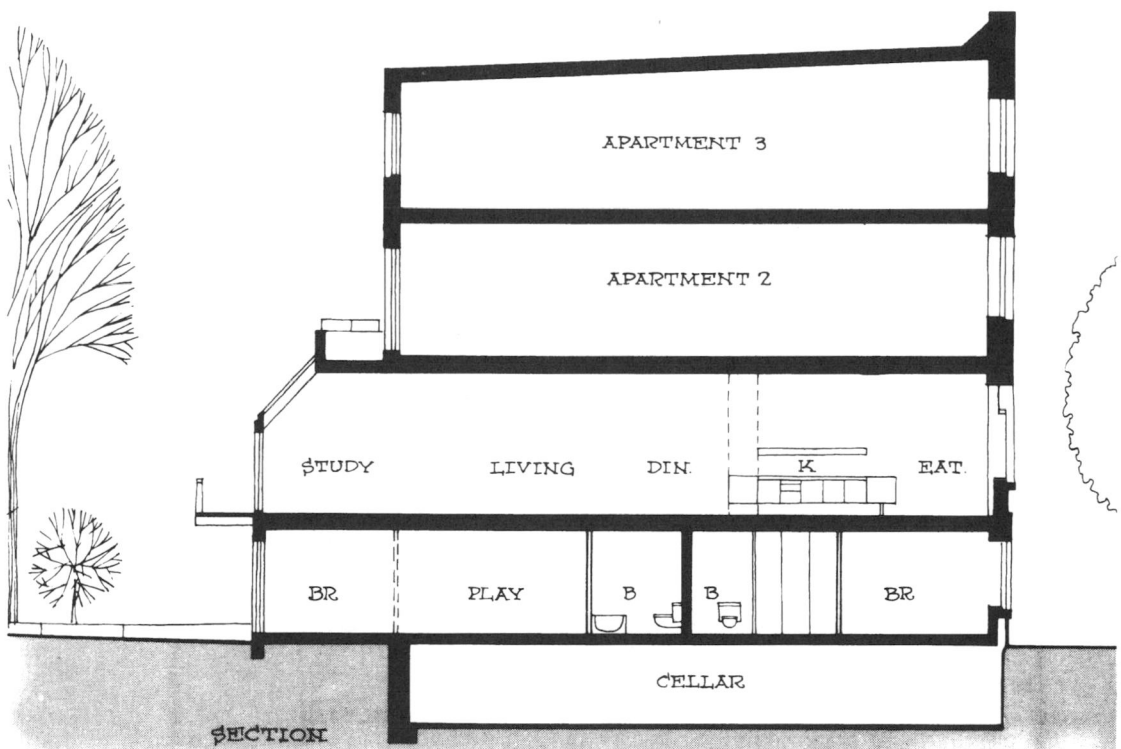

Fig. 1   Brownstone, New York, N.Y. Peter Sampton—Architect.

HISTORIC PRESERVATION

GUIDELINES FOR A SUCCESSFUL LOCAL HISTORIC PRESERVATION PROGRAM

The experience of communities that have undertaken successful programs of historic preservation with urban renewal assistance has shown that a thorough knowledge of a community's historic assets and problems and an understanding of the various programs of federal aid are essential. It is essential, too, that local groups interested in historic preservation participate in urban renewal planning to realize their goals. To do this, local groups:

1. May advise the local planning agency on preparing citywide review and evaluation of buildings and sites of historic importance.

2. May assist the local renewal agency to prepare actual plan proposals important to historic preservation.

3. May help to educate citizens and local offices to the historic preservation values inherent in urban renewal or other project areas and in the tools available for preservation.

4. Must be able to appreciate other renewal objectives besides historic preservation.

5. Must have local public and private funds available to assist in restoration, maintenance, and where necessary, furnishings for structures preserved in urban renewal areas.

To carry out these activities, the local preservation group may establish criteria for preservation, conduct a survey, and prepare an inventory of significant places, form a group to assist preservation activities (a group must be incorporated if it wants to hold title to property), prepare a program of public information, and raise funds to assist in preservation activities.

The National Trust for Historic Preservation has published "Criteria for Evaluating Historic Sites and Buildings" and for use in determining the importance of a property proposed for preservation. The section of the criteria on "Historical and Cultural Significance" states:

"A structure or area should have outstanding historical and cultural significance in the nation or in the state, region, or community in which it exists. Such significance is found in:

"Historic structures or sites in which the broad cultural, political, economic, or social history of the Nation, State or community is best exemplified, and from which the visitor may grasp in three-dimensional form one of the larger patterns of the American heritage.

"Structures or areas that are identified with the lives of historic personages or with important events in the main current of national state or local history.

"Structures or areas that embody the distinguishing characteristics of an architectural type-specimen, inherently valuable for a study of a period-style or method of construction; or a notable work of a master builder, designer or architect whose individual genius influenced his age. Mere antiquity is not sufficient basis for selection of a structure for permanent preservation, but can be a factor if other more significant examples have disappeared or if the building forms a part of an especially characteristic section of a given community. Smaller structures, such as the first squared-log cabins or the sod houses of the pioneers, may be as important relatively as the mansions of the past.

"Structures or sites of archeological interest that contribute to the understanding of aboriginal man in America."

OTHER FEDERAL AIDS FOR PRESERVATION

Other national programs of historic preservation that can offer aid to cities undertaking a preservation program are those of the National Trust for Historic Preservation and the National Park Service. These programs can help to determine and recognize historic properties. Citizen action, however, is needed to prevent deterioration or destruction of a building.

National Trust for Historic Preservation

The National Trust is a privately supported organization chartered by Congress in 1949 to help preserve sites, buildings, and objects significant in American history and culture. The National Trust assists cities by providing information about preservation criteria and techniques. It provides technical advice and counsel for all preservationists. Through its publications, seminars, and conferences it serves as an information center for all preservation activity both in the United States and abroad.

The National Trust has been working with the Urban Renewal Administration to inform both public and private community agencies of historic sites or buildings in urban renewal projects.

National Park Service

The Department of the Interior provides services and assistance for preservation through the National Park Service. Communities may receive technical aid for making a survey and inventory of an area, documenting historical places, judging architectural significance, and undertaking archeological work.

A list of Registered National Historic Landmarks is issued periodically by the Park Service. It enables communities to check quickly those buildings that have exceptional importance. The listing also gives official recognition to historic sites or buildings which may be needed to stimulate preservation interests in the locality.

Another valuable source of information about historic structures is the Historic American Buildings Survey, a long-range program for assembling a national archive of historic American architecture. Photographs and measured drawings of historic architecture are made by the National Park Service and maintained in the Library of Congress. Periodically the Library publishes a list of the buildings which have been measured and for which copies of the drawings may be obtained.

The National Park Service is interested in continuing and expanding the Historic American Buildings Survey and is giving special attention to historic buildings in urban renewal areas. Upon request, the National Park Service makes surveys of buildings and prepares reports of their history along with suggestions for their future use. This report can be especially important if there is any possibility of the building's being demolished or greatly altered. The survey has completed work in urban renewal projects in Portsmouth, N.H.; Boston, Mass.; Providence, R.I.; New York, N.Y.; Cape May, N.J.; Philadelphia, Pa.; Lancaster, Pa.; York, Pa.; Baltimore, Md.; Mobile, Ala.; Chicago, Ill.; and St. Louis, Mo.

Cities may also receive technical advice and assistance for archeological investigations from the National Park Service. Urban renewal can sometimes offer a unique opportunity to recover valuable artifacts from earlier periods of development before they are distributed by new construction.

Providence, R.I.

College Hill, a demonstration study of historic area renewal, was recognized as a classic report in the preservation field soon after its publication in 1959 by the Providence City Plan Commission in cooperation with the Providence Preservation Society. The study, made possible through a grant from the Urban Renewal Administration, is a major contribution to the field of historic preservation. It developed criteria for judging the architectural and historic value of structures and made recommendations for using these criteria in the collecting, scoring, and mapping of data. The College Hill methodology for evaluating structures has been used successfully in several cities, including Wilmington, N.C.; Salem, Mass.; Cape May, N.J.; and New Bedford, Mass.

College Hill, a name given the area of Providence around Brown University, is the site of the original settlement laid out in 1636 by Roger Williams. Religious leader and Indian trader, Williams led his followers into this unmapped land after being expelled from Congregationalist Salem. The new settlement grew rapidly, and in 100 years it was one of the major Colonial towns. By the end of the 18th century, Providence could boast a large number of elegant houses, churches, and public buildings.

Three hundred years after its founding, however, College Hill was plagued with the blight and decay characteristic of the older sections of many cities. Although redevelopment techniques were being worked out for the average city, little was known about renewing a historic area or preserving a large number of historic houses. The Demonstration Study was undertaken to develop renewal techniques that would be sensitive to historic values of a neighborhood as well as to restore College Hill and create a vibrant in-town residential area. The study included:

■ Analysis of architectural styles on College Hill

■ Programs to preserve important buildings of various periods and to encourage a blend of past and present architectural styles

■ Application of the "area concept" of historic preservation; this holds that, while single museum buildings serve an important purpose and isolated outstanding monuments should be preserved, groups of surviving related buildings in their original setting can explain the characteristics of an entire era and give added cultural and historic dimensions to the modern city

■ Techniques for integrating areas of historic architecture into redevelopment plans

■ A master plan for the future growth of College Hill in which plans for the historic areas could take their place in larger neighborhood plans

Several recommendations were made for positive local action. Nearly all have been carried out. For example:

■ In 1959, the proposed Rhode Island State Enabling Act providing for historic area zoning was enacted. In 1960, the Providence City Council passed a Historic District Zoning Ordinance. The Historic District Commission, appointed to administer the ordinance, has set a pattern of careful cooperation with owners to help solve problems of exterior restoration.

■ The proposed federally assisted East Side urban renewal project was enlarged, and planning was undertaken by the city. The College Hill Study made a major contribution to the planning of this project. Throughout the project planning period, liaison was maintained among municipal officials, the Redevelopment Agency and its consultants, College Hill educational institutions, preservationists, and local citizen groups. The demonstration study was reviewed and discussed. With minor exceptions, its objectives were recognized and provided for in the urban renewal plan.

The Redevelopment Agency named the Providence Preservation Society, organized in 1956, as the sponsor for redevelopment of the South Main-South Water Streets historic waterfront area within the urban renewal project.

Fig. 1*a*

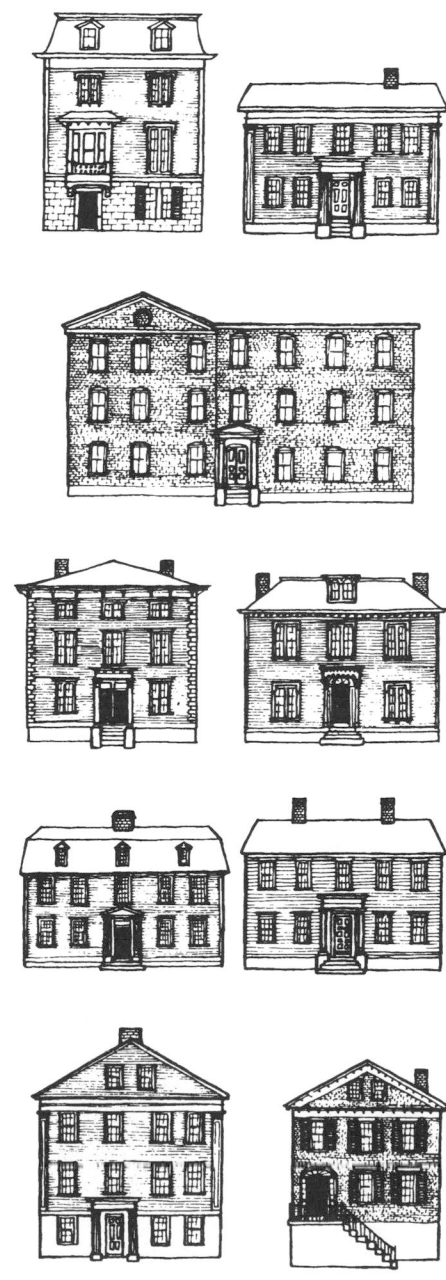

Fig. 1*b* **Buildings from the tourist guide, "Mile of History," which gives visitors a sketch of the historic architecture.**

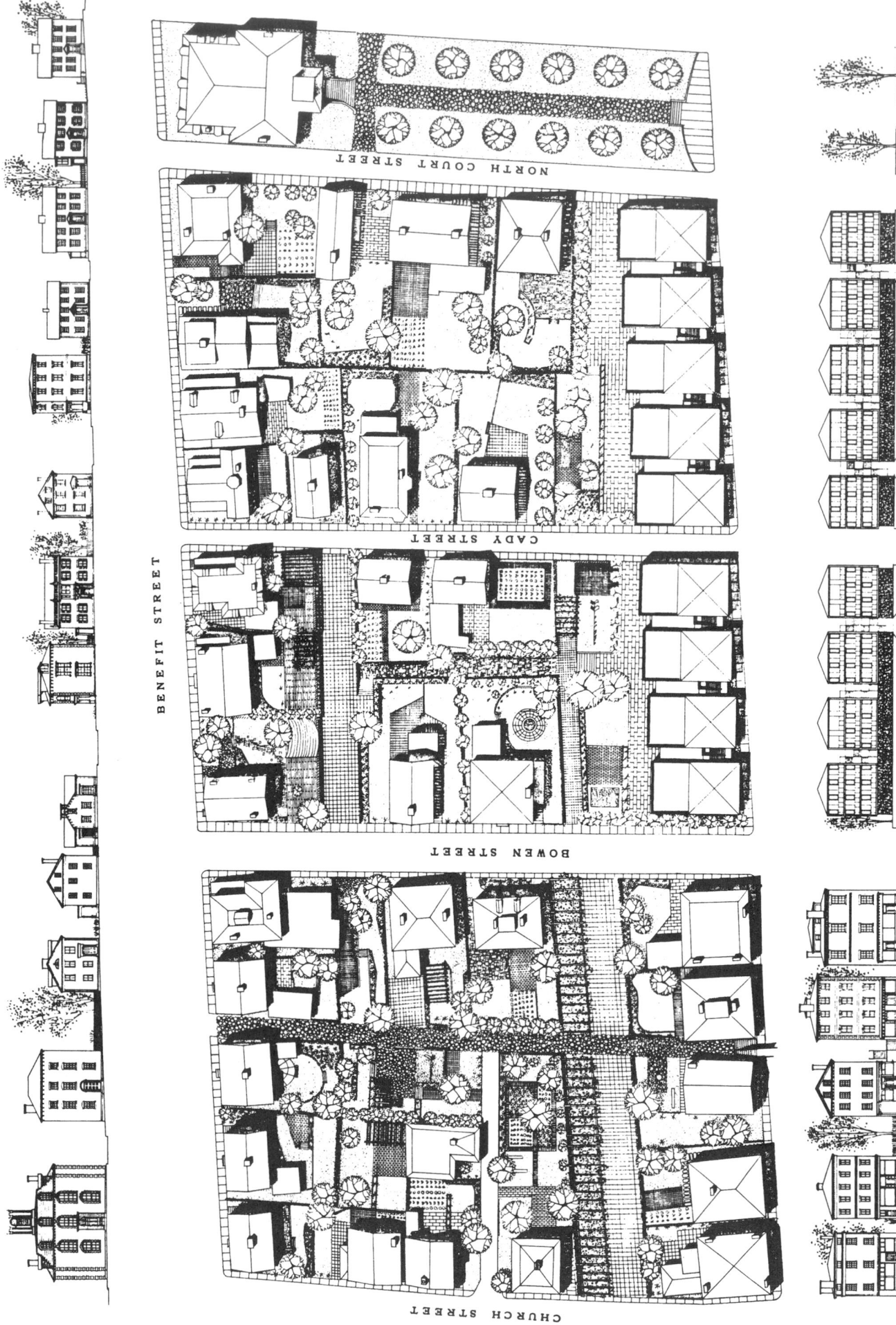

NORTH COURT STREET

CADY STREET

BOWEN STREET

CHURCH STREET

BENEFIT STREET

Fig. 2

Philadelphia, Pa.

Among the many types of treatment for blighted areas included in urban renewal is selective clearance in residential rehabilitation areas. Clearing incompatible industrial and commercial uses and residential buildings that cannot be rehabilitated results in vacant lots that vary in size from parcels for one or two houses up to half a block.

With the assistance of a demonstration grant, the Redevelopment Authority of Phila-

delphia studied this aspect of neighborhood conservation in historic Society Hill. Experience showed that the market for residences in the Society Hill area permits the opportunity for significant improvement and upgrading of existing buildings and for new construction. One phase of the demonstration project dealt with the problem of filling gaps in built-up areas by showing improved techniques for using small land parcels. The report indicates how regeneration of the whole historic area

can be stimulated by well-designed and readily marketable town houses on scattered cleared sites.

The authority analyzed the reasons for new construction in older neighborhoods of Philadelphia in the last decade. It has recommended building-design prototypes that will be compatible with the historic houses in the Washington Square East Unit No. 2 Urban Renewal Area (Society Hill).

Fig. 3b   Proposed street facades of rehabilitated historic houses.

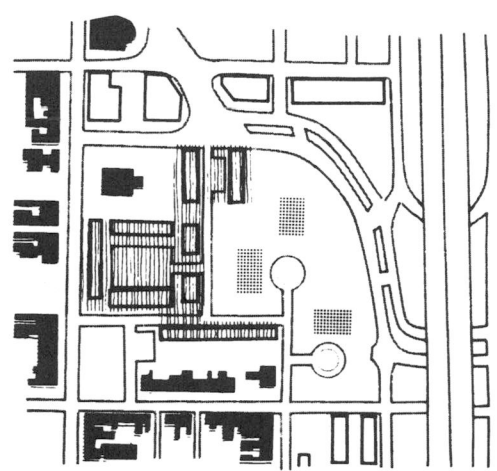

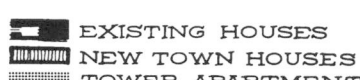
EXISTING HOUSES
NEW TOWN HOUSES
TOWER APARTMENT

Fig. 3a   Dock Street section of the project.

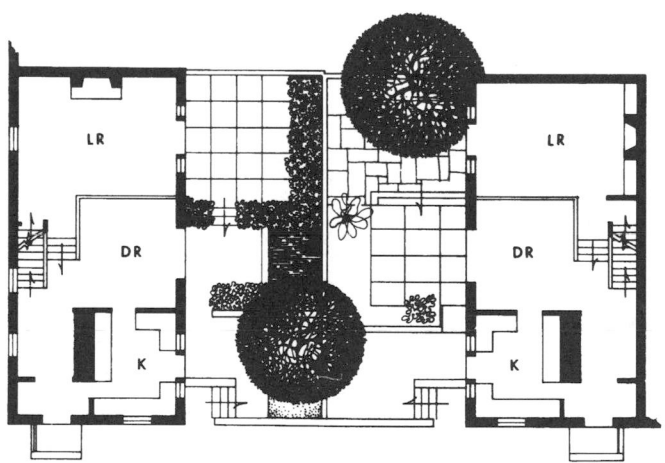

Fig. 4a   Floor plan proposed for new houses at 637 and 639 Pine Street.

Fig. 4b   New and old facades along Pine Street could look like this.

Fig. 4c   Sketches of typical building masses and fenestration of old houses.

HISTORIC PRESERVATION

New Haven, Conn.

Wooster Square in New Haven provides an excellent example of how older areas with character and identity can be preserved through urban renewal conservation treatment. Historically one of the city's best residential districts, with good design and architectural styles, it was able to resist the most serious blighting effects of expanding commercial areas. It suffered some damaging blows but remained a cohesive area, well related to the green space of Wooster Square, with its school and church.

The citizens in Wooster Square became concerned about the plight of their neighborhood and in 1950 asked the city for assistance. At that time, there was little experience anywhere in the field of rehabilitation. Nevertheless, the city went ahead with a plan for renewing the neighborhood. As a first step toward achieving the plan's goals, it persuaded the state highway

department to realign a proposed interstate highway which would have destroyed the cohesiveness of Wooster Square. The state adopted the city's proposal and relocated the highway. Then in 1954, Congress proposed an amendment to the Housing Act of 1949 to provide for rehabilitation as a method for renewal. This permitted federal funds to be used for staff services, project office operation, and other expenses, although not for actual rehabilitation work.

The New Haven City Plan Commission and the Redevelopment Agency prepared an urban renewal plan for the area, and with community approval renewal work was begun. Blighted areas were cleared out and owners helped to rehabilitate their homes. Out of 450 housing units west of the expressway (the major residential portion of the project), all but 50 were left standing for rehabilitation. Project area

work includes improvements in streets and utilities, off-street parking, and school and community facilities.

Special attention was given to preservation of good design and architectural style. Each homeowner was provided with free technical (including architectural) advice from the local renewal agency on bringing the home up to code standards and making it more attractive and harmonious with others in the block. Perspective drawings of each block were prepared, and where appropriate, color perspectives of individual buildings were done for the owners. These drawings emphasize keeping the original appearance of the fine old houses that add richness and variety to the area.

Sketches such as Figs. 5 to 7 were prepared so that owners could visualize how good design could be preserved and enhanced.

Fig. 5

Fig. 6

Fig. 7

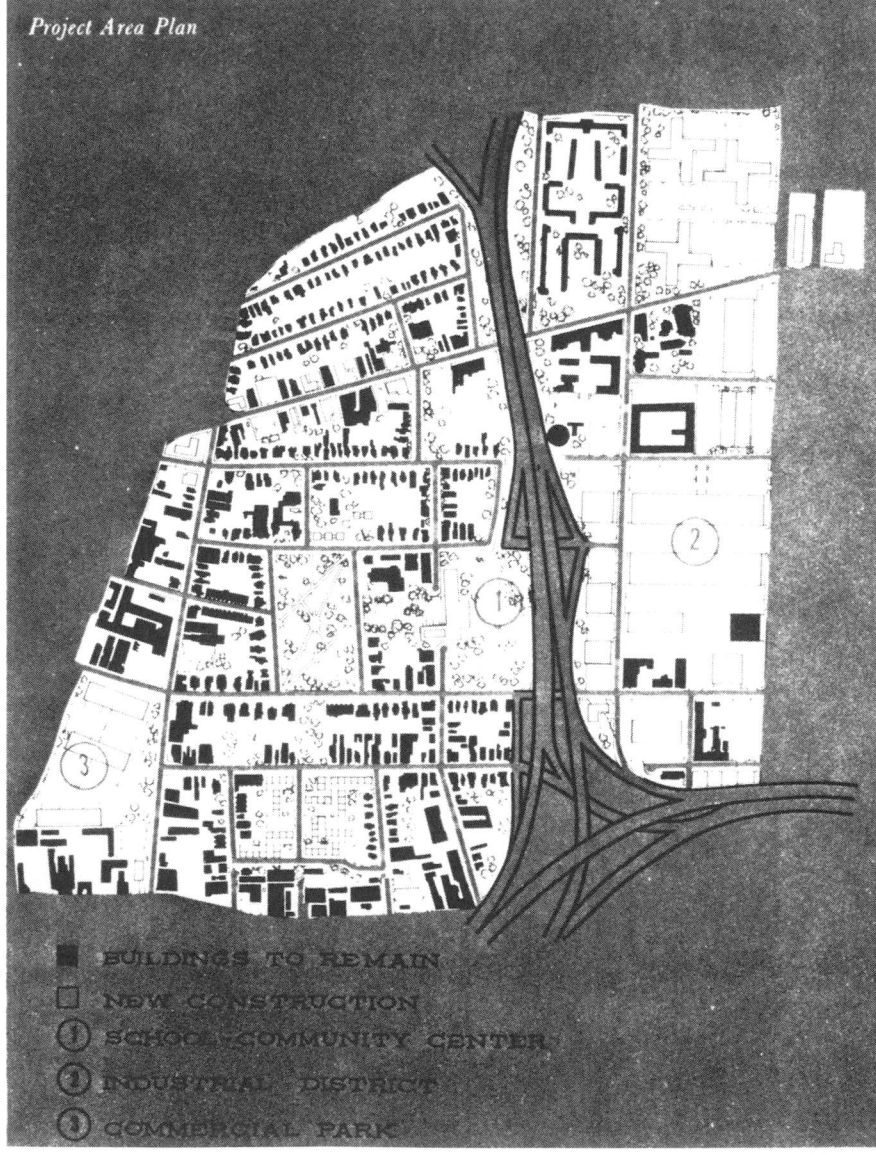

Project Area Plan

■ BUILDINGS TO REMAIN
□ NEW CONSTRUCTION
① SCHOOL-COMMUNITY CENTER
② INDUSTRIAL DISTRICT
③ COMMERCIAL PARK

Fig. 8

13

Housing Controls, Fire Safety, and Security

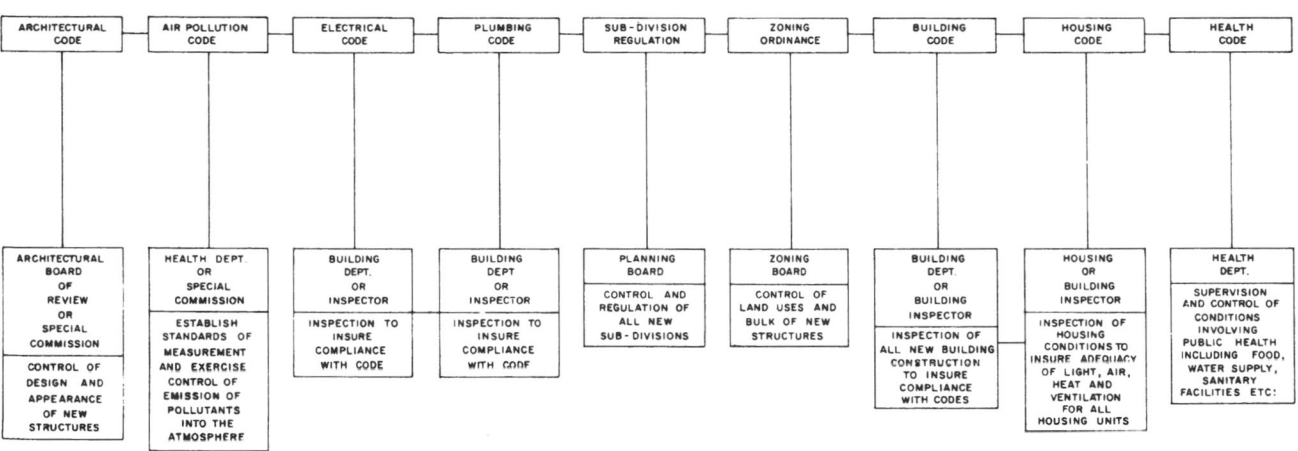

Fig. 1   Codes and enforcement agencies.

Principal participants in developing standards and codes	Building codes	Housing codes	Zoning ordinances	Subdivision regulations
	Engineers-architects	Health specialists	Planners-engineers	Engineers-planners
ELEMENTS REGULATED				
Subject of objective: 1. Natural light (penetration, quality, location).	Windows, yards, courts, light wells, habitable room size, building separations.	Windows, habitable room size	Courts, yards (front, side, rear), building height.	
2. Access and egress	Access to streets, corridors, stairs, doors; exits; access to bathrooms and bedrooms.	Corridors, stairways, doors, exits; access to bathrooms and bedrooms.	Required access to streets	Required access to streets.
3. Occupancy	Room dimensions (area, least dimension, ceiling height); minimum area per person.	Room dimensions (area, least dimension, ceiling height); minimum area per person, minimum area per dwelling unit.	Minimum area per dwelling unit	
4. Air supply	Windows; air conditioning	Windows; air conditioning	Windows, yards, courts	
5. Water supply	Sizes, materials, and construction; fixtures.	Materials, temperature, fixtures, maintenance.	Relation of uses to water supply	Sizes, materials, construction.
6. Air pollution (discharge into air).	Vents and venting systems, blowers and exhaust systems, incinerators.	Vents and venting systems, blowers and exhaust systems, incinerators.	Industrial performance standards; land use locations.	
7. Water pollution	Plumbing systems, septic tanks	Maintenance and functioning of plumbing and fixtures.	Industrial performance standards; land use locations.	Water courses, ground cover, grading.
8. Heating	Design and construction	Design and maintenance		
9. Fire safety	Construction and materials, building separations, access and egress.	Maintenance requirements for interiors, exitways, and heating equipment.	Land use locations and relationship, building separations, access.	Access.

The title of the table above is: IMPLEMENTATION OF OBJECTIVES THROUGH SEPARATE DEVELOPMENT INSTRUMENTS—EXAMPLES OF THE INTERRELATIONSHIP OF STANDARDS AMONG BUILDING CODES, HOUSING CODES, ZONING ORDINANCES, AND SUBDIVISION REGULATIONS

Fig. 2

HOUSING CODE

A housing code is one of a series of ordinances created to protect health, safety, and general welfare through the exercise of the police power. It is primarily a regulatory device designed to effect the preservation of acceptable dwellings and the rehabilitation of salvageable units. It is also used, on occasion, to obtain temporary minimal improvement of severely deteriorated or dilapidated dwellings in an urban renewal area pending their ultimate demolition. Unlike most housing-related legislation, such as building, plumbing, electrical, and fire-prevention codes, which deal with structural or design standards and are prospective in nature, housing codes deal with permissible conditions of occupancy of already existing dwellings. A comprehensive housing code normally specifies the required living space per person, minimum standards for lighting and ventilation, structural soundness and safety, fire protection, basic equipment, and the provision of services, and spells out, preferably in considerable detail, the maintenance responsibilities of owners and occupants. In effect, the housing code attempts to influence the life style of residents by requiring "owners and occupants to conduct themselves and to order their lives so as not to create undue accumulations of filth and trash, to keep their houses in repair, and to keep their walls, floors, and ceilings in a clean condition."[1]

The housing code generally defines the precise conditions for legal occupancy of dwelling units and, in most but not all cases, prescribes the manner in which compliance with locally adopted requirements is to be obtained.

Dwelling Unit Occupancy Requirements

Similarly to building codes, four basic model housing codes have been developed. They are:

1. Recommended Housing Maintenance and Occupancy Ordinance, American Public Health Association (APHA)

2. The Basic Housing Code, The Building Officials' Conference of America (BOCA)

3. The Southern Standard Housing Code, The Southern Standard Building Code Conference (SSBCC)

4. Housing Code, The International Conference of Building Officials (ICBO)

In general there is a reasonable uniformity among most of the codes relative to occupancy requirements as shown in Table 1 in terms of habitable floor area per person.

However, there is a great discrepancy in the current standards for the total floor area requirements (the additional space for halls, foyers, bathrooms, etc.). The APHA committee in 1950 recommended the minimum standards in Table 2.

Misunderstanding of the uses and limitations of code enforcement has led to considerable disillusionment among those who regarded it as a cure-all for urban and suburban housing ills. It is in fact only one of several tools—among them, publicly assisted housing, urban

TABLE 1 Dwelling Unit Occupancy Requirements, Four National Model Housing Codes (Floor Area in Square Feet)

Code	1 person	2 persons	3 persons	4 persons	5 persons
APHA-PHS	150	250	350	450	550
BOCA	150	250	350	450	550
ICBO	200*	200*	290	330	380
SSBCC	150	250	350	450	525

*150 is not prohibited, but the higher standard is recommended.
SOURCE: Model housing codes by organizations named.

TABLE 2 Minimum Total Floor Area (ft²)

For 1 person	400
For 2 persons	750
For 3 persons	1000
For 4 persons	1150
For 5 persons	1400
For 6 persons	1550

renewal, improved services, and enforcement of a full range of codes—that must be employed simultaneously if real progress is to be achieved.

Promulgation and enforcement of a housing code does not automatically ensure the creation of safe and sanitary dwellings, but it may very well prevent the onset or spread of blight in neighborhoods where units were originally of sound construction and, on rare occasions, may even foster the reconstruction of badly deteriorated or dilapidated houses. In a suburban area with manageable housing problems it can help to preserve sound dwellings and to secure the upgrading of all but the poorest units with a minimum expenditure of public funds. By slowing or even halting the unnecessary attrition of sound dwellings, it can help public authorities and nonprofit corporations engaged in the construction of new low- and middle-income units to begin to catch up with the backlog of unfilled housing needs. The prerequisites for the successful use of a housing code appear to be the widespread if not enthusiastic public acceptance of code enactment and enforcement as an appropriate activity of local government, adequate funding of the enforcement operation, the existence of a sufficient supply of housing for temporary relocation, and the availability of financial aids to assist low-income owners and tenants otherwise unable to bear the higher costs of improved housing. Unfortunately, the housing code is least effective when and where it is most needed. A "tight" housing market, with little or no new construction at prices low- and middle-income families can afford, necessarily reduces the options of the enforcing agency. The fact that the housing code applies to existing and already occupied houses, apartments, or rooms means that the forced vacating of a substandard dwelling results from efforts to protect the public health and safety through code enforcement frequently produces increased hardship for those obliged to live in such units.

Housing code enactment and enforcement is a relatively new development in suburban

areas. Although it is somewhat early to judge its effectiveness, it is already evident that legislative changes and administrative modifications must be introduced to permit the housing code to serve the purpose for which it was intended. A major fault is that virtually none of the existing codes established standards for a suitable living environment, i.e., standards for open space, housing density, or schools and their location.

Nothing is said in the typical housing code about private nonresidential properties, such as stores and factories—how they are mixed in with homes; how much noise, smoke, dust, or glare they make, how many signs they put up; how well they are maintained and landscaped. Nothing is said in the typical housing code about the adequacy of public lands such as parks and playgrounds; public structures such as schools, libraries, fire and police stations; or even public facilities such as right of way for access, paving for streets, exclusion of non-neighborhood traffic, sidewalks, storm sewers, street lights, street trees, and street signs.

It is true that many of these elements are provided for by municipalities by inclusion in comprehensive plans, zoning ordinances, subdivision regulations, or through special statutes. However, housing codes differ from the other regulatory procedures in that housing codes apply to already existing structures, whereas the other ordinances and regulations apply at the initiation of new facilities.

An adequate code should also contain sufficient provisions to ensure that implementation of the purpose of the code can be achieved. For example:

1. The power to enter and inspect buildings at reasonable times

2. Written notice served on the owner naming the specific violations found

3. Mechanics for formal conferences with the owner

4. The power to make repairs or demolish the building if the owner is unwilling or otherwise incapable

5. Specified maximum fines and/or prison terms

As increased urbanization takes place, the issues of incompatibility and environmental degradation resulting from nonresidential activities will become more pressing. Noise from aviation activities over residential communities and noise, glare, and fumes from upwind industrial activities can render residential life unpalatable.

[1]Frank P. Grad, *Legal Remedies for Housing Code Violations*, The National Commission on Urban Problems, Research Report 14, U.S. Government Printing Office, Washington, D.C., 1968, p. 3.

MODEL HOUSING CODES

Most housing codes in force in states, counties, cities, and towns have been patterned after one of several model housing codes. There are three fundamental types of model housing codes. They are (1) codes that give particular emphasis to health and safety requirements; (2) codes that are oriented toward the construction and maintenance of building; and (3) codes that stress the legal, administrative, and enforcement provisions. *A Proposed Housing Ordinance* prepared by the American Public Health Association in 1952, is an example of the first fundamental type of model housing code. This suggested legislation embodies the concept that since housing and the housing environment are important factors in the control and prevention of physical disease and injury and are contributory to mental health and social well-being, it is the responsibility of the public health officer to enforce the provisions of the code and to interpret the requirements for the health, safety, and well-being of the residents of the community. It may be said that *A Proposed Housing Ordinance* gives particular emphasis to the *human requirements*, particularly as related to the prevention of disease and injury.

Several organizations of building officials have developed model housing codes. The *BOCA Basic Housing Code,* prepared by the Building Officials Conference of America in 1964, is an example of the second type of model housing code. The underlying principle of this code is that the technical processes which are connected with the construction and maintenance of dwellings and dwelling units as structures are of primary importance. Model housing codes similar to the *BOCA Basic Housing Code* emphasize engineering knowledge of building materials, methods of construction, fire safety standards, and performance requirements for essential equipment and facilities of dwellings. Relatively speaking, model housing codes of this second type place limited emphasis on *human requirements* and stress the technical or *structural requirements* on buildings used for human habitation.

In 1954, the National Institute of Municipal Law Officers prepared and published the *NIMLO Model Minimum Housing Standards Ordinance,* an example of the third type of model housing code. Unlike the model housing codes of the other two types, the *NIMLO Model Minimum Housing Standards Ordinance* contains very few standards. It is more concerned with the legal and administrative aspects of the housing code and emphasizes the importance of uniformity and continuity of this type of legislation with other municipal legislation. (The National Institute of Municipal Law Officers established in 1952 a model ordinance service in order to provide municipalities with a comprehensive code of ordinances on every phase of municipal regulation. New ordinances are developed as need arises, and others are revised from time to time as dictated by experience and court decisions.) The *NIMLO Model Minimum Housing Standards Ordinance* does not specify or suggest standards. Blank spaces are provided for the adopting legislative body to provide its own specification. This model code gives emphasis to the *legal requirements* of housing codes.

SUBSTANTIVE CONTENT OF HOUSING CODES

The substantive content of most housing codes includes three broad subject areas: (1) mini-

mum facilities and installed equipment; (2) maintenance of the dwelling unit and of facilities and equipment; and (3) use, maximum occupancy, and conditions of occupancy. A comprehensive coverage of provisions related to minimum facilities and installed equipment usually includes water supply and wastewater disposal, garbage and rubbish disposal, kitchen and hard-washing sinks, bathing facilities, toilet facilities, means of egress, heating equipment for the dwelling unit and hot water supply, lighting, ventilation, and electrical service.

Most housing codes will contain specific provisions pertaining to maintenance of the dwelling and of the supplied facilities and equipment to include such items as general sanitary conditions; chimneys, flues, and other potential fire hazards; electrical wiring; insect and rodent infestation; internal structural repair; external structural repair; and dampness.

The sections of most housing codes that regulate use, maximum occupancy, and conditions of occupancy usually have provisions pertaining to living space overcrowding; sleeping space overcrowding; doubling of families; separation of sexes; and mixed use of living space for business purposes.

Most, if not all, housing codes tend to rely on general words or phrases without defining them such as *good repair, adequate,* and *safe condition.* These terms are difficult to interpret and sometimes create confusion. Much is left to personal judgment. For example, a cracked wall in a kitchen may be interpreted by the owner as being in good repair because there is no imminent danger of the plaster falling. A housing code inspector who is from the building department may not deem the wall to be in good repair, as the crack suggests that internal structural repair is needed; and a housing code inspector who is from the public health department may classify the wall as not being in good repair, as the crack may provide a harborage for vermin and/or may permit rodents easy access to the dwelling unit.

DWELLING UNIT OCCUPANCY STANDARDS

Of the model housing codes, there is uniformity of requirements for occupancy by one to four persons, of the APHA-PHS Recommended Housing Maintenance and Occupancy Ordinance, the BOCA Basic Housing Code, and the Southern Standard Housing Code of the Southern Building Code Congress. The amount of floor space required in habitable rooms is 150 ft² for the first person and 100 ft² more for each additional person. When the number of occupants is five or more, the SBCC Housing Code reduces the additional space required for each occupant over four to 75 ft².

The ICBO Housing Code does not specify the minimum amount of habitable space in the dwelling unit on an occupancy basis. By indirect calculation it can be determined that 200 ft² is the minimum amount of habitable space for one- or two-person occupancy of a dwelling unit. This can be derived from the requirement that there be a combination living room-bedroom of 150 ft² and a kitchen having at least 50 ft². A dwelling unit of this size may be occupied by two persons. Under the ICBO Housing Code three persons may occupy a two-room dwelling unit with as little as 290 ft² of habitable floor space provided as follows:

Kitchen-living room	150 ft²
Bedroom (triple occupancy)	140 ft²
	290 ft²

Four persons may occupy a three-room dwelling unit with 330 ft² of habitable floor space distributed as follows:

Kitchen-living room	150 ft²
Bedroom (double occupancy)	90 ft²
Bedroom (double occupancy)	90 ft²
	330 ft²

Five persons may occupy a three-room dwelling unit with 380 ft² of habitable floor space arranged as follows:

Kitchen-living room	150 ft²
Bedroom (triple occupancy)	140 ft²
Bedroom (double occupancy)	90 ft²
	380 ft²

The requirements of the state and city housing codes tend to follow the patterns of the particular model code used as a guide. One city code gives occupancy standards in terms of cubic feet of space per occupant.

In general, housing codes that do not use an allotment of square feet of habitable floor space for each occupant tend to be confusing when one is trying to determine maximum occupancy conditions. Housing codes that have different standards for occupancy on a room-size, dwelling-unit arrangement or age-of-occupant basis are confusing also.

There is considerable difference in the amount of habitable space that constitutes a minimum requirement for occupancy of a dwelling unit, particularly if the dwelling unit is to be occupied by several persons. The APHA-PHS and BOCA Housing Codes requirements are considerably greater than the ICBO Housing Code for more than one-person occupancy. For five-person occupancy, the ICBO Housing Code requires only 69 percent of the habitable space required by the APHA-PHS and BOCA Housing Codes.

SLEEPING ROOM OCCUPANCY STANDARDS

There is some uniformity in the requirements for occupancy by one person, but the uniformity is less pronounced for occupancy by two or more persons.

The APHA-PHS, BOCA, and SBCC codes require 70 ft² of habitable floor space for a sleeping room to be occupied by one person. The ICBO code requires that a room used for sleeping should have not less than 90 ft² of floor space and may be occupied by one or two persons. If the sleeping room is to be occupied by two persons, the APHA-PHS housing code requires at least 120 ft² of floor space, while the BOCA and SBCC codes require only 100 ft². The SBCC code allows two persons under the age of 12 years to occupy a sleeping room with but 70 ft² requiring only 35 ft² per child. If the sleeping room is to be occupied by three persons, the APHA-PHS code requires 170 ft² of habitable floor space, the BOCA and SBCC codes 150 ft² if the occupants are older than 12. The ICBO code permits three persons to occupy a sleeping room with 140 ft² of floor space.

A few state and city housing codes include special requirements if the sleeping room is to be used by children.

The difference in the minimum requirements for sizes of sleeping rooms to be occupied by more than one person should be grounds for concern for some people, particularly those who are interested in respiratory disease prevention. The U.S. Army has found that in barracks 72 ft² of floor space per occupant is the minimum requirement to keep respiratory infections under control.

MODEL HOUSING CODES

BATHROOM FACILITIES

The substantive provision that shows the greatest uniformity in the codes is the bathroom facilities requirement. All the housing codes require a private flush toilet, a lavatory sink, and a bathtub or shower with hot and cold running water under pressure. These facilities, located inside the dwelling unit, constitute minimum requirements for sanitary housing.

KITCHEN FACILITIES

For most housing codes the only required facility for a kitchen or a room to be used for the preparation and service of food is a kitchen sink. Very few housing codes include specific requirements for the cooking and storage of food and the storage of utensils and dishes. Of the four model housing codes, only the APHA-PHS code requires, in addition to the kitchen sink, a stove and a refrigerator or similar device, and cabinets and/or shelves for the storage of dishes, utensils, and food. Several housing codes use ambiguous terms such as "adequate facilities for cooking" without definition or elaboration.

With a few exceptions, most housing codes grossly fail to be specific enough about required kitchen facilities. This may stem from the fact that many housing codes are not oriented toward the fulfillment of "human requirements." If a dwelling unit is to have the minimum facilities necessary for the fulfillment of health objectives, there should be facilities for the cooking, preparing, and serving of food.

HEATING FACILITIES

Most codes stipulate that the facilities be capable of heating the dwelling unit to a specified temperature (usually 67 to 70°F) at a specified height above the floor.

ELECTRICAL SERVICE AND FACILITIES

The requirement that dwelling units be provided with electrical service if a connection is available within a reasonable or specified distance (usually 300 ft) from the dwelling unit is basic.

Some codes require the provision of a given number of electrical fixtures or facilities. A few codes call for a stated capacity of electric power to be provided for each dwelling unit; e.g., the APHA-PHS housing code requires an electric service, outlets, and/or fixtures capable of providing at least 3 W of electrical power per square foot of total floor area.

SOLID AND LIQUID WASTE DISPOSAL

All housing codes have provisions concerning the handling of liquid wastes (sewage). Connection to either a public sewer system or an approved septic tank system is required.

Most housing codes require facilities for the storage and/or disposal of solid wastes (garbage and rubbish). Some of the codes stipulate who must provide the containers for the storage of refuse.

LIGHTING

Daylighting of habitable rooms is a minimum requirement of all codes. Most codes require at least one window (or skylight) per habitable room and stipulate the amount of window area to be provided. This specification is usually expressed as a percent of floor area of the room to be daylighted by the window.

Most housing codes require artificial lighting facilities in all habitable rooms and in some other rooms and areas such as bathrooms and hallways.

VENTILATION

All housing codes require natural ventilation of all habitable rooms. Also, all codes require ventilation of toilet rooms, but some permit the use of mechanical ventilation devices in lieu of natural ventilation. The codes specify the percent of window area that will be openable. The values range from 20 to 50 percent. The most common specification is that 45 percent of the window area should be openable.

MINIMUM STANDARDS FOR BASIC EQUIPMENT AND FACILITIES

No person should occupy as owner or occupant or let to another for occupancy any dwelling or dwelling unit, for the purposes of living, sleeping, cooking, or eating therein, which does not comply with the following requirements:

Every dwelling unit should have a room or portion of a room in which food may be prepared and/or cooked, which shall have adequate circulation area and which shall be equipped with the following:

1. A *kitchen sink* in good working condition and properly connected to a water supply system which is approved by the appropriate authority and which provides at all times an adequate amount of heated and unheated running water under pressure, and which is connected to a sewer system approved by the appropriate authority.

2. *Cabinets and/or shelves* for the storage of eating, drinking, and cooking equipment and utensils and of food that does not under ordinary summer conditions require refrigeration for safe keeping; and a counter or table for food preparation; said cabinets and/or shelves and counter or table should be of sound construction furnished with surfaces that are easily cleanable and that will not impart any toxic or harmful effect to food.

3. A *stove* or similar device for cooking food, and a refrigerator or similar device for the safe storage of food at temperatures less than 45°F but more than 32°F under ordinary maximum summer conditions, which are properly installed with all necessary connections for safe, sanitary, and efficient operation; *provided* that such stove, refrigerator, and/or similar devices need not be installed when a dwelling unit is not occupied and when the occupant is expected to provide same on occupancy, and that sufficient space and adequate connections for the safe and efficient installation and operation of said stove, refrigerator, and/or similar devices are provided.

Within every dwelling unit there should be a nonhabitable room which affords privacy to a person within said room and which is equipped with a *flush water closet* in good working condition. Said flush water closet should be equipped with easily cleanable surfaces, be properly connected to a water system that at all times provides an adequate amount of running water under pressure to cause the water closet to be operated properly, and should be properly connected to a sewer system which is approved by the appropriate authority.

Within every dwelling unit there should be a *lavatory sink*. Said lavatory sink may be in the same room as the flush water closet, or if located in another room, the lavatory sink should be located in close proximity to the door leading directly into the room in which said water closet is located. The lavatory sink should be in good working condition and properly connected to a water supply system which is approved by the appropriate authority and which provides at all times an adequate amount of heated and unheated running water under pressure, and which is properly connected to a sewer system approved by the appropriate authority. Water inlets for lavatory sinks should be located above the overflow rim of these facilities.

Within every dwelling unit there should be a room which affords privacy to a person within said room and which is equipped with a *bathtub* or *shower* in good working condition. Said bathtub or shower may be in the same room as the flush water closet or in another room and

should be properly connected to a water supply system which is approved by the appropriate authority and which provides at all times an adequate amount of heated and unheated water under pressure, and which is connected to a sewer system approved by the appropriate authority. Water inlets for bathtubs should be located above the overflow rim of these facilities.

Every dwelling unit should have at least two means of egress leading to safe and open space at ground level. Every dwelling unit in a multiple dwelling shall have immediate access to two or more approved means of egress leading to safe and open space at ground level, or as required by the laws of the state and the appropriate unit. Bedrooms located below the fourth floor should be provided with an exterior door or window of such dimensions as to be used as a means of emergency egress.

Structurally sound hand rails should be provided on any steps containing four risers or more. Porches, patios, and/or balconies located more than 3 ft higher than the adjacent area should have structurally sound protective guard or hand rails.

Each dwelling unit should have facilities for the safe storage of drugs and household poisons.

Access to or egress from each dwelling unit should be provided without passing through any other dwelling unit.

No person should let to another for occupancy any dwelling or dwelling unit unless all exterior doors of the dwelling or dwelling unit are equipped with functioning locking devices.

MINIMUM STANDARDS FOR LIGHT AND VENTILATION

No person should occupy as owner or occupant or let to another for occupancy any dwelling or dwelling unit, for the purpose of living therein, which does not comply with the requirements of this section:

Every habitable room should have at least one window or skylight facing outdoors provided that if connected to a room or area used seasonally (e.g., porch), adequate daylight must be possible through this interconnection. The minimum total window or skylight area, measured between stops, for every habitable room should be at least 10 percent of the floor area of such room, but if light-obstruction structures are located less than 3 ft from the window and extend to a level above that of the ceiling of the room, such window will not be deemed to face directly to the outdoors and will not be included as contributing to the required minimum total window area.

Every habitable room should have at least one window or skylight facing directly outdoors which can be opened easily, or such other device as will ventilate the room adequately, provided that if connected to a room or area used seasonally, adequate ventilation must be possible through this interconnection. The total of openable window or skylight area in every habitable room should be equal to at least 45 percent of the minimum window area size or minimum skylight-type window size, except where there is supplied some other device affording adequate ventilation and approved by the appropriate authority.

Every bathroom and water closet compartment, and nonhabitable room used for food preparation should comply with the light and ventilation requirement for habitable rooms, except that no window or skylight will be required in such rooms if they are equipped with a ventilation system in working condition,

which is approved by the appropriate authority.

Where there is usable electric service readily available from power lines which are not more than 300 ft away from a dwelling, every dwelling unit and all public and common areas should be supplied with electric service, outlets, and fixtures which should be properly installed, should be maintained in good and safe working condition, and should be connected to a source of electric power in a manner prescribed by the ordinances, rules, and regulations of the appropriate unit. The minimum capacity of such services and the minimum number of outlets and fixtures should be as follows:

1. Every dwelling unit should be supplied with at least one 15-A circuit, and such circuit should not be shared with another dwelling unit.

2. Every habitable room should contain at least two separate wall-type duplex electric convenience outlets or one such duplex convenience outlet and one supplied wall- or ceiling-type electric light fixture. No duplex outlet should serve more than two fixtures or appliances.

3. Temporary wiring or extension cords should not be used as permanent wiring.

4. Every nonhabitable room, including water closet compartments, bathrooms, laundry rooms, furnace rooms, and public halls should contain at least one supplied ceiling or wall-type electric light fixture.

5. All electric lights and outlets in bathrooms should be controlled by switches which are of such design as will minimize the danger of electric shock, and such lights and outlets should be installed and maintained in such condition as to minimize the danger of electrical shock.

Every public hall and stairway in every multiple dwelling should be adequately lighted by natural or artificial light at all times, so as to provide in all parts thereof at least 10 ft of light at the tread of floor level. Every public hall and stairway in structures containing not more than two dwelling units may be supplied with conveniently located light switches controlling an adequate lighting system which may be turned on when needed instead of full-time lighting.

MINIMUM THERMAL STANDARDS

No person should occupy as owner or occupant or let to another for occupancy any dwelling or dwelling unit, for the purpose of living therein, which does not comply with the following requirements.

Every dwelling should have heating equipment and appurtenances which are properly installed, are maintained in safe and good working condition, and are capable of safely and adequately heating all habitable rooms, bathrooms, and water closet compartments in every dwelling unit located therein to a temperature of at least 68°F at a distance of 36 in above floor level under ordinary winter conditions.

No owner or occupant should install, operate, or use a heating device, including hot water heating units, which employs the combustion of carbonaceous fuel, which is not vented to the outside of the structure in an approved manner, and which is not supplied with sufficient air to continuously support the combustion of the fuel. All heating devices should be constructed, installed, and operated in such a manner as to minimize accidental burns.

MAXIMUM DENSITY, MINIMUM SPACE, USE, AND LOCATION REQUIREMENTS

No person should occupy or let to be occupied any dwelling or dwelling unit, for the purpose

MINIMUM HOUSING STANDARDS

of living therein, unless there is compliance with the following requirements.

The maximum occupancy of any dwelling unit should not exceed the lesser value of the following two requirements:

1. For the first occupant there should be at least 150 ft^2 of floor space and there should be at least 100 ft^2 of floor space for every additional occupant thereof; the floor space to be calculated on the basis of total habitable room area.

2. A total number of persons should be less than 2 times the number of habitable rooms within the dwelling unit.

Not more than one family, plus two occupants unrelated to the family, except for guests or domestic employees, should occupy a dwelling unit unless a license for a rooming house has been granted by the appropriate authority.

The ceiling height of any habitable room should be at least 7 ft; except that in any habitable room under a sloping ceiling at least one-half of the floor area should have a ceiling height of at least 7 ft, and the floor area of that part of such a room where the ceiling height is less than 5 ft should not be considered as part of the floor area in computing the total floor area of the room for the purpose of determining the maximum permissible occupancy.

No space located up to 4 ft below grade should be used as a habitable room of a dwelling unit unless approved by the appropriate authority in writing.

No space located more than 4 ft below grade should be used as a habitable room of a dwelling unit.

In every dwelling unit of two or more rooms, every room occupied for sleeping purposes by one occupant should contain at least 70 ft^2 of floor space for the first occupant, and every room occupied for sleeping purposes by more than one occupant should contain at least 50 ft^2 of floor space for each occupant thereof.

No dwelling or dwelling unit containing two or more sleeping rooms should have such room arrangements that access to a bathroom or water closet compartment intended for use by occupants of more than one sleeping room can be had only by going through another sleeping room; nor should room arrangements be such that access to a sleeping room can be had only by going through another sleeping room. A bathroom or water closet compartment should not be used as the only passageway to any habitable room, hall, basement, or cellar or to the exterior of the dwelling unit.

Every dwelling unit should have at least 4 ft^2 of floor-to-ceiling height closet space for the personal effects of each permissible occupant; if it is lacking, in whole or in part, an amount of space equal in square footage to the deficiency should be subtracted from the area of habitable room space used in determining permissible occupancy.

ACCEPTABLE EXIT ARRANGEMENTS

Exits should be remote from each other. In determining the distance between exits, measure the corridor centerline as shown in Fig. 1. The distance between exits should not exceed that fixed by maximum distances from doors of living units to exit doors.

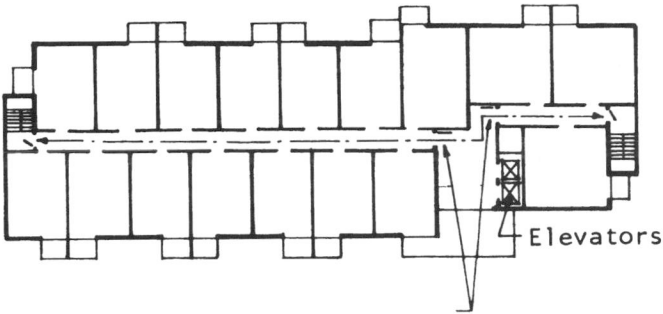

Fig. 1 Fire compartment doors.

Two interior stairs such as the scissor stairs in Fig. 2 may be used when arrangement of the corridor system is similar to the design shown. Since the stair and wall construction separates the two stairs, the construction should not allow the passage of smoke from one to the other.

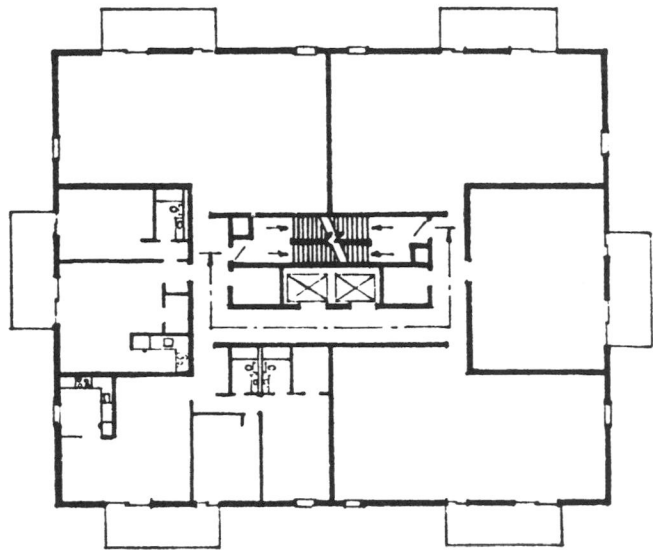

Fig. 2

Dead-end corridors may be used when the length of the corridor is limited. Stair 1 in Fig. 3 is considered a smokeproof tower, because the passageway is open to the outside. It is impossible for smoke to enter the stairs by way of the doors. Wherever possible, such an arrangement is advisable. Stair 2 is an enclosed interior stair and should have positive air pressure relative to the corridor to reduce the possibility of smoke entering the stairs.

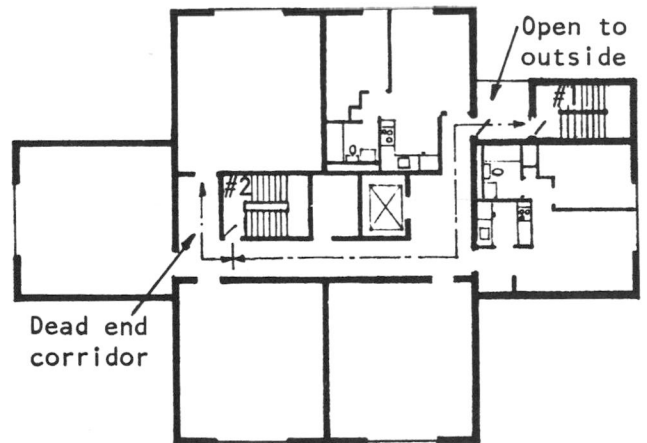

Fig. 3

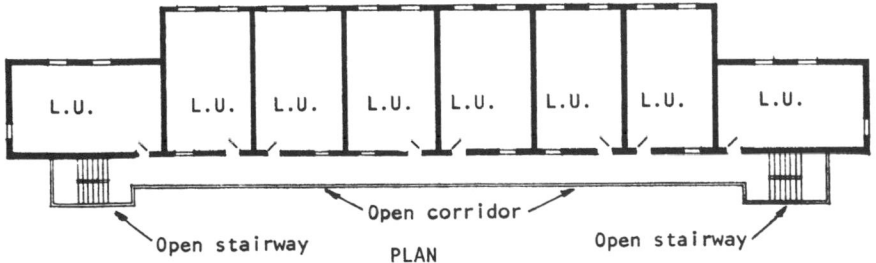

Fig. 4

LOCATION OF EXIT STAIRS

ESCAPE AND REFUGE—BUILDING ESCAPE STRATEGIES

Escape strategies for linear corridors in low-rise buildings are described below. Escape plans should be evaluated to see if alternate routes are available for use in the event that one route becomes untenable due to fire or smoke.

Central Stairs

The central stairs is an undesirable escape strategy because exits cannot be used if the corridor becomes smoke-filled. However, an exterior balcony, as shown on the sketch, could be used to increase chances of rescue.

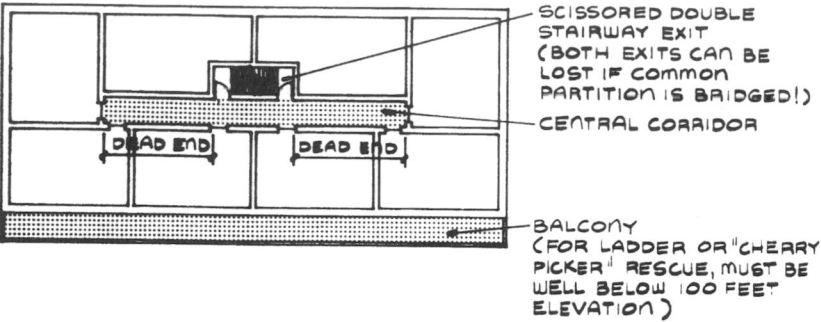

SCISSORED DOUBLE STAIRWAY EXIT (BOTH EXITS CAN BE LOST IF COMMON PARTITION IS BRIDGED!)

CENTRAL CORRIDOR

DEAD END   DEAD END

BALCONY (FOR LADDER OR "CHERRY PICKER" RESCUE, MUST BE WELL BELOW 100 FEET ELEVATION)

Fig. 5

Remote Stairs

Stairway exits at the ends of a corridor can offer alternate escape routes and shorter travel distances to exits for most occupants.

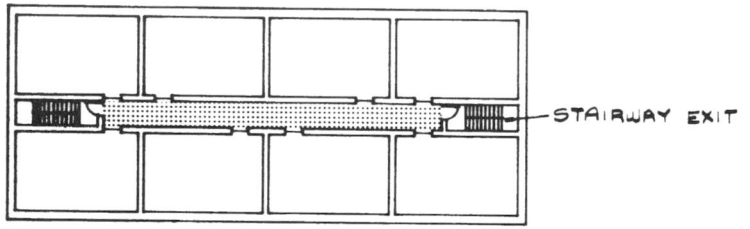

STAIRWAY EXIT

Fig. 6

Exterior Stairs

Protected outdoor stairway exits can provide smoke-free escape routes. There are alternate escape routes to remotely located stairs as shown by the sketch.

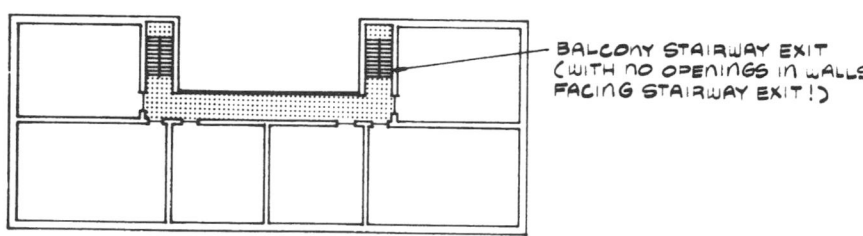

BALCONY STAIRWAY EXIT (WITH NO OPENINGS IN WALLS FACING STAIRWAY EXIT!)

Fig. 7

PLANNING OF FIRE EXITS

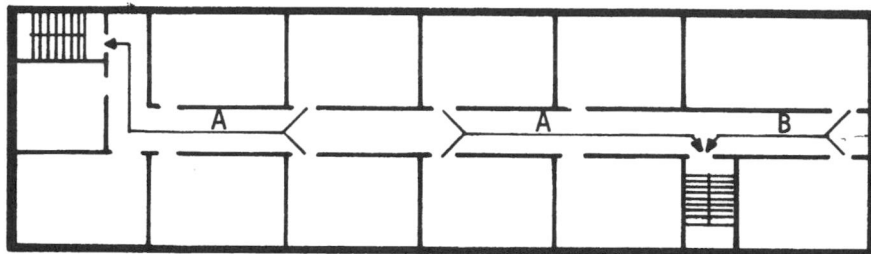

Fig. 1   Plan of residential floor, multifamily housing (*A* = distance between living unit entrance and stairway in corridor affording exit in two directions = 100 ft maximum or 150 ft maximum in buildings with automatic sprinklers).

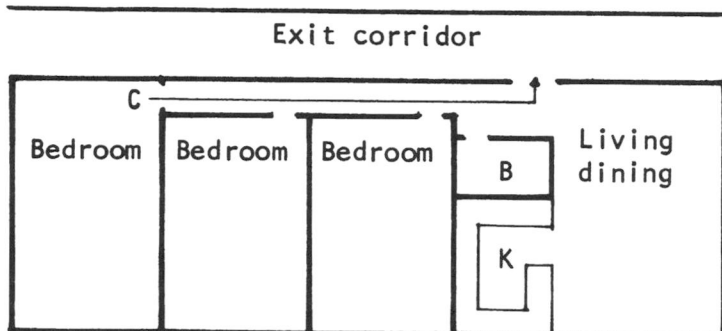

Fig. 2   Plan of living unit, multifamily housing (*C* = distance within a living unit between the door of the most remote room and a doorway to an exit corridor = 50 ft maximum).

EXITS—MULTIPLE DWELLINGS

MAXIMUM DISTANCES OF TRAVEL TO EXITS

Maximum distance of travel (A in Fig. 1) from the door of any room in any dwelling unit to a door opening into an exit passageway on the same story should be 50 ft.

Maximum distance of travel (B) from the main entrance door of any dwelling unit or any room or any part of a fire area not so divided, in a story above the grade story, to a passageway to a door opening into an exit stairway or horizontal exit on the same story: 100 ft in buildings of type 1 or 2 construction; 50 ft in buildings of type 3, 4, and 5 construction. When such buildings have a sprinkler system installed throughout, distance may be increased.

Maximum distance of travel (C) from the door of any room or any point in a fire area not divided, in a basement or below-grade story, to a door opening into an exit stairway or legal open space, or horizontal exit, should be 75 ft.

Maximum distance of travel (D) from doors of below-grade rooms enclosing equipment to a door opening into an exit stairway leading to a legal open space should be 75 ft.

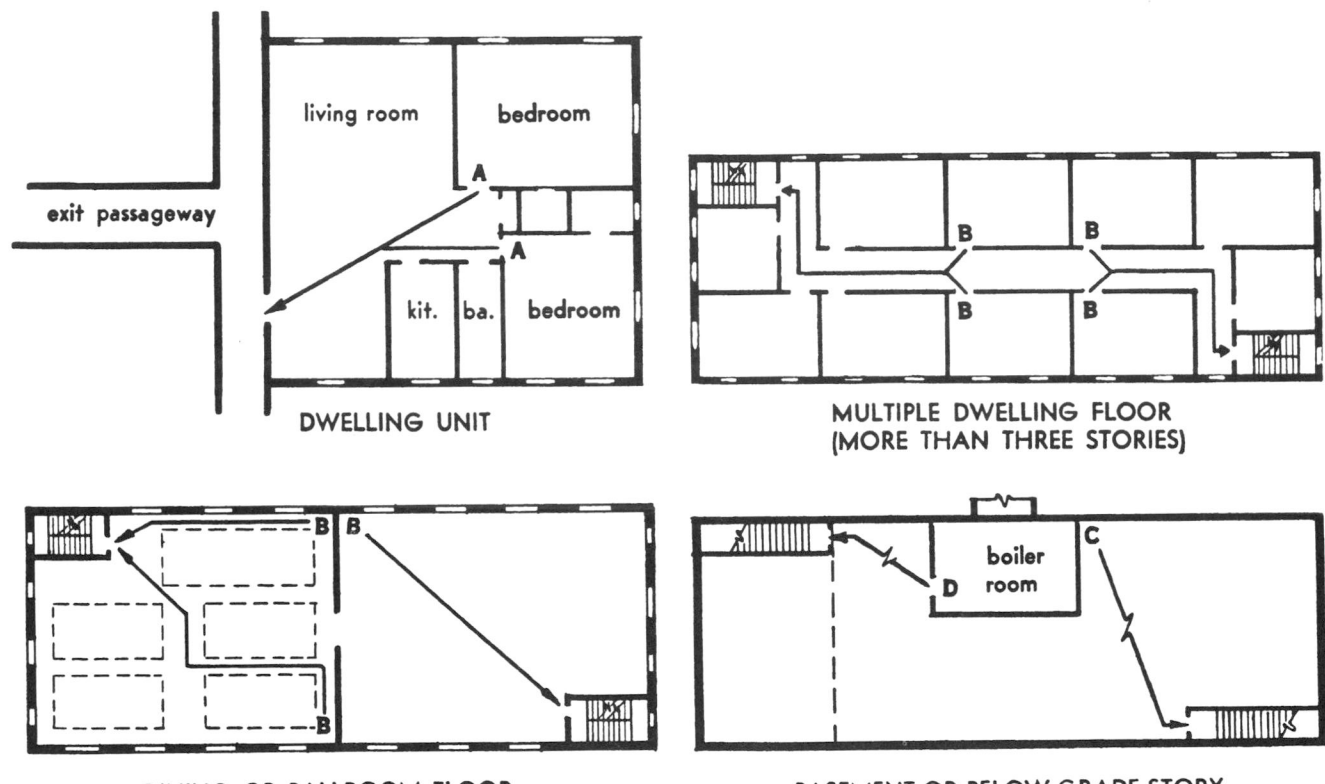

DWELLING UNIT

MULTIPLE DWELLING FLOOR (MORE THAN THREE STORIES)

DINING OR BALLROOM FLOOR

BASEMENT OR BELOW-GRADE STORY

Fig. 1

ACCEPTABLE SCHEMES FOR EXITS

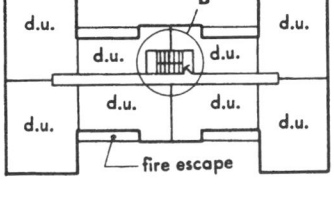

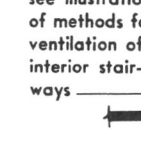

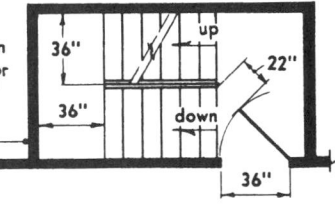

Fig. 2 Two interior stairways 36 in wide located at each end of a public hallway not exceeding 100 ft in length, 36 in in width, throughout the line of travel.

SCHEME

DETAIL

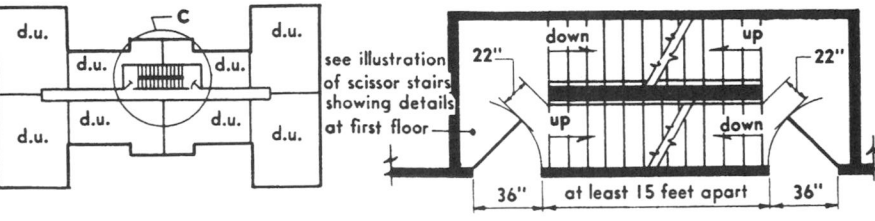

Fig. 3 Two interior stairways 36 in wide, in an enclosed shaft with door to each stairway separated at least 15 ft; public hallway not exceeding 100 ft in length, 36 in in width throughout the line of travel.

SCHEME

DETAIL

TYPICAL ARRANGEMENT OF PUBLIC HALL

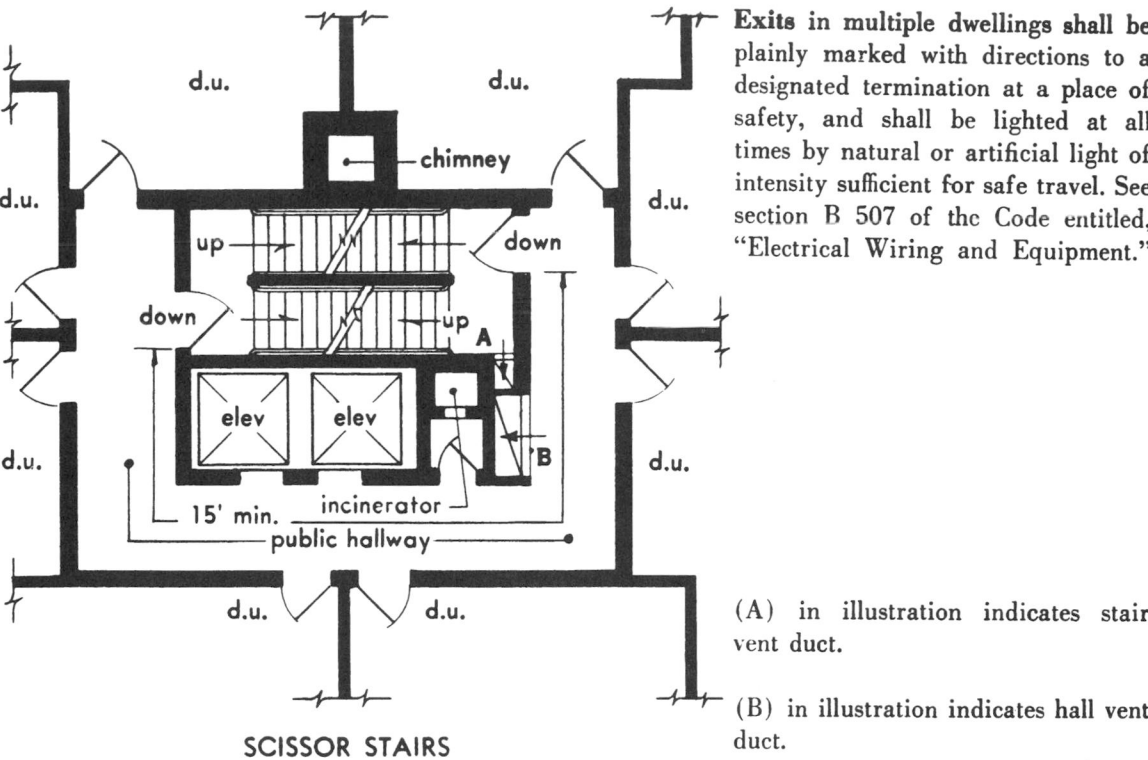

SCISSOR STAIRS

Fig. 4

Exits in multiple dwellings shall be plainly marked with directions to a designated termination at a place of safety, and shall be lighted at all times by natural or artificial light of intensity sufficient for safe travel. See section B 507 of the Code entitled, "Electrical Wiring and Equipment."

(A) in illustration indicates stair vent duct.

(B) in illustration indicates hall vent duct.

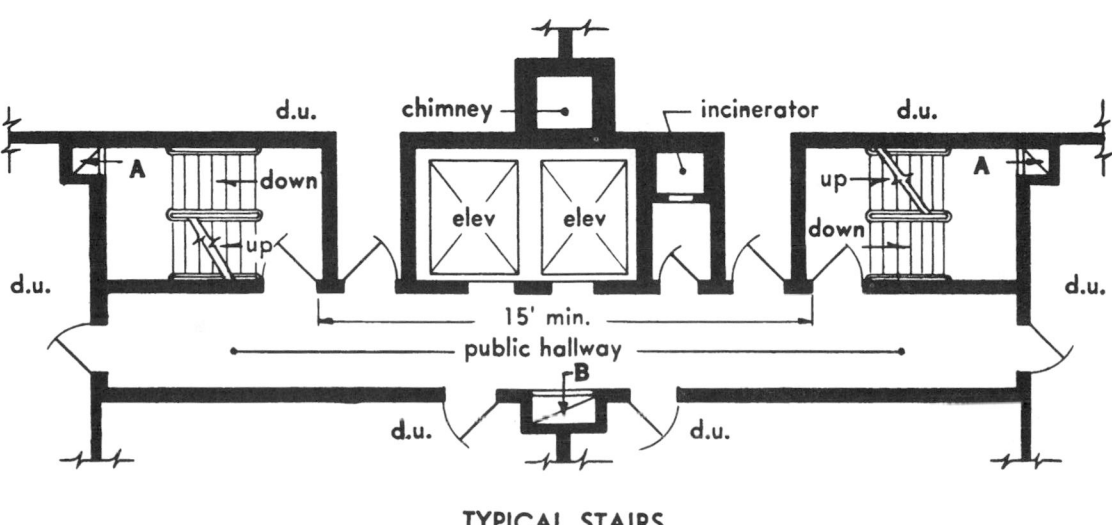

TYPICAL STAIRS

Fig. 5

EXITS—MULTIPLE DWELLINGS

ENCLOSURE REQUIREMENTS

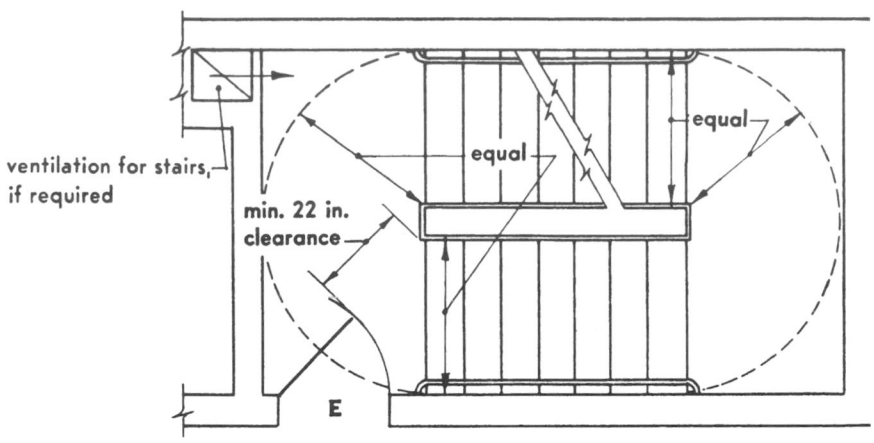

Fig. 6   Plan of typical interior exit stair.

SCISSOR STAIRS

Every passageway and enclosed stairway which serves as an exit or part thereof should be enclosed with fire-resistive construction.

In multiple dwellings more than two stories in height, exit stairways should be separately enclosed. Openings in such construction should be provided with opening protectives.

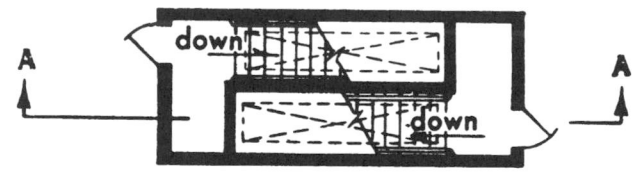

ROOF BULKHEAD PLAN

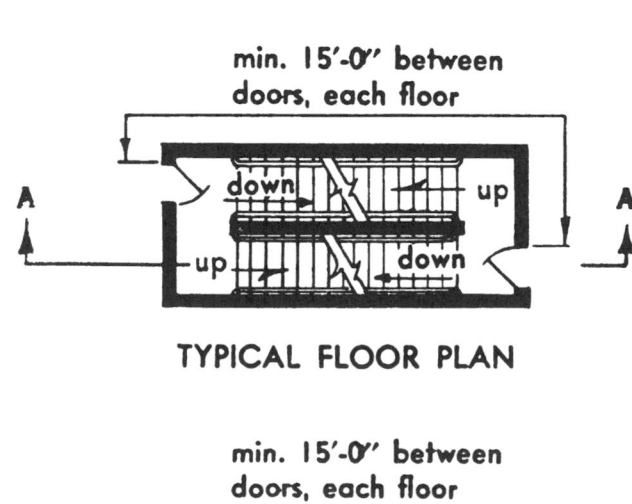

TYPICAL FLOOR PLAN

FIRST FLOOR PLAN

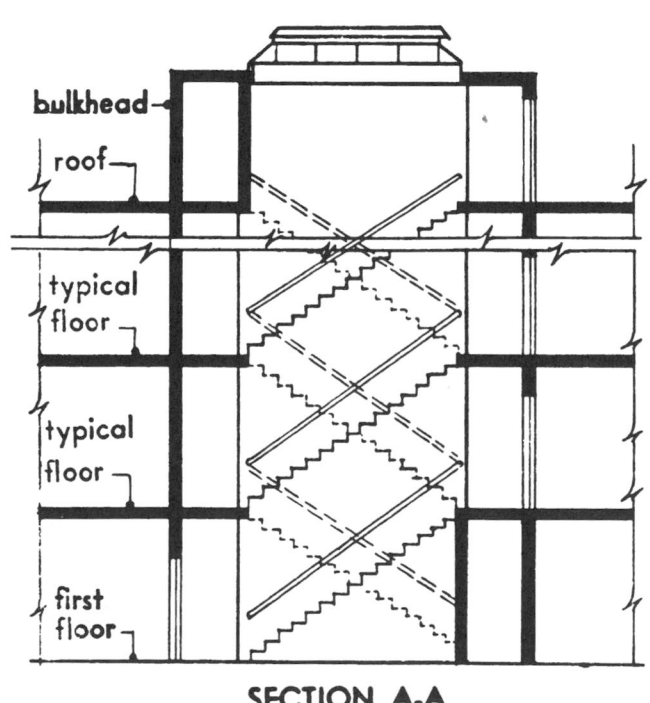

SECTION A-A

Fig. 7a

Fig. 7b

OPEN BALCONY CORRIDOR

A balcony should not be used for any purpose other than entrance or exit and should be no less than 5 ft wide.

Stairs and elevators should be centrally located. Elevators should open into an enclosed area for protection from the weather.

Balcony floors should slope to drains connected to leaders.

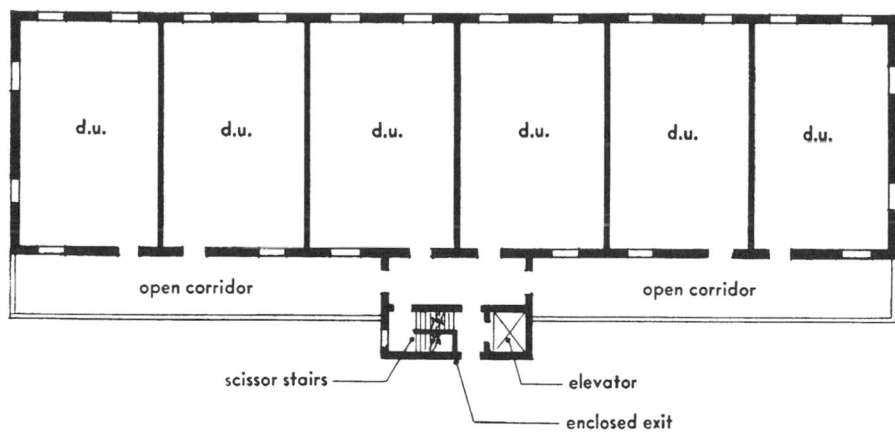

Fig. 8a   Plan.

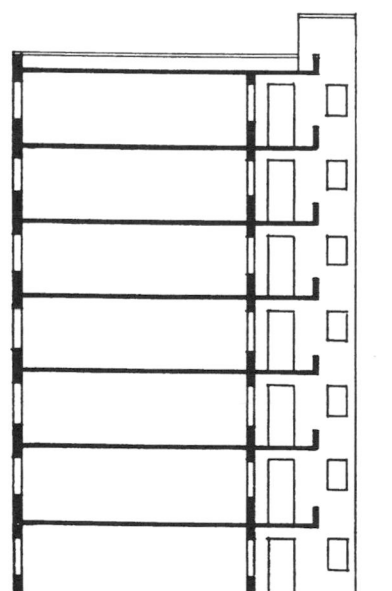

Fig. 8b   Section.

FIRE SAFETY/APARTMENT ALARM SYSTEM

ALARM SYSTEMS

High rates of sickness and accidents among disabled people indicate the need for an alarm system, particularly where they live on their own. The disabled person may easily fall, hurt himself, or become ill and helpless without anyone knowing. An alarm system connecting each dwelling unit to a central control point is therefore recommended in new construction.

There should be two terminals of an alarm system in the dwelling unit, well marked for daytime use and illuminated so that they can be seen at night. One should be located in the bathroom, because many accidents occur there, and the other in the bedroom. If only one alarm button is provided it is preferable to place it in the bathroom. Buttons or pull cords should be placed no more than 600 mm (about 24 in) above floor level, so that they can be reached from a prone position. An alternative to the mechanical pushbutton alarm is one that is triggered by a whistle carried by the disabled person, or by a frequency modulator.

Where the dwelling terminal should be placed is determined by the audible range of the whistle. For deaf persons, a flashing light may operate when the fire alarm system is triggered. For further and more immediate nighttime warning, a vibrator device may be used.

The location of the central terminal varies according to local conditions. If it is placed in a superintendent's apartment it should be heard at any time of the day or night.

Another possible location is the lobby or vestibule, where the system can be combined with the intercom system or perhaps with the fire voice-communications system terminals.

A third alternative is to employ a telephone switchboard, using a telephone alarm system, but this solution is practicable only when staff is provided around the clock.

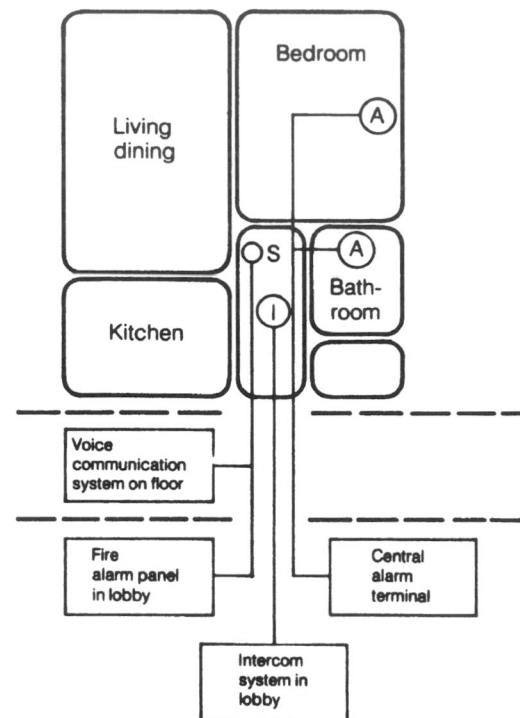

S — Smoke detector

A — Alarm system

I — Intercom

Fig. 1   Alarm systems.

THE SINGLE-FAMILY HOUSE

The basic fire safety principles that should be used in the design of the single-family home are:

1. Two ways out should be provided from all living and sleeping rooms; the second means of escape may be a suitable window.

2. The spatial arrangement should minimize the spread of fire and smoke, particularly from living and utility spaces into the sleeping spaces.

3. The construction should minimize the development and spread of fire.

4. Automatic smoke detectors should be used to safeguard the means of escape from bedrooms.

Spatial Arrangements

Spatial arrangements can improve fire safety by isolating locations where fire is most likely to start and providing uninterrupted paths from basement to the top floor and from living spaces to sleeping spaces. Exiting from second floor windows can be improved by having the windows open onto a porch or garage roof or balcony.

The floor plan in Fig. 1 illustrates an arrangement typical in modern homes. Fire safety is unnecessarily jeopardized because the second floor stairway terminates near the center of the first floor and the basement stairway directly exposes the main exit from the home. In addition, only one second floor window opens onto a roof for easy escape.

Another undesirable spatial layout is the popular open floor plan. The lack of solid partitions between rooms offers no resistance to the spread of fire and smoke throughout the house. With such a floor plan, the strategic installation of smoke detectors is very important.

An improved spatial arrangement for a two-story home is illustrated in Fig. 2. The second floor stairs terminate downstairs adjacent to the outside door, and the basement stairs open away from the entryway. In addition, all second floor bedrooms have a window which opens directly onto a roof.

On a sloping lot, bedrooms can be conveniently located on the lower level. A spatial arrangement such as that shown in Fig. 3 markedly increases the life safety potential for the occupants of the house. The lower-level rooms are only slightly exposed to a fire in the upper, living areas, and every bedroom, except one, has a door leading directly outside. But even for this one bedroom, occupants have excellent potential for escape, through the adjoining rooms or through the grade-level windows.

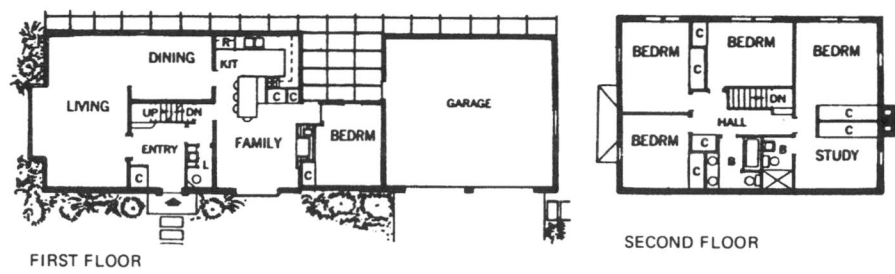

FIRST FLOOR   SECOND FLOOR

Fig. 1   Floor plan for a large two-story residence.

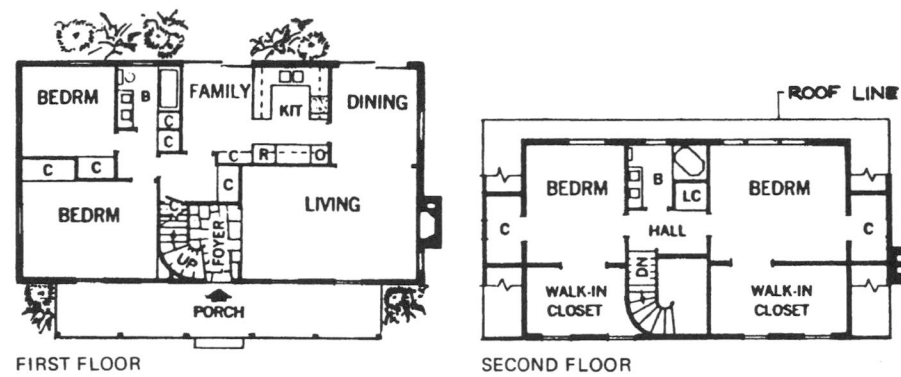

FIRST FLOOR   SECOND FLOOR

Fig. 2   Two-story residence with improved fire safety layout.

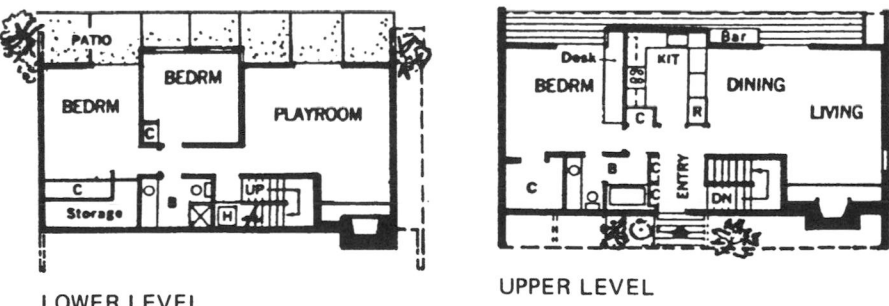

LOWER LEVEL   UPPER LEVEL

Fig. 3   Superior arrangement for fire safety in a two-story home.

FIRE SAFETY PRINCIPLES/SINGLE-FAMILY HOUSE

Additional Examples

The spatial arrangements described and shown on the following pages are merely a few examples of how the floor plan of a home can be varied to obtain improved fire safety. These examples have been presented not as layouts to be followed but as a guide to stimulate the designer and builder.

The ranch-type home (Fig. 4) is easily adapted to a fire-safe arrangement provided the bedroom windows are suitable for emergency escape.

Fire Safety Features

1. Sleeping areas isolated from the living areas and closed off with a door.
2. First-floor windows provide an alternate escape from sleeping areas.
3. Exit from bedroom area is adjacent to outside door.
4. Two ways out from all locations.
5. Smoke detector outside bedrooms.

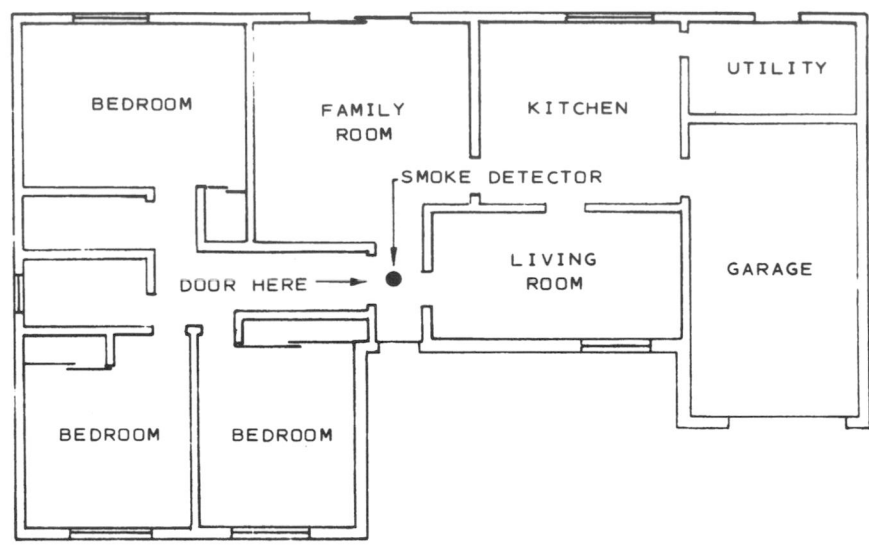

Fig. 4   Application of fire safety principles to a ranch home.

Fire Safety Features

1. Stairs from second floor sleeping area direct to outside door.
2. Living spaces offset from second floor stairway.
3. Two ways out from living spaces.
4. Second floor bedrooms have window opening onto a roof surface.
5. Smoke detector at top of both stairs.

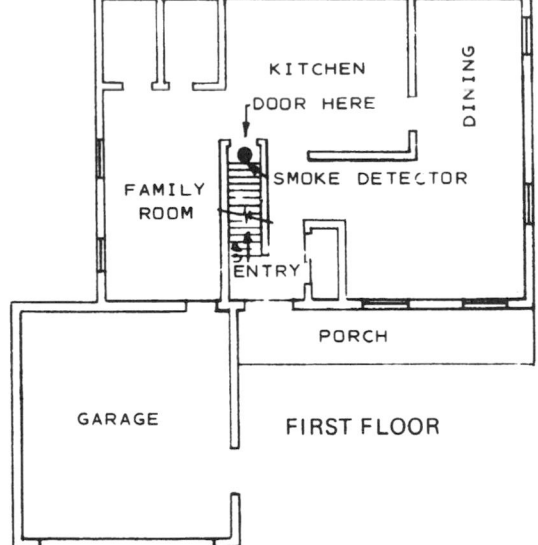

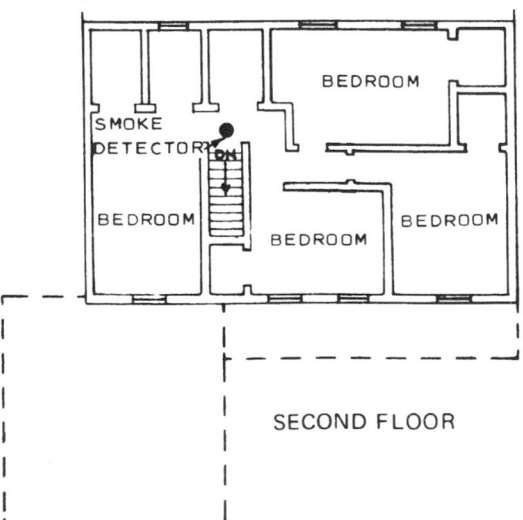

Fig. 5   Fire safety principles applied to a two-story home.

Fire safety features for multifamily housing differ from those for single-family housing primarily in the additional precautions which need to be taken to assure the safety of persons in other family units in the event of fire. The possibility of a fire start in both a family unit and common building areas must be considered. Low-rise housing is differentiated from medium- and high-rise housing principally in the relative ease in evacuating persons from the lower buildings and the comparative difficulty of fire fighting and rescue in the higher buildings.

Thus, the fire safety goals for low-rise multifamily housing are the fire safety principles set forth above as they apply to each family unit, plus the compartmentation needed to assure safe exiting from all parts of the building and to reduce the possibility of severe damage outside the unit of origin. Compartmentation is achieved with construction which will resist the rapid spread of fire, and the degree of such fire resistance, measured in hours and minutes under standard fire test conditions, may vary from 10 min to 2 h under codes applicable to low-rise housing.

Experience has shown that when this type of housing is constructed in conformance with a modern building code, the goal of obtaining a reasonably high degree of life safety for the occupants outside the unit of origin and of limiting fire spread therefrom is achieved. Following are some additional suggestions for increased fire safety within individual living units and for design and construction of techniques which will assist in meeting or exceeding minimum fire safety standards at little or no increase in cost.

SPATIAL ARRANGEMENT

The layout inside each individual family unit should utilize the same principles of fire safety as the ones set forth for the single-family dwelling. Features for the ranch home can be applied to the typical apartment unit, and those for the multistory house can be incorporated into the town house concept. A weakness in many apartment unit designs is the lack of two, separated exits. Often, there are two doors from the apartment to the common corridor, but they are only a few feet apart and do not provide sufficient assurance that one will be available for escape when the other one is blocked by fire or smoke (Fig. 1). If this arrangement is necessary, a door to the corridor should be as close as possible to the bedrooms, but often in the low-rise building it is possible to provide each family unit with a "back" door not opening into the corridor or with a balcony which provides a temporary area of refuge (see Fig. 2).

The interior, enclosed corridor has drawbacks from a fire safety standpoint if it is the only way from which occupants can reach the exterior of the building. Because of its configuration and limited size, it can quickly fill with smoke and heated gases from a fire in one of the adjacent family units if the door to that unit is not closed (see also Corridor Doors, below). In such a situation, an exterior stairway is a good secondary exit. An exterior balcony serving each family unit on that side terminating in a stairway to ground at each end affords superior fire safety even as the sole means of egress, since the possibility of the exterior balcony being completely blocked by smoke and fire so as to prevent escape is remote (Fig. 3).

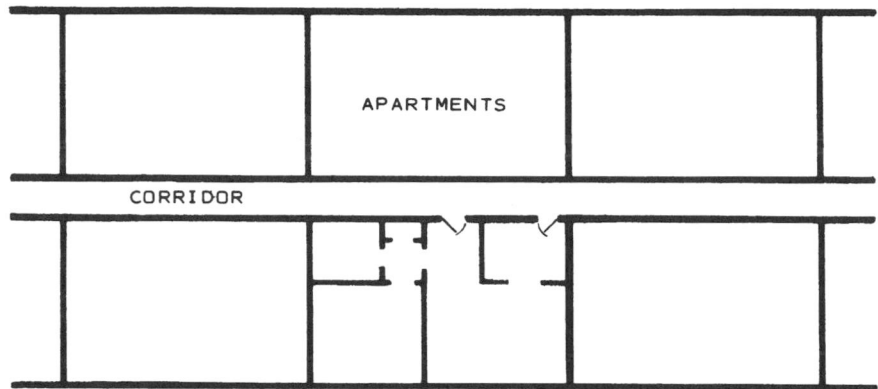

Fig. 1 Typical apartment with single means of egress (second door is too close to be effective as a second exit).

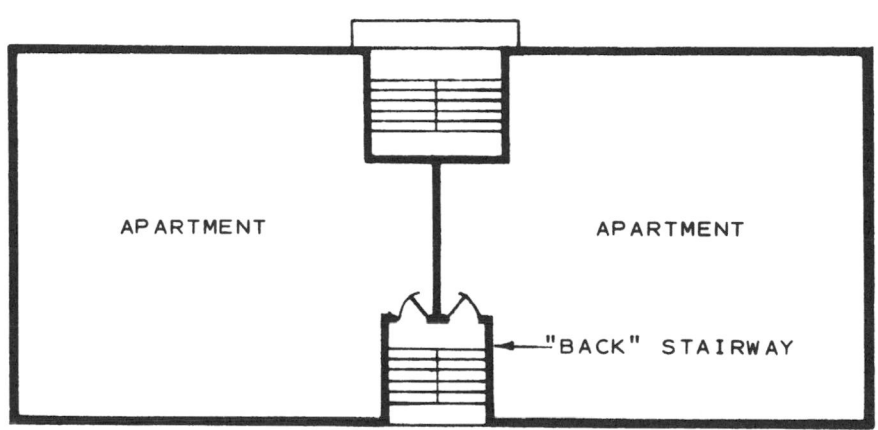

Fig. 2 A door into a second stairway—either enclosed or exterior to the building—is an excellent second exit.

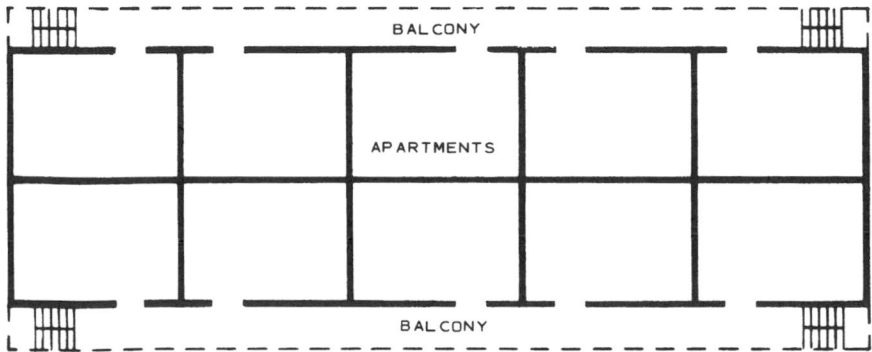

Fig. 3 Exterior balconies and stairways provide smoke-free exiting.

FIRE SAFETY PRINCIPLES/LOW-RISE MULTIFAMILY HOUSING

CORRIDOR DOORS

Many codes require doors between the corridor and individual living units in multifamily residential buildings to have a certain amount of fire resistance, such as is obtained from a solid-core wood or steel door, capable of withstanding a standard test fire for at least 20 min. In addition, to be certain that it will be closed when it is needed as a fire barrier, the door should have a self-closer. Both features have been found to be very effective in confining a fire to the unit of origin and should be provided even if not required by code.

A fire in one living unit can spread smoke into other parts of the building through heating and air-conditioning ducts if these ducts serve more than one apartment. This not only increases smoke damage but may require evacuation of the entire building for even such a minor incident as "burned" food on the kitchen range. Under extreme conditions, fire may spread through such ducts to other apartments. Fortunately at present, central duct systems for heating and air conditioning are not common in multifamily housing, and it is preferable from a fire safety standpoint that each family unit continue to be served only by its own duct system.

PREVENTING SMOKE SPREAD

Placing interior stairways in fire-resistive enclosures is commonly required, both to prevent the vertical spread of smoke and heat in case of fire and to protect the occupants as they leave the building in an emergency. As with smoke barrier partitions and fire walls (see below), stairway doors should be kept closed so that they can perform their function in an emergency. Consequently, they are equipped with self-closers. However, when such doors are frequently used, they tend to be wedged or hooked open, defeating their fire safety function.[1] More and more building codes now permit stairway doors—as well as doors in smoke barriers and fire walls—to be held open by magnetic devices which release the door and allow it to close automatically when a smoke detector at the door is triggered or the fire alarm system operates. Because the magnetic releases are reliable (they are installed on fail-safe circuits) and because they give better assurance that the doors will be closed at the time of need, their installation is recommended on frequently used doors unless other provisions are stipulated by local code.

Smoke barrier partitions in corridors are valuable in preventing the rapid spread of smoke throughout an entire floor. At the same time, they give occupants a place of refuge from which they can slowly exit using the stairways. Whenever long interior corridors are used as the sole means of reaching the exits from the building, smoke barriers lessen the distance a person might have to travel through a smoke-laden atmosphere and they limit the number of units immediately affected by the smoke condi-

tion. As such, they are recommended whenever the corridor length exceeds about 200 ft, which allows a maximum of about 100 ft travel distance in the corridor to a smoke-free area, whereas without the barrier a 150 ft travel distance to the nearest exit is normally permitted by building codes. (HUD Minimum Property Standards call for smoke barriers whenever there are more than eight living units on a floor.) Unless otherwise indicated by local code, smoke barrier partitions should have at least 20 min fire resistance and have self-closing doors which may be held open by magnetic releases (see above). The additional cost of smoke barriers can be minimized by incorporating them as part of fire walls.

Ventilation of corridors in case of fire can be difficult unless a window or other means of removing the contaminated air is provided. Smoke buildup in the corridor can hamper the fire department and increase the smoke damage to the building and living units.

Fire Walls

Improved fire safety combined with reduced construction costs can frequently be achieved by installing a fire wall. In case of fire, the occu-

pants have the option of exiting either down an enclosed stairway or horizontally through a door in the fire wall. The fire wall can often eliminate one or more stairways that would otherwise be required by the building code, since the horizontal exit can qualify for up to 50 percent of the required exits. Also, since the type of construction allowable under a building code depends in part on the area bounded by exterior walls and fire walls, a fire wall reduces the building area from a code standpoint and may allow a less costly construction type. Two examples of the application of fire walls for improved safety and reduced cost are presented below.

Example 1 Figure 4 shows the upper-story floor plan of a low-rise apartment building. Local code had required this building to be of protected noncombustible construction and to have three enclosed stairways and a smoke barrier to subdivide the long corridor. Upgrading the smoke barrier to a fire wall at practically no cost provided the occupants with a horizontal exit and reduced the building area in half for code purposes. The middle stairway could be eliminated and the building built of protected wood-frame construction.

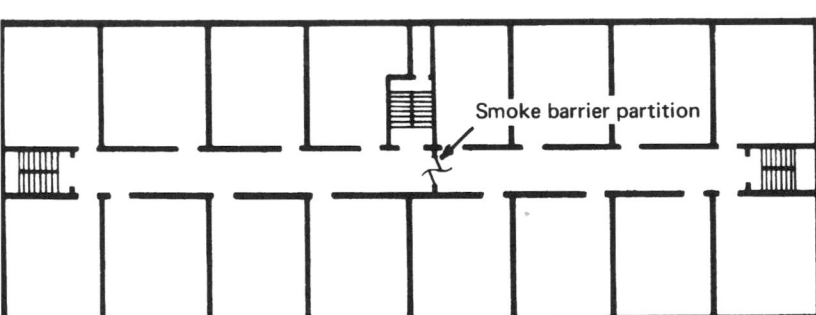

a. Original plan: three stairways and a smoke barrier partition

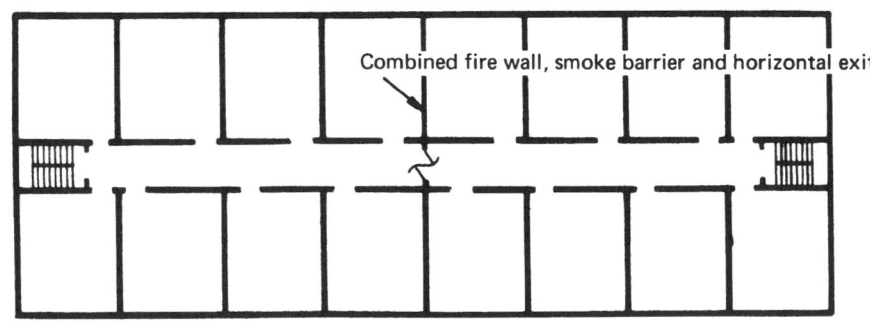

b. Revised plan: two stairways and a fire wall also used as a horizontal exit and a smoke barrier

Fig. 4 Changing the smoke barrier to a fire wall eliminated a stairway in this building and reduced the structural construction costs.

---

[1]Self-closers incorporating a fusible link release used to be popular, but they are of little value for life safety purposes.

Example 2 Two apartment buildings, each having a floor plan as illustrated in Fig. 5a, were required to have two stairways per building, and the facing exterior walls had to be masonry. Combining these into one building separated by a fire wall (Fig. 5b) eliminated the need for one masonry wall and two of the required stairways. This same advantage can often be obtained by connecting two buildings with a passageway to create a horizontal exit and eliminate the need for two stairways.

Firestopping

Firestopping in wood frame walls should be installed in multifamily residences in the same manner as in single-family dwellings. However, multifamily dwellings are required by codes and standards to also have floor-ceiling firestops between living units and in other concealed spaces. Some examples of such firestopping methods are illustrated in Fig. 6.

Other Fire Safety Features

Suggestions for smoke detectors and electrical wiring in single-family dwellings also apply to living units in multifamily dwellings. Fire alarm systems for the remainder of the building (storage and maintenance areas, furnace rooms, garages, etc.), standpipe hose lines, fire extinguishers, main electrical services and feeders, gas piping, etc., should be as required by local codes.

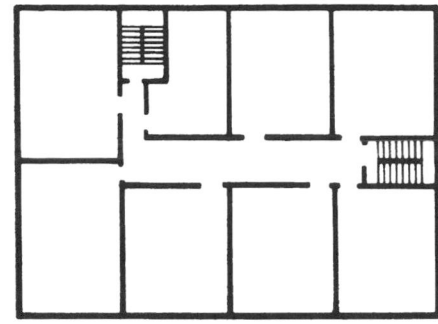

a. Original plan: two buildings, two stairways each

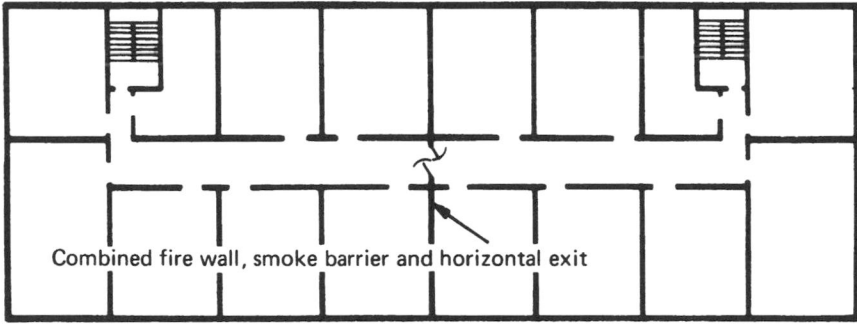

b. Revised plan: one building, two stairways, and a fire wall functioning as horizontal exit and smoke barrier

Fig. 5 Combining two, separated buildings into one, divided by a fire wall-horizontal exit, eliminated two stairways.

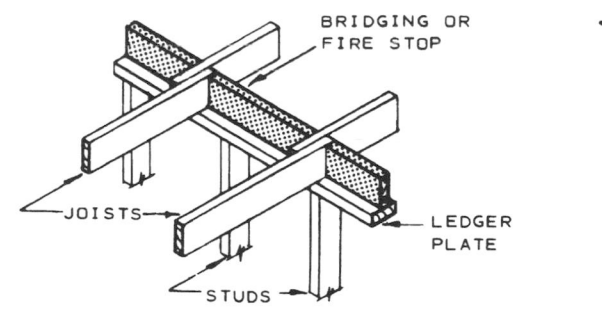

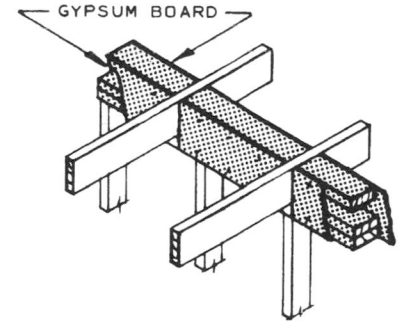

FIRESTOPPING OF SAME CONSTRUCTION AS REQUIRED FOR THE WALL

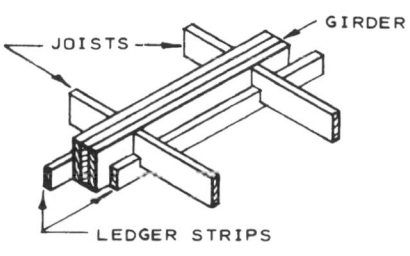

INHERENT FIRESTOPPING

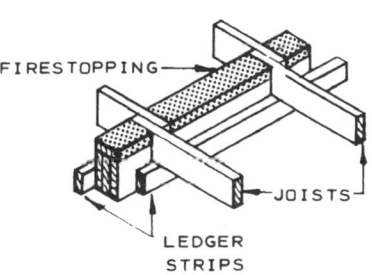

Fig. 6 Firestopping for wood-frame ceiling construction.

FIRE SAFETY PRINCIPLES/MEDIUM- AND HIGH-RISE MULTIFAMILY HOUSING

Medium-rise housing (about four to six or eight stories high) and high-rise housing (more than about six to eight stories high) are distinguished from low-rise housing by the greater need for stringent fire safety regulations as the building becomes larger and higher. As would be expected, it takes longer to evacuate the occupants of higher buildings, up to a point where total evacuation would take so long that it becomes impractical from a fire safety standpoint. In those cases, provisions must be made in the design for areas of safe refuge within the building. Other considerations in the higher buildings are the safety and welfare of the fire fighters who must reach the source of the fire and the structural integrity so that there is no possibility of collapse due to a fire.

All building codes require medium-rise and high-rise buildings to be constructed of materials which have a certain amount of resistance to the heat and flames of a fire. For the lower buildings, depending on the code, wood members, protected by fireproofing materials, may be acceptable, but universally for high-rise buildings, noncombustible materials having a fire resistance rating of not less than 2 h are required for the structural frame. Inasmuch as many other fire safety features are incorporated into building codes as they apply to high-rise buildings, the level of safety of this type of housing is generally very high. However, some incidents in recent years have demonstrated some fire safety weaknesses, and not all building codes have been modified to eliminate these situations. The suggestions here deal to a large extent with these weaknesses and how to avoid them. In all cases, the applicable provisions of local codes should be considered.

COMPARTMENTATION

One of the major differences between low-rise housing and higher buildings is the concept of compartmentation. In the lower buildings, it is generally assumed that the occupants can easily escape—out the window, if necessary. In the high-rise building, particularly, rescue from a window is often impossible, and each apartment is made a place of refuge by constructing it as a fire-resistive compartment in which the occupants can survive a fire elsewhere in the building. The following guidelines are the minimum which should be incorporated in the design in order to have each apartment available as a place of refuge:

1. The apartment should be totally separated from the remainder of the building—adjacent apartments, corridor, and apartments above and below—by construction having a fire resistance rating of not less than 1 h. Doors between the apartment and the corridor should have a fire resistance rating of not less than 45 min, and they should be equipped with a self-closer.

2. Heating and air-conditioning ducts should not serve more than one family unit. (This is not meant to apply to kitchen and bathroom exhaust ducts.)

3. Every bedroom should have an openable window for fresh, smoke-free air.

4. Interior finish in the corridor should be limited to that which is noncombustible or which has a very low fuel contribution. A low flame-spread index is not sufficient in itself.

PREVENTING SMOKE SPREAD

Preventing the spread of smoke, heat, and other products of combustion is critical to fire safety in the high-rise because rapid evacuation is so difficult. If the inside corridor is the only means of reaching stairways and if the occupants must remain in the stairway for many minutes before reaching ground level, both must be kept smoke-free as long as possible. Some building codes now require that every high-rise building be divided vertically into at least two smoke zones, so that on every floor there is an area of temporary refuge. Smoke barrier partitions are used for subdividing the building into smoke zones. Smoke barriers in medium- and high-rise buildings are more critical to life safety than they are in the low-rise because of the difficulty of reaching people trapped in their apartments.

One location for a smoke barrier not always called for by code but which is especially useful is around the elevator lobby, as illustrated in Figs. 1 and 2. Because it can be kept free of smoke, the controlled use of elevators during a fire is possible, thereby aiding fire fighters not only in reaching the fire but in evacuating the elderly and the handicapped who cannot easily use stairways.

As in low-rise housing, a smoke barrier often can be "upgraded" to a fire wall, thereby functioning as a horizontal exit and reducing the number of required stairways.

Stairways are almost always required to be totally enclosed with fire-resistive construction, with self-closing fire doors on openings. Since the stairs are little used for vertical travel, they are more likely not to be wedged or hooked open, and there is usually little need for magnetic hold-open devices.

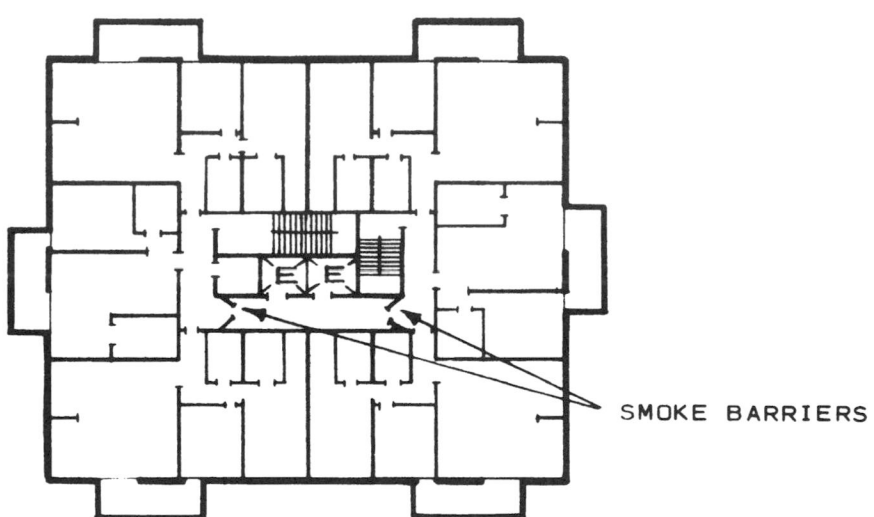

Fig. 1   Smoke barriers for center-core elevator lobby.

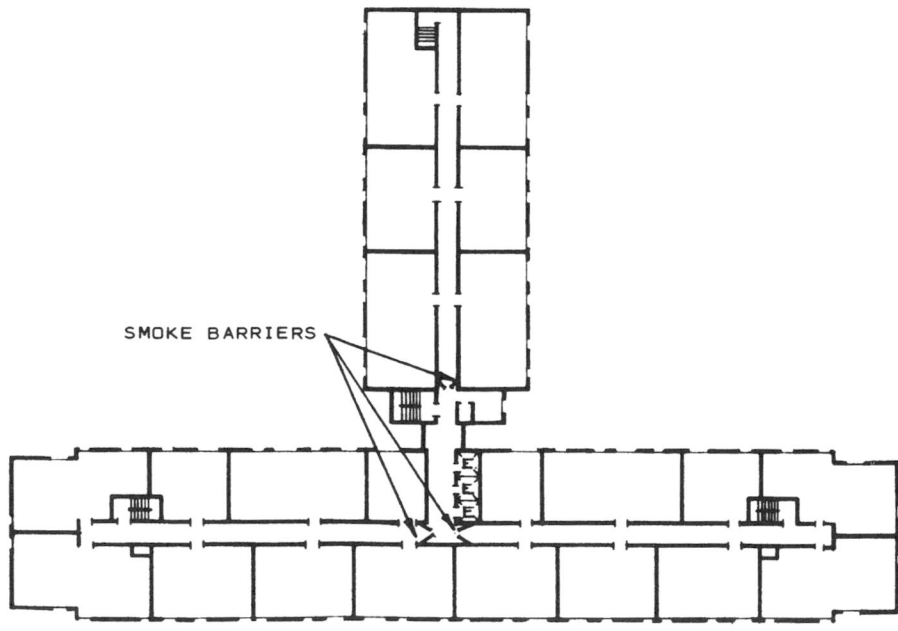

Fig. 2   Smoke barriers for elevator lobby in T-plan high-rise.

DEFENSIBLE SPACE

Defensible space is a term used to describe a series of physical design characteristics that maximize resident control of behavior, particularly crime, within a residential community. A residential environment designed under defensible-space guidelines clearly defines all areas as public, semiprivate, or private. In so doing, it determines who has the right to be in each space and allows residents to be confident in responding to any questionable activity or persons within their complex. The same design concepts improve the ability of police to monitor activities within the community.

Implementation of defensible space utilizes various elements of physical planning and architectural design such as site planning and the grouping and positioning of units, paths, windows, stairwells, doors, and elevators. Provision of defensible-space mechanisms is best achieved in a project's inception, as it involves major decisions with respect to the project.

However, a series of small-scale physical design techniques can be used to create defensible space and consequently to reduce crime in existing residential areas. These techniques consist of subdividing a project or building to limit access and improve neighbors' recognition, thus symbolically defining an area as coming under the sphere of influence of a particular group of inhabitants and improving the inhabitants' surveillance capacity.

The term *limiting access* refers to the use of physical design to prevent a potential criminal from entering certain spaces. Although no barrier is impregnable, physical barriers of this type are real and are relatively difficult to overcome.

In contrast, it is possible to use psychological or *symbolic* barriers that, while presenting no physical restriction, discourage criminal penetration by making an obvious distinction between stranger and intruder and bringing all activity under more intense surveillance. An intruder invading the space defined by such symbolic barriers becomes conspicuous to both residents and police.

Improved neighbor recognition plays a key role in the functional workings of psychological barriers. If, by newly defining areas, neighbors can be made to recognize one another, the potential criminal then cannot only be seen but also be perceived as an intruder. This subdivision of space also reinforces the feeling in residents that they have the right to intervene on their own behalf.

Creating Territorial Areas

Residential developments consisting of large superblocks devoid of interior streets have been found to suffer higher crime rates than projects of comparable size and density in which existing city streets have been allowed to continue through the sites.

Housing sites larger than a city block are best subdivided by through streets. The small scale of neighboring city blocks should be maintained where possible. This directive runs contrary to site planning principles aimed at removing vehicular traffic from the interior of large projects to free areas for recreation. However, large areas of low- and moderate-income projects that have closed off city streets but permitted public access have been considered dangerous by inhabitants and have consequently received minimal use. Through streets bring safety for three reasons:

1. They facilitate direct access to all buildings in the project by car and bus.
2. They bring vehicular and pedestrian traffic into the project and so provide the important measure of safety that comes with the presence of people.
3. They facilitate patrolling by police, provide easy access, and are a means of identifying building locations. Much of the crime deterrence provided by police occurs while they pass through an area in a patrol car.

A project's site should be subdivided so that all of its areas are related to particular buildings or clusters of buildings. No area should be unassigned or simply left "public" (see Fig. 1). Zones of influence should embrace all areas of a project, and the site plan should be so conceived. A *zone of influence* is an area surrounding a building, or preferably an area surrounded by a building, that is perceived by residents as an outdoor extension of their dwellings. As such, it comes under their continued use and surveillance. Residents using these areas should feel that they are under natural observation by other project residents. A potential criminal should equally feel that any suspicious behavior will come under immediate scrutiny.

Grounds should be allocated to specific buildings or building clusters. This practice assigns responsibility and primary claim to certain residents. It also sets up an association between a building resident in his or her apartment and the grounds below.

Residents in projects that are subdivided have the opportunity of viewing a particular segment of the project as their own turf. When an incident occurs there, they are able to determine whether their area or another area is involved. When divisions do not exist within a project plan, an incident in one area is related to the complex and can create the impression of lack of safety in the entire project.

Defining Zones of Transition

Boundaries can be defined by either real or symbolic barriers. Real barriers require entrants to possess a mechanical opening device, a familiar face or voice, or some other means of identification to indicate their belonging prior to entry. That is, access to a residence through a real barrier is by the approval of its occupants only, whether through the issuance of a key or through acceptance by their agents or by electronic signal.

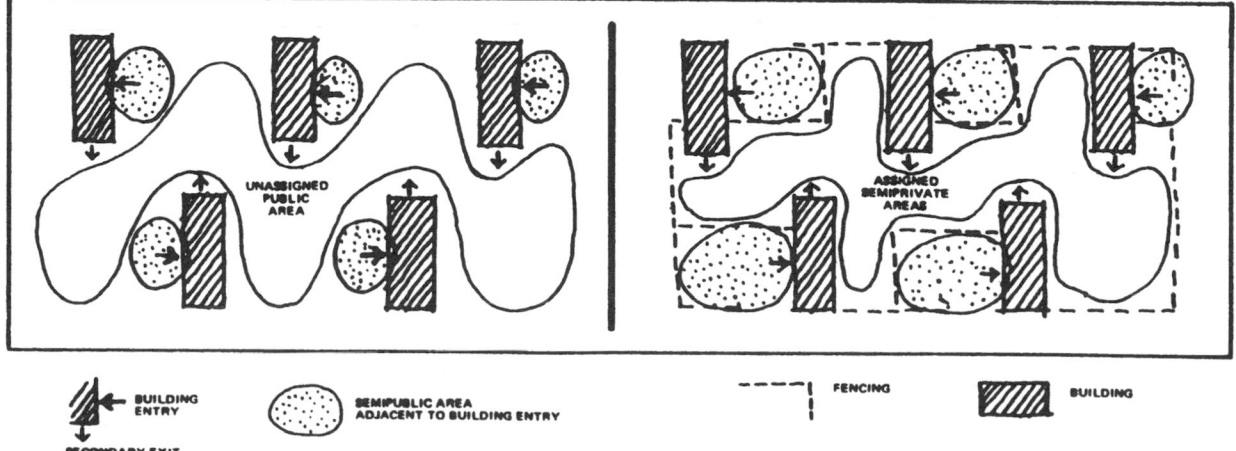

Fig. 1   Alternative site plans with unassigned and assigned areas.

SITE SECURITY

Symbolic barriers define areas or relate them to particular buildings without physically preventing intrusion (see Fig. 2). The success of symbolic versus real barriers in restricting entry rests on four conditions:

1. The capacity of the intruder to read the symbols
2. The capacity of the inhabitants or their agents to maintain controls and reinforce the space definition as symbolically defined
3. The capacity of the defined space to require the intruder to make obvious his or her intentions
4. The capacity of the inhabitants or their agents to challenge the presence of an intruder and to take subsequent action

Since many of these components work in concept, a successful symbolic barrier is one that provides the greatest likelihood of the presence of all these conditions. By employing a combination of symbolic barriers, it is possible to indicate to entrants that they are crossing a series of boundaries without employing literal barriers to define the spaces along the route.

These symbolic tools for restricting space usage assume particular importance in existing projects that cannot be subdivided into territorial areas. When it is still the intent to make space obey semiprivate rules and fall under the influence and control of inhabitants, the introduction of symbolic elements along paths of access can serve this function.

Opportunities for the use of symbolic barriers to define zones of transition are many. As illustrated in Fig. 3, the barriers can occur in moving from the public street to the semipublic grounds of the project, in the transition from outdoors to indoors, and in the transition from the semipublic space of a building lobby to the corridors of each floor.

Symbolic barriers can also be used by residents as boundary lines to define areas of comparative safety. Parents may use symbolic barriers to delimit the areas where young children may play. Similarly, because symbolic barriers force outsiders to realize that they are intruding into a semiprivate domain, they can effectively restrict behavior to that which residents find acceptable.

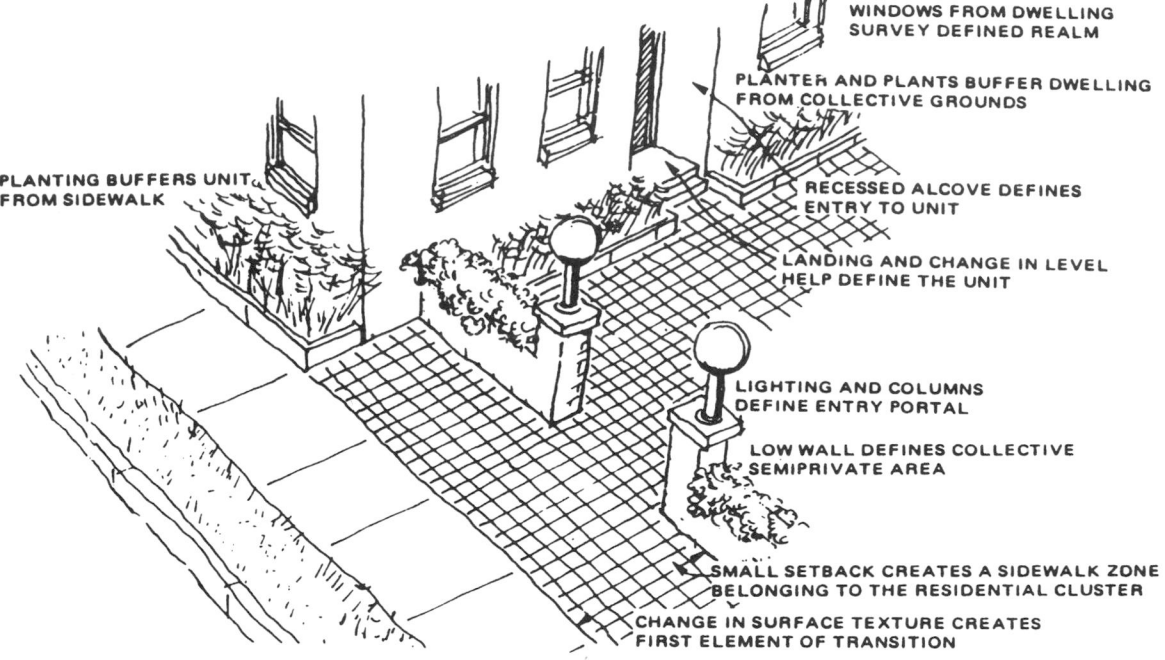

PLANTING BUFFERS UNIT FROM SIDEWALK

WINDOWS FROM DWELLING SURVEY DEFINED REALM

PLANTER AND PLANTS BUFFER DWELLING FROM COLLECTIVE GROUNDS

RECESSED ALCOVE DEFINES ENTRY TO UNIT

LANDING AND CHANGE IN LEVEL HELP DEFINE THE UNIT

LIGHTING AND COLUMNS DEFINE ENTRY PORTAL

LOW WALL DEFINES COLLECTIVE SEMIPRIVATE AREA

SMALL SETBACK CREATES A SIDEWALK ZONE BELONGING TO THE RESIDENTIAL CLUSTER

CHANGE IN SURFACE TEXTURE CREATES FIRST ELEMENT OF TRANSITION

Fig. 2   Symbolic barriers defining zones of transition.

Fig. 3   Zones of transition between public street, project grounds, and building interior.

Locating Amenities

Recreational and open-space areas should serve the needs of different groups. An understanding of what different age groups desire of open-space and recreational facilities is essential to the successful use of such areas. The design and location of these areas within the residential environment should follow the demands, capabilities, and expectations of their eventual users.

All areas of the grounds should be defined for specific uses and designed to suit those uses. Fig. 4 illustrates different uses and users. The areas adjacent to each entry, labeled A, have been allocated for the use of 1- to 5-year-olds, with seating for adults. The larger areas in the center of each entry compound, labeled B, are provided with play facilities serving 6- to 12-year-olds. The areas labeled C are intended for more passive activity and as decorative green areas. The C areas, accessible from the building interiors only, are provided with barbecuing facilities and some seating.

Well-designed recreation facilities improve the security of an area if they provide for activities of a particular group of residents and are adjacent to the residents' interior environs. So designed, the facilities create outdoor zones that are effective extensions of the dwellings. By providing for outdoor activities adjacent to homes, these areas allow residents to assume a further realm of territory and further responsibility.

Children 1 to 5 years in age are most comfortable playing in an outdoor area immediately adjacent to their dwellings, preferably just outside the door in both single-family units and multiple dwellings. Figure 5 illustrates such an area. The location of these facilities adjacent to the entry door to the unit and the inclusion of benches for adults further create a semiprivate buffer zone separating the private zone of the residential interior from the more public zones.

In the design of a multifamily residential complex serving many groups of families, each with its own entry to its own building, the buffers that demarcate each entry zone can also define a larger subcluster within the project.

Figure 6, which extends the concept shown in Fig. 5, illustrates the common entry area of a cluster of buildings. Each entry zone is provided with its own tot play area and surrounding seating. The five entries share a common central play facility for the 5- to 12-year-olds that is large enough to accommodate more active play and sufficiently separated from the dwelling units to reduce noise penetration. The large play area is, however, still very much in view of every dwelling.

Play areas for 12- to 18-year-olds should not be located immediately adjacent to home, but neither should they be too far away. They should be large enough to house activities of interest to this age group: basketball, football, handball, dancing.

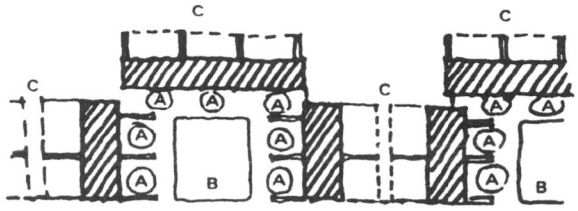

Fig. 4   Ground areas assigned for particular uses.

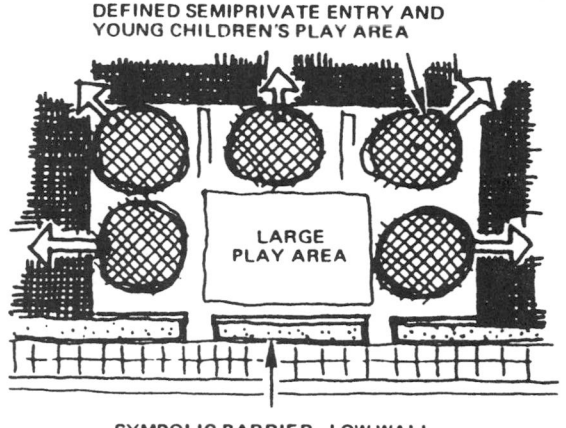

DEFINED SEMIPRIVATE ENTRY AND
YOUNG CHILDREN'S PLAY AREA

LARGE
PLAY AREA

SYMBOLIC BARRIER—LOW WALL

Fig. 5   Play area defining a buffer to a multifamily building entry.

Fig. 6   Common entry to a cluster of buildings.

SITE SECURITY

These teen play areas should not be located in an isolated area of a development, disassociated from dwelling units. This is a common practice (see Fig. 7) that results in the area's neglect, vandalization, or underuse. Rather, teen play areas should be bordered on three or four sides by the dwellings of residents, as illustrated in Fig. 8 and 9.

The teen area should be provided with occasional benches bordering play areas. Benches allow children to gather and watch while only a few play. Children also use the benches for piling extra clothing and for resting after strenuous exercise. Benches give the play area a feeling of stability and containment. When such areas are defined in this way, they frequently are adopted for social uses in the evening.

Green areas unencumbered by play facilities are the pride of the elderly and usually the thorn in the side of 7- to 15-year-olds, who are prevented from using these areas for playfields. It is therefore important to provide such green areas with protection by judicious placement and use of shrubs and fences. However the best guarantee that these green areas will be respected for their decorative purpose is the provision of adjacent and separate play areas and equipment.

Creating Surveillance Opportunities

Surveillance is a major crime deterrent and a major contributor to the image of a safe environment. By allowing tenants to monitor activities in the areas adjacent to their apartment buildings, tenants in areas outside their homes feel that they are observed by other project residents. Surveillance also makes obvious to potential criminals that any overt act or suspicious behavior will come under the scrutiny of project occupants.

The ability to observe criminal activity may not, however, impel an observer to respond with assistance to the person or defense of the property being victimized. The decision to act will depend on the presence of the following conditions:

1. The extent to which the observer has developed a sense of his or her personal and proprietary rights and is accustomed to defending them

2. The extent to which the activity observed is understood to be occurring in an area within the influence of the observer

3. Identification of the observed behavior as being abnormal to the area

4. Identification on the part of the observer with either the victim or the property being vandalized or stolen

5. The extent to which the observer feels that he or she can effectively alter the course of events being observed

Linking opportunities for surveillance to territorially defined areas will go a long way toward ensuring that many of these required conditions will be satisfied.

Designers should position all public paths so that access from public streets to units is as direct as possible. Access arteries should be limited in number to ensure that they are well peopled. They should also be evenly lit. The paths through a project should be designed to allow prescanning before use. There should be no (or few) turns on any artery, and all points along access routes should be observed from the point of origin to the point of destination. When a building is located for the particular use of the elderly, front entrances should face the street and be within 50 ft of the street.

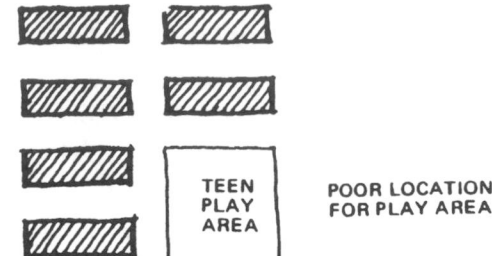

Fig. 7   Teen play area located at the periphery of a project.

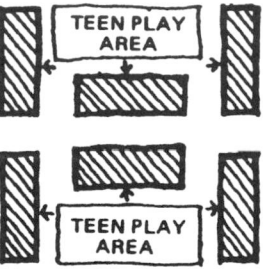

UNITS SURROUND AND LOOK OUT ON TEEN PLAY AREAS, FURTHER ADDING TO THEIR DEFINITION AND ASSOCIATION WITH THE PROJECT

Fig. 8   Teen play area surrounded by buildings and their entries.

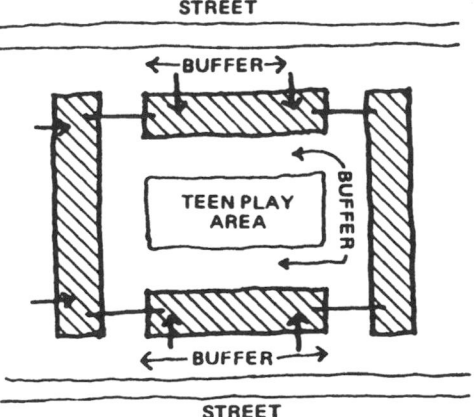

Fig. 9   Teen play area located with a semiprivate zone.

PROTECTING THE INDIVIDUAL DWELLING UNIT

Normally, the first priority for security in multi-family housing is the protection of the individual dwelling unit against burglary. The focus here is on doors and windows, although the existence of other portals (e.g., skylights and attic openings) calls for similar attention to those additional points of vulnerability. In any event, doors, windows, and other means of possible access should be analyzed as total systems, comprised of their frames, locks, and other accessories, as well as their basic components.

1. *Exterior doors.* All exterior doors (including the back door, if any) should be sufficiently secure to withstand the degree of attack anticipated.

   a. *The basic door component* should be of sufficiently heavy construction to withstand the degree of force anticipated. Where security problems are serious, doors with glass panels should be avoided. It is strongly recommended that all exterior wooden doors be of solid-core construction with a minimum thickness of 1¾ in. Both hollow wood doors and thin panel doors are inadequate where serious security problems exist. Although flush doors provide better security, if panel doors are desired, their panels should have a minimum thickness of ½ in. Heavier-duty options are doors with solid wood cores and metal coverings and doors of hollow steel or aluminum construction.

   b. *Door hinges* must also be of heavy-duty construction, and mounted on the inside of the door so that burglars cannot remove the entire door from its hinges. Spring hinges, which close the door automatically, are recommended.

   c. *Door locks* are the one element in the entire security picture about which a standard recommendation can most validly be made for every multifamily housing project: Every exterior dwelling unit door should be equipped with a deadbolt mortise lock with a "throw" of at least one inch, constructed of case-hardened steel, brass, zinc alloy, or bronze. This, if no other, security design feature should be incorporated into every new and existing project. Locks equipped with spring latches only (but not deadbolts) are unsatisfactory, because an intruder can easily push back a spring latch with a celluloid strip. "Key-in-the-knob" locks should not be relied upon for primary exterior lock protection, since a determined burglar can break them with relative ease. The cylinder is a critical element of any lock and must be sufficient to withstand expert lock-picking efforts; it is desirable for a lock cylinder to have at least six pins. Protruding cylinders should be avoided, or protected by a spinner ring, a bevelled ring cylinder guard, or escutcheon plate. There is, however, no such thing as a "burglarproof" lock that can withstand the attack of a skilled burglar with ample time and equipment to practice his or her skills. The value of a good lock is that it can withstand attack by the relatively unskilled burglar, delay the skilled burglar until he or she is driven away or apprehended, or deter either from attempting to break in.

Fig. 1   Door types.

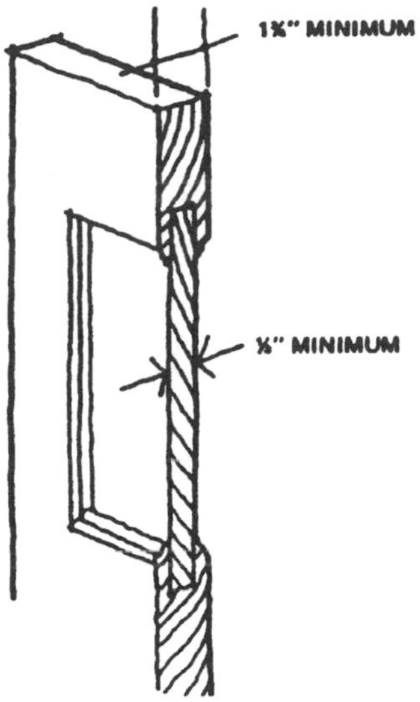

FLUSH   ONE PANEL   TWO PANELS

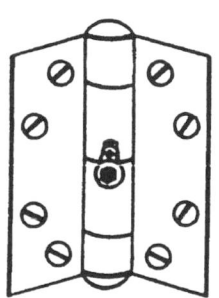

Fig. 2   Panel door.

1¾" MINIMUM

¼" MINIMUM

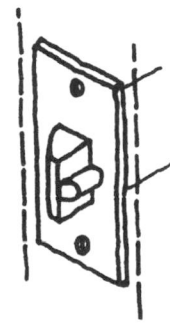

Fig. 3   Nonremovable hinge pin.

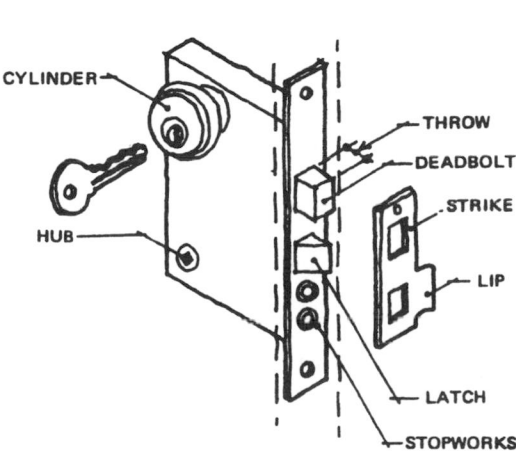

Fig. 4   Deadlatch.

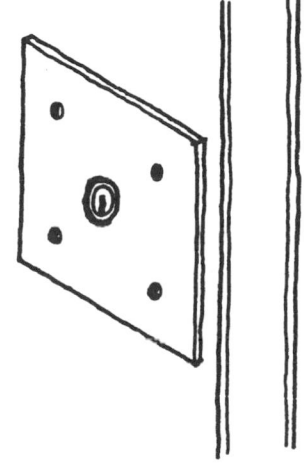

CYLINDER

HUB

THROW

DEADBOLT

STRIKE

LIP

LATCH

STOPWORKS

Fig. 5   Mortise lock.

Fig. 6   Escutcheon plate covering cylinder mortise lock.

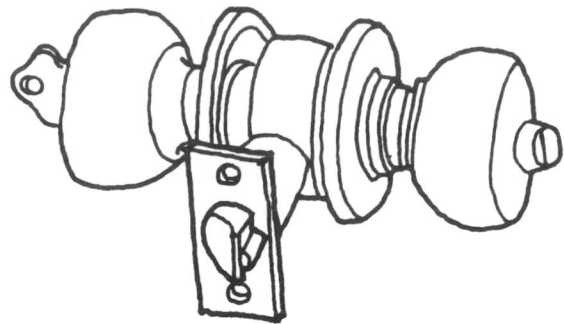

Fig. 7   Key-in-knob lock.

Fig. 8   Spring bolt.

Fig. 9   Horizontal bolt.

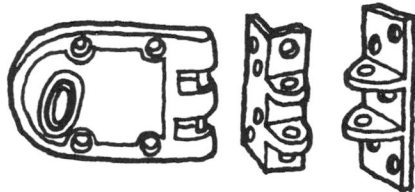

Fig. 10   Vertical bolt.

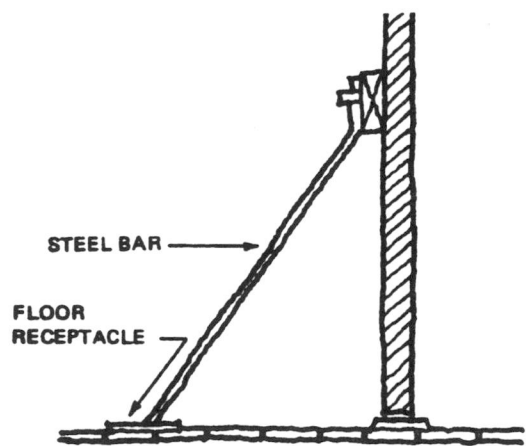

STEEL BAR

FLOOR RECEPTACLE

Fig. 11   Buttress door lock.

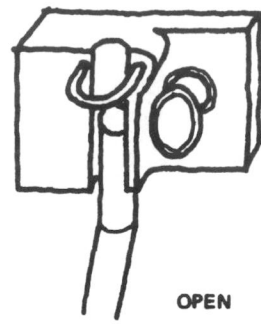

OPEN

Fig. 12   "Magic eye" lock with thumb turn.

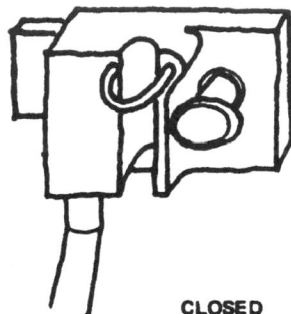

CLOSED

Fig. 13   Buttress door lock with deadbolt.

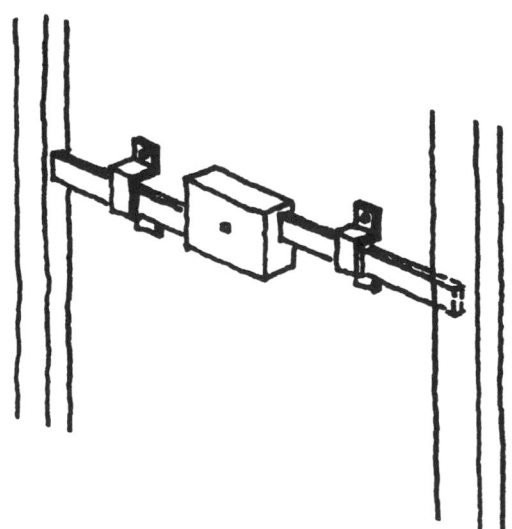

Fig. 14   Double-bar lock.

d. The door frame is often overlooked as a component of the security of a complete door system. All frames should be of heavy-duty construction. Metal-covered wood frames provide optimum cost effectiveness when used in combination with doors of similar construction, but are recommended for use with wooden doors as well. Wooden frames should be at least 2 in thick. If hollow steel frames are used, the air space behind the frame should be filled with crush-resistant material, especially in the area of the strike. For in-swinging doors, rabbeted jambs should be used to prevent tampering in the area of the strike; addition of an L-shaped metal plate in the area of the strike affords extra protection to the lock. For doors opening out, an escutcheon plate, extending beyond the edge of the door and fitting flush with the jamb when the door is closed, will provide similar protection to the lock. All plates mounted on the outsides of doors should be attached with tamper-resistant connectors, such as round-headed carriage bolts or one-way screws.

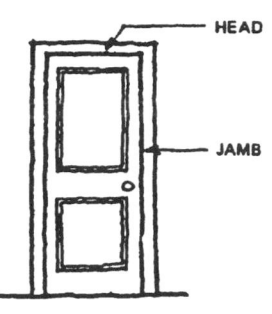

Fig. 15   Door frame.

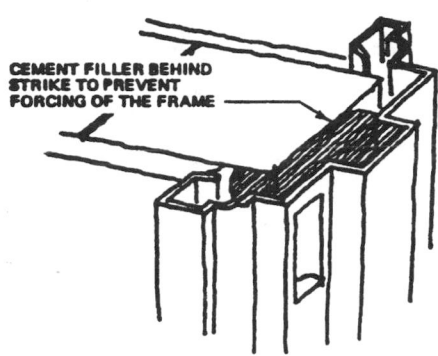

Fig. 16   Hollow metal door frame.

Fig. 17   Door strike.

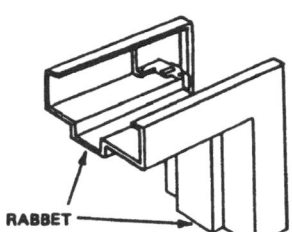

Fig. 18   Rabbeted jamb.

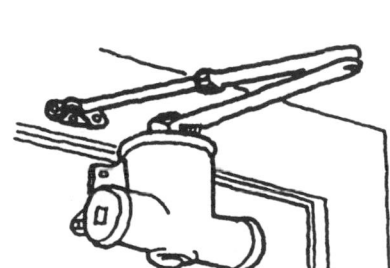

Fig. 19   Door closer.

Fig. 20   Protective angle-iron for doors opening in.

Fig. 21   Escutcheon plate for doors opening out.

BUILDING SECURITY/INDIVIDUAL DWELLING UNIT

e. *Interviewers* are devices installed on opaque doors to allow persons inside the unit to see and hear who is outside without having to unsecure the door. A wide-angle optical interviewer (peephole) should be installed on each exterior door (including the back door, if any). This is a relatively inexpensive measure and should be standard for all multifamily housing projects. The opening of an optical interviewer should be

no more than ¼ in in diameter, and a double glass should be used for safety. Slide-chain interviewers (chain locks) should not be relied upon; they are easily defeated and impart a false sense of security.

f. *Doors with glass panels,* though highly undesirable for exterior use where security problems are serious, must be given special attention wherever they are used. Sliding doors should have

break-resistant glass and should be equipped with a sturdy lock designed specifically for this type of door. For other types of doors with glass panels (e.g., French doors), the deadbolt mortise lock should be key-operated from the inside as well as the outside, in order to prevent the burglar from simply removing a portion of the glass and reaching inside to operate the latch by hand.

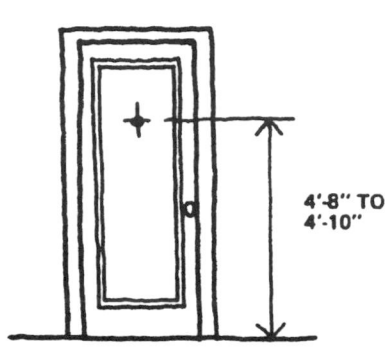

4'-8" TO 4'-10"

Fig. 22   Interviewer location.

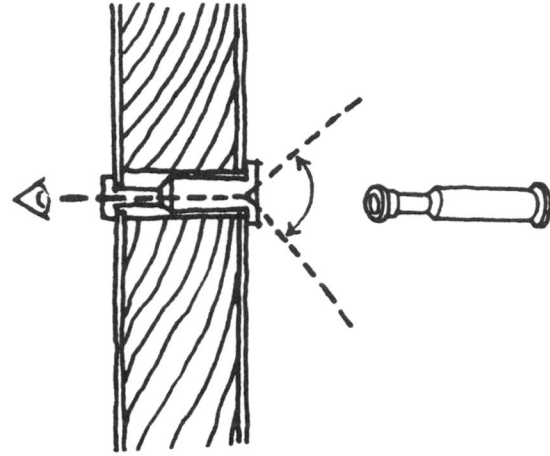

Fig. 23   Interviewer angles.

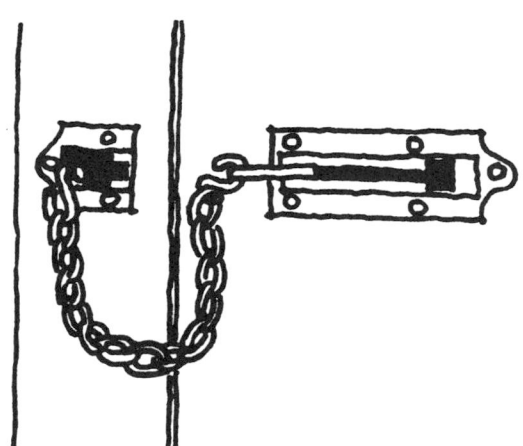

Fig. 24   Chain lock.

2. *Windows.* The dwelling unit windows which are most vulnerable to attack are those situated on the first floor (or otherwise accessible from the ground) and those leading to fire escapes. Also vulnerable, but to lesser degree, are windows located over a canopy (e.g., above a lobby entrance), adjacent stairwell windows or on the top floor. Cornices, ledges or other architectural features can create further vulnerability. In some instances, the threat of determined and resourceful burglars has even extended several stories above the ground or below the roof. Since windows are a prime target for vandalism, that threat, as well as the threat of burglary, should be considered in planning measures for the protection of windows. Careful analysis of experience, trends, and building design will enable management to identify degrees of window vulnerability.

a. *Window glass* can be protected by the use of *unbreakable glass* made of polycarbonate materials, though at relatively high initial purchase cost. Other options which are superior to ordinary window glass include plate glass, tempered glass, and bonded safety glass.

b. *Window locks* are an important element to which little thought is usually given. The only reliable window locks are those of the key-operated variety. However, such locks present problems of fire safety and inconvenience to residents. The standard crescent sash lock, the slide bolt latch, and various friction or pressure devices can easily be overcome, especially if the intruder is willing a break a small section of the glass.

c. *Grilles, bars, and gates* afford reliable protection for vulnerable windows where security problems are great. Such fixtures should be of heavy-duty construction, and should be securely attached to the window frame with machine or roundheaded bolts which cannot be easily removed from the outside. Fire safety requirements must be checked before window grilles, bars, or gates are installed.

Fig. 25   Crescent sash lock.

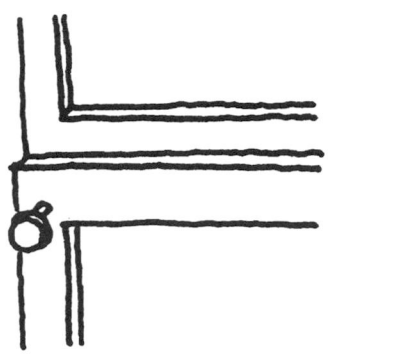

Fig. 26   Thumb screw lock.

Fig. 27   Pin latch.

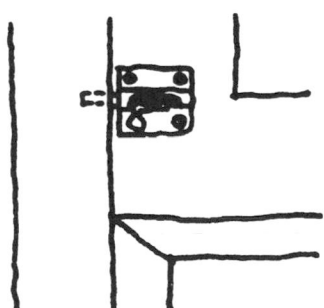

Fig. 28   Slide bolt.

BUILDING SECURITY/INDIVIDUAL DWELLING UNIT

Fig. 29   Keyed window lock.

Fig. 30   Mesh window grille.

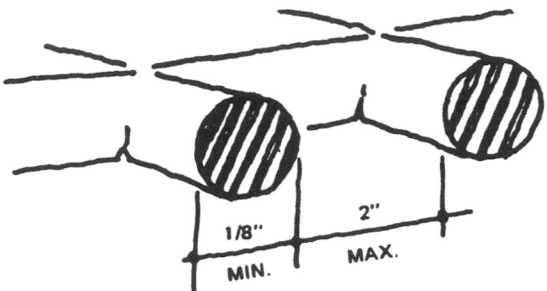

Fig. 31   Wire mesh dimensions.

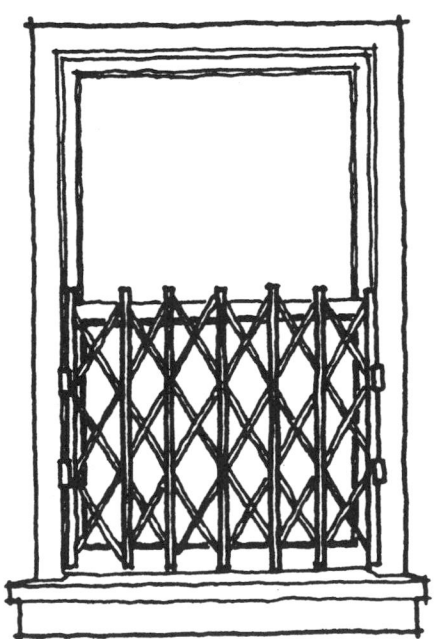

Fig. 32   Window guard.

Fig. 33   Skylight protection.

3. *Electronic alarm systems.* The measures described above for the protection of exterior doors, windows, and other possible points of access to the dwelling unit can be reinforced by electronic alarm systems, although such systems should be used only where necessary and practicable.

    a. *Types of alarm systems.* There are scores of electronic alarm products on the market, varying greatly in price, quality, and complexity. Some are very difficult to install; others need only be plugged into an existing electric outlet. Alarm devices fall roughly into two categories: contact devices and motion detection devices. In the simplest terms, contact devices are mechanical switches which detect the movement of a door or window. Foil strips are a related mechanism used to detect breakage of glass in windows and doors. The second type of alarm system detects the motion of an intruder as he or she moves about the protected space. Motion detection technologies include seismographic devices, photoelectric cells, and ultrasonic detectors. Great caution should be exercised in selecting alarm equipment.

    b. *Alarm reporting systems.* Either a contact or motion detection system may be linked to a local alarm (bell, buzzer, lights on the immediate premises) or to a central alarm (via wires to a security force which is prepared to react when so alerted). Local alarms aim at driving off the burglar or aiding in his or her apprehension, and at alerting residents and neighbors that a break-in is being attempted. The effectiveness of any alarm system depends to a great extent upon the ability to secure a prompt response from the police or other security personnel. False alarms are a major problem, because they diminish the credibility of the system and tend to slow or stop effective response from police, security personnel, and neighbors.

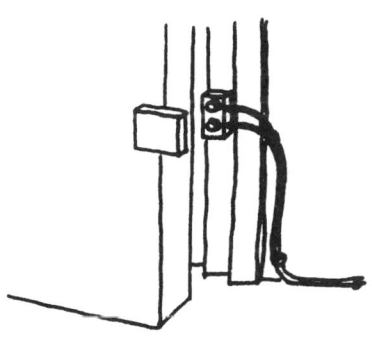

Fig. 34 Contact switch on door.

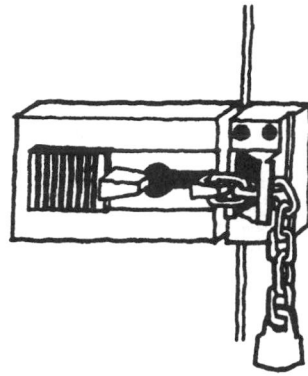

Fig. 35 Lock alarm.

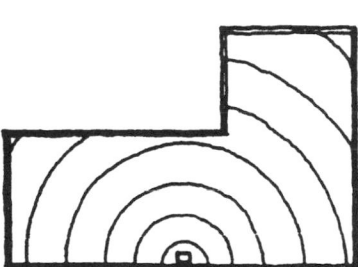

Fig. 36 Ultrasonic detector.

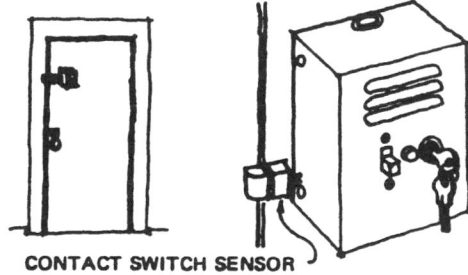

CONTACT SWITCH SENSOR

Fig. 37 Local alarm.

PROTECTING COMMON AREAS
WITHIN MULTIFAMILY BUILDINGS

With regard to protection of common areas within multifamily buildings, a basic choice is the degree to which a "fortress" strategy can and should be adopted. For a project inhabited by families with children, such a strategy may be patently inconsistent with the normal pattern of residents' daily lives. An extreme "fortress" strategy is apt to be most workable in a high-rise building exclusively for the elderly. Acceptability to residents is always a commanding consideration, though it may be possible to obtain resident acceptance through careful educational efforts.

1. *Lobbies* are a first line of building defense. The degree and methods of controlling access must depend upon residents' attitudes and lifestyles and the availability of policing services. Nevertheless, in all instances, the lobby itself and the area immediately outside its doors should be brightly lighted and free of places of concealment. The following additional elements of lobby design merit special attention:

    a. Lobby doors should have large glass panels to facilitate two-way surveillance. Where security problems are great and policing services inadequate to control access, serious consideration must be given to keeping lobby doors locked, especially during evening hours. Where this is done, lobby doors should be equipped with heavy-duty metal frames, a good deadbolt mortise lock set, and a sturdy door closer.

    b. Intercom (annunciator) devices permit residents conveniently to admit callers when lobby doors are locked. The familiar buzzer reply system is satisfactory, but should be installed during initial construction, because costs of wiring installation are very high in existing buildings. A functionally similar alternative is an intercom system utilizing regular telephone wires, instead of separate wiring, so that installation costs are relatively modest. However, a monthly service charge is made for each dwelling unit. If there is a telephone connection in the unit, this system can be used even if the resident does not have a regular telephone. This type of system is available through some local telephone companies. A much less expensive, though also much less satisfactory, method is simply to have the telephone company install a public telephone outside the lobby entrance, so that callers can telephone residents, who can then come to the lobby to open the door. One poten-

tial problem with any of these systems is vandalization of intercom panels or telephones located outside the building's entrance. Difficulties with resident acceptance and vandalism tend to be greatest in buildings with many small children in residence.

2. *Secondary doors* (e.g., emergency exits, delivery doors) each require analysis in terms of ordinary function as well as threat of criminal access. It is sometimes difficult to reconcile these two factors. Fire regulations require that occupants of the building be readily able to open emergency doors from the inside; the best solution here is a vertical-bolt latch or crash bar on the inside, keeping the door locked from the outside at all times. Exit alarms can be installed to alert security personnel upon the opening of emergency doors. All secondary doors should have automatic door closer devices. Glass panels should never be used in such doors, and the construction of the door and its frame should be sufficiently heavy to withstand the degree of attack anticipated. Where secondary doors are continuously used for resident ingress and egress (e.g., doors to garages or parking lots), they should be treated in much the same way as lobby doors.

3. *Garage access* should be controlled, even if doors leading from the garage to other areas of the building are monitored and/or kept locked.

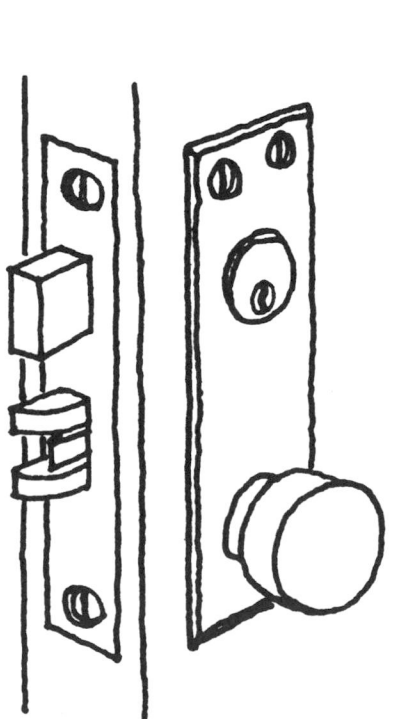

Fig. 1   Antifriction latch bolt.

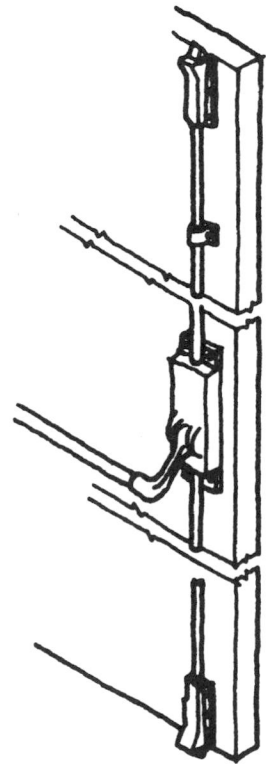

Fig. 2   Vertical bolt on exit door.

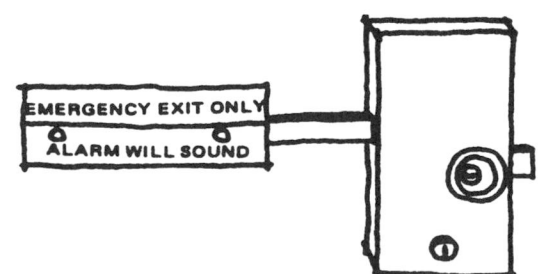

EMERGENCY EXIT ONLY
ALARM WILL SOUND

Fig. 3   Exit alarm.

4. *Elevators* are among the most problematical elements of high-rise multifamily buildings, from the standpoint of vandalism as well as crimes against persons.

    a. Surveillance is a prime factor in elevator security. Buildings should be designed so that the elevator area is fully visible throughout the lobby, and preferably from the area immediately outside the lobby and the street beyond. On levels other than the lobby floor, elevators should open directly on hallways, without recesses or blind corners to restrict two-way visibility. A common and inexpensive device to increase a person's ability to survey the interior of an elevator before entry is a convex mirror placed in the upper back corner of the elevator cab.

    b. Audio-intercom systems permit persons within the elevator to communicate with persons in elevator waiting areas or with security personnel. A continuous audio device is preferable to one which must be activated by pushing a button.

    c. Closed-circuit television is particularly adaptable to elevator security.

    d. Vandalism to elevators can be limited by a variety of measures, including use of stainless steel mushroom buttons, protecting indicator lights with a heavy-duty plastic shield, and use of automatic sliding doors (rather than swinging doors). Door glass, though an element in surveillance, is so susceptible to breakage as to merit avoidance where vandalism is a problem. Where existing elevators have such glass, a piece of metal can be simply welded or bolted over the opening.

5. *Mailboxes and mail rooms* should be located so as to permit maximum surveillance—preferably in or adjacent to the main lobby of the building. The mailboxes themselves should be constructed of heavy metal, with tightly-fitted doors. Locks should be of the cylinder type with at least five pins. Sizes of mailbox doors should be kept at a minimum. A locked mailbox room provides additional security, although it should be subject to full surveillance from the lobby by means of large windows and good lighting. Where back-loading mailboxes are used, a separate mail loading room is often provided. Doors to all such rooms should be of sturdy construction, should be kept locked on a 24-hour basis, and should be equipped with automatic door closers.

6. *Laundry rooms* commonly invite attack on residents or pilferage of coins from laundry machines. A first consideration here is location. Laundry rooms are usually located in basement recesses, and management should carefully consider whether that is the best choice. If acceptable to residents, the laundry room might better be situated in a more active area of the building, adjacent to social rooms or even the main lobby, and fitted with large glass windows to facilitate surveillance. This may accord well with social patterns, and there is no reason why laundry rooms cannot be made attractive. In any event, laundry rooms should be kept locked on a 24-hour basis, with tenants being provided keys. Laundry rooms may be further protected by audio-intercom or closed-circuit television devices.

7. *Social rooms* should provide protection for both people and such valuables as may be kept there. Proximity to other heavily used areas (e.g., the main lobby) can facilitate mutual surveillance by residents in the ordinary course of their activities.

Fig. 4 Elevator mirror.

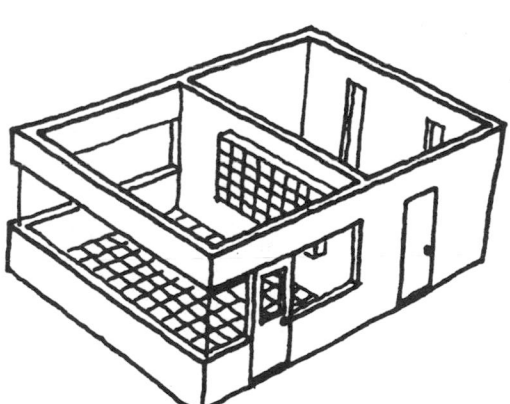

Fig. 5 Mailroom and loading room.

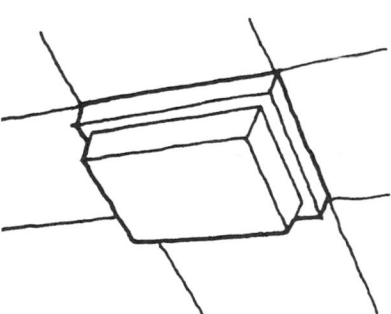

Fig. 6 Unbreakable light fixture.

8. *Storage rooms* for residents' and management's property, merit heavy protective measures. Use of windowless rooms (e.g., in basements) is advisable, and door systems (including their locks and frames) should be of particularly strong construction. Dead bolt locks should always be used. For residents' storage, the additional use of locked bins is recommended.

9. *Management offices* should be protected to the degree they may be attractive targets for burglary or robbery, with particularly strong measures where money, office equipment, or other valuables are kept in such areas. Where significant sums in money or checks are kept in the management office, a strong safe (preferably bolted into the floor) should be used. Collection windows may be further protected by heavy metal grilles or bars and alarms which can be readily activated in the event of robbery. However, one of the best precautions is to avoid keeping large sums in cash or coins on the premises at any time.

10. *Hallways and stairs* should be susceptible to easy surveillance. Open design is generally desirable, and all such areas should be well-lighted at all times. As in the case of elevators, convex mirrors can be used to permit an approaching view of all possible places of concealment. Doors between fire stairwells and other common areas should be kept locked from the outside with hardware to permit emergency egress only. Vertical-bolt latches or crash bars on the inside of such doors and automatic door closers are recommended. Exit alarms provide a local alarm upon the opening of a fire exit door, but have the same weakness as all local alarms—the necessity for prompt response and the problem of nuisance and credibility.

11. *Roofs* merit attention, because of their potential as avenues of escape or access and as isolated areas which may be used for such offenses as drug abuse, assault, and rape. In addition, easy access to the roof presents a safety hazard to children. At the same time, fire safety requirements may demand that access to the roof be available from fire stairs. Where this is the case, doors to the roof should nevertheless be kept locked from the outside, with vertical-bolt latches or crash bar on the inside to permit emergency egress. However, if fire safety codes permit, such doors should also be kept locked from the inside. In either event, but especially where ready egress must be allowed, consideration should be given to installing exit alarms on these doors and means for regular surveillance of roof areas. Rooftop lighting, closed-circuit television, and (to the extent possible) avoidance of structural elements which provide opportunities for concealment on the roof are possible options to promote surveillance.

12. *Interior lighting* for common areas within the building has already been mentioned in connection with several of the specific types of areas discussed above. However, the generally applicable principle merits additional emphasis: all common interior areas should be brightly lighted at all evening hours when they are subject to ordinary use. For lobbies, elevators, hallways, and stairwells, this means 24-hour lighting. Where vandalism is a problem (as is most likely in lobbies, elevators, hallways, and stairwells), vandal-resistant lighting fixtures should be installed. Residents should be requested to make prompt report of inoperative lights, and maintenance staff should be required to make frequent lighting inspections and speedy repairs or replacements. Good lighting for interior common areas is one of the least expensive of security measures, and missing or burned-out bulbs are always a mark of poor management. (Energy conservation measures may impose constraints on lighting.)

DESIGN GUIDELINES FOR LOBBY SECURITY

A number of design criteria have to be met for a controlled entranceway to work. Those discussed below relate to the design of the guard booth, the problem of access to the inner lobby, and the location of monitoring equipment.

The Design of the Guard Booth

The design of the guard booth is critical because it must provide guards with the fullest view possible of the lobby area and at the same time permit them to read the monitoring equipment reporting on other parts of the building. Additionally, guards must have direct access to the lobby to be able to handle any difficulty. They must also be protected from assault while in the booth and still be able to converse with people in the outer lobby area.

It is important to avoid conflicts in the design of guard stations, that is, situations where the guard cannot move from one task to another easily or perform more than one task at a time. For example, it is important that the electronic equipment not block the guard's view of the lobby area. The equipment should be placed so as to permit guards to view and electronically monitor activities while they are carrying out other duties. Many times, this can be accom-

plished by having the television monitor inserted in the guard desk in front of the outer lobby intercom. This permits the guard to check identification and watch the side area while also scanning the monitors.

Audio monitors, if improperly placed and modulated, can also present design conflicts. Ideally, the speakers should be spread apart and coded to the area they are monitoring; and generally they should be placed in the rear of the booth so as to avoid any need to turn them down in order to talk to residents passing by the guard window.

Consideration should also be given to using audio monitors that are activated only by loud noises. The constant din of audio monitors on all the time is extremely tiring for guards—so much so that they can be expected to turn them off at an early opportunity.

Lobby Design

Lobby design must meet several criteria. First, the lobby should be as attractive as possible, just as any lobby area should be whether it is part of a controlled access system or not. It is important not to let the need for access control overwhelm the design. A lobby is an important orienting point in the housing environment. It can shape how people feel about their building,

their home, and even about themselves and where they are in life. To the extent possible, therefore, the lobby should be made as warm and pleasant as possible. At a minimum, it should have a sitting place for people who are waiting for transportation or for someone from the building to meet them.

A second requirement is the need for an outer and inner lobby area with the guard booth established at a central point. People should first enter into an outer lobby and then move into an inner lobby upon approval by the guard who controls the doors from the outer to the inner lobby. It is important in lobby design that guards be able to talk easily from their booth to people requesting entry into the inner lobby.

The third major requirement in lobby design is the need for guards to be able to control people from entering the inner lobby without approval. A number of factors impact on this, but in terms of lobby design, the placement and number of doors is extremely important.

Another issue in lobby design is the location of mailboxes. Ideally, they should be located in full view of the guard station and if possible, they should be positioned in such a way as to prevent people from using their bodies as a shield.

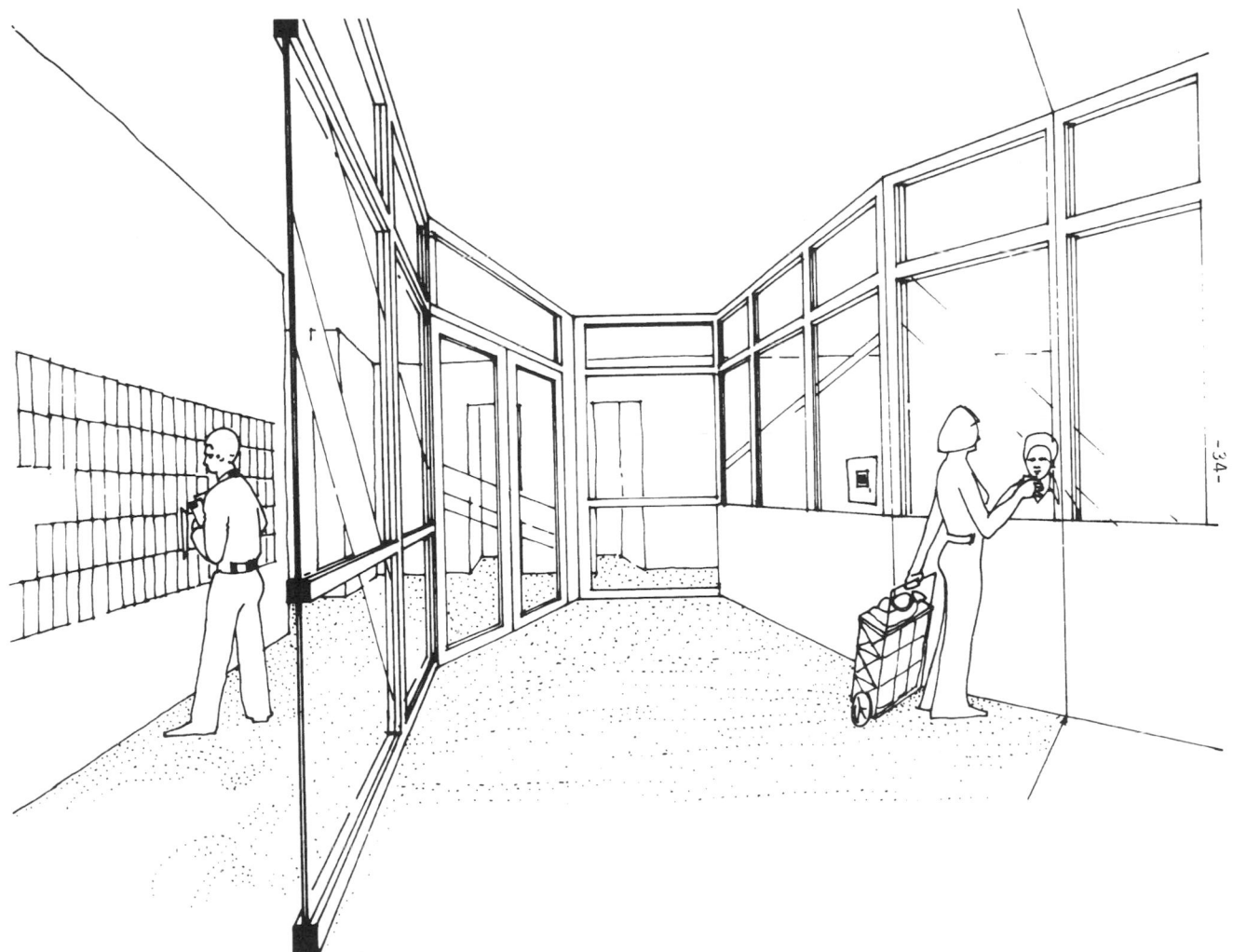

Fig. 7   Location of mailboxes permits perpetrators to use their bodies as masks.

Fire Codes and Security Planning

There is often a built-in conflict between security planning and the need to provide people with safe passage and escape through a building in the case of fire. Security requirements emphasize controlled egress and ingress; fire safety requirements emphasize the need for many exit points.

Adjustments between these frequently competing requirements can be expected to vary somewhat from city to city depending on the fire codes.

OPERATING THE CONTROLLED ENTRANCEWAYS

There are basically three operational requirements of a controlled entranceway. First, there must be physical control over the points of ingress and egress; second, there must be a way to identify residents; and third, the guards must use the physical controls they have available and enforce the identification system to assure that those entering the building have a legitimate reason for being in the building.

These last two elements—the identification system and the conduct of the guards—are as important as the actual physical design of the entranceways. If the guards do not use their control over the entrances or if they do not screen those entering, there will be no control over access.

Identifying Residents

A number of different identification techniques can be used. These include calling on an intercom system directly to the apartment, dialing a special number that opens the door, a key system, the insertion of a card in a card reader that opens the door, and a range of others.

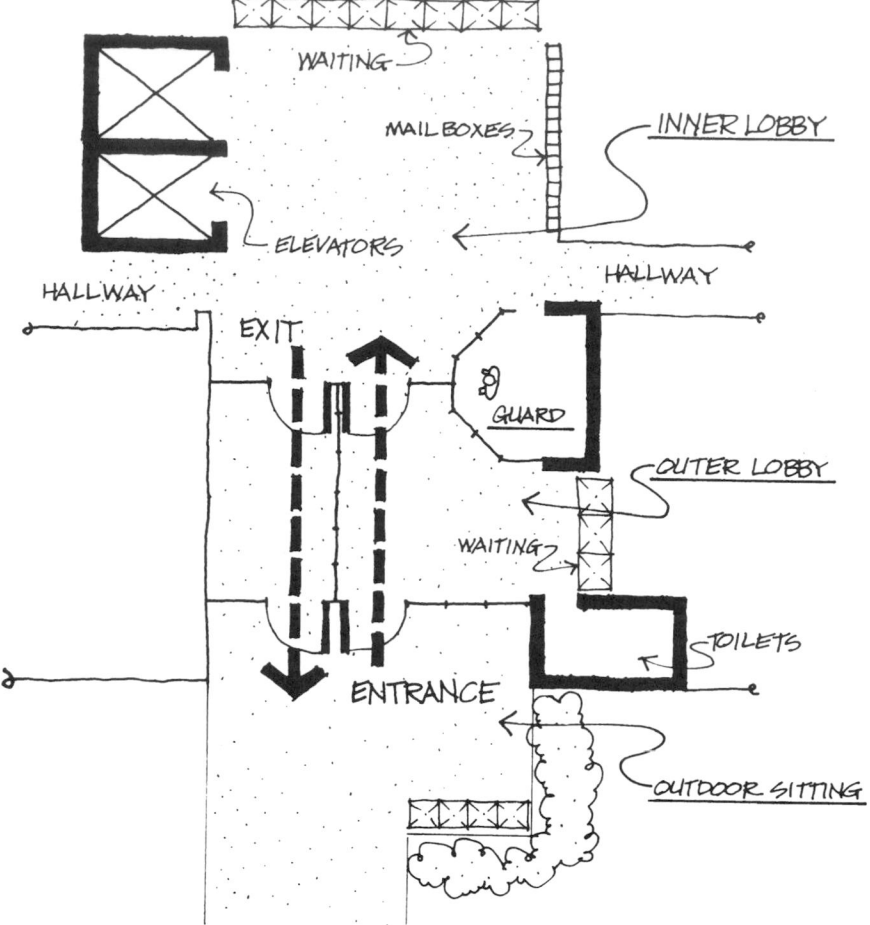

Fig. 8   This design optimizes guard's visibility of lobby and provides for separate entrance and exit areas.

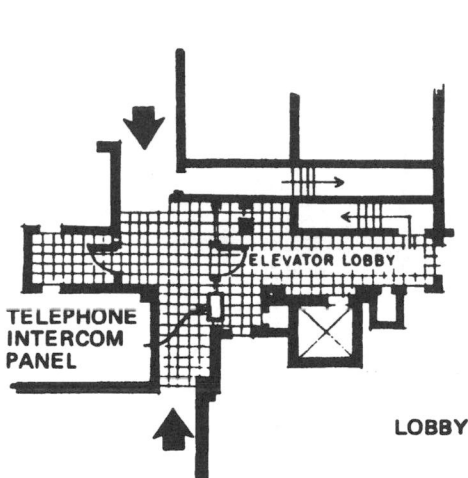

Fig. 9

Fig. 10 Modified entry and TV surveillance of the lobby and elevator.

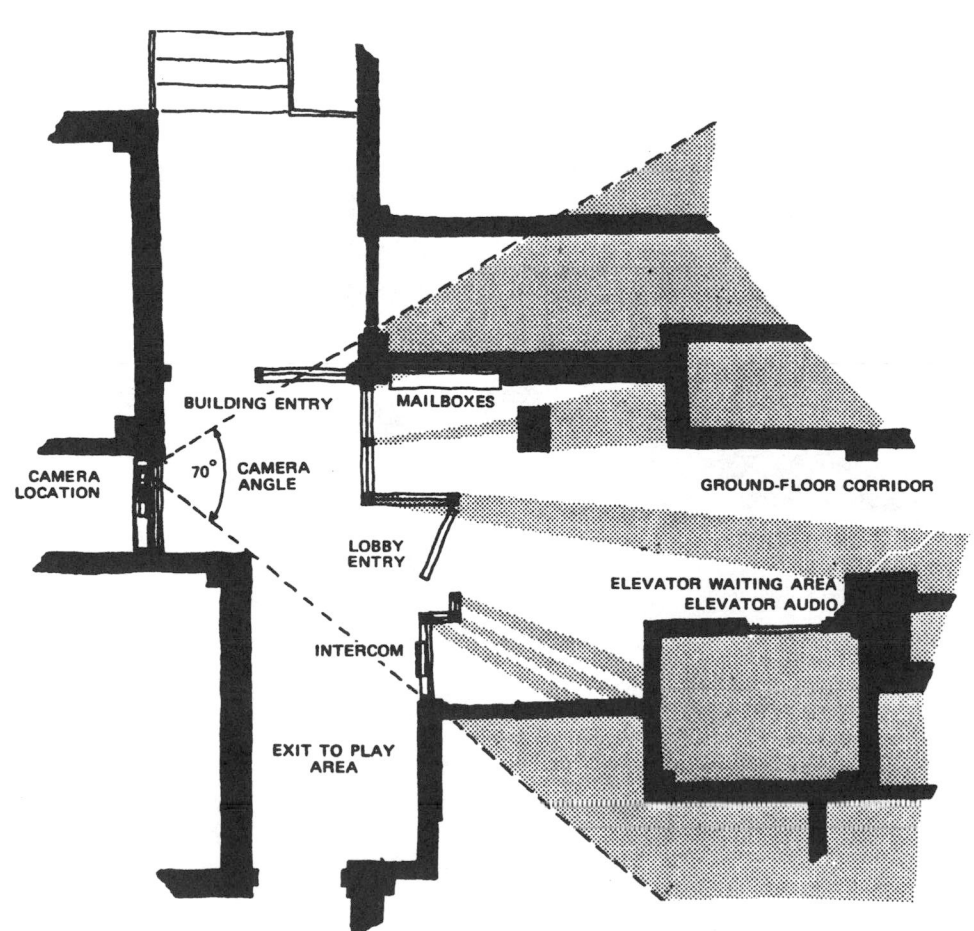

Fig. 11 Study to determine camera position in lobby.

BUILDING SECURITY/EXTERNAL AREA

PROTECTING EXTERIOR AREAS OF THE PROJECT

Normally, at least in housing for families with children, it is infeasible or undesirable to restrict access to the project's grounds to the same degree as with respect to interior common areas. On the other hand, in some instances (e.g., all-elderly projects), it may be both feasible and desirable to extend some degree of a "fortress" strategy to the exterior boundaries of the property. Again, the choice must depend upon a realistic appraisal of all the relevant facts. The following options merit consideration in this connection.

1. *Exterior lighting* should be amply provided for all heavily used areas, such as walkways, entry areas, and parking lots. Lighting levels in projects for the elderly should be well in excess of conventional standards, because light perception declines with advancing age. High placement of lighting fixtures results in wider coverage as well as less susceptibility to vandalism. High-intensity lights are well-suited to large areas, such as parking lots. A variety of vandal-resistant lighting equipment is now being marketed.

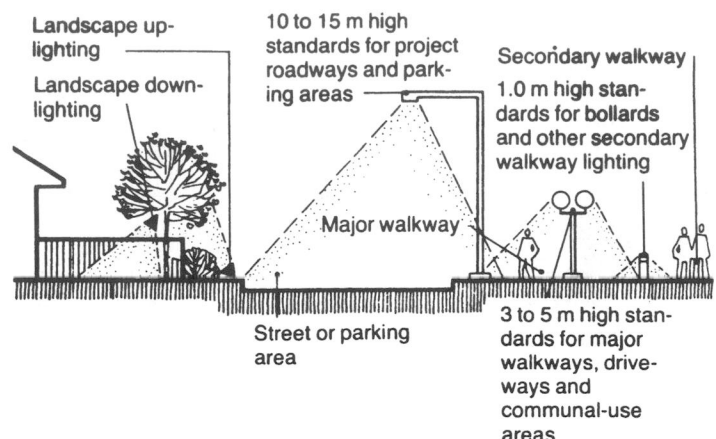

Fig. 1 Purpose, functional intent, and esthetics should be kept in mind when determining the size of lighting fixture and standard (pole).

Fig. 2 Planting and lighting for security.

2. *Recreational areas* for children and adults merit individualized treatment, taking account of the particular use to be made of each, the particular group of users, and the social patterns of the project and the wider neighborhood. Generally, areas designed for use by different age groups (preschool, elementary school, teenagers, adults) should be sufficiently separated by space or other barriers to minimize conflicts. However, all recreational areas should be susceptible to intensive surveillance from streets and sidewalks, and, if possible, from the windows of dwellings. In addition, benches on the perimeter of playgrounds and areas for athletic activities can encourage adults to exercise casual surveillance. While vandalism to playground equipment is a matter of legitimate concern, hard use of such equipment should be no cause for discomfort. Particularly, where their number is high, children "use up" playground equipment during the ordinary course of play, and this is even desirable as a means of diverting normal youthful energies from damage to other elements of the property.

3. *Fences and walls* may be utilized to channel or restrain movement, so as to facilitate surveillance and policing. However, it is seldom feasible to erect complete physical barriers around the property. More widely appropriate is the creation of limited exterior zones for use by elderly residents or small children, with access only by way of an adjacent building. Fences and walls may be objectionable to residents. However, if they are well-designed in both their security and esthetic aspects, it is much easier to gain resident acceptability.

4. *Approaches to building entrances* are particularly prone to crime, and primary entrances should be near the street. Routes from parking lots to building entrances should also be kept short and direct. Dense shrubbery or other possible places of concealment should be avoided in these areas. Every opportunity for casual surveillance should be exploited.

5. *Closed-circuit television* can provide effective surveillance of exterior areas.

CLOSED-CIRCUIT TELEVISION SURVEILLANCE SYSTEMS

Where other means of surveillance are inadequate, the potential and feasibility of closed-circuit television (CCTV) systems should be explored. While initially costly, CCTV may be more economical than such alternatives as design modifications or security patrols, particularly in large projects.

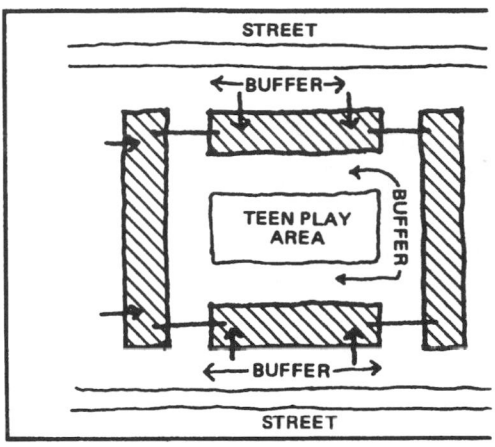

Fig. 3   Teen play area located within a semiprivate zone.

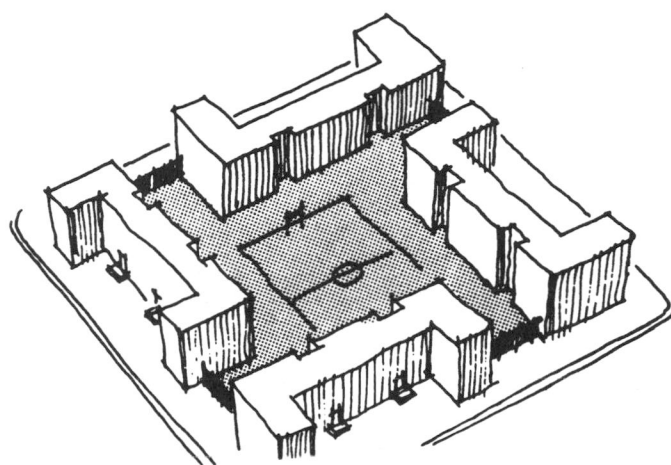

Fig. 4   Use of fencing to define and secure large semiprivate areas.

Fig. 5   Guard booth.

STREET AND ACCESS CRITERIA

The fire department is severely hampered if it cannot gain rapid access to the building in which there is a fire. The following are suggested guidelines for streets and access to buildings:

■ The developed area should be accessible from at least two separate connecting points or one connecting roadway of divided design.

■ Streets should be paved and capable of supporting the heaviest load permitted.

■ Total street width should provide a 22-ft clear width exclusive of parking. Add 7 ft for each lane of parallel parking.

■ Intersections should have a minimum curb radius of 20 ft and meet at approximately a 90° angle.

■ Grades should not exceed 10 percent (8 percent if winter icing is common) except that grades of 15 percent can exist for distances of less than 600 ft.

■ Streets should be clearly marked and houses uniquely numbered in an orderly sequence.

■ Cul-de-sacs should have a clear turning radius of 40 ft exclusive of parking.

■ Fences should have gates allowing access to the rear of the building.

■ Access to buildings should not exceed the following distances from the street or driveway:

100 ft for single-family housing
75 ft for low-rise multifamily housing
50 ft for high-rise multifamily housing

SITE PLANNING—TURNING CLEARANCES FOR NARROW DRIVEWAYS

Table 1 below gives the preferred minimum distance (*d*) in ft for equal legs of the right triangle–shaped open area shown in Fig. 1 for *turning clearance*. Table values are presented for private driveway width (*W*) and vehicle length (*L*).

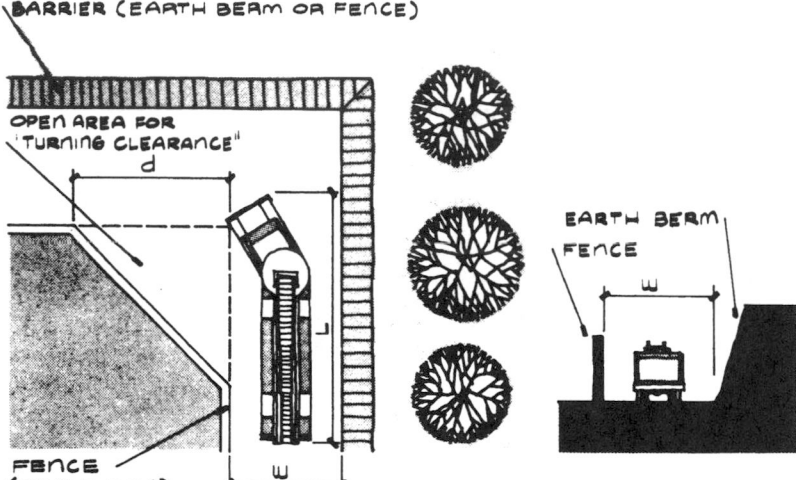

Fig. 1

TABLE 1 Values of *d*

W L	10	12	14	16
35	36	31	25	19
40	38	32	28	25
45	47	37	34	30

*L and W are in feet.

WATER-SUPPLY CRITERIA

An inadequate water supply for fire fighting can result in unnecessary loss of life and property. The following are guidelines for water supplies and hydrants:

■ The following minimum fire flows should be available at a residual pressure of 20 lb/in² for the periods indicated for areas comprised of the following housing types:

Single-family detached (houses separated by more than 30 ft): 500 gal/min for 1 h.

Single-family detached (houses separated less than 30 ft); single-family attached; multifamily low-rise (not over six living units between fire walls if of combustible construction): 750 gal/min for 1½ h.

All other residential buildings: 1000 gal/min for 2 h.

Public buildings: depending on size, height, occupancy, and combustibility of structure and contents: 750 to 2000 gal/min for 1½ h to 4 h.

■ Piping should be lined as required to resist corrosion and tuberculation.

■ Water mains should not be less than 8 in nominal inside diameter.

■ Dead-end mains should not exceed 600 ft in length.

■ Valves should be provided so that not more than 800 ft of piping will be affected by a shutdown.

■ Hydrants should not have less than a 6-in connection to the water main. A gate valve should be installed in the supply connection to each hydrant. Hydrants should be of a style and type in accordance with local practices and regulations. Threads should match those of the fire department.

■ Hydrants should be located at each street intersection with additional hydrants provided at midpoints along all streets, drives, and cul-de-sacs where the distance between intersections exceeds 500 ft. Spacing between hydrants serving multifamily dwellings and public buildings should be reduced to 300 ft.

■ A hydrant should be located within 200 ft of the standpipe siamese connection on high-rise housing.

■ Hydrants should be placed within 5 to 10 ft of street or driveway pavement.

■ Hydrants should not be placed closer than 50 ft to the building being protected unless the building is fire-resistive or the hydrant fronts a blank masonry wall.

QUANTITY OF WATER

With quality water supplies becoming more difficult to find and water demands increasing, this limited resource must be conserved. In order to select a suitable water supply source, the demand that will be placed on it must be known. The elements of water demand include the average daily water use and the peak rate of demand. In the process, the ability of the water source to meet demands during critical periods (when surface flows and ground water tables are low) must be determined. Stored water that would meet demand during these critical periods must also be taken into consideration.

The *peak demand* rates must be estimated in order to determine plumbing and pipe sizing, pressure losses, and storage requirements necessary to supply enough water during periods of peak water demand.

State or local agency requirements may dictate water supply (and component) capacities. Where such agency requirements do not exist, the following discussion of average and peak demands can be used to project water needs.

Average Daily Water Use

Many factors influence water use. For example, the fact that water under pressure is available encourages people to water lawns and gardens, wash automobiles, and perform many other activities at home and on the farm. Modern household appliances such as food waste disposers and automatic dishwashers contribute to a higher total water use and tend to increase peak demands. Since water requirements will influence all features of a water supply under development or improvement, they are very important in planning. Table 1 presents a summary of average water use as a guide in preparing estimates. Local adaptations must be made where necessary, such as in cases where limitations in water supply require active water conservation measures.

The average U.S. family of four uses an estimated 300 gal of water every day. This is water that has been pumped, treated, and distributed or stored for consumption or use. About 95 percent of this water, or 285 gal, ends up as sewage: As much as 120 gal is flushed down toilets, with the remainder going down household drains.

Depending on the type of plumbing fixtures and personal use habits, average household water use is similar to the ranges described in Table 2.

TABLE 1   Planning Guide for Water Use

Types of establishments	Gallons per day
Airports (per passenger)	3–5
Apartments, multiple family (per resident)	60
Bath houses (per bather)	10
Camps	
Construction, semipermanent (per worker)	50
Day with no meals served (per camper)	15
Luxury (per camper)	100–150
Resorts, day and night, with limited plumbing (per camper)	50
Tourist with central bath and toilet facilities (per person)	35
Cottages with seasonal occupancy (per resident)	50
Courts, tourist with individual bath units (per person)	50
Clubs	
Country (per resident member)	100
Country (per nonresident member)	25
Dwellings	
Boardinghouse (per boarder)	50
Luxury (per person)	100–150
Multiple-family apartments (per resident)	40
Rooming houses (per resident)	60
Single family (per resident)	50–75
Estates (per resident)	100–150
Stores (per toilet room)	400
Swimming pools (per swimmer)	10

TABLE 2   Average Household Water Use Activities

Per person per day (indoor use only)

Use	Gallons	Gallons per day	% daily
Toilet (per flush)	1.5–5	25	37
Faucets (per minute use)	3	15	21
Bath/shower (per minute use)	5	15	22
Daily laundry (per load)	25	10	15
Cooking/drinking	3	3	5
Total			100%

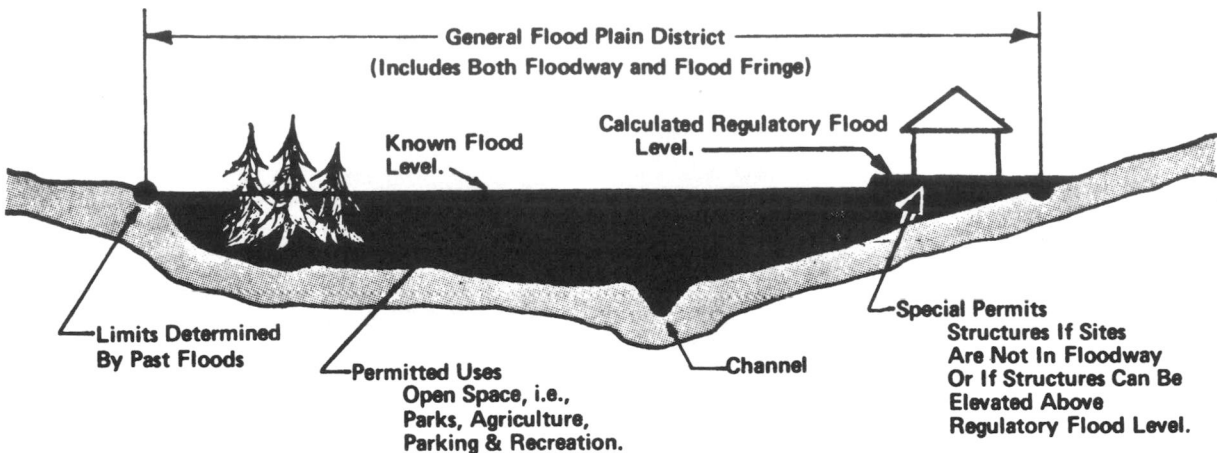

Fig. 1   One floodplain district.

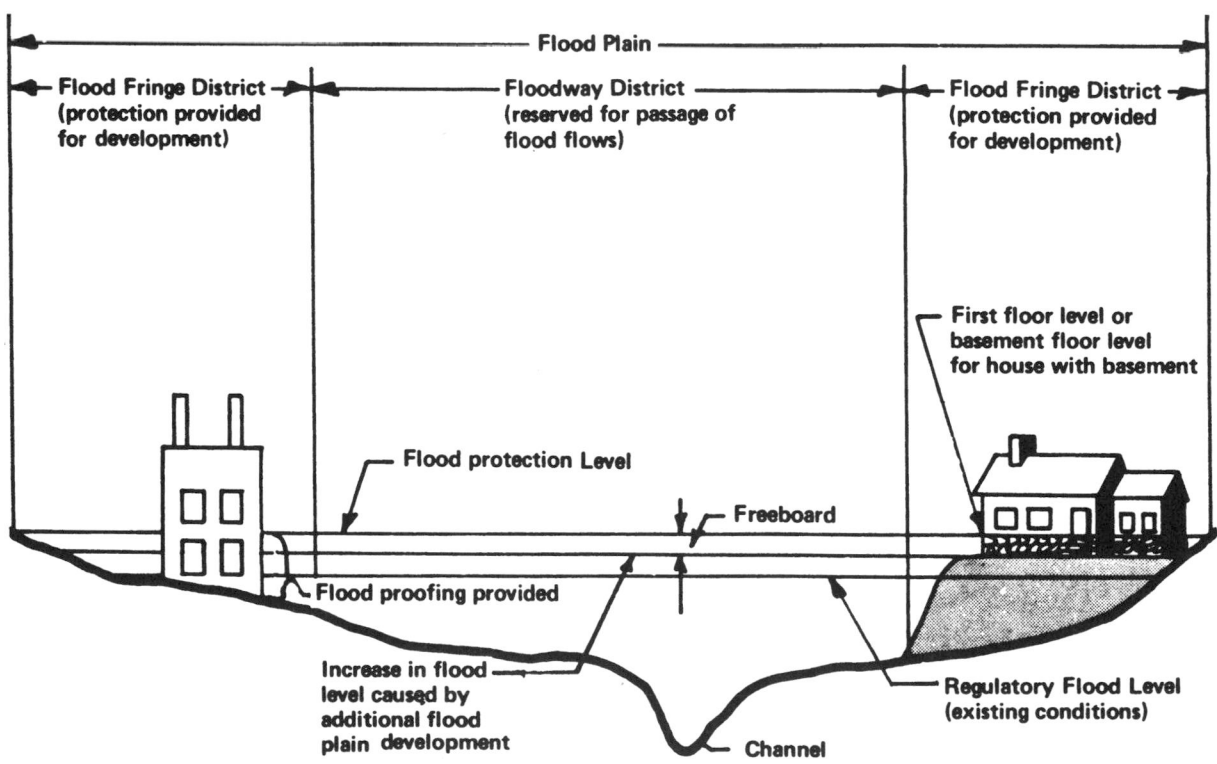

Fig. 2   Two floodplain districts.

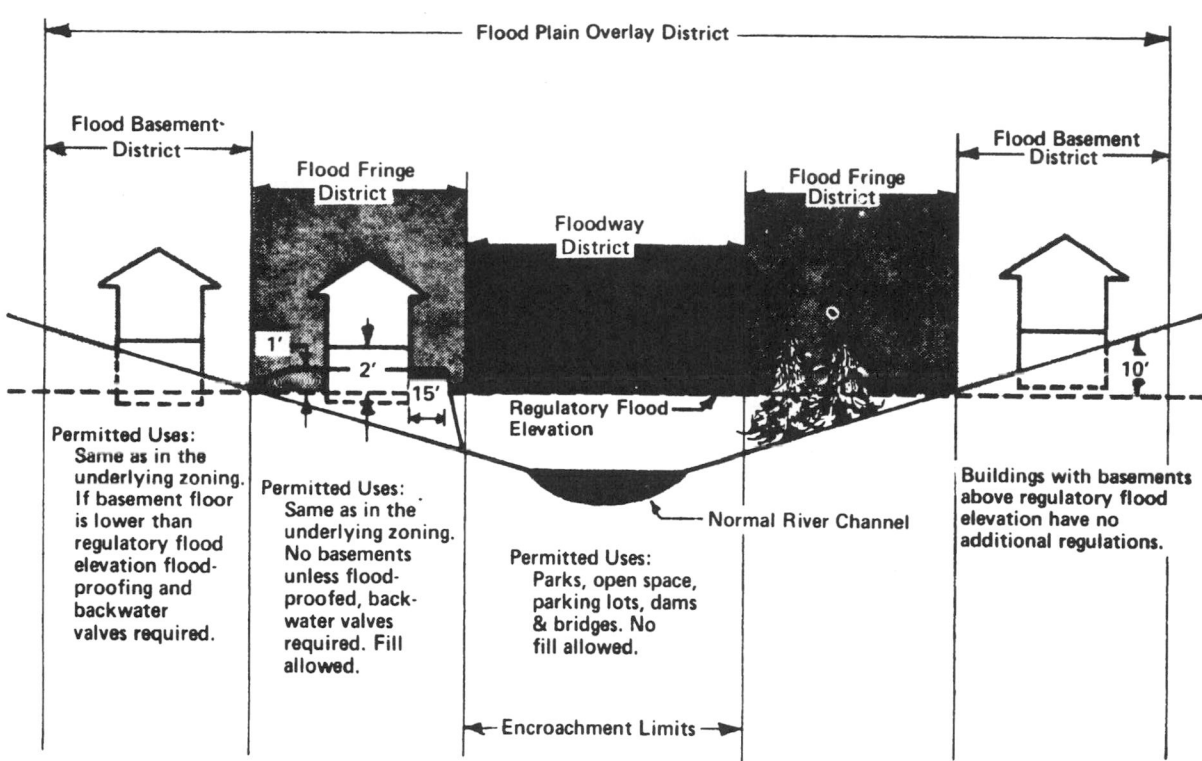

Fig. 3 Addition of flood basement districts to flood fringe and floodway districts.

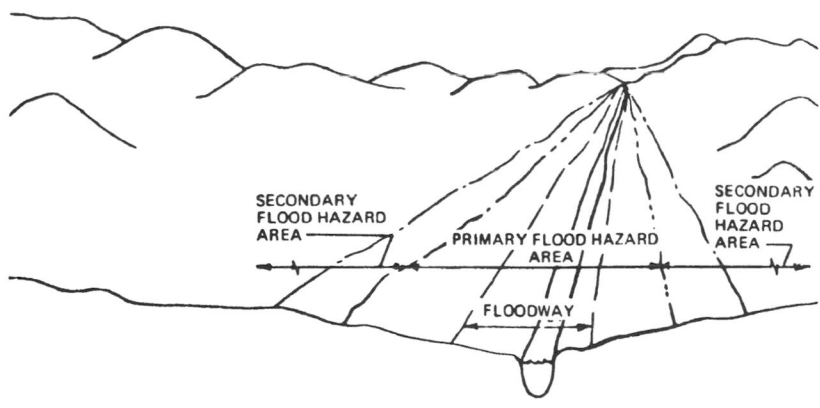

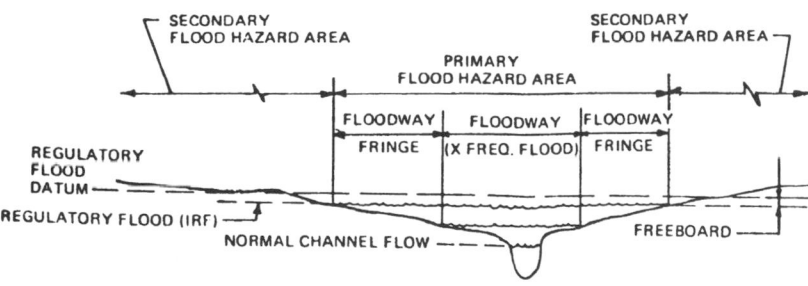

Fig. 4 Flood hazard areas and regulatory flood datum.

COASTAL FLOOD PROTECTION ZONING

Local regulations for dune protection are primarily administered through state agencies or coastal councils. Dune restrictions limit construction to areas a minimum distance from mean high water (MHW). These restrictions varied from a minimum of 200 ft to some local restrictions of 250 ft. Dune restrictions also protect the natural vegetation.

Local zoning regulations limit placement of a structure on a lot by enforcement of side and front yard restrictions and compliance with dune regulations on the ocean side. These restrictions for front and side yards are minimums, allowing smaller lots.

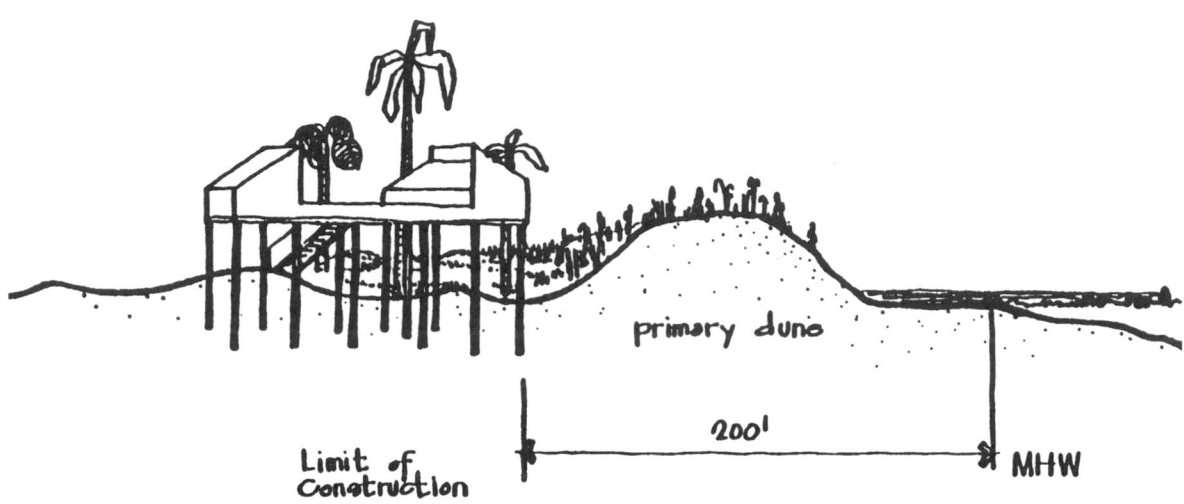

Fig. 1 Example of building setback requirements.

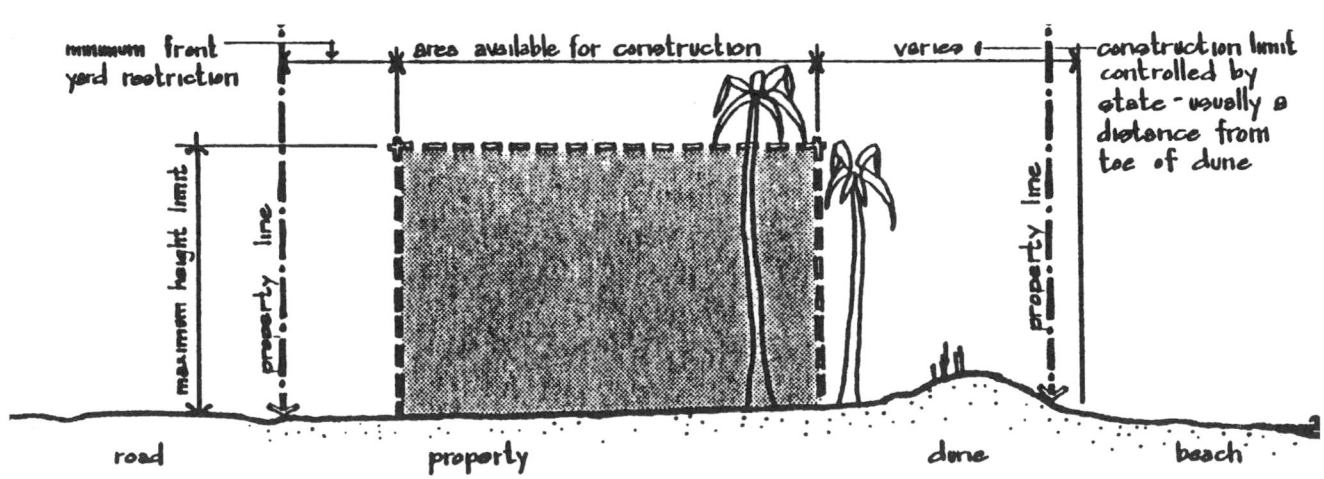

Fig. 2 Zoning restrictions on coastal development.

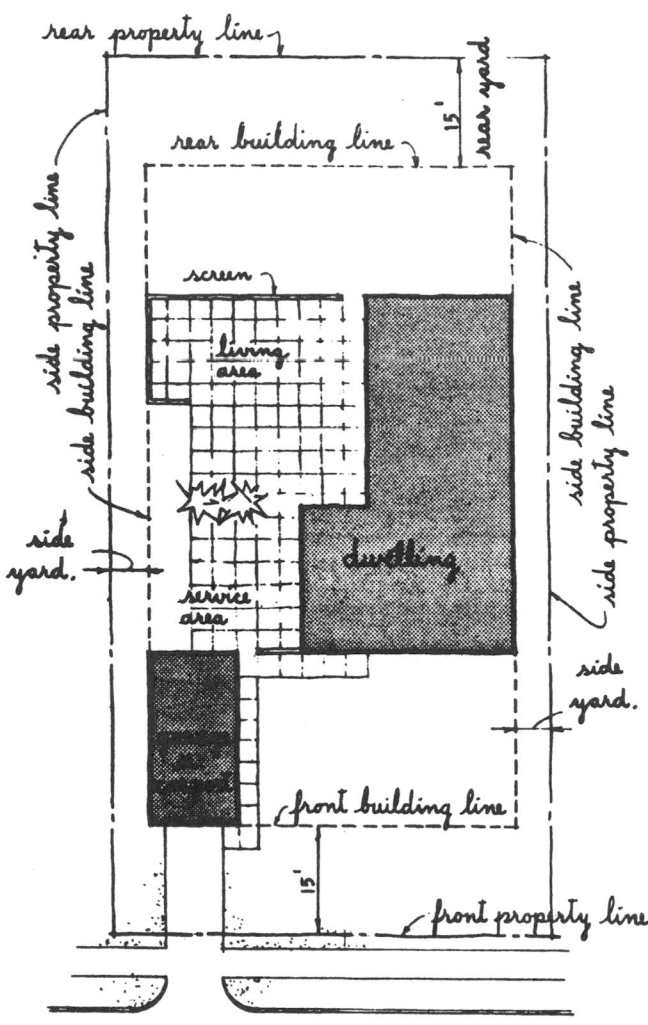

Fig. 1   Interior lot.

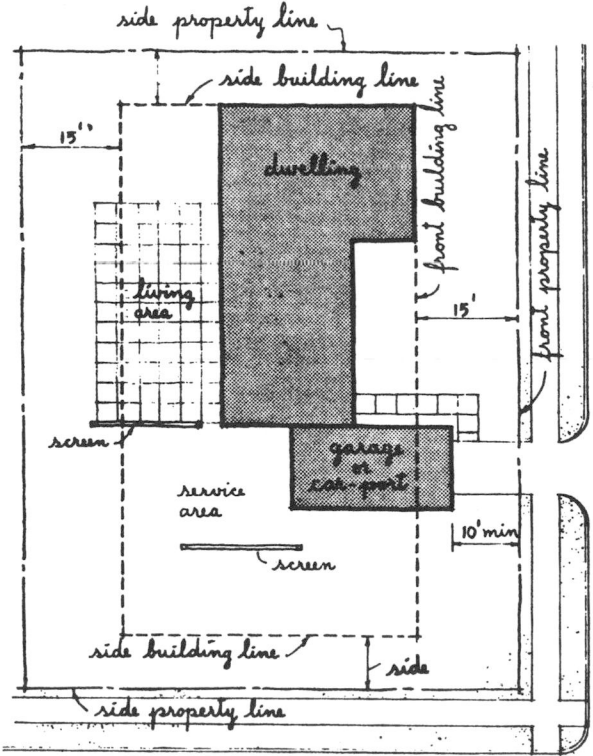

note:
    on a corner lot, either yard facing street may
    be considered *front*. yard opposite selected front
    is rear yard.

Fig. 2   Corner lot.

FLOOR AREA

Floor area is a sum of the gross areas of the several floors of a building or buildings, measured from the exterior faces of exterior walls or from the centerlines of walls separating two buildings. In particular, floor area generally includes:

1. Basement space, except as specifically excluded
2. Elevator shafts or stairwells at each floor
3. Floor space in penthouses
4. Attic space (whether or not a floor has been laid) providing structural headroom of 8 ft or more
5. Floor space in interior balconies or mezzanines
6. Any other floor space used for dwelling purposes, no matter where located within a building
7. Floor space in accessory buildings, except for floor space used for accessory off-street parking
8. Any other floor space not specifically excluded

However, the floor area of a building should not include:

1. Cellar space, except that cellar space used for retailing should be included for the purpose of calculating requirements for accessory off-street parking spaces and accessory off-street loading berths
2. Elevator or stair bulkheads, accessory water tanks, or cooling towers
3. Uncovered steps
4. Attic space (whether or not a floor actually has been laid) providing structural headroom of less than 8 ft
5. Floor space used for mechanical equipment

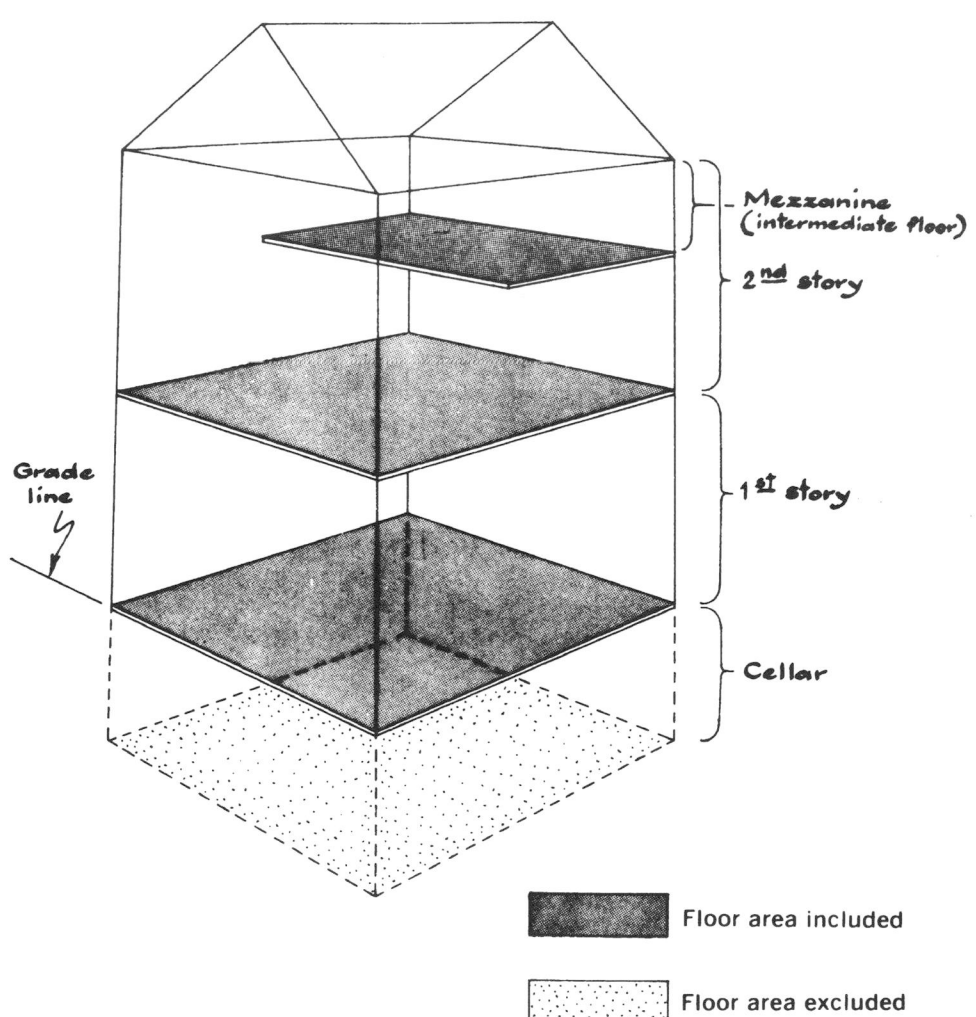

Mezzanine (intermediate floor)

2nd story

Grade line

1st story

Cellar

Floor area included

Floor area excluded

Fig. 1

FLOOR-AREA RATIO (FAR)

Floor-area ratio is the total floor area on a zoning lot, divided by the lot area of that zoning lot.

$$FAR = \frac{\text{total floor area}}{\text{total lot area}}$$

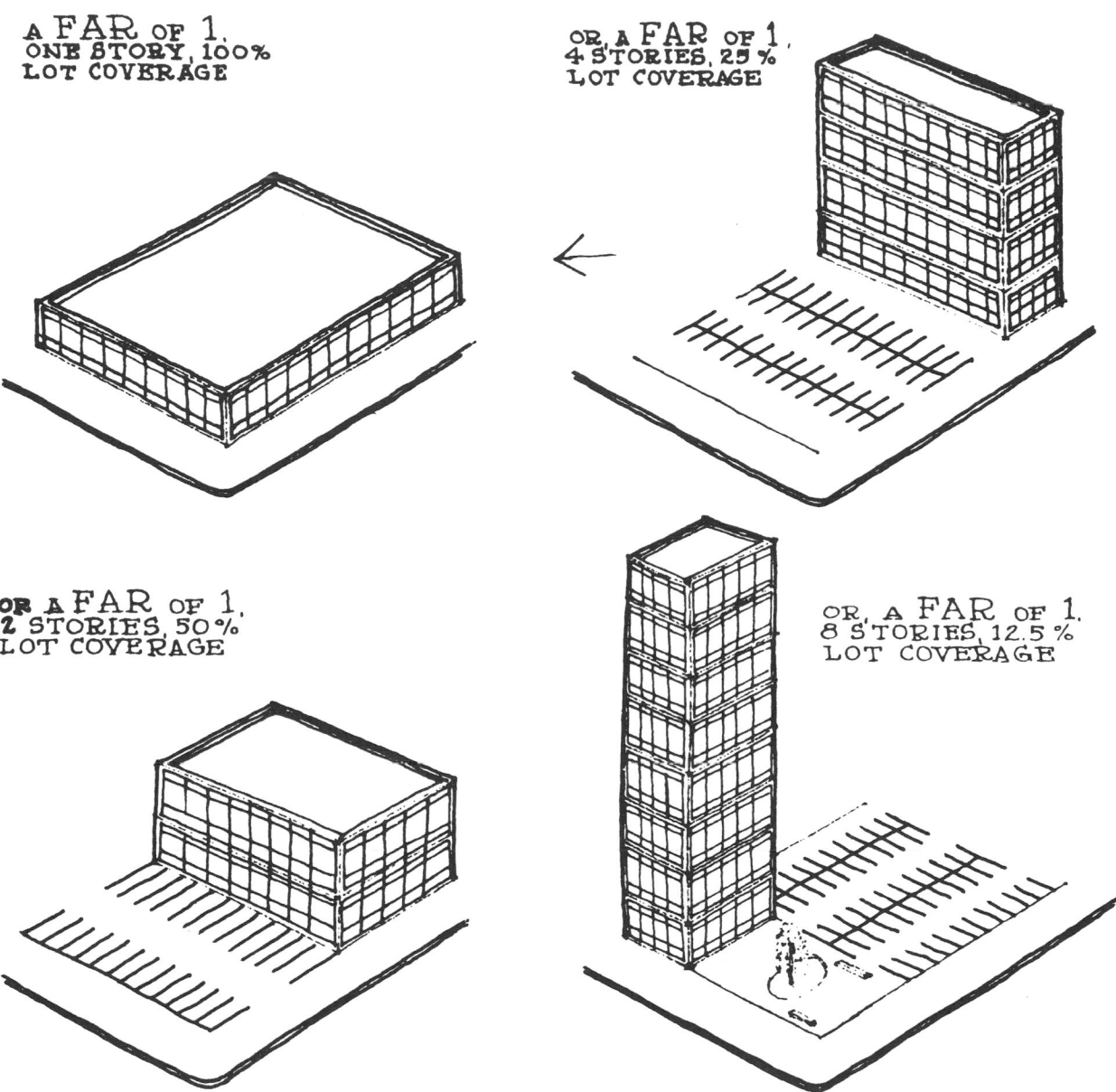

A FAR OF 1.
ONE STORY, 100%
LOT COVERAGE

OR, A FAR OF 1,
4 STORIES, 25%
LOT COVERAGE

OR A FAR OF 1,
2 STORIES, 50%
LOT COVERAGE

OR, A FAR OF 1,
8 STORIES, 12.5%
LOT COVERAGE

Fig. 1

BUILDING HEIGHTS

Building height is the vertical distance measured from the established grade to the highest point of the roof surface for flat roofs; to the deck line of mansard roofs; and to the average height between eaves and ridge for gable, hip, and gambrel roofs.

Story

A story is that part of a building between the surface of a floor (whether or not counted for purposes of computing floor area ratio) and the ceiling immediately above. However, a cellar is not a story.

Basement

A basement is a story (or portion of a story) partly below curb level, with at least one-half of its height (measured from floor to ceiling) above curb level. On through lots the curb level nearest to a story (or portion of a story) should be used to determine whether such story (or portion of a story) is a basement.

Cellar

A cellar is a space wholly or partly below curb level, with more than one-half of its height (measured from floor to ceiling) below curb level. On through lots the curb level nearest to such space should be used to determine whether such space is a cellar.

The total number of stories is counted or measured vertically from the first story.

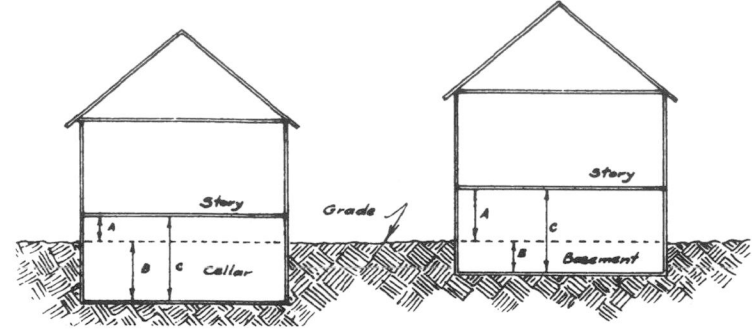

Fig. 1

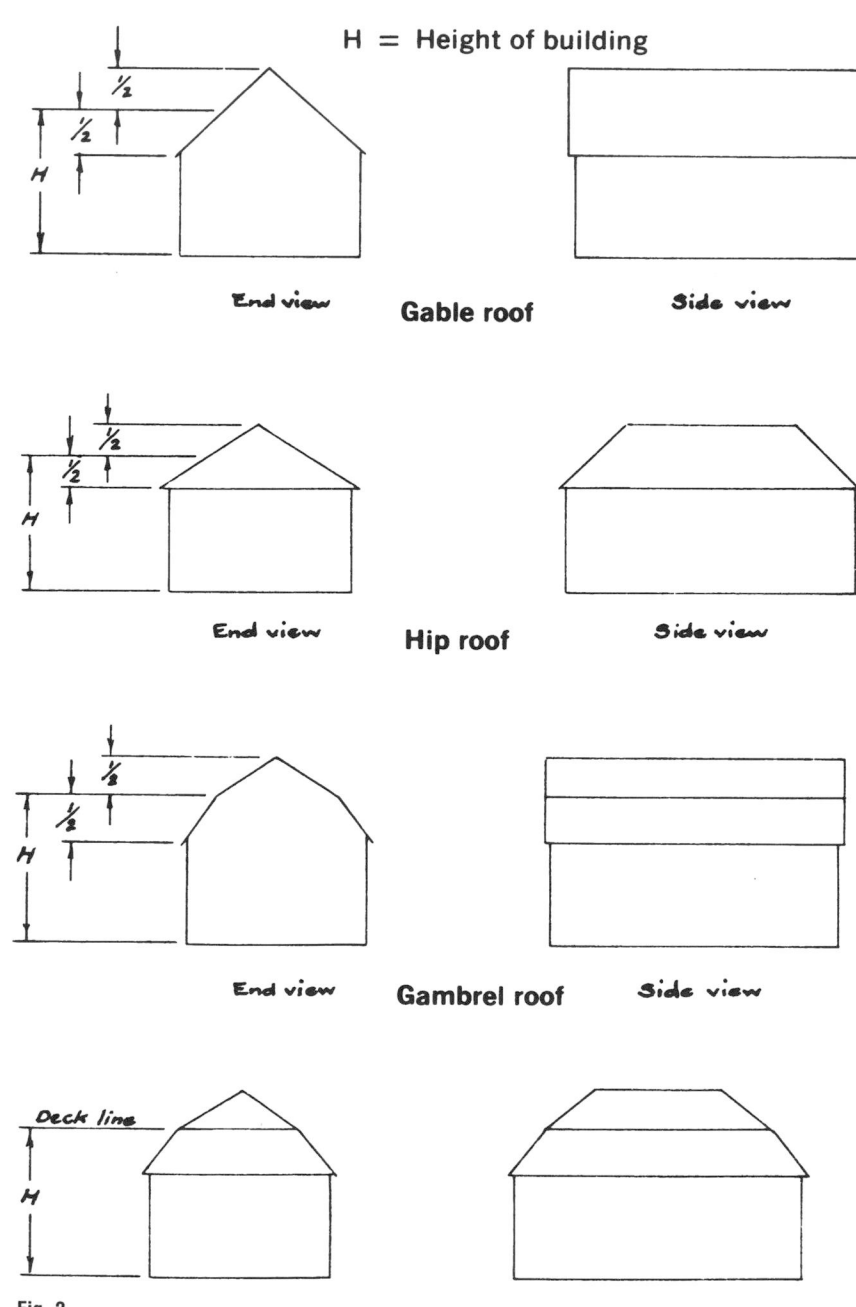

Fig. 2

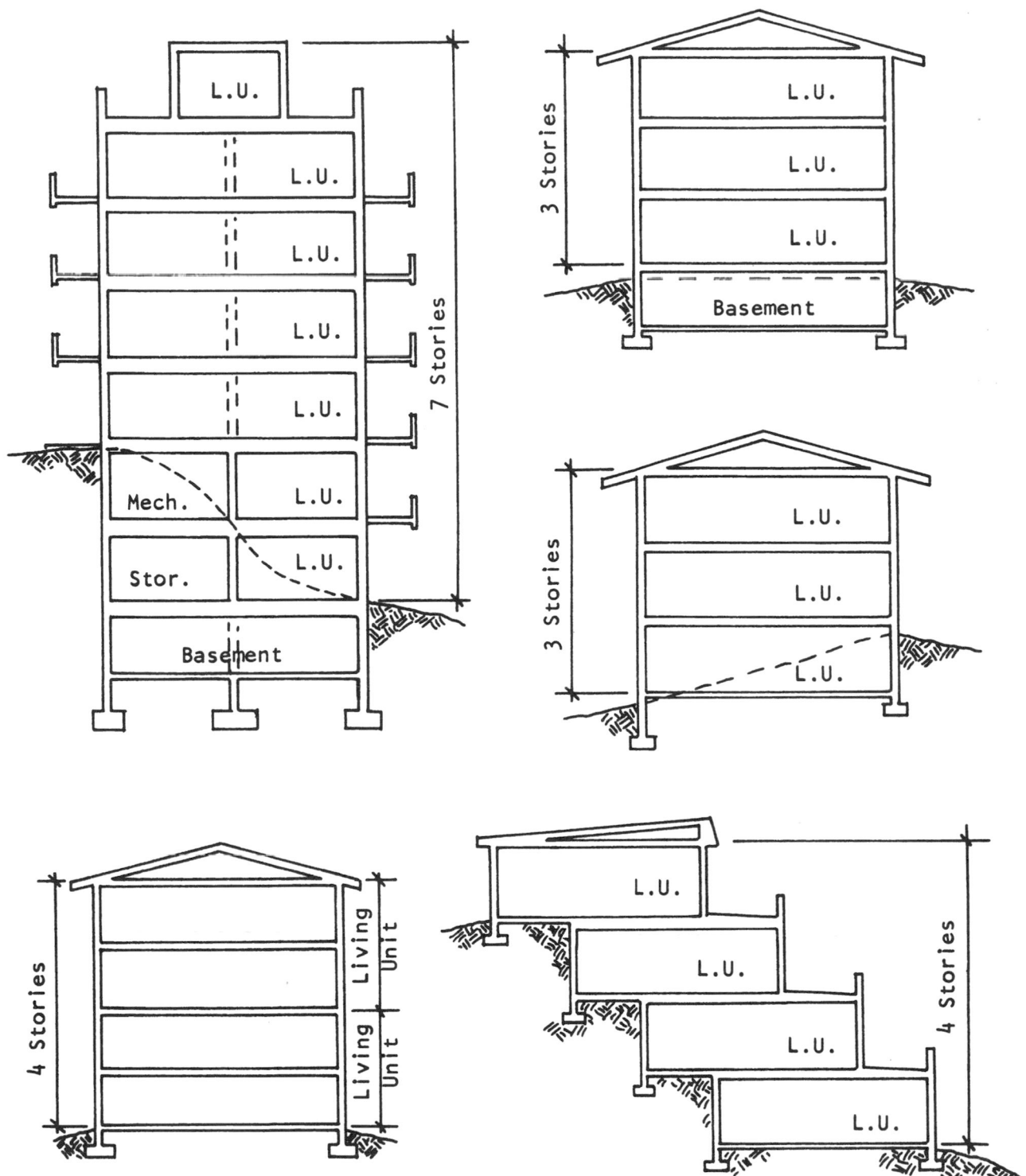

Fig. 3   Determination of number of stories.

SKY EXPOSURE PLANE

A sky exposure plane is an imaginary inclined plane beginning above the street line at a set height and rising over a zoning lot at a ratio of vertical distance to horizontal distance.

On narrow streets, the slope will be less than on wide streets.

The height *h* should relate to the general scale of the neighboring structures.

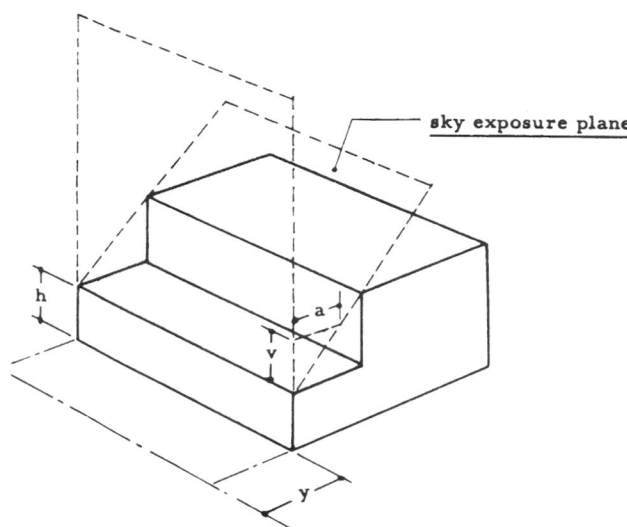

y is the depth of required *front yard*
h is the height of *sky exposure plane* above *front yard line level*
v is the vertical distance
a is the horizontal distance

Fig. 1

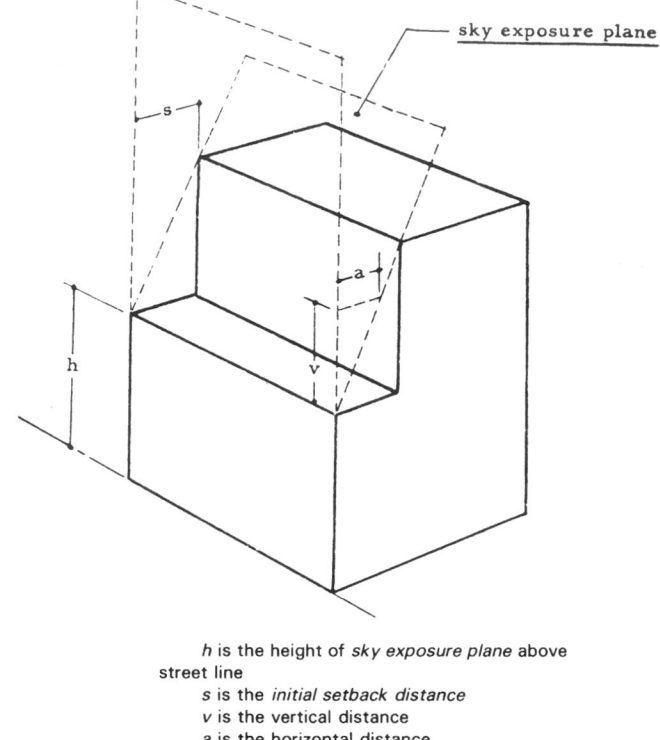

h is the height of *sky exposure plane* above street line
s is the *initial setback distance*
v is the vertical distance
a is the horizontal distance

$$\text{Sky exposure plane} = \frac{\text{vertical distance}}{\text{horizontal distance}}$$

Fig. 2   Illustration of sky exposure plane.

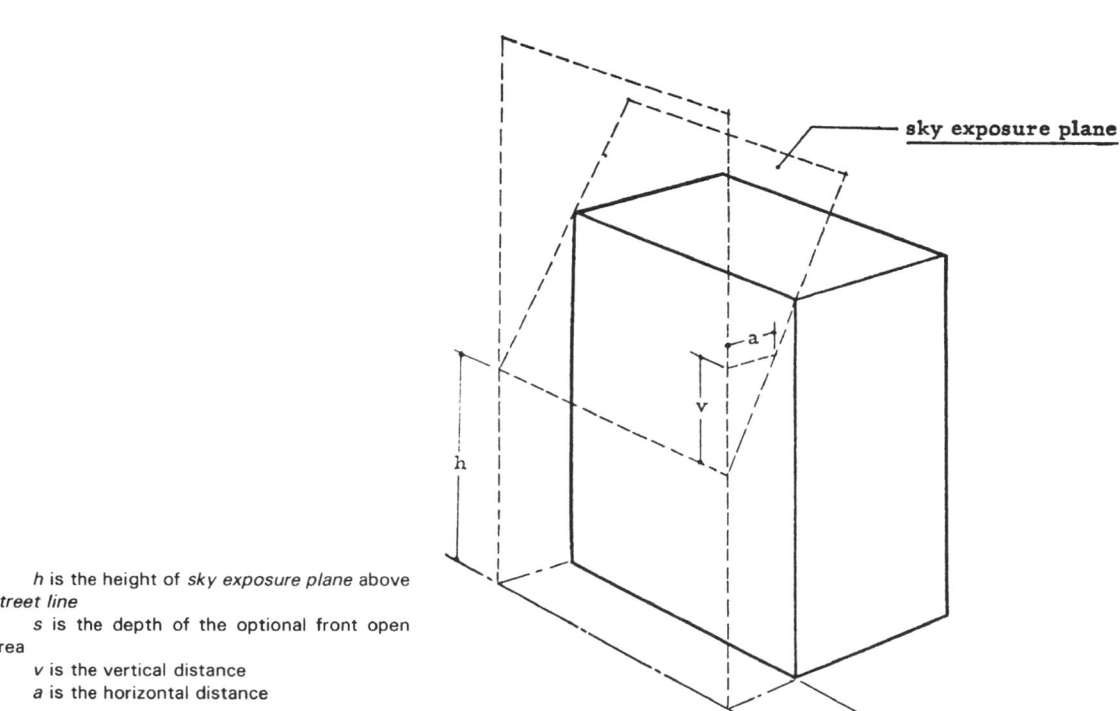

h is the height of *sky exposure plane* above *street line*
s is the depth of the optional front open area
v is the vertical distance
a is the horizontal distance

Fig. 3   Illustration of alternate sky exposure plane.

AREA FOR LIGHT ACCESS

All windows that are needed to satisfy the ventilation requirements of the building code will have to give upon a certain minimum of open space known as the area for light access (ALA). This can be easily and quickly measured with a graphic device marked off in a series of wedge-shaped sections.

The required ALA may be within the lot upon which the building is placed or on the street, or on the required open yard of an adjoining lot. The area for light access is measured by a series of wedges marked out within the segments of a circle.

If an obstruction in front of the window is not higher (from the sill line) than two-thirds the distance from that window, it is not considered an obstruction when checking the window for units of light access.

For residential buildings the wedges are within the band between 40 and 60 ft from the window. Eight wedges (six of them contiguous) must be unobstructed.

For low-bulk commercial buildings the wedges are within the band between 20 and 40 ft from the window. Eight wedges (six of them contiguous) must be unobstructed.

For high-bulk commercial and manufacturing buildings the wedges are within the band between 10 and 20 ft from the window. Eight wedges (all contiguous) must be unobstructed.

The window of a satisfies the requirements for residential buildings: a minimum of eight units of light access, at least six of which are contiguous.

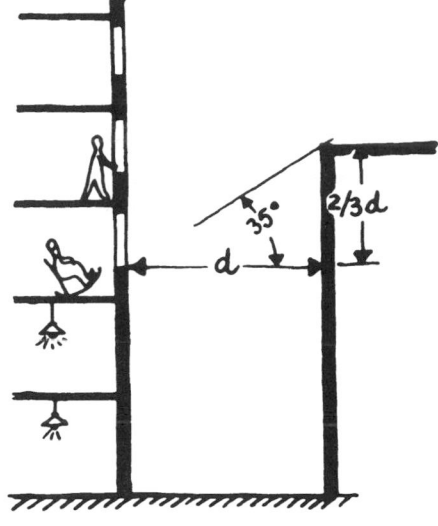

Fig. 1

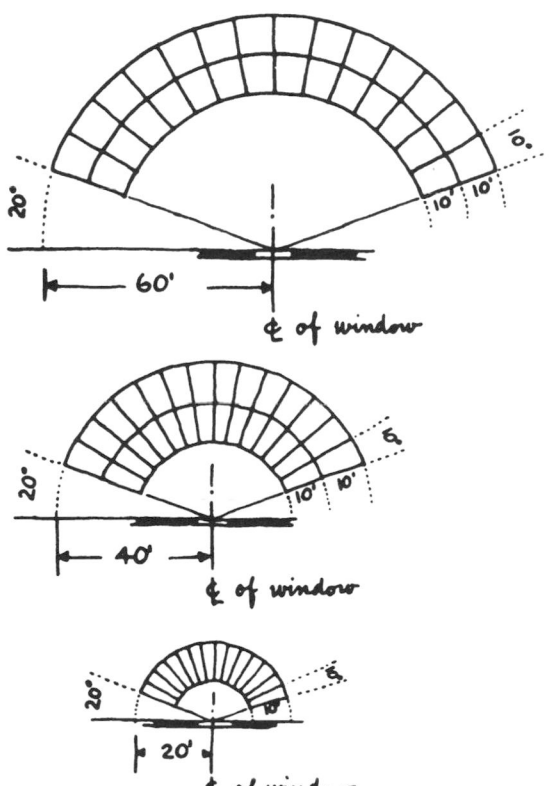

Fig. 2

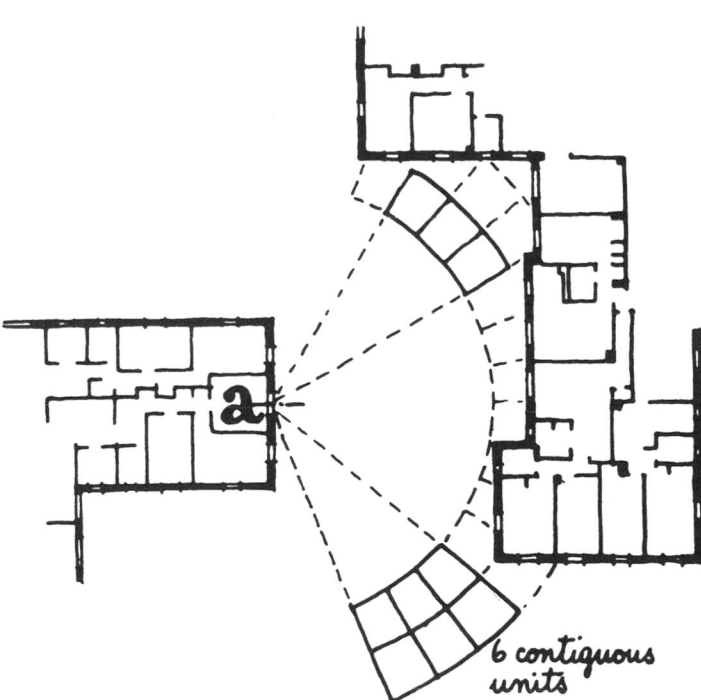

Fig. 3

RESIDENTIAL DENSITY

Density is a measurement of the number of housing units or people for a specific area of land. The term high-density is relative; it differs in meaning from community to community. However, it is used here to refer to housing in excess of approximately 25 dwelling units per net residential acre. This term should not be mistakenly equated with "overcrowding." Indeed, high-density housing has some very positive and unique advantages. If carefully planned and executed, it can offer indoor living space comparable with that of low-density single-family dwellings, and even greater outdoor space. This space, when properly planned and equipped, can serve the outdoor needs of a number of families in a highly satisfactory manner.

Economically, an advantage of high-density housing lies in reduced costs of utility networks and transportation systems. Sewers, water mains, electric power lines, for example, are frequently less costly for clustered, compact housing than for housing scattered at low densities. High-density housing that is concentrated in groups at various places throughout a community can provide a ready, built-in market large enough to support some form of public transportation. The higher the density, presumably the greater the number of potential customers and the more economical the services. As it now stands, urban and suburban sprawl is responsible for rising transportation costs in terms of both dollars and time. The commuter and taxpayer alike bear the brunt of expensive highway networks and mass transit systems required to connect city centers and outlying areas. Another toll is taken through the sacrifice of open space land for parking facilities for private automobiles.

Additional advantages of high-density housing are these: it engenders a community feeling and identity by virtue of the fact that a number of people live in close proximity, and it facilitates the conservation of land resources for use as recreation space. To large segments of the population—the elderly, the childless couple, the single adult, and small groups of working adults—high-density housing has a special appeal. An apartment can be cheaper than a house. It does not require any individual responsibility for maintenance of either the dwelling unit or outdoor grounds. It enhances opportunities for sociability afforded by ease of contact with others of similar age and interests. For the elderly in particular, such accommodations may be preferable to the single-family house which can be physically demanding and lonely. A house emptied of children and activity might be for some a sad reminder of fuller days. Since the American population is a highly mobile one, multiple housing serves another useful purpose. It offers opportunities for attractive rental apartments throughout the country, a real boon to those moving into a new community.

The value of high density as a counterbalance to the spread of American cities is by no means a recent concept. As early as the 1920s, Le Corbusier was advocating increasing net residential densities in his *Plan for a Contemporary City of 3 Million Inhabitants,* with tower apartments of low coverage to allow abundant park and recreation land for the enjoyment of center city residents. Unlike the congested tenements they were intended to replace, these skyscrapers would have light and air, while parks and playgrounds would be close at hand at the base of the apartment towers.

MEASURES OF HOUSING INTENSITY AND RESIDENTIAL DENSITY

A detailed measurement of the intensity of housing development is necessary as a safeguard against overbuilding an area and thereby threatening the health, safety, and general welfare of its residents. Prior to any new construction, an accurate estimate of the existing facilities for specific sites and their populations—sewage-disposal systems, community buildings, for example—is essential if a reasonable intensity is to be maintained.

Housing Intensity

Housing intensity can be measured in a number of ways. For any given building it is possible to measure certain physical characteristics such as height, coverage, total floor space, or bulk. Generally there are local ordinances which impose limits on building size by specifying maximum and minimum building dimensions for different zones throughout a city. A typical zoning ordinance defines maximum height (through number of feet, number of floors, or both), maximum coverage (through percentage of total lot area), and minimum yard sizes (through setbacks from property lines and required areas of usable open space). Building and housing codes and local practice also regulate room sizes and window openings. Additional measures and controls apply to groups of structures. They may stipulate minimum distances between parts of the same building, maximum allowable site coverage, and other bulk specifications which are computed by totaling the individual building areas.

Other controls affecting intensity derive from certain off-site factors. Examples include the following: setback requirements which are figured as an average of the setbacks for other buildings along the same street; height limits which are dictated by the width of an adjoining street (the "angle of light obstruction" measurement); and height limits dictated by special locations—the approach zone to an airport, an area adjacent to a park or other open space, or a district of historic significance.

Combinations of housing intensity measures are also in use. The most significant ones are floor area ratio (FAR) and land use intensity (LUI).

Floor area ratio is determined by a combination of lot area and the total floor space of buildings on a lot. It regulates the total bulk of a building and not coverage or height. The floor area ratio is an index figure which expresses the total permitted floor area as a multiple of the area of the lot. For example, a floor area ratio of 1.0 would apply to a two-story building covering 50 percent of its site, or to a five-story building covering 20 percent.

Land use intensity is also an index figure, but unlike FAR, it is not derived by arithmetic means. Instead it is assigned to a piece of land by local FHA insuring offices on the basis of neighborhood characteristics. For every LUI there are other ratios which apply—FAR, open space ratio, living space ratio, recreation space ratio, occupant car ratio, and total car ratio. In turn each of these ratios is computed by some measure of building bulk, site size, or housing development characteristic.

Residential Density

Residential density is a system of measurement to express in simple mathematical terms the number of people (population density) or amount of housing (accommodation density) in a specified area of land. It can be stated in terms of either gross land area (the entire area of a city precinct or parcel of developed land) or net land area (only that portion specifically devoted to residential uses and excluding areas devoted to major streets, community facilities, and major open spaces).

When employed to measure the degree of crowding within a residential structure, density is expressed in such terms as persons per room or square feet of living space per person. The measure of habitable rooms per acre can be applied to express interior building space on a residential site or within a portion of the community.

Credits

SECTION 1   GENERAL PLANNING AND NEIGHBORHOOD ORGANIZATION

page 14:   Planning Housing Environments for the Elderly, Courtesy The National Council on the Aging, Inc. © 1974.

pages 18–19:   Reprinted from **Fortune Magazine,** September 1969 by special permission © 1969, Time, Inc.

pages 20–22:   Planning the Neighborhood, American Public Health Association, Committee on the Hygiene of Housing, Public Administration Service, Chicago, Ill.

pages 24–28:   HUD Condominium/Cooperative Study, Vol. III, Department of Housing and Urban Development, Washington, D.C., 1975.

pages 30–35:   Manual of Acceptable Practices, Minimum Property Standards, Department of Housing and Urban Development, Washington, D.C., 1973.

pages 36–40:   Noise Assessment Guidelines, Department of Housing and Urban Development, Washington, D.C., 1971.

pages 41–42:   National Park, Recreation and Open Space Standards, National Recreation and Park Association, Washington, D.C., 1971.

page 49:   Ebenezer Howard, **Garden Cities of Tomorrow,** Farber and Farber, London, 1946.

page 50, Fig. 13:   New York Regional Survey of New York and Its Environs—1929.

page 50, Fig. 14:   Clarence Stein in **New Pencil Points,** June 1942.

page 51, Fig. 15:   Architectural Forum.

page 51, Fig. 16; pages 52–53:   Clarence Stein, **Towards New Towns for America,** Reinhold Publishing Corp., New York, 1957.

SECTION 2   SITE CONSIDERATIONS AND SITE PLANNING

page 60:   Know the Soil You Build On, Agriculture Information Bulletin 320, Soil Conservation Service, U.S. Department of Agriculture.

page 61, Fig. 9; pages 75, 102:   Landscape Development, Department of the Interior, Littleton, Colo., 1967.

pages 64–65:   Housing Design and Rehabilitation Guidelines, National Park Service, U.S. Department of the Interior, Washington, D.C., 1988.

pages 67–73:   Allen Carroll, **Developer's Handbook,** Department of Environmental Protection, State of Connecticut, 1972.

page 74:   J. Abel and F. Severud, **Apartment Houses,** Reinhold Publishing Co., New York, 1947.

page 77:   C. Talcott, D. E. Hepler, and P. I. Wallach, **Home Planners' Guide to Residential Design,** McGraw-Hill, New York, 1986.

pages 78–80:   Edward D. Stone, Jr., Associates—Architects, **House and Home,** April 1975.

pages 86, 98:   Minimum Property Standards for One- and Two-Family Houses, FHA, Department of Housing and Urban Development, Washington, D.C., 1973.

pages 91–93:   V. Joseph Kostka, **Planning Residential Subdivisions,** Appraisal Institute of Canada, 1954.

pages 95, 103:   Planning and Design Workbook for Community Participation, prepared for the New Jersey Department of Community Affairs by the Research Center for Urban and Environmental Planning, School of Architecture and Urban Planning, Princeton University, 1969.

pages 105–106:   Design Guide for Reducing Transportation Noise in and around Buildings, National Bureau of Standards, Department of Commerce, Washington, D.C., 1978.

pages 107–108:   Manual of Acceptable Practices, Minimum Property Standards, Department of Housing and Urban Development, Washington, D.C., 1973.

page 109: Bert M. Cohn, **Suggested Guidelines for Fire Protection Criteria for Residential Developments,** Department of Housing and Urban Development, Washington, D.C., 1974.

page 110: M. David Egan, **Concepts in Building Fire Safety,** John Wiley & Sons, New York.

pages 111–120: Solar Dwelling Design Concepts, A.I.A. Research Corp., Department of Housing and Urban Development, Washington, D.C., 1976.

pages 120–125; 126–134: Planning for Housing Security—Site Elements Manual, Department of Housing and Urban Development, Washington, D.C., 1979.

SECTION 3   SUBDIVISIONS AND LAND PLANNING

pages 137–146: Control of Land Subdivision, Office of Planning Coordination, State of New York, Albany, 1968.

pages 147–151: Planning Profitable Neighborhoods—FHA Technical Bulletin No. 7, Federal Housing Administration, Washington, D.C.

page 152: Subdivision Planning Standards, Land Planning Division, Federal Housing Administration, Washington, D.C.

pages 153–158: Cost Effective Site Planning, The National Association of Homebuilders (NAHB), Washington, D.C., 1976.

pages 159–161: Reprinted with permission from Robert Engstrom and Mark Putman from **Planning and Design of Townhouses and Condominiums,** published by the Urban Land Institute, 1979.

pages 162–166: Walter Richardson, FAIA, and Ralph Martin, AIP **House and Home,** January 1975.

page 171, Figs. 4 and 5: House and Home, May 1977.

pages 172–173: Allen Carroll, **Developer's Handbook,** Department of Environmental Protection, State of Connecticut, 1972.

page 173, Fig. 5: J. Rahenkamp and A. Wolffe, **Landscape for Living—The Yearbook of Agriculture,** U.S. Department of Agriculture, Washington, D.C., 1972.

page 174: Open Space Subdivision, New York State Office of Planning Services, Albany, 1961.

pages 175–176: Design Manual—Family Housing, Naval Facilities Engineering Command, Department of the Navy, Alexandria, Va., 1971.

pages 178–181: Planned Unit Development, New York City Planning Commission, 1968.

page 182: Reprinted from the May issue of **Progressive Architecture,** copyright Reinhold Publishing Company, New York.

pages 184–196: Manual—Design and Control of Land Subdivision in Suburban Communities, The Institute of Rational Design, Inc., 1976.

pages 197–201: John Macsai, **Housing,** © 1978 John Wiley & Sons, Inc., reprinted by permission from John Wiley & Sons, Inc., New York.

SECTION 4   COMMUNITY FACILITIES

page 205: J. De Chiara and L. E. Koppelman, **Planning Design Criteria,** Van Nostrand Reinhold Co., New York, 1969.

page 206: Housing Families at High Densities, Vancouver City Planning Department, Vancouver, B.C., Canada, 1978.

page 207; pages 248–250: Operation Breakthrough—An Evaluation Framework for Site Planning, Llewelyn-Davies Associates, Department of Housing and Urban Development, Washington, D.C., 1970.

page 210, Figs. 4 and 5: Housing with Shelter, Department of Housing and Urban Development, Washington, D.C., 1970.

page 210, Fig. 6: Ronald W. Haase, **Designing the Child Development Center,** Office of Economic Opportunity, Washington, D.C.

page 211, Figs. 7 and 8; page 217, Fig. 4; page 218; page 219, Fig. 6; page 241: Definitive Designs, Naval Shore Facilities, Department of the Navy, Washington, D.C., 1972.

page 211, Table 2; page 212, Table 4; pages 214, 256: Planning the Neighborhood, American Public Health Association, Committee on the Hygiene of Housing, Public Administration Service, Chicago, Ill., 1960.

page 213: M. Alexander Gabrielson and Coswell M. Miles, **Sports and Recreation Facilities,** Prentice-Hall, Englewood Cliffs, N.J., 1958.

page 215: Robert C. Hoover and Everett L. Perry, **Church and City Planning,** Bureau of Research and Survey, National Council of the Churches of Christ in the U.S., New York, 1955.

page 217, Table 2: Interim Standards for Small Public Libraries, Public Library Association, A Division of the American Library Association, Chicago, Ill., 1970.

page 221: Guidelines for Developing Public Recreation Facility Standards, Ministry of Culture and Recreation, Toronto, Ontario, Canada.

page 223, Figs. 1 and 2: Wayne R. Williams, **Recreation Places,** Reinhold Publishing Corp., New York, 1958.

page 243, Fig. 22: Ben Chlevin, **Golf Operator's Handbook,** National Golf Foundation, Inc., 1956.

pages 251–253: Reprinted with permission from **The Community Builders Handbook,** The Urban Land Institute, Washington, D.C., 1947, pp. 143–147.

page 254, Fig. 5: **Land Subdivision Regulations,** HHFA, Washington, D.C.

pages 257–261: **Housing Quality—A Program for Zoning Reform,** Urban Design Council of the City of New York, 1975.

SECTION 5 ELEMENTS OF THE DWELLING UNIT

pages 265–266; page 267, Fig. 2; pages 301, 311, 385, 440: **Space Standards for Household Activities,** Illinois Agricultural Experiment Station, 1962.

pages 267–269, 312–313, 325; page 326, Fig. 56; page 337; page 388, Fig. 13; pages 389–390: **The Use and Design of Space in the Home,** Canada Mortgage and Housing Corporation, Ottawa, Ontario, Canada, 1977.

page 270; page 285, Fig. 1; pages 287–289; page 330, Fig. 63; pages 331, 386, 387: **Manual of Acceptable Practices,** Department of Housing and Urban Development, Washington, D.C., 1973.

pages 271–276: **Townhouse Development Process,** Michigan State Housing Development Authority, Lansing, October 1970.

pages 277, 455: **Operation Breakthrough,** Department of Housing and Urban Development, Washington, D.C., 1970.

page 278; page 308, Fig. 7; page 310; page 426, Figs. 5–7: **Housing Families at High Densities,** Vancouver City Planning Department, Vancouver, B.C., Canada, 1978.

pages 279–280: Sandra Honell and Gayle Epp, **Private Space,** Department of Architecture, Massachusetts Institute of Technology, Boston, Mass., 1978.

pages 281, 290, 304, 328–329, 391; page 400, Fig. 4; page 423; page 427, Fig. 1a and 1b; page 451, Fig. 20: Ernest Pickering, **Shelter for Living,** John Wiley and Sons, Inc., New York, 1941.

pages 282–284, 293, 336, 392–398, 402, 410–411: James W. Wentling, **Housing by Lifestyle,** McGraw-Hill, New York, 1990.

page 286, Figs. 4 and 5; page 295, Fig. 7; page 299, Fig. 12; page 308, Fig. 7; page 310; page 330, Fig. 62; page 338, Fig. 6; page 384; page 388, Figs. 11 and 12; page 399: Department of Housing and Urban Development, Washington, D.C.

pages 296–297; page 300, Fig. 16; page 312, Fig. 13; pages 412–417, 421–422, 438–439, 441–443: Julius Panero and Martin Zelnik, **Human Dimension and Interior Space,** Whitney Library of Design/Watson-Guptill Publications, New York, 1979.

page 298, Fig. 13; page 299, Fig. 14; pages 378–383, 418–419, 435: **Planning Guides for Southern Rural Homes,** Southern Cooperative Series Bulletin no. 58, Georgia Agricultural Experiment Station, Experiment, Ga., 1958.

page 307; page 308, Fig. 6: Aristokraft, Inc., Jasper, Ind., 1993.

pages 314–324; page 360, Fig. 29: **Kitchen Design: The New Rules,** National Kitchen and Bath Association, Hackettstown, N.J., 1993.

page 332: **Adaptable Bathrooms and Kitchens,** New York.

pages 333–335; 407: **Housing Ideas for the '80's,** McGraw-Hill, New York, no date.

pages 341–342: **House and Home,** October 1975.

pages 343–344: **House and Home,** June 1976.

pages 345–349, 353–354, 357, 361–362, 364–366, 388–391, 403–406: John Callender, **Time-Saver Standards,** 4th ed., McGraw-Hill Book Co., New York, 1966.

pages 350–351: Elkay Manufacturing Co., Oak Brook, Ill., 1993.

pages 372–377: Merillat Industries, Inc., Adrian, Mich., 1993.

page 408: Waterjet Corp., Canoga Park, Calif.

page 409: Maclevy Products Corp., Queens, N.Y.

page 420: **Architectural Record.**

page 424: **House and Home,** August 1954.

page 425, Fig. 2: C. Talcott, D. E. Hepler, and P. I. Wallach, **Home Planners' Guide to Residential Design,** McGraw-Hill, New York, 1986.

page 425, Fig. 3: Elizabeth Wood, **Housing Design—A Social Theory,** Citizens Housing and Planning Council of New York, Inc., 1961.

page 427, Fig. 1c: **Housing,** December 1979.

pages 428, 431: **Family Housing Handbook,** Midwest Plan Service, Iowa State University, Ames, Iowa, 1971.

pages 429–430: **Maytag Encyclopedia of Home Laundry,** 5th ed., Western Publishing Co., Newton, Iowa, 1982.

pages 432, 437: **Space for Home Sewing,** The Pennsylvania State University, College of Home Economics, Agriculture Experiment Station, 1957.

page 444: J. Abel and F. Severud, **Apartment Houses,** Reinhold Publishing Co., New York, 1947.

page 453, Fig. 1: **A Design Guide for Home Safety,** Department of Housing and Urban Development, Washington, D.C., 1972.

SECTION 6   TYPES OF SINGLE-FAMILY HOUSES

page 461, Figs. 5 and 7; pages 462–463:   Earth-Sheltered Homes, Department of Housing and Urban Development, Washington, D.C., 1980.

page 461, Fig. 8:   Cutting Energy Costs, 1980 Yearbook of Agriculture, U.S. Department of Agriculture, Washington, D.C.

page 475, Figs. 31 and 32:   House and Home, April 1954.

page 478:   House and Home, August 1963.

page 479; page 551, Fig. 8:   C. Talcott, D. E. Hepler, and P. I. Wallach, **Home Planners' Guide to Residential Design,** McGraw-Hill, New York, 1986.

page 480:   House and Home, August 1954.

page 481:   House and Home, August 1954 (Philip Johnson, Architect).

page 482, Fig. 43; page 544:   Robert D. Katz, **Design of the Housing Site,** University of Illinois, 1968.

page 483, Fig. 44:   Genesee River Houses, Rochester, N.Y. Conklin & Rossant, Architects.

page 483, Fig. 45; page 484:   Architectural Record, September 1963.

page 485, Fig. 1:   House and Home, June 1977.

page 485, Fig. 2:   House and Home, May 1977.

pages 488–489, 498:   House and Home, July 1977.

page 490:   House and Home, August 1977.

pages 491–492:   House and Home, November 1977.

page 493, Fig. 1:   Housing with Shelter, Department of Housing and Urban Development, Washington, D.C., 1970.

page 493, Fig. 4:   House and Home, September 1972.

page 494, Figs. 5 and 6:   Housing, May 1978.

page 494, Figs. 7 and 8:   House and Home, March 1977.

page 495, Fig. 8*b*:   Architectural Record.

pages 500–501:   Technical Bulletin, HHFA, Washington, D.C., May–June 1950.

pages 505–506:   Architectural Forum, January 1950.

page 507:   Gilbert Switzer & Associates/Architects, New Haven, Conn.

page 511, Fig. 6; pages 512–513:   House and Home, July 1969.

page 512, Fig. 8:   New York State Urban Development Corp., Charlotte Area Project, Rochester, N.Y., Northrop, Kaelber, & Kopf/Architects.

page 515:   House and Home, September 1973.

page 516:   New Jersey Housing Finance Agency.

pages 517–519:   Architectural Record, June 1968.

page 520:   Architectural Record, September 1972.

page 521:   Architectural Record, March 1975.

page 523:   Reprinted from January 1962 issue of **Progressive Architecture,** © 1962, Reinhold Publishing Co., New York.

page 524:   Reprinted from the March 1976 issue of **Progressive Architecture,** © 1976, Reinhold Publishing Co., New York.

page 525:   House and Home, April 1972.

page 526:   House and Home, July 1972.

pages 527–528:   Housing, June 1979.

page 529:   House and Home, August 1976.

page 530:   House and Home, March 1976.

page 531:   House and Home, February 1976.

page 532, Fig. 52:   House and Home, February 1976.

page 532, Fig. 53:   House and Home, November 1975.

page 533:   House and Home, October 1971.

page 534:   House and Home, September 1977.

page 535:   House and Home, January 1977.

page 536, Fig. 57:   House and Home, June 1977.

page 536, Fig. 58:   Architectural Record, Mid-May 1978.

page 537:   Planned Unit Development with a Home Association, Land Planning Bulletin no. 6, FHA, Washington, D.C., 1964.

page 538, Fig. 62; pages 539, 543:   Design Manual—Family Housing, Naval Facilities Engineering Command, Department of the Navy, Alexandria, Va., 1971.

pages 545–546:   House and Home, January 1974.

pages 547–549: **Farmstead Planning Handbook,** Midwest Plan Service, Iowa State University, Ames, Iowa, 1974.

SECTION 7 TYPES OF APARTMENTS

page 562, Fig. 1; page 564, Fig. 6; page 570, Fig. 1: J. Abel and F. Severud, **Apartment Houses,** Reinhold Publishing Co., New York, 1947.

page 564, Fig. 5; pages 621, 624: **Design Manual—Family Housing,** Naval Facilities Engineering Command, Department of the Navy, Washington, D.C., 1971.

page 565: **Architectural Record,** June 1963.

page 566: **Architectural Record,** Mid-May 1975.

pages 567–568: Reprinted from the March 1976 issue of **Progressive Architecture,** © 1976, Reinhold Publishing Co., New York.

page 569: **Architectural Forum,** January 1952.

page 571, Figs. 2–4; page 577, Figs. 6–9; page 591, Figs. 5 and 6; page 592, Figs. 8–11; page 609, Figs. 5 and 6; page 610; page 622, Figs. 3–5: **Planning and Design Workbook for Community Participation,** prepared for the New Jersey Department of Community Affairs by the Research Center for Urban and Environmental Planning, School of Architecture and Urban Planning, Princeton University, 1969.

page 603, Fig. 35: **Architectural Record,** December 1951.

pages 640–641: **House and Home,** February 1972.

page 642: **House and Home,** July 1973.

page 643: **Architectural Record,** Mid-May 1978.

pages 644–645: **House and Home,** February 1977.

page 646: **House and Home,** November 1973.

SECTION 8 TYPES OF APARTMENT BUILDINGS

page 653: Robert D. Katz, **Design of the Housing Site,** University of Illinois, Urbana, Ill., 1966.

page 654, Fig. 2a: Suffolk County Housing Study, 1968.

page 654, Fig. 2b: **House and Home.**

page 656, Fig. 1: Suffolk County Housing Study.

page 656, Fig. 2: Conklin & Rossant, Architects.

page 659, Fig. 4: **Architectural Forum,** January 1950.

pages 660, 663: Reprinted from the December 1973 issue of **Progressive Architecture,** © 1973, Reinhold Publishing Co., New York.

page 664: **House and Home,** September 1969.

page 665: **Architectural Record,** Mid-May 1976.

page 667: **Progressive Architecture,** March 1976.

page 668: **House and Home,** September 1976.

page 670: Williamsburg, Brooklyn, N.Y., Skidmore, Owens & Merrill, Architects.

pages 672–673: **Architectural Forum,** January 1950.

pages 682, 687, 703: **Architectural Record,** September 1972.

page 685: **Architectural Record,** June 1973.

pages 688–689: **Architectural Record,** September 1977.

page 691: New York City Housing Authority.

pages 692–693: **Architectural Forum,** January 1950.

page 695: **Architectural Record,** January 1969.

pages 696–698: Reprinted from the March 1976 issue of **Progressive Architecture,** © 1976, Reinhold Publishing Co., New York.

page 698, Fig. 9: **Architectural Forum,** November 1966.

pages 700–701: **Architectural Record,** February 1949.

page 702: **Architectural Forum,** January 1952.

pages 704, 718; page 721, Fig. 19; page 728, Fig. 2; pages 742, 759: Samuel Paul, **Apartments,** Reinhold Book Co., New York, 1967.

page 705: **Architectural Record,** February 1949.

page 714, Fig. 7: Reprinted from the February 1966 issue of **Progressive Architecture,** © 1965, Reinhold Publishing Co., New York.

page 717, Fig. 14: Reprinted from the January 1965 issue of **Progressive Architecture,** © 1965, Reinhold Publishing Co., New York.

page 721, Fig. 20: **Architectural Record,** September 1963.

page 722, Fig. 21: **Urban Renewal,** New York City Planning Commission, 1958.

page 722, Fig. 22; page 727, Fig. 5: **Architectural Forum,** March 1964.

page 727, Fig. 6: **Architectural Record,** April 1971.

page 728, Fig. 1: **Architectural Forum,** January 1950.

pages 730–734: New York City Housing Authority.

page 737: **Architectural Record,** September 1963.

page 738: **Architectural Forum,** November 1966 and March 1964.

page 739: **Architectural Forum,** January 1950.

pages 746–747: Reprinted from the January 1966 issue of **Progressive Architecture,** © 1966, Reinhold Publishing Co., New York.

page 748: **Architectural Record,** March 1974.

pages 749–750: **Architectural Forum,** July 1948.

page 751: **Architectural Forum,** June 1948.

page 760: **Architectural Record,** June 1968.

pages 761–766: **Architectural Record,** February 1973.

pages 767–768: **Architectural Record,** January 1969.

pages 769–770: Reprinted from the May 1973 issue of **Progressive Architecture,** © 1973, Reinhold Publishing Co., New York.

page 771: **Architectural Record,** April 1971.

SECTION 9 APARTMENT BUILDING AMENITIES

page 775: Boise Cascade Housing System, **Operation Breakthrough,** Department of Housing and Urban Development, Washington, D.C., 1970.

pages 776–779: **Townhouse Development Process,** Michigan State Housing Development Authority, Lansing, October 1970.

pages 780, 795: **Guidelines for Developing Public Recreation Facility Standards,** Ministry of Culture and Recreation, Toronto, Ontario, Canada, 1976.

pages 781–782: Office of the Chief of Engineers, Department of the Army, Washington, D.C.

page 783, Fig. 2: **Golf Course Developments,** Technical Bulletin no. 70, Urban Land Institute, Washington, D.C., 1974.

page 784; page 785, Fig. 4: **Site Improvement Handbook for Multifamily Housing,** Housing Research and Development, University of Illinois, Urbana, Ill.

page 786, Fig. 1; page 795: **Manual of Acceptable Practices,** Department of Housing and Urban Development, Washington, D.C., 1973.

pages 789–790: **Public Playground Handbook for Safety,** U.S. Consumer Product Safety Commission, Washington, D.C., no date.

pages 791–793, 805, 821–823, 846–847: **Barrier Free Site Design,** U.S. Department of Housing and Urban Development, Washington, D.C., 1977.

page 794; page 796, Fig. 2: Elizabeth Wood, **Housing Design—A Social Theory,** Citizen's Housing and Planning Council of New York, Inc., 1961.

pages 797–798: **Roof Decks Design Guidelines,** Central Mortgage and Housing Corp.

page 799, Figs. 1 and 2; page 800, Figs. 1–3: **Renovating Schools,** New York City Planning Commission, 1974.

page 799, Fig. 3; page 800, Fig. 4; pages 801–802; page 804, Fig. 7: **Learning Environments,** Educational Facilities Lab, New York, no date.

page 801, Fig. 2; pages 834–835; page 842, Fig. 10: **Memo to Architects,** New York City Housing Authority, no date.

pages 806–811: **Suggested Minimum Standards for Residential Pools,** National Swimming Pool Institute, Washington, D.C., 1969.

pages 811–812: **Planning Areas and Facilities for Health, Physical Education, and Recreation,** The Athletic Institute and American Association for Health, Physical Education, and Recreation, 1966.

page 812: Milton Costello, Consulting Engineer, Wantagh, N.Y.

page 813: **A Design Guide for Home Safety,** Department of Housing and Urban Development, Washington, D.C., 1972.

pages 814–816, 818: Maclevy Products Corp., Queens, N.Y.

page 817, Fig. 1: Metos Saunas, Amerec Corp., Bellvue, Washington.

page 817, Fig. 2: Tylo Saunas, Baths International, Inc.

pages 819–820, 824–826, 836, 848: **Housing for the Elderly Development Process,** Michigan State Housing Development Authority, Lansing, Mich., 1974.

pages 827–828: Parking Dimensions, Motor Vehicle Manufacturers Association, 1993.

pages 829, 831: Space and Equipment for Rental Housing, Federal Housing Administration, Washington, D.C., 1950.

pages 832–833: Design Guide for Permanent Parking Areas, National Crushed Stone Association, Washington, D.C., 1970.

pages 836–839: Laundry Guide, Department of Housing and Urban Development in conjunction with National Association of Coin Laundry Equipment Operators, Inc., Baltimore, Md., 1974.

pages 840–841: Design Ideas and Installation Details, Speed Queen Division, McGraw-Edison Co., Ripon, Wisc., no date.

pages 843–844: A Planning Guide for Physicians' Medical Facilities, American Medical Association, no date.

page 851: Operation Breakthrough, Department of Housing and Urban Development, Washington, D.C., 1978.

SECTION 10 SPECIAL TYPES OF HOUSING

pages 855–859: Edward Steinfeld, Associate Professor, Department of Architecture, State University of New York at Buffalo.

pages 860–861, 875–878: Access for All: An Illustrated Handbook of Barrier-Free Design for Ohio, 2d ed., Governor's Committee on Employment of the Handicapped, Columbus, Ohio, 1978.

pages 862–864, 868–872, 874, 887–889, 914–915, 917; page 918, Fig. 11; page 923, Fig. 9: Federal Register—Minimum Guidelines and Requirements for Accessible Design, Architectural and Transportation Barriers Compliance Board, August 4, 1982.

pages 865–867, 879–880; page 908, Figs. 8 and 9; pages 910–911, 920: Handbook for Design, Veterans Administration, Washington, D.C., 1978.

pages 873, 898–899: Georges Selim, **Barrier-Free Design,** The Office of Disabled Student Services, University of Michigan, Ann Arbor, Mich., 1977.

pages 881–886; page 908, Fig. 10; page 909: Adaptable Dwellings, Department of Housing and Urban Development, Washington, D.C., 1979.

page 883, Fig. 4b; pages 900, 904–907: Ronald L. Mace, **An Illustrated Handbook of the Handicapped Section of the North Carolina State Building Code,** Raleigh, N.C., 1974.

pages 890–897: Home in a Wheelchair, Paralyzed Veterans of America, Inc.

page 901: Manual of Acceptable Practices, Department of Housing and Urban Development, Washington, D.C., 1973.

page 912; page 923, Fig. 8: Development of Priority Accessible Networks, U.S. Department of Transportation, Federal Highway Administration, Washington, D.C., 1980.

page 918, Fig. 12; page 921: Barrier Free Site Design, U.S. Department of Housing and Urban Development, Washington, D.C., 1977.

page 919, Fig. 13: Ronald Mace, AIA, **Accessibility Modifications,** North Carolina Department of Insurance, Raleigh, N.C., 1976.

pages 929–932: Congregate Housing for Older People—An Urgent Need, A Growing Demand, Administration of Aging, U.S. Department of Health, Education, and Welfare, Washington, D.C., 1977.

pages 933–936: John Edmonds, Institute of Advanced Architectural Studies, University of York, England, 1977.

pages 948–954: Harold C. Riker, **College Students Live Here,** Educational Facilities Laboratories, Inc., New York, 1961.

pages 959–966: Cornell Misc. Bulletin 15, Cornell Extension Service, no date.

SECTION 11 MOBILE HOMES AND PARKS

pages 969–986: Guidelines for Improving the Mobile Home Environment, Office of Policy Development and Research, Department of Housing and Urban Development, Washington, D.C., 1978.

pages 987–999: Mobile Homes—Alternative Housing for the Handicapped, Saint Andrews Presbyterian College and U.S. Department of Housing and Urban Development, Washington, D.C., 1977.

pages 1000–1007: Site Selection and Design for Disaster Housing Group Sites, Department of Housing and Urban Development, Washington, D.C., 1976.

SECTION 12 REHABILITATION, CONVERSIONS, AND HISTORIC PRESERVATION

pages 1011–1012: Guidelines for Rehabilitating Old Buildings, Departments of Housing and Urban Development and of the Interior, Washington, D.C., 1977.

pages 1013–1015, 1019–1020: Urban Renewal, New York City Planning Commission, 1958.

pages 1017–1018, 1021–1025: Housing in Central Harlem, Architects Renewal Committee in Harlem, Inc. (ARCH), New York City, no date.

page 1026: Reprinted from March 1976 issue of **Progressive Architecture,** © 1976 Reinhold Publishing Co., New York.

page 1027: Architectural Record, February 1978.

page 1031: E. K. Thompson, **Recycling Buildings,** McGraw-Hill, New York, 1977.

pages 1034–1035: House and Home, March 1977.

page 1036; page 1037, Fig. 12: House and Home, February 1974.

page 1037, Fig. 13: House and Home, April 1972.

page 1039: Architectural Record, October 1977.

page 1040: Architectural Record, September 1977.

page 1041: House and Home, November 1972.

pages 1042–1046: Preservation of Historic America, Department of Housing and Urban Development, Washington, D.C., 1966.

SECTION 13 HOUSING CONTROLS, FIRE SAFETY, AND SECURITY

page 1049, Fig. 1: J. De Chiara and L. E. Koppelman, **Urban Planning and Design Criteria,** Van Nostrand Reinhold Co., New York. 1975.

pages 1051–1052: Housing Codes Standards, National Commission on Urban Problems Research Report no. 19, Washington, D.C., 1969.

pages 1053–1054: APHA-PHS Recommended Housing Maintenance and Occupancy Ordinance, Department of Health, Education, and Welfare, Washington, D.C., 1975.

pages 1055, 1057, 1097: Manual of Acceptable Practices, Department of Housing and Urban Development, Washington, D.C., 1973.

pages 1056, 1088: M. David Egan, **Concepts in Building Fire Safety,** John Wiley & Sons, New York.

pages 1058–1061: Code Manual for the State Construction Code, New York State, 1977.

page 1062: Housing Disabled Persons, Canada Mortgage and Housing Corp.

pages 1063–1068: Fire Safety in Housing, Department of Housing and Urban Development, Washington, D.C., 1976.

pages 1069–1072, 1073–1085, 1087: A Design Guide for Improving Residential Security, U.S. Department of Housing and Urban Development, Washington, D.C., 1973.

pages 1083–1085: Controlling Access in Highrise Buildings, U.S. Department of Housing and Urban Development, Washington, D.C., 1978.

page 1086, Fig. 1: Landscape Architectural Design and Maintenance, Canada Mortgage and Housing Corp., 1982.

page 1086, Fig. 2: Housing for the Elderly Development Process, Michigan State Housing Development Authority, Lansing, Mich., 1974.

pages 1090–1092: Regulation for Flood Plains, Report 277, American Society for Planning Officials, Planning Advisory Service, Chicago, 1972.

page 1100: Design of the Housing Site, Department of Urban Planning, University of Illinois, 1966.

Index

About the Editors

JOSEPH DE CHIARA is a practicing architect and city planner in New York City. He has taught at Columbia University, Pratt Institute, Cooper Union, the New York Institute of Technology, and the State University of New York at Farmingdale. He is coauthor of *Time-Saver Standards for Site Planning* and *Time-Saver Standards for Interior Design and Space Planning,* and the author of *Time-Saver Standards for Building Types* and the *Handbook of Architectural Details for Commercial Buildings,* all published by McGraw-Hill.

JULIUS PANERO, AIA, ASID, is principal of the architectural and consulting firm of Panero Zelnik Associates, Architects/Interior Designers, New York City. He is Professor of Interior Design at the Fashion Institute of Technology (FIT), where he has taught interior design for the past 35 years and served as chairperson of the interior design department. He is coauthor of McGraw-Hill's *Time-Saver Standards for Interior Design and Space Planning* and a book entitled *Human Dimension & Interior Space,* for which he was awarded the prestigious ASID Joel Polsky Prize. Mr. Panero is also the author of *Anatomy for Interior Designers.* He holds a Bachelor of Architecture degree from Pratt Institute and a Master of Science degree in Urban Planning from Columbia University.

MARTIN ZELNIK, AIA, ASID, IDEC, is principal of the architectural and consulting firm of Panero Zelnik Associates, Architects/Interior Designers, New York City. He is Professor of Interior Design at FIT, where he has taught interior design for the past 26 years and served as chairperson of the interior design department. He is coauthor of McGraw-Hill's *Time-Saver Standards for Interior Design and Space Planning* and also *Human Dimension & Interior Space,* for which, along with Mr. Panero, he was awarded the ASID Joel Polsky Prize. Mr. Zelnik holds a Bachelor of Fine Arts degree from Brandeis University and a Master of Architecture degree from Columbia University.